ENCYCLOPEDIA *of* MARINE GEOSCIENCES

Encyclopedia of Earth Sciences Series

ENCYCLOPEDIA OF MARINE GEOSCIENCES

Volume Editors

Jan Harff is Professor of Geosciences and Seafloor Geology at the University of Szczecin, Poland. His previous research focused on sedimentary basin analysis at the Central Institute for Physics of the Earth (ZIPE, subsequently the GeoForschungsZentrum, GFZ), Potsdam, and marine geology at the Leibniz Institute for Baltic Sea Research Warnemünde and the University of Greifswald, Germany. He cooperates, on a permanent basis, with marine research institutes of the Chinese Academy of Sciences, Chinese universities, and the Guangzhou Marine Geological Survey, Guangzhou, China. His research interests concern marine geology in general, sedimentology, coastal geology, paleooceanography, paleoclimatology, mathematical geology, and basin modeling. In addition to having authored and coauthored numerous research papers and having served as editor of other scientific publications, he acted as corresponding editor of *Modeling of Sedimentary Systems* (Springer, 1999) and *The Baltic Sea Basin* (Springer, 2011).
Martin Meschede is Professor of Regional and Structural Geology at the Institute of Geography and Geology, University of Greifswald, Germany. His research interests focus on geodynamic processes at plate margins, subduction, large igneous provinces, exhumation, paleogeography, paleoclimatology, basin evolution, and glacial tectonics. He participated in several marine research expeditions, among these are the Joides Resolution of IODP and a diving cruise with *Shinkai 6500*. Besides a number of scientific publications, he is author and coauthor of several textbooks on plate tectonics, structural geology, and regional geology of Germany.
Sven Petersen is a senior researcher at GEOMAR, Helmholtz Centre for Ocean Research Kiel in Germany. His research focuses on understanding the processes that form and change seafloor hydrothermal systems and associated mineral deposits with time. He participated in more than 30 research cruises to submarine hydrothermal systems in the Pacific, Atlantic, and Indian Oceans. Major aims of his research are to understand their chemical variability, the use of mobile drilling techniques and geophysical methods to investigate their subseafloor extent, as well as the use of autonomous underwater vehicles for their exploration.
Jörn Thiede is the leader of the KÖPPEN-Laboratory of the Institute of Earth Sciences of Saint Petersburg State University. He worked during 1967–1987 at the universities of Aarhus (Denmark), Bergen (Norway), Oregon State University in Corvallis (USA), Oslo (Norway), and Kiel (Germany) and learned to sail the world's oceans to understand their history. Afterward, he pursued the foundation of GEOMAR. In 1997, he joined The Alfred Wegener Institute-Helmholtz Centre for Polar and Marine Research. In 2008, he served at the Geocenter Denmark as well as at UNIS (Longyearbyen/Svalbard) and in 2011 was invited to join the St. Petersburg State University (Russia).

Editorial Board

Gerhard Bohrmann
University of Bremen
Bremen, Germany

Charles W. Finkl
The Coastal Education and Research Foundation, Inc.
Coconut Creek, FL, USA
and
Florida Atlantic University
Boca Raton, FL, USA

Mark D. Hannington
University of Ottawa
Ottawa, ON, Canada
and
GEOMAR Helmholtz Centre for Ocean Research
Kiel, Germany

Roland von Huene
GEOMAR Helmholtz Centre for Ocean Research
Kiel, Germany

Carina B. Lange
Universidad de Concepción
Concepción, Chile

Alexander P. Lisitzin
Shirshov Institute of Oceanology, Russian Academy of Sciences
Moscow, Russia

Jason Phipps Morgan
Cornell University
Ithaca, NY, USA

Yujiro Ogawa
Tsukubamirai, Japan

Manik Talwani
Rice University
Houston, TX, USA

Di Zhou
South China Sea Institute of Oceanology, CAS
Guangzhou, P. R. China

Aims of the Series

The *Encyclopedia of Earth Sciences Series* provides comprehensive and authoritative coverage of all the main areas in the Earth Sciences. Each volume comprises a focused and carefully chosen collection of contributions from leading names in the subject, with copious illustrations and reference lists.

These books represent one of the world's leading resources for the Earth Sciences community. Previous volumes are being updated and new works published so that the volumes will continue to be essential reading for all professional earth scientists, geologists, geophysicists, climatologists, and oceanographers as well as for teachers and students. Go to http://link.springer.com to visit the "Encyclopedia of Earth Sciences Series" online.

About the Series Editor

Professor Charles W. Finkl has edited and/or contributed to more than eight volumes in the *Encyclopedia of Earth Sciences Series*. For the past 25 years, he has been the Executive Director of the Coastal Education and Research Foundation and Editor-in-Chief of the international *Journal of Coastal Research*. In addition to these duties, he is Professor at Florida Atlantic University in Boca Raton, Florida, USA. He is a graduate of the University of Western Australia (Perth) and previously worked for a wholly owned Australian subsidiary of the International Nickel Company of Canada (INCO). During his career, he acquired field experience in Australia, the Caribbean, South America, SW Pacific islands, Southern Africa, Western Europe, and the Pacific Northwest, Midwest, and Southeast USA.

Founding Series Editor

Professor Rhodes W. Fairbridge (deceased) has edited more than 24 Encyclopedias in the *Earth Sciences Series*. During his career, he has worked as a petroleum geologist in the Middle East, been a WWII intelligence officer in the SW Pacific, and led expeditions to the Sahara, Arctic Canada, Arctic Scandinavia, Brazil, and New Guinea. He was Emeritus Professor of Geology at Columbia University and was affiliated with the Goddard Institute for Space Studies.

ENCYCLOPEDIA OF EARTH SCIENCES SERIES

ENCYCLOPEDIA of MARINE GEOSCIENCES

edited by

JAN HARFF
University of Szczecin
Szczecin, Poland

MARTIN MESCHEDE
University of Greifswald
Greifswald, Germany

SVEN PETERSEN
GEOMAR Helmholtz Centre for Ocean Research
Kiel, Germany

JÖRN THIEDE
Institute of Earth Science, KÖPPEN-Laboratory
Saint Petersburg State University
St. Petersburg, Russia

Library of Congress Control Number: 2015957581

ISBN: 978-94-007-6237-4
This publication is available also as:
Electronic publication under ISBN 978-94-007-6238-1 and
Print and electronic bundle under ISBN 978-94-007-6239-8

Springer Dordrecht, Heidelberg, New York, London

Printed on acid-free paper

Cover photo: The submersible *DeepSee*, some 600 feet below the surface of the Pacific, descends into a volcanic vent of the seamount Las Gemelas near Costa Rica. Photograph by: Brian J. Skerry/National Geographic Creative

Every effort has been made to contact the copyright holders of the figures and tables which have been reproduced from other sources. Anyone who has not been properly credited is requested to contact the publishers, so that due acknowledgement may be made in subsequent editions.

All rights reserved for the contributions: *Coastal Engineering; Cobalt-rich Manganese Crusts; Hydrothermal Plumes; Marine Evaporites; Sea Walls / Revetments; Terranes*

© Springer Science+Business Media Dordrecht 2016
This work is subject to copyright. All rights are reserved by the Publisher, whether the whole or part of the material is concerned, specifically the rights of translation, reprinting, reuse of illustrations, recitation, broadcasting, reproduction on microfilms or in any other physical way, and transmission or information storage and retrieval, electronic adaptation, computer software, or by similar or dissimilar methodology now known or hereafter developed. The publisher, the authors and the editors are safe to assume that the advice and information in this book are believed to be true and accurate at the date of publication. Neither the publisher nor the authors or the editors give a warranty, express or implied, with respect to the material contained herein or for any errors or omissions that may have been made.

Dedication

We dedicate the *Encyclopedia of Marine Geosciences* to two scientists: Eugen Seibold and Terry Healy. Eugen Seibold has, directly and indirectly, inspired the four of us and supported our enthusiasm for marine geosciences. While a coauthor of the entry "Marine Geosciences: A Short, Eclectic, and Weighted Historic Account," he passed away in 2013 before the *Encyclopedia of Marine Geosciences* was completed. Terry Healy, a marine scientist from New Zealand, had initiated the *Encyclopedia of Marine Geosciences* project before he passed away in 2010. Terry influenced the geoscience community worldwide by his groundbreaking research in coastal erosion, sedimentation, and hazards (including tsunamis). While Eugen Seibold's contributions represent basic sciences, Terry Healy's address applied aspects of marine sciences. In this sense, the two scientists epitomize our intention for the *Encyclopedia of Marine Geosciences* to bridge the basic and the applied compartments of marine geosciences.

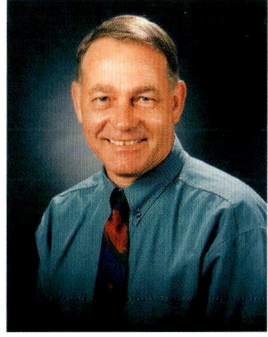

Eugen Seibold (May 11, 1918–October 23, 2013) initiated research on marine geology at the University of Kiel, Germany, with the aim of understanding the fossil depositional environments through the study and comparison of modern sedimentary processes. He realized that marine geology would develop into a fascinating research discipline. He will be remembered as one of the great geoscientists of our time; he was a true man of the *Earth*.

Terry Healy (November 28, 1944–July 20, 2010) was for decades affiliated to the University of Waikato, New Zealand. His special field of research and academic education was the inner shelf oceanography and marine geology, studied on scales from regional to global. Terry Healy's major contribution to marine geosciences involved societal applications of earth science principles.

"Throughout my professional career I have held to a philosophy of demonstrating in my teaching and research, the importance, relevance, and application of earth science principles to the problems of, and ultimate benefit to, the community."

Editors and Editorial Board

Jan Harff
Institute of Marine and Coastal Sciences
University of Szczecin
ul. Mickiewicza 18
70-383 Szczecin
Poland
jan.harff@io-warnemuende.de

Martin Meschede
Institute of Geography and Geology
Ernst-Moritz-Arndt University
Friedrich-Ludwig-Jahn-Str. 17A
17487 Greifswald
Germany
meschede@uni-greifswald.de

Sven Petersen
GEOMAR Helmholtz Centre for Ocean Research
Wischhofstr. 1-3
24148 Kiel
Germany
spetersen@geomar.de

Jörn Thiede
Institute of Earth Sciences
Saint Petersburg State University
V. O., Sredniy prospect, 41
St. Petersburg 199 178
Russia
and
AdW Mainz c/o GEOMAR
Wischhofstr. 1-3
24148 Kiel
Germany
jthiede@geomar.de

Gerhard Bohrmann
MARUM-Center for Marine Environmental Sciences
University of Bremen
P.O. Box 330 440
28334 Bremen
Germany
gbohrmann@marum.de

Charles W. Finkl
Department of Geosciences
Florida Atlantic University
5130 NW 54th Street
Boca Raton 33073 FL
USA
and
The Coastal Education and Research Foundation
Coconut Creek, FL
USA
cfinkl@cerf-jcr.com

Mark D. Hannington
GEOMAR Helmholtz Centre for Ocean Research
Wischhofstr. 1-3
24148 Kiel
Germany
and
Goldcorp Chair in Economic Geology
Department of Earth Sciences
University of Ottawa
K1N 6N5 Ottawa, ON
Canada
mark.hannington@uottawa.ca

Roland von Huene
GEOMAR Helmholtz Centre for Ocean Research
Wischhofstraße 1-3
24148 Kiel
Germany
rhuene@mindspring.com

Carina B. Lange
Centro FONDAP-COPAS
Universidad de Concepción
Casilla 160-C
Concepción
Chile
clange@udec.cl

Alexander P. Lisitzin
P.P. Shirshov Institute of Oceanology
Russian Academy of Sciences (IO RAS)
Nakhimovskii prospekt, 36
Moscow 117851
Russia
lisitzin@ocean.ru

Jason Phipps Morgan
Department of Earth and Atmospheric Science
Cornell University
2122 Snee Hall
Ithaca, NY 14853-1504
USA
jason.phipps.morgan@cornell.edu

Yujiro Ogawa
1-127-2-C-740 Yokodai
Tsukubamirai 300-2358
Japan
fyogawa45@yahoo.co.jp

Manik Talwani
Department of Earth Sciences
MS-126 Rice University
6100 Main Street
Houston, TX 77005
USA
manik.talwani@gmail.com

Di Zhou
South China Sea Institute of Oceanology
Chinese Academy of Sciences
164 West Xingang Road
510301 Guangzhou
China
zhoudiscs@scsio.ac.cn

Contents

Contributors	xv	Bed forms *Miwa Yokokawa*	55
Preface	xxxi		
Acknowledgments	xxxiii	Biochronology, Biostratigraphy *Felix M. Gradstein*	55
Abyssal Plains *David Voelker*	1	Biogenic Barium *Graham Shimmield*	56
Accretionary Wedges *Martin Meschede*	6	Bioturbation *Gerhard Graf*	57
Active Continental Margins *Serge Lallemand*	9	Black and White Smokers *Margaret K. Tivey*	58
Ancient Plate Tectonics *Stephen F. Foley*	13	Bottom Simulating Seismic Reflectors (BSR) *Jürgen Mienert and Stefan Bünz*	62
Anoxic Oceans *Christian März and Hans-Jürgen Brumsack*	20	Bottom-Boundary Layer *Wenyan Zhang*	67
Asphalt Volcanism *Gerhard Bohrmann*	24	Bouma Sequence *Thierry Mulder and Heiko Hüneke*	68
Astronomical Frequencies in Paleoclimates *André Berger*	25	Calcite Compensation Depth (CCD) *Wolfgang H. Berger*	71
Axial Summit Troughs *Samuel Adam Soule and Michael R. Perfit*	33	Carbon Isotopes *Thomas Wagner and Jens O. Herrle*	73
Axial Volcanic Ridges *Isobel Yeo*	36	Carbonate Dissolution *Leif G. Anderson*	78
Beach *Charles W. Finkl*	41	Carbonate Factories *John J.G. Reijmer*	80
Beach Processes *Charles W. Finkl*	47	Chemosynthetic Life *Verena Tunnicliffe*	84

Clay Minerals
Rüdiger Stein	87	Depleted Mantle
Andreas Stracke	182
Coastal Bio-geochemical Cycles
Jing Zhang	93	Diatoms
Arto Miettinen	185
Coastal Engineering
Louise Wallendorf	99	Dinoflagellates
Jens Matthiessen and Michael Schreck	189
Coasts
Charles W. Finkl	103	Driving Forces: Slab Pull, Ridge Push
Carolina Lithgow-Bertelloni	193
Cobalt-rich Manganese Crusts
James R. Hein	113	Dunes
Sara Muñoz-Vallés and Jesús Cambrollé	196
Cold Seeps
Marta E. Torres and Gerhard Bohrmann	117	Dust in the Ocean
*Cécile Guieu and
Vladimir P. Shevchenko*	203
Continental Rise
William W. Hay	122	Earthquakes
Nina Kukowski	209
Continental Slope
William W. Hay	124	El Niño (Southern Oscillation)
Janina Körper	216
Contourites
Heiko Hüneke	127	Energy Resources
Dieter Franke and Christoph Gaedicke	217
Convergence Texture of Seismic Reflectors
Christopher George St. Clement Kendall	132	Engineered Coasts
H. Jesse Walker	226
Crustal Accretion
Benoît Ildefonse	133	Epicenter, Hypocenter
Nina Kukowski	231
Cumulates
Jürgen Koepke	137	Erosion-Hiatuses
William W. Hay	231
Curie Temperature
Hans-Jürgen Götze	138	Estuary, Estuarine Hydrodynamics
*Halina Kowalewska-Kalkowska and
Roman Marks*	235
Currents
Jüri Elken	139		
Data
Michael Diepenbroek	143	Eustasy
Reinhard Dietrich	238
Deep Biosphere
Axel Schippers	144	Events
William W. Hay	239
Deep-sea Fans
Thierry Mulder and Heiko Hüneke	156	Explosive Volcanism in the Deep Sea
Christoph Helo	241
Deep-sea Sediments
Mitchell Lyle	156	Export Production
*Gerold Wefer, Gerhard Fischer and
Morten Iversen*	247
Deltas
Duncan FitzGerald, Ioannis Georgiou and Mark Kulp	171	Fault-Plane Solutions of Earthquakes
Wolfgang Frisch	249

Fjords *Antoon Kuijpers and Camilla S. Andresen*	250	Hurricanes and Typhoons *Don Resio and Shannon Kay*	329
Foraminifers (Benthic) *Anna Binczewska, Irina Polovodova Asteman and Elizabeth J. Farmer*	251	Hydrothermal Plumes *Edward T. Baker*	335
Foraminifers (Planktonic) *Demetrio Boltovskoy and Nancy Correa*	255	Hydrothermal Vent Fluids (Seafloor) *Andrea Koschinsky*	339
Gabbro *Jürgen Koepke*	263	Hydrothermalism *John W. Jamieson, Sven Petersen and Wolfgang Bach*	344
General Bathymetric Chart of the Oceans (GEBCO) *Hans Werner Schenke*	268	Ice-rafted Debris (IRD) *Antoon Kuijpers, Paul Knutz and Matthias Moros*	359
General ocean circulation - its signals in the sediments *Hermann Kudrass*	269	Integrated Coastal Zone Management *Gerald Schernewski*	363
Geochronology: Uranium-Series Dating of Ocean Formations *Vladislav Kuznetsov*	271	International Bathymetric Chart of the Arctic Ocean (IBCAO) *Martin Jakobsson*	365
Geohazards: Coastal Disasters *Gösta Hoffmann and Klaus Reicherter*	276	Intraoceanic Subduction Zone *Hayato Ueda*	367
Geologic Time Scale *Felix M. Gradstein*	283	Intraplate Magmatism *Millard F. Coffin and Joanne M. Whittaker*	372
Glacial-marine Sedimentation *Alexander P. Lisitzin and Vladimir P. Shevchenko*	288	Island Arc Volcanism, Volcanic Arcs *Jim Gill*	379
Glacio(hydro)-isostatic Adjustment *Kurt Lambeck*	294	Lagoons *Andrzej Osadczuk*	385
Gravity Field *Udo Barckhausen and Ingo Heyde*	299	Laminated Sediments *Alan Kemp*	391
Grounding Line *Antoon Kuijpers and Tove Nielsen*	302	Layering of Oceanic Crust *Juan Pablo Canales*	392
Guyot, Atoll *Eberhard Gischler*	302	Lithosphere: Structure and Composition *Martin Meschede*	393
High-pressure, Low-temperature Metamorphism *Hayato Ueda*	311	Lithostratigraphy *Rüdiger Henrich*	397
Hot Spots and Mantle Plumes *William M. White*	316	Magmatism at Convergent Plate Boundaries *Robert J. Stern*	399
Hotspot-Ridge Interaction *Colin Devey*	328	Magnetostratigraphy of Jurassic Rocks *María Paula Iglesia Llanos*	407
		Manganese Nodules *James R. Hein*	408

Mangrove Coasts *Norman C. Duke*	412	Modelling Past Oceans *Jürgen Sündermann*	514
Marginal Seas *Di Zhou*	423	Modern Analog Techniques *Michal Kucera*	514
Marine Evaporites *Walter E. Dean*	427	Mohorovičić Discontinuity (Moho) *Carolina Lithgow-Bertelloni*	515
Marine Gas Hydrates *Gerhard Bohrmann and Marta E. Torres*	433	Moraines *Anders Schomacker*	519
Marine Geosciences: a Short, Ecclectic and Weighted Historic Account *Jörn Thiede, Jan Harff, Wolfgang H. Berger and Eugen Seibold*	437	Morphology Across Convergent Plate Boundaries *Wolfgang Frisch*	523
Marine Geotope Protection *Aarno Kotilainen*	448	Mud Volcano *Achim Kopf*	527
Marine Heat Flow *John G. Sclater, Derrick Hasterok, Bruno Goutorbe, John Hillier and Raquel Negrete*	449	Nannofossils Coccoliths *Hans R. Thierstein*	537
		Nitrogen Isotopes in the Sea *Thomas Pedersen*	539
Marine Impacts and Their Consequences *Henning Dypvik*	460	North Atlantic Oscillation (NAO) *Erik Kjellström*	540
Marine Microfossils *Jakub Witkowski, Kirsty Edgar, Ian Harding, Kevin McCartney and Marta Bąk*	467	Ocean Acidification *Ulf Riebesell*	541
Marine Mineral Resources *Sven Petersen*	475	Ocean Drilling *Kiyoshi Suyehiro*	542
Marine Regression *Jan Harff*	480	Ocean Margin Systems *Gerold Wefer*	552
Marine Sedimentary Basins *Paul Mann*	481	Oceanic Plateaus *Andrew C. Kerr*	558
Marine Transgression *Jan Harff*	489	Oceanic Rifts *Adolphe Nicolas*	567
Mass Wasting *Jan Sverre Laberg, Tore O. Vorren and Matthias Forwick*	490	Oceanic Spreading Centers *Satish C. Singh and Adolphe Nicolas*	571
Metamorphic Core Complexes *Uwe Ring*	491	Oil Spill *Erich Gundlach*	587
Methane in Marine Sediments *Gerhard Bohrmann and Marta E. Torres*	495	Ophiolites *Adolphe Nicolas*	592
Mg/Ca Paleothermometry *Dirk Nürnberg*	499	Organic Matter *Evgeny Romankevich and Alexander Vetrov*	596
Mid-ocean Ridge Magmatism and Volcanism *Ken H. Rubin*	501	Orogeny *Bernhard Grasemann and Benjamin Huet*	602

Editors and Editorial Board

Jan Harff
Institute of Marine and Coastal Sciences
University of Szczecin
ul. Mickiewicza 18
70-383 Szczecin
Poland
jan.harff@io-warnemuende.de

Martin Meschede
Institute of Geography and Geology
Ernst-Moritz-Arndt University
Friedrich-Ludwig-Jahn-Str. 17A
17487 Greifswald
Germany
meschede@uni-greifswald.de

Sven Petersen
GEOMAR Helmholtz Centre for Ocean Research
Wischhofstr. 1-3
24148 Kiel
Germany
spetersen@geomar.de

Jörn Thiede
Institute of Earth Sciences
Saint Petersburg State University
V. O., Sredniy prospect, 41
St. Petersburg 199 178
Russia
and
AdW Mainz c/o GEOMAR
Wischhofstr. 1-3
24148 Kiel
Germany
jthiede@geomar.de

Gerhard Bohrmann
MARUM-Center for Marine Environmental Sciences
University of Bremen
P.O. Box 330 440
28334 Bremen
Germany
gbohrmann@marum.de

Charles W. Finkl
Department of Geosciences
Florida Atlantic University
5130 NW 54th Street
Boca Raton 33073 FL
USA
and
The Coastal Education and Research Foundation
Coconut Creek, FL
USA
cfinkl@cerf-jcr.com

Mark D. Hannington
GEOMAR Helmholtz Centre for Ocean Research
Wischhofstr. 1-3
24148 Kiel
Germany
and
Goldcorp Chair in Economic Geology
Department of Earth Sciences
University of Ottawa
K1N 6N5 Ottawa, ON
Canada
mark.hannington@uottawa.ca

Roland von Huene
GEOMAR Helmholtz Centre for Ocean Research
Wischhofstraße 1-3
24148 Kiel
Germany
rhuene@mindspring.com

Carina B. Lange
Centro FONDAP-COPAS
Universidad de Concepción
Casilla 160-C
Concepción
Chile
clange@udec.cl

Alexander P. Lisitzin
P.P. Shirshov Institute of Oceanology
Russian Academy of Sciences (IO RAS)
Nakhimovskii prospekt, 36
Moscow 117851
Russia
lisitzin@ocean.ru

Jason Phipps Morgan
Department of Earth and Atmospheric Science
Cornell University
2122 Snee Hall
Ithaca, NY 14853-1504
USA
jason.phipps.morgan@cornell.edu

Yujiro Ogawa
1-127-2-C-740 Yokodai
Tsukubamirai 300-2358
Japan
fyogawa45@yahoo.co.jp

Manik Talwani
Department of Earth Sciences
MS-126 Rice University
6100 Main Street
Houston, TX 77005
USA
manik.talwani@gmail.com

Di Zhou
South China Sea Institute of Oceanology
Chinese Academy of Sciences
164 West Xingang Road
510301 Guangzhou
China
zhoudiscs@scsio.ac.cn

Contents

Contributors	xv	Bed forms *Miwa Yokokawa*	55
Preface	xxxi		
Acknowledgments	xxxiii	Biochronology, Biostratigraphy *Felix M. Gradstein*	55
Abyssal Plains *David Voelker*	1	Biogenic Barium *Graham Shimmield*	56
Accretionary Wedges *Martin Meschede*	6	Bioturbation *Gerhard Graf*	57
Active Continental Margins *Serge Lallemand*	9	Black and White Smokers *Margaret K. Tivey*	58
Ancient Plate Tectonics *Stephen F. Foley*	13	Bottom Simulating Seismic Reflectors (BSR) *Jürgen Mienert and Stefan Bünz*	62
Anoxic Oceans *Christian März and Hans-Jürgen Brumsack*	20	Bottom-Boundary Layer *Wenyan Zhang*	67
Asphalt Volcanism *Gerhard Bohrmann*	24	Bouma Sequence *Thierry Mulder and Heiko Hüneke*	68
Astronomical Frequencies in Paleoclimates *André Berger*	25	Calcite Compensation Depth (CCD) *Wolfgang H. Berger*	71
Axial Summit Troughs *Samuel Adam Soule and Michael R. Perfit*	33	Carbon Isotopes *Thomas Wagner and Jens O. Herrle*	73
Axial Volcanic Ridges *Isobel Yeo*	36	Carbonate Dissolution *Leif G. Anderson*	78
Beach *Charles W. Finkl*	41	Carbonate Factories *John J.G. Reijmer*	80
Beach Processes *Charles W. Finkl*	47	Chemosynthetic Life *Verena Tunnicliffe*	84

Clay Minerals *Rüdiger Stein*	87	Depleted Mantle *Andreas Stracke*	182
Coastal Bio-geochemical Cycles *Jing Zhang*	93	Diatoms *Arto Miettinen*	185
Coastal Engineering *Louise Wallendorf*	99	Dinoflagellates *Jens Matthiessen and Michael Schreck*	189
Coasts *Charles W. Finkl*	103	Driving Forces: Slab Pull, Ridge Push *Carolina Lithgow-Bertelloni*	193
Cobalt-rich Manganese Crusts *James R. Hein*	113	Dunes *Sara Muñoz-Vallés and Jesús Cambrollé*	196
Cold Seeps *Marta E. Torres and Gerhard Bohrmann*	117	Dust in the Ocean *Cécile Guieu and Vladimir P. Shevchenko*	203
Continental Rise *William W. Hay*	122	Earthquakes *Nina Kukowski*	209
Continental Slope *William W. Hay*	124	El Niño (Southern Oscillation) *Janina Körper*	216
Contourites *Heiko Hüneke*	127	Energy Resources *Dieter Franke and Christoph Gaedicke*	217
Convergence Texture of Seismic Reflectors *Christopher George St. Clement Kendall*	132	Engineered Coasts *H. Jesse Walker*	226
Crustal Accretion *Benoît Ildefonse*	133	Epicenter, Hypocenter *Nina Kukowski*	231
Cumulates *Jürgen Koepke*	137	Erosion-Hiatuses *William W. Hay*	231
Curie Temperature *Hans-Jürgen Götze*	138	Estuary, Estuarine Hydrodynamics *Halina Kowalewska-Kalkowska and Roman Marks*	235
Currents *Jüri Elken*	139	Eustasy *Reinhard Dietrich*	238
Data *Michael Diepenbroek*	143	Events *William W. Hay*	239
Deep Biosphere *Axel Schippers*	144	Explosive Volcanism in the Deep Sea *Christoph Helo*	241
Deep-sea Fans *Thierry Mulder and Heiko Hüneke*	156	Export Production *Gerold Wefer, Gerhard Fischer and Morten Iversen*	247
Deep-sea Sediments *Mitchell Lyle*	156		
Deltas *Duncan FitzGerald, Ioannis Georgiou and Mark Kulp*	171	Fault-Plane Solutions of Earthquakes *Wolfgang Frisch*	249

Fjords *Antoon Kuijpers and Camilla S. Andresen*	250	Hurricanes and Typhoons *Don Resio and Shannon Kay*	329
Foraminifers (Benthic) *Anna Binczewska, Irina Polovodova Asteman and Elizabeth J. Farmer*	251	Hydrothermal Plumes *Edward T. Baker*	335
Foraminifers (Planktonic) *Demetrio Boltovskoy and Nancy Correa*	255	Hydrothermal Vent Fluids (Seafloor) *Andrea Koschinsky*	339
Gabbro *Jürgen Koepke*	263	Hydrothermalism *John W. Jamieson, Sven Petersen and Wolfgang Bach*	344
General Bathymetric Chart of the Oceans (GEBCO) *Hans Werner Schenke*	268	Ice-rafted Debris (IRD) *Antoon Kuijpers, Paul Knutz and Matthias Moros*	359
General ocean circulation - its signals in the sediments *Hermann Kudrass*	269	Integrated Coastal Zone Management *Gerald Schernewski*	363
Geochronology: Uranium-Series Dating of Ocean Formations *Vladislav Kuznetsov*	271	International Bathymetric Chart of the Arctic Ocean (IBCAO) *Martin Jakobsson*	365
Geohazards: Coastal Disasters *Gösta Hoffmann and Klaus Reicherter*	276	Intraoceanic Subduction Zone *Hayato Ueda*	367
Geologic Time Scale *Felix M. Gradstein*	283	Intraplate Magmatism *Millard F. Coffin and Joanne M. Whittaker*	372
Glacial-marine Sedimentation *Alexander P. Lisitzin and Vladimir P. Shevchenko*	288	Island Arc Volcanism, Volcanic Arcs *Jim Gill*	379
Glacio(hydro)-isostatic Adjustment *Kurt Lambeck*	294	Lagoons *Andrzej Osadczuk*	385
Gravity Field *Udo Barckhausen and Ingo Heyde*	299	Laminated Sediments *Alan Kemp*	391
Grounding Line *Antoon Kuijpers and Tove Nielsen*	302	Layering of Oceanic Crust *Juan Pablo Canales*	392
Guyot, Atoll *Eberhard Gischler*	302	Lithosphere: Structure and Composition *Martin Meschede*	393
High-pressure, Low-temperature Metamorphism *Hayato Ueda*	311	Lithostratigraphy *Rüdiger Henrich*	397
Hot Spots and Mantle Plumes *William M. White*	316	Magmatism at Convergent Plate Boundaries *Robert J. Stern*	399
Hotspot-Ridge Interaction *Colin Devey*	328	Magnetostratigraphy of Jurassic Rocks *María Paula Iglesia Llanos*	407
		Manganese Nodules *James R. Hein*	408

Mangrove Coasts *Norman C. Duke*	412	Modelling Past Oceans *Jürgen Sündermann*	514
Marginal Seas *Di Zhou*	423	Modern Analog Techniques *Michal Kucera*	514
Marine Evaporites *Walter E. Dean*	427	Mohorovičić Discontinuity (Moho) *Carolina Lithgow-Bertelloni*	515
Marine Gas Hydrates *Gerhard Bohrmann and Marta E. Torres*	433	Moraines *Anders Schomacker*	519
Marine Geosciences: a Short, Ecclectic and Weighted Historic Account *Jörn Thiede, Jan Harff, Wolfgang H. Berger and Eugen Seibold*	437	Morphology Across Convergent Plate Boundaries *Wolfgang Frisch*	523
Marine Geotope Protection *Aarno Kotilainen*	448	Mud Volcano *Achim Kopf*	527
Marine Heat Flow *John G. Sclater, Derrick Hasterok, Bruno Goutorbe, John Hillier and Raquel Negrete*	449	Nannofossils Coccoliths *Hans R. Thierstein*	537
		Nitrogen Isotopes in the Sea *Thomas Pedersen*	539
Marine Impacts and Their Consequences *Henning Dypvik*	460	North Atlantic Oscillation (NAO) *Erik Kjellström*	540
Marine Microfossils *Jakub Witkowski, Kirsty Edgar, Ian Harding, Kevin McCartney and Marta Bąk*	467	Ocean Acidification *Ulf Riebesell*	541
Marine Mineral Resources *Sven Petersen*	475	Ocean Drilling *Kiyoshi Suyehiro*	542
Marine Regression *Jan Harff*	480	Ocean Margin Systems *Gerold Wefer*	552
Marine Sedimentary Basins *Paul Mann*	481	Oceanic Plateaus *Andrew C. Kerr*	558
Marine Transgression *Jan Harff*	489	Oceanic Rifts *Adolphe Nicolas*	567
Mass Wasting *Jan Sverre Laberg, Tore O. Vorren and Matthias Forwick*	490	Oceanic Spreading Centers *Satish C. Singh and Adolphe Nicolas*	571
Metamorphic Core Complexes *Uwe Ring*	491	Oil Spill *Erich Gundlach*	587
Methane in Marine Sediments *Gerhard Bohrmann and Marta E. Torres*	495	Ophiolites *Adolphe Nicolas*	592
Mg/Ca Paleothermometry *Dirk Nürnberg*	499	Organic Matter *Evgeny Romankevich and Alexander Vetrov*	596
Mid-ocean Ridge Magmatism and Volcanism *Ken H. Rubin*	501	Orogeny *Bernhard Grasemann and Benjamin Huet*	602

Oxygen Isotopes *Mark Maslin and Alexander J. Dickson*	610	Radiolarians *Kjell R. Bjørklund*	700
Paired Metamorphic Belts *Wolfgang Frisch*	619	Reef Coasts *Charles W. Finkl*	710
Paleoceanographic Proxies *Gerold Wefer*	622	Reefs (Biogenic) *Christian Dullo*	718
Paleoceanography *Jörn Thiede*	628	Reflection/Refraction Seismology *Christian Hübscher and Karsten Gohl*	721
Paleomagnetism and Jurassic Paleogeography *María Paula Iglesia Llanos*	632	Regional Marine Geology *Martin Meschede*	731
Paleophysiography of Ocean Basins *R. Dietmar Müller and Maria Seton*	638	Relative Sea-Level (RSL) Cycle *Jan Harff*	735
Paleoproductivity *Paul Loubere*	648	Salt Diapirism in the Oceans and Continental Margins *Sergey S. Drachev*	741
Palynology (Pollen, Spores, etc.) *Anne de Vernal*	653	Sapropels *Rüdiger Stein*	746
Passive Plate Margins *Paul Mann*	659	Sclerochronology *Christian Dullo*	747
Peridotites *Eric Hellebrand*	665	Sea Walls/Revetments *Louise Wallendorf*	747
Phosphorites *Hermann Kudrass*	666	Sea-Level *Reinhard Dietrich*	748
Placer Deposits *Hermann Kudrass*	668	Seamounts *David M. Buchs, Kaj Hoernle and Ingo Grevemeyer*	754
Plate Motion *R. Dietmar Müller and Maria Seton*	669	Sediment Dynamics *Wenyan Zhang*	761
Plate Tectonics *Martin Meschede*	676	Sediment Transport Models *Wenyan Zhang*	764
Pteropods *Annelies C. Pierrot-Bults and Katja T. C. A. Peijnenburg*	680	Sedimentary Sequence *Christopher George St. Clement Kendall*	768
Pull-apart Basins *Alper Gürbüz*	687	Seismogenic Zone *Eli Silver*	773
Push-up Blocks *Alper Gürbüz*	692	Sequence Stratigraphy *Christopher George St. Clement Kendall*	773
Radiocarbon: Clock and Tracer *Pieter M. Grootes*	695	Serpentinization *Niels Jöns and Wolfgang Bach*	779
Radiogenic Tracers *Anton Eisenhauer*	699	Shelf *William W. Hay*	787

Silica *Paul Treguer*	792	Triple Junctions *A. M. Celâl Şengör*	876
Source Rocks, Reservoirs *Rüdiger Stein*	792	Tsunamis *Jose C. Borrero*	884
Subduction *Serge Lallemand*	793	Turbidites *Thierry Mulder and Heiko Hüneke*	888
Subduction Erosion *Martin Meschede*	803	Underwater Archaeology *Geoffrey N. Bailey*	893
Submarine Canyons *William W. Hay*	807	Upwelling *Colin Summerhayes*	900
Submarine Lava Types *Michael R. Perfit and Samuel Adam Soule*	808	Volcanism and Climate *Olav Eldholm and Millard F. Coffin*	913
Submarine Slides *Roland von Huene*	817	Volcanogenic Massive Sulfides *John W. Jamieson, Mark D. Hannington, Sven Petersen and Margaret K. Tivey*	917
Subsidence of Oceanic Crust *David Voelker*	821	Wadati-Benioff-Zone *Nina Kukowski*	925
Technology in Marine Geosciences *Sven Petersen*	825	Wadden Sea *Achim Wehrmann*	933
Terranes *Maurice Colpron and JoAnne Nelson*	835	Waves *Tarmo Soomere*	940
Tethys in Marine Geosciences *A. M. Celâl Şengör*	838	Wilson Cycle *Martin Meschede*	944
Tidal Depositional Systems *Achim Wehrmann*	849	Author Index	947
Transform Faults *A. M. Celâl Şengör*	859	Subject Index	949

Contributors

Leif G. Anderson
Department of Marine Sciences
University of Gothenburg
Kemigården 4
412 96 Gothenburg
Sweden
leif.anderson@gu.se

Camilla S. Andresen
Geological Survey of Denmark and Greenland (GEUS)
Øster Voldgade 10
1350 Copenhagen
Denmark
csa@geus.dk

Wolfgang Bach
Department of Geosciences
University of Bremen
Klagenfurter Str. (GEO)
28359 Bremen
Germany
and
MARUM-Center for Marine Environmental Sciences
University of Bremen
P.O. Box 330 440
28334 Bremen
Germany
wbach@marum.de

Geoffrey N. Bailey
Department of Archaeology
University of York
The King's Manor
York YO1 7EP
UK
geoff.bailey@york.ac.uk

Marta Bąk
Faculty of Geology, Geophysics and Environmental Protection
AGH University of Science and Technology
Al. Mickiewicza 30
30-059 Kraków
Poland
martabak@agh.edu.pl

Edward T. Baker
Joint Institute for the Study of Atmosphere and Ocean – NOAA/PMEL
University of Washington
7600 Sand Point Way NE
Seattle, WA 98115-6349
USA
edward.baker@noaa.gov

Udo Barckhausen
Federal Institute for Geosciences and Natural Resources
Stilleweg 2
30655 Hannover
Germany
udo.barckhausen@bgr.de

André Berger
Georges Lemaître Center for Earth and Climate Research
Catholic University of Louvain
3, Place Louis Pasteur
B-1348 Louvain-la-Neuve
Belgium
andre.berger@uclouvain.be

Wolfgang H. Berger
Scripps Institution of Oceanography UCSD
9500 Gilman Drive
La Jolla, CA 92093
USA
wberger@ucsd.edu

Anna Binczewska
Institute of Marine and Coastal Sciences
University of Szczecin
ul. Mickiewicza 18
70-383 Szczecin
Poland
anna.binczewska@gmail.com

Kjell R. Bjørklund
Museum of Natural History
University of Oslo
P.O. Box 1172
NO-0318 Blindern, Oslo
Norway
k.r.bjorklund@nhm.uio.no

Gerhard Bohrmann
MARUM-Center for Marine Environmental Sciences
University of Bremen
P.O. Box 330 440
28334 Bremen
Germany
gbohrmann@marum.de

Demetrio Boltovskoy
Institute of Ecology, Genetics and Evolution of Buenos Aires (IEGEBA)
University of Buenos Aires – CONICET
C1428EHA Buenos Aires
Argentina
boltovskoy@gmail.com
demetrio@bg.fcen.uba.ar

Jose C. Borrero
eCoast Ltd., Marine Consulting and Research
47 Cliff Street
3225 Raglan
New Zealand
jose@ecoast.co.nz
jborrero@usc.edu

Hans-Jürgen Brumsack
Institute for Chemistry and Biology of the Marine Environment
University of Oldenburg
P.O. Box 2503
26111 Oldenburg
Germany
brumsack@icbm.de

David M. Buchs
School of Earth and Ocean Sciences
Cardiff University
CF10 3AT Cardiff
UK
buchsd@cardiff.ac.uk

Stefan Bünz
Centre for Arctic Gas Hydrate, Environment and Climate (CAGE)
UiT The Arctic University of Norway
Dramsveien 201
9037 Tromsø
Norway
stefan.buenz@uit.no

Jesús Cambrollé
Department of Plant Biology and Ecology
University of Sevilla
Avda. de la Reina Mercedes, S/N
41012 Seville
Spain
cambrolle@us.es

Juan Pablo Canales
Woods Hole Oceanographic Institution
266 Woods Hole Rd.
Woods Hole, MA 02543-1050
USA
jcanales@whoi.edu

Millard F. Coffin
Institute for Marine and Antarctic Studies
University of Tasmania
Private Bag, 129
7001 Hobart, TAS
Australia
mike.coffin@utas.edu.au

Maurice Colpron
Yukon Geological Survey
P.O. Box 2703 (K-14)
Whitehorse, Yukon Y1A 2C6
Canada
Maurice.Colpron@gov.yk.ca

Nancy Correa
Department of the Argentine Naval Hydrographic Service and School of Marine Sciences
Ministry of Defence and Marine Institute
Av. Montes de Oca 2124
C1270ABV Buenos Aires
Argentina
ncorrea@hidro.gov.ar

Anne de Vernal
Centre GEOTOP, University of Québec in Montréal
CP 8888, succ. Centre-Ville
H3C 3P8 Montréal, QC
Canada
devernal.anne@uqam.ca

Walter E. Dean
U.S. Geological Survey
P.O.Box 25046 MS980 Federal Center
Denver, CO 80225
USA
dean@usgs.gov

Colin Devey
GEOMAR Helmholtz Centre for Ocean Research
Wischhofstr. 1-3
24148 Kiel
Germany
cdevey@geomar.de

Alexander J. Dickson
Department of Earth Sciences
University of Oxford
South Parks Road
Oxford OX1 3AN
UK
alex.dickson@earth.ox.ac.uk

Michael Diepenbroek
MARUM-Center for Marine Environmental Sciences
University of Bremen
P.O. Box 330 440
28334 Bremen
Germany
mdiepenbroek@pangaea.de

Reinhard Dietrich
Institute of Planetary Geodesy
Technical University of Dresden
Helmholtzstr. 10
01062 Dresden
Germany
reinhard.dietrich@tu-dresden.de

Sergey S. Drachev
ExxonMobil House, ExxonMobil International Ltd.
Ermyn Way
KT22 8UX Leatherhead
UK
sergey.s.drachev@exxonmobil.com
sdrachev@gmail.com

Norman C. Duke
James Cook University
TropWATER – Centre for Tropical Water and Aquatic Ecosystem Research
4811 Townsville, QLD
Australia
norman.duke@jcu.edu.au

Christian Dullo
GEOMAR Helmholtz Centre for Ocean Research
Wischhofstr. 1-3
24148 Kiel
Germany
cdullo@geomar.de

Henning Dypvik
Department of Geosciences
University of Oslo
P.O. Box 1047 Blindern
0316 Oslo
Norway
henning.dypvik@geo.uio.no

Kirsty Edgar
School of Earth Sciences
University of Bristol
Wills Memorial Building, Queen's Road
Clifton BS8 1RJ
UK
kirsty.edgar@bristol.ac.uk

Anton Eisenhauer
GEOMAR Helmholtz Centre for Ocean Research
Wischhofstr. 1-3
24148 Kiel
Germany
aeisenhauer@geomar.de

Olav Eldholm
Department of Earth Sciences
University of Bergen
Realfagbygget, Allégate 41
P.O. Box 7800 5020 Bergen
Norway
olav.eldholm@geo.uib.no

Jüri Elken
Marine Systems Institute
Tallinn University of Technology
Akadeemia 15a
12618 Tallinn
Estonia
Juri.elken@msi.ttu.ee

Elizabeth J. Farmer
Department of Earth Science
University of Bergen
P.O. Box 7803
Realfagbygget, Allégaten 41
5020 Bergen
Norway
elizabeth.farmer@uib.no

Charles W. Finkl
Department of Geosciences
Florida Atlantic University
5130 NW 54th Street
Boca Raton 33073 FL
USA
and
The Coastal Education and Research Foundation
Coconut Creek, FL
USA
cfinkl@cerf-jcr.com

Gerhard Fischer
MARUM-Center for Marine Environmental Sciences
University of Bremen
P.O. Box 330 440
28334 Bremen
Germany
gfischer@marum.de

Duncan FitzGerald
Department of Earth and Environment
Boston University
685 Commonwealth Avenue
Boston, MA 02215
USA
dunc@bu.edu

Stephen F. Foley
Department Earth and Planetary Sciences
ARC Centre of Excellence for Core to Crust Fluid Systems
Macquarie University
Building E7A, Room 503
2109 North Ryde, NSW
Australia
stephen.foley@mq.edu.au

Matthias Forwick
Department of Geology
University of Tromsø
Realfagbygget, Dramsveien 201
9037 Tromsø
Norway
matthias.forwick@uit.no

Dieter Franke
Federal Institute for Geosciences and Natural Resources
Stilleweg 2
30655 Hannover
Germany
dieter.franke@bgr.de

Wolfgang Frisch
Department of Geosciences
University of Tübingen
Hölderlinstr. 12
72074 Tübingen
Germany
frisch@uni-tuebingen.de

Christoph Gaedicke
Federal Institute for Geosciences and Natural Resources
Stilleweg 2
30655 Hannover
Germany
christoph.gaedicke@bgr.de

Ioannis Georgiou
Department of Earth and Environmental Sciences
University of New Orleans
1065 Geology and Psychology Building
2000 Lakeshore Drive
New Orleans, LA 70148
USA
igeorgio@uno.edu

Jim Gill
Earth and Planetary Sciences
University of California
1156 High Street
Santa Cruz, CA 95064
USA
jgill@es.ucsc.edu

Eberhard Gischler
Institute of Geosciences
Johann Wolfgang Goethe University
Altenhöferallee 1
60438 Frankfurt am Main
Germany
gischler@em.uni-frankfurt.de

Karsten Gohl
Alfred Wegener Institute, Helmholtz-Centre for Polar and Marine Research
Am Alten Hafen 26
27568 Bremerhaven
Germany
karsten.gohl@awi.de

Hans-Jürgen Götze
Institute of Geosciences
Christian-Albrechts-University Kiel
Otto-Hahn-Platz 1
24118 Kiel
Germany
hajo@geophysik.uni-kiel.de

Bruno Goutorbe
Institute of Geosciences
State University Fluminense
Niteroi
Brazil
brunog@id.uff.br
goutorbe@hotmail.com

Felix M. Gradstein
Museum of Natural History, University of Oslo
P.O. Box 1172 Blindern
0318 Oslo
Norway
felix.gradstein@gmail.com

Gerhard Graf
Institute for Biological Sciences
University of Rostock
Albert-Einstein-Strasse 3
18059 Rostock
Germany
gerd.graf@uni-rostock.de

Bernhard Grasemann
Department for Geodynamics and Sedimentology
University of Vienna
Althanstrasse 14
1090 Wien
Austria
bernhard.grasemann@univie.ac.at

Ingo Grevemeyer
GEOMAR Helmholtz Centre for Ocean Research
Wischhofstr. 1-3
24148 Kiel
Germany
igrevemeyer@geomar.de

Pieter M. Grootes
Institute for Ecosystem Research
Christian-Albrechts-University Kiel
Olshausenstrasse 100
24106 Kiel
Germany
pgrootes@ecology.uni-kiel.de

Cécile Guieu
Villefranche sur Mer Oceanographic Laboratory
Chemin du Lazaret 181
06230 Villefranche sur Mer
France
guieu@obs-vlfr.fr

Erich Gundlach
E-Tech International Inc.
P.O. Box 4321
Boulder, CO 80306
USA
ericheti@cs.com

Alper Gürbüz
Department of Geological Engineering
University of Ankara
06100 Tandoğan, Ankara
Turkey
and
Department of Geological Engineering
University of Niğde
51240 Niğde
Turkey
agurbuz@eng.ankara.edu.tr
agurbuz@nigde.edu.tr

Mark D. Hannington
GEOMAR Helmholtz Centre for Ocean Research
Wischhofstr. 1-3
24148 Kiel
Germany
and
Goldcorp Chair in Economic Geology, Department of Earth Sciences
University of Ottawa
K1N 6N5 Ottawa, ON
Canada
mark.hannington@uottawa.ca

Ian Harding
Ocean and Earth Science, National Oceanography Centre
University of Southampton, Waterfront Campus
European Way
Southampton SO14 3ZH
UK
ich@noc.soton.ac.uk

Jan Harff
Institute of Marine and Coastal Sciences
University of Szczecin
ul. Mickiewicza 18
70-383 Szczecin
Poland
jan.harff@io-warnemuende.de

Derrick Hasterok
Department of Geosciences
The University of Adelaide
SA 5005 Adelaide
Australia
derrick.hasterok@adelaide.edu.au
dhasterok@gmail.com

William W. Hay
Department of Geological Sciences
University of Colorado at Boulder
2045 Windcliff Dr.
Estes Park, CO 80517
USA
whay@gmx.de

James R. Hein
U.S. Geological Survey
400 Natural Bridges Dr.
Santa Cruz, CA 95060
USA
jhein@usgs.gov

Eric Hellebrand
Department of Geology and Geophysics, SOEST
University of Hawaii
1680 East West Rd
Honolulu, HI 96822
USA
ericwgh@hawaii.edu

Christoph Helo
Institute of Geosciences
Johannes Gutenberg-University Mainz
J.-J.-Becher-Weg 21
55128 Mainz
Germany
helo@uni-mainz.de

Rüdiger Henrich
Department of Geosciences
University of Bremen
Klagenfurter Street
28359 Bremen
Germany
henrich@uni-bremen.de

Jens O. Herrle
Institute of Geosciences
Goethe University Frankfurt
Altenhoeferallee 1
60438 Frankfurt
Germany
jens.herrle@em.uni-frankfurt.de

Ingo Heyde
Federal Institute for Geosciences and Natural Resources
Stilleweg 2
30655 Hannover
Germany
ingo.heyde@bgr.de

John Hillier
Department of Geography
Loughborough University
LE11 3TU Leicestershire
UK
j.hillier@lboro.ac.uk

Kaj Hoernle
GEOMAR Helmholtz Centre for Ocean Research
Wischhofstr. 1-3
24148 Kiel
Germany
khoernle@geomar.de

Gösta Hoffmann
Department of Geosciences and Geography
RWTH Aachen University
Lochnerstr. 4-20
52056 Aachen
Germany
goesta.hoffmann@gutech.edu.om
g.hoffmann@nug.rwth-aachen.de

Christian Hübscher
Institute of Geophysics, Center for Earth System Research and Sustainability
University of Hamburg
Bundesstraße 55
20146 Hamburg
Germany
Christian.huebscher@zmaw.de

Benjamin Huet
Department for Geodynamics and Sedimentology
University of Vienna
Althanstrasse 14
1090 Wien
Austria
benjamin.huet@univie.ac.at

Heiko Hüneke
Institute of Geography and Geology
Ernst Moritz Arndt University
Friedrich-Ludwig-Jahn-Str. 17A
17487 Greifswald
Germany
hueneke@uni-greifswald.de

María Paula Iglesia Llanos
IGEBA, National Scientific and Research Council
University of Buenos Aires
Pab. 2, Ciudad Universitaria
C1428EHA Buenos Aires
Argentina
mpiglesia@gl.fcen.uba.ar

Benoît Ildefonse
Geosciences Montpellier
CNRS and University of Montpellier
CC 60 34095, Cedex 5 Montpellier
France
ildefonse@um2.fr

Morten Iversen
Alfred Wegener Institute, Helmholtz Centre for Polar and Marine Research
Am Handelshafen 12
27570 Bremerhaven
Germany
miversen@marum.de

Martin Jakobsson
Department of Geological Sciences
Stockholm University
Svante Arrhenius väg 8
SE-106 91 Stockholm
Sweden
martin.jakobsson@geo.su.se

John W. Jamieson
GEOMAR Helmholtz Centre for Ocean Research
Wischhofstr. 1-3
24148 Kiel
Germany
jjamieson@geomar.de

Niels Jöns
Department of Geology, Mineralogy and Geophysics
Ruhr-University Bochum
Universitätsstr. 150
44801 Bochum
Germany
Niels.Joens@ruhr-uni-bochum.de

Shannon Kay
Coastal Projects
Manson Construction Co.
4309 Pablo Oaks Court, Suite 1
Jacksonville, FL 32202
USA
s.kay.98096@ospreys.unf.edu

Alan Kemp
Ocean and Earth Science, National Oceanography Centre Southampton
University of Southampton Waterfront Campus
European Way
SO14 3ZH Southampton
UK
aesk@noc.soton.ac.uk

Christopher George St. Clement Kendall
Earth and Ocean Sciences
University of South Carolina
2800 Gervais Street
Columbia, SC 29204
USA
kendall@geol.sc.edu

Andrew C. Kerr
School of Earth and Ocean Sciences
Cardiff University
Main Building, Park Place
Cardiff, Wales CF10 3AT
UK
kerra@cardiff.ac.uk

Erik Kjellström
Swedish Meteorological and Hydrological Institute
601 76 Norrköping
Sweden
erik.kjellstrom@smhi.se

Paul Knutz
Geological Survey of Denmark and Greenland (GEUS)
Øster Voldgade 10
1350 Copenhagen
Denmark
pkn@geus.dk

Jürgen Koepke
Institute of Mineralogy
Leibniz University Hannover
Callinstr. 3
30167 Hannover
Germany
koepke@mineralogie.uni-hannover.de

Achim Kopf
MARUM-Center for Marine Environmental Sciences
University of Bremen
P.O. Box 330 440
28334 Bremen
Germany
akopf@marum.de

Janina Körper
Institute of Meteorology
Free University of Berlin
Carl-Heinrich-Becker-Weg 6-10
12165 Berlin
Germany
janina.koerper@met.fu-berlin.de

Andrea Koschinsky
Earth and Environmental Sciences
School of Engineering and Science
Jacobs University
Campus Ring 1
28759 Bremen
Germany
a.koschinsky@jacobs-university.de

Aarno Kotilainen
Geological Survey of Finland
Betonimiehenkuja 4
P.O. Box 96
02151 Espoo
Finland
aarno.kotilainen@gtk.fi

Halina Kowalewska-Kalkowska
Institute of Marine and Coastal Sciences
University of Szczecin
ul. Mickiewicza 18
70-383 Szczecin
Poland
halkalk@univ.szczecin.pl

Michal Kucera
MARUM-Center for Marine Environmental Sciences
University of Bremen
P.O. Box 330 440
28334 Bremen
Germany
michal.kucera@marum.de

Hermann Kudrass
MARUM-Center for Marine Environmental Sciences
University of Bremen
P.O. Box 330 440
28334 Bremen
Germany
kudrass@gmx.de

Antoon Kuijpers
Geological Survey of Denmark and Greenland (GEUS)
Øster Voldgade 10
1350 Copenhagen
Denmark
aku@geus.dk

Nina Kukowski
Institute of Geosciences
Friedrich-Schiller-University
Burgweg 11
07749 Jena
Germany
nina.kukowski@uni-jena.de

Mark Kulp
Department of Earth and Environmental Sciences
University of New Orleans
1065 Geology and Psychology Building
2000 Lakeshore Drive
New Orleans, LA 70148
USA
coastalsrg@gmail.com

Vladislav Kuznetsov
Institute of Earth Sciences
Saint Petersburg State University
Sredniy prospect, 41
199178 St. Petersburg
Russia
v.kuznetsov@spbu.ru

Jan Sverre Laberg
Department of Geology
University of Tromsø
Realfagbygget, Dramsveien 201
9037 Tromsø
Norway
jan.laberg@uit.no

Serge Lallemand
Geosciences Montpellier
University of Montpellier
CC. 60, place E. Bataillon
34095 Montpellier
France
serge.lallemand@gm.univ-montp2.fr
lallem@gm.univ-montp2.fr

Kurt Lambeck
Research School of Earth Sciences
Australian National University
Building 142 Mills Road
Acton, ACT 2601
Australia
kurt.lambeck@anu.edu.au

Alexander P. Lisitzin
P.P. Shirshov Institute of Oceanology
Russian Academy of Sciences (IO RAS)
Nakhimovskii prospekt, 36
Moscow 117851
Russia
lisitzin@ocean.ru

Carolina Lithgow-Bertelloni
Department of Earth Sciences
University College London
Gower Street
WC1E 6BT London
UK
c.lithgow-bertelloni@ucl.ac.uk

Paul Loubere
Department of Anthropology
Northern Illinois University
DeKalb, IL 60115
USA
ploubere@niu.edu
paul@geol.niu.edu

Mitchell Lyle
College of Earth, Ocean, and Atmospheric Sciences
Oregon State University
104 CEOAS Administration Building
Corvallis, OR 97331-5503
USA
mlyle@coas.oregonstate.edu

Paul Mann
Department of Earth and Atmospheric Sciences
University of Houston
312 Science and Research 1
Houston, TX 77204-5007
USA
pmann@uh.edu

Roman Marks
Institute of Marine and Coastal Sciences
University of Szczecin
ul. Mickiewicza 18
70-383 Szczecin
Poland
marks@univ.szczecin.pl

Christian März
School of Civil Engineering and Geosciences (CEGS)
Newcastle University
Drummond Building, Room 4.02
NE1 7RU Newcastle upon Tyne
UK
christian.maerz@ncl.ac.uk

Mark Maslin
Department of Geography
University College London
Pearson Building, Gower Street
WC1E 6BT London
UK
m.maslin@geog.ucl.ac.uk

Jens Matthiessen
Alfred Wegener Institute, Helmholtz Centre for Polar and Marine Research
Am Alten Hafen 26
27568 Bremerhaven
Germany
Jens.Matthiessen@awi.de

Kevin McCartney
Department of Environmental Studies and Sustainability
University of Maine at Presque Isle
302 Folsom Hall
Presque Isle, ME 04769
USA
kevin.mccartney@umpi.edu

Martin Meschede
Institute of Geography and Geology
Ernst-Moritz-Arndt University
Friedrich-Ludwig-Jahn-Str. 17A
17487 Greifswald
Germany
meschede@uni-greifswald.de

Jürgen Mienert
Centre for Arctic Gas Hydrate, Environment and Climate (CAGE)
UiT The Arctic University of Norway
Dramsveien 201
9037 Tromsø
Norway
jurgen.mienert@uit.no

Arto Miettinen
Norwegian Polar Institute
Fram Centre
Postbox 6606 Langnes
NO-9296 Tromsø
Norway
arto.miettinen@npolar.no

Matthias Moros
Leibniz Institute for Baltic Sea Research Warnemünde
Seestrasse 15
18119 Rostock
Germany
matthias.moros@io-warnemuende.de

Thierry Mulder
University of Bordeaux
UMR CNRS 5805 EPOC, Avenue des facultés
33185, Cedex Talence
France
t.mulder@epoc.u-bordeaux1.fr

R. Dietmar Müller
School of Geosciences
The University of Sydney
Madsen Building F09, R. 406
2006 Sydney, NSW
Australia
dietmar.muller@sydney.edu.au

Sara Muñoz-Vallés
Department of Plant Biology and Ecology
University of Sevilla
Avda. de la Reina Mercedes, S/N
41012 Seville
Spain
saramval@us.es

Raquel Negrete
Ensenada Center for Scientific Research and Higher Education
Department of Geology
CICESE
Carretera Tijuana-Ensenada 3918
Zona Playitas
22860 Baja California
Mexico
rnegrete@cicese.mx

JoAnne Nelson
British Columbia Geological Survey
P.O. Box 9333
Victoria, BC V8W 9N3
Canada
JoAnne.Nelson@gov.bc.ca

Adolphe Nicolas
Geosciences Montpellier
University of Montpellier 2
CC. 60, place E. Bataillon
34095, cedex 5 Montpellier
France
adolphe.nicolas@um2.fr

Tove Nielsen
Geological Survey of Denmark and Greenland (GEUS)
Øster Voldgade 10
1350 Copenhagen
Denmark
tni@geus.dk
toveniel@gmail.com

Dirk Nürnberg
GEOMAR Helmholtz Centre for Ocean Research
Wischhofstr. 1-3
D-24148 Kiel
Germany
dnuernberg@geomar.de

Andrzej Osadczuk
Institute of Marine and Coastal Sciences
University of Szczecin
ul. Mickiewicza 18
70-383 Szczecin
Poland
andros@univ.szczecin.pl

Thomas Pedersen
School of Earth and Ocean Sciences
University of Victoria
P.O. Box 1700
Victoria, BC V8W 2Y2
Canada
tfp@uvic.ca
picsdir@uvic.ca

Katja T. C. A. Peijnenburg
Institute for Biodiversity and Ecosystem Dynamics
University of Amsterdam
Mauritskade 61
P.O. Box 94216
1090 Amsterdam
The Netherlands
and
Naturalis Biodiversity Center
Darwinweg 2
2333 CR Leiden
The Netherlands
a.c.pierrot-bults@uva.nl

Michael R. Perfit
Department of Geological Sciences
University of Florida
241 Williamson Hall
P.O. Box 112120
Gainesville, FL 32611-2120
USA
perfit@geology.ufl.edu
mperfit@ufl.edu

Sven Petersen
GEOMAR Helmholtz Centre for Ocean Research
Wischhofstr. 1-3
24148 Kiel
Germany
spetersen@geomar.de

Annelies C. Pierrot-Bults
Institute for Biodiversity and Ecosystem Dynamics
University of Amsterdam
Mauritskade 61
P.O. Box 94216
1090 Amsterdam
The Netherlands
and
Naturalis Biodiversity Center
Darwinweg 2
2333 CR Leiden
The Netherlands
a.c.pierrot-bults@uva.nl

Irina Polovodova Asteman
Bjerknes Centre for Climate Research
Allégaten 55
5007 Bergen
Norway
irina.asteman@uni.no

Klaus Reicherter
Department of Geosciences and Geography
RWTH Aachen University
Lochnerstr. 4-20
52056 Aachen
Germany
k.reicherter@nug.rwth-aachen.de

John J.G. Reijmer
Department of Sedimentology and Marine Geology
VU University Amsterdam
De Boelelaan 1085
1081 Amsterdam
The Netherlands
j.j.g.reijmer@vu.nl
john.reijmer@falw.vu.nl

Don Resio
Taylor Engineering Research Institute
University of North Florida
1 UNF Drive
Jacksonville, FL 32224-7699
USA
don.resio@unf.edu

Ulf Riebesell
GEOMAR Helmholtz Centre for Ocean Research
Düsternbrooker Weg 20
24105 Kiel
Germany
uriebesell@geomar.de

Uwe Ring
Department of Geological Sciences
Stockholm University
Svante Arrhenius väg 8
106 91 Stockholm
Sweden
uwe.ring@geo.su.se

Evgeny Romankevich
P.P. Shirshov Institute of Oceanology
Russian Academy of Sciences
Nakhimovsky prospekt, 36
117997 Moscow
Russia
and
National Research Tomsk Polytechnic University
634050 Tomsk
Russia
Romankevich@mail.ru

Ken H. Rubin
Department of Geology and Geophysics, SOEST
University of Hawaii
1680 East West Rd
Honolulu, HI 96822
USA
krubin@hawaii.edu

Hans Werner Schenke
Alfred-Wegener-Institute, Helmholtz-Centre for Polar and Marine Research
Am Handelshafen 12
27570 Bremerhaven
Germany
hans-werner.schenke@awi.de

Gerald Schernewski
Leibniz Institute for Baltic Sea Research Warnemünde
Seestrasse 15
18119 Rostock
Germany
gerald.schernewski@io-warnemuende.de

Axel Schippers
Federal Institute for Geosciences and Natural Resources
Stilleweg 2
30655 Hannover
Germany
axel.schippers@bgr.de

Anders Schomacker
Department of Geology and Mineral Resources Engineering
Norwegian University of Science and Technology
Sem Sælands Veg 1
7491 Trondheim
Norway
anders.schomacker@ntnu.no

Michael Schreck
Arctic Research Centre
Korea Polar Research Institute
26 Songdomirae-ro
Incheon 406-840
Korea
Michael.Schreck@kopri.re.kr

John G. Sclater
Scripps Institution of Oceanography
University of California San Diego
La Jolla, CA 92093
USA
jsclater@ucsd.edu

Eugen Seibold
(deceased)

A. M. Celâl Şengör
Faculty of Mines, Department of Geology and Eurasia Institute of Earth Sciences
Istanbul Technical University
34469 Ayazaga, Istanbul
Turkey
sengor@itu.edu.tr

Maria Seton
School of Geosciences
The University of Sydney
Madsen Building F09, R. 406
2006 Sydney, NSW
Australia
maria.seton@sydney.edu.au

Vladimir P. Shevchenko
P.P. Shirshov Institute of Oceanology
Russian Academy of Sciences (IO RAS)
Nakhimowskii prospekt, 36
117218 Moscow
Russia
vshevch@ocean.ru

Graham Shimmield
Bigelow Laboratory for Ocean Sciences
60 Bigelow Drive
East Boothbay, ME 04544
USA
gshimmield@bigelow.org

Eli Silver
Earth and Planetary Sciences Department
University of California
1156 High Street
Santa Cruz, CA 95064
USA
esilver@ucsc.edu

Satish C. Singh
Laboratory of Marine Geosciences
Institute of Earth Physics of Paris
Bureau 364 – 1, rue Jussieu
75238, cedex 05 Paris
France
singh@ipgp.fr

Tarmo Soomere
Institute of Cybernetics at Tallinn University of Technology
Akadeemia tee 21
12618 Tallinn
Estonia
and
Estonian Academy of Sciences
Kohtu 6
10130 Tallinn
Estonia
tarmo.soomere@cs.ioc.ee

Samuel Adam Soule
Woods Hole Oceanographic Institution
266 Woods Hole Rd.
MS# 24
Woods Hole, MA 02543-1050
USA
ssoule@whoi.edu

Rüdiger Stein
Alfred Wegener Institute, Helmholtz Centre for Polar and Marine Research (AWI)
Am Alten Hafen 26
27568 Bremerhaven
Germany
ruediger.stein@awi.de

Robert J. Stern
Department of Geosciences
University of Texas at Dallas
800 West Campbell Road, ROC 21
Richardson, TX 75080-3021
USA
rjstern@utdallas.edu

Andreas Stracke
Westphalian Wilhelms-University
Institute of Mineralogy
Corrensstrasse 24
48149 Münster
Germany
stracke.andreas@uni-muenster.de

Colin Summerhayes
Scott Polar Research Institute (SPRI)
Cambridge University
Lensfield Road
CB2 1ER Cambridge
UK
cps32@cam.ac.uk

Jürgen Sündermann
Institute of Oceanography (IFM)
University of Hamburg
Bundesstrasse 53
20146 Hamburg
Germany
juergen.suendermann@ifm.uni-hamburg.de

Kiyoshi Suyehiro
Integrated Ocean Drilling Program
Management International
2-1-6, Etchujima, Koto-ku OLCR 3F
135-8533 Tokyo
Japan
ksuyehiro@iodp.org

Jörn Thiede
Institute of Earth Sciences
Saint Petersburg State University
V. O., Sredniy prospect, 41
St. Petersburg 199 178
Russia
and
AdW Mainz c/o GEOMAR
Wischhofstr. 1-3
24148 Kiel
Germany
jthiede@geomar.de

Hans R. Thierstein
Department of Earth Sciences
Swiss Technical University
Sonneggstrasse 5
8092 Zürich
Switzerland
thierstein@erdw.ethz.ch

Margaret K. Tivey
Marine Chemistry and Geochemistry, Woods Hole
Oceanographic Institution
266 Woods Hole Rd., MS# 08
Woods Hole, MA 02543-1050
USA
mktivey@whoi.edu

Marta E. Torres
College of Earth, Ocean, and Atmospheric Sciences
Oregon State University
104 CEOAS Administration Building
Corvallis, OR 97331-5503
USA
mtorres@coas.oregonstate.edu

Paul Treguer
European Institute for Marine Studies (IUEM)
University of Bretagne Occidentale (UBO)
Technopole Brest-Iroise, Place Copernic
29280 Plouzané
France
paul.treguer@univ-brest.fr

Verena Tunnicliffe
Department of Biology and School of Earth and Ocean
Sciences
University of Victoria
P.O. Box 3020
Victoria, BC V8W 2Y2
Canada
verenat@uvic.ca

Hayato Ueda
Department of Geology
Niigata University
8050 Ikarashi-2-nocho, Nishi-ku
950-2181 Niigata
Japan
ueta@mvf.biglobe.ne.jp

Alexander Vetrov
P.P.Shirshov Institute of Oceanology
Russian Academy of Sciences
Nakhimovsky prospekt, 36
117997 Moscow
Russia
vetrov@ocean.ru

David Voelker
MARUM-Center for Marine Environmental Sciences
University of Bremen
P.O. Box 330 440
28334 Bremen
Germany
dvoelker@marum.de

Roland von Huene
GEOMAR Helmholtz Centre for Ocean Research
Wischhofstr. 1-3
24148 Kiel
Germany
rhuene@mindspring.com

Tore O. Vorren
(deceased)

Thomas Wagner
School of Civil Engineering and Geosciences
Newcastle University
Drummond Building
NE1 7RU Newcastle upon Tyne
UK
thomas.wagner@ncl.ac.uk

H. Jesse Walker
(deceased)

Louise Wallendorf
Hydromechanics Laboratory
United States Naval Academy
590 Holloway Rd.
Annapolis, MD 21402-5042
USA
lou@usna.edu

Gerold Wefer
MARUM-Center for Marine Environmental Sciences
University of Bremen
P.O. Box 330 440
28334 Bremen
Germany
gwefer@marum.de

Achim Wehrmann
Marine Research Department
Senckenberg am Meer
Südstrand 40
26382 Wilhelmshaven
Germany
achim.wehrmann@senckenberg.de

William M. White
Department of Earth and Atmospheric Sciences
Cornell University
4112 Snee Hall
Ithaca, NY 14853
USA
wmw4@cornell.edu

Joanne M. Whittaker
Institute for Marine and Antarctic Studies
University of Tasmania
Private Bag 129 Hobart, TAS 7001
Australia
jo.whittaker@utas.edu.au

Jakub Witkowski
Institute of Marine and Coastal Sciences
University of Szczecin
ul. Mickiewicza 18
70-383 Szczecin
Poland
jakub.witkowski@univ.szczecin.pl

Isobel Yeo
GEOMAR Helmholtz Centre for Ocean Research
Wischhofstr. 1-3
24148 Kiel
Germany
iyeo@geomar.de

Miwa Yokokawa
Laboratory of Geoenvironment
Faculty of Information Science and Technology
Osaka Institute of Technology
Osaka
Japan
miwa@is.oit.ac.jp

Jing Zhang
State Key Laboratory of Estuarine and Coastal Research
East China Normal University
3663 Zhongshan Road North
200062 Shanghai
China
jzhang@sklec.ecnu.edu.cn

Wenyan Zhang
MARUM-Center for Marine Environmental Sciences
University of Bremen
P.O. Box 330 440
28334 Bremen
Germany
wzhang@marum.de

Di Zhou
South China Sea Institute of Oceanology
Chinese Academy of Sciences
164 West Xingang Road
510301 Guangzhou
China
zhoudiscs@scsio.ac.cn

Preface

The globally growing demand for energy and mineral resources, the need for reliable projections of future climate processes, and the necessity for coast protection to mitigate threats and hazards of disasters require a comprehensive understanding of the genesis and structure of the marine geosphere and processes involving it. The "classical" research fields in marine geology are being currently supplemented by the development of more general concepts which strive to integrate marine geophysics, hydrography, climatology, marine biology, and ecology. The term "marine geosciences" has been broadly accepted to serve as an "umbrella" for this new complex field of research and practical solutions in the marine realm.

We have reflected this development by putting together the *Encyclopedia of Marine Geosciences* which appears as a volume of the Encyclopedia in Earth Sciences Series. When collecting entries for the *Encyclopedia of Marine Geosciences*, we intended to provide the current state of knowledge in marine geosciences and to cover their theoretical, applied, and technical aspects. However, as we were working on the Encyclopedia, it became clear to us that one printed volume cannot accommodate all the disciplinary basics of marine geosciences. Therefore, the reader seeking explanation of a specific scientific term is advised to consult not only the list of entries, but also to take a look on the back-of-the-book index. The index can be used as a search engine for text paragraphs in this volume dealing with the words of interest. Because of the complex nature of marine geosciences, we do realize that some overlap with the "classical" research geosciences fields, already covered by other volumes of the *Encyclopedia in Earth Sciences Series*, must have occurred in the present volume. We did our best to avoid repetitions in this volume and we advise the reader to consult, whenever necessary, other *Encyclopedias in the Earth Science Series* volumes such as *Encyclopedia of Coastal Science, Encyclopedia of Geochemistry, Encyclopedia of Paleoclimatology and Ancient Environments*, or *Encyclopedia of Sediments and Sedimentary Rocks*.

With this volume, we hope to reach out to a broad audience of users concerned with marine sciences and technology, from students and scholars in academia to engineers in the industry to decision makers in administration and politics.

The editorial team of the *Encyclopedia of Marine Geosciences* reflects the diversity of disciplines covered: Martin Meschede took over responsibility for manuscripts related to plate tectonics, Sven Petersen dealt with those focusing on magmatism, Jörn Thiede was responsible for entries on deep ocean processes, and Jan Harff edited the entries related to ocean margins and associated processes.

Jan Harff
Martin Meschede
Sven Petersen
Jörn Thiede

Acknowledgments

The work on an encyclopedia covering a wide disciplinary field such as marine geosciences requires a well-established teamwork. First of all, we would like to express our thanks to the members of the Editorial Board who have helped to set up the list of entries that means to define the scientific area to be covered by the encyclopedia. But members of the board also assisted in finding the best authors for the topics defined and also to foster the review process. Thanks also go to our authors for their engagement in the project but also for their patience because of the longer time needed finally for the completion than we expected at the beginning of the project. An excellent cooperation took place with the reviewers and we have to convey our gratitude explicitly, among others, to Roger Anderson, Henning Bauch, Karl-Heinz Baumann, Thorsten Becker, Wolfgang Berger, Jörg Bialas, Svante Björck, Bodo von Bodungen, Peter Bormann (†), Johannes Brumme, Hans Burchard, Suzanne Carbotte, Rebecca Carey, Ronald Clowes, Eduardo Contreras-Reyes, David Cronan, Colin Devey, Tim Dooley, Joanna Dudzinska-Nowak, Christian Dullo, Malte Elbrächter, Bob Embley, Wolfgang Fennel, Charles W. Finkl, Burghard Flemming, Nicolas Flemming, Armin Freundt, Peter Fröhle, Gerald Gabriel, Dennis Geist, Jörg Geldmacher, Bruce Gemmell, Chris German, Daniel Gibson, Pieter M. Grootes, Marc Gutscher, Matthias Haekel, Peter Halbach, William W. Hay, James D. Hays, George Ross Heath, Jim Hein, Christoph Hemleben, Rüdiger Henrich, Sebastian Hölz, Kaj Hoernle, Albrecht Hofmann, Martin Hovland, Roland von Huene, Kim Juniper, Heidemarie Kassens, Christian Koeberl, Jürgen Koepke, Heidrun Kopp, Martin Van Kranendonk, Hermann-Rudolf Kudraß, Vladyslav Kuznetsov, Carina Lange, Gert de Lange, John Lupton, Gianreto Manatschal, Roman Marks, Jean Mascle, Philip A. Meyers, Paul Morgan, Johannes Mrazek, Yujiro Ogawa, Andrzej Osadczuk, Mike Perfit, Jörg Pfänder, Teresa Radzie-Jewska, Mark Reagan, Eoghan Reeves, Klaus Ricklefs, Karol Rotnicki, Ralph Schneider, Michael Schnabel, B. Charlotte Schreiber, Detlef Schulz-Bull, Wolfgang Schlager, Klaus Schwarzer, Andrew Short, Roger Searle, Monika Sobiesiak, Robert Spielhagen, Tarmo Soomere, Karl Stattegger, Rüdiger Stein, Andreas Stracke, Kurt Stüwe, Colin Summerhayes, Adam D. Switzer, Fred Taylor, Frederik Tilmann, Barbara Teichert, Tony Watts, Gordon Webster, Joanne Whittaker, Philip Woodworth, Jiaxue Wu, Isobel Yeo, and Anastasia Zhuravleva.

Last, but not least, we have to mention that the work would not have been possible without supervision and active cooperation of Petra van Steenbergen, Sylvia Blago, and Johanna Klute from Springer's publishing house.

Jan Harff
Martin Meschede
Sven Petersen
Jörn Thiede

A

ABYSSAL PLAINS

David Voelker
MARUM-Center for Marine Environmental Sciences,
University of Bremen, Bremen, Germany

Abyssal plains are flat areas of the ocean floor in a water depth between 3,500 and 5,000 with a gradient well below 0.1°. They occupy around 28 % of the global seafloor. The thickness of the sediment cover seldom exceeds 1,000 m, and the sediments consist of fine-grained erosional detritus and biogenic particles. The first global map of seafloor physiography, published by Heezen and Tharp in 1977, illustrated that the bottom of the ocean is bordered by continental margins of variable width and hosts large submarine mountain chains, the mid-ocean ridges. It also illustrated that between the continental margins and mid-ocean ridges, there exist vast flat and almost featureless regions, the so-called abyssal plains (Figure 1). The global seafloor map has been refined in recent years (e.g., by Smith and Sandwell, 1995; Becker et al., 2009), in a way that has allowed to perform quantitative analyses of the global distribution of specific seafloor features (Harris et al., 2014).

Bathymetry

Earth is a sphere with an equatorial radius of 6,378 km and a surface area of $\sim 5.1 \times 10^{14}$ m^2. Roughly 71 % or $\sim 3.58 \times 10^{14}$ m^2 of that surface is covered by oceans. The largest part of the ocean floor is at a depth between 3,500 and 5,000 m (with a mean ocean depth of 3,795 m), in the depth range of abyssal plains. Exceptions from this rule are (1) continental margins and epeiric seas, (2) mid-ocean ridges, (3) seamounts and basaltic plateaus that are shallower, as well as (4) submarine trenches at subduction zones that are deeper. These areas are exceptional because plate collision forces govern the shape of the seafloor (at subduction zone trenches), because they form a transition between continental and oceanic crust (continental margins) or because of atypical structure of the ocean crust (basaltic plateaus and seamounts). Mid-ocean ridges stand out as the crust that is formed at these plate boundaries is young and hot. Cooling of the plate leads to crustal subsidence to abyssal depth within ~ 12 Ma.

Apart from those areas, the seafloor is relatively monotonous in its depth and also very flat. Those regions that typically stretch from the foot of the continental slopes to the mid-ocean ridges have been classified as a specific physiographic entity and termed abyssal plains. The word abyss comes from the Greek word abyssos "bottomless (pool)" indicating the enormous lightless depths, whereas plain refers to the morphological notion of vast monotonously flat areas. Strictly speaking, the term is a kind of contradiction because a plain cannot be bottomless, but it is meant in the same way as its correspondent "deep-sea plain."

The depth distribution of the world oceans is shown in Figure 2 in the form of a hypsometric curve. The curve shows a bimodal distribution with one peak between 0 and 500 m and another broader peak between 3,500 and 6,000 m. This distribution reflects the double nature of the seafloor: around the continents, large areas of the seafloor are actually drowned parts of the continents with a crustal structure that is similar to the crust underlying the continents. The deeper part of the ocean is in contrast underlain by oceanic crust that is formed at mid-ocean ridges in a similar composition on global scale. The lifespan of a portion of oceanic crust from its creation at a spreading center to its destruction at a subduction zone ranges seldom exceeds 120 Ma (Müller et al., 2008).

J. Harff et al. (eds.), *Encyclopedia of Marine Geosciences*, DOI 10.1007/978-94-007-6238-1,
© Springer Science+Business Media Dordrecht 2016

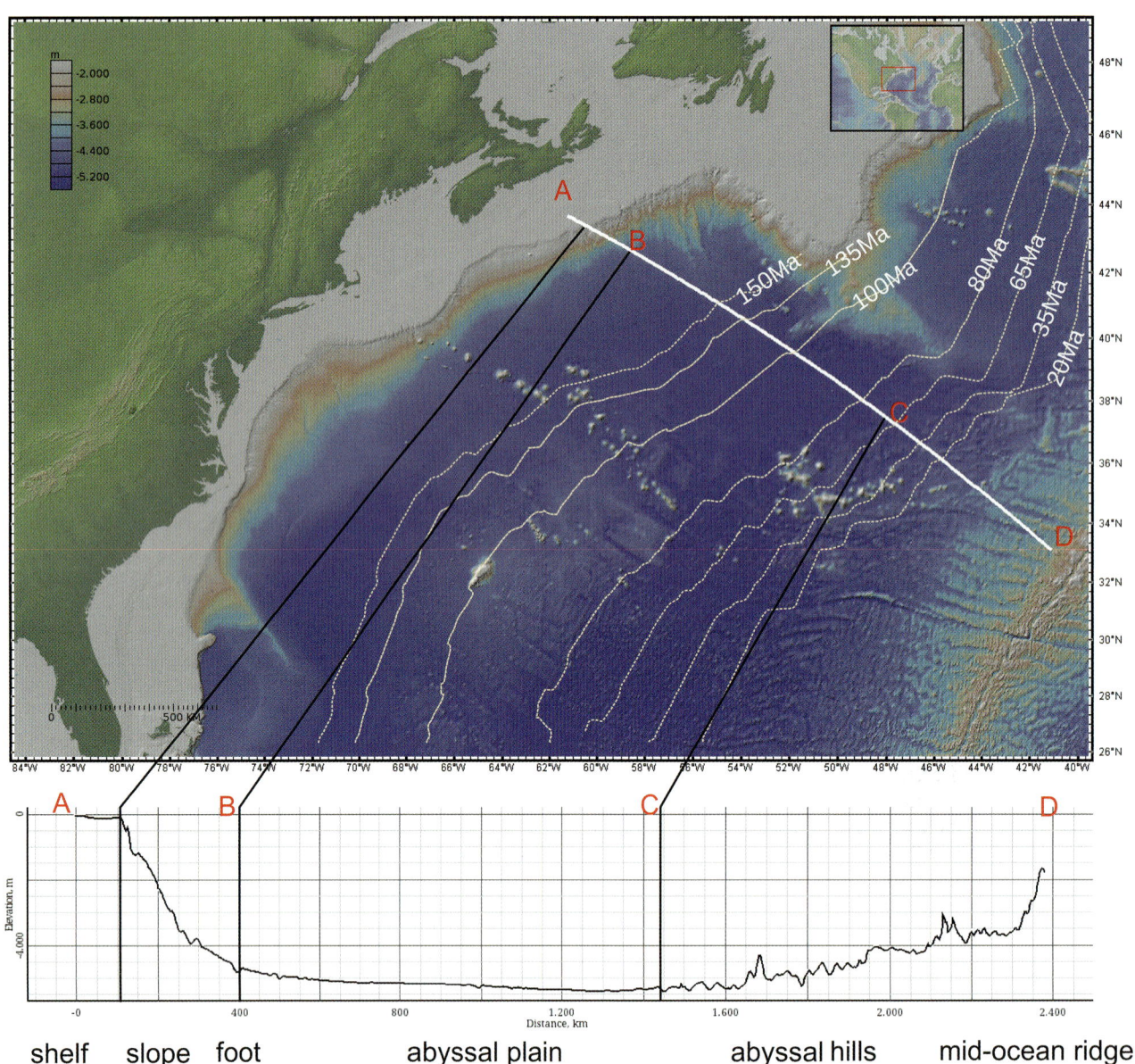

Abyssal Plains, Figure 1 (a) Bathymetric map of the western North Atlantic, showing a broad continental shelf and slope, a prominent abyssal plain (the Hatteras Abyssal Plain), and the Mid-Atlantic Ridge. The map was created with the GeoMapApp (www.geomapapp.org); the underlying bathymetric data were published as Global Multi-Resolution Topography (GMRT) by Ryan et al. (2009). (b) shows a depth profile from the continental shelf offshore Newfoundland to the Mid-Atlantic Ridge (A–D). In this profile, the continental platform and the continental slope are clearly depicted. Towards the Mid-Atlantic Ridge, the roughness of the seafloor increases with decreasing thickness of the sediment cover. Stippled lines indicate seafloor bedrock age according to Müller et al. (2008).

Over that time span, the ocean floor cools, contracts, and sinks, rapid first and slower later (see "Subsidence of Oceanic Crust"), and this subsidence controls the depth of the largest part of the seafloor.

Abyssal plains grade into the foot regions of continental margins towards the land masses and into areas of increasing topographical roughness towards mid-ocean ridges (Figure 1a, b). In order to delimit this physiographic domain, the International Hydrographic Organization (IHO) agreed upon a definition that includes water depth and the variation in relief over a certain radius (e.g., the variation in relief has to be <300 m over a radius of 25 cells of 30 arc sec). Following this definition, Harris et al. (2014) calculated that globally, abyssal plains make up 27.9 % of the seafloor or 19.8 % of the entire surface of the Earth.

The flatness of abyssal plains is exemplified in Figure 1b. It shows a bathymetric profile of the western North Atlantic Ocean from offshore Newfoundland to

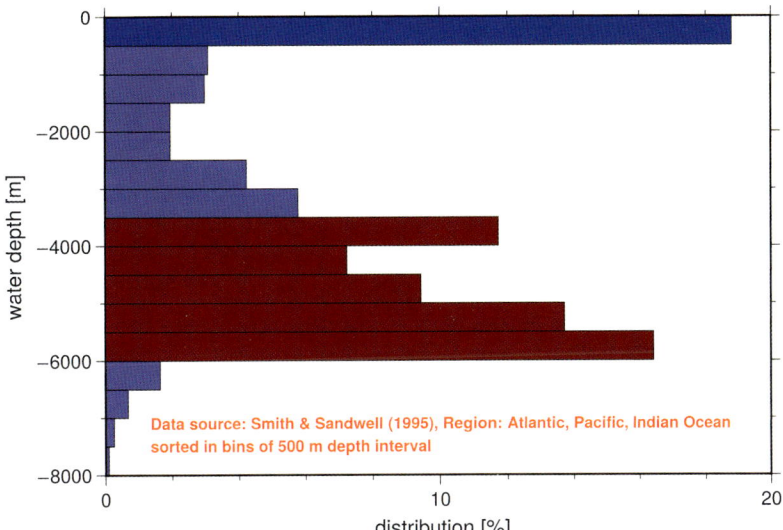

Abyssal Plains, Figure 2 Hypsometric curve of the seafloor of the Atlantic, Pacific, and Indian Oceans. The curve shows the percentage of areas of the seafloor that lie within a certain depth interval and exhibits a bimodal distribution: First maximum lies within the 0–500 m range (continental shelves, *blue bar*), while the largest portion and second maximum lies at water depths between 3,500 and 6,000 m (oceanic seafloor, *red bars*). The second maximum, indicated by *red bars*, also is the typical depth range of abyssal plains. The map is based on the global bathymetric data set of Smith and Sandwell (1995).

the Mid-Atlantic Ridge. The profile across the abyssal plain between points B and C is 1,100 km long. Over this distance, the difference in elevation is 600 m, corresponding to a mean slope gradient of less than 0.04°, similar to shelf areas. In contrast, the continental margins stand out with slope gradients in the range of 1–10° (in places up to 30°). The flanks of the mid-ocean ridges show gradients of 10–20°; some seamounts stand out as steep edifices with slopes as steep as 40°. The profile across the abyssal plain between points B and C is 1,100 km long. Over this distance, the difference in elevation is 600 m, corresponding to a mean slope gradient of less than 0.04°. Figure 3 summarizes this information in the form of a slope gradient map.

Sedimentation

Abyssal plains receive sediments in the form of erosional detritus of the continents and as shell fragments of planktonic animals and algae that live in the upper water column of the open sea. These components form the main constituents of the typical deep-sea or pelagic sediments in varying relative proportions. The growth of the sediment cover in time (the rate of sedimentation) is generally low over abyssal plains (on the order of some cm growth in thickness/ka), as compared to continental margins where sedimentation rates are generally higher by a factor of 100–1,000. Given such low sedimentation rates, on abyssal plain accumulates a sediment cover of some hundred meters. The mean thickness of the sediment cover of abyssal plains is 450 m (Whittaker et al., 2013).

Two reasons account for the comparatively low sedimentation rate: first, the abyssal plains are largely cut off from the main source of sedimentary particles, the erosion of the continents. A large part of sediments that are transported to the sea in rivers are deposited on the shelf regions of the continents during global high stands of the sea-level (when the shelf regions around continents are wide). River-transported sediments surpass this "sediment trap" easily only when rivers directly connect to submarine canyon systems, or when the shelf is narrow, or episodically at sea-level low stands or when turbidites are shaken off. Wind transport of very fine-grained erosional detritus (dust) is an important constituent of abyssal plain sedimentation to the leeward side of arid regions with a constant wind regime (such as offshore the Saharan coast of Western Africa (Morocco, Mauretania)). Erosional detritus of larger grain size (gravel) is brought into the oceans at high latitudes released by melting icebergs (dropstones).

Second, the bioproductivity (density of primary production) of the open seas is low in general (with the exception of zones of equatorial upwelling) because of limited nutrient availability. Low bioproductivity results in little and seasonally variable fallout of shell fragments and organic material from the photic zone. In addition, only a fraction of that export of material from the photic zone ever reaches the seafloor, as the major part gets dissolved or recycled in the water column. This is specifically true for calcareous fragments that become chemically unstable in the deep sea. The water depth underneath of which no calcareous material is preserved in marine sediments is called the carbonate compensation depth (CCD).

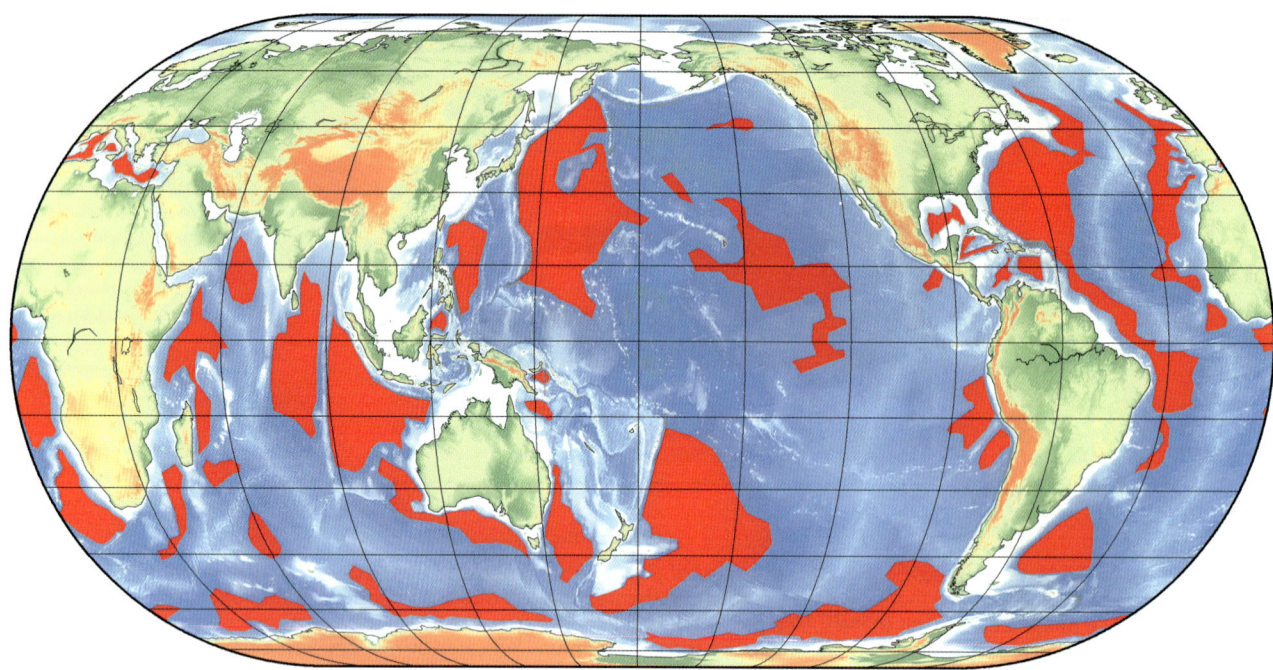

Abyssal Plains, Figure 3 Global distribution of abyssal plains and deep-sea basins.

In short, deep-sea sedimentation is sparse and relatively evenly distributed (no local sources), depending mostly on the thin fallout of particles from the photic zone, the so-called pelagic rain of particles, wind transport of dust, and distal turbidites from higher areas. Over time, this kind of sedimentation blankets the oceanic crust and levels its topography resulting in the very smooth and even abyssal plains. The thickness of sediments that rest on the oceanic basement can be mapped in the form of contour lines (isopachs). Global isopach maps (Whittaker et al., 2013) show that the major trend in sediment thickness is from absent close to the mid-ocean ridges where the oceanic crust is young and the distance to the continental sources of sediments is large to some hundred meters close to the foot of continental slopes, where the oceanic crust is oldest and continents are close. This trend is locally overprinted by other effects such as high bioproductivity belts around the equator and bottom current transport of sediments but general to all oceans (Figure 4).

Life in the abyssal plains

New observation systems employed in the last decades give evidence for numerous life forms in the deep sea and on abyssal plains. One of the fundamental conditions for higher life forms, the constant supply of oxygen is provided for by deep ocean currents that are part of the so-called global conveyor belt of ocean currents. Still, life in the abyssal plains meets very specific challenges, because of which sometimes they are termed "ocean deserts," indicating the low diversity and low density of life forms. First, autotrophic (plant) life is impossible in the lightless depth of the oceans. Therefore practically every life form here depends on the "import" of nutrients from the photic (uppermost) zone of the water column in the form of a rain of particles (dead plants and animals, fecal pellets) from above. Second, this rain is very thin over most of the abyssal plains. Only few ocean surface regions (the so-called upwelling zones) have a high primary production and associated food web to produce a constant export of nutrients to the deep sea that can nourish a dense population of animals on the deep seafloor. Third, this rain is variable in time and space. This is exemplified by whale falls. Dead whale bodies that sink to the bottom of the ocean provide food for years for a very specialized group of bacteria and animals. Yet whale falls are so irregular in time and the cadavers are so small in relation to the vast dimensions of abyssal plains that it is a question yet unresolved, how those animals actually find the prey and what they do "in between" the next lucky strike. Fourth, the abyssal plain is covered by relatively soft mud. This is an unfavorable situation for animals that are attached to the ground, preventing them to settle and build colonies. This fact becomes striking when seamounts are considered. Seamounts are solid rock bodies that rise from the abyssal plains. If abyssal plains are termed deserts in terms of biological activity, then seamounts are the oasis.

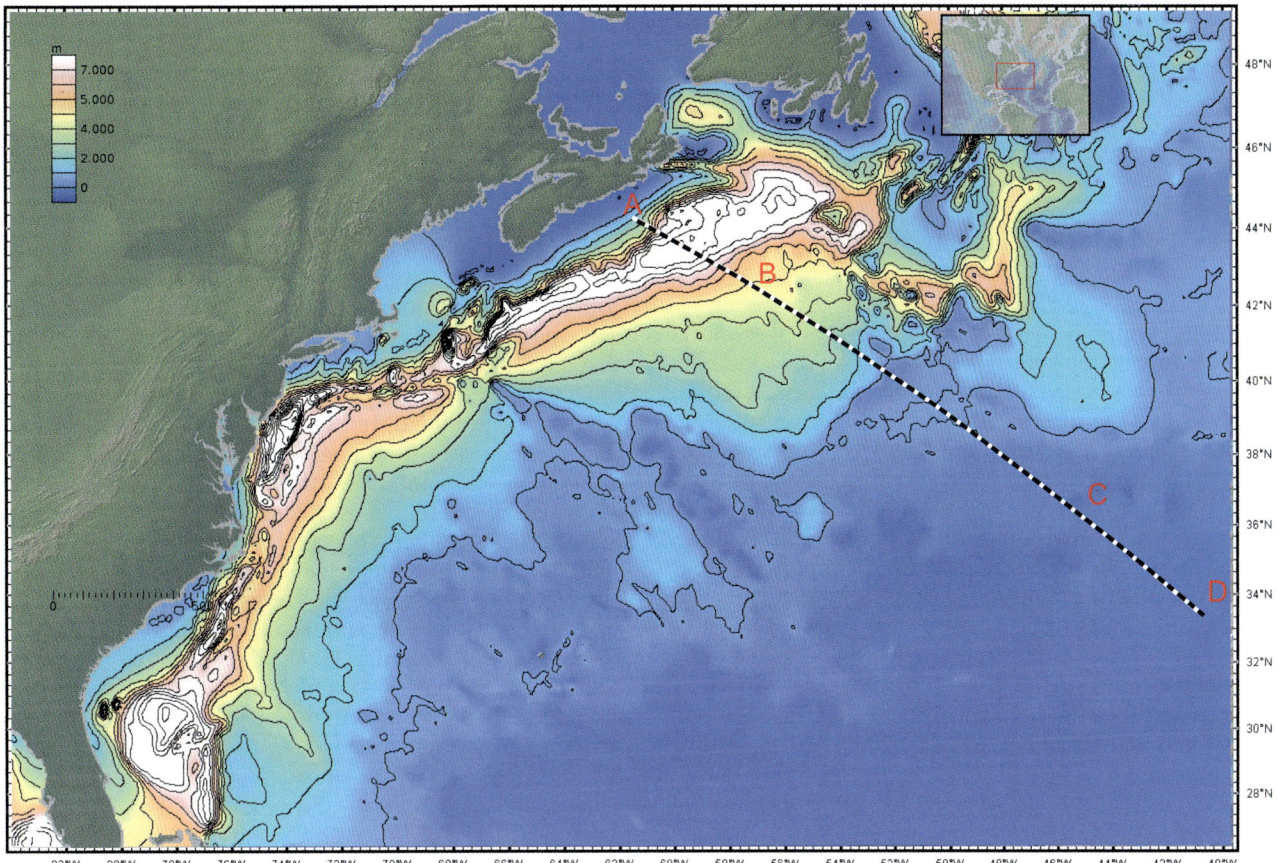

Abyssal Plains, Figure 4 Sediment thickness (isopach) map of the western North Atlantic, based on a global data set of Whittaker et al. (2013). Towards the American continent, the sediment thickness of the abyssal plain exceeds 1,000 m which partly explains the flatness of that region. Towards the Mid-Atlantic Ridge, sediment thickness decreases while the ruggedness of the ocean floor increases. The depth profile of Figure 1 is superimposed.

Bibliography

Becker, J. J., Sandwell, D. T., Smith, W. H. F., Braud, J., Binder, B., Depner, J., Fabre, D., Factor, J., Ingalls, S., Kim, S. H., Ladner, R., Marks, K., Nelson, S., Pharaoh, A., Trimmer, R., Von Rosenberg, J., Wallace, G., and Weatherall, P., 2009. Global bathymetry and elevation data at 30 arc seconds resolution: SRTM30_PLUS. *Marine Geodesy*, **32**, 355–371.

Harris, P. T., Macmillan-Lawler, M., Rupp, J., and Baker, E. K., 2014. Geomorphology of the oceans. *Marine Geology*, **352**, 4–24.

Heezen, B.C., Tharp, M., 1977. World ocean floor panorama. In full color, painted by H. Berann, Mercator Projection, scale 1:23,230,300, 1168 × 1930mm, New York.

IHO, 2008. *Standardization of Undersea Feature Names: Guidelines Proposal form Terminology*, 4th edn. Monaco: International Hydrographic Organisation and Intergovernmental Oceanographic Commission.

Müller, D., Sdrolias, M., Gaina, C., and Roest, W. R., 2008. Age, spreading rates, and spreading asymmetry of the world's ocean crust. *Geochemistry, Geophysics, Geosystems*, **9**(4), doi:10.1029/2007GC001743.

Ryan, W. B. F., Carbotte, S. M., Coplan, J. O., O'Hara, S., Melkonian, A., Arko, R., Weissel, R. A., Ferrini, V., Goodwillie, A., Nitsche, F., Bonczkowski, J., and Zemsky, R., 2009. Global - multi-resolution topography synthesis. *Geochemistry, Geophysics, Geosystems*, **10**, Q03014.

Smith, W. H. F., and Sandwell, D. T., 1995. Bathymetric prediction from dense satellite altimetry and sparse shipboard bathymetry. *Oceanographic Literature Review*, **42**, 409.

Whittaker, J. M., Goncharov, A., Williams, S. E., Müller, R. D., and Leitchenkov, G., 2013. Global sediment thickness data set updated for the Australian-Antarctic Southern Ocean. *Geochemistry, Geophysics, Geosystems*, **14**, 3297–3305.

Cross-references

Calcite Compensation Depth (CCD)
Intraoceanic Subduction Zone
Mid-ocean Ridge Magmatism and Volcanism
Oceanic Plateaus
Seamounts
Subsidence of Oceanic Crust
Turbidites
Upwelling

ACCRETIONARY WEDGES

Martin Meschede
Institute of Geography and Geology, Ernst-Moritz-Arndt University, Greifswald, Germany

Synonyms

Accretionary complex; Accretionary prism

Definition

Accretion defines a process at a convergent plate margin above a subduction zone where material of the subducting lower plate is scraped off and transferred to the overriding upper plate. The offscraped material is accumulated in a wedge-shaped stack of sedimentary layers sometimes containing also offscraped material from the oceanic crust of the subducting plate. It is located directly at the boundary between the two converging plates. This region is called the forearc region of the convergent plate boundary (see entry "Morphology Across Convergent Plate Boundaries" and Figure 3 therein, this volume).

Formation

Accretionary wedges essentially develop as compressional fold-and-thrust belts which are composed primarily of oceanic-plate deposits and, in many cases, continentally derived trench-floor sediment from a nearby continental plate (Figure 1). Sometimes scraped-off parts of the subducted oceanic lithosphere are added to the accretionary complex, which then form ophiolitic rocks in the succession of the accreted material.

Accretionary wedges or prisms typically have wedge-shaped cross sections. They are characterized by one of the most complex internal structures of any tectonic element known on Earth caused by imbricate thrusting and folding of the incoming material. Some parts of accretionary wedges are composed of numerous thin rock layers that are repeated by thrust faults (stacking by duplexing); other parts of wedges or even entire wedges are characterized by large-folded and partly brecciated packages of rocks. Frequently they comprise tectonic mélanges that are composed of mixtures of blocks and thrust slices of many rock types (e.g., basalt, sandstone, limestone, chert, graywacke, and others) that are incorporated in a matrix of fine rock material such as shale or serpentinite. The faults and folds, in general, verge toward the subducting plate (Figure 2). However, anticlinal structures that include some landward-verging reverse faults, the so-called backthrusts, are produced, which create tectonic ridges that are expressed as positive morphological structures at subduction zones (Figure 1). Deformation and sedimentation

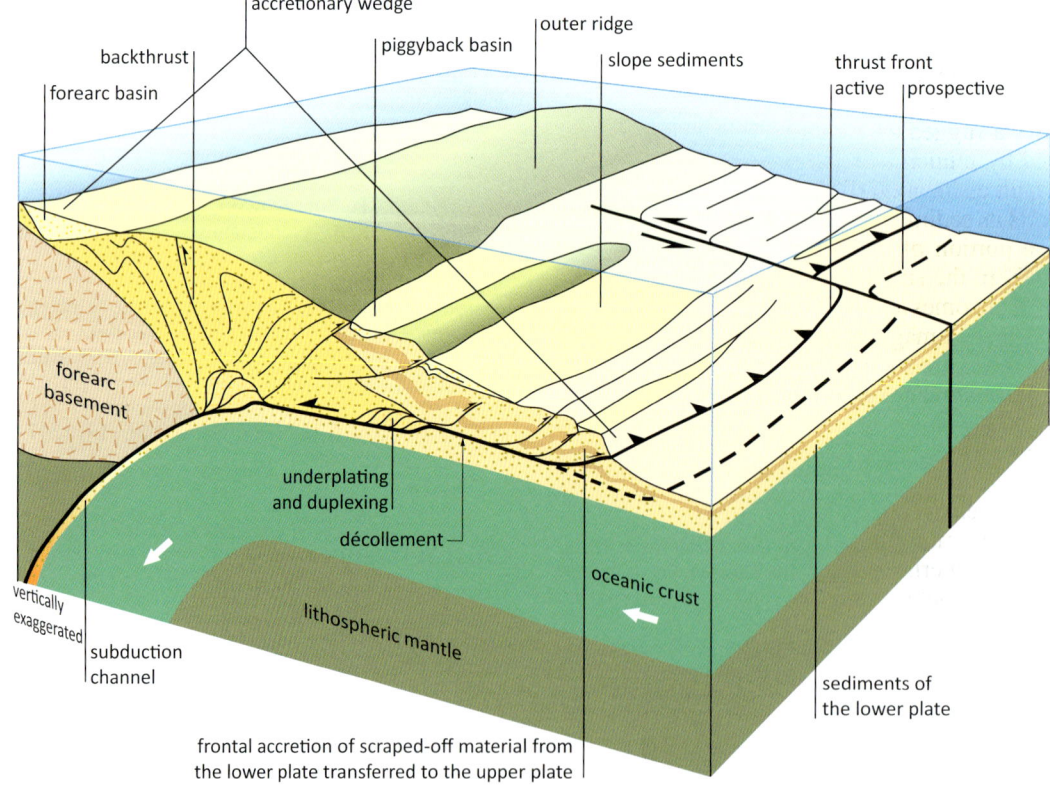

Accretionary Wedges, Figure 1 3D sketch of an accretionary wedge with fold and thrust structures and their morphological expression in the forearc region with basins and trenches filled with sediments (Modified after Fisher, 1996).

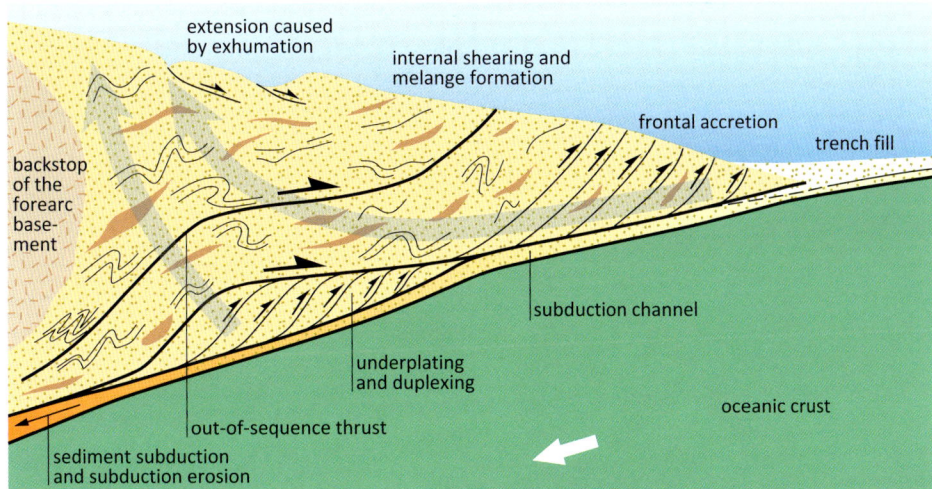

Accretionary Wedges, Figure 2 Schematic section of an accretionary wedge showing its internal structure with frontal and basal accretion and internal deformation caused by contraction and extension (Modified from Cawood et al., 2009). The transparent arrows indicate particle paths within the accretionary wedge during the accretion process.

occur concurrently and incrementally throughout the evolution of the system. The ongoing process causes the most devastating earthquakes on Earth, in some cases of magnitudes more than 9.0 (moment magnitude scale). The earthquakes are concentrated in the seismogenic zone, which in general is situated in the accretionary wedge.

A detachment surface, or décollement, separates the upper part of the accreted section (i.e., zone of offscraping) from material that is underthrust beyond the base of the slope. Above the décollement, scraped-off sediment is transferred to the accretionary prism, and this prism displays a rugged and irregular seafloor morphology governed by numerous tectonic ridges that form by folding and fault dislocation (Figure 1). As the subducting plate transports its sedimentary fill from the trench toward the arc, some portion of the sediment is subducted and transported within the subduction channel down to great depths where it becomes an important factor in the feeding of subduction-related magmas. The remaining portion, or in some cases the entire sedimentary layer and parts of the oceanic crustal basement, can be scraped off forming the accretionary wedge on the upper plate.

An accretionary wedge grows from below. The scraped-off sedimentary layers are stacked and continuously uplifted by renewed underplating from below, a process that results in morphological elevation of the outer ridge. The more material that is scraped off, the higher the elevation of the outer ridge. The process of accretion has been modeled in sandbox experiments so the evolution of the accretionary wedge is well understood (e.g., Gutscher et al., 1996; Dominguez et al., 2000). The underplating process resembles large-scale nappe thrusts in mountain ranges. Previously juxtaposed layers of sediment are stacked during the shortening process, and at each overthrust, older sediments are placed on top of younger ones. The process and sequence of events can also be viewed from the opposite perspective – younger units are forced below older ones by underthrusting.

Occurrences

About half of the convergent plate boundaries on Earth are dominated by the process of accretion in an accretionary wedge (von Huene and Scholl, 1991), which is the opposite process characterized by subduction erosion (see entry "Subduction Erosion" and Figure 1 therein, this volume). Numerous examples of studies exist that examine accretionary wedges from around the world (e.g., Scholl et al., 1980, Silver and Reed, 1988, Westbrook et al., 1988; Kukowski et al., 2001; Gulick et al., 2004, and others). Large accretionary wedges are represented in southwest Japan, Sumatra, large portions of the Gulf of Oman (Makran subduction zone), in the Lesser Antilles, along the Aleutians, and in smaller areas of western North and South America.

Eight overthrust planes that display repetitions of the sedimentary layers have been drilled at the Vanuatu accretionary wedge in the Southwest Pacific (Ocean Drilling Program, ODP Leg 134; Meschede and Pelletier, 1994). Because the sedimentary layer at the subducting plate has a thickness of slightly more than 100 m, the much larger thickness of the accretionary wedge is a result of intense tectonic stacking. In contrast, south of Japan, where the Philippine Sea Plate subducts beneath the Eurasian Plate, a thick sedimentary layer of more than 1000 m is entering the Nankai subduction zone (Gulick et al., 2004). Here, the décollement zone remains in the sedimentary layer and does not cut through the underlying oceanic crust as is the case in Vanuatu. Approximately the lower third of the sedimentary layer is being subducted and does not contribute to the growth of the accretionary wedge.

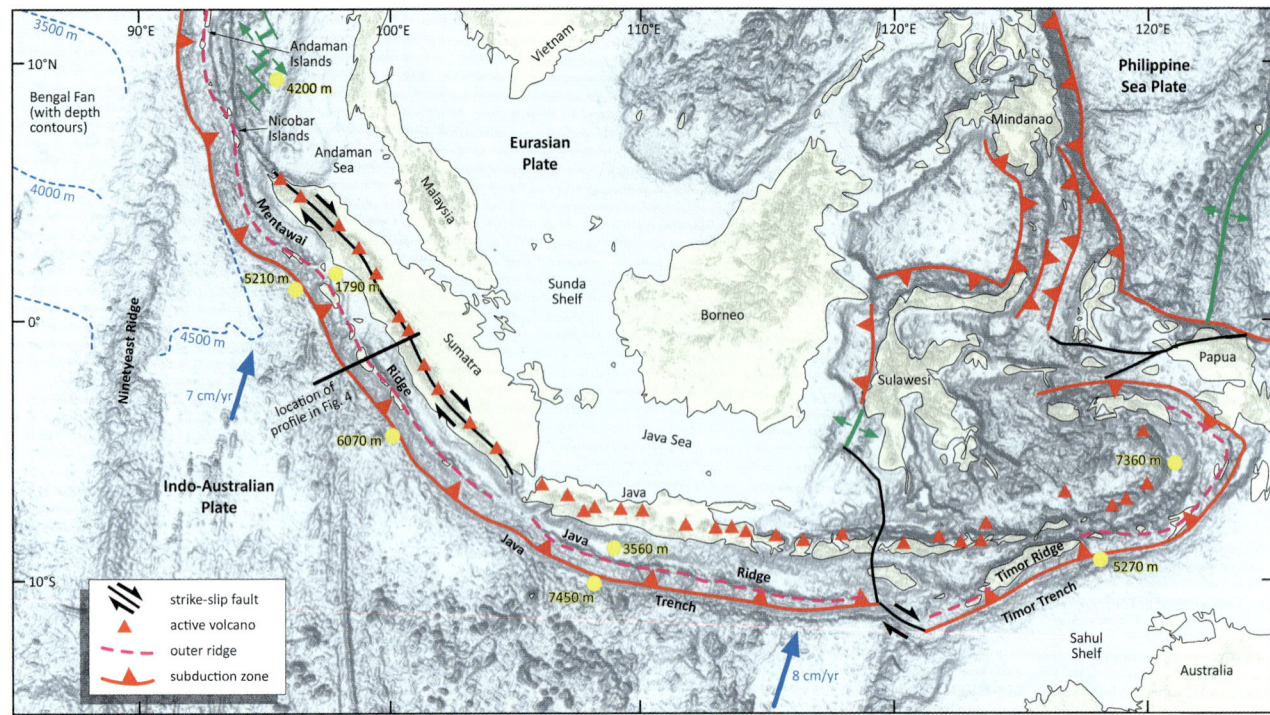

Accretionary Wedges, Figure 3 Map showing contrasting plate-tectonic conditions along the Sunda Arc (Modified from Frisch et al., 2011). In front of Sumatra, sediment of the thick Bengal fan are scraped off and incorporated into the accretionary wedge. This causes the outer ridge to emerge from the sea at this location (Mentawai Ridge; see Figure 4). In front of Java, the deep sea trench and the outer ridge are significantly deeper. In front of Australia, the continental crust of the Sahul Shelf is being subducted beneath the Sunda Arc; this causes a particularly strong uplift of the outer ridge (Timor Ridge) and marks the initial stage of orogenesis.

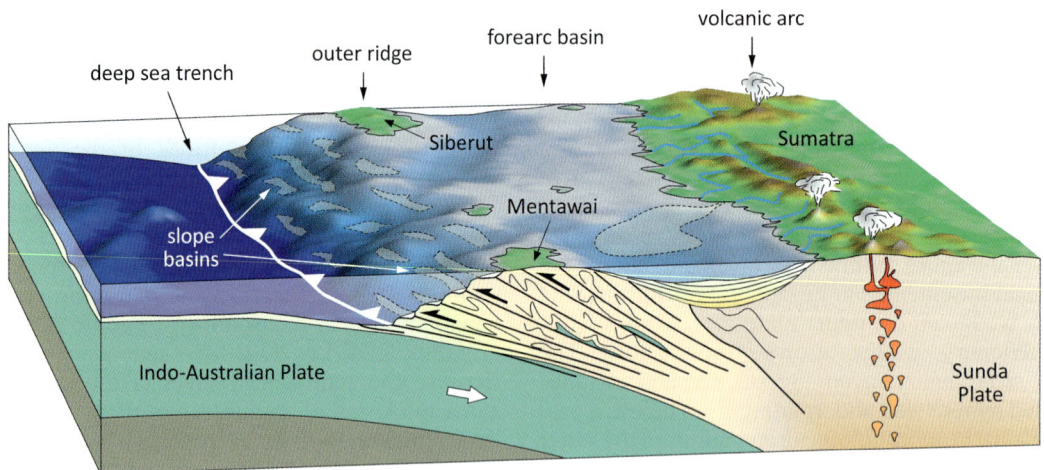

Accretionary Wedges, Figure 4 Example for an accretionary wedge with ridges uplifted above the sea-level (islands of Mentawai and Siberut) and numerous slope basins at the subduction zone off the island of Sumatra (Modified from Frisch et al., 2011).

The Sunda Arc provides an instructive example demonstrating how the shape of the accretionary wedge depends on the amount of sediments transported into the subduction zone (Figure 3). Fed by an enormous monsoon-controlled supply of sedimentary material, the Ganges and Brahmaputra rivers accumulated the huge several-kilometer-thick submarine Bengal fan that extends to the southernmost point of Sumatra. The sedimentary fan rests on the Indo-Australian Plate and is transported toward the NNE where it is being subducted in the Sunda Arc. Therefore, the deep sea trench along the northwestern part of this island-arc system is mostly masked by the high sedimentation rate, while the outer ridge, built by the tectonic stacking of fan-supplied sediments, is emerged several hundred meters above sea-level. It forms islands such as the Andaman and Nicobar Islands and the Mentawai Ridge in front of Sumatra (Figs. 3 and 4). Farther southwest away from the influence of the Bengal fan and adjacent to Java, the trench is distinctive with depths of almost 7500 m. Accordingly, the outer ridge is not well developed and lies mostly below a water depth of 2000 m.

Along Sumatra, the slope of the accretionary wedge falling from the outer ridge to the deep sea trench typically displays distinctive subdivisions. Here active thrusts produce elongated flat areas and depressions – the so-called slope basins (Figure 4). They are common along the Mentawai Ridge where some are rising above sea-level. During their complex history, these basins served as sediment traps recording the uplift history of the outer ridge. Analyses of the microfauna indicate uplift from deep to shallow water conditions. The youngest sediments contain reefs having formed in very shallow water before uplifting above sea-level (Moore et al., 1980).

Bibliography

Dominguez, S., Malavieille, J., and Lallemand, S. E., 2000. Deformation of accretionary wedges in response to seamount subduction: insights from sandbox experiments. *Tectonics*, **19**, 182–196.

Fisher, D. M., 1996. Fabrics and veins in the forearc: a record of cyclic fluid flow at depths of <15 km. In Bebout, G. E., Schell, D. W., Kirby, S. H., and Platt, J. P. (eds.), *Subduction Top to Bottom*. Washington, DC: American Geophysical Union. Geophysics Monograph, Vol. 96, pp. 75–89.

Frisch, W., Meschede, M., and Blakey, R. C., 2011. *Plate Tectonics*. Heidelberg/Dordrecht/London/New York: Springer, 212 pp.

Gulick, S.P.S., Bangs, N.L.B., Shipley T.H., Nakamura, Y., Moore, G., and Kuramoto, S., 2004. Three-dimensional architecture of the Nankai accretionary prism's imbricate thrust zone off Cape Muroto, Japan: Prism reconstruction via en echelon thrust propagation. *Journal of Geophysical. Research*, **109**(B02105). doi:10.1029/2003JB002654.

Gutscher, M. A., Kukowski, N., Malavieille, J., and Lallemand, S., 1996. Cyclical behavior of thrust wedges: insights from high-basal friction sandbox experiments. *Geology*, **24**, 135–138.

Kukowski, N., Schillhorn, T., Huhn, K., von Rad, U., Husen, S., and Flueh, E. R., 2001. Morphotectonics and mechanics of the central Makran accretionary wedge off Pakistan. *Marine Geology*, **173**, 1–19.

Meschede, M., and Pelletier, B., 1994. Structural style of the accretionary wedge in front of the North d'Entrecasteaux Ridge (ODP Leg 134). *Proceedings of the Ocean Drilling Program, Scientific Results*, **134**, 417–429.

Moore, G. F., Billman, H. G., Hehanussa, P. E., and Karig, D. E., 1980. Sedimentology and paleobathymetry of trench-slope deposits, Nias Island, Indonesia. *Journal of Geology*, **88**, 161–180.

Scholl, D. W., von Huene, R., Vallier, T. L., and Howell, D. G., 1980. Sedimentary masses and concepts about tectonic processes at underthrust ocean margins. *Geology*, **8**, 564–568.

Silver, E., and Reed, E., 1988. Backthrusting in accretionary wedges. *Journal of Geophysical Research*, **93**(B4), 3116–3126.

von Huene, R., and Scholl, D. W., 1991. Observations at convergent margins concerning sediment subduction, subduction erosion, and the growth of continental crust. *Review of Geophysics*, **29**, 279–316.

Westbrook, G. K., Ladd, J. W., Buhl, P., Bangs, N., and Tiley, G. J., 1988. Cross section of an accretionary wedge: Barbados Ridge complex. *Geology*, **16**, 631–635.

Cross-references

Earthquakes
Island Arc Volcanism, Volcanic Arcs
Lithosphere: Structure and Composition
Morphology Across Convergent Plate Boundaries
Ophiolites
Seismogenic Zone
Subduction
Subduction Erosion

ACTIVE CONTINENTAL MARGINS

Serge Lallemand
Geosciences Montpellier, University of Montpellier, Montpellier, France

Synonyms

Convergent boundary; Convergent margin; Destructive margin; Ocean-continent subduction; Oceanic subduction zone; Subduction zone

Definition

An active continental margin refers to the submerged edge of a continent overriding an oceanic lithosphere at a convergent plate boundary by opposition with a passive continental margin which is the remaining scar at the edge of a continent following continental break-up. The term "active" stresses the importance of the tectonic activity (seismicity, volcanism, mountain building) associated with plate convergence along that boundary. Today, people typically refer to a "subduction zone" rather than an "active margin."

Generalities

Active continental margins, i.e., when an oceanic plate subducts beneath a continent, represent about two-thirds of the modern convergent margins. Their cumulated length has been estimated to 45,000 km (Lallemand et al., 2005). Most of them are located in the circum-Pacific (Japan, Kurils, Aleutians, and North, Middle, and South America), Southeast Asia (Ryukyus, Philippines, New Guinea), Indian Ocean (Java, Sumatra, Andaman, Makran), Mediterranean region (Aegea, Calabria), or Antilles. They are generally "active" over tens (Tonga, Mariana) or hundreds (Japan, South America) of millions

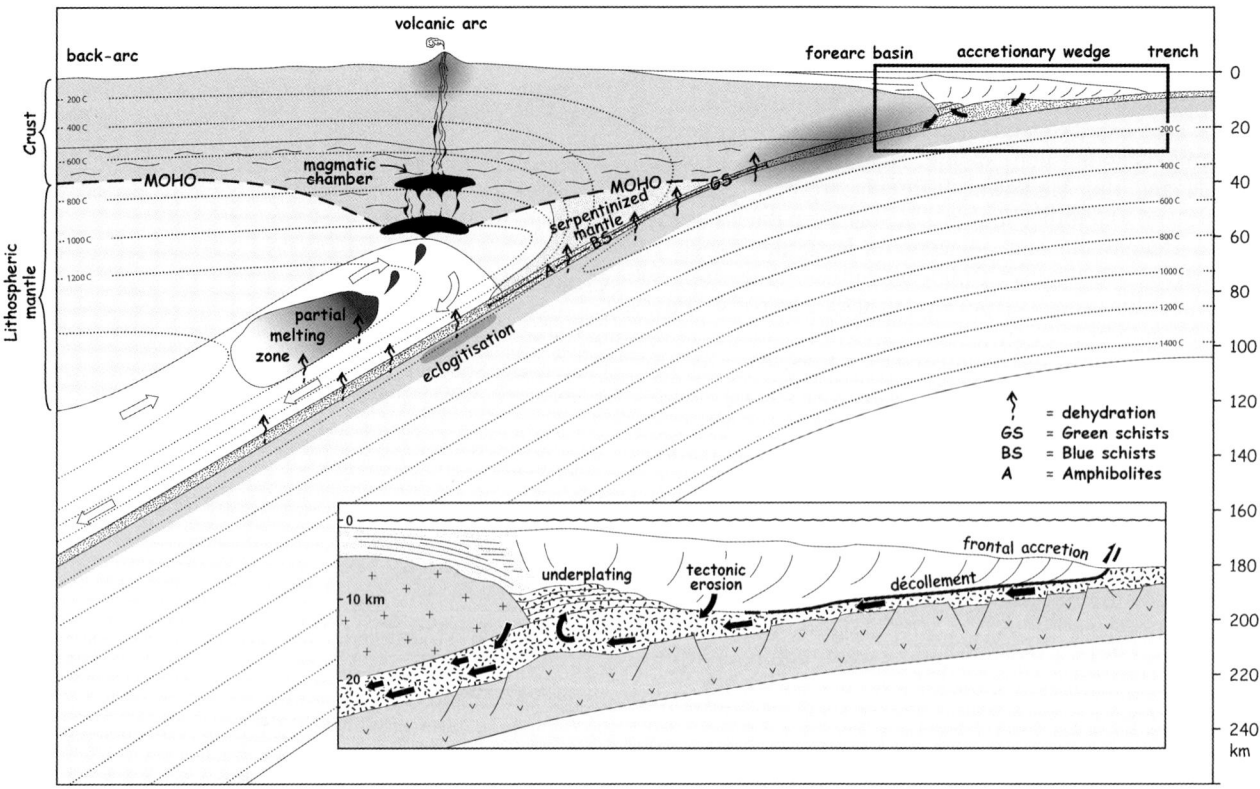

Active Continental Margins, Figure 1 Schematic view of an active continental margin after Lallemand et al. (2005).

of years. This longevity has consequences on their internal structure, especially in terms of continental growth by tectonic accretion of oceanic terranes, or by arc magmatism, but also sometimes in terms of continental consumption by tectonic erosion.

Morphology

A continental margin generally extends from the coast down to the abyssal plain (see Figure 1 and entry "Morphology Across Convergent Plate Boundaries"). It typically includes a continental platform gently dipping seaward and a talus with a steeper slope down to the trench. In detail, active continental margins offer a wide spectrum of morphologies from narrow and steep ones (e.g., Porto Rico) to wide and flat ones (e.g., Mediterranean "Ridge" which is a huge evaporitic accretionary wedge). In general, the continental shelves bordering active margins are narrower than those along passive margins. The trench marks the deepest seaward termination of the active continental margin. It coincides with the surface trace of the subduction fault marking the boundary between the converging plates. Its depth generally ranges between 2 and 7 km depending on the age of the subducting plate (young plate like Juan de Fuca – shallow trench) and the amount of trench fill sediment (thick trench fill like Makran – shallow trench).

Birth, life, and death

The discovery of microdiamonds in two billion-year-old rocks in Canada attests to ultrahigh-pressure metamorphism compatible with subduction in the early Proterozoic (Cartigny et al., 2004). The ongoing processes responsible for the sinking of the lithosphere at that time probably differed from those at present since the Earth was hotter than today. For modern subduction zones, Seiya Uyeda, in 1984, has proposed two modes of subduction mechanisms: a forced one (Chilean type) in contrast with a spontaneous one (Mariana type), further used by many authors like Stern and Bloomer (1992). These notions should now be replaced by compressional versus extensional subduction, based on the observation of the dominant strain within the overriding plate and not on the cause of the subduction itself (Heuret and Lallemand, 2005). Indeed, the observation of nascent or young (less than 10 million years old) subduction zones in the western Pacific has shown that all of them result from ongoing collisions and plate boundaries reorganization (Lallemand et al., 2005). None of them can be simply explained by the spontaneous sinking of an old and dense oceanic lithosphere under its own weight. The main driving force for subduction, after it initiates and develops down to a depth of about 200 km, is the slab pull exerted by the excess mass of the slab with respect to the surrounding mantle (Turcotte and Schubert, 1982; Hassani et al., 1997).

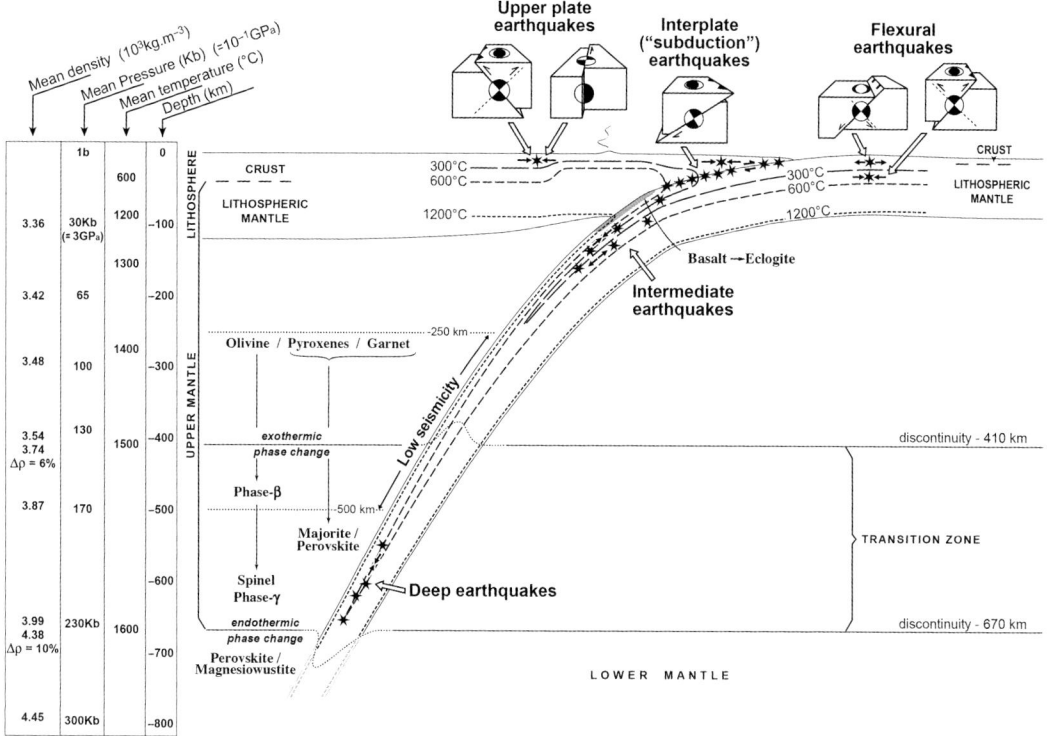

Active Continental Margins, Figure 2 Global view of an oceanic subduction zone with an old slab sinking at a high rate in order to get deep earthquakes after Lallemand et al. (2005). The isotherms are deflected along the top of the cold descending slab. Phase changes occur at various temperature/pressure (depth) conditions. This is why they occur at different depths in the slab. Different types of earthquakes (shallow, intermediate, and deep) and stable minerals are described. Variations in the mean density of the slab is indicated in the left column as the main phase transitions occur.

The mantle convection around a subduction mainly results from the dynamics of the plate subduction, not the opposite (Kincaid and Sacks, 1997). Once the subduction process is launched, it becomes stable over millions of years unless an oceanic plateau, an island arc, or a continental block reaches the subduction zone. If this occurs, subduction may stop in the collision zone and jump to another place where a new oceanic subduction may develop (Taiwan, New Guinea, Philippines, Indonesia, etc.). In a few cases, continental subduction prevails for tens of millions of years before plate reorganization (Himalaya, Weissel et al., 1980; Shemenda, 1992).

Tectonic activity
Earthquakes

The subduction process generates large stresses within both converging plates as well as along the subduction interface. Part of the stress is intermittently released seismically (seismic cycle). The largest earthquakes are produced by the so-called subduction earthquakes, i.e., the seismic expression of mega-ruptures along the frictional part of the plate interface, also called "seismogenic zone" (M9.0 2011 Tohoku earthquake, M9.2 2004 Sumatra earthquake, etc.). These particular earthquakes dissipate at least 85 % of the total seismic energy released in the world (Scholz, 1990). The reason is that the subduction interfaces are the largest faults on Earth (up to several hundred km, even 1,000 + km long) accommodating the largest slips (up to several tens of meters). Kanamori (1977) has shown that the seismic moment for great earthquakes is proportional to the faulted area and the slip on the fault. In addition to the subduction earthquakes, intraplate earthquakes may also occur either in the overriding plate, especially if compressive or extensive stress is transmitted from the subduction zone, or the downgoing plate (commonly called slab when it subducts into the mantle). The intraslab earthquakes include (Figure 2) the flexural earthquakes caused by plate bending near the trench, the intermediate earthquakes (down to about 300 km) resulting from the down-dip stress (often extensional as a result of the slab pull) or the unbending processes necessary to unfold the slab as it penetrates into the mantle, and the deep earthquakes (down to the discontinuity between the upper and lower mantle at 660–670 km). For further details, see also the following entries: "Subduction," "Earthquakes," and "Seismogenic Zone."

Volcanism

Most subduction zones are marked by a volcanic arc located at a distance between 100 and 300 km landward from the trench and 110 ± 20 km above the subducting slab. Arc magmatism results from the dehydration of the crust and overlying sediment carried by the downgoing oceanic plate. Such dehydration occurs at various depths depending on pressure/temperature conditions, but those occurring below the convective mantle wedge trigger melting and metasomatism of the mantle and subsequent rise of magmas. The explosivity of the volcanism is attributed to the high volatile content in the magma, especially H_2O. Calc-alkaline, potassic calc-alkaline, and shoshonitic series characterize the arc volcanoes lying on continental crust. For further details, see entry "Island Arc Volcanism, Volcanic Arcs."

Terrane collision/accretion and mountain building

Oceanic plates may carry seamounts, plateaus, active or fossil island arcs, and active or fossil spreading centers. Moreover, they may be attached to continents. Thus, it follows logically that during long period of activity of an oceanic subduction, collisions with buoyant features carried by the subducting plate occur. Collisions may be frontal like the Ontong-Java plateau with the Solomon arc or oblique like the Philippine Mobile Belt (Lallemand, 1999). Collisions along active continental margins often give rise to the accretion of crustal slivers (e.g., Izu-Bonin arc accretion in Japan; Tamura et al., 2010) or even large-size exotic terranes of oceanic or continental affinities such as the Caribbean (Antilles) or the Okhotsk (Russia) plateaus, the Panama-Choco (Colombia, Panama) or the Halmahera (Philippines) arcs, and the Qiantang, the Lhasa, or the South China continental blocks (e.g., Lallemand et al., 1998; Taboada et al., 2000; Konstantinovskaia, 2001; Roger et al., 2003; Kroehler et al., 2011). All these accreted blocks contribute to continental growth. When the colliding blocks are large enough, they contribute to mountain building such as the Kunlun Mountain Range north of Tibet or the Eastern Cordillera in Colombia.

Summary

Active continental margins are the most common convergent plate boundaries. They represent one class of subduction zones where an oceanic plate subducts beneath a continental plate. Since their tectonic activity commonly lasts tens of millions of years, they are the locus of continental growth and consumption. Most of the seismic energy is released along these margins. They often concentrate seismic, tsunami, and volcanic hazards.

Bibliography

Cartigny, P., Chinn, I., Viljoen, K. S., and Robinson, D., 2004. Early proterozoic ultrahigh pressure metamorphism: evidence from microdiamonds. *Science*, **304**, 853–855.

Hassani, R., Jongmans, D., and Chéry, J., 1997. Study of plate deformation and stress in subduction processes using two-dimensional numerical models. *Journal of Geophysical Research*, **102**(B8), 17951–17965.

Heuret, A., and Lallemand, S., 2005. Plate motions, slab dynamics and back-arc deformation. *Physics of the Earth and Planetary Interiors*, **149**, 31–51.

Kanamori, H., 1977. The energy release in great earthquakes. *Journal of Geophysical Research*, **82**(20), 2981–2987.

Kincaid, C., and Sacks, I. S., 1997. Thermal and dynamical evolution of the upper mantle in subduction zones. *Journal of Geophysical Research*, **102**(B6), 12295–12315.

Konstantinovskaia, E. A., 2001. Arc-continent collision and subduction reversal in the Cenozoic evolution of the Northwest Pacific: an example from Kamchatka (NE Russia). *Tectonophysics*, **333**, 75–94.

Kroehler, M. E., Mann, P., Escalona, A., and Christeson, G. L., 2011. Late Cretaceous-Miocene diachronous onset of backthrusting along the South Caribbean deformed belt and its importance for understanding processes of arc collision and crustal growth. *Tectonics*, **30**, TC6003.

Lallemand, S., 1999. *La subduction océanique*. Amsterdam: Gordon and Breach Science Publishers, 208 pp (in french).

Lallemand, S. E., Popoff, M., Cadet, J.-P., Deffontaines, B., Bader, A.-G., Pubellier, M., and Rangin, C., 1998. Genetic relations between the central & southern Philippine Trench and the Philippine Trench. *Journal of Geophysical Research*, **103**(B1), 933–950.

Lallemand, S., Huchon, P., Jolivet, L., and Prouteau, G., 2005. In Vuibert (ed.), *Convergence lithosphérique*, 182 pp (Paris).

Roger, F., Arnaud, N., Gilder, S., Tapponnier, P., Jolivet, M., Brunel, M., Malavieille, J., Xu, Z., and Yang, J., 2003. Geochronological and geochemical constraints on Mesozoic suturing in east central Tibet. *Tectonics*, **22**(4), 1037.

Scholz, C. H., 1990. *The Mechanics of Earthquakes and Faulting*. New York: Cambridge University Press. 400 pp.

Shemenda, A. I., 1992. Horizontal lithosphere compression and subduction: constraints provided by physical modeling. *Journal of Geophysical Research*, **97**(B7), 11097–11116.

Stern, R. J., and Bloomer, S. H., 1992. Subduction zone infancy: examples from the Eocene Izu-Bonin-Mariana and Jurassic California arcs. *Geological Society of America Bulletin*, **104**, 1621–1636.

Taboada, A., Rivera, L. A., Fuenzalida, A., Cisternas, A., Philip, H., Bijwaard, H., Olaya, J., and Rivera, C., 2000. Geodynamics of the northern Andes: subductions and intracontinental deformation (Colombia). *Tectonics*, **19**(5), 787–813.

Tamura, Y., Ishizuka, O., Aoike, K., Kawate, S., Kawabata, H., Chang, Q., Saito, S., Tatsumi, Y., Arima, M., Takahashi, M., Kanamaru, T., Kodaira, S., and Fiske, R. S., 2010. Missing Oligocene crust of the Izu-Bonin arc: consumed or rejuvenated during collision ? *Journal of Petrology*, **51**(4), 823–846, doi:10.1093/petrology/egq002.

Turcotte, D. L., and Schubert, G., 1982. *Geodynamics: Applications of Continuum Physics to Geological Problems*. New York: Wiley. 450 pp.

Uyeda, S., 1984. Subduction zones: their diversity, mechanism and human impacts. *GeoJournal*, **8**(4), 381–406.

Weissel, J., Anderson, R., and Geller, C., 1980. Deformation of the Indo-Australian plate. *Nature*, **287**, 284–291.

Cross-references

Accretionary Wedges
Crustal Accretion
Driving Forces: Slab Pull, Ridge Push
Earthquakes
Geohazards: Coastal Disasters
Island Arc Volcanism, Volcanic Arcs
Magmatism at Convergent Plate Boundaries
Morphology Across Convergent Plate Boundaries

Ocean Margin Systems
Orogeny
Seamounts
Seismogenic Zone
Subduction
Subduction Erosion
Wadati-Benioff-Zone
Wilson Cycle

ANCIENT PLATE TECTONICS

Stephen F. Foley
Department Earth and Planetary Sciences, ARC Centre of Excellence for Core to Crust Fluid Systems, Macquarie University, North Ryde, NSW, Australia

Definition
Plate tectonics

Plate tectonics is a theory which attempts to explain all forms of geological and geophysical observations of the Earth through time in terms of the movement of rigid lithospheric plates that form the upper few tens of kilometers of the Earth. These observations include rock types and their spatial and temporal relationships to each other, their structures and the formation of mountains, the chemical and physical properties of rocks, as well as seismicity, volcanism, and even the distribution of fauna and flora. These lithospheric plates are made up of the crust and the uppermost layers of the mantle (the mantle lithosphere) that are mechanically coupled and can therefore move as a unit on top of convection currents within the mantle. Although moving at rates of only a few centimeters per year, these convection currents explain the movement of continents across the surface over geological time. The crust involved varies from thin ocean crust (around 7 km) to thick continental crust (25–100 km, averaging 35 km). Many plates, such as the South American plate, are topped by extensive tracts of both. When mantle lithosphere is included, some plates may reach total thicknesses of 250 km in the older cores of the continents.

The fact that all geological processes on the modern Earth are thought to be explainable by plate tectonics does not necessarily mean that this was the case for earlier periods in Earth history when the planet was significantly hotter. In deciding whether this was the case, we have a major problem concerning the preservation of rocks from the first half of Earth history. The production and consumption of ocean crust is axiomatic to the operation of plate tectonics, and yet the oldest ocean crust on the Earth's surface today is less than 180 million years old, which is equivalent to only 4 % of the age of the Earth. Linked to this uncertainty is the growth rate of the continental crust, which is thought by most geoscientists to have been gradual or episodic over the 4.57 Ga age of the Earth, with perhaps less than half of it in existence at the end of the Archean period 2,500 myr ago.

History of plate tectonics

The term *plate tectonics* first emerged at the end of the 1960s, following a turbulent decade in which many branches of the geological sciences reorganized their understanding of geological processes to fit this single all-encompassing theory. Its forerunner was continental drift, which followed centuries-old observations such as the similarity of coastline form between Africa and South America, and hypothesized the lateral movement of continents over thousands of kilometers. Continental drift, most commonly associated with the work of Alfred Wegener, remained controversial for several decades owing to the lack of an explanatory mechanism. It received support from the then new science of paleomagnetism in the 1950s, which demonstrated the movement of continents around rotation poles (Runcorn, 1959), but it was the introduction of marine geology and geophysics in the late 1950s that led to the breakthrough. Magnetic stripes were discovered on the ocean floor (Raff and Mason, 1961) and eventually proved to be symmetrically distributed across mid ocean ridges (Pitman and Heirtzler, 1966). The concept of sea-floor spreading was born (Vine and Matthews, 1963), once again linked to rotation around poles that also explained the newly discovered transform faults between ridge segments (Wilson, 1965). However, if sea-floor spreading occurs commonly on a planet of constant size (there were a few who denied this; Carey, 1975), then the ocean crust produced must also disappear again, and this led to the concept of subduction zones (Le Pichon 1968), where sinking ocean crust is delineated by seismically active zones below island arcs (Isacks et al., 1968). Once the mathematical explanation of the rotation of rigid lithospheric plates around poles and the catalog of possible triple junctions was in place (McKenzie and Parker, 1967; McKenzie and Morgan, 1969), all that remained was to fit regional geology including mountains, rifts, and continental margins into the model to explain the geological history of Earth through time (Dickinson, 1970; Dewey and Bird, 1970). For the last 40 years, plate tectonics has been the single complete geological theory into which all geological and geophysical observations are expected to fit.

Mechanisms of plate tectonics

In order to decide whether plate tectonics functioned either differently or even at all early in Earth's history, it is necessary to consider how plate tectonics is thought to work and what characteristics lead us to decide in favor of it. Some of the features described earlier in favor of continental drift, such as the similarity of coastlines between continents, the movement of continents across the surface of the globe, and polar wander curves defined by paleomagnetism, are compatible with plate tectonics but do not require it. The underlying mechanism lies in the distinction between the rigid upper lithosphere and the ability of the asthenosphere below it to flow and requires convection currents in the mantle below the lithosphere. Forces

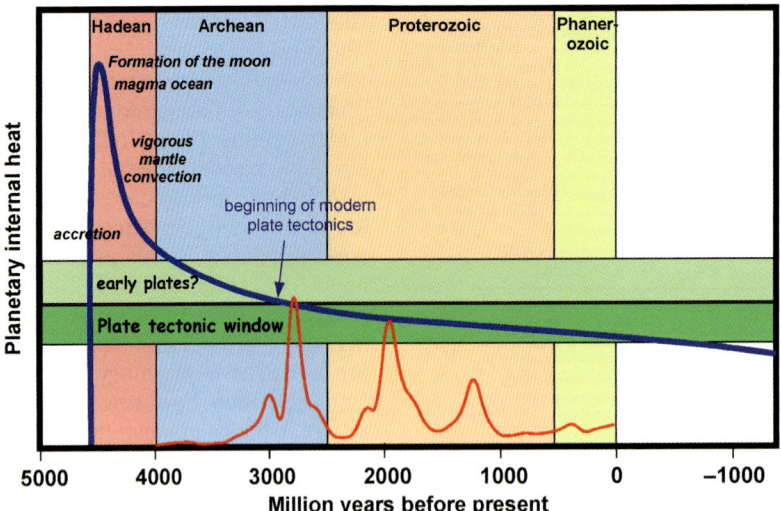

Ancient Plate Tectonics, Figure 1 Evolution of planetary temperature (*blue line*) indicating entry into the plate tectonic window (*green* box) in the mid Archean from hotter conditions resulting from planetary accretion. Crustal formation in the period labeled "early plates?" may have occurred by melting of the base of thick crust without modern-style subduction. The *red line* indicates major episodes of continental crust formation deduced from the ages of igneous rocks. See text for further explanation.

that act to move the plates are (1) the drag caused by partial coupling of the asthenosphere to the lithospheric plates above it, so that the plates follow the convection movements; (2) phase transformations in the subducting crust and lithosphere, which increase the density of the rocks and therefore cause them to sink through the mantle (known as "slab pull"); and (3) a lateral "ridge push" force imparted by the mountains of the mid ocean ridges where ocean crust is formed; new crust is formed at the top of the ridge and so causes a gravitational force pushing the slightly older crust to both sides. The slab pull force is due mainly to two reactions; the conversion of basalt and gabbro to eclogite (about 20 % denser) in the upper 30–40 km and the transformation of olivine to a high-pressure spinel structure at around 400 km depth (12 % density increase).

For the discussion of whether plate tectonics operated earlier in Earth history, mantle convection will certainly have occurred and volcanism may have been more abundant at any particular time, but the second-phase transformation of the slab pull force will only have functioned if subduction to depths of more than 400 km occurred.

The evolution of temperature in the Earth: the background to global tectonic regimes

Since mantle convection is an essential prerequisite for plate tectonics, and convection will occur once a given threshold is reached (a Rayleigh number of more than 2,380), the state of internal heat of the Earth as a function of time is the deciding factor. The blue curve in Figure 1 shows the approximate expected heat evolution of the Earth. The colored background panels refer to the four Eons, the Hadean before the first preserved rocks, the Archean until 2,500 Ma, the Proterozoic, and the Phanerozoic, which started with the massive preservation of skeletal fossils at 542 Ma. The blue line begins with the growth of the planet during the accretion of the solar nebula at around 4,567 Ma. At least 85 % of the internal heat of the planet is derived from accretion energy (collisions between small planetary bodies and planetesimals) and the separation of the metallic core: if the moon was formed by a major impact at about 4,500 Ma as considered likely by most scientists (Lee et al., 1997), then this figure is much more than 90 %. The gradual loss of heat after planet formation and the end of collisions is due to two factors: the loss of heat to space due to conduction through the crust and the gradual radiogenic decay of isotopes, particularly K, U, and Th, which decreases with time leading to a flattening of the curve as the parent isotopes become rarer. The green bar shows a postulated "plate tectonic window" (Condie, 1989) in which plate tectonics may operate. Vigorous convection would follow the accretion process and the great loss of heat at this time is unlikely to have allowed the survival and orderly movement of plates. At the other end of the timescale, continued loss of heat and diminishing heat production from radioactive decay will cause the Earth to drop out of the plate tectonic window and form a one-plate planet sometime in the future; the time at which this may happen is not known.

Note that there are many unknowns in the form of the curve and that it should only be taken as a conceptual illustration. Two important unknowns are the rate of decrease after the accretion peak and the degree to which episodic processes may cause humps in the curve (Davies, 1995). The height of the peak is poorly known, but accepting the giant impact model for the origin of the moon (Canup and Asphaug, 2001), the age of the moon at

4,500 Ma must correspond to the existence of a magma ocean in the Earth that may have been hundreds of kilometers thick, resulting in extremely hot conditions near the surface of the early Earth. However, this state may not have lasted for very long, as investigations of the oxygen isotope compositions of the earliest rocks and minerals (zircons) find that they formed in cool conditions in siliceous crustal rocks (that have since been destroyed) near the Earth's surface (Valley et al., 2002). Since the oldest of these zircons are 4,404 Ma (Wilde et al., 2001), there appears to be only 100 million years for the Earth to progress from extremely hot to cool conditions. The other major simplification in the curve is that it represents the whole Earth and implies a gradual, smooth evolution without any episodic or catastrophic developments. This contrasts with the red line superimposed on Figure 1, which shows the rate of production of new continental crust as gauged by the ages of igneous rocks on the same timescale (Hawkesworth and Kemp, 2006). Although there is some discussion as to whether this really indicates ages of crustal growth rather than merely its survival, there appear to be major periods of crustal formation in the late Archean (3,000–2,500 Ma) and in the early Proterozoic (2,000–1,750 Ma), with an additional pulse at around 1,100 Ma. There may have been a similar episodism in plate movements and subduction (O'Neill et al., 2007). These two uncertainties emphasize that the slope of the curve and its exact form are debatable, and therefore, the point marked for entry into the plate tectonic window at around 3,000 Ma can only be approximate, although there is increasing evidence for the onset of modern-style (i.e., steep and deep) plate tectonics at around 3,200 Ma (Shirey and Richardson, 2011; Van Kranendonk, 2011; Dhuime et al., 2012).

The other open question is that if plate tectonics did begin at about this time, then what happened before that? Was there some other form and movement of lithospheric plates that is distinguishable from the modern style (indicated by the light green band in Figure 1; Brown, 2007), or could small plates tip towards each other like ice floes in the Antarctic, or was there a thick lid of crust that was recycled into the mantle by eclogite dripping from its base?

An outline of Archean geology

To adequately assess whether plate tectonics could have operated during the first half of Earth history, it is important not just to ascertain the presence or absence of indicators for plate tectonics but also to explain the Archean geological record. Archean rocks consist of about 80 % high-grade gneisses, mostly summarized as the tonalite-trondhjemite-granodiorite (or TTG) suite, and 20 % greenstone belts with lower metamorphic grade. The gneisses have igneous origins and can be explained by melting of basaltic protolith material (Foley et al., 2002; Rapp et al., 2003). Their deformation often indicates lateral compression, although in some regions, most notably Western Australia, vertical movements may have been important (Chardon et al., 1996). On the modern Earth, tonalitic melts are formed by melting of basaltic material at subduction zones, and so a subduction environment is the most commonly invoked option to explain them in the Archean. Greenstone belts consist principally of volcanic rocks and sediments wedged between domes of high-grade gneisses. Their most famous constituents are komatiites, which are magnesium-rich mafic volcanics that are almost exclusively Archean in age (Arndt and Nisbet, 1982). However, the majority of volcanics in greenstone belts are more "normal" basaltic to felsic rocks, and not komatiites. The sediments are dominantly texturally immature, whereas shelf sediments such as carbonates and mature sandstones, although present back to 3.5 Ga, are much less common (Eriksson et al., 1994). This would be expected on a planet on which island arc environments and/or volcanic plateaux are common but continents are rare; the first continents would be formed by the amalgamation of island arcs, and all sediments that are preserved would be formed close to these arcs. The rarity of shelf sediments may correspond to the rarity of continents. The relationship between the high-grade gneisses and greenstone belts ought, therefore, to be of paramount interest, but unfortunately, most contacts between them are the locations of later tectonic movements. Examples exist, however, that show basal unconformities, leading to the conclusion that greenstones commonly do not represent ocean crust but were deposited on continents (Bickle et al., 1994; Buick et al., 1995).

There are also changes in rock types towards the end of the Archean; following the peak in continental crust formation at 2.7 Ga (Figure 1), more potassic granites that are typical for later periods in Earth history become most common, often appearing to seal large sections of the newly formed crust together into stable continents. Other rocks remain notably absent – eclogites and blueschists – which are indicators of high-pressure metamorphism at low pressures, and do not appear until around 2,100 Ma.

The characteristics of plate tectonics

From theoretical expectations, we may consider how we can decide if plate tectonics operates on a planet and then deliberate about whether these features are relevant for the early Earth. Table 1 lists the main features by which the operation of either continental drift or plate tectonics can be recognized on the modern Earth. The problem in ascertaining the relevance of plate tectonics for the Archean and Proterozoic Eras is that most of these criteria depend either on geophysical measurements that only provide a picture of the Earth today or on good preservation of rock associations and geological structures. Much of the acceptance and detail of plate tectonics is due to seismological and other geophysical surveys that provide information on all levels of the deep Earth, documenting seismic discontinuities that delineate the size of core and mantle, that the outer core is molten (Fowler, 2005), and

Ancient Plate Tectonics, Table 1 Criteria for continental drift or plate tectonics. Due to the young age of preserved ocean crust, the effects of metamorphism and deformation, and the restriction of geophysics to the present, most evidence for ancient plate tectonics is indirect

Feature	Evidence for	Applicable to Archean?
Geophysical, geomorphological, and geological indicators on the modern Earth		
Matching coastlines	Continental drift	No
Same fauna and flora on different continents	Continental drift	No
Polar wander curves	Continental drift	Maybe, but very limited
Magnetic stripes/reversals	Sea-floor spreading	No
Earthquake zones under arcs	Subduction	No
Seismic tomography	Deep subduction	No
Geological indicators for plate tectonics in the past		
Hadean zircons	Continental crust	Yes
Linear mountain ranges	Continental collision	Yes
Shelf sediments	Continents and their margins	Yes, but restricted
Continental rift rock associations	Continents and lateral movements	Yes, but continents restricted
Large transcurrent faults	Lateral plate movements	Yes, but controversial
Ophiolites	Ocean crust	Yes, but controversial
Geochemical signature of subduction in igneous rocks	Subduction	Yes, but controversial
Arc igneous rock associations	Subduction	Yes, but restricted
Eclogite	Subduction	Yes, but missing at the surface
Blueschist rock belts	Subduction (cold)	Yes, but missing
Sodic granites	Subduction	Yes, but controversial
Paired metamorphic belts	Subduction and collision	Yes, but restricted
Accretionary prisms	Subduction	Yes, but continents restricted

the densities of rocks that constitute these layers (Anderson, 1989).

However, few of the criteria for either continental drift or plate tectonics can be successfully used for the Archean or early Proterozoic, which according to Figure 1 is the time period where plate tectonics is questionable. Of geophysical measurements, seismological records are little more than 100 years old and cannot be carried out in retrospect; only the science of paleomagnetism, which first documented continental drift (Runcorn, 1959), can be used in the distant geological past (Strik et al., 2003), but even here, the accuracy and certainty of interpretations diminishes as we go back into the Archean and requires coupling to accurate age determinations for earlier times (Evans and Pisarevsky, 2008).

We are left with the indirect geological indicators in the second part of Table 1 that rely on the interpretation of the rock record. A cursory look shows that most of these are potentially applicable, but either their interpretation is to some extent controversial or their application is restricted. This restriction is mostly temporal, meaning that there is a time limit for each before which they cannot be unambiguously recognized. This may be because they did not exist or because of the effects of deformation and metamorphism (the large majority of early Archean rocks have been highly deformed and metamorphosed at medium or high temperatures) or even their complete destruction. Many make a first appearance between 3.3 and 2.6 Ga, but a first appearance must be taken to be a minimum age as they may have existed beforehand but no longer be recognizable. An example of differential preservation is the earliest eclogites, which are evidence for metamorphism of basalts or ocean crust at very high pressures. If we were to restrict our attention to those preserved in collision zones on the continents, then we would conclude that the oldest are younger than 2.9 Ga (Mints et al., 2010), whereas they are common as xenoliths trapped in the mantle lithosphere from 3.3 Ga, where they are interpreted to represent pieces of subducted ocean crust (Jacob, 2004).

The criteria in Table 1 address continental crust, ocean crust, and the subduction process, which raises an important point for discussion of plate tectonics. The hypsometric curve, which shows the distribution of crustal levels above and below sea-level on the Earth, shows a bimodal distribution on the Earth, with most crust around 200 m above or 3–4 km below sea-level (Figure 2). This is equivalent to the distinction between continental and oceanic crust and can be taken to indicate the operation of plate tectonics over several hundreds of millions of years. Before plate tectonics began to function, the term "ocean crust" may have little meaning: there may have been just "crust," broadly of mafic composition, and the huge variety of crustal elevations may not have been present.

The time and rate of formation of the *continental* crust is a controversial topic, but currently most geologists accept episodic formation and accumulation of the continental crust through time (red curve in Figure 1). This would mean that there was less than half the current volume of continental crust until a major phase of production between 3.0 and 2.5 Ga at the close of the Archean. The main controversy lies in the degree to which former continental crust may have been recycled into the mantle (Scholl and von Heune, 2007), such that the alleged crustal formation rate really illustrates the rate of crustal survival. Added to this uncertainty is the high degree of deformation and metamorphism in many Archean rocks, which hampers the interpretation of their formation. The existence of zircons dating from the Hadean

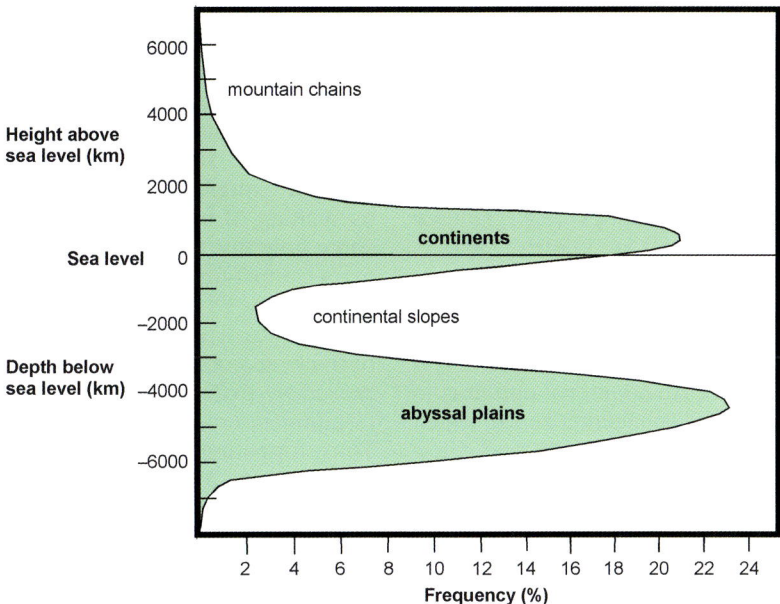

Ancient Plate Tectonics, Figure 2 The hypsometric curve for the Earth today, indicating that most crust is slightly above sea-level (continental crust) or 3–4 km below sea-level (oceanic crust). This bimodality of elevations is a logical consequence of the operation of plate tectonics for several hundred million years.

(4.4–4.0 Ga), although preserved in younger rocks (Wilde et al., 2001), is thought to demonstrate the existence of some siliceous crust at the time, but does that have to be continental crust, or could it be merely siliceous scum on pre-plate tectonic crust? Small sections of felsic crust on large rafts of mafic crust may have been too insignificant volumetrically to prevent wholesale recycling of the crust and its remixing into the mantle.

In terms of structures, plate tectonics results in linear collisional belts of mountains and large strike-slip faults, but taken alone these are indications of lateral crustal movements, but not necessarily plate tectonics. Shelf sediments and rock associations typical of continental rifts are rare until the late Archean, but one cannot expect these before continental crust itself becomes abundant.

The best evidence for ocean crust takes the form of ophiolites, which are rock packages consisting of basalts and gabbros of the ocean crust underlying marine sediments and overlying ultramafic rocks of the uppermost mantle. Opinions differ greatly as to the first appearance of ophiolites; some scientists require the complete package to be present, whereas basalts with an oceanic geochemical signature suffice for others. The first relatively complete ophiolites date from around 2.0 Ga (Scott et al., 1991; Peltonen and Kontinen, 2004), whereas partial ones are claimed for the earliest Archean (de Wit, 1998). Due to the ephemeral nature of ocean crust, it is much more poorly preserved than continental crust, and some sections of mafic crust interpreted as ocean crust may be from the pre-plate tectonic mafic crust. The eclogite xenoliths already mentioned contain clear evidence of having been at the surface and were later returned to the mantle – but by which process? This may not be clear evidence for subduction in the modern plate tectonic style.

Most entries in Table 1 consider the subduction process, as this appears to be the best key to indicating that plate tectonics functioned. All of these either suffer from the preservation problem or are only indirect indictors. The high-pressure, low-temperature metamorphism characteristic of modern subduction produces eclogites and blueschists, and these ideally occur as part of paired metamorphic belts together with low-pressure, high-temperature metamorphism formed at the other side of the subduction suture (Miyashiro, 1961). Blueschists are not known from before 1 Ga (Stern, 2005). However, this time limit addresses large occurrences that can be recognized in the field. Evidence is found for much earlier high-pressure, low-temperature metamorphism as relict features preserved on a microscopic scale in rocks that have been re-equilibrated at lower pressures. These can be used to show that the duality of thermal indicators across subduction sutures can be recognized as far back as 3.26 Ga (Moyen et al., 2006), but at this time the temperature difference between the two sides was considerably smaller than in Phanerozoic times. Before 3.26 Ga, the pair of thermal indicators has not been found, and no relicts for high-pressure, low-temperature metamorphism are known (Brown, 2007; van Kranendonk, 2011).

The most widely used evidence for tectonic settings in old rocks is indirect and concerns the recognition of trace element geochemical signatures reminiscent of modern plate tectonic environments such as island arcs, mid ocean ridges, oceanic plateaux, and back arc basins (Smithies et al., 2005; Pearce, 2008). However, these are prone to overinterpretation, as they are not foolproof on the modern Earth: for example, continental flood basalts are known to have low-Ti characteristics normally associated with island arc volcanism and calc-alkaline rocks that are characteristic of magmatic arcs above subduction zones also been found in non-subduction settings (e.g., Hooper et al., 2002; Willbold et al., 2009) are known. There is a tendency to engrave interpretations from modern settings onto ancient rocks, although most Archean rocks give mixed signals (Pearce, 2008). But here again, the oldest arc-like greenstones appear to be between 3.2 and 3.1 Ga (Smithies et al., 2005). An additional option is to use rock types characteristic for subduction zones, which include andesites, boninites, shoshonites, and adakites. Most of these become frequent in the late Archean (Condie and Kroner, 2008), whereas adakite is a rock type from modern subduction zones (island arcs and Andean-type subduction beneath continental crust) that is similar to tonalitic and trondhjemitic gneisses that make up large tracts of the Archean continental crust. These are thought to indicate a subduction environment because they do not form in modern plume-related environments (Martin, 1999).

A different type of plate tectonics or no plate tectonics at all?

The Archean rock record consists of small continent-like blocks of high-grade gneisses of broadly tonalitic composition and subordinate greenstone belts that are composed mostly of volcanic rocks and texturally immature sedimentary rocks. These rocks are quite common from 3 Ga onwards, but similar rocks are present from the beginning of the Archean: tonalitic gneisses make up the bulk of the Itsaq complex of western Greenland, dated at 3,800 Ma (Nutman et al., 1999), and supracrustal rocks of similar age occur both here and in northern Quebec (Mloszewska et al., 2012). As we trace them back into the Archean, several of the best plate tectonic indicators, such as paired metamorphic belts, high-pressure, low-temperature metamorphism, and greenstone volcanics with arc-like geochemistry, seem to "pinch out" at about 3.1–3.3 Ga. For the beginning of plate tectonics, no single indicator is as convincing as the testimonial given by the accumulation of these lines of evidence at this time. Even so, the lack of eclogites and blueschists leads some to distinguish between *Proterozoic plate tectonics* (approximately 3,300–700 Ma) and *modern plate tectonics* (after 700 Ma; Brown, 2007). Before 3.3 Ga, we have only a limited number of indicators, but these tell us that TTG magmatism occurred from the beginning of the Archean, although with a tendency for MgO poorer, SiO_2-richer compositions earlier on in time. This possibly indicates a type of shallow subduction, in which the subducting crustal plate directly underlies the overriding plate so that reaction of TTG melts with the mantle wedge could not occur. As time passed and the Earth cooled, subduction to deeper levels occurred and reaction with the mantle became more common.

However, the production of TTG melts does not necessarily denote subduction. The requirement from high-pressure experiments is that these melts must be derived by melting of basalt, but these experiments say nothing about how this basalt reaches its melting conditions. The main competitor hypotheses to subduction for explaining the origin of TTG melts are oceanic plateaux and stagnant lid tectonics. The oceanic plateau model uses the analogy of thickened ocean crust on the modern Earth, which are caused by extra heating from below by mantle plumes, resulting in crust up to 35 km thick (Neal et al., 1997). Thus, formation of some crust as oceanic plateaux could apply to an Archean Earth with plate tectonics, but a major uncertainty in this case becomes the behavior of this thick mafic crust in subduction zones. Can such thick crust be subducted as a unit, or does just part of it return to the mantle and the upper parts accumulate at the surface? In this scenario, the basaltic source rocks for TTG melts are derived from thick piles of volcanics that simply become buried to lower crustal levels because of continued volcanism.

The ocean plateau model is generally thought to work within plate tectonics and does not seem adequate for the early Archean if complete recycling of crust commonly occurred. Here a stagnant lid scenario may be more appropriate, in which the thick crust is not subducted, but the lower levels delaminate and melt back into the mantle. Only the upper sections of crust are preserved, a scenario known as "flake tectonics" (Hoffman and Ranalli, 1988). In the earliest Archean or Hadean, blocks of crust may have tipped like ice flows, thus returning extensive blocks of crust to the mantle without the mechanism of subduction as it occurs on the modern Earth.

It is probable that the large degree of crustal recycling implied by the paucity of surviving crust from the early Archean is incompatible with plate tectonics in its modern form. The transition to plate tectonics may have occurred during the period 3.3–3.1 Ga, followed by an immense increase in the survival of continental crust (Figure 1). The Archean-Proterozoic boundary at 2.5 Ga is the much more publicized major disjuncture in the geological record, but these changes in geological style could only have occurred after abundant continental crust had collected for long periods. The middle Archean may have been a time in which plate tectonics operated on some parts of the Earth, but not on others, and it may have been intermittent (O'Neill et al., 2007).

Summary

Taking multiple indicators into account together, there is good evidence for the operation of plate tectonics on the Earth as far back as 3.3 Ga, but it may have been a modified form until the late Proterozoic without deep subduction and the production of ultrahigh pressure metamorphic belts. In the early Archean, the production of early SiO_2-rich crust may have been caused by melting at the base of a uniformly thick global crust or oceanic plateaux, with or without subduction as on the modern Earth. In the past few years, the number of lines of evidence for cool conditions at the end of the Hadean is increasing, such as the stable isotope signatures of zircons and the composition of TTG gneisses. The importance and abundance of komatiites and exceptional mantle temperatures in the Archean may have been overestimated in the past.

Bibliography

Anderson, D. L., 1989. *Theory of the Earth*. Boston: Blackwell, p. 366.

Arndt, N. T., and Nisbet, E. G., 1982. *Komatiites*. London: Allen & Unwin, p. 526.

Bickle, M. J., Nisbet, E. G., and Martin, A., 1994. Archean greenstone belts are not oceanic crust. *Journal of Geology*, **102**, 121–138.

Brown, M., 2007. Metamorphic conditions in orogenic belts: a record of secular change. *International Geology Review*, **49**, 193–234.

Buick, R., Thornett, J. R., McNaughton, N. J., Smith, J. B., Barley, M. E., and Savage, M., 1995. Record of emergent continental crust 3.5 billion years ago in the Pilbara Craton of Australia. *Nature*, **375**, 574–577.

Canup, R. M., and Asphaug, E., 2001. Origin of the moon in a giant impact near the end of the Earth's formation. *Nature*, **412**, 708–712.

Carey, S. W., 1975. The expanding earth: an essay review. *Earth Science Reviews*, **11**, 105–143.

Chardon, D., Choukroune, P., and Jayananda, M., 1996. Strain patterns, decollement and incipient sagducted greenstone terrains in the Archaean Dharwar craton (southern India). *Journal of Structural Geology*, **18**, 991–1004.

Condie, K. C., 1989. *Plate Tectonics and Crustal Evolution*. Oxford: Pergamon Press, p. 476.

Condie, K. C., and Kroner, A., 2008. When did plate tectonics begin? Evidence from the geological record. *Geological Society of America, Special Paper*, **440**, 281–294.

Davies, G. F., 1995. Punctuated tectonic evolution of the Earth. *Earth and Planetary Science Letters*, **36**, 363–380.

De Wit, M. J., 1998. On Archean granites, greenstones, cratons and tectonics: does the evidence demand a verdict? *Precambrian Research*, **91**, 181–226.

Dewey, J. F., and Bird, J. M., 1970. Mountain belts and the new global tectonics. *Journal of Geophysical Research*, **75**, 2625–2647.

Dhuime, B., Hawkesworth, C. J., Cawood, P. A., and Storey, C. D., 2012. A change in the geodynamics of continental growth 3 billion years ago. *Science*, **355**, 1334–1336.

Dickinson, W. R., 1970. Relations of andesites, granites, and derivative sandstones to arc-trench tectonics. *Reviews of Geophysics and Space Physics*, **8**, 813–860.

Eriksson, K. A., Krapez, B., and Fralick, P. W., 1994. Sedimentology of Archean greenstone belts – signatures of tectonic evolution. *Earth-Science Reviews*, **37**, 1–88.

Evans, D. A. D., and Pisarevsky, S. A., 2008. Plate tectonics on early Earth? Weighing the paleomagnetic evidence. *Geological Society of America, Special Paper*, **440**, 249–263.

Foley, S., Tiepolo, M., and Vannucci, R., 2002. Growth of early continental crust controlled by melting of amphibolite in subduction zones. *Nature*, **417**, 837–840.

Fowler, C. M. R., 2005. *The Solid Earth: An Introduction to Global Geophysics*. New York: Cambridge University Press, p. 685.

Hawkesworth, C. J., and Kemp, A. I. S., 2006. Evolution of the continental crust. *Nature*, **443**, 811–817.

Hoffman, P. F., and Ranalli, G., 1988. Archean oceanic flake tectonics. *Geophysical Research Letters*, **15**, 1077–1080.

Hooper, P. R., Binfer, G. B., and Lees, K. R., 2002. Ages of the Steens and Columbia River flood basalts and their relationship to extension-related calc-alkalic volcanism in eastern Oregon. *Geological Society of America Bulletin*, **114**, 43–50.

Isacks, B., Oliver, J., and Sykes, L. R., 1968. Seismology and the new global tectonics. *Journal of Geophysical Research*, **73**, 5855–5899.

Jacob, D. E., 2004. Nature and origin of eclogite xenoliths from kimberlites. *Lithos*, **77**, 295–316.

Le Pichon, X., 1968. Sea-floor spreading and continental drift. *Journal of Geophysical Research*, **73**, 3661–3696.

Lee, D. C., Halliday, A. N., Snyder, G. A., and Taylor, L. A., 1997. Age and origin of the moon. *Science*, **278**, 1098–1103.

Martin, H., 1999. Adakitic magmas: modern analogues of Archaean granitoids. *Lithos*, **46**, 411–429.

McKenzie, D. P., and Morgan, W. J., 1969. Evolution of triple junctions. *Nature*, **224**, 125–133.

McKenzie, D. P., and Parker, R. L., 1967. The North Pacific: an example of tectonics on a sphere. *Nature*, **216**, 1276–1280.

Mints, M. V., Belousova, E. A., Konilov, A. N., Natapov, L. M., Shchipansky, A. A., Griffin, W. L., O'Reilly, S. Y., Dokukina, K. A., and Kaulina, T. V., 2010. Mesoarchean subduction processes: 2.87 Ga eclogites from the Kola Peninsula, Russia. *Geology*, **38**, 739–742.

Miyashiro, A., 1961. Evolution of metamorphic belts. *Journal of Petrology*, **2**, 277–311.

Mloszewska, A. M., Pecoits, E., Cates, N. L., Mojzsis, S. J., O'Neil, J., Robbins, L. J., and Konhauser, K. O., 2012. The composition of Earth's oldest iron formations: the Nuvvuagittuq Supracrustal Belt (Quebec, Canada). *Earth and Planetary Science Letters*, **317–318**, 331–342.

Moeller, A., Appel, P., Mezger, K., and Schenk, V., 1995. Evidence for a 2 Ga subduction zone: eclogites in the Usagaran Belt of Tanzania. *Geology*, **23**, 1067–1100.

Moyen, J.-F., Stevens, G., and Kister, A., 2006. Record of mid-Archean subduction from metamorphism in the Barberton terrain, South Africa. *Nature*, **442**, 559–562.

Neal, C. R., Mahoney, J. J., Kroenke, L. W., Duncan, R. W., and Petterson, M. G., 1997. The Ontong-Java plateau. *American Geophysical Union Monograph*, **100**, 183–216.

Nutman, A. P., Bennett, V. C., Friend, C. R. L., and Norman, M. D., 1999. Meta-igneous (non-gneissic) tonalites and quartz diorites from an extensive ca. 3800 Ma terrain south of the Isua supracrustal belt, southwest Greenland: constraints on early crust formation. *Contributions to Mineralogy and Petrology*, **137**, 364–388.

O'Neill, C., Lenardic, A., Moresi, L., Torsvik, T. H., and Lee, C. T. A., 2007. Episodic Precambrian subduction. *Earth and Planetary Science Letters*, **262**, 552–562.

Pearce, J. A., 2008. Geochemical fingerprinting of oceanic basalts with applications to ophiolite classification and the search for Archean oceanic crust. *Lithos*, **100**, 14–48.

Peltonen, P., and Kontinen, A., 2004. The Jormua Ophiolite: a mafic-ultramafic complex from an ancient ocean-continent transition zone. In Kusky, T. M. (ed.), *Precambrian Ophiolites and Related Rocks*. Amsterdam: Elsevier. Developments in Precambrian Geology 13, pp. 35–72.

Pitman, W., and Heirtzler, J., 1966. Magnetic anomalies over the Pacific-Antarctic ridge. *Science*, **154**, 1164–1171.

Raff, A. D., and Mason, R. G., 1961. Magnetic survey off the west coast of North America, 40°N latitude to 50°N longitude. *Geological Society of America Bulletin*, **72**, 1267–1270.

Rapp, R. P., Shimizu, N., and Norman, M. D., 2003. Growth of early continental crust by partial melting of eclogite. *Nature*, **425**, 605–609.

Runcorn, S. K., 1959. Rock magnetism. *Science*, **129**, 1002–1012.

Scholl, D. W., and von Heune, R., 2007. Crustal recycling at modern subduction zones applied to the past – issues of crustal growth and preservation of continental basement crust, mantle geochemistry, and supercontinent reconstruction. In Hatcher, R. D., Jr., Carlson, M. P., McBride, J. H., and Martínez Cataln, J. R. (eds.), *4-D Framework of Continental Crust*. Boulder: Geological Society of America, Memoir, Vol. 200, pp. 9–32.

Scott, D. J., St-Onge, M. R., Lucas, S. B., and Helmstaedt, H., 1991. Geology and chemistry of the Early Proterozoic Purtuniq ophiolite, Cape Smith Belt, northern Quebec, Canada. In Peters, T., et al. (eds.), *Ophiolite Genesis and Evolution of the Oceanic Lithosphere*. Sultanate of Oman: Ministry of Petroleum and Minerals, pp. 817–849.

Shirey, S. B., and Richardson, S. H., 2011. Start of the Wilson Cycle at 3 Ga shown by diamonds from subcontinental mantle. *Science*, **333**, 434–436.

Smithies, R. H., van Kranendonk, M. J., and Champion, D. C., 2005. It started with a plume – early Archaean basaltic proto-continental crust. *Earth and Planetary Science Letters*, **238**, 284–297.

Stern, R. J., 2005. Evidence from ophiolites, blueschists, and ultrahigh-pressure metamorphic terranes that the modern episode of subduction tectonics began in Neoproterozoic time. *Geology*, **33**, 557–560.

Strik, G. M. H. A., Blake, T. S., Zegers, T. E., White, S. H., and Langereis, C. G., 2003. Paleomagnetism of flood basalts in the Pilbara Craton, Western Australia: late Archaean continental drift and the oldest known reversal of the geomagnetic field. *Journal of Geophysical Research*, **108**(B12), 2551, doi:10.1029/2003JB002475.

Valley, J. W., Peck, W. H., King, E. M., and Wilde, S. A., 2002. A cool early Earth. *Geology*, **30**, 351–354.

Van Kranendonk, M. J., 2011. Onset of plate tectonics. *Science*, **333**, 413–414.

Vine, F. J., and Matthews, D. M., 1963. Magnetic anomalies over ocean ridges. *Nature*, **199**, 947–949.

Wilde, S. A., Valley, J. W., Peck, W. H., and Graham, C. M., 2001. Evidence from detrital zircon for the existence of continental crust and oceans on the Earth 4.4 Gyr ago. *Nature*, **409**, 175–178.

Willbold, M., Hegner, E., Stracke, A., and Rocholl, A., 2009. Continental geochemical signatures in dacites from Iceland and implications for models of early Archaean crust formation. *Earth and Planetary Science Letters*, **279**, 44–52.

Wilson, J. T., 1965. A new class of faults and their bearing on continental drift. *Nature*, **207**, 343–347.

Cross-references

Driving Forces: Slab Pull, Ridge Push
Hot Spots and Mantle Plumes
Intraplate Magmatism
Plate Tectonics
Subduction
Wilson Cycle

ANOXIC OCEANS

Christian März[1] and Hans-Jürgen Brumsack[2]
[1]School of Civil Engineering and Geosciences (CEGS), Newcastle University, Newcastle upon Tyne, UK
[2]Institute for Chemistry and Biology of the Marine Environment, University of Oldenburg, Oldenburg, Germany

Definition

A large body of open marine water free of dissolved molecular oxygen and potentially containing dissolved hydrogen sulfide.

Introduction

Today's world oceans are dominated by oxic water masses, except for some marginal basins like the Black Sea, fjords, upwelling areas, and so-called coastal "anoxic dead zones." However, since the formation of oceans on Earth, anoxic (oxygen-free) conditions have occurred frequently in parts of the global ocean, leading to the deposition of organic-rich sediments that subsequently generated oil or gas, the main energy source for our modern civilization. Here we will focus on deep ocean anoxia, their environmental controls, and their geological expressions.

The Proterozoic

Throughout large parts of the Precambrian, the oceans were anoxic. Oxygen-producing photosynthetic organisms like algae were only beginning to evolve, and atmospheric oxygen levels were orders of magnitude lower than at present. Due to this oxygen-poor atmosphere, sulfate was not produced in significant quantities by chemical weathering of magmatic rocks on land, and sulfate concentrations in the early oceans were low. In contrast, dissolved Fe concentrations were very high due to hydrothermal input related to ocean crust formation, reductive weathering on land, and possibly diagenetic Fe release from sediments. These oxygen-free, sulfate-poor but iron-rich (ferruginous) oceans persisted over much of the Proterozoic until the Great Oxidation Event (Canfield, 1998), which initiated the global deposition of banded iron formations (BIFs). When atmospheric oxygen concentrations reached a threshold, the oxidative weathering of sulfide minerals led to the establishment of the marine sulfate pool. Consequently, organic matter degradation by microbial sulfate reduction started to release hydrogen sulfide into the water column. As the dissolved Fe in the water column reacted with hydrogen sulfide to form pyrite, ferruginous conditions were gradually replaced by sulfidic (euxinic) conditions in the marine environment. With increasing atmospheric oxygen, the spatial extent of marine anoxia/euxinia was reduced, only persisting in certain ocean regions (Poulton et al., 2010). While the expansion and duration of Proterozoic deep ocean anoxia were never reached again, there were still periods in the Phanerozoic with widely distributed deep ocean anoxia.

The Phanerozoic

Several intervals of deep ocean anoxia occurred throughout the Phanerozoic (see review by Meyer and Kump, 2008). Their most prominent geological expressions include organic-rich deposits ("black shales") from the Cambrian of China, the Ordovician of North America, the Silurian of North Africa, and the Devonian-Lower Carboniferous of the USA and Europe. Most of these Paleozoic deposits represent shallow marine environments, and synchronous records from the deep oceans are largely missing. In contrast, there is evidence for widespread deepwater anoxia around the Permian-Triassic boundary, coinciding with the most severe extinction event in Earth history (Meyer and Kump, 2008). In the Mesozoic, a number of rather short-termed but severe and widely recognized intervals of ocean anoxia occurred, the so-called oceanic anoxic events (OAEs; Schlanger and Jenkyns, 1976). Although their full global expansion is debated (e.g., Trabucho Alexandre et al., 2010), there is evidence that at least the North Atlantic and Tethys were periodically affected by intense deepwater anoxia to euxinia, whereas comparable records from the Pacific and Indian Oceans are scarce. The Toarcian Oceanic Anoxic Event (T-OAE) and the Cenomanian-Turonian Boundary Event (CTBE or OAE2) are considered to be the most widespread occurrences of Mesozoic marine anoxia (Jenkyns, 2010). The Cenozoic sediment record is comparably poor in black shale-type deposits. Organic-rich sediments related to the Paleocene-Eocene Thermal Maximum (PETM) are mostly limited to shallow marine environments (Cohen et al., 2007). However, the deep Arctic Ocean record shows clear signs of Paleocene and Eocene deepwater anoxia, most likely related to its restricted circulation and stable water column stratification. The most recent witnesses of short-termed deepwater anoxia are Pliocene-Pleistocene sapropels of the Eastern Mediterranean (Emeis et al., 2000; De Lange et al., 2008).

"Modern analogues"

In the modern ocean, there are no true analogues for open marine black shale formation. However, a number of "near-analogue" environments can serve as natural laboratories to study some of the physical, chemical, and biological processes that interacted in past anoxic oceans (Demaison and Moore, 1980). The most prominent example is the Black Sea, Earth's largest permanently euxinic marine basin (Degens and Ross, 1974). It is characterized by moderate primary productivity but strong salinity stratification and a very restricted connection to the Mediterranean. A similarly well-studied but less restricted and less euxinic marine environment is the Cariaco Basin, and also a number of fjords and bays around the world (e.g., Framvaren Fjord, Saanich Inlet, Kau Bay) exhibit multiannual deepwater anoxia to euxinia (Richards, 1965; Tissot and Welte, 1984). Also coastal upwelling areas (e.g., off Namibia, Chile, Peru, Arabian Sea) share common features with past anoxic oceans as they display high seasonal primary productivity, leading to deepwater oxygen consumption and occasional development of sulfidic water masses (Copenhagen, 1953). Over the past decades, the study of geochemically similar but spatially restricted analogues has advanced our understanding of past anoxic oceans. A wide range of geochemical proxies is now at hand to unravel the paleoenvironmental setting from sedimentary archives.

Environmental controls: organic matter export, water column stratification, plate tectonics, and astronomical forcing

Two fundamental environmental parameters are traditionally suggested as drivers of oceanic anoxia: high primary productivity inducing enhanced organic matter export to the seafloor and enhanced organic matter preservation by limited deepwater ventilation, i.e., low oxygen (Demaison and Moore, 1980). Modern end-members of this "productivity versus preservation" debate are the Black Sea, on the one hand (moderate productivity, severe basin restriction, salinity stratification, absence of molecular oxygen below redoxcline), and upwelling areas, on the other hand (full connection to the open ocean, seasonally very high productivity). Recent studies suggest that most black shale deposits do not allow a strict differentiation between organic matter productivity and preservation as dominant cause for organic carbon accumulation and anoxia. Past examples of restricted anoxic basins include the mid-Cretaceous proto-North Atlantic and the Paleogene Arctic Ocean. Past high-productivity regions with anoxic water masses developed in the mid-Cretaceous equatorial Atlantic as nutrients were "trapped" due to specific circulation patterns (Trabucho Alexandre et al., 2010). To create and maintain marine anoxia, a range of important biogeochemical feedback mechanisms needs to be considered. One well-known "anoxia-productivity feedback" is the efficient recycling of phosphate from marine sediments if bottom-water oxygen is strongly depleted (Ingall et al., 1993; Slomp et al., 2004; März et al., 2008). Recycled phosphate can then be reintroduced into the photic zone and sustain high-productivity and organic matter export.

Although the onsets and terminations of widespread ocean anoxia are still debated, they appear to be related to plate tectonics and volcanism (Meyer and Kump, 2008). Astronomical forcing can regulate the intensity of anoxia once favorable conditions have been created (De Lange et al., 2008). Plate tectonic constellations are responsible for the restriction of ocean basins (e.g., the mid-Cretaceous North and South Atlantic), reducing the renewal and oxygenation of deepwater masses (Erbacher et al., 2001). Volcanic activity introduces large amounts of the greenhouse gas CO_2 into the atmosphere, leading to both global warming and enhanced chemical weathering (Turgeon and Creaser, 2008). Warming reduces the oxygen solubility in the oceans and the vertical mixing of its water masses, and chemical weathering enhances terrestrial nutrient input into the oceans (Adams et al., 2010). Global warming leads to eustatic sea-level rise, concomitant widespread flooding, and an expansion of shelf sea areas. Besides the increased potential for shelf anoxia, these shallow marine areas also trap a large fraction of siliciclastic detritus before it reaches the deep basins. Therefore, organic-rich deepwater deposits are often characterized by reduced terrestrial dilution (Arthur et al., 1988).

Recognizing anoxia in sedimentary archives

As most multicellular organisms (such as benthic foraminifera) require oxygen to survive, anoxic environments are usually strongly deprived in macroscopic life. The sedimentological expressions of this lack of burrowing macrobenthos are finely laminated deposits without any bioturbation.

Both organic and inorganic geochemistry provides a number of proxies to detect anoxia/euxinia in different parts of a water column. For instance, a biomarker to detect euxinic conditions in the photic zone is isorenieratene, an organic pigment exclusively produced by phototrophic green and purple sulfur bacteria (Sinninghe Damsté et al., 1993).

A first-order geochemical approach to reconstruct marine anoxia was the organic carbon to total sulfur (C/S) ratio of Black Sea deposits (Berner, 1984). While normal, oxic marine sediments exhibited C/S ratios around 3, the production of H_2S by organic matter degradation in the Black Sea, and its subsequent precipitation as metal sulfides (mostly pyrite, FeS_2), resulted in C/S ratios <3. A better representation of the C-S-Fe interrelationship is a ternary diagram with the end-members organic C, total S, and reactive Fe, as it allows a distinction between anoxic systems where pyrite formation was limited by H_2S or reactive Fe (Dean and Arthur, 1989).

Under anoxic ferruginous conditions, the sulfate depletion of the water column prevents any enrichment of sedimentary S and hence the recognition of water column anoxia using bulk sedimentary C-S-Fe proxies (Poulton and Canfield, 2011). To distinguish non-sulfidic from sulfidic conditions, a sequential Fe extraction scheme distinguishes between the relative amounts of the Fe pool that can potentially react, or already have reacted, with H_2S (Poulton and Canfield, 2005). In addition to C-S-Fe systematic, the selective removal of sedimentary P relative to organic C under anoxic conditions results in increasing sedimentary C/P ratios under increasingly oxygen-depleted bottom-water conditions (Algeo and Ingall, 2007).

Among the most powerful redox proxies are trace metal enrichments/depletions relative to their lithogenic background contents (Brumsack, 2006). Redox-sensitive metals change their redox states under oxic, suboxic, and anoxic conditions (e.g., Mn, U, V); sulfide-forming metals precipitate sulfide phases/coprecipitate with pyrite if H_2S is available (e.g., Mo, Cd, Zn). While questions remain in our understanding of natural trace metal systematics, several metals have emerged over the years as reliable redox indicators. While Mn tends to be depleted from sediments already under suboxic conditions, Re becomes enriched within sulfidic microenvironments in the sediment. Uranium is enriched if full anoxia is reached in the sediments, while Cd and Zn precipitate as sulfides under weakly sulfidic conditions. Molybdenum requires stronger euxinia in the water column to be transferred to particle-reactive thiomolybdates. The interpretation of trace metal records is complicated by the fact that some of these elements serve as micronutrients and are pre-concentrated via plankton (Böning et al., 2004) before burial. They may as well respond to specific redox conditions in the surface sediment, while others respond to redox changes in the water column. Early diagenetic processes pose another challenge, as trace metals may be redistributed within the sediment, biasing the primary redox record.

Beyond element contents and speciation in marine sediments, the stable isotopic signatures of metals like Mo and Fe are increasingly used as redox indicators (Anbar and Rouxel, 2007). The specific value of Mo isotopes is based on the quantitative removal of Mo from a water column under fully euxinic conditions without any isotope fractionation. Through isotope and mass balance calculations, this ultimately gives the percentage of the global oceanic water mass that was fully euxinic at a given point in time. The stable Fe isotope composition of marine sediments, in conjunction with total Fe enrichment patterns, is controlled by the input of isotopically light Fe from suboxic shelf sediments and the shuttling and distribution of this excess Fe within a suboxic redoxcline. Therefore, the depth of the redoxcline in a water column and the extent of underlying euxinic water masses can be reconstructed (Eckert et al., 2013). However, the full application of these isotope systems as paleo-redox proxies will require a better understanding of these trace metals and their isotopes in the marine environment under a variety of boundary conditions.

Conclusions

Our understanding of anoxic conditions in the oceans has greatly advanced over the past decades. The classic "productivity versus preservation" discussion sparked many of these advances and has motivated scientists to take a closer look at the underlying mechanisms causing, sustaining, and terminating ocean anoxia. This was achieved by combining studies of modern and ancient anoxic marine environments and by working across scientific disciplines. Nowadays, geoscientists can take advantage of a large set of analytical tools and paleoenvironmental proxies that led to a better understanding of, e.g., the causes for the development of anoxic conditions, their effects on organisms, global element cycles and the climate system, duration and spatial extent of anoxia and euxinia, the degree of anoxia or euxinia, and the reasons for their termination. Still, many questions remain about marine anoxia. Facing a warming ocean and intensifying oxygen depletion in coastal and open oceans, these questions are certainly relevant for the development and understanding of our future ocean.

Bibliography

Adams, D. D., Hurtgen, M. T., and Sageman, B. B., 2010. Volcanic triggering of a biogeochemical cascade during oceanic anoxic event 2. *Nature Geosciences*, **3**, 201–204.

Algeo, T. J., and Ingall, E., 2007. Sedimentary C_{org}:P ratios, paleocean ventilation, and Phanerozoic atmospheric pO_2. *Palaeogeography, Palaeoclimatology, Palaeoecology*, **256**, 130–155.

Anbar, A. D., and Rouxel, O., 2007. Metal stable isotopes in paleoceanography. *Annual Reviews of Earth and Planetary Sciences*, **35**, 717–746.

Arthur, M. A., Brumsack, H.-J., Jenkyns, H. C., and Schlanger, S. O., 1988. Stratigraphy, geochemistry, and paleoceanography of organic carbon-rich Cretaceous sequences. In Ginsburg, R. N., and Beaudoin, B. (eds.), *Cretaceous Resources, Events and Rhythms: Background and Plans for Research. Proceedings of ARW*. Digne: Kluwer Academic Publishers.

Berner, R. A., 1984. Sedimentary pyrite formation: an update. *Geochimica et Cosmochimica Acta*, **48**, 605–615.

Böning, P., Brumsack, H.-J., Böttcher, M. E., Schnetger, B., Kriete, C., Kallmeyer, J., and Borchers, S. L., 2004. Geochemistry of Peruvian near-surface sediments. *Geochimica et Cosmochimica Acta*, **68**, 4429–4451.

Brumsack, H.-J., 2006. The trace metal content of recent organic-rich sediments: implications for Cretaceous black shale formation. *Palaeogeography, Palaeoclimatology, Palaeoecology*, **232**, 344–361.

Calvert, S. E., and Price, N. B., 1970. Minor metal contents of recent organic-rich sediment off South West Africa. *Nature*, **227**, 593–595.

Canfield, D. E., 1998. A new model for Proterozoic ocean chemistry. *Nature*, **396**, 450–453.

Cohen, A. S., Coe, A. L., and Kemp, D. B., 2007. The Late Palaeocene-Early Eocene and Toarcian (Early Jurassic) carbon isotope excursions: a comparison of their time scales, associated environmental changes, causes and consequences. *Journal of the Geological Society of London*, **164**, 1093–1108.

Copenhagen, W. J., 1953. *The Periodic Mortality of Fish in the Walvis Region*. Investigational Report 14, Division of Fisheries Union of South Africa.

De Lange, G. J., Thomson, J., Reitz, A., Slomp, C. P., Speranza Principato, M., Erba, E., and Corselli, C., 2008. Synchronous basin-wide formation and redox-controlled preservation of a Mediterranean sapropel. *Nature Geosciences*, **1**, 606–610.

Dean, W. E., and Arthur, M. A., 1989. Iron-sulfur-carbon relationships in organic-carbon-rich sequences I: cretaceous western interior seaway. *American Journal of Science*, **289**, 708–743.

Degens, E. T., and Ross, D. A., 1974. *The Black Sea – Geology, Chemistry, and Biology*. American Association of Petroleum Geologists Memoir 20, Tulsa, Oklahoma.

Demaison, G. J., and Moore, G. T., 1980. Anoxic environments and oil source bed genesis. *Organic Geochemistry*, **2**, 9–31.

Eckert, S., Brumsack, H.-J., Severmann, S., Schnetger, B., März, C., and Fröllje, H., 2013. Establishment of euxinic conditions in the Holocene Black Sea. *Geology*, **41**, 431–434.

Emeis, K. C., Sakamoto, T., Wehausen, R., and Brumsack, H.-J., 2000. The sapropel record of the eastern Mediterranean Sea – results of Ocean Drilling Program Leg 160. *Palaeogeography, Palaeoclimatology, Palaeoecology*, **158**, 371–395.

Erbacher, J., Huber, B. T., Norris, R. D., and Markey, M., 2001. Increased thermohaline stratification as a possible cause for an ocean anoxic event in the Cretaceous period. *Nature*, **409**, 325–327.

Ingall, E. D., Bustin, R. M., and Van Cappellen, P., 1993. Influence of water column anoxia on the burial and preservation of carbon and phosphorus in marine shales. *Geochimica et Cosmochimica Acta*, **57**, 303–316.

Jenkyns, H. C., 2010. Geochemistry of oceanic anoxic events. *Geochemistry, Geophysics, Geosystems*, **11**, Q03004, doi:10.1029/2009GC002788.

Lyons, T. W., Anbar, A. D., Severmann, S., Scott, C., and Gill, B. G., 2009. Tracking euxinia in the ancient ocean: a multiproxy perspective and Proterozoic case study. *Annual Reviews in Earth and Planetary Sciences*, **37**, 507–534.

März, C., Poulton, S. W., Beckmann, B., Küster, K., Wagner, T., and Kasten, S., 2008. Redox sensitivity of P cycling during marine black shale formation: dynamics of sulfidic and anoxic, non-sulfidic bottom waters. *Geochimica et Cosmochimica Acta*, **72**, 3703–3717.

Meyer, K. M., and Kump, L., 2008. Oceanic euxinia in Earth history: causes and consequences. *Annual Reviews in Earth and Planetary Sciences*, **36**, 251–288.

Poulton, S. W., and Canfield, D. E., 2005. Development of a sequential extraction procedure for iron: implications for iron partitioning in continentally derived particulates. *Chemical Geology*, **214**, 209–221.

Poulton, S. W., and Canfield, D. E., 2011. Ferruginous conditions: a dominant feature of the ocean through Earth's history. *Elements*, **7**, 107–112.

Poulton, S. W., Fralick, P. W., and Canfield, D. E., 2010. Spatial variability in oceanic redox structure 1.8 billion years ago. *Nature Geosciences*, **3**, 486–490.

Richards, F. A., 1965. Anoxic basins and fjords. In Riley, J. P., and Skirrow, G. (eds.), *Chemical Oceanography*. London: Academic, pp. 611–643.

Schlanger, S. O., and Jenkyns, H. C., 1976. Cretaceous oceanic anoxic events: causes and consequences. *Geologie en Mijnbouw*, **55**, 179–184.

Sinninghe Damsté, J. S., Wakeham, S. G., Kohnen, M. E. L., Hayes, J. M., and De Leeuw, J. W., 1993. A 6,000-year sedimentary molecular record of chemocline excursions in the Black Sea. *Nature*, **362**, 827–829.

Slomp, C. P., Thomson, J., and De Lange, G. J., 2004. Controls on phosphorus regeneration and burial during formation of eastern Mediterranean sapropels. *Marine Geology*, **203**, 141–159.

Tissot, B. P., and Welte, D. H., 1984. *Petroleum Formation and Occurrence*. Berlin: Springer.

Trabucho Alexandre, J., Tuenter, E., Henstra, G. A., Van der Zwan, K., Van de Wal, R. S. W., Dijkstra, H. A., and De Boer, P. L., 2010. The mid-Cretaceous North Atlantic nutrient trap: black shales and OAEs. *Paleoceanography*, **25**, PA4201, doi:10.1029/2010PA001925.

Turgeon, S. C., and Creaser, R. A., 2008. Cretaceous oceanic anoxic event 2 triggered by a massive magmatic episode. *Nature*, **454**, 323–326.

Cross-references

Biogenic Barium
Deep-sea Sediments
Geologic Time Scale
Laminated Sediments
Marginal Seas
Organic Matter
Paleoproductivity
Sapropels
Source Rocks, Reservoirs
Upwelling

ASPHALT VOLCANISM

Gerhard Bohrmann
MARUM-Center for Marine Environmental Sciences, University of Bremen, Bremen, Germany

Definition

Asphalt volcanism is a type of hydrocarbon seepage associated with submarine mounds found in the Gulf of Mexico and the Santa Barbara Basin. The term asphalt volcanism was introduced because of the lavalike appearance of the asphalt flows and several other indications, which resemble magmatic eruptions.

Asphalt seepage versus asphalt volcanism

Although asphalt deposits have been described from several places of the ocean floor, the term asphalt volcanism has been introduced as a novel type of hydrocarbon seepage by MacDonald et al. (2004) after an area of approximately 1 km^2 solidified asphalt was found on top of one of the Campeche Knolls in the southern Gulf of Mexico. The knoll was subsequently named Chapopote which is the Aztec word for tar. Those knolls in the Gulf of Mexico are clearly associated with salt tectonism which controls the development of hydrocarbon reservoirs and faults that allow oil and gas to escape at the seafloor. Guided by satellite data that showed evidence of persistent oil seepage in the region, the seafloor of Chapopote was mapped and investigated by MacDonald et al. (2004). Visual surveys revealed extensive surface deposits of solidified asphalt and light crude oil, emanating from sites along the southern rim of a crater-like structure in 3.000 m water depth. Some of the asphalt flows were measured to be at least 15 m across and extended far down the slope. In some places the surface appearance of the asphalt deposits was blocky or ropy (Figure 1) similar to pa'hoehoe lava flows of basalt. Furthermore, large areas of the asphalt deposits were colonized by vestimentiferan tubeworms, bacterial mats, and other biological communities. Also discovered alongside the asphalt were locations of sediment/gas hydrate interlayering associated with emanating gas and oil bubbles from the seafloor.

Based on the collected data and observations, MacDonald et al. (2004) postulated repeated, extensive eruptions of molten asphalt under conditions which could destabilize gas hydrates on the seafloor. A violent destabilization of hydrates could contribute to slope failures and mass wasting mapped on Chapopote Knoll, as well as documented on other Campeche Knolls. Based on the idea that the asphalt on Chapopote was molten during extrusion, Hovland et al. (2005) argued for a model that relies on supercritical water being transported vertically upward through a suspected internal conduit within the salt diapir, from near the base of the sedimentary column at perhaps 13 km depth. Organic material including bitumen should have been transported upward together with "hydrothermal-like" components as a hot substance. At the summit of the Chapopote structure, a hot slurry flowed out onto the seafloor where bitumen and asphalt devolatilized rapidly, eventually building up the asphalt volcano's structure.

The Chapopote Knoll was surveyed in greater detail by Brüning et al. (2010) using the MARUM ROV QUEST. The results support the concept that the asphalt deposits on Chapopote originate from seepage of heavy oil with a density slightly greater than water, which leads to remaining petroleum and oil residues on the seafloor. During extrusion of the heavy petroleum, the viscosity increased due to the loss of volatiles, and the heavy petroleum forms the lavalike flow structures along the distance where continuous solidification occurs. The investigations

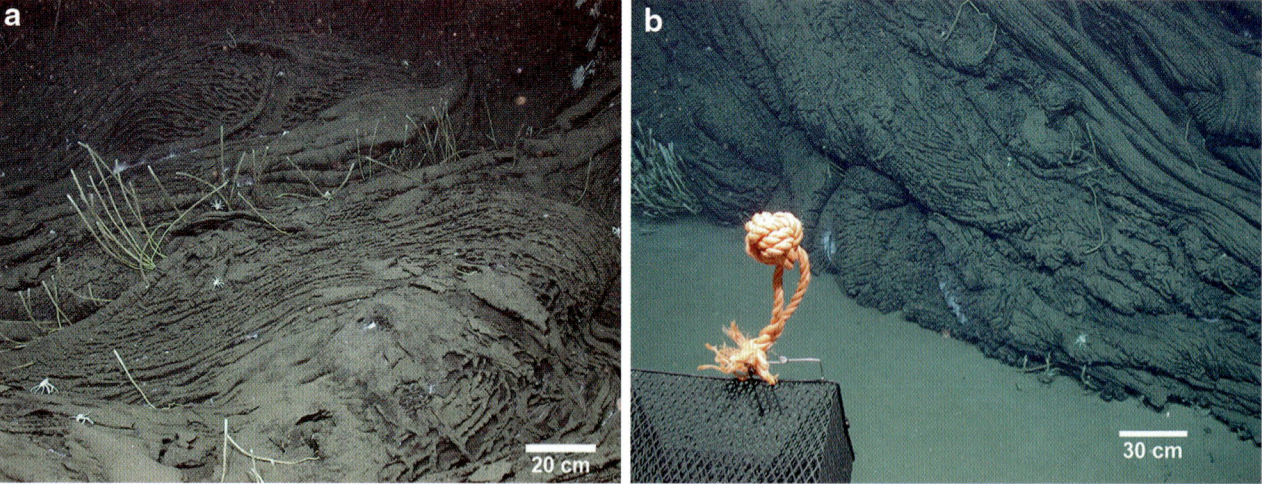

Asphalt Volcanism, Figure 1 Seafloor images from asphalt flows of Chapopote Knoll/Southern Gulf of Mexico taken by MARUM ROV QUEST. (**a**) Vestimentiferan tubeworms and galatheid crabs colonizing the surface of the asphalt. (**b**) Several asphalt layers of ropy surface piled up above the seafloor.

of Brüning et al. (2010) documented that the asphalt is subject to sequential alteration. While fresh asphalt was gooey, older asphalt appeared fragmented and brittle. Highly altered asphalt was often colonized by further chemosynthetic fauna like mytilid clams and others. The change in the consistency of the asphalts goes along with a change in the geochemical composition and microbial signatures (Schubotz et al., 2011). Besides the unusual asphalt formation, the putative "volcanic structure" is representing a very interesting seepage area which extended our knowledge about the broad spectrum of seafloor venting phenomena.

Beside the Gulf of Mexico, asphalt volcanoes are known from the Santa Barbara Basin, California, in much shallower water depths close to the coast. Seven of those morphological structures were described as extinct asphalt volcanoes by Valentine et al. (2010). Radiocarbon dating of carbonate layers intercalated with the asphalt deposits indicated formation of two of the volcanoes between 44 and 31 kyr ago. Based on quantitative assumptions and the geochemistry of samples taken from the volcanoes, the authors estimated the amount of oil and accompanied methane gas, which are emitted at the sites where the residues of the hydrocarbon seepage (i.e., the asphalt) currently are deposited. Since the amount of greenhouse gas (in this case methane) emissions is not known during former times, the study is of great value to reveal estimates of former seepage rates.

Bibliography

Brüning, M., Sahling, H., MacDonald, I. R., Ding, F., and Bohrmann, G., 2010. Origin, distribution, and alteration of asphalts at the Chapopote Knoll, Southern Gulf of Mexico. *Marine and Petroleum Geology*, **27**(5), 1093–1106, doi:10.1016/j.marpetgeo.2009.09.005.

Hovland, M., MacDonald, I. R., Rueslatten, H., Johnsen, H. K., Naehr, T., and Bohrmann, G., 2005. Chapopote asphalt volcano may have been generated by supercritical water. *EOS, Transactions*, **86**(42), 397–402, doi:10.1029/2005EO420002.

MacDonald, I. R., Bohrmann, G., Escobar, E., Abegg, F., Blanchon, P., Blinova, V. N., Brueckmann, W., Drews, M., Eisenhauer, A., Han, X., Heeschen, K. U., Meier, F., Mortera, C., Naehr, T., Orcutt, B., Bernard, B., Brooks, J., and de Farágo, M., 2004. Asphalt volcanism and chemosynthetic life, Campeche Knolls, Gulf of Mexico. *Science*, **304**(5673), 999–1002, doi:10.1126/science.1097154.

Schubotz, F., Lipp, J. S., Elvert, M., Kasten, S., Mollar, X. P., Zabel, M., Bohrmann, G., and Hinrichs, K. U., 2011. *Geochimica et Cosmochimica Acta*, **75**(16), 4377–4398, doi:10.1016/j.gca.2011.05.025.

Valentine, D. L., Reddy, C. M., Farwell, C., Hill, T. M., Pizzarro, O., Yoerger, D. R., Camilli, R., Nelson, R. K., Peacock, E. E., Bagby, S. C., Clarke, B. A., Roman, C. N., and Soloway, M., 2010. Asphalt volcanoes as a potential source of methane to late Pleistocene coastal waters. *Nature Geosciences*, **3**, 345–348, doi:10.1038/NGEO848.

Cross-references

Cold Seeps
Marine Gas Hydrates
Mud Volcano

ASTRONOMICAL FREQUENCIES IN PALEOCLIMATES

André Berger
Georges Lemaître Center for Earth and Climate Research, Catholic University of Louvain, Louvain-la-Neuve, Belgium

Definition

The long-term variations of climate display periods characteristic of three astronomical parameters which are the eccentricity (which fixes the shape of the Earth's orbit), obliquity (the tilt of the equatorial plane on the plane of the Earth's orbit around the Sun), and climatic precession (a measure of the distance from the Earth to the Sun at the summer solstice). Their main periods of variations are 400 and 100 kyr for eccentricity, 41 kyr for obliquity, and 23 and 19 kyr for precession.

Introduction

As in the astronomical theory of paleoclimates the glacial-interglacial cycles are of primary interest, this entry focuses on the long-term variations of the astronomical parameters which are involved in the calculation of the energy received by the Earth from the Sun (here called incoming solar radiation or insolation) at time scales of tens to hundreds of thousands of years.

These are the eccentricity, e, obliquity, ε, and climatic precession, $e \sin \tilde{\omega}$, $\tilde{\omega}$ being the longitude of the perihelion.

The full spectral characteristics of these astronomical elements and the resulting insolation changes date back only to the 1970s. Although the precessional period of 21,000 years was well known since Adhémar (1842) at least, Milankovitch and his contemporaries did not seem to be much interested in these astronomical periods (Berger, 2012). Milankovitch (1920), like Emiliani (1955) 35 years later, estimated only their mean values by counting the number of peaks from the curves that Milankovitch calculated numerically, leading to about 92,000, 40,000, and 21,000 years for e, ε, and $e \sin \tilde{\omega}$, respectively. These periods were confirmed 20 years later when Berger (1973) completed his calculation of the long-term variations of precession, obliquity, and eccentricity. Besides its high accuracy over the Quaternary, the Berger calculation provided, for the first time, a full list as well as the origin of the periods characterizing the theoretical expansion of e (with periods of 413,000, 95,000, 123,000, 99,000, 131,000, and 2,305,000 years in decreasing order of the term's amplitudes), of ε (with periods of 41,000, 53,600, and 29,700 years), and of $e \sin \tilde{\omega}$ (with periods of 23,700, 22,400, 18,900, and 19,200 years) (see Berger, 1978, and for a slightly improved solution Berger and Loutre, 1991). Among these periods, those around 400,000, 2,300,000, and 54,000 and mainly of around 23,000 and 19,000 years

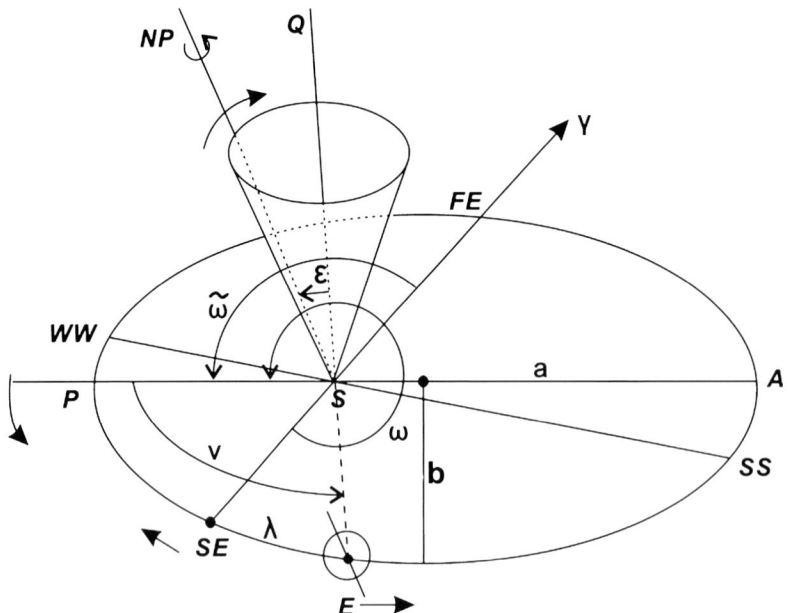

Astronomical Frequencies in Paleoclimates, Figure 1 The Earth (E)'s orbit around the Sun (S). The astronomical elements are defined in the manuscript. a is the semimajor axis of the elliptical orbit and b its semiminor axis, P is the perihelion, A is the aphelion, SE is spring equinox, SS is summer solstice, FE is fall equinox, WW is winter solstice, SQ is the perpendicular to the plane of the ecliptic, S NP is a parallel to the axis of rotation of the Earth describing the cone of precession, ω is the longitude of the perigee, $\tilde{\omega}$ is the longitude of the perihelion relative to the moving vernal point γ, λ is the longitude of the Earth in its orbit, v is the true anomaly, and ε is the obliquity.

were new, their existence not having been previously suspected.

In their science paper, Hays et al. (1976) used a spectral analysis technique which they applied to geological records and to the numerical values of the astronomical parameters calculated by Brouwer and Van Woerkom (1950) and Vernekar (1972) and found also 125,000 and 96,000 years for e, 41,000 years for ε, and 23,000 and 19,000 years for precession.

Orbit of the Earth around the Sun and its axis of rotation

The plane of the Earth's elliptical orbit around the Sun is called the ecliptic. The Sun is not located at the center of the ellipse but rather at one of the foci, a fact that was demonstrated in the seventeenth century by the German astronomer Johannes Kepler (1571–1630) whose three laws are:

- Kepler's first law. Each planet moves in an ellipse with the Sun at one focus.
- Kepler's second law. The radius vector from the Sun to any one planet describes equal areas in equal times.
- Kepler's third law. For any two planets, the squares of the periods are proportional to the cubes of the semimajor axes of their orbits.

As the Earth travels counterclockwise around its orbit each year, it is sometimes nearer to and sometimes farther away from the Sun. Today, the Earth reaches the point in its orbital path known as perihelion, the point at which it is the closest to the Sun, on or about 3 January. On or about 4 July, it reaches aphelion, the point farthest from the Sun. At aphelion the distance between the Earth and the Sun, which is in average roughly 150 million kilometers, is about 5 million kilometers greater than it is at perihelion.

Figure 1 displays the various elements of the Earth's orbit. The ellipse drawn through the perihelion, P, the Earth, E, and the aphelion, A, represents the orbit of the Earth around the Sun.

The Earth's eccentricity, e, which is a measure of its elliptical shape, is given by $\sqrt{a^2 - b^2}/a$, where a is the semimajor axis and b the semiminor axis. The vernal point, γ, also called the First Point of Aries, is the position of the Sun in the sky seen from the Earth at the time of the spring equinox, SE, i.e., in Figure 1 when the Earth crosses the celestial equator from the austral to the boreal hemisphere. It is the origin from which the ecliptical longitudes are measured. The winter (ww) and summer (ss) solstices and γ are shown at their present-day positions. In such a heliocentric system, γ is in the direction of the fall equinox, FE. In practice, observations are made from the Earth and the Sun is considered as if it were revolving around the Earth. In such a case, the system of reference becomes geocentric instead of heliocentric, and the vernal point is in the direction of the spring equinox.

The principal cause of the seasons is not the lack of uniformity of the apparent annual motion of the Sun but rather the tilting of the Earth's axis at an angle, ε, presently $23°27'$ away from a vertical drawn to the plane of the orbit. ε is therefore also the inclination (tilt) of the equator on the ecliptic and is called the obliquity. Seasons occur because the orientation of that axis of rotation remains approximately fixed in space as the Earth revolves about the Sun in 1 year. Each season begins at a particular point in the Earth's orbit. Today these points are reached roughly on 20 or 21 March (spring equinox also called vernal or March equinox), 21 or 22 June (summer solstice also called June solstice), 22 or 23 September (fall equinox also called autumnal or September equinox), and 21 or 22 December (winter solstice also called December solstice). Their respective longitudes differ by a multiple of $90°$ with the longitude at spring equinox being zero by definition.

In the Northern Hemisphere, winter solstice marks the beginning of winter because the North Pole is tipped farthest away from the Sun on that day, making it the shortest day of the year in the entire Northern Hemisphere. Six months later, at summer solstice, the Earth reaches the point at which summer begins in the Northern Hemisphere. At this point, the North Pole is tipped toward the Sun, making the day of the summer solstice the longest day of the year in the Northern Hemisphere. From geometrical consideration, it can be shown that the Sun reaches the zenith at the June-summer solstice at noon of the true solar time (a time similar to the time read on a sundial) at the latitude now of $23°27'$ which defines the tropics of Cancer; it reaches the December-winter solstice at the latitude now of $-23°27'$ which defines the tropics of Capricorn. Between the tropics, the Sun reaches the zenith at noon of the true solar time twice a year. North of the northern polar circle (which latitude is $90° - 23°27' = 66°33'$), at the summer (winter) solstice, the day (night) is 24-h long. At the winter solstice, the Earth is today near perihelion (this approximate coincidence has nothing to do with the beginning of the calendar year).

At the spring and fall equinoxes, the two poles are equidistant from the Sun. On these dates, the number of daylight hours equals the number of hours of darkness at every point on the globe. These two points on the Earth's orbit are therefore known as the equinoxes (the term equinox is indeed derived from the Latin *aequus* (equal) and *nox* (night)). In the Northern Hemisphere, the spring/vernal/March equinox marks the beginning of spring and the fall/autumnal/September equinox the beginning of fall. In the Southern Hemisphere the seasons are reversed.

Orbital elements of a planet

In astronomy, it is usual to define an orbit and the position of the body describing that orbit by six quantities called elements. As shown in Figure 2, three of these elements

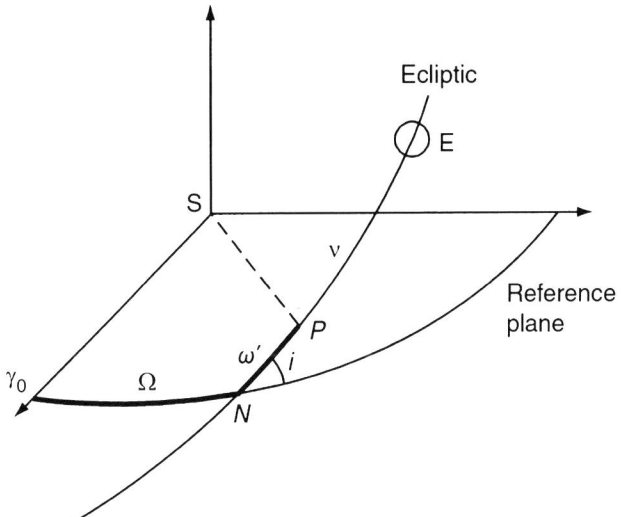

Astronomical Frequencies in Paleoclimates, Figure 2 The position of the Earth E around the Sun S given by six quantities called the elements (see the manuscript for definitions). N is the ascending node, γ_0 is the vernal point at the epoch of reference, Ω (the angle $\gamma_0 SN$) is the longitude of the ascending node, ω' (the angle NSP) is the argument of the perihelion P, i is the inclination of the ecliptic on the plane of reference, and v is the true anomaly.

(Ω, ω', i) define the orientation of the orbit with respect to a set of axes, two of them define the size and the shape of the orbit (a and e, respectively), and the sixth, v (with time), defines the position of the body within the orbit at that time. In the case of a planet moving in an elliptical orbit around the Sun, it is convenient to take a set of orthogonal axes in and perpendicular to the plane of reference, xoy, with the origin, o, at the center of the Sun or at the barycenter of the planetary system. The z-axis is taken to be perpendicular to this reference plane, so that the three axes form a right-handed orthogonal coordinate system. The reference point from which the angles are measured is labeled γ_0. As the reference plane is usually chosen to be the ecliptic at a particular fixed date of reference (called the epoch of reference, to distinguish it from any date of the past or the future), γ_0 is, in such a case, the vernal point at the epoch of reference which is taken to be 1950 CE in the Berger calculation. The point where the orbit cuts the reference plane with an increasing z-coordinate is called the ascending node, N; Ω, the longitude of that ascending node, is measured in the reference plane from γ_0; P is the perihelion; ω' is the argument of the perihelion, an angle measured anticlockwise along the orbit from the ascending node; π is the longitude of the perihelion measured from the reference vernal point, γ_0, and is equal to the sum of Ω and ω'; i is the inclination of the orbital plane relative to the fixed reference plane; v is the true anomaly, the angle measured anticlockwise between the perihelion and the Earth's position; and λ is the longitude of the Earth in its orbit measured from the spring equinox at a specific date. Note that v and λ fix the Earth on its orbit, respectively, from the perihelion and from the spring equinox at a specific date (Figure 1). In the orbital plane, the distance, r, from the Earth to the Sun, normalized to a, is given by the equation of the ellipse:

$$\rho = \frac{r}{a} = \frac{1-e^2}{1+e\cos v}$$

The perihelion distance is consequently given by $a(1-e)$ and the aphelion distance by $a(1+e)$, which leads to the difference between the two being proportional to $2ae$.

Long-term variations of the orbital elements

Due to the mutual attraction of the Sun, the planets, and the Moon, both the orientation and eccentricity of the ecliptic are changing with time. So is the axis of rotation of the Earth and therefore the equator. As a consequence, both the positions of the perihelion and of the vernal point relative to a fixed frame of reference are changing with time. The longitude of the perihelion, $\tilde{\omega}$, relative to the moving vernal point, γ, is shown in Figure 1. It is equal to $\pi + \psi$, where the annual general precession in longitude, ψ, describes the clockwise absolute motion of γ along the Earth's orbit relative to the fixed stars, and the longitude of the perihelion, π, describes the anticlockwise absolute motion of the perihelion relative to the fixed stars.

To compute the long-term variations of the orbital elements, Newton's law of gravitational attraction is applied to the planetary system. This leads to the Lagrange equations, i.e., a system of equations the number of which is equal to six times the number of planets. These equations provide the time evolution of the six orbital elements of each planet. They relate all the orbital elements of the planets between them together and describe their motion around the Sun. However, these equations possess some inconvenient features for orbits with small eccentricities and inclinations, both of which appearing in the denominators of some terms. It is therefore desirable to use a modified form of these equations by setting:

$$h = e \sin \pi$$
$$k = e \cos \pi$$
$$p = \sin i \sin \Omega$$
$$q = \sin i \cos \Omega$$

If approximations are introduced in the development of the disturbing function (which expresses the mutual attraction of the Sun and the planets), the long-term behavior of h, k, p, q are given by

$$h = \sum_k M_k \sin(g_k t + \beta_k)$$
$$k = \sum_k M_k \cos(g_k t + \beta_k) \quad (1)$$

$$p = \sum_k N_k \sin(s_k t + \delta_k)$$
$$q = \sum_k N_k \cos(s_k t + \delta_k) \quad (2)$$

For the Berger (1978) solution, the amplitude M_k and N_k, the mean rates g_k and s_k, and the phases β_k and δ_k were calculated from Bretagnon (1974). Their numerical values are available in Berger (1973 and 1978) and also in Tables 1 and 2 of Berger and Loutre (1990). For the Berger and Loutre (1991) solution, Laskar (1988) was used leading to slightly different values of the orbital elements for periods of time longer than 1 million years (Berger and Loutre, 1992) but not changing our line of argument. Tables 3 and 4 of Berger and Loutre (1991) give the numerical values of the amplitudes, mean rates, and phases in decreasing order of the magnitude of the five most important terms for the two solutions Berger (1978) and Berger and Loutre (1990).

Table 1 provides the periods associated with the five largest amplitudes in decreasing order of magnitude for Eqs. 1 and 2. The values of i for Eq. (1) and j for Eq. (2) are those corresponding to the ordering traditionally used in celestial mechanics. Although there is not a one-to-one relationship between these periods and the individual planets, it can be shown that their origin is, to some extent, associated to one particular planet.

Because of the importance of the mass of Jupiter, the frequency (s_5) associated with that planet is equal to 0. This explains why the spectrum of i is dominated by two periods around 70 kyr and two of about 210 kyr, those characterizing Eq. 2. If the invariable plane (plane perpendicular to the total angular momentum) is taken instead of the ecliptic of the epoch, the term involving s_5 is excluded and the spectrum of the inclination on that plane is dominated by periods of 98, 107, and 1,300 kyr coming from the resonances between the Earth and Mercury, Mars and Venus, and Earth and Mars respectively. The periods of about 100 kyr are close to the periods of 100 kyr of eccentricity but totally different, and they are not associated with the 100 kyr of the geological records (Berger et al., 2005).

Astronomical Frequencies in Paleoclimates, Table 1 Periods related to the frequencies g and s of the most important terms in the series expansion of (e, π) and (i, Ω)

	Frequencies g			Frequencies s	
i	Period	Planet	j	Period	Planet
5	308,043	Jupiter	5		
2	176,420	Venus	3	68,829	Earth
4	72,576	Mars	1	230,977	Mercury
3	75,259	Earth	4	72,732	Mars
1	249,275	Mercury	2	191,404	Venus
			6	49,339	Saturn

Long-term variations of obliquity and precession

After having calculated the motion of the planetary point masses around the Sun, the Poisson equations for the Earth-Moon system were used to compute the long-term variations of two astronomical parameters which, with the eccentricity, play a fundamental role in the long-term seasonal and latitudinal variations of insolation. These two parameters are obliquity, ε and $\tilde{\omega}$ the longitude of the perihelion relative to the moving vernal point.

Although the hypothesis that the planets attract each other as if the mass of every one of them was concentrated in their respective center of mass is valid for the calculation of the orbital elements, the flattening of the planets has a perceptible effect on their rotation. The Moon's gravitational attraction on the Earth causes two small bulges directed toward the Moon and away from it in the ocean and on the solid Earth. Dissipative processes cause a lag in the tidal response, and a torque is exerted that does not vanish when averaged over an orbital period of the Moon. The consequence of this torque is a change in the Earth's angular momentum or, equivalently, to an increase of the length of the day by about one thousandth of a second in 100 years. At the same time, the bulges slow the Moon down in its orbital motion and lead to an increase in the Earth-Moon distance of the order of a few centimeters per year (400 millions years ago, the length of the day was about 22 h, the year consisted of about 400 days, and the Moon must have been 4 % closer to the Earth).

Viewed by an observer on the Earth near the North Pole, the stars appear to trace out concentric circles, the center of which defines the celestial North Pole, the extension of the Earth's rotational axis in the sky. This celestial North Pole currently lies close to the star Polaris. As already noted by Hipparchus in about 120 BCE, the rotational axis is observed to move slowly and to trace out a cone, clockwise, with a half-angle of about 23.5° around the pole of the ecliptic, a motion which takes about 25,700 years to go full circle around the heavens. This steady motion of the rotational axis in space is called the astronomical precession of the Earth (or general precession in longitude). This is due to the inclination of the major axis of the oblate Earth to the ecliptic. Consequently, the net gravitational force on the Earth due to the Sun exerts a torque which attempts to draw the equator into the plane of the ecliptic, but the spinning of the Earth resists this; instead the torque causes motion of the spin axis about the pole of the ecliptic. This observed precession results from the sum of the solar and lunar torques (because of the large mass of the Sun and the proximity of the Moon) plus a rather minor contribution arising from the other planets. In addition, the complex interplay of the solar and lunar orbits induces small oscillations in the secular precessional motion of the rotational axis; these oscillations are known as forced nutations. The principal nutation term arises from a 19-year periodicity in the inclination of the Moon's orbit, but these nutations are not considered in the long-term variations of the astronomical parameters discussed here because of their small magnitude.

These long-term variations of ε and ψ can be expressed analytically as is the case for h, k, p, q:

$$\varepsilon = \varepsilon^* + \sum_i A_i \cos(\gamma_i t + \varsigma_i) \qquad (3)$$

$$\psi = kt + \alpha + \sum_i S_i \sin(\xi_i t + \sigma_i) \qquad (4)$$

ε^* and α are constant of integration and k is the precessional constant. For the Berger (1978) solution, their numerical values are

$$\varepsilon^* = 23°320556$$
$$\alpha = 3°392506$$
$$k = 50''439273$$

The amplitudes, mean rates, and phase of Eqs. 3 and 4 are given in Berger (1978). Some more detailed analytical developments are given in Berger and Loutre (1991) with for the most important terms:

$$\gamma_i = \xi_i = s_j + k$$
$$\varsigma_i = \sigma_i = \delta_j + \alpha$$

Long-term variations of eccentricity, obliquity, and climatic precession

The incoming solar radiation changes from day to day due to the Earth's elliptical motion around the Sun. But there are other changes of interest related to the planetary system and the Sun's interior. In particular, the total solar energy received by the whole Earth over one full year varies, by a very small amount (Berger, 1977; Berger and Loutre, 1994), because the mean Earth-Sun distance varies in relation to changes in the shape of the Earth's orbit around the Sun (the eccentricity, e). The solar output (the so-called solar constant) and the opacity of the interplanetary medium are also changing, but their effects remain difficult to prove at our time scales.

In addition, the seasonal and latitudinal distributions of insolation have also long-term variations which are related to the orbit of the Earth around the Sun and to the inclination of its axis of rotation. These involve three well-identified astronomical parameters (Figure 1): the eccentricity, e; the obliquity, ε; and the climatic precession, $e \sin \tilde{\omega}$, a measure of the Earth-Sun distance at the summer solstice. $\tilde{\omega}$, the longitude of the perihelion, is a measure of the angular distance between the perihelion and the vernal equinox that are both in motion. In a geocentric system, the angle ω is the longitude of the perigee and its numerical values are obtained by adding 180° to $\tilde{\omega}$ (Figure 1). The present-day value of e is 0.016. As a consequence, although the Earth's orbit is very close to a circle, the Earth-Sun distance, and consequently the insolation, varies by as much as 3.2 % and 6.4 %, respectively, over

Astronomical Frequencies in Paleoclimates, Table 2 Amplitudes, mean rates, phases, and periods of the five largest amplitude terms in the trigonometrical expansion of climatic precession, obliquity, and eccentricity (BER78 refers to Berger, 1978 and is based upon Bretagnon, 1974 and BER90 refers to Berger and Loutre, 1991 and is based on Laskar, 1988)

	Amplitudes		Mean rate (″/year)		Phase (°)		Period (years)	
	BER78	BER90	BER78	BER90	BER78	BER90	BER78	BER90
Climatic precession								
1	0.018608	0.018970	54.64648	54.66624	32.0	32.2	23,716	23,708
2	0.016275	0.016318	57.78537	57.87275	197.2	201.3	22,428	22,394
3	0.013007	0.012989	68.29654	68.33975	131.7	153.4	18,976	18,964
4	0.009888	0.008136	67.65982	67.79501	323.6	311.4	19,155	19,116
5	0.003367	0.003870	67.28601	55.98574	102.8	78.6	19,261	23,149
Obliquity								
1	−2462.22	−1969.00	31.60997	31.54068	251.9	247.14	41,000	41,090
2	−857.32	−903.50	32.62050	32.62947	280.8	288.79	39,730	39,719
3	−629.32	−631.67	24.17220	32.08588	128.3	265.33	53,615	40,392
4	−414.28	−602.81	31.98378	24.06077	292.7	129.70	40,521	53,864
5	−311.76	−352.88	44.82834	30.99683	15.4	43.20	28,910	41,811
Eccentricity								
1	0.011029	0.011268	3.13889	3.20651	165.2	169.2	412,885	404,178
2	0.008733	0.008819	13.65006	13.67352	99.7	121.2	94,945	94,782
3	0.007493	0.007419	10.51117	10.46700	294.5	312.0	123,297	123,818
4	0.006724	0.005600	13.01334	13.12877	291.6	279.2	99,590	98,715
5	0.005812	0.004759	9.87446	9.92226	126.4	110.1	131,248	130,615

the course of 1 year. The obliquity, which defines our tropical latitudes and polar circles, is presently 23°27′. The longitude of the perihelion is 102°, which confirms that the Northern Hemisphere winter occurs when the Earth is about the closest to the Sun.

The long-term variations of obliquity are given by Eq. 3

$$\varepsilon = \varepsilon^* + \sum_i A_i \cos(\gamma_i t + \varsigma_i)$$

For eccentricity, its analytical development can be calculated from

$$e = \sqrt{(e \sin \widetilde{\omega})^2 + (e \cos \widetilde{\omega})^2} \quad \text{or from}$$

$$\sqrt{(e \sin \pi)^2 + (e \cos \pi)^2}$$

which leads to

$$e = e^* + \sum_i E_i \cos(\lambda_i t + \phi_i) \tag{5}$$

with $e^* = 0.0287069$.

From the definition of $\widetilde{\omega} = \pi + \psi$ and using Eqs. 1, 4, and 5, the climatic precession parameter $e \sin \widetilde{\omega}$ can be expressed into the following trigonometrical expansion as a quasiperiodic function of time:

$$e \sin \widetilde{\omega} = \sum_i P_i \sin(\alpha_i t + \eta_i) \tag{6}$$

The amplitudes P_i, A_i, and E_i; the frequencies α_i, γ_i, and λ_i; and the phases η_i, ς_i, ϕ_i in Eqs. 3, 5, and 6 (Table 2) have been computed by Berger (1978) and Berger and Loutre (1991). These analytical expansions can be used over 1–2 millions years (Berger and Loutre, 1992), but for more remote times, numerical solutions are necessary (Laskar, 1999).

The sign of the amplitude of terms 3 and 5 in climatic precession and of terms 2 and 3 in eccentricity in column BER78 has been changed relatively to the values given in Berger (1978) in agreement with the change of the phase by 180°. This has been done to allow an easier comparison between the solutions. The five terms given in this table do not allow an accurate computation of the astronomical parameters. More terms are requested. They are available with a computer program in http://www.astr.ucl.ac.be and http://www.elic.ucl.ac.be/modx/elic/index.php?id=83.

Equations 3 and 5 show that ε and e vary quasiperiodically around constant values ε^* (23.32°) and e^* (0.0287). This implies that, in the estimation of the order of magnitude of the terms in insolation formulae where ε and e occur, they may be considered as a constant to a first approximation. On the other hand, in the insolation formulas used to study past and future astronomical forcing of climate, the amplitude of $\sin \widetilde{\omega}$ is modulated by eccentricity in the term $e \sin \widetilde{\omega}$. The envelope of $e \sin \widetilde{\omega}$ is therefore given exactly by e.

Besides their simplicity and practicability for easy computation, Eqs. 3, 5, and 6 and their derivation allow to explain the origin of the main periods associated with the astronomical theory of paleoclimates (for full details see Berger and Loutre, 1990). The periods calculated in

Berger (1978) (but also in Berger and Loutre, 1991) come from the fundamental periods of the orbital elements associated with the g_k in Eq. 1 and s_k in Eq. 2 and from the period of the astronomical precession associated with k.

The main periods of eccentricity in Eq. 5 are 413, 95, 123, 100, and 131 kyr coming, respectively, from the following relationships where the subscripts of g (and s) are the classical ones reported in Table 1:

$$\lambda_1 = g_2 - g_5$$
$$\lambda_2 = g_4 - g_5$$
$$\lambda_3 = g_4 - g_2$$
$$\lambda_4 = g_3 - g_5$$
$$\lambda_5 = g_3 - g_2$$

For obliquity, the main periods are 41 and 54 kyr, coming, respectively, from

$$\gamma_1 = s_3 + k$$
$$\gamma_3 = s_6 + k$$

For climatic precession, the main periods are 23.7, 22.4, 18.98, and 19.16 kyr with

$$\alpha_1 = g_5 + k$$
$$\alpha_2 = g_2 + k$$
$$\alpha_3 = g_4 + k$$
$$\alpha_4 = g_3 + k$$

This leads to conclude that the periods characterizing the expansion of e are nonlinear combinations of the precessional periods and, in particular, that the eccentricity periods close to 100 kyr are originating from the periods close to 23 and 19 kyr in precession:

$$\lambda_1 = \alpha_2 - \alpha_1$$
$$\lambda_2 = \alpha_3 - \alpha_1$$
$$\lambda_3 = \alpha_3 - \alpha_2$$
$$\lambda_4 = \alpha_4 - \alpha_1$$
$$\lambda_5 = \alpha_4 - \alpha_2$$

As a consequence,

$$\lambda_3 = \lambda_2 - \lambda_1$$
$$\lambda_5 = \lambda_4 - \lambda_1$$

which clearly shows that not all the periods of eccentricity are independent. Using the periods associated to the largest amplitude terms in Eqs. 1 and 2, we would conclude that the periods of 413, 95, and 100 kyr are the most fundamental ones (123 and 131 kyr deriving directly from them) with:

- 413 kyr coming from the resonance between Venus and Jupiter
- 95 kyr coming from the resonance between Mars and Jupiter
- 100 kyr coming from the resonance between Earth and Jupiter

For the obliquity, Mars, Earth, and Moon explain the 40-kyr period, Saturn being related to the 54 kyr one. For climatic precession, in addition to the Moon effect, the two periods close to 23 kyr come from Jupiter and Venus and those close to 19 kyr from Mars and Earth.

Instability of the astronomical periods

Over the last few million years, eccentricity varies between 0 and 0.06 with an average period of 95 kyr, being slightly shorter over the future (91 kyr). Correlatively, the line of apses (line through P and A in Figure 1) made a revolution in a fixed reference frame with an average period of 125 kyr but largely varying between 20 and 250 kyr. This dispersion around the average value is related to the existence of very short periods related to the very low values of eccentricity. If these short periods are ignored, the minimum value of the period becomes more realistic (60 kyr).

Figure 3 shows the long-term variations of these three astronomical parameters over the past 400 ka and into the future for the next 100 ka. It shows in particular that the 100-kyr period is not stable in time (Berger et al., 1998), being remarkably shorter over the present day. Actually, the most important theoretical period of eccentricity, 400 kyr, is weak before 1 Ma BP and is particularly strong over the next 400 ka, with the strength of the components in the 100-kyr band changing in the opposite way. It is worth pointing out that this weakening of the 100-kyr period started about 900 ka ago when this period began to appear very strongly in paleoclimate records. This implies that the 100-kyr period found in paleoclimatic records is definitely not linearly related to eccentricity. We are now approaching a minimum of e at the 400-kyr time scale: at 27 ka AP, the Earth's orbit will be circular. Actually, transitions between successive strong 400-kyr cycles (as it is the case now) are characterized by very small eccentricity and short eccentricity cycles with a low amplitude of variation.

For precession, the average period is roughly 21.5 kyr with large dispersion varying between 14 and 30 kyr. At the 400-kyr time scale, the amplitude and frequency modulations are inversely related: when the amplitude of precession is small (large), the period is short (long). It is the reverse at the 100-kyr time scale.

Obliquity varies between 22° and 24°5 with a very stable period of 41 kyr, but there is an amplitude modulation with a time duration of about 1.3 Myr. The spectra of both the amplitude and frequency modulations of obliquity display significant power at 171 and 97 kyr. Although this last might look close to the so-called 100-kyr eccentricity period, they are not related (Mélice et al., 2001). At the 1.3-Myr time scale, a large amplitude corresponds to a short period, the reverse being true at the 170-kyr time scale.

While today the winter solstice occurs near perihelion, at the end of the deglaciation, roughly 10 ka BP, it occurred near aphelion. Moreover, because the length of

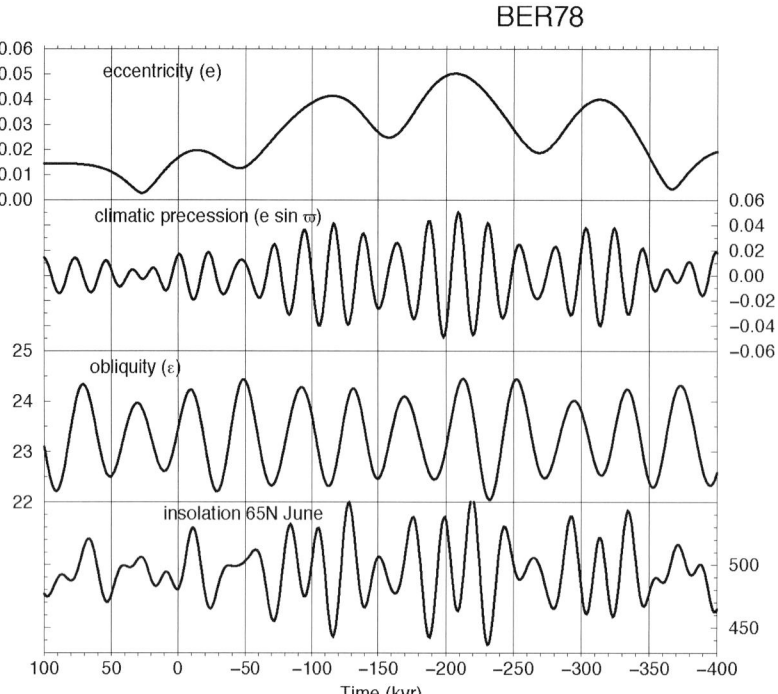

Astronomical Frequencies in Paleoclimates, Figure 3 Long-term variations of eccentricity, climatic precession, obliquity, and insolation at 65°N at the summer solstice from 400 ka BP to 100 ka AP. Time is progressing from *right* to *left* (The data are derived from the formula given in Berger (1978)).

the seasons varies in time according to Kepler's second law, the solstices and equinoxes occur at different calendar dates during the geological past and in the future. Presently in the Northern Hemisphere, the longest seasons are spring (92.8 days) and summer (93.6 days), while autumn (89.8 days) and winter (89 days) are notably shorter. In 1250 CE, spring and summer have had the same length (as did autumn and winter) because the winter solstice occurred at perihelion. About 4,500 years into the future, the Northern Hemisphere spring and winter will have the same short length and consequently summer and fall will be equally long.

Summary and conclusions

The most important periods of the three astronomical parameters which drive the long-term variations of climate are 400 and 100 kyr for eccentricity, 41 kyr for obliquity, and 23 and 19 kyr for precession. Although they are often called Milankovitch periods, they are not originating from his work. For example, the 41-kyr obliquity cycle and the average period of climatic precession, 21 kyr, date from the mid-nineteenth century. Among the properties of these astronomical parameters, some are less known and are stressed here again.

The term climatic precession has been introduced in the 1970s to avoid the too often confusion made at these early times with the astronomical precession.

The first theoretical period of eccentricity, about 400 kyr, and the double precessional peaks, 23 and 19 kyr, were only discovered in the early 1970s and play presently a more and more important role in paleoclimatology. The 400-kyr period is particularly strong over the next 400 kyr, whereas the 100 kyr is very weak. Actually, transitions between successive strong 400-kyr cycles are characterized by very small eccentricity. This is happening now and at 27 ka AP, the Earth's orbit will be circular.

The analytical calculation of the trigonometrical expansions of the astronomical elements shows that the eccentricity periods close to 100 kyr are originating from the periods close to 23 and 19 kyr in precession, and the 400-kyr period is a combination of the two first precessional periods. The periods of eccentricity are therefore not all independent, those of 413, 95, and 100 kyr being the most fundamental ones.

For precession, the average period is roughly 21.5 kyr with a large dispersion varying between 14 and 30 kyr. At the 400-kyr time scale, when the amplitude of precession is small (large), the period is short (long). It is the reverse at the 100-kyr time scale.

Obliquity varies between 22° and 24°5 with a very stable period of 41 kyr, but there is an amplitude modulation with a period of about 1.3 Myr. At this 1.3-Myr time scale, large amplitude corresponds to a short period, the reverse being true at the 170-kyr time scale.

As obliquity and eccentricity vary around constant values (23.32° for obliquity and 0.0287 for eccentricity), these elements may be considered as a constant in a first very rough approximation as compared to the variations of climatic precession. As the amplitude of $e \sin \tilde{\omega}$ is modulated by eccentricity, the envelope of $e \sin \tilde{\omega}$ is given exactly by e.

Because the length of the seasons varies in time according to Kepler's second law, the solstices and equinoxes occur at different calendar dates through time. The meteorological seasons are therefore continuously shifting through the astronomical ones and do not correspond to the same configuration of the Earth relative to the Sun.

Bibliography

Adhémar, J. A., 1842. *Révolution des mers, Déluges périodiques*. Première édition: Carilian-Goeury et V. Dalmont, Paris. Seconde édition: Lacroix-Comon, Hachette et Cie, Dalmont et Dunod, Paris, 359 p.

Berger, A., 1973. *Théorie Astronomique des Paléoclimats*. Dissertation doctorale, Belgium, Université catholique de Louvain, 2 Vol.

Berger, A., 1977. Long-term variations of the Earth's orbital elements. *Celestial Mechanics*, **15**, 53–74.

Berger, A., 1978. Long-term variations of daily insolation and Quaternary climatic changes. *Journal of the Atmospheric Sciences*, **35**(12), 2362–2367.

Berger, A., 2012. A brief history of the astronomical theories of paleoclimates. In Berger, A., Mesinger, F., and Sijacki, D. (eds.), *Climate change at the eve of the second decade of the century. Inferences from Paleoclimates and regional aspects. Proceedings of Milankovitch 130th Anniversary Symposium*. Wien: Springer, pp. 107–129. doi:10.1007/978-3-7091-0973-1.

Berger, A., and Loutre, M. F., 1990. Origine des fréquences des éléments astronomiques intervenant dans le calcul de l'insolation. *Bulletin de la Classe des Sciences de l'Académie royale de Belgique*, **I**(1–3), 45–106. 6ème série.

Berger, A., and Loutre, M. F., 1991. Insolation values for the climate of the last 10 million years. *Quaternary Science Review*, **10**(4), 297–317.

Berger, A., and Loutre, M. F., 1992. Astronomical solutions for paleoclimate studies over the last 3 million years. *Earth Planetary Science Letters*, **111**(2/4), 369–382.

Berger, A., and Loutre, M. F., 1994. Precession, eccentricity, obliquity, insolation and paleoclimates. In Duplessy, J. C., and Spyridakis, M. T. (eds.), *Long-Term Climatic Variations: Data and Modling*. Berlin: Springer. NATO ASI Series, Vol. I22, pp. 107–151.

Berger, A., Loutre, M. F., and Mélice, J. L., 1998. Instability of the astronomical periods from 1.5 Myr BP to 0.5 Myr AP. *Palaeoclimates Data and Modelling*, **2**(4), 239–280.

Berger, A., Mélice, J. L., and Loutre, M. F., 2005. On the origin of the 100-kyr cycles in the astronomical forcing. *Paleoceanography*, **20**, PA4019.

Bretagnon, P., 1974. Termes à longues périodes dans le système solaire. *Astronomy and Astrophysics*, **30**, 141–154.

Brouwer, D., and Van Woerkom, A. J. J., 1950. Secular variations of the orbital elements of principal planets. *Astronomical Papers Prepared for American Ephemeris and Nautical Almanac, US Naval Observatory*, **13**(2), 81–107.

Emiliani, C. R. W., 1955. Pleistocene temperatures. *Journal of Geology*, **63**(6), 538–578.

Hays, J. D., Imbrie, J., and Shackleton, N. J., 1976. Variations in the Earth's orbit: pacemaker of the ice ages. *Sciences*, **194**, 1121–1132.

Laskar, J., 1988. Secular evolution of the solar system over 10 million years. *Astronomy and Astrophysics*, **198**, 341–362.

Laskar, J., 1999. The limits of Earth orbital calculations for geological time scale use. In Shackleton, N. J., McCave, I. N., and Weedon, G. P. (eds.), *Astronomical (Milankovitch) Calibrations of the Geological Time Scale*. London: Philosophical Transactions of the Royal Society, Vol. 357, pp. 1735–1759.

Mélice, J. L., Coron, A., and Berger, A., 2001. Amplitude and frequency modulations of the Earth's obliquity for the last million years. *Journal of Climate*, **14**(6), 1043–1054.

Milankovitch, M., 1920. *Théorie Mathématique des Phénomènes Thermiques Produits par la Radiation Solaire. Académie Yougoslave des Sciences et des Arts de Zagreb*. Paris: Gauthier Villars.

Vernekar, A. D., 1972. Long-period global variations of incoming solar radiation. *Meteorology Monographs*, **12**(34), 130.

Cross-references

Paleoceanography
Relative Sea-level (RSL) Cycle

AXIAL SUMMIT TROUGHS

Samuel Adam Soule[1] and Michael R. Perfit[2]
[1]Woods Hole Oceanographic Institution, Woods Hole, MA, USA
[2]Department of Geological Sciences, University of Florida, Gainesville, FL, USA

Synonyms

Axial summit collapse trough; Axial summit graben; Cleft

Definition

An axial summit trough (AST) is a narrow trough or volcanically modified graben that develops at the crest of a midocean ridge and typically is the locus of volcanic and hydrothermal activity.

Introduction

At the summit of magmatically robust midocean ridges (MORs), a narrow trough may develop that marks the location of the divergent plate boundary and is the locus of most magmatic and hydrothermal activity (Macdonald and Fox, 1988). This axial summit trough (AST) is typically less than 500 m in width and 50 m in depth and may be discontinuous along the ridge in terms of its width, depth, strike, and continuity. ASTs are common along fast- and intermediate-spreading rate ridges (e.g., East Pacific Rise, Juan de Fuca Ridge, Galapagos Spreading Center), and magma-rich, inflated portions of the slow-spreading Mid-Atlantic Ridge and Lau Basin back-arc spreading center.

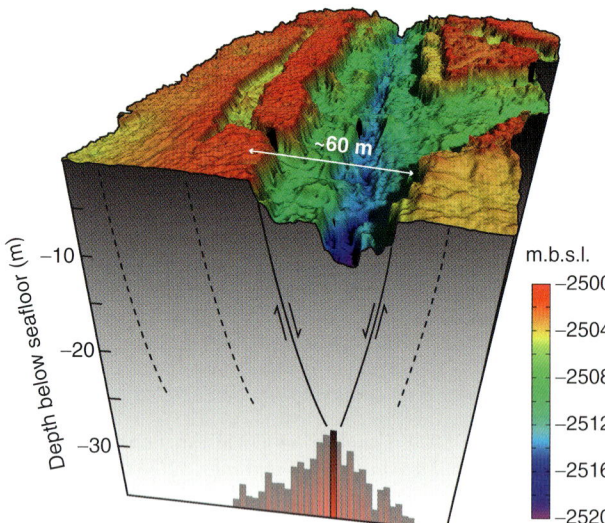

Axial Summit Troughs, Figure 1 Perspective view of the AST at the northern EPR with schematic representation of normal faults between the seafloor and intruding dike (After Soule et al., 2009).

Development of the axial summit trough

The AST is a volcanically modified tectonic graben. The graben forms in response to deformation above magmatic dikes originating at the axial magma lens located 1–3 km beneath the ridge crest (Chadwick and Embley, 1998; Fornari et al., 1998). Models of dike intrusion into a homogeneous elastic medium with physical properties similar to oceanic crust produce horizontal stresses at the seafloor, which reach their maximum at distances from the dike centerline of ~1.5 times the depth to the dike tip (Rubin and Pollard, 1988; Figure 1). Slips along normal faults between these symmetric zones of dilation and the dike tip produce grabens with widths many times greater than depths. The typical depth and width of ASTs on ridges are larger than could be expected for the intrusion of a single 1-m wide dike, the nominal dike width on midocean ridges (Qin and Buck, 2008). Thus, it is assumed that the AST represents accumulated deformation over many dike intrusion cycles. In some instances, an AST may evolve into a much wider (~1–2 km) and deeper (~50–100 m) axial valley by this process (Carbotte et al., 2006; Soule et al., 2009). Other contributions to graben subsidence may include magma withdrawal from subridge melt lenses (Carbotte et al., 2003).

The dimensions of the AST are established by tectonic processes but modified by volcanic processes (Figure 2). Volcanic overprinting during a single eruptive event can range from complete infilling of the AST to minor narrowing and shallowing of the trough based on the style (e.g., low or high eruption rate) and frequency of eruptions (Chadwick and Embley, 1998; Fornari et al., 1998; Soule et al., 2009). At low eruption rates, pillow lavas can fill the trough and completely obscure the AST. At high eruption rates, lava filling the trough can drain back into eruptive fissures after the eruption has ceased. In the latter case, the solidified upper crust of the lava flow that filled the AST founders and leaves only thin fragmental remnants of the flow on the trough floor (Fornari et al., 1998). Some portions of the upper crust may remain intact, supported by lava pillars – hollow conduits composed of solidified lava – that form between the base and upper surface of the flow (Francheteau et al., 1979; Chadwick and Embley, 1998; Gregg et al., 2000; Chadwick, 2003). This can result in an apparent narrowing of the trough although the open space of the graben remains beneath a thin lid of solidified lava.

In addition to hosting the majority of eruptive fissures, the AST hosts the bulk of hydrothermal vents along the ridge crest. High- and low-temperature hydrothermal venting occurs within the trough, commonly colocated with volcanic fissures as well as along the walls of the AST (e.g., Haymon et al., 1991; Fornari et al., 1998). The association of hydrothermal vents with the AST reflects the location of the graben directly over the shallowest axial melt lenses along the ridge. High permeability in the vertical direction reflects the presence of steeply dipping faults and dikes within what is the weakest and thinnest portion of the ocean crust.

The AST as a record of volcanic-tectonic history

Discontinuities in the AST occur in the form of abrupt changes in the depth and/or width of the trough, physical breaks in the continuity of the trough, or changes in trough orientation. These discontinuities, referred to as devals (deviations in axial linearity) (Langmuir et al., 1986), reflect the finest degree of segmentation of the ridge axis (Macdonald et al., 1988; Haymon et al., 1991; White et al., 2000; Haymon and White, 2004; White et al., 2006) and in many cases correlate with other indicators of segmentation such as lava geochemistry, ridge-crest water depth, presence and continuity of the subridge melt lenses, and volcanic deposition processes (Langmuir et al., 1986; Carbotte et al., 2000; Soule et al., 2007, 2009; Smith et al., 2001).

As an example, the AST along the well-studied EPR between 9 °N and 10 °N varies in width from 20 to 300 m and depth from 2 to 20 m over 60 km of ridge length. The AST is divided into roughly eight segments that range in length from 1 to 25 km. The most pronounced change in AST properties occurs at ~9°44′N, coincident with the onset of significant shoaling in the ridge-crest depth (Figure 3). In addition, this marks the southernmost extent of two recent eruptions in 1991–1992 and 2005–2006. The AST maintains average widths of ~180 and ~75 m to the north and south of this discontinuity, respectively. The greater width and depth of the AST across this discontinuity would suggest that a greater

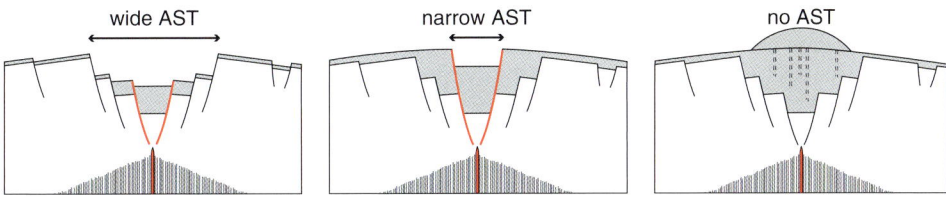

Axial Summit Troughs, Figure 2 Examples of tectonically defined graben (*white*) overprinted by recent volcanic deposition (*gray*) (Soule et al., 2009).

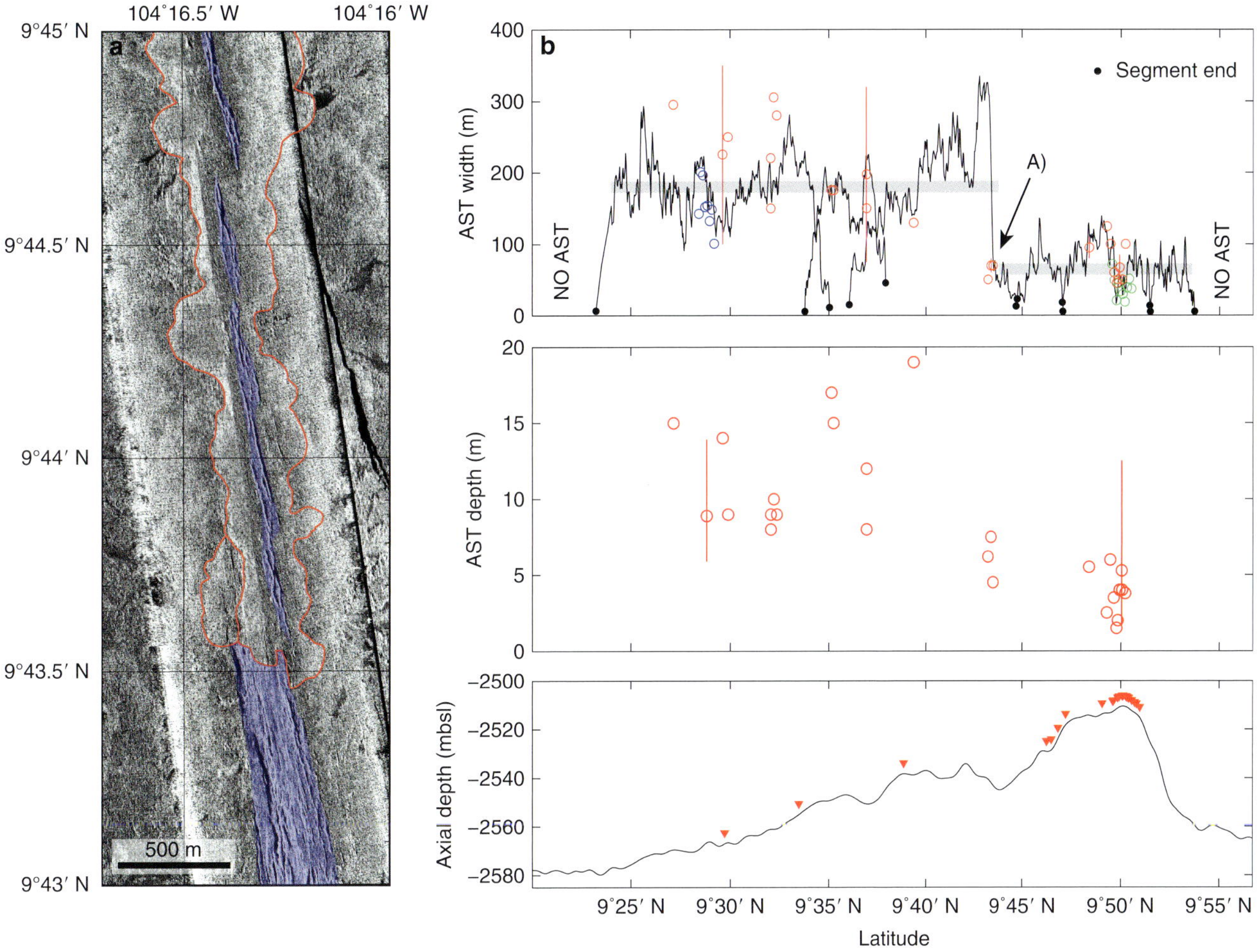

Axial Summit Troughs, Figure 3 (**a**) Example of AST discontinuity in width (*blue shading*) along the East Pacific Rise (*EPR*) as imaged by 120-kHz side-scan sonar backscatter. The AST narrows north of 9°43.5 due to recent volcanic deposition. (**b**) Compilation of AST properties along the EPR shows that this transition in width (measured from side-scan sonar) persists for tens of kilometers north and south and is consistent with AST depth (measured from towed camera crossings) and ridge-crest depth (measured from ship-based bathymetry) as well as geophysical and geochemical indicators of melt supply (Soule et al., 2009).

proportion of dikes reach the seafloor and erupt to the north resulting in greater magmatic overprinting. This is consistent with a break in the axial magma lens and magma compositions that become more mafic to the north of this discontinuity. Thus, it appears that the AST provides a visible record of the frequency of eruptions, which tend to infill a consistently growing graben.

Summary

The axial summit trough is a common feature along the ridge crest of fast- and intermediate-spreading rate ridges and is present along slower-spreading but magma-rich ridges. The trough forms from deformation due to shallowly intruded dikes, with deformation accumulating over many years, but the graben dimensions are modified by volcanic deposition from eruptive fissures located within the AST to degrees depending on the frequency and effusion rates of eruptions. The AST, where present, hosts the majority hydrothermal vents along the ridge axis due to its proximity to crustal heat sources (e.g., magma chambers) and anisotropic permeability within the sheeted dikes and tectonically disrupted upper crust. Discontinuities within the AST are coincident with discontinuities in other ridge-crest properties such as depth, axial magma chamber continuity and melt content, and lava geochemistry.

Bibliography

Carbotte, S. M., Solomon, A., and Ponce-Correa, G., 2000. Evaluation of morphological indicators of magma supply and segmentation from a seismic reflection study of the East Pacific Rise 15°30′–17°N. *Journal of Geophysical Research*, **105**, 2737.

Carbotte, S. M., Ryan, W. B., Jin, W., Cormier, M. H., Bergmanis, E., Sinton, J., and White, S., 2003. Magmatic subsidence of the East Pacific Rise (EPR) at 18°14′ S revealed through fault restoration of ridge crest bathymetry. *Geochemistry, Geophysics, Geosystems*, **4**, doi:10.1029/2002GC000337.

Carbotte, S. M., Detrick, R. S., Harding, A., Canales, J. P., Babcock, J., Kent, G., Ark, E. V., Nedimovic, M., and Diebold, J., 2006. Rift topography linked to magmatism at the intermediate spreading Juan de Fuca Ridge. *Geology*, **34**, 209.

Chadwick, W. W., 2003. Quantitative constraints on the growth of submarine lava pillars from a monitoring instrument that was caught in a lava flow. *Journal of Geophysical Research*, **108**, 2534.

Chadwick, W. W., Jr., and Embley, R. W., 1998. Graben formation associated with recent dike intrusions and volcanic eruptions on the mid-ocean ridge. *Journal of Geophysical Research*, **103**, 9807.

Fornari, D. J., Haymon, R. M., Perfit, M. R., Gregg, T. K. P., and Edwards, M. H., 1998. Axial summit trough of the East Pacific Rice 9°–10°N: geological characteristics and evolution of the axial zone on fast spreading mid-ocean ridge. *Journal of Geophysical Research*, **103**, 9827–9855.

Francheteau, J., Juteau, T., and Rangin, C., 1979. Basaltic pillars in collapsed lava-pools on the deep ocean floor. *Nature*, **281**, 209–211.

Gregg, T. K. P., Fornari, D. J., Perfit, M. R., Ridley, W. I., and Kurz, M. D., 2000. Using submarine lava pillars to record mid-ocean ridge eruption dynamics. *Earth and Planetary Science Letters*, **178**, 195–214.

Haymon, R. M., and White, S. M., 2004. Fine-scale segmentation of volcanic/hydrothermal systems along fast-spreading ridge crests. *Earth and Planetary Science Letters*, **226**, 367–382.

Haymon, R. M., Fornari, D. J., Edwards, M. H., Carbotte, S., Wright, D., and Macdonald, K. C., 1991. Hydrothermal vent distribution along the East Pacific Rise crest (9°09′–54′ N) and its relationship to magmatic and tectonic processes on fast-spreading mid-ocean ridges. *Earth and Planetary Science Letters*, **104**, 513–534.

Langmuir, C. H., Bender, J. F., and Batiza, R., 1986. Petrological and tectonic segmentation of the East Pacific Rise, 5°30′–14°30′ N. *Nature*, **322**, 422–429.

Macdonald, K. C., and Fox, P. J., 1988. The axial summit graben and cross-sectional shape of the East Pacific Rise as indicators of axial magma chambers and recent volcanic eruptions. *Earth and Planetary Science Letters*, **88**, 119–131.

Macdonald, K. C., Fox, P. J., Perram, L. J., Eisen, M. F., Haymon, R. M., Miller, S. P., Carbotte, S. M., Cormier, M. H., and Shor, A. N., 1988. A new view of the mid-ocean ridge from the behaviour of ridge-axis discontinuities. *Nature*, **335**, 217–225.

Qin, R., and Buck, W. R., 2008. Why meter-wide dikes at oceanic spreading centers? *Earth and Planetary Science Letters*, **265**, 466–474.

Rubin, A. M., and Pollard, D. D., 1988. Dike-induced faulting in rift zones of Iceland and Afar. *Geology*, **16**, 413–417.

Smith, M. C., Perfit, M. R., Fornari, D. J., Ridley, W. I., Edwards, M. H., Kurras, G. J., and Von Damm, K. L., 2001. Magmatic processes and segmentation at a fast spreading mid-ocean ridge: detailed investigation of an axial discontinuity on the East Pacific Rise crest at 9°37′N. *Geochemistry, Geophysics, Geosystems*, **2**, doi:10.1029/2000GC000134.

Soule, S. A., Fornari, D. J., Perfit, M. R., and Rubin, K. H., 2007. New insights into mid-ocean ridge volcanic processes from the 2005–2006 eruption of the East Pacific Rise, 9°46′N–9°56′N. *Geology*, **35**, 1079–1082.

Soule, S. A., Escartín, J., and Fornari, D. J., 2009. A record of eruption and intrusion at a fast spreading ridge axis: axial summit trough of the East Pacific Rise at 9–10°N. *Geochemistry, Geophysics, Geosystems*, doi:10.1029/2008GC002354.

White, S. M., Macdonald, K. C., and Haymon, R. M., 2000. Basaltic lava domes, lava lakes, and volcanic segmentation on the southern East Pacific Rise. *Journal of Geophysical Research*, **105**, 23519–23536.

White, S. M., Haymon, R. M., and Carbotte, S., 2006. A new view of ridge segmentation and near-axis volcanism at the East Pacific Rise, 8–12 N, from EM300 multibeam bathymetry. *Geochemistry, Geophysics, Geosystems*, **7**, doi:10.1029/2006GC001407.

Cross-references

Axial Volcanic Ridges
Hydrothermalism
Mid-ocean Ridge Magmatism and Volcanism
Oceanic Spreading Centers

AXIAL VOLCANIC RIDGES

Isobel Yeo
GEOMAR Helmholtz Centre for Ocean Research, Kiel, Germany

Synonyms

Neo-volcanic ridge

Definition

Axial volcanic ridges (AVRs). Composite volcanic edifices, comprising an elongate, typically spreading-normal orientated topographic high, produced within the inner valleys of mid-ocean ridges, usually those that are slow spreading.

Introduction

The surface expression of volcanic activity is extremely variable at different spreading rates (see "Morphology Across Convergent Plate Boundaries"), as a result of the differing eruption styles associated with each. Axial volcanic ridges (AVRs), sometimes called neo-volcanic ridges, have been recognized at many slow-spreading Mid-Atlantic Ridge (MAR) spreading segments (e.g., Ballard and Van Andel, 1977; Karson et al., 1987; Smith and Cann, 1992; Parson et al., 1993; Sempere et al., 1993; Lawson et al., 1996; Bideau et al., 1998; Gracia et al., 1998; Navin et al., 1998; Briais et al., 2000; Peirce and Sinha, 2008; Searle et al., 2010) and some ultraslow-spreading segments elsewhere (Mendel et al., 2003). AVRs are elongate, composite volcanoes, typically with a ridge parallel orientation (Figure 1). They vary in size, although are typically a few kilometers wide, are tens of kilometers long, and reach heights of several hundreds of meters above the surrounding seafloor.

AVRs are usually found in the middle of spreading segments and are often associated with hourglass-shaped axial valleys. They may extend all the way from the center of the valley to the base of the bounding axial valley wall faults or be surrounded by areas of flatter seafloor. Where present, an AVR usually represents the largest volume magmatic structure on the segment.

AVR eruption style and volcanic architecture

AVRs themselves are built almost entirely of agglomerations of volcanic hummocks (Smith and Cann, 1990; Yeo et al., 2012), which are circular or subcircular, probably monogenetic, volcanic cones, or domes 50–500 m in diameter with heights of tens to hundreds of meters (Figure 1 inset). Hummocks are constructed of a combination of pillow, elongate pillow, and lobate lavas (see "Submarine Lava Types"), which are erupted from a central vent and flow outwards and down the sides of the hummock. These hummocky structures are responsible for the rough, lumpy surface texture of AVRs in multibeam data (Figure 1). Several hummocks may be produced in one eruption, often, but not always, distributed along an eruptive fissure forming a hummocky lineament on the seafloor (Searle et al., 2010). The feeder dykes for these eruptions are unlikely to be active for more than one eruption as the typical mid-ocean ridge dyke thickness of less than 2 m (Qin and Buck, 2008) is such that it will solidify before the next predicted dyke emplacement (Head et al., 1996). Such predominantly ridge parallel fissure eruptions are thought to be the dominant eruption style on AVRs. The individual hummocks are formed as a result of point focusing down to a number of individual vents that feed discrete, round edifices (Smith and Cann, 1992; Head et al., 1996; Smith and Cann, 1999). Hummocks may coalesce together to form larger hummocky ridges or mounds (Smith and Cann, 1993; Smith et al., 1995; Head et al., 1996; Lawson et al., 1996; Briais et al., 2000), which are similar to large pillow mound eruptions on intermediate-spreading rate ridges (Yeo et al., 2013). Hummocks commonly collapse down the AVR flanks, converting around 12 % of the lavas erupted on the AVR to talus, which probably also form a component of the AVR structure (Yeo et al., 2012). Rare flat-topped seamounts and small areas of smoother lava flows may also form small parts of the AVR structure.

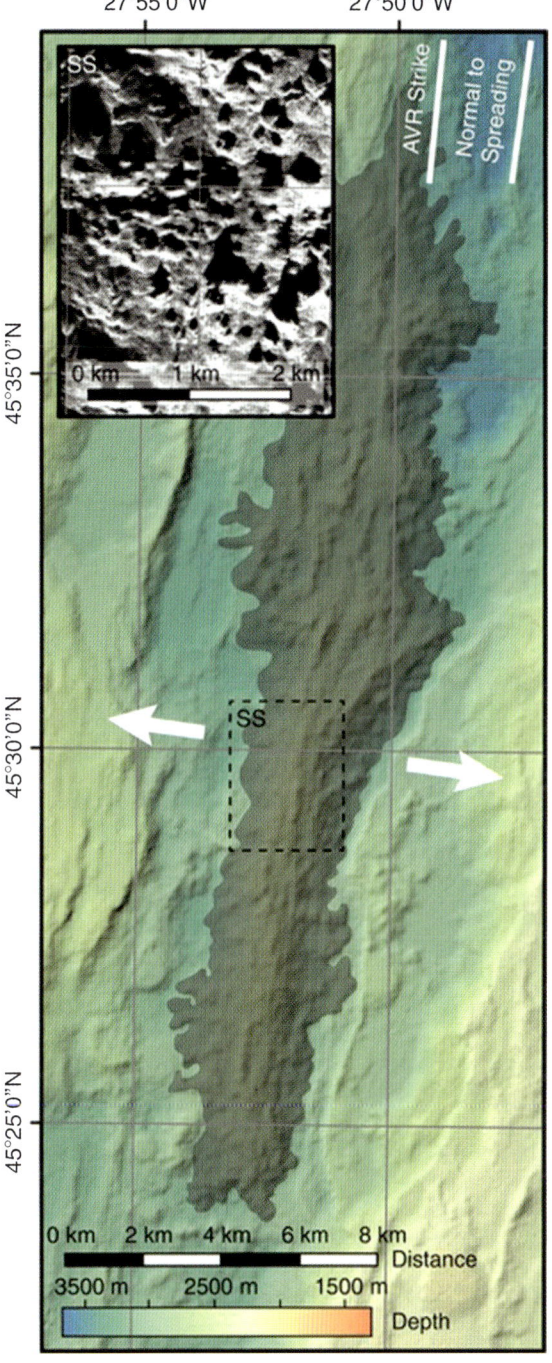

Axial Volcanic Ridges, Figure 1 EM120 bathymetry gridded at 50 m showing the axial volcanic ridge at 45°N on the Mid-Atlantic Ridge as surveyed by Searle et al. (2010). The prominent 22-km-long AVR is shown by the *gray-shaded* area and lies almost parallel to the ridge strike in the center of the hourglass-shaped inner valley. The spreading direction is shown by the *white arrows*. The northern end (north of 45°33) appears more tectonized than the southern end. The rough surface texture is a result of the hummocks that cover its surface. *Inset*: TOBI side-scan mosaic of the hummocky terrain in the area covered by the *dashed box* labeled SS on the main figure (Data is insonified to the north).

The dominant rock type found on AVRs is normal mid-ocean ridge basalt (N-MORB), with variations due to differences in the degree of partial melting or heterogeneities in the source (see "Mid-ocean Ridge Magmatism and Volcanism").

Formation and growth

Melt supply at slow-spreading mid-ocean ridges is irregular in space and time, and, at a typical slow-spreading ridge, melt production is too low to sustain large, steady-state magma chambers anywhere along the segment (Forsyth, 1992; Lin and Morgan, 1992; Sinton and Detrick, 1992; Magde et al., 2000). Therefore at slow- and ultraslow-spreading ridges, volcanism must be episodic. AVRs lie entirely within the Brunhes chron and therefore are difficult to date; however, a number of AVR life cycles as a result of such episodic magmatism have been proposed. Estimates of the lengths of these cycles are highly variable, ranging from several tens of thousands of years (e.g., 10 kyr (Bryan and Moore, 1977), 20 kyr (Sinha et al., 1998), and 25 kyr (Ballard and Van Andel, 1977)) to much longer periods (e.g., 600 kyr (Searle et al., 1998)) on the Mid-Atlantic Ridge and 400 kyr–2.4 Myr on the Southwest Indian Ridge (Mendel et al., 2003). Additionally, where available, the ages measured for AVRs – 10 kyr (Sturm et al., 2000) and ~12 kyr (Searle et al., 2010) – are much younger than the age of the crust calculated based on spreading rate. This, combined with the similarity of estimated ages for lava flows all over an AVR (Yeo and Searle, 2013), suggests that AVRs are the product of episodes of higher than normal volcanic activity.

Such a life cycle is probably comprised of at least one volcanic phase, followed by an amagmatic phase in which the AVR is broken apart and possibly rifted off axis by tectonic activity (Parson et al., 1993; Mendel et al., 2003; Peirce and Sinha, 2008). The length of these various phases and the extent to which rejuvenation may occur during periods of predominantly tectonic extension are poorly constrained. In the extreme, this could, if periods of tectonic extension were insufficient to destroy the AVR between rejuvenation episodes, actually result in an almost steady-state AVR, where a bathymetric high is present nearly all the time, maintained by regular episodic volcanism. However, evidence from the RAMASSES experiment conducted on the Reykjanes Ridge (Sinha et al., 1998) suggests that magma chambers may only exist beneath an AVR on a slow-spreading ridge for around 10 % of the cycle.

Summary

AVRs are large, constructional, volcanic features formed predominantly of volcanic hummocks that are very commonly found on slow- and ultraslow-spreading ridges. They typically lie in the middle of a segment and are the focus of volcanic activity and therefore probably upper crustal construction. Due to the irregular magma supply to slow- and ultraslow-spreading ridges, volcanism on AVRs is almost certainly episodic although the timings and durations of magmatic episodes are currently poorly constrained.

Bibliography

Ballard, R. D., and Van Andel, T. H., 1977. Morphology and tectonics of the inner rift valley at lat 36°50′N on the Mid-Atlantic Ridge. *Geological Society of America Bulletin*, **88**(4), 507–530, doi:10.1130/0016-7606.

Bideau, D., Roger, H., Sichler, B., Bollinger, C., and Guivel, C., 1998. Contrasting volcanic-tectonic processes during the past 2 Ma on the Mid-Atlantic Ridge: submersible mapping, petrological and magnetic results at lat. 34°52 N and 33°55 N. *Marine Geophysical Researches*, **20**(5), 425–458, doi:10.1023/A:1004760111160.

Briais, A., Sloan, H., Parson, L. M., and Murton, B. J., 2000. Accretionary processes in the axial valley of the Mid-Atlantic Ridge 27 degrees N – 30 degrees N from TOBI side-scan sonar images. *Marine Geophysical Researches*, **21**, 87–119, doi:10.1023/A:1004722213652.

Bryan, W. B., and Moore, J. G., 1977. Compositional variations of young basalts in the Mid-Atlantic Ridge rift valley compositional variations of young basalts in the Mid-Atlantic Ridge rift valley near lat 36°49′N. *Geological Society of America Bulletin*, **88**(4), 556–570, doi:10.1130/0016-7606(1977)88<556.

Forsyth, D. W., 1992. Geophysical constrains on mantle flow and melt generation beneath Mid-Ocean Ridges. In Morgan, J. P., Blackman, D. K., and Sinton, J. M. (eds.), *Mantle Flow and Melt Generation and Mid-Ocean Ridges*. Washington, DC: American Geophysical Union, pp. 1–65.

Gracia, E., Parson, L., Bideau, D., and Hekinian, R., 1998. Volcano-tectonic variability along segments of the Mid-Atlantic Ridge between Azores Platform and the Hayes Fracture zone: evidence from submersible and high resolution sidescan data. *Special Publication Geological Society of London*, **148**, 1–15, doi:10.1016/0040-1951(91)90352-S.

Head, W., Wilson, L., and Smith, D. K., 1996. Mid-ocean ridge eruptive vent morphology and substructure: evidence for dike widths, eruption rates, and evolution of eruptions and axial volcanic ridges. *Journal of Geophysical Research*, **101**(B12), 28265–28280, doi:10.1029/96JB02275.

Karson, J. A., Thompson, G., Humphris, S. E., Edmond, J. M., Bryan, W. B., Brown, J. R., Winters, A. T., Pockalny, R. A., Casey, J. F., Campbell, A. C., Klinkhammer, G., Palmer, M. R., Kinzler, R. J., and Sulanowska, M. M., 1987. Along-axis variations in seafloor spreading in the MARK area. *Nature*, **328**, 681–685, doi:10.1038/328681a0.

Lawson, K., Searle, R. C., Pearce, J. A., Browning, P., and Kempton, P., 1996. Detailed volcanic geology of the MARNOK area, Mid-Atlantic Ridge north of Kane transform. *Geological Society, London, Special Publications*, **118**, 61–102, doi:10.1144/GSL.SP.1996.118.01.05.

Lin, J., and Morgan, J. P., 1992. The spreading rate dependence of three-dimensional mid-ocean ridge gravity structure. *Geophysical Research Letters*, **19**(1), 13–16, doi:10.1029/91GL03041.

Magde, L. S., Barclay, A. H., Toomey, D. R., Detrick, R. S., and Collins, J. A., 2000. Crustal magma plumbing within a segment of the Mid-Atlantic Ridge 35°N. *Earth and Planetary Science Letters*, **175**(1–2), 55–67, doi:10.1016/S0012-821X(99)00281-2.

Mendel, V., Sauter, D., Rommevaux-Jestin, C., Patriat, P., Lefebvre, F., and Parson, L. M., 2003. Magmato-tectonic cyclicity at the ultra-slow spreading Southwest Indian Ridge: evidence from variations of axial volcanic ridge morphology and abyssal hills pattern. *Geochemistry, Geophysics, Geosystems*, **4**(5), 1–23, doi:10.1029/2002GC000417.

Navin, D. A., Peirce, C., and Sinha, M. C., 1998. The RAMESSES experiment – II. Evidence for accumulated melt beneath a slow spreading ridge from wide-angle refraction and multichannel reflection seismic profiles. *Geophysical Journal International*, **135**(3), 746–772, doi:10.1046/j.1365-246X.1998.00709.x.

Parson, L. M., Murton, B. J., Searle, R. C., Booth, D., Keeton, J., Laughton, A., Mcallister, E., Millard, N., Redbourne, L., Rouse, I., Shor, A., Smith, D., Spencer, S., Summerhayes, C., et al., 1993. En echelon axial volcanic ridges at the Reykjanes Ridge: a life cycle of volcanism and tectonics. *Earth and Planetary Science Letters*, **117**, 73–87, doi:10.1016/0012-821X(93)90118-S.

Peirce, C., and Sinha, M. C., 2008. Life and death of axial volcanic ridges: segmentation and crustal accretion at the Reykjanes Ridge. *Earth and Planetary Science Letters*, **274**(1–2), 112–120, doi:10.1016/j.epsl.2008.07.011.

Qin, R., and Buck, W. R., 2008. Why meter-wide dikes at oceanic spreading centers? *Earth and Planetary Science Letters*, **265**, 466–474, doi:10.1016/j.epsl.2007.10.044.

Searle, R. C., Keeton, J. A., Owens, R. B., White, R. S., Mecklenburgh, R., Parsons, B., and Lee, S. M., 1998. The Reykjanes Ridge: structure and tectonics of a hot-spot-influenced, slow-spreading ridge, from multibeam bathymetry, gravity and magnetic investigations. *Earth and Planetary Science Letters*, **160**(3–4), 463–478, doi:10.1016/S0012-821X(98)00104-6.

Searle, R. C., Murton, B. J., Achenbach, K., LeBas, T., Tivey, M., Yeo, I., Cormier, M. H., Carlut, J., Ferreira, P., Mallows, C., Morris, K., Schroth, N., van Calsteren, P., and Waters, C., 2010. Structure and development of an axial volcanic ridge: Mid-Atlantic Ridge, 45°N. *Earth and Planetary Science Letters*, **299**, 228–241, doi:10.1016/j.epsl.2010.09.003.

Sempere, J. C., Lin, J., Brown, H. S., Schouten, H., Purdy, G. M., and Oceanography, I., 1993. Segmentation and morphotectonic variations along a slow-spreading center: the Mid-Atlantic Ridge (24°00′N-30°40′N). *Marine Geophysical Researches*, **15**(3), 61–102, doi:10.1007/BF01204232.

Sinha, M. C., Constable, S. C., Peirce, C., White, A., Heinson, G., MacGregor, L. M., and Navin, D. A., 1998. Magmatic processes at slow spreading ridges: implications of the RAMESSES experiment at 57° 45′N on the Mid-Atlantic Ridge. *Geophysical Journal International*, **135**(3), 731–745, doi:10.1046/j.1365-246X.1998.00704.x.

Sinton, J. M., and Detrick, R. S., 1992. Mid-Ocean Ridge magma chambers. *Journal of Geophysical Research*, **97**(B1), 197–216, doi:10.1029/91JB02508.

Smith, D. K., and Cann, J. R., 1990. Hundreds of small volcanoes on the median valley floor of the Mid-Atlantic Ridge. *Nature*, **348**, 152–155, doi:10.1038/348152a0.

Smith, D. K., and Cann, J. R., 1992. The role of seamount volcanism in crustal construction and the Mid-Atlantic Ridge. *Journal of Geophysical Research*, **97**(B2), 152–155, doi:10.1029/91JB02507.

Smith, D. K., and Cann, J. R., 1993. Building the crust at the Mid-Atlantic Ridge. *Nature*, **365**, 707–715, doi:10.1038/365707a0.

Smith, D. K., and Cann, J. R., 1999. Constructing the upper crust of the Mid-Atlantic Ridge: a reinterpretation based on the Puna Ridge, Kilauea Volcano. *Journal of Geophysical Research*, **104**(B11), 25379–25399, doi:10.1029/1999JB900177.

Smith, D. K., Cann, J. R., Dougherty, M. E., Lin, J., Spencer, S., Macleod, C., Keeton, J., Mcallister, E., Brooks, B., Pascoe, R., and Robertson, W., 1995. Mid-Atlantic Ridge volcanism from deep-towed side-scan sonar images, 25 degrees-29 degrees-N. *Journal of Volcanology and Geothermal Research*, **67**, 233–262, doi:10.1016/0377-0273(94)00086-V.

Sturm, M. E., Goldstein, S. J., Klein, E. M., Karson, J. A., and Murrell, M. T., 2000. Uranium-series age constraints on lavas from the axial valley of the Mid-Atlantic Ridge, MARK area. *Earth and Planetary Science Letters*, **181**(1–2), 61–70, doi:10.1016/S0012-821X(00)00177-1.

Yeo, I. A., and Searle, R. C., 2013. High resolution ROV mapping of a slow-spreading Ridge: Mid-Atlantic Ridge 45°N. *Geochemistry, Geophysics, Geosystems*, **14**(6), 1693–1702, doi:10.1002/ggge.20082.

Yeo, I., Searle, R. C., Achenbach, K. L., Bas, L., Tim, P., and Murton, B. J., 2012. Eruptive hummocks: building blocks of the upper ocean crust. *Geology*, **40**(1), 91–94, doi:10.1130/G31892.1.

Yeo, I. A., Clague, D. A., Martin, J. F., Paduan, J. B., and Caress, D. W., 2013. Pre-eruptive flow focusing in dikes feeding historical pillow Ridges on the Juan de Fuca and Gorda Ridges. *Geochemistry, Geophysics, Geosystems*, **14**(9), 3586–3599, doi:10.1002/ggge.20210.

Cross-references

Crustal Accretion
Mid-ocean Ridge Magmatism and Volcanism
Oceanic Rifts
Oceanic Spreading Centers
Submarine Lava Types

B

BEACH

Charles W. Finkl
Department of Geosciences, Florida Atlantic University, Boca Raton, FL, USA
The Coastal Education and Research Foundation, Coconut Creek, FL, USA

Synonyms

Bank; Beachfront; Coast; Lakeshore; Lakeside; Littoral; Margin; Oceanfront; Seaboard; Seafront; Seashore; Seaside; Shingle; Shore; Strand; Waterfront

Definition

In the most rudimentary sense, a beach may be considered a shore (see Coasts) covered by sand (e.g., Shepard, 1973), gravel (e.g., Carter and Orford, 1984; Jennings and Shulmeister, 2002), or larger rock fragments (but lacking bare rock surfaces). More specifically, Jackson (1997) defines a beach as a relatively thick and transitory accumulation of loose waterborne materials that are mostly well-sorted sand and pebbles, but contains admixtures of mud, cobbles, boulders, smoothed rock, and shell fragments. The term was originally used in a scientific sense to designate loose waterworn shingle or pebbles on English shores (Johnson, 1919). Today, beaches are defined in a wider geomorphological sense as having subaerial and submarine components that are systematically interrelated. The beach system includes the dry (subaerial) beach, the wet beach (swash or intertidal zone), the surf zone, and the nearshore zone lying beyond the breakers (Short, 2006). Mostly the nearshore zone is defined as the area beginning from the low-water line extending seaward. The surf zone is part of the nearshore zone. The subaerial beach has a gentle seaward slope, typically with a concave profile, and extends from the low-water mark landward to a place where there is a change in material or physiography (such as a dune, bluff, or cliff, or even a seawall) or to the line of permanent vegetation marked by the limit of highest storm waves or surge. If the area beginning from the low-water line is regarded, the profile is not concave when looking to a beach with a certain tidal range. The dry beach is concave, but at low water when including the wet beach, the morphology switches from a concave beach (dry part) to a convex beach (wet part). Nearshore bars and troughs (in the subtidal domain) are often present in the surf zone, but are obscured by waves and surf, as is the always submerged nearshore zone (Short, 2006). Beach morphology refers to the shape of the beach, surf, and nearshore zone. The beach per se contains numerous morphological–processual subunits such as berms, storm ridges (e.g., Zenkovich, 1967; Pethick, 1984; Komar, 1998; Davis and FitzGerald, 2004), cusps, beach face, shoreface (e.g., Swift et al., 1985), plunge step (Davis and Fox, 1971), scarp, etc.

Related features include beach plains, beach ridges, and beachrock. Beach plains are level or gently undulating areas formed by closely spaced successive embankments of wave-deposited materials added to a prograding shoreline (Jackson, 1997). Beach ridges are low mounds of beach or beach-and-dune materials (sand, gravel, shingle) accumulated by waves on the backshore beyond the present limit of storm waves or ordinary tides. Beach ridges are roughly parallel to the shoreline and mark prior positions of an advancing shoreline (Jackson, 1997). Beachrock is a friable to well-cemented sedimentary rock formed in the intertidal zone in a tropical or subtropical region. Beachrock formations, which slope gently seaward on the beach face, consist of sand or gravel (detrital and skeletal) cemented with calcium carbonate.

Etymology and usage

According to Merriam-Webster (1994), the first known use of the term beach was perhaps in 1535. The term beach was used to describe "loose, water-worn pebbles of the seashore," probably from Old English bæce, bece "stream," from Proto-Germanic *bakiz. Extended to loose, pebbly shores (1590s), and in dialect around Sussex and Kent, beach still has the meaning "pebbles worn by the waves." French grève shows the same evolution (http://www.etymonline.com/index.php?term=beach). Thus, early on the term *beach* did not refer to the expanse of sand normally thought of today. Expansion of term to cover the whole shore in the late sixteenth century may have been related to a popular misunderstanding of the "pebbles" connotation of "beach" in phrases such as "walk on the beach" (http://www.word-detective.com/2007/12/strand-beach/).

Geographic occurrence

Beaches are characteristic of almost all coasts where sediments accumulate alongshore. Sandy beaches occur in all latitudes where there are waves, appropriate types of sediments that are available for accumulation, and a favorable littoral geomorphology (e.g., Davies, 1980; Swift et al., 1985; Pilkey, 2003). Long straight beaches tend to occur along gently seaward-sloping continental shelves with abundant sediment supply, whereas curved beaches in coastal embayments are typically associated with steep continental shelves and a limited sediment supply (Masselink et al., 2011). Although most commonly associated with sedimentary coasts such as occur on barrier islands, for example, beach–dune systems may be quite extensive as major coastal features. It is estimated by various researchers that beaches occur along about a third of the world's coastline (Bird, 1984; Bird and Schwartz, 1985; van der Maarel, 1993), depending on the definitions used for compound coasts. Some coasts, for example, may be dominantly cliffy or rocky, but at the same time they contain beaches in front of the cliffs or between rocky promontories (Figure 1). That is to say, beaches may be small disjunctive pocket accumulations along rocky coasts or they may be the dominant coastal type. Australia, for example, has some 11,761 dominant beaches that make up half of the 30,000-km-long coast (Short and Farmer, 2012). Most Australian beaches, like many others throughout the world, are bordered by some kind of natural feature such as a dune field, cliff, headland, rocks, coral reef, or inlet. According to Dolan et al. (1972), 33 % of the North American shoreline is beach, which further breaks down to 23 % barrier-beach islands, 8 % pocket beaches, and 2 % associated with rock headlands. Of the 19,550-km-long shoreline of the contiguous USA, the US Army Corps of Engineers (COE, 1984) estimates that about 76,100 km (i.e., excluding Alaska, Hawaii, Puerto Rico, and US territories) is beach.

Some of the longest beaches in the world include, for example, Brazil's Praia do Cassino Beach (over 250 km

Beach, Figure 1 Carbonate beach on Long Island, Bahamas, showing biogenic carbonate sand derived from shells and coral reef materials that have been comminuted by wave action to form sand-sized grains. Note the soft pink color of the sand seaward of the wetted perimeter. The coral-rich beach sand contains admixtures of species of Foraminifera or forams (microscopic animals with bright red shells), making these pink beaches some of the best in the world. These beaches, bounded by eolianite headlands, occur on the open Atlantic Ocean side of the island and are also well known on Harbour Island near Eleuthera (Photo: C.W. Finkl).

long), India's Cox's Bazar (about 240 km long), Texas' Padre Island (about 210 km long), Australia's Coorong Beach (194 km), New Zealand's Ninety Mile Beach (each about 145 km long), Mexico's Playa Novillero (about 80 km long), and America's Virginia Beach (about 55 km long).

Grain size, shape, chemical composition, and color

Beaches are the product of wave-deposited sediment, but the materials comprising the beach have an almost unlimited range in size, shape, and composition. In practice, however, each beach area has a particular texture and composition with great variation occurring from place to place. Most beaches are sandy (composed predominantly of sand-sized grains, according to the Udden-Wentworth scale), but the term also includes shingle, gravel (pebbles and cobbles), and boulders that are deposited between the upper swash limit and wave base (e.g., Hardisty, 1990; Short and Woodroffe, 2009). Coarse-grained (cobble, boulder) beaches are prevalent in high latitudes and typify many coastal segments in Canada and the southern Baltic Sea coastline, for example, where glacial materials and bedrock are eroded by coastal processes. Coarse-grained beaches and barriers may also occur along many high-energy coasts, as seen along the west coast of North America, and in low latitudes where they are composed of shell or coral fragments (Figure 2).

Very fine-grained (high percentage of silt and clay) beaches may exceptionally occur under specialized conditions where there is an abundance of fine-grained sediments deposited by wave action as, for example, along the Red Sea and in the vicinity of large deltas and estuaries. A well-known example of a muddy coast with a potential for muddy "beaches" (mud flats) is provided by Anthony et al. (2010) for the 1,500-km-long coast of South America between the Amazon and the Orinoco river mouths. Ephemeral mud "beaches" also occur around mud-lump islands of the Mississippi Delta (Morgan et al. 1963). Similar "beaches" occur along the deltaic plains of the Brazos and Colorado rivers of the Texas coast and the eastern shore of Virginia (Davis, 1978).

Beaches tend to be comprised of well-sorted and generally rounded grains. Exceptions include disk, blade, and roller shapes that are so characteristic of gravel beaches where the larger clasts slide over sand and fine gravel (Pilkey et al., 2011). Biogenic beach gravels are also rather common and may comprise the entire beach sediment, as along coral reef beaches, or may be mixed with sand (cf. Figures 1 and 2). The abundance of biogenic debris (shell, algae, coral fragments) on beaches reflects the provenance of materials and the processes acting on the beach to produce a range of irregular shapes such as the pure *Acropora* coral stick beaches (Davis, 1978).

The composition of beach sediments depends on initial materials where terrigenous (or terrestrial, derived from the erosion of other rocks on land), volcanic (directly derived from volcanic activity and volcanic rocks), and biogenic (composed of shells and skeletons of dead marine organisms) lithologies contribute to the clastic pool (Pilkey et al., 2011). Siliciclastic (quartzose) beaches

Beach, Figure 2 Coarse-grained carbonate beach composed of shell hash on Captiva Island, southwest coast of the Florida Peninsula, USA, facing the Gulf of Mexico. Note the cuspate planform of the beach and a series of beach ridges containing shell fragments heaped up by wave action and swash (Photo: C.W. Finkl).

Beach, Figure 3 Siliciclastic beach sands on Amagansett Beach, Long Island, New York, USA, showing a reflective morphodynamic beach state. As swash runs up the steep beach face, it carries sand grains that overtop the berm to the back beach area. Interdigitating swash marks on the berm mark the landward most transport of sand grains that are dominantly quartz but which may contain minor admixtures of biogenic (shell) fragments and organic matter (Photo: C.W. Finkl).

tend to be more common in middle and high latitudes (Figure 3), whereas carbonates (aragonite and calcite) dominate tropical and subtropical zones. Although a small feldspar fraction is usually present in terrigenous sands, those few beaches that are rich in feldspar tend to be close to the source rock from which the feldspar is derived, such as glacial deposits derived from shield (cratonic) areas containing feldspar-rich granites and gneisses. Beach sands derived from granitic source areas sometimes contain heavy minerals (density > 2.9 g cm^3) that tend to be concentrated in the swash zone by the placering effect to form dark-colored streaks that are sorted out from the rest of the lighter-colored grains.

Beach color is determined by grain size and mineralogical composition so that silicates (derived from igneous granitic rocks, metamorphic rocks such as gneisses and schists, and preexisting sandstones) impart a light color to beaches (Figure 3), whereas more mafic compositions imbue black (Figure 4) to dark gray to dark green colors in fragments of fine-grained rocks such as metamorphic slates or volcanic ricks such as andesite and basalt (Pilkey et al., 2011). The inherent color of the beach's grains does not always determine beach color because grains may be stained by iron oxides, tannic acids, or microorganisms that may produce distinctly red, yellow, or green beaches, for example.

Types of beaches

Because beaches occur in all climatic zones, there are some obvious morphological differences related to severe conditions. In very high latitudes, for example, the water is frozen but for a few weeks when beaches may be affected by waves (Davis, 1978) and consequently beach morphology and texture is somewhat different from those in low latitudes. Beaches in arid climates depend almost solely on wave action to provide sediment from the bedrock coast to form a beach. The best developed beaches are associated with low-lying coasts where large quantities of sediment are available. Beach development requires an abundant sediment supply to produce characteristic morphologies and environmental zonations (Masselink et al., 2011).

The term *beach type* refers to the dominant nature of a beach based on tidal, wave, and current regimes, spatiotemporal extent of the nearshore zone, morphodynamics (beach width, shape, and processes) of the surf zone including bars and troughs, and the subaerial beach (Short, 1993). Irrespective of the specific beach type, most beaches contain several morphogenetic zones: (1) the backshore (nearly horizontal to gently landward-sloping area called the berm), (2) inner swash zone (upper limit of swash to the shoreline), (3) surf zone (from the shoreline to where waves break, the breaker zone), (4) nearshore zone (from the breaker zone, commonly with sandbars, to the wave base), and (5) wave base (depth where waves begin to interact with the seabed to transport sand to the beach and seaward to where sand is transported by large waves that cause beach erosion) (Davis, 1978; Short and Woodroffe, 2009). These zones are greatly variable, depending on the processes and materials that affect

Beach, Figure 4 Black sand beach composed of basaltic grains on the south coast of Iceland west of the Dyrhólaey promontory (headland or cape), near Vik. The mafic composition of the basaltic grains does not make the clasts particularly durable, and they are worn down by wave and surge action to finer-grained particles, as seen in the reticulate pattern on this extremely wide berm and beach plain where there are darker-colored coarse-grained ridges and lighter-colored finer-grained flats. This black basaltic beach and beach plain front a large sandur plain (Photo: C.W. Finkl).

beach type. In addition to these major zonations of beach environments, there are numerous smaller, but important, morphological features such as the plunge step (Davis and Fox, 1971), which is a small and commonly subtle shore-parallel depression in the foreshore that is caused by the final plunge of waves as they break for the last time before surging up the beach face (Davis, 1978).

Classification of beaches

Many factors need to be considered in the classification of beaches (e.g., Finkl, 2004) as they are among the most dynamic features on earth, but even so they retain certain overriding characteristics that facilitate generalization and categorization. The overall gradient of the beach and nearshore influences the amount of wave energy that reaches the beach, giving it its configuration. Beach materials, slope, and exposure interact with waves to produce the morphodynamic beach state that is constantly adjusting to new environmental conditions. Although beaches are one of the most dynamic morphosystems on earth (literally changing every day), they exhibit a range of characteristic morphologies that have been intensively studied in Australia. The analysis of Australian beach characteristics and shoreface dynamics (e.g., Short, 1993; Short, 1999) has led to a beach classification system that is now used internationally.

According to the classification scheme for Australian beaches (see discussion in Short and Woodroffe, 2009), there are 15 major beach types that are derived from three major beach systems: wave dominated, tide modified, and tide dominated. An overview of the beaches of Australia, which occupy half the 29,900-km-long coast (including Tasmania), is provided by Short (2006) in his discussion of the roles of waves, sediment, and tide range that contribute to beach type, particularly through the dimensionless fall velocity and relative tide range. Short's comprehensive study of Australian beach types includes descriptions of their regional distribution, together with the occurrence of rip currents, multibar beach systems, and the influence of geological inheritance and marine biota, a natural progression of comprehensive observational collages and models stemming from seminal works (e.g., Wright and Short, 1984) commonly referred to as the "Australian Beach Model" (Short, 2006).

Recognition of the 15 beach types occurring around the Australian coast provides a basis for identifying similar wave–tide–sediment environments throughout the world and classification of many of the world's beaches. Although applied internationally, the Australian Beach Model is not universal because it does not include tide-modified beaches exposed to higher ocean swell and storm seas, resulting in similar though higher-energy beaches, gravel, and cobble beaches (few occur in Australia), nor ice-affected beaches (because they do not occur in Australia). Nevertheless, this system finds wide application throughout the world as, for example, in the classification and study of Florida east coast beaches (e.g., Benedet et al., 2004), eastern Brazil (e.g., Klein and Menezes, 2001), India (Saravanan et al., 2011), Portugal (e.g., Coelho, Lopes, and Freitas, 2009), France (Sabatier et al., 2009), and so on.

Artificial beaches

When beaches erode along developed shorelines, they are often replenished by sand dredged from offshore and pumped back onto the shore as remediation. Efforts are made to ensure that the dredged sediment closely approximates the nature (size, shape, density, durability, composition, and color) of native beach sands onshore (e.g., Finkl and Walker, 2005). Replenished beaches are engineered to approximate natural morphodynamic beach states so that they perform as sacrificial sand deposits over desired time frames. Some artificial beaches may erode very quickly if impacted by storms and thus fall short of the design life, while others, having greater durability, may exceed the design life, as in the case of Miami Beach, Florida, which has the longest half-life of any renourished beach in America (e.g., Finkl and Walker, 2005). The practice of building artificial beaches, as a soft-engineering shore protection measure, is now so widespread worldwide that many renourished beaches are perceived as natural formations. The practice is, however, not without environmental concerns as borrow sources do not always closely match native beach sands to cause unwanted impacts in the coastal zone (e.g., Bonne, 2010).

Summary

Beaches occur in all latitudes, from tropical to polar regions, and lie on the interface between land and water to form ocean, bay (sound, estuary), and river beaches. They are mostly composed of sand (silicates and carbonates), but there is great compositional and morphological variation between sites. Beaches have great economic and environmental value, serving as tourist destinations and providing habitat as well as buffering impacts from storm waves and surges. Some beaches are artificial, and although their placement is controversial among certain special interest groups, their benefits generally outweigh disadvantages when properly engineered and placed. The morphodynamic classification of beach state finds global application, provides greater understanding of Beach Processes, and contributes to beach safety education programs and surf life saving to reduce public risk.

Bibliography

Anthony, E. J., Gardel, A., Gratiot, N., Proisy, C., Allison, M. A., Dolique, F., and Fromard, F., 2010. The Amazon-influenced muddy coast of South America: a review of mud-bank-shoreline interactions. *Earth-Science Reviews*, **103**(2010), 99–121.

Benedet, L., Finkl, C. W., and Klein, A. H. F., 2004. Morphodynamic classification of beaches on the Atlantic coast of Florida: geographical variability of beach types, beach safety, and Coastal Hazards. *Journal of Coastal Research* (Special Issue No. 39), 360–365.

Bird, E. C. F., 1984. *Coasts: An Introduction to Coastal Geomorphology*. Oxford: Blackwell, 320 p.

Bird, E. C. F., and Schwartz, M. L. (eds.), 1985. *The World's Coastline*. New York: Van Nostrand Reinhold, 1071 p.

Bonne, W. M. I., 2010. European marine sand and gravel resources: evaluation and environmental impacts of extraction – an introduction. In Lancker, V. V.; Bonne, W., Uriarte, A., and Collins, M (eds.), *EUMARSAND: European Marine Sand and Gravel Resources. Journal of Coastal Research*, Special Issue #51, i–vi.

Carter, R. W. G., and Orford, J. D., 1984. Coarse clastic barrier beaches: a discussion of the distinctive dynamic and morphosedimentary characteristics. In Greenwood, B., and Davis, R. A. (eds.), *Hydrodynamics and Sedimentation in Wave-Dominated Coastal Environments. Marine Geology*, **60**, 377–389.

COE (Corps of Engineers), (1984). *Shore Protection Manual*. Vicksburg, MS: Coastal Engineering Research Center, 2 Vols. Available in electronic form: http://openlibrary.org/books/OL3001149M/Shore_protection_manual; and as the *Coastal Engineering Manual* at http://chl.erdc.usace.army.mil/cem

Coelho, C., Lopes, D., and Freitas, P., 2009. Morphodynamics classification of Areão Beach, Portugal. *Journal of Coastal Research*, (Special issue No. 56), 34–38.

Davies, J. L., 1980. *Geographical Variation in Coastal Development*. London: Longman, 204 p.

Davis, R. A., 1978. *Coastal Sedimentary Environments*. New York: Springer, 420 p.

Davis, R. A., and FitzGerald, D. M., 2004. *Beaches and Coasts*. Malden, MA: Blackwell, 419 p.

Davis, R. A., and Fox, W. T., 1971. *Beach and Nearshore Dynamics in Eastern Lake Michigan*. Kalamazoo, MI: Western Michigan University. Technical Report No. 4 (ONR Contract 388–092), 145 p.

Dolan, R., Hayden, B., Hornberger, G., Zieman, J., and Vincent, M., (1972). *Classification of the Coastal Environments of the World. Part I, The Americas*. Charlottesville, VA: University of Virginia. Technical Report No. 1 (ONR Contract 389–158), 13 p.

Finkl, C. W., 2004. Coastal classification: systematic approaches to consider in the development of a comprehensive scheme. *Journal of Coastal Research*, **20**(1), 166–213.

Finkl, C. W., and Walker, H. J., 2005. Beach nourishment. In Schwartz, M. (ed.), *The Encyclopedia of Coastal Science*. Dordrecht, The Netherlands: Kluwer Academic (now Springer), pp. 147–161.

Hardisty, J., 1990. *Beaches: Form and Processes*. London: Unwin Hyman, 324 p.

Jackson, J. A., 1997. Glossary of geology. Denver, Colorado: American Geological Institute, 769 p.

Jennings, R., and Shulmeister, J., 2002. A field based classification scheme for gravel beaches. *Marine Geology*, **186**(3–4), 211–228.

Johnson, D. W., 1919. Shore processes and shoreline development. New York: Wiley, 584 p.

Klein, A. H. F., and Menezes, J. T., 2001. Beach morphodynamic and profile sequence for a headland bay coast. *Journal of Coastal Research*, **17**, 812–835.

Komar, P. D., 1998. *Beach Processes and Sedimentation*. Upper Saddle River, NJ: Prentice Hall.

Masselink, G., Hughes, M. G., and Knight, J., 2011. *Introduction to Coastal Processes and Geomorphology*. London: Arnold.

Merriam-Webster, 1994. *Merriam-Webster's Dictionary of English Usage*. Springfield, MA: Merriam-Webster, 989 p.

Morgan, J. O., Coleman, J. M., and Gagliano, S. W., 1963. *Mudlumps at the Mouth of South Pass, Mississippi River; Sedimentology, Paleontology, Structures, Origin and Relation to Deltaic Processes*. Baton Rouge, LA: Louisiana State University Studies. Coastal Studies Series No. 10.

Pethick, J., 1984. *An Introduction to Coastal Geomorphology*. London: Edward Arnold, 260 p.

Pilkey, O. H., 2003. *A Celebration of the World's Barrier Islands*. New York: Columbia University, 309 p.

Pilkey, O. H., Neal, W. J., Kelley, J. T., and Cooper, A. G., 2011. *The World's Beaches*. Berkeley, CA: University of California Press, 283 p.

Sabatier, F., Anthony, E. J., Héquette, A., Suanez, S., Musereau, J., Ruz, M.-H., and Regnauld, H., 2009. Morphodynamics of beach/dune systems: examples from the coast of France. *Geomorphologie*, **1**(2009), 3–22.

Saravanan, S., Chandrasekar, N., Sheik Mujabar, P., and Hentry, C., 2011. An overview of beach morphodynamic classification along the beaches between Ovari and Kanyakumari, southern Tamil Nadu coast, India. *Physical Oceanography*, **21**(2), 57–71.

Shepard, F. P., 1973. *Submarine Geology*. New York: Harper and Row, 517 p.

Short, A. D., 1993. *Beaches of the New South Wales Coast: A Guide to the Nature, Characteristics, Surf and Safety*. Sydney, NSW: Australian Beach Safety and Management Program, 358 p.

Short, A. D., 1999. *Handbook of Beach and Shoreface Morphodynamics*. Chichester: Wiley, 379 p.

Short, A. D., 2006. Australian beach systems – nature and distribution. *Journal of Coastal Research*, **22**(1), 11–27.

Short, A. D., and Farmer, B., 2012. *101 Best Australian Beaches*. Sydney, NSW: NewSouth Publishing, 222 p.

Short, A. D., and Woodroffe, C. D., 2009. *The Coast of Australia*. Cambridge: Cambridge University Press, 288 p.

Swift, D. J. P., Niederoda, A. W., Vincent, C. E., and Hopkins, T. S., 1985. Barrier island evolution, middle Atlantic shelf, U.S.A. Part I: shoreface dynamics. *Marine Geology*, **63**, 331–361.

van der Maarel, E. (ed.), 1993. *Dry Coastal Ecosystems: Polar Regions and Europe*. Amsterdam, The Netherlands: Elsevier, Ecosystems of the World, Vol. 2A, 600 p.

Wright, L. D., and Short, A. D., 1984. Morphodynamic variability of surf zones and beaches: a synthesis. *Marine Geology*, **56**, 93–118.

Zenkovich, V. P., 1967. *Processes of Coastal Development*. Edinburgh: Oliver and Boyd, 73 p.

Cross-references

Beach Processes
Coastal Engineering
Dunes
Engineered Coasts
Integrated Coastal Zone Management
Placer Deposits
Reef Coasts
Sea Walls/Revetments
Sediment Transport Models
Shelf
Waves

BEACH PROCESSES

Charles W. Finkl
Department of Geosciences, Florida Atlantic University, Boca Raton, FL, USA
The Coastal Education and Research Foundation, Coconut Creek, FL, USA

Synonyms

Beach morphodynamics; Coastal processes; Cusp formation; Littoral processes; Surf zone processes

Definition

The term *beach processes* refers to the collection of processes that are responsible for the formation and maintenance of morphological features and materials that make up all facets of units comprising the beach per se. Because the beach has subaerial and submarine components, processes range from the effects of wind to waves, currents, and tides acting on sediment supply and redistribution to form beach types. Important processes include swash run-up and rundown, overtopping, infiltration into the beachface, step generation, bar formation, cusp development, and so on. Processes and materials interact in a dynamic manner in continual adjustment to changing environmental conditions associated with wind regimes, wave climates, tides, and extreme events such as storm surge and tsunamis.

Introduction

Beaches are complex systems that at first glance appear to be simple enough, sediment piled up along the shore. But, nothing could be farther from the truth as these dynamic systems have historically defied articulation and comprehension because of their continual short-term modification in terms of minutes to hours (e.g., Bascom, 1980; Carter, 1988). High-energy events can induce large-scale beach change that is immediately apparent to an observer. But, even during quiescent periods, the beach is continually changing; literally with every wave grains are moved about the beachface in laminar up- or downrush or by saltation on the berm. Over a longer decadal time frame, some beach materials in tropical and subtropical regions may become indurated by calcium carbonate to form beachrock that in turn provides long-term stability to the beach. Beach processes are similar in general but very different in detail, depending on the ambient energy conditions and materials present. Much is known about beach processes, but some exact modes of formation for some beach features such as beach cusps remain elusive (Guza and Inman, 1975; Komar, 1997; Davidson-Arnott, 2009; Pilkey et al., 2011).

The main agents that affect beaches include waves, tides, currents, and wind (on the dry backbeach area), all of which interact with the sediment to produce distinctive beach morphologies (e.g., Bascom, 1954; Pethick, 1984; Woodroffe, 2002; Davis and FitzGerald, 2004). Breaking waves obviously interact with the sediment, and their position along the beachface depends on tidal ranges and setup or surge. Currents produced by the waves are among the most important processes that change the beach. These important sediment-transporting currents include: (1) combined flow currents, (2) longshore currents, (3) rip currents, and (4) onshore-offshore (cross-shore) currents produced in the swash zone (Davis and FitzGerald, 2004).

Beaches contain noncohesive materials that accumulate in areas affected by wave action along the perimeter of a body of water (whether fresh, brackish, marine, or hypersaline). The material is commonly sand-sized grains

Beach Processes, Figure 1 Dissipative beach along the Oregon coast showing the low-gradient swash that moves up the beachface as laminar flow. Successive incursions of swash move up the beachface as hydraulic jumps until they lose forward momentum and infiltrate into the beach. The high beach groundwater table close to the surface allows swash to run up the beachface for long distances. Pebble and gravel-sized clasts are heaped farther up the beach by storm waves. The backbeach contains coarse-grained lag that has been winnowed by currents when setup temporarily superelevates sea-level during storms (Photo by C.W. Finkl).

(0.06–2.0 mm), but may range through gravel (2–60 mm), cobbles (60–200 mm), and boulders (over 200 mm). Figure 1 shows a gradation in grain size from fine sand on the beachface to cobbles in the backbeach area at the foot of a cliff along the Oregon coast. In very specialized environmental settings, some beaches contain fine-grained materials that are admixed with sand. Intertidal beaches permanently rich in mud (10–20 %) on wave-dominated coasts are, for example, extremely rare because of hydrodynamically controlled nondeposition of mud and lack of substantial nearby mud supplies (Anthony et al., 2011). Mud-rich beaches are generally found associated with high fine-grained discharge deltas such as the Amazon (Anthony and Dolique, 2004) and Mekong (Tamura et al., 2010).

The beach zone is typically subdivided into nearshore, foreshore, and backshore; boundaries between these zones are elastic, responding to cyclic migrations and geo-hydro-meteodynamic events (e.g., Davidson-Arnott, 2009). The nearshore is the submarine part of the beach that lies seaward of the low-tide limit in the region of shoaling waves. The foreshore extends from the low-tide limit to the high-tide upper swash limit that is often marked by a line of flotsam and jetsam. The backshore typically lies landward of the high-tide swash limit and exceptionally lies along the landward limit of storm surge that may include cliffs, dunes, or marshes. The crest of the backshore zone, above the uppermost limit of normal swash, is called the berm. A beach typically has one berm, but there may be several berms depending on cycles and severity of storminess (e.g., Hardisty, 1990; Short, 1999). The example of the Brazilian beach shown in Figure 2 contains a gently seaward-sloping foreshore, a wide berm, and a narrow backbeach fronting eroded dune fronts.

Study of beach processes

Beach processes are studied in many different ways, as, for example, in the field, in wave tanks, or by numerical simulation on computers (e.g., Hardisty, 1990; Dean and Dalrymple, 2002; Davidson-Arnott, 2005; Dingler, 2005). Whatever approach is used, it is necessary to be cognizant of morphodynamic features such as the cross-shore (equilibrium) profile, presence and interaction of rock alongshore as backbeach cliffs, promontories, beachface beachrock, tidal or subtidal hardgrounds, skerries, or coral reefs offshore. The nature of sedimentary materials that is available for beach processes to act upon, the alongshore supply of sediments, offshore biogenic supply of sediments, and thickness of unconsolidated sediments along the continental shelf are all important parameters that influence beach type (Pilkey et al., 2011). It is also not a matter of presence or absence of any of these features, but the degree to which they are present, or not. With so many cross feedback mechanisms, it is generally not possible to isolate discrete processes that do not interact with other processes or which are modified by different kinds of materials that they act upon.

Beach Processes, Figure 2 Intermediate beach morphotype on the coast of Santa Caterina, Brazil. A wide berm lies landward of a low-tide terrace with poorly developed ridges and runnels. Storm surges have nipped away the front of the backbeach dunes bringing dune sand back into the beach system. The maximum extent of inland penetration of swash is marked by the change in the color of the sand on the berm from darker yellow to cream colored on the backbeach (Photo by C.W. Finkl).

Beaches are thus complex processual systems that require much tenacity and expertise to tease out the secrets of beach formation and modification. In spite of the complexity involved, it is possible to outline some of the main processes that are responsible for major beach features. It is also perhaps worth mentioning at this point that it is not profitable to attempt discussion of beach processes without considering the morphological properties of beach features. The association of process and form (Hardisty, 1990) thus leads to appreciation of beach morphodynamics (e.g., Short, 2006), which includes the interaction between breaking waves, currents, rocky headland or submerged hardgrounds, and the sediments that compose the beach. Beaches are dynamic features where composition, morphology, and orientation are spatio-temporally variable as, for example, in the case of headland bay beaches that rotate between promontories (Klein et al., 2002; Short and Trembanis, 2004).

Small-scale beach processes

Waves and currents sort the sediment and move it alongshore and cross-shore. The concept of a "river of sand" is sometimes promoted, but it is a misnomer because there can be many interruptions to the alongshore sediment flow as in the case of capes, promontories (headlands), submarine canyons, deeply dredged inlets, or other engineering structures such as groins and jetties. Alongshore drift may be predominantly in one direction, but there can be seasonal reversals (e.g., during monsoons) or interruptions due to storms, making the concepts of gross and net sediment transport viable (Dean and Dalrymple, 2002). As a general rule in sandy environments, coarser materials remain closer to shore, but along high-energy coasts coarse materials may be heaped up on the beachface and berm. Below mobile surface sediments, bedrock platforms or consolidated materials slope gently seaward, serving as a foundation for the beach. Some beach sediments are heaped up on marine benches during storms and because they lie beyond normal waves and tidal cycles, they are referred to as *perched beaches* being the result of special extreme processes.

Beaches exist in a state of dynamic equilibrium with tides and waves. Storm waves tend to move the sand out to deeper water and flatten the beach cross-shore profile while less energetic waves in quiescent periods tend to move sediment shoreward to build up the beach. Some beaches retain sand that is trapped between headlands, while in other settings sand transits the beach by entering in one end and exiting in the other, somewhat in the context of a conveyor belt. The beach therefore exists when processes and materials balance each other over time (e.g., Dean and Dalrymple, 2002; Davis and FitzGerald, 2004).

The surf zone, the most dynamic part of the beach, extends from the breaker zone to the shore (Wright and Short, 1984). Waves theoretically break when the slope of the advancing face is steeper than 1:0.78 but may be closer to 1:1 based on field data. Waves typically start to interact with the seafloor when the water depth is about 75–80 % of the wave height (Bascom, 1980). Frictional forces then slow the bottom part of the wave while the top continues to move at the same speed, eventually over reaching the slower moving wave base. Waves may break as spilling breakers on low-gradient slopes, as plunging waves on moderate gradients, or as surging waves on steep slopes. Breaking waves transform their potential energy into kinetic energy in the form of broken waves of translation, or wave bores that move shoreward as turbulent white water. Breaking waves dissipate most of their forward-moving energy, which is lost in turbulence. Turbulent flow is an important process because it stirs up large quantities of bottom sediment that will remain in suspension for variable amounts of time depending on the grain size, larger particles settling faster than smaller ones according to Stoke's law and measured in numerous ways in the lab and in the field. This entrained sediment is transported more or less parallel to the shore by the longshore current, which typically reaches from the shoreline through the breaker zone. Under high-energy conditions, breaking storm waves advance across the upper shoreface, sending surge and fast-moving swash up the beachface to erode the berm (Figure 3). Under these conditions when the surf zone temporarily moves shoreward due to increased water depths resulting from surge and setup, beach sands are entrained in cross-shore currents leaving a beach scarp cut into the berm. Beaches equilibrate to a morphodynamic state that is more in balance with quiescent conditions after the storm passes.

Saltation on the berm and backbeach

Once sandy particles are deposited on the berm, wind is the predominant force that moves them inland. Wave-deposited materials are dried by solar radiation, wind, and drainage prior to transport by wind from the berm to backbeach. Wind velocity controls the rate of sand transport, but spatio-temporal variation in sediment deposition depends on beach width (Pethick, 1984; Davidson-Arnott and Law, 1996).

Three main eolian processes transport sediment (Bagnold, 1941). Wind erosion of surface particles on the berm is initiated when air velocities reach about $4.5 \, \text{m s}^{-1}$. Initial particle movement takes place as a rolling motion, termed traction or creep, where particles as large as small pebbles can be tracted by strong winds. About 20–25 % of wind erosion is by traction. When particles are lifted off the ground, becoming suspended in the air, and then return to the ground surface several centimeters downwind, the process is called saltation (Davis, 1985; Carter, 1988). Saltation on the dry berm accounts for 75–80 % of total sediment transport by wind, with beach sand ultimately ending up in dunes behind the beach (Cf. Figure 2). It provides momentum that drives the other two sand transport modes because when a falling particle strikes the berm surface, part of its impact force is transferred to another particle causing it to become airborne.

Beach Processes, Figure 3 Reflective beach type now adjusting to an intermediate morphotype after the passage of storm conditions along the Florida Atlantic coast near Palm Beach. During the high-energy storm conditions, sand was eroded from the berm forming a new steeper beachface than the more gently sloping one associated with the formerly wider beach. The beach scarp, now a hazard to beachgoers, will slump by uprush undercutting during high tides to form a new more gently seaward-sloping beachface (Photo by C.W. Finkl).

Swash run-up and rundown

When beachfaces are steep, waves can transit the shoreface without breaking, and when the broken wave reaches the base of the wet beach, it collapses and runs up the beachface as swash or uprush in the swash zone (e.g., Davis and FitzGerald, 2004). Under normal or quiescent wave conditions, the laminar-flow uprush stops near the top of the slope, with some of the water infiltrating the surface of the beachface per se and then percolating into the beach until it reaches the water table (Cf. Figure 1). If the water table is close to the surface, the remainder flows back down the beachface as backwash. If the water table does not lie close to the surface, the uprush may completely disappear into the beachface with no back flow. These actions produce a relatively steep seaward-sloping swash zone on the beachface, the slope of which can range from 1° to 20°. If waves approach the beach obliquely, and although sediment in this case is transported in a zigzag pattern across the

Beach Processes, Figure 4 Current ripples along the foreshore on a beach in Long Island Sound near Amagansett, Long Island, New York, USA. Shore-parallel ripples in this intertidal zone are developed on both sides (landward and seaward) of a hardground containing algal overgrowth. A hydraulic jump is visible in the swash seaward of the algae poking through the sand cover. The complexity of these small-scale beach processes is shown by superposition of fine-grained sand in a micro-delta deposited on top of the rippled sand bed, as noted by the lighter color of the sand splay and lack of closely spaced ripples. In addition to physical processes, beaches are modified by biological processes as, for example, seen here with mounds (that look like micro-volcanoes approximately 15 cm in diameter) produced by Atlantic jackknife clams (*Ensis directus*) (Photo by C.W. Finkl).

beachface, the net result is sediment transport in the direction of the waves. As sediment accumulates in a seaward-building swash zone, it constructs a berm, the nearly horizontal to slightly landward-dipping sand surface that merges with the backbeach. This is the so-called dry beach where most people sit or lay down when they go to the beach. The swash zone may contain small-scale features such as beach cusps, spaced about every 20–30 m apart, and which are thought to be produced by edge waves (Guza and Inman, 1975). On intertidal flat-lying parts of the foreshore, swash currents may produce various types of ripples. The ripples shown in Figure 4 are typical of many low-gradient intertidal beaches with fine-grained sediments, as seen on this beach on Long Island, New York, where the ripples formed by progressively dissipating swash currents during a falling tidal sequence.

Sheetwash, runoff channel, and sapping-seepage erosion

Rain-induced beach processes (Otvos, 1999) produce interesting minor features on beaches. Subtle features on sandy beach foreshores include swash grooves (braided scour marks), shore normal, flutes, dendritic rills, rhomboidal microrills, and box canyon valleys. These features are related to a range of processes such as water infiltration (in the infiltration zone) and percolation from the beach (in the effluent zone) at low tide or following a storm (e.g., Duncan, 1964; Davis, 1985; Komar, 1997), backwash deflection around obstacles (e.g., Otvos, 1964, 1965; Komar, 1997), backwash flow dissipating into the beach and emerging on the foreshore (Higgins, 1984), sheetwash and cross-beach channeled flow (Otvos, 1999), foreshore dissection by headward eroding box canyons and rill channels (Otvos, 1999), etc. Sheetwash, channeled surface flow, and fresh groundwater driven spring sapping processes produce a wide range of micro-morphodynamic features on the beachface and foreshore. Although appearing to be morphologically minor, these processes play an important subsidiary role in seasonal beach degradation and operate on most beaches of the world. These processes are significant in sediment redistribution on coasts with high rainfalls and wide beaches that are backed by permeable bluffs.

Plunge step formation

The plunge step occurs between the base of the foreshore and the top of the shoreface (Davis, 1985). Wave-dominated shorelines often show a sharp break in grain size at the plunge step where coarser clasts tend to accumulate. The plunge step marks a subtle decline in the foreshore profile that is produced by the final plunge (wave break) before a wave runs up the beachface (Pilkey et al., 2011). The last wave break converts potential energy into kinetic energy that is used up in turbulent flow.

Continual wave breaks in the same location alongshore suspends beach sediments that are transported by currents. The continuous turbulent action and cross-shore movement of sand grains creates a sedimentation deficit that produces the small alongshore trough. The penultimate forward movement of the wave thus moves up the beachface as laminar flow, often with several hydraulic jumps as successive waves break and send new swash on top of the preceding swash flow, in piggyback fashion. Bathers entering the surf are often surprised by the presence of the plunge step, especially when the water is murky, because it may be of decimeter scale, the deeper parts of the trough collecting organic debris that is unpleasant to walk through. The plunge step is best developed in low-tidal range beaches with a steep foreshore slope (Davis, 1985), such as commonly seen along Florida Atlantic coast beaches but is also common along gravel beaches. When setup is very high, the plunge step site may transfer up the beachface to the swash berm or even the storm berm. Direct measurement of step morphology is rare on account of the high-energy conditions localized along this zone.

Bar formation and migration

Surf zone currents transport sediment onshore, alongshore, and offshore to build (sand) wave-formed nearshore bars and troughs occupying the surf zone (e.g., Davis, 1985; Komar, 1997; Greenwood, 2005). Wave-formed bars occur as symmetrical or asymmetrical undulations along the upper shoreface profile in intertidal and subtidal environments (Greenwood, 2005). Barred profiles are in general associated with large values of wave steepness and wave height-to-grain size ratios and are associated with waning stages of shoaling and dissipation of wave energy (Wright et al., 1979; Greenwood, 2005). Bar formation has been related to a number of specific hypotheses that involve convergence of sediment transport, viz., (1) breakpoint hypotheses, (2) infragravity wave hypotheses, and (3) self-organization hypotheses. Wave breaking, for example, is thought to induce a seaward transport of sediment that is, respectively, entrained by roller or helical vortices under plunging or spilling breakers (e.g., Zhang, 1994). Alternatively, convergence of sediment at the breakpoint may be related to onshore transport associated with increasing asymmetry and skewness of high-frequency incident waves and offshore transport through setup-induced undertow (e.g., Thornton et al., 1996). Bar frequency of occurrence and geomorphic position may be produced by the interaction of sandy sediments with infragravity waves that are low frequency (greater 30 s period) waves produced by sets of higher and lower waves that are enhanced by wave breaking across the surf zone. These waves can be standing or progressive-produced as a result of energy dissipation during breaking and frequently related to groupiness (Roelvink and Broker, 1993; Reussink, 1998), amplitude modulation of the incident wave field. As a rule, the longer the infragravity wave period, the more widely spaced the bar(s).

Rip currents and channels

Rip currents are narrow, usually fast seaward-flowing currents that penetrate the surf zone, often in a rip channel that flows to deeper water. Rips are a mechanism for returning water back to the sea that piles up alongshore in setup (e.g., Pilkey et al., 2011; Leatherman, 2013). These fast-moving shore-perpendicular currents transport eroded beach sediment seaward to deep water during high seas. These fast-moving cross-shore currents often occur through gaps in offshore bars but are also caused by depressions in the beach, irregular bottom topography, obstructions in the surf zone such as rock outcrops or groins, or longshore currents in circulating cells (e.g., Pilkey et al., 2011). Rip currents, a major hazard to swimmers, are responsible for most beach rescues and drownings (Short, 1999; Leatherman, 2013). Edge waves, a form of infragravity wave (Guza and Inman, 1975; Reussink, 1998), influence the alongshore spacing of rip currents and channels, which are typically 200–300 m apart on open ocean beaches. Bars and troughs that form on the intertidal part of the beach are known as ridges and runnels, with sometimes as many as a dozen or more occurring on intertidal beaches with high tidal ranges. Individual bars may have their own swash zones and dry beach on the crests until they are covered by the rising tide.

Large-scale morphodynamic beach processes

Equilibrium beach (cross-shore) profiles are produced by steady wave forcing during seasonal cycles. Accretionary (summer) and erosional (winter) beach profiles are expressions of seasonal cycles of wave energy, as elucidated in the classic works of Shepard (1950) and Bascom (1954). Accretionary beaches, which are produced by swell waves with a low wave height (generally <1 m) and a period of 8–12 s, produce a wide and well-developed backbeach with a relatively narrow foreshore (e.g., Davis and FitzGerald, 2004).

Large storm events result in a disequilibrium profile where sand may be permanently lost to deep water (Dean and Dalrymple, 2002). Erosional or storm beaches, on the other hand, are temporary morphodynamic states that are characterized by a profile that is generally flat and featureless with a narrow or nonexistent backbeach (Shepard, 1950; Hallermeier, 1981). Most or all of the beach thus occur in the foreshore position (e.g., Davis and FitzGerald, 2004). Due to storms, waves are larger and more energetic in winter than summer. Long periods of stormy weather, such as El Niño winters, erode beaches to the underlying cobbles or bedrock and deposit sand far offshore in deep water, leaving the beach in disequilibrium, as seen, for example, along the California coast. The winter beach is denuded of sand by storm waves, leaving heavier cobbles behind as lag deposits. Wave-cut platforms underlying the mobile sediments are often exposed during high-energy conditions. Summer wave conditions return sand back to the beach so that the summer beach is covered with a layer of sand that is moved south by alongshore currents and onshore by low-energy wave climate. Other researchers have noted associations between waves, breakers, and beach gradients – the relationship occurring because breaker type is a function of wave characteristics (Pethick, 1984). High values of the surf-scaling factor (e.g., Guza and Inman, 1975) are associated, for example, with wide flat beaches with spilling breakers, while low values are associated with steep narrow beaches and surging breakers (e.g., Wright et al., 1979; Wright et al., 1982). Although the relationship between wave steepness and beach gradient is pronounced, other variables such as the angle of the wave approach (e.g., Sonu and van Beek, 1971) and tides and longshore currents (e.g., Davis, 1985) may also be important.

Morphodynamic beach types

Beach morphotypes range through a sequence of process-form-materials systems between end points marked by low-energy systems with small beach-building waves to high-energy systems with large waves breaking across several hundred meters of surf zone (Wright and Short, 1984). Tides are an additional consideration in beach formation as they range from micro (<2 m), meso (2–4 m), macro (4–8 m), to mega tides (>8 m). Beaches are divided into the three basic morphodynamic types in order to accommodate tidal ranges, wave heights, and variations in grain size (sand to boulders), viz., wave dominated, tide modified, and tide dominated, based on the relative tide range (RTR) (Short, 2006). This process-based grouping of beach morphotypes, as initially conceived by Wright et al. (1982) and expanded to beaches around the Australian continent by Short (1999, 2006, 2012), provides a basis for cataloguing a range of beach processes associated with beaches in low latitude and mid-latitude. Ice-coast beach processes are not yet comprehensively included in the system. Some essential facets of the morphodynamic classification of beaches are briefly summarized here from Short (2012) and Short and Woodroffe (2009).

Wave-dominated beaches are characterized by a RTR tide range that is less than three times the average wave height (RTR < 3). These beaches occur in three morphodynamic states that are referred to as (1) reflective, (2) intermediate, and (3) dissipative.

- Reflective beaches are the result of process-material interactions that are associated with lower waves (H < 0.5 m), longer wave periods, and coarser sediment. As a result of these process-form relationships, all coarse sand and cobble-boulder beaches exhibit a reflective beach state. Reflective beaches contain a relatively steep beachface (5–20°) that reflects backwash. Reflective beach processes do not produce a bar or surf zone.
- Intermediate beaches are common along open coasts and are produced by moderate waves (H = 0.5–2.5 m) interacting with fine- to medium-grained sand. These types of beaches are characterized by a surf zone with one or two bars up to 100 m wide. The bar is typically traversed by regularly spaced shore-perpendicular rip channels and currents.
- Dissipative beaches are produced by high waves (H > 2.5 m) interacting with fine-grained sand. These morphotypes are characterized by gently seaward-sloping beachfaces with low-gradient swash (~1°) and 300–500 m wide surf zone that contains at least two bars. Waves sequentially break on the outer bar first and then on inner bars and in the process dissipate their energy as they move through the surf zone.

Tide-modified beaches are characterized by a RTR that lies between 3 and 10, which implies that the tidal range is increasing and/or wave height is decreasing. This morphodynamic beach type usually has a steeply sloping, cusped, and reflective beachface that is developed in coarser-grained sediments to produce a so-called high-tide beach. Tide-modified morphotypes are fronted by a wide, finer-grained, low-gradient, often featureless, intertidal zone (up to 200 m wide), seaward of which lies a low-tide surf zone that may contain bars and rip channels.

Tide-dominated beaches are produced when the RTR lies between 10 and approximately 50, implying high tides and/or very low waves (H << 0.3 m). These morphotypes are characterized by a low elevation, coarse-grained sediments, irregular, high-tide beach that is fronted by low-gradient (<<1°) intertidal sand or mud flats that may be hundreds of meters wide. Beyond an RTR of 50, tidal flats typically prevail.

An additional morphodynamic beach type is associated with a high-tide reflective beachface that is fronted by intertidal rock flats in middle latitudes. In tropical zones, the high-tide beach may be fronted by a fringing coral reef flat. High latitude morphotypes are seasonally exposed to freezing air and water temperatures that lead to the development of sea ice, shoreface ice, and a frozen snow-covered beach (Short and Woodroffe, 2009).

Summary

Although seemingly simple in appearance, beaches are complex systems where many aspects of their formation are incompletely understood. The morphological results of interactions between processes and materials are categorized in a range of processes-based beach types that are easily recognizable on beaches of the world. Beach processes are observed to operate over myriametric (macro-, meso-, micro-) scales and a range of time frames with grains on the beachface changing by the second to some sand bars that may persist for hundreds of years. Complete understanding of how beaches evolve (accrete) and devolve (erode) over time in different kinds of sedimentary environments provides scope for future studies that involve observation by various means in the field and laboratory.

Bibliography

Anthony, E. J., and Dolique, F., 2004. The influence of Amazon-derived mud banks on the morphology of sandy, headland-bound beaches in Cayenne, French Guiana: a short- to long-term perspective. *Marine Geology*, **208**, 249–264.

Anthony, E. J., Dolique, F., Gardel, A., and Marin, D., 2011. Contrasting sand beach morphodynamics in a mud-dominated setting: Cayenne, French Guiana. *Journal of Coastal Research*, SI 64 (*Proceedings of the 11th International Coastal Symposium*), pp. 30–34.

Bagnold, R. A., 1941. *The Physics of Blown Sand and Desert Dunes*. New York: Chapman and Hall, 265 p.

Bascom, W. H., 1954. Characteristics of natural beaches. In *Proceedings 4th Conference of Coastal Engineering*. Chicago: Illinois, pp. 163–180.

Bascom, W. H., 1980. *Waves and Beaches*. New York: Doubleday.

Carter, R. W. G., 1988. *Coastal Environments*. San Diego, CA: Academic.

Davidson-Arnott, R. A., 2005. Beach and nearshore instrumentation. In Schwartz, M. L. (ed.), *Encyclopedia of Coastal Science*. Dordrecht: Springer, pp. 130–138.

Davidson-Arnott, R. A., 2009. *Introduction to Coastal Processes and Geomorphology*. Cambridge, UK: Cambridge University Press.

Davidson-Arnott, R. A., and Law, M. N., 1996. Measurement and prediction of long-term sediment supply to coastal foredunes. *Journal of Coastal Research*, **12**(3), 654–663.

Davis, R. A., 1985. Beach and nearshore zone. In Davis, R. A. (ed.), *Coastal Sedimentary Environments*. New York: Springer, pp. 379–444.

Davis, R. A., and FitzGerald, D. M., 2004. *Beaches and Coasts*. Oxford: Blackwell, 419 p.

Dean, R. G., and Dalrymple, R. A., 2002. *Coastal Processes with Engineering Applications*. Cambridge: Cambridge University Press, p. 475.

Dingler, J. R., 2005. Beach processes. In Schwartz, M. L. (ed.), *Encyclopedia of Coastal Science*. Dordrecht, The Netherlands: Springer, pp. 161–168.

Duncan, J. R., 1964. The effects of water table and tide cycle on swash-backwash sediment distribution on beach profile development. *Marine Geology*, **2**, 186–197.

Greenwood, B., 2005. Bars. In Schwartz, M. L. (ed.), *Encyclopedia of Coastal Science*. Dordrecht, The Netherlands: Springer, pp. 120–129.

Guza, R., and Inman, D., 1975. Edge waves and beach cusps. *Journal of Geophysical Research*, **80**(21), 2997–3012.

Hallermeier, R. J., 1981. A profile zonation for seasonal sand beaches from wave climate. *Coastal Engineering*, **4**, 253–277.

Hardisty, J., 1990. *Beaches: Form and Processes*. London: Unwin Hyman, 324 p.

Higgins, C. G., 1984. Piping and sapping: development of landforms by groundwater outflow. In Lafleur, R. G. (ed.), *Groundwater as a Geomorphic Agent*. Boston: Allen and Unwin, pp. 18–58.

Klein, A. H., Filho, L. B., and Schumacher, D. H., 2002. Short-term beach rotation processes in distinct headland bay beach systems. *Journal of Coastal Research*, **18**, 442–458.

Komar, P. D., 1997. *Beach Processes and Sedimentation*. Upper Saddle River, NJ: Prentice Hall, 544 p.

Leatherman, S. P., 2013. Rip currents. In Finkl, C. W. (ed.), *Coastal Hazards*. Dordrecht, The Netherlands: Springer, pp. 811–831.

Otvos, E. G., 1964. Observations on rhomboid beach marks. *Journal of Sedimentary Petrology*, **34**, 683–687.

Otvos, E. G., 1965. Type of rhomboid beach surface patterns. *American Journal of Science*, **263**, 271–276.

Otvos, E. G., 1999. Rain-induced beach processes; landforms from ground water sapping and surface runoff. *Journal of Coastal Research*, **15**(4), 1040–1054.

Pethick, J., 1984. *An Introduction to Coastal Geomorphology*. London: Edward Arnold.

Pilkey, O. H., Neal, W. J., Kelley, J. T., and Cooper, A. G., 2011. *The World's Beaches*. Berkeley, CA: University of California Press, 283 p.

Reussink, B. G., 1998. *Infragravity waves in a dissipative multiple bar system*. Doctoral dissertation, Utrecht, The Netherlands: University of Utrecht, 243 p.

Roelvink, J. A., and Broker, I., 1993. Cross-shore profile models. *Coastal Engineering*, **21**, 163–191.

Shepard, F. P., 1950. *Beach Cycles in S. California*. U.S. Army Corps of Engineers BEB Technical Memo 20, 26 p.

Short, A. D. (ed.), 1999. *Handbook of Beach and Shoreface Morphodynamics*. Chichester, UK: Wiley.

Short, A. D., 2006. Australian beach systems – nature and distribution. *Journal of Coastal Research*, **22**, 11–27.

Short, A. D., 2012. Coastal processes and beaches. *Nature Education Knowledge*, **3**(10), 15.

Short, A. D., and Trembanis, A. C., 2004. Decadal scale patterns in beach oscillation and rotation Narrabeen Beach, Australia – time series, PCA and wavelet analysis. *Journal of Coastal Research*, **20**, 523–532.

Short, A. D., and Woodroffe, C. D., 2009. *The Coast of Australia*. Melbourne: Cambridge University Press.

Sonu, C. J., and van Beek, J. L., 1971. Systematic beach changes in the outer banks. N. Carolina. *Journal of Geology*, **79**, 416–425.

Tamura, T., Horaguchi, K., Saito, Y., Nguyen, V. L., Tateishi, M., Ta, T. K. O., Nanayama, F., and Watanabe, K., 2010. Monsoon-influenced variations in morphology and sediment of a mesotidal beach on the Mekong River delta coast. *Geomorphology*, **116**, 11–23.

Thornton, E. B., Humiston, R. T., and Birkemeier, W., 1996. Bar-trough generation on a natural beach. *Journal of Geophysical Research*, **101**, 12097–12110.

Woodroffe, C. D., 2002. *Coasts — Form and Processes*. Cambridge, UK: Cambridge University Press.

Wright, L. D., and Short, A. D., 1984. Morphodynamic variability of surf zones and beaches: a synthesis. *Marine Geology*, **56**, 93–118.

Wright, L. D., Chappell, J., Thom, B., Bradshaw, M., and Cowell, O., 1979. Morphodynamics of reflective and dissipative beach and inshore systems. South Australia. *Marine Geology*, **32**, 105–140.

Wright, L. D., Guza, R. T., and Short, A. D., 1982. Dynamics of a -high-energy dissipative surf-zone. *Marine Geology*, **45**, 41–62.

Zhang, D. P., 1994. Wave flume experiments on the formation of longshore bars produced by breaking waves. *Science Report, Institute Geoscience, University of Tsukuba*, **15**, 47–105.

Cross-references

Beach
Bed Forms
Coastal Engineering
Coasts
Currents
Dunes
Engineered Coasts
Laminated Sediments
Placer Deposits
Sea Walls/Revetments
Sediment Transport Models
Waves

BED FORMS

Miwa Yokokawa
Laboratory of Geoenvironment, Faculty of Information Science and Technology, Osaka Institute of Technology, Osaka, Japan

The term "bedforms" is used for to describe rhythmic topographic features on the surface of granular beds. A wide variety of such features form in response to specific ranges of hydraulic conditions and grain size. Apart from their intrinsic scientific interest, bedforms are important in both geology and engineering. Large subaqueous bedforms can be obstacles to navigation, and their migration can be a threat to submarine structures. Bedforms also play an important role in determining the resistance to flow. Bedforms are also one of the most useful tools available for interpreting ancient sedimentary environments from outcrops.

When unidirectional flow operates on relatively fine-grained sand (less than 0.7 mm), the following bedforms appear in order as flow "strength" is increased: current ripples, dunes, upper-regime plane bed (absence of bedforms), antidunes, and cyclic steps. If the bed material is coarser than 0.7 mm, the ripple regime is replaced by lower-regime plane bed.

Under purely oscillatory flows over fine-grained sediments, small symmetrical regular straight-crested (2D) ripples, less regular 3D ripples, and large 3D ripples appear with increasing oscillatory velocity magnitude. In the case of long-period oscillatory flows, large dome-like mounds with gentle slopes appear; these are known as "hummocky beds." In the case of coarser sediment, ripples tend to stay straight-crested. When the velocity magnitude becomes large, ripples are washed out to plane bed.

When unidirectional flow is superimposed on oscillatory flow, i.e., combined wave-current flow, bedforms change corresponding to the sum of unidirectional flow velocity and oscillatory flow velocity magnitude. When the unidirectional flow velocity is smaller than a threshold, bedforms are the same as those under purely oscillatory flow. When unidirectional flow velocity increases beyond this threshold, bedforms show characteristic rounded crests and asymmetrical profiles. The threshold at which unidirectional flow affects bedform shape decreases with increasing oscillatory period. When the unidirectional component dominates, large, asymmetric dune-like bedforms appear. On the other hand, when the oscillatory component dominates, hummocky beds appear. These bedforms wash out to plane bed at sufficiently high velocities.

Various bed phase diagrams have been proposed to categorize the relationship between bedforms and hydraulic conditions. In order to construct such diagrams, the relevant parameters which govern the relationships must be specified. In general, bed state is a function of seven parameters. Five parameters are common to all cases: grain size, sediment density, fluid density, fluid viscosity, and sediment submerged specific weight. For unidirectional flow, two more parameters, such as flow velocity and flow depth, are needed. For oscillatory flow, two of the following three parameters must be specified: oscillation period, orbital diameter, and maximum orbital velocity. Dimensional analysis allows the seven variables to be grouped into four dimensionless variables. A large body of experimental research on bedforms under unidirectional flows and short-period oscillatory flows is available. Our knowledge of long-period oscillatory flows and multidirectional combinations of combined waves and currents, however, remains limited.

Bibliography

Bridge, J., and Demicco, R., 2008. *Earth Surface Processes Landforms and Sediment Deposits.* Cambridge, UK/New York: Cambridge University Press, pp. 121–254. Chap. 5, 6, 7.

Garcia, M. H., 2008. Chapter 2 Sediment transport and morphodynamics. In: Garcia, M. H. (ed.), *Sedimentation Engineering.* ASCE (American Society of Civil Engineers), pp. 21–163.

Leeder, M., 2011. Chapter 7 Bedforms and sedimentary structures in flows and under waves. In Leeder, M. (ed.), *Sedimentology and Sedimentary Basins*, 2nd edn. Willey-Blackwell, pp. 132–170.

Southard, J.B., Chapter 12 Bed configurations. In Southard, J. B., 12.090 Introduction to Fluid Motions, Sediment Transport, and Current-Generated Sedimentary Structures, Fall 2006. (MIT OpenCourseWare: Massachusetts Institute of Technology), http://ocw.mit.edu/courses/earth-atmospheric-and-planetary-sciences/12-090-introduction-to-fluid-motions-sediment-transport-and-current-generated-sedimentary-structures-fall-2006, 350-443.

BIOCHRONOLOGY, BIOSTRATIGRAPHY

Felix M. Gradstein
Museum of Natural History, University of Oslo, Oslo, Norway

Biochronology attempts to organize the fossil record in linear time, using the notion that fossil events and fossil zones are organized according to the irreversible process of evolution.

The key to biochronology and its building blocks is fossil events. A fossil or paleontological event is the presence of a fossil taxon in its time context, derived from its position in a rock sequence. Most commonly used are first appearance and last appearance datums (FAD and LAD). Since the first or last appearance datum may be difficult to recognize or distinguish where specimen numbers dwindle or get obscured by "noise" (like reworking of fossil tests very common with the tiny nannofossils), it can be advantageous to substitute with first and last consistent (or common) appearances. A first or last appearance datum is called consistent when such stratigraphic range end points are part of an observed continuous stratigraphic range.

If the fossil record encountered in stratigraphic sections that we want to correlate and calibrate in time would be ubiquitous and perfect, i.e., if only time would control the appearance, range, and disappearance of taxa, then biostratigraphy would be a straightforward exercise. The science of biochronology, as developed for the evolutionary first and last occurrence datums of ocean plankton, would be a matter of systematic bookkeeping on a worldwide scale, only constrained by taxonomic deliberations. Unfortunately, the paleontological record is highly imperfect.

Uncertainty factors may be summarized as follows:

1. Quality and quantity of sampling
2. Specimen frequency of fossil taxa
3. Confidence of taxonomic identification
4. Influence of environmental change on the stratigraphic range of taxa
5. Differential rate of taxon evolution in different parts of the world
6. Time lag in migration of taxa, where correlation is over large distances or across major environmental barriers

Hence, biochronology almost always requires careful calibration with independent geomagnetic reversals and stable isotope correlation frameworks.

Biochronology is reaching a pinnacle in the massive fossil event datasets generated with TimeScale Creator software (http://www.tscreator.org). The method and data system confidently link all events in the evolutionary "organic continuum" in an elegant linear time framework, calibrated to Geologic Time Scale 2012 (see Geologic Time Scale). A recent calibration of Cretaceous and Cenozoic planktonic foraminifera and calcareous nannofossils of the temperate to tropical marine realm is by Anthonissen and Ogg (2012). Syntheses for siliceous, organic, and other microfossil biostratigraphy are summarized in the appropriate chapters in *The Geologic Time Scale* (Gradstein et al., 2012), and detailed and updated versions of all stratigraphic scales can be accessed as datasets and graphics at the TimeScale Creator website.

Special mention is made of the detailed and high-resolution (deep time) conodont-foraminifera-ammonoid composite zonation for the Carboniferous, with over 35–40 zones constrained by 36 radiometric dates employed in timescale building for this period (Davydov et al., 2012). Particularly in younger parts of the Carboniferous zonal event, resolution is resolved at a biochronologic scale that provides insight in biota migrations due to the waxing and waning of massive Gondwana supercontinent glaciations.

Summary

Biochronology attempts to order and scale fossil events and fossil ranges in linear time.

Bibliography

Anthonissen, D. E., and Ogg, J. G., 2012. Cenozoic and cretaceous biochronology of planktonic foraminifera and calcareous nannofossils. In Gradstein, F. M. (ed.), *The Geologic Time Scale 2012*. Amsterdam: Elsevier, pp. 1083–1128.

Davydov, et al., 2012. for reference see The Geologic Time Scale.

Gradstein, F. M., et al., 2012. *The Geologic Time Scale*. Amsterdam: Elsevier.

BIOGENIC BARIUM

Graham Shimmield
Bigelow Laboratory for Ocean Sciences, East Boothbay, ME, USA

Biogenic barium usually occurs as discrete microcrystals of the refractory mineral, barite ($BaSO_4$). It may be found in the water column (in the tests of both live and dead planktonic species), in benthic foraminifera, in coral skeletons, and in the underlying sediment. The earliest observations of enriched barium (usually identified as barium concentrations exceeding typical shale or sediment concentrations), and attributed to biological processes, are the work of Revelle et al. (1955) working in the equatorial divergence of the Pacific Ocean. Dehairs et al. (1980) and Bishop (1988) showed that barite ($BaSO_4$) was precipitated in decaying suspended marine particulate matter (particularly diatoms) in oceanic waters. Some studies have suggested that biogenic barium may occur in heavy mineral granules functioning as statoliths in statocyst organs and within protozoans such as Xenophyophoria and Loxodes. Biogenic barium distribution and concentration have been studied in benthic foraminifera and corals as a tracer of bottom water nutrients and upwelling, respectively.

Biogenic barium in sediments is often found in water underlying areas of high productivity. Bishop (1988) has studied the barium content of large and small particles in the Gulf Stream, and Schmitz (1987) has illustrated the use of Ba as a tracer of Indian Ocean plate movement beneath the equatorial upwelling zone on a timescale of millions of years. Virtually, all ocean basins display enrichment of biogenic barium where productivity is elevated and with time (paleoproductivity). Shimmield (1992) suggested that barite-secreting organisms may be confined to a rather discrete zone within the coastal upwelling productivity belt, seaward of the shelf break (under a different nutrient regime), and as a consequence shallow-water, organic-rich sediments may receive little biogenic barium, or the sedimentary barite undergoes diagenesis during sulfate reduction (see below). A similar distribution of biogenic barium was noted by Calvert and Price (1983) in their work off Namibia.

The refractory nature of barite was remarked on in the earliest work by Dymond (1981), something he called a "dissolution residue." The association of biogenic Ba, opal, and biogenic sedimentation has been described by many authors (see reviews in Schmitz, 1987; Gingele et al., 1999). These observations have opened the potential to use biogenic barium downcore distributions as an important proxy for paleoproductivity, given that both organic carbon and opal may suffer from remineralization and dissolution. An important step in quantifying past productivity is to establish the relationship between biogenic barium and organic carbon in sediment traps (Dymond et al., 1992; Francois et al., 1995). Using algorithms developed from empirical observations, downcore variations in biogenic barium flux have been converted to paleoprimary production. This approach has a number of assumptions and potential drawbacks. In particular, it is necessary to calculate the biogenic component of the total barium content in the sediment. This is usually achieved by normalization to lithogenic metals such as aluminum or titanium. There may be additional sources of barium to the sediment from hydrothermal systems or benthic organisms such as xenophyophores. Finally, although barite is very refractory under oxic conditions, during sulfate reduction in anaerobic sediments, barite dissolution may occur.

Bibliography

Bishop, J. K. B., 1988. The barite-opal-organic-carbon association in oceanic particulate matter. *Nature*, **332**, 341–343.

Calvert, S. E., and Price, N. B., 1983. Geochemistry of Namibian shelf sediments. In Suess, E., and Thiede, J. (eds.), *Coastal Upwelling, Part A*. New York: Plenum, pp. 337–375.

Dehairs, F., Chesselet, R., and Jedwab, J., 1980. Discrete suspended particles of barite and the barium cycle in the open ocean. *Earth and Planetary Science Letters*, **49**, 528–550.

Dymond, J., 1981. Geochemistry of Nazca plate surface sediments: an evaluation of hydrothermal, biogenic, detrital, and hydrogenous sources. *Memoirs of the Geological Society of America*, **154**, 133–173.

Dymond, J., Suess, E., and Lyle, M., 1992. Barium in deep-sea sediment: a geochemical proxy for paleoproductivity. *Paleoceanography*, **7**(2), 163–181.

Francois, R., Honjo, S., Manganini, S. J., and Ravizza, G. E., 1995. Biogenic barium fluxes to the deep-sea: implications for paleoproductivity reconstruction. *Global Biogeochemical Cycles*, **9**, 289–303.

Gingele, F. X., Zabel, M., Kasten, S., Bonn, W. J., and Nurnberg, C. C., 1999. Biogenic barium as a proxy for paleoproductivity: methods and limitations of application. In Fischer, G., and Wefer, G. (eds.), *Use of Proxies in Paleoceanography*. Berlin: Springer, pp. 345–364.

Revelle, R., Bramelette, M., Arrenhius, G., and Goldberg, E. D., 1955. Pelagic sediments of the Pacific. *Geological Society of America, Special Paper*, **62**, 221–235.

Schmitz, B., 1987. Barium, high productivity, and northward wandering of the Indian continent. *Paleoceanography*, **2**, 63–77.

Shimmield, G. B., 1992. Can sediment geochemistry record changes in coastal upwelling paleoproductivity? Evidence from northwest Africa and the Arabian Sea. In Summerhayes, C., Prell, W., and Emeis, K. (eds.), *Upwelling Systems Since the Early Miocene*. London: Geological Society. Geology Society Special Publication, Vol. 63, pp. 29–46.

BIOTURBATION

Gerhard Graf
Institute for Biological Sciences, University of Rostock, Rostock, Germany

Definition

Bioturbation is the mixing and displacement of particles in marine and freshwater sediments caused by benthic fauna mainly during foraging and the construction of burrows. As a result particulate proxies and microfossils such as tests of foraminifera deposited at the seafloor may not be found in the time slice corresponding to its deposition, and the interpretation of the geological record can be hampered. Other effects of animal activity, such as biodeposition and bioresuspension, but also fluid transport will not be considered in this section although they are essential for biogeochemical cycles and are used as subprocesses of bioturbation in recent biological literature (cf. Kristensen et al. 2012).

Measurement and modeling

In geological sciences, mainly natural radioactive tracers like ^{210}Pb, ^{234}Th, or ^{7}Be but also chlorophyll a or stained sand grains have been used for a quantitative description. A basic equation for sediment mixing was provided by Berner (1980).

$$\frac{\partial C}{\partial t} = \frac{\partial}{\partial x}\left(D_b(x)\frac{\partial C}{\partial x}\right) - \omega\frac{\partial C}{\partial x} \pm R(C,x,t)$$

C = concentration of tracer, t = time, x = depth, D_b = mixing coefficient, ω = burial velocity, R = reaction term for the tracer

It describes particle mixing in analogy to molecular diffusion. This assumption holds true in the case of many animals moving particles in a stochastic way in small steps (local mixing). Statistically this creates a transport of tracers along the concentration gradient, as does the Brownian movement on the molecular level, and can be calculated via Fick's law of diffusion. The mixing coefficient D_b provides a quantitative measure.

Some animals transport food particle directly in one step from the surface to depth, or particles may drop into open burrows. In these cases the diffusion analogy is inadequate, and this advective transport (nonlocal mixing) has to be considered separately, i.e., the step length and frequency of such events have to be determined and to be included into the model, as, for example, in the gallery-diffusion model by Francois et al. (2002).

Order of magnitude

From published data, Boudreau (1994) calculated a worldwide mean mixing depth of 9.8 ± 4.5 cm. It is, however, well documented that animal burrows may reach as deep as 2 m (Thalassinidea). Mixing coefficients D_b may range from 0.0002 to 370 cm^2 years^{-1}. The latter results are highly dependent on the habitat and the tracers used. The shorter the half-life of the tracer, the higher is D_B. This result is mainly explained by the fact that some animals are highly selective and prefer fresh organic matter.

So far attempts to find general relationships between sediment mixing and biomass or organic carbon content failed, indicating the strong species-specific effect of bioturbation.

Bibliography

Berner, R. A., 1980. *Early Diagenesis: A Mathematical Approach.* Princton: Princton University Press.

Boudreau, B. P., 1994. Is burial velocity a master parameter for bioturbation? *Geochimica et Cosmochimica Acta*, **58**, 1243–1249.

Francois, F., Gerino, M., Stora, G., Durbec, J.-P., and Poggiale, J.-C., 2002. Functional approach to sediment reworking by gallery-forming macrobenthic organisms: modeling and application with the polychaete Nereis diversicolor. *Marine Ecology Progress Series*, **229**, 127–136.

Kristensen, E., Penha-Lopes, G., Delefosse, M., Valdemarsen, T., Quintana, C. O., and Banta, G. T., 2012. What is bioturbation? The need for a precise definition for fauna in aquatic sciences. *Marine Ecology Progress Series*, **446**, 285–302.

BLACK AND WHITE SMOKERS

Margaret K. Tivey
Marine Chemistry & Geochemistry, Woods Hole Oceanographic Institution, Woods Hole, MA, USA

Synonyms

Active vent deposit; Black chimney, White chimney; Black smoker chimney, White smoker chimney; Hydrothermal chimney

Definition

Black and white smokers. Chimney-like edifices composed of mixtures of copper-, iron-, and zinc-sulfide minerals and calcium- and barium-sulfate minerals. They form as very hot (up to ~400 °C or 750 °F) fluids exit very young seafloor and mix with cold seawater, with the high-temperature fluids passing through channels within the edifices into the deep ocean.

Introduction

Black and white smokers are the portions of seafloor hydrothermal vent deposits through which ~200 °C–400 °C hydrothermal fluids travel and exit into the deep ocean, <1,000–5,000 m below sea-level. The hot fluids form as cold seawater percolates down into young, still hot seafloor near the spreading axes of the mid-ocean ridges and spreading centers in back-arc basins; during its transit, the seawater exchanges heat and undergoes chemical reactions with the young oceanic crust. The seawater loses oxygen and sulfate, becomes more acidic (pH decreases), and becomes enriched in metals (e.g., Fe, Mn, Cu, Zn, Pb) and hydrogen sulfide, and its temperature increases from ~2 °C to >400 °C (German and Von Damm, 2006). The hot fluid is very buoyant (because its density decreases to approximately two-thirds of the original density as it is heated to >400 °C) and rises rapidly to the seafloor, exiting at meter-per-second flow rates (Bischoff and Rosenbauer, 1985; Spiess et al., 1980; see "Hydrothermal Vent Fluids (Seafloor)"). When the hot, metal- and sulfide-rich, oxygen-poor fluid exits and mixes with cold, sulfate-rich, metal-poor seawater, minerals precipitate rapidly as chimney-like edifices and as particles within plumes (with a smokelike appearance) above the edifice – see Figure 1.

Discovery

Venting of hydrothermal fluids from the youngest portions of the seafloor along the mid-ocean ridges was first observed in 1977 along the Galapagos Rift where unusual biological communities were found associated with warm vent fluids (17 °C, much warmer than the surrounding 2 °C seawater); the presence of these warm fluids had been predicted based on measurements of heat flow (see "Marine Heat Flow") and bottom seawater thermal anomalies close to the mid-ocean ridge (Corliss et al., 1979). In 1978, massive sulfide deposits that likely formed from much higher temperature fluids were found near 21°N latitude on the East Pacific Rise about 650 m west of the spreading axis (Francheteau et al., 1979). The first actively venting black and white smokers were subsequently discovered in 1979, along the spreading axis of the East Pacific Rise (Spiess et al., 1980) not far from where the deposits were found a year earlier.

The hot fluids were observed exiting "stacks" or "chimneys" that were 1–5 m tall, composed of copper-, iron-, and zinc-sulfide minerals and the mineral anhydrite (calcium sulfate) (Spiess et al., 1980). The observed black chimneys, or black smokers, resembled organ pipes ≤30 cm in diameter and emitted hot (>350 °C) fluid with dark-colored (black) precipitates suspended within the exiting fluid (Spiess et al., 1980). The white chimneys, or white smokers, were covered with worm tubes (making them light-colored) and emitted cooler (<330 °C) fluids at slower flow rates with light-colored precipitates suspended within exiting waters (Spiess et al., 1980; Haymon and Kastner, 1981).

Formation of black smoker chimneys

Black smoker chimneys form in two stages (Haymon, 1983; Goldfarb et al., 1983): (1) deposition of an anhydrite ($CaSO_4$)-dominated wall as hot, calcium-rich vent fluid mixes turbulently with cold, sulfate- and calcium-rich seawater, followed by (2) deposition of a layer of sulfide minerals against the inner side of the anhydrite layer as the Stage 1 anhydrite wall prevents rapid mixing between the hot fluid flowing inside the chimney and the cold

Black and White Smokers, Figure 1 Hot (347 °C) vent fluid exits multiple black smokers on the chimney edifice "Homer" near 17.5°S latitude on the southern East Pacific Rise (Courtesy Woods Hole Oceanographic Institution; M. Lilley and K. Von Damm chief scientists).

seawater present on the outside of the chimney. Chalcopyrite ($CuFeS_2$) is dominant if temperature is >330 °C and pyrite (FeS_2) and wurtzite ($(Zn, Fe)S$) occur at lower temperatures. Infiltration of seawater and hydrothermal fluid components across the porous wall also results in the deposition of sulfide and sulfate minerals and silica in the interstices of the wall, which gradually makes the chimney less porous and more metal-rich – see Figure 2. Interaction of fluids and biota on chimney exteriors can also result in the formation of an iron-sulfide-rich outermost layer on the chimney (Juniper et al., 1992).

Rapid formation of the Stage 1 chimney wall (it can form at rates up to 30 cm per day; Goldfarb et al., 1983) is in part a consequence of anhydrite being an unusual mineral that is more soluble at low temperatures than at high temperatures; if seawater is heated to ~150 °C or greater, anhydrite precipitates (Bischoff and Seyfried, 1978). When the very hot vent fluid exits at meter-per-second velocities into seawater, mixtures above ~150 °C will be saturated in anhydrite, with the sulfate coming from seawater and the calcium from both the vent fluid and seawater (Styrt et al., 1981; Albarède et al., 1981). Metal sulfides and oxides (zinc sulfide, iron sulfide, copper-iron sulfide, manganese oxide, and iron oxide) also precipitate from the vent fluid and vent fluid/seawater mixtures as fine-grained particles. Some of these particles become trapped within and between grains of anhydrite within the Stage 1 chimney walls, giving the anhydrite, which is clear to white in its pure form, a gray to black color (Goldfarb et al., 1983; Haymon, 1983). The remaining particles form a plume of "smoke" above the chimney. Because bottom seawater is denser than the mix of seawater and hydrothermal fluid in the plume, the plume rises a few 100 m above the seafloor to a depth where it is of the same buoyancy as the surrounding ocean water (see "Hydrothermal Plumes").

Once the initial anhydrite-dominated framework is in place, chalcopyrite (or chalcopyrite and pyrite or chalcopyrite and wurtzite for lower temperature black smokers) precipitates on the inner surface of the chimney. Observed young chimney walls are thin, less than a centimeter to a few centimeters, with one side of the wall very hot and one side much colder. The porous chimney wall is subject to steep gradients of temperature and concentrations of elements. As the chimney evolves, the innermost layer thickens (recovery of a chimney known to be only 1 year old had an innermost chalcopyrite layer that was ~1 cm thick; Koski et al., 1994). At the same time that the innermost layer is thickening, aqueous ions, including copper, iron, hydrogen, oxygen, sulfide, sulfate, zinc, sodium, chloride, and magnesium, are transported from areas of high to low concentrations (by diffusion). These elements also are carried by fluids flowing across the wall from areas of high to low pressure (by advection). As a result of these processes, minerals become saturated and precipitate in the pore spaces within the chimney walls, and favorable conditions for microorganisms are established in the outer parts of the chimney walls. In particular, within the chimney walls at temperatures less than ~120 °C, the steep temperature and concentration gradients provide combinations of reducing chemicals from vent fluids (hydrogen, hydrogen sulfide, ferrous iron) and oxidizing chemicals from seawater (oxygen, sulfate, ferric iron) that can be used by microorganisms (bacteria and archaea) as sources of energy (see "Chemosynthetic Life"; Jannasch, 1995). Larger organisms (e.g., alvinellids and paralvinellids; Haymon and Kastner, 1981; Juniper et al., 1992) also reside on the exteriors of chimneys, and their tubes can become incorporated into chimney walls – see Figure 3. As the chimney grows, earlier-formed minerals, as they are exposed to hotter fluids, can be replaced by later-formed minerals.

Formation of white smoker chimneys

The style of mixing between vent fluid and seawater differs in chimneys that emit lower-temperature, white to clear fluids (~200–330 °C). One major reason for this is that the vent fluid is flowing more slowly. Because the

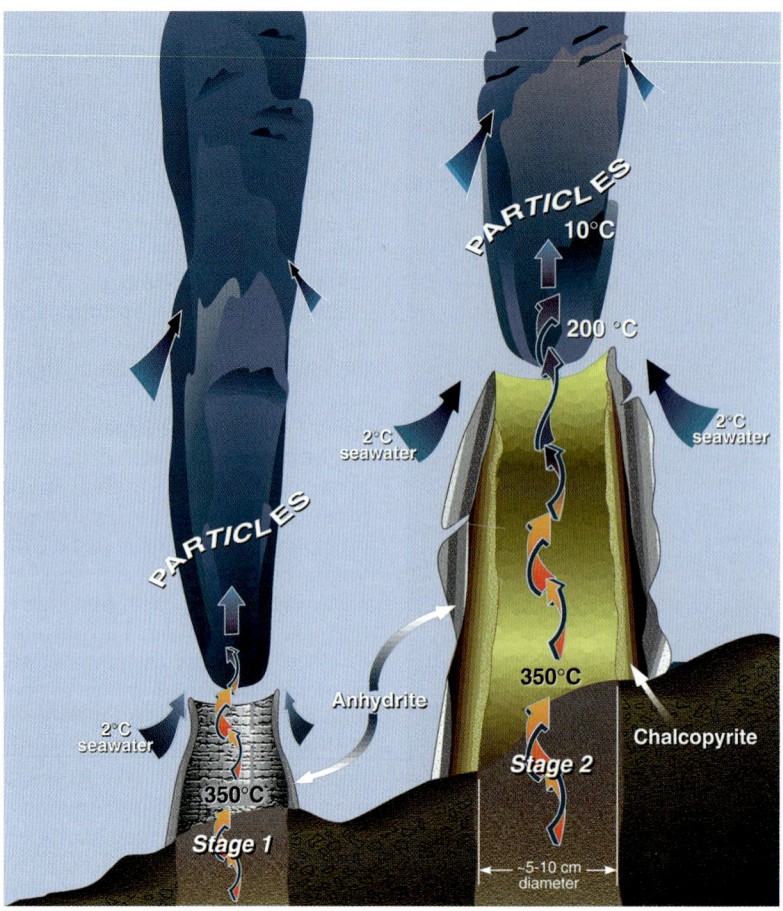

Black and White Smokers, Figure 2 Schematic drawing showing Stage 1 and Stage 2 growth of black smoker chimneys (After Haymon, 1983 and Goldfarb et al., 1983; figure from Tivey, 1998, courtesy Woods Hole Oceanographic Institution).

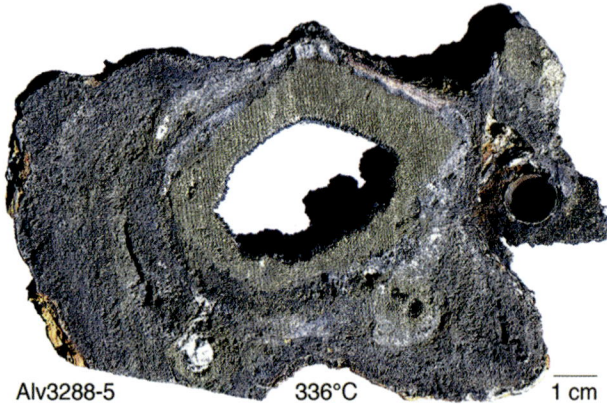

Black and White Smokers, Figure 3 Photograph of a slab taken perpendicular to the open channel across a black smoker chimney that was venting 336 °C fluid, from the southern East Pacific Rise near 21.5°S. The innermost layer (~1 cm thick) is composed of chalcopyrite (*gold* in color), and the outer 0.5–3 cm layer is composed of a mixture of anhydrite and sulfide minerals (*gray* in color). The fossilized tube of an alvinellid is embedded in the chimney wall (*right* side of image) (courtesy Woods Hole Oceanographic Institution).

fluid percolates less vigorously through the porous spires, a greater percentage of metal from the fluid precipitates within the deposit instead of being lost to the ocean within particle-laden plumes (e.g., Haymon and Kastner, 1981; Koski et al., 1994). Some of the white smoker chimneys are dominated by zinc- and iron-sulfide minerals, which form both an initial framework and infilling material (Koski et al., 1994 – see Figure 4). Others contain abundant barite and silica, both as initial framework material and as infilling material, with barite then co-precipitating with silica and sulfide minerals (Hannington and Scott, 1988). Many white smokers lack anhydrite, consistent with a lack of entrained seawater as they form. In contrast to black smokers, where flow is rapid through open conduits of >1 cm diameter, flow in white smokers is through narrow, anastomosing conduits that seal over time, commonly diverting flow horizontally (Fouquet et al., 1993; Koski et al., 1994). Beehive structures or diffusers are another type of smoker found at some vent fields. They have a bulbous morphology, interior temperatures only slightly less than for black smoker chimneys, highly porous interiors, and high-temperature (relative to black

Black and White Smokers, Figure 4 *Left image* is of white smoker chimneys (venting 250–300 °C fluids) from the Kremlin area on the TAG active hydrothermal mound at 26°N on the Mid-Atlantic Ridge (Courtesy of Woods Hole Oceanographic Institution; Peter Rona chief scientist); *right image* is a photograph of a slab taken perpendicular to the axis of one of the Kremlin chimneys, composed dominantly of zinc-sulfide – note the absence of large conduits and prevalence of very narrow channelways (courtesy of Woods Hole Oceanographic Institution).

and white smoker chimneys) exteriors (up to 70 °C; Fouquet et al., 1993; Koski et al., 1994). They form from less focused high-temperature fluids that exhibit slower flow rates than those from black smokers; as in white smoker chimneys, some flow is diverted and occurs through sides of these smokers (Fouquet et al., 1993; Koski et al., 1994). Within some vent fields, hot fluids are trapped beneath overhanging ledges, termed flanges; fluids percolate up through the porous flange layers or flow horizontally and "waterfall" upwards over the lip of the flange (Delaney et al., 1992).

Carbonate-rich chimneys

Very different types of chimneys are found at the Lost City vent field, located 15 km from the axis of the Mid-Atlantic Ridge. The Lost City hydrothermal system is hosted in mantle rocks (peridotite and serpentinite – see "Peridotites"), and the venting fluids have a higher pH than seawater, low metal and sulfide concentrations, and low temperature (<100 °C) relative to black smoker and white smoker fluids (Kelley et al., 2005); the very tall spires being deposited from these fluids are composed of calcium carbonate and magnesium hydroxide minerals (calcite and/or aragonite ($CaCO_3$) and brucite ($MgOH_2$)) that are saturated in the high pH, carbonate- and hydroxide-rich fluids.

Summary and conclusions

The morphologies of actively venting seafloor hydrothermal deposits and the composition and appearance of their hydrothermal plumes reflect the compositions of the fluids (hydrothermal fluid and seawater) from which they form, and the styles of flow and mixing of these fluids. "Black smoker" chimneys with conduit diameters of <2–10 cm form from very high-temperature (>330–400 °C) fluids that flow at rates of meters per second, and the turbulent mixing that results when this fluid exits into seawater results in the formation of particle-laden plumes reminiscent of black smoke. "White smoker" chimneys form from lower-temperature (~200–330 °C) vent fluids that flow less vigorously through anastomosing narrow conduits that can become blocked resulting in diversion of flow; a greater percentage of metals are trapped and deposited within these chimneys than within black smokers, and they tend to be richer in zinc because of the lower temperatures that result in saturation and precipitation of zinc-sulfide minerals.

Over time, collapse and incorporation of these different types of chimneys and flanges onto and into mounds, and subsequent reworking of this material as hot fluids flow through the mounds, forms larger deposits. These larger deposits (see "Volcanogenic Massive Sulfides" and "Marine Mineral Resources") found along the mid-ocean ridges and along the spreading centers within back-arc basins are analogs for some types of ore deposits found on land (e.g., Cyprus-type massive sulfide ore deposits; Francheteau et al., 1979).

Bibliography

Albarède, F., Michard, A., Minster, J.-F., and Michard, G., 1981. $^{87}Sr/^{86}Sr$ ratios in hydrothermal waters and deposits from the East Pacific Rise at 21°N. *Earth and Planetary Science Letters*, **55**, 229–236.

Bischoff, J. L., and Rosenbauer, R. J., 1985. An empirical equation of state for hydrothermal seawater (3.2 percent NaCl). *American Journal of Science*, **285**, 725–763.

Bischoff, J. L., and Seyfried, W. E., Jr., 1978. Hydrothermal chemistry of seawater from 25° to 350°C. *American Journal of Science*, **278**, 838–860.

Corliss, J. B., Dymond, J., Gordon, L. I., Edmond, J. M., von Herzen, R. P., Ballard, R. D., Green, K., Williams, D.,

Bainbridge, A., Crane, K., and van Andel, T. H., 1979. Submarine thermal springs on the Galapagos Rift. *Science*, **203**, 1073–1083.

Delaney, J. R., Robigou, V., McDuff, R. E., and Tivey, M. K., 1992. Geology of a vigorous hydrothermal system on the Endeavour Segment, Juan de Fuca Ridge. *Journal of Geophysical Research*, **97**, 19663–19682.

Fouquet, Y., Wafik, A., Cambon, P., Mevel, C., Meyer, G., and Gente, P., 1993. Tectonic setting, mineralogical and geochemical zonation in the Snake Pit sulfide deposit (Mid-Atlantic Ridge at 23°N). *Economic Geology*, **88**, 2018–2036.

Francheteau, J., Needham, H. D., Choukroune, P., Juteau, T., Seguret, M., Ballard, R. D., Fox, P. J., Normark, W., Carranza, A., Cordoba, D., Guerrero, J., Rangin, C., Bougault, H., Cambon, P., and Hekinian, R., 1979. Massive deep-sea sulphide ore deposits discovered on the East Pacific Rise. *Nature*, **277**, 523–528.

German, C., and Von Damm, K., 2006. Hydrothermal processes. In Holland, H. D., and Turekian, K. K. (eds.), *Treatise on Geochemistry, volume 6: The Oceans and Marine Chemistry*. London: Elsevier, pp. 181–222.

Goldfarb, M. S., Converse, D. R., Holland, H. D., and Edmond, J. M., 1983. The genesis of hot spring deposits on the East Pacific Rise, 21N. In Ohmoto, H., and Skinner, B. J. (eds.), *The Kuroko and Related Volcanogenic Massive Sulfide Deposits*. New Haven, Conn: Economic Geology Publication. Economic Geology Monograph, Vol. 5, pp. 184–197.

Hannington, M. D., and Scott, S. D., 1988. Mineralogy and geochemistry of a hydrothermal silica- sulfide-sulfate spire in the caldera of Axial Seamount, Juan de Fuca Ridge. *Canadian Mineralogist*, **26**, 603–625.

Haymon, R. M., 1983. Growth history of hydrothermal black smoker chimneys. *Nature*, **301**, 695–698.

Haymon, R. M., and Kastner, M., 1981. Hot spring deposits on the East Pacific Rise at 21°N: preliminary description of mineralogy and genesis. *Earth and Planetary Science Letters*, **53**, 363–381.

Jannasch, H., 1995. Microbial interactions with hydrothermal fluids. In Humphris, S. E., Zierenberg, R. A., Mullineaux, L. S., and Thomson, R. E. (eds.), *Seafloor Hydrothermal Systems*. Washington, DC: American Geophysical Union. American Geophysical Union Monograph, Vol. 91, pp. 273–296.

Juniper, S. K., Jonasson, I. R., Tunnicliffe, V., and Southward, A. J., 1992. Influence of a tube-building polychaete on hydrothermal chimney mineralization. *Geology*, **20**, 895–898.

Kelley, D. S., Karson, J. A., Fruh-Green, G. L., Yoerger, D. R., Shank, T. M., Butterfield, D. A., Hayes, J. M., Schrenk, M. O., Olsen, E. J., Proskurowski, G., Jakuba, M., Bradley, A., Larson, B., Ludwig, K., Glickson, D., Buckman, K., Bradley, A. S., Brazelton, W. J., Roe, K., Elend, M. J., Delacour, A., Bernasconi, S. M., Lilley, M. D., Baross, J. A., Summons, R. E., and Sylva, S. P., 2005. A serpentinite-hosted ecosystem: the Lost City hydrothermal field. *Science*, **307**, 1428–1434.

Koski, R. A., Jonasson, I. R., Kadko, D. C., Smith, V. K., and Wong, F. L., 1994. Compositions, growth mechanisms, and temporal relations of hydrothermal sulfide-sulfate-silica chimneys at the northern Cleft segment, Juan de Fuca Ridge. *Journal of Geophysical Research*, **99**, 4813–4832.

Spiess, F. N., Macdonald, K. C., Atwater, T., Ballard, R., Carranza, A., Cordoba, D., Cox, C., Diaz Garcia, V. M., Francheteau, J., Guerrero, J., Hawkins, J., Haymon, R., Hessler, R., Juteau, T., Kastner, M., Larson, R., Luyendyk, B., Macdougall, J. D., Miller, S., Normark, W., Orcutt, J., and Rangin, C., 1980. East pacific rise: hot-springs and geophysical experiments. *Science*, **207**, 1421–1433.

Styrt, M. M., Brackman, A. J., Holland, H. D., Clark, B. C., Pisutha-Arnold, V., Eldridge, C. S., and Ohmoto, H., 1981. The mineralogy and the isotopic composition of sulfur in hydrothermal sulfide/sulfate deposits on the East Pacific Rise, 21°N latitude. *Earth and Planetary Science Letters*, **53**, 382–390.

Tivey, M. K., 1998. How to build a black smoker chimney: the formation of mineral deposits at mid-ocean ridges. *Oceanus*, **41**, 68–74.

Cross-references

Chemosynthetic Life
Hydrothermal Plumes
Hydrothermal Vent Fluids (Seafloor)
Hydrothermalism
Marine Heat Flow
Marine Mineral Resources
Marginal Seas
Oceanic Spreading Centers
Peridotites
Volcanogenic Massive Sulfides

BOTTOM SIMULATING SEISMIC REFLECTORS (BSR)

Jürgen Mienert and Stefan Bünz
Centre for Arctic Gas Hydrate, Environment and Climate (CAGE), UiT The Arctic University of Norway, Tromsø, Norway

Definition

A seismic reflection occurring in the upper few hundred meters of marine sediments mimicking the seafloor, crosscutting sediment layers, and showing a phase reversal is known as a "bottom-simulating reflector." Such a gas hydrate-related BSR originates from a large impedance contrast between a layer of gas-hydrated sediment above and a free gas layer below. A diagenetic-related BSR occurs at the opal-A/opal-CT transition zone, lies often deep and outside the base of the gas hydrate stability zone, shows no phase reversal, and does not always mimic the seafloor.

Introduction

The intent of this article is to describe the two most commonly observed bottom-simulating reflectors (BSRs). The term BSR stems from their principal characteristic that these reflectors mimic the seafloor topography in marine seismic reflection data thereby crosscutting sedimentary strata. BSRs are known to occur in continental margin sediments in regions of gas hydrate and free gas (Shipley et al., 1979) and/or in siliceous ooze (diatoms, radiolaria, silica sponges, silicoflagellates) bearing sedimentary formations (Hein et al., 1978). The largest silica contribution comes from diatoms (Holland and Turekian, 2003), while the largest methane contribution derives from methane-producing Archaea in sub-seafloor sediments (e.g., Kotelnikova, 2002).

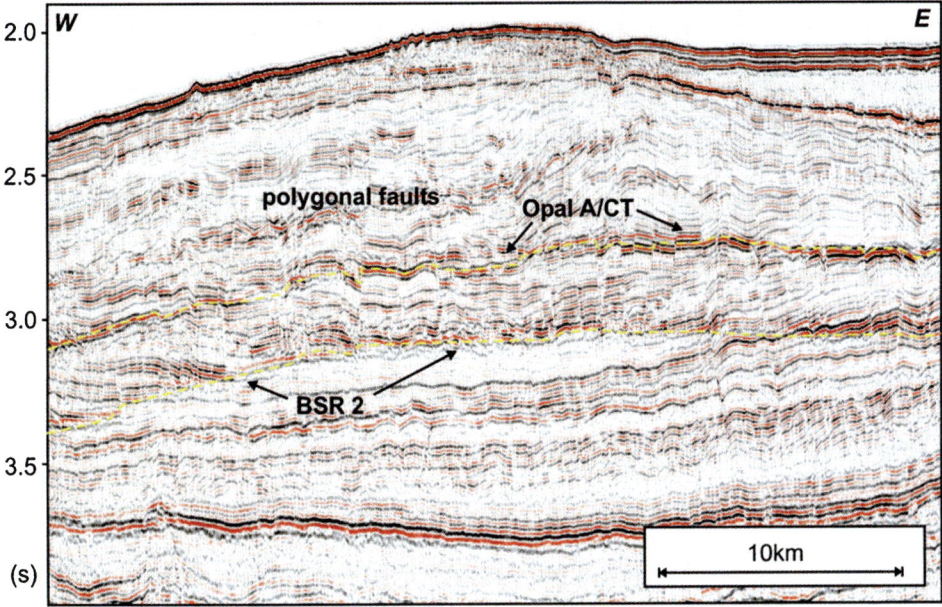

Bottom Simulating Seismic Reflectors (BSR), Figure 1 Reflection seismic profile from the mid-Norwegian margin siliceous sedimentary formation showing an example of an opal-A/opal-CT BSR. The BSR and amplitudes are offset by polygonal faults. The origin of BSR 2 is speculative and may be related to transformation from smectite to illite at higher temperatures (Figure is from Berndt et al., 2004, Figure 4).

Diagenesis-related BSR

Siliceous ooze occurs in a wide range of sedimentary basins indicating times of higher ocean productivity, for example, along equatorial, polar, or coastal upwelling regions (Holland and Turekian, 2003). With increasing burial depth, the temperature and pressure increases, resulting in the dissolution of the siliceous skeletons. The dissolution process causes a collapse of the siliceous framework and a geochemical reaction that converts amorphous opal-A into opal-CT (e.g., Hesse, 1989; Knauth, 1994). Deep-sea drilling project studies allowed documenting the chemical, mineralogical, and structural changes occurring during this process with increasing depth (e.g., Hurd and Birdwhistell, 1983). As a consequence of the skeleton dissolution, an interface develops where the opal-CT formation causes an increase in density and compressional-wave velocity and a decrease in porosity and permeability (e.g., Tribble et al., 1992). If the seismic impedance contrast between opal-A (lower density and velocity and higher porosity and permeability) and opal-CT becomes large enough, a seismic reflector with a positive polarity occurs (Figure 1) (Berndt et al., 2004). It is believed that the temperature increase with burial depth is the main parameter controlling the opal-A/opal-CT transition aside from the time since burial, type of surrounding sediment material, and interstitial waters (e.g., Hein et al., 1978). The opal-A/opal-CT diagenesis causes a volume reduction of as much as 30–40 % (Davies and Cartwright, 2002). Since large areas of siliceous ooze exist in sedimentary formations, a diagenetic BSR (Figure 1) often shows a very large lateral extent, which is uncommon for gas hydrate-/free gas-related BSRs. Moreover, diagenetic BSRs are believed to develop at greater depth below the seafloor, at temperatures (35–50 °C) where hydrate is normally no longer stable (Berndt et al., 2004).

Opal-A/opal-CT BSRs have been reported from many areas containing siliceous sediments such as from the mid-Norwegian margin (e.g., Brekke, 2000; Berndt et al., 2004) (Figure 1). Often, polygonal faults occur in conjunction with diagenetic BSRs as they preferably form in similar types of sediments (Figure 1) (Cartwright and Dewhurst, 1998; Davies and Cartwright, 2002). Polygonal faults show an interruption and vertical offset of continuous reflections leading to short reflection segments. The BSR may be difficult to identify if it runs parallel to the strata, because normally it also does not show a reduced instantaneous frequency (Berndt et al., 2004). The BSR has a strong amplitude, has an apparent polarity that is positive, lies deeper than a gas hydrate-related BSR, and is often interrupted by polygonal faults.

Gas hydrate-related BSR

Gas hydrates occur as an icelike substance composed of water molecules forming a rigid lattice of cages that trap a guest molecule (Sloan, 2003). The predominant guest

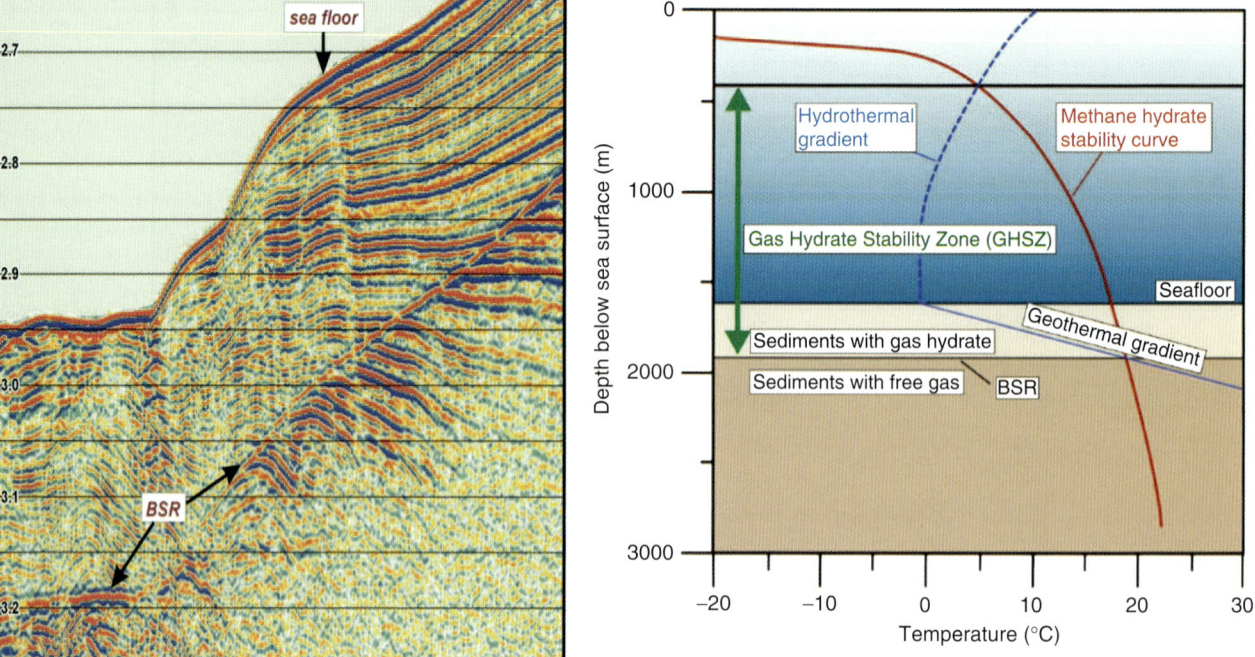

Bottom Simulating Seismic Reflectors (BSR), Figure 2 Reflection seismic profile of a hydrate/free gas BSR that follows the sub-bottom depths and coincides with the predicted depth of the base of the gas hydrate stability zone (schematic diagram, *right*). The gas hydrate stability zone is shown as a function of water temperature, pressure, and geothermal gradient.

molecule in the submarine environment is methane, but also hydrates containing high-order hydrocarbons, carbon dioxide, hydrogen sulfide, or other gas may exist. Gas hydrates occur naturally in the pore space of different types of marine sediments, where appropriate high-pressure and low-temperature conditions exist and there is an adequate supply of gas and water (Figure 2) (Kvenvolden, 1993; Rempel and Buffett, 1997; Sloan, 2003). Those requirements confine marine gas hydrates to the upper few hundred meters of the shallow geosphere of continental margins, where biogenic processes produce sufficient amounts of methane gas. The gas hydrate-related BSR detected on marine seismic reflection data commonly corresponds to the base of the gas hydrate stability zone (GHSZ, Figure 2). It is the result of an acoustic impedance contrast between hydrate-bearing sediments (increase in compressional-wave velocity) and free gas trapped in the sediments underneath (decrease in velocity) gas hydrates (Hyndman and Spence, 1992; Bünz et al., 2003). As a consequence, the hydrate-related BSR has reversed polarity (compared to the seafloor reflection) and is often accompanied by high-reflection amplitudes. The gassy sediments beneath the hydrate-bearing sediments also show on the instantaneous frequency attribute as the free gas predominantly attenuates the high-frequency content of the seismic signal (Berndt et al., 2004).

The BSR shows not always as a reflection proper because gas beneath the hydrated sediments accumulates only in places where rock properties of the host rock have high enough permeability. The seismic reflection shows higher amplitudes preferentially in areas where appreciable amounts of gas accumulate beneath the GHSZ (Figs. 3 and 4). Thus, whether the BSR is a true reflection in its own right on the seismic data is mainly the result of the frequency bandwidth of the acquisition system (Wood et al., 2002). High-frequency seismic acquisition systems often image gas accumulations along layers. The BSR is then identified as the envelope of amplitude increases that crosscuts stratigraphic boundaries (Figs. 2, 3, and 4). The BSR generally lies shallower than a diagenetic BSR, has a smaller areal extent, and might often show in patches over a larger area.

Assumptions on the possible presence of gas hydrate-/free gas-related BSRs along continental margins are based largely on modeling the GHSZ (Figure 2) (Dickens and Quinby-Hunt, 1997; Zatsepina and Buffett, 1998). The modeling and thus the theoretical potential for the existence of a BSR are mainly based on water depth (pressure), seafloor temperature, and the geothermal

Barents Sea – UiT P-Cable 3D seismic

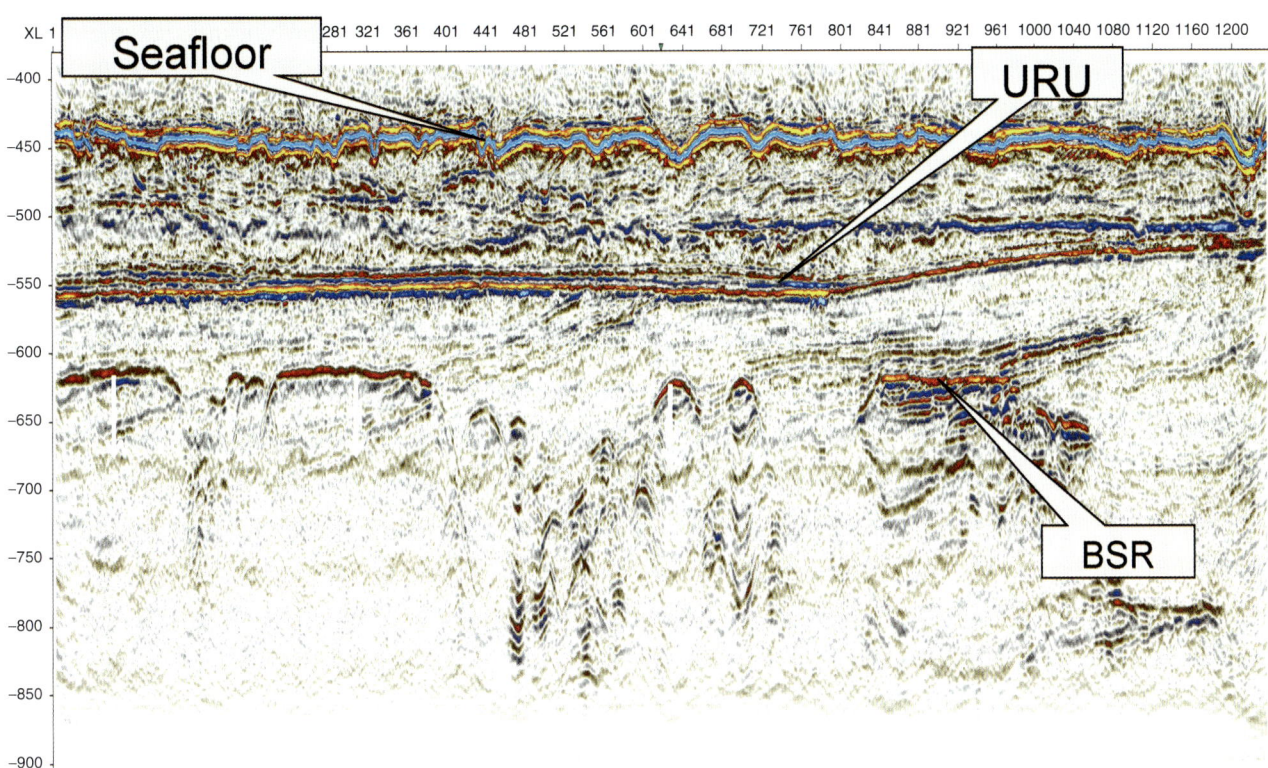

Bottom Simulating Seismic Reflectors (BSR), Figure 3 High-resolution 3D P-Cable seismic data showing higher amplitudes beneath the upper regional unconformity (*URU*) in the SW-Barents Sea. This high-amplitude reflection crosscuts sedimentary reflections with reversed polarity and is interpreted as BSR at the base of the GHSZ.

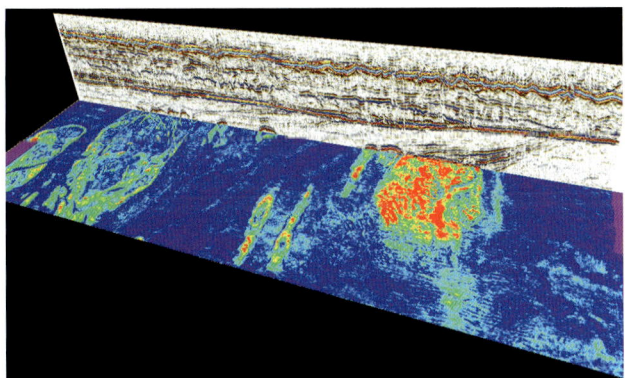

Bottom Simulating Seismic Reflectors (BSR), Figure 4 High-resolution 3D P-Cable seismic data showing higher amplitudes preferentially in areas where appreciable amounts of gas accumulate beneath the GHSZ.

gradient (Figure 2). As these parameters are largely controlled by water depth from the shelf to the deep sea, the methane hydrate BSR shows two pinch-out zones, one at the shallow water depth where pressure becomes too low for hydrates to be stable and one toward the mid-ocean ridge where heat flow becomes too high. Thus, if one describes these two end members, one may still use the terminology BSR but should be clear that the BSR does not parallel the seafloor.

Seismic character of BSRs

BSRs related to hydrate/free gas phase transitions can be distinguished from opal-A/opal-CT phase transitions based on their phase polarity. While diagenetic BSRs have the same positive phase as the seafloor reflection (Figure 1), the hydrate/free gas BSRs show a phase reversal (Figure 2). This reversed polarity is due to the negative impedance contrast between hydrated sediments above (higher density and velocity) and gas-saturated sediments below (Hyndman and Spence, 1992). However, caution is necessary if the phase of the BSR is the only criterion used. Gas trapped beneath a diagenetic BSR may also result in a phase reversal. As a consequence one should use several criteria such as instantaneous frequency, phase reversal, and GHSZ modeling to predict the BSR depth. Frequencies are useful as additional criteria for gas hydrate BSRs. A shift from higher frequencies in the gas hydrate zone to lower frequencies in the free gas zone (Berndt et al., 2004) beneath the BSR may be used.

Dynamics of BSRs

Both the diagenetic- and the hydrate-related BSR may be used to evaluate the thermal state or reconstruct the thermal development of a sedimentary basin (Grevemeyer and Villinger, 2001; Nouzé et al., 2009). Particularly, the hydrate-related BSR is widely used as a proxy for heat flow in marine sediments. On a small scale, BSR-derived heat flow changes may often be associated with structures that focus warm fluids from deep sediments like mud volcanoes (Depreiter et al., 2005) or chimneys (Rajan et al., 2013). In such instance, the BSR may not mimic the seafloor if lateral variations in heat flow or changes in gas compositions exist. Generally, an increase in heat flow causes a shoaling of a BSR, whereas an increase of higher-order hydrocarbons causes a deepening. Hence, the depth of the BSR may vary greatly as, for example, in the Barents Sea depending on the contribution and thus the amount of thermogenic gases migrating into the GHSZ (Chand et al., 2008; Rajan et al., 2013).

More recently, BSR observations and hydrate stability zone modeling of the upper pinch-out zone on continental margins have been used to assess past and contemporary changes in hydrate stability through warming of ocean bottom water (Vogt and Jung, 2002; Mienert et al., 2005; Biastoch et al., 2011; Ferré et al., 2012; Phrampus and Hornbach, 2012).

Inferred former positions of a BSR are often referred to as paleo-BSR. One of the best examples for a paleo-BSR can be found on the Blake Ridge, approx. 450 km offshore Georgia on the East Coast of the United States (Hornbach et al., 2003). This BSR formed when erosion by strong contour currents on the eastern flank of the Blake Ridge removed the top sediments of a hydrated formation causing an adjustment of the hydrate/free gas boundary by moving it deeper. The free gas layer beneath the former BSR crystallized into a newly formed concentrated layer of hydrates causing both a density and velocity increase. Beneath this paleo-BSR, the new BSR formed with free gas underneath the base of the gas hydrate stability zone (BGHSZ). Though the timing of the readjustment is unknown, it presents one good example for the dynamic behavior (in this case deepening) of a BSR due to erosion of sediments and a drop in seafloor temperature.

Summary and conclusions

Bottom-simulating reflectors (BSRs) occur in a wide range of sediments in the world oceans. Such creation of BSRs involves the existence of free gas and water in the pore space of sediments under low temperature and high pressure, forming hydrates beneath the ocean floor. The depth of the BSR defines the base of the gas hydrate stability zone (BGHSZ) under which free gas accumulates. Free gas becomes trapped beneath the hydrate-charged layer causing a distinct impedance contrast and a seismic reflection of reversed (negative) polarity if compared to the seafloor. The GHSZ depends on temperature and pressure and to a lesser degree on salinity and gas composition (thermogenic, biogenic). The thickness of the GHSZ decreases toward the upper continental margins (lower pressure) and sedimented ocean ridges (higher temperature). The second type of BSRs concentrates in regions of siliceous ocean sediments. Here, increases in temperature with burial depth result in dissolution of siliceous skeletons, which in turn creates an interface with higher porosity and permeability above and lower values beneath the interface. If the contrast becomes large enough, a seismic reflector occurs but with positive polarity (no phase reversal). As a consequence, diagenetic BSRs occur commonly deeper, show no phase reversal, and exist over large areas at the opal-A/opal-CT interface in ocean sediments.

Bibliography

Berndt, C., Bünz, S., Clayton, T., Mienert, J., and Saunders, M., 2004. Seismic character of bottom simulating reflectors: examples from the mid-Norwegian margin. *Marine and Petroleum Geology*, **21**, 723–733.

Biastoch, A., Treude, T., Rüpke, L. H., Riebesell, U., Roth, C., Burwicz, E. B., Park, W., Latif, M., Böning, C. W., Madec, G., and Wallmann, K., 2011. Rising Arctic Ocean temperatures cause gas hydrate destabilization and ocean acidification. *Geophysical Research Letters*, **38**, L08602.

Brekke, H., 2000. The tectonic evolution of the Norwegian Sea continental margin with emphasis on the Vøring and More basins. In Nottvedt, A. (ed.), *Dynamics of the Norwegian Margin*. Geological Society of London Special Publication 167. London: Geological Society, pp. 327–378.

Bünz, S., Mienert, J., and Berndt, C., 2003. Geological controls on the Storegga gas-hydrate system of the mid-Norwegian continental margin. *Earth and Planetary Science Letters*, **209**(3–4), 291–307.

Cartwright, J. A., and Dewhurst, D. N., 1998. Layer-bound compaction faults in fine-grained sediments. *Bulletin of the Geological Society of America*, **110**(10), 1242–1257.

Chand, S., Mienert, J., Andreassem, K., Knies, J., Plassen, L., and Fotland, B., 2008. Gas hydrate stability zone modelling in areas of salt tectonics and pockmarks of the Barents Sea suggest an active hydrocarbon venting system. *Marine and Petroleum Geology*, **25**, 625–636.

Davies, R. J., and Cartwright, J. A., 2002. A fossilized opal-A to opal C/T transformation on the northeast Atlantic margin: support for a significantly elevated paleogeothermal gradient during the Neogene? *Basin Research*, **14**, 467–486.

Depreiter, D., Poort, J., Van Rensbergen, P., and Henriet, J. P., 2005. Geophysical evidence of gas hydrates in shallow submarine mud volcanoes on the Moroccan margin. *Journal of Geophysical Research*, **110**, B10103, doi:10.1029/2005JB003622.

Dickens, G. R., and Quinby-Hunt, M. S., 1997. Methane hydrate stability in pore water: a simple theoretical approach for geophysical applications. *Journal of Geophysical Research*, **102**, 773–783.

Ferré, B., Mienert, J., and Feseker, T., 2012. Ocean temperature variability for the past 60 years on the Norwegian-Svalbard margin influences gas hydrate stability on human time scales. *Journal of Geophysical Research, Oceans*, **117**, C10017.

Grevemeyer, I., and Villinger, H., 2001. Gas hydrate stability and the assessment of heat flow through continental margins. *Geophysical Journal International*, **145**, 647–660.

Hein, J. R., Scholl, D. W., Barron, J. A., Jones, M. G., and Miller, J. J., 1978. Diagenesis of Late Cenozoic diatomaceous deposits and formation of the bottom simulating reflector in the southern Bering Sea. *Sedimentology*, **25**, 155–181.

Hesse, R., 1989. Silica diagenesis: origin of inorganic and replacement cherts. *Earth-Science Reviews*, **26**, 253–284.

Holland, H. D., and Turekian, K. K., 2003. *Treatise on Geochemistry*. Elsevier Pergamon, Elsevier Ltd. The Boulevard, Langford Lane, Kidlington, Oxford, QX5 1GB, UK, ISBN 978-0-08-043751-4.

Hornbach, M. J., Holbrook, W. S., Gorman, A. R., Hackwith, K. L., Lizarralde, D., and Pecher, I., 2003. Direct seismic detection of methane hydrate on the Blake Ridge. *Geophysics*, **68**(1), 92–100.

Hurd, D. C., and Birdwhistell, S., 1983. On producing a more general model for biogenic silica dissolution. *American Journal of Science*, **283**, 1–28.

Hyndman, R. D., and Spence, G. D., 1992. A seismic study of methane hydrate marine bottom simulating reflectors. *Journal of Geophysical Research – Solid Earth*, **97**, 6683–6698.

Knauth, L. P., 1994. Petrogenesis of chert. *Reviews of Mineralogy*, **29**, 233–258.

Kotelnikova, S., 2002. Microbial production and oxidation of methane in deep subsurface. *Earth-Science Reviews*, **58**, 367–395.

Kvenvolden, K. A., 1993. Gas hydrates – geological perspective and global change. *Reviews of Geophysics*, **31**, 173–187.

Mienert, J., Vanneste, M., Bunz, S., Andreassen, K., Haflidason, H., and Sejrup, H. P., 2005. Ocean warming and gas hydrate stability on the mid-Norwegian margin at the Storegga Slide. *Marine and Petroleum Geology*, **22**, 233–244.

Nouzé, H., Cosquer, E., Collot, J., Foucher, L. P., Klingelhoefer, F., Lafoy, Y., and Géli, L., 2009. Geophysical characterization of bottom simulating reflectors in the Fairway Basin (off New Caledonia, Southwest Pacific), based on high resolution seismic profiles and heat flow data. *Marine Geology*, **266**(1–4), 80–90.

Phrampus, B. J., and Hornbach, M. J., 2012. Recent changes to the Gulf Stream causing widespread gas hydrate destabilization. *Nature*, **490**(7421), 527–530.

Rajan, A., Bünz, S., Mienert, J., and Smith, A. J., 2013. Gas hydrate in petroleum provinces of the SW-Barents Sea. *Marine and Petroleum Geology*, **46**, 92–106.

Rempel, A. W., and Buffett, B. A., 1997. Formation and accumulation of gas hydrate in porous media. *Journal of Geophysical Research*, **102**, 10151–10164.

Shipley, T. H., Houston, M. H., Buffler, R. T., Shaub, F. J., McMillen, K. J., Ladd, J. W., and Worzel, J. L., 1979. Seismic reflection evidence for the widespread occurrence of possible gas-hydrate horizons on continental slopes and rises. *American Association of Petroleum Geologists Bulletin*, **63**, 2204–2213.

Sloan, D. R., 2003. Fundamental principles and applications of natural gas hydrates. *Nature*, **426**, 353–359.

Tribble, J. S., Mackenzie, F. T., Urmos, J., O'Brien, D. K., and Manghnani, M. H., 1992. Effects of biogenic silica on acoustic and physical properties of clay-rich marine sediments. *American Association of Petroleum Geologists Bulletin*, **76**, 792–804.

Vogt, P. R., and Jung, W. Y., 2002. Holocene mass wasting on upper non-Polar continental slopes – due to post-Glacial ocean warming and hydrate dissociation? *Geophysical Research Letters*, **29**, 55-1-55-4.

Wood, W. T., Gettrust, J. F., Chapman, N. R., Spence, G. D., and Hyndman, R. D., 2002. Decreased stability of methane hydrates in marine sediments owing to phase-boundary roughness. *Nature*, **420**, 656–660.

Zatsepina, O. Y., and Buffett, B. A., 1998. Thermodynamic conditions for the stability of gas hydrate in the seafloor. *Journal of Geophysical Research*, **103**, 24127–24139.

Cross-references

Cold Seeps
Deep-sea Sediments
Marine Gas Hydrates
Methane in Marine Sediments
Silica

BOTTOM-BOUNDARY LAYER

Wenyan Zhang
MARUM-Center for Marine Environmental Sciences, University of Bremen, Bremen, Germany

Definition

In marine geosciences, the bottom boundary layer (BBL) refers to a layer of flow in the immediate vicinity of the solid sea bottom where the effects of viscosity are significant in determining the characteristics of the flow. The BBL was first discovered by Prandtl (1905) in aerodynamics and subsequently applied to other fluids moving on the surface of a solid body.

Starting upwards from the sea bed, the total thickness of the BBL is defined as the distance above the bottom at which the mean flow velocity equals to 0.99 U_∞, where U_∞ is the free-stream velocity of a layer that is in a geostrophic balance overlying the BBL. On top of the geostrophically balanced layer is a surface layer subjected to wind-wave mixing. When both the bottom micro-topography is uniform and the overlying flow is steady, the BBL can be easily quantified from the vertical flow structure. Various ways exist to estimate the thickness δ of the BBL under neutral conditions (e.g., Grant and Madsen, 1986; Nielsen, 1992). In general, the BBL thickness at continental margins is at the order of 5–50 m.

Theoretically the BBL can be classified into three different sub-layers:

(1) A thin inner layer just above the bottom where turbulence is inhibited by the presence of the solid boundary. The flow is controlled by molecular viscosity and the shear stress is consistent with the bottom shear stress.
(2) An outer layer where turbulence shear dominates and viscous shear can be neglected.
(3) A transitional layer where both the viscous shear and the turbulence shear are important.

As the shear stress is almost constant and fulfills Newton's law of viscosity in the inner layer, flow velocity can thus be approximated by a linear form. However, this only applies to a hydraulically smooth bottom where bed roughness is too small to affect the velocity distribution. In hydraulically rough bottom, bed roughness is large enough to produce eddies close to the bottom and the inner layer may not be detectable. Upwards from the inner layer, the importance of molecular viscous decreases and turbulence gradually dominates the flow. Mean flow velocity in this transitional layer obeys the law of the wall and can be approximated by a logarithmic function. The outer turbulent layer takes up a majority (80–90 %) of the BBL. Flow characteristics of this layer mainly depend on the velocity difference with the external free flow (i.e., velocity defect) and the overall scale of the boundary layer.

Wind waves affect the BBL by imposing a wave boundary layer wherever the water depth is less than half of the wave length. The wave-induced oscillatory water motion is affected by the sea bottom within the wave boundary layer.

However, in a natural continental shelf, any definition of the BBL structure is not straightforward due to the influences of many factors (e.g., density stratification, internal waves, seabed topography). A practical indicator for the BBL is a thermohaline pycnocline observed in the water column (e.g., Stips et al., 1998; Perlin et al., 2005). Just above the sea bottom, there is a homogenous layer of temperature, salinity, and density, which indicates the inner layer and the transitional layer.

Bibliography

Grant, W. D., and Madsen, O. S., 1986. The continental-shelf bottom boundary layer. *Annual Review of Fluid Mechanics*, **18**, 265–305.

Nielsen, P., 1992. Coastal bottom boundary layers and sediment transport. In Series Editor-in-Chief Liu, Philip L-F. (ed.), *Advanced Series on Ocean Engineering*. World Scientific Publishing Co. Pte. Ltd., Singapore, Vol. 4.

Perlin, A., Moum, J. N., and Klymak, J. M., 2005. Response of the bottom boundary layer over a sloping shelf to variations in alongshore wind. *Journal of Geophysical Research*, **110**, C10S09, doi:10.1029/2004JC002500.

Prandtl, L., 1905. *Verhandlungen des dritten internationalen Mathematiker-Kongresses in Heidelberg 1904*, Krazer, A. (ed.), Leipzig: Teubner, p. 484. English trans. Ackroyd, J. A. K., Axcell, B. P., Ruban, A. I. (eds.) 2001. *Early Developments of Modern Aerodynamics*. Oxford: Butterworth-Heinemann, p. 77.

Stips, A., Prandke, H., and Neumann, T., 1998. The structure and dynamics of the Bottom Boundary Layer in shallow sea areas without tidal influence: an experimental approach. *Progress in Oceanography*, **41**, 383–453.

Cross-references

Sediment Dynamics
Sediment Transport Models

BOUMA SEQUENCE

Thierry Mulder[1] and Heiko Hüneke[2]
[1]University of Bordeaux, Talence, France
[2]Institute of Geography and Geology, Ernst Moritz Arndt University, Greifswald, Germany

The Bouma sequence (named after Arnold H. Bouma, 1932–2011) is a characteristic set of sedimentary structures typically preserved within positively graded sand or silt-mud couplets. From base to top, Bouma (1962) differentiated the following intervals above an erosion surface or sharp boundary: (Ta) massive to graded sand, (Tb) plane-parallel laminated sand, (Tc) cross-laminated sand and silt, (Td) parallel-laminated sand to silt, and (Te) laminated to homogeneous mud (Figure 1). Because of nonuniform grain size distribution and flow transformations (Fisher, 1983), the complete sequence is rare. Turbidite beds represent the typical deposit of low-concentration turbidity flows and related non-cohesive density flows.

The Bouma sequence is the first model of sediment-laden gravity flows and represents the first predictive model in sedimentology. It is a facies model of combined suspension fallout and traction deposition by a bipartite density flow (see Mulder, 2011, for details). The successive divisions with typical sedimentary structures display a bottom-to-top decline in energy consistent with the grading.

The basal surface of many turbidites, i.e., the lower bounding surface of Ta, commonly displays erosional features produced by turbulent scouring (Lanteaume et al., 1967).

Turbidity flows commonly develop from stratified density flows with a strong vertical velocity gradient. Within such bipartite flows (basal laminar flow and a top turbulent flow), the high particle concentration of its basal parts hinders suspension fallout (Mulder, 2011). Consequently, a poorly graded Ta division can be interpreted as being deposited from a concentrated-flow basal part. Rapid deposition and resulting unstable initial grain packing are also indicated by the common occurrence of water-escape structures such as dish or pillar structures or dewatering pipes.

The Bouma divisions Tb to Te record the passage of the flow body with a fully turbulent regime and reflect flow deceleration (Walker, 1965). The parallel lamination (Tb division) results from plane-bed transport of sand in the upper flow regime. The ripple cross-lamination (Tc division) reflects settling of sand and silt from suspension while lower-flow-regime current ripples migrate on the seabed. Climbing-ripple cross-lamination and convolute lamination would indicate rapid fallout and short-lived liquefaction, respectively. The Tc division is the most common structure in turbidite beds because ripples are the easiest structures to form for a given grain size and velocity/flow energy.

The uppermost divisions (Td and Te) are mainly products of settling from suspension. The pelitic top (Te division) represents the interaction of ongoing pelagic production with the fine terrigenous particle fallout from the turbulent tail of the flow.

Bouma Sequence, Figure 1 Well-developed Bouma sequence (divisions Ta to Te from base to top) in a turbidite bed of the Carboniferous Crackington Formation at the sea cliff near Crackington Haven, Cornwall, England.

Bibliography

Bouma, A. H., 1962. *Sedimentology of Some Flysch Deposits. A Graphic Approach to Facies Interpretation*. Amsterdam: Elsevier.

Fisher, R. V., 1983. Flow transformations in sediment gravity flows. *Geology*, **11**, 273–274.

Lanteaume, M., Beaudoin, B., and Campredon, R., 1967. *Figures sédimentaires du flysch 'Grès d'Annot' du synclinal de Peira-Cava*. Paris: Editions du CNRS.

Mulder, T., 2011. Gravity processes on continental slope, rise and abyssal plains. In Hüneke, H., and Mulder, T. (eds.), *Deep-Sea Sediments*. Amsterdam: Elsevier, pp. 25–148.

Walker, R. G., 1965. The origin and significance of internal sedimentary structures of turbidites. *Proceedings of the Yorkshire Geological Society*, **35**, 1–32.

Cross-references

Deep-sea Fans
Turbidites

C

CALCITE COMPENSATION DEPTH (CCD)

Wolfgang H. Berger
Scripps Institution of Oceanography UCSD, La Jolla, CA, USA

The carbonate compensation depth (CCD) is the dominant facies boundary on the deep-sea floor. It separates calcareous from noncalcareous sediments, with the calcareous deposits ("carbonate ooze") restricted roughly to the shallower half of the deep-sea floor. The link of the ooze boundary to elevation and the fact that deep-sea calcareous sediments are light gray in color, albeit with a buff tint (Figure 1), have invited comparison to the snow line on land. The ooze consists of calcareous skeletal parts and shells of coccolithophorids and foraminifers (or "nannofossils" and foraminifers). Discoasters (symmetrical ray-bearing nannofossils, Figure 1) are extinct.

Calcite Compensation Depth (CCD), Figure 1 View of calcareous ooze from Ontong Java Plateau, western equatorial Pacific, in a partially opened box core. Contents of calcareous ooze (nannofossils, planktonic foraminifers) to the right (Sources: SIO (box core photo, coccosphere, foraminifers) and Bukry and Bramlette (1969), Leg 1 of DSDP (nannofossils. The discoasters are extinct)).

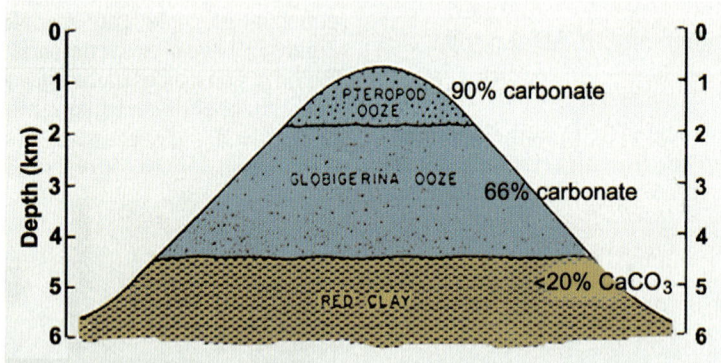

Calcite Compensation Depth (CCD), Figure 2 The general distribution pattern of deep-sea sediments was first explored by the British Challenger Expedition (1872–1876). Based on samples from that expedition, John Murray postulated elevation-linked facies boundaries between *pteropod ooze* and *globigerina ooze* and between *globigerina ooze* and *red clay*. The latter boundary is now known as the "carbonate compensation depth" or "CCD".

Like snowflakes on land, the skeletal parts and empty shells (produced in surface waters) fall to the seafloor, where they either pile up (at the shallower depths) or dissolve (close to and below the CCD). The CCD was discovered and first described by the naturalist John Murray of the Challenger Expedition (Figure 2).

Several important questions arose in the middle of the twentieth century, with respect to the nature of the CCD (e.g., Bramlette, 1961). The central one concerned the pattern of dissolution: did it start above the CCD, or did it set in, abruptly, at the facies boundary (denoting a change from saturation to undersaturation there)? The answer was provided by Peterson's experiment (Peterson, 1966) and by extensive studies of deep-sea sediments (e.g., Berger, 1970a; van Andel et al., 1975; Peterson and Prell, 1985). Dissolution starts well above the CCD; that is, the CCD marks the level where the rate of input of calcareous sediment is balanced by the rate of dissolution. Other questions concerned the position of the CCD, which differs between the Atlantic and Pacific. What processes set the typical depth, why is it regionally different, and why does it change through geologic time? The global mean presumably is a result of a need for balance between total availability of carbonate for deposition and the rate of production by planktonic organisms (the latter being controlled by availability of nutrients, rather than by that of carbonate). Deep circulation greatly affects deviations of the local CCD position from the global mean

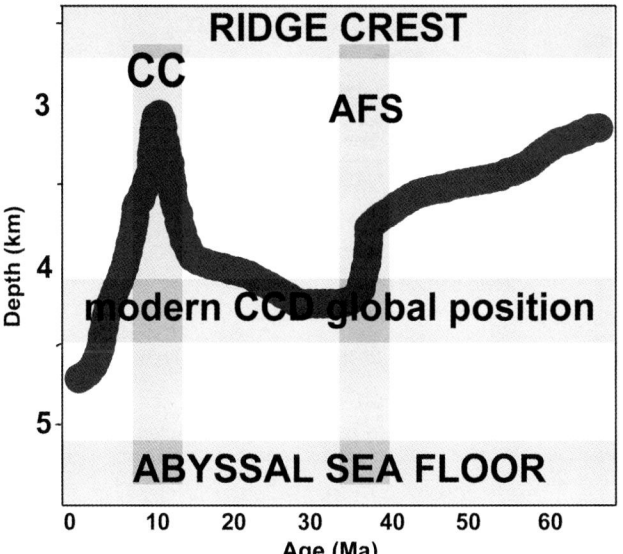

Calcite Compensation Depth (CCD), Figure 3 Schematic representation of CCD fluctuations in the South Atlantic (Simplified from Hsü and Wright (1985) and van Andel et al. (1977)). Approximate modern global position of the CCD after Berger and Winterer (1974). CC "carbonate crash" (slightly later in the Pacific), AFS "Auversian facies shift" (larger step in the Pacific).

(Berger, 1970b). Changes through time (Figure 3) depend on a number of factors, including changes in production of sediment, in rates of dissolution, in deep circulation, and in the sizes of reservoirs containing carbonate and organic carbon. The two most striking CCD events in the Cenozoic (the last 65 million years) are the abrupt appearance of deep-sea carbonate in the earliest Oligocene (the great facies shift at the end of the Eocene) and the "carbonate crash" at the end of the middle Miocene, with the event in the Caribbean leading that of the eastern Pacific by a million years or so. Neither event is fully understood, although there are many propositions regarding possible causes (for review, see Preiss-Daimler, 2011).

Bibliography

Berger, W. H., 1970a. Planktonic foraminifera: selective solution and the lysocline. *Marine Geology*, **8**, 111–138.

Berger, W. H., 1970b. Biogenous deep-sea sediments: fractionation by deep-sea circulation. *Bulletin of the Geological Society of America*, **81**, 1385–1402.

Berger, W. H., and Winterer, E. L., 1974. Plate stratigraphy and fluctuating carbonate line. In Hsü, K. J., and Jenkyns, H. C. (eds.), *Pelagic Sediments on Land and Under the Sea*. Oxford: International Association of Sedimentologists Special Publication, Vol. 1, pp. 11–48.

Bramlette, M. N., 1961. Pelagic sediments. In Sears, M. (ed.), *Oceanography*. Washington, DC: American Association for the Advancement of Science, Vol. 67, pp. 345–366.

Bukry, D., and Bramlette, M. N., 1969. Coccolith age determinations – leg 1, Deep Sea Drilling Project. *Initial Reports*, **1**, 369–387.

Hsü, K. J., and Wright, R., 1985. History of calcite dissolution of the South Atlantic Ocean. In Hsü, K. J., and Weissert, H. J. (eds.), *South Atlantic Paleoceanography*. Cambridge, UK: Cambridge University Press, pp. 149–187.

Murray, J., and Hjort, J., 1912. *The Depths of the Ocean*. London: Macmillan. 821 pp.

Peterson, M. N. A., 1966. Calcite rates of dissolution in a vertical profile in the central Pacific. *Science*, **154**, 1542–1544.

Peterson, L. C., and Prell, W. L., 1985. Carbonate dissolution in recent sediments of the eastern equatorial Indian Ocean: preservation patterns and carbonate loss above the lysocline. *Marine Geology*, **64**, 259–290.

Preiss-Daimler, I., 2011. *The Miocene Carbonate Crash: Shifts in Carbonate Preservation and Contribution of Calcareous Plankton*. PhD thesis, Germany, University of Bremen.

van Andel, T. H., Heath, G. R., and Moore, T. C., Jr., 1975. Cenozoic history and paleoceanography of the central equatorial Pacific Ocean. *Geological Society of America Memoir*, **143**, 1–134.

van Andel, T. H., Thiede, J., Sclater, J. G., and Hay, W. W., 1977. Depositional history of the South Atlantic Ocean during the last 125 million years. *Journal of Geology*, **85**, 651–698.

Cross-references

Marine Microfossils
Paleoceanography

CARBON ISOTOPES

Thomas Wagner[1] and Jens O. Herrle[2]
[1]School of Civil Engineering and Geosciences, Newcastle University, Newcastle upon Tyne, UK
[2]Institute of Geosciences, Goethe University Frankfurt, Frankfurt, Germany

Synonyms

$^{13}C/^{12}C$ isotopes; Carbon isotope stratigraphy; Stable carbon isotopes

Definition

The carbon isotopes with masses ^{12}C and ^{13}C comprise 98.89 % and 1.11 % of the stable carbon on Earth, respectively (Craig, 1953). Because of the low abundances of the rare ^{13}C stable carbon isotope, measurements are expressed as ratios to the more common ^{12}C in a sample ($^{13}C/^{12}C$) and reported in the $\delta^{13}C$ notation relative to the Vienna Pee Dee Belemnite (VPDB) standard in per mil (Coplen, 1996).

Introduction and main applications

The stable carbon isotopes measured on the carbonate and organic carbon fractions in sediments or carbonate tests from planktic and benthic organisms and molecular compounds (biomarkers) isolated from them are classical

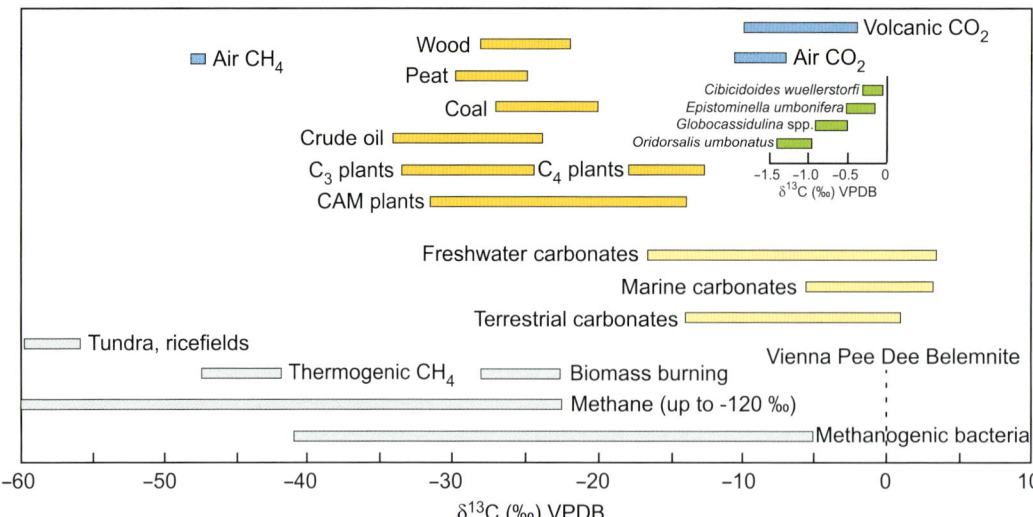

Carbon Isotopes, Figure 1 Variations in stable carbon isotope ratios for different organic and inorganic sources in the modern environment and in sediments (modified from Wefer and Berger (1991), Schidlowski and Aharon (1992)).

proxies in marine sciences. The main applications of carbon isotopes in marine and climate sciences are broad including reconstructions of the global carbon cycle in relation to carbon sources and burial, chemostratigraphy, tracing of water masses, and surface water productivity in the modern and past ocean.

Carbon sources

Stable carbon isotopes are widely used geochemical proxy tools to reconstruct changes in the global carbon cycle. Two main carbon reservoirs exist, the reduced biogenic organic carbon reservoir (organic matter) and the oxidized carbonate reservoir, covering a wide range of stable carbon isotopic signatures from below $-100‰$ to around $5‰$, depending on the source of the carbon (Figure 1). Differences in the organic carbon isotopic composition ($\delta^{13}C_{org}$), both at the bulk and the molecular level, are controlled by fractionation during photosynthesis in the marine and the terrestrial biosphere (e.g., Degens et al., 1968). The most negative (light) carbon isotope values are related to microbiologically produced methane (CH_4) measured in, e.g., gas-hydrate-bearing sediments along continental slopes and related carbonate bioherms, marine seep systems, and Arctic permafrost (e.g., Saltzman and Thomas, 2012; Wendler, 2013; and references therein). The organic carbon from terrestrial C3 plants, the most common type of vegetation on Earth, is isotopically less negative with mean $\delta^{13}C_{org}$ values of about $-27‰$. Corresponding $\delta^{13}C_{org}$ values for marine algae scatter around $-21‰$. Terrestrial vegetation from arid climate zones utilizing the C4 and CAM photosynthetic pathway, mainly Savannah grasslands, has mean $\delta^{13}C_{org}$ values of about $-12‰$. The most positive values are from inorganic carbonate precipitates ($\delta^{13}C_{carb}$), which range from about $0‰$ to $5‰$ (e.g., Hoefs, 2009).

Carbon isotopes in shallow seawater

The average dissolved inorganic carbon (DIC) $\delta^{13}C$ of seawater ($\delta^{13}C_{DIC}$) is dynamically coupled to the global carbon cycle via the partitioning of the main carbon reservoirs between the ocean, atmosphere, and terrestrial biosphere. $\delta^{13}C_{DIC}$ is not uniform in the modern ocean nor constant over time (e.g., Sundquist and Visser, 2004). The global seawater $\delta^{13}C_{DIC}$ composition is mainly controlled by two processes: photosynthesis and microbial decay of algal organic matter and physical fractionation during gas exchanges at the air to sea interface (Broecker and Maier-Reimer, 1992). Marine surface water is generally enriched in ^{13}C, because photosynthesis of marine phytoplankton favors the light ^{12}C over the heavy ^{13}C, leading to more negative $\delta^{13}C$ of organic matter ($\delta^{13}C_{org}$) relative to ambient seawater $\delta^{13}C_{DIC}$ (Garlick, 1974). This process is limited by the availability of nutrients, in particular nitrate and phosphate (Broecker and Peng, 1982).

Most marine organisms exhibit $\delta^{13}C$ values in their carbonate shells, which were formed not in full equilibrium relative to ambient seawater (Figure 1). The reason for this is manifold, but metabolic effects, including photosynthetic activity of algal symbionts, growth rate, and carbonate ion concentrations, are critical (McConnaughy, 1989; Rohling and Cook, 1999). These effects, together with species-specific differences, are summarized as "vital effects" (e.g., Wefer and Berger, 1991).

Modern sea surface water $\delta^{13}C_{DIC}$ varies between $+0.7‰$ in the northern Pacific and $+2.5‰$ in the midlatitude Atlantic (Kroopnick, 1985). Remineralization of organic matter and nutrients within thermocline subsurface water recycles isotopically light ^{12}C. Due to upwelling and wind mixing of shallow waters, this light ^{12}C and nutrient-rich subsurface water can be remobilized back into the surface waters, thereby stimulating primary productivity and influencing the carbon isotopic composition of calcareous shells (e.g., Broecker and Peng, 1982).

Tracing of seawater masses, surface water productivity, and ecology

The $\delta^{13}C_{DIC}$ in water is widely used as a tracer for seawater masses (e.g., Ravelo and Hillarie-Marcel, 2007). Although changes in surface water productivity and mixing influence $\delta^{13}C_{DIC}$, other processes, including water mass formation and ocean circulation, determine the $\delta^{13}C_{DIC}$ composition and therefore can be used as tracers for water masses. Deep water captures the $\delta^{13}C$ signature from its surface water sources. North Atlantic Deep Water (NADW) has therefore relatively high $\delta^{13}C$ values (~1.11‰), whereas Antarctic Bottom Water (ABW) is relatively low (~0.4‰), reflecting the major differences in surface water $\delta^{13}C_{DIC}$ in both high-latitude source regions (Kroopnick, 1985). Variations in deep water $\delta^{13}C$ can therefore be used to trace the history of deep water currents. This approach is also referred to as "aging" of deep water (Rohling and Cook, 1999), where the relative age of deep water provides a measure for ocean circulation.

The $\delta^{13}C$ difference between shallow and deep-dwelling planktic foraminifera ($\Delta\delta^{13}C$) provides a well-established proxy for thermocline productivity and the depth of the shallow-water mixed layer (e.g., Wefer et al., 1999). Increasing $\Delta\delta^{13}C$ values reflect more stratified surface water masses, due to increasing contrasts between shallow and deep waters and vice versa. A comparable approach is used to reconstruct productivity via changes in benthic foraminifera $\Delta\delta^{13}C$ of epifaunal and infaunal species (Zahn et al., 1986). Here, the $\delta^{13}C$ differences are mainly controlled by pore water geochemical gradients and the intensity of organic matter degradation, the latter being a function of organic matter supply (flux) to the sea floor (McCorkle et al., 1990). Large $\Delta\delta^{13}C$ values represent less organic matter flux and thus less primary productivity and vice versa. Benthic foraminifera $\Delta\delta^{13}C$ gradients can, however, be biased by the variability in new production of organic matter (fresh phytodetritus), which influences the stable isotope signal. This effect, constrained by variable primary productivity and organic matter fluxes at seasonal time scales, is referred to as the "phytodetritus" or "Mackensen effect" (Mackensen et al., 1993). Furthermore, the carbon and oxygen isotopic composition of calcite tests of planktic foraminifera is used for reconstructing depth habitats and thus the ecology of extant and extinct species (e.g., Pearson et al., 1993).

Chemostratigraphy and carbon isotope excursions (CIE)

The history of the global carbon cycle and climate through time is preserved in the carbon isotope record of marine and terrestrial sediments (Figure 2). The application of organic and inorganic carbon isotope ratios as chemical fingerprints for stratigraphic correlations was pioneered in the late 1970s with the studies of Berger et al. (1978), Weissert et al. (1979), and Scholle and Arthur (1980), more recently summarized in Weissert et al. (2008), Saltzman and Thomas (2012), and Wendler (2013). The most robust Phanerozoic archives of marine carbon isotope signatures are from individual pristine preserved fossils and bulk carbonate from marine pelagic sediments (e.g., Joachimski and Buggisch, 1993; Veizer et al., 1999; Saltzman and Thomas, 2012; and references therein). Carbon isotope fluctuations of several per mil are also reported from hemipelagic sediments, shallow-water carbonates, and terrestrial sediments. Due to multiple local influences, however, they may differ in their absolute isotope values and ranges from their corresponding pelagic sediment records, complicating their use in stratigraphic correlation across wider areas.

The long-term trend in Earth's climate is interrupted by series of transient events, identified by distinct positive or negative carbon isotope excursions, CIEs (Figure 2). These CIEs, combined with longer trends in the global carbon isotope record, are used in chemostratigraphy to identify chemical events (e.g., Weissert et al., 1998; Zachos et al., 2001), which can be correlated within and across ocean basins and the terrestrial environment (e.g., Gröcke et al., 1999; Herrle et al., 2004). Well-documented examples of CIEs representing major perturbations of the global carbon cycle are reported from, e.g., the Paleogene-Mesozoic greenhouse (Paleocene-Eocene thermal maximum PETM, Cretaceous and Early Jurassic oceanic anoxic events, OAEs, Jenkyns, 2003) and the Paleozoic (Saltzman and Thomas, 2012; Figure 2). Many CIEs coincide with widespread environmental perturbations including severe and short-term global warming (hyperthermal events), ocean acidification and shoaling of the carbonate compensation depth (CCD), enhanced marine organic carbon burial linked to widespread ocean anoxia, major crisis or extinction of biota on land and in the ocean, and sudden shifts in Earth's hydrological cycle and climate. Understanding how the Earth system responded to and recovered from these past extreme events remains a major focal point of Earth sciences, adding to the discussion on anthropogenically induced global warming.

The magnitude and rapidity of CIEs in the Phanerozoic, combined with evidence for massive environmental change, have stimulated intense interdisciplinary research about the underlying trigger and feedback mechanisms. Positive CIEs are reported for multimillion to orbital timescales and are, among other possible mechanisms (Figure 2), commonly associated with periods of globally enhanced marine carbon burial, leading to the concept of oceanic anoxic events (Schlanger and Jenkyns, 1976; Scholle and Arthur, 1980) and global cooling (e.g., Kuypers et al., 1999). Recent studies, combining geochemical with biotic data and biogeochemical modeling, have shown that global cooling can cause perturbations to marine ecosystems and biogeochemical cycles at scales comparable to those associated with global warming (McAnena et al., 2013). Sharp negative CIEs, either isolated within the chemostratigraphic record or in combination with positive CIEs, have been recorded throughout

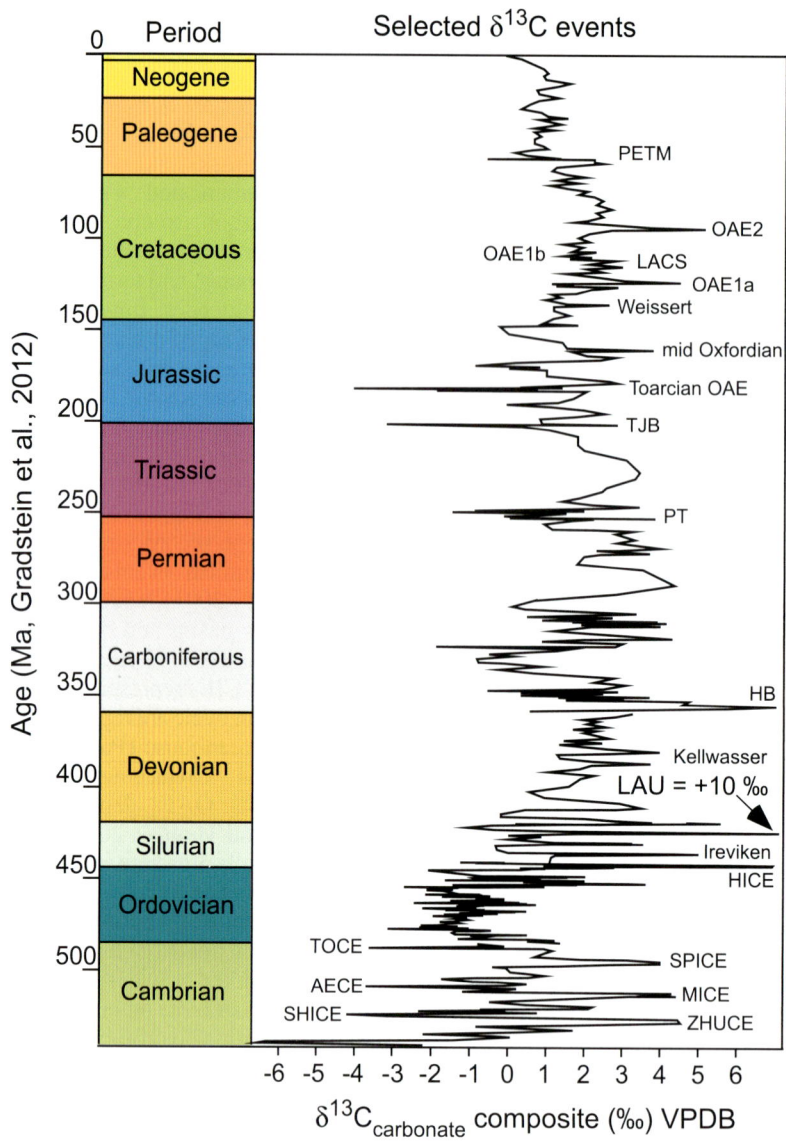

Carbon Isotopes, Figure 2 Phanerozoic carbon isotope record and selected global events (*CIEs*) based on Gradstein et al. (2012) and references therein, Caplan and Bustin (1999), Munnecke et al. (2003), Wagner et al. (2007), Korte and Kozur (2010) and McAnena et al. (2013). *PETM* Paleocene-Eocene thermal maximum, *OAE* oceanic anoxic events, *LACS* late Aptian cold snap, *TJB* Triassic-Jurassic boundary, *PT* Perm-Trias event, *HB* Hangenberg event, *LAU* Lau event, *HICE* Hirnantian event, *TOCE* Top of Cambrian Excursion, *SPICE* Steptoean event, *MICE* Mingxinsi event, *AECE* Archaeocyathids event, *SHICE* Shiyantou event, *ZHUCE* Zhujiaqing event.

the Phanerozoic sedimentary record (e.g., Saltzman and Thomas, 2012) and have become a focal point for Earth system research. Negative CIEs have been related to a large number of mechanisms (Figure 2), including rapid dissociation of metastable marine methane hydrate, buildup and release of isotopically "light" carbon in stratified epicontinental seas, thermogenic release of methane related to the placement of volcanic dykes into coal-bearing strata or thermal maturation of other organic-rich sediments, extensive biomass burning, mobilization of labile carbon pools from terrestrial permafrost, ventilation of marine dissolved organic carbon, and others (for further information on mechanisms, see, e.g., Cohen et al., 2007; deConto et al., 2012).

Conclusions

Stable carbon isotopes are commonly used for a wide range of applications in marine sciences and climate research. Applications include reconstructions of the global carbon cycle including short-term perturbations, chemostratigraphy, and tracing of water masses and surface water productivity in the modern and past ocean.

The development of carbon isotopes at the bulk sediment, foraminifera shell, and compound-specific (molecular) level, especially when combined with complementary evidence on biota, nutrients, ocean chemistry and redox, and modeling, has opened new pathways to unravel the complex interplay between fluctuations of the main carbon pools between the atmosphere, the living biosphere, and carbon buried in the sediments and their interactions with and impact on global climate at multiple temporal and spatial scales.

Bibliography

Berger, W. H., Killingley, J. S., and Vincent, E., 1978. Stable isotopes in deep-sea carbonates: box core ERDC-92, West equatorial Pacific. *Oceanologica Acta*, **1**, 203–216.

Broecker, W. S., and Maier-Reimer, E., 1992. The influence of air and sea exchange on the carbon isotope distribution in the sea. *Global Biogeochemical Cycles*, **6**, 315–320.

Broecker, W. S., and Peng, T.-H., 1982. *Tracers in the Sea*. Palisades: Eldigio Press.

Caplan, M. L., and Bustin, R. M., 1999. Devonian-carboniferous Hangenberg mass extinction event, widespread organic-rich mudrock and anoxia: causes and consequences. *Palaeogeography Palaeoclimatology Palaeoecology*, **148**, 187–207.

Cohen, A. S., Coe, A. L., and Kemp, D. B., 2007. The late Paleocene-early Eocene and Toarcian (early Jurassic) carbon-isotope excursions: a comparison of their timescales, associated environmental changes, causes and consequences. *Journal of the Geological Society (London)*, **164**, 1093–1108.

Coplen, T. B., 1996. New guidelines for the reporting of stable hydrogen, carbon and oxygen isotope ratio data. *Geochimica et Cosmochimica Acta*, **60**, 3359–3360.

Craig, H., 1953. The geochemistry of the stable carbon isotopes. *Geochimica et Cosmochimica Acta*, **3**, 53–92.

DeConto, R., Galeotti, S., Pagani, M., Tracy, D., Schaefer, K., Zhang, T., Pollard, D., and Beerling, D. J., 2012. Past extreme warming events linked to massive carbon release from thawing permafrost. *Nature*, **484**, 87–92.

Degens, E. T., Guillard, R. R. L., Sackett, W. M., and Hellebust, J. A., 1968. Metabolic fractionation of carbon isotopes in marine plankton. I. Temperature and respiration experiments. *Deep Sea Research*, **15**, 1–9.

Garlick, G. D., 1974. The stable isotopes of oxygen, carbon, and hydrogen in the marine environment. In Goldberg, E. D. (ed.), *The Sea*. New York: Wiley, Vol. 5, pp. 393–425.

Gradstein, F. M., Ogg, J. G., Schmitz, M., and Ogg, G., 2012. *The Geological Time Scale*. Amsterdam: Elsevier.

Gröcke, D. R., Hesselbo, S. P., and Jenkyns, H. C., 1999. Carbon-isotope composition of lower Cretaceous fossil wood: ocean–atmosphere chemistry and relation to sea-level change. *Geology*, **27**, 155–158.

Herrle, J. O., Kößler, P., Friedrich, O., Erlenkeuser, H., and Hemleben, C., 2004. High-resolution carbon isotope records of the Aptian to Lower Albian from SE France and the Mazagan Plateau (DSPD site 545): a stratigraphic tool for paleocanographic and paleobiologic reconstruction. *Earth and Planetary Science Letters*, **218**, 149–161.

Hoefs, J., 2009. *Stable Isotope Geochemistry*. Berlin: Springer, Vol. 6.

Jenkyns, H. C., 2003. Evidence for rapid climate change in the Mesozoic-Palaeogene greenhouse world. *Philosophical Transactions of the Royal Society of London, Series A*, **361**, 1885–1916.

Joachimski, M. M., and Buggisch, W., 1993. Anoxic events in the late Frasnian – causes of the Frasnian-Famennian faunal crisis? *Geology*, **21**, 675–678.

Korte, C., and Kozur, H. W., 2010. Carbon-isotope stratigraphy across the Permian-Triassic boundary: a review. *Journal of Asian Earth Sciences*, **39**, 215–235.

Kroopnick, P., 1985. The distribution of ^{13}C of ΣCO_2 in the world oceans. *Deep Sea Research*, **32**, 5784.

Kuypers, M. M. M., Pancost, R., and Shinninghe Damsté, J. S., 1999. A large and abrupt fall in atmospheric CO_2 concentration during Cretaceous times. *Nature*, **399**, 342–345.

Mackensen, A., Hubberten, H. W., Bickert, T., Fischer, G., and Fütterer, D. K., 1993. The ^{13}C in benthic foraminiferal tests of Fontbotia wuellerstorfi (Schwager) relative to the ^{13}C of dissolved inorganic carbon in southern ocean deep water: implications for glacial ocean circulation models. *Paleoceanography*, **8**, 587–610.

McAnena, A., Flogel, S., Hofmann, P., Herrle, J. O., Griesand, A., Pross, J., Talbot, H. M., Rethemeyer, J., Wallmann, K., and Wagner, T., 2013. Atlantic cooling associated with a marine biotic crisis during the mid-Cretaceous period. *Nature Geoscience*, **6**, 558–561.

McConnaughy, T., 1989. ^{13}C and ^{18}O disequilibrium in biological carbonates: I. Patterns. *Geochimica et Cosmochimica Acta*, **53**, 151–162.

McCorkle, D. C., Keigwin, L. D., Corliss, B. H., and Emerson, S. R., 1990. The influence of microhabitats on the carbon isotopic composition of deep-sea benthic foraminifera. *Paleoceanography*, **5**, 161–185.

Munnecke, A., Samtleben, C., and Bickert, T., 2003. The Ireviken event in the lower Silurian of Gotland, Sweden – relation to similar Palaeozoic and Proterozoic events. *Palaeogeography Palaeoclimatology Palaeoecology*, **195**, 99–124.

Pearson, P. N., Shackleton, N. J., and Hall, M. A., 1993. Stable isotope paleoecology of middle Eocene planktonic foraminifera and integrated isotope stratigraphy, DSDP Site 523, South Atlantic. *Journal of Foraminiferal Research*, **23**, 123–140.

Ravelo, A., and Hillarie-Marcel, C., 2007. The use of oxygen and carbon isotopes of foraminifera in paleoceanography. In Hillaire-Marcel, C., and de Vernal, A. (eds.), *Proxies in Late Cenozoic Paleoceanography*. Amsterdam: Elsevier. Developments in Marine Geology, Vol. 1, pp. 735–764.

Rohling, E. J., and Cook, S., 1999. Stable oxygen and carbon isotopes on foraminiferal carbonate shells. In Gupta, B. K. S. (ed.), *Modern Foraminifera*. Dordrecht: Kluwer, pp. 239–258.

Saltzman, M. R., and Thomas, E., 2012. Carbon isotope stratigraphy. In Gradstein, F. M., Ogg, J. G., Schmitz, M., and Ogg, G. (eds.), *The Geological Time Scale*. Amsterdam: Elsevier, Vol. 1, pp. 207–232.

Schidlowski, M., and Aharon, P., 1992. Carbon cycle and carbon isotopic record: geochemical impact of life over 3.8 Ga of Earth history. In Schidlowski, M., et al. (eds.), *Early Organic Evolution: Implications for Energy and Mineral Resources*. Berlin: Springer, pp. 147–175.

Schlanger, S. O., and Jenkyns, H. C., 1976. Cretaceous oceanic anoxic events: causes and consequences. *Geologie en Mijnbouw*, **55**, 179–184.

Scholle, P. A., and Arthur, M. A., 1980. Carbon-isotope fluctuations in Cretaceous pelagic limestones: potential stratigraphic and petroleum exploration tool. *AAPG Bulletin*, **64**, 67–87.

Sundquist, E. T., and Visser, K., 2004. The geologic history of the carbon cycle. *Treatise on Geochemistry*, **8**, 425–472.

Veizer, J., Ala, D., Azmy, K., Bruckschen, P., Buhl, D., Bruhn, F., Carden, G. A. F., Diener, A., Ebneth, S., Godderis, Y., Jasper, T., Korte, C., Pawellek, F., Podlaha, O., and Strauss, H., 1999. $^{87}Sr/^{86}Sr$, $\delta^{13}C$ and $\delta^{18}O$ evolution of Phanerozoic seawater. *Chemical Geology*, **161**, 37–57.

Wagner, T., Wallmann, K., Herrle, J. O., Hofmann, P., and Stüsser, I., 2007. Consequences of moderate 25,000 year lasting emission of light CO2 into the mid-Cretaceous ocean. *Earth and Planetary Science Letters*, **259**, 200–211.

Wefer, G., and Berger, W. H., 1991. Isotope paleontology: growth and composition of extant calcareous species. *Marine Geology*, **100**, 207–248.

Wefer, G., Berger, W. H., Bijma, J., and Fischer, G., 1999. Clues to ocean history: a brief overview of proxies. In Fischer, G., and Wefer, G. (eds.), *Use of Proxies in Paleoceanography*. Berlin: Springer, pp. 1–68.

Weissert, H., McKenzie, J. A., and Hochuli, P., 1979. Cyclic anoxic event in the early Cretaceous Tethys ocean. *Geology*, **7**, 147–151.

Weissert, H., Lini, A., Föllmi, K. B., and Kuhn, O., 1998. Correlation of early Cretaceous carbon isotope stratigraphy and platform drowning events: a possible link? *Palaeogeography Palaeoclimatology Palaeoecology*, **137**, 189–203.

Weissert, H., Joachimski, M., and Sarnthein, M., 2008. Chemostratigraphy. *Newsletters on Stratigraphy*, **42**, 145–179.

Wendler, I., 2013. A critical evaluation of carbon isotope stratigraphy and biostratigraphic implications for late Cretaceous global correlation. *Earth Science Reviews*, **126**, 116–146.

Zachos, J., Pagani, M., Sloan, L., Thomas, E., and Billups, K., 2001. Trends rhythms, and aberrations in global climate 65 Ma to present. *Science*, **292**, 689–693.

Zahn, R., Winn, K., and Sarnthein, M., 1986. Benthic foraminiferal ^{13}C and accumulation rates of organic carbon: *Uvigerina* peregrine group and Cibicidoides wuellerstorfi. *Paleoceanography*, **1**, 27–42.

Cross-references

Anoxic Oceans
Biochronology, Biostratigraphy
Cold Seeps
Currents
Deep-sea Sediments
Marine Gas Hydrates
General Ocean Circulation - its Signals in the Sediments
Marine Microfossils
Modelling Past Oceans
Organic Matter
Oxygen Isotopes
Paleoceanographic Proxies
Paleoproductivity
Sapropels
Source Rocks, Reservoirs

CARBONATE DISSOLUTION

Leif G. Anderson
Department of Marine Sciences, University of Gothenburg, Gothenburg, Sweden

Definition

Calcium carbonate is a chemical compound with the formula $CaCO_3$. It is the main component of shells of marine organisms. Calcium carbonate can dissolve in seawater to form calcium and hydrogen carbonate ions.

$$CaCO_3 + CO_2 + H_2O \rightarrow Ca^{2+} + 2HCO_3^-$$

Several crystal structures of calcium carbonate are found in the marine environment. The most common crystal structures in the ocean are calcite and aragonite, but also ikaite occurs in cold-water environments.

The crystal structure determines how easy these crystals can dissolve, a property that is expressed as the solubility product, K_{sp},

$$K_{sp} = [Ca^{2+}] \times [HCO_3^{2-}]$$

where $[Ca^{2+}]$ and $[HCO_3^{2-}]$ equal the concentration of these properties in the surrounding media. The value of K_{sp} is dependent on temperature, salinity, and pressure. The temperature (T in K) and salinity (S) dependence of the stoichiometric solubility product was determined by Mucci (1983).

$$\log K_{sp} = (b_0 + b_1 T + b_2/T) S^{0.5} + c_0 S + d_0 S^{1.5}$$

The constants for calcite and aragonite are given in Table 1.

Few studies have been made to estimate the pressure effect on K_{sp}. The one most used is by Ingle (1975) who formulated the following expression for calcite based on high-pressure experiments:

$$\log\left(K_{sp}^p/K_{sp}^1\right) = \{(48.8 - 0.53t)(z-10)/10$$
$$+ (-5.88 \times 10^{-3} + 0.1845 \times 10^{-3}t)(z-10)^2/100\}/$$
$$188.93\ (t+273.15)$$

where K_{sp}^p is the solubility product at the depth z (meters), K_{sp}^1 is the solubility product at the surface, and t is the in situ temperature in °C.

In order to express if a water is super- or undersaturated, the term saturation state is used, which is expressed as

$$\Omega = [Ca^{2+}] \times [HCO_3^{2-}]/K_{sp}$$

Consequently, a Ω value above 1 means that the water is supersaturated, and a value below 1 expresses undersaturation. In the global oceans, most surface waters are supersaturated with the highest degrees in the warm tropical areas. In the waters of the warm-water corals, Ω is typically above 3.

Calcite is the most stable calcium carbonate mineral, with the crystals being trigonal-rhombohedral. Aragonite is the next common mineral with its crystal being orthorhombic. Aragonite is thermodynamically unstable at standard temperature and pressure and tends to alter to calcite on scales of 10^7–10^8 years.

Both calcite and aragonite form naturally in many marine mollusk shells and as the calcareous endoskeleton of warm- and cold-water corals. Often, magnesium is incorporated in the crystal structure by taking the place of the calcium ion. Typically, this will increase the solubility, i.e., the more percent magnesium a crystal contains, the more soluble it gets.

Ikaite is the mineral name for the hexahydrate of calcium carbonate, $CaCO_3 \cdot 6H_2O$. It is only found in a metastable state and decomposes rapidly once removed from near-freezing water (below 6 °C). It was first discovered in nature in the bottom of the Ikka Fjord in SW Greenland. Here, it is believed to be created when carbonate ion-rich groundwater seeps out and meets the seawater that is rich in calcium ions. During the last 10 years, its occurrence has been found in sea ice, first reported from the Weddell Sea (Dieckmann et al., 2008), and has been observed as grain sizes of hundreds of micrometers.

The solubility of calcite and aragonite is mainly determined by the carbonate ion concentration in the deep ocean as both the salinity and the calcium ion concentration are quite constant. The carbonate ion concentration is impacted by the formation of carbon dioxide through mineralization of sinking organic matter. This also means that the longer the water has been exposed to sinking organic matter, the higher the carbon dioxide content gets, and the concentration of carbonate ions decreases.

$$CO_2 + H_2O + CO_3^{2-} \rightleftarrows 2HCO_3^-$$

One consequence of this is that the depth at which one finds the water at saturation, i.e., $\Omega = 1$, is deepest in the regions of deepwater formation and shallowest in the regions where the waters have not been in contact with the surface for a long time. This means that the deepest saturation depths are found in the Atlantic Ocean and the shallowest in the Pacific Ocean.

When investigating calcium carbonate in marine sediments, three depth expressions are used. The depth where $\Omega = 1$ is called the calcite (or aragonite) saturation horizon. Deeper than this is the lysocline, which is the horizon where dissolution starts to become noticeable. Finally, we have the carbon compensation depth (CCD) which is defined as the depth where the rain of calcium carbonate is balanced by the rate of dissolution. Operationally, the CCD has been defined as the depth where a fixed content of calcium carbonate (e.g., 10 %) is found in the sediments or alternatively the depth at where the calcium carbonate accumulation rate is 0. A schematic illustration of these three horizons is shown in Figure 1.

Carbonate Dissolution, Table 1 Parameters for the temperature and salinity dependence of the stoichiometric solubility product of calcite and aragonite in seawater

Solid	b_0	$b_1 \, 10^3$	b_2	c_0	$d_0 \, 10^3$
Calcite	−0.77712	2.8426	178.34	−0.07711	4.1249
Aragonite	−0.068393	1.7276	88.135	−0.10018	5.9415

Summary or conclusions

The dissolution of metal carbonates is in the ocean mainly depending on the concentration of carbonate ions. In relatively young seawater, the carbonate ion concentration is

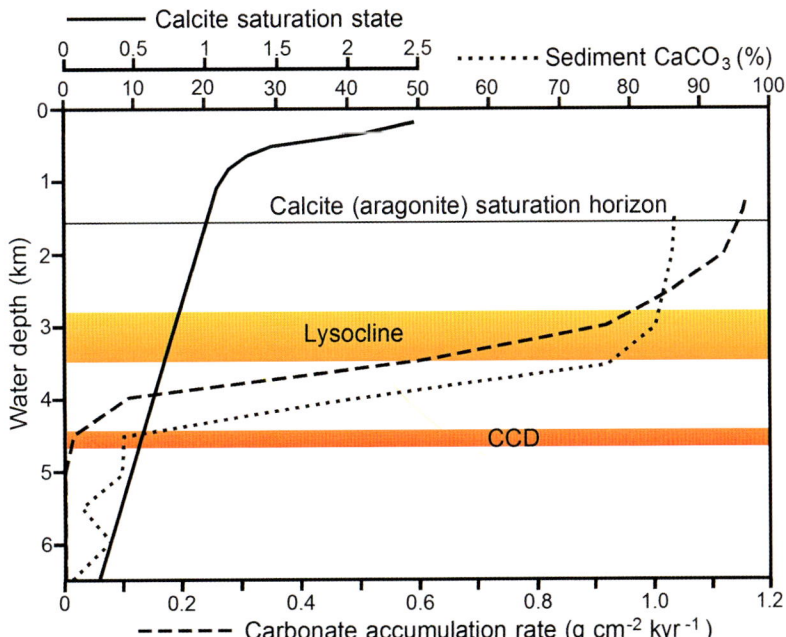

Carbonate Dissolution, Figure 1 Schematic illustration of the position of the calcite saturation horizon, CCD, and lysocline and their relationship to calcium saturation state (Ω), carbonate accumulation rate, and $CaCO_3$ content (modified after Pälike et al., 2012).

high enough to prevent most crystal forms of calcium carbonate to dissolve. However, carbon dioxide is produced when organic matter sediments out of the photic zone and mineralizes in the deep ocean. The carbon dioxide reacts with the carbonate ion and forms hydrogen carbonate which thus contributes to the dissolution of carbonates. Consequently, the sediments at several km depth of the North Atlantic contain carbonates, while sediments at 1 km in the Pacific Ocean might not.

Bibliography

Dieckmann, G. S., et al., 2008. Calcium carbonate as ikaite crystals in Antarctic sea ice. *Geophysical Research Letters*, **35**, L08501, doi:10.1029/2008GL033540.
Ingle, S. E., 1975. Solubility of calcite in the ocean. *Marine Chemistry*, **3**, 301–319.
Mucci, A., 1983. The solubility of calcite and aragonite in seawater at various salinities, temperatures and 1 atmosphere total pressure. *American Journal of Science*, **238**, 780–799.
Pälike, H., et al., 2012. A Cenozoic record of the equatorial Pacific carbonate compensation depth. *Nature*, **488**, 609–614, doi:10.1038/nature11360.

Cross-references

Calcite Compensation Depth (CCD)
Cold Seeps
Marine Microfossils

CARBONATE FACTORIES

John J.G. Reijmer
Department of Sedimentology and Marine Geology,, VU University Amsterdam, Amsterdam, The Netherlands

Synonyms
Carbonate production systems

Definition
Carbonate factories, or production systems, are benthic carbonate associations that show variations in their dominant precipitation mode, mineral composition, and depth range of production as well as growth potential (Schlager, 2000).

Introduction
The term "carbonate factory" was introduced to define the narrow depth zone where tropical reefs and detrital carbonates are produced (e.g., Tucker and Wright, 1990; James and Kendall, 1992; Reading and Levell, 1996). Based on the carbonate factory principle, Schlager (2000) proposed a threefold subdivision of the benthic carbonate production systems, with the planktonic carbonate factory as a fourth system. The latter is traditionally dealt with in the context of paleo-oceanography. The Schlager (2000) carbonate factory concept is based on the style of carbonate precipitation in aquatic realms: (1) abiotic, (2) biotically induced, or (3) biotically controlled (Lowenstam, 1981). In the latter category, a distinction can be made between sunlight-controlled organisms (phototrophic) and nutrient-controlled organisms (heterotrophic).

Environmental parameters
A series of environmental parameters steer the different modes of carbonate precipitation. The abiotic mode is normally encountered in marine and freshwater aquatic settings, although increasing evidence suggests microbes also play a mediating role in the formation of whitings (e.g., Yates and Robbins, 1998; Thompson, 2000) and ooids (Pacton et al., 2012). The differentiation found in the biotically controlled precipitation mode is directed through a series of environmental factors that drive environmental variations, the most important factors being: (1) light, (2) temperature, (3) nutrients, and (4) salinity. These factors set the boundaries for styles of carbonate precipitation and sediment production profiles but also for sediment production and distribution. As a result, the overall morphological development of the carbonate system relates to the dominance of specific factors:

1. Light. This is considered as one of the most important environmental controls. The depth of the photic column, light penetration, varies around the present-day globe with a maximum of almost 150 m for the Pacific atolls (Schlager, 2005).

    The light-saturated zone and euphotic zone not only regulate the growth forms of corals but most importantly the growth rates of the photosynthetic, carbonate-secreting benthos (Schlager, 2003).
2. Temperature. More or less equal to light is temperature, as it regulates the diversity of the biotic association. Each carbonate-secreting species has its own optimum growth window along the temperature scale (Lees, 1975).

    Ocean circulation patterns and latitudinal positions relate to the distribution and occurrence of specific water temperatures and hence determine the distribution of specific organisms across the globe.
3. Nutrients. Nutrient variations play an important role as high nutrient levels reduce calcification rates, but also stimulate the development of filamentous algae, bryozoans, and barnacles (e.g., Halfar et al., 2004, 2006; Reijmer et al., 2012); increase bio-erosion rates (Chazottes et al., 2008); and may hamper coral recruitment (Smith and Buddemeier, 1992; Atkinson et al., 1995).
4. Salinity. Coral communities are tolerant to long-term and short-term salinity variations (Muthiga and Szmant, 1987). The same holds for *Mytilus* shells (Malone and Dodd, 1967) whose calcification rates are not influenced by salinity thresholds. Algal communities exist that have a greater tolerance for high salinities, and they occur in large terminal lakes like

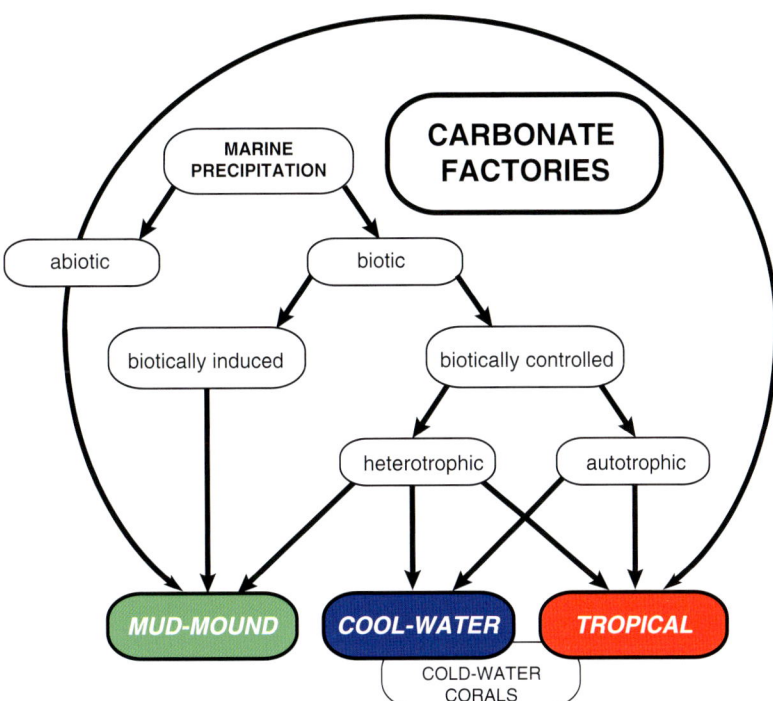

Carbonate Factories, Figure 1 Classification scheme of carbonate factories related to their precipitation modes (see text for details) (modified from Schlager (2003)).

the Great Salt Lake (Stephens, 1990; Harris et al., 2013). Porter et al. (1999) showed, for the Florida Keys coral reef system, that elevated salinities could diminish the negative effects of elevated temperature and conclude that temperature and salinity have opposing effects on coral photosynthesis. In restricted environments, however, biotic diversity might be reduced (Miocene, SE Spain; e.g., Braga and Martin, 1996).

The influence of the aforementioned factors on the growth strategies of skeletal and microbial precipitation and on the formation of non-skeletal particles differs for each component. Various strategies will result in different products at different times. Schlager (2000, 2003, 2005) distinguished three end-members of carbonate sedimentation systems or the so-called carbonate factories. The specific factories possess different sedimentation modes that reflect different growth strategies with different overall geometries, carbonate mineralogy, and grain sizes.

Carbonate factories

The three main factories (Figure 1) that were distinguished are:

(1) T-factory, in which the T is derived from tropical or "top-of-the-water-column" (Schlager, 2005); (2) the C-factory, in which the C stands for cool-water or controlled precipitation; and (3) the M-factory, in which M represents microbial, micrite, or mud-mound (Schlager, 2003, 2005). A fourth factory might be distinguished, the cold-water reef systems, which share characteristics with the T-factory through the type of dominant skeletal builder, e.g., scleractinian corals, and with the C-factory type, because of the production-depth profile and the nutrient-steered and light-independent carbonate production mode.

In the *T-factory*, light and water temperature steer the production profiles ensuring high production rates through biota living in the photic zone (Figure 2). The occurrence of this factory is mostly limited to the tropical zone between 30°N and 30°S, with modification through surface currents related to ocean gyres and upwelling areas.

The light and water temperature steered production mode results in a carbonate platform morphology, with a rim, reef barrier, at the edge of the platform (Figure 2). This barrier protects the shallow-water lagoon environments and forms the upper part of the steep slopes surrounding the platform. These slopes are mostly coarse grained and show fast cementation (Grammer et al., 1999).

The major mineralogy here is aragonite explaining the fast cementation rates found on the slopes (Grammer et al., 1999), reef systems, platform interior, and exposed sediments (Dravis, 1996).

This factory is very sensitive to relative changes in sea-level as it is closely tied to the light-saturated zone, and shows a flat top and steep slopes. A relatively small drop

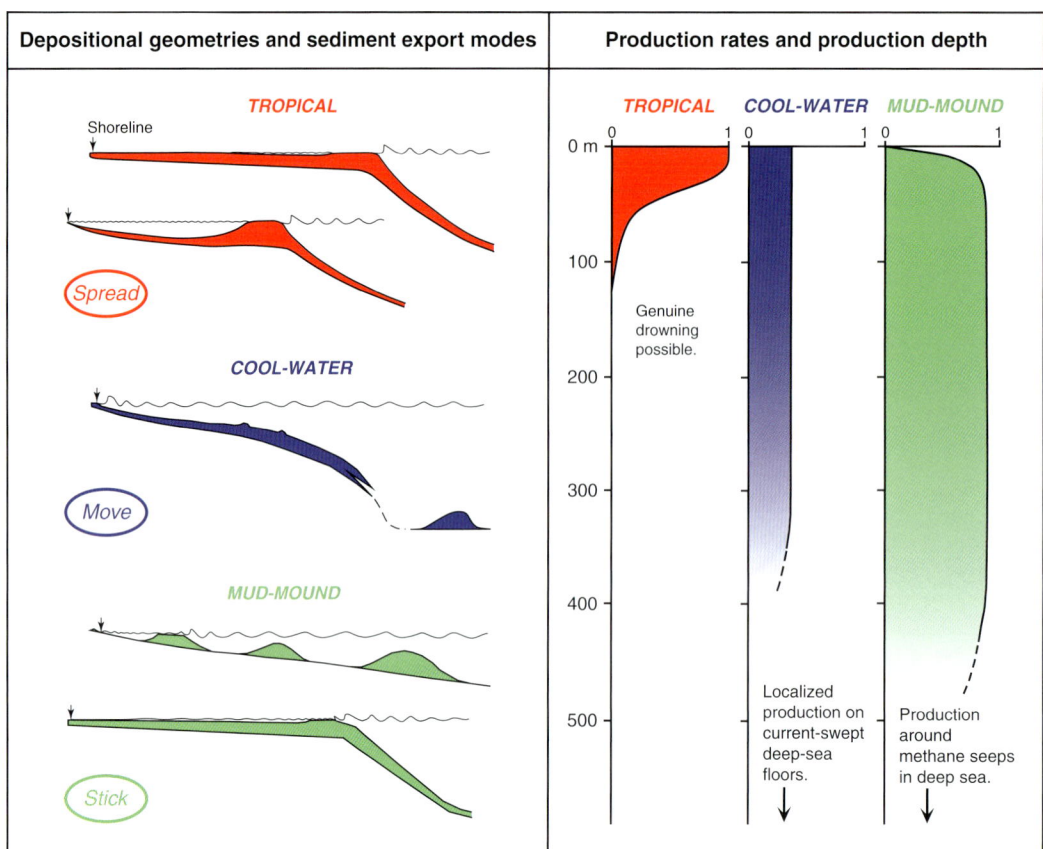

Carbonate Factories, Figure 2 Platform morphologies, production rates, and sediment export mode of the individual carbonate factories (modified from Schlager (2003)).

in sea-level might eliminate most of the sediment production zone. This sedimentation pattern with sea-level-related sharp changes in sediment production is called highstand shedding (e.g., Schlager et al., 1994). One could tag the overall sedimentation pattern as "Spread" as the systems export abundant sediment from shallow-water areas into surrounding basins (Figure 2).

Nutrients and relatively low temperatures steer the biologically controlled production of carbonate sediments in the *C-factory*. In surface waters, the C-factory operates at higher latitudes than the T-factory, northward of 30°N and southward of 30°S. Other realms are upwelling areas and low-temperature waters below the thermocline.

The production profile shows moderate production rates over a large water depth range (Figure 2), so production is not limited to shallow waters (Schlager, 2003). Another characteristic is the open sedimentary system without any shallow-water barriers as found in the other factories. Low cementation rates combined with the absence of a shallow-water barrier result in a sedimentation regime in which waves and currents control sediment transport and redistribution. One could label the overall sedimentation pattern as "Move" as changes in currents and waves result in a shift of the depositional sediment loci. In addition, the sedimentary response to sea-level changes shows large similarities to that seen in clastic systems (Figure 2).

The *M-factory* is characterized by the microbial-mediated, e.g., bacteria and cyanobacteria, precipitation of mud (Schlager, 2000, 2003). The most prominent features of this factory are the microbial mats, which usually contain several microbial species from various key groups, and their organic EPS (extracellular polymeric substances) matrix steering different organo-mineralization processes. The production profile of this factory shows a fairly steady production rate (Figure 2), which can reach tropical rates (Kenter et al., 2005), within a wide range of environments and water depths. Very characteristic for this factory is the formation of mound buildups, e.g., the Carboniferous of Belgium (Lees and Miller, 1995) and Devonian of Algeria (Wendt et al., 1997), albeit large flat-topped platforms are also very frequent, e.g., the Carboniferous of Kazakhstan (Kenter et al., 2005) and Triassic of the Dolomites (Keim and Schlager, 2001). The production profile shows high production rates down to 500 m water depth or even deeper, e.g., around methane seeps (Figure 2). Production in the shallow-water areas is affected by the waves due to the sensitivity of the biota to these physical processes.

The term "Stick" describes the overall sedimentation pattern very well, because of the ability of the system to glue (through microbial activity) sediments and build sturdy systems (Figure 2).

Relative changes in sea-level do not have a large impact on the production rates of this system as the main sediment production loci are situated on the upper slope, below the wave base, and extend over a fairly extensive water depth range. As a result, the production rates of this system are fairly constant and can be described by slope shedding (Kenter et al., 2005), slope-derived sediment production not being affected by sea-level changes.

As a fourth system, one could add the cold-water coral reefs (CWCR), first described from the Northwestern European margin by Teichert (1958). These systems share the stony corals with the T-factory and the nutrient-dependent growth strategy with the C-factory. Characteristic for this *CWCR-factory* is the dominance of ahermatypic corals, e.g., *Lophelia* and *Madrepora*, in a sedimentary system with very high biodiversity. The nutrient dependency and occurrence throughout a wide range of water depths below the photic zone result in the worldwide distribution of this sedimentary system, with new discoveries being added daily (Correa et al., 2012; Mienis et al., 2012).

Production profiles

The production profiles of the different factories not only reflect the water depths in which the factories reach their optimum growth and sediment production, but also the environmental processes that steer the types and amounts of sediment produced. Various environmental factors drive the development of specific carbonate producing organisms and associated modes of carbonate production, which results in differences in the grain-size spectra produced, as well as dominant carbonate mineralogies. These factors then relate to the efficiency of the system to build sturdy structures, fixing the sediments within depositional realms.

So the different styles of carbonate platform development are related to production profiles and thus the sensitivity of the system as a whole to environmental changes. These differences are very well expressed in the response of the T-, C-, and M-factories to relative changes in sea-level. Sharp contrasts are found in the T-factory, with high production during highstands in sea-level when the flat tops of the platforms are flooded. Sharply reduced production occurs during times when the tops become exposed and sediment production is restricted to small surfaces on the steep platform slopes. In the C-factory, relative changes in sea-level have a minor effect on the overall production rates, but do cause re-sedimentation of sediments during these changes. The M-factory also shows fairly stable sediment production and export rates during sea-level changes, as the main sediment production sites are situated on the upper slope. Minor differences might occur during phases of progradation and aggradation (Della Porta et al., 2004). The CWCR-factory strongly depends on the steady influx of nutrients either through slope currents or pelagic input. The CWCR-factories along the Irish margin as well as the communities in the Mediterranean show variations in their occurrence related to glacial and interglacial time intervals.

Summary and conclusions

The carbonate factory concept provides a subdivision of marine aquatic benthic carbonate sediment production systems based on their styles of carbonate precipitation. The planktonic carbonate factory is classified as an additional production system. The overall carbonate sediment production for the individual systems (tropical, cool-water, microbial, and cold-water coral reef systems) depends on specific factory-dependent sediment production profiles.

Bibliography

Atkinson, M. J., Carlson, B., and Crow, G. L., 1995. Coral growth in high-nutrient, low-ph seawater: a case study of corals cultured at Waikiki Aquarium, Honolulu, Hawaii. *Coral Reefs*, **14**(4), 215–223.

Braga, J. C., and Martín, J. M., 1996. Geometries of reef advance in response to relative sea-level changes in a Messinian (uppermost Miocene) fringing reef (Cariatiz reef, Sorbas Basin, SE Spain). *Sedimentary Geology*, **107**, 61–81.

Chazottes, V., Reijmer, J. J. G., and Cordier, E., 2008. Sediment characteristics in reef areas influenced by eutrophication-related alterations of benthic communities and bioerosion processes. *Marine Geology*, **250**(1–2), 114–127.

Correa, T. B. S., Grasmueck, M., Eberli, G. P., Reed, J. K., Verwer, K., and Purkis, S., 2012. Variability of cold-water coral mounds in a high sediment input and tidal current regime, Straits of Florida. *Sedimentology*, **59**(4), 1278–1304.

Della Porta, G., Kenter, J. A. M., and Bahamonde, J. R., 2004. Depositional facies and stratal geometry of an Upper Carboniferous prograding and aggrading high-relief carbonate platform (Cantabrian Mountains, N. Spain). *Sedimentology*, **51**, 267–295.

Dravis, J. J., 1996. Rapidity of freshwater calcite cementation – implications for carbonate diagenesis and sequence stratigraphy. *Sedimentary Geology*, **107**, 1–10.

Grammer, G. M., Crescini, C. M., McNeill, D. F., and Taylor, L. H., 1999. Quantifying rates of syndepositional marine cementation in deeper platform environments – new insight into a fundamental process. *Journal of Sedimentary Research*, **69**(1), 202–207.

Halfar, J., Godinez-Orta, L., Mutti, M., Valdez-Holguín, J. E., and Borges, J. M., 2004. Nutrient and temperature controls on modern carbonate production. An example from the Gulf of California, Mexico. *Geology*, **32**(3), 213–216.

Halfar, J., Godinez-Orta, L., Mutti, M., Valdez-Holguin, J. E., and Borges, J. M., 2006. Carbonates calibrated against oceanographic parameters along a latitudinal transect in the Gulf of California, Mexico. *Sedimentology*, **53**, 297–320.

Harris, P. M., Ellis, J., and Purkis, S. J., 2013. Assessing the extent of carbonate deposition in early rift settings. *American Association of Petroleum Geologists Bulletin*, **97**(1), 27–60.

James, N. P., and Kendall, A. C., 1992. Introduction to carbonate and evaporite facies models. In Walker, R. G., and James, N. P. (eds.), *Facies Models – Response to Sea Level Change*. St. John's: Geological Society of Canada, pp. 265–275.

Keim, L., and Schlager, W., 2001. Quantitative compositional analysis of a Triassic carbonate platform (Southern Alps, Italy). *Sedimentary Geology*, **139**, 261–283.

Kenter, J. A. M., Harris, P. M., and Della Porta, G., 2005. Steep microbial boundstone-dominated platform margins – examples and implications. *Sedimentary Geology*, **178**, 5–30.

Lees, A., 1975. Possible influence of salinity and temperatures on modern shelf carbonate sedimentation. *Marine Geology*, **19**, 159–198.

Lees, A., and Miller, J., 1995. Waulsortian banks. In Monty, C. L. V., Bosence, D. W. J., Bridges, P. H., Pratt, B. R. (eds.), *Carbonate Mud-Mounds – Their Origin and Evolution*. International Association of Sedimentologists, Oxford-London (UK), pp. 191–271.

Lowenstam, H. A., 1981. Minerals formed by organisms. *Science*, **211**, 1126–1131.

Malone, P. G., and Dodd, J. R., 1967. Temperature and salinity effects on calcification rate in Mytilus edulis and its paleoecological implicatons. *Limnology and Oceanography*, **12**(3), 432–436.

Mienis, F., Duineveld, G. C. A., Davies, A. J., Ross, S. W., Seim, H., Bane, J., and Van Weering, T. C. E., 2012. The influence of near-bed hydrodynamic conditions on cold-water corals in the Viosca Knoll area, Gulf of Mexico. *Deep Sea Research, Part I*, **60**, 32–45.

Muthiga, N. A., and Szmant, A. M., 1987. The effect of salinity stress on the rates of aerobic respiration and photosynthesis in the hermatypic coral *Siderastrea Siderea*. *The Biological Bulletin*, **173**, 539–551.

Pacton, M., Ariztegui, D., Wacey, D., Kilburn, M. R., Rollion-Bard, C., Farah, R., and Vasconcelos, C., 2012. Going nano: a new step toward understanding the processes governing freshwater ooid formation. *Geology*, **40**, 547–550.

Porter, J. W., Lewis, S. K., and Porter, K. G., 1999. The effect of multiple stressors on the Florida keys coral reef ecosystem: a landscape hypothesis and a physiological test. *Limnology and Oceanography*, **44**(3), 941–949.

Reading, H. G., and Levell, B. K., 1996. Controls on the sedimentary rock record. In Reading, H. G. (ed.), *Sedimentary Environments: Processes, Facies and Stratigraphy*, 3rd edn. Oxford, UK: Blackwell, pp. 5–52.

Reijmer, J. J. G., Bauch, T., and Schäfer, P., 2012. Carbonate facies patterns in surface sediments of upwelling and non-upwelling shelf environments (Panama, East Pacific). *Sedimentology*, **59**(1), 32–56.

Schlager, W., 2000. Sedimentation rates and growth potential of tropical, cool-water and mud-mound carbonate systems. In Insalaco, E., Skelton, P. W., and Palmer, T. J. (eds.), *Carbonate Platform Systems: Components and Interactions*. London: The Geological Society, pp. 217–227.

Schlager, W., 2003. Benthic carbonate factories of the Phanerozoic. *International Journal of Earth Sciences*, **92**, 445–464.

Schlager, W., 2005. *Carbonate Sedimentology and Sequence Stratigraphy*. Tulsa: SEPM (Society for Sedimentary Geology). SEPM Concepts in Sedimentology and Paleontology, Vol. 8. 200 pp.

Schlager, W., Reijmer, J. J. G., and Droxler, A. W., 1994. Highstand shedding of carbonate platforms. *Journal of Sedimentary Research*, **B64**(3), 270–281.

Smith, S. V., and Buddemeier, R. W., 1992. Global change and coral reef ecosystems. *Annual Review of Ecology and Systematics*, **23**, 89–118.

Stephens, D. W., 1990. Changes in lake levels, salinity and the biological community of Great Salt Lake (Utah, USA), 1847–1987. *Hydrobiologia*, **197**, 139–146.

Teichert, C., 1958. Cold- and deep-water coral banks. *American Association of Petroleum Geologists Bulletin*, **42**(5), 1064–1082.

Thompson, J. B., 2000. Microbial whitings. In Riding, R. E., and Awramik, S. M. (eds.), *Microbial Sediments*. Berlin/Heidelberg: Springer, pp. 250–260.

Tucker, M. E., and Wright, V. P., 1990. *Carbonate Sedimentology*. Oxford, UK: Blackwell. 482 pp.

Wendt, J., Belka, Z., Kaufmann, B., Kostrewa, R., and Hayer, J., 1997. The world's most spectacular carbonate mud mounds (Middle Devonian, Algerian Sahara). *Journal of Sedimentary Research*, **67**(3), 424–436.

Yates, K. K., and Robbins, L. L., 1998. Production of carbonate sediments by a unicellular green alga. *American Mineralogist*, **83**, 1503–1509.

Cross-references

Chemosynthetic Life
Eustasy
Export Production
Foraminifers (Benthic)
Guyot, Atoll
Lagoons
Lithostratigraphy
Marine Sedimentary Basins
Reef Coasts
Reefs (Biogenic)
Sea-Level
Sediment Transport Models
Sedimentary Sequence
Sequence Stratigraphy

CHEMOSYNTHETIC LIFE

Verena Tunnicliffe
Department of Biology and School of Earth & Ocean Sciences, University of Victoria, Victoria, BC, Canada

Synonyms

Chemoautotrophy; Chemolithoautotrophy; Chemosynthesis; Life supported by chemosynthesis

Definition

Chemosynthesis is the process that some microbes use to transform CO_2 into organic molecules. Energy to fuel this synthesis is gained from reduction-oxidation (redox) reactions involving inorganic compounds. The process is analogous to photosynthesis. Chemolithoautotrophy is "self-feeding using chemical energy from inorganic sources."

Chemoautotrophic metabolism

It is very likely that the first microbes on Earth able to fix carbon dioxide into organic carbon compounds did so using chemosynthesis (Martin and Russell, 2003). A major requirement of carbon fixation is energy to fuel the conversion process inside the cell. In plants and some microbes, sunlight supplies that energy but other microbes can capture energy from redox reactions mediated inside the cell. Most redox reactions that are key to energy transformation are aerobic using oxygen (either free or bound in another molecule) as the

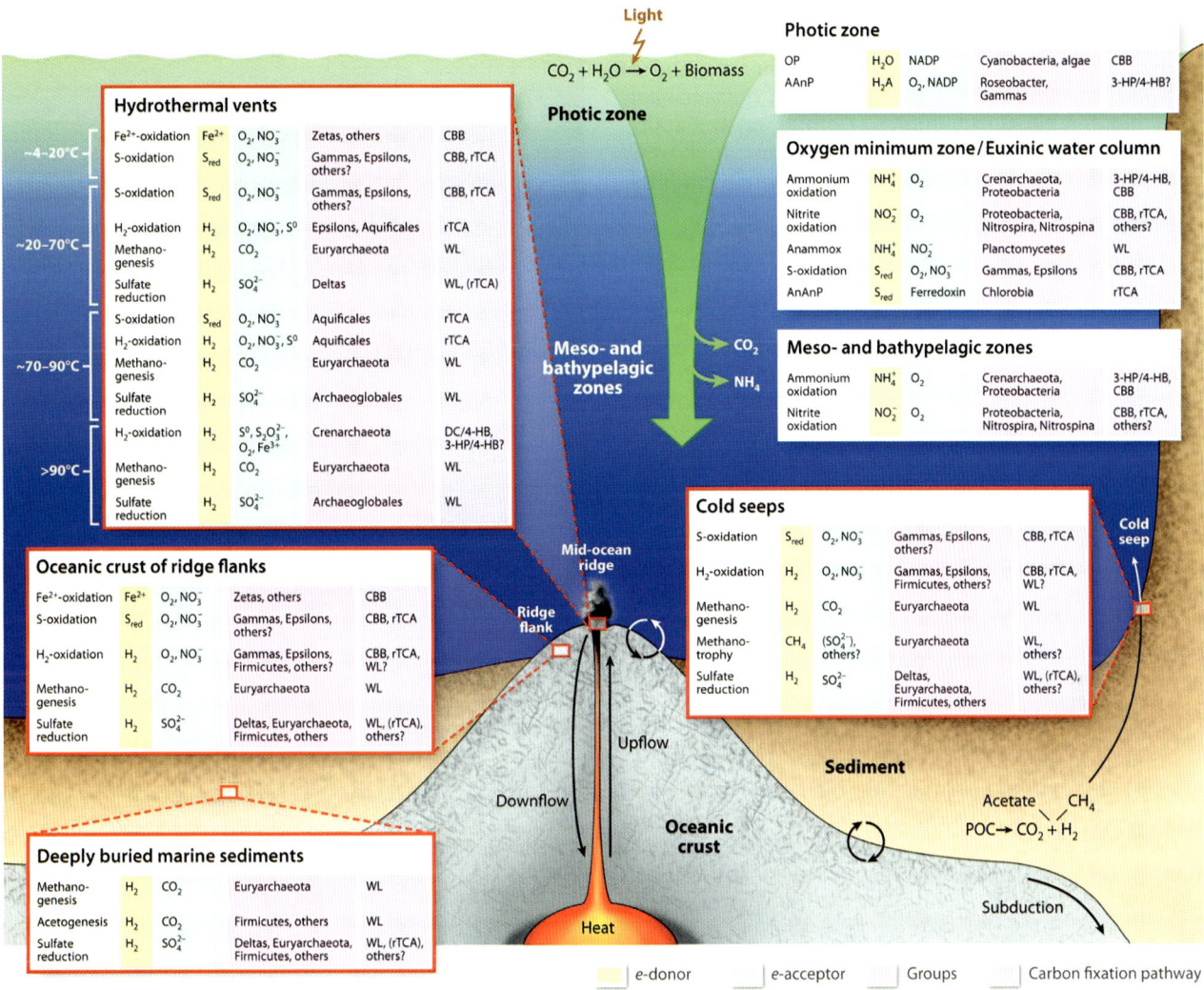

Chemosynthetic Life, Figure 1 Locations and reactants supporting in carbon fixing processes in the ocean. Boxes list the reduction-oxidation reactions and the dominant microbial groups engaged in chemoautotrophy (and photoautotrophy in the upper ocean). The *right column* indicates the biochemical pathway inside the cell involved in creating the organic carbon molecule. *CBB* Calvin-Benson-Bassham cycle, *rTCA* reductive tricarboxylic acid cycle, *WL* reductive acetyl-CoA or Wood-Ljungdahl pathway, *3-HP/4-HB* 3-hydroxypropionate/4-hydroxybutyrate cycle, *DC/4-HB* dicarboxylate/4-hydroxybutyrate cycle (from Hügler and Sievert, 2011).

electron acceptor: sulfur-oxidizing microbes transform hydrogen sulfide to sulfate using oxygen ($H_2S + 2O_2 = SO_4^{2-} + 2H^+$), while consortia of microbes can use sulfate to oxidize methane, thereby producing sulfide ($CH_4 + SO_4^{2-} = HCO_3^- + HS^- + H_2O$). Many compounds act as electron donor/acceptor pairs in a wide variety of chemosynthetic reactions (Figure 1). Microbiologists are discovering many energy favorable differences between compounds in which some microbial groups have evolved to mediate the reaction (Hügler and Sievert, 2011). There are also "mixotrophs" that use both organic and inorganic compounds as electron donors or carbon sources. In shallow water, a mixture of photoautotrophy and chemoautotrophy is possible. Aerobic sulfide oxidation is the reaction that yields the highest amount of energy available to cellular metabolism (McCollom and Shock, 1997). Free energy yield diminishes from O_2 to NO_3^- and NO_2^-, followed by manganese and iron, then SO_4^{2-}, and, lastly, CO_2.

Habitats of chemosynthetic life

Chemosynthesis occurs at interfaces of chemical environments where compounds mingle in disequilibrium (Figure 1). For example, abundant seawater oxygen spontaneously reacts with hydrogen sulfide yet a microbe on an anoxic interface can access both molecules and mediate the reaction inside the cell. Many such settings are fostered by geochemical processes under the seafloor that form emergent fluids rich in reduced compounds

(Tunnicliffe et al., 2003). At hydrothermal vents, high rates of fluid flux form strong gradients of compounds. Within smoker chimneys, zonation of minerals and of microbial diversity reflects a wide array of microhabitats available for chemoautotrophy (Kormas et al., 2006). The geological setting of high temperature rock/water interactions influences the composition of emergent fluids and the resulting microbial community (Amend et al., 2011). Oxidizing habitats on the periphery of vents, and on particulates in hydrothermal plumes, support chemolithoautotrophs that influence metal cycles in the ocean (Holden et al., 2012). Oxidation of reduced minerals within inactive sulfide deposits provides chemical energy to microbes that undergo ecological succession that reflects changing metabolic pathways as the sulfides age (Sylvan et al., 2012). Many similar sources of lithotrophic energy exist in the seafloor, not the least of which includes microbial weathering of basalts (Edwards et al., 2005).

At subduction zones, the pressure of sediment accumulation and mineral dehydration forces fluids upwards through organic matter-rich sediments along faults or bedding planes of the accretionary prism. Seepage that supports chemosynthesis can occur in association with gas hydrate or hydrocarbon deposits and also with brines near salt domes (Tunnicliffe et al., 2003; Levin, 2005). Rising bubbles, carbonate structures, or pockmarks may occur at seeps (Figure 2). Porewater methane generated from organic matter decomposition is the main source of carbon for chemosynthetic processes often through anaerobic oxidation using sulfate oxidation although other reduced compounds (e.g., H_2, NH^{4+}, Fe^{2+}, Mn^{2+}) are also present. The role of microbes in controlling the flux of compounds and behavior of methane reserves is a current interest (Jörgensen and Boetius, 2007).

Chemosynthesis plays an important role in oxygen minimum zones (OMZ) in many enclosed marine basins and in upwelling margins where abundant organic matter sinks into deep water. Microbial decomposition depletes dissolved oxygen often with sulfide and methane forming in and above the seafloor sediments, a feature of increasing concern as oxygen content decreases in the modern ocean (Helm et al., 2011). A complex array of biogeochemical cycles in OMZs reflects the redox reactions that support a diverse chemoautotrophic community.

Another notable site of chemosynthesis is associated with organic remains on the seafloor: large animal carcasses and wood (and sunken organic cargo). Whale skeletons release lipids over many decades creating anoxic conditions from microbial oxidation of organics. In these sulfide-rich bones, microbial chemoautotrophic production supports a community of metazoans for many years (Smith and Baco, 2003).

Microbial diversity and carbon fixation

The diversity of Bacteria and Archaea that use chemoautotrophic pathways continues to grow as new techniques of

Chemosynthetic Life, Figure 2 Representation of multiple habitats. Chemoautotrophy supports diverse animal communities in the deep ocean. White bacterial mats (*lower middle*) are often the first visual indication of reducing conditions. At hydrothermal vents (*upper left*), high biomass of animals develops in a food web dependent on both free-living and symbiotic microbes. Symbioses develop in most of these settings such as endosymbioses in cells of vestimentiferans and clams (*left*) or episymbioses as illustrated in alvinellid polychaetes (*top*) and on the yeti crab claws (*middle*). Various species of clams, mussels, and tubeworms also inhabit cold seeps (background), while highly specific "shipworms" (bivalves) use sunken wood and attract other species. The carcasses of whales (*lower right*) foster another community that includes the "bone-eating worms" depicted (Image credit: Amy Scott-Murray (amy@amyscottmurray.com); use terms under Creative Commons license).

detection emerge. Many microbes remain uncultivated but molecular approaches and genomic techniques to examine functional genes reveal a wide diversity. The Euryarchaeota tend to dominate at vents, especially at higher temperatures, while, among bacterial groups, the epsilonproteobacteria and Aquificales are abundant (Figure 1). Among the alpha- and gammaproteobacteria are the strains that function as symbionts in many animals at reducing habitats such as tubeworms and bivalves (Figure 2). Some microbes form obligate interactions such as the archaeal/bacterial consortia at cold seeps that mediate the anaerobic oxidation of methane with sulfate (Jørgensen and Boetius, 2007). Metabolite exchange among diverse microbes using different redox reactions also occurs in open-ocean low-oxygen zones (Wright et al., 2012). As research explores the metabolic functions expressed by these microbes, discoveries of new biochemical pathways for CO_2 fixation are emerging. In addition to the classic Calvin-Benson cycle, at least five additional pathways are known (Hügler and Sievert, 2011; Figure 1).

Future directions

Chemosynthesis affects the chemical environment in fluid and rocks. How microbes mediate mineral transformations is a major field of biogeochemical research (Holden et al., 2012). Deep in the subseafloor, the minerals olivine and pyroxene react with water releasing hydrogen in a process called serpentinization. The potential for life in these zones is reflected in uplifted ophiolites where microbes are recorded; the metabolic pathways and consequences of the alkaline habitat are far from understood (Schrenk et al., 2013). Measurement of production rates in most marine-reducing habitats is difficult and especially challenging at hydrothermal vents where strong chemical gradients exist. The degree to which microorganisms are interdependent within a community that adapts a variety of metabolisms to changing chemical conditions is unclear (Sievert and Vetriani, 2012) and suggests that studies need to address the entire metabolome of a chemoautotrophic setting. More accessible are open-ocean hypoxic zones where the complexity of microbial interactions in a cascade of redox reactions is revealed by new genomic approaches. The development of modeling tools (e.g., Amend et al., 2011) to assess bioenergetics of redox/synthesis reactions can help formulate testable ideas of metabolic and microbial diversity in many settings. Relationships of these microbes with eukaryotes and metazoans are complex, often supporting communities with the highest natural biomass of the ocean. Symbiotic associations continue to contribute information on cell-cell signaling and evolutionary patterns. Exploration of deep-sea chemosynthetic habitats has led to many fundamental discoveries in basic life processes and redirected our considerations of the origin of life and life beyond Earth. The potential remains to discover energy sources not linked to processes on or above the seafloor. Exploration, both on the seafloor and in laboratories, continues to challenge our concepts of basic life processes.

Bibliography

Amend, J. P., McCollom, T. M., Hentscher, M., and Bach, W., 2011. Catabolic and anabolic energy for chemolithoautotrophs in deep-sea hydrothermal systems hosted in different rock types. *Geochimica et Cosmochimica Acta*, **75**, 5736–5748.

Edwards, K. J., Bach, W., and McCollom, T. M., 2005. Geomicrobiology in oceanography: microbe–mineral interactions at and below the seafloor. *Trends in Microbiology*, **13**, 449–456.

Helm, K. P., Bindoff, N. L., and Church, J. A., 2011. Observed decreases in oxygen content of the global ocean. *Geophysical Research Letters*, **38**, L23602.

Holden, J. F., Breier, J. A., Rogers, K. L., Schulte, M. D., and Toner, B. M., 2012. Biogeochemical processes at hydrothermal vents: microbes and minerals, bioenergetics, and carbon fluxes. *Oceanography*, **25**, 196–208.

Hügler, M., and Sievert, S. M., 2011. Beyond the Calvin Cycle: autotrophic carbon fixation in the ocean. *Annual Review of Marine Science*, **3**, 261–289.

Jørgensen, B. B., and Boetius, A., 2007. Feast and famine – microbial life in the deep-sea bed. *Nature Reviews Microbiology*, **5**, 770–781.

Kormas, K. A., Tivey, M. K., Von Damm, K., and Teske, A., 2006. Bacterial and archaeal phylotypes associated with distinct mineralogical layers of a white smoker spire from a deep-sea hydrothermal vent site (9°N, East Pacific Rise). *Environmental Microbiology*, **8**, 909–920.

Levin, L. A., 2005. Ecology of cold seep sediments: interactions of fauna with flow, chemistry and microbes. *Oceanography and Marine Biology: An Annual Review*, **43**, 1–46.

Martin, W., and Russell, M. J., 2003. On the origins of cells: a hypothesis for the evolutionary transitions from abiotic geochemistry to chemoautotrophic prokaryotes, and from prokaryotes to nucleated cells. *Philosophical Transactions of the Royal Society of London, Series B: Biological Sciences*, **358**, 59–83.

McCollom, T. M., and Shock, E. L., 1997. Geochemical constraints on chemolithoautotrophic metabolism by microorganisms in seafloor hydrothermal systems. *Geochimica et Cosmochimica Acta*, **61**, 4375–4392.

Schrenk, M. O., Brazelton, W. J., and Lang, S. Q., 2013. Serpentinization, carbon, and deep life. *Reviews in Mineralogy and Geochemistry*, **75**, 575–606.

Sievert, S. M., and Vetriani, C., 2012. Chemoautotrophy at deep-sea vents: past, present, and future. *Oceanography*, **25**, 218–233.

Smith, C. R., and Baco, A. R., 2003. Ecology of whale falls at the deep-sea floor. *Oceanography and Marine Biology: An Annual Review*, **41**, 311–354.

Sylvan, J. B., Toner, B. M., and Edwards, K. J., 2012. Life and death of deep-sea vents: bacterial diversity and ecosystem succession on inactive hydrothermal sulfides. *MBio*, **3**, e00279-11, doi:10.1128/mBio.00279-11.

Tunnicliffe, V., Juniper, S. K., and Sibuet, M., 2003. Reducing environments of the deep-sea floor. In Tyler, P. A. (ed.), *Ecosystems of the World: The Deep-Sea*. Amsterdam: Elsevier Press, pp. 81–110. Ch. 4.

Wright, J. J., Konwar, K. M., and Hallam, S. J., 2012. Microbial ecology of expanding oxygen minimum zones. *Nature Reviews Microbiology*, **10**, 381–394.

Cross-references

Anoxic Oceans
Black and White Smokers
Deep Biosphere
Hydrothermalism
Ocean Acidification

CLAY MINERALS

Rüdiger Stein
Alfred Wegener Institute, Helmholtz Centre for Polar and Marine Research (AWI), Bremerhaven, Germany

Definition

Clay minerals are hydrous aluminum phyllosilicates. The main representatives are kaolinite, illite, chlorite, and smectite.

Structure and composition

The structure of clay minerals is characterized by alternation of sheets, which yield the layer structure. The composition and configuration of these sheets are different for different clay minerals. There are, however, two basic types of sheets, composing any given clay mineral (Grim, 1962; Bridley and Brown, 1984; Chamley, 1989): (i) tetrahedral sheets with one silicon atom surrounded by four oxygen atoms in a tetrahedral configuration and (ii) octahedral sheets with an aluminum, magnesium, or iron atom surrounded by hydroxyl groups and oxygen in a sixfold coordinated configuration (Figure 1a–d). Alternation of one tetrahedral and one octahedral sheet yields the 1:1 structure, whereas alternation of two tetrahedral sheets with one octahedral sheet results to the 2:1 structure. In most clay minerals, substitution in tetrahedral and octahedral sites creates a charge deficit known as layer charge, which is balanced by potassium, sodium, calcium, or iron and/or hydrated cations. These cations are hosted between the layers in the interlayer space and therefore are called interlayer cations. During hydration, the interlayer cations may cause expansion or swelling of the interlayer space of some clay minerals (e.g., montmorillonite; a member of the smectite family; Figure 1b). In general, clay minerals are classified by the differences in their layered structures, the type of cations in the structure, the magnitude of layer charge, and the type amounts and kinds of exchangeable ions within the interlayers (Meunier, 2005).

Kaolinite is the most important 1:1 clay mineral (Figure 1a), whereas illite and smectites have 2:1 structures composed of two tetrahedral and one octahedral sheet (Figure 1b, d). Chlorite is characterized by a 2:1 structure with an interlayered sheet of brucite (hydroxide interlayer) (Figure 1c). In addition to these main types of clay minerals, there are clay minerals that display a mixed-layered configuration composed of different basic structures and are classified as mixed-layer clay minerals. The most common mixed-layer clay mineral present in marine sediments is illite/smectite (e.g., Fütterer, 2006).

Formation of clay minerals

Clay minerals are the product of physical and chemical weathering of primary, rock-forming aluminum silicates (feldspars, amphiboles, pyroxenes, etc.) and hydrothermal alteration. Different clay minerals are formed under different climate-dependent weathering conditions. Thus, clay minerals determined in surface sediments and sediment cores may provide information on present and past, respectively, weathering conditions and climate of the source areas as well as transport processes of the terrigenous sediments (see examples below and synthesis by Chamley, 1989). The clay minerals illite and chlorite, for example, are common weathering products of igneous and metamorphic rocks. When mobilized during physical weathering, these minerals are typically found in high-latitude marine sediments. Kaolinite, on the other hand, forms under warm and humid conditions by intensive chemical weathering of feldspars in tropical soils. Thus, kaolinite is often referred to as a low-latitude mineral. Kaolinite, however, is also found in polar regions in sedimentary deposits that were formed either under past warmer and wetter climatic conditions that currently exist or at low latitudes and later displaced northward through plate motion. Derived from the alteration of volcanic rocks, smectite is a good indicator of volcanic sediment sources.

Determination of clay minerals

Clay minerals are commonly determined both accurately and precisely in the <2 μm clay fraction by X-ray powder diffraction (XRD) (Bridley and Brown, 1984; Moore and Reynolds, 1997). For qualitative and semiquantitative estimates of clay mineral composition of marine sediments, the Biscaye method (Biscaye, 1965) is widely used in Marine Geosciences. The method uses oriented samples of the clay fractions prepared by usual sedimentation techniques (Moore and Reynolds, 1997) and subsequent ethylene glycol (EG) solvation and assumes that the sum of the four clay minerals is 100 %.

For example the abundance of illite, is given by the following equation:

$$\text{Illite}(\%) = \frac{4I \times 100}{S + 4I + 2K + 2C}$$

where the weighting factors to convert peak areas to relative weight fractions for smectite, illite, kaolinite, and chlorite are 1, 4, 2, and 2, respectively (Biscaye, 1965). For smectite the 17 Å peak, after removal of the chlorite 14 Å peak, and for illite the 10 Å peak are used in the calculations. Kaolinite is distinguished from chlorite from the peaks at 3.57–3.58 Å (kaolinite) to 3.53–3.54 Å (chlorite) peaks using the slow-scan XRD traces. The relative amount of each mineral is estimated from the 7 Å kaolinite-chlorite peak (Biscaye, 1965). Further details about the X-ray diffraction technique and different evaluation approaches can be found in Heath and Pisias (1979), Bridley and Brown (1984), and Moore and Reynolds (1997).

Significance of clay minerals in marine geosciences

In Marine Geosciences, clay minerals represent a major to dominant proportion of fine-grained deep-sea sediments and thus are important proxies for reconstruction of present and past environmental and depositional as well as diagenetic conditions from studies of surface sediments and sediment cores. Some examples are presented and discussed below. For more details the reader is referred to the literature listed at the end.

The distribution and abundance of clay minerals within the different ocean basins (e.g., Griffin and Goldberg, 1963; Biscaye, 1965; Venkatarathnam and Biscaye, 1973; Kolla et al., 1976; Heath and Pisias, 1979; Naidu

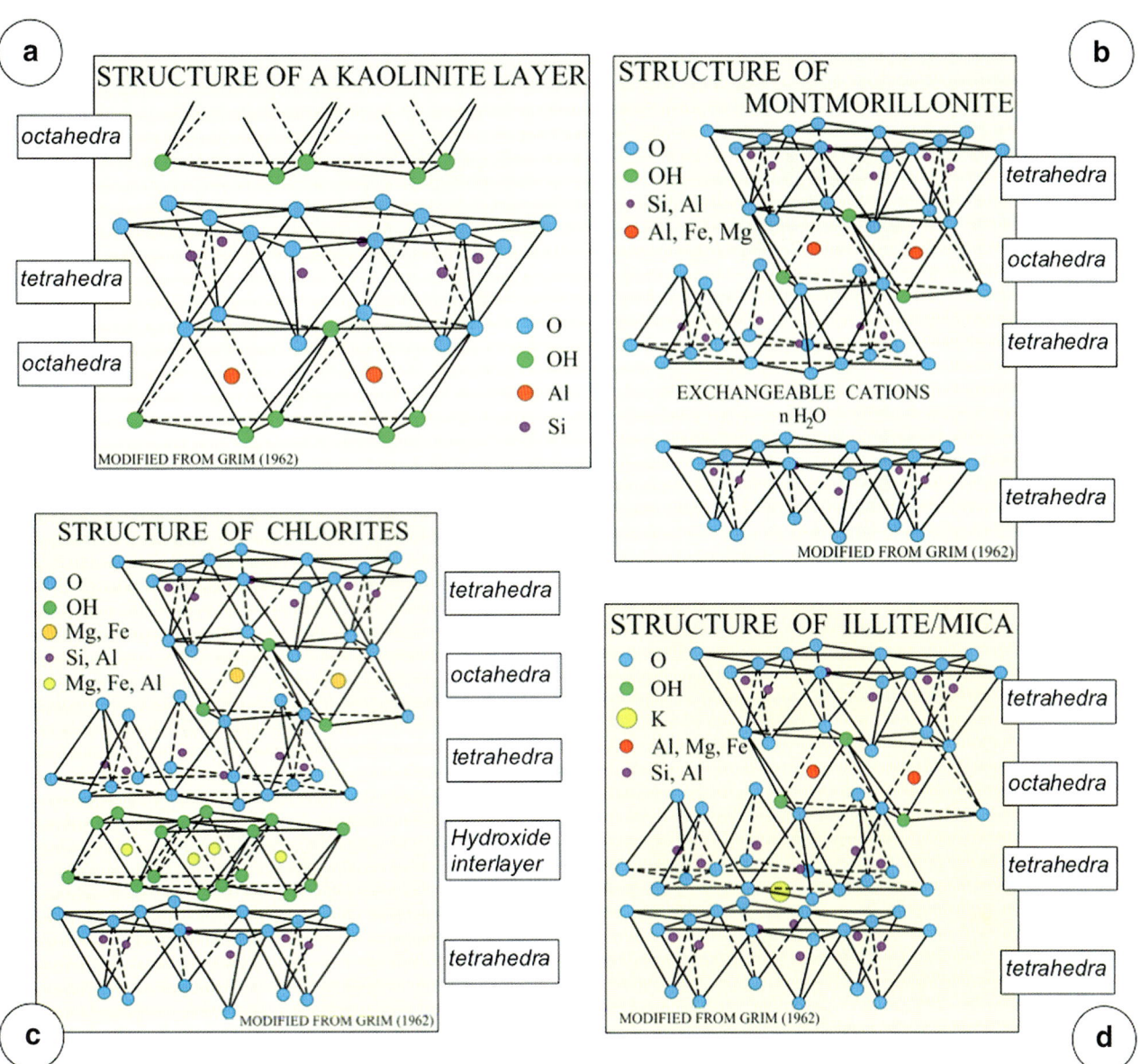

Clay Minerals, Figure 1 Diagrammatic sketch of kaolinite, montmorillonite, chlorite, and illite/mica (from Poppe et al., 2010: (http://pubs.usgs.gov/of/2001/of01-041/htmldocs/clays/kaogr.htm; Webpage 2010; supplemented)).

et al., 1982; Wahsner et al., 1999; Stein, 2008 among others) as well as the world ocean (Griffin et al., 1968; Rateev et al., 1969; Windom, 1976; Rateev et al., 2008) has been studied thoroughly over the past decades. The distribution patterns of the four main clay minerals have been related to latitudinal climate regimes, weathering conditions, wind regimes, river discharge, and ocean currents. Kaolinite concentrations display prominent maxima in low latitudes characterized by maximum chemical weathering (Figure 2a). Illite reaches maximum values in the mid to higher latitudes, especially in the northern oceans (Figure 2b) surrounded by large land masses, and is indicative of terrigenous origin and fluvial and eolian supply. Maximum concentrations of chlorite are observed in the high northern and southern latitudes (Figure 2c), i.e., the subpolar regions characterized by predominantly physical weathering processes. Smectite indicative for a volcanic origin displays maximum abundance in the South Pacific (Figure 2d), i.e., a region of intense volcanic activity and low sedimentation rates due to the large distance to the continents and, thus, limited dilution by detrital clay minerals.

Based on the relationship between clay mineral distribution pattern in surface sediments and modern climatic

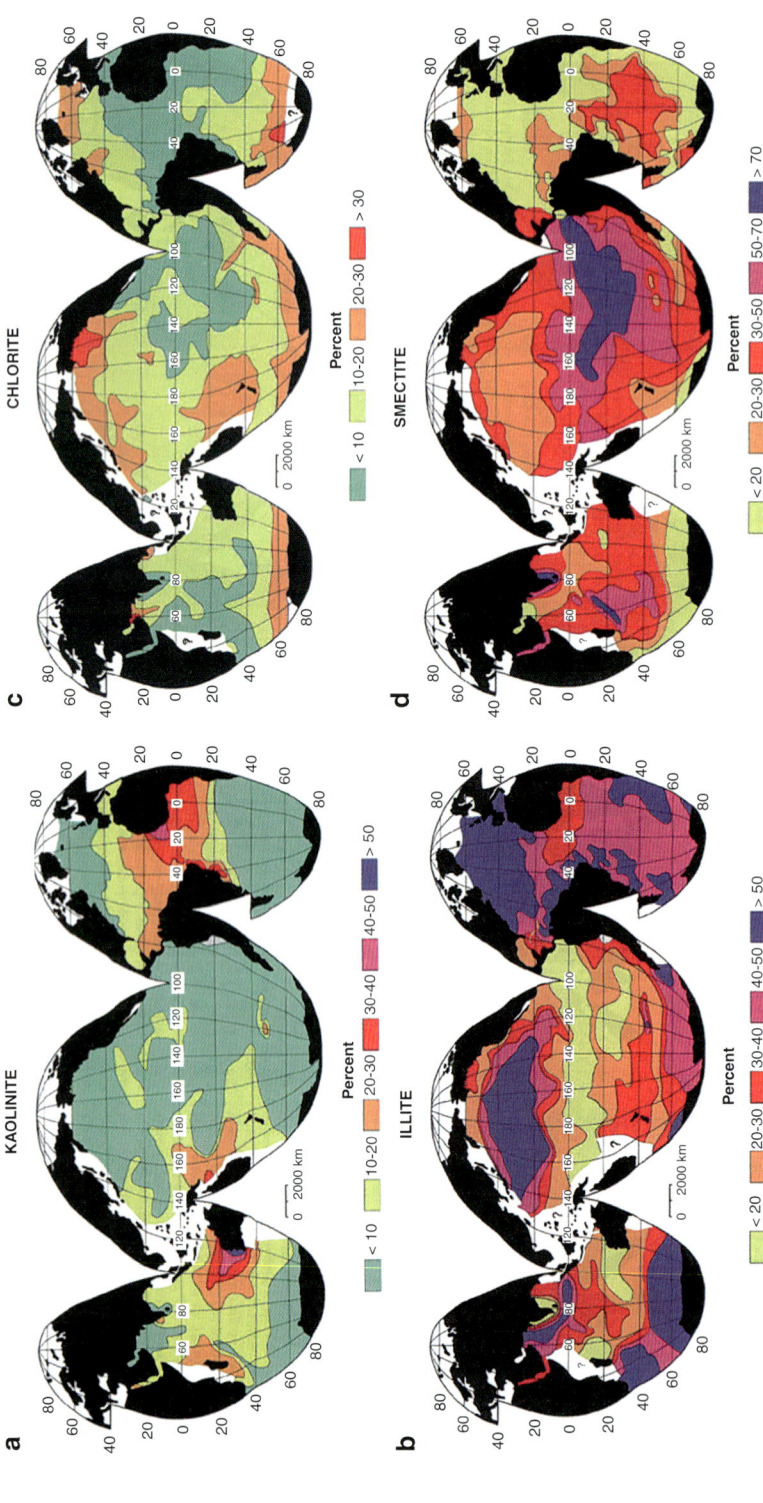

Clay Minerals, Figure 2 Relative distribution of kaolinite (**a**), illite (**b**), chlorite (**c**), and smectite (**d**) in surface sediments from the world ocean as percentage in the carbonate-free clay fraction <2 μm (from Windom, 1976, color version extracted from Fütterer, 2006).

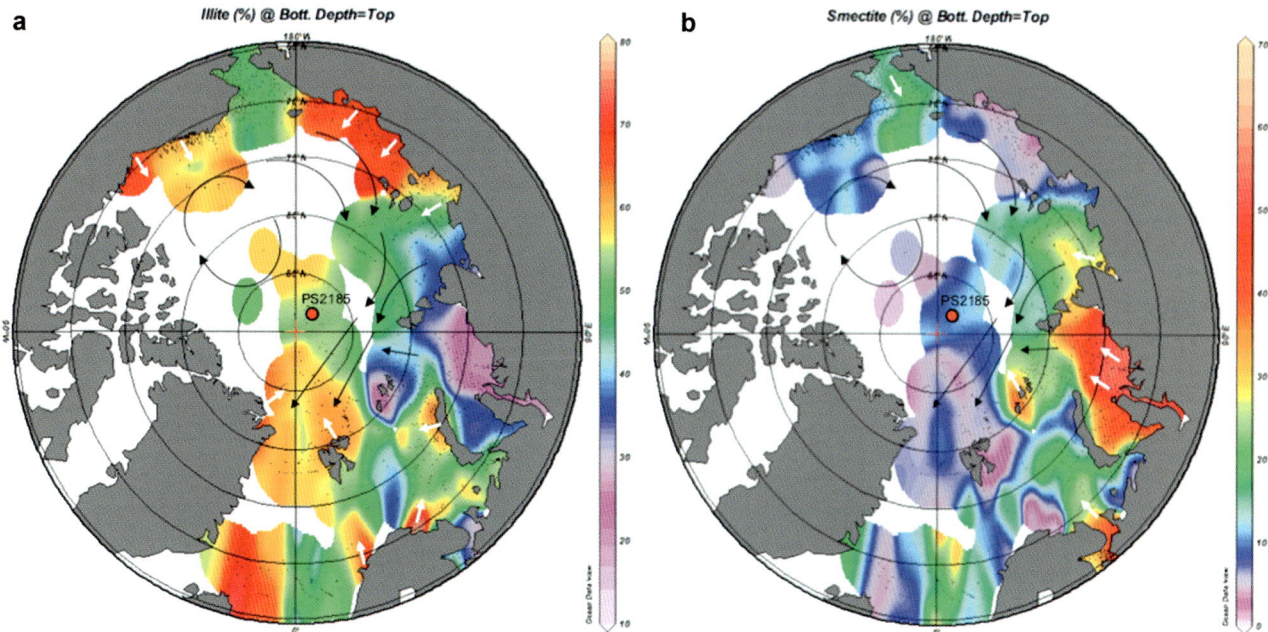

Clay Minerals, Figure 3 Relative distribution of illite (**a**) and smectite (**b**) in Arctic Ocean surface sediments as percentage in the carbonate-free clay fraction <2 μm (from Stein, 2008). The location of Core PS2185 is shown.

conditions, clay mineral data determined in marine sediment cores were used to reconstruct the paleoceanographic and paleoclimatic changes through time, as shown in numerous studies from different parts of the world ocean (e.g., Janecek and Rea, 1983; Stein and Robert, 1985; Stein, 1985; Ehrmann et al., 1992; Robert and Kennett, 1994; Stein, 2008).

In polar and subpolar regions, with predominantly cold climate at least during Neogene and Quaternary times, physical weathering processes dominate and chemical and diagenetic alterations are negligible. Therefore, the clay mineral association in marine sediments is a valuable indicator to identify origin and transport pathways of terrigenous sediments, and clay mineral data were used to distinguish among the shelf areas by the amounts of smectite, illite, and kaolinite in surface sediments (see Stein, 2008 for review). The Kara Sea is characterized by highest smectite values and the East Siberian Sea with lowest smectite and kaolinite values and highest illite concentrations (Figure 3). The smectite maxima in the Kara Sea result from erosion and weathering of extensive flood basalts of the Putoran Massif of the Siberian Hinterland (Duzhikov and Strunin, 1992; Vyssotski et al., 2006). The Yenisei River with its tributaries drains this area and transports smectite into the Kara Sea. The Central Arctic sediments are characterized by a mixture of these three clay minerals but are distinctly depleted in smectite compared to the Kara and Laptev seas.

In the Arctic Ocean, different processes influence the transport of sediment from the shelves into the deep Arctic Basin. An important mechanism responsible for the clay mineral dispersal pattern is current systems. Clay minerals are extremely fine-grained and therefore can be transported over large distances within the water column. In addition, sediment transport with drifting sea ice is of importance in the Arctic Ocean, and clay minerals might be useful proxies to identify source areas of the sea ice (e.g., Pfirman et al., 1997; Dethleff, 2005).

An example for using clay minerals for reconstruction of paleoenvironments in polar regions, records from Core PS2185 recovered from Lomonosov Ridge/Central Arctic Ocean and representing glacial-interglacial variability of the last 200,000 years BP, is shown in Figure 3 (for location of core see Figure 2). Intervals with increased coarse-grained ice-rafted debris (IRD) were recorded in uppermost MIS 7 to MIS 6 (190–130 ka), upper part of MIS 5 (substage 5.2, about 90–80 ka), near the MIS 5/4 boundary (around 75 ka), and in the late MIS 4/early MIS 3 time interval (65–50 ka), indicating major continental glaciations during those times. Concerning the provenance of the IRD and its variability through time in the Eurasian Arctic, bulk, clay, and heavy-mineral associations of Core PS2185 can be used to identify source areas of the terrigenous (IRD) fractions (e.g., Spielhagen et al., 1997; Behrends et al., 1999; Wahsner et al., 1999; Stein, 2008; and references therein). Here, elevated smectite and kaolinite concentrations as well as high clinopyroxene/amphibole ratios during MIS 6, upper MIS 5, and late MIS 4/early MIS 3, mostly coinciding with IRD maxima (Figure 4), serve as a tracer for an IRD origin from the western Laptev/southeastern Kara Sea/Franz Josef Land and central Barents Sea/Franz Josef Land, respectively. High smectite concentrations, however, do not always coincide with high coarse-fraction

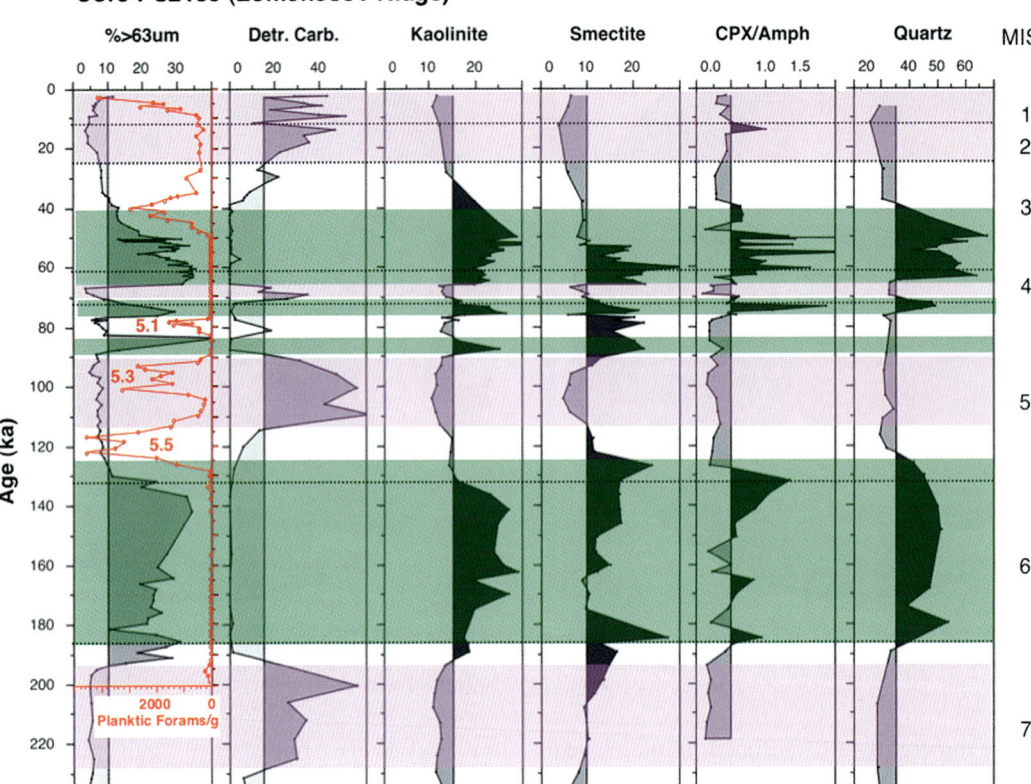

Clay Minerals, Figure 4 Summary plots showing coarse-fraction content >63 μm (wt%), detrital carbonate (% of the coarse fraction >500 μm), and kaolinite and smectite (% of clay minerals in the clay fraction <2 μm) (Data from Spielhagen et al., 1997) as well as the clinopyroxene/ amphibole (CPX/Amph) ratio (Data from Behrends, 1999) and quartz content (Data from Vogt, 1997, 2004) in Core PS2185 for the last 240 ka (MIS 7 to MIS 1), using the age model of Spielhagen et al. (2004). Marine isotope stages MIS 1–7 are indicated. In addition, the concentration of planktic foraminifers per gram sediment is shown (Data from Spielhagen et al., 2004). *Green* and *pink* bars indicate sediment source areas in Eurasia and northern Canada, respectively (Figure from Stein (2008)).

content. In some intervals, e.g., in the uppermost and lowermost MIS 5 (5.1 and 5.5), smectite enrichments correlate with very low terrigenous coarse fraction which may suggest transport by sea ice (or currents) rather than icebergs. For more detailed discussion see Stein (2008) and references therein.

Summary and conclusions

Clay minerals, hydrous aluminum silicates which comprise mainly kaolinite, illite, chlorite and smectite, are major to dominant constituents of fine-grained deep-sea sediments. Clay mineral composition of marine sediments may give important information about source areas and transport mechanisms of terrigenous material as well as weathering processes and climate of the source areas. Thus, clay minerals are widely used in Marine Geosciences as proxies for reconstruction of present and past environmental, depositional, and diagenetic conditions from the studies of surface sediments and sediment cores.

Bibliography

Behrends, M., 1999. Reconstruction of sea-ice drift and terrigenous sediment supply in the Late Quaternary: heavy-mineral associations in sediments of the Laptev-Sea continental margin and the central Arctic Ocean. *Reports on Polar Research*, **310**, 167.

Behrends, M., Hoops, E., and Peregovich, B., 1999. Distribution patterns of heavy minerals in Siberian rivers, the Laptev Sea and the eastern Arctic Ocean: an approach to identify sources, transport and pathways of terrigenous matter. In Kassens, H., Bauch, H., Dmitrenko, I., Eicken, H., Hubberten, H. W., Melles, M., Thiede, J., and Timokhov, L. (eds.), *Land-Ocean Systems in the Siberian: Dynamics and History*. Heidelberg: Springer, pp. 265–286.

Bergaya, F., Theng, B. K. G., and Lagaly, G., 2006. *Handbook of Clay Science*. Amsterdam: Elsevier. Developments in Clay Science, Vol. 1, p. 1224.

Biscaye, P. E., 1965. Mineralogy and sedimentation of recent deep-sea clay in the Atlantic Ocean and adjacent seas and oceans. *Geological Society of America Bulletin*, **76**, 455–486.

Bridley, G. W., and Brown, G., 1984. Crystal structures of clay minerals and their X-ray identification. *Mineralogical Society Monograph*, **5**, 248.

Chamley, H., 1989. *Clay Sedimentology*. Heidelberg: Springer, p. 623.

Dethleff, D., 2005. Entrainment and export of Laptev Sea ice sediments, Siberian Arctic. *Journal Geophysical Research*, **110** (C07009), doi:10.1029/2004JC002740.

Duzhikov, O. A., and Strunin, B. M. (eds.), 1992. *Geology and Metallogeny of Sulfide Deposits, Norilisk Region*. Moscow: USSR SEG Special Publication, p. 60.

Ehrmann, W. U., Melles, M., Kuhn, G., and Grobe, H., 1992. Significance of clay mineral assemblages in the Antarctic Ocean. *Marine Geology*, **107**, 249–273.

Fütterer, D. K., 2006. The solid phase of marine sediments. In Schulz, H. D., and Zabel, M. (eds.), *Marine Geochemistry*, 2nd edn. Heidelberg: Springer, p. 574.

Griffin, J. J., and Goldberg, E. D., 1963. Clay-mineral distribution in the Pacific Ocean. In Hill, M. N. (ed.), *The Sea*. New York: Interscience, Vol. 3, pp. 728–741.

Griffin, J. J., Windom, H., and Goldberg, E. D., 1968. The distribution of clay minerals in the World Ocean. *Deep-Sea Research*, **15**, 433–459.

Grim, R. E., 1962. Clay mineralogy. *Science, New Series*, **135**, 890–898.

Heath, G. R., and Pisias, N. G., 1979. A method for the quantitative estimation of clay minerals in North Pacific deep-sea sediments. *Clays and Clay Minerals*, **27**, 175–184.

Janecek, T. R., and Rea, D. K., 1983. Eolian deposition in the northeast Pacific Ocean: Cenozoic history of atmospheric circulation. *Geological Society of America Bulletin*, **94**, 730–738.

Kolla, V., Henderson, L., and Biscaye, P. E., 1976. Clay mineralogy and sedimentation in the western Indian Ocean. *Deep-Sea Research*, **23**, 949–961.

Meunier, A., 2005. *Clays*. Berlin: Springer, p. 472.

Moore, D. M., and Reynolds, R. C., Jr., 1997. *X-Ray Diffraction and the Identification and Analysis of Clay Minerals*, 2nd edn. Oxford: Oxford University Press, p. 378.

Naidu, A. S., Creager, J. S., and Mowatt, T. C., 1982. Clay mineral dispersal patterns in the North Bering and Chukchi seas. *Marine Geology*, **47**, 1–15.

Pfirman, S. L., Colony, R., Nürnberg, D., Eicken, H., and Rigor, I., 1997. Reconstructing the origin and trajectory of drifting Arctic sea ice. *Journal of Geophysical Research*, **102**(C6), 12575–12586.

Poppe, L. J., Paskevich, V. F., Hathaway, J. C., and Blackwood, D. S., 2010. *A Laboratory Manual for X-Ray Powder Diffraction*. Woods Hole: U. S. Geological Survey Open-File Report 01-041. http://pubs.usgs.gov/of/2001/of01-041/htmldocs/clays/kaogr.htm; Webpage 2010.

Rateev, M. A., Gorbunova, Z. N., Lisitzyn, A. P., and Nosov, G. L., 1969. The distribution of clay minerals in the oceans. *Sedimentology*, **13**, 21–43.

Rateev, M. A., Sadchikova, T. A., and Shabrova, V. P., 2008. Clay minerals in recent sediments of the World Ocean and their relation to types of lithogenesis. *Lithology and Mineral Resources*, **43**, 125–135.

Robert, C., and Kennett, J. P., 1994. Antarctic subtropical humid episode at the Paleocene-Eocene boundary: clay mineral evidence. *Geology*, **22**, 211–214.

Shichi, T., and Takagi, K., 2000. Clay minerals as photochemical reaction fields. *Journal of Photochemistry and Photobiology, C, Photochemistry Reviews*, **1**, 113–130.

Spielhagen, R. F., Bonani, G., Eisenhauer, A., Frank, M., Frederichs, T., Kassens, H., Kubik, P. W., Mangini, A., Nøgaard-Pedersen, N., Nowaczyk, N. R., Schäper, S., Stein, R., Thiede, J., Tiedemann, R., and Wahsner, M., 1997. Arctic Ocean evidence for late Quaternary initiation of northern Eurasian ice sheets. *Geology*, **25**, 783–786.

Spielhagen, R. F., Baumann, K.-H., Erlenkeuser, H., Nowaczyk, N. R., Nørgaard-Pedersen, N., Vogt, C., and Weiel, D., 2004. Arctic Ocean deep-sea record of northern Eurasian ice sheet history. *Quaternary Science Reviews*, **23**, 1455–1483.

Stein, R., 1985. The post- Eocene sediment record of DSDP-Site 366: implications for African climate and plate tectonic drift. *Geological Society of America Memoir*, **163**, 305–315.

Stein, R., 2008. *Arctic Ocean Sediments: Processes, Proxies, and Palaeoenvironment*. Amsterdam: Elsevier. Developments in Marine Geology, Vol. 2, p. 587.

Stein, R., Robert, C., et al., 1985. Siliciclastic sediments at Sites 588, 590, and 591: Neogene and Paleogene evolution in the Southwest Pacific and Australian climate. In Kennett, J. P., and van der Borch, C. (eds.), *Initial Reports of the Deep Sea Drilling Project*. Washington: U.S. Govt. Printing Office, Vol. 90, pp. 1437–1455.

Venkatarathnam, K., and Biscaye, P. E., 1973. Clay mineralogy and sedimentation in the eastern Indian Ocean. *Deep-Sea Research*, **20**, 727–738.

Vogt, C., 1997. Regional and temporal variations of mineral assemblages in Arctic Ocean sediments as climatic indicator during glacial/interglacial changes. *Reports on Polar Research*, **251**, 309.

Vogt, C., 2004. Mineralogy of sediment core p S2185–3. *Data Report*, doi:10.1594/PANGAEA.138269.

Vyssotski, A. V., Vyssotski, V. N., and Nezhdanov, A. A., 2006. Evolution of the West Siberian basin. *Marine and Petroleum Geology*, **23**, 93–126.

Wahsner, M., Müller, C., Stein, R., Ivanov, G., Levitan, M., Shelekova, E., and Tarasov, G., 1999. Clay mineral distributions in surface sediments from the Central Arctic Ocean and the Eurasian continental margin as indicator for source areas and transport pathways: a synthesis. *Boreas*, **28**, 215–233.

Windom, H. L., 1976. Lithogeneous material in marine sediments. In Riley, J. P., and Chester, R. (eds.), *Chemical Oceanography*. New York/London: Academic, pp. 103–155.

Cross-references

Deep-sea Sediments

COASTAL BIO-GEOCHEMICAL CYCLES

Jing Zhang
State Key Laboratory of Estuarine and Coastal Research, East China Normal University, Shanghai, China

Definition

The "Coastal Biogeochemical Cycles" is dealing with the transformation of chemical speciation of elements (i.e., chemical form and valence) and flow (e.g., flux) of materials between biotic and abiotic compartments of coastal and marine environments (here after the coastal ocean) that is defined by landmass (e.g., continents) on one side and the open ocean on another side. The study of biogeochemical cycles in this domain needs an approach that provides an integrated view of the physical, biological, geological, and chemical aspects of materials in the changing marine environment.

Introduction

Our understanding of the ocean was built up along with the ambition of navigation and exploration of marine resources by human beings that started more than 2000 years ago. The coastal ocean, because of its proximity to the land, is the marine environment that has been extensively explored by our human societies. For instance, about 1,500 years ago, our ancestors from Asia started to reclaim the coastal areas for salt production and for agriculture, and the early history of marine trade in the Middle East can be tracked back to the epoch far before the tenth century.

The modern knowledge of biogeochemical cycles started to be established in the late nineteenth century in Northern Europe where scientists began to determine the chemical composition of seawaters from coastal and open oceans. However, the process studies of the dynamics of the biogeochemical cycles only began in the second half of the twentieth century, benefiting from the technical innovations in clean sample collection in seagoing observations and instrumental analysis in the laboratory, particularly when trace elements were of concern.

Besides, the rapid increase and change of human being activities in the environment are a great challenge for the study of biogeochemical cycles. For instance, the enormous increase in agricultural production after the 1930s–1940s by the application of chemically synthesized fertilizers has considerably modified the seaward fluxes of some macronutrients (i.e., N and P), with a deterioration of adjacent coastal environments, such as eutrophication and hypoxia. More recently, the worldwide dam construction over the watersheds has dramatically reduced the sediment loads to the sea, and in some of the rivers, there has been a reduction of flux of dissolved silicate that induces a change in the phytoplankton community structure (e.g., change in biomass from diatoms to flagellates) with a profound impact on the marine food web.

In this work, we briefly review the major aspects of the coastal biogeochemical cycles and address a number of key issues that affect the biogeochemistry of coastal ocean and link to the health of marine ecosystems and sustainability of their exploitation.

External driving forces

Different from the open ocean, the ecosystems of the coastal environment can be taken as overstressed, because of the combined effects of climate change and anthropogenic perturbations. Although either natural forcing or human being activity can be reasonably characterized with state-of-art observational techniques and simulations, the combined effect of these two categories of driving forces is hardly predictable with our up-to-date knowledge and existing data of ocean sciences. This can be exemplified by the comparison with the interaction of different waves in physics: the interaction of individual waves (e.g., in a pond) is nonlinear mode and can produce a new wave spectrum with different period/frequency and amplification.

The natural driving force of the coastal ocean is related to the global and climate changes, with a number of basin-wide persistent processes and systematic trends, e.g., global warming, the sea-level rise, and ocean acidification, as well as by processes with relatively fast and episodic characters with annual and interannual variability, e.g., tropical cyclones and storms, and change in open boundary circulation (Table 1). It should be kept in mind, however, that studying local and regional

Coastal Bio-geochemical Cycles, Table 1 Characteristics of some of the important driving forces in the biogeochemical cycles of coastal oceans

Driving forces	Temporal and spatial dimensions
Tides (e.g., M_2, S_2, K_1, and O_1)	Semidiurnal, daily, monthly, and seasonal variations at basin scale, particularly in the coastal zone with strong effect of mixing and transportation
Ekman pumping and suction	Related to the wind-induced circulation in low- and midlatitude regions, e.g., monsoon areas, with strong seasonal nature at the mesoscale
Riverine influx and groundwater discharge	Daily and seasonal variations with strong impact from human beings in the watersheds, with change in fluxes of water, sediment, and pollutants
Atmospheric wet and dry depositions	Daily and seasonal variations with significant fingerprints from human emission from upwind side, either from land sector or from the seaside
Incursion of open boundary currents	Seasonally and interannually, with significant character linked to climate variability like ENSO, PDO, NAO, etc. and having an impact on basin-wide scale
Eutrophication	At mesoscale in coastal environment, an ongoing decadal process, usually the phenomena is related to the over-enrichment of nutrients from land sources
Oxygen depletion and hypoxia	Can be local or basin-wide phenomena with seasonal and/or annual persistence with negative impact to the food webs (e.g., in particular benthic organisms)
Reclamation	Usually happens in coastal areas with impacts leading to the loss of habitats because of change in land use for the purpose of economics (e.g., aquaculture and urbanization)
Fishery	Including removal of top predators or even also removal of lower food-web levels and negative effect of by-catch, with a strong "top-down" effect on the ecosystem
Coastal engineerings	Events related to the economic activities of human society, for example, the construction of harbors, oil platforms, and sediment dredging; damage to habitats on land and on the sea bottom

changes has to take into account the effects of distant forcing, i.e., global climate change, the interaction of external forcing and geographic settings can derive a different and sometimes unforeseen driving force at local and/or regional scale. The natural forcings should also include change in atmospheric depositions and riverine influx, however, with consideration of the respective effects of the anthropogenic perturbations. In temperate and high-latitude regions, the recurring seasonal changes and their considerable interannual variability are the most important factors in shaping the biogeochemical cycles of materials.

Since the coastal ocean is adjacent to the continental and/or landmass (e.g., islands), the "biogeochemical cycles" in this region is also modulated by the human being activities in the watersheds and in the ocean-side (LOICZ, 2005). Change in land-use and hydraulic engineering (e.g., damming the river) over the watersheds can dramatically modify the seaward transport of weathering products and pollutants, including riverine flux as well as groundwater discharge and their variability at seasonal and interannual scales. For instance, the terrestrial sediment load has been reduced by more than 50 % and the nutrient fluxes increased by up to twofold for some of the top world rivers over the last several decades (Boesch, 2002; Syvitski et al., 2005). Other examples include the reclamation for arable land in the coastal zone for aquaculture, e.g., shrimp ponds that destroy the wetland, and because of expansion of urbanization, particularly so in the developing countries. The global overfishing problem can have a "top-down" effect on the energy and material flow in ecosystems by removing top predators as well as herbivores thus disturbing interactions in the entire food web (GLOBEC, 1997). Moreover, eutrophication in the coastal environment can exert a negative impact on the ecosystem through "bottom-up" effects that modify macronutrients and their molar ratios and hence alter the community structure of phytoplankton (GEOHAB, 2001).

How and to what extent the natural and anthropogenic forcings interact have a determining effect on the biogeochemical cycles in the coastal ocean, which in turn modulates the structure and function of ecosystems. This is probably why the coastal biogeochemical cycles have significant varying temporal and spatial dimensions and accordingly the research and results often have a case-specific nature.

Machinery of the coastal biogeochemical cycles

The coastal environment exhibits the transit zone between the landmass and the open ocean where the properties of land and ocean are strongly modulated. Thus, strong gradients exist for hydrographic parameters (i.e., salinity and temperature), chemical properties (e.g., major and trace element concentrations), turbidity and profiles of photosynthetically available radiation, as well as for the composition of organisms.

Dynamics of biogeochemical cycles of the coastal ocean

The coastal biogeochemical cycles incorporate dynamic processes that include the major pathways of chemical elements between biotic and abiotic compartments and variability of source versus sink terms in the euphotic waters, where the tiny phytoplankton and marine algae use the energy from solar radiation to fix carbon dioxide from atmosphere and change it to chemical energy of organisms, releasing oxygen:

$$CO_2 + H_2O \rightarrow CH_2O + O_2$$

where the symbol "CH_2O" stands for a simplified composition of organic matter. This photosynthesis fuels the biogeochemical cycles and energy flow in the entire water column and through the entire food web. Along with photosynthesis, macronutrients such as N, P, Ca, and Si, and about further 30 minor and trace elements such as As, Cd, Co, Cu, Fe, Mn, Mo, Se, V, Zn, etc. that are biologically essential, are also taken up by phytoplankton (Figure 1). Particularly C, N, P, and O_2 are incorporated and remineralized in a relatively fixed molar quota, the so-called Redfield ratio: $C:N:P:O = 106:16:1:138$ (Redfield et al., 1963; Moral et al., 2003). Chemical elements can be either incorporated as components for soft organic matter or used to form skeletons and cell wall of organisms (GEOTRACES, 2006). Nutrients and trace elements used by the organisms in photosynthesis can be supplied from external sources, like land-source input via rivers and groundwater discharge, atmospheric wet and dry depositions, release through the sediment-water interface, advective transport, and upwellings from open ocean. Generally speaking, all nutrient supply from outside the euphotic zone, which includes nutrients enriched in surface waters during the nongrowth period of phytoplankton, is defined as "new production," which maintains the sustainable food production of ecosystem (e.g., fishery); in contrast phytoplankton uptake of nutrients stemming from remineralization in the food webs of euphotic zone is referred to as "regenerated production" (Dugdale and Goering, 1967).

The organic matter formed in photosynthesis is in part transferred to zooplankton and then through herbivores and carnivores up to the top predators in the food web. Organic matter escaping from water column remineralization sinks eventually to the seafloor as also do inorganic biogenic materials (e.g., carbonate skeletons and silicate shells) that survive dissolution in the water column. The whole journey from the carbon fixation by photosynthesis in surface waters until the organic matter deposition to the sea bottom is referred to as "biological pump" in Global Change Science.

Since the mid-1970s, the significance of microbial loop has been recognized in the ecosystem functioning that complements the traditionally defined "food chain," i.e., phytoplankton via zooplankton to fish, playing an important qualitative and quantitative role in the

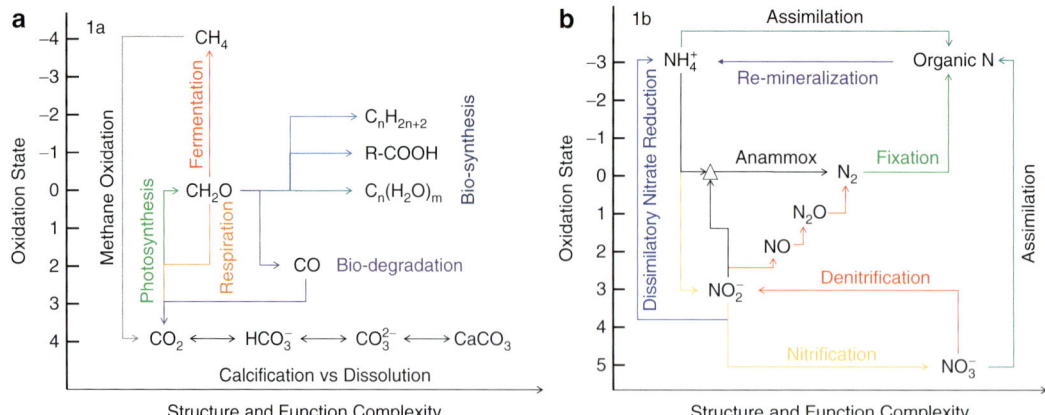

Coastal Bio-geochemical Cycles, Figure 1 Oxidation states of chemical elements and relative structure and function complexities of molecules for carbon (**a**) and nitrogen (**b**) in the biogeochemical cycles.

biogeochemical cycles in the ocean. Within the microbial loop, organic matter is used by bacteria in the water column and pore waters to form new biomass that can be grazed by Protozoa, including flagellates and ciliates, which is then fed by zooplankton and linked to the higher trophic levels of the food web. The microbial loop also comprises pico-phytoplankton and autotrophic bacteria (Azam and Malfatti, 2007). In the ocean, functioning of the microbial loop is similar to a filtration system, by selective use of readily decomposed molecules (i.e., labile fraction) of organic matter, leaving the residual fraction being more refractory in nature and thus not readily used by organisms (Hopkinson and Vallino, 2005). The microbial loop in the coastal environment is also fueled with organic matter from the terrestrial sources (e.g., river) and from atmospheric wet (e.g., rainfall) and dry (i.e., aerosols and gases) depositions.

In Figure 2, the major biogenic functional groups that link the biogeochemical cycles with the structure of food web of ecosystems in the coastal ocean are summarized as a conceptual model. Due to the diverse topography and hydrographical forcing, the biogeochemical cycles in shallow coastal seas exhibit considerably stronger spatial and temporal variations when compared to the open ocean. For instance, coastal systems are exposed to tidal forcing and constrained in semi-enclosed basins landward from the coastline or connected to the open ocean through channels/straits between islands, whereas systems affected by the large rivers can extend seaward to the shelf break and even beyond. Thus, horizontal and vertical gradients in hydrographic and chemical forcings induce that biogeochemical cycles have a zonal nature. Briefly, the coastal waters under influence from terrestrial influx can suffer strongly from eutrophication, where photosynthesis (p) can be exceeded by community respiration (r) (i.e., $p - r < 0$), and the ecosystem becomes heterotrophic with occurrence of hypoxia and/or anoxia in near-bottom waters; in the areas closer to the open ocean on the other side, the environment becomes more oligotrophic and the photosynthesis is higher than respiration (i.e., $p - r > 0$), and the ecosystem turns to be autotrophic (Bauer and Bianchi, 2011).

Biogeochemical modeling and budgets of the coastal ocean

The quantitative description of coastal biogeochemical cycles integrates our understandings of interactions of chemical species between biotic and abiotic compartments and the dynamic processes of biogeochemistry that regulate the transfer of elements from the sources to the sinks including the fluxes across the boundaries between the coastal ecosystems and adjacent marine and/or land provinces (Figure 3).

The early studies of biogeochemical budgets relied on the application of a "box model," owing to the limitation of available data sets and the lack of the ability of mathematic calculations. In the box model, the system of interest is considered well mixed and at the steady state, that is, the concentration and fluxes of a given chemical species do not change with time, and the principle of energy and/or mass balances can be used to solve the problems; usually the calculation of ordinary differential equation(s) is incorporated.

In the more sophisticated studies, the equations that describe the chemical reactions and/or biogeochemical pathways (i.e., parameterization) are coupled with the 1D to 3D numerical models that simulate the hydrodynamic processes and sedimentary dynamics (e.g., transport of the total suspended matter). The output data can be plotted against timescales, showing the variability in spatial and temporal dimensions. In the numerical simulations, the change of concentration (C) for a volume of water parcel at a given location can be described as

$$\frac{\partial C}{\partial t} = [\text{AdvectionTerm}] + [\text{DiffusionTerm}] + [\text{Source} - \text{Sink}]$$

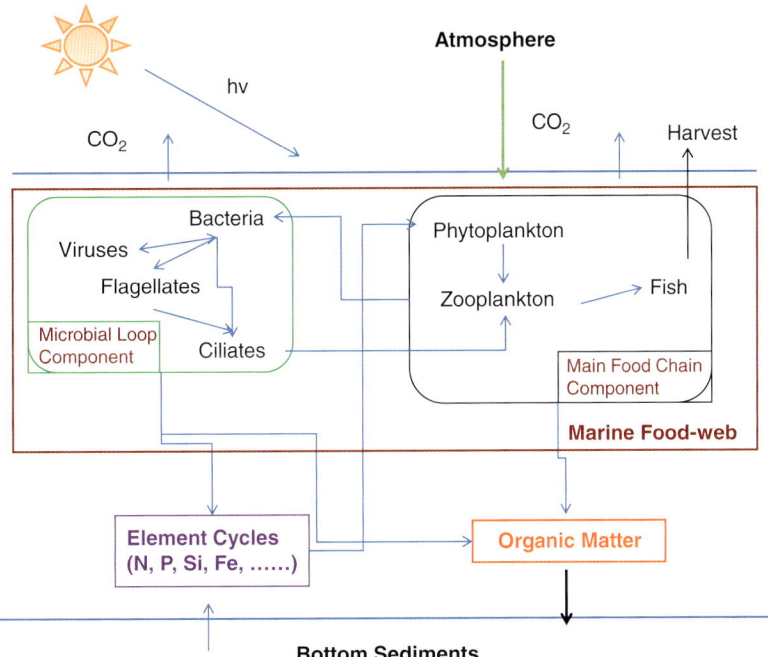

Coastal Bio-geochemical Cycles, Figure 2 Interlinkages of biogeochemical cycles with traditional food chain and microbial loop in the coastal ocean; collectively they make the food web.

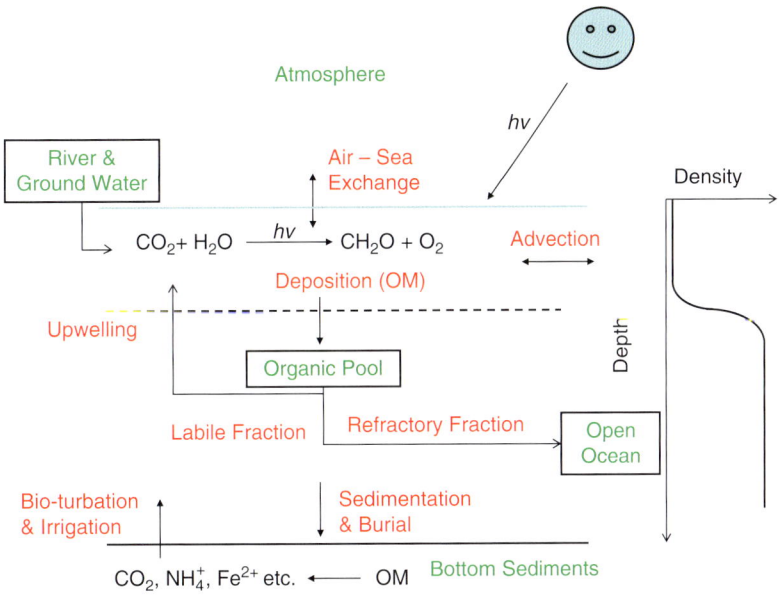

Coastal Bio-geochemical Cycles, Figure 3 Major budgetary pathways for the modeling of coastal biogeochemical cycles, including external forcing terms (i.e., new production) and internal recycle of chemical species (i.e., regenerated production). The extent of seasonal changes is decisive for the spatial and temporal decoupling of new versus regenerated productions, as well as between production and export of organic matter (OM).

The "AdvectionTerm" can be expressed as $-U \cdot \nabla C$, and "DiffusionTerm" by $\nabla \cdot (D \cdot \nabla C)$, where U and D represent the velocity vector and diffusivity tensor, respectively, and $\nabla = i\frac{\partial ()}{\partial x} + j\frac{\partial ()}{\partial y} + k\frac{\partial ()}{\partial z}$, that is, the Hamiltonian. The "Source-Sink" term deals with the chemical reaction(s) that alters the element concentration at molecular level. In the simulations with the 3D model, usually the analytical solutions are not valid anymore; however, the fine and heterogeneous structures of the system of interest can be analyzed through numerical solutions. Moreover, the model outputs of a numerical simulation can be displayed in a biogeochemical field showing the distribution of species of interest in three dimensions with temporal variability, which can be used for the diagnostic analysis of biogeochemical cycles and perspective scenarios of system behavior in the background of climate change (Sarmiento and Gruber, 2006).

Based on the synthesis of data from the Joint Global Ocean Flux Study (JGOFS Web-site: http//ijgofs.whoi.edu/) in the 1990s, it has been indicated that primary production amounts to 0.5×10^{15} mol year^{-1} of organic carbon for the global continental marginal seas; though continental margins occupy only 5–10 % of ocean surface, ca. 15 % of total primary production of global ocean takes place in the coastal oceans (Chen et al., 2003). Whereas seasonal cycles and interannual to decadal variability in a given latitudinal area are similar in the marginal seas and the open ocean at surface, it should be noted, however, that owing to the proximity to the landmass, the coastal biogeochemical cycles have a stronger character of seasonality as well as interannual variability; the dimensions of spatial nature differ considerably between these two realms, revealed, for example, by the change in biogeochemical budgets (Zhang et al., 2007). For instance, in the coastal region of the tropical East Pacific Ocean, the upwelling affects the dissolved iron profiles in the water column, with elevated *Chl-a* in the wide shelf corresponding to high Fe concentrations in surface waters and depletion of dissolved Fe in the narrow shelf inducing low *Chl-a* values (Bruland et al., 2005).

Feedbacks to the earth system and human society

In the coastal as well as the open oceans, there exist close interactions between biogeochemical cycles and ecosystems from end to end (i.e., from microbial to top predators) (IMBER, 2005), with profound impact on the human society, such as the water quality (GEOHAB, 2001) and fishery (GLOBEC, 1997).

As the coastal ocean covers a transitional area from an often more heterotrophic system with large addition of allochthonous organic matter in coast to a more autotrophic system with dominance of autochthonous organic matter at open boundary, the nearshore waters are usually regarded as a CO_2 source to atmosphere, while the offshore waters become the sink of atmospheric CO_2 (Thomas et al., 2004). Hence coastal and marginal seas can play an important role in the global and oceanic carbon cycles by bridging the land, ocean, and atmospheric carbon reservoirs.

Humans are part of the global ecosystem, and hence the future of human society is affected by the interactions between land, ocean, and atmosphere. Human activities on land have a profound impact on the coastal biogeochemical cycles and thus on the interactions between biogeochemistry and ecosystem from end to end, which can change the composition of material fluxes from the coastal environment to the open ocean and to the atmosphere. This in turn feeds back on the sustainability of human society as well as on the marine life (Bauer and Druffel, 1998; Fasham et al., 2001). For instance, coastal eutrophication can induce a change in phytoplankton community structure (e.g., diatoms relative to dinoflagellates), which in turn affects the food web structure as well as fish-catch through transfer of organic matter and chemical energy between trophic levels. It has been indicated that the emission of some of greenhouse gas species (e.g., CH_4 and N_2O) from the coastal ocean accounts for an important part of global budgets considering its relatively small spatial proportion of world ocean (Bange et al., 2005; Naqvi et al., 2010). The predicted impacts of climate change in the future (e.g., global warming and ocean acidification) have been shown to have irreversible negative impacts on the marine ecosystems in low (e.g., coral reefs) as well as in high latitudes (Hoegh-Guldberg et al., 2007; Arrigo et al., 2008).

Summary and concluding remarks

Coastal biogeochemical cycles exhibit interlinks of different mechanisms that regulate the transfer of chemical elements in the marine ecosystem (e.g., food web) and encompass rapid and complicated processes that bridge the material transport between the various compartments of the Earth system, land, ocean, and atmosphere as well as human society. The external forcing factors on coastal biogeochemical cycles can be either natural or anthropogenic, as well as the combination of these two.

Biogeochemistry in the coastal ocean is composed of cycles of macronutrients and trace elements between biotic and abiotic compartments driven by formation of organic matter fueled by solar energy and the metabolisms (i.e., anabolic and catabolic processes) based on the chemical energy stored in the organisms.

Since the late 1970s, it has been recognized that the microbial loop can be an important component of biogeochemical cycles in the ocean, which is an integral part of the entire food web and affects considerably the pathways of biogeochemical cycles in the coastal ocean. It is known that coastal biogeochemical cycles have strong feedbacks to the atmosphere (e.g., emission of greenhouse gases) as well as the open ocean (e.g., lateral export of particulate and dissolved organic carbon) through change in material fluxes.

Acknowledgments

The author expresses gratitude to Profs. Jan Harff and Bodo von Bodungen for their review comments to improve the original manuscript of this work.

Bibliography

Arrigo, K. R., van Dijken, G., and Pabi, S., 2008. Impact of a shrinking Arctic ice cover on marine primary production. *Geophysical Research Letters*, **35**, L19603, doi:10.1029/2008GL035028.

Azam, F., and Malfatti, F., 2007. Microbial structuring of marine ecosystems. *Nature Reviews Microbiology*, **5**, 782–791.

Bange, H. W., Naqvi, S. W. A., and Codispoti, L. A., 2005. The nitrogen cycle in the Arabian Sea. *Progress in Oceanography*, **65**, 145–158.

Bauer, J. E., and Bianchi, T. S., 2011. Dissolved organic carbon cycling and transformation. In Wolanski, E., and McLusky, D. S. (eds.), *Treatise on Estuarine and Coastal Science*. Waltham: Academic Press, Vol. 5, pp. 7–67.

Bauer, J. E., and Druffel, E. R. M., 1998. Ocean margins as a significant source of organic matter to the deep open ocean. *Nature*, **392**, 482–485.

Boesch, D. F., 2002. Challenges and opportunities for science in reducing nutrient over-enrichment of coastal ecosystems. *Estuaries*, **25**, 886–900.

Bruland, K. W., Rue, E. L., Smith, G. J., and Ditullio, G. R., 2005. Iron, macronutrients and diatom blooms in the Peru upwelling regime: brown and blue waters of Peru. *Marine Chemistry*, **93**, 81–103.

Chen, C. T. A., Liu, K. K., and Macdonald, R., 2003. Continental margin exchanges. In Fasham, M. J. R. (ed.), *Ocean Biogeochemistry*. Berlin: Springer, pp. 53–97.

Dugdale, R. C., and Goering, J. J., 1967. Uptake of new and regenerated forms of nitrogen in primary productivity. *Limnology and Oceanography*, **12**, 196–206.

Fasham, M. J. R., Balino, B. M., and Bowles, M. C., 2001. A new vision of ocean biogeochemistry after a decade of the Joint Global Ocean Flux Study (JGOFS). *AMBIO Special Report*, **10**, 4–31.

GEOHAB, 2001. *Global Ecology and Oceanography of Harmful Algal Blooms, Science Plan*. Baltimore and Paris: SCOR and IOC, 87 pp.

GEOTRACES, 2006. *GEOTRACES Science Plan*. Baltimore, MD: Scientific Committee on Oceanic Research (SCOR). 79 pp.

GLOBEC, 1997. *Global Ocean Ecosystem Dynamics Science Plan*. IGBP Report No. 40. Stockholm: IGBP Secretariat, 83 pp.

Hoegh-Guldberg, O., Mumby, P. J., Hooten, A. J., Steneck, R. S., Greenfield, P., Gomez, E., Harvell, C. D., Sale, P. F., Edwards, A. J., Caldeira, K., Knowlton, N., Eakin, C. M., Iglesias-Prieto, R., Muthiga, N., Bradbury, R. H., Dubi, A., and Hatziolos, M. E., 2007. Coral reefs under rapid climate change and ocean acidification. *Science*, **318**, 1737–1742.

Hopkinson, C. S., and Vallino, J. J., 2005. Efficient export of carbon to the deep ocean through dissolved organic matter. *Nature*, **433**, 142–145.

IMBER, 2005. *IMBER Science Plan and Implementation Strategy*. IGBP Report No. 52. Stockholm: IGBP Secretariat, 76 pp.

IMBER, 2010. *Supplement to the IMBER Science Plan and Implementation Strategy*. IGBP report no. 52A. Stockholm: IGBP Secretariat, 36 pp.

LOICZ, 2005. *LOICZ Science Plan and Implementation Strategy*. IGBP report 51/IHDP report no. 18. Stockholm: IGBP Secretariat, 60 pp.

Morel, F. M. M., Milligan, A. J., and Saito, M. A., 2003. Marine bioinorganic chemistry: the role of trace metals in the oceanic cycles of major nutrients. In Elderfield, H. (ed.), *The Oceans and Marine Geochemistry, Treatise on Geochemistry*. Oxford: Elsevier, pp. 113–143.

Naqvi, S. W. A., Bange, H. W., Farias, L., Monterio, P. M. S., Scranton, M. I., and Zhang, J., 2010. Marine hypoxia/anoxia as a source of CH_4 and N_2O. *Biogeosciences*, **7**, 2159–2190.

Redfield, A. C., Ketchum, B. H., and Richards, F. A., 1963. The influence of organisms on the composition of sea-water. In Hill, M. N. (ed.), *The Sea*. New York: Interscience, pp. 26–77.

Samiento, J. L., and Gruber, N., 2006. *Ocean Biogeochemical Dynamics*. Princeton: Princeton University Press.

Syvitski, J. P. M., Vorosmarty, C., Kettner, A. J., and Green, P., 2005. Impacts of humans on the flux of terrestrial sediment to the global coastal ocean. *Science*, **308**, 376–380.

Thomas, H., Bozec, Y., Elkalay, K., and de Baar, H. J. W., 2004. Enhanced open ocean storage of CO_2 from shelf sea pumping. *Science*, **304**, 1005–1008.

Zhang, J., Liu, S. M., Ren, J. L., Wu, Y., and Zhang, G. L., 2007. Nutrient gradients from the eutrophic Changjiang (Yangtze River) Estuary to the oligotrophic Kuroshio waters and re-evaluation of budgets for the East China Sea. *Progress in Oceanography*, **74**, 449–478.

Cross-references

Coasts
Diatoms
Dinoflagellates
Export Production
Ocean Acidification
Ocean Margin Systems
Shelf
Upwelling

COASTAL ENGINEERING

Louise Wallendorf
Hydromechanics Laboratory, United States Naval Academy, Annapolis, MD, USA

Definition

Coastal Engineering is the civil engineering design of structures to protect or restore a shoreline from the effects of erosion.

Introduction

Coastal engineering was formally recognized as a specialty of civil engineering with the birth of the International Conference of Coastal Engineering in 1950. Informally man has been building structures at the border of the land and sea for centuries; the history and heritage of coastal engineering through 1990 is documented for 15 countries in the world in the hope that the evolution of knowledge can be transferred to future generations (Kraus, 1996). The world population has grown from 2.6 billion in 1950 and to over seven billion in 2014 (United Nations, 2010; NOAA, 2013). In addition, populations are concentrated in coastal watersheds; for example, more than 40 % of cities with populations over 500,000 are located on the world's coastlines (NOAA, 2013). The increased population density and associated shoreline development are interfering with natural coastal processes and ecology.

Coastal engineers are now incorporating design elements that both prevent coastal erosion and create a more natural environment sensitive to the site. Coastal engineers work with biologists, geologists, coastal managers, lawyers, oceanographers, and the local population to create designs.

In order to develop a design to protect infrastructure or restore an eroding shoreline, a coastal engineer must consider the site environment, available tools, customers, and regulations.

Site environment

Before developing a design the project site is investigated. Site characteristics of interest to a coastal engineer are:

- Sediment type (rock, gravel, sand, clay, or muddy soils) and distribution of particle sizes and unit weights are determined. Ideally core samples are taken along the cross section of the shoreline. Sediments are classified by the distribution of particle diameters; clays and muddy soils are treated differently than sand or gravel as their properties are affected by water (Healy et al., 2000).
- Shoreline/beach equilibrium profile – what is the natural profile of the shoreline and are there historical records showing how it has changed over time? Shorelines can be surveyed by LIDAR, aerial photography, and field surveys to determine present conditions.
- Site topography and bathymetry – geographic maps and charts provide general information. Surveys of the land and shallow water areas can show the detailed topography of specific shoreline areas especially when the site is changing rapidly.
- Sediment transport rates and direction – erosion and accretion of shorelines in the site area, predominant wave direction, and currents along with numerical models of the site can contribute to sediment transport rate estimates.
- Coastal inlets – the water flow in and out of inlets to marinas/ports and inland waterways for the coastal population can severely affect the natural sediment processes in an area, and special considerations must be used in their design or presence near a project site (Kraus, 2005)
- Marine habitat – a site can have both native and invasive species of flora and fauna. How can the design enhance the local habitat?
- Hydrodynamics – waves, ship wakes, currents, and low-frequency motions: tides, seiches, and storm surge – regional records of the magnitude, and frequency of atmospheric conditions, winds, waves, and tides exist internationally. Coastal engineers can estimate winds, wave, tide, seiche, and storm water levels from available site data. A major consideration is the wind fetch or maximum distance that the wind blows over the body of water and the angle of incidence to the shoreline. Of particular concern is the prediction of wave run-up and wave overtopping on the shoreline (Jones et al., 2005). Wave run-up is the elevation above the still water on a beach or structural slope due to tide, surge, and wave setup. Wave overtopping is the volumetric rate at which run-up flows over the top or crest of a slope of a beach, dune, or structure. Wave diffraction, the bending of wave fronts around the ends of coastal structures, must be considered.
- Weather and extreme events – hurricanes/cyclones, tropical storms, earthquakes, and tsunamis – what are the natural hazards in the site area? With increased coastal populations, hazard maps have been developed to show seismic, extreme wind and flooding and wave inundation areas. Has a risk assessment been performed for the site evaluating the vulnerability of the population to the hazards?
- Sea-level rise/subsidence – what is the forecast rate of change of the mean sea-level? The rates of change of sea-level and its causes are continuously documented and updated by scientists (IPCC, 2014).
- Proximity/access to site – road, railway, marina, and ports – how close is the site to major infrastructure? Coastal shorelines naturally fluctuate over time. Is there a room at the site to allow for fluctuation or does the shoreline protection require a stationary profile to protect infrastructure?

Available tools

Design manuals

Design manuals exist to aid the coastal engineer in design (USACE, 2002; CIRIA, 2007). These manuals detail design guidance for all types of coastal structures. Within these manuals, design methodologies are recommended. Simple formulas and techniques based on empirical field data or laboratory model tests are recommended for preliminary design estimates. The engineer must be familiar with the assumptions made in these formulas to evaluate their suitability for a site.

Numerical tools

When there are gaps in environmental site information, numerical models have been developed to predict meteorological and oceanographic conditions, wave run-up, and overtopping. Increasingly climate change/hazard risk models are being developed for project sites. Coastal scientists and engineers must be familiar with how the models operate and the assumptions used to develop them, before applying them to an individual site design.

When field data of weather, oceanographic topography, and bathymetric conditions exists for a location, the time period of data recording should be compared to the proposed life of the structure. Trends can be forecast or extrapolated using both statistics and modeling.

Physical modeling

For large coastal construction projects, physical model testing can be performed on a scaled model of the design. Coastal model testing techniques are provided in detail (Hughes, 1993). Model tests can confirm and enhance numerical predictions. They are also used when the design is too complicated for a numerical model.

Local knowledge

What is the existing local body of knowledge from coastal protection projects near the site? What types of designs have been used over time in the area and in what ways have they succeeded or failed?

Construction resources

What is the availability and cost of both natural and man-made construction materials (soil and beach fill, stone, timber, concrete, steel)? There is a unique set of concrete construction components for shore protection (Reef balls, Core-loc units, Coastal Haven WAD® (Wave Attenuation Device), etc.) which can be used in certain site conditions with design guidance. How long will it take to obtain the materials for construction and what are the transportation costs?

Often specialized construction equipment (dredges, barges, tugs, workboats, divers) is necessary to complete a coastal construction project. Is the equipment required available locally and is there local infrastructure to get it to the site? Are their local contractors with experience with the equipment?

Increasingly coastal projects are completed with the contribution of community volunteers, particularly with landscaping and project monitoring. How can these resources be efficiently combined with the commercial contractors work?

Customer

Coastal engineering projects are initiated by a customer's desire to protect/enhance their coastal property. The customer can be an individual, business, nonprofit, or public entity. The property may be residential, recreational, commercial, or coastal infrastructure (roads, bridges, marinas, ports, railroads). The primary controls are the budget and the proposed life of project and maintenance intervals. What is the customer's permissible level of damage due to an extreme storm/seismic event? How is the project being funded? If public funds are being used, can the economic return to the community from the contribution be justified? The use, the life, the budget, and an acceptable level of risk are balanced in the final design.

Regulation

Every design project is subject to both engineering codes and permitting for zoning and environmental review. Engineers must maintain knowledge of design and construction codes. Permitting is done at the local, regional, and national levels and varies between jurisdictions. As complex as it has become, obtaining permitting for a project has become a special area of legal and policy expertise separate from the engineering. An engineer must continuously communicate and solicit input from all involved parties to insure the coastal project is built to everyone's satisfaction.

Coastal engineering designs

Coastal engineering design solutions can vary from letting nature take its course to any of the following man-made alterations to the shoreline:

Beach fill or nourishment

Beach fill or nourishment is the restoration of a natural beach with sand fill and reconstructed dunes to recreate a historical beach profile. Dunes protect the backshore of the coast in an extreme storm. Models exist to predict two-dimensional coastal erosion due to extreme storm conditions (Kriebel and Dean, 1985). Correct estimates of wave run-up and overtopping are necessary (Jones et al., 2005). Beach fill must be selected that matches the color, weight, median particle diameter, d_{50}, and distribution at the location.

Living shorelines

The concept of "living shorelines" involves the use of native vegetation and low-lying structures to provide shoreline stabilization while attempting to mimic the natural landscape and preserve the intertidal habitat at a site location (Walker et al., 2011). In low-wave energy locations or within estuaries, shoreline marshes can be constructed with toe stabilization provided by rock or timber sills with low crest elevations that provide water exchange, habitat, and protection especially while the vegetation is regrowing. Openings in the sills are provided for wildlife.

Groins

A groin is a low wall or sturdy timber barrier built out into the sea from a beach to check erosion. They are constructed perpendicular or at a slight angle to the shoreline, as shown in Figure 1a. Groins are intended to interrupt the longshore transport and deposit sand adjacent to the structure; sometimes excessive deposition can occur on the updrift and downdrift sides of the groin, so their length and porosity need to be designed carefully.

Breakwaters

Breakwaters are structures designed to provide shelter from the waves and/or manipulate the littoral transport conditions to trap sand in their lee. Breakwaters transform or reduce the height of waves in their lee. They can cause wave setup or an increase in the water level due to the waves trapped behind them. They can be composed of concrete units or layers of rock with varying weights and sizes. Breakwaters can extend at any angle to the shoreline to create a sheltered area depending on wave conditions, as shown in Figure 1b. Rock or concrete breakwaters are constructed with front and back slopes. Rock breakwaters can be composed of a single homogeneous layer of rock or layers of graded rock. A layered breakwater has a central core of finer stone to improve wave transmission. The central core is covered with a level cap and front and back slopes composed of layers of larger heavier stones to

Coastal Engineering, Figure 1 Examples of shoreline protection: (**a**) Bay Ridge Beach, Chesapeake Bay, MD, USA – timber groin and rock berm breakwater system protecting recreational beach and coastal roadway (photo: Louise Wallendorf). (**b**) Shangri-La Center for Islamic Art, Oahu, Hawaii, USA – breakwater creating small boat refuge and beach (photo: Rob Walker). (**c**) San Diego, CA, USA – community bank protection and public walkway using naturally finished concrete seawall and a rock revetment – extreme wave climate (photo: Walter Crampton, TerraCosta Consulting Group).

prevent movement of the stone in storm wave conditions. Filter cloth is often used on the base of the foundation to prevent erosion.

Berm breakwaters

Berm breakwaters are detached breakwater constructed in series parallel to the shore, either submerged or with a low elevation above the still water level. Sited at the proper distance offshore and with the correct gaps between them, they can create a salient or widening of the beach directly in their lee. When the salient gets too large, a tombolo is formed; the collected sand connects the beach to the breakwater structure. Generally this type of beach is used to grow and protect the shoreline of a recreational beach; Figure 1a shows a groin and berm breakwater system designed for this purpose. Sand collected in this type of design can serve as a reservoir of material when the beach is eroded during a storm (Dean and Dalyrmple, 2002).

Floating breakwaters

Floating breakwaters have been tested periodically in deeper water sites to provide a portable or low-cost alternate to breakwaters constructed on the seafloor. The keys to a successful floating breakwater design are the structural design of the connections between the units and the moorings.

Seawalls/revetments

Seawalls are vertical walls built to delineate the border between sea and land, in an area where the upland contains infrastructure that requires protection from storm surge and wave overtopping during an extreme storm event. Seawalls are often constructed in regions that have experienced high erosion and have limited or no beachfront. Revetments are sloped surface layers designed to protect the shoreline from erosion from waves and currents. They can be as simple as grass on packed clay, as is in river levees, or as more complex armored layers of rock or interlocking man-made elements on top of a soil slope. Figure 1c shows a seawall creatively camouflaged as a natural rock along with a rock revetment which is protecting community housing.

Self-closing flood control structures

To prevent damages to large coastal populations when high river or sea water levels can inundate large areas, self-closing flood control structures are designed. The Netherlands has extensive experience in this specialty of coastal engineering with the design and construction of the 13-segment dam and flood barrier system instituted with the Deltaworks plan in 1953. Both the size of the protected population and the community's contribution to a nation's economy must be evaluated and compared to the high cost of this type of protection (Hatheway, 2008).

Setback or relocation

When a community has experienced high damages from multiple extreme coastal events within one generation, a community can choose to relocate structures or infrastructure inland from the shoreline. Hatteras Lighthouse, a historic coastal structure subject to repeated erosion detaching it from the natural North Carolina shoreline, was moved 1.1 km inland (Tice and Knott, 2000). The downtown area of Hilo, HI, was relocated and replaced with a planned tsunami inundation area after the population experienced the tsunamis of 1946 and 1960 (Hwang, 2011).

Summary and conclusions

A good coastal engineer balances the customer's requirements to prevent coastal erosion, using sound science and practical engineering with a sensitivity to coastal ecology.

While the specific techniques used to forecast environmental conditions and develop components of a coastal design are continuously updated (US Army Corps of Engineers, 2002; CIRIA, 2007), a coastal engineer should have a flexible design philosophy balancing the known and unknown scientific data for a site with engineering analysis under an experienced hand (Dean and Dalrymple, 2002).

Bibliography

Construction Industry Research & Information Association (CIRIA), CUR, and CETMEF, 2007. *The Rock Manual. The Use of Rock in Hydraulic Engineering*. London: CIRIA.

Dean, R. G., and Dalrymple, R. A., 2002. *Coastal Processes with Engineering Applications*. Cambridge: Cambridge University Press.

Jones, C., Broker, I., Coulton, K., Gangai, J., Hatheway, D., Lowe, J., Noble, R., and Srinivas, R., 2005. *Wave Run-up and Overtopping FEMA Coastal Flood Analysis and Mapping Guidelines Focused Study Report*. Washington, DC: FEMA.

Hatheway, D.,, 2008. Assessment of international coastal flood protection levels for disaster mitigation. In *Solutions to Coastal Disasters 2008*. Reston, VA: American Society of Civil Engineers.

Healy, T., Yang, W., and Healy, J.-A. (eds.), 2000. *Muddy Coasts of the World: Processes, Deposits and Function*. Amsterdam: Elsevier Science B.V.

Hughes, S. A., 1993. *Physical Models and Laboratory Techniques in Coastal Engineering*. Singapore: World Scientific.

Hwang, D., 2011. Disaster recovery planning: lessons learned from past events. In *Proceedings, Solutions to Coastal Disasters 2011*. Reston, VA: American Society of Civil Engineers.

Intergovernmental Panel on Climate Change (IPCC), 2014. *Assessment Report on Climate Change 2014 Synthesis (AR5)*. Geneva: IPCC.

Kraus, N. C. (ed.), 1996. *History and Heritage of Coastal Engineering*. Reston, VA: American Society of Civil Engineers.

Kraus, N. C., 2005. Coastal inlet functional design: anticipating morphologic response. In *Coastal Dynamics*. Reston, VA: American Society of Civil Engineers.

Kriebel, D. L., and Dean, R. G., 1985. Numerical simulation of time-dependent beach and dune erosion. *Coastal Engineering*, **9**(3), 221–245. Reston, VA: American Society of Civil Engineers.

NOAA, U.S. Census Bureau, 2013. *National Coastal Population Report*. Washington, DC: National Oceanographic and Atmospheric Association (NOAA).

Tice, J., and Knott, R., 2000. Relocation of the Cape Hatteras Light station: move route design and construction. In *Soil-Cement and Other Construction Practices in Geotechnical Engineering*. Reston, VA: American Society of Civil Engineers, pp. 51–66.

United Nations Population Division, 2010. *World Population Project*. New York: Department of Economic & Social Affairs.

U.S. Army Corps of Engineers, (USACE). 2002. *Coastal Engineering Manual*. Engineer Manual 1110-2-1100. Washington, DC: U.S. Army Corps of Engineers. (in 6 volumes).

Walker, R., Bendell, B., and Wallendorf, L., 2011. Defining engineering guidance for living shoreline projects. In *Coastal Engineering Practice*. Reston, VA: American Society of Civil Engineers, pp. 1064–1077.

Cross-references

Beach Processes
Coasts
Engineered Coasts
Sea Walls/Revetments

COASTS

Charles W. Finkl
Department of Geosciences, Florida Atlantic University, Boca Raton, FL, USA
The Coastal Education and Research Foundation, Coconut Creek, FL, USA

Synonyms

Bank; Beach; Coastline; Coastal area; Coastal zone; Littoral; Margin; Seaboard; Seacoast; Seashore; Seaside; Shore; Shoreline; Strand

Definition

When used as a noun, the coast is part of the land that is next to or adjoining the shore, a sea, or a large body of water such as an inland sea or lake. The definition of coast depends on the jurisdiction or intended use. When used as "the Coast," it refers to a specific geographic coastal segment such as the Pacific coast of North America or the Dutch coast of the Netherlands or even more restrictively as the Coromandel Coast of southeastern India.

The term *coast* resides in the vernacular as a generic reference to the strip of land adjacent to large water bodies, most commonly the sea. While the term covers a broad sweep between land and water, it at the same time becomes more complicated and convoluted when applied in adjectival form as *coastal* (pertaining to the coast) or conjoined with the suffix *line*, making the term *coastline*, or preceding a noun as in *coastal zone* or *coastal area* (see below). Associated with coast, coastal, and coastline are various nuances, inferences, and interdigitations that come into play depending on the use by geoscientists, engineers, surveyors, lawyers, and resource managers, among others. Some of the legion applications of coast terms and salient types of coasts are indicated in what follows.

Origins of the term, variable meanings, and related usage

The term *coast* is a general expression that, in its simplest sense, refers to the contact between land and water. The origin of the term stems from Middle English *cost*, from Anglo-French *coste*, and from Latin *costa* rib, side, akin to Old Church Slavic *kostĭ* bone, with the first known use in the fourteenth century (Merriam-Webster, 1994). Defined as the land near a shore, it is implied to be more or less equivalent to seashore. Other definitions say "...a strip of land that meets the sea or some other large expanse of water... or which extends from the low tide line inland to the first major change in landform features..." (e.g., Jackson, 1997; Oertel, 2005) or, according to the FAO, "The geographical area of contact between the terrestrial and marine environments, a boundary area of indefinite width, appreciably wider than the shore" (Cullinan, 2006).

A slightly more rigorous definition of the term *coast* involves a relationship with seashore, viz., "A strip of land of indefinite length and width (may be tens of kilometers) that extends from the seashore inland to the first major change in terrain features" (Voigt, 1998). The vernacular related to the concept of coast is often somewhat confusing because the terms are indefinite, imprecisely defined, or partially overlap. In the example abstracted from Jackson (1997), the term *shore* is used to help constrain the definition of coastline, viz., technically the line that forms the boundary between the coast and the shore, the shore being the most seaward part of the coast, but further begs the question regarding the difference between coastline and shoreline. Coast, seaside, coastline, and seashore are all words for the land beside or near to the sea, a river, or a lake.

Although usually associated with the marine realm of oceans, seas, gulfs, and fjords, for example, the term may also be applied to large inland freshwater lakes (e.g., Great Lakes in North America, Lake Baikal in southern Siberia, Lake Tanganyika [an African Great Lake], Lake Balaton in the Transdanubian region of Hungary) and to saline or hypersaline bodies of water such as the Dead Sea in the Middle East, Great Salt Lake in the western United States, the Aral Sea between Kazakhstan and Uzbekistan, and so on. The term coast is also often used in a very general way to refer to some specific geographic region, physiographic feature, or anthropogenic influence as, for example, appreciated by the terms Barbary Coast, Coast Mountains (Canada), Coromandel Coast, Gold Coast, Ivory Coast, rocky coast, reef coast, muddy coast, mountain coast, developed coast, Engineered Coasts (q.v.), and so on.

Coastal zone

The coastal zone is a geographic entity including both terrestrial and submerged areas of the coast, defined legally or administratively for coastal zone management (Cullinan, 2006). According to many definitions, the coastal zone includes parts of landmasses that are under the direct influence of maritime conditions where the weather, wildlife, and soils are obviously different from those of noncoastal areas (Clark, 1996). The coastal zone, in general, is usually obvious to decipher but when freedoms are restricted and legal issues are involved, an administrative definition is needed, and this is where things start to get complicated. Typically, the extent of coastal waters is defined first, and then a strip of shoreland is added to interlock land and sea elements (Clark, 1996). Nicholls and Small (2002), for example, also consider coastal populations and define the coastal zone as the land margin within 100 km of the coastline or less than 100 m above mean low tide, whichever comes first.

The coastal zone can thus be defined in different ways. In most cases, some combination of distance from the coast and elevation is used. The Millennium Ecosystem Assessment (MEA, 2005), for example, used 100 km from the coast as a distance threshold and 50 m as an elevation limit, depending on which was closer to the sea. McGranahan, Balk, and Anderson (2007), on the other hand, used 10 m elevation contiguous with the coast with no distance threshold; this procedure mostly delineated areas closer than 100 km from the sea, but some areas extended farther inland to define the coastal zone.

Coastal area

From an engineering point of view, the term *coastal area* refers to areas of land and sea bordering the shoreline and extending seaward through the breaker zone (CERC, 1966). The coastal area is also a technical term for a geographic entity of land and water affected by the biological and physical processes of both terrestrial and marine environments. In this sense it is broadly defined for the purpose of natural resources management (Cullinan, 2006). It should be noted that coastal area boundaries change over time without regard to enabling legislation.

Determination of coastline length: the coastline paradox

Application of the term *coast* is conditioned by scale, purpose of the designation, and the audience or user. The so-called coastline paradox states that a coastline does not have a well-defined length because its measurement is scale dependent. Measurements of the length of a coastline behave like a fractal, being different at varying scale intervals. The smaller the scale interval (the shorter in length, the more detailed the measurement), the longer the coastline. This "magnifying" effect occurs because it takes many more shorter scalar lengths to measure the coastline, and consequently, coastline length becomes greater for convoluted coastlines than for relatively smooth ones. Convoluted coastlines tend to obtain for rocky coasts, and determination of coastline length in these situations can be especially problematic. One solution, as proposed by Spagnolo et al. (2008), is the *indentation index*, which is defined as the ratio between the real length of a coast and its Euclidean length. Application of this method involves digitizing the coastline in a GIS platform to test several Euclidean length values on the same coastline, thereby obtaining a different spatial variability of the indentation index with each trial. The best length values that maximize the spatial variability of the indentation index determine an indentation index pattern of high variance and low spatial autocorrelation. This procedure has applicability to some rocky coasts of limited extent but is limited by the scale of operation in a GIS where it would not be practical to determine world rocky coast lengths.

The ten countries with the longest coastlines, regardless of specific coastal types, are listed in Table 1. The data were derived from the World Resources Institute, based on data calculated in 2000 from the World Vector Shoreline, United States Defense Mapping Agency, 1989 (Soluri and Woodson, 1990). The coastline length was retrieved from the World Vector Shoreline database at a scale of 1:250,000. The estimates were calculated in a Geographic Information System (GIS) platform using an underlying database consistent for the entire world. Coastline length was estimated based on the following criteria: (1) use of individual lines that have two or more vertices and/or nodes, (2) length between two vertices was calculated on the surface of a sphere, (3) the lengths of pairs of vertices were summed and aggregated for each individual line, and (4) the lengths of individual lines were summed for each country. Long coastline lengths thus accrued where landmasses were large (i.e., Canada, United States, Russia, and Australia) and for countries that

Coasts, Table 1 Top ten countries with the longest coastline lengths in the world, as abstracted from http://en.wikipedia.org/wiki/List_of_countries_by_length_of_coastline. Coastline lengths were determined at a nominal scale of 1:250,000 as described above

1	Canada	265,523
2	United States	133,312
3	Russia	110,310
4	Indonesia	95,181
5	Chile	78,563
6	Australia	66,530
7	Norway	53,199
8	Philippines	33,900
9	Brazil	33,379
10	Finland	31,119

had convoluted coastlines formed by many islands, peninsular headlands, gulfs, estuaries, and fjords (i.e., Canada [British Columbia, Nunavut, Newfoundland and Labrador], United States [Alaska], Indonesia, Chile, Australia, Norway, Philippines, Brazil, and Finland). Discrepancies in coastline lengths can be illustrated for Australia (number 6 in Table 1), the only one of the top six completely surrounded by water. The coastline length abstracted from Geoscience Australia's GEODATA Coast 100 K 2004 database (2010), for example, gives a total of 59,736 km. Compared to the total length given in Table 1 (66,530 km), there is a difference of 6,794 km, being accounted for by exclusion of thousands of small fringing islands, many of which occur around Tasmania, among other factors. The nature of the measuring conundrum has been reviewed by Galloway and Bahr (1979) for the Australian coastline and it should be noted that this same problem also exists for other countries, where coastline length may be indeterminate but varies consistently with the accuracy of measurement. Galloway and Bahr (1979) illustrate the magnitude of the coastline paradox by the fact that the total length for the mainland, including islands greater than 12 ha, ranges from 24,330 km when measured on 1:2,500,000 maps with a divider intercept of 100–69,630 km when measured on the same 1:250,000 maps with an intercept of 0.1 km. Coastline lengths should thus be taken with a grain of salt, and the reader should be cognizant of the means by which length was measured.

Classification of coasts: types of coasts

To the layman, coasts are thought of in very general terms such as sandy coast, rocky cost, muddy, Mangrove Coasts (q.v.), Reef Coast (q.v.), and so on. Related terms such as *seaside* (especially British English) refer to an area that is by the sea, especially one where people go for a day or a holiday: a trip to the seaside. It is always the seaside, except when it is used before a noun: a seaside resort. The seaside is British English; in American English seaside is only used before a noun. The term coastline broadly classifies the land along a coast, especially when thinking of its shape or appearance, as, for example, in British Columbia's rugged coastline around Puget Sound. Classification of coast in terms of materials such as sand to a layman refers to a large area of sand on a beach. Although these concepts of coast are useful to laypersons, more rigorous definitions are required for technical applications where scientists and engineers need more detailed information about materials making up the coast (types of rocks and sediments), shape and configuration of landforms along the coast, natural processes operating around the coast (climate, waves, currents), tectonics, sea-level change, coastal-marine ecology, etc.

Organization of different types of coasts into a classification system thus seems simple enough, but as it turns out, the pigeon holing of variants has been elusive (Finkl, 2004). Although it is a natural tendency to try and classify coasts, no universal classification of coasts has yet been put forth. Problems thwarting attempts to classify coasts are legion, but they mostly arise from scalar considerations, the multidimensionality of coasts, and polymorphism that is associated with the concurrence of paleo- and neomorphs along the same coastal segment. The compound nature of coasts often makes it difficult to classify coast according to one type or another, as, for example, when beaches occur seaward of cliffs or as pocket beaches along rocky coasts. Percentage frequencies of occurrence of a particular coastal type may thus overlap if more than one type occurs along a coastal segment, as indicated previously where beaches may occur in front (seaward) of rocky cliffs. Estimates of worldwide coastal types thus differ, and percentage frequency of occurrence overlaps due to the compound nature of many coastal segments. Different types of coasts result from a wide range of processes (e.g., Zenkovich, 1967; Bird, 1985; Fairbridge, 1989; Short, 1999; Woodroffe, 2002) and exhibit distinctive spatial distribution patterns of landforms (e.g., Davies, 1964, 1980) that give character to coastal segments to identify them as sandy, muddy, cliffy, rocky coast, etc.

Previous investigations propose that several salient factors should be taken into account when devising a coastal classification (cf. King, 1966; Davies, 1964, 1980; Bird, 1976; Fairbridge, 1989, 1992; Finkl, 2004): (1) the shape or form (morphology) of the land surface (above and below sea-level), (2) the movement of sea-level relative to the land and vice versa (e.g., change in relative sea-level, RSL), (3) modifying effects of marine processes, (4) climatic influences on process and form, and (5) age and durability of coastal materials. The following examples illustrate the point of proposed terminologies, viz., Atlantic and Pacific type coasts (Suess, 1888); subduction, taphrogenic (rifted), and sediment-loading (alluvial) coasts (Fairbridge, 1992); coastlines of submergence and emergence (Johnson, 1919); primary and secondary coasts (Shepard, 1973); advancing and retreating coasts (Valentin, 1952); soil and unconsolidated coast forms (e.g., cliff, platform, reef); solid and unconsolidated coast materials (e.g., beach, delta, channel, flats, scree/talus) (Owens, 1994); coasts of plate boundary islands and intraplate islands (Nunn, 1994); coasts of shelf coral reefs (patch reefs, crescentic reefs, lagoonal reefs, planar reefs) (Hopley, 1988); coast of unvegetated solitary island, vegetated solitary island, multiple islands, and complex low wooded islands (Stoddart and Steers, 1978); reef coasts (fringing reefs, bank reefs, barrier reefs, atolls, ridge reefs, Bahamian reefs) (Guilcher, 1988); beach coast (dissipative, intermediate, reflective beaches) (Short, 1999); Type A coast platforms (e.g., benches (Zenkovitch, 1967), platforms (Trenhaile, 1987), wave-cut terraces (Leet and Judson, 1958), wave-cut platforms (King, 1963)) and Type B coast platforms (e.g., coastal platforms (Guilcher, 1958), marine benches (Cotton, 1963), wave-cut platforms (Short, 1982)); sea cliffs (plunging, composite) (Davis, 1898; Orme, 1962); etc.

All of a sudden the concept of coast becomes very complicated and far-removed from the realm of laypersons as geoscientific terminologies make headway into the literature. The jargon surrounding the term *coast* is necessary because the term, which at first sight seems simple enough, is actually complex. This is because the perception of coast is not only scale dependent but time dependent as well where old and new landforms are juxtaposed along many coasts. An additional classificatory complication is that coasts are dynamic and constantly changing to the point that the study of coastal (coastline) change takes on a wide purview among researchers, especially in regard to identification and evaluation of evident and potential change associated with variations in sea-level (e.g., Bird, 1985; Fairbridge, 1989; Woodroffe, 2002). New technologies associated with remote sensing capabilities now facilitate the recognition and classification of coasts in terms of natural landform features, ecosystems, and processes (e.g., Shaw et al. 2008; Wang et al., 2010; and Klemas, 2011).

Coastal landforms

Landforms and ecosystems often provide specificity to different types of coasts, and it is perhaps useful to indicate some notable examples. The modifier before the noun *coast* (e.g., cliffy coast, sandy coast, mangrove coast, skerry coast) or as a noun by itself to signify a specific kind of coast (e.g., estuary, rocky headland, spit, tombolo) helps project an image of what is being described. There is, however, great variation in scale within a single coast designation as in the example of estuarine coasts where the St. Lawrence Estuary (Canada) is the longest (1,197 km) in the world followed by Chesapeake Bay (United States) (322 km) to very small estuaries along many coasts of the world.

Rocky coasts

Whereas rocky coasts in general make up about 80 % of the world coast length (Emery and Kuhn, 1980), sea cliffs (Figure 1) are distinctive and dramatic forms of rocky coasts that make up about half of the world's coasts, but

Coasts, Figure 1 Example of rocky coast on the eastern shore of Eleuthra Island, Bahamas, showing eolianite cliffs that are undercut at present sea-level to form a notch. Note the height of these sea cliffs (about 30 m) compared to the man standing on top of the cliff (Photo: C.W. Finkl).

Coasts, Figure 2 Sea stack on the northern Oregon coast (United States), which is predominantly a rocky coast with stretches of beach interspersed between rocky headlands (photo: C.W. Finkl).

there is debate as to the location of the highest sea cliffs depending on the slope of the cliff face and actual free fall distance. The location of the world's highest sea cliffs thus depends on the definition of *cliff*. Kalaupapa, Hawaii (1,010 m high), and the north face of Mitre Peak, Milford Sound, New Zealand, which drops 1,694 meters (Seddon and Dadelszen, 1898), are extreme examples. The average slope of these cliffs is roughly 60 degrees, but a more vertical drop of 1,560 m into the sea occurs at Maujit Qaqarssuasia in the Torssuktak (Tunoq) Fjord (Sound), southeast Greenland (Beckwith, 1989). Considering a truly vertical drop, Mount Thor on Baffin Island in Arctic Canada is among the highest at about 1,370 m for the longest purely vertical drop on Earth (Simon, 2012). These spectacular sea cliffs are tourist attractions around the world for a very specialized kind of coast. Many cliffy coasts have gently sloping seaward ramps (Type A platforms) or subhorizontal surface terminating in a seaward or low-tide scarp (Type B platforms) (Trenhaile, 1987). These shore platforms (wave-cut platforms) often contain shallow pools and are subject to subaerial weathering at low tide. Lithology and structure are important to the types of rocky coasts with volcanic rocks and limestones forming some of the most spectacular and scenic coasts. Many terraces and benches are uplifted above present sea-level, while others are below the water. Sea caves, arches (bridges), and stacks (Figure 2) are additional types of erosional landforms that are characteristic of rocky coasts (Woodroffe, 2002). More subtle forms of rocky coasts are associated with drowned glacial landscapes with drumlins or other glaciated terrains with exposed bedrock to form skerries or partially submerged whalebacks.

Beach and barrier coasts

Beaches (q.v.), which are normally backed by Dunes (q.v.), make up about 70 % of the world's coasts (Bird, 1976) and are well-known types of sedimentary coast. In Australia, where Short and Farmer (2012) report no fewer than 11,761 beaches, dunes back 80 % of the beaches and about 40 % of the coast. Beach sediment is extremely variable in terms of size, shape, color, and chemical

Coasts, Figure 3 Segment of a barrier island along the Florida east coast showing the typical open ocean to mainland shore progression of coastal environments: beach, dunes, back-barrier marshland, lagoon, and mainland. An old inlet (*left* hand side of the photo) is closed off from the ocean by the beach-dune system (photo: C.W. Finkl).

(mineralogical) composition to provide diversity in the appearance of the world's beaches (Pilkey et al., 2011). There are many examples of the world's best beaches, longest beaches, widest beaches, and so on. In the United States, there are about 650 major public recreational beaches, and they have been ranked annually according to fifty criteria since 1991 to determine the top ten beaches (Leatherman, 1997). A range of selected examples of coasts with the longest beaches in the world include Brazil's Praia do Cassino Beach (over 250 km long), India's Cox's Bazar (about 240 km long), Texas' Padre Island (about 210 km long), Australia's Coorong Beach (194 km), New Zealand's Ninety Mile Beach (each about 145 km long), Mexico's Playa Novillero (about 80 km long), and America's Virginia Beach (about 55 km long). Most beaches are sandy (Figure 3), but very fine-grained (high percentage of silt and clay) beaches may occur under specialized conditions where there is an abundance of fine-grained sediments deposited by wave action, as, for example, along the Red Sea and in the vicinity of large deltas and estuaries such as the Bay of Fundy as well as along the northeast coast of South America (Anthony et al., 2010). Coarse-grained (cobble, boulder) beaches are prevalent in high latitudes and typify many coastal segments in Canada, for example, where glacial materials and bedrock are eroded by coastal processes. Cobble beaches and barriers may also occur along many high-energy coasts, as seen along the west coast of North America. Pilkey (2003) estimates that there are about 2,200 barrier coasts (Figure 3) in the world with about 12 % of all open-ocean coasts fronted by these islands. About 73 % of barrier islands occur in the Northern Hemisphere, with about 25 % of all barrier islands occurring in the United States (Pilkey, 2003).

Muddy (mangrove, deltas, estuaries, marshes, and tidal flats) coasts

Muddy coasts are defined by the SCOR Working Group 106 as "a sedimentary-morphodynamic type characterized by fine-grained sedimentary deposits – predominantly silts and clays – within a coastal sedimentary environment" (Wang and Healy, 2002). These fine-grained sedimentary coasts contain rather flat surfaces that are usually associated with intertidal flats. The term *muddy coast* encompasses a range of physiographic types, viz., tidal flats, enclosed sheltered bay deposits, estuarine drowned river valley deposits, inner deposits in barrier-enclosed lagoons, supra-tidal (storm surge) deposits, swamp marsh and wetland deposits, chenier plains, mud deposits veneering eroded shore platforms, ice-deposited mud veneer, and subtidal littoral mud deposits. The geographic distribution of muddy coasts is a complicated matter, as indicated above by the wide range of conditions under which fine-grained sediment may accumulate along the shore. The largest expanses of muddy coasts are associated with tropical mangroves (Figure 4) or temperate salt marshes (Figure 5) (Flemming, 2002). Tropical mangrove systems (see Mangrove Coast) occupy about 75 % of the world's coasts between 25°N and 25°S (Chapman, 1974).

Coasts, Figure 4 Muddy coast along the southeast Florida shore where red mangroves (*Rhizophora mangle*) grow in carbonate muds in sheltered coves shoreward of barrier islands and keys (photo: C.W. Finkl).

Coasts, Figure 5 Example of a middle latitude coastal marsh on Cape Cod, Massachusetts, showing a tidal creek in the foreground, marsh grasses in photo center, and fringing upland vegetation in the background. Smaller tidal channels may occur throughout the interior of these salt marshes (photo: C.W. Finkl).

Discussions of the geographic distributions of muddy coasts may be found in Bird and Schwartz (1985), Davies (1980), and Flemming (2002) where regional details are provided. A well-known example of a muddy coast is provided by Anthony et al. (2010) for the 1,500 km-long coast of South America between the Amazon and the Orinoco river mouths where this coast is the world's muddiest. This is due to the huge suspended-sediment discharge of the Amazon River ($10^6 \times 754$ t year^{-1} ± 9 %), part of which is transported alongshore as mudbanks.

Reef coasts

Coral reef coasts (see "Reef Coasts"), which contain some of the largest diversity of life in the world, are sometimes referred to as the rainforest of the sea. The three widely recognized coral reef regions include the Indo-Pacific Region, wider Caribbean (including the Bahamas to the north and northeastern coasts of South America), and the Red Sea. Because of restrictive requirements for the growth of coral reef ecosystems (Figure 6), coral reef coasts occur along much of the Central and South Pacific Ocean as far north as Hawaii and southern Japan. Farther

Coasts, Figure 6 Coral reefs along the eastern shore of Long Island, Bahamas, fronting the Atlantic Ocean. This eolianite coast is bordered by nearshore fringing coral reefs and a series of shore-parallel *en echelon* deeper reefs that stair step into deeper waters, each reef tract is separated by interreefal sand flats (photo: C.W. Finkl).

south, coral reef coasts occur around Taiwan, most of Southeast Asia, the islands of the Philippines, and along coasts of the northern half of the Australian continent. Coasts of Indonesia and eastern Africa also contain reefs.

Coral reefs, the largest biological constructions on Earth, occur in many different forms (e.g., barrier [or shelf reefs], fringing, atolls) making up distinctive coasts. Barrier reefs occur along continental, mainland, and island coasts and are separated (not attached) from the land by a lagoon of 10 m or more water depth. One of the largest examples is the Great Barrier Reef, which stretches more than 2,000 km along the northeastern coast of Australia (offshore Queensland) (Woodroffe, 2002). Additional examples of large barrier-type reef coasts include the Belizean Barrier Reef and the Florida Reef Tract. Barrier reefs may also occur on mid-ocean islands, viz., Moorea in French Polynesia, Tahiti, Fiji Islands, and Bahamian Archipelago. Fringing reefs (Kennedy and Woodroffe, 2002) are associated with mainland coasts where there is shallow water between the reef flat and the land or where they are partly attached to the land as seen at Ningaloo Reef along the coast of Western Australia and in the Caribbean, Red Sea, and elsewhere. Atolls contain an annular or irregular reef around a central lagoon that lies above a submerged volcanic peak.

Ice (Polar) coasts

An ice (bordered) coast (so-called cold coast or high-latitude coast) forms where a glacier extends to the sea. Perhaps the most spectacular examples occur in Antarctica where continental glaciers form a wall of ice that is in direct contact with the water. About one-third of Antarctica's coastline (i.e., 11,000 km) is occupied by ice cliffs at the ocean interface (Walker, 2005). Water-terminating glaciers also occur in the form of mountain glaciers that flow down to the sea in fjords, as seen, for example, in Greenland and in nonpolar tidewater areas such as southern Alaska and Chile as well as Ellesmere, Baffin Island, Bylot Island, and Devon Island. A major characteristic of ice coasts is presence of ice shelves, icebergs, and sea ice.

Coastal zone (management)

The term *coastal* refers to regional mosaics of habitats including intertidal habitats (mangroves, marshes, mud flats, rocky shores, sandy beaches), semi-enclosed bodies of water (estuaries, sounds, bays, fjords, fjards, gulfs, seas), benthic habitats (coral reefs, sea grass beds, kelp forests, hard and soft bottoms), and the open waters of the coastal ocean to the seaward limits of the exclusive economic zone (EEZ), i.e., from the head of the tidal waters to the outer limits of the EEZ (Christian and Mazzilli, 2007). This definition was adopted for the study of large-scale regional change for the Global Observing Systems of the United Nations. Because coastal areas (*q.v.*) are particularly sensitive to global change, sentinel ecosystems are used to evaluate terrestrial, wetland, and freshwater ecosystems along the coast.

The coastal zone or coast areas are terms that attempt to better define *coast*. The point of view here is in relation to the legalities of management. Such definitions are required to provide order to human use of the coast, which can be quite deleterious unless properly managed and constrained in this ecologically sensitive environment. The main considerations of coast in this sense relate to landward and seaward boundaries or jurisdictions, usually based on tidal lines. As pointed out by Quadros and Collier (2008), the rigorous and consistent realization of the line of intersection between a nominated tidal datum and the land mass is notoriously difficult. This difficulty in turn creates spatial uncertainty and the potential for conflict.

The location of seaward boundaries shows great variation between countries due to different sets of criteria: (1) political, legal, or administrative criteria, such as the outward limit of the national territorial sea (e.g., New Zealand, France (implicitly), Spain, Singapore), or the boundary between provincial or state jurisdiction and

Coasts, Table 2 Examples of legal seaward boundaries of the coast as defined by the coastal zone or coastal area (Based on Cullinan, 2006)

Costa Rica	Mean sea-level
India	Low tide line
Israel	500 m from mean low tide
New Zealand	12 nautical miles of the territorial sea
Sri Lanka	2 km seaward from the mean low water line
United States	Limit of the three-nautical-mile territorial sea

Coasts, Table 3 Examples of landward boundaries of coastal areas showing the variability of jurisdictional zones based on arbitrary lines and physical features (Based on Cullinan, 2006)

Costa Rica	200 m from mean high tide
Cuba	Limits for each of 6 coastal types
India	Up to 500 m from the high tide line
New Zealand	Line of the mean high water springs, or where this line crosses a river, the lesser of the point situated at 1 km upstream or the point obtained by multiplying the width of the river mouth by 5
Sri Lanka	300 m inland from the mean high water line and 2 km inland measured perpendicular to the straight baseline between the natural entrance points for rivers, streams, lagoons, or other bodies of water connected to the sea permanently or periodically
United States	Inland from the shorelines (not defined) of states "only to the extent necessary to control shorelands, the uses of which have a direct and significant impact on the coastal waters, and to control those geographical areas which are likely to be affected by or vulnerable to sea-level rise"
	Examples include
	California: variable line depending on the issues (since 1977), formerly highest elevation of nearest mountain
	Washington: 200 ft from mean high water mark for regulation purposes (planning controls extend to the inland boundaries of coastal counties)
	Florida, American Samoa, Guam, Northern Mariana Islands: entire state or territory

national jurisdiction (USA); (2) physical landmarks such as shoreline, continental shelf, tidal marks (Costa Rica and India), and depth or isobaths (e.g., China and Indonesia); and (3) a combination of arbitrary distances and physical marks such as an arbitrary oceanward distance in kilometers from a tidal mark (e.g., Brazil, Israel, Sri Lanka).

As appears from comparisons in Table 2, several national coastal area management laws take as the maximum seaward boundary the outer limit of the nation's territorial sea. This is a logical choice since under international law the coastal state exercises full sovereignty over this area, and in most countries the majority of the productive and extractive activities that have an impact on the coast take place within 12 nautical miles of the coast.

There is less uniformity with respect to the landward boundaries of coastal areas in national legislation than the seaward boundaries, as shown in Table 3. Some researchers favor extending the landward boundary of the coast inland to include areas that are influenced by the sea in an ecological sense. This procedure may include an entire watershed, as was done in California. This approach may make sense from an ecological perspective, but it is too unwieldy as a basis for controlling activities such as construction on the coast. Most ICM (Integrated Coastal Management) legislation thus prefers greater legal certainty by stipulating a specified distance inland from a tidal baseline.

Coastal management is not only important but necessary to control access, use, and exploitation of natural resources. About sixty per cent of the world's population already lives in coastal areas, while 65 per cent of cities with populations above 2.5 million are located along the world's coasts (Agenda 21 Staff, 1992). With overpopulated and overdeveloped coasts, there is an urgent need for better understanding of jurisdictions and effective management for sustainability.

Coast and ocean jurisdictions

For purposes of both international and domestic law, the boundary line dividing the land from the ocean is called the *baseline*. The seaward boundary of the exclusive economic zone (EEZ) extends 200 nautical miles from the baseline (based on United Nations Convention on the Law of the Sea, UNCLOS). The baseline is determined according to principles described in the 1958 United Nations Convention on the Territorial Sea and the Contiguous Zone and the 1982 United Nations Convention on the Law of the Sea (LOS Convention, also UNCLOS) and is normally the low water line along the coast, as marked on charts officially recognized by the coastal nation. In the United States, the definition has been further refined based on federal court decisions; the US baseline is the mean lower low water line along the coast, as shown on official US nautical charts. The baseline is drawn across river mouths and the opening of bays and along the outer points of complex coastlines. Water bodies inland of the baseline – such as bays, estuaries, rivers, and lakes – are considered "internal waters" subject to national sovereignty (U.S. Commission on Ocean Policy). There are, however, problems associated with the baseline concept due to coastal erosion along unstable coasts. By way of one example, as reported by McGlashan, Duck, and Reid (2008), under British property laws, it is possible that a small area of the upper beach (which regularly changes in shape and size on the foreshore) can be owned by adjacent landowners (technically under their control) despite being regularly inundated by the tides, which normally would be thought of as being owned by the Crown.

Summary

Although coasts account for only 10 % of Earth's land surface, they serve as home to two-thirds of the world's human population. As population density and economic activity in the coastal zone increases, pressures on coastal ecosystems increase. Among the most important pressures are habitat conversion, land cover change, pollutant loads, and introduction of invasive species. It is thus relevant and important to have clear concepts of definitions of coast that can be applied in a range of disciplines.

Bibliography

Agenda 21 Staff, 1992. *United Nations Sustainable Development*. United Nations Conference on Environment & Development, Rio de Janeirol, 3–14 June 1992, v.p.

Anthony, E. J., Gardel, A., Gratiot, N., Proisy, C., Allison, M. A., Dolique, F., and Fromard, F., 2010. The Amazon-influenced muddy coast of South America: a review of mud-bank-shoreline interactions. *Earth-Science Reviews*, **103**(2010), 99–121.

Beckwith, C. (ed.), 1989. *1989 American Alpine Journal*. Golden, CO: The American Alpine Club.

Bird, E. C. F., 1976. *Coasts*. Canberra, BC: Australian National University Press. 282 p.

Bird, E. C. F., 1985. *Coastline Changes: A Global Review*. Chichester: Wiley. 219.

Bird, E. C. F., and Schwartz, M. L. (eds.), 1985. *The World's Coastline*. New York: Van Nostrand Reinhold. 1071 p.

CERC (Coastal Engineering Research Center) Staff, 1966. *Shore Protection, Planning and Design*. Washington, DC: Government Printing Office. 580 p.

Chapman, V. J., 1974. *Salt Marshes and Salt Deserts of the World*. Lehre, Germany: Cramer. 494 p.

Christian, R. R., and Mazzilli, S., 2007. Defining the coast and sentimental ecosystems for coastal observations of global change. *Hydrobiologia*, **577**, 55–70.

Clark, J. R., 1996. *Coastal Zone Management Handbook*. Boca Raton, FL: Lewis Publishers. 928 p.

Cotton, C. A., 1963. Levels of planation of marine benches. *Zeitschrift für Geomorphologie, N.F.***7**, 97–111.

Cullinan, C., 2006. *Integrated Coastal Management Law: Establishing and Strengthening National Legal Frameworks for Integrated Coastal Management*. Rome: FAO. 262 p.

Davies, J. L., 1964. A morphogenic approach to world shorelines. *Zeitschrift für Geomorphologie (Mortensen Sonderheft)*, **8**, 127–142.

Davies, J. L., 1980. *Geographical Variation in Coastal Development*. London: Longman. 204 p.

Davis, W. M., 1898. *Physical Geography* (assisted by W.H. Snyder). Boston: Ginn and Co., 428 p.

Emery, K. O., and Kuhn, G. G., 1980. Erosion of rock coasts at La Jolla, California. *Marine Geology*, **37**, 197–208.

Fairbridge, R. W., 1989. Crescendo events in sea-level changes. *Journal of Coastal Research*, **5**(1), i–vi.

Fairbridge, R. W., 1992. Holocene marine coastal evolution of the United States. In Fletcher, C. H., III, and Wehmiller, J. F. (eds.), *Quaternary Coasts of the United States: Marine and Lacustrine Systems*. Tulsa: SEPM (Society for Sedimentary Geology). SEPM Special Publication No. 4, pp. 9–20.

Finkl, C. W., 2004. Coastal classification: systematic approaches to consider in the development of a comprehensive scheme. *Journal of Coastal Research*, **20**(1), 166–213.

Flemming, B. W., 2002. Geographic distribution of muddy coasts. In Healy, T., and Wang, Y. (eds.), *Muddy Coasts of the World: Processes, Deposits and Function*. Amsterdam: Elsevier, pp. 99–201.

Galloway, R. W., and Bahr, M. E., 1979. What is the length of the Australian coast? *Australian Geographer*, **14**, 244–247.

Guilcher, A., 1958. *Coastal and Submarine Morphology*. London: Methuen. 274 p.

Guilcher, A., 1988. *Coral Reef Geomorphology*. New York: Wiley. 228 p.

Hopley, D., 1988. *The Geomorphology of the Great Barrier Reef*. New York: Wiley. 453 p.

Jackson, J. A. (ed.), 1997. *Glossary of Geology*. Alexandria, VI: American Geological Institute. 769 p.

Johnson, D. W., 1919. *Shore Processes and Shoreline Development*. New York: Wiley. 584 p.

Kennedy, D. M., and Woodroffe, C. D., 2002. Fringing reef growth and morphology: a review. *Earth-Science Reviews*, **57**, 255–277.

King, C. A. M., 1963. Some problems concerning marine planation and the formation of erosion surfaces. *Transactions Papers Institute British Geographers*, **33**, 29–43.

King, C. A. M., 1966. *Beaches and Coasts*. London: Arnold. 403 p.

Klemas, V., 2011. Remote sensing technologies for studying coastal ecosystems: an overview. *Journal of Coastal Research*, **27**(1), 2–17.

Leatherman, S. P., 1997. Beach ratings: a methodological approach. *Journal of Coastal Research*, **13**, 253–258.

Leet, L. D., and Judson, S., 1958. *Physical Geology*. Englewood Cliffs, NJ: Prentice Hall. 502 p.

McGlashan, D. J., Duck, R. W., and Reid, C. T., 2008. Unstable boundaries on a cliffed coast: geomorphology and British Laws. *Journal of Coastal Research*, **24**(1A), 181–188.

McGranahan, G., Balk, D., and Anderson, B., 2007. The rising tide: assessing the risks of climate change and human settlements in low elevation coastal zones. *Environment and Urbanization*, **19**(1), 17–37.

MEA (Millennium Ecosystem Assessment), 2005. *Ecosystems and Human Well-Being: Biodiversity Synthesis*. Washington, DC: World Resources Institute. 86 p.

Merriam-Webster, 1994. *Merriam-Webster's Dictionary of English Usage*. Springfield, MA: Merriam-Webster. 989 p.

Nicholls, R. J., and Small, C., 2002. Improved estimates of coastal population and exposure to hazards. *EOS, Transactions American Geophysical Union*, **83**, 301.

Nunn, P. D., 1994. *Oceanic Islands*. Oxford: Blackwell. 413 p.

Oertel, G. F., 2005. Coasts, coastlines, shores, and shorelines. In Schwartz, M. L. (ed.), *The Encyclopedia of Coastal Science*. Dordrecht, The Netherlands: Springer, pp. 323–327.

Orme, A. R., 1962. Abandoned and composite sea cliffs in Britain and Ireland. *Irish Geography*, **4**, 279–291.

Owens, E. H., 1994. *Canadian Coastal Environments, Shoreline Processes, and Oil Spill Cleanup*. Ottawa, ON: Environment Canada, Environmental Emergency Branch. Report EPS 3/SP/5, 328 p.

Pilkey, O. H., 2003. *A Celebration of the World's Barrier Islands*. New York: Columbia University. 309 p.

Pilkey, O. H., Neal, W. J., Kelley, J. T., and Cooper, A. G., 2011. *The World's Beaches*. Berkeley, CA: University of California Press. 283 p.

Quadros, N. D., and Collier, P. A., 2008. A new approach to delineating the littoral zone for an Australian marine cadastre. *Journal of Coastal Research*, **24**(4), 780–789.

Seddon, R. J., and von Dadelszen, E. J., 1898. *The New Zealand Official Year-Book 1989*. Wellington, New Zealand: John Mackey, Government Printer. 684 p.

Shaw, J. B., Wolinsky, M. A., Paola, C., and Voller, V. R., 2008. An image-based method for shoreline mapping on complex coasts. *Geophysical Research Letters*, **35**, 12.

Shepard, F. P., 1973. *Submarine Geology*. New York: Harper and Row. 519 p.

Short, A. D., 1982. Wave-cut bench and wave-cut platform; Erosion ramp, wave ramp. In Schwartz, M. L., (ed.), *The Encyclopedia of Beaches and Coastal Environments*. Stroudsburg: Hutchinson and Ross, pp. 856–857; 393.

Short, A. D., 1999. *Handbook of Beach and Shoreface Morphodynamics*. Chichester: Wiley. 379 p.

Short, A. D., and Farmer, B., 2012. *101 Best Australian Beaches*. Sydney, Australia: NewSouth Publishing. 222 p.

Simon, S., 2012. *Seymour Simon's Extreme Earth Records*. San Francisco, CA: Chronic Books. 56 p.

Soluri, E. A., and Woodson, V. A., 1990. World vector shoreline. *International Hydrographic Review*, **LXVII**(1), 27–36.

Spagnolo, M., Llopis, A., Pappalardo, M., and Feferici, P. R., 2008. A new approach for the study of the coast indentation index. *Journal of Coastal Research*, **24**(6), 1459–1468.

Stoddart, D. R., and Steers, J. A., 1978. The nature and origin of coral reef islands. In Jones, O. A., and Endean, R., (eds.), *Biology and Geology of Coral Reefs*. Geology II, Vol. IV, New York: Academic Press, pp. 59–105.

Suess, E., 1888. *The Face of the Earth (Das Antlitz der Erde)*, Vol. 2. London: Oxford University Press (English translation 1906 by Sollas, H. B.), 5 Vols.

Trenhaile, A. S., 1987. *The Geomorphology of Rock Coasts*. Oxford: Clarendon Press. 384 p.

Valentin, H., 1952. *Die Küsten der Erde. Petermanns Geographisches Mitteilungen Ergänzsungheft*, 246, 118 p.

Voigt, B., 1998. *Glossary of Coastal Terminology*. Olympia, WA: Washington Department of Ecology, Publication, pp. 98–105 (Modified 2012).

Walker, H. J., 2005. Ice-bordered coasts. In Schwartz, M. L. (ed.), *The Encyclopedia of Coastal Science*. Dordrecht, The Netherlands: Springer, pp. 542–545.

Wang, Y., and Healy, T., 2002. Definition, properties, and classification of muddy coasts. In Healy, T., Wang, Y., and Healy, J.-A. (eds.), *Muddy Coasts of the World: Processes, Deposits and Function*. Amsterdam: Elsevier, pp. 9–18.

Wang, C., Zhang, J., and Ma, Y., 2010. Coastline interpretation from multispectral remote sensing images using an association rule algorithm. *International Journal of Remote Sensing*, **31**(24), 6409.

Woodroffe, C. D., 2002. *Coasts: Form, Process and Evolution*. Cambridge, UK: Cambridge University Press. 623 p.

Zenkovich, V. P., 1967. *Processes of Coastal Development*. Edinburgh: Oliver and Boyd. 73 p.

Cross-references

Beach
Beach Processes
Coastal Engineering
Engineered Coasts
Reef Coasts

COBALT-RICH MANGANESE CRUSTS

James R. Hein
U.S. Geological Survey, Santa Cruz, CA, USA

Synonyms

Cobalt crusts; Cobalt-rich crusts; Cobalt-rich ferromanganese crusts; Fe–Mn crusts; Ferromanganese crusts; Iron–manganese crusts; Polymetallic crusts

Definition

Layered manganese oxide and iron oxyhydroxide (ferromanganese) deposits formed from direct and very slow precipitation of minerals from seawater onto hard substrates; crusts contain minor but significant concentrations of cobalt, titanium, nickel, platinum, molybdenum, tellurium, zirconium, and other metallic and rare-earth elements sorbed from seawater by the particulate manganese and iron minerals.

Introduction

The existence of iron (Fe)–manganese (Mn) crusts (Figure 1) has been known for more than a century; however, during early explorations, crusts were not distinguished from manganese nodules. It was not until the early 1980s that a research cruise was dedicated to the investigation of Fe–Mn crusts and their resource potential for cobalt (Halbach et al., 1982). From then until the end of the twentieth century, many research cruises were undertaken to study crusts; Hein et al. (2000) summarized much of that work. Although the initial attraction for crusts was their high cobalt (Co) content, they have also been considered to be a potential ore for nickel (Ni) and Mn. Subsequent work has shown that crusts host large quantities of many critical metals needed for high-tech, green-tech, and energy applications (Hein et al., 2013; Hein and Koschinsky, 2014). In addition, because crusts precipitate from seawater layer by layer over millions of years, and therefore represent condensed stratigraphic sections, they have been used to study changes in the global ocean and climates that occurred over the past 70 Ma (e.g., Frank et al., 1999).

Occurrence and distribution

Fe–Mn crusts form pavements on rock surfaces and coat talus (pebbles and cobbles) throughout the global ocean (from Antarctica to the North Pole) where the seabed was kept clear of sediment for millions of years. The crusts obtain thicknesses up to 260 millimeter (mm) and grow (accrete) at rates of generally 1–5 mm per million years. The maximum thickness of crusts on seamounts increases with increasing age of the seamount. Fe–Mn crusts form at water depths of 400–7,000 meters (m) on the flanks of extinct submarine volcanoes (seamounts, guyots), ridges, and plateaus, although crusts of more economic interest occur at depths of about 800–2,500 m because Ni, Co, and Mo contents are higher and crusts are thicker. More than 100,000 seamounts occur in the global ocean, many within the Exclusive Economic Zone (EEZ) of Pacific Island nations (Figure 2). The same volcanism that formed the islands or volcanic edifices below carbonate islands and atolls also formed the submerged seamounts on which crusts grow. Abundant seamounts also occur in many areas beyond national jurisdictions (Figure 2). Fewer seamounts occur in the Atlantic and Indian Oceans, and Fe–Mn crusts there are commonly associated with oceanic spreading centers. Those crusts commonly obtain

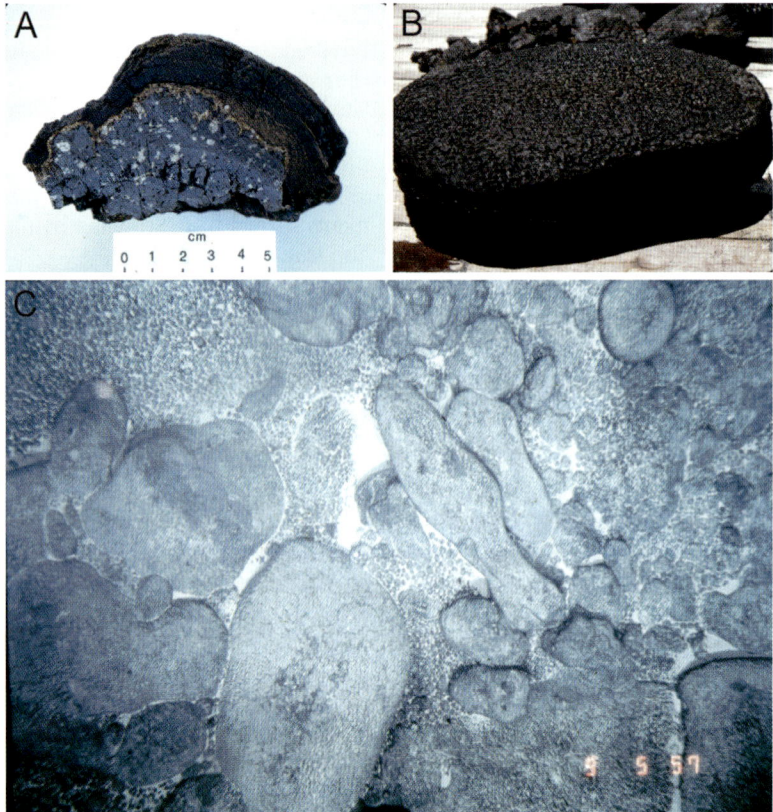

Cobalt-rich Manganese Crusts, Figure 1 Fe–Mn crust photographs: (a) cross section of a Fe–Mn crust showing growth layers and substrate basalt, from Gorda Ridge, NE Pacific, 1,512 m water depth. (b) Fe–Mn crust collected from a seamount in the Marshall Islands EEZ, NW Pacific, water depth 1,780 m; long dimension is about 1 m. (c) Seafloor rocks coated by Fe–Mn crust, Horizon Seamount, Johnston Island EEZ, central Pacific, 2,000 m water depth; about 3×4 m of seabed is displayed.

a contribution of Fe or Mn, mostly to the lowermost layers, from hydrothermal sources associated with the spreading centers; this hydrothermal contribution can dilute the metals in crusts of economic interest. Fe–Mn crusts of economic interest occur distant from hydrothermal sources and from continental margins and thus are not associated with direct volcanic, hydrothermal, or river input. Based on these criteria, coupled with the oldest seamounts occurring in the equatorial NW Pacific, the area of greatest economic interest occurs in the equatorial central-NW Pacific (Figure 2), which is called the prime crust zone (PCZ; Hein et al., 2009).

Formation

Mn oxide ($MnO_2 \cdot xH_2O$) and Fe oxyhydroxide (FeO(OH)) are the dominant minerals that compose Fe–Mn crusts, which precipitate from cold seawater onto rock surfaces, called hydrogenetic precipitation. Seawater is the source of Mn, Fe, and all the metals associated with those main components (Table 1). Elements dissolved in seawater are derived predominantly from the continents via rivers and windblown (eolian) dust. The mineral particles delivered to the ocean by those sources dissolve to various degrees in seawater, and the released elements form ionic and element complexes in seawater. Another source of elements to seawater is hydrothermal venting that occurs along the 89,000 kilometers (km) of oceanic spreading centers and volcanic arcs. However, the continents are the dominant source for all the elements dissolved in seawater except Mn and Fe, which come from hydrothermal input at oceanic spreading centers. Once the elements are dissolved in seawater, a first-order electrochemical model indicates that positively charged ions will sorb onto the surface of the Mn oxide, which has a strong negative surface charge at seawater pH. In contrast, the negative and neutral element complexes in seawater will sorb onto the slightly positively charged surface of Fe oxyhydroxide (Koschinsky and Hein, 2003). Elements may be weakly sorbed, called outer-sphere sorption, or strongly sorbed, called inner-sphere sorption, where covalent bonds usually form that tightly hold the metals to the main Fe and Mn minerals. Many of the metals that are highly enriched in Fe–Mn crusts are oxidized from a mobile state to an immobile state on the surface of the Fe and Mn minerals after inner-sphere absorption (Hein and Koschinsky, 2014). For example, cobalt (Co^{2+}), cerium (Ce^{3+}), and thallium (Tl^+) oxidize to Co^{3+}, Ce^{4+}, and Tl^{3+}, respectively, on the surface of the Mn oxide, and tellurium

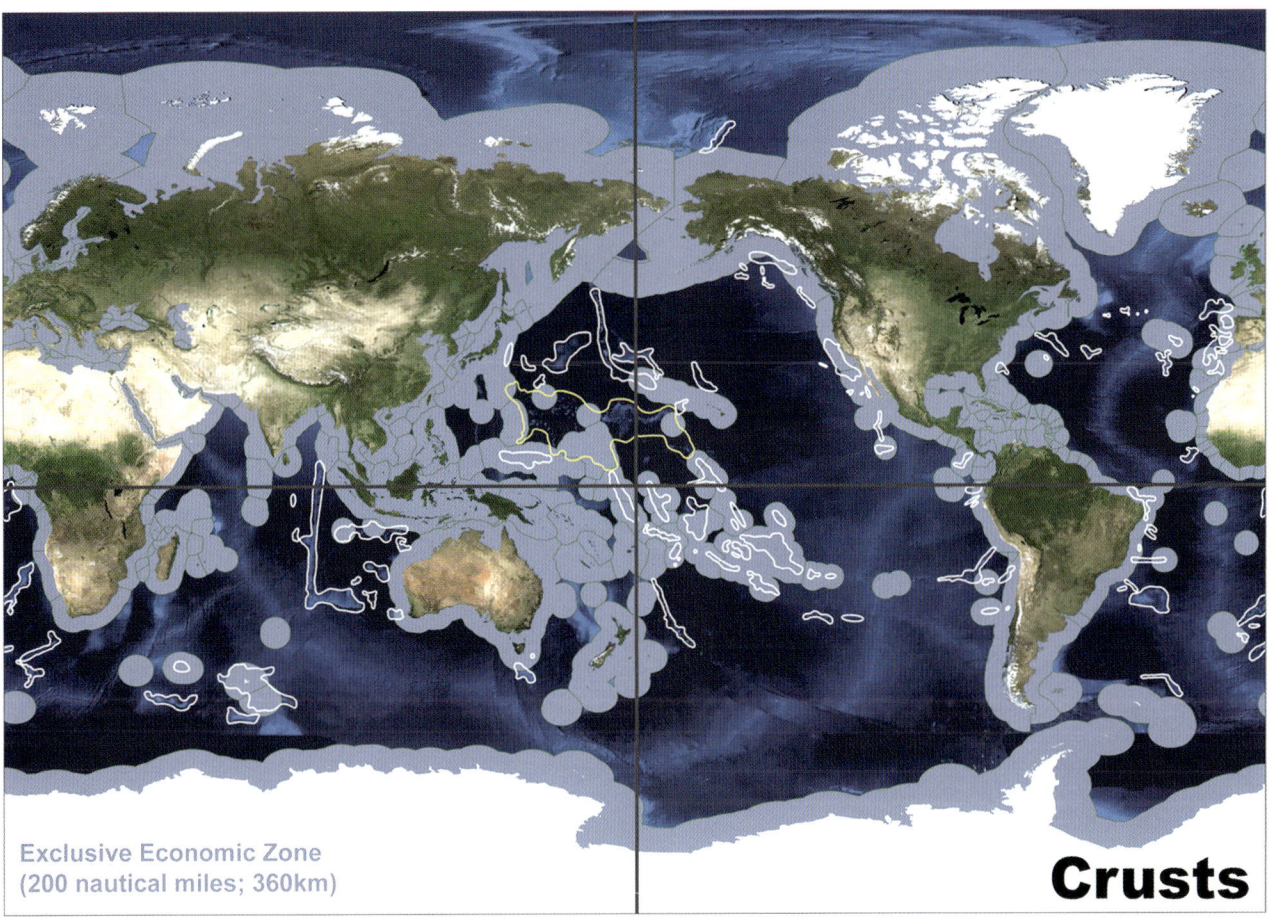

Cobalt-rich Manganese Crusts, Figure 2 Map of global distribution of Exclusive Economic Zones (EEZs, *gray shading*), areas beyond national jurisdictions (*dark black blue* to *pale gray blue*), and global permissive areas for cobalt-rich crust development; the central Pacific prime crust zone (*PCZ*) is the zone of greatest economic interest and is enclosed by a *yellow line*; all other areas are marked with a white line; a permissive area does not mean that economic crust deposits will be found in that area. Small occurrences of crusts will occur in other areas. The equator and 180th parallel are marked by *gray lines*.

(Te^{4+}) and platinum (Pt^{2+}) oxidize to Te^{6+} and Pt^{4+} (e.g., Hein and Koschinsky, 2014). Fe–Mn crusts with slower growth rates also have higher concentrations of metals sorbed onto the main mineral phases. Crusts that form in the oxygen-minimum zone (OMZ) have the slowest growth rates and contain the highest concentrations of metals of economic interest. The OMZ is created by upwelling of nutrient-rich waters along the flanks of seamounts, which promotes productivity in surface waters over the seamounts. As the plankton die and sink through the water column, the organic matter oxidizes and depletes the seawater of oxygen, creating the OMZ. In the central Pacific Ocean, the OMZ commonly occurs from about 500 m to 2,000 m water depths.

Composition

Fe–Mn crusts have a simple mineralogy and are composed predominantly of vernadite (Mn oxide) and noncrystalline Fe oxyhydroxide. They contain minor amounts of detrital minerals such as quartz and feldspar derived from weathering of seabed outcrops and blown by winds from the continents. Older layers of thick crusts contain a phosphate mineral called carbonate fluorapatite, which is a secondary mineral that formed long after the Mn and Fe minerals precipitated from seawater.

Fe and Mn occur in subequal amounts (Fe/Mn ~0.7–1.4) in Fe–Mn crusts, and Co, the trace metal of most economic interest, can be up to 2 % but usually averages 0.33–0.67 % by weight (0.1 % = 1,000 parts per million, ppm) for large areas of the global ocean (Table 1); for smaller mine-site size areas (500 km^2), Co can average 0.8 %. Fe–Mn crusts have the highest concentrations of Te compared to other rock types (global mean about 50 ppm); Te is a rare metal that is needed by the solar-cell industry for thin-film photovoltaics and is the best material for converting sunlight into electricity. Fe–Mn crusts are also significantly enriched in bismuth (Bi), molybdenum (Mo), niobium (Nb), nickel (Ni), platinum (Pt),

Cobalt-rich Manganese Crusts, Table 1 Compiled chemical composition of crusts from selected areas of the global ocean; see Figure 1 for location of PCZ

Element	Atlantic Ocean Mean	N	Indian Ocean Mean	N	Prime crust zone (PCZ) Mean	N	South Pacific Mean	N	California margin Mean	N
Fe (wt%)	20.9	43	22.3	23	16.8	368	18.1	286	23.5	167
Mn	14.5	43	17.0	23	22.8	368	21.7	321	18.2	167
Si	5.21	43	6.82	23	4.04	309	4.75	255	11.2	167
Al	2.20	43	1.83	23	1.01	357	1.28	241	1.84	167
Ti	0.92	43	0.88	23	1.16	351	1.12	230	0.66	167
Bi (ppm)	19	38	30	22	42	40	22	46	16	105
Co	3,608	43	3,291	23	6,655	368	6,167	321	2,977	167
Cu	861	43	1,105	23	982	368	1,082	321	438	167
Li	33	42	8.3	22	3.3	38	3.5	36	15	3
Mo	409	43	392	23	463	334	418	67	354	167
Nb	51	43	61	23	54	49	59	46	31	105
Ni	2,581	43	2,563	23	4,216	368	4,643	321	2,299	167
Pb	1,238	43	1,371	23	1,636	332	1,057	113	1,541	167
Pt	0.57	2	0.21	6	0.48	66	0.47	15	0.07	23
Te	43	37	31	22	60	49	38	38	11	101
Tl	104	38	95	22	160	40	154	46	41	105
Th	52	42	56	18	12	46	15	67	53	105
V	849	43	634	23	642	334	660	177	613	167
W	79	35	80	18	89	42	97	56	59	105
Zn	614	43	531	23	669	331	698	181	561	167
Zr	362	38	535	22	559	49	754	46	473	105
TREE	2,402	20–43	2,541	12–21	2,454	89–300	1,634	17–75	2,352	115

Modified from Hein et al. (2013) and Hein and Koschinsky (2014)
TREE total rare-earth elements including yttrium

thallium (Tl), thorium (Th), titanium (Ti), vanadium (V), tungsten (W), zinc (Zn), zirconium (Zr), and total rare-earth elements plus yttrium (TREE + Y; Table 1) relative to their average contents in the Earth's lithosphere and in seawater. Thorium is one of the few elements that is more abundant in the Atlantic and Indian Ocean crusts than in PCZ Pacific crusts. On average, Fe–Mn crusts have three times more Co, 10 times more Te, three times more TREEs, and 3–14 times more Pt than manganese nodules. In contrast, nodules contain more Ni and copper (Cu) and significantly more lithium (Li) than crusts, whereas both have about equal amounts of Mo.

Paleoceanography

Growth rates and ages of Fe–Mn crusts were first determined using uranium (U)-series isotope and beryllium (Be) isotope ratios. These two techniques provide reliable ages but are limited by the thickness of crust that they can date, the outer ~2 mm (<500,000 years) using U-series and the outer ~20 mm (<12 Ma) using Be isotopes. Extrapolation of those ages to the base of thicker crusts can result in large errors because of potential changes in growth rates. A new technique compares Fe–Mn crust osmium (Os) isotope ratios to those that define a Cenozoic seawater curve; crusts as old at 70 Ma have been dated using Os isotopes (Klemm et al., 2005). Two other techniques that have been employed successfully include nannofossil biostratigraphy and paleomagnetic stratigraphy, but both require lengthy sample preparations. In addition, several empirical equations have been developed that estimate growth rates and therefore ages of crusts and can be used if the more accurate isotopic techniques are not available.

Because Fe–Mn crusts occur throughout the ocean basins at a wide range of water depths, they are ideal for paleoceanographic studies. Textural and geochemical changes in Fe–Mn crusts have been related to the history of seamount subsidence, plate tectonic migration of seamounts, primary productivity, changes in the equator-to-pole thermal gradient and associated ocean mixing, and the extent and intensity of the OMZ, among others. Based on temporal changes in trace-metal isotope distributions (such as lead (Pb), neodymium (Nd), hafnium (Hf), and Be), Fe–Mn crusts have been used to reconstruct past circulation patterns of the oceans and erosion rates of the continents on timescales of millions of years (e.g., Frank et al., 1999). Because these elements reflect different sources and different residence times in seawater, they complement each other as tracers of paleoceanographic events for the past 70 Ma. Other metal isotopes in Fe–Mn crusts that have come into play more recently for paleoceanographic studies include Tl, Mo, Fe, and cadmium (Cd).

Resource consideration

A wide array of trace metals (Co, Ni, Ti, Cu), rare metals (Te, Pt, Zr, Nb, W, Bi, Mo, Tl, Th, V), and REEs are sorbed in large quantities onto the Mn and Fe oxides making crusts a potential resource for many metals used in emerging high-tech, green-tech, and energy applications. Such high concentrations of metals are sorbed because of the very slow growth rates, extreme specific surface area (average 325 m^2/cm^3 of crust), and high porosity (average 60 %) of crusts (Hein et al., 2000). The tonnage of Fe–Mn crusts in most areas of the global ocean is poorly known. However, a rough estimate for the PCZ area is about 7,533 million dry tonnes (Hein and Koschinsky, 2014). Using this tonnage estimate, Fe–Mn crusts in the PCZ are calculated to contain about four times more Co, three and a half times more Y, and an incredible nine times more Te than the entire land-based reserve base for those metals; these crusts also contain half the Bi and a third of the Mn that make up the entire land-based reserve base (Hein et al., 2013). The reserve base is defined as land-based deposits that are currently economically viable (reserves), marginally economic, and subeconomic.

There are two technological challenges to overcome before Fe–Mn crust mining can become viable. A deep-towed or autonomous underwater vehicle-mounted instrument must be developed that can measure Fe–Mn crust thicknesses in situ in real time. This measurement gives the tonnage of crusts per square meter of seabed. This is a difficult challenge because there are a great variety of rocks on which crusts grow that have variable physical properties, many of which overlap with those of the crusts. The second technological challenge is developing a mining tool that can remove the Fe–Mn crust from the substrate rock without collecting any substrate rock, to which crusts can be attached tightly to weakly. That separation of crust from substrate will have to be done on an uneven and commonly rough seabed. These challenges will require significant engineering innovations.

Summary

Fe–Mn crusts grow on nearly all rocks exposed at the seabed throughout the global ocean where sediment does not accumulate. Crusts are composed predominantly of Fe and Mn oxide minerals that precipitate from cold seawater and acquire abundant metals and other elements by sorption from seawater. Metals essential to many emerging technologies are enriched in crusts to the extent that they are considered a potential resource for mining in the near future. Four contracts for exploration for crusts have been taken out with the International Seabed Authority, one each by Japan, China, Russia, and Brazil. Fe–Mn crusts also have a unique potential as recorders of oceanographic events that have occurred over the past 70 Ma.

Bibliography

Frank, M., O'Nions, R. K., Hein, J. R., and Banakar, V. K., 1999. 60 Myr records of major elements and Pb-Nd isotopes from hydrogenous ferromanganese crusts: reconstruction of seawater paleochemistry. *Geochimica et Cosmochimica Acta*, **63**, 1689–1708.

Halbach, P., Manhein, F. T., and Otten, P., 1982. Co-rich ferromanganese deposits in the marginal seamount regions of the Central Pacific Basin—results of the Midpac'81. *Erzmetall*, **35**, 447–453.

Hein, J. R., and Koschinsky, A., 2014. Deep-ocean ferromanganese crusts and nodules, Chapter 11. In Holland, H. D., and Turekian, K. K. (eds.), *Treatise on Geochemistry*. Elsevier: Oxford, Vol. 13, pp. 273–291.

Hein, J. R., Koschinsky, A., Bau, M., Manheim, F. T., Kang, J.-K., and Roberts, L., 2000. Cobalt-rich ferromanganese crusts in the Pacific. In Cronan, D. S. (ed.), *Handbook of Marine Mineral Deposits*. Boca Raton: CRC Press, pp. 239–279.

Hein, J. R., Conrad, T. A., and Dunham, R. E., 2009. Seamount characteristics and mine-site model applied to exploration- and mining-lease-block selection for cobalt-rich ferromanganese crusts. *Marine Georesources and Geotechnology*, **27**, 160–176.

Hein, J. R., Mizell, K., Koschinsky, A., and Conrad, T. A., 2013. Deep-ocean mineral deposits as a source of critical metals for high- and green-technology applications: comparison with land-based deposits. *Ore Geology Reviews*, **51**, 1–14.

Klemm, V., Levasseur, S., Frank, M., Hein, J. R., and Halliday, A. N., 2005. Osmium isotope stratigraphy of a marine ferromanganese crust. *Earth and Planetary Science Letters*, **238**, 42–48.

Koschinsky, A., and Hein, J. R., 2003. Uptake of elements from seawater by ferromanganese crusts: solid phase association and seawater speciation. *Marine Geology*, **198**, 331–351.

Cross-references

Deep-sea Sediments
Dust in the Ocean
Energy Resources
Geochronology: Uranium-Series Dating of Ocean Formations
Guyot, Atoll
Manganese Nodules
Marine Mineral Resources
Paleoceanographic Proxies
Paleoceanography
Phosphorites
Radiogenic Tracers
Seamounts
Technology in Marine Geosciences

COLD SEEPS

Marta E. Torres[1] and Gerhard Bohrmann[2]
[1]College of Earth, Ocean, and Atmospheric Sciences, Oregon State University, Corvalis, OR, USA
[2]MARUM-Center for Marine Environmental Sciences, University of Bremen, Bremen, Germany

Synonyms

Cold vents; Methane seeps

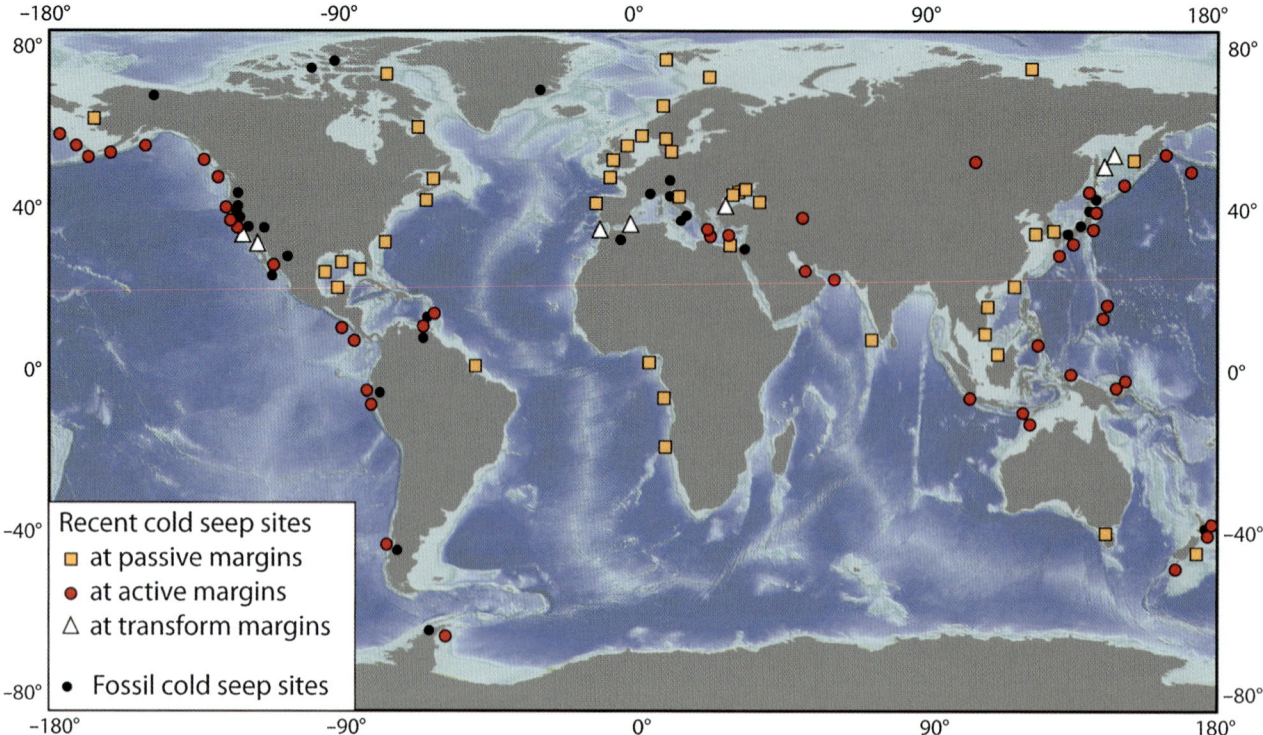

Cold Seeps, Figure 1 Seep locations at active (recent) locations at passive (*orange squares*) and active (*red circles*) margin sites. Locations along transform faults are denoted by white triangles. Fossil sites are shown by black circles. Seep locations based on compilations from Suess (2010), Campbell (2006) and Römer (2011).

Definition

Cold seepage denotes the emission at the seafloor of deep-sourced fluids enriched in methane and hydrogen sulfide, which support a characteristic biome based on chemosynthetic organisms.

Basic process

Oxygen enters the ocean surface water through contact with the atmosphere. From there, dissolved oxygen is brought to greater depths by sinking and circulation of water masses, so that in the majority of the deep oceans, there is enough oxygen to support aerobic decomposition of organic matter. Against this well-oxygenated hydrosphere, discrete locations on the seafloor experience localized discharge of highly reducing fluids from the underlying sediments and igneous crust. Along continental margins, methane emanates from the sediments at locations known as cold seeps (Figure 1).

At some of the cold seeps, methane emission rates exceed mid-ocean ridge values by more than an order of magnitude. Methane plumes are such a characteristic feature of cold seep areas that dissolved methane in the water column was originally used to prospect for cold seep activity (Heeschen et al., 2005). Acoustic imaging techniques have also been used to identify locations of macroseepage, where methane concentration is so large that it exceeds its solubility and bubble plumes can be observed emanating out of the seafloor (Figure 2). Recent acoustic surveys have led to the discovery and detailed mapping of methane gas plumes along continental margins worldwide (e.g., Westbrook et al., 2009; Faure et al., 2010; Römer et al., 2012).

Source, transport, and fate of methane

Sources

Within continental margin sediments, rapid burial of organic matter tied to a limited supply of oxygen from the overlying seawater results in the development of highly reducing environments that favor methane generation by a group of microorganisms called methanogens (see review by Valentine, 2011). Deeper in the sediment, thermal alteration of organic matter generates methane and higher hydrocarbons within the temperature range of 80–200 °C, which typically occur at burial depths greater than 2 km. The combined action of these two processes leads to high levels of methane, which at high-pressure and low-temperature conditions will combine with water to form a solid structure known as gas hydrates (Sloan and Koh, 2008). Because of the pressure, temperature, and methane concentration needed to form gas hydrate, these deposits are commonly found in continental margin sediments at depths larger than 350 m and are stable to a few hundred of meters in the sediment column, depending on the attendant geothermal gradient.

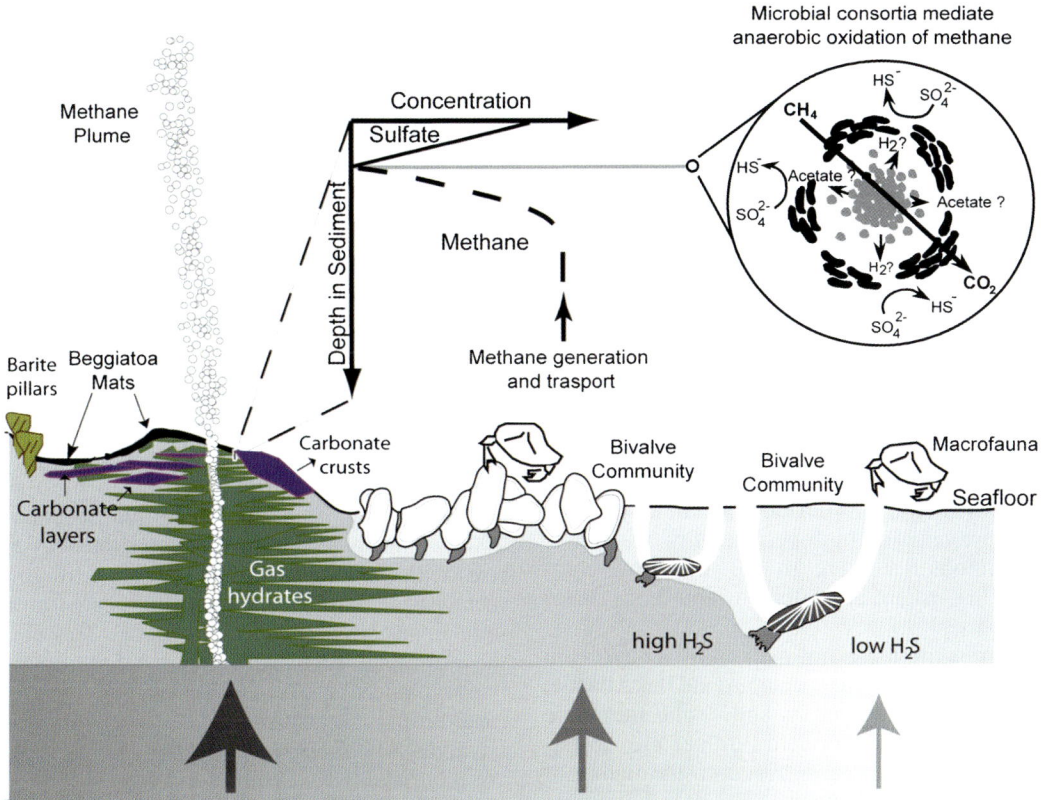

Cold Seeps, Figure 2 Schematic representation of cold seeps, showing methane generation in the sediments, transport and discharge at the seafloor. Carbonate and barite deposits form at cold seeps. The reducing fluids rich in methane and hydrogen sulfide support complex biomes that include microbial mats, chemosynthetic bivalves and macrofauna. *Arrows* denote relative magnitude of upward methane flux.

Transport

Sediments in continental margins are hydrologically active, but the processes that drive fluid flow vary depending on the tectonic and geologic characteristics of each site (Figure 1).

Convergent margins – The best-studied cold seeps to date are located along margins where continental and oceanic plates converge, such as those found landward of the subduction trenches that rim the Pacific Ocean. Cold seeps in these settings were first reported in the early 1980s by LaVerne Kulm and Erwin Suess along the Oregon-Washington coast (Kulm et al., 1986). To date, cold seepage has been widely studied from Alaska to southern Chile in the Eastern Pacific margin, with similar level of research targeting seeps along the western Pacific offshore Russia, Japan, and New Zealand as well as the Mediterranean and Black Seas. Additional seeps have been reported along convergent margins offshore Makran, Indonesia, and Colombia (review in Suess, 2010).

Pore fluids in the sediments account for as much as 50–70 % of their volume, and these fluids are squeezed out from the sediment by the lateral compression of plate movement. As porosity decreases, the rate of compaction-driven dewatering decreases, and thus water contribution from intergranular and fracture porosity is generally largest within the first ~3–7 km of burial (see review by Saffer and Tobin, 2011). With increasing burial, water is released from hydrous minerals. Common in these systems is the smectite to illite conversion, which occurs at temperatures ranging from 60°C to 200 °C (Colten-Bradley, 1987). Fluid flow in convergent margins appears to control seismogenic behavior, and the extensive research targeting the processes involved in fluid generation has revealed that at ~ 10–40 km from the trench, fluid sources from smectite and, to a lesser extent, opal dehydration overtake those generated from compaction (Saffer and Tobin, 2011). The excess water accumulates in the pore space, where it creates overpressures that drive flow toward the seafloor. Whether dewatering occurs by compaction, dehydration, or a combination of these processes, water in convergent margins migrates from deep horizons through high-permeability zones such as faults, fractures, or coarse-grained sediment layers.

When methane accumulates as a gas, it can also generate fractures in response to gas overpressures. Gas migration through gas-induced hydrofracture networks or via structural (e.g., faults) and lithologic (e.g., sand horizons) pathways has been documented through acoustic imaging in seismic profiles and by analyses of sediment cores collected under in situ pressure (e.g., Tréhu et al., 2004; Torres et al., 2011).

In *transform margins*, such as the one generated by the lateral movement between the Pacific and the North American plates along the San Andreas fault zone, thick sediment sequences may be exposed by the sideways plate motion along the fault planes (Legg et al., 1989). These cold seeps have received less attention than those in convergent or passive margins. However, a good example of transform-margin seeps is those found along the San Clemente fault, approximately 100 km southwest of San Diego (Torres et al., 2002).

In this transform margin, the turbidite sand layers of the Navy Fan deposits support migration of subsurface fluids, which vent at the locations where the sediment package is cut by fault-induced escarpments. Both methane anomalies in the water column, as well as the presence of chemosynthetic organisms along the fault traces, delineate the zones of active methane venting. Fluid flow in these settings is driven primarily by the development of a hydraulic head and the presence of high-permeability pathways, similar to conditions that drive groundwater flow commonly studied in onshore and offshore aquifers (Torres et al., 2002).

Passive margins are those not marked by a strike-slip fault or a subduction zone. They occur around the Atlantic, Arctic, and western Indian Oceans and define the entire coasts of Africa, Greenland, northern Europe, and Australia. Because of their extensive distribution, the nature of the seeping fluids and mechanisms that control flow are quite large. Mass transport, which is common in these margins, provides an overriding compressional component to the normal faulting regime typical of passive-margin tectonics. In addition, dehydration reactions and gas generation all create overpressures that drive flow. Flow is channeled along highly permeable lithologies, such as ash layers and turbidite sands through normal faulting.

The first cold seep on a passive margin was discovered by Charles Paull along the Florida Escarpment (Paull et al., 1992). Here, groundwater migrates through carbonate aquifers and discharges in the Gulf of Mexico at a depth of 3,200 m. Since then a large number of seep localities have been identified on the northern margin of the gulf. Fluid flow is associated with salt dome intrusions within these (>10 km) thick sediment packages, from which the thermogenic oil and gas accumulations are currently targeted for oil and gas production. In this margin, migration conduits supply hydrocarbon materials that migrate ~6–8 km toward the sediment surface. More recently, cold seeps were discovered along the Mexican margin of the gulf, which are characterized by the presence of thick oil coatings and lavalike asphalt flow structures on the seafloor (McDonald et al., 2004).

Because of the thermogenic nature of these seeps, it is possible to image the naturally created oil slicks at the sea surface, using satellite data. Ian MacDonald and collaborators have analyzed images from space that reveal the presence of oil slicks across the north-central Gulf of Mexico in water depths greater than 1,000 m (McDonald et al., 2000).

Fate of methane and cold seep biomes

If all methane out of the sedimentary reservoir were to reach the atmosphere, this could have significant impact on the Earth's climate, because methane is a strong greenhouse gas. Fortunately, microorganisms consume a very large fraction of the sedimentary methane under anaerobic conditions. If released to the water column, methane is further consumed by aerobic methane oxidation. This combined microbial filter, reviewed by William Reeburgh (2007), is key in regulating methane flux from the sediments to the ocean and potentially the atmosphere.

Within the sediments, a metabolic pathway whereby methane was oxidized by sulfate was originally proposed based on the chemical composition of the pore water (Reeburgh, 1976; Barnes and Goldberg, 1976). This postulate was later confirmed by observations of a microbial consortium that oxidizes methane to bicarbonate while reducing sulfate to hydrogen sulfide (Boetius and Suess, 2004). The hydrogen sulfide generated in this process rises with the ascending fluids to the seafloor, where it is metabolized by *Beggiatoa* that aggregate forming the bacterial mats typically found at cold seeps (Figure 2). In these complex ecosystems, there are multiple interactions between microbes and macrobiota that involve symbiosis, heterotrophic nutrition, and geochemical feedbacks. Bivalves (*mytilids, vesicomyids, lucinids,* and *thyasirids*) are nourished by symbiotic bacteria that derive their energy from methane and hydrogen sulfide, and constitute prominent members of cold seep fauna. *Vestimentiferan* tube worms, also common at cold seep locations, extract hydrogen sulfide through their roots, which is also metabolized by symbionts. *Pogonophorans, cladorhizid* sponges, gastropods, shrimp, and crabs are also abundant at some locations (Levin, 2005).

Authigenic minerals

In addition to hosting unique communities, chemical reactions between the highly reduced seeping fluids and the oxidant-rich seawater lead to precipitation of characteristic authigenic minerals. Calcium carbonate ($CaCO_3$) is formed as a by-product of methane oxidation, which provides a bicarbonate source characterized by enrichment in ^{12}C. These methane-derived carbonates are ubiquitous near cold vent sites throughout the globe and constitute a significant carbon sink. The dominant mineral phases are aragonite and high-magnesium calcite (Bohrmann et al., 1998).

Microbially mediated carbonate formations have been observed to rise above the seafloor by up to 90 m in height. In the Cascadia margin, for example, carbonate chemoherms display a pinnacle-shaped morphology with steep flanks and are known to have been active for at least 270 ky (Teichert et al., 2005). Their irregular structure and high porosity reflect vigorous outflow of methane-rich fluids. In anoxic bottom waters, such as those found in the Black Sea, microbial consortia fed by methane bubble emissions at the seafloor support the buildup of towerlike carbonate structures several meters in height. Other locations of extensive authigenic carbonate formation are found offshore Costa Rica, South China Sea, New Zealand, and in the Gulf of Cadiz (reviewed by Suess, 2010).

Although not as common as the carbonate occurrences, barite ($BaSO_4$) deposits have been observed along structurally controlled sites of cold-fluid seepage in several continental margins. Barite pillars along the San Clemente fault can reach 10 m in height, and similar deposits have been reported on the Sea of Okhotsk. Less massive but distinct barite occurrences have also been reported in the Alaska margin, Monterey Canyon, Peru margin, and the Gulf of Mexico (Torres et al., 2003). The formation of these deposits is intimately linked with the sulfate concentrations, which in the highly reduced environments that characterize cold seeps is depleted at shallow depths within the sediments. At low sulfate, any barite in the sediment will dissolve, leading to high levels of dissolved barium in the upwardly migrating fluids. Upon discharging at the seafloor, barium quickly reacts with seawater sulfate and precipitates in the newly formed barite deposits.

Paleoseeps

The occurrence of carbonates in the geologic record enriched in the light ^{12}C isotope has been commonly attributed to authigenic formation at paleo-methane seeps, which indicate significant methane release from the Proterozoic to the present (Campbell, 2006). The cold seep barite deposits likely constitute the modern analogues to Paleozoic stratiform barite deposits, now being mined in Nevada, Arkansas, Mexico, and South China. These barite deposits represent large-scale submarine methane venting, which could have affected the Paleozoic carbon cycle and perhaps modified ancient climate (Torres et al., 2003).

Summary

Along the ocean margins, sediments experience chemical, microbial, and physical transformations that result in the generation and expulsion of reduced gases and water at the seafloor, in what is known as cold seepage.

Methane is generated by microbial or thermochemical processes in the organic-rich marine sediments. Hydrogen sulfide is a by-product of microbial oxidation of methane under anaerobic conditions. Water is produced at greater depths by dehydration reactions of smectite and to a lesser extent opal. Accumulation of the excess water and gas creates overpressures that can fracture the overlying geological formations and drive fluid flow. Sediment compaction, generation of hydraulic heads, and tectonic compression also act as driving forces for fluids that eventually discharge at the seafloor.

Whereas the majority of life in the deep ocean floor rely upon low levels of sinking organic matter, the chemosynthetic microbial communities at cold seeps take advantage of high fluxes of bioreactive reductants to support unique, oasis-type cold seep biomes in which the density of organisms is several orders of magnitude greater than in the surrounding regions. Authigenic carbonate and barite deposits, which can reach extremely large accumulations, characterize areas of cold seepage and constitute a geologic record of methane discharge at paleo-seep sites.

Bibliography

Barnes, R. O., and Goldberg, E. D., 1976. Methane production and consumption in anoxic marine sediments. *Geology*, **4**, 297–300.

Boetius, A., and Suess, E., 2004. Hydrate Ridge: a natural laboratory for the study of microbial life fueled by methane from near-surface gas hydrates. *Chemical Geology*, **205**, 291–310.

Bohrmann, G., Greinert, J., Suess, E., and Torres, M., 1998. Authigenic carbonates from the Cascadia subduction zone and their relation to gas hydrate stability. *Geology*, **26**, 647–650.

Campbell, K. A., 2006. Hydrocarbon seep and hydrothermal vent paleoenvironments: past developments and future research directions. *Palaeogeography, Palaeoclimatology, Palaeoecology*, **232**, 362–407.

Colten-Bradley, V. A., 1987. Role of pressure in smectite dehydration–effects on geopressure and smectite-to-illite transformation. *AAPG Bulletin*, **71**, 1414–1427.

Faure, K., Greinert, J., Schneider von Deimling, J., McGinnis, D. F., Kipfer, R., and Linke, P., 2010. Methane seepage along the Hikurangi Margin of New Zealand: geochemical and physical data from the water column, sea surface and atmosphere. *Marine Geology*, **272**, 170–188.

Heeschen, K. U., Collier, R. W., de Angelis, M. A., Suess, E., Rehder, G., Linke, P., and Klinkhammer, G. P., 2005. Methane sources, distributions, and fluxes from cold vent sites at Hydrate Ridge, Cascadia Margin. *Global Biogeochemical Cycles*, **19**. GB2016. doi:10.1029/2004GB002266.

Kulm, L. D., Suess, E., Moore, J. C., Carson, B., Lewis, B. T., Ritger, S. D., Kadko, D. C., Thornburg, T. M., Embley, R. W., Rugh, W. D., Massoth, G. J., Langseth, M. G., Cochrane, G. R., and Scamman, R. L., 1986. Oregon subduction zone: venting, fauna, and carbonates. *Science*, **231**, 561–566.

Legg, M. R., Luyendyk, B. P., Mammerickx, J., Moustier, C., and Tyce, R. C., 1989. Sea Beam survey of an active strike-slip fault: the San Clemente fault in the California Continental Borderland. *Journal of Geophysical Research, Solid Earth*, **94**(B2), 1727–1744.

Levin, L. A., 2005. Ecology of cold seep sediments: interactions of fauna with flow, chemistry, and microbes. *Oceanography and Marine Biology an Annual Review*, **43**, 1–46.

MacDonald, I. R., Buthman, D. B., Sager, W. W., Peccini, M. B., and Guinasso, N. L., 2000. Pulsed oil discharge from a mud volcano. *Geology*, **28**, 907–910.

MacDonald, I. R., Bohrmann, G., Escobar, E., Abegg, F., Blanchon, P., Blinova, V., and De Farago, M., 2004. Asphalt volcanism and chemosynthetic life in the Campeche Knolls, Gulf of Mexico. *Science*, **304**, 999–1002.

Paull, C. K., Chanton, J. P., Neumann, A. C., Coston, J. A., Martens, C. S., and Showers, W., 1992. Indicators of methane-derived carbonates and chemosynthetic organic carbon deposits: examples from the Florida Escarpment. *Palaios*, **7**, 361–375.

Reeburgh, W. S., 1976. Methane consumption in Cariaco Trench waters and sediments. *Earth and Planetary Science Letters*, **28**, 337–344.

Reeburgh, W. S., 2007. Oceanic methane biogeochemistry. *Chemical Reviews*, **107**, 486–513.

Römer, M. 2011., Gas bubble emissions at continental margins: detection, mapping, and quantification. University of

Bremen. Dissertation. http://nbn-resolving.de/urn:nbn:de:gbv:46-00102438-18

Römer, M., Sahling, H., Pape, T., Bohrmann, G., and Spieß, V., 2012. Quantification of gas bubble emissions from submarine hydrocarbon seeps at the Makran continental margin (offshore Pakistan). *Journal of Geophysical Research*, **117**, C10015, doi:10.1029/2011JC007424.

Saffer, D. M., and Tobin, H. J., 2011. Hydrogeology and mechanics of subduction zone forearcs: fluid flow and pore pressure. *Annual Review of Earth and Planetary Sciences*, **39**, 157–186.

Sloan, E. D., and Koh, C. A., 2008. *Clathrate Hydrates of Natural Gases*, 3rd edn. Boca Raton, FL: CRC Press. 752pp.

Suess, E., 2010. *Marine cold seeps Handbook of Hydrocarbon and Lipid Microbiology*. Berlin: Springer, pp. 188–203, doi:10.1007/978-3-540-77587-4_12. ISBN 978-3-540-77587-4.

Teichert, B. M. A., Bohrmann, G., and Suess, E., 2005. Chemoherms on Hydrate Ridge – unique microbially-mediated carbonate build-ups growing into the water column. *Paleogeography, Paleoclimatology, Paleoecology*, **227**, 68–85.

Torres, M. E., McManus, J., and Huh, C.-A., 2002. Fluid seepage along the San Clemente fault scarp: impact on the geochemical barium cycles on a basin-wide scale. *Earth and Planetary Science Letters*, **203**, 181–194.

Torres, M.E., J-H. Kim, J. Choi, J-B Ryu, J-J Bahk, M. Riedel, T. Collett, W-L. Hong and M. Kastner. 2011. Occurrence of high salinity fluids associated with massive near-seafloor gas hydrate deposits. In: Proceedings of the 7th International Conference on Gas Hydrates (ICGH 2011), Edinburgh, Scotland, United Kingdom, July 17–21.

Torres, M. E., Bohrmann, G., Dubé, T. E., and Poole, F. G., 2003. Barite deposition at modern cold seeps: analogue to paleozoic bedded barite. *Geology*, **31**, 897–900.

Tréhu, A. M., Flemings, P. B., Bangs, N. L., Chevalier, J., Gracia, E., Jonhson, J., Liu, C. S., Liu, X., Riedel, M., and Torres, M. E., 2004. Feeding methane vents and gas hydrate deposits at south hydrate ridge. *Geophysical Research Letters*, **31**, L23310.

Valentine, D. L., 2011. Emerging topics in marine methane biogeochemistry. *Annual Review of Marine Science*, **3**, 147–171.

Westbrook, G. K., Thatcher, K. E., Rohling, E. J., Piotrowski, A. M., Pälike, H., Osborne, A. H., and Aquilina, A., 2009. Escape of methane gas from the seabed along the West Spitsbergen continental margin. *Geophysical Research Letters*, **36**, L15608, doi:10.1029/2009GL039191.

Cross-references

Intraoceanic Subduction Zone
Marine Gas Hydrates
Methane in Marine Sediments
Mid-ocean Ridge Magmatism and Volcanism
Mud Volcano
Passive Plate Margins
Transform Faults

CONTINENTAL RISE

William W. Hay
Department of Geological Sciences, University of Colorado at Boulder, Estes Park, CO, USA

Definition

The continental rise is the gently inclined slope between the base of the continental slope and the deep ocean floor.

The expression "continental rise" was first used by Bruce Heezen and Maurice Ewing in their account of the effects of the 1929 Grand Banks earthquake. It was formally defined in 1959 by Heezen et al. in GSA Special Paper 65. The Floors of the Ocean: I, The North Atlantic: "Since we have limited the continental slope to gradients greater than 1:40, we split off this lower portion of the continental margin into a separate province, the continental rise" (p. 19).

In many areas, local morphologic features interfere with the general slopes so that neither the upper nor lower limits of the continental rise are well defined (Figure 1).

The geologic basis for the continental rise

The concept of the continental rise arose in the classic passive margin region of the western North Atlantic. There the continental rise overlies the ocean crust bordering the faulted and fractured continental margin. It is the site where sediment shed from the continent into the deep sea accumulates.

Along active margins where ocean crust is being subducted beneath continental crust, the margin is often marked by an ocean trench; there is no continental rise. However in some areas where there is a great thickness of sediment on the ocean floor entering the subduction zone, or where the supply of sediment from the continent into the subduction zone completely fills the trench, a narrow continental rise may develop.

What controls the configuration of the continental rise

Nature of the crust

Where the continents have broken up, in the Atlantic, Indian, and Arctic oceans, the continental rise overlies the oldest ocean crust. Because the ocean crust subsides with age, this is the region where the differences in elevation between continental and ocean crust are maximal.

Detrital sediment supply

The continental rise is the ultimate site of deposition of sediment eroded off the continents. The current rate of detrital sediment eroded from land and delivered to the sea is estimated to be about 24×10^{12} kg/year. However, the continental shelves can only accommodate about 10 % of this load at high sea-level stands and almost none at sea-level lowstands. This means that most of the detrital sediment load must be passed on to the continental slope, where again accommodation space is very limited, and to the deep sea. The continental rise represents the site of accumulation of most of the sediment eroded off the continental blocks. These conditions have changed over time. The amount of material eroded off the continental blocks depends on topographic relief and changing climatic conditions. The amount of sediment that can accumulate on shelves has decreased with time as subsidence of the shelf areas has slowed. Major erosion of the continental shelves has occurred since the lowering of sea levels associated

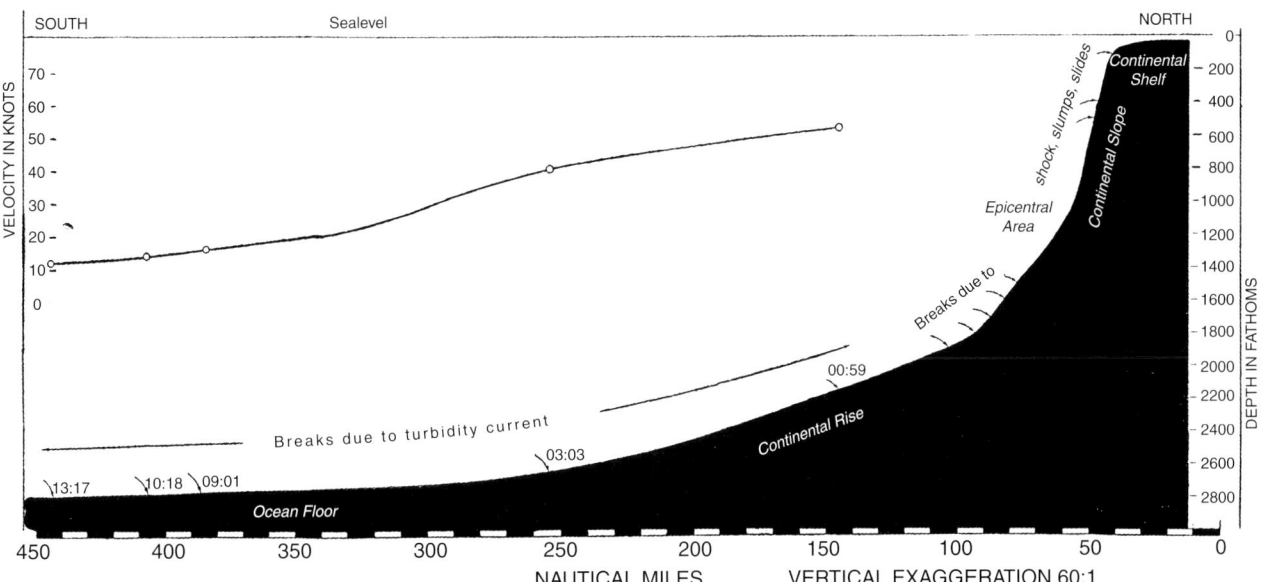

Continental Rise, Figure 1 Heezen and Ewing's (1952) diagram of the continental margin off the Grand Banks, showing the times of cable breaks after the 1929 earthquake and debris flow and turbidity current velocities.

with the buildup of Antarctic and then northern hemisphere ice sheets since the end of the Eocene. Once deposited on the continental rise, there is no place else to go for the sediment except further out onto the abyssal plains.

Detrital sediment reaches the continental rise by three mechanisms. First is simply settling through the water column. The oceanic frontal systems separating the brown turbid waters of the continental shelves from the blue waters of the open ocean generally prevent riverine-suspended sediment from being fed into the surface ocean. However, the supply of dust from land is also large, and this settles into the ocean and becomes a part of the rain of pelagic sediment to the deep sea. There is no special supply of dust to the continental rise.

The second mechanism is through turbidity currents. Fresh riverine waters with a significant suspended sediment load have a density greater than seawater and can flow across the shelf and down onto the slope and rise as turbidity currents. Because the density of seawater increases with the depth of these turbidity currents, upon reaching seawater of equal density the turbid flow may mix with the seawater or flow out on a density surface.

The third mechanism is through failures of the sediments of the continental slope as slumps or catastrophic slides into the deep sea. Gradual slumping appears to be a common process where there is a large sediment supply to a steep slope. Landslides from the slope to the deep sea are much more common than had originally been thought. They seem to occur especially in regions where gas clathrates that acted as cement stabilizing the slope decompose as warming of the waters occurs, as during glacial terminations (Bryn et al., 2006). The usual triggering mechanism is thought to be earthquakes, but large storms may also be a cause. Probably half of the sediment in the continental rise reached its present location through mass movements.

The classic study was that of the Grand Banks' Earthquake of 1929 by Heezen and Ewing. Because the slide and subsequent turbidity current cut submarine telegraph cables, their velocities down the continental rise are known, ranging from 50 to 10 knots (=25.7–10.1 m/s).

Finally, ocean bottom currents redistribute the sediments of the continental rise.

Carbonate deposition

Carbonate deposition onto the continental rises occurs in two ways. First, the production of shallow water carbonate on banks and atolls far exceeds the rate at which it can be accommodated through subsidence. Most of it is swept off into the deep sea during major storms. Much of that is dissolved by the corrosive waters of the deep sea, but because it may be delivered in large quantities in very short intervals of time, it can accumulate along bank margins (e.g., Bahamas, Maldives). The second source is the rain of carbonate from calcareous nannoplankton, planktonic foraminifera, and pteropods that occurs throughout the oceans. This can be incorporated into the continental rise sediments above the compensation depths for the different mineral phases. The proportions of detrital sediment and pelagic carbonate are an important clue to the relative rates of accumulation from these two sources.

Currents

Originally thought to be a region of very quiet water, the deep sea has been found to be the site of many active currents. Most of these are directly associated with the formation of dense waters that sink into the ocean depths and are redistributed throughout the ocean basins. Their velocities are typically in the range of 0.5–5 cm/s. Flows with the higher velocities can transport and even erode unconsolidated sediment.

Sea-level

The lowering of sea-level associated with the buildup of Oligocene and Neogene ice sheets has caused erosion of a very large amount of material off the continental shelves and ultimately deposited on the continental rises.

Summary

The continental rise is the gently inclined slope between the base of the continental slope and the deep ocean floor. It overlies the ocean crust bordering the faulted and fractured continental margin. It is the ultimate site of accumulation of sediment shed from the continent into the deep sea. A small amount of sediment reaches the continental rise by settling through the water column. Much of the detrital sediment is brought down by turbidity currents, but a very large amount of the material is delivered through failures of the sediments of the continental slope descending as slumps or catastrophic slides. Deep sea bottom currents move and redistribute the sediments of the continental rise. Sea-level fluctuations associated with the buildup of ice sheets have caused a very large amount of material from the continental shelves and slopes to be deposited on the continental rises.

Bibliography

Bryn, P., Berg, K., Forsberg, C. F., Solheim, A., and Kvalstad, T. J., 2006. Explaining the Storegga Slide. *Marine and Petroleum Geology*, **22**, 11–19.

Heezen, B. C., and Ewing, M., 1952. Turbidity currents and submarine slumps and the 1929 Grand Banks earthquake. *American Journal of Science*, **250**, 849–873.

Heezen, B., Tharp, M., and Ewing, M., 1959. The floors of the oceans: I. The North Atlantic. *Geological Society of America, Special Paper*, **65**, 1–126.

Cross-references

Continental Slope
Ocean Margin Systems
Shelf

CONTINENTAL SLOPE

William W. Hay
Department of Geological Sciences, University of Colorado at Boulder, Estes Park, CO, USA

Definition

The first use of the term "slope," referring to the edge of the continental block, appears to have been by Rollin D. Salisbury in his book *Physiography*, published in 1907. "The continental platforms and the ocean basins are topographic features of the first order. The contrast between them is emphasized by the fact that there is almost everywhere a rather steep slope from the one to the other, – a steep descent from the continental platforms to the ocean basins, or looked at from the other point of view, a steep ascent from the ocean basins to the continental platforms" (pp. 5–6). Later in the book, he presented the idea that the junction of the continent and the ocean basin was bordered by a belt that represented a concave-upward part of the seafloor "100–300 miles wide" as indicated in Figure 681: *Map of the world showing in black the portions of the sea bottom which are concave upwards.* The term "continental slope" only appears in the book's index. Although it was obvious Salisbury considered the top of the continental slope to be demarcated by the shelf break, a lower boundary was defined by Heezen et al. (1959) as the point at which the gradient becomes less the 1:40 (about 1°25').

History

The discovery of the continental slope goes back to the time of laying the first telegraph cables across the Atlantic (see SHELF), but the detailed bathymetry of continental margins did not become known until after World War II when the United States and USSR began to make extensive measurements of the seafloor depth using SONAR (Doel et al., 2006).

The geologic basis for the continental slope

It was recognized in the late nineteenth century that the margins between the continents and ocean basins were a fundamental geologic boundary, and in the twentieth century, it became evident that this was due to a fundamental difference between the rocks forming the floors of the continental blocks and ocean basins. However, it was not until the development of the theory of plate tectonics that the nature of the slope became fully understood. There are two kinds of continental slope, one associated with passive continental margins, the other with active margins.

On passive margins, the continental slope overlies the region where the continental margin was thinned by faulting during the process of continental breakup. Along active margins, the continental slope marks the boundary between a subducting oceanic plate and the continental crust on the overriding plate. In some instances, the subducting oceanic plate is physically eroding the continental crust.

What controls the configuration of the continental slope

Crustal thinning

Along classic passive margins, it is the original width of thinning of continental crust by faulting and intrusion of ocean margin basalts, an area with a width typically of 100–300 km. Where the margin coincided with a transform fault, as on the E-W Guinea margin of West Africa, the continental slope can be very narrow and abrupt. Along active margins, where the continent is bordered by an oceanic trench, the edge of the continent may be undergoing abrasion, and the slope may be very steep.

Detrital sediment supply

As the available accommodation space on the shelf is reduced, detrital sediment may accumulate on the continental slope. Although initially unconsolidated, it may be held in place by the cementing effect of gas hydrates, especially where significant amounts of organic carbon are present. However, sea-level fluctuations and currents may destabilize slope sediments and landslide scars are common.

Carbonate deposition

The production of carbonate occurs in shallow water, but even minor sea-level fluctuations will expose the original aragonitic sediment to fresh water and cause its conversion to well-cemented calcite or dolomite. As crustal subsidence occurs, carbonate banks may build upward very rapidly, and the marginal slopes may be almost vertical.

Waves

Today, other than tsunamis, wave action has little effect on slope deposits as they are below wave base of even long swell. However, during glacial sea-level lowstands, when the ocean surface was near the shelf breaks, waves actively eroded the upper part of the continental slope.

Currents

Many continental slopes are sculpted by contour currents flowing along them. These currents have velocities of up to 40 cm/s and can easily erode and move sediment. They generally flow in a direction opposite to the surface gyres. Along the western margins of the ocean basins, the currents of the tropical-subtropical surface gyres (e.g., Gulf Stream, Kuroshio) are narrow and deep and can sculpt the upper parts of the continental slopes.

Sea-level

The large sea-level fluctuations associated with the development of high latitude ice sheets lowers the surface of the ocean to the top of the continental slope. During lowstands, rivers may cut into the edge of the continental shelf and carve canyons in the upper part of the continental slope. Their turbid waters can then flow down the slope and their sediment load deposited on the lower slope and continental rise.

Mass wasting

In many areas the continental slope is at the maximum angle of repose for sediments, and very large mass wasting events often triggered by earthquakes can occur. One of the largest was the three Storegga Slides on the Norwegian margin (Bondevik et al., 2005; Haflidason et al., 2005). Their heads were at the edge of the continental shelf (Storegga is Norwegian for "the Great Edge"). The landslides occurred along 290 km of the continental margin with an estimated volume of 3,500 km^3 of debris. The last of the slides occurred about 6,100 BCE and generated a large tsunami, over 20 m high in Northern Scotland.

Continental slope profiles

While the continental slope forms the steeper part of the transition from the shelf to the ocean basin, the profiles differ greatly from place to place as shown in Figures 1 and 2.

Figure 1a shows the steep slope south of Vladivostok into the Japan Sea which occupies a pull-apart basin behind the Japanese archipelago. Figure 1b shows one of the steepest slopes along the edge of a continent, near Accra, Ghana, where the continent-ocean transition is along a transform fault. Figures 1c and d show two very tall, steep continental slopes on active margins associated with oceanic trenches. Figure 1c shows that on the western South American margin at 25°S, from the Andes into the Peru-Chile Trench; note that a continental shelf is lacking. Figure 1d shows the transition from the eastern Aleutians across a shelf into the Aleutian trench. Figure 1e is an E-W section from the Hebrides across the shelf with a slope into Rockall Trough and then up again onto Rockall Bank, a continental fragment. This is the area in which the term "continental shelf" was first used. Figures 1f and g show two continental margins with "borderlands." Figure 1f shows the complex bathymetry of the Southern California Borderland with ridges and basins terminating oceanward in a steep escarpment. Figure 1g is an E-W section off northern Florida showing the transition from the shelf down a slope onto the broad Blake Plateau and finally down a steep slope into the western Atlantic Basin. Figures 1h and i are profiles along the classic passive margin of the northeastern United States. Figure 1h is a NW-SE transect from Cape Cod into the deep Atlantic with a very steep upper slope and a classic concave-upward lower slope and rise. It is evident that there is no distinctive morphological feature that separates the slope from the rise. Figure 1i shows the broad shelf of the New York Bight, the slope, the Hudson Submarine Fan, and the transition onto the Hatteras Abyssal Plain. The trace of the Hudson Canyon is indicated by a dashed line.

Figure 2a shows a typical Antarctic profile, with a broad deep shelf generally sloping toward the ice-laden continent, and a relatively gentle slope into the deep ocean. Figure 2b is an W-E section from the Gulf of Mexico across south Florida to the Great Bahama Bank, all carbonate terrains. Off western Florida, the shelf is very broad and ends at a steep escarpment into the depths of the Gulf of Mexico basin. The shelf off Miami is very narrow, and the slope down into the Florida Strait is interrupted by the Pourtalés Terrace. The slope up to the Bahama Banks is almost vertical and swept clean of sediment by the Florida Current/Gulf Stream.

Figure 2c shows the transition from continent to ocean basin along an N-S section from the Ganges-Brahmaputra Delta into the Bay of Bengal. This is the most gradual

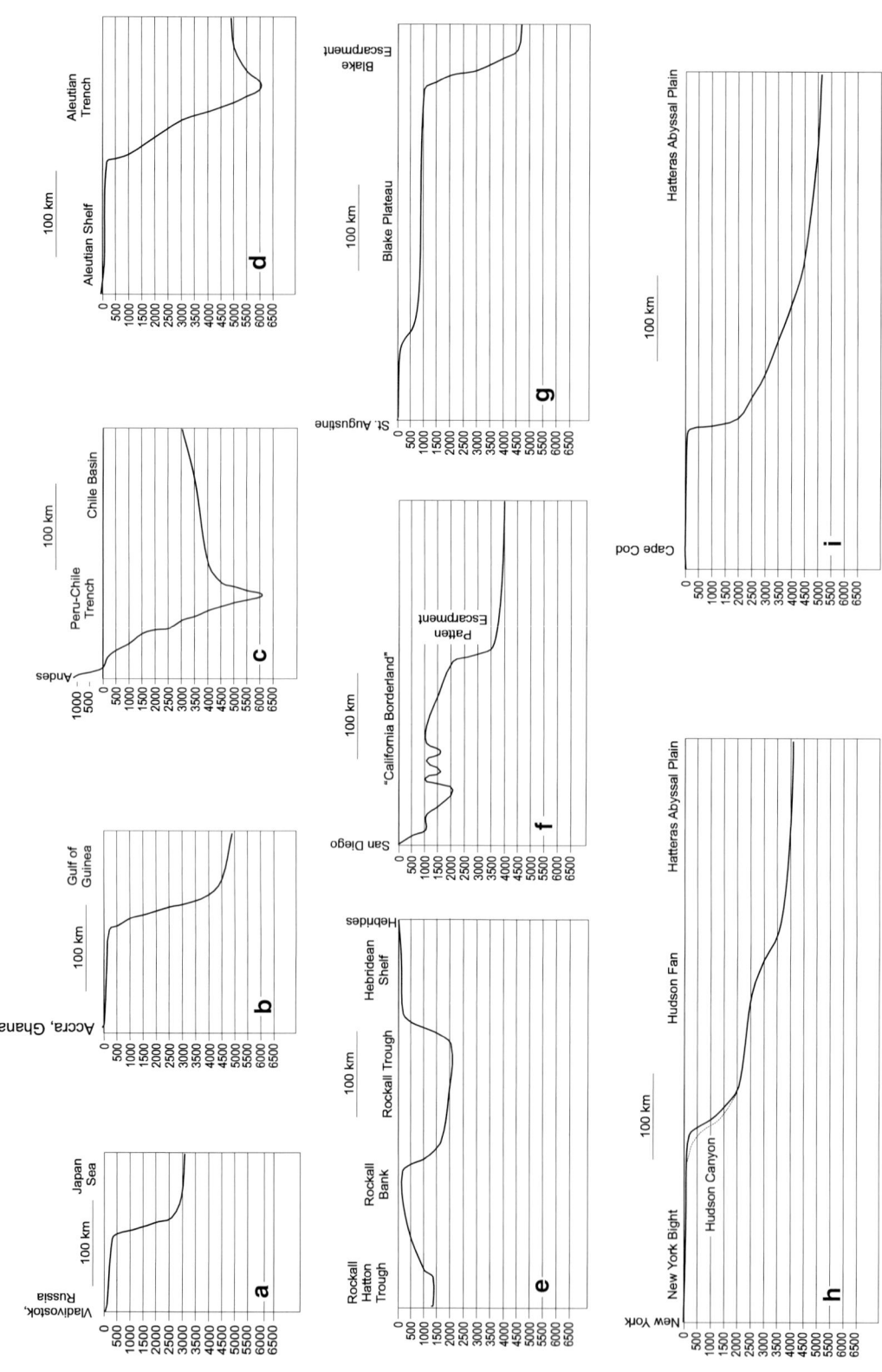

Continental Slope, Figure 1 Continental slope profiles on active and passive margins based on GEBCO (1984) charts.

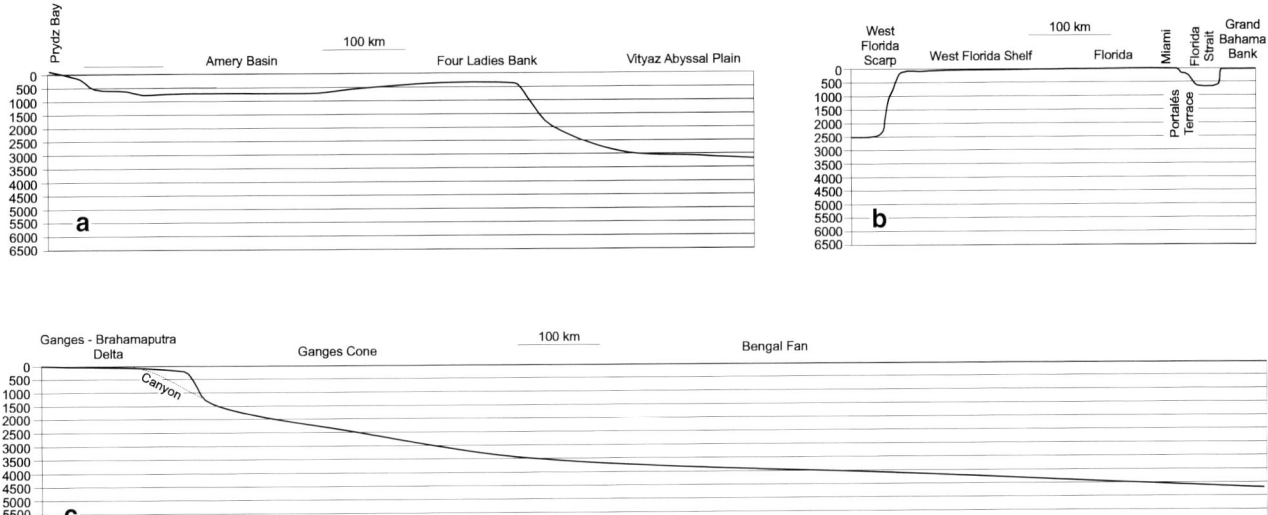

Continental Slope, Figure 2 Profiles of some unusual continental margins based on GEBCO (1984) charts.

transition on Earth, reflecting the huge supply of sediment from the erosion of the Himalaya. The trace of the canyon cut during sea-level lowstands is indicated as a dashed line.

Summary

The continental slope is the steep part of the transition from continent to ocean basin. However, the form of the transition is highly variable from place to place. The "classic" transition is that of the passive continental margin off the northeastern border of the United States, with a steep slope starting at the shelf break and a concave-upward profile. Continental slopes on active margins may descend directly from the shoreline into an oceanic trench. The complexity of the continental breakup process has resulted in a variety of forms of continental slope, many with interruptions that differ from the classic profile.

Bibliography

Bondevik, S., Løvholt, F., Harbitz, C., Mangerud, J., Dawson, A., and Svendsen, J. I., 2005. The Storegga Slide tsunami – comparing field observations with numerical simulations. *Marine and Petroleum Geology*, **22**, 195–208.

Doel, R. E., Levin, T. J., and Marker, M. K., 2006. Extending modern cartography to the ocean depths: military patronage, Cold War priorities, and the Heezen-Tharp mapping project, 1952–1959. *Journal of Historical Geography*, **32**, 605–626.

Haflidason, H., Lien, R., Sejrup, H. P., Forsberg, C. F., and Bryn, P., 2005. The dating and morphometry of the Storegga Slide. *Marine and Petroleum Geology*, **22**, 123–126.

Heezen, B., Tharp, M., and Ewing, M., 1959. The floors of the oceans: I. The North Atlantic. *Geological Society of America, Special Paper*, **65**, 1–126.

Salisbury, R. D., 1907. *Physiography*. New York: Henry Holt, p. 770.

Searle, R. C., Monahan, D., and Johnson, G. L., 1984. *GEBCO General Bathymetric Chart of the Oceans*, 5th edn. Ottawa: Canadian Hydrographic Service. 19 sheets.

Cross-references

Continental Rise
Ocean Margin Systems
Shelf

CONTOURITES

Heiko Hüneke
Institute of Geography and Geology, Ernst Moritz Arndt University, Greifswald, Germany

Synonyms

Bottom-current deposit

Definition

Contourites are the sediments deposited or significantly affected by the action of bottom currents (Stow et al., 2002a; Rebesco, 2005; Stow and Faugères, 2008). They comprise a group of essentially deepwater facies, typically formed below 300 m water depth under the influence of semipermanent currents (Stow et al., 2008).

Bottom currents

Any semipermanent water current that affects the seafloor by resuspending, transporting, and/or controlling the deposition of sediments may be called a bottom current (Rebesco et al., 2008; Stow et al., 2008). They are persistent features within specific regions of the ocean basin (detailed below), more or less continuously affecting the pattern of sedimentation over periods of hundreds to up to millions of years.

In modern oceans, contourites are formed mainly under the influence of bottom currents driven by the major thermohaline and wind-induced circulation systems (Faugères and Stow, 2008). In addition, bottom currents can operate as part of upwelling and downwelling currents, up- and down-canyon currents, internal tides and waves, and seafloor polishing and spillover (see Faugères and Stow, 2008; Shanmugam, 2008; and references therein). These processes produce rather isolated contourite facies commonly interbedded with other deep-sea sediments.

Bottom currents are predominantly unidirectional subsurface currents controlled by the ocean-basin configuration. When their flow direction largely parallels the submarine topography, they are called contour currents. Although bottom currents show a quasi-steady flow, they are highly variable in location, direction, and velocity over relatively short time scales (hours to months). Deepwater tidal effects, seasonal changes in water mass properties, and laterally migrating eddies modulate the velocity of the main bottom-current flow and even flow reversals may occur (Shanmugam, 2008).

The mean velocity of a bottom current and its ability to redistribute sediment are controlled by the Coriolis force and by the topography of the ocean basin (Stow et al., 2002a; Faugères and Mulder, 2011). Stow et al. (2008) differentiate between (i) low-velocity flows (mean velocity <10 cm s^{-1}), which typically move as a broad sluggish water body over low-gradient slopes and abyssal plains; (ii) intermediate-velocity flows (10–30 cm s^{-1}), which become intensified over steeper slopes and around topographic obstacles; and (iii) high-velocity flows (>30 cm s^{-1}), which are constricted and accelerated in narrow passages and over shallow sills. Flow velocities may exceed 100 cm s^{-1} where the flow is particularly constricted, such as through gateways on the deep-sea floor and within shallower straits, where flow velocities in excess of 200 cm s^{-1} have been recorded (Stow et al., 2002a).

Episodic downslope flows, which are not in equilibrium conditions with the oceanic water masses, do not represent bottom currents; for example, turbidity currents are short-lived (in the range of hours to months) and dense (more heavily loaded with suspended sedimentary particles) currents driven by gravity.

Contourite drifts

Bottom current-induced erosion and redeposition produce large-scale accumulations of sediment termed contourite drifts, which may be part of an even larger contourite depositional system, comprising several related drifts and associated erosional elements (Faugères et al., 1993; Stow et al., 2002b; Rebesco, 2005; Faugères and Stow, 2008). Contourite drifts were first described in the North Atlantic Ocean, where they were called "outer ridges" or "sediment drifts" (e.g., McCave and Tucholke, 1986). Present at most of modern continental margins, they are as frequent as deep-sea turbidite systems, occurring anywhere from the abyssal seafloor to slope settings. Sediment bodies deposited on the uppermost part of the continental slope and across the outer shelf edge are named "shallow-water contourite drifts."

Contourite drifts range in size from about 100 km^2 (small patch drifts) to larger than 100,000 km^2 (giant elongated drifts) and occasionally even approaching 1,000,000 km^2 (abyssal sheet drifts). They may form a positive relief of up to 2,000 m height. Larger contourite drifts may record long-term continuity of deposition of over several millions of years, resulting in the accumulation of several hundreds of meters of contourite sediments. Contourite drifts are mainly composed of the sediments deposited by bottom currents, but they may also enclose associated deepwater sediments, particularly pelagites, hemipelagites, and glaciomarine sediments, as well as being locally intercalated with turbidites and other density-flow deposits. They are often closely associated with erosional areas that display coeval current-induced winnowing of the seafloor, sediment bypassing, and the formation of coarser-grained (gravel-lag) contourites (Stow et al., 2008).

The principal types of contourite drifts that have been identified include sheeted, elongate-mounded, channel-related, confined, infill, and mixed systems (McCave and Tucholke, 1986; Faugères et al., 1999; Stow et al., 2002a; Rebesco, 2005). Case studies can be found in Stow and Faugères (1993, 1998), Stow et al. (2002c), Viana and Rebesco (2007), and Rebesco and Camerlenghi (2008). The classification summarized below (Figure 1) is based on a combination of criteria, such as drift distribution (i.e., location) and morphology, and furthermore illustrates the drift development in the context of a particular hydrological background (Faugères and Stow, 2008; Faugères and Mulder, 2011):

(1) Contourite-sheeted drifts accumulate over a broad area with very low relief and form a sediment body of more or less constant thickness. They represent relatively slow rates of deposition of fine-grained contourites (Stow et al., 2008). Extensive, impressive fields of sediment waves may cover their surface. Abyssal-sheeted drifts can extend across large basin plains with deposits of up to a few hundreds of meters thick, while slope-sheeted drifts are plastered against the continental margin and cover smaller areas.

(2) Mounded drifts show an elevated relief with thicker accumulation over a narrower, elongate region. They represent relatively enhanced rates of deposition of fine- to medium-grained contourites, commonly focused into slope-parallel, elongate sediment bodies over moderate to large areas of the deep seafloor (Stow et al., 2008). Along one or both lateral margins of mounded drifts, the flow tends to be markedly intensified, in many cases causing distinct zones that experience intermitted erosion and bypassing (contourite channels, moats, and marginal valleys).

(3) Elongate-mounded drifts are the archetypical contourite accumulations because of their distinctly mounded and elongated geometry. Giant elongated

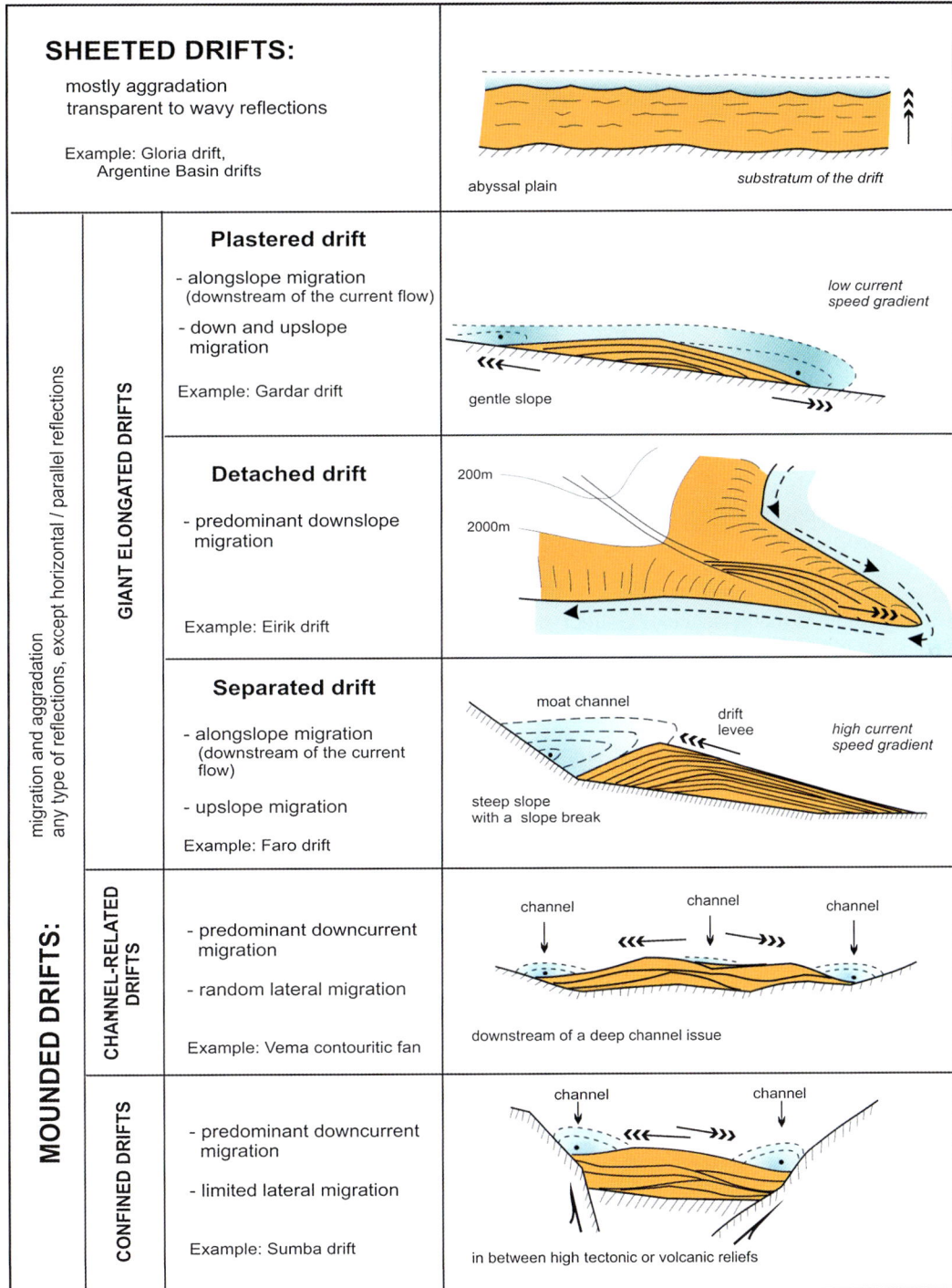

Contourites, Figure 1 Different types of contourite drifts (based on classifications of McCave and Tucholke, 1986; Faugères et al., 1993, 1999; Stow et al., 2002a), showing the general drift geometry and the trend of migration-aggradation (*triple black arrow*), as well as inferred *bottom*-current pathways (*broken-line black arrow*) (Redrawn and modified from Faugères and Mulder, 2011). Position of the main *bottom*-current flow (highest velocities) is indicated by *dashed* contour lines and *blue* color.

drifts are well recognizable as being of contourite origin, and they are the most spectacular because of their size (dimensions ranging from a few tens of kilometers to over 1,000 km). They are characterized by length–width ratios varying from 2:1 to 10:1 and thicknesses reaching several hundreds of meters. Elongate-mounded drifts have been documented anywhere from the outer shelf and upper slope down to the abyssal plains. McCave and Tucholke (1986) identified plastered, separated, and detached elongate-mounded drifts.

(4) Drifts related to deep channels or other oceanic passages (gateways, seaways, gaps, and straits), where the bottom currents are constricted and flow velocities are substantially increased, occur as axial and lateral patch drifts (either mounded or sheet form). Within the channel, the localized deposition of finer-grained contourite facies is closely related to significant erosion of the passage seafloor, widespread sediment bypassing, and the formation of coarse-grained lag deposits. Near the channel exit region, where the flow is expanded and decelerated, sheetlike contourite fans form, typically with a downflow decrease in contourite grain size. The channel-related patch drifts are typically small, a few kilometers in diameter, and up to 150 m thick. The contourite fans are much larger cone-shaped accumulations, up to 100 km or more in radius and 300 m in thickness.

(5) Confined contourite drifts are characterized by an elongate-mounded geometry aligned parallel to the axis of a relatively small confining basin. A significant feature is the presence of distinct moats along both sides of the drift, suggesting that the fairly slow contour current is confined on both margins or that it develops into some kind of circulatory pattern within the basin (Faugères and Stow, 2008).

(6) Mixed drift systems are those that involve significant interaction of *along-slope* bottom currents with other depositional processes in the buildup of the drift body. These drifts include substantial deepwater facies deposited either by episodic *downslope* processes (e.g., turbidites, debrites, and hyperpycnites) or by continuous *vertical* settling (e.g., pelagites, hemipelagites, periplatform carbonates, ice-rafted debris). The regular contourite-drift morphologies may be markedly modified by significant downslope sediment transport, and vice versa, bottom currents crossing deep-sea turbidite systems may induce a lateral shift of the channel-levee complexes. Mulder et al. (2008) outline the variable aspects of these alternating and interacting processes and the resulting drift geometries.

Contourite sediment facies

Depending on the flow characteristics and on local sediment sources, a large variety of contourite facies occurs in modern oceans (Stow et al., 1996, 2002a; Stow and Faugères, 2008). The grain size of contourites varies from fine muds, through silts and sands, to sand and gravel-lag components. The composition of the sediment can be terrigenous siliciclastic, biogenic (calcareous and siliceous), volcaniclastic, and chemogenic (manganiferous, phosphatic). Biogenic components typically comprise whole tests or shell fragments of foraminiferas, radiolarians, diatoms, spicules, and coccoliths. The different grain-size fractions and diverse materials are often poorly sorted mixtures. The rare contourite deposits identified in the fossil record are mainly carbonate-rich contourites (Hüneke and Stow, 2008; e.g. Hüneke 2013).

It is the bottom-current velocity that mainly governs the depositional processes and resulting sediment facies (Faugères and Mulder, 2011). Low-velocity flows allow vertical setting of fine-grained suspended particles from the nepheloid layer, whereas intermediate-velocity flows induce more bed load transport and deposition of silt and sand. High-velocity flows can affect still larger grain sizes and generate large-scale winnowing and erosion, resulting in coarse-grained lag deposits. Dissolution and authigenesis can accompany the current-induced processes, generating very low sedimentation rates and sediment bypassing.

The large variety of contourite facies typically displays (a) abundant bioturbational sedimentary structures, (b) bad preservation of hydrodynamic sedimentary structures, and (c) an irregular vertical variation in grain size. Burrowing animals intensely colonize contourite sediments due to their low accumulation rates and to the generally well-oxygenated (cold) water masses prevailing during bottom-current-induced deposition. Several types of ichnofabrics can be distinguished, depending on the prevailing current intensity (Wetzel et al., 2008). Hydrodynamic (small-scale) sedimentary structures are scarce or completely lacking, being obliterated by the permanent bioturbation, except when the sedimentation rate exceeds the bioturbation rate (Martin-Chivelet et al., 2008). Consequently, the vertical variation in grain size and sediment facies, which typically displays a repeated superposition of coarsening-upward and fining-upward units, may be difficult to recognize.

The facies model for contourites typifies the vertical grain-size variation of a standard mud-silt-sand contourite sequence, which results from variation in contour-current velocity (Faugères et al., 1984; Stow et al., 2002a; Stow and Faugères, 2008). A complete sequence of any particle composition consists of five principal divisions: (C1) lower muddy contourite division, (C2) lower mottled silty contourite division, (C3) middle sandy contourite division, (C4) upper mottled silty contourite division, and (C5) upper muddy contourite division. Traction structures, including rare cross-lamination, are more evident in the coarse silts and sands than in finer-grained facies. There may be an indistinct and discontinuous parallel lamination, and lenses of coarser materials may occur in all intervals.

The centimeter- to decimeter-thick successions of the standard sequence may be truncated. In that case, some divisions are missing, and erosional-sharp contacts or nondeposition surfaces are present between the facies units (typically at division C3). In fact, partial sequences of different thickness are at least as common as the full sequence. The complete absence of certain divisions can be related to an increase in bottom-current velocity, resulting in a prolonged phase of nondeposition and/or erosion. Consequently, gravel-lag deposits, manganiferous horizons, and phosphatized hardgrounds can be produced.

The standard contourite sequence is interpreted to reflect at least sub-regular periodicity in mean bottom-current velocity (and/or in variation in sediment supply), related to climate variation or other external factors. Based on modern age data, Stow et al. (2002a) calculate approximate sequence periodicities for a variety of marginal drifts of terrigenous to mixed composition (Gulf of Cadiz, Rockall Margin, West Shetlands, and Norwegian Margin). In each case, they obtained a periodicity of between 5,000 and 20,000 years (Stow and Faugères, 2008).

Summary

Contourites are generally not easily recognized because of the lack of unambiguous diagnostic criteria. Their deposition involves multiphase entrainment of sedimentary particles and long-distance transport and furthermore depends on the interaction among the depositional processes induced by the various types of bottom currents operating in the deep sea. Hydrodynamic sedimentary structures are usually obliterated by coeval, permanent bioturbational mottling. Therefore, a composite triple-stage approach is recommended for the identification of sediments deposited by bottom currents (Stow et al., 1998). The analysis must include small-scale criteria (sedimentary facies and sequences), medium-scale criteria (drift geometry and internal architecture, facies distribution and trends, bottom-current activity, unconformities), and large-scale criteria (oceanographic features and continental reconstruction, basin configuration, drift location) in order to unambiguously identify a contourite.

Bibliography

Faugères, J. C., and Mulder, T., 2011. Contour currents and contourite drifts. In Hüneke, H., and Mulder, T. (eds.), *Deep-Sea Sediments*. Amsterdam: Elsevier, pp. 149–214.

Faugères, J.-C., and Stow, D. A. V., 2008. Contourite drifts: nature, evolution and controls. In Rebesco, M., and Camerlenghi, A. (eds.), *Contourites*. Amsterdam: Elsevier, pp. 257–288.

Faugères, J.-C., Gonthier, E., and Stow, D. A. V., 1984. Contourite drift moulded by deep Mediterranean outflow. *Geology*, **12**, 296–300.

Faugères, J.-C., Mezerais, M. L., and Stow, D. A. V., 1993. Contourite drift types and their distribution in the North and South Atlantic Ocean basins. *Sedimentary Geology*, **82**, 189–203.

Faugères, J.-C., Stow, D. A. V., Imbert, P., and Viana, A. R., 1999. Seismic features diagnostic of contourite drifts. *Marine Geology*, **162**, 1–38.

Hüneke, H., 2013. Bioclastic contourites: depositional model for bottom-current redeposited pelagic carbonate ooze (Devonian, Moroccan Central Massif). Z. dt. Ges. Geowiss. (German Jour. Geosci.), 164, 253–277.

Hüneke, H., and Stow, D. A. V., 2008. Identification of ancient contourites: problems and palaeoceanographic significance. In Rebesco, M., and Camerlenghi, A. (eds.), *Contourites*. Amsterdam: Elsevier, pp. 323–344.

Martín-Chivelet, J., Fregenal-Martínez, M. A., and Chacón, B., 2008. Traction structures in contourites. In Rebesco, M., and Camerlenghi, A. (eds.), *Contourites*. Amsterdam: Elsevier, pp. 159–182.

McCave, I. N., and Tucholke, B. E., 1986. Deep current-controlled sedimentation in the western North Atlantic. In Vogt, P. R., and Tucholke, B. E. (eds.), *The Geology of North America, The Western North Atlantic Region. Decade of North American Geology*. Boulder: Geology Society of America, pp. 451–468.

Mulder, T., Faugères, J.-C., and Gonthier, E., 2008. Mixed turbidite-contourite systems. In Rebesco, M., and Camerlenghi, A. (eds.), *Contourites*. Amsterdam: Elsevier, pp. 435–456.

Rebesco, M., 2005. Contourites. In Selley, R. C., Cocks, L. R. M., and Plimer, I. R. (eds.), *Encyclopedia of Geology*. Oxford: Elsevier, Vol. 4, pp. 513–527.

Rebesco, M., and Camerlenghi, A. (eds.), 2008. *Contourites*. Amsterdam: Elsevier.

Rebesco, M., Camerlenghi, A., and VanLoon, A. J., 2008. Contourite research: a field in full development. In Rebesco, M., and Camerlenghi, A. (eds.), *Contourites*. Amsterdam: Elsevier, pp. 1–10.

Shanmugam, G., 2008. Deep-water bottom currents and their deposits. In Rebesco, M., and Camerlenghi, A. (eds.), *Contourites*. Amsterdam: Elsevier, pp. 59–81.

Stow, D. A. V., and Faugères, J.-C. (eds.), 1993. *Contourites and Bottom-Currents*. Amsterdam/New York: Elsevier. Sedimentary Geology, Vol. 82, pp. 1–310.

Stow, D. A. V., and Faugères, J.-C. (eds.), 1998. *Contourites, Turbidites and Process Interaction*. Amsterdam: Elsevier Science. Sedimentary Geology, Vol. 115, pp. 1–384.

Stow, D. A. V., and Faugères, J. C., 2008. Contourite facies and the facies model. In Rebesco, M., and Camerlenghi, A. (eds.), *Contourites*. Amsterdam: Elsevier, pp. 223–256.

Stow, D. A. V., Reading, H. G., and Collinson, J. D., 1996. Deep sea. In Reading, H. G. (ed.), *Sedimentary Environments: Processes, Facies and Stratigraphy*. London: Blackwell Science, pp. 395–453.

Stow, D. A. V., Faugères, J.-C., Viana, A. R., and Gonthier, E., 1998. Fossil contourites: a critical review. *Sedimentary Geology*, **115**, 3–31.

Stow, D. A. V., Faugères, J. C., Howe, J. A., Pudsey, C. J., and Viana, A. R., 2002a. Bottom currents, contourites and deep-sea sediment drifts: current state-of-the-art. In Stow, D. A. V., Pudsey, C. J., Howe, J. A., Faugères, J.-C., and Viana, A. R. (eds.), *Deep-Water Contourite Systems: Modern Drifts and Ancient Series, Seismic and Sedimentary Characteristics*. London: Geological Society. Memoir (Geological Society of London), Vol. 22, pp. 7–20.

Stow, D. A. V., Faugères, J.-C., Gonthier, E., Cremer, M., Llave, E., Hernandez-Molina, F. J., Somoza, L., and Diaz-Del-Rio, V., 2002b. Faro-Albufeira drift complex, northern Gulf of Cadiz. In Stow, D. A. V., Pudsey, C. J., Howe, J. A., Faugères, J.-C., and Viana, A. R. (eds.), *Deep-Water Contourite Systems: Modern Drifts and Ancient Series, Seismic and Sedimentary Characteristics*. London: Geological Society. Memoir (Geological Society of London), Vol. 22, pp. 137–154.

Stow, D. A. V., Pudsey, C. J., Howe, J. A., Faugères, J.-C., and Viana, A. R. (eds.), 2002c. *Deep-Water Contourite Systems: Modern Drifts and Ancient Series, Seismic and Sedimentary Characteristics*. London: Geological Society. Memoir (Geological Society of London), Vol. 22, pp. 1–464.

Stow, D. A. V., Hunter, S., Wilkinson, D., and Hernández-Molina, F. J., 2008. Contourite facies and the facies model. In Rebesco, M., and Camerlenghi, A. (eds.), *Contourites*. Amsterdam: Elsevier, pp. 143–156.

Viana, A. R., and Rebesco, M. (eds.), 2007. *Economic and Palaeoceanographic Significance of Contourite Deposits*. London: Geological Society. Geological Society Special Publication, Vol. 276, pp. 1–350.

Wetzel, A., Werner, F., and Stow, D. A. V., 2008. Bioturbation and biogenic sedimentary structures in contourites. In Rebesco, M., and Camerlenghi, A. (eds.), *Contourites*. Amsterdam: Elsevier, pp. 183–202.

Cross-references

Continental Rise
Continental Slope
Deep-sea Sediments

CONVERGENCE TEXTURE OF SEISMIC REFLECTORS

Christopher George St. Clement Kendall
Earth and Ocean Sciences, University of South Carolina, Columbia, SC, USA

Depositional and erosional bounding surfaces define sedimentary sequences, systems tracts, and parasequences (see section on the topics of Sedimentary Sequence in the encyclopedia). They match surfaces that coincide with the convergence and termination of seismic reflectors but are best displayed in outcrops and are often easily identified in well logs and cores.

In seismic reflection data the convergence and termination patterns and their continuity are used to identify sequence stratigraphic surfaces. On seismic data the most prominent of these surfaces are subaerial unconformities marking sequence boundaries (SB) eroded during sea-level fall (Sloss et al., 1949) and time-equivalent submarine correlative conformities (SB-CC) (Hunt and Tucker, 1992; Posamentier et al., 1988; Figure 1).

SB and geometries of sediment fill as they respond to changes in relative sea-level

- An upper sequence boundary:
  - Toplap: termination of strata against an overlying surface, as the result of nondeposition and/or minor erosion
  - Truncation: removal by erosion (often after subsequent tilting; representing the most reliable top-discordant criterion of a sequence boundary) or with a channel termination against erosional surface
- A lower sequence boundary:
  - Onlap – A base-discordant relationship in which initially horizontal strata progressively terminate against an initially inclined surface or in which initially inclined strata terminate progressively updip against a surface of greater initial inclination.
  - Downlap: a relationship in which seismic reflections of inclined strata terminate downdip against an inclined or horizontal surface. Examples of downlap surfaces include a top basin-floor fan surface, a top slope fan surface, and a maximum flooding surface.

Subsequent deformation of the sedimentary section can make onlap indistinguishable from downlap, so in this case the term baselap is used; in contrast offlap refers to en echelon progradation of (outward building) strata often expressed in seismic and other data.

NB: *Visit the SEPM STRATA website* (http://www.sepmstrata.org/) *for explanations of the sequence stratigraphic terminology used here, which is linked to pop-up boxes containing information that clarify the understanding and use of this discipline of stratigraphy.*

Bibliography

Catuneanu, O., 2002. Sequence stratigraphy of clastic systems: concepts, merits, and pitfalls. *Journal of African Earth Sciences*, **35**(1), 1–43.

Hunt, D., and Tucker, M. E., 1992. Stranded parasequences and the forced regressive wedge systems tract: deposition during base-level fall. *Sedimentary Geology*, **81**, 1–9.

Sloss, L. L., 1963. Sequences in the cratonic interior of North America. *Geological Society of America Bulletin*, **74**, 93–114.

Cross-references

Geological Time Scale
Relative Sea-level (RSL) Cycle
Sea-Level
Sedimentary Sequence
Sequence Stratigraphy

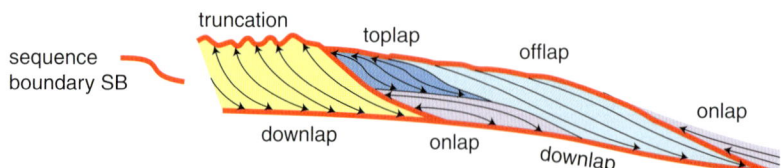

Convergence Texture of Seismic Reflectors, Figure 1 Convergence and termination of seismic reflectors define sequence boundaries (Christopher Kendall, 2013 (after Catuneanu, 2002)).

CRUSTAL ACCRETION

Benoît Ildefonse
Geosciences Montpellier, CNRS and University of Montpellier, Montpellier, France

Synonyms

Formation of ocean crust

Definition

Crustal Accretion. Formation of ocean crust at mid-ocean ridges, by a combination of magmatic, tectonic, and hydrothermal processes.

Introduction

Since the early 1970s, the classical model of a uniformly layered ocean crust, with from top to bottom basaltic lava flows, basaltic sheeted dikes, and gabbros, known as the "Penrose" model (Anonymous, 1972), has considerably evolved, with complementary contributions from ophiolite studies, seafloor geology, marine geophysics, and scientific ocean drilling. At the global scale, there is substantial variability of the mid-ocean ridge morphology and hence of crustal architecture. This results from various modes of accretion that are controlled by magma supply to the ridge, which itself primarily depends on the spreading rate. This article briefly describes the current end-member models for crustal accretion at fast-spreading (>~80 mm/year) and slow-spreading (<~40 mm/year) ridges.

Fast-spreading ridges: a nearly continuous magma chamber

At fast-spreading (and part of intermediate-spreading) ridges, the ocean crust is assumed to be relatively simple and continuously layered, in accordance with the "Penrose" model, as a result from the presence of nearly continuous and steady-state axial magma chamber, along the axis. Most of the first models for mid-ocean ridge magma chambers, in the 1970s and early 1980s, proposed rather large magma bodies, ~10–20 km large across the ridge axis and ~4 km thick at the axis, lying below the sheeted dike complex (e.g., Cann, 1974; Pallister and Hopson, 1981). Using thermal constraints, Sleep (1975) proposed that most of the chamber was filled with magmatic mush and that melt-rich magma would be limited to the top, below the sheeted dikes. This prediction was confirmed about 10 years later, with the first seismic reflection images of an axial magma chamber beneath the Valu Fa Ridge in the Lau Basin (Morton and Sleep, 1985) and beneath the East Pacific Rise (e.g., Hale et al., 1982; Detrick et al., 1987; Kent et al., 1994). This shallow and thin magma lens then became a key element of all models for accretion at fast-spreading ridges. Details of the relationships between the morphologic segmentation of the ridge, the along-axis variability (depth and melt content) of the magma lens, volcanic activity, lava chemistry, and hydrothermal cooling in the upper crust are discussed in many studies published over the last three decades (e.g., Macdonald et al., 1988; Hooft et al., 1997; Singh et al, 1998; Goss et al., 2010; Carbotte et al., 2013; Xu et al., 2014). In intact fast-spread crust, the deepest stratigraphic level reached by drilling is the transition zone between the sheeted dike complex and the gabbros of the lower crust, in the ODP (Ocean Drilling Program) Hole 1256D on the Cocos Plate (Wilson et al., 2006; Teagle et al., 2012; Ildefonse et al., 2014). This zone, also described in ophiolites (e.g., Gillis, 2008; France et al., 2009), marks the thermal boundary layer between the magmatic and hydrothermal systems at the ridge axis. It displays a complex petrological record of the interplay between temporally and spatially intercalated magmatic, hydrothermal, partially melting, and metamorphic processes.

Below the axial magma lens is a seismically attenuated domain, which is believed to correspond to a melt-poor magmatic mush (e.g., Sinton and Detrick, 1992; Dunn et al., 2000). Two competing models are proposed for building the lower, gabbroic crust in this domain. The "gabbro glacier" models (e.g., Henstock et al., 1993; Phipps Morgan and Chen, 1993; Figure 1a) postulate that the entire lower crust is formed by the flow of mushy material downward and outward from the single, shallow axial magma lens. In contrast, "sheeted sill models," largely based on observations made in the Oman ophiolite (e.g., Kelemen et al., 1997; Figure 1b), involve in situ formation of the lower crust by sill intrusions. Boudier et al. (1996) proposed a combination of these two end-members. Deep, lower crustal seismic reflectors, interpreted as magma sills accreting the lower igneous crust, have now been imaged below the axial magma lens at the Juan de Fuca Ridge (Canales et al., 2009) and at the East Pacific Rise (Marjanović et al., 2014). Accretion models can be used for predicting contrasted vertical trends of chemical composition, deformation, and alteration in the gabbroic crust, which still remain to be fully tested in ophiolites and by deep drilling in intact fast-spread crust (e.g., Teagle et al., 2012; Ildefonse et al., 2014). While the gabbro glacier model does not require deep hydrothermal circulation close to the axis, the latter is needed to sustain a sheeted sill model (Figure 1b; e.g., Maclennan et al., 2005). The IODP (Integrated Ocean Drilling Program) Expedition 345 recovered, for the first time, significant sections of layered gabbros from disrupted lower crust exposed at Hess Deep in the equatorial Eastern Pacific (Gillis et al., 2014), where young (~1 Ma) lower fast-spread crust is tectonically exposed at the tip of the Cocos-Nazca rift.

Slow-spreading ridges: a variable, composite crust

In contrast to fast-spread crust, ocean crust created at slow- and ultraslow-spreading ridges is spatially heterogeneous, both along and across the ridge axis. Along parts of slow-spreading ridges (e.g., the centers of ridge segments

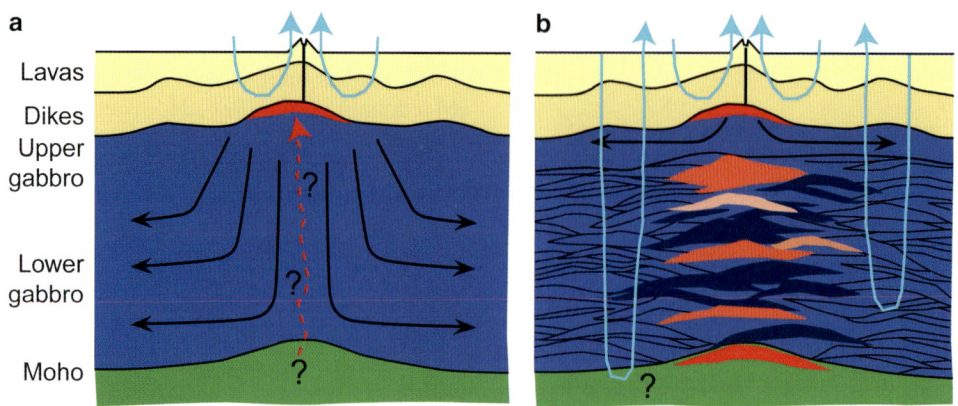

Crustal Accretion, Figure 1 Sketches of end-member crustal accretion models (modified from Kelemen et al., 1997). (a) Gabbro glacier model. (b) Sheeted sill model of in situ formation of the lower crust by on-axis sill intrusions.

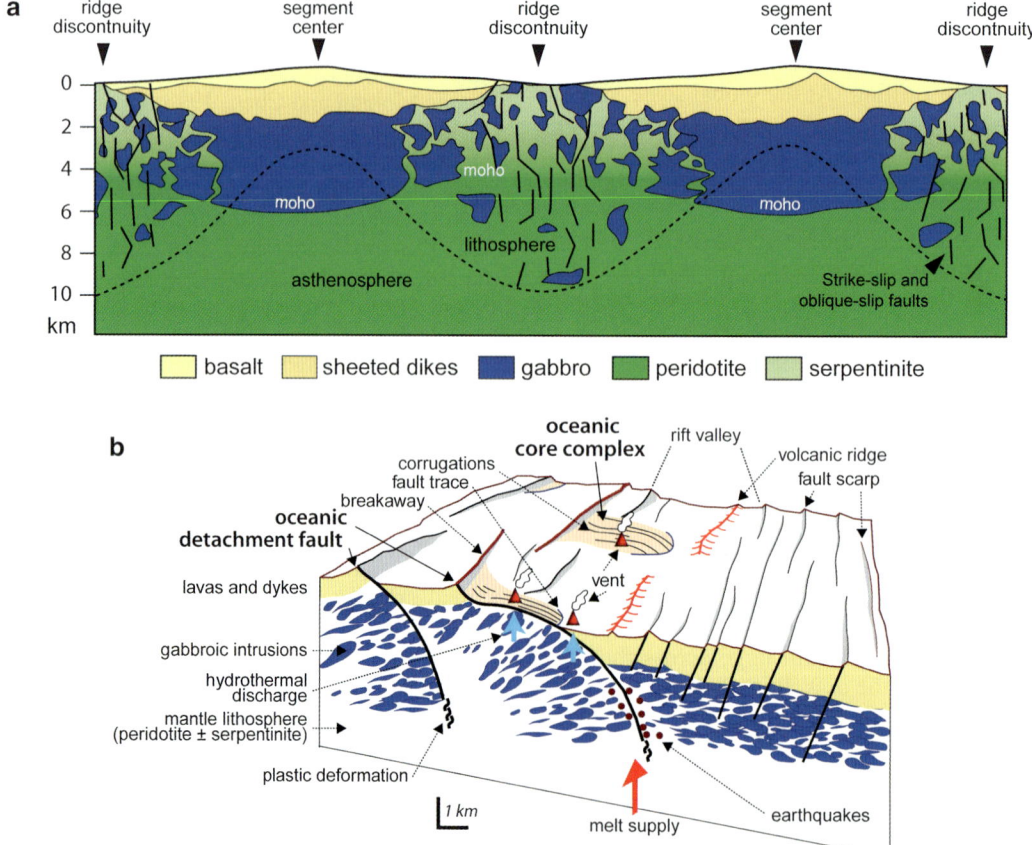

Crustal Accretion, Figure 2 (a) Simplified, interpreted, axis-parallel section through slow-spread crust (modified from Cannat et al., 1995). (b) Sketch of crustal accretion controlled by detachment faulting (modified from Escartín and Canales, 2011).

in the northern Atlantic), magmatic processes dominate. Seismic imaging has revealed an axial magma chamber beneath the Lucky Strike Volcano (37°18′N) at the Mid-Atlantic Ridge (Singh et al., 2006), similar to that imaged along fast-spreading ridges but limited in space to 7 km along the axis. The architecture of slow-spread crust typically changes along axis toward segment ends, where it is more heterogeneous, with a mixture of serpentinized mantle peridotite and gabbroic intrusions locally capped by lavas with or without intervening sheeted dikes (e.g., Cannat, 1993; Figure 2a). This lateral and vertical variability of slow-spread crust has been well documented through ophiolite studies, seafloor geological studies, geophysical surveys, and scientific ocean drilling expeditions (e.g., Karson et al., 1987; Dick, 1989; Cannat et al., 1995; Canales et al., 2000; Kelemen et al., 2007; Blackman et al., 2011; Sauter et al., 2013; Ildefonse et al., 2014; Lagabrielle et al., 2015). In this type of crust, tectonics play a key role in crustal construction, through detachment faulting (e.g., Escartín and Canales, 2011; Figure 2b), which dominates the style of accretion over ~50 % of the Mid-Atlantic Ridge between 12.5°N and 35°N (Escartín et al., 2008). Long-living (~1–2 Ma) detachment faults cap domal features, known as "oceanic core complexes" or "megamullions," where gabbroic rocks, serpentinites, and fault-related material (e.g., talc-amphibole schists) are typically sampled (e.g., Cann et al., 1997; Tucholke et al., 1998; Escartín et al., 2003; Ildefonse et al., 2007; MacLeod et al., 2009; Dick et al., 2010). The magma-poor end-member for this style of accretion is expressed at the ultraslow-spreading Southwest Indian Ridge by the nonvolcanic "smooth seafloor," where almost exclusively serpentinized mantle peridotites are sampled (Cannat et al., 2006; Sauter et al., 2013). Details of the detachment-controlled accretion mode remain poorly constrained. In oceanic core complexes, drill cores document a complex interplay between magmatism, faulting, and hydrothermal cooling, which is yet to be fully resolved (e.g., Blackman et al., 2011; McCaig and Harris, 2012; Ildefonse et al., 2014).

Bibliography

Anonymous, 1972. Penrose field conference on ophiolites. *Geotimes*, **17**, 24–25.

Blackman, D. K., Ildefonse, B., John, B. E., Ohara, Y., Miller, D. J., Abe, N., Abratis, M., Andal, E. S., Andreani, M., Awaji, S., Beard, J. S., Brunelli, D., Charney, A. B., Christie, D. M., Collins, J., Delacour, A. G., Delius, H., Drouin, M., Einaudi, F., Escartín, J., Frost, B. R., Früh-Green, G., Fryer, P. B., Gee, J. S., Godard, M., Grimes, C. B., Halfpenny, A., Hansen, H. E., Harris, A. C., Tamura, A., Hayman, N. W., Hellebrand, E., Hirose, T., Hirth, J. G., Ishimaru, S., Johnson, K. T. M., Karner, G. D., Linek, M., MacLeod, C. J., Maeda, J., Mason, O. U., McCaig, A. M., Michibayashi, K., Morris, A., Nakagawa, T., Nozaka, T., Rosner, M., Searle, R. C., Suhr, G., Tominaga, M., von der Handt, A., Yamasaki, T., and Zhao, X., 2011. Drilling constraints on lithospheric accretion and evolution at Atlantis Massif, Mid-Atlantic Ridge 30°N. *Journal of Geophysical Research*, **116**, B07103, doi:10.1029/2010JB007931.

Boudier, F., Nicolas, A., and Ildefonse, B., 1996. Magma chambers in the Oman ophiolite: fed from the top and the bottom. *Earth and Planetary Science Letters*, **144**, 239–250, doi:10.1016/0012-821X(96)00167-7.

Canales, J. P., Collins, J. A., Escartín, J., and Detrick, R. S., 2000. Seismic structure across the rift valley of the Mid-Atlantic ridge at 23°20′N (MARK area): implications for crustal accretion processes at slow-spreading ridges. *Journal of Geophysical Research*, **105**, 28411–28425, doi:10.1029/2000JB900301.

Canales, J. P., Nedimović, M. R., Kent, G. M., Carbotte, S. M., and Detrick, R. S., 2009. Seismic reflection images of a near-axis melt sill within the lower crust at the Juan de Fuca ridge. *Nature*, **460**, 89–93, doi:10.1038/nature08095.

Cann, J. R., 1974. A model for oceanic crystal structure developed. *Geophysical Journal International*, **39**, 169–187, doi:10.1111/j.1365-246X.1974.tb05446.x.

Cann, J. R., Blackman, D. K., Smith, D. K., McAllister, E., Janssen, B., Mello, S., Avgerinos, E., Pascoe, A. R., and Escartín, J., 1997. Corrugated slip surfaces formed at ridge-transform intersections on the Mid-Atlantic Ridge. *Nature*, **385**, 329–332, doi:10.1038/385329a0.

Cannat, M., 1993. Emplacement of mantle rocks in the seafloor at mid-ocean ridges. *Journal of Geophysical Research*, **98**, 4163–4172, doi:10.1029/92JB02221.

Cannat, M., Mével, C., Maia, M., Deplus, C., Durand, C., Gente, P., Agrinier, P., Belarouchi, A., Dubuisson, G., Humler, E., and Reynolds, J., 1995. Thin crust, ultramafic exposures, and rugged faulting patterns at the Mid-Atlantic Ridge (22°–24°N). *Geology*, **23**, 49–52, doi:10.1130/0091-7613(1995)023<0049:TCUEAR>2.3.CO;2.

Cannat, M., Sauter, D., Mendel, V., Ruellan, É., Okino, K., Escartín, J., Combier, V., and Baala, M., 2006. Modes of seafloor generation at a melt-poor ultraslow-spreading ridge. *Geology*, **34**, 605–608, doi:10.1130/G22486.1.

Carbotte, S. M., Marjanović, M., Carton, H., Mutter, J. C., Canales, J. P., Nedimović, M. R., Han, S., and Perfit, M. R., 2013. Fine-scale segmentation of the crustal magma reservoir beneath the East Pacific Rise. *Nature Geoscience*, **6**, 866–870, doi:10.1038/ngeo1933.

Detrick, R. S., Buhl, P., Vera, E., Mutter, J., Orcutt, J., Madsen, J., and Brocher, T., 1987. Multi-channel seismic imaging of a crustal magma chamber along the East Pacific Rise. *Nature*, **326**, 35–41, doi:10.1038/326035a0.

Dick, H. J. B., 1989. Abyssal peridotites, very slow spreading ridges and ocean ridge magmatism. In Saunders, A. D., and Norry, M. J. (eds.), *Magmatism in the Ocean Basins*. London: Geological Society. Geological Society Special Publications, Vol. 42, pp. 71–105.

Dick, H. J. B., Lissenberg, C. J., and Warren, J. M., 2010. Mantle melting, melt transport, and delivery beneath a slow-spreading ridge: the Paleo-MAR from 23°15′N to 23°45′N. *Journal of Petrology*, **51**, 425–467, doi:10.1093/petrology/egp088.

Dunn, R. A., Toomey, D. R., and Solomon, S. C., 2000. Three-dimensional seismic structure and physical properties of the crust and shallow mantle beneath the East Pacific Rise at 9°30′N. *Journal of Geophysical Research*, **105**, 23537–23555, doi:10.1029/2000JB900210.

Escartín, J., Mével, C., MacLeod, C.J., and McCaig, A.M., 2003. Constraints on deformation conditions and the origin of oceanic detachments: The Mid-Atlantic Ridge core complex at 15°45′N. *Geochem Geophys Geosyst*, **4**, doi:10.1029/2002GC000472.

Escartín, J., and Canales, J. P., 2011. Detachments in oceanic lithosphere: deformation, magmatism, fluid flow, and ecosystems. *Eos, Transactions of the American Geophysical Union*, **92**, 31, doi:10.1029/2011EO040003.

Escartín, J., Smith, D. K., Cann, J., Schouten, H., Langmuir, C. H., and Escrig, S., 2008. Central role of detachment faults in accretion of slow-spreading oceanic lithosphere. *Nature*, **455**, 790–794, doi:10.1038/nature07333.

France, L., Ildefonse, B., and Koepke, J., 2009. Interactions between magma and hydrothermal system in Oman ophiolite and in IODP Hole 1256D: fossilization of a dynamic melt lens at fast spreading ridges. *Geochemistry, Geophysics, Geosystems*, **10**, Q10O19, doi:10.1029/2009GC002652.

Gillis, K. M., 2008. The roof of an axial magma chamber: a hornfelsic heat exchanger. *Geology*, **36**, 299–302, doi:10.1130/G24590A.1.

Gillis, K. M., Snow, J. E., Klaus, A., Abe, N., Adrião, Á. B., Akizawa, N., Ceuleneer, G., Cheadle, M. J., Faak, K., Falloon, T. J., Friedman, S. A., Godard, M., Guerin, G., Harigane, Y., Horst, A. J., Hoshide, T., Ildefonse, B., Jean, M. M., John, B. E., Koepke, J., Machi, S., Maeda, J., Marks, N. E., McCaig, A. M., Meyer, R., Morris, A., Nozaka, T., Python, M., Saha, A., and Wintsch, R. P., 2014. Primitive layered gabbros from fast-spreading lower oceanic crust. *Nature*, **505**, 204–207, doi:10.1038/nature12778.

Goss, A. R., Perfit, M. R., Ridley, W. I., Rubie, K. H., Kamenov, G. D., Soule, S. A., Fundis, A., and Fornari, D. J., 2010. Geochemistry of lavas from the 2005–2006 eruption at the East Pacific Rise, 9°46′N-9°56′N: implications for ridge crest plumbing and decadal changes in magma chamber compositions. *Geochemistry, Geophysics, Geosystems*, **11**, Q05T09, doi:10.1029/2009GC002977.

Hale, L. D., Morton, C. J., and Sleep, N. H., 1982. Reinterpretation of seismic reflection data over the East Pacific Rise. *Journal of Geophysical Research*, **87**, 7707–7717, doi:10.1029/JB087iB09p07707.

Henstock, T. J., Woods, A. W., and White, R. S., 1993. The accretion of oceanic crust by episodic sill intrusion. *Journal of Geophysical Research*, **98**, 4143–4161, doi:10.1029/92JB02661.

Hooft, E. E. E., Detrick, R. S., and Kent, G. M., 1997. Seismic structure and indicators of magma budget along the southern East Pacific Rise. *Journal of Geophysical Research*, **102**, 27319–27340, doi:10.1029/97JB02349.

Ildefonse, B., Blackman, D. K., John, B. E., Ohara, Y., Miller, D. J., MacLeod, C. J., and IODP Expeditions 304/305 Science Party, 2007. Oceanic core complexes and crustal accretion at slow-spreading ridges. *Geology*, **35**, 623–626, doi:10.1130/G23531A.1.

Ildefonse, B., Abe, N., Godard, M., Morris, A., Teagle, D.A.H., and Umino, S., 2014. Formation and evolution of oceanic lithosphere: new insights on crustal structure and igneous chemistry from ODP/IODP Sites 1256, U1309 and U1415. In Stein, R., Blackman, D., Inagaki, F., and Larsen H.-C. (eds.), Earth and life processes discovered from subseafloor environments, a decade of science achieved by the Integrated Ocean Drilling Program (IODP). *Developments in Marine Geology*, **7**, 449–505. doi:10.1016/B978-0-444-62617-2.00017-7.

Karson, J. A., Thompson, G., Humphris, S. E., Edmond, J. M., Bryan, W. B., Brown, J. R., Winters, A. T., Pockalny, R. A., Casey, J. F., Campbell, A. C., Klinkhammer, G., Palmer, M. R., Kinzler, R. J., and Sulanowska, M. M., 1987. Along-axis variations in seafloor spreading in the MARK area. *Nature*, **328**, 681–685, doi:10.1038/328681a0.

Kelemen, P. B., Koga, K., and Shimizu, N., 1997. Geochemistry of gabbro sills in the crust-mantle transition zone of the Oman ophiolite: implications for the origin of the oceanic lower crust. *Earth and Planetary Science Letters*, **146**, 475–488, doi:10.1016/S0012-821X(96)00235-X.

Kelemen, P. B., Kikawa, E., Miller, D. J., and Shipboard Scientific Party, 2007. Leg 209 summary: processes in a 20-km-thick conductive boundary layer beneath the Mid-Atlantic Ridge, 14°–16°N. In Kelemen, P. B., Kikawa, E., and Miller, D. J. (eds.), *Proceedings of the ODP, Science Results*. College Station (Ocean Drilling Program), Vol. 209, pp. 1–33, doi:10.2973/odp.proc.sr.209.001.2007.

Kent, G. M., Harding, A. J., Orcutt, J. A., Detrick, R. S., Mutter, J. C., and Buhl, P., 1994. Uniform accretion of oceanic crust south of the Garrett transform at 14°15′S on the East Pacific Rise. *Journal of Geophysical Research*, **99**, 9097–9116, doi:10.1029/93JB02872.

Lagabrielle, Y., Vitale-Brovarone, A., and Ildefonse, B., 2015. Fossil oceanic core complexes recognized in the blueschist metaophiolites of Western Alps and Corsica. *Earth-Science Reviews*, **141**, 1–26, doi:10.1016/j.earscirev.2014.11.004.

Macdonald, K. C., Fox, P. J., Perram, L. J., Eisen, M. F., Haymon, R. M., Miller, S. P., Carbotte, S. M., Cormier, M. H., and Shor, A. N., 1988. A new view of the mid-ocean ridge from the behaviour of ridge-axis discontinuities. *Nature*, **335**, 217–225, doi:10.1038/335217a0.

Maclennan, J., Hulme, T., and Singh, S. C., 2005. Cooling of the lower oceanic crust. *Geology*, **33**, 357–366, doi:10.1130/G21207.1.

MacLeod, C. J., Searle, R. C., Murton, B. J., Casey, J. F., Mallows, C., Unsworth, S. C., Achenbach, K. L., and Harris, M., 2009. Life cycle of oceanic core complexes. *Earth and Planetary Science Letters*, **287**, 333–344, doi:10.1016/j.epsl.2009.08.016.

Marjanović, M., Carbotte, S. M., Carton, H., Nedimović, M. R., Mutter, J. C., and Canales, J. P., 2014. A multi-sill magma plumbing system beneath the axis of the East Pacific Rise. *Nature Geoscience*, **7**, 825–829, doi:10.1038/ngeo2272.

McCaig, A. M., and Harris, M., 2012. Hydrothermal circulation and the dike-gabbro transition in the detachment mode of slow seafloor spreading. *Geology*, **40**, 367–370, doi:10.1130/G32789.1.

Morton, J. L., and Sleep, N. H., 1985. Seismic reflections from a Lau Basin magma chamber. In Schol, D. W., and Vallier, T. L. (eds.), *Geology and Offshore Resources of Pacific Island Arcs-Tonga Region*. Houston: Circum-Pacific Council for Energy and Mineral Resources. Earth Science Series, Vol. 2, pp. 441–453.

Pallister, J. S., and Hopson, C. A., 1981. Samail ophiolite plutonic suite: field relations, phase variation, cryptic variation and layering, and a model of a spreading ridge magma chamber. *Journal of Geophysical Research*, **86**, 2593–2644, doi:10.1029/JB086iB04p02593.

Phipps Morgan, J., and Chen, Y. J., 1993. The genesis of oceanic crust: magma injection, hydrothermal circulation, and crustal flow. *Journal of Geophysical Research*, **98**, 6283–6297, doi:10.1029/92JB02650.

Sauter, D., Cannat, M., Rouméjon, S., Andreani, M., Birot, D., Bronner, A., Brunelli, D., Carlut, J., Delacour, A., Guyader, V., MacLeod, C. J., Manatschal, G., Mendel, V., Ménez, B., Pasini, V., Ruellan, É., and Searle, R., 2013. Continuous exhumation of mantle-derived rocks at the Southwest Indian Ridge for 11 million years. *Nature Geoscience*, **6**, 314–320, doi:10.1038/ngeo1771.

Singh, S. C., Kent, G. M., Collier, J. S., Harding, A. J., and Orcutt, J. A., 1998. Melt to mush variations in crustal magma properties along the ridge crest at the southern East Pacific Rise. *Nature*, **394**, 874–878, doi:10.1038/29740.

Singh, S. C., Crawford, W. C., Carton, H., Seher, T., Combier, V., Cannat, M., Pablo Canales, J., Dusunur, D., Escartín, J., and Miguel Miranda, J., 2006. Discovery of a magma chamber and faults beneath a Mid-Atlantic Ridge hydrothermal field. *Nature*, **442**, 1029–1032, doi:10.1038/nature05105.

Sinton, J. M., and Detrick, R. S., 1992. Mid-ocean ridge magma chambers. *Journal of Geophysical Research*, **97**, 197–216, doi:10.1029/91JB02508.

Sleep, N. H., 1975. Formation of oceanic crust: some thermal constraints. *Journal of Geophysical Research*, **80**, 4037–4042, doi:10.1029/JB080i029p04037.

Teagle, D. A. H., Ildefonse, B., Blum, P., and the Expedition 335 Scientists, 2012. *Proceeding of the IODP*. Integrated Ocean Drilling Program Management International, Inc., Tokyo, Vol. 335, doi:10.2204/iodp.proc.335.2012.

Tucholke, B. E., Lin, J., and Kleinrock, M. C., 1998. Megamullions and mullion structure defining oceanic metamorphic core complexes on the Mid-Atlantic Ridge. *Journal of Geophysical Research*, **103**, 9857–9866, doi:10.1029/98JB00167.

Wilson, D. S., Teagle, D. A. H., Alt, J. C., Banerjee, N. R., Umino, S., Miyashita, S., Acton, G. D., Anma, R., Barr, S. R., Belghoul, A., Carlut, J., Christie, D. M., Coggon, R. M., Cooper, K. M., Cordier, C., Crispini, L., Durand, S. R., Einaudi, F., Galli, L., Gao, Y. J., Geldmacher, J., Gilbert, L. A., Hayman, N. W., Herrero-Bervera, E., Hirano, N., Holter, S., Ingle, S., Jiang, S. J., Kalberkamp, U., Kerneklian, M., Koepke, J., Laverne, C., Vasquez, H. L. L., Maclennan, J., Morgan, S., Neo, N., Nichols, H. J., Park, S. H., Reichow, M. K., Sakuyama, T., Sano, T., Sandwell, R., Scheibner, B., Smith-Duque, C. E., Swift, S. A., Tartarotti, P., Tikku, A. A., Tominaga, M., Veloso, E. A., Yamasaki, T., Yamazaki, S., and Ziegler, C., 2006. Drilling to gabbro in intact ocean crust. *Science*, **312**, 1016–1020, doi:10.1126/science.1126090.

Xu, M., Pablo Canales, J., Carbotte, S. M., Carton, H., Nedimović, M. R., and Mutter, J. C., 2014. Variations in axial magma lens properties along the East Pacific Rise ($9°30'$N-$10°00'$N) from swath 3-D seismic imaging and 1-D waveform inversion. *Journal of Geophysical Research*, **119**, 2721–2744, doi:10.1002/2013JB010730.

Cross-references

Axial Summit Troughs
Axial Volcanic Ridges
Cumulates
Gabbro
Layering of Oceanic Crust
Lithosphere: Structure and Composition
Ocean Drilling
Oceanic Spreading Centers
Peridotites
Serpentinization

CUMULATES

Jürgen Koepke
Institute of Mineralogy, Leibniz University Hannover, Hannover, Germany

Cumulate is a name for plutonic rocks which were formed either by the process of accumulation of crystals (crystal settling) or by direct crystallization (in situ crystallization) in a magma chamber. Since "cumulate" is clearly a genetic term, it should not be used as rock name (qv. "Gabbro," and the chapter on rock classification of gabbroic rocks therein), but can be used as additional qualifier, in order to express that rocks were formed by a distinct process within a magma chamber.

For the oceanic crust, this term is directly related to the accretion of oceanic gabbros, forming the lower part of the oceanic crust (layer 3; qv. "seismic layers"). The corresponding magma chambers are located directly under the spreading centers of mid-ocean ridges or back-arc systems (qv. "axial magma chamber"). Here, primitive MORB (mid-oceanic ridge basalt; qv. "Mid-ocean Ridge Magmatism and Volcanism") melts differentiate by fractionation of mainly olivine, plagioclase, and clinopyroxene, which may accumulate at the floor of the magma chamber to form a dense crystal mush, resulting in the oceanic gabbros after cooling. Thus, the vast majority of oceanic gabbros can be regarded as cumulate rocks, formed by the process of crystal sedimentation in axial magma chambers.

Traditionally, the term cumulate was used for layered intrusions on continents (e.g., Skaergaard Layered Intrusion, Greenland; Rum Layered Suite, Scotland). Based on the work of Wager et al. (1960), a scheme of textures for cumulate formation was introduced which has been a major contribution to the understanding of igneous textures. Reviews on that are given by Shelly (1993) or Hunter (1996). "Cumulus phases" or "cumulus crystals" are those phases which are accumulated (sedimented) in a magma chamber, while the liquid around is called "intercumulus" liquid. "Adcumulate" textures represent an accumulation of crystals from which the intercumulus liquid was pressed out, possibly driven by compaction. The cumulate phases occupy the entire volume of the rock, resulting in crystals with subhedral to anhedral shapes which are defined by the jointly merged boundaries of adjacent crystals. "Orthocumulate" textures represent accumulated crystals around which a more evolved intercumulus liquid crystallized often resulting in typical poikilitic growth, meaning that relatively large, late crystallized minerals enclose earlier, smaller crystals.

However, the detailed cumulate terminology as established by Wager et al. (1960) is generally not applied to oceanic gabbros, due to several reasons. First, the terms are rather specific, often with a strong genetic aspect, which makes its application to oceanic gabbros difficult, since the mechanisms by which the lower, gabbroic crust forms are poorly constrained (qv. "crustal accretion"; "gabbro"). Moreover, primary textures may be obscured by later strain which may result in magmatic foliation (especially in gabbros from fast-spreading systems) or in crystal-plastic deformation (especially in gabbros from slow-spreading systems). Thus, for the petrographic characterization of oceanic gabbros, common petrographic description terms for mineral structures and rock textures are commonly used. For example, the record of crystallization of a late, evolved melt in the interstices of earlier crystallized minerals, producing minerals like amphibole, oxide, orthopyroxene, apatite, and zircon occurring in many gabbros from slow-spreading ridges (qv. "gabbro"), can simply be described as "late assemblages crystallized interstitially," instead of using a term like "orthocumulate" or "mesocumulate" which might be appropriate in some cases.

Bibliography

Hunter, R. H., 1996. Texture development in cumulate rocks. In Cawthorn, R. G. (ed.), *Layered Intrusions*. Amsterdam: Elsevier, pp. 77–101.

Shelly, D., 1993. *Igneous and Metamorphic Rocks Under the Microscope*. London: Chapman and Hall. 445 p.

Wager, L. R., Brown, G. M., and Wadsworth, W. J., 1960. Types of igneous cumulate. *Journal of Petrology*, **1**, 73–85.

Cross-references

Crustal Accretion
Gabbro
Mid-ocean Ridge Magmatism and Volcanism

CURIE TEMPERATURE

Hans-Jürgen Götze
Institute of Geosciences, Christian-Albrechts-University Kiel, Kiel, Germany

In a physical sense the Curie temperature (CT or T_c) is defined as the temperature that marks the reversible phase crossing of ferromagnetic or ferrimagnetic materials in its high paramagnetic temperature form:

- Above the CT, spontaneous or directed magnetization disappears from crystal domains.
- Below the CT, magnetic materials get magnetic behavior back, i.e., the spontaneous magnetization and magnetic domains without any effect of an external magnetic field.

Therefore, CT is a material-specific temperature, above which magnetic properties of matter, solids, or rocks will change. Also remanence of magnetized ferromagnets is removed above CT (Blakely, 1995).

"CT" is named in honor of the eminent French Physicist *Pierre Curie* (1859–1906), who was a pioneer in crystallography, magnetism, piezoelectricity, and radioactivity and won the Nobel Prize in Physics in 1903.

A certain level in the Earth's crust is called "Curie depth" in which temperatures are so high that the Curie temperature is reached. In the continental crust this temperature is usually achieved at about 20 km depth, while in the oceanic crust, the Curie temperature lies, depending on temperature, pressure, and rock properties, at greater depths.

When rocks are in a ferromagnetic state, exposure to a magnetic field can align their atoms and create an attraction. If the field is taken away, the magnetism remains, as the material has a form of "memory." This can be used to make permanent magnets and demonstrate a variety of interesting physical phenomena. Paramagnetic materials, however, require the maintenance of an external magnetic field to remain magnetized. At the Curie temperature, the heat agitates the atoms inside the material so much that they cannot align, and it loses its magnetism. At Curie temperature rocks lose their spontaneous magnetization (e.g., 580 °C for magnetite; 769 °C for iron (Fe); 1,400 °C for Cobalt (Co); 631 °C for Nickel (Ni); e.g., Buschow, 2001; Bouligand et al., 2009).

This can have important implications. In geology, for example, high temperatures can occur in lightning strikes and volcanic eruptions and are capable of causing the properties of minerals in the Earth's crust to change. Observers looking at magnetic minerals need to consider their history and what may have influenced them. One can map the depth to the Curie temperature isotherm from *magnetic anomalies* in an attempt to provide a measure of crustal temperatures. Such methods are based on the estimation of the depth to the bottom of magnetic sources, which is assumed to correspond to the temperature at which rocks lose their spontaneous magnetization (see above). Therefore, depths to the bottom of magnetic sources show several features correlated well with prominent heat-flow anomalies. Bouligand et al. (2009) used a method based on the spectral analysis of magnetic anomalies. It incorporates a representation where magnetization has a fractal distribution defined by three independent parameters: the depths to the top and bottom of magnetic sources and a fractal parameter related to the geology. The Curie temperature has also practical implications in marine geosciences. For instance, the oceanic crust can be altered by high-temperature fluids related to hydrothermal processes, thereby losing its magnetization. The resulting magnetic lows in basaltic crust can be used to detect fossil zones of hydrothermal upflow and associated volcanogenic massive sulfides (Zhu et al., 2010).

As a practical example the table below (after Bouligand et al., 2009) shows heat-flow values, depths to the Curie temperature isotherm, and the Moho depth and expected basal depth of magnetic sources for an area in the Western United States. For comparison, results from survey in the South China Sea and the Caribbean are included in the table.

Area	Heat flow (mW/m^2)	Curie depth (km)	Moho depth (km)	Basal depth (km)
Great Basin (typical values)	75–95	17–23	25–35	17–23
High Cascades	~100	~16	~45	~16
West Cascades	40–50	41–63	~45	41–45
Great Valley	25–55	36–250	~35	~35
Colorado Plateau	~60	~32	40–50	~32
Eastern Caribbean (Arnaiz-Rodríguez and Nuris, 2013)	40–80	23	20	
South China Sea (Li et al., 2010)	~80	12–22	<15	

Bibliography

Arnaiz-Rodríguez, M. S., and Nuris, O., 2013. Curie point depth in Venezuela and the Eastern Caribbean. *Tectonophysics*, **590**, 38–51, doi:10.1016/j.tecto.2013.01.004.

Blakely, R. J., 1995. *Potential Theory in Gravity and Magnetic Applications.* Cambridge, UK: Cambridge University Press.

Bouligand, C., Glen, J. M. G., and Blakely, R. J., 2009. Mapping Curie temperature depth in the western United States with a fractal model for crustal magnetization. *Journal of Geophysical Research*, **114**, B11104, doi:10.1029/2009JB006494.

Buschow, K. H. J., 2001. *Encyclopedia of Materials: Science and Technology.* Amsterdam: Elsevier. ISBN 0-08-043152-6.

Li, C.-F., Shi, X., Zhou, Z., Li, J., Geng, J., and Chen, B., 2010. Depths to the magnetic layer bottom in the South China Sea area and their tectonic implications. *Geophysical Journal International*, **182**, 1229–1247, doi:10.1111/j.1365-246X.2010.04702.x.

Zhu, J., Lin, J., Chen, Y. J., Tao, C., German, C. R., Yoerger, D. R., and Tivey, M. A., 2010. A reduced crustal magnetization zone near the first observed active hydrothermal vent field on the Southwest Indian Ridge. *Geophysical Research Letters*, **37**, L18303, doi:10.1029/2010GL043542.

Cross-references

Hydrothermalism
Hot Spots and Mantle Plumes
Regional Marine Geology

CURRENTS

Jüri Elken
Marine Systems Institute, Tallinn University of Technology, Tallinn, Estonia

Synonyms

Not to be confused with electrical current; Ocean currents; Water flow

Definition

Currents are large-scale (mainly horizontal) translational motions of seawater, as opposed to the wave motions where water particles perform periodic back-and-forth excursions relative to the state of rest. Ocean currents can flow at great distances; they carry seawater and its inherent properties (heat and salt content, dissolved and particulate matter, etc.) from one region to another.

Ocean currents are expressed in terms of horizontal velocity (m/s) of fluid motion as a function of space and time, given either by the velocity vector components or by the direction and speed of the current. This is the Eulerian description of currents. Current velocity components are often presented as (eastward) zonal current and (northward) meridional current. When using the current presentation by flow direction, zero direction has been agreed for the current flowing to the north; the direction angle is further counted positive in clockwise rotation. In Lagrangian description of currents, trajectories of fluid particles are followed over time.

Development of observation techniques

The first ancient tracks of ocean currents were obtained by investigating the ship logs. As an example, historical cross-Atlantic voyages discovered regions of very strong currents like Gulf Stream, where the sailing of ships was slowed down or speeded up (depending on the sailing direction) because "the currents are more powerful than the wind"; the first Gulf Stream map was drawn in the 1770s by Benjamin Franklin. Drift data of objects floating on the sea surface (boats, message bottles, etc.) carry information about an object-dependent combination of currents, waves, and wind in creating the drag forces.

Modern current-tracking drifters as Lagrangian observing platforms have specially designed drogues to minimize the wind and wave effects on the surface; the drifters are satellite tracked (e.g., WOCE-SVP), and the recorded data are transferred over satellite communication links to the shore stations. Currents and stratification of deeper ocean layers are effectively observed using the profiling ARGO drifters, which move most of the time, neutrally buoyant in deep layers at about 1,000-m depth, but rise periodically to the sea surface for the profile measurement and data communication.

Direct water flow measurements at fixed locations are technically complicated. The flowmeter needs to be fixed in the dynamic ocean environment; or alternatively, the sensor motions due to waves and currents need to be determined with high precision. Since the flow field is usually fluctuating, averaging of current data over time is important. The earliest mechanical rotor-based current meter capable of recording direction and speed from an anchored ship on different depths has been introduced in the 1890s by John Elliott Pillsbury. Vagn Walfrid Ekman introduced in the 1930s reliable current meter design which included mechanical averaging of speed and direction data. This instrument was in service until the 1960s, when Ivar Aanderaa introduced modern electronically recording current meters. Although concepts for acoustic current measurement methods were elaborated already before the 1950s, reliable technologies for replacement of mechanical rotor-dependent speed counters with acoustic point sensors were developed by Neil Brown at the end of the 1970s. A breakthrough of current measurements was the introduction of acoustic Doppler current profilers (ADCP) in the 1980s that allow self-recording measurements of full current profiles from the instruments mounted either on the bottom or on the moving ship.

Complementary to the direct current measurements, indirect dynamic method based on charting of the density fields has been widely used since the beginning of the 1900s to identify the main large-scale current patterns. The method is based on geostrophic relations. Relative currents of one vertical level in reference to another level can be easily calculated from temperature and salinity profiles. Historically absolute currents were calculated with the assumption that deep levels around 1,500-m depth were motionless. In the recent decades, satellite-based altimeters measure sea-level topography (determines the pressure gradients at the sea surface) with high precision; therefore pressure gradients forming the geostrophic currents throughout the water column can be calculated also with high accuracy.

In the coastal areas, near-shore currents, including the tidal streams, can be well detected by modern coastal maritime radars with advanced signal processing algorithms.

Further details of current observing methods can be found in the books by Emery and Thomson (2001) and Talley et al. (2011).

Basic theoretical concepts

Motions of seawater follow the laws of hydrodynamics written in partial differential equations. The flows are generated mainly by the pressure gradients, forming due to the gravity and acting through the whole water column, and by the frictional forces acting on the sea surface due to the wind stress. The balance of forces (momentum) follows Newton's Second Law for the continuum fluid in the system of partial differential equations: the sum of forces due to the gravity, pressure gradients, and friction per unit mass is equal to the acceleration of the water parcel. Acceleration consists from local time rate of momentum velocity change (absent for stationary flows), nonlinear momentum advection per unit mass (can be omitted for small flow speeds and/or curvature), and Coriolis acceleration (can be omitted for short-period motions).

Coriolis acceleration appears due to the rotation of the Earth and deflects in the Northern (Southern) Hemisphere the currents to the right (left) from the original direction. Its magnitude is described by the product of Coriolis parameter with the current speed in rotated, perpendicular direction. Coriolis parameter is zero at the equator and reaches twice the Earth's angular speed on the poles.

In an idealized case of boundless ocean and vanishing horizontal flow variations, the wind stress acting on the sea surface and frictional drag of the underlying layers are balanced by the Coriolis force. This flow is called "Ekman spiral," according to the theoretical work by Vagn Walfrid Ekman published in 1905. In such conditions, the surface drift currents are deflected by 45° to the right (left) from the wind direction in the Northern (Southern) Hemisphere. When going deeper from the surface, the flow speed decays exponentially, but the direction deflects further in linear dependence from the depth. At the Ekman depth (depends on the turbulent viscosity and Coriolis parameter), usually from a few tens of meters to hundred meter, the drift currents become negligibly small. Total water transport in the upper layer (Ekman transport) is perpendicular to the wind direction.

Besides the momentum equations, the flow must follow also the continuity equation. In a simple explanation, when the currents are converging in horizontal direction relative to the bottom, then isopycnals (surfaces of constant water density) and sea-level will rise; in case of diverging currents, the vertical motion is of opposite direction. When the wind blows parallel to the coast, then Ekman transport is offshore or onshore, depending on the wind direction. Ocean waters are usually stratified; warm surface waters are separated from deeper colder

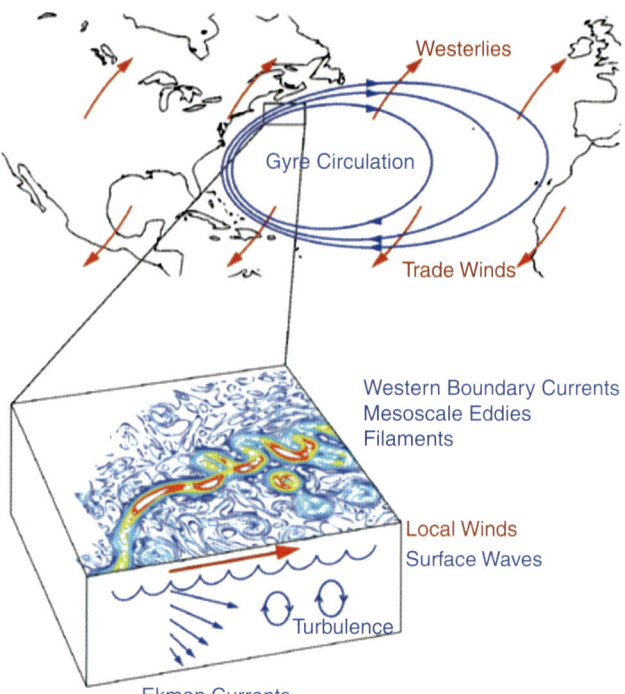

Currents, Figure 1 Schematic of the hierarchy of near-surface currents focused on the Gulf Stream region in the Northern Atlantic Ocean. The gyre circulation consists of a broad eastern component and a narrow, fast, western boundary current (*WBC*). The WBC is enlarged to highlight the mesoscale (100 km and larger) features and the submesoscale eddies, filaments, and turbulence. Global winds drive basin circulation, while local winds force local Ekman currents, surface waves, and turbulence (Dohan and Maximenko, 2010).

waters by a thermocline where temperature drops rapidly. By the continuity arguments, offshore Ekman transport causes upwelling of cold and nutrient-rich deeper waters to the sea surface; downwelling occurs in case of onshore Ekman transport.

The pressure gradients are formed by sea surface topography and spatial variations in water density; the pressure at actual point can be usually calculated by hydrostatic relation. Ocean currents are quite well described by the geostrophic relations where the pressure gradient force is balanced by the Coriolis force. Geostrophic flow is directed along the lines of constant pressure (isobars) in a direction that looking from the higher pressure toward lower pressure, the flow is directed perpendicularly to the right (left) on the Northern (Southern) Hemisphere.

Tilting of sea surface and isopycnals is easily formed by horizontally diverging or converging flows due to the continuity; pressure gradients formed by the flow in such a way have feedback to the flow itself in the momentum equations. As a result, different wave motions are created. In the time scale of days to months, planetary and/or topographic Rossby waves appear. These waves are nearly in the geostrophic balance like are also the mesoscale eddies which have typical size of the order of baroclinic Rossby

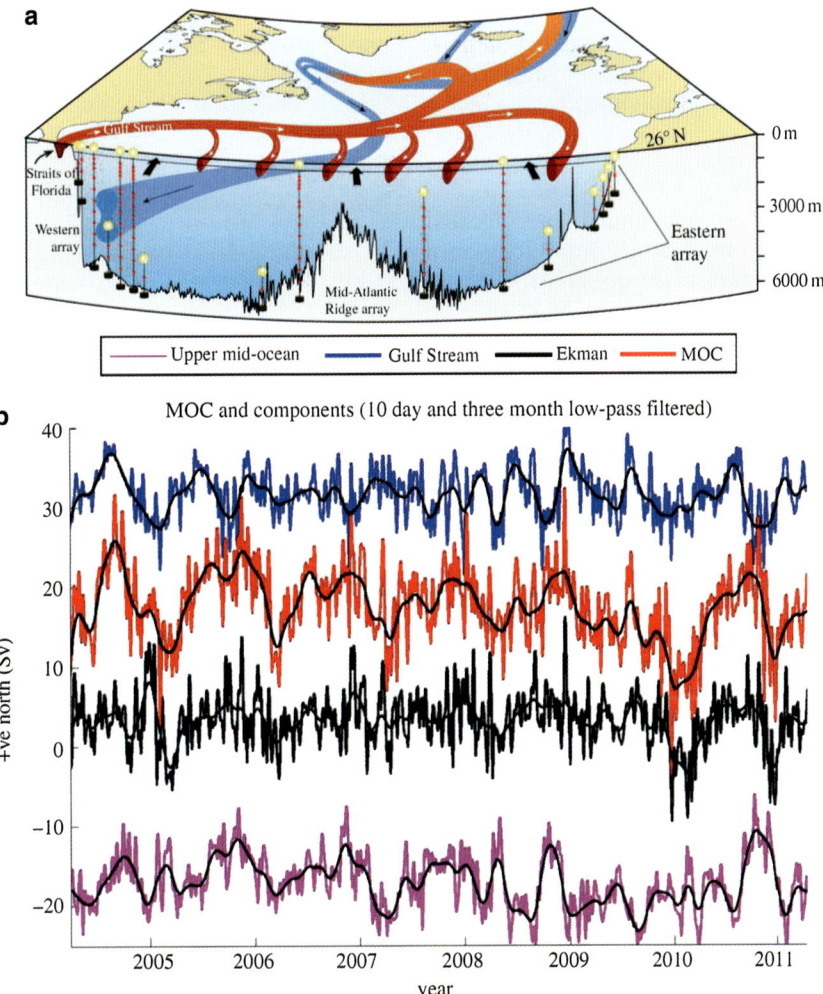

Currents, Figure 2 (a) Schematic of Northern Atlantic circulation with the rapid monitoring array at 26°N. From http://www.noc.soton.ac.uk/rapidmoc/. (b) Gulf Stream (*blue curves*), meridional overturning circulation (MOC; *red curves*), Ekman (*black curves*) and upper mid-ocean (*purple curves*) transports (10 days and 3 months, low-pass filtered) April 2004 to March 2011 as measured by the rapid array. Transports are in Sverdrups (1 Sv = 10^6 m^3 s^{-1}). From http://www.noc.soton.ac.uk/rapidmoc/ (Bryden et al., 2012).

deformation radius (a length measure derived from the depth, stratification strength, and Coriolis parameter). Eddies are mainly generated by the instability of large-scale currents, and these motions follow the law of potential vorticity conservation that counts variations of Coriolis parameter with latitude (beta effect) and depth variation (topographic effect). Vorticity balance as derived in special cases by Sverdrup and Stommel explains also existence of strong narrow-band currents near the western coasts of the oceans: Gulf Stream in the Atlantic, Kuroshio in the Pacific, etc.

Surface and deep currents in the ocean

Surface currents in the ocean follow the global wind patterns, forming the basin-wide gyres with stronger western boundary currents. Many descriptions of surface currents can be found in the literature (e.g., Talley et al., 2011). Modern views have been created with the help of remote sensing and numerical modeling. We refer to the recent paper by Dohan and Maximenko (2010), from where we have adopted the schematic of circulation in the Northern Atlantic (Figure 1). In the Southern Hemisphere, the continents do not cut the ocean flow path; Antarctic Circumpolar Current flows clockwise through all the oceans and provides the global link of water masses of different oceans.

When warm surface waters move toward the polar regions, they cool down. As the water density increases, the waters sink toward greater depths and can move toward tropical regions. Sinking waters are modified by freshwater flux. When the latter exceeds some critical level, the "new" water in the polar region may not be dense enough for sinking; then blocking of warm surface currents may occur. This abyssal flow is also called thermohaline circulation. The overall water parcel cycle through all the oceans – along the thermohaline conveyor belt – takes about 1,000 years.

Changing ocean currents

Ocean circulation varies on interannual and decadal scales, as seen from the observations from long-term arrays installed in different parts of the ocean. An example of water transport variations in different layers of the Northern Atlantic is presented in Figure 2 (Bryden et al., 2012). Such variations have impact to weather patterns and climate, fluxes of dissolved and particular matter, ocean productivity, fish resources, etc.

Conclusions

Currents are large-scale translational motions of seawater. Surface currents in the ocean follow the global wind patterns, forming the basin-wide gyres with stronger western boundary currents. When warm surface waters move toward the polar regions, they cool down. As the water density increases, the waters sink toward greater depths and can move back toward tropical regions. This abyssal flow is also called thermohaline circulation. Ocean circulation varies on interannual and decadal scales, having impact to weather patterns and climate, fluxes of dissolved and particular matter, ocean productivity, and fish resources.

Bibliography

Bryden, H. L., Robinson, C., and Griffiths, G., 2012. Changing currents: a strategy for understanding and predicting the changing ocean circulation. *Philosophical Transactions of the Royal Society A: Mathematical, Physical and Engineering Sciences*, **370**, 5461–5479.

Dohan, K., and Maximenko, N., 2010. Monitoring ocean currents with satellite sensors. *Oceanography*, **23**(4), 94–103.

Emery, W. J., and Thomson, R. E., 2001. *Data Analysis Methods in Physical Oceanography*. Amsterdam/New York: Elsevier.

Talley, L. D., Pickard, G. L., Emery, W. J., and Swift, J. H., 2011. *Descriptive Physical Oceanography: An Introduction*. Amsterdam/Boston: Elsevier - Academic Press.

Cross-references

Anoxic Oceans
Beach Processes
Bottom-Boundary Layer
Estuary, Estuarine Hydrodynamics
Sediment Dynamics
Waves

D

DATA

Michael Diepenbroek
MARUM-Center for Marine Environmental Sciences, University of Bremen, Bremen, Germany

Data are an integral part of geo-scientific work. Full and open access to data is a good scientific practice. It allows for the verification of research findings and is a prerequisite for large-scale and complex research approaches. The management of data covers the entire life cycle of research data from field or real-time data acquisition to long-term archiving and publication as well as analysis and reuse of these data. Reliable usage of data implies that accuracy, consistency, and authenticity of data are maintained and assured during ingest, archiving, and dissemination (data integrity). The Open Archival Information System (OAIS) provides a corresponding reference model.

Production and preparation of data includes control and assessment of data quality (QA/QC), documentation of lineage and process steps, as well as harmonization of data and metadata according to international standards, whereby the heterogeneity of data types and used methods imposes a particular challenge. To ensure the consistency of data and metadata, most of the different science fields maintain specific metadata standards. Consistency is supported by development and usage of common ontologies and vocabularies.

Within data centers, ingest and archiving of data are mostly based on well-defined procedures and supported through corresponding software frameworks and editorial systems operated by professional data managers. The technical development allows storage of data at nearly any scale and complexity. Relational Database Management Systems (RDBMS) are mostly used as storage back ends. Other technologies like the Resource Description Framework (RDF) are available for contextual data such as ontologies.

Increasingly important is the publication of archived data. The publication process involves persistent identification of data (e.g., through Digital Object Identifier – DOI), registration (e.g., using DataCite), license protection, and cataloguing. In this way, data are citable, can be cross-referenced with literature, and are thus part of the scholarly publishing system.

For long-term archiving and publication, a multitude of good quality data centers covering the whole range of earth and environmental sciences are available. Prominent are the data holdings of space agencies (e.g., NASA, ESA), the World Meteorological Organization Information System (WIS), data centers linked to the genome community (EMBL-EBI, GenBank), the Global Biodiversity Information Facility (GBIF), the International Oceanographic Data and Information Exchange (IODE), and multidisciplinary data centers like PANGAEA (Diepenbroek et al., 2002). A long-standing nongovernmental organization comprising many of these facilities is the World Data System (WDS) of the International Council for Science (ICSU). The WDS is certifying and monitoring their members on the base of international standards and policies thus ensuring high quality of supplied data and services.

Data centers are part of the globally evolving information infrastructures. Efficient usage of data from a network of providers requires a high level of data consistency as well as interoperable and generally available added value services for data integration and analysis. Technically, a multitude of content and interoperability standards, as well as warehouse, cloud, and other systems, are available for this purpose. Integration of multidisciplinary data raises particular new research opportunities.

Bibliography

Diepenbroek, M., Gobe, H., Reinke, M., Schindler, U., Schlitzer, R., Sieger, R., and Wefer, G., 2002. PANGAEA-an information system for environmental sciences. *Computers and Geosciences*, **28**, 1201–1210.

DEEP BIOSPHERE

Axel Schippers
Federal Institute for Geosciences and Natural Resources, Hannover, Germany

Synonyms

Deep subseafloor biosphere; Deep subsurface biosphere

Definition

Deep biosphere. Jørgensen and Boetius (2007) define the seafloor as the top meter layer of the seabed that is bioturbated by animals and porous ocean crust that is penetrated by seawater, whereas the deep subsurface, which harbors the deep biosphere, comprises the sediment and rock that is deeper than 1 m below the seafloor (mbsf).

Introduction

The Earth's deep biosphere includes a variety of subsurface habitats, such as mines and deep aquifer systems in the continental realm and sediments and igneous rock in the marine realm. The terrestrial deep subsurface biosphere has been studied in various geological formations (e.g., Gold, 1992; Pedersen, 1993, 1997; Stevens and McKinley, 1995; Amy and Haldeman, 1997; Chapelle et al., 2002; Moser et al., 2003; Lin et al., 2006; Fry et al., 2009; Borgonie et al., 2011; Breuker et al., 2011; Itävaara et al., 2011; Edwards et al., 2012) but is not in the scope of this entry. Jørgensen and Boetius (2007) and Edwards et al. (2012) defined the seafloor as the top meter layer of the seabed that is bioturbated by animals and porous ocean crust that is penetrated by seawater, whereas the deep subsurface, which harbors the deep biosphere, comprises the sediment and rock that is deeper than 1 m below the seafloor. Several review papers have been published about the marine deep biosphere (Jørgensen, 2000; Parkes et al., 2000; Parkes and Wellsbury, 2004; Teske, 2005; Teske, 2006a; Teske, 2006b; Jørgensen et al., 2006; Jørgensen and Boetius, 2007; Fry et al., 2008; Mascarelli, 2009; Schrenk et al., 2010; Edwards et al., 2012; Hoehler and Jørgensen, 2013; Lever, 2013; Orcutt et al., 2013a; Teske et al., 2013; Parkes et al., 2014). This entry gives an overview about the marine deep biosphere in deeply buried sediments and in the ocean crust considering the deep biosphere biomass, the predominant microorganisms and their diversity and activity, as well as implications for biogeochemistry. The marine deep biosphere has been explored at various ocean sites mainly within the framework of the Ocean Drilling Program (ODP) and the Integrated Ocean Drilling Program (IODP).

The deep biosphere in marine sediments

Biomass

Organic carbon from marine primary production is usually the energy source feeding the deep biosphere. Only about 1 % of the total marine primary production of organic carbon is available for deep-sea sediment microorganisms (Hedges, 1992). For the largest part of the ocean seafloor, the organic carbon amount available on the ocean seafloor from sedimentation is on average 1 g of carbon per m^2 per year (Jahnke and Jackson, 1992). Most of the organic carbon is degraded within the top 1-m sediment, and the organic carbon content in the deep subsurface comprises approximately 0.1–1 % of sediment by dry weight (Jørgensen and Boetius, 2007). In deep sediments with extremely low organic carbon flux, radiolysis of water may provide hydrogen and thus an energy source for autotrophic microorganisms independent of photosynthesis (Jørgensen and D'Hondt, 2006; D'Hondt et al., 2009).

A fundamental question in deep biosphere research is its size or the biomass of living cells. This topic was first comprehensively addressed by the research group of Parkes et al. (1994, 2000). For counting of microbial cells under a fluorescence microscope, the cells in a sediment sample were stained with the DNA-intercalating dye acridine orange, and the number of total cells counted is given as acridine orange direct counts (AODC). Using this method for building up an AODC database, the first report about an existence of a deep biosphere in several hundred meters deeply buried sediments was published (Parkes et al., 1994). Studies of contamination potential have demonstrated that it is possible to obtain core samples without outside contamination (Smith et al., 2000), and the enumeration of prokaryotes is now a standard protocol onboard the IODP drillships JOIDES Resolution (Lever et al., 2006) and Chikyu (Yanagawa et al., 2013). The total number of prokaryotes (*Bacteria* and *Archaea*) in the subsurface sediments decreases with sediment depth mainly because it is controlled by the organic carbon content of the sediment as it is the main microbial substrate (Parkes et al., 2000; D'Hondt et al., 2004; Roussel et al., 2008). This data was then further extrapolated to calculate a global biomass of the deep biosphere, and it was estimated that marine sediments harbor over half of all prokaryotic cells on Earth (Whitman et al., 1998). However, these estimates were mainly derived from organic carbon-rich, meso- to eutrophic sediments from continental margins and/or upwelling areas. A recent study included organic carbon-lean, oligotrophic subsurface sediments which contain much less cells and showed that the total cell counts varied between ocean sites by ca. five orders of magnitude (Figure 1; Kallmeyer et al., 2012). They estimated Earth's total number of microbes and living biomass to be 50–78 % and 10–45 %, respectively, lower

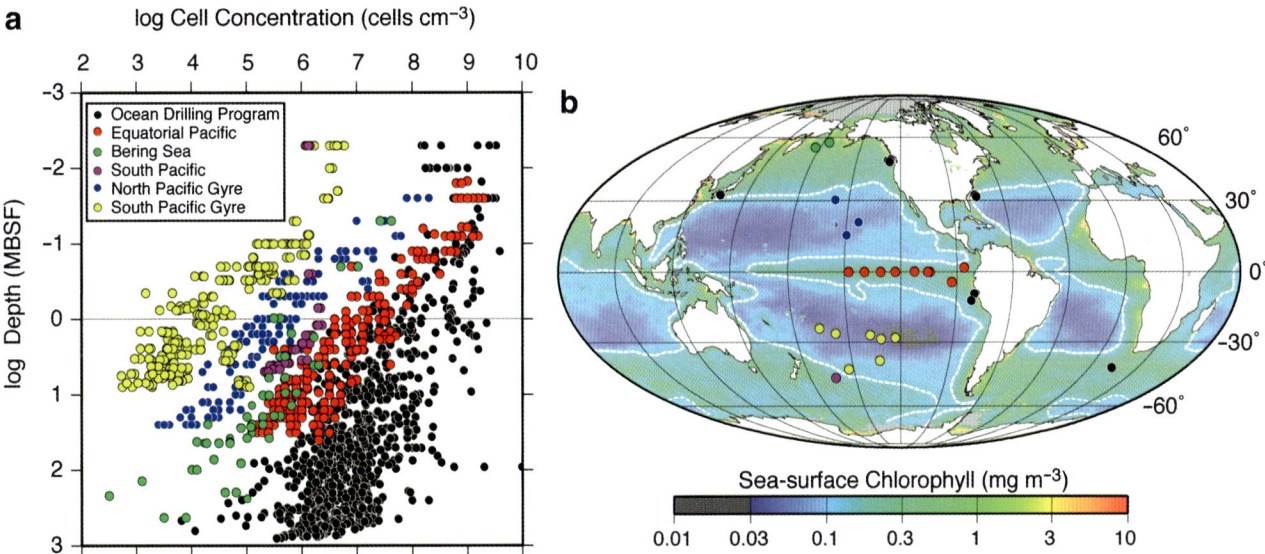

Deep Biosphere, Figure 1 Subseafloor sedimentary cell counts (From Kallmeyer et al., 2012): (**a**) cell counts versus depth (mbsf) for various ocean sites. (**b**) Site locations overlain on a map of time-averaged sea surface chlorophyll (chl-a; Gregg et al., 2005).

than estimates by Whitman et al. (1998). Therefore, Kallmeyer et al. (2012) estimated the values for the global biomass in marine sediments to be now 2.9×10^{29} cells, corresponding to 4.1 petagram (Pg) C and ~0.6 % of Earth's total living biomass. The total microbial abundance in subseafloor sediments (2.9×10^{29} cells) is now similar to the estimates for the total number of prokaryotes in seawater (1.2×10^{29}) and in soil (2.6×10^{29}).

Kallmeyer et al. (2012) were able to count very low cell numbers by using a protocol in which the cells were counted after being detached from the sediment particles which allowed a concentration of more cells on a filter (Kallmeyer et al., 2008). For cell staining the brighter fluorescent dye SYBR Green was used instead of acridine orange. SYBR Green is nowadays more frequently used for total cell counting in sediments than any other dye (Figure 2; Weinbauer et al., 1998; Engelen et al., 2008; Morono et al., 2009; Schippers et al., 2010, 2012). Total cell counts do not provide information about the viability of the cells and their taxonomy. Using the cellular ribosomal RNA-targeting molecular technique catalyzed reporter deposition-fluorescence in situ hybridization (CARD-FISH or FISH), the first direct quantification of living cells – defined by the presence of ribosomes – was provided for deep sediment samples from the equatorial Pacific and the Peru margin (ODP Leg 201, Schippers et al., 2005; Teske, 2005; Biddle et al., 2006). A striking finding was that a large fraction of the subseafloor prokaryotes is alive, even in very old (16 Ma) and deep (>400 mbsf) sediments (Schippers et al., 2005). However, critical for a successful application of CARD-FISH or FISH seems to be the cell permeabilization protocol (Lloyd et al., 2013b).

Microbial activity and biogeochemistry in deep subsurface sediments

Results of the ODP program demonstrated that subseafloor microbial communities are active in biogeochemical cycling (e.g., Parkes et al., 2000). Within sediments and on rock surfaces, bacteria interact with minerals, thus catalyzing dissolution, precipitation, and other surface reactions that have been traditionally viewed as abiotic (Parkes et al., 2007, 2011). Pore water data from a large number of Deep Sea Drilling Project (DSDP) and ODP sites (e.g., Whelan et al., 1986; Kastner et al., 1990; Borowski et al., 1997; D'Hondt et al., 2004), in particular, decreases in sulfate and increases in methane, and the presence of hydrogen sulfide has long been identified by geochemists as manifestations of deep microbial activity. More recently, also stable sulfur isotope measurements have shown to reflect the activity of deep biosphere sulfate-reducing microbial communities (Wortmann et al., 2001). Anaerobic processes, in particular sulfate, Fe(III), and Mn(IV) reduction and methanogenesis, characterize subsurface marine sediments (D'Hondt et al., 2004; Schippers et al., 2010). In continental margin sediments, bacterial sulfate reduction can be responsible for 50 % or more of the organic matter degradation (Jørgensen, 1982; Ferdelman et al., 1999; Fossing et al., 2000). Consequently, sulfate-reducing bacteria (SRB) are among the most numerous microorganisms (Sahm et al., 1999; Ravenschlag et al., 2000; Schippers et al., 2010, 2012). From deep sediment layers in the Pacific Ocean, sulfate-reducing bacteria (SRB) have been enriched with different organic compounds at different temperatures (Barnes et al., 1998), and a novel barophilic sulfate-reducing bacterium, *Desulfovibrio profundus*, has

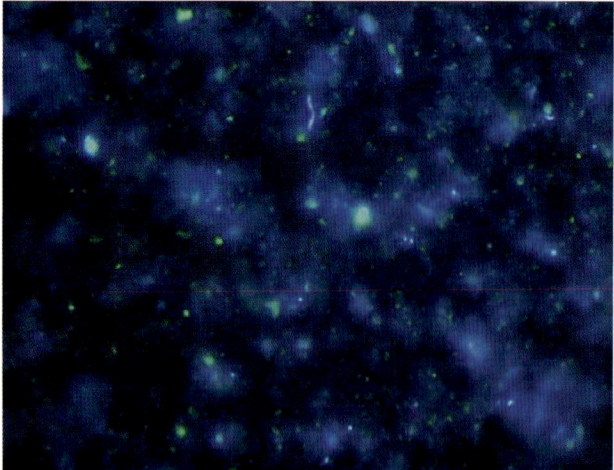
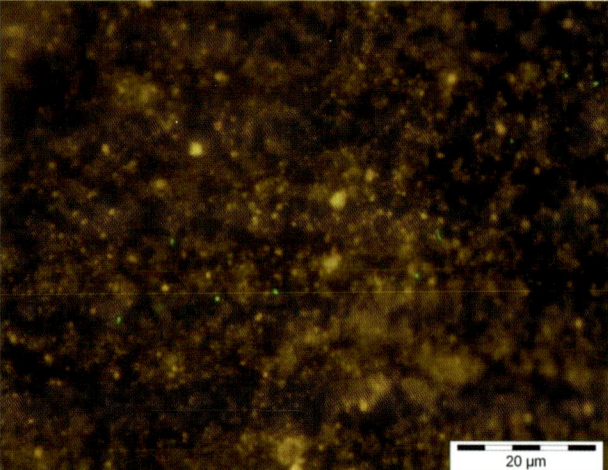

Deep Biosphere, Figure 2 Staining of cells in marine sediment samples with the DNA-intercalating fluorescent dyes DAPI (cells in bright *blue, top*) and SYBR Green (cells in *green, bottom*).

been isolated (Bale et al., 1997). Furthermore, a consortium of *Archaea* and sulfate-reducing bacteria catalyzes anaerobic methane oxidation, an important process in marine sediments regulating the sedimentary flux of methane to the ocean water column (Hoehler et al., 1994; Boetius et al., 2000; Jørgensen and Boetius, 2007; Milucka et al., 2012). In deeper sediments, where sulfate becomes depleted, bacterial methanogenesis and acetogenesis become quantitatively more important (Parkes et al., 2000; Heuer et al., 2009; Lever et al., 2010). Direct evidence for sulfate reduction and methanogenic activity comes from radiotracer experiments performed on sediments from various ODP Legs (Cragg et al., 1996; Parkes et al., 2000, 2005; Hoehler and Jørgensen, 2013).

Organic-lean, oligotrophic, and oxic sediments, with penetration of molecular oxygen to several meters, host oxygen-respiring prokaryotes, which likely dominate organic carbon oxidation (Røy et al., 2012; Ziebis et al., 2012; Orcutt et al., 2013b). In such sediments radiolysis of water may provide hydrogen, and thus an energy source for autotrophic microorganisms independent of photosynthesis (Jørgensen and D'Hondt, 2006; D'Hondt et al., 2009).

Microbial cells in deeply buried marine sediments catabolize 10^4–10^6-fold more slowly than model organisms in nutrient-rich cultures, representing turnover biomass on timescales of centuries to millennia rather than hours to days, and subsist with energy fluxes that are 1,000-fold lower than the typical culture-based estimates of maintenance requirements (Hoehler and Jørgensen, 2013). Furthermore, it has been demonstrated that the microbial communities and their activities change at sediment interfaces over geological times (Coolen et al., 2002; Inagaki et al., 2003; Parkes et al., 2005; Schippers et al., 2012). For example, bacterial populations in subseafloor sediments from the Sea of Okhotsk, composed of pelagic clays with several volcanic ash layers containing fine pumice grains, were approximately two to ten times larger in the ash layers than those in the clays (Inagaki et al., 2003).

Microbial diversity and quantification in deep subsurface sediments

The biomass of the deep subsurface biosphere comprises the three domains of life *Archaea*, *Bacteria*, and *Eukarya*, as well as spores and viruses (Schippers et al., 2005, 2012; Edgcomb et al., 2011; Lomstein et al., 2012; Breuker et al., 2013; Engelhardt et al., 2013; Orsi et al., 2013a; Ciobanu et al., 2014). Total cell numbers alone do not provide information about the microbial diversity and the physiology of the microorganisms that are critical to understanding deep biosphere biogeochemical processes. It is important to ask what types of microorganisms are present and in what abundance and which of these microorganisms are truly active (i.e., not dormant) and participating in deep sedimentary geochemical processes. In the early days of deep biosphere research, classical cultivation techniques were applied, while nowadays molecular techniques are more frequently used (see below).

Using classical cultivation techniques, i.e., the most probable number (MPN) cultivation method, various physiological types of microorganisms have been enriched from deep sediments and their numbers determined (Cragg et al., 1990, 1996; Barnes et al., 1998; Parkes et al., 2000, 2009; D'Hondt et al., 2004; Biddle et al., 2005; Batzke et al., 2007). These types include aerobic ammonifiers, nitrate reducers, fermentative anaerobic heterotrophs, sulfate reducers, methanogens, acetogens, and anaerobic hexadecane oxidizers. MPN population counts ranged from 0 to 10^5 cells/cm^3 and generally decreased with increasing depth. Most of the prokaryotes in natural environments do not grow on standard laboratory media since the complex conditions of the natural habitat are difficult to reproduce. Thus, generally less than 0.6 % of the total cell numbers in deep sediments were detected via cultivation (Cragg et al., 1990, 1996; Barnes et al., 1998; Parkes et al., 2000). Bacteria have been enriched and isolated with different organic compounds at different temperatures (Bale et al., 1997; Barnes et al., 1998; Inagaki et al., 2003; Mikucki et al., 2003; Toffin et al., 2004, 2005; Biddle et al., 2005; Lee et al., 2005; Kobayashi et al., 2008; Parkes et al., 2009). Isolates belonged to the alpha-, gamma-, deltaproteobacteria, Firmicutes, Actinobacteria, and Bacteroidetes (D'Hondt et al., 2004; Batzke et al., 2007). For example, a barophilic sulfate-reducing bacterium, *Desulfovibrio profundus* (Bale et al., 1997), and a methanogenic archaeum, *Methanoculleus submarinus*, have been isolated from deep marine sediments that contain methane hydrates (Mikucki et al., 2003). Cultivation and isolation of microorganisms is still the only way to get novel organisms from the environment; to study the properties for the description of new species, understanding their role in the environment; and to use them for biotechnological applications. However, molecular methods provide a more comprehensive picture about the diversity and abundance of microbial communities in the environment.

In recent years, nucleic acid-based techniques (i.e., DNA and RNA extraction, PCR amplification, and sequencing) have been developed to identify subsurface microorganisms without cultivation. Most of these techniques make it possible to phylogenetically identify specific groups of microorganisms and can highlight microbial diversity and detect new sequences. The analyses of 16S rRNA gene sequences showed that microbial communities of deep marine sediments harbor members of distinct, uncultured bacterial and archaeal phylogenetic lineages (Teske, 2006a; Teske, 2006b; Biddle et al., 2008; Teske and Sørensen, 2008; Fry et al., 2008; Durbin and Teske, 2012; Kubo et al., 2012; Orsi et al., 2013b). Among the *Bacteria*, 16S rRNA gene sequences belonging to the JS-1 group and the *Chloroflexi* were frequently found (Webster et al., 2004, 2007, 2011; Blazejak and Schippers, 2010). Typical groups for the *Archaea* include the Marine Benthic Group B (MBG-B), a deeply branching phylum-level lineage; the Miscellaneous Crenarchaeotal Group (MCG), a frequently detected crenarchaeotal lineage with high intragroup diversity; the South African Gold Mine Euryarchaeotal Group (SAGMEG); and the Marine Benthic Group D (MBG-D), an euryarchaeotal group affiliated with the *Thermoplasmatales* (Durbin and Teske, 2012). The metabolic capabilities of these uncultivated organisms have started to be explored using metagenomic approaches (Biddle et al., 2008, 2011; Rinke et al., 2013; Wasmund et al., 2014). For example, recent metagenomic data indicate that uncultured archaea such as MCG and MBG-D may have a role in protein remineralization in anoxic marine sediments (Lloyd et al., 2013a). The microbial ecology of the deep biosphere has been mainly studied in organic-rich, meso- to eutrophic sediments, and relatively few studies focused on organic carbon-lean, oligotrophic subsurface sediments (Inagaki et al., 2001; Sørensen et al., 2004; Nunoura et al., 2009; Roussel et al., 2009; Durbin and Teske, 2011; Breuker and Schippers, 2013). Since oligotrophic sediments exhibit specific archaeal diversity patterns, the organic carbon content is obviously extremely relevant for the natural selection of distinct *Archaea* (Durbin and Teske, 2012). However, the organic carbon concentration is not a directly proportional index of the sediment trophic state, as substrate availability and organic carbon residence time can vary between sediments with similar organic carbon contents (Durbin and Teske, 2012). Consequently, Durbin and Teske also defined other parameters such as the sedimentation rate, the penetration depth of electron acceptors such as sulfate, and the ammonium concentration to characterize the trophic state of sediments.

Sulfate reduction and methanogenesis are relevant biogeochemical processes in deeply buried sediments (e.g., D'Hondt et al., 2004). However, sulfate-reducing *Bacteria* or methanogenic *Archaea*, which are frequently found in near-surface sediments, were rarely detected in deep sediments (Parkes et al., 2005; Biddle et al., 2006; Inagaki et al., 2006; Teske, 2005; Teske, 2006a; Teske, 2006b; Webster et al., 2006; Fry et al., 2008; Teske and Sørensen, 2008). In addition to 16S rRNA genes, sulfate reducers and methanogens have been detected and quantified via their functional genes encoding dissimilatory sulfite reductase (*dsr*), adenosine 5'-phosphosulfate reductase (*apr*), and methyl-coenzyme M reductase (*mcr*), respectively (Parkes et al., 2005; Schippers and Neretin, 2006; Webster et al., 2006, 2009; Wilms et al., 2007; Engelen et al., 2008; Nunoura et al., 2009; Blazejak and Schippers, 2011; Breuker et al., 2013; Lever, 2013; Ciobanu et al., 2014).

As mentioned above CARD-FISH (and FISH) has been applied for the quantification of living *Bacteria* and *Archaea* in deeply buried subsurface sediments. Another technique with high sensitivity is quantitative real-time polymerase chain reaction (qPCR). It has been used frequently for the enumeration of phylogenetic 16S rRNA genes as well as functional genes coding for enzymes particularly involved in biogeochemical processes. qPCR is based on the online fluorescence detection of PCR products and allows the rapid detection and quantification of gene sequences without the need for labor-intensive post-PCR processing (Heid et al., 1996). For application, DNA is quantitatively extracted from sediment samples, purified, and specifically amplified with a thermocycler using sequence-specific fluorescently labeled probes. There are different formats of the used probes, but the most common are sequence-specific TaqMan probes (Heid et al., 1996) and the intercalating nonspecific SYBR Green dye (Wittwer et al., 1997). The detection limit of the method depends on the target of interest, sample purity, PCR conditions, and other factors but theoretically allows to detect a single DNA molecule (Lockey et al., 1998). Additionally to DNA, RNA can be quantified after application of an additional reverse transcription step.

A quantification of particular prokaryotic groups (i.e., *Bacteria* and *Archaea*) in deep subsurface sediments has

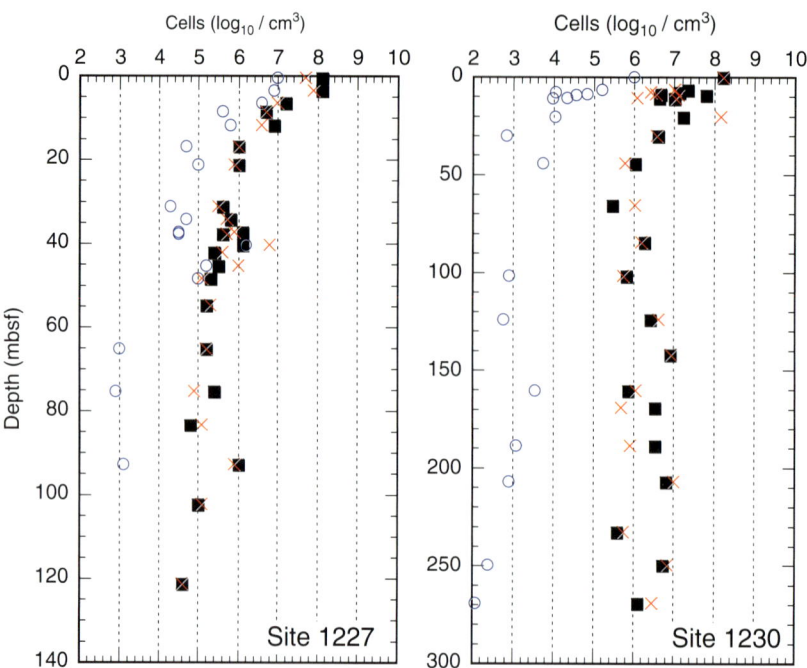

Deep Biosphere, Figure 3 Depth profiles of total prokaryotes (*squares*), *Bacteria* (*crosses*), and *Archaea* (*circles*) determined by qPCR for two Peru continental margin sites (ODP Leg 201; from Schippers et al., 2005).

been done by qPCR in several studies (Figures 3 and 4; Inagaki et al., 2003; Schippers et al., 2005; Inagaki et al., 2006; Schippers and Neretin, 2006; Wilms et al., 2007; Engelen et al., 2008; Nunoura et al., 2009; Webster et al., 2009; Schippers et al., 2010, 2012; Breuker et al., 2013; Breuker and Schippers, 2013; Ciobanu et al., 2014). Eukaryotic 18S rRNA genes were orders of magnitude less abundant than prokaryotic 16S rRNA genes (Schippers and Neretin, 2006; Schippers et al., 2010, 2012; Ciobanu et al., 2014). Published qPCR data on the abundance of *Bacteria* and *Archaea* of several sediment studies show that the ratio of *Archaea* versus *Bacteria* seems to be variable depending on the type of sediment (Table 1; Breuker and Schippers, 2013; Breuker et al., 2013) and/or the qPCR protocols applied in different laboratories (Lloyd et al., 2013b). Using qPCR, an almost equal abundance of *Bacteria* and *Archaea* has been found for the Porcupine Seabight (IODP Exp. 307; Webster et al., 2009), the northeast Pacific ridge flank (IODP Exp. 301; Engelen et al., 2008), Sumatra forearc basins (Schippers et al., 2010), and sediments of the Black Sea and the Benguela upwelling system of the Atlantic coast of Namibia (Figure 4; Schippers et al., 2012). By contrast, *Bacteria* dominated other sediments such as the Sea of Okhotsk (Inagaki et al., 2003), the Gulf of Mexico (IODP Exp. 308; Nunoura et al., 2009), the Peru continental margin, and the equatorial Pacific sediments (Figure 3; ODP Leg 201; Schippers et al., 2005; Inagaki et al., 2006), as well as gas hydrate-bearing sediments from the Cascadia margin (ODP Leg 204; Inagaki et al., 2006) as well as very deep sediments of the Canterbury basin (IODP Exp. 317; Ciobanu et al., 2014). These data on the abundance of *Bacteria* and *Archaea* in deeply buried marine sediments originate from qPCR analysis of rather organic carbon-rich, eutrophic sediments. There, *Bacteria* either dominated or an overall equal portion of *Bacteria* and *Archaea* were determined in all qPCR studies. *Archaea* dominated only clearly in the oligotrophic (total organic carbon ~0.15 ± 0.07 %) and oxic sediments from the North Pond area in 7 Ma western flank of the Mid-Atlantic Ridge 23°N (Breuker and Schippers, 2013). It is not understood which factors control the *Bacteria*/*Archaea* ratio but likely the organic carbon content is an important factor. In addition, a potential underestimation of *Archaea* depending on the applied qPCR protocols has to be considered in future studies (Lloyd et al., 2013b).

The data for the Peru continental margin and the equatorial Pacific sediments (ODP Leg 201) as well as gas hydrate-bearing sediments from the Cascadia margin (ODP Leg 204) gave conflicting results depending on the quantification method. Nucleic acid-based methods (CARD-FISH and qPCR) showed as mentioned above a dominance of *Bacteria* (Schippers et al., 2005; Schippers and Neretin, 2006; Inagaki et al., 2006), while the analysis of intact polar lipids (IPL) of prokaryotic cell membranes determined *Archaea* as major prokaryotes in deeply buried sediments (Biddle et al., 2006; Lipp et al., 2008). These conflicting results may be explained by a potential underestimation of qPCR-determined *Archaea* (Lloyd et al., 2013b) but more likely by a different preservation

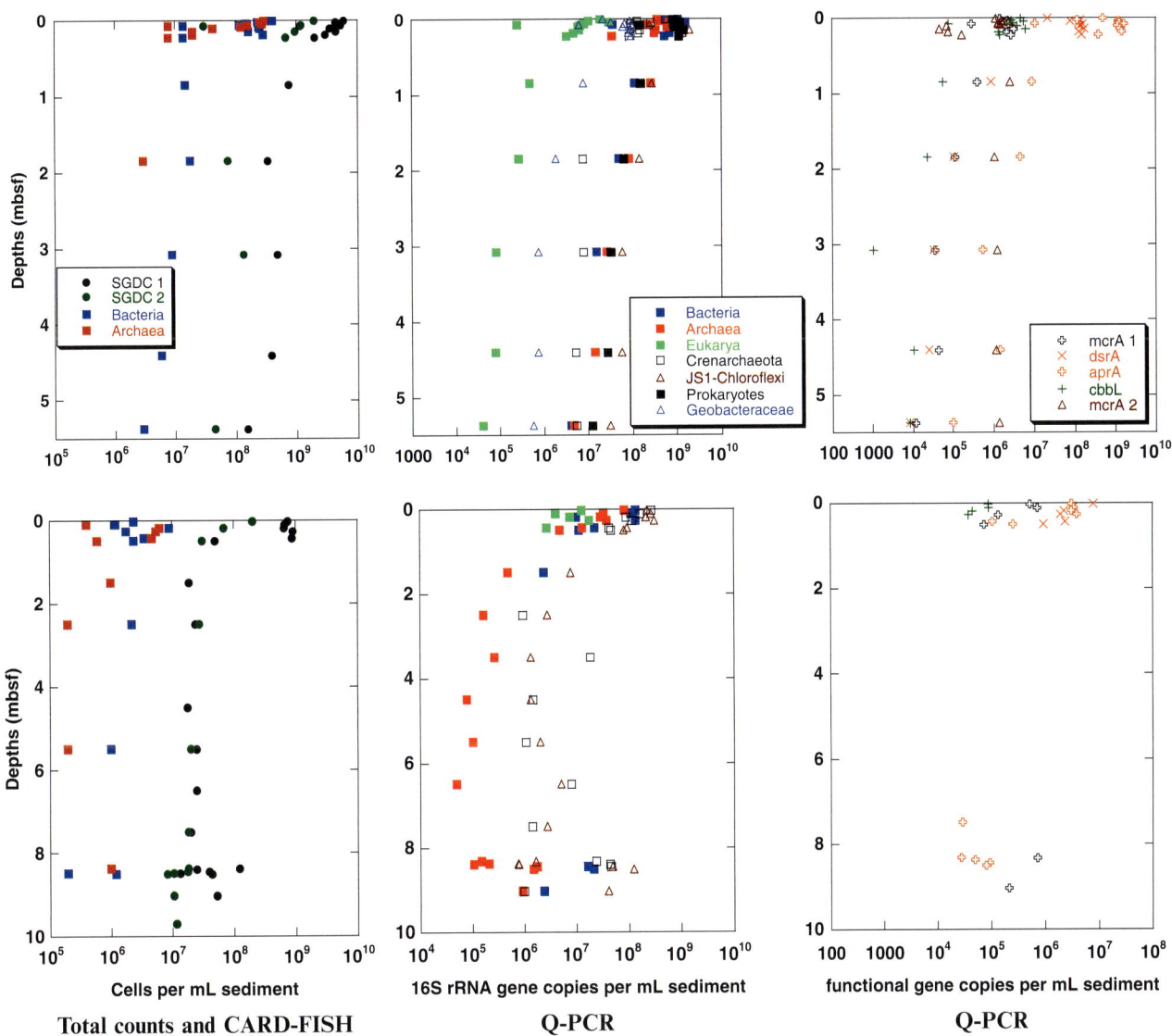

Deep Biosphere, Figure 4 Sediment samples from Benguela upwelling system (station 3, *top*) and from the Black Sea (stations 22, *bottom*; from Schippers et al., 2012). Total cell counts obtained with two different methods (SGDC1 after Weinbauer et al., 1998, SGDC2 after Kallmeyer et al., 2008) and CARD-FISH numbers for *Bacteria* and *Archaea* (*left*); qPCR quantification of 16S rRNA genes of Prokaryotes, *Bacteria*, *Archaea*, *Eukarya*, *Crenarchaeota*, *Geobacteraceae*, and JS-1-*Chloroflexi* (*middle*); and the functional genes *mcrA* 1 and 2, *dsrA*, *aprA*, and *cbbL* (*right*).

of DNA and IPL in deeply buried sediments (e.g., stabilized on clay surfaces or organic matter; Coolen et al., 2006; Inagaki and Nealson, 2006; Schippers and Neretin, 2006). The preservation of archaeal IPL biomarkers in marine sediments has been confirmed based on experimental and modeling approaches (Schouten et al., 2010; Logemann et al., 2011; Xie et al., 2013). These indicate that due to their different chemical nature, a faster degradation of ester-bound bacterial than of ether-bound archaeal IPLs leads to an increasing proportion of archaeal IPL with sediment depth. The fossilized archaeal IPLs mainly derived from sedimented *Archaea* from the water column were masking the much smaller pool of IPLs from living cells in the sediments. As a consequence, the published IPL data do not reflect the abundance of living *Archaea*, and the conclusion that *Archaea* are dominant in deeply buried sediments is therefore questionable (Lipp et al., 2008; Xie et al., 2013).

Life in the ocean crust

Some recent review papers give an overview about life in the ocean crust (Edwards et al., 2005, 2012; Schrenk et al., 2010; Orcutt et al., 2011). Nearly 70 % of the Earth's surface is covered by ocean crust of which the major part is covered by sediments. The marine deep biosphere in the

Deep Biosphere, Table 1 Compilation of published mean total cell counts and qPCR abundance of *Bacteria* and *Archaea* in the depth range of 1–10 and 10–200 mbsf (meter below seafloor) in subsurface marine sediments (*nd* not determined; from Breuker et al., 2013, modified)

Expedition/area	1–10 mbsf			10–200 mbsf			References
	Total counts	*Bacteria*	*Archaea*	Total counts	*Bacteria*	*Archaea*	
ODP Leg 201 Peru margin	10^7–10^8	10^7	10^4–10^7	10^7	10^6	10^3–10^5	Schippers et al. (2005)
ODP Leg 201 Peru margin	10^7–10^8	>90 %	<10 %	10^7	>99 %	<1 %	Inagaki et al. (2006)
ODP Leg 204 Cascadia margin	10^7	>70 %	<30 %	10^6	>70 %	<30 %	Inagaki et al. (2006)
ODP/IODP	nd	~60 %	~40 %	nd	~60 %	~40 %	Lipp et al. (2008)
IODP Exp. 301 Juan de Fuca	10^8–10^9	10^6–10^8	10^5–10^6	10^8	10^6	10^6	Engelen et al. (2008)
IODP Exp. 307 Porcupine Seamount	nd	nd	nd	10^6–10^7	10^5–10^6	10^4–10^5	Webster et al. (2009)
IODP Exp. 308 Gulf of Mexico	10^5–10^6	10^5–10^6	10^5	10^4–10^5	10^4	<10^2	Nunoura et al. (2009)
IODP Exp. 313 New Jersey shallow shelf	10^6	10^6	10^6	10^6	10^6	10^5–10^6	Breuker et al. (2013)
Sea of Okhotsk	10^6–10^7	10^4–10^5	<10^4	10^6–10^7	10^4–10^5	<10^4	Inagaki et al. (2003)
North Sea tidal flat	10^7–10^8	10^7	10^6	nd	nd	nd	Wilms et al. (2007)
SO189 Forearc of Sumatra	10^7–10^8	10^7–10^8	10^7–10^8	nd	nd	nd	Schippers et al. (2010)
M72-5 Black Sea	10^7–10^8	10^5–10^6	10^5–10^6	nd	nd	nd	Schippers et al. (2012)
M76-1 Benguela upwelling	10^7–10^9	10^6–10^8	10^6–10^9	nd	nd	nd	Schippers et al. (2012)
MSM11-1 "North Pond"	10^5–10^6	10^4	10^5–10^6	nd	nd	nd	Breuker and Schippers (2013)

ocean crust has been less intensively studied than that in subsurface sediments. Unlike sediments in which organic carbon, derived from photosynthesis in the oceans, is the major energy source for microbial life, inorganic processes in the ocean crusts likely provide the main energy for the rock-hosted deep biosphere.

Prokaryotic cell abundances on seafloor-exposed basalts are three to four orders of magnitude greater than in overlying deep-sea water. Microbiological analyses of basaltic lavas revealed that the basalt-hosted biosphere harbor is a substantial community of phylogenetically and physiologically diverse microorganisms. This community is dominated by *Bacteria* and comprises at least 16 different taxonomic groups, including all subdivisions of the proteobacteria (Santelli et al., 2008). As energy source for these microorganisms, the alteration of basalt, in particular the oxidation of reduced iron and sulfur in the basalt, has been suggested (Bach and Edwards, 2003). Evidence for this suggestion comes from textural observations in altered glass of oceanic pillow basalts (Fisk et al., 1998), the fact that iron-oxidizing bacteria can live on basalt glass and increase basalt-weathering rates (Edwards et al., 2004) and that microbial biomass and community composition correlate with the age and alteration state of seafloor basalts (Santelli et al., 2009). As a competing hypothesis, iron- and manganese-oxidizing bacteria living as biofilms on basalt rock are rather fed by material from hydrothermal venting than by basalt-weathering processes explaining the formation of iron-manganese crusts on basalt surfaces (Templeton et al., 2009).

Support for microbial life in ocean crust came from the microbial analyses of borehole fluids (reviewed in Edwards et al., 2012). For example, ribosomal RNA gene sequence data of 65 °C fluids from a 300-m-deep borehole indicated the presence of diverse *Bacteria* and *Archaea*, including gene clones of varying degrees of relatedness to known nitrate reducers (with ammonia production), thermophilic sulfate reducers, and thermophilic fermentative heterotrophs, all consistent with the fluid chemistry (Cowen et al., 2003).

Evidence for microbial carbon and sulfur cycling in deeply buried ridge-flank basalts has been recently given by the presence of functional genes of methane-cycling and sulfate-reducing microorganisms, their successful cultivation, as well as signatures of carbon and sulfur stable isotopes in the drilled basalt core. Hydrogen-producing abiotic serpentinization reactions have been discussed to be relevant for providing the energy for these microorganisms (Lever et al., 2013).

In the deepest gabbroic layer of ocean crust, a low diversity of proteobacterial lineages was observed. These were related to *Bacteria* from hydrocarbon-dominated environments and to known hydrocarbon degraders. Archaeal 16S rRNA genes were not detected; however, the functional gene *mcr* of methanogens was detected as well as functional genes coding for enzymes involved in nitrate, sulfate, and metal reduction, carbon and nitrogen fixation, and ammonium, methane, and toluene oxidation (Mason et al., 2010). Since hydrocarbons were detected in the same borehole, probably originating from abiotic serpentinization reactions, the microbial community is likely fed by hydrocarbons formed independently of photosynthesis in the ocean (Mason et al., 2010).

The size of the biomass in oceanic crust is unknown. Ocean crust shall harbor the largest aquifer on Earth (Fisher and Becker, 2000), which may transport molecular oxygen and nitrate for basalt-weathering reactions (Expedition 329 Scientists, 2011; Ziebis et al., 2012; Orcutt et al., 2013b). This provides an enormous porous space and chemical redox gradients as energy source for a potentially huge deep biosphere. It has been suggested based on model estimates that the size of the biomass of the deep biosphere in oceanic crust is at least as big as that in marine sediments (Heberling et al., 2010). Future research will reveal this estimate.

Conclusions

The marine deep biosphere has been explored at various ocean sites mainly within the framework of the Ocean Drilling Program (ODP) and the Integrated Ocean Drilling Program (IODP). The deep biosphere in the ocean crust has been less intensively studied than that in subsurface marine sediments. Unlike sediments in which organic carbon, derived from photosynthesis in the oceans, is the major energy source for microbial life, inorganic processes in the ocean crusts likely provide the main energy for the rock-hosted deep biosphere. The size of the biomass in oceanic crust is unknown. Kallmeyer et al. (2012) estimated the values for the global biomass in marine sediments being now 2.9×10^{29} cells, corresponding to 4.1 petagram (Pg) C and ~0.6 % of Earth's total living biomass. The total microbial abundance in subseafloor sediments (2.9×10^{29} cells) is now similar to the estimates for the total number of prokaryotes in seawater (1.2×10^{29}) and in soil (2.6×10^{29}). The biomass of the deep subsurface comprises the three domains of life *Archaea*, *Bacteria*, and *Eukarya*, as well as spores and viruses. The analyses of 16S rRNA gene sequences showed that microbial communities of deep marine sediments harbor members of distinct, uncultured bacterial and archaeal phylogenetic lineages. Living cells, microbial activity, and the impact of the deep biosphere for biogeochemical processes have been demonstrated.

Bibliography

Amy, P. S., and Haldeman, D. L., 1997. *The Microbiology of the Terrestrial Deep Subsurface*. New York: CRC.

Bach, W., and Edwards, K. J., 2003. Iron and sulfide oxidation within the basaltic ocean crust: implications for chemolithoautotrophic microbial biomass production. *Geochimica et Cosmochimica Acta*, **67**, 3871–3887.

Bale, S. J., Goodman, K., Rochelle, P. A., Marchesi, J. R., Fry, J. C., Weightman, A. J., and Parkes, R. J., 1997. *Desulfovibrio profundus* sp. nov., a novel barophilic sulfate-reducing bacteria from deep sediments layers in the Japan Sea. *International Journal of Systematic Bacteriology*, **47**, 515–521.

Barnes, S. P., Bradbrook, S. D., Cragg, B. A., Marchesi, J. R., Weightman, A. J., Fry, J. C., and Parkes, R. J., 1998. Isolation of sulfate-reducing bacteria from deep sediment layers of the Pacific Ocean. *Geomicrobiology Journal*, **15**, 67–83.

Batzke, A., Engelen, B., Sass, H., and Cypionka, H., 2007. Phylogenetic and physiological diversity of cultured deep-biosphere bacteria from equatorial Pacific Ocean and Peru margin sediments. *Geomicrobiology Journal*, **24**, 261–273.

Biddle, J. F., House, C. H., and Brenchley, J. E., 2005. Enrichment and cultivation of microorganisms from sediment from the slope of the Peru Trench (ODP Site 1230). In Jørgensen, B. B., D'Hondt, S. L., and Miller, D. J. (eds.), *Proceedings of the ODP, Science Results, 201*. Ocean Drilling Program, College Station, pp. 1–19. doi:10.2973/odp.proc.sr.201.107.2005.

Biddle, J. F., Lipp, J. S., Lever, M., Lloyd, K. G., Sørensen, K. B., Anderson, R., Fredricks, H. F., Elvert, M., Kelly, T. J., Schrag, D. P., Sogin, M. L., Brenchley, J. E., Teske, A., House, C. H., and Hinrichs, K.-U., 2006. Heterotrophic archaea dominate sedimentary subsurface ecosystems off Peru. *Proceedings of the National Academy of Sciences of the United States of America*, **103**, 3846–3851.

Biddle, J. F., Fitz-Gibbon, S., Schuster, S. C., Brenchley, J. E., and House, C. H., 2008. Metagenomic signatures of the Peru margin subseafloor biosphere show a genetically distinct environment. *Proceedings of the National Academy of Sciences of the United States of America*, **105**, 10583–10588.

Biddle, J. F., White, J. R., Teske, A. P., and House, C. H., 2011. Metagenomics of the subsurface Brazos-Trinity Basin (IODP site 1320): comparison with other sediment and pyrosequenced metagenomes. *ISME Journal*, **5**, 1038–1047.

Blazejak, A., and Schippers, A., 2010. High abundance of JS-1- and *Chloroflexi*-related *Bacteria* in deeply buried marine sediments revealed by quantitative, real-time PCR. *FEMS Microbiology Ecology*, **72**, 198–207.

Blazejak, A., and Schippers, A., 2011. Real-time PCR quantification and diversity analysis of the functional genes *aprA* and *dsrA* of sulfate-reducing prokaryotes in marine sediments of the Peru continental margin and the Black Sea. *Frontiers in Microbiology*, **2**, 253.

Boetius, A., Ravenschlag, K., Schubert, C. J., Rickert, D., Widdel, F., Gieseke, A., Amann, R., Jorgensen, B. B., Witte, U., and Pfannkuche, O., 2000. A marine microbial consortium apparently mediating anaerobic oxidation of methane. *Nature*, **407**, 623–626.

Borgonie, G., García-Moyano, A., Litthauer, D., Bert, W., Bester, A., van Heerden, E., Möller, C., Erasmus, M., and Onstott, T. C., 2011. Nematoda from the terrestrial deep subsurface of South Africa. *Nature*, **474**, 79–82.

Borowski, W. S., Paull, C. K., and Ussler, W., III, 1997. Carbon cycling with the upper methanogenic zone of continental rise sediments: an example from the methane-rich sediments overlying the Blake Ridge gas hydrate deposits. *Marine Geology*, **157**, 299–311.

Breuker, A., and Schippers, A., 2013. Data report: total cell counts and qPCR abundance of *Archaea* and *Bacteria* in shallow subsurface marine sediments of North Pond: gravity cores collected on site survey cruise prior to IODP Expedition 336. In Edwards, K. J., Bach, W., Klaus, A., and Expedition 336 Scientists (eds.), *Proceedings of the IODP 336*. Integrated Ocean Drilling Program Management International, Inc., Tokyo.

Breuker, A., Köweker, G., Blazejak, A., and Schippers, A., 2011. The deep biosphere in terrestrial sediments of the Chesapeake Bay impact structure, Virginia, USA. *Frontiers in Microbiology*, **2**, 156.

Breuker, A., Stadler, S., and Schippers, A., 2013. Microbial community analysis of deeply buried marine sediments of the New Jersey shallow shelf (IODP Expedition 313). *FEMS Microbiology Ecology*, **85**, 578–592.

Chapelle, F. H., O'Neill, K., Bradley, P. M., Methé, B. A., Ciufo, S. A., Knobel, L. L., and Lovley, D. R., 2002. A hydrogen-based subsurface microbial community dominated by methanogens. *Nature*, **415**, 312–315.

Ciobanu, M.-C., Burgaude, G., Dufresne, A., Breuker, A., Redou, V., Maamar, S. B., Gaboyer, F., Vandenabeele-Trambouze, O., Lipp, J., Schippers, A., Vandenkoornhuyse, P., Barbier, G., Jebbar, M., Godfroy, A., and Alaine, K., 2014. Microorganisms persist at record-depths in the subseafloor of the Canterbury basin. *ISME Journal*, **8**, 1370–1380.

Coolen, M. J. L., Cypionka, H., Sass, A. M., Sass, H., and Overmann, J., 2002. Ongoing modification of Mediterranean Pleistocene sapropels mediated by prokaryotes. *Science*, **296**, 2407–2410.

Coolen, M. J. L., Muyzer, G., Schouten, S., Volkman, J. K., and Sinninghe Damsté, J. S., 2006. Sulfur and methane cycling during the Holocene in Ace Lake (Antarctica) revealed by lipid and DNA stratigraphy. In Neretin, L. N. (ed.), *Past and Present Water Column Anoxia*. Dordrecht: Springer. NATO Science Series, IV. Earth and Environmental Sciences, Vol. 64, pp. 41–65.

Cowen, J. P., Giovannoni, S. J., Kenig, F., Johnson, H. P., Butterfield, D., Rappé, M. S., Hutnak, M., and Lam, P., 2003. Fluids from aging ocean crust that support microbial life. *Science*, **299**, 120–123.

Cragg, B. A., Parkes, R. J., Fry, J. C., Herbert, R. A., Wimpenny, J. W. T., and Getliff, J. M., 1990. Bacterial biomass and activity profiles within deep sediment layers. *Proceedings of the Ocean Drilling Program, Scientific Results*, **112**, 607–619.

Cragg, B. A., Parkes, R. J., Fry, J. C., Weightman, A. J., Rochelle, P. A., and Maxwell, J. R., 1996. Bacterial populations and processes in sediments containing gas hydrates (ODP Leg 146: Cascadia margin). *Earth and Planetary Science Letters*, **139**, 497–507.

D'Hondt, S. L., Jørgensen, B. B., Miller, D. J., Batzke, A., Blake, R., Cragg, B. A., Cypionka, H., Dickens, G. R., Ferdelman, T. G., Hinrichs, K.-U., Holm, N. G., Mitterer, R., Spivack, A., Wang, G., Bekins, B., Engelen, B., Ford, K., Gettemy, G., Rutherford, S. D., Sass, H., Skilbeck, C. G., Aiello, I. W., Guèrin, G., House, C. H., Inagaki, F., Meister, P., Naehr, T., Niitsuma, S., Parkes, R. J., Schippers, A., Smith, D. C., Teske, A., Wiegel, J., Padilla, C. N., and Solis Acosta, J. L., 2004. Distributions of microbial activities in deep subseafloor sediments. *Science*, **306**, 2216–2221.

D'Hondt, S., Spivack, A. J., Pockalny, R., Ferdelman, T. G., Fischer, J. P., Kallmeyer, J., Abrams, L. J., Smith, D. C., Graham, D., Hasiuk, F., Schrum, H., and Stancin, A. M., 2009. Subseafloor sedimentary life in the South Pacific gyre. *Proceedings of the National Academy of Sciences of the United States of America*, **106**, 11651–11656.

Durbin, A. M., and Teske, A., 2011. Microbial diversity and stratification of oligotrophic abyssal South Pacific sediments. *Environmental Microbiology*, **13**, 3219–3234.

Durbin, A. M., and Teske, A., 2012. Archaea in organic-lean and organic-rich marine subsurface sediments: an environmental gradient reflected in distinct phylogenetic lineages. *Frontiers in Microbiology*, **3**, 168.

Edgcomb, V. P., Beaudoin, D., Gast, R., Biddle, J. F., and Teske, A., 2011. Marine subsurface eukaryotes: the fungal majority. *Environmental Microbiology*, **13**, 172–183.

Edwards, K. J., Bach, W., McCollom, T. M., and Rogers, D. R., 2004. Neutrophilic iron-oxidizing bacteria in the ocean: their habitats, diversity, and roles in mineral deposition, rock alteration, and biomass production in the deep-sea. *Geomicrobiology Journal*, **21**, 393–404.

Edwards, K. J., Bach, W., and McCollom, T. M., 2005. Geomicrobiology in oceanography: mineral-microbe interactions at and below the seafloor. *Trends in Microbiology*, **13**, 449–459.

Edwards, K. J., Becker, K., and Colwell, F., 2012. The deep, dark energy biosphere: intraterrestrial life on Earth. *Annual Review of Earth and Planetary Sciences*, **40**, 551–568.

Engelen, B., Ziegelmüller, K., Wolf, L., Köpke, B., Gittel, A., Cypionka, H., Treude, T., Nakagawa, S., Inagaki, F., Lever, M. A., and Steinsbu, B. O., 2008. Fluids from the oceanic crust support microbial activities within the deep biosphere. *Geomicrobiology Journal*, **25**, 55–66.

Engelhardt, T., Sahlberg, M., Cypionka, H., and Engelen, B., 2013. Biogeography of *Rhizobium radiobacter* and distribution of associated temperate phages in deep subseafloor sediments. *ISME Journal*, **7**, 199–209.

Expedition 329 Scientists, 2011. South Pacific Gyre subseafloor life. *IODP Preliminary Report*, **329**, p. 108, doi:10.2204/iodp.pr.329.2011.

Ferdelman, T. G., Fossing, H., Neumann, K., and Schulz, H. D., 1999. Sulfate reduction in surface sediments of the southeast Atlantic continental margin between 15°38′ S and 27°57′S (Angola and Namibia). *Limnology and Oceanography*, **44**, 650–661.

Fisher, A. T., and Becker, K., 2000. Channelized fluid flow in oceanic crust reconciles heat-flow and permeability data. *Nature*, **403**, 71–74.

Fisk, M. R., Giovannoni, S. J., and Furnes, H., 1998. Alteration of oceanic volcanic glass textural evidence of microbial activity. *Science*, **281**, 978–980.

Fossing, H., Ferdelman, T. G., and Berg, P., 2000. Sulfate reduction and methane oxidation in continental margin sediments influenced by irrigation (South-East Atlantic off Namibia). *Geochimica et Cosmochimica Acta*, **64**, 897–910.

Fry, J. C., Parkes, R. J., Cragg, B. A., Weightman, A. J., and Webster, G., 2008. Prokaryotic diversity and activity in the deep subseafloor biosphere. *FEMS Microbiology Ecology*, **66**, 181–196.

Fry, J. C., Horsfield, B., Sykes, R., Cragg, B. A., Heywood, C., Kim, G. T., Mangelsdorf, K., Mildenhall, D. C., Rinna, J., Vieth, A., Zink, K. G., Sass, H., Weightman, A. J., and Parkes, R. J., 2009. Prokaryotic populations and activities in an interbedded coal deposit, including a previously deeply buried section (1.6–2.3 km) above ~150 Ma basement rock. *Geomicrobiology Journal*, **26**, 163–178.

Gold, T., 1992. The deep, hot biosphere. *Proceedings of the National Academy of Sciences of the United States of America*, **89**, 6045–6049.

Gregg, W. W., Casey, N. W., and McClain, C. R., 2005. Recent trends in global ocean chlorophyll. *Geophysical Research Letters*, doi:10.1029/2004gl021808.

Heberling, C., Lowell, R. P., Liu, L., and Fisk, M. R., 2010. Extent of the microbial biosphere in the oceanic crust. *Geochemistry, Geophysics, Geosystems*, **11**(Q08003), 1–15.

Hedges, J. I., 1992. Global biogeochemical cycles: progress and problems. *Marine Chemistry*, **39**, 67–93.

Heid, C. A., Stevens, J., Livak, K. J., and Williams, P. M., 1996. Real time quantitative PCR. *Genome Research*, **6**, 986–994.

Heuer, V. B., Pohlman, J. W., Torres, M. E., Elvert, M., and Hinrichs, K. U., 2009. The stable carbon isotope biogeochemistry of acetate and other dissolved carbon species in deep subseafloor sediments at the northern Cascadia margin. *Geochimica et Cosmochimica Acta*, **73**, 3323–3336.

Hoehler, T. M., and Jørgensen, B. B., 2013. Microbial life under extreme energy limitation. *Nature Reviews Microbiology*, **11**, 83–94.

Hoehler, T. M., Alperin, M. J., Albert, D. B., and Martens, C. S., 1994. Field and laboratory studies of methane oxidation in an anoxic marine sediment: evidence for a methanogen-sulfate reducer consortium. *Global Biogeochemical Cycles*, **8**, 451–463.

Inagaki, F., and Nealson, K. H., 2006. The Paleome: letters from ancient earth. In Neretin, L. N. (ed.), *Past and Present Water Column Anoxia*. Dordrecht: Springer. NATO Science Series, IV. Earth and Environmental Sciences, Vol. 64, pp. 21–39.

Inagaki, F., Takai, K., Komatsu, T., Kanamatsu, T., Fujioka, K., and Horikoshi, K., 2001. Archaeology of *Archaea*: geomicrobiological record of Pleistocene thermal events concealed in a deep-sea subseafloor environment. *Extremophiles*, **5**, 385–392.

Inagaki, F., Suzuki, M., Takai, K., Oida, H., Sakamoto, T., Aoki, K., Nealson, K. H., and Horikoshi, K., 2003. Microbial communities associated with geological horizons in costal subseafloor sediments from the Sea of Okhotsk. *Applied and Environmental Microbiology*, **69**, 7224–7235.

Inagaki, F., Nunoura, T., Nakagawa, S., Teske, A., Lever, M., Lauer, A., Suzuki, M., Takai, K., Delwiche, M., Colwell, F. S., Nealson, K. H., Horikoshi, K., D'Hondt, S. L., and Jørgensen, B. B., 2006. Biogeographical distribution and diversity of microbes in methane hydrate-bearing deep marine sediments on the Pacific Ocean margin. *Proceedings of the National Academy of Sciences of the United States of America*, **103**, 2815–2820.

Itävaara, M., Nyyssönen, M., Kapanen, A., Nousiainen, A., Ahonen, L., and Kukkonen, I., 2011. Characterization of bacterial diversity to a depth of 1500 m in the Outokumpu deep borehole, Fennoscandian Shield. *FEMS Microbiology Ecology*, **77**, 295–309.

Jahnke, R. A., and Jackson, G. A., 1992. The spatial distribution of sea floor oxygen consumption in the Atlantic and Pacific oceans. In Rowe, G. T., and Pariente, V. (eds.), *Deep-Sea Food Chains and the Global Carbon Cycle*. Dordrecht: Kluwer Academic, pp. 295–308.

Jørgensen, B. B., 1982. Mineralization of organic matter in the sea bed – the role of sulphate reduction. *Nature*, **296**, 643–645.

Jørgensen, B. B., 2000. Bacteria and marine biogeochemistry. In Schulz, H. D., and Zabel, M. (eds.), *Marine Geochemistry*. Berlin: Springer, pp. 173–207.

Jørgensen, B. B., and Boetius, A., 2007. Feast and famine – microbial life in the deep-sea bed. *Nature Review Microbiology*, **5**, 770–783.

Jørgensen, B. B., and D'Hondt, S. L., 2006. A starving majority deep beneath the seafloor. *Science*, **314**, 932–934.

Jørgensen, B. B., D'Hondt, S. L., and Miller, D. J., 2006. Leg 201 synthesis: controls on microbial communities in deeply buried sediments. In Jørgensen, B. B., D'Hondt, S. L., Miller, D. J. (eds.), *Proceedings of ODP, Science Results, 201*. Ocean Drilling Program, College Station. http://www-odp.tamu.edu/publications/201_SR/201sr.htm

Kallmeyer, J., Smith, D. C., D'Hondt, S. L., and Spivack, A. J., 2008. New cell extraction procedure applied to deep subsurface sediments. *Limnology and Oceanography: Methods*, **6**, 236–245.

Kallmeyer, J., Pockalny, R., Adhikari, R. R., Smith, D. C., and D'Hondt, S., 2012. Global distribution of microbial abundance and biomass in subseafloor sediment. *Proceedings of the National Academy of Sciences of the United States of America*, **109**, 16213–16216.

Kastner, M., Elderfield, H., Martin, J. B., Suess, E., Kvenvolden, K. A., and Garrison, R. E., 1990. Diagenesis and interstitial-water chemistry at the Peruvian continental margin-major constituents and strontium isotopes. In Suess, E., von Huene, R., et al. (eds.), *Proceedings of the ODP, Scienctific Results*. Vol. 112, Chap. 25, pp. 413–440. http://www-odp.tamu.edu/publications/112_SR/112sr.htm

Kobayashi, T., Koide, O., Mori, K., Shimamura, S., Matsuura, T., Miura, T., et al., 2008. Phylogenetic and enzymatic diversity of deep subseafloor aerobic microorganisms in organics- and methane-rich sediments off Shimokita Peninsula. *Extremophiles*, **12**, 519–527.

Kubo, K., Lloyd, K., Biddle, J., Amann, R., Teske, A., and Knittel, K., 2012. Archaea of the miscellaneous crenarchaeotal group are abundant, diverse and widespread in marine sediments. *ISME Journal*, **6**, 1949–1965.

Lee, Y. J., Wagner, I. D., Brice, M. E., Kevbrin, V. V., Mills, G. L., Romanek, C. S., and Wiegel, J., 2005. *Thermosediminibacter oceani* gen. nov., sp. nov. and *Thermosediminibacter litoriperuensis* sp. nov., new anaerobic thermophilic bacteria isolated from Peru margin. *Extremophiles*, **9**, 375–383.

Lever, M. A., 2013. Functional gene surveys from ocean drilling expeditions – a review and perspective. *FEMS Microbiology Ecology*, **84**, 1–23.

Lever, M. A., Alperin, M., Engelen, B., Inagaki, F., Nakagawa, S., Steinsbu, B. O., Teske, A., and IODP Expedition 301 Scientists, 2006. Trends in basalt and sediment core contamination during IODP Expedition 301. *Geomicrobiology Journal*, **23**, 517–530.

Lever, M. A., Heuer, V. B., Morono, Y., Masui, N., Schmidt, F., Alperin, M. J., Inagaki, F., Hinrichs, K. U., and Teske, A., 2010. Acetogenesis in deep subseafloor sediments of the Juan de Fuca Ridge Flank: a synthesis of geochemical, thermodynamic, and gene-based evidence. *Geomicrobiology Journal*, **27**, 183–211.

Lever, M. A., Rouxel, O., Alt, J. C., Shimizu, N., Ono, S., Coggon, R. M., Shanks, W. C., III, Lapham, L., Elvert, M., Prieto-Mollar, X., Hinrichs, K. U., Inagaki, F., and Teske, A., 2013. Evidence for microbial carbon and sulfur cycling in deeply buried ridge flank basalt. *Science*, **339**, 1305–1308.

Lin, L. H., Wang, P. L., Rumble, D., Lippmann-Pipke, J., Boice, E., Pratt, L. M., Sherwood Lollar, B., Brodie, E. L., Hazen, T. C., Andersen, G. L., DeSantis, T. Z., Moser, D. P., Kershaw, D., and Onstott, T. C., 2006. Long term sustainability of a high energy, low diversity crustal biome. *Science*, **314**, 479–482.

Lipp, J. S., Morono, Y., Inagaki, F., and Hinrichs, K. U., 2008. Significant contribution of archaea to extent biomass in marine subsurface sediments. *Nature*, **454**, 991–994.

Lloyd, K. G., Schreiber, L., Petersen, D. G., Kjeldsen, K. U., Lever, M. A., Steen, A. D., Stepanauskas, R., Richter, M., Kleindienst, S., Lenk, S., Schramm, A., and Jørgensen, B. B., 2013a. Predominant archaea in marine sediments degrade detrital proteins. *Nature*, **496**, 215–218.

Lloyd, K. G., May, M. K., Kevorkian, R. T., and Steen, A. D., 2013b. Meta-analysis of quantification methods shows that *Archaea* and *Bacteria* have similar abundances in the subseafloor. *Applied and Environmental Microbiology*, **79**, 7790–7799.

Lockey, C., Otto, E., and Long, Z., 1998. Real-time fluorescence detection of a single DNA molecule. *BioTechniques*, **24**, 744–746.

Logemann, J., Graue, J., Köster, J., Engelen, B., Rullkötter, J., and Cypionka, H., 2011. A laboratory experiment of intact polar lipid degradation in sandy sediments. *Biogeosciences Discussions*, **8**, 3289–3321.

Lomstein, B. A., Langerhuus, A. T., D'Hondt, S., Jørgensen, B. B., and Spivack, A. J., 2012. Endospore abundance, microbial growth and necromass turnover in deep sub-seafloor sediment. *Nature*, **484**, 101–104.

Mascarelli, A. L., 2009. Low life. *Nature*, **459**, 770–773.

Mason, O. U., Nakagawa, T., Rosner, M., van Nostrand, J. D., Zhou, J., Maruyama, A., Fisk, M. R., and Giovannoni, S. J., 2010. First investigation of the microbiology of the deepest layer of ocean crust. *PloS One*, **5**, e15399.

Mikucki, J. A., Liu, Y., Delwiche, M., Colwell, F. S., and Boone, D. R., 2003. Isolation of a methanogen from deep marine sediments that contain methane hydrates, and description of *Methanoculleus submarines*. sp. nov. *Applied and Environmental Microbiology*, **69**, 3311–3316.

Milucka, J., Ferdelman, T. G., Polerecky, L., Franzke, D., Wegener, G., Schmid, M., Lieberwirth, I., Wagner, M., Widdel, F., and Kuypers, M. M. M., 2012. Zero-valent sulphur is a key intermediate in marine methane oxidation. *Nature*, **491**, 541–546.

Morono, Y., Terada, T., Masui, N., and Inagaki, F., 2009. Discriminative detection and enumeration of microbial life in marine subsurface sediments. *ISME Journal*, **3**, 503–511.

Moser, D. P., Onstott, T. C., Fredrickson, J. K., Brockman, F. J., Balkwill, D. L., Drake, G. R., Pfiffner, S. M., White, D. C., Takai, K., Pratt, L. M., Fong, J., Sherwood Lollar, B., Slater, G., Phelps, T. J., Spoelstra, N., Deflaun, M., Southam, G., Welty, A. T., Baker, B. J., and Hoek, J., 2003. Temporal shifts in the geochemistry and microbial community structure of an ultradeep mine borehole following isolation. *Geomicrobiology Journal*, **20**, 517–548.

Nunoura, T., Soffientino, B., Blazejak, A., Kakuta, J., Oida, H., Schippers, A., and Takai, K., 2009. Subseafloor microbial communities associated with rapid turbidite deposition in the Gulf of Mexico continental slope (IODP Expedition 308). *FEMS Microbiology Ecology*, **69**, 410–424.

Orcutt, B. N., Sylvan, J. B., Knab, N. J., and Edwards, K. J., 2011. Microbial ecology of the dark ocean above at and below the seafloor. *Microbiology and Molecular Biology Reviews*, **75**, 361–422.

Orcutt, B. N., LaRowe, D. E., Biddle, J. F., Colwell, F. S., Glazer, B. T., Reese, B. K., Kirkpatrick, J. B., Lapham, L. L., Mills, H. J., Sylvan, J. B., Wankel, S. D., and Wheat, C. G., 2013a. Microbial activity in the marine deep biosphere: progress and prospects. *Frontiers in Microbiology*, **4**, 189.

Orcutt, B. N., Wheat, C. G., Rouxel, O., Hulme, S., Edwards, K. J., and Bach, W., 2013b. Oxygen consumption rates in subseafloor basaltic crust derived from a reaction transport model. *Nature Communications*, **4**, 2539.

Orsi, W. D., Biddle, J. F., and Edgcomb, V., 2013a. Deep sequencing of subseafloor eukaryotic rRNA reveals active fungi across marine subsurface provinces. *PloS One*, **8**, e56335.

Orsi, W. D., Edgcomb, V., Christman, G. D., and Biddle, J. F., 2013b. Gene expression in the deep biosphere. *Nature*, **499**, 205–208.

Parkes, R. J., and Wellsbury, P., 2004. Deep biospheres. In Bull, A. T. (ed.), *Microbial Diversity and Bioprocessing*. Washington, DC: ASM press, pp. 120–129.

Parkes, R. J., Cragg, B. A., Bale, S. J., Getliff, J. M., Goodman, K., Rochelle, P. A., Fry, J. C., Weightman, A. J., and Harvey, S. M., 1994. Deep bacterial biosphere in Pacific Ocean sediments. *Nature*, **371**, 410–413.

Parkes, R. J., Cragg, B. A., and Wellsbury, P., 2000. Recent studies on bacterial populations and processes in subseafloor sediments: a review. *Hydrogeology Journal*, **8**, 11–28.

Parkes, R. J., Webster, G., Cragg, B. A., Weightman, A. J., Newberry, C. J., Ferdelman, T. G., Kallmeyer, J., Jørgensen, B. B., Aiello, I. W., and Fry, J. C., 2005. Deep sub-seafloor prokaryotes stimulated at interfaces over geological time. *Nature*, **436**, 390–394.

Parkes, R. J., Wellsbury, P., Mather, I. D., Cobb, S. J., Cragg, B. A., Hornibrook, E. R. C., and Horsfield, B., 2007. Temperature activation of organic matter and minerals during burial has the potential to sustain the deep biosphere over geological timescales. *Organic Geochemistry*, **38**, 845–852.

Parkes, R. J., Sellek, G., Webster, G., Martin, D., Anders, E., Weightman, A. J., and Sass, H., 2009. Culturable prokaryotic diversity of deep, gas hydrate sediments: first use of a continuous high-pressure, anaerobic, enrichment and isolation system for subseafloor sediments (DeepIsoBUG). *Environmental Microbiology*, **11**, 3140–3153.

Parkes, R. J., Linnane, C. D., Webster, G., Sass, H., Weightman, A. J., Hornibrook, E. R. C., and Horsfield, B., 2011. Prokaryotes stimulate mineral H_2 formation for the deep biosphere and subsequent thermogenic activity. *Geology*, **39**, 219–222.

Parkes, R. J., Cragg, B., Webster, G., Roussel, E. G., Weightman, A. J., and Sass, H., 2014. A review of prokaryotic populations and processes in sub-surface sediments, including biosphere: geosphere interactions. *Marine Geology*, **352**, 409–425.

Pedersen, K., 1993. The deep subterranean biosphere. *Earth Science Reviews*, **34**, 243–260.

Pedersen, K., 1997. Microbial life in deep granitic rock. *FEMS Microbiology Review*, **20**, 399–414.

Ravenschlag, K., Sahm, K., Knoblauch, C., Jorgensen, B. B., and Amann, R., 2000. Community structure, cellular rRNA content, and activity of sulfate-reducing bacteria in marine Arctic sediments. *Applied and Environmental Microbiology*, **66**, 3592–3602.

Rinke, C., Schwientek, P., Sczyrba, A., Ivanova, N. N., Anderson, I. J., Cheng, J. F., Darling, A., et al., 2013. Insights into the phylogeny and coding potential of microbial dark matter. *Nature*, **499**, 431–437.

Roussel, E. G., Bonavita, M. A. C., Querellou, J., Cragg, B. A., Webster, G., Prieur, D., and Parkes, R. J., 2008. Extending the sub-sea-floor biosphere. *Science*, **320**, 1046.

Roussel, E. G., Sauvadet, A. L., Chaduteau, C., Fouquet, Y., Charlou, J. L., Prieur, D., and Bonavita, M. A. C., 2009. Archaeal communities associated with shallow to deep subseafloor sediments of the New Caledonia Basin. *Environmental Microbiology*, **11**, 2446–2462.

Røy, H., Kallmeyer, J., Adhikari, R. R., Pockalny, R., Jørgensen, B. B., and D'Hondt, S., 2012. Aerobic microbial respiration in 86-million-year-old deep-sea red clay. *Science*, **336**, 922–925.

Sahm, K., MacGregor, B. J., Jørgensen, B. B., and Stahl, D. A., 1999. Sulfate-reduction and vertical distribution of sulphate-reducing bacteria quantified by rRNA slot-blot hybridization in a coastal marine sediment. *Environmental Microbiology*, **1**, 65–74.

Santelli, C. M., Orcutt, B. H., Banning, E., Bach, W., Moyer, C. L., Sogin, M. L., Staudigel, H., and Edwards, K. J., 2008. Abundance and diversity of microbial life in ocean crust. *Nature*, **453**, 653–657.

Santelli, C. M., Edgcomb, V. P., Bach, W., and Edwards, K. J., 2009. The diversity and abundance of bacteria inhabiting seafloor lavas positively correlate with rock alteration. *Environmental Microbiology*, **11**, 86–98.

Schippers, A., and Neretin, L. N., 2006. Quantification of microbial communities in near-surface and deeply buried marine sediments on the Peru continental margin using real-time PCR. *Environmental Microbiology*, **8**, 1251–1260.

Schippers, A., Neretin, L. N., Kallmeyer, J., Ferdelman, T. G., Cragg, B. A., Parkes, R. J., and Jørgensen, B. B., 2005. Prokaryotic cells of the deep sub-seafloor biosphere identified as living bacteria. *Nature*, **433**, 861–864.

Schippers, A., Köweker, G., Höft, C., and Teichert, B., 2010. Quantification of microbial communities in three forearc sediment basins off Sumatra. *Geomicrobiology Journal*, **27**, 170–182.

Schippers, A., Kock, D., Höft, C., Köweker, G., and Siegert, M., 2012. Quantification of microbial communities in subsurface marine sediments of the Black Sea and off Namibia. *Frontiers in Microbiology*, **3**, 16.

Schouten, S., Middelburg, J. J., Hopmans, E. C., and Sinninghe Damste, J. S., 2010. Fossilization and degradation of intact polar lipids in deep subsurface sediments: a theoretical approach. *Geochimica et Cosmochimica Acta*, **74**, 3806–3814.

Schrenk, M. O., Huber, J. A., and Edwards, K. J., 2010. Microbial provinces in the subseafloor. *Annual Review of Marine Science*, **2**, 279–304.

Smith, D. C., Spivack, A. J., Fisk, M. R., Haveman, S. A., and Staudigel, H., 2000. Tracer-based estimates of drilling-induced microbial contamination of deep sea crust. *Geomicrobiology Journal*, **17**, 207–219.

Sørensen, K. B., Lauer, A., and Teske, A. P., 2004. Archaeal phylotypes in a metal-rich and low-activity deep subsurface sediment of the Peru Basin, ODP Leg 201, Site 1231. *Geobiology*, **2**, 151–161.

Stevens, T. O., and McKinley, J. P., 1995. Lithoautotrophic microbial ecosystems in deep basalt aquifers. *Science*, **270**, 450–454.

Templeton, A. S., Knowles, E. J., Eldridge, D. L., Arey, B. W., Dohnalkova, A. C., Webb, S. M., Bailey, B. E., Tebo, B. M., and Staudigel, H., 2009. A seafloor microbial biome hosted within incipient ferromanganese crusts. *Nature Geoscience*, **2**, 872–876.

Teske, A. P., 2005. The deep subsurface biosphere is alive and well. *Trends in Microbiology*, **13**, 402–404.

Teske, A. P., 2006a. Microbial communities of deep marine subsurface sediments: molecular and cultivation surveys. *Geomicrobiology Journal*, **23**, 357–368.

Teske, A. P., 2006b. Microbial community composition in deep marine subsurface sediments of ODP Leg 201: sequencing surveys and cultivations. In Jørgensen, B. B., D'Hondt, S. L., and Miller, D. J. (eds.), *Proceedings of the ODP, Scientific Results, 201*. Ocean Drilling Program, College Station. http://www-odp.tamu.edu/publications/201_SR/VOLUME/CHAPTERS/120.PDF

Teske, A. P., and Sørensen, K. B., 2008. Uncultured archaea in deep marine subsurface sediments: have we caught them all? *ISME Journal*, **2**, 3–18.

Teske, A. P., Biddle, J. F., Edcomb, V. P., and Schippers, A., 2013. Deep subsurface microbiology: a guide to the research topic papers. *Frontiers in Microbiology*, **4**, 122.

Toffin, L., Webster, G., Weightman, A. J., Fry, J. C., and Prieur, D., 2004. Molecular monitoring of culturable bacteria from deep-sea sediment of the Nankai Trough. Leg 190 Ocean Drilling Program. *FEMS Microbiology Ecology*, **48**, 357–367.

Toffin, L., Zink, K., Kato, C., Pignet, P., Bidault, A., Bienvenu, N., et al., 2005. *Marinilactibacillus piezotolerans* sp. nov., a novel marine lactic acid bacterium isolated from deep subseafloor sediment of the Nankai Trough. *International Journal of Systematic and Evolutionary Microbiology*, **55**, 345–351.

Wasmund, K., Schreiber, L., Lloyd, K. G., Petersen, D. G., Schramm, A., Stepanauskas, R., Jørgensen, B. B., and Adrian, L., 2014. Genome sequencing of a single cell of the widely distributed marine subsurface *Dehalococcoidia*, phylum *Chloroflexi*. *ISME Journal*, **8**, 383–397.

Webster, G., Parkes, R. J., Fry, J. C., and Weightman, A. J., 2004. Widespread occurrence of a novel division of bacteria identified by 16S rDNA gene sequences originally found in deep marine sediments. *Applied and Environmental Microbiology*, **70**, 5708–5713.

Webster, G., Parkes, R. J., Cragg, B. A., Newberry, C. J., Weightman, A. J., and Fry, J. C., 2006. Prokaryotic community composition and biogeochemical processes in deep subseafloor sediments from the Peru margin. *FEMS Microbiology Ecology*, **58**, 65–85.

Webster, G., Yarram, L., Freese, E., Köster, J., Sass, H., Parkes, R. J., and Weightman, A. J., 2007. Distribution of candidate division JS1 and other Bacteria in tidal sediments of the German Wadden Sea using targeted 16S rRNA gene PCR-DGGE. *FEMS Microbiology Ecology*, **62**, 78–89.

Webster, G., Blazejak, A., Cragg, B. A., Schippers, A., Sass, H., Rinna, J., Tang, X., Mathes, F., Ferdelman, T. G., Fry, J. C., Weightman, A. J., and Parkes, R. J., 2009. Subsurface microbiology and biogeochemistry of a deep, cold-water carbonate mound from the Porcupine Seabight (IODP Expedition 307). *Environmental Microbiology*, **11**, 239–257.

Webster, G., Sass, H., Cragg, B. A., Gorra, R., Knab, N. J., Green, C. J., Mathes, F., Fry, J. C., Weightman, A. J., and Parkes, R. J., 2011. Enrichment and cultivation of prokaryotes associated with the sulphate-methane transition zone of diffusion-controlled sediments of Aarhus Bay, Denmark, under heterotrophic conditions. *FEMS Microbiology Ecology*, **77**, 248–263.

Weinbauer, M. G., Beckmann, C., and Höfle, M. G., 1998. Utility of green fluorescent nucleic acid dyes and aluminium oxide membrane filters for rapid epifluorescence enumeration of soil and sediment bacteria. *Applied and Environmental Microbiology*, **64**, 5000–5003.

Whelan, J. K., Oremland, R., Tarafa, M., Smith, R., Howarth, R., and Lee, C., 1986. Evidence for sulfate-reducing and methane-producing microorganisms in sediments from Sites 618, 619, and 622. In Bouma, A. H., Coleman, J. M., et al. (eds.), *Initial Reports of DSDP*, Vol. 96, pp. 767–775.

Whitman, W. B., Coleman, D. C., and Wiebe, W. J., 1998. Prokaryotes: the unseen majority. *Proceedings of the National Academy of Sciences of the United States of America*, **95**, 6578–6583.

Wilms, R., Sass, H., Köpke, B., Cypionka, H., and Engelen, B., 2007. Methane and sulfate profiles within the subsurface of a tidal flat are reflected by the distribution of sulfate-reducing bacteria and methanogenic archaea. *FEMS Microbiology Ecology*, **59**, 611–621.

Wittwer, C. T., Herrmann, M. G., Moss, A. A., and Rasmussen, R. P., 1997. Continuous fluorescence monitoring of rapid cycle DNA amplification. *BioTechniques*, **22**, 130–138.

Wortmann, U. G., Bernasconi, S. M., and Böttcher, M. E., 2001. Hypersulfidic deep biosphere indicates extreme sulfur isotope fractionation during single-step microbial sulfate reduction. *Geology*, **29**, 647–650.

Xie, S., Lipp, J. S., Wegener, G., Ferdelman, T. G., and Hinrichs, K. U., 2013. Turnover of microbial lipids in the deep biosphere and growth of benthic archaeal populations. *Proceedings of the National Academy of Sciences of the United States of America*, **110**, 6010–6014.

Yanagawa, K., Nunoura, T., McAllister, S. M., Hirai, M., Breuker, A., Brandt, L., House, C. H., Moyer, C. L., Birrien, J. L., Aoike, K., Sunamura, M., Urabe, T., Mottl, M. J., and Takai, K., 2013. The first contamination assessment for microbiological study in deep-sea drilling and coring by the D/V Chikyu at the Iheya North hydrothermal field in the Mid-Okinawa Trough (IODP Expedition 331). *Frontiers in Microbiology*, **4**, 327.

Ziebis, W., McManus, J., Ferdelman, T., Schmidt-Schierhorn, F., Bach, W., Muratli, J., et al., 2012. Interstitial fluid chemistry of sediments underlying the North Atlantic gyre and the influence of subsurface fluid flow. *Earth and Planetary Science Letters*, **323–324**, 79–91.

Cross-references

Cold Seeps
Deep Biosphere
Deep-sea Sediments
Marine Sedimentary Basins
Methane in Marine Sediments
Ocean Margin Systems
Paleoceanographic Proxies
Regional Marine Geology

DEEP-SEA FANS

Thierry Mulder[1] and Heiko Hüneke[2]
[1]University of Bordeaux, Talence, France
[2]Institute of Geography and Geology, Ernst Moritz Arndt University, Greifswald, Germany

Synonyms
Single-point-sourced turbidite system; Submarine fan

Definition
The term "deep-sea fan" has been coined by Normark (1970) to describe modern turbidite systems with a point source that typically display a fan-shaped outline. This morphological analysis was based on the Pleistocene Californian fans San Lucas and Astoria. Mutti and Ricci Lucchi (1972) used a similar model to describe the fan-shaped facies association of Cretaceous and Tertiary turbidite systems in the Apennines and Spanish Pyrenees.

The depositional system
Models that are more recent try to make a synthesis between morphology, sedimentary processes, depositional environment, and sedimentary facies (Walker, 1978; Mutti, 1979; Barnes and Normark, 1985). They include control factors such as sedimentary source, tectonic context, and eustatic changes.

The term "turbidite system" was thus preferentially used to embrace all deepwater clastic depositional systems. The environmental model of Reading and Richards (1994) includes (a) submarine fans controlled by single-point-sourced supply, (b) submarine ramps displaying multiple-point-sourced delivery, and (c) slope aprons with line-sourced supply. Based on the lithology, mud-rich, mud-/sand-rich, sand-rich, or gravel-rich systems are differentiated.

Mud-rich submarine fans are the typical turbidite systems of modern passive continental margins. The largest fans (10^5–10^6 km^2) are the Bengal, Indus, Amazon, Mississippi, and Zaire fans. Much of the clastic material of these high-efficiency, open-ocean fans is typically supplied during floods from major rivers.

The modern mud-rich submarine fans are subdivided into canyon, channel-levee, and lobe complexes. The canyon usually largely incises the continental slope and shelf, sometimes up to river mouth. On the continental rise, it opens on a channel-levee complex including several generations of stacked channel-levees. The channel is usually meandering and aggradational and bordered with bird wing-shaped (in a transversal cross section) levees built above the fan surface. Levees progressively diminish in height and grain size and pass into lobes extending in the abyssal plain. Lobes are first channeled and become unchanneled with increasing distance.

Today, classification of deep-sea turbidite systems has been officially abandoned (Walker, 1992). Sedimentologists prefer the "do-it-yourself" toolbox using the architectural element concept (Miall, 1985).

Bibliography
Barnes, P. M., and Normark, W. R., 1985. Diagnostic parameters for comparing modern submarine fans and ancient turbidite systems. In Bouma, A., Normark, W., and Barnes, N. (eds.), *Submarine Fans and Related Turbidite Systems*. New York: Springer, pp. 13–14.

Miall, A. D., 1985. Architectural-element analysis: a new method of facies analysis applied to fluvial deposits. *Earth-Science Reviews*, **22**, 261–308.

Mutti, E., 1979. Turbidites et cônes sous-marins profonds. In Homewood, P. (ed.), *Sédimentation Détritique (Fluviatile, Littorale et Marine)*. Fribourg: Institut de Géologie de l'Université de Fribourg, pp. 353–419. Short Course.

Mutti, E., and Ricci Lucchi, F., 1972. Le torbiditi dell'Appennino settentrionale: introductione all'analisi di facies. *Memorie della Societa' Geologica Italiana*, **11**, 161–199.

Normark, W. R., 1970. Growth patterns of deep-sea fans. *American Association of Petroleum Geologists Bulletin*, **54**, 2170–2195.

Reading, H. G., and Richards, M. T., 1994. The classification of deep-water siliciclastic depositional systems by grain size and feeder systems. *American Association of Petroleum Geologists Bulletin*, **78**, 792–822.

Richards, M., Bowman, M., and Reading, H., 1998. Submarine-fan systems I: characterization and stratigraphic prediction. *Marine and Petroleum Geology*, **15**, 689–717.

Walker, R. G., 1978. Deep-water sandstone facies and ancient submarine fans: models for exploration for stratigraphic traps. *American Association of Petroleum Geologists Bulletin*, **62**, 932–966.

Walker, R. G., 1992. Turbidites and submarine fans. In Walker, R. G., and James, N. P. (eds.), *Facies Models: Response to Sea Level Changes*. St. John's: Geological Association of Canada, pp. 1–14. GEOtext 1.

Cross-references
Bouma Sequence
Continental Rise
Continental Slope
Deep-sea Sediments
Submarine Canyons
Turbidites

DEEP-SEA SEDIMENTS

Mitchell Lyle
College of Earth, Ocean, and Atmospheric Sciences, Oregon State University, Corvallis, OR, USA

Synonyms
Pelagic sediments

Definition
The term "deep-sea sediments" or the interchangeable term "pelagic sediments" refers to sediments that deposit

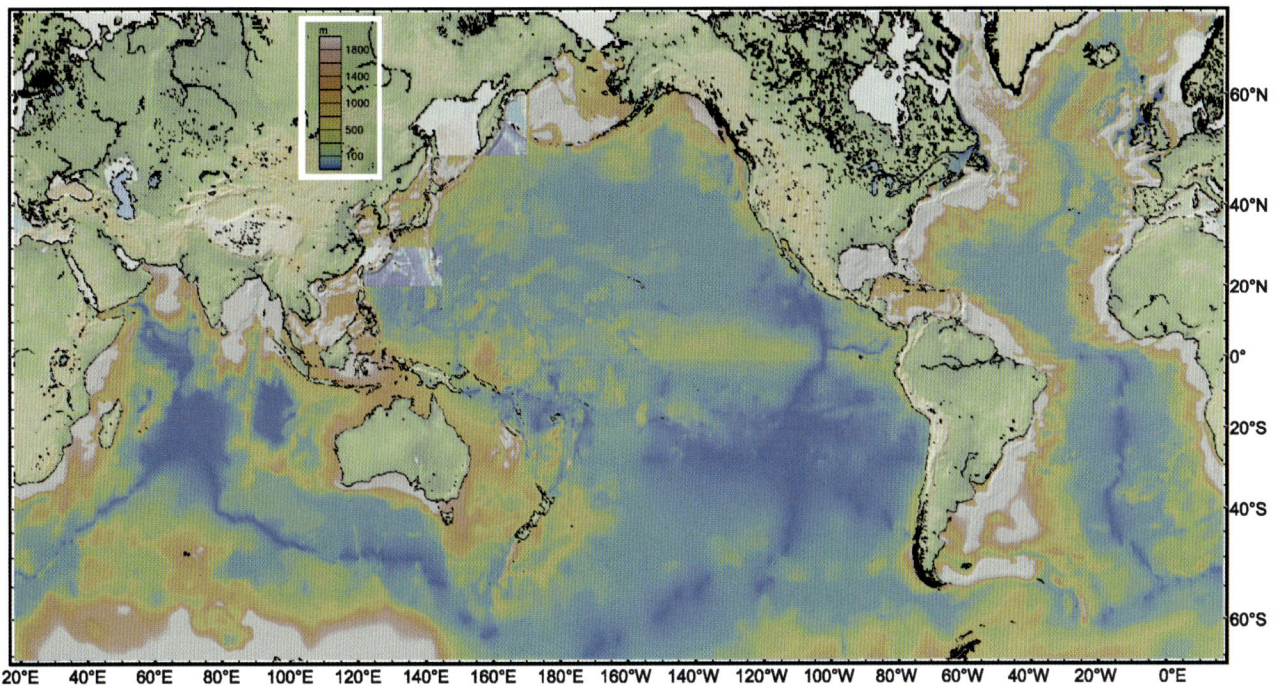

Deep-sea Sediments, Figure 1 Sediment thickness in the world's oceans (Whittaker et al., 2013; http://www.ngdc.noaa.gov/mgg/sedthick). The thickest sediments (*beige*) are along continental margins where turbidites accumulate. The thinnest sediments (*dark blue*) are on areas far from land or on zero age crust along the mid-ocean ridges where there has not been time for sediments to accumulate.

slowly in the abyssal ocean beyond the continental margins.

What are deep-sea sediments?

Deep-sea sediments typically have sedimentation rates less than $30 \text{ m}/10^6$ years, and rates as low $0.1 \text{ m}/10^6$ years have been reported. The slow sedimentation rates and unusual sediment compositions reflect the low fluxes of aluminosilicates eroded from continents. The terrigenous material that does deposit is often windblown dust. Other solids produced through biological activity, through hydrothermal leaching of basalts, or even by earth's bombardment by meteorites can make up large fractions of a deep-sea sediment deposit.

Not all sediments in the deep ocean are deep-sea sediments depositing slowly. A significant portion of the deep ocean is filled by turbidites, which are gravity flow deposits that typically originate from the continental margins. These sediments will be discussed briefly in section "Turbidites and Hemipelagic Sediment Components" but are really extensions of continental margin sedimentation.

Typically, the total sediment accumulation in the deep sea that is on basalt crust is less than 250 m (Figure 1; Whittaker et al., 2013). For most ocean regions, sediment production is small compared to erosion of mountains and river transport. Because of plate tectonics, there is little time to accumulate a thick sediment pile because the ocean crust is young – the seafloor is totally resurfaced in about 160 million years. Thickest accumulations outside of the turbidite regimes are on older shallower crust or underneath zones of high biological productivity. Sediments are thicker under regions of high productivity because the $CaCO_3$ and SiO_2 tests of plankton are a major deep-sea sediment component (section "Biogenic Sediment Components") and in shallow plateaus because there is better preservation (section "Seafloor Processes Affecting Deep-Sea Sediment Composition: Dissolution, Early Diagenesis, and Sediment Movement"; Berger, 1970).

Why study deep-sea sediments?

Deep-sea sediments are the main means to study environmental change within the marine world. Because they accumulate slowly but steadily, analysis of sediment cores can provide environmental records that span millions of years. Components of deep-sea sediments record changes in ocean temperature and salinity and track the growth and decay of ice sheets. They can be used to track the change of winds and aridity on continents in response to climate cycles. They are also used to study ocean biogeochemical cycles and how they are changed by natural variation in weathering or greenhouse gas emissions. The strong imprint of Milankovitch solar insolation cycles in the sediments is an important indication of the sensitivity of the earth and oceans to minor changes in climate drivers (Hays et al., 1976; Hodell et al., 2001; Lisiecki and Raymo, 2005).

Regions with coherent sedimentation patterns within the oceans are extremely large compared with those on land. Patterns of change in the oceans can thus be elucidated from relatively few sediment cores. Large regions in the oceans have similar environmental conditions (e.g., temperature, nutrients, and salinity), and most of the ocean basins are being covered with a continual rain of sediment. Deep-sea sediment regimes are also large because they reflect the relative size of the depositional processes, whether it is the region underneath continental dust plumes or regions with coherent production of planktonic hard parts. On land, the primary variables (precipitation, temperature, seasonality) vary on a much smaller scale and thus must be sampled more densely. Terrestrial basins that systematically collect sediments are also more rare than in the oceans.

Geophysical and geologic study of deep-sea sediments

Marine geologists have difficulty accessing deep-sea sediments, because the average ocean depth is ~3,800 m and recovery of ocean bottom samples is technically challenging and time consuming. While sampling is hard, geophysical methods that use acoustic pulses are much easier to perform at sea than on land. Sound propagates well through water and it is much easier to produce and receive an acoustic pulse through the ocean than through the land surface. Marine geologists are much more dependent on geophysical methods than continental geologists.

Marine geologists use acoustic methods to map the topography of the seafloor (echo sounding or multibeam bathymetry) and to map sediment layers beneath the seafloor (reflection seismology). Multibeam mapping uses acoustic technology to emit a focused array of acoustic pulses and uses the time for each pulse to return to calculate ocean floor depths in a swath perpendicular to the ship. The travel time data (time for the echo to hit the bottom and return) can be converted into a topographic map of the seafloor by using measured ocean sound velocity to convert from time to depth (e.g., Caress and Chayes, 1996; Mosher, 2011). An example of the resultant map is shown at the top of Figure 2, a survey around the IODP drill site U1335 (Expedition 320/321 Scientists, 2010).

Seismic reflection measures the time for an echo to be reflected back from sediment layers. The echoes occur where sediment density and/or acoustic velocity change and are typically caused by a change of sediment lithology. Rarely, acoustic interference causes composite echoes from closely spaced beds (Mayer, 1979). The echoes from sediment layers produce a cross section of sediment deposition through the sediments (Figure 2 bottom; Mayer et al., 1986). The combination of swath mapping and seismic reflection allows marine geologists to interpret how sediments have accumulated, allowing them to understand the long-term development of a marine sediment sequence (e.g., Sacchetti et al., 2013; Brothers et al., 2013).

Such information has limited use without ground truth from sediment samples. Marine geologists use a variety of techniques to sample near-surface sediments by lowering coring devices to the seafloor on a wire rope. Where the ocean is greater than 3 km deep, it takes 2–6 h to lower a corer to the bottom, collect a sample, and bring it back to the ship. Sediment cores taken by wireline range in length from 0.3 m to about 20 m long, sufficient to study relatively recent geologic events. Because of the effort needed to collect these cores, the sediment cores are archived in core repositories around the world (e.g., Lamont-Doherty Earth Observatory core repository, http://www.ldeo.columbia.edu/core-repository/collections). Locations of many of these geologic samples can be plotted using the online mapping and database tool GeoMapApp (http://www.geomapapp.org/).

While wireline coring can collect near-surface sediments, drilling technology must be used to sample the entire sediment column and the ocean crust (Duce et al., 2011). The time needed to sample the entire sediment column may take days or weeks, depending on the thickness of the sediments and the care taken to recover a full sediment column for scientific analysis. Once recovered, the drilled sediment column can be correlated to the seismic reflection profile to identify and date specific seismic reflection horizons. With dated seismic horizons, it is possible to study regional variations in sedimentation by using seismic reflection profiles (Mayer et al., 1986; Tominaga et al., 2011).

Types of deep-sea sediments

Tens of thousands of archived deep-sea sediment samples have made it possible to map different types of sediments, linked to regional environmental conditions (Menard, 1964; Lisitzin, 1972; Dymond, 1981). Deep-sea sediments are often classified by the source of their major components. For example, sediments consisting of ferromanganese oxyhydroxides precipitated from mid-ocean ridge hydrothermal plumes are referred to as hydrothermal sediments.

Based upon a source definition, there are four major classes of deep-sea sediment components: (a) *terrigenous* sediments, aluminosilicates from the continents, divided into *hemipelagic* (water-transported continental debris) and *aeolian* (windblown dust); (b) *biogenic* sediments, primarily calcareous or siliceous hard parts of organisms; (c) *hydrothermal* sediments, derived from mid-ocean ridge hot springs; and (d) *authigenic* sediments, precipitated directly out of seawater. Most deep-sea sediments are mixtures of the different components. For example, hemipelagic sediments will also have a small percentage of biogenic-sourced plankton tests.

The sediment types have typical sedimentation rates, reflecting their relative supply to the pelagic environment (Figure 3). Sedimentation rates are plotted against sediment C_{org} content because the amount of C_{org} is a primary indicator of the rate of primary productivity in the surface

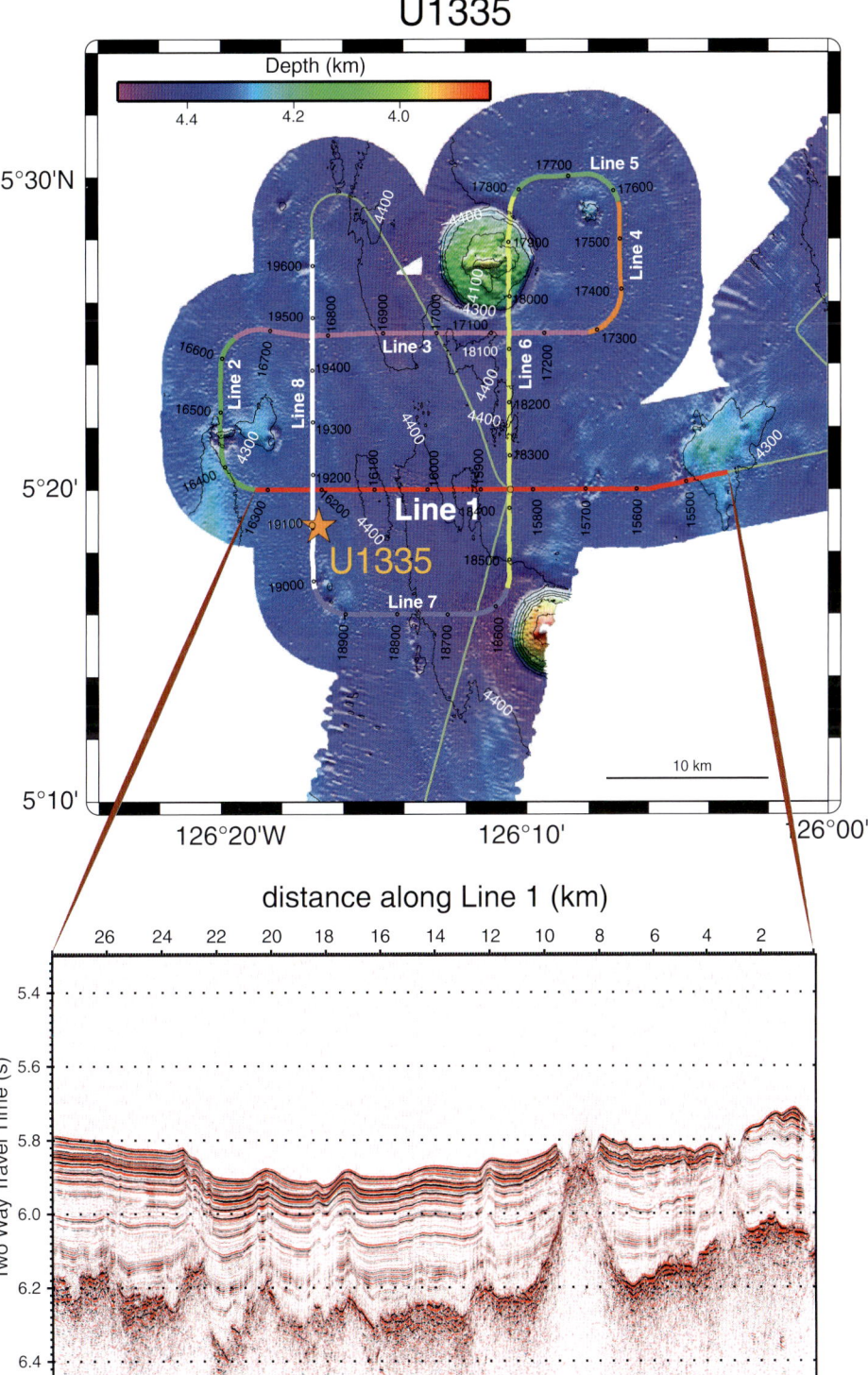

Deep-sea Sediments, Figure 2 An example of a swath map bathymetry map (*top*) and a seismic reflection profile (*bottom*) from a site survey for IODP Site U1335, in the central tropical Pacific Ocean (Expedition 320/321 Scientists, 2010). The bathymetric map shows areas of current scour near seamounts and areas of disturbed sediment along Line 1. The seismic reflection profile along Line 1 shows that the sediment disturbances found by the swath map survey are where small seamounts still stick up through the sediment column. The parallel seismic horizons in the seismic reflection profile mark changes in lithology. In the tropical Pacific, these normally result from changes in sediment $CaCO_3$ content (Mayer et al., 1986).

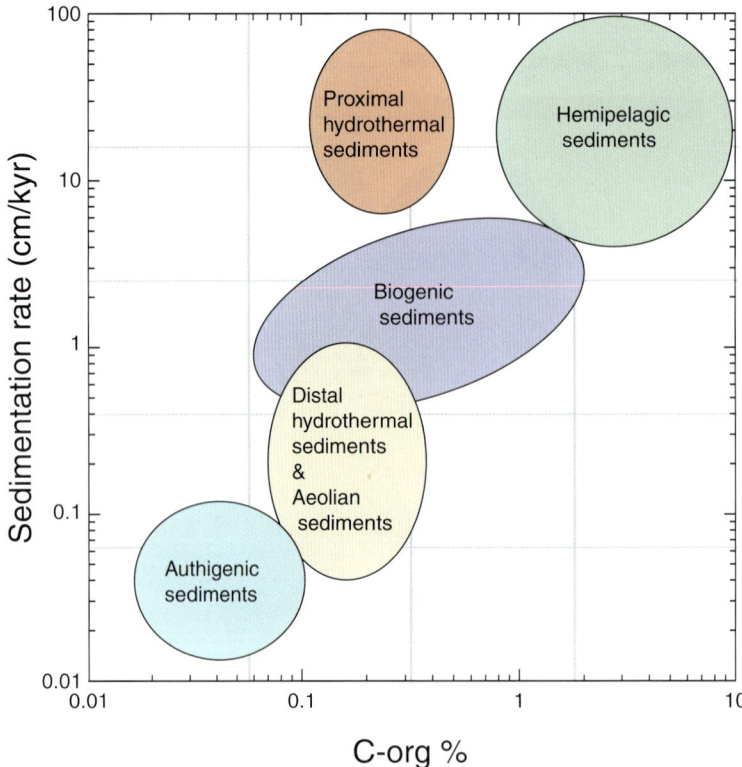

Deep-sea Sediments, Figure 3 Ranges of sedimentation rates versus organic carbon content for different types of deep-sea sediments. Highest sedimentation rates and C_{org} contents are found in hemipelagic sediments, because of high terrigenous sediment deposition and high biological productivity along continental margins. Authigenic sediments have the lowest rates of deposition and low C_{org} % because the rate of precipitation of these oxides from seawater is so slow.

waters and the intensity of early diagenesis (chemical and physical changes as sediment is converted to rock; section "Seafloor Processes Affecting Deep-Sea Sediment Composition: Dissolution, Early Diagenesis, and Sediment Movement") depends upon C_{org} degradation.

Turbidites and hemipelagic sediment components

The most abundant deep-sea sediments are terrigenous sediments derived from rivers or glaciers and which may temporarily reside on continental shelves before being deposited on abyssal plains. Turbidites are terrigenous in origin but not considered pelagic sediments. They are rapidly deposited from turbidity currents, relatively thick sediment-water slurries that flow down from continental shelves into the adjacent deep ocean. They are a type of continental margin sediment that happens to lie on the deep-ocean floor.

Turbidite-filled abyssal plains are common along continental margins (Figure 4) and are most commonly associated with passive continental margins, e.g., most of the Atlantic Basin, especially where there is high flux of sediments carried to continental margins by rivers or glaciers. Turbidites mostly fill basins at the toe of continental margins, and because they are deposited from dense sediment suspensions, they tend to ignore topography and form a relatively flat deposit surrounding existing ridges or other upraised topography. An example of turbidite deposition can be found on the right-hand side of Figure 5, nearest to the North American continent. Figure 5 shows a seismic reflection profile across the Gorda Ridge, a slow-spreading mid-ocean ridge near the California-Oregon border.

Hemipelagic sediments, in contrast to turbidites, are fine-grained clay sediments that fall out of the water column and drape over the seafloor topography (Damuth, 1977; Gorsline et al., 1984). The sediment drape across topography can be found on the higher topography on Figure 5. Ocean crust is being formed at the Gorda Ridge axis and is transported east and west from the ridge axis. As the crust moves, it gradually gets covered with a thick layer of fine-grained sediment. Sediment at 2203z on Figure 5, for example, is about 150 m thick and is on crust about two million years old, indicating an average sedimentation rate of $70-80 \text{ m}/10^6$ years.

The fine-grained sediments that make up hemipelagic sediments originate from turbid plumes from rivers or from resuspension of fine-grained shelf sediments by internal waves or storms. The fine-grained sediments are then carried by currents and eventually deposited (Karlin, 1980; Krissek, 1984; Biscaye et al., 1988).

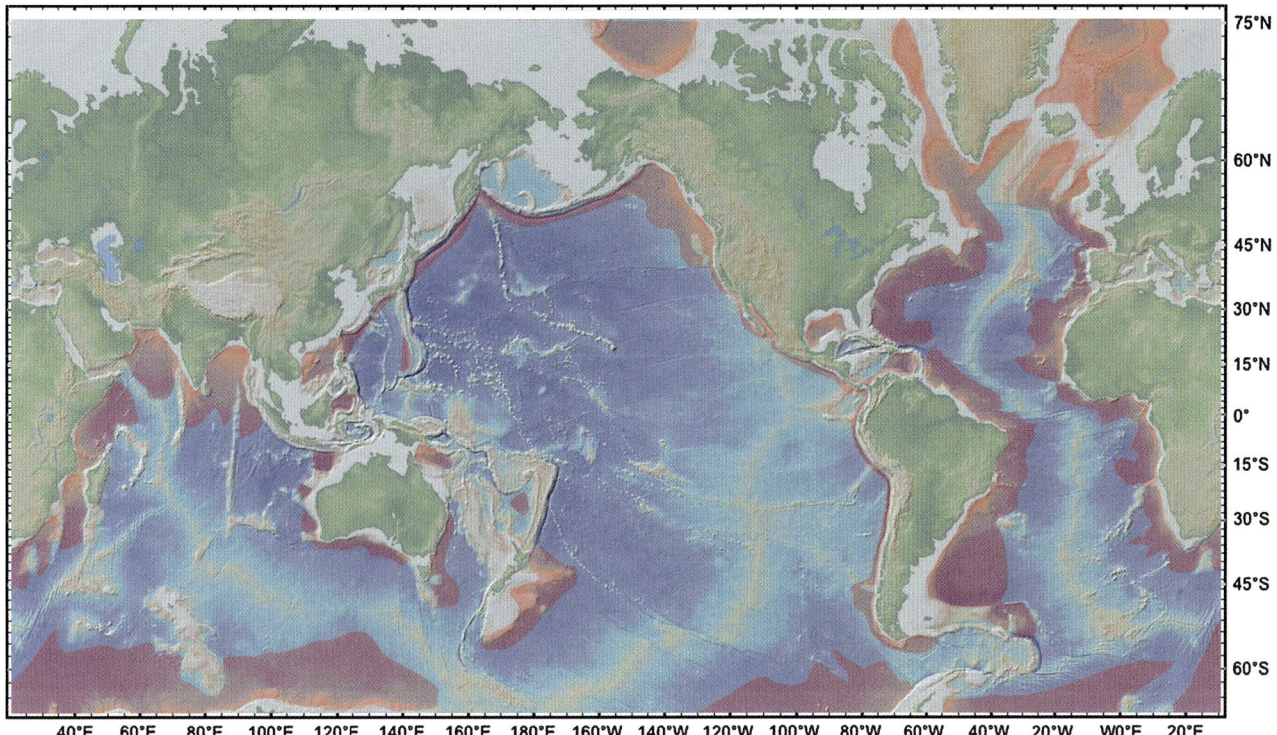

Deep-sea Sediments, Figure 4 *Red* shading shows locations of turbidite abyssal plains and deep-sea fans in the world ocean, assigned by using GeoMapApp (http://www.geomapapp.org/) to identify areas of flat topography near continental margins and to access geo-referenced analog seismic reflection data. Within the abyssal plain region, one often finds fine-grained hemipelagic sediments draping over higher topography above turbidite deposition. Largest turbidite plains are associated with passive continental margins and large river drainages.

Sediment composition can be used to track average current flow along the continental margins and how sediments move from the rivers to the deep sea. Krissek (1984), for example, estimated that 50–90 % of the river sediment flux to the Oregon/northern California margin escaped to deposit in the deep sea.

The Atlantic Ocean has a higher percentage of its total floor covered by turbidites than other ocean basins because it is surrounded by passive continental margins where continental sediment accumulates. The Pacific Ocean has a much higher percentage of its basin covered with deep-sea sediments because the sediment load from rivers is smaller and because much of the sediment is trapped in ocean trenches or back-arc basins. The Indian Ocean is home to two major deep-sea fans associated with sediments draining from the continental collision of Asia and India (Clift and Gaedicke, 2002; Bastia et al., 2010).

Aeolian and volcanic ash sediment components

Windblown dust is an important terrigenous component in slowly accumulating sediments of the deep sea, downstream from major dust sources on land (Figure 6; Rea, 1994; Mahowald et al., 1999, 2006). Not only is dust important as a sediment component, but dust deposition can also be critical to deliver micronutrients like Fe to the oceans (Martin 1990; Maher et al., 2010). Dust originates from the arid regions of the earth, found near 30° N or S of the equator. Most dust is produced from regions that are periodically wet rather than the hyperarid centers of deserts where there is not enough precipitation to weather the rocks quickly (Rea, 1994).

Windblown dust is a major component of deep-sea sediments where other sources are rare, e.g., the northwest Pacific Ocean (Figure 6). In the North Pacific, ocean crust is old and deep, and biogenic sediments dissolve before burial (see section "Seafloor Processes Affecting Deep-Sea Sediment Composition: Dissolution, Early Diagenesis, and Sediment Movement"). There is little left after dissolution except dust and the fraction that precipitates from seawater (*authigenic* sediments, section "Authigenic and Cosmogenic Sediment Components and Ferromanganese Nodules"). The highest dust flux into the oceans is from North African and Middle Eastern dust sources into the Arabian Sea and from North Africa westward into the subtropical Atlantic Ocean (Figure 6). Because these regions also have high biogenic sediment flux, the sediments are not as distinctly aeolian as sediments from the North Pacific.

Lowest dust fluxes are in the South Pacific, where the lowest sedimentation rates in the world ocean are also

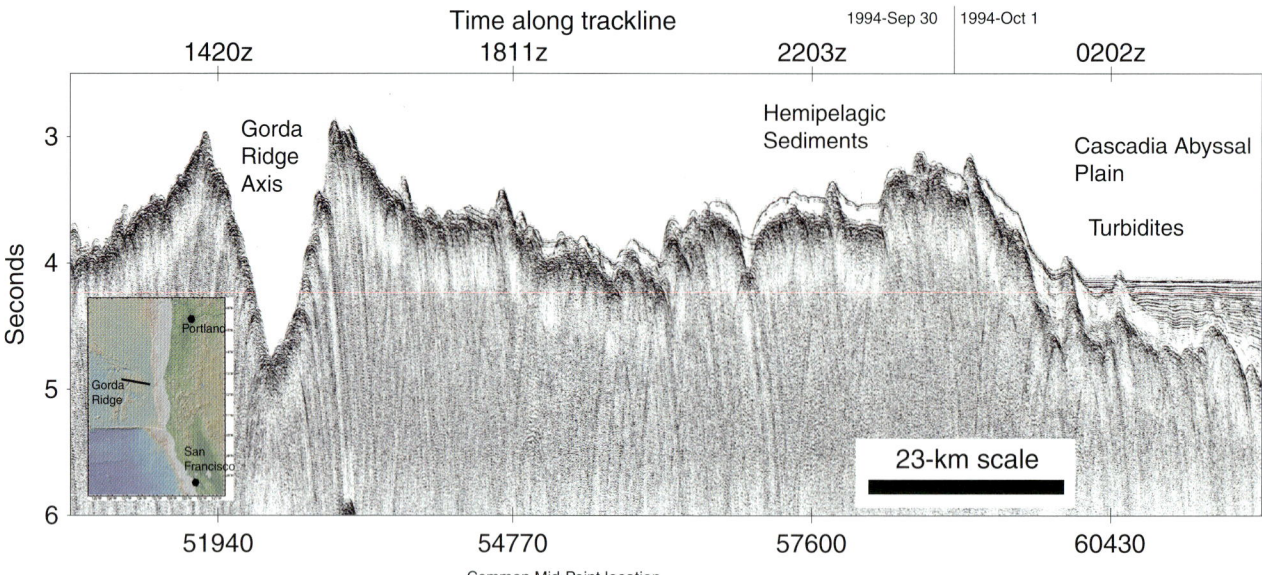

Deep-sea Sediments, Figure 5 Seismic reflection profile across the Gorda Ridge, near the California-Oregon border. The location of the profile is shown on the inset map. The first reflection is the reflection from the seafloor, while the dark reflection is the echo from basaltic basement. Depth scale is in two-way travel time, where 1 s is equivalent to 750 m depth in water. Turbidite sediments (*right side*) fill in topography and typically have horizontal layering. Hemipelagic sediments are fine-grained continental sediments carried out in the water column and drape upon the topography. New ocean crust at a mid-ocean ridge is typically bare, but sediments thicken away from the mid-ocean ridge as the crust ages and there is time for sediments to deposit. Seismic profile is from cruise EW9413, from the UTIG seismic database (http://www.ig.utexas.edu/sdc/cruise.php?cruiseIn=ew9413). Processing: band-pass filter, 20–450 Hz, 1 s AGC.

found (Rea et al., 2006). Typically, dust fluxes into the Southern Hemisphere oceans are 1–2 orders of magnitude lower than those in the North Pacific (Rea, 1994). In the Southern Hemisphere, there are only two major dust sources, one in South America and one in Australia (Figure 6). In contrast, there are extensive northern arid regions, including northern Africa, Asia, and North America.

Volcanic ash is also a significant air-transported sediment component that can be found a surprisingly long distance from explosive volcanoes (Scheidegger et al., 1980; Olivarez et al., 1991). It also can be a significant source of Fe to the oceans (Duggen et al., 2010). Volcanic ash is most often recognized as discrete coarse layers of volcanic glass (e.g., Drexler et al., 1980), but dispersed ash can also be found using its distinctive chemical composition (Finney et al., 1988; Olivarez et al., 1991).

Biogenic sediment components

Biogenic sediments are primarily composed of mineralized hard parts produced by a variety of marine organisms. They also include organic remains (the C_{org} fraction) and elements like Ba that precipitate within C_{org}-rich particle rain as it falls from the euphotic zone to the seafloor (Griffith and Paytan, 2012). In shallow water, corals and calcareous algae are major producers of biogenic sediments, but in the pelagic realm, the production of mineral tests by protist plankton is the major source of this deep-sea sediment component (Kennett, 1982). Production by plankton is often large enough to produce sediment accumulations that are essentially pure biogenic tests (Berger, 1973; Burckle and Cirilli, 1987; Lyle et al., 2010).

Biogenic sediments are divided into siliceous and calcareous oozes based on the composition of the plankton test. Tests are made of either biogenic SiO_2 (biogenic opal or biogenic silica) or biogenic $CaCO_3$ (biogenic carbonate). Both phytoplankton and zooplankton contribute to biogenic sediment production (Lipps, 1993). Single-celled photosynthetic algae are a major source of both biogenic silica (diatoms, with a minor contribution by silicoflagellates) and biogenic carbonate (coccolithophorids). The larger zooplankton tests are also significant fractions of biogenic sediment: foraminifera produce sand-sized tests of biogenic carbonate, whereas radiolaria produce silt- to fine-sand-sized tests of biogenic silica.

Biogenic carbonates are commonly used for paleoceanographic reconstructions because the tests of foraminifera are large enough to be easily separated from sediments for analysis. The chemistry of the test gives quantitative data about the environmental conditions at the time they were formed (e.g., Dekens et al., 2007). The preserved microfossil assemblages (census of different species preserved in a layer) can also be used to infer environmental conditions (e.g., Le and Shackleton, 1994) because plankton species live within characteristic

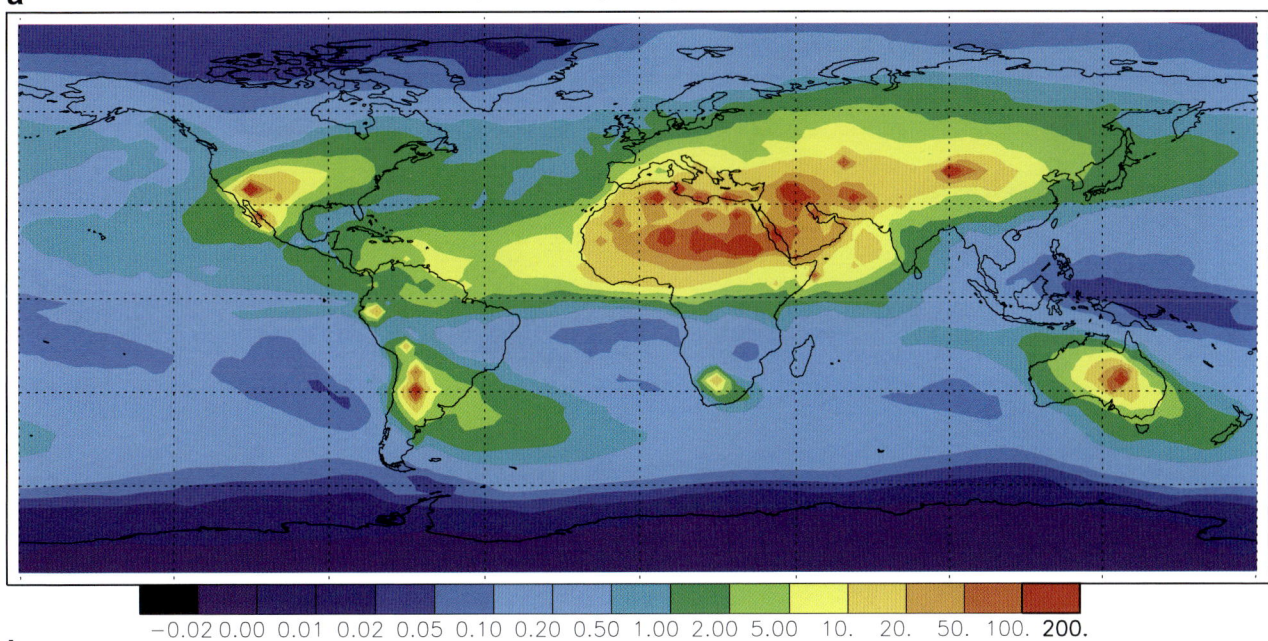
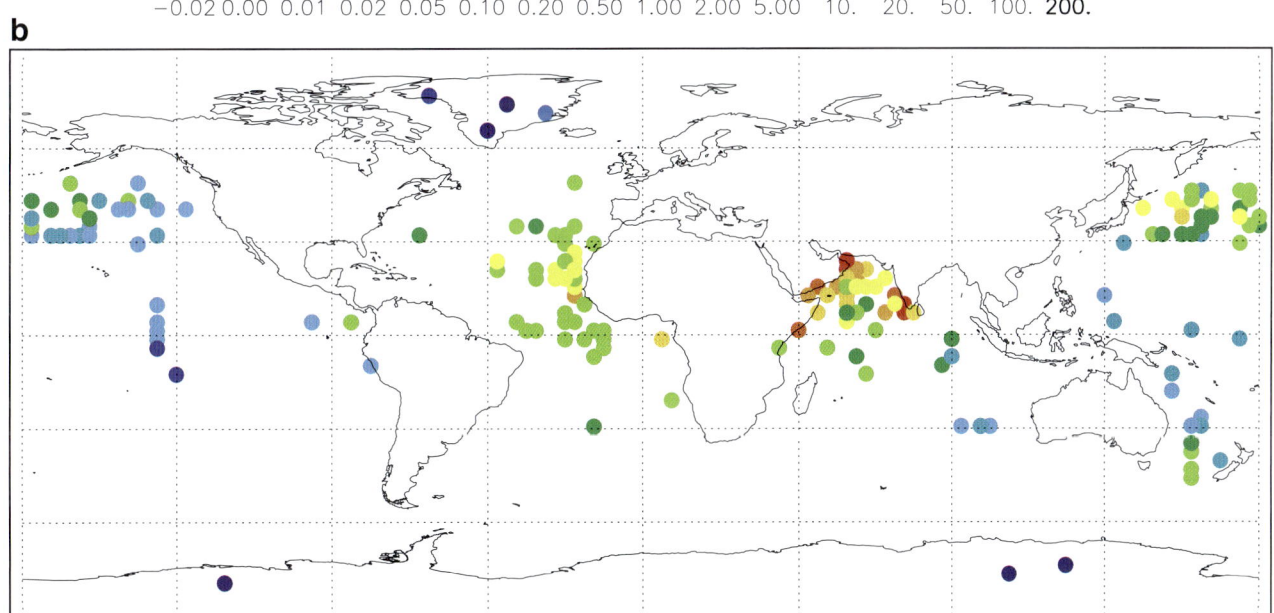

Deep-sea Sediments, Figure 6 Modeled modern global dust deposition compared to measured Holocene dust mass accumulation rates (MAR) in sediment cores from DIRTMAP (From Mahowald et al. (2006)). The highest dust deposition into the oceans is in the northern tropical Atlantic and the Arabian Sea. Extremely low dust fluxes are found in the Southern Hemisphere because there are few sources for dust.

environmental conditions. Equations can be made to relate the amount of each assemblage from a past time to the environmental conditions at that time. The assemblage techniques can be used to assess past ocean temperatures, e.g., CLIMAP project members (1976).

Pelagic sediments rich in biogenic components can be mapped by their opal and carbonate content (Figure 7; Archer, 1996, Leinen et al., 1986). $CaCO_3$ is prominent in the Atlantic Ocean, because the ocean chemistry in the basin leads to better preservation (section "Seafloor Processes Affecting Deep-Sea Sediment Composition: Dissolution, Early Diagenesis, and Sediment Movement"). In the Pacific Ocean, biogenic sediments are primarily found in the south Pacific because the crust is shallower

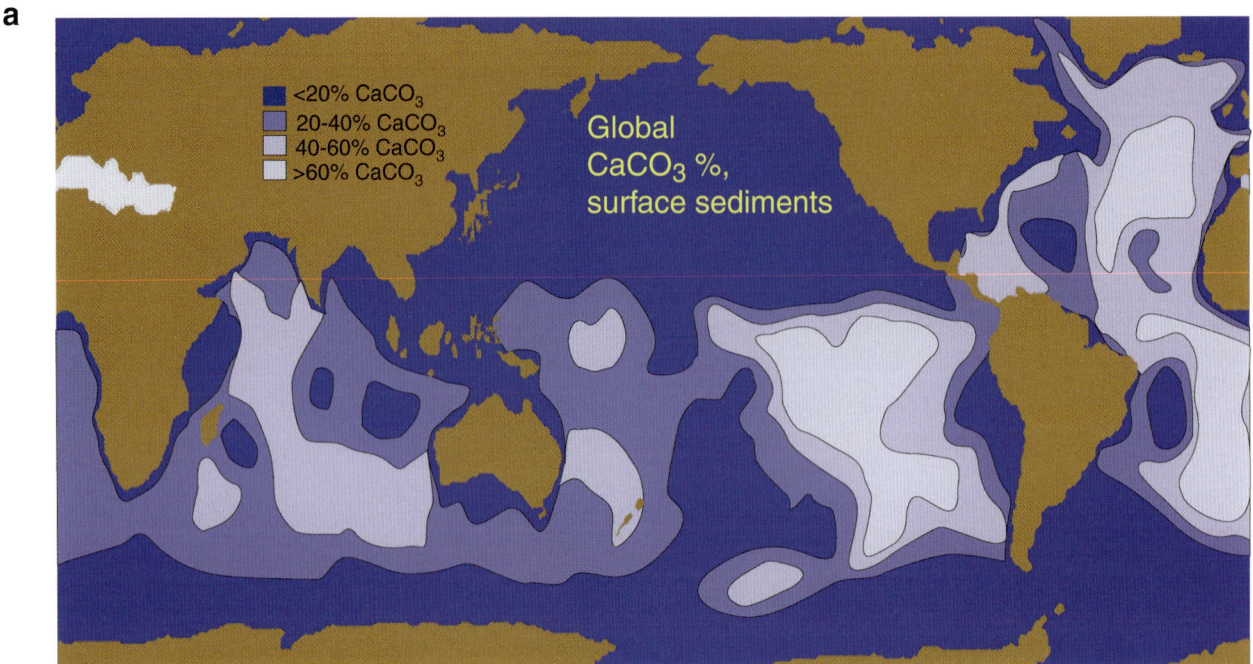

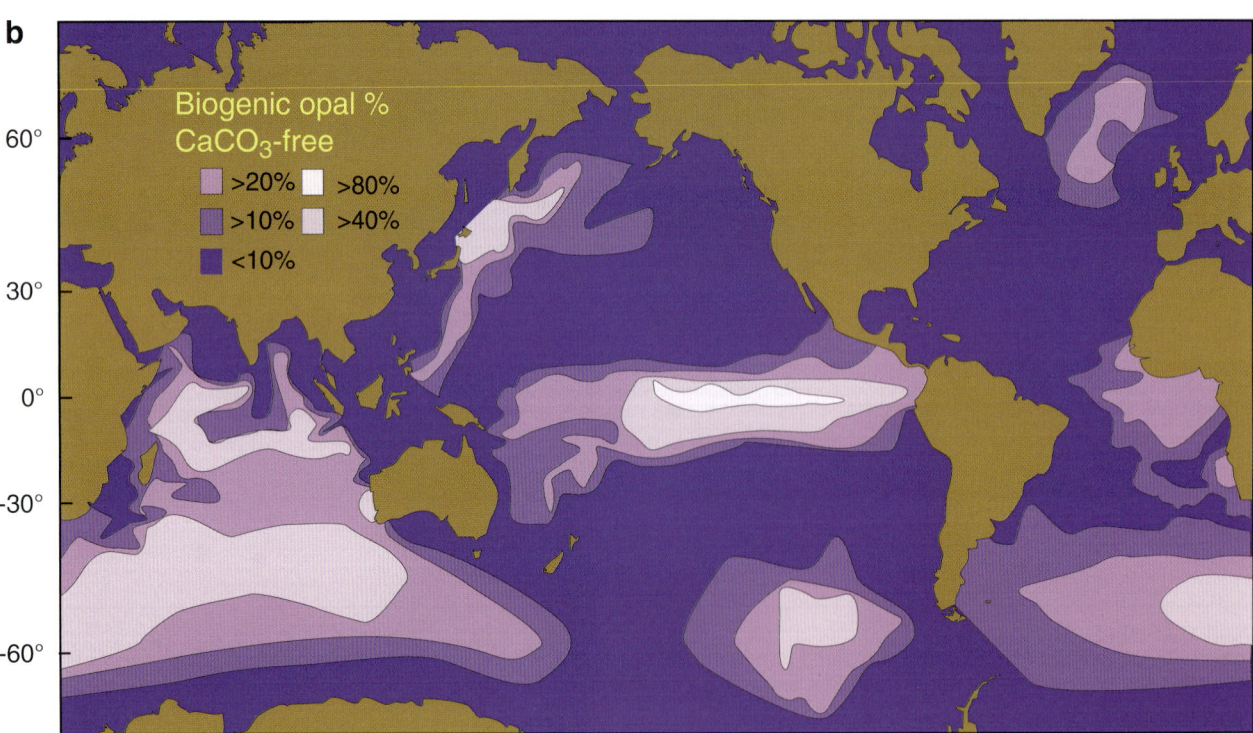

Deep-sea Sediments, Figure 7 (a) Distribution of $CaCO_3$ in the world ocean after Archer (1996). The highest carbonate contents are in the Atlantic and in the Southern Hemisphere. The North Pacific is too deep to preserve $CaCO_3$. (b) Distribution of biogenic opal on a carbonate-free basis (as a percentage of the noncarbonate sediment fraction) from Leinen et al. (1986). Equatorial Pacific sediments are rich in both biogenic opal and $CaCO_3$, while sediments near the Antarctic polar front are typically diatom-rich.

and above the calcite compensation depth (CCD). Because $CaCO_3$ solubility is pressure dependent, the CCD represents where all $CaCO_3$ dissolves and none has been preserved. $CaCO_3$ will not accumulate in the oldest and deepest parts of the ocean basin like the North Pacific, even though plankton produce $CaCO_3$ and it rains to the seafloor (e.g., Zaric et al., 2005). High production of $CaCO_3$ occurs in pelagic, moderately strong upwelling regimes, especially in the tropics (Figure 7a). Because the CCD represents the difference between production and dissolution, high $CaCO_3$ production forces down the CCD under the equatorial upwelling region resulting in higher $CaCO_3$ content in equatorial sediments (van Andel and Moore, 1974).

Biogenic opal distribution in Figure 7b is presented on a carbonate-free basis, as a percentage of the noncarbonate fraction of the sediments. Because biogenic opal is undersaturated in ocean waters, it only accumulates where sedimentation rates are reasonably high. For the most part, opal is an important but secondary biogenic sediment component in the pelagic regime. However, diatom oozes can be found where upwelling brings nutrient-rich waters to the euphotic zone and high diatom production results. Diatom oozes are characteristic of the Southern Ocean, for example (Burckle and Cirilli, 1987). A zone rich in radiolarian opal has often been noted along the north edge of the equatorial Pacific upwelling zone, but this primarily represents the outcropping of Eocene and older biogenic radiolarian oozes (Riedel, 1967). Eocene radiolarian oozes result from a shallow CCD prior to 34 million years ago and conditions favorable to radiolarian production of biogenic opal. Some researchers have proposed that the rich deposits of Eocene biogenic opal were caused by elevated dissolved Si in the oceans delivered by a very active silica weathering cycle, but Moore et al. (2008) showed that, at least in the equatorial Pacific, biogenic opal burial was no greater in the Eocene than the Holocene. Additional Si inputs to the oceans are not needed to produce the ooze deposits.

Both C_{org} and Ba are also biogenic sediment components. Near continental margins, C_{org} sediment contents are typically 1–2 % and occasionally can reach as high as ~20 % where hydrodynamically light sediments are deposited near coastal upwelling zones (Müller and Suess, 1979). In most of the pelagic realm, however, C_{org} contents are less than 0.25 % (Pedersen and Calvert, 1990). C_{org} burial is enhanced where primary productivity and sedimentation rates are high and secondarily where bottom water oxygen contents are low (Pedersen and Calvert, 1990). Enhanced C_{org} burial strongly affects the subseafloor redox environment and causes mobilization of some elements as well as enhanced burial of others (section "Seafloor Processes Affecting Deep-Sea Sediment Composition: Dissolution, Early Diagenesis, and Sediment Movement"; Froelich et al., 1979; Calvert and Pedersen, 1993; Morford and Emerson, 1999). Preservation of C_{org} has changed with time, and C_{org} is more poorly preserved when Earth climates were warm, perhaps because heterotrophs have higher metabolic demand under warm conditions (Olivarez Lyle and Lyle, 2006).

Ba is enriched in biogenic sediments as well as in hydrothermal sediments (Griffith and Paytan, 2012). As the remains of plankton fall through the water column, barite ($BaSO_4$) precipitates in microenvironments within the falling debris (Dehairs et al., 1980; Griffith and Paytan, 2012). Biogenic Ba is the most consistently preserved of the biogenic sediment components in deep-sea sediments. In the Holocene, less than 5 % of the C_{org} particulate rain is typically buried in deep-sea sediments, versus 5–10 % of biogenic opal, and 30 % biogenic Ba (Dymond et al., 1992; Dymond and Lyle, 1994). $CaCO_3$ has highly variable preservation depending on water depth and ocean chemistry. Where biogenic sediments are severely dissolved after deposition, a barite ($BaSO_4$) residue can be the last remaining marker of biogenic deposition (Dymond, 1981). Barite is preserved in well-oxygenated sediments but will dissolve under sulfate-reducing conditions and can be remobilized (Torres et al., 2002).

Hydrothermal sediment components

The mid-ocean ridges are a world-circling volcanic mountain chain, creating ~3 km^2 of new hot basalt crust each year (Williams and von Herzen, 1974). Hot springs that cool the crust are ubiquitous along mid-ocean ridges, form rich environments for benthic life (Kelley et al., 2002), and form distinctive sediments that precipitate around the vents and are dispersed in hydrothermal plumes. Unusual sediment composition found on all mid-ocean ridges led to the hypothesis that hydrothermal systems must be common in the oceans (Boström and Peterson, 1966, 1969).

Near hydrothermal vents, where large-scale precipitation of minerals from solution results from mixing with normal ocean water, massive sulfide deposits and oxidized remains of massive sulfides and other minerals are found (Haymon, 2005). Proximal hydrothermal deposits can be highly varied because of the variability of temperature and mixing conditions. The deposits themselves are scattered and uncommon because active vent fields are small relative to the total ridge length and are spaced ~20–50 km apart along the ridge crest (Baker and Urabe, 1996).

Farther away from the actual vents, hydrothermal plumes consist of particles with much more uniform composition. The plumes distribute particles rich in amorphous Fe-Mn oxyhydroxides and poor in Al that form the hydrothermal sediment component (Boström and Peterson, 1969; Dymond, 1981). Hydrothermal sediments are most apparent in the South Pacific along the East Pacific Rise, partly because little besides biogenic carbonates deposit there and partly because highest ocean crustal formation occurs there. As the hydrothermal plume material drifts away from mid-ocean ridges, the Fe-Mn oxyhydroxide particles scavenge other elements from seawater, like P and the rare earth elements, and form a significant sink for these elements (Wheat et al., 1996; Ruhlin and Owen, 1986). Downcore variation in the rate that hydrothermal sediments accumulate has been used to track changes in hydrothermal activity through time. Lyle et al. (1987) showed that hydrothermal activity at the East Pacific Rise increased dramatically during ridge reorganizations, presumably because fracturing allowed significantly better seawater access to hot rock.

Authigenic and cosmogenic sediment components and ferromanganese nodules

A fraction of deep-sea sediments are transition metal-rich oxyhydroxides precipitated directly from seawater, referred to as the authigenic or hydrogenous sediment component (Goldberg, 1954). "Authigenic" has also been used to refer to aluminosilicate minerals that are formed as other unstable components dissolve within deep-sea sediments, e.g., authigenic smectite clays (Cole and Shaw, 1983) or authigenic phillipsite (Stonecipher, 1976; Bernat and Church, 1978). However, this is more properly considered early diagenesis (section "Seafloor Processes Affecting Deep-Sea Sediment Composition: Dissolution, Early Diagenesis, and Sediment Movement").

The authigenic ferromanganese oxyhydroxides are only a significant sediment component when sedimentation rates are very low, because of the low concentrations of transition metals in seawater and slow precipitation rates. Ferromanganese nodules, an authigenic concretion, typically have growth rates of $1-10$ mm/10^6 years, for example, versus typical sedimentation rates in the surrounding sediments that are $\sim$1,000 times faster (Heath, 1981; Dymond et al., 1984).

Authigenic oxyhydroxides include both ferromanganese nodules and the dispersed ferromanganese oxyhydroxides in sediments. Ferromanganese nodules are large concretions of Fe-Mn oxyhydroxides ranging in size from about 0.5 to 10 cm (Figure 8) and are found in all low-sedimentation-rate ocean basins (Glasby, 2006). Typical authigenic oxyhydroxides contain nearly equal amounts of Fe and Mn, unlike hydrothermal plume sediments which have Fe:Mn weight ratios near 3.5 (Dymond, 1981). More Mn is in the authigenic fraction because Mn has a longer residence time than Fe in the oceans, and because Mn is preferentially reduced and remobilized from continental margin sediments (Lyle, 1981). High concentrations of co-precipitated transition metals such as Co, Ni, and Cu have made ferromanganese nodules a potentially attractive mining target, although only trial mining has yet been carried out (Glasby, 2006). Detailed chemical analyses of tops and bottoms of nodules have shown that the chemistry and even growth rates of ferromanganese nodules have been influenced by significant interactions with the sediments (Dymond et al., 1984). The side of the nodule buried within sediments is enriched in Mn, Ni, and Cu and has a faster growth rate than the top of the nodule exposed to seawater. Lyle et al. (1984) suggested that metals sorbed on sediment grains provide the additional source of metals to nodule bottoms.

In very slowly accumulating sediments, it is possible to separate out a cosmogenic sediment component of small meteorite grains that have slowly accumulated (Kyte, 2002). The amount of cosmogenic debris is estimated by measuring elements or isotopes that are highly enriched in meteorites like Ir or ^{3}He. There have been rare large meteor impact layers, like the Cretaceous/Paleogene event that caused the extinction of dinosaurs, but for the most part, there has been a relatively constant background deposition through time (Kyte, 2002). Assuming a constant cosmogenic deposition rate, it is possible to estimate sedimentation rates in cores since the sedimentation rate will be inversely proportional to the concentration of the cosmogenic debris (Marcantonio et al., 1996).

Deep-sea Sediments, Figure 8 Ferromanganese nodule collected at the top of a sediment core taken from the Southwest Pacific gyre in 2005 (R/V Melville TUIM-03 expedition). Ferromanganese nodules are very common in this region. The nodule is nearly the same diameter as the coring device ($\sim$10 cm).

Seafloor processes affecting deep-sea sediment composition: dissolution, early diagenesis, and sediment movement

Particles that have fallen through the water column are transformed before they become sediment. Unstable, mainly biogenic, components dissolve, new minerals are formed by early diagenesis, and particles may be fractionated and moved about by seafloor currents before burial.

The composition of deep-sea sediments is strongly affected by dissolution of particles at the seafloor and by early diagenesis. There is a significant flux of biogenic debris to the seafloor even in low-productivity regions like ocean gyres, but most of the particulate rain that lands on the seafloor dissolves or degrades before burial. For example, at 40°N in the Pacific Ocean, 1.8 g/cm^2/kyr of CaCO$_3$ and a similar amount of biogenic opal flux were measured in a sediment trap at 5,016 m (Honda et al., 2002), but only clays are found in sediments below because of dissolution at the seafloor. Nevertheless, elements scavenged on the surface of biogenic particles that rain to the seafloor may be carried to the seafloor and buried (Balistrieri et al., 1981; Fischer et al., 1986).

Sediment dissolution and the carbonate compensation depth

Biogenic silica must be buried relatively quickly to prevent dissolution, since all ocean waters are undersaturated with respect to biogenic opal (Ragueneau et al., 2000). Slowly accumulating sediments do not preserve biogenic opal.

The solubility of calcite increases with increasing pressure, decreasing temperature, and decreasing dissolved carbonate ion, $[CO_3]^=$ (Boudreau et al., 2010). Deep-ocean water temperature is low and relatively constant, but pressure causes the calcite solubility product constant to double between 2,000 and 5,000 m water depth (Boudreau et al., 2010). Deeper parts of the ocean lack $CaCO_3$ because of elevated $CaCO_3$ dissolution. The calcite compensation depth (CCD) is defined as the depth where the rate of dissolution matches the $CaCO_3$ particulate rain, and all $CaCO_3$ is dissolved before burial. Since the CCD depends upon both production and dissolution, the CCD has changed as environmental conditions have changed (Figure 9; Pälike and Expedition 320/321 Shipboard Scientists, 2012).

Shallow ocean waters are supersaturated with respect to $CaCO_3$, but there is a depth, referred to as the lysocline, where the increasing effects of $CaCO_3$ dissolution become apparent, e.g., the depth below which dissolution-susceptible foraminifera rapidly disappear. This depth occasionally matches the depth where undersaturation occurs, but it may also show where the cumulative effects of dissolution are easily measured. The lysocline and CCD are at different depths in different ocean basins because they experience different levels of $CaCO_3$ production and because the different basins have different deep water $[CO_3]^=$ levels. In the Atlantic, the lysocline ranges from 4 to 5 km deep, and the CCD is 5–6 km deep (Figure 7; Biscaye et al., 1976). In contrast, the Pacific Ocean lysocline ranges from <3 to 4 km deep, and the CCD is typically about 4,500 m deep (Berger et al., 1976). The Atlantic has much better $CaCO_3$ preservation than the Pacific or Indian Ocean because deep waters in the basin are flushed by high $[CO_3]^=$ water North Atlantic Deep Water sourced from the North Atlantic surface ocean rather than from lower $[CO_3]^=$ water sourced from the Antarctic.

Early diagenesis

The chemical environment within sediments is significantly different from that of the abyssal ocean, so that some minerals will dissolve and new minerals can form. The changes are referred to as early diagenesis. Oxidation of organic matter is a major driver of early diagenesis, leading to reduction and remobilization of Fe and Mn and the formation of oxide-rich surface sediment layers (Froelich et al., 1979; Finney et al., 1988). Other early diagenetic reactions include the formation of calcite overgrowths on foraminifera and formation of diagenetic dolomite ($CaMgCO_3$). Diagenetic calcite overgrowths can severely affect the chemical composition of foraminiferal tests and thus cause major errors in proxy estimates of past temperatures and other properties (e.g., Pearson et al., 2001).

Near-bottom sediment movement and sediment focusing

The cycle of elements to the seafloor is further complicated by near-seafloor particle dynamics. The particle rain from surface waters can fall to the seafloor but then be resuspended back into the lower water column before final incorporation into sediments (McCave, 2010). In the abyssal realm, Walsh et al. (1988) found that there was a "rebound" sediment flux caught in sediment traps up to 500 m above the bottom. Chemical analysis of the near-bottom sediment traps showed that the additional flux is intermediate in composition between particle rain and the surface sediment, suggesting that the recycled particulate flux includes degrading particles resuspended before they were completely incorporated into the sediment.

During the resuspension process, currents and tides can cause sediments to be transported laterally, and sediments will preferentially fill in low topography. Over time, in depositional environments, the infill will tend to flatten seafloor topography because sedimentation rates are higher in abyssal valleys (Tominaga et al., 2011). The amount of drape versus sediment infill appears to depend upon topography and rate of sedimentation. The high topography of the slow-spreading mid-Atlantic Ridge results in bare ridges with little sediment drape and much basin infill (Ruddiman, 1972). In contrast, the subdued abyssal hill topography in the tropical Pacific results in high amounts of sediment drape and only small differences between sediment cover on abyssal hills versus valleys (Tominaga et al., 2011). As can be seen in Figure 2, seafloor topography still mimics the topography of basaltic basement, even though the sediment column is many times thicker than the original basement topography.

Summary and conclusions

Deep-sea sediments deposit slowly and are distinctive chemically from typical terrestrial or continental margin sediments. Because such a small fraction of clays is deposited in the deep ocean compared to continental margins, deep-sea sediments can have large fractions of biogenic debris or metal oxides precipitated from hydrothermal plumes or directly from seawater. Deep-sea sediments also steadily accumulate, making them an excellent archive of Earth's environmental change.

Acknowledgments

I thank G. Ross Heath for his review of the manuscript. This study was supported in part by NSF grant OCE-0962184.

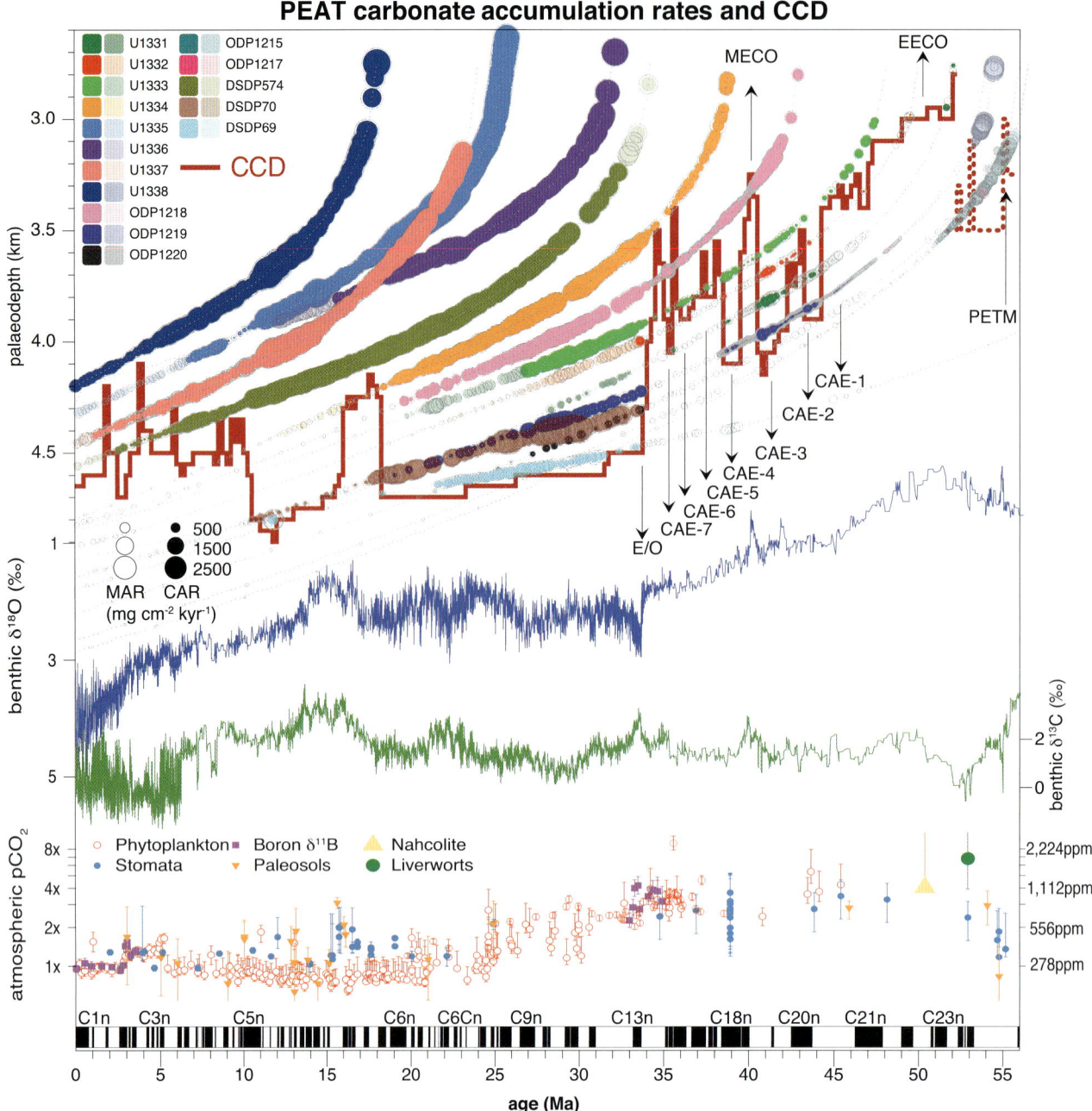

Deep-sea Sediments, Figure 9 Changes in the equatorial Pacific CCD (carbonate compensation depth) for the last 50 million years, from Pälike and Expedition 320/321 Shipboard Scientists (2012). The *red line* marks the depth of the CCD over time, while the bubbles mark $CaCO_3$ mass accumulation rate plotted versus paleodepth of different drill sites from the equatorial Pacific. A shallow CCD means that there is little $CaCO_3$ preserved in the sediments. The major change in CCD occurs at about 34 million years ago (the Eocene-Oligocene boundary), where the CCD permanently drops by about 1 km. Also shown are the benthic oxygen isotope profile (*blue*), an indicator for temperature and ice volume, the carbon isotope profile (*green*), and a compilation of estimates for atmospheric CO_2.

Bibliography

Archer, D. E., 1996. A data-driven model of the global calcite lysocline. *Global Biogeochemical Cycles*, **10**(3), 511–526.

Baker, E. T., and Urabe, T., 1996. Extensive distribution of hydrothermal plumes along the superfast spreading East Pacific Rise, 13°30′–18°40′S. *Journal of Geophysical Research, Solid Earth*, **101**, 8685–8695.

Balistrieri, L., Brewer, P. G., and Murray, J. W., 1981. Scavenging residence times of trace metals and surface chemistry of sinking particles in the deep ocean. *Deep Sea Research*, **28A**, 101–121.

Bastia, R., Das, S., and Radhakrishna, M., 2010. Pre- and post-collisional depositional history in the upper and middle Bengal fan and evaluation of deepwater reservoir potential along the northeast Continental Margin of India. *Marine and Petroleum Geology*, 27, 2051–2061.

Berger, W. H., 1970. Biogenous deep-sea sediments: fractionation by deep-sea circulation. *Geological Society of America Bulletin*, 81, 1385–1402.

Berger, W. H., 1973. Cenozoic sedimentation in the eastern tropical Pacific. *Geological Society of America Bulletin*, 84, 1941–1954.

Berger, W. H., Adelseck, C. G. J., and Mayer, L. A., 1976. Distribution of carbonate in surface sediments of the Pacific Ocean. *Journal of Geophysical Research*, 81, 2617–2627.

Bernat, M., and Church, T. M., 1978. Deep-sea phillipsites: trace geochemistry and mode of formation. In Sand, L. B., and Mumpton, F. A. (eds.), *Natural Zeolites: Occurrence, Properties, Use*. Oxford/New York: Pergamon Press, pp. 259–267.

Biscaye, P. E., Kolla, V., and Turekian, K. K., 1976. Distribution of calcium carbonate in surface sediments of the Atlantic Ocean. *Journal of Geophysical Research*, 81, 2595–2603.

Biscaye, P. E., Anderson, R. F., and Deck, B. L., 1988. Fluxes of particles and constituents to the eastern United States continental slope and rise: SEEP-I. *Continental Shelf Research*, 8, 855–904.

Boström, K., and Peterson, M. N. A., 1966. Precipitation from hydrothermal exhalations on the East Pacific Rise. *Economic Geology*, 39, 1258–1265.

Boström, K., and Peterson, M. N. A., 1969. The origin of aluminum-poor ferromanganoan sediments in areas of high heat flow on the East Pacific Rise. *Marine Geology*, 7, 427–447.

Boudreau, B. P., Middelburg, J. J., and Meysman, F. J. R., 2010. Carbonate compensation dynamics. *Geophysical Research Letters*, 37, L03603.

Brothers, D. S., ten Brink, U. S., Andrews, B. D., and Chaytor, J. D., 2013. Geomorphic characterization of the U.S. Atlantic continental margin. *Marine Geology*, 338, 46–63.

Burckle, L. H., and Cirilli, J., 1987. Origin of diatom ooze belt in the Southern Ocean: implications for late quaternary paleoceanography. *Micropaleontology*, 33, 82–86.

Calvert, S. E., and Pedersen, T. F., 1993. Geochemistry of recent oxic and anoxic marine sediments: implications for the geological record. *Marine Geology*, 113, 67–88.

Caress, D. W., and Chayes, D. N., 1996. Improved processing of Hydrosweep DS multibeam data on the R/V Maurice Ewing. *Marine Geophysical Research*, 18, 631–650.

Clift, P., and Gaedicke, C., 2002. Accelerated mass flux to the Arabian Sea during the middle to late Miocene. *Geology*, 30, 207–210.

CLIMAP Project Members, 1976. The surface of the ice-age earth. *Science*, 191, 1131–1137.

Cole, T. G., and Shaw, H. F., 1983. The nature and origin of authigenic smectites in some recent marine sediments. *Clay Minerals*, 18, 239–552.

Damuth, J. E., 1977. Late quaternary sedimentation in the western equatorial Atlantic. *Geological Society of America Bulletin*, 88, 695–710.

Dehairs, F., Chesselet, R., and Jedwab, J., 1980. Discrete suspended particles of barite and the barium cycle in the open ocean. *Earth and Planetary Science Letters*, 49, 528–550.

Dekens, P. S., Ravelo, A. C., and McCarthy, M. D., 2007. Warm upwelling regions in the Pliocene warm period. *Paleoceanography*, 22, PA3211, p. 12.

Drexler, J. W., Rose, W. I., Jr., Sparks, R. S. J., and Ledbetter, M. T., 1980. The Los Chocoyos Ash, Guatemala: a major stratigraphic marker in middle America and in three ocean basins. *Quaternary Research*, 13, 327–345.

Duce, R. A., et al., 2011. *Scientific Ocean Drilling: Accomplishments and Challenges*. Washington, DC: National Academy, p. 146.

Duggen, S., Olgun, N., Croot, P., Hoffmann, L., Dietze, H., Delmelle, P., and Teschner, C., 2010. The role of airborne volcanic ash for the surface ocean biogeochemical iron-cycle: a review. *Biogeosciences*, 7, 827–844.

Dymond, J., 1981. Geochemistry of Nazca plate surface sediments: an evaluation of hydrothermal, biogenic, detrital, and hydrogenous sources. *Geological Society of American Memoir*, 154, 133–173.

Dymond, J., and Lyle, M., 1994. Particle fluxes in the ocean and implications for sources and preservation of ocean sediments. In Hay, W. W., et al. (eds.), *Material Fluxes on the Surface of the Earth*. Washington, DC: National Academy, pp. 125–143.

Dymond, J., Lyle, M., Finney, B., Piper, D. Z., Murphy, K., Conard, R., and Pisias, N., 1984. Ferromanganese nodules from MANOP Sites H, S, and R; control of mineralogical and chemical composition by multiple accretionary processes. *Geochimica et Cosmochimica Acta*, 48, 931–949.

Dymond, J., Suess, E., and Lyle, M., 1992. Barium in deep-sea sediment: a geochemical proxy for paleoproductivity. *Paleoceanography*, 7, 163–181.

Expedition 320/321 Scientists, 2010. Site U1335. In Pälike, H., Lyle, M., Nishi, H., Raffi, I., Gamage, K., Klaus, A., and the Expedition 320/321 Scientists (eds.), *Proceedings of the IODP, 320/321*. Tokyo: Integrated Ocean Drilling Program Management International, doi:10.2204/iodp.proc.320321.107.2010.

Finney, B. P., Lyle, M. W., and Heath, G. R., 1988. Sedimentation at MANOP Site H (eastern equatorial Pacific) over the past 400,000 years: climatically induced redox variations and their effects on transition metal cycling. *Paleoceanography*, 3, 169–189.

Fischer, K., Dymond, J., Lyle, M., Soutar, A., and Rau, S., 1986. The benthic cycle of copper; evidence from sediment trap experiments in the eastern tropical North Pacific Ocean. *Geochimica et Cosmochimica Acta*, 50, 1535–1543.

Froelich, P. N., Klinkhammer, G. P., Bender, M. L., Luedtke, N. A., Heath, G. R., Cullen, D., Dauphin, P., Hammond, D., Hartman, B., and Maynard, V., 1979. Early oxidation of organic matter in pelagic sediments of the eastern equatorial Atlantic: suboxic diagenesis. *Geochimica et Cosmochimica Acta*, 43, 1075–1090.

Glasby, G. P., 2006. Manganese: predominant role of nodules and crusts. In Schuz, H. D., and Zabel, M., (eds.), *Marine Geochemistry*, 2nd rev. Berlin/Heidelberg: Springer, pp. 371–427.

Goldberg, E. D., 1954. Marine geochemistry 1. Chemical scavengers of the sea. *Journal of Geology*, 62, 249–265.

Gorsline, D. S., Kolpack, R. L., Karl, H. A., Drake, D. E., Fleischer, P., Thornton, S. E., Schwalbach, J. R., and Savrda, C. E., 1984. Studies of fine-grained sediment transport processes and products in the California Continental Borderland. *Geological Society of London Special Publications*, 15, 395–415.

Griffith, E. M., and Paytan, A., 2012. Barite in the ocean-occurrence, geochemistry and palaeoceanographic applications. *Sedimentology*, 59, 1817–1835.

Haymon, R. M., 2005. TECTONICS/hydrothermal vents at mid-ocean ridges. In Selley, R. C., Cocks, L. R. M., and Plimer, I. R. (eds.), *Encyclopedia of Geology*. Oxford: Elsevier, pp. 388–395, doi:10.1016/B0-12-369396-9/00448-2.

Hays, J. D., Imbrie, J., and Shackelton, N. J., 1976. Variations in the earth's orbit: pacemaker of the ice ages. *Science*, 194, 1121–1131.

Heath, G. R., 1981. Ferromanganese nodules of the deep sea. *Economic Geology*, 75, 736–756.

Hodell, D. A., Charles, C. D., and Sierro, F. J., 2001. Late Pleistocene evolution of the ocean's carbonate system. *Earth and Planetary Science Letters*, 192, 109–124.

Honda, M. C., Imai, K., Nojiri, Y., Hoshi, F., Sugawara, T., and Kusakabe, M., 2002. The biological pump in the northwestern North Pacific based on fluxes and major components of particulate matter obtained by sediment-trap experiments, 1997-2000. *Deep Sea Research II*, **49**, 5595–5625.

Karlin, R., 1980. Sediment sources and clay mineral distributions off the Oregon coast. *Journal of Sedimentary Petrology*, **50**, 543–560.

Kelley, D. S., Baross, J. A., and Delaney, J. R., 2002. Volcanoes, fluids, and life at mid-ocean ridge spreading centers. *Annual Reviews of Earth and Planetary Science*, **30**, 385–491.

Kennett, J. P., 1982. *Marine Geology*. Englewood Cliffs: Prentice-Hall, p. 813.

Krissek, L. A., 1984. Continental source area contributions to fine-grained sediments on the Oregon and Washington continental slope. *Geological Society of London Special Publications*, **15**, 363–375.

Kyte, F. T., 2002. Tracers of the extraterrestrial component in sediments and inferences for Earth's accretion history. *Geological Society of America Special Papers*, **356**, 21–38.

Le, J., and Shackleton, N. J., 1994. Reconstructing paleoenvironment by transfer function: model evaluation with simulated data. *Marine Micropaleontology*, **24**, 187–189.

Leinen, M., Cwienk, D., Heath, G. R., Biscaye, P., Kolla, V., Thiede, J., and Dauphin, J. P., 1986. Distribution of biogenic silica and quartz in recent deep-sea sediments. *Geology*, **14**, 199–203.

Lipps, J. H., 1993. *Fossil Prokaryotes and Protists*. Oxford: Blackwell Scientific, p. 342.

Lisiecki, L. E., and Raymo, M. E., 2005. A Pliocene-Pleistocene stack of 57 globally distributed benthic d18O records. *Paleoceanography*, **20**, PA1003.

Lisitzin, A. P., 1972. Sedimentation in the world ocean. *SEPM Special Publication*, **17**, 218.

Lyle, M., 1981. Formation and growth of ferromanganese oxides on the Nazca Plate. *Geological Society of American Memoir*, **154**, 269–294.

Lyle, M., Heath, G. R., and Robbins, J. M., 1984. Transport and release of transition elements during early diagenesis: sequential leaching of sediments from MANOP sites M and H. Part I. pH 5 acetic acid leach. *Geochimica et Cosmochimica Acta*, **48**, 1705–1715.

Lyle, M., Leinen, M., Owen, R. M., and Rea, D. K., 1987. Late Tertiary history of hydrothermal deposition at the East Pacific Rise, 19° S; correlation to volcano-tectonic events. *Geophysical Research Letters*, **14**, 595–598.

Lyle, M., Pälike, H., Nishi, H., Raffi, I., Gamage, K., Klaus, A., and Expedition 320/321 Scientific Party, 2010. The Pacific equatorial age transect, IODP expeditions 320 and 321: building a 50-million-year-long environmental record of the equatorial Pacific Ocean. *Scientific Drilling*, **9**, 4–15.

Maher, B. A., Prospero, J. M., Mackie, D., Gaiero, D., Hesse, P. P., and Balkanski, Y., 2010. Global connections between aeolian dust, climate and ocean biogeochemistry at the present day and at the last glacial maximum. *Earth-Science Reviews*, **99**, 61–97.

Mahowald, N., Kohfeld, K., Hansson, M., Balkanski, Y., Harrison, S. P., Prentice, I. C., Schulz, M., and Rodhe, H., 1999. Dust sources and deposition during the last glacial maximum and current climate: a comparison of model results with paleodata from ice cores and marine sediments. *Journal of Geophysical Research, Atmosphere*, **104**, 15895–15916.

Mahowald, N. M., Muhs, D. R., Levis, S., Rasch, P. J., Yoshioka, M., Zender, C. S., and Luo, C., 2006. Change in atmospheric mineral aerosols in response to climate: last glacial period, preindustrial, modern, and doubled carbon dioxide climates. *Journal of Geophysical Research, Atmosphere*, **111**, D10202, p. 22.

Marcantonio, F., Anderson, R. F., Stute, M., Kumar, N., Schlosser, P., and Mix, A., 1996. Extraterrestrial 3-He as a tracer of marine sediment transport and accumulation. *Nature*, **383**, 705–707.

Martin, J. H., 1990. Glacial-interglacial CO2 change: the iron hypothesis. *Paleoceanography*, **5**(1), 1–13.

Mayer, L. A., 1979. The origin of fine scale acoustic stratigraphy in deep-sea carbonates. *Journal of Geophysical Research*, **84**, 6177–6184.

Mayer, L. A., Shipley, T. H., and Winterer, E. L., 1986. Equatorial Pacific seismic reflectors as indicators of global oceanographic events. *Science*, **233**, 761–764.

McCave, I. N., 2010. Nepheloid layers. In Steele, J. H., Thorpe, S. A., and Turekian, K. K. (eds.), *Marine Ecological Processes: A Derivative of the Encyclopedia of Ocean Sciences*. London: Academic, pp. 1861–1870.

Menard, H. W., 1964. *Marine Geology of the Pacific*. New York: McGraw-Hill, p. 271.

Moore, T. C., Jr., Jarrard, R. D., Olivarez Lyle, A., and Lyle, M., 2008. Eocene biogenic silica accumulation rates at the Pacific equatorial divergence zone. *Paleoceanography*, **23**, PA2202, p. 22.

Morford, J. L., and Emerson, S., 1999. The geochemistry of redox sensitive trace metals in sediments. *Geochimica et Cosmochimica Acta*, **63**, 1735–1750.

Mosher, D. C., 2011. Cautionary considerations for geohazard mapping with multibeam sonar: resolution and the need for third and fourth dimensions. *Marine Geophysical Research*, **32**, 25–35.

Müller, P. J., and Suess, E., 1979. Productivity, sedimentation rate, and sedimentary organic matter in the oceans -I. Organic carbon preservation. *Deep-Sea Research*, **26A**, 1347–1362.

Olivarez Lyle, A., and Lyle, M., 2006. Organic carbon and barium in Eocene sediments: is metabolism the biological feedback that maintains end-member climates? *Paleoceanography*, **21**, 1–13, doi:10.1029/2005PA001230.

Olivarez, A. M., Owen, R. M., and Rea, D. K., 1991. Geochemistry of eolian dust in pacific pelagic sediments: implications for paleoclimatic interpretations. *Geochimica et Cosmochimica Acta*, **55**, 2147–2158.

Pälike, H., and Expedition 320/321 Shipboard Scientists, 2012. A Cenozoic record of the equatorial Pacific carbonate compensation depth. *Nature*, **488**, 609–614.

Paytan, A., and Griffith, E. M., 2007. Marine barite: recorder of variations in ocean export productivity. *Deep Sea Research II*, **54**, 687–705.

Pearson, P. N., Ditchfield, P. W., Singano, J., Harcourt-Brown, K. G., Nicholas, C. J., Olsson, R. K., Shackleton, N. J., and Hall, M. A., 2001. Warm tropical sea surface temperatures in the late Cretaceous and Eocene epochs. *Nature*, **413**, 481–487.

Pedersen, T. F., and Calvert, S. E., 1990. Anoxia versus productivity: what controls the formation of organic-carbon-rich sediments and sedimentary rocks? *American Association of Petroleum Geologists Bulletin*, **74**, 454–466.

Ragueneau, O., Tréguer, P., Anderson, R. F., Brzezinski, M. A., DeMaster, D. J., Dugdale, R. C., Dymond, J., Fischer, G., François, R., Heinze, C., Leynaert, A., Maier-Reimer, E., Martin-Jézéquel, V., Nelson, D. M., and Quéguiner, B., 2000. A review of the Si cycle in the modern ocean; recent progress and missing gaps in the application of biogenic opal as a paleoproductivity proxy. *Global and Planetary Change*, **26**, 317–365.

Rea, D. K., 1994. The paleoclimatic record provided by eolian deposition in the deep sea: the geologic history of wind. *Reviews of Geophysics*, **32**, 159–195.

Rea, D. K., et al., 2006. Broad region of no sediment in the Southwest Pacific Basin. *Geology*, **34**, 873–876.

Riedel, W. R., 1967. Radiolarian evidence consistent with spreading of the pacific floor. *Science*, **157**, 540–542.

Ruddiman, W. F., 1972. Sediment redistribution on the Reykjanes Ridge: seismic evidence. *Geological Society of America Bulletin*, **83**, 2039–2062.

Ruhlin, D. E., and Owen, R. M., 1986. The rare earth geochemistry of hydrothermal sediments from the East Pacific Rise: examination of a seawater scavenging mechanism. *Geochimica et Cosmochimica Acta*, **50**, 393–400.

Sacchetti, F., Benetti, S., Quinn, R., and Cofaigh, C. O., 2013. Glacial and post-glacial sedimentary processes in the Irish Rockall Trough from an integrated acoustic analysis of near-seabed sediments. *Geo-Marine Letters*, **33**, 49–66.

Scheidegger, K. F., Corliss, J. B., Jezek, P. A., and Ninkovich, D., 1980. Compositions of deep-sea ash layers derived from North Pacific volcanic arcs: variations in time and space. *Journal of Volcanology and Geothermal Research*, **7**, 107–137.

Stonecipher, S. A., 1976. Origin, distribution, and diagenesis of phillipsite and clinoptilolite in deep-sea sediments. *Chemical Geology*, **17**, 307–318.

Tominaga, M., Lyle, M., and Mitchell, N. C., 2011. Seismic interpretation of pelagic sedimentation regimes in the 18-53 Ma eastern equatorial Pacific: basin-scale sedimentation and infilling of abyssal valleys. *Geochemistry, Geophysics, Geosystems*, **12**, Q03004, p. 22.

Torres, M. E., McManus, J., and Huh, C.-A., 2002. Fluid seepage along the San Clemente Fault scarp: basin-wide impact on barium cycling. *Earth and Planetary Science Letters*, **203**, 181–194.

van Andel, T. H., and Moore, T. C., Jr., 1974. Cenozoic calcium carbonate distribution and calcite compensation depth in the central equatorial Pacific Ocean. *Geology*, **2**, 87–92.

Walsh, I., Fischer, K., Murray, D., and Dymond, J., 1988. Evidence for resuspension of rebound particles from near-bottom sediment traps. *Deep Sea Research*, **35**, 59–70.

Wheat, C. G., Feeley, R. A., and Mottl, M. J., 1996. Phosphate removal by oceanic hydrothermal processes: an update of the phosphorus budget of the oceans. *Geochimica et Cosmochimica Acta*, **60**, 3593–3608.

Whittaker, J., Goncharov, A., Williams, S., Müller, R. D., and Leitchenkov, G., 2013. Global sediment thickness dataset updated for the Australian-Antarctic Southern Ocean, Geochemistry, Geophysics. *Geosystems*, **14**, 3297–3305, doi:10.1002/ggge.20181.

Williams, D. L., and Von Herzen, R. P., 1974. Heat loss from the earth: new estimate. *Geology*, **2**, 327–328.

Zaric, S., Donner, B., Fischer, G., Mulitza, S., and Wefer, G., 2005. Sensitivity of planktonic foraminifer to sea surface temperature and export production as derived from sediment trap data. *Marine Micropaleontology*, **55**, 75–105.

Cross-references

Biogenic Barium
Black and White Smokers
Calcite Compensation Depth (CCD)
Carbonate Dissolution
Clay Minerals
Deep-sea Fans
Diatoms
Dust in the Ocean
Export Production
Foraminifers (Planktonic)
Hydrothermal Plumes
Hydrothermalism
Manganese Nodules
Marine Microfossils
Marine Mineral Resources
Mid-ocean Ridge Magmatism and Volcanism
Modelling Past Oceans
Nannofossils Coccoliths
Ocean Drilling
Paleoceanography
Radiolarians
Reflection/Refraction Seismology
Sediment Dynamics
Upwelling

DELTAS

Duncan FitzGerald[1], Ioannis Georgiou[2] and Mark Kulp[2]
[1]Department of Earth and Environment, Boston University, Boston, MA, USA
[2]Department of Earth and Environmental Sciences, University of New Orleans, New Orleans, LA, USA

Synonyms
River-mouth deposits

Definition

As defined by Moore and Asquith (1971), a delta is the subaerial and submerged contiguous sediment mass deposited in a body of water (ocean or lake) primarily by the action of a river. Wright (1985) added that deltas included secondary riverine-derived deposits that had been reworked and molded by waves, currents, or tides.

The term *delta*, for the sedimentary accumulation at a river mouth, was coined by the Greek historian Herodotus who in the fifth century recognized a similarity in shape of the Greek letter Δ to the tract of land at the mouth of the Nile River (*The Histories*, 450–420 AD). Although the distinctive morphology is lacking in many river-mouth land tracts, the term has become accepted to describe both the geographical region near a river mouth and the sedimentary package deposited by fluvial processes into a depositional basin. Although sediment delivery to a delta is by a fluvial system, the dynamic interaction of riverine and ocean processes controls morphologic, stratigraphic, and sedimentologic variability of deltaic environments. Deltaic environments exist throughout the world and have been widely recognized in the sedimentary rock record. Buried deltaic sandstones are associated with reservoir rocks containing many of the world's major natural gas and oil resources (Morse, 1994). Modern deltas have abundant food sources derived from diverse flora and fauna, extensive cultivable land, and navigable waterways reaching from the coast to interiors of continents. Thus, these settings became the centers of numerous formative cultures and today are the home to some of the densest populations in the world. It is estimated that approximately 61 % of the world's population live on deltas and that percentage continues to grow (Bianchi and Allison, 2009). Deltaic systems are also recognized as important carbon sinks due to burial of vegetative detritus coming down rivers as well as phytoplankton blooms spawned by nutrient-rich waters. Thus, the progradation or

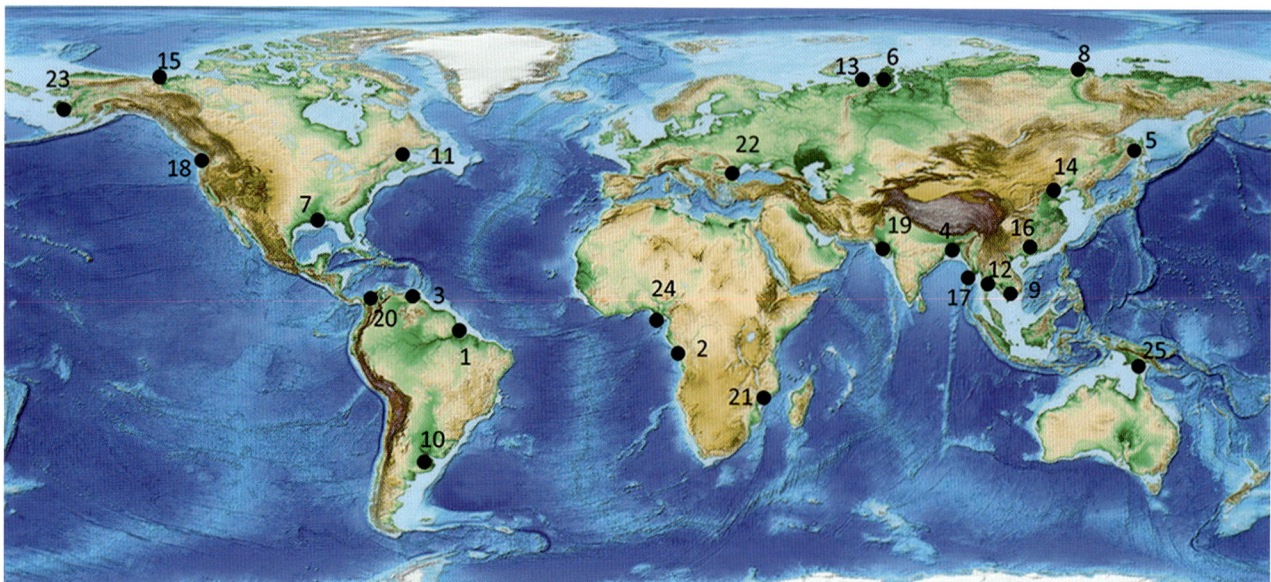

Deltas, Figure 1 Map showing the global distribution of the 25 largest sediment discharge rivers (refer to Table 1). These rivers roughly coincide with the major deltas of the world. Note that most of these river systems are located along trailing-edge coastlines close to the equator (modified from Coleman and Prior, 1980). Background image from ETOPO7, NOAA database.

erosional reworking and retreat of deltas affects the carbon budget and climate change.

Globally, deltas are found on all continents and in all climates (Figure 1). In a general sense, the locations of deltas are similar, at the terminus of a catchment basin that provides sediment load into an ocean, gulf, lagoon, estuary, or lake. Although a delta may form regardless of the size of the fluvial system or receiving basin, some tectonic settings are more conducive than others to the development of major deltaic landscapes. Using the Inman and Nordstrom (1971) tectonic classification of coasts, trailing-edge coasts typically have the largest drainage basins followed by marginal-sea coasts, and the smallest are common to leading-edge coasts. Of the 58 major river systems in the world with drainage areas greater than 10^5 km^2, 56.9 % occur on trailing-edge coasts, 34.5 % along marginal-sea coasts, and 8.6 % on collision (leading-edge) coasts.

Trailing-edge coasts provide a geologic setting favorable to delta development because of their tectonic stability and geologically old, low-relict terrains with extensive river systems that provide an abundant supply of sediment. Moreover, trailing-edge coasts often border broad continental shelves, which provide shallow-water platforms for the formation of deltas. Marginal-sea coasts typically provide relatively low-energy settings with many trailing coast characteristics that are conducive to delta development. Mountainous and immature drainage systems of tectonically active, leading-edge coasts (Pacific coast of South America) with small catchment basins and low sediment supply generally limit the likelihood of extensive deltaic deposition. However, it is the mountains of tectonically active coastal margins (e.g., Andes) or former coastal margins and present subduction zones (Himalayas) that can supply vast amounts of sediment to trailing-edge margin deltas (e.g., Amazon and Ganges-Brahmaputra, respectively). Figure 1 shows the locations and Table 1 lists the sizes of the world's 25 largest deltaic systems, according to McKee (2005) and Meade (1996).

Delta types and formation

Previous summaries of the world's deltas have focused primarily on the major large deltas (e.g., Coleman and Wright, 1975). A review of the recent literature on Late Quaternary deltaic systems reveals that deltas can be classified into four basic types depending on sea-level position (highstand or lowstand) and sediment supply (high or low) (Kindinger, 1988; Boyd et al., 1989; Nichol et al., 1996). The four basic types of deltas include: (1) lacustrine, (2) bayhead (or lagoonal), (3) continental shelf, and (4) continental margin. In a geomorphic and evolutionary sense, these four types of deltas represent end members and transitional forms describe many deltas. For example, during a sufficiently long sea-level stillstand, a lacustrine deltaic system having an adequate sediment supply and wave and tidal setting could evolve into a continental shelf margin deltaic system through progradation.

Lacustrine deltas

A lacustrine delta consists of sedimentary deposits located landward of the coast or paralic zone within an inland lake or small inland sea. In these cases, the environmental

Deltas, Table 1 Top 25 largest rivers based on freshwater and sediment discharges (modified from McKee, 2005, after Milliman and Meade, 1983; Meade, 1996)

River	Sediment discharge (10^6 t year^{-1})	Sediment discharge rank	Water discharge (10^9 m^3 year^{-1})	Water discharge rank	Drainage basin area (10^6 km^2)
Amazon	1,150	1	6,300	1	6.15
Zaire	43	22	1250	2	3.82
Orinoco	150	11	1200	3	0.99
Ganges-Brahmaputra	1,050	3	970	4	1.48
Yangtze (Changjiang)	480	4	900	5	1.94
Yenisey	5		630	6	2.58
Mississippi	210	7	530	7	3.27
Lena	11		510	8	2.49
Mekong	160	9	470	9	0.79
Parana/Uruguay	100	14	470	10	2.83
St Lawrence	3		450	11	1.03
Irrawaddy	260	5	430	12	0.43
Ob	16		400	13	2.99
Amur	52	20	325	14	1.86
Mackenzie	100	13	310	15	1.81
Pearl (Xi Jiang)	80	16	300	16	0.44
Salween	100	15	300	17	0.28
Columbia	8		250	18	0.67
Indus	50	21	240	19	0.97
Magdalena	220	6	240	20	0.24
Zambezi	20		220	21	1.2
Danube	40	24	210	22	0.81
Yukon	60	19	195	23	0.84
Niger	40	25	190	24	1.21
Purari/Fly	110	12	150	25	0.09
Yellow (Huang He)	1,100	2	49		0.77
Godavari	170	8	92		0.31
Red (Hunghe)	160	10	120		0.12
Copper	70	17	39		0.06
Choshui	66	18			0.003
Liao He	41	23	6		0.17

setting is typically a shallow-water basin with a low-energy regime. Along the Gulf of Mexico, the best example of a lacustrine delta is the Grand Lake delta within the Atchafalaya drainage basin of south-central Louisiana. At the start of the nineteenth century, Grand Lake, within the Atchafalaya basin, was a shallow, low-energy lake (Tye and Coleman, 1989). However, by the 1950s a large lacustrine delta had filled this lake and started conveying its sediment load into the Atchafalaya Bay. This rapid infilling largely resulted from the progressively increasing Mississippi River flow into the Atchafalaya basin (van Heerden and Roberts, 1988). Historically, if left uncontrolled, the Mississippi River would have avulsed from its modern course into the Gulf of Mexico past New Orleans to the Atchafalaya course located farther west. Other lacustrine delta examples include deltas in Lake Geneva and Lake Constance (Switzerland) (Müller, 1966; Reineck and Singh, 1973) and Lake Ayakum Delta, Tibet (Figure 2).

Bayhead/lagoonal deltas

Bayhead/lagoonal deltas are located in the upper reaches of the coastal zone. In many cases they represent an evolutionary step from a lacustrine delta during a sea-level stillstand or withdrawal, which allows the delta to build into coastal environments. Bayhead-delta settings are typically protected by barrier-island systems or remnants of preexisting antecedent topography, resulting in a low-energy environment. Examples of bayhead deltas are found within the northern part of Mobile Bay (USA, Alabama; Kindinger et al., 1994), Trinity Bay and San Antonio Bay (USA, Texas; McEwen, 1969; Donaldson et al., 1970; van Heerden and Roberts, 1988), and Merrymeeting Bay, Maine (Fenster et al., 2005). One of the largest bayhead deltas in the world exists in the Rio de la Plata at the border between northern Argentina and Uruguay where the Parana and Uruguay Rivers have combined to build a bayhead delta 280 km long and 40 km wide.

Deltas, Figure 2 Lake Ayakum Delta in Tibet (Astronaut photograph ISS027-E-16922, acquired on 25 April 2011).

Continental shelf deltas

Continental shelf deltas develop during prolonged sea-level stillstands and a moderate sea-level fall. Trailing-edge or marginal-sea continental shelf deltas typically are characterized by an abundant sediment supply, and their geomorphology reflects the wave and tidal conditions in which they develop. Examples of continental shelf deltas include those associated with rivers of the Lena along the Russian Siberian coast (Figure 3), the Mississippi in Louisiana (USA) (Figure 4), the Ebro in Spain, the Colville in Alaska (USA) (Frazier, 1967; Maldonado, 1975; Naidu and Mowatt, 1975; Alonso et al., 1990), the Ganges-Brahmaputra (Goodbred and Kuehl, 2000), and those along the East China Sea (Zhou et al., 2014).

Continental shelf margin deltas

Continental shelf margin deltas are typically most well developed during sea-level lowstands or formed by rivers with large sediment loads during prolonged sea-level stillstands. The majority of shelf margin deltas in the recent geologic record have developed during falling sea-level or sea-level lowstand conditions as river valleys lengthen, incise across subaerially exposed continental shelf sediments, and discharge their sedimentary load at the shelf margin (Suter, 1994). With sufficient sediment load or during lowstand conditions, these deltas continue to build seaward, creating an apron of deltaic sediment along the shelf edge that extends the preexisting continental margin into deeper basinal waters. Numerous lowstand shelf margin deltas, as listed above, formed during the

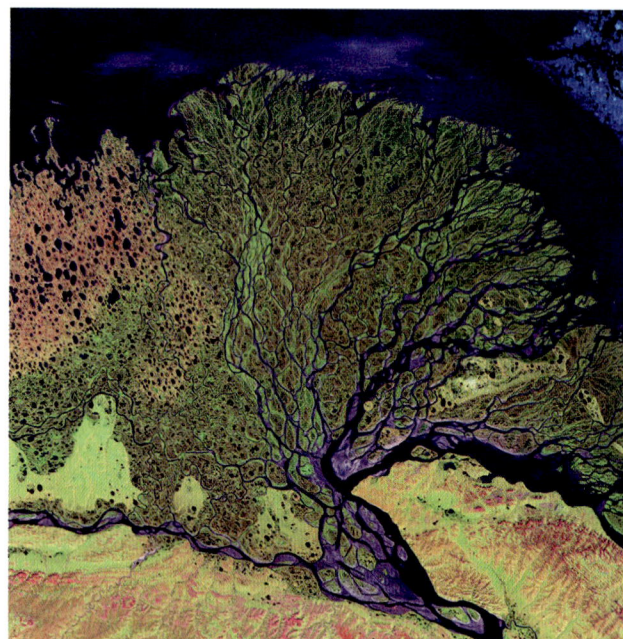

Deltas, Figure 3 Lena Delta (from: earthobservatory.nasa.gov).

Late Pleistocene sea-level fall and lowstand culminating at approximately 18,000 year BP. Some specific examples include those found seaward of the Rio Grande and the Mississippi River deltas (Morton and Price, 1987; Suter et al., 1987; Suter, 1994). Sufficiently large rivers carrying substantial sediment loads may also construct shelf margin delta deposits during sea-level highstand conditions. The main stem of the modern Mississippi River delta represented by the "bird-foot" delta is an example of a shelf margin highstand delta (Boyd et al., 1989).

Delta components

The delta plain contains a subaerial and subaqueous zone where sediments are discharged from the alluvial valley and accumulate in the receiving basin (Figure 5). Typically, a delta plain is subdivided into an upper delta plain and a lower delta plain, each of which is characterized by different vegetation, morphology, and depositional processes (Reineck and Singh, 1973; Coleman, 1981; Elliot, 1986).

Upper delta plain Many of the features of the upper delta plain are inherited from its alluvial valley. In the upper delta plain, deposition is primarily fluvial as the main stem of the river leaves the alluvial valley discharging sediment onto the upper delta plain. The vegetative landscape of the upper delta plain is a freshwater environment, generally beyond the influence of marine incursions created by the tidal regime of the receiving basin. The major landform includes distributary channels with their subaerial natural levees, point bars, and crevasse

Deltas, Figure 4 Mississippi bird-foot delta in Louisiana, USA (from: http://eoimages.gsfc.nasa.gov).

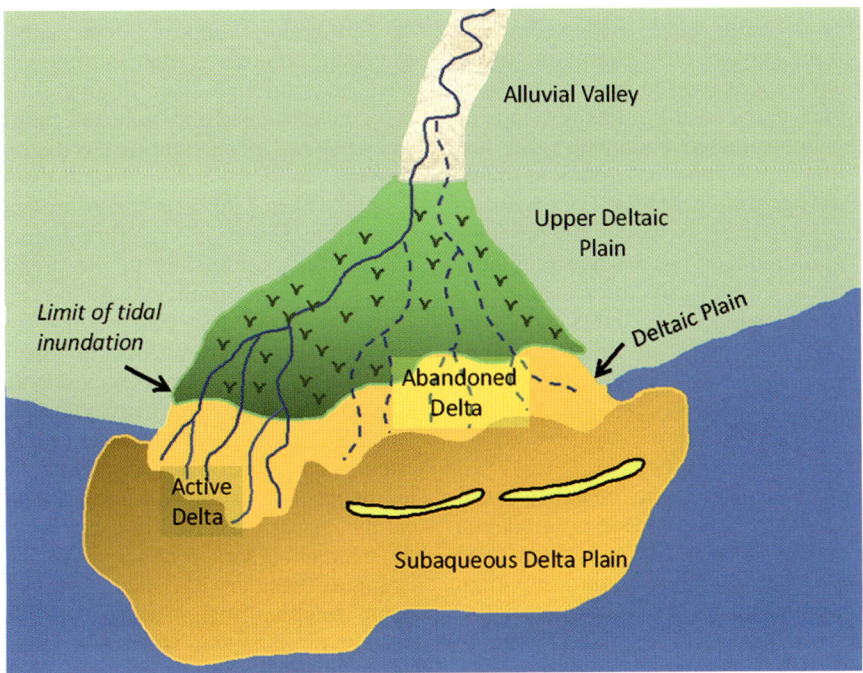

Deltas, Figure 5 Schematic diagram showing the primary physiographic components and depositional environment of a delta (modified from Coleman and Prior, 1980).

splays, as well as interdistributary environments characterized by low-lying swampy areas.

Lower delta plain The lower delta plain is low relief and characterized by tidal incursion and greater sedimentation compared to the upper delta plain. Elevations of the natural levees are relatively low, producing frequent overbank flooding, which results in channel crevassing and avulsion. Distributary shifting is a common dynamic process, resulting in a complex pattern of multi-aged distributary channels in various stages of evolution (Figure 5). The lower delta plain contains active and abandoned distributary channels leading to different stages of regression and transgression within a single delta complex (Penland et al., 1988). Because the lower delta plain is a transitional environment influenced by freshwater fluvial and saline marine waters, the vegetative landscape generally contains species tolerant of both brackish and saline water. For example, the lower modern Mississippi River delta is characterized by *Phragmites australis*, an invasive species that tolerates fluctuating salinities, river stage, and storms (White, 1993). The resiliency of this plant community is largely responsible for stabilizing deltaic deposits and greatly reducing erosion and reworking during frequent hurricane impacts. Mangroves serve a similar function in low-latitude deltas, such as Ganges-Brahmaputra.

Subaqueous delta plain This is the shallow, seaward sloping part of the delta that is characterized by sedimentation at the ends of distributary channels and the reworking of these deposits onshore and laterally by waves and tides. The subaqueous delta plain forms a broad apron of interfingering deltaic and marine sediments that transitions seaward into continental shelf sediments. Major components of the subaqueous delta plain consist of distributary bar mouth sands incised by channels bordered by subaqueous levees. Seaward of this region is an area of increasingly finer-grained sediments and less marine reworking as the depositional environment transitions from distributary bar mouth sand to sandy silt of the delta front and finally to the silty clay of the prodelta. At some deltas where progradation has reached at or close to the shelf edge, sediment from distributaries is conveyed directly into submarine canyons (e.g., Indus Goodbred and Kuehl, 2000; Giosan et al., 2006; Ganges-Brahmaputra).

Deltaic sediments

In general, deltaic sedimentary lithosomes contain a coarsening upward sequence that reflects seaward progradation of the delta into its receiving basin. At any given time, there is typically a well-developed fining grain size pattern extending seaward from the subaerial to subaqueous portions of a delta. The relatively coarsest-grained material is deposited proximal to the river mouth where transport competence is highest, whereas the finest-grained material is carried farther seaward and deposited in more distal locations. Thus, as sediments are delivered to the receiving basin from the river mouth, they accumulate in a subaqueous, fine-grained depositional zone referred to as the *prodelta*. The prodelta forms the platform across which the delta progrades and subsequently aggrades. The prodelta sedimentary package is a widespread laterally continuous interval composed primarily of the finest sediment fraction transported by the fluvial and ocean system. Thus, the overall finest sediment is found at the base of the prodelta sequence and the whole sequence coarsens upward. Bioturbation of prodelta sediments is a function of the rate of deposition.

The delta front is located between the fine-grained prodelta deposition and the landward, coarser-grained distributary mouth deposits of a progradational deltaic system. The relatively coarser-grained sediment of the delta front generally consists of interbedded clays, silts, and sands. This zone is dominated by the interaction of fluvial and marine processes, resulting in sand-rich accumulations landward of the advancing prodelta that are commonly a result of high-stage fluvial discharge or storm reworking. They are usually highly bioturbated. The delta front also represents a zone of transition between deposits representative of progradation and aggradation.

In a progradational deltaic sequence, distributary channel deposits overlie the relatively finer-grained delta front and prodelta deposits. The framework of distributary channels consists of distributary mouth bars overlain and bordered by natural levee deposits. Distributary mouth bar deposits are primarily subaqueous deposits that grade laterally into relatively finer-grained deposits; locally, mouth bar deposits may contain fine-grained beds within the generally sandy matrix of the mouth bar. The presence of fine-grained deposits represents deposition during low-stage conditions when current velocities are relatively weak. During high-stage flood conditions, natural levees can be overtopped and breached creating crevasse splays. These splays occur within the distributary network and provide conduits for sediment dispersal into interdistributary bays. Crevasse splays are one of the major landform processes in riverine-dominated deltas (e.g., Mississippi) contrasting to the preponderance of beach ridges at wave-dominated deltas (e.g., Ebro, Danube). However, crevassing can produce loss of stream power downstream of the cut due to loss of flow, leading to decreased channel competency and sedimentation within the channel. A case study in South Pass, one of the Mississippi River distributaries, in Louisiana, USA, reported that lateral flow loss from crevassing, including flow into a relict distributary channel, diminished stream power significantly such that South Pass aggraded immediately downstream (Clark et al., 2013).

Interdistributary bay and marsh sediments are generally as volumetrically significant as prodelta sediments. The interdistributary area is a low-energy depositional environment and less dynamic than areas of the delta plain characterized by multiple channels, channel bifurcations, and channel avulsions. Interdistributary deposits characteristically consist of fining upward deposits consisting of clay-rich bay sediments overlain by organic-rich marsh deposits. This fine-grained depositional environment is punctuated by sandy depositional events associated with overbank flooding and crevasse splay events. Channels within the crevasse splay also create elongated sand deposits.

Delta processes

Deltas develop when sediment-laden river water decelerates upon entering a receiving basin and sediment transport competence of the river is lost. Consequently, the interplay between fluvial and river-mouth processes constitutes an important controlling factor in the nature of deltaic deposition (Bates, 1953). Fluvial processes are particularly important in deltaic environments where wave and tide regimes are less important. In such environments the primary riverine factors, influencing the nature of deltaic deposition, are river discharge, sediment load, and textural character of the sediment. Generally, the primary forces controlling the character of deltaic sedimentation are (1) the riverine effluent inertia and diffusion, (2) friction between the sediment load and the floor of the receiving basin, and (3) density contrasts between the effluent and the receiving basin waters (Bates, 1953; Wright, 1977; Orton and Reading, 1993).

Homopycnal conditions Homopycnal flow describes flow conditions where the density contrast between riverine and basinal waters is small. Discharge patterns are influenced primarily by inertia of the riverine water, which allows the effluent to radially spread into the receiving basin. As a result of deltaic progradation, fine-grained bottomset sediments are overlain by relatively coarser-grained topset beds and river-mouth bar deposits, creating a coarsening upward depositional sequence common to deltaic systems. Gilbert (1885) first described this type of deltaic system associated with the paleo-lakes of the western USA; hence, they are termed Gilbert-type deltas.

Hyperpycnal conditions Hyperpycnal conditions occur when the riverine discharge is denser than the waters of the receiving basin. Consequently, the sediment-laden effluent moves along the floor of the receiving basin as a density current. This type of discharge condition is rare and typically restricted to deltaic systems rich in silt and coarse-grained sediments discharging into freshwater lakes, such as the Rhone River entering Lake Geneva in Switzerland and Bella Coola fjord delta of British Columbia (Kostashcuk, 1985). Although not common, hyperpycnal conditions may exist in some deltas for short periods during early spring, when river water temperatures are appreciably lower than those in the receiving basin, contributing to the formation of density currents. This is usually a seasonal phenomenon associated with spring and summer meltwater discharges from glaciers.

Hypopycnal conditions Hypopycnal discharge conditions occur when riverine inflow is less dense than the waters of the receiving basin, a condition commonly present where rivers debouch into marine waters. Most of the world's deltas form under hypopycnal conditions. In hypopycnal settings, as the riverine inflow enters the receiving basin, it undergoes spreading and expansion facilitating turbulent mixing resulting in the deposition of the sediment load. Finer-grained sediments are transported distally into the receiving basin (prodelta) as the freshwater sediment plume disperses above relatively more dense basinal waters. The relatively coarser-grained sediments are deposited in close proximity to the river mouth (distributary mouth bars and delta front). In shallow-water hypopycnal discharge conditions, friction between the receiving basin bottom and the riverine leads to multiple bifurcations of the distributary system. Conversely, in deepwater during hypopycnal discharge conditions, buoyancy dominates during the dispersal of the effluent, producing elongated distributaries such as those present at the modern Mississippi River depocenter (Suter, 1994).

Channel processes

Deltas grow by utilizing their distributary channel networks through a series of successive bifurcations, whereby a channel splits into two. Bifurcations are caused by sediment deposition at the mouths of distributary channels resulting in flow around a bar. Each bifurcation can be stable, whereby both channels receive the same flow, or unstable, where one channel receives greater flow than the other. At each bifurcation, sedimentation produces a distributary mouth bar that eventually becomes subaerial, thereby stabilizing the channel position. With each successive bifurcation, channels become smaller in both width and depth as they approach the delta front. In a study of 11 deltas worldwide, Edmonds and Slingerland (2007) found that channel widths decrease nonlinearly with bifurcation order, suggesting that across the delta there is continuous reduction in channel geometry. However, this should not be interpreted as loss of stream power, because deltas can maintain channel efficiency such that their ability to transport sediment downstream remains unaltered. A case study by Esposito et al. (2013) showed that the loss in dimensions with each bifurcation is met with an increase in water surface slope, resulting in unchanged velocity in the channels. This evolution can be significant in the delta growth cycle, because efficient channels can extend via depositional processes throughout the change in stage of the river (Esposito et al., 2013) or can extend during low flow conditions by incising through the delta front. Shaw and Mohrig (2014) reported that the channels of the Wax Lake Delta in Louisiana, USA, have extended seaward primarily through channel incision during low flow conditions, rather than solely by deposition.

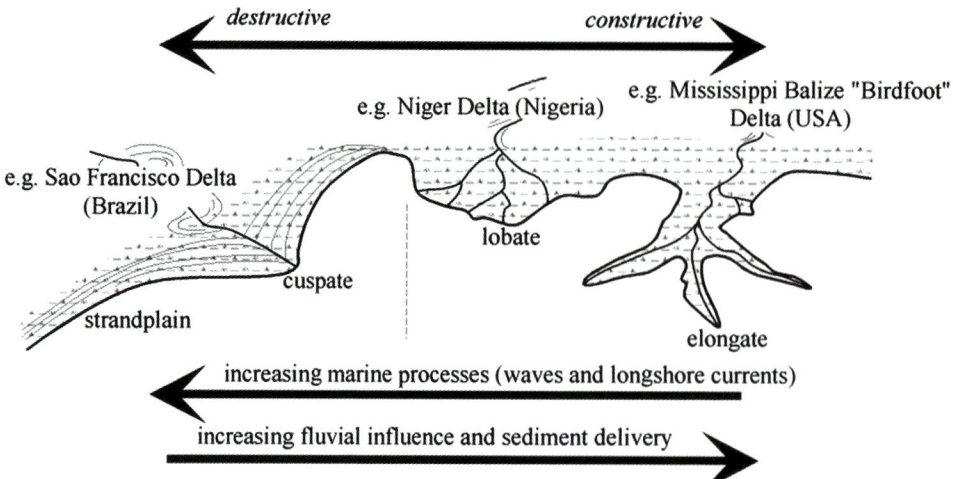

Deltas, Figure 6 Schematic diagram showing the variation in deltaic geomorphology as a function of the relative influences of fluvial and marine processes (from Fisher et al., 1969).

Delta morphology

Although sediment delivery to a delta is via a fluvial system, it is ultimately the dynamic interaction of riverine and marine basinal processes that control the morphologic, sedimentologic, and stratigraphic variability of a deltaic accumulation. Water depth and basinal configuration, tidal range, wave climate, and coastal currents are the primary basinal processes that control delta morphology. The dynamics of these processes and complex interaction between them lead to a highly variable array of deltaic configurations (Coleman and Wright, 1975).

One of the first and simplest attempts to describe deltaic variability as a function of processes depicts the morphology of river deltas as a function of sediment influx and marine processes such as waves and longshore currents (Fisher et al., 1969; Figure 6). The tidal regime of the receiving basin was not considered in the formulation of this model. Fundamentally, this model describes the processes leading to deltaic deposition as constructive or destructive (Davis, 1983). Deltas dominantly influenced by the nearshore wave climate are termed destructional, characteristically cuspate shaped with poorly developed distributaries, such as well-developed strand plains of the Sao Francisco delta in Brazil. Intermediate between destructional and constructional forms is the lobate-shaped delta, typified by the Niger delta of Africa. Lobate deltas possess well-developed distributary networks and a smooth coastal outline. Deltas that are strongly influenced by sediment deposition are essentially constructional and are characterized by distinctive elongate distributaries. An example of this type of delta is the modern "bird-foot" depocenter of the Mississippi River (Figure 4).

In the 1970s, a process-response approach to understanding morphologic variability of deltas incorporated features of the constructional/destructional approach but also considered the effects of the receiving basin tidal regime. In Galloway's (1975) scheme, a variety of

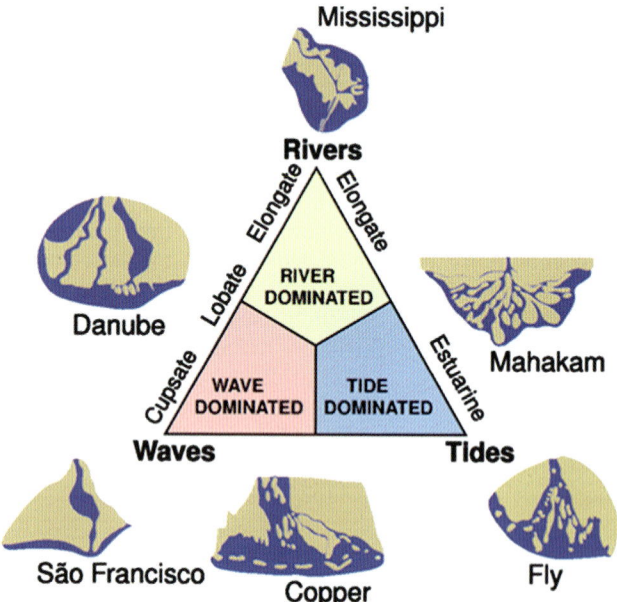

Deltas, Figure 7 Ternary classifications of deltas on the basis of dominate processes and the resulting morphology (after Galloway, 1975).

previously documented delta morphologies (e.g., Fisher et al., 1969; Wright and Coleman, 1973) were classified within the framework of a ternary diagram that differentiated delta morphology as a function of sediment input, wave energy flux, and tidal energy flux (Figure 7).

End member deltas (Figure 7) dominated by sediment input are supplied by a large well-developed drainage system, capable of transporting large volumes of sediment to the deltaic coast. Distributary switching is an important process, resulting in highly constructive elongate to lobate deltas with straight to sinuous active and abandoned

Deltas, Table 2 Characteristics of different delta types

Characteristics	Fluvial dominated	Wave dominated	Tide dominated
Geomorphology	Elongate to lobate	Arcuate	Estuarine to irregular
Channels distributaries	Straight to sinuous distributaries Flaring straight to sinuous distributaries	Meandering	
Sediments Framework facies	Muddy to mixed Distributary mouth bar and channel-fill sands, delta-margin sand sheet	Sandy Coastal barrier and beach-ridge sands	Variable Estuary fill and tidal sand ridges
Framework geometry depositional strike	Parallels depositional slope Parallels depositional slope	Parallels	

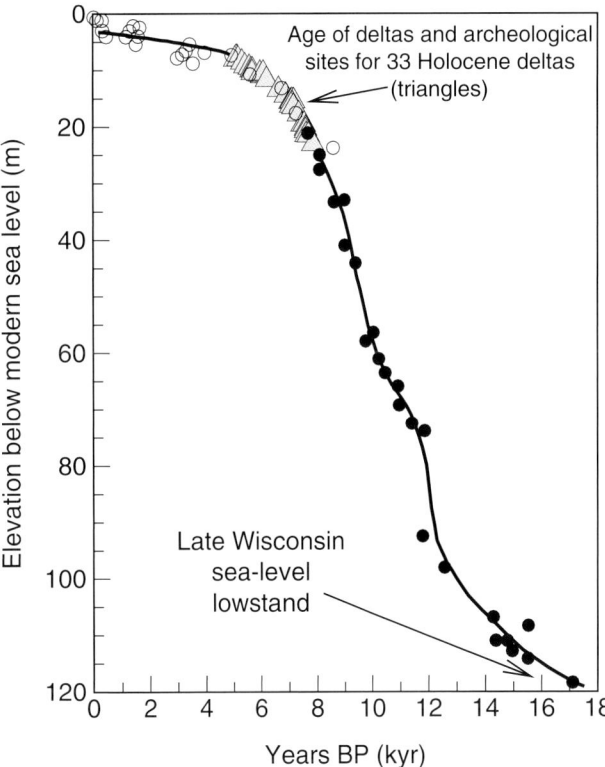

Deltas, Figure 8 Graph showing the time of formation for latest Quaternary deltaic systems and ages of deltaic archaeological sites in relation to deceleration in the rate of sea-level rise following the late Wisconsin sea-level lowstand (ages (*triangles*) from Stanley and Warne, 1997; sea-level curve (*solid line*) from Fairbanks, 1989, derived from radiocarbon dates from Barbados (*open circles*) and other Caribbean sites (*closed circles*)).

channels. The bulk composition of the sediment supply is muddy to mixed, and the framework facies include distributary mouth bar, channel-fill sands, and delta-margin sand sheets (Table 2).

Wave-dominated cuspate and lobate deltas are usually characterized by small sediment discharge relative to the volume of sediment reworking by wave energy. The high-wave-energy regime leads to winnowing of fine-grained sediments leaving behind a relatively coarser-grained sediment fraction. Consequently, the bulk composition of wave-dominated deltas is typically sandy. The primary distributary channels are meandering, and the framework facies are coastal barrier and beach-ridge sands (Table 2).

Tide-dominated deltas are shaped by onshore-offshore sediment transport caused by reversing tidal currents. They occur primarily in funnel-shaped, macro-tidal embayments, such as the Kuskokwim Delta in the Alaskan Bering Sea or the mouth of the Elbe River in the German Bight. Fine-grained sediment that enters the embayment is deposited offshore and along the borders of the estuary forming broad tidal flats transitioning into wetland areas (2–10s km wide) (Table 2). Riverine-derived coarse sediment is transported offshore, reworked by tidal currents, and deposited as linear sand shoals, also called tidal ridges. These features are usually subtidal with a relief of several meters, greater than 0.5 km wide, and 5–20 km in length. It appears that under very high sediment discharge conditions, such as at the mouth of the Fly River in Papua New Guinea, elongated channel-parallel islands have developed that are up to 60 km long and 5–10 km wide (Canestrelli et al., 2010).

A fundamental sedimentologic difference between wave- and tide-dominated deltas is that tide-dominated systems formed sand bodies oriented shore normal, whereas wave-dominated deltas contain sand bodies trending parallel to shore. The bulk of sand contained within river-dominated deltas is associated with distributary channels, and thus, these sand bodies are sinuous and oriented normal to oblique to shore (Rich, 1923).

Modern deltas

Formation, sea-level, and sedimentation Most of the world's deltas began forming 6–8,000 years ago (Figure 8; Stanley and Warne, 1997) during a period when the rate of sea-level rise began to slow (Ibanez et al., 2014; Maselli and Trincardi, 2013). During this initial period, estuaries filled with sediment and were gradually transformed into prograding deltaic shorelines. In the present regime of accelerating sea-level rise, the survival of deltas is related to their ability to accrete vertically through overbank flooding, crevassing, and distributary switching. Because these sedimentary systems are composed of varying percentages of mud, they are susceptible to compaction and subsidence, which creates additional accommodation space and need for additional sediment to maintain their areal footprint. In a study of basal marsh peats, Tornqvist et al. (2008) showed that Mississippi delta subsidence was primarily due to compaction of Holocene muds. Contrastingly, Goodbred and Kuehl (2000) suggested that subsidence of the Ganges-Brahmaputra Delta was primarily a product of tectonics and not compaction. Given the proximity of the Indian-Eurasian plate collision, this condition is not surprising.

Deltas and human impacts Deltaic settings continue to play an important role in global society. As mentioned earlier, deltaic environments served as the culture hearth for early civilizations throughout the world because of their tremendous variety of food resources such as fish and wildlife. Moreover, the rich alluvial soils of deltaic landscapes allowed for the establishment of bountiful crops vital to the establishment and expansion of early cultures (Stanley and Warne, 1997). The historic impact of human development on delta evolution is revealed in a comprehensive study of European deltas by Maselli and Trincardi (2013). They showed that the Ebro, Rhone, Po, and Danube deltas underwent rapid progradation during the Roman Empire as a result of widespread deforestation and extensive agricultural use of the land, leading to greater sediment contribution to rivers and delta growth. They also reported widespread delta erosion and shoreline retreat coincided with the collapse of the Roman Empire in 400 AD when lands were reforested leading to lower sediment influx to rivers. A second period of broad delta erosion was caused by extensive dam building between 1960–1990 (Maselli and Trincardi, 2013). Similar comprehensive studies and findings have also been reported for the Mississippi River delta (Kessel, 2003; Tweel and Turner, 2012) and Ebro (Guillen and Palanques, 1997).

The extent of delta erosion on a global scale is summarized in a thoughtful paper by Giosan et al., 2014. They show that dam construction, levee building, and other human impacts have led to a drastic reduction in sediment conveyance to many of the world's largest deltas including the Nile, Indus, Mississippi, Danube, Yangtze, and Rhone with sediment loads having decreased by 60–98 % during the past century. They state that most of the world's largest and moderately sized deltas have experienced significant erosion and are predicted to succumb to rising sea-level during the next century, as they do not receive sufficient sediment to maintain their elevation and areal extent (Giosan et al., 2014). The demise of deltaic lands can be hastened by subsidence brought on by the subsurface withdrawal of hydrocarbons (e.g., Mississippi; Morton and Bernier, 2010) or water (e.g., Po; Teatini et al., 2011). A detailed analysis by Blum and Roberts (2009) of the Mississippi River sediment load indicates that the amount of sediment coming down the river is vastly inadequate to maintain the existing delta plain above sea-level, and therefore, they predict that most of the delta will drown by 2100. Contributing to erosion and drowning, it also should be noted that 11 of the largest 25 deltas are impacted by hurricanes (cyclones, typhoons) (Bianchi and Allison, 2009) and that 85 % of the world's largest deltas have experienced severe flooding during the last two centuries (Syvitski et al., 2009).

The partial solution to erosion and the eventual drowning of deltas will require a multiple-faceted approach undoing certain human alterations and implementing new ones. For example, dam removal will allow sediment behind dams to be transported to the coast, reestablishing natural sediment conveyance systems. Since 1912, 1,150 dams have been removed along American rivers, including 51 dams in 2013 (American Rivers, 2013). Other solutions to delta land loss might involve: (1) initiating crevasse splays in the lower delta plain, (2) creating channel diversions to fill large open water areas (e.g., Atchafalaya delta formation, Mississippi delta), or (3) diverting the entire river flow to form new distributary delta lobes.

Summary and conclusions

Deltas are sedimentary deposits that exist at the mouths of rivers and whose morphology and facies architecture are a product of riverine sediment discharge (type and load) and processes within the receiving basin, including waves, tides, storms, and tectonic regime. Modern deltas began forming 6–8,000 years ago coincident with a slowing in the rate of SLR when river-derived sediment filled estuaries producing shoreline progradation. Riverine-dominated deltas are characterized by elongate sandy distributary deposits with intervening muddy bays and wetlands; wave-dominated deltas tend to be cuspate to lobate in shape having numerous prograding chenier/beach ridges; tide-dominated deltas are characterized by funnel-shaped embayments with linear islands and/or subtidal sand shoals and onshore mud flats and wetlands. Deltas of the world are under siege due to their overall low-lying elevation and high rates of relative SLR brought on by subsidence and accelerating global SLR. In addition, dams have drastically reduced sediment contribution to most deltas, leading to widespread shoreline erosion. The fact that deltas coincide with dense human population centers means that they are particularly susceptibility to RSLR, especially because many deltas are impacted by periodic cyclones and severe flooding.

Bibliography

Alonso, B., Field, M. E., Gardner, J. V., and Maldonado, A., 1990. Sedimentary evolution of the pliocene and pleistocene Ebro margin, northeastern Spain. *Marine Geology*, **95**, 313–331.

American Rivers, 2013. *American Rivers Strategic Plan 2014–2018*, Washington, DC, 13 pp. Accessed at http://www.americanrivers.org/assets/pdfs/AmericanRiversStrategicPlan2014-2018.pdf?506914.

Bates, C. C., 1953. Rational theory of delta formation. *American Association of Petroleum Geologists Bulletin*, **37**, 2119–2161.

Bianchi, T. S., and Allison, M. A., 2009. Large-river delta-front estuaries as natural "recorders" of global environmental change. *Proceedings of the National Academy of Sciences*, **106**, 8085–8092.

Blum, M. D., and Roberts, H. H., 2009. Drowning of the Mississippi Delta due to insufficient sediment supply and global sea-level rise. *Nature Geoscience*, **2**, 488–491.

Boyd, R., Suter, J. R., and Penland, S., 1989. Sequence stratigraphy of the Mississippi Delta. *Gulf Coast Association of Geological Societies Transactions*, **39**, 331–340.

Canestrelli, A., Fagherazzi, S., Defina, A., and Lanzoni, S., 2010. Tidal hydrodynamics and erosional power in the Fly River delta, Papua New Guinea. *Journal of Geophysical Research, Earth Surface*, **115**, F04033. 14 pp.

Clark, R., Georgiou, I. Y., FitzGerald, D. M., 2013. An evolutionary model of a retrograding subdeltaic distributary of a river-dominated system. In *Association for the Sciences of Limnology and Oceanography*, Abstracts with program, 12107, New Orleans, LA.

Coleman, J. M., 1981. *Deltas: Processes of Deposition and Models for Exploration*. Minnesota: Burgess.

Coleman, J. M., and Prior, D. B., 1980. Deltaic sand bodies. In *American Association of Petroleum Geologists Continuing Education Course*. Tulsa, OK: American Association of Petroleum Geologists. 171 p.

Coleman, J. M., and Wright, L. D., 1975. Modern river deltas: variability of processes and sand bodies. In Broussard, M. L. (ed.), *Deltas: Models for Exploration*. Houston, TX: Houston Geological Society, pp. 99–146.

Davis, R. A., Jr., 1983. *Depositional Systems: A Genetic Approach to Sedimentary Geology*. Englewood Cliffs: Prentice Hall.

Donaldson, A. C., Martin, R. H., and Kanes, W. H., 1970. Holocene Guadalupe delta of Texas gulf coast. In Morgan, J. P. (ed.), *Deltaic Sedimentation: Modern and Ancient*. Tulsa, OK: Society of Economic Paleontologists and Mineralogists Special Publication, Vol. 15, pp. 107–137.

Edmonds, D. A., and Slingerland, R. L., 2007. Mechanics of river mouth bar formation: implications for the morphodynamics of delta distributary networks. *Journal of Geophysical Research*, **112**, F02034, doi:10.1029/2006JF000574.

Elliot, T., 1986. Deltas. In Reading, H. G. (ed.), *Sedimentary Environments*. Oxford: Blackwell Scientific Publications, pp. 113–154.

Esposito, C. R., Georgiou, I. Y., and Kolker, A. K., 2013. Hydrodynamic and geomorphic controls on mouth bar evolution. *Geophysical Research Letters*, **8**, 1540–1545.

Fairbanks, R. G., 1989. A 17,000-year glacio-eustatic sea level record: influence of glacial melting rates on the younger dryas event and deep-ocean circulation. *Nature*, **342**, 637–642.

Fenster, M. S., FitzGerald, D. M., Belknap, D. F., Knisley, B. A., Gontz, A., and Buynevich, I. V., 2005. Controls on estuarine sediment dynamics in Merrymeeting Bay, Kennebec River estuary, Maine, U.S.A. In FitzGerald, D. M., and Knight, J. (eds.), *High Resolution Morphodynamics and Sedimentary Evolution of Estuaries*. New York: Springer, pp. 173–194.

Fisher, W. L., Brown, L. F. Jr., Scott, A. J., and McGowen, J. H., 1969. *Deltas Systems in the Exploration for Oil and Gas: A Research Colloquium*. University of Texas/Bureau of Economic Geology. Austin, Texas

Frazier, D. E., 1967. Recent deltaic deposits of the Mississippi River: their development and chronology. *Gulf Coast Association of Geological Societies Transactions*, **27**, 287–315.

Galloway, W. E., 1975. Process framework for describing the morphologic and stratigraphic evolution of deltaic depositional systems. In Broussard, M. L. (ed.), *Deltas: Models for Exploration*. Houston, TX: Houston Geological Society, pp. 87–96.

Gilbert, G. K., 1885. The topographical features of lake shores. *United States Geological Survey Annual Report*, **5**, 104–108.

Giosan, L., Constantinescu, S., Clift, P. D., Tabrez, A. R., Danish, M., and Inam, A., 2006. Recent morphodynamics of the Indus delta shore and shelf. *Continental Shelf Research*, **26**, 1668–1684.

Giosan, L., Syvitski, J., Constantinescu, S., and Day, J., 2014. Climate change: protect the world's deltas. *Nature*, **516**, 31–33.

Goodbred, S. L., Jr., and Kuehl, S. A., 2000. The significance of large sediment supply, active tectonism, and eustasy on margin sequence development: late quaternary stratigraphy and evolution of the Ganges–Brahmaputra delta. *Sedimentary Geology*, **133**, 227–248.

Guillén, J., and Palanques, A., 1997. A historical perspective of the morphological evolution in the lower Ebro river. *Environmental Geology*, **30**, 174–180.

Ibáñez, C., Day, J. W., and Reyes, E., 2014. The response of deltas to sea-level rise: natural mechanisms and management options to adapt to high-end scenarios. *Ecological Engineering*, **65**, 122–130.

Inman, D. L., and Nordstrom, C. E., 1971. On the tectonic and morphologic classification of coasts. *Journal of Geology*, **97**, 1–21.

Kessel, R. H., 2003. Human modifications to the sediment regime of the lower Mississippi River flood plain. *Geomorphology*, **56**, 325–334.

Kindinger, J. L., 1988. Seismic stratigraphy of the Mississippi-Alabama shelf and upper continental slope. *Marine Geology*, **83**, 79–94.

Kindinger, J. L., Balson, P. S., and Flocks, J. G., 1994. Stratigraphy of the mississippi-alabama shelf and the mobile river incised-valley system. In Dalrymple, R. W., Boyd, R., and Zaitlin, B. A. (eds.), *Society of Economic Paleontologists and Mineralogists*. Special Publication, Tulsa, Oklahoma, 51, pp. 83–95.

Kostaschuk, R. A., 1985. River mouth processes in a fjord delta, British Columbia Canada. *Marine Geology*, **69**, 1–23.

Maldonado, A., 1975. Sedimentation, stratigraphy and development of the Ebro Delta, Spain. In Broussard, M. L. (ed.), *Deltas: Models for Exploration*. Houston, TX: Houston Geological Society, pp. 311–338.

Maselli, V., and Trincardi, F., 2013. Man made deltas. *Scientific Reports*, **3**, 1926, doi:10.1038/srep01926.

McEwen, M. C., 1969. Sedimentary facies of the modern Trinity delta. In Lanlford, R. R., and Rogers, J. J. W. (eds.), *Holocene Geology of the Galveston Bay Area*. Houston: Houston Geological Society, pp. 53–77.

McKee, B. A., Cohen, A. S., Dettman, D. L., Palacios-Fest, M. R., Alin, S. R., and Ntungumburanye, G., 2005. Paleolimnological investigations of anthropogenic environmental change in Lake Tanganyika: II. Geochronologies and mass sedimentation rates based on 14C and 210Pb data. *Journal of Paleolimnology*, **34**, 19–29.

Meade, R. H., 1996. River-sediment inputs to major deltas. In *Sea-Level Rise and Coastal Subsidence, Coastal Systems and Continental Margins*. Dordrecht: Springer, Vol. 2, pp. 63–85.

Milliman, J. D., and Meade, R. H., 1983. World-wide delivery of river sediment to the oceans. *The Journal of Geology*, **91**, 1–21.

Moore, G. T., and Asquith, D. O., 1971. Delta: term and concept. *Geological Society of America Bulletin*, **82**, 2563–2568.

Morse, D. G., 1994. Siliciclastic reservoir rocks. In Magoon, L. B., and Dow, W. G. (eds.), *The Petroleum System – From Source to Trap*. AAPG Memoir #60, Tulsa, OK, pp. 121–139.

Morton, R. A., and Bernier, J. C., 2010. Recent subsidence-rate reductions in the Mississippi Delta and their geological implications. *Journal of Coastal Research*, **26**, 555–561.

Morton, R. A., and Price, W. A., 1987. Late quaternary sea-level fluctuations and sedimentary phases of the Texas coastal plain and shelf. In Nummedal, D., Pilkey, O. H., and Howard, J. D. (eds.), *Sea-Level Fluctuation and Coastal Evolution*. Society of Economic Paleontologists and Mineralogists. Special Publication, Tulsa, OK, 41, pp. 181–198.

Müller, G., 1966. The new Rhine delta in Lake Constance. In Broussard, M. L. (ed.), *Deltas: Models for Exploration*. Houston, TX: Houston Geological Society, pp. 107–124.

Naidu, A. S., and Mowatt, T. C., 1975. Depositional environments and sediment characteristics of the Colville and adjacent deltas, Northern Arctic Alaska. In Broussard, M. L. (ed.), *Deltas: Models for Exploration*. Houston, TX: Houston Geological Society, pp. 283–307.

Nichol, S. L., Boyd, R., and Penland, S., 1996. Sequence stratigraphy of a coastal-plain incised valley estuary: Lake Calcasieu Louisiana. *Journal of Sedimentary Research*, **66**(4), 847–857.

Orton, G. J., and Reading, H. G., 1993. Variability of deltaic processes in terms of sediment supply, with particular emphasis on grain size. *Sedimentology*, **40**, 475–512.

Penland, S., Boyd, R., and Suter, J. R., 1988. Transgressive depositional systems of the Mississippi delta plain: a model for barrier shoreline and shelf sand development. *Journal of Sedimentary Research*, **58**, 932–949.

Reineck, H. E., and Singh, I. B., 1973. *Depositional Sedimentary Environments with Reference to Terrigenous Clastics*. New York: Springer.

Rich, J. L., 1923. Shoestring sands of eastern Kansas. *American Association of Petroleum Geologists Bulletin*, **7**, 103–113.

Shaw, J. B., and Mohrig, D., 2014. The importance of erosion in distributary channel network growth, Wax Lake Delta, Louisiana, USA. *Geology*, **42**, 31–34.

Stanley, D. J., and Warne, A. G., 1997. Holocene sea-level change and early human utilization of deltas. *Geological Society of American Today*, **7**, 1–7.

Suter, J. R., 1994. Deltaic coasts. In Carter, R. W. G., and Woodroffe, C. D. (eds.), *Coastal Evolution: Late Quaternary Shoreline Morphodynamics*. Cambridge: Cambridge University Press, Tulsa, pp. 87–114.

Suter, J. R., Berryhill, H. L. Jr., and Penland, S., 1987. Late quaternary sea-level fluctuations and depositional sequences, southwest louisiana continental shelf. In Nummedal, D., Pilkey, O. H., and Howard, J. D. (eds.), *Sea-Level Fluctuation and Coastal Evolution*. Society of Economic Paleontologists and Mineralogists. Special Publication, Tulsa, OK, 41, pp. 199–219.

Syvitski, J. P., Kettner, A. J., Overeem, I., Hutton, E. W., Hannon, M. T., Brakenridge, G. R., and Nicholls, R. J., 2009. Sinking deltas due to human activities. *Nature Geoscience*, **2**, 681–686.

Teatini, P., Castelletto, N., Ferronato, M., Gambolati, G., Janna, C., Cairo, E., and Bottazzi, F., 2011. Geomechanical response to seasonal gas storage in depleted reservoirs: a case study in the Po River basin, Italy. *Journal of Geophysical Research, Earth Surface*, **116**(F2), 2003–2012.

Törnqvist, T. E., Wallace, D. J., Storms, J. E., Wallinga, J., Van Dam, R. L., Blaauw, M., and Snijders, E. M., 2008. Mississippi Delta subsidence primarily caused by compaction of Holocene strata. *Nature Geoscience*, **1**, 173–176.

Tweel, A. W., and Turner, R. E., 2012. Watershed land use and river engineering drive wetland formation and loss in the Mississippi River birdfoot delta. *Limnology and Oceanography*, **57**, 18–28.

Tye, R. S., and Coleman, J. M., 1989. Depositional processes and stratigraphy of fluvially dominated lacustrine deltas: Mississippi Delta plain. *Journal of Sedimentary Petrology*, **59**, 973–996.

van Heerden, I., and Roberts, H. H., 1988. Facies development Atchafalaya delta, Louisiana: a modern bayhead delta. *American Association of Petroleum Geologists*. **72**, 439–453.

White, D., 1993. Vascular plant community development on mudcats in the Mississippi River delta, Louisiana, USA. *Aquatic Botany*, **45**, 171–194.

Wright, L. D., 1977. Sediment transport and deposition at river mouths: a synthesis. *Geological Society of America Bulletin*, **88**, 857–868.

Wright, L. D., 1985. River deltas. In Davis, R. A. (ed.), *Coastal Sedimentary Environments*. New York: Springer, pp. 1–76.

Wright, L. D., and Coleman, J. M., 1973. Variations in morphology of major river deltas as functions of ocean wave and river discharge regimes. *American Association of Petroleum Geologists Bulletin*, **57**, 370–398.

Zhou, L., Liu, J., Saito, Y., Zhang, Z., Chu, H., and Hu, G., 2014. Coastal erosion as a major sediment supplier to continental shelves: example from the abandoned Old Huanghe (Yellow River) delta. *Continental Shelf Research*, **82**, 43–59.

Cross-references

Beach Processes
Coasts
Estuary, Estuarine Hydrodynamics
Lagoons
Marine Regression
Marine Sedimentary Basins
Sediment Transport Models
Shelf
Waves

DEPLETED MANTLE

Andreas Stracke
Westphalian Wilhelms-University, Institute of Mineralogy, Münster, Germany

Definition

Depleted mantle: The depleted mantle is the part of Earth's mantle from which basaltic melt has been extracted in one or multiple melting events at, for example, mid-ocean ridges, hot spots, or island arcs. Elements that do not fit into the crystal lattice of mantle minerals, the so-called incompatible elements, are preferentially incorporated into the melt and thus removed from the mantle, leaving the latter *depleted* with respect to these elements. This residual depleted mantle amounts to anywhere between 30 % and 100 % of Earth's mantle.

Basic facts about Earth's mantle

Earth's mantle extends from the core-mantle boundary at 2900 km depth, the Gutenberg discontinuity, to the base of Earth's crust, the Mohorovičić discontinuity. With about two-thirds of Earth's mass, the mantle is the largest silicate reservoir of our planet. Earth's most abundant elements, O, Mg, Si, Fe, Ca, and Al are the main constituents of Earth's mantle (Palme and O'Neill, 2014). The mantle is mostly peridotitic in composition, with varying mineral assemblage that adapts to the large range of pressure-temperature conditions in the mantle. At upper mantle pressures, a typical mantle peridotite with a density of 3300 kg per m^3 is made up of about 55 % olivine, 35 % ortho- and clinopyroxene, and 5–10 % Al-bearing phase such as plagioclase, spinel, or garnet. At the core-mantle boundary, the mantle is mostly composed of the high-pressure minerals perovskite and ferropericlase and has a density of about 5600 kg per m³.

Although the mantle is solid over its entire depth range, it convects in response to density gradients between Earth's cold surface and hot interior. Thermal mantle convection is the primary agent of cooling of the Earth and transports hot mantle from the deep interior into shallower regions. At mid-ocean ridges, shallow mantle rises passively in response to plate spreading and melts partially through adiabatic decompression. The latter process generates the oceanic crust at mid-ocean ridges and is responsible for about 80 % of Earth's volcanism. At hot spots, where hot buoyant mantle rises from the deeper mantle, some melting occurs because temperatures exceed those of the ambient shallow mantle. At island arcs, basaltic and andesitic crust is generated by fluid-induced melting of the mantle wedge overlying the subducted slab. The newly generated basaltic crust is eventually subducted at convergent plate margins, establishing a large-scale plate tectonic cycle of ocean crust generation and its return into the mantle. In this way, the mantle has evolved continuously since the onset of plate tectonics.

Formation and evolution of the "depleted" mantle

During partial mantle melting at mid-ocean ridges, hot spots, and island arcs, elements that do not fit into the crystal lattices of mantle minerals, the so-called incompatible elements (e.g., Cs, Rb, Ba, Th, U, Nb, Ta, K, Sr, Zr, Pb, Hf, Ti, and rare earth elements), are extracted preferentially from the mantle and incorporated into the newly formed basaltic crust. Left behind is a residual mantle whose mineralogical and chemical composition has changed by melting at shallow depths (usually <100 km depth), that is, it has become *depleted* in incompatible elements. This residual mantle is therefore referred to as the "depleted mantle."

The depleted mantle (DM) has undergone a complex geological evolution. Depleted mantle that has formed at mid-ocean ridges, hot spots, or island arcs is transported with the overlying oceanic crust and eventually subducted at convergent plate margins. Although part of the oceanic crust, consisting of igneous rocks and their sedimentary cover, is removed during subduction, most of the oceanic crust is recycled back into the mantle. Once introduced into the deeper mantle, the subducted oceanic crust and mantle become stretched and intermingled by mantle convection. In this way, crustal material that is enriched in incompatible elements is reincorporated into the DM. These processes counteract depletion by melt extraction and re-enrich the DM in incompatible elements.

Over geologic time, the DM is thus expected to have evolved into an assemblage of materials that is mineralogically and chemically heterogeneous, consisting of variably depleted peridotitic mantle and some remnants of recycled crustal rocks (e.g., van Keken et al., 2014). Owing to the different composition, size, and distribution of its individual components (e.g., residual peridotitic mantle, oceanic crust, continental crust, and marine sediment), the DM is heterogeneous on a range of scales, from the 1000-km scale of ocean basins (Dupré and Allègre, 1983; Hart, 1984) down to the cm scale of single hand specimens. Each individual component of the DM can also vary compositionally: the composition of the residual peridotitic mantle varies because of differing degrees of depletion (melt extraction) or refertilization (metasomatism) (e.g., Johnson et al., 1990; Bodinier and Godard, 2014). Crustal components (oceanic crust, continental crust, and marine sediment) are compositionally variable as a result of different conditions of formation (e.g., Plank, 2014; Rudnick and Gao, 2014; White and Klein, 2014) and varying extents of hydrothermal alteration (e.g., Staudigel, 2014) and modification during subduction (e.g., Bebout, 2014; Ryan and Chauvel, 2014).

Composition of the depleted mantle

Direct investigation of mantle composition is a complicated task because mantle rocks are rarely exposed at the surface. Mantle peridotites are sporadically found on the ocean floor, in ophiolite complexes and peridotite massifs on the continents, and as xenoliths in lavas (e.g., Bodinier and Godard, 2014). Fragments of the DM recovered from the ocean floor, abyssal peridotites, are often highly altered through interaction with extracted melt and seawater. Peridotites from ophiolites, peridotite massifs, and mantle xenoliths are also generally affected by melt-rock interaction (metasomatism) and weathering and are sometimes metamorphosed (Bodinier and Godard, 2014). Owing to their scarcity, small scale, and prevalent modification by secondary processes (metamorphism, metasomatism, alteration), it is difficult to derive a representative composition of the DM by direct investigation of the available mantle samples. Current estimates of average DM composition (Salters and Stracke, 2004; Workman and Hart, 2005) are therefore mostly indirectly inferred from mantle-derived melts: basalts generated at mid-ocean ridges (MORB).

Mantle melting integrates over different components of the DM, and one strategy of constraining DM composition is to use compositional parameters that are invariant in MORB (Salters and Stracke, 2004; Workman and Hart, 2005). MORBs, however, have highly heterogeneous chemical compositions (Hofmann, 2014; Arevalo and McDonough, 2010; Jenner and O'Neill, 2012; Gale et al., 2013) that reflect the different conditions during melting and the compositional diversity of their DM source. Estimates of DM composition thus span a range of compositions, mirroring the compositional variability observed in MORB. In detail, there may be a systematic bias between MORB and the DM because some components melt preferentially and may thus become overrepresented in the melt (MORB) relative to their mantle source, the DM (Stracke, 2012).

Although estimating the composition of the DM is thus complex, current estimates should adequately account for the compositional range of the DM. Even more difficult than estimating its composition is assessing the proportion of the DM relative to the total mantle. Current estimates are based mainly on geochemical mass balances or "box" models and range anywhere from 30 % to 100 % (e.g., Allègre et al., 1983; Boyet and Carlson, 2006; DePaolo, 1980; Hofmann, 1988; Jacobsen and Wasserburg, 1979; O'Nions et al., 1979; Zindler and Hart, 1986). Constraining which fraction of the mantle corresponds to DM hence remains a considerable challenge.

The composition and relative proportion of DM, however, are key parameters for inferring Earth's bulk composition, for assessing global geochemical cycles, and for understanding the compositional evolution and differentiation of our planet and will thus continue to be a focus of geoscience research.

Summary

The depleted mantle forms by partial melting of the upper mantle at mid-ocean ridges and hot spots. Partial melting "depletes" the residual mantle in incompatible elements and preferentially incorporates them into the generated melt. Hence mantle from which basaltic melt has been extracted in one or multiple melting events is defined as "depleted mantle." The large-scale plate tectonic cycle of mantle melting, ocean crust generation, and subsequent subduction at convergent margins forms the backbone of silicate Earth evolution. Knowledge of the composition and mass fraction of the depleted mantle is therefore key for assessing global geochemical cycles and for understanding the compositional evolution and differentiation of our planet.

Bibliography

Allègre, C. J., Hart, S. R., and Minster, J. F., 1983. Chemical structure and evolution of the mantle and the continents determined by inversion of Nd and Sr isotopic data. II. Numerical experiments and discussion. *Earth and Planetary Science Letters*, **66**, 191–213.

Arevalo, R., and McDonough, W. F., 2010. Chemical variations and regional diversity observed in MORB. *Chemical Geology*, **271**, 70–85.

Bebout, G. E., 2014. 4.20 – Chemical and isotopic cycling in subduction zones. In Holland, H. D. and Turekian, K. K. (ed.), *Treatise on Geochemistry*, 2nd edn. Oxford: Elsevier, pp. 703–747.

Bodinier, J. L., and Godard, M., 2014. 3.4 – Orogenic, ophiolitic, and abyssal peridotites. In Holland, H. D. and Turekian, K. K. (ed.), *Treatise on Geochemistry*, 2nd edn. Oxford: Elsevier, pp. 103–167.

Boyet, M., and Carlson, R. W., 2006. A new geochemical model for the Earth's mantle inferred from ^{146}Sm-^{142}Nd systematics. *Earth and Planetary Science Letters*, **262**, 505–516.

DePaolo, D. J., 1980. Crustal growth and mantle evolution: inferences from models of element transport and Nd and Sr isotopes. *Geochimica et Cosmochimica Acta*, **44**, 1185–1196.

Dupré, B., and Allègre, C. J., 1983. Pb-Sr isotope variation in Indian Ocean basalts and mixing phenomena. *Nature*, **303**, 142–146.

Gale, A., Dalton, C. A., Langmuir, C. H., Su, Y., and Schilling J.-G., (2013), The mean composition of ocean ridge basalts, Geochem. Geophys. Geosyst., **14**, doi:10.1029/2012GC004334.

Hart, S. R., 1984. A large-scale isotope anomaly in the southern-hemisphere mantle. *Nature*, **309**, 753–757.

Hofmann, A. W., 1988. Chemical differentiation of the Earth – the relationship between mantle, continental-crust, and oceanic-crust. *Earth and Planetary Science Letters*, **90**, 297–314.

Hofmann, A. W., 2014. 3.3 – Sampling mantle heterogeneity through oceanic basalts: isotopes and trace elements. In Turekian, H. D. H. K. (ed.), *Treatise on Geochemistry*, 2nd edn. Oxford: Elsevier, pp. 67–101.

Jacobsen, S. B., and Wasserburg, G. J., 1979. Mean age of mantle and crustal reservoirs. *Journal of Geophysical Research*, **84**, 7411–7427.

Jenner, F. E., and H. St. C. O'Neill (2012), Analysis of 60 elements in 616 ocean floor basaltic glasses, Geochem. Geophys. Geosyst., 13, Q02005, doi:10.1029/2011GC004009.

Johnson, K. T. M., Dick, H. J. B., and Shimizu, N., 1990. Melting in the oceanic upper mantle – an ion microprobe study of diopsides in abyssal peridotites. *Journal of Geophysical Research*, **95**, 2661–2678.

O'Nions, R. K., Evensen, N. M., and Hamilton, P. J., 1979. Geochemical modeling of mantle differentiation and crustal growth. *Journal of Geophysical Research*, **84**, 6091–6101.

Palme, H., and O'Neill, H. S. C., 2014. 3.1 – Cosmochemical estimates of mantle composition. In Holland, H. D. and Turekian, K. K. (ed.), *Treatise on Geochemistry*, 2nd edn. Oxford: Elsevier, pp. 1–39.

Plank, T., 2014. 4.17 – The chemical composition of subducting sediments. In Holland, H. D. and Turekian, K. K. (ed.), *Treatise on Geochemistry*, 2nd edn. Oxford: Elsevier, pp. 607–629.

Rudnick, R. L., and Gao, S., 2014. 4.1 – Composition of the continental crust. In Holland, H. D. and Turekian, K. K. (ed.), *Treatise on Geochemistry*, 2nd edn. Oxford: Elsevier, pp. 1–51.

Ryan, J. G., and Chauvel, C., 2014. 3.13 – The subduction-zone filter and the impact of recycled materials on the evolution of the mantle. In Holland, H. D. and Turekian, K. K. (ed.), *Treatise on Geochemistry*, 2nd edn. Oxford: Elsevier, pp. 479–508.

Salters, V. J. M., and A. Stracke (2004), Composition of the depleted mantle, Geochem. Geophys. Geosyst., 5, Q05004, doi:10.1029/2003GC000597.

Staudigel, H., 2014. 4.16 – Chemical fluxes from hydrothermal alteration of the oceanic crust. In Holland, H. D. and Turekian, K. K. (ed.), *Treatise on Geochemistry*, 2nd edn. Oxford: Elsevier, pp. 583–606.

Stracke, A., 2012. Earth's heterogeneous mantle: a product of convection-driven interaction between crust and mantle. *Chemical Geology*, **330–331**, 274–299.

van Keken, P. E., Ballentine, C. J., and Hauri, E. H., 2014. 3.14 – Convective mixing in the Earth's mantle. In Holland, H. D. and Turekian, K. K. (ed.), *Treatise on Geochemistry*, 2nd edn. Oxford: Elsevier, pp. 509–525.

White, W. M., and Klein, E. M., 2014. 4.13 – Composition of the oceanic crust. In Holland, H. D. and Turekian, K. K. (ed.), *Treatise on Geochemistry*, 2nd edn. Oxford: Elsevier, pp. 457–496.

Workman, R. K., and Hart, S. R., 2005. Major and trace element composition of the depleted mantle. *Earth and Planetary Science Letters*, **231**, 53–72.

Zindler, A., and Hart, S., 1986. Chemical geodynamics. *Annual Review of Earth and Planetary Sciences*, **14**, 493–571.

Cross-references

Active Continental Margins
Ancient Plate Tectonics
Gabbro
Hot Spots and Mantle Plumes
Intraoceanic Subduction Zone
Intraplate Magmatism

Island Arc Volcanism, Volcanic Arcs
Layering of Oceanic Crust
Magmatism at Convergent Plate Boundaries
Mid-ocean Ridge Magmatism and Volcanism
Oceanic Plateaus
Oceanic Rifts
Oceanic Spreading Centers
Ophiolites
Peridotites
Subduction
Subduction Erosion

DIATOMS

Arto Miettinen
Norwegian Polar Institute, Fram Centre, Tromsø, Norway

Definition
Diatoms are single-celled, eukaryotic, photosynthetic algae of the class Bacillariophyceae.

Introduction
Diatoms are microscopic (usually 2–200 μm) algae characterized by a compound silica cell wall, called a frustule composed of two interlocking thecae, each of which consists of a valve and a series of girdle bands. Diatoms are the most species-rich group of algae (more than 200 genera and up to 100,000 species) and are found in almost every aquatic environment. The diatom taxonomy is based on the shape and ornamentation of the valves including a variety of pores, processes, ribs, spines, marginal ridges, elevations, and other distinguishable features. Both planktonic and benthic forms exist. Further details of diatom biology can be found in, e.g., Round et al. (1990) and Not et al. (2012).

Diatoms play a significant role in global climate via the global carbon cycle. They constitute the major part of phytoplankton and are major primary producers in the oceans producing as much as 40 % of the annual organic carbon in the ocean. They also contribute a major fraction of the downward flux of particulate organic carbon and, thus, a major fraction of the export of CO_2 to deep seawater (Hopkinson et al. 2011).

In the oceans, diatoms occur from the tropics to the poles, but are most abundant in polar to temperate, nutrient-rich regions, where silicic acid is not a limiting factor. Sea surface temperature (SST), sea-ice conditions, light and nutrient levels, stability of the surface water layer, and grazing are the most important factors determining the distribution and abundance of diatoms in the ocean surface waters. Diatoms live in the photic zone in the uppermost surface waters (commonly 0–50 m, max. 200 m), because they are dependent on light for photosynthesis. Therefore, mostly planktonic species occur in the open ocean and benthic species occur only in the nearshore environments.

Diatom applications
Di/atoms are used extensively in paleoenvironmental studies, especially in paleoceanography and paleolimnology. For example, diatoms have proved to be an excellent paleoclimatic tool for SST reconstructions in the Nordic Seas and the North Atlantic. Diatoms are good paleoindicators as they are a species-rich group of algae and living diatoms have specific tolerances to temperature, salinity and pH levels, and other environmental parameters. The high diversity of diatoms makes them particularly useful in high-latitude oceans where calcareous microfossils are often poorly preserved. Therefore, diatoms are one of the foremost tools available in the biostratigraphic studies and paleoclimatic reconstructions. In this entry, modern diatom-based reconstruction methods and topical results are summarized from the North Atlantic region.

Diatom biostratigraphy
Fossil diatom records reveal that the first diatoms, radial centrics, appeared in the Jurassic, multipolar centrics in the Early Cretaceous, pennates in the Late Cretaceous, and the first raphid pennates at 55 Ma in the Paleogene (Not et al. 2012). Since diatom species have undergone rapid evolution through the Cenozoic, it is possible to establish a diatom biostratigraphy for the North Atlantic; eight Pleistocene diatom datum events were identified and tied to the oxygen isotope record and paleomagnetic stratigraphy in studies on sites ODP 919 and ODP 983 (Koç and Flower 1998; Koç et al. 1999) in the subpolar North Atlantic:

- The first occurrence (FO) of *Pseudoeunotia doliolus* at 1.89 Ma
- The FO of *Proboscia curvirostris* at 1.53 Ma
- The FO of *Neodenticula seminae* at 1.25 Ma
- The last occurrence (LO) of *Neodenticula seminae* at 0.84 Ma
- The LO of *Nitzschia fossilis* at 0.68 Ma
- The LO of *Nitzschia reinholdii* at 0.6 Ma
- The LO of *Proboscia curvirostris* and *Thalassiosira jouseae* at 0.3 Ma

This biostratigraphy can be updated by the ninth event, "the second occurrence (SO) of *Neodenticula seminae* in the 1990s." *N. seminae* is a marine planktonic diatom which belongs to the modern assemblage of the subarctic North Pacific and its high-latitude marginal seas. The FO of *N. seminae* in the North Atlantic is an indicator of the cooling, which started at 1.26 Ma, leading to the transition from the dominance of 41 ka cycles in the climate record to the dominance of 100 ka cycles and intensified Northern Hemisphere glaciations (Koç et al. 1999). Conditions in the North Atlantic were too warm for subarctic *N. seminae* before 1.26 Ma, after which conditions cooled to subarctic environment favorable for the species. *N. seminae* thrived in the North Atlantic until conditions turned too severe with perennial sea-ice cover leading to the disappearance of the species at 0.84 Ma (Koç et al. 1999). After an absence of more than 0.8 Ma, *N. seminae* reappeared in the Labrador Sea in the late 1990s (Reid et al. 2007) and was found for the first time in the northern Nordic Seas with a widespread modern distribution in the mid-2000s (Miettinen et al. 2013). The introduction of *N. seminae* in the Nordic Seas can be associated with an increased influence of Pacific waters via the Arctic, probably due to diminished Arctic sea ice and/or changed ocean circulation in the Arctic Ocean. This might suggest the initiation of a unique climatic transition of the scale seen during the mid-Pleistocene transition though now the appearance may indicate a transition to warmer conditions (Miettinen et al. 2013).

Diatom-based sea-ice reconstructions

Sea ice has a prominent effect on climate and thus paleo-sea-ice reconstructions are important for climate modeling and for setting the recent sea-ice variability into a geological context. For the sea-ice reconstructions, three diatom-based methods are available.

The qualitative method is based on the statistically (Imbrie-Kipp method and a Q-mode factor analysis) defined occurrence of diatom assemblages according to their relation to modern surface hydrography (Andersen et al. 2004). The dominant marine diatom species in the specific sea-ice assemblage are *Fragilariopsis cylindrus*, *F. oceanica*, *Bacterosira bathyomphala*, *Thalassiosira hyalina*, *T. nordenskioeldii*, and resting spores of *T. gravida* (Koç-Karpuz and Schrader 1990). For example, the sea-ice cover of the Nordic Seas for the last 14,000 years was reconstructed by this method (Koç et al. 1993). In addition to the previous marine assemblage, some brackish-water diatom species indicating sea ice occur in the Arctic, such as *Melosira arctica* (Boetius et al. 2013 and references therein).

The transfer function for May sea-ice cover is based on diatoms preserved in surface sediment samples from the North Atlantic and the associated modern sea-ice concentrations, using the maximum likelihood method (Justwan and Koç 2008). The results attained by this method are comparable to the ones obtained by other paleo-proxies (e.g., dinoflagellates) indicating the reliability of this new technique for the quantitative reconstruction of May sea ice. The method has been used, e.g., for sea-ice reconstruction from the Norwegian Sea showing a marked decrease in May sea-ice concentration at the transition between the Younger Dryas and the Holocene (Justwan and Koç 2008).

Recently, the highly branched isoprenoid lipid, IP_{25}, is an approach based on long carbon chains produced by diatoms living in association with seasonal sea ice (Belt et al. 2007). IP_{25} has emerged as a potential sea-ice-specific proxy for the past sea-ice cover as it seems to reliably reconstruct trends in observed sea-ice concentrations (Weckström et al. 2013).

Diatom-based sea surface temperature (SST) reconstructions

Diatom SST transfer functions

Diatom-based quantitative SST reconstructions from the North Atlantic have been generated using data sets consisting of modern diatom assemblages of the surface sediment samples and modern SST data from the North Atlantic (e.g., Andersen et al. 2004; Jiang et al. 2005; Miller and Chapman 2013) and the transfer function methods, such as weighted averaging partial least square (WA-PLS; ter Braak and Juggins 1993), Imbrie-Kipp transfer function method (IKM; Imbrie and Kipp 1971), maximum likelihood (ML), the modern analog technique (MAT; Hutson 1980), and the artificial neural network approach (ANN; Malmgren and Nordlund 1997).

Recently, Miettinen et al. (unpublished) expanded the most commonly used diatom SST calibration data set of 139 surface sediment samples with 52 diatom species (Andersen et al. 2004) to 184 surface samples and refined the SST equation. This new WA-PLS diatom transfer function has a root mean square error (RMSE) of 0.8 °C, a coefficient of determination between observed and inferred values (r^2) of 0.96, and a maximum bias of 0.6 °C. These statistical criteria indicate that diatoms show good analogue relations to modern oceanic conditions and are linearly related to SST.

The WA-PLS method is a commonly used transfer function method for SST reconstructions because it shows the best statistical fit between observed SST and estimated SST through the temperature range of modern calibration data. The reconstructions indicate summer temperatures; as for diatom transfer functions, August SST gives the best match (Berner et al. 2008). The WA-PLS method can be regarded as the unimodal-based equivalent of multiple linear regression, i.e., a diatom species has an optimal abundance along the environmental gradient being investigated. The method uses several components in the final transfer function. These components are selected to maximize the covariance between the environmental variables to be reconstructed to improve predictive power of the method (ter Braak and Juggins 1993).

Late Holocene SSTs in the North Atlantic and the Nordic seas

Recently, several quantitative high-resolution SST reconstructions have been generated from the North Atlantic in order to investigate the variability of the SST pattern, the surface currents, and the forcing factors behind the variability. For example, the most recent high-resolution diatom August SST records show persistent opposite climate trends toward warming in the subpolar North Atlantic (Miettinen et al. 2011, 2012) and cooling in the Norwegian Sea (Berner et al. 2011) during the late Holocene (Figure 1). An apparent tendency to coherent antiphased SST variations between the regions is also revealed for the multicentennial time scales implying an SST seesaw

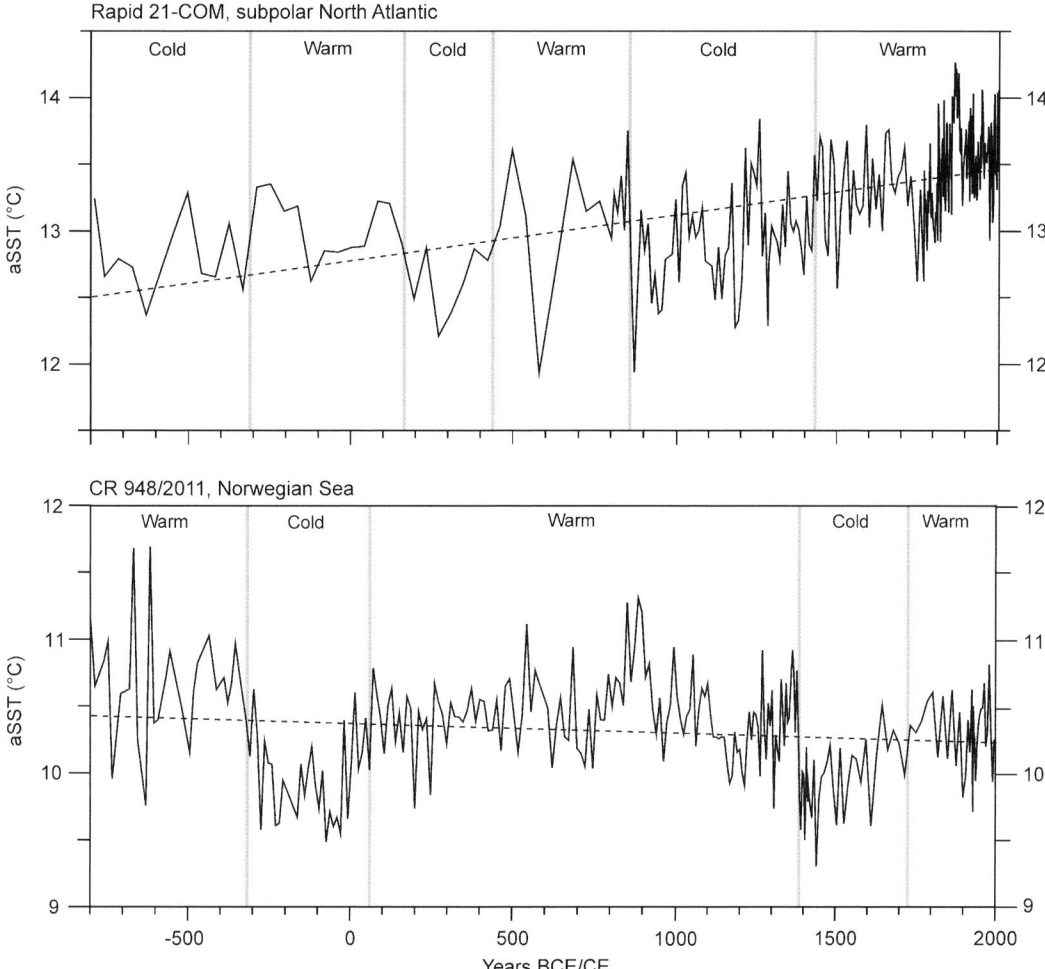

Diatoms, Figure 1 Sea surface temperature records (aSST) from the subpolar North Atlantic (core Rapid 21-COM) and the Norwegian Sea (core CR948/2011) for the last 2,800 years (adapted from Miettinen et al. 2012). *Stippled lines* refer to linear trends of the aSST records.

between the northern subpolar North Atlantic and the Norwegian Sea (Miettinen et al. 2012). This seesaw might have had a strong effect on two major climate anomalies in northwest Europe during the past millennium: the Medieval Warm Period (MWP) and the Little Ice Age (LIA). During the MWP, warming of the sea surface in the Norwegian Sea occurred in parallel with cooling in the northern subpolar North Atlantic, whereas the opposite pattern emerged during the LIA. Coupled changes in SST between the subpolar North Atlantic and the Norwegian Sea indicate common driving forces behind the observed variability. The emerging spatial pattern of changes resembles the one predicted by modeling studies and is associated with rapid changes in the regimes of the North Atlantic overturning circulation (AMOC) (Hofer et al. 2011). The observed aSST seesaw between the subpolar North Atlantic and the Norwegian Sea could be a surface expression of the variability of the eastern and western branches of the AMOC with a possible amplification through atmospheric feedback (the North Atlantic Oscillation, NAO).

In the subpolar North Atlantic, the last 200 years represent the warmest SST period of the late Holocene. The last three decades since the 1970s show a strong decadal SST variability with several years among the warmest of the record but also the coldest years since the 1820s. These cold years correspond with the documented great salinity anomalies in the North Atlantic suggesting that the fluxes of cold, low-salinity waters from the Arctic have been increased recently. Furthermore, August SST and the August NAO index (Luterbacher et al. 2002) show similar but reversed multidecadal-scale variability indicating a close coupling between the oceanic and atmospheric patterns corresponding with the general NAO/SST pattern in the North Atlantic (Wanner et al. 2001; Miettinen et al. 2011).

Summary and conclusions

In marine geosciences, diatoms are one of the foremost micropaleontological tools available for biostratigraphic, paleoceanographic, and climatic studies. Their high sensitivity to surface ocean conditions makes them invaluable especially for the quantitative reconstructions of SST, sea ice, sea-ice margin, distribution of surface water masses, variability of the surface currents, and ocean–atmosphere interactions. Although quantitative reconstructions are the topical methods, highly valuable environmental data can be extracted from the diatom stratigraphic data as the statistical errors are always minimized in the qualitative data.

The diatom research has progressively concentrated on high-resolution studies and reconstructions providing the most useful data for the climate research; the numerical SST reconstructions are routinely archived into the scientific databases (e.g., NOAA) with open access to the scientific community. In extending the record of climate variability beyond the era of instrumental measurements, diatom records help to infer information about the mechanisms, forcing factors, and spatial and temporal ranges of climate variations.

Bibliography

Andersen, C., Koç, N., Jennings, A., and Andrews, J. T., 2004. Nonuniform response to the major surface currents in the Nordic Seas to insolation forcing: implications for the Holocene climate variability. *Paleoceanography*, **19**, PA2003, doi:10.1029/2002PA000873.

Belt, S. T., Massé, G., Rowland, S. J., Poulin, M., Michel, C., and LeBlanc, B., 2007. A novel chemical fossil of palaeo sea ice: IP$_{25}$. *Organic Geochemistry*, **38**, 16–27.

Berner, K. S., Koç, N., Divine, D., Godtliebsen, F., and Moros, M., 2008. A decadal-scale Holocene sea surface temperature record from the subpolar North Atlantic constructed using diatoms and statistics and its relation to other climate parameters. *Paleoceanography*, **23**, PA2210, doi:10.1029/2006PA001339.

Berner, K. S., Koç, N., Godtliebsen, F., and Divine, D., 2011. Holocene climate variability of the Norwegian Atlantic Current during high and low solar insolation forcing. *Paleoceanography*, **26**, PA2220, doi:10.1029/2010PA002002.

Boetius, A., Albrecht, S., Bakker, K., Bienhold, C., et al., 2013. Export of algal biomass from the melting Arctic sea ice. *Science*, **339**, 1430–1432.

Hofer, D., Raible, C. C., and Stocker, T. F., 2011. Variations of the Atlantic meridional overturning circulation in control and transient simulations of the last millennium. *Climate of the Past*, **7**, 133–150.

Hopkinson, B. M., Dupont, C. L., Allen, A. E., and Morel, M. M., 2011. Efficiency of the CO_2-concentrating mechanism of diatoms. *Proceedings of the National Academy of Sciences*, **108**, 3830–3837.

Hutson, W. H., 1980. The Agulhas current during the Late Pleistocene: analysis of modern faunal analogs. *Science*, **207**, 64–66.

Imbrie, J., and Kipp, N. G., 1971. A new micropaleontological method for quantitative micropaleontology: application to a late Pleistocene Caribbean core. In Turekian, K. (ed.), *Late Cenozoic Glacial Ages*. New Haven: Yale University Press, pp. 71–181.

Jiang, H., Eiriksson, J., Schultz, M., Knudsen, K.-L., and Seidenkrantz, M.-S., 2005. Evidence for solar forcing of sea-surface temperature on the North Icelandic Shelf during the late Holocene. *Geology*, **33**, 73–76.

Justwan, A., and Koç, N., 2008. A diatom based transfer function for reconstructing sea ice concentrations in the North Atlantic. *Marine Micropaleontology*, **66**, 264–278.

Koç, N., and Flower, B., 1998. High-resolution Pleistocene diatom biostratigraphy and paleoceanography of Site 919 from the Irminger Basin. In Saunders, A. D., Larsen, H. C., and Wise, S. W., Jr. (eds.), *Proceedings of the Ocean Drilling Program, Scientific Results*. College Station: Ocean Drilling Program, Vol. 152, pp. 209–219.

Koç, N., Jansen, E., and Haflidason, H., 1993. Paleoceanographic reconstruction of surface ocean conditions in the Greenland, Iceland and Norwegian Seas through the last 14 ka based on diatoms. *Quaternary Science Reviews*, **12**, 115–140.

Koç, N., Hodell, H., Kleiven, D. A., and Labeyrie, L., 1999. High-resolution Pleistocene diatom biostratigraphy of Site 983 and correlations to isotope stratigraphy. In Jansen, E., Raymo, M., and Blum, P. (eds.), *Proceedings of the Ocean Drilling Program, Scientific Results*. College Station: Ocean Drilling Program, Vol. 162, pp. 51–62.

Koç-Karpuz, N., and Schrader, H., 1990. Surface sediment diatom distribution and Holocene paleo-temperature variations in the Greenland, Iceland and Norwegian Seas through the last 14 ka based on diatoms. *Paleoceanography*, **5**, 557–580.

Luterbacher, J., Xoplaki, E., Dietrich, D., Jones, P. D., Davies, T. D., Portis, D., Gonzalez-Rouco, J. F., von Storch, H., Gyalistras, D., Casty, C., and Wanner, H., 2002. Extending North Atlantic oscillation reconstructions back to 1500. *Atmospheric Science Letters*, **2**, 114–124.

Malmgren, B. A., and Nordlund, U., 1997. Application of artificial neural networks to paleoceanographic data. *Palaeogeography, Palaeoclimatology, Palaeoecology*, **136**, 359–373.

Miettinen, A., Koç, N., Hall, I. R., Godtliebsen, F., and Divine, D., 2011. North Atlantic sea surface temperatures and their relation to the North Atlantic Oscillation during the last 230 years. *Climate Dynamics*, **36**, 533–543.

Miettinen, A., Divine, D., Koç, N., Godtliebsen, F., and Hall, I. R., 2012. Multicentennial variability of the sea surface temperature gradient across the subpolar North Atlantic over the last 2.8 kyr. *Journal of Climate*, **25**, 4205–4219.

Miettinen, A., Koç, N., and Husum, K., 2013. Appearance of the Pacific diatom *Neodenticula seminae* in the northern Nordic Seas – an indication of changes in Arctic sea ice and ocean circulation. *Marine Micropaleontology*, **99**, 2–7.

Miller, K. R., and Chapman, M. R., 2013. Holocene climate variability reflected in diatom-derived sea surface temperature records from the subpolar North Atlantic. *The Holocene*, **23**, 882–887.

Not, F., Siano, R., Kooistra, W. H. C. F., Simon, N., Vaulot, D., and Probert, I., 2012. Diversity and ecology of eukaryotic marine phytoplankton. In Piganeau, G. (ed.), *Advances in Botanical Research, Genomic insights into the Biology of Algae*. Amsterdam: Elsevier, Vol. 64, pp. 1–53.

Reid, P. C., Johns, D. G., Edwards, M., Starr, M., Poulin, M., and Snoeijs, P., 2007. A biological consequence of reducing Arctic ice cover: arrival of the Pacific diatom *Neodenticula seminae* in the North Atlantic for the first time in 800 000 years. *Global Change Biology*, **13**, 1910–1921.

Round, F. E., Crawford, R. M., and Mann, D. G., 1990. *The Diatoms. Biology & Morphology of the Genera*. Cambridge: Cambridge University Press.

ter Braak, C. J. F., and Juggins, S., 1993. Weighted averaging partial least squares regression (WA-PLS): an improved method for reconstructing environmental variables from species assemblages. *Hydrobiologia*, **269**(270), 485–502.

Wanner, H., Bronnimann, S., Casty, C., Gyalistras, D., Luterbacher, J., Schmutz, C., Stephenson, D. B., and Xoplaki, E., 2001. North Atlantic oscillation – concepts and studies. *Surveys in Geophysics*, **22**, 321–382.

Weckström, K., Massé, G., Collins, L. G., Hanhijärvi, S., Bouloubassi, I., Sicre, M.-A., Seidenkrantz, M.-S., Schmidt, S., Andersen, T. J., Andersen, M. L., Hill, B., and Kuijpers, A., 2013. Evaluation of the sea ice proxy IP_{25} against observational and diatom proxy data in the SW Labrador Sea. *Quaternary Science Reviews*, **79**, 53–62.

Cross-references

Marine Microfossils
Modelling Past Oceans
North Atlantic Oscillation (NAO)
Paleoceanographic Proxies
Paleoceanography

DINOFLAGELLATES

Jens Matthiessen[1] and Michael Schreck[2]
[1]Alfred Wegener Institute, Helmholtz Centre for Polar and Marine Research, Bremerhaven, Germany
[2]Arctic Research Centre, Korea Polar Research Institute, Incheon, Korea

Definition

Dinoflagellates (Greek, δινη, *dino*, "whirl" and Latin *flagellum*, "whip, scourge") are unicellular protists that have two distinctive flagella during at least part of their life cycle.

Introduction

Dinoflagellates are a biologically complex group of protists that comprise planktonic, meroplanktonic, and benthic species. They have different modes of nutrition making it difficult to attribute the group as a whole to animals or plants. Some species produce toxins that impact human health through consumption of contaminated seafood or water or aerosol exposure (Hackett et al., 2004). Toxic algae blooms are increasingly documented over the past decades (http://oceanservice.noaa.gov/hazards/hab) and may have a considerable economic impact. Extensive blooms of dinoflagellates may cause a coloration of water known as red tide. Some species are an important source of bioluminescence (Hackett et al., 2004). Since it has been finally accepted more than 50 years ago that the fossil hystrichospheres in Mesozoic and Cenozoic sediments are cysts of dinoflagellates (e.g., Dale, 1983), dinoflagellate cysts have become important in stratigraphy and understanding past environments.

General characteristics

Dinoflagellates ($c.$ 2 to 2,000 μm) are primarily unicellular eukaryotes but some species are colonial, and chain formation is common (Taylor et al., 2008). Organisms are assigned to the division Dinoflagellata (kingdom Alveolata) based on possession of one or more of a suite of characters including an amphiesma, two dissimilar flagella, and a unique type of nucleus (Taylor, 1987). The amphiesma is the complex outer region of the cell wall, usually containing a single layer of flattened vesicles. These amphiesmal vesicles may contain thecal plates usually composed of cellulose (thecate or armored forms). Six basic types of tabulation, i.e., arrangement of amphiesmal vesicles, are known (Fensome et al., 1993). Dinoflagellates without thecal plates are called athecate, naked, or unarmored.

A distinctive flagellar apparatus consisting of a coiled transverse flagellum within a cingular groove and a posterior flagellum within a sulcal groove enables a spiral motion and to move freely in the water column (Taylor, 1987). Swimming speeds range from centimeters to a few meters per hour. Vertical migration is a result of endogenous rhythms. This motility permits to optimize position in the euphotic zone to a limited extent to take full advantage of light and nutrients and avoids sinking under very stable water conditions.

The unique type of nucleus, the dinokaryon, is characterized by chromosomes that remain condensed between cell divisions, and a lack of histones. Dinoflagellates may have special vacuole-like structures of unknown function called pusules (usually two per cell). The accessory pigment peridinin that enables energy transfer may be present in photosynthetic cells.

Ecology of dinoflagellates

Dinoflagellates live in all aquatic environments and have been observed both in snow and sea ice (Taylor et al., 2008). They are most abundant in shallow marine settings but also occur in fully oceanic environments. The biogeographic distribution is primarily determined by temperature, and the same species occur within similar climatic zones in both hemispheres. True endemism is rare, and some species have a bipolar distribution. More than 2,300 species have been described (Gómez, 2012) of which more than 180 are marine benthic (Hoppenrath et al., 2014) and 350 freshwater species (Mertens et al., 2012).

Dinoflagellates have diverse feeding mechanisms and utilize various modes of nutrition: they may be phototrophic, heterotrophic, and mixotrophic and may be free living, endosymbionts, or parasites (Jeong et al., 2010). Most species are probably mixotrophic or heterotrophic feeding on diverse preys such as bacteria, picoeukaryotes, nanoflagellates, diatoms, other dinoflagellates, heterotrophic protists, and metazoans or ingest particulate matter or dissolved substances. They are important in planktonic marine food webs since they may have both a considerable grazing impact on natural populations and are excellent prey for mixotrophic protists and metazoans. Together with diatoms and coccolithophores, dinoflagellates are among the most prominent marine primary producers today, thus playing an important role in the global carbon cycle.

Dinoflagellate cysts (dinocysts)

Dinoflagellates may form different types of cysts during various stages of their complex life cycle that involve asexual and sexual and motile and nonmotile stages (Taylor, 1987). Resting cysts represent a dormant stage in which normal life processes are greatly reduced. They are part of the sexual reproduction cycle (hypnozygotes) but may also be formed asexually (Kremp, 2013). Vegetative cysts are metabolically and/or reproductively active nonmotile cells. Temporary cysts are formed asexually as a result of adverse conditions. Digestion cysts that form after feeding are rare. Dale (1983) suggests that resting cysts may have three possible functions: protection, propagation, and dispersion. The latter may be extremely effective in introducing viable dinocysts into new geographic areas via transport in ships' ballast water (Taylor et al., 2008). Resting cysts may remain viable in sediments for centuries (Ribeiro et al., 2011).

Formation of resting cysts is a complex process and may be induced by various biotic and abiotic factors but is often related to peak abundances of the vegetative cells occurring at various times of the year (e.g., Matthiessen et al., 2005). After a mandatory dormancy period of variable length, excystment is triggered by different environmental factors. The cytoplast excysts through an opening in the cell wall, the archeopyle, which is an important feature for taxonomic definition of cyst genera. Only a minority of living dinoflagellates produce resting cysts (less than 20 %, Head, 1996). Establishing cyst-theca relations are complicated by the fact that a single dinoflagellate species may produce cyst morphotypes attributable to different cyst species (Rochon et al., 2009).

Fossil record of dinoflagellates

Dinoflagellates are preserved in the fossil record predominantly through their resting cysts. Micropaleontologists mainly focus on organic-walled cysts (i.e., consisting of a refractory biomacromolecule called dinosporin, Fensome et al., 1993) but calcified cysts are increasingly recognized in tropical to temperate environments (Zonneveld et al., 2005). Siliceous skeletons are rare. Taphonomic processes that alter dinocyst assemblages while sinking through the water column are relatively little known (Matthiessen et al., 2005), but species-selective aerobic degradation at the seafloor is an important process (Zonneveld et al., 2008).

Fossil cysts first occurred in the Triassic with a subsequent major radiation from late Triassic to mid-Jurassic, but molecular biomarkers indicate that ancestors of dinoflagellates originated in the Proterozoic (Hackett et al., 2004). Species diversity was highest in the Cretaceous declining throughout the Cenozoic and followed the global sea-level record with high diversity corresponding to intervals of high sea-level and large shelf seas (Pross and Brinkhuis, 2005). To date more than 4,000 fossil cyst species have been described.

Separate classification schemes have been developed by biologists and paleontologists for living dinoflagellates and fossil cysts before their natural relationship was discovered. Therefore, the resting cysts are often attributed to a different genus and species than their motile stage. Due to their nutritional strategies, dinoflagellates have been handled under the International Code either of Botanical or Zoological Nomenclature. Based on morphological characteristics, a phylogenetic classification at suprageneric level including both extant and fossil vegetative cells and cysts has been proposed by Fensome et al. (1993). Cyst species are generally described based on morphology, but molecular genetic studies become increasingly important to unravel the intricate phylogenetic relationship between taxa difficult to distinguish by morphology (Matsuoka and Head, 2013 and references therein). The database dinoflaJ2 comprises the classification of fossil and living dinoflagellates down to the generic rank; an index of fossil dinoflagellates at generic, specific, and intraspecific rank; and the references of original descriptions (Fensome et al., 2008b).

Ecology of extant dinoflagellate cysts

Like dinoflagellates, their cysts are found in all aquatic environments and occur even in regions with a seasonal sea-ice cover (e.g., Dale, 1996; Matthiessen et al., 2005; Mertens et al., 2012; Zonneveld et al., 2013). In general, diversity is highest in shallow marine settings (continental shelf and rise) and decreases toward the poles as a function of annual mean sea-surface temperature (Chen et al., 2011). Apart from changes in assemblage composition in relation to environmental gradients, cyst morphology (e.g., process morphology and length) may be affected by environmental stress such as temperature and salinity variability (Dale, 1996; Rochon et al., 2009; Jansson et al., 2014). The assemblage composition generally depends on both water mass properties and surface water circulation pattern. Application of multivariate ordination methods (canonical correspondence, detrended correspondence, and regression analysis) on regional and global data sets confirms a relationship to different physical (e.g., mean annual and seasonal surface temperature, salinity, upwelling intensity, sea-ice cover), biological (e.g., chlorophyll-a concentration, primary productivity), and chemical (e.g., phosphate, nitrate, and bottom water oxygen concentration) water mass properties. Ecological preferences are relatively well defined for a number of extant species (Zonneveld et al., 2013). The sensitivity for nutrient availability makes them ideal to identify areas of high productivity such as polynyas and upwelling regions and also of human-induced pollution and eutrophication if these signals can be differentiated from climate change (Dale, 2009). Biogeographic distributions of assemblages on regional and hemispheric scale have been widely used to develop transfer functions (using primarily the modern analogue technique) in order to quantitatively reconstruct sea-surface temperature and salinity, seasonal extent of sea-ice cover, and primary productivity in Quaternary sediments (e.g., de Vernal and Marret, 2007; de Vernal et al., 2007; Bonnet et al., 2012).

Paleoecology of extinct dinoflagellate cysts

Dinocysts are increasingly used for paleoenvironmental reconstructions (Pross and Brinkhuis, 2005) but the definition of ecological preferences of extinct species remains a challenge. Various combinations of actuo-paleontological, empirical, and statistical approaches including comparison with the morphology and ecology of co-occurring extant genera and species, the identification of latitudinal and onshore-offshore gradients from paleobiogeographic data, the interpretation of statistical analyses (e.g., correspondence analysis) on dinocyst distribution in relation to independent paleoenvironmental information, and the relation between dinocyst assemblages and geochemical proxies for water mass properties may yield qualitative and quantitative ecological information on, e.g., temperature, salinity, onshore-offshore gradients, bottom water oxygenation, and productivity (e.g., Versteegh and Zonneveld, 1994; Pross and Brinkhuis, 2005; De Schepper et al., 2011; Bijl et al., 2011; Masure et al., 2013; Schreck and Matthiessen, 2013). These ecological parameters are qualitatively known for some extinct species, groups of taxa or complexes of genera (e.g., Pross and Brinkhuis, 2005), and the correlation of species abundance to geochemical proxies (e.g., Mg/Ca temperatures on co-occurring planktonic foraminifera, De Schepper et al., 2011) is promising for providing quantitative data.

Biostratigraphy

Since the middle of the twentieth century, palynostratigraphy has emerged as a routine tool in both hydrocarbon exploration and academic research in Mesozoic and Cenozoic sediments, and numerous biostratigraphic zonations have been erected for Triassic to Neogene sediments (Stover et al., 1996). Dinocysts typically exhibit high abundances in neritic settings; thus, the derived stratigraphic information is complementary to that obtained from typically more offshore groups such as planktonic foraminifers, coccolithophores, and radiolarians (Pross and Brinkhuis, 2005). Significant progress has been made during the past four decades of scientific ocean drilling (DSDP, ODP, IODP) by assessing stratigraphic ranges against independent chronostratigraphic information. Recently, the focus is slowly moving from defining new zonations toward calibrating bioevents to the geological time scale on both regional (De Schepper and Head, 2008; Fensome et al., 2008a; Schreck et al., 2012) and global scales (Williams et al., 2004). This avoids the inherent problem of zonations that zones named after the same species may have different age ranges. However, few studies illustrate that dinocyst bioevents are rarely synchronous worldwide, and low-, mid-, and high-latitude bioevents should be distinguished to account for the observed latitudinal control on species ranges. Nonetheless, some bioevents are useful on regional and/or supraregional scale and enable stratigraphic correlations between different basins in the mid- and high latitudes (Schreck et al., 2012).

Conclusions

In recent years, biological and paleontological studies have provided a wealth of new information relevant for the application of recent and fossil dinocysts in marine geosciences. However, our knowledge of their ecology is still biased to coastal and shelf environments, and many open ocean regions such as the Pacific yet remain largely unexplored. The phylogenetic relationship to the motile form of many extant species is unknown, and molecular genetic studies will be particularly useful to address the long-standing question whether a single dinoflagellate species forms different cyst species. In the geological record, dinocysts are of eminent importance for paleoenvironmental interpretation and biostratigraphy in high latitudes where preservation of calcareous and biosiliceous microfossils is poor, but data from lower latitudes are required to inevitably improve independent age calibration of bioevents. This will also provide new data on the temporal and spatial distribution of fossil dinocysts, which together with calibration of species abundances to geochemical proxies for, e.g., surface temperature, will lead to a better understanding of cyst paleoecology.

Bibliography

Bijl, P. K., Pross, J., Warnaar, J., Stickley, C. E., Huber, M., Guerstein, R., Houben, A. J. P., Sluijs, A., Visscher, H., and Brinkhuis, H., 2011. Environmental forcings of Paleogene Southern Ocean dinoflagellate biogeography. *Paleoceanography*, **26**, PA1202, doi:10.1029/2009PA001905.

Bonnet, S., de Vernal, A., Gersonde, R., and Lembke-Jene, L., 2012. Modern distribution of dinocysts from the North Pacific Ocean (37–64°N, 144°E–148°W) in relation to hydrographic conditions, sea-ice and productivity. *Marine Micropaleontology*, **84–85**, 87–113.

Chen, B., Irwin, A. J., and Finkel, Z. V., 2011. Biogeographic distribution of diversity and size-structure of organic-walled dinoflagellate cysts. *Marine Ecology Progress Series*, **425**, 35–45.

Dale, B., 1983. Dinoflagellate resting cysts: "benthic plankton". In Fryxell, G. A. (ed.), *Survival Strategies of the Algae*. Cambridge: Cambridge University Press, pp. 69–144.

Dale, B., 1996. Dinoflagellate cyst ecology: modelling and geological applications. In Jansonius, J., and McGregor, D. C. (eds.), *Palynology: Principles and Applications*. Dallas: American Association of Stratigraphic Palynologists Foundation, pp. 1249–1275.

Dale, B., 2009. Eutrophication signals in the sedimentary record of dinoflagellate cysts in coastal waters. *Journal of Sea Research*, **61**, 103–113.

De Schepper, S., and Head, M. J., 2008. Age calibration of dinoflagellate cyst and acritarch events in the Pliocene–Pleistocene of the eastern North Atlantic (DSDP Hole 610A). *Stratigraphy*, **5**, 137–161.

De Schepper, S., Fischer, E. I., Groeneveld, J., Head, M. J., and Matthiessen, J., 2011. Deciphering the palaeoecology of Late Pliocene and Early Pleistocene dinoflagellate cysts. *Palaeogeography Palaeoclimatology Palaeoecology*, **309**, 17–32.

de Vernal, A., and Marret, F., 2007. Organic-walled dinoflagellates: tracers of sea-surface conditions. In Hillaire-Marcel, C., and de Vernal, A. (eds.), *Proxies in Late Cenozoic Paleoceanography*. New York: Elsevier, pp. 371–408.

de Vernal, A., Rochon, A., and Radi, T., 2007. Dinoflagellates. In Elias, S. A. (ed.), *Encyclopedia of Quaternary Science*. Amsterdam: Elsevier, pp. 1652–1667.

Fensome, R. A., Taylor, F. J. R., Norris, G., Sarjeant, W. A. S., Wharton, D. I., and Williams, G. L., 1993. A classification of living and fossil dinoflagellates. *Micropaleontology*, Special Publication Number, 7, pp. 1–351.

Fensome, R. A., Crux, J. A., Gard, I. G., MacRae, R. A., Williams, G. L., Thomas, F. C., Fiorini, F., and Wach, G., 2008a. The last 100 million years on the Scotian margin, offshore eastern Canada: an event-stratigraphic scheme emphasizing biostratigraphic data. *Atlantic Geology*, **44**, 93–126.

Fensome, R. A., MacRae, R. A., and Williams, G. L., 2008b. *DINOFLAJ2, Version 1*. American Association of Stratigraphic Palynologists, Data series no. 1, http://dinoflaj.smu.ca/wiki

Gómez, F., 2012. A checklist and classification of living dinoflagellates (Dinoflagellata, Alveolata). *CICIMAR Oceánides*, **27**, 65–140.

Hackett, J. D., Anderson, D. M., Erdner, D. L., and Bhattachary, D., 2004. Dinoflagellates: a remarkable evolutionary experiment. *American Journal of Botany*, **91**, 1523–1534.

Head, M. J., 1996. Modern dinoflagellate cysts and their biological affinities. In Jansonius, J., and McGregor, D. C. (eds.), *Palynology: Principles and Applications*. Dallas: American Association of Stratigraphic Palynologists Foundation, pp. 1197–1248.

Hoppenrath, M., Murray, S. A., Chomérat, N., and Horiguchi, T., 2014. *Marine Benthic Dinoflagellates – Unveiling Their Worldwide Biodiversity*. Stuttgart: Schweizerbart. Kleine Senckenberg-Reihe, Vol. 54.

Jansson, I.-M., Mertens, K. N., Head, M. J., de Vernal, A., Londeix, L., Marret, F., Matthiessen, J., and Sangiorgi, F., 2014. Statistically assessing the correlation between salinity and morphology in cysts produced by the dinoflagellate *Protoceratium reticulatum* from surface sediments of the North Atlantic Ocean, Mediterranean–Marmara–Black Sea region, and Baltic–Kattegat–Skagerrak estuarine system. *Palaeogeography Palaeoclimatology Palaeoecology*, **399**, 202–213.

Jeong, H. J., Yoo, Y. D., Kim, J. S., Seong, K. A., Kang, N. S., and Kim, T. H., 2010. Growth, feeding, and ecological roles of the mixotrophic and heterotrophic dinoflagellates in marine planktonic food webs. *Ocean Science Journal*, **45**, 65–91.

Kremp, A., 2013. Diversity of dinoflagellate life cycles: facets and implications of complex strategies. In Lewis, J. M., Marret, F., and Bradley, L. (eds.), *Biological and Geological Perspectives of Dinoflagellates*. London: The Micropalaeontological Society, Special Publications, Geological Society, pp. 197–205.

Masure, E., Aumar, A.-M., and Vrielynck, B., 2013. Worldwide palaeogeography of Aptian and Late Albian dinoflagellate cysts: implications for sea-surface temperature gradients and palaeoclimate. In Lewis, J. M., Marret, M., and Bradley, L. R. (eds.), *Biological and Geological Perspectives of Dinoflagellates*. London: The Micropalaeontological Society, Special Publications, Geological Society, pp. 97–125.

Matsuoka, K., and Head, M. J., 2013. Clarifying cyst-motile relationships in dinoflagellates. In Lewis, J. M., Marret, M., and Bradley, L. R. (eds.), *Biological and Geological Perspectives of Dinoflagellates*. London: The Micropalaeontological Society, Special Publications, Geological Society, pp. 325–350.

Matthiessen, J., de Vernal, A., Head, M. J., Okolodkov, Y., Zonneveld, K. A. F., and Harland, R., 2005. Modern organic-walled dinoflagellate cysts and their (paleo-)environmental significance. *Paläontologische Zeitschrift*, **79**, 3–51.

Mertens, K. N., Rengefors, K., Moestrup, Ø., and Ellegaard, M., 2012. A review of recent freshwater dinoflagellate cysts: taxonomy, phylogeny, ecology and palaeocology. *Phycologia*, **51**, 612–619.

Pross, J., and Brinkhuis, H., 2005. Organic-walled dinoflagellate cysts as paleoenvironmental indicators in the Paleogene; a synopsis of concepts. *Paläontologische Zeitschrift*, **79**, 53–59.

Ribeiro, S., Berge, T., Lundholm, N., Andersen, T. J., Abrantes, F., and Ellegaard, M., 2011. Phytoplankton growth after a century of dormancy illuminates past resilience to catastrophic darkness. *Nature Communications*, doi:10.1038/ncomms1314.

Rochon, R., Lewis, J., Ellegaard, M., and Harding, I. C., 2009. The *Gonyaulax spinifera* (Dinophyceae) "complex": perpetuating the paradox? *Review of Palaeobotany and Palynology*, **155**, 52–60.

Schreck, M., Matthiessen, J., and Head, M. J., 2012. A magnetostratigraphic calibration of Middle Miocene through Pliocene dinoflagellate cyst and acritarch events in the Iceland Sea (Ocean Drilling Program Hole 907A). *Review of Palaeobotany and Palynology*, **187**, 66–94.

Schreck, M., and Matthiessen, J., 2013. *Batiacasphaera micropapillata*: palaeobiogeographic distribution and palaeocological implications of a critical Neogene species complex. In Lewis, J. M., Marret, M., and Bradley, L. R. (eds.), *Biological and Geological Perspectives of Dinoflagellates*. London: The Micropalaeontological Society, Special Publications, Geological Society, pp. 301–314.

Stover, L. E., Brinkhuis, H., Damassa, S. P., de Verteuil, L., Helby, R. J., Monteil, E., Partridge, A. D., Powell, A. J., Riding, J. B., Smelror, M., and Williams, G. L., 1996. Mesozoic-Tertiary dinoflagellates, acritarchs and prasinophytes. In Jansonius, J., and McGregor, D. C. (eds.), *Palynology: Principles and Application*. Dallas: American Association of Stratigraphic Palynologists Foundation, pp. 641–750.

Taylor, F. J. R., 1987. *The Biology of Dinoflagellates*. Oxford: Blackwell Scientific Publications, p. 785.

Taylor, F. J. R., Hoppenrath, M., and Saldarriaga, J. F., 2008. Dinoflagellate diversity and distribution. *Biodiversity and Conservation*, **17**, 407–418.

Versteegh, G. J. M., and Zonneveld, K. A. F., 1994. Determination of (paleo-)ecological preferences of dinoflagellates by applying detrended and canonical correspondence analysis to Late Pliocene dinoflagellate cyst assemblages of the south Italian Singa section. *Review of Palaeobotany and Palynology*, **84**, 181–199.

Williams, G. L., Brinkhuis, H., Pearce, M. A., Fensome, R. A., and Weegink, J. W., 2004. Southern Ocean and global dinoflagellate cyst events compared: index events for the late Cretaceous – Neogene. In Exon, N. F., Kennett, J. P., and Malone, M. J. (eds.), *Proceedings of the Ocean Drilling Program, Scientific Results 189*. College Station: Ocean Drilling Program, doi:10.2973/odp.proc.sr.189.107.2004.

Zonneveld, K. A. F., Meier, K. J. S., Esper, E., Siggelkow, D., Wendler, I., and Willems, H., 2005. The (palaeo)environmental significance of modern calcareous dinoflagellate cysts: a review. *Paläontologische Zeitschrift*, **79**, 61–77.

Zonneveld, K. A. F., Versteegh, G., and Kodrans-Nsiah, M., 2008. Preservation and organic chemistry of late Cenozoic organic-walled dinoflagellate cysts: a review. *Marine Micropaleontology*, **68**, 179–197.

Zonneveld, K. A. F., Marret, F., Versteegh, G. J. M., Bogus, K., Bonnet, S., Bouimetarhan, I., Crouch, E., de Vernal, A., Elshanawany, R., Edwards, L., Esper, O., Forke, S., Grøsfjeld, K., Henry, M., Holzwarth, U., Kielt, J.-F., Kim, S.-Y.,

Ladouceur, S., Ledu, D., Chen, L., Limoges, A., Londeix, L., Lu, S.-H., Mahmoud, M. S., Marino, G., Matsouka, K., Matthiessen, J., Mildenhal, D. C., Mudie, P., Neil, H. L., Pospelova, V., Qi, Y., Radi, T., Richerol, T., Rochon, A., Sangiorgi, F., Solignac, S., Turon, J.-L., Verleye, T., Wang, Y., Wang, Z., and Young, M., 2013. Atlas of modern dinoflagellate cyst distribution based on 2405 data points. *Review of Palaeobotany and Palynology*, **191**, 1–197.

Cross-references

Deep-sea Sediments
Marine Microfossils
Paleoceanography
Paleoceanographic Proxies

DRIVING FORCES: SLAB PULL, RIDGE PUSH

Carolina Lithgow-Bertelloni
Department of Earth Sciences, University College London, London, UK

Definition

Plate Driving Forces: The forces that drive the motions of tectonic plates at the surface.

Slab Pull: The force exerted by the weight of the subducted slab on the plate it is attached to.

Ridge Push: The pressure exerted by the excess height of the mid-ocean ridge.

Introduction

The history of the development of plate tectonics is centrally tied to the question of what drives plate motions. This has been the case since the failure of Wegener's ideas about polflucht to explain continental drift to the seminal papers by Elsasser (1969), Solomon and Sleep (1974), and Forsyth and Uyeda (1975) on slab pull and ridge push. This contribution cannot possibly review all the seminal and historical publications that led to the establishment of the terms and concepts of *slab pull* and *ridge push* as major plate driving forces; instead, it gives a brief historical introduction and then focuses on modern views of the plate-mantle system and what remains to be understood.

Historical overview

Beginning studies on plate driving forces all used parameterizations of forces acting at plate boundaries (e.g., Solomon and Sleep, 1974; Harper, 1975; Forsyth and Uyeda, 1975; Chapple and Tullis, 1977). This choice was partly out of necessity and partly a historical remnant of the development of plate tectonics with a contraposition of surface and interior. The perspective necessarily was a geological one, focusing on what could be seen and observed. Seminal early work focused on mantle convection and showed that the top thermal boundary layer of the convecting system could have plate-like properties and velocities (Turcotte and Oxburgh, 1967). This division between studies of plate driving forces from the plate or mantle perspective remains today. Early studies were partly influenced by the need to explain the apparent contradiction between a fluid, weak mantle, and transmitting stresses to strong plates, a corollary to Harold Jeffreys' objections to continental drift. These were the days before global seismic tomography (Dziewonski et al., 1977) and there was an understandable reluctance to correlate deep and unseen mantle structure with the surface. The existence of a low-velocity zone (the asthenosphere), which coincided (Anderson, 2007) with a very weak, very low-viscosity region, furthered the notion that stresses from mantle convection could not be transmitted to the lithosphere and effectively decoupled plates from the mantle. All return flow generated by the subduction of slabs was limited to convection in the asthenosphere (Forsyth and Uyeda, 1975). As global tomographic models emerged (Dziewonski et al., 1977), the suggestion that global geoid anomalies were correlated with past subduction (Chase and Sprowl, 1983) opened the door for studies that linked interior flow and surface motions. These ideas culminated in the classic papers of Hager and O'Connell (1981) on plate motions and driving forces in the context of mantle flow. The clear correlation between the long-wavelength pattern of subduction, mantle flow, and the geoid (Richards and Engebretson, 1992) further motivated the link between convection and the interior, leading to integrated studies of mantle flow to predict plate motions incorporating ever larger complexity in rheology and boundary conditions (Ricard and Vigny, 1989; Lithgow-Bertelloni and Richards, 1995; Becker and O'Connell, 2001; Conrad and Lithgow-Bertelloni, 2002; Becker, 2006; Stadler et al., 2010; van Summeren et al., 2012).

Basic principles

Plate driving forces in the traditional formulation are driving and resisting forces acting at the edges and the base of lithospheric plates, balanced so that there is no-net torque on any given plate or the lithosphere as a whole (Solomon and Sleep, 1974). In other words, plates are in dynamical equilibrium. The assumption of no-net torque is justified by the high viscosity of mantle and lithosphere, precluding acceleration and inertial forces. The balance of forces is best illustrated in the formulation of Forsyth and Uyeda (1975) shown in Figure 1. The main driving forces are slab pull (F_{SP}) and ridge push (F_{RP}), and forces like mantle drag (F_{DF} and F_{CD}) can both oppose and drive plate motions. Suction (F_{SU}), first identified by Elsasser (1971), pulls the overriding plate toward the trench. There are significant resisting forces, slab resistance (F_{SR}), transform resistance (F_{TF}), and collisional resistance (F_{CR}). For the sake of brevity, we only focus on slab pull and ridge push.

Slab pull is defined as $F_{SP} = \Delta\rho g \sin\theta V_s$ where $\Delta\rho$ is the density contrast between the slab and the surrounding

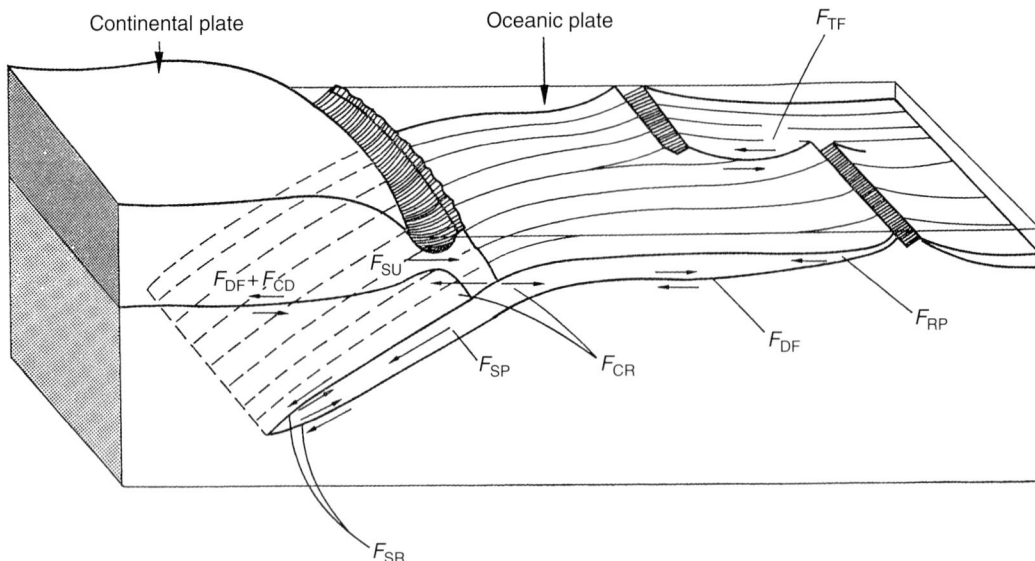

Driving Forces: Slab Pull, Ridge Push, Figure 1 Cartoon of plate driving forces illustrating the parameterization of forces at the edges and base of the plates (reproduced from Forsyth and Uyeda (1975)).

mantle, g the gravitational acceleration, θ is the dip angle of the slab, and V_S the volume of the slab. The slab pulls on the plate and is resisted by viscous drag along the interface (F_{SR}). This definition implies a strong slab capable of transmitting stresses to the plate and therefore acting as a stress guide (Elsasser, 1969).

Ridge push is induced by the pressure gradient at the ridge crest due to its higher elevation with respect to the surrounding oceanic lithosphere. It is approximately given by $\Delta P = 1/2\ g(\rho_o - \rho_w)l$ where ρ_O and ρ_W are the densities of the oceanic lithosphere and water, respectively, and l the elevation of the ridge (McKenzie, 1972). It acts at the plate boundary and is resisted by mantle drag (F_{DF}) and transform resistance (F_{TF}).

Mantle drag (F_{DF}) is the shear exerted by the flowing mantle at the base of the lithosphere and considered as largely opposing plate motions as the plate drags across the surface of the mantle.

Minor forces such as collisional resistance also act to impede plate motions. Explicitly parameterizing these forces is a way to account for the differences in composition and rheology of the continental crust (Meade and Conrad, 2008).

These basic definitions of the force balance acting on the plates view the plate-mantle system as largely decoupled. This view is strongly prevalent in the geological and tectonophysics community. Many modern studies focus on individual plates and forces at the edges of plates (Coblentz et al., 1998; Bird, 1998; Govers and Meijer, 2001, among many others). In geodynamics, there is a different point of departure. Plates are not seen as rafts on a convecting fluid mantle and hence decoupled; instead, plates are mantle convection (Bercovici, 2003). In this view, plates and the mantle are inextricably linked and the same thing as plates represent the upper thermal boundary layer of the convecting system. In this framework, the only relevant physical force driving plate motions is gravity acting on lateral density variations, and the only resisting forces are viscoelastic stresses. Plate velocities are the result of the integration of the net shear tractions induced by mantle flow acting on the base of the lithosphere. The net tractions are those resulting from the driving buoyancy as well as the resistance-induced viscous flow in the mantle from plates moving at the surface. Therefore, net tractions can be both in the direction of plate motions or against it and any other direction. It is the integral of these tractions over the area of the plate that generates the plate velocity vector. This is already clear when comparing textbook estimates of slab pull (10^{13} N/m) and ridge push (10^{12} N/m) with the gravitational body force of the top thermal boundary layer of the mantle convecting system (Turcotte and Schubert, 2014).

The difficulty lies not in understanding the individual balance of driving and resisting forces but how to physically understand the interplay between the buoyancy and rheology in light of our incomplete knowledge of both: specifically, (a) complete catalog and understanding of the energy sources available to generate buoyancy (e.g., radioactive decay, cooling from the core, crustal composition) and (b) the rheology of mantle and plates. The latter, the ignorance of rheology as a function of composition, volatile content, stress, deformation, temperature, and pressure, is the most serious obstacle to a full theory of mantle convection and plate tectonics.

In this light, force balancing on individual plates by a catalog of forces acting at the edge is one way to parameterize the strength of plates, the rheology of plate boundaries, and the available buoyancy.

State of current knowledge

The importance of rheology for plates, plate boundaries, and slabs in generating realistic plate tectonics is illustrated by Stadler et al. (2010) and Alisic et al. (2012). These studies match plate motions globally and locally in the no-net-rotation reference frame with very high-resolution simulations that incorporate lithospheric thickening, upper-mantle slab temperature profiles defined by seismicity, and lower-mantle buoyancy from seismic tomography and complex nonlinear rheologies. The inclusion of nonlinear rheology and variations in rheology at the plate boundary are important for matching other observables, such as the state of stress of the slab defined by deep earthquakes. The buoyancy of the slab remains the most important driving factor, but the details of the rheology of the plate boundary, slab, and mantle are critical for matching plate velocities and defining the slab-plate coupling.

At present, the slab pull force is understood as the weight of the slab exerting a pulling stress on the plate it is attached to. The buoyancy of the slab induces flow, but the slab is strong enough to support its own weight and transmit all the stresses to its attached plate, in other words act as a stress guide (Elsasser, 1969). However, once the slab gets into the lower mantle and warms up in a more viscous lower mantle, it is no longer able to support its own weight and is supported by viscous stresses in the mantle (Conrad and Lithgow-Bertelloni, 2002; Stadler et al., 2010). This view reconciles a mantle convection view of plate driving forces with the more traditional approach of Forsyth and Uyeda (1975), where the slab pull force pulls at the edge of the slab and is resisted by viscous drag along the slab interface.

Ridge push is just another source of buoyancy acting over the entire surface area of oceanic plates. It arises from the cooling and thickening of the oceanic lithosphere as it ages (Lister, 1975). At the ridge or for young oceanic lithosphere (<20–30 My), the density contrast with respect to the mantle is negative and no flow is induced. This driving force is again balanced by viscous stresses acting at the bottom of the oceanic lithospheric plates.

The magnitude of individual forces depends on the tectonics of each plate and the local viscosity structure. Globally, the ratio of forces due to the excess buoyancy of slabs including those in the lower mantle (slab pull) and those due to lithospheric thickening (ridge push) is 70–80 % to 30–20 % (Lithgow-Bertelloni and Richards, 1995; Becker and O'Connell, 2001).

The effect of lateral variations in viscosity, deep continental roots, and parameterized slab strength (van Summeren et al., 2012) shows that the effect of roots is minimal and does not affect plate driving forces substantially, but the presence of a laterally varying asthenosphere is critical and slab buoyancy and strength (slab pull) remain the most important contribution to the force budget. Locally, however, the tractions exerted by large-scale mantle upwellings can contribute significantly to the force balance of individual plates (Lithgow-Bertelloni and Silver, 1998).

Summary and conclusions

The plate-mantle system is one and indivisible. Plates are part of mantle convection and interact with all parts of the mantle through their buoyancy and rheology. The classic view of plates decoupled from the mantle by a very low-viscosity asthenospheric layer does not conform to the present understanding. While the presence of the asthenosphere is critical to making plates and matching plate velocities (Tackley, 2000; Richards et al., 2001), it does not preclude interaction with the deeper mantle as far as the core-mantle boundary. The terms slab pull and ridge push are parameterizations of the buoyancy and strength of plates and slabs. Matching plate motions and the state of stress of plates and slabs is critically dependent on the rheology of the plates, slab, and mantle as well as the plate boundary, which remains the most uncertain physical property affecting the plate-mantle system.

Bibliography

Alisic, L., Gurnis, M., Stadler, G., Burstedde, C., and Ghattas, O., 2012. Multi-scale dynamics and rheology of mantle flow with plates. *Journal of Geophysical Research: Solid Earth (1978–2012)*, **117**, B10.

Anderson, D. L., 2007. *The New Theory of the Earth*. Cambridge: Cambridge University Press.

Becker, T. W., 2006. On the effect of temperature and strain-rate dependent viscosity on global mantle flow, net rotation, and driving forces. *Geophysical Journal International*, **167**, 943–957.

Becker, T. W., and O'Connell, R. J., 2001. Predicting plate velocities with mantle circulation models. *Geochemistry, Geophysics, Geosystems*, **2**, 1060.

Bercovici, D., 2003. The generation of plate tectonics from mantle convection. *Earth and Planetary Science Letters*, **205**, 107–121.

Bird, P., 1998. Testing hypotheses on plate-driving mechanisms with global lithosphere models including topography, thermal structure, and faults. *Journal of Geophysical Research: Solid Earth (1978–2012)*, **103**, 10115–10129.

Chapple, W. M., and Tullis, T. E., 1977. Evaluation of the forces that drive the plates. *Journal of Geophysical Research*, **82**, 1967–1984.

Chase, C. G., and Sprowl, D. R., 1983. The modern geoid and ancient plate boundaries. *Earth and Planetary Science Letters*, **62**, 314–320.

Coblentz, D. D., Zhou, S., Hillis, R. R., Richardson, R. M., and Sandiford, M., 1998. Topography, boundary forces, and the Indo-Australian intraplate stress field. *Journal of Geophysical Research: Solid Earth (1978–2012)*, **103**, 919–931.

Conrad, C. P., and Lithgow-Bertelloni, C., 2002. How mantle slabs drive plate tectonics. *Science*, **298**, 207–209.

Dziewonski, A. M., Hager, B. H., and O'Connell, R. J., 1977. Large-scale heterogeneities in the lower mantle. *Journal of Geophysical Research*, **82**, 239–255.

Elsasser, W. M., 1969. Convection and stress propagation in the upper mantle. In Runcorn, S. K. (ed.), *The Application of Modern Physics to the Earth and Planetary Interiors*. Hoboken: Wiley-Interscience, pp. 1–41.

Elsasser, W. M., 1971. Sea-floor spreading as thermal convection. *Journal of Geophysical Research*, **76**, 1101–1112.

Forsyth, D. W., and Uyeda, S., 1975. On the relative importance of driving forces of plate motions. *Geophysical Journal of the Royal Astronomical Society*, **43**, 163–200.

Govers, R., and Meijer, P. T., 2001. On the dynamics of the Juan de Fuca plate. *Earth and Planetary Science Letters*, **189**, 115–131.

Hager, B. H., and O'Connell, R. J., 1981. A simple global model of plate dynamics and mantle convection. *Journal of Geophysical Research*, **86**, 4843–4867.

Harper, J. F., 1975. On the driving forces of plate tectonics. *Geophysical Journal of the Royal Astronomical Society*, **40**, 465–474.

Lister, C. R. B., 1975. Gravitational drive on oceanic plates caused by thermal contraction. *Nature*, **257**, 663–665.

Lithgow-Bertelloni, C., and Richards, M. A., 1995. Cenozoic plate driving forces. *Geophysical Research Letters*, **22**, 1317–1320.

Lithgow-Bertelloni, C., and Silver P. G., 1998. Dynamic topography, plate driving forces and the African Superswell, Nature, **395**, 269–272.

McKenzie, D. P., 1972. Plate tectonics. In Robertson, E. C. (ed.), *The Nature of the Solid Earth*. New York: McGraw Hill, pp. 323–360.

Meade, B. J., and Conrad, C. P., 2008. Andean growth and the deceleration of South American subduction: time evolution of a coupled orogen-subduction system. *Earth and Planetary Science Letters*, **275**, 93–101.

Ricard, Y., and Vigny, C., 1989. Mantle dynamics with induced plate tectonics. *Journal of Geophysical Research*, **94**, 17543–17560.

Richards, M. A., and Engebretson, D. C., 1992. Large-scale mantle convection and the history of subduction. *Nature*, **355**, 437–440.

Richards, M. A., Yang, W. S., Baumgardner, J. R., and Bunge, H. P., 2001. Role of a low-viscosity zone in stabilizing plate tectonics: implications for comparative terrestrial planetology. *Geochemistry, Geophysics, Geosystems*, **2**, 2000GC000115

Solomon, S. C., and Sleep, N. H., 1974. Some simple physical models for absolute plate motions. *Journal of Geophysical Research*, **79**, 2557–2567.

Stadler, G., Gurnis, M., Burstedde, C., Wilcox, L. C., Alisic, L., and Ghattas, O., 2010. The dynamics of plate tectonics and mantle flow: from local to global scales. *Science*, **329**, 1033–1038.

Tackley, P. J., 2000. Mantle convection and plate tectonics: towards and integrated physical and chemical theory. *Science*, **288**, 2002–2007.

Turcotte, D. L., and Oxburgh, E. R., 1967. Finite amplitude convective cells and continental drift. *Journal of Fluid Mechanics*, **28**, 29–42.

Turcotte, D. L., and Schubert, G., 2014. *Geodynamics*. Cambridge: Cambridge University Press.

van Summeren, J., Conrad, C. P., and Lithgow-Bertelloni, C. R., 2012. The importance of slab pull and a global asthenosphere to plate motions. *Geochemistry, Geophysics, Geosystems*, **13**, Q0AK03.

Cross-references

Marine Heat Flow
Plate Motion
Plate Tectonics
Subduction

DUNES

Sara Muñoz-Vallés and Jesús Cambrollé
Department of Plant Biology and Ecology, University of Sevilla, Seville, Spain

Definition

Dunes are sedimentary landforms originated by Aeolian processes. More than simple accumulations of loose sand, dunes show particular shapes with defined windward and slip face slopes and an internal laminated structure due to their formation dynamics of sand deposition. Adequate supplies of well-sorted sands and winds capable of moving these sands for at least part of the year are required to their formation, both in coastal and continental areas. In the case of continental dunes, availability of sandy sediments originated by desertification in emerged lacustrine deposits or due to alteration of rocks can provide considerable sandy supplies. Under arid or semiarid climate conditions, where vegetation is sparse or absent, such deposits are mobilized by wind, resulting in the origination and development of dune fields such as the existing in the Sahara, Arabian, Australian, Kalahari, or Mohave deserts (García Mora, 2000). In the case of coastal dunes, they are relatively complex phenomena that form at the interface of terrestrial, oceanic, and atmospheric systems, where the influence from the sea and the vegetation cover are relevant factors to their formation and development. In addition, due to the joint action of tides and wind-driven currents, submarine sand dunes are also originated in continental shelves (Le Bot and Trentesaux, 2004), while the existence of sandy bed material in rivers also promotes the formation of fluvial sand dunes (Carling et al., 2000).

This entry focuses on coastal dunes as sedimentary landforms and ecosystem and describes their occurrence, typology, formation dynamics, and main ecological traits. In addition, some information about environmental services and conservation is included.

Occurrence of coastal dunes

Coastal dunes are distributed worldwide, from polar to tropical latitudes, comprising very diverse climates and biomes, but covering a really limited global area (Figure 1). They mainly develop in association with dissipative coasts dominated by marine winds, with an ample supply of loose, sand-sized sediment (i.e., continental shores, sandy barriers and spits), but they also occur on lake and estuary shorelines (Carter, 1988; Martínez et al., 2004). Coastal dunes present a large variety of forms (see the subsection "Dune Typology" in this entry) and dimensions (they range in size from less than 1 m to several kilometers) related with spatial and temporal variations in sediment supply and wind regime (Hesp, 2004; Martínez et al., 2004; Pye and Tsoar, 2009). For instance, dunes in geographical areas such as Australia show greater developments and sizes due to the enormous volumes of sand supply from the continental shelf, in contrast with areas from USA or Europe, where dune systems show in general smaller sizes and longitudinal developments due to the action of the littoral drift currents (Sanjaume et al., 2011). On the other hand, dune fields migrating obliquely onshore to onshore (not alongshore), in very general terms, are more common in the tropics and adjacent humid subtropics than in temperate areas, where tabular dune fields are more common (Hesp, 2004).

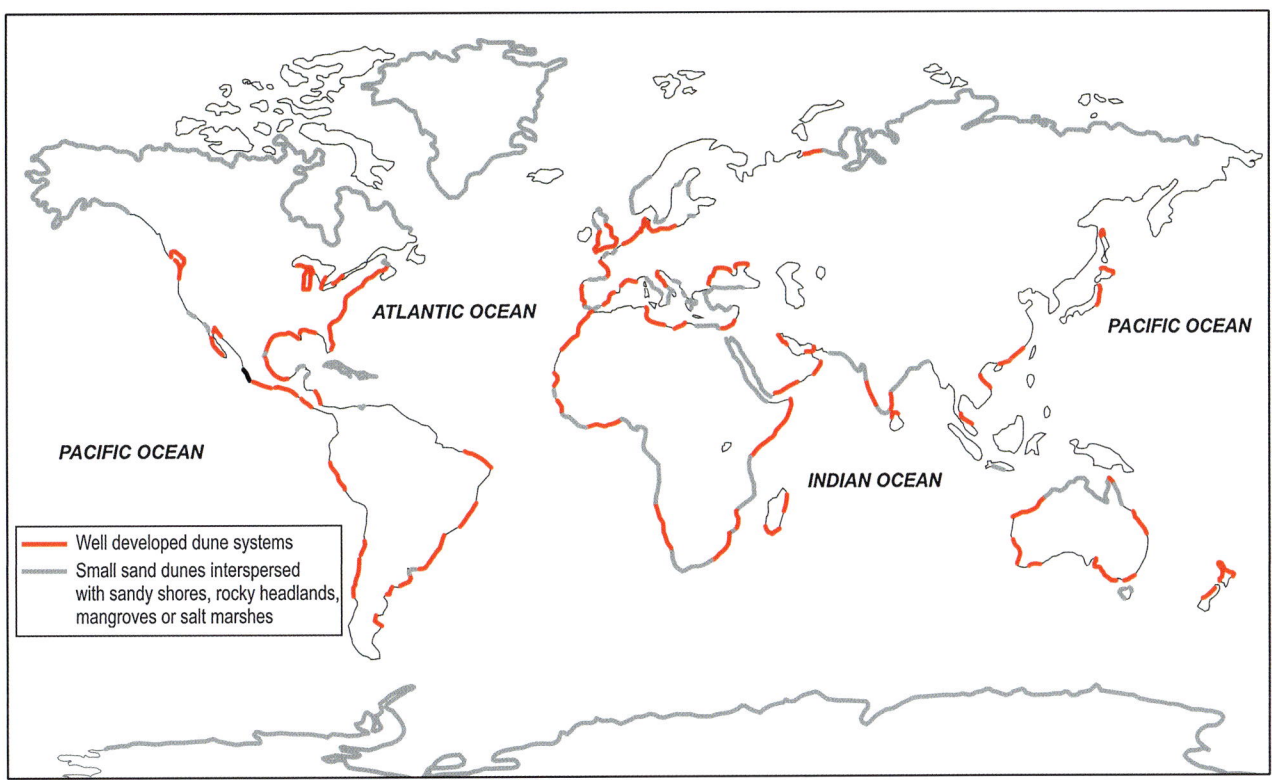

Dunes, Figure 1 Worldwide distribution of coastal dunes (Adapted from Martínez et al., 2004).

Dunes typology

The different factors involved in the formation and development of coastal dunes result in a wide variety of forms and dimensions. During the last decades, different authors have proposed diverse dune classifications based on different criteria (e.g., morphogenetic, geometric, morphoecological) (see Sanjaume et al., 2011), since the study of sand dunes is a multidisciplinary matter, carried out by researchers from different science branches such as physical (e.g., engineering, physics), life (ecology, botany, zoology), or earth (geomorphology, sedimentology, hydrology) sciences (Pye and Tsoar, 2009).

In general, the most widely spread and applied coastal dune classification has been based on morphogenetic criteria, sometimes considering vegetation traits. Here we summarize the most common dune types found in coastal environments (based on the descriptions by Goldsmith, 1985; Hesp, 2002; Pye and Tsoar, 2009; otherwise cited):

- **Embryo dunes**: also named *shadow dunes*; they are the most elementary and small coastal dunes that form in the wind shadows of clumps of vegetation or other small obstacles to sand transport by the wind. Sometimes, the term *shadow dune* is used to name the smaller forms, with decimetrical or centimetrical lengths.
- **Foredunes**: dune ridges that run parallel to the shore, more or less perpendicular to the prevailing wind. Foredunes are usually developed from several former *embryo dunes* that grow and coalesce to form an incipient *foredune*, corresponding to a higher evolutionary dune stage (see the subsection "Dune Formation" in this entry).
- **Transverse dunes**: alignments or longitudinal ridges with a gentle slope in the windward face but more pronounced in the slip face, more or less parallel to the coastline and always arranged perpendicular to the prevailing wind that originates them. It includes other types of dunes if they are arranged perpendicular to the prevailing winds, such as foredunes, barchans, parabolic, or precipitation dunes.
- **Barchans**: dunes with crescent shape and wings developed downwind.
- **Parabolic dunes**: relatively stable, U-shaped dunes with lateral lobes anchored by vegetation and the crest migrating downwind.
- **Reversing dunes**: a special type of transverse dune, showing similar slopes both in the windward and slip faces due to winds acting in opposite directions.
- **Transgressive or precipitation dunes**: mobile dunes with high dynamism and absence of vegetation cover due to high burial rates, which rush on the existing tree line and may become a burial hazard for roads and towns.
- **Climbing dunes**: dunes originated by a sand-laden wind or a migratory dune that bump into a topographic barrier – a hill or a cliff – and climb it totally or partially.

- **Cliff-top dunes**: *climbing dunes* that have remained at the top of a cliff due to erosion, usually disconnected from the original source of sand. *Falling dunes* are *cliff-top dunes* that are provided with excess of sand supply so that the dunes keep moving down the side of the cliff lee.
- **Echo dunes**: elongated mounds with a more or less linear ridge, generated near a subvertical topographic barrier and parallel to this barrier.
- **Wrap-around dunes**: dunes that surround the barrier and their shape is adapted to the obstacle. *Nebkha* – or *nabkha*, also named *hummock* or *coppice dune* – is a particular type formed in and around clumps of grass, herbs, bushes, etc. (Figure 2).

Other classifications are based on biological, ecological, and morphological traits, where different dune areas are related with different habitats colonized by characteristic plant species in each case, which often coincide with particular geomorphological situations. Corresponding to different stages of dune maturity and/or evolution, *white* (carbonate rich sands due to shell content) and *yellow* (siliceous sand) *dunes* are referred to mobile dunes with scarce or null vegetation cover, where sand surface is clearly visible, while *grey dunes* are referred to dunes that are semistabilized and stabilized (fixed) with vegetation (mainly grass), where a major organic matter content in the soil would provide a darker color to the sands (Sanjaume et al., 2011). On the other hand, the dune classification carried out by Cooper (1967) is based on factors driving the dune genetics and development, and it can be sometimes related with the dune conservation status and vulnerability. It groups dunes according to factors such as the sediment exchange with the beach: *primary dunes* derive directly from the beach sand transported by the wind and maintain an active sediment exchange with the beach while *secondary dunes* are derived from the erosion of *primary dunes*.

Dune formation

The origination and development of coastal dunes are determined by a complex interaction of factors, the main ones being wind velocity and direction, sand supply availability, influence from marine processes, moisture content, vegetation, and topography of the beach and its adjacent continental zone (Ranwell, 1972; Klijn, 1990; Hesp, 2002; Sanjaume et al., 2011). Coastal dunes are originated when the amount of sand that reaches the beach from the sea is greater than the amount that the beach loses, for instance, by erosion during storms, and this sand surplus is transported inland by the wind toward a dry zone of the beach, out of the reach of normal waves and tides (Olson, 1997).

Once the sand is deposited in a wide and dry enough area of the beach, a selection of grain sizes occurs in relation to the wind intensity and grain traits. Four types of Aeolian transport of particles take place: very coarse sand (1–2 mm diameter) and granules (2–4 mm) are rolled or slid forward (***traction*** or *creep*) and generally remain in the beach and absorb part of the wind transport energy. On the other hand, very fine grains (silt and clay) that have not been washed away by waves during the transport of beach sediments tend to be carried upward in turbulent eddies (***suspension***) and finally winnowed out of the beach and dune. As a result, middle size-range grains consisting mostly of fine to coarse sand, ranging, respectively, from 0.125 to 0.25 mm (fine sand), 0.25–0.5 mm (medium sand), and 0.5–1 mm diameter (coarse sand) remain to the formation and development of the dunes. These sand grains are transported by wind by *saltation* or by *reptation*. During **saltation**, the sand grains are lifted up into the air and describe a parabolic trajectory for a short distance before falling back to the surface. During **reptation**, particles on the ground may set in motion at wind velocities lower than those required to move them by wind alone, due to collisions with grains that are already in motion by saltation (Bagnold, 1941; Olson, 1997). This transport by wind takes place almost entirely within 0.5 m of the ground surface, with nearly 90 % of this movement within 2.5 cm of the surface (Bagnold, 1941).

Dune formation begins with the deposition of sand transported by the onshore winds, to the lee of the roughness elements by reverse flows. It is therefore necessary to set some obstacles in the beach surface, natural or artificial, such as vegetation, pebbles and stones, shells, marine debris, or some irregularities existing out of the reach of waves and tides (Ranwell, 1972; Olson, 1997). More usually, tidal litter initiates sand accumulation at the top of the backshore. This is firstly colonized by annual plant species and subsequently by perennial grasses capable of growing up through sand accretion, forming *shadow dunes*, *mounds*, and *nebkhas*. Although other inert obstacles are capable of producing sand deposition, these incipient dunes are not able to grow once the obstacle is buried. The development of *embryo dunes* and their transformation from ephemeral to permanent features is therefore favored by vegetation establishment and succession (Ranwell, 1972; Hesp, 2002). Scattered embryo dunes may coalesce to form a continuous band parallel with the strandline and would grow upward and in width to form a *foredune*. In situations of high availability of sand supply, the resulting morphology is consistent with a series of parallel foredunes of increased age from seaward to landward, separated by interdune depressions or *slacks* (Figure 3). Otherwise, a foredune regression and a dune topography dominated by washovers would occur.

Rates of beach-dune sediment exchange (particularly the balance between the beach and frontal dune sediment budgets) strongly influence the frontal dune development and morphology. In this regard, Hesp (1988) differed 5 foredune types (Fa – Fe), corresponding to gradual situations associated to the loss of vegetation cover and foredune disintegration. Under conditions of positive balance of beach sediment supply, the formation of the coastal dune continues until a reduction in sand supply

Dunes, Figure 2 Dune morphologies (*arrows* indicates wind direction): (**a**) *Shadow dunes, embryo dune,* and *foredune* in Isla Cristina Coast, Huelva (Spain); (**b**) *Barchans* in Jericoacoara Beach, Ceara State (Brazil); (**c**) *Parabolic dune* (also named *blowout*) in Doñana National Park (Spain); (**d**) *Parabolic dunes* near Foxton Beach, North Island (New Zealand); (**e**) *Transgressive* (also *climbing*) *dune* in Bolonia dune Natural Monument, Cádiz (Spain); (**f**) *Transgressive dune field* in Myall Lakes National Park (Australia); (**g**) *Cliff-top dunes* in El Asperillo Cliff Natural Monument, Doñana Natural Park (Spain); (**h**) *Hummocks* (also named *nebkhas* or *coppice dunes*) in Valdevaqueros dune Natural Monument, Cádiz (Spain) (Images **a–g** have been extracted from Google Earth, 2014).

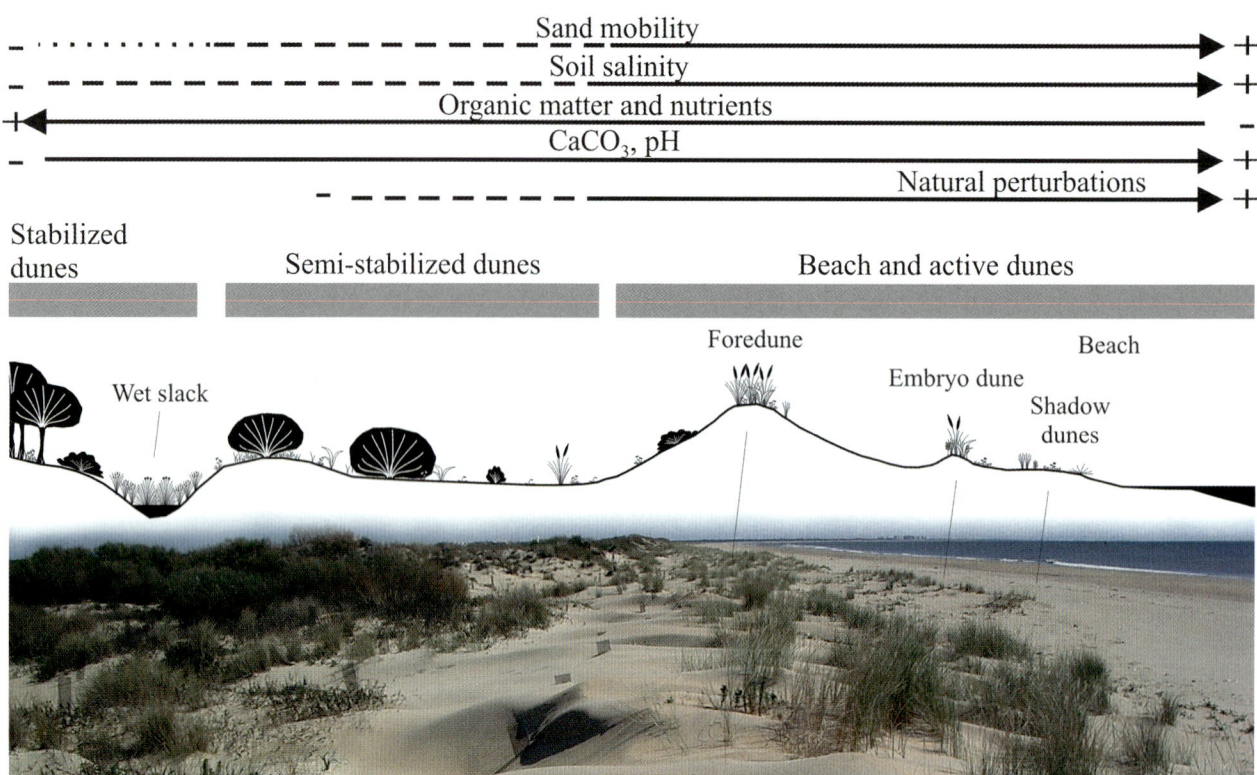

Dunes, Figure 3 Main environmental gradients on coastal dunes (adapted from Carter, 1988; Brown and McLachlan, 1990; Muñoz-Vallés, non published data).

occurs, due to a change in the beach sedimentary balance, from prograding to erosive, or to the formation of new seaward embryo dunes (Hesp and Martínez, 2007). Therefore, the frontal dunes will either grow vertically to their maximum height and then erode moving landwards (e.g., in beaches with narrow backshore and prevailing and intense enough winds on the shore) or prograde seawards through the development of new embryo dunes and foredune ridges (in beaches with broad and high backshore levels and moderate onshore winds). Under generally negative balance of beach sediment supply, embryo dunes will not form and the foredune will undergo net erosion. In extreme circumstances, the foredune will become disrupted by blowouts, leading to the development of transgressive sand sheets and dunes (Ranwell, 1972; Hesp, 2002).

Dunes as ecosystems

Coastal dunes support unique and complex ecosystems of high ecological value. They represent a harsh environment for the survival of many living beings; hence, species of sand flora and fauna are specifically adapted to such stressful conditions, mainly related to sand properties, sedimentary dynamics, the proximity to the sea, and the influence of wind (Ranwell, 1972; Chapman, 1976; Carter, 1988; Maun, 2009). One of the main traits that make these ecosystems so particular is the nature of the substrate, i.e., sands, onto which the biocenosis is developed. Apart from its mobility (burial, but also bypassing and erosion), fine to coarse siliceous sands that commonly comprise the dunes are characterized by a considerable pore size, which confers these substrates a free drainage and a pursuant low capacity to retain water or nutrients, such as phosphorous, nitrogen, or potassium (Maun, 2009). Severe leaching often occurs, which leads to substantial losses of nutrient and organic matter in the soil. This leads to a general low vegetation coverage, which is related to high air and soil temperatures and light intensities in low and medium latitudes. In addition, this characteristic low plant abundance is usually unable to develop a real soil, and only a partially formed surface horizon (*arenosol*), with poor levels of humus, is noticeable in the more mature stages of the dune development. On the other hand, the influence of the sea determines a series of environmental and biological gradients from the shoreline and landwards (Figure 3). In this regard, factors such as incidence of *salt spray*, wind and sand blasting and saltwater flooding (in the strandline), tend to decrease or disappear inland while nutrient concentration, sands stability, vegetation cover, and, in general, the ecosystem maturity increase (Ranwell, 1972; Carter, 1988; Doody, 2012). In this regard, coastal dunes have served scientific research on ecological succession through dune age gradients (*chronosequences*), and some of these studies have relevantly contributed to the elaboration of ecological theories and to a major understanding of the general functioning of ecosystems (e.g., Cowles, 1899; Clements, 1916; Salisbury, 1922; Olson, 1958; Yarranton and Morrison, 1974).

A characteristic vegetation type including unique species with specific adaptations to the harsh environmental conditions establishes on coastal dunes (Ranwell, 1972; Maun, 2009). Such vegetation shows a wide distribution through different climatic and geographic areas, both at the species level (particularly those established on the beach and active dunes) and at the genus level or their growth forms, so these ecosystems have relatively similar vegetation worldwide in temperate latitudes (Doing, 1985). Coastal dunes also support a high number of rare and endemic species (Martínez et al., 1993; Heslenfeld et al., 2004). Other plants from inland ecosystems, ruderal, and/or alien species may be also present, mainly associated with the degradation status of the dune. Vegetation is closely related to the sediment dynamics and geomorphological processes of the dune, and a continuous interaction is maintained between both elements (Ranwell, 1972; Hesp, 2002). On the other hand, the environmental gradients previously referred determine the establishment of the different plant species, particularly in the youngest areas, nearest to the sea, while biological interactions (mainly positive for the interacting species, although they may be also negative) play a relevant role in determining the structure and composition of vegetation, particularly in more stabilized and mature situations of the dune, toward the inland areas (Lichter, 1998; Maun, 1998). As a result, vegetation becomes ordered in bands from the strandline and landward, more or less parallel to the coastline and originating a particular type of landscape composed of a mosaic of habitats, both in dune systems developed on continental coasts, spits, or barrier islands.

Environmental services and conservation of coastal dunes

Coastal dune systems are not only relevant from the geomorphological, geological, or ecological point of view. They also have a recognized historical, archaeological, scenic, cultural, and socioeconomic interest (Nordstrom et al., 2000; Heslenfeld et al., 2004; Martínez et al., 2004; van der Meulen et al., 2004). In addition, they provide us with different direct and indirect benefits and resources (which are known as *environmental services*), acting as zones for livestock grazing, agriculture, sand exploitation for construction purposes and water extraction, or for human settlement, urban development, tourism, and recreational use (Carter, 1988; van der Meulen and Salman, 1996; Williams et al., 1997; Kutiel et al., 1999; Baeyens and Martínez, 2004; Heslenfeld et al., 2004). Furthermore, coastal dune systems play a relevant role as natural barriers that preserve other inland natural areas and human settlements by buffering the effects of storm waves, tides, and wind and preventing groundwater salinization by intrusion of sea water, among others (Carter, 1988; Hesp, 2000; Kidd, 2001; Gómez-Pina et al., 2002).

Nevertheless, due to an excessive exploitation of the resources and services they offer, the human demographic expansion and the relatively recent urban, engineering and industrial development, many coastal dune systems are being seriously degraded at present, some of them irreversibly altered and lost (Carter, 1988; Heslenfeld et al., 2004; Doody, 2012). Only in Europe, it is estimated that around 25 % of the coastal dune surface has been lost since 1900, while around 55 % of the remaining surface has lost its natural character, and approximately 85 % of the currently existing dunes are endangered (Heslenfeld et al., 2004). Among the main damages that coastal dunes suffer are the net loss or fragmentation of natural habitats, loss of landscape scenic quality, alteration of soil and aquifers, water and air pollution, hazards for endangered species, global biodiversity decline, and increase of the ecosystem vulnerability. On the other hand, the anthropogenic habitat degradation and destruction (mainly associated to urban development and tourism), dune stabilization and eutrophication, expansion of invasive species, and the effects of global warming and the progressive sea-level rise are identified as the main current threats to coastal dunes (Carter, 1988; Kutiel et al., 1999; Nordstrom et al., 2000; Heslenfeld et al., 2004; Martínez et al., 2004; Doody, 2012; Muñoz-Vallés and Cambrollé, 2013).

Currently, many coastal dunes are being protected under regional, national, and international initiatives, devoting a great effort in the conservation, management, and restoration of such ecosystem from different disciplines and government levels, in pursuit of a sustainable development that takes into account both the ecosystem health and the use of their services (Doody, 2012).

Summary

Coastal dunes are Aeolian landforms that form at the interface of terrestrial, oceanic, and atmospheric systems, mainly associated with dissipative coasts dominated by marine winds with an ample supply of well-sorted sands and occurring on ocean, lake, and estuary shorelines. Distributed worldwide, they comprise very diverse climates and biomes and present a large variety of forms and dimensions related with sediment supply and wind regime, where vegetation plays a relevant role. The influence of the sea determines a series of environmental and biological gradients from the shoreline and landwards. The harsh environment characterized by the scarcity of water and nutrients, substrate mobility and sand burial, incidence of salt spray and sand blasting, high air and soil temperatures and light intensities in low and medium latitudes, and saltwater flooding in the strandline determines the establishment of species of singular sand flora and fauna that are specifically adapted to such stressful conditions.

Coastal dune systems have a recognized geomorphological, geological, ecological, historical, archaeological, scenic, cultural, and socioeconomic interest and provide us with different direct and indirect benefits and resources including their role as natural barriers that buffer the effects of storms, sea, and wind. Nevertheless, they cover a really limited global area that is largely endangered at present, mainly due to human direct and indirect impacts. This has led to the development of initiatives and efforts focused on dune conservation, management, and restoration from different disciplines and government levels.

Bibliography

Baeyens, G., and Martínez, M. L., 2004. Animal life in sand dunes: from exploitation and prosecution to protection and monitoring. In Martínez, M. L., and Psuty, N. P. (eds.), *Coastal Dunes, Ecology and Conservation*. Berlin: Springer. Ecological Studies, Vol. 171, pp. 279–296.

Bagnold, R. A., 1941. The physics of blown sand and desert dunes. *Geomorphology*, **17**, 339–350.

Brown, A. C., and McLachlan, A., 1990. *Ecology of Sandy Shores*. Amsterdam: Elsevier. 328 pp.

Carling, P. A., Gölz, E., Orr, H. G., and Radecki-Pawlik, A., 2000. The morphodynamics of fluvial sand dunes in the River Rhine, near Mainz, Germany. I. Sedimentology and morphology. *Sedimentology*, **47**, 227–252.

Carter, R. W. G., 1988. *Coastal Environment. An Introduction to the Physical, Ecological and Cultural Environment*. London: Academic. 617 pp.

Chapman, V. J., 1976. *Coastal Vegetation*. Oxford: Pergamon Press. 292 pp.

Clements, F. E., 1916. *Plant Succession: An Analysis of the Development of Vegetation*. Carnegie Institute (ed.), Washington DC: Washington Publishing, Vol. 242, 512 pp

Cooper, W. S., 1967. *Coastal Sand Dunes of California*. Washington, DC: Geological Society of America. Geological Society of America Memoir, Vol. 104.

Cowles, H. C., 1899. The ecological relations of the vegetation of the sand dunes of Lake Michigan. *Botanical Gazette* **27**, 95–117, 167–202, 281–308, 361–391

Doing, H., 1985. Coastal fore-dunes zonation and succession in various parts of the World. *Vegetatio*, **61**, 65–75.

Doody, J. P., 2012. *Sand Dune Conservation, Management and Restoration*. Dordrecht: Springer. Coastal Research Library, Vol. 4.

García Mora, M. R., 2000. *Vulnerabilidad de los ecosistemas dunares costeros del Golfo de Cádiz. Tipos funcionales y estructura de la vegetación*. Doctoral thesis. Universidad de Sevilla, Sevilla

Goldsmith, V., 1985. Coastal dunes. In Davis, R. A., Jr. (ed.), *Coastal Sedimentary Environments*. New York: Springer.

Gómez-pina, G., Muñoz-Pérez, J. J., Ramírez, J. L., and Ley, C., 2002. Sand dune management problems and techniques, Spain. *Journal of Coastal Research*, **36**, 325–332.

Heslenfeld, P., Jungerius, P. D., and Klijn, J. A., 2004. European coastal dunes: ecological values, threats, opportunities and policy development. In Martínez, M. L., and Psuty, N. P. (eds.), *Coastal Dunes, Ecology and Conservation*. Berlin: Springer. Ecological Studies, Vol. 171, pp. 335–351.

Hesp, P. A., 1988. Morphology, dynamics and internal stratification of some established foredunes in southeast Australia. In Hesp, P., and Fryberger, S. G. (eds.), *Aeolian Sediments*. Sedimentary Geology, Vol. 55, pp. 17–41.

Hesp, P. A., 2000. *Coastal Sand Dunes. Form and Function*. Massey University. Rotorua: Rotorua Printers, 24 pp.

Hesp, P. A., 2002. Foredunes and blowouts: initiation, geomorphology and dynamics. *Geomorphology*, **48**, 245–268.

Hesp, P. A., 2004. Coastal dunes in the tropics and temperate regions: locations, formation, morphology and vegetation processes. In Martínez, M. L., and Psuty, N. P. (eds.), *Coastal Dunes, Ecology and Conservation*. Berlin: Springer. Ecological Studies, Vol. 171, pp. 29–49.

Hesp, P. A., and Martínez, M. L., 2007. Disturbance processes and dynamics in coastal dunes. In Johnson, E. A., and Miyanishi, K. (eds.), *Plant Disturbance Ecology*. Burlington: Academic, pp. 215–248.

Kidd, R., 2001. *Coastal Dune Management: A Manual of Coastal Dune Management and Rehabilitation Techniques*. Newcastle: NSW Department of Land and Water Conservation Coastal Unit. 96 pp.

Klijn, J. A., 1990. Dune forming factors in a geographical context. In Bakker, T. W., et al. (eds.), *Dunes of the European Coasts*. Cremlingen-Destedt: Catena. Catena Supplement, Vol. 18.

Kutiel, P., Zhevelev, H., and Harrison, R., 1999. The effect of recreational impacts on soil and vegetation of stabilized coastal dunes in the Sharon Park, Israel. *Ocean and Coastal Management*, **42**, 1041–1060.

Le Bot, S., and Trentesaux, A., 2004. Types of internal structure and external morphology of submarine dunes under the influence of tide- and wind-driven processes (Dover Strait, northern France). *Marine Geology*, **211**, 143–168.

Lichter, J., 1998. Primary succession and forest development on coastal Lake Michigan sand dunes. *Ecological Monographs*, **68**, 487–510.

Martínez, M. L., Moreno-Casasola, P., and Castillo, S., 1993. Biodiversidad Costera: Playas y Dunas. In Salazar Vallejo, S. I., and González, N. E. (eds.), *Biodiversidad Marina y Costera de México*. Chetumal: Centro de Investigaciones de Quintana Roo y CONABIO, pp. 160–181.

Martínez, M. L., Psuty, N. P., and Lubke, R. A., 2004. A perspective on coastal dunes. In Martínez, M. L., and Psuty, N. P. (eds.), *Coastal Dunes, Ecology and Conservation*. Berlin: Springer. Ecological Studies, Vol. 171, pp. 3–10.

Maun, M. A., 1998. Adaptations of plants to burial in coastal sand dunes. *Canadian Journal of Botany*, **76**, 713–738.

Maun, M. A., 2009. *The Biology of Coastal Sand Dunes*. Oxford: Oxford University Press. 265 pp.

Muñoz-Vallés, S., and Cambrollé, J., 2013. Coastal dune hazards. In Finkl, C. W. (ed.), *Coastal Hazards*. Dordrecht: Springer. The Coastal Research Library, Vol. 6, pp. 491–510.

Nordstrom, K. F., Lampe, R., and Vandemark, L. M., 2000. Reestablishing naturally functioning dunes on developed coasts. *Environmental Management*, **25**, 37–51.

Olson, J. S., 1958. Rates of succession and soil changes on southern Lake Michigan sand dunes. *Botanical Gazette*, **119**, 125–170.

Olson, J. S., 1997. Organic and physical dune building. In Van der Maarel, E. (ed.), *Dry Coastal Ecosystems: General Aspects*. Amsterdam: Elsevier, pp. 63–91.

Pye, K., and Tsoar, H., 2009. *Aeolian Sand and Sand Dunes*. Berlin: Springer.

Ranwell, D. S., 1972. *Ecology of Salt Marshes and Sand Dunes*. London: Chapman and Hall.

Salisbury, E. J., 1922. The soils of Blakeney Point: a study of soil reaction and succession in relation to the plant covering. *Annals of Botany*, **36**, 391–431.

Sanjaume, E., Gracia, F. J., and Flor, G., 2011. Introducción a la geomorfología de sistemas dunares. In Sanjaume, E., and Gracia, F. J. (eds.), *Las dunas en España*. Cádiz: Sociedad Española de Geomorfología.

Van der Meulen, F., and Salman, A. H. P. M., 1996. Management of Mediterranean coastal dunes. *Ocean & Coastal Management*, **30** (2–3), 177–195.

Van der Meulen, F., Bakker, T. W. M., and Houston, J. A., 2004. The costs of our coasts: examples of dynamic dune management from Western Europe. In Martínez, M. L., and Psuty, N. P. (eds.), *Coastal Dunes, Ecology and Conservation*. Berlin: Springer. Ecological Studies, Vol. 171, pp. 259–277.

Williams, A. T., Randerson, P., and Sothern, E., 1997. Trampling and vegetation response on sand dunes in South Wales, UK. In García Novo, F., Crawford, R. M. M., and Díaz Barradas, M. C. (eds.), *The Ecology and Conservation of European Dunes*. Sevilla: Universidad de Sevilla, pp. 287–300.

Yarranton, G. A., and Morrison, R. G., 1974. Spatial dynamics of a primary succession: nucleation. *Journal of Ecology*, **62**, 417–428.

Cross-references

Beach
Beach Processes
Bed Forms
Coastal Engineering
Coasts
Engineered Coasts
Integrated Coastal Zone Management
Laminated Sediments
Marine Sedimentary Basins
Ocean Margin Systems
Sediment Dynamics
Sediment Transport Models
Tidal Depositional Systems

DUST IN THE OCEAN

Cécile Guieu[1] and Vladimir P. Shevchenko[2]
[1]Villefranche sur Mer Oceanographic Laboratory, Villefranche sur Mer, France
[2]P. P. Shirshov Institute of Oceanology, Russian Academy of Sciences (IO RAS), Moscow, Russia

Definition

Mineral particles, mainly composed of clay and quartz, are emitted from arid regions of the continents. The presence of these nonspherical particles in the atmosphere influences the radiative budget of the Earth. Mainly under the form of pulsed events, much of this mineral material enters the open ocean with high spatial and temporal variability. Following a settling of the largest particles during transport, dust deposition on the ocean surface consists mostly of particles of a few microns in size. These particles bring to the ocean surface new nutrients (e.g., phosphorus) and metals (e.g., iron) that are essential to life, impacting marine biogeochemistry and thus ocean uptake of carbon dioxide and deep-sea sedimentation.

The fate of dust in the ocean

Despite observations and references to dust events dating back to ancient times, such as "blood rain" or "red rain" (e.g., in *Homer's Iliad*), the significance of desert dust and its emission, transport in the atmosphere, and deposition in large areas of the ocean has been demonstrated only very recently. We know now that dust particles emitted in the atmosphere and deposited at the surface of the ocean (Table 1) do impact the global climate as they influence the radiative budget of the Earth (e.g., Miller and Tegen, 1998), the marine biogeochemistry in various areas of the ocean (e.g., Martin et al., 1991; Jickells et al., 2005), and the sedimentation in the deep ocean (e.g., Windom, 1975; Lisitzin, 2011). However, the linkage of dust to ocean productivity and climate is far from being fully understood (Schulz et al., 2012).

Emissions: Today, the major regional sources are concentrated in a broad "dust belt" (Prospero et al., 2002) in the northern hemisphere (Table 1, Figure 1a). The chemical composition of desert dust aerosol particles reflects that of the average Earth surface rocks with a dominance of SiO_2 (~60 %) and Al_2O_3 (10–16 %) resulting from the dominance of quartz and clay minerals (Goudie and Middleton, 2006). From estimates of mass concentrations, desert dust and sea spray aerosol are the largest aerosol contributions on a global scale (Jickells et al., 2005; Andreae and Rosenfeld, 2008).

Dust transport: Vertical and horizontal scales of dust transport from arid areas are clearly defined (Lisitzin, 2011). Local transport includes movements of sand and gravel by rolling and saltation over the desert surface within 0–10 km from the source. Regional transport (10–1,000 km) occurs at heights of 5–7 km (to the cloud top and above). Finer particles (silt and clay) are transported over longer distances. The long-range global transport (more than 1,000 km) may occur during the emission of finer-grained pelitic material above the in-cloud scavenging height (5–7 km). Mineral crustal particles in the atmosphere have a mean size around 2 μm (diameter range 0.1–10 μm), and the largest particles fall out quickly during their long-range transport (thousands of kilometers, e.g., Ginoux et al., 2001). Together with lithogenic particles, biogenic components and nutrients, black carbon, and different pollutants enter the dust load transported to the ocean from industrial centers and areas of biomass burning, for example, from Southeast Asia (e.g., Chin et al., 2007; Guieu et al., 2010; Lisitzin, 2011). This mixing with, for example, anthropogenic acids (such as HNO_3) between emission and deposition regions may result in deposition of dust enriched in nitrogen (Geng et al., 2009).

Deposition: One of the main characteristics of dust deposition is its very high spatial and temporal variability (e.g., Jickells et al., 2005). A few, intense events may account for the bulk of the annual deposition (e.g., Loÿe-Pilot and Martin, 1996). The largest contribution is from arid North Africa (58 % of the total emissions, Table 1), which feeds large areas of the North Atlantic where high deposition dust fluxes occur. Saharan dust has also strong deposition in regional seas such as the Mediterranean Sea (Guerzoni et al., 1999). Deposition occurs as wet or dry

Dust in the Ocean, Table 1 Dust emissions from various source areas and deposition in the ocean

Dust emissions				Dust deposition in the ocean		
	Mt. year^{-1}	%	References		Mt. year^{-1}	References
North Africa	1,367	58	Mahowald et al. (2010)	North Atlantic	202	Jickells et al. (2005)
Middle East/Central Asia	760	20	Mahowald et al. (2010)	Indian Ocean	118	Jickells et al. (2005)
Australia	120.3	5	Mahowald et al. (2010)	North Pacific	72	Jickells et al. (2005)
North America	121.9	7	Mahowald et al. (2010)	Mediterranean Sea	40	Guerzoni et al. (1999)
East Asia	100.6	2	Mahowald et al. (2010)	South Pacific	29	Jickells et al. (2005)
South America	98.5	6	Mahowald et al. (2010)	South Atlantic	17	Jickells et al. (2005)
South Africa	6.25	1	Mahowald et al. (2010)	Arctic Ocean	5.7	Shevchenko and Lisitzin (2004)
Total emission	2,575			**Total deposition to the ocean**	**478**	

deposition. Our knowledge on this partitioning both from models and observation is still quite fragmented (Mahowald et al., 2005). The few observationally based estimates suggest that wet deposition is more important than dry deposition over ocean regions (Hand et al., 2004), although observations are limited in both space and time.

Variability of the emissions/deposition in geological history and present time: Desert dust emission being very sensitive to climate, dust deposition is not constant over time. For example, during glacial periods, dust deposition rates were three to four times higher globally (Rea, 1994; Kohfeld and Harrison, 2001) and 2–20 times higher at high latitude (e.g., Fisher, 1979; Petit et al., 1990; Steffensen, 1997) compared to interglacial periods. Dust deposition varied by a factor of 4 regionally in the later part of the twentieth century (Prospero and Lamb, 2003) and doubled in the twentieth century over much, but not the entire globe (Mahowald et al., 2010).

Contribution to ocean sedimentation: The regions where "arid marine (oceanic) sedimentation" occurs are located in the tropical zone and cover the same climatic zones as arid continental regions. The arid oceanic regions – where evaporation is higher than atmospheric precipitation – account for about one third of the modern ocean surface. They are by far larger than the total area of arid continental sedimentogenesis (Lisitzin, 1996; Lisitzin, 2011). In North Atlantic, Pacific, and Indian Ocean deep-sea sediments, dust particles account for up to 10–50 % or more of the noncarbonate fraction (e.g., Chavagnac et al., 2008; Lisitzin, 2011). The sedimentation rate in these areas is very low (<1 mm/1,000 year). Red clays and other abyssal sediments contain fine detrital quartz (diameter <1 μm). Volcanic ashes that are transformed to authigenic minerals (phillipsite, montmorillonite, clinoptilolite, etc.) play an important role in Pacific Ocean sediments, suggesting the influence of a tectonic factor, i.e., active margins (Lisitzin, 2011). Using sediment traps, studies of airborne dust settling in the water column (e.g., Romero et al., 1999; Lee et al., 2009; Brust and Waniek, 2010; Ternon et al., 2010) attempted to link the atmospheric dust input with interior ocean processes while also providing data for selected time intervals. It was demonstrated that the lithogenic particle flux in sediment traps differs seasonally and interannually with no systematic variations between lithogenic and biogenic flux. In the Mediterranean Sea, Ternon et al. (2010) revealed a series of "lithogenic events" (corresponding to both high particulate organic carbon (POC) and high lithogenic marine fluxes), likely resulting from interactions between lithogenic particles and dissolved organic matter present at the time of deposition. Such strong and rapid POC exports are thus not directly related to a fertilization effect (Bressac and Guieu, 2013) but rather mediated by aggregation processes and dust ballasting, which can account for ~45 % of the total annual POC export following an extreme dust event (Ternon et al., 2010).

Dust and the ocean biogeochemistry: This link has been mostly explored in iron-limited, high-nutrient low-chlorophyll (HNLC) regions (i.e., Boyd et al., 2007) after John Martin formulated his "iron hypothesis" linking stimulation of new production and atmospheric CO_2 drawdown during the last glacial maximum to higher atmospheric dust iron inputs at that time. Although a number of in situ mesoscale experiments and microcosm and mesocosm approaches since the 1990s have improved our understanding of the actual role of dust input in ocean

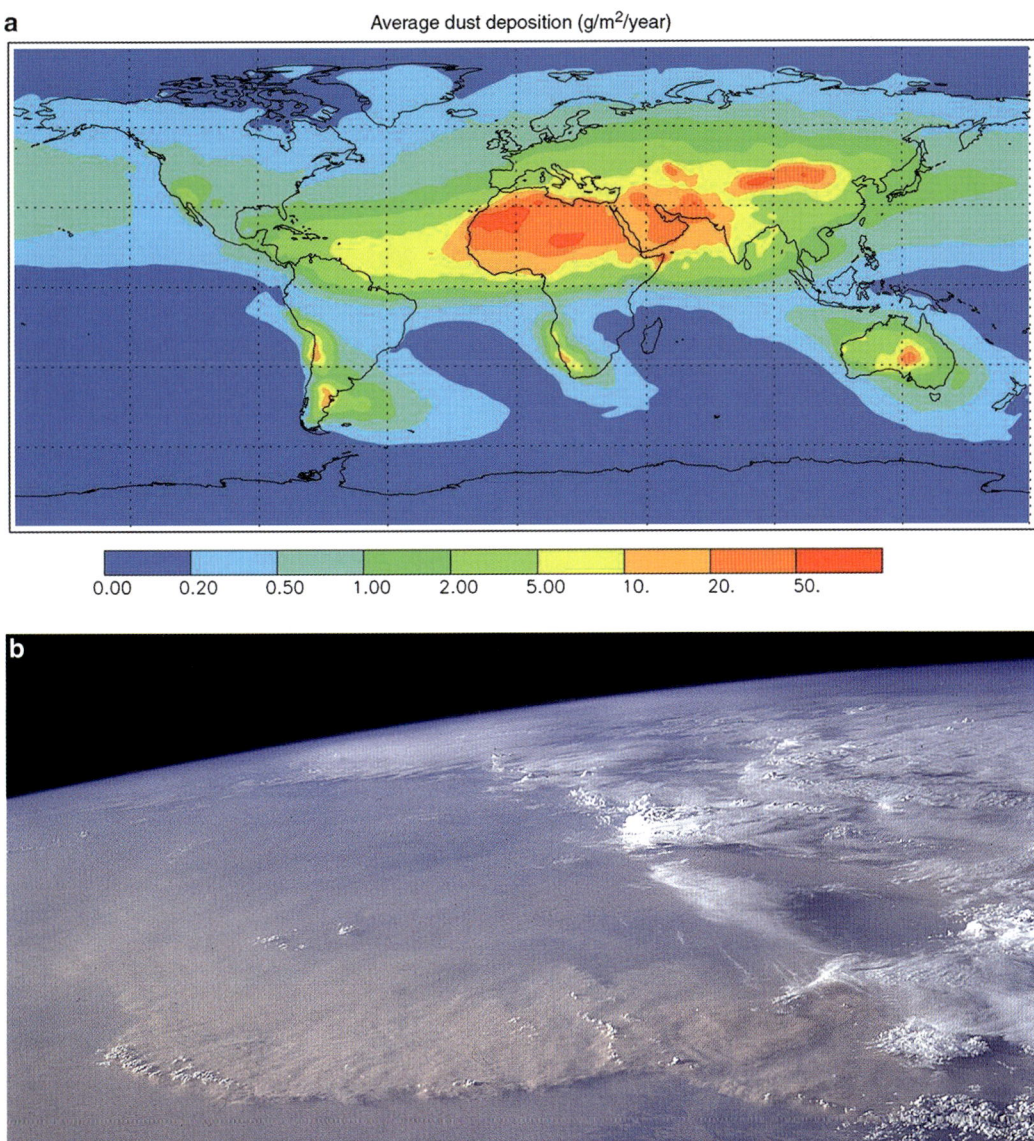

Dust in the Ocean, Figure 1 (a) Map of average dust deposition (g m^{-2} year^{-1}) to the world oceans (Jickells et al., 2005) showing the "dust belt" in the northern hemisphere (reproduced by permission of *Science*). (b) Desert dust emission over several 100 km^2 in the Sahara Desert. A portion of the particles produced eventually reaches the ocean surface (leading to complex biogeochemical processes) and then the seafloor where they contribute to deep-sea sedimentation. Image courtesy of Image Science and Analysis Laboratory, NASA Johnson Space Center (Astronaut Photograph STS049-92-71, http://eol.jsc.nasa.gov/).

biogeochemistry, our knowledge of the impact of dust deposition on the biogeochemistry of different areas of the ocean is still fragmented (de Leeuw et al., 2014). Dust dissolution experiments have shown the potential of dust to release nutrients such as phosphorus (e.g., Ridame and Guieu, 2002) and iron (e.g., Baker and Croot, 2010). Large dust deposition events may also be a sink for dissolved iron in the surface ocean via scavenging processes occurring during the sinking of the particles. The importance of this scavenging depends on the iron-binding capacity of seawater where the deposition occurs (Wagener et al., 2010). The impact on biota varies and depends on the trophic status of the regions affected by the deposition. In high-nutrient low-chlorophyll (HNLC) regions where phytoplankton growth is limited by the availability of iron, additional iron supply from dust will directly affect primary production and species composition (de Baar et al., 2005; Boyd et al., 2007). In oligotrophic regions limited and/or co-limited by nitrogen, phosphorus, and iron availability, the impact of dust (providing both P and Fe) will stimulate nitrogen fixation organisms (Mills et al., 2004). Enrichment of dust in anthropogenic nitrogen (see above) can stimulate biological activity when the dust deposition relieves both P and N limitation and co-limitation (Ridame et al., 2013). Toxic elements such as Cu contained in transported dust can also negatively impact phytoplankton (Paytan et al., 2009). Different phytoplankton groups show differential responses to dust addition (e.g., Paytan et al., 2009; Giovagnetti et al., 2013). Heterotrophic bacteria are stimulated also by dust inputs in oligotrophic regions such as the Mediterranean Sea (Pulido-Villena et al., 2008). The dominant response (bacterial production vs. primary production) has been shown to be dependent on the degree of oligotrophy, with heterotrophic bacteria having a superior ability to take up nutrients under ultraoligotrophic conditions (Marañón et al., 2010). Such a competition for the additional nutrients between heterotrophic bacteria and phytoplankton will result in opposite effects on CO_2 drawdown, in particulate organic carbon export, and thus on the carbon budget (Marañón et al., 2010; Guieu et al., 2014).

Conclusions

Recent studies on the continental sources of dust aerosol production, transport, and deposition in the ocean increased the understanding of the importance of long-range transport of dust. The dust consists of fine (mostly less than 2 μm) particles originating mainly in the arid tropical and subtropical regions extending into the oceans and is similar in composition to deep-sea (pelagic) red clays. The total contribution of dust sedimentation is almost equal to the net influx of the riverine terrigenous material to the pelagic zones of the ocean (Windom, 1975; Lisitzin, 2011). By bringing to the ocean surface iron and other essential nutrients, dust creates an impact on the ocean's biogeochemistry, affecting carbon fixation, respiration, and nitrogen fixation, depending on the ongoing limitation and co-limitation where the deposition occurs (Moore et al., 2013). A number of important issues, relevant to the International Surface Ocean-Lower Atmosphere Study (SOLAS) research community, have recently been raised to better link atmospheric deposition of nutrients, ocean productivity, and feedbacks to climate (Law et al., 2013). In this complex picture, dust is nodal, in particular because of the dominant role of strong-pulse events in atmospheric deposition (Guieu et al., 2014). A better understanding of these links is crucial to accurately represent the present role of atmospheric deposition of new nutrients and particles in marine biogeochemical models and to be able to predict its evolution in a context of global and anthropogenic changes.

Bibliography

Andreae, M. O., and Rosenfeld, D., 2008. Aerosol-cloud-precipitation interactions. Part 1. The nature and sources of cloud-active aerosols. *Earth-Science Reviews*, **89**, 13–41.

Baker, A. R., and Croot, P. L., 2010. Atmospheric and marine controls on aerosol iron solubility in seawater. *Marine Chemistry*, **120**, 4–13.

Boyd, P. W., Jickells, T., Law, C. S., Blain, S., et al., 2007. A synthesis of mesoscale iron-enrichment experiments 1993–2005: key findings and implications for ocean biogeochemistry. *Science*, **315**, 612–617.

Bressac, M., and Guieu, C., 2013. Post-depositional processes: what really happens to new atmospheric iron in the ocean surface? *Global Biogeochemical Cycles*, **27**, 859–870, doi:10.1002/gbc.20076.

Bressac, M., et al., 2011. A mesocosm experiment coupled with optical measurements to observe the fate and sinking of atmospheric particles in clear oligotrophic waters. *Geo-Marine Letters*, doi:10.1007/s00367-011-0269-4.

Brust, J., and Waniek, J. J., 2010. Atmospheric dust contribution to deep-sea particle fluxes in the subtropical Northeast Atlantic. *Deep-Sea Research Part I*, **57**, 988–998.

Chavagnac, V., Lair, V., Milton, J. A., Lloyd, A., Croudace, I. W., Palmer, M. R., Green, D. R. H., and Cherkashev, G. A., 2008. Tracing dust input to the Mid-Atlantic Ridge between 14°45′N and 36°14′N: geochemical and Sr isotope study. *Marine Geology*, **247**, 208–225.

Chin, M., Diehl, T., Ginoux, P., and Malm, W., 2007. Intercontinental transport of pollution and dust aerosols: implications for regional air quality. *Atmospheric Chemistry and Physics*, **7**, 5501–5517.

de Baar, H. J. W., Boyd, P. W., Coale, K. H., Landry, M. R., Tsuda, A., Assmy, P., Bakker, D. C. E., Bozec, Y., Barber, R. T., Brzezinski, M. A., Buesseler, K. O., Boyé, M. P., Croot, L., Gervais, F., Gorbunov, M. Y., Harrison, P. J., Hiscock, W. T., Laan, P., Lancelot, C., Law, C. S., Levasseur, M., Marchetti, A., Millero, F. J., Nishioka, J., Nojiri, Y., van Oijen, T., Riebesell, U., Rijkenberg, M. J. A., Saito, H., Takeda, S., Timmermans, K. R. T., Veldhuis, M. J. W., Waite, A. M., and Wong, C.-S., 2005. Synthesis of iron fertilization experiments: from the iron age in the age of enlightenment. *Journal of Geophysical Research*, **110**, C09S16, doi:10.1029/2004JC002601.

de Leeuw, G., Guieu, C., Arneth, A., Bellouin, N., Bopp, L., Boyd, P. W., Denier van der Gon, H. A. C., Desboeufs, K. V., Dulac, F., Facchini, M. C., Gantt, B., Langmann, B., Mahowald, N. M., Marañon, E., O'Dowd, C., Olgun, N., Pulido-Villena, E., Rinaldi, M., Stephanou, E. G., and Wagener, T., 2014. Ocean–atmosphere interactions of particles. In Liss, P. S., and Johnson, M. T. (eds.), *Ocean–Atmosphere Interactions of Gases and Particles*. Heidelberg: Springer, pp. 171–246.

Fisher, D. A., 1979. Comparison of 105 years of oxygen isotope and insoluble impurity profiles from the Devon Island and Camp Century ice cores. *Quaternary Research*, **11**, 299–305.

Geng, H., Park, Y., Hwang, H., Kang, S., and Ro, C. U., 2009. Elevated nitrogen-containing particles observed in Asian dust aerosol samples collected at the marine boundary layer of the Bohai Sea and the Yellow Sea. *Atmospheric Chemistry and Physics*, **9**, 6933–6947.

Ginoux, P., Chin, M., Tegen, I., Prospero, J. M., Holben, B., Dubovik, O., and Lin, S.-J., 2001. Sources and distributions of dust aerosols simulated with the GOCART model. *Journal of Geophysical Research*, **106**, 20255–20273.

Giovagnetti, V., Brunet, C., Conversano, F., Tramontano, F., Obernosterer, I., Ridame, C., and Guieu, C., 2013. Assessing the role of dust deposition on phytoplankton ecophysiology and succession in a low-nutrient low-chlorophyll ecosystem: a mesocosm experiment in the Mediterranean Sea. *Biogeosciences*, **10**, 2973–2991.

Goudie, A. S., and Middleton, N. J., 2006. *Desert Dust in the Global System*. Berlin/NewYork: Springer. 296 pp.

Guerzoni, S., Chester, R., Dulac, F., Herut, B., Loÿe-Pilot, M. D., Measures, C., Migon, C., Molinaroli, E., Moulin, C., Rossini, P., Saydam, C., Soudine, A., and Ziveri, P., 1999. The role of atmospheric deposition in the biogeochemistry of the Mediterranean Sea. *Progress in Oceanography*, **44**, 147–190.

Guieu, C., Loÿe-Pilot, M.-D., Benyaya, L., and Dufour, A., 2010. Spatial and temporal variability of atmospheric fluxes of metals (Al, Fe, Cd, Zn and Pb) and phosphorus over the whole Mediterranean from a one-year monitoring experiment; biogeochemical implications. *Marine Chemistry*, **120**, 164–178.

Guieu, C., Ridame, C., Pulido-Villena, E., Bressac, M., Desboeufs, K., and Dulac, F., 2014a. Impact of dust deposition on carbon budget: a tentative assessment from a mesocosm approach. *Biogeosciences*, **11**, 5621–5635.

Guieu, C., Aumont, O., Paytan, A., Bopp, L., Law, C. S., Mahowald, N., Achterberg, E. P., Marañón, E., Salihoglu, B., Crise, A., Wagener, T., Herut, B., Desboeufs, K., Kanakidou, M., Olgun, N., Peters, F., Pulido-Villena, E., Tovar-Sanchez, A., and Völker, C., 2014. Significant biological responses to pulsed atmospheric deposition in Low Nitrate Low Chlorophyll regions of the ocean. *Global Biogeochemical Cycles*, 28(11), 1179–1198.

Hand, J. L., Mahowald, N. M., Chen, Y., Siefert, R. L., Luo, C., Subramaniam, A., and Fung, I., 2004. Estimates of atmospheric processed soluble iron from observations and a global mineral aerosol model: biogeochemical implications. *Journal of Geophysical Research: Atmospheres (1984–2012)*, **109**(D17), D17205.

Jickells, T. D., An, Z. S., Andersen, K. K., Baker, A. R., Bergametti, G., Brooks, N., et al., 2005. Global iron connections between dust, ocean biogeochemistry and climate. *Science*, **308**, 67–71.

Kohfeld, K. E., and Harrison, S. P., 2001. DIRTMAP: the geological record of dust. *Earth-Science Reviews*, **54**, 81–114.

Law, C. S., Brévière, E., de Leeuw, G., Garçon, V., Guieu, C., Kieber, D., Kontradowitz, S., Paulmier, A., Quinn, P., Saltzman, E., Stefels, J., and von Glasow, R., 2013. Evolving research directions in surface ocean-lower atmosphere (SOLAS) science. *Environmental Chemistry*, **10**, 1–16.

Lee, C., Peterson, M. L., Wakeham, S. G., Armstrong, R. A., Cochran, J. K., Miquel, J. C., Fowler, S. W., Hirschberg, D., Beck, A., and Xue, J., 2009. Particulate organic matter and ballast fluxes measured using time-series and settling velocity sediment traps in the northwestern Mediterranean Sea. *Deep Sea Research, Part II*, **56**, 1420–1436.

Lisitzin, A. P., 1996. *Oceanic Sedimentation. Lithology and Geochemistry*. Washington, DC: American Geophysical Union. 400 pp.

Lisitzin, A. P., 2011. Arid sedimentation in the oceans and atmospheric particulate matter. *Russian Geology and Geophysics*, **52**, 1100–1133.

Loÿe-Pilot, M.-D., and Martin, J. M., 1996. Saharan dust input to the western Mediterranean: an eleven years record in Corsica. In Guerzoni, S., and Chester, R. (eds.), *The Impact of Desert Dust Across the Mediterranean*. Dordrecht: Kluwer, pp. 191–199.

Mahowald, N. M., Baker, A. R., Bergametti, G., Brooks, N., Duce, R. A., Jickells, T. D., Kubilay, N., Prospero, J. M., and Tegen, I., 2005. Atmospheric global dust cycle and iron inputs to the ocean. *Global Biogeochemical Cycles*, **19**(4), GB4025.

Mahowald, N. M., Kloster, S., Engelstaedter, S., Moore, J. K., Mukhopadhyay, S., McConnell, J. R., Albani, S., Doney, C., Bhattacharya, A., Curran, M. A. J., Flanner, M. G., Hoffman, F. M., Lawrence, D. M., Lindsay, K., Mayewski, P. A., Neff, J., Rothenberg, D., Thomas, E., Thornton, P. E., and Zender, C. S., 2010. Observed 20th century desert dust variability: impact on climate and biogeochemistry. *Atmospheric Chemistry and Physics*, **10**, 10875–10893.

Marañón, E., et al., 2010. Degree of oligotrophy controls the response of microbial plankton to Saharan dust. *Limnology and Oceanography*, **55**, 2339–2352.

Martin, J., Gordon, R. M., and Fitzwater, S. E., 1991. The case for iron. *Limnology and Oceanography*, **36**, 1793–1802.

Miller, R. L., and Tegen, I., 1998. Climate response to soil dust aerosols. *Journal of Climate*, **11**, 3247–3267.

Mills, M. M., Ridame, C., Davey, M., La Roche, J., and Geider, R. J., 2004. Iron and phosphorus co-limit nitrogen fixation in the eastern tropical North Atlantic. *Nature*, **429**, 292–294.

Moore, C. M., Mills, M. M., Arrigo, K. R., Berman-Frank, I., Bopp, L., Boyd, P. W., Galbraith, E. D., Geider, R. J., Guieu, C., Jaccard, S. L., Jickells, T. D., La Roche, J., Lenton, T., Mahowald, N. M., Marañón, E., Marinov, I., Moore, J. K., Nakatsuka, T., Oschlies, A., Saito, M. A., Thingstad, T. F., Tsuda, A., and Ulloa, O., 2013. Processes and patterns of oceanic nutrient limitation. *Nature Geoscience*, doi:10.1038/ngeo1765.

Paytan, A., Mackey, K. R. M., Chen, Y., Lima, I. D., Djney, S. C., Mahowald, N., Laniosa, R., and Post, A. F., 2009. Toxicity of atmospheric aerosols on marine phytoplankton. *PNAS*, **106**, doi:10.1073/pnas.08114868106.

Petit, J. R., Mounier, L., Jouzel, J., Korotkevich, Y. S., Kotlyakov, V. I., and Lorius, C., 1990. Palaeoclimatological and chronological implications of the Vostok core dust record. *Nature*, **343**, 56–58.

Prospero, J., and Lamb, P., 2003. African droughts and dust transport to the Caribbean: climate change implications. *Science*, **302**, 1024–1027.

Prospero, J. M., Ginoux, P., Torres, O., Nicholson, S. E., and Gill, T. E., 2002. Environmental characterization of global sources of atmospheric soil dust identified with the nimbus 7 total ozone mapping spectrometer (TOMS) absorbing aerosol product. *Reviews of Geophysics*, **40**, doi:10.1029/2000RG000095.

Pulido-Villena, E., Wagener, T., and Guieu, C., 2008. Bacterial response to dust pulses in the western Mediterranean: implications for carbon cycling in the oligotrophic ocean. *Global Biogeochemical Cycles*, **22**, GB1020, doi:10.1029/2007GB003091.

Rea, D. K., 1994. The Paleoclimatic record provided by eolian deposition in the deep sea: the geologic history of wind. *Reviews of Geophysics*, **32**, 159–195.

Ridame, C., and Guieu, C., 2002. Saharan input of phosphorus to the oligotrophic water of the open western Mediterranean. *Limnology and Oceanography*, **47**, 856–869.

Ridame, C., Guieu, C., and L'Helguen, S., 2013. Strong stimulation of N2 fixation in oligotrophic Mediterranean Sea: results from dust addition in large in situ mesocosms. *Biogeosciences*, **10**, 7333–7346.

Romero, O. E., Lange, C. B., Swap, R., and Wefer, G., 1999. Eolian-transported freshwater diatoms and phytoliths across the equatorial Atlantic record: temporal changes in Saharan dust transport patterns. *Journal of Geophysical Research*, **104**, 3211–3222.

Schulz, M., Prospero, J. M., Baker, A. R., Dentener, F., et al., 2012. The atmospheric transport and deposition of mineral dust to the ocean – implications for research needs. *Environmental Science and Technology*, **46**, 10390–10404.

Shevchenko, V. P., and Lisitzin, A. P., 2004. Aeolian input. In Stein, R., and Macdonald, R. W. (eds.), *The Organic Carbon Cycle in the Arctic Ocean*. Berlin/Heidelberg/New York: Springer, pp. 53–54.

Steffensen, J. P., 1997. The size distribution of microparticles from selected segments of the Greenland ice core project ice core representing different climatic periods. *Journal of Geophysical Research*, **102**, 26755–26763.

Ternon, E., Guieu, C., Loye-Pilot, M.-D., Leblond, N., Bosc, E., Gasser, B., Miquel, J.-C., and Martin, J., 2010. The impact of Saharan dust on the particulate export in the water column of the North Western Mediterranean Sea. *Biogeosciences*, **7**, 809–826.

Wagener, T., Guieu, C., and Leblond, N., 2010. Effects of dust deposition on iron cycle in the surface Mediterranean Sea: results from a mesocosm seeding experiment. *Biogeosciences*, **7**, 3769–3781.

Windom, H. L., 1975. Eolian contribution to marine sediments. *Journal of Sedimentary Petrology*, **45**, 520–529.

Cross-references

Clay Minerals
Deep-sea Sediments
Export Production
Nitrogen Isotopes in the Sea
Organic Matter

E

EARTHQUAKES

Nina Kukowski
Institute of Geosciences, Friedrich-Schiller-University, Jena, Germany

Synonyms
Seismic event

Definition
The term earthquake describes ground shaking and radiation of seismic energy, which is caused, e.g., by slip on a fault, magmatic activity, or cavity collapse.

Introduction
Earthquakes are events of seismic ground shaking caused, e.g., by the release of elastic stress accumulated along a previously locked fault zone (Figure 1), by magmatic activity, or by the collapse of caves. Further, human activity such as mining and hydrocarbon exploitation, underground nuclear explosions, geothermal fluid injection, or rapid water level changes in dam lakes may cause earthquakes. Mostly, stress release occurs suddenly and unpredicted. In the case of large earthquakes, ground shaking may be considerable and cause devastation and fatalities.

By far the most earthquakes, about 90 % of all natural seismic events, are of tectonic origin, and in the following sections only these will be dealt with. Earthquakes caused by magmatic or human activity have characteristics different from those of tectonic events and therefore in most cases can be identified by experienced scientists.

Before instruments were developed to detect and record ground motion related to earthquakes, knowledge about earthquakes having occurred only came from human observations and reporting. Therefore, historically only earthquakes sufficiently large to be felt by humans were identified. Instrumentally it is possible to detect very small earthquakes, which remain unfelt. The basic principle of a seismometer (from the Greek "seismos," quake, and "meter," measure), as it was firstly invented by J. Milne (1850–1913) in 1882, is a sensor that detects and amplifies ground vibration, which may be recorded as a seismogram either in a directly visible form or is converted to an analogue or digital electronic signal before display (Figure 1b).

Generation, travel, and recording of seismic waves
Faulting and earthquakes mechanisms
Tectonic earthquakes are related to motion along faults occurring in either the three basic (Figure 1c) or mixed types of faults. These reflect the variable orientations of the principal axes of stress with respect to the horizontal plane. When motion on the fault occurs as a pure up or down slip in the slip direction of the fault plane, then it is called a dip-slip fault. In case of upward motion this corresponds to a thrust or reverse fault, and when the motion is downward to a normal fault. In case of horizontal motion, parallel to the strike of the fault, this is called a strike-slip fault. Thrust faulting, normal faulting, strike-slip faulting, or any combination of them will occur on a fault, depending on whether the stress regime in the corresponding region is compressional, extensional or in a state of shear, and on how the usually preexisting fault strike and dip are oriented with respect to the direction of maximum compression and tension.

Faulting within the Earth occurs under high confining pressure. At depth, the minimum tensional stress corresponds to the maximum compressional stress, and vice versa.

Failure or faulting in case of brittle deformation, respectively, occurs on the plane of maximum shear stress. After

J. Harff et al. (eds.), *Encyclopedia of Marine Geosciences*, DOI 10.1007/978-94-007-6238-1,
© Springer Science+Business Media Dordrecht 2016

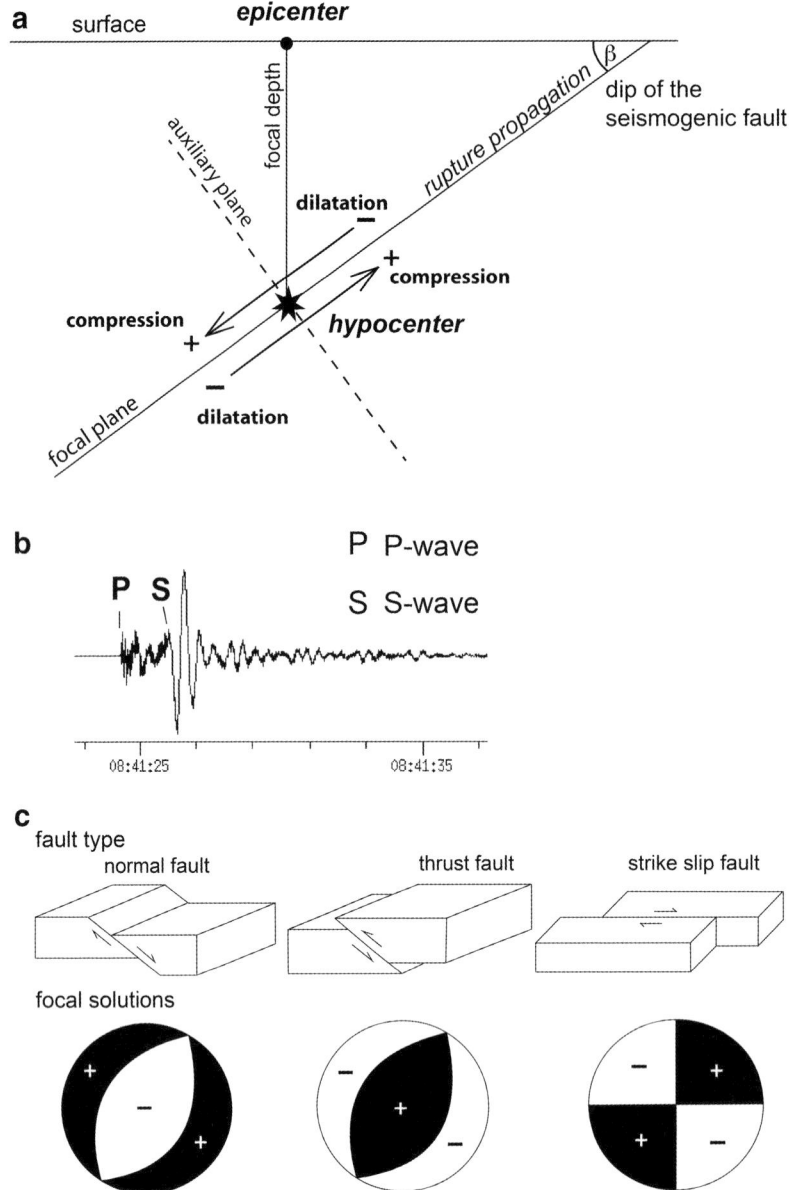

Earthquakes, Figure 1 (a) Hypocenter position and rupture propagation along a fault. The epicenter is directly located above the hypocenter from where the rupture initiates. (b) Seismogram of a near earthquake recorded by the Thuringian Seismic Network (TSN) in central Germany at a distance of about 15 km. P marks the first arrival of longitudinal P-waves, S marks the first arrival of transversely oscillating S-waves. (c) Three basic types of tectonic faults and illustration (beach balls) of the related focal mechanisms derived from the azimuth- and distance-dependent P-wave first motion polarity pattern. + refers to compressional first motions, − refers to extensional first motions. See text for more explanation and Figure 2b, c for examples of focal solutions.

a fault has been initiated, further motion is controlled by friction along the fault planes. In the fifteenth century, Leonardo da Vinci experimentally discovered the two main laws of friction and also observed that rough surfaces exert higher friction than smooth ones (Scholz, 2002). However, this knowledge remained largely forgotten until in 1699, Amontons described the two main laws of friction, namely that frictional force is independent of the size of the surface in contact but is proportional to normal load.

As friction along faults is often unstable, slip usually occurs rapidly as a rupture dynamically propagating along a fault surface, leading to sudden motion that generates seismic waves. Thus, seismicity is the short-timescale phenomenon of brittle tectonics, whereas faulting is its long-timescale expression. As faults move through

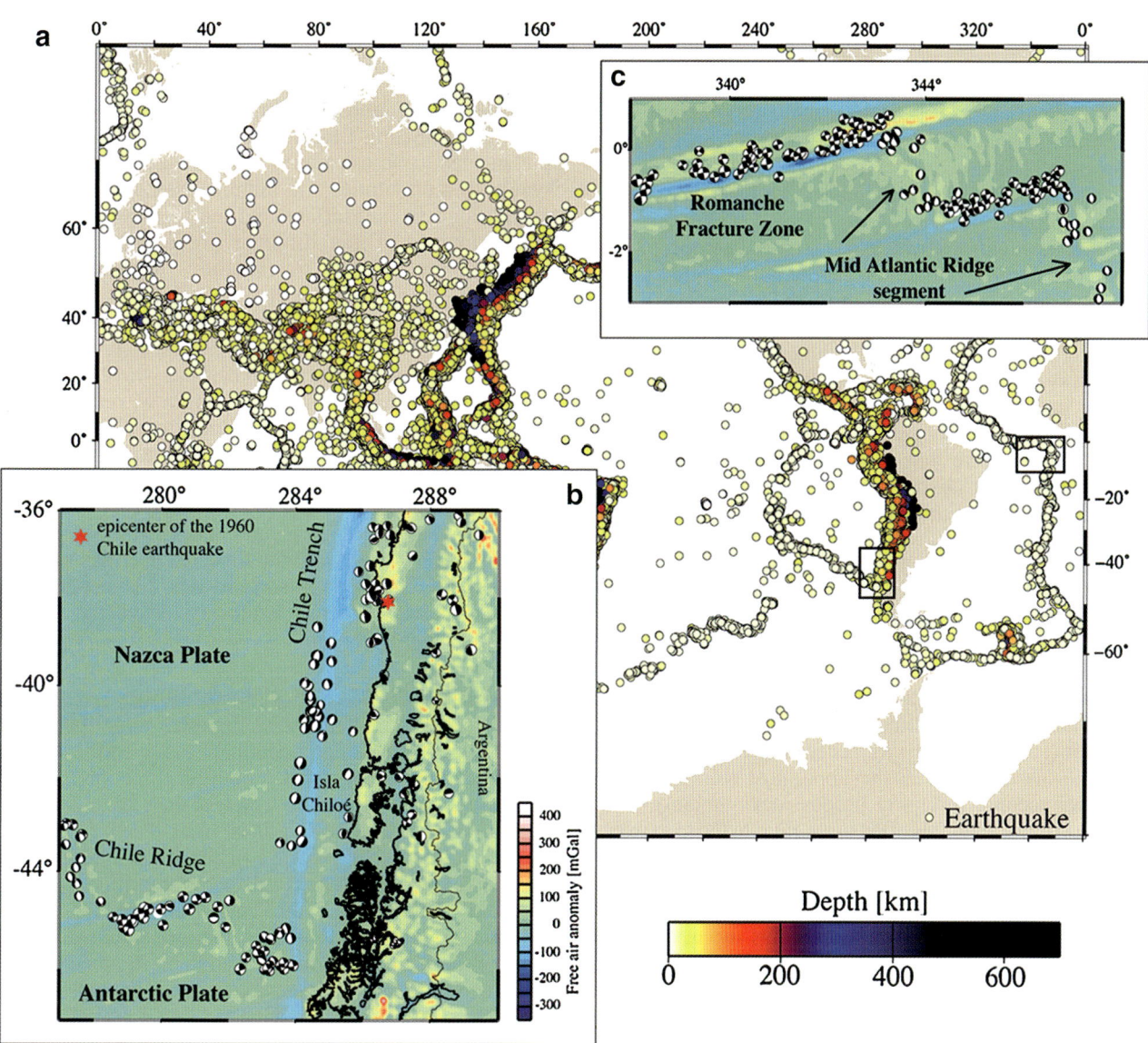

Earthquakes, Figure 2 Global seismicity: (**a**) earthquake location of ≥5 earthquakes from 1978 to 2012 (data from USGS website). Earthquake occurrence mirrors plate boundaries as the region of highest seismicity, but significant earthquakes also occur in the interior of plates (map generated using the Submap-tool 4.0 described in Heuret and Lallemand, 2005). (**b**) Focal solutions for a region around the Chile triple junction. (**c**) Focal solutions for a segment of the Mid-Atlantic Ridge.

repeated earthquakes, their record of faulting mirrors the history of past seismicity (Scholz, 2002).

Shear stress builds up on a locked fault as a result of tectonic motions, until the failure stress is reached, and the accumulated stress is completely, or mostly partially, relieved in a seismic event. Most earthquakes are associated with recurring displacements on preexisting faults. During an earthquake the accumulated elastic energy is released by physical displacement of the ground. In the case of shallow earthquakes, this rupture most often propagates along the fault at velocities of about 75–95 % of that of shear waves (S-waves) (Kanamori and Brodsky, 2004) and generates seismic waves that propagate through surrounding rocks and the entire Earth. However, minor fractions of the stored energy may be dissipated as heat by friction on the fault.

As the presence of overpressured fluid reduces friction, and therefore rock strength, fluids may play an important role in faulting and therefore also in earthquake processes (e.g., Hickman et al., 1995 and articles therein). There is worldwide evidence for fluid driven seismicity for all types of tectonic earthquakes; important examples are the Parkfield earthquake along the San Andreas Fault zone, swarm earthquakes, or megathrust events in the seismogenic zone of subduction zones.

Seismic waves

Elastic seismic energy released during an earthquake travels away from the hypocenter as body waves through the Earth and also, as a result of wave interference near the Earth's surface, as surface waves. On the seismic timescale, usually deciseconds up to almost 1 h for the longest free oscillations of the whole planet, the Earth behaves as an elastic body.

Compressional or P-waves (P for primary) propagate by stretching and contracting the medium parallel to the direction the wave travels through the entire Earth, whereas in the case of shear or S-waves (S for secondary), swinging is perpendicular to the direction of propagation. Shear waves cannot travel through fluids. The fact that already in early world-wide records of P- and S-waves after the great 1897 magnitude 8.1 Assam earthquake no direct S-waves were found that had traveled through the Earth's center led Dixon Oldham (1858–1936) to infer the existence of an Earth core and to conclude that it was liquid. This, however, relates only to the outer core, as the existence of a solid inner core was later proved by Inge Lehmann (1888–1993) in 1936.

Seismic P-waves travel at sound velocities of about 1.5 km per second in water and faster, i.e., several km per second, in most rocks. In sedimentary rocks, P-waves are as fast as between about 1.5 and 6 km per second, whereas in crystalline, igneous, and metamorphic crustal rocks they usually are faster, between about 4 and 7 km per second, and in the uppermost mantle around 8 km per second. In highly porous rocks or loose sediments, seismic velocities often are similar to those in water or even lower. S-waves travel about 1.7 times more slowly than P-waves. The velocity ratio between P- and S-waves depends, e.g., on fluid content and therefore differs slightly between various tectonic settings and in the presence of significant amounts of fluid or major fluid motion. Therefore, it can be used to study the role of fluids for earthquake processes (e.g., Schurr et al., 2003; Eberhart-Phillips et al., 2008).

There are also two types of surface waves, Rayleigh waves (LR) and Love waves (LQ). While Rayleigh waves swing in the vertical plane with particles moving along a retrograde ellipse, particle motion of Love-waves is horizontal and transverse to the travel direction. Love waves are the fastest surface waves and travel with velocities close to the subsurface velocity of S-waves, whereas Rayleigh waves travel slightly slower, with about 90 % of the shear wave velocity. Although surface waves generated by crustal earthquakes are associated with generally larger amplitudes of ground motion than body waves, their periods are also significantly longer (see Figure 1b). Therefore, in the local distance of several tens to hundred kilometers, the largest destructions are usually associated with shear waves and Love waves, and not with true surface waves.

Seismic waves are reflected and refracted at lithological, rheological, diagenetic, or other boundaries with a contrast in acoustic impedance, i.e., the product of seismic velocity and density. Besides reflection and refraction, seismic waves can also undergo conversion, e.g., from a P- to an S-wave, and vice versa. Thus, the body waves P and S originating from the earthquake source will be split in many phases, due to the interaction with many velocity discontinuities and inhomogeneities along their way through the subsurface and the deeper parts of the Earth's interior. These different phases usually arrive along different and sometimes rather complicated travel paths at different times at a distant seismograph. The different wave onsets can be identified one after the other; however, one has to keep in mind that the arrival of some theoretically possible but weaker later waves may be masked through the coda of other, earlier waves. Because of damping and geometric spreading, but also focusing effects due to the complicated velocity structure and inhomogeneities of the subsurface and the deep interior of the Earth, the amplitudes of P- and S-waves vary with distance from their source in a complicated way despite the overall tendency to decay with distance. Besides this the amplitude ratio between P- and S-waves also depends on the angle at which their seismic rays leave the source (Aki and Richards, 2002).

Seismic records and fault plane solutions

High precision modern seismographs most basically consisting of an inertial pendulum mass and auxiliary electronic transducers and amplifiers record ground motion, which may be as low as less than 10^{-10} m for a magnitude 2 earthquake and as large as 10 cm in case of a large seismic event. Because of the small size of most earthquake signals, these need to be amplified. This is commonly done by attaching a coil to the mass that moves through the magnetic field of a magnet attached to the frame of the instrument, thus transducing the mechanical movement into an equivalent electronic signal, which can be then amplified and converted into a digital signal. Further, highly accurate synchronized timing between the seismic records at many stations is crucial for precisely locating seismic sources. Whereas older instruments had provided only analogue records on slowly moving paper, modern seismometers provide digital records in high temporal resolution.

Seismic records may be used to identify the focal mechanism of the earthquake and to reconstruct motion along a seismic fault. The oldest and simplest way is to compare P-wave first-motion polarity recorded for a given event by several seismic stations that lie in different directions and at different distances from the focus. Consider motion along a dipping fault such that the material above the fault, the so-called hanging-wall, moves upwards. Then, this material is subject to compression and the region beneath the fault undergoes dilatation, i.e., extension. With the compensatory downward motion of the region beneath the fault, there will be two regions of compression and extension, separated by the fault plane and an auxiliary plane orthogonal to it (Figure 1b). A positive (+; upward) P-wave first motion observed in vertical component records corresponds to a push away from the hypocenter and thus to a seismic ray that has left the source in an extensional quadrant, and vice versa an observed dilatational first motion (−; downward) to a compressional segment in the source volume. Thus, by plotting the azimuth of the station to the hypocenter and its distance together with the polarity of first motion (dark shaded area for compression, light colored areas for dilatational first motions, cf. Figure 1c) on a stereogram, the two orthogonal great circle lines that separate best the + − and −polarity data points can be considered as the fault plane and its orthogonal auxiliary plane, from which both the strike and dip of these planes but also the slip direction can be derived. However, which of these two planes relates to the acting fault plane cannot be inferred from polarity readings alone but requires additional amplitude and frequency observations or independent field evidence. Moreover, existing faults constitute planes of weakness that may be reactivated in stress fields that are not optimally oriented to the direction of maximum pressure and tension in the source area, and as frictional strength of faults is less than the stress necessary to form them, inferring the real stress-field orientation and the size of the deviatoric stresses from the orientation of active faults or earthquake focal mechanism is also ambiguous without additional information (Scholz, 2002).

The accuracy of the estimation of both the hypocenter and the focal plane of an earthquake depends on the number of records studied from stations in different directions and distances to the hypocenter. Therefore, since the beginning of the twentieth century, global and local, permanent and temporary seismic networks have been installed. This need made seismology play an important role in leading the way for open data distribution. Nowadays, the scientific community does not only share seismograms, but more and more permits also open access in real time to digital recordings from stations worldwide via the Internet.

Earthquake size and frequency

Earthquakes size and frequency of occurrence are unevenly distributed in space and time, with small events being much more common than the larger ones.

Earthquake size is both estimated in a qualitative way based on observations like the amount of damage, felt shaking or vibrating glasses, rattling doors, etc., and in a more objective instrumentally measured quantitative way in order to get a measure of seismic moment released or seismic wave energy radiated by the seismic source independent on the distance of observations. The classification of earthquake shaking effects is done by means of intensity scales, and that of the size of seismic sources by magnitude scales, respectively.

There exist several intensity scales, among them the Modified Mercalli (MM) scale, which is mainly used in

Earthquakes, Table 1 Modified Mercalli intensity scale (After Stein and Wysession, 2003; Turcotte and Schubert, 2014)

I	Usually not felt at all
II	Just a few people felt the earthquake, mainly on upper floors of buildings
III	Lamps and similar objects swing, vibrations like from a passing truck
IV	Glasses in a cupboard and windows clink, felt by many indoors
V	Felt by nearly everyone with people sleeping mostly always waking up
VI	Glassware and windows break, felt by all
VII	People have difficulties to stand upright, minor damage
VIII	Branches break from trees and many buildings partially collapse
IX	Most people panic, cracks occur in the ground, solid buildings partially collapse
X	Landslides occur, large buildings fall completely, rails bend
XI	Few buildings remain standing, bridges destroyed
XII	Complete damage, objects thrown in the air

the USA, and the European Macroseismic Scale (EMS). Common to these scales is the definition of 12 degrees through which increasing amounts of felt ground-shaking-related phenomena and the degree of devastation is described. Although these degrees and their description are based on human observation and judgement, and thus are somewhat subjective, the proper application of intensity scales yields reasonable results how to take into account, e.g., different percentages of damage or different amounts of vulnerability. Table 1 is intended to provide an idea of the character of each intensity degree, but is far from the detailed description of the MM or EMS (for more details see, e.g., modern textbooks on geophysics like Fowler, 2005, or Clauser, 2014, and references therein). Moreover, macroseismic studies are the only way to estimate not only the size but also the epicenter and depth of historical earthquakes, which occurred prior to the instrumental record. Therefore, the evaluation of epicentral intensity and location of preinstrumental earthquakes yields not only important data for more reliable long-term seismic hazard estimates but related seismic intensity distribution maps for assessing differences in vulnerability and seismic risk of settlement areas exposed to earthquake shaking. Still older paleoearthquakes, which took place before any human reporting can be identified through, e.g., tsunami deposits, sand intrusions, fallen stalagmites, and other co-called seismites (Goldfinger, 2011). If dated, the area, which was affected by the same earthquake, can be used to estimate its size. However, mostly, only major and great earthquakes can be detected in these ways. If several paleoearthquakes were identified and dated, e.g., through isotopic methods, in the same area, recurrence intervals can be estimated (e.g., Cisternas et al., 2005).

In order to describe the size of an earthquake in a physically more meaningful way, modern magnitudes are calculated from instrumental measurements of either the

Earthquakes, Table 2 Earthquake parameters (After Stein and Wysession, 2003; Turcotte and Schubert, 2014). Magnitude does not refer to a specific magnitude

Size	Magnitude	Rupture length	Annual number	Distance up to which the earthquake is felt [km]	Energy released (10^{15} J/year)
Great	≥ 8	> 200 km	1	> 600	0–1000
Major	7–7.9	70 km	18	400	100
Strong	6–6.9	20 km	120	220	30
Moderate	5–5.9	ca. 3 km	800	150	5
Light	4–4.9		6,000	80	1
Minor	3–3.9		50,000	15	0.2
Micro	0–2.9		Many thousands per day, mostly not detected	0	

scalar seismic moment M_0, a tensorial property related to fault area, slip vector, and shear modulus (e.g., Hanks and Kanamori, 1979; Scholz, 2002) or of the radiated energy E_s by integration of the squared P-wave velocity records (e.g., Choy and Boatwright, 1995). The latter is suggested to be superior to traditional seismic energy estimates from empirical Gutenberg–Richter magnitude–energy relationships (e.g., Gutenberg and Richter, 1956). However, also these modern moment (M_w) and energy magnitudes (M_e) are semiempirical as they have been scaled to the empirical 20 s M_s surface-wave magnitude and the classical empirical relationship $\log E_s = 1.5 M_s + 4.8$ according to Richter and Gutenberg.

The scalar seismic moment M_0 is a relation between shear modulus μ, rupture area A, and average displacement D ($M_0 = \mu A D$). It is a static measure of earthquake size, but does not say anything about, e.g., rupture velocity or stress drop, which strongly control the energy radiated by an earthquake of a given seismic moment. Therefore, M_0 is only a rough average estimate of the released seismic energy and thus the moment magnitude M_w ($M_w = 2/3 \log_{10} M_0 - 6.0$ with M_0 in Nm; cf. Fowler, 2005) is less well suited for a realistic assessment of the seismic shaking hazard than the energy magnitude M_e ($M_e = 2/3 \log_{10} E_s - 2.9$; e.g., Choy and Boatwright, 1995). However, both modern magnitudes should be used complementary, as they are related to different properties of the source in order to achieve a reliable assessment of important issues related to earthquakes such as hazard or fault maturity. Table 2 intends to provide a brief overview of some earthquake parameters and terms discussed without referring to a specific magnitude.

Large earthquakes occur rarely, but release most seismic energy, e.g., 27 % of the entire seismic energy released between 1904 and 1986 was that released during the M_w 9.5 Valdivia, Chile (Figure 2), earthquake, the largest earthquake recorded so far instrumentally. The number of earthquakes occurring in a certain period of time is a function of their size (Table 2). This is expressed in the logarithmic frequency–magnitude relationship ($\log N = a - bM$ with N being the number of earthquakes with a magnitude greater than M occurring in a certain time span, a being the intercept, and b the slope). Richter and Gutenberg first introduced this relationship in 1954.

Global occurrence of earthquakes

The majority of tectonic earthquakes occur in the Earth's crust with 95 % of the global seismic moment released along plate boundaries (Scholz, 2002) with convergent margins and continental collision zones exhibiting largest seismic events (Figure 2). Large, often destructive earthquakes also occur in plate interiors, however with a much lower frequency. Also here, earthquakes seem to cluster in certain regions. Earthquakes with hypocenters deeper than some tens of kilometers almost only are found along the Wadati-Benioff zones (see Wadati-Benioff-Zone) of subduction zones.

A typical displacement in a very large earthquake is 10 m. If the relative velocity across a plate boundary would be 50 mm per year, it would take 200 years to accumulate this displacement. Large earthquakes at subduction zones and major transform faults such as the San Andreas recur in about such periods of time. Since regular displacements do not need to be accommodated in plate interiors, the period of time between major intraplate earthquakes is much longer (Turcotte and Schubert, 2014).

Tectonic earthquakes in the shallow brittle part of the crust and along the seismogenic zone of convergent plate boundaries mainly are caused when stress accumulated along a locked fault exceeds rock strength. As rock strength is dependent on the tectonic regime, stresses necessary to generate earthquakes may differ in different tectonic environments such as compressional or tensile tectonic regimes (Scholz, 2002). However, independent of the tectonic regime, shallow earthquakes seem to be limited by the brittle behavior of rock and do not seem to occur at temperatures exceeding about 600 °C (McKenzie et al., 2005). However, in subduction zones, earthquakes occur down to depths of more than 600 km, asking for other potential mechanisms causing those deep earthquakes (*see* Wadati-Benioff-Zone).

Unusual earthquakes

Immediately after a large earthquake, the so-called main shock, usually, many smaller events, the so-called aftershocks, occur in the same region, i.e., along parts of the fault plane that had ruptured before. This seismic activity often decays over some weeks or months, depending on the size of the main earthquake.

If seismic activity is observed over weeks or months, but without any major event, this is called an earthquake swarm. Further, there are also unusual types of seismic events like those causing a tsunami larger than expected from their magnitude (Okal and Newman, 2001). Since about 15 years, another type of seismic events, called slow slip events, has been identified along certain plate boundaries.

Earthquake swarms

Earthquake swarms are periods of seismic activity of more or less equal size without dominating events, an abrupt onset and ending of the seismic activity. Individual magnitudes scarcely exceeding three or four, which are occurring over weeks or months in a certain region limited in space (e.g., Scholz, 2002; Hainzl et al., 2012). Earthquake swarms usually consist of several hundreds to several thousands detected individual events. Often, they start and cease with gradually increasing and decreasing magnitudes, respectively. Whereas earthquake swarms were first mainly detected in volcanic regions, they now have been identified in nearly all seismotectonic settings, such as subduction zones close to the seismogenic thrust, transform faults, or intraplate regions (Holtkamp and Brudzinski, 2011; Hainzl et al., 2012). B-values of earthquake swarms often are between 1.5 and 2, and therefore considerably larger than those obtained from global or regional earthquake data sets (Scholz, 2002; Holtkamp and Brudzinski, 2011). Fluids have been proposed to play an important role for earthquake swarms, e.g., pore pressure changes or magma transport. Further, earthquake swarms may be potentially related to slow slip events.

Slow earthquakes

Recently, a special type of release of seismic energy has been identified from GPS records in several subduction zones around the Pacific, like Cascadia offshore the USA and Canada west coast (Dragert et al., 2001) or Hikurangi offshore the east coast of New Zealand's north island (Wallace et al., 2012). Well-studied regions with SSEs recurring in relatively regular intervals, like Cascadia or Hikurangi show that these events may nucleate at very shallow depths, about 5–6 km beneath the seafloor, but also at depths of 30–40 km indicating they can occur in very different thermal regimes. Several types of slow earthquakes are known, e.g., tremors, slow slip events, or very low-frequency earthquakes. Mostly, they seem to occur in subduction zones with a young downgoing plate, but have also been detected beneath the San Andreas fault system (Beroza and Ide, 2011).

It seems that fluids are invoked in their occurrence, possibly at very high fluid pressures. However, SSEs' mechanisms so far are largely not understood (Schwartz and Rokosky, 2007). An important role of Love waves also has been hypothesized. Slow earthquakes seem to occur episodically, often quite frequent, with recurrence times of only a few years or even less at the same location and typically show durations of hours to several months. Slow earthquakes cannot be detected with seismometers and so far only have been identified from geodetic (GPS) records. Slip caused by slow earthquakes suggests that usually their size is comparable to that of usual earthquakes with magnitudes around 6–6.5.

Seismic hazard

Most seismic hazard is related to major and great earthquakes, as one magnitude >8 earthquake releases more energy than all other earthquakes during an entire year. It describes the probability of a certain level of ground shaking caused by recurring earthquakes (Shedlock et al., 2000; Stein and Wysession, 2003).

Damage caused by an earthquake mainly is the consequence of ground acceleration, which describes the amount of shaking or strong motion, and is used as a measure for seismic risk. Other hazardous phenomena occurring in relation to earthquakes are, e.g., slumping in regions of steep slope, liquefaction, or tsunamis (see Wadati-Benioff-Zone).

The size of an earthquake does not necessarily directly relate to damage and fatalities caused by it as the depth of the hypocenter, and consequently ground motion as well as bedrock in the region affected by an earthquake largely control destruction. When nonconsolidated, water-saturated sand, which is typical, e.g., for river estuaries or shores, is repeatedly shaken, especially at resonance period, e.g., during ground motion caused by a large earthquake, it may behave like a liquid, i.e., it completely looses its shear strength and therefore no longer is able to carry, e.g., buildings. As a consequence of this liquefaction, buildings may sink several meters and become severely damaged, like during the 1906 San Francisco, 1964 Alaska, 2010 Canterbury, or 2011 Christchurch earthquakes (Quigley et al., 2013). Strike-slip earthquakes seem to be especially prone to a high degree of damage relative to their magnitudes as their hypocenters usually are quite shallow. If sediments redistributed by liquefaction keep their geotechnical properties, they may be indicative of paleoearthquakes (Green et al., 2005).

Therefore, globally, seismic hazard differs considerably between different tectonic settings and to a large degree mirrors the distribution of brittle strain rate, i.e., deformation (Triep and Sykes, 1997). To quantify seismic hazard, data of historical and instrumentally recorded earthquakes, their source characteristics, and related strong ground motions have been compiled to compute global seismic hazard and assign a level of seismic hazard to every location worldwide (Giardini et al., 1999).

Bibliography

Aki, K., and Richards, P. G., 2002. *Quantitative Seismology*, 2nd edn. New York: Freeman.

Beroza, G. C., and Ide, S., 2011. Slow earthquakes and non-volcanic tremor. *Annual Review of Earth and Planetary Sciences*, **39**, 271–296, doi:10.1146/annurev-earth-040809-152531.

Choy, G. L., and Boatwright, J. L., 1995. Global patterns of radiated seismic energy and apparent stress. *Journal of Geophysical Research*, **100**(B9), 18,205–18,228.

Cisternas, M., et al., 2005. Predecessors of the giant 1960 Chile earthquake. *Nature*, **437**, 404–407, doi:10.1038/nature03943.

Clauser, C., 2014. *Einführung in die Geophysik*. Berlin: Springer, p. 407 pp.

Dragert, H., Wang, K., and James, T. S., 2001. A silent slip event on the deeper Cascadia subduction interface. *Science*, **292**, 1525–1528.

Eberhart-Phillips, D., Reyners, M., Chadwick, M., and Stuart, G., 2008. Three-dimensional attenuation structure of the Hikurangi subduction zone in the central North Island, New Zealand. *Geophysical Journal International*, **174**, 418–434, doi:10.1111/j.1365-246X.2008.03816.

Fowler, M., 2005. *The Solid Earth*, 2nd edn. Cambridge: Cambridge University Press, p. 685 pp.

Giardini, D., Grüntha, G., Shedlock, K.M., and Zhang, P., 1999. The GSHAP global seismic hazard map. *Annali di Geofisica* **42**, 1225–1230.

Goldfinger, C., 2011. Submarine paleoseismology based on turbidite records. *Annual Review of Marine Science*, **3**, 35–66, doi:10.1146/annurev-marine-120709-142852.

Green, R. A., Obermeier, S. F., and Olson, S. M., 2005. Engineering geologic and geotechnical analysis of paleoseismic shaking using liquefaction effects: field examples. *Engineering Geology*, **76**, 263–293, doi:10.1016/j.enggeo.2004.07.026.

Gutenberg, B., and Richter, C. F., 1954. *Seismicity of the Earth and Associated Phenomena*, 2nd edn. Princeton: Princeton University Press, p. 310 pp.

Gutenberg, B., and Richter, C. F., 1956. Magnitude and energy of earthquakes. *Annali di Geofisica*, **9**, 1–15.

Hainzl, S., Fischer, T., and Dahm, T., 2012. Seismicity-based estimation of the driving fluid pressure in the case of swarm activity in Western Bohemia. *Geophysical Journal International*, **191**, 271–281, doi:10.1111/j.1365-246X.2012.05610.x.

Hanks, T. C., and Kanamori, H., 1979. A moment magnitude scale. *Journal of Geophysical Research*, **84**, 2348–2350.

Heuret, A., and Lallemand, S., 2005. Plate motions, slab dynamics and back-arc deformation. *Physics of the Earth and Planetary Interiors*, **149**, 31–51.

Hickman, S., Sibson, R., and Bruhn, R., 1995. Introduction to special section: mechanical involvement of fluids in faulting. *Journal of Geophysical Research*, **100**, 12 831–12 840.

Holtkamp, S. G., and Brudzinski, M. R., 2011. Earthquakes swarms in circum-Pacific subduction zones. *EPSL*, **305**, 205–215, doi:10.1016/j.epsl.2011.03.004.

Kanamori, H., and Brodsky, E. E., 2004. The physics of earthquakes. *Reports on Progress in Physics*, **67**, 1429–1496, doi:10.1088/0034-4885/67/8/R03.

McKenzie, D., Jackson, J., and Priestley, K., 2005. Thermal structure of oceanic and continental lithosphere. *Earth and Planetary Science Letters*, **233**, 337–349, doi:10.1016/j.epsl.2005.02.005.

Okal, E. A., and Newman, A. V., 2001. Tsunami earthquakes: the quest for a regional signal. *Earth and Planetary Science Letters*, **124**, 45–70.

Okal, E. A., and Sweet, J. R., 2007. Frequency-size distributions for intraplate earthquakes. *Geological Society of America Special Papers*, **425**, 59–71, doi:10.1130/2007.2425(05).

Pacheco, J. F., and Sykes, L. R., 1992. Seismic moment catalog of large shallow earthquakes, 1900 to 1989. *Bulletin of the Seismological Society of America*, **82**, 1306–1349.

Quigley, M. C., Bastin, S., and Bradley, B. A., 2013. Recurrent liquefaction in Christchurch, New Zealand, during the Canterbury earthquake sequence. *Geology*, **41**, 419–422, doi:10.1130/G33944.1.

Scholz, C. H., 2002. *The Mechanics of Earthquakes and Faulting*. Cambridge: Cambridge University Press, p. 471 pp.

Schurr, B., Asch, G., Rietbrock, A., Trumbull, R., and Haberland, C., 2003. Complex patterns of fluid and melt transport in the central Andean subduction zone revealed by attenuation tomography. *Earth and Planetary Science Letters*, **215**, 105–119, doi:10.1016/S0012-821X(03)00441-2.

Schwartz, S. Y., and Rokosky, J. M., 2007. Slow slip events and seismic tremor at circum-Pacific subduction zones. *Reviews of Geophysics*, **45**, RG3004. 32 pp.

Shedlock, K. M., Giardini, D., Grünthal, G., and Zhang, P., 2000. The GSHAP global seismic hazard map. *Seismologica Research Letters*, **71**, 679–686.

Stein, S., and Wysession, M., 2003. *An Introduction to Seismology, Earthquakes, and Earth Structure*. Malden: Blackwell, p. 498 pp.

Triep, E. O., and Sykes, L. R., 1997. Frequency of occurrence of moderate to great earthquakes in intracontinental regions: Implications for changes in stress, earthquake prediction, and hazards assessments. *Journal of Geophysical Research*, **102**, 9923–9948.

Turcotte, D. L., and Schubert, G., 2014. *Geodynamics*, 3rd edn. Cambridge: Cambridge University Press, p. 623 pp.

Wallace, L. M., Beavan, J., Bannister, S., and Williams, C., 2012. Simultaneous long-term and short-term slow slip events at the Hikurangi subduction margin, New Zealand: implications for processes that control slow slip event occurrence, duration, and migration. *Journal of Geophysical Research*, **117**, B11402, doi:10.1029/2012JB009489.

Wells, D. L., and Coppersmith, K. J., 1994. New empirical relationships among magnitude, rupture length, rupture width, rupture area, and surface displacement. *Bulletin of the Seismological Society of America*, **84**, 974–1002.

Cross-references

Epicenter, Hypocenter
Geohazards: Coastal Disasters
Plate Motion
Seismogenic Zone
Subduction
Tsunamis
Wadati-Benioff-Zone

EL NIÑO (SOUTHERN OSCILLATION)

Janina Körper
Institute of Meteorology, Free University of Berlin, Berlin, Germany

Synonyms

El Niño-Southern Oscillation (ENSO) warm event

Antonym

La Niña

The term El Niño (Spanish for "the Child Christ") was initially used by Peruvian fishermen to describe a warm-water current that periodically flows along the coast of Ecuador and Perú about Christmas time. It is an oscillation associated with the unusually large basin-wide warmings of the eastern tropical Pacific Ocean that occur every few years (about 2–7 years) and change the local and regional ecology and have global impacts. The oceanic event is connected to the atmospheric component termed "Southern Oscillation." The coupled atmosphere-ocean phenomenon is collectively known as the El Niño-Southern Oscillation (ENSO). While El Niño refers to the warm phase of ENSO, La Niña (Spanish for "the girl") refers to the cold phase of ENSO. See Trenberth (1997) for a review on definitions of El Niño.

An initial positive sea surface temperature (SST) anomaly intensifies due to a positive feedback first hypothesized by Bjerknes (1969). The anomaly reduces the east-west SST gradient, which leads to reduced equatorial westerlies of the Walker circulation and inherent weakening of equatorial upwelling. This reinforces the positive SST anomaly. To explain the determination of an El Niño event and its quasiperiodic reoccurrence, two mechanisms are still discussed. ENSO is explained by either a self-sustained, naturally oscillatory mode of the coupled ocean atmosphere system or a stable mode triggered by stochastic forcing. See Wang and Fiedler (2007) and Wang et al. (2012) for a review of these mechanisms.

There are two types of El Niño that differ with respect to their spatial structure, evolution, underlying mechanisms, and their global impacts. The eastern-Pacific type is characterized by SST anomalies centered in the eastern Pacific cold tongue region. The SST anomalies of the central-Pacific type, also referred to as date line El Niño or warm pool El Niño, are centered near the International Date Line (Wang et al., 2012).

Bibliography

Bjerknes, J., 1969. Atmospheric teleconnections from the equatorial pacific. *Monthly Weather Review*, **97**, 163–172.
Trenberth, K. E., 1997. The definition of El Niño. *Bulletin of the American Meteorological Society*, **78**, 2771–2777.
Wang, C., and Picaut, J., 2004. Understanding ENSO physics – a review. In Wang, C., Xie, S. P., and Carton, J. A. (eds.), *Earth's Climate: The Ocean-Atmosphere Interaction*. Washington, DC: AGU, pp. 21–48.
Wang, C., Deser, C., Yu, J.-Y., DiNezio, P., Clement, A., 2012. El Niño and Southern Oscillation (ENSO): a review. In Glymn, P., Manzello, D., Enochs, I. (eds.), *Coral Reefs of the Eastern Pacific*. Springer.

Cross-references

Coasts
Currents
Sea-Level

ENERGY RESOURCES

Dieter Franke and Christoph Gaedicke
Federal Institute for Geosciences and Natural Resources, Hannover, Germany

Definition

Energy resources. Energy resources are used to satisfy the energy consumption of the global economy. Energy in general is the capacity of a system to perform work. The unit for measurement of this scalar physical quantity is the joule. A resource is defined as a source of supply or support that is ready to use if or when it is needed. Energy derived directly from a natural resource is called primary energy, i.e., burning coal for heating or collecting solar energy with solar cells to generate electricity. When primary energy is transferred into another form of energy, i.e., burning coal in a power plant to generate electric power, this energy is called secondary energy.

Energy resources are generally divided into (1) renewable energy sources and (2) fossil energy resources. The renewable sources are continually replenished from sunlight, wind, rain, water, waves, tides, geothermal heat, and biomass. Fossil energy resources consist of coal, uranium, natural gas, and oil. The most important *offshore* energy resources are oil and gas. Renewable offshore energy is mainly derived from wind, waves, and tides.

Introduction

Energy resources from offshore areas are of increasing importance. This is valid for both renewable and fossil energy resources: On the one hand, new exploration methods and technical innovations enlarge the volumes and accessible areas for the exploitation of fossil energy resources. On the other hand, technological advancements and the increasing efficiency in using the renewable energy resources, combined with the benefits of mass production and market competition, are continuously improving the competitiveness of renewable technologies.

World primary energy consumption grew by more than 40 % over the past 20 years and is likely to grow by a similar amount over the next decades. The world's most important energy resource is oil. With an annual worldwide consumption of about 34 billion (10^9) barrels, it accounts for more than 30 % of the world's primary energy consumption. The global oil supply is continuously growing and rose to up to 90 million barrels/day in 2011, but further growth is uncertain. In fact there is considerable controversy about the date when the maximum rate of oil production is reached. Without doubt the amount of oil on earth is limited. However, there is an ongoing discussion about the percentage that has already been extracted and mainly about the date when a further increase in production will no longer be possible ("peak oil"). Peak oil must not be confused with oil depletion. Even when reaching the point of maximum production, there will be continuous oil supply on high levels; only a further increase in production from thereon is not possible. Optimistic scenarios suggest that additional reserves can be obtained by a more effective oil recovery from producing fields, the usage of unconventional resources, and that significant additional oil – and also natural gas – is found in the frontier areas of the Arctic and the deepwater areas of the continental margins. It is not only the geology and reservoir performance that defines if the targeted resources are exploited, it also depends on future demand and prices, the availability of new technologies, the infrastructure, environmental issues, and political and regulatory considerations. The impact of carbon dioxide on the climate may result in efforts to limit the extensive use of coal, oil, and natural gas.

At present, renewable energy represents only a minor percentage of global primary energy consumption, but it is growing very rapidly. Renewable energy costs continue to drop as a result of technological innovation and mass production. A main driving force to expand the use of renewable energy is its relatively low impact on climate and on the environment.

Offshore renewable energy

In order to decrease emissions of carbon dioxide into the atmosphere, serious efforts are undertaken to develop renewable energy sources.

Offshore wind farms boast some advantages in comparison to onshore installations: the wind blows more constantly most of the time, and the demands on "land" use are less competitive offshore. Technical challenges include the foundations of the wind turbines installed in shallow water. This requires detailed information of the sediment distribution not only on the surface but also some tens of meters below the seafloor, where peat horizons or shallow gas may occur. Sediment movement may restrict the foundations of wind turbines. Many shallow areas of the sea are important habitats for the reproduction of marine mammals, fishes, and benthic invertebrates. These biological communities are not only disturbed during the construction but also may be perturbed during operation.

As the number of offshore wind turbines installed in the marine realm increases, integrating them into the existing power supply structures will become increasingly important. While the present total capacity of offshore wind farms is in the megawatt range, plans call for much larger offshore units. A substantial increase in the amount of wind power will have to be integrated into national and international electricity grids. The next step in offshore wind farming will lead to installation in deeper and more remote areas far away from the coast. Since the wind blows more strongly and comparatively smaller areas are needed, electricity generation with wind turbines out at sea is particularly attractive. The organizational and technological requirements are however significantly higher than on land, i.e., great distances to the coast, ocean depths of up to 40 m, and the harsh sea climate with humid, salty air, strong temperature fluctuations, severe storms, squalls, and high wave loads (see Ocean Margin Systems).

The development of wave power devices is still in progress and so far only prototypes are being tested at sea. With the recent worldwide growth in interest in renewables, wave energy may also become more important. Wave power devices harness the continuous energy of waves to generate electricity. These may be installed at the shoreline, nearshore, or offshore. Tidal power plants convert the potential energy or the power of currents between low and high tide into electricity. The major advantage of this renewable energy is that tides are more predictable than other renewable energy sources. However, regions with sufficiently high tidal ranges or flow velocities are limited. Further technological improvements are necessary to bring production costs of tidal energy down to competitive levels.

Other ocean energy sources that are at present not widely used include ocean thermal energy, making use of the small temperature difference between deep and shallow water, salinity gradient energy which becomes available from the difference in the salt concentration between seawater and river water, and the energy of marine currents.

Fossil energy resources

The most important offshore energy resources are oil and gas. Given the huge onshore resources, coal and uranium are not expected to be produced in the coming decades from offshore areas.

Petroleum classifications, reserves, and resources

Oil and gas are trapped in subsurface geological structures and the volumes cannot be physically examined or inspected. Regional estimates are based on data that provide indirect evidence of the scale of the resource base. Classification usually starts with a petroleum play, which is a concept to recognize patterns in petroleum occurrence that can help to predict the results of future exploration. Plays are characterized by the presence or prediction of source rocks, reservoirs, and seals. Hydrocarbon generation in sedimentary basins (see Marine Sedimentary Basins) is controlled by the source rock, its lithology, facies, distribution, and organic content and by the tectonic evolution of the basin. The amount of organic matter in the source rock depends on the production and input of organic matter during sedimentation (see Paleoceanography) but also on preservation or degradation during diagenesis. The increasing heat and pressure during subsidence and continuing sedimentation control the generation of hydrocarbons in the source rock, while porosity and permeability control the migration of hydrocarbons from the source rock into a reservoir. The potentiality of sedimentary basins thus depends also on subsidence, burial, and physical conditions such as pressure and heat flow. Porous structures for accumulations and seals trapping the generated hydrocarbons are prerequisites for petroleum reservoirs.

The amount of the producible and usable energy resources in a specific setting depends on the geological conditions of the deposits, the state of the scientific and technical knowledge, the technological potential of development and production, as well as the economic and political requirements. Correspondingly, the intentions and methods used to assess the amounts of energy resources considerably vary. There is one general distinction that applies to all energy resource statistics: the distinction between those energy resources whose exploitation is regarded as proven (reserves) and those energy resources whose existence is only assumed or whose production is currently not considered economically feasible (resources) (Andruleit et al., 2013). *Oil and gas reserves* are defined as the volumes that have been accurately recorded and can be commercially recovered in the future using the current technical possibilities. *Resources* are those amounts of oil and gas which have been geologically proven, but which at the present time cannot be economically recovered, and the amounts which have not been proven, but which can be expected for geological reasons in the concerned area. There is a dynamic boundary between resources and reserves and, e.g., variations in the price or technical advances will convert resources to reserves and vice versa.

There is no common definition to subdivide oil and gas into *conventional and unconventional occurrences*. Probably most commonly classified are conventional occurrences as the ones which can be exploited by standard exploration and production methods. According to this definition, the development and use of unconventional occurrences requires alternative technologies. Aspects of economic efficiency, or whether the individual deposit is already used for production, are not considered in this definition. However, once an alternative technology becomes established, earlier unconventional resources may be classified as conventional.

Giant oil and gas fields (those with 79,000,000 m^3 (500 million barrels) of ultimately recoverable oil or gas equivalent) account for about 40 % of the world's petroleum reserves (Halbouty, 2001). The majority of the world's giant oil and gas fields exist in two characteristic tectonic settings – passive continental margin and rift structures – highlighting the potential impact of future offshore resources. Passive continental margins are the outcome of a "successful rift" which resulted in continental separation and the formation of new oceanic crust, while in failed rifts, extension terminated at some stage during the evolution to leave an aborted rift (see "Plate Tectonics").

Offshore petroleum exploration and production
Evolution of offshore petroleum exploration

Offshore drilling for oil in seawater began off the US West Coast, just south of Santa Barbara, California, as early as 1896. Significant offshore exploration, however, only started in the middle of the twentieth century. In the Gulf of Mexico, off the Louisiana coast, the first productive wells were installed in the 1940s. However, water depth was still far less than 100 m. 1947 is commonly used as the year defining the initiation of offshore drilling beyond the sight of land. From then on, exploration also began in the Caspian Sea, the Persian Gulf, the Arabian Sea off India, and the North Sea. Some discoveries were made in China and Australia. Exploration peaked in 1982 when nearly 200 new fields were discovered. In 1986, the first deepwater field (>1,500 m) was discovered in the Gulf of Mexico (Mensa gas field). Based on the dominant discovery pattern of the 1990s, the deep water became the primary new exploration target for petroleum companies (Esser, 2001). Several new discoveries were made up until the twenty-first century, but the number of new discoveries never again reached the high numbers of 1982. In 2000, when the oil price dropped to about 12$/barrel, almost all coastal countries bordering passive continental margins were involved in deepwater exploration. Until this time, offshore exploration had been concentrated in shallow waters. Only about 40 production rigs were installed in waters exceeding 500 m depth, accounting for about 3 % of global oil production. More than 2000 new discoveries were made since that time, the most important being in the Gulf of Mexico, offshore Brazil, Iraq, and Australia. Global deepwater production has more than tripled since 2000, and in 2010, about 10 % of the global oil production came from deepwater wells. The fact that more than 50 % of newly discovered reserves are found in deep waters over the last few years indicates that the proportion of deepwater oil and gas production will grow further in the near future.

Deepwater petroleum

There is a wide range of definitions for deep water. The most widespread definition refers to water depths greater than 500 m. That is about the depth at which traditional development platforms cannot be implemented. Deepwater exploration came along with major improvements of exploration and production methods. Major challenges include 3-D seismic data acquisition (see Technology in Marine Geosciences) and advanced seismic processing integrated with interpretation workstations capable of handling and visualizing large 3-D volumes, the use of seismic amplitudes for hydrocarbon and reservoir prediction, methodologies for basin analysis and predictive stratigraphy (see Deep-sea Sediments), and drilling and logging tools (dynamically positioned drill ships, automated pipe handling systems, horizontal drilling and well completions, and subsea systems).

Offshore oil production represents about one-third of present world oil output. Globally, there are more than 150 major deepwater fields that have the potential to come on-stream over the next few years. Drilling activity has particularly increased in the "golden triangle" of Brazil,

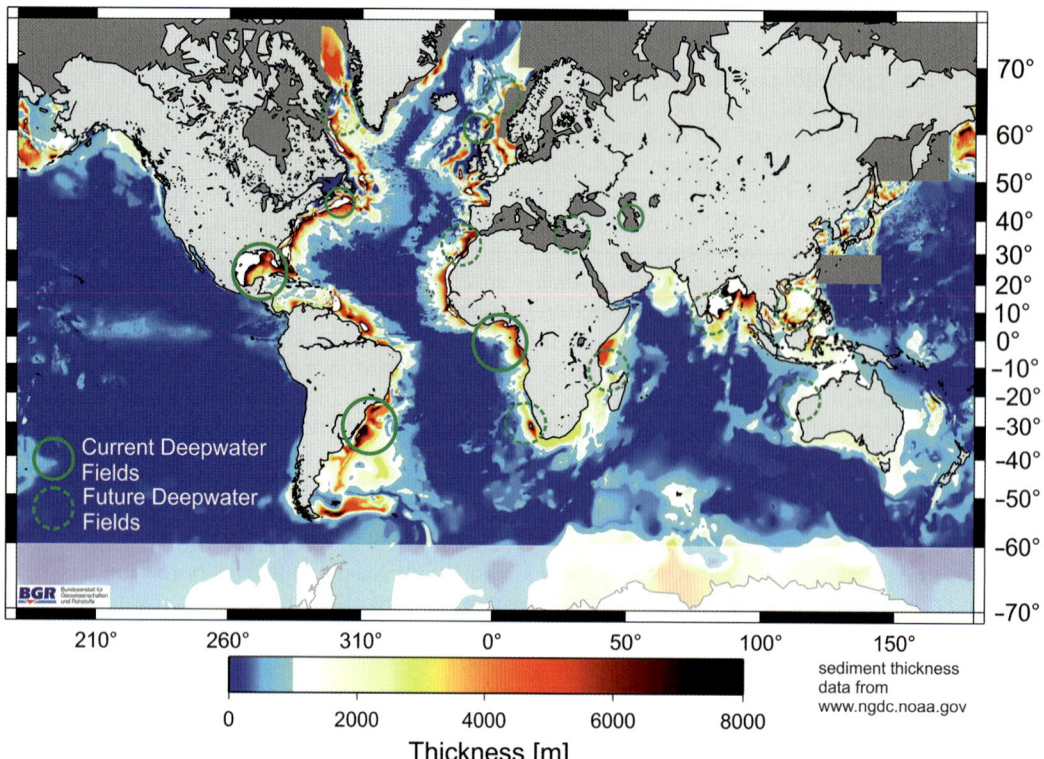

Energy Resources, Figure 1 Sediment thickness exceeding 1 km along the continental margins (data from ngdc.nooa.gov). Present (*solid*) and future (*dashed*) deepwater fields are indicated by *green* circles. Drilling activity has particularly increased in the "golden triangle" of Brazil, the Gulf of Mexico, and West Africa, but in the future, many more potentially productive areas are expected.

the Gulf of Mexico, and West Africa (Figure 1). The exact size of the hydrocarbon resources at the continental margins is unknown, but recent estimates indicate there could be as much as 3.8 billion barrels of oil and 609 billion (10^9) cubic meters of natural gas in the eastern Gulf of Mexico alone (Hughes and Kinnersley, 2010). Of particular interest are basins with evaporite deposits, especially those with pre-salt oil and gas accumulations. In Brazil, almost 20 billion barrels of oil equivalent were reported discovered in subsalt Cretaceous deposits.

Exploration and production in the Arctic realm

Exploration and production of oil and gas in the Arctic realm has been taking place for several decades. At present, the activities focus on northern Alaska, the Barents Sea, and on West Siberia. The Prudhoe Bay and Kuparuk fields in Alaska and the Urengoy field in the Russian Federation are prime examples of the giant field potential of the region. The first Arctic offshore oil platform is located in the Pechora Sea, in the southeastern Barents Sea. Production with the first ice-resistant platform "Prirazlomnaya" started in 2014.

The Canadian Arctic has been intensively explored over the past decades and more than 100 wells have been drilled. Despite considerable gas potential, the economic risks and the complex geology resulted in no production so far. Offshore Greenland is considered by some researchers as the region with the largest undiscovered resources. In both East and West Greenland, a couple of exploration licenses have been sold, but so far there is no known noteworthy discovery, and there is no production at the moment. The land and sea areas of the north of the European continent mainly belong to the exclusive economic zone of Norway and Russia (Figure 2). The first licenses for oil and gas exploration in the Norwegian Barents Sea were awarded in 1980, leading to the discovery of the Snøhvit gas field in 1984. In recent years, several oil discoveries were made in the Norwegian sector of the Barents Sea. On the Russian side, seismic surveying started in the 1970s, leading to the discovery of giant fields like Shtokmanovskoye (estimated to hold 3.8 trillion m^3 of recoverable gas), as well as Ledovoye and Ludovskoye. Further south in the Pechora Sea, many smaller fields were identified. More than 100 wells have been drilled in total, and the current assessment is that there is some 42,000–50,000 million barrels of oil equivalents in the Barents Sea, some 80 % of this on the Russian side. In the Kara Sea, to the east of Novaya Zemlya, Russians have discovered two other giant gas fields, Leningradskoye and Rusanovskye. Oil production in Russia shows a continuous increase. In 2005, it reached some 110 million barrels/day, and since then Russia was able to expand

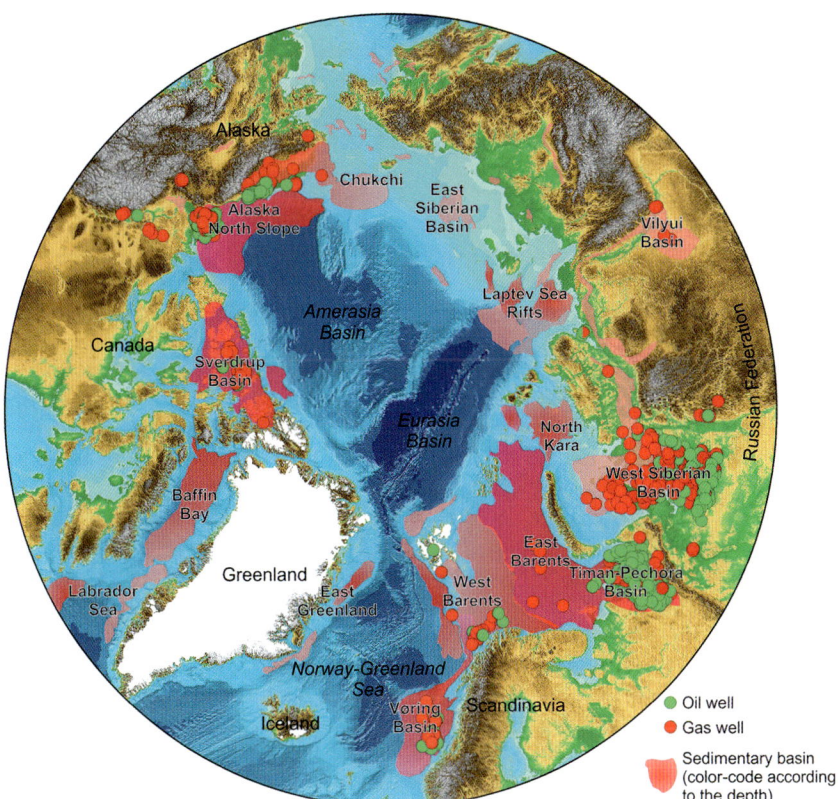

Energy Resources, Figure 2 Geography and bathymetry of the Arctic region. Major sedimentary basins with hydrocarbon potential (*red shaded*) are located within the exclusive economic zones of the Arctic countries. The sedimentary basins with hydrocarbon potential are commonly located in less than 500 m water depth (modified after Grantz et al. 2011).

its oil production still further. Oil reserve figures indicate that Russia can be one of the world's key oil producers for at least the next 40 years.

Among the greatest uncertainties in future energy supply and a subject of considerable environmental concern is the amount of oil and gas yet to be found in the Arctic (Gautier et al. 2009). Further research and a better understanding of the complex geological configuration of the Arctic is a prerequisite for reliable estimates of petroleum resources. However, the geological conditions for petroleum generation are favorable in the Arctic realm (Figure 2). A large majority of the undiscovered petroleum resources are expected to be found on the shallow Arctic shelves, well within the 200 nm exclusive economic zones of the countries bordering the Arctic Ocean. As the Arctic becomes more accessible, interest in its potential resources is increasing – but as well the public concern for the environmental consequences of petroleum exploration and production. When it becomes necessary to drill in harsh and sensitive environments such as the Arctic region, the costs of producing oil and gas are very high. However, interest for petroleum exploration is enhanced by estimates about how much oil and gas could be found in the Arctic region. The most reliable results in forecasting offshore large fields are obtained when the geological environment of a well-studied region onshore is extrapolated to the adjacent shelf area. By using a probabilistic geology-based methodology, the United States Geological Survey has assessed the entire area north of the Arctic Circle and concluded that about 30 % of the world's undiscovered gas and 13 % of the world's undiscovered oil may be found there, mostly offshore under less than 500 m of water (Gautier et al., 2009).This assertion of undiscovered Arctic reserves of 90 billion barrels of oil, 47 trillion (10^{12}) m^3 of natural gas, and 44 billion barrels of natural gas liquids (NGLs) still remains unproven. But there is no doubt that the resources are substantial in a global context, and there is consensus that some ¾ of these are expected to be natural gas.

Environmental impact of offshore petroleum production

The production of natural resources is inherently associated with environmental risks. Minor oil spills are quite common during oil production. On the other hand, there are natural oil and gas seeps all over the ocean seafloor. There is not a single oil field in the world that would be completely leak-proof. In the Gulf of Mexico, naturally seeping oil partly reaches the sea surface but is mainly deposited on the seafloor. This enables the growth of three kinds of naturally occurring oil-consuming bacteria: sulfate-reducing bacteria and acid-producing bacteria are anaerobic, while general bacteria are aerobic. These bacteria act to remove oil from an ecosystem, and their biomass tends to replace other populations in the food chain.

Most marine oil spills are the result of accidents during transportation; however, deep-sea accidents during production or exploration activities also occur. Although they are rare, they may have a significant impact on the environment, such as the accident with the Deepwater Horizon platform in 2010. Spilt oil penetrates the plumage of birds and the fur of mammals, reducing its insulating ability and making them more vulnerable to temperature fluctuations and much less buoyant in the water. Cleanup and recovery from an oil spill is difficult and depends upon many factors including the type of oil spilled, the temperature of the water, weather condition, and the types of shorelines and beaches involved.

The worst oil spills recorded in history include the crude oil carrier Amoco Cadiz accident in 1978, about 5 km off the coast of Brittany, France. The vessel contained 1857 barrels of crude oil when it ran aground. Severe weather resulted in the complete breakup of the vessel and the entire cargo was spilled into the sea. The fuel oil was swept onto the French shoreline, and the total extent of the spill includes more than 300 km of coastline, where the oil penetrated the sand on several beaches to a depth of half a meter. The resultant environmental impact was massive due to delayed cleanup operations owing to rough seas. In 1979, the Ixtoc I exploration well in the Gulf of Mexico suffered a blowout in waters 50 m deep. Due to a loss of hydrostatic pressure, the drilling rig lost mud circulation. Without the counterpressure from the circulating mud, the pressure in the oil reservoir forced oil up the well resulting in a disastrous blowout. Large quantities of oil and gas reached the surface and caught fire. The fire and the eventual collapse of the drilling rig caused damage to the underlying foundations of the well structures, which led to significant amounts of oil leaking into the Gulf. The incident led to the loss of 84,500 barrels of oil that spread to beaches. The water contamination directly affected food fish and octopus species. In early 1991, a disastrous environmental impact resulted from the war in Kuwait. The retreating Iraqi military forces set 700 oil wells on fire in Kuwait. This severely affected the Gulf economy and environment and destroyed several tankers and terminals that resulted in the loss of close to 8.5 million barrels of oil each day. One of the most devastating manmade environmental disasters occurred in 1989. The Exxon Valdez, an oil tanker that was carrying about 200,000 m^3 of oil, hit Prince William Sound's Bligh Reef on the south coast of Alaska and released 42,000–120,000 m^3 of crude oil into the sea. An inaccessible location made it difficult to deploy rescue and clean up operations quickly. The oil spilled retains its toxicity to this day putting a large amount of marine life and vegetation at risk. In 2009 an oil and gas leak occurred in the Montara well in the Timor Sea, off the northern coast of Western Australia in shallow waters. The oil spill was released following a blowout and continued leaking for 74 days. The oil leak could have been as high as 2,000 barrels. BP's Deepwater Horizon explosion in 2010 killed 11 people and injured several others, as well as resulted in a severe oil spill. The oil spill flowed for 3 months and spewed out 5.8 million barrels of crude oil. The spill is considered one of the worst accidents in US history and caused grave damage to the surrounding wildlife and biosphere, putting at risk many endangered species. The Gulf's fishing and tourism industry were also seriously affected. A 6-month offshore drilling moratorium below 150 m of water was enforced.

The Arctic area with its cold climate and waters represents a particularly vulnerable environment with respect to wildlife, biodiversity, fisheries, and nature. Environmental regulations in the Arctic realm are typically stricter than further south. Accidents analogous to BP's Deepwater Horizon disaster in the Gulf of Mexico are expected to have a much stronger impact on the highly sensitive Arctic environment. The risk resulting from transportation at sea has been shown quite plainly from the accident of the Exxon Valdez tanker in 1989 in offshore southern Alaska. There is already a risk of oil spills from the increased oil tanker traffic, and there are additional concerns raised by the increased petroleum exploration and production activities. The conditions in Arctic environments are less favorable for removing oil spills than in moderate or warm climate conditions. The water temperature is close to zero degrees Celsius, slowing down microbial processes and reducing the dissolution and evaporation of oil compounds. Because oil-degrading microbes are usually found in environmental samples, including relatively pristine Arctic environments, it is likely that the potential for oil degradation is already present in the ocean around Northern Greenland, but it is not known how fast and to what extent the degraders will respond to increased oil concentrations in waters and sediments. A better understanding of the potential of the indigenous microbial communities to react to naturally or accidentally spilled hydrocarbons is necessary to delineate risks and develop possible measures for remediation in the future.

Future targets for the petroleum exploration

One of the open questions is how much oil and gas will be discovered in previously unexplored offshore areas. Potential targets include shallow shelf areas that in the past were not accessible, because of, e.g., the harsh environmental conditions in the Arctic, unconventional resources, and deepwater areas (see Regional Marine Geology). In well-explored basins (see Marine Sedimentary Basins), further deepwater exploration will probably include more giant gas discoveries in relation to oil giants and a move to deposits in deeper water along passive continental margins (Chakhmakhchev and Rushworth, 2010).

The deep water is where many of the largest finds of recent times have been made – but also poses challenges,

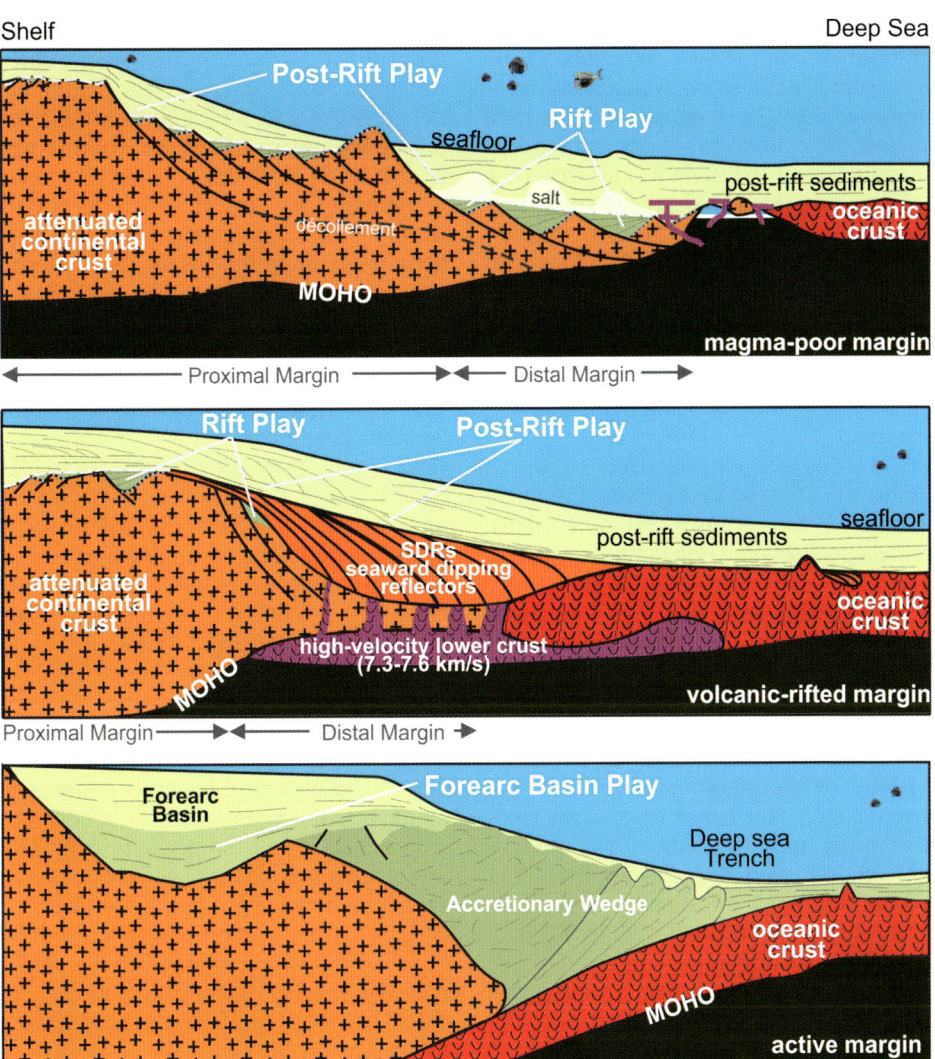

Energy Resources, Figure 3 Schematic sketch of the end-member extremes of continental margins. *Top*: The magma-poor margin is defined by a wide area of highly attenuated continental crust where the upper crust is deformed by deep-reaching listric faults that may sole out on a common detachment surface, the proximal margin. In the distal margin, the listric faults may cut across the entire crust leading to a detachment at the Mohorovičić (MOHO) discontinuity. Further seaward, extensional allochthons may be situated on exhumed mantle before relatively thin oceanic crust is reached. *Middle*: Volcanic rifted margins show a comparatively narrow proximal margin with considerable crustal thinning over a short distance, thick wedges of syn-rift volcanic flows manifest in seismic reflection data as seaward dipping reflectors (SDRs), and a wide high-velocity (Vp > 7.3 km/s) lower-crust seaward of the continental rifted margin. The oceanic crust is comparatively thick at these margins, especially close to the continent-ocean transition. *Bottom*: Currently not in the focus are hydrocarbons in forearc basins at active margins. These basins develop along active continental margin by the subsidence induced by the subducted oceanic crust.

as seen at BP's Deepwater Horizon disaster. Besides subsalt plays, advances in deep water have been made in exploiting stratigraphic traps, deltas, thrust structures, and carbonate reservoirs. Exploration success often is driven by a better understanding of the geology. Our present knowledge about the structure and architecture of passive continental margins (see Regional Marine Geology) is mainly derived from the interpretation of seismic data, supported by other geophysical data (see Technology in Marine Geosciences), and at times controlled by drilling. Such interpretation underpins exploration concepts, and future exploration is expected to be challenged by the maturation of conceptual models.

There are two end-member extremes of passive rifted margins: (1) magma-poor rifted margins and (2) volcanic rifted margins (Whitmarsh et al., 2001; Menzies et al., 2002; Franke, 2013; Peron-Pinvidic et al., 2013). Magma-poor rifted margins typically evolve along wide rifts that may be divided into a proximal and a distal part (Figure 3). The proximal margin is characterized by high-angle listric faults related to fault-bounded rift basins, while the distal margin is characterized by extremely thinned continental crust that is potentially separated from the oceanic crust by a domain of exhumed subcontinental mantle. The boundary between the two domains is at the point where listric faults cut across the entire crust reaching the mantle (see Lithosphere: Structure and Composition). In the proximal margin, typically a detachment is interpreted between the brittle upper crust and the mantle, while at the distal margin, brittle upper crust and upper mantle are separated by only a thin lower crustal layer or are juxtaposed (Figure 3). This leads to coupling and the development of a detachment at the crust-mantle boundary, allowing the exhumation of the mantle further seaward.

Volcanic rifted margins are characterized by up to 15 km thick wedges of volcanic flows (see "Intraplate Magmatism") (Hinz, 1981; Mutter et al., 1982), manifested in multichannel seismic reflection data as seaward dipping reflectors (SDRs), and high-velocity ($Vp > 7.3$ km/s) lower-crust seaward of the rifted continental margin (Figure 3). Because thinning of the continental crust typically occurs over a comparably short distance (50–100 km), the continent-ocean transition is abrupt and located in the vicinity of the SDRs (Figure 3). Various models place the continent-ocean transition, and thus the seaward limit of the presence of potential rift basins, at the seaward edge, in the center of, or at the landward edge of the seaward dipping reflectors. The effusive magmatism of volcanic margins and the corresponding lavas represented by the seaward dipping reflections obscure the seismic images of potential underlying fault blocks that undoubtedly accompanied the rifting. The major challenge will be a better imaging of sub-basalt structures to understand the complexity in the rift development, the heat flow history, and hydrocarbon maturation and thus to identify prospective plays (Skogseid, 2001).

Forearc basins (see Subduction) are only by way of an exception in the focus of petroleum companies. Such basins are located at almost all convergent continental margins between the deep-sea trench and the magmatic arc (Figure 3). Active margins stretch for a total length of some 44,000 km (von Huene and Scholl, 1991), mainly around the Pacific Ocean, but also along the Indonesian Island Arc, in the Caribbean, the Mediterranean, and the Arabian Sea. The widespread occurrence of this basin type makes it worth reviewing its hydrocarbon potential. Forearc basins may overlie either continental crust or accreted rocks that became attached to the continental plate during the collision of the involved plates. Commonly, forearc basins are bounded seaward by an outer-arc high that is part of the accretionary prism. The tectonic evolution of forearc basins is mainly controlled by the subduction of an oceanic plate. Important factors controlling the basin dynamics are convergence rate, lateral movements of the bending axis, dip angle of the oceanic plate, obliquity of convergence, roughness and sedimentary cover of the oceanic crust, and the sedimentary input of the landward magmatic arc and continent. Forearc basins typically show an elongated shape, extending parallel to the trench-arc system. Their width commonly ranges between 25 and 125 km, but their length may reach 500 km. Forearc basins are widespread; however, their hydrocarbon potential is believed to be low. This is due to generally low heat flow and the complicated structural history with long-lasting and usually ongoing tectonic activity affecting the basins, mainly terrigenous sediment input from the volcanic arc, and thus the questionable presence of prolific source rocks. Nevertheless, a tectonic basin classification alone is not sufficient to evaluate the petroleum potential of a sedimentary basin. The heat flow values of forearc basins range between 20 and 45 mW/m^2, with a typical value of 40 mW/m^2 (Dickinson 1995). In contrast, passive margins typically exhibit a heat flow between 30 and 50 mW/m^2, or even more in cases of forced fluid flow. Relatively cold oceanic crust is subducted at continental margins, and this reduces the heat flow beneath the forearc basin. Heat flow is typically further reduced by high sediment input from the volcanic arc and the adjacent continent.

Natural gas hydrate is an unconventional energy resource that is currently unexploited. Gas hydrate is a potentially vast source of hydrocarbon energy; the dissociation of 1 m^3 of gas hydrate can release 164 m^3 of gas. Gas hydrates occur under conditions of high pressure and low temperature and are abundantly found in marine environments and the permafrost environments of polar regions. However, the exact volumes are unknown and it is unclear how the gas may be produced. Some estimation predict that the carbon bound in gas hydrates is twice as high as in all other recoverable and non-recoverable energy resources such as oil, gas, and coal (Kvenvolden, 1993). Not only gas hydrates, but also the free gases trapped below the hydrates can be seen as potential energy sources.

Example: Simeulue forearc basin of the Sunda Arc, Indonesia

The Sumatra-Java area is part of the Sunda Arc (see Subduction) that stretches from the Andaman Sea in the northwest to the Banda Sea in the east. Along the Sunda Arc, the Indo-Australian Plate subducts under the Eurasian Plate. The western Indonesian forearc basins extend for more than 1,800 km from northwest of Aceh to southwest Java. The width of the basins varies from less than 70 km to the south of the Sunda Strait to about 120 km to the west of northern Sumatra. The basins form a strongly subsiding belt between the elevated Sumatra Paleozoic-Mesozoic arc massif cropping out along Sumatra and Java and the rising outer-arc high (Berglar et al., 2010).

The Simeulue forearc basin extends between Simeulue Island and northern Sumatra. It matches all features of a typical forearc basin (e.g., low heat flow, tectonic setting controlled by subduction). It is a frontier shallow shelf area with few wells and no wells in the basin center:

therefore it is studied using geophysical data and geological surface samples. Multichannel seismic data show bright spots above potential hydrocarbon reservoirs in carbonate buildups. Amplitude versus offset analysis indicates the presence of gas, and surface geochemical prospecting indicates thermal hydrocarbon generation. Heat flow in the Simeulue Basin ranges between 37 and 74 mW/m^2, as deduced from 1-D petroleum system modeling. Two possible source rocks (Eocene and Early-Middle Miocene; see Geological Time Scale) were assigned for 3-D petroleum system modeling in the Simeulue Basin. Due to a similar pre-Miocene geological evolution in the present-day back-arc and the forearc, it can be assumed that the back-arc source rocks also occur in the forearc.

Modeling based on two heat flow scenarios (40 and 60 mW/m^2) reveals that oil and gas generation is possible within and below the main depocenters of the central and southern Simeulue basin (Lutz et al., 2011). Deep burial (>6 km) of source rocks can compensate for low heat flow. This is an example of a prolific forearc basin and highlights the necessity to carefully evaluate the hydrocarbon potential of each basin individually.

Example: gas hydrates: an unconventional energy resource

Gas hydrates are drawing attention because of their contribution to geological hazards, to climate change, and because of their potential as a future energy supply. From the stability diagrams and the phase boundary of methane hydrate, it can be deduced that elevated pressure and reduced temperatures are necessary for the formation of gas hydrates. Additionally, sufficiently high concentrations of gases are essential. These conditions limit the occurrence of gas hydrates to marine environments (sediments of continental slopes, rises, and in the deep seas) and to permafrost environments (sediments in the Arctic region and in continental ice sheets). At continental margins, the formation of gas hydrates is primarily controlled by the sedimentation rate and the organic matter content. Therefore, the gases involved in the gas hydrates are mostly of biogenic origin. Gas hydrates in permafrost regions can also be the result of the migration of gases from deeper hydrocarbon reservoirs into the Gas Hydrate Stability Zone (Kvenvolden, 1993).

Most of the marine gas hydrates are detected based on the occurrence of bottom simulating reflectors (BSRs) in reflection seismic data. The BSR indicates the base of the Gas Hydrate Stability Zone and mirrors the seafloor. A BSR marks the phase boundary between free gas in the deeper sediments and the gas hydrates of the stability zone, which is determined by temperature, pressure, gas composition, and pore water salinity. Therefore, BSRs are not restricted to layer boundaries and may thus intersect them. In reflection seismic data, a BSR is generated by the inversion in the P-wave velocity from the hydrates to unconsolidated sediments. However, although the BSR is often cited as evidence of gas hydrates, the magnitude of the BSR amplitudes primarily depends on the amount of free gas below the BSR. There are examples of BSRs with no associated gas hydrates, or BSRs within the Gas Hydrate Stability Zone. Nevertheless, the existence of gas hydrates in sediments decreases, or even hinders the further migration of gases, so that in marine sediments, the highest concentration of the gas hydrates is generally just above the BSRs (Dillon and Max, 1998).

Much research is still needed to determine the geological and economic feasibility of extracting gas hydrates. Also the environmental implications of gas hydrate exploration must be considered thoroughly for safe and environmentally friendly exploration. Until now, the exploration of gas hydrates has not been successful due to the complex technical requirements for economic extraction, despite existing offshore and onshore know-how from the oil and gas industry. This is because, in contrast to conventional gas deposits, gas hydrate reservoirs react largely unpredictably to changes in physical and chemical environmental parameters. Therefore, better understanding of the reservoir behavior of gas hydrates is required before significant commercial exploration could take place. However, even if only a small fraction of the estimated gas hydrates could be produced, the resource could be very important (Collet, 2000).

Potential options for the recovery of gas hydrates are based on changing the factors that control their phase stability. This can be achieved by three principal methods. Producing the free gas below the Gas Hydrate Stability Zone reduces the local pressure so that some of the gas hydrates at the base of the Gas Hydrate Stability Zone dissociates and can be produced (depressurization). In the case of thermal stimulation, the dissociation of gas hydrates is achieved by increasing the temperature either by heating or by injecting warm surface water into the Gas Hydrate Stability Zone. Injection of inhibitors such as methanol into the gas hydrate-containing sediments changes the chemical composition so that some of the gas hydrates dissociate and release gas.

Summary

Energy resources from marine environments are of increasing importance. This is valid for both renewable and fossil energy resources: renewable energy represents at present only a minor percentage of global primary energy consumption but is growing very rapidly. Natural oil and gas are the most important offshore energy resources. Information concerning the amount of the producible and usable oil and gas resources depends on the geological conditions of the deposits, the state of the scientific and technical knowledge, the technological potential of the development, and production as well as on the economic and political requirements.

Offshore and deepwater petroleum exploration and exploration in the Arctic are ongoing since decades. At present drilling activity has particularly increased in the "golden triangle" of Brazil, the Gulf of Mexico, and West Africa. The greatest potential of undiscovered oil and gas is assumed in deepwater settings at continental margins and in the Arctic realm.

While it is believed that renewable energy has minor environmental impact, the production of oil and gas resources inherently comes along with a risk for the environment. Best practice regulations are necessary to reduce risks in the future.

Bibliography

Andruleit, H., Bahr, A., Babies, H. G., Franke, D., Meßner, J., Pierau, R., Schauer, M., Schmidt, S., and Weihmann, S., 2013. Energiestudie 2013 Reserven, Ressourcen und Verfügbarkeit von Energierohstoffen. Bundesanstalt für Geowissenschaften und Rohstoffe (BGR) für die Deutsche Rohstoffagentur (DERA), 112p.

Berglar, K., Gaedicke, C., Franke, D., Ladage, S., Klingelhoefer, F., and Djajadihardja, Y. S., 2010. Structural evolution and strike-slip tectonics off north-western Sumatra. *Tectonophysics*, **480** (1–4), 119–132.

Chakhmakhchev, A., and Rushworth, P., 2010. Global overview of recent exploration investment in deepwater – new discoveries, plays and exploration potential. Paper presented at the *AAPG Convention September 12–15, 2010*, Calgary, AB, Search and Discovery Article #40656.

Collet, T. S., 2000. Natural gas hydrates as a potential energy resource. In Max, M. D. (ed.), *Natural Gas Hydrate in Oceanic and Permafrost Environments*. Dordrecht: Kluwer Academic, pp. 123–136.

Dickinson, W. R., 1995. Forearc basins. In Busby, C. J., and Ingersoll, R. V. (eds.), *Tectonics of Sedimentary Basins*. Oxford: Blackwell Science, pp. 221–261.

Dillon, W. P., and Max, M. D., 1998. Oceanic methane hydrate: the characters of the Blake Ridge hydrate stability zone, and the potential for methane extraction. *Journal of Petroleum Geology*, **21**(3), 343–357.

Esser, R., 2001. Discoveries of the 1990s. In Downey, M. W., Threet, J. C., and Morgan, W. A. (eds.), *Petroleum Provinces of the Twenty-First Century*. Tulsa: American Association of Petroleum Geologists. AAPG Memoir 74, pp. 35–43.

Franke, D., 2013. Rifting, lithosphere breakup and volcanism: comparison of magma-poor and volcanic rifted margins. *Marine and Petroleum Geology*, **43**, 63–87, doi:10.1016/j.marpetgeo.2012.11.003.

Grantz, A., Scott, R. A., Drachev, S. S., Moore, T. E., and Valin, Z. C., 2011. Sedimentary successions of the Arctic Region (58–64° to 90°N) that may be prospective for hydrocarbons. *Geological Society, London, Memoirs*, **35**, 17–37.

Gautier, D. L., Bird, K. J., Charpentier, R. R., Grantz, A., Houseknecht, D. W., Klett, T. R., Moore, T. E., Pitman, J. K., Schenk, C. J., Schuenemeyer, J. H., Sørensen, K., Tennyson, M. E., Valin, Z. C., Wandrey, C. J., 2009. Assessment of undiscovered oil and gas in the arctic. *Science*, **324**, 1175–1179. doi:10.1126/science.1169467

Halbouty, M., 2001. Giant oil and gas fields of the decade 1990–1999: an introduction. Paper presented at the *AAPG Convention*, Denver, CO. Search and Discovery Article #20005.

Hinz, K., 1981. A hypothesis on terrestrial catastrophes: wedges of very thick oceanward dipping layers beneath passive continental margins – their origin and paleoenvironmental significance. *Geologisches Jahrbuch Reihe*, **E22**, 3–28.

Hughes, C., and Kinnersley, D., 2010. *Developments in Deepwater. Energy Briefings Series 2010 Deepwater.* Energy Institute. www.deloitte.co.uk/energybriefings

Kvenvolden, K. A., 1993. Gas hydrates- geological perspective and global changes. *Reviews of Geophysics*, **31**, 173–187.

Lutz, R., Gaedicke, C., Berglar, K., Schloemer, S., Franke, D., and Djajadihardja, Y. S., 2011. Petroleum systems of the Simeulue forearc basin, offshore Sumatra, Indonesia. *AAPG Bulletin*, **95**(9), 1589–1616.

Menzies, M. A., Klemperer, S. L., Ebinger, C. J., and Baker, J., 2002. Characteristics of volcanic rifted margins. In Menzies, M. A., Klemperer, S. L., Ebinger, C. J., and Baker, J. (eds.), *Volcanic Rifted Margins*. Boulder: Geological Society of America. Geological Society of America Special Paper Boulder, Vol. 362, pp. 1–14.

Mutter, J. C., Talwani, M., and Stoffa, P. L., 1982. Origin of seaward-dipping reflectors in oceanic crust off the Norwegian margin by "subaerial sea-floor spreading". *Geology*, **10**(7), 353–357, doi:10.1130/0091-7613(1982)10<353:oosrio>2.0.co;2.

Peron-Pinvidic, G., Manatschal, G., and Osmundsen, P. T., 2013. Structural comparison of archetypal Atlantic rifted margins: a review of observations and concepts. *Marine and Petroleum Geology*, **43**, 21–47, doi:10.1016/j.marpetgeo.2013.02.002.

Skogseid, J., 2001. Volcanic margins: geodynamic and exploration aspects. *Marine and Petroleum Geology*, **18**(4), 457–461.

von Huene, R., and Scholl, D. W., 1991. Observations at convergent margins concerning sediment subduction, subduction erosion, and the growth of continental crust. *Reviews of Geophysics*, **29**(3), 279–316, doi:10.1029/91rg00969.

Whitmarsh, R. B., Manatschal, G., and Minshull, T. A., 2001. Evolution of magma-poor continental margins from rifting to sea-floor spreading. *Nature*, **413**(6852), 150–154.

Cross-references

Bottom Simulating Seismic Reflectors (BSR)
Deep-sea Sediments
Geologic Time Scale
Intraplate Magmatism
Lithosphere: Structure and Composition
Marine Sedimentary Basins
Ocean Margin Systems
Oceanic Spreading Centers
Paleoceanography
Regional Marine Geology
Subduction
Technology in Marine Geosciences

ENGINEERED COASTS

H. Jesse Walker
Department of Geography and Anthropology, Louisiana State University, Baton Rouge, LA, USA

Definition

Engineered coasts, as used in this essay, refer to all coastal zones that have consciously modified by humans from their natural state. Emphasized are those that have been reclaimed, converted into ports, urban, and recreational

---

H. Jesse Walker is deceased.

centers, and those protected from atmospheric and oceanographic processes.

Introduction

Although humans did not originate in a coastal environment, once they "discovered" it, they entered an ecological niche that contained numerous opportunities and challenges. However, long before humans arrived on the scene, the coastal zone's gross character had been established by the constructive and destructive forces that accompanied the movement of Earth's several migrating plates, forces that are still occurring and that may cause havoc with human endeavors and often tax the expertise of coastal engineers. With the more localized actions of wave, current, tide, wind, glacier, biota, and sea-level change occurring along the world's 10^6 km long coastline (Bird and Schwartz, 1985), numerous highly varied forms developed. These forms, which are not necessarily mutually exclusive, include barrier islands, beaches, cliffs, coral reefs, deltas, dunes, estuaries and lagoons, mangrove swamps, marshes, and mudflats, among others. Because of location, structure, configuration, and composition, some coastal types are more amenable to human utilization and modification than others.

Occupational history and coastal impact

Although humans came in contact with the oceanic shores of Africa, Europe, and Asia and even Australia quite early, it was not until near the end of the Pleistocene that they moved into the Americas. It was not long thereafter that most of the world's habitable coastal zones were occupied even if sparsely. Morphological impact was mainly limited to the construction of middens while exploiting the zone's biological resources.

With the coming of agriculture, deltas were colonized, political organizations founded, and trade utilizing coastal locations developed. As trade expanded, coastal cities prospered. This trend continued to such an extent that today some 60 % of the world's population is coastal as are two-thirds of the world's largest cities (Viles and Spencer, 1995). Further, the pressures exerted by such concentrations are resulting in a coastal zone that is being humanly modified more rapidly than at any time in the past.

Recognizing that not all human modification is direct or even intentional, the discussion that follows, nonetheless, places emphasis on those changes that are engineered, namely, reclamation, harbors and ports, shoreline protection, and soft engineering.

From tide pools to polders

Just who was the first coastal engineer is unknown. However, if engineering includes all conscious modification of the environment, then he or she might have been the person who first rearranged boulders along a rocky shore in order to aid in harvesting tidal pools. The objective was to enhance food supply, an objective that has dominated much of coastal engineering. Such a simple structure that takes advantage of tides and waves has evolved into concrete tidal ponds in which the Japanese cultivate abalone, rock-bound *loko* (fish ponds) in aboriginal Hawaii, and solar salt pans in many countries around the world (Walker, 2002). Although only a few of the Hawaiian fish ponds are in use today, their structures still mark the shore line, and, for the most part, solar salt production has expanded from small pans to very large fields in France, Korea, Italy, the USA (Ver Plank, 1958), and elsewhere.

The structures associated with such endeavors capitalized on the physical and chemical processes of the near-shore waters. In contrast, as Chapman noted, "...reclamation usually signifies exclusion of marine or estuarine waters from littoral or riparian lands..." (1982, 514). Such exclusion can be achieved by raising the level of the land or by damming and then draining the area being reclaimed. Both practices have a long history. In China, a well-documented history of the Shijiang (Pearl River) delta traces reclamation through hundreds of years and for scores of kilometers. By 1950, it had more than 1,300 km of dykes and levees. Reclamation has been so extensive in China that more than half of its mudflats have disappeared (Halvany, 2009). Similar projects have been undertaken elsewhere. In Korea a plan, which is still ongoing, called for reclaiming 7,270 km^2 of tidal flat and shallow water areas.

Although reclamation has been mainly in the name of agriculture and mariculture, other objectives are gaining in importance. For example, in Singapore, it is being done in the name of industry, urban renewal, shipping, recreation, housing, and transportation with each reflecting its own engineering challenges (Walker and McGraw, 2010).

A somewhat different but more "sophisticated" (according to Volker, 1982, 2) type of reclamation is that generally known as *poldering*. Volker defines polder as a "...reclaimed level area having a naturally high watertable but where the surface and groundwater levels can be controlled" (1982, 2). Low-lying coastal areas, coastal marshes, and tidal embayments have all been polderized. Because polders are isolated from surrounding water bodies, elaborate drainage systems are necessary and excess water must be transferred through sluice gates or by pumps – achieved in part, and especially in the past, by picturesque Dutch windmills. The key to poldering is the dyke that surrounds the reclaimed land (Figure 1a). There is some question as to whether the first dykes in the Netherlands were defensive structures aimed at protecting the *terpen* (dwelling mounds) from floods or offensive structures that helped create new land (Harris, 1957). Whereas poldering is usually associated with the Netherlands, it is a practice that probably goes back at least 6,000 years. In more modern times (often with the help of Dutch engineers), it has been used in countries such as the USA, Venezuela, Colombia, India, Vietnam, and Russia.

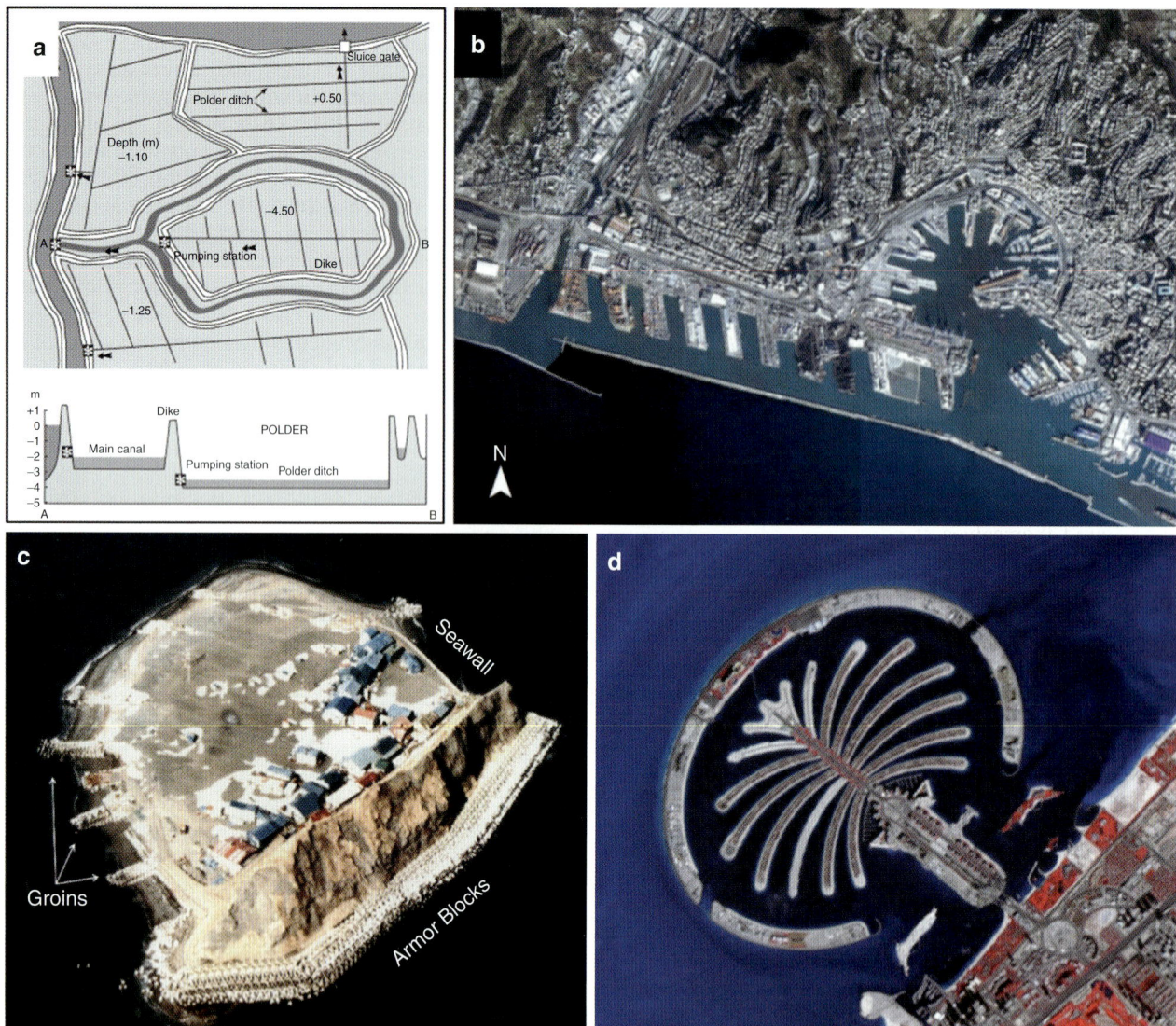

Engineered Coasts, Figure 1 Examples of engineered coasts: (a) the polder system (modified from Stive and Waterman, 2002); (b) the port of Genoa: note the protected natural harbor and the engineered structures (screenshot used by permission. Copyright@Esri. ArcGIS Data Providers); (c) Kojima, Japan; ~5 ha in size (from Walker, 2012); (d) Palm Jumeirah, a 5.72^2 km artificial residential and recreational island in Dubai (USGS/NASA Landsat).

Harbors and ports

When sailors began to ply the seas, they used irregularities in the shoreline as harbors. Eventually they were modified converting them into ports. The Mediterranean, with its very irregular shorelines, became extensively used by various trading and warring cultures such as the Minoans, Phoenicians, Greeks, and Romans. The Phoenicians, for example, had more than 75 ports between the Atlantic Ocean and the Black Sea (Sarton, 1960).

Although port development in the so-called Dark Ages and for a few centuries after was limited, during the periods of exploration/discovery/colonization and industrialization, ports increased in number and complexity (Figure 1b) and today are common along most of the coastlines of the world. Their number, size, and complexity vary with many factors including the nature of the coastline, location, population, and function.

Because natural harbors are rare along some coasts, ports are constructed by enclosing open ocean areas with offshore breakwaters and in some locales by dredging behind natural coastal barriers. Another type of modern facility is the deep-water mooring buoy from which pipelines transfer materials to onshore facilities. Ports are classified in a variety of ways, e.g., by size, location, and function. The World Port Index (NGIA, 2012) divides ports into eight types including those on rivers, lakes, and canals as well as on ocean shorelines. Of the eight types, the largest number falls in the coastal breakwater category, suggesting extensive shoreline engineering (Figure 1b). The WPI, which provides information on only 3,700 ports worldwide, excludes many, mostly small fishing ports. For example, Japan, with a coastal length of about 30,000 km, has more than 4,000, 28 % listed as commercial and 72 % as fishing ports (Shapiro, 1984). The WPI lists only 292 for Japan.

The structures that are found in ports include piers, docks, warehouses, and other support facilities. The type and number of which vary greatly from port to port depending mainly on their function.

Shoreline protection

"The need for coastal protection only becomes evident as human activities become established...and fixed structures and buildings are constructed..." (Gourlay, 1996, 12). Once humans began to fill the coast with dwellings, roads, industries, and other infrastructure elements, they often found themselves at odds with nature. According to Horikawa, beach erosion is "...one of the largest world problems from the perspective of land preservation" (1978, 327), and Bird (1996) adds emphasis by noting that 75 % of the world's beaches are in retreat. Human reaction to coastal erosion has mostly been "stabilized" so that today artificial structures dominate extensive sections of the world's shorelines.

This "hard" or "static" engineering approach centers around structures that are fixed in position and that have durability. They include those that are land/sea interface structures such as seawalls, bulkheads, and revetments. The seawall is the one "...most generally regarded by the public as representing the best form" (French, 2001, 51). Designed to separate land from sea, it in essence becomes an artificial cliff. Some, like the one along a typhoon-prone area on the Pacific coast of Japan, are more than 12 m high. Japan's coastline, much of which is also threatened by tsunami, has more than 8,000 km of dykes and seawalls.

Breakwaters, designed to eliminate or reduce wave action, are used to protect the shore behind or adjacent to them. Most early breakwaters were attached to the shore, especially in connection with harbors. Today they are also constructed off shore. They are especially abundant in Italy and Japan. In the USA, along the southern coast of California, detached breakwater construction began in 1899 and eventually extended along 14 km of shoreline (U.S. Army, 1977). Other hard structures include jetties and groins (Figure 1c). They are nonparallel structures designed to protect river entrances from silting (jetties) and to trap long-shore drift and combat shore erosion (groins).

The materials of which such structures are made are highly varied. Natural rock as rubble and riprap was the traditional material, although timber, steel piling, and, more recently, manufactured armor blocks are also used. The first fabricated armor unit was a simple concrete block that served as a substitute for natural boulders. In 1950 the French designed an interlocking unit called "tetrapod" which subsequently led to numerous designs including dolos (South Africa), akmon (the Netherlands), trumpets (Spain), and more than 30 others in Japan with such names, when translated from the Japanese, as spindle, turtle, and igloo (Walker, 1988).

The extent and character of an engineered shoreline varies greatly from one country or locale to another and depends not only on the nature of the coast itself but also on how humans, with their varied population densities, histories, cultural preferences, and technological expertise, have adapted to it. For example, less than 2 % of Finland, with its hard-rock, rebounding coast, is bordered by artificial structures, whereas more than 85 % of the coastline of Belgium is so bordered.

Soft engineering

Hard engineering has been the most common method of combating coastal erosion despite the fact that it is expensive and often unsightly and frequently aggravates erosion. Recently, a technique that is more compatible with natural processes has been utilized. Basically, it is the replenishment of the sand that has been removed from a beach by erosion. The first example of such an engineering feat may have been the creation of a beach in Turkey with sand shipped from Egypt more than 2,000 years ago for Cleopatra (El-Sammak and Tucker, 2002). Today, that beach (on the island of Sedir) is a popular tourist destination but one with very strict preservation restrictions.

Beach nourishment, proposed in the United States in 1916, was done sparingly for decades; the first large scale beach nourishment in Europe was at Norderney, Germany, in 1951. Beach nourishment mainly elevates the beach and advances the shoreline, thereby increasing its effectiveness against erosion and, of course, its desirability for recreation. Sediment sources for beach nourishment projects are varied including from offshore; transport is by truck, bulldozer, dredge, and pipeline (Finkl and Walker, 2002). Durability also varies; effectiveness averages between 5 and 10 years, although some, such as Miami Beach, Florida, last much longer.

Another soft structural procedure is the construction of dunes which serve as barriers to wave action, locales for vegetative growth, and reservoirs of sand during storms (National Research Council, 1990).

Other types of engineered coasts

Although reclamation, ports, and shoreline protection dominate coastal engineering, there are other modifications that have altered the natural coastal zone. Some examples include the construction of highways and railways along the top, the bottom, and on the slopes of cliffy coasts; the creation of flood gates as on the Thames, in Venice lagoon, and in the Netherlands; the damming of straits creating fresh water bodies as in Hong Kong; the construction of offshore islands for drilling platforms as in the Beaufort and Kara Seas; and the construction of marine parks in a number of countries. Marinas are a specialized type of port that takes many forms primarily dedicated to recreation. Recently the concept has been expanded with the use of Belgian and Dutch contractors in the construction of a 5.72^2 km residential and recreational complex in the form of a palm off the shore of Kuwait (Figure 1d).

Summary, conclusions, and the future

The time between when humans first utilized the varied resources of the coastal zone and when they first consciously began its modification (engineering) consumed most of human history. However, once such modification began, change progressed rapidly. Today, coastal reclamation projects dominate many coastal sites, ports are numerous in coastal countries, and protective structures front many shorelines. Further, urban waterfronts which reflect a great variety of modern-day activities – industry, commerce, housing, recreation – also demonstrate the complexity of modern-day engineered coasts.

Although most coastal engineering has been in support of such normal endeavors, much of it is in response to catastrophic events – e.g., landslides, hurricanes and typhoons, earthquakes, and tsunami. Recent examples include the hurricane/flooding reconstructions from Katrina in New Orleans and Sandy in the northeastern United States and the reconstructions resulting from the tsunamis in Southeast Asia and Japan. The style of engineering in response to coastal disasters also has a cultural component that differs among countries (Bijker, 2007).

Because this summary paper about coastal engineering has only been able to touch on selected characteristics of the topic, it is recommended that the reader consult other, more definitive, treatments such as Artificial Structures and Shorelines (Walker, 1988) and, especially, History and Heritage of Coastal Engineering (Kraus, 1996). The Artificial Structures volume demonstrates how coastal engineers have impacted the shorelines of 48 countries around the world, whereas the History and Heritage volume provides detailed and well-illustrated histories of coastal engineering for 15 major countries.

Engineering and coastal modification in the future will continue to evolve as environmental conditions such as sea-level and storminess change, as technology develops, and as the attitudes of humans vis-à-vis the coastal zone modify. The recent acceptance of the importance of coastal wetlands impact on reclamation and the trend toward soft engineering are but two examples.

Bibliography

Bijker, W., 2007. American and Dutch coastal engineering: differences in risk conception and differences in technological culture. *Social Studies of Science*, **37**, 143–151.

Bird, E. C. F., 1996. *Beach Management*. Chichester: Wiley.

Bird, E. C. F., and Schwartz, M. L. (eds.), 1985. *The World's Coastline*. New York: Van Nostrand Reinhold Company.

Chapman, D. M., 1982. Land reclamation. In Schwartz, M. L. (ed.), *Encyclopedia of Beaches and Coastal Environments*. Stroudsburg: Hutchinson & Ross, pp. 513–516.

El-Sammak, A., and Tucker, M., 2002. Ooids from Turkey and Egypt in the Eastern Mediterranean and a love-story of Antony and Cleopatra. *Facies*, **46**(1), 217–227.

Finkl, C. W., and Walker, H. J., 2002. Beach nourishment. In Chen, J., Eisma, D., Hotta, K., and Walker, H. (eds.), *Engineered Coasts*. Dordrecht: Kluwer Academic Publishers, pp. 1–22.

French, P. W., 2001. *Coastal Defences: Processes, Problems and Solutions*. London: Routledge.

Gourlay, M. R., 1996. History of coastal engineering in Australia. In Kraus, N. C. (ed.), *History and Heritage of Coastal Engineering*. New York: ASCE, pp. 1–88.

Halvany, M. G., 2009. Wetlands and reclamation. In *International Encyclopedia of Human Geography*. Amsterdam: Elsevier, pp. 241–246. Downloaded on December 27, 2012 from http://www.sciencedirect.com.lib.lsu.edu

Harris, L. E., 1957. Land drainage and reclamation. In Singer, C., Holmyard, E., Hall, A., and Williams, T. (eds.), *A History of Technology*. London: Oxford University Press, pp. 300–323.

Horikawa, K., 1978. *Coastal Engineering*. New York: Wiley.

Kraus, N. C. (ed.), 1996. *History and Heritage of Coastal Engineering*. New York: American Society of Civil Engineers, 603 pp.

National Research Council, 1990. *Managing Coastal Erosion*. Washington, DC: National Academy Press.

NGIA, 2012. *World Port Index*. Springfield, VA: National Geospatial Intelligence Agency. Pub. 150.

Sarton, G., 1960. *A History of Science*. Cambridge: Harvard University Press.

Shapiro, H. A., 1984. Coastal area management in Japan: an overview. *Coastal Zone Management Journal*, **12**, 19–56.

Stive, M. J. F., and Waterman, R. E., 2002. The Netherlands: the Zuyder Zee project. In Chen, J., Eisma, D., Hotta, K., and Walker, H. (eds.), *Engineered Coasts*. Dordrecht: Kluwer Academic Publishers, pp. 279–308.

U. S. Army, 1977. *Shore Protection Manual*. Washington, DC: U.S. Government Printing Office.

Ver Plank, W. E., 1958. *Salt in California*. San Francisco: Division of Mines, State of California. Bulletin 175.

Viles, H., and Spencer, T., 1995. *Coastal Problems*. London: Edward Arnold.

Volker, A., 1982. Polders: an ancient approach to land reclamation. *Natural Resources*, **18**, 2–13.

Walker, H. J. (ed.), 1988. *Artificial Structures and Shorelines*. Dordrecht: Kluwer Academic Publishers, 708 pp.

Walker, H. J., 2002. Marinas, sea-level reservoirs, solar salt pans and other artificial shorelines. In Chen, J., Eisma, D., Hotta, K., and Walker, H. (eds.), *Engineered Coasts*. Dordrecht: Kluwer Academic Publishers, pp. 151–183.

Walker, H. J., 2012. In defense of an island: Kojima's fight against the sea. *Focus*, **55**, 11–18.

Walker, H. J., and McGraw, M., 2010. Reclamation and the coastal zone with emphasis on the United States of America. In Piastra, S. (ed.), *Land Reclamations: Geo-Historical Issues in a Global Perspective*. Bologna: Pàtron Editore, pp. 137–167.

Cross-references

Beach
Beach Processes
Coastal Engineering

Coasts
Deltas
Hurricanes and Typhoons
Plate Motion
Relative Sea-level (RSL) Cycle
Sea Walls/Revetments
Tsunamis
Wadden Sea
Waves

EPICENTER, HYPOCENTER

Nina Kukowski
Institute of Geosciences, Friedrich-Schiller-University, Jena, Germany

Synonyms
Focus (hypocenter)

Definition
The *hypocenter* of an earthquake is the subsurface location at which energy stored along a locked fault is first released. The *epicenter* is that location at the surface, which is closest to the hypocenter.

During an earthquake, rupture propagates from the hypocenter to both sides along the originally locked fault (see "Earthquakes," Figure 1). The hypocenter is directly located beneath the epicenter, and the distance or depth of the hypocenter, respectively, is the focal depth. Seismic waves initially propagate spherically away from the hypocenter and will be reflected and refracted at, e.g., the boundaries of geological layers, which show a contrast in impedance, which is the product of seismic velocity and rock density. This leads to complicated travel paths of seismic waves arriving at seismic stations.

One of the most important inverse problems of seismology is to find the hypocenter, i.e., the location, where the earthquake nucleates, from the arrival times of the different seismic waves at seismic stations (Stein and Wysession, 2003). Whereas the distance to the epicenter, the epicentral distance (Lowrie, 2007), can be estimated from the time difference of the arrivals of p- and s-waves at one seismic station, to localize the epicenter, distances determined from three stations are necessary, from which it can be found through triangulation. As Earth's velocity structure is not homogenous, there will be an error in the epicenter estimate, which will be quite small, if data from many seismic stations are available and if these stations are not too far away.

Focal depth can range from close to the surface of the Earth to several hundred kilometers depth (see "Earthquakes," and "Wadati-Benioff-Zone"). It can be estimated from the epicentral distance and the distance traveled by the direct wave. Again, combining estimates from several seismic stations improves the accuracy of determining focal depth.

In case of shallow earthquakes, the epicenter usually also is within the area of greatest damage. However, in case of deeper earthquakes, when rupture is propagating all along the fault to the surface, damage usually is greatest where the fault reaches the surface. This is especially true for seismogenic zone (see "Seismogenic Zone") earthquakes along subduction zones (see "Subduction"), which can reach the largest magnitudes observed. For such seismic events, the seismogenic fault often reaches the seafloor, which may lead to destructive tsunamis (see "Tsunamis").

Bibliography
Lowrie, W., 2007. *Fundamentals of Geophysics*, 2nd edn. New York: Cambridge University Press. 38199.
Stein, S., and Wysession, M., 2003. *An Introduction to Seismology, Earthquakes and Earth Structure*. Malden, MA: Blackwell, 498 pp.

Cross-references
Earthquakes
Fault-Plane Solutions of Earthquakes
Seismogenic Zone
Subduction
Wadati-Benioff-Zone

EROSION-HIATUSES

William W. Hay
Department of Geological Sciences, University of Colorado at Boulder, Estes Park, CO, USA

Definition
Erosion is the processes whereby the detritus and solutes resulting from the weathering of sediments and rock are transported to a site of deposition. *Hiatus* derives from the Latin, meaning a gap in something. In its geological sense, a *hiatus* represents either a cessation in deposition of sediments, resulting in an interval during which no strata form, or an erosional surface within the stratigraphic sequence representing a gap in the rock record.

History
Prior to the introduction of the ideas of seafloor spreading and plate tectonics, documented by the recovery of long sequences of ocean basin sediments by the Deep Sea Drilling Project (DSDP) in the late 1960s and 1970s, it was thought that the deep sea was the repository of an essentially undisturbed monotonous but continuous multibillion-year record of Earth history. It was thought that sedimentation had been slow but continuous in a

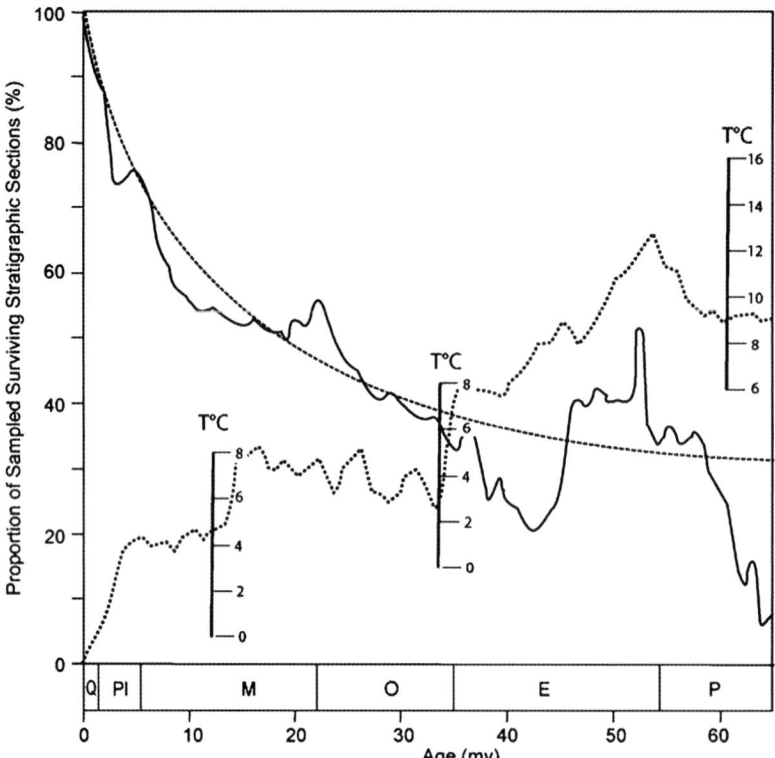

Erosion-Hiatuses, Figure 1 Moore and Heath's (1977) compilation of the proportion of surviving stratigraphic sections for the Cenozoic (*solid line*), a decay curve assuming an erosion factor of 0.07 %/my (*dashed curve*) and a plot of ocean bottom water temperatures based on oxygen isotopes in benthic foraminifera after Savin (1977) (*dotted line*), with temperature scales reflecting the changing oxygen isotope ratios in response to the buildup of ice on land.

sluggish environment that had always been cold. The first estimates of the velocities of currents in the deep sea came in the 1950s through analysis on the data collected by the German "Meteor" expedition of 1925–1927, revealing that current velocities of the deep western boundary currents ranged between 2 and 18 cm/s.

It was only after the recovery of extensive sediment cores by the DSDP that it became evident that the record was one not only of changing conditions over the past 180 million years, but that there were many gaps in the record caused by erosion and redeposition by bottom currents. As sonic swath mapping of the seafloor proceeded, it was realized that there are many bottom features, such as sediment drifts, sand sheets, dunes, gravel bars, and scour marks that reflect the activity of bottom currents.

The overall pattern of deep-sea erosion and deposition

Moore and Heath (1977) compiled data on the amount of the sampled chronostratigraphic section preserved in cores taken during the first 32 legs (313 sites) of the DSDP.

It was evident that there were many gaps in the record and that the missing parts of the record increased with age. This was an expression of the phenomenon that had been discovered a few years earlier by James Gilluly based on studies of the age distribution of sediments on the continents: that younger sediments are largely the product of erosion and redeposition of older sediments. A much more complete discussion of the distribution of hiatuses in deep-sea sediments, both in time and space, was presented by Moore et al. (1978). The first 32 legs of the DSDP were close to being a random sample of the sediments on the deep ocean floor. Subsequently the selection of sites for sediment drilling and coring was adjusted to prefer areas where loss of section due to hiatuses would be minimal.

It was evident that the widely held assumption that the deep sea was the ultimate repository of sedimentary material and was unaffected by the erosion characteristic of the land, continental shelf, and slope was incorrect (Figure 1).

A modern interpretation of the data would be that the initiation of intensive bottom water formation as a result

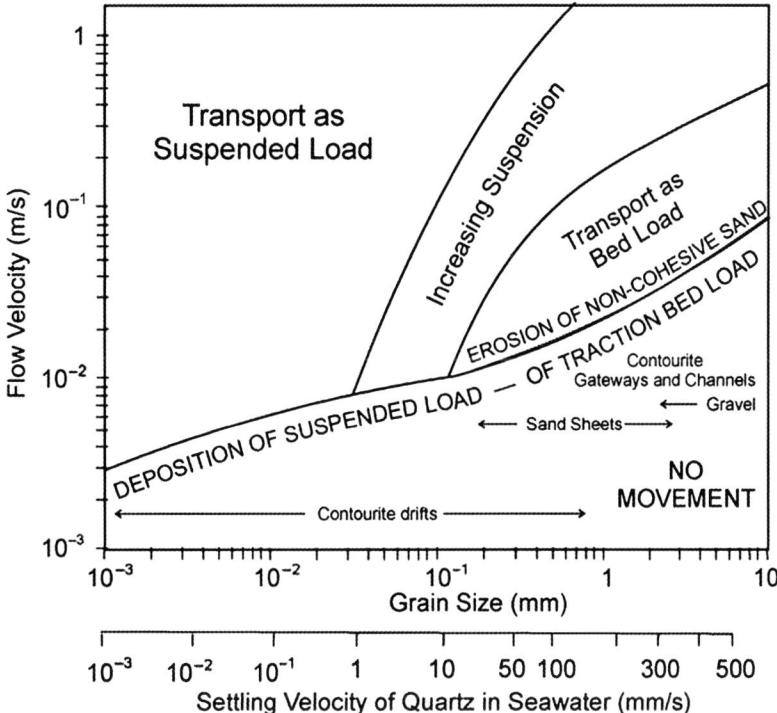

Erosion-Hiatuses, Figure 2 Relation between seawater flow velocity and erosion and deposition of sediments with some major current-generated bottom features (after Brown et al. 1993).

of global cooling caused erosion and redeposition of unconsolidated sediments on the deep ocean floor. Because the deep ocean waters are corrosive to the $CaCO_3$ skeletal materials of oceanic plankton, their erosion and resuspension often resulted dissolution, destroying the fossil record of their reworking into younger sediments.

Erosion, suspension, transport, and deposition of sediment on the ocean floor

Figure 2 shows the relation between current flow velocity and sediment grain size in terms of erosion, transport, and deposition of unconsolidated sediment.

It has become evident that ocean bottom waters are everywhere in motion. Velocities of the order of 2 cm/s are common, but significant variations are introduced by the tides and eddies. These can result in "benthic storms" when flow velocities reach 20 cm/s or more. Local flow velocities are controlled by the bottom roughness and topography. The currents suspend, scour, transport, and redeposit sediment, leaving hiatuses in the geologic record.

Extent of hiatuses

Hiatuses can be recognized as unconformities in seismic profiles. Figure 3 shows interpretation of a profile in the Atlantic off Florida by Vail et al. (1980). Unconformities are common and extensive. This is a region that is affected by the north to south-flowing western boundary current of the North Atlantic Basin.

Summary

Erosion, once thought to be rare or impossible in the deep sea, has turned out to be a common phenomenon. It produces hiatuses (gaps in the record of geologic time) in the stratigraphic record. Ocean bottom current velocities vary with time but can exceed 20 cm/s and easily erode and transport bottom sediment either as suspended load or along the bottom by traction. Carbonate skeletal elements of nannoplankton and foraminifera, eroded and suspended in the water column, are subject to dissolution by ocean deep waters so that reworked fossils are relatively rare.

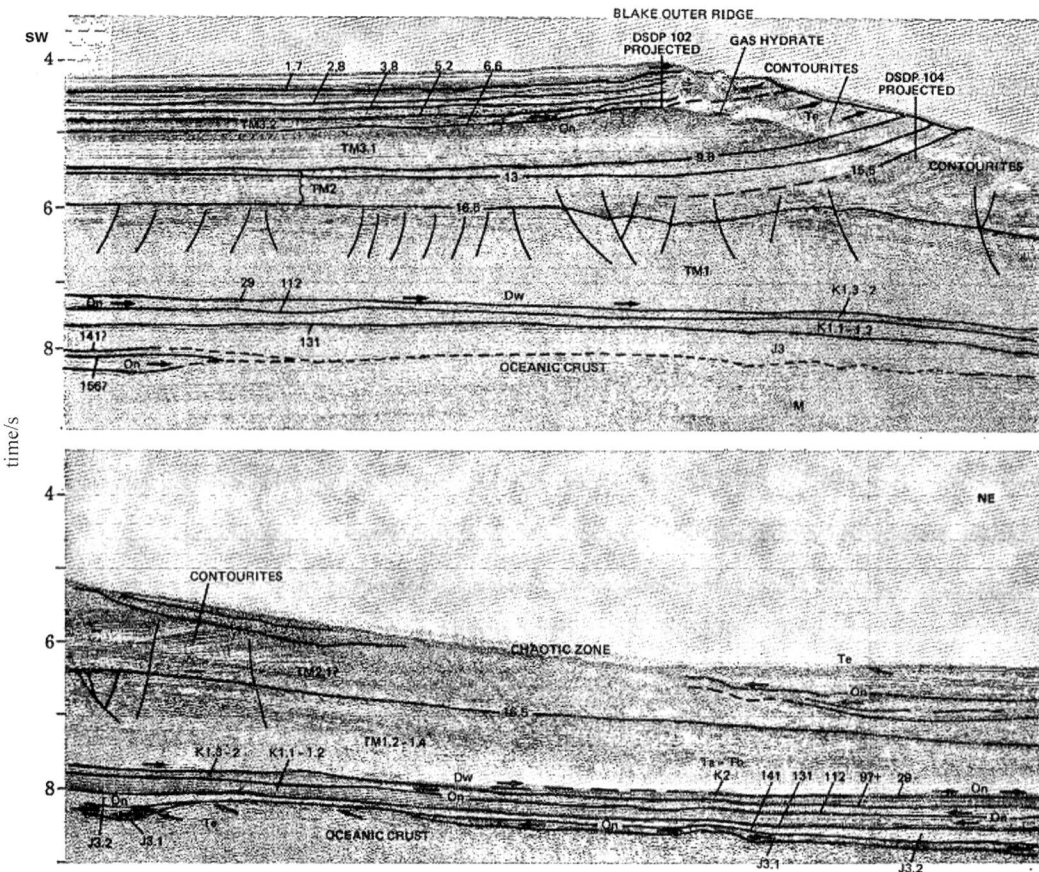

Erosion-Hiatuses, Figure 3 Unconformities (hiatuses) in the Blake Outer Ridge region of the western North Atlantic (Vail et al. 1980). Unconformities are marked as heavy lines. Numbers associated with them are estimates of the ages of the sediment below the unconformity.

Bibliography

Brown, J., Colling, A., Park, D., Phillips, J., Rothery, D., and Wright, J., 1993. *Waves, Tides, and Shallow Water Processes*. Milton Keynes: The Open University, p. 187.

Gilluly, J., 1969. Geological perspective and the completeness of the geologic record. *Geological Society of America Bulletin*, **80**, 2303–2312.

Gross, T. F., and Williams, A. J., 1991. Characterization of deep-sea storms. *Marine Geology*, **99**, 81–301.

Moore, T. C., Jr., and Heath, C. R., 1977. Survival of deep-sea sedimentary sections. *Earth and Planetary Science Letters*, **37**, 71–80.

Moore, T. C., Jr., van Andel, T. H., Sancetta, C., and Pisias, N., 1978. Cenozoic hiatuses in pelagic sediments. *Micropaleontology*, **24**, 113–138.

Savin, S., 1977. The history of the Earth's surface temperature during the past 100 million years. *Annual Review of Earth and Planetary Sciences*, **5**, 319–355.

Stow, D. A. V., Faugères, J.-C., Howe, J. A., Pudsey, C. J., and Viana, A. R., 2002. Bottom currents, contourites and deep-sea sediment drifts: current state-of-the-art. In Stow, D. A. V., Pudsey, C. J., Howe, J. A., Faugères, J.-C., and Viana, A. R. (eds.), *Deep-Water Contourite Systems: Modern Drifts and Ancient Series, Seismic and Sedimentary Characteristics*. London: Geological Society. Memoirs, Vol. 22, pp. 7–20.

Vail, P. R., Mitchum, R. M., Shipley, T. H., Buffler, R. T., and Matthews, D. H., 1980. Unconformities of the North Atlantic. *Philosophical Transactions of the Royal Society of London A*, **294**, 137–155, doi:10.1098/rsta.1980.0021.

Wüst, G., 1958. Über Stromgeschwindigkeiten und Strommengen in der Atlantischen Tiefsee. *Geologische Rundschau*, **47**, 187–195.

Cross-references

Contourites
Deep-sea Sediments
Ocean Drilling

ESTUARY, ESTUARINE HYDRODYNAMICS

Halina Kowalewska-Kalkowska and Roman Marks
Institute of Marine and Coastal Sciences, University of Szczecin, Szczecin, Poland

Synonyms
River mouth area

Definition
According to the definition given by Cameron and Pritchard (1963), an estuary is a semi-enclosed coastal body of water which has a free connection with the open sea and within which seawater is measurably diluted with freshwater derived from land drainage. This definition applies to the so-called classical (positive) estuaries of the Earth's temperate latitudes in which freshwater input is the main driver of the long-term circulation. Freshwater inflow establishes longitudinal density gradients resulting in long-term surface outflow and net inflow underneath. In some arid basins, called inverse (negative) estuaries, where evaporation exceeds the sum of precipitation and runoff, the loss of freshwater becomes a main force shaping long-term circulation and the development of longitudinal density gradients, in analogy to classical estuaries (Valle-Levinson, 2010).

Introduction
Estuaries constitute unique ecosystems influenced by freshwater input and tide currents which establish the estuarine circulation often called the gravitational circulation. This steady-state circulation may be modified by tidal currents, the Earth's rotation, atmospheric forcing, topographic features, and river discharge (MacCready and Geyer, 2010; Geyer and MacCready, 2014). The water circulation controls many estuarine processes and phenomena. One of peculiar features in estuaries is the occurrence of turbidity maxima in the area of interface of fresh- and salt waters. These zones of high suspended sediment concentrations influence primary production, contaminant flushing, fish migration, and dredging (Mitchell et al., 1998). Estuaries are regions extensively exploited by man since they serve as locations for ports, urban and industry development, sewage discharge systems, recreational areas, farming, as well as fishing and hunting grounds (Wolanski and McLusky, 2012). It should be kept in mind, however, that any pollutant in an estuary, i.e., pollution loading, oil spills, thermal discharges, coastal dredging, and offshore drilling, may significantly affect its environment.

Origin of estuaries
Considering their geological origin, estuaries are divided into four types (Pritchard, 1967; Valle-Levinson, 2010):

- **Drowned river valleys (coastal plain estuaries)**: originally rivers, formed as a result of the Pleistocene increase in sea-level, typically funnel shaped, several kilometers wide, and shallow (of about 10 m depth) with large width/depth discrepancy. The salinity decreases steadily from about 30 ‰ (at the mouth) to about 0.1 ‰ (at the head). In this type of estuary the stretch of the freshwater river above the upper limit of intrusion of sea-derived salt is still subject to tidal action. Examples of drowned river valleys are Chesapeake Bay, Delaware Bay, and the Elbe Estuary.
- **Fjords**: located in high-latitude regions, where glaciation has been a major factor in shaping the land, characterized by an elongated, deep channels with steep sides and a sill. The sill is related to a moraine of either a currently active glacier or an extinct glacier. Fjords are typical in Scandinavia, Greenland, and Antarctica as well as in New Zealand, Canada, and Patagonia.
- **Bar-built estuaries**: formed when offshore barrier sand islands and sandpits build above sea-level and extend between headlands in a chain, broken by one or more inlets, usually shallow, oriented parallel to the coast. In these estuaries the circulation is provided by the wind, as tides are considerably choked and the river discharge is of low volume. Bar-built estuaries are found in northern Mexico or in southern Portugal.
- **Tectonic estuaries**: formed by earthquakes or by the Earth's crust fractures and creases that generated faults in regions adjacent to the ocean, typically developed after sinking a part of the crust and forming a hollow basin. An example of this type of estuary is San Francisco Bay.

Water balance of estuaries
In terms of water balance estuaries may be classified according to Pritchard (1952) as:

- **Positive estuaries**: in which freshwater input from rivers, rainfall, groundwater, and ice melting exceeds evaporation. The longitudinal density gradient drives a net volume outflow to the ocean; surface outflow is stronger than near-bottom inflow.
- **Inverse (negative) estuaries**: in which evaporation exceeds the combined sources of freshwater. They show stronger surface inflow than near-bottom outflow.
- **Neutral estuaries**: in which evaporation and freshwater input are in balance.

With the exceptions of lagoons in arid or semiarid regions, most estuaries are positive. As reported by Valle-Levinson et al. (2001), some estuaries could change seasonally from one type to another (e.g., they are positive during rainy seasons, whereas negative during dry periods). The estuary during the neutral stadium represents a temporal transition between positive and negative stages (Kjerfve, 1989).

Salinity vertical structure and mixing processes

Estuaries may be divided based on the vertical stratification and the extent of lateral homogeneity (Bowden, 1967; Pritchard, 1989). According to Valle-Levinson (2010), this classification includes:

- **Salt-wedge estuary**: caused by weak tidal forcing and large river discharge, the tidally averaged salinity profiles show the sharp pycnocline, with mean flow dominated by outflow throughout most of the water column and weak inflow in a near-bottom layer; mixing between fresh- and salt waters is negligible (e.g., the Mississippi River, Ebro River, Rio de la Plata).
- **Strongly stratified (fjord-type) estuaries**: formed as a result of moderate to large river discharge and weak to moderate tidal forcing. The tidally averaged salinity profiles have a well-developed pycnocline with weak vertical variations above and below it. The inflow is weak because of weak mixing with freshwater and weak horizontal density gradients. An example of this type of estuary is the almost tideless Baltic Sea.
- **Weakly (partially) stratified estuaries**: resulted from moderate to strong tidal forcing and weak to moderate river discharge. A weak pycnocline or continuous stratification from surface to bottom, except near the bottom mixed layer, is observed. Stronger tidal currents induce mixing between fresh- and salt waters (e.g., Chesapeake Bay, Delaware Bay, James River).
- **Well-mixed estuaries**: produced by strong tidal forcing and weak river discharge, in which mean salinity profiles are practically uniform and mean flows are unidirectional with depth. In wide estuaries inflow may develop on one side across the estuary and outflow on the other side (e.g., the lower Chesapeake Bay in early autumn).

Estuaries would tend to shift from strongly stratified to well mixed with (a) decreasing river flow, (b) increasing tidal velocities, (c) increasing width, and (d) decreasing depth (Pritchard, 1989). They respond to consecutive tidal cycles, meteorological forcing, and topographic features. For instance, the Hudson River Estuary changes from highly stratified during neap tides to weakly stratified during spring tides (Valle-Levinson, 2010).

Principal forcing variables

The tidal velocity and the freshwater flow constitute the basis of the prognostic approach to classification of estuaries. Geyer and MacCready (2014) mapped the estuarine parameter space using two nondimensional parameters Fr_f and M, where Fr_f is the freshwater Froude number, the net velocity due to river flow scaled by the maximum possible frontal propagation speed, with a tidal Froude number as the other axis (Geyer, 2010) and M is the mixing number which is based on the ratio of the tidal timescale to the vertical mixing timescale (Geyer and MacCready, 2014). Estuaries with similar hydrodynamic conditions would be expected to appear at the similar place in the Fr_f-M parameter space. They are depicted as not points in the parameter space but rectangles owing to the spring-neap variations in tidal velocity as well as changes in river discharge and variations in depth. Salt-wedge estuaries such as the Mississippi and the Ebro River (with higher values of the Fr_f number) are near the top of the Fr_f-M space. Partially mixed estuaries fall in the middle (e.g., the Hudson and the James Rivers). Fjords appear in the lower-left corner, whereas well-mixed estuaries (the M number higher than 1) appear in the lower-right part of the estuarine parameter space.

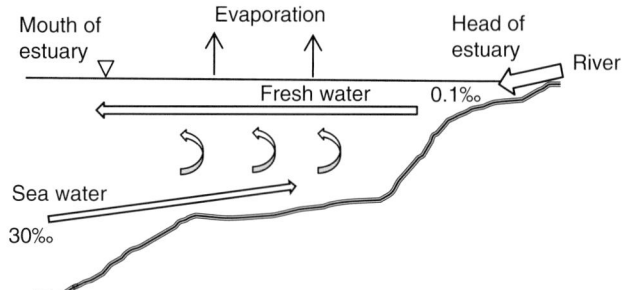

Estuary, Estuarine Hydrodynamics, Figure 1 Scheme of gravitational circulation in a partially mixed estuary.

Estuarine hydrodynamics

Currents in the most common partially stratified estuaries are primarily induced by density and elevation differences between fresh- and salt waters. They establish two-layer gravitational circulation, which is maintained by dynamic balance between advective and diffusive processes. Primarily less dense freshwater has a tendency to remain in the surface layer of estuary (Figure 1). However, due to the impact of tide and wind, a vertical exchange between fresh and salt layers takes place. That process explains the existence of longitudinal and vertical salinity gradients in the estuary (Kjerfve, 1989). The time-average pressure surfaces tilt seaward in less dense surface layer forcing a net outflow of freshwater. In the bottom layer they tilt upstream, driving the salty and dense water toward land. At a certain depth in mid-water column, the pressure surfaces become horizontal and a level of no net motion is observed. The net outflow from the estuary can have a much greater volume than the river discharge as it carries some of the seawater back toward the sea.

Tidal straining is one of effects forcing water circulation within estuaries and refers to variations in stratification that may not reach the well-mixed limit (MacCready and Geyer, 2010). Due to the convective instability of

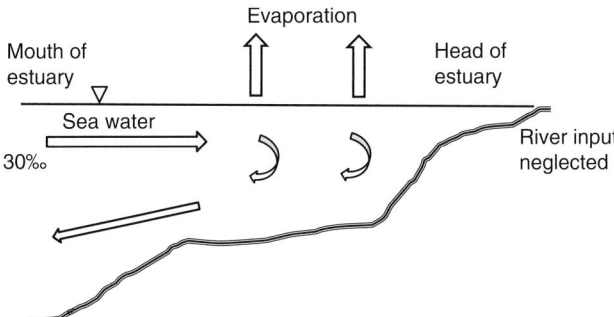

Estuary, Estuarine Hydrodynamics, Figure 2 Scheme of inverse estuarine circulation.

the near-bottom flow at the flood tide, the vertical velocity profile is much more uniform at the flood than at the ebb tide. As a consequence, velocity profiles are bottom intensified during flood and surface intensified during ebb (Burchard and Baumert, 1998). Burchand and Hetland (2010) revealed that tidal straining is the governing process in generating estuarine circulation for periodically stratified estuaries. The process may not only be driven by tidal asymmetries resulting from horizontal buoyancy gradients but from other asymmetric process as well (e.g., wind straining). Tidal straining is the main mechanism for generation of estuarine turbidity maxima.

Spring-neap modulation of tidal forcing has been found in many estuaries. In these estuaries during neap tides, potential energy input due to straining by the mean gravitational circulation exceeds the buoyancy flux due to tidal mixing and results in increase in stratification (Geyer and MacCready, 2014).

Lateral flows in estuaries are usually much smaller than the dominant tidal currents directed along the channel (Lerczak and Geyer, 2004). In many systems differential advection drives secondary flows, with the exception in the vicinity of channel beds where flow curvature is the main factor (Chant, 2010). The Coriolis acceleration is the next important forcing variable influencing the estuarine secondary circulation. Lerczak and Geyer (2004) found that the lateral flow tends to be stronger during flood than ebb tides. Moreover, Scully et al. (2009) demonstrated that tidal rectification of lateral advection may act as a driving force for the residual estuarine circulation.

In some enclosed water bodies located in arid or semi-arid regions, inverse current system may emulate as a result of negligible river discharge and low precipitation but high evaporation. In those negative estuaries, the high evaporation rates cause increase in salinity and density of a surface layer, which sinks deeper into the ocean (Figure 2). Then it outflows seaward at the near-bottom layer, while less dense ocean water inflows landward as a surface layer. Reverse estuaries present the longitudinal density gradient of the opposite sign as compared with positive estuaries, i.e., water density increases landward (Valle-Levinson, 2010). Examples of reverse estuaries are the Bay of Guaymas in Mexico in the late spring (Valle-Levinson et al., 2001) or Shark Bay in Australia (Hetzel et al., 2013). In some estuaries with freshwater input from rivers but very small, during hot and dry season strong evaporation may cause the occurrence of salt plug near the river mouth. That salinity maximum zone serves as a buffer between the fresh to brackish waters and the offshore waters. That phenomenon was recorded, i.e., in the South Alligator River or the Escape River in Australia (Wolanski, 1986).

Investigations of estuarine processes

Initially the processes taking place in estuaries were mainly investigated during field experiments. Since the 1950s estuarine modeling has emerged, starting from studies by Ketchum (1951) and Stommel (1951). Subsequently, mathematical models describing mixing and circulation in estuaries have been devised by, i.e., Stommel and Farmer (1952), Pritchard (1952), and Hansen and Rattray (1965). In the recent years the extended capacity of numerical models allowed to simulate physical processes in estuaries with a high accuracy. Burchard et al. (2004) applied General Estuarine Transport Model (GETM) for simulating dynamics of the estuarine turbidity maxima in the Elbe Estuary. The GETM was also successfully applied for studying the water dynamics in the Baltic Sea (Holtermann et al., 2014). Scully et al. (2009) explored the mechanisms driving the estuarine circulation in the Hudson River Estuary utilizing the Regional Ocean Modeling System (ROMS). An estuarine and coastal version of Princeton Ocean Model was efficacious for the simulations of water circulation in the Delaware Bay (Schmalz, 2009), the Hudson Estuary (Hellweger et al., 2004), or the tideless Oder Estuary in the Baltic Sea (Kowalewska-Kalkowska and Kowalewski, 2006). The recent study by Burchard et al. (2011) allowed to quantify the contribution of the gravitational circulation, tidal straining, advectively driven circulation, and horizontal mixing circulation to longitudinal and lateral residual circulation in tidally energetic estuaries.

Summary

Estuaries are complex systems, with nonlinear cross-coupling and feedback between the circulation and density structure (Geyer and MacCready, 2014). The water circulation within them is controlled by river discharge, tidal currents, the Earth's rotation, atmospheric forcing, and bathymetry impacts. The classifications of estuarine systems show the great diversity in their regimes. Many estuaries cross boundaries between different estuarine types as a result of the numerous processes. Recent advances in the understanding of estuarine circulation allowed to decompose the residual estuarine circulation and study the contributions from various processes taking place in tidally energetic estuaries (Burchard et al., 2011).

Estuaries, extensively exploited by man, need careful management and continuous monitoring of environmental processes occurring there (Wolanski and McLusky, 2012). The nowadays research is based on the joint applications of field measurements and advanced numerical models. It is expected that gained knowledge will prevent the deterioration of the estuarine natural environments and preserve them as biodiverse-rich and healthy marine ecosystems for future generations.

Bibliography

Bowden, K. F., 1967. Circulation and diffusion. In Lauff, G. H. (ed.), *Estuaries*. Washington: American Association for the Advancement of Science, Vol. 83, pp. 15–36.

Burchard, H., and Baumert, H., 1998. The formation of estuarine turbidity maxima due to density effects in the salt wedge. A hydrodynamic process study. *Journal of Physical Oceanography*, **28**, 209–321.

Burchard, H., and Hetland, R. D., 2010. Quantifying the contribution of tidal straining and gravitational circulation to residual circulation in periodically stratified tidal estuaries. *Journal of Physical Oceanography*, **40**, 1243–1262.

Burchard, H., Bolding, K., and Villarreal, M. R., 2004. Three-dimensional modelling of estuarine turbidity maxima in a tidal estuary. *Ocean Dynamics*, **54**, 250–265.

Burchard, H., Hetland, R. D., Schulz, E., and Schuttelaars, H. M., 2011. Drivers of residual estuarine circulation in tidally energetic estuaries: straight and irrotational channels with parabolic cross section. *Journal of Physical Oceanography*, **41**, 548–570.

Cameron, W. M., and Pritchard, D. W., 1963. Estuaries. In Hill, M. N. (ed.), *The Sea*. New York: John Wiley and Sons, Vol. 2, pp. 306–324.

Chant, R. J., 2010. Estuarine secondary circulation. In Valle-Levinson, A. (ed.), *Contemporary Issues in Estuarine Physics*. New York: Cambridge University Press, pp. 100–124.

Geyer, W. R., 2010. Estuarine salinity structure and circulation. In Valle-Levinson, A. (ed.), *Contemporary Issues in Estuarine Physics*. New York: Cambridge University Press, pp. 12–26.

Geyer, W. R., and MacCready, P., 2014. The estuarine circulation. *Annual Review of Fluid Mechanics*, **46**, 175–197.

Hansen, D. V., and Rattray, M., 1965. Gravitational circulation in straits and estuaries. *Journal of Marine Systems*, **23**, 104–122.

Hellweger, F. L., Blumberg, A. F., Schlosser, P., Ho, D. T., Caplow, T., Lall, U., and Li, H. H., 2004. Transport in the Hudson estuary: a modeling study of estuarine circulation and tidal trapping. *Estuaries*, **27**, 527–538.

Hetzel, Y., Pattiaratchi, C., and Lowe, R. J., 2013. Intermittent dense water outflows under variable tidal forcing in Shark Bay, Western Australia. *Continental Shelf Research*, **66**, 36–48.

Holtermann, P. L., Burchard, H., Gräwe, U., Klingbeil, K., and Umlauf, L., 2014. Deep-water dynamics and boundary mixing in a nontidal stratified basin: a modeling study of the Baltic Sea. *Journal of Geophysical Research, Oceans*, **119**, 1465–1487.

Ketchum, B. H., 1951. The exchanges of fresh and salt waters in tidal estuaries. *Journal of Marine Research*, **10**, 18–38.

Kjerfve, B., 1989. Estuarine geomorphology and physical oceanography. In Day, J. W., Jr., Hall, C. A. S., Kemp, W. M., and Yáñez-Arancibia, A. (eds.), *Estuarine Ecology*. New York: John Wiley & Sons, pp. 47–78.

Kowalewska-Kalkowska, H., and Kowalewski, M., 2006. Hydrological forecasting in the Order estuary using a three-dimensional hydrodynamic model. *Hydrobiologia*, **554**, 47–55.

Lerczak, J. A., and Geyer, W. R., 2004. Modeling the lateral circulation in strait, stratified estuaries. *Journal of Physical Oceanography*, **34**, 1410–1428.

MacCready, P., and Geyer, W. R., 2010. Advances in estuarine physics. *Annual Review of Marine Science*, **2**, 35–58.

Mitchell, S. B., West, J. R., Arundale, A. M. W., Guymer, I., and Couperthwaites, J. S., 1998. Dynamics of the turbidity maxima in the upper Humber estuary system, UK. *Marine Pollution Bulletin*, **37**, 190–205.

Pritchard, D. W., 1952. Estuarine hydrography. *Advances in Geophysics*, **1**, 243–280.

Pritchard, D. W., 1967. What is an estuary: physical viewpoint. In Lauff, G. H. (ed.), *Estuaries*. Washington: American Association for the Advancement of Science, Vol. 83, pp. 3–5.

Pritchard, D. W., 1989. Estuarine classification – a help or a hindrance. In Neilson, B. J., Kuo, A., and Brubaker, J. (eds.), *Estuarine Circulation*. Clifton: Humana Press, pp. 1–38.

Schmalz, R. A., Jr., 2009. Comparison of the Princeton Ocean Model and the Regional Ocean Modeling System hindcasts in the Delaware River and Bay. In Starrett, S. (ed.), *Proceedings of World Environmental and Water Resources Congress 2009: Great Rivers*. Reston, Va: American Society of Civil Engineers, pp. 1–15.

Scully, M. E., Geyer, W. R., and Lerczak, J. A., 2009. The influence of lateral advection on the residual estuarine circulation: a numerical modeling study of the Hudson River estuary. *Journal of Physical Oceanography*, **39**, 107–124.

Stommel, H., 1951. *Recent Developments in the Study of Tidal Estuaries*. Technical report, Woods Hole: Woods Hole Oceanographic Institution, Ref. No. 51–33, pp. 1–18.

Stommel, H., and Farmer, H. G., 1952. *On the Nature of Estuarine Circulation*. Technical report, Woods Hole: Woods Hole Oceanographic Institution, Ref. No. 52–51, pp. 1–172.

Valle-Levinson, A., 2010. Definition and classification of estuaries. In Valle-Levinson, A. (ed.), *Contemporary Issues in Estuarine Physics*. New York: Cambridge University Press, pp. 1–11.

Valle-Levinson, A., Delgado, J. A., and Atkinson, L. P., 2001. Reversing water exchange patterns at the entrance to a semiarid coastal lagoon. *Estuarine, Coastal and Shelf Science*, **53**, 825–838.

Wolanski, E., 1986. An evaporation-driven salinity maximum zone in Australian tropical estuaries. *Estuarine, Coastal and Shelf Science*, **22**, 415–424.

Wolanski, E., and McLusky, D. (eds.), 2012. *Treatise on Estuarine and Coastal Science*. London: Academic.

Cross-references

Currents
Deltas
Geohazards: Coastal Disasters
Lagoons
Sea-Level

EUSTASY

Reinhard Dietrich
Institute of Planetary Geodesy, Technical University of Dresden, Dresden, Germany

A volume change of the global ocean divided by the area of the global ocean is called eustatic effect. The eustatic effect is one number representing the average value of global sea-level change (Pugh, 1987, p. 460; Emery and

Aubrey, 1991, p. 165). It may be caused by mass exchange between land and sea (built up or melting of continental ice and snow, changes in hydrological conditions on land) or by a density change (temperature, salinity) of the ocean water. The latter one is also called steric effect.

If the contribution of land ice is considered only, the resulting effect is called glacial eustasy. However, the built up or melting of ice masses will never result in a uniform sea-level change. Here, the changes of the gravity field due to the mass changes ("geoid effect") and the response of the solid earth due to changing mass loads ("glacio-isostatic adjustment") have to be considered as well.

For the time period 1993–2010, a value of 2.8 ± 0.5 mm/year for the eustatic effect has been determined (IPCC, 2013). The main contributions come from thermal expansion (1.1 mm/year), glaciers and ice caps (0.76 mm/year), and ice sheets (0.8 mm/year).

Bibliography

Emery, K. O., and Aubrey, D. G., 1991. *Sea Levels, Land Levels, and Tide Gauges*. New York/Berlin/Heidelberg/London/Paris/Tokyo/Hong Kong/Barcelona: Springer.

IPCC, 2013. Summary for policymakers. In Stocker, T. F., Qin, D., Plattner, G.-K., Tignor, M., Allen, S. K., Boschung, J., Nauels, A., Xia, Y., Bex, V., and Midgley, P. M. (eds.), *Climate Change 2013: The Physical Science Basis. Contribution of Working Group I to the Fifth Assessment Report of the Intergovernmental Panel on Climate Change*. Cambridge, UK/New York: Cambridge University Press.

Pugh, D. T., 1987. *Tides, Surges and Mean-Sea-Level*. Chichester/New York/Brisbane: Wiley.

Cross-references

Glacio(hydro)-isostatic Adjustment
Relative Sea-level (RSL) Cycle
Sea-Level

EVENTS

William W. Hay
Department of Geological Sciences, University of Colorado at Boulder, Estes Park, CO, USA

Definition

Events are relatively short-lived phenomena recorded in the geologic record by depositional, erosional, or geochemical features. They may be of local significance (e.g., a storm deposit (tempestite), debris flow, submarine landslide, tsunami deposit) or more extensive (e.g., a volcanic ash deposit, changes in the distribution of fossil species, changes in ocean water properties such as anoxia reflected by sapropels) or even global (eustatic sea-level changes, variations in isotopic ratios, changes in seawater composition).

Major events in ocean history

Knowledge of the history of the oceans depends in part on how much of the ocean floor is still available for an examination. As shown in Figure 1a, only half of the ocean crust with its sediment that existed 180 million years ago is still present. The oldest fragment of the ocean crust in the ocean basins is between 180 and 200 million years old, but other bits of the older ocean crust are preserved as ophiolites in mountain ranges. Subduction and consequent loss of the geologic record primarily affect the Pacific and Indian Oceans. Subduction in the Atlantic Ocean basins is limited to the eastern peripheries of the Caribbean and Scotia Plates.

The salinity of the ocean has generally declined during the Phanerozoic as salt has been removed to form evaporite deposits. Erosion of older evaporites on the continental blocks slowly returns salt to the ocean, but overall, the trend has been toward freshening of the ocean. Salinities in the Precambrian are thought to have been generally above 50 ‰. During the Paleozoic, they were of the order of 48 ‰ to 42 ‰. Figure 1b shows major salt extractions and changes in the average ocean salinity since 180 Ma.

The breakup of Pangaea produced several isolated deep basins on ocean crust, particularly the opening Gulf of Mexico and southern South Atlantic, that accumulated large amounts of salt, and each significantly reduced ocean salinity. The deposition of salt in the isolated Mediterranean and Red Sea basins and the Persian Gulf lowered salinity to its modern value of about 34.72 ‰. The evaporites in these deposits on ocean crust will not return to the ocean until continental collisions or other plate tectonic processes bring them back above sea-level. The buildup of ice on land during the Pleistocene glacials increased ocean salinity to as high as 35.90 ‰, indicated by the asterisk at the top of the figure.

Ocean anoxic events (OAEs), shown in Figure 1c, were episodes of extensive anoxia in the intermediate and/or deep ocean. Most of these are known only from the Tethys and North Atlantic, as indicated by the width of the gray bars in the figure. Durations of these events, as estimated by Erba (2004) but here expressed in millions of years, were as follows: the Toarcian OAE – ?1? myr, Weissert OAE – 2 myr, OAE 1a – 1.250 myr, OAE 1b 0.046 myr, OAE 1c – 1 myr, OAE 1 day – 0.3 myr, and OAE 2 (the Bonarelli event) – 0.3–0.4 myr. Each of the OAEs occurred during very warm greenhouse conditions.

During the Cretaceous and Paleocene, the ocean's deep waters were warm, containing only about half as much dissolved oxygen as cold deep waters. This was undoubtedly one of the factors contributing to the development of OAEs.

Also indicated in Figure 1c is the Cretaceous-Tertiary boundary event, probably caused by the impact of an asteroid in Yucatan. The global effect of this event is marked in the ocean by the extinction of most calcareous nannoplankton and planktonic foraminifera. The recovery of the diversity in the plankton took several million years.

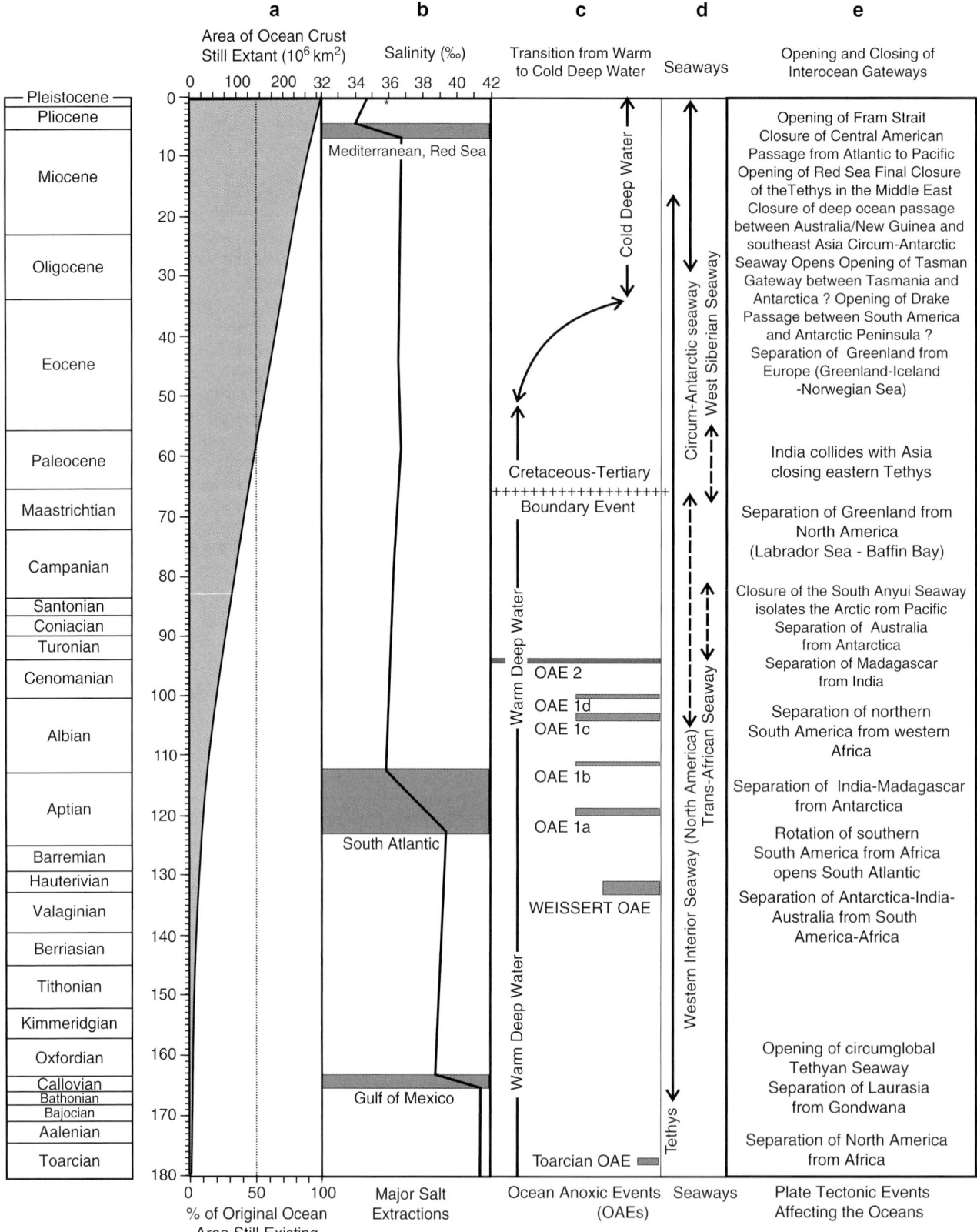

Events, Figure 1 Major events in the history of the oceans since 180 Ma.

A major shift in circumglobal circulation occurred in the mid-to late Tertiary. From about 165 to 8 Ma, circumglobal ocean circulation was possible through the Tethys, located about 20–30° north of the equator. However, the Tethys consisted of many basins separated by swells, both of which were constantly changing, affecting the width and depth of shallow water passages. Connections at depth were even more variable, and it is possible that a deep passage through the entire Tethys never existed. As the Tethys closed, circumglobal circulation shifted to the high latitudes of the Southern Hemisphere. A continuous path for circulation around the Antarctic continent became possible with the opening of the Drake Passage and Tasman Gateway during the Eocene.

Seaways across the continental blocks were associated with the high sea levels of the Cretaceous. These are shown as dotted lines in Figure 1d. The Western Interior Seaway of North America connected the Gulf of Mexico with the Arctic Ocean, but probably had an arm via the present Hudson Bay region across Greenland to northern Europe. Finally, during the middle of the Cretaceous, a seaway extended across Africa from the northern South Atlantic to the Mediterranean. In the latest Cretaceous and early Tertiary, a very shallow seaway extended across western Siberia periodically connecting the Tethys with the Arctic.

A shallow seaway from the North Atlantic to the Arctic between Greenland and Europe (not shown in Figure 1d) existed from the time of the separation of North America and Africa until spreading began to separate Greenland from Europe to form the Greenland-Iceland-Norwegian (GIN) Sea. However, a deep connection between the Arctic and North Atlantic is blocked by the Greenland-Iceland-Scotland Ridge.

The opening and closing of major interconnections between the ocean basins are shown in Figure 1e. Although very important over the long term, these were gradual "events" that occurred over intervals of a few million years. Opening passageways were initially shallow and became deeper with time, just as closures initially restricted deep water circulation and ended by shutting off surface circulation.

Summary

Events are episodes of change recorded by depositional, erosional, or geochemical features. They may be only local or regional in extent or global, reflecting significant trends in ocean evolution.

Bibliography

Dercourt, J., Ricou, L. E., Adamia, S. A., Csaszar, G., Funk, H., Lefeld, J., Rakus, M., Sandulescu, M., Tollman, A., and Tchoumatchenko, P., 1990. *Northern Margin of Tethys; Paleogeographical Maps*. Bratislava: Statny Geologicky Ustav Dionyza Stura. 11 sheets.

Dercourt, J., Ricou, L.-E., and Vrielynck, B. (eds.), 1993. *Atlas of Tethys Paleoenvironmental Maps*. Paris: Gauthier-Villars. 14 maps + explanatory notes.

Dercourt, J., Gaetani, M., Vrielynck, B., Barrier, E., Biju-Duval, B., Brunet, M. F., Cadet, J. P., Crasquin, S., and Sandulescu, M., 2000. *Peri-Tethys Paleogeographical Atlas*. Paris: Gauthier-Villars. 24 maps + explanatory notes.

Erba, E., 2004. Calcareous nannofossils and Mesozoic oceanic anoxic events. *Marine Micropaleontology*, **52**, 85–106.

Haq, B. U., Hardenbol, J., and Vail, P. R., 1987. Chronology of fluctuating sea levels since the Triassic. *Science*, **235**, 1156–1167.

Hay, W. W., Migdisov, A., Balukhovsky, A. N., Wold, C. N., Flögel, S., and Söding, E., 2006. Evaporites and the salinity of the ocean during the Phanerozoic: implications for climate, ocean circulation and life. *Palaeogeography Palaeoclimatology Palaeoecology*, **240**, 3–46.

Hohbein, M. W., Sexton, P. F., and Cartwright, J. A., 2012. Onset of North Atlantic deep water production coincident with inception of the Cenozoic global cooling trend. *Geology*, **40**, 255–258.

Holser, W. T., 1977. Catastrophic chemical events in the history of the ocean. *Nature*, **267**, 403–498.

Holser, W. T., 1985. Gradual and abrupt shift in ocean chemistry during Phanerozoic time. In Holland, H. D., and Trendall, A. F. (eds.), *Patterns of Change in Earth Evolution, Dahlem Konferenzen 1984*. Berlin: Springer, pp. 123–143.

Jenkyns, H. C., 1999. Mesozoic anoxic events and paleoclimate. Zentralblatt für Geologie und Paläontologie, Teil I 1998. *Heft*, **5–6**, 943–949.

Kazmin, V. G., and Napatov, L. M. (eds.), 1998. *The Paleogeographic Atlas of Northern Eurasia*. Moscow: Institute of Tectonics of the Lithospheric Plates, Russian Academy of Natural Sciences. 26 maps.

Knauth, L. P., 2005. Temperature and salinity history of the Precambrian ocean: implications for the course of microbial evolution. *Palaeogeography Palaeoclimatology Palaeoecology*, **219**, 53–69.

Müller, R. D., Sdrolias, M., Gaina, C., and Roest, W. R., 2008. Age, spreading rates, and spreading asymmetry of the world's ocean crust. *Geochemistry, Geophysics, Geosystems*, **9**, Q04006, doi:10.1029/2007GC001743. 19 pp.

Schlanger, S. O., and Jenkyns, H. C., 1976. Cretaceous oceanic anoxic events: causes and consequences. *Geologie en Mijnbouw*, **55**, 179–184.

Sclater, J. G., Jaupart, C., and Galson, D., 1980. The heat flow through oceanic and continental crust and the heat loss of the Earth. *Reviews of Geophysics and Space Physics*, **18**, 269–311.

Cross-references

Anoxic Oceans
Calcite Compensation Depth (CCD)
Erosion-Hiatuses
Marine Evaporites
Sedimentary Sequence

EXPLOSIVE VOLCANISM IN THE DEEP SEA

Christoph Helo
Institute of Geosciences, Johannes Gutenberg-University Mainz, Mainz, Germany

Definition

Explosive eruption	Eruption characterized by magma fragmentation at the vent, producing pyroclasts.
Effusive eruption	Extrusion of magma as lava flow.

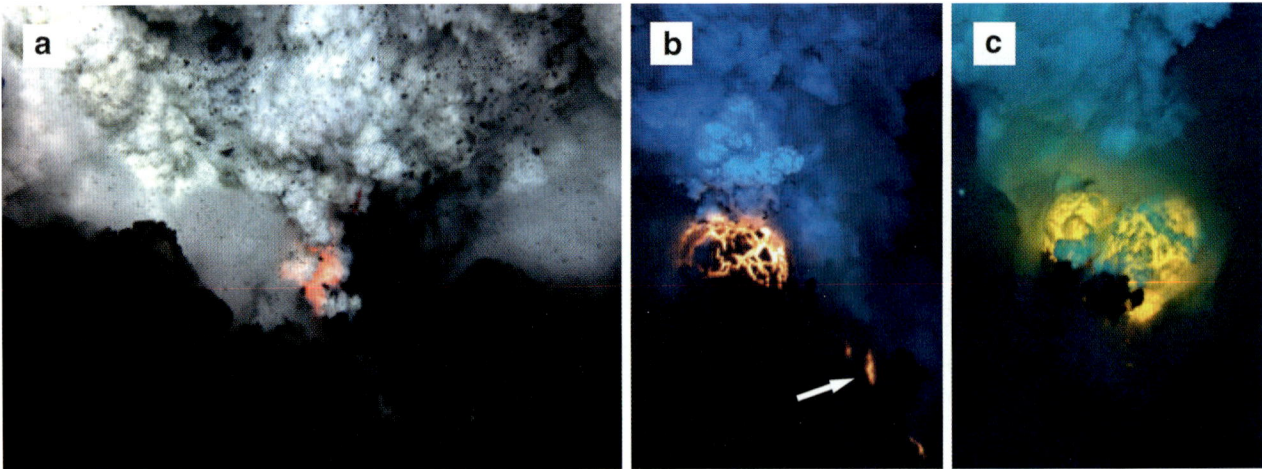

Explosive Volcanism in the Deep Sea, Figure 1 Explosive deep-sea eruption at West Mata, Lau back-arc basin. (**a**) Strong active degassing and pyroclasts formation. Field of view is ~2.5 m. (**b**), (**c**) Magma bubble bursts, in (**b**) visibly accompanied by effusion of degassed lava (white arrow). (**b**) Field of view is ~1.25 m; (**c**) base of bubble is ~0.5–0.8 m (Resing et al. (2011), adapted by permission from *Macmillan Publishers Ltd*: Nature Geoscience, *copyright* (2011)).

Pyroclast	Rock or glass fragment expelled from volcanic vent during the eruption.
Volcaniclast	Nongenetic term for any fragment of volcanic origin.
Magma	Multiphase system containing silicate melt ± crystals ± bubbles.
Melt, silicate	Liquid phase of a molten rock.
Volatile	Chemical compound of the melt that can form a free gas phase (mostly water, carbon dioxide, and sulfur, minor chlorine and fluorine).
Solubility	Maximum concentration of a volatile phase dissolvable in the melt; decreases strongly with declining pressure.
Exsolution	Formation of a free magmatic gas phase (commonly as bubbles), as the volatile concentration exceeds solubility.

Introduction

Within the past decades, it has been realized that volcanoes in the deep sea (i.e., water depths in excess of a few hundreds of meter) compare in their general spectrum of eruption styles to those on land and explosive activity is common to both environments (Figure 1). Either setting covers the full compositional width from basaltic to rhyolitic. The seafloor is volcanically the most active place on Earth (roughly 75 % output volume of terrestrial volcanism), with mid-ocean ridge basalts (MORB) accounting for most of the eruptive output. While explosive eruptions of rhyolitic volcanoes in the deep sea may reveal themselves prominently by so-called floating pumices (buoyant, highly porous pyroclasts), ongoing basaltic eruptions commonly remain undetected as monitoring of the deep sea is still sparse and technically challenging.

Our current understanding of deep-sea volcanism is vitally owed to indirect information sources: (1) ophiolite sequences and other uplifted blocks of ancient seafloor (e.g., Moores et al., 1984; Staudigel and Schmincke, 1984), (2) dredging and ocean drilling of the ocean seafloor (e.g., Fox and Hezeen, 1965), and (3) tedious mapping and sampling using underwater vehicles (cf. Yoerger et al, 2007; Rubin et al., 2012). The abundance of fine-grained volcaniclastic material in addition to lava flows was recognized early on. Initial models on the formation of volcaniclastites at water depths below 500 m focused largely on nonexplosive mechanisms (Fisher and Schmincke, 1984), mainly lava granulation due to cooling contraction and spalling of glassy lava rinds (Schmincke et al., 1978; Bonatti and Harrison, 1988).

Clastic deposits from what is more recently interpreted as basaltic *pyroclastic* eruptions have been recovered from all major oceanic settings: mid-ocean ridge (MOR), subduction arc, back-arc basin, and intraplate environments. Although explosive eruptions seem volumetrically minor compared to their effusive counterparts, in particular along MOR (Clague et al., 2009a), they too bear implications on the behavior of volatile phases within magmatic reservoirs and likely impact on the properties and evolution of the volcanic edifices. (This chapter focuses on basaltic volcanoes, due to their vast abundance on the seafloor. Key to their general eruption behavior is the low viscosity of basaltic magmas, opposing to the high viscosity common to rhyolitic systems.)

Classification and terminology of eruption styles and products

Explosive subaerial eruptions are often classified as one type of either steady activity, ranging from low-energy

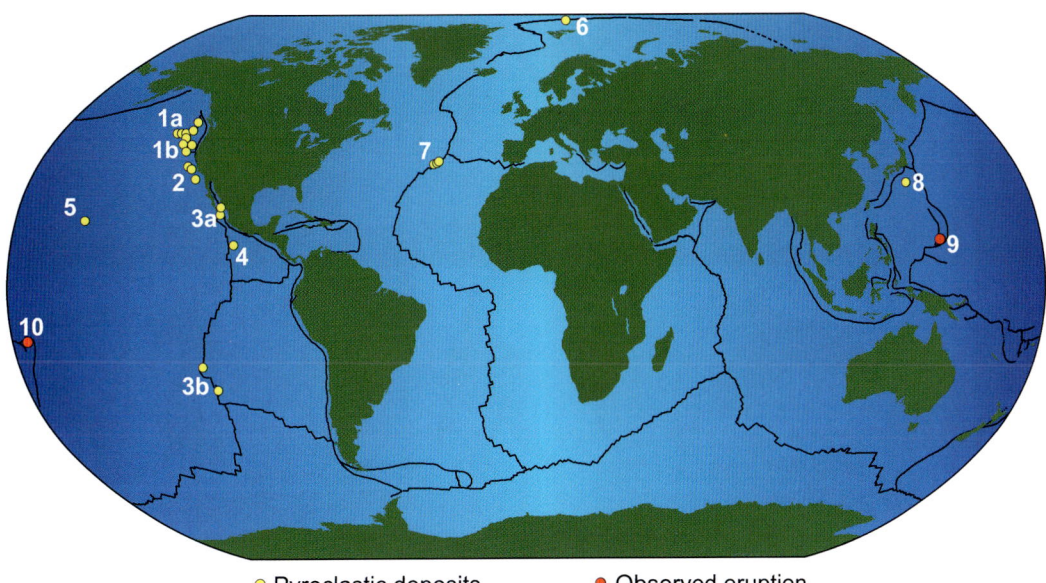

Explosive Volcanism in the Deep Sea, Figure 2 Known locations of basaltic explosive volcanism in the deep sea: (*1a*) Juan de Fuca Ridge, Axial Volcano and Vance Seamounts, (*1b*) Gorda Ridge and President Jackson Seamounts, (*2*) Taney Seamounts and Ben Seamount, (*3a, b*) East Pacific Rise, (*4*) Seamount Six, (*5*) Hawaii, (*6*) Gakkel Ridge, (*7*) Mid-Atlantic Ridge south of the Azores, (*8*) Sumisu basin, Izu-Bonin back-arc, (*9*) NW Rota-1, Mariana arc, (*10*) W-Mata, Lau back-arc basin. See text for references.

Hawaiian fire fountains to violent *ultra-Plinian* eruptions, or transient activity classified as *Strombolian* to *Vulcanian*. Discrete, short-lived burst events are characteristic for the latter, whereas steady eruptions display sustained ejection of fragmented magma-gas mixtures. Principally, the same categories can be applied to submarine explosive activity (Head and Wilson, 2003).

Explosive activity of basaltic low-viscosity systems is generally driven by expanding magmatic gas bubbles *(magmatic eruption)*. When ground or seawater is mingled with the magma prior to the eruption, rapid water-steam expansion drives extremely violent *phreato-* or *hydromagmatic* activity (cf. White et al., 2003). However, the vigorousness of phreatomagmatic events ceases rapidly at water depths beyond 100 m and should be strongly suppressed in the deep sea (Zimanowski and Büttner, 2003).

Magma fragmentation is at last responsible for the formation of the clasts observed in primary deposits. Fragmentation maybe magmatic, that is, due to the expansion of magmatic gas bubbles, caused by water-magma interaction or both. In either case, the term *pyroclastic* applies. (Various, slightly different definitions exists for the terminology of eruption products and deposits (cf., Fisher and Schmincke, 1984; Cas and Wright, 1987; White and Houghton, 2006).) More specifically, particles maybe referred to as *hydroclastic* if derived from the second mechanism. This does not equate to the term *hyaloclastic* that typically bears connotation to effusive activity when flowing lava is quenched in contact with water.

Occurrence

Pyroclastic deposits and explosive activity is documented at various MOR in the Pacific and Atlantic Ocean, near-ridge seamounts and intraplate seamounts on the Pacific Plate, as well as the Izu-Bonin-Mariana arc and Lau back-arc basin, Western Pacific Ocean (Figure 2) (e.g., Gill et al., 1990; Fouquet et al., 1998; Maicher et al., 2000; White et al., 2003 and chapters within; Davis and Clague, 2006; Chadwick, 2008; Sohn et al., 2008; Clague et al., 2009a and references within; Resing et al., 2011). This distribution of eruption sites is most certain strongly biased by, if not purely reflecting the pattern of scientific sampling surveys. In fact, the presence of pyroclastic fragments appears to be ubiquitous at most MOR (Clague et al., 2009a). MOR known for widespread pyroclastic deposits include the *East Pacific Rise*, the *Juan de Fuca Ridge*, the *Gorda Ridge*, the central *Mid-Atlantic Ridge*, and the *Gakkel Ridge*.

Explosive activity is thus documented, (1) over nearly the entire depth range of the worldwide MOR system, (2) for a wide array of spreading rates, and (3) for the entire range of MORB geochemical compositions. Individual eruptions at MOR are thought to be small in volume, depositing only a thin cover of pyroclasts onto the adjacent lava flows (Sohn et al., 2008).

More prominent pyroclastic deposits or clastic sections containing abundant pyroclastic fragments are known for many near-ridge seamounts. These often host one or multiple calderas (collapse craters) at their summit and volcaniclastic sections are frequently found as lithified

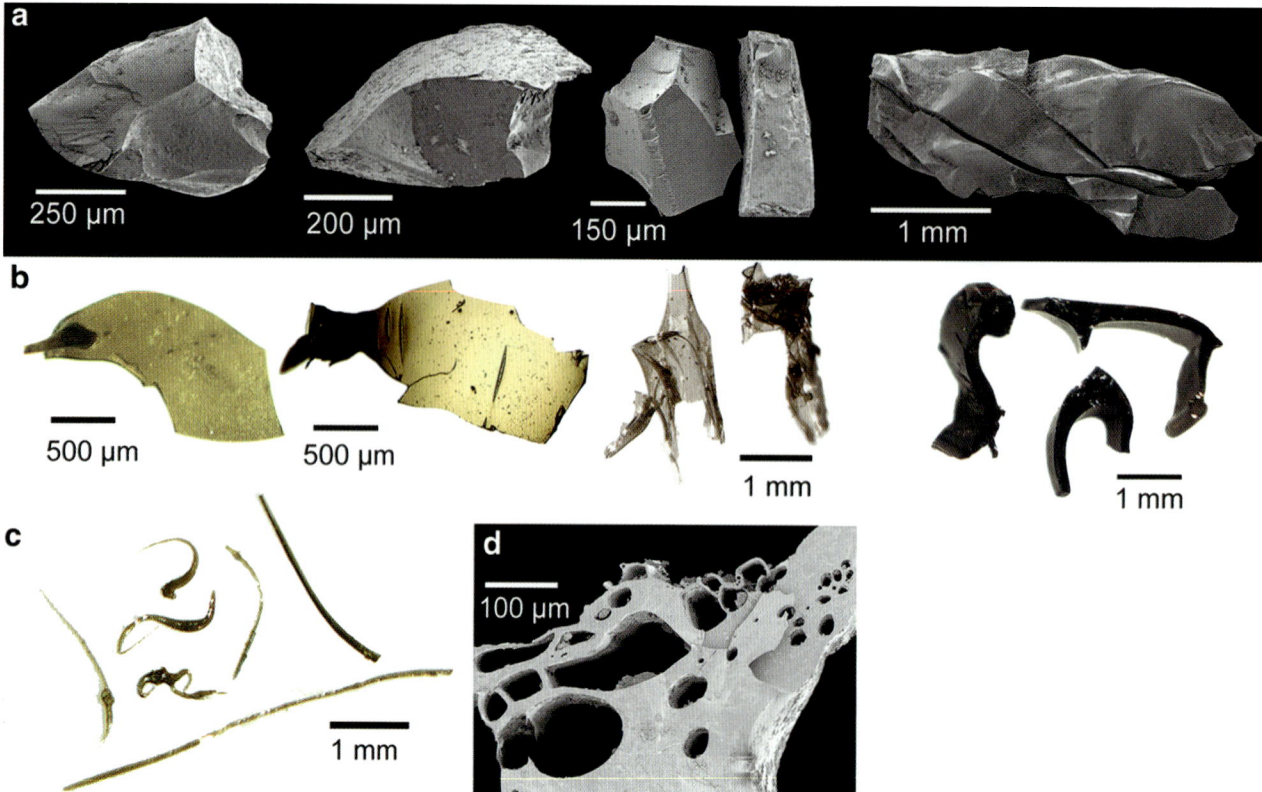

Explosive Volcanism in the Deep Sea, Figure 3 Common morphologies of basaltic deep-sea pyroclasts. (**a**) Angular fragments, (**b**) Limu o Pele, (**c**) Pele's hair, and (**d**) clasts with tube vesicles. Typical order of abundance: (**a**) > (**b**) >> (**c**), (**d**). Samples collected during dives T1009 and T1010 on Axial Seamount, by D. Clague at Monterey Bay Aquarium Research Institute.

units exposed within the caldera wall. Well-studied examples include the *Vance Seamount chain*, west of the Juan de Fuca ridge; the *President Jackson Seamounts*, west of the Gorda ridge; or *Seamount Six*, east to the Northern East Pacific Rise. The intraplate seamount *Loihi*, Hawaii, exhibits unconsolidated volcaniclastic sequences of several meters thickness, derived from Strombolian and Hawaiian activity (Clague et al., 2003a; Schipper et al., 2010). For wide parts of the deposits, the accumulation was found to be fairly rapid at an average rate of 0.37 cm/year (Clague, 2009).

Along the subduction zone of the Western Pacific Plate, submarine explosive activity is known for the Izu-Bonin-Mariana arc and the Lau back-arc basin. Explosive submarine activity in this setting has been recognized as early as 1990 (Gill et al., 1990) and directly observed at *NW Rota-1*, Mariana arc (Chadwick, 2008), and *West Mata*, Lau Basin (Resing et al., 2011) – the first deep-sea eruptions ever to be witnessed by eye. Detailed mapping of the morphology and structure of the volcanic terrain and deposits in the Lau back-arc basin suggests pyroclastic eruptions to be a prevalent phenomenon in this environment rather than the exception.

Eruption products

Produced fragments are often glassy and of low microlite content due to fast quenching. A core of morphological characteristics is common to many pyroclastic glasses in the deep sea and can be used for a rough subdivision as illustrated in Figure 3.

1. *Angular, blocky* clasts (Figure 3a), with sharp edges. This is the most abundant form, at MOR typically dense, and stronger vesiculated in seamount deposits. The dense character may appear counterintuitive for explosively derived deposits, but is consistent with the inferred eruption conditions (see below).
2. *Limu o Pele* (Figure 3b). Thin rapidly quenched melt films, largely described as broken bubble walls. They come in manifold morphologies, ranging from platy to bended and complexly folded varieties, with thicknesses usually between 10 and 200 μm. Limu o Pele is typically seen as a characteristic clast form indicative of explosive activity of low-viscosity magmas.
3. A minor component of fragments may be present as fluidal forms like *Pele's hair* or small *ribbons* (Figure 3c).

4. Fragments with prominent *tubular vesicles* loosely resembling tube pumice (Figure 3d).
5. Various *lithics* ranging from basaltic rock clasts to mineral fragments from a hydrothermal stockwerk may occur.

Deep-sea pyroclastic deposits are in tendency finer grained than subaerial Strombolian deposits. Although larger fragments and spatter are known, grain sizes typically range between <63 μm and 2–4 mm. This is diagnostic of increased fragmentation efficiency.

General eruption mechanism

Although much if not most about the nature of explosive submarine eruptions is still puzzling, a general and simplistic view may be described by the following: Hydromagmatic activity is largely ruled out as the main force to trigger fragmentation and formation of pyroclastic deposits, in particular as the violence of water-magma interaction diminishes with increasing water pressure (cf. Zimanowski and Büttner, 2003). Instead the volcaniclastic record is often considered to represent explosive volcanic activity and magma fragmentation driven primarily by exsolved magmatic gas. Further fragmentation due to rapid quenching upon contact with seawater is thought to reduce the primary clast size (Potuzak et al., 2008; Helo et al., 2013).

At MOR, pyroclastic eruptions are widely envisioned to occur as Strombolian bubble bursts accompanying effusive activity. In exceptional cases, more energetic eruptions and fragmentation levels deeper within the conduit seem possible. The extent of pyroclastic discharge has been proposed to depend broadly on the rise rate of the magma within the conduit. Sheet lavas forming under high extrusion rates have been associated with intensive formation of pyroclastic fragments, whereas the opposite appears to hold true for the slow extrusion of pillow lavas. Direct observations at *NW Rota-1* and *W-Mata* (Figure 1) have confirmed both the transient eruption character, i.e., discrete eruption pulses, and the coeval occurrence of explosive and effusive activity. The eruptive events at *NW Rota-1* were Strombolian in style and characterized by the repeated ascent of gas-rich pockets. This gave rise to a cycle of individual eruptive bursts sustained for several minutes with short response intervals (Chadwick et al., 2008). Explosive activity on *W-Mata* ranges from lava fountaining to bubble bursting and vigorous active degassing. Its eruptions produced a variety of Limu o Pele and Pele's hair (Clague et al., 2009b).

After fragmentation, pyroclasts may be entrained into an eruption plume that mainly rises by thermal buoyancy (Deardorff et al., 2011). Constraints on how effective the entrainment of clasts into a rising plume aids to their dispersal are still poor. Field observations and modeling suggest that small Limu o Pele might advect up to few hundreds of meters within buoyant plumes and can be dispersed up to a few kilometers from the vent (e.g., Barreyre et al., 2011).

Role of magmatic volatiles

The principle role of volatiles is to form gas bubbles that drive the eruption and ultimately cause fragmentation. Formation of bubbles typically happens when magma rises from deep, high-pressure levels to shallow, low-pressure levels. Since volatile solubility decreases as pressure is reduced, the concentration of one or more volatile compounds will at some point exceed the solubility and starts to exsolve.

At MOR, CO_2 is the crucial volatile to drive explosive eruptions (Clague et al., 2003b; Helo et al., 2011; Pontbriand et al., 2012). Due to its low solubility (e.g., Dixon et al., 1995; Newman and Lowenstern, 2002; Papale et al., 2006; Witham et al., 2012), it starts exsolving from the magma at early stages during the ascent from the mantle into a shallow magmatic reservoir. This situation is broadly analogous to subaerial volcanoes. When the magma eventually starts to erupt from the shallow reservoir, the two cases differ. Under subaerial pressure condition, vigorous exsolution of various volatiles and fast bubble growth is characteristic for this stage, while high ambient water pressure at the seafloor hampers syn-eruptive CO_2 and H_2O exsolution (Shaw et al., 2010; Helo et al., 2011). Eruptions are then merely driven by gas bubbles that formed in the magmatic reservoir prior to the onset of eruption. This is the key to resolve the apparent paradox of dense pyroclasts forming during explosive activity: at the onset of eruption, the free gas phase already present becomes concentrated in larger potentially coalesced bubbles rising in the conduit, with the interstitial melt remaining fairly dense as only marginal amounts of new bubbles grow at this point. In the pyroclastic record, the interstitial dense melt is then represented by angular clasts, while Limu o Pele fragments represent remnants of the larger pockets of bubbles. The bulk vesicularity (effectively the large bubbles) becomes thereby completely obscured in the deposits.

Magmatic H_2O is considerably more soluble in basaltic melts (cf. Moore, 2008) than CO_2. Exsolution of H_2O therefore requires very low pressure and/or a high amount of H_2O dissolved in the magma. Consequently magmatic H_2O qualifies as a driving force for submarine explosive eruptions merely at volcanic settings with sufficiently high concentrations of dissolved H_2O (e.g., Schipper et al., 2010).

The initial volatile budget of many deep-sea volcanic systems, in particular MOR, may be too low or just adequate to drive efficiently explosive activity at the given water depth (Head and Wilson, 2003; Shaw et al., 2010). This obstacle is solved effectively when magmatic foam collapse models are invoked (e.g., Vergniolle and Jaupart, 1990). Magmatic gas that exsolves from magma arriving in the reservoir progressively accumulates at the top section of the magma body. This way, a high degree of vesicularity can build up, culminate in a sudden collapse of the foam, and trigger explosive activity. Hence, the capacity to erupt explosively becomes less dependent on the intrinsic volatile content but rather on the potential to sequester gas bubbles within the shallow reservoir. Such mechanisms have been argued for to facilitate deep-sea explosive activity, specifically at MOR systems (e.g., Clague et al., 2009a; Helo et al., 2011; Pontbriand et al., 2012).

Concluding remarks

Deep-sea volcanoes in all principle oceanic settings display the capability to erupt explosively. Volcanism in this environment is therefore more vigorous than previously thought. Notably, this includes MOR, the most volatile-poor, but in total the most productive volcanoes on Earth. Many of the observed features and relationships that led to our current concept of explosive deep-sea volcanism are still based on a limited number of cases, and much remains unknown about the nature of submarine explosive eruptions and the phenomenon is still enigmatic.

Bibliography

Barreyre, T., Soule, S. A., and Sohn, R. A., 2011. Dispersal of volcaniclasts during deep-sea eruptions: settling velocities and entrainment in buoyant seawater plumes. *Journal of Volcanology and Geothermal Research*, **205**, 84–93.

Bonatti, E., and Harrison, C. G. A., 1988. Eruption styles of basalt in oceanic spreading ridges and seamounts: effect of magma temperature and viscosity. *Journal of Geophysical Research*, **93** (B4), 2967–2980.

Cas, R. A. F., and Wright, J. V., 1987. *Volcanic Successions: Modern and Ancient*. London: Chapman & Hall.

Chadwick, W. W., Jr., Cashman, K. V., Embley, R. W., Matsumoto, H., Dziak, R. P., de Ronde, C. E. J., Lau, T. K., Deardorff, N. D., and Merle, S.G., 2008. Direct video and hydrophone observations of submarine explosive eruptions at NW rota-1 volcano, Mariana arc. *Journal of Geophysical Research*, **113**, B08S10.

Clague, D. A., 2009. Accumulation rates of volcaniclastic deposits on loihi seamount, Hawaii. *Bulletin of Volcanology*, **71**, 705–710.

Clague, D. A., Batiza, R., Head, J. W. I., and Davis, A. S., 2003a. Pyroclastic and hydroclastic deposits on loihi seamount, Hawaii. In White, J. D. L., Smellie, J. L., and Clague, D. A. (eds.), *Explosive Subaqueous Volcanism*. Washington, DC: AGU, pp. 73–95.

Clague, D. A., Davis, A. S., and Dixon, J. E., 2003b. Submarine eruptions on the gorda Mid-Ocean ridge. In White, J. D. L., Smellie, J. L., and Clague, D. A. (eds.), *Explosive Subaqueous Volcanism*. Washington, DC: AGU, pp. 111–125.

Clague, D. A., Paduan, J. B., and Davis, A. S., 2009a. Widespread strombolian eruptions of mid-ocean ridge basalt. *Journal of Volcanology and Geothermal Research*, **180**, 171–188.

Clague, D.A., Rubin, K.H., Keller, N.S., 2009b. *Products of Submarine Fountains and Bubble-burst Eruptive Activity at 1200 m on West Mata Volcano, Lau Basin*. Eos Trans. AGU Fall Meet. Suppl., Abstract V43I-02.

Davis, A. S., and Clague, D. A., 2006. Volcaniclastic deposits from the North Arch volcanic field, Hawaii: explosive fragmentation of alkalic lava at abyssal depths. *Bulletin of Volcanology*, **68**, 294–307.

Deardorff, N. D., Cashman, K. V., and Chadwick, W. W., Jr., 2011. Observations of eruptive plume dynamics and pyroclastic deposits from submarine explosive eruptions at NW Rota-1, Mariana arc. *Journal of Volcanology and Geothermal Research*, **202**, 47–59.

Dixon, J. E., Stolper, E. M., and Holloway, J. R., 1995. An experimental study of water and carbon dioxide solubilities in mid ocean ridge basaltic liquids. 1. Calibration and solubility models. *Journal of Petrology*, **36**, 1607–1631.

Fisher, R. V., and Schmincke, H. U., 1984. *Pyroclastic Rocks*. Berlin: Springer.

Fouquet, Y., et al., 1998. Extensive volcaniclastic deposits at the Mid-Atlantic ridge axis: results of deep-water basaltic explosive volcanic activity? *Terra Nova*, **10**, 280–286.

Fox, P. J., and Heezen, B. C., 1965. Sands of the Mid-Atlantic ridge. *Science*, **149**, 1367–1370.

Gill, J., et al., 1990. Explosive deep-water basalt in the sumisu backarc rift. *Science*, **248**, 1214–1217.

Head, J. W., and Wilson, L., 2003. Deep submarine pyroclastic eruptions: theory and predicted landforms and deposits. *Journal of Volcanology and Geothermal Research*, **121**, 155–193.

Helo, C., Longpré, M.-A., Shimizu, N., Clague, D. A., and Stix, J., 2011. Explosive eruptions at mid-ocean ridges driven by CO_2-rich magmas. *Nature Geosciences*, **4**, 260–263.

Helo, C., Clague, D. A., Dingwell, D. B., and Stix, J., 2013. High and highly variable cooling rates during pyroclastic eruptions on axial seamount, Juan de Fuca ridge. *Journal of Volcanology and Geothermal Research*, **253**, 54–64.

Maicher, D., White, J. D. L., and Batiza, R., 2000. Sheet hyaloclastite: density-current deposits of quench and bubble-burst fragments from thin, glassy sheet lava flows, seamount Six, eastern pacific ocean. *Marine Geology*, **171**, 75–94.

Moore, G., 2008. Interpreting H_2O and CO_2 contents in melt inclusions: constraints from solubility experiments and modeling. In Putirka, K.D., Tepley III, F.J. (eds.), *Minerals, Inclusions and Volcanic Processes*. Reviews in Mineralogy and Geochemistry. Chantilly, Virgina, USA: Mineralogical Society of America, Geochemical Society, pp. 333–361.

Moores, E. M., Robinson, P. T., Malpas, J., and Xenophonotos, C., 1984. Model for the origin of the troodos massif, Cyprus, and other Mideast ophiolites. *Geology*, **12**, 500–503.

Newman, S., and Lowenstern, J. B., 2002. VOLATILECALC: a silicate melt-H2O-CO2 solution model written in visual basic for excel. *Computational Geosciences*, **28**, 597–604.

Papale, P., Moretti, R., and Barbato, D., 2006. The compositional dependence of the saturation surface of $H_2O + CO_2$ fluids in silicate melts. *Chemical Geology*, **229**, 78–95.

Pontbriand, C. W., et al., 2012. Effusive and explosive volcanism on the ultraslow-spreading Gakkel Ridge, 85 degrees E. *Geochemistry, Geophysics, Geosystems*, **13**, Q10005.

Potuzak, M., Nichols, A. R. L., Dingwell, D. B., and Clague, D. A., 2008. Hyperquenched volcanic glass from Loihi Seamount, Hawaii. *Earth and Planetary Science Letters*, **270**, 54–62.

Resing, J. A., et al., 2011. Active submarine eruption of boninite in the northeastern Lau Basin. *Nature Geoscience*, **4**, 799–806.

Rubin, K. H., Soule, S. A., Chadwick, W. W., Jr., Fornari, D. J., Clague, D. A., Embley, R. W., Baker, E. T., Perfit, M. R., Caress, D. W., and Dziak, R. P., 2012. Volcanic eruptions in the deep sea. *Oceanography*, **25**, 142–157.

Schipper, C. I., White, J. D. L., Houghton, B. F., Shimizu, N., and Stewart, R. B., 2010. Explosive submarine eruptions driven by volatile-coupled degassing at Lo'ihi Seamount, Hawai'i. *Earth and Planetary Science Letters*, **295**, 497–510.

Schmincke, H.-U., Robinson, P.T., Ohnmacht, W. and Flower, M. FJ., 1978. Basaltic hyaloclastites from hole 396B, DSDP Leg 46. In: Dmitriev, L., Heirtzler, J., et al., (eds.), *Initial Reports Deep Sea Drilling Project*. Vol. 46: Washington, U.S. Government Printing Office, pp. 341–348.

Shaw, A. M., Behn, M. D., Humphris, S. E., Sohn, R. A., and Gregg, P. M., 2010. Deep pooling of low degree melts and volatile fluxes at the 85 degrees E segment of the gakkel ridge: evidence from olivine-hosted melt inclusions and glasses. *Earth and Planetary Science Letters*, **289**, 311–322.

Sohn, R. A., et al., 2008. Explosive volcanism on the ultraslow-spreading gakkel ridge, arctic ocean. *Nature*, **453**, 1236–1238.

Staudigel, H., and Schmincke, H.-U., 1984. The Pliocene seamount series of La Palma/Canary Islands. *Journal of Geophysical Research*, **89**(B13), 11195–11215.

Vergniolle, S., and Jaupart, C., 1990. Dynamics of degassing at Kilauea Volcano, Hawaii. *Journal of Geophysical Research*, **95**, 2793–2809.

White, J. D. L., and Houghton, B. F., 2006. Primary volcaniclastic rocks. *Geology*, **34**, 677–680.

White, J. D. L., Smellie, L. L., and Clague, D. A., 2003. *Explosive Subaqueous Volcanism*. Washington, DC: American Geophysical Union.

Witham, F., Blundy, J., Kohn, S. C., Lesne, P., Dixon, J., Churakov, S. V., and Botcharnikov, R., 2012. SolEx: a model for mixed COHSCl-volatile solubilities and exsolved gas compositions in basalt. *Computational Geosciences*, **45**, 87–97.

Yoerger, D. R., et al., 2007. Autonomous and remotely operated vehicle technology for hydrothermal vent discovery, exploration, and sampling. *Oceanography*, **20**, 152–161.

Zimanowski, B., and Büttner, R., 2003. Phreatomagmatic explosions in subaqueous volcanism. In White, J. D. L., Smellie, J. L., and Clague, D. A. (eds.), *Explosive Subaqueous Volcanism*. Washington, DC: AGU, pp. 51–60.

Cross-references

Axial Volcanic Ridges
Intraoceanic Subduction Zone
Intraplate Magmatism
Island Arc Volcanism, Volcanic Arcs
Marginal Seas
Mid-ocean Ridge Magmatism and Volcanism
Ocean Drilling
Ophiolites
Seamounts
Submarine Lava Types

EXPORT PRODUCTION

Gerold Wefer[1], Gerhard Fischer[1] and Morten Iversen[2]
[1]MARUM-Center for Marine Environmental Sciences, University of Bremen, Bremen, Germany
[2]Alfred Wegener Institute, Helmholtz Centre for Polar and Marine Research, Bremerhaven, Germany

Introduction

Export production (EP) is the fraction of organic carbon formed by primary production in the photic zone (photosynthesis) that is not degraded before it sinks below 100–150 m. Typically, EP is measured in units of carbon due to its importance for the ocean's biological carbon pump. Under steady-state conditions and on longer timescales (years), EP is assumed to be in proportion to the part of primary production that is fed by upwelled nitrate (Eppley and Peterson, 1979), named new production (Dugdale and Goering, 1967). A fraction of EP is then transferred to the deeper ocean (below 1,000 m) and only less than 1 % of the primary produced organic carbon is finally stored in the sediment. This transfer of organic matter from surface waters to the deep sea is known as the biological pump (Neuer et al., 2014). The final sequestration of organic carbon in the sediments, which stores CO_2 from the upper ocean and atmosphere over longer timescales (>1,000 years), is controlled by the strength of the biological pump.

Primary production, export production, and deep ocean flux

EP generally mirrors surface distribution of phytoplankton standing stock and primary production. Usually, this fraction is below 25 % but can be higher in coastal upwelling systems and areas with pulsed production such as the polar environments. Primary production and export production are low in the so-called ocean deserts such as the great oligotrophic gyres. Often sediment traps are used to measure flux and flux attenuation in the deeper water column. Typically, the flux attenuation is described by a power function with high attenuation at shallow depths (Suess, 1980).

Aggregation, marine snow, and fecal pellets

Particles of a few μm in size are not able to sink and constitute the sediment. Therefore, the EP (export flux) mainly consists of larger particles that are formed in the surface waters by physical and biological processes such as coagulation and collision (marine snow) and feeding activity by zooplankton (fecal pellets). Marine snow and fecal pellets consist of organic matter from living and dead organisms and of mineral particles produced both biogenic and non-biogenic. The minerals act as ballast for larger particles and may enhance their settling velocities (Iversen and Ploug, 2010), which have an important influence on the strength of the biological pump.

Sediment traps

Sediment traps collect sinking particles over time and area at a certain water depth, mostly around 1,000 m. The collected material is used to estimate vertical fluxes of organic carbon and other components such as carbonate, biogenic opal, or minerals (Buesseler et al., 2007). Sediment traps can be anchored to the seafloor and have multiple time-controlled collectors to reveal annual patterns of particle sedimentation/flux with a temporal resolution of a few weeks. Alternatively, sediment traps can also drift freely with the water body. These drifting traps sample sinking particles in the upper few hundred meters (export production/flux) over hours to days. Recently, drifting traps have been equipped with gels to collect intact marine snow particles or fecal pellets. Often these particles fall apart when collected with conventional sediment traps.

Bibliography

Buesseler, K. O., Antia, A. N., Chen, M., Fowler, S. W., Gardner, W. D., Gustafsson, Ö., Harada, K., Michaels, A. F., Rutgers van der Loeff, M., Sarin, M., Steinberg, D. K., and Trull, T., 2007.

An assessment of the use of sediment traps for estimating upper ocean particle fluxes. *Journal of Marine Research*, **65**(3), 345–416.

Dugdale, R. C., and Goering, J. J., 1967. Uptake of new and regenerated forms of nitrogen in primary productivity. *Limnology and Oceanography*, **12**, 196–206.

Eppley, R. W., and Peterson, B. J., 1979. Particulate organic matter flux and planktonic new production in the deep ocean. *Nature*, **282**, 677–680.

Iversen, M. H., and Ploug, H., 2010. Ballast minerals and the sinking carbon flux in the ocean: carbon-specific respiration rates and sinking velocities of marine snow aggregates. *Biogeosciences*, **7**, 2613–2624.

Neuer, S., Iversen, M., and Fischer, G., 2014. The Ocean's Biological Carbon pump as part of the global carbon cycle. Limnology and Oceanography, e-Lectures, doi:10.4319/lol.2014.sneuer.miversen.gfischer.9. http://aslo.org/lectures/14_009/14_009_neuer_iversen_fischer.html

Suess, E., 1980. Particulate organic carbon flux in the oceans–surface productivity and oxygen utilization. *Nature*, **288**, 260–263, doi:10.1038/288260a0.

F

FAULT-PLANE SOLUTIONS OF EARTHQUAKES

Wolfgang Frisch
Department of Geosciences, University of Tübingen, Tübingen, Germany

Definition

Earthquakes are caused by ruptures (faults) in solid rock with instantaneous movement, during which seismic body waves are emanated by elastic rebound. By evaluation of the seismic waves, the focus (location) of the earthquake, the spatial orientation of the fault plane, and the mode of movement can be determined. The result is called "fault-plane solution."

Fault-plane solutions

The orientation of planes of movement at plate boundaries or within any block of solid rock can be deduced from earthquake data. In the case of an earthquake triggered at a fault plane, the two blocks move by creating an instantaneous offset up to several meters. This results in the generation of the two types of seismic body waves. Primary (P-) waves oscillate in the longitudinal direction of propagation. They are faster than secondary (S-) waves that oscillate transversally. If all of the seismic data from a given earthquake collected around the Earth are put into a diagram, four quadrants and the two separating planes (Figure 1) can be determined with their spatial orientation. In the two quadrants that are in the direction of movement of each block, the first motion of the primary waves is away from the earthquake focus, and an observer on the Earth's surface first receives a push; the wave starts with a compressive movement (compressive first motion shown as black quadrants in Figure 1). First motion in the other two quadrants, shown in white, is in the opposite direction; it starts with a tension and is dilatational. Each of these motions is registered by seismographs.

One of the separating planes represents the slip plane generated by the earthquake; the other one is an aiding plane that has no use in nature. However, initially it is not possible to decide which one of these two planes was the slip plane (Figure 2). Commonly, this can be deduced from geological observations if the approximate orientation of a fracture zone is known. On the other hand, careful analysis of seismic data generated by the aftershock activity following every large earthquake provides the opportunity to identify the slip plane because of the shift of the seismic centers. If the slip plane is known, the sense of movement is easily detected (Figure 2). Direction of movement in the slip plane is orthogonal to the aiding plane. A process similar to that used for analyzing P-waves can also be used for the analysis of S-waves.

Conclusions

Using this method, which is called fault-plane solution, the orientation of a slip plane and the sense of movement can be determined with high accuracy. Fault-plane solutions allow for a reconstruction of plate boundaries and their movement patterns. The analysis of earthquake first motion data impressively confirmed the concept of three different types of plate boundaries.

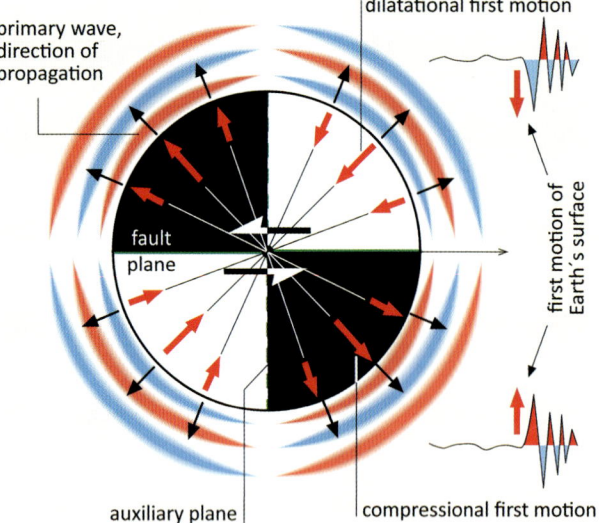

Fault-Plane Solutions of Earthquakes, Figure 1 Principle of fault-plane solutions of an earthquake hypocenter. The fault plane of the earthquake (orthogonal to the paper plane) and an orthogonal virtual aiding plane define four quadrants. First motions of the primary waves oscillating in the propagation direction and expressed by vertical motions in the soil indicate the sense of movement of the blocks displaced during the earthquake. Seismograms subdivide two quadrants with compressive and two quadrants with dilatative first motion.

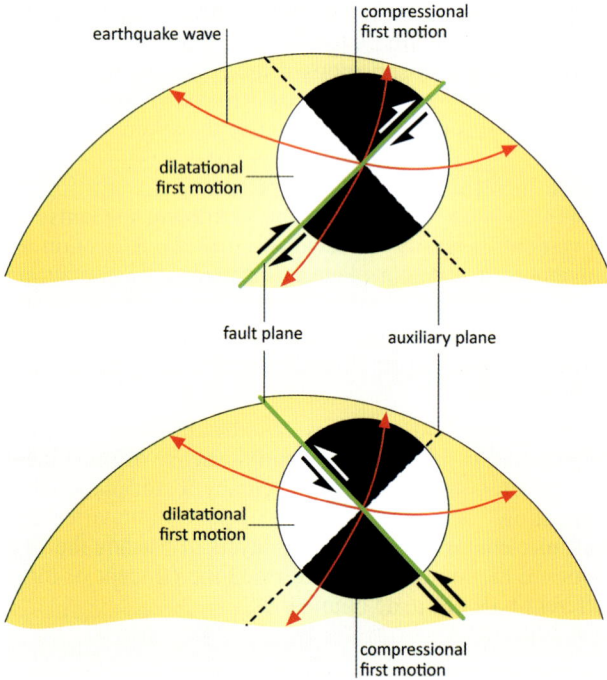

Fault-Plane Solutions of Earthquakes, Figure 2 Ambiguity of fault-plane solutions illustrated by an earthquake that produces an overthrust. Illustration is a schematic vertical section through the Earth.

Bibliography

Frisch, W., Meschede, M., and Blakey, R., 2011. *Plate Tectonics. Continental Drift and Mountain Building*. Berlin/Heidelberg: Springer. 212 pp.

Green, H. W., 1994. Solving the paradox of deep earthquakes. *Scientific American*, **271**(3), 50–57.

Isacks, B., and Molnar, P., 1969. Mantle earthquake mechanisms and the sinking of the lithosphere. *Nature*, **223**, 1121–1124.

Okomura, K., Yoshioka, T., and Kuscu, I., 1993. *Surface faulting on the North Anatolian Fault in these two millennia*. U.S. Geological Survey Open-File Report, 94–568, pp. 143–144.

Sieh, K. E., 1978. Prehistoric large earthquakes produced by slip on the San Andreas fault at Pallett Creek, California. *Journal of Geophysical Research*, **83**, 3907–3939.

Sykes, L. R., 1967. Mechanism of earthquakes and nature of faulting on the mid-ocean ridges. *Journal of Geophysical Research*, **72**, 2131–2153.

Cross-references

Earthquakes
Epicenter, Hypocenter

FJORDS

Antoon Kuijpers and Camilla S. Andresen
Geological Survey of Denmark and Greenland (GEUS), Copenhagen, Denmark

Definition

Fjords are defined as deep, elongated arms of the sea that have been (or are presently being) excavated or modified by land-based glaciers (Syvitski et al., 1987).

Summary

The term "fjord" has its origin in the Old Norse word "fjorthr" which also included freshwater lakes. However, according to above definition, we will deal here exclusively with deep, semi-enclosed coastal inlets that generally are long relative to their width. Globally, almost all fjords are found at higher latitude, i.e., north of 43° and south of 42°. The longest (>200 km) fjords are Scoresby Sund, East Greenland, and Sognefjord in Norway. The inland termination and seaward opening of the fjord are referred to as head and mouth, respectively. Fjord topography is often characterized by one or more submarine sills representing relict moraines or bedrock ridges. Sediments from fjord basins may provide high-resolution records of both marine and terrestrial environmental changes (Gilbert, 2000), with maximum accumulation rates amounting to more than 1 m per year (Syvitski and Farrow, 1989). Sedimentation patterns in the fjord are controlled by glacial and fluvial conditions, topography, sea-level, and hydrography of coastal waters (Syvitski et al., 1987). Usually, an estuarine circulation regime with a saline, occasionally poorly oxygenated deep layer and a low-salinity surface layer marks the hydrography. In fjords where a tidewater glacier is present, turbid meltwater from beneath the glacier rises as a plume into the surface layer, where flocculation of silt and clay particles leads to settling of these fine-grained particles contributing to the bulk mass of the fjord deposits (Mugford and Dowdeswell, 2011). Lamination may occur due to the effect of seasonal melting on the depositional process. These fine-grained sediments do, however, often display intervals with coarser ice-rafted debris which may reflect changes in iceberg calving activity of the local glacier (Andresen et al., 2012). A common feature of fjord basins with a high sediment input is a wide variety of slope failures along side-wall margins or basement highs and frequent occurrence of turbidites and mass-flow deposits in the adjacent basin. This slope instability implies a potential geohazard risk in the form of a local tsunami.

Bibliography

Andresen, C. S., Straneo, F., Ribergaard, M. H., Bjørk, A. A., Andersen, T. J., Kuijpers, A., Nørgaard-Pedersen, N., Kjær, K. H., Schjøth, F., Wecklström, K., and Ahlstrøm, A. P., 2012. Rapid response of Helheim Glacier in Greenland to climate variability over the past century. *Nature Geoscience*, **5**, 37–41.

Gilbert, R., 2000. Environmental assessment from the sedimentary record of high-latitude fjords. *Geomorphology*, **32**, 295–314.

Mugford, R. I., and Dowdeswell, J. A., 2011. Modelling glacial meltwater plume dynamics and sedimentation in high-latitude fjords. *Journal of Geophysical Research*, **116**, F01023, doi:10.1029/2010JF001735.

Syvitski, J. P. M., Burrel, D. C., and Skei, J. M., 1987. *Fjords, Processes and Products*. New York: Springer.

Syvitski, J. P. M., and Farrow, G. E., 1989. Fjord sedimentation as an analogue for small hydrocarbon-bearing fan deltas. In Whateley, M. K. G., and Pickering, K. T. (eds.), *Deltas: Sites and Traps for Fossil Fuels*. Geological Society of London. Special Publication, Vol. 41, Department of Geology, The University Leicester, pp. 21–43.

Cross-references

Clay Minerals
Contourites
Currents
Grounding Line
Ice-rafted Debris (IRD)
Laminated Sediments
Layering of Oceanic Crust
Marine Sedimentary Basins
Moraines
Paleoceanography
Regional Marine Geology
Sea-level
Sediment Dynamics
Sediment Transport Models
Sedimentary Sequence
Shelf
Tsunamis
Turbidites

FORAMINIFERS (BENTHIC)

Anna Binczewska[1], Irina Polovodova Asteman[2] and Elizabeth J. Farmer[3]
[1]Institute of Marine and Coastal Sciences, University of Szczecin, Szczecin, Poland
[2]Bjerknes Centre for Climate Research, Bergen, Norway
[3]Department of Earth Science, University of Bergen, Bergen, Norway

Synonyms

Benthic foraminifera; Benthic foraminifers; Informally benthic forams

Definition

Benthic foraminifera are unicellular, aquatic (marine and brackish) eukaryotic organisms. They are benthic bottom dwellers characterized by high diversity and abundance. In benthic foraminifera, the cytoplasmic body is encased in organic or mineralized test (shell), which provides a fossil record (Cambrian to recent). Due to the presence of the test, benthic foraminifera are useful indicators of past and present-day environmental changes. Foraminiferal shells consist of single or multiple chambers interconnected with each other by a channel, called *foramen*, which gives the name to the entire Foraminifera group.

Introduction

One of the first findings on benthic foraminifera dates back to the seventeenth century, and a drawing by Antonie van Leeuwenhoek of a specimen nowadays recognized as the genus *Elphidium*. Following numerous studies by, e.g., A. d'Orbigny (1826), H.B. Brady, J.A Cushman and A. R. Loeblich, and Jr. H. Tappan made a substantial contribution to foraminiferal research (Sen Gupta, 1999a). These and subsequent investigations resulted in benthic foraminiferal classification system consisting of 16 orders, Allogromiida, Astrorhizida, Litoulida, Trochamminida, Textulariida, Fusulinida, Miliolida, Carterinida, Spirillinida, Lagenida, Rotaliida, Buliminida, Globigerinida, Involutinida, Robertinida, and Silicoloculinida (Loeblich and Tappan, 1988), which have been recently placed in 3 to 4 major classes (Pawlowski et al., 2013; Mikhalevich, 2013). Based on shell morphology, about 5 000 modern and 40 000 fossil species are described (Sen Gupta, 1999). Recent molecular studies reveal a much higher genetic diversity due to numerous cryptic species and genetically distinctive phylotypes (Pawlowski et al., 2014).

Morphology and taxonomy

Foraminifera belong to the kingdom Protista because their body consists of a single cell filled with cytoplasm. Cellular structures include mitochondria, vacuoles, and nuclei placed in a cytoplasmic body and encased in a shell (test). Foraminiferal shells can be calcareous

Foraminifers (Benthic), Figure 1 Scanning electronic microscope images of selected benthic foraminifera: 1, *Reophax sp.* with a shell of *Cassidulina laevigata* (*arrow*); 2, *Elphidium excavatum*; 3, *Lagena sulcata*; 4, *Bolivinellina pseudopunctata*; 5, *Melonis barleeanus*, 6a,b, *Ammonia* sp. (a, spiral side; b, umbilical side); 7, *Eggerelloides scaber*; 8, *Pyrgo williamsoni*.

(precipitated $CaCO_3$), arenaceous (built of sediment particles), or organic (made of tectin).

Calcareous forms (Figure 1: 2–6, 8) are the most abundant, due to their greater diversity and better potential for preservation. The calcareous test wall is divided into two main types: porcellaneous (e.g., suborder Miliolina), which are milky to white in color, translucent to opaque under light, and with no pores (Figure 1: 8), and hyaline (e.g., Rotaliina), which are transparent with a glassy, shiny appearance and always contain pores (Figure 1: 2, 4–6) (Murray, 1979).

Agglutinated forms (e.g., order Astrorhizida) have an organic membrane to which they glue foreign particles – mineral grains (e.g., quartz, rutile), volcanic ash, or biogenic debris (e.g., sponge spicules, foraminifera, diatoms, coccoliths) with an organic and/or calcareous cement (Kuhnt et al., 2005; Thomsen and Rasmussen, 2008). Test surfaces are smooth to very rough and often irregular (Figure 1: 1, 7) and brownish in color and may be brittle and fragile, causing difficulties in taxonomic identification of some species (Schröder, 1988).

The organic-walled foraminifera (e.g., order Allogromida) secrete an organic membrane made of tectin. Their soft and elastic body readily changes in shape but has low or no fossilization potential.

Test size depends largely on environmental conditions prevailing during growth and ranges from < 0.01 mm in very small individuals to several centimeters (in tropical latitudes). Larger benthic foraminifera (e.g., *Nummulites gizehensis*) are often disc-shaped, contain many chambers and can grow up to 15 cm, being still a single-celled organism.

The generally accepted classification of benthic foraminifera is based on characteristic test morphologies (morphotypes). While more recent molecular studies have added greater detail to taxonomic classifications (e.g., Pawlowski et al., 2013), we focus on the morphological aspects here.

Enormous diversity of test appearance allows foraminiferal classification from the order, family, or genus level to the species level, frequently including several morphotypes. Tests of benthic foraminifera consisting of a single chamber are called *unilocular* (Figure 1: 3); of two chambers, *bilocular*; or of several chambers, *multilocular* (Figure 1: 2, 4–8). While unilocular species grow by increasing the single chamber size, multilocular individuals grow by adding new chambers. The addition of new chambers usually occurs according to a well-defined pattern: uniserial, whereby chambers are added linearly (e.g., *Reophax*); biserial (Figure 1: 4) or triserial (Figure 1: 7), where chambers are added in two or three parallel series (e.g., *Bolivina* or *Eggerelloides*); and coiled (e.g., *Elphidium, Ammonia*). Coiling also follows strict patterns as each chamber is added in a spiral – either planispiral when chambers are added along a single plane (e.g., *Haynesina, Elphidium*) (Figure 1: 2, 5) or trochospiral when chambers spiral along more than one plane, up into a spire (e.g., *Ammonia, Trochammina)* (Figure 1: 6a, b).

The aperture (opening in the shell) can assume a variety of shapes, positions, and forms and may include additional structures such as teeth, tubercles, and flaps.

Taxonomically important are also other shell features such as the shell margin (e.g., keeled or rounded) and surface ornamentation (e.g., bosses, costae, ridges, sutures). For more detailed descriptions of shell morphology, see Boltovskoy and Wright (1976) and Haynes (1981).

Biology

Pseudopodia: The foraminiferal shell functions as protection for the soft internal body. Cell interaction with the external environment occurs via extensions of cytoplasm called *granuloreticulopodia (pseudopodia)*, which extend from the test through the aperture. Granuloreticulopodia branch, creating a network around the test, and carry out many of the fundamental biological functions in foraminifera. They take part in growth, reproduction, respiration, metabolism, test construction, locomotion, feeding, and protection.

Trophic relationships: Benthic foraminifera have various feeding strategies including symbiosis (e.g., *Alveolinidae*), parasitism (e.g., *Hyrrokkin sarcophaga*), suspension, and deposit feeding (e.g., *Ammotium cassis, Elphidium* spp.) (Pawlowski, 2012). Foraminiferal diets may include diatoms, dinoflagellates, macroalgae, organic detritus, and bacteria. Some species are carnivorous and ingest small copepods, shrimp larvae, and even other foraminifera (cannibalism). For some, the sediment surface provides a sufficient amount of nourishment, while other species dig into the sediment to find food. During feeding, pseudopodia build a feeding cyst surrounding the shell and provide a safe space for extra- or intracellular digestion (Murray, 1979). Pseudopodia often collect food and transport it toward the aperture and inside the cell. In addition, some carnivorous species use pseudopodia to catch prey and/or keep it in the net (e.g., *Astrammina rara*).

Locomotion: Some benthic foraminifera live attached to the seabed by extending pseudopodia to anchor themselves to pebbles, macrophytes, mussel shells, or worm tubes, thereby avoiding displacement by water masses. Pseudopodia are also used to burrow into the sediments to avoid predation or destruction and to change location to find better food sources.

Growth and Reproduction: During growth, pseudopodia build a growing cyst, outlining a new chamber frame and creating the base for calcification in calcareous group and for gathering and attaching foreign particles to in agglutinated group (Haynes, 1981). Benthic species follow a great variety of reproductive models, among which a life cycle with two (sexual and asexual) generations is well understood and described (Goldstein, 1999).

Ecology and geographic distribution

Though benthic foraminifera mainly occur in marine habitats, some families (e.g., Allogromidae and Lagenidae) have been reported in freshwater or even terrestrial environments (Pawlowski, 2012). Benthic foraminifera interact with other components of meio- and macrofauna and are influenced by a combination of abiotic and biotic factors. Abiotic factors include temperature, salinity, currents, light, dissolved oxygen, nutrients, pH, trace elements, and substrate. Interaction with other organisms, e.g., competition for space and food, symbiosis, grazing, and parasitism, constitutes biotic factors (Murray, 2006). Some species are more sensitive to ecological factors and prefer narrow factor ranges, while others have wider tolerance limits (Murray, 1991).

On the global scale, water depth, temperature, and salinity are, perhaps, the leading factors determining distribution, diversity, and abundance of benthic foraminifera (Boltovskoy and Wright, 1976).

The distribution of foraminifera within the water column is controlled by seawater saturation with respect to $CaCO_3$. Occurrence of calcareous assemblages at deep-sea settings depends on the carbonate compensation depth (CCD). The CCD in the central Pacific occurs between 4000–5000 m, while in the Atlantic Ocean, it lies deeper at around 5000–6000 m (Boltovskoy and Wright, 1976). Thus, ocean depths located above the CCD are characterized by a highly diverse and rich calcareous benthic fauna. However, with increasing water depth, calcareous foraminifera decrease in abundance and disappear. Simultaneously, an increase in proportion of agglutinated and organic-walled taxa takes place. Surface ocean waters are generally saturated with $CaCO_3$; however, lower temperatures and higher atmospheric CO_2 may cause $CaCO_3$ dissolution in shallow areas at high latitudes, resulting in rich agglutinated and organic-walled assemblages. Consequently, sediments from the Antarctic shelf (e.g., Weddell and Ross Sea) where the CCD lies at 400-500 m water depth are carbonate impoverished. However, there are a few calcareous species adapted to such extreme conditions, e.g., *Nuttallides umboniferus* (Murray, 2006).

Salinity is of great importance to foraminifera in general. While planktonic species are more sensitive to salinity changes, most benthic foraminifera have broad salinity tolerance ranges. Fully marine conditions (30–40 psu) favor many benthic species and promote their reproduction, survival, and growth. Therefore, open oceans have rich and well-preserved benthic foraminiferal assemblages. In environments with salinities <30 psu, foraminiferal reproduction decreases (Bradshaw, 1955; Bradshaw, 1957) and tests have thinner walls, often loose ornamentation, and become reduced in size or deformed (e.g., Almogi-Labin et al., 1992; Stouff et al., 1999; Polovodova and Schönfeld, 2008).

Coastal habitats with brackish waters often host benthic foraminiferal species from the agglutinated *Trochammina*, *Jadammina*, and *Miliammina* and calcareous *Elphidium*, *Miliolidae*, and *Ammonia* genera (Boersma et al., 1998). In the brackish Baltic Sea, carbonate preservation is limited and foraminiferal assemblages are largely dominated by the calcareous *Elphidium* and *Ammonia* and agglutinated *Miliammina*, *Trochammina*, and *Reophax* genera. Intertidal-subtidal lagoons and estuaries often host foraminifera adapted to extreme daily salinity fluctuations from 35 to 0.5 psu (Murray, 2006). In the Mediterranean Sea, during dry seasons brackish lagoons become hypersaline but still hold foraminiferal species, which tolerate salinities of >45 psu (Murray, 2006).

Substrate determines foraminiferal microhabitats, and, based on substrate preferences, benthic foraminifera can be epifaunal (e.g., *Cibicidoides wuellerstorfi*) and shallow to deep infaunal (e.g., *Melonis barleeanum, Uvigerina peregrina, Globobulimina affinis, Chilostomella ovoidea*). Epifaunal species live at the sediment surface, while shallow and deep infaunal taxa dwell at 2–4 cm and below 4 cm in the sediments, respectively (Gooday, 2001).

Shell geochemistry

Calcareous benthic foraminifera incorporate aspects of the chemical elements prevailing in seawater as they precipitate their shells. Hence, these organisms record environmental information – being preserved as fossils serves as a signal carrier for climate and ocean changes taking place during their life cycle. This makes foraminifera important *paleoproxies* (indicators of past changes).

The pioneering work of Cesare Emiliani, using stable oxygen isotopes from planktonic foraminifera as a proxy for temperature change, fundamentally changed our view on past climates (Emiliani, 1955). This paleothermometer can be applied to benthic foraminifera: depending on bottom water temperatures and glacial-interglacial cycles, benthic foraminifera incorporate more or less ^{16}O or ^{18}O into their shells, in much the same way as planktonic species in the upper water column. Changes in $^{18}O/^{16}O$ ratios have been widely considered as indices of climate cycles during the Pleistocene and Quaternary (Ravelo and Hillaire-Marcel, 2007). Oxygen isotopic records also mirror ice volume fluctuations and, hence, allow estimation of past sea-level changes (Mix and Ruddiman, 1984; Shackleton, 1987). The application of oxygen isotopes improved our understanding of glacial-interglacial stages in the North Atlantic during the Pleistocene (Shackleton et al., 1988).

Another approach increasingly used in paleoclimate studies is magnesium/calcium ratios measured in foraminiferal shells. The Mg/Ca is used as a paleotemperature proxy and can be used to constrain the salinity and temperature components of the oxygen isotope records with respect to seawater (Kristjansdottir et al., 2007). The Mg/Ca ratios measured on benthic foraminifera proved glacial-interglacial temperature shifts in the deep ocean during the Quaternary, as well as deep water mass temperature oscillations (Martin et al., 2002). Additionally, other trace element ratios (Cd/Ca and Ba/Ca) in benthic foraminiferal shells can be analysed to study seawater nutrient content (Lea, 1999).

Conclusions

Benthic foraminifera are an important component of marine habitats. Due to their high fossilization potential, worldwide occurrence, high abundances, high reproduction rates and rapid response to environmental changes, benthic foraminifera are used in numerous studies from biostratigraphy and reconstructions of past climate, ocean and sea-level changes, to bio-monitoring and ocean acidification (Horton et al., 1999; Murray, 2006; Schönfeld et al., 2012; Haynert et al., 2012; Dolven et al., 2013). The uniformitarian principle "the present is a key to the past" has been widely applied in paleoenvironmental interpretations based on foraminiferal assemblages, with reference to their present ecological preferences. These marine organisms show enormous potential - providing interdisciplinary knowledge linking biology, geosciences and industry. This potential will undoubtedly expand to other subjects in years to come.

Bibliography

Almogi-Labin, A., Perelis-Grossovicz, L., and Raab, M., 1992. Living Ammonia from a hypersaline inland pool, Dead Sea area, Israel. *Journal of Foraminiferal Research*, **22**, 257–266.

Boersma, A., 1998. Foraminifera. In Haq, B. U., and Boersma, A. (eds.), *Introduction to Marine Micropaleontology*. Singapore: Elsevier, pp. 19–78.

Boltovskoy, E., and Wright, R., 1976. *Recent Foraminifera*. The Hague: W. Junk.

Bradshaw, J. S., 1955. Preliminary laboratory experiments on ecology of foraminiferal populations. *Micropaleontology*, **1**, 351–358.

Bradshaw, J., 1957. Laboratory studies on the rate of growth of the foraminifera *Streblus beccarii* (Linne) var. tepida (Cushman). *Journal of Paleontology*, **31**, 1138–1147.

Dolven, J., Alve, E., Rygg, B., and Magnusson, J., 2013. Defining past ecological status and in situ reference conditions using benthic foraminifera: a case study from Oslofjord, Norway. *Ecological Indicators*, **29**, 219–223.

Emiliani, C., 1955. Pleistocene temperatures. *Journal of Geology*, **63**, 538–578.

Frew, R. D., Dennis, P. F., Heywood, K. J., Meredith, M. P., and Boswell, S. M., 2000. The oxygen isotope composition of water masses in the northern North Atlantic. *Pergamon*, I **47**, 2265–2286.

Goldstein, S. T., 2003. Foraminifera: a biological overview. In Sen Gupta, B. K. (ed.), *Modern Foraminifera*. London: Kluwer, pp. 37–55.

Gooday, A. J., 2001. Benthic foraminifera. In Turekian, K. K. (ed.), *Climate and Oceans*. Princeton: Elsevier, pp. 425–436.

Haynert, K., Schönfeld, J., Polovodova Asteman, I., and Thomsen, J., 2012. The benthic foraminiferal community in a naturally CO_2-rich coastal habitat of the Southwestern Baltic Sea. *Biogeosciences*, **9**, 4421–4440.

Haynes, J. R., 1981. *Foraminifera*. London: Macmillan.

Hayward, B.W., Cedhagen, T., Kaminski, M., and Gross, O. 2015. World Foraminifera Database. Accessed through: Hayward, B. W., Cedhagen, T., Kaminski, M., and Gross, O. (2015) World Foraminifera Database at http://www.marinespecies.org/foraminifera/aphia.php?p=taxdetails&id=114000. 02 May 2015.

Horton, B., Edwards, R. J., and Lloyd, J. M., 1999. UK intertidal foraminiferal distributions: implications for sea-level studies. *Marine Micropaleontology*, **36**, 205–223.

Kristjansdottir, G. B., Lea, D. W., Jennings, A. E., Pak, D. K., and Belanger, C., 2007. New spatial Mg/Ca-temperature calibrations for three Arctic, benthic foraminifera and reconstruction of north Iceland shelf temperature for the past 4000 years. *Geochemistry, Geophysics, Geosystems*, **8**, 1–27.

Kuhnt, W., Hess, S., Holbourn, A., Paulsen, H., and Salomon, B., 2005. The impact of the 1991 Mt. Pinatubo eruption on deep-sea foraminiferal communities: a model for the Cretaceous-Tertiary (K/T) boundary? *Palaeogegraphy, Palaeoclimatology, Palaeoecology*, **224**, 83–107.

Lea, D. W., 1999. Trace elements in foraminiferal calcite. In Sen Gupta, B. K. (ed.), *Modern Foraminifera*. London: Kluwer, pp. 259–277.

Loeblich, A. R., and Tappan, H., Jr., 1988. *Foraminiferal Genera and Their Classification*. New York: Van Nostrand Reinhold.

Martin, P. A., Lea, D. W., Rosenthal, Y., Shackleton, N. J., Sarnthein, M., and Papenfuss, T., 2002. Quaternary deep sea temperature histories derived from benthic foraminiferal Mg/Ca. *Earth and Planetary Science Letters*, **198**, 193–209.

Mikhalevich, V. I., 2013. New insight into the systematics and evolution of the foraminifera. *Micropaleontology*, **59**, 493–527.

Mix, A. C., and Ruddiman, W. F., 1984. Oxygen-isotope analyses and Pleistocene ice volumes. *Quaternary Science*, **21**, 1–20.

Murray, J. W., 1979. *British Nearshore Foraminiferids*. London: Academic.

Murray, J. W., 1991. *Ecology and Palaeoecology of Benthic Foraminifera*. Harlow: Longman.

Murray, J. W., 2006. *Ecology and Applications of Benthic Foraminifera*. New York: Cambridge.

Pawlowski, J., 2012. Foraminifera. In Schaechter, M. (ed.), *Eukaryotic Microbes*. Amsterdam: Elsevier.

Pawlowski, J., Holzmann, M., Berney, C., Fahrni, J., Gooday, A. J., Cedhagen, T., Habura, A., and Bowser, S. S., 2003. The evolution of early foraminifera. *Proceeding of the National Academy of Science of the USA*, **100**, 11494–11498.

Pawlowski, J., Holzmann, M., and Tyszka, J., 2013. New supraordinal classification of foraminifera: molecules meet morphology. *Marine Micropaleontology*, **100**, 1–10.

Pawlowski, J., Lejzerowicz, F., and Esling, P., 2014. Next-Generation environmental diversity surveys of foraminifera > preparing the future. *Biological Bulletin*, **227**, 93–106.

Polovodova, I., and Schönfeld, J., 2008. Foraminiferal test abnormalities in the western Baltic Sea. *Journal of Foraminiferal Research*, **38**, 318–336.

Ravelo, A. Ch., and Hillaire-Marcel, C., 2007. The use of oxygen and carbon isotopes of foraminifera in Paleoceanography. In Hillaire-Marcel, C., and De Vernal, A. (eds.), *Developments in Marine Geology*. London: Elsevier, Vol. 1, pp. 735–760.

Rohling, E. J., and Cooke, S., 1999. Stable oxygen and carbon isotopes in foraminiferal carbonate shells. In Sen Gupta, B. K. (ed.), *Modern Foraminifera*. London: Kluwer, pp. 239–258.

Schönfeld, J., Alve, E., Geslin, E., Jorissen, F., Korsun, S., Spezzaferri, S., and Members of the FOBIMO Group, 2012. *Marine Micropaleontology*, **94–95**, 1–13.

Schröder, C.J., 1988. Subsurface preservation of agglutinated foraminifera in the Northwest Atlantic Ocean. *Abhandlungen der Geologischen Bundesanstalt-A*, **41**, 325–336.

Sen Gupta, B. K., 1999a. Introduction to modern foraminifera. In Sen Gupta, B. K. (ed.), *Modern Foraminifera*. London: Kluwer, pp. 3–6.

Sen Gupta, B. K., 1999b. Systematics of modern foraminifera. In Sen Gupta, B. K. (ed.), *Modern Foraminifera*. London: Kluwer, pp. 7–36.

Shackleton, N. J., 1987. Oxygen isotopes, ice volume and sea level. *Quaternary Science Reviews*, **6**, 183–190.

Shackleton, N. J., Imbrie, J., Pisias, N. G., and Rose, J., 1988. The evolution of oceanic oxygen-isotope variability in the North Atlantic over the past three million years. *The Royal Society Publishing*, **318**, 679–688.

Stouff, V., Debenay, J.-P., and Lesourd, M., 1999. Origin of double and multiple tests in benthic foraminifera: observations in laboratory cultures. *Marine Micropaleontology*, **36**, 189–204.

Thomsen, E., and Rasmussen, T. L., 2008. Coccolith-agglutinating foraminifera from the early Cretaceous and how they constructed their tests. *Journal Foraminiferal Research*, **38**, 193–214.

Cross-references

Biochronology, Biostratigraphy
Carbonate Dissolution
Calcite Compensation Depth (CCD)
Foraminifers (Planktonic)
Glacial-marine Sedimentation
Marine Microfossils
Mg/Ca Paleothermometry
Oxygen Isotopes
Paleoceanographic Proxies
Paleoceanography

FORAMINIFERS (PLANKTONIC)

Demetrio Boltovskoy[1] and Nancy Correa[2]
[1]Institute of Ecology, Genetics and Evolution of Buenos Aires (IEGEBA), University of Buenos Aires – CONICET, Buenos Aires, Argentina
[2]Department of the Argentine Naval Hydrographic Service (Ministry of Defence), and School of Marine Sciences (Marine Institute), Buenos Aires, Argentina

Synonyms

Planktic foraminifers; Planktonic foraminifera; Planktonic foraminifers

Definition

Planktonic foraminifers (from the Latin "foramen," hole or orifice, and "ferre," to bear) are exclusively marine, open-ocean, single-celled, eukaryotic protists that secrete a multichambered calcitic shell. The group comprises 45–50 living morphospecies, but the actual number of genetically different forms is probably much higher. They inhabit all latitudes, chiefly above 100 m water depth. Planktonic foraminifers are very important in the biogeochemistry of calcite in the oceans. Their rich fossil record

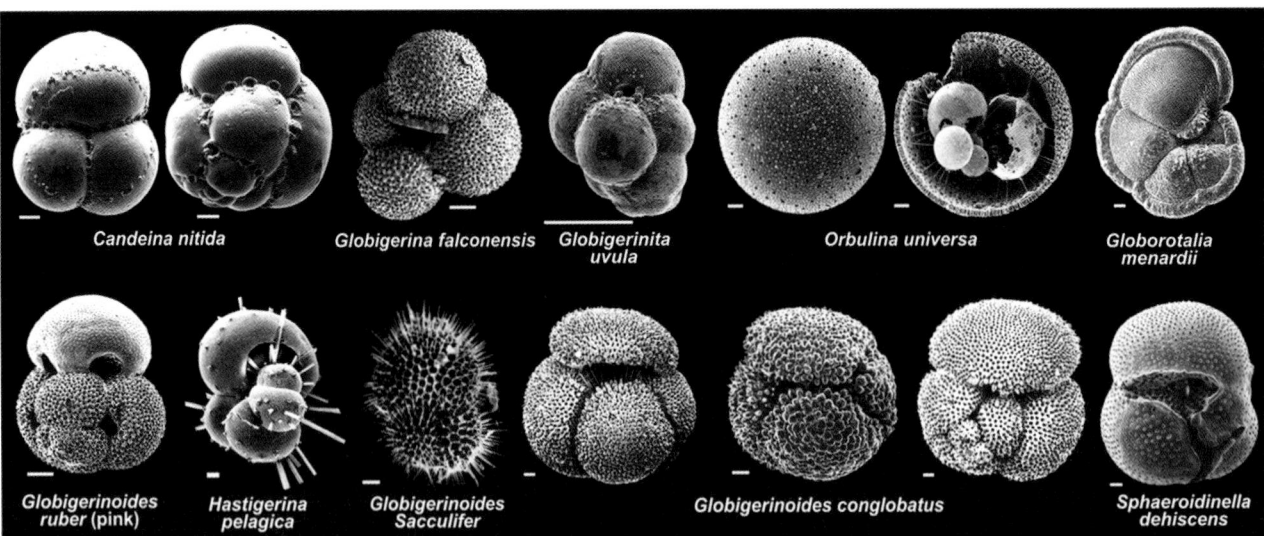

Foraminifers (Planktonic), Figure 1 Examples of representative foraminiferal morphospecies. Notice that *Globigerina falconensis*, *Orbulina universa*, *Globigerinoides ruber*, *Globigerinoides conglobatus*, and *Sphaeroidinella dehiscens* are spinose species (like *Globigerinoides sacculifer*), but spines are usually absent in specimens retrieved from bottom sediments (like most of those figured here). Scale bar equals 50 μm. *G. sacculifer* is from Bé (1968); all others are from Kemle-von Mücke and Hemleben (1999).

is widely used for stratigraphic and paleoenvironmental studies.

Shell and cell

All planktonic foraminifers build shells (up to 1.5 mm, mean around 0.3 mm in size) composed of mostly more or less globular chambers arranged in a streptospiral (where each chamber is half a whorl), planispiral (coiled in a single horizontal plane), trochospiral (helical), or, very rarely, triserial coil. An exception is the genus *Orbulina*, which builds a complete sphere around a small trochospiral test which may be resorbed prior to gametogenesis (Figure 1). Biserial arrangement of chambers is present in some fossil species. The shell is made of calcium carbonate (CO_3Ca) with hexagonal crystals (calcite), usually formed by the addition of lamellae separated by organic material. The outer lamella of successive chambers may also cover previous chambers, in which case, the wall of earlier parts of the test becomes considerably thicker. In most species, prior to gametogenesis the test is covered with a more or less continuous additional calcite layer with coarse crystalline structure. Calcite comprises around 80–90 % of the shell material, the remaining being represented by $MgCO_3$, $FeCO_3$, SiO_2, Sr, and several amino acids (Boltovskoy and Wright, 1976; Hemleben et al., 1989). The shell wall comprises poreless areas (such as the keel), but most of it is perforated by pores, which are obliterated by an organic, sometimes porous, layer in living specimens. In some species the shell surface may bear pustules or a cancellate, honeycomb-like, ornamentation. Spinose species have long calcitic spines, circular, triangular, or three bladed in cross section, which are shed during gametogenesis.

The cytoplasm typically fills all shell chambers (occasionally the last chamber may be only partly filled) and also extends as a thin layer covering the shell surface, where it forms a dense array of very thin filopodia or reticulopodia. This outer layer may be foamy ("bubble capsule"), or finely reticulate, or smooth and sheathlike.

Organelles present in the cytoplasm comprise a single nucleus (usually in one of the inner chambers, with granular nucleoplasm and strands of chromatin and heterochromatin), small mitochondria, peroxisomes (involved in the synthesis of carbohydrates and metabolism of waste products), endoplasmic reticulum, Golgi complex, vacuoles (digestive, waste, etc.), and fibrillar bodies (which probably aid in flotation) (Hemleben et al., 1989). In addition to these, the cytoplasm contains various inclusions, such as lipid droplets and pigment granules.

Reproduction and growth

As opposed to benthic species, which show a range of reproductive modes, in planktonic foraminifers, the only process that has been observed to date is the release of thousands of very small (3–5 μm), free-swimming, biflagellate "swarmers," which fuse producing a zygote. Subsequently, this zygote grows into an adult individual (Hemleben et al., 1989).

Most of the spinose, symbiont-bearing species, as well as some non-spinose ones have a life span of 2–4 weeks and a lunar or semilunar cycle, reproducing at either full moon or new moon, while other non-spinose forms, in particular the deeper-living ones, seem to follow a yearly cycle (Kemle-von Mücke and Hemleben, 1999).

All planktonic Foraminifera perform ontogenetic vertical migrations. The deep-living species reproduce in surface waters and then migrate to layers as deep as 1,000 m or more (Hemleben et al., 1989). Surface-dwelling species reproduce in the pycnocline or chlorophyll maximum layer, and the offspring migrate to more surficial waters.

Growth of the shell starts with a small spherical chamber ("proloculum") and the subsequent addition of progressively larger chambers around it (roughly at the rate of one chamber every 48 h; Hemleben et al., 1989). Successive stages (five, according to Brummer et al. (1987)) are characterized by marked morphological changes, including inflation of chambers, development of pores, and development of spine collars, ridges, secondary apertures, etc. Prior to reproduction, the shell wall thickens, spines are shed, and internal septa may disappear.

Trophic relationships

Most species are fairly omnivorous, feeding on a large spectrum of organisms including diatoms, dinoflagellates, coccolithophorids, radiolarians, ciliates, pteropods, various invertebrate larvae, copepods, etc. Cannibalism has been observed in non-spinose species, but not in spinose forms. Food items can be several times larger than the foraminifer itself. Spinose species tend to prefer animal prey, whereas non-spinose ones feed chiefly on phytoplankton. The prey is surrounded by the protist's rhizopodia which eventually engulf the organism and transfer ruptured pieces of its body to digestive vacuoles in the extra- or intrashell cytoplasm by protoplasmic streaming (Hemleben et al., 1989).

In addition to heterotrophic feeding, many planktonic foraminifers possess algal symbionts (dinoflagellates and chrysophytes) in their cytoplasm (Figure 2). Symbiotic algae are common in most (but not all) spinose species (where their presence often seems obligatory) and almost always absent in the non-spinose ones. These algae are normally located in the outer cell layer (where lighting is best), but have also been observed to perform diel migrations within the cell, withdrawing to the inner cytoplasm at night and dispersing around the periphery during the day (Bé et al., 1977). A single foraminifer may host as many as 10,000 symbiotic algae (Spero and Parker, 1985), which are either digested or released into the environment when the host reproduces (thus, offspring acquire their symbionts from the medium rather than from their parents). For the foraminifer, the advantages of this symbiotic relationship are the energy supply (either as free extracellular organic matter or when symbionts are digested by the host), enhancement of calcification, and intracellular elimination of host waste metabolites (Goldstein, 2003).

Taxonomy

During recent years, the taxonomy of most protists has undergone profound changes in association with the widespread use of molecular and biochemical data (e.g., rRNA gene sequences). A subject of particular interest has been

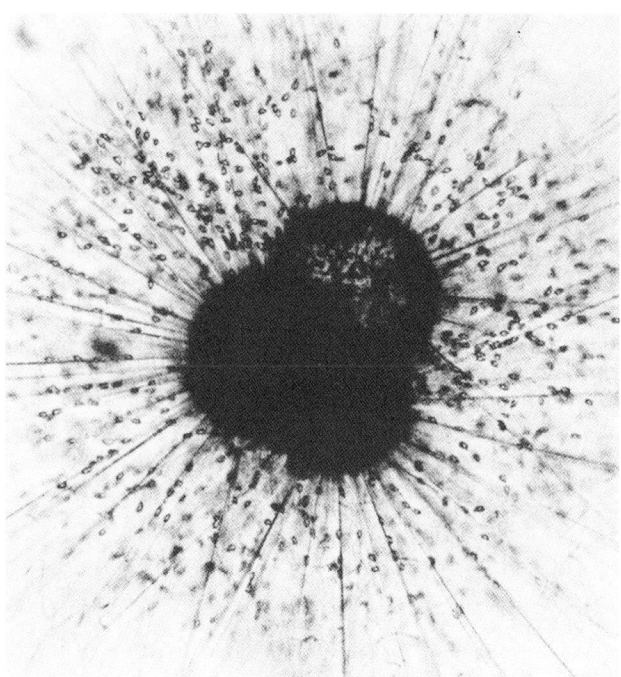

Foraminifers (Planktonic), Figure 2 Zooxanthellae (symbiotic algae) dispersed along the spines and rhizopodia of a specimen of *Globigerinoides ruber* (from Bé et al. (1977)).

the elucidation of phylogenetic relationships between higher-rank taxa, for which reason familiar and suprafamiliar assignments have changed repeatedly. According to Lee et al. (2000), planktonic foraminifers are distributed in two superfamilies and nine families included in the order Globigerinida of the class Foraminifera, phylum Granuloreticulosa. A few years later, Adl et al. (2005) placed the Foraminifera in the "supergroup" Rhizaria but refrained from further subdividing them because existing morphology-based schemes are not fully consistent with molecular phylogenetic data.

At the genus and species level, the classification of Foraminifera has traditionally been based on features of the test, such as spines, coiling, size and arrangement of chambers, ornamentation, etc. (see Hottinger, 2006, for a complete glossary of terms used in foraminiferal research). As opposed to other protozooplanktonic groups, like the radiolarians, the low diversity of planktonic Foraminifera and their comparatively large size have contributed toward developing a reasonably stable classification system. Most researchers recognize between 45 and 50 living morphospecies (Hemleben et al., 1989), although some stretch this number to 64 (Saito et al., 1981). Molecular data, which are presently available for a number of species, never suggest the need to lump existing morphospecies, but often indicate that traditional taxonomy has included several genetically different organisms under a single name (cryptic species). De Vargas et al. (2004) noticed that the eight morphological

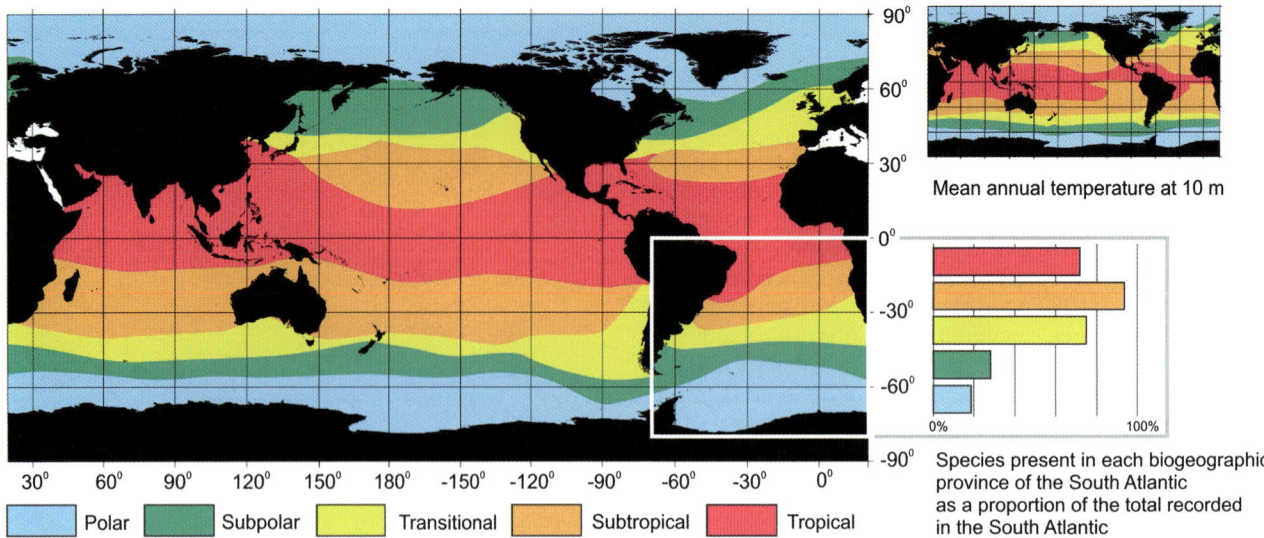

Foraminifers (Planktonic), Figure 3 Major foraminiferal biogeographic provinces (slightly modified from Bé (1977)); notice resemblance with pattern of mean annual temperature at 10 m (based on data from Locarnini et al. (2006)); temperature intervals are <2 °C, 2–8 °C, 8–16 °C, 16–26 °C, and >26 °C. *Bar graph* at the *bottom right* shows numbers of foram species recorded in each biogeographic province in the South Atlantic (*highlighted* on main map), as a proportion of the overall total (39) for this area (based on data from Kemle-von Mücke and Hemleben (1999)).

species that have been genetically analyzed (DNA sequence coding for the small subunit and internal transcribed spacer of the nuclear ribosomal RNA genes) until that year contain between three and six distinct genetic entities ("sister species") each. Interestingly, upon closer examination, most of the new, genetically identified species within traditional morphological species turn out to show minor but recognizable morphological differences (De Vargas et al., 2004; Aurahs et al., 2009). It seems plausible that niche partitioning (driven by regional differences in food availability, salinity, temperature), rather than allopatric speciation through vicariance, is responsible for this diversification (Darling and Wade, 2008; Aurahs et al., 2009; Seears et al., 2012).

Geographic patterns of abundance and species composition

Data on the distribution patterns of living planktonic Foraminifera come from three sources: plankton tows, sediment traps, and surface sediment samples. Each has advantages and drawbacks. Plankton samples retrieve whole assemblages (unless the mesh size used is too large) unbiased by postmortem processes (see below), but the sample size is normally limited and the sample is but a snapshot representative of a very restricted time offset. Sediment traps are particularly suitable for investigating temporal cycles and sedimentation processes, but their yields are subject to several distorting mechanisms, such as lateral advection, selective dissolution, and fragmentation due to grazing. Surface sediment samples are by far the most widely available and used for biogeographic purposes, but as proxies of the distribution of the living assemblages, they are also the most biased. The sedimentary remains of planktonic foraminifers can differ strongly from the corresponding planktonic assemblages due to a number of mechanisms, including selective dissolution of the less resistant shells (both on the way to the seafloor and after settling), reworking by bottom fauna, winnowing by bottom currents, lateral advection by subsurface and deep currents, submersion and extended survival of colder water forms under warmer water areas ("equatorward shadows"), fragmentation due to grazing, vertical integration of shallow- and deep-living species, different reproduction modes, and integration of seasonally dissimilar abundance patterns (Vincent and Berger, 1981; Boltovskoy, 1994; Schiebel and Hemleben, 2005). Many of these modifications tend to enhance the proportions of cold water species, resulting in assemblages indicative of colder waters than those overlying the corresponding sediments.

Despite these shortcomings, the worldwide biogeographic pattern originally proposed by Bé (1977), based chiefly on sedimentary samples, is the most widely accepted today and probably realistic in general terms (Figure 3).

As with most other zooplanktonic organisms, the two major attributes of foraminiferal distribution patterns respond to different constraints: species compositions depend mainly on temperature, whereas abundance depends on primary production. Within normal open-ocean values, other variables have little influence on foraminiferal biogeography. For example, planktonic Foraminifera are scarce or absent altogether in low- or high-salinity waters (i.e., below 30 and above 45 PSU: Boltovskoy and Wright, 1976; Hemleben et al., 1989),

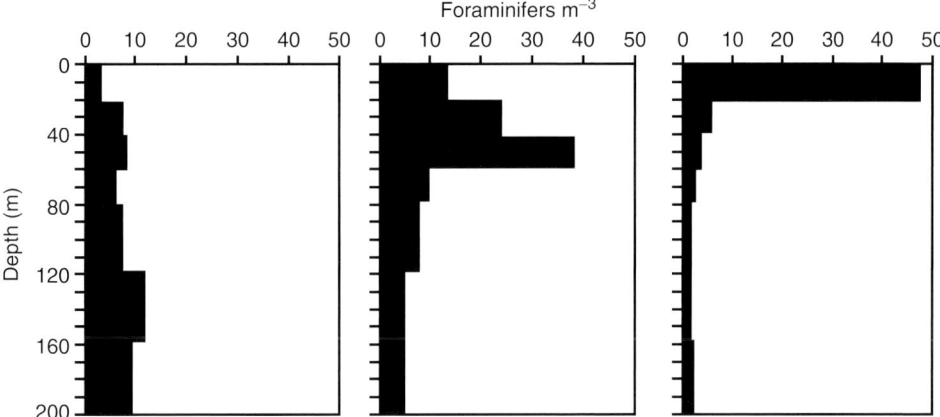

Foraminifers (Planktonic), Figure 4 Vertical distribution of planktonic Foraminifera at three stations in the Sea of Japan in fall (from Kuroyanagi and Kawahata (2004)).

but within these values, the influence of salinity is rather limited (Hemleben et al., 1989; Bijma et al., 1990).

The worldwide biogeography of planktonic foraminifers is characterized by a system of latitudinally oriented belt-like provinces, punctuated by conspicuous north-south drifts where ocean currents distort the east-west orientation of isotherms (e.g., the Gulf Stream in the North Atlantic or the Humboldt Current in the South Pacific; Figure 3). The limits of these provinces are strikingly similar to near-surface isotherms (Figure 3), which reinforces the notion of the overwhelming importance of temperature for determining specific makeups. Each province is characterized by a particular combination of co-occurring species, but individual species ranges seldom coincide with province boundaries, so that specific ranges overlap extensively and species restricted to any one province are very few (normally below 10 % of those present anywhere in the province). Diversity drops conspicuously from the equator to the poles, but the highest numbers of species are often found in the subtropics (Figure 3). With a few exceptions, known morphospecies inhabit all three major oceans. Most species dwell in the surface waters, either during their entire life (all species of *Globigerinoides*, *Globigerina*) or as juveniles, while the adults migrate below 100 m (*Globigerinella adamsi*, *Sphaeroidinella dehiscens*, *Neogloboquadrina pachyderma*, etc.), but a number of forms are usually found at 50–100 m (*Globigerina bulloides*, *Hastigerina pelagica*, *Orbulina universa*, *Globigerinella siphonifera*, *Globigerina calida*, *Pulleniatina obliquiloculata*, *Neogloboquadrina dutertrei*, *Candeina nitida*, *Globigerina glutinata*, and others). Deep-water species (below 1,000 m) are *Hastigerinella digitata* and *Globorotalia truncatulinoides*.

Foraminiferal abundances normally range around 0.001 to >1 individual per L of water, with oligotrophic central oceanic gyres hosting the lowest densities and high-productivity upwelling regions the highest. Peak densities, however, have been recorded in sea-ice communities (ca. 500 individuals per L; Arnold and Parker, 2003). Although vertical abundance profiles depend on the vertical migrations (chiefly ontogenetic) which most species perform, normally they are characterized by peak abundances in the euphotic layer, around 10–50 m, and steeply decreasing values below this depth (Figure 4).

The annual flux of calcitic tests at 100 m water depth is around 1.3–3.2 Gt, which represents 23–56 % of the global, open marine $CaCO_3$ flux. Although much of the calcium carbonate dissolves before reaching the seafloor, the number of tests that do settle on the bottom is so high that the resulting carbonate oozes cover almost 50 % of the seafloor (Lisitzyn, 1974; Kennett, 1982; see "Deep-sea Sediments"), with remains of planktonic foraminifers largely dominating these deposits, especially in the open ocean. The distribution of this so-called *Globigerina* ooze, however, is uneven. Because the dissolution of calcite increases with depth (see "Carbonate Dissolution"), down to about 3,500 m (depending on ocean basin and area), preservation of foraminiferal tests is excellent or good; between this depth (known as the lysocline) and ca. 4,500 m, dissolution is considerably faster and preservation deteriorates rapidly. Below 4,500, which is known as the calcite compensation depth (see "Calcite Compensation Depth (CCD)"), the rate of supply of calcium carbonate is balanced by the rate of dissolution, so there is no net accumulation (Seibold and Berger, 1996).

Stratigraphic and paleoecologic applications

The geologic record of planktonic foraminifers dates back to the late Triassic or Jurassic (BouDagher-Fadel, 2012), becoming particularly abundant since the mid-Cretaceous. They have been used extensively for defining zonal stratigraphies from the upper Valanginian to the recent (Bolli et al., 1985), proving of great value for oil exploration purposes (see "Biochronology, Biostratigraphy").

Recent assemblages from the water column have been used in biogeographic surveys and as tracers of currents and water masses (e.g., Boltovskoy and Wright, 1976). Paleoceanographic studies with Foraminifera are based on several approaches (see "Paleoceanography" and "Paleoceanographic Proxies"). Past changes in the proportions of species as compared with present-day specific makeups (usually aided by more or less complex mathematical manipulations of the data, like the transfer function techniques proposed by Imbrie and Kipp (1971)) allow reconstructing past ecological and oceanographic settings (temperature, fronts, primary productivity; e.g., CLIMAP (1976)). Downcore changes in the relationship between right-coiling versus left-coiling specimens of some species (e.g., *Neogloboquadrina pachyderma*, whose proportions of right-coiling individuals increase with increasing temperature) have been used to reconstruct the water temperature at which these foraminifers lived. Right- versus left-coiling ratios, however, may also depend on other conditions (e.g., Thiede, 1971; Bolli et al., 1985). Stable isotopes of oxygen and carbon in foraminiferal tests are also used to determine shifts in the temperature, salinity, oxygen content, and fertility of the ocean over the past hundreds to millions of years (see Rohling and Cooke, 2003, for a review) (see "Carbon Isotopes" and "Oxygen Isotopes").

Conclusions

Planktonic foraminifers are among the most important marine microfossils. Their widespread occurrence and high abundances, their excellent stratigraphic record, and the comparatively good knowledge of their biology and ecology make them excellent tools for stratigraphic, paleoecologic, and biogeographic work. Their limited diversity and relatively stable taxonomy have been major advantages for their use as proxies of present and past oceanographic settings as well as for tackling distributional and evolutionary problems. However, recent changes in our understanding of the taxonomy of planktonic foraminifers brought about by molecular studies pose new challenges for the use of these protists in applied surveys.

Bibliography

Adl, S. M., Simpson, A. G. B., Farmer, M. A., Andersen, R. A., Anderson, O. R., Barta, J. R., Bowser, S. S., Brugerolle, G., Fensome, R. A., Fredericq, S., James, T. Y., Karpov, S., Kugrens, P., Krug, J., Lane, C. E., Lewis, L. A., Lodge, J., Lynn, D. H., Mann, D. G., Mccourt, R. M., Mendoza, L., Moestrup, Ø., Mozley-Standridge, S. E., Nerad, T. A., Shearer, C. A., Smirnov, A. V., Spiegel, F. W., and Taylor, M. F. J. R., 2005. The new higher level classification of eukaryotes with emphasis on the taxonomy of protists. *The Journal of Eukaryotic Microbiology*, **52**, 399–451.

Arnold, A. J., and Parker, W. C., 2003. Biogeography of planktonic foraminifera. In Sen Gupta, B. K. (ed.), *Modern Foraminifera*. New York: Kluwer, pp. 103–122.

Aurahs, R., Grimm, G. W., Hemleben, V., Hemleben, C., and Kučera, M., 2009. Geographic distribution of cryptic genetic types in the planktonic foraminifer *Globigerinoides ruber*. *Molecular Ecology*, **18**, 1692–1706.

Bé, A. W. H., 1968. Shell porosity of recent planktonic foraminifera as a climatic index. *Science*, **161**, 881–884.

Bé, A. W. H., 1977. An ecological, zoogeographic and taxonomic review of recent planktonic foraminifera. In Ramsay, A. T. S. (ed.), *Oceanic Micropaleontology*. London: Academic, pp. 1–101.

Bé, A. W. H., Hemleben, C., Anderson, O. R., Spindler, M., Hacunda, J., and Tuntivate-Choy, S., 1977. Laboratory and field observations of living planktonic foraminifera. *Micropaleontology*, **23**, 155–179.

Bijma, J., Faber, W. W., Jr., and Hemleben, C., 1990. Temperature and salinity limits for growth and survival of some planktonic foraminifers in laboratory cultures. *Journal of Foraminiferal Research*, **20**, 95–116.

Bolli, H. M., Saunders, J. B., and Perch-Nielsen, K. (eds.), 1985. *Plankton Stratigraphy*. London: Cambridge University Press, pp. 1–1032.

Boltovskoy, D., 1994. The sedimentary record of pelagic biogeography. *Progress in Oceanography*, **34**, 135–160.

Boltovskoy, E., and Wright, R., 1976. *Recent Foraminifera*. The Hague: W. Junk, pp. 1–515.

BouDagher-Fadel, M. K., 2012. Biostratigraphy and geological significance of planktonic foraminifera. In Wignall, P. B. (ed.), *Developments in Paleontology and Stratigraphy*. New York: Elsevier, pp. 1–301.

Brummer, G. J. A., Hemleben, C., and Spindler, M., 1987. Ontogeny of extant spinose planktonic foraminifera (Globigerinidae), a concept exemplified by *Globigerinoides sacculifer* (Brady) and *G. ruber* (d'Orbigny). *Marine Micropaleontology*, **12**, 357–381.

CLIMAP Project Members, 1976. The surface of the ice age earth. *Science*, **191**, 1131–1137.

Darling, K. F., and Wade, C. M., 2008. The genetic diversity of planktic foraminifera and the global distribution of ribosomal RNA genotypes. *Marine Micropaleontology*, **67**, 216–238.

de Vargas, C., Saez, A. G., Medlin, L., and Thierstein, H. R., 2004. Super-species in the calcareous plankton. In Thierstein, H. R., and Young, J. R. (eds.), *Coccolithophores: From Molecular Processes to Global Impact*. Berlin: Springer, pp. 251–298.

Goldstein, S. T., 2003. Foraminifera: a biological overview. In Sen Gupta, B. K. (ed.), *Modern Foraminifera*. New York: Kluwer, pp. 37–55.

Hemleben, C., Spindler, M., and Anderson, O. R., 1989. *Modern Planktonic Foraminifera*. New York: Springer, pp. 1–363.

Hottinger, L., 2006. Illustrated glossary of terms used in foraminiferal research. http://paleopolis.rediris.es/cg/CG2006_M02/index.html

Imbrie, J., and Kipp, N. G., 1971. A new micropaleontological method for quantitative paleoclimatology: application to a late Pleistocene Caribbean core. In Turekian, K. K. (ed.), *The Late Cenozoic Glacial Ages*. New Haven: Yale University Press, pp. 71–181.

Kemle-Von Mücke, S., and Hemleben, C., 1999. Foraminifera. In Boltovskoy, D. (ed.), *South Atlantic Zooplankton*. Leiden: Backhuys, pp. 43–73.

Kennett, J. P., 1982. *Marine Geology*. Englewood Cliffs: Prentice Hall, pp. 1–813.

Kuroyanagi, A., and Kawahata, H., 2004. Vertical distribution of living planktonic foraminifera in the seas around Japan. *Marine Micropaleontology*, **53**, 173–196.

Lee, J. J., Pawlowski, J., Debenay, J.-P., Whittaker, J., Banner, F., Gooday, A. J., Tendal, O., Haynes, J., and Faber, W. W., 2000. Class foraminifera. In Lee, J. J. (ed.), *The Illustrated Guide to the Protozoa*, 2nd edn. Lawrence: Society of Protozoologists, pp. 872–951.

Lisitzyn, A. P., 1974. *Osadkoobrazovanie v Okeanakh*. Moskva: Nauka, pp. 1–438.

Locarnini, R. A., Mishonov, A. V., Antonov, J. I., Boyer, T. P., and Garcia, H. E., 2006. World ocean atlas 2005, volume 1: temperature. In Levitus, S. (ed.), *NOAA Atlas NESDIS 61*. Washington, DC: U.S. Government Printing Office, pp. 1–182.

Rohling, E. J., and Cooke, S., 2003. Stable oxygen and carbon isotopes in foraminiferal carbonate shells. In Sen Gupta, B. K. (ed.), *Modern Foraminifera*. New York: Kluwer, pp. 239–258.

Saito, T., Thompson, P. R., and Breger, D., 1981. *Systematic Index of Recent and Pleistocene Planktonic Foraminifera*. Tokyo: University of Tokyo Press, pp. 1–190.

Schiebel, R., and Hemleben, C., 2005. Modern planktic foraminifera. *Paläontologische Zeitschrift*, **79**, 135–148.

Seears, H. A., Darling, K. F., and Wade, C. M., 2012. Ecological partitioning and diversity in tropical planktonic foraminifera. *BMC Evolutionary Biology*, **12**, 54.

Seibold, E., and Berger, W. H., 1996. *The Sea Floor*. Berlin: Springer, pp. 1–356.

Sen Gupta, B. K. (ed.), 2002. *Modern Foraminifera*. New York: Kluwer, pp. 1–371.

Spero, H. J., and Parker, S. L., 1985. Photosynthesis in the symbiotic planktonic foraminifer *Orbulina universa* and its potential contribution to oceanic primary productivity. *Journal of Foraminiferal Research*, **15**, 273–281.

Thiede, J., 1971. Variations in coiling ratios of Holocene planktonic foraminifera. *Deep-Sea Research*, **18**, 823–831.

Vincent, E., and Berger, W. H., 1981. Planktonic foraminifera and their use in paleoceanography. In Emiliani, C. (ed.), *The Oceanic Lithosphere, The Sea*. New York: Wiley, Vol. 7, pp. 1025–1119.

Cross-references

Biochronology, Biostratigraphy
Calcite Compensation Depth (CCD)
Carbonate Dissolution
Carbon Isotopes
Deep-sea Sediments
Foraminifers (Benthic)
Marine Microfossils
Oxygen Isotopes
Paleoceanographic Proxies
Paleoceanography

G

GABBRO

Jürgen Koepke
Institute of Mineralogy, Leibniz University Hannover, Hannover, Germany

Definition

Gabbro is a medium- to coarse-grained, mafic intrusive igneous rock, mostly grayish to greenish in color, and is the most abundant rock in the oceanic crust. Oceanic gabbros are mainly composed of three minerals: olivine, plagioclase, and clinopyroxene, sometimes associated with orthopyroxene, amphibole, and Fe-Ti oxides. They are formed in magma chambers (qv. "Mid-ocean Ridge Magmatism and Volcanism") under the mid-ocean ridges, where primitive MORB melts (mid-oceanic ridge basalt; qv. "mid-ocean ridge magmatism") differentiate by fractional crystallization. The accumulated minerals in dense crystal mushes and subsequent slow cooling lead to the formation of gabbros in the lower oceanic crust.

Introduction

Gabbro is the most common rock of the oceanic crust and forms its lower part (for review see Coogan, 2014). The name "gabbro" was suggested from the German geologist Leopold von Buch after a town in the Italian Tuscany region. Gabbros from the oceanic crust are named "oceanic gabbros" in order to distinguish them from gabbros from continents. Oceanic gabbros mostly consist of three minerals: forsterite-rich olivine, calcium-rich plagioclase feldspar (usually labradorite or bytownite), and clinopyroxene, which has the composition of augite. Orthopyroxene, amphibole, and Fe-Ti oxides might also occur, but only in small amounts, and usually crystallized late forming interstitial crystals.

Oceanic gabbros are regarded as the frozen crystal mushes accumulated during fractional crystallization of primitive MORB melts in the magma chambers under the mid-ocean ridges, forming – after subsequent cooling – the layer 3 of the oceanic crust (qv. "Layering of Oceanic Crust"). In contrast, the corresponding differentiated melts rise upward to erupt as sheeted dikes (layer 2b) or as basaltic sheet flows or pillow basalts (layer 2a).

Owing to the scarcity of natural exposures of the deep gabbroic basement, especially from fast-spreading oceanic crust, complementary studies on ophiolites (qv. "Ophiolites") have helped establish models for the structure of the oceanic crust.

Classification of oceanic gabbros

Gabbroic rocks can be classified using the modal abundances of the constituent minerals olivine, plagioclase, clinopyroxene, and orthopyroxene, based on the International Union of Geological Sciences (IUGS) system shown in Figure 1 (Le Maitre, 1989). Figure 1 is composed of four individual triangular plots, the three outer ones define gabbroic rocks (or mafic plutonic rocks), while the inner triangle defines ultrabasic (or ultramafic) rocks. The presence of more than 5 vol% plagioclase in Figure 1 divides ultrabasic from gabbroic rocks. In addition to the major rock names given in Figure 1, the plutonic part of the oceanic crust may contain:

Diorite: in the IUGS classification, diorite is distinguished from gabbro by the anorthite content (An) of plagioclase. Diorites have plagioclase containing <50 mol% An, while gabbros have plagioclase containing >50 mol% An. The vast majority of oceanic gabbroic rocks are gabbros.

Oxide gabbros: gabbros with Fe-Ti oxides >2 vol%

Dolerite: is applied to fine- or medium-grained gabbroic rocks with dominant ophitic or subophitic textures

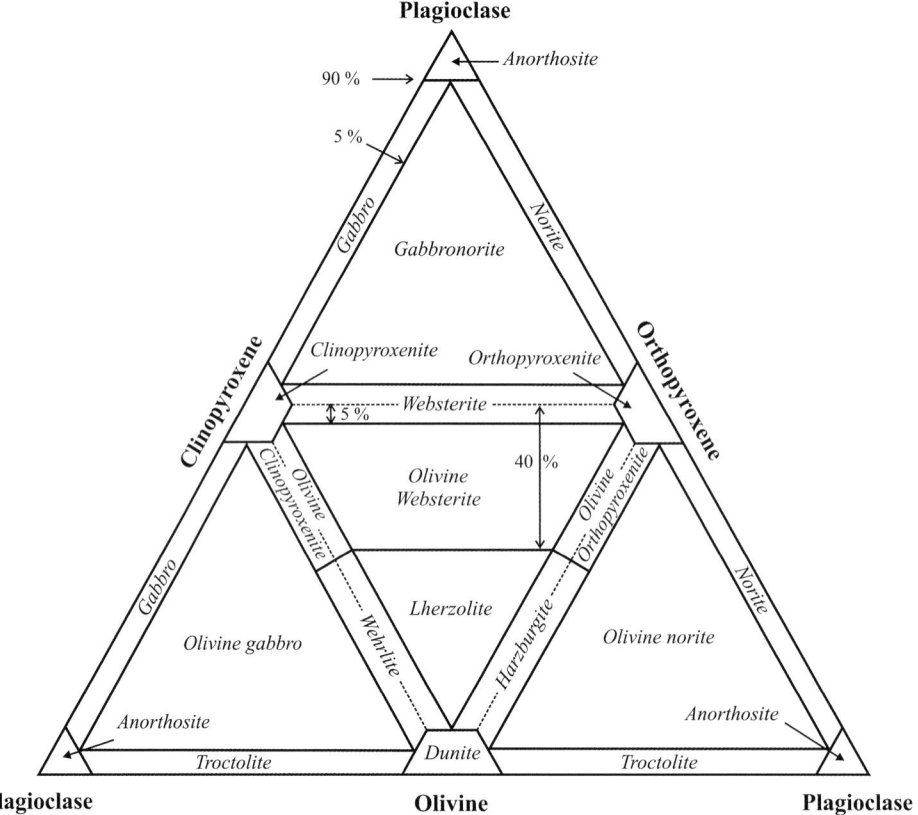

Gabbro, Figure 1 Modal classification scheme for mafic to ultramafic plutonic igneous rocks (after Le Maitre, 1989).

(larger clinopyroxene crystals enclose smaller plagioclase crystals).

Oceanic plagiogranite: the term "oceanic plagiogranite" is an often used field term for light-colored, felsic plutonic rocks within the oceanic crust (e.g., Coleman and Donato, 1979; Koepke et al., 2007), but its exact definition is often questioned. Its use is in accord with IUGS classification, defining the term "oceanic plagiogranite" as a "series of plutonic rocks of the oceanic crust consisting of plagioclase, ranging in composition from oligoclase to anorthite, quartz, and minor amounts of hornblende and pyroxene" (Le Maitre, 1989).

General characteristics of oceanic gabbros

Oceanic gabbros are cumulate (qv. "cumulate") rocks formed by accumulation of fractionated minerals during MORB differentiation in shallow magma chambers. Owing to the relatively slow cooling in the lower crust, these are mostly medium-grained (grain sizes, 1–5 mm) to coarse-grained (5–30 mm) rocks. Typical cooling rates for deeper gabbros from fast-spreading ridges (>500 m below the sheeted dike complex) are 100–10,000 °C per million years (between 1,000 °C and 600 °C, derived from Ca-in-olivine geospeedometry; for details see Coogan, 2014). Crystal settling in magma chambers results in monotonous equigranular, granular textures, mostly with subhedral major minerals such as prismatic olivine and clinopyroxene and tabular plagioclase in concordance with the formation as cumulate (see microphotographs in Figure 2).

Ophiolites show that deep gabbros from fast-spreading ridges exhibit a characteristic layering, which was recently confirmed by direct drilling into the lower crust of the fast-spreading East Pacific Rise (EPR) at Hess Deep (Gillis et al., 2014). A special gabbro type, so-called vari-textured gabbro, occurs at fast-spreading ridges at high level directly below the sheeted dike complex and is quite different in structure and appearance compared to the deeper cumulate gabbro. Vari-textured gabbros are highly heterogeneous, often showing domains with contrasting grain size and varying mode, frequently with poikilitic crystal growth and even pegmatitic growth (e.g., Coogan et al., 2002; Koepke et al., 2011). In contrast to the lower gabbros, which are regarded as cumulates, these gabbros are assumed to correspond more to frozen melts. Gabbros from slow-spreading ridges are in general isotropic and do not show any pronounced layering. A characteristic feature of gabbros from slow-spreading ridges is the abundant crystallization of a late, evolved melt producing

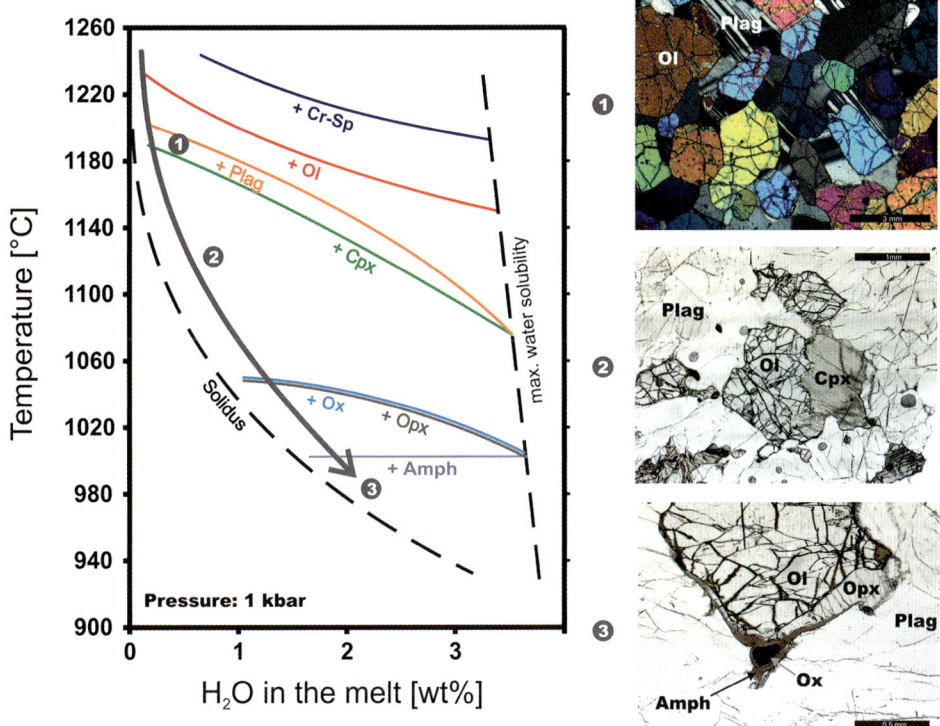

Gabbro, Figure 2 Phase diagram with temperature vs. water content in the melt for a primitive MORB and microphotographs of oceanic gabbros reflecting different stages of differentiation as indicated in the phase diagram. The phase diagram is for 1 kbar (corresponding to a crustal depths of ~3km), modified after Feig et al., (2006). The oxygen fugacity varies as a function of water content from QFM (corresponding to quartz-magnetite-fayalite oxygen buffer) for low water activities to QFM + 4 for high water activities. The phase boundaries correspond to the appearance (+) of the mineral phases. For details see Feig et al., (2006). The gray arrow corresponds to the evolution trend for MORB crystallization in typical shallow magma chambers, with numbers indicating specific stages of magma crystallization/evolution. The microphotographs show corresponding oceanic gabbros which crystallized at these stages. *1*: Early extensive crystallization of olivine and plagioclase leading to (spinel-bearing) troctolites. *2*: Main (cotectic) crystallization of olivine, plagioclase, and clinopyroxene leading to olivine gabbro. *3*: Late crystallization of orthopyroxene, amphibole and Fe-Ti oxide leading to interstitial growth between earlier phases formed during the main crystallization. Abbreviations: *Ol* olivine, *Cr-Sp* chromium-rich spinel, *Cpx* clinopyroxene, *Opx* orthopyroxene, *Plag* plagioclase, *Ox* Fe-Ti oxide, *Amph* amphibole. Samples: *1* troctolite from the Mid-Atlantic Ridge (MAR, 305-U1309D-248R-2, 7–9 cm; image from Expedition Scientific Party, 2005). *2* olivine gabbro from Southwest Indian Ridge (SWIR, 176-735B-166-R-2, 61–71 cm). *3* olivine gabbro from SWIR (176-735B-121-R-3, 65–75 cm). Microphotograph 1 under crossed, 2 and 3 under plane-polarized polars.

interstitial mineral assemblages composed of amphibole oxides, orthopyroxene ± apatite ± zircon.

Many oceanic gabbros show signs of deformation with specific characteristics for gabbros from fast- and from slow-spreading systems. Gabbros from fast-spreading ridges (including the gabbros from the Oman ophiolite) were deformed primarily in a partly molten regime as shown by the abundance of magmatic foliation. Subsequently, no or only minor sub-solidus crystal plastic deformation took place. In contrast, in gabbros from slow-spreading ridges, magmatic foliation is rare, but crystal plastic deformation in a melt-absent regime is common, indicated by the presence of high-temperature shear zones. These contrasting modes of deformation reflect the significant differences in formation/emplacement of gabbros of the lower crust between fast- and slow-spreading ridges (see next chapter).

Gabbro accretion within the oceanic crust

In contrast to the upper oceanic crust, which consists generally of effusive MORBs, the structure, composition, mineralogy, and in situ physical properties of the lower gabbroic crust are poorly constrained. The mode of gabbro accretion (qv. "Crustal Accretion (incl. Gabbro Glacier Model)") is significantly different between fast- and slow-spreading ridges (qv. "Spreading Rates and Ridge Morphology").

At fast-spreading ridges, seismic tomography revealed that much of the lower oceanic crust represents a crystal mush and that a melt lens tens of meters thick exists near the top of the lower crust (e.g., Detrick et al., 1987). Conceptual models for the formation of the deep, fast-spread crust include the "gabbro-glacier" model (e.g., Phipps Morgan and Chen, 1993), where the lower crust is formed in the melt lens, and the "sheeted sill" model

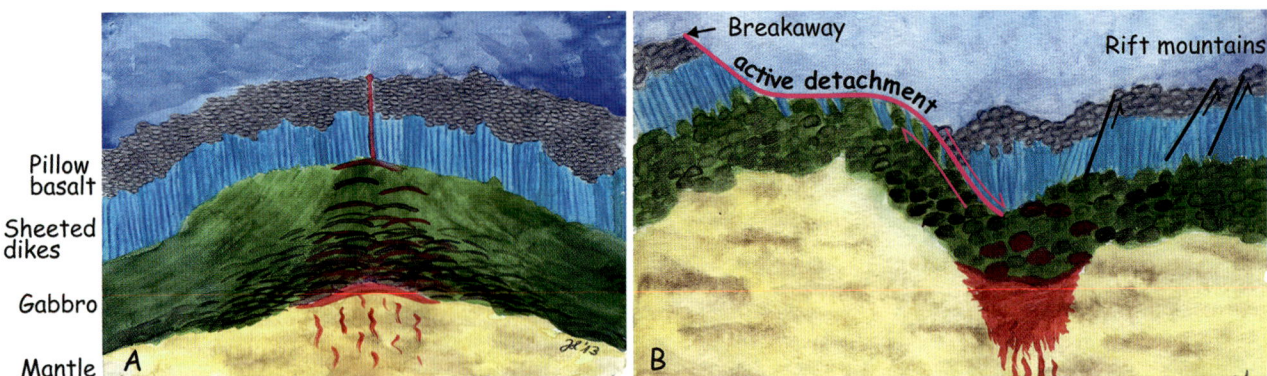

Gabbro, Figure 3 Sketches of gabbro accretion models. (**a**). Intermediate model for fast-spreading oceanic crust, where parts of the lower crust are formed in the shallow melt lens, while other parts in lower crustal sills (modified after Natland and Dick, 2009). Gabbro sills in red are formed by actual magmatic activity; sills in darker color represent the record of earlier intrusions. (**b**). Gabbro accretion and emplacement via long-lived, low-angle detachment faults for slow-spreading crust after an example for an oceanic core complex at 23°N at MAR (after Dick et al., 2008). Gabbro bodies in red are formed by actual magmatic activity; bodies in darker color represent the record of earlier intrusions (drawings by Janna Lehmann).

(e.g., Bedard, 1991; Kelemen et al., 1997), where the lower gabbros are generated by the intrusion of sills of gabbroic mushes. Intermediate models, where part of the lower crust forms in the shallow melt lens and the rest in lower crustal sills, are also discussed (e.g., Boudier et al., 1996; Natland and Dick, 2009). A sketch of such an intermediate model is presented in Figure 3a. The gabbro layer at fast-spreading ridges is usually covered by a 1–2 km thick sequence of basaltic rocks. Thus, there are only three locations where gabbros at fast-spread oceanic crust were studied directly, all formed at the EPR. Two of them are complex tectonic windows where gabbros are exposed as a consequence of ridge/ridge interactions: Hess Deep and Pito Deep. The third location is the Site 1256 located in ~15 Ma-year-old crust formed at the EPR under superfast spreading conditions (for more details on these locations see Coogan, 2014).

While at fast-spreading ridges, axial magma chambers were seismically imaged at many locations, for slow-spreading ridges, this is the case only for one locality (at MAR, 9°N; Singh et al., 2006). The reason therefore is that at slow-spreading ridges extension often takes place in a regime where the magma flux is low. As a consequence considerable amounts of gabbros of the deeper part of the oceanic crust can be emplaced on the seafloor via long-lived, low-angle detachment faults (Figure 3b) (e.g., Cannat, 1993; Ildefonse et al., 2007). The tectonic extension produces Oceanic Core Complexes (OCC) which include smooth gabbro domes perpendicular to the mid-ocean ridge (e.g., Ildefonse et al., 2007). Thus, OCC provide access to the gabbro layer and underlying mantle rocks at slow-spreading oceanic ridges. Owing to the lack of magmatic activity at such localities, individual gabbro bodies may intrude directly into the shallow mantle at a range of depths (Figure 3b) – a model sometimes referred to as a "plum pudding." Many gabbros from oceanic core complexes show characteristic deformations both in the plastic and in the brittle regime, which is a direct expression of tectonic shearing in a temperature regime varying from magmatic down to the sub-greenschist facies. Owing to the relatively easy access to oceanic gabbros in OCC, there are more detailed investigations of gabbros from slow-spreading ridges than from fast-spreading ridges. A detailed compilation on key locations is provided by Coogan (2014 and references therein).

Formation of oceanic gabbros

Experimental studies of MORB crystallization at low pressures reveal the characteristic crystallization sequence confirmed by the petrography of basalts and gabbros which is illustrated in the phase diagram of Figure 2. Since primitive MORBs are practically dry, containing only a very small content of water, the path of crystallization (gray arrow) starts at the left side of Figure 2. Chromium spinel is the first phase crystallizing at very high temperatures ($>1,250\ °C$) but disappears at lower temperatures. The first silicate phase to crystallize is olivine followed by plagioclase. The cotectic crystallization and subsequent accumulation of olivine and plagioclase lead to the formation of troctolitic cumulates. Correspondingly, Cr-spinel-bearing troctolites are the most primitive gabbroic rocks. At lower temperatures, clinopyroxene is saturated leading to the co-crystallization of olivine–plagioclase–clinopyroxene, which correspond to the main stage of crystallization, taking place over a temperature interval $>100\ °C$, explaining that the vast majority of oceanic gabbros correspond to olivine gabbro. At the pressure of 1 kbar, plagioclase is always crystallizing before clinopyroxene. This characteristic feature changes with pressure, and higher-pressure plagioclase crystallization is suppressed, stabilizing the co-crystallization of olivine–clinopyroxene (Gaetani et al., 1993; Feig et al., 2006). About 200 °C below the liquidus of the system, orthopyroxene and Fe-Ti oxides crystallize (the latter strongly controlled by the prevailing oxygen fugacity), followed by amphibole, indicating that the water content in the residual melt has been raised to significant amounts, due to the crystallization of only water-free minerals in the main stage crystallization. At this stage, most of the melts are already crystallized and these minerals form interstitial parageneses.

Geochemistry of oceanic gabbros

Students' textbooks often state that gabbros are equivalent in composition to basalts and that the difference between the two rock families is their grain size, which is a direct function of the cooling rate (high cooling rates produce basalts; low cooling rates gabbros). However, oceanic gabbros are cumulates (=accumulated crystals; qv "cumulate"), and thus their bulk composition is strongly dependent of the presence and the modal amount of accumulated crystals. For example, if oceanic gabbros are strongly enriched in oxides ("oxide gabbro"; see above), the silica content of the bulk rock can be <40 wt% SiO_2, due to the high concentration of the heavy element oxide FeO. If crystallization of plagioclase dominates, anorthosites (Figure 1) may be formed with very high Al_2O_3 contents (>20 wt%), and abundant crystallization of olivine produces olivine-rich troctolites with MgO contents >25 wt%. In contrast, MORBs are frozen melts with more homogeneous major element composition, which varies only within a few wt% for the major element oxides as SiO_2, Al_2O_3, and MgO. Another consequence of the cumulate nature of oceanic gabbros is that the concentrations of incompatible trace elements are generally lower than those of MORBs. For example, the average content of K_2O and P_2O_5 in typical gabbros from the drilled core from Hole 735B at SWIR is 0.04 and 0.01 wt%, respectively (data set of Dick et al., 2000), while the corresponding value for typical primitive MORB is 0.07 and 0.09 (e.g., average from the data base PETD, Lehnert et al., 2000), respectively. This depleted nature of typical oceanic gabbros is a consequence of the accumulation of the cotectic assemblage olivine + plagioclase + clinopyroxene (±orthopyroxene) resulting in rocks which often represent nearly pure cumulates with very low proportions of residual melt (~1.4 % as average value for the gabbros from the SWIR drill core, Natland and Dick, 2001). Hence, the bulk content of incompatible elements of oceanic gabbros is more or less a direct function of the amount of interstitial late crystallization where incompatible elements are enriched in phases like amphibole and apatite, and which is in general decoupled from the main stage of crystallization.

Summary

Gabbros are the most common rocks in the oceanic crust and form its lower part (layer 3). They represent slowly cooled crystal mushes composed mainly of olivine, plagioclase, and clinopyroxene, corresponding to the accumulated phases during differentiation of primitive MORB in shallow magma chambers under the ridges. Their mode of accretion is strongly controlled by the spreading rate, with the formation of layered gabbros at fast-spreading ridges and more isotropic gabbros at slow-spreading ridges. The latter often show signs of plastic deformation and vast crystallization of evolved interstitial parageneses.

Bibliography

Bedard, J. H., 1991. Cumulate recycling and crustal evolution in the Bay of islands ophiolite. *Journal of Geology*, **99**, 225–249.

Boudier, F., Nicolas, A., and Ildefonse, B., 1996. Magma chambers in the Oman ophiolite: fed from the top and the bottom. *Earth and Planetary Science Letters*, **144**, 239–250.

Cannat, M., 1993. Emplacement of mantle rocks in the seafloor at mid-ocean ridges. *Journal of Geophysical Research-Solid Earth*, **98**, 4163–4172.

Coleman, R. G., and Donato, M. M., 1979. Oceanic plagiogranite revisited. In Barker, F. (ed.), *Trondhjemites, Dacites, and Related Rocks*. Amsterdam: Elsevier, pp. 149–167.

Coogan, L. A., 2014. The lower oceanic crust. In Turekian, K., and Holland, H. D. (eds.), *Treatise on Geochemistry*, 2nd edn. Amsterdam: Elsevier, doi:10.1016/B978-0-08-095975-7.00316-8.

Coogan, L. A., Thompson, G., and MacLeod, C. J., 2002. A textural and geochemical investigation of high level gabbros from the Oman ophiolite: implications for the role of the axial magma chamber at fast-spreading ridges. *Lithos*, **63**, 67–82.

Detrick, R. S., Buhl, P., Vera, E., Mutter, J., Orcutt, J., Madsen, J., and Brocher, T., 1987. Multichannel seismic imaging of a crustal magma chamber along the East Pacific Rise. *Nature*, **326**, 35–41.

Dick, H. J. B., Natland, J. H., Alt, J. C., Bach, W., Bideau, D., Gee, J. S., Haggas, S., Hertogen, J. G. H., Hirth, G., Holm, P. M., Ildefonse, B., Iturrino, G. J., John, B. E., Kelley, D. S., Kikawa, E., Kingdon, A., LeRoux, P. J., Maeda, J., Meyer, P. S., Miller, D. J., Naslund, H. R., Niu, Y. L., Robinson, P. T., Snow, J., Stephen, R. A., Trimby, P. W., Worm, H. U., and Yoshinobu, A., 2000. A long in situ section of the lower ocean crust: results of ODP Leg 176 drilling at the Southwest Indian Ridge. *Earth and Planetary Science Letters*, **179**, 31–51.

Dick, H. J. B., Tivey, M. A., and Tucholke, B. E., 2008. Plutonic foundation of a slow-spreading ridge segment: oceanic core complex at Kane megamullion, 23 degrees 30 ' N, 45 degrees 20 ' W. *Geochemistry, Geophysics, Geosystems*, **9**.

Expedition Scientific Party, 2005. Oceanic core complex formation, Atlantis Massif – oceanic core complex formation, Atlantis Massif, Mid-Atlantic Ridge: drilling into the footwall and hanging wall of a tectonic exposure of deep, young oceanic lithosphere to study deformation, alteration, and melt generation. Integrated Ocean Drilling Program Preliminary Reports, 305, http://iodp.tamu.edu/publications/PR/305PR/305PR.PDF.

Feig, S., Koepke, J., and Snow, J., 2006. Effect of water on tholeiitic basalt phase equilibria: an experimental study under oxidizing conditions. *Contributions to Mineralogy and Petrology*, **152**, 611–638.

Gaetani, G. A., Grove, T. L., and Bryan, W. B., 1993. The influence of water on the petrogenesis of subduction-related igneous rocks. *Nature*, **365**, 332–334.

Gillis, K. M., Snow, J. E., Klaus, A., Abe, N., Adriao, A. B., Akizawa, N., Ceuleneer, G., Cheadle, M. J., Faak, K., Falloon, T. J., Friedman, S. A., Godard, M., Guerin, G., Harigane, Y.,

Horst, A. J., Hoshide, T., Ildefonse, B., Jean, M. M., John, B. E., Koepke, J., Machi, S., Maeda, J., Marks, N. E., McCaig, A. M., Meyer, R., Morris, A., Nozaka, T., Python, M., Saha, A., and Wintsch, R. P., 2014. Primitive layered gabbros from fast-spreading lower oceanic crust. *Nature*, **505**, 204–207.

Ildefonse, B., Blackman, D. K., John, B. E., Ohara, Y., Miller, D. J., Macleod, C. J., and Integrated Ocean Drilling Program Expeditions 304/305 Science Party, 2007. Oceanic core complexes and crustal accretion at slow-spreading ridges. *Geology*, **35**, 623–626.

Kelemen, P. B., Koga, K., and Shimizu, N., 1997. Geochemistry of gabbro sills in the crust-mantle transition zone of the Oman ophiolite: implications for the origin of the oceanic lower crust. *Earth and Planetary Science Letters*, **146**, 475–488.

Koepke, J., Berndt, J., Feig, S. T., and Holtz, F., 2007. The formation of SiO_2-rich melts within the deep oceanic crust by hydrous partial melting of gabbros. *Contributions to Mineralogy and Petrology*, **153**, 67–84.

Koepke, J., France, L., Müller, T., Faure, F., Goetze, N., Dziony, W., and Ildefonse, B., 2011. Gabbros from IODP Site 1256 (Equatorial Pacific): insight into axial magma chamber processes at fast-spreading ocean ridges. *Geochemistry, Geophysics, Geosystems*, **12**, doi:10.1029/2011GC003655.

Le Maitre, R. W., 1989. *A Classification of the Igneous Rocks and Glossary of Terms*. Oxford: Blackwell. 193 p.

Lehnert, K., Su, Y., Langmuir, C. H., Sarbas, B., and Nohl, U., 2000. A global geochemical database structure for rocks. *Geochemistry, Geophysics, Geosystems*, **1**, 1999GC000026.

Natland, J. H., and Dick, H. J. B., 2001. Formation of the lower ocean crust and the crystallization of gabbroic cumulates at a very slowly spreading ridge. *Journal of Volcanology and Geothermal Research*, **110**, 191–233.

Natland, J. H., and Dick, H. J. B., 2009. Paired melt lenses at the East Pacific Rise and the pattern of melt flow through the gabbroic layer at a fast-spreading ridge. *Lithos*, **112**, 73–86.

Phipps Morgan, J. P., and Chen, Y. J., 1993. The genesis of oceanic-crust – magma injection, hydrothermal circulation, and crustal flow. *Journal of Geophysical Research-Solid Earth*, **98**, 6283–6297.

Singh, S. C., Crawford, W. C., Carton, H., Seher, T., Combier, V., Cannat, M., Canales, J. P., Dusunur, D., Escartin, J., Miranda, J. M., 2006. Discovery of a magma chamber and faults beneath a Mid-Atlantic Ridge hydrothermal field. Nature, **442**, 1029–1032

Cross-references

Crustal Accretion
Cumulates
Layering of Oceanic Crust
Mid-ocean Ridge Magmatism and Volcanism
Ophiolites

GENERAL BATHYMETRIC CHART OF THE OCEANS (GEBCO)

Hans Werner Schenke
Alfred-Wegener-Institute, Helmholtz-Centre for Polar and Marine Research, Bremerhaven, Germany

Definition

GEBCO, the General Bathymetric Chart of the Oceans, is a global seafloor mapping program under the auspices of the International Hydrographic Organization (IHO) and the Intergovernmental Oceanographic Commission (IOC) of UNESCO.

History

GEBCO was established in 1903 in Wiesbaden, Germany, in the course of the committee meeting on ocean nomenclature and charting, chaired by Prince Albert I of Monaco (Carpine-Lancre et al., 2003).

Five editions of GEBCO were published until 1983. The 5th edition was digitized and used for the development of the GEBCO Digital Atlas (GDA). With the latest version, published at the event of the GEBCO Centenary 2003 (GDA-CE), GEBCO entered the era of digital bathymetry. The GDA-CE includes also undersea feature names and is continuously updated with new single and multibeam soundings.

Structure

GEBCO is directed by a Guiding Committee which organizes the work of the subcommittees and acts for GEBCO toward the parent organizations. It supervises the development, maintenance, and updating of all GEBCO products and fosters the cooperation with related marine scientific associations.

GEBCO runs three subcommittees (SC):

1. Technical SC on Ocean Mapping (TSCOM): The main task is the development of modern gridding techniques and the continuous updating of the global bathymetric grid.
    Products: global grid, *GEBCO CookBook* (IHO Publication B-11).
2. SC on Undersea Feature Names (SCUFN): The function of SCUFN is to evaluate name proposals considering the IHO Publication B-6 "Standardization of Undersea Feature Names" and to maintain and update the IHO Publication B-8 GEBCO Web Gazetteer (http://www.ngdc.noaa.gov/gazetteer/).
3. SC on Regional Undersea Mapping (SCRUM): SCRUM was established in 2009 and tasked to initiate and support regional ocean mapping projects and coordinate the International Bathymetric Chart (IBC) programs, like the IBC of the Arctic Ocean (http://www.ibcao.org) and the Southern Ocean (http://www.ibcso.org). A GEBCO world map was compiled and published on http://www.gebco.net/data_and_products/gebco_world_map/.

The following international agencies and universities contribute to GEBCO:

1. The National Geophysical Data Center (NGDC) of the US National Oceanic and Atmospheric Administration (NOAA) operates and maintains the IHO Data Center for Digital Bathymetry (http://www.ngdc.noaa.gov/mgg/bathymetry/iho.html).
2. The British Oceanographic Data Centre (BODC) played a key role in GEBCO's transition from analog

to digital products. BODC maintains the GDA-CE and continuously updates the global bathymetric grid (https://www.bodc.ac.uk/data/online_delivery/gebco/gebco_08_grid/).
3. The **IHO** has embedded GEBCO in the Inter-Regional Coordination Program. IHO operates and maintains the GEBCO Gazetteer of Undersea Feature Names and publishes on behalf of GEBCO bathymetric publications (Ward, 2010) (http://www.iho.int/srv1/).
4. The **University of New Hampshire (UNH)** conducts since 2004 in cooperation with GEBCO the "Nippon Foundation/GEBCO Postgraduate Certificate in Ocean Bathymetry Training Program" in order to train academics and hydrographers in ocean mapping (http://ccom.unh.edu/gebco-students-scholars).

Bibliography

Carpine-Lancre, J., et al., 2003. *The History of GEBCO 1903–2003*. Lemmer: GITC bv. ISBN 90-806205-4-8.
Ward, R., 2010. General bathymetric chart of the oceans. *Hydro International*, **14**(5), 45.

Cross-references

General Bathymetric Chart of the Oceans (GEBCO)
Plate Tectonics

GENERAL OCEAN CIRCULATION - ITS SIGNALS IN THE SEDIMENTS

Hermann Kudrass
MARUM-Center for Marine Environmental Sciences, University of Bremen, Bremen, Germany

Definition

The worldwide pattern of ocean circulation is governed by a balance of surface and deep and bottom currents.

Ocean circulation

Ocean circulation is driven by the higher solar radiation at low latitudes compared to the radiation at high latitudes and the resulting differences in evaporation. The transport routes and the velocity of currents are determined by the Earth's rotation, the atmospheric circulation, and the configuration of the ocean basins. The meridional heat and salt transfer by warm tropical and subtropical water is mostly confined to the western side of the ocean basins and cold-water flow towards the equator along the eastern side. A considerable part of the poleward warm-water flow at the surface is balanced by cold bottom water flowing towards the equator. At a global scale, this flow pattern is simplified by the salt conveyor belt or the thermohaline circulation (Figure 1). On geological time scales, this circulation pattern considerably varied, mainly driven by changes of solar insolation and plate tectonic movements. The reconstruction of these changes, the main subject of paleoceanography, entirely relies on the record of ocean currents preserved in marine sediments.

Surface currents and bottom currents produce different sedimentary records. Surface currents are mainly recorded in the underlying sediments by contrasting regional distributions of specific proxy parameters. In contrast, the signals of the subsurface currents, which flow in direct contact with the seafloor, can be more directly deduced from the sediment properties and composition.

Surface currents

A great variety of proxy parameters in the sedimentary archives is used to decode the extent and intensity of surface currents. First of all, the remains of surface-dwelling organisms of coccolithophorids, diatoms, radiolaria, planktonic foraminifera, pteropods, and dinoflagellates are good indicators for characterizing temperature, salinity, and fertility of the surface water. Geochemical composition of calcareous and siliceous tests (e.g., Mg/Ca, P, Ba, Cd) and their isotopic compositions (C, O, Si, Nd), as well as organic carbon and biomarkers (e.g., alkenones) and their isotopic compositions (C, N, H), are successfully used to identify origin and transport paths of the surface waters. In case that the surface currents are in contact with the seafloor, benthic communities also contain a record of changes, like corals for the ENSO signals (Cobb et al., 2003) or benthic foraminifera for the Indonesian Throughflow (Holbourn et al., 2011). In special cases, the spread of debris from drifting icebergs can define ocean-wide transport paths. A successful strategy to decipher changes in global oceanic circulation is the drilling of sediment core transects, e.g., by the International Ocean Drilling Program (IODP). The transects usually compromise drill holes along the postulated transport path. In special cases, the opening or closing of gateways by plate tectonic movements was investigated (e.g., Tasmanian Gateway for the Circum-Antarctic Current (Exon et al., 2004), mid-American uplift for the initiation of the Gulf Current (Pisias et al., 1995; Haug and Tiedemann, 1998), or the latitudinal shifts and intensity changes of the Equatorial Current in the Pacific Ocean (Lyle and Wilson, 2006)).

Bottom currents

As for the surface currents, the properties of bottom waters are documented by benthic biological communities of mollusks, ostracods, deep-water corals, and epi- and endobenthic foraminifera, as well as the isotopic and elemental composition of their shells, skeletons, and tests. Thus, changes in bottom currents can be reconstructed from fossilized remains of these organisms. In addition, bottom currents generate a much more direct sedimentary record than the surface currents, as they directly impact the sediment. First of all, the fine-scale structure and the composition of the sediments are shaped by the flow, which generates typically silty laminated sediments in extended

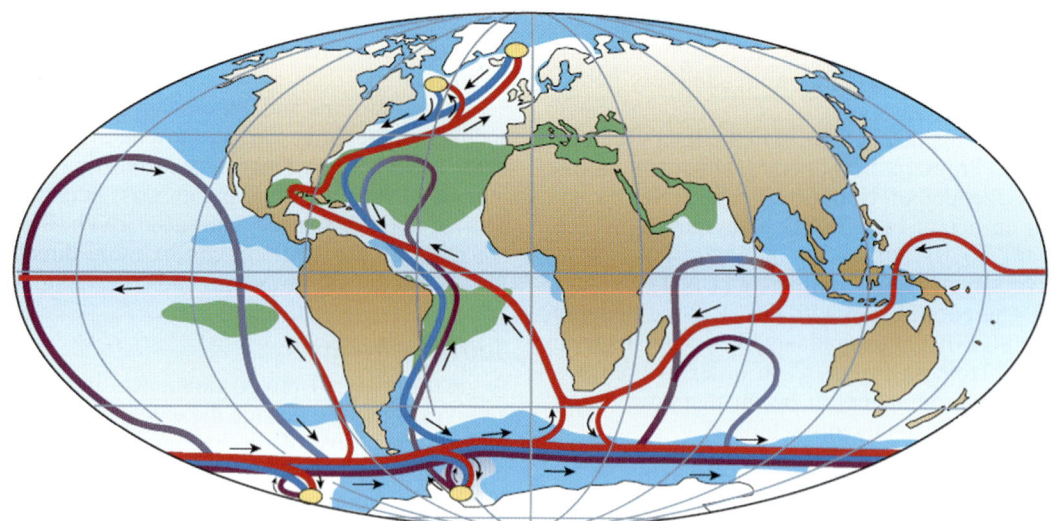

General ocean circulation - its signals in the sediments, Figure 1 Global thermohaline circulation and surface salinity (from Rahmstorf, 2002). *Yellow ovals* deep-water formation, *blue lines* deep currents, *purple lines* bottom currents, *red lines* surface currents, *green shading* salinity >36 ‰, *blue shading* salinity <34 ‰.

sediment waves and drift mounds. These drifts are aligned along the contour-following bottom currents and are called contourites. The variation of current velocities is recorded by bedding and grain-size distribution, the latter one mostly relying on the sortable silt. On a larger scale, the internal architecture and the horizontal distribution of these contourite beds can be studied by high-resolution seismics. The internal structure of the accumulation patterns of sediment waves and drift mounds indicates the direction of sediment transport (Stow et al., 2002). Especially the contourites generated by the Mediterranean Outflow (Roque et al., 2012), the Antarctic Circumpolar Current (Michels et al., 2001), and the western boundary currents of the North Atlantic Deep Water (Müller-Michaelis et al., 2013) contain a long-term record of bottom-water characteristics and variations of the flow intensity.

Summary

A high variety of physical and geochemical parameters of the circulating water masses are recorded by biological, geochemical, and petrophysical proxies, which are preserved in the bottom sediments and can be used to monitor changes of the general ocean circulation pattern.

Bibliography

Cobb, K. M., Charles, C. D., Cheng, H., and Edwards, R. L., 2003. El Nino/southern oscillation and tropical pacific climate during the last millennium. *Nature*, **424**(6946), 271–276.

Exon, N. F., Kennett, J. P., and Malone, M. J., 2004. Leg 189 synthesis: cretaceous-Holocene history of the Tasmanian Gateway. In Exon, N. F., Kennett, J. P., and Malone, M. J. (eds.), *Proceedings of the ODP Scientific Results*, Vol. 189, pp. 1–37.

Haug, G. H., and Tiedemann, R., 1998. Effect of the Isthmus of Panama on Atlantic Ocean thermohaline circulation. *Nature*, **393**, 673–676.

Holbourn, A. E., Kuhnt, W., and Xu, J., 2011. Indonesian throughflow variability during the last two glacial cycles: the Timor Sea outflow. In *The SE Asian Gateway: History and Tectonics of Australia-Asia Collision*. Geological Society, London, Special Publications No. 355, pp. 283–303, doi:10.1144/SP355.14.

Lyle, M., and Wilson, P. A., 2006. Leg 199 synthesis: evaluation of the equatorial Pacific in the early Cenozoic. In Lyle, M., Wilson, P. A., and Firth, J. V. (eds.), *Proceedings of ODP Scientific Results*, Vol. 199, pp. 1–39.

Michels, K. H., Rogenhagen, J., and Kuhn, G., 2001. Recognition of contour-current influence in mixed contour-turbidite sequences of the western Weddell Sea, Antarctica. *Marine Geology*, **22**, 465–485.

Müller-Michaelis, A., Uenzelmann-Neben, G., and Stein, R., 2013. A revised early Miocene age for the instigation of the Eirik Drift, off Greenland: evidence from high-resolution records. *Marine Geology*, **340**, 1–15.

Pisias, N. G., Mayer, L. A., and Mix, A. C., 1995. Paleoceanography of the eastern Pacific during the Neogene: synthesis of Leg 138 drilling results. In Pisias, N. G., et al. (eds.), *Proceedings of ODP Scientific Results*, Vol. 138, pp. 5–21.

Rahmstorf, S., 2002. Ocean circulation and climate during the past 120,000 years. *Nature*, **419**, 207–214.

Rebesco, M., and Camerlenghi, A. (eds.), 2008. *Contourites*. Amsterdam: Elsevier Science, pp. 1–688.

Roque, C., Duarte, H., Terrinha, P., Valadares, V., Noiva, J., Cachao, M., Ferreira, J., Legoinha, P., and Zitellini, N., 2012. Pliocene and Quaternary depositional model of the Algarve margin contourite drifts (Gulf of Cadiz, SW Iberia): seismic architecture, tectonic control and paleoceanographic insights. *Marine Geology*, **303**, 42–62.

Stow, D. A. V., Faugères, J.-C., Pudsey, C. J., and Viana, A. R., 2002. Bottom currents, contourites and deep-sea sediment drifts: current state-of-the-art. In *Deep-Water Contourite Systems: Modern Drifts and Ancient Series, Seismic and Sedimentary Characteristics*. London: Geological Society. Geological Society of London, Memoirs, Vol. 22, pp. 7–20.

Cross-references

Biogenic Barium
Contourites
Currents
Deep-sea Sediments
Diatoms
Dinoflagellates
Foraminifers (Benthic)
Foraminifers (Planktonic)
Ice-rafted Debris (IRD)
Paleoceanographic Proxies
Paleoceanography
Pteropods
Radiogenic Tracers
Radiolarians

GEOCHRONOLOGY: URANIUM-SERIES DATING OF OCEAN FORMATIONS

Vladislav Kuznetsov
Institute of Earth Sciences, Saint Petersburg State University, St. Petersburg, Russia

Definition

Uranium-series methods allow estimating quantitatively an age of various types of ocean and terrestrial formations and are based on phenomenon of radioactive decay/accumulation of isotopes from two naturally occurring U-series, with parent isotopes being ^{238}U and ^{235}U.

Introduction

Deep-sea sediments are deposited under a thick water layer and protected from external influences such as climate change, wind erosion, and other destruction processes. Therefore, deep-sea sediments represent unique well-preserved record of climatic and geological events that occurred during the Quaternary period. Hence, they can be used for reconstruction of these events in time applying the U-series dating methods. Theory says that in natural objects in which the activities of all of the nuclides are equal, all isotopes from the same decay chain are in radioactive secular equilibrium (Figure 1). In most environmental materials, however, a break in the decay chain and a state of radioactive disequilibrium are observed, and it is the main prerequisite for the use of U-series dating methods.

The chemical behavior of U-series nuclides, their contents, and mechanisms of accumulation in the ocean environment were widely researched in the second half of the twentieth century. As a result, the basic principles and methods of radioisotope geochronology of ocean sediments are established and well summarized (Kuznetsov, 1976; Smart, 1991; Ivanovich and Harmon, 1992; Cochran, 1992; Wagner, 1998; Cochran and Masque, 2003; Edwards et al., 2003; Henderson and Anderson, 2003; Kuznetsov, 2008; Kuznetsov and Maksimov, 2012).

For dating different ocean formations, two types of the U-series disequilibrium methods are available which are based on radioactive (1) decay or (2) accumulation of daughter isotope (Ku, 1976; Ivanovich and Harmon, 1992).

At the present time, the most widespread and well-founded methods are $^{230}Th_{excess}$, $^{231}Pa_{excess}$, $^{230}Th/^{234}U$, and $^{231}Pa/^{235}U$ dating methods of different ocean formations. This entry further discusses the use of such short-living U-series nuclides as ^{234}Th (half-life = 24.1 day), ^{224}Ra (3.64 day), etc., allowing assess rates of processes in the modern ocean, as well as ^{210}Pb (22 years), allowing estimates of accumulation rates of the marine shelf and lacustrine sediments.

$^{230}Th_{excess}$ and $^{231}Pa_{excess}$ dating methods

In oxidizing aqueous conditions typical for most seawater, U forms the soluble uranyl carbonate species, and the most part of this element content is held in the ocean waters. ^{230}Th and ^{231}Pa are produced by decay of ^{238}U (and ^{234}U) and ^{235}U, respectively. Both elements form the positively charged or neutral hydroxide species which are strongly absorbed onto suspended particles and rapidly removed from the water to the ocean floor (Sackett, 1960; Scott et al., 1972; Kuznetsov, 1976; Anderson et al., 1983a; Anderson et al., 1983b; Taguchi et al., 1989; Cochran, 1992). Thus, significant concentrations/activities of ^{230}Th, as well as ^{231}Pa, are accumulated on the surface of the ocean floor, much more than in seawater (Moore and Sackett, 1964; Kuznetsov, 1976). ^{230}Th (and ^{231}Pa) in the sediments consists of both the ^{230}Th (^{231}Pa) derived from ^{234}U (^{235}U) decay in the seawater and small portion of ^{230}Th (^{231}Pa) contained in mineral detrital component. Only the unsupported ^{230}Th or ^{231}Pa, the so-called $^{230}Th_{excess}$ (or $^{230}Th_{xs}$) and $^{231}Pa_{excess}$ (or $^{231}Pa_{xs}$), derived directly from seawater ^{234}U (^{235}U) decay can be used to calculate the sediment age. Therefore, both isotope activities measured in the sediment samples must be corrected for this small portion of detrital ^{230}Th (and ^{231}Pa) to estimate $^{230}Th_{xs}$ ($^{231}Pa_{xs}$). The correction technique is described (see Henderson and Anderson, 2003 for details).

There are a number of ocean objects suitable for $^{230}Th_{xs}$ (or $^{231}Pa_{xs}$) dating: (1) deep-sea sediments of different origin (foraminiferal or metalliferous sediments, for instance) and (2) ferromanganese formations (Fe-Mn nodules and crusts). The theoretical distribution of $^{230}Th_{xs}$ or $^{231}Pa_{xs}$ radioisotopes in sediments and Fe-Mn formations reflects exponential decay due to their half-life time.

Formerly, the alpha-emitting nuclides, such as ^{230}Th and ^{231}Pa, were measured by alpha spectrometry following analytical procedures needed to extract the isotopes from a sample. Effective age ranges, depending on the isotope's half-life ($T_{1/2}$ = 75.2 kyr for ^{230}Th and $T_{1/2}$ = 34.3 kyr for ^{231}Pa) and activity level in a sample, are ca. 3–350 kyr for the $^{230}Th_{xs}$ method and ca. 5–150 kyr

Geochronology: Uranium-Series Dating of Ocean Formations, Figure 1 Simplified scheme of the ^{238}U-series (**a**) and ^{235}U-series (**b**) chains.

for the ^{231}Pa$_{xs}$ method. Modern mass spectrometric analysis allowed extending the age ranges for both methods. It is now possible to determine ^{230}Th$_{xs}$ ages from several decades to ca. 500 kyr and ^{231}Pa$_{xs}$ ages up to 250 kyr (Chen et al., 1986; Edwards et al., 1986; Edwards et al., 1987; Edwards et al., 1997). Mass spectrometric methods for measuring Th, Pa, and U isotopes have greatly reduced sample size requirements and improved analytical precision.

Theoretical prerequisites for the application of both the methods are:

1. ^{230}Th and ^{231}Pa as well as parent uranium may not migrate in the sedimentary stratum. A lot of data support the point of view that Th and Pa are always immobile in aqueous conditions (Kuznetsov, 1976; Huh and Ku, 1984; Cochran, 1992; Henderson and Anderson, 2003). Possible migration of U as a mobile element may not significantly affect the vertical distribution of the daughter nuclides due to its low concentration in sediments compared with ^{230}Th and ^{231}Pa (Kuznetsov, 1976).
2. The concentration of U producing ^{230}Th and ^{231}Pa in ocean water must remain constant during aging, i.e., for the last 300–350 kyr. Seawater U concentration does not change during at least for the last ca. 400 kyr according to Kuznetsov (1976), Chen et al. (1986), Rosenthal et al. (1995), and Henderson and Anderson (2003).
3. The ^{230}Th and ^{231}Pa sedimentation rate must be constant through time. However, fluctuations in sedimentation rate are often observed due to biogenic and/or terrigenous supply changes. These variations can be corrected by applying normalization of ^{230}Th$_{ex}$ (^{231}Pa$_{ex}$) specific activities to carbonate-free material of sediments (Kuznetsov, 1976, 2008; Ivanovich and Harmon, 1992).
4. There are no any disturbances in sediment stratigraphy.

If these assumptions are fulfilled, the vertical distribution of both nuclides in sediments becomes close to their theoretical exponential decrease from the surface layers to the lower layers.

After deposition, the ^{230}Th$_{xs}$ and ^{231}Pa$_{xs}$ decay and tend to come to equilibrium with the parent uranium. The age of individual sediment layer (i) can be calculated from ^{230}Th$_{xs}$ and ^{231}Pa$_{xs}$ values as

$$^{230}\text{Th}_{xs}(0) = {}^{230}\text{Th}_{xs}(i) \cdot e^{-\lambda(230)t} \text{ and}$$

$$^{231}\text{Pa}_{xs}(0) = {}^{230}\text{Pa}_{xs}(i) \cdot e^{-\lambda(231)t}, \qquad (1)$$

where:

- ^{230}Th$_{xs}$(0), ^{231}Pa$_{xs}$(0) = specific activities of the nuclides in the outer surface layer
- ^{230}Th$_{xs}$(i), ^{231}Pa$_{xs}$(i) = specific activities of the nuclides in the underlying layer i
- $\lambda = \ln 2 / T_{1/2}$, where $T_{1/2}$ is the nuclide half-life

These two methods are widely used in the determination of deep-sea sedimentation rates during the Holocene and Late and Middle Neopleistocene as well as age accumulation rates of Fe-Mn nodules and crusts (Ku and Broecker, 1967; Ku and Broecker, 1969; Kuznetsov, 2008). For example, the recent determinations of Fe-Mn nodule and crust time formation showed their ancient ages of several Myr and accumulation rates being in the wide range from millimeters to centimeters per million years (Finney et al., 1984; Huh and Ku, 1984; Claude-Ivanaj et al., 2001; Kuznetsov, 2008). Eisenhauer et al. (1992) used ^{230}Th$_{xs}$ method to determine the growth rates of Mn crusts from the Pacific Ocean. The study showed that the Mn crusts grew faster during the warm climatic periods compared to the cold periods. Both the ^{230}Th$_{xs}$ and ^{231}Pa$_{xs}$ are applied to determine the age boundaries of marine isotope stages (MIS) reflecting global climate changes in the past. These data have played an important role in building the timescale of changes in the isotopic composition of oxygen in the foraminiferal shells from the ocean sedimentary columns (Broecker and Van Donk, 1970). For instance, the boundary between MIS 6 and MIS 5 detected in the foraminiferal shells by δ^{18}O analysis shows astronomical date about 127 Kyr and agrees well with the age range of 127–128 Kyr obtained by the U-series dating methods (Edwards et al., 1986; Edwards et al., 1987; Zhu et al., 1993). Currently, both the methods are widely used in the study of the sedimentation processes and paleoclimate changes in the Arctic (Not and Hillaire-Marcel, 2010).

^{230}Th/^{234}U and ^{231}Pa/^{235}U dating methods

Both the ^{230}Th/U and ^{231}Pa/U methods play an important role in dating marine carbonates (corals, mollusk shells) as well as the ^{230}Th/U method which is most widely applied in dating seafloor massive sulfide (SMS) deposits in the ocean hydrothermal zones. ^{230}Th/U dating involves calculating ages from radioactive decay and accumulation relationships among ^{238}U, ^{234}U, and ^{230}Th, whereas ^{231}Pa/U dating involves calculating ages from the ingrowth of ^{231}Pa from its grandparent ^{235}U. The age range for these methods is the same as for ^{230}Th$_{xs}$ and ^{231}Pa$_{xs}$ (see above).

The ^{230}Th (and ^{231}Pa) content in seawater is negligible, while the average concentration of U is about ~ 3 μg/l (Chen et al., 1986). The biogenic precipitation of calcium carbonate from seawater (e.g., when coralline skeleton is formed) is accompanied by the incorporation of uranium, which occurs in the soluble uranyl carbonate complexes in seawater, while the ^{230}Th (and ^{231}Pa) is strongly absorbed onto suspended matter. As a result, the carbonate radiometric system contains a deficiency of ^{230}Th (and ^{231}Pa) in comparison with parent ^{234}U (^{235}U). The first applications of both the ^{230}Th/U and ^{231}Pa/U methods in dating marine carbonates were demonstrated in the second half of the last century by Barnes et al. (1956), Rosholt and Antal (1962), Veeh (1966), Sakanoue et al. (1967), Ku (1968), Broecker et al. (1968), Kaufman et al. (1971), and others.

The SMS deposits (or sulfide ores, elsewhere in the publications) were formed from hydrothermal fluids originated from the seawater circulated in the basalts of the oceanic crust. The reduced hydrothermal fluids contain two orders of magnitude less uranium than the seawater (Michard et al., 1983). The hydrothermal fluids mix with seawater which results locally in reducing conditions. By this, easily soluble uranyl carbonate complexes dissolved in seawater become transformed into absorbable uranyl or poorly soluble U^{IV} ions. The latter coprecipitates with transitional sulfides discharged with the fluid. As a result, uranium (without its daughter nuclide ^{230}Th) is accumulated in the SMS deposits on the seafloor. Common uranium concentrations in the SMS deposits range up to 10 ppm (parts per million) or more (Lalou et al., 1996; Kuznetsov et al., 2002; Kuznetsov et al., 2006; Kuznetsov et al., 2007). First ^{230}Th/U ages of sulfide ores from the East Pacific Rise (EPR) and Mid-Atlantic Ridge (MAR) were obtained at the end of the last century by Lalou and Brichet (1982, 1987) and Lalou et al. (1988, 1993, 1995, 1996, 1998).

There are two main prerequisites for ^{230}Th/U (and ^{231}Pa/U) dating of both marine carbonates and sulfide ores (Ivanovich and Harmon, 1992; Lalou et al., 1996; Geyh, 2001; Kuznetsov, 2008; Kuznetsov et al., 2011): (1) sulfides/carbonates contain uranium without thorium immediately after deposition, and (2) during their aging, sulfides/carbonates were under conditions of chemically closed system with regard to uranium and thorium.

A thorough check whether the ^{230}Th/U (and ^{231}Pa/U) method is applicable is necessary in order to obtain reliable ^{230}Th/U (and ^{231}Pa/U) ages of both marine carbonates and sulfide ores. Different approaches to testing the theoretical positions of these methods are described by Edwards et al. (2003) for dating carbonates and by Lalou et al. (1996) and Kuznetsov et al. (2011) for dating SMS deposits.

The ^{230}Th/U age of a sample is derived from Eq. 2 (Ivanovich and Harmon, 1992):

$$\frac{^{230}\text{Th}}{^{234}\text{U}} = \frac{^{238}\text{U}}{^{234}\text{U}}\left(1 - e^{-\lambda_0 t}\right) + \left[\left(1 - \frac{\lambda_0}{\lambda_0 - \lambda_4}\right)\left(1 - \frac{^{238}\text{U}}{^{234}\text{U}}\right)\left(1 - e^{(\lambda_4 - \lambda_0)t}\right)\right], \quad (2)$$

where:

λ_0, λ_4 – decay constants for the ^{230}Th and ^{234}U; ^{230}Th/^{234}U and ^{238}U/^{234}U – AR; ^{234}U, ^{238}U, ^{230}Th – specific activities; and t – age of a sample

Both the ^{230}Th/U and ^{231}Pa/U methods play an important role in the study of sea-level changes in the past. Paired ^{230}Th/U and ^{14}C dating of coral sequences from a number of sites (such as the Barbados, Tahiti, Papua New Guinea) has allowed reconstructing the deglacial sea-level history (Bard et al., 1990; Edwards et al., 1993; Stein et al., 1993; Bard et al., 1996; Hanebuth et al., 2000; Cutler et al., 2003). For example, it was established that during the Last Glacial Maximum between 20 and 22 kyr ago, the minimum sea level was about 120 m lower than at the present time (Bard et al., 1990; Hanebuth et al., 2000). Sea-level reconstructions over the timescale of a full glacial-interglacial cycle during past ca. 140 kyr, as well as earlier interglacial periods, were also determined using a large number of ^{230}Th/U and ^{231}Pa/U datings of corals, carbonate bank sediments, and speleothems (e.g., see recent publications by Arslanov et al., 2002; Cutler et al., 2003; Frank et al., 2006; Andersen et al., 2008; Waelbroeck et al., 2008; Kuznetsov, 2008).

The discovery of massive sulfide deposits within the East Pacific Rise and Mid-Atlantic Ridge in the 1970s to 1980s has gathered great scientific and economic interest due to high concentration of Cu, Zn, Pb, Fe, Mn, Au, Ag, and a number of rare chemical elements as well as huge size of dozens of million tons each. Applying the ^{230}Th/U dating of sulfide ore deposits, the chronology of ore-forming processes having the episodic or pulse-type origin can be evaluated (Lalou et al., 1995; Lalou et al., 1998; You and Bickle, 1998; Kuznetsov et al., 2006; Kuznetsov et al., 2007; Kuznetsov, 2008; Kuznetsov et al., 2011; Kuznetsov and Maksimov, 2012). The recent dating results of SMS samples from the "Semenov" hydrothermal district (located at 13°31′ N, 44°59′ W in a depth between 2,360 and 2,580 m b.s.l.) at the MAR allowed evaluating the total ore formation period of ca. 124 kyr (Kuznetsov et al., 2011). It was shown that the hydrothermal activity and related ore formation had a pulse pattern marked by the certain number of episodes with duration of up to several 1,000 years.

Summary

Two types of the U-series disequilibrium methods are available for dating different ocean formations and are based on radioactive (1) decay or (2) accumulation of daughter isotope. At the present time, the most widespread and well-founded methods are (1) $^{230}Th_{excess}$, $^{231}Pa_{excess}$, and (2) $^{230}Th/^{234}U$ and $^{231}Pa/^{235}U$ dating methods of different ocean formations.

A number of ocean objects are suitable for $^{230}Th_{xs}$ and $^{231}Pa_{xs}$ dating: deep-sea sediments of different origin such as carbonate (foraminiferal), silicate (radiolarian), or metalliferous sediments and ferromanganese nodules and crusts. Both the $^{230}Th/U$ and $^{231}Pa/U$ methods play an important role in dating corals and mollusk shells as well as the $^{230}Th/U$ method which is most widely applied in dating seafloor massive sulfide deposits in the ocean hydrothermal zones.

Modern mass spectrometric analysis allows to determine $^{230}Th_{xs}$ and $^{230}Th/U$ ages between several decades to ca. 500 kyr and $^{231}Pa_{xs}$ and $^{231}Pa/U$ ages up to ca. 250 kyr. Mass spectrometric measurement of Th, Pa, and U isotopes requires only tens or hundreds of milligrams of sample and provides high analytical precision.

Both the $^{230}Th_{xs}$ and $^{231}Pa_{xs}$ dating methods are widely used in the determination of sedimentation rates of deep-sea formations; they play an important role in building the timescale of global climate changes in the past as well. Both the $^{230}Th/U$ and $^{231}Pa/U$ methods are widely used in the study of sea-level changes in the past.

Bibliography

Andersen, M. B., Stirling, C. H., Potter, E.-K., Halliday, A. N., Blake, S. G., McCulloch, M. T., Ayling, B. F., and O'Leary, M., 2008. High-precision U-series measurements of more than 500,000 year old fossil corals. *Earth and Planetary Science Letters*, **265**, 229–245.

Anderson, R. F., Bacon, M. P., and Brewer, P. G., 1983a. Removal of ^{230}Th and ^{231}Pa from the open ocean. *Earth and Planetary Science Letters*, **62**, 7–23.

Anderson, R. F., Bacon, M. P., and Brewer, P. G., 1983b. Removal of ^{230}Th and ^{231}Pa at ocean margins. *Earth and Planetary Science Letters*, **66**, 73–90.

Arslanov, K. A., Tertychny, N. I., Kuznetsov, V. Y., Chernov, S. B., Lokshin, N. V., Gerasimova, S. A., Maksimov, F. E., and Dodonov, A. E., 2002. $^{230}Th/U$ and ^{14}C dating of mollusc shells from the coasts of the Caspian, Barents, White and Black Seas. *Geochronometria*, **21**, 49–56.

Bard, E., Hamelin, B., Fairbanks, R. G., and Zindler, A., 1990. Calibration of the ^{14}C timescale over the last 30,000 years using mass spectrometric U-Th ages from Barbados corals. *Nature*, **345**, 461–468.

Bard, E., Hamelin, B., Arnold, M., Montaggioni, L., Cabioch, G., Faure, G., and Rougerie, F., 1996. Deglacial sea level record from Tahiti corals and the timing of global meltwater discharge. *Nature*, **382**, 241–244.

Barnes, J. W., Lang, E. J., and Portratz, K. A., 1956. Ratio of ionium to thorium in coral limestone. *Science*, **124**, 175–176.

Broecker, W. S., and Van Donk, J., 1970. Insolation changes, ice volumes and the ^{18}O record in deep-sea cores. *Reviews of Geophysics and Space Physics*, **8**(1), 169–198.

Broecker, W. S., Thurber, D. L., Goddard, J., Ku, T. K., Matthews, R. K., and Mesolella, K. J., 1968. Milankovitch hypothesis supported by precise dating of coral reefs and deep sea sediments. *Science*, **159**, 297–300.

Chen, J. H., Edwards, R. L., and Wasserburg, G. J., 1986. U-238, U-234, and Th-232 in seawater. *Earth and Planetary Science Letters*, **80**, 241–251.

Claude-Ivanaj, C., Hofmann, A. W., Vlastelic, I., and Koschinsky, A., 2001. Recording changes in ENADW composition over the last 340 ka using high-precision lead isotopes in a Fe-Mn crust. *Earth and Planetary Science Letters*, **188**, 73–89.

Cochran, J. K., 1992. The oceanic chemistry of the uranium and thorium series nuclides. In Ivanovich, M., and Harmon, R. S. (eds.), *Uranium-Series Disequilibrium: Applications to Earth, Marine, and Environmental Sciences*, 2nd edn. Oxford: Clarendon, pp. 334–395.

Cochran, J. K., and Masque, P., 2003. Short-lived U/Th-series radionuclides in the ocean: tracers for scavenging rates, export fluxes and particle dynamics. *Reviews in Mineralogy and Geochemistry*, **52**, 461–492.

Cutler, K. B., Edwards, R. L., Taylor, F. W., Cheng, H., Adkins, J., Gallup, C. D., Cutler, P. M., Burr, G. S., Chappell, J., and Bloom, A. L., 2003. Rapid sea-level fall and deep-ocean temperature change since the last interglacial. *Earth and Planetary Science Letters*, **206**, 253–271.

Edwards, R. L., Chen, J. H., and Wasserburg, G. J., 1986. ^{238}U-^{234}U-^{230}Th-^{232}Th systematics and the precise measurement of time over the past 500,000 years. *Earth and Planetary Science Letters*, **81**, 175–192.

Edwards, R. L., Chen, J. H., Ku, T.-L., and Wasserburg, G. J., 1987. Precise timing of the last interglacial period from mass spectrometric analysis of ^{230}Th in corals. *Science*, **236**, 1547–1553.

Edwards, R. L., Beck, J. W., Burr, G. S., Donahue, D. J., Druffel, E. R. M., and Taylor, F. W., 1993. A large drop in atmospheric $^{14}C/^{12}C$ and reduced melting during the Younger Dryas, documented with ^{230}Th ages of corals. *Science*, **260**, 962–968.

Edwards, R. L., Cheng, H., Murrell, M. T., and Goldstein, S. J., 1997. Protactinium-231 dating of carbonates by thermal ionization mass spectrometry: implications for Quaternary climate change. *Science*, **276**, 782–786.

Edwards, R. L., Gallup, C. D., and Cheng, H., 2003. Uranium-series dating of marine and lacustrine carbonates. *Uranium-Series Geochemistry*, **52**, 363–405.

Eisenhauer, A., Gogen, K., Pernicka, E., and Mangini, F., 1992. Climatic influences on the growth rates of Mn crusts during the late Quaternary. *Earth and Planetary Science Letters*, **109**, 25–36.

Finney, B., Heath, G. R., and Lyle, M., 1984. Growth rates of manganese-rich nodules at MANOP Site H (Eastern North Pacific). *Geochimica et Cosmochimica Acta*, **48**(5), 911–919.

Frank, N., Turpin, L., Cabioch, G., Blamart, D., Tressens-Fedou, M., Colin, C., and Jean-Baptiste, P., 2006. Open system U-series ages of corals from a subsiding reef in New Caledonia: implications for sea level changes and subsidence rate. *Earth and Planetary Science Letters*, **249**, 274–289.

Geyh, M. A., 2001. Reflections on the $^{230}Th/U$ dating of dirty material. *Geochronometria*, **20**, 9–14.

Hanebuth, T., Stattegger, K., and Grootes, P. M., 2000. Rapid flooding of the Sunda Shelf: a late-glacial sea-level record. *Science*, **288**, 1033–1035.

Henderson, G. M., and Anderson, R. F., 2003. The U-series toolbox for paleoceanography. *Reviews in Mineralogy and Geochemistry*, **52**(1), 493–531.

Huh, C. A., and Ku, T. L., 1984. Radiochemical observation on manganese nodules from three sedimentary environments in the North Pacific. *Geochimica et Cosmochimica Acta*, **48**(5), 951–963.

Ivanovich, M., and Harmon, R. S. (eds.), 1992. *Uranium-Series Disequilibrium: Applications to Earth, Marine and Environmental Sciences*, 2nd edn. Oxford: Clarendon, p. 902.

Kaufman, A., Broecker, W. S., Ku, T.-L., and Thurber, D. L., 1971. The status of U-series methods of mollusk dating. *Geochimica et Cosmochimica Acta*, **35**, 1155–1183.

Ku, T.-L., 1968. Protactinium-231 method of dating coral from Barbados Island. *Journal of Geophysical Research*, **73**, 2271–2276.

Ku, T. L., 1976. The uranium-series methods of age determination. *Annual Review of Earth and Planetary Sciences*, **4**, 347–380.

Ku, T. L., and Broecker, W. S., 1967. Uranium, thorium and protactinium in a manganese nodule. *Earth and Planetary Science Letters*, **2**, 317–320.

Ku, T. L., and Broecker, W. S., 1969. Radiochemical studies of manganese nodules of deep-sea origin. *Deep Sea Research*, **16**, 625–637.

Kuznetsov, Y. V., 1976. *Radiokhronologia okeana (Radiochronology of Ocean)*. Moscow: Atomizdat, p. 279 (in Russian).

Kuznetsov, V. Y., and Maksimov, F. E., 2012. *Metody chetvertichnoy geokhronometrii v paleogeografii I morskoy geologii (Methods of Quaternary Geochronometry in Palaeogeography and Marine Geology)*. Saint-Petersburg: Nauka, p. 191 (in Russian).

Kuznetsov V. Y., 2008. *Radiokhronologia chetvertichnikh otlozheniy (Radiochronology of Quaternary Deposits)*. Saint-Petersburg, p. 312 (in Russian).

Kuznetsov, V. Y., Arslanov, K. A., Shilov, V. V., and Cherkashev, G. A., 2002. ^{230}Th-excess and ^{14}C dating of pelagic sediments from the hydrothermal zone of the North Atlantic. *Geochronometria*, **21**, 33–40.

Kuznetsov, V., Cherkashev, G., Lein, A., Shilov, V., Maksimov, F., Arslanov, K., Stepanova, T., Baranova, N., Chernov, S., and Tarasenko, D., 2006. ^{230}Th/U dating of massive sulfides from the Logatchev and Rainbow hydrothermal fields (Mid-Atlantic Ridge). *Geochronometria*, **25**, 51–56.

Kuznetsov, V. Y., Cherkashev, G. A., Bel'tenev, V. E., Lein, A. Y., Maximov, F. E., Shilov, V. V., and Stepanova, T. V., 2007. The ^{230}Th/U dating of sulfide ores in the ocean: methodical possibilities, measurement results and perspectives of application. *Doklady Earth Sciences*, **417**(8), 1202–1205.

Kuznetsov, V., Maksimov, F., Zheleznov, A., Cherkashov, G., Bel'tenev, V., and Lazareva, L., 2011. ^{230}Th/U chronology of ore formation within the Semyenov hydrothermal district (13°31′ N) at the Mid-Atlantic Ridge. *Geochronometria*, **38**, 72–76.

Lalou, C., and Brichet, E., 1982. Ages and implications of East Pacific Rise sulphide deposits at 21°N. *Nature*, **300**, 169–171.

Lalou, C., and Brichet, E., 1987. On the isotopic chronology of submarine hydrothermal deposits. *Chemical Geology*, **65**, 197–207.

Lalou, C., Reyss, L. G., Brichet, E., Krasnov, S., Stepanova, T., Cherkashev, G., and Markov, V., 1988. Chronology of a recently discovered hydrothermal field at 14°45′ N, Mid Atlantic Ridge. *Earth and Planetary Science Letters*, **144**, 483–490.

Lalou, C., Reyss, J. L., Brichet, E., Arnold, M., Thompson, G., Fouquet, Y., and Rona, P. A., 1993. New age data for MAR hydrothermal sites: TAG and Snakepit chronology revisited. *Journal of Geophysical Research*, **98**, 9705–9713.

Lalou, C., Reyss, J.-L., Brichet, E., Rona, P. A., and Thompson, G., 1995. Hydrothermal activity on a 10^5-year scale at a slow-spreading ridge. TAG hydrothermal field, Mid-Atlantic Ridge 26°N. *Journal of Geophysical Research*, **100**, 17855–17862.

Lalou, C., Reyss, J. L., Brichet, E., Krasnov, S., Stepanova, T., Cherkashev, G., and Markov, V., 1996. Initial chronology of a recently discovered hydrothermal field at 14°45′ N, Mid- Atlantic Ridge. *Earth and Planetary Science Letters*, **144**, 483–490.

Lalou, C., Reyss, J.-L., and Brichet, E., 1998. Age of sub-bottom sulfide samples at the TAG active mound. In Herzig, P. M., Humphris, S. E., Miller, D. J., and Zierenberg, R. A. (eds.), *Proceedings of the Ocean Drilling Program*. College Station: Ocean Drilling Program. Scientific Results, Vol. 158, pp. 111–117.

Michard, A., Albarede, F., Michard, G., Minster, J. F., and Charlou, J. L., 1983. Rare-earth elements and uranium in high-temperature solutions from East Pacific Rise hydrothermal vent field (13° N). *Nature*, **303**(5920), 795–797.

Moore, W. S., and Sackett, W. M., 1964. Uranium and thorium series inequilibrium in sea water. *Journal of Geophysical Research*, **69**(24), 5401.

Not, C., and Hillaire-Marcel, C., 2010. Time constraints from ^{230}Th and ^{231}Pa data in late Quaternary, low sedimentation rate sequences from the Arctic Ocean: an example from the northern Mendeleev Ridge. *Quaternary Science Reviews*, **29**, 3665–3675.

Rosenthal, Y., Boyle, E. A., Labeyrie, L., and Oppo, D., 1995. Glacial enrichments of authigenic Cd and U in subantarctic sediments: a climatic control on the elements' oceanic budget? *Paleoceanography*, **10**(3), 395–413.

Rosholt, J. N., and Antal, P. S., 1962. Evaluation of the Pa231/U-Th230/U method for dating Pleistocene carbonate rocks. *U.S. Geological Survey Professional Paper*, **450-E**, 108–111.

Sackett, W. M., 1960. Protactinium-231 content of ocean water and sediments. *Science*, **132**, 1761.

Sakanoue, M., Konishi, K., and Komura, K., 1967. Stepwise determinations of thorium, protactinium, and uranium isotopes and their application in geochronological studies. In *Proceedings of the Symposium on Radioactive Dating and Methods of Low-Level Counting,* Monaco, pp. 313–329.

Scott, M. R., Osmond, T. K., and Cochran, T. K., 1972. Sedimentation rates and sediment chemistry in the South Indian Basin. In Hayes, D. E. (ed.), *Antarctic Oceanology II, the Australian-New-Zealand Sector*. Washington, DC: American Geophysical Union. Antarctic Research Series, Vol. 19, pp. 317–334.

Smart, P. L., 1991. Uranium series dating. In Smart, P. L., and Frances, P. D. (eds.), *Quaternary Dating Methods*. London: Quaternary Research Association. Technical Guide, Vol. 4, pp. 45–83.

Stein, M., Wasserburg, G., Aharon, P., Chen, J. H., Zhu, Z. R., Bloom, A. L., and Chapell, J., 1993. TIMS U series dating and stable isotopes of the last interglacial event at Papua New Guinea. *Geochimica et Cosmochimica Acta*, **57**, 2541–2554.

Taguchi, K., Harada, K., and Tsunogai, S., 1989. Particulate removal of ^{230}Th and ^{231}Pa in the biologically productive northern North Pacific. *Earth and Planetary Science Letters*, **93**, 223–232.

Veeh, H. H., 1966. 230Th/238U and 234U/238U ages of Pleistocene high sea level stand. *Journal of Geophysical Research*, **71**, 3379–3386.

Waelbroeck, C., Frank, N., Jouzel, J., Parrenin, F., Masson-Delmotte, V., and Genty, D., 2008. Transferring radiometric dating of the last interglacial sea level high stand to marine and ice core records. *Earth and Planetary Science Letters*, **265**, 183–194.

Wagner, G. A., 1998. *Age Determination of Young Rocks and Artifacts*. Berlin/Heidelberg: Springer, p. 467.

You, C.-F., and Bickle, M., 1998. Evolution of an active sea-floor massive sulphide deposit. *Nature*, **394**, 668–671.

Zhu, Z. R., Wyrwoll, K. H., Chen, L., Chen, J., Wasserburg, G. J., and Eisenhauer, A., 1993. High-precision U-series dating of last interglacial events by mass spectrometry: Houtman Abrolhos Islands, Western Australia. *Earth and Planetary Science Letters*, **118**, 281–293.

Cross-references

Geologic Time Scale
Marine Microfossils
Radiocarbon: Clock and Tracer
Sclerochronology

GEOHAZARDS: COASTAL DISASTERS

Gösta Hoffmann and Klaus Reicherter
Department of Geosciences and Geography, RWTH Aachen University, Aachen, Germany

Definition

We define the term "hazard" as a potential source of harm, where "harm" is defined as the "injury or damage to the health of people, or damage to property or the environment" following the definition as given by ISO/IEC (2014). The term "risk" is defined as a "combination of the probability of occurrence of harm and the severity of that harm" (ISO/IEC, 2014). Furthermore, the term "risk" may be expressed as a function of hazard, exposure, and vulnerability. A geohazard is a geological state that may lead to widespread damage presenting severe threats to humans, property, and the natural and built environment. A coastal geohazard is a natural physical phenomenon usually associated with seacoasts but to a limited extent also along lakeshores, where the initiating process of the disaster may originate at great distances from the point of impact.

Introduction

There are numerous geohazards which affect coastal regions, including beach erosion, landslide/cliff collapse, wave attack, flooding, and tsunamis. They are always related to the landward movement of water. These processes can be either slow or fast and may be temporary or permanent in nature. Some hazards have meteorological causes (climatic hazards, e.g., storms, Nicholls, 2004), whereas others may be driven by Earth's endogenous forces such as seismic events (earthquakes and tsunamis, e.g., Atwater, 1987) and volcanic eruptions (e.g., Firth et al., 1996). Furthermore, coastal hazards can be gravity induced (landslide/cliff collapse, e.g., Dawson et al., 2009; Del Río and Gracia, 2009). Relative sea-level changes may pose a hazard to many low-lying countries as do subsurface changes induced by humans (e.g., due to groundwater extraction and associated settlement). As the coastal zone is a preferred area of human settlement, there is always some exposure to coastal hazards. Coastal cities host many different kinds of economically relevant infrastructure (e.g., harbors) and are often used for recreational purposes. Two percent of the world's surface is defined as low elevation coastal zone (LECZ). McGranahan et al. (2007) defined the LECZ as a contiguous area along the coast lower than 10 m above sea-level that is inhabited by 10 % of the world's population. Small and Nicholls (2003) estimate that around 1.2 billion people are living within 100 km of the coast throughout the globe. In addition to urbanization, population growth within the coastal zone is expected to be more rapid than in other areas in the near future. For example, Adger et al. (2005) calculate that more than 50 % of the world's population is likely to live in coastal areas by 2030. The areas affected by coastal hazards directly depend on the topography; low-lying countries such as the Netherlands or Bangladesh are particularly prone to the hazards caused by rising sea levels.

Cliff collapse and other types of mass movements are hazards when they occur close to settlements along cliff edges (Dewez et al., 2013). Furthermore, beaches at the base of a cliffed coastline might be used for recreational activities (Günther and Thiel, 2009). Mass movements occur if stability thresholds are crossed. Critical parameters are related to meteorological conditions and to tide- and wave-controlled shoreface processes. Cliff collapse is a form of episodic coastal retreat, and important parameters controlling the processes are mainly the lithology (including water content) and structure. Frost action, in particular the number of freeze-thaw cycles, plays an important role in the weathering of semiconsolidated rocks; the chalk cliffs of northern Europe is a good example of this (Duperret et al., 2002; Kuhn and Prüfer, 2014). Bioerosion becomes an important agent, especially in the retreat of limestone-dominated coasts within the tropics (Vita-Finzi and Cornelius, 1973; Taboroši and Kázmér, 2013).

Despite the immediate destruction by shock waves, earthquake-related coastal hazards are mainly attributed to coseismic (crustal) uplift or subsidence as witnessed, e.g., during the 1964 Alaska earthquake (Plafker, 1969).

A further criterion to distinguish coastal hazards is the duration of time, which ranges from seconds for tsunami landfall to thousands of years for relative sea-level changes. Another important issue is to separate coastal hazards into local, regional, and global hazards. All geohazards may cause significant loss of life and may have severe economic impacts. In the following, extreme wave events caused by storm surges and tsunamis are discussed.

Extreme wave events

Extreme wave events can develop into a coastal hazard, and they primarily result from two fundamentally different phenomena: storm surges and tsunamis. Whereas the intensity of such events is high, the frequency is usually low. The major difference between these two natural phenomena lies in the energy source responsible for the piling up of water. An earthquake-generated tsunami is the result of endogenous forces with the energy derived from the Earth's interior. A storm surge, however, is fueled by exogenous forces; the energy that drives this process, which is ultimately derived from the sun, is transmitted via the climate system onto the coast. In contrast to normal ocean waves, tsunami waves are characterized by a longer wavelength (200–300 km) with periods from minutes to hours. The speed of a tsunami wave within the deep sea reaches up to 800 km/h with a very low amplitude of less than 1 m. The amplitude increases in shallow water (shoaling) as the wave gets compressed. A tsunami wave moves the entire water column, which causes refraction effects. Bryant (2005) notes 124 tsunami events as a result of 15,000 earthquakes for the period 1861–1948. In general, storm surges occur much more frequently than tsunami waves.

Storm surges

Storm surges are long-wavelength, low-amplitude sea surface displacements resulting in superelevated water levels that are over and above the regular astronomically forced highest water levels (high tides). Storm surges are a climatically driven phenomenon, accompanied with reduced atmospheric pressure. They are therefore surface gravity waves. The low pressure results in the reduction of the vertical force on the sea surface and hence leads to a rise of the water level. This phenomenon is also known as the "inverted barometer effect" (Wunsch and Stammer, 1997). A linear relationship exists between atmospheric pressure and sea-level. A sea-level drop of 1 cm is observed with an increase of 1 mbar of atmospheric pressure (Ross, 1854). Storm surges are short-term changes in water level. This is due to the presence of wind which moves the low-pressure system. The variations in sea-level can be about ±15 cm.

The coastal hazards associated with storm surge impacts are coastal flooding, erosion, and salinization. The flooding of a normally dry coastal zone occurs where there are large low-lying areas; hence, flooding is especially severe in estuaries and other coastal areas characterized by geologically young and unconsolidated sedimentary strata. The situation is enhanced in cases of local land subsidence which may be induced by humans, e.g., due to excessive groundwater extraction (Galloway and Burbey, 2011). Artificial drainage of swampy coastal areas, as practiced for centuries in the Netherlands, leads to peat compaction and oxidization. As a consequence, the surface is lowered and the flooding potential amplified.

Wave impacts can result in significant erosion along the shore, a process which is sometimes amplified by additional erosional agents such as floating debris (e.g., trees) and sea ice. In addition to the effects along the coastline, the water level in rivers may rise. This is because severe weather conditions are often accompanied by extraordinary precipitation events, which can cause a noncoastal flooding hazard. Coastal flooding in river mouths (especially in backwater areas) may also occur without additional water from precipitation. This is because the base level changes during times of storm-induced higher sea level. Coastal storms have an "event" character as they result in comparatively sudden coastal changes and are the main reason for coastal retreat. This retreat is not a gradual and/or linear process but the sum of sporadic events and the intermittent movement of sediments. At least 70 % of sandy beaches around the world are recessional (Bird, 1985; Schlacher et al., 2007).

Coastal storms are generated by low-pressure weather systems such as tropical cyclones (synonyms: hurricanes, typhoons) as well as extratropical storms (low-pressure systems including blizzards). Extratropical cyclones form over both land and sea, in contrast to tropical cyclones that only form over the sea. Cyclones rotate anticlockwise in the Northern Hemisphere and clockwise in the Southern Hemisphere due to the Coriolis effect. A sea surface temperature above 26 °C is one of the most important requirements for a tropical cyclone to form. Energy in the form of latent heat is transferred from the ocean into the climate system by the condensation of water vapor. Frontal activity associated with the interaction of warm and cold air masses is the main driving force in the formation of extratropical cyclones.

The severity of a storm surge is influenced by the wind conditions, the drag on the sea surface due to the wind fetch, the timing of the tides, the wind speed, and also the duration of the event. Of equal importance is the orientation of the coast relative to the storm path, as wind directed at a right angle causes the highest rise in water level. Furthermore, the storm surge extent depends on the speed of movement of the pressure system, the topography, the shape of the coastline, and the bathymetry.

The passing of an atmospheric pressure system and the associated storm surge activity may last from hours to days. Extratropical cyclones are usually larger and longer lasting than tropical cyclones, which are always nonfrontal storms (Gray, 1979). However, tropical cyclones are more severe as they contain more potential energy. The severity of the storm surge hazard increases if the situation persists over several tidal cycles, because the risk of flooding increases significantly if high tide and storm surge coincide. Furthermore, the maximum height of the water level is controlled by instantaneous water level changes caused by waves.

Tsunami

Tsunamis are the result of a large mass of water being displaced by different processes. Earthquake-induced tsunamis have small wave heights offshore and very long wavelengths which can reach several hundred kilometers (line source tsunamis). This is in contrast to tsunamis caused by mass movements (subaerial or underwater origin), volcanic eruptions and explosions, glacier calvings, or meteorite impacts, which have very high waves and very short wavelengths close to the origin (point source tsunamis). More than 85 % of tsunamis have tectonic causes such as fault movements; 5 % are caused by volcanic activity, 5 % by landslides, and 5 % by a combination

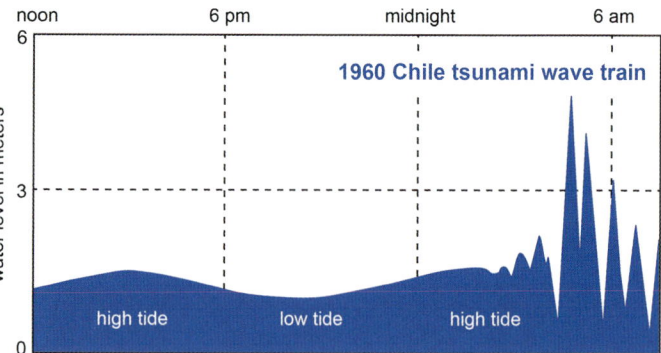

Geohazards: Coastal Disasters, Figure 1 Tsunami wave train approaching the Japanese coastline after the 2011 Tohoku earthquake; note that tsunami has already made landfall (*left*; photo by Douglas Sprott is licensed under CC BY-NC 2.0). Gauge data showing the wave train from the 1960 Chilean earthquake tsunami along the Japanese coast (Onagawa) with tide data superimposed (*right*; modified from Atwater et al., 1999).

of all other causes – impact tsunamis are very rare (Rhodes et al., 2006). Eighty percent of all tsunamis occur in the Pacific Ocean, but other oceans and seas (like the Mediterranean Sea) have also been affected by tsunamis. Tsunami evidence has also been documented on the shorelines of large and deep lakes, such as Lake Geneva in Switzerland (Kremer et al., 2012). The speed of a tsunami wave is calculated by the square root of the product of acceleration by gravity (9.8 m/s^2) and the water depth; in deep water, the wavelength of the tsunami must also be taken into account. Extreme waves can also be induced by storms and become tsunamis when approaching the coastline due to shoaling. These are then referred to as meteotsunamis (Monserrat et al., 2006). Shoaling can result in the superposition of waves and a dramatic reduction in wave velocity. The plural form is "tsunamis" or "tsunami."

A tsunami train is a series of waves, which are destructive when they make landfall. The first wave in the tsunami wave train is often not the largest one and therefore leaves almost no deposits (or only in favorable setting) as the deposits are usually reworked by following waves. The time interval between waves hitting the coastline varies (Figure 1) but can be around 10 min or more. Generally, three to four major wave landfalls can be documented in the sedimentary record. The ecology and the geomorphology of coastlines are severely affected and modified during earthquake-induced tsunami landfall.

Reconstructing paleostorm and tsunami events

Coastal zone management and improved decision-making in coastal planning and development depend on mitigation and adaptation strategies. These should be applied to reduce the risks associated with geohazards because the hazards themselves cannot be prevented (they are unavoidable by nature). The design of flood protection measures relies on inundation statistics on frequency and duration of water levels above a specified elevation threshold (exceedance probability statistics on extreme water levels). The risks can only be statistically assessed if recurrence intervals (probability that harm will occur) are known and if the severity, as well as magnitude, of potential events can be quantified (severity of harm). Knowledge of past extreme wave events is therefore essential for risk assessment, which also forms the base of multihazard early warning systems. In the ideal case, statistical analyses rely on observed events and are based on instrumental measurements. However, the period covered by instrumental data may not be statistically significant, as extreme wave events typically have a low frequency (Wolman and Miller, 1960). Consequently, other archives need to be made accessible and summarized in the form of catalogs and databases in order to extend the record into the past. Historical data are useful, especially if the flooding level can be correlated to a reference level such as high water marks (Brázdil et al., 2006). Basic sources of documented data on floods might be found in written description (e.g., Satake et al., 1996; Lau et al., 2010; Atwater et al., 2014). Uncertainties concerning the timing of the event are usually low when using these descriptions; however, limitations arise because the maximum flooding level may only be reconstructed if the landmarks that are described still exist and are identifiable. Hence, precision of dating and magnitude commonly get more vague and uncertain with increasing event age in the preinstrumental period. The historical period is also often too short (Switzer et al., 2014). Therefore, for most areas, the archaeological and geological records prove to be the only archives reaching back sufficiently long enough into the past to cover the worst-case scenario which will have very long return periods. Event deposits can only be expected from the second half of the Holocene as older deposits would likely have been eroded and reworked with fast sea-level rise (Hoffmann and Lampe, 2007).

Paleostorm deposits

The sedimentary evidence left behind by a storm surge is referred to as "tempestite." Nott (2004) introduced the term "paleotempestology" as the study of prehistoric storms. The identification of paleostorm deposits along

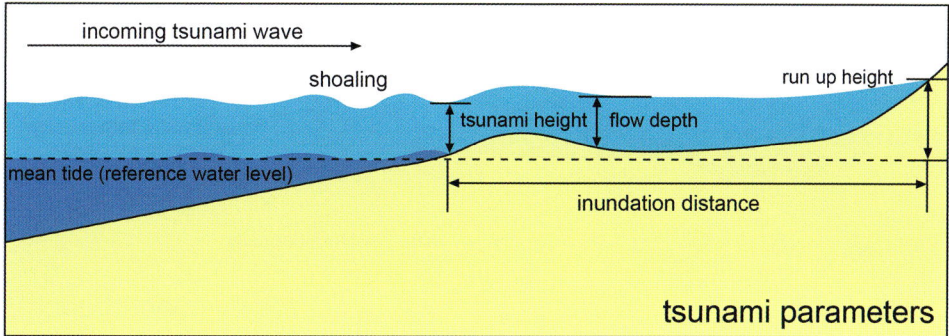

Geohazards: Coastal Disasters, Figure 2 Tsunami parameters (*top*).

coastlines is based on geomorphological and sedimentological work. There is the potential for both sedimentary and erosional evidence. Paleostorm deposits may form individual layers which differ from the coastal facies. Such sandy event deposits are found as stratified, often shelly strata in back barrier lagoons and swamps where fine-grained sediments usually accumulate (e.g., Switzer and Jones, 2008). The deposits are commonly tens of centimeters thick and usually laminated due to individual wave pulses. Paleosoils are useful indicators in coastal profiles when overlain by a suspicious deposit. This indicates a hiatus that covers at least the time of soil formation and underlines the event character of the deposit. Geomorphological evidence of paleostorms can be seen in coastal ridges and splays. Landward thinning sheets, which appear as fan-shaped tongues in the coastal geomorphology, form as a consequence of the overtopping of dune morphology and the breaching of barriers and berms along sandy coasts. Besides these fine-grained sandy deposits, coarse-grained coastal deposits such as boulders have been described as built by tropical cyclones (e.g., Scheffers and Scheffers, 2006). Cox et al. (2012) convincingly demonstrate extratropical storm activation of boulder deposits along the west coast of Ireland. Another deposit associated with storm surges is wrack lines. These are composed of floatable material. Although the preservation potential might be low, Hoffmann and Reicherter (2014) recently documented how these deposits may be used to precisely date the flooding event. In general, reworking of older material along the shore is common during storm surge impact, and this may hamper dating of the event deposits.

Paleotsunami

Tsunami action is mirrored in geological records and archives as phases of erosion, reworking, and redeposition leaving unconformities and unusual sedimentary layers. Tsunamites or tsunami deposits are proof that past or prehistoric tsunami events have occurred along coastlines or lakeshores. Several parameters obtained from recent or historical tsunamis are used in paleotsunami research, the most important of which are the inundation distance and run-up height (Figure 2). The maximum run-up and inundation are delineated by the location of the wrack line or wrack line deposits (Figure 3a). These are then used as basic parameters for coastal evacuation and emergency planning. Local effects of coastal morphology such as bays, fjords, and rias can amplify wave heights and inundations. Refraction of tsunami waves (bending) may occur and mainly depends on the morphology and shape of the seabed in front of the coast. Wave reflection also occurs, which describes the wave bouncing back after striking the shoreline.

Sedimentary evidence for paleotsunamis is found along the shorelines of marine and lacustrine environments. There are clear sedimentological characteristics to identify such sediments as high-energy deposits. However, the distinction between storm- and tsunami-related deposits is almost impossible (e.g., Goff et al., 2004; Switzer et al., 2005; Kortekaas and Dawson, 2007; Morton et al., 2007; Bahlburg and Spiske, 2012; Shanmugam, 2012 and references therein) as the sediments have numerous similarities and a substantial lack of unequivocal diagnostic criteria.

The shorelines of the Earth vary from place to place and can be simplified into two characteristic environments: flat sandy coasts or marshes and cliffed rocky coasts (Figure 3a, b). The formation of a coast with sediments at the seashore depends on several parameters including the position of the sea-level, the tidal influence, and the presence of river inlets (e.g., sculpted into bays, cuspate forelands with mudflats, marshes, sandy beaches, or dunes). The geomorphology of a coast plays a major role in preserving tsunamigenic deposits.

Many shorelines have developed relatively recently due to Holocene sea-level rise. Other shorelines have been affected by tectonic movements (uplift or subsidence). Due to the glacial sea-level low stands, some tsunamis older than 5,000 years may not be preserved along the seashores as they are now drowned or have been reworked. However, interglacial/interstadial tsunamigenic deposits (during a high stand, e.g., MIS 5) have the potential to be preserved.

Paleotsunami research focuses on suitable archives along affected coastlines. To have a suitable archive, the deposits must firstly have had the potential to be preserved, which means subsidence and sedimentation

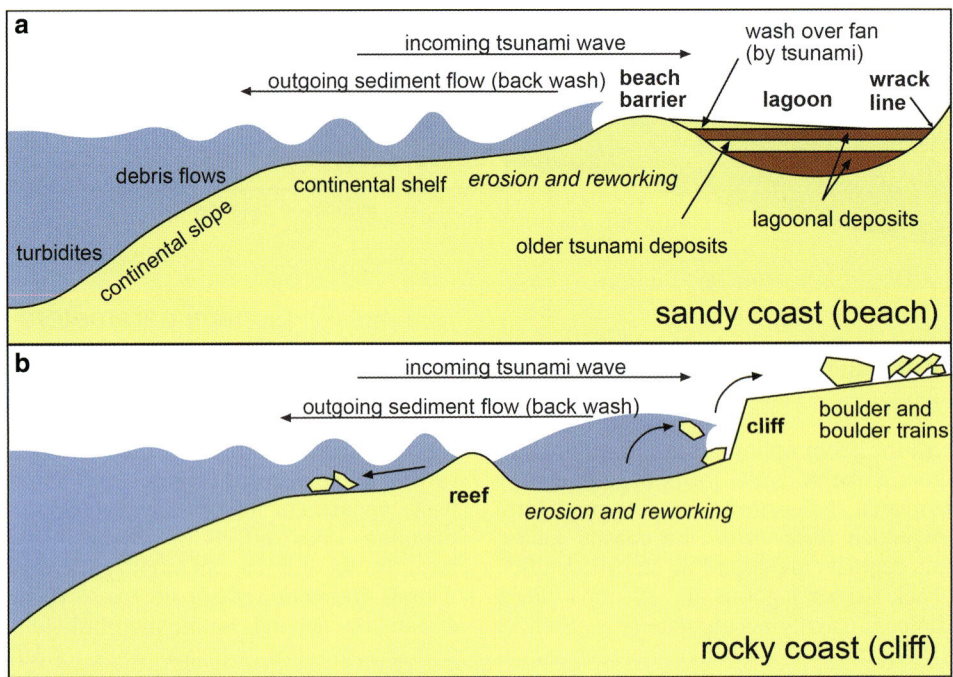

Geohazards: Coastal Disasters, Figure 3 (a) Tsunami landfall and deposits along sandy shores and various types of backwash deposits (*middle*). (b) Tsunami landfall at rocky coasts (*bottom*).

should prevail and not uplift and erosion; secondly, the deposits should be distinguishable from storm layers (tempestites); and thirdly, the deposits should yield datable material. The distinction of storm and tsunami deposits proves to be particularly difficult. Hence, many researchers (e.g., Lario et al., 2010) refer to extreme wave events (EWE). Lagoons or marshy flatlands have been widely used as archives for investigating tsunami deposits (Figure 3a). In these settings, tsunami landfall has a huge effect on the local sedimentary environment and ecology. Freshwater, brackish, or hypersaline lagoons will be affected by short durations of marine ingressions containing characteristic marine flora and fauna. Furthermore, coarse-grained deposits (sand sheets and washover fans) will be deposited in fine-grained lagoonal clays or evaporites. After landfall, the lagoonal environment may then reestablish itself including sedimentation, and hence sand sheets can be preserved (Figure 3a).

Along steep rocky shores, the locations of huge blocks and the arrangement of smaller blocks are used to delineate tsunami landfall. Tsunami boulders or "boulder trains" are chains of imbricated blocks interpreted as remnants of tsunami action (Figure 3b). These "boulder trains" have been found all around the world (Scheffers, 2008). However, Williams and Hall (2004) and Cox et al. (2012) describe the same type of deposits and clearly attribute storm waves as depositional agent.

Offshore records of paleotsunamis are rare, and related deposits are ambiguous; therefore, they cannot be reliably associated with tsunami action. Backwash deposits into the sea/lake after tsunami landfall by the backflow of the water mass are also hard to identify. For relatively recent tsunamis, backwash deposits can be identified mainly by the presence of anthropogenic pollution as evidenced by some chemical proxies; however, older tsunamigenic deposits are complex and hard to recognize as they may not yield these characteristic constituents.

Tsunami sedimentology has advanced considerably in the past 25 years since Brian Atwater in 1987 associated sand layers on the Pacific Coast of Washington State with the occurrence of major tsunami-producing subduction earthquakes (Atwater, 1987). The 2004 Indian Ocean and 2011 Tohoku-oki tsunamis and their deposits significantly changed our understanding of sedimentary features, sediment architecture, and the processes of tsunamigenic deposition, preservation, and postdepositional changes. Prior to 2004, catastrophic events in the historical record were too infrequent to be satisfactorily studied, and the identification of tsunami remains was ambiguous. Paleotsunami sedimentology is complex, but more and more researchers have integrated sedimentological, geochemical, and paleontological analyses from post-tsunami field surveys and laboratory analyses. This multidisciplinary approach allows a datum to be assigned to the event layers using a combination of several dating methods. To some extent, a clear assignment of an event layer to tsunami action will always remain equivocal, but large storms have different wavelengths, have a major peak of wave action, and have no train of several major waves. Consequently, tsunami deposits should be different to those caused by storms. Some authors also refer to the recurrence periods of, e.g., tropical storms, which is an order of magnitude higher than that of major (paleo)tsunamis (e.g., Nanayama et al., 2003). The tsunami database has been expanded during the last decades and, in combination with an improved understanding of tsunami depositional mechanisms (Shiki et al., 2008), has helped to identify paleotsunami deposits in many parts of the world.

Recognizing and identifying paleotsunami deposits and assigning the magnitude and extent of (pre)historic tsunamis are important parts of paleoseismological studies. An outstanding example of this is from Japan, where tsunami deposits of a historical event in AD 869 were identified on the Sendai plain by Minoura et al. (2001). The 2011 Tohoku-oki tsunami deposits now cover the AD 869 deposits in the same area. The presence of the older deposits was used to determine/verify the occurrence of a previous large earthquake and tsunami event and proved that this area is susceptible to large-scale tsunami inundation. The world then realized that this part of Japan had experienced magnitude 9 earthquakes with associated tsunamis in the past. Sedimentary, geochemical (inorganic, organic, isotopic), and paleontological proxies are usually used for the identification of tsunamites.

Field surveys aim to map the three-dimensional distribution of a sand sheet and to find evidence for a regional continuity of the layer. Washover fans, typically taper and fine landward, have a common height above mean sea-level and contain marine fossils. These observations can easily be achieved by shallow drillings and GPS control of the drilling locations for large strips along a coast. Ground-penetrating radar surveys as a shallow, nondestructive geophysical tool have been proven to produce reliable images of the facies architecture of fine-grained tsunami sediments (Koster et al., 2014).

When dating event deposits, the same challenges are faced as for paleostorm deposits. This is because reworking is a common process during deposition. Any dates obtained from the event deposits thus give a maximum age, and the youngest date would ideally give an indication of the timing.

Conclusions

Coastal hazards have the potential to cause havoc, especially where the coastline is densely populated. As the hazards involved are the consequence of natural processes, they cannot be avoided. Hence, adaptation strategies have to be implemented. A proper risk assessment should be based on knowledge of recurrence intervals as well as the definition of the worst-case scenario. As the recurrence interval might be rather long and historical information may not be available, this information can only be derived from deposits left behind by the impact of high-energy events. However, at the moment, there is no clear sedimentological fingerprint to discriminate storm and tsunami deposits; for more information, the reader is referred to the benchmark papers of Kortekaas and Dawson (2007), Morton et al. (2007), Switzer and Jones (2008), and Chagué-Goff et al. (2011). Mapping of high-energy deposits along the world's coastal zones requires an interdisciplinary (geoarchaeology) multiproxy and holistic approach. Transdisciplinary studies between geoscientists and historians are necessary as catalogs of past extreme wave events should include instrumental, historical, and geological data. Based on the results of these investigations, coastal engineers should design defense structures accordingly.

Bibliography

Adger, W. N., Hughes, T. P., Folke, C., Carpenter, S. R., and Rockström, J., 2005. Social-ecological resilience to coastal disasters. *Science*, **309**(5737), 1036–1039.

Atwater, B. F., 1987. Evidence for great Holocene earthquakes along the outer coast of Washington state. *Science*, **236**(4804), 942–944.

Atwater, B. F., Cisternas, M., Bourgois, J., Dudley, W. C., Hendley, J. W. II., and Stauffer, P. H., 1999. *Surviving a Tsunami – Lessons from Chile, Hawaii and Japan*. US Geological Survey Circular 1187. This report and any updates to it are available online at: http://pubs.usgs.gov/circ/c1187/

Atwater, B. F., Fuentes, Z., Halley, R. B., Ten Brink, U. S., and Tuttle, M. P., 2014. Effects of 2010 Hurricane Earl amidst geologic evidence for greater overwash at Anegada, British Virgin Islands. *Advances in Geosciences*, **38**(38), 21–30.

Bahlburg, H., and Spiske, M., 2012. Sedimentology of tsunami inflow and backflow deposits: key differences revealed in a modern example. *Sedimentology*, **59**(3), 1063–1086.

Bird, E. C. F., 1985. *Coastline Changes*. New York: Wiley. 219 pp.

Brázdil, R., Kundzewicz, Z. W., and Benito, G., 2006. Historical hydrology for studying flood risk in Europe. *Hydrological Sciences Journal*, **51**(5), 739–764.

Bryant, E., 2005. *Natural Hazards*. Cambridge: Cambridge University Press.

Chagué-Goff, C., Schneider, J. L., Goff, J. R., Dominey-Howes, D., and Strotz, L., 2011. Expanding the proxy toolkit to help identify past events – lessons from the 2004 Indian Ocean Tsunami and the 2009 South Pacific Tsunami. *Earth-Science Reviews*, **107**(1), 107–122.

Cox, R., Zentner, D. B., Kirchner, B. J., and Cook, M. S., 2012. Boulder ridges on the Aran Islands (Ireland): recent movements caused by storm waves, not tsunamis. *The Journal of Geology*, **120**(3), 249–272.

Dawson, R. J., Dickson, M. E., Nicholls, R. J., Hall, J. W., Walkden, M. J., Stansby, P. K., and Watkinson, A. R., 2009. Integrated analysis of risks of coastal flooding and cliff erosion under scenarios of long term change. *Climatic Change*, **95**(1–2), 249–288.

Del Río, L., and Gracia, F. J., 2009. Erosion risk assessment of active coastal cliffs in temperate environments. *Geomorphology*, **112**(1), 82–95.

Dewez, T. J., Rohmer, J., Regard, V., and Cnudde, C., 2013. Probabilistic coastal cliff collapse hazard from repeated terrestrial laser surveys: case study from Mesnil Val (Normandy, northern France). *Journal of Coastal Research*, **65**, 702–707.

Duperret, A., Genter, A., Mortimore, R. N., Delacourt, B., and De Pomerai, M. R., 2002. Coastal rock cliff erosion by collapse at Puys, France: the role of impervious marl seams within chalk of NW Europe. *Journal of Coastal Research*, **18**, 52–61.

Firth, C. A. L. L. U. M., Stewart, I., McGuire, W. J., Kershaw, S., and Vita-Finzi, C., 1996. *Coastal Elevation Changes in Eastern Sicily: Implications for Volcano Instability at Mount Etna*. London: Geological Society, Special Publications, Vol. 110, No. 1, pp. 153–167.

Galloway, D. L., and Burbey, T. J., 2011. Review: regional land subsidence accompanying groundwater extraction. *Hydrogeology Journal*, **19**(8), 1459–1486.

Goff, J., McFadgen, B. G., and Chagué-Goff, C., 2004. Sedimentary differences between the 2002 Easter storm and the 15th-century Okoropunga tsunami, southeastern North Island, New Zealand. *Marine Geology*, **204**(1), 235–250.

Gray, W. M., 1979. Hurricanes: their formation, structure and likely role in the tropical circulation. In Shaw, D. B. (ed.), *Meteorology Over the Tropical Oceans*. Bracknell, Berks: Royal Meteorological Society, J. Glaisher House, Grenville place, pp. 155–218.

Günther, A., and Thiel, C., 2009. Combined rock slope stability and shallow landslide susceptibility assessment of the Jasmund cliff area (Rügen Island, Germany). *Natural Hazards and Earth System Science*, **9**(3), 687–698.

Hoffmann, G., and Lampe, R., 2007. Sediment budget calculation to estimate Holocene coastal changes on the southwest Baltic Sea. *Marine Geology*, **243**(2007), 143–156.

Hoffmann, G., and Reicherter, K., 2014. Reconstructing Anthropocene extreme flood events by using litter deposits. *Global and Planetary Change*, **122**, 23–28.

ISO/IEC, 2014. *Safety Aspects – Guidelines for Their Inclusion in Standards*. London: International Organization for Standardization/The International Electrotechnical Commission Guide 51. British Standards Institution. http://shop.bsigroup.com/ProductDetail/?pid=000000000030299269

Kortekaas, S., and Dawson, A. G., 2007. Distinguishing tsunami and storm deposits: an example from Martinhal, SW Portugal. *Sedimentary Geology*, **200**(3), 208–221.

Koster, B., Hoffmann, G., Grützner, C., and Reicherter, K., 2014. Ground penetrating radar facies of inferred tsunami deposits on the shores of the Arabian Sea (northern Indian Ocean). *Marine Geology*, **351**, 13–34, doi:10.1016/j.margeo.2014.03.002.

Kremer, K., Simpson, G., and Girardclos, S., 2012. Giant Lake Geneva tsunami in AD 563. *Nature Geoscience*, **5**, 756–757.

Kuhn, D., and Prüfer, S., 2014. Coastal cliff monitoring and analysis of mass wasting processes with the application of terrestrial laser scanning: a case study of Rügen, Germany. *Geomorphology*, **213**, 153–165.

Lario, J., Luque, L., Zazo, C., Goy, J. L., Spencer, C., Cabero, A., and Alonso-Azcárate, J., 2010. Tsunami vs. storm surge deposits: a review of the sedimentological and geomorphological records of extreme wave events (EWE) during the Holocene in the Gulf of Cadiz, Spain. *Zeitschrift für Geomorphologie, Supplementary Issues*, **54**(3), 301–316.

Lau, A. Y. A., Switzer, A. D., Dominey-Howes, D., Aitchison, J. C., and Zong, Y., 2010. Written records of historical tsunamis in the northeastern South China Sea: challenges associated with developing a new integrated database. *Natural Hazards and Earth System Sciences*, **10**, 1793.

McGranahan, G., Balk, D., and Anderson, B., 2007. The rising tide: assessing the risks of climate change and human settlements in low elevation coastal zones. *Environment and Urbanization*, **19**(1), 17–37.

Minoura, K., Imamura, F., Sugawara, D., Kono, Y., and Iwashita, T., 2001. The 869 Jogan tsunami deposit and recurrence interval of large-scale tsunami on the Pacific coast of northeast Japan. *Journal of Natural Disaster Science*, **23**(2), 83–88.

Monserrat, S., Vilibić, I., and Rabinovich, A. B., 2006. Meteotsunamis: atmospherically induced destructive ocean waves in the tsunami frequency band. *Natural Hazards and Earth System Science*, **6**(6), 1035–1051.

Morton, R. A., Gelfenbaum, G., and Jaffe, B. E., 2007. Physical criteria for distinguishing sandy tsunami and storm deposits using modern examples. *Sedimentary Geology*, **200**(3), 184–207.

Nanayama, F., Satake, K., Furukawa, R., Shimokawa, K., Atwater, B. F., Shigeno, K., and Yamaki, S., 2003. Unusually large earthquakes inferred from tsunami deposits along the Kuril trench. *Nature*, **424**, 660–663.

Nicholls, R. J., 2004. Coastal flooding and wetland loss in the 21st century: changes under the SRES climate and socio-economic scenarios. *Global Environmental Change*, **14**(1), 69–86.

Nott, J., 2004. Paleotempestology: the study of prehistoric tropical cyclones – a review and implications for hazard assessment. *Environment International*, **30**(3), 433–447.

Plafker, G., 1969. *Tectonics of the March 27, 1964, Alaska Earthquake*. Washington, DC: US Government Printing Office, p. 74.

Rhodes, B., Tuttle, M., Horton, B., Doner, L., Kelsey, H., Nelson, A., and Cisternas, M., 2006. Paleotsunami research. *Eos, Transactions American Geophysical Union*, **87**(21), 205–209.

Ross, J. C., 1854. On the effect of the pressure of the atmosphere on the mean level of the ocean. *Philosophical Transactions of the Royal Society of London*, **144**, 285–296.

Satake, K., Shimazaki, K., Tsuji, Y., and Ueda, K., 1996. Time and size of a giant earthquake in Cascadia inferred from Japanese tsunami records of January 1700. *Nature*, **379**(6562), 246–249.

Scheffers, A., 2008. Tsunami boulder deposits. In Shiki, T., Tsuji, Y., Yamakazi, T., and Minoura, K. (eds.), *Tsunamiites – Features and Implications*. Amsterdam: Elsevier Scientific, pp. 299–318.

Scheffers, A., and Scheffers, S., 2006. Documentation of the impact of Hurricane Ivan on the coastline of Bonaire (Netherlands Antilles). *Journal of Coastal Research*, **22**, 1437–1450, doi:10.2112/05-0535.1.

Schlacher, T. A., Dugan, J., Schoeman, D. S., Lastra, M., Jones, A., Scapini, F., and Defeo, O., 2007. Sandy beaches at the brink. *Diversity and Distributions*, **13**(5), 556–560.

Shanmugam, G., 2012. Process-sedimentological challenges in distinguishing paleo-tsunami deposits. *Natural Hazards*, **63**(1), 5–30.

Shiki, T., Tachibana, T., Fujiwara, O., Goto, K., Nanayama, F., and Yamazaki, T., 2008. Characteristic features of tsunamiites. In Shiki, T., Tsuji, Y., Yamakazi, T., and Minoura, K. (eds.), *Tsunamiites – Features and Implications*. Amsterdam: Elsevier Scientific, pp. 319–336.

Small, C., and Nicholls, R. J., 2003. A global analysis of human settlement in coastal zones. *Journal of Coastal Research*, **19**, 584–599.

Switzer, A. D., and Jones, B. G., 2008. Large-scale washover sedimentation in a freshwater lagoon from the southeast Australian coast: sea-level change, tsunami or exceptionally large storm? *The Holocene*, **18**(5), 787–803.

Switzer, A. D., Pucillo, K., Haredy, R. A., Jones, B. G., and Bryant, E. A., 2005. Sea level, storm, or tsunami: enigmatic sand sheet deposits in a sheltered coastal embayment from southeastern New South Wales, Australia. *Journal of Coastal Research*, **21**, 655–663.

Switzer, A. D., Yu, F., Gouramanis, C., Soria, J. L. A., and Pham, D. T., 2014. Integrating different records to assess coastal hazards at multi-century timescales. *Journal of Coastal Research, SI*, **70**, 723–728.

Taboroši, D., and Kázmér, M., 2013. Erosional and depositional textures and structures in coastal karst landscapes. In *Coastal Karst Landforms*. Dordrecht: Springer, pp. 15–57.

Vita-Finzi, C., and Cornelius, P. F. S., 1973. Cliff sapping by molluscs in Oman. *Journal of Sedimentary Research*, **43**, 31–32.

Williams, D. M., and Hall, A. M., 2004. Cliff-top megaclast deposits of Ireland, a record of extreme waves in the North Atlantic – storms or tsunamis? *Marine Geology*, **206**(1), 101–117.

Wolman, M. G., and Miller, J. P., 1960. Magnitude and frequency of forces in geomorphic processes. *The Journal of Geology*, **68**, 54–74.

Wunsch, C., and Stammer, D., 1997. Atmospheric loading and the oceanic "inverted barometer" effect. *Reviews of Geophysics*, **35**(1), 79–107.

Cross-references

Beach Processes
Coastal Engineering
Coasts
Earthquakes
Erosion-Hiatuses
Hurricanes and Typhoons
Integrated Coastal Zone Management
Lagoons
Oil Spill
Radiocarbon: Clock and Tracer
Submarine Slides
Tsunamis
Waves

GEOLOGIC TIME SCALE

Felix M. Gradstein
Museum of Natural History, University of Oslo, Oslo, Norway

Introduction

The geologic time scale (GTS) is the principal tool for deciphering and understanding the long and complex history of our planet, Earth. As Arthur Holmes, the father of the geologic time scale, once wrote (Holmes, 1965, p. 148): "To place all the scattered pages of earth history in their proper chronological order is by no means an easy task." Ordering these scattered and torn pages and understanding the physical, chemical, and biological processes that acted on them since Earth appeared and solidified require a detailed and accurate time scale.

This calibration to linear time of the succession of events recorded in the rocks on Earth has three components (Figure 1):

1. The international stratigraphic divisions and their correlation in the global rock record
2. The means of measuring linear time or elapsed durations from the rock record
3. The methods of joining the two scales, the stratigraphic one and the linear one

For clarity and precision in international communication, the rock record of Earth's history is subdivided into a "chronostratigraphic" scale of standardized global stratigraphic units, such as "Devonian," "Miocene," "*Zigzagiceras zigzag* ammonite zone," or "polarity Chron C25r." Unlike the continuous ticking clock of the "chronometric" scale (measured in years before the year AD 2000), the chronostratigraphic scale is based on relative time units in which global reference points at boundary stratotypes define the limits of the main formalized units, such as "Permian." The chronostratigraphic scale is an agreed-upon convention, whereas its calibration to linear time is a matter for discovery or estimation.

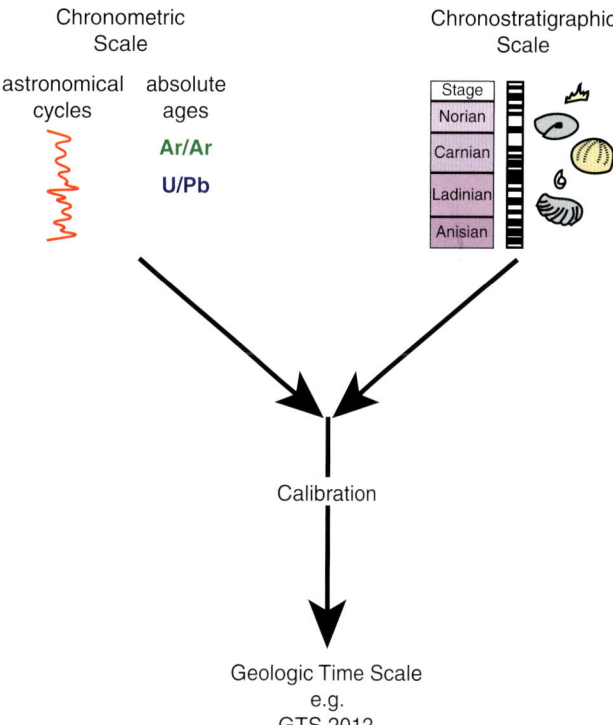

Geologic Time Scale, Figure 1 The construction of a geologic time scale is the merger of a chronometric scale, measured in years, and a chronostratigraphic scale, consisting of formalized definitions of geologic stages, biostratigraphic zonation units, magnetic polarity zones, and other subdivisions of the rock record.

By contrast, Precambrian stratigraphy is formally classified chronometrically, i.e., the base of each Precambrian eon, era, and period is assigned an arbitrary numerical age (Van Kranendonk, 2012). The reason for this is the difficulty to actually define globally correlative Precambrian rock units and date them. An exception is the final period of the Precambrian and the Ediacaran that now has a chronostratigraphic definition, with a GSSP (see below) for its base (Narbonne et al., 2012).

Under the auspices of the International Commission on Stratigraphy (ICS), over 65 % of the international stratigraphic divisions of the Phanerozoic and their correlative events are now (January 2013) standardized, using the GSSP (Global Boundary Stratotype Section and Point) concept. This concept identifies "globally" correlative levels in continuous (marine) sedimentary sections where stage boundaries are defined and (literally) pinned down, under state protection. Correlations from and to the rockbound pin level facilitate extra-regional and/or global stage boundary recognition. On the Internet site https://engineering.purdue.edu/Stratigraphy/, the GSSP record is summarized and kept track of.

Conventions and standards

Ages are given in years before "Present" (BP). To avoid a constantly changing datum, "Present" was fixed as AD 1950 (as in ^{14}C determinations), the date of the beginning of modern isotope dating research in laboratories around the world. For most geologists, this offset of official "Present" from "today" is not important. However, for archeologists and researchers into events during the Holocene (the past 11,500 years), the offset between the "BP" convention from radiogenic isotope laboratories and actual total elapsed calendar years becomes significant. The offset between the current year and "Present" has led many Holocene specialists to use a "2000 BP," which is relative to the year AD 2000.

For clarity, the linear age in years is abbreviated as *a* (for annum), and ages are measured in *ka*, *Ma*, or *Ga* for thousands, millions, or billions of years before present. Elapsed time or duration is often abbreviated as *yr* (for year) and longer durations in *kyr* or *myr*. Therefore, the Cenozoic began at 66 Ma and spans 66 myr (to the present day, defined as the year AD 2000).

The uncertainties in computed ages or durations are expressed as standard deviation with 2-sigma (95 %) confidence. The uncertainty is indicated by "±" and will have implied units of thousands or millions of years as appropriate to the magnitude of the age.

Historical overview of geologic time scales

Stitching together the many data points on the loom of time requires an elaborate combination of earth science and mathematical/statistical methods. Hence, the time and effort involved in constructing a new geologic time scale and assembling all relevant information is considerable. Because of this and because continuous updating in small measure with new information is not advantageous to the stability of any common standard, new geologic time scales come out sparsely.

Since 1981, six successive Phanerozoic GTSs have been published, each new one achieving higher resolution, better error analysis, and more users worldwide (Harland et al., 1982, 1990; Gradstein and Ogg, 1996; Gradstein et al., 2004 = GTS2004; Ogg et al., 2008; Gradstein et al., 2012 = GTS2012).

Geologic time scale (GTS2012)

The current standard geologic time scale is GTS2012 (Gradstein et al., 2012), constructed between 2004 and 2012 with a team of over 65 earth science and other specialists. Many of its coauthors are officers of ICS. Philosophy, methodology, chronostratigraphy, and geochronology are laid out in 32 chapters. Three data appendices detail stages of color coding, all radiometric dates, and Cretaceous through Cenozoic calcareous microfossils biochronology.

Construction of GTS2012 may be summarized in four steps:

Step 1. Construct an updated global chronostratigraphic scale for the Earth's rock record.
Step 2. Identify key linear-age calibration levels for the chronostratigraphic scale using radiogenic isotope age dates, and/or apply astronomical tuning to cyclic sediment, or scale and interpolate (near) linear segments of stable isotope sequences.
Step 3. Interpolate the combined chronostratigraphic and chronometric scale, for example, with a smoothing cubic spline, where direct information in specific stratigraphic intervals is wanting.
Step 4. Calculate or estimate error bars on the combined chronostratigraphic and chronometric information to obtain a geologic time scale with estimates of uncertainty on boundary ages and on unit durations.

The first step, integrating multiple types of stratigraphic information in order to construct the chronostratigraphic scale, is the most time-consuming; it summarizes and synthesizes centuries of detailed geological research while reconciling it with the most up-to-date information. The second step, identifying which radiogenic isotope and cycle-stratigraphic studies would be used as the primary constraints for assigning linear ages, is the one that is evolving most rapidly since the last decade. Historically, time scale building went from an exercise with very few and relatively inaccurate radiogenic isotope dates, as used by Holmes (1960), to one with many dates with greatly varying analytical precision (like GTS89).

The new philosophy for Step 2 that was started in GTS2004 and expanded in GTS2012 is to select stratigraphically meaningful radiogenic isotope dates with high analytical precision. More than 260 radiogenic isotope dates were thus selected for their reliability and stratigraphic importance to calibrate the geologic record in linear time. All 260+ GTS2012 age dates are detailed in Appendix 2 of GTS2012 by Schmitz (2012).

In addition to selecting radiogenic isotope ages based upon their stratigraphic control and analytical precision, GTS2012 also applied the following criteria or corrections:

1. Stratigraphically constrained radiogenic isotope ages with the U-Pb method on zircons were accepted from the isotope dilution mass spectrometry (IDMS) method, but not from the high-resolution ion microprobe (SHRIMP).
2. ^{40}Ar-^{39}Ar radiogenic isotope ages were recomputed to be in accord with the revised ages for laboratory monitor standards: 527.0 ± 2.6 Ma for MMhb-1 (Montana hornblende), 28.51 ± 0.06 Ma for TCR (Taylor Creek sanidine), and 28.201 ± 0.046 Ma for FCT (Fish Canyon sanidine). Systematic ("external") errors and uncertainties in decay constants are partially incorporated (Kuiper et al., 2008; Schmitz, 2012). As in GTS2004, no glauconite-based dates are used.

The bases of Paleozoic, Mesozoic, and Cenozoic are bracketed by analytically precise ages at their GSSP or

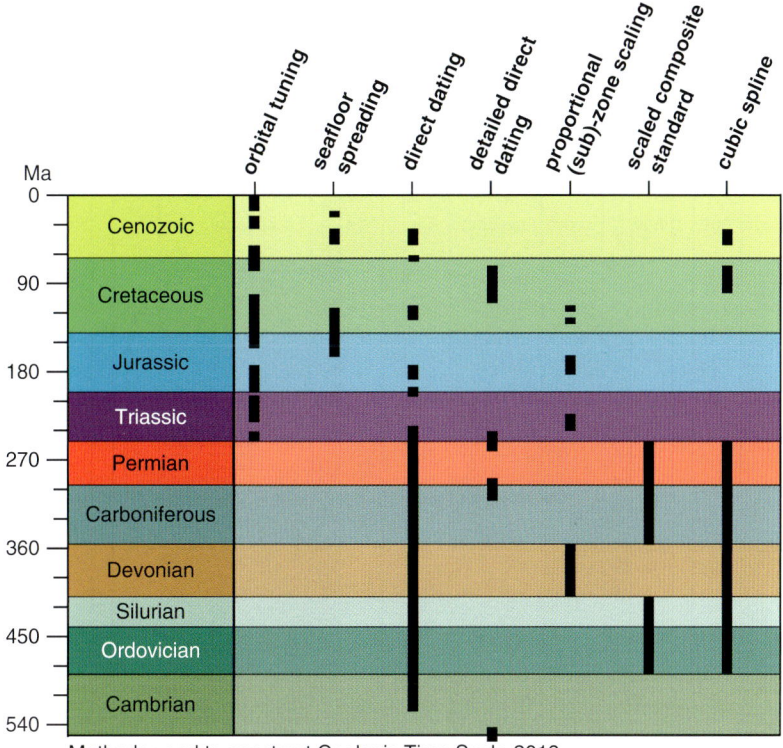

Geologic Time Scale, Figure 2 Methods used to construct Geologic Time Scale GTS2012 integrate different techniques depending on the quality of data available within different intervals.

primary correlation markers, respectively, 541 ± 0.63 Ma, 252.16 ± 0.5 Ma, and 65.95 ± 0.05 Ma, and there also are direct age dates on base Carboniferous, base Permian, base Jurassic, base Cenomanian, and base Eocene, but most other period or stage boundaries prior to the Neogene lack direct age control. No radiometric age dates were available to GTS2012 for Barremian, Bathonian, Norian, Kungurian through Wordian, Givetian, Pragian, Pridoli, Gorstian, Aeronian, Floian, and large parts of Cambrian. The Callovian through Hauterivian and Upper Triassic are particularly low in stratigraphically meaningful radiometric dates. Therefore, the third step, linear interpolation, still plays a key role for most of GTS2012, as it did in GTS2004. This detailed and high-resolution interpolation process incorporates several techniques, depending upon the available information (Figure 2):

1. A composite standard of graptolite zones spanning the uppermost Cambrian, Ordovician, and Silurian interval was derived from hundreds of sections in oceanic and slope environment basins using the Constrained Optimization (CONOP) method (Cooper and Sadler, 2012). With average zone thickness from many sections taken as directly proportional to zone duration, the detailed composite sequence was scaled using high-precision zircon and sanidine age dates.

2. High-precision ID-TIMS U-Pb zircon ages of a great many interstratified tuffs and tonsteins were used to calibrate the detailed litho-, cyclo-, and biostratigraphic framework of Carboniferous basins of Eastern Europe. Age resolution of <0.05 % or ca 100 kyr was obtained for these Carboniferous volcanics. This precision allows the resolution of time in the Milankovitch frequency band and confirms the long-standing hypothesis that individual high-frequency Pennsylvanian cyclothems and bundles of cyclothems into fourth-order sequences are the eustatic response to orbital eccentricity (ca 100 and 400 kyr) forcing. Combining the new datings with orbital tuning of fourth-order sequences in the basins scales Carboniferous stages to a new level of precision (Davydov et al., 2012).

3. Only three Cretaceous stages have ratified GSSPs, but reasonable stable consensus definitions exist for the other nine ones (Ogg and Hinnov, 2012). The age model for the Berriasian through Barremian ammonite zones is mainly derived from correlations to the M-Sequence of marine magnetic anomalies plus intervals with cycle stratigraphy and/or linearization of strontium isotope trends. The Aptian-Albian ammonite zones are scaled according to their correlation to microfossil and nannofossil data, which, in turn, are scaled by cycle stratigraphy. Base Cenomanian has a direct age date, using the Japanese sediment record. A spline fit of numerous radioisotopic dates from

Geologic Time Scale, Figure 3 The new Geologic Time Scale 2012.

volcanic ash horizons interbedded with US Western Interior ammonites with adjustments for cycle stratigraphy of some intervals provides a high-resolution numerical scale for the Cenomanian through early Maastrichtian. The late Maastrichtian correlations rely on microfossil data calibrated to a spline and cycle fit of C-Sequence marine magnetic anomalies.

4. Astronomical tuning of cyclic sediments was used for Neogene, Oligocene, Paleocene, and most of Eocene and portions of the Cretaceous, Jurassic, and Triassic. The Oligocene through Neogene astronomical scale is directly tied to the Present (Hilgen et al., 2012; Vanden Berghe et al., 2012); the older astronomical scale provides linear duration constraints on polarity chrons, biostratigraphic zones, and entire stages.

5. Proportional scaling was undertaken relative to component biozones or subzones. This procedure was necessary in portions of the Triassic and Jurassic. Devonian stages were scaled from approximate equal duration of a set of high-resolution subzones of ammonoids and conodonts and tentative estimates of sedimentary cycles' duration, fitted to an array of dates.

The actual statistical procedures and error analysis employed for the above proportionally scaled zones and stages in major parts of GTS2012 were programmed and executed by O. Hammer (in Agterberg et al., 2012). The uncertainties on older stage boundaries systematically increase owing to potential systematic errors in the different radiogenic isotope methods, rather than to the analytical precision of the laboratory measurements. In this connection we mention that biostratigraphic error is fossil event and fossil zone dependent, rather than age dependent.

Ages and durations of Neogene stages derived from orbital tuning are considered to be accurate to within a precession cycle (~20 kyr) assuming that all cycles are correctly identified and that the theoretical astronomical tuning for progressively older deposits is precise. Paleogene dating combines tuning, radiometrics, and C-sequence splining; hence, stage age uncertainty is larger and varies between 0.2 and 0.5 Ma.

The Geologic Time Scale GTS2012 is displayed in Figure 3 (from Gradstein et al., 2012).

Stratigraphic charts and tables

The plethora of names for time and time-rock units in local regions lends itself to the production of wall charts and stratigraphic lexicons that visualize the links between regional schemes and the standard scale. The international standard is developed in collaboration with officers of the International Commission on Stratigraphy (ICS). The Commission for the Geological Map of the World (CGMW) in Paris and ICS closely collaborate on the map and color coding of chronostratigraphic units on the standard chart. The updated charts in PDF format are freely available from websites https://engineering.purdue.edu/stratigraphy/ and www.nhm2.uio.no/stratlex and also are distributed by the CGMW.

A sophisticated, albeit easy time scale, graphics program for GTS2012 is called *TSCreator @*. This popular software, with its large time scale dataset, is freely available from https://engineering.purdue.edu/Stratigraphy/tscreator/. Now there are more than 25,000 Cambrian through Holocene biostratigraphic, sea-level, magnetic, and geochemical events in the public software database, all calibrated to GTS2012. Cross-correlations are in place for trilobites, conodonts, graptolites, ammonoids, fusulinids, chitinozoans, megaspores, nannofossils, foraminifers, dinoflagellates, radiolarians, diatoms, strontium isotope, C-org and oxygen curves, eustatic sea-level curves, etc. Scalable vector graphics output of any time scale interval is an easy option. A majority of linear scale drawings, with its calibrated event and zonal data in the GTS2012 book, were initiated in TSCreator, before being drafted in a final format.

Summary

The geologic time scale (GTS) provides the framework for the physical, chemical, and biological processes in time on Earth. The most up-to-date GTS is available since October 2012. Geoscientists can easily create and print time scale charts, utilizing the freely available software package "Time Scale Creator @."

Bibliography

Agterberg, F. P., Hammer, O., and Gradstein, F. M., 2012. Statistical procedures. In Gradstein, R., et al. (eds.), *The Geologic Time Scale 2012*. Amsterdam: Elsevier, pp. 269–275.

Cooper, R. A., and Sadler, P. M., 2012. The Ordovician period. In Gradstein, R. M., et al. (eds.), *The Geologic Time Scale 2012*. Amsterdam: Elsevier, pp. 489–525.

Davydov, V. I., Korn, D., and Schmitz, M. D., 2012. The Carboniferous period. In Gradstein, R. M., et al. (eds.), *The Geologic Time Scale 2012*. Amsterdam: Elsevier, pp. 603–653.

Gradstein, F. M., and Ogg, J. G., 1996. A Phanerozoic time scale. *Episodes*, **19**, 3–5. with insert.

Gradstein, F. M., Ogg, J. G., Smith, A. G., Agterberg, F. P., Bleeker, W., Cooper, R. A., Davydov, V., Gibbard, P., Hinnov, L. A., House, M. R., Lourens, L., Luterbacher, H.-P., McArthur, J., Melchin, M. J., Robb, L. J., Shergold, J., Villeneuve, M., Wardlaw, B. R., Ali, J., Brinkhuis, H., Hilgen, F. J., Hooker, J., Howarth, R. J., Knoll, A. H., Laskar, J., Monechi, S., Powell, J., Plumb, K.A., Raffi, I., Röhl, U., Sanfilippo, A., Schmitz, B., Shackleton, N. J., Shields, G. A., Strauss, H., Van Dam, J., Veizer, J., van Kolfschoten, T. H., and Wilson, D., 2004. *A Geologic Time Scale 2004*. Cambridge University Press, 589 pp.

Gradstein, F. M, Ogg, J. G., Schmitz, M. D., Ogg, G. M., Agterberg, F. P., Anthonissen, D. E., Becker, T. R., Catt, J. A., Cooper, R. A., Davydov, V. I., Gradstein, S. R., Henderson, C. M., Hilgen, F. J., Hinnov, L. A., McArthur, J. M., Melchin, M. J., Narbonne, G. M., Paytan, A., Peng, S., Peucker-Ehrenbrink, B., Pillans, B., Saltzman, M. R., Simmons, M. D., Shields, G. A., Tanaka, K. L.,Vandenberghe, N., Van Kranendonk, M. J., Zalasiewicz, J., Altermann, W., Babcock, L. E., Beard, B. L., Beu, A. G., Boyes, A. F., Cramer, B. D., Crutzen, P. J., van Dam, J. A., Gehling, J. G., Gibbard, P. L., Gray, E. T., Hammer, O., Hartmann, W. K., Hill, A. C., Paul F. Hoffman, P. F., Hollis, C. J., Hooker, J. J., Howarth, R. J., Huang, C., Johnson,

C. M., Kasting, J. F., Kerp, H., Korn, D., Krijgsman, W., Lourens, L. J., MacGabhann, B. A., Maslin, M. A., Melezhik, V. A., Nutman, A. P., Papineau, D., Piller, W. E., Pirajno, F., Ravizza, G. E., Sadler, P. M., Speijer, R. P., Steffen, W., Thomas, E., Wardlaw, B. R., Wilson, D. S., and Xiao, S., 2012. *The Geologic Time Scale 2012*. Boston: Elsevier, 1129 pp.

Harland, W. B., Cox, A. V., Llewellyn, P. G., Pickton, C. A. G., Smith, A. G., and Walters, R., 1982. *A Geologic Time Scale*. Cambridge: Cambridge University Press, 131 pp.

Harland, W. B., Armstrong, R. L., Cox, A. V., Craig, L. A., Smith, A. G., and Smith, D. G., 1990, *A Geologic Time Scale 1989*. Cambridge: Cambridge University Press, 263 pp.

Hilgen, F. J., Lourens, L. J., and VanDam, J. A., 2012. The Neogene Period. In Gradstein, R. M., et al. (eds.), *The Geologic Time Scale 2012*. Amsterdam: Elsevier, pp. 923–979.

Holmes, A., 1960. A revised geological time-scale. *Transactions of the Edinburgh Geological Society*, **17**, 183–216.

Holmes, A., 1965. *Principles of Physical Geology*. London: Nelson Printers, 1288 p.

Kuiper, K. F., Deino, A., Hilgen, F. J., Krijgsman, W., Renne, P. R., and Wijbrans, J. R., 2008. Synchronizing rock clocks of Earth history. *Science*, **320**(5875), 500–504.

Narbonne, G., Xiao, S., and Shields, G. A., 2012. The Ediacaran period. In Gradstein, R. M. (ed.), *The Geologic Time Scale 2012*. Amsterdam: Elsevier, pp. 413–437.

Ogg, J. G., Ogg, G., and Gradstein, F. M., 2008. *The Concise Geologic Time Scale*. Cambridge: Cambridge University Press, 177 pp.

Ogg, J. G., and Hinnov, L. A., 2012. The cretaceous period. In Gradstein, R. M., et al. (eds.), *The Geologic Time Scale 2012*. Amsterdam: Elsevier, pp. 793–855.

Schmitz, M. D., 2012. Radiogenic isotopes geochronology. In Gradstein, R. M. (ed.), *The Geologic Time Scale 2012*. Amsterdam: Elsevier, pp. 115–127.

Vanden Berghe, N., Hilgen, F. J., and Speijer, R. P., 2012. The Paleogene Period. In Gradstein, R. M., et al. (eds.), *The Geologic Time Scale 2012*. Amsterdam: Elsevier, pp. 855–923.

Van Kranendonk, M., 2012. A chronostratigraphic division of the Precambrian. In Gradstein, R. M., et al. (eds.), *The Geologic Time Scale 2012*. Amsterdam: Elsevier, pp. 299–393.

Cross-references

Biochronology, Biostratigraphy
Geochronology: Uranium-series Dating of Ocean Formations
Radiocarbon: Clock and Tracer

GLACIAL-MARINE SEDIMENTATION

Alexander P. Lisitzin and Vladimir P. Shevchenko
P.P. Shirshov Institute of Oceanology, Russian Academy of Sciences (IO RAS), Moscow, Russia

Definition

Glacial-marine sedimentation is the process of sediment deposition in the sea after release from ice shelves, tidewater ice fronts, icebergs, or sea ice.

Introduction

Glacial type of sedimentation dominates in polar and subpolar parts of the World Ocean and continues the regions of glacial sedimentation both on the land (in region land ice sheets, mountain glaciation, and other types of inland ice) and in seas and oceans (Lisitzin, 1972, 2002, 2010). Water catchments are changed here by ice catchments and annually are replenished by new portions of snow. In this kingdom of snow, ice and cold minimal values of winter temperatures reach $-71\ °C$ in the Northern Hemisphere and $-89\ °C$ in the Southern Hemisphere. It influences on the development of life in the ocean and on the land and residual soils at the stage of preparation of sedimentary matter on the land. It determines at a large extent both quantity, composition, and properties of marine bottom sediments and suspended particulate matter in water column.

Glacial-marine dynamics and associated sedimentary processes are closely tied to glacial regime and reflect dominant climate conditions (Trusel et al., 2010). Glacial-marine processes such as sediment flux and yield, terminus fluctuations, calving rates, freshwater flux, and mass balance are useful for comparison among differing glacial and climatic regimes. Understanding the relationship of complex modern processes to those found in glacial-marine sediment record can enhance interpretation of past glacial-marine processes and changes in biological productivity (Powell, 1984; Trusel et al., 2010).

According to the composition and genesis of ice and sedimentary material, glacial type of sedimentation could be divided into two subtypes:

1. Sedimentation from glacier ice, including iceberg sedimentation (defined by land glaciers reaching the sea)
2. Sea-ice sedimentation (defined by sea ice)

Sedimentation from glacier ice

Icebergs are released where glaciers, ice sheets, and ice shelves terminate in the ocean (Warren, 1992). Often icebergs released in fjords (Mugford and Dowdeswell, 2010). The dominant sedimentary process in most fjords is sedimentation from the brackish plumes rich in suspended particulate matter that emerges from either glacier-fed river mouths or tidewater cliff. Modern glacial-marine environments are widely spread in the Antarctic, Alaska, Greenland, Baffin Island, British Columbia, Svalbard, Norway, Novaya Zemlya, and Severnaya Zemlya (Drewry and Cooper, 1981; Elverhøi et al., 1983; Molnia, 1983; Cowan and Powell, 1991; Syvitski et al., 1996; Gilbert et al., 2003; Kehrl et al., 2011; Politova et al., 2012; Szczucinski and Zajaczkowski, 2012; Chewings et al., 2014; Miller et al., 2015).

The vertical sediment fluxes in these areas are relatively high. For example, maximum sediment flux from temperate glaciers in southeastern Alaska was measured in McBride Inlet; in average it was 53 $kg\ m^{-2}\ day^{-1}$ (Cowan and Powell, 1991). In Kongsfjorden, Svalbard, near Kronebreen, glacier sediment flux reached

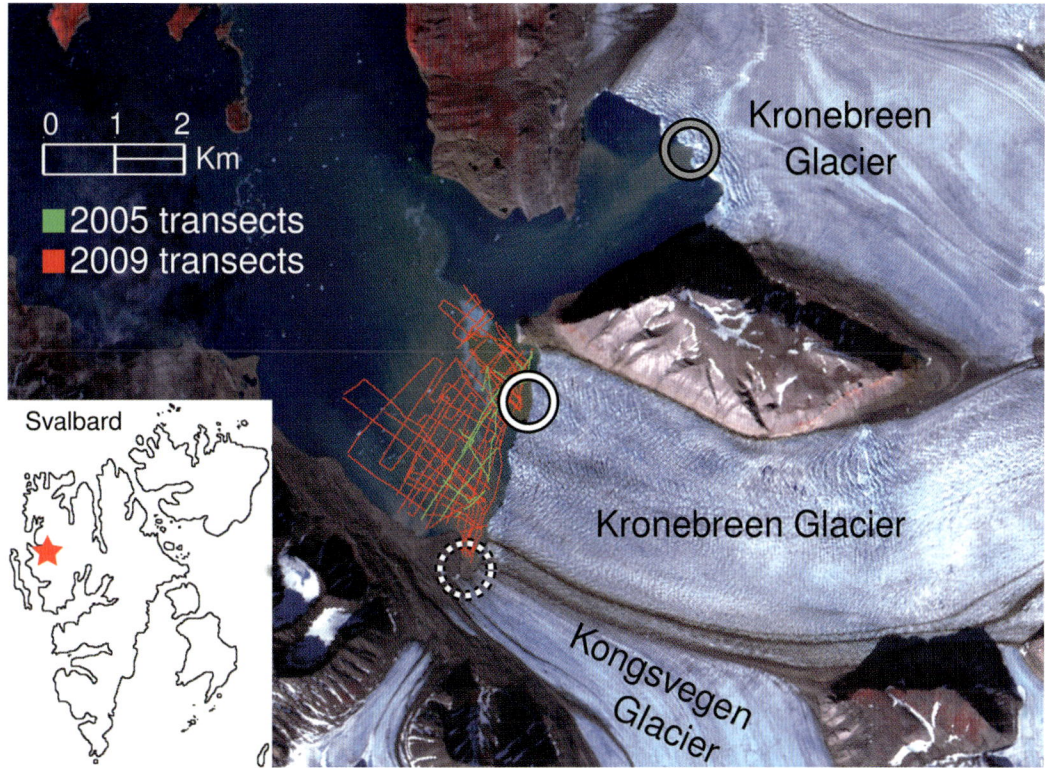

Glacial-marine Sedimentation, Figure 1 Kronebreen and Kongsvegen terminate at the head of Kongsfjorden, western Svalbard (Kehrl et al., 2011). Two point sources deliver sediment to the complex's terminus: an ice-marginal stream (*dashed white circle*) and a subglacial stream (*solid white circle*). The *gray circle* indicates a subglacial stream that emerged between Landsat images taken on 30 June and 7 July 2002. All three sediment sources form plumes that extend into the fjord. *Green* and *red lines* indicate 2005 and 2009 echo-sounding transects, respectively.

18.6 kg m^{-2} day^{-1} (Trusel et al., 2010). Studies in Billefjorden, a subpolar fjord in Svalbard, demonstrate that vertical downward particulate matter fluxes can reach up to several kg m^{-2} d^{-1} in subpolar non-glacier contact and glacier contact setting during the short summer season (Szczucinski and Zajaczkowski, 2012).

Studies in Kongsfjorden, western Svalbard, demonstrated that glacial-marine sedimentation processes continue to shape the fjord floor (Kehrl et al., 2011). Two point sources supply sediment to the fjord floor adjacent to the glacier complex: (1) a subglacial stream that exits Kronebreen just north of the centerline and (2) an ice-marginal stream that flows alongside the southern margin of Kongsvegen (Figure 1). Sedimentation rates are greatest near the ice cliff (>100 mm a^{-1}) and decrease to 50–100 mm a^{-1} ~10 km away from the terminus (Elverhøi et al., 1983). About 90 % of the total sediment input is deposited within 400 m of the ice front, thereby decreasing water depth at the grounding line (Elverhøi et al., 1983; Svendsen et al., 2002).

Icebergs transport sediments far from glaciers, in some cases up to thousands of kilometers, releasing them gradually as they melt (Dowdeswell and Murray, 1990; Syvitski et al., 1996). There are no rivers and multiyear pack ice in the Antarctic. The distance of one-year ice field transport, having a small reserve of cold, is small. Polar regions of the Southern Ocean are the main area of iceberg sedimentation now. In the Northern Hemisphere, Greenland and Canadian archipelago are the main areas of iceberg sedimentation. Load-carrying ability of icebergs in contrast to sea-ice fields reaches tens and hundreds of tons. Their large sizes (up to hundreds of km, generally less) ensure large reserve of cold; in some years icebergs cross the Southern Ocean and reach the southern ends of continents.

The area of distribution of icebergs reaches 62 * 10^6 km^2 in the Southern Hemisphere and only 3.3 * 10^6 km in the Northern Hemisphere (Lisitzin, 2002). More often (according to satellite data) they are found not only a continent together with one-year ice but far out of limits of sea-ice distribution, up to 26–30°S; i.e., they dominate in the Southern Hemisphere, crossing temperate zone and reaching arid zone.

A belt of icebergs girdles the Antarctica. Its width ranges from 400 to 1,200 km, averaging 500–700 km (Lisitzin, 2002). The zones of their distribution and concentrations of coarse material in bottom sediments have been established during expeditions. Near the coasts concentration of ice-rafted material exceeds 100 kg/m^3 of sediments and gradually decreases to values less than

$1~kg~m^{-3}$. Most Antarctica icebergs discharge their load near the place of their origin; here submarine moraines are formed.

Fine material of icebergs is formed in bottom layers of glaciers as a result of the detachment of rock fragments from the glacier bed and their breaking up and abrasion (up to glacier milk).

Sea-ice sedimentation

The average time of life of sea ice in the Northern Hemisphere is 1.3 years (here the center of existence of multiyear ice is situated). There is no multiyear ice in the Southern Hemisphere. Sea ice is developed at the area of $26 \pm 3 * 10^6~km^2$; $13.1 * 10^6~km^2$ of them are in the Arctic Ocean (Lisitzin, 2002). The average thickness of sea ice is about 1.5 m, though in the area of multiyear ice, it reaches 4–6 m. The thickness of ice in water determines its carrying capacity – it is sufficient to carry coarse material including boulder with weight of tens to hundreds of kg (for icebergs the weight of boulders can be many tons).

In the main area of modern multiyear ice distribution in the Arctic, the average area of ice cover changes during a year from $11.4 * 10^6~km^2$ in March to $7 * 10^6~km^2$ in September. In the last years the area of ice cover considerably decreased as a result of climate warming.

Sea ice could be divided into one-year ice (which melts in summer and loses sedimentary matter included in it) and multiyear (packs) that could exist in a few years (up to 10–15 years in the Beaufort Gyre).

Correspondingly, one-year ice is the agent of nearby and middle-range transport of sedimentary matter, while multiyear ice is the agent of long-range transport (for 100–1,000 km). Major routes of their drift and horizontal fluxes according to satellite data and drifting station observations are demonstrated in Figure 2. One-year ice fields and pack ice fields discharge sedimentary matter differently. For one-year ice fields the spring melting along the outer periphery takes place at large area and sedimentary matter deposits as a result of the so-called carpet-like discharge. The discharge runs differently for pack ice. The melting occurs along the contact with warm waters (temperature from $0~°C$ to $1–2~°C$ above zero) in the Fram Strait (meeting with the North Atlantic Current) and in confluence of the Kuroshio and Oyashio currents (northwestern Pacific Ocean) (Lisitzin, 1968, 1996).

The distribution of ice-rafted material is analyzed in comparison with data on ice exposition (time of ice covering at this bottom part). After that the granulometric composition of ice-rafted material is studied by sieve method, roundness index is determined in comparison with standard samples, and fouling by organisms and concretions (for ancient samples) is estimated.

The most important part of the study of ice-rafted stones is petrographic analysis with selection of typical samples of standard collection and with looking through thin sections under a microscope. It allows to judge about complexes of rocks and to sort out provinces of ice-rafted material distribution in bottom sediments (the same as sorting out results of mineralogical analysis of aleuritic fraction).

In some cases, for example, in the Kara Sea, an unusual wide distribution of basalt stones is registered (Lisitsin et al., 2004; Lisitsyn et al., 2004). Their source is situated in the catchment area at the Putorana Plateau. Comparative petrochemical analyses and determination of age confirmed marker value of this type of rocks. The preparation of maps of distribution of different types of rocks (generally few tens) in the upper layer of bottom sediments in the Sea of Okhotsk, Bering Sea, and northern Pacific Ocean made possible to establish areas of each rock type distribution and places of their supply from the continent (according to growth of concentrations). It became possible to determine the sources of these rocks as a result of comparison of collected rocks with rocks in the catchment area and to reveal ways of migration of ice fields marked by rock material (according to their contents in bottom sediments) (Lisitsin, 1968, 1972, 1996, 2001, 2002, 2010; Lisitsin et al., 2004; Lisitsyn et al., 2004).

Coarse waste material is an impressive indication of ice rafting, but it is not sole. The study of finer fractions (sand-aleuritic and politic) and biogenic material gives important data. Analysis of sedimentary matter obtained by melting of sea ice gives important materials also. These studies made possible to reveal one unexpected source of glacial sediments – aeolian (Darby et al., 1974; Lisitzin, 2002; Shevchenko et al., 2002; Shevchenko, 2010; Chewings et al., 2014; Miller et al., 2015). The main feature of aeolian supply here is that aeolian material is deposited mostly during polar winter at low temperatures. In summer aeolian material is washed out rapidly from the atmosphere by rains. In winter most part of catchment area is covered by snow and frozen. The sedimentary aeolian matter in winter is supplied mainly as a result of long-range (often more than 1,000 km) transport of matter in the atmosphere at levels higher than clouds.

Lastly, it is important to say about the new mechanism of fine sediment supply by sea-ice fields, which originated in polynyas along the boundary between fast ice and drift ice. Polynyas formation is related to strong winds separating the drift ice from the fast ice; the winds are particularly strong along the periphery of the east Siberian anticyclone (Kara, East Siberian, and Laptev seas) and accompanied by very low temperatures (-30 to $-40~°C$). The extent of some polynyas from time to time reaches 1–3,000 km. This giant freezing plant combines open water of polynyas with strong wind and very low temperature. This leads to overcooling of both the surface layer and a significant part of the water column. Frazil ice and sometimes bottom ice are formed around centers of crystallization represented by suspended particles. The tiny freshwater ice crystals become even less dense compared to the host marine water (the process of cold distillation) and appear as hanging to the suspended particles; like floats, they

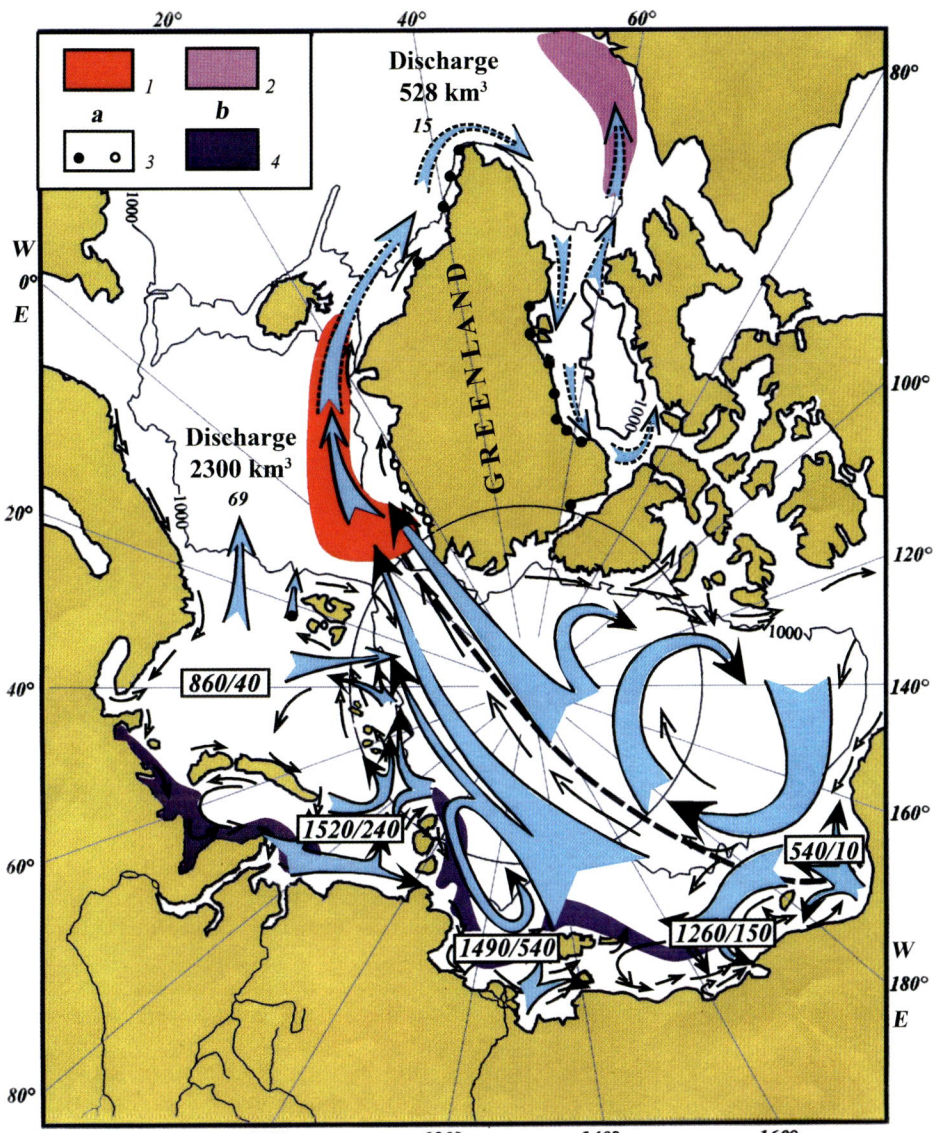

Glacial-marine Sedimentation, Figure 2 Major routes of ice drift and ice-rafted sedimentary material transport (Lisitzin, 2002, 2010). Horizontal fluxes of sedimentary matter, entrainment areas, transportation dynamics, and major areas of discharge, cryodepocenters (fronts of contact with warm waters): *1*, discharge zones, basins, and cryodepocenters of ice-rafted sedimentation; *2*, areas adjacent to Greenland where ice is supplemented by icebergs; *3*, major (*a*) and minor (*b*) areas of iceberg supply; *4*, areas of fast ice development (entraining shallow water sediments). *Arrows*, major routes of ice drift; *dashed arrows*, iceberg drift; *dashed line*, major Arctic ice divide separating Eurasian and North American sedimentary matter transport. Numbers in rectangles, ice volume (km^3): first number, at the end of winter; second number, ice carried in the central part of the Arctic Ocean. Numbers in cryodepocenters: volume of annual ice melt (km^3/year); semibold, ice-rafted and iceberg-rafted material supply (10^6 tons/year).

quickly rise to the surface and aggregate to form shuga. This phenomenon, called as ice pump (Lisitzin, 2010), is widely spread, but studied insufficiently due to operation in the polar night. Ice pump transfers fine suspended particulate matter from water to sea ice of the shelf. It is this type of processes that facilitates the sedimentary matter transportation from the major suppliers of newly formed ice (the Laptev, Kara, East Siberian, Chukchi, Beaufort seas) into the Central Arctic and after that to cryodepocenters (Nürnberg et al., 1994; Reimnitz et al., 1994; Lindemann et al., 1999; Eicken et al., 2005; Dethleff and Kuhlmann, 2010; Darby et al., 2011, 2015). Sediments can be entrained in sea ice during anchor ice formation (Reimnitz et al., 1987; Darby et al., 2011). Polar researches studying ice flows leaving the Arctic Ocean through the Fram Strait have noticed that the ice in general becomes darker and contains more dark spots of sedimentary matter ("dirty ice") (Figure 3).

Glacial-marine Sedimentation, Figure 3 Dirty ice in the Yermak Plateau area.

Glacial-marine Sedimentation, Table 1 Annual sediment flux in the ice melting areas and ice-covered areas in the Arctic

Position	Years	Flux, g $m^{-2} yr^{-1}$	Reference
Arctic			
Eastern Fram Strait	1987–1990	122–231	Hebbeln, 2000
Eastern Fram Strait	2000–2005	13–32	Bauerfeind et al., 2009
Fram Strait	2007	489	Cámara-Mor et al., 2011
Lomonosov Ridge	1995–1996	10.3	Fahl and Nöthig, 2007
Northeast Water Polynya	1992–1993	4.2–9.8	Ramseier et al., 1997
Peary Channel	1989–1990	1.1	Hargrave et al., 1994
Antarctic			
Bransfield Strait	1983–1984	107.7	Fischer et al., 1988
Weddell Sea	1985	0.37	Fischer et al., 1988

[7]The activity in the ice recently was used to estimate that the annual transport and release of sediments to the ablation area of the Fram Strait is about 500 g m^{-2} (Cámara-Mor et al., 2011), a value comparable to previously measured fluxes in sediment traps deployed in this area and larger than in the areas mostly covered by ice (Table 1).

Conclusions

Thus, the zone of modern glacial sedimentation, connected with sea ice and icebergs, integrally occupies about 20 % of the World Ocean surface, i.e., this type of sedimentary process is very important in the ocean both now and in the past (Lisitzin, 1968, 1996, 2002, 2010; Powell, 1984; Dowdeswell and Scourse,1990; Stein, 2008; Levitan and Lavrushin, 2009). During of glacial periods it became the major not only on continents but in World Ocean. Thus, the studies of the last few decades convincingly proved the existence of particular material-genetic type of bottom sediments, which occupied during the interglacials about 20 % of the surface of Cenozoic oceans and during glaciations (now we know about 6 of them) about 30 %, i.e., the global material-genetic type of marine and continental sedimentation, discovered and studied in the last century, is very important.

Bibliography

Bauerfeind, E., Nöthig, E.-M., Beszczynska, A., Fahl, K., Kaleschke, L., Kreker, K., Klages, M., Soltwedel, T., Lorenzen, C., and Wegner, J., 2009. Particle sedimentation patterns in the eastern Fram Strait during 2000–2005: results from the Arctic long-term observatory HAUSGARTEN. *Deep-Sea Research Part I*, **56**, 1471–1487.

Cámara-Mor, P., Masque, P., Garcia-Orellana, J., Kern, S., Cochran, J. K., and Hanfland, C., 2011. Interception of atmospheric fluxes by Arctic sea ice: evidence from cosmogenic [7]Be. *Journal of Geophysical Research*, **116**, C12041, doi:10.1029/2010JC006847.

Chewings, J. M., Atkins, C. B., Dunbar, G. B., and Golledge, N. R., 2014. Aeolian sediment transport and deposition in a modern high-latitude glacial marine environment. *Sedimentology*, **61**, 1535–1557.

Cowan, E. A., and Powell, R. D., 1991. Ice-proximal sediment accumulation rates in a temperate glacial fjord, southeastern Alaska. In Anonymous (ed.), *Glacial Marine Sedimentation: Paleoclimatic Significance. Geological Society of America*, Special Paper. Boulder, Colorado. 261, pp. 61–73.

Darby, D. A., Burcle, L. H., and Clark, D. L., 1974. Airborne dust on the Arctic pack ice, its composition and fallout rate. *Earth and Planetary Science Letters*, **24**, 166–172.

Darby, D. A., Myers, W. B., Jakobsson, M., and Rigor, I., 2011. Modern dirty sea ice characteristics and sources: the role of anchor ice. *Journal of Geophysical Research*, **116**, C09008, doi:10.1029/2010JC006675.

Darby, D. A., Myers, W., Herman, S., and Nicholson, B., 2015. Chemical fingerprinting, a precise and efficient method to determine sediment sources. *Journal of Sedimentary Research*, **85**, 247–253.

Dethleff, D., and Kuhlmann, G., 2010. Fram Strait sea-ice sediment provinces based on silt and clay composition identify Siberian Kara and Laptev seas as main source regions. *Polar Research*, **29**, 265–282.

Dowdeswell, J. A., and Murray, T., 1990. Modelling rates of sedimentation from icebergs. In Dowdeswell, J.A., and Scourse, J.-D. (eds.), *Glacial Environments: Processes and Sediments*. Geological Society of London, Special Publications. 53, pp. 121–137.

Dowdeswell, J. A., and Scourse, J. D., 1990. On the description and modeling of glacimarine sediments and sedimentation. In Dowdeswell, J.A., and Scourse, J.D. (eds.), *Glacial Environments: Processes and Sediments*. Geological Society of London, Special Publications, 53, pp. 1–13.

Drewry, D. J., and Cooper, A. P. R., 1981. Processes and models of Antarctic glaciomarine sedimentation. *Annals of Glaciology*, **2**, 117–122.

Eicken, H., Gradinger, R., Gaylord, A., Mahoney, A., Rigor, I., and Melling, H., 2005. Sediment transport by ice in the Chulchi and Beaufort Seas: increasing importance due to changing ice conditions? *Deep-Sea Research II*, **52**, 3281–3302.

Elverhøi, A., Lønne, Ø., and Seland, R., 1983. Glaciomarine sedimentation in a modern-fjord environment, Spitsbergen. *Polar Research*, **1**(2), 127–149.

Fahl, K., and Nöthig, E.-M., 2007. Lithogenic and biogenic particle fluxes on the Lomonosov Ridge (central Arctic Ocean) and their relevance for sediment accumulation: vertical vs. lateral transport. *Deep-Sea Research Part I*, **54**, 1256–1272.

Fischer, G., Fütterer, D., Gersonde, R., Honjo, S., Ostermann, D., and Wefer, G., 1988. Seasonal variability of particle flux in the Weddell Sea and its relation to sea ice cover. *Nature*, **335**, 426–428.

Gilbert, R., Chong, Å., Dunbar, R. B., and Domack, E. W., 2003. Sediment trap records of glacimarine sedimentation at Müller Ice Shelf, Lallemand Fjord, Antarctic Peninsula. *Arctic, Antarctic, and Alpine Research*, **35**, 24–33.

Hargrave, B. T., Bodungen von, B., Stoffyn-Egli, P., and Mudie, P. J., 1994. Seasonal variability in particle sedimentation under permanent ice cover in the Arctic Ocean. *Continental Shelf Research*, **14**, 279–293.

Hebbeln, D., 2000. Flux of ice-rafted detritus from sea ice in the Fram Strait. *Deep-Sea Research Part II*, **47**, 1773–1790.

Kehrl, L. M., Hawley, R. L., Powell, R. D., and Brigham-Grette, J., 2011. Glacimarine sedimentation processes at Kronebreen and Kongsvegen, Svalbard. *Journal of Glaciology*, **57**(205), 841–847.

Levitan, M. A., and Lavrushin, Y. A., 2009. Sedimentation history in the Arctic Ocean and subarctic seas for the last 130 kyr. *Lecture Notes in Earth Sciences*, **118**, 1–416.

Lindemann, F., Hölemann, J. A., Korablev, A., and Zachek, A., 1999. Particle entrainment in newly forming sea ice – freeze-up studies in October 1995. In Kassens, H., Bauch, H. A., Dmitrenko, I. A., Eicken, H., Hubberten, H. W., Melles, M., Thiede, J., and Timokhov, L. A. (eds.), *Land-ocean systems in the Siberian arctic: dynamics and history*. Berlin: Springer, pp. 113–125.

Lisitsin, A. P., Kharin, G. S., and Chernyshova, E. A., 2004. Coarse ice-rafted debris in the bottom sediments of the Kara Sea. *Oceanology*, **44**(3), 412–427.

Lisitsyn, A. P., Kharin, G. S., and Chernysheva, E. A., 2004. Basalts from coarse ice-rafted detritus in the bottom sediments of the Kara Sea. *Oceanology*, **44**(4), 554–563.

Lisitzin, A. P., 1968. *Recent Sedimentation Processes in the Bering Sea*. Jerusalem: Israel Program for Scientific Translation. 614 pp.

Lisitzin, A. P., 1972. Sedimentation in the World Ocean, Society of Economic Paleontologists and Mineralogists. Tulsa, SEPM Special Publication 17, 128 pp.

Lisitzin, A. P., 1996. *Oceanic Sedimentation: Lithology and Geochemistry*. Washington, DC: American Geophysical Union Press. 400 pp.

Lisitzin, A. P., 2001. Lithology of lithosphere plates. *Russian Geology and Geophysics*, **42**(4), 522–559.

Lisitzin, A. P., 2002. *Sea-Ice Sedimentation in the Ocean. Recent and Past*. Berlin/Heidelberg/New York: Springer-Verlag. 563 pp.

Lisitzin, A. P., 2010. Marine ice-rafting as a new type of sedimentogenesis in the Arctic and novel approaches to studying sedimentary processes. *Russian Geology and Geophysics*, **51**(1), 12–47.

Miller, M. F., Fan, Z., and Bowser, S., 2015. Sediment beneath multi-year sea ice: delivery by deltaic and eolian processes. *Journal of Sedimentary Research*, **85**, 301–314.

Molnia, B. F., 1983. Subarcic glacial-marine sedimentation: a model. In Molnia, B. F. (ed.), *Glacial-Marine Sedimentation*. New York: Plenum, pp. 95–144.

Mugford, R. I., and Dowdeswell, J. A., 2010. Modeling iceberg-rafted sedimentation in high-latitude fjord environments. *Journal of Geophysical Research*, **115**, F03024, doi:10.1029/2009JF001564.

Nürnberg, D., Wollenburg, I., Dethleff, D., Eicken, H., Kassens, H., Letzig, T., Reimnitz, E., and Thiede, J., 1994. Sediments in Arctic sea ice: implications for entrainment, transport and release. *Marine Geology*, **119**, 185–214.

Politova, N. V., Shevchenko, V. P., and Zernova, V.V., 2012. Distribution, composition, and vertical fluxesw of particulate matter in bays of Novaya Zemlya Archipelago, Vaigach Island at the end of summer. *Advances of Meteorology*, Article ID 259316, 15 p.

Powell, R. D., 1984. Glacimarine processes and inductive lithofacies modeling of ice shelf and tidewater glaciers sediments based on Quaternary examples. *Marine Geology*, **57**, 1–52.

Ramseier, R. O., Bauerfeind, E., Garrity, C., and Walsh, I. D., 1997. Seasonal variability of sediment trap collections in the Northeast Water polynya. Part I: sea-ice parameters and particle flux. *Journal of Marine Systems*, **10**, 359–369.

Reimnitz, E., Kempema, E. W., and Barnes, P. W., 1987. Anchor ice, sea-bed freezing, and sediment dynamics in shallow arctic seas. *Journal of Geophysical Research*, **92**, 14671–14678.

Reimnitz, E., Dethleff, D., and Nürnberg, D., 1994. Contrasts in Arctic shelf sea-ice regimes and some implications: Beaufort Sea versus Laptev Sea. *Marine Geology*, **119**, 215–225.

Shevchenko, V. P., 2010. Aerosols over the Russian Arctic seas. In Nikiforov, S. (ed.), *Seabed Morphology of the Russian Arctic Shelf*. New York: Nova Science Publishers, Inc, pp. 87–92.

Shevchenko, V. P., Lisitsyn, A. P., Polyakova, E. I., Dethleff, D., Serova, V. V., and Stein, R., 2002. Distribution and composition of sedimentary material in the snow cover of arctic drift ice (Fram Strait). *Doklady Earth Sciences*, **383**(3), 385–389.

Stein, R., 2008. *Arctic Ocean Sediments: Processes, Proxies and Paleoenvironment. Developments in Marine Geology*, 2nd edn. Amsterdam: Elsevier. 592 pp.

Svendsen, H., Beszczynska-Møller, A., Hagen, J. O., Lefauconnier, B., Tverberg, V., Gerland, S., Ørbæk, J. B., Bischof, K., Papucci, C., Zajaczkowski, M., Azzolini, R., Bruland, O., Wiencke, C., Winther, J.-G., and Dallmann, W., 2002. The physical environment of Kongsfjorden–Krossfjorden, an Arctic fjord system in Svalbard. *Polar Research*, **21**(1), 133–166.

Syvitski, J. P. M., Andrews, J. T., and Dowdeswell, J. A., 1996. Sediment deposition in an iceberg-dominated glacimarine environment, east Greenland: basin fill implications. *Global and Planetary Change*, **12**, 251–270.

Szczucinski, W., and Zajaczkowski, M., 2012. Factor controlling downward fluxes of particulate matter in glacier-contact and non-glacier contact settings in a subpolar fjord (Billefjorden, Svalbard). *International Association of Sedimentologists. Special Publication*, **44**, 369–386.

Trusel, L. D., Powell, R. D., Cumpston, R. M., and Brigham-Grette, J., 2010. Modern glacimarine processes and potential future behaviour of Kronebreen and Kongsvegen polythermal tidewater glaciers, Kongsfjorden, Svalbard. In Howe, J. A., Austin, W. E. N., Forwick, M., and Paetzel, M. (eds.), *Fjord Systems and Archives*. Geological Society, London, Special Publications, 344, pp. 89–102.

Warren, C. R., 1992. Iceberg calving and the glacioclimatic record. *Progress in Physical Geography*, **16**(3), 253–282.

Cross-references

Energy Resources
Ice-rafted Debris (IRD)

GLACIO(HYDRO)-ISOSTATIC ADJUSTMENT

Kurt Lambeck
Research School of Earth Sciences, Australian National University, Acton ACT, Australia

Synonyms

Glacial rebound; Isostasy

Definition

Isostasy refers to the response of the Earth to surface loading. Glacio-hydro-isostasy refers to the specific case of surface ice and water loads during glacial cycles. The most important observation of this process is the change in sea levels around the world.

Introduction

Isostatic adjustment of the Earth refers to the response of the Earth to changes in surface loading. The classic examples are the Airy or Pratt models (Watts, 2001) in which the surface load (of, for example, sediments) is supported locally either by a deflection of the crust (Airy) or by a change in crustal density (Pratt) beneath the load such that, at some deeper and constant depth in the mantle, pressures are constant. In these models, the crust (or lithosphere) does not support shear stresses. In models of regional isostatic adjustment (e.g., the Vening Meinesz model), the load is supported by the elastic strength of the crust or lithosphere and by the buoyancy forces acting at the base of the layer (Heiskanen and Vening-Meinesz, 1958). In these regional elastic models, small-amplitude, short-wavelength bulges form in the crustal deformation around the load, and in the geological literature, reference is sometimes made to a lithospheric forebulge. In both cases, the mantle is represented as a zero-viscosity fluid and the isostatic state is indicative of the Earth's deformational response only when the loading histories are longer than the relaxation time of the mantle. This relaxation time is characteristically of the order of 10^4 years, and the models are not appropriate for analyzing the response of the Earth-ocean system to changing glacial loads where both the mantle flow and gravity field changes are dominant factors in determining the shape of the solid Earth and ocean surface.

The Earth's response to changing ice and water loads

The glacio-hydro-isostatic adjustment refers to the Earth's response to the changing ice and water surface loads during glacial cycles, particularly to the last phase of deglaciation, that have characteristic load-cycle time scales of 10^3–10^5 years, of similar magnitude as the mantle relaxation times. Hence response models must consider the flow induced in the mantle by the surface loading cycle, and in consequence, the response continues for some thousands of years after the ice loads have stabilized.

When ice sheets grow, the crust and lithosphere beneath the ice subsides as mantle materials flow away from the stressed regions. At the same time, water is taken out of the oceans to feed the growing ice sheets, unloading the underlying mantle and resulting in ocean floor uplift relative to the center of mass of the Earth. The reverse occurs during the decay phase of the ice cycle. The glacio-hydro-isostasy is therefore a global phenomenon, and models of which will have to consider both the ice and water loads through time.

Several observational measures of the Earth's response exist: (i) the deformation of the land surface and this is measured by geodetic methods, most notably by GPS, (ii) changes in the planet's gravity field as both internal and surface mass redistribution occurs and this is measured with gravity meters as well as inferred from perturbations in Earth-satellite orbits, (iii) changes in the Earth's rotation measured by astronomical and satellite methods, and (iv) changes in sea-level as measured by tide gauges and as preserved in the geological record. The instrumental records extend back for decades to at most about 100 years and measure only the response long after the end of the last major deglaciation phase. They are, therefore, likely to also include the Earth's response to any recent changes in the residual ice sheets.

The geological evidence for sea-level change is particularly important since it extends back in time, primarily to the time of the Last Glacial Maximum (LGM) but with scattered evidence for earlier periods. The observation is of the change in sea-level relative to the land surface and is, therefore, a relative measurement and includes both the crustal deformation associated with the changes in surface loads and the changes in the shape of gravitational equipotential surfaces. A raised shoreline could mean either that there has been land uplift, that there has been a reduction in ocean volume, or that there has been a redistribution of water within the ocean basin. The glacio-hydro-isostatic effect is a combination of all three. Observations of sea-level therefore contain information on the Earth's rheology as well as on the ice history, and on any other processes, principally of tectonic and geodynamic origin.

Models for predicting the Earth's response to glacial cycles and the concomitant water loading have been well developed over recent decades with a focus on linear viscoelastic models for the mantle rheology and in particular on a Maxwell rheology with realistic radial elastic and density stratification (Peltier, 1974; Cathles, 1975). The rheological parameters are usually expressed as depth-dependent effective viscosities and as an effective elastic thickness of the lithosphere that provide a formulation that appear to work remarkably well in explaining much of the observational evidence even though the relations between these parameters and laboratory evidence for mantle behavior under stress remains unclear. A critical element in these models is the treatment of the water loading on a deforming Earth such as to be gravitationally consistent and conserve ice-water mass in which the water load itself is a function of the deformation of the Earth and of the redistribution of the water (Farrell and Clark, 1976; Mitrovica and Milne, 2003).

The pattern of global sea-level change

The pattern of sea-level induced by the growth and decay of ice sheets is conveniently divided into three zones: areas within or near the margins of the former ice sheets (near-field), areas far from these ice sheets (far-field), and an intermediate-field. During deglaciation, the dominant contribution to sea-level change in the near-field is the crustal rebound as mantle materials flow back beneath once ice-loaded continents. This is the glacio-isostatic component. For the largest ice sheets, perhaps up to 4,000 m thick and 1,000 km or more radius, the total rebound since the LGM approaches the local isostatic limit of 800–1,000 m but only the rebound that occurred after the region became ice-free is preserved in the record. Rates of crustal uplift of up to 1 cm/year still occur in areas such as the Hudson Bay (Canada) or the Gulf of Bothnia (Fennoscandia). The other not insignificant contributions are from changes in the gravitational attraction in the Earth-ocean-ice system, the change in the ocean volume and the deformation of the Earth caused by the changing water load. The net result is that where there once were large ice sheets, sea-level has been falling and continues to fall long after the last of the ice has vanished. Toward the edge of the ice sheet, the crustal rebound component is reduced, and the other contributions begin to dominate the signal such that the relative sea-level signal may oscillate in time between rising and falling.

In far-field regions, it is the change in the ocean volume and the deformation of the Earth caused by the water loads that are the dominant contributors. With increasing water, the ocean floor subsides, approaching the local isostatic limit in mid-oceans of about 25 % of the change in the ocean depth. Along continental margins, because of the elastic properties of the lithosphere and the viscous flow in the mantle, the subsidence will be about half this. This is the hydro-isostatic component. Since the LGM, $\sim 50 \times 10^6$ km^3 of ice has been added into the oceans raising globally averaged sea-level by ~135 m but by less along the continental far-field margins because of the isostatic effects. At the end of glaciation, the ocean volume remains constant but the isostatic effects continue and sea levels fall, producing small-amplitude (up to ~3 m) highstands peaking at around 6,000 years ago that will be strongly dependent on nearby coastline geometry.

In the intermediate-field, the deformational, gravitational, and ocean volume contributions are often of similar magnitudes but opposing signs, such that the sea-level signal can be spatially complex. But the dominant feature of this response is the combination of the subsidence of a broad zone of crustal uplift that developed around the ice sheet during the loading phase and the change in the gravitational attraction between ice and ocean. Together, these factors result in sea levels outside of the ice margins rising throughout the post-glacial phase at rates that are a function of the ice sheet dimensions and of the distance from the ice margin. This pattern will extend out to several thousand kilometers beyond the center of the largest ice sheets.

Viewed globally, the glacio-hydro-isostatic response can therefore be expected to present a complex spatial and temporal variability, and this is indeed observed (Figure 1). For Hudson Bay and the Gulf of Bothnia, near the centers of the North American and Fennoscandian ice sheets respectively, sea-level has been falling over past millennia, and observations of the timing and rates of this fall provide constraints on both the mantle response function and local ice thickness. In the northwestern Norway example of Andøya in Figure 1, sea levels since the respective areas became ice-free have fallen and risen in time due to competing contributions from the crustal rebound and ocean volume increase, and observations from these locations provide insight into the local details of the ice sheet. The Andøya record, for example,

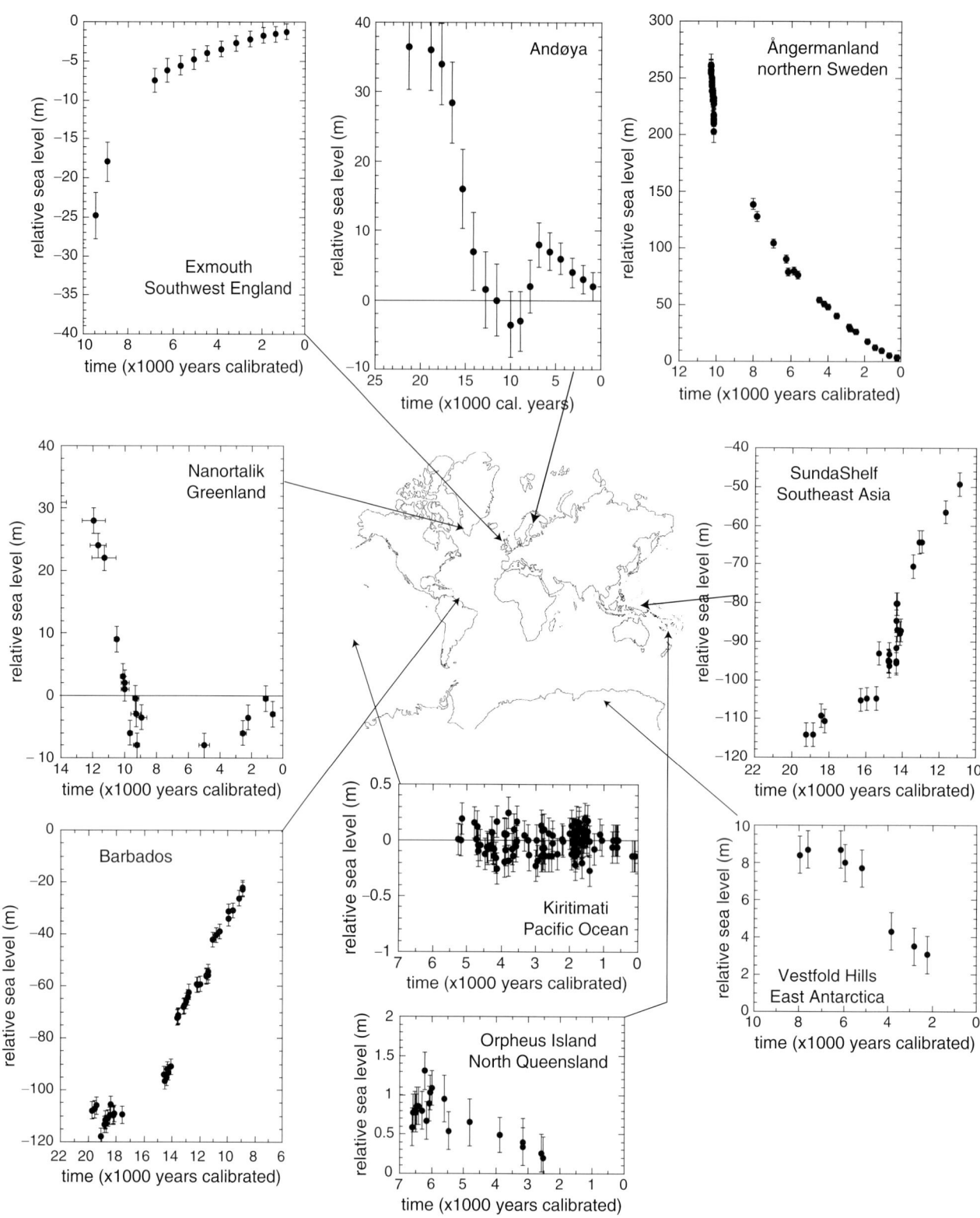

Glacio(hydro)-isostatic Adjustment, Figure 1 (Continued)

indicates that thick ice must have extended out to the edge of the continental shelf before the formation of the oldest sea-level signals at ~20,000 years ago. The record from the Vestfold Hills, East Antarctica, is of shorter duration than that of Andøya, but the fall in level during the past 6,000 years is similar and indicative of this area having been extensively glaciated. Further outside the ice margins as in southern England, sea-level has continued to rise up to the present as a result of the subsidence of the crust beyond the ice margins, and the spatial variability of this rise provides information on the mantle response function. At the far-field continental margin site of Orpheus Island, Australia, sea-level has been falling for the past 6,000 or so years, reflecting the Earth's response to the water loading, and the spatial variability of this highstand provides a measure of the Earth's response as well as any changes in ocean volume during this period. Far-field observations from the time of the LGM, such as the Bonaparte Gulf (Australia) or Sunda Shelf (Southeast Asia), provide a first approximation estimate of the total change in ice volume that has occurred since the LGM, and records from the late glacial period, such as from Barbados, constrain in a first approximation the timing and rate of melting of the global ice sheets. Second approximation solutions require that these observations be corrected for the isostatic (and, where relevant, the tectonic) contributions to the local sea-level change.

Inversion of sea-level data for geophysical and glaciological parameters

From this brief global survey, it should be evident that geophysical inversion of global data sets should be able to provide constraints on both the Earth's rheology and on ice sheet histories. There are often a number of caveats to such models concerning the distribution of the sea-level data, the parameterization of the rheology function, assumptions about the role of tectonics, or the nature of independent constraints on the location of former ice margins and on ice thickness through time.

The resolution for the radial dependence of mantle viscosity is limited in these inversions (Paulson et al., 2007) in part because of the limiting nature of the distribution of field data and in part because of uncertainties in the knowledge of the ice sheets, but solutions do point to an increase in mantle viscosity from the average upper mantle value to the average lower mantle value by a factor of 50–100 or more, consistent with independent – geodynamically reasoned – estimates (e.g., Mitrovica and Forte, 2004; Čížková et al., 2012).

Limitations in the knowledge of the past ice sheets are particularly severe in their impact on the inferred rheologies. The usual assumption is that the ice margin history is known, and the ice thickness through time is treated as a partially known function. Of the major past ice sheets, the Fennoscandian ice models is best constrained from rebound analyses because of an extensive and reliable database of both sea-level change within its former maximum margin and of the ice margin history. The North American ice sheet is less well constrained, and the Antarctic ice sheet remains largely unconstrained. Recent ice models obtained from sea-level data inversions include Peltier (2004) for a global model, Lambeck et al. (2010) for northern Europe, Fleming and Lambeck (2004) for Greenland, and Whitehouse et al. (2012) for Antarctica.

Generally, the models provide a good description of the observational evidence (Figure 2) and a basis for predictive models for the evolution of coastlines and water depths from the LGM to the present, provided that the model parameters are from consistent Earth-ice solutions because of the trade-offs that can occur between them through incomplete a priori information of the ice sheets. Figure 3 provides one such example for the Mediterranean where the contours provide lines of constant sea-level change from the LGM to the present. Of note is that these contours follow closely the coastline, indicative of the importance of the water loading within the Mediterranean basin. The influence of the local Alpine ice sheet is restricted to the northern Adriatic and Gulf of Genoa. Less visibly evident is the longer wavelength influence of the North American and Eurasian ice sheets consisting of a north–south gradient across the region. Results such as these are model-parameter dependent and can be, and have been, tested against local independent sea-level data, but they provide useful starting points for discussing

Glacio(hydro)-isostatic Adjustment, Figure 1 Representative sea-level curves from different localities around the world from the time of the Last Glacial Maximum to the present. Note the different time and sea-level change scales. The Ångerman result, from central Sweden, is representative of a location near the center of a former ice sheet, with crustal rebound exceeding the increase in ocean volume throughout the melting and post-glacial phase. The Andøya result, from the northwest coast of Norway, and the Nanortalik result, from southern Greenland, are representative of sites near the edge of an ice sheet. Initially, the rebound dominates, followed by a period when the increase in ocean volume dominates and a later phase when the rebound is again dominant. In the case of the Greenland signal, a major contributor to the sea-level change is from the North American deglaciation. The East Antarctic site of Vestfold Hills is similar to that of Andøya indicating that substantial changes in ice volume occurred here before the area became ice-free. The southwest England result is representative of locations outside of the ice sheet where the crustal rebound after the onset of melting is one of subsidence. The signal is initially dominated by the increase in ocean volume and later by the crustal response. The Sunda Shelf and Barbados locations are far-field sites, and the sea-level signal is dominated by the change in ocean volume, modulated by glacio-hydro-isostatic contributions. The North Queensland site is representative of far-field continental margins where a small-amplitude highstand forms in Late Holocene time due to the water-loading component. At the mid-ocean island site of Kiritimati, the change in sea-level is virtually constant for 5,000 years due to a cancellation out of the various isostatic contributions to sea-level change.

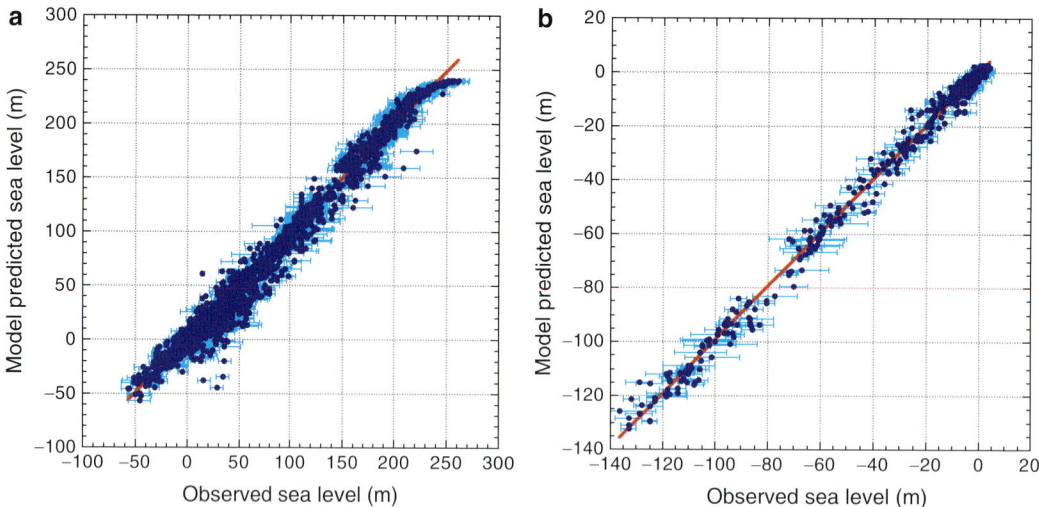

Glacio(hydro)-isostatic Adjustment, Figure 2 Two examples of comparison of sea-level observations from the past 20,000 years with predictions based on glacio-hydro-isostatic models. The red lines are linear regressions, in both cases with gradients that depart from unity by less than 1 % and with correlation coefficients greater than 99 %. (**a**) For a near-field solution of Fennoscandia, with nearly 3,300 data points. The anomalous points below the linear regression line identify sites where the ice model is either deficient in ice or where the retreat occurred earlier than assumed in the ice model. These sites are all from the Norway coast and indicate that thick ice covered the continental shelf during the LGM. (**b**) For a global far-field solution consisting of some 990 data points. Systematic departures from the regression line, such as occurring at observed depths between −70 and −80 m and corresponding to ages around 14,000 years ago, point to a need to modify the global ice volume function. By combining far- and near-field analyses of sea-level data, it becomes possible to improve the ice models for the individual ice sheets as well as for the total changes in ice volume from the time of maximum glaciation to the present.

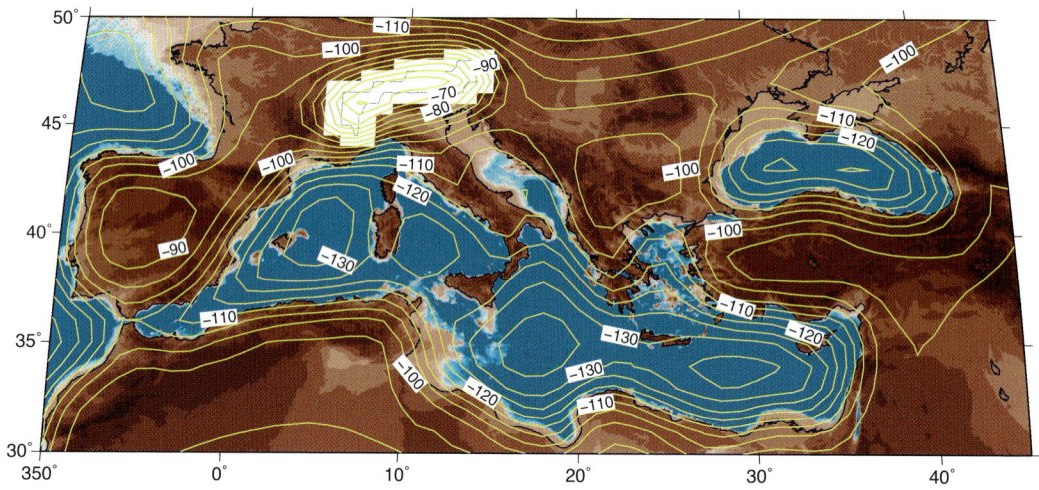

Glacio(hydro)-isostatic Adjustment, Figure 3 Reconstruction of the difference in sea-level between the Last Glacial Maximum and today for the Mediterranean basin, based on the glacio-hydro-isostatic model whose ice sheet and Earth-rheology parameters have been constrained by sea-level data from both far- and near-field sites. The contours (at 5 m intervals) are of equal change in sea-level between the LGM and present. Such models can be compared against archaeological and geological sea-level indicators from across the region to test the models themselves or to infer tectonic contributions to sea-level change.

geological and archaeological data from coastal areas or as a reference surface for evaluating tectonic rates of vertical motion. At the LGM and early late glacial period, much of the northern Adriatic is predicted to be subaerial, as are other shelf areas, most notably the Gulf of Gabes. Land bridges are predicted between mainland Italy and Malta and the crossing between Corsica-Sardinia, and continental Europe is reduced to little more than 10 km.

Glacial rebound before the last glacial maximum

While most discussions on glacio-hydro-isostatic adjustment have focused on the last phase of deglaciation, it is important to note that this process is important also on longer time scales when it comes to interpreting interglacial sea levels in terms of differences in ice volumes relative today (Lambeck et al. 2012; Raymo et al. 2011. First, the response to the glacial history immediately before the interglacial will determine the nature of the interglacial sea-level function in the same way as the LGM determined the character of Holocene sea levels. Second, the observation of the past interglacial sea levels are with respect to present sea-level which is still evolving in response to the last glacial loading-unloading phase. This latter part can be predicted with some confidence from analyses of the recent interval, but the accuracy of the first part will be very much limited by the knowledge of ice extent during the earlier glacial maxima. Hence, estimates of the differences in ice volumes between present and earlier interglacials remain uncertain.

Summary

Glacio-hydro isostasy is an important physical process describing the Earth-ocean response to changes in ice sheets during glacial cycles. It is a global phenomenon and operates long after the surface loads have stabilized. It manifests itself in a wide range of geophysical and geological observations of which the most important is sea-level change. Observations of the isostatic response provide an important input for understanding the evolution of past ice sheets and ocean volumes.

Bibliography

Cathles, L. A., 1975. *The Viscosity of the Earth's Mantle*. Princeton: Princeton University Press, p. 403.
Čížková, H., van den Berg, A. P., Spakman, W., and Matyska, C., 2012. The viscosity of the Earth's lower mantle inferred from sinking speed of subducted lithosphere. *Physics of the Earth and Planetary Interiors*, **200–201**, 56–62.
Farrell, W. E., and Clark, J. A., 1976. On post-glacial sea level. *Geophysical Journal of the Royal Astronomical Society*, **46**, 647–667.
Fleming, K., and Lambeck, K., 2004. Constraints on the Greenland ice sheet since the last glacial maximum from observations of sea-level change and glacial-rebound modelling. *Quaternary Science Reviews*, **23**, 1053–1077.
Heiskanen, H., and Vening-Meinesz, F. A., 1958. *The Earth and Its Gravity Field*. New York: McGraw-Hill, p. 470.
Lambeck, K., and Chappell, J., 2001. Sea level change through the last glacial cycle. *Science*, **292**, 679–686.
Lambeck, K., Purcell, A., Zhao, J., and Svensson, N.-O., 2010. The Scandinavian ice sheet: from MIS4 to the end of the Last Glacial Maximum. *Boreas*, **39**, 410–435.
Lambeck, K., Purcell, A., and Dutton, A., 2012. The anatomy of interglacial sea levels. The relationship between sea levels and ice volumes during the Last Interglacial. *Earth and Planetary Science Letters*, **315–316**, 4–11.
Mitrovica, J. X., and Forte, A. M., 2004. A new inference of mantle viscosity based upon joint inversion of convection and glacial isostatic adjustment. *Earth and Planetary Science Letters*, **225**, 177–189.
Mitrovica, J. X., and Milne, G. A., 2003. On post-glacial sea level: I. General theory. *Geophysical Journal International*, **154**, 253–267.
Paulson, A., Zhong, S., and Wahr, J., 2007. Limitations on the inversion for mantle viscosity from postglacial rebound. *Geophysical Journal International*, **168**, 1195–1209.
Peltier, W. R., 1974. The impulse response of a Maxwell Earth. *Reviews of Geophysics*, **12**, 649–669.
Peltier, W. R., 2004. Global glacial isostasy and the surface of the Ice-Age Earth: the ICE-5G (VM2) model and GRACE. *Annual Reviews in Earth and Planetary Science*, **32**, 111–149.
Raymo, M., Mitrovica, J. X., O'Leary, M. J., DeConto, R. M., and Hearty, P. J., 2011. Departures from eustasy in Pliocene sea-level records. *Nature Geoscience*, **4**, 328–332.
Watts, A. B., 2001. *Isostasy and Flexure of the Lithosphere*. Cambridge: Cambridge University Press.
Whitehouse, P. L., Bentley, M. J., and Le Brocq, A. M., 2012. A deglacial model for Antarctica. *Quaternary Science Reviews*, **32**, 1–24.

Cross-references

Relative Sea-level (RSL) Cycle
Sea-Level

GRAVITY FIELD

Udo Barckhausen and Ingo Heyde
Federal Institute for Geosciences and Natural Resources, Hannover, Germany

Definition

The gravity field in marine geosciences is the gravitational force that the Earth's mass exerts on objects on or near its surface.

Introduction

The basic principle for the Earth's gravity field in classical mechanics is Newton's law of universal gravitation. It states that every massive body exerts an attractive force on all other massive objects. The force is proportional to the product of the bodies' masses and inversely proportional to the square of the distance between them. For extended bodies like the Earth, their mass can be assumed to be concentrated as a point mass in their center (strictly this is only true for spheres with a symmetrical mass distribution). In geophysics, Newton's law is used in the form

$$\underline{K} = G \frac{mM}{r^2} \underline{r^0}$$

where $G = 6.672 * 10^{-11} \mathrm{Nm^2 kg^{-2}}$ is the gravitational constant, m the mass of a body on or near Earth's surface, M Earth's mass, r the distance between m and the Earth's center, and $\underline{r^0}$ the unit vector giving the direction from m to the Earth's center.

Since the Earth is rotating, centrifugal acceleration has to be taken into account when calculating resulting gravity:

$$z = \Omega^2 R \cos \varphi$$

z points vertically away from Earth's rotational axis with Ω being the angular velocity, φ the geographic latitude, and R Earth's radius.

The resulting acceleration of gravity is

$$g = (982{,}037 - 3{,}389 \cos^2 \varphi)\,\text{mGal} \quad \text{with mGal} = 10^{-5}\,\text{m/s}^2$$

for the simplifying assumption that Earth has the shape of a sphere. As a result of the latitude dependence, g is smaller at the equator than at the poles.

Another effect of the centrifugal forces is that Earth's body is not a perfect sphere, but has a bulge at the equator and in first approximation has the shape of a spheroid. The commonly used model for the reference spheroid is the WGS84 (Stacey, 1992) which uses the following values:

Equatorial radius 6,378,136 m
Polar radius 6,356,751 m
Flattening 1/298.257

The International Union for Geodesy and Geophysics (IUGG) has set the following formula for the gravity on the spheroid (Stacey, 1992):

$$\gamma_0 = 978{,}032.7\bigl(1 + 0.0053024\,\sin^2\varphi \\ - 0.0000058\,\sin^2 2\varphi\bigr)\,\text{mGal}$$

where φ is the geographical latitude.

An even better approximation of the Earth is the so-called geoid which reflects the deviations from the rotationally symmetric shape of the reference spheroid. The deviations are on the order of a few tens of meters and only in the area south of India reach a value of more than 100 m. The geoid can be described with spherical functions, which has recently been done to degree and order 2,159 (Pavlis et al., 2012). Over the oceans, the surface of the geoid is equivalent to the mean sea surface.

Measurements

The classical way of measuring gravity on the sea surface is by using straight line gravity meters which are installed on board of ships to measure the vertical component of the gravity field. A marine gravity meter typically consists of a gyrostabilized platform and a gravity sensor. Different types of sensors are in use, most of which have a mass that is suspended on a metal spring to measure local differences in gravity. Thus, they measure relative differences which have to be tied to a point of known absolute gravity in a harbor. Similar instruments are being used on aircrafts as well.

As an example, the Sea-Air-Gravimeter System KSS32M (BGGS GmbH, Meersburg, Germany) consists of a gravity sensor, a gyrostabilized platform containing the power supply and all electronics for gravimeter and gyro, and a standard laptop for controlling and data acquisition. The gravity sensor GSS30 is based on a non-astatized axially symmetric spring-mass system (Figure 1). The mass consists of a vertical tube which is held by spring-suspended wires such that a friction-free movement is limited to the vertical direction. The measuring spring inside the tube generates an equilibrium with respect to gravity. Displacements are sensed by a capacitive pickoff and after processing voltage changes serve to adjust the mass to the zero position (Torge, 1989).

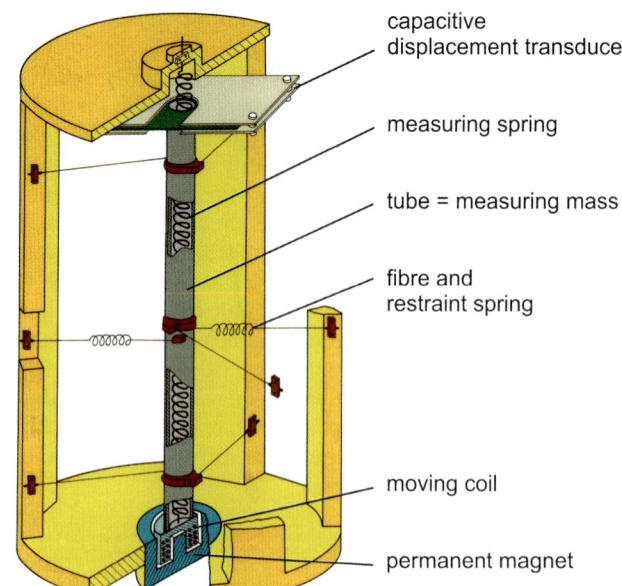

Gravity Field, Figure 1 Gravity sensor GSS30 principle, BGGS GmbH, Meersburg.

Due to the installation of the sea-air-gravity meter on a platform moving relative to the Earth, a centrifugal acceleration (the Earth being assumed at rest) and a Coriolis acceleration (rotating Earth) occur. The vertical components of these inertial accelerations affect the measured gravity value (Eötvös effect, von Eötvös, 1919). At normal ship speeds, the Eötvös correction, E_c, can be approximated by

$$E_c = 7.503\,v\,\sin\alpha\,\cos\varphi + 0.004154\,v^2\,\text{mGal}$$

where v is the platform velocity in knots, α is the course, and φ is the latitude.

Correcting the measured gravity for the Eötvös effect and the normal gravity of the reference spheroid, free-air gravity anomalies are obtained. They reflect the gravitational attraction due to submarine topography and geological structure.

A dramatically improved view of the marine gravity field came with Earth-orbiting satellites which carried radar altimeters able to measure the height of the sea surface with great accuracy (Tapley et al., 1982). Since the gradient of the sea surface at short wavelengths depicts the gravitational effect of the density boundary between the water and the rocks of sea bottom, it was possible to calculate a predicted topography of the seafloor which revealed the large-scale tectonic fabric of the ocean floor

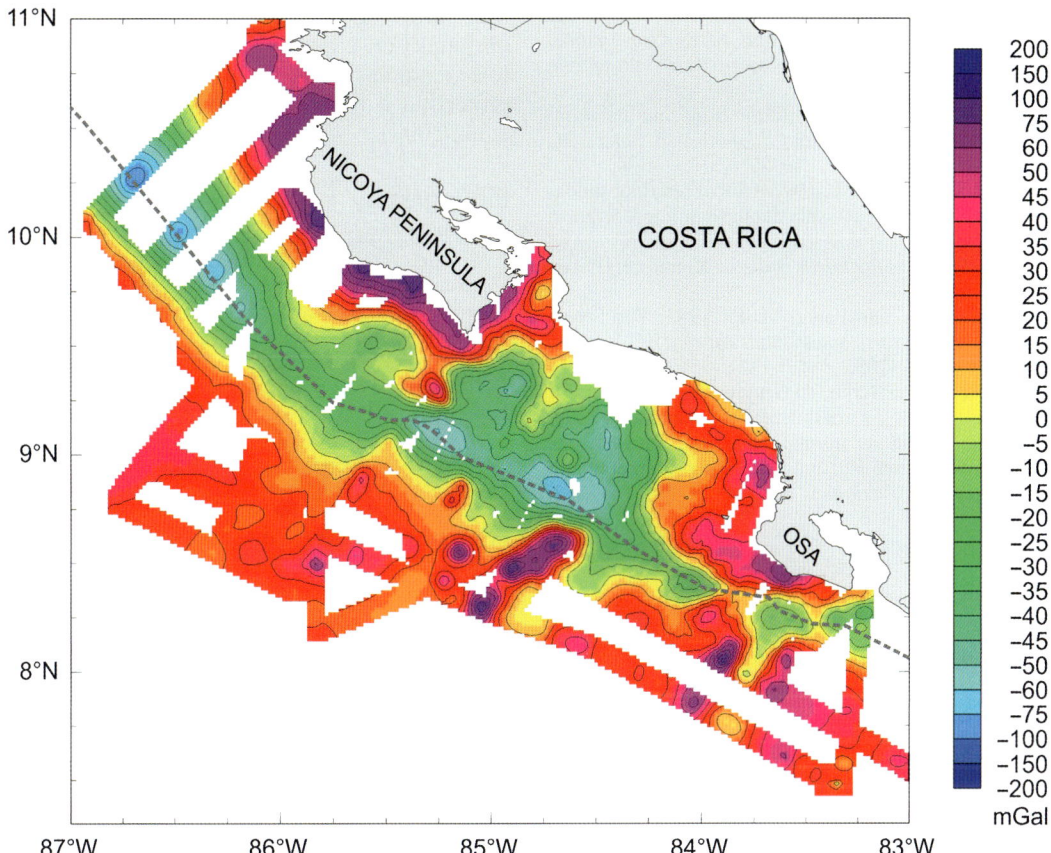

Gravity Field, Figure 2 Gravity map of an area offshore the Pacific coast of Costa Rica. The map was compiled from data acquired with a sea-air-gravity meter on a ship. The *dashed line* indicates the position of the deep-sea trench where the Pacific plate subducts beneath Central America. Seaward of the trench, local high gravity values are related to seamounts on the Pacific plate. Landward of the trench, the complex geology of the subduction zone is reflected by the large number of local gravity highs and lows.

(Smith and Sandwell, 1997). Today the spatial resolution of the satellite altimetry-derived gravity maps of the world's oceans as well as the constructed bathymetry is better than 20 km. Ship-based measurements reach significantly higher resolutions; however, they are only available along ship tracks and small areas which have been surveyed systematically but cover only a fraction of the oceans. Since to date all satellite missions measuring the gravity field have been flown at an orbital inclination, gaps still exist in high polar regions where little or no gravity data are available.

Marine gravity and tectonics

For marine geosciences, the gravity field provides a wealth of information related to tectonics regionally and globally (Figure 2). The large structures of the mid-ocean ridges became visible in the gravity maps for the first time thus confirming the concept of plate tectonics (Haxby et al., 1983; Smith, 1998). Thousands of seamounts could be detected by their local gravity anomalies and reveal plate boundaries, magmatic processes associated to hot spots, plate motions, and others (Craig and Sandwell, 1988). Where both gravity and independently measured bathymetry data exist, gravity anomalies not directly related to the seafloor topography indicate density contrasts within the Earth's crust. These can be modeled in order to get insight into geologic structures at plate boundaries and other places where the geology is more complex. Areas which are out of isostatic equilibrium due to uplift or depression also become visible in the gravity field. Typically a combination of gravity, magnetic, seismic, and bathymetric data is used for integrated studies in marine geosciences. In the case of the gravity field, satellite-derived data with full spatial coverage of the longer wavelengths and high-resolution shipboard data supplement each other.

Bibliography

Craig, C. H., and Sandwell, D. T., 1988. Global distribution of seamounts from Seasat profiles. *Journal of Geophysical Research*, **93**, 10408–10420.

Haxby, W. F., Karner, G. D., LaBrecque, J. L., and Weissel, J. K., 1983. Digital images of combined oceanic and continental data sets and their use in tectonic studies. *EOS Transactions American Geophysical Union*, **64**(52), 995–1004.

Pavlis, N. K., Holmes, S. A., Kenyon, S. C., and Factor, J. K., 2012. The development and evaluation of the Earth Gravitational Model 2008 (EGM2008). *Journal of Geophysical Research*, **117**, B04406.

Smith, W. H. F., 1998. Seafloor tectonic fabric from satellite altimetry. *Annual Reviews of Earth and Planetary Science*, **26**, 697–737.

Smith, W. H. F., and Sandwell, D. T., 1997. Global sea floor topography from satellite altimetry and ship depth soundings. *Science*, **277**, 1956–1962.

Stacey, F. D., 1992. *Physics of the Earth*, 3rd edn. Kenmore: Brookfield Press.

Tapley, B. D., Born, G. H., and Parke, M. E., 1982. The Seasat altimeter data and its accuracy assessment. *Journal of Geophysical Research*, **87**, 3179–3188.

Torge, W., 1989. *Gravimetry*. Berlin: de Gruyter.

von Eötvös, R., 1919. Experimenteller Nachweis der Schwereänderung, die ein auf normal geformter Erdoberfläche in östlicher und westlicher Richtung bewegter Körper durch diese Bewegung erleidet. *Annalen der Physik*, **59**, 743–752, Ser. 4.

Cross-references

Plate Tectonics
Technology in Marine Geosciences

GROUNDING LINE

Antoon Kuijpers and Tove Nielsen
Geological Survey of Denmark and Greenland (GEUS), Copenhagen, Denmark

Definition

The grounding line, or more properly the grounding zone, of glaciers and ice sheets is where their marine margins cease to be in contact with the seafloor (Dowdeswell and Fugelli, 2012).

A variety of features indicative of ice sheet and glacier dynamics marks the grounding zone. These grounding zone features indicate episodic rather than catastrophic ice sheet retreat (Dowdeswell and Fugelli, 2012), but can still yield important information on ice sheet and glacier response to changes in climate and ocean temperature. Observations of the Antarctic ice sheet show grounding line retreat and ice shelf collapse at the Antarctic Peninsula associated with warming in the Amundsen Sea (Payne et al., 2004). Enhanced submarine melting at the grounding line has also been proposed as trigger mechanism for recent acceleration of large, marine-based glaciers in Greenland (Rignot et al., 2010). Modeling results confirm how ocean warming and enhanced subglacial drainage lead to increases in melting near the grounding line (Jenkins, 2011). Ice sheet and glacier retreat leave grounding zone depositional systems on the seabed which include ice-contact deltas and moraines and fans associated with subglacial and basal meltwater discharge (Powell, 1990). At quasi-stable grounding lines, the fans may grow to sea-level and form ice-contact deltas (Seramur et al., 1997). Formation of the grounding line features helps stabilization of the ice sheet in protecting it from retreat driven by sea-level rise (Alley et al., 2007).

Bibliography

Alley, R. B., Anandakrishnan, S., Dupont, T. K., Parizek, B. R., and Pollard, D., 2007. Effect of sedimentation on ice sheet grounding-line stability. *Science*, **315**, 1838–1841.

Dowdeswell, J. A., Fugelli, E. M. G., 2012. The seismic architecture and geometry of grounding-zone wedges formed at the marine margins of past ice sheets. *Geological Society of America Bulletin*, doi:10.1130/B30628.1.

Jenkins, A., 2011. Convection-driven melting near the grounding lines of ice shelves and tidewater glaciers. *Journal of Physical Oceanography*, **41**, 2279–2294.

Payne, A. J., Vieli, A., Sheperd, A. P., Wingham, D. J., and Rignot, E. J., 2004. Recent dramatic thinning of the largest West Antarctic ice stream triggered by oceans. *Geophysical Research Letters*, **31**, L23401, doi:10.1029/2004GL021284.

Powell, R. D., 1990. Glacimarine processes at grounding-line fans and their growth to ice-contact deltas. In *Special Publications*. London: Geological Society, Vol. 53, Glaciomarine Environments: Processes and Sediments (J.A. Dowdeswell, J.D. Scourse, Eds.), pp. 53–73.

Rignot, E. J., Xu, Y., Koppes, M. N., Menemenlis, D., Schodlok, M., Spreen, G., 2010. Submarine melting at the grounding line of Greenland's tidewater glaciers: observations and implications. In Abstracts AGU Fall Meeting 2010. American Geophysical Union, Abstract #C12B-07R.

Seramur, K. C., Powell, R. D., and Carlson, P. R., 1997. Evaluation of conditions along the grounding line of temperate marine glaciers: an example from Muir Inlet, Glacier Bay, Alaska. *Marine Geology*, **140**, 307–328.

Cross-references

Currents
Eustasy
Fjords
Ice-rafted Debris (IRD)
Moraines
Paleoceanography
Regional Marine Geology
Sea-Level
Sediment Dynamics
Sediment Transport Models
Shelf

GUYOT, ATOLL

Eberhard Gischler
Institute of Geosciences, Johann Wolfgang Goethe University, Frankfurt am Main, Germany

Definition

Atolls are circular- to irregular-shaped, isolated oceanic reef structures at or near sea-level that enclose a lagoon

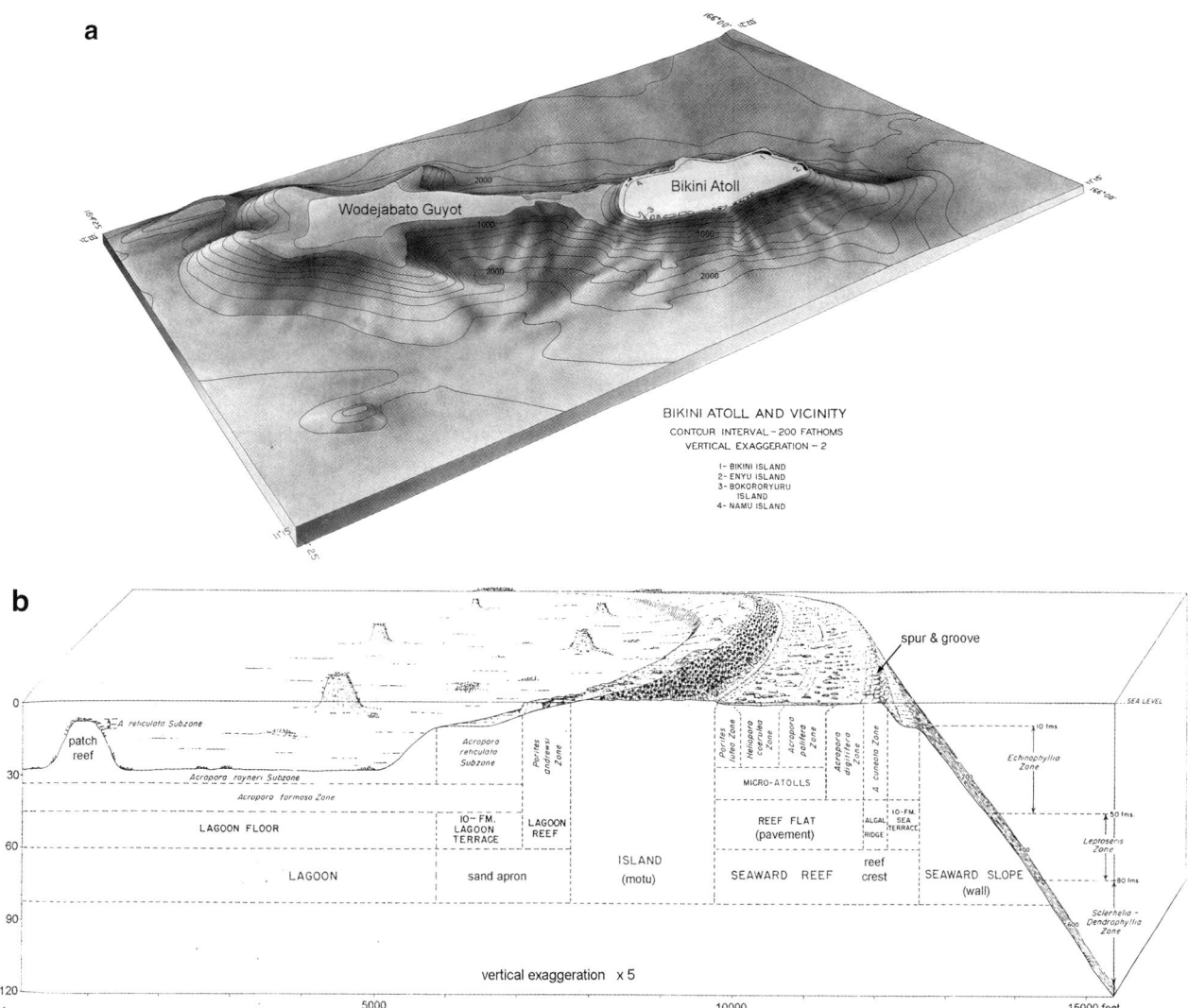

Guyot, Atoll, Figure 1 (**a**) Bikini Atoll and Wodejabato Guyot (*top* at 1.3 km water depth) in the western Pacific (Modified from Ladd et al., 1950). (**b**) Geomorphology and zonation at Bikini Atoll (modified from Wells, 1954). 1,000 ft = 300 m; 1 fathom = 1.85 m.

usually tens of meters deep. The term atoll derives from the Divehi (Maldivian) word atolu. Atoll reefs occur in all oceans; they are most common in the Indo-Pacific realm.

Guyots, a term coined by Hess (1946), are flat-topped seamounts located in several hundred to several thousand meters water depth. They are most common in the northwestern Pacific Ocean. Guyots usually are former volcanoes that are in many cases capped by drowned atoll reefs and pelagic sediments.

Occurrence, geomorphology, and sediments
Atolls

Atolls are circular to irregular, narrow, shallow-water reef structures that enclose a deeper lagoon (Figure 1). Therefore, their shape has been likened to that of a bucket (Purdy and Gischler, 2005; and references therein). There are some 400 atolls worldwide (Stoddart, 1965). The majority occurs in the open Pacific and Indian oceans where they typically have a volcanic basement. There are also some atolls that rest on continental crust, for instance, on the topographic and structural highs of fault blocks at passive margins, e.g., in offshore Belize, in the southern Great Barrier Reef, and in the Red Sea. Tiladummati-Miladummadulu in the Maldives (Indian Ocean) covers 3,800 km^2 and is the largest atoll on Earth. Kwajalein (Marshall Islands) covers 2,200 km^2 and is the largest Pacific atoll. Great Bahama Bank (ca. 78,000 km^2) and Great Chagos Bank (12,650 km^2) are very large isolated carbonate platforms in the Atlantic and Indian Oceans, respectively, which are not considered atolls. They lack continuous surface-breaking reefs at their margins and a deeper interior lagoon in the case of the Bahamas.

The shallow and partly emergent margin of atolls includes reef crest, reef flat (pavement), sand apron, and marginal islands (Figure 1b). It is relatively narrow (up to several hundred meters) as compared to the much larger lagoon area. The margin is characterized by coral reefs, which exhibit geomorphological and ecological zonations (Ladd et al., 1950; Wiens, 1962, 234–243). On the ocean side of the reef crest, which is often composed of an algal ridge in the Pacific, there is a fore-reef slope that in many cases dips about 30°. The shallow fore-reef slope is usually characterized by the spur and groove system, a comb-shaped structure of ridges perpendicular to the reef crest that may have either constructional or erosional origins (Gischler, 2010). The fore-reef slope transitions to a break in slope and a steep wall several hundreds of meters high (James and Ginsburg, 1979). On the lagoon side of the reef crest, a cemented pavement, occasionally with islands, and a sand apron occur. It is not uncommon that a distinct sandy slope transitions to the lagoon proper. Sand aprons are significant in that they contribute to the lateral infilling of atoll lagoons by lagoonward progradation (Purdy and Gischler, 2005). Atoll margins are usually not continuous but feature several passes that connect the open ocean with the interior lagoon. In many Pacific atolls, the passes are located on the leeward sides (Wiens, 1962).

Atoll lagoons get up to 100 m deep; the average depth amounts to some 36 m based on an analysis of the geomorphology of 295 atolls (Purdy and Winterer, 2001). The same authors showed that atoll lagoon depths covary significantly with precipitation rates, thereby supporting the contention that inherited karst relief from solution during glacial sea-level lowstands is crucial in atoll development. As compared to the atoll margin that exhibits accretion rates of several meters per kyr, background sedimentation within Quaternary atoll lagoons averages decimeters per kyr (Purdy and Winterer, 2001; Purdy and Gischler, 2005). Atoll lagoons usually accommodate numerous coral patch reefs. In the Maldives, lagoonal reefs as well as some marginal reefs may form characteristic ringlike structures called faroes. Their formation is largely attributed to the changing wind and current regime of the Indian monsoon (Purdy and Bertram, 1993).

Sand and rubble islands termed cays are common on both marginal and lagoonal atoll reefs. Beachrock is a very common characteristic of sandy beaches on reef islets (Gischler, 2007). Elongated cays on Pacific reef margins are called motu, and the channels between them hoa. Motus are in many cases relatively stable due to underpinning by mid- to late Holocene (ca. 4–2 kyrs BP) reef flat deposits that formed during higher-than-present sea-level (Dickinson, 2009). Human colonization in the Pacific only started after high tide levels had fallen below the upper level of these older Holocene deposits, some 2,000 years ago (Dickinson, 2009).

Modern atoll sediments are usually coralgal grainstones on marginal and lagoonal reefs. Atoll lagoon sediments often include mollusk-rich and foraminifer-rich wackestones and packstones as well as nonskeletal facies (Gischler and Lomando, 1999; Gischler, 2006; and references therein). Atoll lagoon deposits have also been successfully used as archives of event sedimentation (Klostermann et al. 2014).

Guyots

Guyots are seamounts with more or less flat tops that in many cases lie in about 1–2 km water depth (Figure 1). The tops typically are 3.0–4.5 km above the surrounding seafloor and may exhibit reliefs of tens of meters. The large majority can be found in the northwestern Pacific at 8–27°N and 146–165°E (Hess, 1946; Hamilton, 1956; Menard, 1984; Wessel, 1997; Flood, 1999). There are some 50,000 seamounts, many of them guyots, in the Pacific Ocean, about 50 % of them not charted and <1 % sampled for dating (Wessel, 1997). Some guyots are also known from other oceans (e.g., Pratt, 1963, 1967). Their shape is circular to oval. Guyot size varies and diameters may reach >200 km (Davies et al., 1972). Slopes of guyots are concave and reach 20° inclination toward the top. Guyots have been repeatedly investigated by dredging, seismics, and drilling. ODP cruises 143 and 144 were especially devoted to the Pacific guyots (Winterer et al., 1995; Haggerty and Premoli-Silva, 1995).

Ideal successions of guyot tops include the volcanic (basaltic) edifice, a shallow-water atoll reef section with elevated rims, and a package of pelagic ooze, usually capped by a manganese-iron or phosphorite crusts (Schlanger et al., 1987; Grötsch and Flügel, 1992; Jenkyns and Wilson, 1999; Flood, 1999, 2001) (Figure 2). However, there are also a number of guyots that are purely volcanic, lack a reefal succession, or feature mixed volcanic and limestone breccias (Jenkyns and Wilson, 1999).

The shallow-water reef deposits often are Cretaceous to the Eocene in age. They may get as thick as 1.6 km as on Resolution Guyot and contain clear indicators of a reefal origin such as corals, rudist mollusks, calcareous sponges (stromatoporoids), calcareous algae, larger foraminifera, and nonskeletal grains including ooids (Grötsch and Flügel, 1992; Jenkyns and Wilson, 1999; Flood, 1999, 2001). The majority of shallow-water reefs on guyots drowned in the Cretaceous (Matthews et al., 1974; Wilson et al., 1998; Jenkyns and Wilson, 1999). The age range of pelagic oozes comprises the Cretaceous to the Quaternary. The oozes get as thick as 150 m as on top of Allison Guyot. They are characterized in many cases by stratigraphic gaps and the so-called ghost faunas, i.e., mixtures of reworked foraminifera that comprise many biostratigraphical zones. The formation of ferromanganese crusts is likely the result of microbial activity. Additional processes include phosphatization and aragonite dissolution (Grötsch and Flügel, 1992). Collectively, these features are characteristic of very low sedimentation rates and/or nondeposition (omission), erosion, condensation of sediment, and dissolution (Hamilton, 1956; Lonsdale et al.,

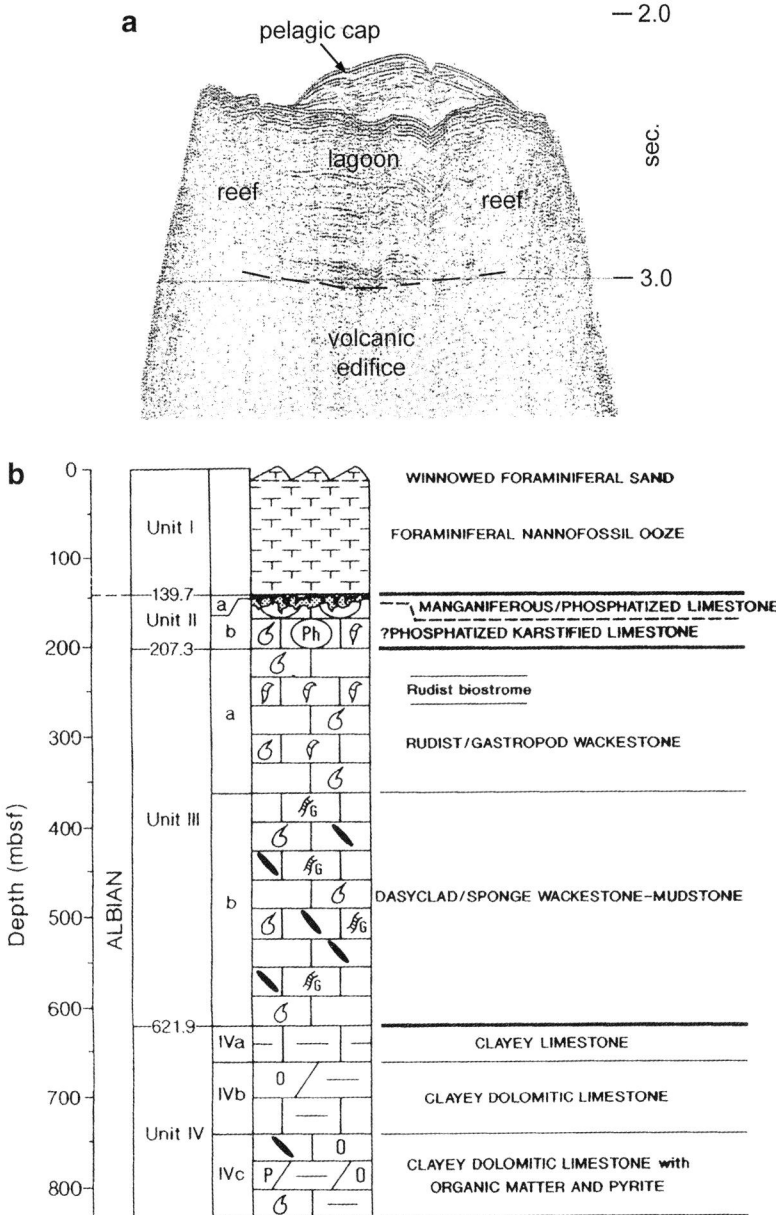

Guyot, Atoll, Figure 2 (a) Seismic cross section and (b) sedimentary succession on top of Allison Guyot (modified from Winterer and Metzler, 1984; Jenkyns and Wilson, 1999).

1972; Grötsch and Flügel, 1992; Gischler, 1996). Lonsdale et al. (1972) recorded current speeds of up to 15 cm/s on the top of Horizon Guyot and observed substantial erosion due to winnowing of sediment. On the top of guyots, flow acceleration and eddies, internal waves, and upwelling, called Taylor columns (Boehlert and Genin, 1987), apparently increase nutrient concentrations, enhance primary productivity, and sustain filter-feeding organisms such as soft corals, crinoids, or sponges.

Genetic connection between atolls and guyots

In his universally known subsidence theory, Darwin (1842) genetically connected fringing, barrier, and atoll reefs. During the subsidence of volcanic islands, fringing reefs develop into barrier reefs and eventually to atolls due to vertical reef accretion. For example, the distribution of reef types in the Society Islands (south Pacific) may be used in support of Darwin's theory. Away from the hot spot toward the west, volcanic islands are progressively older (Guillou et al., 2005) and reef types change from

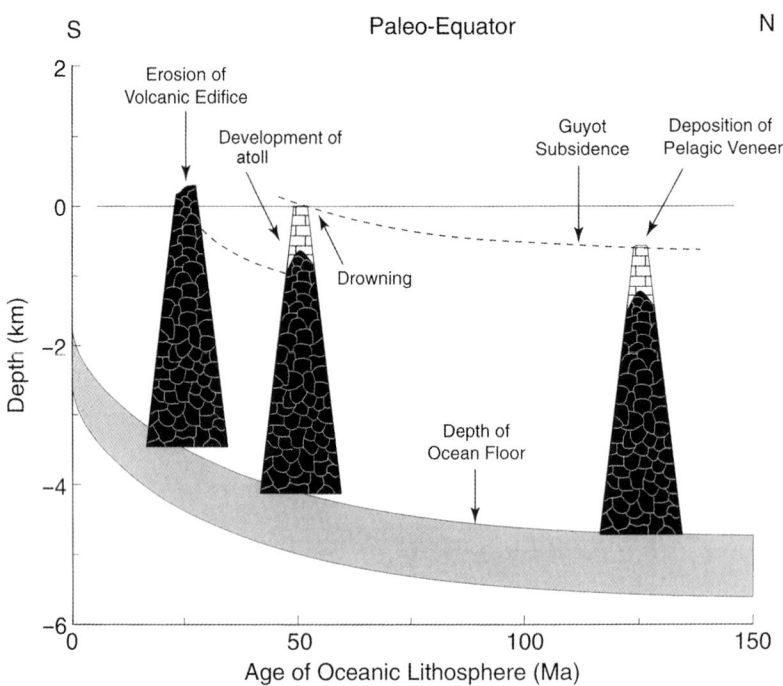

Guyot, Atoll, Figure 3 Idealized development of a Pacific guyot (modified from Jenkyns and Wilson, 1999).

fringing to barrier to atoll reefs (Gischler, 2011). During continued subsidence during cooling and transport of oceanic lithosphere away from the area of active volcanism, atoll reefs eventually drown and guyots develop (Figure 3). Impressive manifestations of this development are reef and seamount chains (Grigg, 1982; Konishi, 1989; Grigg, 1997). The several thousand km-long Hawaiian reef chain and the Emperor seamounts are imposing examples of northwestward and northward movement of reefs and seamounts away from a hot spot location (Greene et al., 1978; Grigg, 1982, 1997; Flood, 1999, 2001). Further examples include the Line and Marshall Island chains (Schlanger and Premoli-Silva, 1986). Wave planation of the volcanic island plays a central role in the glacial-control theory of reef development (Daly, 1915). Island erosion has subsequently been used when explaining guyot development (Hess, 1946; Hamilton, 1956). However, results of ODP legs 143 and 144 support the contention that the more or less flat summit of guyots is constructional and a consequence of carbonate accretion (Flood, 1999, 2001).

The significance of subsidence of a volcanic island for the development of isolated reefs in the open ocean was proven for the first time by drilling on Enewetak Atoll, Marshall Islands (Ladd et al., 1953), which recovered 1.25–1.4 km of reefal limestone overlying Eocene basalt. Darwin's theory may indeed be applied to the long-term development of oceanic reef structures; however, it fails when being applied to short-term and especially to Quaternary reef development because glacio-eustatic sea-level fluctuations are not considered. For example, Saller and Koepnick (1990) showed that reef facies in Enewetak had not simply aggraded and/or slightly retrograded as predicted by Darwin's theory, but significantly prograded; furthermore, major unconformities were identified in the drill cores. Similar observations were made in Mururoa (Camoin et al., 2001). The former observation is a consequence of falling sea-level trends; the latter results from repeated subaerial exposure during sea-level lowstands.

Sea-level fluctuations have been proven to be a major environmental factor of reef growth and, hence, of atoll formation (e.g., Camoin et al., 2001; Montaggioni, 2005). Reef and carbonate platform architecture, i.e., aggradation, progradation, or retrogradation and backstepping of reef sequences as well as drowning, has been largely attributed to the course of sea-level in combination with sediment supply, i.e., the productivity of the carbonate factory (e.g., Schlager, 1993). There are atolls which have a very complex geological history of separation and coalescence of platforms during aggrading, prograding, retrograding, and drowning episodes, largely controlled by sea-level change and subsidence, e.g., the Maldives archipelago that is composed of 22 major atolls since the Neogene (Purdy and Bertram, 1993; Belopolsky and Droxler, 2003).

Meteoric dissolution (karst) during sea-level lowstands may even produce atoll and barrier reef geomorphologies, as shown during experiments with limestone blocks and acid and in the examples of uplifted carbonate islands with erosional atoll and barrier reef shapes, e.g., Kita Daito Jima (Japan), Makatea (French Polynesia), or Mangaia

(Cook Islands) (Purdy, 1974; Purdy and Winterer, 2001, 2006). Purdy and Winterer (2001, 2006) also found a significant correlation between atoll and barrier reef lagoon depths and precipitation rates. They showed statistically that barrier reef lagoon depths are not lower as compared to atoll lagoon depths as would be expected when applying Darwin's model of continued subsidence. Purdy and Winterer (2006) also argued that fringing reefs located within barrier reefs also do not fit the idea of a progressive development of fringing and barrier reefs. Based on results obtained during ocean drilling project legs 143 and 144 (Winterer et al., 1995; Haggerty and Premoli-Silva, 1995), Winterer and Metzler (1984) and Winterer (1998) concluded that many guyots with Cretaceous reef caps located in the western Pacific experienced a subaerial karst phase before drowning, which enhanced their atoll geomorphology.

The mechanisms of reef drowning, however, still remain somewhat enigmatic and controversial. Grigg (1982, 1997) suggested that movement of the lithospheric plate has been transporting reefs of the Hawaiian chain out of the warm tropical-subtropical realm toward higher latitudes where reef growth rates are progressively slower, thereby causing reef demise and drowning. Hallock and Schlager (1986) and Rougerie and Fagerstrom (1994) argued that drowning of many mid-Cretaceous reefs, including those in the Pacific, coincided with oceanic anoxic events (OAEs), i.e., during the upwelling of nutrient-rich, oxygen-depleted water masses. Wilson et al. (1998) and Jenkyns and Wilson (1999) were able to prove that numerous Cretaceous reefs, now forming the tops of western Pacific guyots, were exterminated when the reefs were situated in equatorial latitudes. Possibly, very high temperatures caused reef decline, as observed today, e.g., during strong El Niño years. Reef-building scleractinian corals may lose their photosymbionts, bleach, and eventually die off (Hoegh-Guldberg, 1999). Likewise, equatorial upwelling of nutrient-rich waters might have been detrimental to coral and rudist reef growth. Based on the observation that the growth potential of corals and coral reefs usually exceeds the rate of rise of sea-level, Schlager (1981) concluded that only very rapid pulses of sea-level rise and/or environmental deterioration may cause reef drowning. While some late Quaternary reefs in the Indo-Pacific region have indeed been accreting at rates of up to 20 m/kyr (Montaggioni, 2005), the majority of Holocene Atlantic reefs accumulated with average rates of 3–5.5 m/kyr (Gischler, 2008; Hubbard, 2009). These rates are well below rapid rates of glacio-eustatic sea-level rise, e.g., during the last post-glacial and that predicted by the ICDP for the twenty-first century. Thus, rapid sea-level rise alone is a potential cause of reef drowning. Interestingly, guyots, i.e., drowned atolls, exist adjacent to living atolls, e.g., in the Marshall Islands. An instructive example is Wodejabato (Sylvania) Guyot (Figure 1), which is positioned next to Bikini Atoll (Ladd et al., 1950; Hamilton, 1956; Schlanger et al., 1987). Wodejabato is a former atoll reef that drowned in the latest Cretaceous and the top of which is now 1,335 m below sea-level (Camoin et al., 1998). In this case, environmental deterioration such as nutrient excess cannot be used as an explanation for drowning. It is highly unlikely that nutrient excess would cause the demise of some reefs in the middle of the Pacific Ocean while adjacent ones would not be harmed. A more plausible explanation could be differences in the elevation of the reef foundation. Elevation controls the timing and, hence, the rate of initial eustatic sea-level rise over a platform. The latter is crucial as full carbonate accretion may be delayed initially due to the time necessary for biotic colonization (Kim et al., 2012). Likewise, the relief of the inundated reef pedestal during sea-level rise, providing a variety of habitats of reef builders, might be of significance (Camoin et al., 1998).

Summary

Many atolls and guyots in the Pacific are genetically connected by the processes of subsidence of volcanic islands, reef accretion, and reef drowning. Still, subsidence alone may not satisfactorily explain reef development and drowning. Sea-level is another crucial factor of reef development. Sea-level rise results in reef aggradation and retrogradation and/or backstepping. Sea-level stalling and fall may cause progradation and karstification during subaerial exposure. Possible causes of drowning include rapid pulses of sea-level rise, reef elevation and relief, and environmental stress such as low sea-surface temperatures as well as high sea-surface temperatures and upwelling of nutrient-rich and oxygen-depleted waters.

Bibliography

Belopolsky, A. V., and Droxler, A., 2003. Imaging Tertiary carbonate systems – the Maldives, Indian Ocean: insights into carbonate sequence interpretation. *Leading Edge*, **22**, 646–652.

Boehlert, G. W., and Genin, A., 1987. A review of the effects of seamounts on biological processes. *Geophysical Monograph*, **43**, 319–334.

Camoin, G. F., Arnaud-Vanneau, A., Bergersen, D. D., Enos, P., and Ebren, P., 1998. Development and demise of mid-oceanic carbonate platforms, Wodejabato Guyot (NW Pacific). *International Association of Sedimentologists Special Publication*, **25**, 39–67.

Camoin, G., Ebren, P., Eisenhauer, A., Bard, E., and Faure, G., 2001. A 300,000 years record of sea-level changes, Mururoa Atoll (French Polynesia). *Palaeogeography Palaeoclimatology Palaeoecology*, **175**, 325–341.

Daly, R. A., 1915. The glacial-control theory of coral reefs. *Proceedings of the American Academy of Arts and Sciences*, **51**, 157–251.

Darwin, C. R., 1842. *The Structure and Distribution of Coral Reefs*. London: Smith Elder, 214 p.

Davies, T. A., Wilde, P., and Clague, D. A., 1972. Koko Seamount: a major guyot at the southern end of the Emperor Seamounts. *Marine Geology*, **13**, 311–321.

Dickinson, W. R., 2009. Pacific atoll living: how long already and until when? *GSA Today*, **19**, 4–10.

Flood, P., 1999. Development of northwest Pacific guyots: general results from Ocean Drilling Program legs 143 and 144. *Island Arc*, **8**, 92–98.

Flood, P. G., 2001. The 'Darwin Point' of Pacific Ocean atolls and guyots: a reappraisal. *Palaeogeography Palaeoclimatology Palaeoecology*, **175**, 147–152.

Gischler, E., 1996. Late Devonian-early Carboniferous deep-water coral assemblages and sedimentation on a Devonian seamount: Iberg Reef, Harz Mts, Germany. *Palaeogeography Palaeoclimatology Palaeoecology*, **123**, 297–322.

Gischler, E., 2006. Sedimentation on Rasdhoo and Ari atolls, Maldives, Indian Ocean. *Facies*, **52**, 341–360.

Gischler, E., 2007. Beachrock and intertidal precipitates. In Rigby, D., and MacLaren, S. (eds.), *Chemical Sediments and Landscapes*. Oxford: Blackwell, pp. 365–390.

Gischler, E., 2008. Accretion patterns in Holocene tropical coral reefs: do massive coral reefs with slowly growing corals accrete faster than branched coral (acroporid) reefs with rapidly growing corals? *International Journal of Earth Sciences*, **97**, 851–859.

Gischler, E., 2010. Indo-Pacific and Atlantic spurs and grooves revisited: the possible effects of different Holocene sea level history, exposure, and reef accretion rate in the shallow fore reef. *Facies*, **56**, 173–177.

Gischler, E., 2011. Sedimentary facies of Bora Bora, Darwin's type barrier reef (Society Islands, south Pacific): the unexpected occurrence of non-skeletal grains. *Journal of Sedimentary Research*, **81**, 1–17.

Gischler, E., and Lomando, A. J., 1999. Recent sedimentary facies of isolated carbonate platforms, Belize-Yucatan system, Central America. *Journal of Sedimentary Research*, **69**, 747–763.

Greene, H. G., Dalrymple, G. B., and Clague, D. A., 1978. Evidence for northward movement of the Emperor Seamounts. *Geology*, **6**, 70–74.

Grigg, R. W., 1982. Darwin point: a threshold for atoll formation. *Coral Reefs*, **1**, 29–34.

Grigg, R. W., 1997. Paleoceanography of coral reefs in the Hawaiian-Emperor Chain – revisited. *Coral Reefs*, **16**, S33–S38.

Grötsch, J., and Flügel, E., 1992. Facies of sunken Early Cretaceous atoll reefs and their capping Late Albian drowning succession (Northwestern Pacific). *Facies*, **27**, 153–174.

Guillou, D., Maury, R. C., Blais, S., Cotten, J., Legendre, C., Guille, G., and Caroff, M., 2005. Age progression along the Society hotspot chain (French Polynesia) based on new unspiked K-Ar ages. *Société Géologique France Bulletin*, **176**, 135–150.

Haggerty, J. A., Premoli-Silva, I. (eds.), 1995. Scientific results. Northwest Pacific atolls and guyots. *Proceedings of Ocean Drilling Program*, **144**, p. 1059.

Hallock, P., and Schlager, W., 1986. Nutrient excess and the demise of coral reefs and carbonate platforms. *Palaios*, **1**, 289–298.

Hamilton, E. L., 1956. Sunken islands of the Mid-Pacific Mountains. *Geological Society of America Memoir*, **64**, p. 97.

Hess, H. H., 1946. Drowned ancient islands of the Pacific Basin. *American Journal of Science*, **244**, 772–791.

Hoegh-Guldberg, O., 1999. Climate change, coral bleaching and the future of the world's coral reefs. *Marine and Freshwater Research*, **50**, 839–866.

Hubbard, D. K., 2009. Depth and species-related patterns of Holocene reef accretion in the Caribbean and western Atlantic: a critical assessment of existing models. *International Association of Sedimentologists Special Publication*, **40**, 1–18.

James, N. P., and Ginsburg, R. N., 1979. The seaward margin of Belize barrier and atoll reefs. *International Association of Sedimentologists Special Publication*, **3**, p. 191.

Jenkyns, H. C., and Wilson, P. A., 1999. Stratigraphy, paleoceanography, and evolution of Cretaceous Pacific guyots: relics from a greenhouse earth. *American Journal of Science*, **299**, 341–392.

Kim, W., Fouke, B. W., Petter, A. L., Quinn, T. M., Kerans, C., and Taylor, F., 2012. Sea-level rise, depth-dependent carbonate sedimentation and the paradox of drowned platforms. *Sedimentology*, **59**, 1677–1692.

Klostermann, L., Gischler, E., Storz, D., and Hudson, J. H., 2014. Sedimentary record of late Holocene event beds in a mid-ocean atoll lagoon, Maldives, Indian Ocean: potential for deposition by tsunamis. *Marine Geology*, **348**, 37–43.

Konishi, K., 1989. Limestone of the Daiichi Kashima seamount and the fate of a subducting guyot: fact and speculation from the Kaiko "Nautile" dives. *Tectonophysics*, **160**, 249–265.

Ladd, H. S., Tracey, J. I., Wells, J. W., and Emery, K. O., 1950. Organic growth and sedimentation of an atoll. *Journal of Geology*, **58**, 410–425.

Ladd, H. S., Ingerson, E., Townsend, R. C., Russell, M., and Stephenson, H. K., 1953. Drilling on Eniwetok Atoll, Marshall Islands. *American Association of Petroleum Geologists Bulletin*, **37**, 2257–2280.

Lonsdale, P., Normark, W. R., and Newman, W. A., 1972. Sedimentation and erosion on Horizon Guyot. *Geological Society of American Bulletin*, **83**, 289–316.

Matthews, J. L., Heezen, B. C., Catalano, R., Coogan, A., Tharp, M., Natland, J., and Rawson, M., 1974. Cretaceous drowning of reefs on mid-Pacific and Japanese guyots. *Science*, **184**, 462–464.

Menard, H. W., 1984. Origin of guyots: the Beagle to Seabeam. *Journal of Geophysical Research*, **B89**, 11117–11123.

Montaggioni, L. F., 2005. History of Indo-Pacific coral reef systems since the last glaciation: development patterns and controlling factors. *Earth Science Reviews*, **71**, 1–75.

Pratt, R. M., 1963. Great Meteor seamount. *Deep Sea Research*, **10**, 17–25.

Pratt, R. M., 1967. Photography of seamounts. *Johns Hopkins Oceanography Study*, **3**, 145–158.

Purdy, E. G., 1974. Reef configuration: cause and effect. *Society of Economic Paleontologists and Mineralogists, Special Publication*, **18**, 9–76.

Purdy, E. G., and Bertram, G. T., 1993. Carbonate concepts from the Maldives, Indian Ocean. *American Association of Petroleum Geologists, Studies in Geology*, **34**, 1–56.

Purdy, E. G., and Gischler, E., 2005. The transient nature of the empty bucket model of reef sedimentation. *Sedimentary Geology*, **175**, 35–47.

Purdy, E. G., and Winterer, E. L., 2001. Origin of atoll lagoons. *Geological Society of America Bulletin*, **113**, 837–854.

Purdy, E. G., and Winterer, E. L., 2006. Contradicting barrier reef relationships for Darwin's evolution of reef types. *International Journal of Earth Sciences*, **95**, 143–167.

Rougerie, F., and Fagerstrom, J. A., 1994. Cretaceous history of Pacific basin guyot reefs: a reappraisal based on geothermal endo-upwelling. *Palaeogeography Palaeoecology Palaeoclimatology*, **112**, 239–260.

Saller, A. H., and Koepnick, R. B., 1990. Eocene to early Miocene growth of Enewetak Atoll: insight from strontium-isotope data. *Geological Society of America Bulletin*, **102**, 381–390.

Schlager, W., 1981. The paradox of drowned reefs and carbonate platforms. *Geological Society of America Bulletin*, **92**, 197–211.

Schlager, W., 1993. Accommodation and supply – a dual control on stratigraphic sequences. *Sedimentary Geology*, **86**, 111–136.

Schlanger, S. O., and Premoli-Silva, I., 1986. Oligocene sea-level falls recorded in mid-Pacific atoll and archipelagic apron settings. *Geology*, **14**, 392–395.

Schlanger, S. O., Campbell, J. F., and Jackson, M. W., 1987. Post-Eocene subsidence of the Marshall Islands recorded by drowned atolls on Harriuer and Sylvania guyots. *Geophysical Monograph*, **43**, 165–174.

Stoddart, D. R., 1965. The shape of atolls. *Marine Geology*, **3**, 369–383.

Wells, J. W., 1954. Recent corals of the Marshall Islands, Bikini and nearby atolls, II, oceanography. *United States Geological Survey. Professional Paper*, **260-I**, 385–486.
Wessel, P., 1997. Sizes and ages of seamounts using remote sensing: implications for intraplate volcanism. *Science*, **277**, 802–805.
Wiens, H. J., 1962. *Atoll Environment and Ecology.* New Haven: Yale University Press, p. 532.
Wilson, P. A., Jenkyns, H. C., Elderfield, H., and Larson, R. L., 1998. The paradox of drowned carbonate platforms and the origin of Cretaceous Pacific guyots. *Nature*, **392**, 889–894.
Winterer, E. L., 1998. Cretaceous karst guyots: new evidence for inheritance of atoll morphology from subaerial erosional terrain. *Geology*, **26**, 59–62.
Winterer, E. L., and Metzler, C. V., 1984. Origin and subsidence of guyots in mid-Pacific mountains. *Journal of Geophysical Research*, **B89**, 9969–9979.
Winterer, E. L., Sager, W. W., Firth, J. V., Sinton, J. M. (eds.), 1995. Scientific results. Northwest Pacific atolls and guyots. Proceedings of Ocean Drilling Program, **143**, p. 629

Cross-references

Carbonate Dissolution
Carbonate Factories
El Niño (Southern Oscillation)
Erosion-Hiatuses
Foraminifers (Benthic)
Foraminifers (Planktonic)
Hot Spots and Mantle Plumes
Lagoons
Plate Motion
Reefs (Biogenic)
Relative Sea-level (RSL) Cycle
Seamounts
Tsunamis

H

HIGH-PRESSURE, LOW-TEMPERATURE METAMORPHISM

Hayato Ueda
Department of Geology, Niigata University, Niigata, Japan

Synonyms

High-pressure/temperature (high-P/T) metamorphism; High-pressure metamorphism

Definition

Metamorphism: Changes of mineral textures and phases in rocks at depths without significant change of bulk-rock chemical composition except for vapor components.

High-pressure/low-temperature metamorphism: Metamorphism that undergoes at high pressure with low-temperature conditions than usual, typically with thermal gradients as low as 10 °C/km.

High-pressure/temperature metamorphism: Metamorphism under conditions of high pressure versus temperature ratios, i.e., of low thermal gradients.

Introduction

In Earth's interior, temperature generally increases as the depth increases as a result of heat conduction. Because rocks as well as melts and fluids sometimes migrate fast enough for evident heat advection, geothermal fields are modified by regional or local geodynamics. In *subduction and collision zones*, lower temperatures than normal geotherms are maintained with thermal gradients as low as 10 °C/km, because the slabs cooled on the surface are constantly supplied to the depths. High-pressure/low-temperature (HPLT) metamorphism is believed almost exclusively to occur at such cool conditions inside the convergent plate margins. Although only a few cases of seafloor exposures of HPLT metamorphic rocks are so far known, HPLT metamorphism might be ongoing inside every subduction zone corresponding to deep-sea trenches. HPLT rocks occasionally record physical and chemical conditions, dynamics and element transport, and their historical changes inside deep subduction or collision zones as mineral phases and textures. HPLT metamorphic rocks commonly contain significant amounts of hydrous and/or carbonate minerals. When these rocks are further subducted down to deeper portions with increasing temperature and pressure, they release volatile fluids such as water and carbon dioxide as well as other elements as solutes via dehydration or decarbonation. The released volatiles affect chemistry, rheology, and melting conditions of the upper mantle. Therefore, HPLT metamorphic rocks act as containers which deliver volatiles from the surface to the mantle.

Geodynamic settings and source rocks

Source rocks (protoliths) of HPLT metamorphic rocks differ between two contrasting convergent settings of collision and subduction zones (Maruyama et al., 1996). In collision zones, materials being subducted are leading edges of *passive continental margins* consisting of granitic basements, clastic and calcareous sediments of *continental shelves and slopes*, and rift-related basaltic rocks. Aluminous and/or iron-rich metapelites originated from mud and soil affected by subaerial weathering are also occasionally accompanied. Materials input into subduction zones consist of *oceanic crust* and overlying pelagic sediments, as well as trench-fill deep-sea *turbidites* and mudstone, and *seamounts*. This subduction zone rock assemblage is comparable to rocks of accretionary complexes. Basaltic rocks have chemical compositions correlative to *mid-ocean ridge* and oceanic island basalts. Metachert characteristically occur, and some of them are occasionally manganese. HPLT metamorphic rocks are

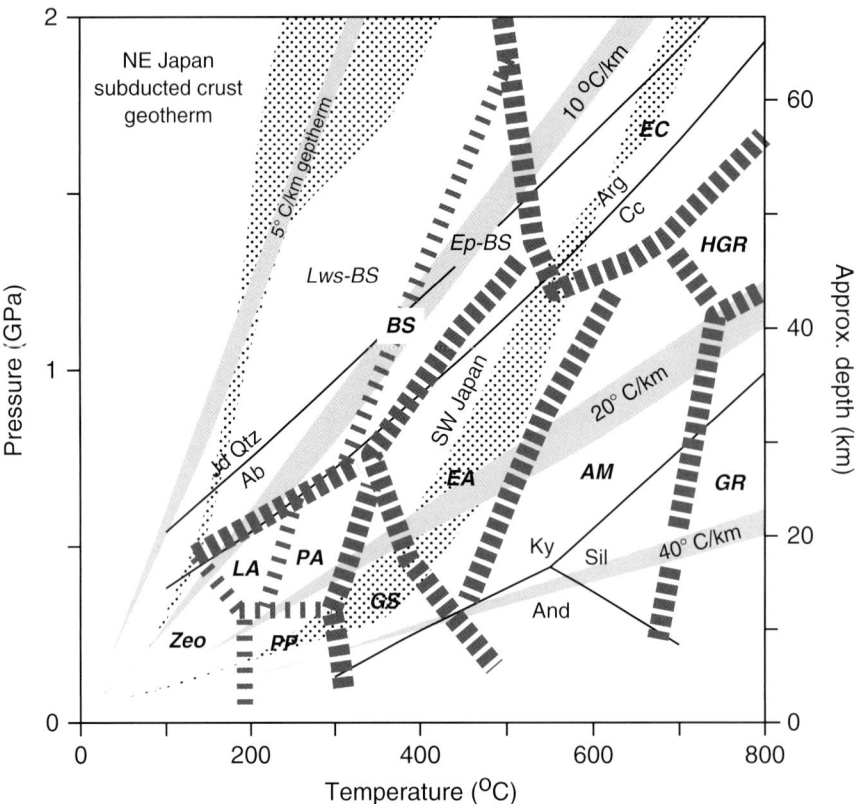

High-pressure, Low-temperature Metamorphism, Figure 1 Metamorphic facies in a pressure versus temperature coordinate (compiled from petrogenetic grids by Oh and Liou, 1988; Evans, 1990; Frey et al., 1991) with references of geothermal gradients (*gray* bands) and geotherms of subducted cold (*NE Japan*) and warm (*SW Japan*) oceanic crusts (sand hatches: after Peacock and Wang, 1999). Abbreviations: *AM* amphibolite facies, *BS* blueschist facies, *EA* epidote amphibolite facies, *EC* eclogite facies, *GR* granulite facies, *GS* greenschist facies, *HGR* high-pressure granulite facies, *LA* lawsonite-albite facies, *PA* prehnite-actinolite facies, *PP* prehnite-pumpellyite facies, and *Zeo* zeolite facies.

commonly associated with *serpentinites* especially abundant on seafloor exposures in intraoceanic subduction zones and on-land regional metamorphic belts accompanied by *ophiolites*.

Mineralogy and metamorphic facies

A chemically equilibrated rock consists of the most stable assemblage and compositions of minerals for a given pressure, temperature, and chemical compositions of bulk-rock and interstitial fluids. When these conditions change, chemical reactions occur consuming some mineral components or phases and producing others in order to maintain the stability. As pressure increases, sodic plagioclase diminishes and finally disappears, generating sodic pyroxene (jadeite or omphacite) and sodic amphibole (e.g., glaucophane), both diagnostic of HPLT metamorphism. Lawsonite and aragonite are formed exclusively in HPLT conditions. In mafic rocks, Fe-Mg garnet tends to develop at high pressure as well as high temperature.

As a result of changes above, blueschist (glaucophane schist) at lower temperatures and eclogite (garnet-omphacite rock) at higher temperatures represent HPLT metamorphic rocks of basaltic compositions. Typical mineral assemblages of both the rocks define the blueschist and eclogite facies, respectively (Figure 1).

The blueschist facies mafic rocks characteristically contain sodic amphiboles such as glaucophane, occasionally with lawsonite, sodic (jadeitic) pyroxene, aragonite, epidote, pumpellyite, chlorite, phengite, and titanite. Plagioclase (albite) occurs at relatively lower pressures but disappears as pressure increases. Pelitic rocks are typically phengite-chlorite schist. Winkler (1967) set the lawsonite-albite facies, in which lawsonite coexists with neither sodic pyroxenes nor sodic amphiboles. It suggests lower pressure conditions than the typical blueschist facies and can be regarded as sub-facies of the latter. The typical blueschist facies is subdivided into lawsonite-blueschist and epidote-blueschist sub-facies (Evans, 1990) where sodic pyroxenes and sodic amphiboles coexist with lawsonite in the former and epidote in the latter.

The eclogite facies is characterized by the occurrence of Fe-Mg garnet and omphacite in the absence of plagioclase. Epidote or lawsonite, hornblende or barroisite, phengite, paragonite or kyanite, talc, and rutile are

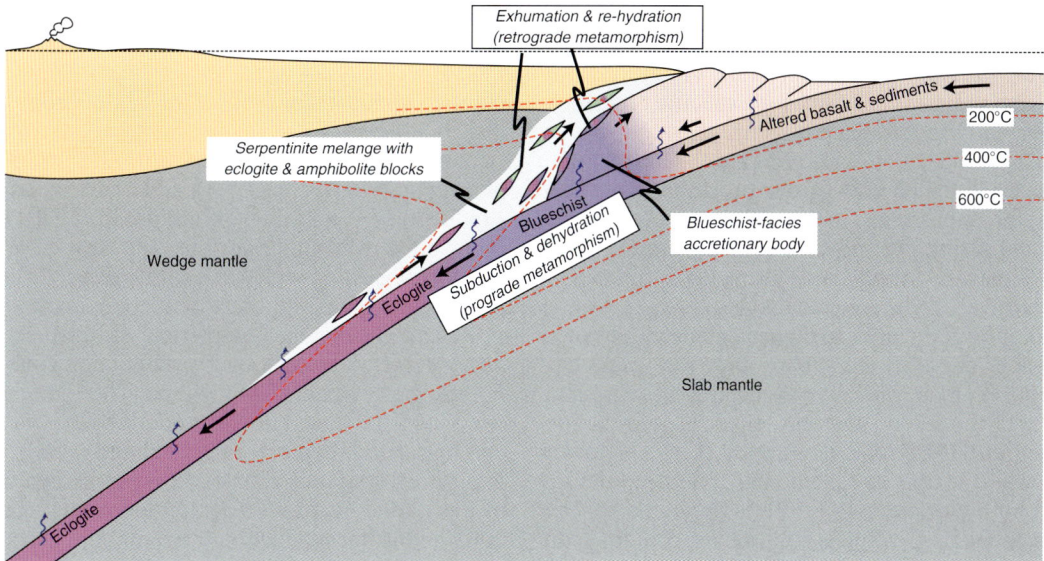

High-pressure, Low-temperature Metamorphism, Figure 2 A sketch of HPLT metamorphism in a conceptual subduction zone system. *Red dashed lines* show schematic isothermal contours.

occasionally accompanied. Lawsonite occurs at lower temperatures than epidote; and kyanite + omphacite occurs at higher pressures than paragonite. Amphiboles disappear as pressure increases. Equivalent pelitic rocks are typically garnet-phengite schist occasionally with talc + chloritoid and talc + kyanite associations (Mottana et al., 1990). Stability of the eclogite facies extends to the wet solidus temperatures, so that merely "high-pressure (HP) metamorphism" is often applied to such rocks. At higher pressure than ~2.7 GPa (ca. 100 km deep), coesite occurs instead of quartz, and such rocks are referred to as ultrahigh-pressure (UHP) rocks.

Metamorphic facies suggesting warmer conditions than typical HPLT metamorphism are also not uncommon in subduction and collision zones. These are the pumpellyite-actinolite facies, high-pressure parts of the greenschist and amphibolite facies (epidote amphibolites and epidote-garnet amphibolites). Sodic-calcic amphiboles such as winchite and barroisite are common in these rocks.

Prograde and retrograde metamorphism

Many of HPLT metamorphic rocks originated from volcanic and sedimentary rocks once on the surface and then were subducted to depths and finally returned back to the surface again. Such transports result in a general metamorphic history with increase and subsequent decrease of pressure and temperature. Recrystallization at the highest temperature and/or pressure which a rock experienced is assigned to the peak metamorphism, and transitional processes toward and after the peak are prograde and retrograde metamorphism, respectively.

Starting materials of HPLT metamorphic rocks are rich in hydrous minerals (clays in mudstone and alteration minerals in basalt, etc.). Increasing temperature and pressure induce dehydration reactions. Prograde dehydration reactions usually well proceed with scarce obstructing factors. It is thus sometimes difficult to find traces of prograde stage, except for mineral inclusions and relic cores capsuled by progressively grown minerals.

Conversely, retrograde metamorphism is governed by hydration reactions, and its extent is controlled by supply of water fluids from outside the rocks. In cases of limited water supply, rocks are incompletely recrystallized preserving mineralogy of the peak condition. However, in subduction zones, rocks may migrate upward above the subducting slab, which constantly releases water fluids upward (Figure 2). Probably by this reason, retrograde recrystallization is sometimes quite extensive in HPLT rocks. It is not uncommon that the peak-stage mineralogy had been almost completely erased by retrograde mineralization.

Pressure-temperature paths and dynamics

Pressure and temperature of each metamorphic stage can be better estimated when geobarometers and geothermometers are applicable. These estimates are based on chemical equilibria including solid solution minerals, and pressure and temperature dependence of each equilibrium constant is calibrated by experimental results or thermodynamic data for minerals. Major geothermometry for HP rocks (e.g., Carswell and Harley, 1990; Ravna and Paquin, 2003) uses Fe-Mg exchange reactions between garnet and other ferromagnesian minerals (e.g., Ellis and Green, 1979) and net transfer reactions (e.g., Ravna and Terry, 2004), in which reactant components are consumed for products.

Net transfer reactions result in appearance and disappearance of minerals, as well as their compositional changes. The pressure and temperature conditions for these limits and mineral chemistry for a specific bulk-rock composition can be simulated using thermodynamic dataset of minerals (e.g., Holland and Powell, 1998). Such petrogenetic grids, called as P-T pseudosections, have also become useful approaches to investigate metamorphic conditions and paths. Thermo-Calc (Powell and Holland, 1998) and PerpleX (Connolly and Petrini, 2002) are examples of the software for these calculations.

A transition from the peak to the retrograde conditions, with prograde ones if available, draws a pressure-temperature (P-T) path, which we expect to reflect thermal structures and modes of material flow in subduction zones. When an HPLT rock returned back to shallower levels, nearly adiabatic, isothermal decompression represented by a clockwise P-T path is predicted for the case of fast ascending, whereas significant cooling with decompression with a hairpin path for sluggish ascending (Ernst, 1988). Such contrasting exhumation rates has been supported by geochronology (Rubatto and Hermann, 2001; Anczkiewicz et al., 2004; Baldwin et al., 2010). On the other hand, numerical simulations commonly predict characteristic inverted thermal structure in subduction zones with deeply subducted but still cool slab overlain by the hot wedge mantle (e.g., Peacock and Wang, 1999; Figure 2). When an oceanic crustal rock is detached from the subducted slab, it is expected to ascend via a higher-temperature conduit than descending one on the slab, resulting in clockwise P-T paths (Gerya et al., 2002). Counterclockwise P-T paths with isobaric cooling known in the Franciscan Complex in California (Wakabayashi, 1990) are difficult to explain by ordinary thermal structures in subduction zones. They are attributed to transient cooling of the subduction zone just after its inception.

Exhumation tectonics

Exhumation, the processes in which HPLT metamorphic rocks ascend from depths to the surface, is one of the controversial subjects on subduction zones dynamics. There are many ideas for the driving forces to lift up the HPLT rocks. Buoyancy of the subducting edifices (continents, oceanic plateaus, mid-oceanic ridges, etc.: Maruyama et al., 1996), of subducted crustal materials (Chemanda, 1995), or of serpentinized wedge mantle (Guillot et al., 2001) is one of the likely candidates. Isostatic rebound of the subducted buoyant slab is expected when the dense precursor parts of the slab is broken off at depths (Davis and von Blanckenburg, 1995). Traction force of the subducting slab could induce a counter flow above the subduction interfaces (Cloos, 1982). Underplating accretion at depths could jack up the HPLT rocks (Platt, 1986).

Because HPLT rocks are exposed not at trenches but inside the hanging-wall crusts, unroofing mechanisms to remove the overlying crustal materials are also necessary for HPLT rocks to reach to the surface. Erosion plays an important role for unroofing on land (Chemanda et al., 1995), whereas its effect might be minimal in cases of marine subduction settings. One of the proposed ideas is gravity collapse of an *accretionary wedge* when it is too steepened by underplating accretion (Platt, 1986). Upper plate rifting (Hill et al., 1993), and transtension by oblique subduction (Ave Lallemant and Guth, 1990) also, could remove the overburden for the HPLT rocks. When HPLT rocks uprise along with serpentinite diapir, it would penetrate the hanging-wall crust by its buoyancy.

Recently developed numerical simulation techniques provide useful deductions for subduction zone dynamics. For example, Gerya and Stöckhert (2006) modeled at least two kinds of induced flow in a subduction zone depending on contrasting rheology of crustal and mantle rocks. One is a set of drag and counter flows in serpentinized parts of the wedge mantle (serpentinite channel: Guillot et al., 2001) along the subduction interface, and the counter flow carries up higher-grade HP and UHP rocks from greater depths to the base of the hanging-wall crust. The others are shallower intra-crustal flows, which lift up lower-grade HPLT rocks with fragments of higher-grade rocks. This generally agrees with natural occurrences that low-grade HPLT accretionary rocks tend to occur as coherent masses (Kimura et al., 1996) whereas higher-grade blocks of eclogites and amphibolites are occasionally hosted in serpentinite mélanges (Coleman and Lanphere, 1971) in ancient subduction complexes.

Occurrences in the sea

Except for submerged extensions of landmass terrains, sea-bottom occurrences of HPLT metamorphic rocks have been so far recognized only in the Philippine Sea Plate.

The first finding (Maekawa et al., 1992, 1993; Maekawa, 1995) was from ODP Leg125 drill core (Hole 778A), which penetrated the Conical Seamount, a forearc serpentinite *seamount* ~50 km arc-ward from the Mariana Trench. HPLT metamorphic minerals, such as lawsonite, pumpellyite, sodic pyroxene, aragonite, and blue amphibole (winchite), were found in pebble-sized metabasite clasts hosted in serpentinite. Mineral chemistry and occurrences suggest the fragments underwent low-grade blueschist facies, in which lawsonite does not coexist with sodic amphibole and sodic pyroxene, with the metamorphic conditions of 150–200 °C 0.5–0.6 GPa (Maekawa et al., 1993, 1995). These results first confirmed that HPLT rocks are really formed in modern subduction zone. The serpentinite seamount itself is build up of highly fractured and sheared serpentinites, and groundwater vents and serpentine mudflows were also observed (Fryer et al., 1990). Based on these occurrences, the HPLT metabasites are considered to have carried up from the subduction interface of ~20 km depths by diapir or mud volcanism. Fragments of HPLT rocks with serpentinites are also found in other serpentinite seamounts and inner trench slopes in the Izu-Bonin-Mariana forearc (Tararin, 1995; Fryer et al., 2000; Maekawa et al., 2004).

Relics of HP metamorphism were found in amphibolites collected from the Ohmachi (Omachi) Seamount near the volcanic front of the Izu-Bonin arc (Ueda et al., 2004). These rocks are extensively recrystallized during the retrograde stage, resulting in garnet amphibolite and epidote-albite amphibolite mineralogy. Eclogite and blueschist facies minerals such as omphacite, kyanite, and glaucophane occur as micro-inclusions in zoisite, epidote, hornblende, and garnet. The highest grade sample underwent the peak metamorphism at 600–700 °C and 2–2.5 GPa (ca. 60–80 km deep), followed by a retrograde overprint at 550–600 °C and < ~1 GPa (shallower than approx. 30 km deep). Although the age of these HP rocks has not been published, pre-Eocene exhumation was suggested by Ueda et al. (2004) based on the age of cover volcano-sedimentary sequence. These HP metamorphic rocks are hosted in part of a serpentinite body consisting of antigorite schist and massive antigorite serpentinite. These serpentinites comprise a coherent body without pulverization, and schistose parts exhibit lateral extension in ductile manners (Hirauchi et al., 2010; Ueda et al., 2011). These modes of occurrence differ from the forearc serpentinite seamounts, and Ueda et al. (2011) considered that the serpentinite body preserves the structure of deep serpentinite channel in an *intraoceanic subduction zone*.

Summary

Petrology, structural geology, and numerical modeling of HPLT rocks commonly depict dynamic flows beneath forearcs with great vertical and lateral displacements of regional rock masses, which might be commonly ongoing in subduction zones. Such deductions have, however, not been well linked with seafloor geology of forearc areas. For example, stacks of thrust sheets (nappe piles) and hanging-wall low-angle normal faults (detachment) are ones of the typical structures of HPLT terranes exposed on land. However, such structures have not yet been detected in forearcs on the seafloors. Serpentinite diapirs and mud volcanoes that penetrate the Mariana forearc crusts are in turn not so suitable with dominantly flat-lying structures of HPLT terranes. Filling these gaps and comprehension of subduction zone dynamics will greatly owe to further investigations by marine geoscience.

Bibliography

Anczkiewicz, R., Platt, J. P., Thirlwall, M. F., and Wakabayashi, J., 2004. Franciscan subduction off to a slow start: evidence from high-precision Lu–Hf garnet ages on high grade-blocks. *Earth and Planetary Science Letters*, **225**, 147–161.

Ave Lallemant, H. G., and Guth, L. R., 1990. Role of extensional tectonics in exhumation of eclogites and blueschists in an oblique subduction setting: northeastern Venezuela. *Geology*, **18**, 950–953.

Baldwin, S., Monteleone, B. D., Webb, L. E., Fitzgerald, P. G., Grove, M., and Hill, E. J., 2004. Pliocene eclogite exhumation at plate tectonic rates in eastern Papua New Guinea. *Nature*, **431**, 263–267.

Carswell, D. A., and Harley, S. L., 1990. Mineral barometry and thermometry. In Carswell, D. A. (ed.), *Eclogite Facies Rocks*. Glasgow/London: Blackie, pp. 83–110.

Chemenda, A. I., Mattauer, M., Malavieille, J., and Bokun, A. N., 1995. A mechanism for syn-collisional rock exhumation and associated normal faulting: results from physical modeling. *Earth and Planetary Science Letters*, **132**, 225–232.

Cloos, M., 1982. Flow melanges: numerical modeling and geologic constraints on their origin in the Franciscan subduction complex, California. *Geological Society of America Bulletin*, **93**, 330–345.

Connolly, J. A. D., and Petrini, K., 2002. An automated strategy for calculation of phase diagram sections and retrieval of rock properties as a function of physical conditions (0.4 Mb). *Journal of Metamorphic Geology*, **20**, 697–708.

Davis, J. H., and von Blanckenburg, F., 1995. Slab breakoff: A model of lithosphere detachment and its test in the magmatism and deformation of collisional orogens. *Earth and Planetary Science Letters*, **129**, 85–102.

Ellis, D. J., and Green, D. H., 1979. An experimental study of the effect of Ca upon the garnet-clinopyroxene Fe-Mg exchange equilibria. *Contributions to Mineralogy and Petrology*, **71**, 13–22.

Evans, B. W., 1990. Phase relations of epidote-blueschists. *Lithos*, **25**, 3–23.

Frey, M., de Capitani, C., and Liou, J. G., 1991. A new petrogenetic grid for low-grade metabasites. *Journal of Metamorphic Geology*, **9**, 497–509.

Fryer, P., Saboda, K. L., Hohnson, L. E., Mackey, M. E., Moore, G. F., and Stoffers, P., 1990. Conical seamount: SeaMARC II, Alvin submersible, and seismic reflection studies. In Fryer, P., Pearce, J. A., Stokking, L. B., et al. (eds.), *Proceedings of the Ocean Drilling Program Initial Reports, 125*. College Station: Ocean Drilling Program, pp. 69–80.

Fryer, P., Lockwood, J. P., Becker, N., Phipps, S., and Todd, C. S., 2000. Significance of serpentinite mud volcanism in convergent margins. In Dilek, Y., Moores, E. M., Elthon, D., and Nicolas, A. (eds.), *Ophiolites and Oceanic Crust: New Insights from Field Studies and the Ocean Drilling Program*. Boulder: Geological Society of America, pp. 35–51.

Gerya, T., and Stöckhert, B., 2006. Two-dimensional numerical modeling of tectonic and metamorphic histories at active continental margins. *International Journals of Earth Science (Geologische Rundschau)*, **95**, 250–274.

Gerya, T., Stöckhert, B., and Perchuk, A. L., 2002. Exhumation of high-pressure metamorphic rocks in a subduction channel: a numerical simulation. *Tectonics*, **21**, 1056, doi:10.1029/2002TC001406.

Guillot, S., Hatori, K., de Sigoyer, J., Nägler, T., and Auzene, A.-L., 2001. Evidence of hydration of the mantle wedge and its role in the exhumation of eclogites. *Earth and Planetary Science Letters*, **193**, 115–127.

Hill, E. J., Baldwin, S. L., and Lister, G. S., 1993. Unroofing of active metamorphic core complexes in the D'Entrecasteaux Islands, Papua New Guinea. *Geology*, **20**, 907–910.

Hirauchi, K., Michibayashi, K., Ueda, H., and Katayama, I., 2010. Spatial variations in antigorite fabric across a serpentinite subduction channel: insights from the Ohmachi Seamount, Izu-Bonin frontal arc. *Earth and Planetary Science Letters*, **299**, 196–206.

Holland, T. J. B., and Powell, R., 1998. An internally-consistent thermodynamic dataset for phases of petrological interest. *Journal of Metamorphic Geology*, **16**, 309–343.

Kimura, G., Maruyama, S., Isozaki, Y., and Terabayashi, M., 1996. Well-preserved underplating structure of the jadeitized Franciscan complex, Pacheco Pass, California. *Geology*, **24**, 75–78.

Maekawa, H., Shozui, M., Ishii, T., Saboda, K. L., and Ogawa, Y., 1992. Metamorphic rocks from the serpentinite seamounts in the Mariana and Izu-Ogasawara forearcs. In Fryer, P., Pearce, J. A., Stokking, L. B., et al. (eds.), *Proceedings of the Ocean Drilling Program Scientific Results 125*. College Station: Ocean Drilling Program, pp. 415–430.

Maekawa, H., Shozui, M., Ishii, T., Fryer, P., and Pearce, J. A., 1993. Blueschist metamorphism in an active subduction zone. *Nature*, **364**, 520–523.

Maekawa, H., Fryer, P., and Ozaki, A., 1995. Incipient blueschist-facies metamorphism in the active subduction zone beneath the Mariana forearc. In Talor, B., and Natland, J. (eds.), *Active Margins and Marginal Basins of the Western Pacific*. Washington, DC: American Geophysical Union, pp. 281–289.

Maekawa, H., Yamamoto, K., Ueno, T., Osada, Y., and Nogami, N., 2004. Significance of serpentinites and related rocks in the high-pressure metamorphic terranes, circum-Pacific regions. *International Geology Review*, **46**, 426–444.

Maruyama, S., Liou, J. G., and Terabayashi, M., 1996. Blueschists and eclogites of the world and their exhumation. *International Geology Review*, **38**, 485–594.

Mottana, A., Carswell, D. A., Chopin, C., and Oberhänsli, R., 1990. Eclogite facies mineral paragenesis. In Carswell, D. A. (ed.), *Eclogite Facies Rocks*. New York: Chapman & Hall, pp. 14–52.

Oh, C. W., and Liou, J. G., 1998. A petrogenetic grid for eclogite and related facies under high-pressure metamorphism. *Island Arc*, **7**, 36–51.

Peacock, S. M., and Wang, K., 1999. Seismic consequences of warm versus cool subduction metamorphism: examples from Southwest and Northeast Japan. *Science*, **286**, 937–939.

Platt, J. P., 1986. Dynamics of orogenic wedges and the uplift of high-pressure metamorphic rocks. *Geological Society of America Bulletin*, **97**, 1037–1053.

Ravna, E. J. K., and Paquin, J., 2004. Thermobarometric methodologies applicable to eclogites and garnet ultrabasites. In Carswell, D. A., and Compagnoni, R. (eds.), *Ultrahigh Pressure Metamorphism*. Budapest: Eötvös University Press, pp. 229–259.

Ravna, E. J. K., and Terry, M. P., 2004. Geothermobarometry of UHP and HP eclogites and schists –an evaluation of equilibria among garnet – clinopyroxene – kyanite – phengite – coesite / quartz. *Journal of Metamorphic Geology*, **22**, 579–592.

Rubatto, D., and Hermann, J., 2001. Exhumation as fast as subduction? *Geology*, **29**, 3–4.

Tararin, L. A., 1995. Blueschist facies metamorphic rocks from the basement of the Izu-Bonin inner trench wall. In *Proceedings of the 8th International Symposium on Water–Rock Interaction*. Vladivostok: Balkema, pp. 611–614.

Ueda, H., Usuki, T., and Kuramoto, Y., 2004. Intra-oceanic unroofing of eclogite-facies rocks in the Omachi Seamount, Izu-Bonin frontal arc. *Geology*, **32**, 849–852.

Ueda, H., Niida, K., Usuki, T., Hirauchi, K., Meschede, M., Miura, R., Ogawa, Y., Yuasa, M., Sakamoto, I., Chiba, T., Izumino, T., Kuramoto, Y., Azuma, T., Takeshita, T., Imayama, T., Miyajima, Y., and Saito, T., 2011. Seafloor geology of the basement serpentinite body in the Ohmachi Seamount (Izu-Bonin arc) as exhumed parts of a subduction zone within the Philippine Sea. In Ogawa, Y., Anma, R., and Dilek, Y. (eds.), *Accretionary Prisms and Convergent Margin Tectonics in the Northwest Pacific Basin*. Dordrecht: Springer, pp. 97–128.

Wakabayashi, J., 1990. Counterclockwise P-T-t paths from amphibolites, Franciscan complex, California: relics from the early stages of subduction zone metamorphism. *The Journal of Geology*, **98**, 657–680.

Winkler, H. G. F., 1967. *Petrogenesis of Metamorphic Rocks*, 2nd edn. New York: Springer, p. 237.

Cross-references

Accretionary Wedges
Continental Slope
Contourites
Geochronology: Uranium-Series Dating of Ocean Formations
Intraoceanic Subduction Zone
Lithosphere: Structure and Composition
Mid-ocean Ridge Magmatism and Volcanism
Ophiolites
Passive Plate Margins
Seamounts
Serpentinization
Shelf
Subduction
Turbidites

HOT SPOTS AND MANTLE PLUMES

William M. White
Department of Earth and Atmospheric Sciences, Cornell University, Ithaca, NY, USA

Definition

Hot Spot. Persistent volcanism over many millions or tens of millions of years at fixed location on the Earth's surface in an absolute reference frame. This leads to a chain of volcanic islands and seamounts that get progressively older in the direction of *lithospheric plate motion*. The Hawaiian Islands provide the classic example.

Mantle Plume. Mantle plumes are relatively narrow columns of hot, buoyant rock rising from the deep mantle, probably the core–mantle boundary in many cases, and partially melting in the uppermost mantle. The magma produced in this way is responsible for hot spot volcanism and oceanic volcanic islands and *seamounts*.

Introduction

Wilson (1963) pointed out the existence of chains of volcanic islands in the Pacific whose alignment was nearly parallel and whose age increased with distance from the East Pacific Rise. He proposed that the volcanoes formed as convection currents in the shallow mantle dragged oceanic floor over a fixed melting region in the deeper mantle. He also pointed out that ridges in the Atlantic seemed to form Vs centered on the Mid-Atlantic Ridge and argued that these stationary "hot spots" (he did not use the term) provided compelling evidence of the combination of continental drift and seafloor spreading that, by the end of the decade, would come to be known as plate tectonics.

Among the great successes of plate tectonics is its explanation of most magmatism on Earth. The largest fraction occurs at mid-ocean ridges, where every year some 20 km^3 of magma solidifies into new oceanic crust as a direct consequence of the mantle convection that drives *plate motion*. As mantle rises beneath *mid-ocean ridges*, it partially melts (e.g., Klein and Langmuir, 1987).

The resulting magma then rises to fill the gap between the spreading plates. The second most voluminous type of magmatism is *island arc volcanism*, which occurs at *subduction zones*, where oceanic crust returns into the mantle. The primary cause of melting there is water released by dehydration of the subducting oceanic crust, which lowers the melting point of the overlying mantle wedge (e.g., Grove et al., 2006). However, plate tectonics sensu stricto fails to provide an explanation for *intraplate volcanism*, such as hot spot volcanoes, even though hot spots inspired the early evolution of plate tectonic theory. This group includes some of the most active volcanoes on the planet, such as Kilauea in Hawaii and Piton de la Fournaise on Reunion.

Wilson did not attempt to explain why stationary melting regions in the mantle should occur. It was Morgan (1971) who proposed the now widely accepted explanation: Wilson's "hot spots" were the surface manifestation of convection plumes of solid rock rising from the lower mantle (Figure 1). Morgan described these mantle plumes as similar to convection cells in the atmosphere that give rise to thunderstorms. Like air, hot mantle is buoyant and consequently can rise in a more or less vertical column. The analogy can be taken one step further as well. The rising air in a thunder cell eventually undergoes a phase change with water vapor partially condensing. Similarly, rising solid mantle eventually undergoes a phase change and partially melts. While gravity causes the raindrops to fall, it causes the buoyant melt to rise.

Early laboratory and numerical experiments revealed that plumes could readily arise at a hot thermal boundary layer (Whitehead and Luther, 1975). As seismic studies revealed that no thermal boundary layer exists at the 670 km seismic discontinuity or elsewhere in the mid-mantle (e.g., Jordan et al., 1993), a thermal boundary layer at the base of the mantle, across which temperature may increase as much as 1,000 °C, seemed to be the only possible source of mantle plumes. Hence, the idea that plumes are generated at the core–mantle boundary became an integral part of the mantle plume theory. The idea that plumes are generated at the core–mantle boundary found further support from the calculations of Davies (1988) and Sleep (1990) showing the total heat flux of mantle plumes matched the expected heat flux from the core (as it is only about 10 % of the heat flux associated with spreading centers). More recent numerical simulations of mantle convection, which take advantage of higher computational power and contain more realistic physics, suggest this need not be the case. Farnetani and Samuel (2005) showed that thermochemical plumes (plumes that differ in density from surrounding mantle because of both temperature and chemistry) may come in "a great variety of plume shapes and sizes" and may neck to the point where they are no longer visible seismically. This is particularly true as they pass through the 670 km discontinuity. Davies and Bunge (2006) found that cold mantle downwellings plunging into regions of hot mantle can initiate upwelling plumes on the fringes

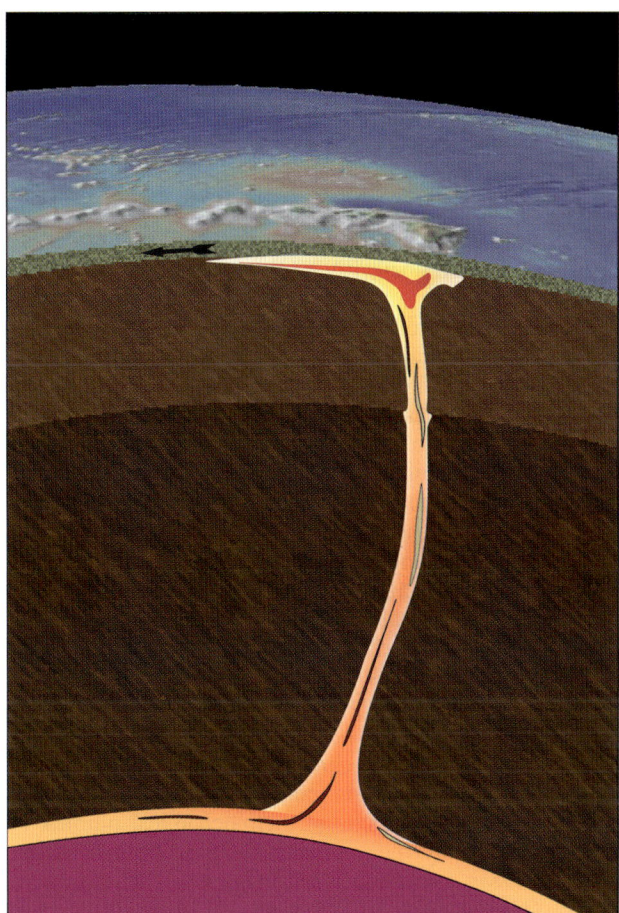

Hot Spots and Mantle Plumes, Figure 1 Cartoon illustrating the Hawaiian mantle plume. Plumes are thought to arise from the hot thermal boundary layer (D″) at the core–mantle boundary in most cases and can incorporate heterogeneities present in that layer. They may be deflected by mantle convection currents and encounter resistance at the upper–lower mantle boundary due to a phase change. Melting is concentrated in the uppermost 100 or 200 km of the plume "stem" and shield volcanoes develop it. As it flattens against the lithosphere, the plume is drawn out in the direction of the plate motion. Smaller extents of melting extend downstream and may be responsible for post-shield and rejuvenescent magmatism.

of the hot regions. The deformed hot regions and upwellings look much like the splash of a water droplet, so Davies and Bunge called them "splash plumes."

At first controversial, Morgan's plume theory is now widely, but not universally (e.g., Anderson, 2000), accepted as the explanation for most intraplate volcanism. It does not, however, explain all intraplate volcanism, which can have a variety of causes. These include tensional rifting and lithospheric extension (e.g., Van Wijk and Blackman, 2005), plate flexure (Hirano et al., 2006), shear-driven upwelling (e.g., Conrad et al., 2010), and small-scale convection features immediately beneath the lithosphere (Ballmer et al., 2007; King, 2007). Indeed, there are hundreds of thousands of small volcanic seamounts, most unrelated to mantle plumes. This entry will focus on only hot spots related to mantle plumes. Campbell (2007) and White (2010) provide additional reviews.

Plumes and large igneous provinces

Early laboratory and numerical simulations of plumes (e.g., Whitehead and Luther, 1975; Campbell and Griffiths, 1990; Griffiths and Campbell, 1990; Olsen, 1990) suggested that plumes necessarily start with a large voluminous head. As Morgan (1972) noted, a number of oceanic island and seamount chains extrapolate to large igneous provinces, or LIPs, which include flood basalts or "traps" on continents and oceanic plateaus in the oceans and result from massive eruptions of basalt over a relatively short period of time. For example, the Reunion plume, currently beneath volcanically active Reunion Island, can be connected through Mauritius, the Mascarene Plateau, and the Chagos-Laccadive Ridge to the Deccan Traps flood basalt province of India, produced 65 Ma ago by the eruption of more than 500,000 km^3 of basalt in roughly a million-year period. These observations led to the hypothesis that LIPs are the products of plume heads and oceanic islands are the product of plume tails (Richards et al., 1989). This idea became an essential component of the plume theory. However, relatively few oceanic island chains can be confidently associated with LIPs. Subsequent numerical simulations indicate that the surface expression of thermochemical plumes, as opposed to purely thermal plumes, "may be a headless, age-progressive volcanic chain" (Farnetani and Samuel, 2005). *Oceanic plateaus*, a variety of LIP, are the subject of a separate entry and will not be further discussed here.

How many plumes?

Table 1 lists 29 oceanic island chains/hot spots possibly produced by mantle plumes, but just how may mantle plumes are presently active is unclear and debated. Morgan suggested there were about 20 mantle plumes. Sleep (1990) estimated buoyancy fluxes for 37 plumes. Courtillot et al. (2003) considered 49 "hot spots" and concluded that only 7–10 were products of deep mantle plumes based on five criteria: (1) a long-lived track, (2) a LIP at initiation, (3) a topographic anomaly indicative of a buoyancy flux in excess of 10^3 kg/s, (4) high ^{3}He/^{4}He and ^{20}Ne/^{22}Ne ratios, and (5) negative shear-wave anomalies indicative of hot mantle. Only those hot spots meeting at least three or more of these criteria were considered to be of "primary plumes" of likely deep mantle origin. They suggested that many of the remaining hot spots were "secondary," perhaps triggered by large low shear-wave velocity provinces (LSSVPs) located at the base of the mantle and extend up to 1,000 km above it beneath Africa and the central equatorial Pacific.

As noted earlier, it is possible that plumes might begin without large buoyant heads and some may be short lived or young. In addition, Courtillot's noble gas isotope criterion assumes mantle plumes always consist of relatively undegassed material, whereas other chemical evidence discussed below suggests recycled oceanic crust may be the dominant component in many cases. Thus, of the Courtillot's 5 criteria, the last, seismic evidence of hot mantle, is perhaps the most important. Subsequent seismic studies by Montelli et al. (2004, 2006) found slow seismic wave velocity anomalies under 30 of 33 hot spots that extended at least to depths greater than the upper–lower mantle boundary at 660 km. Of these, the velocity anomalies extend to depths of greater than 1,450 km in 23 of the hot spots examined. In addition, Montelli et al. (2006) identified three "plumes" in the lowermost mantle that did not extend into the upper mantle (and are not associated with surface hot spots).

Ebinger and Sleep (1998) suggested that ponding and flow of the "Ethiopian" or Afar plume beneath the African lithosphere could account not only for the Mid-Tertiary Ethiopian flood basalt province and modern Afar volcanism but also hot spots in central North Africa such as Tibesti, the Cameroon line volcanism, and the oceanic island volcanism on the Comoros. Furthermore, some of the plumes imaged by Montelli et al. appear to merge at depth, such as the Ascension and St. Helena plumes. Thus, a single plume in some cases might be responsible for more than one hot spot.

Hot spots aren't entirely stationary

Both Wilson (1963) and Morgan (1971) noted the apparently fixed nature of hot spots, and this became integral to mantle plume theory and was often used to reconstruct plate motions. Subsequent work showed that hot spots do move relative to one another (e.g., Molnar and Stock, 1987), although the velocities are much lower than those of lithospheric plates. In addition, the relative motion between Pacific hot spots and between Indo-Atlantic hot spots is much smaller than the relative motion between a Pacific hot spot and an Atlantic or Indian hot spot (e.g., Courtillot et al., 2003). It was perhaps always overly simplistic to expect hot spots to remain stationary; after all, convection plumes in the atmosphere are not stationary: thunderstorms move and can be bent by atmospheric shear. We might expect similar behavior in the mantle. Indeed, based on a mantle convection model, Steinberger et al. (2004) were able to explain nearly all the apparent relative motion between hot spots. The surprising thing is how little, not how much, hot spots move relative to one another. The most likely reason is that three-fourths of a plume's journey to the surface takes place through the lower mantle, whose viscosity is probably one to two orders of magnitude greater than that of the upper mantle and consequently flow is sluggish.

Petrology and geochemistry of oceanic island basalts and mantle plumes

Morgan (1971) pointed out that his model was compatible with the compositional difference between oceanic island basalts (OIB) and mid-ocean ridge basalts (MORB)

Hot Spots and Mantle Plumes, Table 1 Mantle plumes and oceanic island chains

Chain	Location	Oceanic crust age	Youngest volcanism	Oldest volcanism	Flux (Mg s^{-1})	Seismic depth[c] (km)	Large igneous province
Amsterdam–St. Paul	39°S 77°E	3–5 Ma	~0 Ma	>18 Ma			Rajmahal(?)
Ascension	8°S 14°W	7 Ma	2 Ma	6–7 Ma		~2,800	No
Austral–Cook	29°S 140°W	40–90 Ma	0 Ma	30 Ma	3.3[a], 3.9[b]	~2,800	No
Azores	38°N 28°W	0 Ma	0 Ma	36 Ma	1.1[a]	~2,800	Azores Plateau?
Balleny	65°S 167°E	10–20[7]	0?[7]	2.6			Farrar and Tasmania?[8]
Bouvet	54°S 3°E	<9.6 Ma	0 Ma	2.5 Ma	0.4[a]	>1,900	Karoo?[8]
Canary Is	28°N 18°W	155 Ma	0 Ma	68 Ma	1[a]	~2,800	No
Cape Verde Is	15°N 24°W	140 Ma	0 Ma	>20 Ma	1.6[a], 0.5[b]	~2,800	No
Comoros	11°S 43°E	128 Ma	0 Ma	7.7 Ma			Madagascar?[9]
Crozet Is	46°S 50°E	80 Ma	0 Ma	183? Ma	0.5[a]	~2,800	Karoo?[10]
San Felix–San Ambrosio	26°S 80°W	20–33 Ma	0 Ma	3.5 Ma	1.6[a], 2.3[b]		No
Easter Is	27°S 108°W	4.5 Ma	0.11 Ma	>26 Ma	3.3[a]	~2,800	Mid-Pacific mountains[10]
Galapagos	1°S 92°W	5 Ma	0 Ma	~100 Ma	1[a]	~2,800	The Caribbean[11]
Tristan da Cunha–Gough	37°S 12°W	16 Ma	0 ma	130 Ma	1.7[a], 0.5[b]		Etendeka and Parana
Hawaii	19°N 155°W	93 Ma	0 Ma	>80 Ma	8.7[a], 6.2[b]	≥2,350	?
Heard–Kerguelen	49°S 69°E	118 Ma	0 Ma	118 Ma	0.5[a], 0.2[b]	~2,800	Kerguelen Plateau, Rajmahal?[12]
Iceland	65°N 10°W	0 Ma	0 Ma	61 Ma	1.4[a]	~2,350	North Atlantic Tertiary
Jan Mayen	80°N 9°W	0	0	>0.7 Ma			Siberia?[13]
Juan Fernandez Is	33°S 80°W	20–33 Ma	1 Ma	4.2 Ma	1.6[a], 1.7[b]	≥2,350	No
Louisville	50°S 139 W	~66 Ma	1 Ma	>80 Ma	≤0.9[a], 3[b]	≥1,450	Ontong Java
Madeira	33°N 17°W	129 Ma	1 Ma	70 Ma			No
Marion–Prince Edward	47°S 38°E	0	0 Ma	88 Ma?			Madagascar?[14]
The Marquesas	9°S 138°W	63 Ma	0.5 Ma	43 Ma (?)	3.3[a], 4.6[b]		Shatsky Rise?[10]
Martin Vaz/Trindade	29°S 19°W	90 Ma	1.5 Ma	84 Ma	0.5[a], 0.8[b]		Poxoreu?[16]
Reunion	21°S 56°E	64 Ma	0	65 Ma	1.9[a], 0.9[b]	≥1,900	Deccan
Pitcairn–Gambier	25°S 130°W	23 Ma	0[17]	8 Ma	3.3[a], 1.7[b]		No
Samoa	15°S 168°W	100 Ma	0	24 Ma	1.6[a]		No
The Society Is	18°S 149°W	45–63 Ma	0	4.3 Ma	3.3[a], 5.8[b]	~2,800	No
St. Helena	16°S 6°W	41 Ma	2.5 Ma[19]	>81 Ma	0.5[a], 0.3[b]	~2,800	No?

Modified from White (2010)
[a]Plume flux in 10^6 g/s from Sleep (1990)
[b]Plume flux in 10^6 g/s from Davies (1988)
[c]Depth of the s- or p-wave seismic velocity anomaly from Montelli et al. (2006)

established by earlier studies, such as those of Gast (1969) and Hedge (1966). There are several reasons for this difference. First, Morgan predicted that plumes should bring "relatively primordial material" to the shallow mantle, and hence basalts produced by melting of plumes should be chemically distinct from the upper mantle. Second, Morgan predicted that plumes should bring "heat" as well, this too should influence the composition of melts; all other things being equal, one would expect the extent of melting in plumes to be greater than beneath mid-ocean ridges. However, all other things are not always equal, as will be explained below. Finally, if plumes are buoyant, the rate of melting and extents of melting should be different than beneath mid-ocean ridges. Each of these will be considered in the following sections.

Plumes consist of a diverse mix of primitive and recycled materials

Melting of the mantle preferentially removes elements that, because of their ionic size or charge, are not easily accommodated in mantle minerals. These so-called

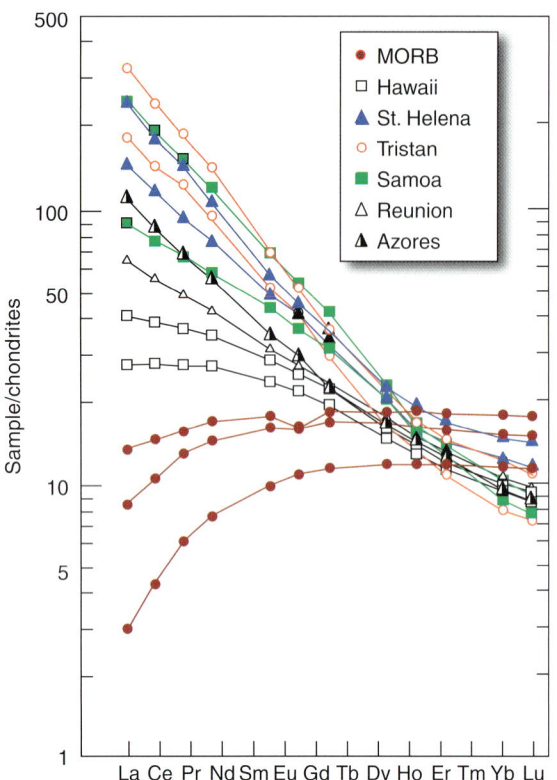

Hot Spots and Mantle Plumes, Figure 2 Rare earth patterns in selected oceanic island and mid-ocean ridge (*MORB*) basalts. Oceanic island basalts are relatively enriched in incompatible elements such as the light rare earth elements, while mid-ocean ridge basalts are relatively depleted in such elements (modified from White (2013)).

incompatible elements, including K, Rb, Sr, Ba, Nb, Ta, Hf, Zr, U, Th, Pb, and the light rare earths, are relatively depleted in MORB but are enriched in OIB (e.g., Figure 2). Schilling (1973) demonstrated a progressive change in the La/Sm ratio of basalts erupted along the Reykjanes Ridge as Iceland was approached and argued this reflected chemical variation in the mantle source of these basalts (see "Hotspot-Ridge Interaction"). Changes in elemental ratios such as La/Sm in basalts can, however, result from differences in magma generation and evolution, such as the extent of partial melting and fractional crystallization, as O'Hara (1973) argued at the time. In contrast, radiogenic isotope ratios are insensitive to these processes. Consequently, the study by Hart et al. (1973), which showed that $^{87}Sr/^{86}Sr$ also increased along the Reykjanes Ridge toward Iceland, clearly demonstrated a difference between the mantle beneath Iceland and the Mid-Atlantic Ridge to the south. Subsequent studies documented similar variations along mid-ocean ridges in the vicinity of other hot spots, including the Azores (White et al., 1976), the Galapagos (Schilling et al., 1976, 1982), Afar (Schilling et al., 1992), Tristan da Cunha (Hanan et al., 1986), etc.

Schilling (1973) followed Morgan and referred to the Iceland mantle plume as "primordial," in the sense that it consisted of mantle from which little or no melt had been previously extracted. A natural evolution of this idea was a two-layered mantle, in which the upper mantle source of MORB had been incompatible element-depleted (see "Depleted Mantle") by melt extraction to form the crust and the lower mantle consisted of primitive, unprocessed material (e.g., DePaolo and Wasserburg, 1976; O'Nions et al., 1979; Jacobsen, 1988). Relatively high $^{3}He/^{4}He$ ratios in Icelandic and other OIB seemed consistent with this idea (Kurz et al., 1982). Subsequent work has shown that plumes cannot consist entirely of primitive or primordial mantle (see below). First, Pb isotope ratios of OIB cannot be explained through mixing of primordial material and incompatible element-depleted upper mantle. Second, Nd isotope ratios in OIB extend well below any plausible value for primitive mantle, which likely lies between ε_{Nd} of 0 (DePaolo and Wasserburg, 1976) and +7 (Caro and Bourdon, 2010). Furthermore, as Figure 3 illustrates, Sr and Nd isotope ratios in OIB exhibit dispersion well beyond what can be explained by simple two-component mixing of depleted and primitive mantle (White and Hofmann, 1982). Finally, mantle seismic tomography reveals whole mantle convection that is inconsistent with the idea of a simple two-layered mantle initially envisioned by geochemists (Grand et al., 1997).

It became subsequently clear that mantle plumes come in a variety of geochemical "flavors" implying the deep mantle consists of material that has evolved through a variety of processes (White, 1985). Perhaps the most significant of these is subduction of oceanic crust and sediment into the deep mantle (Hofmann and White, 1982). The signature of ancient crustal material is particularly apparent in the so-called EM II OIB (Figure 3), notably Samoa and the Society Islands, which have non-mantle *oxygen isotope ratios* (Eiler et al., 2005), extreme radiogenic isotopic compositions (Jackson et al., 2007), and crustal-like trace element ratios (White and Duncan, 1996). More recently, Cabral et al. (2013) have found mass-independent-fractionated sulfur isotope ratios in sulfide inclusions in a lava from Mangaia of the Cook Islands, which belong to the HIMU group (Figure 3). Mass-independent fractionation of sulfur isotopes occurred only in the atmosphere prior to the great oxidation event 2.3 billion years ago. Thus the source of at least some HIMU lavas also includes anciently recycled surface material. Willbold and Stracke (2010) have argued based on trace element and isotope systematics that the EM I group (Figure 3), which includes islands such as Kerguelen and Tristan da Cunha, carry the chemical signature of lower continental crust, which returns to the mantle either through lower crustal foundering/delamination or *subduction erosion*. Finally, Blichert-Toft et al. (1999) documented a strong, curved correlation between Hf and Pb isotope ratios in Hawaiian basalts that almost certainly reflects mixing between mantle and an ancient crustal component, most likely sedimentary. Thus, crustal material subducted into the deep mantle may be pervasive in mantle plumes.

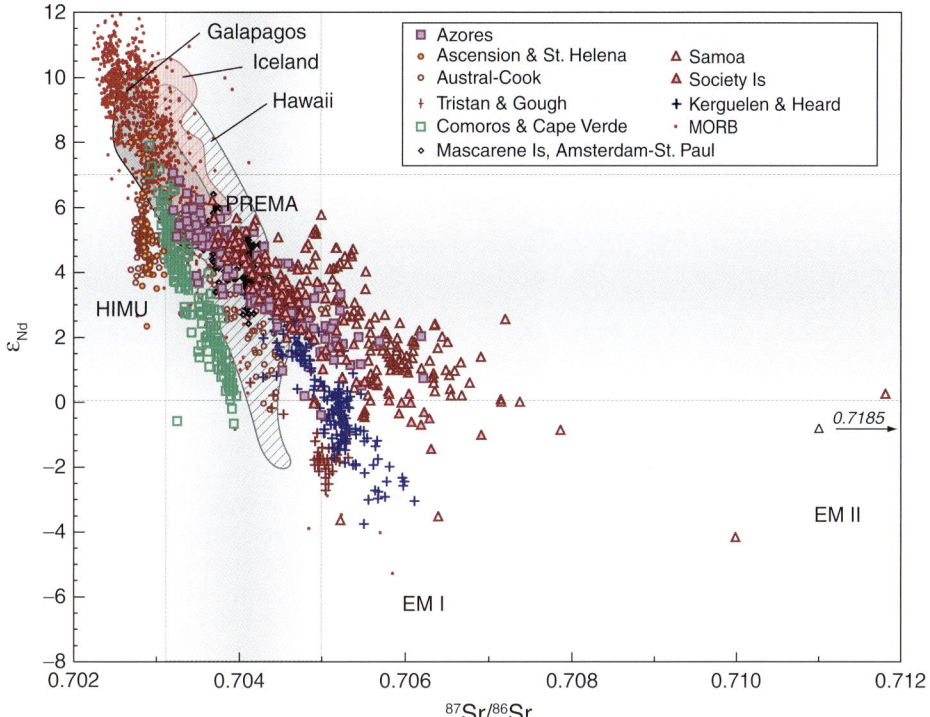

Hot Spots and Mantle Plumes, Figure 3 Sr and Nd isotope ratios in selected oceanic island and mid-ocean ridge basalts. ε_{Nd} is the deviation in parts in 10,000 of the $^{143}Nd/^{144}Nd$ ratio from the chondritic value (=0.512638). Island chains form a variety of distinct trends on this diagram, indicating their sources have evolved in a variety of ways (White, 1985). HIMU, EM I, EM II, and PREMA are the end-member names of Zindler and Hart (1986). *Gray areas with lines* indicate the likely composition of primitive mantle or bulk silicate Earth.

In contrast, there is no evidence in any OIB of a component produced by melting in the deep mantle, although there is seismic evidence that localized melting might occur in small regions near the core–mantle boundary. Experiments show that partitioning of elements between deep mantle minerals (those in which silicon is octahedrally coordinated by oxygen) and liquids is quite different from solid–liquid partitioning in the upper mantle (e.g., Corgne et al., 2005). So far, element abundances in all OIB are consistent with their distinct signatures having formed in the upper mantle or crust and not in the lower mantle.

Noble gas isotope ratios, however, indicate that a near-primitive component is present in many mantle plumes. As noted above, high $^3He/^4He$ ratios in many OIB indicate they are derived from mantle that has been substantially less degassed than the *depleted* upper *mantle*. Decay of U and Th constantly produces ^4He in the mantle so that the $^3He/^4He$ ratio decreases through time. When mantle melts, He is efficiently degassed from the mantle into the atmosphere and lost to space. Consequently, processed mantle as well as recycled crustal material should have low $^3He/^4He$ (Porcelli and Elliott, 2008). Some OIB do, but high ratios predominate. Hanan and Graham (1996) noted the association of high $^3He/^4He$ ratios with intermediate Pb isotope ratios and argued that a primitive component, which they termed "C," is present in all or most plumes. Thus, while recycled material dominates the geochemical signature of plumes just as strawberries, vanilla, and chocolate dominate the flavor of ice cream, this may not be the main ingredient.

Recent studies of other noble gas isotope ratios, particularly of xenon, support the notion that at least some plumes contain a very primitive component indeed. Xenon isotopic composition varies because of decay of two extinct radionuclides: ^{129}I, which decays to ^{129}Xe with a 15.7 Ma half-life, and ^{244}Pu, which has a 82 Ma half-life and whose principal fission products include the heavy Xe isotopes (Xe isotopic composition also varies due mass-dependent fractionation and fission of ^{238}U). $^{129}Xe/^{130}Xe$ ratios in OIB are higher than in the atmosphere and are higher still in MORB. These differences, as well as those in heavy fissogenic Xe isotopes must have been established within the first 100 Ma of Earth's history, require that the plume source to less degassed than the MORB source and that subsequent mixing between them over the next 4.3 billion years must have been limited (Mukhopadhyay, 2012). The primitive signal appears to be largely confined to noble gases; concentrations and isotope ratios of other elements indicate plumes consisting primarily of processed material.

Mantle plumes can also be heterogeneous, exhibiting systematic geographic variation in composition. The best-documented examples are Hawaii and the Galapagos. Tatsumoto (1978) and Stille et al. (1986) showed that the Hawaiian hot spot consists of two volcanic chains: Loa and Kea, whose Pb isotopic compositions are distinct and have remained so for about 5 Ma. Subsequent studies (e.g., Abouchami et al., 2005; Bryce et al., 2005; Weis et al., 2011) fully support that early work and show the difference extends to other isotope ratios as well. Parallel chains having distinct radiogenic isotopic characteristics have also been identified in the Marquesas, the Society Islands, Samoa, and the Tristan da Cunha–Gough hot spot (Desonie et al., 1993; Huang et al., 2011; Chauvel et al., 2012; Payne et al., 2013; Rohde et al., 2013). In the Galapagos, four distinct plume components, northern, central, southern, and a depleted eastern one, can be distinguished in the mixing trends (White et al., 1993; Harpp and White, 2001). Hoernle et al. (2000) showed that these same domains occur in the same relative positions in a geochemical profile across the Cocos Ridge (the Galapagos hot spot track on the Cocos Plate) indicating that the spatial zonation of the Galapagos hot spot has continued for at least 14 Ma. As they point out, this implies "that plume material can ascend from the lower mantle, possibly from the core-mantle boundary, with little stirring occurring during ascent, and that zonation in hotspot lavas may in some cases reflect spatial heterogeneity within the lower mantle source." These zonations may in turn reflect chemical zonation and structure in the lowermost mantle (Weis et al., 2011; Farnetani et al., 2012; Harpp et al., 2014).

In summary, plumes carry a variety of material from the deep mantle to the surface, including anciently subducted oceanic and continental crust and very primitive noble gas signal preserved from the earliest days of Earth's history.

Lithological composition of plumes

The lithology of plumes, which is their mineralogical makeup, is more difficult to assess because it cannot be directly observed, only inferred. Without question, the mantle consists predominately of *peridotite*. However, populations of mantle xenoliths always contained a small but significant fraction of rocks, such as eclogite and pyroxenite, similar in composition to basalt. Allègre and Turcotte (1986) pointed out that continuous production and subduction of oceanic crust over billions of years was likely to produce a "marble-cake" mantle with ubiquitous stringers of mafic material embedded in an ultramafic matrix. Indeed, recent numerical simulations suggest 97 % of the mantle has been processed through the oceanic crust cycle (Huang and Davies, 2007). If, as isotopic and trace element studies suggest, plumes are particularly rich in recycled crustal materials, one might expect them to be lithologically distinct from asthenospheric mantle and particularly rich in mafic materials such as eclogite and pyroxenite. Since this material melts at lower temperature than peridotite, might the mantle plume melts be derived predominantly from pods and stringers of mafic material rather than the more refractory peridotite in which they are embedded?

This question remains unresolved and debated. Stracke et al. (1999) argued against the existence of garnet-pyroxenite or eclogite in the Hawaiian source on the basis of combined Hf-Nd-Th isotope and trace element systematics. On the other hand, Reiners (2002) argued that "temporal-compositional trends in the form of decreasing incompatible elements and MgO, with increasing SiO_2, as the eruption proceed"... "reflect sequential eruption of melt from increasing depths in mantle melting regions that are compositionally zoned because of different solidi and melt productivities of distinct peridotite and pyroxenite lithologic domains."

A particularly interesting, provocative, and consequently controversial idea was proposed by Sobolev et al. (2005), who argue on the basis of SiO_2, Ni, and Mn concentrations that part of the Hawaiian source is olivine-free. They suggested that eclogite, which originates as recycled oceanic crust, composes up to 30 % of the Hawaiian plume. As a consequence of its much lower solidus, the eclogite melts at greater depth than peridotite. This melt then reacts with surrounding peridotite to form pyroxenite, which subsequently melts at shallower depth. Sobolev et al. (2007) argue that recycled oceanic crust in the form of eclogite stringers, and consequently this two-stage melting process, is pervasive in the mantle; according to them even MORB are composed of 10–30 % pyroxenite-derived melts. A number of other studies have also concluded that pyroxenites play an important role in the genesis of Hawaiian magmas. Putirka et al. (2011), however, argue that high Ni contents and high FeO/MnO ratios in Hawaiian basalts compared to MORB reflect higher temperatures (by 300°) and pressures (4–5 GPa vs. 1–3 GPa) of melting under Hawaii than beneath mid-ocean ridges and that both Hawaiian basalts and MORB are generated by melting of peridotite rather than pyroxenite or eclogite. They suggest that subducted material is thoroughly mixed back into the peridotite matrix and does not survive "high temperatures and potentially long residence times" in the deep mantle as a distinct lithological component.

Evidence that plumes are hot and buoyant

If plumes are hotter than the depleted upper mantle, this should produce petrologic and chemical distinctions. To begin with, one would expect them to melt more. Because melting in both plumes and at mid-ocean ridges results from adiabatic decompression of rising mantle, the extent of melting depends both on temperature, which determines the depth at which melting begins, and lithospheric thickness, which determines the height to which mantle can rise, and consequently the total amount of melting for mantle of a given temperature and composition. At mid-ocean ridges, the lithosphere is thin to nonexistent and melting continues to shallow depth (<30 km). Older lithosphere, for example, beneath Hawaii, may approach

90 km. Thus, the extent of melting alone is not a good indicator of mantle temperature.

A useful comparison is between melting in ridge-centered plumes and mid-ocean ridges distant from them. Dick et al. (1984) found that abyssal peridotites in the vicinity of hot spots had mineral abundances and compositions suggesting they had melted to greater degree than had peridotites away from hot spots. They noted that the peridotite mineralogy was correlated regionally with basalt chemistry and concluded that hot spots were indeed hotter and hence likely products of mantle plumes. Michael and Bonatti (1985) reached similar conclusions from a study of peridotite compositions. Klein and Langmuir (1987) concluded from a comprehensive study of MORB chemistry that mantle temperatures beneath the mid-ocean ridges varied by as much as 250 °C. The highest temperatures, and highest extents of melting, invariably occurred near hot spots such as Iceland. This, of course, makes some sense because Iceland crust is far thicker (~30 km) than normal oceanic crust (~6–7 km), requiring much greater amounts of melt for its creation.

Putirka et al. (2007) used a refined version of the Roeder and Emslie (1970) olivine geothermometer to calculate melt temperatures of magmas from mid-ocean ridges and oceanic islands and from them derived mantle potential temperatures. They calculated mantle potential temperatures of $1,454 \pm 81$ °C for mantle beneath mid-ocean ridges, 1,722 °C for mantle beneath the Hawaiian and Samoan chains, and 1,616 °C for mantle beneath Iceland. These translate to excess temperatures of 268 °C and 162 °C, respectively, in overall good agreement with earlier studies. Consistent with this, Putirka et al. (2011) calculated that the Hawaiian magmas are generated by up to 20 % melting, compared to 8 % for MORB.

Potential temperatures calculated by Herzberg et al. (2007) for ambient mantle beneath mid-ocean ridges were lower: in the range of 1,280–1,400 °C, but they nevertheless concluded that plumes are typically 200–300 °C hotter. Herzberg and Asimow (2008) estimated mantle excess potential temperatures for a variety of other plumes and found that they scattered between those for Hawaii (~200 °C) and for Iceland (~100 °C). These estimates are also in good agreement with studies that use entirely different approaches, such as the width of the geochemical anomaly along ridges (Schilling, 1991) and excess topography and geoid height (Watson and Mckenzie, 1991). Finally, Courtier et al. (2007) found a correlation between petrologically determined temperatures in OIB and MORB and the seismically determined thickness of the underlying mantle transition zone. Beneath hot spots, the transition zone is thinner, precisely the expected result if hot spots result from hot mantle rising through the transition zone. There is, however, not a complete consensus on this point as Green et al. (2001) and Falloon et al. (2007), also using the olivine geothermometer, found no difference in potential temperatures between the sub-ridge and the sub-Hawaiian mantle. Possible explanations for this discrepancy are discussed by Putirka et al. (2007).

The melt production rate is related to the upwelling rate. The extent of radioactive disequilibrium between ^{230}Th and its parent, ^{238}U, should depend on these rates (e.g., Spiegelman and Elliot, 1993). Consequently, the degree to which the ^{230}Th/^{238}U activity ratio (activity ratios will hereafter be designated by parentheses) exceeds 1 should be a measure of the rate of mantle upwelling. Indeed, Chabaux and Allegre (1994) demonstrated that (^{230}Th/^{238}U) ratios in OIB are inversely correlated with buoyancy flux as estimated by Sleep (1990). Bourdon et al. (2006) demonstrated that the relationship observed by Chabaux and Allègre is precisely that of the expected based on simple assumptions about melt porosity and partition coefficients and the relationship between melt production and upwelling rate. They further showed that (^{230}Th/^{238}U) and (^{231}Pa/^{235}U) ratios vary with distance from the assumed center of plumes in a manner that is consistent with buoyancy fluxes and excess temperatures in the range of 100–300 °C.

Rejuvenescent volcanism

One aspect of oceanic island volcanism that remains somewhat puzzling is the so-called "post-erosional" or "rejuvenescent" volcanism: eruptions that occur after a long hiatus, sometimes several million years. The Honolulu volcanic series of Oahu, which includes the iconic Diamond Head cone, is the classic example. This phenomenon was first recognized in Hawaii (e.g., Sterns, 1946; MacDonald, 1968) and has since been recognized in at least Samoa, the Society Islands, the Marquesas, Madeira, the Canaries, the Louisville Seamounts, Mauritius, and, most recently, Jasper Seamount (Konter et al., 2009). Usually these cases mimic the Hawaiian pattern in which the post-erosional phase involves eruption of alkaline to highly alkaline, low-silica lavas enriched in incompatible elements (all of which indicate low extents of melting) but with a more depleted isotopic signature than lavas of the main shield-building phase (e.g., Chen and Frey, 1983). The eruption rates in the post-erosional phase are, however, far lower than in the shield-building phase. Garcia et al. (2010) estimated the rejuvenescent Koolau Series on Kaua'i comprising about 0.1 % of the volcano volume, while Moore et al. (2011) estimate the rejuvenescent phase on Mauritius comprising about 0.05 % of the volume, even though these lavas cover most of the island's surface. Rejuvenescent volcanism has not been identified in on- or near-ridge hot spots such as the Galapagos and Azores.

In Samoa, the rejuvenescent volcanism is so extensive on the westernmost island of Savai'i that it was initially thought that age progressed from east to west, opposite of the prediction of the hot spot model (Hawkins and Natland, 1975). Natland (1980) argued that Samoan volcanism resulted from lithospheric flexure as the Pacific plate passes north of the Tonga–Kermadec subduction zone rather than a mantle plume. However, Koppers et al. (2008) obtained ^{40}Ar–^{39}Ar ages of 5 Ma on basalts dredged from the flanks of Savai'i, which agree well with the predicted age assuming a lithospheric motion of 7.1 cm/year and a current location of the hot spot beneath the young, volcanically active seamount Vailulu'u. The volume of Samoan rejuvenescent volcanism does seem unusually large; Konter and Jackson (2012) estimate that it is responsible for up to 2 % of the volume of Savai'i. It is also unusual in that the post-erosional volcanics do not have a depleted radiogenic isotope signature (Wright and White, 1987; Konter and Jackson, 2012).

The two main questions are why does volcanism flair up again after the volcano has moved off the plume and what is melting to produce post-erosional magmas? Several ideas have been put forth as to the cause. The numerical model of plume dynamics of Ribe and Christensen (1999) predicts small secondary melting zone downstream from the hot spot as a consequence of upwelling and decompression of the mantle plume as it interacts with the overlying lithosphere. Jackson and Wright (1970), ten Brink and Brocher (1987), and Bianco et al. (2005) have proposed it results from lithospheric flexure caused by the growth of the next volcano in the chain. Liu and Chase (1991) proposed that rejuvenescent magmas are derived from the base of the lithosphere which is conductively heated by the plume sufficiently to melt to a small degree.

Because rejuvenescent lavas are isotopically distinct from the shield phase lavas, they must have a different source. Chen and Frey (1983) suggested the source was oceanic lithosphere, but in detail the isotope and trace element systematics are inconsistent with this hypothesis (Yang et al., 2003; White, 2010). White and Duncan (1996) and Fekiacova et al. (2007) suggested it was deep mantle viscously entrained by the plume, and geodynamic models do indeed show a region of melting deep in the plume extending well downstream of the principal melting regime (e.g., Hofmann and Farnetani, 2013; Figure 1). Bianco et al. (2005) and Paul et al. (2005) proposed that the Hawaiian and Reunion mantle plumes, respectively, are lithologically heterogeneous with a relatively incompatible-element-enriched mafic component embedded within a depleted peridotite. According to these models, the pyroxenite melts more or less completely in the plume stem as it rises beneath the growing shield volcano. Subsequently, due either to lithospheric flexure or buoyant lateral spreading of the plume, small degree melts of the depleted peridotite are generated and erupt as the post-erosional lavas. Moore et al. (2011) proposed a variant of the Chen and Frey hypothesis in which pyroxenite veins form in the lowermost lithosphere as early shield-building stage melts react with it when it is still cold. Eventually, conductive heating by the plume causes these veins to melt to a small degree (Gurriet, 1987), producing the rejuvenescent magmas. Except for Samoa, where lithospheric flexure appears to be the cause, no consensus has formed around any of these hypotheses.

Summary

Hot spots are the surface manifestation of mantle plumes. This hypothesis, proposed by Morgan (1971) to explain Wilson's "hot spots," was initially controversial, but over the last 40 years, a wealth of geophysical, geochemical, and petrological data now place it on firm ground. Oceanic island chains are located on bathymetric "swells," which allow estimates of the buoyancy fluxes of the plumes (Davies, 1988; Sleep, 1990). The total heat flux carried by plumes is about 10 % of the total mantle heat flux, indicating that plumes are a secondary feature of mantle convection, with plate tectonics being the dominant one. Estimated buoyancy fluxes correlate with upwelling ratios estimated from (^{230}Th/^{238}U) ratios in the lavas (Bourdon et al., 2006). Regions of slow seismic velocity, indicative of high temperatures and extending into the deep mantle and the base of the mantle in many cases, have been found under most of the oceanic hot spots (Montelli et al., 2006). Consistent with this, petrological studies demonstrate that the potential temperatures of the mantle from which OIB are generated are some 100–300 °C hotter than ambient asthenosphere (e.g., Putirka et al., 2007). Isotope and trace element geochemistry reveal that plumes contain a variety of materials originating on or near the Earth's surface that have been recycled into the deep mantle through subduction, subduction erosion, or lower crustal foundering. Many plumes also appear to sample very primitive material, which may have remained essentially unprocessed since the Earth's earliest days.

Bibliography

Abouchami, W., Hofmann, A. W., Galer, S. J. G., Frey, F. A., Eisele, J., and Feigenson, M., 2005. Lead isotopes reveal bilateral asymmetry and vertical continuity in the Hawaiian mantle plume. *Nature*, **434**(7035), 851–856.

Allègre, C. J., and Turcotte, D. L., 1986. Implications of a two component marble-cake mantle. *Nature*, **323**, 123–127.

Anderson, D. L., 2000. The thermal state of the upper mantle; No role for mantle plumes. *Geophysical Research Letters*, **27**(22), 3623–3626, doi:10.1029/2000gl011533.

Ballmer, M. D., van Hunen, J., Ito, G., Tackley, P. J., and Bianco, T. A., 2007. Non-hotspot volcano chains originating from small-scale sublithospheric convection. *Geophysical Research Letters*, **34**(23), L23310, doi:10.1029/2007gl031636.

Bianco, T. A., Ito, G., Becker, J. M., and Garcia, M. O., 2005. Secondary Hawaiian volcanism formed by flexural arch decompression. *Geochemistry, Geophysics, Geosystems*, **6**(8), Q08009, doi:10.1029/2005gc000945.

Blichert-Toft, J., Frey, F., and Albarede, F., 1999. Hf isotope evidence for pelagic sediments in the source of Hawaiian basalts. *Science*, **285**, 879–882.

Bourdon, B., Ribe, N., Stracke, A., Saal, A., and Turner, S. P., 2006. Insights into the dynamics of mantle plumes from uranium-series geochemistry. *Nature*, **444**, 713–717, doi:10.1038/nature05341.

Bryce, J. C., DePaolo, D. J., and Lassiter, J. C., 2005. Geochemical structure of the Hawaiian plume: Sr, Nd, and Os isotopes in the 2.8 km HSDP-2 section of the Mauna Kea volcano. *Geochemistry, Geophysics, Geosystems*, **6**, doi:10.1029/2004GC000809.

Cabral, R. A., Jackson, M. G., Rose-Koga, E. F., Koga, K. T., Whitehouse, M. J., Antonelli, M. A., Farquhar, J., Day, J. M. D., and Hauri, E. H., 2013. Anomalous sulphur isotopes in plume lavas reveal deep mantle storage of Archaean crust. *Nature*, **496**(7446), 490–493, doi:10.1038/nature12020.

Campbell, I. H., 2007. Testing the plume theory. *Chemical Geology*, **241**(3–4), 153–176, doi:10.1016/j.chemgeo.2007.01.024.

Campbell, I. H., and Griffiths, R. W., 1990. Implications of mantle plume structure for the evolution of flood basalts. *Earth and Planetary Science Letters*, **99**, 79–93.

Caro, G., and Bourdon, B., 2010. Non-chondritic Sm/Nd ratio in the terrestrial planets: consequences for the geochemical evolution of the mantle crust system. *Geochimica et Cosmochimica Acta*, **74**(11), 3333–3349.

Chabaux, F., and Allegre, C., 1994. ^{238}U-^{230}Th-^{226}Ra disequilibrium in volcanics: a new insight into melting conditions. *Earth and Planetary Science Letters*, **126**, 61–74.

Chauvel, C., Maury, R. C., Blais, S., Lewin, E., Guillou, H., Guille, G., Rossi, P., and Gutscher, M. A., 2012. The size of plume heterogeneities constrained by Marquesas isotopic stripes. *Geochemistry, Geophysics, Geosystems*, **13**(7), Q07005, doi:10.1029/2012gc004123.

Chen, C.-Y., and Frey, F. A., 1983. Origin of Hawaiian tholeiite and alkalic basalt. *Nature*, **302**, 785–789, doi:10.1038/302785a0.

Conrad, C. P., Wu, B., Smith, E. I., Bianco, T. A., and Tibbetts, A., 2010. Shear-driven upwelling induced by lateral viscosity variations and asthenospheric shear: a mechanism for intraplate volcanism. *Physics of the Earth and Planetary Interiors*, **178**(3'4), 162–175, doi:10.1016/j.pepi.2009.10.001.

Corgne, A., Liebske, C., Wood, B. J., Rubie, D. C., and Frost, D. J., 2005. Silicate perovskite-melt partitioning of trace elements and geochemical signature of a deep perovskitic reservoir. *Geochimica et Cosmochimica Acta*, **69**(2), 485–496.

Courtier, A. M., Jackson, M. G., Lawrence, J. F., Wang, Z., Lee, C.-T. A., Halama, R., Warren, J. M., Workman, R., Xu, W., Hirschmann, M. M., Larson, A. M., Hart, S. R., Lithgow-Bertelloni, C., Stixrude, L., and Chen, W.-P., 2007. Correlation of seismic and petrologic thermometers suggests deep thermal anomalies beneath hotspots. *Earth and Planetary Science Letters*, **264**(1–2), 308–316, doi:10.1016/j.epsl.2007.10.003.

Courtillot, V., Davaille, A., Besse, J., Stock, J., 2003. Three distinct types of hotspots in the Earth's mantle. *Earth and Planetary Science Letters*, **205**, 295–308, doi:10.1016/S0012-821X(02)01048-8.

Davies, G. F., 1988. Ocean bathymetry and mantle convection, 1, large-scale flow and hotspots. *Journal of Geophysical Research*, **93**, 10467–10480.

Davies, J. H., and Bunge, H.-P., 2006. Are splash plumes the origin of minor hotspots? *Geology*, **34**(5), 349–352, doi:10.1130/G22193.1.

DePaolo, D., and Wasserburg, G., 1976. Inferences about magma sources and mantle structure from variations of ^{143}Nd/^{144}Nd. *Geophysical Research Letters*, **3**, 743–746, doi:10.1029/GL003i012p00743.

Desonie, D. L., Duncan, R. A., and Natland, J. H., 1993. Temporal and geochemical variability of volcanic products of the Marquesas Hotspot. *Journal of Geophysical Research, Solid Earth*, **98**(B10), 17649–17665, doi:10.1029/93JB01562.

Dick, H. J. B., Fisher, R. L., and Bryan, W. B., 1984. Mineralogic variability of the uppermost mantle along mid-ocean ridges. *Earth and Planetary Science Letters*, **69**, 88–106.

Ebinger, C. J., and Sleep, N. H., 1998. Cenozoic magmatism throughout east Africa resulting from impact of a single plume. *Nature*, **395**(6704), 788–791.

Eiler, J. M., Carr, M. J., Reagan, M., and Stolper, E., 2005. Oxygen isotope constraints on the sources of Central American arc lavas. *Geochemistry, Geophysics, Geosystems*, **6**(7), Q07007, doi:10.1029/2004gc000804.

Fallon, T. J., Danyushevsky, L. V. A. A., Green, D. H., and Ford, C. E., 2007. The application of olivine geothermometry to infer crystallization temperatures of parental liquids; implications for the temperature of MORB magmas. *Chemical Geology*, **241**(3–4), 207–233, doi:10.1016/j.chemgeo.2007.01.015.

Farnetani, C. G., and Samuel, H., 2005. Beyond the thermal plume paradigm. *Geophysical Research Letters*, **32**, L0731, doi:10.1029/2005GL022360.

Farnetani, C. G., Hofmann, A. W., and Class, C., 2012. How double volcanic chains sample geochemical anomalies from the lowermost mantle. *Earth and Planetary Science Letters*, **359–360**(0), 240–247, doi:10.1016/j.epsl.2012.09.057.

Fekiacova, Z., Abouchami, W., Galer, S. J. G., Garcia, M. O., and Hofmann, A. W., 2007. Origin and temporal evolution of Koolau Volcano, Hawaii: inferences from isotope data on the Koolau Scientific Drilling Project (KSDP), the Honolulu Volcanics and ODP Site 843. *Earth and Planetary Science Letters*, **261**(1–2), 65–83.

Garcia, M. O., Swinnard, L., Weis, D., Greene, A. R., Tagami, T., Sano, H., Gandy, C. E., 2010. Petrology, Geochemistry and Geochronology of Kaua'i Lavas over 4.5 Myr: Implications for the Origin of Rejuvenated Volcanism and the Evolution of the Hawaiian Plume. *Journal of Petrology*, **51**(7), 1507–1540, doi:10.1093/petrology/egq027.

Gast, P. W., 1969. The isotopic composition of lead from St. Helena and Ascension Islands. *Earth and Planetary Science Letters*, **5**, 253–259.

Grand, S. P., van der Hilst, R. D., and Widiyantoro, S., 1997. Global seismic tomography: a snapshot of convection in the mantle. *Geological Society of America: GSA Today*, **7**(4), 1–7.

Green, D. H., Fallon, T. J., Eggins, S., and Yaxley, G., 2001. Primary magmas and mantle temperatures. *European Journal of Mineralogy*, **13**, 437–451.

Griffiths, R. W., and Campbell, I. H., 1990. Stirring and structure in mantle starting plumes. *Earth and Planetary Science Letters*, **99**(1–2), 66–78.

Grove, T. L., Chatterjee, N., Parman, S. W., and Médard, E., 2006. The influence of H_2O on mantle wedge melting. *Earth and Planetary Science Letters*, **249**(1–2), 74–89, doi:10.1016/j.epsl.2006.06.043.

Gurriet, P., 1987. A thermal model for the origin of post-erosional alkalic lava, Hawaii. *Earth and Planetary Science Letters*, **82**(1–2), 153–158, doi:10.1016/0012-821X(87)90115-4.

Hanan, B. B., and Graham, D. W., 1996. Lead and helium isotope evidence from oceanic basalts for a common deep source of mantle plumes. *Science*, **272**(5264), 991–995.

Hanan, B. B., Kingsley, R. H., and Schilling, J.-G., 1986. Pb isotope evidence in the South Atlantic for migrating ridge-hotspot interactions. *Nature*, **322**(6075), 137–144.

Harpp, K. S., and White, W. M. 2001. Tracing a mantle plume: Isotopic and trace element variations of Galápagos seamounts. *Geochemistry, Geophysics, Geosystems*, **2**(1042), doi:10.1029/2000GC000137.

Harpp, K. S., Hall, P. S., and Jackson, M. G., 2014. Galápagos and Easter: a tale of two hotspots. In *The Galapagos: A Natural Laboratory for the Earth Sciences*. Washington, DC: AGU, Vol. 204, pp. 27–40.

Hart, S. R., Schilling, J. G., and Powell, J. L., 1973. Basalts from Iceland and along the Reykjanes Ridge: Sr isotope geochemistry. *Nature Physical Science*, **246**, 104–107.

Hawkins, J. W., and Natland, J. H., 1975. Nephelinites and basanites of the Samoan linear volcanic chain: their possible tectonic significance. *Earth and Planetary Science Letters*, **24**, 427–439.

Hedge, C. E., 1966. Variations in radiogenic strontium found in volcanic rocks. *Journal of Geophysical Research*, **71**, 6119–6126, doi:10.1029/JZ071i024p06119.

Herzberg, C., and Asimow, P. D., 2008. Petrology of some oceanic island basalts: PRIMELT2.XLS software for primary magma calculation. *Geochemistry, Geophysics, Geosystems*, **9**(9), Q09001, doi:10.1029/2008GC002057.

Herzberg, C., Asimow, P. D., Arndt, N., Niu, Y., Lesher, C. M., Fitton, J. G., Cheadle, M. J., and Saunders, A. D., 2007. Temperatures in ambient mantle and plumes: constraints from basalts, picrites, and komatiites. *Geochemistry, Geophysics, Geosystems*, **8**, doi:10.1029/2006gc001390.

Hirano, N., Takahashi, E., Yamamoto, J., Abe, N., Ingle, S. P., Kaneoka, I., Hirata, T., Kimura, J.-I., Ishii, T., Ogawa, Y., Machida, S., and Suyehiro, K., 2006. Volcanism in response to plate flexure. *Science*, **313**(5792), 1426–1428, doi:10.1126/science.1128235.

Hoernle, K., Werner, R., Morgan, J. P., Garbe-Schoenberg, D., Bryce, J., and Mrazek, J., 2000. Existence of complex spatial zonation in the Galapagos Plume for at least 14 m.y. *Geology*, **28**(5), 435–438, doi:10.1130/0091-7613(2000)28<435:EOCSZI>2.0.CO;2.

Hofmann, A. W., and Farnetani, C. G., 2013. Two views of Hawaiian plume structure. *Geochemistry, Geophysics, Geosystems*, **14**(12), 5308–5322, doi:10.1002/2013GC004942.

Hofmann, A. W., and White, W. M., 1982. Mantle plumes from ancient oceanic crust. *Earth and Planetary Science Letters*, **57**, 421–436.

Huang, J., and Davies, G. F., 2007. Stirring in three-dimensional mantle convection models and implications for geochemistry; 2, Heavy tracers. *Geochemistry, Geophysics, Geosystems*, **8**, Q03017, doi:10.1029/2006GC001312.

Huang, S., Hall, P. S., and Jackson, M. G., 2011. Geochemical zoning of volcanic chains associated with Pacific hotspots. *Nature Geoscience*, **4**(12), 874–878, doi:10.1038/ngeo1263.

Jackson, E. D., and Wright, T. L., 1970. Xenoliths in the Honolulu volcanic series. *Journal of Petrology*, **11**, 405–430.

Jackson, M. G., Hart, S. R., Koppers, A. A. P., Staudigel, H., Konter, J., Blusztajn, J., Kurz, M., and Russell, J. A., 2007. The return of subducted continental crust in Samoan lavas. *Nature*, **448**(7154), 684–687, doi:10.1038/nature06048.

Jacobsen, S. B., 1988. Isotopic and chemical constraints on mantle-crust evolution. *Geochimica et Cosmochimica Acta*, **52**, 1341–1350.

Jordan, T. H., Puster, P., Glatzmeyer, G. A., and Tackley, P. J., 1993. Comparisons between seismic earth structures and mantle flow models based on radial correlation functions. *Science*, **261**, 1427–1431, doi:10.1126/science.261.5127.1427.

King, S. D., 2007. Hotspots and edge-driven convection. *Geology*, **35**(3), 223–226, doi:10.1130/g23291a.1.

Klein, E. M., and Langmuir, C. H., 1987. Ocean ridge basalt chemistry, axial depth, crustal thickness and temperature variations in the mantle. *Journal of Geophysical Research*, **92**, 8089–8115.

Konter, J. G., and Jackson, M. G., 2012. Large volumes of rejuvenated volcanism in Samoa: evidence supporting a tectonic influence on late-stage volcanism. *Geochemistry, Geophysics, Geosystems*, **13**(6), Q0AM04, doi:10.1029/2011GC003974.

Konter, J. G., Staudigel, H., Blichert-Toft, J., Hanan, B. B., Polvé, M., Davies, G. R., Shimizu, N., and Schiffman, P., 2009. Geochemical stages at Jasper Seamount and the origin of intraplate volcanoes. *Geochemistry, Geophysics, Geosystems*, **10**, doi:10.1029/2008gc002236.

Koppers, A. A. P., Russell, J. A., Jackson, M. G., Konter, J., Staudigel, H., and Hart, S. R., 2008. Samoa reinstated as a primary hotspot trail. *Geology*, **36**(6), 435–438, doi:10.1130/g24630a.1.

Kurz, M. D., Jenkins, W. J., Schilling, J. G., and Hart, S. R., 1982. Helium isotopic variations in the mantle beneath the central North Atlantic Ocean. *Earth and Planetary Science Letters*, **58**, 1–14.

Liu, M., and Chase, C. G., 1991. Evolution of Hawaiian basalts: a hotspot melting model. *Earth and Planetary Science Letters*, **104**(2'Äi4), 151–165, doi:10.1016/0012-821X(91)90201-R.

MacDonald, G. A., 1968. Composition and origin of Hawaiian lavas. In Coats, R. R., Hay, R. N., and Anderson, C. A. (eds.), *Studies in Volcanology*. Boulder: GSA, Vol. 116, pp. 82–133.

Michael, P. J., and Bonatti, E., 1985. Peridotite composition from the North Atlantic: regional and tectonic variations and implications for partial melting. *Earth and Planetary Science Letters*, **73**, 91–104.

Molnar, P., and Stock, J., 1987. Relative motions of hotspots in the Pacific, Atlantic and Indian oceans since late Cretaceous time. *Nature*, **327**, 587–591.

Montelli, R., Nolet, G., Dahlen, F. A., Masters, G., Engdahl, R., and Hung, S.-H., 2004. Finite-frequency tomography reveals a variety of plumes in the mantle. *Science*, **303**(5656), 338–343, doi:10.1126/science.1092485.

Montelli, R., Nolet, G., Dahlen, F. A., and Masters, G., 2006. A catalogue of deep mantle plumes: new results from finite-frequency tomography. *Geochemistry, Geophysics, Geosystems*, **7**(11), Q11007, doi:10.1029/2006gc001248.

Moore, J., White, W. M., Paul, D., Duncan, R. A., Abouchami, W., Galer, S. J. G., 2011. Evolution of shield-building and rejuvenescent volcanism of Mauritius. *Journal of Volcanology and Geothermal Research*, **207**(1–2), 47–66, doi:10.1016/j.jvolgeores.2011.07.005.

Morgan, W. J., 1971. Convection plumes in the lower mantle. *Nature*, **230**, 42–43.

Morgan, W. J., 1972. Plate motions and deep mantle convection. In Shagam, R., Hargraves, R. B., Morgan, W. J., Van Houten, F. B., Burk, C. A., Holland, H. D., and Hollister, L. C. (eds.), *Studies in Earth and Space Sciences*. Boulder: Geological Society of America, Vol. 132, pp. 7–22.

Mukhopadhyay, S., 2012. Early differentiation and volatile accretion recorded in deep-mantle neon and xenon. *Nature*, **486**(7401), 101–104.

Natland, J. H., 1980. The progression of volcanism in the Samoan linear volcanic chain. *American Journal of Science*, **280-A**(Part 2), 709–735.

O'Hara, M. J., 1973. Non-primary magmas and dubious mantle plume beneath Iceland. *Nature*, **243**(5409), 507–508, doi:10.1038/243507a0.

O'Nions, R. K., Evensen, N. M., and Hamilton, P. J., 1979. Geochemical modelling of mantle differentiation and crustal growth. *Journal of Geophysical Research*, **84**, 6091–6101.

Olsen, P., 1990. Hot spots, swells and mantle plumes. In Ryan, M. P. (ed.), *Magma Transport and Storage*. New York: Wiley, pp. 33–51.

Paul, D., White, W. M., and Blichert-Toft, J., 2005. Geochemistry of Mauritius and the origin of rejuvenescent volcanism on oceanic island volcanoes. *Geochemistry, Geophysics, Geosystems*, **6**, doi:10.1029/2004gc000883.

Payne, J. A., Jackson, M. G., and Hall, P. S., 2013. Parallel volcano trends and geochemical asymmetry of the Society Islands hotspot track. *Geology*, **41**(1), 19–22, doi:10.1130/g33273.1.

Porcelli, D., and Elliott, T., 2008. The evolution of He Isotopes in the convecting mantle and the preservation of high $^3He/^4He$ ratios. *Earth and Planetary Science Letters*, **269**(1'2), 175–185, doi:10.1016/j.epsl.2008.02.002.

Putirka, K. D., Perfit, M., Ryerson, F. J., and Jackson, M. G., 2007. Ambient and excess mantle temperatures, olivine thermometry, and active vs. passive upwelling. *Chemical Geology*, **241**(3–4), 177–206, doi:10.1016/j.chemgeo.2007.01.014.

Putirka, K., Ryerson, F. J., Perfit, M., and Ridley, W. I., 2011. Mineralogy and composition of the oceanic mantle. *Journal of Petrology*, **52**(2), 279–313, doi:10.1093/petrology/egq080.

Reiners, P. W., 2002. Temporal-compositional trends in intraplate basalt eruptions: implications for mantle heterogeneity and melting processes. *Geochemistry, Geophysics, Geosystems*, **3**(2), 1–30, doi:10.1029/2001GC000250.

Ribe, N., and Christensen, U., 1999. The dynamical origin of Hawaiian volcanism. *Earth and Planetary Science Letters*, **171**, 517–531, doi:10.1016/S0012-821X(99)00179-X.

Richards, M. A., Duncan, R. A., and Courtillot, V. E., 1989. Flood basalts and hot-spot tracks: plume heads and tails. *Science*, **246**, 103–107.

Roeder, P. L., and Emslie, R. F., 1970. Olivine-liquid equilibrium. *Contributions to Mineralogy and Petrology*, **29**, 275–289.

Rohde, J., Hoernle, K., Hauff, F., Werner, R., O'Connor, J., Class, C., Garbe-Schönberg, D., and Jokat, W., 2013. 70 Ma chemical zonation of the Tristan-Gough hotspot track. *Geology*, **41**(3), 335–338, doi:10.1130/g33790.1.

Schilling, J.-G., 1973. Iceland mantle plume: geochemical study of the Reykjanes Ridge. *Nature*, **242**, 565–571.

Schilling, J.-G., 1991. Fluxes and excess temperatures of mantle plumes inferred from their interaction with migrating mid-ocean ridges. *Nature*, **352**, 397–403.

Schilling, J.-G., Anderson, R. N., and Vogt, P., 1976. Rare earth, Fe and Ti variations along the Galapagos spreading centre, and their relationship to the Galapagos mantle plume. *Nature*, **261**, 108–113.

Schilling, J.-G., Kingsley, R. H., and Devine, J. D., 1982. Galapagos hot spot-spreading center system 1. spatial petrological and geochemical variations (83 W-101 W). *Journal of Geophysical Research*, **87**, 5593–5610.

Schilling, J.-G., Kingsley, R. H., Hanan, B. B., and McCully, B. L., 1992. Nd-Sr-Pb isotopic variations along the Gulf of Aden: evidence for Afar mantle plume-continental lithosphere interaction. *Journal of Geophysical Research*, **97**, 10927–10996.

Sleep, N. H., 1990. Hotspots and Mantle Plumes: some phenomenology. *Journal of Geophysical Research*, **95**, 6715–6736.

Sobolev, A. V., Hofmann, A. W., Sobolev, S. V., and Nikogosian, I. K., 2005. An olivine-free mantle source of Hawaiian shield basalts. *Nature*, **434**(7033), 590–597, doi:10.1038/nature03411.

Sobolev, A. V., Hofmann, A. W., Kuzmin, D. V., Yaxley, G. M., Arndt, N. T., Chung, S.-L., Danyushevsky, L. V., Elliott, T., Frey, F. A., Garcia, M. O., Gurenko, A. A., Kamenetsky, V. S., Kerr, A. C., Krivolutskaya, N. A., Matvienkov, V. V., Nikogosian, I. K., Rocholl, A., Sigurdsson, I. A., Sushchevskaya, N. M., and Teklay, M., 2007. The amount of recycled crust in sources of mantle-derived melts. *Science*, **316**(5823), 412–417, doi:10.1126/science.1138113.

Spiegelman, M., and Elliot, T., 1993. Consequences of melt transport for uranium series disequilibrium in young lavas. *Earth and Planetary Science Letters*, **118**, 1–20.

Steinberger, B., Sutherland, R., and O'Connell, R. J., 2004. Prediction of Emperor-Hawaii seamount locations from a revised model of global plate motion and mantle flow. *Nature*, **430**(6996), 167–173.

Sterns, H. T., 1946. *Geology of the Hawaiian Islands*. Honolulu: Hawaii Division of Hydrology.

Stille, P., Unruh, D. M., and Tatsumoto, M., 1986. Pb, Sr, Nd, and Hf isotopic constraints on the origin of Hawaiian basalts and evidence for a unique mantle source. *Geochimica et Cosmochimica Acta*, **50**, 2303–2320.

Stracke, A., Salters, V. J. M., and Sims, K. W. W., 1999. Assessing the presence of garnet-pyroxenite in the mantle sources of basalts through combined hafnium-neodymium-thorium isotope systematics. *Geochemistry, Geophysics, Geosystems*, **1**, doi:10.1029/1999gc000013.

Tatsumoto, M., 1978. Isotopic composition of lead in oceanic basalt and its implication to mantle evolution. *Earth and Planetary Science Letters*, **38**, 63–87.

ten Brink, U. S., and Brocher, T. M., 1987. Multichannel seismic evidence for a subscrustal intrusive complex under Oahu and a model for Hawaiian volcanism. *Journal of Geohysical Research*, **92**(B13), 13687–13707.

Van Wijk, J. W., and Blackman, D. K., 2005. Dynamics of continental rift propagation: the end-member modes. *Earth and Planetary Science Letters*, **229**(3'4), 247–258, doi:10.1016/j.epsl.2004.10.039.

Watson, S., and Mckenzie, D., 1991. Melt generation in plumes: a study of Hawaiian volcanism. *Journal of Petrology*, **32**, 501–537.

Weis, D., Garcia, M. O., Rhodes, J. M., Jellinek, M., and Scoates, J. S., 2011. Role of the deep mantle in generating the compositional asymmetry of the Hawaiian mantle plume. *Nature Geoscience*, **4**(12), 831–838, doi:10.1038/ngeo1328.

White, W. M., 1985. The sources of ocean basalts: radiogenic isotopic evidence. *Geology*, **13**, 115–118.

White, W. M., 2010. Oceanic Island basalts and mantle plumes: the geochemical perspective. *Annual Review of Earth and Planetary Sciences*, **38**(1), 133–160, doi:10.1146/annurev-earth-040809-152450.

White, W. M., 2013. *Geochemistry*. Oxford: Wiley-Blackwell.

White, W. M., and Duncan, R. A., 1996. Geochemistry and geochronology of the Society Islands: new evidence for deep mantle recycling. In Hart, S. R., and Basu, A. (eds.), *Earth Processes: Reading the Isotope Code*. Washington, DC: AGU, Vol. 95, pp. 183–206.

White, W. M., and Hofmann, A. W., 1982. Sr and Nd isotope geochemistry of oceanic basalts and mantle evolution. *Nature*, **296**, 821–825.

White, W. M., Schilling, J.-G., and Hart, S. R., 1976. Evidence for the Azores mantle plume from strontium isotope geochemistry of the Central North Atlantic. *Nature*, **263**, 659–663.

White, W. M., McBirney, A. R., and Duncan, R. A., 1993. Petrology and geochemistry of the Galapagos: portrait of a pathological mantle plume. *Journal of Geophysical Research*, **98**(B11), 19533–19563.

Whitehead, J. A., and Luther, D. S., 1975. Dynamics of laboratory diapir and plume models. *Journal of Geophysical Research*, **80**(5), 705–717.

Willbold, M., and Stracke, A., 2010. Formation of enriched mantle components by recycling of upper and lower continental crust. *Chemical Geology*, **276**(3'4), 188–197, doi:10.1016/j.chemgeo.2010.06.005.

Wilson, J. T., 1963. A possible origin of the Hawaiian Islands. *Canadian Journal of Physics*, **41**, 863–870.

Wright, E., and White, W. M., 1987. The origin of Samoa: new evidence from Sr, Nd, and Pb isotopes. *Earth and Planetary Science Letters*, **81**(2–3), 151–162, doi:10.1016/0012-821X(87)90152-X.

Yang, H. J., Frey, F. A., and Clague, D. A., 2003. Constraints on the source components of lavas forming the Hawaiian North Arch and Honolulu Volcanics. *Journal of Petrology*, **44**(4), 603–627.

Zindler, A., and Hart, S. R., 1986. Chemical geodynamics. *Annual Review of Earth and Planetary Sciences*, **14**, 493–571.

Cross-references

Depleted Mantle
Hotspot-Ridge Interaction
Intraplate Magmatism
Mid-ocean Ridge Magmatism and Volcanism
Oceanic Plateaus
Oceanic Spreading Centers
Oxygen Isotopes
Peridotites
Plate Motion
Plate Tectonics
Seamounts
Subduction
Subduction Erosion

HOTSPOT-RIDGE INTERACTION

Colin Devey
GEOMAR Helmholtz Centre for Ocean Research,
Kiel, Germany

Synonyms
Plume-ridge interaction

Definition
The exchange of material (magma, mantle rock) between an intraplate mantle melting anomaly (hotspot, thought to be caused in places by the presence of a mantle plume) and the global spreading ridge system. Evidence for the interaction is found in the depth of the spreading axis, its morphology, the chemistry of the lavas (both on the spreading axis and possibly at the hotspot), and sometimes by the presence of linear volcanic ridges between spreading axis and hotspot. These linear volcanic ridges generally do not show clear age-progressive volcanism, in contrast to the volcanoes of the hotspot itself.

Description
As a result of the relative fixity of hotspots with respect to the moving plates, their distance from active spreading centers changes over geological time. Depending on the plate tectonic situation, hotspots can both migrate toward and away from ridges over time, respectively, weakening and strengthening the hotspot-ridge interaction. At some limiting minimum separation (which may be several hundred km and is dependent on whether the two are approaching or separating from each other), the two begin to interact. Evidence for hotspot-ridge interaction comes from both present-day systems and from fossil traces preserved on the seafloor.

In some cases, such as Iceland, the hotspot lies directly beneath the spreading center (the Mid-Atlantic Ridge), and there is evidence that the spreading axis regularly relocates via ridge jumps to maintain this situation despite plate motion ("ridge capture"). In others (e.g., Galapagos hotspot and Galapagos Spreading Center) there is evidence that the spreading axis is currently moving away from the hotspot (e.g., Ito and Lin, 1995). It has been proposed (Mittelstaedt et al., 2011) that ridge jumps occur more regularly at slow spreading rates which may explain, for example, the tendency for Atlantic hotspots (e.g., Azores, St. Helena, Tristan, Bouvet, as well as Iceland) to cluster near the axis. When the hotspot lies directly on the spreading axis, mantle upwelling can extend to relatively shallow depths, producing a large volume of magma. This is at least one factor (in addition to plume temperature and composition) for Iceland's large land surface compared to other volcanic ocean islands and has been suggested as a mechanism by which large oceanic igneous provinces (such as the Ontong Java-Hikurangi-Manihiki cluster) could have attained their exceptional size (e.g., Taylor, 2006). Ridge-centered hotspots produce volcanic trails on both plates on either side of the ridge, as is seen, for example, in the Greenland-Faroe Ridge, which the Iceland hotspot has produced since opening of the North Atlantic, and the dual Carnegie-Cocos Ridges produced by the Galapagos hotspot (e.g., Harpp et al., 2005).

Off-axis hotspots produce variable effects on the bathymetry of the adjacent axis, related to both the hotspot-ridge separation distance (the effect falls off with increasing separation) and the intensity of hotspot volcanism (more productive hotspots have larger effects on the ridge). The Galapagos hotspot, for example, produces a large bathymetric anomaly (e.g., Christie et al., 2005) on the Galapagos Spreading Center (hotspot-ridge separation 150 km). In the past, this hotspot lay directly on the spreading axis, which led to the construction of a volcanic ridge (the Cocos Ridge) on the Cocos Plate. The Easter hotspot, in contrast, has only very limited effect on the bathymetry of the East Pacific Rise bathymetry (separation 350 km), and no trace is visible on the opposite plate (see, e.g., Haase et al., 1996).

The hotspot supplies material to the spreading axis. Whether this is in the form of direct magma transport or the flow of mantle material is still unclear (c.f., Braun and Sohn, 2003; Kokfelt et al., 2005; Stroncik et al., 2008; Villagomez et al., 2014). The geochemical influence of the hotspot on the ridge often leads to the axis erupting magmas with compositions more enriched in mantle-incompatible elements than surrounding segments of the ridge, although evidence for the eruption of extremely depleted magmas at both hotspots and their adjacent axes has also been found (Portnyagin et al., 2009; Stroncik and Devey, 2011). Several hotspot-influenced ridges have been found to erupt magmas more fractionated than basalt – in some cases, these magmas have compositions very similar to early continental crust (Haase et al., 2005). Hotspot-ridge interaction also clearly influences the composition of the hotspot magmas – the extent to which this influence results from (a) the effects of lithospheric thickness on depth and degree of melting, (b) the flow of material from the ridge to the hotspot, or (c) the consequence of the removal of hotspot material toward the ridge is hotly debated (Haase et al., 1996; Keller et al., 2000; Stroncik and Devey, 2011)

When the hotspot is close to the ridge, the seafloor separating them is also the site of intense volcanism (e.g., Christie et al., 2005; Fretzdorff et al., 1996; Maia et al., 2000). Sometimes this is manifested as linear volcanic ridges running between hotspot and ridge (e.g., Kopp et al., 2003; Maia et al., 2000) which show no systematic age-distance relationship, in contrast to the hotspot volcanoes themselves (O'Connor et al., 2001).

Summary
Hotspot-ridge interaction affects large areas of the seafloor, resulting in bathymetric, structural, and geochemical anomalies on both ridge and hotspot.

Bibliography

Braun, M. G., and Sohn, R. A., 2003. Melt migration in plume-ridge systems. *Earth and Planetary Science Letters*, **213**, 417–430.

Christie, D. M., Werner, R., Hauff, F., Hoernle, K., and Hanan, B. B., 2005. Morphological and geochemical variations along the eastern Galapagos Spreading Center. *Geochemistry, Geophysics, Geosystems*, **6**(1), Q01006. doi:10.1029/2004GC000714

Fretzdorff, S., Haase, K. M., and Garbe-Schönberg, C.-D., 1996. Petrogenesis of lavas from the Umu Volcanic Field in the young hotspot region west of Easter Island, southeastern Pacific. *Lithos*, **38**, 23–40.

Haase, K. M., Devey, C. W., and Goldstein, S. L., 1996. Two-way exchange between the Easter mantle plume and the Easter microplate spreading axis. *Nature*, **382**, 344–346.

Haase, K. M., Stroncik, N. A., Hékinian, R., and Stoffers, P., 2005. Nb-depleted andesites from the Pacific-Antarctic Rise as analogs for early continental crust. *Geology*, **33**(12), 921–924.

Harpp, K. S., Wanless, V. D., Otto, H. R., Hoernle, K., and Werner, R., 2005. The Cocos and Carnegie aseismic ridges: a trace element record of long-term plume-spreading center interaction. *Journal of Petrology*, **46**(1), 109–133.

Ito, G., and Lin, J., 1995. Oceanic spreading center-hotspot interactions: constraints from along-isochron bathymetric and gravity anomalies. *Geology*, **23**, 657–660.

Keller, R. A., Fisk, M. R., and White, W. M., 2000. Isotopic evidence for Late Cretaceous plume-ridge interaction at the Hawaiian hotspot. *Nature*, **405**, 673–676.

Kokfelt, T., Lundstrom, C., Hoernle, K., Hauff, F., and Werner, R., 2005. Plume–ridge interaction studied at the Galápagos spreading center: evidence from 226Ra–230Th–238U and 231Pa–235U isotopic disequilibria. *Earth and Planetary Science Letters*, **234**(1–2), 165–187.

Kopp, H., Kopp, C., Morgan, J., Flueh, E., Weinrebe, W., and Morgan, W., 2003. Fossil hot spot-ridge interaction in the Musicians Seamount Province: geophysical investigations of hot spot volcanism at volcanic elongated ridges. *Journal of Geophysical Research: Solid Earth*, **108**(B3), 2160.

Maia, M., Ackermand, D., Dehghani, G. A., Gente, P., Hekinian, R., Naar, D., O'Connor, J., Perrot, K., Morgan, J. P., Ramillien, G., Revillon, S., Sabetian, A., Sandwell, D., and Stoffers, P., 2000. The Pacific-Antarctic ridge-foundation hotspot interaction: a case study of a ridge approaching a hotspot. *Marine Geology*, **167**(1–2), 61–84.

Mittelstaedt, E., Ito, G., and van Hunen, J., 2011. Repeat ridge jumps associated with plume-ridge interaction, melt transport, and ridge migration. *Journal of Geophysical Research: Solid Earth*, **116**, B01102.

O'Connor, J. M., Stoffers, P., and Wijbrans, J. R., 2001. En echelon volcanic elongate ridges connecting intraplate Foundation Chain volcanism to the Pacific-Antarctic spreading center. *Earth and Planetary Science Letters*, **192**, 633–648.

Portnyagin, M., Hoernle, K., and Savelyev, D., 2009. Ultra-depleted melts from Kamchatkan ophiolites: evidence for the interaction of the hawaiian plume with an oceanic spreading center in the Cretaceous? *Earth and Planetary Science Letters*, **287**(1–2), 194–204.

Stroncik, N. A., and Devey, C. W., 2011. Recycled gabbro signature in hotspot magmas unveiled by plume-ridge interactions. *Nature Geoscience*, **4**(6), 393–397.

Stroncik, N., Niedermann, S., and Haase, K., 2008. Plume–ridge interaction revisited: evidence for melt mixing from He, Ne and Ar isotope and abundance systematics. *Earth and Planetary Science Letters*, **268**(3–4), 424–432.

Taylor, B., 2006. The single largest oceanic plateau: Ontong Java–Manihiki–Hikurangi. *Earth and Planetary Science Letters*, **241**, 372–380.

Villagomez, D. R., Toomey, D. R., Geist, D. J., Hooft, E. E. E., and Solomon, S. C., 2014. Mantle flow and multistage melting beneath the Galapagos hotspot revealed by seismic imaging. *Nature Geoscience*, **7**(2), 151–156.

Cross-references

Hot Spots and Mantle Plumes
Oceanic Plateaus
Oceanic Spreading Centers
Seamounts

HURRICANES AND TYPHOONS

Don Resio[1] and Shannon Kay[2]
[1]Taylor Engineering Research Institute, University of North Florida, Jacksonville, FL, USA
[2]Coastal Projects, Manson Construction Co., Jacksonville, FL, USA

Definition

Hurricanes and typhoons are atmospheric circulation systems of tropical origin characterized by low pressure at the center and near surface winds spiraling inward around this center, typically storm size ranges from 10 to 80 km for the radius to maximum wind speeds with cloud cover extending from about 150 to 1,500 km. In meteorological terms, hurricanes and typhoons are low pressure, warm-core cyclones, originating in warm waters with closed surface winds rotating about an eye. Table 1 shows that several different names are used in different areas around the world to refer to these storms. The origins of the local names for these storms, hurricanes and typhoons, come from the languages of people who lived in areas most affected by tropical cyclones.

- "Hurricane" was derived from the ancient Mayan God of Storms, "Hurakan," known for causing the great flood which destroyed humanity, based on ancient Carib God, "Hurican," the god of evil (Mythica, 2013).
- "Typhoon" originates from the Chinese word "*dàfēng (tai fung)*" meaning big wind.
- "Willy Willy" is a local term that originated along west coast of Australia.

For simplicity, the generic meteorological term, tropical cyclone, will be used this entry.

Introduction

As shown in Figure 1, tropical cyclones represent a serious, persistent threat to coastal communities around the world. Given the existence of ongoing, possibly accelerating, global sea-level rise along with the increasing importance of ports and harbors in global trade, it is very likely

Hurricanes and Typhoons, Table 1 Common terms used to reference tropical cyclones around the World

Region	Term Definition
North America	Tropical depression: maximum 1-min wind speed < 39 mph (62 km/h) Tropical storm: maximum 1-min wind speed >39mph (62 km/h) Hurricane: maximum 1-min winds speed > 74 mph (120 km/h)
Southern Hemisphere and Fiji	Tropical low: maximum sustained wind speed < 90 km/h Tropical cyclone: maximum sustained wind speed > 90 km/h Severe tropical cyclone: > 165 km/h
Western Pacific	Tropical depression: maximum sustained wind speed < 61 km/h Tropical storm: maximum sustained wind speed > 61 km/h Severe tropical storm: maximum sustained wind speed > 89 km/h Typhoon: maximum sustained wind speed > 118 km/h Super typhoon: maximum sustained wind speed > 252 km/h
India	Tropical depression: maximum sustained wind speed < 62 km/h Tropical storm: maximum sustained wind speed > 63 km/h Tropical cyclone: maximum sustained wind > 118 km/h Intense tropical cyclone: maximum sustained wind speed > 166 km/h

that the cost of coastal disasters in terms of both economic impacts and loss of life will continue to rise, making the study of these storms a critical priority within scientific, engineering, and planning communities.

Formation of a tropical cyclone

Tropical cyclones derive their energy from rising air above warm ocean waters, coupled with the earth's rotation. Key factors contributing to cyclone formation include (Gray, 1968, 1979; Emanuel, 1991):

1. Ocean water must contain abundant heat to compensate for heat loss via evaporation at the air-sea interface (typically above 26 °C, to a depth of about 50 m).
2. The atmosphere must cool sufficiently with height to retain potential instability in the rising air, otherwise the density of the rising air will approach that of the surrounding air and its vertical motion will dissipate.
3. Relatively high atmospheric water content in the mid-tropospheric level is needed since condensation of water vapor fuels continued development of thunderstorms at this level and condensation rates depend on relative humidity.
4. A horizontal displacement of the circulation center at least 500 km from the equator is needed to enable Coriolis acceleration (the effect of the earth's rotation on motions in an inertial frame of reference) to force horizontal air paths toward a closed circulation via a quasi-gradient balance before the winds and pressure differential within the storm become sufficiently large to shift toward a cyclostrophic balance.

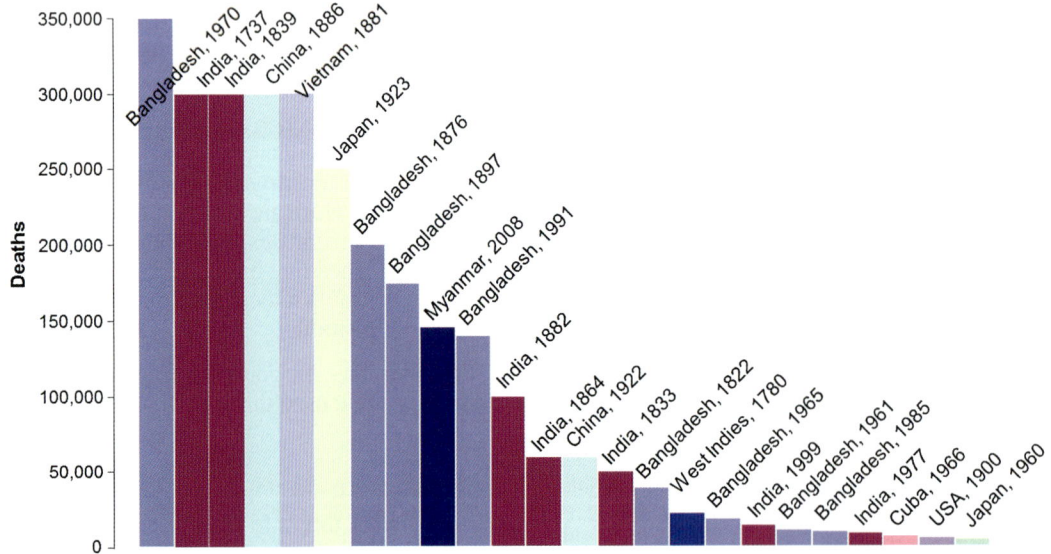

Hurricanes and Typhoons, Figure 1 Highest ranking estimated mortality due to tropical cyclone generated storm surges (Dube, 2007).

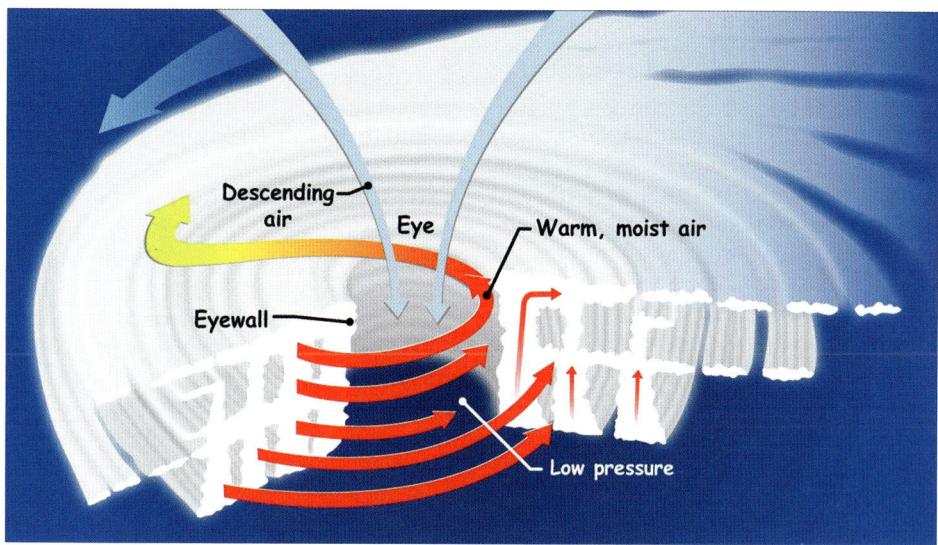

Hurricanes and Typhoons, Figure 2 Schematic diagram of airflow through a tropical cyclone (NASA, 2013).

5. A disturbance must exist over a sufficiently broad ocean surface to allow Coriolis acceleration to produce horizontal trajectories that move toward a closed circulation.
6. Since vertical wind shear can disrupt the organized energy flow that drives a hurricane (heat taken from the ocean surface by evaporation carried vertically where it is released by condensation), such shear can inhibit the formation and/or strengthening of tropical cyclones.

Anatomy of a tropical cyclone

Tropical cyclogenesis is initiated by large areas of convection developing in warm ocean regions around the world. Organized synoptic-scale motions (spatial scales in the range of 10–100 s of km and temporal scales in the range of hours to days) are typically required to initiate cyclogenesis. Features such as easterly waves, stationary fronts, or broad areas of atmospheric divergence represent types of disturbances which can create such organized convection. Rising air within a large area of convection begins to create generally decreased pressures, which, in turn, draws near-surface air from surrounding areas into the low-pressure region. The speed of air moving over the ocean surface directly influences the transfer rate of moisture; thus the vertical motions create a feedback which increases the amount of heat transferred into the overriding air passing into the region of low pressure, i.e., the faster the air moves the more moisture it accumulates. The air trajectories spiral inward toward the center of the low pressure system due to Coriolis acceleration and contribute additionally to the convection in this region. The heat in the rising air is released when it condenses in the vicinity of the top of the frictional boundary layer within the atmosphere (about 1–2 km above the water surface), further adding to the convective instability and contributing to a continued pressure reduction. In developing storms a stable layer in the atmosphere, located about 15 km above sea-level, creates a region of high pressure above the storm center and causes the air at this level to spiral outwards with an opposite rotation from the lower trajectories, as shown in Figure 2.

Emanuel (1986, 1991) provides an excellent treatise on the formation of tropical cyclones in terms of a Carnot engine that converts the heat from evaporation at the ocean surface into mechanical energy. This energy cycle can continue to grow in strength as long as the rate of gain of heat (from the ocean into the atmosphere) from the evaporation at the surface and condensation in the cloud layer exceeds the sum of energy losses in the system, loss of kinetic energy due to drag at the air-sea interface, loss of heat energy due to upwelling of cooler water from subsurface regions, etc. As an interesting consequence of this theoretical work, parametric models for the maximum potential intensity (MPI) of a hurricane have been formulated (Emanuel, 1988; Holland, 1997; Tonkin et al., 2000). These models indicate that the MPI increases as a function of sea surface temperate (SST); and since SST is known to be increasing around the world (Trenberth et al., 2007), recent work (Knutson et al., 2010) has suggested that the frequency of intense tropical cyclones can be expected to rise in the future.

- **Effects of storm translation on wind fields**: As might be expected from the conceptual framework presented, there is no mechanism which will produce strong asymmetry in a stationary storm; hence such storms tend to have approximately symmetric wind fields around their centers, often termed the eye of these storm. However, storms are imbedded within steering currents produced by large-scale motions (spatial scales greater than about 5,000 km and temporal scales greater than five days) within the atmosphere. These steering currents roughly coincide with many of the average seasonal circulation features around the globe, producing east to west motions at latitudes between 30° latitude north and south and curving toward west to east motions at latitudes poleward of 35° latitude north and south. Tropical cyclones move with the general speed of these large-scale currents in which they are imbedded. The winds in the tropical cyclone and large-scale wind system are superposed vectorially. Relative to the earth's surface, this creates higher wind speeds on the side of the storm moving in the same direction as the large-scale motion and lower wind speeds on the side moving against the large-scale motion. At the same time, the vector addition changes the wind directions slightly (typically less than 10°). The overall effect is to create an asymmetric wind field with generally stronger(weaker) winds on the right (left) side of a tropical cyclone in the northern hemisphere and stronger (weaker) winds on the left side of a tropical cyclone in the southern hemisphere.

Features within a tropical cyclones

- **The eye and eye wall**: As shown in Figure 2, there is an often cloudless region of relatively calm winds, termed the "eye" of the storm, around which the rest of the storm rotates. In the eye, the atmospheric pressure typically reaches its minimum values accompanied by the warmest temperatures. The warming in this region is caused by slowly sinking air from the high pressure region above. In well-developed storms, the eye is surrounded by an eyewall, as illustrated in Figure 2. The eyewall is where the strongest winds and highest rainfall rates are usually located. The existence of the eye and eyewall is related to the dynamics of flow within tropical cyclones. As air nears the center of the storm, the centrifugal acceleration would become very large if wind speeds around the eye remained constant. The pressure gradient required to maintain a cyclostrophic balance cannot be maintained inside the eyewall area and the slowly sinking air inside the eyewall slowly merges with the air in the eyewall, where it rises to complete the cycle.
- **Rain bands**: Although many well-developed tropical cyclones are relatively symmetric, lower intensity or storms during the early formative stage of developments can be much less symmetric and contain more irregular organized motions. This creates regions of higher and lower winds throughout the storm, leading to somewhat unique patterns of clouds and winds in tropical cyclones. One prevalent feature related to these organized variations is termed a rain band. Rain bands are heavily convective regions containing high rainfall rates and typically higher wind speed than adjacent portions of the storm. These bands form long, narrow cloud systems which are oriented in the direction of the wind. Along the lower level of these bands, convergence is a maximum, leading to some low-level subsidence in adjacent areas. In many cases the spiral bands converge into the storm center merging into the eyewall.

Extratropical transition

When tropical cyclones encounter pronounced horizontal temperature gradients in the atmosphere, as typical of many ocean areas poleward of 35° north or south, they begin to undergo extratopical transition (Jones et al., 2003). This introduces a new source of energy into the energy balance, termed a baroclinic energy source, which is the same mechanism that supplies energy to extratropical storms. Most hurricanes along the US East Coast are undergoing extratropical transition by the time they get into the New York Bight; consequently, it should not be a surprise that Hurricane Sandy was deriving a substantial part of its energy from baroclinic sources (Shen et al., 2013) when in made landfall. This also means that the MPI is not a viable concept for quantifying the maximum storm intensity at this latitude.

Impacts of tropical cyclones

Tropical cyclones represent a major threat to property, livelihood and safety around the world. Critical threats are underestimated in many areas due to their relative rarity; and even in areas frequently exposed to tropical cyclones, an extreme cyclone can produce inundation, inland flooding, mudslides and wave run-up which significantly exceed the experience of local inhabitants. For this reason, such storms often produce catastrophic, unanticipated impacts, even when appropriate warnings are issued. The principal contributors to damages from tropical cyclones are storm surge, damaging waves, strong winds and inland flooding.

- **Storm surge**: The rise of water level generated by the storm winds and waves is termed the storm surge and is responsible for the majority of deaths in most areas. The superposition of storm surge on local tides is sometimes termed the storm tide. Inundation due to the storm tide, or total water level, is driven by three main processes: (1) lower atmospheric pressures at the water surface, termed the "inverted barometer effect" (roughly 1 m of sea-level rise for a 100 mb lowering of the atmospheric pressure at the surface), (2) direct momentum transfer from wind to water, and (3) fluxes into the water column from breaking waves approaching the coast. Additional, factors affecting surge include: storm size, forward velocity, angle of coastal incidence, wave stress (Irish et al., 2009), coastal bathymetry (Resio and Westerink, 2008) and interactions with rivers (Kerr et al., 2013; Resio et al., 2013). In areas with broad continental shelves, direct wind forcing is usually the dominant surge generation mechanism, while on steep coasts, typical of many island areas around the world, waves and wave run-up are often the dominant flooding

Tracks and Intensity of Tropical Cyclones, 1851-2006

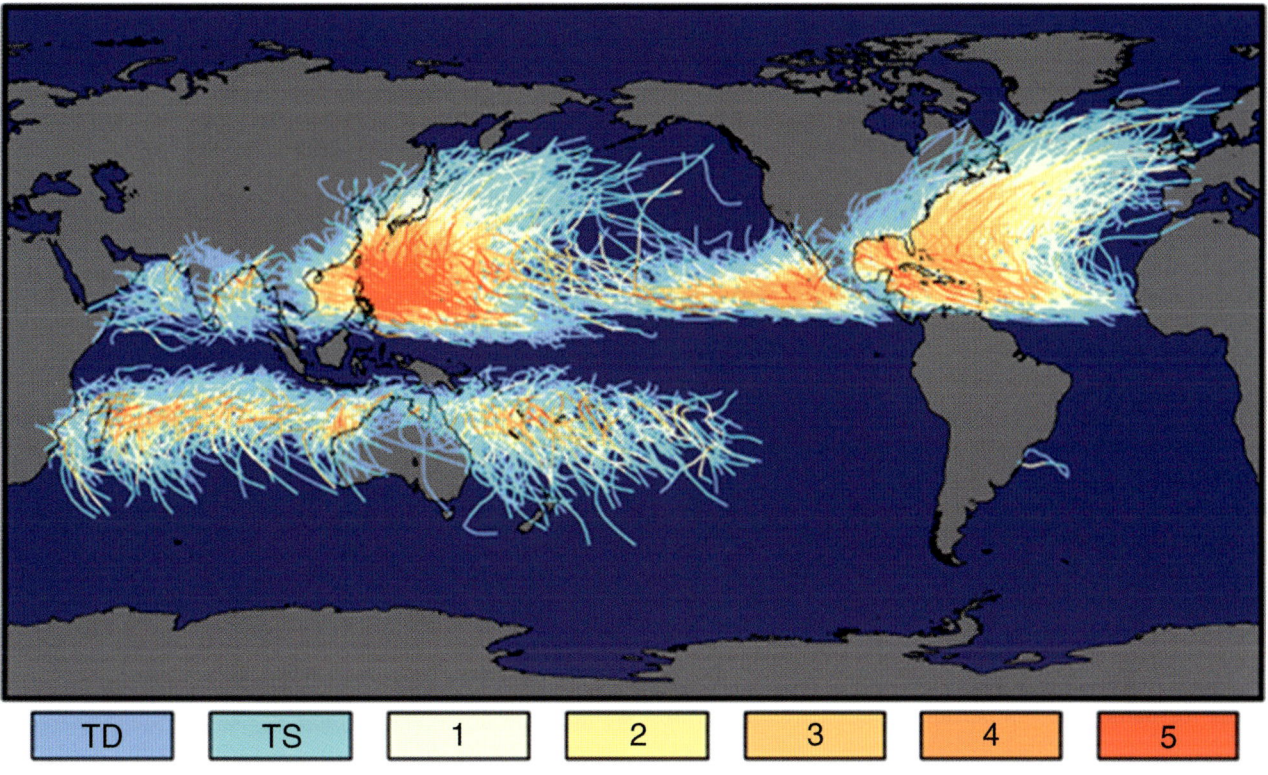

Saffir-Simpson Hurricane Intensity Scale

Hurricanes and Typhoons, Figure 3 Worldwide distribution of combined tropical cyclone frequency and intensity. The lack of tropical cyclones near the equator can be clearly seen in this analysis by NASA available online at www.goes-r.gov/users/comet/tropical/textbook_2nd_Edition/. The Saffir-Simpson scale referenced here was developed initially as a 1–5 scale as an indicator of wind damage to structures on land and should not be used as a scale for surge, wave or inland flooding hazards (Kay, 2010).

mechanism. There is typically a pronounced asymmetry in surge heights along the coast, related to the wind speed asymmetry previously noted and the direction of the wind, with higher surges to the right of landfall (relative to direction of storm heading) in the northern hemisphere and to the left in the southern hemisphere.
- **Rough seas**: Large waves pose a major threat to offshore structures, such as oil platforms and transshipment facilities, as well as to commercial shipping and recreational boating. Besides the loss of life and property when a vessel is damaged or sinks, delays in the transport of goods and the impact of increased ship time related to rough seas can dramatically influence shipping costs. In addition to the effects of waves offshore, large waves along coasts, particularly when superposed upon the elevated water levels, are also responsible for substantial loss of life and property in coastal communities.
- **Strong winds**: Strong winds associated with tropical cyclones also pose a major threat to coastal communities. Roofs can be lifted, trees uprooted, and power lines torn down by the magnitude of force from the wind. In some areas of the world with minimal building codes, damage from winds alone can produce massive devastation.
- **Inland flooding**: Coastal areas of many islands around the world are bounded by steep topography. When the moisture laden air in tropical cyclones is forced up these slopes, orographic lifting can produce extreme rainfall rates, which often produce extreme flooding in coastal rivers and streams. The effects of such high precipitation rates can be compounded by mudslides on water-soaked slopes; and in many areas where the mountain rivers lead into coastal ports and harbors, the coupling of elevated coastal water levels and the very high river discharges can combine to inundate entire coastal towns.

A global perspective

Tropical cyclones play a vital role in the global atmospheric circulation and provide vital rain to many areas of the world. However, when these storms exceed local threshold values related to wind speed, precipitation or surge generation, their impacts are extremely devastating. Recent events, such as Hurricane Katrina and the hybrid cyclone Sandy in the United States, show that even highly industrialized countries continue to be very vulnerable to these storms. The relative rarity of intense tropical cyclones makes it very difficult to develop, execute, and maintain effective mitigation plans. Even areas experiencing frequent many tropical cyclones can be devastated by an unusually intense storm, as was the case recently with Typhoon Haiyan in the Philippines in November 2013 which set a new world record as the most intense landfalling tropical cyclone. Figure 3 (www.goes-r.gov/users/comet/tropical/textbook_2nd_Edition/) gives an overview of areas around the world that are most likely to experience tropical cyclones; however, as shown by the hybrid cyclone Sandy, even if an area is only infrequently affected, these storms still pose a very real and increasing threat.

Summary

Tropical cyclones represent one of nature's most destructive forces and the effects of climate variability and continued coastal development are likely to exacerbate these impacts. Disasters and disaster recovery will depend heavily on improved predictions over short time intervals for evacuation decision-making and over many years for community planning and development of resilient coastal areas for natural and urban landscapes (Brantley et al., 2014; Sealza and Sealza, 2014).

These storms are fundamentally a heat engine taking energy from the sea via the latent heat of evaporation and depositing the energy at mid-tropospheric levels via condensation. This engine converts the heat energy directly into mechanical energy (winds, waves and currents). Potential positive feedbacks, related to the dependence of evaporation rates on wind speed and storm intensity and mid-tropospheric heating rates, exist in this system. Therefore, tropical cyclones will continued to strengthen in ocean areas with high water temperatures, minimal vertical shear, and moist air aloft until the mechanical and heat energy losses equal the heat sources.

References

Brantley, S. T., Bissett, S. N., Young, D. R., Wolner, C. W., and Moore, L. J., 2014. Barrier Island morphology and sediment characteristics affect the recovery of Dune building grasses following storm-induced overwash. *PloS One*, 9(8), e104747.

Dube, S., 2007. Storm surge force associated with sudden changes in intensity and/or track in the Bay of Bengal. Retrieved from http://www.wmo.int/pages/prog/arep/wwrp/pdf%20files/6%20Nov/SDube.pdf. Accessed November 20, 2013.

Emanuel, K., 1986. An air-sea interaction for tropical cyclones, Part I: steady-state maintenance. *Journal of the Atmospheric Sciences*, 43, 585–604.

Emanuel, K., 1988. The maximum intensity of hurricanes. *Journal of the Atmospheric Sciences*, 45, 1143–1155.

Emanuel, K., 1991. The theory of hurricanes. *Fluid Mechanics*, 23, 179–196.

Encyclopedia Mythica, 2013. Hurakan. Retrieved from http://www.pantheon.org/articles/h/hurakan.html. Accessed November 17, 2013.

Gray, W. M., 1968. A global view of the origin of tropical disturbances and storms. *Monthly Weather Review*, 96, 669–700.

Gray, W. M., 1979. Hurricanes: their formation, structure and likely role in the tropical circulation. In Shaw, D. B. (ed.), *Meteorology Over Tropical Oceans*. James Glaisher House, Berkshire: Royal Meteorological Society, pp. 155–218.

Holland, G. J., 1997. The maximum potential intensity of tropical cyclones. *Journal of the Atmospheric Sciences*, 54, 2519–2541.

Irish, J. L., Resio, D. T., and Cialone, M. A., 2009. A surge response function approach to coastal hazard assessment: Part 2, quantification of spatial attributes of response functions. *Nat Hazards*, 51, 183–205, doi:10.1007/s11069-9381-4.

Jones, S. C., Harr, P. A., Abraham, J., Bosart, L. F., Bowyer, P. J., Evans, J. L., Hanley, D. E., Hanstrum, B. N., Hart, R. E., Laurette, F. O., Sinclair, M. R., Smith, R. K., and Thorncroft, C., 2003. The extratropical transition of tropical cyclones: forecast challenges, current understanding, and future directions. *Weather and Forecasting*, 18, 1052–1092.

Kay, J., 2010. Storm surge removed from Saffir-Simpson scale. *USA Today*, Retrieved February 13, 2010.

Kerr, P. C., Westerink, J. J., Dietrich, J. C., Martyr, R. C., Tanaka, S., Resio, D. T., Smith, J. M., Westerink, H. J., Westerink, L. G., Wamsley, T., van Ledden, M., and de Jong, W., 2013. Surge generation mechanisms in the lower Mississipi river and discharge dependency. *Journal of Waterway, Port, Coastal, and Ocean Engineering*, 139, 326–335.

Knutson, T. R., McBride, J. L., Chan, J., Emanuel, K., Holland, G., Landsea, C., and Sugi, M., 2010. Tropical cyclones and climate change. *Nature Geoscience*, 3(3), 157–163.

NASA, 2013. How do hurricanes form. Retrieved from http://pmm.nasa.gov/education/articles/how-do-hurricanes-form. Accessed November 20, 2013.

Resio, D. T., and Westerink, J. J., 2008. Modeling the physics of hurricane storm surges. *Physics Today*, September, 33–38.

Resio, D. T., Irish, J. L., Westerink, J. J., and Powell, N. J., 2013. The effect of uncertainty on estimates of hurricane surge hazards. *Natural Hazards*, 66, 1443–1459, doi:10.1007/s11069-012-0315-1.

Sealza, I. S., and Sealza, L. P., 2014. Recovering from the effects of natural disaster: the case of urban Cagayan de Oro, Philippines. *European Journal of Sustainable Development*, 3(3), 103–110.

Shen, B. W., DeMaria, J. L., Li, F., and Cheung, S., 2013. The genesis of Hurricane Sandy (2012) simulated with a global mesoscale model. *Geophysical Research Letters*, 40, 4944–4950.

Tonkin, H., Holland, G. J., Holbrook, N., and Henderson-Sellers, A., 2000. An evaluation of thermodynamic estimates of climatological maximum potential cyclone intensity. *Monthly Weather Review*, 138, 746–762.

Trenberth, K. E., Jones, P. D., Ambenje, P., Bojariu, R., Easterling, D., Tank, A. K., Parker, D., Rahimzadeh, F., Renwick, J. A., Rusticucci, M., Soden, B., and Zhai, P., 2007. Observations: surface and atmospheric climate change. In Solomon, S., Qin, D., Manning, M., Chen, Z., Marquis, M., Averyt, K. B., Tignor, M., and Miller, H. L. (eds.), *Climate Change 2007: The Physical Science Basis. Contribution of Working Group I to the Fourth Assessment Report of the Intergovernmental Panel on Climate Change*. Cambridge, UK/New York: Cambridge University Press.

Cross-references

Beach Processes
Coastal Engineering
Coasts
Geohazards: Coastal Disasters
Integrated Coastal Zone Management
Waves

HYDROTHERMAL PLUMES

Edward T. Baker
Joint Institute for the Study of Atmosphere and Ocean – NOAA/PMEL, University of Washington, Seattle, WA, USA

Definition

Hydrothermal plumes. Near-bottom oceanic layers, with distinct chemical and physical signatures, formed by the mixing of ambient ocean water and discharging hydrothermal vent fluids.

Introduction

Hydrothermal circulation is a primary agent for the exchange of heat and chemicals between the Earth's crust and ocean (see "Hydrothermalism"). Hydrothermal discharge occurs on the seafloor where crustal fluids have access to magmatic heat sources, such as along mid-ocean ridges on divergent plate boundaries, volcanic arcs at convergent plate boundaries, and intraplate hot spots (see "Mid-ocean Ridge Magmatism and Volcanism," "Hot Spots and Mantle Plumes"). These hydrothermal vents occupy only a tiny fraction of the seafloor, but the hydrothermal plumes formed by the mixing of vent fluids and seawater can be traced even across ocean basins. Mapping and sampling hydrothermal plumes locates vent fields, describes the dispersal of chemicals from vent fields, estimates hydrothermal fluxes of heat and chemicals, traces ocean circulation, and provides information on the transfer of heat and chemicals to the ocean during eruptions that construct the Earth's crust.

Hydrothermal plume characteristics

Hydrothermal plumes form above the sites of venting as hot hydrothermal fluids rise, entraining ambient seawater with a consequent continuous increase in plume volume, until nonbuoyancy is achieved and the plume disperses laterally. Models of plume formation are based on observations of atmospheric plumes and the principles of turbulent entrainment described by Morton et al. (1956). Plume rise height, which depends on the buoyancy flux of the discharge (a function of temperature and volume flux) and the ambient vertical density gradient, typically ranges from 10s to 100s of meters. Entrainment ultimately dilutes high-temperature ($\sim$150–400 °C) vent fluids by roughly a factor of 10^4 or more. At the level of lateral spreading, hydrothermal plumes have different hydrographic characteristics depending on the local salinity profile (Speer and Rona, 1989). In the deep Pacific, where salinity increases with depth, the plume is warm and salty compared to ambient seawater. In the deep Atlantic, where salinity decreases with depth, the plume is relatively cool and fresh. Using these models and local hydrographic profiles, it is possible to calculate the true hydrothermal temperature anomaly within the laterally spreading plume. Several detailed papers on plume formation can be found in Humphris et al. (1995).

Hydrothermal fluids are enriched, relative to typical oceanic deep waters, by up to a factor of $\sim 10^7$ in several key chemical tracers found in vent fluids (see "Hydrothermal Vent Fluids (Seafloor)") (e.g., ^{3}He, Mn, Fe, CH_4, H_2, H_2S, suspended particles, and other trace species). These tracers allow hydrothermal plumes to be detected at significant distances away from hydrothermal vent sites (Figure 1). Hydrothermal species can be classified as conservative or nonconservative. Conservative species include heat and ^{3}He, and their concentration in plumes changes only by dilution. Nonconservative species, such as suspended hydrothermal precipitates, Mn, Fe, CH_4, H_2, and H_2S, are also lost by chemical and biological degradation or by deposition. Their residence time in the plume varies widely, from several years for dissolved Mn and Fe and Mn oxyhydroxides to months or days for suspended particles and CH_4 and to days or hours for H_2 and H_2S. These differences in residence time make it possible to roughly estimate the age of a plume and thus its distance from the source.

Hydrothermal plumes can be categorized on the basis of their discharge and temporal characteristics. Discrete vent sources are individual orifices that discharge high-temperature fluids ($\sim$150–400 °C). These include the familiar "black smokers" emitting opaque smokestack-like plumes with high concentrations of precipitated hydrothermal metal sulfides and oxyhydroxides (see "Black and White Smokers"). Diffuse vent sources leak low-temperature (a few 10s of °C) fluids over a broadly distributed area. Plumes from diffuse vent sources have few suspended particles and have insufficient rise height to form a distinct above-bottom layer. Plumes from stable vents that discharge for years are termed chronic plumes. Plumes formed in association with a magma eruption or intrusion are termed event (or mega-) plumes. Event plume discharge lasts only minutes to days yet forms large, eddy-like plume features extending up to a kilometer above the seafloor (Baker et al., 2011).

Hydrothermal plume applications

Exploration

The earliest hydrothermal plume sampling occurred during a near-bottom sensor tow that detected a temperature spike (Weiss et al., 1977) and ^{3}He anomalies (Lupton et al., 1977) on the Galapagos Rift. This observation led directly to the discovery of seafloor vents at the same location, the first ever found on a mid-ocean ridge (Corliss et al., 1979; see "Mid-ocean Ridge Magmatism and Volcanism"). This success demonstrated the power of using hydrothermal plumes as a tool for finding the vents themselves. Researchers quickly recognized that "tow-yoing" a sensor package through near-bottom water to construct a two-dimensional transect of plumes is an efficient and high-resolution search and sampling strategy (Figure 1; Baker et al., 1985).

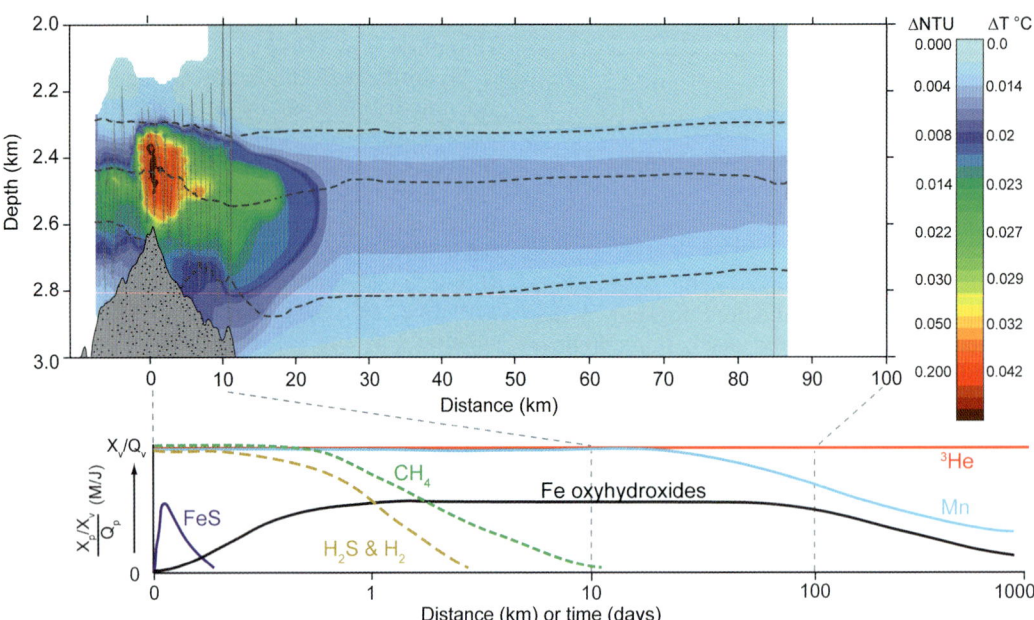

Hydrothermal Plumes, Figure 1 (*Top*) A hydrothermal plume formed on the East Pacific Rise ridge axis rises ~300 m above the seafloor before local currents advect it westward into the ocean interior (see Figure 2). A CTD tow-yo (*gray sawtooth line*) maps the plume at its origin, supplemented by two off-axis vertical casts (*gray vertical lines*) through the distal plume. The plume spreads along density contours (*dashed lines*). Color scale is applicable to hydrothermal anomalies of both temperature (ΔT °C) and suspended particle concentration (estimated by Nephelometric Turbidity Units (ΔNTU)). Plume data from Baker and Urabe (1996). (*Bottom*) An idealized drawing of how plume chemistry changes over time and distance. Each curve shows the change in the relative concentration of a tracer ((X_p/X_v)) to hydrothermal heat (Q_p) in the plume (p) relative to the same ratio in undiluted vent (v) fluids (X_v/Q_v). All tracers are dissolved species except FeS and Fe oxyhydroxides, which precipitate from dissolved species as vent fluids are diluted. About half the hydrothermal Fe is deposited close to the vent as sulfides, while the other half more slowly precipitates and settles as fine-grained Fe oxyhydroxides. Manganese precipitates and settles similarly to Fe. Only ^{3}He is conservative throughout the dispersal.

These hydrothermal plume mapping strategies pursued over the last 30 years have allowed researchers to build maps of vent distributions on scales from local to global. As of 2011, a quarter of all oceanic spreading centers have been systematically explored for hydrothermal plumes, more than 11,000 km in detail and another 6,500 km with sparser sampling by vertical profiles alone (http://www.pmel.noaa.gov/eoi/PlumeStudies/global-vents/index.html). This effort has been instrumental in accumulating a global inventory of almost 600 active vent fields (http://www.interridge.org/IRvents). This distribution demonstrates that the spatial density of vent fields on mid-ocean ridges increases roughly linearly with increasing spreading rate. Because the rate of magma supply also increases with spreading rate, the presence of a magmatic heat source, not tectonically controlled crustal permeability, is thus shown to be the first-order control on the global distribution of vents (Baker and German, 2004).

More recently, the advent of Autonomous Underwater Vehicles (AUVs) has improved synchronous sampling of hydrothermal plumes even further. Hydrothermal plumes mapped by sensor tow-yos can typically locate seafloor vent sites to within a few kilometers to a few hundred meters. That precision can then be improved by a detailed AUV search, using observations of near-bottom maxima in physical and chemical anomalies to pinpoint a vent site (German et al., 2008).

Ocean chemistry

Because buoyant hydrothermal plumes entrain a volume of seawater ~10^4 times greater than the discharging vent fluids, a volume equivalent to the entire ocean cycles through hydrothermal plumes every 4–8 kyr (Elderfield and Schultz, 1996), similar to the ~1 kyr mixing time of the global deep ocean. Thus vent fluids, with concentrations of Fe and Mn often enriched ~10^6:1 relative to normal seawater, influence ocean chemistry to a surprising degree (German and Von Damm, 2003). In buoyant plumes, metals such as Cu, Zn, and Pb coprecipitate with Fe as sulfide phases and settle rapidly to the seafloor. In the near-vent nonbuoyant plume, P, K, V, As, Cr, U, and others are rapidly scavenged by fine-grained Fe oxyhydroxides and, under oxic conditions, maintain constant Fe/metal ratios. The Fe/metal ratios reflect the concentration of the dissolved species in local seawater. The Fe/P ratios in near-ridge sediments, for example, might prove useful in paleoceanographic studies of ancient ocean P concentrations (Feely et al., 1998). Elements such as the rare earths show a different behavior, progressively

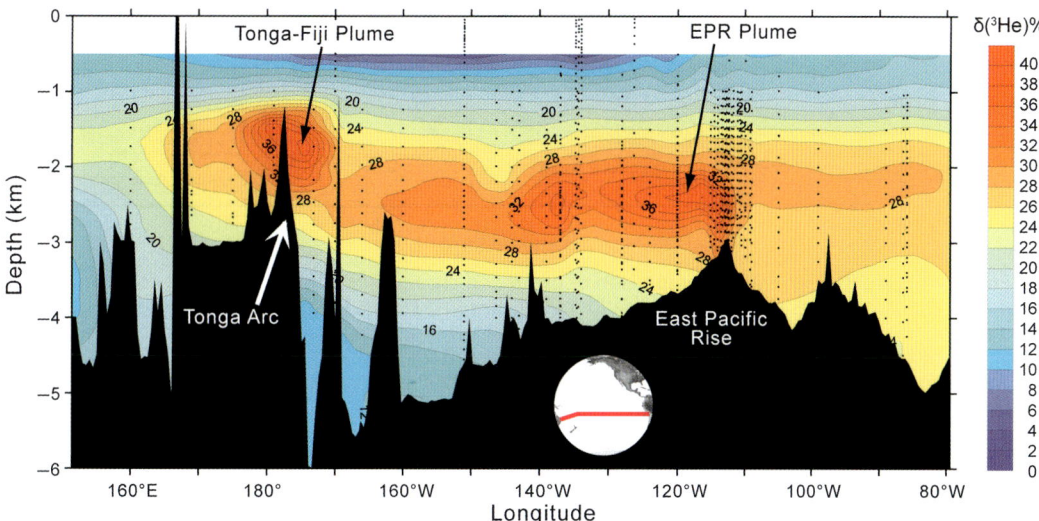

Hydrothermal Plumes, Figure 2 Plumes of $\delta(^3He)\%$ along 15°S in the Pacific Ocean ($\delta(^3He\%)$ are the percentage deviation of the $^3He/^4He$ ratio in water from the ratio in air). A plume originating from hydrothermal venting along the axis of the East Pacific Rise (113°W at this latitude) advects westward almost to Australia (see Figure 1). A second major plume, centered near 180° at a depth of ~1,700 m, apparently has a local origin near the northern end of the Tonga Arc (reprinted with permission from Lupton et al. (2003)).

increasing their concentration on Fe and Mn oxyhydroxides as scavenging continues during plume dispersal. For many of the above elements, hydrothermal removal rates exceed or roughly equal their oceanic input by rivers (Rudnicki and Elderfield, 1993).

Hydrothermal flux measurements

Estimates of hydrothermal heat and chemical fluxes from entire vent fields are important for geophysical, chemical, and ecological studies. Such measurements are difficult, and only about 30 have been reported (Ramondenc et al., 2006). Almost all have used some form of plume observations, either of the rising buoyant plume or of the dispersing nonbuoyant plume. At an individual vent, the heat flux has been calculated by measuring the temperature, salinity, and velocity of the rising plume at multiple depths above the seafloor (e.g., Bemis et al., 1993) and by direct measurements of flow at the vent orifice (e.g., Ramondenc et al., 2006). The heat flux from high-temperature vent structures typically ranges between ~1 and ~50 MW. Heat flux estimates from an entire vent field are commonly made by measuring the net current flow through a two-dimensional transect of the hydrothermal heat anomaly down current from a vent field (e.g., Thomson et al., 1992). A geochemical approach, determining the plume inventory of a short-lived isotope, such as ^{222}Rn, emitted at the plume source, has also been employed (e.g., Rosenberg et al., 1988). Published estimates of vent field heat fluxes vary widely, ranging from ~0.1 to 4 GW.

Global ocean tracers

Understanding patterns of mixing and transport in the deep ocean requires knowledge of the circulation patterns on the scale of decades to centuries. Measurements at this scale are impractical, but hydrothermal venting provides a tracer that can image flow at scales from regional to ocean basin (Lupton et al., 2003). Helium is stable, completely conservative, and injected into the deep ocean only by hydrothermal venting along mid-ocean ridges, volcanic arcs, and a few intraplate volcanoes. Hydrothermal plumes are highly enriched in the isotope ^3He, which can be very precisely measured. Ocean-basin scale sampling, such as by the Geochemical Ocean Section Study (GEOSECS) and World Ocean Circulation Experiment (WOCE) programs, has produced images of ^3He-enriched plumes, typically centered between 2,000 and 3,000 m, that span the meridional and zonal extents of the Pacific, Indian, and Atlantic Oceans (http://cchdo.ucsd.edu/). For example, the abundance of fast-spreading ridges in the Pacific creates intense ^3He plumes that illustrate a varied zonal flow pattern. Westward deep flow is centered along 40°N and in two subequatorial jets at 10°N and 10°S (Figure 2). Compensating eastward flows occur at 20°N and 30°S.

The hydrothermal plume distribution can also be used to judge the performance of global circulation and mixing models. On the global scale, most models can adequately reproduce the differences in ^3He concentration between the major ocean basins. Relative to the observed ^3He distribution, however, existing models tend to underestimate the ventilation of the deep ocean and are too coarse to reproduce the strong and meridionally narrow currents implied by the ^3He data in the Pacific.

Event plumes

The creation of ocean crust by lava eruptions is a fundamental Earth process, involving immediate and

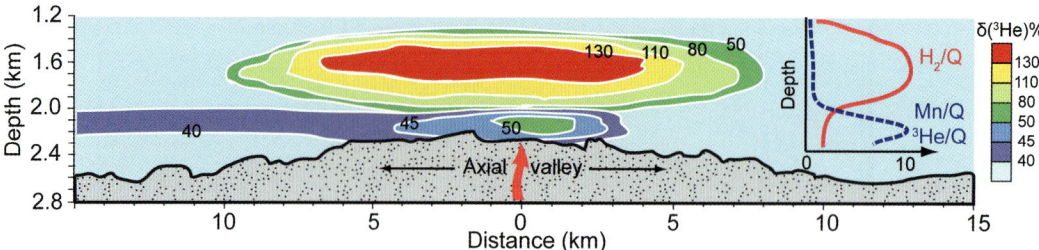

Hydrothermal Plumes, Figure 3 A schematic comparison of event and chronic hydrothermal plumes; contours are $\delta(^3He)\%$ (compare to Figure 2). Chronic plumes closely overlie seafloor vent sites (*red arrow*), typically near the spreading axis of a mid-ocean ridge. Because the vents can be active for years, the plume is quasi-steady state and spreads under the influence of local currents. Event plumes form from brief and massive releases of fluids during a seafloor eruption, forming three-dimensionally symmetrical vortices rising much higher than chronic plumes. Using the hydrothermal heat anomaly (Q) as a reference, Mn/Q (nM/J) and $^3He/Q$ (fM/J) are high in chronic plumes, while H_2/Q (nM/J) is high in event plumes.

immense transfers of heat and chemicals from crust to ocean (see "Explosive Volcanism in the Deep Sea"). This transfer creates unique hydrothermal plumes called "event plumes" (or "megaplumes"). First observed in 1986, event plumes are massive ellipsoidal eddies with distinctive and consistent chemical signatures (reviewed in Baker et al., 2011). Confirmed diameters range from ~1 to 20 km (and perhaps up to 70 km) and thicknesses from ~0.05 to 1.2 km (Figure 3). These characteristics indicate they form from a linear source, such as a dike eruption, rather than a vent-like point source (Lavelle, 1995). Concentrations (in terms of the species/heat ratio) of diagnostic magmatic tracers such as ^3He and dissolved Mn are nearly uniform among all event plumes, implying the same formation process for each.

Confirmed eruption events at five separate locations have each created from one to at least eight individual event plumes. At three other locations where event plumes were found, no information on recent eruptions is available. Possible formation processes include the release of preformed hydrothermal fluids during crustal rupturing, the rapid cooling of an intruded dike, and the rapid cooling of an erupting lava flow. Observations favoring the preformed fluids hypothesis include the similarity between ^3He/heat ratios in event plumes and mature vent fluids and the discovery of thermophilic, presumably crustal dwelling, microbes in one event plume. No convincing mechanism for the storage and instant release of such crustal fluids has yet been proposed. A primary role for dikes or lava is supported by extremely high concentrations of H_2 in event plumes, proof of extensive interaction between molten lava and event plumes during their formation. However, theoretical models of lava cooling and studies of cooled lava on the seafloor indicate that any transfer of heat from molten lava to seawater during an eruption occurs far too slowly to generate the enormous buoyancy flux required to form an event plume (see "Submarine Lava Types"). Thus, the physical and chemical processes that create event plumes remain elusive, and we have only a poor understanding of a fundamental Earth process: the interactions that occur between the solid and liquid Earth as new magma builds the ocean basins.

Conclusions

Hydrothermal plumes disperse heat, chemicals, and organisms from the solid Earth throughout the deep ocean. Their broad extent and varied chemistry facilitate the search for seafloor vent sites and the quantification of vent discharge. Although the search for plumes has been intensive over the last 30 years, over three quarters of spreading center length, and a similar proportion of volcanic arc length, remain unexplored.

Bibliography

Baker, E. T., and German, C. R., 2004. On the global distribution of hydrothermal vent fields. In German, C. R., Lin, J., and Parson, L. M. (eds.), *Mid-Ocean Ridges: Hydrothermal Interactions Between the Lithosphere and Oceans*. Washington, DC: American Geophysical Union. Geophysical Monograph Series, Vol. 148, pp. 245–266.

Baker, E. T., and Urabe, T., 1996. Extensive distribution of hydrothermal plumes along the superfast-spreading East Pacific Rise, 13°50′-18°40′S. *Journal of Geophysical Research*, **101**, 8685–8695.

Baker, E. T., Lavelle, J. W., and Massoth, G. J., 1985. Hydrothermal particle plumes over the southern Juan de Fuca Ridge. *Nature*, **316**, 342–344.

Baker, E. T., Lupton, J. E., Resing, J. A., Baumberger, T., Lilley, M., Walker, S. L., and Rubin, K., 2011. Unique event plumes from a 2008 eruption on the Northeast Lau Spreading Center. *Geochemistry, Geophysics, Geosystems*, **12**, Q0AF02, doi:10.1029/2011GC003725.

Bemis, K. G., Von Herzen, R. P., and Mottl, M. J., 1993. Geothermal heat flux from hydrothermal plumes on the Juan de Fuca Ridge. *Journal of Geophysical Research*, **98**, 6351–6365.

Corliss, J. B., Dymond, J., Gordon, L. I., Edmond, J. M., von Herzen, R. P., Ballard, R. D., Green, K., Williams, D., Bainbridge, A., Crane, K., and van Andel, T. H., 1979. Submarine thermal springs on the Galapagos Rift. *Science*, **203**, 1073–1083.

Elderfield, H., and Schultz, A., 1996. Mid-ocean ridge hydrothermal fluxes and the chemical composition of the ocean. *Annual Review of Earth and Planetary Science*, **24**, 191–224.

Feely, R. A., Trefry, J. H., Lebon, G. T., and German, C. R., 1998. The relationship between P/Fe and V/Fe ratios in hydrothermal precipitates and dissolved phosphate in seawater. *Geophysical Research Letters*, **25**, 2253–2256.

German, C. R., and Von Damm, K. L., 2003. Hydrothermal processes. In Elderfield, H. (ed.), *The Oceans and Marine Geochemistry*, Vol. 6 *Treatise on Geochemistry* (eds. Holland, H. D., and Turekian, K. K.). Oxford: Elsevier-Pergamon, pp. 181–222.

German, C. R., Yoerger, D. R., Jakuba, M., Shank, T. M., Langmuir, C. H., and Nakamura, K., 2008. Hydrothermal exploration with the Autonomous Benthic Explorer. *Deep-Sea Research I*, **55**, 203–219.

Humphris, S., Zierenberg, R., Mullineaux, L. S., and Thomson, R. (eds.), 1995. *Seafloor Hydrothermal Systems: Physical, Chemical, Biological, and Geological Interactions*. Washington, DC: American Geophysical Union. Geophysical Monograph Series, Vol. 91.

Lavelle, J. W., 1995. The initial rise of a hydrothermal plume from a line segment source – results from a three-dimensional numerical model. *Geophysical Research Letters*, **22**, 159–162.

Lupton, J. E., Weiss, R. F., and Craig, H., 1977. Mantle helium in hydrothermal plumes in the Galapagos Rift. *Nature*, **267**, 603–604.

Lupton, J. E., Pyle, D. G., Jenkins, W. J., Greene, R., and Evans, L., 2003. Evidence for an extensive hydrothermal plume in the Tonga-Fiji region of the south Pacific. *Geochemistry, Geophysics, Geosystems*, **5**, Q01003, doi:10.1029/2003GC000607.

Morton, B. R., Taylor, G. I., and Turner, J. S., 1956. Turbulent gravitational convection from maintained and instantaneous sources. *Proceedings of the Royal Society of London*, **A234**, 1–23.

Ramondenc, P., Germanovich, L. N., Von Damm, K. L., and Lowell, R. P., 2006. The first measurements of hydrothermal heat output at 9°50′N, East Pacific Rise. *Earth and Planetary Science Letters*, **245**, 487–497.

Rosenberg, N. D., Lupton, J. E., Kadko, D., Collier, R., Lilley, M., and Pak, H., 1988. A geochemical method for estimating the heat and mass flux from a seafloor hydrothermal system. *Nature*, **334**, 604–607.

Rudnicki, M. D., and Elderfield, H., 1993. A chemical model of the buoyant and neutrally buoyant plume above the TAG vent field, 26°N, Mid-Atlantic Ridge. *Geochimica et Cosmochimica Acta*, **57**, 2939–2957.

Speer, K. G., and Rona, P. A., 1989. A model of an Atlantic and Pacific hydrothermal plume. *Journal of Geophysical Research*, **94**, 6213–6220.

Thomson, R. E., Delaney, J. R., McDuff, R. E., Janecky, D. R., and McClain, J. S., 1992. Physical characteristics of the Endeavour Ridge hydrothermal plume during July 1988. *Earth and Planetary Science Letters*, **111**, 141–154.

Weiss, R. F., Lonsdale, P., Lupton, J. E., Bainbridge, A. E., and Craig, H., 1977. Hydrothermal plumes in the Galapagos Rift. *Nature*, **267**, 600–602.

Cross-references

Black and White Smokers
Explosive Volcanism in the Deep Sea
Hot Spots and Mantle Plumes
Hydrothermal Vent Fluids (Seafloor)
Hydrothermalism
Mid-Ocean Ridge Magmatism and Volcanism
Oceanic Spreading Centers
Submarine Lava Types
Volcanogenic Massive Sulfides

HYDROTHERMAL VENT FLUIDS (SEAFLOOR)

Andrea Koschinsky

Earth and Environmental Sciences, School of Engineering & Science, Jacobs University, Bremen, Germany

Synonyms

Black-smoker fluids; Hydrothermal solutions; Seafloor hot springs

Definition

Hydrothermal vent fluid (seafloor): a hot (up to >400 °C) aqueous solution discharging at the seafloor that typically originates from the reaction of seawater with oceanic crust under high-temperature high-pressure conditions, leading to enrichments in dissolved components such as metals and gases. Hydrothermal vent fluids may also originate from discharges of magma-derived waters from beneath submarine volcanoes, which typically mix with seawater or other vent fluids prior to venting.

Introduction

Hydrothermal fluids on the seafloor typically form by circulation of seawater in fractured oceanic crust at volcanically active sites in the ocean (Figure 1) (Von Damm, 1995; German and Seyfried, 2014). These places include mid-ocean ridges, back-arc spreading centers, and hot-spot or arc-related submarine volcanoes. Circulating seawater is heated by a heat source such as a magma chamber or associated hot rock and, during heating and chemical reaction with the surrounding rock, undergoes a suite of chemical modifications. These include acidification, leaching of major and trace components (including trace metals such as copper, zinc, and gold) from the rock, precipitation of specific mineral phases, and redox reactions including the formation of reduced gases such as methane, hydrogen sulfide, and hydrogen. In the reaction zone in the subsurface, the fluids can reach temperatures of several hundred degrees Celsius, which makes them very buoyant. Furthermore, if fluid temperatures reach the boiling point under the respective pressure conditions during their circulation and ascent back to the seafloor, the fluids can boil and phase-separate into a vapor phase and a residual brine phase. This process is a major cause of salinity variations in vent fluids. After ascent the hot, low pH, and reduced fluids discharge at the seafloor through larger channels or finer cracks, reacting with the cold seawater and forming hot (up to >400 °C) focused black-smoker fluids and, in most cases, sulfide edifices or "chimneys." Mixing with seawater and the associated cooling and chemical changes induce the precipitation of sulfide minerals either forming sulfide structures or "black smoke" particles. When the fluid is cooled or diluted deeper in the subsurface prior to discharging at the seafloor, diffuse and translucent fluid emissions at lower

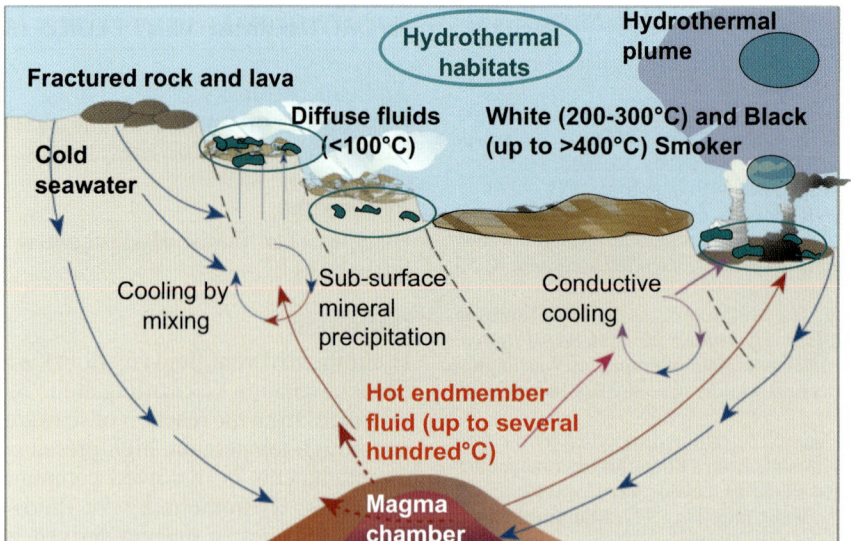

Hydrothermal Vent Fluids (Seafloor), Figure 1 Simplified sketch showing the evolution of hydrothermal fluids in a geologically active site; cold seawater entrains the fractured crust, heats up and reacts with the rock in the reaction zone, and then flows up again as hot hydrothermal fluid to the seafloor. If it is not significantly cooled and remains largely undiluted, it forms hot black smokers or white smokers (with whitish minerals forming at slightly lower temperatures) during mixing with cold ambient seawater. If mixing and cooling already takes place in the subsurface, part of the mineral load will precipitate here, and the fluids will be emitted as mostly transparent, cooler diffuse fluids. *Green ovals* show sites where hydrothermal habitats can be found.

temperatures seep through cracks in the seafloor. The metals discharged and minerals formed by the vent fluids are important for the formation of potentially valuable ore deposits at the seafloor and contribute to the chemical composition of the ocean. The reduced gases in the fluids also nurture rich specially adapted hydrothermal ecosystems that are based on chemosynthetic primary production (i.e., by gaining energy and chemical components directly from the fluids) rather than photosynthesis (see Figure 1).

Occurrence of hydrothermal vent fluids

The majority of hydrothermal circulation producing vent fluids occurs along the mid-ocean ridges, which span about 60,000 km through the global oceans (see, e.g., http://vents-data.interridge.org/ventfields and Figure 2). The first low-temperature (up to 17 °C) hydrothermal vents had been discovered at the Galapagos Spreading Center in 1977 (Corliss et al., 1979; Edmond et al., 1979), while the first black smokers with temperatures around 380 °C had been found at 21°N on the East Pacific Rise (EPR) in 1979 (Spiess et al., 1980). While hydrothermal fluid emanations seem to be more abundant on fast-spreading ridges such as the EPR, dozens of active hydrothermal vent sites have also been discovered on slow-spreading ridges such as the Mid-Atlantic Ridge (MAR) or Central Indian Ridge (see compilation by Edmonds (2010)). Even a few off-axis systems with fluids of a very distinct composition, such as the Lost City field on the MAR (Kelley et al., 2001), have been discovered. While hydrothermal activity and fluid composition seem to be highly variable on fast-spreading ridges on timescales of months to years, due to frequent volcanic activities (e.g., Baker et al., 1998), vent fluid composition has been shown to be rather stable over many years in systems on the MAR (e.g., Edmonds, 2010; Schmidt et al., 2011). Hydrothermal fluid venting is also found associated with hot-spot-related intraplate volcanism, such as at some Pacific islands like Hawaii, as well as at arc volcanoes and back-arc spreading centers such as the Manus Basin and North Fiji Basin in the Western Pacific.

Composition of hydrothermal vent fluids

Hydrothermal fluids from different locations or even within discrete vent fields can span a wide range of chemical composition, with individual parameters often varying over several orders of magnitude (see, e.g., German and Seyfried, 2014). The fluids are usually acidic, with pH values down to ~2, major ions (such as Na, Ca, Cl, etc.) either increased or decreased relative to seawater, and the majority of elements including Si and most trace metals significantly enriched relative to seawater. The reducing nature of the hydrothermal fluids is reflected by the absence of oxygen and the presence of reduced gases such as hydrogen, methane, and hydrogen sulfide. Parameters determining the chemical composition of hydrothermal vent fluids include pressure (both the pressure of the sub-seafloor reaction zone and the emanation site at the seafloor), temperature during fluid evolution and emanation, the mineralogical and chemical composition of the host rock, and the reaction time. In back-arc hydrothermal systems, magmatic fluids (i.e., waters exsolved from water-rich magmas) also can affect hydrothermal

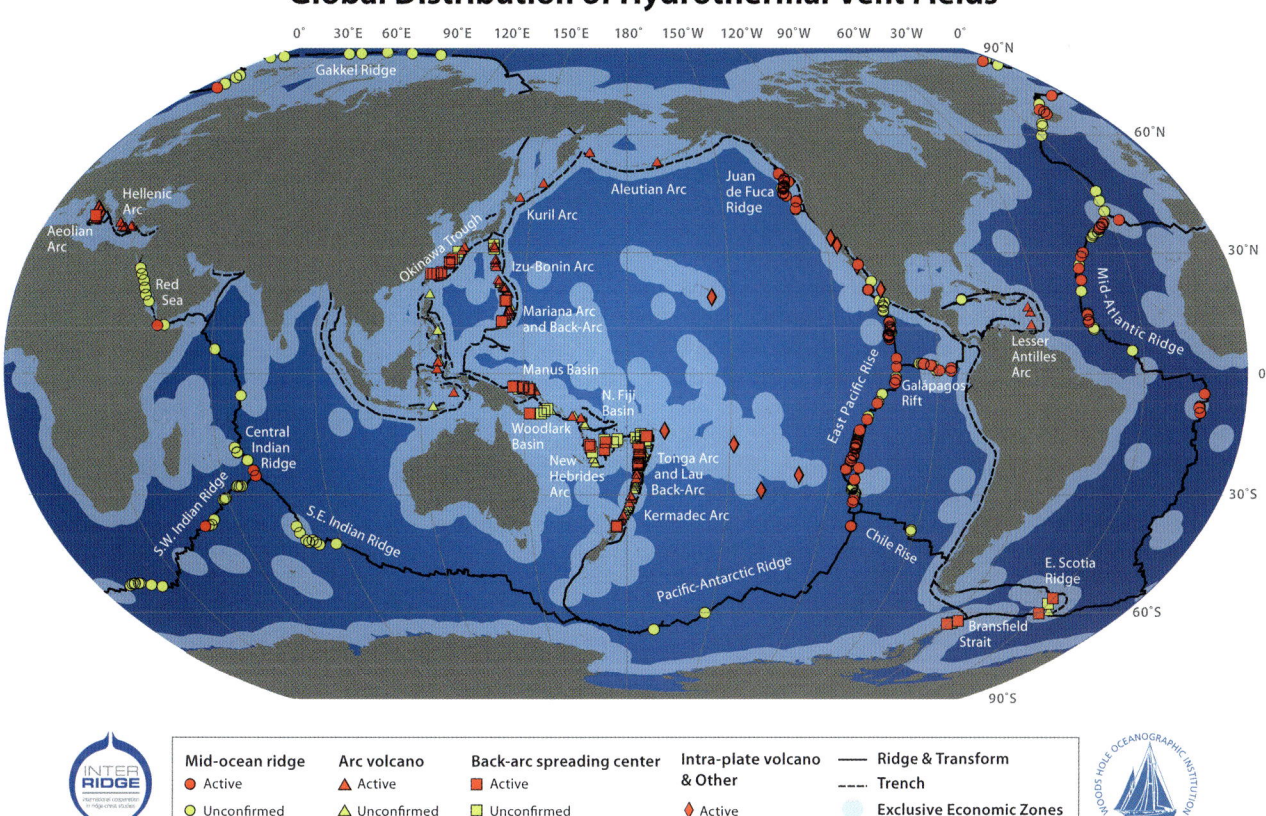

Hydrothermal Vent Fluids (Seafloor), Figure 2 Global distribution of seafloor hydrothermal systems and related mineral deposits, with about 300 sites of high-temperature hydrothermal venting (Beaulieu, 2010).

fluids if they are added at depth within the reaction zone (Gamo et al., 2006; Reeves et al., 2011). These magmatic fluids may also discharge directly toward the seafloor forming unique acidic hydrothermal vents. Apart from in situ studies and direct sampling and analysis, laboratory experimental studies and thermodynamic calculations (e.g., Bischoff and Rosenbauer, 1985) have also helped to understand factors controlling the composition of hydrothermal fluids.

High temperatures, as well as acidity, are necessary to leach large amounts of metals such as Fe and Cu from the rock and keep them in solution. Fluids emanating with very high temperatures and very rich in metals (up to millimolar amounts of Fe and Mn and micromolar amounts of Cu and other metals) can be found in deep ($\geq$3,000 m water depth) hydrothermal systems, where the pressure-dependent boiling point allows the fluids to reach temperatures of $\geq$400 °C (e.g., Koschinsky et al., 2008). Fluids reaching the boiling point phase-separate into a vapor phase rich in gases and a residual brine phase being rich in major ions and trace metals. Emanation of low-chlorinity vapor and high-chlorinity brine phases can be observed both spatially and temporally segregated, with the denser brine phase often emanating subsequent to the vapor phase (Butterfield et al., 1997; von Damm et al., 1997).

The difference in composition for fluids from different types of hydrothermal systems is also related to the different host rocks that react with the entraining seawater. At fast-spreading centers, the fluids are characterized by reaction with basalt, resulting in high concentrations of S, Si, and many metals. At slow-spreading ridges such as the MAR, fluids often carry a pronounced ultramafic signature from reactions with mantle rocks, such as very high hydrogen and methane concentrations due to serpentinization reactions (e.g., Kelley et al., 2001; Charlou et al., 2002; Kelley et al., 2005; Schmidt et al., 2011). The few available fluid data from back-arc basins and island arcs indicate typically very low pH values and a strong enrichment of trace metals including As, Au, Hg, Pb, Sn, and Sb originating potentially from magmatic sources as well as leaching of the rocks (e.g., andesites) (e.g., Hannington et al., 2005; Yang and Scott, 2006; De Ronde et al., 2011; Reeves et al., 2011).

Role of hydrothermal vent fluids for the formation of ore deposits

When the hot metal-rich hydrothermal fluids mix with ambient seawater, either in the sub-seafloor or when they emanate at the seafloor, they precipitate large amounts of its metal and sulfur load as minerals, forming massive sulfide deposits, mineral chimneys, and black smoke in the rising plume (e.g., Hannington et al., 2005). The steep temperature and geochemical gradients in the mixing zones lead to a sequence of precipitation of sulfide minerals and others, such as anhydrite, carbonates, silicates, and finally, in more oxidized zones, oxyhydroxides of Fe and Mn. These deposits are an important sink of metals mobilized from the oceanic crust by the hydrothermal fluids and thus limit the amount of material that is transported into the oceanic water column. They are very rich in many valuable metals such as Au, Ag, Cu, and Zn (e.g., de Ronde et al., 2011), depending on the composition of the ore-forming fluid. Hence, they may represent important metal resources for our future needs in modern technologies, although estimated overall quantities of deposits are small in scale relative to terrestrial metal resources (Hannington et al., 2011; Hein et al., 2013).

Role of hydrothermal vent fluids for supporting deep-sea vent communities and the origin of life

Hydrothermal vent fluids do not only feed hydrothermal ore deposits with precious metals but also nurture rich hydrothermal ecosystems with their chemical energy and material released in the fluids mixing with seawater (e.g., van Dover, 2000). The members of these ecosystems are both adapted to extreme conditions in the fluids, such as low pH, high temperatures, and high metal contents, and are able to sustain bioproductivity without any light but based on the material and energy available in the vent fluids (e.g., Tunnicliffe, 1992). During chemosynthesis, microorganisms gain energy from redox reactions in the mixing zone to build up biomass and feed a rich community of grazers, symbiotic mussels, shrimps, tube worms, and other organisms. While the type and composition of the fluid has an impact on the evolution of the ecosystem, at the same time the organisms change the fluid composition in the mixing zone. They mediate redox reactions, take up chemical components, and excrete others that may influence the (bio)chemical reactivity and fluxes of these metals. Hydrothermal fluids are also discussed to be potential sites where life on Earth may have evolved. While in early Earth history the surface of the Earth was a rather hostile place, the seafloor was a more protected place with hydrothermal fluids providing energy and chemical components suitable to sustain microbial life (Martin et al., 2008). However, also in the search for extraterrestrial life, the potential existence of hydrothermal fluids plays a major role. Astrobiological research focuses on geothermally active bodies in our solar system in which water exists or may have existed, such as Mars and Venus, or the icy satellites of the giant planets in the outer solar system, such as Titan and Europa (e.g., Vance et al., 2007). Hence, knowledge on the functions of hydrothermal fluids for life on Earth is an important prerequisite for the search for life elsewhere in our solar system.

Role of hydrothermal vent fluids for heat and energy transfer from the oceanic crust into the ocean

Hydrothermal fluids are a crucial medium for the exchange of heat and matter between the oceanic crust and the ocean (Kadko et al., 1995; Baker, 2007). The globally widespread occurrence of hydrothermal fluid emanations on the seafloor, which has been operating for most of Earth's history, makes hydrothermal input one of the main, but poorly quantified, sources of components such as trace metals into the global oceans. The magnitude of hydrothermal input seems to be on the order of riverine input for some elements, such as Mn (Edmond et al., 1979; Elderfield and Schulz, 1996). Among the many problems that exist for a more precise assessment of hydrothermal element fluxes to the ocean are the unknown total number and size of hydrothermal vent fields and their chemical variability, the role of poorly quantified or undiscovered off-axis flow, and an incomplete understanding of the transformation processes of fluid components during mixing of fluid and ambient seawater and within the rising and dispersing hydrothermal plume. Formation of mineral deposits, oxidation of Fe and Mn in the plume, and scavenging of trace metals on Mn and Fe oxides in the hydrothermal plumes (German et al., 1997) significantly reduce the hydrothermal metal fluxes of many elements, for example. However, modeling and chemical speciation studies have shown that for Fe, the hydrothermal metal flux must be significantly higher than previously assumed, probably contributing to the biogeochemical cycling of Fe in the oceans (Tagliabue et al., 2010). While the exact mechanisms of metal transport and transformation from the vent sites into the ocean are still unclear, several studies hint at a strong role of colloidal sulfides or hydroxides and organic metal complexes (Sander and Koschinsky, 2011).

Summary and conclusions

Hydrothermal fluid emanations at the seafloor are a widespread phenomenon in volcanic, magmatic, or tectonically active areas in the world oceans, such as mid-ocean ridges and submarine volcanoes. Their wide range in emanation temperatures (up to >400 °C), chemical composition (up to millimolar concentrations in reduced gases and metals), and temporal variability reflect the specific nature (such as water depth and host rock) of the site location. Hydrothermal fluids are important on a global scale for the formation of metal ore deposits on the seafloor, for nurturing chemosynthetic ecosystems specifically adapted to extreme conditions, and for metal fluxes into the water column and the chemical budget of the ocean. However, the exact nature and quantity of the related processes is still not fully understood. It can be expected that each expedition discovering new hydrothermal vent systems will broaden our view on the diversity of hydrothermal vent fluids and help understand their relevance for the exchange of heat and material between the oceanic crust and the ocean.

Bibliography

Baker, E. T., 2007. Hydrothermal cooling of midocean ridge axes: do measured and modeled heat fluxes agree? *Earth and Planetary Science Letters*, **263**, 140–150.

Baker, E. T., Massoth, G. J., Feely, R. A., and Cannon, G. A., 1998. The rise and fall of the CoAxial hydrothermal site, 1993–1996. *Journal of Geophysical Research*, **103**(B5), 9791–9806.

Beaulieu, S. E., 2010. InterRidge Global Database of Active Submarine Hydrothermal Vent Fields, Version 2.0: http://www.interridge.org/irvents. July 2011.

Bischoff, J. L., and Rosenbauer, R. J., 1985. An empirical equation of state for hydrothermal seawater (3.2 percent NaCl). *American Journal of Science*, **285**, 725–763.

Butterfield, D. A., Jonasson, I. R., Massoth, G. J., Feely, R. A., Roe, K. K., Embley, R. E., Holden, J. F., McDuff, R. E., Lilley, M. D., and Delaney, J. R., 1997. Seafloor eruptions and evolution of hydrothermal fluid chemistry. *Philosophical Transactions of the Royal Society of London*, **A355**, 369–386.

Charlou, J. L., Donval, J. P., Fouquet, Y., Jean-Baptiste, P., and Holm, N., 2002. Geochemistry of high H_2 and CH_4 vent fluids issuing from ultramafic rocks at the Rainbow hydrothermal field (36°14'N, MAR). *Chemical Geology*, **191**, 345–359.

Connelly, D. P., et al., 2011. Hydrothermal vent fields and chemosynthetic biota on the world's deepest seafloor spreading centre. *Nature Communications*, **3**, doi:10.1038/ncomms1636.

Corliss, J. B., et al., 1979. Submarine thermal springs on the Galapagos Rift. *Science*, **203**, 1073–1083.

De Ronde, C. E. J., et al., 2011. Submarine hydrothermal activity and gold-rich mineralization at Brothers Volcano, Kermadec Arc, New Zealand. *Mineralium Deposita*, **46**, 541–584.

Edmond, J. M., Measures, C. I., McDuff, R. E., Chan, L. H., Collier, R., Grant, B., Gordon, L. I., and Corliss, J. B., 1979. Ridge-crest hydrothermal activity and the balances of the major and minor elements in the ocean: the Galapagos data. *Earth and Planetary Science Letters*, **46**, 1–18.

Edmonds, H. N., 2010. Chemical signatures from hydrothermal venting on slow spreading ridges. In Rona, P. A., Devey, C. W., Dyment, J., and Murton, B. J. (eds.), *Diversity of Hydrothermal Systems on Slow Spreading Ocean Ridges*. Washington, DC: American Geopysical Union. Geophysical Monograph, Vol. 188.

Elderfield, H., and Schultz, A., 1996. Mid-ocean ridge hydrothermal fluxes ad the chemical composition of the ocean. *Annual Review of Earth and Planetary Sciences*, **24**, 191–224.

Gamo, T., Ishibashi, J., Tsunogai, U., Okamura, K., and Chiba, H., 2006. Unique geochemistry of submarine hydrothermal fluids from arc–back-arc settings of the western Pacific. In Christie, D. M., Fisher, C. R., Lee, S.-M., and Givens, S. (eds.), *Back-Arc Spreading Systems: Geological, Biological, Chemical, and Physical Interactions*. Washington, DC: American Geophysical Union. AGU Monograph, Vol. 166, pp. 147–161.

German, C. R., and Seyfried, W. E., Jr., 2014. Hydrothermal processes. In *Treatise on Geochemistry*, 2nd edn. Oxford, Elsevier, Vol. 8, pp. 191–233.

German, C. R., Campbell, A. C., and Edmond, J. M., 1997. Hydrothermal scavenging at the Mid-Atlantic Ridge: modification of trace element dissolved fluxes. *Earth and Planetary Science Letters*, **107**, 101–114.

Hannington, M. D., de Ronde, C. D., and Petersen, S., 2005. Sea-floor tectonics and submarine hydrothermal systems. *Economic Geology*, **100**, 111–141.

Hannington, M., Jamieson, J., Monecke, T., Petersen, S., and Beaulieu, S., 2011. The abundance of seafloor massive sulfide deposits. *Geology*, **39**, 1155–1158.

Hein, J. R., Mizell, K., Koschinsky, A., and Conrad, T. A., 2013. Deep-ocean mineral deposits as a source of critical metals for high- and green-technology applications: comparison with land-based resources. *Ore Geology Reviews*, **51**, 1–14.

Kadko, D., Baross, J., and Alt, J., 1995. The magnitude and global implications of hydrothermal flux. In Humphris, S. E., Zierenberg, R. A., Mullineaux, L. S., and Thomson, R. E. (eds.), *Seafloor Hydrothermal Systems: Physical, Chemical, Biological, and Geological Interactions*. Washington, DC: American Geophysical Union. Geophysical Monograph, Vol. 91, pp. 446–466.

Kelley, D. S., Karson, J. A., Blackman, D. K., Fruh-Green, G. L., Butterfield, D. A., Lilley, M. D., Olson, E. J., Schrenk, M. O., Roe, K. K., Lebon, G. T., Rivizzigno, P., and The AT3–60 Shipboard Party, 2001. An off-axis hydrothermal vent field near the Mid-Atlantic Ridge at 30°N. *Nature*, **412**, 145–149.

Kelley, D., et al., 2005. A serpentinite-hosted ecosystem: the Lost City hydrothermal field. *Science*, **307**, 1428–1434.

Koschinsky, A., Garbe-Schönberg, D., Sander, S., Schmidt, K., Gennerich, H. H., and Strauß, H., 2008. Hydrothermal venting at pressure-temperature conditions above the critical point of seawater, 5°S on the Mid-Atlantic Ridge. *Geology*, **36**, 615–618.

Martin, W., Baross, J., Kelley, D., and Russel, M. J., 2008. Hydrothermal vents and the origin of life. *Nature Reviews. Microbiology*, **6**, 805–814.

Reeves, E. P., Seewald, J. S., Saccocia, P., Bach, W., Craddock, P. R., Shanks, W. C., Sylva, S. P., Walsh, E., Pichler, T., and Rosner, M., 2011. Geochemistry of hydrothermal fluids from the PACMANUS, northeast Pual and Vienna Woods hydrothermal fields, Manus Basin, Papua New Guinea. *Geochimica et Cosmochimica Acta*, **75**, 1088–1123.

Sander, S., and Koschinsky, A., 2011. Metal flux from hydrothermal vents increased by organic complexation. *Nature Geoscience*, **4**, 145–150.

Schmidt, K., Garbe-Schönberg, D., Koschinsky, A., Strauss, H., Jost, C. L., Klevenz, V., and Königer, P., 2011. Fluid elemental and stable isotope composition of the Nibelungen hydrothermal field (8°18'S, Mid-Atlantic Ridge): constraints on fluid–rock interaction in heterogeneous lithosphere. *Chemical Geology*, **280**, 1–18.

Spiess, F. N., et al., 1980. East Pacific Rise: hot springs and geophysical experiments. *Science*, **207**, 1421–1433.

Tagliabue, A., et al., 2010. Hydrothermal contribution to the oceanic dissolved iron inventory. *Nature Geoscience*, **3**, 252–256.

Tunnicliffe, V., 1992. The nature and origin of the modern hydrothermal vent fauna. *Palaios*, **7**, 338–350.

Van Dover, C. L., 2000. *The Ecology of Deep-Sea Hydrothermal Vents*. Princeton: Princeton University Press, p. 424.

Vance, S., Harnmeijer, J., Kimura, J., Hussmann, H., Demartin, B., and Brown, M., 2007. Hydrothermal systems in small ocean planets. *Astrobiology*, **7**, 987–1005.

Von Damm, K. L., 1995. Controls on the chemistry and temporal variability of seafloor hydrothermal fluids. In Humphris, S. E., Zierenberg, R. A., Mullineaux, L. S., and Thomson, R. E. (eds.), *Seafloor Hydrothermal Systems: Physical, Chemical, Biological, and Geological Interactions*. Washington, DC: American Geophysical Union. Geophysical Monograph, Vol. 91, pp. 222–247.

Von Damm, K. L., Buttermore, L. G., Oosting, S. E., Bray, A. M., Fornari, D. J., Lilley, M. D., and Shanks, W. C., III, 1997. Direct observation of the evolution of a seafloor black smoker from vapor to brine. *Earth and Planetary Science Letters*, **149**, 101–112.

Yang, K., and Scott, S. D., 2006. Magmatic fluids as a source of metals in seafloor hydrothermal systems. In Christie, D. M., Fisher, C. R., Lee, S.-M., and Givens, S. (eds.), *Back-Arc Spreading Systems: Geological, Biological, Chemical, and Physical Interactions*. Washington, DC: American Geophysical Union. Geophysical Monograph Series, Vol. 166, pp. 163–184.

Cross-references

Black and White Smokers
Hydrothermal Plumes
Hydrothermalism
Marine Mineral Resources
Oceanic Spreading Centers
Serpentinization
Volcanogenic Massive Sulfides

HYDROTHERMALISM

John W. Jamieson[1], Sven Petersen[1] and Wolfgang Bach[2,3]
[1]GEOMAR Helmholtz Centre for Ocean Research, Kiel, Germany
[2]Department of Geosciences, University of Bremen, Bremen, Germany
[3]MARUM-Center for Marine Environmental Sciences, University of Bremen, Bremen, Germany

Definition

Hydrothermalism refers to the convective circulation of seawater through oceanic lithosphere with possible contributions of volatiles and metals from underlying magma chambers.

Introduction

The circulation of seawater through oceanic lithosphere is the principal mechanism for the transfer of heat and chemicals between the ocean and underlying crust and has a direct effect on the composition and heat budget of oceanic lithosphere and the composition of seawater (e.g., German and Seyfried, 2014). Along submarine tectonic plate boundaries (see "Mid-ocean Ridge Magmatism and Volcanism," "Magmatism at Convergent Plate Boundaries," "Intraoceanic Subduction Zone," "Hydrothermal Plumes") and hot-spot chains, the heat associated with crystallization in shallow magma reservoirs and cooling of hot rock drives the circulation of seawater through the permeable crust. As the fluid circulates, it is heated, reacts chemically with the surrounding rock, and ultimately discharges back at the seafloor (see "Hydrothermal Vent Fluids (Seafloor)"). To date, over 500 hydrothermal discharge sites have been discovered along submarine tectonic boundaries, at an average spacing of about one major discharge site per 100 km of ridge axis (http://www.interridge.org/IRvents). Cooling of the crust by hydrothermal circulation accounts for about 30 % of the heat lost from newly formed (<1 million year old) oceanic lithosphere (Wolery and Sleep, 1976; German and Seyfried, 2014).

The circulation of seawater and the fluid-rock interactions (both chemical and thermal) that occur are a direct result of magmatic heat input into the system and cooling of the newly accreted lithosphere. Our understanding of these processes derives mainly from observations of alteration minerals within oceanic crust exposed on large-scale oceanic faults and from Ocean Drilling Program drill cores (Alt and Teagle, 2003; Teagle et al., 2003). Additional information has been gained from slices of oceanic crust that have been thrust onto land by plate tectonic processes (i.e., ophiolites; (Gillis, 2002). As seawater flows downward into the crust and reacts with the surrounding rock, progressive chemical alteration of both the rock and fluid takes place. As a consequence of this process, some dissolved components within the original seawater are transferred to the rock, whereas chemical constituents within the host rock may be leached by the fluid (Figure 1). Evidence from seismic studies indicates that the depths within the crust to which fluids circulate can vary from less than 1 km to greater than 4 km, depending on factors such as the depth of the axial magma chamber and the depth of the brittle-ductile transition zone. Temperatures of the circulating seawater-derived fluids can reach more than 500 °C in the reaction zone at the base of the circulating cell. Here, the intensity of fluid-rock interactions reaches a maximum, and the high temperatures cause the fluid, which is now rich in dissolved metals and reduced sulfur leached from the surrounding rock, to become positively buoyant. Seawater is converted to variable salinity fluids in the course of supercritical phase separation, and these fluids rapidly ascend back to the surface along focused, deep-seated permeable pathways such as fault planes (Figure 1). Where hot, mineral-laden fluid reaches the seafloor, it mixes with cold ambient seawater, causing a sudden change in temperature, redox conditions, and the precipitation of sulfide minerals from the hydrothermal fluid. The precipitated minerals both accumulate at the vent site, forming massive sulfide spires and mounds, and are vented into the overlying water column as "smoke," which has led to such vents being referred to as "black or white smokers" or "chimneys." Several (tens to hundreds) individual vents often occur in a cluster, or vent field, that can be spread over hundreds of m above a hydrothermal upflow zone. As the chimneys grow, they often collapse, and a new chimney begins to grow on the talus of the collapsed chimney. Over time, the sulfide material from neighboring vents coalesces to form massive sulfide mounds (Figure 1). Most of the dissolved material, however, is ejected into the water column, forming a *hydrothermal plume* that can rise to several hundred meters above the seafloor (Figure 1). The thermal and chemical anomalies associated with these plumes can be traced over several kilometers from the vent site, and the

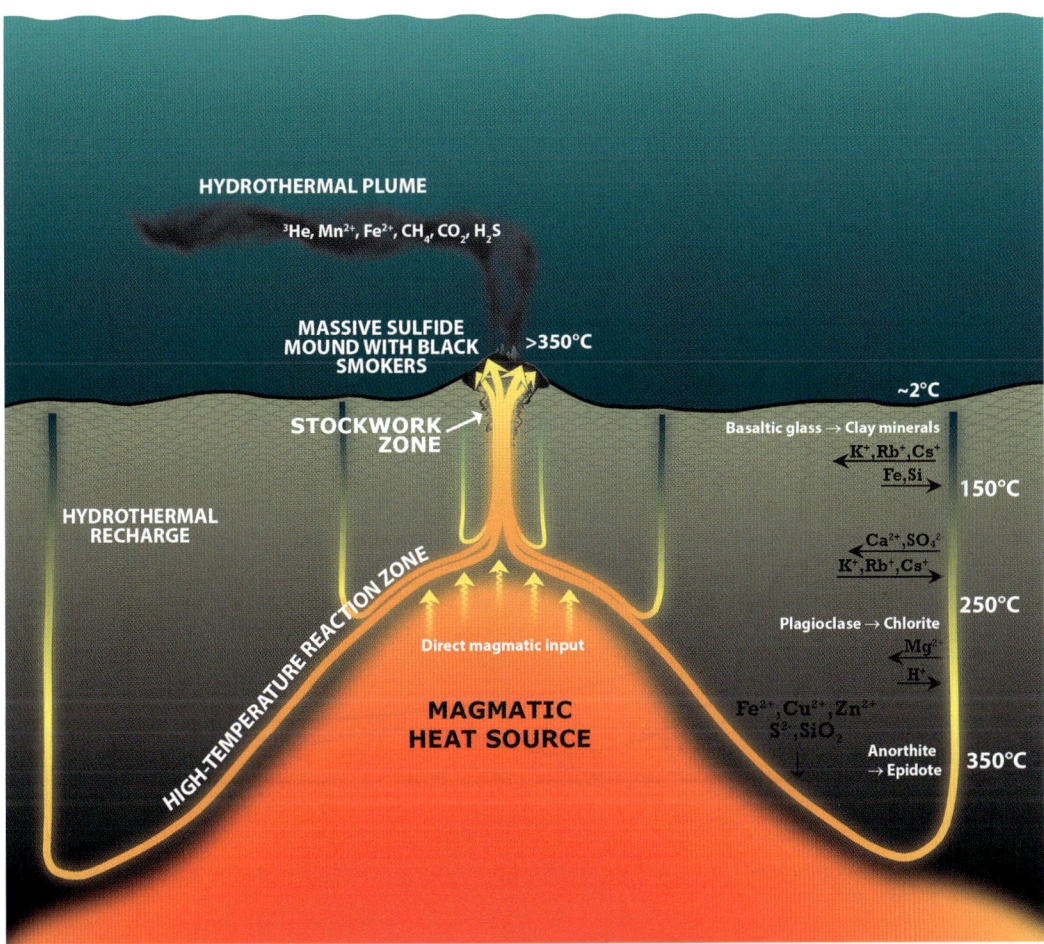

Hydrothermalism, Figure 1 Generalized schematic of a hydrothermal system at a fast-spreading mid-ocean ridge showing fluid temperature gradients and fluid-rock chemical exchange.

detection of hydrothermal plumes, historically, has been the primary method for the discovery of hydrothermal venting on the seafloor.

The massive sulfide deposits that form on the seafloor are composed primarily of sulfide minerals, as well as sulfates (anhydrite and barite) and amorphous silica. The deposits can contain economically significant concentrations of Cu, Zn, Pb, Au, and Ag, and massive sulfide deposits that have been discovered in ancient oceanic crust now obducted on land (referred to as *volcanogenic massive sulfide* deposits) have been mined for their precious and base metal content and continue to be an important source for these metals. With the increasing global demand for metal resources and the decreasing rate of discovery of large ore deposits on land to satisfy this demand, modern hydrothermal sulfide deposits on the seafloor are increasingly being considered as a viable future source for metals (see "Marine Mineral Resources"). Although, at the time of writing, no modern seafloor massive sulfide deposits have been exploited, several companies and national governments are actively exploring the seafloor for massive sulfide deposits that are sufficiently large and exhibit sufficiently high metal concentrations to be exploited.

The following sections provide an overview of the physical and chemical characteristics of typical submarine hydrothermal systems. Many aspects of hydrothermal systems (e.g., *black and white smokers*, *hydrothermal fluids*, *hydrothermal plumes*, *volcanogenic massive sulfides*, *chemosynthetic life*) are covered in more detail in their own separate entries, and the reader is directed to those entries for further details.

Discovery

The discovery of active hydrothermal venting on the seafloor, and especially their link to chemosynthetic life, was among the most significant scientific advances of the twentieth century. The link between seafloor volcanism and hydrothermalism in the oceans was first articulated as an outgrowth of plate tectonic theory in the 1960s, about 20 years before the discovery of black

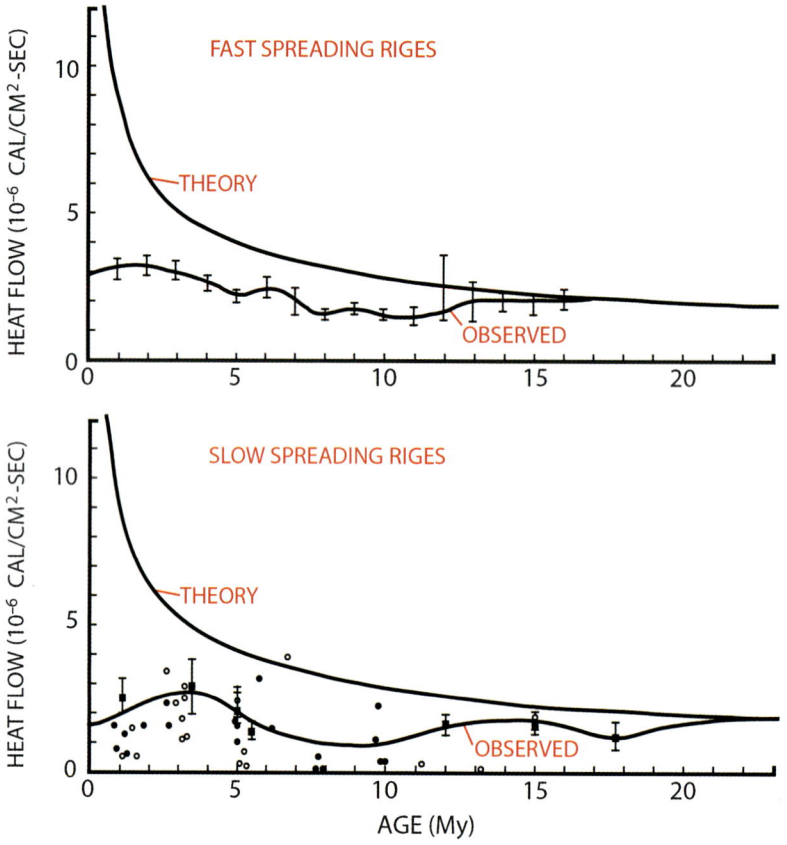

Hydrothermalism, Figure 2 Theoretical versus observed heat flow profiles as a function of crustal age for fast- and slow-spreading ridges (modified from Wolery and Sleep, 1976).

smokers, although the notion of massive sulfide deposits forming at submarine volcanoes existed among economic geologists long before then. By the late 1970s, many of the details of ore formation in the submarine environment had already been established from the study of fossil hydrothermal sulfide deposits on land. A significant metal enrichment had also been identified in sediments flanking the East Pacific Rise mid-ocean ridge, which was attributed to, as yet undiscovered, hydrothermal activity along the nearby ridge (Boström, 1966). Thus, the discovery of seafloor hydrothermal vents and the associated massive sulfide deposits came about not only because of technological advances that allowed deep diving to the mid-ocean ridges but also because of existing ideas about the likelihood that some form of hydrothermal activity should exist on the seafloor. However, those who participated on the discovery cruises did not foresee the spectacular "black smoker" chimneys that formed at the vents or the unusual biological communities that were found thriving in such extreme environments.

The first clue to the magnitude of seafloor hydrothermal activity along the ridges came from heat flow measurements in young oceanic crust adjacent to the East Pacific Rise. The observed heat flow was found to be much lower than that predicted for conductive cooling of the new crust forming along the ridge, and a number of researchers postulated that hydrothermal convection of seawater through the ridges must account for this discrepancy (Figure 2) (Lister, 1972; Wolery and Sleep, 1976). Shortly thereafter, evidence for the injection of mantle helium into the oceans was discovered above the Galapagos Rift, and the first warm-water vents and vent animals were found at the Galapagos hot springs in 1977 (Lupton et al., 1977; Weiss et al., 1977; Corliss et al., 1979). Two years later, black smoker vents and chemosynthetic tube worms were discovered at 21°N on the East Pacific Rise, providing the first direct evidence of vigorous hydrothermal circulation and the formation of sulfide deposits on the seafloor (Francheteau et al., 1979; Spiess et al., 1980; Hekinian et al., 1983). This discovery began more than two decades of intensive exploration of the oceans for additional evidence of seafloor hydrothermal activity. Within the next 5 years, more than 50 sites of hydrothermal venting and related mineral deposits were found on the mid-ocean ridges; by the early 1990s there were 150. At the time of writing, more than 500 sites of seafloor hydrothermal activity and related mineral deposits have been found (Figure 3; Beaulieu et al., 2013).

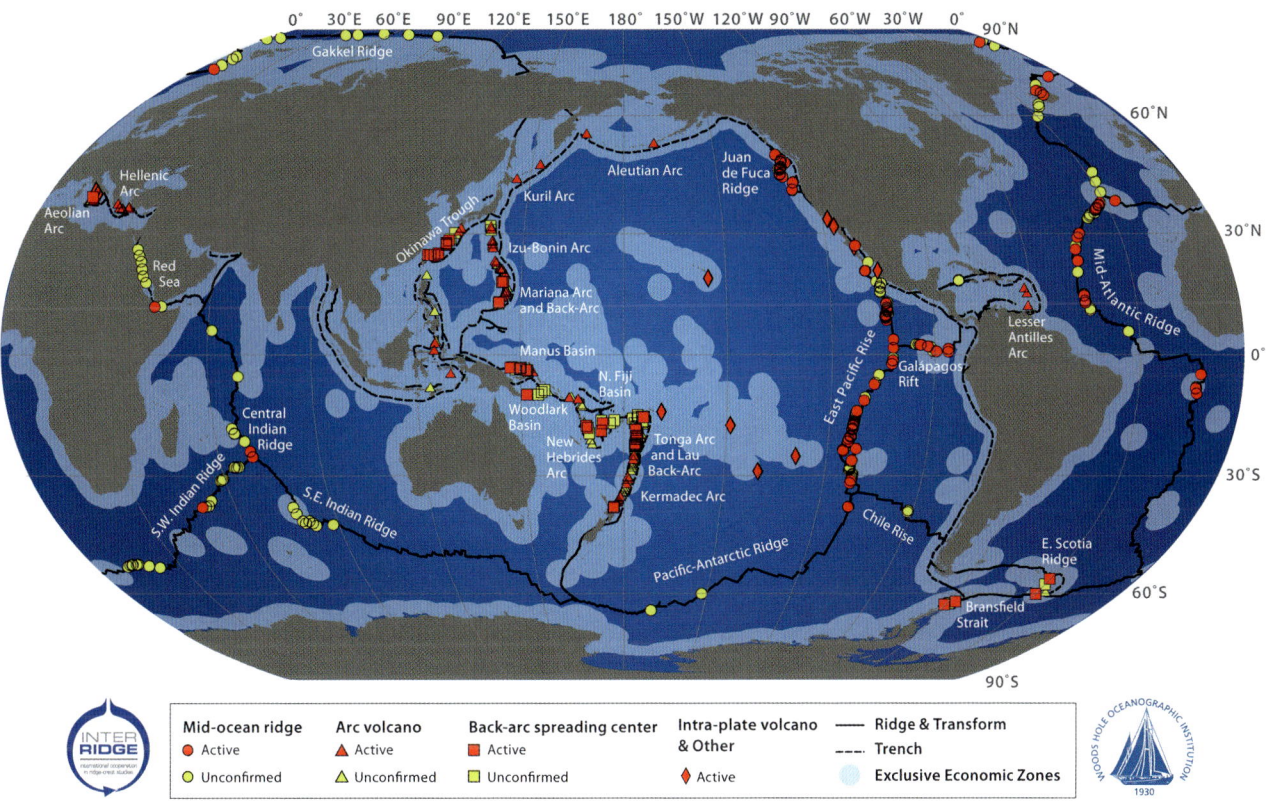

Hydrothermalism, Figure 3 Global distribution of confirmed and inferred seafloor hydrothermal systems (from InterRidge at http://www.interridge.org/irvents/). About 300 are sites of active high-temperature hydrothermal venting; 165 are confirmed sites of volcanogenic massive sulfide accumulation.

Besides mid-ocean ridges, vent sites are now known at a variety of submarine tectonic settings, including island arcs, and back-arc basins in both oceanic and continental crust and associated with more isolated hot-spot-related seafloor volcanic chains. Large differences are seen in the compositions of the hydrothermal fluids and in the compositions of the deposits, reflecting variations in both the depth of venting (hence, pressure) but more significantly, variations in the host rock lithologies and makeup of the basement in the different settings.

Hydrothermal circulation

With the exception of well-defined discharge sites, the parameters that describe the geometry of hydrothermal systems (e.g., depth, fluid volumes, and recharge sites) are poorly defined. Insights can come from direct observations of the seafloor, seismic tomography, or analysis of bathymetric data. However, these observations are limited to only the surface expressions of hydrothermal systems. For example, although seafloor discharge sites are often observed to occur on or in close proximity to exposed faults, the relationship between hydrothermal upflow, discharge, and the permeable fault planes remains unclear (Tivey and Johnson, 2002). Much of our understanding of the subsurface geometry of hydrothermal systems has come from studies of ancient hydrothermal systems preserved in ophiolites (Hannington et al., 1995). For example, Richards et al. (1989) showed that hydrothermal upflow zones occur as discrete pipes beneath vent sites. Further evidence for a narrow, focused upflow zone has come from studies of thermally induced crustal demagnification (or "burn holes") beneath hydrothermal discharge sites (Tivey and Johnson, 2002) and numerical fluid flow simulations (Coumou et al., 2008).

Hydrothermal recharge zones are especially enigmatic, since recharge is effectively invisible on the seafloor, especially compared to the dynamic activity at high-temperature discharge sites. Numerical models and heat flow measurements indicate that hydrothermal recharge can occur at scales ranging from tens of meters to several kilometers from the discharge area (Figure 1) (Coumou et al., 2008; Johnson et al., 2010; Hasenclever et al., 2014). Recharge that occurs near the discharge site can include a component of recirculated, hot, hydrothermal fluid that did not vent at the surface, which results in a significant increase in the temperature of the circulating fluids in proximity to the discharge site. Recent numerical modeling indicates that about two thirds of the recharge occurs on-axis by warm (up to 300 °C) recharge close to the discharge sites, while one third occurs as colder and broader recharge up to several kilometers away from the axis (Hasenclever et al., 2014).

The following is a brief description of the evolution of hydrothermal fluid as it circulates through the oceanic lithosphere and the fluid-rock interactions that occur. The reader is directed to the entry on *hydrothermal vent fluids* for a more detailed description of the physical and chemical properties and evolution of hydrothermal fluids.

Hydrothermal recharge

Seawater undergoes important chemical changes when heated. These changes were first established experimentally by Bischoff and Seyfried (1978) and Bischoff and Rosenbauer (1988) and are in agreement with the first measurements of hydrothermal fluids (Edmond et al., 1979). At ~130 °C, anhydrite, which has retrograde solubility, will begin to precipitate from dissolved Ca^{2+} and SO_4^{2-} in the seawater, removing almost all of the Ca^{2+} and about one third of the sulfate. Beginning at about 25 °C, magnesium begins to precipitate from seawater in a brucite-like molecule [$Mg(OH)_2$] as part of the smectite structure. When seawater is heated slowly in the recharge zone, Ca^{2+} is released at the same rate that Mg^{2+} is taken up (Elderfield et al., 1999) and the pH remains near neutral. At temperatures in excess of 150–250 °C, calcium is then lost from the fluid, when Ca silicates (prehnite, clinozoisite, tremolite) become stable. If seawater is heated rapidly (i.e., in chimney walls), protons may be released in exchange for Mg when magnesium hydroxide sulfate hydrate (MHSH) may precipitate and the pH drops. Considerable acidity is also generated when protons are released into solution in exchange for Ca^{2+}. The released Ca^{2+} also combines with the remaining sulfate to form more anhydrite. Leftover sulfate is thermochemically reduced to hydrogen sulfide by reaction with ferrous iron in the host rock. As a result, SO_4^{2-} concentrations decrease to nearly zero by the time fluids reach their maximum temperature.

Sequential water-rock reactions along the hydrothermal fluid flow path are responsible for the exchange of different elements between the fluid and substrate at progressively higher temperatures. At temperatures of about 40–60 °C, basaltic glass in the substrate is altered to clay minerals, zeolites, and Fe oxyhydroxides. Alkali elements such as K, Rb, and Cs are precipitated out of seawater and fixed in the rock at temperatures up to about 100 °C. Initially, this exchange is balanced by the leaching of Fe and Si from volcanic glass. At 150 °C, the alkali elements that were originally fixed in the rocks at low temperatures are leached, and Ca, Mg, and Na are added to the rock to maintain charge balance. Magnesium is fixed first as Mg-rich smectite (<200 °C) and then as chlorite (>200 °C). The H^+ released during Ca fixation is partly consumed by the hydrolysis of the igneous minerals, but contributes to the overall acidity of the fluids and promotes metal leaching from the rock. At about 250 °C, Na^+ is fixed as albite, partly replacing the remaining plagioclase. In the highest-temperature parts of the hydrothermal system (>350 °C), all Mg is lost from the hydrothermal fluid, and Ca fixation causes the formation of secondary plagioclase and epidote and the release of even more H^+, which is then buffered in a system that has become rock-dominated (e.g., Seyfried, 1987). At the same time, reaction of water with Fe-bearing minerals in the rock (e.g., olivine, pyroxene, Fe sulfides) liberates metals and H_2 and results in major reduction of the fluids. Thus, for a system hosted in basaltic/gabbroic oceanic crust, the highest-temperature fluids are in equilibrium with an assemblage of plagioclase-epidote-quartz-magnetite-anhydrite-pyrite, referred to as the PEQMAP buffer (Pester et al., 2012). The major sources of Fe, Mn, Zn, and Cu in hydrothermal fluids are immiscible sulfides and ferromagnesian minerals in basalt. Feldspar is a major source of Pb and Ba especially in more felsic volcanic rocks associated with hydrothermal systems related to volcanic arcs and in back-arc regions.

High-temperature reaction zones in different geological environments

The sum of the reactions occurring along the path of hydrothermal circulation at basaltic mid-ocean ridge segments results in a fluid that is slightly acidic, reduced, silica- and alkali-rich, and Mg-poor relative to seawater. Metals and reduced sulfur reach maximum concentrations in the high-temperature "reaction zone," at temperatures in excess of ~400 °C and pressures of 400–500 bars (~2 km depth). Here, fluids in equilibrium with the PEQMAP buffer assemblage have a narrow range of aH_2S and aH_2, and elements such as Si and Fe become so enriched that they are major constituents of the hydrothermal fluids. Under these conditions, SiO_2 concentrations quickly rise to near quartz saturation. Because quartz solubility is pressure dependent, silica concentrations measured in the highest-temperature fluids at the seafloor may provide an indication of the depth of the reaction zone, which commonly reaches the top of the crystallizing subvolcanic intrusions.

In subduction-related environments, the more felsic composition of the melts results in higher concentrations of alkali elements and other trace elements such as Ba and Pb (Hannington et al., 2005). Higher volatile contents of the fluids (e.g., CO_2 and SO_2) are also common, relative to mid-ocean ridge fluids, and likely reflect direct contributions from volcanic degassing.

Where ultramafic rocks are exposed to hydrothermal circulation of seawater, a number of different reactions lead to fluid compositions that are dramatically different from those of ordinary mid-ocean ridges, including both low-pH, metal-rich fluids and high-pH, metal-depleted fluids over a wide range of temperatures (Allen and Seyfried, 2004; Seyfried et al., 2011). Very high concentrations of H_2 (up to 16 mmol/kg) are mainly produced by oxidation of ferrous iron in olivine during serpentinization. Large methane plumes are also common in the water column above the exposed peridotite. The abundant CH_4 is produced during serpentinization through a series of redox reactions that convert carbon dioxide to CH_4 (i.e., $CO_2 + 4H_2 = CH_4 + 2H_2O$).

$$18(Mg_{1.67}Fe_{0.33})SiO_4 + 23H_2O \rightarrow 9Mg_3Si_2O_5(OH)_4$$
Olivine solid solution　　　　　　　　Serpentin
$$+ 3Mg(OH)_2 + 2Fe_3O_4 + 2H_2$$
　Brucite　　　Magnetite

The compositions of the hydrothermal fluids also appear to be sensitive to small differences in the mineralogy of the ultramafic rocks (e.g., different proportions of olivine and pyroxene). For example, the low pH of the vent fluids at Rainbow is interpreted to result mainly from Ca-Mg exchange with pyroxene and the precipitation of tremolite and talc in the gabbroic rocks (Allen and Seyfried, 2004). Unusually high Fe concentrations in the end-member fluids are believed to result from leaching of magnetite by the low-pH fluids.

Hydrothermal discharge

Hydrothermal venting is the most recognizable feature of hydrothermal systems. High-temperature, focused venting produces *black and white smokers* and *volcanogenic massive sulfide* deposits that are often associated with diverse *chemosynthetic ecosystems*.

The hot hydrothermal fluid that formed at depth has a low density relative to cold seawater and thus rises buoyantly to the seafloor in a matter of minutes (at meters per second) along focused zones of high permeability (e.g., deep normal faults). Whereas convecting seawater has a residence time in the crust of several years from the point of recharge to the reaction zone, the transit from the base of the hydrothermal system to the seafloor is rapid (Kadko and Moore, 1988; Kadko and Butterfield, 1998). Observations of the rock record indicate that the rate of ascent is so fast that the fluid does not reequilibrate with the surrounding rock during ascent, although there may be some reaction (e.g., silicification of the wall rock and precipitation of metals). Minor amounts of sulfide are precipitated, but the major subseafloor deposition of sulfide occurs only in well-developed "stockwork zones" or feeder systems immediately beneath the seafloor vents where entrainment of seawater (which reduces the fluid temperature) and possibly boiling contribute to the precipitation of sulfide minerals (see below). Significant mixing and precipitation of sulfides below the seafloor result in the venting of clear fluids and a distinct lack of black or white "smoke."

Measured concentrations of most elements in fluids that are discharged at the seafloor are somewhat lower than in the reaction zone because of mineral precipitation and dilution caused by mixing with seawater. However, the concentrations in the unmixed fluids can be calculated by assuming that Mg and SO_4 were quantitatively removed from hydrothermal seawater at depth and any Mg or SO_4 in the sample fluids must therefore be a result of mixing with seawater at or below the seafloor. This is the basis for the "end-member" fluid model (Edmond et al., 1979; Von Damm et al., 1985). In this model, the concentrations of major elements in samples collected at different temperatures or degrees of mixing can be extrapolated back to their original values along mixing lines between seawater and an undiluted, zero-Mg or zero-SO_4 fluid. Under some conditions, the uptake of Mg by altered rocks can be reversed, and SO_4 may be added to the fluids from other sources, so the zero-Mg, zero-SO_4 end-member model cannot be universally applied.

Venting of hydrothermal fluids results in the precipitation and accumulation of sulfide-rich chimneys and mounds at the vent site. Chimneys that form at ~350 °C "black smoker" vents on mid-ocean ridges are composed of an assemblage of Cu, Fe, and Zn sulfide minerals including pyrite (±marcasite), pyrrhotite, chalcopyrite, Fe-rich sphalerite, and minor isocubanite and bornite, together with anhydrite. Lower-temperature vents (so-called white smokers) are dominated by pyrite, marcasite, sphalerite, and minor galena and sulphosalts, together with abundant amorphous silica, barite, and minor anhydrite; pyrrhotite and chalcopyrite are rare or absent. Mineral assemblages from extinct chimneys can be readily ascribed to a high- or low-temperature origin by analogy with the mineralogy of active vents. Mineralogical zoning in the chimneys reflects strong gradients in fluid temperatures (e.g., Tivey, 2007). High-temperature fluid conduits in black smokers commonly are lined by chalcopyrite together with pyrrhotite, isocubanite, and locally bornite; the outer portions of chimneys are commonly made up of the minerals deposited at lower temperature, such as sphalerite, pyrite, and marcasite. However, detailed examination reveals complex intergrowths, replacements, and recrystallization of the minerals resulting from a dynamic and locally chaotic environment of sulfide precipitation within the chimneys.

Subseafloor stockwork mineralization

The feeder zones of most hydrothermal vent sites contain vertically extensive networks of sulfide veins and disseminated sulfides, referred to as the "stockwork zone," and intensely altered rocks (i.e., the "alteration pipe"). Stockwork mineralization can contain as much as 30–40 % of the total metal transported by hydrothermal fluids. The main alteration minerals are quartz, chlorite, illite, and kaolinite. These are produced by reaction of the hydrothermal fluids with the host rocks and by entrainment of local seawater into the upflow zone. Pyrite, which occurs as a replacement of Fe-bearing igneous phases, is commonly associated with the intense clay alteration in the central portion of the stockwork. Intense silicification adjacent to sulfide veins typically destroys any primary igneous textures in the volcanic rocks and converts them to a dominantly quartz-chlorite assemblage. Alteration temperatures in the cores of the stockwork were likely close to the exit temperatures of black smoker vents at the surface, as indicated by fluid inclusion homogenization temperatures in quartz and anhydrite in the veins (Petersen et al., 2000). In most cases, the alteration also causes the rocks surrounding the discharge zone to become insulated, sealing the system against the ingress of seawater and allowing higher-temperature fluids to occupy the alteration pipe (Lowell et al., 1993). However, chloritized pillow rims are also common beyond the main stockwork zone, indicating fluid flow between pillows even where sulfide veins are not present. Much of this chlorite alteration is likely caused by entrainment of seawater into the hydrothermal upflow rather than precipitation from the high-temperature fluids (Janecky and Shanks, 1988). Abundant anhydrite is also found within some feeder zones, precipitated from the progressive heating of seawater that is drawn into the mound beneath black smoker vents. Lenses of massive sulfide are also found within permeable zones in the volcanic rocks below the seafloor, such as interflow breccias, hyaloclastite, or sediments.

The role of phase separation

In the high-temperature reaction zone, some fluids have temperatures and pressures higher than the critical point of seawater (i.e., 407 °C and 298 bars, Figure 4) and a small amount of high-salinity brine condenses from the bulk fluid (Bischoff and Rosenbauer, 1988). Dissolved gases such as CO_2, H_2, and H_2S partition strongly into the less dense phase, whereas chloride partitions into the denser condensate. The majority of monovalent cations, such as Na^+ are not fractionated relative to chloride, but divalent cations like Ca^{2+} and the metals become highly concentrated in the brine phase (Butterfield et al., 2003). The large differences in chloride concentration control the metal concentrations in the fluids, as most cations are carried in solution at high temperatures as aqueous chloride complexes (e.g., $FeCl_2$) (Helgeson et al., 1981). Evidence of high-temperature phase separation has been found in numerous studies of fluid inclusions from rocks in the reaction zone (i.e., fluids with salinities both greater and less than seawater) (Delaney et al., 1987; Vanko et al., 1992; Kelley and Fruh-Green, 2001). However, active venting of supercritical fluids has been documented only recently along the southern Mid-Atlantic Ridge (e.g., Koschinsky et al., 2008).

At lower pressures and temperatures, fluids rising to the seafloor may intersect the two-phase curve for seawater at temperatures below the critical point and will begin to boil. At the depth ranges typical for the majority of mid-ocean ridges (>2,500 m), most fluids at the point of discharge are well below the two-phase curve for seawater. However, at water depths of less than about 1,500 m,

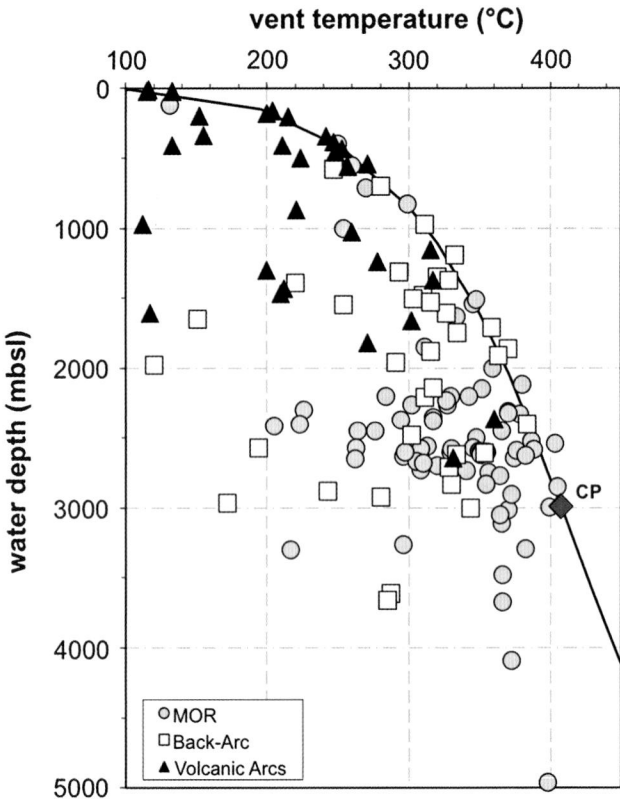

Hydrothermalism, Figure 4 Two-phase curve for seawater (after Bischoff and Rosenbauer, 1987) and temperature/depth information of known active vent fields in various geological settings (data from Beaulieu et al. (2013) with additions for systems discovered after 2012). The critical point of seawater (CP, red diamond) is at 407 °C and 298 bars. A 350 °C ascending hydrothermal fluid will undergo subcritical phase separation and cooling at venting depths of < ~1,500 mbsl.

for example, at locations where hot-spot and mid-ocean ridges interact (Iceland, Azores, Ascension, Axial Seamount), the near-vertical adiabats of typical black smoker fluids at 350 °C will intersect the two-phase boundary more than 100 m below the seafloor (Figure 4). These fluids will cool dramatically as they rise along the boiling curve. At these low pressures, the density difference between the vapor and liquid is much greater than in the reaction zone, resulting in the rapid separation of the vapor phase.

As a result, there are important physical limitations to the depth at which high-temperature massive sulfide deposits can form if the pressure at the seafloor is insufficient to prevent boiling (Ohmoto, 1996). In particular, it is unlikely that a Cu-rich massive sulfide deposit could develop at water depths of less than 500 m (i.e., a boiling temperature of <265 °C), unless the composition of the hydrothermal fluids is very different from that of typical black smokers. Metals such as Cu, which are most effectively transported at high temperatures, will be deposited in a subseafloor boiling zone; metals that are soluble at lower temperatures (e.g., Au, Ag, As, Sb, Hg, Tl, Pb, and Zn) will be transported to the seafloor above the boiling zone and deposited in the seafloor precipitates (Hannington et al., 1999). The extent of boiling is therefore a major control on the bulk composition of the hydrothermal deposits forming at the seafloor and could lead to abundant subseafloor vein mineralization over a large vertical depth range coincident with the boiling zone.

Global distribution

Mid-ocean ridges

On mid-ocean ridges spreading at between 50 and 150 mm/year, the occurrence and distribution of hydrothermalism is roughly proportional to the spreading rate (Baker and German, 2004). The frequency of hydrothermal vents along the magmatically robust fast-spreading ridges is high, relative to the more magma-starved slow-spreading ridges (Beaulieu et al., 2013). However, associated hydrothermal sulfide deposits on the fast-spreading ridges tend to be small, and the fewer deposits on slow-spreading ridges are commonly larger, reflecting longer-lived hydrothermal deposition. The incidence of large hydrothermal plumes is unexpectedly high at ultraslow-spreading ridges. Although most high-temperature hydrothermal systems occur in the axial zones of the ridges, a number of important hydrothermal systems also have developed at ridge-axis volcanoes and at the intersections of ridges with major transform faults or hot spots (mantle plumes). Widespread but typically low-temperature hydrothermal activity also has been found on off-axis volcanoes. Curiously, although intraplate hotspot volcanoes account for a large proportion of seafloor volcanism (about 12 % of the Earth's magmatic budget (Schmidt and Schmincke, 2000), almost no high-temperature hydrothermal activity has been reported that is associated with these volcanoes, except in proximity to ridges. This apparent contradiction may be due to the great depth of the heat source and the narrow magma conduits beneath hotspot volcanoes (Hannington et al., 2005). The lack of documented hydrothermalism in other parts of the oceans (e.g., the polar regions and the Southern Ocean, Figure 3) mainly reflects the difficulties of marine research at these latitudes. Discoveries of hydrothermal plumes and massive sulfide deposits in the high Arctic (e.g., Gakkel Ridge) and in Antarctica (e.g., Bransfield Strait, East Scotia Ridge) confirm that seafloor hydrothermal activity in remote parts of the oceans is little different from that observed elsewhere. Along slow- and ultraslow-spreading ridges (<20 mm/year full spreading rate), at high latitudes and elsewhere, hydrothermal activity appears to be demonstrably more abundant than predicted based on the relation between magma budget and hydrothermal activity established for ridges spreading at faster rates.

A variety of geophysical measurements indicate that about 3 to 6×10^{13} kg/year of seawater must be circulated through the axial zones of the world's mid-ocean ridges and heated to a temperature of at least 350 °C to remove the heat from newly formed crust (Schultz and Elderfield, 1997; Alt, 2003; Mottl, 2003). Other estimates, based on geochemical mass balances, have ranged as high as 1.5×10^{14} kg/year (Baker et al., 1996) and as low as 7×10^{12} kg/year (Nielsen et al., 2006). These calculations assume that the component of lower-temperature diffuse flow (responsible for up to 90 % of the total heat discharge) represents end-member fluid in which the elements have been conservatively diluted with seawater. The combined flux of low- and high-temperature fluids is $>10^{14}$ kg/year, enough to cycle the entire mass of the world's oceans (1.37×10^{21} kg) through the axis of the mid-ocean ridges every 5–10 Ma (Elderfield and Schultz, 1996). If the flux of heat from the flanks of the ridges is included in this calculation (crust older than 1 Ma but <65 Ma), the cycling time for the worlds' oceans through the ridges is less than 1 Ma.

Based on an assumption that fast-ridge EPR-style venting was globally representative, Mottl (2003) and Sinha and Evans (2004) estimated that the heat transport associated with high-temperature convection at ridge axes is $1.8 + 0.3 \times 10^{12}$ W. As noted above, only about 10 % of this heat is discharged at black smoker temperatures because of mixing and conductive cooling of fluids on their way to the seafloor (Ginster et al., 1994). Conversely, German et al. (2010) argue that closer to 50 % of the total on-axis heat flow on slow-spreading ridges may be in large black smoker systems. Assuming a heat flux of 2–5 MW for a single black smoker vent (Converse et al., 1984; Bemis et al., 1993; Ginster et al., 1994), between 50,000 and 100,000 black smokers would be required to account for the high-temperature flux (10 % of $1.8 + 0.3 \times 10^{12}$ W) or a density of at least one black smoker for every 1 km of ridge axis. Because their distribution is far from uniform, it is more likely that this heat is removed by fewer larger-scale hydrothermal systems. For example, some large vent fields contain as many as 100 black smoker vents with a heat output equivalent to 200–500 MW (Becker and von Herzen, 1996; Kelley et al., 2002). Only 1,000 high-temperature vent fields of this magnitude would be required to account for the high-temperature flux, equivalent to one large hydrothermal system every 100 km along the ridges. Fewer than 100 vent fields of this size are presently known, but only about 20 % of the mid-ocean ridge system has been explored in detail so far. The annual flux of metals and sulfur from these vents is estimated to be on the order of 10^6 tonnes per year, assuming a discharge rate of 200–500 kg/s and total combined metal and sulfur concentrations of about 250 mg/kg. This may be only a small fraction of the metal and sulfur originally mobilized by high-temperature fluids at depth, if a large proportion is retained in the crust as a result of cooling, mixing, and water-rock interaction in the subseafloor (Alt, 2003).

Independent estimates based on the actual distribution of known vent sites suggest that the spacing of sulfide occurrences is quite regular. Hannington and Monecke (2009) examined the spacing of large vent fields on

32 ridge segments and calculated an average spacing between sulfide deposits of 98 km. The spacing is greater on the slow-spreading ridges (167 km) than on the fast-spreading ridges (46 km), but the individual sulfide occurrences, and therefore the cumulative heat output, on the slow-spreading ridges are larger on average (Bach and Frueh-Green, 2010; Hannington et al., 2011). The spacing of known mineral deposits is similar to the global incidence of venting detected remotely as hydrothermal plumes (Baker and German, 2004; Baker, 2007).

Ultramafic environments

At some slow-spreading ridges, tectonic processes locally expose gabbroic and ultramafic rocks. Near large faults, vertical tectonics and especially low-angle detachment faulting expose large sections of the underlying oceanic lithosphere, referred to as oceanic core complexes. At several locations on the Mid-Atlantic Ridge, hydrothermal systems have been located on top of oceanic core complexes (e.g., Rainbow, Logatchev, Ashadze, Nibelungen, Semyenov, Irinovskoe, Von Damm fields). Although it has been suggested that some hydrothermal circulation at these sites is driven by the heat of serpentinization of the exposed ultramafic rocks, high-temperature black smoker activity in these settings is considered to reflect a deep magmatic heat source (Kelley et al., 2002; Mevel, 2003; Melchert et al., 2008). Gabbro has been shown to intrude the lithospheric mantle prior to uplift and exposure (Cannat, 1996), which may drive high-temperature hydrothermal circulation in the ultramafic rocks, and the hydrothermal fluids are interpreted as products of high-temperature reactions with both gabbroic subvolcanic intrusions and peridotite.

The Lost City vent field, located at the intersection of the Mid-Atlantic Ridge and the Atlantis Fracture Zone, is an example of a rare hydrothermal system that may be driven primarily by the heat of serpentinization. Here, peridotite exposed at the seafloor is reacting directly with cold seawater, releasing significant quantities of a Mg-poor, CH_4- and H_2-rich fluid but no metals (Kelley et al., 2001). The fluids are buffered by equilibrium with diopside, brucite, and serpentine produced by the hydration of peridotite, which results in pH values of ~10–11. As the warm, up to 91 °C, high-pH (thus OH-rich) fluids mix with Mg-rich seawater, brucite is formed directly on the seafloor and in subseafloor fractures (Kelley et al., 2005). The formation of serpentine and other Mg-silicates also buffers Si to very low concentrations, and only traces of silica are deposited at the seafloor. The high pH of the vent fluids causes the conversion of HCO_3^- in the seawater to carbonate ions and triggers carbonate precipitation. Towering carbonate (aragonite) chimneys, 10–30 m in height (up to 60 m), and travertine-like deposits form (Kelley et al., 2001; Palandri and Reed, 2004). Polymetallic sulfides are absent because of a lack of metals and H_2S in the vent fluids, and the low temperatures prevent anhydrite saturation. The serpentinization of olivine is exothermic and generates enough heat to raise the subseafloor temperatures to as much as 150 °C. However, theoretical calculations based on major cation concentrations in the fluids and heat balance suggest that exothermic reactions cannot be the only source of heat (Allen and Seyfried, 2004).

Suprasubduction zones

Hydrothermal systems in subduction-related environments are broadly similar to those at the mid-ocean ridges, and features typical of mid-ocean ridge hydrothermal systems (e.g., black smokers, hydrothermal plumes) are also observed in arc and back-arc settings. In contrast to the mid-ocean ridges, however, convergent margins are characterized by a range of different crustal thicknesses, heat flow regimes, and magma compositions. The compositions of the volcanic rocks vary from mid-ocean ridge basalt (MORB) to more felsic lavas (andesite, dacite, and rhyolite) with geochemical affinities to island arcs. This variability leads to major differences in the composition of the hydrothermal fluids and the mineralogy and bulk composition of the sulfide deposits. For instance, many of the arc-related systems are directly affected by influx of magmatic SO_2, which disproportionates to native sulfur and sulfuric acid. These seafloor solfataras are characterized by large accumulations of sulfur and fields of white smokers venting fluids with pH <1.5. About one third of the known high-temperature seafloor hydrothermal systems occur at submarine volcanic arcs or in back-arc basins. Although there are no estimates of the total hydrothermal flux, distributions along axis, as determined from overlying hydrothermal plume incidence, appears directly proportional to the lengths of the arc and back-arc segments, just as seen along mid-ocean ridges (Beaulieu et al., 2013) hence, accounting for about 20 % of the mid-ocean ridge flux. The spacing of hydrothermal systems along the volcanic fronts of island arcs remain poorly known, as these volcanoes are still being discovered and explored. On the Mariana and Kermadec arcs, which are the most completely surveyed, hydrothermal plumes have been found above one third of the submarine volcanoes, which occur at intervals of ~70 km along the lengths of the arcs (de Ronde et al., 2003). However, the incidence and intensity of hydrothermal activity associated with these volcanoes are highly variable and not yet well characterized. Some plumes detected within or above the summit calderas of arc volcanoes are related to high-temperature hydrothermal venting and black smoker activity, but many of the plume signals appear to be due to passive degassing or diffuse venting rather than high-temperature hydrothermal vents (Embley et al., 2006).

The major controls on vent fluid compositions in arc and back-arc settings are the same as those at mid-ocean ridges, although there is abundant evidence that magmas directly supply a number of components, including CO_2 and metals, to the hydrothermal fluids and boiling affects many of the hydrothermal systems at the volcanic fronts of the arcs. Nearly all of the volcanoes at the fronts of the arcs have summit calderas at water depths of less than 1,600 m. This has a major impact on the temperatures of the hydrothermal fluids due to boiling, and there is abundant visual evidence of phase separation at many of the vents (Stoffers et al., 1999) (see section on phase separation for further details). Higher-temperature hydrothermal vents are found mainly in the back-arc rifts or on volcanoes with large and deep calderas, which are below the boiling depth of the hydrothermal fluids. The Lau Basin was among the first of the Western Pacific arc- and back-arc systems to be recognized as a young ocean basin formed by the splitting of an island arc (Karig, 1970). And, one of the first known occurrences of seafloor hydrothermal mineralization anywhere in the oceans was recovered by accident from the northern part of the Lau Basin in 1975, long before the discovery of black smoker deposits on the East Pacific Rise (Bertine and Keene, 1975). Black smoker activity has since been found on all of the spreading segments of the Lau Basin and in all of the other back-arc spreading centers of the Western Pacific region.

Massive sulfide deposits in volcanic arc/back-arc environments also may contain abundant native gold, Ag-rich galena (PbS), Pb-As-Sb sulphosalts (including tennantite and tetrahedrite), stibnite (Sb_2S_3), orpiment (As_2S_3), realgar (AsS), and locally cinnabar (HgS). These differences reflect the composition of the underlying volcanic rocks, which commonly included more fractionated suites of lava (andesite, dacite, rhyolite) and possibly direct contributions of metals from magmatic volatiles.

Sedimentary environments

About 5 % of the world's active spreading centers are covered by sediment from nearby continental margins (Hannington et al., 2005). These include sediment-covered ridges such as Middle Valley and Escanaba Trough on the Juan de Fuca and Gorda Ridges and rift grabens that have become major depocenters for sediment (e.g., Guaymas Basin in the Gulf of California). Subseafloor intrusions (e.g., sill-sediment complexes) are common and are indicated by high heat flow in the sedimented valleys, where high-temperature hydrothermal venting is locally focused around the margins of buried sills (Einsele, 1985; Zierenberg et al., 1993). The highest-temperature fluids originate in the volcanic basement where the upper few hundred meters of fractured basalt is several orders of magnitude more permeable than the overlying sediments and can accommodate large-scale fluid flow.

Although the hydrothermal fluids originate in the basement, extensive interaction and buffering by sediments occur en route to the seafloor. The fluids are generally depleted in metals and enriched in the alkalies, boron, ammonia, and organic-derived hydrocarbons compared to seawater-basalt systems. They have a higher pH and are more reduced than fluids that reacted only with basalt. Sill intrusions into the sediments produce organic matter derived hydrocarbons, mainly methane, as well as CO_2 and H_2. Lower venting temperatures compared to sediment-free mid-ocean ridges result from subseafloor mixing and interaction with the sediments; together with the higher pH, this accounts for the low metal concentrations at the seafloor (German and Von Damm, 2004). The high pH and alkalinity of the vent fluids, combined with high SiO_2 concentrations, results in the precipitation of abundant Mg-silicates (e.g., talc and stevensite) and carbonate during mixing with seawater, and chimneys composed only of anhydrite, barite, amorphous silica, smectite, or carbonate are common. Calcite, dolomite, and manganiferous carbonates are also important constituents of the hydrothermal precipitates. Products of the thermal degradation of organic matter, including liquid petroleum and solid bitumens, are commonly preserved in the hydrothermal precipitates (Peter et al., 1991), and higher hydrocarbons (e.g., C1 through C6) are also present in the vent fluids (Simoneit, 1988). The high concentrations of ammonia in the vent fluids, which are several orders of magnitude higher than in typical mid-ocean ridge black smokers (e.g., 10–15 mmol/kg at Guaymas Basin compared to <0.01 mmol/kg NH_4 at 21°N EPR), originate from the breakdown of N-bearing organic compounds in the sediments, and this produces a strong pH buffer via the reaction $NH_3 + H^+ = NH_4^+$.

The massive sulfide deposits that form at sediment-covered mid-ocean ridges (e.g., Escanaba Trough, Southern Gorda Ridge, and Guaymas Basin, Gulf of California) are locally more complex than the black smoker chimneys at sediment-starved mid-ocean ridges. Hydrothermal fluids that interact with continent-derived turbiditic sediments leach Pb, Ba, and other elements from feldspar and other detrital components in the sediments, and the sulfide deposits include such minerals as arsenopyrite (FeAsS), tetrahedrite ($Cu_{12}Sb_4S_{13}$), loellingite ($FeAs_2$), stannite (Cu_2FeSnS_4), jordanite ($Pb_{14}As_7S_{24}$), native bismuth, and other Pb-As-Sb sulphosalts (Koski et al., 1988; Zierenberg et al., 1993). Due to the strongly reduced nature of hydrothermal fluids that have reacted with organic material in the sediment, pyrrhotite is also a common constituent of these deposits.

Many of the characteristics of hydrothermal systems within sedimented ridge environments are also observed in sedimented back-arc basins, such as in the Okinawa Trough (Ishibashi et al., 2015). The basin sediments are dominated by terrigenous material, commonly with a high organic matter content. This can lead to concentrations of hydrocarbons, including gas hydrates, in the sediment-filled basins (Glasby and Notsu, 2003). Abundant carbonate minerals, similar to those found on sediment-covered mid-ocean ridges, also occur in sedimented back-arc environments (Nakashima et al., 1995).

Plumes

A characteristic feature of high-temperature hydrothermal vents is hydrothermal plumes that form above the active vent-sources. The plumes, consisting of a mixture of hydrothermal fluid diluted by seawater, rise several hundred meters above the seafloor until they attain neutral buoyancy, at which point the plume spreads laterally along the direction of the local currents (see "Hydrothermal Plumes"). Plumes can readily be detected for several kilometers away from their source, from measurements of physical and chemical parameters in the water column (e.g., positive temperature, ^{3}He, CH_4 and turbidity anomalies, and negative Eh anomalies). The detection of plumes in the water column has been the primary method for detecting active hydrothermal venting on the seafloor. While it was long believed that most transition metals from hydrothermal vents are deposited near the vent sites, hydrothermal plumes have recently been recognized as a potentially important source of dissolved iron throughout the oceans (Tagliabue et al., 2010).

The minerals that precipitate out of hydrothermal fluids at a vent site and are entrained into the buoyant ascending plume (i.e., the "smoke") are commonly dispersed over a wide area by near-bottom and mid-depth currents and do not contribute to the growth of sulfide deposits, except in rare cases. At typical black smoker vents, at least 90 % of the metals and sulfur discharged at the seafloor is lost to the hydrothermal plume. Sulfide particles larger than about 50 μm in size settle out of the plume within a kilometer of the vent, but finer particles, typically less than 20 μm in size, are dispersed (Baker et al., 1985; Mottl and McConachy, 1990; Feely et al., 1994). The finer particles account for most of the particle load of the plumes, but a large proportion of these may oxidize and dissolve before settling, releasing metals back into seawater (Feely et al., 1987; Metz and Trefry, 1993). However, up to 10 % of total Fe within a plume may be in the form of nanoparticles (<200 nM size fraction) that are resilient to both oxidation and particle settling and can therefore be dispersed significant distances off-axis (Yucel et al., 2011).

The metalliferous sediments in the Atlantis II Deep of the Red Sea are an exceptional example of metal deposition from hydrothermal fluids vented into the water column. In the Atlantis II Deep, metal-rich fluids are injected into the base of a highly stratified brine pool by hydrothermal vents located within the deep (Zierenberg and Shanks, 1983; Scholten et al., 2000). These metal-bearing brines have salinities, which are many times greater than vent fluids on the mid-ocean ridges, most likely as a consequence of seawater circulation through Miocene evaporites on the flanks of the rift. The brines sink, rather than rise as buoyant hydrothermal plumes, settling into several deep anoxic basins and depositing the metals as thick horizons of oxides and sulfides on the basin floor.

Summary

The circulation of seawater through oceanic lithosphere and the associated fluid-rock thermal and chemical exchange that occurs in the subsurface has a major effect on the composition of both seawater and oceanic lithosphere. The wide range of different mineralogical and geochemical characteristics of hydrothermal systems is a result of a combination of source rock composition, pressure- and temperature-dependent solubility controls, and possible direct contributions of metals and gases from subvolcanic magmas. These characteristics provide clues to the geology and chemical processes that are occurring in the underlying crust. The heating and subsequent venting of circulated seawater is the major mechanism for cooling hot, newly formed oceanic crust. The resulting chemical exchange includes the leaching and transport of base and precious metals from the underlying rock to the seafloor. The mixing of hot, metal-rich hydrothermal fluids with cold seawater at sites of focused black or white smoker discharge on the seafloor result in the formation of volcanogenic massive sulfide deposits that often host chemosynthetic-based ecosystems. The massive sulfide deposits that formed in ancient oceans and are now exposed on land are an important source for Cu, Zn, Pb, Au, and Ag and are mined for their metal contents. In the future, hydrothermal sulfide deposits on the modern seafloor may similarly be exploited for their rich metal content.

Bibliography

Allen, D. E., and Seyfried, W. E., 2004. Serpentinization and heat generation: constraints from Lost City and rainbow hydrothermal systems. *Geochimica Et Cosmochimica Acta*, **68**, 1347–1354.

Alt, J. C., 2003. Hydrothermal fluxes at mid-ocean ridges and on ridge flanks. *Comptes Rendus Geoscience*, **335**, 853–864.

Alt, J. C., and Teagle, D. A. H., 2003. Hydrothermal alteration of upper oceanic crust formed at a fast-spreading ridge: mineral, chemical, and isotopic evidence from ODP Site 801. *Chemical Geology*, **201**, 191–211.

Bach, W., and Frueh-Green, G. L., 2010. Alteration of the oceanic lithosphere and implications for seafloor processes. *Elements*, **6**, 173–178.

Baker, E. T., 2007. Hydrothermal cooling of midocean ridge axes: do measured and modeled heat fluxes agree? *Earth and Planetary Science Letters*, **263**, 140–150.

Baker, E. T., and German, C. R., 2004. On the global distribution of hydrothermal vent fields. *Mid-Ocean Ridges*, **148**, 245–266.

Baker, E. T., Lavelle, J. W., and Massoth, G. J., 1985. Hydrothermal particle plumes over the southern Juan de Fuca Ridge. *Nature*, **316**, 342–344.

Baker, E. T., Chen, Y. J., and Morgan, J. P., 1996. The relationship between near-axis hydrothermal cooling and the spreading rate of mid-ocean ridges. *Earth and Planetary Science Letters*, **142**, 137–145.

Beaulieu, S. E., Baker, E. T., German, C. R., and Maffei, A., 2013. An authoritative global database for active submarine hydrothermal vent fields. *Geochemistry Geophysics Geosystems*, **14**, 4892–4905.

Becker, K., and von Herzen, R. P., 1996. *Pre-Drilling Observations of Conductive Heat Flow at the TAG Active Mound Using Alvin*. Texas A & M University, Ocean Drilling Program, College Station, TX, USA.

Bemis, K. G., Vonherzen, R. P., and Mottl, M. J., 1993. Geothermal heat-flux from hydrothermal plumes on the Juan de Fuca Ridge. *Journal of Geophysical Research-Solid Earth*, **98**, 6351–6365.

Bertine, K. K., and Keene, J. B., 1975. Submarine barite-opal rocks of hydrothermal origin. *Science*, **188**, 150–152.

Bischoff, J. L., and Rosenbauer, R. J., 1987. Phase-separation in seafloor geothermal systems - an experimental-study of the effects on metal transport. *American Journal of Science*, **287**, 953–978.

Bischoff, J. L., and Seyfried, W. E., 1978. Hydrothermal chemistry of seawater from 25-degrees-C to 350-degrees-C. *American Journal of Science*, **278**, 838–860.

Bischoff, J. L., and Rosenbauer, R. J., 1988. Liquid–vapor relations in the critical region of the system NaCl-H_2O from 380-degrees-C to 415-degrees-C – a refined determination of the critical-point and 2-phase boundary of seawater. *Geochimica Et Cosmochimica Acta*, **52**, 2121–2126.

Butterfield, D. A., Seyfried, J. W. E., and Lilley, M. D., 2003. *Composition and Evolution of Hydrothermal Fluids*. Berlin: Dahlem University Press.

Cannat, M., 1996. How thick is the magmatic crust at slow spreading oceanic ridges? *Journal of Geophysical Research-Solid Earth*, **101**, 2847–2857.

Converse, D. R., Holland, H. D., and Edmond, J. M., 1984. Flow-rates in the axial hot springs of the East Pacific Rise (21-degrees-N) – implications for the heat budget and the formation of massive sulfide deposits. *Earth and Planetary Science Letters*, **69**, 159–175.

Corliss, J. B., Dymond, J., Gordon, L. I., Edmond, J. M., Herzen, R. P. V., Ballard, R. D., Green, K., Williams, D., Bainbridge, A., Crane, K., and Vanandel, T. H., 1979. Submarine thermal springs on the Galapagos Rift. *Science*, **203**, 1073–1083.

Coumou, D., Driesner, T., and Heinrich, C. A., 2008. The structure and dynamics of mid-ocean ridge hydrothermal systems. *Science*, **321**, 1825–1828.

de Ronde, C. E. J., Faure, K., Bray, C. J., Chappell, D. A., and Wright, I. C., 2003. Hydrothermal fluids associated with seafloor mineralization at two southern Kermadec arc volcanoes, offshore New Zealand. *Mineralium Deposita*, **38**, 217–233.

Delaney, J. R., Mogk, D. W., and Mottl, M. J., 1987. Quartz-cemented breccias from the Mid-Atlantic Ridge – samples of a high-salinity hydrothermal upflow zone. *Journal of Geophysical Research-Solid Earth and Planets*, **92**, 9175–9192.

Edmond, J. M., Measures, C., McDuff, R. E., Chan, L. H., Collier, R., Grant, B., Gordon, L. I., and Corliss, J. B., 1979. Ridge crest hydrothermal activity and the balances of the major and minor elements in the ocean – Galapagos data. *Earth and Planetary Science Letters*, **46**, 1–18.

Einsele, G., 1985. Basaltic sill-sediment complexes in young spreading centers – genesis and significance. *Geology*, **13**, 249–252.

Elderfield, H., and Schultz, A., 1996. Mid-ocean ridge hydrothermal fluxes and the chemical composition of the ocean. *Annual Review of Earth and Planetary Sciences*, **24**, 191–224.

Elderfield, H., Wheat, C. G., Mottl, M. J., Monnin, C., and Spiro, B., 1999. Fluid and geochemical transport through oceanic crust: a transect across the eastern flank of the Juan de Fuca Ridge. *Earth and Planetary Science Letters*, **172**, 151–165.

Embley, R. W., Chadwick, W. W., Jr., Baker, E. T., Butterfield, D. A., Resing, J. A., De Ronde, C. E. J., Tunnicliffe, V., Lupton, J. E., Juniper, S. K., Rubin, K. H., Stern, R. J., Lebon, G. T., Nakamura, K.-i., Merle, S. G., Hein, J. R., Wiens, D. A., and Tamura, Y., 2006. Long-term eruptive activity at a submarine arc volcano. *Nature*, **441**, 494–497.

Feely, R. A., Lewison, M., Massoth, G. J., Robertbaldo, G., Lavelle, J. W., Byrne, R. H., Vondamm, K. L., and Curl, H. C., 1987. Composition and dissolution of black smoker particulates from active vents on the Juan de Fuca Ridge. *Journal of Geophysical Research-Solid Earth and Planets*, **92**, 11347–11363.

Feely, R. A., Gendron, J. F., Baker, E. T., and Lebon, G. T., 1994. Hydrothermal plumes along the East Pacific Rise, 8-degrees-40′ to 11-degrees-50′N - Particle distribution and composition. *Earth and Planetary Science Letters*, **128**, 19–36.

Francheteau, J., Needham, H. D., Choukroune, P., Juteau, T., Seguret, M., Ballard, R. D., Fox, P. J., Normark, W., Carranza, A., Cordoba, D., Guerrero, J., Rangin, C., Bougault, H., Cambon, P., and Hekinian, R., 1979. Massive deep-sea sulfide ore-deposits discovered on the East Pacific Rise. *Nature*, **277**, 523–528.

German, C., and Seyfried, W., 2014. Hydrothermal processes. In Turekian, H. (ed.), *Treatise on Geochemistry*, 2nd edn. Amsterdam: Elsevier.

German, C. R., and Von Damm, K. L., 2004. *Hydrothermal Processes*. London, UK: Elsevier.

German, C. R., Thurnherr, A. M., Knoery, J., Charlou, J. L., Jean-Baptiste, P., and Edmonds, H. N., 2010. Heat, volume and chemical fluxes from submarine venting: a synthesis of results from the Rainbow hydrothermal field, 36 degrees N MAR. *Deep-Sea Research Part I-Oceanographic Research Papers*, **57**, 518–527.

Gillis, K. M., 2002. The rootzone of an ancient hydrothermal system exposed in the Troodos ophiolite, Cyprus. *Journal of Geology*, **110**, 57–74.

Ginster, U., Mottl, M. J., and Vonherzen, R. P., 1994. Heat-flux from black smokers on the Endeavour and Cleft Segments, Juan de Fuca Ridge. *Journal of Geophysical Research-Solid Earth*, **99**, 4937–4950.

Glasby, G. P., and Notsu, K., 2003. Submarine hydrothermal mineralization in the Okinawa Trough, SW of Japan: an overview. *Ore Geology Reviews*, **23**, 299–339.

Hannington, M., and Monecke, T., 2009. Global exploration models for polymetallic sulphides in the area: an assessment of lease block selection under the draft regulations on prospecting and exploration for polymetallic sulphides. *Marine Georesources & Geotechnology*, **27**, 132–159.

Hannington, M., Jonasson, I., Herzig, P., and Petersen, S., 1995. Physical and chemical processes of seafloor mineralization at mid-ocean ridges. *Geophysical Monograph*, **11**, 115–157.

Hannington, M., Bleeker, W., and Kjarsgaard, I., 1999. Sulfide mineralogy, geochemistry, and ore genesis of the kid creek deposit: part i. North, Central, and South orebodies. In Hannington, M., and Barrie, T. (eds.), *The Giant Kidd Creek Volcanogenic Massive Sulfide Deposit*, Society of Economic Geologists Monograph Series. Littleton: Society of Economic Geologists, Vol. 10, pp. 163–224.

Hannington, M., De Ronde, C., and Petersen, S., 2005. Sea-floor tectonics and submarine hydrothermal systems. In Hedenquist, J., Thompson, J., Goldfarb, R., and Richards, J. (eds.), *Economic Geology One Hundredth Anniversary Volume 1905–2005*. Littleton: Society of Economic Geologists, pp. 111–141.

Hannington, M., Jamieson, J., Monecke, T., Petersen, S., and Beaulieu, S., 2011. The abundance of seafloor massive sulfide deposits. *Geology*, **39**, 1155–1158.

Hasenclever, J., Theissen-Krah, S., Ruepke, L. H., Morgan, J. P., Iyer, K., Petersen, S., and Devey, C. W., 2014. Hybrid shallow on-axis and deep off-axis hydrothermal circulation at fast-spreading ridges. *Nature*, **508**, 508–+.

Hekinian, R., Francheteau, J., Renard, V., Ballard, R. D., Choukroune, P., Cheminee, J. L., Albarede, F., Minster, J. F., Charlou, J. L., Marty, J. C., and Boulegue, J., 1983. Intense hydrothermal activity at the axis of the East Pacific Rise near 13-degrees-N – submersible witnesses the growth of sulfide chimney. *Marine Geophysical Researches*, **6**, 1–14.

Helgeson, H. C., Kirkham, D. H., and Flowers, G. C., 1981. Theoretical prediction of the thermodynamic behavior of aqueous-electrolytes at high-pressures and temperatures. 4. Calculation of activity-coefficients, osmotic coefficients, and apparent molal and standard and relative partial molal properties to 600-degrees-C and 5 KB. *American Journal of Science*, **281**, 1249–1516.

Ishibarshi, J., Ikegami, F., Takeshi, T., and Urabe, T., 2015. Hydrothermal activity in the okinawa trough back-arc basin: geological background and hydrothermal mineralization. In Ishibashi, J., Okino, K., and Snuamura, M. (eds.), *Subseafloor Biosphere Linked to Hydrothermal Systems: TAIGA Concept*. Tokyo: Springer, pp. 337–360.

Janecky, D. R., and Shanks, W. C., 1988. Computational modeling of chemical and sulfur isotopic reaction processes in seafloor hydrothermal systems – chimneys, massive sulfides, and subjacent alteration zones. *Canadian Mineralogist*, **26**, 805–825.

Johnson, H. P., Tivey, M. A., Bjorklund, T. A., and Salmi, M. S., 2010. Hydrothermal circulation within the Endeavour Segment, Juan de Fuca Ridge. *Geochemistry Geophysics Geosystems*, **11**, 1–13.

Kadko, D., and Butterfield, D. A., 1998. The relationship of hydrothermal fluid composition and crustal residence time to maturity of vent fields on the Juan de Fuca Ridge. *Geochimica Et Cosmochimica Acta*, **62**, 1521–1533.

Kadko, D., and Moore, W., 1988. Radiochemical constraints on the crustal residence time of submarine hydrothermal fluids - endeavour ridge. *Geochimica Et Cosmochimica Acta*, **52**, 659–668.

Karig, D. E., 1970. Ridges and basins of Tonga-Kermadec island arc system. *Journal of Geophysical Research*, **75**, 239.

Kelley, D. S., and Fruh-Green, G. L., 2001. Volatile lines of descent in submarine plutonic environments: insights from stable isotope and fluid inclusion analyses. *Geochimica Et Cosmochimica Acta*, **65**, 3325–3346.

Kelley, D. S., Karson, J. A., Blackman, D. K., Fruh-Green, G. L., Butterfield, D. A., Lilley, M. D., Olson, E. J., Schrenk, M. O., Roe, K. K., Lebon, G. T., Rivizzigno, P., and Party, A. T. S., 2001. An off-axis hydrothermal vent field near the mid-Atlantic Ridge at 30 degrees N. *Nature*, **412**, 145–149.

Kelley, D. S., Baross, J. A., and Delaney, J. R., 2002. Volcanoes, fluids, and life at mid-ocean ridge spreading centers. *Annual Review of Earth and Planetary Sciences*, **30**, 385–491.

Kelley, D. S., Karson, J. A., Fruh-Green, G. L., Yoerger, D. R., Shank, T. M., Butterfield, D. A., Hayes, J. M., Schrenk, M. O., Olson, E. J., Proskurowski, G., Jakuba, M., Bradley, A., Larson, B., Ludwig, K., Glickson, D., Buckman, K., Bradley, A. S., Brazelton, W. J., Roe, K., Elend, M. J., Delacour, A., Bernasconi, S. M., Lilley, M. D., Baross, J. A., Summons, R. T., and Sylva, S. P., 2005. A serpentinite-hosted ecosystem: the lost city hydrothermal field. *Science*, **307**, 1428–1434.

Koschinsky, A., Garbe-Schoenberg, D., Sander, S., Schmidt, K., Gennerich, H.-H., and Strauss, H., 2008. Hydrothermal venting at pressure-temperature conditions above the critical point of seawater, 5 degrees S on the Mid-Atlantic Ridge. *Geology*, **36**, 615–618.

Koski, R. A., Shanks, W. C., Bohrson, W. A., and Oscarson, R. L., 1988. The composition of massive sulfide deposits from the sediment-covered floor of Escanaba Trough, Gorda Ridge – implications for depositional processes. *Canadian Mineralogist*, **26**, 655–673.

Lister, C. R. B., 1972. Thermal balance of a mid-ocean ridge. *Geophysical Journal of the Royal Astronomical Society*, **26**, 515.

Lowell, R. P., Vancappellen, P., and Germanovich, L. N., 1993. Silica precipitation in fractures and the evolution of permeability in hydrothermal upflow zones. *Science*, **260**, 192–194.

Lupton, J. E., Weiss, R. F., and Craig, H., 1977. Mantle helium in hydrothermal plumes in Galapagos Rift. *Nature*, **267**, 603–604.

Melchert, B., Devey, C. W., German, C. R., Lackschewitz, K. S., Seifert, R., Walter, M., Mertens, C., Yoerger, D. R., Baker, E. T., Paulick, H., and Nakamura, K., 2008. First evidence for high-temperature off-axis venting of deep crustal/mantle heat: the Nibelungen hydrothermal field, southern Mid-Atlantic Ridge. *Earth and Planetary Science Letters*, **275**, 61–69.

Metz, S., and Trefry, J. H., 1993. Field and laboratory studies of metal uptake and release by hydrothermal precipitates. *Journal of Geophysical Research-Solid Earth*, **98**, 9661–9666.

Mevel, C., 2003. Serpentinization of abyssal peridotites at mid-ocean ridges. *Comptes Rendus Geoscience*, **335**, 825–852.

Mottl, M. J., 2003. *Partitioning of Energy and Mass Fluxes Between Mid-ocean Ridge Axes and Flanks at High and Low Temperature*. Berlin: Dahlem University Press.

Mottl, M. J., and McConachy, T. F., 1990. Chemical processes in buoyant hydrothermal plumes on the East Pacific Rise near 21-degrees-N. *Geochimica Et Cosmochimica Acta*, **54**, 1911–1927.

Nakashima, K., Sakai, H., Yoshida, H., Chiba, H., Tanaka, Y., Gamo, T., Ishibashi, J., and Tsunogai, U., 1995. *Hydrothermal Mineralization in the Mid-Okinawa Trough*. Tokyo: Terra Science.

Nielsen, S. G., Rehkamper, M., Teagle, D. A. H., Butterfield, D. A., Alt, J. C., and Halliday, A. N., 2006. Hydrothermal fluid fluxes calculated from the isotopic mass balance of thallium in the ocean crust. *Earth and Planetary Science Letters*, **251**, 120–133.

Ohmoto, H., 1996. Formation of volcanogenic massive sulfide deposits: the Kuroko perspective. *Ore Geology Reviews*, **10**, 135–177.

Palandri, J. L., and Reed, M. H., 2004. Geochemical models of metasomatism in ultramafic systems: serpentinization, rodingitization, and sea floor carbonate chimney precipitation. *Geochimica Et Cosmochimica Acta*, **68**, 1115–1133.

Pester, N. J., Reeves, E. P., Rough, M. E., Ding, K., Seewald, J. S., and Seyfried, W. E., Jr., 2012. Subseafloor phase equilibria in high-temperature hydrothermal fluids of the Lucky Strike Seamount (Mid-Atlantic Ridge, 37 degrees 17 ' N). *Geochimica Et Cosmochimica Acta*, **90**, 303–322.

Peter, J. M., Peltonen, P., Scott, S. D., Simoneit, B. R. T., and Kawka, O. E., 1991. C-14 ages of hydrothermal petroleum and carbonate in Guaymas Basin, Gulf of California – implications for oil generation, expulsion, and migration. *Geology*, **19**, 253–256.

Petersen, S., Herzig, P. M., and Hannington, M. D., 2000. Third dimension of a presently forming VMS deposit: TAG hydrothermal mound, Mid-Atlantic Ridge, 26 degrees N. *Mineralium Deposita*, **35**, 233–259.

Richards, H. G., Cann, J. R., and Jensenius, J., 1989. Mineralogical zonation and metasomatism of the alteration pipes of Cyprus sulfide deposits. *Economic Geology*, **84**, 91–115.

Seyfried, W. E., 1987. Experimental and theoretical constraints on hydrothermal alteration processes and mid-ocean ridges. *Annual Review of Earth and Planetary Sciences*, **15**, 317–335.

Schmidt, R., and Schmincke, H. U., 2000. Seamounts and island building. In Sigurdsson, H. (ed.), *Encyclopedia of Volcanoes*. Sand Diego, CA: Academic Press, pp. 383–402.

Scholten, J. C., Stoffers, P., Garbe-Schonberg, D., and Moammar, M., 2000. *Hydrothermal Mineralization in the Red Sea*. Boca Raton: CRC press.

Schultz, A., and Elderfield, H., 1997. Controls on the physics and chemistry of seafloor hydrothermal circulation. *Philosophical Transactions of the Royal Society a-Mathematical Physical and Engineering Sciences*, **355**, 387–425.

Seyfried, W. E., Pester, N. J., Ding, K., and Rough, M., 2011. Vent fluid chemistry of the Rainbow hydrothermal system (36 degrees N, MAR): phase equilibria and in situ pH controls on subseafloor alteration processes. *Geochimica Et Cosmochimica Acta*, **75**, 1574–1593.

Simoneit, B. R. T., 1988. Petroleum generation in submarine hydrothermal systems – an update. *Canadian Mineralogist*, **26**, 827–840.

Sinha, M. C., and Evans, R. L., 2004. *Geophysical Constraints Upon the Thermal Regime of the Ocean Crust*. Washington DC, USA: American Geophysical Union.

Spiess, F. N., Macdonald, K. C., Atwater, T., Ballard, R., Carranza, A., Cordoba, D., Cox, C., Diazgarcia, V. M., Francheteau, J., Guerrero, J., Hawkins, J., Haymon, R., Hessler, R., Juteau, T., Kastner, M., Larson, R., Luyendyk, B., Macdougall, J. D., Miller, S., Normark, W., Orcutt, J., and Rangin, C., 1980. East Pacific Rise – hot springs and geophysical experiments. *Science*, **207**, 1421–1433.

Stoffers, P., Hannington, M., Wright, I., Herzig, P., de Ronde, C., and Shipboard Sci, P., 1999. Elemental mercury at submarine hydrothermal vents in the Bay of Plenty, Taupo volcanic zone, New Zealand. *Geology*, **27**, 931–934.

Tagliabue, A., Bopp, L., Dutay, J.-C., Bowie, A. R., Chever, F., Jean-Baptiste, P., Bucciarelli, E., Lannuzel, D., Remenyi, T., Sarthou, G., Aumont, O., Gehlen, M., and Jeandel, C., 2010. Hydrothermal contribution to the oceanic dissolved iron inventory. *Nature Geoscience*, **3**, 252–256.

Teagle, D. A. H., Bickle, M. J., and Alt, J. C., 2003. Recharge flux to ocean-ridge black smoker systems: a geochemical estimate from ODP Hole 504B. *Earth and Planetary Science Letters*, **210**, 81–89.

Tivey, M. A., and Johnson, H. P., 2002. Crustal magnetization reveals subsurface structure of Juan de Fuca Ridge hydrothermal vent fields. *Geology*, **30**, 979–982.

Tivey, M. K., 2007. Generation of seafloor hydrothermal vent fluids and associated mineral deposits. *Oceanography*, **20**, 50–65.

Vanko, D. A., Griffith, J. D., and Erickson, C. L., 1992. Calcium-rich brines and other hydrothermal fluids in fluid inclusions from plutonic rocks, Oceanographer Transform, Mid-Atlantic Ridge. *Geochimica Et Cosmochimica Acta*, **56**, 35–47.

Von Damm, K. L., Edmond, J. M., Grant, B., and Measures, C. I., 1985. Chemistry of submarine hydrothermal solutions at 21-degrees-N, East Pacific Rise. *Geochimica Et Cosmochimica Acta*, **49**, 2197–2220.

Weiss, R. F., Lonsdale, P., Lupton, J. E., Bainbridge, A. E., and Craig, H., 1977. Hydrothermal plumes in Galapagos Rift. *Nature*, **267**, 600–603.

Wolery, T. J., and Sleep, N. H., 1976. Hydrothermal circulation and geochemical flux at mid-ocean ridges. *Journal of Geology*, **84**, 249–275.

Yucel, M., Gartman, A., Chan, C. S., and Luther, G. W., III, 2011. Hydrothermal vents as a kinetically stable source of iron-sulphide-bearing nanoparticles to the ocean. *Nature Geoscience*, **4**, 367–371.

Zierenberg, R. A., and Shanks, W. C., 1983. Mineralogy and geochemistry of epigenetic features in metalliferous sediment, Atlantis-II Deep, Red Sea. *Economic Geology*, **78**, 57–72.

Zierenberg, R. A., Koski, R. A., Morton, J. L., Bouse, R. M., and Shanks, W. C., 1993. Genesis of massive sulfide deposits on a sediment-covered spreading center, Escanaba Trough, Southern Gorda Ridge. *Economic Geology and the Bulletin of the Society of Economic Geologists*, **88**, 2069–2098.

Cross-references

Black and White Smokers
Chemosynthetic Life
Hydrothermal Plumes
Hydrothermal Vent Fluids (Seafloor)
Intraoceanic Subduction Zone
Island Arc Volcanism, Volcanic Arcs
Magmatism at Convergent Plate Boundaries
Marine Mineral Resources
Mid-ocean Ridge Magmatism and Volcanism
Oceanic Spreading Centers
Subduction
Volcanogenic Massive Sulfides

ICE-RAFTED DEBRIS (IRD)

Antoon Kuijpers[1], Paul Knutz[1] and Matthias Moros[2]
[1]Geological Survey of Denmark and Greenland (GEUS), Copenhagen, Denmark
[2]Leibniz Institute for Baltic Sea Research Warnemünde, Rostock, Germany

Definition

Ice-rafted debris (IRD) is a terrigenous material transported within a matrix of ice and deposited in marine or lake sediments when the ice matrix melts (US National Climatic Data Center).

History of observations

Coarse-grained clasts interpreted as ice-rafted debris were first recognized in seabed samples collected during a 1928 expedition of the US Coast Guard vessel "Marion" in the northern Labrador Sea and Baffin Bay (Ricketts and Trask, 1932). A decade later, Bramlette and Bradley (1940) described glacially transported striated clasts and erratics from North Atlantic seabed sediments. The more common use of IRD as a proxy of glacial variability commenced with the systematic sampling of deep-sea sediments in the early 1970s, notably by the international Deep Sea Drilling Programme (DSDP) and the "Vema" cruises operated from Lamont-Doherty Earth Observatory in the USA. Initially, IRD analyses were mainly applied to understand the long-term Cenozoic evolution of ice sheets (Berggren, 1972). This was highlighted by the study of Shackleton et al. (1984), who traced the development of major northern hemisphere glaciations back to about 2.5 Ma ago. A different approach demonstrated by, among others, Ruddiman (1977) was the application of ice-rafted sand-sized material measured in multiple core records to investigate late Quaternary ice dispersal patterns in the North Atlantic. In 1988, a milestone article by Hartmut Heinrich, reporting on the origin and consequences of late Quaternary cyclic ice rafting, set the stage for a new understanding of rapid ice sheet–ocean interactions during the last glacial in the North Atlantic region (Heinrich, 1988). Subsequent work by Andrews and Tedesco (1992) and Broecker et al. (1992) identified these ice rafting cycles (so-called Heinrich, H-events) as distinct layers rich in detrital carbonate derived from Paleozoic limestone formations in the Hudson Bay region, thus pointing to the Laurentide Ice Sheet (LIS) as the main iceberg source. Heinrich (1988) documented that these cyclic layers were marked by their high lithic-to-foraminifera ratio and increased relative abundance of the polar planktic foraminifera $N.\ pachyderma$ sinistral with low $\delta^{18}O$ values indicative of low-salinity polar water. Bond et al. (1997) noted the carbonate-rich layers to have a sharp basal contact suggesting short-lived, catastrophic discharges of icebergs from the LIS. These IRD horizons are widely distributed in late Quaternary North Atlantic sediments between 40° and 55° N, occurring with intervals of about 7,000 years that partly correspond to shorter (<1,000 year), but intense cooling periods in Greenland ice core records (Bond and Lotti, 1995). The geochemical composition of the Heinrich layers shows a Hudson Strait source for H1, H2, H4, and H5, but chemical and mineralogical analyses suggest another source for H3 (Gwiazda et al., 1996). Since the mid-1990s, numerous paleoceanographic studies have utilized IRD to document regional and temporal ice sheet variability and the effect of meltwater on ocean thermohaline circulation (e.g., Vidal et al., 1997; Knutz et al., 2002). In the past decade, focus on the topic has further increased due to the ongoing debate on climate warming, Antarctic and Greenland Ice Sheet stability, and implications for sea-level rise and ocean circulation.

Ice rafting processes and IRD analysis

Icebergs form the primary transport agent for IRD (Figure 1), but sediments can also be carried by sea ice drift. Even when using detailed IRD analytical techniques, it appears difficult to distinguish between these two transport mechanisms (Tantillo et al., 2012). The sediment load of individual icebergs depends on the basal (cold-based versus warm-based) thermal regime of the glacier from which the iceberg has calved (Drewry, 1986). Aeolian deposition and rock fall from exposed, ice-free cliffs are other mechanisms supplying IRD load to glaciers. Sea ice-derived IRD originates from seabed grounding and beach contact processes as well as wind-blown depositions (Gilbert, 1990). Ice melting and rafting processes and subsequent deposition of IRD is largely a function of ambient ocean water temperatures. Although there is no standard technique for IRD analyses, the proxy is commonly measured as the number of detrital grains in the >150 μm size fraction per gram bulk sediment (dry weight). Another approach is simply to define IRD as weight percentage of material with grain sizes larger than coarse silt (>63 μm). These thresholds are arbitrary, since all grain sizes, from clay to boulder size, may potentially be ice rafted. In proximal glacimarine environments, the finer grain sizes generally make up the largest amount of glacially derived sediments. X-ray analysis of intact sediment cores is often used for counts of larger-sized (e.g., gravel) IRD components.

Sources and distribution of IRD

IRD is not only present close to formerly glaciated continental margins, but has been found widespread in the subpolar North Atlantic, and occasionally even further south, i.e., in the eastern North Atlantic off the Strait of Gibraltar. Studies of North Atlantic glacial IRD provenance have provided important information on the relative contributions of the Laurentide, Greenland, and NW European ice sheets. During H-events 1 and 2, ice discharge from Hudson Strait contributed with large amounts of detrital carbonate, at approx. 14,500 and 20,000 ^{14}C year BP, respectively. (Andrews and Tedesco, 1992; Broecker et al., 1992). Furthermore, H-event layers contain igneous fragments of hornblende and feldspar displaying a dominant Paleoproterozoic (1,600–1,800 Ma) provenance age (Hemming et al., 1998), which is consistent with a Laurentide Ice Sheet (LIS) origin. On the other hand, IRD showing younger isotopic ages and containing chalk coccoliths are inferred to have been derived from NW European glacial outlets (Scourse et al., 2000). These isotopic and mineralogical fingerprints of European/Scandinavian derived icebergs appear to have formed precursors to the main LIS response, thus adding to the complexity of H-events. A possible IRD contribution to H-events by icebergs from the Greenland Ice Sheet has been a matter of discussion. Despite the scarcity of high-resolution records from the Greenland margin and lack of consistent chronological framework, some studies show that Greenland Ice Sheet iceberg calving events occurred more frequently than H-events (Stein et al., 1996; Andrews, 2000). An independent behavior of the Greenland Ice Sheet during the last deglaciation has been shown by Knutz et al. (2011) and is illustrated by the vast amount of continental ice still present in Greenland today. Timing of iceberg surging stages on the NW European margin reflects also here different ice sheet behaviors before and during Heinrich events (Scourse et al., 2000; Knutz et al., 2002). The arrival of European and Icelandic

Ice-rafted Debris (IRD), Figure 1 Example of drifting ice transporting a significant load of land-derived sediments which is being deposited on the seabed as "IRD" as a result of the ice melting process (photo courtesy J.T. Andrews, INSTAAR, Boulder, Colorado).

detritus preceding the deposition of detrital carbonate-rich debris from the LIS (Bond et al., 1999; Grousset et al., 2001) further support the conclusions by Dowdeswell et al. (1999) that the dynamics of Quaternary ice sheets surrounding the Nordic Seas and North Atlantic were asynchronous. This implies that each glaciated margin may have behaved differently in response to external ocean–climate forcing.

Iceberg surging mechanisms

Since the early 1990s, several possible mechanisms responsible for forcing large-scale IRD events have been proposed (e.g., Alley and Clark, 1999). The binge–purge hypothesis of MacAyeal (1993) invoked internal ice sheet dynamics as a driver for H-events. However, this mechanism is not supported by the detailed structure of H-events and cannot explain North Atlantic records showing millennial and centennial scale IRD fluctuations that appear to be closely coupled to the ocean–atmosphere climate system (e.g., Bond and Lotti, 1995). Apart from internal ice sheet dynamics, ocean circulation changes, sea-level fluctuations, variations in solar parameters, as well as ice-load-induced earthquakes have been proposed being responsible for iceberg surging. In addition, Hulbe et al. (2004) postulate catastrophic ice shelf breakup induced by climate-controlled meltwater infilling of surface crevasses as recently witnessed along the Antarctic Peninsula. Observations of recent iceberg discharge processes and ice rafting in an east Greenland fjord reported by Reeh et al. (1999) demonstrated that the transport and deposition of iceberg-derived IRD increased in periods of enhanced (subsurface) advection of "warm" Atlantic water. These findings were used in a study by Moros et al. (2002) and Kuijpers et al. (2005) who concluded that large-scale IRD events in the North Atlantic most likely were triggered by enhanced northward ocean heat transport which caused bottom melting of floating outlet glaciers and ice shelves, leading to ice sheet destabilization and iceberg surging. More recently, a sudden acceleration, thinning, and retreat of the Jakobshavn Isbræ, west Greenland, were observed, which have been attributed to the warming of the subsurface ocean currents off west Greenland (Holland et al., 2008). Studies of sedimentary records of the past ca. 100 years from this area and from another active glacier calving site on the southeast Greenland coast (Lloyd et al., 2011; Andresen et al., 2012) have meanwhile confirmed a correlation between enhanced ocean subsurface warming and associated glacier bottom melting and increase in glacier acceleration and iceberg IRD production. It is without doubt that also the other factors referred to above as, for instance, surface (air) temperature warming and ice sheet dynamics, influence iceberg calving activity. However, increasing evidence has demonstrated the important role of ocean (sub)surface warming when trying to explain iceberg surging over larger areas, such as during the Heinrich events (e.g., Alvares-Solas et al., 2010), or more locally, as recently observed in South Greenland fjords. Within this context, one should keep in mind that ocean heat transport pathways are influencing the temperature regime of various parts of the North Atlantic not everywhere in the same way, which may be one of the reasons for reported asynchronous behavior of glacial ice sheets. The occurrence of iceberg-derived IRD in sedimentary records thus not only provides a tool to reconstruct former ice sheet dynamics, glacier calving activity, and iceberg drift but also yields information on ocean current patterns. In addition, having a different origin and a distribution pattern more depending on atmospheric circulation, sea ice-derived IRD can provide additional information for assessing prevailing wind directions and albedo conditions.

Summary

IRD has been found to be widely distributed in the entire subpolar North Atlantic. Glacial sedimentary records from this region display discrete IRD layers at time intervals of about 7,000 years named "Heinrich" layers that witness large-scale iceberg surging of the North American Laurentide Ice Sheet. This conclusion was made based on the lithology of the IRD involved, showing, among others, a large contribution of detrital dolomitic carbonate derived from the sedimentary rocks around Hudson Bay and Strait. Mineral studies have provided also evidence for IRD originating from European and Greenland glaciers. Asynchronous deposition of IRD from these various sources suggests a different stability regime of ice sheets west and east of the North Atlantic. Although several mechanisms may have played a role, increasing evidence arises which demonstrates that (sub)surface ocean warming and associated bottom melting of floating glaciers and ice shelves have been an important mechanism triggering large-scale iceberg calving ("Heinrich") events, both under glacial climate and under present-day warming conditions.

Bibliography

Alley, R. B., and Clark, P. U., 1999. The deglaciation of the Northern Hemisphere: a global perspective. *Annual Review of Earth and Planetary Sciences*, **27**, 149–182.

Alvares-Solas, J., Charbit, S., Ritz, C., Paillard, D., Ramstein, G., and Dumas, C., 2010. Links between ocean temperature and iceberg discharge during Heinrich events. *Nature Geoscience*, **3**, 122–126.

Andresen, C. S., Straneo, F., Ribergaard, M. H., Bjørk, A. A., Andersen, T. J., Kuijpers, A., Nørgaard-Pedersen, N., Kjær, K., Schjøth, F., Weckström, K., and Ahlstrøm, A. P., 2012. Rapid response of Helheim Glacier in Greenland to climate variability over the past century. *Nature Geoscience*, **5**, 37–41.

Andrews, J. T., 2000. Icebergs and iceberg rafted detritus (IRD) in the North Atlantic: facts and assumptions. *Oceanography*, **13**(3), 100–108.

Andrews, J. T., and Tedesco, K., 1992. Detrital carbonate-rich sediments, northwest Labrador Sea: implications for ice-sheet dynamics and iceberg rafting (Heinrich) events in the North Atlantic. *Geology*, **20**, 1087–1090.

Berggren, W. A., 1972. Late Pliocene-Pleistocene glaciation. In Laughton, A. S., Berggren, W. A., et al. (eds.), *Initial Reports of the Deep-Sea Drilling Programme*. Washington, DC: US Government Printing Office, Vol. 12, pp. 953–963.

Bond, G. C., and Lotti, R., 1995. Iceberg discharges into the North Atlantic on millennial time scales during the Last Glaciation. *Science*, **267**, 1005–1009.

Bond, G., Showers, W., Cheseby, M., Lotti, R., Almasi, P., deMenocal, P., Priore, P., Cullen, H., Hajdas, I., and Bonani, G., 1997. A pervasive millennial-scale cycle in North Atlantic Holocene and glacial climates. *Science*, **278**, 1257–1266.

Bond, G., Showers, W., Elliot, M., Evans, M., Lotti, R., Hadjas, I., Bonani, G., and Johnson, S., 1999. The North Atlantic's 1–2 kyr climate rhythm: relation to Heinrich events, Dansgaard/Oeschger and the Little Ice Age. In Clark, P. U., et al. (eds.), *Mechanisms of Global Climate Changes at Millennial Timescales*. Washington, DC: AGU. Geophysical Monograph Series, 112, pp. 35–58.

Bramlette, M. N., and Bradley, W. H., 1940. *Geology and biology of North Atlantic deep-sea cores between Newfoundland and Ireland. Part I. Lithology and geological interpretations*. United States Geological Survey, Professional Paper 196A: 1–34.

Broecker, W. S., 1994. Massive iceberg discharges as triggers for global climate change. *Nature*, **372**, 421–424.

Broecker, W. S., Bond, G., McManus, J., Klas, M., and Clark, M., 1992. Origin of the Northern Atlantic's Heinrich events. *Climate Dynamics*, **6**, 265–273.

Dowdeswell, J. A., Maslin, M. A., Andrews, J. T., and McCave, I. N., 1995. Iceberg production, debris rafting, and the extent and thickness of Heinrich layers (H-1, H-2) in North Atlantic sediments. *Geology*, **23**, 301–304.

Dowdeswell, J. A., Elverhøj, A., Andrews, J. T., and Hebbeln, D., 1999. Asynchronous deposition of ice-rafted layers in the Nordic seas and North Atlantic Ocean. *Nature*, **400**, 348–351.

Drewry, D., 1986. *Glacial Geologic Processes*. London: Edward Arnold.

Gilbert, R., 1990. Rafting in glacimarine environments. In Dowdeswell, J. A., and Scource, J. D. (eds.), *Glacimarine Environments: Processes and Sediments*. London: Geological Society. Special Publications of the Geological Society of London, Vol. 53, pp. 105–120.

Grousset, F. E., Cortijo, E., Huon, S., Herve, L., Richter, T., Burdloff, D., Duprat, J., and Weber, O., 2001. Zooming in on Heinrich layers. *Paleoceanography*, **16**, 240–259.

Gwiazda, R. H., Hemming, S. R., and Broecker, W. S., 1996. Provenance of icebergs during Heinrich event 3 and their contrast to their sources during other Heinrich episodes. *Paleoceanography*, **11**, 371–378.

Heinrich, H., 1988. Origin and consequences of cyclic ice rafting in the Northeast Atlantic Ocean during the past 130,000 years. *Quaternary Research*, **29**(2), 143–152.

Hemming, S. R., Broecker, W. S., Sharp, W. D., Bond, G. C., Gwiazda, R. H., McManus, J. F., Klas, M., and Hajdas, I., 1998. Provenance of Heinrich layers in core V28-82, northwestern Atlantic: $^{40}Ar/^{39}Ar$ ages of ice-rafted hornblende, Pb isotopes in feldspar grains, and Nd-Sr-Pb isotopes in the fine sediment fraction. *Earth and Planetary Science Letters*, **164**, 317–333.

Holland, D., Thomas, R. H., de Young, B., Ribergaard, M. H., and Lyberth, B., 2008. Acceleration of Jakobshavn Isbrae triggered by warm subsurface ocean waters. *Nature Geoscience*, **1**, 659–664.

Hulbe, C. L., MacAyeal, D. R., Denton, G. H., Kleman, J., and Lowell, T. V., 2004. Catastrophic ice shelf breakup as the source of Heinrich event icebergs. *Paleoceanography*, **19**, PA1004, doi:10.1029/2003PA000890.

Knutz, P. C., Hall, I. R., Zahn, R., Rasmussen, T. L., Kuijpers, A., Moros, M., and Shackleton, N. J., 2002. Multidecadal ocean variability and NW European ice sheet surges during the last deglaciation. *Geochemistry, Geophysics, Geosystems*, **3**(12), 1–9.

Knutz, P. C., Ebbesen, H., Christiansen, S., Sicre, M.-A., and Kuijpers, A., 2011. A triple stage deglacial retreat of the southern Greenland Ice Sheet driven by vigorous Irminger Current. *Paleoceanography*, **26**, PA3204, doi:10.1029/2010PA002053.

Kuijpers, A., Heinrich, H., and Moros, M., 2005. Climatic warming: a trigger for glacial iceberg surges ("Heinrich events") in the North Atlantic? Review of Survey Activities. *Geological Survey of Denmark and Greenland Bulletin*, **7**, 53–56.

Linthout, K., Troelstra, S. R., and Kuijpers, A., 2000. Provenance of coarse Ice Rafted Detritus near the SE Greenland Margin. *Netherlands Journal of Geosciences*, **79**(1), 109–121.

Lloyd, J. M., Moros, M., Perner, K., Telford, R., Kuijpers, A., Jansen, E., and McCarthy, D., 2011. A 100 year record of ocean temperature control on the stability of Jakobshavn Isbrae, West Greenland. *Geology*, doi:10.1130/G32076.1.

MacAyeal, D. R., 1993. Binge/purge oscillations of the Laurentide ice sheet as a cause of the North Atlantic's Heinrich events. *Paleoceanography*, **8**, 775–784.

Moros, M., Kuijpers, A., Snowball, I., Lassen, S., Bäckström, D., Gingele, F., and McManus, J., 2002. Were glacial iceberg surges in the North Atlantic triggered by climatic warming? *Marine Geology*, **192**, 393–417.

Reeh, N., Mayer, C., Miller, H., Højmark-Thomsen, H., and Weidick, A., 1999. Present and past climate control on fjord glaciations in Greenland: implications for IRD deposition in the sea. *Geophysical Research Letters*, **26**, 1039–1042.

Ricketts, N. G., and Trask, P. D., 1932. *The "Marion" expedition to Davis Strait and Baffin Bay, 1928 – Scientific Results, Part 1, The Bathymetry and Sediments of Davis Strait*. Washington, DC: United States Government Printing Office. U.S. Treasury Department, Coast Guard Bulletin, Vol. 19, pp. 1–81.

Ruddiman, W. F., 1977. Late Quaternary deposition of ice-rafted sand in the sub-polar North Atlantic (40–60° N). *Geological Society of America Bulletin*, **88**, 1813–1827.

Scourse, J. D., Hall, I. R., McCave, I. N., Young, J. R., and Sudgon, C., 2000. The origin of Heinrich layers: evidence from H2 for European precursor events. *Earth and Planetary Science Letters*, **181**, 187–195.

Shackleton, N. J., Backman, J., Zimmerman, H., Kent, D. V., Hall, M. A., Roberts, D. G., Schnitker, D., Baldauf, J. G., Desprairies, A., Homrighausen, R., Huddlestun, P., Keene, J. B., Kaltenback, A. J., Krumsiek, K. A. O., Morton, A. C., Murray, J. W., and Westberg-Smith, J., 1984. Oxygen isotope calibration of the onset of ice rafting and history of glaciation in the North Atlantic region. *Nature*, **307**, 620–623.

Stein, R., Nam, S.-I., Grobe, H., and Hubberten, H., 1996. Late Quaternary glacial history and short-term ice-rafted debris fluctuations along the East Greenland continental margin. In Andrews, J. T., Austin, W. E. N., Bergsten, H., and Jennings, A. E. (eds.), *Late Quaternary Palaeoceanography of the North Atlantic Margins*. London: Geological Society. Geological Society Special Publication, Vol. 111, pp. 135–151.

Tantillo, B., St. John, K., Passchier, S., and Kearns, L., 2012. *Can sea ice-rafted debris be distinguished from iceberg-rafted debris based on grain surface features? Analysis of quartz grains from modern Arctic Ocean sea ice floes*. In GSA Annual Meeting Abstracts.

Vidal, L., Labeyrie, L., Cortijo, E., Arnold, M., Duplessy, J. C., Michel, E., and Becqué, S., 1997. Evidence for changes in the North Atlantic Deep Water linked to meltwater surges during the Heinrich events. *Earth and Planetary Science Letters*, **146**, 13–27.

Cross-references

Currents
Deep-sea Sediments
Fjords
Glacial-marine Sedimentation
Grounding Line
Paleoceanographic Proxies
Paleoceanography
Radiocarbon: Clock and Tracer
Regional Marine Geology
Sea-Level
Sediment Transport Models
Sequence Stratigraphy
Shelf

INTEGRATED COASTAL ZONE MANAGEMENT

Gerald Schernewski
Leibniz Institute for Baltic Sea Research Warnemünde, Rostock, Germany

Synonyms

Coastal area planning or sustainable coastal zone development; Integrated coastal area management (ICAM); Integrated coastal resources management; Several terms are used for concepts and approaches that share similar objectives with ICZM

Definition

A large number of ICZM definitions are available (e.g., Sorensen, 1997; Cicin-Sain and Knecht, 1998; Salomons et al., 1999). A comprehensive international definition is provided by the European Commission (1999): "ICZM is a dynamic, continuous and iterative process designed to promote sustainable management of coastal zones. ICZM seeks, over the long-term, to balance the benefits from economic development and human uses of the Coastal Zone, the benefits from protecting, preserving, and restoring Coastal Zones, the benefits from minimizing loss of human life and property, and the benefits from public access to and enjoyment of the Coastal Zone, all within the limits set by natural dynamics and carrying capacity." The term "integrated" refers to the integration of objectives; multiple instruments needed to meet these objectives; relevant policy areas, sectors, and levels of administration; the terrestrial and marine components; time and space; as well as different disciplines. Although referring to "management," ICZM covers an iterative cyclic process of information collection, planning, decision making, management, and monitoring of implementation. The term "planning" is meant in a broad sense as strategic policy development. Participation of all interested and affected parties (stakeholders) is a core element in ICZM.

A general definition of the "coastal zone" does not exist. It has to cover a band of land and sea that can be defined based on, e.g., functional relationships, geographic properties, or administrative boundaries. Depending on national definitions, the seaward boundary can be several 100 m or far offshore up to the outer limit of the exclusive economic zone (EEZ). The same is true for the landward boundaries. In general, a trade-off exists between a comprehensive large-scale definition of the coastal zone and a narrow, small-scale approach with high practical, administrative, and political acceptance.

Background and history

Coastal regions are among the most populated and productive areas, with outstanding economic and ecologic value. Increasing competition for maritime and coastal space and increasing pressures on resources lead to a deterioration of natural, socioeconomic, and cultural resources. The impacts of climate change are expected to further increase the exposure of the coast. In the past, coastal planning activities or development decisions took place in a sectorial way, hardly being linked to each other. This fragmented approach to planning and management leads to inefficient use of resources, conflicting claims on space, and missed opportunities for more sustainable coastal development.

The United States Coastal Zone Management Program (CZMP), established under the Coastal Zone Management Act (CZMA) of 1972, can be regarded as one of the first policy frameworks for coordinated coastal zone planning and management (Knecht and Archer, 1993). During the last 40 years, ICZM as a concept evolved from nonintegrated sectorial toward integrated approaches (Sorensen, 1997; Cicin-Sain and Knecht, 1998; Salomons et al., 1999). Initially the emphasis of ICZM was on the environment, later economic aspects as well as social and ethical components were included (Turner, 2000). ICZM received worldwide recognition after the United Nations Conference on Environment and Development (UNCED) in 1992 in Rio de Janeiro, Brazil. The resulting Agenda 21, Chapter 17.1, called for "new approaches to marine and coastal area management and development, at the national, sub-regional, regional and global levels, approaches that are integrated in content and are precautionary and anticipatory in ambit."

State and challenges

Many ICZM policies, projects, initiatives, and activities have developed during the last decades at local, regional, and national level (e.g., Sorensen, 1997; Shipman and Stojanovic, 2007). The exchange of experiences and learning from best practice examples plays an important role in ICZM (Olsen et al., 1998). Against this background the European Commission maintains the OURCOAST online database with about 350 case studies on major themes like adaptation to coastal risks and climate change, planning and land management instruments, institutional coordination mechanisms, as well as information and communication (OurCoast, 2011).

Despite all efforts, ICZM still suffers from weaknesses, e.g., insufficient political and legal status or the lack of a consistent and applicable process for practitioners and policy makers (e.g., Shipman and Stojanovic, 2007). Shipman (2012) outlined future demand: ICZM has to be politically robust and legitimate, coherent in its outcomes and benefits, and with a clear guidance of how to achieve them. Recently, ICZM agreements have been developed and adopted for international, large-scale regional sea frameworks. For example, the "Protocol on Integrated Coastal Zone Management in the Mediterranean," the first supranational (21 countries located in Europe, Africa, and Asia), legally binding ICZM agreement entered into force in 2011. The ratification includes the European Union and is legally binding for all member states. Formally, it addresses recent ICZM demands, provides a vision of sustainable development, defines the coastal zone, and has legitimacy, a hierarchy of strategies and plans, as well as a coherence of governance and actions. It is outcome focused, provides process guidelines for each stage, and is interactive and expandable. The future will show if it can serve as model for regional seas around the world (Rochette and Billé, 2012; Shipman, 2012).

Some major future challenges are the integration of ICZM with maritime spatial planning (Smith et al., 2011), maritime policy and marine environment protection (Cicin-Sain and Belfiore, 2005), the integration of land and marine spatial planning, and the further development and application of tools, like systems approach frameworks (SAF) for coastal zones (Hopkins et al., 2011).

Measuring success

Several indicator sets have been suggested to measure the state of and the progress toward sustainability in coastal zones on a local (Henocque, 2003; Hoffman, 2009), European (Burbridge, 1997; Pickaver et al., 2004), and worldwide level (Olsen, 2003; Belfiore et al., 2006). Indicators provide a simplified view on complex phenomena, quantify information, and make it comparable. Many application exercises (e.g., Pickaver, 2009) and critical evaluations (Wallis, 2006; Bell and Morse, 2008) took place. Indicator sets are regarded as important tools, e.g., in European coastal and maritime policy (Meiner, 2010), and are used to monitor the EU Sustainable Development Strategy. On a local level, ICZM indicator sets are still poorly accepted. Recent approaches provide indicator sets, a scoring system, and participative self-assessment approaches to comprehensive policy tools for local coastal authorities (SUSTAIN partnership, 2012). It not only allows the evaluation of the sustainability performance but can be used as a local planning and management tools.

Conclusions

Existing conflicts in the coastal zone as well as future challenges, like climate change, call for an ICZM. ICZM principles and concepts are not new but still need to evolve and to prove its applicability and practical relevance.

Bibliography

Belfiore, S., Barbiere J., Bowen R., Cicin-Sain B., Ehler C., Mageau C., McDougall D., and Siron, R., 2006. *A Handbook for Measuring the Progress and Outcomes of Integrated Coastal and Ocean Management*. IOC Manuals and Guides No. 46, ICAM Dossier, Vol. 2, UNESCO.

Bell, S., and Morse, S., 2008. *Sustainability Indicators – Measuring the Immeasurable?* 2nd edn. London: Earthscan.

Burbridge, P. R., 1997. A generic framework for measuring success in integrated coastal management. *Ocean and Coastal Management*, **37**(2), 175–189.

Cicin-Sain, B., and Belfiore, S., 2005. Linking marine protected areas to integrated coastal and ocean management: a review of theory and practice. *Ocean and Coastal Management*, **48**(11–12), 847–868.

Cicin-Sain, B., and Knecht, R. W., 1998. *Integrated Coastal and Ocean Management: Concepts and Practices*. Washington, DC: Island Press.

European Commission, 1999. *Towards a European Integrated Coastal Zone Management (ICZM) Strategy: General Principles and Policy Options*. Luxembourg: Office for Official Publications of the European Communities. ISBN 92-828-6463-4.

Henocque, Y., 2003. Development of process indicators for coastal zone management in France. *Ocean and Coastal Management*, **46**, 363–379.

Hoffmann, J., 2009. Indicators for an ICZM. Experience with a problem-oriented approach. *Journal of Coastal Conservation*, **13**(2–3), 141–150.

Hopkins, T. S., Bailly, D., and Støttrup, J. G., 2011. A systems approach framework for coastal zones. *Ecology and Society*, **16**(4), 25, doi:10.5751/ES-04553-160425.

Knecht, R. W., and Archer, J., 1993. Integration in the US coastal management program. *Ocean and Coastal Management*, **21**, 183–200.

Meiner, A., 2010. Integrated maritime policy for the European Union – consolidating coastal and marine information to support maritime spatial planning. *Journal of Coastal Conservation*, **14**, 1–11.

Olsen, S., Tobey, J., and Kerr, M., 1998. A common framework for learning from ICZM experience. *Ocean and Coastal Management*, **37**(2), 155–174.

Olsen, S. B., 2003. Frameworks and indicators for assessing progress in integrated coastal management initiatives. *Ocean and Coastal Management*, **46**(3–4), 347–361.

OurCoast. 2011. Comparative analyses of the OURCOAST cases. http://ec.europa.eu/ourcoast

Pickaver, A. H., Gilbert, C., and Breton, F., 2004. An indicator set to measure the progress in the implementation of integrated coastal zone management in Europe. *Ocean and Coastal Management*, **47**(9–10), 449–462.

Pickaver, A. H., 2009. Further Testing of the Approved EU Indicator to Measure the Progress in the Implementation of Integrated

Coastal Zone Management in Europe. In Moksness, E., Dahl, E., and Støttrup, J. (eds.), *Integrated Coastal Zone Management*. Oxford: Wiley-Blackwell.

Rochette, J., and Billé, R., 2012. ICZM protocols to regional seas conventions: what? why? how? *Marine Policy*, **36**(5), 977–984.

Sorensen, J., 1997. National and international efforts at integrated coastal management: definitions, achievements, and lessons. *Coastal Management*, **25**(1), 3–41.

Salomons, W., Turner, R. K., Lacerda, L. D. de, and Ramachandran, S. (eds.) 1999. *Perspectives on Integrated Coastal Zone Management*. Springer Series: Environmental Science and Engineering Subseries: Environmental Science, XVIII. New York: Springer Publ.

Shipman, B., and Stojanovic, T., 2007. Facts, fictions, and failures of integrated coastal zone management in Europe. *Coastal Management*, **35**(2–3), 375–398.

Shipman, B., 2012. *ICZM2.0 – A New ICZM for an Era of Uncertainty*. VLIZ Special Publication, 61, p. 11.

Smith, H. D., Maes, F., Stojanovic, T. A., and Ballinger, R. C., 2011. The integration of land and marine spatial planning. *Journal of Coastal Conservation*, **15**, 291–303.

SUSTAIN Partnership. 2012. Measuring coastal sustainability. A guide for the self-assessment of sustainability using indicators and a means of scoring them. http://www.sustain-eu.net/what_are_we_doing/measuring_coastal_sustainability.pdf

Turner, R. K., 2000. Integrating natural and socio-economic science in coastal management. *Journal of Marine Systems*, **25**(3–4), 447–460.

Wallis, A. M., 2006. Sustainability indicators: is there consensus among stakeholders? *International Journal of Environment and Sustainable Development*, **5**(3), 287–296.

Cross-references

Beach Processes
Coastal Bio-geochemical Cycles
Coastal Engineering
Coasts
Dunes
Engineered Coasts
Fjords
Lagoons
Mangrove Coasts
Sediment Dynamics
Wadden Sea

INTERNATIONAL BATHYMETRIC CHART OF THE ARCTIC OCEAN (IBCAO)

Martin Jakobsson
Department of Geological Sciences, Stockholm University, Stockholm, Sweden

Definition

The International Bathymetric Chart of the Arctic Ocean (IBCAO). Mapping project focused on assembling and compiling all available depth information from the Arctic Ocean to portray the shape of its seafloor.

Introduction

IBCAO was initiated as a bathymetric mapping project in St. Petersburg, Russia, in 1997 (Macnab and Grikurov, 1997). The goal was to develop a digital database containing all available bathymetric data north of 64°N and from this compile a bathymetric map and digital bathymetric model (DBM) representing the shape of the Arctic seafloor. DBM is a digital terrain model (DTM) representing the bathymetry of the seafloor. The most common format of a DBM is a regular grid with defined cell spacing. Each grid cell is represented by a bathymetric value, i.e., a depth. IBCAO builds on the volunteer efforts of investigators and data contributions from countries that carry out mapping in the Arctic Ocean. An established editorial board comprises its main organizational structure. The activity has been endorsed by the Intergovernmental Oceanographic Commission (IOC), the International Arctic Science Committee (IASC), the International Hydrographic Organization (IHO), and the General Bathymetric Chart of the Oceans (GEBCO). IBCAO products are available for download from http://www.ibcao.org/.

IBCAO digital bathymetric models

The first IBCAO compilation was released to the public in 2000 (Jakobsson et al., 2000). This "beta version" consisted of a DBM with grid cell spacing of 2.5 × 2.5 km on a polar stereographic projection. After feedback from the Arctic research community resulting in corrections of artifacts and incorporation of additional data, Version 1 was completed and released in 2001 (Jakobsson and IBCAO Editorial Board Members, 2001).

In 2008, a finer grid spacing of 2 × 2 km was implemented in Version 2.0 when several large multibeam surveys became available (Jakobsson et al., 2008). The subsequent Version 3.0, completed in 2012, represents the largest improvement so far taking advantage of new data sets collected by Arctic nations to substantiate an extension of the continental shelf under the United Nations Convention on the Law of the Sea (UNCLOS) Article 76, opportunistic data collected from fishing vessels, and data acquired from submarines and from research ships of various nations (Jakobsson et al., 2012). Version 3.0 has grid spacing of 500 × 500 m, and the area covered by multibeam surveys has increased from ~6 % in Version 2.0 to ~11 % in Version 3.0. A shaded relief map and a 3D portrayal based on Version 3.0 are shown in Figure 1.

Compilation of the IBCAO DBM is based on gridding the bathymetric database with the algorithm continuous curvature splines in tension (Smith and Wessel, 1990) available in the Generic Mapping Tools (GMT) (Wessel and Smith, 1991). Soundings always have the priority in the gridding procedure before isobaths digitized from published maps. Compilation of Version 3.0 involved an additional final gridding step where data with a spatial

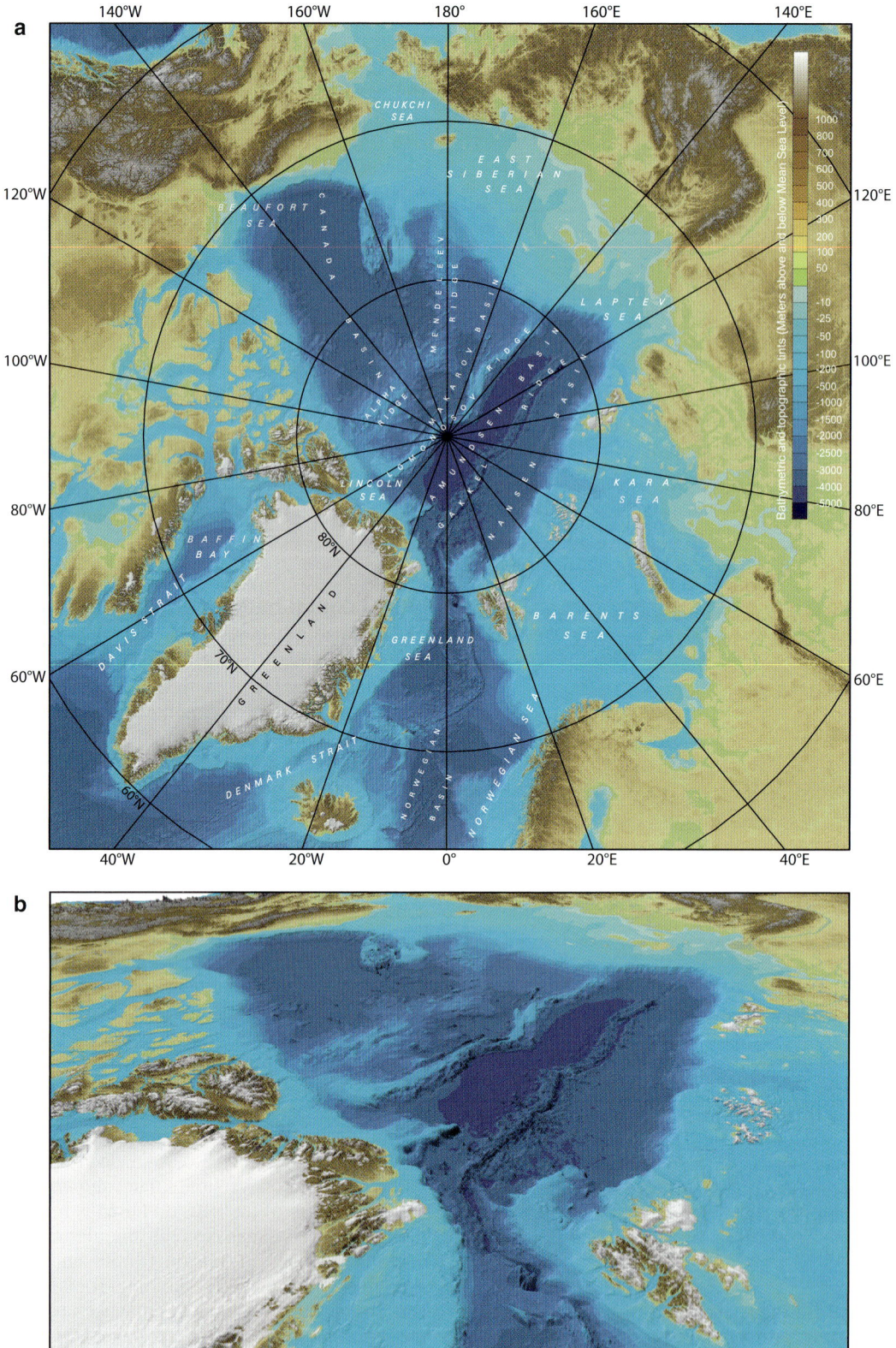

International Bathymetric Chart of the Arctic Ocean (IBCAO), Figure 1 Shaded relief map (**a**) and 3D portrayal (**b**) of IBCAO Version 3.0.

density of 500 m between points or higher were added using the remove-restore method (e.g., Smith and Sandwell, 1997; Hell and Jakobsson, 2011).

IBCAO printed map

In 2004, a printed map at a scale of 1:6,000,000 was published based on the IBCAO Version 1.0 DBM. Based on a shaded relief generated from the DBM and overlaid depth contours, this map portrayed the shape of the Arctic Ocean seafloor north of 64°N. A polar stereographic map projection was used with the true scale set at 75°N and the WGS 84 horizontal datum. The map limit, scale, and projection were taken from the GEBCO Arctic Ocean Sheet 5.17 published in 1979 (Johnson et al., 1979).

Bibliography

Hell, B., and Jakobsson, M., 2011. Gridding heterogeneous bathymetric data sets with stacked continuous curvature splines in tension. *Marine Geophysical Research*, **32**, 493–501.

Jakobsson, M., and IBCAO Editorial Board Members, 2001. Improvement to the International Bathymetric Chart of the Arctic Ocean (IBCAO): updating the data base and the grid model. In Union, A. G. (ed.), *American Geophysical Union Fall Meeting*. San Francisco: EOS Transactions, pp. OS11B–OS0371B.

Jakobsson, M., Cherkis, N., Woodward, J., Macnab, R., and Coakley, B., 2000. New grid of Arctic bathymetry aids scientists and mapmakers. *EOS, Transactions American Geophysical Union*, **81**, 89, 93, 96.

Jakobsson, M., Macnab, R., Mayer, L., Anderson, R., Edwards, M., Hatzky, J., Schenke, H. W., and Johnson, P., 2008. An improved bathymetric portrayal of the Arctic Ocean: implications for ocean modeling and geological, geophysical and oceanographic analyses. *Geophysical Research Letters*, **35**, L07602.

Jakobsson, M., Mayer, L., Coakley, B., Dowdeswell, J. A., Forbes, S., Fridman, B., Hodnesdal, H., Noormets, R., Pedersen, R., Rebesco, M., Schenke, H. W., Zarayskaya, Y., Accettella, D., Armstrong, A., Anderson, R. M., Bienhoff, P., Camerlenghi, A., Church, I., Edwards, M., Gardner, J. V., Hall, J. K., Hell, B., Hestvik, O., Kristoffersen, Y., Marcussen, C., Mohammad, R., Mosher, D., Nghiem, S. V., Pedrosa, M. T., Travaglini, P. G., and Weatherall, P., 2012. The International Bathymetric Chart of the Arctic Ocean (IBCAO) version 3.0. *Geophysical Research Letters*, **39**, L12609.

Johnson, G. L., Monahan, D., Grönlie, G., and Sobczak, L., 1979. Sheet 5.17. In Canadian Hydrographic Service, O (ed.), *General Bathymetric Chart of the Oceans (GEBCO)*, 5th edn. Ottawa: Canadian Hydrographic Service.

Macnab, R., and Grikurov, G., 1997. *Report: Arctic Bathymetry Workshop*. St. Petersburg: Institute for Geology and Mineral Resources of the Ocean (VNIIOkeangeologia), p. 38.

Smith, W. H. F., and Sandwell, D. T., 1997. Global seafloor topography from satellite altimetry and ship depth soundings. *Science*, **277**, 1957–1962.

Smith, W. H. F., and Wessel, P., 1990. Gridding with continuous curvature splines in tension. *Geophysics*, **55**, 293–305.

Wessel, P., and Smith, W. H. F., 1991. Free software helps map and display data. *EOS Transactions, American Geophysical Union*, **72**, 441, 445–446.

Cross-references

Ocean Margin Systems
Plate Tectonics

INTRAOCEANIC SUBDUCTION ZONE

Hayato Ueda
Department of Geology, Niigata University, Nishi-ku, Niigata, Japan

Synonyms

Oceanic subduction zone

Definition

Intraoceanic subduction zone. A subduction zone developed beneath the leading edge of an oceanic lithosphere.

Introduction

Intraoceanic subduction zones were developed beneath leading edges of *oceanic crusts*. Although the accompanied *volcanic arcs* (intraoceanic island arcs) have thickened crusts compared with normal oceanic crusts, they are still immature and thinner than continental crusts. It is thus distinguished from continental arcs (Active Continental Margins). An intraoceanic subduction zone is associated with an oceanic basin, i.e., *back-arc basin*, behind the volcanic arc. Continental island arcs such as Northeast Japan and Kurile arcs are also accompanied by back-arc basins (marginal basins) at least partly of oceanic crusts. These arcs developed upon old continental crusts split from continents behind and can also be distinguished from intraoceanic arcs consisting essentially of arc magmatic rocks developed in intraoceanic settings.

Modern intraoceanic subduction zones cluster in the western Pacific and also occur in the western Atlantic (Figure 1). They contrast with eastern Pacific margins bounded almost entirely by continental subduction zones. Among them, the Izu-Bonin-Mariana, Lesser Antilles (Barbados), and South Sandwich subduction zones developed along margins of independent small oceanic plates mechanically decoupled with continents. The Tonga-Kermadec and Aleutian subduction zones are situated along margins of well-developed oceanic basin connected with continents. In Southeast Asia to Melanesia regions, shorter subduction zones are complicatedly arranged. Some arcs are bounded by subduction zones of opposite polarities or being switched from one to another (e.g., Philippine and Solomon arcs). There are also many indistinct trenches with poorly developed volcanic arcs and abandoned trenches in these areas.

Intraoceanic subduction zones are the major sites for descending of the Earth's surface materials (oceanic crust, sediments, and fluids), seismic and volcanic activities, *hydrothermalism*, and growth of sialic continental crusts. They could be modern analogues for origins of *ophiolites* on land and can also give hints to consider tectonics and environments in the early Earth's history when continental crusts had not yet been well-developed.

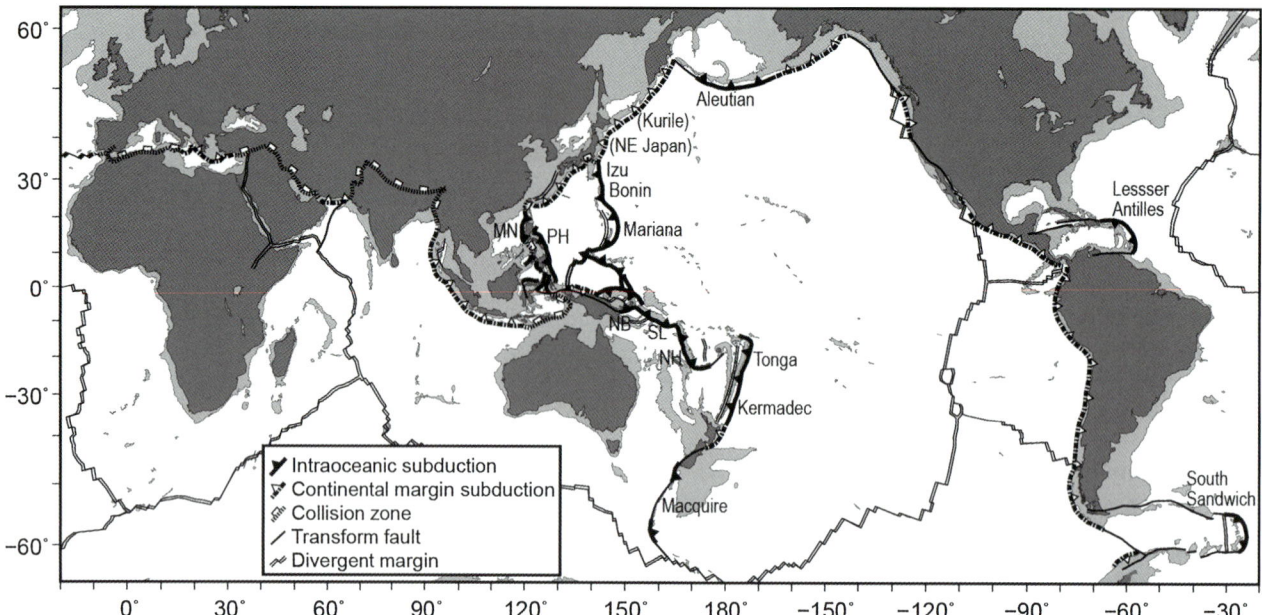

Intraoceanic Subduction Zone, Figure 1 Distribution of trenches of intraoceanic subduction zones. MN Manila, P Philippine, NB New Britain, NH New Hebrides (Vanuatu), and SL Solomon.

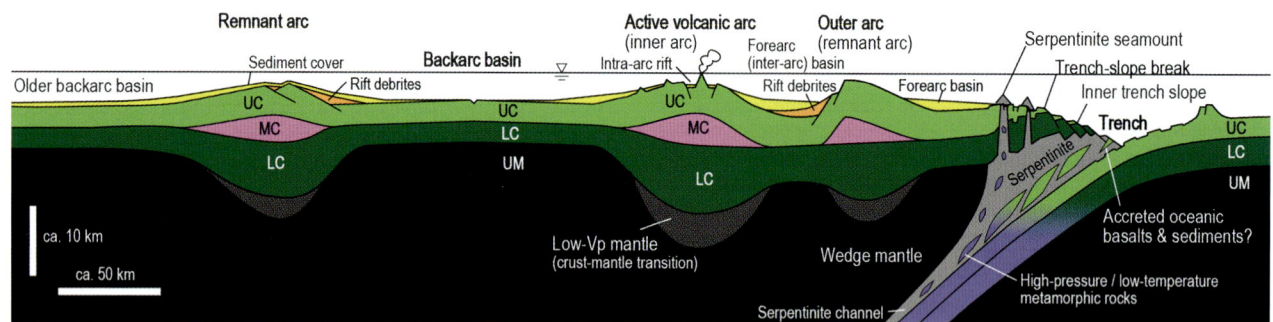

Intraoceanic Subduction Zone, Figure 2 A cross-sectional picture for the features of intraoceanic subduction zone. Schematic crust and mantle structures refer to seismic profiles across the Izu-Bonin-Mariana arc back-arc system (Suyehiro et al., 1996; Takahashi et al., 2008, 2009), except for speculative illustration of forearc parts close to the trench concerning geological evidences. UC upper crust, MC middle crust, LC lower crust, and UM upper mantle.

General architecture

Features of topography and crustal structures resulting from intraoceanic *subduction* are illustrated in Figure 2. The plate boundary on the surface is marked by a trench, and a volcanic arc with thickened crust is arranged parallel to the trench. Typical back-arc basins are formed by *seafloor spreading*, which initiates from intra-arc rift splitting the volcanic arc. A half of the split and inactivated arc, referred to as a remnant arc (Karig, 1972), is left behind the back-arc basin far from the trench comprising a non-seismic ridge. The counterpart of the remnant arc may comprise the basement of new volcanic arc after back-arc spreading (daughter arc) or is occasionally found as the outer arc, a currently non-volcanic structural high to the trench side of the active volcanic arc. In a well-developed intraoceanic subduction system, there are thus multiple arrays of arcs.

Intraoceanic trench

Because remote from continental margins, trenches of typical intraoceanic subduction zones are starved with terrigenous clastic sediments except for the trench off Barbados and Aleutian as shown later. Trench floors are marked by rough topography with horst and graben resulting from slab bending and with subducting *seamounts* (Figure 2) and are accompanied by thin pelagic sediment covers and local landslide debris. The subducting plate in the Western Pacific are dominantly old, cool, and thus dense. Scarce sediment fill over dense subducting plate created the Earth's deepest trenches there. Limited sediment supply and fast convergence result in poor development or absence of *accretionary wedges* and, instead, facilitate *subduction erosion* removing hanging-wall materials around trenches. On the contrary, abundant clastic turbidites from adjacent continents bury trenches of the Lesser Antilles (Barbados) and eastern Aleutian subduction zones, where sediments are being scraped off to comprise *accretionary wedges*.

Forearc

Forearc regions between the trench and the volcanic front consist of inner trench slope, forearc basin, and outer arc. Surface slope dips greatly change between the steeper inner trench slope and the flat forearc basin, and the boundary between them is called as a trench-slope break.

Inner trench slopes to outer margins of forearc basins of intraoceanic non-accretionary margins are dominated by normal faults with stepping fault scarps and/or horsts and grabens (Hussong and Uyeda, 1981). In Izu-Bonin-Mariana and Tonga-Kermadec forearcs, the basement rocks there show ophiolitic assemblage of serpentinized *peridotites*, *gabbro*, and early arc volcanic rocks including *boninites* (Ishii, 1985; Bloomer and Fisher, 1987; Bloomer et al., 1995). Minor occurrences of mid-ocean ridge basalts (MORB: see "Mid-ocean Ridge Magmatism and Volcanism") and alkali basalts of *seamount* origins, chert, and limestone were presumably accreted from the subducted Pacific plate (Johnson et al., 1991).

In outer margins of the Mariana forearc basin, serpentinite seamounts of conical shapes are aligned parallel to the trench, in addition to serpentinite horsts of tectonic features (Fryer, 1992). Similar serpentinite seamounts also occur in the inner trench slope of the Bonin forearc (Fryer, 1992). The Mariana forearc serpentinite seamounts are *mud volcanoes* of highly fractured and clayey serpentinites, occasionally with serpentine mud flows and cold-water vents with carbonate chimneys (Fryer et al., 1990). These serpentinites originated from residual peridotites after extensive extraction presumably of arc magmas and boninites (Ishii et al., 1992) and are thought to represent forearc mantle rocks.

Forearc basins are filled by thick sediments comprising of flat and smooth surfaces. Tectonic deformation is less extensive compared with trench-side areas, suggesting more rigid basements. The Moho surfaces beneath forearcs become indistinct as seismic velocity of the mantle is reduced toward the trench (Suyehiro et al., 1996). If successfully imaged, forearc crust is thinner than volcanic arc crust and partly as thin as normal oceanic crust (Kodaira et al., 2010).

Subduction interface

There has been no drill hole which penetrated intraoceanic subduction boundaries of non-accretionary type. However, it is probable that the wedge mantle extends to or very close to the trenches (Bloomer and Fisher, 1987; Kamimura et al., 2002) in the cases of the Tonga-Kermadec and Izu-Bonin-Mariana subduction zones, taking into account thin forearc crusts and common occurrences of peridotite and serpentinites on inner trench slopes. Reduction of seismic velocity in the wedge mantle toward the trench suggests *serpentinization* of peridotites (Hyndman and Peacock, 2003). These lines of evidence imply that serpentinites are in contact with the subducting oceanic crust since shallow depths of these intraoceanic subduction zones and that crust-crust interface is limited. This may result in narrow zones of seismic coupling, in contrasts to continental subduction zones, where contacts of thick continental crusts and the subducted slabs comprise extensive *seismogenic zones* (Hyndman et al., 1997).

Serpentinized subduction interface is also supported by the occurrences of serpentinite seamounts in the Mariana forearc. Basaltic clasts recording *high-pressure/low-temperature metamorphism* enclosed in the serpentinite suggest the origin of the mud volcanoes from the subduction interface (Maekawa et al., 1993). Mixed occurrences of sheared serpentinite matrix and clasts and blocks of ophiolitic and metamorphic rocks are correlative to serpentinite melange.

Because water fluids might be supplied from the underlying slab, a wedge mantle is expected to be serpentinized from its base, where mechanical strength is significantly reduced. This predicts a structure that the weak serpentinite zone is sandwiched between more rigid layers: the slab and the unhydrated mantle wedge (Figure 2). Strain resulted from plate convergence concentrate to this serpentinite zone, where material flows are channelized. The concept *serpentinite channel* (Guillot et al., 2001) is applied to the zone and is regarded as a conduit for upwelling of deeply subducted rocks.

Extension and weak seismic coupling

Uyeda and Kanamori (1979) classified subduction zones into Mariana and Chilean types in terms of stress regime in the hanging-wall plate. The Mariana type is characterized by extensional stress evidenced by active back-arc rifting or spreading. It is associated with steep dip subduction and tends to occur with west-dipping subduction polarity. Large inter-plate *earthquakes* are scarce, suggesting weak coupling between the two converging plates. The Chilean type in contrast is characterized by compression without active back-arc spreading, gently dipping slabs dominantly of eastward subduction, and common large inter-plate earthquakes with strong coupling. As evident from its nomenclature, typical intraoceanic subduction zones such as Izu-Bonin-Mariana and Tonga belong to the Mariana type. Extension and weak inter-plate coupling is attributed to relative retreat of the trench off the hanging-wall plate. Uyeda and Kanamori (1979) argued several ideas to explain the trench retreat: by sinking of the slab (trench rollback); migration of the hanging-wall plate away from the trench fixed to the deep mantle, and eastward migration of trenches reflecting the global asthenosphere flow.

Crustal growth

Magma supply from the mantle in subduction zones (see "Island Arc Volcanism, Volcanic Arcs" and "Magmatism at Convergent Plate Boundaries") progressively thickens arc crusts. Continental arcs develop overprinting the preexisting continental crust of long histories and complex structures, whereas intraoceanic arcs are more juvenile (less than several tens m.y. old) with simpler starting materials of hanging-wall oceanic *lithosphere*. They can thus provide more elementary models for magma genesis and crustal growth, whose processes are referred to as subduction factory.

Intraoceanic arc crusts are 10–30 km thick with significant variation even in a single arc (e.g., Kodaira et al., 2007). They are intermediate between those of oceanic and continental crusts in terms of thickness and composition. The Izu-Bonin-Mariana arc characteristically has the middle crust layer of 6–6.5 km/s (Suyehiro et al., 1996) not found in oceanic crusts. It is assigned to felsic plutonic rocks such as tonalite (low-K variety of granitoids), compared with crustal section exposed in the Izu collision zone (Tanzawa Mountains) scraped off the Izu arc (Kawate and Arima, 1998).

Magma input from the mantle is dominantly basaltic, whereas the bulk continental crusts are estimated as andesitic (Turcotte, 1989). The transformation mechanism from basaltic ingredients to more felsic products is a major subject on subduction factory. Removal of mafic components from the differentiated crusts via delamination is one of the convincing candidates (Turcotte, 1989; Kay and Kay, 1993). Tatsumi et al. (2008) tested several hypotheses for crustal differentiation with volumetric simulation comparing to the Izu-Bonin-Mariana arc. They showed that remelting (anatexis) of basaltic crust producing a tonalitic middle crust and a high-density basal layer of mafic restites and *cumulates* is the most compatible model with the present crustal structure, and the low-Vp mantle (crust-mantle transition zone) observed in seismic refraction profiles (e.g., Kodaira et al., 2007; Takahashi et al., 2008) is assigned to the basal layer. Comparable restites occur as mafic garnet granulite at the base of the Kohistan arc crustal section (Garrido et al., 2006), a fossil intraoceanic arc exposed on land. If such basal layer is delaminated from the overlying crust, average composition of the rest crust could be more felsic.

Earliest stage of subduction

It is generally accepted that the an intraoceanic subduction begins at a *fracture zone* or a *transform fault* among preexisting oceanic crust, since Uyeda and Ben-Avraham (1972) considered for the genesis of the Philippine Sea plate by an idea that an oceanic transform boundary in the Pacific switched to be convergent when Pacific plate motion changed in the Eocene. Stern (2004) synthesized possible processes of subduction initiation into two major types. One is induced by compression as exampled by the Solomon subduction system, where the collided Ontong Java Plateau jammed the previous trench to the north of the arc and new subduction initiated facing the south. Another is spontaneous slab sinking resulting from a high-density contrast between the adjacent oceanic lithosphere and characterized by extension and eruption of *boninites* in the forearc as exampled in the Izu-Bonin-Mariana subduction zone.

Boninites are high-magnesian varieties of basaltic andesite characteristically occurring in not all but some intraoceanic subduction zones and ophiolites. It suggests anomalously high temperature at shallow parts of the mantle above the slab (Crawford et al., 1989; Pearce et al., 1992). Stern and Bloomer (1992) assumed the spontaneous sinking model of the colder oceanic lithosphere at a transform boundary, initially without convergence. They argued that the asthenosphere might inject to fill the vacancy above the sinking slab, promoting a preferable condition for the boninitic genesis. This model is supported by a numerical simulation by Hall et al. (2003), in which however initial push to help bend the thick lithosphere was necessary. Recently, Reagan et al. (2010) and Ishizuka et al. (2011) found occurrences of MORB-like basalts (FAB: forearc basalts) underlying boninites in the Mariana and Bonin forearcs, respectively. They attributed them to the first injection of asthenosphere prior to the boninite stage with major supply of water from the slab. They also explained a common ophiolite succession, where boninite magmatism followed just after formation of MORB-like oceanic crusts, by analogy of the Mariana and Bonin forearc succession.

Summary

Although most of the system is immersed in deep water, an intraoceanic subduction zone may be one of the most dynamic regions on the Earth's surface. Lateral convergence and divergence, sinking and uprising, and accretion and removal concurrently occur in a single system of adjoining areas. Intraoceanic subduction zone could also give hints to consider how Earth's simplest layering of the mantle, oceanic crusts, marine water, and atmosphere interacted and evolved to more diverse and complicated structure and chemistry as seen in the present. Comprehensive understanding of such dynamics and evolution is still in progress, and further investigations are necessary heavily depending on oceanic surveys, as well as prediction and testing by numerical modeling recently significantly developed (Gerya, 2011).

Bibliography

Bloomer, S. H., and Fisher, R. L., 1987. Petrology and geochemistry of igneous rocks from the Tonga Trench – a non-accreting plate boundary. *Journal of Geology*, **95**, 469–495.

Bloomer, S. H., Taylor, B., MacLeod, C. J., Stern, R. J., Fryer, P., Hawkins, J. W., and Johnson, L., 1995. Early arc volcanism and the ophiolite problem: a perspective from drilling in the Western Pacific. In Taylor, B., and Natland, J. (eds.), *Active Margin and Marginal Basins of the Western Pacific*. Washington, DC: AGU, pp. 1–30.

Crawford, A. J., Faloon, T. J., and Green, D. H., 1989. Classification, petrogenesis and tectonic setting of boninites. In Crawford, A. J. (ed.), *Boninites and Related Rocks*. London: Unwin Hyman, pp. 1–49.

Fryer, P., 1992. A synthesis of Leg 125 drilling of serpentinite seamounts on the Mariana and Izu-Bonin forearcs. In Fryer, P., Pearce, J. A., Stokking, L. B., et al. (eds.), *Proceedings of the Ocean Drilling Project Scientific Results, 125*. TX, College Station, pp. 593–614.

Fryer, P., Saboda, K. L., Hohnson, L. E., Mackey, M. E., Moore, G. F., and Stoffers, P., 1990. Conical seamount: SeaMARC II, Alvin submersible, and seismic reflection studies. In Fryer, P., Pearce, J. A., Stokking, L. B., et al. (eds.), *Proceedings of the Ocean Drilling Program Initial Reports, 125*. TX, College Station, pp. 69–80.

Garrido, C. J., Bodinier, J.-L., Burg, J.-P., Zeilinger, G., Hussain, S. S., Dawood, H., Chaudhry, M. N., and Gervilla, F., 2006. Petrogenesis of mafic garnet granulite in the lower crust of the Kohistan paleo-arc complex (Northern Pakistan): Implications for intra-crustal differentiation of island arcs and generation of continental crust. *Journal of Petrology*, **47**, 1873–1914.

Gerya, T. V., 2011. Intra-oceanic subduction zones. In Brown, D., and Ryan, P. D. (eds.), *Arc-Continent Collision*. Berlin/Heidelberg: Springer, pp. 23–51.

Guillot, S., Hattori, K. H., De Sigoyer, J., Nägler, T., and Auzende, A.-L., 2001. Evidence of hydration of the mantle wedge and its role in the exhumation of eclogites. *Earth and Planetary Science Letters*, **193**, 115–127.

Hall, C. E., Gurnis, M., Sdrolias, M., Lavier, L. L., and Müller, R. D., 2003. Catastrophic initiation of subduction following forced convergence across fracture zones. *Earth and Planetary Science Letters*, **212**, 15–30.

Hussong, D. M., and Uyeda, S., 1981. Tectonic processes and the history of the mariana arc: a synthesis of the results of deep sea drilling project leg 60. In Hussong, D. M., Uyeda, S., et al. (eds.), *Initial Reports of Deep Sea Drilling Project*. Washington, DC: U.S. Govt. Printing Office, Vol. 60, pp. 909–929.

Hyndman, R. D., and Peacock, S. M., 2003. Serpentinization of the forearc mantle. *Earth and Planetary Science Letters*, **212**, 417–432.

Hyndman, R. D., Yamano, M., and Oleskevich, D. A., 1997. The seismogenic zone of subduction thrust faults. *Island Arc*, **6**, 244–260.

Ishii, T., 1985. Dredged samples from the Ogasawara fore-arc seamount or "Ogasawara paleoland"-"fore-arc ophiolite". In Nasu, N., et al. (eds.), *Formation of Active Ocean Margins*. Tokyo: Terrapub, pp. 307–342.

Ishii, T., Robinson, P. T., Maekawa, H., and Fiske, R., 1992. Petrological studies of peridotites from diapiric serpentinite seamounts in the Izu-Ogasawara-Mariana forearc, Leg 125. In Fryer, P., Pearce, J. A., Stokking, L. B., et al. (eds.), *Proceedings of Ocean Drilling Project Scientific Results, 125*. TX, College Station, pp. 445–485.

Ishizuka, O., Tani, K., Reagan, M., Kanayama, K., Umino, S., Harigane, Y., Sakamoto, I., Miyajima, Y., Yuasa, M., and Dunkley, D. J., 2011. The timescales of subduction initiation and subsequent evolution of an oceanic island arc. *Earth and Planetary Science Letters*, **306**, 229–240.

Johnson, L. E., Fryer, P., Taylor, B., Silk, M., Jones, D. L., Sliter, W. V., and Ishii, T., 1991. New evidence for crustal accretion in the outer Mariana fore arc: Cretaceous radiolarian cherts and mid-ocean ridge basalt-like lavas. *Geology*, **19**, 811–814.

Kamimura, A., Kasahara, J., Shinohara, M., Hino, R., Shiobara, H., Fujie, G., and Kanazawa, T., 2002. Crustal structure study at the Izu-Bonin subduction zone around 31°N: implications of serpentinized materials along the subduction plate boundary. *Physics of the Earth and Planetary Interiors*, **132**, 105–129.

Karig, D. E., 1972. Remnant arcs. *Geological Society of America Bulletin*, **83**, 1057–1068.

Kawate, S., and Arima, M., 1998. Petrogenesis of the Tanzawa plutonic complex, central Japan: exposed felsic middle crust of the Izu–Bonin – Mariana arc. *Island Arc*, **7**, 342–358.

Kay, R. W., and Kay, S. M., 1993. Delamination and delamination magmatism. *Tectonophysics*, **219**, 177–189.

Kodaira, S., Sato, T., Takahashi, N., Ito, A., Tamura, Y., Tatsumi, Y., and Kaneda, Y., 2007. Seismological evidence for variable growth of crust along the Izu intraoceanic arc. *Journal of Geophysical Research*, **112**, B05104, doi:10.1029/2006JB004593.

Kodaira, S., Noguchi, N., Takahashi, N., Ishizuka, O., and Kaneda, Y., 2010. Evolution from fore-arc oceanic crust to island arc crust: a seismic study along the Izu-Bonin fore arc. *Journal of Geophysical Research*, **115**, B09102, doi:10.1029/2009JB006968.

Maekawa, H., Shozui, M., Ishii, T., Fryer, P., and Pearce, J. A., 1993. Blueschist metamorphism in an active subduction zone. *Nature*, **364**, 520–523.

Pearce, J. A., van der Laan, S. R., Arculus, R. J., Murton, B. J., Ishii, T., Peate, D. W., and Parkinson, I. J., 1992. Boninite and harzburgite from Leg 125 (Bonin-Mariana forearc): a case study of magma genesis during the initial stages of subduction. In Fryer, P., Pearce, J. A., Stokking, L. B., et al. (eds.), *Proceedings of Ocean Drilling Project Scientific Results, 125*. TX, College Station, pp. 623–659.

Reagan, M. K., Ishizuka, O., Stern, R. J., Kelley, K. A., Ohara, Y., Blichert-Toft, J., Bloomer, S. H., Cash, J., Fryer, P., Hanan, B. B., Hickey-Vargas, R., Ishii, T., Kimura, J., Peate, D. W., Rowe, M. C., and Woods, M., 2010. Fore-arc basalts and subduction initiation in the Izu-Bonin-Mariana system. *Geochemistry, Geophysics, and Geosystems*, **11**, Q03X12, doi:10.1029/2009GC002871.

Stern, R. J., 2004. Subduction initiation: spontaneous and induced. *Earth and Planetary Science Letters*, **226**, 275–292.

Stern, R. J., and Bloomer, S. H., 1992. Subduction zone infancy: examples from the Eocene Izu-Bonin-Mariana and Jurassic California. *Geological Society of America Bulletin*, **104**, 1621–1636.

Suyehiro, K., Takahashi, N., Ariie, Y., Yokoi, Y., Hino, R., Shinohara, M., Kanazawa, T., Hirata, N., Tokuyama, H., and Taira, A., 1996. Continental crust, crustal underplating and low-Q upper mantle beneath an oceanic island arc. *Science*, **272**, 390–392.

Takahashi, N., Suyehiro, K., and Shinohara, M., 1998. Implications from the seismic crustal structure of the northern Izu-Bonin arc. *Island Arc*, **7**, 383–394.

Takahashi, N., Kodaira, S., Tatsumi, Y., Kaneda, Y., and Suyehiro, K., 2008. Structure and growth of the Izu-Bonin-Mariana arc crust: 1. Seismic constraint on crust and mantle structure of the Mariana arc–back-arc system. *Journal of Geophysical Research*, **113**, B01104, doi:10.1029/2007JB005120.

Takahashi, N., Kodaira, S., Tatsumi, Y., Yamashita, M., Sato, T., Kaiho, Y., Miura, S., No, T., Takizawa, K., and Kaneda, Y., 2009. Structural variations of arc crusts and rifted margins in the southern Izu-Ogasawara arc–back arc system. *Geochemistry, Geophysics, and Geosystems*, **10**, Q09X08, doi:10.1029/2008GC002146.

Tatsumi, Y., Shukuno, H., Tani, K., Takahashi, N., Kodaira, S., and Kogiso, T., 2008. Structure and growth of the Izu-Bonin-Mariana arc crust: 2. Role of crust-mantle transformation and the transparent Moho in arc crust evolution. *Journal of Geophysical Research*, **113**, B02203, doi:10.1029/2007JB005121.

Turcotte, D. L., 1989. Geophysical processes influencing the lower continental crust. In Mereu, R. F., et al. (eds.), *Properties and Processes of Earth's Lower Crust*. Washington, DC: AGU, pp. 321–329.

Uyeda, S., and Ben-Avraham, Z., 1972. Origin and development of the Philippine Sea. *Nature*, **240**, 176–178.

Uyeda, S., and Kanamori, H., 1979. Back-arc opening and the mode of subduction. *Journal of Geophysical Research*, **84**, 1049–1061.

Cross-references

Accretionary Wedges
Active Continental Margins
Cumulates
Earthquakes
Gabbro
Hydrothermalism
Island Arc Volcanism, Volcanic Arcs
Magmatism at Convergent Plate Boundaries
Marginal Seas
Mud Volcano
Oceanic Spreading Centers
Ophiolites
Peridotites
Seamounts
Seismogenic Zone
Serpentinization
Subduction
Subduction Erosion
Transform Faults

INTRAPLATE MAGMATISM

Millard F. Coffin and Joanne M. Whittaker
Institute for Marine and Antarctic Studies, University of Tasmania, Hobart, TAS, Australia

Definition

Intraplate magmatism constitutes igneous activity distal from the boundaries of the tectonic plates and is thus considered to be unrelated to the processes of seafloor spreading, subduction, and transform faulting.

Introduction

Intraplate magmatism is pervasive within both oceanic and continental crust. Expansive spatially and across geologic time, it is complex because of both its causative magmatic processes as well as its interplay with Earth's mantle geodynamics and plate tectonic processes. The scale of intraplate magmatism ranges from large igneous provinces (LIPs) encompassing millions of cubic kilometers of igneous rock to small individual volcanoes. Similarly, compositions of intraplate igneous rock span the spectrum of extrusive compositions between highly mafic and highly silicic. However, intraplate volcanic rocks are dominantly basaltic and have compositions distinctly different from those found at "normal" mid-ocean ridges and in arc-trench systems. Herein we focus on intraplate magmatism in the oceans that cover 71 % of Earth's surface. We examine intraplate magmatism both present and past in the oceans, which has occurred predominantly within oceanic crust. The geological record of oceanic crust flooring much of the world's oceans extends back to ~200 million years ago, or slightly less than 5 %, of Earth history.

Tectonic setting

Tectonic setting determines whether or not magmatism is considered intraplate. Although it may appear relatively straightforward to distinguish between intraplate magmatism and plate boundary magmatism, this is certainly not the case. Diffuse zones of plate deformation, forebulges (flexural bends on subducting plates), and zones of intraplate extension and convergence complicate the present-day picture, together with magmatism exploiting ancient plate boundaries and other zones of weakness. Moving back through geologic time, identification of plate boundaries and the times when they were active becomes increasingly more challenging, as does our knowledge of the geochronology of candidate intraplate igneous rocks, so we focus on the present, augmented by our knowledge of the past 200 million years in the oceans.

Currently, 25 tectonic plates collectively occupy 97 % of Earth's surface (DeMets et al., 2010). Boundaries of the majority of these are relatively distinct; however, diffuse boundaries characterize portions of the Capricorn, Caribbean, Lwandle, Macquarie, and Sur plates (Figure 1). Forebulges, in this case flexural lithospheric bulges seaward of subduction zone trenches, are common features throughout the global ocean and are considered to be intraplate rather than part of a plate boundary. Intraplate stress can result in zones of extension and convergence, examples of which are found in the global ocean; these may or may not develop into plate boundaries.

Morphology and characteristics of oceanic intraplate magmatism

The scale of intraplate volcanic features in the oceans ranges over many orders of magnitude, from small volcanoes on the seafloor that rise <100 m above surrounding seafloor to massive oceanic plateaus with areas exceeding 10^6 km^2 and volumes of igneous rock greater than 10^7 km^3. Within this continuum are volcanoes rising hundreds to thousands of meters above the ambient seafloor, linked chains of volcanoes, and submarine ridges, all of which can have subaerial expressions. Below, we examine these various features, including key examples, in more detail.

Individual undersea volcanoes, known as seamounts, sea knolls, and/or guyots (hereafter all referred to as "seamounts"), are among the most ubiquitous landforms on Earth (Wessel et al., 2010) and are found across the entire range of ages of oceanic crust, i.e., from zero age to approaching 200 million years. Excluding subduction-related island arc volcanoes but including volcanoes formed near seafloor spreading centers, Wessel et al. (2010) predict the population of seamounts in the

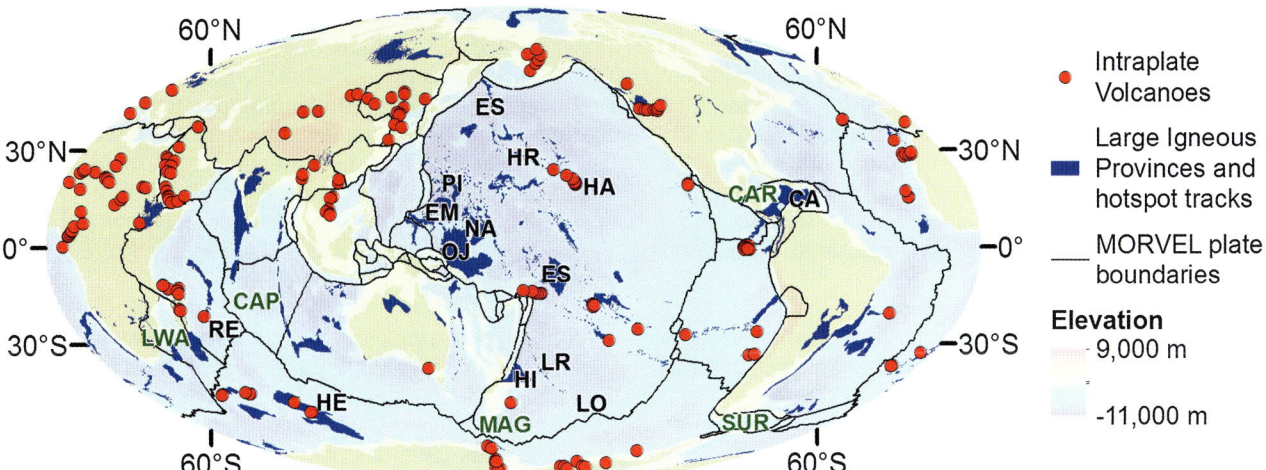

Intraplate Magmatism, Figure 1 Global distribution of intraplate magmatism. Holocene intraplate volcanoes in the oceans and on the continents are depicted in *red* (modified from Siebert et al., 2010). Current "MORVEL" (Mid-Ocean Ridge VELocity) plate boundaries are indicated in *black* (DeMets et al., 2010). Large igneous provinces, mostly younger than 250 million years, but some of which were emplaced at plate boundaries, are shown in *blue* (Coffin and Eldholm, 1994). Intraplate magmatism abbreviations: *CA* Caribbean, *EM* East Mariana, *ES* Emperor Seamount Chain, *HA* Hawaii (island), *HR* Hawaiian Ridge, *HE* Heard Island, *HI* Hikurangi, *LO* Louisville, *LR* Louisville Ridge, *MA* Manihiki, *NA* Nauru, *OJ* Ontong Java, *PI* Pigafetta, *RE* Réunion Island. Plate abbreviations: *CAP* Capricorn, *CAR* Caribbean, *LWA* Lwandle, *MAC* Macquarie, *SUR* Sur.

global ocean >1,000 m in height (above surrounding seafloor) to be ~125,000, but it could range from 45,000 to 350,000. For seamounts >100 m in height, they predict a population of 25,000,000, but it could range from 8,000,000 to 80,000,000. However, only a minuscule fraction of these features have been sampled and dated (e.g., the online Seamount Catalog (http://earthref.org/SC/) includes approximately 1,800 seamounts).

The most comprehensive database of volcanoes active during the past ~10,000 years (Holocene) includes 1,559 volcanoes, of which 215 are classified as intraplate (Siebert et al., 2010; http://www.volcano.si.edu/). However, 25 of these are situated on or near plate boundaries (DeMets et al., 2010). Of the remaining 190 considered intraplate, 47 are in the ocean (Figure 1). This number is similar to estimates of active hot spots in the ocean (e.g., Courtillot et al., 2003). While it is likely that most, if not all, active volcanoes that form islands in the ocean are included in the Holocene database, volcanic islands constitute only a small fraction of volcanic features in the ocean. Among the most active intraplate volcanoes on Earth currently are the islands of Hawaii in the Pacific Ocean, Réunion in the Indian Ocean, and Heard in the Southern Ocean (Figure 1).

Hawaii, on which Mauna Kea rises >9,000 m above surrounding Pacific Ocean seafloor forming Earth's tallest volcano and mountain (Figure 1), is the largest volcano in what is arguably the best-studied intraplate volcanic chain – the Hawaiian-Emperor hot spot track – on the planet. Analyses of extensive subaerial and submarine surficial lavas on Hawaii, and of buried lavas obtained by the Hawaii Scientific Drilling Project, indicate a shield-building phase of dominantly tholeiitic magmas and high eruption rates followed by a major decrease in eruption rates coinciding with a shift to alkaline magmas (e.g., Clague and Dalrymple, 1987; Stolper et al., 2009). Tholeiitic shield volcanism followed by alkaline post-shield volcanism is a general characteristic of many intraplate volcanoes. Hawaiian volcanoes form two parallel geographic chains with distinct isotopic compositions (Loa and Kea trends) believed to reflect distinct mantle sources for the two chains.

Piton de la Fournaise on Réunion ascends >7,000 m above adjacent Indian Ocean seafloor (Figure 1). Although it shares many characteristics with Hawaii, it displays both early and waning stages of alkaline volcanism bracketing a steady-state stage of tholeiitic activity (e.g., Albarède et al., 1997). The relatively simple elemental and isotopic geochemistry of Piton de la Fournaise lavas suggests a rather homogeneous mantle source.

Heard surmounts the Kerguelen Plateau in the Southern Ocean (Figure 1), and the Big Ben massif reaches a maximum elevation of 2,745 m above sea-level at Mawson Peak. Because the island is 80 % glaciated, knowledge of Heard Island lava petrology and geochemistry is quite limited compared to Hawaii and Réunion. However, the elemental and isotopic geochemistry of Heard rocks studied to date displays among the largest range of compositions of any oceanic island lavas (e.g., Barling et al., 1994). This heterogeneity is attributed to continent-derived material in the mantle source, shallow-level contamination of magmas ascending through continental crust of the Kerguelen Plateau, or both (Barling et al., 1994).

A small subset of seamounts are sea knolls confined to the forebulge of oceanic plates seaward of trenches; these have been termed petit spots (e.g., Hirano et al., 2001). These features rise 100–200 m above surrounding seafloor, with a volume typically <1 km^3. Lavas from petit spots are alkaline, and their elemental and isotopic compositions suggest that the magmas originate in the asthenosphere (Hirano et al., 2006).

Little differentiates age-progressive chains of volcanoes from submarine ridges, e.g., the Hawaiian-Emperor seamount chain consists of the younger Hawaiian Ridge and the older Emperor seamount chain (Figure 1). Similarly, an age-progressive chain of seamounts in the South Pacific is known as the Louisville Ridge (Figure 1). While both the Hawaiian-Emperor seamount chain and Louisville Ridge are intraplate, the tectonic settings of emplacement of other prominent age-progressive submarine ridges (e.g., Walvis/Rio Grande and Greenland-Iceland-Faroes in the Atlantic Ocean, Ninetyeast and Chagos-Laccadive/Mascarene in the Indian Ocean) appear to be on, or proximal to, plate boundaries. The crustal nature – oceanic or continental – of still other submarine ridges (e.g., Alpha-Mendeleev in the Arctic Ocean) remains controversial.

The archetypal Hawaiian-Emperor chain of volcanoes extends for more than 5,800 km from the active intraplate submarine volcano of Loihi southeast of the island of Hawaii to Meiji Seamount on the outer slope of the Kuril-Kamchatka Trench in the northwest (Figure 1). The chain includes more than 80 islands, atolls, reefs, banks, and seamounts. Overall, lavas from the Hawaiian-Emperor chain show significant heterogeneity, including tholeiitic and alkalic basalts and their differentiates, and silica-undersaturated lavas (e.g., Regelous et al., 2003).

In contrast to the Hawaiian-Emperor chain, lavas of the entirely submarine Louisville seamount chain ("Louisville Ridge," Figure 1) are relatively homogeneous. Only alkaline rocks have been dredged and drilled from Louisville seamounts, and it may be that Louisville shield volcanoes are produced by massive alkaline volcanism, as opposed to the tholeiitic shield volcanism characterizing Hawaiian-Emperor chain volcanoes (e.g., Koppers et al., 2013).

On the massive end of the scale of intraplate magmatism lie some oceanic plateaus, one category of large igneous province (LIP, e.g., Coffin and Eldholm, 1994). Oceanic plateaus constitute a subset of submarine plateaus; the latter may be continental (e.g., Campbell), oceanic (e.g., Ontong Java), or hybrid (e.g., Kerguelen) in crustal nature. The sole oceanic plateau forming today is Iceland astride the Mid-Atlantic Ridge, so it does not meet the intraplate criterion. Some ancient oceanic plateaus, e.g., Shatsky in the northwest Pacific Ocean and the conjugate Ceara/Sierra Leone provinces in the central Atlantic Ocean, also formed at or near plate boundaries (e.g., Seton et al., 2012).

The volcanic rocks of many oceanic and hybrid plateaus formed during the Cretaceous Normal Superchron (~121 to ~83 million years ago), which makes identification of tectonic setting challenging due to the absence of marine magnetic anomalies in adjacent ocean basins. These Cretaceous oceanic – Caribbean, Conrad, Crozet, East Mariana, Hess, Hikurangi, Manihiki, Nauru, Ontong Java, Pigafetta – and hybrid – Broken, Kerguelen, Madagascar, Naturaliste, Wallaby – plateaus may have formed proximal or distal to plate boundaries.

Among the best studied of these plateaus, i.e., those with multiple samples of igneous basement rocks, are greater Ontong Java (Ontong Java, East Mariana, Nauru, Pigafetta), the conjugate Manihiki/Hikurangi, and the Caribbean (Figure 1). Lying in the western equatorial Pacific Ocean, the greater Ontong Java Plateau, including the Plateau proper as well as the neighboring East Mariana, Nauru, and Pigafetta basin flood basalts, encompasses ~4.27 × 10^6 km^2, or ~0.8 %, of the Earth's surface (Ingle and Coffin, 2004) (Figure 1). Emplaced ~120 million years ago (e.g., Neal et al., 1997; Fitton et al., 2004), the Ontong Java Plateau appears to have formed within ~130- to ~160-million-year-old oceanic lithosphere (e.g., Ishikawa et al., 2005), thus qualifying as intraplate magmatism. The elemental and isotopic geochemistry of greater Ontong Java Plateau tholeiitic lavas is remarkably homogeneous, suggesting a homogeneous mantle source (e.g., Fitton et al., 2004).

The conjugate Manihiki and Hikurangi plateaus in the southwest Pacific Ocean (Figure 1) also formed ~120 million years ago (e.g., Hoernle et al., 2010; Timm et al., 2011), and it has been proposed that they formed adjacent to the Ontong Java Plateau (Taylor, 2006). Plate tectonic reconstructions suggest a plate boundary proximal to Manihiki and Hikurangi when they formed (e.g., Seton et al., 2012), casting some doubt as to whether they represent intraplate magmatism. Nevertheless, tholeiitic lavas of igneous basement of both plateaus show strong elemental and isotopic geochemical similarities to Ontong Java basalts, suggesting a common mantle source (Hoernle et al., 2010).

The Caribbean flood basalt province (Figure 1), consisting of basalts, picrites, and komatiites, formed over a relatively long time span from 139 to 69 million years ago, with a major portion emplaced ~90 million years ago and a secondary peak of volcanism at ~75 million years ago (e.g., Kerr et al., 1997; Sinton et al., 1998; Hoernle et al., 2004). Elemental and isotopic geochemistry shows that the two major pulses are largely compositionally distinct, indicating at least two mantle sources (Hoernle et al., 2004).

Causes of intraplate magmatism

The concept of intraplate magmatism developed at the same time as the plate tectonic paradigm. Such magmatism requires an explanation other than the plate tectonic processes of seafloor spreading and subduction.

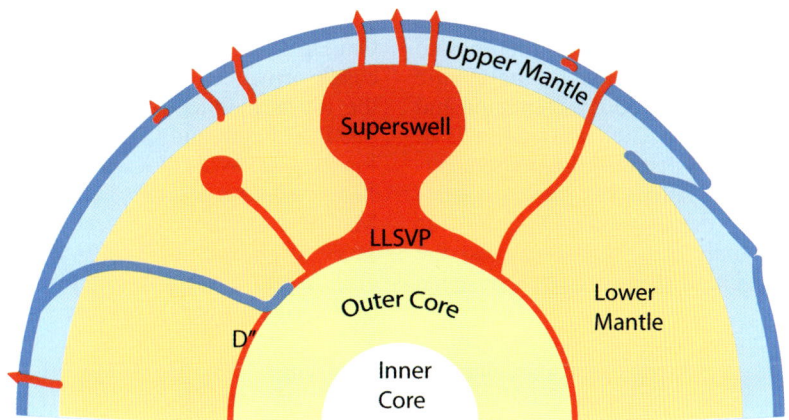

Intraplate Magmatism, Figure 2 Cartoon depicting source regions of intraplate magmatism: (i) core-mantle boundary/D"/LLSVP, (ii) lower mantle-upper mantle boundary/670-km discontinuity, and (iii) upper mantle. LLSVP, large low shear velocity provinces (after Coffin and Eldholm 1994 and Courtillot et al. 2003).

Age-progressive volcanism along the Hawaiian island chain (McDougall, 1964) provided the first insights into potential causes of intraplate magmatism. The age of volcanism systematically increases to the northwest from the island of Hawaii, a "hot spot" generating significant magmatism that is distal from any current plate boundary (Wilson, 1963, 1965). Another explanation for intraplate magmatism is mantle ascending and partially melting via fissures and faults where the lithosphere is extending (e.g., Turcotte and Oxburgh, 1973; Anderson, 1995). The trend of the Hawaiian-Emperor seamount chain, however, is oblique to preexisting fracture zones, which show no evidence for extension.

Observations of multiple age-progressive volcanic island chains led to the interpretation that hot spots are surface expressions of deep mantle plumes (Morgan, 1971, 1972), to which large igneous provinces have also been attributed, both on the continents (e.g., Richards et al., 1989) and in the oceans (e.g., Coffin and Eldholm, 1994). Subsequently, other interpretations of hot spot islands arose, e.g., channeled asthenospheric flow from a hot spot to a seafloor spreading center (Morgan, 1978; Sleep, 2002). Estimates of the depth of the source of hot spots vary considerably, from the core-mantle boundary to the upper mantle (Figure 2; e.g., Anderson, 1998; Clouard and Bonneville, 2001; Courtillot et al., 2003), with Morgan's (1971, 1972) original deep mantle interpretation being the deep end-member. Buoyancy fluxes of hot spots also vary considerably – by a factor of 20 – with Hawaii being the largest (Sleep, 1990).

Hot spots can be categorized into three different types, depending on which mantle boundary layer is the source, according to a set of five criteria (Courtillot et al., 2003). The five criteria are (1) long-lived tracks, (2) flood basalts at their initiation, (3) large buoyancy flux, (4) high He or Ne ratio, and (5) underlying low shear velocities. In a constellation of 49 active hot spots globally, seven meeting at least three of these criteria were proposed to originate deep in the mantle (core-mantle boundary): Hawaii, Easter, and Louisville in the Pacific hemisphere and Iceland, Afar, Réunion, and Tristan in the Indo-Atlantic hemisphere. Five of these "primary" hot spots are intraplate, with Iceland and Afar being the exceptions. Another ~20 "secondary" hot spots, including Caroline, McDonald, Pitcairn, Samoa, and Tahiti, may originate at the boundary between the upper and lower mantle, the 670-km discontinuity. The remaining ~20 may originate entirely within the upper mantle, possibly passive responses to lithospheric extension (e.g., Natland and Winterer, 2005; Foulger, 2007). In alignment with the different types of hot spots, finite-frequency tomography of the Earth's mantle reveals plumes traceable to different depths in the mantle (Montelli et al., 2004, 2006), although discrepancies exist between these and the Courtillot et al. (2003) results.

Conceptually, mantle plumes are columns of hot, solid material that ascend buoyantly from the core-mantle boundary or the 670-km discontinuity toward the Earth's surface (Figure 2; e.g., Campbell and Griffiths, 1990; Griffiths and Campbell, 1990, 1991; Farnetani and Richards, 1994; Campbell, 2005). New plumes consist of a large head followed by a small tail. When the head of a plume reaches the base of the lithosphere, it flattens and spreads laterally to a diameter exceeding 1,000 km. The temperature excess of a plume head is highest at the center of the head and decreases toward the margin. Both plume heads and plume tails should produce high-temperature picrites, which erupt early and be most abundant near the center of the plume head and less abundant toward the margin. Flood volcanism resulting from decompressional melting should be preceded by domal uplift of hundreds of meters at the center of the dome. The lithosphere acts as a lid on the ascending plume and the extent of decompressional melting, with the thickness of the lithosphere being a key factor in controlling the amount of melt produced.

The concept of predominantly thermal mantle plumes has evolved to thermochemical plumes that incorporate both thermal and chemical heterogeneities in the mantle (e.g., Davaille et al., 2002, 2003, 2005; Farnetani and Samuel, 2005; Lin and van Keken, 2006a, b). Models of thermochemical plumes suggest that upwelling characteristics depend on the buoyancy ratio, which is the ratio of the stabilizing chemical density anomaly to the destabilizing thermal anomaly. Low ratios suggest the absence of a plume tail, and high ratios suggest long-lived hot spot tracks.

In contrast to solely deep Earth processes accounting for intraplate magmatism, an exogenous cause, namely, impacts of large bolides, has also been postulated to instigate such magmatism (e.g., Rogers, 1982; Glikson, 1999; Abbott and Isley, 2002; Jones et al., 2002, 2005; Muller, 2002; Ingle and Coffin, 2004). Large impacts in the ocean basins can excavate some or all of the lithosphere and potentially the uppermost portions of the asthenosphere, thereby inducing voluminous decompressional melting of mantle material. Oblique impacts may impart high shear to the core-mantle boundary, triggering one or more mantle avalanches and mantle plumes.

Thus, whether caused by endogenous or exogenous processes, and whether originating at the core-mantle boundary, the 670-km discontinuity, or within the upper mantle, intraplate magmatism results from decompressional melting of mantle material as it ascends toward the Earth's crust through the asthenosphere and lithosphere.

Nature and source of intraplate magmatism type examples

The three highly active intraplate hot spots described earlier – Hawaii, Réunion, and Heard islands (Figure 1) – are classified by Courtillot et al. (2003) either as originating (Hawaii, Réunion) or possibly originating, with more data (Heard), at the core-mantle boundary (Figure 2). By implication, this implies that the entire Hawaiian-Emperor chain (Figure 1) was created by a hot spot with its origin at the core-mantle boundary. Louisville Ridge (Figure 1) is classified similarly.

In stark contrast, petit spots appear to be sourced entirely within the upper mantle (Figure 2). They form where the outer rise of a subducting plate flexes and develops normal faults, providing pathways for mantle to ascend and melt via decompression (e.g., Hirano et al., 2008). Melt volumes are orders of magnitude less than hot spots such as Hawaii, Réunion, and Heard.

The upper mantle is also likely to be the source for the vast majority of the tens of millions of intraplate seamounts rising >100 m above surrounding seafloor that aren't petit spots (Figure 2). Fissures and faults allow anomalously warm and/or wet mantle to ascend and melt via decompression, producing these features (e.g., Natland and Winterer, 2005). Faults may also influence where seamounts form that possibly originate from deeper in the mantle (e.g., McNutt et al., 1989).

The sheer volumes of igneous rock comprising the Ontong Java, Manihiki, Hikurangi, and Caribbean flood basalt provinces (Figure 1) would suggest an origin in the lower mantle (Figure 2; e.g., Coffin and Eldholm, 1994; Eldholm and Coffin, 2000). Evidence is equivocal for the Louisville hot spot being the source of Ontong Java, Manihiki, and Hikurangi flood basalts (e.g., Mahoney et al., 1993; Neal et al., 1997; Richards et al., 1989; Riisager et al., 2003; Antretter et al., 2004; Chandler et al., 2012). The Caribbean flood basalts have been linked to the Galapagos hot spot (e.g., Duncan and Hargraves, 1984; Kerr et al., 1997; Sinton et al., 1998), which has been classified as possibly originating, with more data, at the core-mantle boundary (Figure 2; Courtillot et al., 2003).

Models for the formation of the Ontong Java Plateau include various plume incarnations (e.g., Bercovici and Mahoney, 1994; Farnetani and Richards, 1994; Larson, 1997; Neal et al., 1997; Ito and Clift, 1998; Tejada et al., 2004; Hoernle et al., 2010), bolide impact (e.g., Rogers, 1982; Jones et al., 2002, 2005; Ingle and Coffin, 2004), and upwelling eclogite (Korenaga, 2005). None of the models to date satisfy all of the observations (e.g., Kerr and Mahoney, 2007). Plume models have also been invoked for the Manihiki and Hikurangi plateaus (e.g., Larson, 1997; Ito and Clift, 1998; Hoernle et al., 2010). In contrast to Ontong Java, plume models seem to satisfy most observations from these two plateaus. The dismembered nature of the Caribbean flood basalt province means that it doesn't lend itself readily to modeling; however, to date only the plume model has been invoked for its formation (e.g., Duncan and Richards, 1991; Kerr et al., 1997).

Summary

Intraplate magmatism in the ocean occurs over many orders of magnitude in scale, and the predominant volcanic rock type is basalt. Type examples of robust active intraplate magmatism – the islands of Hawaii, Réunion, and Heard, all probably sourced from the core-mantle boundary – show a general pattern of alkaline pre-shield, tholeiitic shield, and alkaline post-shield volcanism. Classic seamount chains such as the Hawaiian-Emperor and Louisville, also probably sourced from the core-mantle boundary, display differing degrees of elemental and isotopic compositional variations, with the Hawaiian-Emperor chain quite heterogeneous and Louisville rather homogeneous. The massive oceanic plateaus (Ontong Java, Manihiki, Hikurangi) and ocean basin flood basalts (Caribbean, East Mariana, Nauru, Pigafetta), all likely sourced from the core-mantle boundary, although perhaps instigated by exogenous mechanisms, range from homogeneous (e.g., Ontong Java) to heterogeneous (e.g., Caribbean) in their elemental and isotopic compositions. Finally, the tens of millions of seamounts, including petit spots, that rise >100 m above surrounding seafloor probably originate overwhelmingly in the upper mantle and mostly exploit fissures and faults in the oceanic lithosphere.

Bibliography

Abbott, D. H., and Isley, A. E., 2002. Extraterrestrial influences on mantle plumes. *Earth and Planetary Science Letters*, **205**, 53–62.

Albarède, F., Luais, B., Fitton, G., Semet, M., Kaminski, E., Upton, B. G. J., Bachèlery, P., and Cheminée, J.-L., 1997. The geochemical regimes of Piton de la Fournaise volcano (Réunion) during the last 530,000 years. *Journal of Petrology*, **38**(2), 171–201.

Anderson, D. L., 1995. Lithosphere, asthenosphere, and perisphere. *Reviews of Geophysics*, **33**, 125–149.

Anderson, D. L., 1998. The scales of mantle convection. *Tectonophysics*, **284**, 1–17.

Antretter, M., Riisager, P., Hall, S., Zhao, X., and Steinberger, B., 2004. Modelled paleolatitudes for the Louisville hot spot and the Ontong Java Plateau. In Fitton, J. G., Mahoney, J. J., Wallace, P. J., and Saunders, A. D. (eds.), *Origin and Evolution of the Ontong Java Plateau*. London: Geological Society. Special publication, 229, pp. 21–30.

Barling, J., Goldstein, S. L., and Nicholls, I. A., 1994. Geochemistry of Heard Island (southern Indian Ocean): characterization of an enriched mantle component and implications for enrichment of sub-Indian ocean mantle. *Journal of Petrology*, **35**(4), 1017–1053.

Bercovici, D., and Mahoney, J., 1994. Double flood basalts and plume head separation at the 660-kilometer discontinuity. *Science*, **266**, 1367–1369.

Campbell, I. H., 2005. Large igneous provinces and the mantle plume hypothesis. *Elements*, **1**, 265–269.

Campbell, I. H., and Griffiths, R. W., 1990. Implications of mantle plume structure for the evolution of flood basalts. *Earth and Planetary Science Letters*, **99**, 79–93.

Chandler, M. T., Wessel, P., Taylor, B., Seton, M., Kim, S.-S., and Hyeong, K., 2012. Reconstructing Ontong Java Nui: Implications for Pacific absolute plate motion, hotspot drift and true polar wander. *Earth and Planetary Science Letters*, **331–332**, 140–151.

Clague, D. A., and Dalrymple, G. B., 1987. The Hawaiian-Emperor volcanic chain, Part 1: geological evolution. *United States Geological Survey Professional Paper*, **1350-1**, 5–54.

Clouard, V., and Bonneville, A., 2001. How many Pacific hotspots are fed by deep-mantle plumes? *Geology*, **29**, 695–698.

Coffin, M. F., and Eldholm, O., 1994. Large igneous provinces: crustal structure, dimensions, and external consequences. *Reviews of Geophysics*, **32**, 1–36.

Courtillot, V., Davaille, A., Besse, J., and Stock, J., 2003. Three distinct types of hotspots in the Earth's mantle. *Earth and Planetary Science Letters*, **205**, 295–308.

Davaille, A., Girard, F., and Le Bars, M., 2002. How to anchor hotspots in a convecting mantle? *Earth and Planetary Science Letters*, **203**, 621–634.

Davaille, A., Le Bars, M., and Carbonne, C., 2003. Thermal convection in a heterogeneous mantle. *Comptes Rendus Geoscience*, **335**, 141–156.

Davaille, A., Stutzmann, E., Silveira, G., Besse, J., and Courtillot, V., 2005. Convective patterns under the Indo-Atlantic <<box>> *Earth and Planetary Science Letters*, **239**, 233–252.

DeMets, C., Gordon, R. G., and Argus, D. F., 2010. Geologically current plate motions. *Geophysical Journal International*, **181**, 1–80, doi:10.1111/j.1365-246X2009.04491.x.

Duncan, R. A., and Hargraves, R. B., 1984. Plate tectonic evolution of the Caribbean region in the mantle reference frame. In Bonini, W. E., Hargraves, R. B., and Shagam, R. (eds.), *The Caribbean-South American Plate Boundary and Regional Tectonics*. Geological Society of America Memoir, Geological Society of America, Denver, Colorado, USA, Vol. 162, pp. 81–93.

Duncan, R.A., and Richards, M.A., 1991. Hotspots, mantle plumes, flood basalts, and true polar wander. *Reviews of Geophysics*, **29**, 31–50.

Eldholm, O., and Coffin, M. F., 2000. Large igneous provinces and plate tectonics. In Richards, M. A., Gordon, R. G., and van der Hilst, R. D. (eds.), *The History and Dynamics of Global Plate Motions*. American Geophysical Union Geophysical Monograph, American Geophysical Union, Washington, DC, USA, Vol. 121, pp. 309–326.

Farnetani, C. G., and Richards, M. A., 1994. Numerical investigations of the mantle plume initiation model for flood basalt events. *Journal of Geophysical Research*, **99**(B7), 13813–13833.

Farnetani, C.G., and Samuel, H., 2005. Beyond the thermal plume paradigm. *Geophysical Research Letters*, **32**, doi: 10.1029/2005GL022360

Fitton, J. G., Mahoney, J. J., Wallace, P. J., and Saunders, A. D., 2004. Origin and evolution of the Ontong Java Plateau: introduction. In Fitton, J. G., Mahoney, J. J., Wallace, P. J., and Saunders, A. D. (eds.), *Origin and Evolution of the Ontong Java Plateau*. London: Geological Society. Special publication, 229, pp. 1–8.

Foulger, G.R., 2007. The "plate" model for the genesis of melting anomalies. In Foulger, G. R., and Jurdy, D. M. (eds.), *Plates, Plumes, and Planetary Processes*. Geological Society of America Special paper, Geological Society of America, Denver, Colorado, USA, 430, 1–28

Glikson, A. Y., 1999. Oceanic mega-impacts and crustal evolution. *Geology*, **27**(5), 387–390.

Griffiths, R. W., and Campbell, I. H., 1990. Stirring and structure in mantle starting plumes. *Earth and Planetary Science Letters*, **99**, 66–78.

Griffiths, R. W., and Campbell, I. H., 1991. Interaction of mantle plume heads with the Earth's surface and onset of small-scale convection. *Journal of Geophysical Research*, **96**(B11), 18295–18310.

Hirano, N., Kawamura, K., Hattori, M., Saito, K., and Ogawa, Y., 2001. A new type of intra-plate volcanism; young alkali-basalts discovered from the subducting Pacific plate, northern Japan Trench. *Geophysical Research Letters*, **28**, 2719–2722.

Hirano, N., Takahashi, E., Yamamoto, J., Abe, N., Ingle, S. P., Kaneoka, I., Hirata, T., Kimura, J.-I., Ishii, T., Ogawa, Y., Machida, S., and Suyehiro, K., 2006. Volcanism in response to plate flexure. *Science*, **313**, doi: 10.1126/science.1128235.

Hirano, N., Koppers, A. A. P., Takahashi, A., Fujiwara, T., and Nakanishi, M., 2008. Seamounts, knolls and petit-spot monogenetic volcanoes on the subducting Pacific plate. *Basin Research*, **20**, 543–553.

Hoernle, K., Hauff, F., and van den Bogaard, P., 2004. 70 m.y. history (139–69 Ma) for the Caribbean large igneous province. *Geology*, **32**, 697–700.

Hoernle, J., Hauff, F., van den Bogaard, P., Werner, R., Mortimer, J., Garbe-Schönberg, D., and Davy, B., 2010. Age and geochemistry of volcanic rocks from the Hikurangi and Manihiki oceanic Plateaus. *Geochimica et Cosmochimica Acta*, **74**, 7196–7219.

Ingle, S., and Coffin, M. F., 2004. Impact origin for the greater Ontong Java Plateau? *Earth and Planetary Science Letters*, **218**, 123–134.

Ishikawa, A., Nakamura, E., and Mahoney, J. J., 2005. Jurassic oceanic lithosphere beneath the southern Ontong Java Plateau: evidence from xenoliths in alnöite, Malaita, Solomon Islands. *Geology*, **33**(5), 393–396.

Ito, G., and Clift, P. D., 1998. Subsidence and growth of Pacific Cretaceous plateaus. *Earth and Planetary Science Letters*, **161**, 85–100.

Jones, A. P., Price, G. D., Price, N. J., DeCarli, P. S., and Clegg, R. A., 2002. Impact induced melting and the development of large igneous provinces. *Earth and Planetary Science Letters*, **202**, 551–561.

Jones, A. P., Wünemann, K., and Price, G. D., 2005. Modeling impact volcanism as a possible origin for the Ontong Java Plateau. In Foulger, G. R., Natland, J. H., Presnall, D. C., and Anderson, D. L. (eds.), *Plates, Plumes, and Paradigms*. Geological Society of America. Special paper, Geological Society of America, Denver, Colorado, USA, Vol. 388, pp. 711–720.

Kerr, A. C., and Mahoney, J. J., 2007. Oceanic plateaus: problematic plumes, potential paradigms. *Chemical Geology*, **241**, 332–353.

Kerr, A. C., Tarney, J., Marriner, G. F., Nivia, A., and Saunders, A. D., 1997. The Caribbean-Colombian cretaceous igneous province: the internal anatomy of an oceanic plateau. In Mahoney, J. J., and Coffin, M. F. (eds.), *Large Igneous Provinces: Continental, Oceanic, and Planetary Flood Volcanism*. American Geophysical Union Geophysical Monograph, American Geophysical Union, Washington, DC, USA, Vol. 100, pp. 123–144.

Koppers, A. A. P., Yamazaki, T., Geldmacher, J., and The IODP Expedition 330 Scientific Party, 2013. IODP expedition 330: drilling the Louisville seamount trail in the SW Pacific. *Scientific Drilling*, **15**, doi: 10.2204/iodp.sd.15.02.2013.

Korenaga, J., 2005. Why did not the Ontong Java Plateau form subaerially? *Earth and Planetary Science Letters*, **234**, 385–399.

Larson, R. L., 1997. Superplumes and ridge interactions between Ontong Java and Manihiki Plateaus and the Nova-Canton Trough. *Geology*, **25**, 779–782.

Lin, S.-C., and van Keken, P. E., 2006a. Dynamics of thermochemical plumes: 1. Plume formation and entrainment of a dense layer. *Geochemistry, Geophysics, Geosystems*, **7**(2), doi:10.1029/2005GC001071.

Lin, S.-C., and van Keken, P. E., 2006b. Dynamics of thermochemical plumes: 2. Complexity of plume structures and its implications for mapping mantle plumes. *Geochemistry, Geophysics, Geosystems*, **7**(2), doi:10.1029/2005GC001072.

Mahoney, J. J., Storey, M., Duncan, R. A., Spencer, K. J., and Pringle, M. S., 1993. Geochemistry and age of the Ontong Java Plateau. In Pringle, M. S., Sager, W. W., Sliter, W. V., and Stein, S. (eds.), *The Mesozoic Pacific: Geology, Tectonics, and Volcanism*. American Geophysical Union Geophysical Monograph, American Geophysical Union, Washington, DC, USA, Vol. 77, pp. 233–262.

McDougall, I., 1964. Potassium-Argon Ages from Lavas of the Hawaiian Islands. *Geological Society of America Bulletin*, **75**, 107–128.

McNutt, M., Fischer, K., Kruse, S., and Natland, J., 1989. The origin of the Marquesas fracture zone ridge and its implications for the nature of hot spots. *Earth and Planetary Science Letters*, **91**, 381–393.

Montelli, R., Nolet, G., Dahlen, F. A., Master, G., Engdahl, E. R., and Hung, S.-H., 2004. Finite-frequency tomography reveals a variety of plumes in the mantle. *Science*, **303**, 338–343.

Montelli, R., Nolet, G., Dahlen, F. A., and Masters, G., 2006. A catalogue of deep mantle plumes: New results from finite-frequency tomography. *Geochemistry, Geophysics, Geosystems*, **7**(11), doi:10.1029/2006GC0012.

Morgan, W. J., 1971. Convection plumes in the lower mantle. *Nature*, **230**, 42–43.

Morgan, W. J., 1972. Deep mantle convection plumes and plate motions. *American Association of Petroleum Geologists Bulletin*, **56**(2), 203–213.

Morgan, W. J., 1978. Rodriguez, Darwin, Amsterdam, a second type of hotspot island. *Journal of Geophysical Research*, **83**(B11), 5355–5360.

Muller, R. A., 2002. Avalanches at the core-mantle boundary. *Geophysical Research Letters*, **29**, doi:10.1029/2002GL015938.

Natland, J. H., and Winterer, E. L., 2005. Fissure control on volcanic action in the Pacific. In Foulger, G. R., Natland, J. H., Presnall, D. C., and Anderson, D. L. (eds.), *Plates, Plumes, and Paradigms*. Geological Society of America. Special paper, Geological Society of America, Denver, Colorado, USA, Vol. 388, pp. 687–710.

Neal, C. R., Mahoney, J. J., Kroenke, L. W., Duncan, R. A., and Petterson, M. G., 1997. The Ontong Java Plateau. In Mahoney, J. J., and Coffin, M. F. (eds.), *Large Igneous Provinces: Continental, Oceanic, and Planetary Flood Volcanism*. American Geophysical Union Geophysical Monograph, American Geophysical Union, Washington, DC, USA, Vol. 100, pp. 183–216.

Regelous, M., Hofmann, A. W., Abouchami, W., and Galer, S. J. G., 2003. Geochemistry of lavas from the Emperor Seamounts, and the geochemical evolution of Hawaiian magmatism from 85 to 42 Ma. *Journal of Petrology*, **44**, 113–140.

Richards, M. A., Duncan, R. A., and Courtillot, V. E., 1989. Flood basalts and hot-spot tracks: plume heads and tails. *Science*, **246**, 103–107.

Riisager, P., Hall, S., Antretter, M., and Zhao, X., 2003. Paleomagnetic paleolatitude of Early Cretaceous Ontong Java Plateau basalts: implications for Pacific apparent and true polar wander. *Earth and Planetary Science Letters*, **208**, 235–252.

Rogers, G. C., 1982. Oceanic plateaus as meteorite impact signatures. *Nature*, **299**, 341–342.

Seton, M., Müller, R. D., Zahirovic, S., Gaina, C., Torsvik, T., Shephard, G., Talsma, A., Gurnis, M., Turner, M., Maus, S., and Chandler, M., 2012. Global continental and ocean basin reconstructions since 200 Ma. *Earth-Science Reviews*, **113**, 212–270.

Siebert, L., Simkin, T., and Kimberley, P., 2010. *Volcanoes of the world*, 3rd edn. Berkeley: University of California Press. 568 pp.

Sinton, C. W., Duncan, R. A., Storey, M., Lewis, J., and Estrada, J. J., 1998. An oceanic flood basalt province within the Caribbean plate. *Earth and Planetary Science Letters*, **155**, 221–235.

Sleep, N. H., 1990. Hotspots and mantle plumes: some phenomenology. *Journal of Geophysical Research*, **95**(B5), 6715–6736.

Sleep, N. H., 2002. Ridge-crossing mantle plumes and gaps in tracks. *Geochemistry, Geophysics, Geosystems*, **3**(12), doi:10.1029/2001GC000290.

Stolper, E. M., DePaolo, D. J., and Thomas, D. M., 2009. Deep drilling into a mantle plume volcano: The Hawaii scientific drilling project. *Scientific Drilling*, **7**, doi: 10.2204/iodp.sd.7.02.2009.

Taylor, B., 2006. The single largest oceanic plateau: Ontong Java-Manihiki-Hikurangi. *Earth and Planetary Science Letters*, **241**, 372–380.

Tejada, M. L. G., Mahoney, J. J., Castillo, P. R., Ingle, S. P., Sheth, H. C., and Weis, D., 2004. Pin-pricking the elephant: evidence on the origin of the Ontong Java Plateau from Pb-Sr-Hf-Nd isotopic characteristics of ODP Leg 192 basalts. In Fitton, J. G., Mahoney, J. J., Wallace, P. J., and Saunders, A. D. (eds.), *Origin and Evolution of the Ontong Java Plateau*. London: Geological Society. Special publication, 229, pp. 133–150.

Timm, C., Hoernle, K., Werner, R., Hauff, F., van den Bogaard, P., Michael, P., Coffin, M. F., and Koppers, A., 2011. Age and geochemistry of the oceanic Manihiki Plateau, SW Pacific: new evidence for a plume origin. *Earth and Planetary Science Letters*, **304**, 135–146.

Turcotte, D. L., and Oxburgh, E. R., 1973. Mid-plate tectonics. *Nature*, **244**, 337–339.

Wessel, P., Sandwell, D. T., and Kim, S.-S., 2010. The global seamount census. *Oceanography*, **23**(1), 24–33.

Wilson, J. T., 1963. A possible origin of the Hawaiian Islands. *Canadian Journal of Physics*, **41**, 863–870.

Wilson, J. T., 1965. Evidence from oceanic islands suggesting movement in the Earth. *Philosophical Transactions of the Royal Society of London, Mathematical, Physical and Engineering Sciences*, **A258**, 145–167.

Cross-references

Active Continental Margins
Anoxic Oceans
Black and White Smokers
Chemosynthetic Life
Events
Explosive Volcanism in the Deep Sea
Gravity Field
Guyot, Atoll
Hot Spots and Mantle Plumes
Hydrothermal Plumes
Hydrothermal Vent Fluids (Seafloor)
Hydrothermalism
Island Arc Volcanism, Volcanic Arcs
Magmatism at Convergent Plate Boundaries
Marginal Seas
Marine Impacts and Their Consequences
Mid-ocean Ridge Magmatism and Volcanism
Ocean Drilling
Oceanic Plateaus
Oceanic Rifts
Oceanic Spreading Centers
Ophiolites
Paleophysiography of Ocean Basins
Plate Motion
Plate Tectonics
Pull-apart Basins
Reflection/Refraction Seismology
Subduction
Submarine Lava Types
Transform Faults
Triple Junctions
Volcanism and Climate
Volcanogenic Massive Sulfides
Wilson Cycle

ISLAND ARC VOLCANISM, VOLCANIC ARCS

Jim Gill
Earth & Planetary Sciences, University of California,
Santa Cruz, CA, USA

Synonyms

Arc magmatism; Magmatism at convergent plate boundaries; Subduction zone magmatism

Definition

Island arcs are the marine subset of currently active or formerly active volcanoes located near the boundary between two converging tectonic plates (see Figure 2 in Stern, Magmatism at Convergent Plate Boundaries). Convergence results in the denser plate being thrust (subducted) beneath the rigid overlying plate creating an arcuate-shaped plate boundary on a spherical Earth that is usually marked by an oceanic trench. The volcanoes are located 150–350 km away from the trench on the overlying plate and define an arc in plan view (Frank, 1968). When the crust of the overlying plate is less than ~25 km thick, most of the volcanoes lie below sea-level. Only the largest volcanoes form islands, which, therefore, constitute an island arc. A similar distribution of subduction-related volcanoes also characterizes active continental margins where the crust of the overlying plate is thicker and subaerial so that the volcanoes are not islands (e.g., in western North, Central, and South America; Kamchatka, Russia), but the general processes are similar. "Volcanic arc" refers to the volcanoes produced by subduction in both settings.

Characteristics of island arcs

Almost 80 % of the active subaerial volcanoes on Earth lie within 500 km of a convergent plate boundary and, therefore, are "island arc volcanoes." Their lavas range from basalt to rhyolite. The rock type "andesite," named for the Andes mountains in South America, occurs at about 80 % of these volcanoes. Therefore, volcanism in island arcs is more andesitic than in other tectonic environments on Earth, which are mostly basaltic. Because average continental crust is broadly andesitic in composition, continental crust is either created or recycled or both by island arc magmatism. And because modern plate tectonics is unique to Earth within our solar system, so also are arc volcanism and continental crust of andesitic composition.

Most active island arc volcanoes lie 50–100 km apart and 125 ± 30 km above the underlying earthquakes that occur near the top of the subducting plate (the Wadati-Benioff-Zone). Most individual arcs (see Figure 2 in Stern, Magmatism at Convergent Plate Boundaries, for their names and distribution) differ from one another by the pairing of the converging plates, although some continuous plate boundaries are conventionally divided into two or more arcs by a morphological contrast in the upper plate (e.g., Kermadec vs. Tonga, Mariana vs. Bonin and Izu, Sunda vs. Banda). Within any one arc, the volcanoes are remarkably linear over distances of several hundred kilometers subparallel to, but displaced from, the plate boundary. This line often is called the volcanic or magmatic front, which is the locus of volcanoes closest to the convergent plate boundary.

All of the magma types in island arcs are characterized by distinctive ratios of trace elements and isotopes as the result of recycling material from the underthrust plate. More-fluid-mobile elements (e.g., Sr and Pb) are enriched in arc magmas relative to less-fluid-mobile elements (e.g., REE), and the isotopic compositions of Sr and Pb are more radiogenic, especially relative to the isotopic composition of Nd, a REE.

Arc crust

The seismically best-imaged island arc crust and upper mantle are at the Izu-Bonin-Mariana and Aleutian arcs

(Takahashi et al., 2007; Kodaira, 2007; Calvert, 2011). In Izu, the crust varies sinusoidally from <25 to ~35 km thick along the length of the arc with a wavelength of 80–100 km. Crust is thicker where the Quaternary arc front volcanoes have erupted mostly basaltic magma and thinner where the magmas are mostly rhyolitic in composition. The velocity structure is similar in both cases, with a $Vp = 6.0$–6.5 km/s middle crust at 5–10 km depth that is characteristic of continents and consistent with an andesitic composition and a lower crust that is consistent with plagioclase-bearing mafic cumulates ($Vp = 6.8$–7.2 km/s) overlying plagioclase-free mafic rocks (olivine, pyroxene, amphibole: $Vp = 7.2$–7.6 km/s). This highest-velocity layer is sometimes assigned to the lowermost crust (Kodaira, 2007) and sometimes to the uppermost mantle (Takahashi et al., 2007; Tatsumi et al., 2008). The Aleutian arc differs in that its $Vp = 6.0$–6.5 layer is at a shallower depth (<5 km) and velocities generally are higher at all depths, indicating a more mafic composition. The overall velocity structure of arcs is consistent with a two-stage process of arc crustal production and internal differentiation whereby juvenile magma from the underlying mantle is basaltic and the crustal structure reflects in situ magmatic differentiation and anatexis (e.g., Tatsumi et al., 2008). The crust in all modern island arcs has higher velocity at depths >10 km than old continental interiors.

Across-arc variations

All island arcs include volcanoes farther from the plate boundary than the volcanic front in what is called the "rear arc." Rear arc volcanoes are usually smaller and more likely to be submarine seamounts rather than subaerial islands. Consequently, they are less well known. There can be hundreds of them that extend up to 100 km behind the volcanic front. The volcanic front and rear arc volcanoes together constitute the island arc or volcanic arc. The thickness and velocity structure of the crust are roughly the same for ~100 km across unrifted volcanic arcs (Kodaira, 2007). Therefore, although the volcanoes are smaller in modern rear arcs, the majority of arc crust may be created in the rear arc because it is wider than the volcanic front.

Volcanic rocks in the rear arc usually differ in chemical composition from those at the front. They have higher concentrations of K and most incompatible trace elements (those favoring melt over crystals) but less enriched Sr, Nd, Pb, and Hf isotope ratios. They are more likely to have amphibole phenocrysts in mafic rocks. The differences usually are attributed to smaller degrees of mantle melting beneath the rear arc in response to less fluxing by liquids from the subducting plate that are themselves less water rich than beneath the volcanic front (e.g., Hochstaedter et al., 2001).

Backarc magmas gradually lose their subduction-related character altogether by several hundred kilometers behind the volcanic front and become like those at mid-ocean ridges (MORB) where there is active seafloor spreading in the backarc (e.g., Central Lau, North Fiji, and Manus Basins) or like those at ocean islands (OIB) in broad extensional zones without spreading. The distinction between rear arc and backarc, and consequently the width of volcanic arcs, is somewhat subjective depending on what evidence of "subduction-related character" is used. The following concentration ratios are indicative of subduction-related magmatism: $U/Nb > 0.03$, $Ba/La > 20$, $Pb/Ce > 0.08$, $H_2O/Ce > 250$, and $La/Nb > 1$. However, enrichment in these elements derived from the subducting plate can postdate active subduction by as much as a few million years.

Along-arc variations

As noted above, the volcanic fronts of island arcs are characterized by along-arc spacing of both volcanoes and the thickness of the middle crust at 50–100 km intervals. In addition, there are similarly spaced chains of volcanoes across the width of many volcanic arcs (e.g., across the Kermadec, Tonga, Mariana, Izu, Kurile, and Sangihe arcs; e.g., Hochstaedter et al., 2001). In NE Honshu, Japan, the along-arc spacing of these volcanic chains correlates with the spacing of low-gravity anomalies at the surface and low seismic velocities (especially Vs/Vp) in the upper mantle (Tamura et al., 2002). This along-strike three-dimensional structure of island arcs suggests spatial periodicity of magma production and mantle upwelling. This spacing has been attributed directly to the rheology of melt from the subducting plate (Marsh, 1974; Behn et al., 2011) or indirectly to the rheology of flux-melted mantle in the overlying wedge (Honda and Yoshida, 2005).

Relationship of island arcs to subduction parameters

Island arcs differ most among themselves in the age and velocity of the subducting plate and, therefore, slab surface temperatures and in the nature and thickness of subducting sediment. The location of volcanic fronts is thought to be controlled by the thermal structure of the subducting plate or the overlying mantle or both. A generic arc cross section is shown in Figure 2 of Stern (Magmatism at Convergent Plate Boundaries). P-T paths of the slab surface for different subducting plates, and for different layers within the subducting crust of any one arc, are shown in Figs. 1 and 2 (Syracuse et al., 2010; van Keken et al., 2011). The location of the volcanic front sometimes corresponds to the depth of maximum dehydration or the onset of melting, in one or more layers of subducted crust. Alternatively, the volcanic front may overlie where the anhydrous solidus of the convecting mantle comes closest to the plate boundary (England and Katz, 2010). The across-arc width of island arcs may be related to the loci of slab melting. Subducting sediment is expected to melt beneath all volcanic fronts, and subducting basalt beneath most volcanic fronts, if they are water saturated (Figure 1). Because the P-T path of

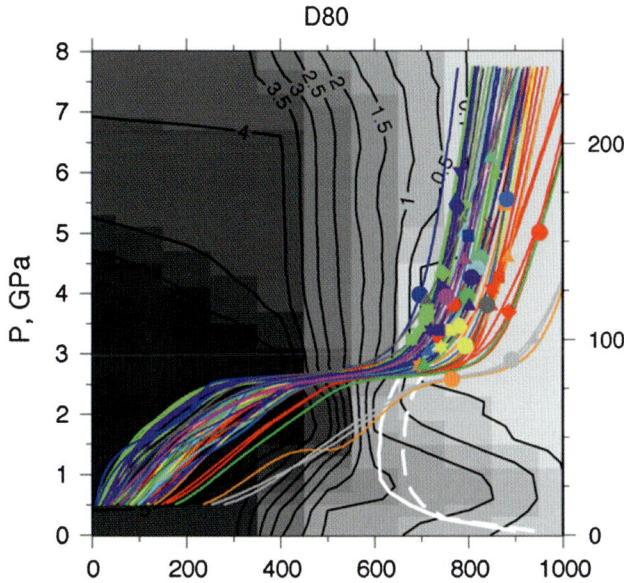
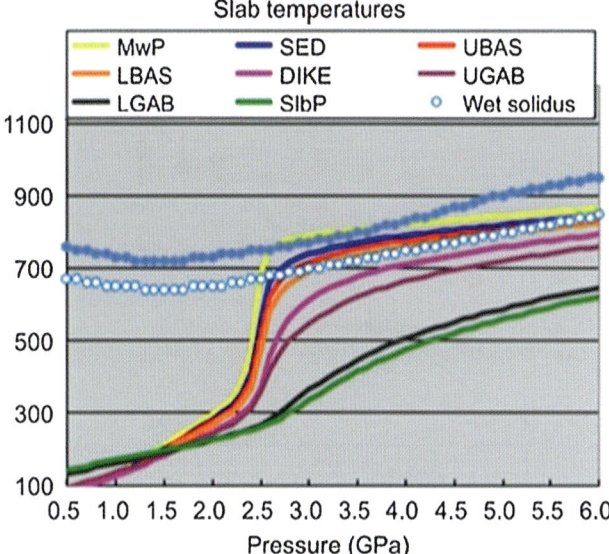

Island Arc Volcanism, Volcanic Arcs, Figure 1 Thermal models for slab surface temperatures beneath 30 different island arcs from Syracuse et al. (2010). Each *colored line* is a thermal model for a different arc. Although the calculated slab surface temperature varies by 200 °C beneath different arcs, in all cases it crosses the water-saturated solidi of sediment (*solid white line*) and basalt (*dashed white line*) slightly trenchward of the volcanic front that typically is ~125 ± 20 km above the slab. The labeled contours, and the shades of gray, show the maximum calculated amount of water in wt% bound in minerals in the slab.

Island Arc Volcanism, Volcanic Arcs, Figure 2 Slab temperatures beneath the Izu island arc where the slab below the volcanic front is at 3.2 GPa (125 km) and the slab below the rearmost arc volcanism is at about 6 GPa. Figure is from Version 4 of the Arc Basalt Simulator (Kimura et al., 2010). Each color is a different portion of subducting oceanic crust: SED (sediment), UBAS and LBAS (upper and lower basalt; oceanic crust layer 2), and UGAB and LGAB (upper and lower gabbro; oceanic crust layer 3). The overlying and underlying mantles are MwP and SlbP, respectively. The wet (*open circle*) and dehydration (*closed circle*) solidi apply to both sediment and basalt. The slight difference in wet solidi from Figure 1 below 3 GPa reflects the range of experimental results chosen in the papers cited. The thermal model for Izu is from Figure 1 (Syracuse et al., 2010). Note that both the sediment and uppermost basalt are predicted to start melting slightly trenchward of the volcanic front and to remain partially molten beneath the full width of the arc.

the slab surface is predicted to be steeper than the water-saturated solidi, the width of arcs corresponds roughly to where the slab surface temperature exceeds the solidi (Figure 1). Moreover, slab surface temperatures calculated from geodynamic and geochemical models agree ±100 °C for volcanic front magmas (Cooper et al., 2012). For this to be other than coincidence, water-saturated melting near the slab surface may be ubiquitous, minerals like monazite and phlogopite must be residual during sediment melting for the geochemical models to work, and arc magmas must rise quasi-vertically from the slab.

Arc magmas differ from one another in composition for many reasons, but chief among them are differences in locally subducting sediment. Figure 3 shows that the Th/La ratio of the slab-derived component in arc magmas correlates well with that ratio in locally subducting sediment on a global basis (Plank, 2005). That is, what goes in largely controls what comes out in many cases. Because isotope ratios are unaffected by melting processes, they would be an even better test of this conclusion, but no similar global synthesis is available.

Temporal evolution

It has been thought that individual arc volcanoes and entire arcs become more enriched in incompatible elements and less enriched in iron through time (Gill, 1981, Chap. 7). That is, arc magmas seemed to become higher K, more calcalkaline, and less tholeiitic with time. However, there are many exceptions to this generalization at all time scales. Moreover, recent studies of the oldest volcanic rocks in the Mariana, Bonin, Izu, and Tongan arcs (Reagan et al., 2010, Ishizuka et al., 2011, Todd et al., 2012), and of supra-subduction zone ophiolites, suggest a more complex evolution during the initiation of an island arc. The magmatic sequence starts with tholeiitic basalt generated by decompression melting with little or no contribution from the subducted plate, followed by boninite (high-Mg, low-Ti andesite) generated by flux melting of the shallow mantle and then by low-K arc tholeiite with more of a recycled component than during the initial phase. In some locations, two or more of these rock types erupt simultaneously in the same place. Higher-K and less-Fe-enriched magmas appear occasionally later in arc history, especially in the rear arc, and sometimes in response to special tectonic events like arc rifting or

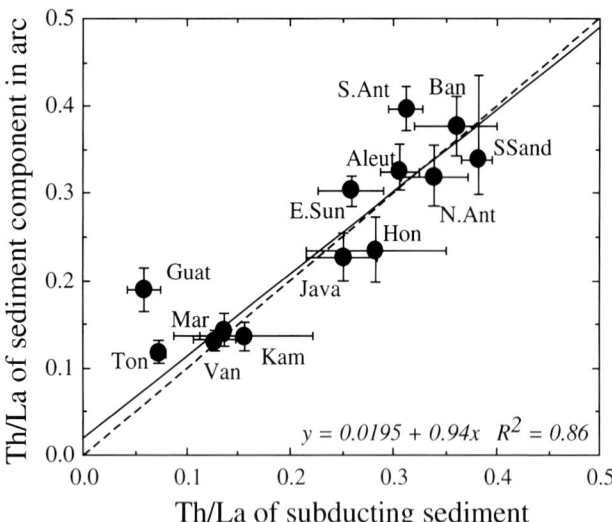

Island Arc Volcanism, Volcanic Arcs, Figure 3 Th/La of the slab-derived component in arc magmas compared to the Th/La ratio of sediment subducting beneath the same arcs. See Plank (2005) for details about both axes and the error bars.

seamount collision. These calcalkaline magmas are more common where crust is thicker or the subducting plate is younger.

Ore deposits

Island arcs host many economically important ore deposits. For example, venting of high-temperature hydrothermal fluids into ocean water by submarine arc volcanoes causes Fe-Pb-Cu-Zn-Ag sulfides to precipitate and accumulate as volcanogenic massive sulfide deposits (Hannington 2014). Hydrothermal systems that are thermally and chemically driven by plutons associated with continental andesitic to rhyolitic volcanic centers in unusually compressional settings in produce the Cu and Mo porphyry deposits that are the principal sources for these and many other societally important metals (e.g., Richards and Mumin, 2013). Much of the world's Au precipitated in shallower portions of arc magmatic systems where hydrothermal fluids boil (e.g., Cook et al., 2011).

Summary

Terrestrial island arcs or volcanic arcs are regions up to 100 km wide of subaerial and submarine volcanoes whose trenchward limit is ~120 km above the subducting plate. Their magmas are more andesitic than elsewhere on Earth or other planets and contribute to continent formation. Their volcanoes are often explosive, and their magmatism creates important metal ore deposits. There are 3-D variations in the spacing of volcanoes and the composition of magmas both of which are related to the thermal structure of the subducting plate and overlying convecting mantle.

Bibliography

Behn, M., Keleman, P., Hirth, G., Hacker, B., and Massone, H., 2011. Diapirs as the source of the sediment signature in arc lavas. *Nature Geoscience*, **4**, 641–646.

Calvert, A. J., 2011. The seismic structure is island arc crust. In Brown, D., and Ryan, P. (eds.), *Arc-Continent Collision*. New York: Springer. Frontiers in Earth Sciences, pp. 87–119.

Cook, D. R., Deyell, C. L., Waters, P. J., Gonzales, R. I., and Zawi, K., 2011. Evidence for magmatic-hydrothermal fluids and ore-forming processes in epithermal and porphyry deposits of the baguio district, Philippines. *Economic Geology*, **106**, 1399–1424.

Cooper, L., Ruscitto, D., Plank, T., Wallace, P., Syracuse, E., and Manning, C., 2012. Global variations in H_2O/Ce: 1. Slab surface temperatures beneath volcanic arcs. *Geochemistry Geophysics Geosystems*, **13**, Q03024, doi:10.1029/2011GC003902.

England, P., and Katz, R., 2010. Melting above the anhydrous solidus controls the location of volcanic arcs. *Nature*, **467**, 700–703.

Frank, F., 1968. Curvature of island arcs. *Nature*, **220**, 363.

Gill, J., 1981. *Orogenic Andesites and Plate Tectonics*. New York: Springer, p. 390pp.

Hannington, M. D., 2014. Volcanogenic massive sulfide deposits. In *Treatise on Geochemistry*, 2nd edn. Amsterdam: Elsevier, Vol. 13, pp. 463–488.

Hochstaedter, A., Gill, J., Peters, R., Broughton, P., Holden, P., and Taylor, B., 2001. Across-arc geochemical trends in the Izu-Bonin arc: contributions from the subducting slab. *Geophysics, Geochemistry, Geosystems*, Paper No. 2000GC000105.

Honda, S., and Yoshida, T., 2005. Application of the model of small-scale convection under the island arc to the NE Honshu subduction zone. *Geochemistry Geophysics Geosystems*, **6**, Q01002, doi:10.1029/2004GC000785.

Ishizuka, O., Tani, K., Reagan, M., Kanayama, K., Umino, S., Harigane, Y., Sakamoto, I., Miyajima, Y., Yuasa, M., and Dunkley, D., 2011. The timescales of subduction initiation and subsequent evolution of an oceanic island arc. *Earth and Planetary Science Letters*, **306**, 229–240.

Kimura, J.-I., Adam, J. R. K., Rowe, M., Nakano, N., Katakuse, M., van Keken, P., Hacker, B., and Stern, R. J., 2010. Origin of cross-chain geochemical variation in Quaternary lavas from northern Izu arc: a quantitative mass balance approach on source identification and mantle wedge processes. *Geochemistry, Geophysics, Geosystems*, doi:10.1029/2010GC003050.

Kodaira, S., 2007. Seismological evidence for variable growth of continental crust in the Izu-Bonin intra-oceanic arc. *Journal of Geophysical Research*, **112**, B05104, doi:10.1029/2006JB004593.

Marsh, B., and Carmichael, I.S.E., 1974. Benioff zone magmatism. *Journal of Geophysical Research*, **79**, 1196–1206.

Plank, T., 2005. Constraints from Thorium/Lanthanum on sediment recycling at subduction zones and the evolution of continents. *Journal of Petrology*, **46**, 921–944.

Reagan, M., Ishizuka, O., Stern, R., Kelley, K., Ohara, Y., Blichert-Toft, J., Bloomer, S., Cash, J., Fryer, P., Hanan, B., Hickey-Vargas, R., Ishii, T., Kimura, J., Peate, D., Rowe, M., and Woods, M., 2010. Fore-arc basalts and subduction initiation in the Izu-Bonin-Mariana system. *Geochemistry Geophysics Geosystems*, **11**, Q003X12, doi:10.1029/2009GC002871.

Richards, J. P., and Mumin, A. H., 2013. Magmatic-hydrothermal processes within an evolving Earth: Iron oxide-copper-gold and porphyry $Cu \pm Mo \pm Au$ deposits. *Geology*, **41**, 767–770, doi:10.1130/G34275.1.

Syracuse, E., van Keken, P., and Abers, G., 2010. The global range of subduction zones thermal models. *Physics of the Earth and Planetary Interiors*, **51**, 1761–1782.

Takahashi, N., Kodaira, S., Klemperer, S., Tatsumi, Y., Kaneda, Y., and Suyehiro, K., 2007. Crustal structure and evolution of the Mariana intra-oceanic island arc. *Geology*, **35**, 203–206.

Tamura, Y., Tatsumi, Y., Zhao, D., Kido, Y., and Shukuno, H., 2002. Hot fingers in the mantle wedge: new insights into magma genesis in subduction zones. *Earth and Planetary Science Letters*, **197**, 105–116.

Tatsumi, Y., Shukuno, H., Tani, K., Takahashi, N., Kodaira, S., and Kogiso, T., 2008. Structure and growth of the Izu-Bonin-Mariana arc crust: 2. Role of crust-mantle transformation and the transparent Moho in arc crust evolution. *Journal of Geophysical Research*, **113**, B02203, doi:10.1029/2007JB005121.

Todd, E., Gill, J., and Pearce, J., 2012. A variably enriched mantle wedge and contrasting melt types during arc stages following subduction initiation in Fiji and Tonga, southwest Pacific. *Earth and Planetary Science Letters*, **335–336**, 180–194.

Van Keken, P., Hacker, B., Syracuse, E., and Abers, G., 2011. Subduction factory: 4. Depth-dependent flux of H2O from subducting slabs worldwide. *Journal of Geophysical Research*, **116**, B01401, doi:10.1029/2010JB0079.

Cross-references

Magmatism at Convergent Plate Boundaries
Marginal Seas
Plate Tectonics
Subduction
Volcanogenic Massive Sulfides

L

LAGOONS

Andrzej Osadczuk
Institute of Marine and Coastal Sciences, University of Szczecin, Szczecin, Poland

Definition

The term *lagoon* is derived from the Latin *lacun* and the later Italian *laguna*, which originally referred to the shallow water body around Venice (Skeat, 1893). Although there are many definitions of lagoon, they differ only in details. A lagoon is defined as a *body of brackish, marine or hypersaline water impounded by a sandy barrier and having an inlet connecting with the open ocean* (Phleger, 1981) or *water areas separated from the sea by narrow and low-laying land strips and usually differing from the sea in salinity* (Kaplin, 1982) or a *shallow coastal water body separated from the ocean by a barrier, connected at least intermittently to the ocean by one or more restricted inlets, and usually oriented shore-parallel* (Kjerfve, 1994). A full overview of the definitions of the term "lagoon" in relation to various coastal environments has been given by Tagliapietra et al. (2009).

Key features and variety of lagoons

Coastal lagoons represent about 11 % of the world's coastlines and are widespread all along the world's ocean coasts (Nichols and Boon, 1994). However, lagoons develop on aggrading or formerly aggraded coastal plains which are gently sloping seaward. They are most common in microtidal environments, although many examples of lagoons can be found in mesotidal and even macrotidal environments (Hayes, 1979; Nichols and Boon, 1994). Many lagoons are well developed on coasts of mid-latitudes, especially along the Atlantic coast of the North and South America (Figure 1). Also, many lagoons occur on southern coast of Europe, on southern shores of the North Sea, and in some enclosed seas, such as Black Sea, Caspian Sea, and southern Baltic. Lagoons can also be found in low (tropical) and high (polar) latitude zones. In the first case, lagoons in particular are common along both coasts of Middle America (Gulf of Mexico, Caribbean Sea). On the other hand, along the shores of Alaska and the Chukchi Peninsula, we can find many examples of arctic lagoons (e.g., Kasegaluk Lagoon, Elson Lagoon, Kunergvin Lagoon). The size of coastal lagoons varies greatly, from about a hectare to more than 10,000 km^2 (Bird, 2008). Some lagoons extend over 200 km, e.g., Lagoa dos Patos (Brazil) and Laguna Madre (Gulf of Mexico).

With respect to hydrodynamics and water exchange with the adjacent open sea, coastal lagoons are often classified as *closed, partly closed, open*, and *estuarine* (Nichols and Allen, 1981) (Figure 2) or *choked, restricted*, and *leaky* (Kjerfve, 1986). A *closed lagoon* is produced by intermediate to high waves, and strong longshore drift. Such lagoon has no entrance channel, but that does not mean that there is no water exchange with the sea. Water flow, although very limited, can take place through the permeable sediments of the barrier. Washovers or winds across the barrier are the leading processes of sediment influx. Organic production or chemical precipitation is often a chief sediment-producing process in such conditions. In a *partly closed lagoon*, high wave energy acting on the nearshore bed or headlands and strong longshore currents build an extensive barrier island with prominent flood delta and poorly developed ebb delta. Over long periods, tidal currents and wave power may reach a dynamic balance. Ocean-lagoon exchange is poor, and entrapment of both sandy and muddy sediments is good. An *open lagoon* is a result of wave action in combination with moderate tidal action and river inflow. High waves and longshore drift build short barrier islands, while tidal

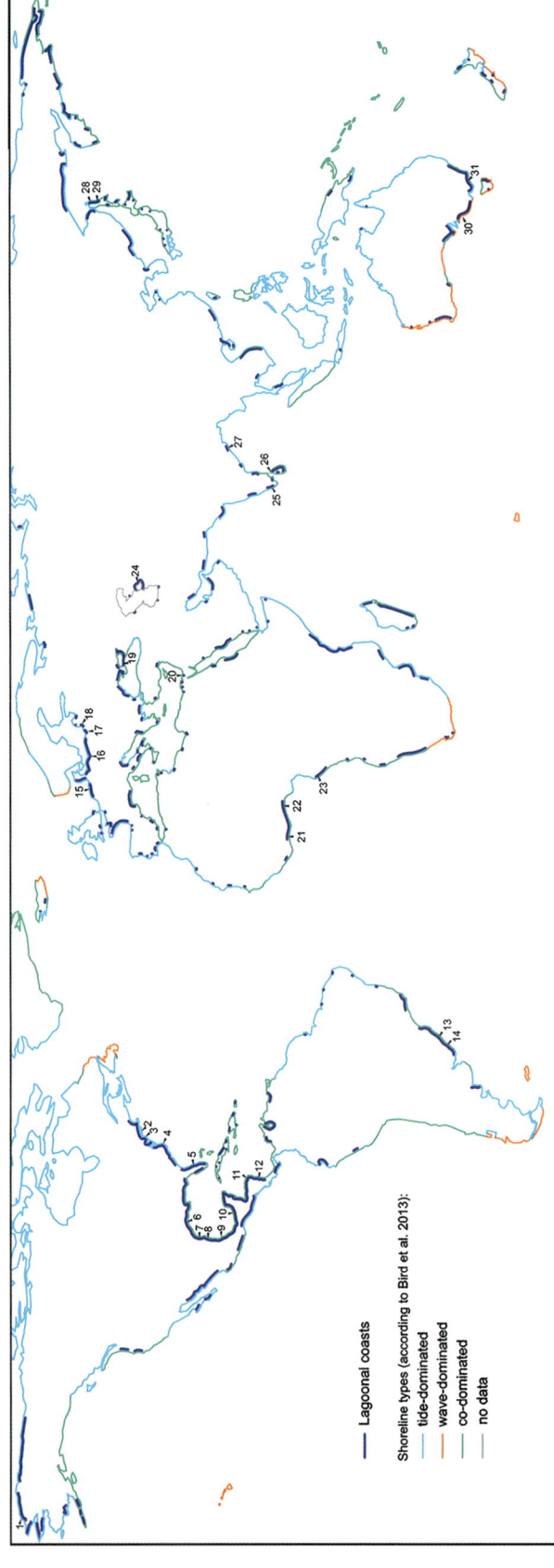

Lagoons, Figure 1 Worldwide distribution of modern coastal lagoons (*medium blue*). Shoreline classification according to relative wave and tidal ranges after Bird et al. (2013), modified (tide dominated, *sky blue*; wave dominated, *orange*; co-dominated, *green*; no data, *gray*). Numbers indicate the location of the largest lagoons (longer than 40 km). *1*, Kasegaluk Lagoon; *2*, Great South Bay; *3*, Barnegat Bay; *4*, Pamlico Sound; *5*, Indian River; *6*, Galveston Bay; *7*, Laguna Madre (USA); *8*, Laguna Madre (Mexico); *9*, Laguna de Tamiahua; *10*, Laguna de Terminos; *11*, Laguna de Caratasca; *12*, Laguna de Perlas; *13*, Lagoa dos Patos; *14*, Lagoa Mirim; *15*, Wadden Sea; *16*, Szczecin Lagoon; *17*, Vistula Lagoon; *18*, Curonian Lagoon; *19*, Sivash Lagoon System; *20*, Lake Bardawil; *21*, Ébrié Lagoon; *22*, Lagos Lagoon and Lekki Lagoon; *23*, Lagune Nkomi; *24*, Garabogazköl Aylagy (Kara-Bogaz-Gol); *25*, Vembanad Lake; *26*, Jaffna Lagoon (Sri Lanka); *27*, Chilika Lake; *28*, Zaliv Piltun; *29*, Zaliv Nyyskiy; *30*, Coorong Lagoon; *31*, Gippsland Lakes.

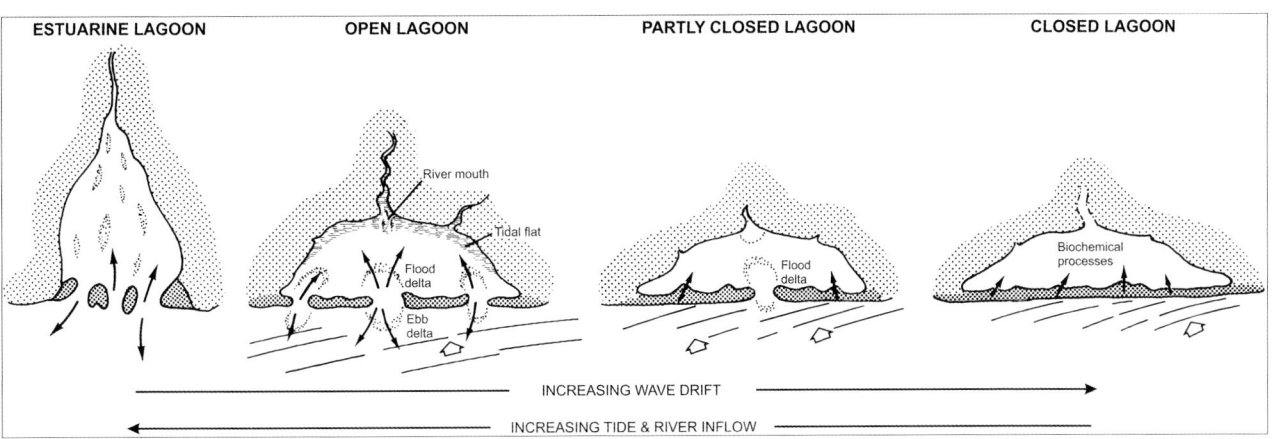

Lagoons, Figure 2 Types of costal lagoons according to Nichols and Allen (1981).

currents maintain several moderate-size entrance channels with well-developed flood and ebb deltas. In such lagoon, muddy sediments tend to accumulate on flats along inner parts, whereas sandy sediments accumulate in the immediate vicinity of entrance channels. An *estuarine lagoon* is characterized by strong tidal action augmented by large river inflow. Wave energy and longshore drift are of little importance in this case. Tidal currents maintain several broad inlets that provide an intensive exchange of water (Nichols and Allen, 1981).

Choked lagoons (according to Kjerfve, 1986) are usually connected to the sea by a single narrow and shallow entrance and occur along coasts affected by high wave energy or significant littoral drift (e.g., Lagoa dos Patos in Brazil, Coorong in Australia). A *restricted* lagoon consists of a large and wide water body, usually parallel to the shore, which has two or more entrance channels or inlets (e.g., Laguna de Terminos in Mexico, Lake Pontchartrain in the USA). *Leaky* lagoons are elongated shore-parallel water bodies with many tidal passes along coasts where tidal currents are sufficient to overcome the tendency to closure by wave action and littoral drift (e.g., the Mississippi Sound, intertidal areas of Wadden Sea) (Kjerfve, 1986; Kjerfve and Magill, 1989; Kjerfve, 1994). Duck and da Silva (2012) have attributed certain dominant hydromorphological conditions and geomorphological features to these three types of lagoons.

Some of geoscientists distinguish between "estuarine lagoons" into which rivers flow and "marine lagoons" without a major freshwater input (Barnes, 1980). "Marine lagoons" are often called "coastal lakes," especially when the connection with the sea is reduced or temporarily obliterated (Tagliapietra et al., 2009). However, the use of the term "estuarine lagoon" should be very prudent, because the word "estuary" is derived from the Latin words "aestus" and "aestuarium," which means "tide" and "tidal place" (tidal inlet). Thus, in the absence of tides, the term "estuarine lagoon" is unsuitable (Osadczuk et al., 2007).

Depending on local conditions, lagoons exhibit salinities which range from almost completely fresh (e.g., Szczecin Lagoon, Poland/Germany; Curonian Lagoon, Lithuania/Russia) or brackish (e.g., Chilika Lake, India; Lagoa dos Patos, Brazil) to hypersaline (e.g., Laguna Madre, USA; Coorong, Australia).

Formation and evolution of lagoons

Contemporary lagoons owe their origin largely to eustatic sea-level rise near the end of the Pleistocene 15,000 year ago. The consequence of a rapid but variable rate of sea-level rise was inundation of river valleys and low-lying coastal depressions (Nichols and Biggs, 1985). Most modern lagoons have developed during the sea-level rising of the last 6–8 thousand year ago (Phleger, 1981). Lagoons develop on aggrading or formerly aggraded coastal plains which are gently sloping seaward. Very often they are associated with modern or ancient river valley. From the geological point of view, lagoons are ephemeral. They form, evolve, and infill within a short timespan and may be considered as effects of events or processes on generally prograding coasts (Phleger, 1981; Bird, 1994; Cooper, 1997).

The emergence of a barrier is a primary requirement for the creation and functioning of a lagoon (Figure 3). The barriers may take a variety of forms, including headland spits (single and double sided), emergent offshore bars, coral reefs, and barrier inland chains. Differences in the genesis and structure of the barriers are a consequence of differences in controlling mechanisms, such as regional topography; the stage in the shore development; the source, type, and rate of sediment supply; the pace and direction of change in sea-level, tidal range; and meteorological and climatic conditions (Fisher, 1982).

There are several models of the development of barrier relating to sea-level changes. Most of them are related to marine transgression. As sea-level rises, a barrier may undergo of three responses: erosional (the Bruun role), translational (rollover model), or stationary (overstepping model) (Carter, 1988; Cooper, 1997). The barrier's response to the sea-level rise has a direct impact

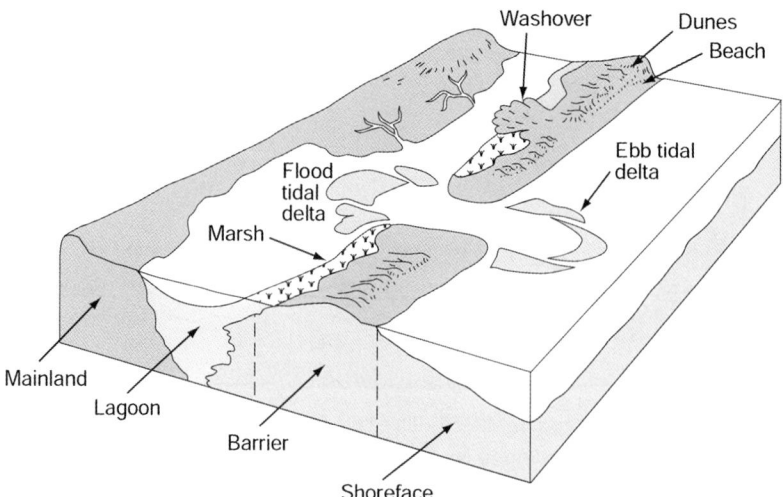

Lagoons, Figure 3 Block diagram depicting major features of lagoon environment (according to Cooper, 2007).

on the evolution of a lagoon. If a transgressive lagoonal barrier stabilizes against a break in slope or topographic high and accretes upward rather than landward, the volume of the lagoon may increase if the sea-level rise exceeds the fluvial sediment supply. If the barrier continues to move landward as the sea-level rises, the actual lagoon volume may remain constant (Cooper, 1997). However, evolutionary models for coarse-grained barriers and associated lagoons in temperate latitudes show that they differ from sandy barriers (Carter et al., 1989).

An important element of each lagoon system is connection to the open sea. Typically, there are one or more entrances or tidal inlets. Most inlets form by the following mechanisms: (1) storm-generated scour channels, (2) spit growth across the entrances to flooded valleys during the mid-Holocene, (3) intersection of major tidal channels by landward-migrating transgressive barrier islands, and (4) tidal prism-controlled, evenly spaced inlets that evolve in regions lacking major former river valleys (Hayes and FitzGerald, 2013).

Based on the morphology of the barrier-inlet elements, Davis and Hayes (1984) distinguished three types of barrier coasts based on the morphology of the barrier-inlet elements: (1) "wave dominated" (the barrier is narrow with abundant washover features and a generally low elevation; inlets are typically widely spaced, small, and unstable); (2) "mixed-energy" (barriers are typically short and drumstick shaped, with one end prograding in the form of beach/dune ridges and the other transgressing landward; inlets are large and stable); (3) "tide dominated" (barriers are absent, and sand bodies tend to be oriented with an onshore-offshore orientation as the result of strong tidal currents).

Gibeaut and Davis (1991), taking into account previous findings relating to the synergies between wave and tidal energy, have identified four categories of barrier inlets: (1) tide dominated, (2) mixed-energy straight, (3) mixed-energy offset, (4) wave dominated. This relationship also affects the shape of the ebb delta. Generally, the "tide-dominated" ebb delta extends well beyond the barriers and is dominated by channel-margin linear bars. The "mixed-energy" ebb delta has a distinct terminal lobe but also with well-defined channel-margin linear bars. The barrier shorelines may be aligned or offset. The "wave-dominated" ebb delta is very small and is dominated by the terminal lobe if the inlet is not completely closed (Davis, 2013). Although the comparative influence of waves and tides is the fundamental relationship in barrier-inlet morphodynamics, in the opinion of Davis (2013), variables such as tidal prism, sediment abundance, and sea-level change are equally important.

In general, coastal lagoons are naturally net sediment sinks. However, sometimes some portion of sediments may escape a lagoon by flushing into the sea, by wind erosion of coastal flats, or by oxidation of organic matter. Sediments can be supplied to lagoons from various sources. In many lagoons, the sediment components are derived locally from shore or barrier erosion or from sources inside lagoon, which are organic production and chemical precipitation. In other lagoons, sedimentary material can come also from distant sources. Very often, a catchment area is a source of significant amounts of fluvial sediments. Some amounts of sediment may reach the lagoon from the sea during tide action (Nichols and Allen, 1981). It should be noted that the supply of sediment is strongly influenced by climate. Comparison of lagoon features and processes in different climatic zones was given by Nichols and Allen (1981).

Lagoons act as sinks of particles through sedimentation, acting as important buffers of the land-sea interface. A large number of processes are responsible for this phenomenon. During transport and recycling, fine particles are modified by aggregation, disintegration, and reforming. Microorganisms and benthic fauna play an important role in these processes. Of particular importance are also wind waves that induce sediment resuspension. Although sediments are extensively reworked, especially by storms, lagoons primarily function as net sediment sinks in which the accumulation rates adjust to submergence (Nichols and Allen, 1981; Boyd et al., 1992; Nichols and Boon, 1994; Brito et al., 2012). On the other hand, submerged aquatic vegetation and salt marsh vegetation as well microphytobenthos may exert a strong influence on the morphology of the lagoons bottom by bio-stabilization of sediment (Widdows and Brinsley, 2002). Generally speaking, lagoonal sediments are a mixture of components, derived either from external sources (river, sea, shore, barrier) or from internal sources (organic production, chemical precipitation, erosion of older deposits) (Nichols and Boon, 1994).

From an ecological perspective, coastal lagoons are highly productive ecosystems, 10–15 times higher than continental shelves (Valiela, 1995), excluding, of course, some hypersaline lagoons. Thus, many of them can also provide a wide range of natural services, very important for the development of society. Anthony et al. (2009) perceive a wide range of social values of lagoons, in such categories as pragmatic, scholarly, aesthetic, and tactic, especially highlighting commercial, recreational, and tourism uses of lagoons. A comprehensive and reliable overview of some European semi-enclosed coastal systems, including lagoons, mainly from the ecological and socioeconomic perspective, has been given by Newton et al. (2014).

Coastal lagoons act as buffers of the land-sea interface, providing valuable ecosystem services such as nutrient recycling, decomposition of organic matter, and removal of pollutants. Lagoons are regions of restricted exchange, subject to anthropogenic pressures that result in problems such as eutrophication. Lagoons are extremely vulnerable to climate changes. Climate change, especially sea-level rise and global warming, is likely to affect shallow coastal lagoons and to increase their vulnerability to eutrophication (Brito et al., 2012).

In arid regions, lagoonal sediments are formed primarily as a result of chemical precipitation and biological extraction of calcium carbonate. Hot and dry climate creates favorable conditions for the deposition of evaporites. Thus, carbonates, chlorides, and sulfates can accumulate in the lagoonal environment. Such processes are observed, for example, on the southern coast of the Persian Gulf, where lagoonal waters have a salinity of up to 67 ‰. In the local lagoons, e.g., the Khor al Bazam, many variations of carbonate sediments occur, such as pelleted lime muds, oolites, and pellet aggregates, as result of high organic production (Alsharhan and Kendall, 2003). Their distribution and the characteristics of individual grain types are controlled largely by their hydrodynamic setting, with interactions between wave and current action, sediment mobility, burial rate, and cementation. Coastal shallows are covered by cyanobacterial mats and mangrove swamps, and the supratidal salt flats (sabkhas) are covered by gypsum-salt deposits (Kendall and Alsharhan, 2011).

There are some conceptual models of lagoon evolution. One of the earliest concepts of lagoon evolution, known as "Lucke model," assumes that coastal lagoons evolve from a marine embayment to a relatively deep, partially enclosed, back-barrier lagoon (Lucke, 1934). In developing this concept, Oertel et al. (1992) point out that at wave-dominated coasts, the floors of barrier lagoons are initially smooth after lagoon formation, and basin infilling is dependent on changes in capacity relative to sea-level rise and sediment input from siltation, runoff, inlets, and cross-island transport. At tide-dominated coasts, the lagoon floors are irregular and reflect the antecedent topography of pre-transgressed landscape. The interfluve areas of the topography produced very shallow areas in the lagoon that form tidal flats or are colonized by marshes. In "Lucke model," the open-water embayment represents the first stage in the evolutionary sequence, while a model proposed by Oertel et al. (1992) assumes that the open-water embayment may become end-member of the evolutionary cycle. However, it should be considered that "Lucke model" was developed for stationary sea-level conditions. A model of transgressive evolution of lagoons was presented by Nichols (1989), based upon the interplay between sedimentation rate and relative sea-level rise. The model assumes three different types of lagoons: a "deficit lagoon," a "surplus lagoon," and an "equilibrium lagoon." In a "deficit lagoon," sea-level rise exceeds sediment supply and lagoon becomes deeper. In a "surplus" lagoon, sediment supply exceeds sea-level rise and lagoon becomes increasingly filled with sediments. In "equilibrium lagoon," the rate of sedimentation is in equilibrium with the rate of submergence from sea-level rise, thus lagoon morphology remains unchanged.

In general, the evolution of many coastal lagoons is a result of the combined effects of marine and fluvial processes leading to trapping and infilling of semi-enclosed coastal systems, including coastal lagoons, and the reshaping of seaward boundaries (Kjerfve and Magill, 1989). However, there is wide variation in the evolution of coastal lagoons. Their development can progress in a variety of ways and with different results. Thus, lagoons can be found at various stages of development in an evolutionary sequence.

Overall, lagoon development is the result of the balance between those processes which act to reduce the size of a lagoon and those which act to increase it. It is mainly related to the balance between the supply of sediment and the rate of sea-level rise. The response of a barrier to sea-level rise has a direct bearing on lagoonal evolution. If a transgressive lagoonal barrier stabilizes against a break in slope or topographic high and accretes upward rather than landward, then the volume of the lagoon may increase if sea-level rise exceeds fluvial sediment supply. If the barrier continues to move landward as sea-level rises, then the actual lagoon volume may remain constant.

In certain cases, the barrier progrades seaward, and this may reduce contact with the sea (Cooper, 1997). However, it should be noted that the rate and direction of the lagoon evolution depend on many factors. According to Cooper (1997), these factors may be divided into macroscale and local or microscale variations. Macroscale controls include sea-level changes, climate, and tectonic stability. Local controls include fluvial sediment supply, longshore drift, salinity, surrounding geology, costal morphology, lagoon orientation, wave energy, and tidal range. Carter et al. (1989) proposed a hierarchy of controls on the effect of sea-level changes on coarse-grained barrier-lagoon evolution. Such hierarchical arrangement of dependencies shows the complexity and interrelationships between various factors.

Very often, lagoon evolves from an estuary or marine embayment to a partially enclosed barrier-lagoon system and then, with progressive infilling, to a marsh or deltaic-filled lagoon, ending the cycle with a depositional plain or with an eventual destruction by marine erosion (Nichols, 1989). For example, the area of the Odra River mouth, southern Baltic coast, is subjected to such an evolution since the late Pleistocene (Borówka et al., 2005; Osadczuk et al., 2007).

Based on detailed studies of sedimentation processes in one of the world's largest lagoons (Laguna Madre, Texas), Morton et al. (2000) have found that the historical average annual accumulation rate in the lagoon (1 mm/year) is substantially less than the historical rate of relative sea-level rise (4 mm/year). Lagoon submergence coupled with erosion of the western shore indicates that Laguna Madre is being submerged slowly and migrating westward rather than filling, as some have suggested.

According to Adlam (2014), the existence of numerous examples of lagoons that remain unfilled suggests that a discontinuity in their geologic evolution occurs before maturity is reached. He has presented a new hypothesis that geologic evolution of lagoons takes place in two phases which are defined by changes in physical processes. The early phase aligns with the traditional view but ends when a depth threshold is reached. In the late phase, sedimentation is inhibited by the local energy regime and can only proceed if the lagoon surface area is reduced. Validation of the existence of these two distinct process-based phases of geologic evolution will improve the reliability of predictive models for lagoon shoreline changes and rates of basin fill as well as reduce the risk of misinterpreting past conditions from the geological record.

Conclusions

The formation and evolution of coastal lagoons is controlled by many factors, including global, regional, and local determinants at different levels of importance. Very important is the hierarchy of the circumstances and processes that control the evolution of this environment. A comprehensive model of lagoons' evolution should take into account as many factors as possible. The model should include, in particular, such variables as geological conditions; neotectonic movements; lagoon morphology and its orientation; connection with the sea; sea-level changes, tidal amplitude, and waves energy; climate and local meteorological conditions; character of sediments supply; sedimentation rate; salinity; primary production; and anthropogenic influences. Although there were several attempts to create a comprehensive and universal model of the lagoons' evolution, unfortunately, it was not created, so far, one that would take into account all variables.

Bibliography

Adlam, K., 2014. Coastal lagoons: geologic evolution in two phases. *Marine Geology*, **355**, 291–296.

Alsharhan, A. S., and Kendall, C. G. St. C., 2003. Holocene coastal carbonates and evaporites of the southern Arabian Gulf and their ancient analogues. *Earth-Science Reviews*, **61**, 191–243.

Anthony, A., Atwood, J., August, P., Byron, C., Cobb, S., Foster, C., Fry, C., Gold, A., Hagos, K., Heffner, L., Kellogg, D. Q., Lellis-Dibble, K., Opaluch, J. J., Oviatt, C., Pfeiffer-Herbert, A., Rohr, N., Smith, L., Smythe, T., Swift, J., and Vinhateiro, N., 2009. Coastal lagoons and climate change: ecological and social ramifications in U.S. Atlantic and Gulf coast ecosystems. *Ecology and Society*, **14**(1), 8.

Barnes, R. S. K., 1980. *Coastal Lagoons; The Natural History of a Neglected Habitat*. Cambridge: Cambridge University Press. 106 pp.

Bird, E. C., 1994. Physical setting and geomorphology of coastal lagoons. In Kjerfve, B. (ed.), *Coastal Lagoon Processes*. Elsevier Oceanography Series 60. Amsterdam: Elsevier Science Publications B.V, **60**, 9–40.

Bird, E. C., 2008. *Coastal Geomorphology: An Introduction*. Chichester: Wiley. 411 pp.

Bird, C. E., Franklin, E. C., Smith, C. M., and Toonen, R. J., 2013. Between tide and wave marks: a unifying model of physical zonation on littoral shores. *PeerJ*, **1**, e154, doi:10.7717/peerj.154.

Borówka, R. K., Osadczuk, A., Witkowski, A., Wawrzyniak-Wydrowska, B., and Duda, T., 2005. Late Glacial and Holocene depositional sequences in the eastern part of the Szczecin Lagoon (Great Lagoon) basin – NW Poland. *Quaternary International*, **130**, 87–96.

Boyd, R., Dalrymple, R., and Zaitlin, B. A., 1992. Classification of clastic coastal depositional environments. *Sedimentary Geology*, **80**, 139–150.

Brito, A., Newton, A., Tett, P., and Fernandes, T., 2012. How will shallow coastal lagoons respond to climate change? A modelling investigation. *Estuarine, Coastal and Shelf Science*, **112**, 98–104.

Carter, R. W. G., 1988. *Coastal Environments: An Introduction to the Physical, Ecological and Cultural Systems of Coastlines*. London: Academic.

Carter, R. W. G., Forbes, D. L., Jennings, S. C., Orford, J. D., Shaw, J., and Taylor, R. B., 1989. Barrier and lagoon coast evolution under differing relative sea level regimes: examples from Ireland and Nova Scotia. *Marine Geology*, **88**, 221–242.

Cooper, J. A. G., 1997. Lagoons and microtidal coasts. In Carter, R. W. G., and Woodroffe, C. D. (eds.), *Coastal Evolution. Late Quaternary Shoreline Morphodynamics*. Cambridge: Cambridge University Press.

Cooper, J. A. G., 2007. Temperate coastal environments. In Perry, C., and Taylor, K. (eds.), *Environmental Sedimentology*. Malden: Blackwell Publishing, pp. 263–301.

Davis, R. A., 2013. A new look at barrier-inlet morphodynamics. In: Kana, T., Michel, J., and Voulgaris, G. (eds.), *Journal of Coastal Research*, **69**(Special Issue), 1–12.

Davis, R. A., and Hayes, M. O., 1984. What is a wave-dominated coast? *Marine Geology*, **60**, 313–329.

Duck, R. W., and da Silva, J. F., 2012. Coastal lagoons and their evolution: a hydromorphological perspective. *Estuarine, Coastal and Shelf Science*, **110**, 2–14.

Fisher, J. J., 1982. Barrier islands. In Schwartz, M. L. (ed.), *The Encyclopedia of Beaches and Coastal Environments*. Hutchinson-Ross Pub. Co., Stroudsberg, PA, **XV**, 124–133.

Gibeaut, J. C., and Davis, R. A., 1991. Computer simulation modeling of tidal inlets. In *Coastal Sediments'91*. Seattle: American Society of Civil Engineers, pp. 1389–1403.

Hayes, M. O., 1979. Barrier island morphology as a function of tidal and wave regime. In Leatherman, S. P. (ed.), *Barrier Islands*. New York: Academic, pp. 1–28.

Hayes, M. O., and FitzGerald D. M., 2013. Origin, evolution, and classification of tidal inlets. *Journal of Coastal Research*, **69** (Special Issue), 14–33. Proceedings, Symposium in Applied Coastal Geomorphology to Honor Miles O. Hayes.

Kaplin, P. A., 1982. Lagoon and lagoonal coasts. In Schwartz, M. L. (ed.), *The Encyclopedia of Beaches and Coastal Environments*. Stroudsburg, PA: Hutchinson Ross Publishing Company. 504.

Kendall, C. G. St. C., and Alsharhan A. S., 2011. Coastal Holocene carbonates of Abu Dhabi, UAE: depositional setting, geomorphology, and role of cyanobacteria in micritization. *International Association of Sedimentologists*. Special Publication, **43**, 205–220.

Kjerfve, B., 1986. Comparative oceanography of coastal lagoons. In Wolfe, D. A. (ed.), *Estuarine Variability*. Elsevier Oceanography Series 60. Amsterdam: Elsevier Science Publications B.V, New York: Academic, pp. 63–81.

Kjerfve, B., 1994. Coastal lagoons. In Kjerfve, B. (ed.), *Coastal Lagoon Processes*. Amsterdam: Elsevier. Elsevier Oceanography Series, **60**, 1–8.

Kjerfve, B., and Magill, K. E., 1989. Geographic and hydrodynamic characteristics of shallow coastal lagoons. *Marine Geology*, **88**, 187–199.

Lucke, J. B., 1934. A theory of evolution of lagoon deposits on shorelines of emergence. *Journal of Geology*, **42**, 561–584.

Morton, R. A., Ward, G. H., and White, W. A., 2000. Rates of sediment supply and sea-level rise in a large coastal lagoon. *Marine Geology*, **167**(3–4), 261–284.

Newton, A., et al., 2014. An overview of ecological status, vulnerability and future perspectives of European large shallow, semi-enclosed coastal systems, lagoons and transitional waters. *Estuarine, Coastal and Shelf Science*, **140**, 95–122.

Nichols, M., 1989. Sediment accumulation rates and relative sea-level rise in lagoons. *Marine Geology*, **88**, 201–219.

Nichols, M., and Allen G., 1981. Sedimentary processes in coastal lagoons. In *Coastal Lagoon Research, Present and Future. Proceedings of an UNESCO, IABO Seminar*. Beaufort, NC: UNESCO. UNESCO Technical Papers in Marine Science, **33**, 27–80.

Nichols, M., and Biggs, R., 1985. Estuaries. In Davis, R. A., Jr. (ed.), *Coastal Sedimentary Environments*. New York: Springer, pp. 94–175.

Nichols, M., and Boon, J. D., 1994. Sediment transport processes in coastal lagoons. In Kjerfve, B. (ed.), *Coastal Lagoon Processes*. Elsevier Oceanography Series 60. Amsterdam: Elsevier Science Publications B.V, **60**, 157–219.

Oertel, G. F., Kraft, J. C., Kearney, M. S., and Woo, H. J., 1992. A rational theory for barrier-lagoon development. In Fletcher, C. H., and Wehmiller, J. F. (eds.), *Quaternary Coasts of the United States*. SEPM (Society for Sedimentary Geology), Tulsa, U.S.A., **48**, 77–87.

Osadczuk, A., Musielak, S., and Borówka, R. K., 2007. Why should the River Odra mouth area not be regarded as an estuary? A geologist's point of view. *Oceanological and Hydrobiological Studies*, **XXXVI**(2), 87–99.

Phleger, F.B., 1981. A review of some general features of coastal lagoons. In *Coastal lagoon research, present and future. Proceedings of an UNESCO, IABO Seminar*. Beaufort, NC: UNESCO. UNESCO Technical Papers in Marine Science, **33**, 7–14.

Skeat, W. W., 1893. *An Etymological Dictionary of the English Language*. Clarendon Press, Oxford, 844 pp.

Tagliapietra, D., Sigovini, M., and Ghirardini, A. V., 2009. A review of terms and definitions to categorise estuaries, lagoons and associated environments. *Marine and Freshwater Research*, **60**, 497–509.

Valiela, I., 1995. *Marine Ecological Processes*, 2nd edn. New York: Springer.

Widdows, J., and Brinsley, M., 2002. Impact of biotic and abiotic processes on sediment dynamics and the consequences to the structure and functioning of the intertidal zone. *Journal of Sea Research*, **48**, 143–156.

Cross-references

Coasts
Estuary, Estuarine Hydrodynamics
Integrated Coastal Zone Management
Marine Evaporites
Marine Sedimentary Basins
Ocean Margin Systems
Reef Coasts
Relative Sea-level (RSL) Cycle
Sea-Level

LAMINATED SEDIMENTS

Alan Kemp
Ocean and Earth Science, National Oceanography Centre Southampton, University of Southampton Waterfront Campus, Southampton, UK

Synonyms

Marine varves

Laminated sediments are characterized by the presence of planar, bed-parallel layers that originate from changes in particle composition or grain size, with thickness of the order 1 cm or less. The physical processes involved in the transport and deposition of sedimentary particles may produce parallel laminae in sands or silts, but the term "laminated sediment" is more commonly associated with the formation of laminae by successive depositional or "flux" events in hemipelagic or pelagic settings. Such laminae may be of biogenic origin, formed by the remains of marine phytoplankton such as diatoms or coccolithophores, or in marine systems nearer land, may comprise terrigenous clay and silt particles.

Since the pioneering bottom camera observations of the deep sea in the early 1980s, it has been understood that pelagic sediments are formed by rapid settling from seasonal productivity events that occur in the surface ocean. Rapid sedimentation to the sea floor, within days to weeks, is mediated by aggregates of phytoplankton, often enhanced by sticky, transparent extrapolymer particles and by zooplankton fecal pellets. Normally this seasonal input is rapidly reprocessed on the sea floor by the feeding activity of benthic surface dwellers and burrowers that effectively homogenize the sediment to depths of several centimeters. When the activity of benthic organisms is suppressed or eliminated, laminated sediments that record the seasonal flux cycle are preserved. The benthos is most commonly suppressed by low levels of bottom water oxygenation. This oxygen depletion generally occurs in basins with sills where the bottom waters are not in direct contact with the open ocean and so are not renewed.

Silled basins in which laminated sediments are preserved range in size from coastal fjords such as the Saanich Inlet in British Columbia to the California borderland basins including the Santa Barbara Basin. On even larger scales preservation of laminated sediments also occurs in marginal seas with shallow sills such as the Baltic Sea or the Black Sea. Bottom water oxygenation may also be sufficiently reduced to preserve laminae beneath highly productive upwelling regimes over open shelves, such as the Gulf of California or the Peru Shelf, where the intense flux of organic matter uses all available water column dissolved oxygen. In many of these marginal settings, there is an alternation of biogenic algal remains, often formed by the spring bloom or "fall dump" with laminae of terrigenous sediment commonly supplied by river runoff from seasonal rains. It is common to observe an annual couplet of laminae (biogenic and terrigenous), but up to 22 individual laminae per year have been identified in the Saanich Inlet which has rates of deposition exceeding 1 cm per year.

A further mechanism for the preservation of laminae is the physical suppression of benthic activity by the rapid deposition of tough mats of diatoms in "laminated diatom mat deposits." These are generated by the rapid flux of diatoms beneath oceanic frontal systems and occur extensively in ancient sediments of the equatorial Pacific and around the Southern Ocean.

Annually laminated marine sediments are also referred to as "marine varves" and may be used to develop marine varve chronologies to form a method of establishing the absolute and relative timing of climatic events. Time series of interannual variability are also derived from lamina sequences and are used to investigate past variation in the frequencies of climatic phenomena such as El Niño.

Bibliography

Hughen, K. A., Overpeck, J. T., Peterson, L. C., and Trumbore, S., 1996. Rapid changes in the tropical Atlantic region during the last deglaciation. *Nature*, **380**, 51–54.

Kemp, A. E. S. (ed.), 1996. Palaeoclimatology and palaeoceanography from laminated sediments. Geological Society of London, Special Publication, 116, 258pp.

Kemp, A. E. S., Pike, J., Pearce, R. B., and Lange, C. B., 2000. The "fall dump": a new perspective on the role of a shade flora in the annual cycle of diatom production and export flux. *Deep-Sea Research II*, **47**(2129), 2154.

Kemp, A. E. S., Pearce, R. B., Grigorov, I., Rance, J., Lange, C., Quilty, P., and Salter, I., 2006. The production of giant marine diatoms and their export at oceanic frontal zones: implications for Si and C flux from stratified oceans. *Global Biogeochemical Cycles*, **20**, GB4S04, doi:10.1029/2006GB002698.

LAYERING OF OCEANIC CRUST

Juan Pablo Canales
Woods Hole Oceanographic Institution, Woods Hole, MA, USA

Magma generated by decompression melting of the upwelling mantle beneath mid-ocean ridges (MORs) rises buoyantly and accumulates in crustal magma chambers (e.g., Forsyth, 1992). The long-held view is that oceanic crust is built from in situ crystallization of melts in these reservoirs, as well as from melts extracted from the magma chamber(s) in the form of dikes that eventually break the seafloor during submarine eruptions (e.g., Sinton and Detrick, 1992). These processes result in oceanic crust that is magmatically layered, with a lower section typically 4–5 km thick composed of mafic plutons (gabbros), a 1–2 km thick layer of sheeted basaltic dikes above it, and a carapace of a few to several hundred meters thick of extrusive basalts (pillow lavas and sheet flows), known as the Penrose model of oceanic crust (Penrose Conference participants, 1972). Although these units all have common parental magmas, they exhibit distinct elastic properties because of differences in mineralogical textures produced by crystallization processes and different porosity structure (i.e., the shallower extrusive section is highly porous, with porosity decreasing with depth throughout the crust). As a result, the speed of seismic waves propagating through the crust increases with depth while at the same time the rate at which seismic velocities increase with depth (velocity gradient) diminishes with depth. These seismic layers have been historically known as seismic Layer 2 (upper and mid-crust, subdivided in 2A and 2B based on changes in velocity gradient) and Layer 3 (lower crust) (e.g., White et al., 1992). To a large degree they correspond to the lithological layering of the oceanic crust, although deviations from this direct correlation are known to exist (Christeson et al., 2007; Detrick et al., 1994; Wilson et al., 2006).

The lithological and seismic layering described above has successfully explained many observations made at ophiolites (sections of oceanic crust exposed on land) as well as observations at crust formed at MORs with fast-spreading rates (>80 mm/year full rate), where magmatism is the dominant mode of lithospheric accretion. However, it is becoming increasingly accepted that the Penrose model fails to explain many features observed at MORs where plates spread apart at slower rates. Along slow- and ultra-slow-spreading MORs, the lithosphere is generally thicker and colder than at fast-spreading MORS, tectonic activity is a dominant process accommodating plate separation, and magmatism is highly variable both in time and space (e.g., Dick et al., 2010; Michael et al., 2003; Schroeder et al., 2007). In these settings, the interplay between magmatic and tectonic extension and their space-time variability result in oceanic lithosphere in which the crust is no longer stratified but disrupted, forming a heterogeneous mixture of magmatic and mantle rocks (Cannat, 1993).

Bibliography

Cannat, M., 1993. Emplacement of mantle rocks in the seafloor at mid-ocean ridges. *Journal of Geophysical Research*, **98**, 4163–4172.

Christeson, G. L., McIntosh, K. D., and Karson, J. A., 2007. Inconsistent correlation of seismic layer 2a and lava layer thickness in oceanic crust. *Nature*, **445**, 418–421.

Detrick, R. S., Collins, J. A., Stephen, R. A., and Swift, S. A., 1994. In situ evidence for the nature of the seismic layer 2/3 boundary in oceanic crust. *Nature*, **370**, 288–290.

Dick, H. J. B., Lissenberg, C. J., and Warren, J. M., 2010. Mantle melting, melt transport, and delivery beneath a slow-spreading ridge: the paleo-MAR from 23°15'N to 23°45'N. *Journal of Petrology*, **51**(1–2), 425–467.

Forsyth, D. W., 1992. Geophysical constraints on mantle flow and melt migration beneath mid-ocean ridges. In Phipps Morgan, J., Blackman, D. K., and Sinton, J. M. (eds.), *Mantle flow and melt generation at mid-ocean ridges*. Washington, DC: AGU, pp. 1–65.

Michael, P. J., Langmuir, C. H., Dick, H. J. B., Snow, J., Goldstein, S. L., Graham, D. W., Lenhert, K., Kurras, G. J., Jokat, W., Mühe, R., and Edmonds, H. N., 2003. Magmatic and amagmatic seafloor generation at the ultraslow-spreading Gakkel ridge, Arctic Ocean. *Nature*, **423**, 956–961.

Penrose Conference participants, 1972. GSA Penrose Field Conference on Ophiolites. *Geotimes*, **17**, 24–25.

Schroeder, T., Cheadle, M. J., Dick, H. J. B., Faul, U. H., Casey, J. F., and Kelemen, P. B., 2007. Nonvolcanic seafloor spreading and corner-flow rotation accommodated by extensional faulting at 15°N on the Mid-Atlantic Ridge: A structural synthesis of ODP Leg 209. *Geochemistry, Geophysics, Geosystems*, **8**(6), Q06015, doi:10.1029/2006GC001567.

Sinton, J. M., and Detrick, R. S., 1992. Mid-ocean ridge magma chambers. *Journal of Geophysical Research*, **97**, 197–216.

White, R. S., McKenzie, D., and O'Nions, R. K., 1992. Oceanic crustal thickness from seismic measurements and rare earth element inversions. *Journal of Geophysical Research*, **97**, 19683–19715.

Wilson, D. S., Teagle, D. A. H., Alt, J. C., Banarjee, N. R., Umino, S., Miyashita, S., Acton, G. D., Anma, R., Barr, S. R., Belghoul, A., Carlut, J., Christie, D. M., Coggon, R., Cooper, K. M., Cordier, C., Crispini, L., Durand, S. R., Einaudi, F., Galli, L., Gao, Y., Geldmacher, J., Gilbert, L. A., Hayman, N. W., Herrero-Bervera, E., Hirano, N., Holter, S., Ingle, S., Jiang, S., Kalberkamp, U., Kerneklian, M., Koepke, J., Laverne, C., Lledo-Vasquez, H. L., Maclennan, J., Morgan, S., Neo, N., Nichols, H. J., Park, S.-H., Reichow, M. K., Sakuyama, T., Sano, T., Sandwell, R., Scheibner, B., Smith-Duque, C. E., Swift, S. A., Tartarotti, P., Tikku, A. A., Tominaga, M., Veloso, E. A., Yamasaki, T., Yamazaki, S., and Ziegler, C., 2006. Drilling to gabbro in intact ocean crust. *Science*, **312**, 1016–1020.

Cross-references

Crustal Accretion
Cumulates
Gabbro
Mid-ocean Ridge Magmatism and Volcanism
Ophiolites
Submarine Lava Types

LITHOSPHERE: STRUCTURE AND COMPOSITION

Martin Meschede
Institute of Geography and Geology, Ernst-Moritz-Arndt University, Greifswald, Germany

Definition

The lithosphere (from the Greek words *lithos* [λίθος] meaning rocky and *sphaira* [σΦαῖρα] meaning sphere) is defined as the outermost rigid shell of the Earth (Skinner et al., 2003). It comprises the crust and the upper portion of the mantle which is stiff and rigid and behaves elastically over millions of years. The mantle part of the lithosphere is named the "lithospheric mantle." Beneath the lithospheric mantle follows the asthenosphere (from the Greek words *asthenés* [ἀσθενής] meaning weak and *sphaira* [σΦαῖρα] meaning sphere). Both the lithospheric mantle and the upper part of the asthenosphere are mainly composed of peridotitic rocks.

Introduction

The crust overlying the lithospheric mantle is either of oceanic type which is mainly composed of gabbroic and basaltic rocks resulting in a mean density of 3.0 g/cm^3, or of continental type with a gabbroic/granulitic layer at its base and a variety of rock types in the upper part formed by magmatic differentiation, metamorphism, and/or sedimentation over long periods of the Earth's history (Figure 1). The mean density of the continental crust is 2.8–2.9 g/cm^3 in the deeper portion and 2.6–2.7 g/cm^3 in the upper portion. The boundary between the Earth's crust and the lithospheric mantle is known as the Mohorovičić discontinuity (short form: Moho; Figure 1), named after the Croatian geophysicist Andrija S. Mohorovičić (1857–1936).

Oceanic lithosphere is continuously formed at mid-oceanic ridges and returned into the mantle at subduction zones. Oceanic lithosphere is, therefore, much younger than continental lithosphere. The oldest oceanic lithosphere is about 185 myrs old (Müller et al. 2008), whereas continental lithosphere comprises remnants of all periods of the Earth's history. The oldest parts of

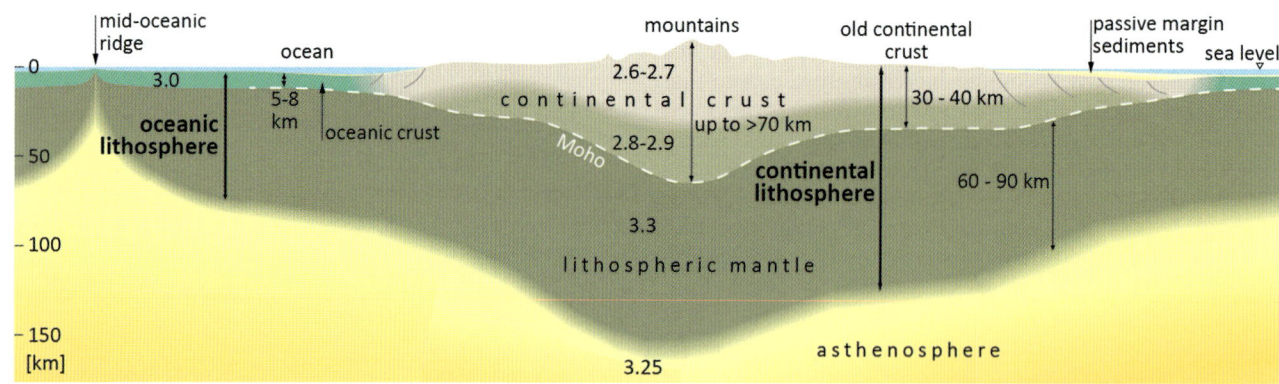

Lithosphere: Structure and Composition, Figure 1 Variations in thickness of continental and oceanic lithosphere. Numbers refer to mean density (g/cm³). Moho, Mohorovičić discontinuity.

continental lithosphere are represented by cratons which are composed of billions of years old mostly metamorphic and/or magmatic rocks. The lithosphere of cratons is thicker and less dense than typical lithosphere and acts as a stabilizer for these regions.

Structure

The oceanic lithosphere is composed of the lithospheric mantle and the oceanic crust with mainly gabbroic and basaltic rocks formed at the oceanic spreading centers. The thickness of the oceanic lithosphere increases with increasing distance to the spreading center because new lithospheric material is continuously added at the base of the lithosphere by conductive cooling (Figure 2). At about 1200 °C, the partially molten asthenospheric material gets completely solid and is converted into lithospheric mantle. The density of the solid material increases a little from 3.25 g/cm³ for the asthenosphere to about 3.30 g/cm³ since thermal contraction makes it denser than the asthenospheric material. As a consequence the mean density of the oceanic lithosphere as an entirety (lithospheric mantle plus oceanic crust) increases with the thickness of the lithospheric mantle. At a certain point depending on the spreading velocity and the distance from the spreading center, the stable layering changes into instability with a slightly denser lithosphere on top of the lighter asthenosphere (Figure 2). This configuration applies for lithosphere older than 80–90 myrs in relation to the onset of spreading at a mid-oceanic ridge. With an instable layering, a new subduction zone may be initiated spontaneously by gravitational downwelling of old oceanic lithosphere.

The continental lithosphere is composed of the lithospheric mantle and the continental crust with a large variety of rocks formed during the entire Earth's history. Continental crust is unsubductable because of its little density. Although the lower part is made up of gabbroic/basaltic and granulitic rocks, the mean density of the entire continental crust is about 2.7 g/cm³ and thus significantly lower than oceanic crust (Figure 1). Rocks of the upper portion of the continental crust were formed mostly by processes that changed the primary composition of the parent material. Basaltic magmas differentiate into lighter melts, forming felsic (word combines letters from feldspar and silica), plutonic, and volcanic rocks (e.g., granite, granodiorite, and, resp., rhyolite, dacite, etc.) that are enriched in silica and light elements (O, Al, Na, K). Erosion, transportation, and sedimentation result in the formation of sedimentary rocks, and, caused by continuous sedimentation and orogenic processes, the rocks may be metamorphosed by elevated pressure and temperatures. During the Earth's history, the amount of continental crust has been increasing, and the growth appears to have occurred in spurts of increased activity corresponding to the preservation of juvenile continental crust due to the formation of supercontinents (Condie, 2000). It is assumed that basaltic eruptions of superplumes and the subsequent magmatic and erosional processing of the basaltic protolith are the most important contribution of crustal growth (Albarède, 1998).

The thickness of the lithosphere depends on the plate tectonic setting. Normally, an evolved continental lithosphere has a thickness of about 90–120 km; oceanic lithosphere is slightly thinner with 70–90 km (Figure 1). This is mainly the result of the different thicknesses of oceanic vs. continental crust. Oceanic crust has a thickness of 5 to mostly 8 km, whereas an evolved continental crust has an average thickness of 30–35 km. Abnormal thin lithosphere exists in continental rifts, where the continental lithosphere may be less than 50 km thick, and at mid-oceanic spreading centers, where new oceanic lithosphere is formed. In these narrow zones, the lithosphere is confined to a thin layer (2–5 km) of oceanic crust only. The oceanic lithosphere is getting thicker with increasing distance from the spreading ridge (see above). Continental lithosphere may be substantially thickened during orogenetic processes. The collision of two continental plates results in the development of large mountain ranges such as the Himalayas or the Alps where overthrusting of large portions of the continental crust occurs. This may result in

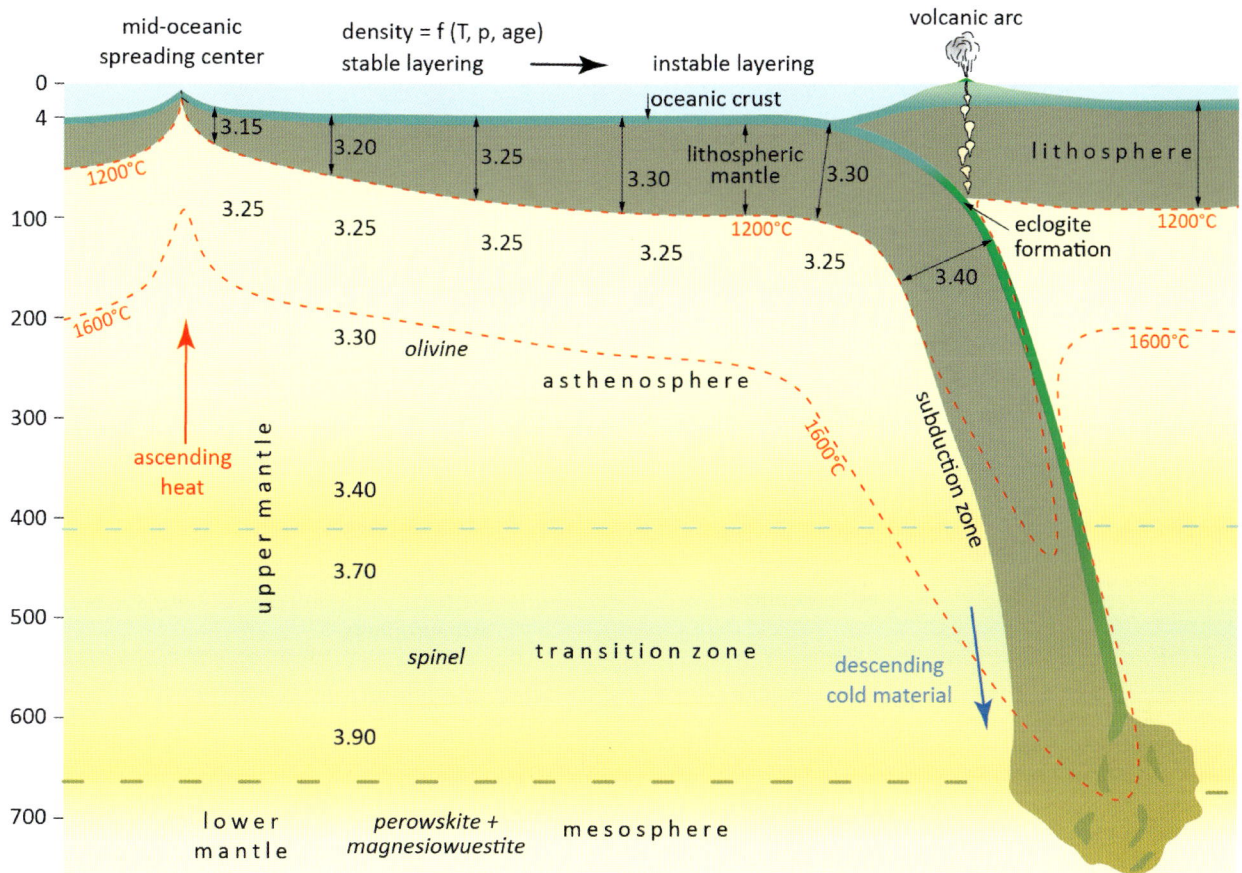

Lithosphere: Structure and Composition, Figure 2 Increasing thickness and mean density of the oceanic lithosphere as a function of temperature (*T*), pressure (*p*), and age (distance to the mid-oceanic spreading center). The stable layering becomes instable with the increasing age of the lithosphere. Numbers refer to mean density of total oceanic lithosphere and upper mantle, respectively.

the thickening of the continental lithosphere to more than 200 km and a doubling of the crustal part of the lithosphere which may sum up to more than 75 km in maximum (Figure 1). This yields an isostatically instable situation which over long periods of time will be compensated by surface erosion and isostatic rebound until the average thickness of continental crust of 30–35 km is restored.

Composition

The lithosphere is the most rigid layer of the Earth. It behaves far less ductile than the asthenosphere, which is partially molten (from 1 % to 5 %, higher percentage underneath mid-oceanic spreading ridges) and thus able to provide the mobility necessary for the tectonic movement of the lithospheric plates. However, the exact composition of the mantle is not known with certainty. For the upper mantle, it is inferred from rock material found as xenoliths in volcanic lavas originating from deep magma sources of up to 300 km depth and from ophiolites which are considered to represent lithospheric mantle and oceanic crust material. The mean composition of the upper mantle rocks is geochemically determined as 45 % SiO_2, 38 % MgO, 8 % FeO, 4.5 % Al_2O_3, 3.5 % CaO, and 1 % other additional compounds (Schubert et al., 2001). In terms of mineralogic composition, it consists mostly of olivine (~58 %), a magnesium-silicate undersaturated in silica ($(Mg,Fe)_2[SiO_4]$), of two different pyroxenes, the orthopyroxene enstatite ($Mg_2[Si_2O_6]$; ~14 %) and the clinopyroxene diopside ($CaMg[Si_2O_6]$; ~16 %), and garnet (~12 %). Because of increasing pressure in deeper parts of the mantle, the olivine is transformed into the denser crystallographic spinel structure in the transition zone below the asthenosphere at about 400–410 km depth (Figure 2), where the seismic velocity strongly increases.

Typical mantle rock types formed by the minerals mentioned above belong to the group of peridotites. The asthenosphere consists on average of a garnet-bearing lherzolite as a specific type of peridotite (type locality at Lac de Lherz, French Pyrenees). It is mostly composed of olivine, orthopyroxene, clinopyroxene, and garnet. The lherzolitic composition is modified when the asthenospheric material ascends into a wedge below a mid-oceanic spreading center. The rise of peridotite causes

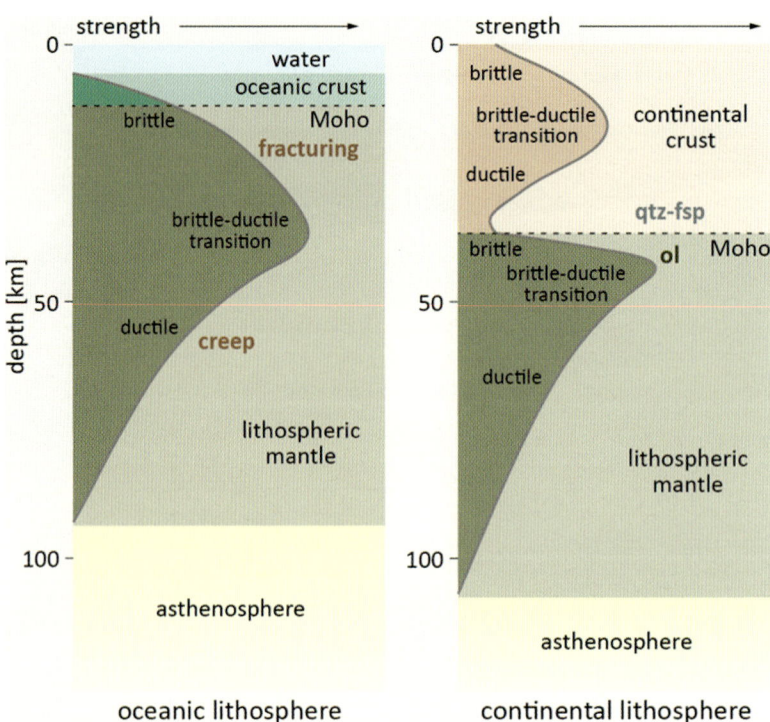

Lithosphere: Structure and Composition, Figure 3 Strength profiles of oceanic vs. continental lithosphere (modified after Kearey et al. 2009). *qtz-fsp* quartzo-feldspathic rocks, *ol* olivine-dominated peridotite.

an increase in partial melting because of pressure reduction. This melt has a basaltic composition and forms the new oceanic crust at spreading centers. The original lherzolitic peridotite of the asthenosphere changes into a harzburgite composition (type locality at Bad Harzburg, Germany) which is mainly composed of olivine and orthopyroxene and dunite (mostly olivine). The clinopyroxene predominantly migrates into the basaltic melt at the spreading centers. Eclogites with omphacite (a pyroxene) and garnet as main compounds develop at subduction zones when gabbroic/basaltic rocks reach high-temperature/low-pressure metamorphic conditions with the descending slab of oceanic lithosphere (Figure 2). All observed mantle material, which is today exposed, for instance, in ophiolitic assemblages, is not primary but has had a liquid fraction removed by partial melting. The removed material is of basaltic composition. To reconstruct the original composition, Ringwood (1962) assumed that the upper part of the lithospheric mantle consists mainly of dunite and peridotite. More primitive material follows further downward. Chemically, these rocks correspond to a mixture of one part of basalt to four parts of dunite. He named this hypothetical primitive material "pyrolite" which is taken as an equivalent for primitive fertile mantle.

Strength

Depending on loading conditions and time scale, the lithosphere exhibits elastic, brittle, or ductile properties (Kirby, 1983; Turcotte and Schubert, 2002). At relatively low temperatures near the surface, rocks tend to break under stress. This is called brittle deformation. At greater depth, as temperature and pressure increase, it is more likely that rocks will accommodate stress by changing shape or deforming (compression, stretching, bending), rather than breaking. This is called ductile deformation which means the ability of a solid material to deform under stress without breaking. The strength of the lithosphere (i.e., the total force per unit width necessary to deform a lithospheric section at a given strain rate) is a function of temperature, composition, and crustal thickness. Confining pressure increases the rock strength but, on the other hand, increasing temperature weakens the rocks. At low temperatures frictional processes dominate and brittle deformation occurs. With increasing depth, the temperature overcomes the strengthening effect of confining pressure, leading to a ductile behavior (Figure 3).

The large-scale strength characteristics of oceanic and continental lithosphere differ. The oceanic lithosphere increases in strength down to the brittle-ductile transition in the lithospheric mantle at a depth of about 30–40 km. Beneath the brittle-ductile transition zone, the ductility increases and the strength decreases. The continental lithosphere has a bimodal strength profile. The first maximum in strength occurs at the brittle-ductile transition in the continental crust at a depth of about 20–30 km, where most of the earthquakes nucleate. A second maximum is observed below the Moho in the lithospheric mantle at a depth of about 40–50 km where the rock composition changes from quartzo-feldspathic rocks to olivine-dominated peridotite (Figure 3).

Lithospheric rocks are considered elastic, but not viscous, and exhibit at least a minimum of strength. At some depth the temperature reaches a critical point when rocks start behaving like a viscous fluid. That depth is treated as the boundary between lithosphere and asthenosphere. The asthenosphere is viscous because partial melting occurs in the order of 1–5 %, which allows for the movement of lithospheric plates relative to each other. This process is known as plate tectonics.

Bibliography

Albarède, F., 1998. The growth of continental crust. *Tectonophysics*, **296**, 1–14.
Condie, K., 2000. Episodic continental growth models: afterthoughts and extensions. *Tectonophysics*, **322**, 153–162.
Kearey, P., Klepeis, K. A., and Vine, F., 2009. *Global Tectonics*, 3rd edn. London: Wiley-Blackwell, 496 pp.
Kirby, S. H., 1983. Rheology of the lithosphere. *Reviews Geophysical Space Physics*, **21**, 1458–1487.
Müller, R. D., Sdrolias, M., Gaina, C., and Roest, W. R., 2008. Age, spreading rates and spreading symmetry of the world's ocean crust. *Geochemistry, Geophysics, Geosystems*, **9**, Q04006, doi:10.1029/2007GC001743.
Ranalli, G., 1995. *Rheology of the Earth*, 2nd edn. London: Chapman & Hall, 413 pp.
Ringwood, A. E., 1962. A model for the upper mantle. *Journal of Geophysical Research*, **67**, 857–867.
Schubert, G., Turcotte, D. L., and Olson, P., 2001. *Mantle Convection in the Earth and Planets*. Cambridge: Cambridge University Press, 940 pp.
Skinner, B. J., Porter, S. C., and Park, J., 2003. *The Dynamic Earth: An Introduction to Physical Geology*. London: Wiley, 648 pp.
Turcotte, D. L., and Schubert, G., 2002. *Geodynamics*, 2nd edn. Cambridge: Cambridge University Press, 456 pp.

Cross-references

Gabbro
High-pressure, Low-temperature Metamorphism
Mid-ocean Ridge Magmatism and Volcanism
Mohorovičić Discontinuity (Moho)
Oceanic Spreading Centers
Ophiolites
Peridotites
Silica
Subduction

LITHOSTRATIGRAPHY

Rüdiger Henrich
Department of Geosciences, University of Bremen, Bremen, Germany

Lithostratigraphy comprises the categorization of sediment/rock strata based on lithology (e.g., color, texture, grain size, and composition), physical properties (seismic character, nature of well log response, magnetic susceptibility), and structural sedimentary inventory.

A broad spectrum of methods is applied to describe and correlate marine lithostratigraphic units. This includes visual core description, smear slide analysis, digital imaging, diffuse color reflectance spectrophotometry, magnetic susceptibility, physical properties (gamma ray attenuation, moisture, density, sonic), grain size analysis, and geochemical analysis (e.g., $CaCO_3$, TOC, element composition measured with an XRF sediment scanner).

Fundamental unit of lithostratigraphy is the formation comprising beds as subunits that may reveal similar lithologies or were deposited in a certain environment. The higher hierarchy of classification terms includes groups and supergroups. Sediments deposited under similar conditions revealing distinct lithological, structural, and paleontological characteristics are summarized as facies. Walther's law, the most important principle concerning lateral variation and vertical stacking of facies, implies that facies that occur in conformable vertical succession of strata also occur in laterally adjacent environments. Also, it is essential to realize that the extension of facies bodies is always limited. They laterally and vertically pinch out and/or transit into another facies. Hence, the spatial arrangement of facies is in the form of more or less elongated lens-shaped bodies that are vertically stacked on top of each other. In the marine realm, pelagic facies often develops far extending flat lens-shaped units, whereas continental margin and shelf settings comprise less-extensive and complex-shaped facies bodies often interfingering in complex manner. These aspects bear essential implications when applying lithostratigraphic correlations. The use of bed and facies boundaries as chronostratigraphic time lines is limited and thus should always be calibrated with chrono- and biostratigraphic methods. In open ocean settings, different surface water masses are characterized by typical pelagic communities with distinct feed webs and production patterns (see Hüneke and Henrich (2011) for review). In addition, changes in surface water mass configurations and their properties, in particular their pelagic production rate, often correspond with climate shifts and thus represent time lines that can be traced over far distances. Hence, certain pelagic parameter records, for example, carbonate records, have been successfully applied as stratigraphic tools. However, due to variable lateral advection of terrigenous material (see Henrich and Hüneke 2011), the applicability of carbonate records as stratigraphic tools is rather limited in continental margin and shelf settings. On the other hand, distinct short-term sedimentation events like volcanic eruptions (e.g., volcanic ash layers) or glaciomarine diamictons and IRD layers (i.e., Heinrich events) deposited during the rapid decay of the marine-based parts of huge ice sheets (see review by Henrich and Hüneke (2011)) provide excellent time lines for lithostratigraphic correlations.

Bibliography

Henrich, R., and Hüneke, H., 2011. Hemipelagic advection and periplatform sedimentation. In Hüneke, H., and Mulder, T. (eds.), *Deep-Sea Sediments*. Amsterdam: Elsevier. Developments in Sedimentology, Vol. 63, pp. 215–351.

Hüneke, H., and Henrich, R., 2011. Pelagic sediments: modern and ancient. In Hüneke, H., and Mulder, T. (eds.), *Deep-Sea Sediments*. Amsterdam: Elsevier. Developments in Sedimentology, Vol. 63, pp. 353–396.

M

MAGMATISM AT CONVERGENT PLATE BOUNDARIES

Robert J. Stern
Department of Geosciences, University of Texas at Dallas, Richardson, TX, USA

Synonyms

Arc magmatism; Island arc; Magmatism; Subduction zone magmatism

Definition

Convergent margin magmatism – also called arc magmatism – occurs at straight or curved alignments (arcs) of discrete volcano-plutonic complexes that form above a subduction zone, where oceanic plate material (lithosphere plus sediments) is recycled into Earth's mantle (Figure 1). Convergent plate margins are one of the two types of plate boundaries where abundant igneous activity is caused by plate tectonics (the other is associated with divergent plate boundaries or mid-ocean ridges (MOR); hot-spot or within-plate igneous activity may be related to mantle plumes). Convergent margins and associated igneous activity occur in three regions: (A) around most of the Pacific Ocean (circum-Pacific Ring of Fire), (B) along much of the southern margin of Eurasia; and (C) in two short arcs in the western Atlantic Ocean (Figure 2).

Principal Components: Convergent margins are subdivided into forearc, volcanic front, and backarc (Figure 1) with increasing distance from the plate boundary, marked by the trench. Forearcs are stable regions underlain by cool mantle, in contrast to the volcanic front, which is underlain by hot mantle. Forearc crust is generated when the subduction zone first formed and is often emplaced on land as ophiolite (Stern et al., 2012).

Accretionary Wedges are found only for active continental margins with high sedimentation rates. Island arc magmatic activity is concentrated at the volcanic front (sometimes called the magmatic front, because not all magmas erupt), defined as the locus of the first volcanoes inboard from the trench. Some island arcs are under tension. Extension can be so strong that true seafloor spreading occurs, forming a backarc basin (Taylor, 1995). Backarc basins are entirely submarine, lying ~2–5 km below the sea surface. Actively spreading backarc basins are associated with the Aeolian, S. Sandwich, Tonga-Kermadec, New Hebrides, Andaman-Nicobar, and Mariana arcs, and are also found in the western Mediterranean Sea (Figure 2). Many marginal seas originated as backarc basins. Active backarc basin spreading ridges erupt basalts that are similar in composition to those erupted at mid-ocean ridges except for containing much higher concentrations of magmatic water and fluid mobile trace elements. In contrast to subduction zones and associated island arcs, which remain active for several tens or even hundreds of millions of years, backarc basins stay active for only 10–20 million years, perhaps because continued widening due to seafloor spreading eventually separates the region of mantle upwelling from subduction-related water inputs needed for melt generation (Stern and Dickinson, 2010). Interarc rifts also can be found where island arcs experience extension, but not to the point of seafloor spreading. Interarc rifts are associated with the Ryukyu, Izu, and Aegean arcs.

Subduction zone and mantle wedge processes

Generally built above a dipping zone of earthquakes, convergent margin magmatism demonstrates that a subduction zone operates deep beneath a region. Subduction zones mark the subsurface expression of convergent plate boundaries, one of the three main types of plate tectonic boundaries, the others being divergent and transform plate

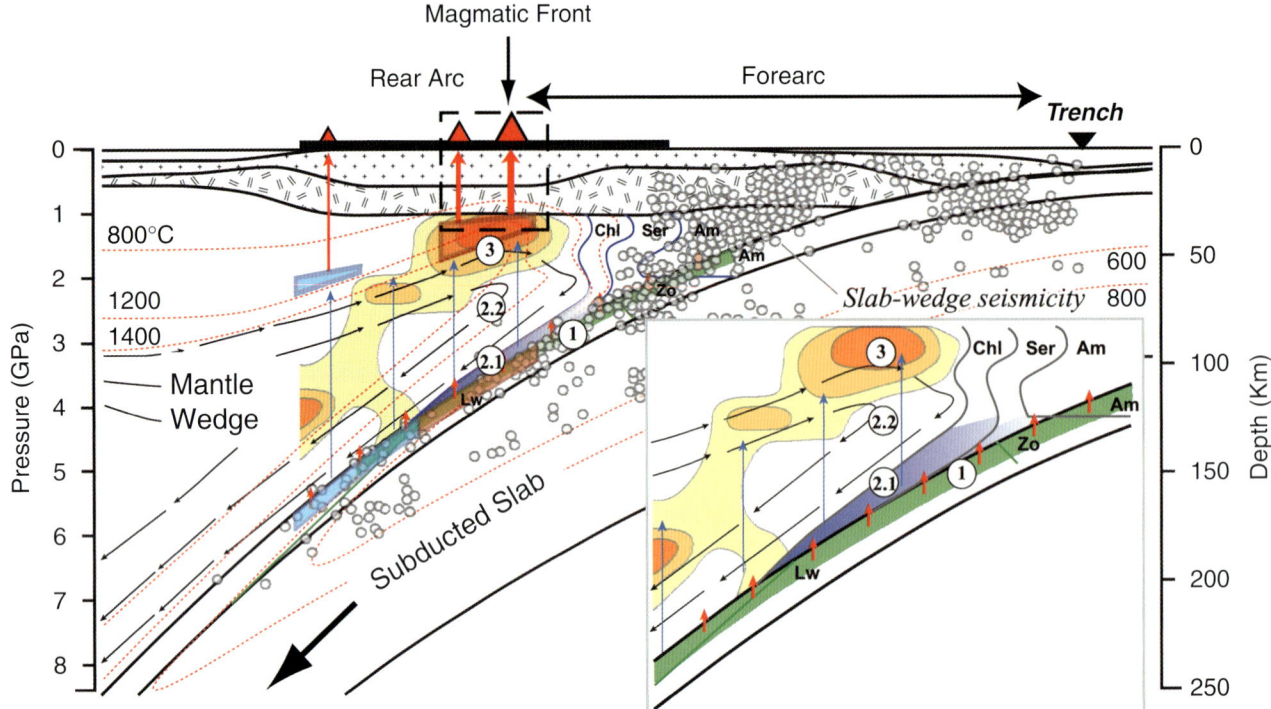

Magmatism at Convergent Plate Boundaries, Figure 1 Simplified cross-section of a convergent plate margin and underlying subduction zone, emphasizing flow of asthenospheric mantle in Mantle Wedge (*thin black arrows*) driven by motion of subducted oceanic plate (Subducted Slab) accompanied by transfer of fluids and melts emanating from subducted oceanic crust and sediments to mantle wedge (*thin blue arrows*). Yellow and red regions indicate seismically slow regions based on seismic tomography beneath NE Japan (Nakajima et al., 2005); these are interpreted as partially melted mantle. Circled numbers indicate loci of important subduction-related magmagenetic processes. Open circles are earthquake locations. *Chl* chlorite, *Ser* serpentine, *Am* amphibole, *Lw* lawsonite, *Zo* zoisite. Important processes in subduction zone and mantle wedge: (1). Dehydration or melting of subducted sediment and altered oceanic crust. (2.1) Alteration of cold mantle just above subducted slab, which also descends and releases fluids. (2.2) Rise of fluids from metasomatized mantle sole up into hotter mantle regions. (3) Melting of the mantle (Kimura et al. 2009). *Dashed box* shows the approximate position of Figure 3 (Reproduced by permission of American Geophysical Union).

boundaries. There are 55,000 km of convergent plate boundaries on Earth (Figure 2; Lallemand, 1999). Subduction zones were first recognized by dipping arrays of earthquakes, called Wadati-Benioff zones (WBZs) after the two Japanese and American seismologists who recognized this geometry (see earthquake foci, Figure 1). Subduction zone earthquakes are unique in that they can occur much deeper (down to 670 km) than in any other tectonic environment, where seismicity is shallower than 30 km. Because earthquakes occur where rocks are cold enough to break, the greater depth indicates that subduction zones are unusually cold regions of Earth's mantle, due to downwelling of cold surface rocks. The somewhat puzzling recognition that convergent margin magmatism is associated with cold regions of mantle downwelling is explained by what happens where cold, wet, subducted seafloor mixes with hot circulating mantle (asthenosphere) above a subduction zone.

Oceanic plate material (slab) is ~100 km thick. It is overwhelmingly composed of relatively cool upper mantle lithosphere, ~6 km of variably altered oceanic crust, and a few hundred meters of sediment. These three components of the subducted slab play different roles in subduction zone processes. The density of the mantle component makes the slab sink, the altered oceanic crust provides most of the water to stimulate the overlying mantle to melt, and the thin sediments (<0.5 % of subducted slab mass) control the geochemically distinctive composition of convergent margin magmas. Because it has been seafloor for tens or even hundreds of millions of years, the upper parts of slabs (sediment, oceanic crust, and uppermost mantle) are cold and wet as the plate arrives at an oceanic trench and starts to sink into a subduction zone. This material is progressively squeezed and baked as it descends (1 in Figure 1), first liberating water and then at greater depth, partial melts of the subducted sediments. Subducted oceanic crust generally does not melt except where very young, hot seafloor or an active mid-ocean ridge is subducted or along subducted plate edges (Drummond and Defant, 1990). Hydrous fluids and melts rise into the overlying wedge of mantle, hydrating and altering (metasomatizing) its base (2.1 in Figure 1).

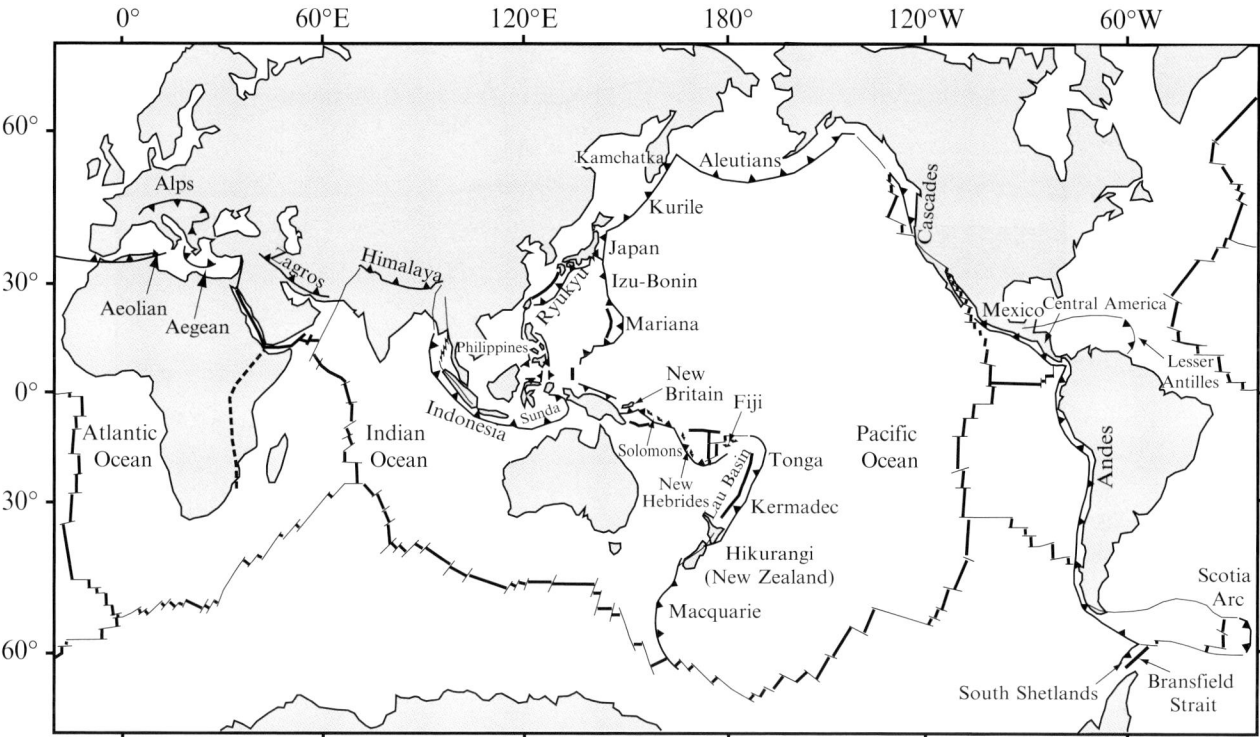

Magmatism at Convergent Plate Boundaries, Figure 2 Global map of major convergent and divergent plate boundaries, also showing the location of oceanic convergent margins (map modified from Stern (2002). Reproduced by permission of American Geophysical Union).

The wet base of the mantle wedge is viscously coupled with the sinking slab and is also carried down with, releasing more fluids as these descend together (2.2 in Figure 1).

Somehow – the general process is clear but the details are still being worked out – hydrous fluids, sediment melts, and (rarely) oceanic crust melts ascend through several tens of kilometers of downward-flowing mantle to reach mantle that is hot enough to melt when fluxed by hydrous fluids or sediment melts (3 in Figure 1). This is because the addition of such liquids reduces the melting temperature of mantle peridotite by several hundred °C (Grove et al., 2012). Partial melting of the mantle generally occurs where the subduction zone lies 100–130 km deep. Mantle melting forms primitive magma, called "basalt." Like basalt melts from other tectonic environments, those of convergent margins have low contents of Si and high contents of Mg. Because of the way they are generated, primitive convergent margin magmas contain much higher concentrations of water than primitive basalts from other tectonic environments, typically 3–4 wt% H_2O, compared to <0.5 wt% H_2O in mid-ocean ridge basalts (Stern, 2002). Arc magmas also are enriched in trace elements that are transported in fluids (especially alkali metals and alkaline earths K, Rb, Ba, and Sr; also U and Pb), sweated out of subducted sediments and altered oceanic crust.

Crustal processes

Although deep subduction zone processes are everywhere similar, eruptive products, eruptive style, and mineralization vary widely, largely depending on the crust on the overriding plate. If this is thick continental crust, then an active continental margin with towering volcanic mountains like the Andes results; if this is thin oceanic crust, then chains of mostly submerged volcanoes known as island arcs develop. These volcanic islands may rise from bases 2–3 km below sealevel and can be wholly submarine seamounts or partially subaerial if the summits of stratovolcanoes are tall enough to emerge from the sea. Whether they rise above sealevel or not, arc volcanoes are typically spaced 20–30 km apart (mean = 27 km; de Ronde et al., 2003). The total length of island arcs is ~22,000 km (de Ronde et al., 2003], about one third of the total length of mid-ocean ridges (Baker et al., 2008). About one third of this is made up of intraoceanic arcs (i.e., built on oceanic crust: Izu-Bonin-Mariana, Tonga-Kermadec, Aleutians, and South Sandwich arcs; Figure 2); the remainder are built on thinned or rifted continental crust, such as Indonesia or Japan Intra-oceanic arcs are particularly difficult to study because these are mostly under water but they are of special interest because constructing formation of juvenile arc crust is thought to be the first step in forming continental crust (Stern 2010).

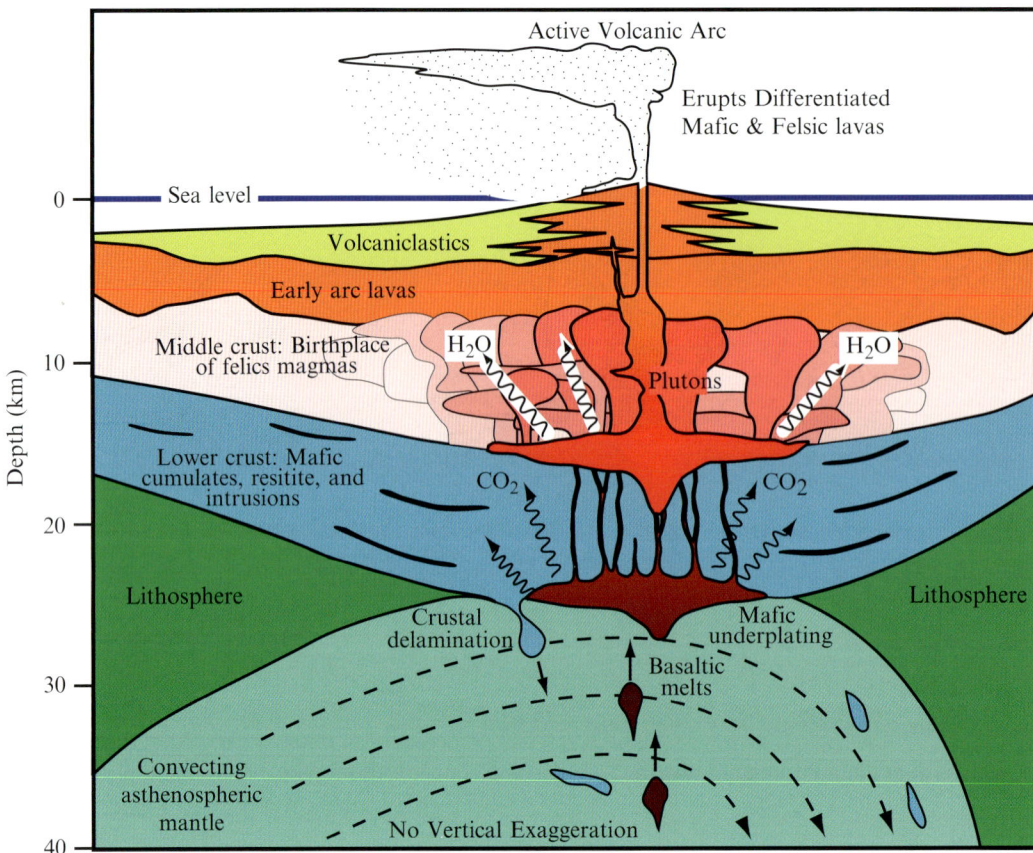

Magmatism at Convergent Plate Boundaries, Figure 3 Idealized convergent margin volcano-plutonic complex, from Stern (2002). Trench is to the right. Only an idealized section through an intraoceanic arc is shown; similar processes are expected beneath Andean-type arcs. Note that the asthenosphere is shown extending up to the base of the crust; delamination or negative diapirism is shown, with blocks of the lower crust sinking into and being abraded by convecting mantle. Regions where degassing of CO_2 and H_2O is expected are also shown (Reproduced by permission of American Geophysical Union).

Primitive basaltic melt is typically ~1,300 °C as it rises out of the mantle wedge but such hot, primitive melt rarely erupts without interacting with and being changed by the crust that it has to pass through. Some rising primitive melts heat the crust and cause it to melt, yielding secondary "anatectic" magmas, which form at much lower temperatures of ~800 °C. Other primitive basaltic melts pool in crustal magma chambers, fractionating as the cooling magma forms crystals that settle out, changing the composition of the residual melt. Both secondary melting and fractional crystallization yield melts that are significantly enriched in silica. These are felsic magmas, akin to granite, and known by their volcanic names: dacite and rhyolite. Felsic igneous rocks are important components of island arcs and are even more important parts of plutonic arc crust. Felsic magmagenesis is complemented by formation of denser lower crust gabbro and pyroxenitic upper mantle cumulates (Figure 3).

The abundance of felsic magmas in convergent margins partly results from the fact that these volcanoes remain over the same crustal tract for unusually long times. In contrast to volcanoes at divergent plate boundaries and hot spots on fast-moving plates, which have short and simple lives, convergent margin volcanoes and the magma plumbing system beneath them are more complex because they grow where magma flows up to the surface for millions of years. Such long volcano lifetimes result because the volcanic front occurs where the subducted plate is ~110 km deep. The volcanic front migrates if the dip of the subducted slab varies much, moving farther away from the trench if slab dip decreases, closer to the trench if slab dip increases. However, subduction zones are massive, slow-moving features that extend deep into the Earth. Unless something happens to disrupt well-established downwelling – such as subduction of buoyant continental crust or break-off of part of the slab – the subduction zone geometry and the location of the volcanic front, with its underlying magmatic plumbing system, will remain approximately in the same position for many millions of years.

Because the location of the volcanic front may not change much, arc volcanoes tend to develop "hot zones" at the base of the crust overlain by variably warm, weak crust surrounding conduits (Annen et al. 2006). This reflects cumulative heating by innumerable pulses of ~1,300 °C magma moving from the upper mantle through the crust towards the surface. The crust is further heated when basaltic magma stalls in it without erupting. Crustal melts and products of fractional crystallization are both felsic combine to generate andesitic and felsic melts. The more felsic the magma, the more polymerized it is, because Si and O link up to form long chains of silicon and oxygen that resist flow. The Si-rich, viscous nature of magmas are half of the reason that convergent margin eruptions are the most violent of among the three great tectonic settings.

The volatile-rich nature (especially H_2O, CO_2, and S) of convergent margin magmas is the other half of the reason that convergent margin magmas can erupt so explosively – and why most ore deposits are associated with ancient arcs. Of the three great magmatotectonic settings, convergent margin magmas contain by far the greatest concentration of magmatic gases and volatiles (Roggensack et al., 1997; Plank et al., 2013). Convergent margin high-Si melts inherit the water- and sulfur-rich nature of the mantle melt, which ultimately result from the fluids released by the subducted slab. The volatiles are dissolved in the magma when it forms in the upper mantle but are released as magma approaches the surface and pressure is greatly reduced. The exsolved volatiles turn into gas bubbles with a tremendous increase in volume. The gas bubbles can escape quietly through low-viscosity mafic magma but must break out violently if they are in viscous felsic magma.

A significant proportion of convergent margin melts crystallize several kilometers deep within the crust, where they form large bodies of crystalline rocks known as plutons (Figure 3). Plutons are similar in scale to the overlying volcano, typically several to a few tens of kilometers in diameter. Although lavas and other products of volcanoes characterize an active island arc, once a subduction zone stops operating (perhaps due to plate collision), the associated magmatic arc shuts down; erosion remove the volcanic rocks and exposes the plutonic roots of the arc. Because island arcs are elongate magmatic systems, an ancient arc can be identified by similarly elongate aggregations of plutons called batholiths. Batholiths, like island arcs, are typically about 100 km wide and thousands of kilometers long. Examples of batholiths include the Sierra Nevada of California, which is spectacularly exposed at Yosemite National Park, USA.

Volcanism

Eruptions from island arc volcanoes can be especially violent and dangerous. This is partly due to high volatile contents (especially water) in both primitive (basaltic) convergent margin magmas and more evolved felsic magmas, and this powers the violence of eruptions from volcanoes at convergent margins. Because the solubility of volatiles in magma decreases as pressure falls during magma ascent, volatiles are progressively released as magma rises towards the surface prior to eruption. This involves a tremendous increase in volume when pressure is low enough for magmatic volatiles to flash into gas. Depending on the viscosity of the magma, the rate of volatile diffusion from the melt into the gas phase varies, which – depending on magma ascent rate – can lead to volatile supersaturation in magmas and dramatic explosive bubble nucleation and greater magma acceleration. Because magmatic volatiles readily exsolve out of low viscosity mafic magmas and the bubbles can rise faster than the melt phase (i.e., are decoupled), basaltic eruptions tend to be less dangerous (but note that there are examples of more violent "Plinian" basaltic eruptions, such as 1886 Mt. Tarawera, New Zealand eruption). In contrast, in high viscosity felsic melt bubbles often remain coupled, driving high magma buoyancy and acceleration, making subaerial felsic eruptions the most violent on Earth.

Today, eruption violence is measured using the Volcanic Explosivity Index (VEI; Newhall and Self, 1982) a logarithmic scale (base 10) for erupted volume. The largest Cenozoic eruptions have VEI = 8: Yellowstone 630 ka; Toba 74 ka; Taupo 26.5 ka. These entail violent and catastrophic eruptions where >1,000 km^3 of magma is erupted over short durations of likely weeks to months.

Submarine arc volcanism is not yet well understood compared to subaerial volcanism but stunning discoveries (Embley et al., 2007) encourage further exploration. In some ways submarine volcanism is very different than subaerial counterparts. In contrast to eruptions into the atmosphere, eruption into water strongly modifies eruption and pyroclast transport processes due to the contrasting properties of water compared to air. Increasing hydrostatic pressure associated with increasing eruption depth influences volatile exsolution and particularly expansion in the shallow conduit during magma ascent. Magma fragmentation and pyroclast transport processes are also strongly modified in the submarine setting due to the greater heat capacity, density and viscosity of water compared with air.

Submarine arc volcanism is not yet well understood, however it is very likely that this mode of eruption shows as much variability in style as subaerial volcanism. Submarine basaltic eruptions range from quiet effusion (including pillow basalts), and low to moderate intensity explosive eruptions. Felsic eruptions are also common along submarine volcanic arcs, from effusive dome growth and spalling of the pumiceous carapace to low – moderate explosive cone-forming eruptions and likely catastrophic caldera-forming eruptions (Fiske et al., 2001; Smith et al., 2003). Subaqueous eruptions are poorly understood for several reasons, most notably the lack of witnessed eruptions, but it is clear from theory and seafloor exploration with submergence vehicles that increasing hydrostatic pressure has a profound effect on limiting explosivity (e.g., Allen and McPhie 2009). In the last decade remote sensing methods together with some good luck has led to the observations of both basaltic and felsic submarine eruptions (e.g., West Mata volcano in

Magmatism at Convergent Plate Boundaries, Figure 4 Schematic illustration of the major geological characteristics of mineral deposit types that typically occur in oceanic arc environment and back-arc spreading center, modified after Figure 14 of Lydon (2007).

the NE Lau Basin (Resing et al., 2011; https://www.youtube.com/watch?v=xRaEcGHHsVY), NW Rota in the Mariana Arc (Chadwick et al., 2008; https://www.youtube.com/watch?v=Cz4niudHwMs); Havre volcano in the Kermadec arc, NZ (Carey et al., 2014; Figure 2). Such events have dramatically increased our knowledge of submarine eruption and pyroclast transport processes, and challenged existing paradigms in volcanology.

Sulfur

Consonant with their volatile-rich nature, convergent margin magmas generally contain more S than those from other tectonic settings (Sharma et al., 2004). This is manifested in several ways. One is that lakes of molten sulfur are sometimes found on convergent margin volcanoes, both above and below sealevel. Sulfur pools at the bottoms of crater lakes exist on subaerial volcanoes in Central America, Indonesia, Andes, and Japan (Kim et al., 2011). Molten sulfur forms as a result of the disproportionation of magmatic sulfur dioxide (SO_2), released from degassing magma, according to the reaction $SO_2 + H_2O = H_2SO_4 + S$. Pools of molten sulfur in equilibrium with acid water is the result. Examples of active disproportionation associated with a submarine eruption has been investigated at NW Rota-1 in the Mariana arc, where yellowish clouds of newly-formed sulfur billowed out of the eruption pit http://www.youtube.com/watch?v=RY3a7rIFlEs. Pools of molten sulfur are also found in craters of some submarine volcanoes, for example the summits of Nikko and Daikoku in the Mariana Arc (Embley et al., 2007) and in Volcano O in the Lau Basin (Kim et al., 2011; Figure 2).

The combined effects of magmatic sulfur and explosive arc volcanism can affect climate. During Plinian eruptions, magmatic SO_2 is converted to small (0.1-1 mm diameter) droplets of sulfuric acid and, if injected into the stratosphere, can remain suspended for many months, encircling the globe. These droplets reflect incoming sunlight, cooling the Earth below and modifying climate. As an example of this, the 1991 eruption of Pinatubo in the Philippines resulted in Northern Hemisphere were lower than normal by up to 3 °C in the winters of 1991–1992 and 1992–1993 (Robock, 2000).

Mineral deposits

Sulfide mineral deposits are another important manifestation of magmatic sulfur in convergent margin magmas. Important mineral deposits are associated with modern and ancient island arcs, including interarc rifts and backarc basins (Figure 4). Arc mineral deposits are dominated by Cu-Pb-Zn-Au sulfides (Hannington et al., 2005). The style of mineralization is related to arc crust composition and thickness (Kessler, 1997). Some deposits form via submarine hydrothermal and magmatic degassing systems on the seafloor (de Ronde et al., 2003) and others are associated with subvolcanic intrusions above sealevel (Lydon, 2007). In both cases, mineralizing fluids transport metals to the site of deposition.

There are two sources of mineralizing fluids in submarine arc volcanoes: (1) metals derived through leaching of pre-existing rock by ambient surface water; and (2) metals derived from a degassing magma body. In the first case, seawater or groundwater is heated and acidified by reaction with degassing sulfur. Fluids associated with convergent margin magmas are very acidic and corrosive because these magmas release tremendous amounts of sulfur; this combines with H_2O to produce H_2SO_4, sulfuric acid. These hot, acid waters are produced when heated by a magma body, resulting in development of a hydrothermal convection cell. This draws in cold water to be heated and acidified above the magma body, which then rises to the surface along faults and fractures, which become upflow channels. Hydrothermal circulation and mineralization are enhanced in calderas, where major faults favor fluid circulation above a shallow magma body. The circulation of hot, acidic solutions through the surrounding rock leaches and breaks down primary igneous minerals into clays – hydrothermal ore deposits are invariably associated with a lot of extremely altered rocks – and such alteration liberates a lot of metals, which are carried away by the acidic fluids. In the second case, metals are released by magmatic degassing of metal-enriched volatiles. Most ore-bearing fluids reflect the combined effects of both hydrothermal leaching and magmatic degassing, and the relative contributions are likely to vary in time and space (Yang and Scott, 1996).

However the metal-bearing acid solutions are generated, the metals these transport will ultimately be deposited where and when the fluid interacts with a different chemical environment, especially where the rising acid fluid mixes with higher pH fluids, such as seawater. This can form massive sulfide deposits associated with backarc basin hydrothermal vents, which are very much like black smokers associated with mid-ocean ridges. Another style of mineralization occurs if the fluid rises into a region where P is low enough that the fluid boils away, leaving the metals behind. This is what happens in shallow subvolcanic intrusions such as porphyry deposits (Sillitoe, 1997). Phase separation (boiling) is also common in arc and even backarc basin hydrothermal systems. Some ore deposits that form in other tectonic environments reflect the generation of immiscible sulfide liquid, which separates from the silicate melt and sinks to the bottom to form mineral deposits, such as the giant Bushveld deposit of S. Africa. This mode of mineralization is not common in island arcs because the magma is too rich in H_2O and O_2 to form much reduced sulfur (sulfide) in the magma body.

The composition of the host magma can also exert important controls on mineralization. For example, sulfide deposits that are commonly associated with felsic igneous activity (both intrusive and extrusive) are typically Pb, In, Bi, and Sb (+/− Au, Ag)-rich, compared to those associated with predominantly mafic igneous rocks, which tend to be more Cu, Zn and Co (+/− Au, Ag)-rich (Timm et al., 2012). Similarly, styles of mineralization vary according to whether igneous activity is submarine (volcanogenic or Kuroko-type massive sulfides) or at shallow depth beneath a large volcano (porphyry and epithermal deposits).

The hydrothermal system associated with Brothers Volcano in the Kermadec arc, NZ (Figure 2) is a good example of submarine arc hydrothermal mineralization. Hydrothermal vents are found at three sites in the caldera at 1,300–1,800 m water depth associated with dacite and rhyodacite domes above a mafic substrate. Hot (max. 305 °C), acidic (pH > 1.9) fluids vent through sulfide chimneys up to 7 m tall (de Ronde et al., 2005). Hydrothermal vents are concentrated along caldera faults. Two main types of chimneys predominate: Cu-rich (up to 28.5 wt% Cu) and more common Zn-rich (up to 48.5 wt% Zn; de Ronde and 18 others 2011). Brothers hydrothermal fields only ~3 km apart provide useful examples of the range of hydrothermal processes and sources seen in submarine arc volcanoes. The Cone hydrothermal site is today more strongly affected by degassed magmatic components than is the NW Caldera site. These two Brothers vent fields represent near end-members of a continuum that spans magma-dominated and seawater-dominated hydrothermal system.

Porphyry Cu-Au deposits are another important style of arc mineralization. This mineralization is found in shallow plutonic rocks affected by rapid uplift, cooling, and depressurization of upper crustal magma chambers, releasing metal-bearing magmatic fluids. Porphyry Cu-Au and epithermal deposits seem to be more common in atypical arc settings and in association with unusual, especially highly potassic, igneous rocks (Sillitoe, 1997).

Summary

Igneous activity at convergent plate boundaries is one of the most important manifestations of plate tectonics. Convergent-margin basalts demonstrate that crust and sediments are being recycled back down into earth's mantle in the underlying subduction zone, and that some of this deeply-recycled material has escaped up into and mixed with overlying mantle and caused it to melt. Basaltic magmas rise from the mantle above the subduction zone towards the surface. Among all Earth magmas, those produced at convergent margins contain the highest concentrations of magmatic volatiles, especially water. Long, curved alignments of volcanoes in the ocean define island arcs, and similar alignments on continental margins mark where the flux of magma out of the mantle above a subduction zone is especially great. Some mantle-derived magmas don't rise to the surface but stall in the crust. The interaction of mantle-derived magma with pre-existing crust sometimes leads to fractional crystallization and secondary melting, forming magmas that are much richer in silica than mantle melting yields. Because of the combination of high silica and water contents, arc volcanic eruptions are the most violent on Earth. Fortunately silica-rich arc magmas often stall in the crust, where they cool and form granitic plutons. Subterranean alignments of plutons in the crust beneath volcanic arcs define batholiths, which may be exposed after subduction stops and the volcanic cover is removed by erosion. Convergent margin igneous activity over Earth history is responsible for forming most continental crust and economic ore deposits.

Bibliography

Allen, S. R., and McPhie, J., 2009. Products of neptunian eruptions. *Geology*, 37, 639–642.

Annen, C., Blundy, J. D., and Sparks, R. S. J., 2006. Genesis of intermediate and Silicic Magmas in deep crustal hot zones. *Journal of Petroleum*, 47, 505–539.

AWST, 1990. Volcanic ash cloud shuts down all four engines of Boeing 747–400, causes $80 million in damage. *Aviation Week and Space Technology* 93, 1 January, 1990.

Baker, E. T., Embley, R. W., Walker, S. L., Resing, J. A., Lupton, J. E., Nakamura, K.-I., de Ronde, C. E. J., and Massoth, G. J., 2008. Hydrothermal activity and volcano distribution along the Mariana arc. *Journal of Geophysical Research*, 113, B08S09, doi:10.1029/2007JB005423.

Carey, R. J., Wysoczanski, R., Wunderman, R., and Jutzeler, M., 2014. Discovery of the largest historic silicic submarine eruption. *Eos, Transactions American Geophysical Union*, 95(19), 157–159.

Chadwick, W. W., Jr., Cashman, K. V., Embley, R. W., Matsumoto, H., Dziak, R. P., de Ronde, C. E. J., Lau, T. K., Deardorff, N. D., and Merle, S. G., 2008. Direct video and hydrophone observations of submarine explosive eruptions at NW Rota-1 volcano, Mariana arc. *Journal of Geophysical Research*, 113, 0810, doi:10.1029/2007JB005215.

de Ronde and 18 others, 2011. Submarine hydrothermal activity and gold-rich mineralization at Brothers Volcano, Kermadec Arc, New Zealand. *Mineralium Deposita*, 46, 541–584.

de Ronde, C. E. J., Hannington, M. D., Stoffers, P., Wright, I. C., Ditchburn, R. G., Reyes, A. G., Baker, E. T., Massoth, G. J., Lupton, J. E., Walker, S. L., Greene, R. R., Soong, C. W. R., Ishibashi, J., Lebon, G. T., Bray, C. J., and Resing, J. A., 2005. Evolution of a submarine magmatic-hydrothermal system: brothers Volcano, Southern Kermadec Arc, New Zealand. *Economic Geology*, 100, 1097–1133.

de Ronde, C. E. J., Massoth, G.J., Baker, E. T., and Lupton, J.E., 2003. Submarine hydrothermal venting related to volcanic arcs. *Society of Economic Geologists*. Special publication number, 10, pp. 91–110.

Drummond, M.S. and M.J. Defant, 1990. A Model for trondhjemite-tonalite-dacite genesis and crystal growth via slab melting: Archean to modern comparisons. Journal of Geophysical Research 95 (B13), 21503–21521

Embley, R. W., Baker, E. T., Butterfield, D. A., Chadwick, W. W., Lupton, J. E., Resing, J. A., de Ronde, C. E. J., Nakamura, K., Tunnicliffe, V., Dower, J. F., and Merle, S. G., 2007. Exploring the submarine ring of fire: Mariana Arc–Western Pacific. *Oceanography*, 20, 68–79.

Fiske, R. S., Naka, J., Iizasa, K., Yuasa, M., and Klaus, A., 2001. Submarine silicic caldera at the front of the Izu-Bonin arc, Japan: voluminous seafloor eruptions of rhyolite pumice. *Geological Society of America Bulletin*, 113, 813–824.

Grove, T. L., Till, C. B., and Krawczynski, M. J., 2012. The role of H_2O in subduction zone magmatism. *Annual Review of Earth and Planetary Sciences*, 40, 413–439.

Hannington, M.D., de Ronde, C.D., and Petersen, S., 2005. Sea-floor tectonics and submarine hydrothermal systems. In: *Economic Geology 100th Anniversary Volume*. In Hedenquist, J. W., Thompson, J. F. H., Goldfarb, R. J. and Richards, J. P. (eds.), Society of Economic Geologists, Littelton, Colorado, USA pp. 111–141.

Kessler, S. E., 1997. Metallogenic evolution of convergent margins: selected deposit models. *Ore Geology Reviews*, 12, 153–171.

Kim, J., Lee, K.-Y., and Kim, J.-H., 2011. Metal-bearing molten sulfur collected from a submarine volcano: implications for vapor transport of metals in seafloor hydrothermal systems. *Geology*, 39, 351–354.

Kimura, J., Hacker, B. R., van Keken, P. E., Kawabata, H., Yoshida, T., and Stern, R. J., 2009. Arc Basalt Simulator version 2, a simulation for slab dehydration and fluid-fluxed mantle melting for arc basalts: modeling scheme and application. *Geochemistry, Geophysics, Geosystems*, 10, Q09004, doi:10.1029/2008GC002217.

Lallemand, S., 1999. *La Subduction Oceanique*. Newark: Gordon and Breach.

Lydon, J. W., 2007. An overview of the economic and geological contexts of Canada's major mineral deposit types. In Goodfellow, W. D. (ed.), *Mineral deposits of Canada a synthesis of major deposit-types, district metallogeny, the evolution of geological provinces, and exploration methods*. St. John's, Newfoundland: Geological Association of Canada Mineral Deposits Division, pp. 3–48.

Nakajima, J., Takei, Y., and Hasegawa, A., 2005. Quantitative analysis of the inclined low-velocity zone in the mantle wedge of northern Japan: a systematic change of melt-filled pore shapes with depth and its implications for melt migration, Earth Planet. *Science Letters*, 234, 59–70.

Newhall, C., and Self, S., 1982. The volcanic explosivity index (VEI): an estimate of explosive magnitude for historical volcanism. *Journal of Geophysical Research, Oceans*, 87(C2), 1231–1238.

Plank, T., Kelley, K. A., Zimmer, M. M., Hauri, E. H., and Wallace, P. J., 2013. Why do mafic magmas contain ~4wt% water on average? Earth Planet. *Science Letters*, 364, 168–179.

Resing, J. A., Rubin, K. H., Embley, R. W., Lupton, J. E., Baker, E. T., Dziak, R. P., Baumberger, T., Lilley, M. D., Huber, J. A., Shank, T. M., Butterfield, D. A., Clague, D. A., Keller, N. S., Merle, S. G., Buck, N. J., Michael, P. J., Soule, A., Caress, D. W., Walker, S. L., Davis, R., Cowen, J. P., Reysenbach, A.-L., and Thomas, H., 2011. Active submarine eruption of boninite in the northeastern Lau Basin. *Nature Geoscience*, 4, 799–806, doi:10.1038/ngeo1275.

Robock, A., 2000. Volcanic eruptions and climate. *Reviews of Geophysics*, 38, 191–219.

Roggensack, K., Hervig, R. L., McKnight, S. B., and Williams, S. N., 1997. Explosive basaltic volcanism from Cerro Negro Volcano: influence of volatiles on eruptive style. *Science*, 277, 1639–1642.

Sharma, K., Blake, S., Self, S., and Krueger, A. J., 2004. SO_2 emissions from basaltic eruptions, and the excess sulfur issue. *Geophysical Research Letters*, 31, L13612, doi:10.1029/2004GL01968.

Sillitoe, R. H., 1997. Characteristics and controls of the largest porphyry copper-gold and epithermal gold deposits in the circum-Pacific region. *Australian Journal of Earth Sciences*, 44, 373–388.

Smith, I.E.M., Worthington, T.J. Stewart, R.B., Price, R.C., and Gamble, J.A. 2003. *Felsic volcanism in the Kermadec arc, SW*

Pacific: *crustal recycling in an oceanic setting*. London: Geological Society. Special publication, 219, pp. 99–118.

Stern, R. J., 2002. Subduction zones. *Reviews of Geophysics*, **40**, doi:10.1029/2001RG000108

Stern, R.J. 2010. The anatomy and ontogeny of modern intraoceanic arc systems. In: Kusky, T. M., Zhai, M.-G., and Xiao, W. (eds.), *The Evolving Continents: Understanding Processes of Continental Growth*. London: Geological Society. Special publication, 338, pp. 7–34.

Stern, R. J., and Dickinson, W. R., 2010. The Gulf of Mexico is a Jurassic back-arc basin. *Geosphere*, **6**, 739–754.

Stern, R. J., Reagan, M., Ishizuka, O., Ohara, Y., and Whattam, S., 2012. To understand subduction initiation, study forearc crust; to understand forearc crust, study ophiolites. *Lithosphere*, **4**, 469–483.

Taylor, B. (ed.), 1995. *Backarc Basins: Tectonics and Magmatism*. New York: Springer. 524 p.

Timm, C., de Ronde, C. E. J., Leybourne, M. I., Layton-Matthews, D., and Graham, I. J., 2012. Sources of chalcophile and siderophile elements in Kermadec arc lavas. *Economic Geology*, **107**, 1527–1538.

Yang, K., and Scott, S. D., 1996. Possible contribution of a metal-rich magmatic fluid to a sea-floor hydrothermal system. *Nature*, **383**, 420–423.

Cross-references

Accretionary Wedges
Active Continental Margins
Black and White Smokers
Chemosynthetic Life
Explosive Volcanism in the Deep Sea
Gabbro
High-pressure, Low-temperature Metamorphism
Hydrothermal Plumes
Hydrothermalism
Intraoceanic Subduction Zone
Lithosphere: Structure and Composition
Marginal Seas
Marine Mineral Resources
Mohorovičić Discontinuity (Moho)
Oceanic Spreading Centers
Ophiolites
Paired Metamorphic Belts
Peridotites
Seamounts
Serpentinization
Subduction
Subduction Erosion
Volcanogenic Massive Sulfides
Wadati-Benioff-Zone

MAGNETOSTRATIGRAPHY OF JURASSIC ROCKS

María Paula Iglesia Llanos
IGEBA, National Scientific and Research Council, University of Buenos Aires, Buenos Aires, Argentina

Rocks can retain a magnetic imprint acquired in the geomagnetic field that existed during their formation. The main goal of magnetic polarity stratigraphy is to document and calibrate the global geomagnetic polarity sequence in stratified rocks and to apply this geomagnetic polarity scale for high-resolution correlation of marine magnetic anomalies and of polarity zones in other sections.

Magnetic polarity scales younger than 160 Ma are derived from seafloor magnetic anomalies. For older times, such scales must be derived from polarity successions obtained in the continents yielding a precise age control such as that provided by ammonites, i.e., 1 myr.

In this study, emphasis is paid to the Jurassic period. Five ammonite-bearing sections located to the north and center of the Neuquén basin, west-central Argentina, have been sampled for paleomagnetic purposes. These sections are made up of ammonite-bearing sedimentary and, subordinately, volcanic rocks of Lower Jurassic age.

The paleomagnetic study showed that there are two magnetic components, one soft that coincides with the dipolar field of the region and another one harder that is dissimilar from any younger magnetization. According to the field tests for paleomagnetic stability and petrographical studies, the latter component has been acquired during or shortly after the deposition (cooling) of sedimentary (volcanic) rocks. Thus, the hard component is interpreted as the original Jurassic.

With polarities isolated in all five sections, the first Lower Jurassic magnetostratigraphic scale of the Southern Hemisphere scale has been constructed. To anchor the polarities to the international geomagnetic scale, we used the ammonite zonation and correlation from Riccardi (2008), and the international geomagnetic polarity scale from Ogg (2004) has been considered. Thus, the resultant magnetostratigraphic scale is made up of 11 dominant reverse (JR_1 to JR_{11}) and 12 dominantly normal (JN_1 to JN_{12}) polarities, included in 19 ammonite zones. They comprise the Andean P. rectocostatum which is correlated with the Tethyan P. planorbis zone to the Andean Phymatoceras that is correlated with the Tethyan G. thouarsense.

With respect to the isolated polarities, a good correlation was achieved with the international scale that helped to constrain some issues that were still at loose ends. In the first place, in the Hettangian-Sinemurian interval, polarities comprised within the international Liasicus and Semicostatum biozones should be displaced down until the dominant normal polarity coincides with the regional JN_3. In the second place, during the Pliensbachian and Toarcian, the frequency of reversals increases, and yet, correlations between the scales are more solid. Results allowed assigning precise ages to those levels devoid of diagnostic fossils. This is the case of the Sinemurian-Pliensbachian boundary as well as the Toarcian in the north and the Pliensbachian-Toarcian in the center of the basin.

Bibliography

Ogg, J., 2004. The jurassic period. In Gradstein, F., Ogg, J., and Smith, A. (eds.), *A Geologic Time Scale*. Cambridge: Cambridge University Press, pp. 307–343.

Riccardi, A. C., 2008. The marine Jurassic of Argentina: a biostratigraphic framework. *Episodes*, **31**, 326–335.

Cross-references

Marine Microfossils
Plate Tectonics

MANGANESE NODULES

James R. Hein
U.S. Geological Survey, Santa Cruz, CA, USA

Synonyms

Fe-Mn nodules; Ferromanganese nodules; Iron-manganese nodules; Polymetallic nodules

Definition

Manganese and iron oxide mineral deposits formed on or just below the sediment-covered surface of the deep-ocean seabed by accretion (precipitation) of oxide layers around a nucleus, thereby forming nodules of various shapes and sizes and which contain minor but significant amounts of nickel, copper, cobalt, lithium, molybdenum, zirconium, and rare earth elements.

Introduction

The existence of manganese (Mn) nodules (Figure 1) has been known since the late 1800s when they were collected during the Challenger expedition of 1873–1876. However, it was not until after WWII that nodules were further studied in detail for their ability to adsorb metals from seawater. Many of the early studies did not distinguish Mn nodules from Mn crusts. Economic interest in Mn nodules began in the late 1950s and early 1960s when John Mero finished his Ph.D. thesis on this subject, which was published in the journal Economic Geology (Mero, 1962) and later as a book (Mero, 1965). By the mid-1970s, large consortia had formed to search for and mine Mn nodules that occur between the Clarion and Clipperton fracture zones (CCZ) in the NE Pacific (Figure 2). This is still the area considered of greatest economic potential in the global ocean because of high nickel (Ni), copper (Cu), and Mn contents and the dense distribution of nodules in the area. While the mining of nodules was fully expected to begin in the late 1970s or early 1980s, this never occurred due to a downturn in the price of metals on the global market. Since then, many research cruises have been undertaken to study the CCZ nodules, and now 15 contracts for exploration sites have been given or are pending by the International Seabed Authority (ISA). Many books and science journal articles have been published summarizing the early work (e.g., Baturin, 1988; Halbach et al., 1988), and research has continued to the present day (e.g., ISA, 1999; ISA, 2010). Although the initial attraction for nodules was their high Ni, Cu, and Mn contents, subsequent work has shown that nodules host large quantities of other critical metals needed for high-tech, green-tech, and energy applications (Hein et al., 2013; Hein and Koschinsky, 2014).

Occurrence and distribution

Mn nodules are concretions that occur on or near the surface of the sediment that covers abyssal plains throughout the global ocean, where sedimentation rates are low, less than 10 centimeters (cm) per thousand years (Figure 2). Low sedimentation rates occur if the abyssal plain is distant from a continental margin or the margin is adjacent to a deep-sea trench, has a wide continental shelf, or is arid. In the open ocean, low sedimentation rates occur away from regions of strong upwelling and high biological productivity, such as the equatorial zone of high productivity. Nodules are generally about golf ball sized, most commonly 1–12 cm, but can vary in diameter from millimeter (mm)-size (micronodules) to as large as 20 cm. The maximum size of nodules increases with increasing diagenetic input of metals (see next section). Nodules of greatest economic interest occur in areas of moderate primary productivity in surface waters (ISA, 2010). The location of nodule growth relative to the calcite compensation depth (CCD), controlled by productivity in surface waters, is important because above the CCD, biogenic calcite increases sedimentation rates and dilutes the organic matter required for diagenetic reactions in the sediment that release Ni and Cu (Verlaan et al., 2004; Cronan, 2006). The most Ni- and Cu-rich nodules form near, but generally below the CCD.

Most Mn nodules form at water depths of about 3,500–6,500 meters (m), although abyssal plains can have hills and ridges several hundred meters high and seamounts up to 2 kilometers (km) high rising above the plains. Nodules can occur on all of those structures if they are covered by sediment. Mn nodules have been collected from many of the large seamounts on which Fe-Mn crusts form, but most of those nodules differ significantly from abyssal plain nodules since the seamount nodules form by hydrogenetic accumulation only (see below) and commonly have a large rock fragment as the nucleus. The seamount nodules are mostly Mn- and iron (Fe)-oxide-coated rock pebbles and cobbles, while abyssal plain nodules usually have a minute nucleus relative to the thickness of the accreted Mn- and Fe-oxide layers.

The greatest concentrations of metal-rich nodules occur in the following regions: in the NE Pacific CCZ (Figure 2) that extends from the Mexico EEZ to as far west as the longitude of Hawaii, in the SE Pacific Peru Basin, in the central South Pacific Penrhyn Basin, and in the Central Indian Ocean Basin. In the CCZ, the Mn nodules lie on abyssal sediments covering an area of about nine million square kilometers with up to 75 kilograms (kg) of nodules per square m of seabed, but more commonly average less than 15 kg per square m wet weight. Their high abundance in this region can be attributed to slow rates of sedimentation, the abundant availability of material for the Mn and Fe to nucleate around, and moderate productivity in

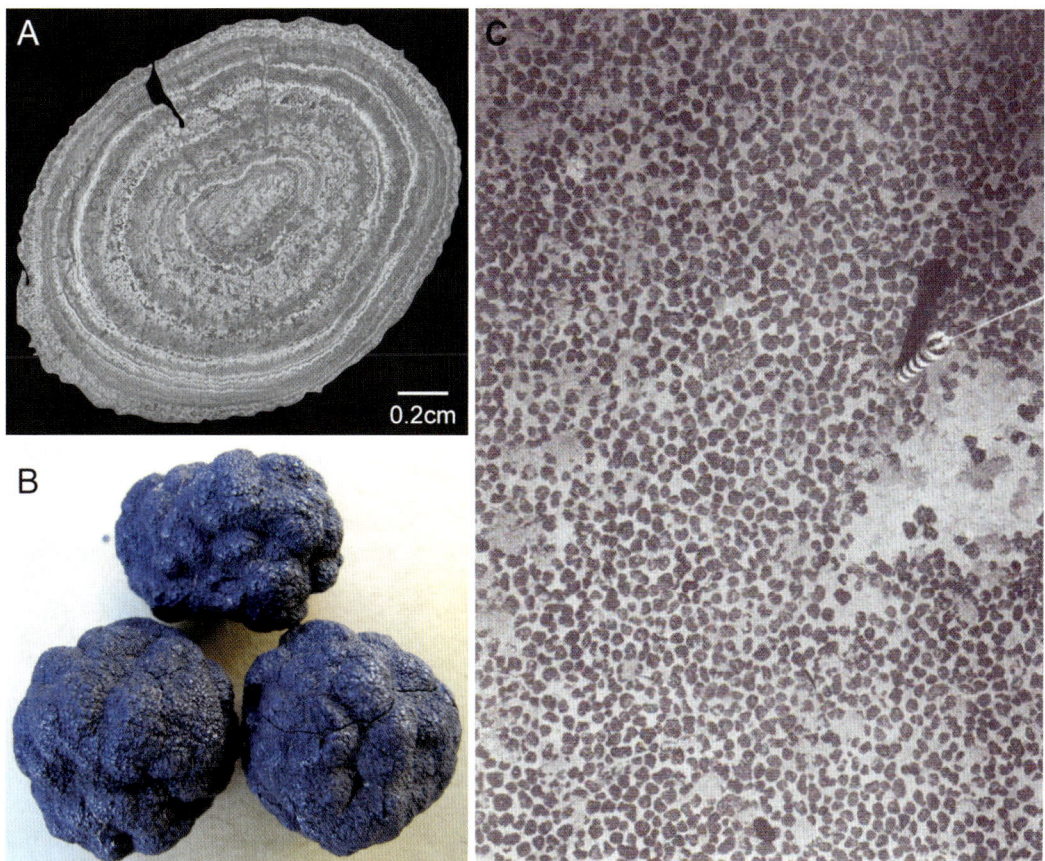

Manganese Nodules, Figure 1 Manganese nodule photographs: (**a**) cross-section of a manganese nodule from the CCZ showing concentric layering (laminations), fractures, and a nucleus composed of an old nodule fragment, 4,353 m water depth; (**b**) diagenetic-hydrogenetic manganese nodules from the CCZ, NE Pacific Ocean; each nodule is 3 cm in diameter; (**c**) seafloor photograph of nodules within the CCZ nodule field and the trigger weight; field of view is about 3 × 4 m.

surface waters. Most Mn-nodule fields are found in areas beyond national jurisdictions, with the exception of the Penrhyn Basin field, which lies predominantly within the exclusive economic zone (EEZ) of the Cook Islands (Figure 2). Fewer large abyssal plains occur in the Atlantic and Indian Oceans compared to the Pacific, which are dominated topographically by the mid-ocean ridge systems. The large Argentine Basin in the SW Atlantic may possess nodule fields, but little exploration has occurred there.

Formation

Mn nodules grow by precipitation of Mn and Fe oxides around a nucleus that is commonly composed of a fragment of an older nodule. Unlike Mn crusts, the Mn and Fe minerals in nodules precipitate from two sources, cold seawater (hydrogenetic) and sediment pore waters (diagenetic), which is seawater modified by chemical reactions within the sediment. Nodules can be solely hydrogenetic (e.g., Penrhyn Basin and Atlantic Ocean; Figure 2) or solely diagenetic (such as Peru Basin; Figure 2), but most nodules acquire metals from both sources. Hydrogenetic nodules grow remarkably slowly, at a rate of about 1 to 5 mm per million years, like Mn crusts, whereas diagenetic nodules grow at rates up to 250 mm per million years. Because most nodules form by both hydrogenetic and diagenetic precipitation, they grow at intermediate rates of several tens of mm per million years. Average nodule growth rates are four orders of magnitude slower than the sedimentation rates, yet the nodules remain at the sediment surface. There have been a number of proposals (e.g., Halbach et al., 1988) as to how this happens, although the actual cause is still a mystery.

Sediment pore fluids are the predominant source of Ni, Cu, and Mn, and seawater is the dominant source of cobalt (Co). Metals in the pore fluid are derived from early diagenetic oxidation-reduction (redox) reactions in upper sediment layers and are incorporated into the Mn-oxide minerals forming at the seabed.

Composition

Mn nodules are composed predominantly of the minerals δ-MnO_2 (hydrogenetic vernadite), 10 Å manganate (diagenetic todorokite, buserite, asbolan), and less

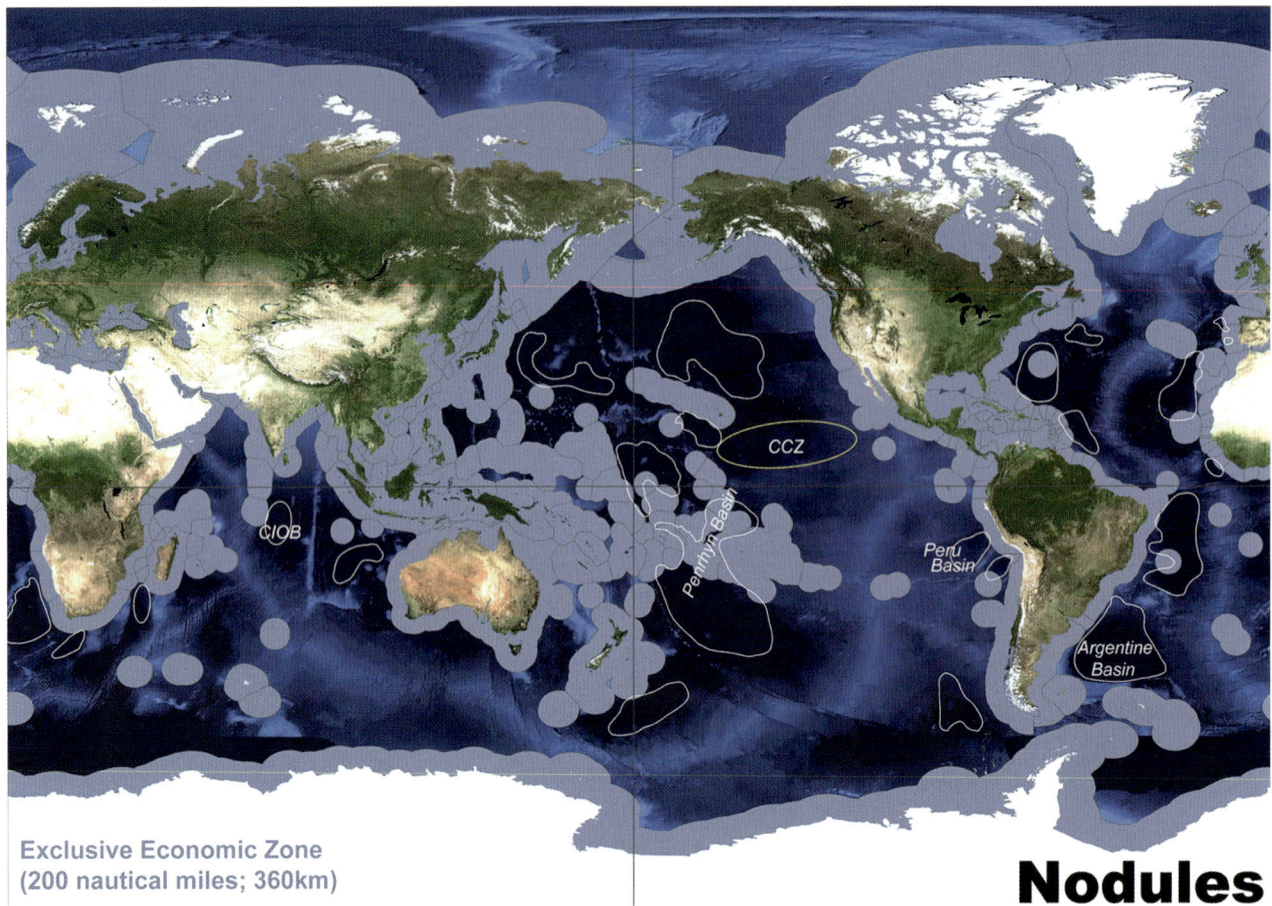

Manganese Nodules, Figure 2 Map of global distribution of Exclusive Economic Zones (EEZ, *gray shading*), areas beyond national jurisdictions (*dark black-blue* to *pale gray-blue*), and global permissive areas for development of abyssal plain manganese nodules; the Clarion-Clipperton Zone (CCZ) is the zone of greatest economic interest and is enclosed by a *yellow line*; all other areas are marked with a *white line*; a permissive area does not mean that economic nodule deposits will be found in that area; small patches of scattered nodules will occur in other areas; the equator and 180th parallel are marked by *gray lines*; CIOB Central Indian Ocean Basin.

commonly diagenetic birnessite; lesser amounts of X-ray amorphous Fe oxyhydroxides also occur. Minor amounts of detrital aluminosilicate minerals and quartz and authigenic minerals (formed in the sediment from alteration of other materials: volcanic glass, minerals, microfossils) are also present. The sheet and tunnel structures of the diagenetic Mn-oxide minerals, which allow for the acquisition from sediment pore fluids of large amounts of Ni, Cu, and other metals that strengthen the mineral structure, are important physical properties of Mn nodules.

Mn nodules typically have three to six times more Mn than Fe (Table 1), in contrast to Fe-Mn crusts, another type of deep-ocean mineral deposit (see entry on "Cobalt-rich Manganese Crusts"). Mn concentrations can be greater than 30 % in diagenetic nodules, such as Peru Basin nodules. Nickel and Cu have been the metals of greatest economic interest and have mean concentrations in four nodule fields that vary from 0.38 % to 1.32 % and 0.23 % to 1.07 %, respectively (Table 1; 1,000 ppm = 0.1 %). These percentages will vary somewhat in mine-site-size areas (about 20,000 square km) within these major nodule fields. Nodules are also more enriched in Ni, Cu, and Li than Fe-Mn crusts, while crusts are more enriched in the other rare metals and the rare earth elements (REEs). The market for Li is growing because of its use in rechargeable batteries. Li in CCZ nodules averages 131 ppm and is especially high in diagenetic nodules, averaging 311 ppm in Peru Basin nodules (Hein and Koschinsky, 2014), the latter of which is comparable to grades in some terrestrial Li ores. REEs in nodules are also of economic interest but are generally two to six times lower than in Mn crusts, with maximum total REEs plus yttrium (Y) in CCZ nodules of about 0.08 %; however, the average for Cook Island nodules is 0.17 %. The main carrier for REEs and Y, in general, is the hydrogenetic component. Platinum (Pt) is very low in nodules compared to crusts and not of economic interest. Mean concentrations of other metals of interest as potential

Manganese Nodules, Table 1 Chemical composition of nodules from selected areas of the global ocean; see Figure 2 for locations

Element	CCZ Nodules Mean	N	Peru Basin nodules Mean	Indian Ocean nodules Mean	N	Cook Island nodules Mean	N
Fe (wt%)	6.16	66	6.12	7.14	1,135	16.1	982
Mn	28.4	66	34.2	24.4	1,135	15.9	982
Si	6.55	12	4.82	10.02	36	8.24	46
Al	2.36	65	1.50	2.92	49	3.25	46
Ti	0.32	66	0.16	0.42	53	1.15	46
Bi (ppm)	8.8	12	3.3	–	–	11	26
Co	2,098	66	475	1,111	1,124	4,116	982
Cu	10,714	66	5,988	10,406	1,124	2,290	982
Li	131	66	311	110	38	61	26
Mo	590	66	547	600	38	262	46
Nb	22	66	13	98	3	92	26
Ni	13,002	66	13,008	11,010	1,124	3,756	982
Pb	338	66	121	731	38	2,000	46
Pt	0.13	12	0.04	–	–	0.21	25
Te	3.6	66	1.7	40	3	23	26
Th	15	66	6.9	76	3	34	26
Tl	199	12	129	347	3	138	26
V	445	66	431	497	16	920	46
W	62	66	75	92	3	59	26
Zn	1,366	66	1,845	1,207	692	516	46
Zr	307	66	325	752	3	468	34
TREE	813	66	403	1,039	11–50	1,707	31–41

Modified from Hein et al. (2013), Hein and Koschinsky (2014)
Dash means no data, *TREE* means total rare earth elements including yttrium

by-products of Ni-Cu-Mn mining include Co (0.05–0.41 %), Mo (262–600 ppm), and zirconium (Zr; 307–752 ppm).

Resource consideration

The tonnages of nodules and therefore of the amount of contained metals are moderately well known for the CCZ and the Central Indian Ocean Basin nodules. A conservative estimate for the CCZ nodules is about 21,100 million dry tonnes. That tonnage of nodules would provide nearly 6,000 million tonnes of Mn, more than the entire land-based reserve base (LBRB) of Mn (Hein et al., 2013; Hein and Koschinsky, 2014). In addition, the tonnage of Ni is twice that of the entire LBRB, and Co is three times greater than the LBRB; the LBRB includes resources that are currently economic (reserves), marginally economic, and subeconomic. The amount of Cu in the CCZ nodules is about a fifth that of the LBRB. In addition, nodules in the CCZ have 1.4 times more tellurium (Te) and 4 times more Y than the entire LBRB. Metals in CCZ nodules as a percent of the LBRB are Mo 63 %, Li 19 %, and total REEs plus Y as oxides 11 %. Peru Basin nodules have the highest Li contents, and nodules there have roughly 2.7 million tonnes of contained Li; this tonnage of Li is the same as in the CCZ nodules, but with less than 42 % of the total dry tonnage of the CCZ nodules.

The high potential ore grades mentioned above in addition to other drivers have sparked an incredible amount of interest in mining Mn nodules. Some of these drivers include (1) the global population increasing beyond seven billion people and a growing middle class in developing countries with increased demands for goods containing many rare metals, (2) decreasing ore grades in existing land-based mines, and (3) the need to remove enormous amounts of barren-rock overburden to access ore bodies, necessitating development of ultra-deep open-pit mines and mega-underground operations.

The CCZ is the area of greatest economic interest and where 14 exploration contracts of 75,000 square km each and one of 58,620 square km have been signed or are pending with the ISA; there is also one nodule contract of 75,000 square km in the Central Indian Ocean Basin, signed with India. The total area of seabed under contract for nodule exploration amounts to 1,183,620 square km. States and state agencies with contracts in the CCZ include China, France, Germany, Japan, Korea, Russia, and Inter-Ocean Metals which is composed of a group of States: Bulgaria, Cuba, The Czech Republic, Poland, The Russia Federation, and The Slovak Republic. The rest of the CCZ contracts are with companies, four of which have partnered with a developing country: Nauru, Tonga, Kiribati, and Cook Islands.

The mining process does not present a major technological challenge because the nodules can be removed fairly easily from the surface of the sediment-covered seabed on which they rest. Trial excavation as early as 1978 was successful in transporting Mn nodules to a ship. New technology for the mining equipment is under development and is being scaled-up to work in larger areas for long-term operations. Riser pipe or other lifting technologies required to transport the ore from the seabed to the ship developed in the 1970s are also being improved with modern technology. The actual area of seabed that is mined will probably not exceed about 25 % of the 75,000 square km contract sites because of topographic obstacles and the lower abundance distribution of nodules. Before large-scale mining of nodules can be carried out, questions related to the environmental impact of mining activities must be addressed. Mining of Mn nodules would considerably disturb parts of the seabed and suspend sediment that will be carried by currents for as yet to be determined distances. The development of mining technology that significantly reduces the sediment plume would be an important advancement.

Summary

Mn nodules are found on the surface of sediment on all abyssal plains throughout the global ocean where sedimentation rates are low, less than 10 cm per thousand years. Nodules generally grow at rates of a few tens of mm per million years, yet they remain at the sediment surface even though sedimentation rates are four orders of magnitude higher, which is still a mystery. Nodules are composed predominantly of Mn minerals that precipitate from seawater and from sediment pore fluids and acquire abundant metals and other elements from those two sources by adsorption. Metals essential to many emerging technologies are enriched in nodules (Ni, Cu, Co, Mo, Li, Zr, REEs) to the extent that they are considered a potential resource for mining in the near future. Fifteen contracts for exploration for nodules have been taken out or are pending with the ISA that cover more than one million square km of seabed.

Bibliography

Baturin, G. N., 1988. *The Geochemistry of Manganese and Manganese Nodules in the Ocean*. Dordrecht: D. Reidel.

Cronan, D. S., 2006. Processes in the formation of central Pacific manganese nodule deposits. *Journal of Marine Science and the Environment*, **C4**, 41–48.

Halbach, P., Friedrich, G., and von Stackelberg, U. (eds.), 1988. *The Manganese Nodule Belt of the Pacific Ocean – Geological Environment Nodule Formation, and Mining Aspects*. Stuttgart: Ferdinand Enke Verlag. 254 pp.

Hein, J. R., and Koschinsky, A., 2014. Deep-ocean ferromanganese crusts and nodules. In Holland, H. D., and Turekian, K. K. (eds.), *Treatise on Geochemistry, Chapter 11*. Oxford: Elsevier, Vol. 13, pp. 273–291.

Hein, J. R., Mizell, K., Koschinsky, A., and Conrad, T. A., 2013. Deep-ocean mineral deposits as a source of critical metals for high- and green-technology applications: comparison with land-based deposits. *Ore Geology Reviews*, **51**, 1–14.

ISA, 1999. *Deep-Seabed Polymetallic Nodule Exploration: Development of Environmental Guidelines, ISA/99/02*. Kingston: International Seabed Authority. 289.

ISA, 2010. *A Geological Model of Polymetallic Nodule Deposits in the Clarion-Clipperton Fracture Zone*. Kingston: International Seabed Authority. ISA Technical Study, Vol. 6, p. 105.

Mero, J. L., 1962. Ocean-floor manganese nodules. *Economic Geology*, **57**, 747–767.

Mero, J. L., 1965. *The Mineral Resources of the Sea*. Amsterdam: Elsevier.

Verlaan, P. A., Cronan, D. S., and Morgan, C. L., 2004. A comparative analysis of compositional variations in and between marine ferromanganese nodules and crusts in the South Pacific and their environmental controls. *Progress in Oceanography*, **63**, 125–158.

Cross-references

Abyssal Plains
Carbonate Dissolution
Calcite Compensation Depth (CCD)
Deep-sea Sediments
Dust in the Ocean
Energy Resources
Geochronology: Uranium-Series Dating of Ocean Formations
Cobalt-rich Manganese Crusts
Marine Mineral Resources
Paleoceanographic Proxies
Paleoceanography
Radiogenic Tracers
Seamounts
Technology in Marine Geosciences

MANGROVE COASTS

Norman C. Duke
James Cook University, TropWATER – Centre for Tropical Water and Aquatic Ecosystem Research, Townsville, QLD, Australia

Synonyms

Coastal swampland; Coastal swamps; Coastal wetland; Mangrove forested coast; Mangrove shoreline; Tidal wetland coast

Definition

Mangrove Forests: Vegetated tidal habitat comprised of saltwater-tolerant trees and shrubs. Mangroves are part of a unique combination of plant types called tidal forests or mangrove forests to distinguish them from individual trees and shrubs, also called mangroves.

Mangrove Plants: Mangroves are trees, shrubs, palms, or ground ferns, generally exceeding one-half meter in height, that normally grow above mean sea-level and below highest tidal levels in soft sediments along the intertidal zone of less exposed marine coastal environments and estuarine margins.

Mangrove Coasts, Figure 1 Luxuriant mangrove vegetation cloaking the shoreline of semiarid regions exemplifies the significance and importance of mangrove coast habitats; as for Badu Island in Torres Strait, Australia.

Tidal Salt Marsh: Shrubbery often associated with mangroves. Tidal salt marsh consists of shrubbery, sedges, and grasses, generally less than one-half meter in height, that normally grow above mean sea-level and below highest tidal levels in soft sediments along the intertidal zone of less exposed marine coastal environments and estuarine margins.

Tidal Salt Pan: Often associated with mangroves and tidal salt marsh, tidal salt pans consist of carpeted surface layers of microphyte vegetation covering flat-pan areas above mean sea-level and below highest tidal levels along the intertidal zone of less exposed marine coastal environments and estuarine margins.

Tidal Wetlands: Tidal wetland habitats are the combination of the three intertidal vegetation assemblages as mangroves, tidal salt marsh, and tidal salt pan. These characteristically occur between mean sea-level and the highest tidal levels. Mangroves dominate in wet tropical conditions. In arid and cooler latitudes, the proportion of mangroves is notably diminished to nonexistent, replaced by the corresponding dominance of tidal salt marsh shrubbery and flat expanses of tidal salt pan.

Introduction

Mangrove coasts are shorelines fringed by mangrove and salt marsh vegetation. They form a significant part of coastal tidal wetlands as distinctive habitats of tropic and temperate shorelines. Tidal wetlands have vegetation of varying complexities from forested mangrove woodlands, thick mangrove and salt marsh shrubbery, and low dense samphire plains to microalgal covered salt pans (Tomlinson, 1994). In the tropics, mangroves are often the dominant shoreline ecosystem comprised chiefly of flowering trees and shrubs uniquely adapted to coastal and estuarine tidal conditions (Duke, 2011). They form distinctly vegetated and often densely structured habitat of verdant closed canopies cloaking coastal margins and tidal waterways of equatorial, tropical, and subtropical regions of the world. Normally, but not exclusively, these vegetation assemblages grow in soft sediments above mean sea-level in the intertidal zone of sheltered coastal environments and estuarine margins (Figure 1).

The plants of mangrove coasts are well known for their morphological and physiological adaptations coping with salt, saturated anoxic soils, and regular tidal inundation; notably with specialized attributes like: exposed air-breathing roots above ground; extra, above-ground stem support structures; salt-excreting leaves; low water potentials and high intracellular salt concentrations to maintain favorable water relations in saline environments; and viviparous water-dispersed propagules. With such attributes, these habitats have essential roles in coastal productivity and connectivity, often supporting high biodiversity and biomass not possible in upland vegetation, especially in more arid regions.

Mangrove coasts are key sources of primary production with highly dependent trophic linkages between plants and animals, as nursery and breeding sites of benthic and arboreal life, as well as physical shelter and protection from severe storms, river flows, and large tsunami waves. Within tropical latitudes, mangrove coasts nestle mostly between two other iconic ecosystems of coral reefs and tropical rainforests. All three are intimately interconnected, providing mutual protection and sustenance. Each of these ecosystems also creates biota-structured environments, where the organisms themselves provide and build the physical structure among which associated life is nurtured and sheltered. Without this living structure, these habitats and the many organisms dependent on them simply would not exist. This essentially identifies how such a large group of plants and animals are so vulnerable.

For example, bordering mangrove coasts, colonial coral reefs often flourish in the shallow warm seas created and protected from land runoff by mangrove vegetation (Duke and Wolanski, 2001). Mangroves absorb unwanted nutrients and sediments of turbid waters to stabilize eroding and depositional shorelines. In modern human times, this buffering role also includes the capture of harmful chemicals in runoff waters from agricultural lands. The specialized plant assemblages of mangrove coasts provide a broad range of essential, and often undervalued, ecosystem services along with their more acknowledged roles as habitats of high productivity and as fishery nursery sites (Robertson and Duke, 1990). In such ways, the consequences of disturbing mangrove coast habitats are expected to have far-reaching implications and impacts on neighboring ecosystems and dependent biota.

Uniquely vulnerable mangroves require special management

Ancestral mangroves are believed to have reinvaded marine environments in multiple episodes from diverse earlier lineages of land-based, flowering plants over the previous 50–100 million years, culminating in today's mangrove flora. The evolution of mangrove plants appears constrained by key functional attributes essential to their survival in characteristically saline, inundated settings where isotonic extremes, desiccation, and hydrologic exposure combine as uniquely harsh restrictions on organisms living in the tidal zone and within estuaries (Duke et al., 1998). Such is the mangrove coast. This land-sea interface is also a highly dynamic environment, where even subtle changes in climate, sea-level, sediment, and nutrient inputs have dramatic consequences for the survival, distribution, and health of organisms living there. Specifically, the predominant direct human disturbances affecting mangrove coasts include: eutrophication, pollution, dredging, landfill, overfishing, and sedimentation. The combination of these pressures, coupled with changes in global climate and sea-level rise, has resulted in once healthy mangrove areas becoming endangered communities with local extinction evident in some places (Duke et al., 2007). Lessons from various rehabilitation projects demonstrate that restoration is costly and unlikely to succeed. This places even greater urgency on managers to develop and apply effective protective measures beforehand, to avoid irreplaceable losses and further degradation of remaining, remnant mangrove coast ecosystems.

Over the past two centuries, human progress and development have increased dramatically. And, human society has prospered and flourished accordingly. These great human achievements have come at a huge environmental cost though, with massive alterations to natural landscapes to accommodate demands of expanding industry, trade, and population. Mangrove coasts are particularly vulnerable because of their naturally highly dynamic geophysical circumstances, coupled with a preference by human communities for development along coastlines and waterways. This explains why there have been significant losses of tidal wetland habitats in most coastal areas. But, the longer-term consequences and implications have only been partially considered. It is reasonable to ask, does it really matter? Are remaining ecosystems sustainable? And, how do we quantify and monitor changes so we might manage both the changes and the consequences better? The answers can only come from knowing more about the factors that affect functioning of habitats along mangrove coasts.

Factors influencing mangrove coasts: a strategy

There is an urgent need for local resource managers and environmental enthusiasts to learn more about the range of factors influencing mangroves and tidal wetlands in their area. Help is also needed for optimization of sustainable management outcomes, with best-practice assessment methods to direct how mangrove coasts might best be managed in view of current sociopolitical, economic, and environmental pressures.

To achieve such outcomes, six key strategies are recommended:

1. Identification and quantification of the type and biota of tidal wetlands
2. Surveys of past changes in extent and character of wetland areas
3. Assessment of ecosystem health and how these ecosystems are changing
4. Identification of the chief drivers of change and assessment of longer-term implications
5. Determination of what will be lost in terms of ecosystem goods and services
6. Decisions on management options based on assessments of extent, change, health, pressures, and values

The far greatest problem facing tidal wetlands is the ever-growing global population with its predictable, continually expanding demand for natural resources and space, including mangrove coasts. One key consequence also is that our global climate is changing dramatically, and the loss of vital natural habitats and biological resources is a direct result (Duke et al., 2007). Where we acknowledge the value of these natural ecosystems for the numerous direct and indirect services they provide, then we must act urgently to become their diligent guardians. It is imperative that communities are aware of the changes taking place, and what is being lost, so they can support more informed management to respond in more effective ways. This awareness starts with a better knowledge and appreciation of the types of changes observed on mangrove coasts and what can be inferred from key indicators of change.

Mangrove coasts as indicators of change

Tidal wetlands of the mangrove coast are ancient ecosystems having evolved over 100 million years. During this time, the earth, sea-level, and climate have changed dramatically. Mangrove coasts of today are comprised of plants that are the survivors of all previous changes through the millennia. These ecosystems consequently have well-developed strategies for survival with a defined capacity for dealing with change. As tidal wetland ecosystems respond to change, they rely on inherent adaptive capacity (Duke et al., 1998). Where changes can be identified, described, measured, and monitored, they form the basis for a more enlightened monitoring and assessment strategy. For example, if a tidal wetland habitat had shifted upland, this might demonstrate and quantify the effects of sea-level rise. Two deductions to be made from such observations are that mangroves responded to sea-level rise and that we might evaluate the rate of net change. The value in this approach in combination with direct instrument measures, like sea/tide level elevation stations, is that mangrove coast plants integrate daily and seasonal term fluctuations. These changes, when viewed from above, are significantly enhanced along characteristically low-profile slopes in numerous locations. Furthermore, the exaggerated shifts can be readily determined retrospectively based on interpretation of vegetative condition and specific location from historical aerial imagery.

There are also questions about what causes such changes. This would be better understood if we were able to expand our general knowledge and monitoring of the full complement of forces or drivers that influence mangrove coasts. Such drivers act at local and global levels, with some delivered directly, others indirectly, and, of course, some are natural. In all situations, tidal wetlands are responding to changing influences in characteristic ways that are useful as indicators of change. With systematic identification of the different types of change in tidal wetlands, we are able to identify the responsible drivers and quantify their importance and anticipated consequences.

Key pressures on mangrove coasts

Pressures on mangrove coasts with their tidal wetlands (Figure 2) can be conveniently grouped under three broad headings (Duke et al., 2003):

- Direct human pressures, as mostly intended and obvious
- Indirect human pressures, as mostly unintended and less obvious
- Natural; being not obviously human influenced, if at all

These classifications are useful when making decisions about sustainable management of coastal and estuarine ecosystems, and they form the basis for classification of all types of observed change. As it happens, the more direct human influences are those that are more easily mitigated, resulting in relatively rapid, improved environmental outcomes. By contrast, it has been inherently difficult to differentiate between human and natural influences on global temperatures. So, it is appropriate at this time to group these more indirect human influences along with the natural pressures. This accepts that some drivers might be somewhat subjective with overlaps and ambiguity.

Indicators of change in mangrove coasts

Changes are usually measured in terms of ecosystem responses to external pressures. Estuarine, coastal, and marine ecosystems are influenced by a number of key factors acting as primary stressors and drivers. The responses of affected habitats are often unique and distinctive making them useful indicators of the key drivers.

The six overall drivers of change for mangrove coasts may be grouped under the following headings:

- Severe Human Disturbance
- Hydrology and Sediments
- Pollutants
- Pests and Pathogens
- Climate Change
- Severe Natural Disturbance

While these drivers often act in combination, their effects on tidal wetland ecosystems however predominantly result in recognizable and distinct responses. These outcomes are often observed in patches of habitat dominated by dramatic and noticeable change. In such instances, the key characteristics and features present can be used as an indicator, or proxy, of each dominant prevailing driver. Based on such information, it is possible to develop more effective management intervention actions taking account of the full array of circumstances. For instance, where pressures are largely natural or globally affected, then a local management response might best be one of adaptation and accommodation. By contrast, if a stressor were predominantly locally derived, then this could be managed and controlled locally to mitigate such effects. The difference between these two management response options is extreme. It emphasizes why prior knowledge of influencing factors and change can greatly facilitate effective and sustainable outcomes.

In general, the many specific types of change can be grouped under each of these headings (Figure 2). Twenty-one types of change are listed. These represent most of all possible changes likely. The following sections describe briefly the drivers of change and the influencing factors affecting mangrove coasts; along with some illustrative examples. Comparative quantification of these pressures essentially can be derived from measures of area

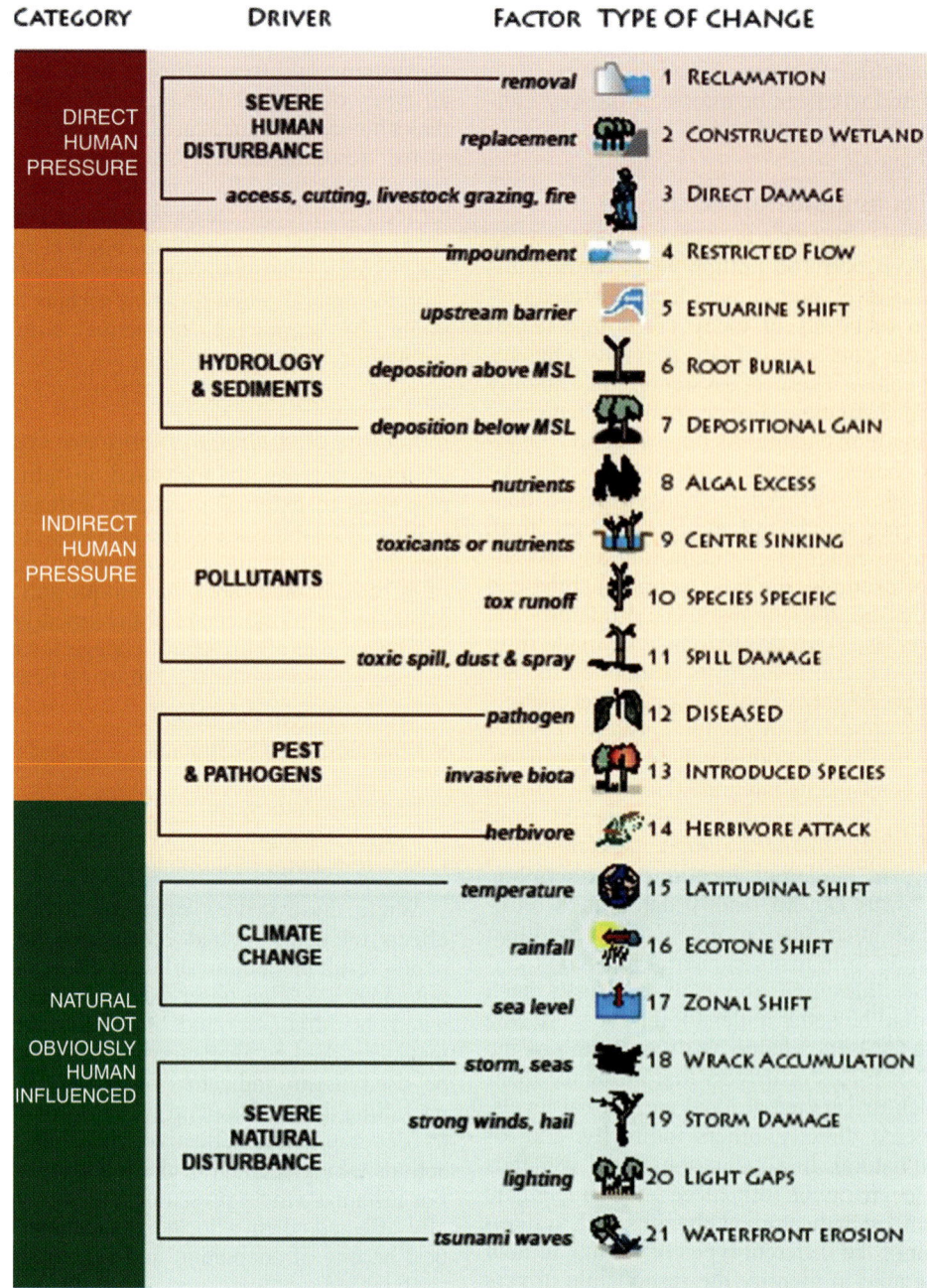

Mangrove Coasts, Figure 2 A listing of the 21 key types of change in tidal wetlands grouped under three pressure categories (human direct, indirect, and mostly natural), for six groupings of drivers of change, along with listings of respective factors that influence coastal and estuarine habitats (Duke et al., 2003).

of each indicator observed across the landscape, as specific estuarine systems or coastal enclaves.

Severe human disturbance
Predominant drivers

The drivers of severe human disturbance are grouped as direct human influences. They mostly include removal and replacement of tidal wetlands with coastal development as shoreline lands for ports and cities (Figure 3). However, with increasing populations in such areas, there are additional more cryptic influences of excessive access, encroachment, and inappropriate use. In rural areas, significant direct damage is caused by livestock grazing by foliage removal and breakage, plus trampling and sediment disturbance. In arid areas, particularly during

Mangrove Coasts, Figure 3 Severe human disturbance with reclamation between 1972 and 2003 of extensive mangrove areas at the mouth of the Brisbane River, Queensland, Australia, for port and coastal industry expansion.

drought periods, fires also severely damage mangroves, further destroying the shallow root systems as well as foliage. These latter drivers often result in the largely unquantified loss of tidal wetlands and their function associated with significant habitat damage, degradation, and disruption. Furthermore, these activities exacerbate negative perceptions of tidal wetlands where many damaged areas become breeding grounds for mosquitoes and other biting insects, as well as becoming a prime source of noxious smells from degrading sulfurous soils.

Factor indicators

Tidal wetlands are removed with landfill and replaced with artificial rock walls, particularly as a direct result of reclamation in areas of expanding human development of coastal and estuarine margins. Remnant intact areas are subject to notable damage and degradation as a direct result of excessive amounts of access and abuse indicated by tracks, paths, wheel ruts, mangrove cutting, inappropriate digging, fires, plus livestock grazing and trampling. Quantification of these effects can be made by measuring the surface area of: altered land use for port, industry, and urban development along coastal and estuarine margins and river mouths; recruitment and expansion of new mangrove area assisted by site hydrological preparation, planting, and sometimes installation of protective devices; and inappropriate human access, curiosity, and recreation activities. Comparatively high turbidity levels are likely to be found in associated estuarine waters since such areas will have lost buffering capacity and fine sediment retention. Turbidity can be measured using automated instruments or by hand.

Change type and corresponding factors

For severe human disturbance (Figure 2), measurable areas of disturbance for respective types of change include: reclamation loss with removal (Duke, 1997; Duke et al., 2003), constructed wetlands with replacement (Spurgeon, 1998; Primavera, 2000; Lewis-III, 2005), and direct damage as excessive access and other damage (Rubin et al., 1999; Duke et al., 2003; Walters, 2005; Jupiter et al., 2007).

Hydrology and sediments
Predominant drivers

Drivers related to hydrology and sediments are often grouped as indirect human influences. While the influences of hydrology and sediment displacement are present naturally in coastal and estuarine waters, they have been greatly increased as a consequence of accidental or unintended human impacts. The disturbance of coastal environments in the name of human progress has led to significant alterations in water flow accompanied by notable increases in coastal sediment loads. These drivers are disturbances that alter flows and mobilize sediments in waterways, including: catchment erosion through removal of vegetation especially riparian vegetation, increased runoff flow rates with urban hardening, straightened and hardened runoff drainage channels, vessel wake and turbulence, and dredging of vessel channels and coastal development projects that generate dust. Alterations to flows will also influence plant distributions sensitive to, and dependent on, upstream salinity gradients.

The impacts of fine sediment deposition are further observed downstream within estuaries of disturbed catchments, indicated by: increased and prolonged turbidity in estuarine coastal waters with enlarged plumes during periods of greater runoff; blocking and choking estuarine waterways with sediments reducing overall channel depths; depositional gains as measured by increased formation of mudbanks downstream, often colonized by mangroves; and the unusually rapid formation of mangrove islands and verges at the mouths of estuaries and surrounding coastal foreshores.

Factor indicators

Changes due to hydrology and sediments are influenced by a number of factors. Constructions, like roads and sea walls, and natural features, like beach berms, all alter water flow and tidal exchange. Tree death is caused mostly by excessive inundation of breathing roots, erosion of sediments, or deposition of sediments. Mangrove occupation of downstream mud banks is the result of clearing catchment vegetation, causing soil erosion coupled with alterations to estuarine margins because of changes in channel hydrology. Raised deposition ridges burying mangroves are caused by the influence of waves and hydrological conditions altered by storms and dredging. Construction of weirs, barrages, and dams reduces freshwater runoff so saline conditions and salt-tolerant mangrove species penetrate further upstream. Comparatively high turbidity levels in estuarine waters are measured using automated instruments and by hand.

Change type and corresponding factors

For hydrology and sediments (Figure 2), measurable areas of disturbance for respective types of change include: restricted flow with impoundment (Olsen, 1983; Gordon, 1988; Jupiter et al., 2007), estuarine shift from barriers to flow from upstream (Duke et al., 2003) and root burial from deposition above mean sea-level (Ellison, 1999, 2009; Duke et al., 2003) depositional gain from deposition below mean sea-level (Duke and Wolanski, 2001; Wolanski and Duke, 2002; Duke et al., 2003).

Pollutants

Predominant drivers

Pollutants in tidal wetlands are mostly the indirect consequence of human activities elsewhere. Pollutants are either excess amounts of naturally occurring substances or foreign and manufactured compounds. Such substances are grouped as two distinct types, namely, those natural ones including those that stimulate plant growth like nutrients and others which inhibit plant growth like toxicants. Two chief nutrients, nitrogen and phosphorus, are essential nutrients for plant and animal growth, but they are derived from two quite different sources. While nitrogen is obtained naturally from the atmosphere by bacteria and microalgae and phosphorus is derived naturally from eroded inorganic sediments, this effectively limits their availability respectively. However, with human development, this has changed dramatically, and there are now two new chief sources, namely, point source sewage outflows from urban and industrial treatment sites and diffuse source runoff from agricultural lands and urban-industrial development. The effects downstream are readily noticeable with unnaturally elevated levels of nutrients. The effects on tidal wetlands are largely unquantified but include abundant growth of epiphytic algae and excess wrack smothering sensitive above-ground breathing roots. Also washed downstream are chemical agents like agricultural herbicides in catchment runoff. Other toxic chemicals, notably petroleum hydrocarbons, are added to estuaries in chronic oil spill incidents in port areas. The fact that these chemicals are not often, if ever, found in natural environments indicates that the biota often have no prior adaptation, or capacity, to respond successfully to some toxic elements. Another natural pollutant are dust clouds although these will have a negative effect on canopy health where photosensitive foliage is covered by fine sediments leading to reduced photosynthetic capacity and loss of foliage density, with possible dieback.

Oil spills are mostly accidental and unintended, but while the application can be direct, it is often delayed by a matter of days. This is significant since delay is usually substantial where petroleum hydrocarbons degrade rapidly in the open with most toxic fractions lost first. Spilled oil on reaching mangrove coasts noticeably smothers plant surfaces resulting in tree death in extreme circumstances. Herbicides, by contrast, are essentially water soluble and invisible to see. Their presence is cryptic but their effect is more obvious, with the notable death of sensitive species. The source of agricultural chemicals is agricultural lands upstream in the catchment.

Factor indicators

Changes due to pollutants, nutrients, and toxicants are influenced by a number of factors. Nutrient-laden fertilizers used on crop lands in surrounding catchment areas plus sewage outflows and septic seepage increase smothering, algal growth on mangrove roots. In Eastern Australia, a currently unknown agent, in combination with high nutrient loads, has led to the rapid decomposition of below-ground roots, resulting in sediment collapse, subsidence and centre sinking. This leads to shallow anoxic pools within mangroves and salt marsh vegetation. In other instances, species specific death of Avicennia trees was correlated with toxic herbicides in runoff from catchment in coastal areas (Duke et al. 2005a). Spillage of toxic chemicals used in local transport causes tree death from toxicity and smothering of breathing surfaces. Dust buildup on foliage notably affects plant health and survival.

Change type and corresponding factors

For pollutants (Figure 2) measurable areas of disturbance for respective types of change include: algal excess with high nutrient levels (Dennison and Abal, 1999), center sinking likely related to a toxicant and/or nutrient (Duke et al., 2010), species specific with toxic pesticides in run-off (Duke et al., 2005a), and spill damage from toxic and suffocating petroleum spills (Duke and Watkinson, 2002; Duke and Burns, 2003; Duke et al., 1997; Duke et al., 1999; Duke et al., 2000) along with dirty foliage from excessive dust (Duke et al., 2003).

Pests and pathogens
Predominant drivers

Pests and pathogens are mostly indirect drivers resulting from the accidental or unintended consequences of human activities elsewhere. Pests are organisms detrimental to other more benign biota. They are often exotic and invasive, but this is not always the case. Local biota may also be pest species. The source of exotic pests includes: constructed wetlands, aquaculture, aquaria, vessel fouling, ballast water, and dredge spoil. Pest species can replace native species, cause alteration to habitat, or result in toxic blooms. Pathogens are bacteria, virus, or fungi which cause disease in biota, including humans. Some pathogens occur naturally in marine and estuarine waters. However, high concentrations of microorganisms are carried downstream in runoff from upstream sources of human and agricultural waste following storm events. In addition to catchment runoff and stormwater flows, further potential sources include: aquaculture cultures, exotic species, imported stock feed, and sewage discharges.

Factor indicators

Changes due to pests and pathogens are influenced by a number of factors. Herbivore life cycles may also be influenced by high nutrient content in foliage and additional habitat stressors. Further indicators of pests and pathogens include: presence and abundance of pest species and pathogens, evidence of harm to native species, loss of native species, and habitat disruption. Indicators of pathogens include: presence of pathogens in water, plant and animal lesions, responses of sensitive species, disrupted reproductive success, evidence of diseases, and mass mortality events.

Change type and corresponding factors

For pests and pathogens (Figure 2), measurable areas of disturbance for respective types of change include: diseased as a result of a pathogen (Fomba and Singh, 1990; Wesre et al., 1991; Duke, 1997), introduced species involving invasive biota (Allen, 1998; Allen et al., 2000), and herbivore attack with an active herbivore (Johnstone, 1981; Robertson and Duke, 1987; Rau and Murphy, 1990; Feller and McKee, 1999; Duke, 2002; Clarke and Kerrigan, 2002).

Climate change
Predominant drivers

Climate change combines both human and natural influences delivered as discrete drivers affecting different types of change. Mangrove coasts have always existed in a state of change. Their efficient and effective responses to altered conditions are the hallmark of their very existence. The tidal zone where they thrive is constantly changing, and it has always been a habitat at considerable risk. Over geological time, for example, inhabitants of the tidal zone have successfully followed sea-level changes ranging over tens to hundreds of meters. Therefore, mangrove coast habitats present today must be appreciated as the survivors of millions of years of adaptation and change. The component organisms have well-developed and specialized attributes to facilitate their survival in response to natural influences like temperature, rainfall, and sea-level. The question is: can they cope now with the accelerated influences of human-induced climate change?

Factor indicators

Changes due to climate change are influenced by a number of factors. Increased air and water temperatures will allow predominantly cold-restricted mangroves to recruit and survive in previously inhospitable higher-latitude locations. Rainfall variability notably affects soil moisture and salinity conditions to influence species at physiological extremes causing species at key ecotones to die or recruit. And, finally with sea-level change, as an added consequence of global warming, the entire tidal zone must shift upward with mangrove encroachment into upland habitat and corresponding losses at the waterfront. This type of change will curiously be similar to the responses of mangrove coasts to more localized changes in tidal amplitude, as observed in Southern Moreton Bay in Eastern Australia (Figure 4).

Change type and corresponding factors

For climate change (Figure 2), measurable areas of disturbance for respective types of change include: latitudinal shift with temperature change (Gilman et al., 2008; Wilson and Saintilan, 2012, Williams et al., 2013), ecotone shift with rainfall variability (Fosberg, 1961, Duke et al., 2003; Meynecke et al., 2006; Gilman et al., 2008), and zonal shift with sea-level change (Duke et al., 2002; Gilman et al., 2008).

Severe natural disturbance
Predominant drivers

The drivers of severe natural disturbance complement and expand on the influences noted above concerning climate change. As noted, mangrove coasts have always existed in a state of change. And, they will continue to do so. Some changes will also add to these influences, based on arguably natural factors, like wrack buildup as a result of higher sea waves. The continued influence of severe storms, is excepted to have worsened with changing

Mangrove Coasts, Figure 4 Severe natural disturbance with localized raised sea-level in Southern Moreton Bay, Queensland, Australia, in 2003. Note tidal salt marsh seedlings among dead and decaying *Casuarina* trees that marked an extensive supra-tidal zone 25 years earlier.

climatic conditions. For instance, more frequent lightning strikes will result in more forest gaps leading to increased forest replacement, recruitment, recovery, and turnover (Duke 2001). And, all extreme events exemplified by large seas, storm surges, and especially tsunami waves have added influences on waterfront stands, ever vulnerable to erosion and severely limited opportunities for recruitment and replacement.

Factor indicators

Changes due to severe natural disturbances are influenced by a number of factors. Following severe storm activity and blooms emanating from deteriorating water quality, bloom and storm debris will accumulate and build up to smother mangrove and salt marsh plants. Affected plants may or may not die. By contrast, severe storms, cyclonic winds, twisters, gusts, hailstorms, strong wave activity, and high flows will usually have notable and severe impacts on mangrove coast plants. This is expected to be exacerbated by climate change with increased severity of storms and other pressures. Forest light gaps will be created more frequently and they may well increase in size. Such impacts will be enhanced by large waves and strong currents from severe storms, large run off events, chronic boat wash, tectonic events and tsunami waves.

Change type and corresponding factors

For severe natural disturbance (Figure 2) easurable areas of disturbance for respective types of change include: wrack accumulation with storm seas (Boston, 1983; Ellison, 2009), storm damage with strong winds and hail (Houston, 1999; Greening et al., 2006; Proffitt et al., 2006), light gaps with lightning (Sousa, 1984; Duke, 2001; Amir and Duke, 2009), and waterfront erosion with large waves and tsunami (Danielsen et al., 2005; Dahdouh-Guebas et al., 2005; Gedan et al., 2011).

Conclusions

Mangroves and tidal wetlands of the mangrove coast are essential to the sustainability of highly productive natural coastal environments. However, these ecosystems and their dependent biota are under serious threat this century from the escalation of large-scale land clearing and conversion of coastal wetlands (including mangrove coasts) associated with the development of coastal lands for agricultural, aquaculture, port, urban, and industrial use. In populated areas, key coastal rivers have become little more than drains transporting eroded mud and effluent to settle in downstream estuarine reaches, as well as on shallow embayments and inshore coral reefs. The mangrove coasts that remain are becoming depauperate, nonfunctional vestiges of their former state. Where healthy, mangrove-lined estuaries had traditionally offered respite and critical dampening of land runoff, in recent years these bastions of coastal buffering – the unique "coastal kidney," mangrove coasts – are succumbing to the increasing and unrelenting pressure of expanding human population into coastal and estuarine tidal wetland habitat of tropical and subtropical shorelines worldwide.

To help allay the worsening situation, a program called MangroveWatch (Duke and Mackenzie 2009; 2010; www.mangrovewatch.org.au) specifically fosters and facilitates partnerships between community volunteers and scientists for raising awareness and for rigorous monitoring and assessment of habitat biodiversity and health. The ready adoption of this program so far raises the possibility of enormous potential benefits; where local community members take responsibility for monitoring the health and well-being of their sections of the mangrove coast.

Bibliography

Allen, J. A., 1998. Mangroves as alien species: the case of Hawaii. *Global Ecology and Biogeography Letters*, **7**, 61–71.

Allen, J. A., Krauss, K. W., Duke, N. C., Herbst, D. R., Björkman, O., and Shih, C., 2000. *Bruguiera* species in Hawai'i: systematic considerations and ecological implications. *Pacific Science*, **54**, 331–343.

Amir, A., and Duke, N.C., 2009. A forever young ecosystem: light gap creation and turnover of subtropical mangrove forests in Moreton Bay, southeast Queensland, Australia. In *Proceedings of the 11th Pacific Science Inter-Congress*, Tahiti, French Polynesia.

Boston, K. G., 1983. The development of salt pans on tidal marshes, with particular reference to south-eastern Australia. *Journal of Biogeography*, **10**, 1–10.

Clarke, P. J., and Kerrigan, R. A., 2002. The effects of seed predators on the recruitment of mangroves. *Journal of Ecology*, **90**, 728–736.

Dahdouh-Guebas, F., Jayatissa, L. P., Nitto, D. D., Bosire, J. O., Seen, D. L., and Koedam, N., 2005. How effective were mangroves as a defence against the recent tsunami? *Current Science*, **15**(12), 443–447.

Danielsen, F., Sorensen, M. K., Olwig, M. F., Selvam, V., Parish, F., Burgess, N. D., Hiraishi, T., Karunagaran, V. M., Rasmussen, M. S., Hansen, L. B., Quarto, A., and Suryadiputra, N., 2005. The Asian Tsunami: a protective role for coastal vegetation. *Science*, **310**, 643.

Dennison, W. C. and E. G. Abal, Eds., 1999. Moreton Bay Study. A scientific basis for the Healthy Waterways Campaign. Brisbane, South East Queensland Regional Water Quality Management Strategy, Brisbane City Council.

Duke, N. C., 1992. Mangrove floristics and biogeography. In Robertson, A. I., and Alongi, D. M. (eds.), *Tropical Mangrove Ecosystems*. Washington, DC: American Geophysical Union. Coastal and Estuarine Studies Series, pp. 63–100. 329 p.

Duke, N. C., 1997. Mangroves in the Great Barrier Reef World Heritage Area: current status, long-term trends, management implications and research. In Wachenfeld, D., Oliver, J., and Davis, K. (eds.), *State of the Great Barrier Reef World Heritage Area Workshop.* Townsville: Great Barrier Reef Marine Park Authority, pp. 288–299.

Duke, N. C., 2001. Gap creation and regenerative processes driving diversity and structure of mangrove ecosystems. *Wetlands Ecology and Management*, **9**, 257–269.

Duke, N. C., 2002. Sustained high levels of foliar herbivory of the mangrove *Rhizophora stylosa* by a moth larva *Doratifera stenosa* (Limacodidae) in north-eastern Australia. *Wetlands Ecology and Management*, **10**, 403–419.

Duke, N. C., 2006. Australia's mangroves. The authoritative guide to Australia's mangrove plants. In *Australia's Mangroves. The Authoritative Guide to Australia's Mangrove Plants*. Brisbane: The University of Queensland and Norman C Duke. 200 p.

Duke, N. C., 2011. Mangrove Islands. Encyclopedia of Modern Coral Reefs. Structure, Form and Process. D. Hopley. Dordrecht, The Netherlands, Springer: 653–655.

Duke, N. C., and Burns, K. A., 2003. Fate and effects of oil and dispersed oil on mangrove ecosystems in Australia. In *Environmental Implications of Offshore Oil and Gas Development in Australia: Further Research. A Compilation of Three Scientific Marine Studies.* Canberra: Australian Petroleum Production and Exploration Association (APPEA), pp. 232–363. 521 p.

Duke, N. C. and J. Mackenzie. 2009. MangroveWatch: a new monitoring program that partners mangrove scientists and community volunteers. Seagrass-Watch(39): 11.

Duke, N. C., and Mackenzie, J., 2010. Pioneering mangrove monitoring program partners experts with the community. *Wetlands Australia*, **18**, 24–25.

Duke, N. C., and Watkinson, A. J., 2002. Chlorophyll-deficient propagules of *Avicennia marina* and apparent longer term deterioration of mangrove fitness in oil-polluted sediments. *Marine Pollution Bulletin*, **44**, 1269–1276.

Duke, N. C., and Wolanski, E., 2001. Muddy coastal waters and depleted mangrove coastlines – depleted seagrass and coral reefs. In Wolanski, E. (ed.), *Oceanographic Processes of Coral Reefs. Physical and Biology Links in the Great Barrier Reef.* Washington, DC: CRC Press, pp. 77–91. 356 p.

Duke, N. C., Pinzón, Z. S., and Prada, M. C., 1997. Large-scale damage to mangrove forests following two large oil spills in Panama. *Biotropica*, **29**(1), 2–14.

Duke, N. C., Ball, M. C., and Ellison, J. C., 1998. Factors influencing biodiversity and distributional gradients in mangroves. *Global Ecology and Biogeography Letters*, **7**, 27–47.

Duke, N. C., Pinzón, Z. S., and Prada, M. C., 1999. Recovery of tropical mangrove forests following a major oil spill: a study of recruitment, growth, and the benefits of planting. In Yáñez-Arancibia, A., and Lara-Domínguez, A. L. (eds.), *Mangrove Ecosystems in Tropical America/ [Ecosistemas de Manglar en América Tropical]*. Silver Spring: Instituto de Ecologia A. C. Mexico, UICN/ORMA Costa Rica, and NOAA/NMFS, pp. 231–254. 380 p.

Duke, N. C., Burns, K. A., Swannell, R. P. J., Dalhaus, O., and Rupp, R. J., 2000. Dispersant use and a bioremediation strategy as alternate means of reducing the impact of large oil spills on mangrove biota in Australia: the Gladstone field trials. *Marine Pollution Bulletin*, **41**, 403–412.

Duke, N. C., E. Y. Y. Lo and M. Sun. 2002. Global distribution and genetic discontinuities of mangroves - emerging patterns in the evolution of Rhizophora. Trees. Structure and Function 16: 65–79.

Duke, N. C., Lawn, P., Roelfsema, C. M., Phinn, S., Zahmel, K. N., Pedersen, D., Harris, C., Steggles, N., and Tack, C., 2003. *Assessing Historical Change in Coastal Environments. Port Curtis, Fitzroy River Estuary and Moreton Bay Regions.* Brisbane: Historical Coastlines Project, Marine Botany Group, Centre for Marine Studies, The University of Queensland. 258 pages plus appendices.

Duke, N. C., Bell, A. M., Pedersen, D. K., Roelfsema, C. M., and Bengtson Nash, S., 2005a. Herbicides implicated as the cause of severe mangrove dieback in the Mackay region, NE Australia – serious implications for marine plant habitats of the GBR World Heritage Area. *Marine Pollution Bulletin*, **51**, 308–324.

Duke, N. C., Lawn, P., Roelfsema, C., Zahmel, K., Pedersen, D., and Tack, C., 2005b. Changing coastlines in the Fitzroy Estuary – assessing historical change in coastal environments. In Noble, B., Bell, A., Verwey, P., and Tilden, J. (eds.), *Fitzroy in Focus*. Brisbane: Coastal CRC – Cooperative Research Centre for Coastal Zone, Estuary and Waterway Management, pp. 6–9, 41–46. 106 p.

Duke, N. C., Meynecke, J.-O., Dittmann, S., Ellison, A. M., Anger, K., Berger, U., Cannicci, S., Diele, K., Ewel, K. C., Field, C. D., Koedam, N., Lee, S. Y., Marchand, C., Nordhaus, I., and Dahdouh-Guebas, F., 2007. A world without mangroves? *Science*, **317**, 41–42.

Duke, N., Haller, A., Brisbane, S., Wood, A., and Rogers, B., 2010. 'Sinking Centres' in Moreton Bay Mangroves. Maps Showing

Areas of Unusual Anoxic Ponds and Mangrove Dieback in Tidal Wetlands of the Bay Area in 2003–08. Brisbane: School of Biological Sciences, The University of Queensland. 223 p.

Ellison, J. C., 1999. Impacts of sediment burial on mangroves. *Marine Pollution Bulletin*, **37**(8–12), 420–426.

Ellison, J. C., 2009. Geomorphology and Sedimentology of Mangroves. In Perillo, E., Wolanski, E., Cahoon, D., and Brinson, M. (eds.), *Coastal Wetlands: An Integrated Ecosystem Approach*. Amsterdam: Elsevier BV, pp. 565–591.

FAO, 2007. *The World's Mangroves 1980–2005, FAO Forestry Paper 153*. Rome: Forest Resources Division, FAO, p. 77.

Feller, I. C. and K. L. McKee. 1999. Small gap creation in a Belizean mangrove forests by a wood-boring insect. Biotropica 31(4): 607–617.

Field, C. D., 1995. *Journey Amongst Mangroves*. Okinawa, Japan: International Society for Mangrove Ecosystems (ISME). 140 p.

Fomba, S. N., and Singh, N., 1990. Crop losses caused by rice brown spot disease in mangrove swamps of northwestern Sierra-Leone. *Tropical Pest Management*, **36**, 387–393.

Fosberg, F. R. 1961. Vegetation-free zone on dry mangrove coastline. U.S. Geol. Soc. Prof. Papers 424(D): 216–218.

Gedan, K. B., Kirwan, M. L., Wolanski, E., Barbier, E. B., and Silliman, B. R., 2011. The present and future role of coastal wetland vegetation in protecting shorelines: answering recent challenges to the paradigm. *Climatic Change*, **106**, 7–29.

Gilman, E. L., Ellison, J., Duke, N. C., and Field, C., 2008. Threats to mangroves from climate change and adaptation options: a review. *Aquatic Botany*, **89**, 237–250.

Gordon, D. M., 1988. Disturbance to mangroves in tropical-arid Western Australia: hypersalinity and restricted tidal exchange as factors to mortality. *Journal of Arid Environments*, **15**, 117–145.

Greening, H., Doering, P., and Corbett, C., 2006. Hurricane impacts on coastal ecosystems. *Estuaries and Coasts*, **29**(6), 877–879.

Hogarth, P. J., 1999. *The Biology of Mangroves*. Oxford: Oxford University Press.

Houston, W. A. 1999. Severe hail damage to mangroves at Port Curtis, Australia. Mangroves and Saltmarshes 3: 29–40.

Johnstone, I. M., 1981. Consumption of leaves by herbivores in mixed mangrove stands. *Biotropica*, **13**, 252–259.

Jupiter, S. D., Potts, D. C., Phinn, S. R., and Duke, N. C., 2007. Natural and anthropogenic changes to mangrove distributions in the Pioneer River Estuary (Queensland, Australia). *Wetlands Ecology and Management*, **15**(1), 51–62.

Lewis-III, R. R., 2005. Ecological engineering for successful management and restoration of mangrove forests. *Ecological Engineering*, **24**, 403–418.

Meynecke, J.-O., Lee, S. Y., Duke, N. C., and Warnken, J., 2006. Effect of rainfall as a component of climate change on estuarine fish production in Queensland, Australia. *Estuarine, Coastal and Shelf Science*, **69**, 491–504.

Mumby, P. J., Edwards, A. J., Arias-Gonzalez, J. E., Lindeman, K. C., Blackwell, P. G., et al., 2004. Mangroves enhance the biomass of coral reef fish communities in the Caribbean. *Nature*, **427**, 533–536.

Olsen, H. F., 1983. *Biological Resources of Trinity Inlet and Bay Queensland*. Brisbane: Queensland Department of Primary Industries. 64 pp.

Primavera, J. H., 2000. The values of wetlands: landscape and institutional perspectives. Development and conservation of Philippine mangroves: institutional issues. *Ecological Economics*, **35**, 91–106.

Proffitt, C. E., Milbrandt, E. C., and Travis, S. E., 2006. Red mangrove (*Rhizophora mangle*) reproduction and seedling colonization after hurricane charley: comparisons of Charlotte Harbor and Tampa Bay. *Estuaries and Coasts*, **29**(6), 972–978.

Rau, M. T., and Murphy, D. H. 1990. Herbivore attack on mangrove plants at Ranong. In Mangrove Ecosystems, UNDP/UNESCO Regional Mangrove Project RAS/86/120. Occasional Papers no. 7, pp. 25–37.

Robertson, A. I., and Duke, N. C., 1987. Insect herbivory and mangrove leaves in North Queensland. *Australian Journal of Ecology*, **12**, 1–7.

Robertson, A. I., and Duke, N. C., 1990. Mangrove fish communities in tropical Queensland, Australia: spatial and temporal patterns in densities, biomass and community structure. *Marine Biology*, **104**, 369–379.

Rubin, J. A., Gordon, C., and Amatekpor, J. K., 1999. Causes and consequences of mangrove deforestation in the Volta Estuary, Ghana: some recommendations for ecosystem rehabilitation. *Marine Pollution Bulletin*, **37**(8), 441–449.

Saenger, P., 2002. *Mangrove Ecology, Silviculture and Conservation*. Dordrecht: Kluwer Academic Publishers. 360 p.

Sousa, W. P., 1984. The role of disturbance in natural communities. *Annual Review of Ecology and Systematics*, **15**, 353–391.

Spalding, M. D., Blasco, F., and Field, C. D. (eds.), 1997. *World Mangrove Atlas*. Okinawa, Japan: International Society for Mangrove Ecosystems. 178 p.

Spurgeon, J., 1998. The socio-economic costs and benefits of coastal habitat rehabilitation and creation. *Marine Pollution Bulletin*, **37**, 373–382.

Tomlinson, P. B., 1994. *The Botany of Mangroves*, 2nd edn. Cambridge: Cambridge University Press. 413 p.

Walters, B. B., 2005. Ecological effects of small-scale cutting of Philippine mangrove forests. *Forest Ecology and Management*, **206**, 331–348.

Wesre, G., Cahill, D., and Stamps, D. J., 1991. Mangrove dieback in north Queensland, Australia. *Transactions of the British Mycological Society*, **79**(1), 165–167.

Williams, A., Eastman, S., Eash-Loucks, W., Kimball, M., Lehmann, M., and Parker, J., 2013. Record Northernmost Endemic Mangroves on the United States Atlantic Coast with a Note on Latitudinal Migration. Southeastern Naturalist in press, 10 p.

Wilson, N. C., and Saintilan, N., 2012. Growth of the mangrove species *Rhizophora stylosa* Griff. at its southern latitudinal limit in eastern Australia. *Aquatic Botany*, **101**, 8–17.

Wolanski, E., and Duke, N. C., 2002. Mud threat to the Great Barrier Reef of Australia (Increasing mud, decreasing mangroves: threats posed by unconsolidated muds to the Great Barrier Reef of Australia). In Healy, T. R., Wang, Y., and Healy, J.-A. (eds.), *Muddy Coasts of the World: Processes, Deposits and Function*. Amsterdam: Elsevier Science B.V, pp. 533–542. 542 p.

Cross-references

Beach
Beach Processes
Bioturbation
Coastal Bio-geochemical Cycles
Coastal Engineering
Coasts
Deltas
Engineered Coasts
Estuary, Estuarine Hydrodynamics
Geohazards: Coastal Disasters
Hurricanes and Typhoons
Integrated Coastal Zone Management
Lagoons
Oil Spill
Organic Matter
Reef Coasts

Sea-Level
Sediment Dynamics
Tidal Depositional Systems
Tsunamis
Waves

MARGINAL SEAS

Di Zhou
South China Sea Institute of Oceanology, Chinese Academy of Sciences, Guangzhou, China

Synonyms
Marginal basin; Back-arc basin; Back-arc sea

Definition
In geology, a marginal sea is the sea bordering continents, semi-isolated from open ocean by *island arc* or land ridge, and underlain by *oceanic crust* (Karig, 1971). Thus a marginal sea is a deep-sea basin with water depth in general over ~3,000 m, while its slopes and shelves are called as its *continental margins*.

As a geographic and oceanographic term, a "marginal sea" represents a partially enclosed sea adjacent to or widely open to open ocean at the surface, but bounded by submarine ridges on the seafloor (Wikipedia). The presence of oceanic crust is not required.

As a geopolitical term, a marginal sea is equivalent to a territorial water, which is important in determining what maritime resources that a state can exploit (Wikipedia).

Geographic distribution
The marginal seas in geological context are distributed mainly in western Pacific Ocean, forming a giant sea chain of ~12,000 km long from the north to the south. A small number of marginal seas are seen in Caribbean, Andaman, and off Antarctica. See Figure 1 for the names and locations of the marginal seas. The Gulf of California is regarded as having oceanic crust in its southern portion and thus may be considered as a marginal sea in geological context (Curry et al., 1982).

The marginal seas in geography and oceanography context include not only the marginal seas in geological context but also their continental margins, as well as other semi-enclosed seas such as the Arabian Sea, Baltic Sea, Bay of Bengal, Bering Sea, Beaufort Sea, Black Sea, Gulf of Mexico, the Mediterranean Sea, Red Sea, Ross Sea, Weddell Sea, all four of the Siberian seas (Barents, Kara, Laptev, and East Siberian) [Water Encyclopedia], Yellow Sea, East China Sea, and Bohai Sea.

Features and formation models of marginal seas in geological context

History of the concept
Although the term "marginal sea" appeared much earlier in context of geography and oceanography, the clear definition of marginal sea or marginal basin in the geological context was given several years after establishment of *plate tectonics* (Karig, 1971). Early researches on marginal seas concentrated to their nature and origin (Matsuda and Uyeda, 1971; Packham and Falvey, 1971; Sleep and Toksöz, 1971). Then there were a long period of extensive geological and geophysical investigations including *ocean drilling*, which revealed basic features and led to various formation models of marginal seas (cf. the summary in Tamaki and Honza (1991)). Recent studies of marginal seas have extended to climate and environmental impacts of the marginal seas in the present and in the geological history (Wang, 1999).

Features
Marginal seas in geological context are of similar water depths as normal oceans, but different from open oceans in several aspects (Tamaki and Honza, 1991; Wang, 1999): (1) Almost all marginal seas are located on the east side of a continent and separated from an open ocean by a *trench-arc system*. (2) Marginal seas have a short life span from their opening by the breakup of *lithosphere* to the stop of *seafloor spreading*, usually less than 25 Ma. After the cessation of spreading, many marginal seas are closed by *subduction* of themselves. (3) Marginal seas may change their spreading directions due to the movements of surrounding tectonic terrains. (4) Compared with open oceans of the same age, marginal seas usually have 600–800 m deeper water depths. (5) The marginal seas in geographic context are the sites for active *land-sea interactions* and have one to two orders of magnitude higher deep-water sedimentation rates, larger *mass flux and energy flux*, and higher *biodiversity* compared with open oceans.

Formation models
The formation of a marginal sea is indicated by the generation of oceanic crust as the consequence of lithosphere breakup and *seafloor spreading*. The models of formation mechanism of marginal seas can be grouped into three categories: *back-arc spreading* models, the slab-pull model, and the pull-apart model. Because most marginal seas are formed behind island arcs, several formation models focus on the mechanism of back-arc spreading (Figure 2), which may be grouped further into active models (Models 1 and 2) in which *asthenosphere* flow provides the driving force for the opening of marginal sea, and passive models where the driving force is slab movement, either the retreat of the overriding back-arc continental slab (Model 3) or the retreat of the subducting oceanic slab (Models 4 and 5).

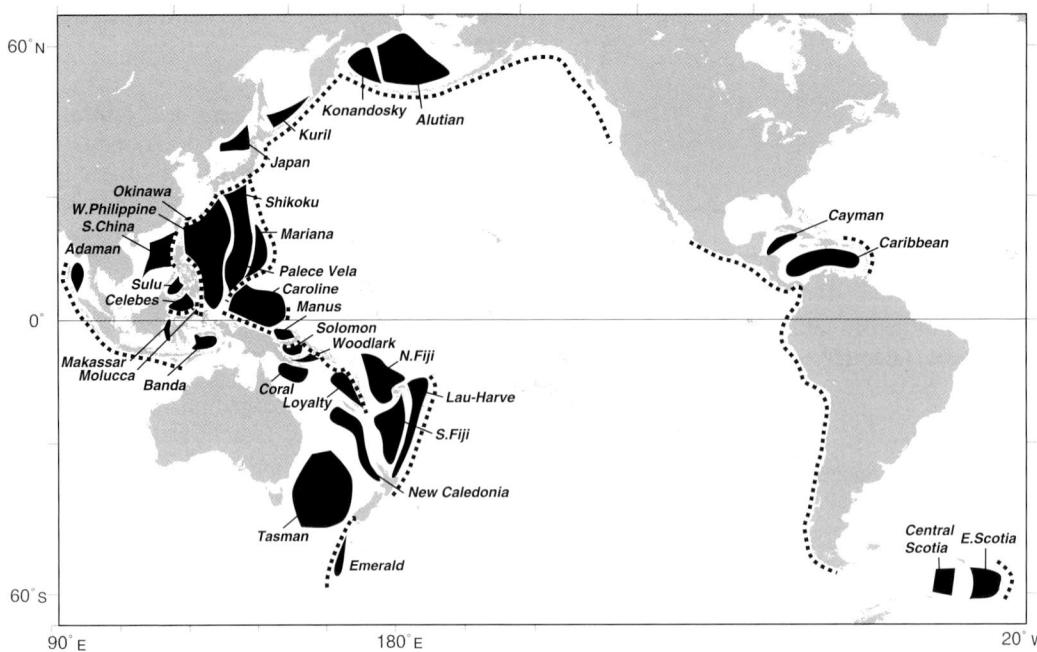

Marginal Seas, Figure 1 Distribution of marginal seas around the Pacific Ocean. *Solid areas* show marginal seas, and *blocked lines* show trenches. Abbreviations: *E* east, *N* north, *S* south, *W* west (from Tamaki and Honza (1991)).

Back-arc spreading models predict the axis of marginal sea opening should be parallel to the subduction trench and island arc; but this is not always seen. For the marginal seas do not have trench-parallel axes, non-back-arc spreading models are proposed. The slab-pull model (Model 6) suggests the marginal sea opened by the pulling force of a frontal subducting oceanic slab. This driving force is caused by the *negative buoyancy*, which occurs when the heavier oceanic slab subducts into the lighter *asthenosphere*. Other non-back-arc spreading models are related to tectonic forces. For example, the pull-apart model (Model 7) suggests that the marginal sea was formed by the *pull-apart* force associated with large *strike-slip fault*(s).

Features of marginal seas in geographic and oceanographic context

Water circulation

Water circulation patterns in marginal seas depend largely on bathymetry, freshwater input (e.g., river runoff and precipitation), evaporation, and connectivity to the open ocean circulations.

As marginal seas are highly isolated from the open ocean, the water circulation is dependent largely on the balance between freshwater input and evaporation. If freshwater input exceeds evaporation, as is the case in the Black and Baltic Seas, the excess freshwater will tend to flow seaward near the sea surface, diluting the marginal sea. If evaporation exceeds freshwater input, as in the Mediterranean Sea, the marginal seawater becomes saltier and then sinks and flows near the bottom toward the less salty open ocean. These general water circulation patterns are often modified by local bathymetry, which in some instances serve to restrict water flow (Water Encyclopedia).

In the marginal seas with relatively good connection with open oceans, the water circulation is significantly influenced by the connectivity to the open ocean circulations, which is in turn influenced by the *sea-level changes*. For example, in the system of western Pacific marginal seas, the water circulation is strongly influenced by the inflow of the western boundary currents (Kuroshio and Oyashio) of the western Pacific Ocean (Figure 3). During glacial periods, the sea-level is low and the boundary currents are flowing only outside the marginal seas, while in interglacial periods, the western boundary currents are partially running through the marginal seas. This change in water circulation might influence the climate system of the Earth. For example, the reduction of Kuroshio warm water in glacial times decreased heat and humidity supply to East Asia (Wang, 1999).

Sedimentation

The marginal seas intercepted the flux of terrigenous sediments from continents and have the sedimentation rates one or two orders of magnitude higher than the sedimentation rates in open oceans. This is responsible for the development of wide continental shelves in marginal seas, such as in the marginal seas between Asia and western Pacific Ocean. Although South and East Asian continent and their

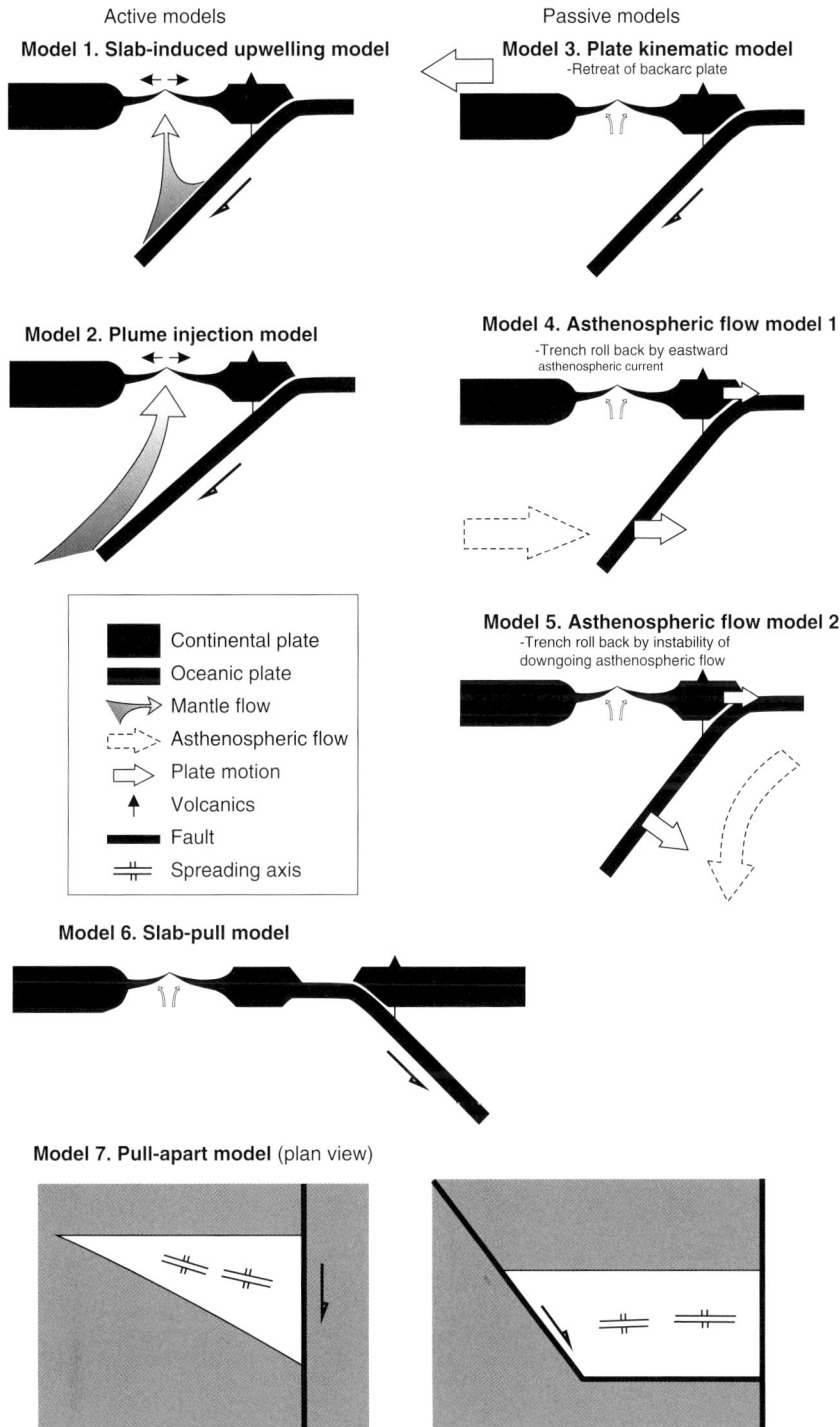

Marginal Seas, Figure 2 Back-arc spreading models for the origin of marginal seas. Models 1–5 are redraw from Tamaki and Honza (1991).

islands provide over 70 % of terrigenous suspended materials to the global ocean, these sediments were intercepted by marginal seas so that there are no large submarine deep-sea fans developed in the western Pacific comparable to Indian Ocean fans (Wang, 1999, 2004). The wide continental shelves of marginal seas are vulnerable to sea-level changes and thus might influence the climate. For example, because the South China Sea is a major source of

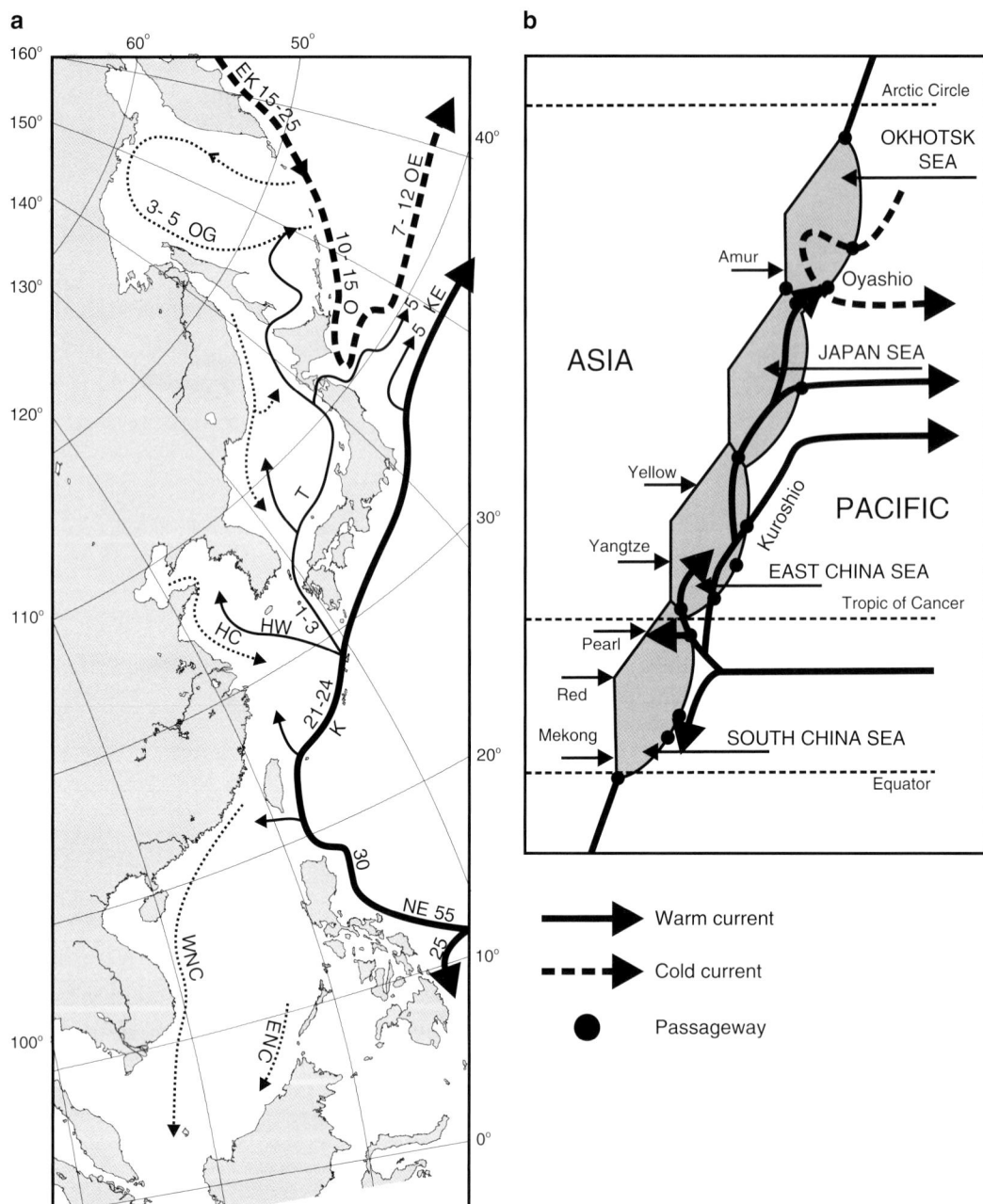

Marginal Seas, Figure 3 The system of marginal seas and the western boundary currents in western Pacific Ocean (redraw from Wang, 2004). (**a**) Western boundary currents flowing through the marginal seas. (**b**) Diagram showing the hydrographic system in the marginal seas. Capital letters are abbreviations of the currents, among which K for Kuroshio and O for Oyashio; numbers give the flow rate in Sv (1 Sv = 10^6 km^3/s).

moisture for the summer monsoon rain in China, the reduction of shelf area in the last glacial maximum has enhanced the aridity in the China hinterland (Wang, 2004).

Primary productivity and biomass production

Marginal seas generally exhibit intermediate levels of primary production and biomass production, in comparison with the highest rates in coastal *upwelling* regions and the lowest rates in open ocean regions. This is because variations in primary productivity in oceans and seas occur as a result of water column chemistry (e.g., nutrient and trace elements concentrations) and dominant physical processes. River runoff and water column mixing introduce dissolved nutrients, trace elements, and suspended particles into the photic (light) zones of nearshore regions and resulted in increased primary production; the addition of suspended particles increases water turbidity, which results in reduced sunlight penetration and decreased primary productivity (Water Encyclopedia).

Importance of marginal seas

The marginal seas are often rich in oil and gas, fishery and biomass, and other resources. They also provide spaces for shipping and other human activities. These seas are the frontal areas of sea-land interactions with extensive mass, energy, and gene fluxes and have significant influences on global climate system. The thick sediments in the marginal seas contain detailed and often magnified records on the geological and environmental evolution of adjacent land masses (Clift et al., 2004; Wang, 1999).

The marginal seas in the geological context are a first-order tectonic feature in the Earth system. Along the boundaries between land masses and open oceans, marginal seas formed only in particular segment mainly in the western Pacific Ocean. The origin of marginal seas is a key to a deep understanding of the Earth's geodynamics. The formation of marginal seas had exerted significant impacts on the Earth system, such as on the geography, environment, and resources, on the ocean circulation and global climate, on the land-sea interactions and sedimentation, etc. These are the current research frontiers in geosciences.

Summary

It should be noticed that the definition of "marginal sea" is slightly different in disciplines of geology, geography and oceanography, and geopolitics. The most restrict definition is in geological context where a marginal sea must be underlain by oceanic crust, which is formed by lithosphere breakup and seafloor spreading. Marginal seas have many features different from those of open oceans. The formation of marginal seas created a space for extensive interactions between land mass and open oceans and has a significant impact on global geography, oceanography, geology, and climate system. The marginal seas also provide abundant natural resources and space for human being. The study of marginal seas is a frontier field in geosciences.

Bibliography

Clift, P. D., Layne, G. D., and Bliesztajn, J., 2004. Marine sedimentary evidence for monsoon strengthening, Tibetan uplift and drainage evolution. In Clift, P., Kuhnt, W., Wang, P., and Hayes, D. (eds.), *Continent-Ocean Interactions within East Asian Marginal Seas*. Washington, DC: American Geophysical Union, pp. 255–282.

Curray, J.R., Moore, D.G., Kelts, K., Einsele, G., 1982. Tectonics and geological history of the passive continental margin at the tip of Baja California, in: Curray, J.R., Moore, D.G. (Ed.), Initial Reports DSDP 64. U.S. Government Printing Office, Washington, p. 1089–1116.

Karig, D. E., 1971. Origin and development of marginal basins in the western Pacific. *Journal of Geophysical Research*, **76**, 2542–2561.

Matsuda, T., and Uyeda, S., 1971. On the pacific-type orogeny and its model – extension of the paired belts concept and possible origin of marginal seas. *Tectonophysics*, **11**, 5–27.

Packham, G. H., and Falvey, D. A., 1971. An hypothesis for the formation of marginal seas in the western pacific. *Tectonophysics*, **11**, 79–109.

Sleep, N., and Toksöz, M. N., 1971. Evolution of marginal basins. *Nature*, **233**, 548–550.

Tamaki, K., and Honza, E., 1991. Global tectonics and formation of marginal basins: role of the western Pacific. *Episodes*, **14**, 225–230.

Wang, P., 1999. Response of Western Pacific marginal seas to glacial cycles: paleoceanographic and sedimentological features. *Marine Geology*, **156**, 5–39.

Wang, P., 2004. Cenozoic deformation and the history of sea-land interactions in Asia. In Clift, P., Kuhnt, W., Wang, P., and Hayes, D. (eds.), *Continent-Ocean Interactions within East Asian Marginal Seas*. Washington, DC: AGU, pp. 1–22.

Cross-references

Lithosphere: Structure and Composition
Marginal Seas
Ocean Drilling
Oceanic Spreading Centers
Plate Tectonics
Subduction
Transform Faults
Upwelling

MARINE EVAPORITES

Walter E. Dean
U.S. Geological Survey, Denver, CO, USA

Definition

Marine evaporites form by extreme evaporation of seawater either directly in an open-water environment with restricted circulation or diagenetically from sediment pore waters.

Introduction

Geology hangs its interpretive powers on the principal of uniformitarianism, i.e., the present is the key to the past. Unfortunately for marine evaporates, there is no present example for the thick, aerially extensive ancient evaporite deposits ("saline giants"). That would require a large, deep basin with a marine connection but with restricted circulation, and a huge influx of seawater for a long time. The marine connection could be surface water or marine-fed groundwater. It would also require a warm climate with extreme net evaporation that existed for a long time. The Quaternary is not a good analog for the extent and distribution of climates that existed throughout most of the Phanerozoic because during the Quaternary the world was in an icehouse mode. However, during most of the Phanerozoic, the world was in a warm, greenhouse mode (Warren, 2010). In addition, the average salinity of the world ocean during most of the Phanerozoic was 40–50‰ in contrast to 35‰ today (Hay et al., 2006).

Large evaporite basins occur on all continents, except Antarctica, and range in age from Late Precambrian to Late Miocene (Warren, 2006, 2010). The most recent saline giant occurred in the Latest Miocene (Messinian) in basins of the Mediterranean region (the so-called Messinian salinity crisis; Ryan et al., 1973; Hsü et al., 1973, 1978). Today, there are many environments with extreme net evaporation, but there are no connected large, deep basins with restricted marine connection. One candidate would be the Red Sea with a little more restriction and possibly a more arid climate. Leg 23 of the Deep Sea Drilling Project (DSDP) recovered bedded middle Miocene evaporates from the Red Sea when the basin was more restricted than today (Stoffers and Ross, 1974). Comprehensive reviews of evaporite deposits, marine and nonmarine, are given by Warren (2006, 2010).

Marine evaporite minerals

The first mineral to crystallize from evaporation of seawater is calcium carbonate ($CaCO_3$), usually as calcite, but at other times aragonite may form. Next is calcium sulfate either as gypsum ($CaSO_4 \cdot 2H_2O$) or anhydrite ($CaSO_4$). The next most abundant evaporite mineral is halite (NaCl), commonly known as rock salt. These two minerals, gypsum/anhydrite and halite, constitute more than 95 % by volume of the world's Phanerozoic evaporates (Warren, 2010). Under conditions of extreme, nearly complete evaporation potassium (K) and/or magnesium (Mg), minerals may form. These, often referred to as bittern or potash salts, are much less common. They include polyhalite ($K_2Ca_2Mg(SO_4)_4 \cdot 2H_2O$, sylvite (KCl), carnallite ($KCl_2 \cdot MgCl_2 \cdot 6H_2O$), langbeinite ($K_2SO_4 \cdot 2MgSO_4$), kainite ($MgSO_4 \cdot KCl \cdot 3H_2O$), glauberite ($Na_2SO_4 \cdot CaSO_4$), kieserite ($MgSO_4 \cdot H_2O$), bischofite ($MgCl_2 \cdot 6H_2O$), epsomite ($MgSO_4 \cdot 7 H_2O$), hexahydrite ($MgSO_4 \cdot 6 H_2O$), loeweite ($Na_4Mg_2 (SO_4)_4 \cdot 5H_2O$), and vanthoffite ($Na_6Mg(SO_4)_4$). Any one bed in an evaporite deposit commonly is monomineral, especially those composed of gypsum, anhydrite, or halite. Many more salts have been recorded from evaporite deposits but the above are by far the most common. Paragenetic mineral pairs with sylvite are common enough that they have been given mine-mineral names. Sylvite combined with halite is called "sylvanite," and sylvite combined with anhydrite and/or kieserite and/or polyhalite is called "hartsalz" by the Germans.

Permian Zechstein formation

The Upper Permian Zechstein Formation underlies most of Poland, Germany, the Netherlands, Denmark, east central England, and the North Sea. It is the most extensively studied evaporite sequence in the world. The Zechstein consists of four depositional cycles. The second oldest depositional cycle is the Stassfurt cycle or series named for its type section at Stassfurt, Germany. The Stassfurt is the most studied and most economically important of the Zechstein series. At Stassfurt, the basal anhydrite is only 50 m thick, overlain by 500 m of halite, topped off by only 50 m of potash salts. Borings into the Zechstein throughout Germany in the early nineteenth century determined the locations of subsurface bedded rock salt and other evaporite minerals that formed a stratigraphic sequence more than 2,300 m thick at Stassfurt. At that time potassium and magnesium deposits (potash) associated with salt were considered contaminants and were discarded. However, it was soon recognized that potassium was necessary for plant growth, and for the rest of the nineteenth century, most of the potash of the world was supplied from mines near Stassfurt. Bischoff, in his book *Elements of Chemical and Physical Geology* (1864), was perhaps the first to recognize that halite and anhydrite (or gypsum) were precipitated from seawater but did not propose an exact method.

The following six zones represent a complete progressive evaporite cycle within the Stassfurt from youngest to oldest:

Hartsalz – sylvite plus anhydrite and/or kieserite, polyhalite, and halite
Carnallite – 55–60 % carnallite, 25 % halite, 17 % kieserite
Kieserite and carnallite – with approximately 65 % halite
Polyhalite – with approximately 93 % halite
Anhydrite plus halite – approximately 96 % anhydrite-laminated halite
Basal anhydrite – with dolomite laminae and some halite

Probably the most significant experiment in understanding the origin of evaporite minerals was the classic work of an Italian chemist named Usiglio (1849). He determined the sequence of precipitation of evaporite minerals from seawater by simply evaporating an aliquot of Mediterranean seawater and noting the sequence and quantities of precipitates with progressive evaporation.

The results of Usiglio's experimental evaporation of seawater (1849; reproduced in Clarke, 1924) show that calcium carbonate begins to precipitate inorganically when seawater has been evaporated to about half of its original volume. The most common carbonate mineral to form from seawater at normal-marine temperatures and pressures is low-Mg calcite, although aragonite, high-Mg calcite, and dolomite may also form if conditions are right, usually at higher salinities and/or in association with biogenic controls. Aragonite is unstable and with time and burial aragonite will recrystallize to form low-Mg calcite.

When the original water volume has been reduced to about 20 %, calcium sulfate begins to precipitate. Under salinity and temperature conditions of most evaporite deposits, anhydrite should theoretically be the most stable calcium sulfate mineral. However, in experimental and observed modern evaporite systems, gypsum is the almost universal calcium sulfate mineral to form. This may or may not have been true in the past. Primary anhydrite has been found in a few Holocene evaporite deposits such as the sabkhas of the Persian Gulf, or associated with the heavy metal deposits in the deep, hot-brine basins of the Red Sea (Miller et al., 1966), and in Clayton Playa, Nevada (Moiola and Glover, 1965).

By the time the solution has been reduced to about 10 % of the original volume, most of the calcium sulfate has been deposited, and halite starts to separate along with gypsum (or anhydrite). When the volume of the solution is less than 5 % of the original, polyhalite takes the place of gypsum or anhydrite. Polyhalite separates together with halite until the solution is saturated with respect to Mg-bearing sulfates free from Ca and K. If the early-formed salts remain in contact with residual liquids at all stages and react freely with them, the final products of crystallization of the bittern salts are:

0° to 30 °C: epsomite + carnallite + bischofite + halite + anhydrite

13° to 17.5 °C: hexahydrite + carnallite + halite + anhydrite

17.5° to 110 °C: carnallite + bischofite + halite + anhydrite

Usiglio's experiment reproduced most of the evaporite mineral sequence found at Stassfurt.

To provide the theoretical or physical chemical explanation of the evaporite sequence at Stassfurt, the Dutch chemist Jacobus van't Hoff dedicated most of his career in the late 1890s and early 1900s to the study of the mineral equilibria of the Stassfurt salts. This was the beginning of experimental petrology. Van't Hoff's calculation produced a sequence of evaporites that was remarkably similar to the Stassfurt sequence. The main differences were with the potash salts. An informative historical summary of van't Hoff's work is provided by Eugster (1971), and an extensive description of the Zechstein is provided by Borchert and Muir (1964).

To explain the sedimentology of the sequence of minerals found at Stassfurt, Carl Ochsenius expanded the earlier model of Bischoff for the formation of evaporite minerals by evaporation of seawater and proposed the barred-basin model whereby a lagoon or seaway was separated from the open ocean by a bar. Ochsenius suggested that the Gulf of Kara Bogaz (Kara-Bogaz-Gol), which is separated from the Caspian Sea by a bar, might serve as an analog for a barred basin but on a smaller scale, both in terms of area and depth, than the large Phanerozoic evaporite basins.

Permian Castile and Salado formations

Another modification of the Bischoff–Ochsenius bar theory was proposed by R. H. King (1947) to explain the thick (1,300–2,000 ft; 400–600 m) anhydrite deposits, with minor halite, in the late Permian (Ochoan; 255 ka) Castile Formation in the Delaware Basin of west Texas and southeastern New Mexico. The Delaware Basin is an intercratonic basin with an area of 25,980 km^2 and was probably about 500 m deep before evaporite deposition (Kirkland, 2003). King proposed a "reflux model" whereby some brine was returned ("refluxed") to the open ocean but a continuous supply of seawater kept the salinity within the anhydrite stability field.

Adams (1944) showed that the "minor" halite in the Castile was not minor in the deeper, eastern part of the basin where it mostly occurs as thick beds of halite interlaminated with anhydrite. He produced the first comprehensive cross section of the basin based on borehole lithologies and defined the boundary between the Castile and the overlying Salado Formation. In the eastern part of the basin, halite makes up about one third of the thickness but represents only one fiftieth of the time of deposition (Anderson, 2011). In the shallower, uplifted western part of the basin, halite beds have been dissolved resulting in beds of dissolution breccias. During the late Permian when the Castile was deposited, the Delaware Basin was an embayment within a region of extreme aridity in the equatorial western margin of the supercontinent Pangaea surrounded by the well-known Permian reef complex (Guadalupian Capitan Formation) with saline groundwater seeping into the basin from the south and west from the giant Panthalassa ocean (Anderson and Dean, 1995; Anderson, 2011).

The Castile anhydrite consists mostly of seasonal laminae couplets of dark, organic-stained calcite and lighter-colored anhydrite, the typical "banded anhydrite" of most of the Castile (Figure 1) (Anderson et al., 1972). Each light–dark couplet is interpreted as an annual increment of sedimentation (a varve) (Anderson and Kirkland, 1966; Anderson et al., 1972). The Castile is divided into four anhydrite members, I–IV (from oldest to youngest), and three halite members, I–III (from oldest to youngest) (Anderson et al., 1972). Each halite member represents the culmination of a major salinity cycle. The halite units consist of cm-thick halite layers and thin anhydrite laminae. At the top of a salinity cycle, the typical calcite-banded anhydrite grades into nodular anhydrite (Figure 1). Dean et al. (1975) illustrate a progression from typical laminated anhydrite to nodular anhydrite but still with laminae of calcite that may be highly distorted to even resemble nodular "chicken wire" anhydrite that is common in supratidal evaporates (discussed below). The laminae are remarkably uniform in composition, thickness, and delicate detail over distances of at least 113 km (Anderson et al., 1972). The calcite–anhydrite varve couplets are repeated with remarkable regularity with more than 209,000 calcite–anhydrite and anhydrite–halite varve couplets in 450 m of stratigraphic section (Anderson, 2011). The evidence supports the deep-basin, deep-water interpretation especially for the typical "banded anhydrite" that represents the preponderance of time of deposition of the Castile.

The calcite–anhydrite varves of the Castile are very similar to the dolomite–anhydrite varves (jahresringe) of the Permian Zechstein Formation of Germany that were correlated for distances of as much as 390 km by Anderson and Kirkland (1966) using measurement by

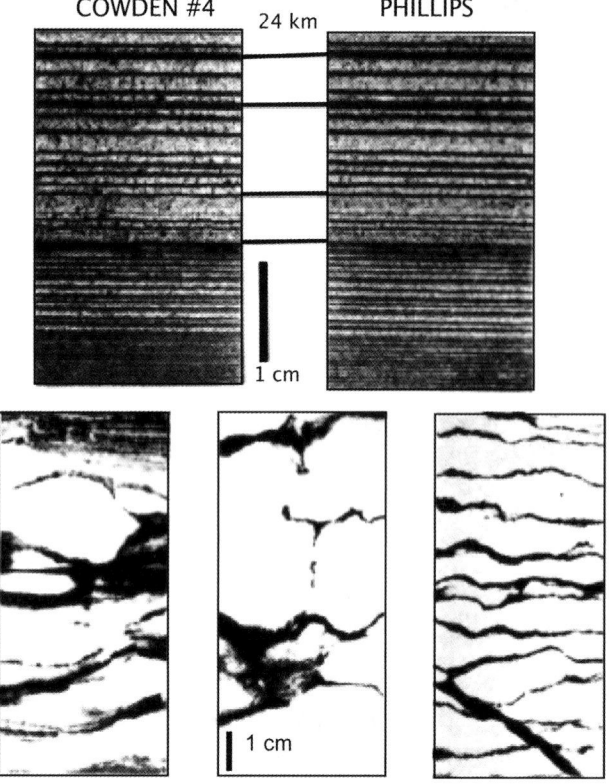

Marine Evaporites, Figure 1 (*Top*) Correlation of typical laminated anhydrite (*light*) and organic-stained calcite (*dark*) varve couplets of the Castile Formation between the Cowden #4 and Phillips cores separated by 24 km in Culberson County, Texas. (*Bottom*) Nodular anhydrite in the Phillips core of the Castile Formation. The nodules started as displacive growth within the anhydrite laminae of normal varve couplets. Dark material is organic-stained calcite.

Richter-Bernburg (1957). The stratigraphic regularity and lateral continuity of laminated anhydrite, such as those found in the Castile and Zechstein Formations, are the main hallmarks of deep-basin, deep-water evaporates.

The anhydrite–halite laminations in the Castile halite members continue into the base of the overlying Salado Formation, which is dominated by halite and potash. The Salado is as much as 700 m thick and spills out over the Delaware Basin, as defined by the Capitan Reef and enclosed Castile Formation. Most of the Salado was deposited in shallow water over a longer time span than the Castile with much of the time represented by nondeposition. The Salado covers an area of about 150,000 km^2, although the area that includes the Carlsbad Potash District is only 70,000 km^2 (Lowenstein, 1988). The main potash-bearing unit is the McNutt Zone in the middle of the Salado that contains sylvite and langbeinite and minor carnallite, kieserite, and kainite (Barker and Austin, 1993). It is in the thick halite beds of the Salado that the Waste Isolation Pilot Plant (WIPP) site is located east of Carlsbad, New Mexico, for the disposal of low-level radioactive wastes.

Shallow-water evaporites

The deep-basin, deep-water model of evaporite deposition, represented by the Zechstein, Castile, and Salado Formations, prevailed during the first half of the twentieth century, but during the 1950s, examples of shallow-water evaporites began to emerge, interpreted as having been deposited in restricted estuaries or basins ("arms of the sea") (e.g., Scruton, 1953; Masson, 1955; Morris and Dickey, 1957).

Marginal-marine evaporite deposition was further stimulated with the documentation in the 1960s of supratidal ("sabkha") evaporite deposition along the Arabian (Abu Dhabi) margin of the Persian Gulf (e.g., Shearman, 1966; Kinsman, 1966, 1969; Butler, 1969). Here a whole spectrum of shelf carbonates and evaporites has been developed. A typical sediment sequence might include subtidal aragonite muds, pelleted muds, and oolites; intertidal laminated cyanobacterial (blue-green algal) mats with trapped and bound carbonate muds; and supratidal (sabkha) nodular anhydrite or gypsum in a carbonate matrix that may be dolomitized. With no modern analogs for the thick, extensive evaporite sequences of the Phanerozoic, the Holocene occurrences of marginal-marine evaporates have been used, and sometimes overused, to explain ancient evaporite deposits. For example, Shearman (1966) suggested that many ancient evaporite deposits, especially those dominated by anhydrite, but not excluding thick halite and potash deposits, had been formed by diagenesis in a sabkha-like environment. A review of Holocene carbonates and evaporites of the Arabian coast is presented by Alsharhan and Kendell (2003), and a discussion of supratidal mudflat evaporite deposition in general is presented in Kirkland and Evans (1973).

The application of the "sabkha model" was accelerated with the exciting recovery of upper Miocene (Messinian) anhydrite and halite from the floor of the Mediterranean Sea on Leg 13 of the Deep Sea Drilling Project (DSDP; Ryan et al., 1973; Hsü et al., 1973) and cored again in 1975 on DSDP Leg 42A (Garrison et al., 1978; Hsü et al., 1978; Kidd et al., 1978). The sediments that were cored with poor recovery on DSDP Legs 13 and 42A were from the upper ~70 m of the upper Messinian evaporite that may be up to 2 km thick based on seismic profiles and boreholes in the seafloor, outcrops on land, and up to 600 m of halite in the subsurface. The conclusion was that the evaporites from the upper part of the upper evaporite in eastern and western basins of the Mediterranean were deposited in shallow subaqueous to subaerial environments and that the Mediterranean must have dried up and refilled several times with a huge waterfall at the Strait of Gibraltar. This is called the deep-basin, desiccation hypothesis, and the event has been called the "Messinian Salinity Crisis" (MSC). It has been used to explain a wide

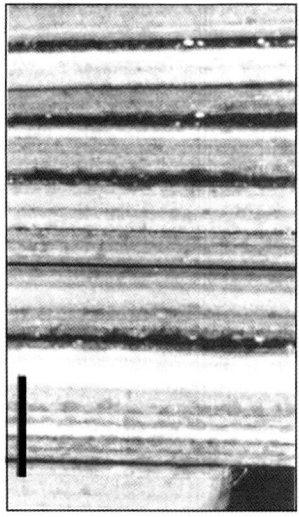
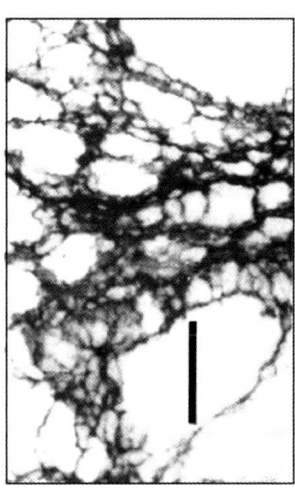

Marine Evaporites, Figure 2 (*Left*) Laminated anhydrite, Miocene, from the Mediterranean. Deep Sea Drilling Project, Site 124, Core 8, Section 1, interval 0–5 cm. (*Right*) Nodular anhydrite, Miocene, from the Mediterranean. Deep Sea Drilling Project, Site 124, Core 8, Section 1, interval 70–74 cm. Bars = 1 cm.

variety of paleoceanographic and paleoclimatic events and has been the subject of numerous articles in both the scientific and popular literature, of books, and even of television shows.

The key features of the evaporites recovered on DSDP Legs 13 and 42A are laminated and nodular anhydrite (Figure 2), which are key fabrics for the interpretative application of the sabkha model. This has led to an uncritical use of the equations: laminated sediments = algal mats = tidal-flat environment and nodular anhydrite = sabkha = subaerial environment. Dean et al. (1975) presented sedimentological criteria to distinguish between deep-basin, deep-water, and shallow-water or sabkha evaporates. Schreiber et al. (1976) and Schreiber and El Tabakh (2000) described deep-water laminated gypsum in the Messinian of Sicily in which the laminae are continuous with carbonate interlaminations, for which the name "balatino" is often used. Ryan (2009) concluded that the finely laminated "balatino" anhydrite in the Mediterranean DSDP cores (e.g., Figure 2, left) when compared to the anhydrite varves of the German Zechstein appears to be "typical of relatively deep and calm waters." Hardie and Lowenstein (2004, 453), after careful examination of cores from both Legs 13 and 42A, concluded: "Several features of these [Miocene] evaporates used by the DSDP workers as evidence of shallow-water conditions are more comparable with deposition under deep-water (below wave base) conditions, and others can only be considered as of uncertain origin."

Since DSDP Leg 42A, there have been a number of symposia and workshops reassessing the Messinian Salinity Crisis. Most notably, 22 papers from a colloquium "The Messinian Salinity Crisis Revisited" held in Palermo, Italy, occupy volumes 188 and 189 (2006) of "Sedimentary Geology" (Rouchy et al., 2006; Rouchy and Caruso, 2006). Most of those papers were based on borehole and outcrop data on land around the Mediterranean. In fact, supratidal evaporite deposition in general and the MSC in particular have dominated the evaporite literature in the decades since the 1960s.

A reassessment of the MSC was made by Ryan (2009). He concluded that a review of the current knowledge of the MSC requires the insertion of an early "deep-water," "deep-basin" stage that was not considered in the formulation of the original desiccation hypothesis in 1973. He states: "The revised model of evaporative concentration now has shallow-margin, shallow-water and deep-basin, deep-water precursors to desiccation."

Summary

Large marine evaporite basins occur on all continents, except possibly Antarctica, and range in age from Late Precambrian to Late Miocene. The progressive evaporation of seawater results in the precipitation of the following succession of evaporite minerals: a carbonate mineral (calcite or dolomite), a calcium sulfate mineral (gypsum or anhydrite), and sodium chloride (halite or rock salt). Anhydrite and halite constitute more than 95 % by volume of the world's Phanerozoic evaporates. Under conditions of extreme evaporation potassium (K) and/or magnesium (Mg), minerals called potash or bittern salts may form, which are much less common. There is no modern analog for the thick, aerially extensive, ancient evaporite deposits ("saline giants").

The most extensively studied "saline giant" in the world is the Upper Permian Zechstein Formation that underlies most of northern Europe. The Zechstein consists of four depositional cycles or series that culminate in halite and, in the extreme, potash. The second oldest depositional cycle of the Zechstein is the Stassfurt series and is the most studied and most economically important of the Zechstein series and comes closest to the experimental and theoretical sequences of evaporite minerals.

Another important "saline giant" consists of the late Permian Castile and Salado Formations of the Delaware Basin of west Texas and southeastern New Mexico. Halite beds occur at the culminations of four major salinity cycles that define four anhydrite and three halite members of the Castile. The anhydrite units consist of interlaminated anhydrite and thin calcite laminae. The halite units consist of cm-thick halite layers and thin anhydrite laminae. The anhydrite–halite laminations in the Castile halite members continue into the overlying Salado Formation, which is dominated by halite and potash. The Salado Formation is as much as 700 m thick and spills out over the Delaware Basin, as defined by the Capitan Reef and the enclosed Castile Formation.

The deep-basin, deep-water model of evaporite deposition, represented by the Zechstein, Castile, and Salado Formations, prevailed during the first half of the twentieth century, but during the 1950s examples of shallow-water evaporites began to emerge, interpreted as having been deposited in restricted marginal-marine estuaries or basins. Marginal-marine evaporite deposition was further stimulated with the documentation in the 1960s of supratidal ("sabkha") evaporite deposition along the Arabian margin of the Persian Gulf. With no modern analogs for the thick, extensive evaporite sequences of the Phanerozoic, the Holocene occurrences of marginal-marine sabkha evaporates have been used to explain ancient evaporite deposits. The application of the "sabkha model" was accelerated with the recovery of upper Miocene (Messinian) anhydrite and halite from the floor of the Mediterranean Sea on Legs 13 and 42A of the Deep Sea Drilling Project (DSDP). The implication was that these evaporates were deposited in shallow subaqueous to subaerial environments and that the Mediterranean must have dried up and refilled several times with a huge waterfall at the Strait of Gibraltar. This has been called the deep-basin, desiccation hypothesis, and the event has been called the "Messinian Salinity Crisis." Supratidal evaporite deposition in general and the Messinian Salinity Crisis in particular have dominated the evaporite literature in the decades since the 1960s.

Bibliography

Adams, J. E., 1944. Upper Permian Ochoan series of Delaware Basin, west Texas and southeastern New Mexico. *American Association of Petroleum Geologists Bulletin*, **28**, 1592–1625.

Alsharhan, A. S., and Kendall, C. G. S. C., 2003. Holocene coastal carbonates and evaporates of the southern Arabian Gulf and their ancient analogues. *Earth-Science Reviews*, **61**, 191–243.

Anderson, R. Y., 2011. Enhanced climate variability in the tropics: a 200,000 yr annual record of monsoon variability from Pangea's equator. *Climate of the Past*, **7**, 757–770.

Anderson, R. Y., and Dean, W. E., 1995. Filling the Delaware Basin: hydrologic and climatic controls on the Upper Permian Castile Formation varved evaporite. In Scholle, P. A., Peryt, T. M., and Ulmer-Scholl, D. S. (eds.), *The Permian of Northern Pangea*. Berlin: Springer. Sedimentary Basins and Economic Resources, Vol. 2, pp. 61–78.

Anderson, R. Y., and Kirkland, D. W., 1966. Intrabasin varve correlation. *Geological Society of America Bulletin*, **77**, 241–256.

Anderson, R. Y., Dean, W. E., Kirkland, D. W., and Snider, H. I., 1972. Permian Castile varved evaporite sequence, West Texas and New Mexico. *Geological Society of America Bulletin*, **83**, 59–86.

Barker, J. M., and Austin, G. S., 1993. Economic geology of the Carlsbad potash district, New Mexico. *New Mexico Geological Society Guidebook, 44th Field Conference, Carlsbad Region, New Mexico and West Texas, 1993*. pp. 283–291.

Borchert, H., and Muir, R. O., 1964. *Salt Deposits, The Origin, Metamorphism and Deformation of Evaporites*. London: D. Van Nostrand.

Butler, G. P., 1969. Modern evaporite deposition and geochemistry of coexisting brines, the sabkha, Trucial Coast, Arabian Gulf. *Journal of Sedimentary Petrology*, **39**, 70–89.

Clarke, F. W., 1924. *The Data of Geochemistry*, 5th edn. U. S. Geological Survey Bulletin, Vol. 770, p. 841.

Dean, W. E., Davies, G. R., and Anderson, R. Y., 1975. Sedimentological significance of nodular and laminated anhydrite. *Geology*, **3**, 367–372.

Eugster, H. P., 1971. The beginnings of experimental petrology. *Science*, **173**, 481–489.

Garrison, R. E., Schreiber, B. C., Bernoilli, D., Fabricus, F. H., Kidd, R. B., and Méliers, F., 1978. In Hsü, K. J., Montadert, L., et al. (eds.), *Initial Reports of the Deep Sea Drilling Project*. Washington, DC: U.S. Government Printing Office, Vol. 42, Part 1, pp. 4571–4612.

Hardie, L. A., and Lowenstein, T. K., 2004. Did the Mediterranean Sea dry out during the Miocene? A reassessment of the evaporite evidence from DSDP Legs 13 and 42A cores. *Journal of Sedimentary Petrology*, **74**, 453–461.

Hay, W. W., Migdisov, A., Balukhovsky, A. N., Wold, C. N., Flögel, S., and Söding, E., 2006. Evaporites and the salinity of the ocean during the Phanerozoic: implications for climate, ocean circulation and life. *Palaeogeography, Palaeoclimatology, and Palaeoecology*, **240**, 3–46.

Hsü, K. J., Ryan, W. B. F., and Cita, M. B., 1973. Late Miocene desiccation of the Mediterranean. *Nature*, **24**(240), 244.

Hsü, K. J., Montadert, L., Bernoulli, D., Cita, M. B., Erickson, A., Garrison, R. E., Kidd, R. B., Méliéres, F., Müller, C., and Wright, R., 1978. History of the Mediterranean salinity crisis. In Hsü, K. J., Montadert, L., et al. (eds.), *Initial Reports of the Deep Sea Drilling Project*. Washington, DC: U.S. Government Printing Office, Vol. 42, Part 1, pp. 1053–1068.

Kidd, R. B., Bernoulli, D., Garrison, R. E., Fabricus, F. H., and Méliéres, F., 1978. Lithologic findings of DSDP Leg 42A, Mediterranean Sea. In Hsü, K. J., Montedert, L., et al. (eds.), *Initial Reports of the Deep Sea Drilling Project*. Washington, DC: U.-S. Government Printing Office, Vol. 42, Part 1, pp. 1079–1094.

King, R. H., 1947. Sedimentation in the Permian Castile sea. *American Association of Petroleum Geologists Bulletin*, **31**, 470–477.

Kinsman, D. J. J., 1966. Gypsum and anhydrite of Recent age, Persian Gulf. In *Second Symposium on Salt*. Cleveland: Northern Ohio Geological Society, pp. 302–326.

Kinsman, D. J. J., 1969. Modes of formation, sedimentary associations, and diagnostic features of shallow-water and supratidal evaporites. *American Association of Petroleum Geologists Bulletin*, **53**, 830–840.

Kirkland, D. W., 2003. An explanation for the varves of the Castile evaporates (Upper Permian), Texas and New Mexico, USA. *Sedimentology*, **50**, 899–920.

Kirkland, D. W., and Evans, R. (eds.), 1973. *Marine Evaporites: Origin, Diagenesis, and Geochemistry*. Stroudsburg: Dowden, Hutchinson, and Ross. Benchmark Papers in Geology.

Lowenstein, T. K., 1988. Origin of depositional cycles in a Permian "saline giant": the Salado (McNutt zone) evaporates of New Mexico and Texas. *Geological Society of America Bulletin*, **100**, 592–608.

Masson, P. H., 1955. An occurrence of gypsum in southwest Texas. *Journal of Sedimentary Petrology*, **25**, 72–79.

Miller, A. R., Densmore, C. D., Degens, E. T., Hathaway, J. C., Manheim, F. T., McFarlin, P. F., Pocklington, R., and Jokela, A., 1966. Hot brines and recent iron deposits in deeps of the Red Sea. *Geochimica et Cosmochimica Acta*, **30**, 341–359.

Moiola, R. J., and Glover, E. D., 1965. Recent anhydrite from Clayton Playa, Nevada. *American Mineralogist*, **50**, 2063–2069.

Morris, R. C., and Dickey, P. A., 1957. Modern evaporite deposition in Peru. *American Association of Petroleum Geologists Bulletin*, **41**, 2467–2474.

Ochsenius, K., 1888. On the formation of rock salt and mother liquor salts. *Philadelphia Academic Society Proceedings*, Part 2, 181–187.

Richter-Bernburg, G., 1957. Isochrone Warven im Anhydrite des Zechstein 2: Germany, *Geol. Landesanst. Geol. Jb. Bd.*, **74**, 601–610.

Rouchy, J. M., and Caruso, A., 2006. The Messinian salinity crisis in the Mediterranean basin: a reassessment of the data and an integrated scenario. *Sedimentary Geology*, **188–189**, 35–67.

Rouchy, J. M., Suc, J. P., Ferrandini, J., and Ferrandini, M., 2006. The Messinian salinity crisis revisited. Editorial. *Sedimentary Geology*, **188–189**, 1–8.

Ryan, W. B. F., 2009. Decoding the Mediterranean salinity crisis. *Sedimentology*, **56**, 95–136.

Ryan, W. B. F., Hsü, K. J., et al., 1973. *Initial Reports of the Deep Sea Drilling Project*. Washington, DC: U.S. Government Printing Office, Vol. 13, Parts 1 and 2, p. 1447.

Schreiber, B. C., and El Tabakht, M., 2000. Deposition and early alteration of evaporates. *Sedimentology*, **47**, 215–238.

Schreiber, B. C., Friedman, G. M., Decima. A., and Schreiber, E., 1976. Depositional environments of Upper Miocene (Messinian) evaporite deposits of the Sicilian Basin. *Sedimentology*, **23**, 729–760.*

Scruton, P. C., 1953. Deposition of evaporates. *American Association of Petroleum Geologists*, **37**, 2498–2512.

Shearman, D. J., 1966. Origin of marine evaporites by diagenesis. *Institute of Mining and Metallurgy Transactions, Section B*, **75**, 208–215.

Stoffers, P., and Ross, D. A., 1974. Sediment history of the Red Sea. In Whitmarsh, R. B., Weser, O. E., Ross, D. A., et al. (eds.), *Initial Reports of the Deep Sea Drilling Project*. Washington, DC: U.S. Government Printing Office, Vol. 23, pp. 849–865.

Usiglio, J., 1849. Analyse de L'eau de la Mediterranee sur les cotes de France. *Annalen der Chemie*, **27**(92–107), 172–191.

Warren, J. K., 2006. *Evaporites: Sediments, Resources and Hydrocarbons*. Berlin: Springer.

Warren, J. K., 2010. Evaporites through time: tectonic, climatic, and eustatic controls on marine and nonmarine deposits. *Earth-Science Reviews*, **98**, 217–268.

Cross-references

Anoxic Oceans
Astronomical Frequencies in Paleoclimates
Events
Laminated Sediments
Marine Mineral Resources
Ocean Drilling
Organic Matter
Paleoceanography
Source Rocks, Reservoirs

MARINE GAS HYDRATES

Gerhard Bohrmann[1] and Marta E. Torres[2]
[1]MARUM-Center for Marine Environmental Sciences, University of Bremen, Bremen, Germany
[2]College of Earth, Ocean, and Atmospheric Sciences, Oregon State University, Corvallis, OR, USA

Synonyms

Clathrates; Gas hydrates; Methane hydrates

Definition

Gas hydrates are non-stoichiometric solid compounds in which low-molecular-weight gases are trapped (guests) within water cavities (hosts). The presence of gas hydrates is controlled by temperature, pressure, and the availability of appropriate gases and water.

Historic background

Gas hydrates were first described in the nineteenth century by Sir Humphry Davy, with his pioneering synthesis of chlorine hydrate. This type of compounds sparked industrial interest in the 1930s, when it became apparent that methane hydrates formed spontaneously within natural gas pipelines in permafrost regions. Research spearheaded by the oil and gas industry at that time created the foundation for our current understanding of the behavior of these compounds (Sloan and Koh, 2008). By the 1970s, Russian scientists had postulated that under certain temperature and pressure conditions, methane hydrates could form in natural systems and indeed observed these natural deposits in the Siberian permafrost. This discovery was soon followed by reports of methane hydrate presence in Canada's MacKenzie Delta and the Caspian and Black Seas. The next important advance in the study of these deposits came with the realization that sediments hosting gas hydrate could be imaged by seismic data (Stoll et al., 1971), thus providing a prospecting tool that fueled exploratory efforts along continental margins worldwide.

Gas hydrates were first cored by the Deep Sea Drilling Project (DSDP) in 1970 on the Blake Ridge (US continental margin), and the first deep-sea gas hydrate specimens were observed in sediments recovered from the Middle America Trench during DSDP Legs 66 and 67 in 1979. Since then, scientific ocean drilling has provided critical data toward the fundamental understanding of the nature, composition, and distribution of naturally occurring gas hydrates (Sloan and Koh, 2008). Large-scale new projects spearheaded by national-industry consortia are targeting gas hydrate deposits in search for new energy resources. Such projects are under way in the Indian Ocean, the South China Sea, the eastern sea off Korea, and both the eastern and western margins of Japan (Collett et al., 2009). In addition to scientific drilling efforts, near-surface gas hydrate deposits have been recovered from continental margins worldwide (Kvenvolden, 1988). These near-surface deposits are usually associated with locations of active cold seepage, where rapid transport of methane allows for massive deposits to form near the seafloor (Suess et al., 1999; Torres et al., 2011).

Gas hydrate stability in marine sediments

The stability of gas hydrate is fundamentally controlled by thermodynamic equilibria (Sloan and Koh, 2008). The pressure/temperature conditions required for pure methane hydrate stability are illustrated in Figure 1. The phase boundary is determined by the gas composition and by the ionic strength of the water (Figure 1a). The presence of CO_2, H_2S, ethane, and/or propane shifts the stability curve to the right, in effect increasing the gas hydrate stability. The presence of dissolved ions in the pore fluids, on the other hand, decreases the gas hydrate stability, so that an increase in salinity of the fluids from which the gas hydrate is forming shifts the phase boundary to the left (Figure 1a).

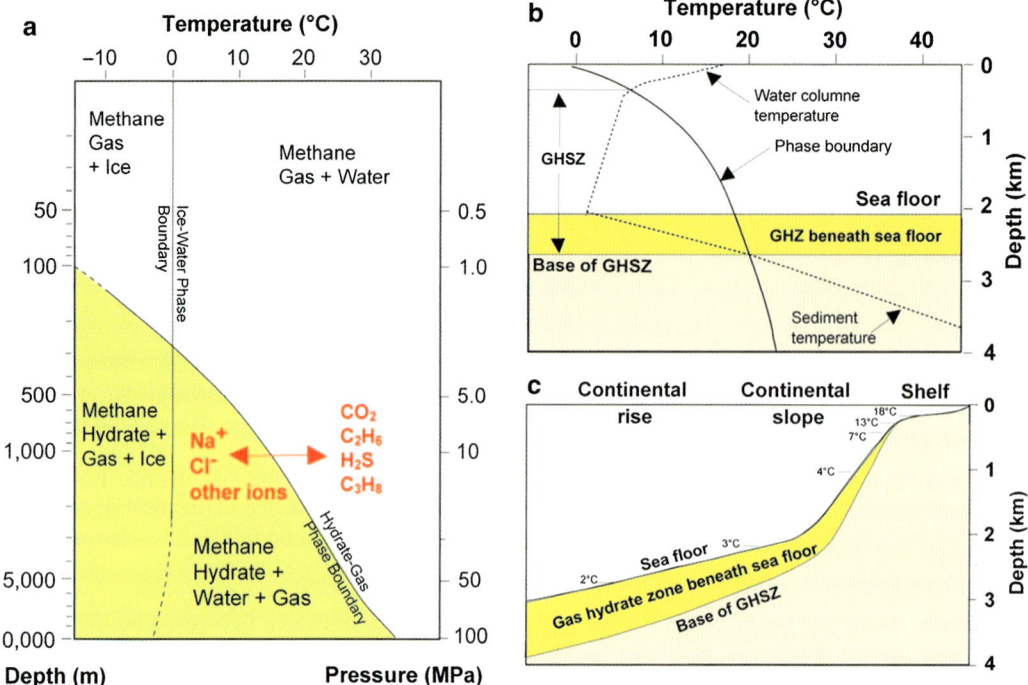

Marine Gas Hydrates, Figure 1 Phase diagram showing the boundary between methane hydrate (*yellow*) and free methane gas (*white*). Additions of ions and the presence of gases other than methane shift this phase boundary and hence gas hydrate stability (after Kvenvolden, 1988).

Temperature and pressure conditions consistent with methane hydrate stability are found at the seafloor nearly everywhere at water depths exceeding 300 m, but may be as deep as 800 m, depending on the regional seawater temperature (Figure 1c). For example, within the Black Sea where the bottom water temperature is 9 °C, the upper limit of hydrate stability is around 700 m water depth, whereas in the Mediterranean Sea, hydrates can be expected at depths >1,200 m with bottom water temperature of 14 °C and higher salinities (Pape et al., 2010). In contrast, in polar oceans, gas hydrate may be stable at depth <300 m (Kvenvolden, 1988).

Below the seafloor, temperatures increase with depth tracking the regional geothermal gradient (Figure 1b). As temperature in the sediments increases, it will eventually get high enough to cross the phase boundary, such that gas hydrate is no longer stable beneath this depth. Since the geothermal gradient is often quite uniform across broad regions beneath the seafloor, the thickness of the gas hydrate stability zone (GHSZ) is quite constant for a given water depth; hence, the thickness of the GHSZ may reach 800–1,000 m below seafloor in deep-water areas and shoals upward as water depth decreases (Figure 1c).

In addition to the P/T considerations described above, gas hydrates will only form if there is enough methane to stabilize the gas hydrate structure. In most of the deep ocean basins, however, methane concentrations in pore water are too low for gas hydrate formation (Claypool and Kaplan, 1974). Such that in general, gas hydrate formation is restricted to continental margins and enclosed seas, where organic matter accumulates rapidly enough to support in situ methane production by bacteria or where existing free or dissolved gas (sometimes from thermogenic sources and petroleum reservoirs) is transported into the gas hydrate stability zone (GHSZ).

Gas hydrate nucleation and growth also depend on sediment grain size, shape, and composition, as predicted by based on capillary pressure models and laboratory experiments using different host matrices (Clennell et al., 1999). These predictions were confirmed by field observations that document the preferential gas hydrate formation in coarse lithologies (Torres et al., 2008). In fine-grained sediments, gas hydrate tends to form discrete veins and lenses and occurs in massive blocks only in the near-sediment surface, where gas hydrate growth can overcome the overburden pressure, in effect displacing the sediment in a manner similar to the well-known ice heaving process (Torres et al., 2004).

Remote sensing techniques

Geophysical surveys are commonly used for detecting the presence of gas hydrate in marine sediments, through the

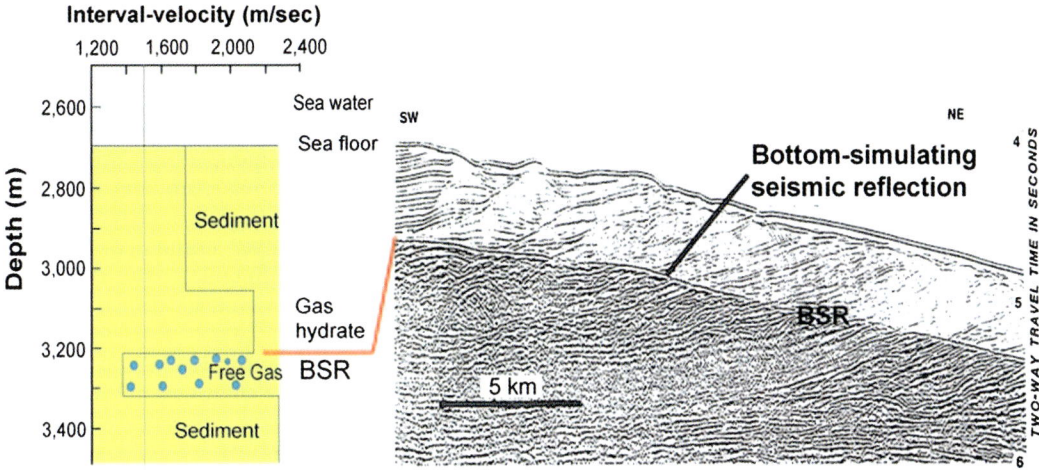

Marine Gas Hydrates, Figure 2 Seismic record from Blake Ridge showing a distinct reflection, known as the bottom-simulating reflection (Shipley et al., 1979). A seismic velocity model (*left*) shows the strong contrast of velocity across the BSR.

observation of a seismic reflection called "bottom-simulating seismic reflection" or BSR, because it approximately mimics the seafloor (Shipley et al., 1979). The negative polarity of this reflection indicates that it results from a decrease in acoustic impedance (defined as a product of density and seismic velocities) with depth (Figure 2). The BSR cuts across reflections of stratigraphic origin, making it readily apparent in marine seismic records. The BSR occurs approximately at the depth where the base of the gas hydrate stability zone is predicted based on thermodynamic equilibria (e.g., Shipley et al., 1979).

Although some details of the seismic reflection properties are not yet fully understood, it appears that the strength and the characteristics of the BSR are determined by the presence of free gas below the gas hydrate zone (Holbrook et al., 1996).

Imaging and quantification of natural gas hydrate

Gas hydrate rapidly dissociates during core recovery when using non-pressurized coring devices. Estimates of gas hydrate abundance and distribution rely by necessity on indirect estimates that include: total gas volumes from pressure-core samples and geochemical measurements of dissolved chloride (Cl$^-$) concentration and water isotopic composition. Like ice, gas hydrate incorporates water but excludes dissolved ions during formation. With time, excess dissolved ions will diffuse away from the location of gas hydrate formation, reestablishing a local background value. However, as gas hydrate dissociates during core retrieval, it adds fresh water to the pore space, thereby diluting dissolved ions. Since chloride in general behaves conservatively at depths in which gas hydrate accumulates, it is used as a good quantitative tracer of gas hydrate concentration (Torres et al., 2008). In addition, the water trapped within the hydrate cage is isotopically heavier than seawater, in both the oxygen and hydrogen atoms; hence, the heavy-oxygen isotopes and the deuterium of the pore fluids sampled from sediments that originally hosted gas hydrate may be used to estimate gas hydrate concentration (Kim et al., 2012).

Geophysical logging records provide the only means of obtaining high-resolution data at in situ conditions from the entire borehole, including those intervals for which sediment is not recovered. The most robust logging indicators of in situ methane hydrate are elevated electrical resistivity, acoustic velocities, and NMR-log data. Estimates of the amount of gas hydrate derived from resistivity log data are perhaps the most commonly used. Resistivity data are used as input to the Archie relation to estimate the fraction of pore space in sediment occupied by gas hydrate, with special considerations for the nature of the reservoir (see Riedel et al., 2010).

Distribution and dynamics

The mode of gas hydrate occurrence can be broadly classified into end-member regimes, which are based on the mechanisms that control gas transport into the GHSZ (Tréhu et al., 2004). The most commonly occurring deposits have been denoted as distributed low-flux systems, where most of the available methane is generated in the vicinity of where gas hydrate is formed, and fluid flow is pervasive. Gas hydrate in the system occurs either as pore filling (commonly within coarse lithologies) or as vein filling in fine-grained sediments. In contrast, in the focused high-flux systems, methane from a large volume of sediment is concentrated within well-defined conduits that channel methane to the seafloor where it forms massive gas hydrate deposits. Torres et al. (2004) argued that to sustain the massive deposits found on Hydrate Ridge, methane must be supplied in the gas phase, since methane solubility is too low for aqueous transport to maintain the observed hydrate accumulations. These deposits are characterized by the presence of brines that result from rapid incorporation of water into the growing gas hydrate structure, with salt remaining in solution and reaching concentrations as much as twice that of seawater (Torres et al., 2011). Massive gas hydrate and chloride brines in near-seafloor sediments along continental margins are not at all uncommon, and may represent a significant carbon reservoir that is susceptible to perturbations driven by changes in bottom water temperature.

Gas hydrates and energy

In several review articles, e.g., Boswell and Collett (2010), four gas hydrate reservoir types are evaluated in terms of their resource potential: sand-dominated reservoirs, clay-dominated fractured reservoirs, seafloor massive deposits, and clay-dominated disseminated reservoirs. Current exploration, production testing, and modeling target arctic and marine hydrate-bearing sand reservoirs in the North Slope of Alaska, Gulf of Mexico, and Japan, with additional research offshore China, Korea, and India. Proposed methods of gas recovery from gas hydrates were reviewed by Boswell and Collett (2010). Traditionally, these were based on: heating the reservoir above the gas hydrate stability, decreasing the reservoir pressure below gas hydrate equilibrium, and injecting a chemical inhibitor into the reservoir to decrease the gas hydrate stability. More recently, laboratory experiments have documented the feasibility of displacing methane molecules in the hydrate structure with carbon dioxide molecules, thus releasing methane and sequestering CO_2. In 2011 a multiyear project began to field-test the CO_2 for methane exchange concept on the Alaska Ignik Sikumi test well. This field test showed that CO_2 can be injected into a gas hydrate sandy reservoir, followed by methane production (Schoderbek et al., 2012).

Gas hydrates and climate

Methane is a powerful greenhouse gas with a greenhouse warming potential (GWP) 23 times that of CO_2 on a per-molecule basis. Sudden release of methane from gas hydrate therefore has the potential to affect global climate, and there are interesting although controversial hypotheses that attribute past climate variations to methane release from marine gas hydrates in response to ocean warming and/or sea-level change (Dickens, 2011). These hypotheses have yet to be confirmed and more research is needed to evaluate hydrate response to environmental change: the fate of methane from marine gas hydrate reservoirs to the seabed, ocean surface, and the atmosphere.

Gas hydrate observatories

Numerous laboratory and field studies at gas hydrate-bearing sites, including several drilling expeditions in the past decades, have provided critical background data on the conditions of gas hydrate stability and have given an overall view of the composition and distribution of gas hydrates in nature. Fundamental questions remain as to the residence time of gas hydrates near the seafloor and deeper within the sediment column, the nature and driving mechanisms for flow and biological interactions in environments where gas hydrates are present and fluid (aqueous and gas) migration occurs, and the role of the ocean in mitigating gas input to the atmosphere from the seafloor. The processes involved in modulating carbon transport within the seabed and in the water column are highly dynamic and can only be understood through monitoring of complementary parameters over decadal or longer timescales. Efforts are under way to collect time series data at several locations, e.g., at the Canadian cabled observatory, NEPTUNE (http://www.neptunecanada.ca/research/research-projects/project.dot?inode=30424), at Hydrate Ridge, offshore Oregon (http://www.oceanobservatories.org/infrastructure/ooi-station-map/regional-scale-nodes/hydrate-ridge/), and in the Gulf of Mexico (http://ieeexplore.ieee.org/xpls/abs_all.jsp?arnumber=4625508&tag=1).

Summary

Methane and other low-molecular-weight gases, such as ethane and carbon dioxide, can combine with water to form ice-like substances at high pressure or low temperature in what are known as gas hydrates. The stability of these compounds is controlled by several factors including pressure, temperature, salinity, and gas concentrations. Gas hydrates are commonly imaged by the presence of a bottom-simulating reflector in the seismic records, which is the most common tool for identifying hydrate-bearing sediments. Other geophysical tools and borehole logging strategies are used to characterize the presence, distribution, and abundance of these deposits. These techniques are calibrated against geochemical proxies that include pore fluid freshening by gas hydrate dissociation during conventional core recovery and controlled depressurization experiments from pressure cores. Interest in gas hydrate rises from their role in carbon cycling, potential climate effects, and energy considerations. They constitute a large carbon reservoir, and sand-hosted deposits are currently being explored for economic uses, with one of the most innovative approaches being the CO_2 for methane exchange.

Bibliography

Bohrmann, G., and Torres, M. E., 2006. Gas hydrates in marine sediments. In Schulz, H. D., and Zabel, M. (eds.), *Marine Geochemistry*. Heidelberg, Germany: Springer, pp. 481–512, doi:10.1007/3-540-32144-6_14.

Boswell, R., and Collett, T. S., 2010. Current perspectives on gas hydrate resources. *Energy and Environmental Science*, **4**, 1206–1215.

Claypool, G. E., and Kaplan, I. R., 1974. The origin and distribution of methane in marine sediments. In Kaplan, I. R. (ed.), *Natural Gases in Marine Sediments*. New York: Plenum Press, pp. 99–139.

Clennell, M. B., Hovland, M., Booth, J. S., Henry, P., and Winters, W. J., 1999. Formation of natural gas hydrates in marine sediments 1. Conceptual model of gas hydrate growth conditioned by host sediment properties. *Journal of Geophysical Research*, **104**(B10), 22985–23003.

Collett, T. S., Johnson, A. H., Knapp, C. C., and Boswell, R., 2009. Natural gas hydrates: A review. In Collett, T., Johnson, A., Knapp, C., and Boswell, R. (eds.), Natural gas hydrates - Energy resource potential and associated geologic hazards: AAPG Memoir, **89**, 146–219.

Dickens, G. R., 2011. Methane release from gas hydrate systems during the Paleocene-Eocene thermal maximum and other past hyperthermal events: setting appropriate parameters for discussion. *Climate of the Past Discussions*, **7**, 1139–1174.

Holbrook, W. S., Hoskins, H., Wood, W. T., Stephen, R. A., and Lizarralde, D., 1996. Methane hydrate and free gas on the Blake Ridge from vertical seismic profiling. *Science*, **273**(5283), 1840–1843.

Hyndman, R. D., and Spence, G. D., 1992. A seismic study of methane hydrate marine bottom-simulating reflectors. *Journal of Geophysical Research*, **97**, 6683–6698.

Kim, J.-H., Torres, M. E., Choi, J., Bahk, J.-J., Park, M.-H., and Hong, W. H., 2012. Inferences on gas transport based on molecular and isotopic signatures of gases at acoustic chimneys and background sites in the Ulleung Basin. *Organic Geochemistry*, **43**, 26–38.

Kvenvolden, K. A., 1988. Methane hydrate – a major reservoir of carbon in the shallow geosphere? *Chemical Geology*, **71**, 41–51.

Pape, T., Kasten, S., Zabel, M., Bahr, A., Abegg, F., Hohnberg, H.-J., and Bohrmann, G., 2010. Gas hydrates in shallow deposits of the Amsterdam mud volcano, Anaximander Mountains, northeastern Mediterranean Sea. *Geo-Marine Letters*, **30**(3–4), 187–206, doi:10.1007/s00367-010-0197-8.

Riedel, M., Willoughby, E., and Chopra, S. (eds.), 2010. *Geophysical Characterization of Gas Hydrates*. Tulsa: Society of Exploration Geophysicists. SEG Geophysical developments series, Vol. 14, p. 392.

Schoderbek, D., Martin, K., Howard, J., Silpngarmlert, S., and Hester, K., 2012. North Slope hydrate field trial: CO_2/CH_4 exchange. In *Proceedings Arctic Technology Conference*, December 3–5, 2012, Houston, TX, 17 p.

Shipley, T. H., Houston, M. H., Buffler, R. T., Shaub, F. J., McMillen, K. J., Ladd, J. W., and Worzel, J. L., 1979. Seismic evidence for widespread possible gas hydrate horizons on continental slopes and rises. *AAPG Bulletin*, **63**, 2204–2213.

Sloan, E. D., and Koh, C. A., 2008. *Clathrate Hydrates of Natural Gases*, 3rd edn. Boca Raton, FL: CRC Press, p. 752.

Stoll, R. D., Ewing, J. I., and Bryan, G. M., 1971. Anomalous wave velocities in sediments containing gas hydrates. *Journal of Geophysical Research*, **76**, 2090–2094.

Suess, E., Torres, M. E., Bohrmann, G., Collier, R. W., Greinert, J., Linke, L., Rehder, G., Tréhu, A., Wallmann, K., Winckler, G., and Zuleger, E., 1999. Gas hydrate destabilization: enhanced dewatering, benthic material turnover and large methane plumes at the Cascadia convergent margin. *Earth and Planetary Science Letters*, **170**, 1–15.

Torres, M. E., Wallmann, K., Tréhu, A. M., Bohrmann, G., Borowski, W. S., and Tomaru, H., 2004. Gas hydrate growth, methane transport, and chloride enrichment at the southern summit of Hydrate Ridge, Cascadia margin off Oregon. *Earth and Planetary Science Letters*, **226**, 225–241.

Torres, M. E., Tréhu, A. M., Cespedes, N., Kastner, M., Wortmann, U. G., Kim, J.-H., Long, P., Malinverno, A., Pohlman, J. W., Riedel, M., and Collett, T., 2008. Methane hydrate formation in turbidite sediments of northern Cascadia, IODP Expedition 311. *Earth and Planetary Science Letters*, **271**(1–4), 170–180.

Torres, M.E., Kim, J-H., Choi, J., Ryu, J-B., Bahk, J-J., Riedel, M., Collett, T., Hong, W-L., and Kastner, M., 2011. Occurrence of high salinity fluids associated with massive near-seafloor gas hydrate deposits. In *Proceedings of the 7th International Conference on Gas Hydrates* (ICGH 2011), July 17–21, 2011, Edinburgh, Scotland.

Tréhu, A.M., Flemings, P.B., Bangs, N.L., Chevallier, J., Gràcia, E., Johnson, J.E., Liu, C.-S., Liu, X., Riedel, M., and Torres, M.E., 2004. Feeding methane vents and gas hydrate deposits at south Hydrate Ridge. *Geophysical Research Letters*, **31**, doi:10.1029/2004GL021286

Cross-references

Bottom Simulating Seismic Reflectors (BSR)
Chemosynthetic Life
Cold Seeps
Marine Mineral Resources
Methane in Marine Sediments
Mud Volcano

MARINE GEOSCIENCES: A SHORT, ECCLECTIC AND WEIGHTED HISTORIC ACCOUNT

Jörn Thiede[1,2], Jan Harff[3], Wolfgang H. Berger[4] and Eugen Seibold
[1]Institute of Earth Sciences, Saint Petersburg State University, St. Petersburg, Russia
[2]AdW Mainz c/o GEOMAR, Kiel, Germany
[3]Institute of Marine and Coastal Sciences, University of Szczecin, Szczecin, Poland
[4]Scripps Institution of Oceanography UCSD, La Jolla, CA, USA

Synonyms

Marine geological sciences, Marine geology

Definition

Marine geosciences are covering all phenomena and processes related to the formations of shallow shelf seas and of the deep ocean. They draw on modern dynamics of seafloor and sediment formation, marine geophysics and tectonics, volcanology, geochemistry, microbiology, biology, and paleontology of marine organisms. They are of great economic importance because of the wealth of nonliving marine resources.

Introduction: why and how to study the global seafloors

In the past, there have been very different perspectives how scientists approached "marine geology" or as we say today "marine geosciences": There were those who studied oceans, shallow seas, and their sediments as products of specific geological processes out of their own right (often at US institutions and closely linked to oceanography), and there were those who came from investigating ancient marine deposits on land but then turned to the marine realm to find modern analogs and thus explanations for their fossil phenomena (as in many European

---

Eugen Seibold is deceased.

institutions). This entry leans heavily on the "Introduction" of Seibold and Berger (1993).

Another trade of the marine geosciences is the fact that technical new, sometimes dramatic developments, often driven by the military, have resulted in new discoveries. We will address some of these in due course. They include the developments of echo sounders, sensitive magnetometers, deep-diving research submarines, deep-sea drilling, ROVs and AUVs, and remote sensing, to name a few.

The history of the marine geosciences is also closely linked to the evolution of suitable and dedicated research vessels. Most of the vessels deployed during the early years of ocean exploration were former war or cargo ships, converted as good as possible into carriers of scientific instrumentation and/or laboratories; hence, they were a compromise between research needs and possibilities. The honor to have actually envisioned the first dedicated ocean research vessel falls on Fridtjof Nansen with the Fram, used for his famous crossing of the ice-infested Arctic Ocean (1893–1896) when he detected that it was a deep-sea basin (Nansen, 1904). It was the first of three ships (the other ones were the Norwegian Maud and the German Gauss) specially constructed and equipped to sail into ice-covered waters, and hence they represented the first generation of dedicated polar research vessels. Despite the great importance of the polar seas for ocean and atmosphere dynamics, strangely enough it should take another seven decades until the next generation of dedicated polar research vessels would be built (the Swedish Oden, the German Polarstern, and the USCG Cutter Healy).

Modern, dedicated research vessels suited for marine geosciences came essentially into service only after the middle of last century. Famous ocean research institutions such as Woods Hole, Lamont, or Scripps in the USA and similar institutions in other countries essentially started their work on the oceans with vessels which had been built for completely different purposes. Only when the needs for having access to modern and well-equipped laboratories at sea, for deploying heavy seismic gear, for sampling long stratigraphic sections of deep ocean crust and sediment cover, as well as for deploying scientific deep-diving submarines entered the marine geosciences, highly specialized vessels were constructed. The development probably culminated when the Deep Sea Drilling Project was founded in 1968 (with the GLOMAR Challenger as its drilling platform), which developed into the International Phase of Ocean Drilling in 1985 (which used the JOIDES Resolution as its main tool) and which in 2003 then widened its scope (IODP, "Integrated Ocean Drilling Program," since 2013 "International Ocean Discovery Program") by employing in addition to JOIDES Resolution the Japanese Chikyu with its capability for riser drilling (since 2005) as well as the MSPs (Mission Specific Platforms, since 2004) of ECORD (European Consortium for Ocean Research Drilling, cf. Kiyoshi, this volume for further details). No question, this development will continue into the future, because marine geosciences have entered the stage of "big science" requiring state-of-the-art, very expensive instruments which then mostly have to be operated through international cooperation.

Pioneers of marine geosciences: identifying the basic questions and the phase of big national expeditions

Marine geology is a young offshoot of geology, a branch of science which begins with James Hutton (1726–1797) and his *Theory of the Earth* (Edinburgh, 1795). Among other things, Hutton studied marine rocks on land. Changes of sea-level ("encroachment of the ocean" and "the placing of materials accumulated at the bottom of the sea in the atmosphere above the surface of the sea") were a central tenet of his "Theory." Thus, the question of what happens at the bottom of the sea was raised at the beginning of systematic geologic investigations. This question had to be attacked if the marine deposits on land were to be understood. Hutton was not alone in these concerns. A few years before the *Theory of the Earth* appeared, the great chemist Antoine Laurent Lavoisier (1743–1794) distinguished two kinds of marine sedimentary layers: those that are formed in the open sea at great depth, which he called *pelagic* beds, and those formed along the coast, which he termed *littoral* beds. "Great depth" for Lavoisier was everything beyond wave base. Lavoisier supposed that sediment particles would settle quietly in deep water, and reworking would be much less in evidence here than near the shore.

Soon true expeditions to explore the oceans were mounted, with initiators from many nations. The first global scientific expedition (1803–1806) of the Russian Admiral A. J. Krusenstern discovered that the deep oceans were cold all over (Krusenstern, 1819). One of the most famous ones was the cruise of HMS Beagle (1831–1836) which transported Charles Darwin around the world (Darwin et al., 1839). Later followed the expedition of HMS Beacon into the Mediterranean and several others. Even though ocean exploration extends into the nineteenth century or even before, geoscientific information about the ocean floors was first limited to water depths (probed by sounding wires at specific locations), seafloor geomorphology, seafloor deposits, as well as rocks. Interest in the properties of the global seafloors arose in North America (Maury, 1855). These basically national efforts culminated in the British circumglobal expedition on HMS Challenger. The seafloor itself became the focus of attention, not for the sake of the clues it would yield for the purposes of land geology but for the clues it contained for its own evolution and its role in the history of Earth. The new emphasis is first evident in the works of the Scotsman, John Murray, naturalist on the HMS Challenger Expedition (1872–1876). This expedition, led by the biologist Charles Wyville Thomson (1830–1882), marked the beginning of modern oceanography. It established the general morphology of the deep-sea floor and the types of sediments covering it. Murray and Renard (1891) laid a foundation for the sedimentology of the deep ocean floor. Murray's studies established a fundamental dichotomy of shallow water and shelf sediments on the one hand and of deep-sea deposits on the other. It became a commonplace textbook truism that not true deep-sea deposits are found on land anywhere (which we now know to be false). The basic morphology of the mid-ocean ridge started to emerge during the later part (Challenger Expedition 1872–1876) and the last years of the nineteenth century (German Valdivia Expedition, cf. Chun, 1903). Substantial work done by Norwegian scientists then allowed Murray and Hjort to write their famous book on the depths of the ocean (Murray and Hjorth, 1912).

Marine geology started in earnest with geologists going to sea and looking for the processes which helped produce the marine rocks with which they were familiar on land. Such work, obviously, began in the intertidal and in easily accessible shallow waters, and it was undertaken by a great many investigators. The German geologist Johannes Walther was very familiar with the observations resulting from this work and was himself a pioneer in these types of studies. Firmly rooted in classical geology, Walther (1894) set an outstanding example of applying *uniformitarianism*, that is, Hutton's doctrine that observable processes are sufficient to explain the geologic record and in 1893 gave an account of marine environments and the ecology of their inhabitants, with emphasis on shell-forming organisms and the deposits they produce.

The German expeditions on RV Meteor into the Atlantic Ocean resulted in substantial progress in marine geosciences. It carried a systematic survey of the morphology of the South Atlantic along traverses between Africa and South America. It was the first one to employ echo sounding devices which had been invented in Kiel, Germany, by Alexander Behm in 1912 and which now allowed to carry out systematic bathymetric surveys of the seafloor; its expeditions into the South Atlantic (1925–1927) defined location and shape of the mid-ocean ridge. This expedition also succeeded to systematically sample deep-sea floor deposits in the South Atlantic (Correns and Schott, 1935, 1937).

A satisfying explanation of the overall morphology of ocean basins and of the various types of continental margins could only come from geophysics, which deals with the motions and forces deep within the Earth (see further below). It is no coincidence that a geophysicist first formulated a global hypothesis for ocean margin morphology which proved to be viable: the meteorologist Alfred Wegener. He became intrigued with the parallelism of the coast lines bordering the Atlantic Ocean (a phenomenon noted already in 1801 by the famous naturalist-explorer Alexander von Humboldt). It seemed to Wegener that the continents looked like puzzle pieces which belong together. He then learned quite by accident that paleontologists had invoked former land bridges between the shores facing each other across the Atlantic to explain striking similarities between the fossil records of both sides of the ocean. After an extensive literature search, he became convinced that the continents had once been joined and had broken up after the Paleozoic. Replacing the concept of land bridges with his hypothesis of *continental drift*, he started the "debate of the century" in geology with a talk and article in 1912 and especially with his book on *The Origin of Continents and Oceans*, first published in 1915 (cf. Wegener, 1966, 2005). He envisioned granitic continents floating in basaltic mantle magma like icebergs in water and drifting about on the surface of Earth in response to unknown forces derived from the rotation of the planet. Wegener's hypothesis, in modified form, is now an integral part of *seafloor spreading and plate tectonics* (see further below), the ruling theories explaining the geomorphology and geophysics of the ocean floor and of the crust of the Earth in general.

Research on marine sedimentation in the following decades is summarized in Trask (1939). This symposium spanned the range of sedimentary environments, from the beach to deep sea, and many of the articles were written by pioneers in marine geology. As marine geologic studies progressed and moved further out to sea, there was a gradual change of emphasis in the set of problems to be solved. The doctrine of a quiet deep ocean (see also further below) was challenged when Ph. H. Kuenen demonstrated by experiment that clouds of sediment could be transported downslope on the seafloor at great speed and to great depth, due to the fact that muddy water is heavier than the clear water surrounding it. Sediment transported in suspension in this fashion settles out at its site of deposition, heavy and large grains first, fine grains last. The resulting layer is *graded*, and such layers are indeed common in the geologic record (e.g., Alpine *flysch* deposits). Kuenen made many other important contributions to marine geology, attacking a broad range of subjects in his book (Kuenen, 1950) and in numerous other publications.

The investigations of the seafloor proceeded parallel to that of marine sedimentation. Coastal landforms were the most accessible, and considerable information had been accumulated early in this century (Johnson, 1919). Much additional work on these topics was done by F. P. Shepard in the USA and also by J. Bourcart in France, who were able to test earlier concepts against field data collected in shallower water. These two pioneers of marine geology especially attacked the problems of the origin of continental margins and of submarine canyons, by studying their morphology and associated sedimentary processes.

F. P. Shepard's wide-flung field areas include the shelf off the US East Coast, shelf and slope off the US West Coast, and the floor of the Gulf of Mexico. Shepard (1948) summarizes the results of this work and presents global statistics on seafloor morphology. In the same year, M. B. Klenova's textbook *Geology of the Sea* was printed. Among other Russian pioneers, N. I. Andrusov (Black Sea) and N. M. Strachov (lithogenesis) should be mentioned. Menard (1964) described the marine geology of the Pacific Ocean. Other representative works of Shepard cover the Northwest of Mexico (Shepard et al., 1960) and submarine canyons (Shepard and Dill, 1966). J. Bourcart carried out similar geomorphologic and sedimentologic studies off the shores of France, especially in the Mediterranean. His concept of continental margin "flexure," with uplift landward of a *hinge line* and downwarp seaward of it, proved useful in explaining the migrations of sea-level across the shelf and in studying the nature of sediment accumulation on the continental slope.

The marine geomorphologist par excellence was B. C. Heezen, whose *physiographic diagrams* of the seafloor (drawn with his collaborator, Marie Tharp) show great insight into the tectonic and sedimentologic processes of the seafloor (Heezen et al., 1959). His graphs now appear in virtually all textbooks of geology and geography. Of B. C. Heezen it has been said (somewhat facetiously, by E. C. Bullard) that he perfected the art of drawing maps of regions where no data are available. Much of his work is summarized in the beautifully illustrated book by Heezen and Hollister (1971). A brilliant summary of our knowledge about the geological properties of sea and ocean floors, seafloor spreading, and plate tectonics has been presented by Kennett (1982), later updated by Seibold and Berger (1993).

Modern marine geosciences in Russia, China, and other countries with similar "science histories" came about much later. Russia woke up after World War II when the cold war generated a fierce competition with western nations. Russian marine institutions acquired ocean-going research vessels and they were suddenly present on all major ocean basins; scientists from the Shirshov Institute of Oceanology and of the Vernadsky Institute of Geochemistry, both belonging to the Russian Academy of Sciences, made important contributions to global ocean atlases of all major ocean basins, and they appeared as authors of important textbooks. These important contributions to marine geosciences were only becoming familiar to western scientists after they had been translated into English (f. e. Lisitzin, 1972, 1996, 2002 or Emelyanov, 2005, see also Lobkovsky et al., 2013, 2014).

The fallacy of the quiet deep ocean

The earliest observers of deep-sea sediments were left with the idea that the deep ocean contained a tranquil depositional environment where fine-grained sediment comprised mainly clay and shell/skeleton remains of the plankton organisms living in the ocean water column. This soon proved to be wrong when Wüst (1936) based on the oceanographic data of the Meteor expedition discovered patterns of deep ocean water movements which transported cold and dense Antarctic bottom waters along their deepwater pathways to the North, in fact across the equator to the Northern Hemisphere. Today we also know that a similar process transports cold and dense Arctic water from the Norwegian-Greenland Sea across the Greenland-Scotland Ridge into the deep basins of the North Ocean to form the North Atlantic Deep Water (for a general discussion, cf. Dietrich et al., 1975). Based on these earlier observations and on dating the times when ocean bottom waters were able to absorb the last time carbon dioxide from the atmosphere, Broecker (1991) developed the concept of the global ocean conveyor belt and its influence on climate change.

Strong circumstantial evidence that Kuenen's concept is important in deep-sea sedimentation was first presented by Heezen and Ewing (1952). In 1929 an important earthquake occurred off the Grand Banks, and at that time mysteriously a series of submarine telegraph cables failed, many in sequence downslope with increasing distance from the epicenter. There was much speculation about the causes, until Heezen and Ewing (1952) determined that turbidity currents, maybe also slumps triggered by the earthquake, had resulted in the cable breaks.

To assume that the deep ocean has a tranquil depositional environment has proven to be completely wrong, and many entries of this volume address phenomena and products of the highly dynamic sedimentary processes which influence (f. e. submarine slides) and in part shape (f. e. abyssal plains) large tracts of the deep-sea floors.

Stratigraphy, biostratigraphy, biochronology, and the dawn of paleoceanography

The first indications of stratigraphic changes in South Atlantic sediment cores were presented by Philippi et al., (1910) who had described the composition of sediment cores taken by the first German Antarctic Expedition. The next major step was achieved by Correns and Schott (1936) who deduced from changes in the composition of planktonic foraminiferal fauna in sediment cores taken by the German Meteor expeditions that one could develop a biostratigraphic correlations of the cores, that these changes were related to the transition of deposits from the last Ice Age to the Holocene, and that one could use these data to calculate oceanic sedimentation rates.

A major leap was achieved by the Swedish Deep-Sea Expedition on Albatross because the scientists from Gothenburg had developed the first piston corer, which provided for the possibility for collecting relatively undisturbed several meter long sediment cores. The Albatross traversed along a transglobal course the tropical-subtropical zones of all major ocean basins (see for a summary: Olausson, 1996). In all basins they detected patterns of a cyclical repetition of more or less calcareous, mainly biogenic deposits which were soon linked to glacial-interglacial changes. The Swedish Albatross cores can be considered to represent the dawn of paleoceanography and marine paleoclimatology which in later decades developed into a very important branch of marine geosciences providing important input data for modeling the evolution of past climates.

The further development of absolute dating techniques by Urey and others, the approaches of Emiliani (cf. 1992, Shackleton and many others learned to use the variability of the fractionation of heavy and light O-isotopes between glacials and interglacials to determine sea surface and bottom temperatures. They also linked them to the transfer processes of large amounts of freshwater from the oceans through evaporation/precipitation to and from the large glacial ice sheets, resulting in isochronous isostatic sea-level changes on the order of 120–140 m. It was soon recognized that ocean sediment cores represented one of the best global archives of past climatic changes. CLIMAP, a large US-led paleoclimate project soon, used this fact to establish the global distribution of climatically important parameters, and they succeeded to publish a map of quantitative ocean surface temperature reconstructions for the time slice of the last glacial maximum (CLIMAP, 1976). At the same time they detected in the time series deduced from some Southern Ocean sediment cores a systematic and regular variability of a number of paleoceanographic parameters (Hays et al., 1976) which could be linked to the Milankovitch frequencies (for eccentricity, obliquity, and precession) which had been postulated in close collaboration with Milankovitch to regulate the regular Quaternary glacial-interglacial alterations already by Köppen and Wegener (1924). Imbrie and Imbrie (1979) thought – at that time – they had solved the mystery of the ice ages.

However, during recent decades, new discoveries in marine geology have provided a deeper insight into the last Glacial period. In particular, the dated and regionalized ice-rafted debris (IRD) in sediments of North Atlantic and the Arctic Sea as well as correlation with the oxygen isotope signature effected in Greenland ice cores have led to the event stratigraphy based on climate cycles (the Dansgaard-Oeschger cycles). These cycles are determined by alteration between warm (Greenland interstadials, GI) and cold (Greenland stadials, GS) events and describe the pacing of climate variation during the last 100,000 years (see Björck et al. 1998 and the "Dansgaard-Oeschger Cycles," "Encyclopedia of Paleoclimate and Ancient Environments"). Some of the cold events are called the "Heinrich Events" after the marine geologist who mapped the extension of sea ice-affected sediments which can be seen distributed basin-wide in North Atlantic (see Heinrich 1988, and "Ice Rafted Debris (IRD)," this volume). For the late Pleistocene to Holocene, Bond et al. (1997) discovered cold events (the "Bond Events"), based mainly on petrologic proxies of drift ice sediments in North Atlantic. Dating of the sediments led to identification of a climate cycle for the time span between the Marine Isotopic Stage 3 (MIS 3) and the Little Ice Age (the "Bond Cycle") with a pacing of $\approx 1{,}470 \pm 500$ years (Bond et al. 1997).

Although the driving mechanisms of this cyclicity is still intensely debated, identification of climatic cycles below the orbitally forced Milankovitch pacing has opened a new wide field for cooperation between geologists, oceanographers, climatologists, and modelers.

In the meantime, numerous paleoclimatic and oceanographic proxies have been developed (Fischer and Wefer, 1999), which allow to reconstruct quantitatively a vast array of ocean and climate properties. Some of them can be applied to the longtime series which are available through the acquisition of the cores obtained through the deep-sea drilling activities, in some areas actually reaching into Jurassic sediments.

Paleoceanography has developed into an important branch of the marine geosciences and by providing quantitative boundary conditions to climate modelers is reaching an ever-increasing importance for establishing possible scenarios for future climates.

Seafloor spreading and plate tectonics: the new paradigm

It was the work of seagoing geophysicists which eventually led to the acceptance of *plate tectonics*. The pioneering efforts of E. C. Bullard on earth magnetism, heat flow, and seismic surveying and of M. Ewing and his associates in all aspects of marine geophysics were of central importance in this development (although M. Ewing did not himself advocate seafloor spreading). These scientists also were instrumental in elucidating the structure of continental margins (Ewing et al., 1950; Bullard and Gaskell, 1941).

The turning point in the scientific revolution which shook the earth sciences and which culminated in *plate tectonics* is generally taken to be the seminal paper of Hess (1962). Hess started his distinguished career working with the Dutch geophysicist, F. A. Vening-Meinesz, on the gravity anomalies of deep-sea trenches. These investigations resulted in the hypothesis that trenches may be surface expressions of the downgoing limbs of mantle convection cells. As a naval officer, Hess discovered and mapped a great number of flat-topped seamounts, whose morphology suggested widespread sinking of the seafloor. Stimulated by subsequent discoveries on the mid-ocean ridge (rift morphology, heat flow, and others), he proposed his idea of the seafloor being generated at the center of the mid-ocean ridge, moving away and downward as it ages, finally to disappear into trenches. The term "seafloor spreading" was introduced by Dietz (in 1961, cf. Vine and Matthews, 1963; Le Pichon, 1968) for this postulated phenomenon. *Seafloor spreading* and its offspring *plate tectonics* have since become the basic framework within which the data of marine geology are interpreted.

As previously mentioned Russian marine geoscientists were slow to follow because their country was dominated by "fixists," and the new ideas about "seafloor spreading" and "plate tectonics" were only slowly accepted. It took eminent scientists such as Khain and Khalilov (2008) and Lev Zoneshain (2013) to push for the acceptance of the new ideas.

Only some 35 years ago, it was still possible for geologists to think that the sediments on the deep-sea floor might contain the entire Phanerozoic record and lead us back even into the Precambrian. Today, there would be few indeed who would harbor such fond hopes. The most ancient sediments recovered from the seafloor are about 180 million years old, which is less than 5 % of the age accorded to fossil-bearing sedimentary deposits on land. Where, then, is the debris which must have washed into the deep sea, for the several billion years that continents have existed? According to the hypothesis of seafloor spreading, all sediments accumulating on the ocean bottom are swept toward the trenches, as on a tectonic conveyor belt. Here, some of the sediments are *subducted* into the Earth's crust; others are scraped off against the inner wall of the trench. Thus, the ocean floor is cleaned of sediments by constant renewal and destruction.

This remarkable idea did not exactly find a warm welcome when first proposed by Hess (1962) and Dietz (1961). Somewhat similar ideas had been advanced much earlier (f. e. by Alfred Wegener) and had likewise been discounted as premature speculations. Gradually, however, as more facts became available to test the hypothesis, opposition weakened, and by 1970 there were only a very few defenders of tradition who challenged the concept of a moving seafloor.

The opposition to large-scale horizontal movement on Earth's surface had a long history. It all began with *continental drift*, an idea that arose early in the present century and challenged the prevailing assumption of *fixism* with the new arguments of *mobilism.* The most serious challenge, as already mentioned, came from the German geophysicist A. Wegener, who proposed drastic changes in the distribution of continents and ocean basins within the last 200 million years. He also proposed that the continents actively plowed through the magma which carries them, an idea that proved to be wrong. In addition, he had a timetable for the drifting of certain land masses which was quite unrealistic. Skeptical geophysicists recognized these weaknesses in Wegener's proposal and argued forcibly for the rejection of the theory of continental drift. Thus, despite the support from geologists familiar with the incredible similarities of ancient rocks and fossils in South America and South Africa, Wegener's hypothesis did not find general acceptance before the 1960s.

In essence, evidence for continental drift was dismissed because the proposed mechanism was wrong. Only in the late 1950s, through the work on *geomagnetism* and *polar wandering* by E. Irving and S. K. Runcorn (Runcorn, 1967), did it become possible again to talk about continental drift without being thought ignorant of physical principles. The most compelling evidence, however, came from the magnetism of the seafloor itself (Vine and Matthews, 1963), and final proof came from deep-sea drilling, almost a decade later.

Toward the end of the 1960s, the new hypothesis of seafloor spreading had metamorphosed into the theory of *plate tectonics*. The theory is based on the mapping of magnetic anomalies and earthquakes, by which it is possible to define large regions of the Earth's surface that move as units (*plates*), with earthquakes occurring at the boundaries of these units. The mathematical tool allowing efficient description of the movement of the plates is a theorem of the Swiss mathematician Leonhard Euler (1707–1783), which states that uniform motion on a sphere is uniquely described by defining rotation about a *pole*. The path of migration of any point on a plate appears as a portion of a circle about that pole. Bullard and associates introduced Euler's theorem to global tectonics, in 1965, to produce a new *fit* of the continents bordering the Atlantic. In the same year, J. T. Wilson explained the nature of what he called *transform faults* as boundaries between "rigid plates," where there is lateral motion generated at a spreading center; in subduction zones they can end at a trench.

For the present, and at any time in the geologic past, plate motions on the surface of Earth are defined by giving the outlines of the dozen or so plates, the geographic location of the poles of rotation, and the associated angular velocities. The principles of this approach were presented by W. J. Morgan early in 1967 and by D. P. McKenzie and R. L. Parker also in 1967. Using this methodology, X. Le Pichon, in 1968, made a global map of the relative motions of the major plates. The same concepts and methods allowed the large-scale reconstruction of the past geographic configurations of continents and ocean basins, back into the Cretaceous period (cf. Le Pichon, 1968; Allégre, 1988; Lobkovsky et al., 2013, 2014).

Deep-sea drilling: discovering new dimensions of ocean history

The emergence of the new paradigm greatly stimulated the beginning of a major venture in marine geology: the systematic exploration of the deep-sea floor by drilling. Up to 1968, before the drilling vessel *GLOMAR Challenger* set out on her first cruise (from Galveston in Texas), knowledge of sediments older than 1 million years was entirely based on cores taken in regions of greatly reduced sedimentation and erosion, where younger sediments had been removed. Such a core would typically represent a short stretch of time somewhere within the long history of the ocean. Detailed comparison with other cores, from other regions, was hardly ever possible, because of the difficulty of exact age assignment. The task of reconstructing a global ocean history for the pre-Quaternary could only be seriously attempted when more or less continuous sequences of samples became available, through drilling, from many different regions. The earliest 23 legs of DSDP were actually dominated by marine geophysicists who spot cored the sites because they wanted mainly to date seismic reflectors. Only during Leg 24 in the NW Indian Ocean, the principle of continuous coring was adopted by the chief scientists. The materials recovered by drilling resulted in a major jump in biostratigraphic resolution – soon it became possible to date samples within sequences with a temporal resolution of less than a million years relative to some standard. The first major result was full confirmation of the theory of seafloor spreading: Sediments overlying the "basement" basalt showed the exact age predicted by the geophysicists from counting magnetic anomalies. Occasionally, sediments were somewhat younger, which means that the basaltic rock was bare for a while, before it collected sediments. Other major results were more subtle but have had far-reaching implications for our understanding of the coevolution of life and climate and for the role of the ocean in climatic change over long periods of time.

In the early years of deep-sea drilling, present sedimentation patterns were assumed to be static, so that downcore changes would be interpreted mainly in terms of motions of the seafloor. It soon became clear, however, that the ocean's productivity changed markedly through geologic time, which produced large changes in sedimentation patterns. Also, it became evident that changes in ocean conditions (as reflected in sediments) were quite sudden at certain times and that such "steps" in climatic change were associated with reorganizations of the marine biosphere, including extinctions and subsequent radiations. The most intriguing period in the ocean's history, in this regard, proved to be the transition between the Cretaceous and the Tertiary, which witnessed large-scale extinction of tropical planktonic organisms (as well as many other terrestrial and oceanic organisms).

The importance of the deep-sea record for the reconstruction of Earth's history for the last 100 million years or so became obvious very quickly. The record on land is patchy and incomplete by nature – land is eroded and delivers sediment to the sea. This renders suspect all arguments about the pace of *evolution* that depend on a continuous record (Darwin pointed this out a long time ago). Only the record in the deep ocean can promise complete sequences, and even here gaps prove to be quite common in many settings. The gaps are not distributed randomly but occur preferentially just where conditions change. This illustrates a certain obstinacy of the ocean in yielding up the secrets of its history, something that many marine geologists have come to appreciate.

Deep-sea drilling has gone through important steps of evolution. While the technique to reenter a drill hole through a cone several times to exchanging the actual drilling tool has already been developed during DSDP and ODP times, the engaged international scientific community is nor striving for further progress. For the years 2003–2023, it has been decided to continue these efforts under the frame of IODP (Integrated Ocean Discovery Program, cf. Kiyoshi, this volume). Instead of using only one drilling platform with a limited scope of the scientific problems to be attacked, IODP is now trying to drill deeper and more dangerous areas than before by employing the Japanese Chikyu which can deploy a riser-protected drill string thus allowing into the seismically active zones of Japan, for example. JOIDES Resolution is still available to conduct drilling in all major ocean basins, with the cold subpolar deep-sea basins being much in the focus.

A system of Mission Specific Platforms has been equipped to drill in shallow shelf seas for determining, for example, timing and extent of global sea-level changes, the history of the Great Barrier Reef, or the central Arctic Ocean, putting the age of the first Northern Hemisphere ice covers back into the Eocene. In addition to drilling, technologies have been developed to install numerous measuring devices or observatories on a permanent or semipermanent basis in some of the drill holes.

One of the most notable aspects of deep ocean drilling – now improved through the deployment of the larger drilling vessels JOIDES Resolution and Chikyu – is the creation of an international community of marine geologists, hundreds of scientists from around the world who were shipmates at one time or another, eagerly listening for the driller's call "core-on-deck" and sharing the thrill of exploring where nobody had gone before.

The treasures of the seafloor

After the discovery of the principles in plate tectonics and mapping of the oceanic spreading centers, it was only a question of time to dive down to the production centers of new seafloor. Here, along the spreading axes, hot mantle material is upwelling and rises as magma and emerges as lava produced by cooling new oceanic crust. In 1977 scientists from the Scripps Institution of Oceanography used the deep submersible Alvin to dive to the East Pacific Rise. These scientists described the first time visually active hydrothermal venting and related processes. To their surprise they found a diverse inventory of venting phenomena such as hot water chimneys separated into black and white smokers along with chemosynthetic life forms which were completely unknown before (Londsdale, 1977). For the history of hydrothermal vent discovery, see "Hydrothermalism" (this encyclopedia). The discovery of chemosynthetic life initiated the development of new biological theory, such as the assumption of Gold (1992) who postulated a phylogenetic pathway of organisms from the "Deep Hot Biosphere" to the surface.

Predicted by economic geologists who had studied hydrothermal sulfides on land, in the course of exploring marine hydrothermal vents, significant polymetallic enrichment was found in the vicinity of the vents which was assigned to hydrothermal activities at or in the vicinity of the spreading centers. The identification of massive sulfide deposits leads to an intense international activity in exploration of marine ore reserves (Francheteau et al., 1979). Saline brines and metalliferous marine muds had been identified in the Red Sea already in the 1960s of the last century (Degens and Ross 1969), but the development of a general genetic concept of marine ores could not be accomplished before the discoveries of the hydrothermal marine sulfides (Hannington et al., 2011). Together with the discovery of hydrothermal sulfides along spreading centers of the oceans, polymetallic (manganese) nodules covering wide areas of the oceanic basins have raised international interest during the last decades of the twentieth century ("Marine Mineral Resources," "Manganese Nodules," this encyclopedia). In a kind of "gold rush," the developed nations spared neither cost nor effort to explore marine ore deposits. National and international consortia (as Interoceanmetal) explored the ocean floor, whereby areas of the Pacific Ocean between Clarion and Clipperton Fracture Zones (CCZ) were – and are still – in the center of interest because of their economic favorability. Despite intense mapping and exploration, an industrial exploitation of polymetallic nodules has never started because of the drop in ore prices and unsolved environmental problems expected to be provoked by an industrial submarine ore mining. In contrast, commercial interests currently are focusing on seafloor massive sulfides (SMS) located on convergent plate boundaries (Rona, 2003), and in 2005 the first commercial SMS survey was launched at the Manus Basin, Papua New Guinea.

During the last third of the twentieth century, the international interest in seabed-bounded mineral resources exploded in the fear of a depletion of onshore deposits. So, extensive exploration programs were launched worldwide for sand and gravel, placer (heavy minerals, diamonds, gold), phosphorites, and carbonates on the continental margins, and most of the adjoining coastal states exploit mineral offshore resources ("Marine Mineral Resources," this encyclopedia) according to international and national laws. A historical pace forward in this respect was the establishment of an international legal regime for the world's oceans and seas by the United Nations Convention on the Law of the Sea in 1982 (United Nations, 1997). Within the frame of this law, the International Seabed Authority (ISA) based at Kingston, Jamaica, was founded, a UN body taking care of the organization, regulation, and control of all mineral-related activities in the international seabed area at the high sea, beyond the limits of national jurisdiction (for the program see United Nations, 2004). In addition to mineral deposits, another resource of the seafloor did raise enormous interest during the second half of the last century: hydrocarbons (oil and gas) and gas hydrates ("Energy Resources," this encyclopedia). After the first discoveries of offshore oil deposits at the US West Coast, during World War II, their systematic exploration and exploitation started in the Gulf of Mexico. After the war, worldwide discoveries followed in the Persian Gulf, Caspian Sea, Arabian Sea and the North Sea, and many other areas. The last and probably very big fossil hydrocarbon province is expected to be found under the Arctic shelf seas and continental margins. Very soon the areas of interest moved from the shelf to deepwater fields at water depths >1,500 m so that at the beginning of the twenty-first century already >50 % of new discoveries are located at deepwater environments.

Target areas for present future exploration and drilling activities are the so-called golden triangle of Brazil, the Gulf of Mexico, and West Africa and the oil and gas in deepwater environments at continental margins of the Arctic Ocean. Drilling in the deep ocean raises the risk of technical accidents and environmental disasters. The Deepwater Horizon oil spill in the Gulf of Mexico ("Oil Spill," this encyclopedia) in 2012 has shown the limits and threats of current exploitation of marine resources.

Another phenomenon of the world ocean related to energy sources has been discovered during the 1970s of the twentieth century: gas hydrates ("Marine Gas Hydrates," this encyclopedia). In 1970 the first time marine gas hydrates have been sampled at the Blake Ridge (Atlantic continental margin of North America). After these first discoveries (Kvenvolden and McMenamin 1980), drilling campaigns and international research programs have helped during the last decades to understand the theoretical background for genesis, structure, and appearance of marine (and continental) methane hydrates (cf. Suess et al., 1999). But, we are still far from the exploitation of gas hydrate reserves as energy source.

Oil and gas deposits and accumulation of gas hydrates are bounded to ocean margins with their sedimentary wedges hosting continental erosional products with relative high proportions of organic matter descending from the high productivity of the ocean margin waters. This organic matter forms the source for hydrocarbons or methane trapped in hydrate cages. For the exploration and balancing of reserves, a stratigraphic framework for the sedimentary successions is of extreme importance.

Deep biosphere

The discoveries of the "Deep Hot Biosphere" along with the investigation of hydrothermalism had raised special interest in the new "exotic" ecosystems. The ODP and later on the IODP drilling campaigns lead to a systematic search for evidences of microbial life in marine sediments and igneous rocks. These systematic discoveries did change completely the understanding of the limits of living organisms on our planet (Parkes et al., 2000; D'Hondt et al., 2004; Jørgensen et al., 2006). For an overview about discoveries and review articles, see "Deep Biosphere" (this encyclopedia). Despite the major progress in research, there are still important questions waiting to be answered. Among these, the balance of the microbial life, the energy source of microbial growth, the role of deep biosphere in the generation and destruction of hydrocarbons and gas hydrates, or in the remobilization of material from descending lithospheric plates is just a selection of important question. Origin and evolutionary history of the deep biosphere and the influence of the marine environments on their abundance is to be addressed for understanding the evolution of ancient life on Earth and will provide analogs for comparative cosmological studies. The crucial scientific problems and pathways to solution have been described as key topics in the science plans of

Integrated Ocean Drilling Program (Moore et al., 2001) and Integrated Ocean Discovery Program (Bickle et al. 2011) for the time span 2003–2023. Besides the scientific questions technological challenges were to be accepted. These challenges are sampling in order to avoid contamination of the microbiological samples during drilling and recovery of the cores.

Sequence stratigraphy

Based on Suess' (1906) fundamental theory of eustacy to link physical stratigraphy of different regions with the frame of global transgressions and regressions, Sloss (1964) and Vail et al. (1977) developed the concept of sequence stratigraphy as a worthwhile tool for a stratigraphic correlation of sedimentary units displayed by seismic images and correlated world-wise. The basic assumption is that at continental margins because of synchronous sea-level change during low-stand phases, wide areas are exposed to subaerial erosion so that the erosional surfaces displayed as reflectors in seismic imaging form spatial/temporal boundaries of "sedimentary sequences" which can be correlated basin wide. The original findings of Sloss (1964) and Vail et al. (1977) were based on data from marine Mesozoic to Cenozoic sequences in an onshore position in relative low resolution. With the beginning of the ocean drilling activities, it became obvious that this program can contribute to test the synchroneity and amplitudes of sea-level events, with the aim to explain the controlling mechanism on margin stratal architecture. In order to answer these questions, it was decided to drill for a global comparison transect offshore the tectonically stable New Jersey siliciclastic margin, on carbonate margins along Australia, at the Bahamas, and isolated northwest Pacific atolls. First results allowed comparisons of slope sequences with the eustatic curve of Haq et al. (1987), the Bahamas records, and benthic foraminiferal oxygen isotope data (Miller et al., 1998). Refinements after the first findings which promised correlation to the Cenozoic stratigraphic architecture on continental slopes and carbonate platforms to the orbital and even suborbital scales were achieved by the deployment of Mission Specific Platforms (MSPs) managed by ECORD to drill at the Jersey shallow shelf (IODP Expedition 313). To correlate marine isotopic stages (MIS) 5–1 of the continental shelf, reliable and high-resolution sea-level data for the last glacial cycle (LGC) are needed. Grant et al. (2012) have developed a probabilistic model of the LGC sea-level change based on a statistical analysis of oxygen isotopes of Red Sea and Mediterranean Sea Foraminifera. Lobo and Ridente (2013) studies give reason to hope that we will succeed to assign even the architecture of the MIS 5–1 shelf sediments to high-frequency Milankovitch cycles.

Sea-level and society

Sea-level and coastline shifts related to climate change and coastal hazards such as tsunamis are threatening increasingly the human society living along the edge of the world's oceans and seas. However, the influence of coastal change on mankind is not a recent phenomenon. Even in their early phases of cultural development, human populations reacted to marine regressions and transgressions and to climatic and environmental changes by migration and adjustment of their socioeconomic systems to the changing natural environment. Therefore, questions of postglacial sea-level change were put on the international research agenda at the end of the last century.

The phenomenon of sea-level change has been tackled in two ways: by empirical data interpretation and by modeling approaches. Pirazzoli (1991) compiled within the frame of an International Geological Correlation Program (IGCP) project postglacial sea-level data that lead the first time to a global overview about quantitative regional trends in coastline change mainly triggered by glacio-isostatic adjustment (GIA). Lambeck and Chappell (2001) used numerical models to describe the deformation of the Earth's crust to the loading and unloading by continental ice sheets and water masses in marine basins. But, besides the interest in historical reconstruction of the paleo-coastlines, increasing interest is paid to the generation of future scenarios. Future projections of sea-level processes are directly related to climate change and the interference of natural and anthropogenic driving forces so that at the beginning of the twenty-first century the door for cooperation between geoscientists, climatologists, and socioeconomists has been widely opened (Rahmsdorf, 2007).

Conclusions. Marine geosciences: quo vadis?

To answer the question of future development of targets and pathways in marine geosciences, it is helpful to go for the IODP Science Plan 2013–2023 (Moore et al., 2001; Bickle et al. 2011). This plan is characterized by a clear commitment to the "System Earth" – an interdisciplinary and integrated approach. The following research themes are on the agenda of the next 10 years:

- Climate and ocean change: reading the past, informing the future. Behaviors of climate, oceans, and ice caps under increasing greenhouse conditions and expansion of the database for climate modeling and prediction
- Biosphere frontiers: deep life, biodiversity, and environmental forcing of ecosystems. Exploration of life forms beneath the seafloor and in extreme environments where microbes live isolated from the photosynthetic worlds at the limits of natural habitats; relation between biodiversity and fast climate change; and the influence of large volcanic eruptions and the impacts extraterrestrial bodies on biodiversity
- Earth connections: deep processes and their impact on earth's surface environment. Geochemical exchange processes between the solid earth, oceans, and atmosphere; drilling into the Earth's mantle; understanding the properties and nature of changes of the Earth's magnetic field; and evolution of mineral deposits resulting

from exchange processes between ocean crust and seawater
- Earth in motion: short-term geodynamic processes of large and immediate relevance for mankind (earthquakes, tsunamis, landslides); understanding of fluid-flow processes in sediments and volcanic crust; formation and stability of gas hydrates; borehole observation laboratories and their networks; and setup of early warning systems for disasters

Despite the fact that drilling to the seafloor using advanced technologies will provide the scientific community with samples of rocks, fluids, and gasses from environments never sampled before, the deep understanding of Earth's systems, but also the societal service of marine geosciences, requires an embedding of new data into an array of monitoring and modeling approaches. Synchronous to the demands of the estimation of risk of hazards and projections of trends to the future, long-term instrumental data series gain increasingly importance. These data series have to be integrated into warning systems of disasters such as tsunamis. Besides the deep oceans, marginal seas and their coastal zones and river mouth systems will come into focus of national and international research projects. Studies of the modification of matter and energy on the pathway from the source at the continents to the sink in marine-receiving basins – including expressively anthropogenic effects – will help to manage processes at the interface between the terrestrial and the marine realm. In addition to this service of marine geosciences to management and planning of natural environments in order to find the balance between their exploitation and sustainable protection, another "classic" task will remain on the agenda: To explore the history of planet Earth, its structure and dynamics, as well as new potential sources for the supply of the society with raw materials and energy.

Bibliography

Allègre, C., 1988. *The Behavior of the Earth – Continental and Seafloor Mobility*. Cambridge: Harvard University Press, Vol. XII, p. 272. ISBN 0-674-06458-5.

Andrée, K., 1920. *Geologie des Meeresbodens*. Leipzig: Gebr. Borntraeger. Beschaffenheit, nutzbare Materialien am Meeresboden, Vol. II. 689 pp.

Baturin, G. N., 1988. *The Geochemistry of Manganese and Manganese Nodules in the Ocean*. Dordrecht: D. Reidel. 342 pp.

Bickle, M. et al. Anonymous, 2011. *Illuminating Earth's Past, Present, and Future – The International Ocean Discovery Program Exploring the Earth Under the Sea – Science Plan for 2013–2023*. Washington, DC: Integrated Ocean Drilling Program Management International, 84 p.

Björck, S., Walker, M. J. C., Cwaynar, L. C., Johnsen, S., Knudsen, K.-L., Lowe, J. J., Wohlfarth, B., and Intimate Members, 1998. An event stratigraphy for the Last Termination in the North Atlantic region based on the Greenland ice-core record: a proposal by the INTIMATE group. *Journal of Quaternary Science*, **13**(4), 283–292.

Bond, G., Showers, W., Cheseby, M., Lotti, R. Almasi, P., deMenocal, P., Priore, P., Cullen, H., Hajdas, I., and Bonani, G., 1997. A pervasive millennial-scale cycle in North Atlantic Holocene and glacial climates. *Science*, **278**(5341), 1257–1266.

Broecker, W. S., 1991. The great ocean conveyor. *Oceanography*, **4**(2), 79–89.

Bullard, E. C., and Gaskell, T. F., 1941. Submarine seismic investigations. *Proceedings Royal Society Series A*, **177**, 476–499.

Chun, C.,1903. *Aus den Tiefen des Weltmeeres*. Jena: Gustav Fischer, 592 p. (with the bathymetric Chart of the German Deepsea-Expedition 1898/99).

CLIMAP Project Members, 1976. The surface of the iceage earth. *Science*, **191**(4232), 1131–1137.

Correns, C. W., and W. Schott, 1935/1937. Die Sediments des Äquatorialen Atlantischen Ozeans. 1. Lfg.: A. Die Verfahren der Gewinnung und Untersuchung der Sediemtns (Correns). B. Die Foraminiferen in dem äquatorialen Teil des Atlantischen Ozeans (Schott).- 2. Lfg.: C. Zusammenstellung der Untersuchungsergebnisse nach Stationen geordnet. D. Die Auswertung der Ergebnisse.- Wiss. Ergebn. der dt. Atlantischen Expedition auf dem Forschungs- und Vermessungsschiff "Meteor", Wiss. Ergebn., Vol. 3.3., Berlin: de Gruyter, 298 p.

Darwin, C., King, P. P., and Fitzroy, R., 1839. *Narrative of the Surveying Voyages of His Majesty's Ships Adventure and Beagle Between the Years 1826 and 1836, Describing Their Examination of the Southern Shores of South America, and the Beagle's Circumnavigation of the Globe*. London: Henry Colburn.

D'Hondt, S., Jørgensen, B. B., Miller, D. J., et al., 2004. Distributions of microbial activities in deep subseafloor sediments. *Science*, **306**, 2216–2221.

Degens, E. T., and Ross, D. A. (eds.), 1969. *Hot Brines and Recent Heavy Metal Deposits in the Red Sea*. New York: Springer. 600 pp.

Dietrich, G., Kalle, K., Krauss, W., and Siedler, G., 1975. *Allgemeine Meereskunde. Eine Einführung in die Ozeanographie: mit 48 Tabellen, 3*, 12th edn. Berlin: Borntraeger, p. 593.

Dietz, R. S., 1961. Continent and ocean basin evolution by spreading of the sea floor. Nature, 4779, 855–857.

Emelyanov, E. M., 2005. *The Barrier Zones in the Ocean*. Berlin: Springer. 632 pp.

Emiliani, C., 1992. *Planet Earth: Cosmology, Geology, and the Evolution of Life and Environment*. Cambridge: Cambridge University Press. 718 pp.

Ewing, M. J. L., Worzel, N. C., Steenland, F., and Press, F., 1950. Geophysical investigations in the emerged and submerged Atlantic coastal plain: PART V: Woods Hole, New York, and Cape May Sections. *Geological Society of America Bulletin*, **61**(9), 877–892, doi:10.1130/0016-7606(1950)61[877:giitea]2.0.co;2.

Fischer, G., and Wefer, G. (eds.), 1999. *Use of Proxies in Paleoceanography: Examples from the South Atlantic*. Berlin: Springer. 735 pp.

Foundation, E. S., 1987. *Report of the Second Conference on Scientific Ocean Drilling (COSOD II), Strasbourg, France, 6–8 Jul 1987*. Strasbourg: European Science Foundation, Vol. VIII. 142 pp.

Francheteau, J., Needham, H. D., Choukroune, P., Juteau, T., Seguret, M., Ballard, R. D., Fox, P. J., Normark, W., Carranza, A., Cordoba, D., Guerrero, J., Rangin, C., Bougault, H., Cambon, P., and Hekinian, R., 1979. Massive deep-sea sulfide ore-deposits discovered on the east Pacific rise. *Nature*, **277**, 523–528.

Gold, T., 1992. The deep, hot biosphere. *Proceedings of the National Academy of Sciences*, **89**(13), 6045–6049.

Grant, K. M., Rohling, E. J., Bar-Matthews, M., Ayalon, A., Medina-Elizalde, M., Bronk Ramsey, C., Satow, C., and Roberts, A. P., 2012. Rapid coupling between ice volume and polar temperature over the past 150,000 years. *Nature*, **491**, 744–747.

Hannington, M., Jamieson, J., Monecke, T., Petersen, S., and Beaulieu, S., 2011. The abundance of seafloor massive sulfide deposits. *Geology*, **39**, 1155–1158.

Haq, B. U., Hardenbol, J., and Vail, P. R., 1987. Chronology of fluctuating sea levels since the Triassic (million years ago to present). *Science*, **235**, 1156–1167.

Hays, J. D., Imbrie, J., and Shackleton, N. J., 1976. Variations in the Earth's orbit: pacemaker of the Ice Ages. *Science*, **194**(4270), 1121–1132.

Heezen, B. C., and Ewing, M., 1952. Turbidity currents and submarine slumps and the 1929 Grand Banks earthquake. *American Journal of Science*, **250**, 849–873.

Heezen, B. C., and Hollister, C. D., 1971. *The Face of the Deep*. New York: Oxford University Press, Vol. VIII. 659 pp.

Heezen, B. C., Tharp, M., and Ewing, M., 1959. The Floors of the Oceans. I. The North Atlantic. *Geological Society of America.*: Spec Pap 65, 122 pp.

Heinrich, H., 1988. Origin and consequences of cyclic ice rafting in the northeast Atlantic Ocean during the past 130,000 years. *Quaternary Research*, **29**, 142–152.

Hess, H. H., 1962. History of ocean basins. In Buddington, F., Engel, A. E. J., and James, H. L. (eds.), *Petrologic Studies: A Volume in Honor of A.* Boulder: The Geological Society of America, pp. 599–620.

Hutton, J., 1795. *Theory of the Earth*. Edinburgh.

Imbrie, J., and Imbrie, K. P., 1979. *Ice Ages. Solving the Mystery*. Hillside: Enslow Publ.. 224 p.

Johnson, D. W., 1919. *Shore Processes and Shoreline Development*. New York: Wiley, Vol. XVII. 584 pp.

Jørgensen, B. B., D'Hondt, S. L., and Miller, D. J., 2006. Leg 201 Synthesis: controls on microbial communities in deeply buried sediments. In Jørgensen, B.B., S.L. D'Hondt and D.J. Miller (eds.), *Proc. ODP, Sci. Res. 201*. College Station: Ocean Drilling Program, Texas A&M, University.

Kennett, J., 1982. *Marine Geology*. Englewood Cliffs: Prentice Hall. 813 pp.

Khain, V. E., and Khalilov, E. N., 2008. *Spatiotemporal Patterns of Seismic and Volcanic Activity*. Burgas: SWB. 304 pp.

Köppen, W., and Wegener, A., 1924. *Die Klimate der geologischen Vorzeit*. Berlin: Borntraeger, Vol. IV. 255 pp.

von Krusenstern, A. J., 1819. Beyträge zur Hydrographie der grössern Ozeane als Erläuterungen zu einer Karte des ganzen Erdkreises nach Mercatos Projection. Leipzig: Kummer, 248 pp.

Kuenen, P. H., 1950. *Marine Geology*. New York: Wiley. 568 p.

Kvenvolden, K. A., and McMenamin, M. A., 1980. Hydrates of natural gas: their geologic occurrence. *U. S. Geologocal Survey*, Menlo Park CA. III.11p., ill., map.

Lambeck, K., and Chappell, J., 2001. Sea level change through the last glacial cycle. *Science*, **292**, 679–686.

Le Pichon, X., 1968. Sea-floor spreading and continental drift. Journal of Geophysical Research, 73, 3661–3697.

Lisitzin, A. P., 1972. Sedimentation in the World Ocean. SEPM/Tulsa/OK, Spec. Publ.: 17, 218 pp.

Lisitzin, A. P., 1996. *Oceanic Sedimentation: Lithology and Geochemistry*. Washington, DC: American Geophysical Union. 400 pp.

Lisitzin, A. P., 2002. *Sea-Ice and Iceberg Sedimentation in the Ocean; Recent and Past*. Berlin: Springer. 563 pp.

Lobkovsky, L. I., Lisitzin, A. P., Dubinin, E. P., Rabinovich, A. B., and Yakovenko, O. I. (eds.), 2013. *The World Ocean*. Moscow: Scientific World. Ocean Geology and Tectonics, Oceanic Catastrophic Phenomena, Vol. 1. 644 pp.

Lobkovsky, L. I., Lisitzin, A. P., Neiman, V. G., Romankevich, E. A., Flint, M. V., and Yakovenko, O. I. (eds.), 2014. *The World Ocean*. Moscow: Scientific World. Physics, Chemistry and Biology of the Ocean, Sedimentation in the Ocean and Earth Geospheres Interaction, Vol. 2. 576 pp.

Lobo, F. J., and Ridente, D., 2013. Milankovitsch cyclicity in modern continental margins: stratigraphic cycles in terrigenous shelf settings. Boletín Geológico y. Minero, 124(2), 169–185.

Lonsdale, P., 1977. Clustering of suspension-feeding macrobenthos near abyssal hydrothermal vents at oceanic spreading centers. *Deep Sea Research*, **24**(9), 857–863.

Maury, M. F., 1855. *The Physical Geography of the Sea*. New York: Harper and Brothers. 209 p.

Menard, H. W., 1964. *Marine geology of the Pacific*. New York: McGraw-Hill, Vol. X. 271 pp.

Miller, K. G., Mountain, G. S., Browning, J. V., Kominz, M., Sugarman, P. J., Christie-Blick, N., Katz, M. E., and Wright, J. D., 1998. Cenozoic global sea level sequences and the New Jersey transect: results from coastal plain and slope drilling. *Reviews of Geophysics*, **36**, 569–601.

Moore, T. C. et al. Anonymous, 2001. *Earth, Oceans and Life – Scientific Investigation of the Earth System Using Multiple Drilling Platforms and New Technologies Integrated Ocean Drilling Program Initial Science Plan, 2003–2013*. Washington, DC: International Working Group Support Office, 110 p.

Murray, J., and Hjort, J., 1912. *The Depths of the Ocean: A General Account of the Modern Science of Oceanography Based Largely on the Scientific Researches of the Norwegian Steamer Michael Sars in the North Atlantic*. London: Macmillan. 821 pp.

Murray, J., Renard, A. F., and Thomson, C. W., 1891. *Deep-Sea Deposits*. London: Eyre & Spottiswoode. 525 pp.

Nansen, F., 1904. The Bathymetric Features of the North Polar Seas, with a Discussion of the Continental Shelves and Previous Oscillations of the Shoreline. Norwegian North Polar Expedition 1893–1896. Scientific Results, Vol. 4, No. 13, 231 p.

Olausson, E., 1996. *The Swedish Deep-Sea Expedition with the "Albatross" 1947–1948. A Summary of the Sediment Core Studies*. Grafiska AB: Novum. 98 p. ISBN 91-971465-6-0.

Parkes, R. J., Cragg, B. A., and Wellsbury, P., 2000. Recent studies on bacterial populations and processes in subseafloor sediments: a review. *Hydrogeology Journal*, **8**, 11–28.

Philippi, E., Reinisch, R., and Gebbing, J., 1910. *Die Grundproben der Deutschen Südpolar-Expedition 1901–1903*. Berlin: G. Reimer, pp. 416–616.

Pirazzoli, P. A., 1991. *World Atlas of Holocene Sea-Level Changes*. Amsterdam: Elsevier. Oceanography Series, Vol. 58. 300 p.

Rahmstorf, S., 2007. A semi-empirical approach to projecting future sea-level rise *Science*, **315**, 368–370.

Rona, P. A., 2003. Resources of the sea floor. *Science*, **299**, 673–674.

Runcorn, S. K., 1967. Polar wandering and continental drift. In Markovitz, W., and Guinot, B. (eds.), International Astronomical Union, Symposium 32. D. Reidel, Dordrecht, pp. 80–85.

Schopf, T. J. M., 1980. *Paleoceanography*. Cambridge: Harvard University Press. 341 p.

Seibold, E., and Berger, W., 1993. *The Seafloor – An Introduction to Marine Geology*. Berlin: Springer. 356 p.

Shepard, F. P., 1948. *Submarine Geology*. New York: Harper, Vol. XVI. 348 pp.

Shepard, F. P., and Dill, R. F. (eds.), 1966. *Submarine Canyons and Other Sea Valleys*. Chicago: Rand McNally College Publishing Company, Vol. XIII. 381 p.

Shepard, F. P., Phleger, F. B., and van Andel, T. H., 1960. *Recent Sediments, Northwest Gulf of Mexico, a Symposium Summarizing the Results of Work Carried on in Project 51 of the American Petroleum Institute 1951–1958*. Tulsa: American Assoc. of Petroleum Geologists. 394 p.

Sloss, L. I., 1964. Tectonic cycles of the North American craton. Kansas Geological Survey Bulletin, 169, 449–459.

Suess, E., 1906. *Das Antlitz der Erde*. Wien: F. Tempsky and G. Freytag. 789 p.

Suess, E., Torres, M. E., Bohrmann, G., Collier, R. W., Greinert, J., Linke, P., Rehder, G., Trehu, A., Wallmann, K., Winckler, G., and Zuelger, E., 1999. Gas hydrate destabilization: enhanced dewatering, benthic material turnover and large methane plumes at the Cascadia convergent margin. *Earth and Planetary Science Letters*, **170**, 1–15.

Trask, P. D. (ed.), 1939. *Recent Marine Sediments*. Tulsa: AAPG. 736 p.

United Nations (UN), 1997. *The Law of the Sea: United Nations Convention on the Law of the Sea*. Division for Ocean Affairs and the Law of the Sea, Office of Legal Affairs United Nations Publication No. E.97.V.10, 294 p.

United Nations (UN), 2004. *Marine Mineral Resources Scientific Advances and Economic Perspectives*. Unites Nations Division for Ocean Affairs and the Law of the Sea, Office of Legal Affairs, and the International Seabed Authority, ISBN: 976-610-712-2, 118 p.

Vail, P. R., Mitchum, R. M. Jr., and Thompson, S., 1977. Relative changes of sea level from coastal onlap. In: Payton, C. E. (ed.), *1977: Seismic Stratigraphy – Application to Hydrocarbon Exploration*.American Association of Petroleum Geologists (AAPG), Tulsa. Mem 26, 63–82.

Vine, F. J., and Matthews, D., 1963. Magnetic anomalies over oceanic ridges. *Nature*, **4897**, 947–949.

Walther, J., 1893. *Bionomie des Meeres. Beobachtungen über die marinen Lebensbezirke und Existenzbedingungen. Einleitung in die Geologie als historische Wissenschaft / Beobachtungen über die Bildung der Gesteine und ihrer organischen Einschlüsse*. Jena: Fischer. 196 p.

Walther, J., 1894. *Lithogenesis der Gegenwart. Beobachtungen über die Bildung der Gesteine an der heutigen Erdoberfläche. Einleitung in die Geologie als historische Wissenschaft/ Beobachtungen über die Bildung der Gesteine und ihrer organischen Einschlüsse, Theil 3*. Jena: Fischer, Vol. VII, pp. 535–1055.

Wegener, A., 1915/1929/ 2005. Die Entstehung der Kontinente und Ozeane. Reprint of the 1st ed. with handwritten comments of him and of the 4th completely revised ed., Berlin: Borntraeger, 481 pp.

Wegener, A., 1966. The Origin of Continents and Oceans, Translation by J. Biram. Mineola: Dover Publications.

Wüst, G.,1936. Schichtung und Zirkulation des Atlantischen Ozeans. Wiss. Ergebn. Dt. Atlant. Expedition VFS Meteor 1925–1927: 6(1), Berlin, 1–288.

Zonenshain, L., 2013. *Paleogeodynamics: The Plate Tectonic Evolution of the Earth*. Washington, DC: American Geophysical Union. 218 pp.

MARINE GEOTOPE PROTECTION

Aarno Kotilainen
Geological Survey of Finland, Espoo, Finland

Synonyms

Marine geosites; Marine geotopes

Marine geotope is (1) an Earth science site at the seafloor (modified after Erikstad, 1994) or a (2) specific site or area at the seafloor that has significant geological and geomorphological value (modified after Stürm, 1994). Marine geosite is a key locality or area, at the seafloor, showing geological (and geomorphological) features of intrinsic scientific interest that allow us to understand the key stages in the evolution of the Earth (modified after Wimbledon and Smith-Meyer, 2012).

Marine geosites can be large or small, e.g., submarine canyons, seamounts, or submarine moraines. Geological and geomorphological features and their diversity (i.e., geodiversity) are of great importance for several reasons, both in terrestrial and marine environments. Values of geodiversity include intrinsic or existence value, cultural value, aesthetic value, economic value, functional value, and research/educational value (Gray, 2004). Geosites, as well as landscapes, shaped and defined by their geology, form the geological heritage, which is a part of the natural heritage of any country (Wimbledon and Smith-Meyer, 2012), on land and at the seafloor. Geomorphological features and rocks are just as much part of natural heritage as animals and plants. Thus conservation of geosites and geodiversity is justified and important also in marine environment.

The conservation of geological and geomorphological features, that is, geoconservation, has a long history. One of the first geoconservation, by legal action, has taken place already in 1819 in Scotland (McMillan et al., 1999). Geoconservation has been a well-established activity in the terrestrial environment, but it has been earlier largely absent in the marine environment. However, at present there are already several marine protected areas that are defined by geology. There are several national and international programs for site inventories and protection of important geological sites, like UNESCO's Global Geoparks and IUGS Geosites Programmes. In addition UNESCO keeps a World Heritage List of certain places on Earth that have outstanding universal value and should form part of the common heritage of humankind. That list includes unique places like the Great Barrier Reef in Australia and Kvarken Archipelago in Finland, among others, that make up the world's marine heritage. Conservation of both biological and geological components of nature forms a basis for the sustainable use of marine resources, sustaining biodiversity and ecosystem functioning in the world's oceans and seas, and is the responsibility of all countries and their governments.

Bibliography

Erikstad, L., 1994. The building of an international airport in an area of outstanding geological diversity and quality. In O'Halloran, D. (ed.), *Geological and Landscape Conservation*. London: Geological Society, pp. 47–51.

Gray, M., 2004. *Geodiversity – Valuing and Conserving Abiotic Nature*. John Wiley & Sons Ltd, The Atrium, Southern Gate, Chichester,West Sussex PO19 8SQ, England. 434 p.

McMillan, A. A., Gillanders, R. J., and Fairhurst, J. A., 1999. *The Building Stones of Edinburgh*, 2nd edn. Edinburgh: Edinburgh Geological Society.

Stürm, B., 1994. The geotope concept: geological nature conservation by town and country planning. In O'Halloran, D. (ed.), *Geological and Landscape Conservation*. London: Geological Society, pp. 27–21.

Wimbledon, W. A. P., and Smith-Meyer, S. (eds.), 2012. *Geoheritage in Europe and its conservation*. Oslo: ProGEO, Norway. 405 pp.

Cross-references

Moraines
Reefs (Biogenic)
Seamounts
Submarine Canyons

MARINE HEAT FLOW

John G. Sclater[1], Derrick Hasterok[2], Bruno Goutorbe[3], John Hillier[4] and Raquel Negrete[5]
[1]Scripps Institution of Oceanography, University of California San Diego, La Jolla, CA, USA
[2]Department of Geosciences, University of Adelaide, Adelaide, Australia
[3]Institute of Geosciences, State University Fluminense, Niteroi, Brazil
[4]Department of Geography, Loughborough University, Leicestershire, UK
[5]Ensenada Center for Scientific Research and Higher Education, Department of Geology, CICESE, Zona Playitas, Baja California, Mexico

Definition

Heat flow is a measure of the rate of thermal energy transferred (power) from one place to another normalized area, commonly denoted q. The units of q are equivalent to energy time^{-1} area^{-1}, which is W m^{-2} in the System International (SI) where a watt (W) is equivalent to a joule second^{-1} (J s^{-1}). Since the heat flow is typically small for most Earth processes, we often report heat flow in mW m^{-2}, where 1 mW = 10^{-3} W. Heat flow is physically defined by Fourier's law, given by the product of the thermal conductivity, k, with a thermal gradient between two points, $\Delta T/\Delta z$, i.e., $q = k\,(\Delta T/\Delta z)$. In marine geophysics steady-state conductive heat flow is the measurement typically made. Thermal gradients are estimated by driving a 3–10 m long, thin steel probe into the soft marine sediments. The probe contains several thermistors, to measure temperature (mK accuracy), distributed along the length at known distances. Such measurements over the short length of such a probe are possible because of the insulating effect of poorly mixed deep-sea water reservoirs. Thermal conductivity is more challenging to estimate but is typically done by (1) modeling the decay of heat due to insertion of the temperature probe or (2) performing measurements on retrieved sediment core. Very few heat flow measurements have been made through or around bare rock, and thus, inherently, the current measurements do not account for the advective heat loss due to the circulation of water at the Earth's surface (Louden and Wright, 1989).

Method of measurement

There are three principal methods of measuring seafloor heat flow. One pioneered by Bullard (1954) (Figure 1a) involves the use of a thin steel probe with the temperature measuring devices located within the steel probe. The second developed by Ewing (Gerard et al., 1962) (Figure 1c) uses a standard core tube and measures the temperature gradient using thermistors within outriggers mounted on the outside of the core tube. The third, the violin-bow probe (Figure 1b), designed by Lister in 1967 but described in Lister (1979), measures the temperature within a thin tube attached to a very strong 3–5 mm diameter steel rod. In this method, the thermal conductivity is determined in situ by pulsing heat into the tube and afterward examining the temperature decay with time from the pulse. The other two instruments rely on known relations between sediment properties and thermal conductivity from a nearby sediment core. In addition to the above methods, a number of measurements have been made using temperatures measured in a core pushed ahead of drill string in the Deep Sea Drilling and Ocean Drilling Programs (Pribnow et al., 2000). Agreement between the shallow measurements and the borehole observations is generally good (Hyndman et al., 1984) and confirm that bottom water temperature does not vary significantly in the eastern north Pacific and the central and eastern Atlantic. However, water temperature variations observed offshore western North America, South Atlantic, and eastern Pacific affect measurements with shallow, <5 m, penetration (Davis et al., 2003).

The above techniques require sediments so that the probe can penetrate below the surface. A thermal blanket (Johnson and Hutnak, 1997), made of a thin (5 cm) urethane foam sheet with known thermal conductivity and thermistors on either side, can be draped directly over a rock outcrop. Thermal blankets have seen very little use as they require placement by an ROV and must be left for approximately 12 h to equilibrate.

Introduction and early work

Measurement of the flow of heat through the Earth is crucial because it is heat and temperature differences that primarily drive motion within the Earth. Heat flow is difficult to measure because it involves measuring a gradient at or near the boundary between two different substances: sediment and water for the oceans and rock/sediment and air on land. Despite these difficulties heat flow measurements have contributed to many major advances in the Earth sciences. This contribution is especially true of the measurements from the ocean floor, and it is these measurements that are the subject of this review.

Essentially one man, Sir Edward Bullard, initiated marine heat flow as a subdiscipline of geophysics at the end of WWII. Given the controversy created by Lord Kelvin (1895) who used temperature gradients from mines on land to compute the age of the Earth, it is truly surprising that it took so long for heat flow measurements to be made

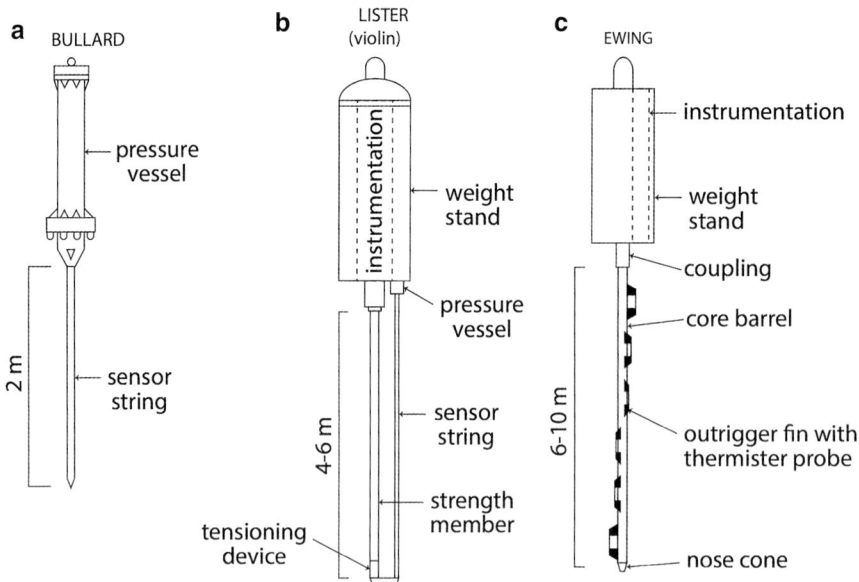

Marine Heat Flow, Figure 1 Mechanical configuration of the three most commonly used marine heat flow probes: (**a**) Bullard probe (see details in text); (**b**) Lister "violin-bow", developed to measure thermal conductivity and temperature gradient in situ at the same location; and (**c**) Ewing outrigger, developed to enable temperature gradient measurements to be taken during a standard Lamont-Doherty Geological Observatory coring station (After Louden and Wright (1989)).

in the oceans. Like almost all other geophysical measurements, the results from the first observations yielded the most startling results and helped change our view of the thermal properties of the Earth.

Bullard (1939) started in this field, before WWII, by measuring the heat flow on land in South Africa. His work there established the standards and techniques that are still currently in use in the global heat flow field. In 1945, he showed that the heat flow through the stable continents is remarkably uniform at ~60 mWm^{-2} (Bullard, 1945). He speculated that the heat was caused by the radioactivity of granite and, since that was missing on the ocean floor, it would be lower. He concluded "it would be of great interest to verify this by heat flow measurements under the oceans."

Hans Pettersson, a Swedish radiochemist, attempted the earliest marine temperature gradient measurements. He reported two measurements from the Pacific in his report on the 1947–1948 Swedish Deep-Sea Expedition (Pettersson, 1949). The temperature gradients were low and implied a heat flow (20–30 mWm^{-2}) much below that through the continents. As this was as expected, if radioactivity were the cause, the incentive to make more measurements evaporated, and little effort was made to continue them (Gustaf Arrhenius, Personal Communication, 2010).

Bullard had questions regarding both the method of measurement and the conclusions of Pettersson. In 1949, on an extended visit to Scripps Institution of Oceanography (SIO) together with colleagues, Revelle, graduate student Maxwell, and the SIO technical staff, he designed an instrument with a long probe to measure the temperature gradient in the soft sediments of the ocean floor (Figure 1a). Multiplying this value by the conductivity measured on sediment recovered in nearby cores gave the heat flow.

After Bullard left SIO to become Director of the UK National Physical Laboratory (NPL), the group perfected the design of the original instrument and made six measurements in the deep North Pacific (Revelle and Maxwell, 1952). The values were close to the mean value for the continents (~60 mW m^{-2}). In his companion paper, Bullard (1952) pointed out that the values should have been much lower. He speculated that the extra heat necessary to elevate the heat flow values came from thermal convection cells in the upper mantle.

On returning to England, Bullard redesigned the probe and took six additional measurements in the North Atlantic from RRS *Discovery*. He combined all the additional measurements taken at NPL and SIO into one publication (Bullard et al., 1956). This paper was the first to point out the correlation, now known to be the feature of all mid-ocean ridges, between very high heat flow (i.e., >100 mWm^{-2}) and the crest of the mid-ocean ridges. This correlation was dramatically confirmed by further work in the Pacific by Von Herzen (1959) and Von Herzen and Uyeda (1963) and in the South Atlantic by Vacquier and Von Herzen (1964).

The theory of seafloor spreading implies the intrusion of hot molten magma at the crest of the mid-ocean ridges. The values reported by Bullard et al. (1956) and Von Herzen (1959) were critical to the development of this theory by Hess (1962). In his seminal paper, Hess wrote: "The Mid-Ocean Ridges are the largest topographic features on the surface of the Earth. Menard has shown that their crests closely correspond to median lines in the oceans and suggests that they may be ephemeral features." Bullard et al. (1956) and Von Herzen (1959) show that they have unusually high heat flow along their crests.

Historical models of plate creation

The major question posed by the papers of Hess (1962) and Dietz (1961) (who expanded on the original idea of Hess) on seafloor spreading was "how did the zones of ocean floor creation relate to the places where the oceanic crust was being destroyed and how thick was the region that was moving?" Wilson (1965) argued that the simple concept that the Earth is made up of rigid plates of greater than 50 km thickness created at spreading centers and destroyed in trenches could explain the huge deep topographic offsets observed on the mid-ocean ridges. Importantly this showed that the region that was moving was much thicker than just the crust. Almost immediately, Ewing et al. (1966) and McKenzie (1967) used heat flow data to test the concept of a thick moving plate. This plate is now called the lithosphere. Ewing et al. (1966) argued that it did not, whereas McKenzie (1967) disagreed and constructed a simple model of plate creation (Figure 2a) that was compatible with the observations.

Sclater and Francheteau (1970) used this plate model to show that the global heat flow data and the subsidence of the ocean floor (Figure 2a), rather than being an argument against the plate theory, as Ewing et al. (1966) had suggested, strongly supported it. Because of the scatter in the heat flow (left-hand side of Figure 2b), it was realized that it was more useful to plot seafloor subsidence against crustal age. Sleep (1969) and Sclater et al. (1971) demonstrated that there was a simple relationship for the entire age of the ocean floor (Figure 2a) that was compatible with the creation of a plate about 100 km thick. Parker and Oldenburg (1973) showed that for the first 80 my in the North Pacific, there was a simple relation between subsidence and the square root of seafloor age (i.e., $z \propto \sqrt{t}$). This is to be expected if the creation of the ocean floor were to be represented by the cooling of a simple thermal boundary layer (Turcotte and Oxburgh, 1967). Davis and Lister (1974) showed that for crust <80 Ma, this relation held for all three oceans considered by Sclater et al. (1971).

Hydrothermal circulation in the oceanic crust

Values from the initial measurements near the ridge axis in the oceans, according to Bullard et al. (1956), were highly scattered. This prevented use of these data for quantitative analysis (left-hand side of Figure 2b). In addition, the averaged heat flow values near the ridge crest were much lower than those predicted by the thermal model that best matched the subsidence of the older ocean floor (see Sclater and Francheteau, 1970). The problem was resolved by Lister (1972) who showed that the observed heat flow followed the values predicted by simple models of fluid flow in a permeable crust where upward flow occurred in narrow vents with the downward flow being broad and diffuse. Immediately, it was realized that this hypothesis explained both the variability in the heat flow values and the low averaged values near the ridge crest. The hydrothermal hypothesis of Lister (1972) predicted the output of hot jets of water on the seafloor, and their discovery (Williams et al., 1974) spawned extensive interdisciplinary studies of seafloor hydrothermal circulation (see Davis and Elderfield, 2004 for a review). Sclater et al. (1976) showed that filtering heat flow data on young seafloor, only to include measurements made in well-sedimented areas with no nearby basement outcrops. This dramatically increased the heat flow to close to that predicted the models that matched the topography. Davis et al. (1999) and Davis and Becker (2004) documented clearly that, over variable but well-sedimented cover, the heat flow is directly related to the depth to basement (Figure 2b). This can only be possible if vigorous water circulation, occuring within the basaltic basement beneath the sediment, maintains a constant temperature at the sediment/crust interface.

Modern thermal models

Using carefully filtered heat flow data (Sclater et al., 1976) and specially selected depth-age provinces, Parsons and Sclater (1977) showed (a) how extraordinarily well the plate model fits the observational data and (b) that various parameters of the plate could be obtained by constructing physical models that fit the data within measurement error (Figure 3a: top and bottom). Out of these analyses came a realization (Parsons and McKenzie, 1978) that the lithosphere, even when represented as a flat plate, was still simply the upper boundary layer of a cooling convection system (Turcotte and Oxburgh, 1967). Parsons and Sclater (1977) fit the data with a plate thickness of 125 km and a basal temperature of 1,330 °C. In an analysis using a much more complete data set, Hillier and Watts (2005) showed that the depth data for the North Pacific filtered to remove the effects of swells gave results very similar to that of Parsons and Sclater (1977). However, both papers matched the heat flow data by adding a 4 mW m^{-2} to account for the effect of radioactivity within the ocean crust. Such a large value is not supported by the overall measured radioactivity of the crust (Staudigal, 2003).

In an alternative observational approach, Stein and Stein (1992) ignored both filtering and the selection of special provinces and plotted all the depth and heat flow data in the Pacific and Northwest Atlantic against age of the ocean floor. They found a model (GDH1) that best fit both data sets that had a plate thickness of 95 km and a bottom temperature of 1,425 °C (Figure 3b). Though the basal temperature is too high to explain the rare Earth distribution in the melted rock found at the ridge axis (McKenzie et al., 2005), GDH1 was the first model that actually had the predicted heat flow high enough to fit the data over ocean floor >130 Ma without assuming the

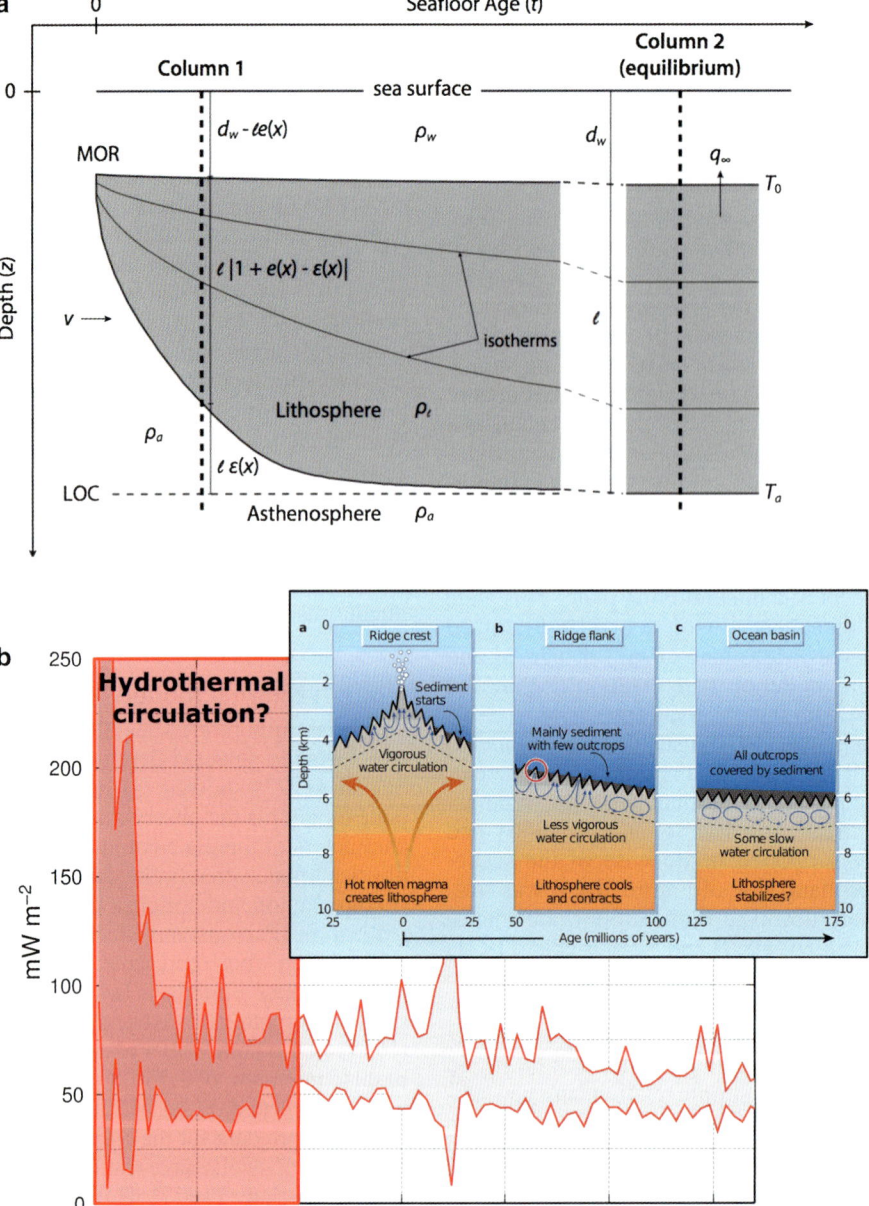

Marine Heat Flow, Figure 2 (**a**) Simple model for the intrusion of the oceanic lithosphere under a mid-ocean ridge (Sclater et al., 1971, based on McKenzie, 1967) that accounts the subsidence with age by balancing the pressure at the base columns 1 and 2 assuming a fluid underlies them. (**b**) (*Left*). The envelope of unfiltered heat flow data versus age marked by *red lines* (from compilation of Francis Lucazeau (personal communication see Goutorbe et al., 2011)). (**b**) (*Inset*) Changes in ocean floor depth, sediment thickness, and hydrothermal circulation plotted against increasing age of the oceanic lithosphere. The ocean floor is depicted in *black*, sediment in *brown*, and hydrothermal circulation in *blue*. In (**b**) (*Inset*): *a*, heat flow is dominantly by advection; in *b*, by advection and conduction; and in *c*, dominantly by conduction. The *red circle* in *b* illustrates the flow pattern observed by Fisher et al. (2003) on the flank of the Juan de Fuca where hot fluid is emanating through a basement outcrop (Sclater 2003).

addition of an unreasonably high value for the crustal radioactivity.

Whether or not the depths increase as the square root of age on the oldest seafloor has proven controversial (e.g., Crough, 1975; Schroder, 1984; Marty and Cazenave, 1989; Stein and Stein, 1992; Carlson and Johnson, 1994; Hillier and Watts, 2005). The distinction is key: a direct continuation would favor the simplest cooling half-space

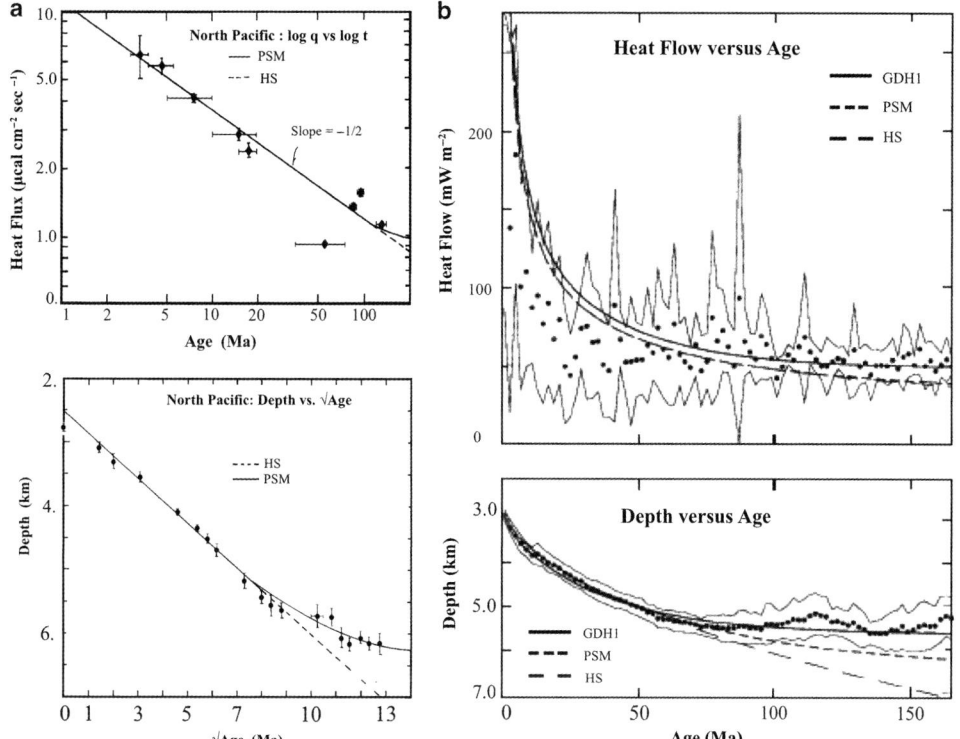

Marine Heat Flow, Figure 3 (a) (*Top*). Heat flow means calculated using data filtered for sediment thickness, plotted on a log/log scale versus age. The theoretical curve has a slope of $-1/2$ until an age of about 120 Ma. The small difference between the plate model and the cooling half-space model (*dashed line*: $-t1/2$ dependence) at older ages is just visible (Parsons and Sclater, 1977). (**a**) (*Bottom*). Mean depths and standard deviations plotted versus the square root of age for the North Pacific (Parsons and Sclater, 1977). (**b**) Data and models for heat flow (*top*) and ocean depth (*bottom*) as a function of age. Depths are an average of the North Pacific and Northwest Atlantic. Heat flow values are from sites in the same regions with data averaged in 2 Myr bins, and the envelope represents one standard deviation about the mean. GDH1, PSM, and HS are, respectively, the best fit model to the data, Parsons and Sclater (1977), and the half-space cooling model (from Stein and Stein (1992)).

model of the lithosphere, while shallower depths, a "flattening," favor a model of the lithosphere with basal heat input (plate, constant basal temperature, or, CHABLIS, constant basal heat flow (Doin and Fleitout, 1996)). Hillier (2010) has shown conclusively that the depths flatten with age on oldest ocean floor strongly favoring a model with basal heat input. This impacts upon expectations for seafloor heat flow.

Under continued scrutiny, how the depth and heat flow data are selected and the above controversy have brought to light two problems with the models. First, the plate thickness using the selected provinces was always thicker than that for the data taken as a whole. Second, Nagihara et al. (1996) summarized a series of carefully selected heat flow surveys on old ocean floor. They pointed out that neither the boundary layer nor GDH1 and the Parson and Sclater (1977) plate models could explain both the heat flow and age and depth and age relations on the oldest (>130 Ma) ocean floor without assuming the addition of an unreasonably high value for the crustal radioactivity. The data selection remains critical and a key to refining models.

Heat flow data

The global heat flow compilation published by the International Heat Flow Commission reported more than 22,000 stations. It incorporated the earlier work of Lee and Uyeda (1965), Simmons and Horai (1968), Jessop et al. (1975), Pollock et al. (1993), and Gosnold and Panda (2002) (http:www.geophysik.rwth.aachen.de/IHFC/heatflow.html). This database has been updated and extended by Laske and Masters (http://igppweb.ucsd.edu/~gabi/ rem.html). It was used in Davies and Davies (2010) and Davies (2013).

The latest papers reference compilations that now include >50,000 unique heat flow sites worldwide. The majority of the data are sparsely but relatively evenly distributed across the globe with a few isolated high-definition studies. For this review, we considered two recent compilations: Hasterok (2010) slightly updated in 2013 and Goutorbe et al. (2011). For the marine realm, both consist of about 14,000 + heat values, if the measurements on the shallow continental shelf are ignored. They differ mostly in the amount of confidential industry data that have been included. All

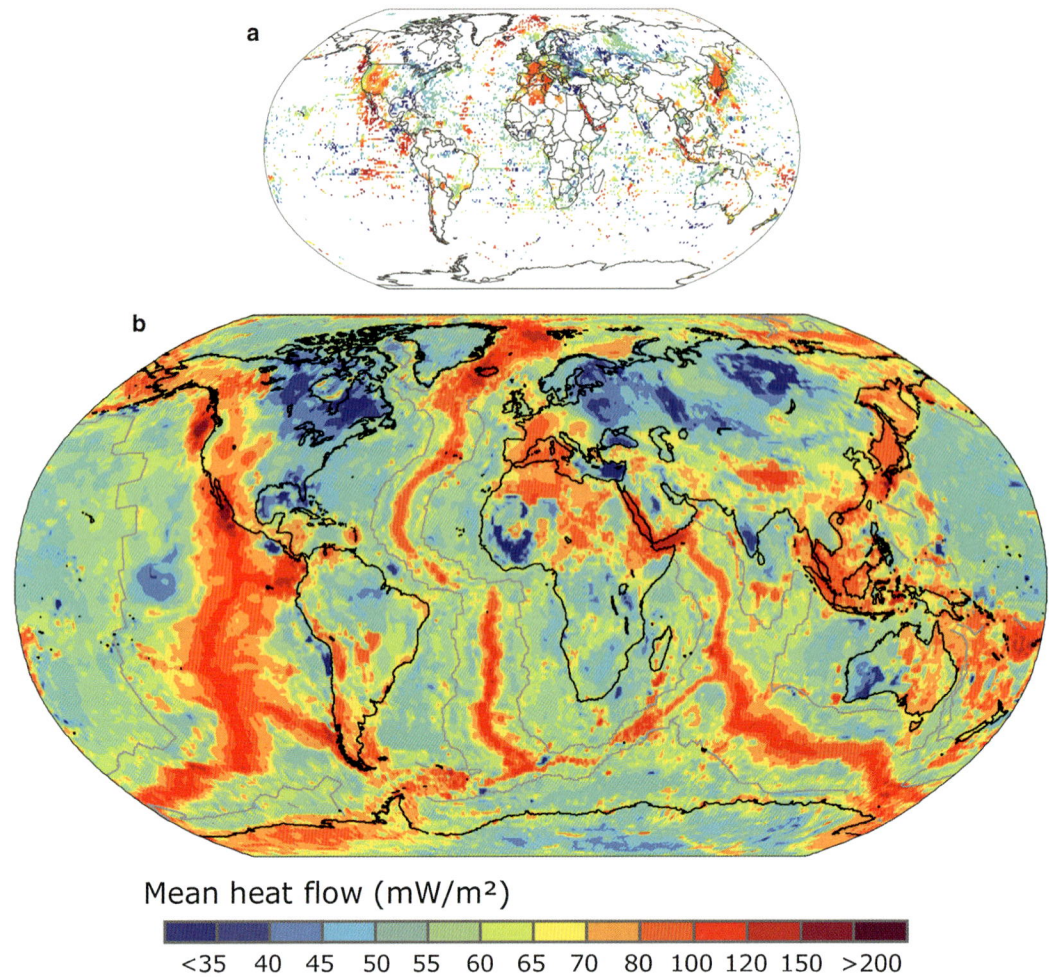

Marine Heat Flow, Figure 4 A 1° × 1° global heat flow grid used for this paper. (**a**) Gaussian filter with a 500 km radius was applied to smooth unexplained small-scale variations. Grid is available in the online supplementary material to Goutorbe et al. (2011). (**b**) A color contour chart of global heat flow that uses a "similarity method" to fill in gaps in coverage (From Goutorbe et al. (2011)).

start from the compilation of Francis Lucazeau (personal communication).

To illustrate the current state of the data, we constructed presentations based on Goutorbe et al. (2011). We present the raw data in 1° by 1° bins (Figure 4a) and a contour plot of the heat flow values based on a "similarity method" (see Goutorbe et al. (2011) for explanation) (Figure 4b). On a global scale, this analysis and a prior, less rigorous contour map, constructed using a similar approach by Shapiro and Rizwoller (2004), show a large area of strikingly high heat flow associated with the plate boundaries in the oceans. Davies (2013, Figure 7), using the theoretical heat flow versus age, presents an even more striking example of this effect. All of these charts illustrate how important the creation of oceanic plate is for the heat loss of the Earth (~60 % Sclater et al., 1981).

Heat flow versus age

The theory of plate tectonics predicts subsidence and decreasing heat flow as young buoyant lithosphere ages and cools. Because of the scatter in heat flow over young ocean floor, for most studies, models derived from the subsidence have dominated. Hasterok (2013) has recalibrated oceanic cooling models to include a filtered, hydrothermally free data set that simultaneously allows for a reasonable mantle temperature (~1,350 °C). He matched the observed heat flow data to a plate model and found that thinner plate cooling models with a thickness of 90 km gave the best fit (Figure 5a).

Hasterok (2013) considered five models (Figure 5a) when plotting heat flow against age of the ocean floor. All models appear to give the same degree of fit for younger ocean floor. However, as a result of the necessity to

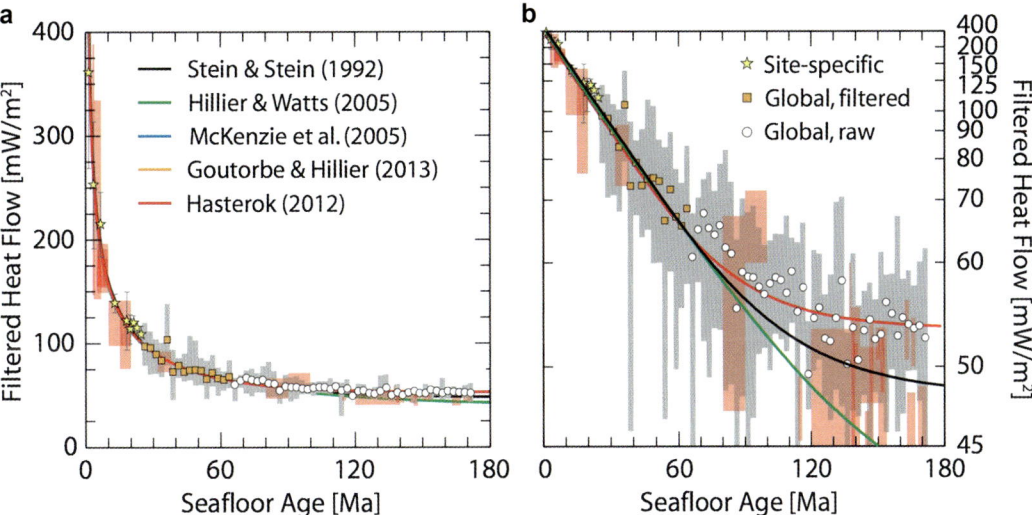

Marine Heat Flow, Figure 5 (a) Worldwide observed heat flow data (*gray rectangles*: interquartile range) in 2.5 Ma bins from well-sedimented areas plotted against age compared with the five different thermal models that match the relation between depth and age (see text for explanation). *Open circles*: Raw global data. *Red rectangles*: detailed heat flow surveys over well-sedimented areas (Sclater et al., 1976, Nagihara et al., 1996) (after Hasterok (2013)). (b) Median heat flow versus age, vertical scale is scaled by heat flow^{-2} to emphasize flattening on older ages. The symbols and curves are the same as for (a) (after Hasterok (2013)).

include the data over the young ocean floor, the linear scale is too small to differentiate between the models at older ages. To enlarge the scale for the older ages, but still retain the young data, he plotted the data with the distance on the vertical axis scaled as 1/heat flow^2 (Figure 5b). In such a plot, only the models of Crosby and McKenzie (2009) and Goutorbe and Hillier (2013) give a good match to his statistical best fit to the filtered heat flow data. However, the model of Goutorbe and Hillier (2013) has a higher basal temperature of 1,390° and a greater plate thickness (106–110 km). Both models match the heat flow relation without adding the effect of crustal radioactivity.

The best-fitting heat flow age model by Hasterok (2013), which fits the heat flow data, can be parameterized by the following simple relationship:

For $t \leq 17.4$ Ma, $q = 506.7\, t^{-1/2}$.
For $t > 17.4$ Ma, $q = 53 + 106\, e^{-0.034607 t}$.

As a result of the reduction in the scatter in heat flow, it is necessary to reconsider the variation in the topography. It is created both by mantle plumes and their topographic expression (Goutorbe and Hiller, 2013) (Figure 6a). In addition, Crosby et al. (2006) have suggested that cells small-scale convection that give rise to dynamic topography also have an effect. To investigate the importance of this, they removed the effects of both mantle plumes and the dynamic topography, both positive and negative, by only considering depths in areas of low free-air gravity. Using these data Crosby and McKenzie (2009) found that the depth-age data in the North Pacific best fit a plate ~90 km thick and 1,315 °C bottom temperature.

This plate is much thinner than the thickness of 125 km given by Parsons and Sclater (1977). The thermal model of Crosby and McKenzie (2009) constructed to fit the depth-age relation also accounts for the heat flow relation for the old ocean floor, thus allowing a single thermal model to match both the heat flow and the depth without assuming significant radioactivity in the ocean crust.

Crosby et al. (2006), Crosby and McKenzie (2009), and Goutorbe and Hillier (2013) start with the assumptions of Parsons and McKenzie (1978) where the cooling oceanic lithosphere consists of a mechanical and thermal boundary layer. At older ages the thermal boundary layer becomes unstable and heat is added to the bottom of the plate (Figure 6a). McKenzie et al. (2005) have pointed out that the creation of a plate of constant thickness is a good approximation to this concept. The bottom of the plate lies within the thermal boundary layer (Figure 6b).

Both the models of Crosby and McKenzie (2009) and Goutorbe and Hillier (2013) consider the effect of dynamic topography within the mantle (Figure 6a). In the case of Goutorbe and Hillier (2013), they remove only the major plumes and their topographic expression. Crosby and McKenzie (2009) assume that all areas where the gravity anomaly is large are associated with small-scale dynamic topography and remove them from their analysis. The model of Crosby and McKenzie (2009) has the advantage of having the lower bottom temperature for the plate fixed at ~1,315 °C, and McKenzie et al. (2005) have argued that this is the required temperature for upwelling to make a crustal thickness of 7 km. However, it has the disadvantage that it is statistical in basis and

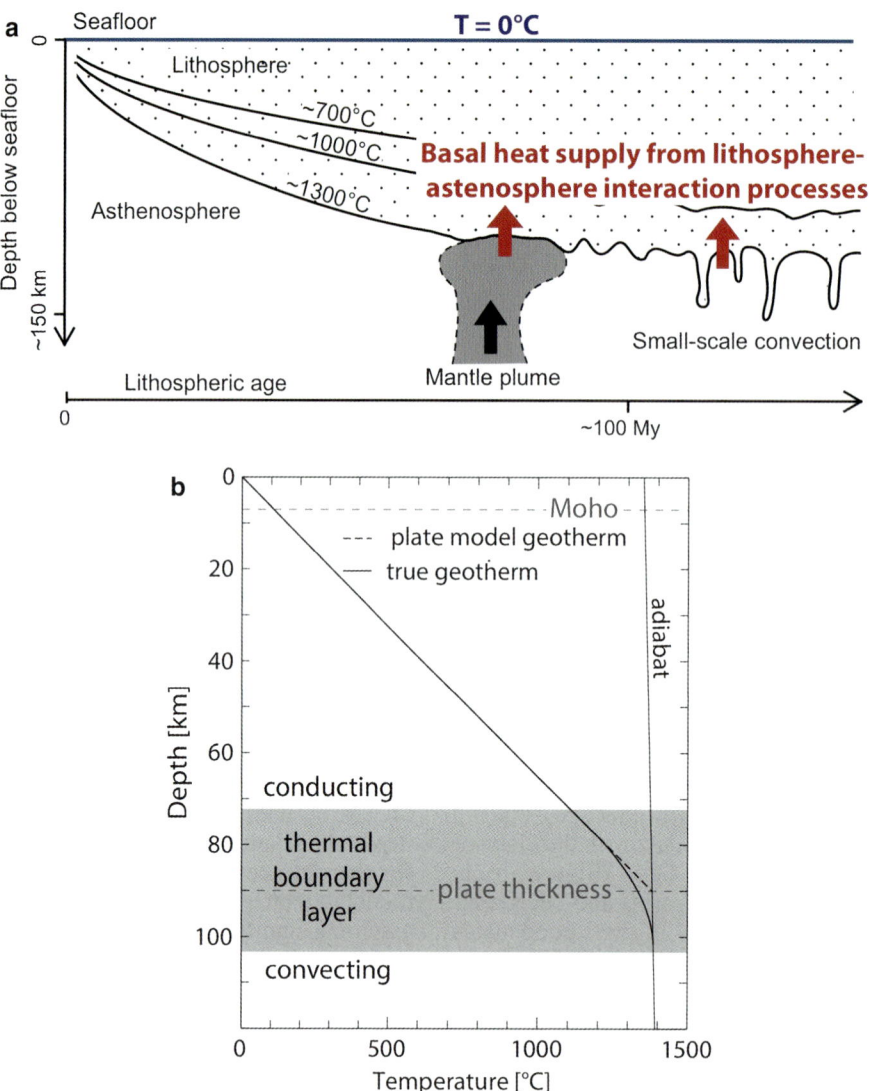

Marine Heat Flow, Figure 6 (a) Conceptual model of Goutorbe and Hillier (2013). The oceanic lithosphere consists of a mechanical and thermal boundary layer that is affected by thermal erosion due to overturn within the thermal boundary layer and by major mantle plumes. For the model of Crosby et al. (2006) and Crosby and McKenzie (2009), there is an additional effect due to small-scale convection in the upper mantle. (b) The base of the plate model lies within the thermal boundary layer. Note that the plate model geotherm lies close to that of a thermal boundary layer (after Crosby et al. (2006)).

the areas of low gravity have not yet been mapped in detail. Though the model of Goutorbe and Hillier (2013) has disadvantage of a high basal temperature, it has the advantages that with a thin plate (100+ km), it still fits the heat flow age relation and has clearly defined areas that have been removed when constructing the model to match the fit between depth and age.

Global heat loss/global heat flow estimates

Following the compilation by Lee and Uyeda (1965), a number of studies have estimated the global heat flow (Figure 7a). Estimates, prior to the widespread acknowledgement of advective heat transport through young seafloor, gave ~30–31 TW (1 TW equals 10^{12} W) (Lee and Uyeda, 1965; Lee, 1970; and Chapman and Pollack, 1975). Approximately 20 TW came from the oceans and ~10 TW from the continents, giving a global heat flow of ~30–31 TW. However, these values are now known to be too low.

Vigorous hydrothermal circulation through young seafloor is now recognized as a major mechanism by which the Earth redistributes heat (Lister, 1972). Hydrothermal circulation near the ridge axis is so vigorous that conductive heat flow values can be reduced by nearly two orders of magnitude and expelled in highly concentrated regions that give rise to black smokers (Williams et al., 1974).

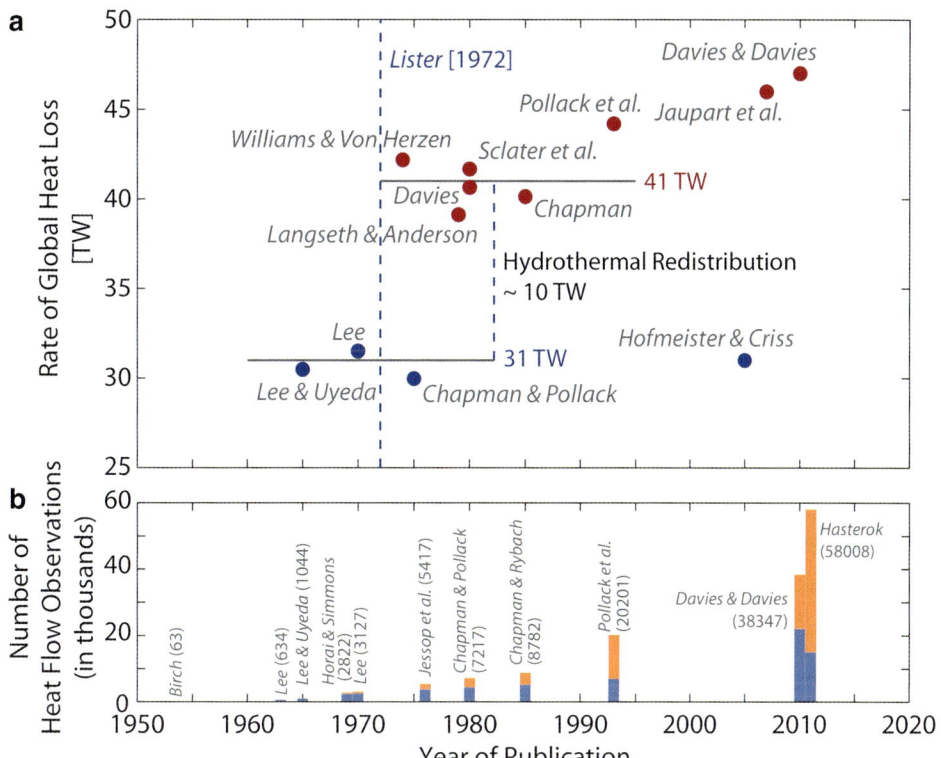

Marine Heat Flow, Figure 7 (a). Estimates of global heat flow of the Earth by date of publication (see references). *Blue circles* represent estimates derived from all heat flow observations, whereas the *red circles* substitute hydrothermally affected estimates with heat flow estimates derived from plate cooling models. Note the abrupt change when it was realized that the scatter in heat flow values was due to hydrothermal circulation. (b) Number of heat flow data within the global heat flow database (see list of references). *Upper bars* are the number of continental measurements, and *lower bars* are the number of oceanic measurements. The decrease in oceanic heat flow observations between Davies and Davies (2010) and Hasterok et al. (2011) is a result of reclassification of marine observations collected on the continental margins as continental in the latter study.

As the igneous crust becomes covered by sediment, direct flow between the oceans and lithosphere is cutoff. In extensively sedimented regions (Figure 2bc), heat flow may recover to near background values (Davis et al., 1999). However, flow may still occur between igneous outcrops, extracting heat beneath regions well covered by sediment for several tens of kilometers (Davis et al., 1999; Hutnak et al., 2008). The first global heat loss estimates adjusted to account for heat flow deficits on young seafloor now give global heat flow at ~41 TW with ~11 TW coming through the continents (Williams and Von Herzen, 1974; Langseth and Anderson, 1979; Davies, 1980; Sclater et al., 1980; Chapman and Pollack, 1980; Chapman (personal communication 1985 in Chapman and Pollack, 1987)) to ~44 TW (Pollack et al., 1993) (Figure 7a).

Even with an approximate doubling (Figure 7b) of the reported measurements, there has been little change in the value of global heat flow. The new data still show a significant observational deficit extending to oceanic crust of <60 Ma. Estimates of the magnitude of this deficit range from 9 to 11 TW (Sclater et al., 1980; Stein and Stein, 1992; Elderfield and Schultz, 1996; Mottl, 2003).

Hasterok (2013) provides the most recent estimate using a plate cooling model calibrated for heat flow at young ages. His estimated global heat flow deficit due to this ventilated flow is 8 (+2,−3) TW. This is ~20 % of the total heat loss of the Earth and >30 % of the heat flow through the deep oceans. Many of the differences in the estimates among older models result from the choice of cooling model parameters. However, adding a predicted heat loss estimate through hotspots rather than a change in the cooling model explains the increase from 44 TW (Pollack et al., 1993) to 46–47 TW (Jaupart et al., 2007; Davies and Davies, 2010). Ignoring hydrothermal circulation in crust <65 Ma and calculating the global heat loss of the Earth from observational data (Hofmeister and Criss, 2005) yields a value that is significantly low.

Jaupart et al. (2007) calculated the heat loss through the oceans by breaking the heat flow into two intervals: the first from 0 to 80 Ma where the simple boundary layer model gives a good match to the data and the second on older seafloor where the heat flow is approximately constant. They find that the heat loss due to plate creation is ~24.5 TW and that due to conductive flow on crust >80 Ma is 4.5 TW. Adding to the amount released by plate creation, the extra heat loss at the swells of ~3 TW raises the nonconductive heat loss to 27.5 TW. This is approximately 85 % of the heat loss through the oceans. On the continents, they estimate that about half of the surface heat flow is due to radioactive decay in the upper continental lithosphere. This leaves a conductive flow from the upper mantle under the continents of ~6 TW, with much of this coming from sedimentary basins and shelves as a result of stretching of the continental crust (Jaupart et al., 2007). Thus, almost all the surface heat flow of the Earth, except that due to radioactivity on the continents, has as its origin convection in the upper mantle. The heat flow from the upper mantle through the cratons and platforms is very low, and these areas act like an isotherm (Jaupart et al., 2007). Thus, the heat loss from the upper mantle shows a huge difference between that occurring beneath the oldest continental lithosphere and that created by plate creation at the mid-ocean ridges. This must have a major effect on convection in the upper mantle.

Conclusions

Many fundamental advances in our understanding of the Earth have come from the study of marine heat flow:

1. The early heat flow values were the same as, if not higher than, the values on continents. Because of the low radioactivity, they implied an additional source of heat to that given by radioactivity to explain them.
2. High values observed at the crest of mid-ocean ridges were a key observation in the development of the theory of seafloor spreading by Hess (1962).
3. Attempts to understand the high scatter in heat flow measurements near the crest of the mid-ocean ridges led Lister (1972) to propose widespread hydrothermal circulation at the crest of mid-ocean ridges.
4. There is general agreement that thermal models requiring input of heat from below best match the relations between the heat flow and subsidence with age.
5. Removing the effect of dynamic topography in determining the relation between depth ages allows the construction of two self-consistent models that match both the heat flow and subsidence on old ocean floor without requiring significant radioactivity in the oceanic crust.
6. A remarkable discovery associated with plate tectonics is that one simple thermal model can explain the general dependence of both marine heat flow and bathymetry on age.
7. Creation of oceanic plate is the major way in which the Earth loses heat. From a global loss of 45–47 TW, three quarters comes through the oceans with one fifth occurring by the advection of seawater at the crest of the mid-ocean ridges. Creation of oceanic plate accounts for ~60 % of the heat loss of the Earth.
8. There is a large difference between the heat loss from plate creation and that occurring under the oldest continents.

Acknowledgments

We thank Paul Morgan for a helpful review.

Bibliography

Birch, F., 1954. The present state of geothermal investigations. *Geophysics*, **19**, 645–659.

Bullard, E. C., 1939. Heat flow in South Africa. *Proceedings of the Royal Society of London Series A*, **173**, 474–502.

Bullard, E. C., 1945. Thermal history of the earth. *Nature*, **156**, 35.

Bullard, E. C., 1952. Heat flow through the floor of the eastern North Pacific Ocean. *Nature*, **170**, 200.

Bullard, E. C., 1954. The flow of heat through the floor of the Atlantic Ocean. *Proceedings of the Royal Society of London Series A*, **222**, 408–429.

Bullard, E. C., Maxwell, A. E., and Revelle, R., 1956. Heat flow through the deep sea floor. *Advances in Geophysics*, **3**, 153–181.

Carlson, R. L., and Johnson, H. P., 1994. On modeling the thermal evolution of the oceanic upper mantle: an assessment of the cooling plate model. *Journal of Geophysical Research*, **99**, 3201–3214.

Chapman, D., and Pollack, H., 1980. Global heat flow: spherical harmonic representation (abstract). *Eos, Transactions American Geophysical Union*, **61**, 383.

Chapman, D., and Pollack, H., 1987. Global heat flow: spherical harmonic representation. In *19th General Assembly of the IUGG*, Abstracts Volume, 1, Vancouver, Canada, p. 50.

Chapman, D., and Pollock, H., 1975. Global heat flow: a new look. *Earth and Planetary Science Letters*, **28**, 23–32.

Chapman, D., and Rybach, L., 1985. Heat flow anomalies and their interpretation. *Journal of Geodynamics*, **4**, 3–37.

Crosby, A., and McKenzie, D., 2009. An analysis of young ocean depth, gravity and global residual topography. *Geophysical Journal International*, **178**, 1198–1219.

Crosby, A., McKenzie, D., and Sclater, J., 2006. The relations between depth, age and gravity in the oceans. *Geophysical Journal International*, **166**, 553–573.

Crough, S., 1975. Thermal model of oceanic lithosphere. *Nature*, **256**, 388–390.

Davies, G., 1980. Review of oceanic and global heat flow estimates. *Reviews of Geophysics*, **18**, 718–722.

Davies, J., and Davies, D., 2010. Earth's surface heat flux. *Solid Earth*, **1**(1), 5–24.

Davies, J., 2013. Global map of solid earth surface heat flow. *Geochemistry, Geophysics, Geosystems*, **14**, 4608–4622.

Davis, E., and Becker, K., 2004. Observations of temperature and pressure: constraints on oceanic crustal hydrologic state, properties and flow. In Davis, E., and Elderfield, H. (eds.), *Hydrogeology of the Oceanic Lithosphere*. Cambridge: Cambridge University Press, pp. 225–227.

Davis, E. E., Wang, K., Becker, K., Thomson, R. E. and Yashayaev, I., 2003. Deep-ocean temperature variations and implications for errors in seafloor heat flow determinations. Journal of Geophysical Research 108(B1), 2034, doi:10.1029/2001JB001695.

Davis, E., and Elderfield, H., 2004. *Hydrology of the Oceanic Lithosphere*. Cambridge: Cambridge University Press.

Davis, E., and Lister, C., 1974. Fundamentals of ridge crest topography. *Earth and Planetary Science Letters*, **21**, 405–413.

Davis, E., Chapman, D., Wang, K., Villinger, H., Fisher, A., Robinson, S., Grigel, J., Pribnow, D., Stein, J., and Becker, K., 1999. Regional heat flow variations across the sedimented Juan de Fuca Ridge eastern flank: constraints on lithospheric cooling and lateral hydrothermal heat transport. *Journal of Geophysical Research*, **104**, 17675–17688.

Dietz, R. S., 1961. Continent and ocean basin evolution by spreading of the seafloor. *Nature*, **190**, 854–857.

Doin, M., and Fleitout, L., 1996. Thermal evolution of the oceanic lithosphere: an alternative view. *Earth and Planetary Science Letters*, **142**, 121–136.

Elderfield, H., and Schultz, A., 1996. Mid-ocean ridge hydrothermal flux and the chemical composition of the ocean. *Annual Review of Earth and Planetary Sciences*, **24**, 191–224.

Ewing, M., LePichon, X., and Langseth, M. G., 1966. Crustal structure of the mid-Atlantic ridge. *Journal of Geophysical Research*, **71**, 1611–1636.

Fisher, A. T., Davis, E. E., Hutnak, M., Spiess, V., Zulsdordorff, L., Cherakaoul, A., Christiansen, L., Edwards, K., Macdonald, R., Vellinger, H., Mottl, M. J., Wheat, C. G., and Becker, K., 2003. Hydrothermal recharge and discharge across 50 km guided by seamounts on a young ridge flank. *Nature*, **421**, 618–621.

Gerard, R., Langseth, M. G., and Ewing, M., 1962. Thermal gradient measurements in the water and bottom sediment of the western Atlantic. *Journal of Geophysical Research*, **67**, 785–803.

Gosnold, W., and Panda, B., 2002. *The Global Heat Flow Database of the International Heat Flow Commission*. www.und/edu/org/ihfc/index2.html.

Goutorbe, B., and Hillier, J. K., 2013. An integration to optimally constrain the thermal structure of oceanic lithosphere. *Journal of Geophysical Research*, **118**, doi:10.1029/2012JB009527

Goutorbe, B., Poort, J., Lucazeau, F., and Raillard, S., 2011. Global heat flow trends resolved from multiple geological and geophysical proxies. *Geophysical Journal International*, **187**, 1405–1419.

Hasterok, D., 2010. *Thermal State of the Oceanic and Continental Lithosphere*. Salt Lake City: University of Utah, PhD thesis, pp. 156

Hasterok, D., 2013. A heat flow based cooling model for tectonic plates. *Earth and Planetary Science Letters*, **361**, 34–43.

Hasterok, D., Chapman, D., and Davis, E., 2011. Oceanic heat flow: implications for global heat loss. *Earth and Planetary Science Letters*, **311**, 386–395.

Hess, H., 1962. History of ocean basins. In Engel, A. E. J., James, H. L., and Leonard, B. F. (eds.), *Petrologic Studies*. Boulder: Geological Society of America, Vol. 4, pp. 599–620.

Hillier, J. K., 2010. Subsidence of 'normal' seafloor: Observations do indicate 'flattening'. *Journal of Geophysical Research*, **115**, doi:10.1029/2008JB005994

Hillier, J. K., and Watts, A., 2005. Relationship between depth and age in the North Pacific Ocean. *Journal of Geophysical Research*, **110**, doi: 10.1029/2004JB003406.

Hofmeister, A., and Criss, R., 2005. Earth's heat flux revised and linked to chemistry. *Tectonophysics*, **395**, 159–177.

Horai, K., and Simmons, G., 1969. Thermal conductivity of rock-forming minerals. *Earth and Planetary Science Letters*, **6**, 359–368.

Hutnak, M., Fisher, A., Harris, R., Stein, C., Wang, K., Spinelli, G., Schindler, M., Villinger, H., and Sliver, E., 2008. Large heat and fluid flux driven through mid-plate outcrops on ocean crust. *Nature Geoscience*, **1**, 611–614.

Hyndman, R. D., Langseth, M. G., and Von Herzen, R. P., 1984. Review of Deep sea Drilling Project geothermal measurements through Leg 71. Initial Report of the Deep Sea Drilling Project, 78B, 813–823.

Jaupart, C., Labrosse, S., and Mareschal, J. C., 2007. Temperature, heat and energy in the mantle of the earth. In Bercovici, D. (ed.), *Treatise on Geophysics, vol 7: Mantle Convection*. Amsterdam: Elsevier, pp. 253–303.

Jessop, A. M., Hobart, M. A., and Sclater, J. G., 1975. *The World Heat Flow Data Collection*. Ottawa: Earth Physics Branch: Energy, Mines and Resources.

Johnson, P., and Hutnak, N., 1997. Conductive heat loss in recent eruptions at mid-ocean ridges. Geophysical Research Letters, 24, 3089–3092.

Johnson, P. H., and Carlson, R. L., 1992. Variations of sea floor depth with age: a test of models based on drilling results. *Geophysical Research Letters*, **19**, 1971–1974.

Kelvin, W. T., 1895. On the age of the earth. *Nature*, **51**, 438–440.

Langseth, M., and Anderson, R., 1979. Correction. *Journal of Geophysical Research*, **84**, 1139–1140.

Lee, W., 1963. Heat flow data analysis. *Reviews of Geophysics*, **1**, 449–479.

Lee, W., 1970. On the global variations in terrestrial heat flow. *Physics of Earth and Planetary International*, **2**, 332–341.

Lee, W., and Uyeda, S., 1965. Review of heat flow data. In Lee, W. (ed.), *Terrestrial Heat Flow*. Washington: American Geophysical Union. Geophysical monograph 8, pp. 87–100.

Lister, C., 1972. On the thermal balance of a Mid-ocean ridge. *Geophysical Journal of the Royal Astronomical Society*, **26**, 515–535.

Lister, C., 1979. The pulse-probe method of conductivity measurement. *Geophysical Journal of the Royal Astronomical Society*, **57**, 451–461.

Louden, K. E., and Wright, J. A., 1989. Marine heat flow data: a new compilation of observation and brief review of its analysis. In Wright, J. A., and Louden, K. E. (eds.), *CRC Handbook of Seafloor Heat Flow*. Boca Raton: CRC Press, pp. 3–72.

Marty, J. C., and Cazenave, A., 1989. Regional variations in subsidence rate of oceanic plates: a global analysis. *Earth and Planetary Science Letters*, **94**(301–315), 1989.

McKenzie, D., 1967. Some remarks on heat flow and gravity anomalies. *Journal of Geophysical Research*, **72**, 6261–6273.

McKenzie, D., Jackson, J., and Priestly, K., 2005. Thermal structure of oceanic and continental lithosphere. *Earth and Planetary Science Letters*, **223**, 337–349.

Menard, H. W., 1969. Elevation and subsidence of oceanic crust. *Earth and Planetary Science Letters*, **6**, 275–284.

Mottl, M., 2003. Partitioning of energy fluxes between midocean ridge axes at high and low temperature. In Halbach, P., Tuncliffe, V., and Hein, J. (eds.), *Energy and Mass Transfer in Marine*. Berlin: Dahlem University Press, pp. 271–286.

Nagihara, S., Lister, C., and Sclater, J., 1996. Reheating of old oceanic lithosphere: deductions from observations. *Earth and Planetary Science Letters*, **139**, 91–104.

Parker, R., and Oldenburg, D., 1973. Thermal model of ocean ridges. *Nature*, **242**, 137–139.

Parsons, B., and McKenzie, D., 1978. Mantle convection and the thermal structure of the lithosphere and the thermal structure of the lithosphere. *Journal of Geophysical Research*, **83**, 4485–4496.

Parsons, B., and Sclater, J., 1977. An analysis of the variation of ocean floor bathymetry and heat flow with age. *Journal of Geophysical Research*, **82**, 803–827.

Pettersson, H., 1949. Exploring the bed of the ocean. *Nature*, **164**, 469–470.

Pollack, H., Hurter, S., and Johnson, J., 1993. Heat flow from the Earth's interior: analysis of the global data set. *Reviews of Geophysics*, **31**, 267–280.

Pribnow, D. F. C., Kinoshita, M., and Stein, C. A., 2000. *Thermal Data Collection and Heat Flow Recalculations for ODP Legs 101–180*. Hanover: Institute for Joint Geoscientific Research, (CGA).

Revelle, R., and Maxwell, A. E., 1952. Heat flow through the floor of the eastern North Pacific Ocean. *Nature*, **170**, 199–200.

Schroeder, W., 1984. The empirical age-depth relation and depth anomalies in the Pacific Ocean. *Journal of Geophysical Research*, **89**(9873–9883), 1984.

Sclater, J. G., 2003. Ins and outs on the ocean floor. *News and Views, Nature*, **421**, 590–591.

Sclater, J. G., and Francheteau, J., 1970. The implications of terrestrial heat flow observations on current tectonic and geochemical models of the crust and upper mantle of the earth. *Geophysical Journal of the Royal Astronomical Society*, **20**, 509–542.

Sclater, J. G., Anderson, R. N., and Bell, M. L., 1971. The elevation of ridges and the evolution of the central Eastern Pacific. *Journal of Geophysical Research*, **76**, 7888–7915.

Sclater, J. G., Crowe, J., and Anderson, R. N., 1976. On the reliability of oceanic heat flow averages. *Journal of Geophysical Research*, **81**, 2997–3006.

Sclater, J. G., Jaupart, C., and Galson, D., 1980. The heat flow through the oceanic and continental crust and the heat loss of the earth. *Reviews of Geophysics and Space Physics*, **18**, 269–311.

Sclater, J. G., Parsons, B., and Jaupart, C., 1981. Oceans and continents: similarities and differences in the mechanisms of heat loss. *Journal of Geophysical Research*, **86**, 11535–11552.

Shapiro, N. M., and Ritzwoller, M. H., 2004. Inferring heat flux distributions guided by a global seismic model: particular application to Antarctica. *Earth and Planetary Science Letters*, **223**, 213–214.

Simmons, G., and Horai, K., 1968. Heat flow data 2. *Journal of Geophysical Research*, **73**, 6608–6609.

Sleep, N., 1969. Sensitivity of heat flow and gravity to the mechanism of sea-floor spreading. *Journal of Geophysical Research*, **76**(542), 1969.

Staudigal, H., 2003. Hydrothermal alteration processes in the oceanic crust. In Rudnick, R. (ed.), *Treatise on Geochemistry*. Amsterdam: Elsevier, Vol. 3, pp. 511–545.

Stein, C., and Stein, S., 1992. A model for the global variation in oceanic depth and heat flow with lithospheric age. *Nature*, **359**, 123–129.

Turcotte, D. L., and Oxburgh, E. R., 1967. Finite amplitude convection cells and continental drift. *Journal of Fluid Mechanics*, **28**, 29–42.

Vacquier, V., and Von Herzen, R., 1964. Evidence for connection between heat flow and the Mid-Atlantic ridge magnetic anomaly. *Journal of Geophysical Research*, **69**, 1093–1101.

Vogt, P. R., 1967. Steady state crustal spreading. *Nature*, **215**, 811–817.

Von Herzen, R., 1959. Heat-flow values from the South-eastern Pacific. *Nature*, **183**, 882–883.

Von Herzen, R., and Uyeda, S., 1963. Heat flow through the eastern Pacific Ocean floor. *Journal of Geophysical Research*, **68**, 4219–4250.

Williams, D., and Von Herzen, R., 1974. Heat loss from the earth: new estimate. *Geology*, **2**, 327–328.

Williams, D., Von Herzen, R. P., Sclater, J. G., and Anderson, R. N., 1974. The Galapagos spreading center: lithospheric cooling and hydrothermal circulation. *Geophysical Journal of the Royal Astronomical Society*, **38**, 609–626.

Wilson, J. T., 1965. A new class of faults and their bearing on continental drift. *Nature*, **207**, 343–347.

Cross-references

Hydrothermalism
Lithosphere: Structure and Composition
Plate Tectonics
Technology in Marine Geosciences
Subsidence of Oceanic Crust
Oceanic Spreading Centers

MARINE IMPACTS AND THEIR CONSEQUENCES

Henning Dypvik
Department of Geosciences, University of Oslo, Oslo, Norway

Introduction

Asteroid and comet impacts (meteorite impacts) are important factors in shaping the surface of the terrestrial planets. The impacts release vast amounts of energy resulting in sudden increase in temperature and pressure, structural deformations, and redistribution of target and bolide (asteroid and comet) material. It is presently well accepted that impacts, especially into marine environments, have had a major influence on the development of the Earth. The Earth is covered by 71 % seawater (7 % shelf seas and 64 % deep oceans), and, consequently, impacts into marine targets should be the most common.

Asteroids and comets (both generally called bolides, projectiles, or impactors) impact regularly on the Earth and other planetary bodies in our solar system, forming impact craters, expelling ejecta material, and triggering earthquakes and tsunamis. The impact craters registered on the Earth range from a few meters to hundreds of kilometers in diameter (http://www.passc.net/EarthImpactDatabase/) mainly determined by bolide size, composition, and velocity as well as the composition of the target area and angle of impact (Figure 1). The most probable impact angle is 45° for asteroids and comets (Gilbert, 1893; Shoemaker, 1962).

The impacts on dry land form subaerial impact craters, while submarine or sublacustrine craters are the result of marine or lacustrine impacts (subaqueous impact craters), respectively (Dypvik and Jansa, 2003). The marine impacts/submarine craters can in general be classified into two broad classes: shallow marine, neritic, and deep marine, bathyal. The shallow marine cases represent impacts and related deposits on continental shelves with water depths down to about 300 m, while the deep marine impacts are found below 300 m, i.e., beyond the shelf edge and into the deep ocean. It should be noted that the critical depth depends on projectile size (D) and water depth (H). The shallow marine case is realized when H < D. Even 1,000 m is shallow water for a 10 km projectile.

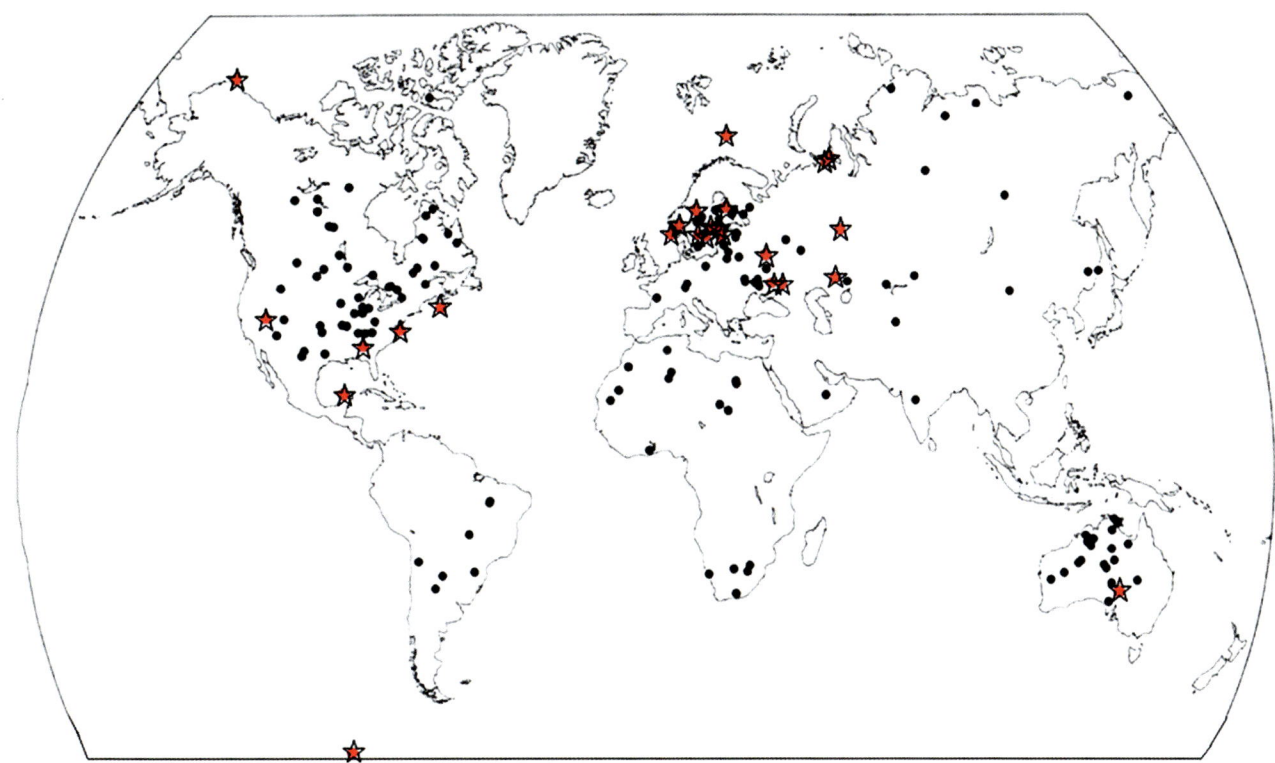

Marine Impacts and Their Consequences, Figure 1 Impact craters on the Earth. Marine craters/impacts are marked by *red stars* (Figure modified from Dypvik and Jansa (2003)).

Impact cratering

According to NASA (2012), several thousand hazardous asteroids and comets exist, but detailed knowledge of their distribution is still lacking. Compositionally, the bolides can be recognized by comparable minerals to terrestrial rocks, often with the addition of enrichments in Pt-group metals (Ru, Rh, Pd, Os, Ir, Pt) and siderophile elements (e.g., Ni, Co, Au). High enrichment of Ir is of special interest and commonly used as an extraterrestrial indicator (French and Koeberl, 2010).

Generally, two main types of impact structure appear: simple and complex (Melosh, 1989; French, 1998). The simple craters are normally bowl shaped and less than 4 km in diameter, while the complex craters (>4 km) can have various shapes, often with a central peak and in some cases internal ring(s) (Figure 2).

Crater formation is completed within a few minutes, the actual time depending mainly on crater size which is controlled by bolide size, target composition, bolide velocity, and composition. The crater development can be subdivided into three main stages: contact/compression, excavation, and modification (French, 1998). In the contact and compression stage, kinetic energy is converted into shock waves that are transmitted into the target material; the duration of this stage is only a few seconds (Melosh, 1989). The result is a fractured and brecciated target area and a totally or partially evaporated projectile.

The succeeding excavation stage lasts longer than the contact/compression stage and is the result of the release waves (rarefaction) when the crater is opened up and a bowl-shaped depression is formed in the target site. Crushed and melted material is ejected, and a so-called transient crater is formed. The transient crater is a bowl-shaped depression with a structurally uplifted rim.

Ejecta deposits have been recognized in association with several impacts, i.e., with wide distributions of crushed target material as well as melted bolide and target material. In the marine impact cases, this phase is highly influenced and volatized by the vaporized water (Figure 2).

The excavation stage is followed by a modification stage dominated by processes remodeling the unstable transient craters, controlled by gravity and rock mechanics and crater sedimentation. The modification stage can last longer than the excavation stage, i.e., a few minutes for the largest impacts (French, 1998).

Marine target sites consist of three layers: from top down, the water column, thick sedimentary successions, and underlying crystalline basement. Their different thicknesses and rheological properties greatly influence cratering developments. First, a crater is generated in the water, then a crater is formed on the sea floor, and finally, in the crystalline basement if enough energy is still left after the bolide's penetration of the water and sedimentary

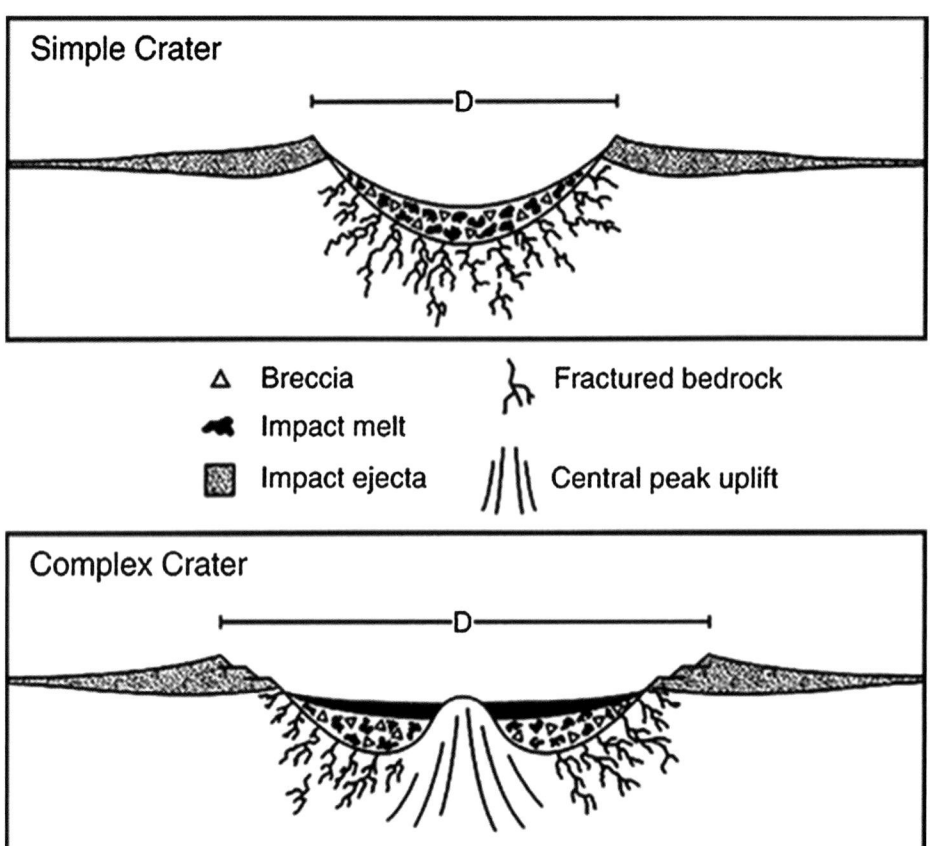

Marine Impacts and Their Consequences, Figure 2 Diagram of the main configuration of simple and complex craters (NASA website 2013; craters.gsfc.nasa.gov).

columns. The interaction of the impacting body and the water causes loss of the body's kinetic energy through oceanic drag prior to the ocean floor collision (Wünnemann and Lange, 2002).

Oberbeck et al. (1993) presented one of the first detailed descriptions of marine impacts and submarine craters, discussing ejecta distribution, tsunami generation, crater formation, and current generation. Focusing on shallow seas, their Figure 6 (Figure 3 here) still gives a good illustration of the marine impacts, crater formation and violent ejecta distribution, tsunami generation, active crater modification, and later postimpact sedimentation filling of the craters.

Water depth will have little effect for the large bolides, but for small- and medium-sized ones, the ocean will function as an efficient filter, with marine impact craters only forming in relatively shallow water (H) compared to the size of the bolide (D) (Artemieva and Shuvalov, 2002); $H < D$.

Impact craters in marine sedimentary basins are filled by postimpact sediments rather quickly after impact. Consequently, submarine craters commonly are very well preserved compared to subaerial ones where erosion-dominated processes are more active. The well-preserved submarine craters of Mjølnir and Chesapeake Bay are good examples (Gohn et al., 2009; Dypvik et al., 2010).

The Earth Impact Database (December 2013) names 183 craters of which only 28 (may be three additional ones, totally 31) have been recognized as marine impacts (Figure 1). They all have been formed by impact into marine basins, but presently, most of them occur on land though a few are still located in a submarine setting. A closer look at the numbers, e.g., simple subtraction $(183 - 28(31))$, gives 155(152) structures which should represent the minimum number of subaerial impacts. 155(152) on land impacts should, taking the land/sea area ratio into consideration, indicate at least 379(372) marine impacts/submarine craters. Since 155(152) clearly is a minimum number, the 379(372) may also be regarded as a minimum total for marine impacts.

The fact that only 28 marine impact structures have been described so far is probably due to:

- The difficulties in searching for marine impact structures and too few engaged impact researchers.

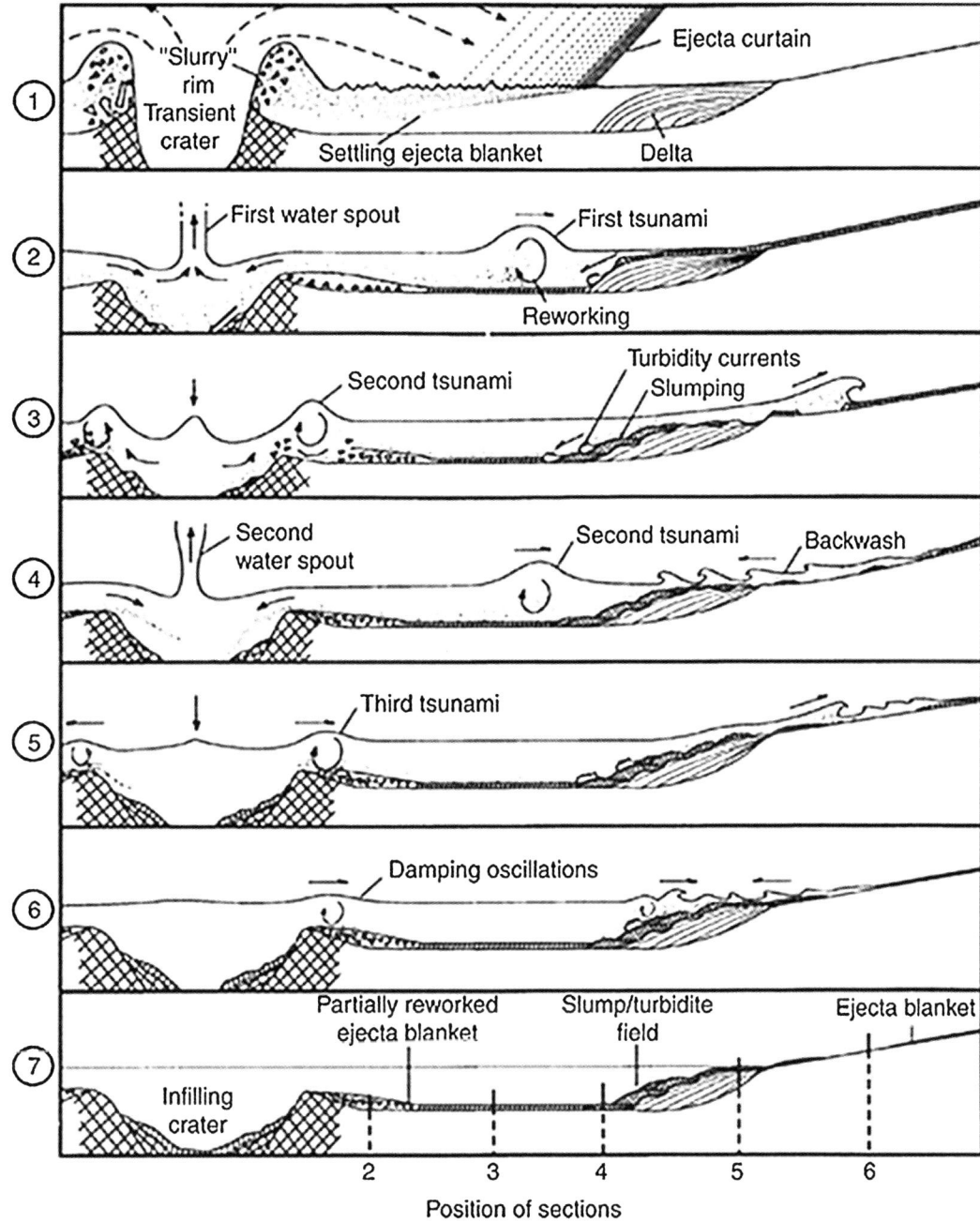

Marine Impacts and Their Consequences, Figure 3 Hypothetical stages of events produced by a bolide impacting a shallow sea (The figure is from Oberbeck et al. (1993) who claims the seven stages will occur over a period of several hours. The vertical scale is exaggerated).

- All oceanic crust is younger than approximately 180 million years, due to plate consumption and generation in the plate tectonic cycle.
- Lack of knowledge of how to recognize deep oceanic impacts, since we know so little about these structures.
- The oceans and epicontinental seas globally cover huge areas of which only a tiny portion have been covered by geological and geophysical analyses.
- A large number of asteroids and comets never reach the sea floor; consequently fewer submarine craters should be expected compared to subaerial ones (Davison and Collins, 2007).

Through the years, crater and impact statistics have been compiled, and, e.g., in the Task Force UK report of 2000 (Table 1), the general impact frequency and crater sizes are listed.

Marine Impacts and Their Consequences, Table 1 Impact cratering bolide size (m and km), crater diameter (km), and impact frequency (year) (Modified from Task Force Group, 2000)

Bolide (m/km)	Crater diameter (km)	Average interval between impacts (year)
75 m	1.5 km	1,000 years
350 m	6 km	16,000 years
700 m	12 km	63,000 years
1.7 km	30 km	250,000 years
3 km	60 km	1 million years
7 km	125 km	10 million years
16 km	250 km	100 million years

Methods of studying marine impacts

Fossil marine impact structures may be found on land and can consequently be as accessible for detailed analyses as the subaerial ones. Modern and fresh submarine craters, present on and below the ocean floor, are more difficult and costly to study, being dependent on geophysical investigations. In particular, seismic, gravimetric, and magnetic surveys are needed to detect and describe these structures and their ejecta distributions, e.g., as done in the Chicxulub, Chesapeake Bay, and Mjølnir cases (Gohn et al., 2009; Dypvik et al., 2010; Schulte et al., 2010). Craters on the Earth traditionally need additional detailed information to be recognized and accepted as impact generated and not circular cavities with different origins. Cores and rock samples are necessary for structural, mineralogical, and geochemical analysis (French and Koeberl, 2010). Coring is normally expensive and rarely done, likely one important reason for our poor understanding of marine impacts.

Various core analyses are needed in order to differentiate crater depressions from other cavities on the seafloor, formed by, e.g., gas discharge, other structuring on the seafloor (faulting/fracturing), and volcanic eruptions. Rocks and sediments affected by impact can mineralogically and geochemically be characterized by the presence of, e.g., shocked altered minerals (shocked quartz), Ni-rich spinels, and geochemical enrichments (Ir in particular) (French and Koeberl, 2010).

Marine impacts

In a marine impact, water absorbs the energy very effectively; consequently, a very large bolide is needed to produce even a modest crater on the abyssal plain. The mobilization of sediment following an impact tends to obscure such a signature except in cases where the bolide diameter is comparable to the ocean depth (Gisler et al., 2010). It should be mentioned that so far very few coastal tsunami deposits related to impacts have been found and described (e.g., Chicxulub; Smit, 1999).

Marine impacts represent an intricate mixture of relatively poorly understood processes compared to subaerial cratering (Ormö and Lindström, 2000; Dypvik and Jansa, 2003; Dypvik and Kalleson, 2010). The relations between bolide size (D) and water depth (H) are crucial in constraining submarine crater formation, as also mentioned above. It has been shown that the ratio H/D is of great importance on the resulting cratering process. No subaqueous crater is formed at the seafloor if $H/D > 10$ (Gault and Sonett, 1982; Artemieva and Shuvalov, 2002), while if $10 < H/D < 1$, the water column plays an important role during the cratering process to the size and morphology of the resulting structure (Shuvalov, 2002). When $H < D$, there is only a minor direct influence of the water column to the cratering process (Shuvalov, 2002).

In shallow water impacts, the crater rims are not normally leveled out (Ormö and Lindström, 2000; Dypvik and Jansa, 2003) and may even be taller than the returning water and consequently even hamper the resurge of water (e.g., the 2.7 km diameter rim of the Ritland structure; Riis et al., 2012; Shuvalov et al., 2012). Already in 1982, Gault and Sonett (1982) carried out experiments to model deep and shallow ocean impacts. They showed that water is remarkably effective in shielding the seafloor against significant excavation. In shallow water craters, their experiments demonstrated that central peaks in smaller craters could extend even above the water surface, as, e.g., in the Mjølnir case (Dypvik et al., 2010). In deep waters, the rims as well as the central peaks may be quickly leveled out by the generated current's reworking of soft sediments, explaining the typical marine crater characteristic of low or absent rims (Dypvik and Jansa, 2003). Resurge gullies may be formed during rather shallow marine impacts, cutting across the rim. Mass and debris deposits, turbidites, and other gravity deposits are produced as results of crater wall and central peak collapse, during and after crater excavation. Such gravity flow-related deposits are found both within and outside the marine impact craters (Dypvik and Kalleson, 2010).

Marine impacts or impact into wet targets can result in ejecta transportation as ground-hugging flows (Melosh, 1982; Masaitis, 2002). Such accretionary impact lapilli may occur both in proximal and distal environments relating to large- and middle-sized impact craters. The Alamo Breccia of Western USA and Albion Formation of Mexico and Belize are possible examples (Warme, 2004; Pope et al., 2005; Morrow et al., 2009).

In marine impacts, the sedimentation of the modification stage includes breccias, slumps, avalanches, mass/debris flows, and turbidity currents. Impact-generated earthquakes may trigger liquefaction reactions in the water-saturated seafloor sediments. The impact-related sediments rich in ejecta may consequently be widely distributed, like the ejecta and tsunamites from the Chicxulub impact (Smit, 1999; Clayes et al., 2002). The Baltic Narva Breccia was possibly formed from slumping, displacements, and wave reworking after the middle Devonian Kaluga impact into an epicontinental, about 300 m deep sea (Masaitis, 2002).

The Chicxulub impact event at the boundary between Cretaceous and Paleogene (K/Pg) took place in shallow water, near the shelf edge at the Yucatan peninsula (Mexico) (Schulte et al., 2010). In that case, a 10 km bolide hit, generated earthquakes, and large parts of the outer shelf/shelf edge collapsed, triggering slumps and slides, avalanches, gravity flows, turbidity currents, and tsunamis.

The 142-million years old Mjølnir structure was found in the Barents Sea during geophysical exploration; later structural, mineralogical, and geochemical analysis proved its meteoritic origin (Dypvik et al., 2010). Mjølnir looks different to subaerial craters with its subdued and possibly gullied rim, chaotic marine succession, and marine fossils in the black shale formations above and below. On top of the folded, slumped, and fractured sediments forming the Mjølnir crater base, there is a sedimentary package about 20 m thick. It consists of breccias, partly from autobrecciation and mass-flow deposits including turbidites, forming the syn- and early postimpact succession. It is all topped by black shales, demonstrating a return of the black clay sedimentation in the basin, which prevailed before impact.

So far, no deep sea impact craters have been found, in spite of the large areas of the Earth which are and have been covered by deep seas. Davison and Collins (2007) claim there is still no definite relationship between impactor properties, water depth, and final crater diameter. Artemieva and Shuvalov (2002) and Gisler et al. (2010) presented similar relations between water, impactor size, and cratering as given by the ratio between bolide size and water depth and (H/D). For shallow craters with a ratio below 3–4, decreasing crater diameters are modeled with increasing water depth, but a ratio larger than 7–8 will result in no crater at all, though possibly some disturbance to the seafloor. Intermediate ratios may, according to Davison and Collins, result in relatively broad, complex craters. Artemieva and Shuvalov (2002) claim water depth/bolide diameter ratios larger than 25 will have no visual bottom effects.

The Eltanin impact (Bellingshausen Sea, SE Pacific) represents the only known large meteorite impact into a deep ocean. This Pliocene event (2.15 million years ago) (Figure 1) involved an approximately 1 km diameter bolide hitting the sea where it was about 4,000 m deep, and no crater was developed on the sea floor, but projectile ejecta and debris (reworked seafloor sediments) were spread in a diameter of at least 600 km across the ocean floor (Kyte, 1988; Kyte et al., 1988; Gersonde et al., 1997; Artemieva and Shuvalov, 2002; Wünnemann and Lange, 2002). The Eltanin-disturbed zone on the seafloor is characterized by a 20–40 m thick unit of chaotically mixed sediments that most probably originated from impact-induced turbulent currents (Gersonde et al., 1997; Gersonde et al., 2002).

Detailed observations of spherule beds in Australia, South Africa, and Greenland are possible remnants of Precambrian (Archean) oceanic impact activity, but without a recognized crater. These possible Precambrian craters may have been eroded or consumed during plate tectonic activity, and presently only the spherule layers have been found (Simonson et al., 2002; Simonson and Glass, 2004). It is speculated whether these spherules could reflect large oceanic impacts, maybe comparable or larger than Eltanin of the SE Pacific.

Consequences of impact

In addition to the impact itself and associated earthquakes, several other related processes can create hazardous situations in the oceans and along the coastlines (Figure 3); here, atmospheric effects, tsunamis, and currents must be mentioned.

Several authors (e.g., Ward and Asphaug, 2000; Artemieva and Shuvalov, 2002; Shuvalov, 2002; Ward and Asphaug, 2002; Gisler et al., 2010; Glimsdal et al., 2010) have stated that tsunamis will be generated, but according to Gisler et al. (2010), the real dangers from wet impacts are mainly atmospheric: blast wave from the explosion, winds, fallout, and high temperatures over small distances. Possible contamination effects of impact should also be mentioned as suggested partly causing the K/Pg extinction tied to the Chicxulub impact (Schulte et al., 2010). The K/Pg extinction was to a large extent controlled by bolide and target area composition, water depth, and impact velocity and direction (Schulte et al., 2010).

In the Chicxulub case, a bolide possibly 10 km in diameter hit a shallow to deep sea with significant lateral bathyal variations from 100 m to probably 2,000 m (Gulick et al., 2008; Schulte et al., 2010), creating an about 180 km diameter crater. The K/Pg extinction was triggered by the Chicxulub impact and its major pollution consequences such as soot and dust distribution partly shutting down photosynthesis, thermal radiation and different aerosols, acid rain, and possible change in the ocean circulation (Pope et al., 1997; Schulte et al., 2010). It should however be noted that several effects contributed to the large biological extinction which has to be evaluated from the background of ambient environmental conditions and the existing ecosystem (Kring, 2003).

The potential hazard of meteorite impact and tsunami generation in the ocean is controversial (Ward and Asphaug, 2000; Ward and Asphaug, 2002; Korycansky and Lynett, 2005; Korycansky and Lynett, 2007; Wünnemann et al., 2007; Gisler et al., 2010) with respect to the destructive power of the large waves (tsunamis) generated. It is important to mention that several types of tsunamis/waves are generated by impacts (Matsui et al., 2002; Glimsdal et al., 2010). Recent studies demonstrate that tsunamis generated by impacts are in general less destructive than the more common tsunamis generated by, e.g., plate tectonic movements. This reflects the nonlinear shoaling of the impact tsunamis (Korycansky and Lynett, 2007; Wünnemann et al., 2007; Glimsdal et al., 2010; Gisler et al., 2010). Impact tsunami may indeed undergo strong diminution of amplitude on continental slopes and shelves (Melosh, 2003). This could be caused by wave breaking or bottom friction, reducing the amplitude of the shoreline run up, maybe to 30 % of the deepwater wave amplitude (Korycansky and Lynett, 2005).

Shuvalov (2002) carried out numerical modeling of tsunami formation, demonstrating, e.g., that the Mjølnir impact was responsible for a major disturbance of the water column. Following the impact-induced outward water surge, large amplitude tsunamis were formed, and the rapid resurge of seawater into the excavated crater transported large amounts of ejecta and crater wall material back into the crater. Glimsdal et al. (2010) modeled more specifically the tsunami generation and propagation from the Mjølnir event (Dypvik et al., 2010). Tsunamis started to develop as an undulatory bore, developing into a train of solitary waves. Waves with amplitudes exceeding 200 m were formed, but during shoaling, the waves broke rather far from the coastlines in relatively deep water. The tsunamis, however, induced strong bottom currents in the 30–90 km/h range, which caused strong reworking of bottom sediments with dramatic consequences for the marine environments.

Acknowledgments

L. Thompson kindly helped with the impact crater statistics (http://www.passc.net/EarthImpactDatabase/), while P. Claeys, V. Shuvalov, and A. J. Read commented on an earlier draft of the manuscript.

Bibliography

Artemieva, N., and Shuvalov, V., 2002. Shock metamorphism on the ocean floor (numerical simulations). *Deep-Sea Research II*, **49**, 959–968.

Clayes, P., Kiessling, W., and Alvarez, W., 2002. Distribution of Chicxulub ejecta at the Cretaceous-Tertiary boundary. In Koeberl, C., and MacLeod, K. G. (eds.), *Catastrophic Events and Mass Extinctions: Impacts and Beyond*. Boulder: Geological Society of America. Special Paper, Vol. 356, pp. 55–68.

Davison, T., and Collins, G. S., 2007. The effect of the oceans on the terrestrial crater size-frequency distribution: insight from numerical modeling. *Meteoritics and Planetary Science*, **42**, 1915–1927.

Dypvik, H., and Jansa, L., 2003. Sedimentary signatures and processes during marine bolide impacts: a review. *Sedimentary Geology*, **161**, 309–337.

Dypvik, H., and Kalleson, E., 2010. Mechanisms of late synimpact to early postimpact crater sedimentation in marine-impact structures. In Reimold, W. U., and Gibson, R. L. (eds.), *Large Meteorite Impacts and Planetary Evolution IV*. Boulder: Geological Society of America. Special Paper, Vol. 465, pp. 301–318.

Dypvik, H., Tsikalas, F., and Smelror, M., 2010. *The Mjølnir Impact Event and Its Consequences*. Heidelberg/Dordrecht/London/New York: Springer. The Springer Series Impact Studies. 318 pp.

French, B. M., 1998. *Traces of Catastrophe*. LPI Contribution, Lunar Planetary Institute, Houston, TX, 954, 120 pp.

French, B. M., and Koeberl, C., 2010. The convincing identification of terrestrial meteorite impact structures: what works, what doesn't and why. *Earth-Science Reviews*, **98**, 123–170.

Gault, D. E., and Sonett, C. P., 1982. Laboratory simulations of pelagic asteroidal impact: atmospheric injection, benthic topography, and the surface wave radiation field. In Silver, L. T., and Schultz, P. H. (eds.), *Geological Implications of Impacts of Large Asteroids and Comets on the Earth*. Boulder: Geological Society of America. Special Paper, Vol. 190, pp. 69–92.

Gersonde, R., Kyte, F. T., Bleil, U., Diekman, B., Flores, J. A., Gohl, K., Grahl, G., Hagen, R., Kuhn, G., Sierro, F. J., Volker, D., Abelmann, A., and Bostwik, J. A., 1997. Geological record and reconstruction of the Late Pliocene impact of the Eltanin asteroid in the Southern Ocean. *Nature*, **390**, 357–363.

Gersonde, R., Deutsch, A., Ivanov, B. A., and Kyte, F. T., 2002. Oceanic impacts – a growing field of fundamental science. *Deep-Sea Research II*, **49**, 951–957.

Gilbert, G. K., 1893. *The Moon's Face. A Study of the Origin of Its Features*. Washington, DC: Bulletin of Philosophical Society, Vol. 12, pp. 241–292.

Gisler, G., Weaver, R., and Gittings, M., 2010. Calculation of asteroid impacts into deep and shallow water. *Pure and Applied Geophysics*, **168**, 1187–1198.

Glimsdal, S., Pedersen, G. K., Langtangen, H. P., Shuvalov, V., and Dypvik, H., 2010. The Mjølnir tsunami. In Dypvik, H., Tsikalas, F., and Smelror, M. (eds.), *The Mjølnir Impact Events and Its Consequences*. Heidelberg/Dordrecht/London/New York: Springer, Vol. The Springer Series Impact Studies, pp. 257–271.

Gohn, G. S., Koeberl, C., Miller, K. G., and Reimold, W. U., 2009. *The ICDP-USGS Deep Drilling Project in the Chesapeake Bay Impact Structure: Results From the Eyreville Core Holes*. Boulder: The Geological Society of America. Special Paper, Vol. 458.

Gulick, S. P., et al., 2008. Importance of pre-impact crustal structure for the asymmetry of the Chicxulub impact crater. *Nature Geoscience*, **1**, 131–135.

Jansa, L., 1993. Cometary impacts into ocean: their recognition and the threshold constraint for biological extinctions. *Palaeogeography Palaeoclimatology Palaeoecology*, **104**, 271–286.

Korycansky, D. G., and Lynett, P. J., 2005. Offshore breaking of impact tsunami: the Van Dorn effect revisited. *Geophysical Research Letters*, **32**, L10608.

Korycansky, D. G., and Lynett, P. J., 2007. Run-up from impact tsunami. *Geophysical Journal International*, **170**, 1076–1088.

Kring, D., 2003. Environmental consequences of impact cratering events as a function of ambient conditions on Earth. *Astrobiology*, **3**, 133–152.

Kyte, F. T., 1988. The extraterrestrial component in marine sediments: description and interpretation. *Paleoceanography*, **3**, 235–247.

Kyte, F. T., Zhou, L., and Wasson, J. T., 1988. New evidence on the size and possible effects of a late Pliocene oceanic asteroid impact. *Science*, **241**, 63–65.

Masaitis, V. L., 2002. The middle Devonian Kaluga impact crater (Russia): new interpretation of marine setting. *Deep-Sea Research II*, **49**, 1157–1169.

Matsui, T., Imamura, F., Tajika, E., Nakano, Y., and Fujisawa, Y., 2002. Generation and propagation of a tsunami from the Cretaceous-Tertiary impact event. In Koeberl, C., and Macleod, K. G. (eds.), *Catastrophic Events and Mass Extinctions: Impacts and Beyond*. Boulder: Geological Society of America. Special Paper, Vol. 356, pp. 69–77.

Melosh, H. J., 1982. *The Mechanics of Large Meteoroid Impacts in the Earth's Oceans*. Boulder: Geological Society of America. Special Paper, Vol. 190, pp. 121–127.

Melosh, H. J., 1989. *Impact Cratering: A Geologic Process*. New York: Oxford University Press, p. 245.

Melosh, H. J., 2003. Impact-generated tsunamis: An overrated hazard. In: *34 Lunar and Planetary Science Conference*, League City, TX, # 2013.

Morrow, J. R., Sandberg, C. A., Malkowski, K., and Joachmiski, M. M., 2009. Carbon isotope chemostratigraphy and precise dating of middle Frasnian (lower Upper Devonian) Alamo Breccia, Nevada, USA. *Palaeogeography Palaeoclimatology Palaeoecology*, **282**, 105–118.

NASA, 2012. http://nssdc.gsfc.nasa.gov/planetary/planets/.

Oberbeck, V. R., Marshall, J. R., and Aggarwal, H., 1993. Impacts, tillites and the breakup of Gondwanaland. *The Journal of Geology*, **101**, 1–19.

Ormö, J., and Lindström, M., 2000. When a cosmic impact strikes the sea bed. *Geological Magazine*, **137**, 67–80.

Pope, K. O., Baines, K. H., Ocampo, A. C., and Ivanov, B. A., 1997. Energy, volatile production, and climatic effects of the Chicxulub Cretaceous/Tertiary impact. *Journal of Geophysical Research*, **102**, 21645–21664.

Pope, K. O., Ocampo, A. C., Fischer, A. G., Vega, F. J., Ames, D. E., King, D. T., Jr., Fouke, B. W., Wachtman, R. J., and Kletetschka, G., 2005. Chicxulub impact ejecta deposits in southern Quintana Roo, Mexico and central Belize. In Kenkmann, T., Hörz, F., and Deutsch, A. (eds.), *Large Meteorite Impacts III*. Boulder: Geological Society of America. Special Paper, Vol. 384, pp. 171–190.

Riis, F., Kalleson, E., Dypvik, H., Krøgli, S. O., and Nilsen, O., 2012. The Ritland impact structure, southwestern Norway. *Meteoritics and Planetary Science*, **46**, 748–761.

Schulte, P., et al., 2010. The Chicxulub asteroid impact and mass extinction at the Cretaceous -Paleogene boundary. *Science*, **327**, 1214–1218.

Shoemaker, E. M., 1962. Interpretation of lunar craters. In Kopal, Z. (ed.), *Physics and Astronomy of the Moon*. New York/London: Academic, pp. 283–359.

Shuvalov, V. V., 2002. Numerical modeling of impacts into shallow seas. In Plado, J., and Pesonen, L. J. (eds.), *Impacts in Precambrian Shields*. Heidelberg/New York: Springer. Impact Studies, pp. 323–336.

Shuvalov, V., Dypvik, H., Kalleson, E., Setså, R., and Riis, F., 2012. Modeling the 2.7 km in diameter, shallow marine Ritland impact structure. *Earth, Moon and Planets*, **108**, 175–188.

Simonson, B. M., and Glass, B. P., 2004. Spherule layers - records of ancient impacts. *Annual Review of Earth and Planetary Science*, **32**, 329–391.

Simonson, B. M., Hassler, S. W., Smit, J., and Sumner, D., 2002. How many late Archean impacts are recorded in the Hamersley basin of Western Australia? In *Lunar and Planetary Science Conference*, League City, TX, 33, #1772.

Smit, J., 1999. The global stratigraphy of the Cretaceous-Tertiary boundary impact ejecta. *Annual Review of Earth and Planetary Science*, **27**, 75–113.

Suuroja, K., Suurjoa, S., All, T., and Floden, T., 2002. Kärdla (Hiiumaa Island, Estonia) – the buried and well preserved Ordovician marine impact structure. *Deep Sea Research II*, **49**, 1121–1144.

Task Force Group, 2000. *Report of the Task Force on Potentially Hazardous Near-Earth Objects*. London: Information Unit, British National Space Centre. 56 pp.

Ward, S. N., and Asphaug, E., 2000. Asteroid impact tsunami: a probabilistic hazard assessment. *Icarus*, **145**, 64–78.

Ward, S. N., and Asphaug, E., 2002. Impact tsunami – Eltanin. *Deep-Sea Research II*, **49**, 1073–1079.

Warme, J. E., 2004. The many faces of the Alamo impact breccia. *Geotimes*, **49**, 26–29.

Wünnemann, K., and Lange, M. A., 2002. Numerical modelling of impact induced modifications of the deep-sea floor. *Deep-Sea Research II*, **49**, 969–981.

Wünnemann, K., Weiss, R., and Hofmann, K., 2007. Characteristics of oceanic impact-induced large water waves – re-evaluation of the tsunami hazard. *Meteoritics and Planetary Science*, **42**, 1893–1903.

Cross-references

Events
Tsunamis

MARINE MICROFOSSILS

Jakub Witkowski[1], Kirsty Edgar[2], Ian Harding[3], Kevin McCartney[4] and Marta Bąk[5]
[1]Institute of Marine and Coastal Sciences, University of Szczecin, Szczecin, Poland
[2]School of Earth Sciences, University of Bristol, Clifton, UK
[3]Ocean and Earth Science, National Oceanography Centre, University of Southampton, Waterfront Campus, Southampton, UK
[4]Department of Environmental Studies and Sustainability, University of Maine at Presque Isle, Presque Isle, ME, USA
[5]Faculty of Geology, Geophysics and Environmental Protection, AGH University of Science and Technology, Kraków, Poland

Synonyms

Fossil microbiota

Definition

Marine microfossils are fossils of small size (usually <1 mm) that need to be studied by microscopy to deduce the morphological characteristics of the organism or skeletal element concerned.

Introduction

Marine microfossils are produced by a variety of microorganisms that may be auto or heterotrophic or planktic or benthic. Although most marine microfossils are body fossils of single-celled organisms (protists), some are small-sized elements of macrobiota, e.g., conodont elements or sponge spicules. The vast majority of marine microfossils possess permineralized skeletons that are composed of one of several common skeletal materials such as calcium carbonate (foraminifera (Figure 1), calcareous nannofossils), silica (radiolarians (Figure 2), diatoms (Figure 3), silicoflagellates, ebridians), organic compounds (acritarchs, chitinozoa, dinoflagellates (Figure 4), scolecodonts), or calcium phosphate (conodonts). Obviously, groups that lack mineralized or refractory skeletal elements have a considerably lower potential for fossilization.

History of study

The discipline of micropaleontology was founded by Christian Gottfried Ehrenberg (1795–1876), a German microscopist who was the first person to systematically describe microfossils (including diatoms, dinoflagellates,

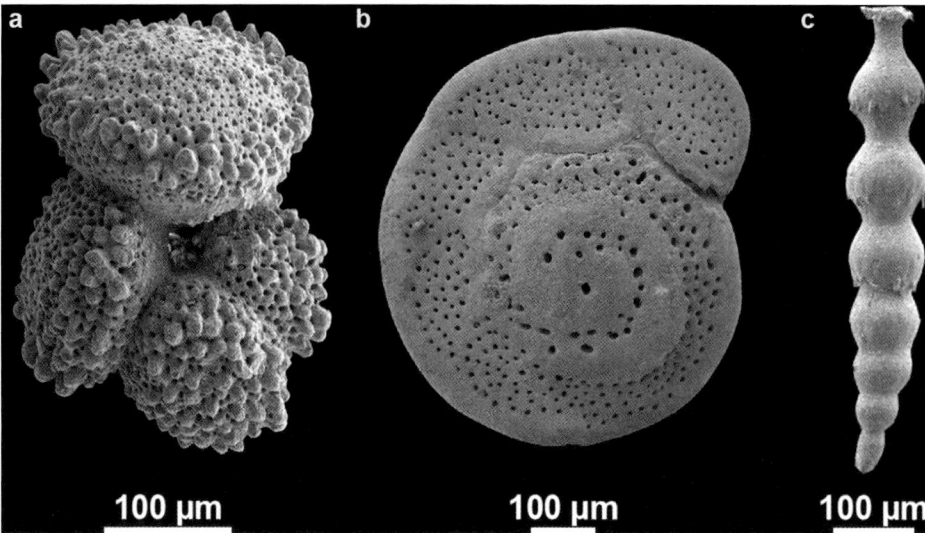

Marine Microfossils, Figure 1 Scanning electron micrographs of the calcite tests of selected planktic (a) and benthic (b, c) foraminifera. (a) *Acarinina*, (b) *Cibicidoides*, (c) *Stilostimella*.

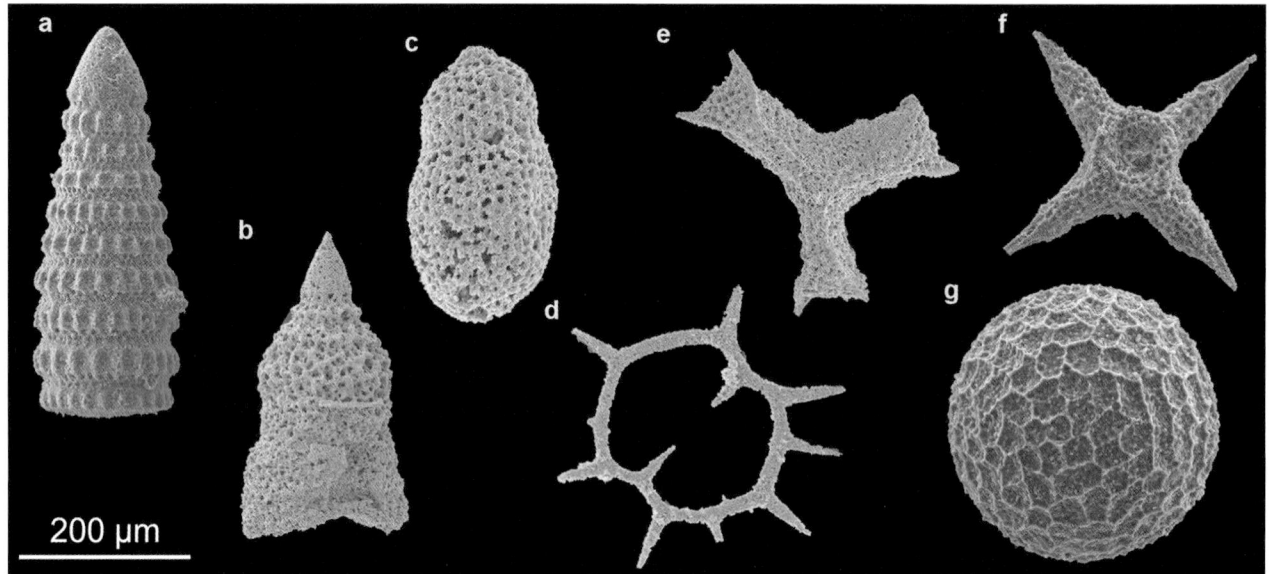

Marine Microfossils, Figure 2 Scanning electron micrographs showing Cretaceous radiolarian skeleton variability. Genera: (a) *Pseudodictyomitra*, (b) *Torculum*, (c) *Diacanthocapsa*, (d) *Acanthocircus*, (e) *Halesium*, (f) *Crucella*, (g) *Archaeocenosphaera*.

foraminifera, radiolarians, and silicoflagellates) publishing nearly 400 scientific papers on fossil and recent microorganisms during his career (Tanimura et al., 2009). Other prominent early workers in the field of marine micropaleontology include Ernst Heinrich Haeckel (1834–1919), a pioneering radiolarian worker and creator of the famous illustrative work "Art Forms in Nature" (see Tanimura et al., 2009), and Georges Deflandre (1897–1973), whose remarkably rich publication record includes papers on nearly every known microfossil group (Evitt, 1975).

Early marine micropaleontologists tended to have a descriptive, taxonomic approach to their study. However, with the realization that fossils have great utility in constraining the relative age of sediments came an increased awareness of possible microfossil applications in biostratigraphy, paleoceanography, geochemistry, and even nanotechnology (e.g., Gebeshuber and Crawford, 2006).

The study of microfossils shifted from the occupation of a narrow group of scholars into one of the basic

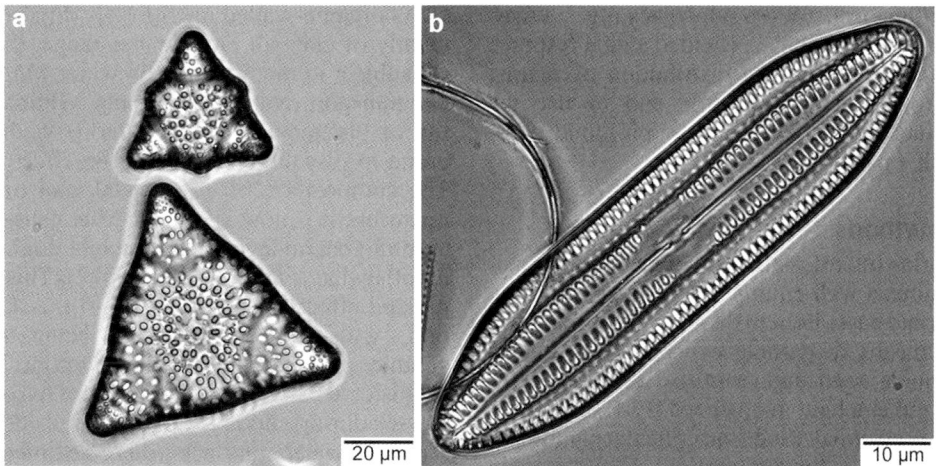

Marine Microfossils, Figure 3 Examples of main morphological groups in diatoms, light micrographs: (a) centric diatom (*Medlinia*); (b) pennate diatom (*Oestrupia*).

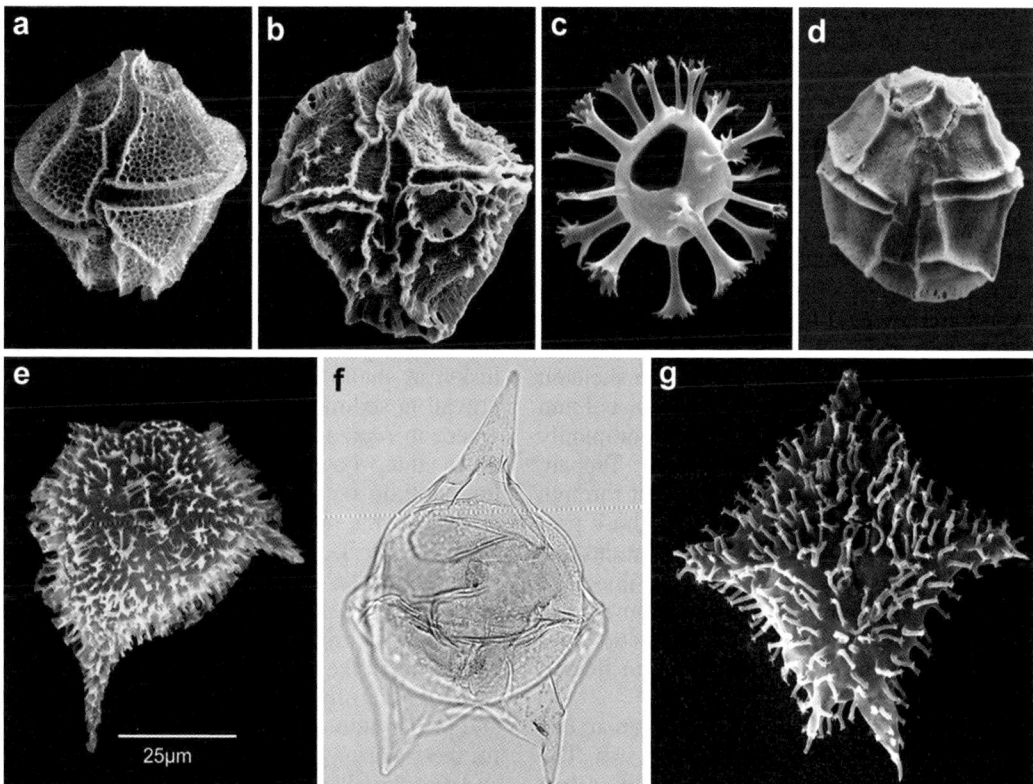

Marine Microfossils, Figure 4 Examples of modern (a) and fossil (b–g) dinoflagellates. (a–e, g) are SEM micrographs; (f) is a light micrograph. All micrographs to approximately the same scale. (a) *Gonyaulax*, (b) *Cribroperidinium*, (c) *Kleithriasphaeridium*, (d) *Meiourogonyaulax*, (e) *Pseudoceratium*, (f) *Deflandrea*, (g) *Wetzelliella*.

disciplines of modern Earth science with the inception of the hugely successful international deep ocean drilling collaborations which began in the 1960s. The Deep Sea Drilling Project (DSDP, 1968–1983) and its successors, the Ocean Drilling Program (ODP, 1985–2003), Integrated Ocean Drilling Program (IODP, 2003–2013), and now the International Ocean Discovery Program (IODP, 2013–present), have collectively recovered thousands of kilometers of sediment cores from the world's oceans (("Ocean Drilling" by Kiyoshi, this volume) see www.deepseadrilling.org, www.odplegacy.org, www.iodp.org). These sediments have provided a vital resource for geologists to understand the distribution of marine microfossils through geological time and to use the remains of marine microbiota to answer profound questions about Earth's past.

Summary of methods

The study of marine microfossils, termed marine micropaleontology, requires a well-equipped laboratory to safely handle the acids and bases used to dissolve large quantities of rock. Some means of particle concentration (microsieves, heavy liquids, centrifuges) are also desirable. Most of all, however, either a light microscope (LM) or a strong binocular scope is essential. Since the optical resolution of both these instruments is limited, marine microfossils are commonly examined by scanning electron microscope (SEM), transmission electron microscope (TEM), or other sophisticated microscopy techniques including computed tomography (CT) scanning and synchrotron analysis that permit 3D internal and external reconstruction of fossils (Cunningham et al., 2014; Schmidt et al., 2013).

Distribution patterns

The occurrence of marine microfossils in sediments is largely dependent on the distribution of living populations and taphonomic factors. The distribution of marine microbiota may be controlled by the availability of nutrients, sunlight, water temperature, oxygen, and other factors. Following the death of a microorganism, a complex series of processes is initiated that may eventually allow its incorporation into sediment. For instance, after the death of a planktic diatom cell (Figure 3), the skeleton (termed "frustule") settles through the water column toward the seabed. Since marine waters are commonly undersaturated with respect to silica (e.g., Tappan, 1980), diatom frustules often dissolve in transit through the water column, and thus only a fraction of those produced reach the deep ocean. Of those diatom frustules that avoid dissolution during settling, a large portion dissolves at the water-sediment interface (see summary in Witkowski et al., 2012). Thus, only a minor fraction of a planktic diatom assemblage is incorporated into marine sediments (e.g., Zielinski and Gersonde, 1997).

Calcareous microfossils (Figure 1) in particular can undergo dissolution at various depths in the ocean. The carbonate compensation depth (CCD) is an important threshold within the ocean where the arrival of carbonate particles such as calcareous microorganism is balanced by dissolution; in practical terms, calcium carbonate is not deposited and/or preserved in large abundances below the CCD, at ~4.6 km in modern oceans (Pälike et al., 2012). Thus, modern benthic calcareous microbiota such as foraminifers cannot dwell in the deep ocean below ~4.6 km, and planktic calcareous microorganisms that inhabit surface waters will not be incorporated into deep-sea sediments deposited below the CCD.

As organic-walled microfossils (Figure 4) are predominantly of clay-silt particle size range, their assemblages are subject to modification by winnowing during sinking and transport by bottom currents. While one of the benefits of organic-walled microfossils is that they are not prone to dissolution, they can be affected by oxidation. The composition of the skeletal wall of organic-walled microfossils is now known to differ markedly, even down to intra-generic level within such biological groups as the dinoflagellates (Bogus et al., 2012). This can have a significant effect on the resistance of the skeleton to oxidation, with most peridinioids being relatively more volatile and the gonyaulacoids more refractory (e.g., Bogus et al., 2014). Thus, exposure to oxygenated water masses during sinking, and/or long residence times at the sediment-water interface where sedimentation rates are low, will bias the composition of an organic-walled microfossil assemblage recovered from a sediment sample (e.g., Zonneveld et al., 1997, 2001).

The interplay between the factors described above controls the distribution of marine microbiota in sediments and the fate of microfossils that underlies the zonal distribution of pelagic sediments in the oceans ("Deep-sea Sediments" by Lyle, this volume). Generally speaking, wherever the terrigenous sediment input is low and nutrient supply high enough to sustain planktic microbiota in ocean surface waters, biogenic pelagic sediments will form on the seabed. At low latitudes, sediments are dominated by calcareous microfossils, predominantly foraminifera (forming the so-called Globigerina ooze that blankets much of the seafloor) and calcareous nannofossils, but potentially also the highly dissolution-susceptible pteropods (a group of tiny aragonitic mollusks) at shallow sites. Siliceous microfossils generally prevail in sediments deposited at high latitudes and elsewhere in regions of enhanced upwelling. This does not mean that diatoms or radiolarians do not occur in low-latitude pelagic sediments but rather that they are diluted by the more abundant calcareous taxa favored in warmer waters. Further, high alkalinity of carbonate-rich sediments ultimately boosts silica dissolution. On the other hand, cooler waters of high-latitude seas are more favorable for silica precipitation, which is why pelagic sediments in the Southern Ocean are impoverished in calcareous microfossils. In locations where the abundance and preservation of mineralized microfossils are depleted for these reasons, organic-walled forms such as acritarchs and dinoflagellates may well be the main groups of marine microfossils preserved in the sediment record and thus available for study in sediments deposited in high-latitude seas (e.g., Eldrett et al., 2004). These patterns are one of the main drivers of microfossil paleobiogeography through Earth history.

Marine microfossils through geological time

Earth's fossil microbiota spans some ~3.2 billion years (gyr) (Javaux et al., 2010) or longer – see the controversy surrounding the purported "microfossils" in the ~3.5 gyr Apex Chert (e.g., Schopf, 1993; Marshall et al., 2011). The acritarchs, an important microfossil group of uncertain and perhaps polyphyletic systematic position, first occur in rocks from the Meso-/Neoproterozoic transition (Knoll et al., 2006) and are the earliest age-diagnostic microfossils, used for correlation in upper Proterozoic and lower Paleozoic sediments (Mendelson, 1993).

Radiolarians and foraminifera (Figures 1 and 2), both extant groups, are known from the Cambrian onward (Casey, 1993; Boersma, 1998) and are important biostratigraphic indicators. For instance, the fusulinid and schwagerinid benthic foraminifera commonly grew to large sizes and underwent a particularly rapid evolution during the Carboniferous and Permian, which makes them useful zonal markers for the younger Paleozoic (Davydov et al., 2012; Henderson et al., 2012). However, the toothlike conodonts, which range from Cambrian through Triassic and are believed by many to represent the feeding apparatus of early chordates (Donoghue et al., 2000), provide the primary means of global age correlation of Paleozoic strata. Simple vaselike chitinozoa, which are presumed to represent an ontogenic stage (i.e., eggs) in the life cycle of a cryptic animal group (Paris and Nolvak, 1999), first appeared in the Ordovician, with morphologically more complex forms evolving through the Paleozoic before the extinction of the group in the late Devonian (Grahn and Paris, 2010). Scolecodonts – fossil worm jaws – are a minor group which occur from the Paleozoic onward but are of little biostratigraphic use despite being found most commonly in lower Paleozoic rocks. Another biostratigraphically important microfossil lineage in Paleozoic and younger marine sediments is the ostracods, first identified in the Ordovician. Both the end-Permian and Triassic brought about major turnovers in oceanic microbiota which led to the development of three prominent phytoplankton groups during the Mesozoic, calcareous nannoplankton, dinoflagellates (Figure 4), and diatoms (Figure 3) (Falkowski et al., 2004), and the rise of the zooplankton – planktic foraminifera (Hart et al., 2003). Although not as significant in today's oceans, other microfossil groups that appeared later in the Mesozoic include silicoflagellates (McCartney et al., 2014) and ebridians (Hoppenrath and Leander, 2006).

Microfossils in biostratigraphy

Microfossils can occur in marine sediments in large abundances (yielding up to several billion individuals per gram of dry mass, e.g., Davies et al., 2009) and have widespread distributions and relatively rapid evolutionary rates making them ideal targets for dividing up geological time and correlating spatially disparate sediments. The advent of deep-sea drilling in recent decades allowed microfossils rather than macrofossils to form the basis for age determination in marine sedimentary sequences. However, since most chronostratigraphic units were defined based on onshore exposures where microfossils are absent or sparse (Bolli et al., 1985), one of the major challenges of modern micropaleontology has been to tie emerging microfossil biostratigraphic zonations to the evolving chronostratigraphic framework (e.g., Cande and Kent, 1995; Gradstein et al., 2012). For planktic foraminifera and calcareous nannofossils, most bioevents are now tied directly to the geomagnetic polarity time scale (GPTS; Berggren et al., 1995; Wade et al., 2011), and many dinoflagellate bioevents are also so calibrated (e.g., Eldrett et al., 2004; Brinkhuis et al., 2003). However, in the longer term the aim is to fully calibrate microfossil-based data to the astronomical time scale, which provides the potential for much higher temporal resolution than that based on magnetics (e.g., Wade et al., 2011).

Here we focus on two examples from the Cenozoic to illustrate how calcareous and siliceous microfossils are used as zonal markers in low- and high-latitude sites.

Low-latitude Eocene zonation (56.0–33.9 million years ago)

Planktic foraminiferal and calcareous nannofossil biostratigraphy provide the basis for age control in most low-latitude Cenozoic marine sediments. There are currently 16 planktic foraminiferal biostratigraphic zones recognized in Eocene low-latitude deep-sea sediments, prefixed "E" by Berggren and Pearson (2005) (Figure 5). Magnetochronologic calibration is available for all of these zones, and the five uppermost zones are also astronomically calibrated (Wade et al., 2011). Low-latitude siliceous microfossil schemes are considerably less advanced than those based on calcareous and other microfossils largely due to a preservational bias acting against siliceous microfossils at low latitudes. However, the Eocene contains 13 radiolarian zones that are tied directly to the GPTS (Nigrini et al., 2005). In contrast, diatom biozones are typically only indirectly calibrated to the GPTS via correlation to planktic foraminifer, calcareous nannofossil, and/or radiolarian zones. Recently, Barron et al. (2015) have provided considerable amendments to the preexisting low-latitude diatom zonation scheme of Fenner (1984, 1985), and 13 zones are identified. But some diatom data remain uncertain, and no information is currently available to evaluate diachrony of key marker species, such as with foraminifera (e.g., Edgar et al., 2010) (Figure 5).

High-latitude zonation for the neogene

Siliceous microfossils offer better age control than their calcareous counterparts in high-latitude marine records, particularly in the Southern and equatorial Pacific Oceans which are characterized by widespread deposition of silica-rich sediments (Ragueneau et al., 2000) throughout the Cenozoic, enabling the development of detailed zonal schemes based on diatom bioevents (e.g., Harwood and Maruyama, 1992; Censarek and Gersonde, 2002; Kamikuri et al., 2012; Moore et al., 2015). Using data from 32 Neogene deep-sea sites in the Southern Ocean, Cody et al. (2008) compiled a computer-aided quantitative synthesis of 116 diatom first and last occurrences relative

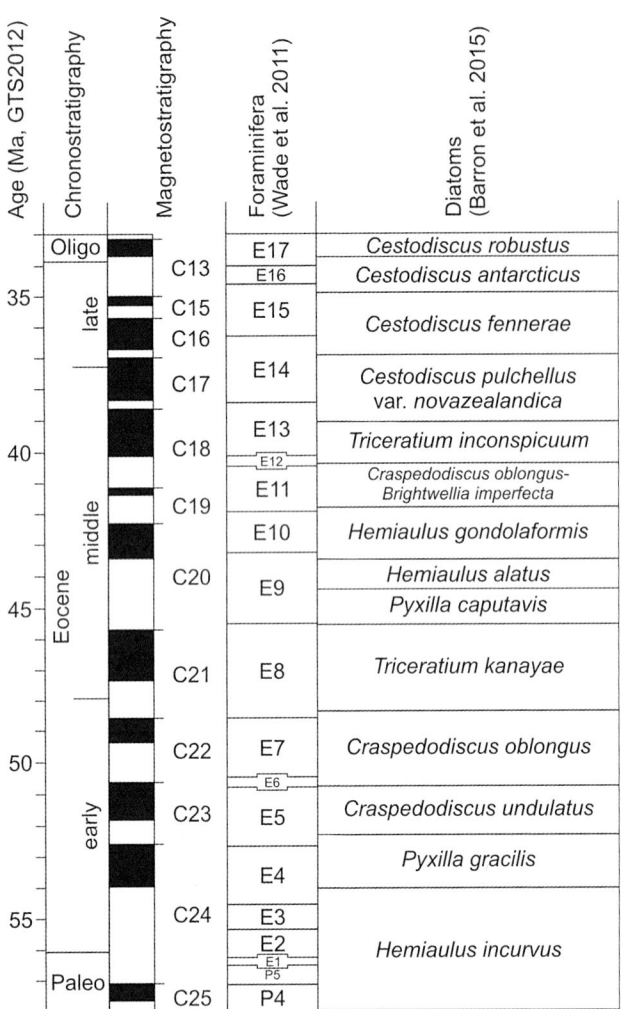

Marine Microfossils, Figure 5 Comparison of Eocene foraminifer- and diatom-based biostratigraphic zonations presented versus the 2012 geological time scale (Gradstein et al., 2012, GTS 2012) and chronostratigraphic and magnetostratigraphic divisions (modified from Barron et al. (2015), using data from Wade et al. (2011) and Barron et al. (2015)).

to magnetostratigraphy and radiometrically dated volcanic ash horizons. The outcome is an improved temporal resolution of the Antarctic diatom biostratigraphy, on the order of ~100,000 years, a resolution inconceivable for the diatoms in the Paleogene low latitudes.

Paleoenvironmental utility of microfossils

Marine microfossils are one of the main tools used to reconstruct Earth's past climate and oceans. This can be achieved in various ways, ranging from simple observations of changes in taxonomic composition or the absolute abundance of microfossil assemblages to infer qualitative environmental changes to more quantitative transfer function approaches or chemical analyses. Below are several notable examples from the recent literature.

Elemental composition and stable isotope records

Currently, one of the major applications of microfossils is to utilize the chemical composition of body fossils to reconstruct various paleoceanographic parameters, e.g., temperature, salinity, etc., and/or the paleoecology of the organisms themselves. $\delta^{13}C$ and $\delta^{18}O$ values are routinely measured in marine carbonates: either from foraminiferal tests (typically from selected specimens of a single species) or from bulk sediment. The $\delta^{13}C$ of carbonates is a function of the $\delta^{13}C$ of dissolved inorganic carbon in seawater and reflects changes in the carbon cycle, e.g., biological pump strength or removal of isotopically light carbon from the ocean-atmosphere system by enhanced organic carbon burial. $\delta^{18}O$ fluctuations can be interpreted in terms of changes in salinity, seawater temperature, or continental ice sheet volume. Long-term compilations of foraminiferal stable isotope records (e.g., Zachos et al., 2001, 2008; Cramer et al., 2009) are standard references for paleoceanographic studies of Cenozoic sedimentary successions. For instance, time series of Mg/Ca ratios can be used in paleothermometry; a recent example is a study by Bohaty et al. (2012), who generated Mg/Ca ratios in conjunction with $\delta^{18}O$ records to deconvolve the relative contribution of ocean cooling and ice growth across the Eocene-Oligocene transition and constrain the timing of the inception of Antarctic ice sheets.

Siliceous microfossils are typically less widely utilized in geochemical studies, but in recent years, a suite of new silica-based proxies have become available for reconstructing past paleoceanographic conditions. Two notable developments are diatom $\delta^{18}O$, typically generated from young, high-latitude sediments that are commonly characterized by pristine diatom preservation (e.g., Swann and Leng, 2009), and $\delta^{30}Si$, which can be generated from diatom valves, radiolarian tests, or sponge spicules (Egan et al., 2012) and provides insight on silicic acid utilization, and data from various levels in the water column, e.g., from sponges and diatoms, can be used to interpret productivity variations (Egan et al., 2013).

Paleoproductivity estimates

Photoautotrophic microfossil groups, e.g., dinoflagellates and diatoms, are well suited to reconstructing paleoproductivity provided that no significant diagenetic alteration of assemblages has taken place in sediments. Aside from being used as salinity and oxygenation indices and for determining onshore-offshore trends (e.g., Sluijs et al., 2005), organic-walled microfossils such as dinoflagellates can also provide a more useful productivity measure in neritic environments where a major portion of modern marine primary productivity originates. For instance, the peridinioid/gonyaulacoid ratio (P/G ratio, the proportion of cyst species produced by photoautotrophic versus heterotrophic dinoflagellates), has been used to reconstruct relative nutrient availability fluctuations across past abrupt global climatic events (see Sluijs et al., 2005; Bijl et al., 2010). Principal component analysis of dinocyst distributions in the Oligocene has also been used to relate productivity changes to the structure of the water column (stratification), eutrophication, and oxygen depletion (e.g., Pross and Schmiedl, 2002). Fluctuations of biogenic silica content in sediment, usually measured spectrophotometrically (e.g., Iwasaki et al., 2014), are also used as a paleoproductivity proxy and comprise photoautotrophic diatoms and silicoflagellates and heterotrophic radiolarians and ebridians that feed on primary producers. Variations in diatom accumulation rates can be indicative of periods of elevated diatom production and flux (e.g., Witkowski et al., 2014). Owing to their unique life cycle characteristics, including resting spore formation in some planktic taxa, diatoms offer a number of ways to infer past variations in nutrient availability. For instance, Davies et al. (2009), based on laminated high-latitude late Cretaceous sediments including alternating resting spore and vegetative valve laminae, have successfully reconstructed seasonal fluctuations in diatom production and flux, thus testifying to ice-free summer conditions in the late Mesozoic Arctic.

Conclusions

Although the study of marine microfossils has so many facets, covering some two thirds of Earth's history and so many systematic groups, its main value is in the applied aspects, particularly in biostratigraphy and paleoceanography. The benefits of marine microfossils have been fully appreciated only recently, but scientific interest in them is certain to continue for many decades to come.

Bibliography

Barron, J. A., Stickley, C. E., and Bukry, D., 2015. Paleoceanographic, and paleoclimatic constraints on the global Eocene diatom and silicoflagellate record. *Palaeogeography Palaeoclimatology Palaeoecology*, **422**, 85–100.

Berggren, W. A., and Pearson, P. N., 2005. A revised tropical to subtropical paleogene planktic foraminiferal zonation. *Journal of Foraminiferal Research*, **35**, 279–298.

Berggren, W. A., Kent, D. V., Swisher, III, C. C., and Aubry, M.-P., 1995. A revised Cenozoic geochronology and chronostratigraphy. Geochronology, time scales, and global stratigraphic correlation. SEPM special publication, 54.

Bijl, P. K., Houben, A. J. P., Schouten, S., Bohaty, S. M., Sluijs, A., Reichart, G.-J., Sinninghe Damsté, J. S., and Brinkhuis, H., 2010. Transient middle Eocene atmospheric CO_2 and temperature variations. *Science*, **330**, 819–821.

Boersma, A., 1998. Foraminifera. In Haq, B. U., and Boersma, A. (eds.), *Introduction to Marine Micropaleontology*. New York: Elsevier, pp. 19–78.

Bogus, K., Harding, I. C., King, A., Charles, A. J., Zonneveld, K., and Versteegh, G., 2012. The composition and diversity of dinosporin in species of the *Apectodinium* complex (Dinoflagellata). *Review of Palaeobotany and Palynology*, **183**, 21–31.

Bogus, K., Mertens, K. N., Lauwaert, J., Harding, I. C., Vrielinck, H., Zonneveld, K. A. F., and Versteegh, G. J. M., 2014. Differences in the chemical composition of organic-walled dinoflagellate resting cysts from phototrophic and heterotrophic dinoflagellates. *Journal of Phycology*, **50**, 254–266.

Bohaty, S. M., Zachos, J. C., and Delaney, M. L., 2012. Foraminiferal Ca/Mg evidence for Southern Ocean cooling across the Eocene-Oligocene transition. *Earth and Planetary Science Letters*, **317–318**, 251–261.

Bolli, H. M., Saunders, J. B., and Perch-Nielsen, K., 1985. Comparison of zonal schemes for different fossil groups. In Bolli, H. M., Saunders, J. B., and Perch-Nielsen, K. (eds.), *Plankton Stratigraphy*. Cambridge: Cambridge University Press, Vol. 1, pp. 3–10.

Brinkhuis, H., Sengers, S., Sluijs, A., Warnaar, J., and Williams, G. L., 2003. Latest Cretaceous–earliest Oligocene and Quaternary dinoflagellate cysts, ODP Site 1172, East Tasman Plateau. In Exon, N. F., Kennett, J. P., Malone, M. J. (eds.), *Proceedings of the Ocean Drilling Program, Scientific Results*. College Station, TX, Vol. 189, pp. 1–48 (Online).

Cande, S. C., and Kent, D. V., 1995. Revised calibration of the geomagnetic polarity timescale for the Late Cretaceous and Cenozoic. *Journal of Geophysical Research*, **100**(B4), 6093–6095.

Casey, R. E., 1993. Radiolaria. In Lipps, J. H. (ed.), *Fossil Prokaryotes and Protists*. Boston: Blackwell Scientific Publications, pp. 249–284.

Censarek, B., and Gersonde, R., 2002. Miocene diatom biostratigraphy at ODP sites 689, 690, 1088, 1092 (Atlantic sector of the Southern Ocean). *Marine Micropaleontology*, **45**, 309–356.

Cody, R. D., Levy, R. H., Harwood, D. M., and Sadler, P. M., 2008. Thinking outside the zone: high-resolution quantitative diatom biochronology for the Antarctic Neogene. *Palaeogeography Palaeoclimatology Palaeoecology*, **260**, 92–121.

Cramer, B. S., Toggweiler, J. R., Wright, J. D., Katz, M. E., and Miller, K. G., 2009. Ocean overturning since the Late Cretaceous: inferences from a new benthic foraminiferal isotope compilation. *Paleoceanography*, **24**, PA4216.

Cunningham, J.A., Rahman, I.A., Lauterschlanger, S., et al., 2014. A virtual world of paleontology. Trends in Ecology & Evolution, 29, 347–357.

Davies, A., Kemp, A. E. S., and Pike, J., 2009. Late Cretaceous seasonal ocean variability from the Arctic. *Nature*, **460**, 254–259.

Davydov, V. I., Korn, D., and Schmitz, M. D., 2012. The Carboniferous period. In Gradstein, F. M., Ogg, J. G., Schmitz, M. D., and Ogg, G. (eds.), *The Geologic Timescale 2012*. Amsterdam: Elsevier, pp. 603–651.

Donoghue, P. C. J., Forey, P. L., and Aldridge, R. J., 2000. Conodont affinity and chordate phylogeny. *Biological Reviews*, **75**, 191–251.

Edgar, K. M., Wilson, P. A., Sexton, P. F., Gibbs, S. J., Roberts, A. P., and Norris, R. D., 2010. New biostratigraphic, magnetostratigraphic and isotopic insights into the Middle Eocene Climatic Optimum in low latitudes. *Palaeogeography Palaeoclimatology Palaeoecology*, **297**, 670–682.

Egan, K. E., Rickaby, R. E. M., Leng, M. J., Hendry, K. R., Hermoso, M., Sloane, H. J., Bostock, H., and Halliday, A. N., 2012. Diatom silicon isotopes as a proxy for silicic acid utilisation: a Southern Ocean core top calibration. *Geochimica et Cosmochimica Acta*, **96**, 174–196.

Egan, K. E., Rickaby, R. E. M., Hendry, K. R., and Halliday, A. N., 2013. Opening the gateways for diatoms primes Earth for Antarctic glaciation. *Earth and Planetary Science Letters*, **375**, 34–43.

Eldrett, J. S., Harding, I. C., Firth, J. V., and Roberts, A. P., 2004. Magnetostratigraphic calibration of Eocene-Oligocene dinoflagellate cyst biostratigraphy from the Norwegian-Greenland Sea. *Marine Micropalaeontology*, **204**, 91–127.

Evitt, W. R., 1975. Memorial to Georges Deflandre, 1897–1973. *Geological Society of America Memorials*, **5**, 1–11.

Falkowski, P. G., Katz, M. E., Knoll, A. H., Quigg, A., Raven, J. A., Schofield, O., and Taylor, F. J. R., 2004. The evolution of modern eukaryotic phytoplankton. *Science*, **305**, 354–360.

Fenner, J., 1984. Eocene-Oligocene planktic diatom stratigraphy in the low latitudes and the high southern latitudes. *Micropaleontology*, **30**, 319–342.

Fenner, J., 1985. Late Cretaceous to oligocene planktic diatoms. In Bolli, H. M., Saunders, J. B., and Perch-Nielsen, K. (eds.), *Plankton Stratigraphy*. Cambridge: Cambridge University Press, pp. 713–762.

Gebeshuber, I. C., and Crawford, R. M., 2006. Micromechanics in biogenic hydrated silica: hinges and interlocking devices in diatoms. *Proceedings of the Institution of Mechanical Engineers*, **220**(Part J), 787–796.

Gradstein, F. M., Ogg, J. G., Schmitz, M. D., and Ogg, G. M. (eds.), 2012. *The Geologic Time Scale 2012*. Amsterdam: Elsevier.

Grahn, Y., and Paris, F., 2010. Emergence, biodiversification and extinction of the chitinozoan group. *Geological Magazine*, **147**, 1–11.

Hart, M. B., Hylton, M. D., Oxford, M. J., Price, G. D., Hudson, W., and Smart, C. W., 2003. The search for the origin of the planktic Foraminifera. *Journal of the Geological Society*, **160**, 341–343.

Harwood, D. M., and Maruyama, T., 1992. Middle Eocene to Pleistocene diatom biostratigraphy of Southern Ocean sediments from the Kerguelen Plateau, Leg 120. In: Wise, S. W., Jr., Schlich, R., et al. (eds.), *Proceedings of the Ocean Drilling Program, Scientific Results*. College Station, TX, Vol. 120, pp. 683–733.

Henderson, C. M., Davydov, V. I., and Wardlaw, B. R., 2012. The Permian period. In Gradstein, F. M., Ogg, J. G., Schmitz, M. D., and Ogg, G. (eds.), *The Geologic Timescale 2012*. Amsterdam: Elsevier, pp. 653–679.

Hoppenrath, M., and Leander, B. S., 2006. Ebriid phylogeny and the expansion of the Cercozoa. *Protist*, **157**, 279–290.

Iwasaki, S., Takahashi, K., Ogawa, Y., Uehara, S., and Vogt, C., 2014. Alkaline leaching characteristics of biogenic opal in Eocene sediments from the central Arctic Ocean: a case study in the ACEX cores. *Journal of Oceanography*, **70**, 241–249.

Javaux, E. J., Marshall, C. P., and Bekker, A., 2010. Organic-walled microfossils in 3.2-billion-year-old shallow-marine siliciclastic deposits. *Nature*, **463**, 934–939.

Kamikuri, S., Moore, T. C., Ogane, K., Suzuki, N., Pälike, H., and Nishi, H., 2012. Early Eocene to early Miocene radiolarian biostratigraphy for the low-latitude Pacific Ocean. *Stratigraphy*, **9**, 77–108.

Kennett, J. P., 1982. *Marine Geology*. New Jersey: Prentice-Hall.

Knoll, A. H., Javaux, E. J., Hewitt, D., and Cohen, P., 2006. Eukaryotic organisms in Proterozoic oceans. *Philosophical Transactions of the Royal Society B*, **361**, 1023–1038.

Marshall, C. P., Emry, J. R., and Olcott Marshall, A., 2011. Haematite pseudomicrofossils present in the 3.5-billion-year-old Apex Chert. *Nature Geoscience*, **4**, 240–243.

McCartney, K., Witkowski, J., and Harwood, D. M., 2014. New insights into skeletal morphology of the oldest known silicoflagellates: *Variramus, Cornua* and *Gleserocha* gen. nov. Revue de Micropaléontologie 57, pp. 75–91.

Mendelson, C. V., 1993. Acritarchs and prasinophytes. In Lipps, J. H. (ed.), *Fossil Prokaryotes and Protists*. Boston: Blackwell Scientific Publications, pp. 77–104.

Moore, T. C., Jr., Kamikuri, S., Erhardt, A. M., Baldauf, J., Coxall, H., and Westerhold, T., 2015. Radiolarian stratigraphy near the Eocene-Oligocene boundary. *Marine Micropaleontology*, **116**, 50–62.

Nigrini, C., Sanfilippo, A., and Moore, Jr., T. J., 2005. Cenozoic radiolarian biostratigraphy: a magnetobiostratigraphic chronology of Cenozoic sequences from ODP Sites 1218, 1219 and 1220, Equatorial Pacific. In: Wilson, P. A., Lyle, M., Firth, J. V. (eds.), *Proceedings of the Ocean Drilling Program, Scientific Results*. College Station, TX, Vol. 199, pp. 1–56 (Online).

Pälike, H., Norris, R. D., Herrle, J. O., et al., 2006. The heartbeat of the Oligocene climate system. *Science*, **314**, 1894–1899.

Pälike, H., Lyle, M. W., Nishi, H., et al., 2012. A Cenozoic record of the equatorial Pacific carbonate compensation depth. *Nature*, **488**, 609–614.

Paris, F., and Nolvak, J., 1999. Biological interpretation and paleobiodiversity of a cryptic fossil group: the "chitinozoan animal". *Geobios*, **32**, 315–324.

Pross, J., and Schmiedl, G., 2002. Early Oligocene dinoflagellate cysts from the Upper Rhine Graben (SW Germany): paleoenvironmental and paleoclimatic implications. *Marine Micropaleontology*, **45**, 1–24.

Ragueneau, O., Tréguer, P., Leynaert, A., et al., 2000. A review of the Si cycle in the modern ocean: recent progress and missing gaps in the application of biogenic opal as a paleoproductivity proxy. *Global and Planetary Change*, **26**, 317–365.

Schmidt, D.N., Rayfield, E.J., Cocking, A., and Marone, F., 2013. Linking evolution and development: synchrotron radiation X-ray tomographic microscopy of planktic foraminifers. Palaeontology, 56, 741–749.

Schopf, J. W., 1993. Microfossils of the Early Archean Apex Chert: new evidence for the antiquity of life. *Science*, **260**, 640–646.

Sluijs, A., Pross, J., and Brinkhuis, H., 2005. From greenhouse to icehouse; organic-walled dinoflagellate cysts as paleoenvironmental indicators in the Paleogene. *Earth Science Reviews*, **68**, 281–315.

Swann, G. E. A., and Leng, M. J., 2009. A review of diatom $\delta^{18}O$ in palaeoceanography. *Quaternary Science Reviews*, **28**, 384–398.

Tanimura, Y., Tuji, A., Aita, Y., Suzuki, N., Ogane, K., and Sakai, T., 2009. Joint Haeckel and Ehrenberg Project "Reexamination of the Haeckel and Ehrenberg Microfossil Collections as a Historical and Scientific Legacy": a summary. *National Museum of Nature and Science Monographs*, **40**, 1–5.

Tappan, H., 1980. *Paleobiology of Plant Protists*. San Francisco: W.H. Freeman.

Wade, B. S., Pearson, P. N., Berggren, W. A., and Pälike, H., 2011. Review and revision of Cenozoic tropical planktic foraminiferal biostratigraphy and calibration to the geomagnetic polarity and astronomical time scale. *Earth-Science Reviews*, **104**, 111–142.

Witkowski, J., Bohaty, S. M., McCartney, K., and Harwood, D. M., 2012. Enhanced siliceous plankton productivity in response to middle Eocene warming at Southern Ocean ODP Sites 748 and 749. *Palaeogeography, Palaeoclimatology, Palaeoecology* **326–328**, 78–94.

Witkowski, J., Bohaty, S. M., Edgar, K. M., and Harwood, D. M., 2014. Rapid fluctuations in mid-latitude siliceous plankton production during the Middle Eocene Climatic Optimum (ODP Site 1051, western North Atlantic). *Marine Micropaleontology*, **106**, 110–129.

Zachos, J. C., Pagani, M., Sloan, L., Thomas, E., and Billups, K., 2001. Trends, rhythms, and aberrations in global climate 65 Ma to present. *Science*, **292**, 686–693.

Zachos, J. C., Dickens, G. R., and Zeebe, R. E., 2008. An early Cenozoic perspective on greenhouse warming and carbon-cycle dynamics. *Nature*, **451**, 279–283.

Zielinski, U., and Gersonde, R., 1997. Diatom distribution in the Southern Ocean surface sediments (Atlantic sector): implications for paleoenvironmental reconstructions. *Palaeogeography Palaeoclimatology Palaeoecology*, **129**, 213–250.

Zonneveld, K. A. F., Versteegh, G. J. M., and de Lange, G. J., 1997. Preservation of organic-walled dinoflagellate cysts in different oxygen regimes: a 10,000 year natural experiment. *Marine Micropaleontology*, **29**, 393–405.

Zonneveld, K. A. F., Versteegh, G. J. M., and de Lange, G. J., 2001. Palaeoproductivity and post-depositional aerobic organic matter decay reflected by dinoflagellate cyst assemblages of the Eastern Mediterranean S1 sapropel. *Marine Geology*, **172**, 181–195.

Websites

Deep Sea Drilling Project. Available online at www.deepseadrilling.org. Accessed 18 July 2015.

Ocean Drilling Program. Available online at www.odplegacy.org. Accessed 18 July 2015.

International Ocean Discovery Program. Available online at www.iodp.org. Accessed 18 July 2015.

Cross-references

Biochronology, Biostratigraphy
Deep-sea Sediments
Diatoms
Dinoflagellates
Foraminifers (Benthic)
Foraminifers (Planktonic)
Geologic Time Scale
Ocean Drilling
Paleoceanography
Palynology (Pollen, Spores, etc.)
Pteropods
Radiolarians

MARINE MINERAL RESOURCES

Sven Petersen
GEOMAR Helmholtz Centre for Ocean Research, Kiel, Germany

Synonyms

Seabed mineral resources

Definition

Marine mineral resources are accumulations of minerals that form at or below the seabed and from which metals, minerals, elements, or aggregate might be extracted as a resource. They are distinguished from energy resources such as oil, gas, or gas hydrates and living resources such as fish, although many of the metals potentially available are essential for green-energy applications.

Introduction

Earth provides natural resources including minerals and metals that are vital for human life. At present almost all of these resources are mined on land with large high-grade deposits becoming more and more difficult to find, driving industry to lower-grade sites where mining has greater environmental impacts or to greater depth. At the same time, the global demand for metals is suspected to rise further due to steady population growth, expected to reach 9 billion by the year 2050 (United Nations, 2014), and strong economic growth of countries such as China and India. The population growth may cause increasing land-use conflicts between the mining industry and the need to feed and house the growing population. In addition to the rising demand for metals, geopolitical issues can also limit the availability of metal resources. This was evident over the past years with China blocking export of "rare earth elements" from global markets awakening the media and policy. There is therefore a foreseeable risk of increasing resource supply shortages for metals that are important to the economy (DOE, 2011; European Commission, 2014). Hence, a number of countries are looking for ways to ensure secure supplies of these critical metals. In this rapidly changing global economic landscape, mining in the deep sea is one of the areas of interest not only for commercial entities but also for governments.

Legal aspects

Marine mineral resources fall under two different legal regimes depending on their location: they either occur within the legal boundaries of national jurisdiction or occur in areas beyond national jurisdictions. In terms of international law, the legal framework for deep-sea mining of mineral resources is set out in the United Nations Convention on the Law of the Sea (UNCLOS) as modified by the Part XI Implementation Agreement. UNCLOS distinguishes between "maritime zones" under the jurisdiction of coastal states (internal and archipelagic waters, territorial sea, exclusive economic zone, and continental shelf) and "areas beyond national jurisdiction," namely, international waters and the seabed beyond the continental shelves of coastal states (called *The Area* in Part XI of UNCLOS). Since it went into effect in 1994, this treaty has formed the basis for the use of marine raw materials from the seabed beyond the limits of national jurisdiction. As of March 2015, UNCLOS has been signed by 166 countries and the European Union.

With regards to marine mineral resources in areas under national jurisdiction, the coastal states have regulatory jurisdiction and can design and adopt their own legislation. The outer limit of the national jurisdiction is defined by the 200 nautical mile exclusive economic zone (EEZ), but individual states can apply to the Commission on the Limits of the Continental Shelf (CLCS) for an extension for up to 350 nautical miles or even beyond that if certain geological conditions are met (extended continental shelf as a juridical term not as a geological term).

All rights to prospect (search without exclusive rights), exploration (search with exclusive rights), and deep-sea mining of mineral resources in The Area are managed and promoted by the International Seabed Authority (ISA) that was established in Kingston (Jamaica) in 1994 based on UNCLOS. The regulatory regime for mineral resources of the deep sea is not complete. Regulations on prospecting and exploration in The Area have been adopted for manganese nodules, Co-rich ferromanganese crusts, and seafloor massive sulfides (SMS) (ISA, 2013a). The regulations for deep-sea mining (exploitation) of manganese nodules are currently being developed (ISA, 2013b).

The seabed: a source of raw materials for humankind?

The seabed is already an important source of commodities for humankind. Sand and gravel as well as oil and gas have been mined from the sea for decades. Moreover, the so-called placer deposits of diamonds have been extracted off the coast of South Africa and Namibia for a long time in addition to deposits of tin, titanium, and gold along the coasts of Africa, Asia, and Australia (Cronan, 1992, 2000; Rona, 2003, 2008). These placer deposits are accumulations of minerals with a high density and chemical resistance to weathering that form along beaches by gravity settling due to wave or current movement.

Obtaining raw materials from the sea is therefore not a new practice. Interest is also growing with respect to *phosphorite* occurrences with projects being considered within the EEZ of New Zealand, Mexico, Namibia, and the USA. Marine phosphorites are sedimentary rocks that contain high concentrations of phosphate and form along continental margins with oceanic upwelling (Burnett and Riggs, 1990).

Raw materials are not confined to shallow water. Potential mineral resources from the deep sea currently investigated are *manganese nodules*, *cobalt-rich manganese crusts*, as well as *massive sulfides* (SPC, 2013; Hein et al., 2013; Fouquet and Lacroix, 2014). Details of the formation, characteristics, and resource potential of all these commodities can be found in the respective entries of this encyclopedia. The raw materials from the deep sea are commonly described as renewable resources. This is misleading since, for instance, the metals in manganese nodules and cobalt-rich ferromanganese crusts take millions of years to accumulate. Even the considerably faster growing SMS take thousands of years to accumulate to economically interesting amounts.

SMS, also known as black smoker deposits, are occurrences of metal-bearing minerals that form on and below the seabed as a consequence of the interaction of seawater with a heat source (magma) in the sub-seafloor region of volcanically active oceanic spreading centers and along volcanic arcs (Hannington et al., 2005). These occurrences are commonly associated with "oasis of life" harboring chemosynthetic faunal communities. By far the majority of SMS occurrences presently known are small, 3-dimensional bodies that can contain metals such as copper, zinc, gold, and silver. Other trace elements that are important for a variety of industry uses (Bi, Ga, Ge, In, Te) can be enriched at certain sites and are commonly considered as possible important by-products. Known SMS deposits at the seafloor rarely exceed a few million tonnes of metal (Hannington et al., 2011) with the exception of the metalliferous muds in the Atlantis II Deep of the Red Sea, which formed under different conditions and was estimated to contain 90+ million tonnes of ore (Nawab, 1984). New technologies to explore for inactive sites away from the ridge axis are being developed in order to investigate the full ocean potential for sulfides (Cathles, 2011; Hannington, 2011). As a result, there is considerable interest in exploration both within the EEZ of coastal states and The Area (see below, ECORYS, 2014). The technology for mining these deposits is currently being built, and first deep-sea tests were performed by the Japanese in their coastal waters (Masuda et al., 2014). The Canadian company *Nautilus Minerals* is constructing deep-sea SMS collectors for its Solwara 1 deposit in the territorial waters of Papua New Guinea. This site could become the first deep-sea mine site for SMS, as a mining license was issued by the PNG government in 2011. Current planning foresees the start of mining operations in 2018.

Manganese nodules occur widely on the vast, sediment-covered abyssal plains at depths of about 4,000–6,500 m. They are mineral concretions made up largely of manganese and iron that form around a hard nucleus and incorporate metals from the sediment and seawater. As manganese nodules form directly on the seafloor, these deposits can be regarded as a 2-dimensional resource. The greatest concentrations of metal-rich nodules occur in the Clarion-Clipperton Zone (CCZ; Hein and Koschinsky, 2014), which extends from Hawaii to Mexico. Nodules are also concentrated in the Peru Basin, near the Cook Islands, and at abyssal depths in the Indian and Atlantic Oceans. Manganese and iron are the principal metals in polymetallic nodules, but the metals of greatest economic interest are nickel, copper, and cobalt that combined can reach between 2.5 and 3 wt% (Hein and Koschinsky, 2014). In addition, there are enrichments of other valuable metals, such as molybdenum, vanadium, titanium, the rare earth elements (REE), and lithium, that have industrial importance in many high-tech and green-tech applications and can possibly be recovered as by-products once appropriate processing techniques are developed. The geochemistry of manganese nodules varies from ocean to ocean, and details are given in the respective entry of this encyclopedia.

Manganese nodules were explored intensely in the 1970s and 1980s mainly for their potential use as a copper resource. An estimate of the in situ value of these metals, published in 1962 by Mero, initiated a "gold rush" mentality that resulted in national exploration programs by numerous countries. At the same time, this "gold rush" was also responsible for setting the stage for international regulations culminating in the development of the United Nations Convention on the Law of the Sea (UNCLOS). During this first period of activities, several hundred tonnes of manganese nodules were collected from the seafloor during several hours of operation showing that mining is, in principle, possible and relatively easy. However, the step from a few hours of operation to reliable mining during 250–300 days a year is challenging.

Cobalt-rich ferromanganese crusts form on nearly all rock surfaces in the ocean that are free of sediment and are composed of manganese oxides and iron oxyhydroxides that precipitate directly from seawater. Their thickness varies from less than 1 mm to about 20 cm. They form at water depths of 600–7,000 m on the flanks of volcanic seamounts, ridges, and plateaus with the thickest and most metal-rich crusts forming in depths between 800 and 2,500 m (Hein et al., 2010). Due to their slow growth rate of only 1–5 mm/million years (Hein and Koschinsky, 2014), economic occurrences are limited to old volcanic edifices. Many of these volcanic seamounts are located within the EEZs of Pacific Island states with fewer seamounts, and hence crusts, in the Atlantic and Indian Ocean. Crusts have, in general, lower copper and nickel concentrations than manganese nodules, and therefore, cobalt is the metal of greatest economic interest commonly exceeding values of 0.5 wt% cobalt. Other potential by-products are the rare earth elements (REE), platinum group elements, and tellurium (Te). Mining of crusts seems technologically more difficult as crusts are attached to a substrate rock, and dilution of the crusts by the substrate needs to be minimized. On the other hand, dispersion of sediments may not be a major issue of crust mining due to the lack of sediments (see below). Technology for crust mining is still in a conceptual state (Hein and Koschinsky, 2014).

Environmental concerns of deep-sea mining

To date no commercial mining activities have occurred anywhere in the deep sea limiting a full assessment of environmental impacts (Boschen et al., 2013). Nevertheless, some of the main impacts seem clear and are bound to the disaggregation of the ore at the mine site, the lifting to the surface vessel, and the dewatering of the ore slurry, which may lead to (1) loss of habitat (possibly with species extinction), (2) the formation of plumes and resedimentation, and (3) return water (discharge) and its effects on pelagic and/or benthic fauna depending on the depth of discharge (van Dover, 2010, 2011; Collins et al., 2013) Pollution from ships, noise from surface or seabed vehicles, and tailings disposal on land that accompanies the processing of all ores may also have negative impacts on the environment.

It is important to note that the impacts will be different for the various mineral types. While some ecosystems may recover quickly (SMS sites), those in the abyssal plains may take thousands of years or more to recover to the original species richness and distribution. The potential impacts of manganese nodule mining have been investigated on various scales by studies such as the DOMES, BIE, INDEX, DISCOL, and ATESEPP projects (see Hein and Koschinsky, 2014 and references therein). During the DISCOL study, an area of 11 km^2 was disturbed and monitored, the largest impacted area so far (Borowski and Thiel, 1998; Thiel et al., 2001). For SMS, the most comprehensive study of potential mining impacts is reported by the environmental impact assessment of Nautilus Minerals for its Solwara 1 deposit (see www.cares.nautilusminerals.com). As deep-sea mining is not yet occurring, the process of defining the regulations for mining by the ISA (Pandey, 2013; ISA, 2013b; Lodge et al., 2014) provides the opportunity for nongovernment organizations and scientists to contribute input that helps shape the regulations beyond the codes of conducts that have been issued by scientists in the past (Devey et al. 2007; IMMS, 2011). Comprehensive environmental impact assessment, mitigation strategies, baseline studies, and long-term monitoring need to be implemented into the regulations. Naturally, the duration of the impact and the size of the impacted area are of great influence. While mining for manganese nodules would cover several hundreds of square kilometers per year and for Co-rich ferromanganese crusts several tens of square kilometers per year for a single mine site, future mining in the Solwara 1 SMS project in Papua New Guinea would cover an area of only 0.112 km^2.

Ongoing exploration activities

For The Area, only exploration contracts have been issued by the ISA (Table 1; Figure 1). Up until March 2015, 26 applications have been approved out of which:

- Sixteen contracts for the exploration of polymetallic nodules, 14 are signed and 2 are pending signature (Table 1). Most of the contract areas (15) are located in the Clarion-Clipperton Zone covering 1.171 million km^2 and only one in the Central Indian Ocean (77,000 km^2). Of the 16 contractors, 8 are government bodies, while the remaining 8 are private companies sponsored by states. It is interesting to note that until 2011 only government bodies applied for exploration contracts in The Area.
- Six exploration contracts, covering 10,000 km^2 each, have been approved for the investigation of SMS in The Area since 2011. All six contractors are government bodies and represent the states of China, Russia, South Korea, France, India, and Germany. Four of the contracts are for the Indian Ocean, the remaining two for the central Atlantic. The contract for India still needs to be signed.
- Four contracted areas for the exploration of cobalt-rich manganese crusts, covering 3,000 km^2 each, have been approved for government entities, signed by China, Japan, Russia and Brazil. The Brazilian application is in the Atlantic (Rio Grande Rise), while the other three contract areas are located in the far Western Pacific.

These 26 approved contracts in The Area cover an area of 1.24 million km^2. This is as big as France, Great Britain, and Germany combined. It should be noted that all contracts are exploration licenses, granted for a period of 15 years, and are subject to relinquishments over time. Regulations for deep-sea mining have not been released, despite the fact that the 6 earliest contracts for manganese nodules will expire in 2016. At this time, the contractors have to decide whether they want to start mining or extend the exploration contract for another 5 years if justifications warrant that extension.

Marine Mineral Resources, Table 1 Status of contracts with the International Seabed Authority

Contractor	Sponsoring state	Location	Size (km^2)	Date of entry into force	Expiry date
Manganese nodules					
Interoceanmetal Joint Organization	Bulgaria, Cuba, Czech Republic, Poland, Russian Federation, and Slovakia	CCZ	75,000	Mar. 29, 2001	Mar. 28, 2016
Yuzhmorgeologiya	Russian Federation	CCZ	75,000	Mar. 29, 2001	Mar. 28, 2016
Government of the Republic of Korea	–	CCZ	75,000	Apr. 27, 2001	Apr. 26, 2016
COMRA	China	CCZ	75,000	May 22, 2001	May 21, 2016
DORD	Japan	CCZ	75,000	June 20, 2001	June 19, 2016
IFREMER	France	CCZ	75,000	June 20, 2001	June 19, 2016
Government of India	–	CIOB	77,000	Mar. 25, 2002	Mar. 24, 2017
BGR	Germany	CCZ	77,271	July 19, 2006	July 18, 2021
Nauru Ocean Resources Inc.	Nauru	CCZ	74,830	July 22, 2011	July 21, 2026
Tonga Offshore Mining Limited	Tonga	CCZ	74,713	Jan. 11, 2012	Jan. 10, 2027
G-TEC Sea Mineral Resources NV	Belgium	CCZ	77,000	Jan. 14, 2013	Jan. 13, 2028
UK Seabed Resources Ltd.	United Kingdom	CCZ	58,620	Feb. 8, 2013	Feb. 7, 2028
Marawa Research and Exploration Ltd.	Kiribati	CCZ	74,990	Jan. 19, 2015	Jan. 18, 2030
Ocean Minerals Singapore	Singapore	CCZ	58,280	Jan. 22, 2015	Jan. 21, 2030
UK Seabed Resources Ltd.	United Kingdom	CCZ	74,920	To be signed	To be signed
Cook Islands Investment Corp.	Cook Islands	CCZ	73,000	To be signed	To be signed
China Minmetals Corporation	China	CCZ	72,740	To be signed	To be signed
Seafloor massive sulfides					
COMRA	China	Southwest Indian Ridge	10,000	Nov. 18, 2011	Nov. 17, 2026
Government of the Russian Federation	–	Mid-Atlantic Ridge	10,000	Oct. 29, 2012	Oct. 28, 2027
Government of the Republic of Korea	–	Central Indian Ocean	10,000	June 24, 2014	June 23, 2029
IFREMER	France	Mid-Atlantic Ridge	10,000	Nov. 18, 2014	Nov. 17, 2029
BGR	Germany	Central Indian Ocean	10,000	May 6, 2015	May 5, 2030
Government of India	–	Central Indian Ocean	10,000	To be signed	To be signed
Co-rich manganese crusts					
JOGMEC	Japan	Western Pacific	3,000	Jan. 27, 2014	Jan. 26, 2029
COMRA	China	Western Pacific	3,000	Apr. 29, 2014	Apr. 28, 2029
Government of the Russian Federation	–	Western Pacific	3,000	Mar. 10, 2015	Mar. 9, 2030
Companhia de Pesquisa de Recursos Minerais S.A.	Brazil	South Atlantic	3,000	Nov. 9, 2015	Nov. 8, 2030

BGR Federal Institute for Geosciences and Natural Resources of Germany, *COMRA* China Ocean Mineral Resources Research and Development Association, *DORD* Deep Ocean Resources Development Co. Ltd. (Japan), *JOGMEC* Japan Oil, Gas, and Metals National Corporation, *IFRMER* Institut français de recherche pour l'exploitation de la mer (France), *CCZ* Clarion-Clipperton Zone, *CIOB* Central Indian Ocean Basin

Another application for exploration work on manganese nodules in the Clarion-Clipperton Zone has been received by the ISA from China Minmetals Corporation (sponsored by China) in 2014 and will be handled during its annual session in July 2015.

Exploration activities in the EEZ of coastal states for mineral resources in shallow water are numerous (e.g., dredging operations for sand and gravel but also for placer deposits). Few large projects for phosphorites are under consideration (see entry "Phosphorites"). For the deep sea, the number of contracts can only be estimated since many countries do not make their contracts public. Nautilus Minerals and Neptune Minerals are the two companies for which some information can be

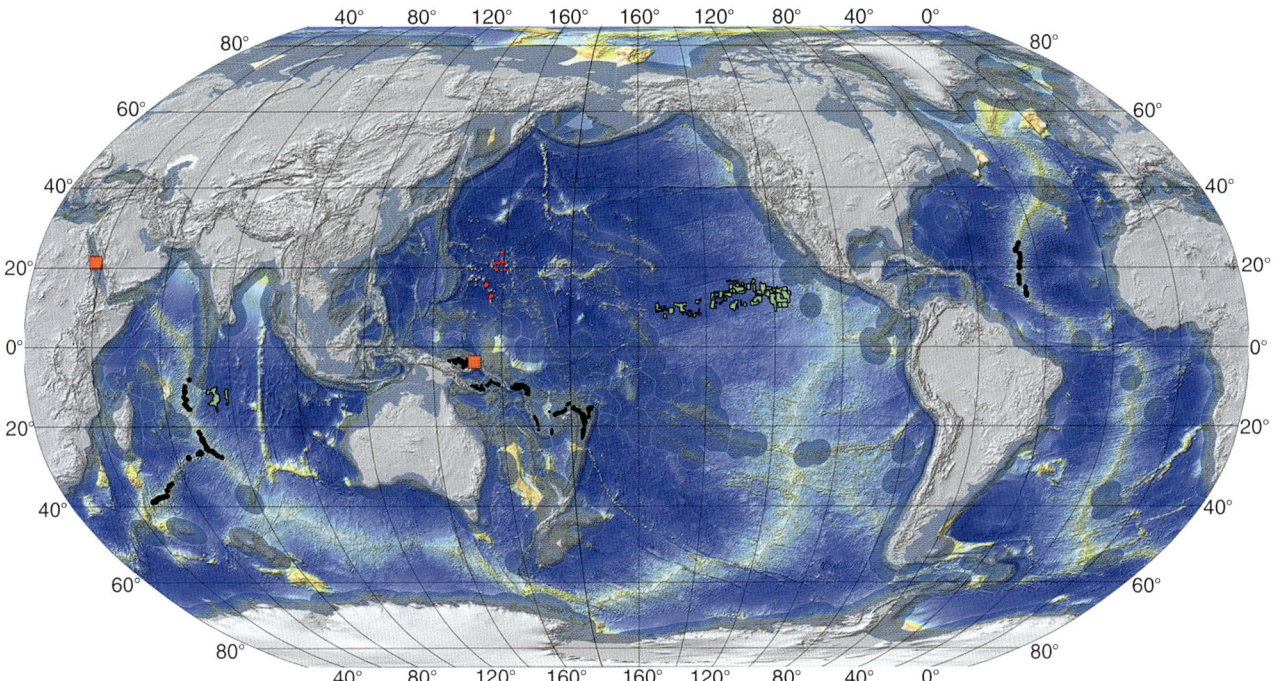

Marine Mineral Resources, Figure 1 Exploration and mining activities for deep-sea resources. Exploration contract areas for manganese nodules (*green*), Co-rich manganese crusts (*red*), and seafloor massive sulfides (*black* in The Area and *brown* within the EEZ of coastal states) are distinguished. Also shown by *orange squares* are the two mining licenses: the Solwara 1 project in Papua New Guinea and the Atlantis II Deep in the Red Sea.

retrieved. Both have extensive exploration contracts in several countries of the Western Pacific, and applications for Italy, New Zealand, Japan, and Portugal are pending. In a 2013 paper, Hein et al. estimated about 900,000 km^2 for exploration contracts in areas of national jurisdiction.

While there are only exploration licenses in The Area, two deep-sea *mining licenses* have been issued by national governments, both for SMS. The government of Papua New Guinea granted a license to Nautilus Minerals Inc. in 2011 (Solwara 1 project in the Bismarck Sea), and the governments of Saudi Arabia and Sudan granted a 30-year mining license to Diamond Fields International in 2010 for the Atlantis II project in the Red Sea. In both projects mining has not yet started.

There are also some strong national exploration programs, namely, by Korea, France, and Japan, within the EEZ of countries in the Western Pacific (Masuda et al., 2014; Fouquet and Lacroix, 2014). Most Pacific Island states regard deep-sea mining as a potential way to generate revenue and are therefore interested to establish regulations and governance structures (SPC, 2013).

Summary

Marine mineral resources such as sand, gravel, and placer deposits are currently being mined in shallow water (<500 m); however, economic, geopolitical, and strategic considerations make deep-sea mineral resources attractive to investors and governments. In contrast to land-based mining, there is only little or no overburden to remove to expose the mineral resource. Also, due to the higher grade of marine resources compared to land deposits, less ore is needed to provide the same amount of metal. Whether deep-sea mining will become viable in the future depends to a large extent on the ability to develop mining systems capable of efficient, reliable operation in deep-sea environments. Due to their abundance, manganese nodules and Co-rich ferromanganese crusts are clearly vast resources, and their mining would augment global metal supply. Since they are 2-dimensional deposits covering large areas of the seafloor, the environmental impact of mining may also be huge. SMS are likely to be the first mineral resource to be mined (within the EEZ of Papua New Guinea); however, most of these deposits are small, and the economic impact of mining them is likely to be small.

Bibliography

Borowski, C., and Thiel, H., 1998. Deep-sea macrofaunal impacts of a large-scale physical disturbance experiment in the Southeast Pacific. *Deep Sea Research, Part II*, **45**, 55–81.

Boschen, R. E., Rowden, A. A., Clark, M. R., and Gardner, J. P. A., 2013. Mining of deep-sea seafloor massive sulfides: a review of the deposits, their benthic communities, impacts from mining, regulatory frameworks and management strategies. *Ocean and Coastal Management*, **84**, 54–67.

Burnett, W. C., and Riggs, S. R., 1990. *Phosphate Deposits of the World. Neogene to Modern Phosphorites*. Cambridge: Cambridge University Press, Vol. 3, 467 pp.

Cathles, L. M., 2011. What processes at mid-ocean ridges tell us about volcanogenic massive sulfide deposits. *Mineralium Deposita*, **46**, 639–657.

Collins, P. C., Croot, P., Carlsson, J., Colaço, A., Grehan, A., Hyeong, K., Kennedy, R., Mohn, C., Smith, S., Yamamoto, H., and Rowden, A., 2013. A primer for the environmental impact assessment of mining at seafloor massive sulfide deposits. *Marine Policy*, **42**, 198–209.

Cronan, D. S., 1992. *Marine Minerals in Exclusive Economic Zones*. Dordrecht, Netherlands: Springer, 209 pp.

Cronan, D. S., 2000. *Handbook of Marine Mineral Deposits*. Boca Raton, FL: CRC Press, 406 pp.

Devey, C. W., Fisher, C. R., and Scott, S., 2007. Responsible science at hydrothermal vents. *Oceanography*, **20**, 162–171.

DOE (U.S. Department of Energy), 2011. *Critical Materials Strategy*. Washington, DC: DOE/PI-009.

ECORYS, 2014. *Study to investigate the state of knowledge of deep-sea mining*. Final Report under FWC MARE/2012/06-SC E2013/04, 196 pp.

European Commission, 2014. Report on critical raw materials for the EU. http://ec.europa.eu/enterprise/policies/raw-materials/files/docs/crm-report-on-critical-raw-materials_en.pdf

Fouquet, Y., and Lacroix, D., 2014. *Deep Marine Mineral Resources*. Heidelberg: Springer, 166 pp.

Hannington, M., 2011. Comments on "What processes at mid-ocean ridges tell us about volcanogenic massive sulfide deposits" by L.M. Cathles. *Mineralium Deposita*, **46**, 659–663.

Hannington, M. D., de Ronde, C. D., and Petersen, S., 2005. Sea-floor tectonics and submarine hydrothermal systems. In *Economic Geology 100th Anniversary Volume*. Littleton, CO: Society of Economic Geologists Inc., pp. 111–141.

Hannington, M., Jamieson, J., Monecke, T., Petersen, S., and Beaulieu, S., 2011. The abundance of seafloor massive sulfide deposits. *Geology*, **39**, 1155–1158.

Hein, J. R., and Koschinsky, A., 2014. Deep-ocean ferromanganese crusts and nodules. In *Treatise on Geochemistry*. Amsterdam: Elsevier, Vol. 2, pp. 273–291, doi:10.1016/B978-0-08-095975-7.01111-6.

Hein, J. R., Conrad, T. A., and Staudigel, H., 2010. Seamount mineral deposits: a source of rare metals for high-technology industries. *Oceanography*, **23**, 184–189.

Hein, J. R., Mizell, K., Koschinsky, A., and Conrad, T. A., 2013. Deep-ocean mineral deposits as a source of critical metals for high- and green-technology applications: comparison with land-based resources. *Ore Geology Reviews*, **51**, 1–14.

IMMS, 2011. *International Marine Minerals Society Code for Environmental Management of Marine Mining*. www.immsoc.org/IMMS_code.htm

ISA, 2013a. *Consolidated Regulations and Recommendations on Prospecting and Exploration*. Kingston, Jamaica: International Seabed Authority, pp. 1–214.

ISA, 2013b. *Technical Study No. 11: Towards the Development of a Regulatory Framework for Polymetallic Nodule Exploitation in the Area*. Kingston, Jamaica: International Seabed Authority, p. 83.

Lodge, M., Johnson, D., Le Gurun, G., Wengler, M., Weaver, P., and Gunn, V., 2014. Seabed mining: International Seabed Authority environmental management plan for the Clarion-Clipperton Zone. A partnership approach. *Marine Policy*, **49**, 66–72.

Masuda, N., Okamoto, N., and Kawai, T., 2014. Sea-floor massive sulfide mining – its possibility and difficulties to emerge as a future business. In Drebenstedt, C., and Singhal, R. (eds.), *Mine Planning and Equipment Selection*. Heidelberg: Springer, pp. 105–112.

Mero, J. L., 1962. Ocean-floor manganese nodules. *Economic Geology*, **57**, 747–767.

Nawab, Z. A., 1984. Red Sea mining: a new era. *Deep Sea Research Part A: Oceanographic Research Papers*, **31**, 813–822.

Pandey, A., 2013. Exploration of deep seabed polymetallic sulphides: scientific rationale and regulations of the International Seabed Authority. *International Journal of Mining Science and Technology*, **23**, 457–462.

Rona, P. A., 2003. Resources of the sea floor. *Science*, **299**, 673–674.

Rona, P. A., 2008. The changing vision of marine minerals. *Ore Geology Reviews*, **33**, 618–666.

SPC, 2013. Deep Sea minerals. Summary and highlights. In Baker, E., and Beaudouin, Y. (eds.), *SPC-EU EDF10 Deep Sea Minerals Project*. Suva, Fiji: Secretariat of the Pacific Community (SPC), Vols. 1 & 2.

Thiel, H., Schriever, G., Ahnert, A., Bluhm, H., Borowski, C., and Vopel, K., 2001. The large-scale environmental impact experiment DISCOL-reflection and foresight. *Deep Sea Research, Part II*, **48**, 3869–3882.

United Nations, 2014. Concise report on the world population situation in 2014. UN Department of Economic and Social Affairs Population Division, ST/ESA/SER.A/354, 30 pp.

Van Dover, C. L., 2010. Mining seafloor massive sulphides and biodiversity: what is at risk? *ICES Journal of Marine Science*, **68**, 341–348.

Van Dover, C. L., 2011. Tighten regulations on deep-sea mining. *Nature*, **470**, 31–33.

Cross-references

Black and White Smokers
Chemosynthetic Life
Cobalt-rich Manganese Crusts
Energy Resources
Hydrothermal Plumes
Hydrothermal Vent Fluids (Seafloor)
Hydrothermalism
Island Arc Volcanism, Volcanic Arcs
Manganese Nodules
Mid-ocean Ridge Magmatism and Volcanism
Phosphorites
Placer Deposits
Plate Tectonics
Seamounts
Volcanogenic Massive Sulfides

MARINE REGRESSION

Jan Harff
Institute of Marine and Coastal Sciences, University of Szczecin, Szczecin, Poland

Synonyms
Seaward shoreline shift

A marine regression occurs either due to relative sea-level fall (forced regression) or to increased sediment supply during a time when the relative sea-level is stable or even rising causing the shoreline to shift seaward (normal regression) (Posamentier and Allen, 1999; Catuneanu, 2002). In case of forced regression former seafloor is emerged – converted into land. The relative sea-level change is an effect of eustatic change and/or tectonic/isostatic vertical movement of the Earth's crust (von Bubnoff, 1954; Seibold and Berger, 1993). The opposite of regression is "transgression," which occurs if the relative sea-level rises and former land is submerged. During regression, the emerged seafloor is exposed to erosional processes (truncation) and incision by fluvial systems. Only minor amounts of terrestrial sediments accumulate so that in the geological record, times of regression are marked by sedimentary hiatuses. A marine transgression following the period of sea-level low-stand hiatus leads to a new accumulation of marine sediments covering the erosional surface discontinuously and forming an unconformity. Because of the petrophysical contrast, the unconformities can be traced within seismic recordings as synchronous reflectors (sequence stratigraphy, seismic stratigraphy). Relative sea-level changes bound cycles in the Earth's history (Einsele, 1992) which can be hierarchically ordered on time spans from hundreds of millions years to millennia (Suess et al., 1906; von Bubnoff, 1954; Posamentier et al., 1988; Vail et al., 1991).

Bibliography

Catuneanu, O., 2002. Sequence stratigraphy of clastic systems: concepts, merits, and pitfalls. *Journal of African Earth Sciences*, 35(1), 1–43.
Einsele, G., 1992. *Sedimentary Basins*. Berlin/Heidelberg: Springer, p. 628.
Posamentier, H. W., and Allen, G. P., 1999. *Siliciclastic Sequence Stratigraphy: Concepts and Applications*. SEPM Concepts Sedimentol. Paleontol. Ser. vol. 7, Tulsa, Oklahoma: Society for Sedimentary Geology.
Posamentier, H. W., Jervey, M. T., and Vail, P. R., 1988. Eustatic controls on clastic deposition. I. Conceptual framework. In: Wilgus, C. K., Hastings, B. S., Kendall, C. G. St. C., Posamentier, H. W., Ross, C. A., Van Wagoner, J. C. (eds.), *Sea Level Changes - An Integrated Approach*. SEPM Special Publication No 42. pp. 110–124, Tulsa, Oklahoma: Society for Sedimentary Geology.
Seibold, E., and Berger, W. H., 1993. *The Sea Floor – An Introduction to Marine Geology*. Berlin/Heidelberg: Springer.
Suess, E., Sollas, W. J., and Sollas, H. B. C., 1906. *The Face of the Earth*. Oxford: Clarendon.
Vail, P. R., Audemard, I. P., Bowman, S. A., Eisner, P. N., and Perez-Cruz, C., 1991. The stratigraphic signatures of tectonics, eustasy and sedimentology – an overview. In Einsele, G. (ed.), *Cycles and Events in Stratigraphy*. Berlin: Springer.
von Bubnoff, S., 1954. *Grundprobleme der Geologie*. Berlin: Akademie Verlag.

Cross-references

Marine Transgression
Relative Sea-level (RSL) Cycle
Sea-Level
Sedimentary Sequence
Sequence Stratigraphy

MARINE SEDIMENTARY BASINS

Paul Mann
Department of Earth and Atmospheric Sciences,
University of Houston, Houston, TX, USA

Synonyms

Active margin basins; Oceanic basins; Passive margin basins; Submarine basins

Definition

Marine sedimentary basins are any modern basin that is currently below sea-level and influenced by marine sedimentary processes and sedimentation. Many ancient marine sedimentary basins are now deformed and uplifted into modern, mountain chains as part of ongoing or ancient orogenic – or mountain-building – processes.

The modern marine sedimentary basins of the world's ocean basins can be subdivided into continental shelves, slopes, rises, abyssal plains, and oceanic trenches in passive margin settings (Figure 1). As these topics have been defined and discussed by other authors of the encyclopedia, I will focus on the origin of the main tectonic classes of marine sedimentary basins in active margin settings that include those produced in rift, collisional, subduction, and strike-slip environments, along with bolide impacts into marine environments (Figure 1).

Actively subsiding environments of modern marine sedimentary basins are commonly found near continent–ocean boundaries and land–sea interfaces, and for that reason, these basins can exhibit rapid alternations between marine and continental depositional systems.

A common sequence of events in the life stages of a basin is to move from initial subsidence with filling by continental sediments to marine incursion as the basin moves below sea-level – often as the consequence of accelerated tectonic subsidence – and back to continental sedimentation as the basin either moves above sea-level as the result of tectonic deformation or eventually fills to capacity as the rate of continental and marine sedimentation exceeds its rate of subsidence.

The examples below illustrate basins near continent–ocean boundaries and near land–sea interfaces that are in the process of sedimentary transitions from nonmarine to marine or vice versa (Figure 1).

Continental and marine rift basins: basins formed by stretching

The East African rift is the archetypal, continuous, and branching continental rift formed by stretching of continental crust extending 4000 km across East Africa (Figure 2a). Extension is linked to the rise of the mantle-derived African superplume that became active in mid-Cenozoic time. GPS results from Saria et al. (2014) show east–west opening of the rifts relative to the African

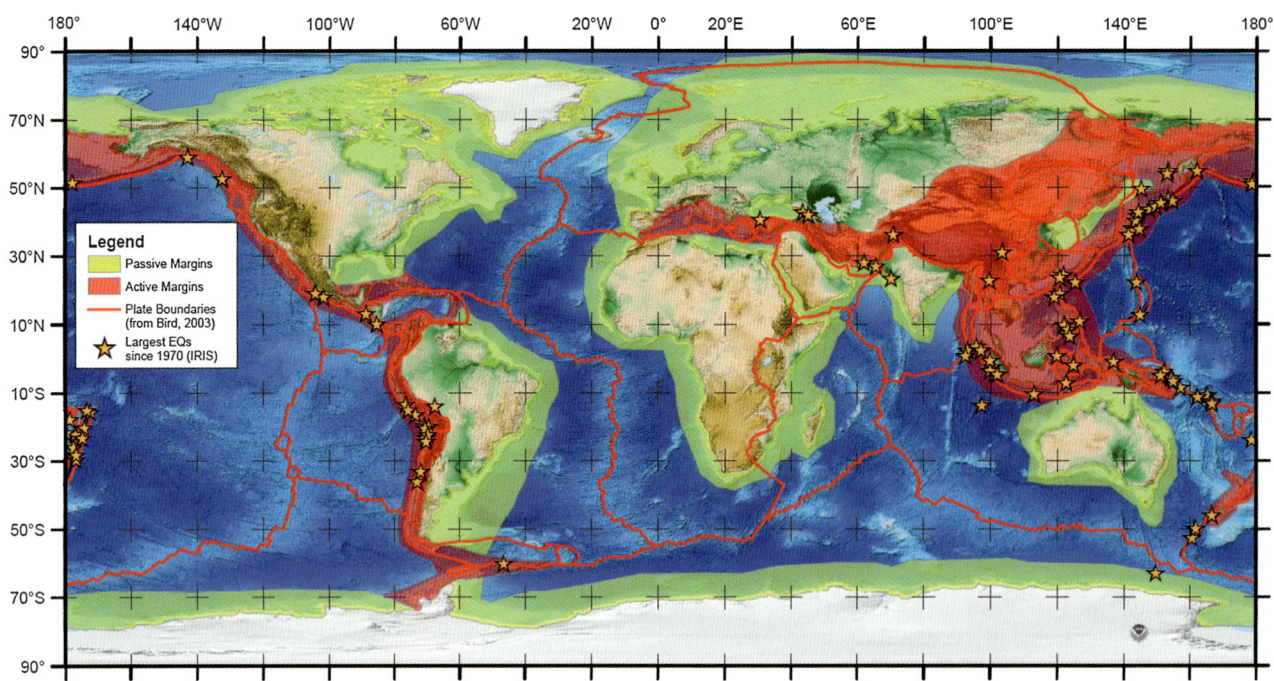

Marine Sedimentary Basins, Figure 1 Map of passive (*green*) and active (*red*) margins of the world modified from passive margin compilation by Bradley (2008). *Stars* show largest earthquakes since 1970 that are confined to the active plate margins shown with *red shading*.

plate to the west along a twin zone of rifting in East Africa (Figure 2b).

The northern terminus of the rift in the Afar area is locally 155 m below sea-level and influenced by periodic marine incursions that have led to evaporite deposits. As the rift progressively opens, marine incursions will progress further and further to the south along the rift axis and converting the rift from a nonmarine to a marine basin (Figure 2b).

The cross section in A and B from Saria et al. (2014) shows the typical structure of East Africa with either a half-graben structure or a full-graben structure. In inland areas of the African continent, the rift fills are removed from marine influence and remain continental.

Failed rift basins: basins formed by stretching

The West Siberian basin is a large failed rift system formed by rifting during the Permian–Triassic period as the result of a failed breakup of the northern Eurasia continent (Mann et al., 2003). As with many failed rifts, the West Siberian basin forms a multi-branched zone of rifting extending hundreds of kilometers into a continental area where fault-bounded rifts narrow and eventually terminate.

Much of the rifted areas are overlain by a thick and less faulted sag basin filled by marine sedimentary rocks. The thickness of the sag basin is controlled by thermal subsidence that produces a large, bowl-shaped marine sedimentary basin overlying the underlying and elongate, parallel rifts (Figure 3a). The sag basin is relatively symmetrical with respect to the location of the underlying rifts that again reflects its origin by thermal subsidence (Figure 3a).

Widespread volcanic rocks related to the lithospheric thinning during the rifting event are present both on the eastern rift flank and underlying the sag basin as shown in the stippled pattern on the cross section in Figure 3b (Reichow et al., 2002).

Collision-related basins: basins formed by flexure

Some of the largest marine and nonmarine basins formed on continental crust are foreland basins produced by the downward flexing of the edge of a continent in response to the overthrusting and loading by a colliding continent, arc, or terrane. The northern margin of the South American continent in the countries of Colombia, Venezuela, and Trinidad provides an excellent regional example of the continental flexure of the continental crust of the South American plate beneath the diachronously colliding "Great Arc of the Caribbean" (Escalona and Mann 2011) (Figure 4). Foreland basins in this area are highly asymmetrical with the deeper edge of the foreland basin controlled by thrust faults of the colliding arc block in the north near the coast of the Caribbean Sea and the more gently sloping edge of the basin in the south controlled by the flexed area of continental crust.

Marine rocks in foreland basins like the eastern Venezuelan basin were deposited during the underfilled stage of basin development when thrusting and depression of

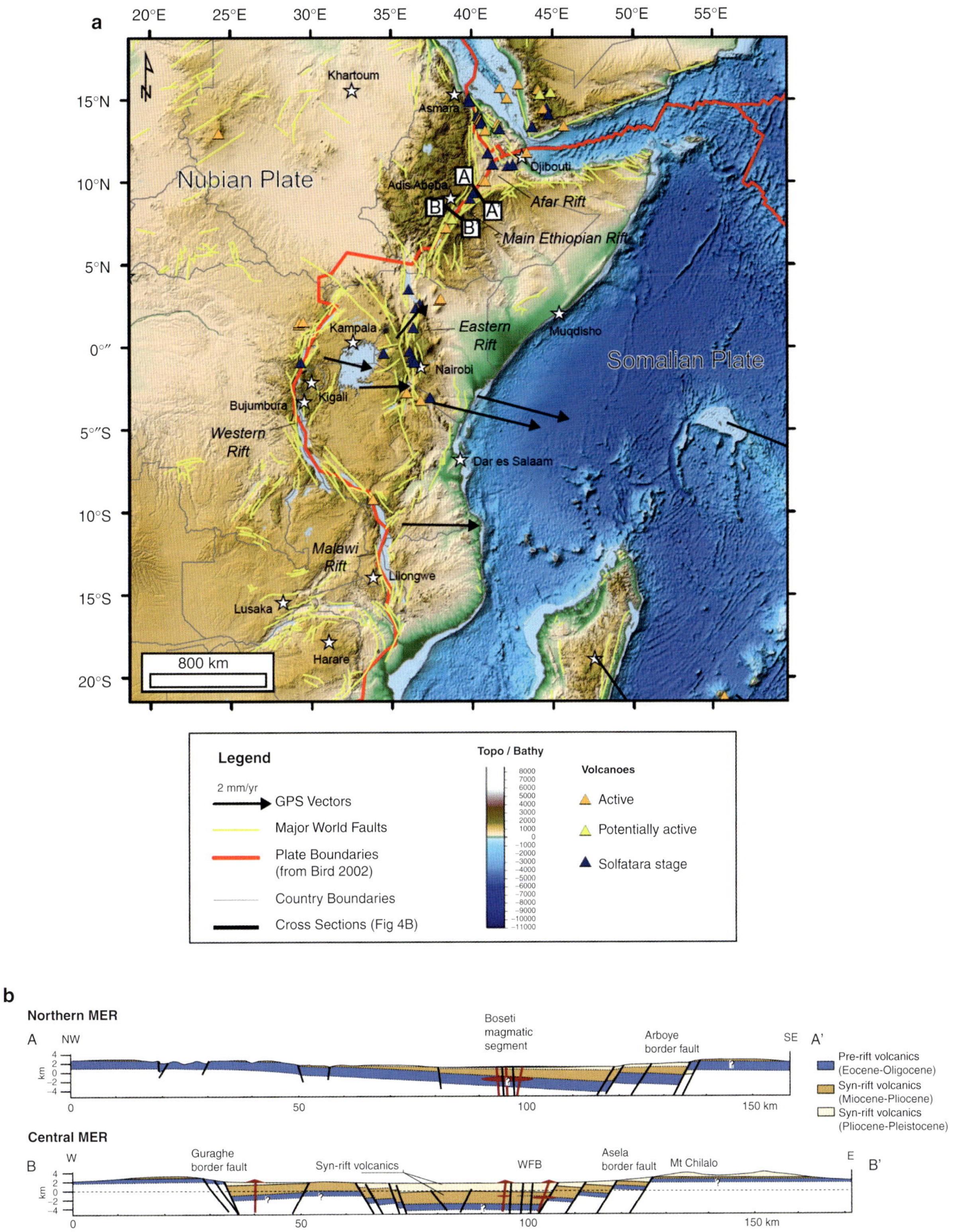

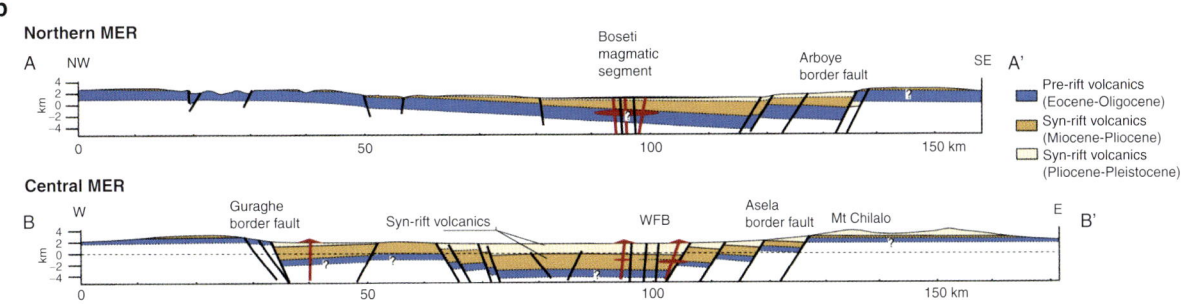

Marine Sedimentary Basins, Figure 2 Tectonic map of the East African rift system modified from Saria et al. (2014). *Yellow lines* indicate active rift faults, and *black arrows* show GPS velocities relative to a fixed Africa plate. The Afar rift at the northern end of the East African rift system is locally 155 m below sea-level and subject to marine incursions. Marine influence will likely expand southward along the topographically depressed rift axis as the rift widens in an east–west direction.

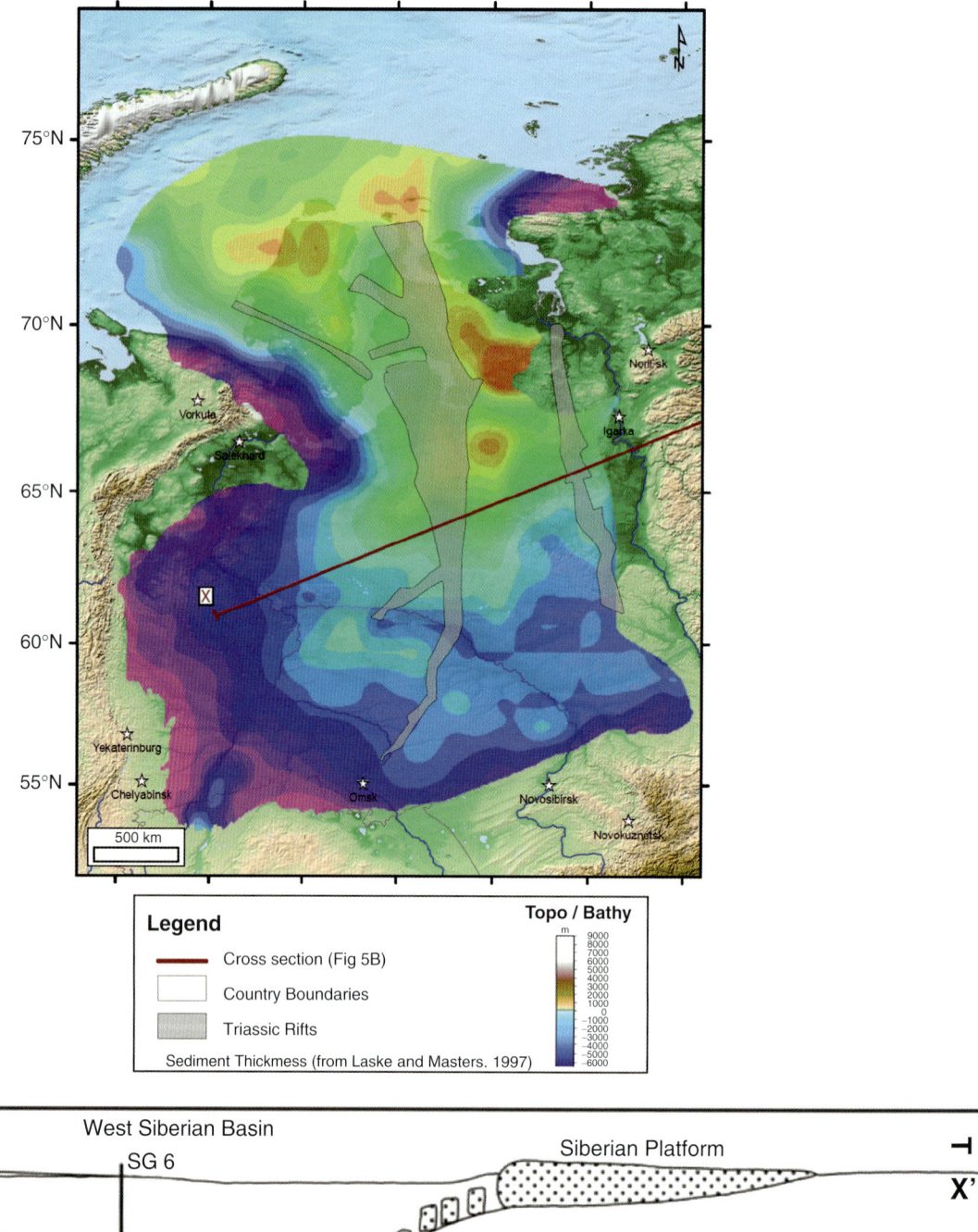

Marine Sedimentary Basins, Figure 3 Tectonic map of the failed West Siberian rift basin modified from Laske and Masters (1997) and Mann et al. (2003) with colors illustrating the thickness variations in the thick sag basin overlying the three elongate rifts shown in *gray*. East–west cross section from Reichow et al. (2002) shows the underlying rifts, an overlying, marine sag basin, and the presence of widespread volcanic rocks in stipple pattern both on the eastern rift flank and beneath its central sag basin.

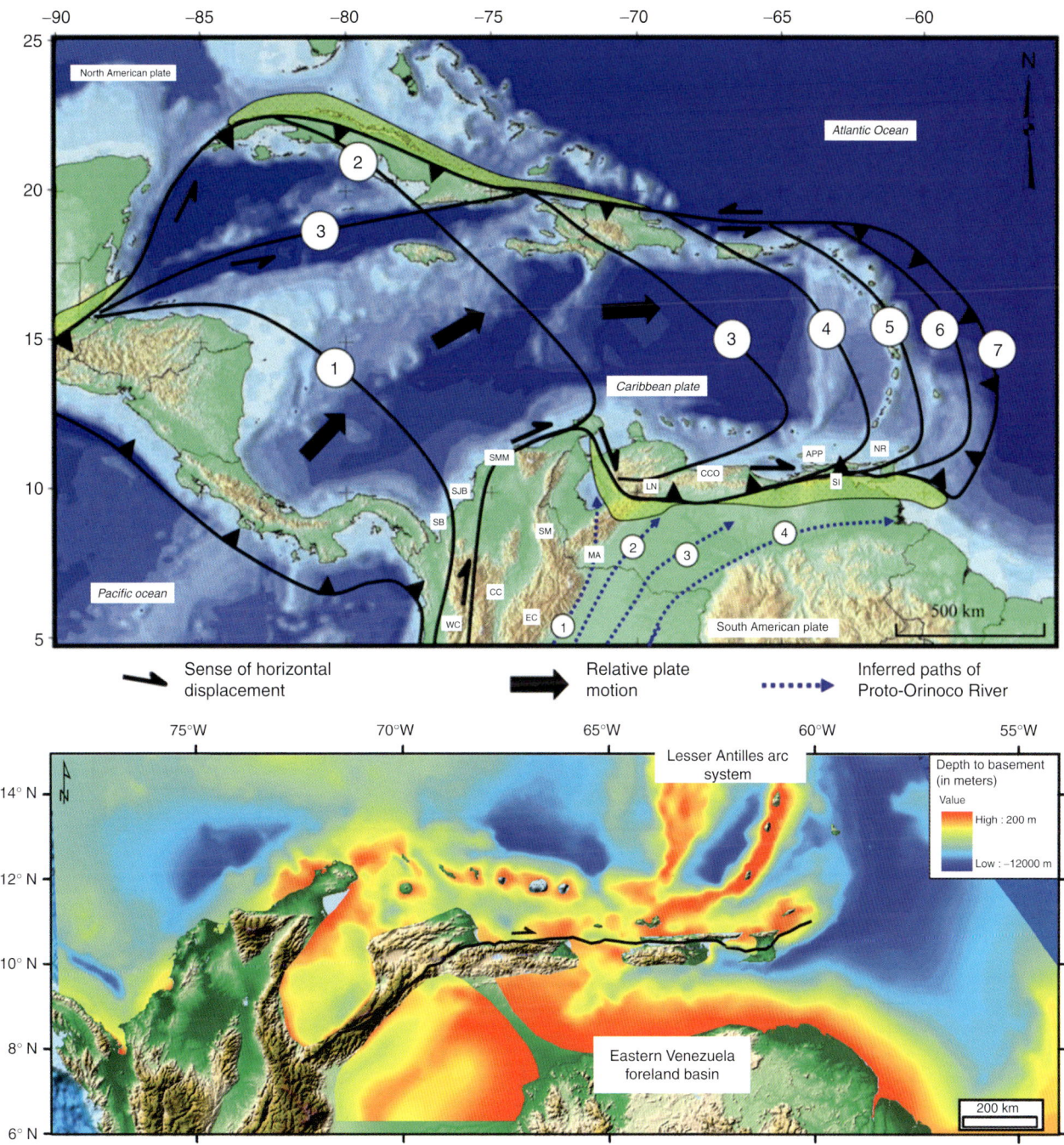

Marine Sedimentary Basins, Figure 4 Map showing the leading edge of the Great Arc of the Caribbean as it progressively and obliquely collided from west to east with the northern, continental margin of South America. Numbers represent the time that the intraoceanic Great Arc of the Caribbean occupied this location: 1, late Cretaceous; 2, Paleocene; 3, Eocene; 4, Oligocene; 5, Miocene; 6, Pliocene; and 7, present day. Depth to basement map for the top of crystalline continental crust of northern South America and the offshore area occupied by the Great Arc from Escalona and Mann (2011). The eastern Venezuelan basin is a large, foreland basin formed by the flexure of the continental edge of northern South America beneath the colliding Great Arc. Marine influence in the foreland basin is greatest during times of basin deepening that accompany periods of maximum thrusting and lithospheric flexure.

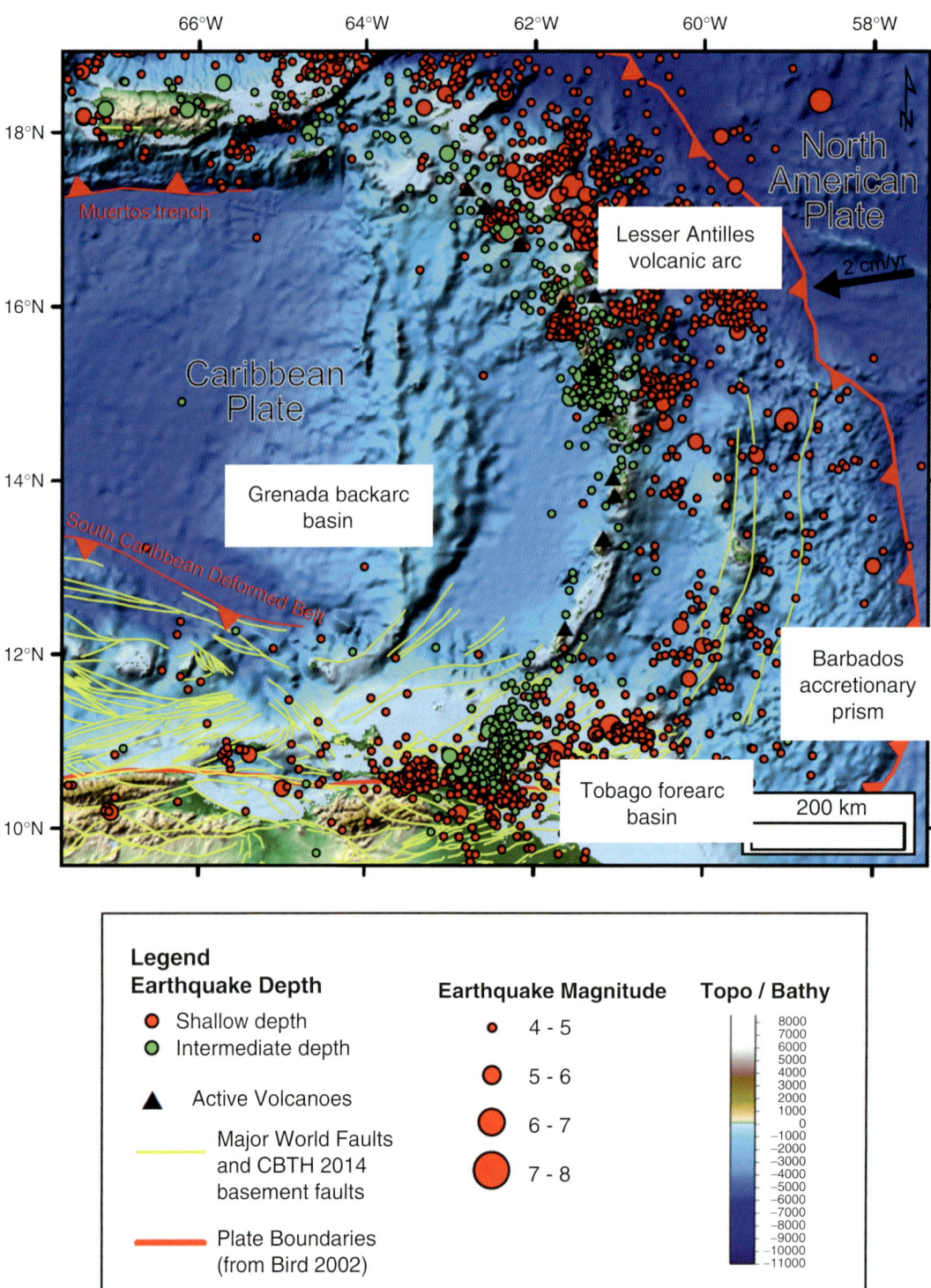

Marine Sedimentary Basins, Figure 5 Map of the Lesser Antilles island arc modified from Aitken et al. (2011) showing major basins formed mainly by the extensional process of "slab rollback" beneath the arc system. Arc-related basins are formed by stretching of the arc lithosphere as the subducting slabs rolls back in an eastward direction.

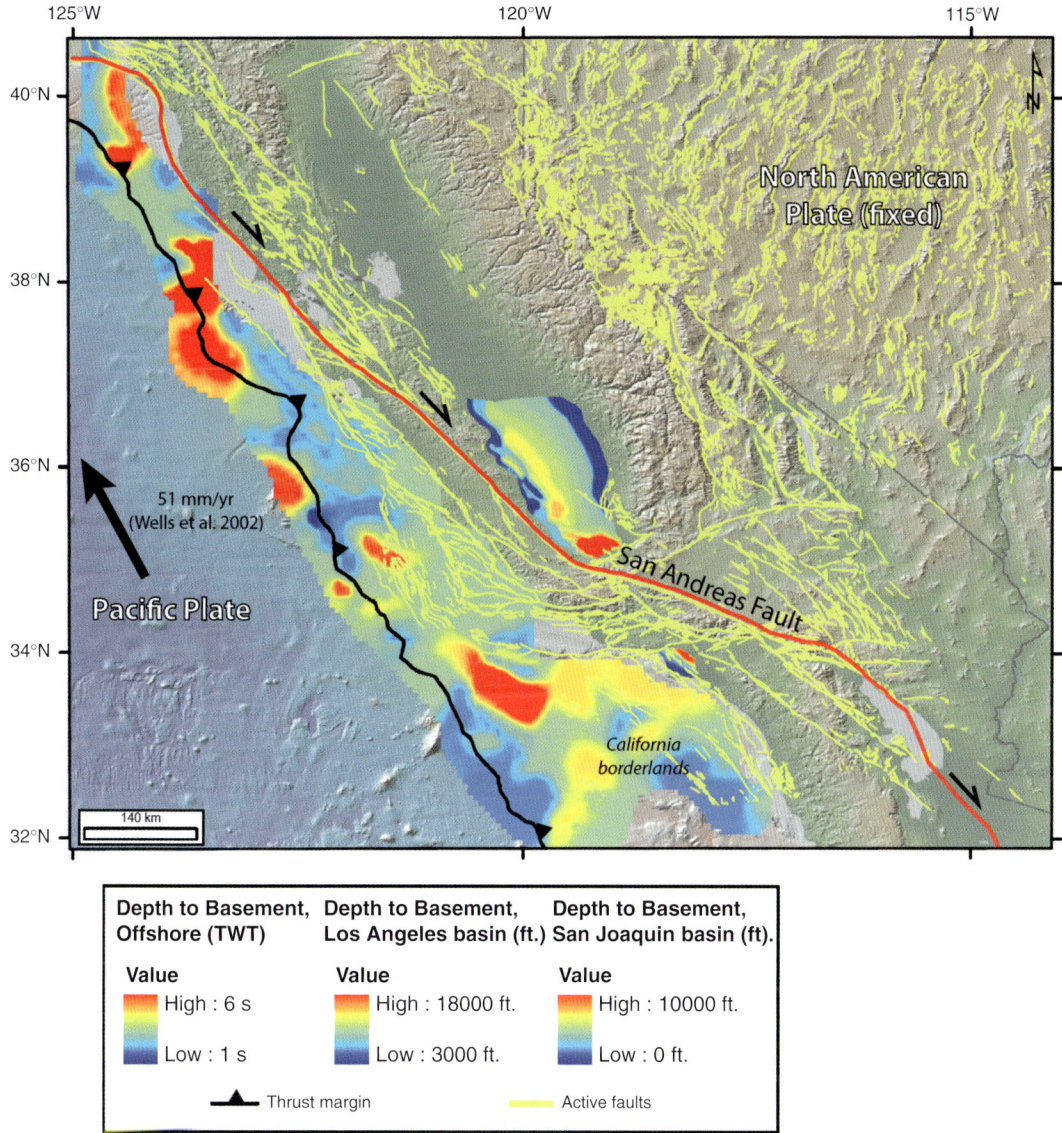

Marine Sedimentary Basins, Figure 6 Map of the active right-lateral, strike-slip San Andreas Fault system of California showing variations in depth to basement along the fault system modified from Mann (2013). Most strike-slip basins are small and formed as a consequence of either localized shortening or extension along strike-slip faults. The deepening of basins in the area of the southern San Andreas Fault is related to shortening-related flexure produced by the prominent restraining bend in that area.

continental crust beneath the basin reached a climax and outpaced the rate of sedimentation (Escalona and Mann 2011). As the zone of deformation between the northern South American continent and the obliquely colliding Great Arc of the Caribbean moved progressively eastward, the tectonically controlled subsidence rate waned, and the basin filled to sea-level with both marine and nonmarine sediments (Figure 4). Presently the central and eastern part of the foreland basin in Venezuela is filled to sea-level and forms a broad, alluvial plain with the Orinoco delta at its eastern end adjacent to the Atlantic Ocean (Figure 4).

Subduction-related basins: basins formed by stretching

Subduction-related basins generally form by processes of extension linked to the underlying, extensional process of "slab rollback" or steepening of the subducting slab that underlies the arc system. Extension produces back-arc basins like the Grenada basin and forearc basins like the Tobago basin that are fully marine in intraoceanic settings as the Lesser Antilles arc shown in Figure 5 and discussed by Aitken et al. (2011). Although occupying a much different tectonic setting, subduction-related basins form by the same processes of stretching that are observed in

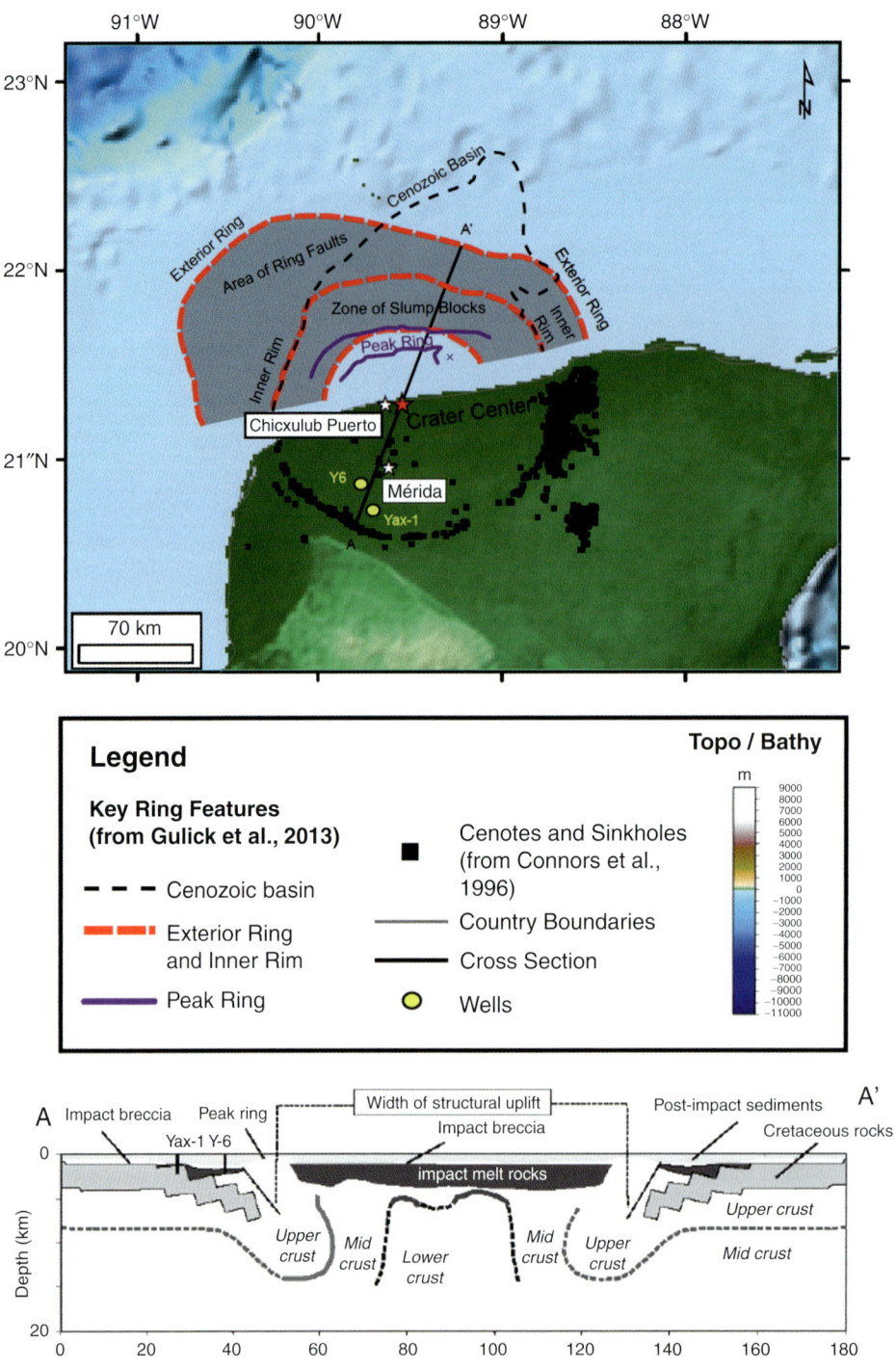

Marine Sedimentary Basins, Figure 7 Map and cross section of Chicxulub impact crater of latest Cretaceous age in the offshore of the Yucatan Peninsula in the Gulf of Mexico modified from Gulick et al. (2013). Marine sedimentary basins produced by bolides are comparable in size to strike-slip basins but smaller than foreland basins.

continental rift basins like the East African rift (Figure 2) and West Siberian basin (Figure 3).

Strike-slip basins: basins formed by stretching and flexure

Strike-slip basins formed in marine and nonmarine basins may form by stretching as in the case of pull-apart basins or can form by flexure in restraining bend settings where thrust faulting and folding dominates. The California strike-slip margin includes examples of both nonmarine basin types in areas of elevated mountains along the controlling and central San Andreas strike-slip fault and deep marine basins of pull-apart origin in the California borderlands offshore of southern California and in the pull-apart basin province of the centerline of the Gulf of California (Mann, 2007) (Figure 6).

Restraining bend tectonics in the Big Bend area of southern California is dominated by thrusting and flexure of adjacent basins beneath uplifted mountain blocks. For this reason, the depth to basement is greater in the restraining bend areas than in the more extensional areas offshore (Figure 6). The line of deeper basins parallel to the coast of California in the Pacific Ocean reflects the trace of a relict subduction margin that preceded the current phase of strike-slip motion along the onland San Andreas strike-slip fault zone (Mann, 2013) (Figure 6).

Basins produced by submarine bolide impacts

Bolide impacts into marine areas can produce deep marine basins as illustrated in the 180–200 km-wide Chicxulub impact crater of latest Cretaceous age in the offshore of the Yucatan Peninsula in the Gulf of Mexico (Figure 7). Cross sections of the crater show a multi-ringed structure with a central structural uplift that is greater than 10 km and a Moho displaced by 1–2 km (Gulick et al., 2013). The outer rim produced during impact produced a belt of large slump blocks. Impact deposits along with normal marine sedimentation eventually filled the crater with marine sedimentary rocks up to 3 km in thickness (Figure 7).

Summary of marine sedimentary basins

Marine sedimentary basins form by a variety of tectonic processes including lithospheric stretching in rift (Figures 2 and 3), strike-slip (Figure 6), and subduction (Figure 5) settings and by lithospheric flexure in collisional (Figure 4) and strike-slip (Figure 6) settings.

The sizes of marine sedimentary basins in foreland or collisional settings are generally much larger than either rift, strike-slip, or bolide impacts in marine settings (Figure 7). The degree of marine fill of these basins is directly related to the proximity of the basin to marine waters. Marine waters can flow into the axis of a juvenile rift or foreland basin during its earlier underfilled stage. Non-marine fill can fill the basin to sea-level during its later, overfilled stage.

Bibliography

Aitken, T., Mann, P., Escalona, A., and Christeson, G., 2011. Evolution of the Grenada and Tobago basins and implications for arc migration. *Marine and Petroleum Geology*, **28**, 235–258.

Bradley, D., 2008. Passive margins through earth history. *Earth-Science Reviews*, **91**, 1–26.

Escalona, A., and Mann, P., 2011. Tectonics, basin subsidence mechanisms, and paleogeography of the Caribbean-South American plate boundary zone. *Marine and Petroleum Geology*, **28**, 8–39.

Gulick, S., Christeson, G., Barton, P., Grieve, R., Morgan, J., and Urrutia-Fucugauchi, J., 2013. Geophysical characterization of the Chicxulub impact crater. *Reviews of Geophysics*, **51**, 31–52.

Laske, G., and Masters, G., 1997. A global digital map of sediment thickness: EOS. *Transactions of the American Geophysical Union*, **78**, 483.

Mann, P., Gahagan, L., and Gordon, M., 2003. Tectonic setting of the world's giant oil and gas fields. In Halbouty, M. T. (ed), *Giant Oil and Gas Fields of the Decade, 1990–1999*. AAPG Memoir 78, Tulsa, Oklahoma, pp. 15–105.

Mann, P., 2007. Global catalogue, classification, and tectonic origins of restraining and releasing bends on active and ancient strike-slip fault systems. In Cunningham, W., and Mann, P. (eds), *Tectonics of Strike-slip Restraining and Releasing Bends*. London: Geological Society. Special Publications, Vol. 290, pp. 13–142.

Mann, P., 2013. Comparison of structural styles and giant hydrocarbon occurrences within four, active strike-slip regions: California, Southern Caribbean, Sumatra, and East China. In Gao, D. (ed), *Dynamic Interplay among Tectonics, Sedimentation, and Petroleum Systems*. AAPG Memoir 100, pp. 43–93.

Reichow, M., Saunders, A., White, R., Pringle, M., Al'Mukhamedov, A., Medvedev, A., and Kirda, N., 2002. $^{40}Ar/^{39}Ar$ dates from the West Siberian basin: Siberian flood basalt province doubled. *Science*, **296**, 1846–1849.

Saria, E., Calais, E., Stamps, D., Delvaux, D., and Hartnady, C., 2014. Present-day kinematics of the East African rift. *Journal of Geophysical Research, Solid Earth*. **119**, doi:10.1002/2013JB010901.

Cross-references

Active Continental Margins
Energy Resources
Marine Impacts and Their Consequences
Morphology Across Convergent Plate Boundaries
Passive Plate Margins
Pull-apart Basins
Shelf
Subduction
Subsidence of Oceanic Crust

MARINE TRANSGRESSION

Jan Harff
Institute of Marine and Coastal Sciences, University of Szczecin, Szczecin, Poland

Synonyms

Landward shoreline shift

A marine transgression occurs if the rate of relative sea-level rise cannot be compensated by sediment supply and the coastline shifts landward (Posamentier et al., 1988; Posamentier and Allen, 1999; Catuneanu, 2002), so that former land is submerged and converted into seafloor. The relative sea-level change is an effect of eustacy modified by tectonic/isostatic vertical displacements of the Earth's crust. The opposite of transgression is "regression", which occurs if the relative sea-level drops, emerging former seafloor as land. During a transgression, the advancing highly dynamic coastal zone reworks the surface of the submerged land, resulting in the formation of coarse coastal sediments. These sediments can be used as "transgression horizons" in the geological record as an indicator of an advancing ancient sea and form ideal markers in stratigraphic correlation of well sections and seismic records (sequence stratigraphy, seismic stratigraphy). In the seismic record, reflectors marking transgressive strata are characterized by "onlapping" at their lower boundary. Relative sea-level changes bound cycles in the Earth's history which can be hierarchically ordered based on the time spans of millions of years to millennia (Suess et al., 1906; von Bubnoff, 1954; Vail et al., 1991; Seibold and Berger, 1993).

Bibliography

Catuneanu, O., 2002. Sequence stratigraphy of clastic systems: concepts, merits, and pitfalls. *Journal of African Earth Sciences*, **35**(1), 1–43.

Posamentier, H. W., and Allen, G. P., 1999. *Siliciclastic Sequence Stratigraphy: Concepts and Applications*. SEPM Concepts Sedimentol. Paleontol. Ser. vol. 7, Tulsa, Oklahoma: Society for Sedimentary Geology.

Posamentier, H. W., Jervey, M. T., and Vail, P. R., 1988. Eustatic controls on clastic deposition. I. Conceptual framework. In: Wilgus, C. K., Hastings, B. S., Kendall, C. G. St. C., Posamentier, H. W., Ross, C. A., and Van Wagoner, J. C. (eds.), *Sea Level Changes – An Integrated Approach*. SEPM Special Publication No 42, Tulsa, Oklahoma: Society for Sedimentary Geology. pp. 110–124.

Seibold, E., and Berger, W. H., 1993. *The Sea Floor – An Introduction to Marine Geology*. Berlin/Heidelberg: Springer.

Suess, E., Sollas, W. J., and Sollas, H. B. C., 1906. *The Face of the Earth*. Oxford: Clarendon.

Vail, P. R., Audemard, I. P., Bowman, S. A., Eisner, P. N., and Perez-Cruz, C., 1991. The stratigraphic signatures of tectonics, eustasy and sedimentology – an overview. In Einsele, G. (ed.), *Cycles and Events in Stratigraphy*. Berlin: Springer.

von Bubnoff, S., 1954. *Grundprobleme der Geologie*. Berlin: Akademie Verlag.

Cross-references

Marine Regression
Relative Sea-level (RSL) Cycle
Sea-Level
Sedimentary Sequence
Sequence Stratigraphy

MASS WASTING

Jan Sverre Laberg, Tore O. Vorren and Matthias Forwick
Department of Geology, University of Tromsø, Tromsø, Norway

Synonyms

Mass movement; Slope failure

Driven by the pull of gravity, mass wasting comprises all the sedimentary processes related to remobilization of sediments deposited on slopes including creep, sliding, slumping, flow, and fall. The initial stage of remobilization is limited to a layer or zone of weakness, i.e., a stratigraphic interval with physical properties allowing liquefaction from an external trigger. This interval may be located from meters to >100 m below the sea floor. Following the initial liquefaction, the overlying succession is mobilized as blocks, slabs, or ridges and deformation mainly occurs in a basal shear zone. During further movement, disintegration into debris flows or mudflows with or without blocks of consolidated sediments may occur. The incorporation of ambient water during disintegration may lead to the development of turbidity currents due to erosion of the frontal part. If the velocity of the flow reaches a critical value, hydroplaning may occur (Mohrig et al., 1998).

The initial stage of mass wasting may be located anywhere on a slope and involve gradients <1°. Many slope failures have a retrogressive development, i.e., that shallower parts of the slope successively get affected due to the loss of support from below (Kvalstad et al., 2005). In such cases, weak layers or zones at different stratigraphic levels may be involved. Such a development may last from hours to days, and individual flows can vary in volume and run-out distance.

The shallowest part of the slide scar is delineated by a slope-parallel or amphitheater-shaped headwall or scarp that may be >100 m high. Sidewalls form by lateral shear in downslope direction. However, they sometimes also involve compressional features. Where weak layers at different stratigraphic depths are involved, prominent escarpments within the slide scar develop. Slide scars may be partly or completely evacuated.

Parts of the evacuated material are frequently deposited in areas with marked reductions in slope gradients, as, e.g., at the transition from the continental slope to the continental rise. Due to the reduction in slope gradient, the shear stress drops below the shear strength of the reworked sediments, leading, e.g., to freezing "en masse" of debris flows/mudflows. However, some debris flows/mudflows may proceed beyond the base of the continental rise and onto the abyssal plain. Run-out distances of debris flows

---

Tore O. Vorren is deceased.

can occasionally exceed 1,500 km from the point of failure and deposition occurs as a result from the transformation of a decelerating turbidity current (Talling et al., 2007).

On glaciated continental margins, fans or delta-like protrusions occur in front of many glacial troughs or channels crossing the continental shelves. Vorren et al. (1989) proposed naming these features "trough mouth fans" (TMF). Debris flows are the main building blocks of TMFs (Vorren et al., 1998). They originated on the upper slope, emplaced during periods when the grounding lines of ice streams were located near or at the shelf breaks at the trough mouths, probably resulting in very high sediment input. This situation caused oversteepened slopes and/or excess pore-fluid pressures in the accumulated sediments, creating instabilities and mass movement.

Bibliography

Kvalstad, T. J., Andresen, L., Forsberg, C. F., Berg, K., Bryn, P., and Wangen, M., 2005. The Storegga slide: evaluation of triggering sources and slide mechanics. *Marine and Petroleum Geology*, **22**, 245–256.

Mohrig, D., Whipple, K. X., Hondzo, M., Ellis, C., and Parker, G., 1998. Hydroplaning of subaqueous debris flows. *Geological Society of America Bulletin*, **110**, 387–394.

Talling, P. J., Wynn, R. B., Masson, D. G., Frenz, M., Cronin, B. T., Schiebel, R., Akhmetzhanov, A. M., Dallmeier-Tiessen, S., Benetti, S., Weaver, P. P. E., Georgiopoulou, A., Zühlsdorff, C., and Amy, L. A., 2007. Onset of submarine debris flow deposition far from original giant landslide. *Nature*, **450**, 541–544, doi:10.1038/nature06313.

Vorren, T. O., Lebesbye, E., Andreassen, K., and Larsen, K.-B., 1989. Glacigenic sediments on a passive continental margin as exemplified by the Barents Sea. *Marine Geology*, **85**, 251–272.

Vorren, T. O., Laberg, J. S., Blaume, F., Dowdeswell, J. A., Kenyon, N. H., Mienert, J., Rumohr, J., and Werner, F., 1998. The Norwegian-Greenland sea continental margins: morphology and late quaternary sedimentary processes and environment. *Quaternary Science Reviews*, **17**, 273–302.

Cross-references

Geohazards: Coastal Disasters
Ocean Margin Systems
Tsunamis

METAMORPHIC CORE COMPLEXES

Uwe Ring
Department of Geological Sciences, Stockholm University, Stockholm, Sweden

Definition

Metamorphic core complexes result from horizontal lithospheric extension and form in low-viscosity lower crust when extension occurs at high rates and deformation within the upper crust becomes localized in detachment faults. They are oval shaped usually updomed structures in which mid-crustal basement rocks of higher metamorphic grade have been tectonically juxtaposed against low-grade upper crustal rocks.

Introduction

Extension of Earth's lithosphere is one of the most fundamental processes that shape the face of our planet. Extension and breakup of continental lithosphere is key to understand the evolution of continents, the origin of sedimentary basins, and their hydrocarbon potential, as well as the thermohaline circulation in the oceans and thus global climate.

The most spectacular form of extension tectonics is the formation of metamorphic core complexes. Metamorphic core complexes mainly develop in continental crust, especially where it has been previously thickened by collisional processes. These processes heated up the thickened crust mainly by radioactive decay thereby weakening it and ultimately causing its failure. In oceans, metamorphic core complexes may form as well near mid-ocean ridges when magma supply is not efficient enough to accommodate extension (e.g., North Atlantic; Tucholke et al., 1998). This entry covers continental core complexes only.

What is a core complex?

Metamorphic core complexes are usually oval-shaped bodies in map view their long axis is typically some 20–50 km long. They result from localized extension, which drags out the middle or lower crust (metamorphic core) from beneath fracturing and extending upper crustal rocks and exposed beneath shallow-dipping extensional faults (detachments) of large areal extent (hundreds to more than one thousand square kilometers) (Lister and Davis, 1989) (Figure 1). As the size of a fault is linked to its displacement, the large areal extent of the detachments indicates considerable tectonic transport along those faults. Displacement along core complex detachments is typically of the order of 20–80 km but may exceed 100 km (Foster and John, 1999; Ring et al., 2001; Brichau et al., 2006) and ultimately juxtaposes rock types with radically different geological histories (Figure 2). The large displacements show that deformation during lithospheric extension is strongly localized during core complex formation.

The detachments separating the metamorphic core from the upper crustal rocks are usually underlain by broader brittle-ductile to ductile extensional shear zones. The deformation in both is kinematically coordinated. In general, the detachments (and shear zones as well) dip at low angles (<30°). There has been some debate in the past whether these low-dip angles represent original values or whether the faults have been rotated from a steep orientation (≈60°) into their present low-angle position. Although it has been demonstrated that some low-angle detachments have been rotated from a higher-angle position (Axen et al., 1995; Gessner et al., 2001), there is incontrovertible evidence that many low-angle faults

Metamorphic Core Complexes, Figure 1 Field photos of detachment faults. (a) Rawhide Buckskin detachment in the Basin and Range province of the western USA. Yellowish-gray rocks in the lower two thirds of the photo form the metamorphic core; the red-brown hill is a tilted block of sediments in the hanging wall directly above the detachment fault, the latter of which is dipping very modestly to the left in this photo. (b) Kuzey detachment in the Menderes Massif of west Turkey. The surface in the upper left represents the detachment plane dipping at about 30° to the right; the hill above this plane in the middle of the photo represents hanging wall sediments. (c) Naxos detachment in the Aegean Sea. The view is onto the detachment plane, which is characterized by a marked stretching lineation representing the tectonic transport direction during extensional deformation. Sedimentary rocks in the foreground are coarse-clastic breccias that make up the basal section of the hanging wall (Olivier Vanderhaeghe provided photos in (a) and (c), Klaus Gessner for the photo in (b)).

originated in a low-angle orientation (Wernicke, 1981; Lister and Davis, 1989; Cowan et al., 2003; Collettini and Holdsworth, 2004; Axen, 2007). A low-angle origin of the detachment faults readily explains the moderate variation in metamorphic grade along the crustal section exposed in the footwalls of core complexes (Lister and Davis, 1989).

The footwalls of core complexes are usually made up of middle/lower crustal metamorphic rocks that may be intruded by syn-extensional plutons (Figure 2). The deeper parts of these plutons are frequently not or very weakly deformed; toward the detachments, they are overprinted by extension-related mylonites. In a number of cases, it has been convincingly shown that high-grade metamorphism and pluton emplacement are a consequence of core complex formation (Ring et al., 2010; Schulte et al., 2014), whereas in some other cases, temporal relationships between metamorphism, magmatism, and core complex formation remain elusive (Reynolds, 1982).

In general, rocks in the footwalls of core complex detachments show a distinct sequence of ductile deformation fabrics that were successively overprinted by brittle structures toward the structurally higher part of the shear zone underlying the detachment (Figure 3). In the simplest case, high-temperature deformation structures may be preserved at the bottom of the shear zone. As the footwall is dragged up toward the surface, deformation becomes more localized, and lower-temperature deformation structures overprint earlier high-temperature structures in the upper portions of the evolving shear zone and are successively overprinted by cataclastic deformation once the footwall has reached the upper crust. The detachment faults at the top of the ductile shear zone are basically upper crustal manifestations of the shallow-dipping, normal-slip shear zones (Reynolds, 1982; Davis, 1983). This simple, idealized overprinting sequence is usually not fully met because deformation is commonly heterogeneously distributed in nature.

The footwalls of metamorphic core complexes cool very rapidly at rates exceeding $50-100\,°C\,Myr^{-1}$ as they are dragged up to the surface (Fitzgerald et al., 1993; Brichau et al., 2006; Thomson and Ring, 2006). Commonly thermochronologic ages that record true cooling through a certain closure temperature (e.g., fission-track and (U-Th)/He ages) are identical with error for minerals with different closure temperatures (Fitzgerald et al., 1993; Foster and John, 1999; Brichau et al., 2007). These cooling ages are also identical, or almost identical, to deformation ages from the extensional mylonites (Kumerics et al., 2005). Because most fault systems in metamorphic core complex are low angle, the amount of footwall exhumation is moderate (Deckert et al., 2002; Ring et al., 2003).

The upper crustal rocks above the detachments are often syn-extensional clastic sediments, which contain detritus of the metamorphic core. Therefore, the sediments provide evidence for syn-extensional exposure of the

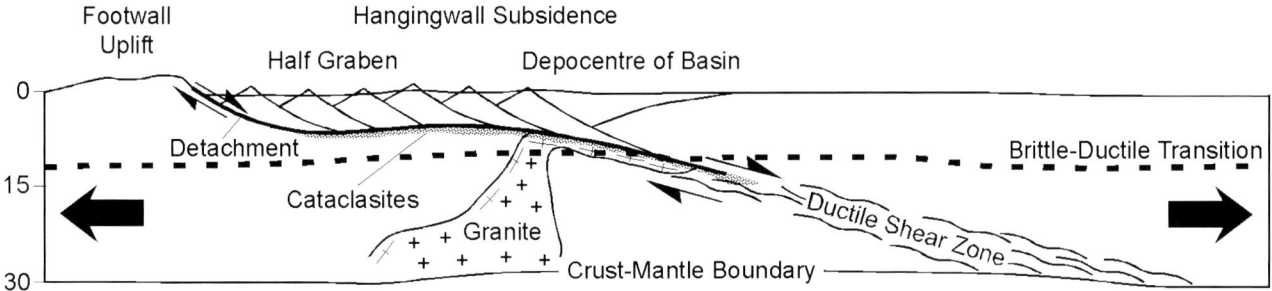

Metamorphic Core Complexes, Figure 2 Idealized geometry of a continental core complex resulting from localized deformation as a response of horizontal lithospheric extension. Shallow-dipping ductile shear zone at depth grades into cataclastic shear zone and detachment fault in the upper crust. Note that in this model the crust-mantle boundary has been slightly exhumed as a result of extensional thinning of the crust. In cases where the lower crust has a lower viscosity, pronounced lower crustal flow may prevent thinning of the lower crust leaving the crust-mantle boundary flat.

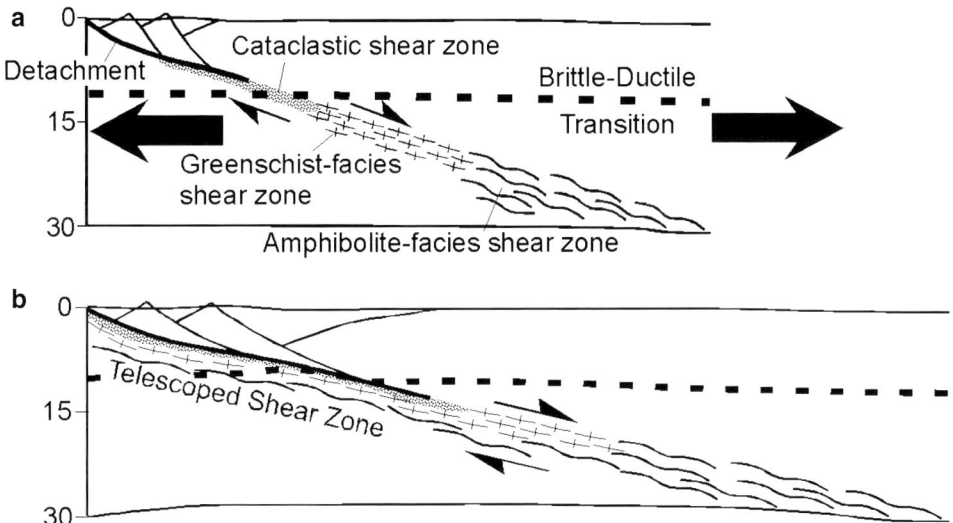

Metamorphic Core Complexes, Figure 3 Idealized scheme for the evolution of ductile-to-brittle deformation in the uppermost footwall of core complexes. (**a**) Onset of core complex formation; extension is accommodated by detachment near surface, narrow cataclastic shear zone in deeper part of brittle crust, and greenschist- to amphibolite-facies shear zone in ductile lower crust below the brittle-ductile transition; note that shear zone is widening with depth (with increasing temperature). (**b**) As the amphibolite shear zone is dragged to the surface, it is overprinted by lower-temperature greenschist-facies deformation, which is in turn cataclastically overprinted as the shear zone moves into the upper crust.

exhuming and uplifting footwall and record the exhumation history of the evolving core complex.

Core complexes also offer an opportunity to look into the problem of absolute versus relative displacement along tectonic faults. Shear-sense indicators in ductile and brittle shear zones provide evidence for the relative displacement of faults; they do not per se provide constraints as to the absolute motion of either hanging- or footwall relative to the Earth surface. However, in a core complex the metamorphic core usually has been juxtaposed against non-metamorphosed sediments in the hanging wall. This provides evidence for the absolute upward motion of the footwall relative to the Earth surface. The sediments of the hanging wall remained more or less stationary with respect to the Earth surface.

Forms of continental extension and thermal stratification of continental crust

The metamorphic grade of high-grade rocks in the footwalls of exposed core complexes commonly shows that the rocks formed under a high thermal gradient exceeding $30\ °C\ km^{-1}$. The high thermal gradient leads to a relatively hot and soft crust with a shallow brittle-ductile transition (≤ 10 km depths) and a relatively high-strength contrast between the ductile lower crust and the brittle upper crust.

It has been shown that the thermal state of the continental lithosphere and its rheological structure control the variation between flow in the ductile middle and lower crust and localized deformation in the brittle upper crust and explains the location and architecture of core complexes (Buck, 1991). Continental rifts (like the East African Rift) form in relatively cold and strong lower crust. In hot, weak, low-viscosity lower crust, core complexes develop (Gessner et al., 2007). It is a general observation that there is a lack of a topographic low above-core complexes and that the crust-mantle boundary is flat in a number of core complexes (Gans, 1987). Both observations suggest that the lower crust is extremely weak and fluid (Buck, 1988). It is assumed that the hot and weak lower crust enables viscous flow to accommodate horizontal pressure gradients laterally across detachments, and this lower crustal flow may be sufficient to accommodate crustal thickness contrasts, thus leaving the crust-mantle boundary flat across the extended lithosphere as observed, for example, in the Basin and Range province (Block and Royden, 1990). In other words, as extension thins the crust, the fluid lower crust is being drawn in from the surrounding areas keeping topographic and crustal thickness gradients small. Because of these processes a hot crust is not thinning significantly during core complex formation, which makes it hard to break it apart to form a new plate boundary.

Examples of regions of extended lithosphere

The classical example of a region that has undergone large-scale continental extension is the Basin and Range province in western North America (Anderson, 1971; Armstrong, 1972, 1982). The Basin and Range records >100 % of extension that has in part been accommodated by the formation of numerous individual core complexes (Coney, 1980). Another well-studied example of large-scale continental extension is the Aegean Sea extensional province in the eastern Mediterranean (Jolivet and Brun 2010; Ring et al., 2010). In both settings, extension by core complex formation was followed by rift-type extension along high-angle normal faulting. The rift-type extension records a stage of much more limited horizontal extension at a later stage of continental extension.

Radiometric dating of extension-related structures shows that there is a pause in tectonic activity between the core complex and the rift stage. It appears that this lull records the time for the hot, formerly overthickened crust to cool down. Core complex formation acts as a mechanism (valve) of how the overheated crust can get rid of its excess energy (heat). The pronounced lower crustal ductile flow causes advective transport of low-viscosity material toward the Earth surface and thereby cools the crust, preconditioning it for rift faulting and eventual continental breakup. This lull in tectonic activity leads to a "messy" split of hot continents. Examples from the Liguro-Provençal basin (Italy/France) and the West Coast of New Zealand show that it took about 25–30 Myr from initial core complex formation to sea floor spreading (Rosenbaum and Lister, 2004; Schulte et al., 2014).

Summary and conclusions

Metamorphic core complexes are spectacular examples of large-scale continental extension, usually juxtaposing metamorphic lower crust against upper crustal rocks. They are bounded by low-angle normal faults that accommodate 10's of kilometers of displacement. Metamorphic core complexes form when the strength contrast between ductile lower crust and brittle upper crust is high, which means when the thermal gradient is relatively high and therefore the brittle-ductile transition shallow (≤ 10 km depths). These hot conditions of the ductile crust make it hard for the lithosphere to fully break apart to form a passive continental margin. Therefore, the lower crust needs to cool down before continental breakup can occur.

Bibliography

Anderson, R. E., 1971. Thin-skin distension in Tertiary rocks of southeastern Nevada. *Geological Society of America Bulletin*, **82**, 43–58.

Armstrong, R. L., 1972. Low-angle (denudation) faults, Hinterland of the Sevier Orogenic Belt, Eastern Nevada and Western Utah. *Geological Society of America Bulletin*, **83**, 1729–1754.

Armstrong, R. L., 1982. Cordilleran metamorphic core complexes—from Arizona to southern Canada. *Annual Reviews of Earth Sciences*, **10**, 129–154.

Axen, G. J., 2007. Research focus: significance of large-displacement, low-angle normal faults. *Geology*, **35**, 287–288.

Axen, G. J., Bartley, J. M., and Selverstone, J., 1995. Structural expression of a rolling hinge in the footwall of the Brenner Line normal fault, eastern Alps. *Tectonics*, **14**, 1380–1392.

Block, L., and Royden, L. H., 1990. Core complex geometries and regional scale flow in the lower crust. *Tectonics*, **9**, 557–567.

Brichau, S., Ring, U., Carter, A., Ketcham, R. A., Brunel, M., and Stockli, D., 2006. Constraining the long-term evolution of the slip rate for a major extensional fault system using thermochronology. *Earth and Planetary Science Letters*, **241**, 293–306.

Brichau, S., Ring, U., Carter, A., Monié, P., Bolhar, R., Stockli, D., and Brunel, M., 2007. Extensional faulting on Tinos Island, Aegean Sea, Greece: how many detachments? *Tectonics*, **26**(4), 19, doi: TC4009, 10.1029/2006TC001969.

Buck, W. R., 1988. Flexural rotation of normal faults. *Tectonics*, **7**, 959–973.

Buck, W. R., 1991. Modes of continental lithospheric extension. *Journal of Geophysical Research*, **96**, 20161–20178.

Collettini, C., and Holdsworth, R. E., 2004. Fault zone weakening and character of slip along low-angle normal faults: insights from the Zuccale fault, Elba, Italy. *Journal of the Geological Society*, **161**, 1039–1051.

Coney, P. J., 1980. Cordilleran metamorphic core complexes: An overview. In Crittenden, M. D., Jr., Coney, P. J., and Davis, G. H. (eds.), *Cordilleran metamorphic core complexes*. Boulder: Geological Society of America. Geological Society of America Memoir 153, pp. 7–31.

Cowan, D. S., Claddouhos, T. T., and Morgan, J., 2003. Structural geology and kinematic history of rocks formed along low-angle normal faults, Death Valley, California. *Geological Society of America Bulletin*, **115**, 1230–1248.

Crittenden, M. D., Coney, P. J., and Davis, G. H. (eds.), 1980. Tectonic significance of metamorphic core complexes of the North American Cordillera. *Memoir Geological Society of America*, **153**.

Davis, G. H., 1983. Shear-zone model for the origin of metamorphic core complexes. *Geology*, **11**, 342–347.
Deckert, H., Ring, U., and Mortimer, N., 2002. Tectonic significance of Cretaceous bivergent extensional shear zones in the Torlesse accretionary wedge, central Otago Schist, New Zealand. *New Zealand Journal Geology and Geophysics*, **45**, 537–547.
Fitzgerald, P. G., Reynolds, S. J., Stump, E., Foster, D. A., and Gleadow, A. J. W., 1993. Thermochronologic evidence for timing of denudation and rate of crustal extension of the south mountains metamorphic core complex and sierra estrella, Arizona. *International Journal of Radiation Applications and Instrumentation*, **21**, 555–563.
Foster, D. A., and John, B. E., 1999. Quantifying tectonic exhumation in an extensional orogen with thermochronology: examples from the southern Basin and Range Province. *Geological Society Special Publication*, **154**, 343–364, doi:10.1144/GSL.SP.1999.154.01.16.
Gans, P., 1987. An open-system two-layer crustal stretching model for the eastern Great Basin. *Tectonics*, **6**, 1–12.
Gessner, K., Ring, U., Johnson, C., Hetzel, R., Passchier, C. W., and Güngör, T., 2001. An active bivergent rolling-hinge detachment system: the Central Menderes metamorphic core complex in western Turkey. *Geology*, **29**, 611–614.
Gessner, K., Wijns, C., and Moresi, L., 2007. A dynamic process model for tectonic denudation of metamorphic core complexes. *Tectonics*, **26**.
John, B. E., and Foster, D. A., 1993. Structural and thermal constraints on the initiation angle of detachment faulting in the southern Basin and Range: the Chemehuevi Mountains study. *Geological Society of America Bulletin*, **105**, 1091–1108.
Jolivet, L., and Brun, J.-P., 2010. Cenozoic geodynamic evolution of the Aegean. *International Journal of Earth Sciences*, **99**, 109–38.
Krabbendam, M., 2001. When the Wilson cycle breaks down: how orogens can produce strong lithosphere and inhibit their future reworking. *Geological Society Special Publication*, **184**, 57–75.
Kumerics, C., Ring, U., Brichau, S., Glodny, J., and Monie, P., 2005. The extensional Ikaria shear zone and associated brittle detachments faults, Aegean Sea, Greece. *Journal of the Geological Society London*, **162**, 701–721.
Lister, G. S., and Davis, G. A., 1989. The origin of metamorphic core complexes and detachment faults formed during Tertiary continental extension in the northern Colorado River region, U.-S. *Journal of Structural Geology*, **11**, 65–94, doi:10.1016/0191-8141(89)90036-9.
Reynolds, S. J., 1982. *Geology and Geochronology of the South Mountains, Central Arizona*. Unpublished Ph.D dissertation, University of Arizona, Tucson, Arizona.
Ring, U., Layer, P. W., and Reischmann, T., 2001. Miocene high-pressure metamorphism in the Cyclades and Crete, Aegean Sea, Greece: evidence for large-magnitude displacement on the Cretan detachment. *Geology*, **29**, 395–398.
Ring, U., Thomson, S. N., and Bröcker, M., 2003. Fast extension but little exhumation: the Vari detachment in the Cyclades, Greece. *Geological Magazine*, **140**, 245–252.
Ring, U., Glodny, J., Thomson, S., and Will, T., 2010. The Hellenic subduction system: High-pressure metamorphism, exhumation, normal faulting and large-scale extension. *Annual Review of Earth and Planetary Sciences*, **38**, 45–76, doi:10.1146/annurev.earth.050708.170910.
Rosenbaum, G., and Lister, G. S., 2004. Neogene and Quaternary rollback evolution of the Tyrrhenian Sea, the Apennines and the Sicilian Maghrebides. *Tectonics*, **23**, doi:10.1029/2003TC001518.
Schulte, D. O., Ring, U., Thomson, S., Glodny, J., and Carrad, H., 2014. Two-stage development of the Paparoa Metamorphic Core Complex, West Coast, South Island, New Zealand: hot continental extension precedes seafloor spreading by ~25 Myr. *Lithosphere*, **6**, 177–194, doi:10.1130/L348.1.
Thomson, S. N., and Ring, U., 2006. Thermochronologic evaluation of post-collision extension in the Anatolide Orogen, western Turkey. *Tectonics*, **25**, TC3005, 20 p, doi:10.1029/2005TC001833.
Tucholke, B. E., Lin, J., and Kleinrock, M. C., 1998. Megamullions and mullion structure defining oceanic metamorphic core complexes on the Mid-Atlantic Ridge. *Journal of Geophysical Research, B: Solid Earth*, **103**, 9857–9866.
Wernicke, B. P., 1981. Low-angle normal faults in the Basin and Range Province-Nappe tectonics in an extending orogen. *Nature*, **291**, 645–648.

Cross-references

Magmatism at Convergent Plate Boundaries
Marginal Seas
Oceanic Rifts
Passive Plate Margins
Plate Tectonics

METHANE IN MARINE SEDIMENTS

Gerhard Bohrmann[1] and Marta E. Torres[2]
[1]MARUM-Center for Marine Environmental Sciences, University of Bremen, Bremen, Germany
[2]College of Earth, Ocean, and Atmospheric Sciences, Oregon State University, Corvallis, OR, USA

Definition

Methane is the simplest hydrocarbon with the chemical formula CH_4. At room temperature and standard pressure, methane is a colorless, odorless gas. It is the main component of natural gas and thus important as an energy source. Due to its ability to absorb energy on the infrared band, methane plays a direct role in the Earth's greenhouse effect.

Methane distribution

Methane represents a key component of the carbon cycle in marine sediments. The amount of carbon that presently occurs as methane in marine sediments is thought to be ~500–10,000 Gt (Kvenvolden and Lorenson, 2001; Milkov, 2004; 1 Gt = 10^{15} g). The size of this reservoir is ultimately determined by a balance of organic carbon sources and sinks, which vary in time and space. The loci of large methane deposits are further controlled by thermal, lithological, and structural characteristics of the sediments. Methane production is fundamentally controlled by organic carbon rain to the sediments, which in turn depends on ocean productivity and water depths (Suess, 1980). Only a fraction of the organic carbon that reaches the seafloor is available for methane generation at greater depth, because a significant part of the organic carbon rain is already remineralized within the top meters of the sediment column (review by Zonneveld et al., 2010). Early organic matter decomposition is partially controlled by the rate of sedimentation and is higher in regions of low sediment accumulation. Consequently, large methane concentrations are usually found in the rapidly accumulating sediments that underlie the high-productivity regions of most continental margins. The sediments beneath the low-productivity waters of the ocean gyres and those overlying recently formed oceanic crust do not harbor significant amounts of this simple hydrocarbon.

Sources

Our understanding of the origin and distribution of methane in marine sediments was pioneered by the work of Claypool and Kaplan (1974), who described microbial methane generation as the terminal catabolic product of organic matter degradation. As such, it follows an ecological succession of microbial ecosystems, driven mainly by the redox condition of the sediment. The remineralization of organic material proceeds in a continuous sequence of redox reactions that cascade from high to low free energy yield until the electron acceptor source is depleted. The sediment depths that harbor the various stages of this redox ladder depend on the nature and magnitude of the organic loading, sedimentation rate, seawater oxygenation, and sediment porosity. Methane production, commonly referred to as methanogenesis, is only significant in environments where burial of organic matter is rapid enough to survive remineralization by the more energy favorable electron acceptors (O_2, NO_3, Fe, Mn, and SO_4). Globally, microbial methanogenesis in marine and freshwater systems is thought to generate ~85 Tg CH_4/year (1 Tg = 10^{12} g; Reeburgh et al., 1993).

Microbial methanogenesis is a multistep process, performed exclusively by a diverse group of obligate anaerobic archaea, generally known as methanogens. These microorganisms can live in a wide range of temperature, salinity, and pH, but are limited in the substrates they can utilize for growth. A detailed description of the microbial methanogenic pathways in marine sediments is given by Wellsbury et al., (2000). The main pathway for microbial methane generation in organic-rich anoxic sediments is thought to proceed via autotrophic methanogenesis, whereby advantage of the energy released from the reduction of carbon dioxide by molecular hydrogen is taken. This is a multistep process, but the net reaction can be written as

$$CO_2 + 8\,H^+ + 8\,e^- \rightarrow CH_4 + 2\,H_2O \quad (1)$$

Heterotrophic methanogens are known to utilize simple organic acids such as acetic acid to generate methane via acetoclastic fermentation, which releases both methane and CO_2:

$$CH_3COOH \rightarrow CH_4 + CO_2 \quad (2)$$

Temperature is a critical factor in sub-seafloor methane generation, as it may impact enzyme activity and metabolic rates (Colwell et al., 2008). Usually deeper in the sediment, thermal alteration of organic matter generates methane and higher order hydrocarbons within the temperature range of 50–200 °C. At typical oceanic geothermal gradients of 20–50 °C/km, depending on the local geothermal gradient, sediment depths larger than 1 km are required to produce significant amounts of hydrocarbons by thermochemical reaction, with low molecular weight hydrocarbons (methane to butane) being produced at rates that are proportional to temperature.

Even though a few hyperthermophilic methanogens have been shown to produce methane at temperatures of ~122 °C (Takai et al., 2008), the upper limit for subsurface methanogenesis appears to be ~80 °C (Wilhems et al., 2001), which corresponds to depths of 1.500–3.000 m below seafloor, depending on the local geothermal gradient.

Several combinations of molecular and isotopic characteristics are commonly used to identify hydrocarbon sources (Figure 1), as well as their post-genetic alterations. Microbial methanogenesis generates mainly methane (C_1) and traces of ethane (C_2) and propane (C_3), resulting in relatively high $C_1/(C_2 + C_3)$ ratios (commonly > 1,000; Figure 1a). In contrast, thermogenic gases are commonly enriched in higher hydrocarbons, resulting in low $C_1/(C_2 + C_3)$ ratios (<100). The now classic differentiation scheme places microbial and thermogenic methane into distinct regions within the $\delta^{13}C_{CH4}$ and δD_{CH4} field (Figure 1b). In addition, carbon and hydrogen isotopes also aid in elucidating autotrophic vs heterotrophic microbial pathways. Differences in carbon isotopic fractionation arise from the large kinetic effects involved in breaking the C-O bond during CO_2 reduction (Eq. 1), which leads to a more ^{13}C-depleted methane than the methane generated by heterotrophic methanogenesis (Eq. 2), in which the methyl group remains intact. The hydrogen isotope signature may also provide information on the metabolic pathways, as acetate fermentation yields methane with a δD value lower than −250 ‰, whereas carbonate fermentation leads to δD values ranging from −150 ‰ to −250 ‰ (Whiticar et al., 1986).

Sediment reservoirs

Marine sediments contain an estimated 500–10,000 Gt (Kvenvolden and Lorenson, 2001; Milkov, 2004) of methane carbon. Because of its low solubility, methane dissolved in the pore fluids accounts for a negligible fraction of the total sedimentary reservoir. Another small fraction is adsorbed to solid phases (Reeburgh 2007). Traditional natural gas reservoirs are locations characterized by the presence of a source rock (high organic carbon needed for methane generation), a sedimentary trap (normally high-porosity layers where methane gas can accumulate), and a seal rock (which prevents methane leakage from the reservoir). Perhaps the larger methane component of the marine sediment pool occurs in the form of methane hydrates. These compounds are ice-like

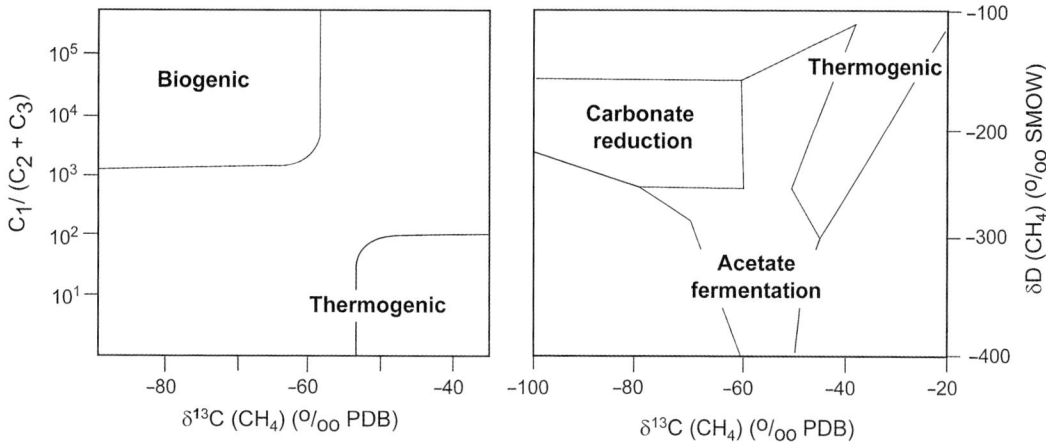

Methane in Marine Sediments, Figure 1 Assignment of microbial and thermogenic methane sources based on (a) the ratio of methane (C_1) to higher hydrocarbons ($C_2 + C_3$) plotted against the stable carbon isotopic composition of methane (After Claypool and Kvenvolden, 1983) and (b) the stable hydrogen and carbon isotopes of methane (After Schoell, 1988). Stable isotope composition expressed in the δ-notation.

solids in which methane molecules are held within a hydrogen-bonded water lattice. These solids are stable at low temperatures and high pressures, as governed by their thermodynamic stability, and necessitate enough methane so to stabilize the hydrate structure (Bohrmann and Torres, 2006). Upon recovery from the seafloor, methane hydrates decompose, and at atmospheric conditions they release approximately 164 m³ of methane per cubic meter of methane hydrate, thus explaining the large fraction of methane that is sequestered within methane hydrate deposits. This reservoir is, however, highly dynamic, and it has been described as a carbon storage capacitor, which stores and releases ^{13}C-depleted carbon at rates linked to external conditions such as deep ocean temperature (Dickens et al., 2011). Based on this concept, the massive negative carbon isotopic excursion recorded in marine and terrestrial sediments during the Paleocene-Eocene Thermal Maximum at ~55 Ma has been interpreted as a widespread release of isotopically light carbon from dissociating marine methane hydrates (Dickens, 2011).

Methane sinks

The fraction of the methane that is not trapped in conventional natural gas reservoirs or within methane hydrates migrates through the sediments via diffusion and/or advection, and in case of passing the seafloor, oxygen depletion in the overlying water column results as a consequence of methane oxidation (Boetius and Wenzhöfer, 2013).

In diffusion-dominated environments, methane is almost completely consumed by the microbially driven anaerobic oxidation of methane (AOM), which acts as a highly efficient methane filter, preventing dissolved methane release to the ocean and atmosphere. Reeburgh et al., (1993) estimated that this filter consumes nearly 90 % of the total methane produced in marine and freshwater systems. Sulfate is the most likely electron acceptor for this reaction, given its high concentration in seawater, which supplies sulfate by diffusion into the sediments. In the sulfate-mediated AOM, several clades of anaerobic methanotrophic archaea (ANMEs) often form synergistic aggregates with sulfate-reducing Deltaproteobacteria (Boetius et al., 2000). Together, these microorganisms are thought to oxidize methane to carbon dioxide while reducing sulfate (SO_4^{2-}) to hydrogen sulfide (HS^-), splitting the energy supplied by this coupled process between them (Hoehler et al., 1994). This process can be summarized by the net reaction:

$$SO_4^{2-} + CH_4 \rightarrow HS^- + HCO_3^- + H_2O \qquad (3)$$

The hydrogen sulfate is feeding chemosynthetic fauna, whereas the hydrogen carbonate is often used for carbonate precipitation. Both processes are typical seafloor manifestations at cold seeps (Bohrmann and Torres, 2006). The sulfate-supported AOM consortia (Boetius et al., 2000) depend on sulfate and methane and thus typically occur at the sulfate-methane interface zone (SMI), a biogeochemical depth level where seawater sulfate and methane from the underlying anoxic sediments converge. The SMI depths in these environments are well defined and typically occur from a few centimeters to several meters below the surface, although deeper SMIs have been observed on the Indian margin (Briggs et al., 2012) and Demerara Rise (Arndt et al., 2009) sediment.

Alternative electron acceptors, which involve the coupling of methane oxidation to the reduction of nitrite, nitrate, and reactive metals, have been recently documented. An uncommon mechanism is mediated by an extremely unusual bacterium, *Methoxymirabilis oxyfera*, that makes its own oxygen to fuel AOM in anoxic environments (Joye 2012). In addition, Milucka et al., (2012) described an ANME-2 clade that uses

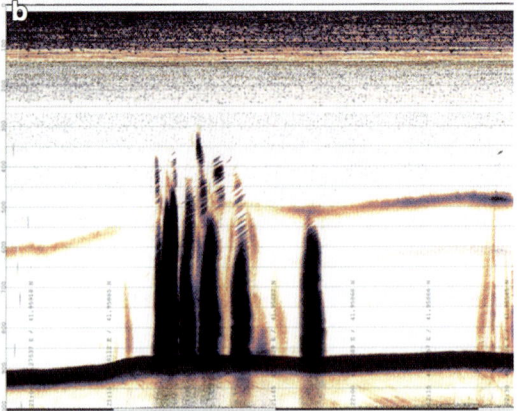

Methane in Marine Sediments, Figure 2 Gas bubble ebullition from seafloor to the water column documented (**a**) (*left*) by visual recording of ROV QUEST at 3,000 m water depth at the Makran continental margin and (**b**) (*right*) by hydroacoustic recording of a sediment echosounder (PARASOUND) at Batumi Seep at 800 m water depth, eastern Black Sea.

sulfate-reduction strategy to single-handedly mediate both AOM and sulfate reduction, and in so, eliminating the need for a microbial partner to shuttle electrons or metabolites. Other substrates of oxidizing power, which include halogenated organics and gradual dissolution of oxidant-bearing minerals, are thought to play a minor role in the overall AOM process and are detailed in a methane biogeochemistry review by Valentine (2011).

In advection-dominated systems, methane flux acts to shoal the SMI to depths of less than a few centimeters. In these environments, the high methane flux often overwhelms or even bypasses the AOM biofilter, leading to methane release to the bottom water in what are commonly known as gas seeps (Figure 2). For example, <40 % of methane is anaerobically consumed at the Håkon Mosby mud volcano (Niemann et al., 2006) leaving a large proportion of the methane to escape into the overlying water.

Methane released from the sediments at individual cold seeps by gas ebullition can be imaged as bubble plumes in the water column using acoustic survey techniques (Figure 2) and reaches values of $0.1–22 \times 10^6$ mol/year (Römer et al., 2012). Notwithstanding this massive release of methane gas, globally it accounts for just a few percent of the global atmospheric methane flux, highlighting the role of microbial activity in marine sediments and the water column in regulating methane concentrations in the atmosphere. As we have stated, most sedimentary methane is oxidized within the sediments by AOM, and the excess methane discharging at the seafloor is consumed aerobically near the sediment-water interface (Sommer et al., 2006) or within the overlying water column (Valentine et al., 2001). There are concerns, however, that methane currently trapped within the extensive Arctic shelves, as either gas hydrate or within "ice-bonded" sediments, can be discharged directly into the atmosphere as comparatively warmer Arctic Ocean waters flood the relatively cold permafrost areas of the shelf, driven by the continuing warming effects associated with the Holocene sea-level rise.

Summary

Methane forms primarily from the breakdown of organic matter, which can be mediated by microbial activity or occur through thermochemical processes. Thus highest methane contents occur in rapidly accumulating sediments that underlie regions of high primary productivity, such as those bordering most continental margins. Methane in marine sediments occurs in dissolved, gaseous, hydrated, or adsorbed states, as determined by thermodynamic constraints as well as by sediment structure and composition. In the presence of geologic traps, methane accumulates in the sediments as natural gas, and these deposits are commonly exploited commercially. However, the biggest methane reservoir in marine sediments is thought to exist in the form of methane hydrate, where methane molecules are trapped inside a water cage.

In the absence of geologic or gas hydrate traps, methane tends to migrate through the sediment via diffusion or advection. During this migration, methane can be effectively consumed by anaerobic microbial oxidation, a process which is thought to remove ~90 % of sedimentary methane, most likely via sulfate reduction mediated by Deltaproteobacteria in conjunction with anaerobic methanotrophic archaea (ANMEs). Methane oxidation in marine sediments is an area of active research, and various alternate electron acceptors and microbial pathways have been identified. At high advective rates, the large flux of methane overwhelms the microbial filter, and it is discharged at the seafloor in what is known as cold seeps, the most impressive of which are imaged as bubble plumes in the water column. Notwithstanding this massive discharge, the global methane input from marine sediments, including that from marine gas hydrate, accounts for a few percent of the global atmospheric methane flux.

Bibliography

Arndt, S., Hetzel, A., and Brumsack, H. J., 2009. Evolution of organic matter degradation in Cretaceous black shales inferred from authigenic barite: a reaction-transport model. *Geochimica et Cosmochimica Acta*, **73**(7), 2000–2022.

Boetius, A., and Wenzhöfer, F., 2013. Seafloor oxygen consumption fuelled by methane from cold seeps. *Nature Geoscience*, doi:10.1038/ngeo1926.

Boetius, A., Ravenschlag, K., Schubert, C., Rickert, D., Widdel, F., Gieseke, A., Amann, R., Jørgensen, B. B., Witte, U., and Pfannkuche, O., 2000. A marine microbial consortium apparently mediating anaerobic oxidation of methane. *Nature*, **407**, 623–626.

Bohrmann, G., and Torres, M. E., 2006. Gas hydrates in marine sediments. In Schulz, H. D., and Zabel, M. (eds.), *Marine Geochemistry*. Berlin/Heidelberg: Springer, pp. 481–512.

Briggs, B. R., Inagaki, F., Morono, Y., Futagami, T., Huguet, C., Rosell-Mele, A., Lorenson, T., and Colwell, F. S., 2012. Bacterial dominance in subseafloor sediments characterized by gas hydrates. *FEMS Microbiology Ecology*, **81**, 88–98.

Claypool, G. E., and Kaplan, I. R., 1974. The origin and distribution of methane in marine sediments. In Kaplan, I. R. (ed.), *Natural Gases in Marine Sediments*. New York: Plenum Press, pp. 99–139.

Colwell, F. S., Boyd, S., Delwiche, M. E., Reed, D. W., Phelps, T. J., and Newby, D. T., 2008. Estimates of biogenic methane production rates in deep marine sediments at Hydrate Ridge, Cascadia Margin. *Applied and Environmental Microbiology*, **74**, 3444–3452.

Dickens, G. R., 2011. Down the Rabbit Hole: toward appropriate discussion of methane release from gas hydrate systems during the Paleocene-Eocene thermal maximum and other past hyperthermal events. *Climate of the Past*, **7**, 831–846.

Hoehler, T., Alperin, M. J., Albert, D. B., and Martens, C., 1994. Field and laboratory studies of methane oxidation in an anoxic marine sediment: evidence for a methanogen-sulfate reducer consortium. *Global Biogeochemical Cycles*, **8**, 451–463.

Joye, S. B., 2012. A piece of the methane puzzle. *Nature*, **491**, 538–539.

Kvenvolden, K. A., and Lorenson, T. D., 2001. The global occurrence of natural gas hydrate. *Geophysical Monograph*, **124**, 87–98.

Milkov, A. V., 2004. Global estimates of hydrate-bound gas in marine sediments: how much is really out there? *Earth-Science Reviews*, **66**, 183–197.

Milucka, J., Ferdelmann, T. G., Polerecky, L., Franzke, D., Wegener, G., Schmid, M., Lieberwirth, I., Wagner, M., Widdel, F., and Kuypers, M. M. M., 2012. Zero-valent sulphur is a key intermediate in marine methane oxidation. *Nature*, **491**, 541–546.

Niemann, H., Lösekann, T., de Beer, D., Elvert, M., Nadalig, T., Knittel, K., Amann, R., Sauter, E. J., Schlüter, M., Klages, M., Foucher, J. P., and Boetius, A., 2006. Novel microbial communities of the Haakon Mosby mud volcano and their role as methane sink. *Nature*, **443**, 854–858.

Reeburgh, W. S., 2007. Oceanic methane biogeochemistry. *Chemical Reviews*, **107**, 486–513.

Römer, M., Sahling, H., Pape, T., Spieß, V., and Bohrmann, G., 2012. Quantification of gas bubble emissions from submarine hydrocarbon seeps at the Makran continental margin (offshore Pakistan). *Journal of Geophysical Research, Oceans*, **117**, C10015, doi:10.1029/2011JC007424.

Schoell, M., 1988. Multiple origins of methane in the Earth. *Chemical Geology*, **71**(1–3), 1–10.

Sommer, S., Pfannkuche, O., Linke, P., Luff, R., Greinert, J., Drews, M., Gubsch, S., Pieper, M., Poser, M., and Viergutz, T., 2006. Efficiency of benthic filter: biological control of the emission of dissolved methane from sediments containing shallow gas hydrates at Hydrate Ridge. *Global Biogeochemical Cycles*, **20**, doi:10.1029/2004GB002389.

Suess, E., 1980. Particulate organic carbon flux in the oceans – surface productivity and oxygen utilization. *Nature*, **288**, 260–263.

Takai, K., Nakamura, K., Toki, T., Tsunogai, U., Miyazaki, M., et al., 2008. Cell proliferation at 122 degrees C and isotopically heavy CH4 production by a hyperthermophilic methanogen under high-pressure cultivation. *Proceedings of the National Academy of Sciences of the United States of America*, **105**, 10949–10954.

Valentine, D. L., 2011. Emerging topics in marine methane biogeochemistry. *Annual Review of Marine Science*, **3**, 147–171.

Valentine, D. L., Blanton, D. C., Reeburgh, W. S., and Kastner, M., 2001. Water column methane oxidation adjacent to an area of active hydrate dissociation, Eel River Basin. *Geochimica et Cosmochimica Acta*, **65**, 2633–2640.

Wellsbury, P., Goodman, K., Cragg, B.A., and Parkes, R.J., 2000. The geomicrobiology of deep marine sediments from Blake Ridge containing methane hydrate (Sites 994, 995 and 997). In Paull, C., Matsumoto, R., Wallace, P.J., and Dillon, W.P. (eds.), *Proceeding of ODP*, Vol. 164, pp. 379–391.

Whiticar, M. J., 1999. Carbon and hydrogen isotope systematics of bacterial formation and oxidation of methane. *Chemical Geology*, **161**, 291–314.

Whiticar, M. J., Faber, E., and Schoell, M., 1986. Biogenic methane formation in marine and freshwater environments. CO_2 reduction vs. Acetate fermentation – Isotope evidence. *Geochimica et Cosmochimica Acta*, **50**, 693–709.

Wilhelms, A., Larter, S. R., Head, I., Farrimond, P., di Primio, R., and Zwach, C., 2001. Biodegradation of oil in uplifted basins prevented by deep-burial sterilization. *Nature*, **411**, 1034–1037.

Zonneveld, K. A. F., Versteegh, G. J. M., Kasten, S., Eglinton, T. I., Emeis, K. C., Huguet, C., Koch, B. P., de Lange, G. J., de Leeuw, J. W., Middelburg, J. J., Mollenhauer, G., Prahl, F., Rethemeyer, J., and Wakeham, S., 2010. Selective preservation of organic matter in marine environments; processes and impact on the fossil record. *Biogsciences*, **7**, 483–511.

Cross-references

Chemosynthetic Life
Cold Seeps
Deep Biosphere
Energy Resources
Marine Gas Hydrates

Mg/Ca PALEOTHERMOMETRY

Dirk Nürnberg
GEOMAR Helmholtz Centre for Ocean Research, Kiel, Germany

Definition

Mg/Ca paleothermometry: Geochemical proxy used in paleoceanography to reconstruct past temperatures of the ocean surface.

Reliable sea surface temperature (SST) estimates are crucial to the reconstruction and modeling of past climate change. Reconstructions of the thermal state of the ocean, in turn, may help to assess the significance of instrumentally observed climate variability and may validate global circulation models used to predict future climate change. Most promising in this respect is the Mg/Ca paleothermometry, which is based on the temperature dependence of the substitution of magnesium into biogenic calcite. It has long been known that tropical calcitic shells of marine organisms are generally more enriched in magnesium than subpolar shells (e.g., Savin and Douglas, 1973). But in particular for the important group of foraminifera (marine protozoa), a well-defined and species-specific calibration of the shell Mg/Ca ratio on ocean temperature was missing until the mid-1990s (Nürnberg, 1995; Nürnberg et al., 1996). Such species-specific Mg/Ca versus temperature relationships, however, are essential for all further quantitative paleotemperature reconstructions.

Meanwhile, the analytical protocol for Mg analyses has established (Greaves et al., 2008), and the Mg/Ca paleothermometry providing an SST accuracy of ~0.5–1 °C became one of the most important tools in paleoceanography. From cultivation experiments and core-top calibration studies, an exponential relationship between temperature and foraminiferal Mg/Ca was suggested for most planktonic species (e.g., Elderfield and Ganssen, 2000; Anand et al., 2003; Regenberg et al., 2009). Although temperature is the primary control on Mg/Ca, salinity and pH exert additional influence on Mg^{2+} uptake. Mg/Ca positively changes with salinity (Nürnberg et al., 1996; Kisakürek et al., 2008), which requires caution when expecting large-scale salinity variations through time. Seawater pH, instead, is negatively correlated to Mg^{2+} incorporation (Kisakürek et al., 2008). Both effects are assumed to cancel each other (Lea et al., 2000).

The method offers unique advantages compared with other SST proxies. Most importantly, Mg/Ca is measured on the same biotic carrier as stable oxygen isotopes ($\delta^{18}O$), thereby avoiding the seasonality and/or habitat effects that occur when proxy data from different faunal groups are used. Also, the magnitude and timing of SST and $\delta^{18}O$ changes can be separated, allowing to constrain the timing of surface ocean warming with respect to continental ice sheet melting and to approximate changes in ocean salinity (e.g., Nürnberg, 2000).

Most prejudicial to Mg/Ca paleothermometry is that calcite dissolution may cause the preferential removal of relatively soluble magnesium-enriched calcitic parts of the shells (Brown and Elderfield, 1996) and thus may limit the applicability of Mg/Ca as a tracer for SST. Regenberg et al. (2006) defined the oceanic 18–26 μm kg^{-1} ΔCO_3^{2-} threshold of calcite saturation as critical for the removal. Various algorithms to correct for dissolution effects were proposed accordingly (Rosenthal et al., 2000; Dekens et al., 2002; Regenberg et al., 2006). Whether the long-term changes in seawater Mg/Ca (Tyrrell and Zeebe, 2004) need to be considered for the Mg/Ca paleothermometry, it is still deficiently studied.

Bibliography

Anand, P., Elderfield, H., and Comte, M. H., 2003. Calibration of Mg/Ca thermometry in planktonic foraminifera from a sediment trap time series. *Paleoceanography*, **18**, PA1050, doi:10.1029/2002PA000846.

Brown, S. J., and Elderfield, H., 1996. Variations in Mg/Ca and Sr/Ca ratios of planktonic foraminifera caused by postdepositional dissolution: evidence of shallow Mg-dependent dissolution. *Paleoceanography*, **11**(5), 543–551.

Dekens, P. S., Lea, D. W., Pak, D. K., and Spero, H. J., 2002. Core top calibration of Mg/Ca in tropical foraminifera: refining paleotemperature estimation. *Geochemistry, Geophysics, Geosystems*, **3**, 1–29, doi:10.1029/2001GC00.

Elderfield, H., and Ganssen, G., 2000. Past temperature and $\delta^{18}O$ of surface ocean waters inferred from foraminiferal Mg/Ca ratios. *Nature*, **405**, 442–445.

Greaves, M., Caillon, N., Rebaubier, H., Bartoli, G., Bohaty, S., Cacho, I., Clarke, L., Cooper, M., Daunt, C., Delaney, M., deMenocal, P., Dutton, A., Eggins, S., Elderfield, H., Garbe-Schönberg, C. D., Goddard, E., Green, D., Groeneveld, J., Hastings, D., Hathorne, E., Kimoto, N., Klinkhammer, G., Labeyrie, L., Lea, D. W., Marchitto, T., Martínez-Botí, M. A., Mortyn, P. G., Ni, Y., Nürnberg, D., Paradis, G., Pena, L., Quinn, T., Rosenthal, Y., Russell, A., Sagawa, T., Sosdian, S., Stott, L., Tachikawa, K., Tappa, E., Thunell, R., and Wilson, P. A., 2008. Interlaboratory comparison study of calibration standards for foraminiferal Mg/Ca thermometry. *Geochemistry, Geophysics, Geosystems*, **9**(8), 1–27, doi:10.1029/2008GC001974.

Kisakürek, B., Eisenhauer, A., Böhm, F., Garbe-Schönberg, D., and Erez, J., 2008. Controls on shell Mg/Ca and Sr/Ca in cultured planktonic foraminiferan, *Globigerinoides ruber* (white). *Earth and Planetary Science Letters*, **273**, 260–269, doi:10.1016/j.epsl.2008.06.026.

Lea, D. W., Pak, D. K., and Spero, H. W., 2000. Climate impact of later Quaternary equatorial Pacific sea surface temperature variations. *Science*, **289**, 1719–1724.

Nürnberg, D., 1995. Magnesium in tests of Neogloboquadrina pachyderma sinistral from high northern and southern latitudes. *Journal of Foraminiferal Research*, **25**, 350–368.

Nürnberg, D., 2000. Taking the temperature of past ocean surfaces. *Science*, **289**, 1698–1699.

Nürnberg, D., Bijma, J., and Hemleben, C., 1996. Assessing the reliability of magnesium in foraminiferal calcite as a proxy for water mass temperatures. *Geochimica et Cosmochimica Acta*, **60**, 803–814.

Regenberg, M., Nürnberg, D., Steph, S., Groeneveld, J., Garbe-Schönberg, D., Tiedemann, R., and Dullo, W. C., 2006. Assessing the effect of dissolution on planktonic foraminiferal Mg/Ca ratios: evidence from Caribbean core tops. *Geochemistry, Geophysics, Geosystems*, **7**, 1–23, doi:10.1029/2005GC001019002E.

Regenberg, M., Steph, S., Nürnberg, D., Tiedemann, R., and Garbe-Schönberg, D., 2009. Calibrating Mg/Ca ratios of multiple planktonic foraminiferal species with $\delta^{18}O$-calcification temperatures: paleothermometry for the upper water column. *Earth and Planetary Science Letters*, **278**, 324–336, doi:10.1016/j.epsl.2008.12.019.

Rosenthal, Y., Lohmann, G. P., Lohmann, K. C., and Sherrell, R. M., 2000. Incorporation and preservation of Mg in *Globigerinoides sacculifer*: implications for reconstructing the temperature and $^{18}O/^{16}O$ of seawater. *Paleoceanography*, **15**, 135–145.

Savin, S. M., and Douglas, R. G., 1973. Stable isotope and magnesium geochemistry of recent planktonic foraminifera from the South Pacific. *Geological Society of America Bulletin*, **84**, 2327–2342.

Tyrrell, T., and Zeebe, R. E., 2004. History of carbonate ion concentration over the last 100 million years. *Geochimica et Cosmochimica Acta*, **68**(17), 3521–3530.

Cross-references

Foraminifers (Benthic)
Foraminifers (Planktonic)
Paleoceanographic Proxies
Paleoceanography
Reefs (Biogenic)

MID-OCEAN RIDGE MAGMATISM AND VOLCANISM

Ken H. Rubin
Department of Geology & Geophysics, SOEST,
University of Hawaii, Honolulu, HI, USA

Definition

Mid-Ocean Ridge: A linear, narrow volcanic and tectonic region which marks the constructive boundary between two tectonic plates. It is divided into segments by transform faults and other offsets. The global ridge system cuts through every major ocean basin and comes ashore in a few places like Iceland. It forms an approximately 65,000 km long, globe-encircling, largely submarine mountain chain.

Magmatism: Magmatism is the production and migration of magma, which is molten rock produced from the partial or complete melting of solid materials within a planetary body. When magma erupts on the surface, it is known as lava.

Volcanism: Volcanism is the eruption of molten rock, hot gases, or solidified rock fragments from an opening ("vent") in the Earth's crust. Volcanism occurs on Earth and other planets and moons. Most of the volcanism on Earth occurs at mid-ocean ridges, almost always sight unseen.

Mid-Ocean Ridge Magmatism: By far, the dominant type of lava resulting from magmatic activity at mid-ocean ridges is basalt, also called mid-ocean ridge basalt (MORB). However, small amounts of other extrusive magma types (predominantly andesite, dacite, and picrite) also erupt there. Variations in the rate and style of magma production, volcanism, and faulting result from variations in upper mantle temperature, upper mantle composition, and seafloor spreading rate through the ridge system, causing differences in igneous ocean crustal thickness, structure, and composition.

Mid-Ocean Ridge Volcanism: Volcanic eruptions at ocean ridges vary in style, intensity, and duration around the globe as a function of parameters such as magma chamber depth, magma supply rate, and eruption depth beneath sea-level. Nevertheless, the dominant eruption styles either produce fissure-fed lava flows or point-source, single vent volcanic lava mounds. Nearly all volcanic products at ridges exhibit rapid quench textures, formed by the freezing of molten magma upon contact with cold seawater.

Introduction and history

Magmatism at mid-ocean ridges is one of our planet's most important geological processes, forming a dense, low-lying, igneous crust that floors the vast ocean basins, making up the solid rocks over nearly two-thirds of the Earth's surface. For most of geological history, the greatest number and volume of volcanic eruptions on Earth has occurred in the ocean, along mid-ocean ridges. Earth's globe-encircling volcanic system is a truly colossal geological structure (Figure 1). Yet it is one that many people are not aware of because the global mid-ocean ridge system mostly lies beneath the sea (except in rare occurrences, such as in Iceland, where the mid-ocean ridge is exposed on land due to an anomalously thickened crust from the effects of the colocated Iceland Hotspot). Viewed from space and with the world's oceans stripped away, the mid-ocean ridge system would stand out as one of our planet's most prominent surface geological features. The ridge system generally produces a layered crust, with volcanic rocks on top, coarse-grained intrusive rocks at depth, and an intervening layer of dikes. Together, these layers form the igneous basement of the oceanic crust.

Mid-ocean ridges are segmented at various length scales into volcano-scale units as well as longer coherent segments that are separated by major offsets and discontinuities (e.g., Macdonald, 1982). Although at their upper surface mid-ocean ridges are dominated by volcanic rocks, the number and frequency of volcanic eruptions that formed them is not well known (e.g., Rubin et al., 2012). Compared to eruptions on land, we know less about submarine eruptions and the magmas that drive them, especially those that occur in the deep sea (here defined as below 500 m depth) because eruptions in the deep ocean are much more difficult to detect and observe (Figure 1). Our first knowledge of submarine volcanic rocks from mid-ocean ridges probably comes from reports by deep-sea cable repair ships (Hall, 1876) and the Challenger Expedition (Murray and Renard, 1891), both of which recovered volcanic rocks from the seafloor in the 1870s. Thus, while geologists working on land have had thousands of years to develop an understanding of the magmatic processes that drive subaerial volcanism, marine geologists have had only a century or so. Still, concerted multinational efforts in the last 35–40 years have accelerated the pace of learning, with several marine expeditions to map, sample, and observe mid-ocean ridges occurring around the globe each year. Sampling campaigns over the past several decades have resulted in greatly increased, but highly uneven, sampling of the global ridge system. Some areas have few to no known rocks recovered from them, and at the other extreme, several locations have sampling density equivalent to the best-studied volcanoes on land (with 10–100 samples taken per km of ridge; Figure 1).

Historically active volcanoes in the past 500 years

○ Subaerial (n = 497) ● Shallow submarine (n = 23) ● ● Deep submarine (n = 17)

sample density along the mid-ocean ridge (km^{-1})

0.001   0.01   0.1   1   10   100

Mid-ocean Ridge Magmatism and Volcanism, Figure 1 Map of the world's continents (*green*) and oceans (*blue*) with major plate boundaries (*black lines*) and several types of volcano eruption and sampling data depicted. The *white*, *gray*, and *black circles* represent the locations of all documented volcanic eruptions on Earth over the past 500 years (as described in the caption of Rubin et al., 2012, Figure 1), and the colored dots represent sampling density of lavas from the world's mid-ocean ridge system (as described in the caption of Rubin et al., 2009, Figure 3; sample sites were compiled from the petrologic database, http://www.petdb.org). Notice that sample density along mid-ocean ridges varies by a wide range and that most of the ridge system is highly under-sampled (less than 1 sample per 10 km or not sampled at all). Notice also that there have been far more subaerial eruptions than submarine ones detected over the past 500 years globally, which is due to the much greater difficulty of identifying submarine eruptions, especially deep ones, and due to the relative youth of the field of submarine volcano studies. Shallow submarine eruptions leave a surface expression (as volcanic debris or ash/gas plumes) and are somewhat easier to detect and study than deep submarine eruptions. Most of the known deep submarine eruptions have occurred at mid-ocean ridges.

Mid-ocean ridge basalt (MORB) is one of Earth's most common and most intensely studied rock types. Murray and Renard (1891) published an early account of the composition of MORB, demonstrating that it was similar in major element composition to basalts erupted on land. Major progress in understanding the range of compositions of volcanic rocks erupted on mid-ocean ridges began roughly 75 years later, with the exploration and sampling of ridges during the early years of testing and acceptance of the theory of plate tectonics. Initially, samples were collected mostly by dredging, although by the 1970s, scientists had also begun to deploy both human-occupied submersibles, such as *Alvin*, to directly observe and sample the seafloor (e.g., Ballard and Van Andel, 1977), and to use drilling, via the Deep Sea Drilling Project (DSDP) and its successor programs to recover intrusive rocks from beneath the seafloor. Evidence that these mid-ocean ridge volcanoes were underlain by active magmatic systems came via the discovery of warm springs on the Galápagos Spreading Center in 1977 (Corliss et al., 1979) and hot springs on the East Pacific Rise in 1979 (Spiess et al., 1980). Soon thereafter, ever-improving observational technologies paved the way for discoveries of specific eruptions (e.g., Embley et al., 1991; Haymon et al., 1993; see also reviews in Baker et al., 2012; Rubin et al., 2012). Exploration and monitoring of mid-ocean ridges continue to this day as we expand our understanding of magmatism and volcanism there. Current understanding of the recent volcanic history of several submarine mid-ocean ridge volcanoes now rivals that at volcanoes on land (Rubin et al., 2012; Clague et al., 2013).

Magma generation at mid-ocean ridges

Overview

Separation of tectonic plates causes compositionally heterogeneous mantle beneath mid-ocean ridges (e.g., Allègre et al., 1984) to rise and decompress, partially melting the upper mantle (e.g., Klein and Langmuir, 1987; Kelemen et al., 1997) and producing basaltic parent melts (e.g., Elthon, 1979). Magmatic compositions are modified by multiple processes during ascent and storage prior to eruption (e.g., Klein and Langmuir, 1987; Sinton and Detrick, 1992; Rubin and Sinton, 2007). Mid-ocean ridge magma compositions therefore reflect a combination of magmatic processes related to (a) the composition and temperature of mantle beneath the ridge, (b) the volcanic plumbing system within ridge volcanoes, and (c) cooling from the infiltration of overlying seawater along faults and fissures in the upper crust leading to magma differentiation (e.g., Macdonald, 1982; Phipps Morgan and Chen, 1994). Variations in the rate and style of magma production, volcanism, and faulting result from upper mantle temperature and composition variations (e.g., Dalton et al., 2014) and a change in seafloor spreading rate from <1 cm/year to 20 cm/year through the ridge system, causing differences in igneous ocean crustal thickness (e.g., Bown and White, 1994), structure and morphology (e.g., Small, 1998), and composition (e.g., Rubin and Sinton, 2007).

Away from hotspots, magma supply to ridges from mantle melting covaries positively with spreading rate, so that high spreading rate generally equates to high magma supply. Superimposed on this plate-spreading-driven magma supply, differences in mantle composition, melting, and melt transport further modulate melt supply to the ridge axis along its length (e.g., Langmuir et al., 1986; Sinton et al., 1991), in turn producing physical and chemical segmentation of the mid-ocean ridge at 10–100 km length scales. Upward migration of MORB melts through the mantle, followed by accumulation in crustal magma bodies called magma chambers, results in magma differentiation and crystallization (e.g., Klein and Langmuir, 1987; Grove et al., 1992; Herzberg, 2004) and perhaps to melt-rock reactions (e.g., Lissenberg and Dick, 2008; Lissenberg et al., 2013) prior to either eruption or emplacement as a pluton within the oceanic crust. Structural and thermal conditions within the mantle and crust produce magma supply and magmatic differentiation variations at even smaller spatial scales (i.e., 1–10 km along the ridge axis), producing spatial patterns in erupted magma compositions akin to those found within and between neighboring terrestrial volcanoes (e.g., Reynolds et al., 1992; Perfit et al., 1994; Perfit and Chadwick, 1998). Studies of individual eruption deposits (e.g., Rubin et al., 2001; Bergmanis et al., 2007; Goss et al., 2010; Colman et al., 2012) and dikes that feed eruptions (e.g., Pollock et al., 2009) reveal spatial geochemical patterns that record variations in the delivery of individual magma batches to, and magma mixing and differentiation conditions within, subaxial magma reservoirs from which they erupt.

Magma and mantle source compositions

Like all mafic magma generation on Earth, the major element composition of MORB magmas is largely set by the composition of the parent rock in the mantle source, the melting depth, and the degree to which the source rock is partially melted. MORB major element compositions vary less than basalts erupted on land, in part because the aforementioned melting parameters and source compositions are comparatively homogeneous beneath ridges and in part because the thinner and compositionally more uniformly oceanic crust imparts significantly less chemical variation to MORB magmas during magma accumulation and differentiation before eruption. Still, MORB display geographical variations in major element, trace element, and radiogenic isotopic compositions that reflect regional differences in mantle composition and melting conditions around the globe (e.g., Dupre and Allègre, 1980; Hamelin and Allègre, 1985; Klein and Langmuir, 1987; Mahoney et al., 1989; Asimow et al., 2004; Hanan et al., 2004; Niu and O'Hara, 2008), as well as the influence of near-ridge hotspots (e.g., Schilling, 1991; Cushman et al., 2004).

Globally, the trace constituents in MORB (trace elements and radiogenic isotopes) vary more significantly than the major elements. Petrologists use these constituents to understand the mineralogy, chemical composition, and history of a magma's mantle source rock, as well as to help track preeruptive magma differentiation and magma-host rock reactions in the crust. MORB have been widely used to identify variations in upper mantle composition although such mantle signatures in MORB still need to be distinguished from variations imparted to the magmas by melting, magma transport, melt-rock reactions, mixing, and magma differentiation (e.g., see review by Rubin et al., 2009). Of particular importance are the degree to which different types of mantle rocks (i.e., lithologies), also known as "mantle heterogeneities," melt at different rates and proportions (e.g., Ito and Mahoney, 2005; Stracke et al., 2006; Russo et al, 2009) and the extent of subsequent magma mixing and homogenization (Rubin and Sinton, 2007). The latter works to partially or fully obscure inherent mantle source rock variability in aggregated melts of the mantle, becoming more pronounced as magma supply increases.

Magma chamber processes

Nearly all MORB erupt bearing the chemical and mineralogical signatures of magma differentiation (e.g., olivine and plagioclase crystallization) and partial loss of initial melt magmatic gas loads, essentially requiring that these magmas accumulated and partially crystallized in magma bodies at one or more locations in the crust before eruption. There is considerable debate as to the nature of these magma bodies, where they reside, and how they impart

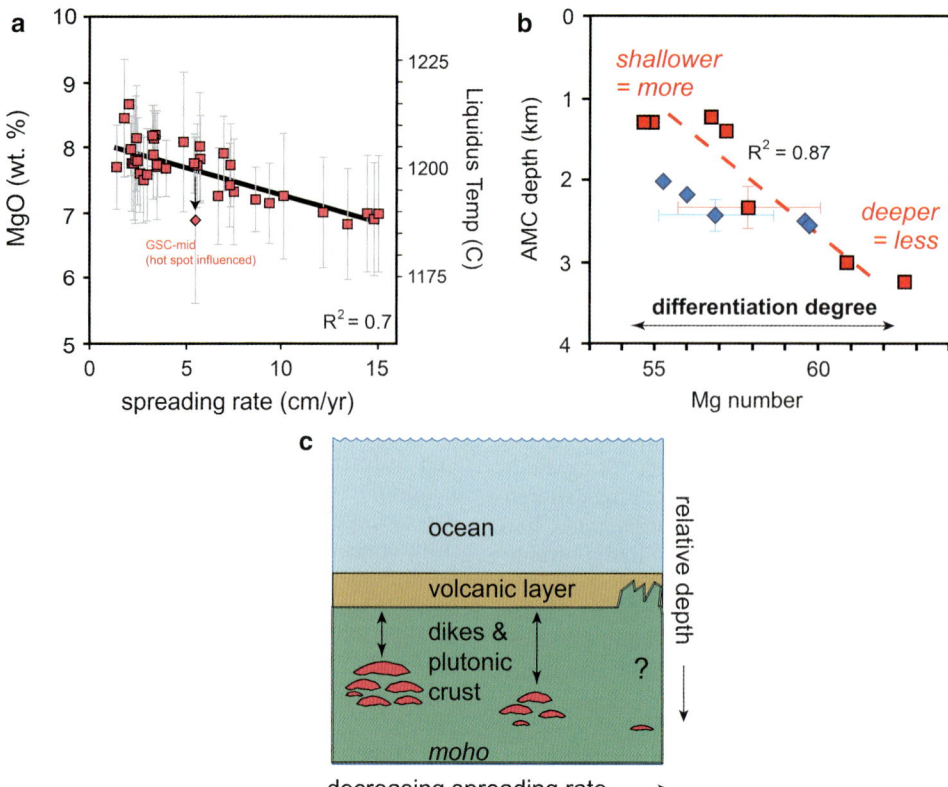

Mid-ocean Ridge Magmatism and Volcanism, Figure 2 Three diagrams depicting compositional attributes of mid-ocean ridge basalts that can be used to infer conditions of magma differentiation, as discussed in the text. *Panel a*, which is modified from Figure 1 of Rubin and Sinton (2007) (see that paper for details of data analysis and spreading rate calculations), shows Mg variations and inferred eruption temperature in ridge section means from different geographic regions of the global mid-ocean ridge system using a >11,000 basalt glass database. Note the strong linear relationships with spreading rate (the *solid black line* is a linear regression to the data). *Gray bars* are 1 s deviations for each mean. Overall, erupted compositions are cooler and more differentiated at faster-spreading rate ridges and also more variable in differentiation-related parameters (see Rubin and Sinton, 2007 for a detailed discussion). The central Galapagos Spreading Center, which is an example of a hotspot-influenced ridge having anomalously high magma supply due to active upwelling, falls off the trend (*diamond symbol*). *Panel b*, axial magma chamber (AMC) depth versus Mg content of erupted MORB showing variations in both means of ridge sections (from *panel a*) also having AMC depth measurements (*red symbols*) and segments of the Juan de Fuca ridge having AMC depth (*blue symbols*). See the caption in Figure 8 in Rubin and Sinton (2007) for details of data sources and handing. This panel shows that magmas erupted from deeper magma chambers are generally less differentiated. Magma chambers are generally shallower at higher spreading rates and deeper at slower-spreading rates, where overall melt supply is high and low, respectively. The linear regression includes the *red* symbols. *Panel c*, A generalized depiction of the greater depth, smaller size, and lower connectivity of AMCs at low spreading rates, as discussed in Rubin et al., (2009). A wide range of geophysical and geochemical parameters were used to infer these variations, as discussed briefly in this article and in detail in Rubin et al. (2009).

structure to the oceanic crust (e.g., see discussion of the "gabbro-glacier" and "sheeted-sill" models in Coogan, 2007 and Keleman and Aharonov, 1998). An important characteristic of many mid-ocean ridges is the axial magma chamber (AMC), discovered in the 1980s by seismic reflection studies. The AMC represents the shallowest level of magma accumulation in the crust (Sinton and Detrick, 1992) and can be segmented at a variety of spatial and temporal scales (e.g., Carbotte et al., 2013). Far fewer AMC reflections have been detected at the ridges with very slow to slow-spreading rates (i.e., < 1–2 cm/year), despite most magmas erupted there showing the chemical and mineralogical signatures of differentiation at crustal pressures, which implies that the AMC bodies there are mostly too small and discontinuous to be seismically imaged. Where they can be detected, the size, continuity, and depth of the AMC vary substantially through the ridge system, but overall these parameters scale with spreading rate: the AMC generally becomes more continuous and shallower (as little as 1.5 or 2 km below the seafloor) at the fastest-spreading rates (Figure 2). This depth variation has been successfully modeled as an interplay between the rate of magma supply from below and the intensity of hydrothermal cooling of the shallowest crust from above (Phipps Morgan and Chen, 1994).

Global MORB magma compositions support the Phipps Morgan and Chen (1994) heat supply model: MORB major element compositions display a general relationship between petrologic indicators of fractionation depth and spreading rate, with evidence for generally higher magma crystallization pressures (Sinton and Detrick, 1992; Michael and Cornell, 1998) and magmatic gas equilibration pressures (Paonita and Martelli, 2007) being more common at slow-spreading ridges. A more recent study found very regular variations in MORB chemistry with spreading rate and AMC depth in a large, 11,000 sample group of MORB from around the globe (Figure 2). Here, the least differentiated MORB were found to occur at slow-spreading rates, becoming progressively more differentiated and erupting from a shallower AMC as spreading rate increases (Rubin and Sinton, 2007). This dichotomy, that cooler magmas generally erupt from ridges with the highest magma supply and hotter magmas from ridges with the lowest magma supply, demonstrates the importance of the AMC and its depth within the crust for modulating MORB composition.

At shorter temporal and spatial scales, chemical variations in MORB within and between successive historical eruptions at sites having an AMC show that magma chamber thermal conditions were relatively constant at these few sites over this time interval, with just 10–20 °C of magmatic heat loss over one to two decades (e.g., at 9° 50′N East Pacific Rise, Goss et al. 2010; 17° S East Pacific Rise, Bergmanis et al., 2007; CoAxial segment of the Juan de Fuca Rise, Embley et al., 2000). Many of these same eruption deposits also preserve magma temperature gradients along the eruptive fissures that indicate thermal variations within the AMC along axis that are not mixed out during eruptions (e.g., Rubin et al., 2001; Sinton et al., 2002; Bergmanis et al., 2007; Goss et al., 2010).

Evolved magma compositions

Small volumes of high-silica lavas (andesites and dacites) occur on mid-ocean ridges, often associated with ridge discontinuities such as propagating rift tips (e.g., Sinton et al., 1983), overlapping spreading centers (e.g., Wanless et al., 2010), and ridge-transform intersections (e.g., Schmitt et al., 2011); however, highly differentiated lavas can also occur in more "normal" ridge settings (e.g., Wanless et al., 2010) and at regions showing ridge-hotspot interaction (Haase et al., 2005; Colman et al., 2012). Many authors have suggested a tectonic control on the occurrence of highly differentiated lavas on mid-ocean ridges, primarily by allowing for extensive cooling of the AMC margins near places where magma supply might be lower. This promotes conditions that allow for extensive fractional crystallization of the magmas. Similar highly differentiated lavas also occur in Iceland, where strong shifts in oxygen isotopic composition relative to mantle values indicate a major role as well for the melting of altered crust (e.g., Martin and Sigmarsson, 2007). The bulk composition of many highly evolved mid-ocean ridge lavas is remarkably similar, and several different occurrences have been successfully modeled by a combination of extensive MORB fractional crystallization plus/minus partial melting and/or assimilation of seawater-altered oceanic crust (e.g., Haase et al., 2005; Wanless et al., 2010). Radiometric dating of zircons from dacite lava domes erupted near the intersection of the Juan de Fuca ridge and the Blanco transform (in the NE Pacific Ocean) indicates that this differentiation can occur rapidly (over just 10 millennia or so; Schmitt et al., 2011).

The plutonic crust

Beneath the volcanic carapace at mid-ocean ridges lies a thick plutonic crust. This intrusive material comprises roughly five times the volume of the extrusive (volcanic) section (White et al., 2006). Direct in situ observation of this crust is difficult, so that much of what we know about this relatively inaccessible material comes from drill hole studies. However, rare plutonic exposures on the seafloor occur in fracture zones and other faulting-exposed "tectonic windows," as well as in ophiolites (pieces of young oceanic crust obducted onto land, e.g., Cann, 1974; Pallister and Hopson,1981). Collectively, studies of these materials indicate a large range of lower crustal rock types, from mafic to evolved compositions, and a range of petrologic textures, some acquired during complex cooling and melt-reaction histories (e.g., Coogan, 2007; Coogan et al., 2007; Keleman and Aharonov, 1998; Lissenberg and Dick, 2008; Lissenberg et al., 2009; Suhr et al., 2008). The extent to which magmas erupted at the surface interact with and reflect this diversity of rocks in the lower crust is an area of active research.

Volcanic eruptions at mid-ocean ridges

Volcanic activity occurs predominantly on the axis of the mid-ocean ridge system with largely unknown frequency. Our understanding of how deep submarine volcanism has built structures on mid-ocean ridges has been greatly aided by sonar and visual mapping of submarine volcanic edifices and by sampling and compositional analysis of rocks there to learn the types and distributions of volcanic products that occur. There is a large literature on this subject, much of which is reviewed in Rubin et al. (2012). In summary, mid-ocean ridges produce low-lying elongate volcanoes from the combined effects of fissure eruptions of generally low viscosity magmas and the constant rafting away of eruption deposits due to plate separation. Bathymetric relief is lower at higher magma supply (e.g., faster-spreading) ridges because volcanic construction dominates over tectonic rifting processes (e.g., Buck et al., 1997; Macdonald, 1998). In fact, the boundaries of individual volcanoes and individual eruptions become more difficult to define at the fastest-spreading ridges because magma is erupted at relatively high frequency (e.g., Sinton et al., 2002; Bergmanis et al., 2007) from nearly continuous along ridge melt axial magma chambers (e.g., Singh et al., 1998) leading to axial high and mound-like topography (e.g., White et al., 2002). At slower-spreading rates, tectonic forces (faulting and plate separation) dominate over volcanic construction, leading to the common axial valley morphology of the ridge, with volcanic accretion commonly localized along smaller ridges within this valley (e.g., Searle et al., 2010; Yeo et al., 2012) and fault block-dominated rift valley walls. At the very slowest-spreading rates, volcanic activity appears to be discontinuous along the highly tectonized ridge axis, leading to exposures of volcanic, plutonic, and even mantle rocks on the seafloor (e.g., Michael et al., 2003; Cannat et al., 2008).

Eruption products

The different styles of submarine volcanic eruptions and the deposits they produce are controlled by magma chemistry and physical properties, magma eruption rate, and physical conditions at the volcanic vent (e.g., Gregg and Fink, 1995; Perfit and Chadwick, 1998). The size, volume, thickness, and dispersal of both effusive and pyroclastic volcanic deposits provide key information about the conditions of the eruption(s) that produced them. Both effusive and explosive eruption styles are observed at mid-ocean ridges, although effusive lava flows are by far the dominant volcanic product of mid-ocean ridge eruptions. Lava flow thickness, run out, and surface morphology reflect variations in magma viscosity, effusion rate, local slope, topographic obstructions, and the sequence of emplacement events during individual eruptions (see discussion in Rubin et al., 2012 and references therein). Submarine lava flow morphologies are typically classified by the length scale of the quenched-crust units that collectively make up a lava flow, which themselves form as a function of flow rate and cooling rate. Common different types of lava deposits are known as pillow, lobate, and sheet lavas (see Figure 3). Common methods of detection and mapping of these flow types involve direct or remote visual observation from submersibles and towed camera systems (see Rubin et al., 2012), although a recent development employs high-resolution sonar data from autonomous underwater vehicles that have been automatically classified by surface geometry and texture to generate detailed lava flow morphology maps over relatively large areas of seabed (McClinton et al., 2013). Volumetrically minor deposits of lapilli- to ash-sized explosive fragments also occur throughout the depth range of observations in the deep sea (e.g., Clague et al., 2009; Helo et al., 2011).

Eruption frequency and duration

Just like volcanoes on land, the periodicity of volcanism at mid-ocean ridges reflects the rate of magma input to the volcano, the buildup of tectonic stress, and conditions within the crustal magma reservoir. Marine scientists currently lack the information to determine eruption frequency, size, and duration for all but a very small number of submarine mid-ocean ridge volcanoes (reviewed in Rubin et al., 2012). Individual eruptions can last just days to perhaps a year. In terms of eruption frequency, many mid-ocean ridge volcanoes likely operate at a quasi-steady state (e.g., Wadge, 1982) in which eruption volume and repose (the time between eruptions) correlate positively as the result of a relatively constant magma flux. Using such a steady-state assumption and calculating magma supply rate from the spreading rate and an assumption of constant crustal thickness, the smallest average eruption repose interval (~10 year) should occur at the fastest-spreading rates (for instance, on the Southern East Pacific Rise; Sinton et al., 2002 or at Axial Volcano, a ridge-centered hotspot on the Juan de Fuca ridge with anomalously high magma supply). An average repose interval of roughly 1,000 years should occur on slower-spreading ridges like the Mid-Atlantic Ridge (see Perfit and Chadwick, 1998; Sinton et al., 2002 and references therein). The small number of direct eruption observations (there are two submarine mid-ocean ridge sites that have been observed to have erupted twice) and studies utilizing volcano-scale geological mapping and deposit dating via high-precision radiometric or paleomagnetic chronological methods indicate that volcanic eruption frequency and repose are likely not steady state at the individual eruption scale, but that averaged over several eruptions, they do occur at roughly the predicted steady-state interval (e.g., Rubin et al., 2012; Bowles et al., 2014).

Eruption dynamics

Our understanding of the dynamics and mechanisms of submarine volcanic eruptions is informed by a relatively recent but growing body of direct and remote sensing observations. Intermediate- to fast-spreading mid-ocean ridges (i.e., the only places we have such observations) are dominated by short duration, high eruption rate events, or clusters of events (see Rubin et al., 2012 and references therein). Eruption rate variations produce a range of lava morphologies. Lava flows will advance for hours or days during each eruption pulse (e.g., Soule et al., 2012). Eruption pulses may continue if the pressure release was incomplete and/or pressure continues to build in the crustal melt lens by recharge. Widespread diffuse venting of hydrothermal fluids continues through the carapace of a new lava flow well after the eruption wanes. Focused flow hydrothermal activity (i.e., from chimneys) can persist at a site over multiple eruptions, although the character of fluids and chimneys at any given vent site is dominated by subterranean plumbing systems that are highly sensitive to changes that occur during eruptions (e.g., Von Damm, 2004).

Decadal-scale studies at several mid-ocean ridge sites with high melt supply reveal that magma chambers are persistent features over eruptive cycles (e.g., Carbotte et al., 2012). Magma chamber pressures increase over time (e.g., Chadwick et al., 2006), triggering increased seismicity rates along the ridge over several years (e.g., Tolstoy et al., 2006). When sufficient pressure has built in the

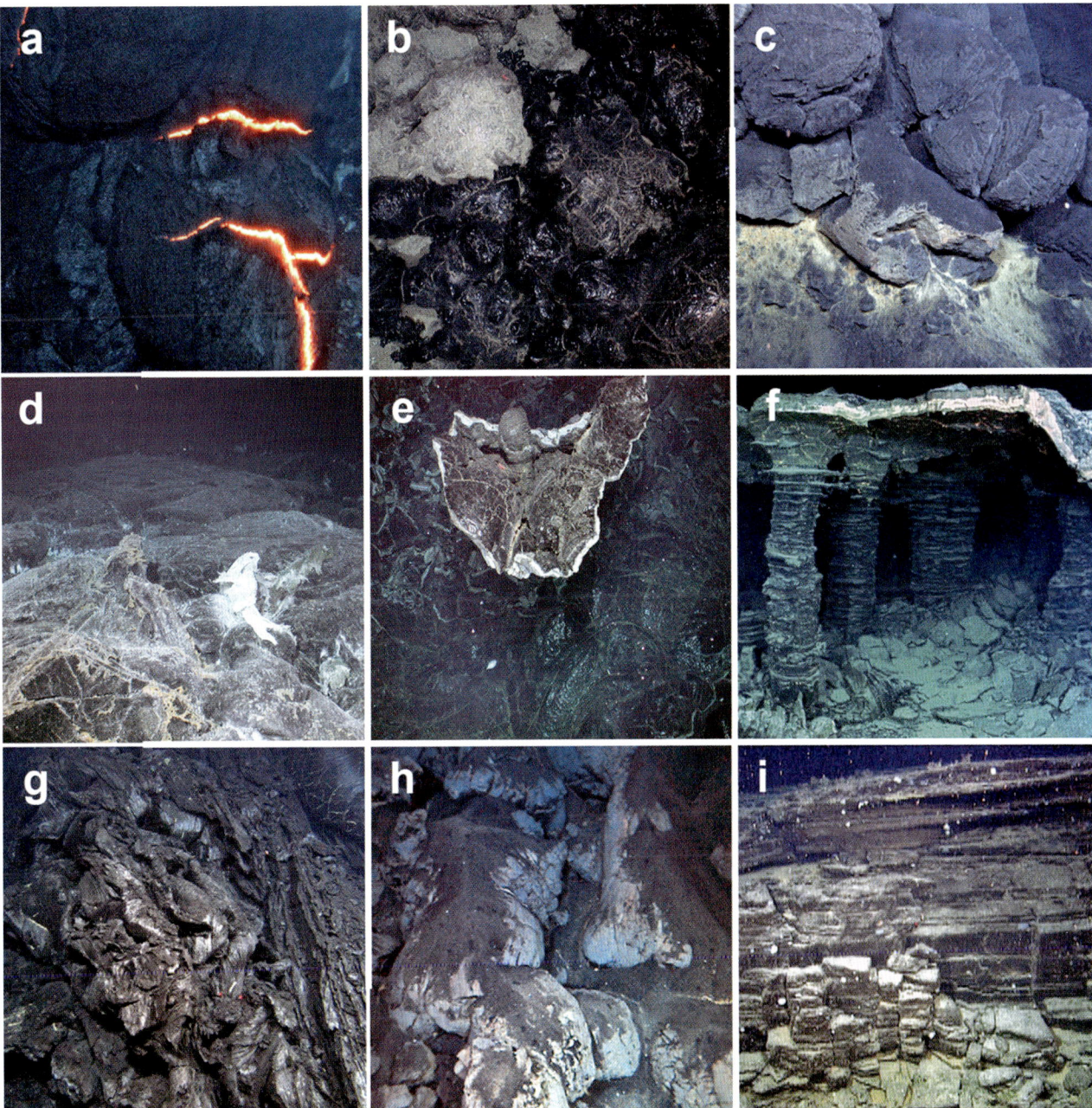

Mid-ocean Ridge Magmatism and Volcanism, Figure 3 Photographs of deep-sea eruption products roughly arranged by volcanic effusion rate (slowest at the *top* of the image) and/or eruption style (effusive or explosive). Panels *a*, *b*, and *c* depict pillow lavas which general extrude slowest and have the smallest cool units; the active pillow lava lobe inflating on a transverse and radial crack in Panel *a* is from West Mata volcano rather than a ridge setting, but the mode of emplacement is the same; Panel *b* shows a downward-looking view of 2005–2006 pillow lavas overlying older pillows at 9° 50′N EPR; Panel *c* shows a very young pillow lava at West Mata volcano colonized by an orange microbial mat. Panels *d*, *e*, and *f* depict lobate lavas, which have larger cooling units that often crust over and continue to flow underneath, leading to drainage of the interior and collapse of the surface crust; Panel *d* is of months old lobate lavas at 10° 45′N EPR covered by various microbial mats; Panel *e* is a downward-looking view of a remnant piece of lobate crust above the collapsed interior of lavas erupted in 2005–2006 at 9° 50′N on the East Pacific Rise; Panel *f* is a side-on view of 2–3 m tall lava pillars in the interior of the partially collapsed lobate lava flow erupted at Axial Seamount in 1998. Panel *g* shows a high-effusion-rate sheet flow from the NE Lau Spreading Center; Panel *h* shows young volcanic pyroclasts overlying pillow lavas from an eruption at Gakkel Ridge; Panel *i* shows a 1 m high section of consolidated pyroclastic deposit at one of the Vance Seamounts, near the JDFR. This figure is a modified subset of images of deep-sea eruptions in Rubin et al. (2012) [Photo sources: (**a**) Jason ROV dive J2-414, 2009; (**b**) WHOI TowCam, 2006; (**c**) Jason ROV dive J2-418, 2009; (**d**) Alvin HOV dive 3935, 2003; (**e**) WHOI TowCam, 2006; (**f**) ROPOS ROV dive R743, 2003 (**h**) Camper camera sled, 2007; (**i**) Tiburon ROV dive T1011, 2006].

magma body, diking initiates (e.g., Dziak et al., 2007) and may reach the surface given sufficient overpressure. Event plumes, which are thermally buoyant volumes of seawater carrying chemical signatures of volcanic and hydrothermal inputs, likely form at this time by advection of magmatic heat to the seafloor and/or discharge of hydrothermal fluids stored in the crust (e.g., Baker et al., 2011, 2012). The amount of magmatic gas within the system and the eruption depth will determine the extent of explosive activity that might accompany the effusive component of the eruption (e.g., Clague et al., 2009; Helo et al., 2011).

Eruption detection

This discipline of submarine eruption detection and study is quite a bit younger than complementary studies of eruptions on land. The first evidence for an eruption at a mid-ocean ridge came in the form of a hydrothermal event plume detected along the Cleft segment of the Juan de Fuca Ridge (JdFR) in 1986 (Baker et al., 1987) – a follow-up study of the seabed led to the discovery of a series of pillow mounds up to 75 m high in the same area, making them the first young deep-sea lava flows of known age (Chadwick et al., 1991; Embley et al., 1991). In 1991, the Alvin submersible happened upon the aftermath of a volcanic eruption at 9°50′N on the East Pacific Rise, finding dead and charred tube worms strewn among and under fresh lava and "snowblower" hydrothermal vents diffusely spewing white sulfur-rich microbial floc (Haymon et al., 1993). Radiometric dating subsequently demonstrated that this discovery occurred just 2–4 weeks after the eruption had stopped (Rubin et al., 1994). The first remotely detected mid-ocean ridge eruption was in 1993, when scientists at NOAA/PMEL detected an earthquake swarm on the CoAxial segment of the Juan de Fuca Ridge within the first month of real-time acoustic monitoring using the US Navy Sound Surveillance System (SOSUS) (Fox et al., 1994). Since then, multiple other eruptions have been detected and described by both remote and direct methods (see Rubin et al., 2012 for a full listing). For instance, the real-time hydroacoustic monitoring capability has led to multiple eruption discoveries and responses in the NE Pacific (e.g., Perfit and Chadwick, 1998; Dziak et al., 2007; Dziak et al., 2011; Baker et al., 2012).

Current investigations and controversies

The scientific community has learned a great deal about mid-ocean ridges over the half century, since the realization that they were centers of volcanism and seafloor spreading. Yet there is a great deal that is not well known about both volcanism and magmatism that occur in the Earth's most active and prolific volcanic province, so research continues on multiple fronts. These include the timescales of magmatic processes, magmatic processes leading to intrusions, and the architecture of the ridge at depth, eruption characteristics, and what are the spatial scales of mantle heterogeneity and how is this heterogeneity sampled and modified by magmatic activity. This section highlights some of these current topics of investigation.

Magmatic timescales

Researchers are actively developing and employing novel methods to unravel the timescales of magmatic processes at mid-ocean ridges, from the duration of melting to the rate of magma transport from mantle depths to the crust and the duration of magma residence in the crust before eruption. The nuclides of the naturally occurring U and Th decay series provide the best tracers for these applications because collectively they have a range of half-lives and chemical characteristics, hence varying recovery rates to secular equilibrium following chemical perturbations (where secular equilibrium is the condition of all isotopes in a chain decaying at the same activity, measured as decay constant multiplied by the number of atoms present). Recent U-series radioactive disequilibria studies have demonstrated that melting, melt extraction, and melt accumulation rates can range from millennial (Elliott and Spiegelman, 2007; Stracke et al., 2006; Sinton et al., 2002) to decadal timescales (Rubin et al., 2001; Bergmanis et al., 2007). For instance, covariation of short-lived ^{210}Pb–^{226}Ra disequilibria (half-life = 22 years.) and other compositional attributes of historically erupted MORB at intermediate to fast-spreading ridges indicate that magmatic conditions can fluctuate rapidly, even between successive eruptions separated by just a decade or two (Rubin et al, 2005; Bergmanis et al., 2007). And eruption of lavas with ^{226}Ra–^{230}Th disequilibria (half-life = 1,600 year.) far away from the axis of spreading implies eruption ages of one to eight thousand years even when located on crust of 75–100 kyr spreading age (i.e., assuming steady-state volcanism and post-eruption movement of lava flows by plate spreading; Waters et al., 2013 Standish and Sims, 2010). Despite these successes, it must be noted that the U-series constraints in the aforementioned studies and others like them are semi-quantitative and highly dependent upon model parameterizations of melting using imprecisely known melting rates, melt fractions, melt porosities, melt flow conditions, lithological diversity, and the spatial scales of mantle compositional and mineralogical variations, all of which are targets for future research.

At even finer temporal scales, the ^{210}Po–^{210}Pb dating method provides ultrahigh-resolution lava eruption ages (±1–3 weeks) by charting the post-eruption in-growth of initially degassed, volatile ^{210}Po from erupting magma. The method has been used on several submarine eruptions since its first use on the East Pacific Rise in 1994 (Rubin et al., 1994) to provide temporal constraint on post-eruption recovery of hydrothermal systems and biological colonization and ecological succession on newly emplaced seafloor (summarized in Rubin et al, 2012). Recent applications using large sample numbers along with detailed lava flow mapping by manned submersible and remotely operated vehicle (e.g., Soule et al, 2007) provided data to reconstruct eruption dynamics and the temporal and spatial evolution of a lava flow field. For example, detailed study of a lava flow erupted at 9° 50′N EPR showed that this long fissure eruption began in summer 2005, with subsequent smaller eruptive pulses from ever-shortening fissures, culminating in a small eruption in mid-late January 2006 (Rubin et al., 2015), with eruption pulses being correlated in time with hydrothermal fluid exit temperatures recorded in situ. Paleomagnetic intensity dating of lavas (e.g., Carlut et al., 2004; Bowles et al., 2006) and radiocarbon dating of planktonic and benthic foraminifera collected from lava flow tops (Clague et al., 2013) are two other methods that both show promise for developing site-specific century to millennial eruption histories.

Mantle compositions and the spatial scales of mantle heterogeneities

Many researchers over the past several decades have been motivated to use MORB magmas to probe upper mantle compositional variations because of several factors that should simplify the connection of erupted magmas to their mantle source, including the common occurrence of relatively undifferentiated magmas, the thin and compositionally less variable oceanic crust (compared to the continents), and the broad spatial coverage of the upper mantle by mid-ocean ridges, which span much of the globe. Such studies have revealed many first-order variations in mantle compositions, as discussed above. More recently, studies addressing the physical nature and distribution of heterogeneities have shown that a few percent of highly fusible mantle pyroxenite veins can dominate aggregate melt composition but not melt flux (e.g., Russo et al, 2009) and that isotopic compositional domains along parts of the ridge system reflect stretching, thinning, and folding of compositionally distinct domains by mantle convection (Graham et al., 2006). Spectral analysis of MORB chemistry along the Mid-Atlantic Ridge has revealed mantle heterogeneity arising from modern hotspots as well as ancient heterogeneities preserved by incomplete mantle mixing during convection (Agranier et al., 2005) and perhaps different mantle-upwelling patterns beneath Indian and Atlantic mid-ocean ridges compared to those in the Pacific (Meyzen et al., 2007). New, high-quality trace element data sets from relatively large numbers of MORB have revealed variations in the degree of depletion of the MORB source mantle around the world (Arevalo and McDonough, 2010; O'Neill and Jenner, 2012) and a more refined definition of and characterization of "normal" MORB (aka N-MORB), as well as depleted and enriched MORB variants in a geographic framework (Gale et al., 2013).

Such studies will help researchers with their goal of understanding the spatial scale of mantle composition and lithology variations and how they are expressed in erupted MORB compositions. However, over the past decade, researchers have also come to realize that magmatic process produces significant overprinting of mantle compositional variations in MORB (e.g., Rubin and Sinton, 2007; Rubin et al., 2009; O'Neill and Jenner, 2012). For instance, the magnitude and temporal variability of melt supply largely affects the degree to which mantle heterogeneity is expressed in MORB magmas (Rubin and Sinton, 2007), which among other things produces reduced variance of mantle parameters in intermediate to high magma supply Pacific ridges compared to other ocean basins, an effect incorrectly assumed to reflect mantle heterogeneity or upwelling patterns in several of the aforementioned studies. The coupling of apparent mantle compositional homogenization with increased magma differentiation, which can be observed over a number of length scales (Rubin et al, 2009), implies that much of this mixing happens at shallow crustal levels, where magmas pool and differentiate in melt-rich lenses prior to eruption. O'Neill and Jenner (2012) argue that disruption of trace element variations in MORB occurs in frequently replenished, open-system magma chambers, although other authors have also argued that magma-rock reactions during intergranular porous flow deeper in the crust also modify the signatures of mantle source characteristics in the erupted MORB (Lissenberg et al., 2013). New advances in this area will likely require more eruption and volcano-scale studies using geological mapping, sampling, and chemical analysis to sort out the full range of processes that operate and under which conditions they occur.

Another way to get at the spatial scale of mantle composition and lithology variations is to look at mantle rocks themselves where they are exposed on the seafloor or in ophiolites. The mantle beneath mid-ocean ridges is thought to be largely depleted by melt prior extraction and magmatism, leading to, for instance, continental growth. But within this depleted matrix likely also lie veins and domains of enriched material injected into the mantle at subduction zones and slowly mixed and spread around by mantle convection. Studies of mantle rocks at ridges have shown that (a) they range in composition, from very melt depleted on ridges near hotspots to less so at locations away from hotspots (e.g., Dick et al., 1984), (b) a large proportion of the upwelling mantle beneath some ridges contributes little magma to the ridge axis because of extreme prior melt extraction (e.g., Liu et al., 2008), and (c) the average mantle beneath ridges may be far more isotopically depleted and compositionally more variable than inferred from erupted oceanic basalts (e.g., Stracke et al., 2011).

Eruption detection and response

Multidisciplinary studies at several focus sites around the world illustrate the important interplays between magmatism, hydrothermal activity, and biological processes at active and dormant mid-ocean ridge volcanoes (e.g., Kelley et al., 2002; Fornari et al., 2012). Mid-ocean ridge eruptions provide windows into rapidly changing magmatic, hydrothermal, and deep-sea biological processes (e.g., Delaney et al., 1998), so there has been great interest in detecting eruptions or intrusions, followed by rapidly organized seagoing response efforts to study these interplays (see reviews in Baker et al., 2012; Rubin et al., 2012). Such field studies, which began in the early 1990s, have largely been in response to remotely detected seismicity or serendipitous discovery. Despite the logistical difficulty of getting research vessels, equipment, and crew together on short notice, thirty-five recognized events have been responded to on submarine volcanoes over the world (i.e., at mid-ocean ridges and arc and intraplate seamounts; Baker et al., 2012). This list includes eleven historical eruptions at mid-ocean ridges (Rubin et al., 2012). As of this writing, the most recent of these eruptions, which occurred at Axial Seamount in 2011 (on the Juan de Fuca), was a triumph of prior planning and deployment of modern instrumentation, so that the geophysical, geodetic, and geological characteristics of the event were captured in exquisite detail (Caress et al., 2012; Chadwick et al., 2012; Dziak et al., 2012). The newest approaches in submarine eruption studies include autonomous detection stations, autonomous vehicle development for response efforts, and the establishment of permanent seafloor observatories. These will provide extraordinary research opportunities at a handful of sites, but active monitoring of most of the global ridge system will take a concerted effort by the international scientific community, as well as many more interested scientists and infrastructure to conduct these rapid response research efforts.

Summary

The global mid-ocean ridge system is a vast, complicated array of volcanoes, the great majority of which remain unvisited and sparsely sampled. Detailed studies at a few dozen sites reveal patterns of volcanic style, magmatic processes, and erupted compositions that generally reflect variations in spreading rate, magma supply, mantle composition, and proximity to ridge discontinuities and hotspots. Such studies also reveal the range of site-specific geological conditions that affect MORB composition and mid-ocean ridge volcanism. These effects are being studied in ever more locales at ever higher resolution to understand the spatial and temporal scales of ocean ridge magmatic processes, revealing how mid-ocean ridge volcanoes operate and how they sample the underlying mantle. Combined petrological, geochemical, geophysical, and geological data about mid-ocean ridges indicate that inferred mantle compositions are significantly modified in range, magnitude, and length scale at mid-ocean ridges by preeruptive magmatic processes such as magma residence and transport through the upper mantle and lower crust. Submarine eruptions construct new ocean crust and are a primary agent for the transfer of heat, chemicals, and microbes from the Earth's mantle or crust into the overlying ocean. Studying these eruptions is therefore important as well for a complete understanding of the chemistry and biology of the deep sea.

Bibliography

Agranier, A., et al., 2005. The spectra of isotopic heterogeneities along the mid-Atlantic Ridge. *Earth and Planetary Science Letters*, **238**, 96–109.

Allègre, C. J., Hamelin, B., and Dupre, B., 1984. Statistical analysis of isotopic ratios in MORB: the mantle blob cluster model and the convective regime of the mantle. *Earth and Planetary Science Letters*, **71**, 71–84.

Arevalo, R., Jr., and McDonough, W. F., 2010. Chemical variations and regional diversity observed in MORB. *Chemical Geology*, **271**, 70–85.

Asimow, P. D., Dixon, J. E., and Langmuir, C. H., 2004. A hydrous melting and fractionation model for mid-ocean ridge basalts: application to the Mid-Atlantic Ridge near the Azores. *Geochemistry, Geophysics, Geosystems*, **5**, Q01E16.

Baker, E. T., Massoth, G. J., and Feely, R.A., 1987. Cataclysmic hydrothermal venting on the Juan de Fuca Ridge. *Nature*, **329** (6135), 149–151.

Baker, E. T., Lupton, J. E., Resing, J. A., Baumberger, T., Lilley, M. D., Walker, S. L., and Rubin, K. H., 2011. Unique event plumes from a 2008 eruption on the Northeast Lau Spreading Center. *Geochemistry, Geophysics, Geosystems*, **12**, Q0AF02.

Baker, E. T., Chadwick, W. W., Jr., Cowen, J. P., Dziak, R. P., Rubin, K. H., and Fornari, D. J., 2012. Hydrothermal discharge during submarine eruptions: the importance of detection, response, and new technology. *Oceanography*, **25**, 128–141.

Ballard, R. D., and van Andel, T. H., 1977. Morphology and tectonics of the inner rift valley at lat $36°50'$ N on the Mid- Atlantic Ridge, *Geol. Soc. Am. Bull.*, **88**, 507–530.

Ballard, R. D., van Andel, T. H., and Holcomb, R. T., 1982. The Galapagos Rift at $86°W$ 5. Variations in volcanism, structure, and hydrothermal activity along a 30-kilometer segment of the rift valley. *Journal of Geophysical Research*, **87**, 1149–1161.

Bergmanis, E., Sinton, J. M., and Rubin, K. H., 2007. Recent eruptive history and Magma reservoir dynamics on the Southern East Pacific Rise at $17.5°S$. *Geochemistry, Geophysics, Geosystems*, **8**, Q12O06.

Bowles, J., Gee, J. S., Kent, D. V., Perfit, M. R., Soule, S. A., and Fornari, D. J., 2006. Paleointensity applications to timing and extent of eruptive activity, $9–10°$ N East Pacific Rise. *Geochemistry, Geophysics, Geosystems*, **7**, Q06006.

Bowles, J. A., Colman, A., McClinton, J. T., Sinton, J. M., White, S. M., and Rubin, K. H., 2014. Eruptive timing and 200 year episodicity at $92°$ W on the hot spot-influenced Galapagos Spreading Center derived from geomagnetic paleointensity. *Geochemistry, Geophysics, Geosystems*, **15**, 2211–2224.

Bown, J. W., and White, R. S., 1994. Variation with spreading rate of oceanic crustal thickness and geochemistry. *Earth and Planetary Science Letters*, **121**, 435–451.

Buck, W. R., Carbotte, S. M., and Mutter, C., 1997. Controls on extrusion at mid-ocean ridges. *Geology*, **25**, 935–938.

Cann, J. R., 1974. A model for oceanic crustal structure developed. *Geophysical Journal of the Royal Astronomical Society*, **39**, 169–187.

Cannat, M., Sauter, D., Bezos, A., Meyzen, C., Humler, E., and Le Rigoleur, M., 2008. Spreading rate, spreading obliquity, and melt supply at the ultraslow spreading Southwest Indian Ridge. *Geochemistry, Geophysics, Geosystems*, **9**, Q04002.

Carbotte, S. M., Canales, J. P., Nedimović, M. R., Carton, H., and Mutter, J. C., 2012. Recent seismic studies at the East Pacific Rise $8°20'–10°10'N$ and endeavour segment: Insights into mid-ocean ridge hydrothermal and magmatic processes. *Oceanography*, **25**, 100–112.

Carbotte, S. M., Marjanović, M., Carton, H., Mutter, J. C., Canales, J. P., Nedimović, M. R., Han, S., and Perfit, M. R., 2013. Fine-scale segmentation of the crustal magma reservoir beneath the East Pacific Rise. *Nature Geoscience*, **6**, 866–870.

Caress, D. W., Clague, D. A., Paduan, J. B., Martin, J. F., Dreyer, B. M., Chadwick, W. W., Jr., Denny, A., and Kelley, D. S., 2012. Repeat bathymetric surveys at 1-metre resolution of lava flows erupted at Axial Seamount in April 2011. *Nature Geoscience*, **5**, 483–488.

Carlut, J., Cormier, M.-H., Kent, D. V., Donnelly, K. E., and Langmuir, C. H., 2004. Timing of volcanism along the northern East Pacific Rise based on paleointensity experiments on basaltic glasses. *Journal of Geophysical Research*, **109**, B04104.

Chadwick, W. W., Jr., Embley, R. W., and Fox, C. G., 1991. Evidence for volcanic eruption on the southern Juan de Fuca Ridge between 1981 and 1987. *Nature*, **350**, 416–418.

Chadwick, W. W., Jr., Nooner, S. L., Zumberge, M. A., Embley, R. W., and Fox, C. G., 2006. Vertical deformation monitoring at axial seamount since its 1998 eruption using deep-sea pressure sensors. *Journal of Volcanology and Geothermal Research*, **150**, 313–32.

Chadwick, W. W., Jr., Nooner, S., Butterfield, D. A., and Lilley, M. D., 2012. Seafloor deformation and forecasts of the April 2011 eruption at Axial Seamount. *Nature Geoscience*, **5**, 474–477.

Clague, D. A., Paduan, J. B., and Davis, A. S., 2009. Widespread strombolian eruptions of mid-ocean ridge basalt. *Journal of Volcanology and Geothermal Research*, **180**, 171–188.

Clague, D. A., Dreyer, B. M., Paduan, J. B., Martin, J. F., Chadwick, W. W., Jr., Caress, D. W., Portner, R. A., Guilderson, T. P., McGann, M. L., Thomas, H., Butterfield, D. A., and Embley, R. W., 2013. Geologic history of the summit of axial seamount, Juan de Fuca Ridge. *Geochemistry, Geophysics, Geosystems*, **14**, 4403–4443.

Colman, A., Sinton, J. M., White, S. M., McClinton, J. T., Bowles, J. A., Rubin, K. H., Behn, M. D., Cushman, B., Eason, D. E., Gregg, T. K. P., Grönvold, K., Hidalgo, S., Howell, J., Neill, O., and Russo, C., 2012. Effects of variable magma supply on mid-ocean ridge eruptions: constraints from mapped lava flow fields along the Galápagos Spreading Center. *Geochemistry, Geophysics, Geosystems*, **13**, Q08014.

Coogan, L. A., 2007. The lower oceanic crust. *Treatise on Geochemistry*, **3**(19), 1–45.

Coogan, L. A., Jenkin, G. R. T., and Wilson, R. N., 2007. Contrasting cooling rates in the lower oceanic crust at fast- and slow-spreading ridges revealed by geospeedometry. *Journal of Petrology*, **48**, 2211–2231.

Corliss, J. B., Dymond, J., Gordon, L. I., Edmund, J. M., von Herzen, R. P., Ballard, R. D., Green, K., Williams, D., Brainbridge, A., Crane, K., and Van Andel, T. J., 1979. Submarine thermal springs on the Galapagos Rift. *Science*, **203**, 1073–1083.

Cushman, B., Sinton, J. M., Ito, G., and Dixon, J. E., 2004. Glass compositions, plume-ridge interaction, and hydrous melting along the Galápagos Spreading Center, 90.5°W to 98°W. *Geophysics Geochemistry Geosystems*, **5**, 8–17.

Dalton, C. A., Langmuir, C. H., and Gale, A., 2014. Geophysical and geochemical evidence for deep temperature variations beneath Mid-Ocean Ridges. *Science*, **344**, 80–83.

Delaney, J. R., Kelley, D. S., Lilley, M. D., Butterfield, D. A., Baross, J. A., Wilcock, W. S. D., Embley, R. W., and Summit, M., 1998. The quantum event of oceanic crustal accretion: Impacts of diking at mid-ocean ridges. *Science*, **281**, 222–230.

Dick, H. J. B., Fisher, R. L., and Bryan, W. B., 1984. Mineralogic variability of the uppermost mantle along mid-ocean ridges. *Earth and Planetary Science Letters*, **69**, 88–106.

Dupre, B., and Allègre, C. J., 1980. Pb-Sr-Nd isotopic correlation and the chemistry of the North Atlantic mantle. *Nature*, **286**, 17–22.

Dziak, R. P., Bohnenstiehl, D. R., Cowen, J. P., Baker, E. T., Rubin, K. H., Haxel, J. H., and Fowler, M. J., 2007. Rapid dike emplacement leads to eruptions and hydrothermal plume release during seafloor spreading events. *Geology*, **35**, 579–582.

Dziak, R. P., Hammond, S. R., and Fox, C. G., 2011. A 20-year hydroacoustic time series of seismic and volcanic events in the Northeast Pacific Ocean. *Oceanography*, **24**, 280–293.

Dziak, R. P., Haxel, J. H., Bohnenstiehl, D., and Matsumoto, H., 2012. Seismic precursors and magma ascent before the April 2011 eruption at Axial Seamount. *Nature Geoscience*, **5**, 478–482.

Elliott, T., and Spiegelman, M., 2007. Melt migration in oceanic crustal production: a U-series perspective. *Treatise on Geochemistry*, **3**(14), 465–510.

Elthon, D., 1979. High magnesia liquids as the parental magma for ocean floor basalts. *Nature*, **278**, 514–518.

Embley, R. W., Chadwick, W. W., Jr., Perfit, M. R., and Baker, E. T., 1991. Geology of the northern Cleft segment, Juan de Fuca Ridge: recent lava flows, sea-floor spreading, and the formation of megaplumes. *Geology*, **19**, 771–775.

Embley, R. W., Chadwick, Jr., W. W., Perfit, M. R., Smith, M. C., and Delaney. J. R., 2000. Recent eruptions on the CoAxial segment of the Juan de Fuca Ridge: Implications for mid-ocean ridge accretion processes, *J. Geophys. Res.*, **105**, 16501–16525.

Fornari, D. J., Von Damm, K. L., Bryce, J. G., Cowen, J. P., Ferrini, V., Fundis, A., et al., 2012. The East Pacific Rise between 9°N and 10°N: twenty-five years of integrated, multidisciplinary oceanic spreading center studies. *Oceanography*, **25**, 18–43.

Fox, C. G., Dziak, R. P., Matsumoto, H., and Schreiner, A. E., 1994. Potential for monitoring low-level seismicity on the Juan de Fuca Ridge using military hydrophone arrays. *Marine Technology Society Journal*, **27**, 22–30.

Gale, A., Dalton, C. A., Langmuir, C. H., Su, Y., and Schilling, J.-G., 2013. The mean composition of ocean ridge basalts. *Geochemistry, Geophysics, Geosystems*, **14**, 489–518.

Goss, A. R., Perfit, M. R., Ridley, W. I., Rubin, K. H., Kamenov, G. D., Soule, S. A., Fundis, A., and Fornari, D. J., 2010. Geochemistry of lavas from the 2005–2006 eruption at the East Pacific Rise, 9°46′N–9°56′N: Implications for ridge crest plumbing and decadal changes in magma chamber compositions. *Geochemistry, Geophysics, Geosystems*, **11**, Q05T09.

Graham, D. W., Blichert-Toft, J., Russo, C. J., Rubin, K. H., and Albarède, F., 2006. Cryptic striations in the upper mantle revealed by hafnium isotopes in Southeast Indian Ridge basalts. *Nature*, **440**, 199–202.

Gregg, T. K. P., and Fink, J. H., 1995. Quantification of submarine lava-flow morphology through analog experiments. *Geology*, **23**, 73–76.

Grove, T. L., Kinzler, R. J., and Bryan, W. B., 1992. Fractionation of mid-ocean ridge basalt (MORB). In Phipps Morgan, J., Blackman, D. K., and Sinton, J. M. (eds.), *Mantle Flow and Melt Generation at Mid-Ocean Ridges*. Washington, DC: American Geophysical Union. American Geophysical Union Monograph, Vol. 71, pp. 281–310.

Haase, K. M., Stroncik, N. A., Hékinian, R., and Stoffers, P., 2005. Nb-depleted andesites from the Pacific-Antarctic rise as analogs for early continental crust. *Geology*, **33**, 921–924.

Hall, M., 1876. Note upon a portion of Basalt from the mid-Atlantic. *The Mineralogical Magazine and Journal*, **1**, 1–3.

Hamelin, B., and Allègre, C. J., 1985. Large-scale regional units in the depleted upper mantle revealed by an isotope study of the South-West Indian Ridge. *Nature*, **315**, 196–199.

Hanan, B. B., Blichert-Toft, J., Pyle, D. G., and Christie, D. M., 2004. Contrasting origins of the upper mantle revealed by hafnium and lead isotopes from the Southeast Indian Ridge. *Nature*, **432**, 91–94.

Haymon, R. M., Fornari, D. J., Von Damn, K. L., Lilley, M. D., Perfit, M. R., Edmond, J. M., Shanks, W. C., Lutz, R. A., Grebmeier, J. M., Carbotte, S., et al., 1993. Volcanic eruption of the mid-ocean ridge along the East Pacific Rise crest at 9° 45–52′N: direct submersible observations of seafloor phenomena associated with an eruption event in April 1991. *Earth and Planetary Science Letters*, **119**, 85–101.

Helo, C., Longpre, M., Shimizu, A. N., Clague, D. A., and Stix, J., 2011. Explosive eruptions at mid-ocean ridges driven by CO_2-rich magmas. *Nature Geoscience*, **4**, 260–263.

Herzberg, C., 2004. Partial crystallization of mid-ocean ridge basalts in the crust and mantle. *Journal of Petrology*, **45**, 2389–2405.

Ito, G., and Mahoney, J. J., 2005. Flow and melting of a heterogeneous mantle: 1. Method and importance to the geochemistry of ocean island and mid-ocean ridge basalts. *Earth and Planetary Science Letters*, **230**, 29–46.

Kelemen, P., and Aharonov, E., 1998. Periodic formation of magma fractures and generation of layered gabbros in the lower crust beneath oceanic spreading ridges. In Buck, W. R., Delaney, P. T., Karson, J. A., and Lagabrielle, Y. (eds.), *Faulting and Magmatism at Mid-Ocean Ridges*. Washington, DC: American Geophysical Union. American Geophysical Union Monograph, Vol. 106, pp. 267–289.

Kelemen, P. B., Hirth, G. B., Shimizu, N., Spiegelman, M., and Dick, H. J. B., 1997. A review of melt migration processes in the adiabatically upwelling mantle beneath oceanic spreading centers. *Philosophical Transactions of The Royal Society A*, **355**, 283–318.

Kelley, D. S., Baross, J. A., and Delaney, J. R., 2002. Volcanoes, fluids, and life at ridge spreading centers. *Annual Review of Earth and Planetary Sciences*. **30**, 385–491.

Klein, E. M., and Langmuir, C. H., 1987. Global correlation of ocean ridge basalt chemistry with axial depth and crustal thickness. *Journal of Geophysical Research*, **92**, 8089–8115.

Langmuir, C. H., Bender, J. F., and Batiza, R., 1986. Petrological and tectonic segmentation of the East Pacific Rise, 5 30′-14 30′N. *Nature*, **322**, 422–429.

Lissenberg, C. J., and Dick, H. J. B., 2008. Melt–rock reaction in the lower oceanic crust and its implications for the genesis of mid-ocean ridge basalt. *Earth and Planetary Science Letters*, **271**, 311–325.

Lissenberg, J. C., Rioux, M., Shimizu, N., Bowring, S. A., and Mével, C., 2009. Zircon dating of oceanic crustal accretion. *Science*, **3**, 1048–1050.

Lissenberg, C. J., MacLeod, C. J., Howard, K. A., and Godard, M., 2013. Pervasive reactive melt migration through fast-spreading lower oceanic crust (Hess Deep, equatorial Pacific Ocean). *Earth and Planetary Science Letters*, **361**, 436–447.

Liu, C.-Z., et al., 2008. Ancient, highly heterogeneous mantle beneath Gakkel ridge, Arctic Ocean. *Nature*, **452**, 312–316.

Macdonald, K. C., 1982. Mid-ocean ridges: fine scale tectonic, volcanic and hydrothermal processes within the plate boundary zone. *Annual Review of Earth and Planetary Science*, **10**, 155–190.

Macdonald, K. C., 1998. Linkages between faulting, volcanism, hydrothermal activity and segmentation on fast spreading centers. In Buck, R. (ed.), *Faulting and Magmatism at Mid-Ocean Ridges*. Washington, DC: AGU. AGU Geophysical Monograph, Vol. 106, p. 27.

Mahoney, J. J., Natland, J. H., White, W. M., Poreda, R., Fisher, R. L., Bloomer, S. H., and Baxter, A. N., 1989. Isotopic and geochemical provinces of the western Indian Ocean spreading centers. *Journal of Geophysical Research*, **94**, 4033–4053.

Martin, E., and Sigmarsson, O., 2007. Crustal thermal state and origin of silicic magma in Iceland: the case of Torfajökull, Ljósufjöll and Snæfellsjökull volcanoes. *Contributions to Mineralogy and Petrology*, **153**, 593–605.

McClinton, T., White, S. M., Colman, A., and Sinton, J. M., 2013. Reconstructing lava flow emplacement processes at the hot spot-affected Galápagos Spreading Center, 95°W and 92°W. *Geochemistry, Geophysics, Geosystems*, **14**, 2731–2756.

Meyzen, C. M., et al., 2007. Isotopic portrayal of the Earth's upper mantle flow field. *Nature*, **447**, 1069–1074.

Michael, P. J., and Cornell, W. C., 1998. Influence of spreading rate and magma supply on crystallization and assimilation beneath mid-ocean ridges: evidence from chlorine and major element chemistry of mid-ocean ridge basalts. *Journal of Geophysical Research*, **103**, 18325–18356.

Michael, P. J., Langmuir, C. H., Dick, H. J. B., Snow, J. E., Goldstein, S. L., Graham, D. W., Lehnert, K., Kurras, G., Jokat, W., Mühe, R., and Edmonds, H. N., 2003. Magmatic and amagmatic seafloor generation at the ultraslow-spreading Gakkel Ridge, Arctic Ocean. *Nature*, **423**, 956–961.

Murray, J., and Renard, A. F., 1891. *Report on Deep-Sea Deposits, Based on Specimens Collected During the Voyage of H.M.S. Challenger in the Years 1872–76*. London: H. M. Stationary Office.

Niu, Y., and O'Hara, M. J., 2008. Global Correlations of Ocean Ridge Basalt Chemistry with Axial Depth: a New Perspective. *Journal of Petrology*, **49**, 633–664.

O'Neill, H. S. C., and Jenner, F. E., 2012. The global pattern of trace-element distributions in ocean floor basalts. *Nature*, **491**, 698–705.

Pallister, J. S., and Hopson, C. A., 1981. Samail ophiolite plutonic suite: field relations, phase variation, cryptic variation and layering, and a model of a spreading ridge magma chamber. *Journal of Geophysical Research*, **86**, 2593–2644.

Paonita, A., and Martelli, M., 2007. A new view of the $He-Ar-CO_2$ degassing at mid-ocean ridges: homogeneous composition of magmas from the upper mantle. *Geochimica et Cosmochimica Acta*, **71**, 1747–1763.

Perfit, M. R., and Chadwick, W. W., Jr., 1998. Magmatism at mid-ocean ridges: constraints from volcanological and geochemical investigations. In Buck, W. R., Delaney, P. T., Karson, J. A., and Lagabrielle, Y. (eds.), *Faulting and Magmatism at Mid-Ocean Ridges*. Washington, DC: American Geophysical Union. American Geophysical Union Monograph, Vol. 106, pp. 59–115.

Perfit, M. R., Fornari, D. J., Smith, M. C., Bender, J. F., Langmuir, C. H., and Haymon, R. M., 1994. Small-scale spatial and temporal variations in mid-ocean ridge crest magmatic processes. *Geology*, **22**, 375–379.

Phipps Morgan, J., and Chen, Y. J., 1994. The genesis of oceanic crust: magma injection, hydrothermal circulation, and crustal flow. *Journal of Geophysical Research*, **98**, 6283–6298.

Pollock, M. A., Klein, E. M., Karson, J. A., and Coleman, D. S., 2009. Compositions of dikes and lavas from the Pito Deep Rift: implications for accretion at superfast spreading centers. *Journal of Geophysical Research*, **114**, B03207.

Reynolds, J. R., Langmuir, C. H., Bender, J. F., Kastens, K. A., and Ryan, W. B. F., 1992. Spatial and temporal variability geochemistry of basalts from the East Pacific Rise. *Nature*, **359**, 493–499.

Rubin, K. H., and Sinton, J. M., 2007. Inferences on mid-ocean ridge thermal and magmatic structure from MORB compositions. *Earth and Planetary Science Letters*, **260**, 257–276.

Rubin, K. H., Macdougall, J. D., and Perfit, M. R., 1994. $^{210}Po-^{210}Pb$ dating of recent volcanic eruptions on the sea floor. *Nature*, **368**, 841–844.

Rubin, K. H., Smith, M. C., Bergmanis, E. C., Perfit, M. R., Sinton, J. M., and Batiza, R., 2001. Geochemical heterogeneity within mid-ocean ridge lava flows: insights into eruption, emplacement and global variations in magma generation. *Earth and Planetary Science Letters*, **188**, 349–367.

Rubin, K. H., van der Zander, I., Smith, M. C., and Bergmanis, E. C., 2005. Minimum speed limit for ocean ridge magmatism from ^{210}Pb-^{226}Ra-^{230}Th disequilibria. *Nature*, **437**, 534–538.

Rubin, K. H., Sinton, J. M., Maclennan, J., and Hellebrand, E., 2009. Magmatic filtering of mantle compositions at mid-ocean ridge volcanoes. *Nature Geoscience*, **2**, 321–328.

Rubin, K. H., Soule, S. A., Chadwick, W. W., Fornari, D. J., Clague, D. S., Embley, R. W., Baker, E. T., Perfit, M. R., Caress, D. W., and Dziak, R. P., 2012. Volcanic eruptions in the deep sea. *Oceanography*, **25**, 142–157.

Rubin, K. H., Soule, Fornari, D. J., Tolstoy, M. Bryce, J. G. Prado, F., Wauldhauser, F., Shank, T. M. Perfit, M. R., and Von Damm, K., 2015. First detailed study of a multiphase deep-sea mid-ocean ridge eruption reveals new volcanic style (2015)

Russo, C. J., Rubin, K. H., and Graham, D. W., 2009. Mantle melting and magma supply to the Southeast Indian Ridge: the roles of lithology and melting conditions from U-series disequilibria. *Earth and Planetary Science Letters*, **278**, 55–66.

Schilling, J.-G., 1991. Fluxes and excess temperatures of mantle plumes inferred from their interaction with migrating mid-ocean ridges. *Nature*, **352**, 397–403.

Schmitt, A. K., Perfit, M. R., Rubin, K. H., Stockli, D. F., Smith, M. C., Cotsonika, L. A., Zellmer, G. F., Ridley, W. I., and Lovera, O. M., 2011. Rapid cooling rates at an active mid-ocean ridge from zircon thermochronology. *Earth and Planetary Science Letters*, **302**, 349–358.

Searle, R. C., Murton, B. J., Achenbach, K., LeBas, T., Tivey, M., Yeo, I., Cormier, M. H., Carlut, J., Ferreira, P., Mallows, C., et al., 2010. Structure and development of an axial volcanic ridge: mid-Atlantic Ridge, 45°N. *Earth and Planetary Science Letters*, **299**, 228–241.

Sims, K. W. W., et al., 2012. Chemical and isotopic constraints on the generation and transport of magma beneath the East Pacific Rise. *Geochimica et Cosmochimica Acta*, **66**, 3481–3504.

Singh, S. C., Kent, G. M., Collier, J. S., Harding, A. J., and Orcutt, J. A., 1998. Melt to mush variations in crustal magma properties along the ridge crest at the southern East Pacific Rise. *Nature*, **394**, 874–878.

Sinton, J. M., and Detrick, R. S., 1992. Mid-ocean ridge magma chambers. *Journal of Geophysical Research*, **97**, 197–216.

Sinton, J. M, Wilson, D., Christie, D. M., Hey, R. N., and Delaney, J. R., 1983. Petrological consequences of rift propagation on oceanic spreading ridges. *Earth and Planetary Science Letters*, **62**, 193–207.

Sinton, J. M., Smaglik, S. M., Mahoney, J. J., and Macdonald, K. C., 1991. Magmatic, processes at superfast spreading mid-ocean ridges: glass compositional variations along the East Pacific Rise, 13-23 S. *Journal of Geophysical Research*, **96**, 6133–6155.

Sinton, J., Bergmanis, E., Rubin, K. H., Batiza, R., Gregg, T. K., Grönvold, P. K., Macdonald, K., and White, S., 2002. Volcanic eruptions on mid-ocean ridges: New evidence from the superfast-spreading East Pacific Rise, 17°-19°S. *Journal of Geophysical Research*, **107**(B6), 2115.

Small, C., 1998. Global systematics of mid-ocean ridge morphology. In Buck, W. R., Delaney, P. T., Karson, J. A., and Lagabrielle, Y. (eds.), *Faulting and Magmatism at Mid-Ocean Ridges*. Washington, DC: American Geophysical Union. American Geophysical Union Monograph, Vol. 106, pp. 59–115.

Soule, S. A., Fornari, D. J., Perfit, M. R., and Rubin, K. H., 2007. New insights into mid-ocean ridge volcanic processes from the 2005–2006 eruption of the East Pacific Rise, 9°46′N–9°56′N. *Geology*, **35**, 1079–1082.

Soule, S. A., Nakata, D. S., Fornari, D. J., Fundis, A. T., Perfit, M. R., and Kurz, M. D., 2012. CO_2 variability in mid-ocean ridge basalts from syn-emplacement degassing: constraints on eruption dynamics. *Earth and Planetary Science Letters*, **327–328**, 39–49.

Spiess, F. N., Macdonald, K. C., Atwater, T., Ballard, R., Carranza, A., Cordoba, D., Cox, C., Diaz Garcia, V. M., Francheteau, J., Guerrero, J., et al., 1980. East pacific rise: hot springs and geophysical experiments. *Science*, **207**, 1421–1433.

Standish, J. J., and Sims, K. W. W., 2010. Young off-axis volcanism along the ultraslow-spreading Southwest Indian Ridge. *Nature Geoscience*, **3**, 286–292.

Stracke, A., Bourdon, B., and McKenzie, D., 2006. Melt extraction in the Earth's mantle: constraints from U–Th–Pa–Ra studies in oceanic basalts. *Earth and Planetary Science Letters*, **244**, 97–112.

Stracke, A., Snow, J. E., Hellebrand, E., von der Handt, A., Bourdon, B., Birbaum, K., and Günther, D., 2011. Abyssal peridotite Hf isotopes identify extreme mantle depletion. *Earth and Planetary Science Letters*, **308**, 359–368.

Suhr, G., Hellebrand, E., Johnson, K. T. M., and Brunelli, D., 2008. Stacked gabbro units and intervening mantle: a detailed look at a section of IODP Leg 305, Hole U1309D. *Geochemistry, Geophysics, Geosystems*, **9**, Q10007.

Tolstoy, M., Cowen, J. P., Baker, E. T., Fornari, D. J., Rubin, K. H., Shank, T. M., Waldhauser, F., Bohnenstiehl, D. R., Forsyth, D. W., Holmes, R. C., et al., 2006. A sea-floor spreading event captured by seismometers. *Science*, **314**, 1920–1922.

Von Damm, K. L., et al., 2004. Evolution of the hydrothermal system at East Pacific Rise 9°50′N: geochemical evidence for changes in the upper oceanic crust. In German, C. R. (ed.), *Mid-Ocean Ridges: Hydrothermal Interactions Between the Lithosphere and Ocean*. Washington, DC: American Geophysical Union. AGU Geophysical Monograph, Vol. 148, pp. 285–304.

Wadge, G., 1982. Steady state volcanism: evidence from eruption histories of polygenetic volcanoes. *Journal of Geophysical Research*, **87**, 4035–4049.

Wanless, V. D., Perfit, M. R., Ridley, W. I., and Klein, E., 2010. Dacite petrogenesis on mid-ocean ridges: evidence for oceanic crustal melting and assimilation. *Journal of Petrology*, **51**, 2377–2410.

Waters, C. L., Sims, K. W. W., Klein, E. M., White, S. M., et al., 2013. Sill to surface: linking young off-axis volcanism with subsurface melt at the overlapping spreading center at 9°03′N East Pacific Rise. *Earth and Planetary Science Letters*, **369–370**, 59–70.

White, S. M., Haymon, R. H., Fornari, D. J., Perfit, M. R., and Macdonald, K. C., 2002. Volcanic structures and lava morphology of the East Pacific Rise, 9-10N: constraints on volcanic segmentation and eruptive processes at fast-spreading ridges. *Journal of Geophysical Research*, **107**, B8, doi:10.1029/2001JB000571.

White, S. M., Crisp, J. A., and Spera, F. J., 2006. Long-term volumetric eruption rates and magma budgets. *Geochemistry, Geophysics, Geosystems*, **7**, Q03010.

Yeo, I., Searle, R. C., Achenbach, K. L., Le Bas, T. P., and Murton, B. J., 2012. Eruptive hummocks: building blocks of the upper ocean crust. *Geology*, **40**, 91–94.

Cross-references

Axial Summit Troughs
Axial Volcanic Ridges
Crustal Accretion
Depleted Mantle
Explosive Volcanism in the Deep Sea

Hydrothermal Plumes
Hydrothermal Vent Fluids (Seafloor)
Layering of Oceanic Crust
Lithosphere: Structure and Composition
Mid-ocean Ridge Magmatism and Volcanism
Oceanic Rifts
Oceanic Spreading Centers
Peridotites
Plate Tectonics
Submarine Lava Types

MODELLING PAST OCEANS

Jürgen Sündermann
Institute of Oceanography (IFM), University of Hamburg, Hamburg, Germany

Understanding the formation of the Earth's surface and the evolution of the Earth–Moon system requires knowledge of the ocean's circulation and tides through geological history. The most important tools to get this information are numerical models simulating the physical and biogeochemical state of the sea. These models are widely tested for the present Earth system. Research on climate, global fluxes of matter, marine ecosystems, as well as operational services like weather forecast are based on them. The models are founded on the laws of nature (e.g., Newton's conservation of momentum) which are formulated as mathematical equations and transformed into complex program codes on major electronic computers. For model validation, measured field data as sea-level elevations, surface temperatures, or ice coverage are used. Past ocean models were based on paleontological data of topography, sediments, and marine organisms which were available for times more than two billion years ago. The models are driven by astronomical reconstructions of the Earth's, the Moon's, and the Sun's ancient orbital motions and scenario-like assumptions on early volcanism, greenhouse gases, sun spots, etc. Coupled ocean–atmosphere–cryosphere–biosphere models simulating ancient circulations and climates, warm and cold periods, extend some 10–100 million years back (Heinemann et al., 2009). They are most reliable for the Holocene with increasing accuracy toward the present (Fischer and Jungclaus, 2010). Tidal models explaining geological sediment patterns or the increases of the length of day and the Moon distance cover the Earth's history back to the Proterozoic (Kagan and Sündermann, 1996). Examples are model calculations of ocean currents and tides for the Gondwana period, for the Tethys ocean, for the early Atlantic, for the world ocean during last glacials and interglacials, and for the recent millennium. From model results, we know today that there exist multiple equilibrium states of the atmospheric and oceanic circulation, e.g., with and without the Gulf stream. We also know that the rate of deceleration of the Earth's rotation by tidal friction through geological history was not uniform due to continental drift and changing ocean resonance.

Bibliography

Fischer, N., and Jungclaus, J. H., 2010. Effects of orbital forcing on atmosphere and ocean heat transports in Holocene and Eemian climate simulations with a comprehensive Earth system model. *Climate of the Past*, **6**, 155–168.

Heinemann, M., Jungclaus, J. H., and Marotzke, J., 2009. Warm Paleocene/Eocene climate as simulated in ECHAM5/MPI-OM source. *Climate of the Past*, **5**, 785–802.

Kagan, B. A., and Sündermann, J., 1996. Dissipation of tidal energy, paleotides and evolution of the Earth-Moon system. *Advances in Geophysics*, **38**, 179–266.

MODERN ANALOG TECHNIQUES

Michal Kucera
MARUM-Center for Marine Environmental Sciences, University of Bremen, Bremen, Germany

Modern analog technique, often abbreviated as MAT, is a popular type of transfer function used in paleoceanography to reconstruct past ocean properties from the composition of fossil assemblages extracted from deep-sea sediments. In the vast majority of cases, MAT is applied to reconstruct past sea-surface temperature from remains of marine unicellular microplankton such as planktonic foraminifera, radiolaria, diatoms, and dinoflagellates.

The method has been introduced to paleoceanography by Hutson (1980). Mathematically, it is akin to k-nearest neighbor regression. Its procedure is computationally and conceptually very simple. First, a similarity coefficient is calculated between the composition of a fossil assemblage in a geological sample and the compositions of corresponding assemblages in a "calibration dataset" of surface-sediment samples. Then, the surface-sediment samples are sorted by their similarity coefficient, and a subset of those with most similar assemblages, termed the "nearest" or "best modern analogs," is used in the second step to calculate an estimate of the desired ocean property. The estimate represents a mean of the present-day values of the ocean property at the site of the deposition of the best modern analogs, typically weighted by the respective similarity coefficient of each sample. Like all paleoceanographic transfer functions, MAT relies on the assumption that the reconstructed ocean property is a statistically important determinant of the assemblage composition of the studied organisms. It also assumes that the fossil assemblage and its covariance with the ocean property are represented in the calibration dataset. If this is not the case, the fossil assemblage is said to represent a no-analog condition, which may render the reconstruction of past ocean properties invalid.

MAT is a strict interpolator; it does not attempt to extract a general relationship between assemblage composition and environmental factors. Its applicability is therefore contingent on appropriate coverage of the environmental gradient and of the associated assemblage compositions. Because of no-analog conditions due to evolution of niche requirements of the constituent species, assemblage-based transfer functions are typically limited in their applicability to late Quaternary sediments. Sea-surface temperature reconstructions via MAT can be obtained to a precision of 1 °C, and the variance among the SST values of the best analogs can be used to assess the uncertainty of a fossil estimate.

Bibliography

Hutson, W. H., 1980. The Agulhas current during the late Pleistocene: analysis of modern faunal analogs. *Science*, **207**, 64–66.

MOHOROVIČIĆ DISCONTINUITY (MOHO)

Carolina Lithgow-Bertelloni
Department of Earth Sciences, University College London, London, UK

Synonyms
Crust-mantle transition; Moho

Definition
Moho: The rapid increase in seismic velocity with which we often identify the boundary between Earth's crust and mantle. More specifically, the Moho is the depth at which the P-wave velocity first increases rapidly or discontinuously to 7.6–8.6 km/s, or if steep velocity gradients are not present, then it is the depth at which the P-velocity exceeds 7.6 km/s (Steinhart, 1967).

Introduction
The Mohorovičić discontinuity, also known as the Moho, is a discrete jump in seismic wave velocities – a seismic discontinuity – that has come to define the boundary between the crust and the mantle. Its significance stems from its clear demarcation of the first boundary of the chemically differentiated solid Earth. This seminal discovery (Mohorovicic, 1910) complemented that of Earth's other major chemical boundaries with the nearly contemporaneous discovery of the outer core (Oldham, 1906) and the inner core (Lehman, 1936) a few decades later.

The seismic discontinuity at the Moho is explained generally as a chemical (compositionally distinct) boundary separating crust (both oceanic and continental) from the mantle, although in detail the Moho is a much more complex boundary. This is in contrast with most other seismic discontinuities in the mantle, which are explained by changes in the structure of minerals with homogeneous composition as a function of pressure. The petrologic Moho is defined as the change from more felsic crust to more mafic mantle, but it does not always coincide with the Mohorovičić discontinuity as determined by refraction velocity studies or reflection studies. The velocity discontinuity may in places occur within the crust as the basalt to eclogite metamorphic transition (O'Reilly and Griffin, 2013) or within the mantle at the base of a serpentinized layer. Depending on tectonic setting its thickness may vary.

Today, mapping of the Moho at global and regional scales with a variety of geophysical probes, from active-source seismic reflection and refraction studies to teleseismic studies to gravity inversions, is a crucial aspect of geological and geophysical research. Drilling through the Mohorovičić discontinuity remains an elusive goal.

In honor of the centenary of its discovery, an excellent and comprehensive review of historical progress and state-of-the-art research was published in December 2013 (*Moho: 100 years after Andrija Mohorovičić, Tectonophysics, 609, 2013*).

Historical overview
Andrija Mohorovičić saw a peculiar signal in seismograms from a shallow earthquake 40 km southeast of Zagreb, Croatia, in the Kupa Valley on 8 October 1909. Mohorovičić obtained records from seismic stations within 800 km of the epicenter for both earthquake and aftershocks. He observed additional primary (P) wave arrivals with different apparent velocities, at distances between 400 and 720 km, which he called $\bar{P}$. The "normal" arrival was the direct travel path of the P-wave from hypocenter to receiver. The other was from a wave traveling along the interface between two media and refracted towards the receiver at the critical angle (Figure 1). The latter headwave (P_n) arrived earlier and its apparent velocity, ~7.8 km/s, was much faster than the direct wave. Mohorovičić surmised that $\bar{P}$ could be from no deeper than ~50 km and it must represent "... a sudden change of the material which makes up the interior of the earth, because there a step in the velocity of the seismic wave must exist" (Mohorovičić, 1910, p. 34). The P_n headwave identifies the seismic discontinuity that marks the compositionally distinct boundary between crust and mantle. Over the following two decades, contributions from Sir Harold Jeffreys, Victor Conrad, and others showed that the Mohorovičić discontinuity was ubiquitous.

However, even by the late 1950s, mapping of the Moho was limited. But from the late 1960s onwards, regional maps were published for Europe (Morelli et al., 1967), the Soviet Union (Belyaevsky et al., 1973), and the USA (Warren and Healy, 1973). Remarkably, these are very similar to today's compilations (Thybo et al., 2013).

Drilling to and through the Moho to understand the nature of the discontinuity and its relation to changes in other physical properties is difficult. The only direct attempt is Project Mohole, which ended in 1967

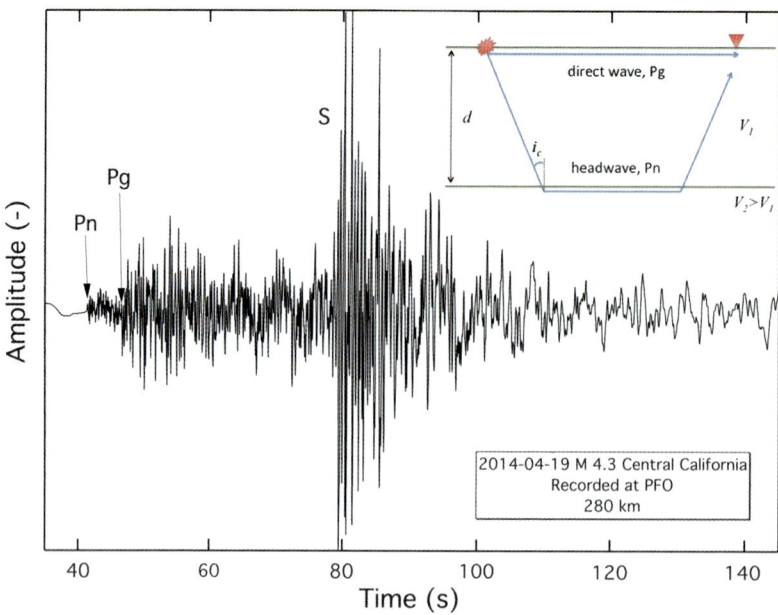

Mohorovičić Discontinuity (Moho), Figure 1 A seismogram from a M 4.3 earthquake in Central California recorded at station PFO, 280 km away, shows the early arrival of P_n. Inset: Cartoon showing the direct P and the headwave, P_n, along the interface.

(Greenberg, 1974). Though unsuccessful it developed critical technology and led directly to the Deep Sea Drilling Project.

Observational techniques

Studies of the Mohorovičić discontinuity have largely focused on seismic techniques. The analysis of seismograms for the P_n and other arrivals was used to map the Moho and the presence of other intracrustal discontinuities (Prodehl et al., 2013), particularly on continents. In the oceans, studies of seismic refraction were pioneered by Ewing and Ewing (1959) with the use of controlled-source seismology (active sources). Following these first studies (e.g., Shor, 1963; Sutton et al., 1971), marine multichannel reflection studies helped define the oceanic Moho (e.g., Hasselgreen and Clowes, 1995). Active-source seismology has proved very effective in continental regions through the use of both deep seismic reflection and refraction experiments. Reflection studies provide the highest resolution images of the Moho, whereas seismic refraction studies provide the best velocity control, and hence relation to the standard Moho definition. More recently the use of earthquake receiver functions has revealed the richness of intracrustal structure such as double and inverted Mohos in subduction zones (Bostock et al., 2002). Receiver functions allow for the removal of earthquake source complexity and image local structure. Stacking and migration produce depth sections and illuminate the variability of the Moho, though generally with lower resolution than active-source studies, except across dense networks. Novel techniques like inversions of ambient noise (Shapiro and Campillo, 2004) and surface waves (Lebedev et al., 2013) further illustrate Moho complexity. Surface wave waveforms and dispersion are strongly influenced by Moho depth, but combined with a priori knowledge of seismic velocities; receiver functions and gravity can be jointly inverted to create regional maps of the Moho.

It is possible to use gravity data to study crustal structure and detect the Moho as the density increases across the interface (e.g., Oldenburg, 1974). However, gravitational inversions are nonunique and require additional information. It is possible to overcome the nonuniqueness by a joint inversion of gravity data with point information on crustal structure from seismology. High-resolution global gravity data from satellite missions (GRACE and GOCE) has led to global and regional Moho models (Reguzzoni et al., 2013; van der Meijde et al., 2013).

Current knowledge

The Mohorovičić discontinuity is not fully mapped in the oceanic basins, and the distribution of observations is irregular (Mutter and Carton, 2013). Often we refer instead to the reflection Moho, the deepest, high-amplitude, laterally extensive reflection or group of reflections present at depths commensurate with other estimates of crustal thickness (Cook et al., 2010). However, current knowledge puts its average depth below the seafloor at 6–7 km reaching thicknesses >25 km at oceanic plateaus like Ontong Java (Gladczenko et al., 1997). Initial observations implied a sharp interface and little change towards the ridge crest. Recent work has shown that the Moho ranges from a sharp interface to a gradual transition (Mutter and Carton, 2013). Moho variability in the oceans

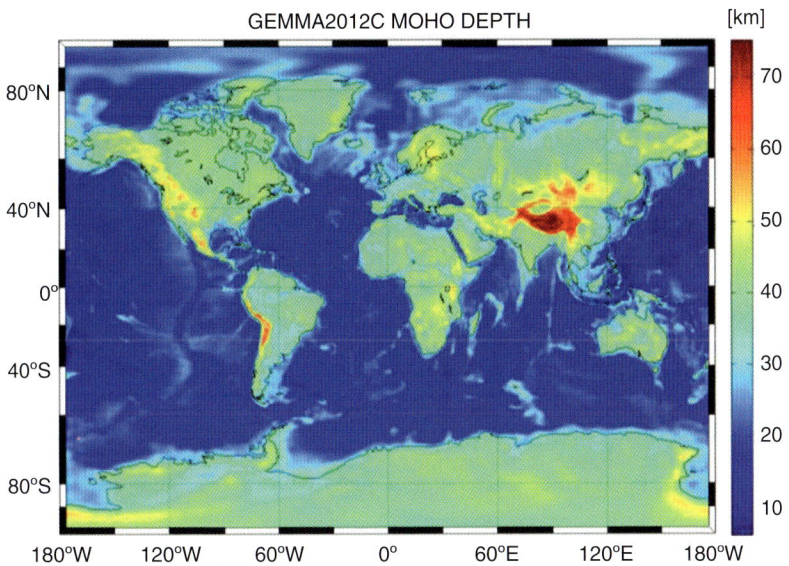

Mohorovičić Discontinuity (Moho), Figure 2 Global crustal thickness model derived from GOCE gravity data (reproduced from Reguzzoni et al., 2013).

does not correlate with spreading rate or age. The Moho appears to form very close to zero age and change little so that it is fully formed 50,000 years from the ridge crest.

On continents the Moho averages ~35 km depth. It is as deep as 65–70 km in the Andes and Himalayas and shallower than 10 km in Afar. Variations in Moho depth and therefore crustal thickness are strongly correlated with tectonic regime. Cratons and platforms average ~40 km thick crust, while mountain belts are above ~45–50 km. Global crustal thicknesses from new inversion of the gravity field measured by GOCE are shown in Figure 2.

The oceanic Moho is exposed on continents where oceanic sections (ophiolites) have been abducted onto the continent as in Oman. In regions of continental collision, the crust-mantle boundary is exposed such as in the Ivrea zone in the Alps. In the oceans, the mantle outcrops at ultra-slow spreading centers like the Gakell ridge in the Arctic.

The Mohorovičić discontinuity is not always indicative of the boundary between crust and mantle. In the oceans, this is known as the difference between the petrologic and seismic Moho. We expect the lower crust to be more mafic and to be derived from partial melting of the mantle below. The velocity of mafic lower crust can be very similar to that of the mantle, complicating observations of the Moho. The Moho may occur within the crust in continental regions where the lower crust has transformed to eclogite, or within the mantle, in oceanic regions at the base of the serpentinized layer (O'Reilly and Griffin, 2013).

Controversies and gaps in current knowledge

The last 105 years since Andrija Mohorovičić's discovery have brought enormous progress, but gaps and controversies remain. A complete mapping of the oceanic Moho from ridge to subduction zone remains elusive, not least because of the sparse coverage of ocean bottom seismometers (Mutter and Carton, 2013). Current outstanding efforts of global compilations of Moho depths include those from active-source experiments (Prodehl and Mooney, 2012), receiver functions, and published information like CRUST1.0 (Laske et al., 2013) and those combining gravity and prior crustal models such as GEMMA2012C (Reguzzoni et al., 2013).

The fact that the seismic Moho and the base of the crust do not always coincide makes the interpretation of the origin of the Moho challenging. In the oceans, the Moho may be relatively simple as a transition from the crust to the mantle. However, on continents, the Moho is generally much more complex, partly due to tectonics (Cook et al., 2010; Carbonell et al., 2013). Possibly the crust-mantle boundary is not sharp but transitional and changing with time (O'Reilly and Griffin, 2013). These complications are particularly evident at subduction zones, where both the Moho of subducted oceanic crust and the overriding plate are visible. The flat slab segment of Argentina provides a beautiful illustration (Gans et al., 2011). The double Moho in subduction zones is often associated with a low velocity region, which may represent a hydrated oceanic layer (Bostock, 2013).

The presence of fluids should have a strong effect on physical properties including electrical conductivity. Measuring the electrical Moho (eMoho) is often complicated by past tectonism and metamorphism though still visible with high-quality data (Jones, 2013). We expect seismic properties such as anisotropy and attenuation to change across the Moho as well as rheological properties, but our knowledge of these changes is limited to a few regional examples (Thybo et al., 2013).

Summary

The Mohorovičić seismic discontinuity, known today as the Moho, was discovered in 1909 by Andrija Mohorovičić. It is defined by a jump in seismic velocities from 5.6 to ~7.8 km/s. It represents the boundary between the more felsic crust and the mafic mantle. Combining seismic, gravity, and other data to map the Moho has led to global compilations (CRUST1.0 and GEMMA2012C) that have achieved impressive resolution and coverage. The petrologic and seismic Moho do not always coincide because of complications due to eclogitisation and serpentinization. Modern seismic techniques have revealed richness in structure including double and inverted Mohos at subduction zones (Bostock, 2013) and great complexity in continental regions (Cook et al., 2010; Carbonell et al., 2013). It is now clear that the Mohorovičić discontinuity is a dynamic feature. Ongoing efforts include detecting and understanding other changes in physical properties across it, such as electrical conductivity, scale changes in heterogeneities across it, and viscosity.

Bibliography

Belyaevsky, N. A., Borisov, A. A., Fedynsky, V. V., Fotiadi, E. E., Subbotin, S. I., and Volvovsky, I. S., 1973. Structure of the earth's crust on the territory of the U.S.S.R. *Tectonophysics*, **20**, 35–45.

Bostock, M. G., 2013. The Moho in subduction zones. *Tectonophysics*, **609**, 547–557.

Bostock, M. G., Hyndman, R. D., Rondenay, S., and Peacock, S. M., 2002. An inverted continental Moho and serpentinization of the forearc mantle. *Nature*, **417**, 536–538.

Carbonell, R., Levander, A., and Kind, R., 2013. The Mohorovičić discontinuity beneath the continental crust: an overview of seismic constraints. *Tectonophysics*, **609**, 353–376.

Cook, F. A., White, D. J., Jones, A. G., Eaton, D. W. S., Hall, J., and Clowes, R. M., 2010. How the crust meets the mantle: lithoprobe perspectives on the Mohorovičić discontinuity and crust-mantle transition. *Canadian Journal of Earth Sciences*, **47**, 315–351.

Ewing, J., and Ewing, M., 1959. Seismic-refraction measurements in the Atlantic Ocean basins, in the Mediterranean Sea, on the mid-Atlantic ridge and in the Norwegian Sea. *Bulletin of the Geological Society of America*, **70**, 291–318.

Gans, C. R., Beck, S. L. G., Zandt, G., Gilbert, H., Alvarado, P., Anderson, M., and Linkimer, L., 2011. Continental and oceanic crustal structure of the Pampean flat slab region, western Argentina, using receiver function analysis: new high-resolution results. *Geophysical Journal International*, **186**, 45–58.

Gladczenko, T., Coffin, M., and Eldholm, O., 1997. Crustal structure of the Ontong Java plateau: modeling of new gravity and existing seismic data. *Journal of Geophysical Research*, **102**, 22711–22729.

Greenberg, D. S., 1974. Mohole: geopolitical fiasco. In Gass, I. G., Smith, P. J., and Wilson, R. C. L. (eds.), *Understanding the Earth*. Maidenhead: Open University Press, pp. 343–349.

Hasselgren, E. O., and Clowes, R. M., 1995. Crustal structure of northern Juan de Fuca plate from multichannel reflection data. *Journal of Geophysical Research*, **100**, 6469–6486.

Jones, A. G., 2013. Imaging and observing the electrical Moho. *Tectonophysics*, **609**, 423–436.

Laske, G., Masters, G., Ma, Z., and Pasyanos, M., 2013. Update on CRUST1.0 – A 1-degree global model of Earth's crust. *Geophysical Research Abstracts*, **15**, EGU2013–EGU2658.

Lebedev, S., Adam, J. M.-C., and Meier, T., 2013. Mapping the Moho with seismic surface waves: a review, resolution analysis, and recommended inversion strategies. *Tectonophysics*, **609**, 377–394.

Lehman, I., 1936. P'. *Travaux Scientifique*, **14**, 88–94.

Mohorovičić, A., 1910. Potres od 8.X 1909. Jahrbuch des meteorologischen observatoriums in Zagreb (Agram) für das Jahr 1909, 1–56. English translation, 1992. Earthquake of 1909 October 8. *Geofizika*, **9**, 3–55.

Morelli, C., Bellemo, S., Finetti, I., and De Visintini, G., 1967. Preliminary depth contour maps for the Conrad and Moho discontinuities in Europe. *Bolletino di Geofisica Teorica ed Applicata*, **9**, 1–48.

Mutter, J. C., and Carton, H. D., 2013. The Mohorovičić discontinuity in ocean basins: some observations from seismic data. *Tectonophysics*, **609**, 314–330.

O'Reilly, S. Y., and Griffin, W. L., 2013. Moho vs crust-mantle boundary: evolution of an idea. *Tectonophysics*, **609**, 535–546.

Oldenburg, D. W., 1974. The inversion and interpretation of gravity anomalies. *Geophysics*, **39**, 526–536.

Oldham, R. D., 1906. Constitution of the interior of the Earth as revealed by earthquakes. *Quarterly Journal of the Geological Society*, **62**, 456–475.

Prodehl, C., and Mooney, W. D., 2012. *Exploring the Earth's Crust—History and Results of Controlled-Source Seismology*. Denver: Geological Society of America Memoir 208.

Prodehl, C., Kennett, B., Artemieva, I. M., and Thybo, H., 2013. 100 years of seismic research on the Moho. *Tectonophysics*, **609**, 9–44.

Reguzzoni, M., Sampietro, D., and Sansò, F., 2013. Global Moho from the combination of the CRUST2.0 model and GOCE data. *Geophysical Journal International*, **195**, 222–237.

Shapiro, N. M., and Campillo, M., 2004. Emergence of broadband Rayleigh waves from correlations of the ambient seismic noise. *Geophysical Research Letters*, **31**, L07614.

Shor, G. G., Jr., 1963. Refraction and reflection techniques and procedure. In Hill, M. N. (ed.), *The Sea*. New York: Interscience, Vol. 3, pp. 19–38.

Steinhart, J. S., 1967. Mohorovicic discontinuity. In Runcorn, S. K. (ed.), *International Dictionary of Geophysics*. Oxford: Pergamon Press, Vol. 2, pp. 991–994.

Sutton, G. H., Maynard, G. L., and Hussong, D. M., 1971. Widespread occurrences of a high velocity basalt layer in the Pacific crust found with repetitive sources and sonobuoys. In Heacock, J. G. (ed.), *The Structure and Physical Properties of the Earth's Crust*. Washington, DC: American Geophysical Union, Vol. 14, pp. 193–209.

Thybo, H., Artemieva, I. M., and Kennett, B., 2013. Moho: 100 years after Andrija Mohorovičić. *Tectonophysics*, **609**, 1–8.

van der Meijde, M., Julià, J., and Assumpção, M., 2013. Gravity derived Moho for South America. *Tectonophysics*, **609**, 456–467.

Warren, D. H., and Healy, J. H., 1973. The crust in the conterminous United States. *Tectonophysics*, **20**, 203–213.

Cross-references

Gravity Field
Layering of Oceanic Crust
Lithosphere: Structure and Composition
Ophiolites
Reflection/Refraction Seismology

MORAINES

Anders Schomacker
Department of Geology and Mineral Resources Engineering, Norwegian University of Science and Technology, Trondheim, Norway

Synonyms

Glacial landform; glacial sediment; till

Definition

"Moraine" is a genetic term for a landform deposited in a glacial environment (Benn and Evans, 2010; Schomacker, 2011). Commonly, "moraine" refers to a ridge-shaped glacial landform such as an end moraine. It also describes large-scale terrestrial glacial landscapes, i.e., hummocky moraines formed by the melt-out of debris-charged dead-ice or ground moraines (till plains) formed in subglacial environments. In marine settings, "moraine" almost entirely refers to ridge- or wedge-shaped landforms deposited at the margin of tidewater glaciers and oriented transverse to the ice-flow direction. Moraines consist of sediments directly deposited by a glacier and/or deformed by a glacier.

Introduction

"Moraine" spans over an extremely wide range of glacial landforms deposited in many different environments. End moraines and other marginal moraines form at the margins of glaciers and both in terrestrial and aquatic environments (e.g., Boulton et al., 1999; Bennett, 2001; Benediktsson et al., 2008, 2009, 2010; Evans et al., 2012). Till plains or ground moraines form in the subglacial environment of terrestrial glaciers (e.g., Dyke and Prest, 1987; Ross et al., 2009; Schomacker et al., 2010). Its analogue in aquatic environments is subglacially streamlined terrains, typically formed by (paleo-) ice streams (Figure 1a; Dowdeswell et al., 2008; Winsborrow et al., 2010; Jakobsson et al., 2011; Howe et al., 2012; Livingstone et al., 2012; Nitsche et al., 2013). On land, hummocky moraines form in dead-ice environments by the melt-out of stagnant, debris-covered glacier ice (e.g., Boulton, 1972; Clayton et al., 2008; Schomacker, 2008). Numerous other types of moraines are known to form in these glacial environments or in a combination of them. They comprise landforms such as ribbed moraines (Figure 1b), controlled moraines, morainal banks, and many more (e.g., Möller, 2006, 2010; Evans, 2009; Benn and Evans, 2010; Goff et al., 2012).

The interior of moraines consists of a variety of sediments. Overall, they can be divided into sediments deposited at the same time as the moraine (e.g., basal till plain, ground moraine, ice-marginal dump moraine) and glacially deformed sediments that were originally deposited prior to the moraine formation (e.g., glaciotectonic end moraines, Figure 2). Sedimentologic and geomorphologic features of moraines formed in glaciomarine environments are mainly described from studies of Pleistocene moraines that have been isostatically uplifted and are now accessible on land (e.g., Nemec et al., 1999; Lindén and Möller, 2005; Evans et al., 2012). Numerous studies have described the processes of terrestrial ice-marginal moraine formation, whereas much less is known of the processes forming moraines at ice margins terminating in the sea due to the inaccessible environment. The morphology and internal architecture of submarine moraine ridges have also been deciphered by geophysical methods such as multibeam and seismic reflection surveys (e.g., Dowdeswell et al., 2008; Shaw et al., 2009; Jakobsson et al., 2011; Goff et al., 2012).

This entry aims to briefly review the most important moraine types formed in marine glacial environments. Focus is on the geomorphology and sedimentology of ice-marginal moraines as well as their implication for glaciation history.

Ice-marginal moraines formed at tidewater glaciers

Moraine ridges formed at the margins of tidewater glaciers reflect deposition in complex environments where the action of meltwater, gravity, and subglacial processes control the morphology and sedimentary facies. The range of "moraines" formed in this environment spans from subaqueous fans, mainly of glaciofluvial sediments, to transverse moraine ridges built up by both subglacial till and glaciofluvial sediments, e.g., so-called De Geer moraines. In paleoglaciological reconstructions, such moraine ridges are interpreted as formed at a quasi-stationary ice margin. During longer glacier stillstands, large subaqueous fans might accumulate and merge or even build up to the water surface of the basin as an ice-contact delta (Fyfe, 1990; Nemec et al., 1999). It may be confusing that these landforms are often referred to as "moraines" although they should be regarded as marginal ice-contact deltas (Fyfe, 1990; Nemec et al., 1999). The use of the term "moraine" to describe these landforms is probably because they are broadly ridge shaped and delineate a former ice margin. Their paleoglaciological significance is therefore similar to terrestrially formed end moraines that mark the extent of a glacier advance or the position of a stationary ice margin (e.g., Bakker, 2005).

Excellent geological sections in gravel pits in the Kregnes "moraine" in Gauldalen in mid-Norway allowed Nemec et al. (1999) to study the sedimentology and architecture of this ice-marginal deposit. The Kregnes moraine was formed during the Younger Dryas cold period along the western margin of the last Scandinavian ice sheet. The majority of the "moraine" is a Gilbert-type delta with 150 m thick gravelly foreset beds and 2–3 m topset beds with beach deposits reaching the local marine limit of 175 m a.s.l. However, the study by Nemec et al. (1999)

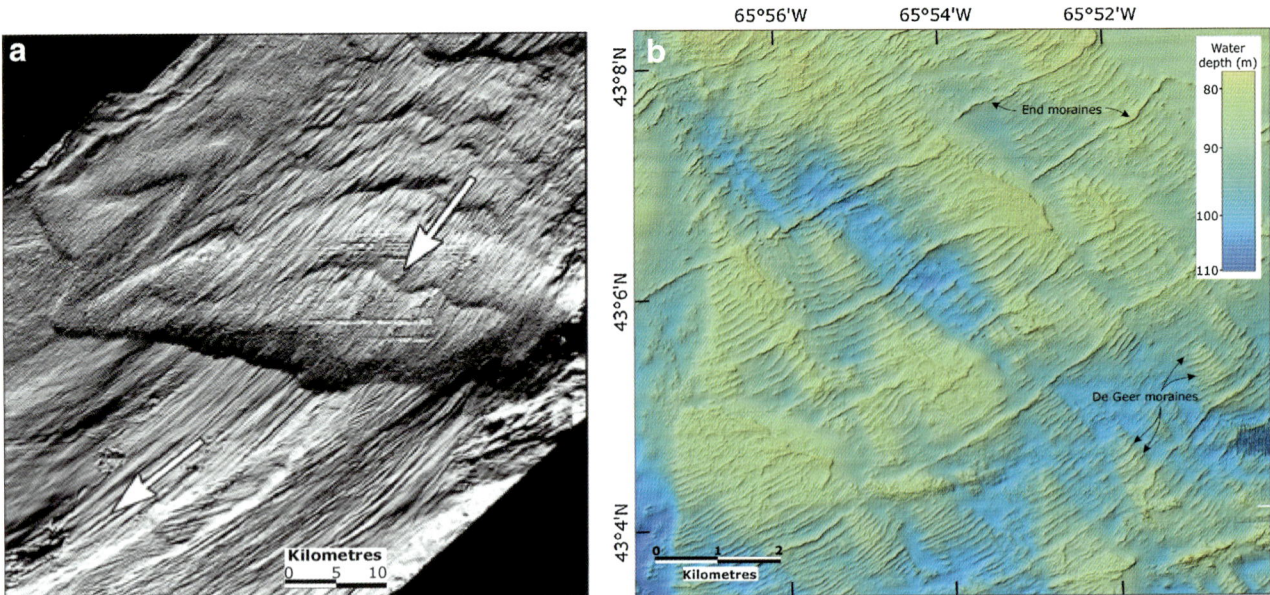

Moraines, Figure 1 Examples of seafloor geomorphology of paleo-ice streams. (**a**) Swath-bathymetric data revealing a grounding-zone wedge and mega-scale glacial lineations in Vestfjorden, north Norway. *Arrows* indicate former ice-flow direction (Modified from Dowdeswell et al. (2008)). (**b**) End moraines and De Geer moraines visible on a multibeam imagery on German Bank, Canadian Shelf (modified from Shaw et al., 2009).

Moraines, Figure 2 Oblique aerial view from 1936 of the prominent push moraine in front of the Holmströmbreen glacier, central Spitsbergen, Svalbard. Ice flow was from left toward right. The end moraine consists of glaciotectonically deformed, fine-grained marine sediments. Subset of aerial photograph number S36 1382, © Norwegian Polar Institute.

suggests that the Kregnes moraine was initially a submarine ice-contact fan that experienced glaciotectonic deformation. As sedimentation in the fjord basin progressed, the deposit evolved into an ice-contact delta and a large glaciofluvial delta. Similar architecture is known from numerous other "marine moraines" (e.g., Lønne, 1995, 2001).

Lindén and Möller (2005) and Lindén et al. (2008) described the geomorphology and sedimentology of transverse moraine ridges (ribbed moraines and De Geer moraines) in north Sweden formed at calving tidewater glacier margins during the retreat of the last Scandinavian ice sheet. They showed that these moraines generally consist of a lower, proximal part of basal till or glacially

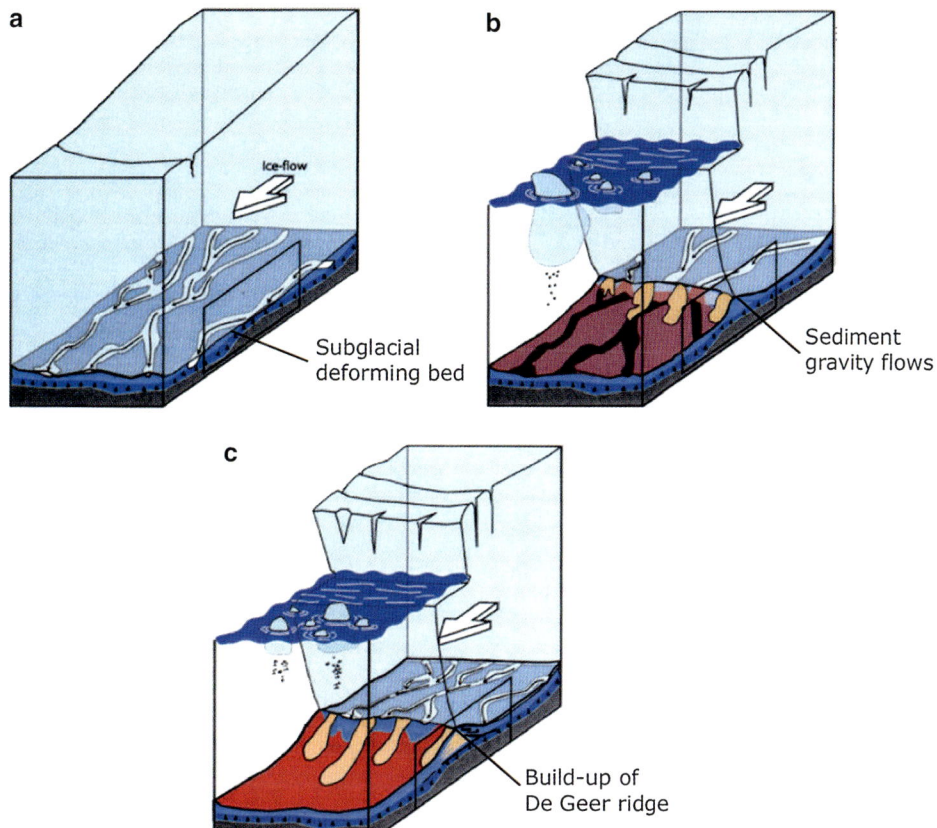

Moraines, Figure 3 Conceptual model for the formation of De Geer moraines at the grounding line of a glacier terminating in deeper water. The model is based on glacial geological investigations in northern Sweden. (**a**). Deposition of subglacial deformation till proximal to the grounding line. This till is the "base" for the subsequently deposited De Geer moraines. A subglacial channel network transports meltwater and sorted sediments toward the grounding line. (**b**). Glaciotectonic deformation at the grounding line and thickening of the subglacial bed initiates the De Geer moraine. Subglacial till accumulation on the proximal side of the ridge and sediment gravity flows on the distal side. (**c**). The ridge buildup and deformation continues until the grounding line moves up-glacier due to a calving event. Subsequently, glacioaquatic fine-grained sediments accumulate in an on-lapping style to the ridge (modified after Lindén and Möller, 2005).

deformed sediments proximal to the grounding line. The distal part of the ridges is characterized by lee-side cavity-deposited sediments or sediment gravity flows distal to the grounding line (Figure 3). The larger ribbed moraines were interpreted to have formed further up-glacier than the smaller De Geer moraines that formed at the grounding line. Transverse moraine ridges, very similar in size and geomorphology to the Swedish De Geer moraines, have been described from Wisconsinan glacial landforms at the seafloor off Nova Scotia (Figure 1b; Shaw et al., 2009). Ottesen et al. (2008) suggest that minor transverse moraine ridges on the sea floor in front of a modern surge-type tidewater glacier on Svalbard might be a modern analogue to the De Geer moraines described by Lindén and Möller (2005).

Conclusions

The term "moraine" describes a glacial landform directly deposited by a glacier and/or deformed by a glacier. Many marine moraines are ridge shaped, transverse to the ice-flow direction, and indicate the extent of a glacier advance or the position of a stationary glacier margin. Moraines formed at the margins of glaciers grounded in marine or large lacustrine basins comprise ridges of mainly till, diamict, and outwash sediments. The term "moraine" is also used for ice-marginal landforms anywhere in the continuum of grounding-line moraines to ice-contact or even glaciofluvial deltas.

Bibliography

Bakker, M. A. J., 2005. *The Internal Structure of Pleistocene Push Moraines. A Multidisciplinary Approach with Emphasis on Ground-Penetrating Radar*. TNO, Geological Survey of the Netherlands, 180 pp.

Benediktsson, Í. Ö., Möller, P., Ingólfsson, Ó., van der Meer, J. J. M., Kjær, K. H., and Krüger, J., 2008. Instantaneous end moraine and sediment wedge formation during the 1890 glacier surge of Brúarjökull, Iceland. *Quaternary Science Reviews*, **27**, 209–234.

Benediktsson, Í. Ö., Ingólfsson, Ó., Schomacker, A., and Kjær, K. H., 2009. Formation of submarginal and proglacial end moraines: implications of ice-flow mechanism during the 1963–64 surge of Brúarjökull, Iceland. *Boreas*, **38**, 440–457.

Benediktsson, Í. Ö., Schomacker, A., Lokrantz, H., and Ingólfsson, Ó., 2010. The 1890 surge end moraine at Eyjabakkajökull, Iceland: a re-assessment of a classic glaciotectonic locality. *Quaternary Science Reviews*, **29**, 484–506.

Benn, D. I., and Evans, D. J. A., 2010. *Glaciers and Glaciation*. London: Hodder Education.

Bennett, M. R., 2001. The morphology, structural evolution and significance of push moraines. *Earth-Science Reviews*, **53**, 197–236.

Boulton, G. S., 1972. Modern Arctic glaciers as depositional models for former ice sheets. *Journal of the Geological Society of London*, **128**, 361–393.

Boulton, G. S., van der Meer, J. J. M., Beets, D. J., Hart, J. K., and Ruegg, G. H. J., 1999. The sedimentary and structural evolution of a recent push moraine complex: Holmströmbreen, Spitsbergen. *Quaternary Science Reviews*, **18**, 339–371.

Clayton, L., Attig, J. W., Ham, N. R., Johnson, M. D., Jennings, C. E., and Syverson, K. M., 2008. Ice-walled-lake plains: implications for the origin of hummocky glacial topography in middle North America. *Geomorphology*, **97**, 237–248.

Dowdeswell, J. A., Ottesen, D., Evans, J., Ó Cofaigh, C., and Anderson, J. B., 2008. Submarine glacial landforms and rates of ice-stream collapse. *Geology*, **36**, 819–822.

Dyke, A. S., and Prest, V. K., 1987. Late Wisconsinan and Holocene history of the Laurentide Ice Sheet. *Géographie physique et Quaternaire*, **41**, 237–263.

Evans, D. J. A., 2009. Controlled moraines: origins, characteristics and palaeoglaciological implications. *Quaternary Science Reviews*, **28**, 183–208.

Evans, D. J. A., Hiemstra, J. F., and Cofaigh, C. O., 2012. Stratigraphic architecture and sedimentology of a Late Pleistocene subaqueous moraine complex, southwest Ireland. *Journal of Quaternary Science*, **27**, 51–63.

Fyfe, G. J., 1990. The effect of water depth on ice-proximal glaciolacustrine sedimentation: Salpausselkä I, southern Finland. *Boreas*, **19**, 147–164.

Goff, J. A., Lawson, D. E., Willems, B. A., Davis, M., and Gulick, S. P. S., 2012. Morainal bank progradation and sediment accumulation in Disenchantment Bay, Alaska: Response to advancing Hubbard Glacier. *Journal of Geophysical Research – Earth Surface*, **117**, F02031.

Howe, J. A., Dove, D., Bradwell, T., and Gafeira, J., 2012. Submarine geomorphology of the Sea of the Hebrides, UK. *Marine Geology*, **315–318**, 64–76.

Jakobsson, M., Anderson, J. B., Nitsche, F. O., Dowdeswell, J. A., Gyllencreutz, R., Kirchner, N., Mohammad, R., O'Regan, M., Alley, R. B., Anandakrishnan, S., Eriksson, B., Kirshner, A., Fernandez, R., Stolldorf, T., Minzoni, R., and Majewski, W., 2011. Geological record of ice shelf break-up and grounding line retreat, Pine Island Bay, West Antarctica. *Geology*, **39**, 691–694.

Lindén, M., and Möller, P., 2005. Marginal formation of De Geer moraines and their implications to the dynamics of grounding-line recession. *Journal of Quaternary Science*, **20**, 113–133.

Lindén, M., Möller, P., and Adrielsson, L., 2008. Ribbed moraine formed by subglacial folding, thrust stacking and lee-side cavity infill. *Boreas*, **37**, 102–131.

Livingstone, S. J., Cofaigh, C. O., Stokes, C. R., Hillenbrand, C. D., Vieli, A., and Jamieson, S. S. R., 2012. Antarctic palaeo-ice streams. *Earth-Science Reviews*, **111**, 90–128.

Lønne, I., 1995. Sedimentary facies and depositional architecture of ice-contact glaciomarine systems. *Sedimentary Geology*, **98**, 13–43.

Lønne, I., 2001. Dynamics of marine glacier termini read from moraine architecture. *Geology*, **29**, 199–202.

Möller, P., 2006. Rogen moraine: an example of glacial reshaping of pre-existing landforms. *Quaternary Science Reviews*, **25**, 362–389.

Möller, P., 2010. Melt-out till and ribbed moraine formation, a case study from south Sweden. *Sedimentary Geology*, **232**, 161–180.

Nemec, W., Lønne, I., and Blikra, L. H., 1999. The Kregnes moraine in Gauldalen, west-central Norway: anatomy of a Younger Dryas proglacial delta in a palaeofjord basin. *Boreas*, **28**, 454–476.

Nitsche, F. O., Gohl, K., Larter, R. D., Hillenbrand, C.-D., Kuhn, G., Smith, J. A., Jacobs, S., Anderson, J. B., and Jakobsson, M., 2013. Paleo ice flow and subglacial meltwater dynamics in Pine Island Bay, West Antarctica. *The Cryosphere*, **7**, 249–262.

Ottesen, D., Dowdeswell, J. A., Benn, D. I., Kristensen, L., Christiansen, H. H., Christensen, O., Hansen, L., Lebesbye, E., Forwick, M., and Vorren, T. O., 2008. Submarine landforms characteristic of glacier surges in two Spitsbergen fjords. *Quaternary Science Reviews*, **27**, 1583–1599.

Ross, M., Campbell, J. E., Parent, M., and Adams, R. S., 2009. Palaeo-ice streams and the subglacial landscape mosaic of the North American mid-continental prairies. *Boreas*, **38**, 421–439.

Schomacker, A., 2008. What controls dead-ice melting under different climate conditions? A discussion. *Earth-Science Reviews*, **90**, 103–113.

Schomacker, A., 2011. Moraine. In Singh, V. P., Singh, P., and Haritashya, U. K. (eds.), *Encyclopedia of snow, ice and glaciers*. Dordrecht: Springer, pp. 747–756.

Schomacker, A., Kjær, K. H., and Krüger, J., 2010. Subglacial environments, sediments and landforms at the margins of Mýrdalsjökull. In Schomacker, A., Krüger, J., and Kjær, K. H. (eds.), *The Mýrdalsjökull Ice Cap, Iceland. Glacial processes, sediments and landforms on an active volcano*. Amsterdam: Elsevier. Developments in Quaternary Science, Vol. 13, pp. 127–144.

Shaw, J., Todd, B. J., Brushett, D., Parrot, D. R., and Bell, T., 2009. Late Wisconsinan glacial landsystems on Atlantic Canadian shelves: new evidence from multibeam and single-beam sonar data. *Boreas*, **38**, 146–159.

Winsborrow, M. C., Andreassen, K., Corner, G. D., and Laberg, J. S., 2010. Deglaciation of a marine-based ice sheet: Late Weichselian palaeo-ice dynamics and retreat in the southern Barents Sea reconstructed from onshore and offshore glacial geomorphology. *Quaternary Science Reviews*, **29**, 424–442.

Cross-references

Deltas
Fjords
Geochronology: Uranium-Series Dating of Ocean Formations
Glacio(hydro)-isostatic Adjustment
Grounding Line
Ice-rafted Debris (IRD)
Regional Marine Geology

MORPHOLOGY ACROSS CONVERGENT PLATE BOUNDARIES

Wolfgang Frisch
Department of Geosciences, University of Tübingen, Tübingen, Germany

Definition

Convergent plate boundaries are plate boundaries where one plate subducts beneath the other due to convergent plate movement. Due to its higher density, mostly oceanic lithosphere is the subducting part.

Four types of convergent plate boundaries

Four types of convergent plate boundaries are recognized (Figures 1 and 2):

The first type occurs when oceanic lithosphere is subducted below another oceanic lithosphere ("Intraoceanic Subduction Zone") to create a volcanic island arc system built on oceanic crust ("ensimatic island arc"; *sima* – artificial word first used by Wegener made from silicon and magnesium to characterize ocean floor and Earth's mantle). Examples for intraoceanic, ensimatic island arc systems include the Mariana Islands in the Pacific and the Lesser Antilles in the Atlantic.

The second type occurs where oceanic lithosphere is subducted beneath continental lithosphere and an island arc underlain by continental crust forms ("ensialic island arc"; *sial* – silicon and aluminum for continental crust). The island arc of this system is separated from the continent by a marine basin underlain by oceanic crust. Examples for island arc systems underlain by continental crust are the Japanese Islands and the eastern Sunda Arc.

The third type of convergent plate boundary represents the active continental margins where oceanic lithosphere is subducted beneath continental lithosphere without a marine basin behind the volcanic arc; rather, the arc is built directly on the adjacent continent. The continental margin is connected directly to the hinterland, although a shallow marine basin may exist behind the volcanic arc. Examples for active continental margins are the Andes, SE Alaska, and the western and central Sunda Arc that includes Sumatra and Java.

The fourth type of convergent margin occurs along zones of continent-continent collision. If two continental

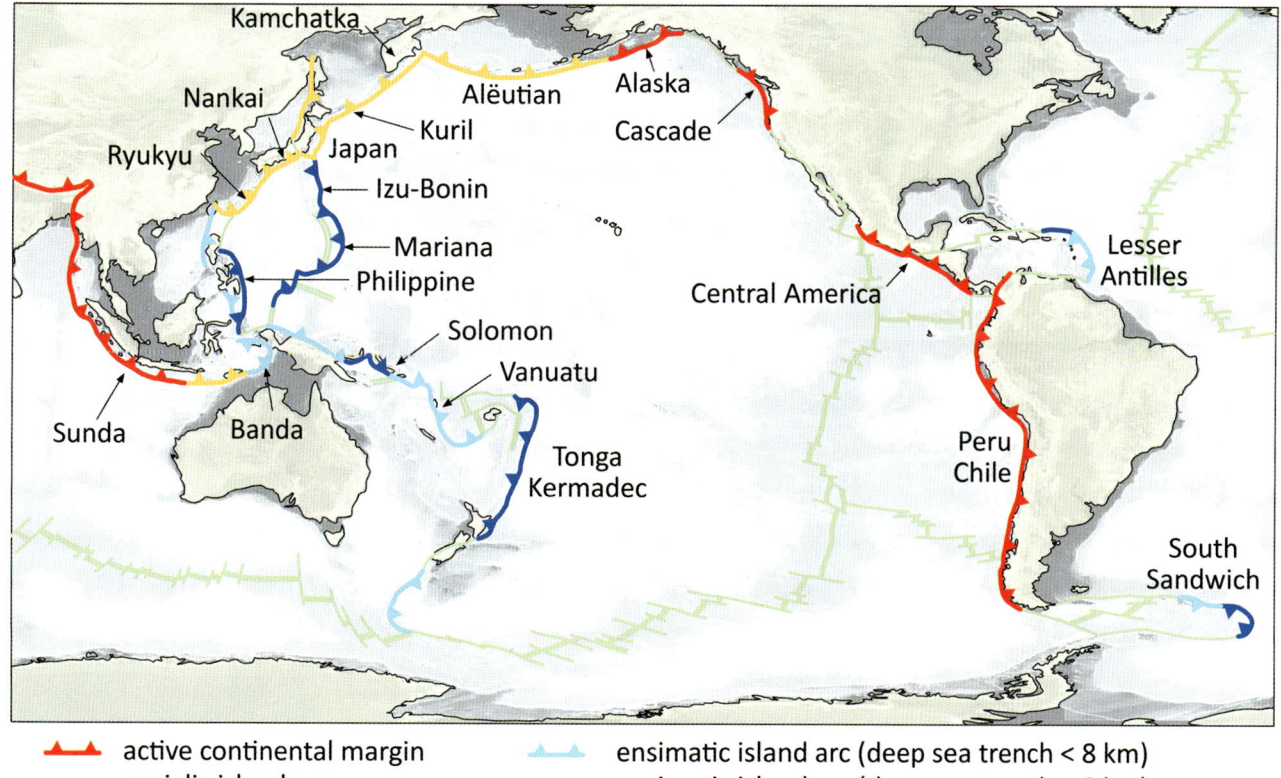

Morphology Across Convergent Plate Boundaries, Figure 1 Convergent plate margins of the Earth characterized by ensimatic island arcs (underlain by oceanic crust), ensialic island arcs (underlain by continental crust), and active continental margins.

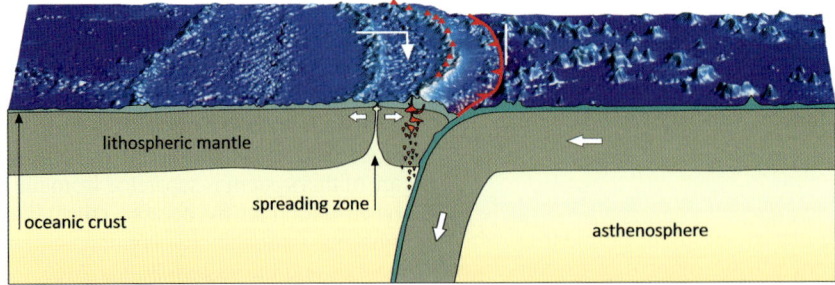
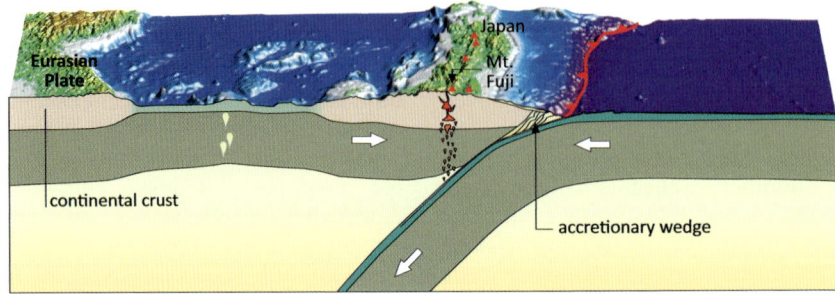
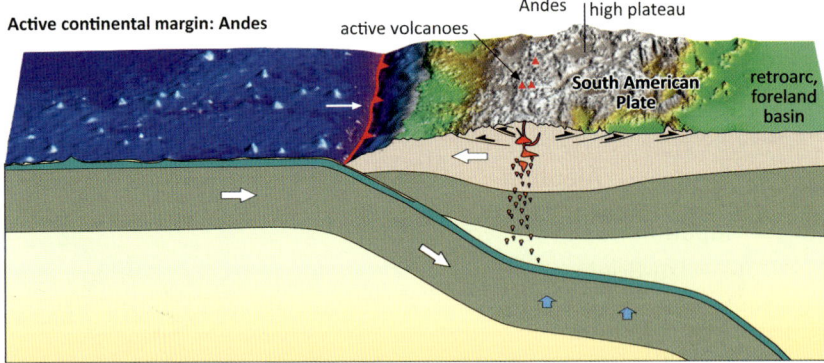
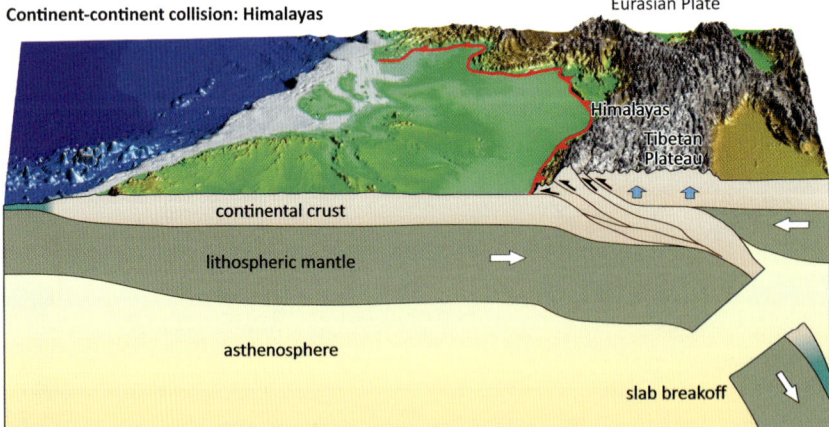

Morphology Across Convergent Plate Boundaries, Figure 2 Examples of different types of plate margins with subduction zones. The island arc of the Marianas developed on oceanic crust, that of Japan on continental crust. The volcanic zone of the Andes is built on the South American continent (active continental margin). The collision of two continents produces a mountain range like the Himalayas – subduction wanes, leading to slab breakoff.

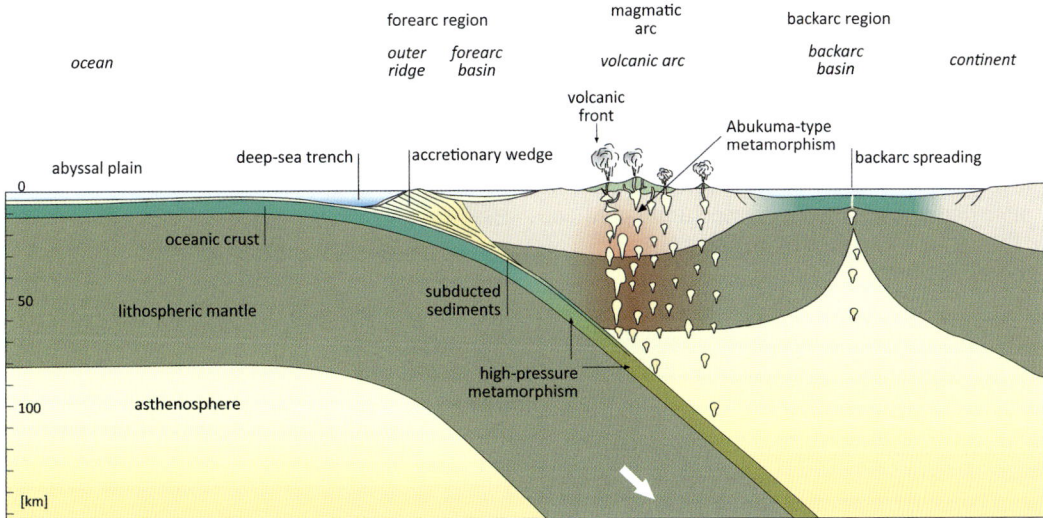

Morphology Across Convergent Plate Boundaries, Figure 3 Structure and morphology of a plate margin system with subduction zone and ensialic island arc.

masses collide during continuous subduction, they eventually merge. Telescoping of the two plates and the buoyancy of the subducting continent eventually leads to a standstill of subduction within the collision zone. The oceanic part of the subducting plate tears off and continues to drop down, a process referred to as "slab breakoff." Continent-continent collisions ultimately result in the formation of mountain ranges like the Himalayas or the Alps.

Structure and topography of plate margin systems with subduction zones

Systems of convergent plate boundaries are characterized by a distinct topographic and geologic subdivision. Although the plate boundary itself is only represented by a line at the surface, commonly within a deep-sea trench, a zone several hundreds of kilometers wide is formed by processes which are related to subduction. The volcanic zone above the subduction zone, in many cases expressed as an island arc, is the dominating element of this plate boundary system. We use arc as shorthand for the terms volcanic arc or magmatic zone. The arc is the point of reference for the convergent boundary that is usually divided into three parallel zones: from the trench to the arc is the forearc zone, the arc zone comprises the magmatic belt, and the region behind the arc is the backarc zone (Figure 3). This generally agreed upon subdivision of three parts turned out to be practical in order to describe the complex structures of convergent plate boundary systems.

Deep-sea trenches form the major topographic expression at convergent plate boundaries. These deep, narrow furrows surround most of the Pacific Rim and small portions of the rims that surround the Indian and Atlantic oceans. Oceanic lithosphere is bent downward under the margin of the "upper plate" and dives into the asthenosphere. The result is an elongated deep trench that is located between the abyssal plains and the border of the upper plate. The deepest trenches, with water depths of about 11,000 m, are known from the Challenger and the Vitiaz Deep (named after a British research vessel and a Russian researcher, respectively) in the southern Mariana trench. Water depths of more than 10,000 m are also known from the Kurile, Izu-Bonin, Philippine, and Tonga-Kermadec trenches (Figure 1).

Convergent plate boundaries are responsible for the greatest differences of relief on the Earth's surface. Differences in altitude of 10 km between the deep-sea trench and volcanoes of the magmatic arc are not unusual. A relief of 14,300 m across a distance of less than 300 km can be observed between Richards Deep (−7,636 m) and Llullaillaca (+6,723 m) in the Chilean Andes, the highest active volcano on Earth.

In a transect from the trench onto the upper plate, the following morphological features generally occur (Figure 3). The landward side of the deep-sea trench is part of the upper plate and consists of a slope with an average steepness of several degrees. In front of the Philippine Islands, the angle exceeds 8° where a rise from −10,500 m to −200 m occurs over a distance of 70 km. The so-called outer ridge follows behind the slope. In most cases, the ridge remains substantially below sea-level; however, in several cases, islands emerge above sea-level (Sunda Arc: Mentawai; Lesser Antilles: Barbados). The outer ridge is not always distinctive. It is prominent in those places where large amounts of sediment are scraped off from the subducting plate along the plate boundary and accreted onto the base of the upper plate. Next in the transect, directly in front of the volcanic arc, lies the forearc basin, another prominent morphological element.

Collectively, the deep-sea trench, outer ridge, and forearc basin comprise the forearc region. The distance between the plate boundary and magmatic zone has a width of between 100 and 250 km. This region is also called the arc-trench gap, a magmatic gap that with very few exceptions is void of magmatic activity. Low temperatures in the crust are caused by the coolness of the subducting plate underneath; the low temperature gradient prevents the formation of magma by melting or the rise of magma from deeper sources.

The volcanic arc, with an average width of 100 km, is the central part of the island arc or the active continental margin system and is characterized by significant magmatic activity. Approximately 90 % of all active volcanoes above the sea-level, most of which are in or around the Pacific Ocean, are subduction-related volcanoes. The zone of volcanism has a sharp boundary on the forearc margin and a gradual one on the backarc margin (Figure 3).

Island arcs are separated from adjacent continents by a marine basin underlain by ocean crust, the backarc basin (Figures 2 and 3). Active continental margins have a backarc region that may have thinned continental crust and/or a zone of compressional structures.

In principle the arcuate shape of island arcs is easy to explain. Before a plate enters a subduction zone, it possesses a curvature according to the curvature of the Earth. When it dives into the subduction zone, this curvature inverts and convex becomes concave. Adjacent arcs commonly display a catenary-like pattern as observed in the Western Pacific where one island arc drapes next to the other. The radius of an arc is a function of the radius of the Earth and the inclination angle of the subduction zone, but is also influenced by structural features of the upper plate. This becomes evident along continental margins where the shape of the continent substantially controls the shape of the plate boundary, e.g., along the continental margin on the western edge of South America.

Morphological differences caused by Mariana- and chile-type subduction

The western rim of the Pacific Plate is characterized by lithosphere with ages greater than 100 Ma. Therefore, spontaneous subduction driven by the density of the cool oceanic lithosphere dominates (Mariana-type subduction). However, younger lithosphere typical of the eastern margin of the Pacific, mostly less than 50 Ma old, as well as that of the Philippine Sea Plate and the western Sunda Arc is too buoyant to subduct spontaneously, but rather is forced underneath the upper plate by compressional forces (Chile-type subduction). Such conditions promote shallow subduction and a strong coupling to the upper plate. In other words, the horizontal compressional force is transferred from the subducting plate to the upper plate and compressional structures such as folding and stacking of crustal units evolve far into the upper plate (Figure 4, lower). In contrast, spontaneous subduction with steeply

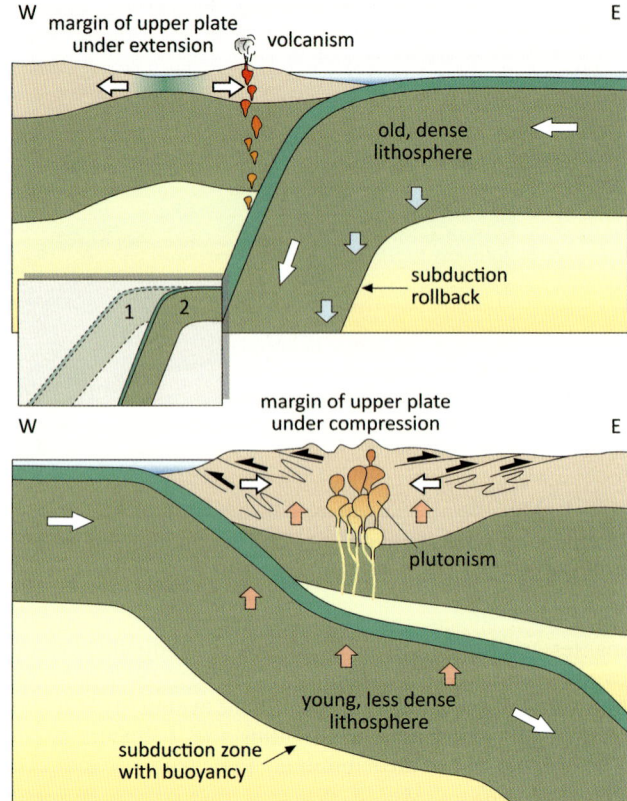

Morphology Across Convergent Plate Boundaries, Figure 4 Mariana-type (*above*) and Chile-type (*below*) subduction zones. At the Marianas, subduction occurs easily and the subduction zone rolls back to the east (insert). At Chile, young and specifically light lithosphere pushes upwards; subduction is forced. The two types of subduction cause highly different morphological features.

dipping subduction zones generates extensive decoupling along the plate boundary so less deformation of the upper plate occurs and extensional structures form (Figure 4, upper). Forced subduction may evolve from spontaneous subduction as increasingly younger portions of the downgoing plate are subducted thus changing the vertical component at a plate boundary to a more strongly coupled horizontal force.

Steeply dipping subduction zones such as the ***Mariana-type subduction*** produce significant consequences. A strong slab pull of the subducting lithosphere forces the subduction zone to roll back oceanward – i.e., towards the subducting plate or easterly into the edge of the Pacific Plate in the case of the Marianas (Figure 4). The locus of the hinge zone, where the ocean floor is bent down into the subduction zone and which marks the plate boundary, migrates backwards, away from the locus of the arc. This process is termed subduction or slab rollback (Figure 4, insert). The rollback causes extensional forces to develop at the plate boundary, and the edge of the upper plate is extended. The rollback causes the island arc to migrate towards the outside of the arc system which in turn results in a strong extension of the backarc region. If the arc was originally built on a continent, this process leads to the separation of the island arc from the upper plate continent. The resulting separation behind the island arc generates a backarc basin and with sufficient extension, new oceanic crust may form (e.g., Sea of Japan). The western Pacific is characterized by numerous island arc–backarc systems (Figure 1). The garland- or drape-like alignment of the island arcs results from their separation from the continent. Typically, the garlands have lengths from 2,000 to 2,500 km. Because the Mariana Arc comprises an intraoceanic subduction zone, the formation of the backarc basin occurred solely within oceanic crust (Figure 2).

Subduction rollback that forms during Mariana-type subduction has an effect on the topography of the plate boundary system. The mean topographic elevation on the edge of the upper plate is low because of the effect of suction and extension. Like in the Marianas, the magmatic arc may only form a chain of small islands representing the peaks of the volcanoes. Also, the deep-sea trenches are significantly deeper on average because the subducting plate bends downward very steeply and the island arcs do not deliver enough sedimentary material to even partially fill the trench. The world's deepest trenches, all with depths greater than 8,000 m, occur along westward-directed subduction zones (including those of the Atlantic Ocean) that involve old oceanic lithosphere (Figure 1). The rollback of the westward-directed subduction zones may be enhanced by an eastward-directed asthenospheric current as calculated from global plate drift compared to flow of the sublithospheric mantle.

A curious situation at subduction zones with backarc basins (Mariana-type subduction) is the juxtaposition of strong convergence and divergence. Although relative plate movement velocities at convergent plate boundaries in the western Pacific approach 9 cm/year, a wide area at the edge of the upper plate is under extension. Compressive structures are restricted essentially to the tip of the upper plate.

The subduction zones and convergent plate boundaries of *Chile-type subduction* are fundamentally different in many ways. The subducting plate is intermittently coupled to the upper plate because of the buoyancy of the former; this causes compression and thickening of the upper plate (Figure 4). Thickening of continental crust leads to the generation of a mountain range. Deep-sea trenches are shallower and volcanic zones are characterized by significantly higher elevations than volcanoes associated with Mariana-type subduction. The highly elevated hinterland is more likely to deliver sedimentary material into the trenches, and the sediment supply is generally not hampered by intervening ridges. Even more consequential than the topographic height is the structural height, the total amount of uplift within the volcanic zone, which may exceed 20 km. Crustal structures originally formed at depth are subsequently uncovered by erosion. In fact, metamorphic and intrusive magmatic rocks at the surface in the Andes can be used to estimate the amount of uplift and subsequent amount of erosion.

Along Chile-type subduction zones, the compressional forces are transferred far into the upper plate. Therefore, earthquakes in this area are particularly strong and frequent. Regionally, compressional structures with overthrusts evolve that are oceanward-directed in the forearc and continentward-directed in the backarc region (Figure 4). Consequently, the backarc area is also under compressional stress. Crustal shortening occurs across the entire magmatic zone, which in turn has an effect on the magmatism itself. Melts rising from the asthenospheric wedge above the subduction zone typically become trapped in the crust and crystallize as intrusions or melt adjacent crustal rocks and feed highly explosive acidic volcanoes. The collective processes of tectonic stacking and magmatic accretion have formed a crustal thickness of 70 km in the Andes, one of the areas of thickest continental crust on Earth.

Bibliography

Bott, M. P. H., 1982. *The Interior of the Earth: Its Structure, Constitution and Evolution*, 2nd edn. London: Edward Arnold. 403pp.

Doglioni, C., Harabaglia, P., Merlini, S., Mongelli, F., Peccerillo, A., and Piromallo, C., 1999. Orogens and slabs vs. their direction of subduction. *Earth Science Reviews*, **45**, 167–208.

Frisch, W., Meschede, M., and Blakey, R., 2011. *Plate Tectonics. Continental Drift and Mountain Building*. Berlin/Heidelberg: Springer. 212pp.

Nicolas, A., 1995. *The Mid-Oceanic Ridges. Mountains Below Sea Level*. Berlin/Heidelberg: Springer. 217pp.

Cross-references

Active Continental Margins
Intraoceanic Subduction Zone
Island Arc Volcanism, Volcanic Arcs
Marginal Seas
Subduction

MUD VOLCANO

Achim Kopf
MARUM-Center for Marine Environmental Sciences, University of Bremen, Bremen, Germany

Synonyms

Diatreme; Mud diapir; Pockmark

Definition

A type of sedimentary volcano (Kopf, 2002; van Loon, 2010) composed of fine-grained mud, in both subaqueous and subaerial settings.

Landforms proposed to be mud volcanoes: Pitted cone, Mound, Mud diapir, Diatreme.

Origin of term

The Greek "diapeirein" means "to pierce," and the first mention of diapiric structures dates back to Leymerie (1881), who suggested the word "tiphon." This nomenclature was adapted by Choffat (1882) when describing sedimentary diapiric manifestations in Portugal, calling these small domes "tiphonique." Mrazec's (1927) definition of diapirism as the process of forceful movement of a more or less plastic body from areas of greater pressure to areas of less pressure has since then widely been accepted. It has often been restricted to salt structures instead of having been applied on mechanical grounds, although some workers used it for effusive or intrusive magmatic processes (e.g., Wegmann, 1930). The term "sedimentary volcanism" supposedly (Saunders, personal communication, 2000) goes back to Kugler (1933).

Introduction and historical background

Mud volcanoes on Earth extrude relatively low temperature slurries of gas, liquid, and rock to the surface from depths of meters to kilometers (Kopf, 2002). Slurries build circular to subcircular deposits of mud and rock breccia that range in diameter from a few meters up to tens of kilometers. They can produce a variety of morphologies, domes, cones, caldera-like forms, or relatively flat structures, reflecting physical properties of the rising material. The depth of origin of the different phases may be variable within the same mud volcano and ranges from a few meters to several kilometers sub-bottom.

Mud intrusion and extrusion are well-known phenomena whereby fluid-rich, fine-grained sediments ascend within a lithologic succession due to their buoyancy. These processes have long been recognized as related to the occurrence of petroleum, regional volcanic and earthquake activity, and orogenic belts (see the enlightening presentation by Ansted (1866)). The abundance of mud volcanoes was first examined systematically on a broad scale by Higgins and Saunders (1974). These authors mostly focused on mud volcanoes on land and used industry drillhole data to establish relationships between mud volcanism, hydrocarbons, and regional tectonics. The onshore mud volcanoes and their important role in predicting petroleum reservoirs have been summarized by Rhakmanov (1987) and were heavily used at times when drilling the subsurface had not been invented yet. Recent improvements in seafloor imagery and seismic exploration led to the discovery of countless mud extrusive provinces all over the world (Figure 1). In fact, compared to the compilation by Higgins and Saunders (1974), more than twice as many occurrences are presently known, with the number gradually increasing through time. Most importantly, mud volcanoes occur along convergent plate margins where fluid-rich sediment is accumulated in deep-sea trenches at high rates. Such deposits then enter the subduction factory, where liquids and volatiles are released due to incipient compactional stress and temperature. Studies of geophysical data and samples of mud volcanoes have considerably improved the understanding of the mechanics, driving forces, and evolution of the features through the most recent Earth history (e.g., Barber et al., 1986; Brown, 1990). In addition, deep ocean drilling and submersible studies shed crucial light on eruptivity, emission of volatiles, and potential hazard originating from violent mud extrusion (e.g., Bagirov et al., 1996; Robertson et al., 1996; Kopf, 1999). The wealth of results attests that mud extrusion predominantly occurs in collisional settings (i.e., zones of convergent plate motion), with mostly pore fluids during early (often marine) stages and hydrocarbons at later stages (often on land; e.g., Jakubov et al., 1971; Tamrazyan, 1972; Speed and Larue, 1992). Although quantification of fluid and mud discharge in mud volcanoes is not easy due to their short-lived nature and inaccessibility on the seafloor, first-order estimates regarding flux rates have recently been attempted for various features and regions (e.g., Henry et al., 1996; Kopf and Behrmann, 2000). When put into a broader context, such estimates indicate that mud extrusion contributes significantly to fluid backflux from the lithosphere to the hydrosphere. Along wide parts of large accretionary prisms (like the Barbados or Mediterranean Ridges), hundreds of features can cause fluid expulsion at rates exceeding those at the frontal part of the prism (see discussion in Kopf et al., 2001). Provided that the process of mud volcanism has been equally common in the past, these features have been major players in fluid and gas budgets and geochemical cycling in collision zones.

Formation, occurrence, and habitus

Mud volcanoes (MVs) are the surface expressions of deeply rooted vertical structures, and they are composed of coarse, poorly sorted "mud breccia" made up of materials which migrated upward from zones of overpressure. Sediment extrusion is driven by a mobile fluid phase, most often a mixture of aqueous and gaseous fluids (on Earth, water, hydrocarbons, and CO_2; on Mars: it is proposed to be partly methane). Mud volcanoe (MV) occurrences have the following aspects in common: (1) an origin from thick, rapidly deposited sequences of marine clays, (2) a tertiary age, (3) a structural association due to tectonic shortening and/or earthquake activity, (4) sediment overpressuring and accompanied fluid emission (free gas, brines, gas hydrate water, rarely oil), and (5) polymictic assemblages of the surrounding rock present in the ejected argillaceous matrix (e.g., Ansted, 1866; Higgins and Saunders, 1974; Fertl, 1976; Yassir, 1989; Kopf, 2002). Less commonly, volcanic and geothermal activity, often associated with ascent of hydrothermal fluids, may form mud domes and mud pools (e.g., at the basal flanks of the Costa Rica volcanic arc onshore). See Figure 2.

The possible causes of undercompaction are numerous, and sediment loading is one mechanism which may apply in any setting. It can be enhanced by rapid plate motion in

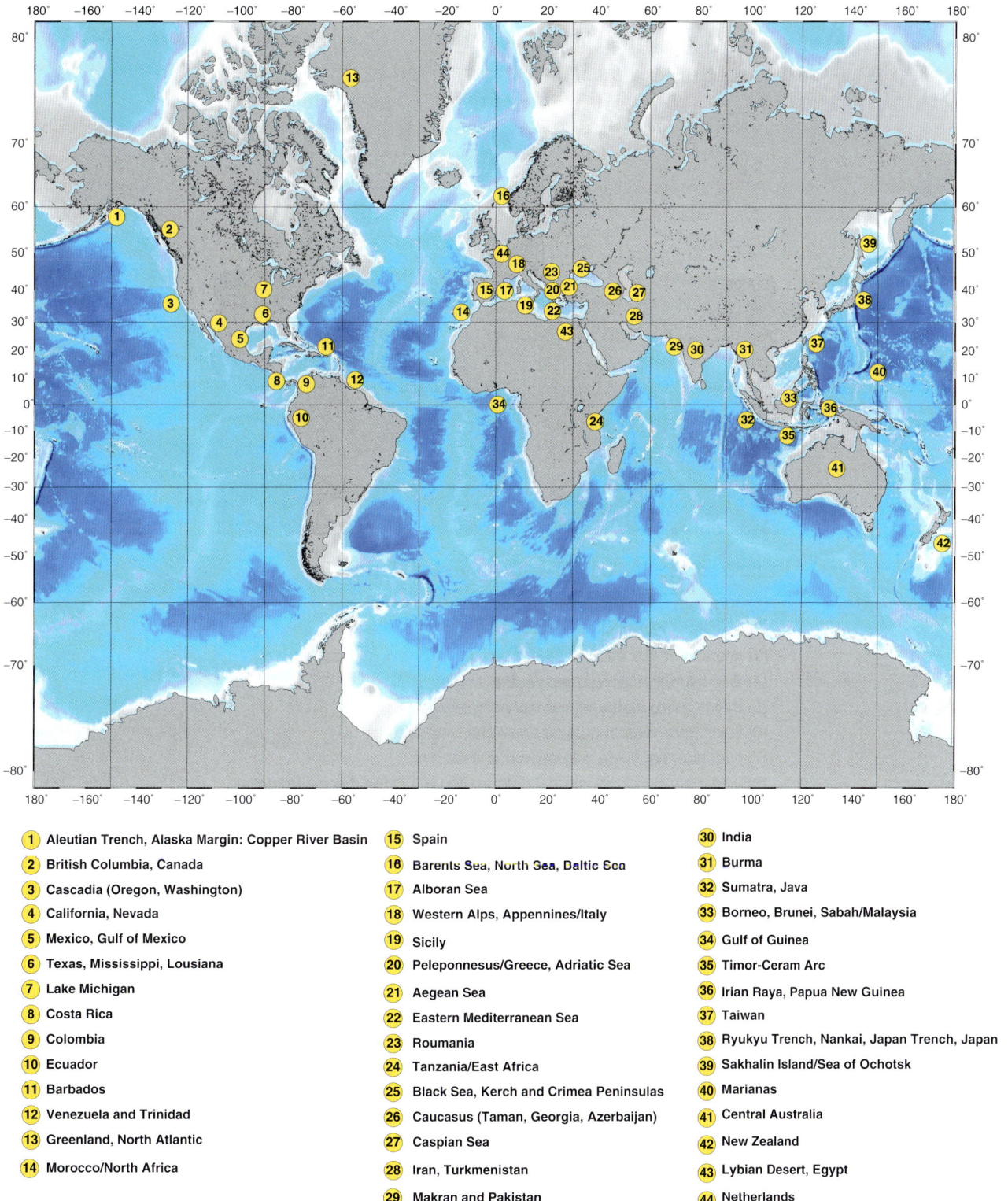

Mud Volcano, Figure 1 Map of global mud volcano occurrence based on compilation by Kopf (2002) and recent publications on individual regions (after Kopf, 2002).

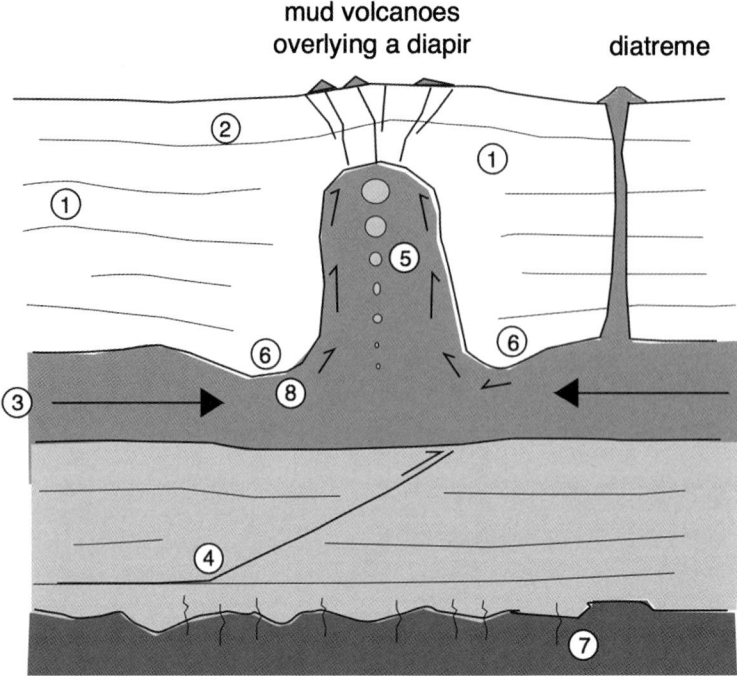

Mud Volcano, Figure 2 Schematic diagram of a mud diapir, MV extrusions (= mud volcanoes) and diatremes, including possible fluid sources (numbered 1–8); geochemically mature fluids may be found among categories *3, 4,* and *7*; while categories *1, 6, 8,* water from dissociated gas hydrates may provide "freshened" fluids (after Kopf, 2002).

a subduction zone setting, because dewatering may not keep pace with the rate the undercompacted material is underthrust beneath the overriding plate. However, among the numerous other possibilities causing undercompaction, the most prominent are gas hydrate dissociation, tectonic loading, diagenesis and mineral dehydration reactions, hydrocarbon generation, aquathermal pressuring, or osmosis (e.g., Yassir, 1989). Several of these processes may act simultaneously in MV areas at depth, so that distinction between them is difficult. A summary of various origins for elevated pore pressures is given in Table 1.

On Earth, MVs are particularly abundant in Azerbaijan, the Caspian Sea, and the Eastern Mediterranean Sea, where their contributions to global budgets and fluid transfer from the geosphere to the hydrosphere/atmosphere are significant (Kopf et al., 2001). Smaller features are typically tens of centimeters of cm to a few meters high and wide, while larger edifices may be up to hundreds of m high and several km wide (Kopf, 2002). In general, the largest features have been described for the Mediterranean Ridge accretionary complex, where seismic reflection data reveal pie-shaped (i.e., tabular), deeply rooting MVs of a few tens of kilometers across (Huguen et al., 2001; Huguen et al., 2004). Evidence for a deep origin of the ejecta is manifold, and mud and embedded clasts may originate from different (often shallower) levels than the aqueous or gaseous fluids (e.g., Robertson and Kopf, 1998; Kopf, 2002; Kopf and Deyhle, 2002; Dupré et al., 2014).

In terms of their morphology, the clast content usually defines the height, while viscosity and flux rate define the overall shape. Little information exists concerning the geometry of the conduit (or feeder channel) for ancient and modern examples. Surface observations indicate small diameters of 15 cm (Taiwan MVs; Yassir

Mud Volcano, Table 1 Causes for overpressuring, distinguished by their origin, mechanism, and corresponding geological setting. A selection of references is provided (Modified from Clennell, 1992)

Origin	Mechanism	Environment	Significance	Selected references
Burial	Sedimentary loading, compaction/settling	Any sedimentary setting, i.e., deltas, active and passive margins	Major in such settings	Braunstein and O'Brien, 1968; Morgan et al., 1968; Moon and Hurst, 1984
	Slumping, sliding	Marine slopes of active and passive margins	Major on slopes	Lewis, 1990
Tectonic	Tectonic loading	Any compressional margin, thrust zones and wedges	Major in such settings	Shipley et al., 1990; Westbrook and Smith, 1983
	Deep level ducting	Accretionary complexes	Major in such settings	Moore, 1989
	Smectite dehydration	Accretionary complexes	Can be major	Fitts and Brown, 1999
Thermogenic	Opal/qtz reactions smectite dehydration	Any setting with biosilica any setting with abundant clay deposition	Usually minor can be major	Kastner, 1981; Schoonmaker, 1987; Colten-Bradley, 1987
	Other diagenesis	Deeper subduction zone	Minor	Moore and Saffer, 2001
	Metamorphism	Deep subduction zones; other collision zones	Usually minor, but locally important	Bebout, 1995
	Methanogenesis/hydrocarbon generation	Any setting; reservoirs	Can be locally important	Ridd, 1970; Hedberg, 1974
	Thermal epansion; hydrothermal pressuring	Magmatic arcs and ridges	Can be locally important	Barker and Horsfield, 1982
Biogenic	Methanogenesis	Shallow marine settings; accretionary prisms	Can be very important	Ritger et al., 1987; Suess et al., 1999
Other	Osmosis	Clay-bearing sedimentary environments	Very minor	Fertl, 1976
	Gas hydrate processes	Arctic margins, accreting subduction zones	Significant	Hensen et al., 2007; Grevemeyer et al., 2004
	Permafrost water	Arctic margins	Locally important	Paull et al., 2007, Paull et al., 2011

(1989) and 30 cm (Greece; Stamatakis et al. (1987) to 1.5 m (Burma; Pascoe (1912) and 2 m (Sakhalin; Gorkun and Siryk (1968). The first values agree well with cemented chimneys in the ancient "Verrua" MVs in Italy (Cavagna et al., 1998). Also, theoretical considerations and calculations based on Stokes' law indicate similar diameters of 2–3 m for the Mediterranean Ridge MVs (Kopf and Behrmann, 2000). However, some controversial data are reported for the feeder geometry at depth. From seismic data, conduits of 1.5–3.5 km width have been proposed for the Black Sea area (Limonov et al., 1997). Similarly, a width of 2 km has been proposed for the bigger of the Barbados Ridge features (Griboulard et al., 1998) and the Alboran Sea mud domes (Perez-Belzuz et al., 1997).

The relationship between the width of the conduit, the associated ascent velocity, and the resulting surface topography is simplistically illustrated in Figure 3 (assuming identical material properties in each case). Given the variable degree of fluid saturation and difference in overlying medium (seawater vs. air), differences in longevity/resistance to erosion and morphology exist between onshore and offshore features, not to mention extraterrestrial mud volcanoes (see next paragraph).

Natural examples of pockmark fields associated with mud volcanoes as well as onshore and offshore examples are presented in Figure 4 and are particularly abundant in thickly sedimented delta fans such as the Mississippi, Nile, or Niger deltas (e.g., Loncke and Mascle, 2004; Judd and Hovland, 2007; Sultan et al., 2014).

Mud volcanoes are commonly associated with neotectonic regions or natural gas fields, but exceptions to this rule are known. Extraterrestrial analogs are, e.g., Martian pitted cones (McGill, 2005). Undersea examples include Haakon Mosby, Napoli and Milano, El Arraiche (Morocco), Gulf of Cadiz, Barbados and Mediterranean Ridges; and Conical Seamount (Marianas Arc), while subaerial examples are Lusi, Dashgil, or Lokhbatan (Kopf, 2002; van Loon, 2010). Mud volcanoes serve as "valves" in many compressional belts, so that their episodic activity and associated transient changes in physical and chemical properties make them suitable indicators for, e.g., seismicity and other deep-seated processes (e.g., Hedberg, 1974; Manga et al., 2009). Next to active margins, MVs are also frequently associated with large deep sea fans such as that of the Nile (Loncke and Mascle, 2004), Niger, or Gulf of Guinea (Sultan et al., 2014). Here, the depth of origin of the ejecta is shallower, and clasts, a prerequisite of tectonic

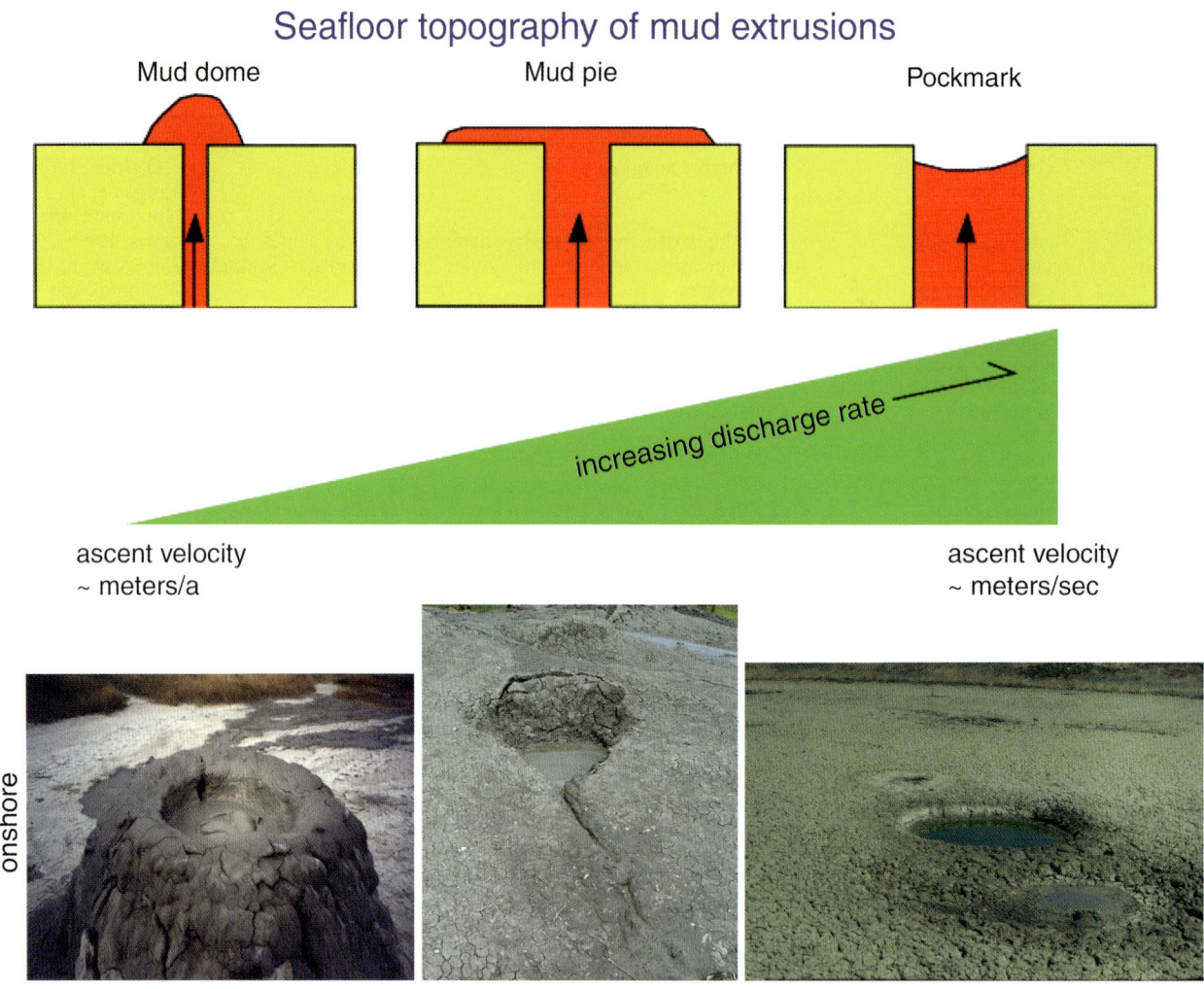

Mud Volcano, Figure 3 Variable feeder geometries and associated surface topography of mud extrusive phenomena, assuming otherwise identical physical properties and buoyancy forces. Note the "negative" topography (i.e., depression) in case of high discharge rates, which often are associated with a depleted (or drained) reservoir at depth. Such mud domes or diatremes are difficult to distinguish from pockmarks and may actually be identical except that pockmarks do not involve upward transport of a solid phase. Figures illustrating the various features onshore are from (*left* to *right*): Appenines/Italy (A. Kopf/Univ. Bremen) 2 x; Georgia (V. Lavrushin/RAS Moscow).

forcing in collisional belts/subduction zones, are scarce or absent; naturally, the height of these features is low and depressions from degassing (formerly organic-rich sediment) are commonly observed.

Landforms derived from the same processes

Onshore examples versus offshore examples, difference in rate of erosion/ longevity between the two; sand volcanoes (Burne, 1970) in delta settings, restricted to rapid sediment accumulation and/or excess overburden stress as a trigger.

Similar landforms derived from different processes

The term "sedimentary volcanism" refers to the similar geometry of many mud extrusions and igneous volcanoes, and only very rarely are MVs connected with igneous activity (e.g., von Gumbel, 1879; White, 1955; Chiodini et al., 1996). Magmatic volcanoes show fairly similar shapes (shield vs. cone), but comprise totally different material. Often the two are associated, e.g., hydrothermal fluids of arc volcanoes feed mud pools/volcanoes in Costa Rica.

Essential concepts and significance

The process of mud volcanism provides possible models for the origin of pitted cones and domes with various morphology on Earth (Kopf, 2002) as well as those on Mars, Venus, and several icy moons (Bradak and Kereszturi, 2003).

Mud volcanoes contribute significantly to the global fluid reflux from the geosphere to the hydrosphere,

Mud Volcano, Figure 4 Examples of pockmark and MV features: (a) active pockmarks 2 days after an EQ in the Hellenic subduction zone, off Greece (Courtesy of George Ferentinos); (b) aerial photograph of a large MV in Pakistan (Courtesy of Peter Clift); (c) crater lake of Dashgil MV, Azerbaijan, with the Caspian Sea in the background. Note heavy CH_4 bubbling (A. Kopf); (d) Fluid-bearing fault feeding a small MV on the Mediterranean Ridge accretionary complex (Modified after Kopf et al., 2011).

namely, in subduction zones (Kopf et al., 2001), while the amount of solid mass is usually not as prominent (Kopf, 1999).

Several studies have attempted to estimate the contribution of MVs to global methane emissions and effectively climate change (Dimitrov, 2002; Kopf, 2003; Milkov et al., 2003; Judd, 2005); however, and despite considerable variations in the database used and methane volumes estimated, it has been accepted that methane release is significant and probably has been over significant periods during Earth's history. The fate of some of the submarine methane remains controversial; however, even if a fraction is fixated as carbonates or oxidized to CO and CO_2 in the water column, there may be a contribution to global warming.

Mud volcanoes provide possible microhabitats in the subsurface, providing pathways for fluid migration and fracturing sediments through which MVs move sediments from depth to surface. This process is significant for deep biosphere research on Earth and may be important for astrobiological studies elsewhere (Oehler and Allen, 2010, 2011).

The first marine drilling campaign into MVs, Ocean Drilling Program Leg 160, attested that the initial deposits in MV formation were clast-supported, coarse sediments, whereas later stages of eruptive activity showed fewer clasts, no clasts, or even mousy (i.e., gas-rich) mud comprising almost pure clay (Robertson and Kopf, 1998). The resulting concept assumes that in fault-bound MVs, the initial deposit represents the fault breccia, and once that is removed, the underlying mud reservoir is depleted.

Current investigations

Microbial turnover rates at active MVs are fuelled by seepage at enhanced rates (e.g., Niemann et al., 2006), and a growing body of work is clustering around the best investigated features such as the Haakon Mosby MV off Norway or features in the Mediterranean, Gulf of Cadiz, or Gulf of Mexico. Given that the flux has been shown to be highly episodic (e.g., MacDonald et al., 2003), current research often involves long-term monitoring of key parameters, e.g., methane bubbling, aqueous fluid seepage, or temperature transients, as has recently been shown for Haakon Mosby MV (de Beer et al., 2006). Similarly, long-term probes were deployed in the Nile deep sea fan MVs, on the Mediterranean Ridge, or the Nankai prism in Japan. These studies build on long-term flux quantification by osmotic samplers, piezometers, or borehole instruments (Kopf et al., 2015) and unravel transient changes in physical parameters that foster MV activity. Other work focuses on onshore settings such as the prominent Lusi mud volcano (Mazzini et al., 2007; Mazzini et al., 2009; Davies et al., 2008), whose triggering mechanism is controversially debated as being somewhere between anthropogenic and seismic. Similarly, other ongoing researches similarly address the seismic triggering of MVs and how groundwater flux in onshore settings (Rojstaczer et al., 1995; Chigira and Tanaka, 1997) as well as methane release in offshore settings (Tsunogai et al., 2012) may be associated with it.

Future prospects/research avenues: gaps in current knowledge

When taking the aforementioned aspect a step further, certain geophysical or geochemical transients can be utilized to identify earthquake precursory phenomena related to MV activity. Recent investigations of active domes in the Kumano Basin, Japan revealed that both methane plumes in the water column (Tsunogai et al., 2012) and/or pore pressure and temperature transients in the formation can be related to seismicity. Historical records further attest that the chemistry of both aqueous and gaseous fluids changes a few hours up to a few days prior to large earthquakes, most likely as a result of stress changes at depth (Kopf et al., 2005). This finding may represent a versatile way to use MVs as stress indicators because of their fluid-rich nature and hydraulic connection to depth. They function similar to a valve; stress changes from deformation before, during, and after EQs may be discernable as transients in pore pressure, fluid chemistry, and mud or gas discharge rates if affordable instruments with standardized data protocols were widely used.

Summary and conclusions

Mud volcanoes represent a global phenomenon where significant amounts of solids and fluids interact between the geosphere and hydrosphere, thereby providing a window to depth otherwise hardly accessible by, e.g., drilling. Ascent is driven by buoyancy and aided by faults and other pathways, with the MVs being hydraulically connected to depth and hence may be capable of recording stress variations and precursors to geohazards. Hydrocarbon emissions from MVs are suspected to not only contribute to greenhouse climate but also to serve as indicators for reservoirs at depth.

Bibliography

Ansted, D. T., 1866. On the mud volcanoes of the Crimea, and on the relation of these and similar phenomena to deposits of petroleum. *Proceedings of Royal Institute*, **IV**, 628–640.

Bagirov, E., Nadirov, R., and Lerche, I., 1996. Flaming eruptions and ejections from mud volcanoes in Azerbaijan: statistical risk assessment from the historical records. *Energy Exploration Exploitation*, **14**, 535–583.

Barber, A. J., and Brown, K. M., 1988. Mud diapirism: the origin of mélanges in accretionary complexes? *Geology Today*, **4**, 89–94.

Barber, A. J., Tjokrosapoetro, S., and Charlton, T. R., 1986. Mud volcanoes, shale diapirs, wrench faults and mélanges in accretionary complexes, eastern Indonesia. *AAPG Bulletin*, **70**, 1729–1741.

Barker, C., and Horsfield, B., 1982. Mechanical versus thermal cause of abnormally high pore pressures in shales. *AAPG Bulletin*, **66**, 99–100.

Bebout, G., 1995. The impact of subduction-zone metamorphism on mantle-ocean chemical cycling. *Chemical Geology*, **126**, 191–218.

Bradak, B., and Kereszturi, A., 2003. Mud volcanism as model for various planetary surface processes. *34th Lunar and Planetary Science Conference*, Houston TX (Copernicus), Abstract 1304.

Braunstein, G., and O'Brien, G. D., 1968. Indexed bibliography. In: Braunstein, G., and O'Brien, G. D. (eds.), *Diapirism and Diapirs*. AAPG Mem., 8, pp. 385–414.

Brown, K. M., 1990. The nature and hydrogeologic significance of mud diapirs and diatremes for accretionary systems. *Journal of Geophysical Research*, **95**, 8969–8982.

Burne, R. V., 1970. The origin and significance of sand volcanoes in teh Bude formation (Cornwall). *Sedimentology*, **15**, 211–228.

Cavagna, S., Clari, P., and Martire, L., 1998. Methane-derived carbonates as an evidence of fossil mud volcanoes: a case history from the Cenozoic of Northern Italy. In *Proceedings V International Conference on Gas in Marine Sed*, Bologna, Italy, September 2003, pp. 106–110.

Chigira, M., and Tanaka, K., 1997. Structural features and the history of mud volcano in southern Hokkaido, northern Japan. *Journal of the Geological Society Japan*, **103**, 781–793.

Chiodini, G., D'Alessandro, W., and Parello, F., 1996. Geochemistry of gases and waters discharged by the mud volcanoes at Paterno, Mt. Etna (Italy). *Bulletin of Volcanology*, **58**, 51–58.

Choffat, P., 1882. Note préliminaire sur les vallés tiphoniques et sur les eruptions d'Ophite et de teschenite en Portugal. *Buletinl de la Societe Géologique de France*, **10**, 3rd ser.

Clennell, M. B., 1992. *The Mélanges of Sabah, Malaysia*. Unpublished PhD thesis, University of London, 483pp.

Colten-Bradley, V. A., 1987. Role of pressure in smectite dehydration – effects on geopressure and smectite-to-illite transformation. *AAPG Bulletin*, **71**, 1414–1427.

Davies, R. J., Brumm, M., Manga, M., Rubiandini, R., Swarbrick, R., and Tingay, M., 2008. The East Java mud volcano (2006 to present): an earthquake or drilling trigger? *Earth and Planetary Science Letters*, **272**, 627–638.

De Beer, D., Sauter, E., Niemann, H., Kaul, N., Foucher, J. P., Witte, U., Schlüter, M., and Boetius, A., 2006. In situ fluxes and zonation of microbial activity in surface sediments of the Håkon Mosby Mud Volcano. *Limnology and Oceanography*, **51**, 1315–1331.

Dimitrov, L. I., 2002. Mud volcanoes – the most important pathway for degassing deeply buried sediments. *Earth Science Reviews*, **59**, 49–76.

Dupré, S., Mascle, J., Foucher, J.-P., Harmegnies, F., Woodside, J., and Pierre, C., 2014. Warm brine lakes in craters of active mud volcanoes, Menes caldera off NW Egypt: evidence for deep-rooted thermogenic processes. *Geo-Marine Letters*, doi:10.1007/s00367-014-0367-1.

Fertl, W. H., 1976. Abnormal formation pressures: implications to exploration drilling and production of oil and gas. In *Developments in Petroleum Science II*. Oxford: Elsevier, p. 382.

Fitts, T. G., and Brown, K. M., 1999. Stress induced smectite dehydration ramifications for patterns of freshening fluid expulsion in the N. Barbados accretionary wedge. *Earth and Planetary Science Letters*, **172**, 179–197.

Gorkun, V. N., and Siryk, I. M., 1968. Calculating depth of deposition and volume of gas expelled during eruptions of mud volcanoes in southern Sakhalin. *International Geology Review*, **10**(1), 4–12.

Grevemeyer, I., Kopf, A. J., Fekete, N., Kaul, N., Villinger, H. W., Heesemann, M., Wallmann, K., Spiess, V., Gennerich, H.-H., Müller, M., and Weinrebe, W., 2004. Fluid flow through active mud dome Mound Culebra offshore Nicoya Peninsula, Costa Rica: evidence from heat flow surveying. *Marine Geology*, **207**, 145–157.

Griboulard, R., Bobier, C., Faugères, J. C., Huyghe, P., Gonthier, E., Odonne, F., and Welsh, R., 1998. Recent tectonic activity in the South Barbados prism: deep-towed side-scan sonar imagery. *Tectonophysics*, **284**, 79–99.

Hedberg, H., 1974. Relation of methane generation to undercompacted shales, shale diapirs and mud volcanoes. *AAPG Bulletin*, **58**, 661–673.

Hensen, C., Nuzzo, M., Hornibrook, E., Pinheiro, L. M., Bock, B., Magalhães, V. H., and Brückmann, W., 2007. Sources of mud volcano fluids in the Gulf of Cadiz – indications for hydrothermal imprint. *Geochimica et Cosmochimica Acta*, **71**, 1232–1248.

Henry, P., Le Pichon, X., Lallemant, S., Lance, S., Martin, J. B., Foucher, J.-P., Fiala-Médioni, A., Rostek, F., Guilhaumou, N., Pranal, V., and Castrec, M., 1996. Fluid flow in and around a mud volcano field seaward of the Barbados accretionary wedge: results from Manon cruise. *Journal of Geophysical Research*, **101**(B9), 20297–20323.

Higgins, G. E., and Saunders, J. B., 1974. Mud volcanoes – their nature and origin. *Verhandlungen der Naturforschenden Gesellschaft Basel*, **84**, 101–152.

Huguen, C., Mascle, J., Chaumillon, E., Woodside, J. M., Benkhelil, J., Kopf, A., and Volkonskaia, A., 2001. Deformational styles of the Eastern Mediteranean Ridge and surroundings from combined swath mapping and seismic reflection profiling. *Tectonophysics*, **343**, 21–47.

Huguen, C., Mascle, J., Chaumillon, E., Kopf, A., and Woodside, J., 2004. Stuctural setting and tectonic control on mud volcanoes: evidence from the central and Eastern Mediterranean Ridge. *Marine Geology*, **209**, 245–263.

Jakubov, A. A., Ali-Zade, A. A., and Zeinalov, M. M., 1971. *Mud Volcanoes of the Azerbaijan SSR*. Baku: Publishing House of the Academy of Sciences of the Azerbaijan SSR, p. 257.

Judd, A., 2005. Gas emissions from mud volcanoes: significance for global change. In Martinelli, G., and Panahi, B. (eds.), *Mud Volcanoes, Geodynamics and Seismicity*. Dordrecht: Springer. NATO Science Series IV, pp. 147–158.

Judd, A., and Hovland, M., 2007. *Seabed Fluid Flow: The Impact on Geology, Biology and the Marine Environment*. Cambridge: Cambridge University Press.

Kastner, M., 1981. Authigenic silicates in deep-sea sediments: formation and diagenesis. In Emiliani, C. (ed.), *The Sea*. New York: Wiley, Vol. 7, pp. 915–980.

Kopf, A., 1999. Fate of sediment during plate convergence at the Mediterranean Ridge accretionary complex: volume balance of mud extrusion versus subduction-accretion. *Geology*, **27**, 87–90.

Kopf, A. J., 2002. Significance of mud volcanism. *Reviews of Geophysics*, **40**(2), 2–52.

Kopf, A., and Behrmann, J.H., 2000 Extrusion dynamics of mud volcanoes on the Mediterranean Ridge accretionary complex. In Vendeville, B., Mart, Y., and Vigneresse, J.-L. (eds.), From the arctic to the mediterranean: Salt, shale, and igneous diapirs in and around Europe. *Journal of Geological Society London, Special Publication*, **174**, 169–204

Kopf, A. J., and Deyhle, A., 2002. Back to the roots: source depths of mud volcanoes and diapirs using boron and B isotopes. *Chemical Geology*, **192**, 195–210.

Kopf, A., Klaeschen, D., and Mascle, J., 2001. Extreme efficiency of mud volcanism in dewatering accretionary prisms. *Earth and Planetary Science Letters*, **189**, 295–313.

Kopf, A. J., 2003. Important global impact of methane degassing through mud volcanoes on past and present Earth climate. *International Journal of Earth Sciences*, **92**(5), 806–816.

Kopf, A., Freudenthal, T., Ratmeyer, V., Bergenthal, M., Lange, M., Fleischmann, T., Hammerschmidt, S., Seiter, C., and Wefer, G., 2015. Simple, affordable and sustainable borehole observatories for complex monitoring objectives. *Geoscientific Instrumentation, Methods and Data Systems*, GI-2014-21.

Kopf, A., Clennell, M. B., and Brown, K. M., 2005. Physical properties of extruded muds and their relationship to episodic extrusion and seismogenesis. In Martinelli, G., and Panahi, B. (eds.), *Mud Volcanoes, Geodynamics and Seismicity*. Dordrecht: Springer. NATO Science Survey IV, pp. 263–283.

Kugler, H. G., 1933. Contribution to the knowledge of sedimentary volcanism in Trinidad. *Journal of the Institution of Petroleum Technologists Trinidad*, **19**(119), 743–760.

Lewis, K. B., 1990. Why does the Hikurangi Channel go east? Program-Abstracts, New Zealand Geological Society Miscellaneous Publication, 50a, p. 80.

Leymerie, A., 1881. *Description géologique et pléontologique des Pyreenées de la Haute Garonne*. Toulouse (private), 1010 pp.

Limonov, A. F., van Weering, T. C. E., Kenyon, N. H., Ivanov, M. K., and Meisner, L. B., 1997. Seabed morphology and gas venting in the Black Sea mud volcano area: observations with the MAK-1 deep-tow side-scan sonar and bottom profiler. *Marine Geology*, **137**, 121–136.

Loncke, L., and Mascle, J., 2004. Mud volcanoes, gas chimneys, pockmarks and mounds on the Nile deep sea-fan (Eastern Mediterranean): geophysical evidences. *Marine and Petroleum Geology*, **21**, 669–689.

Macdonald, I. R., Sager, W. W., and Peccini, M. B., 2003. Gas hydrate and chemosynthetic biota in mounded bathymetry at mid-slope hydrocarbon seeps: Northern Gulf of Mexico. *Marine Geology*, **198**, 133–158.

Manga, M., Brumm, M., and Rudolph, M. L., 2009. Earthquake triggering of mud volcanoes. *Marine and Petroleum Geology*, **26**, 1785–1798.

Mazzini, A., Svensen, H., Akhmanov, G. G., Aloisi, G., Planke, S., Malthe-Sorenssen, A., and Istadi, B., 2007. Triggering and dynamic evolution of the LUSI mud volcano, Indonesia. *Earth and Planetary Science Letters*, **261**, 375–388.

Mazzini, A., Nermoen, A., Krotkiewski, M., Podladchikov, Y., Planke, S., and Svensen, H., 2009. Strike-slip faulting as a trigger mechanism for overpressure release through piercement structures: implications for the Lusi mud volcano, Indonesia. *Marine and Petroleum Geology*, **26**, 1751–1756.

McGill, G. E., 2005. Geologic Map of Cydonia Mensae–Southern Acidalia Planitia, Mars: Quadrangles MTM 40007, 40012, 40017, 450007, 45012, 45017, USGS Geological Investigation Survey. I-2811.

Milkov, A. V., Sassen, R., Apanasovich, T. V., and Dadashev, F. G., 2003. Global gas flux from mud volcanoes: a significant source of fossil methane in the atmosphere and the ocean. *Geophysical Reseasch Letters*, **30**. doi:10.1029/2002GL016358, 4 pp.

Moon, C. F., and Hurst, C. W., 1984. Fabrics of muds and shales: an overview. *Geological Society, London, Special Publications*, **15**, 579–593.

Moore, J. C., 1989. Tectonics and hydrogeology of accretionary prisms: role of the décollement zone. *Journal of Structural Geology*, **11**(1–2), 95–106.

Moore, J. C., and Saffer, D., 2001. Updip limit of the seismogenic zone beneath the accretionary prism of southwest Japan: An effect of diagenetic to low-grade metamorphic processes and increasing effective stress. *Geology*, **29**, 183–186.

Morgan, J. P., Coleman, J. M., and Gagliano, S. M., 1968. Mudlumps: diapiric tructures in the Mississippi delta sediments.

In: Braunstein, G., and O'Brien, G. D. (eds.), *Diapirism and Diapirs*. AAPG Mem. 8, pp. 145–161.

Mrazec, M. L., 1927. *Les plis diapirs et le diapirisme en général*. Comptes-Rendus Inst. Géol. Roumanie, Séances, **VI**.

Niemann, H., Duarte, J., Hensen, C., Omoregie, E., Magalhães, V. H., Elvert, M., Pinheiro, L. M., Kopf, A., and Boetius, A., 2006. Microbial methane turnover at mud volcanoes of the Gulf of Cadiz. *Geochimica et Cosmochimica Acta*, **70**, 5336–5355.

Oehler, D. Z., and Allen, C. C., 2010. Evidence for pervasive mud volcanism in Acidalia Planitia, Mars. *Icarus*, **208**, 636–657.

Oehler, D. Z., and Allen, C. C., 2011. Habitability of a large ghost crater in chryse planitia, Mars. Exploring Mars Habitability, 14 June 2011.

Pascoe, E.H. (1912) The oilfields of Burma. *Memories Geoogical Survey of India*, **40**(pt. 1): 1–54.

Paull, C. K., Ussler III, W., Dallimore, S. R., Blasco, S. M., Lorenson, T. D., Melling, H., Medioli, B. E., Nixon, F. M., and McLaughlin, F. A., 2007. Origin of pingo-like features on the Beaufort Sea shelf and their possible relationship to decomposing methane gas hydrates. *Geophysical Research Letters*, **34**(1), doi:10.1029/2006GL027977.

Paull, C. K., Dallimore, S. R., Hughes Clarke, J. E., Blasco, S., Melling, H., Lundsten, E., Vagle, S., and Collett, T. S., 2011. Degrading permafrost and gas hydrate under the Beaufort Shelf and marine gas hydrate on the adjacent continental slope. American Geophysical Union, Fall meeting 2011, abstract #GC52A-02.

Perez-Belzuz, F., Alonso, B., and Ercilla, G., 1997. History of mud diapirism and trigger mechanisms in the Western Alboran Sea. *Tectonophysics*, **282**, 399–422.

Rhakmanov, R. R., 1987. *Mud Volcanoes and Their Importance in Forecasting of Subsurface Petroleum Potential*. Moscow: Nedra Publication (in Russian).

Ridd, M. F., 1970. Mud volcanoes in New Zealand. *AAPG Bulletin*, **54**, 601–616.

Ritger, S., Carson, B., and Suess, E., 1987. Methane-derived authigenic carbonates formed by subduction-induced pore-water expulsion along the Oregon/Washington margin. *Geological Society of America Bulletin*, **98**, 147–156.

Robertson, A. H. F., and Kopf, A., 1998. Origin of clasts and matrix within milano and napoli mud volcanoes, Mediterranean Ridge accretionary complex. In Robertson, A. H. F., Emeis, K.-C., and Richter, C. (eds.), *Proceedings of the Ocean Driller Program, Scientific Results*. College Station, TX: Ocean Drilling Program, Vol. 160, pp. 575–596.

Robertson, A. H. F., and Scientific Party of ODP Leg 160, 1996. Mud volcanism on the Mediterranean Ridge: initial results of Ocean Drilling Program Leg 160. *Geology*, **24**, 239–242.

Rojstaczer, S. A., Wolf, S., and Michel, R., 1995. Permeability enhancement in the shallow crust as a cause of earthquake-induced hydrological changes. *Nature*, **373**, 237–239.

Schoonmaker, J., 1987. Diagenesis of smectite in an accretionary complex: Implications for the hydrology and dynamics of the subduction process, paper presented at 25th Annual Meeting. Clay Miner. Soc., Grand Rapids, Mich., 18–21.

Shipley, T. H., Stoffa, P. L., and Dean, D. F., 1990. Underthrust sediments, fluid migration paths and mud volcanoes associated with the accretionary wedge off Costa Rica, Middle America Trench. *Journal of Geophysical Research*, **95**, 8743–8752.

Speed, R. C., and Larue, D. K., 1992. Barbados: architecture and implications for accretion. *Journal of Geophysical Research*, **75**, 3633–3643.

Stamatakis, M. G., Baltatzis, E. G., and Skounakis, S. B., 1987. Sulfate minerals from a mud volcano in the Katakolo area, western Peloponnesus, Greece. *American Mineralogist*, **72**, 839–841.

Suess, E., Torres, M. E., Bohrmann, G., Collier, R. W., Greinert, J., Linke, P., Rehder, G., Trehu, A., Wallmann, K., Winckler, G., and Zuleger, E., 1999. Gas hydrate destabilization: enhanced dewatering, benthic material turnover and large methane plumes at the Cascadia margin. *Earth and Planetary Science Letters*, **170**(1–2), 1–15.

Sultan, N., Bohrmann, G., Ruffine, L., Pape, T., Riboulot, V., Colliat, J. L., De Prunele, A., Dennielou, B., Garziglia, S., and Himmler, T., 2014. Pockmark formation and evolution in deep water Nigeria: rapid hydrate growth versus slow hydrate dissolution. *Journal of Geophysical Research, Solid Earth*, **119**(4), 2679–2694.

Tamrazyan, G. P., 1972. Peculiarities in the manifestation of gaseous-mud volcanoes. *Nature*, **240**, 406–408.

Tsunogai, U., Maegawa, K., Sato, S., Komatsu, D. D., Nakagawa, F., Toki, T., and Ashi, J., 2012. Coseimic massive methane release from a submarine mud volcano. *Earth and Planetary Science Letters*, **311–314**, 79–85.

van Loon, A.J., 2010. Sedimentary volcanoes: Overview and implications for the definition of a volcano on Earth. In Cañón-Tapia, E., and Szakács, A., (eds.), *What Is a Volcano?*. Geological Society of America Special Paper 470, pp. 31–41. doi:10.1130/2010.2470(03).

von Gumbel, C. W., 1879. Über das Eruptionsmaterial des Schlammvulkans von Paterno am Ätna und der Schlammvulkane im Allgemeinen. *Sitz k Akad*, **IX**, 270–273.

Wegmann, C. E., 1930. Über Diapirismus. *Bulletin de la Commision Geologique de Finlande*, **92**, 58–76.

Westbrook, G. K., and Smith, M. J., 1983. Long décollements and mud volcanoes: evidence from the Barbados Ridge Complex for the role of high pore-fluid pressure in the development of an accretionary complex. *Geology*, **11**, 279–283.

White, D. E., 1955. Violent mud volcano eruption of Lake City hot springs, Northeast California. *GSA Bulletin*, **66**(9), 1109–1130.

Yassir, N. A., 1989. *Mud Volcanoes and the Behaviour of Overpressured Clays and Silts*. Unpublished PhD thesis, University College London, 249 pp.

Cross-references

Chemosynthetic Life
Cold Seeps
Deep Biosphere
Marine Gas Hydrates
Methane in Marine Sediments

N

NANNOFOSSILS COCCOLITHS

Hans R. Thierstein
Department of Earth Sciences, Swiss Technical University, Zürich, Switzerland

Definition

Nannofossils/coccoliths: Minute (ca. 10^{-6} m) calcite skeletons, produced by marine phytoplankton and found in deep-sea sediments

General significance

Coccolithophorids are a major group of primary producers in the photic zone of the world's oceans today. Their single cells (coccospheres) are covered by coccoliths, which are structured aggregates of calcite crystallites (Figure 1).

The calcite platelets (coccoliths) covering these cells sink to the ocean floor and form the dominant part of the calcareous oozes that are found on most ocean floors at water depths of less than 4–5 km (i.e., above the CCD). Vast amounts of nannofossil chalks have accumulated on the world's ocean floors since the late Jurassic. Nannofossils include coccoliths and a few other biogenic remains of similar size, such as calcitic cysts of dinoflagellates (*Thoracosphaera* sp.), or of unknown taxonomic affinity, such as star-shaped *Discoaster* sp. in Tertiary sediments or rock-forming *Nannoconus* sp. in Cretaceous deposits.

Coccoliths and nannofossils have been used in the dating of sediments (biostratigraphy), in the reconstruction of the history of the world's oceans (paleoceanography), and in the analyses of evolutionary processes (Figure 2).

Biostratigraphy

Nannofossils have become very important in the determination of the geologic age of fine-grained, marine sediments (Perch-Nielsen, 1989; Bown, 1999) both in academic research and in oil exploration. Nannofossils are a major constituent of many marine sediments (up to 90 % in calcareous oozes), they are a highly diverse group (up to 150 known species in the late Cretaceous), and they can be quickly prepared on smear slides and examined in a polarizing light microscope. A quantum step in our knowledge of the taxonomy and biostratigraphy of nannofossils – and of many other marine microfossil groups – was taken with the initiation of systematic drilling of the world's ocean floors by the Deep-Sea Drilling Project in 1967.

Paleoceanography

This field was initiated with the publication of *Studies in Paleo-Oceanography* (Hay, 1974) containing the first integrated attempts to reconstruct aspects of the history of the world's oceans. Since then, various paleoceanographic proxies – prominently among them are nannofossils – were used to reconstruct the history of surface currents, long-term changes in water temperature, carbonate saturation of oceanic water masses, primary productivity, etc. (e.g., Stoll and Ziveri, 2004). Correlations between coccolith sizes of certain taxa and calcification and pCO_2 changes in laboratory experiments and a sediment core may hold potential for a new paleoproxy

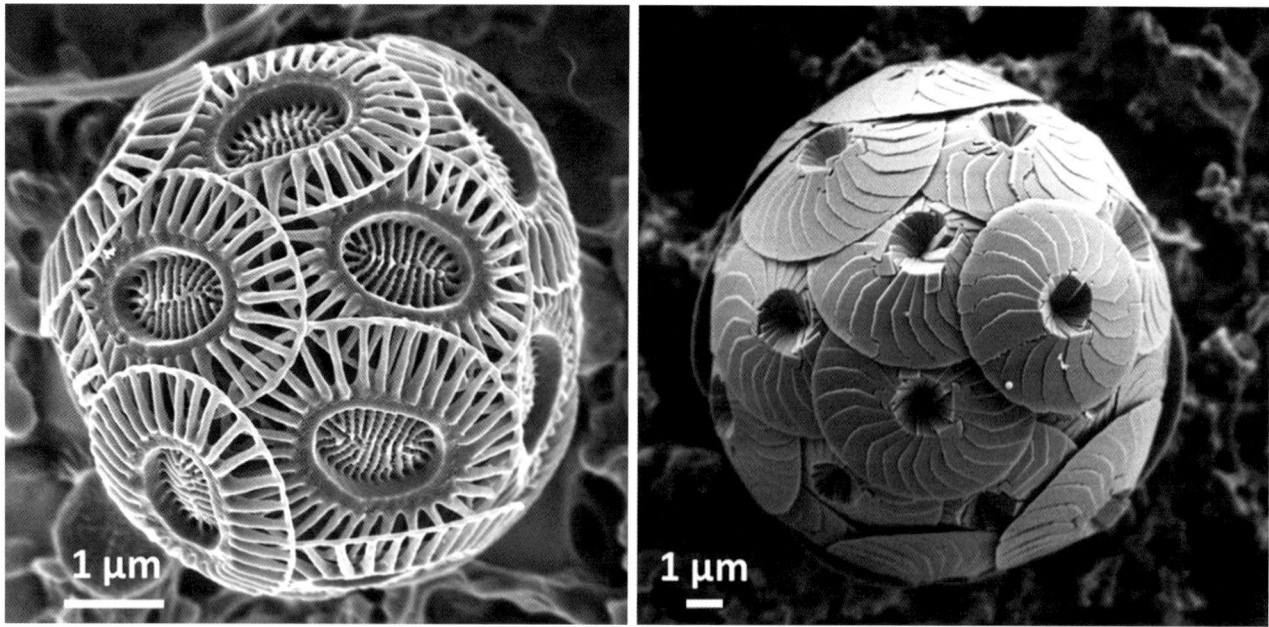

Nannofossils Coccoliths, Figure 1 Coccospheres of *Emiliania huxleyi* (left) and *Calcidiscus leptoporus* (right) as revealed in a scanning electron microscope. The coccospheres sit on a filter through which a water sample was rinsed from 25 m water depth off Bermuda (North Atlantic) on 1 May 1991.

Nannofossils Coccoliths, Figure 2 Scanning electron micrograph of a Cretaceous nannofossil chalk showing a still intact coccosphere of *Watznaueria barnesae* embedded in rock particles which are mostly coccoliths and fragments of them. This sample is from 200 m below seafloor at DSDP Site 258, which was drilled on Exmouth Plateau (off SE Australia) at a water depth of 3 km.

for past global climate changes (e.g., Iglesias-Rodriguez et al., 2008).

Evolution

About 250 coccolithophore species are known to live in today's ocean waters (Young et al., 2003), but coccoliths of only about 50 of them are found in recent deep-sea sediments (Bown et al., 2004). The long-term species diversity history shows a rather gradual increase from 230 Ma (mid-Triassic) to an all-time high of about 140 species at 70 Ma (late Cretaceous), followed by a drastic reduction at the K/T boundary (65 Ma) and numerous fluctuations throughout the Cenozoic (Bown et al., 2004). The causes for these species richness changes or of other macroevolutionary parameters, such as coccolith size of individual species or entire assemblages, however, remain enigmatic (Langer et al., 2006; Herrmann and Thierstein, 2012).

Summary and conclusions

Nannofossils and coccoliths are the microscopic remains of single-celled planktonic algae that have lived in the world's oceans for about 230 Ma. Nannofossils are found in many calcareous rocks and help geologists to tell the age of these deposits and the environmental conditions at that time.

Bibliography

Bown, P. R. (ed.), 1999. *Calcareous Nannofossil Biostratigraphy*. Norwell: Kluwer. 315 p.

Bown, P. R., et al., 2004. Calcareous nannoplankton evolution and diversity through time. In Thierstein, H. R., and Young, J. R. (eds.), *Coccolithophores – From Molecular Processes to Global Impact*. Berlin: Springer, pp. 481–508.

Hay, W. W. (ed.), 1974. *Studies in Paleo-oceanography*. SEPM Special Publication, Vol. 20. 218 p.

Herrmann, S., and Thierstein, H. R., 2012. Cenozoic coccolith size changes – evolutionary and/or ecological controls? *Paleogeography Palaeoclimatology Palaeoecology*, **333–334**, 92–106.

Iglesias-Rodriguez, M. D., et al., 2008. Phytoplankton calcification in a high-CO_2 world. *Science*, **320**, 336–340.

Langer, G., et al., 2006. Species-specific responses of calcifying algae to changing seawater carbonate chemistry. *Geochemistry, Geophysics, Geosystems*, **7**(9), 1–12.

Perch-Nielsen, K., 1989. Mesozoic and Cenozoic calcareous nannofossils. In Bolli, H. M., et al. (eds.), *Plankton Stratigraphy*. Cambridge: Cambridge University Press, Vol. 1, pp. 329–554.

Stoll, H. M., and Ziveri, P., 2004. Coccolithophorid-based geochemical paleoproxies. In Thierstein, H. R., and Young, J. R. (eds.), *Coccolithophores – From Molecular Processes to Global Impact*. Berlin: Springer, pp. 529–562.

Young, J. R., et al., 2003. A guide to extant coccolithophore taxonomy. *Journal of Nannoplankton Research, Special Issue*, **1**, 125.

Cross-references

Biochronology, Biostratigraphy
Calcite Compensation Depth (CCD)
Deep-sea Sediments
Dinoflagellates
Marine Microfossils
Paleoceanographic Proxies
Paleoceanography

NITROGEN ISOTOPES IN THE SEA

Thomas Pedersen
School of Earth and Ocean Sciences, University of Victoria, Victoria, BC, Canada

Definition

Nitrogen has two stable isotopes, ^{14}N and ^{15}N, comprising, respectively, 99.64 % and 0.36 % of the nitrogen atoms on earth. The isotopes are fractionated one from the other in a wide variety of biochemical processes in the sea (see Sigman et al., 2009, for a comprehensive review). Measurements are reported in the delta notation, relative to nitrogen in air (N_2):

$$\delta^{15}N\ (‰) = \left[\frac{(^{15}N/^{14}N)_{sample} - (^{15}N/^{14}N)_{air}}{(^{15}N/^{14}N)_{air}} \right] \times 1,000$$

Understanding of fractionation phenomena has accelerated rapidly in recent decades and has shed important new light on, for example, nutrient cycling in the modern and past ocean (see Galbraith et al., 2008, for a review).

Dissolved nitrogen occurs in the sea in a wide variety of molecules and ions, including N_2 (gas), nitrate (NO_3^-), nitrite (NO_2^-), ammonia (NH_3), ammonium (NH_4^+), and urea. Fractionation occurs primarily during the transformation of one such nitrogen-bearing species to another:

- Nitrogen fixation, or diazotrophy, is the conversion of triply bonded N_2 (gas) to reduced N species (like NH_4^+) by nitrogen-fixing cyanobacteria in the surface ocean, with only modest kinetic discrimination against the heavy isotope (0 to -2 ‰). Such fixation is the principal pathway by which dissolved nitrogen species are introduced to the ocean.
- Denitrification occurs in oxygen-minimum zones in the subsurface below roughly 100–300 m and ~1 km water depth off NW Mexico, Peru, Namibia, and Western India, for example, and refers to the stepwise bacterially mediated conversion of NO_3^- to N_2 (gas). This has a large fractionation factor (~25 ‰; Brandes et al., 1998), producing isotopically light end product N_2 and enriched residual nitrate with $\delta^{15}N$ values as high as +20 ‰ in waters several 100 m below the surface.
- Denitrification also occurs in the pore waters of almost all continental margin sediments, but because NO_3^- is typically completely consumed in such deposits, there is normally no net (overall) isotope fractionation between the N_2 gas product and the nitrate precursor.
- Anaerobic ammonium oxidation (anammox) is a relatively recently discovered process in the sea, first reported in 2002. Nitrite (produced either by nitrate reduction or oxidation of ammonium) serves as an electron acceptor (thus, an oxidant) in the bacterially mediated anaerobic oxidation of NH_4^+. The effect of anammox on nitrogen isotope distributions in the sea is as yet poorly known.

Bibliography

Brandes, J. A., Devol, A. H., Yoshinari, T., Jayakumar, D. A., and Naqvi, S. W. A., 1998. Isotopic composition of nitrate in the central Arabian Sea and eastern tropical North Pacific: a tracer for mixing and nitrogen cycles. *Limnology and Oceanography*, **43**, 1680–1689.

Galbraith, E. D., Sigman, D. M., Robinson, R. S., and Pedersen, T. F., 2008. Nitrogen in past marine environments. In Bronk, D. A., Mulholland, M. R., and Capone, D. G. (eds.), *Nitrogen in the Marine Environment.* Amsterdam: Elsevier, pp. 1535–1597.

Sigman, D. M., Karsh, K. L., and Casciotti, K. L., 2009. Nitrogen isotopes in the ocean. In Steele, J. H., Turekian, K. K., and Thorpe, S. A. (eds.), *Encyclopedia of Ocean Sciences*. London: Academic, pp. 40–54.

NORTH ATLANTIC OSCILLATION (NAO)

Erik Kjellström
Swedish Meteorological and Hydrological Institute, Norrköping, Sweden

The large-scale atmospheric circulation in the North Atlantic sector is characterized by westerly winds that result from the meridional pressure gradient between the Azores High in the south and the Icelandic Low in the north. As the strength of this pressure gradient varies over time, also the strength of the westerlies varies. In periods of strong north-south gradient, the westerlies are more pronounced and vice versa when the gradient is weak. Fluctuations between these different states are known as the North Atlantic Oscillation (NAO).

The strength and phase of the NAO can be summarized by an index (NAOI). There exist a few different NAOI definitions. The most commonly used is likely the one comparing normalized mean sea-level pressure (MSLP) in Iceland and in Portugal (Hurrell, 1995). One of its main strengths is that this type of NAO index can be derived also for historical times as surface pressure readings exist at some locations dating back to the nineteenth century. Other definitions, such as those based on principal component analysis (PCA), are also commonly used (e.g., Barnston and Livezey, 1987). A benefit of the latter type of methods is that they can better depict variability/changes in the center of action of the NAO.

The NAO has a profound impact on the climate in the North Atlantic region. During winter the positive phase is associated with temperatures below average in western Greenland and southern Europe and above average in large parts of the eastern USA and northern Europe. At the same time, the precipitation is above average in northern Europe and in the eastern USA and below average in southern Europe and western Greenland. In the negative phase of the NAO, opposite anomalies are found in temperature and precipitation in these areas. Also in summer, there are impacts on the climate with distinct differences between northern and southern Europe in the respective phases of the NAO (Folland et al., 2009).

As the large-scale pressure pattern varies on different time scales, fluctuations in the NAOI are seen starting already at daily time scales ranging over weeks and months to seasons or years or even decadal time scales. Studies of the temporal behavior of the NAOI show some weakly significant and intermittent oscillations on quasi-biennial and quasi-decadal time scales (see review of Wanner et al., 2001). This shows that the NAO is not a pure random process indicating that there may be predictive potential associated with it.

Bibliography

Barnston, A. G., and Livezey, R. E., 1987. Classification, seasonality and persistence of low frequency atmospheric circulation patterns. *Monthly Weather Review*, **115**, 1083–1126.

Folland, C. K., Knight, J., Linderholm, H. W., Fereday, D., Ineson, S., and Hurrell, J. W., 2009. The summer North Atlantic oscillation: past, present and future. *Journal of Climate*, **22**, 1082–1103.

Hurrell, J. W., 1995. Decadal trends in the North Atlantic oscillation: regional temperatures and precipitation. *Science*, **269**, 676–679.

Wanner, H., Bronnimann, S., Casty, C., Gyalistras, D., Luterbacher, J., Schmutz, C., Stephenson, D. B., and Xoplaki, E., 2001. North Atlantic oscillation – concepts and studies. *Survey in Geophysics*, **22**, 321–382.

OCEAN ACIDIFICATION

Ulf Riebesell
GEOMAR Helmholtz Centre for Ocean Research Kiel, Kiel, Germany

Ocean acidification refers to the process of increasing seawater acidity by dissolving additional carbon dioxide (CO_2) from the atmosphere.

As CO_2 dissolves in seawater, it forms carbonic acid (H_2CO_3), which readily dissociates into bicarbonate (HCO_3^-) and hydrogen (H^+) ions. The hydrogen ion concentration determines the acidity of seawater, expressed by the pH scale. Part of the hydrogen ions released in this process is buffered by the seawater carbonate system by consuming carbonate ions (CO_3^{2-}) and forming additional bicarbonate. As pH is defined as the negative logarithm of the hydrogen ion concentration, pH decreases as the acidity increases (Figure 1).

The increase in atmospheric CO_2 concentration since preindustrial times from 280 ppm (parts per million) to ca. 400 ppm has caused a decline in surface ocean pH by 0.12 units. This corresponds to an increase in seawater acidity (hydrogen ion concentration) of 30 %. If atmospheric CO_2 concentrations continue to increase at present rates, surface ocean pH values will decline by an additional 0.35 units by the year 2100 (Wolf-Gladrow et al., 1999). This corresponds to more than a doubling in seawater acidity and a decline in carbonate ion concentrations by 45 % relative to preindustrial levels. These changes in seawater chemistry are expected to widely affect marine organisms and ecosystems (Raven et al., 2005). Most prominently, calcifying organisms such as corals, bivalves, sea urchins, coralline algae, and a variety of calcareous plankton species have shown to be sensitive to ocean acidification (Gattuso and Hansson, 2011). Equally important as the magnitude of ocean acidification is its rate of change, as this determines to what extent species have the ability to adapt. The current speed of CO_2 increase in the atmosphere and the corresponding rate of ocean acidification are unprecedented in Earth's history for at least 300 million years (Hönisch et al., 2012).

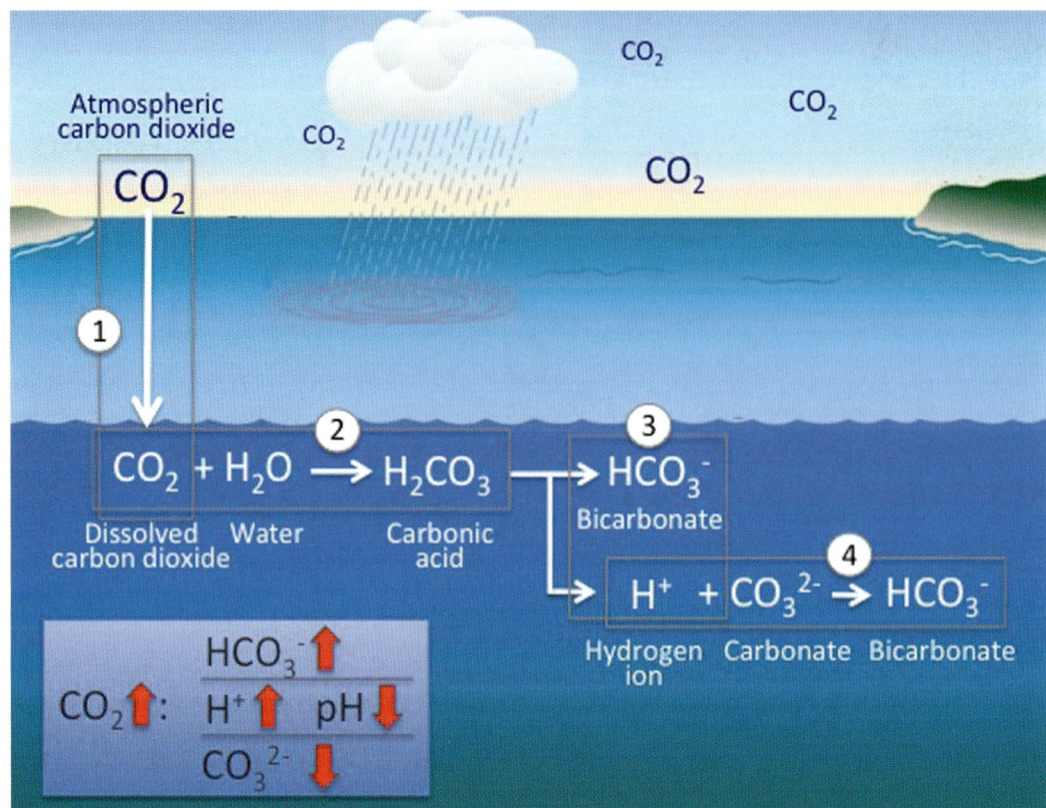

Ocean Acidification, Figure 1 The process of ocean acidification: (*1*) atmospheric carbon dioxide (CO_2) dissolving in seawater; (*2*) dissolved CO_2 reacting with water to form carbonic acid (H_2CO_3); (*3*) carbonic acid dissociating to bicarbonate (HCO_3^-) and hydrogen ion (H^+); and (*4*) hydrogen ion reacting with carbonate (CO_3^{2-}) to form bicarbonate. As a result of ocean acidification, the concentrations of CO_2, HCO_3^-, and H^+ increase, whereas the pH and CO_3^{2-} concentrations decrease.

Bibliography

Gattuso, J.-P., and Hansson, L., 2011. *Ocean Acidification*. Oxford: Oxford University Press.

Hönisch, B., Ridgwell, A., Schmidt, D. N., Thomas, E., Gibbs, S. J., Sluijs, A., Zeebe, R., Kump, L., Martindale, R. C., Greene, S. E., Kiessling, W., Ries, J., Zachos, J. C., Royer, D. L., Barker, S., Marchitto, T. M., Jr., Moyer, R., Pelejero, C., Ziveri, P., Foster, G. L., and Williams, B., 2012. The geological record of ocean acidification. *Science*, **335**, 1058–1063.

Raven, J., Caldeira, K., Elderfield, H., Hoegh-Guldberg, O., Liss, P., Riebesell, U., Shepherd, J., Turley, C., and Watson, A., 2005. *Ocean Acidification Due To Increasing Atmospheric Carbon Dioxide*. Royal Society Report, Policy Document 12/05, London.

Wolf-Gladrow, D., Riebesell, U., Burkhardt, S., and Bijma, J., 1999. Direct effects of CO_2 concentration on growth and isotopic composition of marine plankton. *Tellus*, **51B**, 461–476.

OCEAN DRILLING

Kiyoshi Suyehiro
Integrated Ocean Drilling Program, Management International, Tokyo, Japan

Definition

Ocean drilling broadly refers to science conducted by utilizing ocean drilling platforms (vessels) to retrieve material samples and measurements from boreholes drilled beneath the seafloor in deep seawater not achieved by other shallow sampling methods such as dredging, gravity coring, or shallow piston coring (Figure 1).

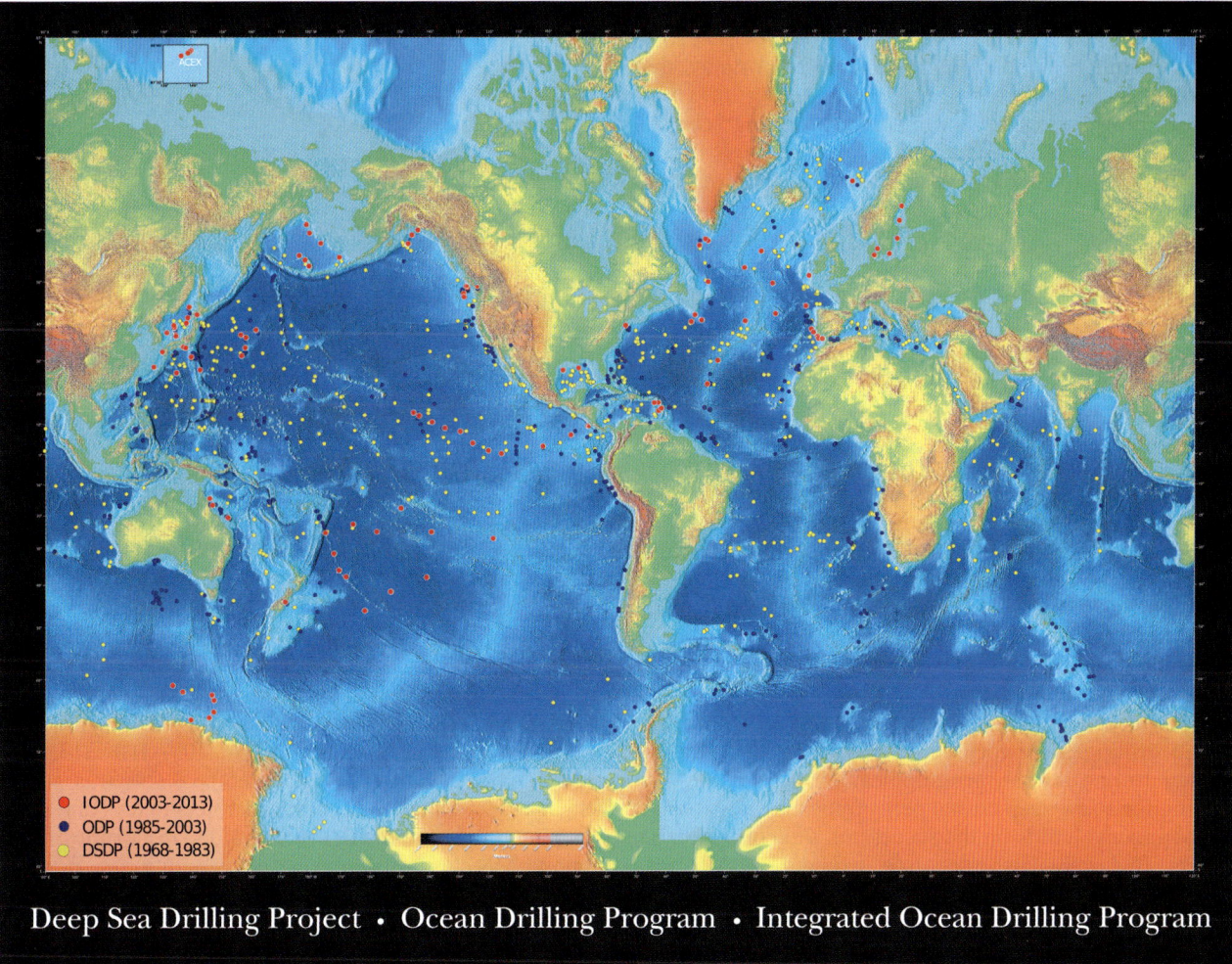

Ocean Drilling, Figure 1 World map view of ocean drilling sites drilled through DSDP (1968–1983, 624 sites), ODP (1984–2003, 669 sites), and IODP (2004–2013, 127 (up to U1427) + 58 (up to M58) + 22 (up to C22) = *207* sites) era (1968–2013). IODP sites are shown with abbreviated expedition names. ACEX sites (Exp. 302, Lomonosov Ridge, M2-4) in small box are at 87°51–56′N latitudes. Southernmost sites (270–272) are at latitudes 77°07–43′S (DSDP Leg 28, Ross Sea). (Seafloor bathymetry deepens with oceanic plate age. High anomalies depict ocean ridges, island arcs, hot spots, large igneous provinces, and continental margins, and lows indicate ocean trenches. These and the configuration of oceans including gateways control the ocean circulation driven by atmospheric motion. The drill sites are not as geographically uniform or dense as compared, e.g., to the Argo float monitoring network (www.argo.ucsd.edu)).

Scientific analyses are applied to the samples (cores with interstitial waters) obtained by wire line coring and downhole measurements made by logging tools. Long-term borehole observatories are established in some cases to obtain time series records. These data and samples are used to describe and understand the Earth's history, structure, and processes that shaped the present-day Earth system and to predict its future from a wide spectrum of disciplines of marine geo-biosciences.

The framework for conducting scientific ocean drilling was established in 1968 and in 1975 became an international scientific program with participating member countries and consortia from the Americas, Asia, Europe, and Oceania to support the operations of drilling vessels and science to be conducted on board. Scientists interested in obtaining samples and data from boreholes in world oceans submit proposals for eventual drilling expeditions or request samples from archives to be evaluated by peer review. Scientists compete to make their cases why drilling in a particular geographical locality can help in solving problems of fundamental or global importance that cannot be achieved otherwise.

The progress of scientific ocean drilling can be described in terms of technological advancements in core sampling and downhole measurements, program management, and scientific achievements.

Ocean drilling history

The history of scientific ocean drilling starts from the Mohole Project Phase 1 (1958–1966), which eventually aimed to penetrate the Mohorovičić discontinuity, which is the seismic velocity discontinuity that separates the crust and the mantle (Bascom, 1961). The Mohole Project started from engineering developments, as scientific ocean drilling from a vessel without anchoring in deep water had not existed. On April 2, 1961, *CUSS I* with a dynamic positioning system to keep its station made its historic achievement of sampling basalt beneath 183 m of sediments in 3,658 m of water depth offshore Guadalupe Island. Although this Phase 1 was a success, technical difficulties and financial constraints prevented the project to advance to the next phase.

The success of Phase 1 opened the era of scientific ocean drilling more focused on sediments and shallow crustal drilling at many geographical locations rather than aiming for a few deep crustal penetrations. Exploration of the sediments builds on the success of Kullenberg piston corers (e.g., Swedish deep-sea expedition Albatross 1947–1948). These led to successful years of scientific programs: Deep Sea Drilling Project (1968–1983), International Phase of Ocean Drilling (1975–1983), Ocean Drilling Program (1985–2003), Integrated Ocean Drilling Program (2003–2013), and International Ocean Discovery Program (2013–2023) (Table 1).

After the retirement of the drilling vessel, *Glomar Challenger* for DSDP (Legs 1–96), *JOIDES Resolution* (*JR*) was introduced from ODP (Legs 100–210; a Leg is typically about 8 weeks long with about 25 scientists on board) and 32 *JR* expeditions for IODP (similar to ODP mode). IODP has been a major upgrade for scientific ocean drilling as the participating countries had agreed to operate three uniquely capable drilling platforms to expand the potential drilling targets to ice-covered, very shallow (a few tens of meters water depth), or very deep water drilling (>~5 km) (Figure 2).

The European Consortium for Ocean Research Drilling (ECORD) introduced chartered drilling vessels for IODP in 2004. This type of platform is called Mission-Specific Platform and is aimed to drill in waters where *JR* is not the optimal platform. The first MSP was a fleet of vessels to drill in ice-covered Arctic Sea (Exp. 302). In 2007, a riser-equipped drilling vessel *Chikyu* joined the program as planned for IODP. *Chikyu* is capable of drilling deeper because the riser pipes that allow mud circulation to prevent hole collapse are normally used by the oil industry.

Ocean Drilling, Table 1 Ocean drilling platforms

Glomar challenger (DSDP, retired 1983)
Built 1968: 123 m length, 20 m breadth, 10,500 t displacement (LT)
Maximum drilling/coring depth: 8,375 m (reentry depth 7,960 m)
Drilling water depth: >6,000 m
JOIDES Resolution (ODP and IODP (overhauls in 1984 and 2009, scientists 25))
Built 1978: 143 m length, 21.3 m breadth, 10,282 gross tonnage
Maximum drilling/coring depth: 8,375 m (reentry depth 7,960 m)
Drilling water depth: 75 m to >6,000 m
Wire line coring retrieval rate: ~200 m/min
Chikyu (IODP, scientists and lab staff 50)
Built 2005: 210 m length, 38 m breadth, 56,752 gross tonnage
Maximum drilling/coring depth: 10,000 m
Drilling water depth: 500–2,500 m (riser drilling mode)
Wire line coring retrieval rate: ~150 m/min
Mission-Specific Platforms (IODP; various chartered platforms outfitted with drill rigs)
Vidar Viking, Oden, and Sovetskiy Soyuz were utilized for drilling through the Arctic ice (Exp. 302). Other expeditions by *DP* (dynamically positioned) *Hunter* (Exp. 310), *L/B Kayd* (liftboat, 313), *Greatship Maya* (325), and *Greatship Manisha* (347)

Currently, *Chikyu* can drill in riser mode to 2,500 m water depth with 10,000 m drill pipes. During IODP, five MSP expeditions were conducted (Exps. 302, 310, 313, 325, 347, Arctic Sea and shallow waters). *Chikyu* was mostly engaged in a multiple-expedition project to drill and study the plate subduction fault off central Japan capable of generating M8-class earthquakes (ten expeditions). *JR* holds the record of penetrating deep in oceanic crust (2,111 m below seafloor (mbsf) as of 2013) at Hole 504B, and *Chikyu* holds the record of deep drilling in thick sedimentary sequence of continental margin (2,466 mbsf as of 2013) at Hole C0020.

The *Glomar Challenger* (*GC*) visited 624 sites during DSDP and cored 97 km total length at 57 % recovery rate. *JR* visited 669 sites during ODP and cored 222 km at 69 % recovery rate.

In a few occasions, *JR* has drilled in shallow water depths at <500 m (38 m water depth test (Leg 143) and other legs such as Leg 194 in 304 m, Exp. 307 in 420 m, Exp. 317 in 84 m, and Exp. 334 in 137 m). Deepwater depth drillings have been successfully made in depths well exceeding the normal ocean basin depths (e.g., Leg 129 in 5,980 m (*JR*), Exp. 343 in 6,898 m (*Chikyu*)).

Scientific ocean drilling owes much to the technology of dynamic positioning and wire line coring. The former allows the drillship to sail almost anywhere in the world oceans in deep waters and drill at the desired location keeping the position. The latter allows continuous coring without retrieving the drill string. The precision of absolute position on Earth and relative location in a particular seafloor locale has improved over the years by GPS positioning and seafloor swath bathymetry. The ability to recover as much as possible from beneath the seafloor owes much to the drilling technology, i.e., cutting into the sediments and rocks by drill bits. The wire line corer (typically each core trip brings on board a cylindrical sample (9.5 m length with 6.2 cm diameter)) runs up and down inside the drill string as the drilling advances the corer length. The actual drilling is made by various designs of corers determined by the physical properties of the cores to be taken. This is why a detailed site survey before drilling is essential to anticipate the samples. In soft sediments, hydraulic piston coring can be employed with minimum disturbance to the obtained cores as it involves a vertical downward thrust motion to penetrate deeper without drill bit rotation (Figure 3).

Once a hard rock stratum is reached or the sediments get too consolidated and hard for piston coring, rotary core barrel (RCB) with a core bit at the bottom is used (Figure 3). This system has been used throughout the history of scientific ocean drilling. See literature for other variations and details (e.g., Huey, 2009). Coring challenges are met at hard rocks, chert layers, alternating soft and hard materials such as sills into sediments, or high-temperature environments such as proximities of hydrothermal vents.

Ocean drilling logistics and technology

Scientific ocean drilling is a basic research and has been open to any scientists to submit drilling proposals with clear targets and objectives. A successful drilling proposal will demonstrate that drilling is a unique method to advance the state of scientific understanding in a significant way. Overall decadal-scale scientific goals are defined through open international debate (ref. in Table 3), by which each proposal is evaluated for merits. Such goals have evolved from understanding the history and structure of the Earth to physical and chemical processes that shaped the present state of the Earth with increasing interest in the role of the sub-seafloor biosphere and relating to the future of the habitable planet Earth.

Drilling is the most expensive part of the scientific quest for new discoveries and verifications (e.g., IODP budgets and expenditures are given in IODP Annual Reports). International participation by national science funding agencies and institutions has supported the ocean drilling since 1975 (Table 2). As of 2013, there are 26 countries participating as 6 countries and 2 consortia (the USA, Japan, PRC, Brazil, India, South Korea), ECORD (Austria, Belgium, Canada, Denmark, Finland, France, Germany, Iceland, Ireland, Israel, Italy, the Netherlands, Norway, Poland, Portugal, Spain, Sweden, Switzerland, the UK), and ANZIC (Australia, New Zealand). The international cost sharing is dedicated to providing the common infrastructure to be enjoyed by all the scientists supported by the participating member countries through international governmental agreements. Each of the drilling expeditions is conducted as an ocean drilling community project with chief scientists (most often cochiefs) nominated to oversee the science. Normally, scientific analyses take place at each scientist's home

Ocean Drilling, Figure 2 Drilling platforms utilized in the current IODP program (see Table 1). *JOIDES Resolution* has been operating since the start of ODP. *Chikyu* operated by Japan Agency for Marine-Earth Science and Technology is a new addition with long-awaited riser drilling capability suitable for deep drilling. Mission-Specific Platform charters are an operationally optimal platform for drilling not suitable for *JR* or *Chikyu*, such as ice-covered or shallow waters.

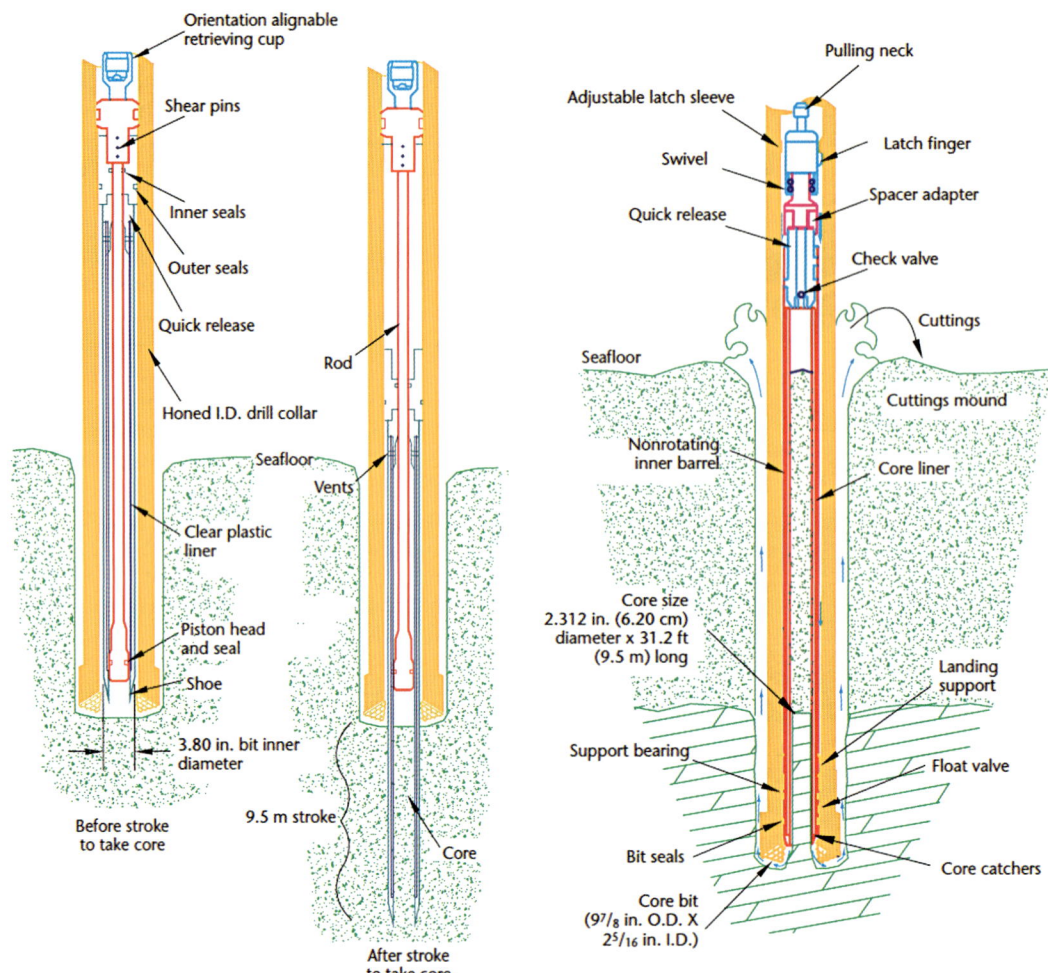

Ocean Drilling, Figure 3 Coring tools. Improvements are continuously made to increase core recovery rates in difficult formations such as hard-soft alternating layers, high temperature, very hard rocks, or chert layers. They allow the core barrel to hydraulic piston coring to be used for *APC* (advanced piston corer for *JR*) and *HPCS* (hydraulic piston coring system for *Chikyu*) to recover undisturbed samples in soft sediments. Beyond the depths of piston coring, *RCB* (rotary core barrel) is utilized, where the drill bit rotates downward to recover core sample into the core liner. Core disturbance is inevitable. Another often-utilized tool is XCB (extended core barrel for *JR*) and *ESCS* (extended shoe coring system for *Chikyu*) (not shown). Its principle is to allow the core barrel to capture the core a few inches ahead of the main bit roller cones (e.g., Huey, 2009).

institution for eventual scientific publications. Each country funds this part. In essence, drilling and sampling are the common infrastructure supported internationally, and all the rest is left to normal scientific competition.

The data and findings are made openly and freely available (www.iodp.org). All obtained core samples are archived at core repositories (core repositories in North America, Europe, and Asia store cores from global oceans according to geographical location; Table 3) ready to be utilized for new hypotheses to be verified or to be scrutinized by advanced analyses. Scientific drilling attempts not to directly seek natural resources for commercial purposes but to look into their scientific significance.

The drill site needs to be characterized by geological and geophysical methods such as the seafloor geomorphology and seismic reflection data to warrant the discovery to be made with good chance in a safe manner. A special international proposal evaluation scheme has been established to allow for international experts in scientific disciplines and drilling technology to advise the proponents and plan for an optimal drilling schedule. This voluntary science advisory system is the core of selecting proposals for scheduling based on scientific merits and technical feasibilities.

Once a drilling proposal is adopted for scheduling, an international science team is assembled to define a scientific project to achieve the proposal objectives considering all the technical aspects of drilling offshore including weather and ocean conditions. The team covers all aspects of required expertise: sedimentology, paleontology, paleomagnetism, petrology, organic and inorganic geochemistry, microbiology, physical properties, etc. Chief

Ocean Drilling, Table 2 A brief history of scientific ocean drilling

1961	US Project Mohole Phase 1: *CUSS I* keeping its position at 3,558 m water depth (dynamic positioning) off Baja California proves deep-ocean core drilling and samples basalt beneath fossil-rich sediments
1963	US Project LOCO: *Submarex* cores tertiary sediments on the Nicaragua Rise in the Caribbean Sea
1964	JOIDES (Joint Oceanographic Institutions for Deep Earth Sampling) is formed: Lamont Geological Observatory (now Lamont-Doherty Earth Observatory), Inst. of Marine Science (now Rosenstiel School of Marine and Atmospheric Science), Scripps Institution of Oceanography, and Woods Hole Institution of Oceanography
1965	US Project CALDRILL: JOIDES conducts a transect drilling on the Blake Plateau off Florida
1968	US Project DSDP (Deep Sea Drilling Project) starts with *Glomar Challenger* as the drillship. DSDP Leg 1 is conducted in the Gulf of Mexico to find a salt dome and traces of petroleum
1969	DSDP Leg 3 in the Atlantic Ocean verifies the seafloor spreading hypothesis
1975	DSDP enters an international phase (International Program for Ocean Drilling, IPOD) with the participation of the Federal Republic of Germany, Japan, the UK, USSR, and France
1979	Hydraulic piston coring becomes available in DSDP. Sediment cores are sampled with little disturbance
1979–1983	A series of experiments of borehole seismometer installations conducted. Ocean sub-bottom seismometer (Sites 482, 494, 543, 581) and marine seismic system (Sites 395, 595)
1981	The first international Conference on Scientific Ocean Drilling (COSOD) held at the University of Texas
1985	Ocean Drilling Program (ODP) is started with *JOIDES Resolution* as the drillship
1987	Second COSOD held in Strasbourg, France, to redefine the future of ocean drilling science
1989	Digital ocean bottom downhole seismometer experiment (Leg 128)
	Oldest Pacific Ocean crust dated at 170 Ma (Leg 129)
1991	Sub-seafloor hydrogeology observatory experiment (CORK) started (Site 858 and subsequently many others)
1993	ODP Leg 148 reaches 2,111 m below seafloor (deepest ocean basin borehole)
1998	Ocean Seismic Network first installation (OSN-1, Site 843, ION prototype)
1999	Conference on Multiple Platform Exploration (COMPLEX) of the Ocean held in Vancouver, Canada
1999–2001	Borehole seismo-geodetic observatories installation in western Pacific (ION Sites 1150, 1151, 1179, 1201)
2004	Integrated Ocean Drilling Program (IODP) is started utilizing three drilling platforms. The *JOIDES Resolution* embarks on the first cruise (Expedition 301)
	The first Mission-Specific Platform expedition of ECORD (European Consortium for Ocean Research Drilling) utilizes the *Vidar Viking* to core at the Lomonosov Ridge in the Arctic Ocean through the ice

Ocean Drilling, Table 2 (Continued)

2007	The first riser-equipped drillship *Chikyu* of JAMSTEC (Japan Agency for Marine-Earth Science and Technology) sails on IODP Expedition 314
2009	IODP New Ventures in Exploring Scientific Targets (INVEST) Conference in Bremen, Germany, to set the course of ocean drilling science in 2013 and beyond
2012	*Chikyu* sets the deepest scientific coring depth at 2,466 m below seafloor (Exp. 337)
	Chikyu installs thermistor string across the fault zone of the 2011 Tohoku earthquake (M9.0). Recovered with data in 2013
2013	The International Ocean Discovery Program succeeds IODP

Ocean Drilling, Table 3 Data and reference publications from ocean drilling programs

Reports from ocean expeditions
Initial Reports of the Deep Sea Drilling Projects, Vol. 1–96, available online www.deepseadrilling.org, Scripps Inst. Oceanography, USA, 1969–1986
Proceedings of the Ocean Drilling Program; Initial Reports, Vol. 101–210, available online www.odp.tamu.edu/publications/pubs_lr.htm, Texas A&M U., USA, 1986–2004
Proceedings of the Ocean Drilling Program; Scientific Reports, Vol. 101–210, available online www.odp.tamu.edu/publications/pubs_lr.htm, Texas A&M U., USA, 1988–2007
Preliminary Report of the Integrated Ocean Drilling Program, Vol. 301–340 (as of 2013), available online publications.iodp.org/proceedings, IODP Management International Inc., USA/Japan, 2004-
Proceedings of the Integrated Ocean Drilling Program, Vol. 301–336 (as of 2013), available online publications.iodp.org/proceedings, IODP Management International Inc., USA/Japan, 2004-
Scientific Drilling Journal, available online (www.iodp.org), 2005-, published by IODP and International Continental Scientific Drilling Program

International planning documents guiding the programs
Earth, Oceans and Life, Integrated Ocean Drilling Program, Initial Science Plan, 2003–2013, pp. 110, IWG Support Office, Japan/USA, 2001
Illuminating Earth's Past, Present, and Future, Science Plan for 2013–2023, pp. 84, IODP Management International, Inc., 2011
Other publications available online at www.odplegacy.org/program_admin/long_range.html

Core repositories
Bremen Core Repository in Bremen, Germany, hosts DSDP, ODP, and IODP cores from the Atlantic Ocean, Mediterranean and Black Seas and Arctic Ocean (http://www.marum.de/en/IODP_Bremen_Core_Repository.html)
Gulf Coast Repository in Texas, USA, stores DSDP, ODP, and IODP cores from the Pacific Ocean, the Caribbean Sea and Gulf of Mexico, and the Southern Ocean (http://iodp.tamu.edu/curation/gcr/)
Kochi Core Center (joint with Japan Agency for Marine Science and Technology and Kochi University) in Kochi, Japan hosts DSDP, ODP, and IODP cores from the Indian Ocean and seas of Asia and Oceania (http://www.kochi-core.jp/en/)

Ocean Drilling, Figure 4 Core sample photographs. A few examples from a vast collection of legacy cores since DSDP openly available to scientists from core repositories. **a** Exp. 302 M0004-11× (297.30–301.35 m; ~49 Ma): there is an extremely dense concentration of remains of the hydropterid (freshwater) fern *Azolla* – unprecedented in other Eocene sections yielding *Azolla*. The event testifies to an extreme change in environmental conditions at the base of the *lower* middle Eocene (Lutetian) (from Proc. IODP 302, 2006). **b** Exp. 312 1256D-223R (1450.78–1452.28 mbsf): gabbro (*lower* crustal rock) section first recovered from the Pacific Ocean after ODP Leg 206 and IODP Exp. 309 at Hole 1256D. **c** Exp. 343 C0019-17R (821.5–822.5 mbsf): strongly deformed clays inferred to be the slip zone (decollement) of the 2011 Tohoku earthquake fault (Proc. IODP 343, 2013) (This photo was taken before cutting into halves (from CDEX, JAMSTEC)).

scientists are required to synthesize the results from different disciplines in order to prove, alter, or disprove the hypotheses they conceived. This characteristic norm of an ocean drilling expedition, in which an international team of scientists works around the clock for about 2 months isolated from land, offers a unique experience to each scientist. The drillships (Chikyu and JOIDES Resolution) are equipped with laboratory equipments for shipboard scientists to conduct core sample analyses to complete a preliminary report before returning ashore. It is a collective, concerted, and concentrated effort for the team on board often having to deal with contingencies on the spot under the pressure of time.

As a drill core (typically 9.5 m long) is recovered from depth on deck, it is cut into 1.5 m sections (Figure 4). The sample from the core catcher at the bottom provides microfossil data (micropaleontology) for dating. An X-ray CT (computed tomography) scanner is a tool on *Chikyu* that offers a 3-D look (X-ray attenuation in response to irregularities inside the core) of the core before dissection. Interstitial water and whole-round core section may be taken for geochemical and microbiological studies. Nondestructive standard measurements of P-wave velocity, gamma-ray attenuation (GRA) density, magnetic susceptibility, electrical resistivity, and natural gamma-ray radiation are made to estimate density, porosity, permeability, or magnetic signatures after the core temperature is stabilized. It is then customary to split the core into archive half and working half. Visual core description is made on the archive half together with other nondestructive measurements (photo, magnetism, etc.) that is subsequently stored in containers. Working halves are to be dissected by scientists as discrete samples for analyses on board and onshore. Geological age determination of obtained samples is an important process and is estimated from paleontological and geomagnetic observations before radiometric age determination is made onshore. Physical properties, X-ray diffraction, and organic and inorganic geochemical parameters are also obtained on board. Microscopes of various resolutions (optical, digital, and scanning electron microscopes) are powerful tools to identify microfossils and determine petrology to elemental composition level. All the initial results from these observations and measurements are described in a single publication for each cruise. Parameters of interest are also searchable across all cruises from the IODP database system (sedis.iodp.org).

Another important source of information of the borehole is to use downhole logging tools to measure physical, chemical, and structural properties through the borehole wall. These tools provide continuous records of in situ data such as sonic velocities, density, or electrical resistivity that complement the core measurements which may be limited by the recovery rate or biased by drilling disturbances.

Ocean drilling science

The human habitable zone, which is the land surface of the Earth, is at the juncture of the fluid sphere (atmosphere and hydrosphere) and the solid sphere. Energy and material movements from above and below (ocean and air circulations and internal mantle/core convections) define the state of the Earth at the surface including the ecosystem. Ocean drilling science contributes in describing these processes from the records of the past in the sediments and igneous rocks below and from in situ borehole measurements. Paleoceanography is a field of science borne out of ocean drilling.

Climate history

For any data of geological past scattered around the globe, time marking is essential in marking precise dates and obtaining time derivatives for the specimens of interest and tying them to plate tectonics, climate, paleoceanography, and life evolution. Biological time markers (planktonic microfossils) are established for the last 150 Ma. These are linked to magnetic polarity reversals and to radiometric absolute age of rocks. The resolution is as fine as 0.1 Ma level and can be further refined by tuning in to the Earth's orbital change (Milankovitch cycles (20–100 kyr)) detectable in sediment cores. Drilling in corals and high-sedimentation-rate sites allows for even higher temporal resolution. Since the ocean floors are destroyed by subduction, the oldest core sample is ~170 myrs old (ice cores <~800 kyr) covering ages since Jurassic. Drilling data contributed in defining the average temperature change over this time period, mostly warmer than the present Earth but covering the period without continental ice ("greenhouse Earth") and the period with ice sheets present ("icehouse Earth").

While the basic geological inferences come from lithostratigraphy, core specimen also provides information on past environments from oxygen isotope ratios and planktonic or benthic microfossils to infer the surface and bottom temperatures, which in turn infer continental ice volumes. The balancing depth of the carbonate supply and its dissolution rates called CCD (calcite compensation depth) was found to have abruptly deepened at about the time of the Antarctic glaciation (Pälike et al., 2012). Recent finding dates the Antarctic glaciation onset to 33 Ma and to be influenced by the atmospheric carbon dioxide; it also coincides with the separation of Australia and South America from Antarctica. Such findings form and advance the field of paleoceanography. It is now possible to infer atmospheric carbon dioxide concentrations to link to past warm climates. These past time series in a short time scale (decadal to millennial) indicate abrupt climate changes likely to be associated with the "flip-flop" of deep-ocean circulation in the North Atlantic. In the longer time scale is the shift from warm climates to the current era with cryosphere. Organic carbon-rich layers from the mid-Cretaceous suggested prevailing anoxic conditions during warm ocean time ("greenhouse"). Drilling in high latitudes elucidated continental glaciation timings as well as developments of ocean current pathways (e.g., Antarctic bottom water or North Atlantic bottom water or isolation of Pacific and Atlantic Oceans at Isthmus of Panama). A significant recent IODP discovery was obtained by drilling in the Arctic Ocean for the first time to go back 55 myrs and to find freshwater fern in 49 Ma when there was no ice (IODP Exp. 302; Moran et al., 2006). The cooling in the Arctic started about 46 Ma much earlier than previously conceived.

There are numerous other key findings that are relevant to shaping the Earth's environment. Drilling across the Cretaceous-Tertiary border time (65 Ma) clearly substantiated the bolide impact causing mass extinction and recovery. The Mediterranean Sea was dried near the end of the Miocene creating salt deposition influenced by the restriction of the water flow into the sea (DSDP Leg 13).

Microbiology

The investigation of the deep biosphere has brought surprising results of finding microbes in excess of 500 m depths into the sediments (Parkes et al., 1994). The sub-seafloor microbes do not depend on photosynthesis. So far, the deepest record is finding active microbes 1,626 mbsf in 111-Ma-old sediments (Leg 210; Roussel et al., 2008). Microbes are found to prevail in organic-rich or hydrate-bearing sediments. Traces of microbes in volcanic rocks are also found (Mason et al., 2010; Exps. 304, 305); thus, the limit is not the sediments, but more likely to be the thermal condition and energy supply (e.g., Bach and Edwards, 2003). Studies of life sub-seafloor have expanded recently with growing interests in defining the habitat, ecosystem, link to fluids, limiting environmental conditions, and the role of these life forms in the Earth-life coevolution and geochemical cycles. Drilling contaminates the core samples, and therefore, technological scrutiny to study only the uncontaminated portion is required.

Solid earth processes

Oceanic bathymetric highs range from scattered seamounts to seamount chains to oceanic plateaus, not directly linked to volcanisms at mid-ocean ridges. One of the major discoveries about the genesis of these features is that it is proven by ocean drilling that the bend between the Emperor seamount chain and the Hawaiian chain is produced not by the plate motion change but by the change in the mantle upwelling motion (e.g., Tarduno et al., 2009). This is an example that working hypotheses will be modified by new findings.

Large igneous provinces (LIPs) are continental flood basalts and oceanic plateaus created by a large supply of mantle-derived magma in a relatively short period (1–10 myrs) (Legs 130, 192, Ontong Java Plateau; Legs 119,120, 183, Kerguelen Plateau; Leg 198, Exp. 324, Shatsky Rise). Mantle plumes are considered the likely source, and their impact on climate (greenhouse Earth) and ocean anoxia (oxygen-depleted ocean) is not well understood, partly because of the difficult core sampling strategy.

Plate subduction occurs both in accretionary tectonics and subduction erosion tectonics around the Pacific Rim. Recent deep drilling in the Nankai subduction zone has revealed the tsunami genesis history in geological perspective (IODP NanTroSEIZE Exps.).

The database can be utilized for objectives that can be achieved through global sampling not possible for a single cruise such as compiling for rare-earth element distribution (Kato et al., 2011).

Ocean igneous crust

Ocean crustal rocks beneath the sediments have not been recovered with significant continuous coring except for recovering the topmost section for obtaining the date of the seafloor formation. The basic structure of an oceanic crust is estimated from geophysical (remote sensing) or geological evidence offshore (exposures of rocks formed at high pressure and temperature) and onshore (ophiolite considered to be obducted oceanic crusts) to be about

6–7 km plus or minus a few kilometers and composed of basalt (pillow basalt, sheeted feeder dikes), gabbro (lower crust), and the mantle rock (variations of peridotite). The Moho discontinuity is seismologically detectable almost ubiquitously in oceans, but its petrologic characteristics dependent on the spreading rate and postulated to be the gabbro to peridotite boundary or an alteration front of serpentinized peridotite remain to be verified.

Since the seafloor spreading hypothesis was verified on Leg 3 of DSDP, the physics and chemistry of how the oceanic crusts are formed at ocean ridges and age and cool away from the ridge axes forming the uppermost layer of the lithosphere have been an important focus of ocean drilling. The difference in the seafloor spreading rate is currently considered a major factor in determining the petrologic structure of oceanic crusts. Where magma supply is sufficient to balance the spreading rate, the crustal layering is expected to be vertically stratified. Crusts formed at fast spreading rate ($>\sim 100$ mm/year) are in this category. Crusts formed at slow spreading rate ($<\sim 50$ mm/year) accompany tectonic adjustments such as exhumation of deeper crustal or mantle rocks, as the magma supply seems insufficient to accommodate the spreading rate and to cause larger lateral heterogeneity than crusts from fast spreading ridges.

Deep penetration coring has been made since DSDP at Sites 332 (Leg 37, mid-Atlantic), 395 and 920–922 (Legs 45, 153, mid-Atlantic), 418 (Legs 52–53, Bermuda Rise), 504 (Legs 69, 70, 83, 111, 137, 140, 148, E. Pacific, Nazca Plate), 735 (Legs 118, 176, SW Indian Ridge), 801 (Legs 129, 185, W. Pacific), 1256 (Leg 206, Exps. 309, 312, 335, E. Pacific, Cocos Plate), and 1309 (Exps. 304, 305, mid-Atlantic, Atlantis Massif). Hole 332B was the first multiple reentry hole to deepen the hole with new bits. Holes 735B and 1309D recovered gabbroic rocks representing the lower crust. Hole 1256D, although not the deepest, reached gabbros at 1,407 m below seafloor (Wilson et al., 2006). These with drilling attempts to shallower penetrations into the igneous crusts contributed in observing the lateral change away from and in parallel to the ridge segmentation (e.g., Leg 209, mid-Atlantic Ridge) or test hypotheses of finding exhumed lower crust or mantle rocks (Leg 147, Exp. 345, Hess Deep). Other attempts include drilling into continental rifted margins (Leg 103, Galicia; Legs 149, 173, Iberia; Leg 210, Newfoundland; Leg 104, Norway; Leg 152, Greenland) to investigate the continental breakup process from continent-ocean transition structure and their relation to magma feed at the time of rifting.

Through ocean drilling combined with surface geophysical surveys, the understanding of petrologic evolution of oceanic crust has progressed to infer its relation to spreading rates, seismic models (e.g., Layer 2/Layer 3 transition; Detrick et al., 1994; Swift et al., 2009), interactions with fluids, and inclusions of microbes (Mason et al., 2010). To further the understanding of oceanic crusts in relation to mantle convection and plate tectonics, the current state of sampling is insufficient depth-wise and also plate age-wise. The feasibility of deep penetration into the lower crust and reaching the mantle across the Moho boundary is being seriously studied with the advent of the riser drillship *Chikyu*.

Fluids

The notion of the dark and motionless world of the seafloor and below is completely revised by numerous seafloor investigations and drilling expeditions. Fluids are seeping up and down not only at hydrothermal vents but also at plate subduction zones and possibly elsewhere providing energy for microbes to thrive and massive sulfide minerals to deposit or serpentinizing mantle peridotites. The physical behavior of the fluid fluxes is being studied at and around ridges by borehole observatories including chemical and biological parameters (Davis et al., 2000, Davis et al., 2000; IODP Exps. 336, 337).

Marine gas hydrates have become important scientific targets in order to understand their formation, distribution, and role in affecting the environment from the viewpoints of natural resource potential, methane storage, and control in relation to sub-seafloor ecosystem and geohazard risk.

Observatories

Ocean-drilled boreholes have long been perceived to provide a stable environment for global seismic observatories (e.g., Byrne et al., 1987; Stephen et al., 2003). The global seismic network has progressively advanced the understanding of earthquakes and the entire internal structure of the Earth as the frequency bandwidth and dynamic range widened. 3-D tomography of the Earth and complex fault motion description are direct products of the network data. The weakness of the global network has been the inhomogeneous distribution of the seismic stations because of the lack of stations in the oceans (71 % of Earth's surface). There have been several attempts to overcome this (Suyehiro et al., 2006) utilizing ocean drilling holes. Technically, a broadband digital seismic observatory with equal fidelity as on-land observatories can be established in the boreholes on solid basement rock that provide more quiet observational environment than on seafloor or in soft sediments. Feeding power for long term and getting data back in real time are the remaining challenges.

While complementing the global resolution of the network, ocean sites uniquely contribute in obtaining new data at upper mantle depths where on-land stations do not have the resolving power. For example, a new implication is made on the thickness of the ocean lithosphere suggesting that the lithosphere-asthenosphere boundary is not a simple rheological change produced by the cooling of the lithosphere (Kawakatsu et al., 2009) or a detailed anisotropic structure of the ocean upper mantle (Shinohara et al., 2008). Regional-scale seafloor cable networks are being established to cover tectonically active areas such as off Japan or off North America to obtain seismic and geodetic data continuously real time from boreholes (e.g., DONET (www.jamstec.go.jp/donet/e/), NEPTUNE Canada (www.neptunecanada.com)).

The main objectives of developing CORK (Circulation Obviation Retrofit Kit) technologies for enabling time series observations have been to identify and quantify sub-seafloor fluid flows and associated processes where physically focused flows may be expected such as in the vicinities of ocean ridges (thermally driven flows) and plate subduction zones (tectonic compaction).

In order to observe in situ fluid flow and estimate flow rates and permeability with minimum disruption, CORK design is intended to seal holes and control perturbation originally at the seafloor and subsequently evolved to enable multilevel isolated zones inside the borehole. The observation parameters expanded from original pressure and temperature to chemical and biological sampling.

It has been proven that these borehole observations not only provide permeability to be estimated from tidal responses but also function as volumetric strainmeters at a microstrain level (Davis et al., 2001). Transients such as associated with seismic events can be registered (Sites 1173, 1255). Furthermore, the temperature and pressure series from multiple CORK observatories where observations can be correlated, combined with the subsurface heterogeneous structure, can provide 4-D flow pattern (Holes 1026B, 1027C).

Conclusions

The ocean floor constitutes the top of the oceanic plates created by seafloor spreading and destroyed by plate subduction as part of the mantle convection. At the same time, the ocean floor is the bottom of the water column that hosts the thermohaline "conveyor belt" ocean circulation system and ocean currents. The chronological records of these processes are kept in the sediments and crust to be explored by ocean drilling. Ocean drilling provides ways to retrieve such information from sub-seafloor core samples and data to better understand how these systems operate and interact with each other.

Ocean drilling has evolved from the early days of exploratory stage to the more recent idea-testing stage over half a century backed with technological advancements for better core recovery and deeper penetration and higher-resolution downhole measurements. Paleoceanography is an Earth science field borne from ocean drilling. Recent new field is the biosphere beneath the oceans.

Ocean drilling boreholes can also be utilized as monitoring observatories for active processes in situ or for remote sensing for hydrology, geophysics, and biology. The framework of conducting ocean drilling has evolved over the last 50 years, but currently the sampling density in space is quite undersampled particularly in high latitudes, deeper crusts, or high-temperature environments.

Operating drilling platforms year-round requires international coordination of funding agencies, scientists, engineers, and platform operators. It has grown from a single non-riser drilling vessel era (DSDP-ODP) to a multiple platform era (IODP) that enables very deep penetrations or drilling in very shallow waters or ice-covered seas. The history of ocean drilling has been to advance for recovery of high-resolution and high-quality samples with long history in all oceans as well as deeper penetrations into igneous rocks and developing capabilities to monitor ongoing geophysical, geochemical, and biological processes. It should be recognized that much of the ocean floor remains undersampled considering the heterogeneous nature of the processes in the oceans and beneath. Ocean drilling is emphasizing its utility to predictions of climate, geohazards, and ecosystem through new discoveries in basic science.

Ocean drilling complements continental scientific drilling and ice core drilling.

Bibliography

Bach, W., and Edwards, K. J., 2003. Iron and sulfide oxidation within basalt ocean crust: implications for chemolithoautotrophic microbial biomass production. *Geochimica et Cosmochimica Acta*, **67**, 3871–3887.

Bascom, W., 1961. *A Hole in the Bottom of the Sea: The Story of the Mohole Project*. Garden City: Doubleday & Co., p. 352.

Becker, K., and Davis, E. E., 2005. A review of CORK designs and operations in the Ocean Drilling Program. In Fisher, A. T., Urabe, T., Klaus, A., and The Expedition 301 Scientists (eds.), *Proceeding of the Integrated Ocean Drilling Program*. College Station, Integrated Ocean Drilling Program Management International, Washington, DC, Vol. 301, doi:10.2204/iodp.proc.301.104.2005.

Byrne, D. A., Harris, D., Duennebier, F. K., and Cessaro, R., 1987. The ocean sub-bottom seismometer system installed in Deep Sea Drilling Project Hole 581C, Leg 88: a technical review. *Initial Reports of the Deep Sea Drilling Project*, **88**, 65–88, doi:10.2973/dsdp.proc.88.106. Washington (U. S. Government Printing Office).

Davis, E. E., Wang, K., Becker, K., and Thomson, R., 2000. Formation scale hydraulic and mechanical properties of oceanic crust inferred from pore-pressure response to periodic seafloor loading. *Journal of Geophysical Research*, **105**, 13,423–13,435.

Davis, E. E., Wang, K., Thomson, R. E., Becker, K., and Cassidy, J. F., 2001. An episode of seafloor spreading and associated plate deformation inferred from crustal fluid pressure transients. *Journal of Geophysical Research*, **106**, 21953–21963.

Davis, E. E., Kinoshita, M., Becker, K., Wang, K., Asano, Y., and Ito, Y., 2013. Episodic deformation and inferred slow slip at the Nankai subduction zone during the first decade of CORK borehole pressure and VLFE monitoring. *Earth and Planetary Science Letters*, **368**, 110–118.

Detrick, R. S., Collins, J., Stephen, R., and Swift, S., 1994. In situ evidence for the nature of the seismic layer 2/3 boundary in oceanic crust. *Nature*, **370**, 288–290, doi:10.1038/370288a0.

Fisher, A. T., Wheat, C. G., Becker, K., Davis, E. E., Jannasch, H., Schroeder, D., Dixon, R., Pettigrew, T. L., Meldrum, R., MacDonald, R., Nielsen, M., Fisk, M., Cowen, J., Bach, W., and Edwards, K., 2005. Scientific and technical design and deployment of long-term, subseafloor observatories for hydrogeologic and related experiments, IODP Expedition 301, eastern flank of Juan de Fuca Ridge. In Fisher, A. T., Urabe, T., Klaus, A., and The Expedition 301 Scientists (eds.), *Proceeding of the Integrated Ocean Drilling Program*. College Station, Integrated

Ocean Drilling Program Management International, Washington, DC, Vol. 301, doi:10.2204/iodp.proc.301.103.2005.

Huey, D. P., 2009. Final report on IODP drilling and coring technology -past and present. IODP Management International, 183 p.

Kato, Y., Fujinaga, K., Nakamura, K., Takaya, Y., Kitamura, K., Ohta, J., Toda, R., Nakashima, T., and Iwamori, H., 2011. Deep-sea mud in the Pacific Ocean as a potential resource for rare-earth elements. *Nature Geoscience*, **4**, 535, doi:10.1038/NGEO1185.

Kawakatsu, H., Kumar, P., Takei, Y., Shinohara, M., Kanazawa, T., Araki, E., and Suyehiro, K., 2009. Seismic evidence for sharp lithosphere-asthenosphere boundaries of oceanic plates. *Science*, **324**, 499–502.

Mason, O., Nakagawa, T., Rosner, M., Van Nostrand, J., Zhou, J., Maruyama, A., Fisk, M., and Giovannoni, S., 2010. First investigation of the microbiology of the deepest layer of ocean crust. *PloS One*, **5**(11), e15399, doi:10.1371/journal.pone.0015399.

Moran, K., Backman, J., et al., 2006. The Cenozoic palaeoenvironment of the Arctic Ocean. *Nature*, **441**, 601–605, doi:10.1038/nature04800.

National Research Council, 2011. *Scientific Ocean Drilling: Accomplishments and Challenges*. Washington, DC: The National Academies Press.

Pälike, H., Lyle, M. W., Nishi, H., et al., 2012. A Cenozoic record of the equatorial Pacific carbonate compensation depth. *Nature*, **488**, 609–614, doi:10.1038/nature11360.

Parkes, R. J., Cragg, B. A., Bale, S. J., Getliff, J. M., Goodman, K., Rochelle, P. A., Fry, J. C., Weightman, A. J., and Harvey, S. M., 1994. Deep bacterial biosphere in Pacific Ocean sediments. *Nature*, **371**, 410–4413, doi:10.1038/371410a0.

Polyak, L., Alley, R. B., Andrews, J. T., et al., 2010. History of sea ice in the Arctic. *Quaternary Science Reviews*, **29**, 1757, doi:10.1016/j.quascirev.2010.02.010.

Pross, J., Contreras, L., Bijl, P. K., et al., 2012. Persistent near-tropical warmth on the Antarctic continent during the early Eocene epoch. *Nature*, **488**, 73–77, doi:10.1038/nature11300.

Roussel, E. G., Bonavita, M. A. C., Querellou, J., Cragg, B. A., Webster, G., Prieur, D., and Parkes, R. J., 2008. Extending the sub-sea-floor biosphere. *Science*, **320**, 1046, doi:10.1126/science.1154545.

Sakaguchi, A., Chester, F., Curewitz, D., Fabbri, O., Goldsby, D., Gaku Kimura, G., Li, C.-F., Masaki, Y., Screaton, E. J., Tsutsumi, A., Ujiie, K., and Yamaguchi, A., 2011. Seismic slip propagation to the up-dip end of plate boundary subduction interface faults: vitrinite reflectance geothermometry on Integrated Ocean Drilling Program NanTroSEIZE cores. *Geology*, **39**, 395–398, doi:10.1130/G31642.1.

Screaton, E., Kimura, G., Curewitz, D., Moore, G., Chester, F., Fabbri, O., et al., 2009. Interactions between deformation and fluids in the frontal thrust region of the NanTroSEIZE transect offshore the Kii Peninsula, Japan: results from IODP Expedition 316 Sites C0006 and C0007. *Geochemistry, Geophysics, Geosystems*, **10**, Q0AD01, doi:10.1029/2009GC002713.

Shinohara, M., Fukano, T., Kanazawa, T., Araki, E., Suyehiro, K., Mochizuki, M., Nakahigashi, K., Yamada, T., and Mochizuki, K., 2008. Upper mantle and crustal seismic structure beneath the Northwestern Pacific Basin using a seafloor borehole broadband seismometer and ocean bottom seismometers. *Physics of the Earth and Planetary Interiors*, **170**, 95–106.

Stephen, R. A., Spiess, F. N., Collins, J. A., Hildebrand, J. A., Orcutt, J. A., Peal, K. R., Vernon, F. L., and Wooding, F. B., 2003. Ocean seismic network pilot experiment. *Geochemistry, Geophysics, Geosystems*, **4**, 1092, doi:10.1029/2002GC000485.

Suyehiro, K., Montagner, J.-P., Stephen, R. A., Araki, E., Kanazawa, T., Orcutt, J., Romanowicz, B., Sacks, S., and Shinohara, M., 2006. Ocean seismic observatories. *Oceanography*, **19**, 144–149.

Swift, S., Reichow, M., Tikku, A., Tominaga, M., and Gilbert, L., 2009. Velocity structure of upper ocean crust at Ocean Drilling Program Site 1256. *Geochemistry, Geophysics, Geosystems*, **9**, Q10O13, doi:10.1029/2008GC002188.

Tarduno, J., Bunge, H.-P., Sleep, N., and Hansen, U., 2009. The bent Hawaiian-Emperor hotspot track: inheriting the mantle wind. *Science*, **324**, 50–53.

The Oceanography Society, 2006. The Impact of the Ocean Drilling program. In Kappel, E., Burger, R., and Fujioka, K. (eds.), *Oceanography*, Rockville, 19(4): 20–177.

Wilson, D. S., Teagle, D. A. H., Alt, J. A., et al., 2006. Drilling gabbro in intact ocean crust. *Science*, **312**, 1016–1020.

Winterer, E. L., 2000. Scientific ocean drilling, from AMSOC to COMPOST. In *50 Years of Ocean Discovery*. Washington, DC: Ocean Studies Board, National Research Council, pp. 117–127.

Cross-references

Deep Biosphere
Deep-sea Sediments
Geologic Time Scale
Marine Geosciences: A Short, Ecclectic and Weighted Historic Account
Paleoceanography
Plate Tectonics
Regional Marine Geology
Technology in Marine Geosciences

OCEAN MARGIN SYSTEMS

Gerold Wefer
MARUM-Center for Marine Environmental Sciences, University of Bremen, Bremen, Germany

Definition

Transitional zones between the oceans and continents, representing dynamic systems in which numerous processes shape the environment.

Introduction

Ocean margins are the transitional zones between the oceans and continents where most of the sediments derived from the land are deposited. The effective processes here are influenced by a variety of steering mechanisms, from mountain building and climate on the land to tectonics and sea-level fluctuations at the margins of the seas. These areas are also of great importance for the global biogeochemical cycles. Although these areas make up only about 20 % of the ocean's surface, 50 % of the global marine production takes place here. Compared to the open ocean, the ocean margins represent more dynamic systems (see Wefer et al., 2002).

The region of the ocean margins extends from the coastal zone across the shelf and the continental slope to the continental rise (Figure 1). About 60 % of all people live in the adjacent coastal land areas, and they have intensively used the coastal waters for the extraction of raw materials and food over a long period. In recent times,

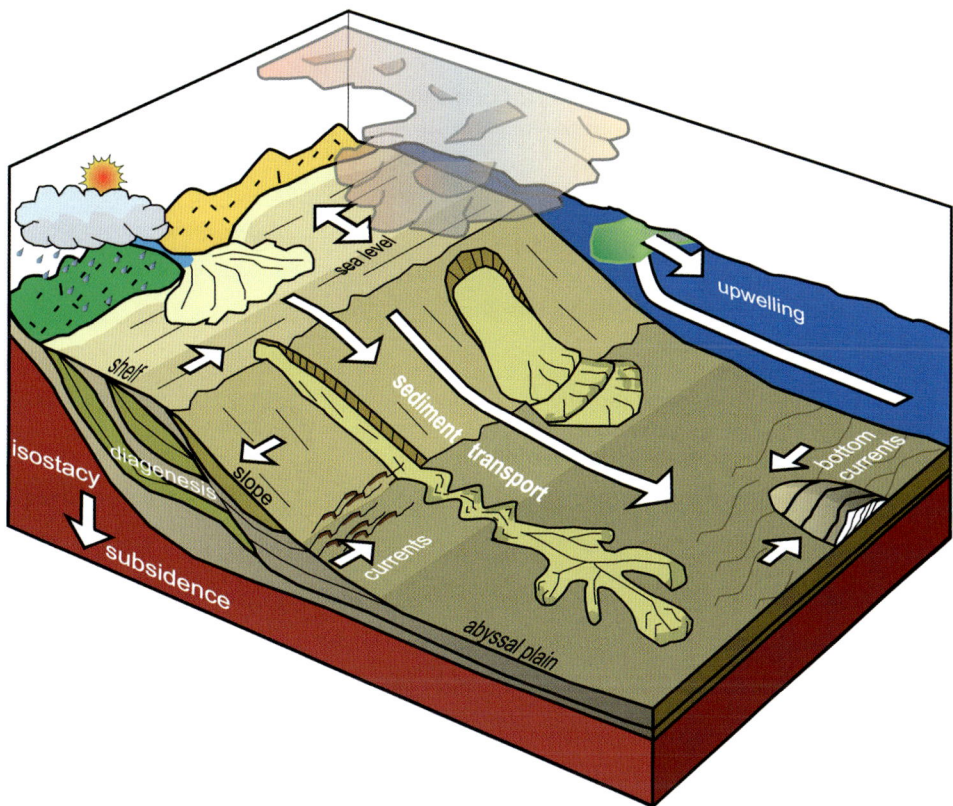

Ocean Margin Systems, Figure 1 Sedimentation processes at ocean margins (MARUM).

human activity has spread farther out into the oceans as the margins receive increasing attention as potential centers for hydrocarbon exploration and industrial fisheries. The great commercial potential of these regions, however, is countered by the presence of high potential hazards, for example, in the form of earthquakes and possible tsunamis triggered by slope instability, which can have a direct impact on the densely populated coastal regions.

The basis for the wealth of natural resources at the ocean margins lies in their high biological productivity. This can lead to the development of large fish populations on the one hand and to the deposition of immense amounts of organic materials on the other, which, through diagenetic processes, can be transformed into energy sources. It is often the case, however, that prerequisite conditions for the formation of resources are first created by large-scale sediment transport processes, i.e., from the shelf into the deep sea. The ocean margins are a dynamic system in which many processes shape the environment and impact the utilization and hazard potentials for humans.

Margin types

The terms Atlantic margin and Pacific margin were first introduced in the nineteenth century (see Seibold and Berger, 1996). Atlantic-type margins are steadily subsiding areas with thick sequences of deposited sediments. In contrast, Pacific-type margins are uplifted, associated with volcanism, faulting, and other mountain-building processes. Atlantic-type margins are also called "passive margins" and Pacific-type margins "active margins." These characterizations are due to tectonics, particularly the occurrence of earthquakes and active volcanism (Figure 2).

Atlantic-type (passive) margins

The Atlantic-type margins originate through the breakup of plates or parts of plates, a process that is still taking place today in the Atlantic at a speed of 2–4 cm per year, in connection with subsidence and deposition of sediments on the ocean margins. A recent model is the Red Sea. Also the upper Rhine valley is an example of the initial stage of uplift and breakup of lithospheric plates. In an early phase of rifting, the deposition of salt is common, e.g., in the South Atlantic. In special areas, e.g., seaward of large deltas, up to 15 km of sediments can be deposited. Atlantic-type margins are also common in East Africa, Australia, off India, and around the Antarctic continent.

Pacific-type (active) margins

On Pacific-type margins three collision types can be distinguished: continent-continent collision, e.g., in the Himalayan region; continent-ocean collision, e.g.,

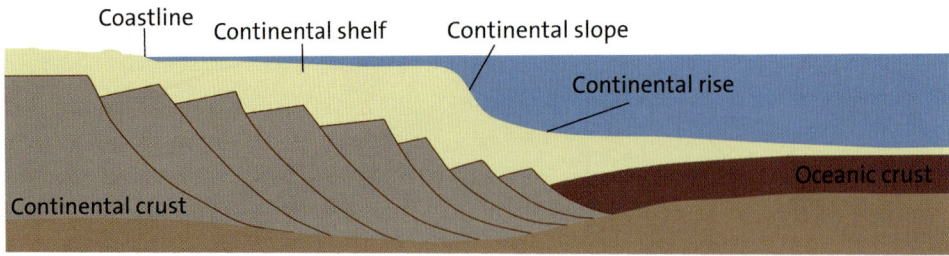
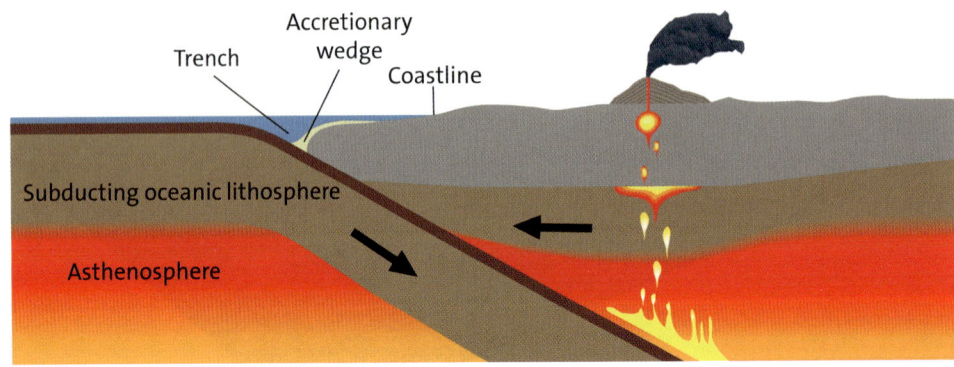

Ocean Margin Systems, Figure 2 Features of a passive ocean margin (*top*) and active margin (*bottom*) (modified after Tarbuck and Lutgens (2009)).

between the Pacific and South America; and subductions in the area of island arcs, e.g., in the Mariana Trench. Faulting, shearing of sediments, volcanism, and melting of the submerging lithosphere are characteristics of colliding margins. Special rocks called ophiolites, a mixture of pelagic sediments and basaltic rocks, are characteristically formed at subduction zones.

In contrast to passive margins, steep slopes are formed on active margins by plate collision, favoring mass transport into the deep-sea trenches. As result of the higher pressures, sediments are subject to dewatering, which is documented by the release of gases and fluid at sites called cold seeps. A special phenomenon of subduction at island arcs is back-arc spreading, the breakup of lithosphere behind the island arc, which is common in the western Pacific.

Morphology

The composition and structure of ocean margins are governed by a wide variety of sedimentation processes along the sedimentary pathway.

Shelf

Shelf regions are characterized by high sedimentation rates because of the input of terrigenous materials from the land and high biological productivity. Massive sediment transport is widespread in these areas due to the high energy levels caused by tides, waves, and the dissipation of energy from internal ocean waves at the shelf edges.

Continental margin

Continental margins are also characterized by high sedimentation rates. Short-term mass transport of sediments is common, depending on the steepness of the slope. These episodic transport events occur at various intensities, ranging from slow-moving slope slides of intact layered packages to stirred-up, disorderly debris flows and further to turbidites that can extend for distances up to 1,000 km. Sedimentation patterns can thus be modified by sediment redistribution.

Through exploration for gas and oil on continental margins, large infrastructures have been installed in recent years, leading to increased concern regarding stability of the margins. In history, large submarine landslides are known to have occurred with a volume of more than 20,000 km^3 of material transported, within an area of more than 100,000 km^2. These events can be triggered, for example, by increased sediment overload, by expulsion of gas hydrates, or through an external trigger such as an earthquake. In addition to the few known very large events that have influenced areas of up to 100,000 km², there have probably been a large number of smaller slides.

Climate coupling

Climate plays a prominent role in driving the processes that shape the environment, and it also interacts closely with ocean circulation and productivity. The climate-induced fluctuations of sea-level alone, with swings of over 100 m during the Late Quaternary glacial cycles causing broad shelf regions to alternately be dry and part of the land and then flooded and covered by the sea, highlight the sensitivity of the ocean margins to climate. In addition to the fact that these areas represent extremely active transitional areas between the oceans and continents, the presence of high sedimentation rates due to the input of terrigenous sediments is a very important characteristic.

Direct climatological measurements, which are critical for a basic understanding of such processes, only go back to the year 1850. Records extending farther back in time are documented primarily in the ocean margin sediments. The ocean margin sediments clearly provide an impressive archive of past climatic, productivity, current, and sedimentation conditions.

The past thousand years represent a particularly interesting time interval. In addition to historical records, there are also numerous high-resolution climate archives available from this period (e.g., coral skeletons, tree rings, continental slope sediments) that make it possible to compare the effects of relatively small-scale natural climate anomalies (e.g., the Middle Ages warm period and the "Little Ice Age") to recent variations caused by humans under somewhat similar boundary conditions.

Another time scale of interest includes the past 30,000 years, which incorporate the transition from the last ice age to the present warm period, the Holocene, which began about 10,000 years ago. Another important time interval encompasses the past 500,000 years (the last glacial-interglacial cycles). During this period, processes leading to variability of the global ocean in the Late Quaternary can be identified. These processes, in turn, are closely coupled to the drastic climate fluctuations over the course of the glacial-interglacial cycles. The ocean is a central component of the global carbon and hydrological cycles, and it also has a critical influence on the other two major components of the climate system, the atmosphere and continental ice.

Productivity

Organic matter deposition

The deposition of large amounts of organic material resulting from high-productivity conditions along the ocean margins leads to very high turnover rates at the seafloor, so these areas can also be characterized as efficient biogeochemical reactors. These high turnover rates of organic carbon, along with associated elements such as nitrogen, phosphorus, sulfur, and iron, play a critical role in the global mass cycles, but above all in the global carbon cycle.

In high-productivity regions, elevated flux rates of organic substance often lead to the formation of an oxygen-minimum zone. In the highly productive ocean margins, these oxygen-minimum zones are in direct contact with the seafloor, strongly affecting the organism communities and the biogeochemical processes. The oxygen deficit varies with the dynamic balance between the flux of degradable organic detritus, stratification stability, and exchange time of the water masses. Interactions between the seafloor and the oxygen-depleted water column are much more intense than in the deep sea, leading to a close coupling between the bottom-water processes and those in the sediment. Oxygen concentration is one of the most important biological-chemical steering parameters for these processes.

Sediments at the seafloor are the habitat for countless, so far only partly known, microorganisms that drive these cycles and contribute significantly to the conversion of both organic and inorganic materials. Along with the normal flux of materials from the water column into the sediments, these microorganisms contribute to an intense return of nutrients from the sediments back into the water column.

Gas hydrates

Gas hydrates are firm, icelike associations of gas molecules and water, present in water depths between 300 and 700 m (see "Marine Gas Hydrates"). Besides methane, other hydrocarbons and other gases like carbon dioxide and nitrogen can also form gas hydrates at higher pressures and lower temperatures. Besides specific pressure and temperature conditions, an additional requisite is a sufficient amount of gas. In ocean margin sediments, the gas can be the product of microbial decomposition of organic material. Gas can also be produced by "thermocatalytic" processes in deep sediment columns. Because abundant organic material is commonly available on ocean margins, gas hydrates are common there. High plankton production and high sedimentation rates are favorable for the deposition of large amounts of organic material on the seafloor (see "Sapropels").

There are concerns regarding the instability of gas hydrates concentrated at the ocean margins as it relates to the warming of the oceans, which is in part caused by humans. The temperature increase could, on the one hand, lead to a wide-scale destabilization of the continental slopes resulting in large slides (see "Submarine Slides") with possible tsunamis and, on the other hand, release immense amounts of methane, which would contribute to a significant strengthening of the greenhouse effect.

Methane hydrates are also often discussed as a potential source of energy for the future. Although estimates are highly variable, the resources might be comparable to today's known fossil fuel (see "Energy Resources"). If they are eventually exploited, the exploration for methane hydrates will probably start in permafrost areas.

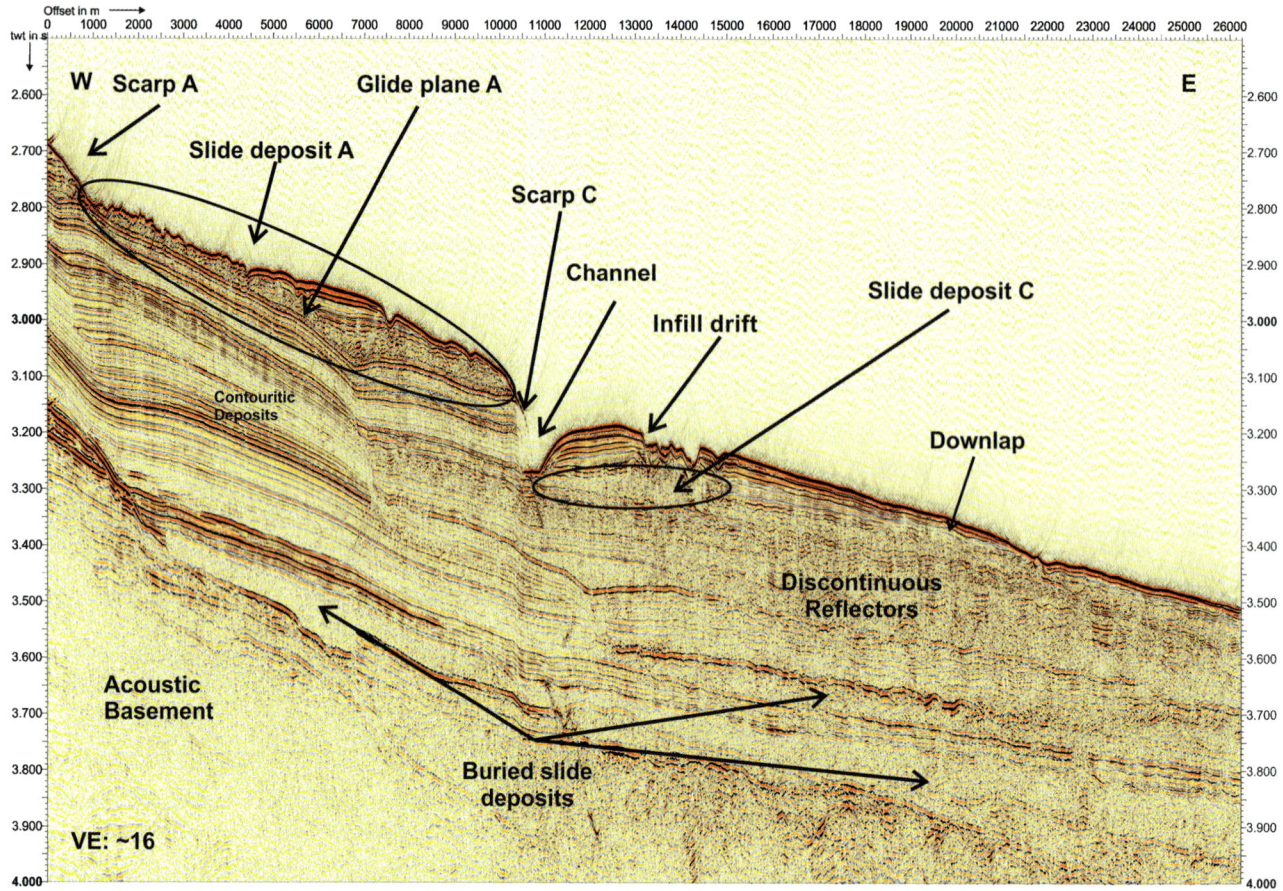

Ocean Margin Systems, Figure 3 High-resolution multichannel seismic profiles off Argentine (Krastel et al., 2014).

Sedimentological features

Benthic boundary layer

An important transition area between water and sediment is called the benthic boundary layer (BBL), where the exchange of material takes place. Processes at the BBL determine the deposition of terrigenous material and organic substances, as well as the release of nutrients into the water column. Furthermore, the BBL is important for the lateral transport of particles at ocean margins, which is controlled by hydrodynamics within the BBL (see "Bottom-Boundary Layer"). Within the BBL, but also deep in the sediments, intensive early diagenetic processes take place, which are significant for the carbon, nitrogen, and oxygen cycles in the ocean.

Turbidites

Turbidites often reach the deep ocean regions through various types of submarine canyons. Some canyons are formed initially in connection with slope slides and then gradually cut their way upslope, over time, into the shelf region (see "Turbidites"). Another type originates near the mouths of glacial rivers (i.e., during low sea-level stands, especially at the shelf edge) or during high stand has cut its way back to the inner shelf.

These sedimentation processes, oriented toward the open ocean, are overprinted by transport processes parallel to the slope that are driven by bottom currents and can be either erosive or accumulative, forming contourites or drift sediment bodies. Because ocean margins include the active interfaces in such balances, these regions are critical for understanding of the global mass transport and transformation of material in the land-atmosphere-ocean system.

Contourites

Contourites are deepwater sedimentary deposits influenced by bottom currents, seafloor morphology, and sediment supply (Figure 3) (see "Contourites"). Bottom currents are driven by thermohaline circulation flowing parallel to the continental margin or along the deep-sea floor. These currents may also be influenced by wind or tidal forces. Sediment for contourite deposits is supplied by erosion processes at the seafloor. Transport to deeper water occurs mostly through by turbidites.

Cold seeps

Besides the transport of dissolved substances by rivers and through the atmosphere, an important transport of dissolved components and gases takes place on vents and seeps, as well as diffusive transport between the seafloor and water column. This also includes submarine releases of groundwater. Seeps are characterized by a special community of benthic organisms, the formation of massive authigenic carbonates, gas releases, and occurrence of gas hydrates. Chemosynthetic communities are common, dominated by giant tube worms and large bacterial mats (see "Cold Seeps").

The return of materials from the sediments into the water is not limited to the microbiological and diagenetic processes in the uppermost sediment layers. Similar to the hydrothermal ("hot") vents at the mid-oceanic ridges, recent discoveries of cold seeps along the ocean margins have revealed ascending fluids flowing into the seawater, in part originating at great depths in the sediments. These fluids, often enriched in methane, sporadically maintain self-sufficient ecosystems based on chemosynthesis such as those known from the hydrothermal vents.

Recently discovered numerous large carbonate mounds (see "Carbonate Factories") located on the continental slope are associated with the discharge openings of such fluids. These carbonate deposits are often occupied by a fascinating ecosystem.

Biology

At ocean margins a large variety of habitats and biological communities are present, e.g., coral mounds, sediment slopes and gravel banks, canyons, and mud volcanoes. With new technology, especially using remotely operated vehicles (ROVs), new habitats have been discovered in recent decades. One example is illustrated by deepwater corals providing a habitat for a great number of other organisms. These communities are in danger as a result of increased commercial use of living and nonliving deepwater resources on the ocean margins, particularly deep-sea fishing and oil and gas production. Recently discussions have begun regarding the exploration of metals from the deep sea. This planned exploration must be combined with careful monitoring of the deep-sea environment.

Cold-water corals

Over the past several decades, reef-forming cold-water corals have been intensively studied (see "Reefs (Biogenic)"). They are comparable in extent and diversity to tropical coral reefs. Together with the two dominant species, *Lophelia pertusa* and *Madrepora oculata*, more than 1,000 other species were identified living in the cold-water coral environments (Freiwald and Roberts, 2005). Ocean margins contain a large number of different habitats and biological environments. Because ocean margins are increasingly being used for fisheries, as a biological resource, and for the exploitation of oil, gas, and minerals, management of the environment and establishment of conservation areas are under discussion (e.g., coral reefs off Norway).

Deep biosphere

In addition to these previously unknown ecosystems, significantly more important systems, quantitatively, have been discovered within the depths of the sediments (see "Deep Biosphere"). The presence of a rich bacterial life below the seafloor has been identified, reaching a sediment depth of more than 1,000 m and penetrating even into the volcanic rocks below the sediment cover. It is estimated that this newly discovered ecosystem accounts for up to 10 % of the active global biomass.

Human usage

The most obvious impacts of human usage on the oceans are naturally seen in the coastal regions. As the transition area between land and sea, the coastal zones are becoming more and more threatened by increasing intensive use as residential or industrial areas and as agri- and aquacultural areas. Added to this are the pressures of harbor sites and tourism. Already today, 60 % of the world's population lives in this zone, and 75 % of all cities with populations over 2.5 million are located in estuarine regions (see "Integrated Coastal Zone Management").

In addition to the input of pollutants and nutrients to the coastal waters, they are increasingly being directly used for human activities in which clear traces of the use remain. The shore, banks and shallow-water areas are being altered by construction (e.g., dikes, harbors, and pipelines, as well as river-bottom deepening and river correction and gravel and sand removal), which has a lasting effect on the current conditions, sea-level change (absolute level and storm-flood frequency), and rates of erosion and deposition (see "Coastal Engineering" and "Engineered Coasts").

On the seaside of the coastal zone, the wide continental shelf is exploited by fisheries and marine mining (see "Marine Mineral Resources"). The shelf regions represent high-productivity regions for fisheries, with 95 % of the total global catch coming from this zone. They are also easily accessible for marine mining, especially for the production of oil and gas, as well as gravel and other building materials. Oil exploration and especially the fisheries have also begun to advance out onto the continental slope. Fishery activity particularly contributes to large-scale changes on the seafloor through intensive bottom trawling. This leads, e.g., to the endangerment of unique ecosystems such as the cold-water coral reefs that occur on the continental slopes in water depths up to 1,000 m.

Conclusions

Ocean margins are extremely important for mankind due to their great commercial potential for hydrocarbon exploration and fisheries. On the other hand, these areas possess a high risk of hazards in the form of earthquakes and tsunamis. Increasing human activities, including construction, oil and gas exploration, fisheries, and possible deep-sea mining for metals, can have a great impact on sediment mobilization and transport and on the ecosystems in general. Our present understanding of the complex sedimentation processes and ecosystems on continental margins is not sufficient, and integrated studies using modern technologies for observation and modeling are needed.

Bibliography

Freiwald, A., and Roberts, J. M. (eds.), 2005. *Cold-Water Corals and Ecosystems*. Berlin/Heidelberg: Springer.

Krastel, S., Lehr, J., Winkelmann, D., Schwenk, T., Preu, B., Strasser, M., Wynn, R. B., Georgiopoulou, A., and Hanebuth, T. J. J., 2014. Mass wasting along Atlantic continental margins: a comparison between NW-Africa and the de la Plata River region (northern Argentina and Uruguay). In Krastel, S., et al. (eds.), *Submarine Mass Movements and Their Consequences*. Advances in Natural and Technological Hazards Research. Dordrecht: Springer, Vol. 37, pp. 459–469.

Seibold, E., and Berger, W., 1996. *The Sea Floor*. Berlin/Heidelberg: Springer.

Tarbuck, E. J., and Lutgens, F. K., 2009. *Earth Science*. Upper Saddle River: Pearson Prentice-Hall.

Wefer, G., Billet, D., Hebbeln, D., Jørgensen, B. B., and van Weering, T. (eds.), 2002. *Ocean Margin Systems*. Berlin/Heidelberg: Springer.

Cross-references

Active Continental Margins
Coasts
Cold Seeps
Contourites
Deep-sea Sediments
Marine Gas Hydrates
Mass Wasting
Methane in Marine Sediments
Passive Plate Margins
Reefs (Biogenic)
Turbidites

OCEANIC PLATEAUS

Andrew C. Kerr
School of Earth and Ocean Sciences, Cardiff University, Cardiff, Wales, UK

Definition

Oceanic plateaus are large areas of elevated, over-thickened basaltic ocean floor ($>5 \times 10^5$ km^3) which have formed throughout most of Earth's history, and, unlike most oceanic crust, they are not primarily the result of seafloor spreading processes and melting of ambient upper mantle but rather are widely regarded to have been formed by decompression melting of hot mantle plumes (Kerr, 2014).

Introduction

The vast erupted and intruded volume of oceanic plateaus means that they are classed as large igneous provinces (LIPs), along with continental flood basalts and volcanic rifted margins. The term large igneous provinces was originally proposed by Coffin and Eldholm (1992) as a generic term for igneous provinces with a volume exceeding 0.1×10^6 km^2 (see also Coffin and Eldholm 2005). More recently, this definition has been revised by Bryan and Ernst (2008), who proposed a classification scheme that includes giant radiating dike swarms and silicic LIPs. Bryan and Ernst's formal definition is as follows

> LIPs are magmatic provinces with areal extents $>0.1 \times 10^6$ km^2, igneous volumes $>0.1 \times 10^6$ km^3 and maximum lifespans of ~50 m.y. that have intraplate tectonic settings or geochemical affinities, and are characterised by igneous pulse(s) of short duration (~1–5 m.y.), during which a large proportion (>75 %) of the total igneous volume has been emplaced.

In the early 1970s, new evidence from seismic surveys revealed that several large portions of the ocean floor were much thicker than "normal" oceanic crust (6–7 km) (see Layering of Oceanic Crust). One of the first areas of such over-thickened crust to be identified was the Caribbean Plate which Donnelly (1973) proposed to be an "oceanic flood basalt province." The term "oceanic plateau" was first used by Kroenke (1974) to describe the large area of thickened crust (>30 km) and then recently discovered in the Western Pacific and now known as the Ontong Java Plateau.

The thick crustal sections and elevated topography of oceanic plateaus (generally 2–3 km above the surrounding ocean floor) result in them being more buoyant than normal oceanic crust. Thick oceanic plateaus and those that are relatively young when they interact with subduction zones are therefore difficult to subduct (e.g., Cloos, 1993; Tetreault and Buiter, 2012). The resultant collision of thick, buoyant oceanic crust with subduction zones results in extensive oceanic plateau fragments being preserved in accreted terrains (see below). These accreted oceanic plateaus are more likely to be preserved in the geological record than continental flood basalts that are more easily eroded by subaerial processes, and so oceanic plateaus could have been major contributors to the growth of continental crust throughout Earth's history (Kerr and Mahoney, 2007).

As our knowledge of the ocean basins and accreted terrains has improved over the last 30 years, many more oceanic plateaus have been identified, both relatively recent "in situ" plateaus (Figure 1) and older accreted examples.

Characteristics of oceanic plateaus

The three best-known extant (i.e., still predominantly in the oceanic realm and not accreted) oceanic plateaus are the Ontong Java, Kerguelen, and Caribbean–Colombian plateaus (Figure 1). The following sections will mostly

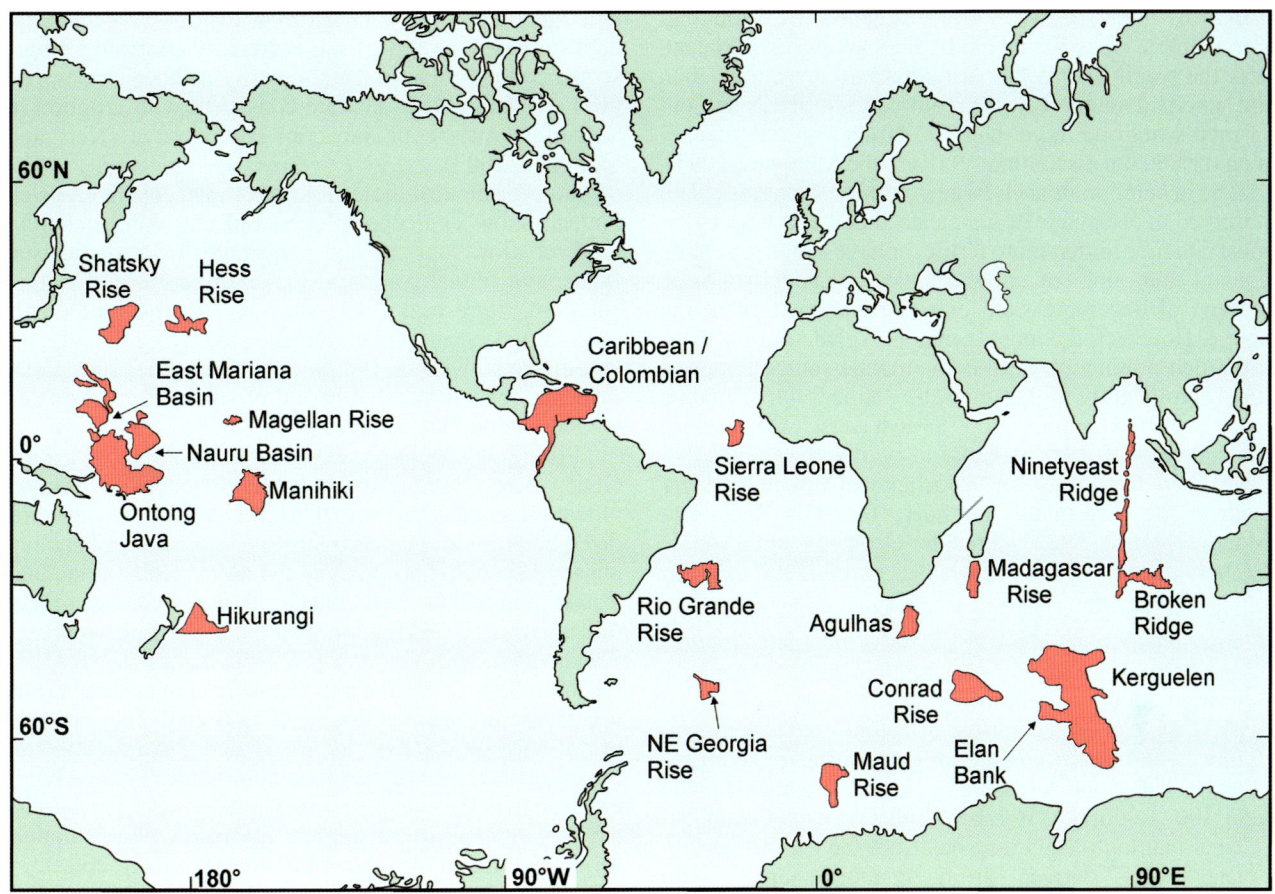

Oceanic Plateaus, Figure 1 Map showing all major oceanic plateaus formed within the last 150 Ma (Modified from Kerr, 2014).

draw on these examples to illustrate the salient features of oceanic plateaus and their formation.

Duration of formation

Both the Ontong Java and Caribbean plateaus have been formed during more than one phase of eruption, but geochronological data suggest that the vast bulk of the volcanism occurred in one short (<3 m.y.) main initial episode (like many similarly derived continental flood basalts), with subsequent magmatism in both plateaus being volumetrically minor. The main phase of Ontong Java eruption occurred within a few million years around 120 Ma, whereas the bulk of the Caribbean Plateau was formed around 93–89 Ma (see reviews in Kerr et al., 2003 and Fitton and Godard, 2004). Although recent $^{40}Ar-^{39}Ar$ dates have widened the age span of magmatism on the Kerguelen Plateau, very large volumes of magma were nevertheless generated in several pulses, 120–110 Ma and 105–100 Ma, over geologically short timescales (Duncan, 2002).

Structure

The crustal structure of oceanic plateaus can be deciphered from accreted plateau fragments (e.g., Kerr et al., 1997) and from seismic and gravity surveys on extant plateaus (e.g., Miura et al., 2004). In general, oceanic plateaus comprise a lowermost layer of predominantly olivine cumulates, overlain by isotropic gabbros. The base of the extrusive sequence is dominated by geochemically heterogeneous high-MgO lavas that are in turn succeeded upward by relatively homogeneous basaltic lavas (a similar structure has been proposed for continental flood basalt provinces).

Accreted oceanic plateau sections generally do not possess a sheeted feeder-dike complex, unlike many ocean ridge-type and supra-subduction zone ophiolites. Instead, the lava succession of oceanic plateaus is intruded by a significant number of thick sills or sheets (e.g., Kerr et al., 1997). In the accreted exposures of Ontong Java Plateau crust in the Solomon Islands, the lack of dikes has been interpreted either as evidence that most lavas have traveled a considerable distance or that the volume of magma being emplaced was too great to be accommodated by extension alone and that most magma was emplaced laterally in sills (Kerr and Mahoney, 2007).

Temperature of the source

The high rate and large volume of magma production in these plateaus and in many large continental flood basalt provinces imply relatively deep asthenospheric melting. For example if, as seems likely, the Ontong Java Plateau was the result of ~25 % peridotite partial melting, then the spherical mantle volume required to form this amount of melt would be more than 700 km across (i.e., deeper than the 660 km discontinuity) (Kerr and Mahoney, 2007).

The generation of such large volumes of magma as are observed in oceanic plateaus means that the source region must either be hotter, more fertile, or have a higher volatile content than ambient upper mantle, or undergo a large amount of decompression, or some combination of the above. However, the most common explanation for these high melt production rates is the mantle source temperatures significantly higher than the ambient upper mantle (see review in Campbell, 2007). Most geoscientists regard the source of this hot material to be mantle plumes that have risen from a thermo-geochemical boundary layer such as the core mantle boundary (see Hot Spots and Mantle Plumes). The evidence for elevated mantle source temperatures for oceanic plateaus is extensive:

Firstly, as reviewed by Kerr and Mahoney (2007), the eruption of high-MgO (>15 wt%) magmas (picrites and komatiites) in some oceanic plateaus, particularly in the Caribbean Plateau and in Precambrian accreted plateaus, provides evidence of mantle source temperatures well in excess of the pressure-corrected temperature of the normal upper asthenosphere (1,300–1,400 °C). Herzberg and Gazel (2009) calculated that high-MgO magmas from the Caribbean Plateau were derived from mantle sources with temperatures (corrected for pressure) up to 1,620 °C.

Secondly, Herzberg (2004) has used mantle melting phase relations to calculate that the Ontong Java primary magmas contained 17–19 wt% MgO and were derived from a mantle source with a temperature range of 1,500–1,560 °C. These figures are confirmed by trace element-based estimates of the mean degree of melting in the source region of the Ontong Java Plateau that range from 25 % to 30 %, and these are consistent with derivation from a mantle source with a temperature greater than >1,465 °C (Fitton and Godard, 2004).

Thirdly, numerical modeling of the petrological characteristics of large igneous provinces (e.g., Farnetani and Richards, 1994) indicates that mantle source temperatures well in excess of ambient upper mantle are required in order to generate the large melt volumes of these oceanic plateaus from peridotite.

Furthermore, in addition to these petrological constraints, several Cretaceous oceanic plateaus display significant evidence for subaerial or shallow marine eruption, resulting from seafloor uplift. This uplift is due to the thermal buoyancy of the plume and is consistent with sources hotter than the ambient upper mantle. In the Caribbean Plateau, these include shallow water sedimentary sequences with carbonized tree trunks, overlying accreted Caribbean Plateau basalts in Colombia (Kerr and Mahoney, 2007). In addition to this, volcaniclastic deposits near the top of the accreted Caribbean Plateau sequence on Aruba, in the southern Caribbean, comprise tuffs and ignimbrites that are consistent with eruption in a shallow marine or subaerial environment (Kerr and Mahoney, 2007).

Many of the drill holes into the Cretaceous Kerguelen Plateau show evidence of subaerial eruption, including oxidized flow tops, aa and pahoehoe flows, pyroclastic flows, and fluvial conglomerates. Furthermore, the presence of terrestrial and shallow marine sediments containing terrestrial plant remains, overlying the igneous basement at ODP Leg 183, Site 1138, provides substantial evidence for a prolonged period of subaerial exposure of the Kerguelen Plateau (Frey et al., 2003).

Drilling on the Ontong Java Plateau at Site 1184 during ODP Leg 192 discovered extensive basaltic phreatomagmatic deposits that, in conjunction with oxidized horizons and wood fragments, indicate a substantial phase of subaerial eruption on the eastern lobe of the plateau (Fitton and Godard, 2004), although evidence of significant uplift is lacking for other parts of the plateau (e.g., Tejada et al., 2002).

This evidence of subaerial exposure of oceanic plateaus is entirely consistent with upwelling from the deep mantle of hot buoyant material and associated dynamic uplift of the lithosphere above the ascending plume. This, in combination with the evidence from petrology and geochemistry, provides significant support for mantle plume models for the formation of oceanic plateaus (see Kerr, 2003).

Geochemical features of oceanic plateaus

Geochemically, most oceanic plateau rocks can be subdivided into three main groups:

1. Basalts and dolerites with a narrow compositional range and flat to slightly light rare earth element-enriched chondrite-normalized patterns. Most igneous rocks from Cretaceous oceanic plateaus have MgO contents ranging from ~6 to 11 wt%, with most lavas possessing a relatively restricted MgO range of 6.5–8.5 wt%. Concentrations of incompatible elements are similar to those in mid-ocean-ridge basalts, whereas radiogenic isotope ratios are broadly ocean-island-like (see reviews in Kerr and Mahoney, 2007; Kerr, 2014). Virtually all sampled Ontong Java Plateau lavas are of this type, as are many of the lavas of the Caribbean and Kerguelen plateaus.
2. Basalts with high La/Nb ratios and high initial $^{87}Sr/^{86}Sr$ (>0.705), along with low $^{143}Nd/^{144}Nd$ (<0.5127). Many early Kerguelen Plateau lavas possess this signature that is consistent with contamination by continental crust (see review in Kerr, 2014).
3. Picrites and komatiites with high-MgO contents (>12 wt%), which have a wider range of isotope ratios and incompatible trace element concentrations and

ratios than group 1. Such rocks are widespread in the Caribbean Plateau (e.g., Kerr et al., 2003).

Explaining compositional differences between the groups

High-MgO rocks have been found in accreted sections of the Caribbean Plateau but not (thus far) in the Ontong Java Plateau. Furthermore, the incompatible-element and isotopic compositions of high-MgO Caribbean Plateau lavas are more heterogeneous than those of lower-MgO basalts from both the Caribbean and Ontong Java plateaus.

It has been proposed that oceanic plateau source regions are heterogeneous and that the homogeneity of many plateau lavas and intrusions is a result of magma mixing, either in the melting column or in magma chambers (Kerr et al., 1997). The source region of the Caribbean Plateau was clearly markedly heterogeneous over much of its area and on a relatively small scale (Kerr et al., 2003). Such widespread and small-scale heterogeneity is also consistent with evidence from present-day lavas found at mid-ocean ridges and oceanic islands (e.g., some oceanic islands display greater isotopic and elemental heterogeneity than observed in the entire Ontong Java Plateau).

Why compositionally heterogeneous, high-MgO lavas have not been found on the Ontong Java Plateau but are relatively common in the Caribbean Plateau may well be related to differences in the thickness of the plateaus and in the depth to which the crust has been exhumed during accretion (Kerr and Mahoney, 2007). High-MgO, heterogeneous lavas are most likely to be erupted where there are few magma chambers and the magmatic plumbing system is poorly developed; in such areas, individual magma batches are more liable to ascend quickly from the source to the surface with little mixing in magma chambers. Therefore, in large igneous provinces, high-MgO lavas are most commonly found at the base of the succession. Later magmas are more likely to become trapped in magma chambers and so mix and fractionate (e.g., Kerr et al., 2003).

Accreted crustal sections provide the best way of studying the internal structure of oceanic plateaus, because they can expose deeper levels of the crust than those reachable by drilling,. However, due to the thicker lithosphere and the length of time between formation and accretion (Kerr and Mahoney, 2007), only the topmost 3–4 km of the Ontong Java Plateau basement crop out in large accreted thrust slices in the Solomon Islands (Petterson et al., 1997). Therefore, the deeper plateau sections that, according to the model of Kerr (2003), will be markedly heterogeneous, and the more Mg-rich lavas are not exposed on the Ontong Java Plateau.

Tejada et al. (2004) has proposed an alternative model to explain the homogeneity of the basalts of the Ontong Java Plateau, which suggests that the mantle source of the Ontong Java Plateau was essentially homogeneous at the very large scale and extent of melting. Tejada et al. (2004) argued that although mixing of more heterogeneous magmas in large open-system magma chambers could partly explain this geochemical homogeneity, mixing would always have to result in essentially the same composition and so reasoned that this was unlikely given that there is little evidence for magma mixing in the isotope and trace element data.

Given the great volume of melting and the large fractions of partial melting involved, it is probable that the melting process played a key role in averaging out smaller-scale source heterogeneity to produce the remarkable isotopic and incompatible-element homogeneity of basalts from the Ontong Java Plateau. Perhaps significantly, the estimated extents of melting for most of the Caribbean Plateau are lower than for the Ontong Java, a feature that would have resulted in less homogenization of melts in the mantle (Kerr and Mahoney, 2007). A similar conclusion was also reached by Heydolph et al. (2014) to explain the geochemical homogeneity of initial volcanism at the Shatsky Rise LIP.

As noted above, some of the oldest lavas drilled on the Kerguelen Plateau possess geochemical features consistent with contamination by continentally derived material. These continental fragments are likely to have been incorporated within the Indian Ocean lithosphere during the breakup of Gondwana prior to plateau formation. During formation of the Kerguelen Plateau, partial melts from these fragments would have contaminated the plateau magmas. This model has been confirmed by the discovery at ODP Site 1137 of a conglomerate containing Precambrian garnet–biotite gneiss cobbles (see review in Frey et al., 2002). Furthermore, seismological evidence also indicates the presence of continental crust in parts of the plateau, although the overall amount is not well established (e.g., Borissova et al., 2003).

Buoyancy and accretion of oceanic plateaus

The main reason for the elevated nature of oceanic plateaus (~2–3 km above the abyssal ocean floor) is the thickness of their basaltic crustal sections, which is generally much greater than the 6–7 km of normal oceanic crust generated at spreading centers. For instance, the Ontong Java Plateau has an average crustal thickness of ~33 km (e.g., Richardson et al., 2000), whereas the Caribbean Plateau crust varies from 8 to 15 km thick (see review in Kerr et al., 2003).

This thicker basaltic crust results in oceanic plateaus being more buoyant than the surrounding ocean floor and is the key reason why plateaus often resist subduction and so are often accreted to continental margins, thereby increasing their likelihood of preservation in the geological record. The extent of the buoyancy depends on both crustal thickness and plateau age. Plateaus that collide with a subduction zone only a few million years after they form will be more likely to resist subduction than a plateau of similar thickness that encounters a subduction zone many millions of years later (e.g., Cloos, 1993). This is related to the fact that oceanic crust becomes less buoyant as it cools.

The Caribbean Plateau largely resisted subduction because it collided with an arc at the entrance to the proto-Caribbean seaway only a few million years after the plateau formed at ~93–89 Ma (Kerr et al., 2003). In contrast, the Ontong Java Plateau, most of which formed at ~120 Ma, collided with the Solomon Islands subduction zone ~100 m.y. after its formation (e.g., Petterson et al., 1997). In this case, the plateau has resisted subduction because of a combination of very thick crust (Cloos, 1993), an anomalously small amount of post-emplacement subsidence, and a >300 km-thick mantle "root" that has helped to choke subduction (Wessel and Kroenke, 2000). Thus, only the topmost 3–4 km of the Ontong Java Plateau are uplifted and exposed on the Solomon Islands (Petterson et al., 1997), whereas the much deeper crustal regions of the more buoyant Caribbean Plateau are exposed in accreted sections around its margins and in Colombia (Kerr et al., 2003).

When an oceanic plateau collides with a subduction zone, various things can happen (Figure 2):

1. At an ocean–ocean convergent margin, the subduction direction (polarity) can reverse (the so-called subduction flip; Figure 2a). This happened in the Solomon Islands subduction zone, following the collision with the Ontong Java Plateau (e.g., Petterson et al., 1997).
2. In addition to subduction polarity reversal at its leading edge, the collision of a plateau with an island arc can result in a "back-stepping of subduction" behind the accreting plateau where a new subduction zone is initiated (Figure 2a). This occurred as the Caribbean Plateau was tectonically emplaced between North and South America and has resulted in the present-day Central American arc and the formation of the Caribbean Plate (Kerr et al., 2003).
3. Collision of an oceanic plateau with an active continental margin also results in subduction back-stepping, but without subduction polarity reversal (Figure 2b). This happened in the Late Cretaceous when part of the Caribbean Plateau collided with the northwestern margin of South America, leading to the preservation of oceanic plateau crust in accreted blocks (Kerr et al., 2003). Indeed, it is clear that oceanic plateau accretion has occurred several times along this margin; portions of at least two or three plateaus appear to comprise accreted terrains that are commonly grouped with the Caribbean Plateau (see Kerr and Tarney, 2005).

Plateaus that collide with intra-oceanic island arcs can also eventually be accreted onto and incorporated into continental crust. Consequently, it has been suggested that the accretion of oceanic plateaus may have contributed to continental crustal growth throughout much of geological time (e.g., Abbott and Mooney, 1995; Tetreault and Buiter, 2012). Sequences interpreted to be of oceanic

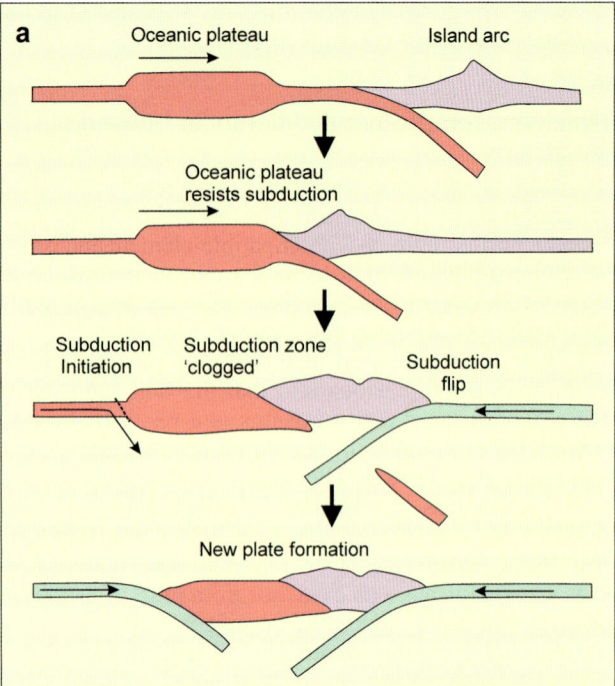

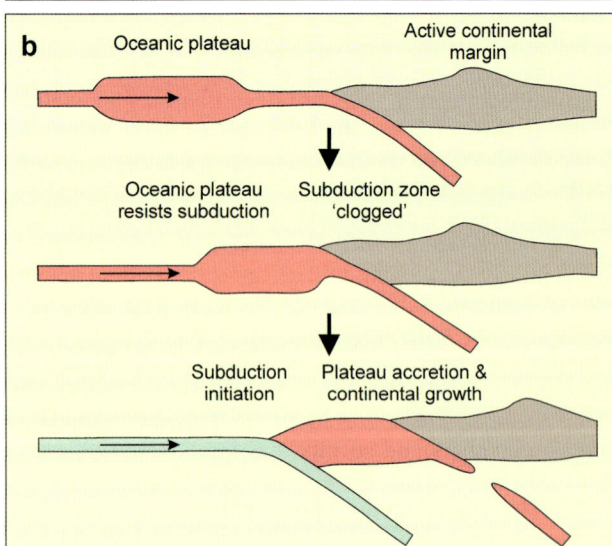

Oceanic Plateaus, Figure 2 Idealized cross sections illustrating the likely effects of the collision of an oceanic plateau with (**a**) an island arc and (**b**) an active continental margin (Modified from Kerr and Mahoney, 2007).

plateau origin have been argued to represent an important component both in the upper and lower crust of Precambrian greenstone belts (see review in Kerr, 2003)

Identifying oceanic plateaus in the geological record

Oceanic plateaus are therefore likely to have been accreted to the continents throughout much of geological time.

The diagnostic geochemical and geological features of Cretaceous oceanic plateaus can be used to help distinguish plateau sequences in the geological record from those originating in other tectonic settings. These diagnostic features are reviewed below and more fully discussed in Kerr et al. (2000).

Igneous rocks formed in any subduction zone setting are relatively easily distinguished from oceanic plateau sequences because arcs possess generally more evolved lavas, with high La/Nb ratios, and only occasionally high-MgO lavas. Additionally, oceanic plateaus do not possess the abundant volcanic ash layers present in volcanic arc sequences. However, a La/Nb ratio ~1 is not an entirely robust discriminant of an oceanic plateau sequence, given that the Kerguelen oceanic plateau often possesses low La/Nb values, due to magma interaction with the continental lithosphere beneath the plateau. This illustrates the point that geochemical discriminants of tectonic setting should not be used in isolation and that other geological evidence should also be considered. In the case of Kerguelen, the lack of volcaniclastic horizons and the thickened oceanic crust help confirm its oceanic plateau affinity.

Most mid-ocean ridge basalts possess light rare earth element-depleted chondrite-normalized patterns in contrast to the generally flat patterns of oceanic plateau lavas. Furthermore, high-MgO lavas are much more common in oceanic plateaus than at mid-ocean ridges.

The geochemistry of igneous rocks is only of limited use in distinguishing between volcanic successions formed in back-arc basins from those formed in oceanic plateaus. However, the mantle temperature below a back-arc basin (i.e., ambient mantle) is lower than a mantle plume (200–300 °C hotter than ambient mantle), and this results in the eruption of relatively few high-MgO lavas. Furthermore, because of their proximity to active subduction zones, back-arc basin sequences will also be more likely to contain abundant volcaniclastic horizons than oceanic plateaus.

Examples of accreted oceanic plateaus

On the basis of many of the criteria outlined above, a significant number of accreted oceanic plateaus have been identified in the geological record (Kerr, 2014). Examples of some of the more significant plateaus are briefly reviewed below:

Wrangellia oceanic plateau

The accreted Wrangellia plateau, comprising basalts, with minor picrites, is discontinuously exposed along the continental margin of Alaska and Canada over a distance of ~2,300 km. The plateau formed in the Panthalassa Ocean ca. 230–225 Ma, in a progressively shallowing marine environment, which eventually became subaerial (see reviews in Greene et al., 2008; Greene et al., 2009a; Greene et al., 2009b). The plateau accreted to western North America in the Late Jurassic–Early Cretaceous.

Most of the Wrangellia plateau basalts contain 5–9 wt% MgO and are light rare earth element enriched, unlike the Ontong Java Plateau and most of the Caribbean Plateau. In contrast, many of the lower basalts are light rare earth element depleted with MgO contents that extend to more magnesian (6–12 wt% MgO) compositions (Greene et al., 2008; Greene et al., 2009a; Greene et al., 2009b). Modeling suggests that the magmas were derived by 23–27 % partial melting of a mantle source ~200 °C hotter than ambient mantle, thus supporting a hot mantle plume origin for the Wrangellia plateau (Greene et al., 2009b).

Carboniferous to Cretaceous oceanic plateaus accreted in Japan

Japan comprises a series of terrains that have been accreted to the Eurasian plate over the past 400 Ma. The Chugoku and Chichibu belts in southwest Japan contain up to 30 % basaltic material. Geochemical data suggests that these basaltic assemblages are remnants of a Carboniferous plume-derived oceanic plateau that formed in the Panthalassa Ocean (Tatsumi et al., 2000).

Permian greenstones of the Mino–Tamba accretionary complex found in southwest Japan comprise pillowed and massive basalt flows with minor picrites, with similar isotopic and trace element compositions (Ichiyama et al., 2008) to the Ontong Java and Caribbean plateaus. Ichiyama et al. (2008) proposed that these basalts were formed by partial melting of a shallow mantle plume head below the thick oceanic lithosphere in the Early Permian.

In contrast to the other accretionary belts in Japan, the Sorachi–Yezo terrain is dominated by oceanic crust and lithosphere (Kimura et al., 1994; Ichiyama et al., 2012). It comprises pillowed basalts with occasional picrites and dolerite sills of Tithonian age (150–145 Ma), which have been interpreted to represent an accreted oceanic plateau. This Jurassic–Early Cretaceous plateau (named the Sorachi Plateau; Kimura et al., 1994) is the same age as the Shatsky Rise (Figure 1), suggesting that the Sorachi Plateau and the Shatsky Rise were originally a single plateau that formed near the Kula–Pacific–Farallon triple junction at ~150 Ma (Kimura et al., 1994; Ichiyama et al., 2014). The Sorachi part of the plateau was carried northwestward on the Kula Plate and collided with the Eurasian continental margin in the mid-Cretaceous.

Precambrian oceanic plateaus

As already noted, the accreted oceanic plateau material has been a significant contributor to crustal growth. Some of the oldest oceanic plateau sequences comprise pillow basalts and komatiites found in the ~3.5 Ga greenstone belts of the Kaapvaal Craton of Southern Africa. The geochemical signatures of these rocks suggest an oceanic plateau origin (Chavagnac, 2004) and, along with mafic rocks found in the Pilbara Craton of Australia, they represent some of the oldest oceanic plateau material on Earth (see review in Kerr, 2003).

Greenstone belts of the Canadian Superior province and the Baltic Shield (3.0–2.7 Ga) consist of many lava formations that have been interpreted to be remnants of accreted oceanic plateaus. The evidence for an oceanic plateau origin is based on the occurrence of pillow basalts and komatiites with no terrestrial sedimentary intercalations or sheeted dike swarms, possessing broadly chondritic La/Nb ratios and low positive ε_{Nd} (see reviews in Ernst and Buchan, 2001; Kerr, 2003).

Proterozoic oceanic plateau terrains have also been identified in the Birimian Province of Western Africa, the Arabian–Nubian Shield, the Flin Flon Belt in Canada, and the Iron King Volcanics in Arizona. More detailed summaries of these accreted terrains and other Precambrian oceanic plateaus are given in Ernst and Buchan (2001) and Kerr (2014).

Environmental impact of oceanic plateau formation

Many periods of oceanic environmental crisis in Earth's history are marked by significant black shale deposits, which indicate oxygen-poor conditions in the deep ocean, i.e., oceanic anoxic events (OAEs). In the Cretaceous, these events are associated with periods of mass extinction and oceanic plateau formation particularly around the Cenomanian–Turonian boundary (93.5 Ma) and in the Aptian (124–112 Ma).

Cenomanian–Turonian Boundary (CTB) event

As reviewed in detail by Kerr (1998, 2005), the CTB event is characterized by the worldwide deposition of organic-rich black shales, indicating widespread anoxia in the oceans at this time. The CTB is also marked by a rise in sea-level and significantly higher average global surface temperatures than the present day. These higher temperatures are most likely due to an increase in global atmospheric CO_2 to 3–5 times higher than preindustrial levels (Fletcher et al., 2008). These phenomena were accompanied by an extinction event that resulted in the demise of 26 % of all known genera (Raup and Sepkoski, 1986).

The CTB event has been linked to the formation of oceanic plateaus (Kerr, 1998, 2005; Kuroda et al., 2007). An oceanic plateau trace element signature in black shales was first noted by Kerr (1998) and was subsequently reported by Snow et al. (2005). More recent isotopic studies (Kuroda et al., 2007) on CTB black shales from Italy showed that their Pb isotope signatures are significantly shifted toward the geochemical composition of the Caribbean plateau, thus implying a substantial input from the plateau.

Links between CTB oceanic plateau volcanism and environmental perturbation

Most plume-related volcanism around the CTB occurred in the oceans, with the formation of significant portions of the Caribbean, Ontong Java, and Kerguelen plateaus. The probable worldwide environmental effects of such volcanism are reviewed below and summarized in Figure 3.

The net result of the emplacement of millions of cubic kilometers of magma both into and onto the preexisting ocean crust in combination with the uplift associated with buoyant plume head impingement on the base of the oceanic lithosphere will be to raise the sea-level, due to the displacement of seawater. The progressive rise in global sea-level throughout the mid-Cretaceous (Haq et al., 1987) may signal the gradual impingement of the Caribbean, Ontong Java, and Kerguelen plume heads below the oceanic lithosphere. This uplift of oceanic lithosphere and subsequent magma emplacement are also likely to have resulted in the disruption of oceanic circulation systems such that cool, polar (oxygenated) water was not circulated to lower latitudes, leading to increased anoxia in the oceans (Kerr, 1998). In addition to the CO_2-induced, elevated, global oceanic temperatures, hydrothermal fluids from oceanic plateau volcanism may also have contributed to warmer oceans and, thus, to anoxia, since oxygen is significantly less soluble in warmer water (see review in Kerr, 2005).

Calculations suggest that increased volcanic activity around the CTB is likely to have added ~10^{17} kg of CO_2 to the atmosphere (Kerr, 1998). This volcanic activity probably also released substantial amounts of SO_2, chlorine, fluorine, and H_2S into the seawater, resulting in the CTB oceans being more acidic. This could have resulted in the dissolution of oceanic carbonate and may help to explain the lack of carbonate at the CTB (Kerr, 1998, 2005), which would also release more CO_2 to the atmosphere. Higher atmospheric CO_2 levels and nutrient upwelling from the deep ocean is likely to have resulted in increased productivity in ocean surface waters (Figure 3), leading to the widespread deposition of black shales and, so, ultimately a reduction in CO_2 levels (Jarvis et al., 2011).

Other links between oceanic plateau formation and black shale formation

Black shales are relatively common in the geological record and are generally associated with, sometimes severe, environmental disruption (Wignall, 1994; Bond and Wignall, 2014). It is highly significant that these black shale events also correlate with the formation of oceanic plateaus or plume-related volcanic rifted margins (Kerr, 2014).

During the Aptian–Albian (125–100 Ma), three distinct oceanic anoxia events (with associated black shales) occurred, indicating a prolonged period of global environmental disturbance (Wignall, 1994). The Aptian–Albian was also a major period of oceanic plateau formation in the Pacific and Indian Oceans (Coffin and Eldholm, 2005). Although many of the geochemical features of the Aptian–Albian OAE are very similar to those of the CTB, one of the black shale horizons in the mid-Aptian shows sharp decrease in $\delta^{13}C$. This decrease is consistent

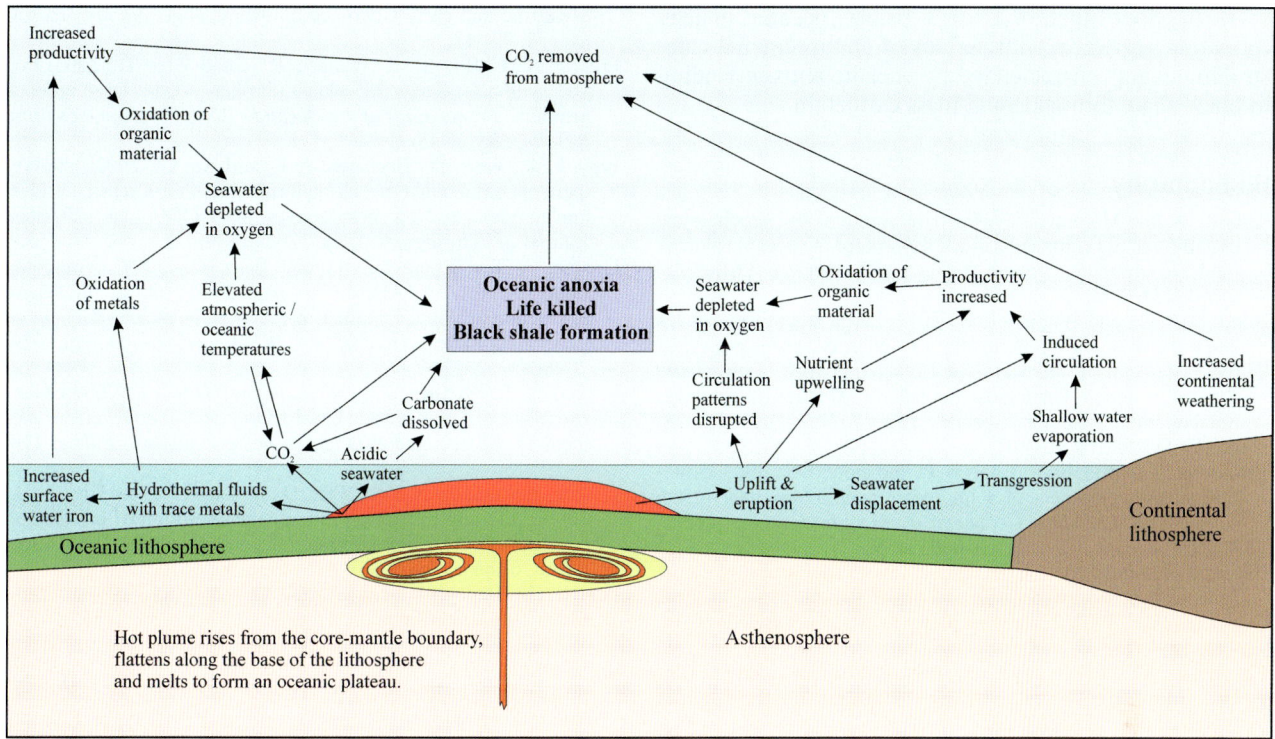

Oceanic Plateaus, Figure 3 Diagram summarizing the possible physical and chemical environmental effects of the formation of large igneous provinces around the Cenomanian–Turonian boundary (Modified from Kerr, 2005).

with the catastrophic release of methane hydrates from ocean floor sediments and would have substantially contributed to global warming and oceanic anoxia in the mid-Aptian (Jahren et al., 2005).

Significant black shale formations are found throughout the Mesozoic and these usually correlate with oceanic plateau (or rifted margin) volcanism, e.g., Kimmeridgian to Tithonian (155–146 Ma) black shales formed at the same time as the Sorachi Plateau in the Western Pacific. Furthermore, Toarcian (187–178 Ma) black shales are coeval with the eruption of the Karoo, Ferrar, and Weddell Sea large igneous province during the breakup of Gondwana (Kerr, 2003). Furthermore, Condie et al. (2001) have proposed that the correlation between black shales, paleoclimatic disturbance, and mantle superplume events extends into the Precambrian and have noted that there is a particularly good correspondence between these events at 1.9 and 2.7 Ga.

The recognition that oceanic anoxia and black shale deposition may have been a result of oceanic plateau formation has important implications for the location of potential oil source rocks in the geological record. It is probable that many of the world's Cretaceous oil source rocks owe their existence to the emplacement of oceanic plateaus that caused widespread global anoxia (Kerr et al., 2003).

Thus, the formation of oceanic plateaus can have far-reaching environmental effects and can set in motion a chain of events that often leads to global warming, oceanic anoxia, black shale deposition, and, ultimately, mass extinction.

Summary

Oceanic plateaus are large areas of over-thickened oceanic crust that are generally regarded to have formed by decompression melting of hot mantle plumes. Oceanic plateaus are found throughout the geological record. The thick crustal sections of oceanic plateaus are difficult to subduct. As a result the upper crustal portions of oceanic plateaus are frequently accreted to continental margins and so have been an important contributor to crustal growth throughout Earth's history.

Much of our knowledge of the composition, origin, and structure of oceanic plateaus has come from our study of Cretaceous oceanic plateaus such as the Ontong Java, Caribbean, and Kerguelen plateaus. The homogeneous basalts of the Ontong Java Plateau either indicate a homogeneous mantle source region or, more likely, homogenization of heterogeneous magmas during mantle melting. In contrast, the mantle source region of the Caribbean Plateau is markedly heterogeneous, and high-MgO lavas are abundant. Initial magmatism on the Kerguelen Plateau is associated with the breakup of Gondwana in the Early Cretaceous, and as a result the early plateau, lavas are contaminated with continental crust.

Throughout much of Earth's history, oceanic plateau formation correlates with periods of environmental catastrophic events characterized by oceanic anoxia, leading to black shale formation and mass extinction.

Bibliography

Abbott, D., and Mooney, W., 1995. The structural and geochemical evolution of the continental crust: support for the oceanic plateau model of continental growth. *Reviews of Geophysics*, **33**(Supplement), 231–242.

Bond, D. P. G., and Wignall, P. B., 2014. Large igneous provinces and mass extinctions: an update. In Keller, G., and Kerr, A. C. (eds.), *Volcanism, Impacts, and Mass Extinctions: Causes and Effects*. Boulder: Geological Society of America. Geological Society of America Special Paper, 505, pp. 29–55, doi:10.1130/2014.2505(02).

Borissova, I., Coffin, M. F., Charvis, P., and Operto, S., 2003. Structure and development of a microcontinent: Elan Bank in the southern Indian Ocean. *Geochemistry, Geophysics, Geosystems*, **4**, 1071, doi:10.1029/2003GC000535, 9.

Bryan, S. E., and Ernst, R. E., 2008. Revised definition of Large Igneous Provinces (LIPs). *Earth Science Reviews*, **86**, 175–202.

Campbell, I. H., 2007. Testing the plume theory. *Chemical Geology*, **241**, 153–176.

Chavagnac, V., 2004. A geochemical and Nd isotopic study of Barberton komatiites (South Africa): implication for the Archean mantle. *Lithos*, **75**, 253–281.

Cloos, M., 1993. Lithospheric buoyancy and collisional orogenesis: subduction of oceanic plateaus, continental margins, island arcs, spreading ridges, and seamounts. *Geological Society of America Bulletin*, **105**, 715–737.

Coffin, M. F., and Eldholm, O., 1992. Volcanism and continental break-up: a global compilation of large igneous provinces. In Storey, B. C., Alabaster, T., and Pankhurst, R. J. (eds.), *Magmatism and the Causes of Continental Breakup*. London: Geological Society of London. Special publication, pp. 17–30.

Coffin, M. F., and Eldholm, O., 2005. Large igneous provinces. In Selley, R. C., Cocks, R., and Plimer, I. R. (eds.), *Encyclopedia of Geology*. Oxford: Elsevier, pp. 315–323.

Condie, K. C., Marais, D. J. D., and Abbott, D., 2001. Precambrian superplumes and supercontinents: a record in black shales, carbon isotopes, and paleoclimates? *Precambrian Research*, **106**, 239–260.

Donnelly, T. W., 1973. Late Cretaceous basalts from the Caribbean, a possible flood basalt province of vast size. *Eos*, **54**, 1004.

Duncan, R. A., 2002. A time frame for construction of the Kerguelen Plateau and Broken Ridge. *Journal of Petrology*, **43**, 1109–1119.

Ernst, R. E., and Buchan, K. L., 2001. Large mafic magmatic events through time and links to mantle plume heads. In Ernst, R. E., and Buchan, K. L. (eds.), *Mantle Plumes: Their Identification Through Time*. Boulder: Geological Society of America. Geological Society of America, Special Paper, 352, pp. 483–575.

Farnetani, C. G., and Richards, M. A., 1994. Numerical investigations of the mantle plume initiation model for flood basalt events. *Journal of Geophysical Research – Solid Earth*, **99**(B7), 13813–13833.

Fitton, J. G., and Godard, M., 2004. Origin and evolution of magmas on the Ontong Java Plateau. In Fitton, J. G., Mahoney, J. J., Wallace, P. J., and Saunders, A. D. (eds.), *Origin and Evolution of the Ontong Java Plateau*. London: Geological Society. Geological Society of London, Special publication, 229, pp. 151–178.

Fletcher, B. J., Brentnall, S. J., Anderson, C. W., Berner, R. A., and Beerling, D. J., 2008. Atmospheric carbon dioxide linked with Mesozoic and early Cenozoic climate change. *Nature Geoscience*, **1**, 43–48.

Frey, F. A., Coffin, M. F., Wallace, P. J., and Weis, D., 2003. Leg 183 synthesis: Kerguelen plateau-broken ridge-a large igneous province. *Proceedings of the Ocean Drilling Program, Scientific Results*, **183**, 1–48.

Frey, F. A., Weis, D., Borisova, A.Y., and Xu, G., 2002, Involvement of continental crust in the formation of the Cretaceous Kerguelen Plateau: New perspectives from ODP Leg 120 sites: Journal of Petrology, v. 43, p. 1207–1239.

Greene, A. R., Scoates, J. S., and Weis, D., 2008. Wrangellia flood basalts in Alaska: a record of plume-lithosphere interaction in a Late Triassic accreted oceanic plateau. *Geochemistry, Geophysics, Geosystems*, **9**(12), Q12004, doi:10.1029/2008GC002092.

Greene, A. R., Scoates, J. S., Weis, D., and Israel, S., 2009a. Geochemistry of Triassic flood basalts from the Yukon (Canada) segment of the accreted Wrangellia oceanic plateau. *Lithos*, **110**, 1–19.

Greene, A. R., Scoates, J. S., Weis, D., Nixon, G. T., and Kieffer, B., 2009b. Melting history and magmatic evolution of basalts and picrites from the accreted Wrangellia oceanic plateau, Vancouver Island, Canada. *Journal of Petrology*, **50**, 467–505.

Haq, B. U., Hardenbol, J., and Vail, P. R., 1987. Chronology of fluctuating sea levels since the Triassic. *Science*, **235**, 1156–1167.

Herzberg, C., 2004. Partial melting below the Ontong Java Plateau. In Fitton, J. G., Mahoney, J. J., Wallace, P. J., and Saunders, A. D. (eds.), *Origin and Evolution of the Ontong Java Plateau*. London: Geological Society. Geological Society of London, Special publication, 229, pp. 179–184.

Herzberg, C., and Gazel, E., 2009. Petrological evidence for secular cooling in mantle plumes. *Nature*, **458**, 619–622.

Heydolph, K., Murphy, D. T., Geldmacher, J., Romanova, I. V., Greene, A., Hoernle, K., Weis, D., and Mahoney, J., 2014. Plume versus plate origin for the Shatsky Rise oceanic plateau (NW Pacific): insights from Nd, Pb and Hf isotopes. *Lithos*, **200**, 49–63.

Ichiyama, Y., Ishiwatari, A., and Koizumi, K., 2008. Petrogenesis of greenstones from the Mino-Tamba belt, SW Japan: evidence for an accreted Permian oceanic plateau. *Lithos*, **100**, 127–146.

Ichiyama, Y., Ishiwatari, A., Kimura, J.-I., Senda, R., Kawabata, H., and Tatsumi, Y., 2012. Picrites in central Hokkaido: evidence of extremely high temperature magmatism in the Late Jurassic ocean recorded in an accreted oceanic plateau. *Geology*, **40**, 411–414.

Ichiyama, Y., Ishiwatari, A., Kimura, J.-I., Senda, R., and Miyamoto, T., 2014. Jurassic plume-origin ophiolites in Japan: accreted fragments of oceanic plateaus. *Contributions to Mineralogy and Petrology*, **168**, 1–24.

Jahren, A. H., Conrad, C. P., Arens, N. C., Mora, G., and Lithgow-Bertelloni, C., 2005. A plate tectonic mechanism for methane hydrate release along subduction zones. *Earth and Planetary Science Letters*, **236**, 691–704.

Jarvis, I., Lignum, J. S., Gröcke, D. R., Jenkyns, H. C., and Pearce, M. A., 2011. Black shale deposition, atmospheric CO_2 drawdown, and cooling during the Cenomanian-Turonian Oceanic Anoxic Event. *Paleoceanography*, **26**, PA3201, doi:10.1029/2010PA002081.

Kerr, A. C., 1998. Oceanic plateau formation: a cause of mass extinction and black shale deposition around the Cenomanian-Turonian boundary. *Journal of the Geological Society (London)*, **155**, 619–626.

Kerr, A. C., 2003. Oceanic plateaus. In Rudinck, R. L. (ed.), *The Crust. Treatise on Geochemistry* (Holland, H. G., Turekian, K. K. (eds.)). Oxford: Elsevier-Pergamon, Vol. 3, pp. 537–566.

Kerr, A. C., 2005. Oceanic LIPs: the kiss of death. *Elements*, **1**, 289–292.

Kerr, A. C. 2014. Oceanic plateaus. In: Rudnick, R. (ed.), The Crust. Chapter 18. *Treatise on Geochemistry*, 2nd edn (Holland, H. C., and Turekian, K. (Series eds.)). Elsevier, Amsterdam, Vol. 4, pp. 631–667.

Kerr, A. C., and Mahoney, J. J., 2007. Oceanic plateaus: problematic plumes, potential paradigms. *Chemical Geology*, **241**, 332–353.

Kerr, A. C., and Tarney, J., 2005. Tectonic evolution of the Caribbean and northwestern South America: the case for accretion of two Late Cretaceous oceanic plateaus. *Geology*, **33**, 269–272.

Kerr, A. C., Tarney, J., Marriner, G. F., Nivia, A., and Saunders, A. D., 1997. The Caribbean-Colombian Cretaceous igneous province: the internal anatomy of an oceanic plateau. In Mahoney, J. J., and Coffin, M. (eds.), *Large Igneous Provinces: Continental, Oceanic and Planetary Flood Volcanism*. Washington, DC: American Geophysical Union. American Geophysical Union Monograph, 100, pp. 45–93.

Kerr, A. C., White, R. V., and Saunders, A. D., 2000. LIP reading: recognizing oceanic plateaux in the geological record. *Journal of Petrology*, **41**, 1041–1056.

Kerr, A. C., White, R. V., Thompson, P. M. E., Tarney, J., and Saunders, A. D., 2003. No oceanic plateau – no Caribbean plate? The seminal role of an oceanic plateau in Caribbean plate evolution. In Bartolini, C., Buffler, R. T., and Blickwede, J. (eds.), *The Gulf of Mexico and Caribbean Region: Hydrocarbon Habitats, Basin Formation and Plate Tectonics*. Tulsa: AAPG. Memoir, 79, pp. 126–168.

Kimura, G., Sakakibara, M., and Okamura, M., 1994. Plumes in central Panthalassa? Deductions from accreted oceanic fragments in Japan. *Tectonics*, **13**(4), 905–916.

Kroenke, L. W., 1974. Origin of continents through development and coalescence of oceanic flood basalt plateaus. *Eos*, **55**, 443.

Kuroda, J., Ogawa, N. O., Tanimizu, M., Coffin, M. F., Tokuyama, H., Kitazato, H., and Ohkouchi, N., 2007. Contemporaneous massive subaerial volcanism and late Cretaceous Oceanic Anoxic Event 2. *Earth and Planetary Science Letters*, **256**, 211–223.

Miura, S., Suyehiro, K., Shinohara, M., Takahashi, N., Araki, E., and Taira, A., 2004. Seismological structure and implications of collision between the Ontong Java Plateau and Solomon Island Arc from ocean bottom seismometer-airgun data. *Tectonophysics*, **389**, 191–220.

Petterson, M. G., Neal, C. R., Mahoney, J. J., Kroenke, L. W., Saunders, A. D., Babbs, T. L., Duncan, R. A., Tolia, D., and McGrail, B., 1997. Structure and deformation of north and central Malaita, Solomon Islands: tectonic implications for the Ontong Java Plateau Solomon arc collision, and for the fate of oceanic plateaus. *Tectonophysics*, **283**, 1–33.

Raup, D. M., and Sepkoski, J. J., 1986. Periodic extinction of families and genera. *Science*, **231**, 833–836.

Richardson, W. P., Okal, E. A., and VanderLee, S., 2000. Rayleigh-wave tomography of the Ontong-Java Plateau. *Physics of the Earth and Planetary Interiors*, **118**, 29–51.

Snow, L. J., Duncan, R. A., and Bralower, T. J., 2005. Trace element abundances in the Rock Canyon Anticline, Pueblo, Colorado, marine sedimentary section and their relationship to Caribbean plateau construction and oxygen anoxic event 2. *Paleoceanography*, **20**(4), PA3005, doi:10.1029/2004PA001093.

Tatsumi, Y., Kani, T., Ishizuka, H., Maruyama, S., and Nishimura, Y., 2000. Activation of Pacific mantle plumes during the Carboniferous: evidence from accretionary complexes in southwest Japan. *Geology*, **28**, 580–582.

Tejada, M. L. G., Mahoney, J. J., Neal, C. R., Duncan, R. A., and Petterson, M. G., 2002. Basement geochemistry and geochronology of central Malaita, Solomon islands, with implications for the origin and evolution of the Ontong Java Plateau. *Journal of Petrology*, **43**, 449–484.

Tejada, M. L. G., Mahoney, J. J., Castillo, P. R., Ingle, S. P., Sheth, H. C., and Weis, D., 2004. Pin-pricking the elephant: evidence on the origin of the Ontong Java Plateau from Pb-Sr-Hf-Nd isotopic characteristics of ODP Leg 192 basalts. In Fitton, J. G., Mahoney, J. J., Wallace, P. J., and Saunders, A. D. (eds.), *Origin and Evolution of the Ontong Java Plateau*. London: Geological Society. Geological Society of London, Special publication, 229, pp. 133–150.

Tetreault, J. L., and Buiter, S. J. H., 2012. Geodynamic models of terrane accretion: testing the fate of island arcs, oceanic plateaus, and continental fragments in subduction zones. *Journal of Geophysical Research*, **117**(B8), B08403, doi:10.1029/2012JB009316.

Wessel, P., and Kroenke, L. W., 2000. Ontong Java Plateau and Late Neogene changes in Pacific plate motion. *Journal of Geophysical Research*, **105**, 28255–28278.

Wignall, P. B., 1994. *Black Shales*. Oxford: Oxford University Press. Geology and Geophysics Monographs, 30. 130pp.

Cross-references

Anoxic Oceans
Hot Spots and Mantle Plumes
Intraplate Magmatism
Mid-ocean Ridge Magmatism and Volcanism
Volcanism and Climate

OCEANIC RIFTS

Adolphe Nicolas
Geosciences Montpellier, University of Montpellier 2, Montpellier, France

Definition and historical review

Rifts share comparable morphologies whether they start as continental rifts, evolve to oceanic rifts, or, successfully, open to become a mid-oceanic ridge. They are elongated with an axial valley a few tens of kilometers wide, below symmetrical shoulders some 2,000 m higher than the regional elevation. This relief is thermally supported by the upwelling of the asthenospheric mantle through the thinning lithosphere. However, it is now realized that due to evolution of tectonically hyperextended systems, such architectures may be absent even in extensive continental rifts such as the North Sea. The rift upwelling triggers partial melting in the upper mantle and, in turn, volcanism in the rift. The rift itself results from collapsing of the shoulders' vault, along normal faults implying some tectonic extension.

These attributes are common to two modes of mantle behavior during rifting, being "passive" or "active," respectively. "Passive" rifting is induced by stretching and rupturing of the continental lithosphere that triggers the upwelling of the underlying asthenosphere (Mckenzie, 1978) (Figure 1a). Passive rifting has been successfully modeled both numerically (Huismans and Beaumont, 2007; Sachau and Koehn, 2010) and by analogic experiments (Brun and Beslier, 1996). It has received the support of field

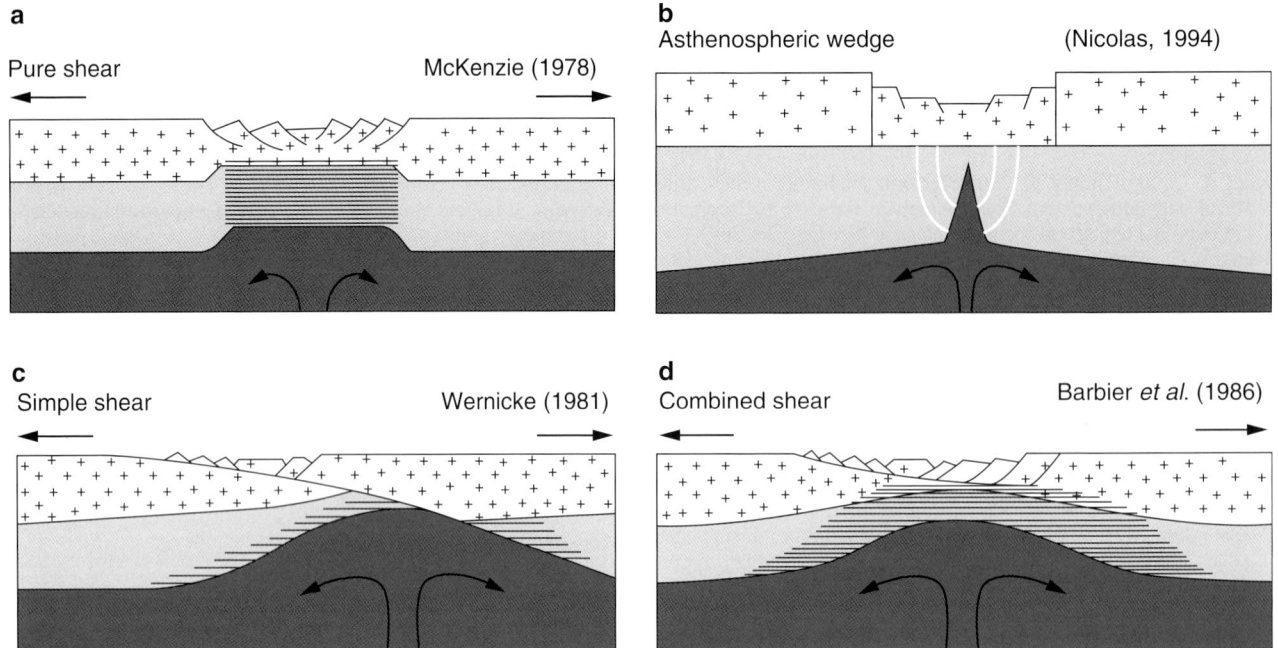

Oceanic Rifts, Figure 1 Theoretical models of rifting. (a) and (b) symmetrical rifting, (a) passive with homogeneous stretching of lower crust and upper lithosphere, inducing an asthenospheric upwelling. (b) Active rifting above a mantle plume. (c) Asymmetrical simple shear along a lithospheric weak zone, also inducing passive asthenosphere upwelling. (d) Combination of simple shear and stretching of lower crust and upper lithosphere, also with passive asthenosphere upwelling.

observations such as in the Rio Grande Rift where stretching preceded volcanic activity in the rift, in contrast with the East African Rift where asthenosphere upwelling preceded stretching (Allen and Allen, 2005). However, in a "normal," homogeneous, continental lithosphere, with a temperature at Moho depth of only ~500 °C, it is not expected that mantle peridotites would yield to the horizontal far-field tensile stress applied to the lithosphere (Nicolas et al., 1994). "Active" rifting is initiated with a local upwelling of an underlying, hot mantle plume (Figure 1b). Another attribute of active rifting should be a copious magmatic activity in the rift because twice more melt is generated by a 100 °C overheated mantle plume compared to normal rising asthenosphere (White and McKenzie, 1989). Provided that this upwelling occurred in an area of tensile stress, rifting can occur (otherwise, the plume builds a major volcanic edifice, such as the Hoggar). Rift initiation may postdate an episode of thermal erosion of the lithosphere by the mantle plume. Spohn and Schubert (1982) predict that a plume can reduce the lithosphere by a factor of two in 30 Myr. Rupture in this thinned and heated lithosphere is now possible and a wedge of asthenospheric mantle can rise, resulting in rift opening (McMullen and McMorhaz, 1989; Nicolas et al., 1994).

If the affected lithosphere is homogeneous, the rifting should be symmetrical, although symmetry versus asymmetry depends also on other factors such as the rates of extension. If the lithosphere was heterogeneous, for instance, guided in continental lithosphere by mechanical or thermal weakness lines, rifting should be asymmetrical (Wernicke, 1981) (Figure 1c). In marine environments, such as the slow-spreading Atlantic Ocean, the analogues are the oceanic core complexes (Tucholke et al., 2008). Rifting may be more complex, combining pure and simple shear (Figure 1d).

We describe briefly below the Northern Red Sea, illustrating the situation of an active and symmetrical rifting, and next, the situation of a passive and asymmetrical rifting with the Galicia Bank where the Atlantic Ocean first opened. The western margins of the Atlantic Ocean in Newfoundland have been classically considered as illustrating a magma-rich system, contrasting with the magma-poor eastern margin of the Galicia Bank, rifting by a major tectonic extension and thinning. Recently, it has been recognized that such relatively simple models evolve in time, with hyperextension affecting the ocean-continent margins. Finally, the views developed here by the author may be biased because of a better knowledge of the deep mantle and lower crust structures of rifts.

The Northern Red Sea: an active rift

The Red Sea oceanic rift opens and propagates northward between thinned continental crusts, since ~11 Ma (Hempton, 1987; Izzeldin, 1987). In its northern part considered here, at its initial stage, rifting was symmetrical (Buck et al., 1988). The rift is narrow with small oceanic basins filled by hot metalliferous brines, between

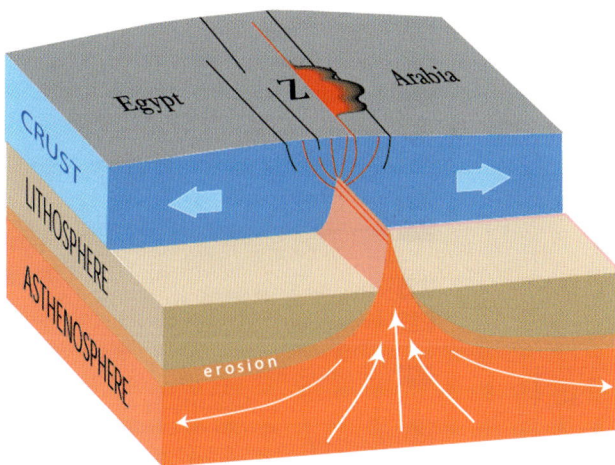

Oceanic Rifts, Figure 2 Early lithospheric thermal erosion above a mantle plume. At that time, the plume was located below Egypt. Lithospheric fracturing opened conduits for rising basaltic melt from the melting asthenospheric intrusion (*red*), beneath the lithosphere (*brown*) and continental crust (*blue*), stretched and invaded by basaltic dikes. This intrusion penetrated the crust and opened a failed rift, the Zabargad Miocene marine basin (sediments in *yellow*, basalt sills in *deep blue*). Opening of the Red Sea occurred 10 to 15 My later, with stretching and fracturing of the preheated continental crust (Nicolas et al., 1994).

attenuated continental crust shoulders. Nowadays, this Red Sea rift expands northwards in a continental lithosphere, being fed by a hot spot located in the Southern Red Sea, at the triple junction with Aden Gulf and the Afar continental rift. The hot spot volcanic activity is 30 Ma old (Jones, 1976). In the Southern Red Sea, rifting is more complex with asymmetrical features (Buck et al., 1988; Coleman and McGuire, 1988; Dixon et al., 1989).

In the Northern Red Sea, a first episode of failed rifting opened small, elongated basins during Miocene, 23 Myr ago (Cochran and Martinez, 1984). Such an early episode of rifting is also recorded in powerful injections of N-S diabase dikes and intrusions (Coleman and McGuire, 1988; Baldridge et al., 1991). Narrow basins, locally opening within the attenuated continental margin of the Red Sea, anticipate the subsequent and successful Red Sea northwards rifting. One of those basins opened in the western margin of the Red Sea is now the Zabargad Island standing over a large gravity high. It displays a peridotite massif, a piece of former and melting asthenosphere that intruded and heated the lower continental crust. It was accompanied by copious diabase diking, parallel to dikes in the Northern Red Sea, with an early age (Sebai et al., 1991). The peridotite massif is a pristine and fertile plagioclase lherzolite, hardly molten during its ascent (Styles and Gerdes, 1983; Bonatti and Seyler, 1987; Nicolas et al., 1994).

Figure 2 proposes a scenario where the lithosphere needs to be thermally eroded before rifting occurs because the tensile stress of the normal continental lithosphere is too strong to yield mechanically. The Miocene event prepares the following event dated early Pliocene, 10–15 My later, with major stretching of the continental crust to its rupture, generating the first oceanic lithosphere. Heating by a mantle plume should be accompanied by a regional doming in the kilometer range of the Northern Red Sea. However, thermal relaxation during 10–15 My and necking of the weakened upper lithosphere have limited the uplift.

Passive rifting

Traditionally, opening of the Atlantic has been presented as showing a striking contrast between the American continental margin, mainly along Newfoundland, that is intruded and covered by thick basalts and the European side of this ocean, a tectonically denudated margin exposing the underlying mantle that is best observed in Galicia (Figure 3b). These contrasted features have been studied along the Iberia-Newfoundland transect (location in Figure 3 of ODP Legs 149, 173 and 210). These features have been explained by an initial asymmetry with a detachment occurring along the European margin (Boillot et al., 1988) and, along the American margin, upwelling of mantle associated with a massive basaltic activity. This simple model is now disputed on the basis of intense field research introducing the concept of hyperextension. Oil companies also conducted research on both margins of the Atlantic, with scientific drilling and many seismic reflection and refraction surveys. The discovery of hydrocarbons in deep margins in the South Atlantic has promoted a new interest in rifted margins. One result has been a major change in the way of thinking about evolution of rift systems with new questions such as how hyperextension relates to magmatism or thermal and isostatic behaviors. The earlier contrast between the opposed margins is now attenuated with complex detachment tectonics being active on both margins. This is illustrated in Figure 3 by the schemes of successive ocean-continent transitions (OCT). In some way, the present day model (Figure 3d) becomes symmetrical. This points out that symmetry depends on the scale: rift systems may be symmetric on the scale of the lithosphere but not necessarily on the scale of the crust.

Conclusions

Historically in the 1970s, two views were prevailing on the origin of rifts. In the model of passive rifting, extension forces applied to a continental lithosphere would stretch and thin this lithosphere with, consequently, uprise of the underlying asthenosphere (Figure 1a). This uprise is responsible for two main attributes of rifts: volcanism and thermally supported reliefs. In the active model, rifting is primarily triggered and localized by a mantle plume (Figure 1b). The asthenosphere rises, either by thermal erosion or fracturing of the lithosphere (Figure 2). In this active model, crustal extension may be limited.

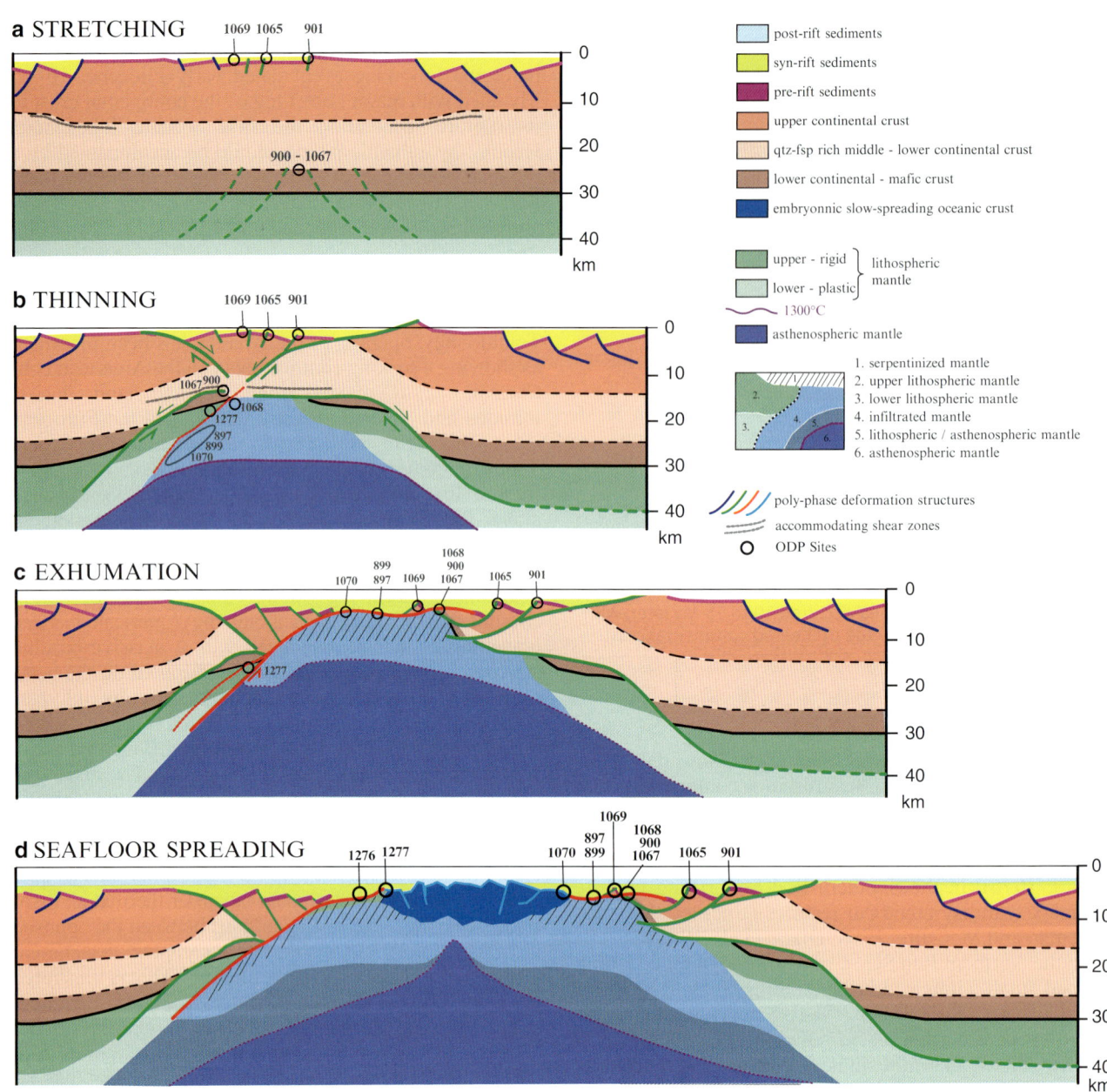

Oceanic Rifts, Figure 3 A new model for the mid-ocean rifting of the Atlantic Ocean, (a) and (b) from continental rifting, with development of conjugate detachments and mantle uprise, (c) OCT between European margin in Galicia and American margin in Newfoundland, with a new major detachment from the western margin and exhumation of lithospheric mantle, and (d) seafloor spreading of the Atlantic Ocean (Péron-Pinvidic and Manatschal 2009).

Thermal erosion alone would only create a volcanic dome, such as Hoggar. We conclude that active rifting requires also tensional forces in the lithosphere. Passive or active rifting reflects some of the observed variety of rifts. Another major factor is whether the lithosphere subjected to rifting was homogeneous or already bearing scars, namely, major shear bands, either thermal or geological, for instance, the trace of a former subduction zone. With respect to symmetrical rifting (Figure 1a, b), this would favor asymmetrical rifting (Figure 1c). In this simple model, the strength of a normal, homogeneous continental lithosphere would be too high to yield, unless a flat-lying weak zone is present. This would generate a cold detachment fault (Figure 1c). However, these models remain very simple. There is also an evolution in time. As illustrated by the Red Sea evolution, the initiation of an oceanic rift in the North tends to be symmetrical, whereas in the Southern Red Sea that is in a further stage of oceanic rifting, heterogeneous structures appear such as detachment faults that increase the asymmetry. Despite 40 years of active research, there is still much to learn about rifting as illustrated by recent developments.

Acknowledgments

This manuscript has benefited from Françoise Boudier's and Gianreto Manatschal's reviews.

Bibliography

Allen, P. A., and Allen, J. R., 2005. *Basin Analysis*. Malden: Blackwell, p. 537.
Baldridge, W. S., Eyal, Y., Bartov, Y., Steinitz, G., and Eyal, L., 1991. Miocene magmatism of Sinai related to the opening of the Red Sea. *Tectonophysics*, **197**, 181–201.
Boillot, G., Girardeau, J., and Kornprobst, J., 1988. The rifting of the Galicia margin: crustal thinning and emplacement of mantle rock on the seafloor. In *Proceeding ODP, Scientific Results*, Vol. 103, pp. 741–756
Bonatti, E., and Seyler, M., 1987. Crustal underplating and evolution in the Red Sea Rift: uplifted gabbro/gneiss crustal complexes on Zabargad and Brothers Islands. *Journal of Geophysical Research*, **92**, 12803–12821.
Brun, J. P., and Beslier, M. O., 1996. Mantle exhumation at passive margins, 1996. *Earth and Planetary Science Letters*, **142**, 161–17.
Buck, W.R., Martinez, F., Steckler, M.S., and Cochran, J.R., 1988. Thermal consequences of lithospheric extension; pure and simple. Tectonics, **7**(2), 213–234.
Burrus, J., 1989. Review of geodynamic model for extensional basins: the paradox of stretching in the Golfe de Lions (Northern Mediterranean). *Bulletin Société géologique de France*, **8**, 377–393.
Cloetingh, S., and Ziegler, P. A., 2007. Tectonic models for the evolution of sedimentary basins. *Treatise on Geophysics*, **6**(11), 485–611.
Cochran, J. R., and Martinez, F., 1988. Evidence from the northern Red Sea on the transition from continental to oceanic rifting. *Tectonophysics*, **153**, 25–53.
Coleman, R. G., and McGuire, A. V., 1988. Magma systems related to the Red Sea opening. *Tectonophysics*, **150**, 77–100.
Hempton, M. R., 1987. Constraints on Arabian plate motion and extensional history of the Red Sea. *Tectonics*, **6**, 687–705.
Huismans, R. S., and Beaumont, C., 2003. Symmetric and asymmetric lithospheric extension: relative effects of frictional-plastic and viscous strain softening. *Journal of Geophysical Research*, **108**, 111–137, doi:10.1029/2002jb002026.
Huismans, R. S., and Beaumont, C., 2007. Roles of lithospheric strain softening and heterogeneity in determining the geometry of rifts and continental margins. In *Imaging, Mapping and Modeling Continental Lithosphere Extension and Breakup*. London: Geological Society of London, Vol. 282, pp. 111–131.
Izzeldin, A. Y., 1987. Seismic, gravity and magnetic surveys in the central part of the Red Sea: their interpretation and implications for the structure and evolution of the Red Sea. *Tectonophysics*, **143**, 269–306.
Jones, P. W., 1976. Ages of the lower flood basalts of the Ethiopian plateau. *Nature*, **261**, 567–569.
McKenzie, D. P., 1978. Some remarks on the development of sedimentary basins. *Earth and Planetary Science Letters*, **40**, 25–32.
McMullen, R. J., and McMorhaz, B., 1989. An interactive thermoelastic rift mechanism. *Journal of Geophysical Research*, **94**, 13951–13960.
Nicolas, A., Achauer, U., and Daignières, L., 1994. Rift initiation by lithospheric rupture. *Earth and Planetary Science Letters*, **123**, 281–298.
Péron-Pinvidic, G., and Manatschal, G., 2009. The final rifting evolution at deep magma-poor passive margins from Iberia-Newfoundland: a new point of view. *International Journal of Earth Science*, **98**, 1581–1597, doi:10.1007/s00531-008-0337-9.
Sachau, T., and Koehn, D., 2010. Faulting of the lithosphere during extension and related rift-flank uplift: a numerical study. *International Journal of Earth Sciences*, **99**, 1619–1632.
Sebai, A., Zumbo, V., Féraud, G., Bertrand, H., Hussain, H. G., Giannerini, G., and Campredon, R., 1991. Ar40/Ar39 dating of alkaline and tholeitic magmatism of Saudi Arabia related to early Red Sea rifting. *Earth and Planetary Science Letters*, **104**, 473–487.
Spohn, T., and Schubert, G., 1982. Convective thinning of the lithosphere: a mechanism for the initiation of continental rifting. *Journal Geophysical Research*, **87**, 4669–4681.
Styles, P., and Gerdes, K. D., 1983. St John's Island (Red Sea): a new geophysical model and implications for the emplacement of ultramafic rocks in fracture zones and at continental margins. *Earth and Planetary Science Letters*, **65**, 353–368.
Tucholke, B. E., Behn, M. D., Buck, W. R., and Lin, J., 2008. Role of melt supply in oceanic detachment faulting and formation of megamullions. *Geology*, **36**(6), 455–458, doi:10.1130/G24639A.
Wernicke, B., 1981. Low-angle normal faults in the Basin and Range Province: nappe tectonics in an extending orogen. *Nature*, **291**, 645–648.
White, R., and McKenzie, D., 1989. Magmatism at rift zones: the generation of volcanic continental margins and flood basalts. *Journal Geophysical Research*, **94**, 7685, doi: 10.1029/JB94iB06p07685

Cross-references

Crustal Accretion
Layering of Oceanic Crust
Metamorphic Core Complexes
Oceanic Spreading Centers
Plate Tectonics

OCEANIC SPREADING CENTERS

Satish C. Singh[1] and Adolphe Nicolas[2]
[1]Laboratory of Marine Geosciences, Institute of Earth Physics of Paris, Paris, France
[2]Geosciences Montpellier, University of Montpellier 2, Montpellier, France

Synonyms

Back-arc spreading center, Mid-ocean ridge, Oceanic ridge

Definition

Seafloor spreading is a process on the earth's surface where oceanic lithosphere is created.

Full spreading rate is the rate (mm/year) at which two plates diverge from each other.

Ridge axis and plane define the position of the spreading center.

Neovolcanic zone lies on the ridge axis where melt is delivered to the surface and lavas erupt.

Hydrothermal circulation is a process where seawater is circulated and heated in the crust.

Hydrothermal vent field (also known as black smokers) is the site on the seafloor where mineral-laden hot water exits the seafloor.

Axial melt lens (AML) is a thin body of magma seismically imaged beneath oceanic spreading centers.

Magma chamber is a domain beneath ridge crests between melt lens and Moho comprised mixtures of melt and crystals.

Layer 2A is the seismically defined top layer of oceanic basaltic crust made of lava flows.

Layer 2B is the seismically defined layer between layer 2A and layer 3, the gabbroic layer, consisting of a diabase dike sequence.

Lid comprises the basaltic layers 2A and 2B above the axial melt lens.

Layer 3 is the seismically defined layer comprised of gabbro unit between the lid and upper mantle.

Moho is the seismically defined boundary between the crust and mantle.

Lithosphere is the upper layer of the earth comprised of the uppermost mantle and crust that moves as a rigid layer over the asthenosphere.

Asthenosphere is the deformable part of the earth mantle over which the lithosphere may move independently.

Seismic reflection is a geophysical technique to image to sharp physical boundary in the crust.

Seismic refraction (also referred as tomography) is a geophysical technique to image large-scale velocity structure of the earth.

Introduction

Plate tectonic theory defines the movement of the upper part of the earth's surface where plates (lithosphere) move over the asthenosphere. Different plates interact with one another along boundaries that are classified as (1) divergent plate boundaries (spreading centers), (2) convergent plate boundaries (subduction and collision zones), and (3) transform plate boundaries along which two plates slip in opposite directions (Figure 1). At oceanic spreading centers, two plates move apart due to the upwelling of the asthenosphere near the surface and due to the pull of plates as they are subducted at convergent boundaries. As the hot asthenospheric mantle rises beneath ridges,

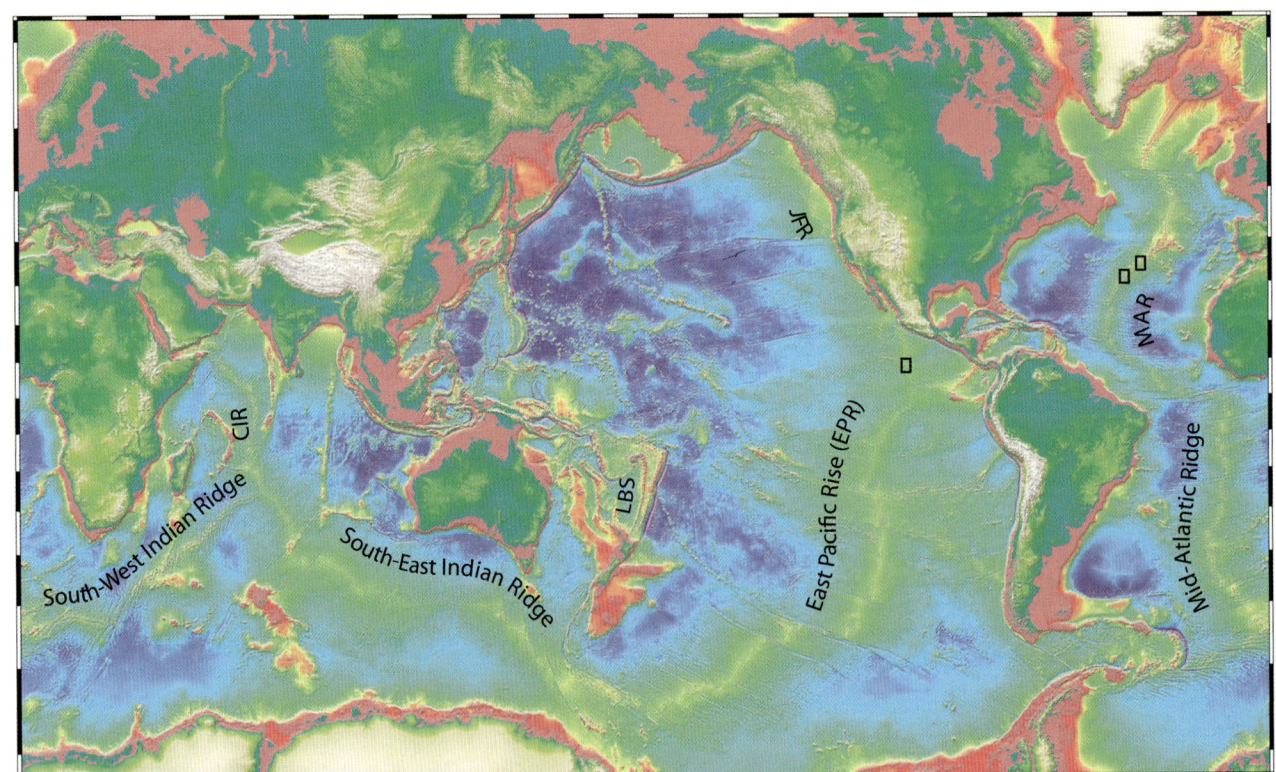

Oceanic Spreading Centers, Figure 1 Global map of the Oceans showing the main spreading centers. *CIR* central Indian ridge, *LBS* Lau Basin spreading center, *JFR* Juan du Fuca Ridge, *MAR* Mid-Atlantic Ridge. *Black rectangles* indicate location where the images have been used in this entry.

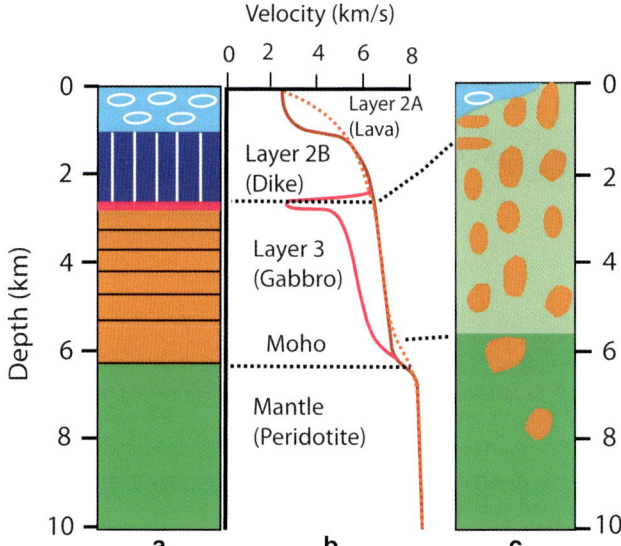

Oceanic Spreading Centers, Figure 2 Simple conceptual models of oceanic crust based on a combination of seismic results and petrological observations. (**a**) A layered cake model beneath a fast spreading center containing lava flow at the *top* (*white ellipses* in *light blue* background), dike sequences (*vertical white streaks* in *dark blue* background), and thin melt lens (*pink*), gabbroic lower crust (*black horizontal streaks* in *brown* background) above mantle peridotite (*green*). (**b**) Representative velocity models as a function of depth beneath a fast spreading center (*pink*), off axis (*brown*), and beneath a slow spreading center (*dashed brown*) defining different seismologically defined layers 2A, 2B, and 3. (**c**) Complex geological model at slow spreading ridges. The definition of all the background colors are the same as that in (**a**), except the *light green color* that represents serpentinite.

the pressure decreases leading to partial melting of the mantle, yielding basaltic melt rising towards the surface of the earth to form the oceanic crust. Part of the melt erupts on the seafloor forming extrusive lava flows at the ridge axis where it solidifies. Laboratory measurements of seismic velocities of these lava flows correspond to the seismically defined layer 2A (Figure 2). Lavas are fed by intrusive dikes whose velocities correspond to the seismically defined layer 2B. Below the dikes that feed the overlying lava flows is a small melt lens (AML), molten at ~1,200 °C. Crystals forming in this magma body feed a gabbroic mush subsiding in a magma chamber below. Away from the ridge axis, these melt lenses and the magma chambers cool and crystallize to form the gabbroic lower crust, whose velocities correspond to the seismically defined layer 3. The cooling and crystallization is achieved by efficient hydrothermal circulation of the seawater throughout the crust. The heated seawater returns to the seafloor axis as hydrothermal vents, the black smokers. Below the crust, the mantle consists of high-density peridotites. The boundary between the crust and mantle is called Mohorovičić seismic discontinuity or Moho for short. Thus, based on a combination of seismic velocities and the petrological descriptions, the oceanic crust can be divided into three layers (Figure 2a): (1) layer 2A consisting mainly of lava flows, (2) layer 2B consisting mainly of dike sequences, and (3) layer 3 containing mainly gabbroic rocks. However, as we shall discuss later, there are significant deviation from this simple model of the oceanic crust (e.g., Figure 2b).

Diverging plates at spreading centers exert horizontal stresses leading to fracturing of the cold and brittle part of the crust and upper mantle (lithosphere) causing earthquakes and faulting. With increasing spreading rate, the amount of melting in the mantle increases leading to increased crustal thicknesses. As presented below, the various physical features and differences in crustal structure can be correlated with spreading rate, rates of cooling, and faulting of the crust. However, other parameters, the nature of the rising mantle and the extent that it melts, may play a more important role. The mantle is unlikely to be homogeneous and could be more or less fertile, warmer or colder, and dry or hydrated. This is known from inferences made from the chemistry of basalts derived from the mantle and the study of ophiolites where different types of mantle have been sampled (Nicolas, 1989). However, a "standard" mantle may be defined. The various physical attributes of spreading centers are therefore largely controlled by the mantle upwelling beneath them. Consequently, the ocean spreading ridges can be classified on the basis of their spreading rates and this in turn can provide a clearer understanding of seafloor spreading processes.

There are four main types of spreading centers based on spreading rates: (1) fast spreading center with a spreading rate greater than 70 mm/year and superfast ones, say more than ~120 mm/year; (2) intermediate spreading center with spreading rate between 40 and 69 mm/year; and (3) slow spreading center with spreading rate between 20 and 39 mm/year and (4) ultraslow spreading center with spreading less than 20 mm/year. Figure 1 shows the location of the main spreading centers on earth. The East Pacific Rise is a fast spreading ridge, Juan du Fuca and Lau Spreading Centers are intermediate spreading ridges, Mid-Atlantic and Central Indian Ridges are slow spreading ridges, and Southwest Indian ridge is an ultraslow spreading ridge.

A simple conceptual model for oceanic ridges

In order to address how rates of seafloor spreading are related to most of the features of fast, intermediate, slow, and ultraslow spreading ridges, as well as oceanic rifts (see entries "Oceanic Rifts" and "Ophiolites" in this volume), a good approach is to address the basic question of how melting operates in the rising mantle below these different spreading systems and how it generates distinct oceanic spreading centers (Figure 3).

Figure 3a illustrates a pressure-temperature diagram showing a trajectory of rising mantle within an

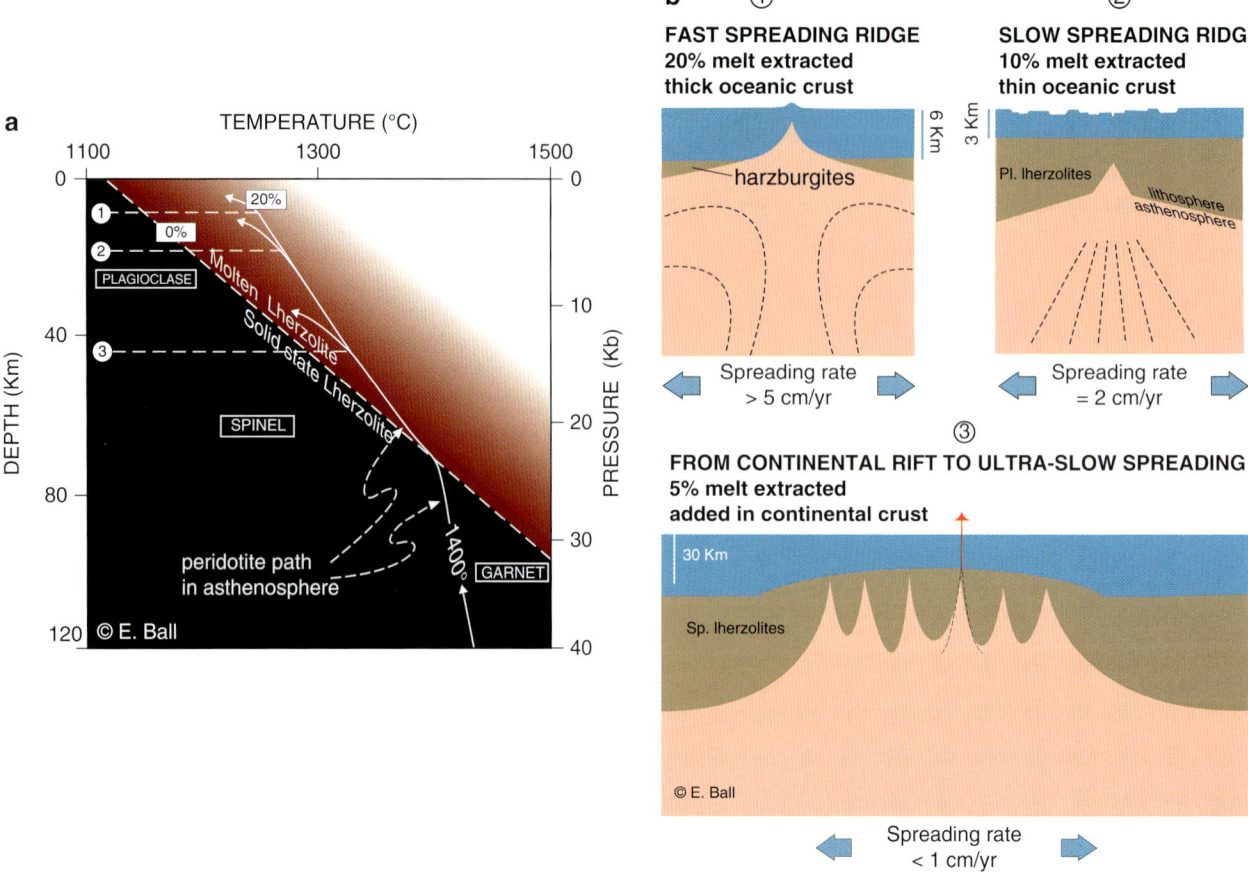

Oceanic Spreading Centers, Figure 3 (**a**) Simplified sketch of melting in a rising mantle along a steep pressure-temperature adiabatic path (in *black* field), before it reaches the lithosphere and follows shallower paths due to conductive cooling. Partial melting occurs in the *red* domain, above the dry mantle solidus, where the adiabatic curve flattens due to the enthalpy of melting. Above the solidus, the three main trajectories depend on the depth of first melting (here 80 km), the melting fraction per Gp), and the meeting point of each trajectory with the lithosphere: Case (1) corresponds to the trajectory for a thin lithosphere, (2) for a thicker lithosphere, and (3) for a very thick lithosphere. (**b**) The corresponding transverse structures of ridges and rifts: case (1) for a fast spreading rate with a thin lithosphere and large melt delivery, (2) for a slow spreading ridge with a thicker lithosphere and less melt delivery, and (3) for a ultraslow spreading ridge with a very thick lithosphere and associated little or no melting (Nicolas, 1989, 1992).

asthenospheric upper mantle where the thermal regime is convective (adiabatic gradient ~0.3 °C/km). At a depth of ~80 km, this mantle trajectory crosses the steeper fertile lherzolite mantle melting curve above which the melt fraction increases linearly with a slope largely smaller than that in the dry adiabatic domain below. Due to enthalpy of melting, a shallower slope results from the heat used for further melting. Melting ceases depending on the depth where the uprising mantle meets the solid lithosphere. However, magmas keep rising, following the conductive gradient in the static lithosphere. This new gradient has a flatter trajectory up to the earth's surface. In this simple model, the nature of any spreading center is controlled only by the lithospheric thickness that is itself dependent on the spreading rate. In a fast spreading situation (Figure 3b1) at the ridge axis, the asthenosphere rises up to ~2 km beneath the lithospheric lid, and ~20 % melt is generated. Lithosphere would be thicker (<10 km) at a slow spreading ridge (Figure 3b2); the melting column is shorter near the central rift valley, thus generating ~10 % basaltic melt. Finally, a ~60 km-thick lithosphere would create a continental rifting opening at ~0.5 cm/y as in the French Massif Central (Figure 3b3). In an oceanic rift such as Red Sea, or in an ultraslow oceanic ridge, lithosphere would remain thicker than that beneath slow spreading ridges. Below we discuss the general features of the two main spreading systems, fast and slow with brief comments on ultraslow ridges and with some comments on back-arc spreading centers.

Fast spreading center

The East Pacific Rise (EPR) is the simplest and most stable geological system in space and time and, thus, the best adapted to understand the process of seafloor accretion.

Morphology

The morphology of fast spreading centers is characterized by an axial high, which can be up to 15–20 km wide and ~500 m high relative to nearby abyssal plain with the shallowest points around neovolcanic zone where volcanism and hydrothermal venting are concentrated and often have a region of volcanic collapse at the center (Figure 4). The presence of axial highs is due to the upwelling of hot mantle beneath the spreading centers. The lithosphere exactly beneath the ridge axis is limited to the lid itself, and therefore, very few faults are observed, but as the lithosphere drifts and thickens, some faults develop away from the ridge that are expressed as abyssal hills. On the larger scale of several hundreds of kilometers, fast spreading ridges are segmented by large transform faults and overlapping spreading centers that offset the spreading center axis. In a few places rotating microplates ranging from ~500 km across at the Easter microplate to ~50 km at the Wilkes nanoplate also offset the spreading center. The scale of individual segments could be from 50 to 150 km. Two segments along a ridge may overlap, developing basins between overlapping spreading centers or duelling, one propagating forward against the other that is receding (MacDonald et al., 1988) (Figure 4). Within a larger ridge segment, a smaller change in the linear nature of the axis is known as a DEViation in Axial Linearity (DEVAL). Because the lithosphere remains thin even off axis, segmentation reflects some independence between lithospheric and asthenospheric tectonic activity (Toomey et al., 2007).

Structure

The layer 2A, the extrusive volcanic unit, is either defined by a change in P-wave velocity gradient around 4.5 km/s contour or by seismic image (Harding et al., 1993) (Figure 2). The thickness of layer 2A varies from 150 to 200 m on the ridge axis to 300–700 m away from the ridge axis due to accumulation of lava (Kent et al., 1993). The low velocity of layer 2A relative to the velocity measured in basaltic lava is generally due to the presence of pores (fluids), where the porosity decreases with depth from 30 % near the surface to 5–6 % at the base of layer 2A (Collier and Singh, 1998).

Layer 2B, the sheeted dike layer, is defined by a low-velocity gradient layer, where the P-wave velocity ranges from 4.5 to 6 km/s (Figure 2). The presence of high velocity suggests a further decrease of porosity down to ~1 %, but a low-velocity hydrothermal-rich layer has also been imaged at the base of layer 2B (Singh et al., 1999). The thickness of layer 2B varies from 1.5 to 2 km thick, depending upon the depth of the axial melt lens (AML) that defines the base of dike layer above and gabbro below.

The P-wave velocity in the AML has been measured to be between 2.5 and 4 km/s (Murase and McBirney, 1973). The large decrease in velocity leads to a strong seismic

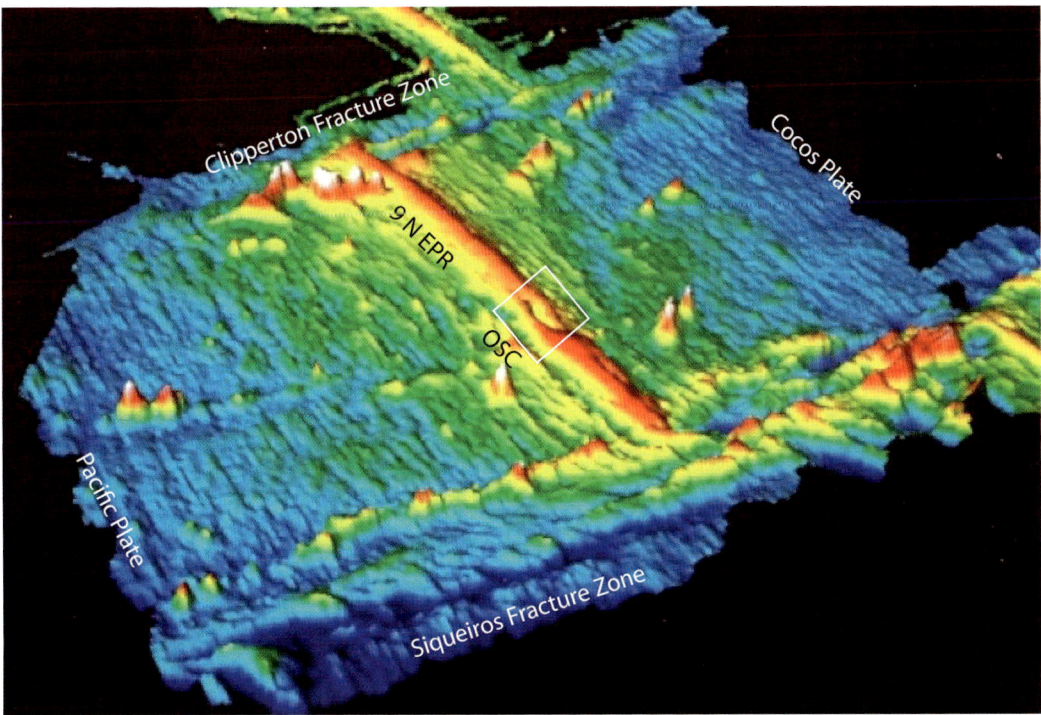

Oceanic Spreading Centers, Figure 4 The seafloor morphology of the 9° N EPR, showing the 9° N overlapping spreading center (OSC in *white box*) and the Clipperton and Siqueiros transforms and associated fracture zones.

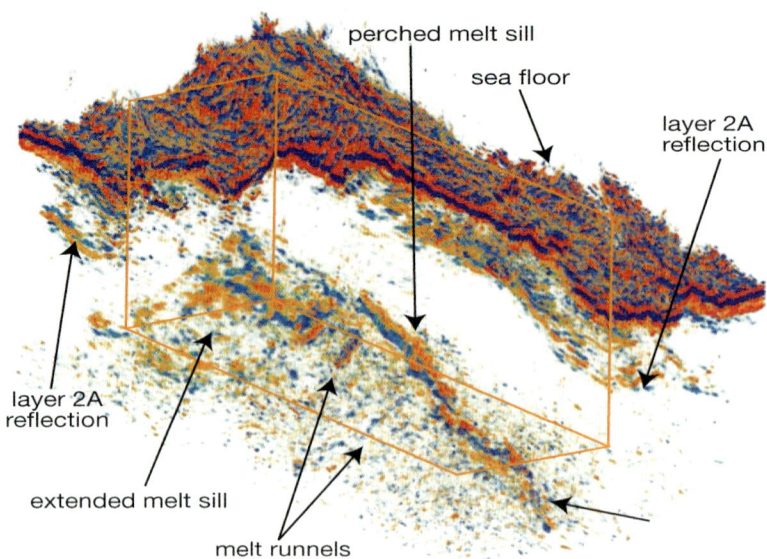

Oceanic Spreading Centers, Figure 5 3D view of AML beneath 9° N overlapping spreading center (Kent et al., 2000).

reflection, generally with a negative polarity as compared to the seafloor (Detrick et al., 1987; Kent et al., 1993). The velocity within the AML could be estimated using seismic full waveform inversion (Singh et al., 1998). The AMLs could be 200–4,500 m wide and 50–100 m thick, although thicker melt lenses (300 m) have been observed beneath the 9°N overlapping spreading center (Kent et al., 2000) (Figure 5). AMLs are present along a significant part (70 %) of the ridge axis beneath fast spreading centers. AMLs are not only present at the center of the ridge axes but also at segment ends (Kent et al., 2000) (Figure 5). Melt lenses have also been observed off axis as well (Canales et al., 2011). Along the ridge axis AMLs can be segmented into areas of pure-melt and crystal-melt mush zones that can be distinguished from zero and non-zero S-wave velocity (Singh et al., 1998). Areas of pure melt seem to be responsible for vigorous hydrothermal circulation and the presence of hydrothermal vent fields observed on the seafloor (Singh et al., 1999). The cooling and crystallization of melt from the AMLs are responsible for the formation of the upper part of layer 3 (upper gabbro).

On the ridge axis, a low-velocity layer is observed beneath the AML, where the P-wave velocity varies from 3.5 to 6.0 km/s, suggesting the presence of partial melt (5–30 %) (Dunn et al., 2000). This region is known as the axial magma chamber (AMC) and is thought to consist of thin melt sills or melt pockets rather than large regions of magma (Figure 6). The width of the AMC is generally 5–8 km, which could up to ~10 km wide at the Moho depths as noted in ophiolites such as in Oman where the internal nature and structure of the AMC has been studied (see Figure 8). The cooling and crystallization of melt in the AMC and AML forms the gabbroic layer 3. Layer 3 is characterized by a low-velocity gradient where the P-wave velocity varies from 6.5 to 7 km/s over a depth range of ~3 km (Harding et al., 1989).

The P-wave velocity in the mantle beneath the ridge axis is about 7.6–7.8 km/s, which increases to 8.0–8.1 km/s away from the ridge axis (Dunn et al., 2000). The P-wave velocity in the mantle is determined by seismic refraction methods. The strong velocity contrast at the crust-mantle boundary could produce Moho reflections. The Moho is observed beneath the ridge axis at zero age, suggesting that the crust at fast spreading centers is formed at zero age (Singh et al., 2006a) (Figure 7a). The Moho reflection can have different characters: sharp, shingled, or diffuse (Kent et al., 1994) (Figures 7b–d). A reflection from the crust and mantle would not be observed if the boundary is thick (~1 km) where the velocity gradually increases with depth. In both cases, wide-angle reflections are observed on seismic data, allowing crustal thickness estimation. However, from wide-angle reflection data alone, it is uncertain if the crust-mantle boundary is sharp or has a velocity gradient (Moho Transition Zone, MTZ).

The mantle beneath the spreading center is characterized by a low-velocity anomaly that extends down to 70 km in depth over a 50 km-wide zone (Toomey et al., 1998), suggesting the presence of partial melt. Some of these melts could get trapped below the Moho in a form of a sill, similar to those interpreted to exist as the MTZ in ophiolites (Crawford and Webb, 2002; Nedimovic et al., 2005).

The lithosphere-asthenosphere boundary lies at the top of the magma chamber and the level of the axial melt lens, rapidly deepening away from the ridge axis. The precise nature of this boundary away from the ridge axis remains

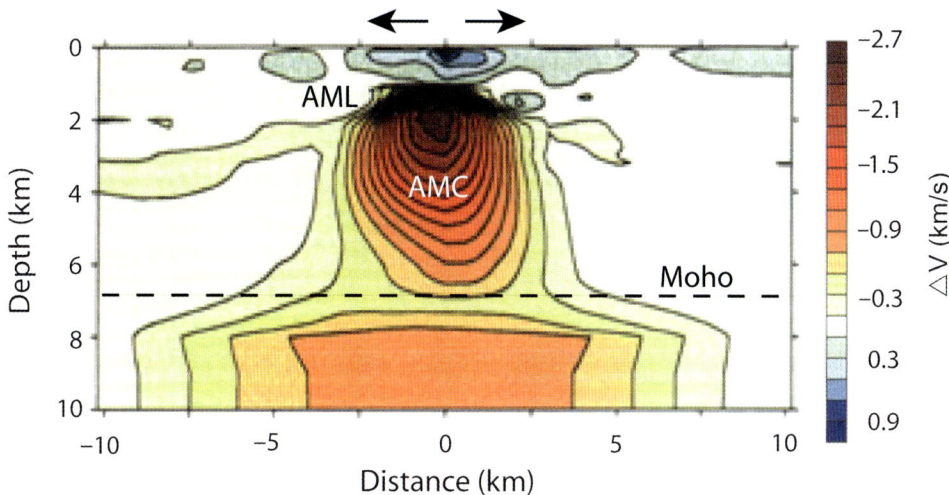

Oceanic Spreading Centers, Figure 6 2D P-wave velocity anomaly beneath 9° 30′ N East Pacific Rise (Dunn et al., 2000). AML stands for axial melt lens where seismic velocities are remarkably low because of large melt fractions, and AMC is axial magma chamber where melt fraction would be in the range of 20 % (Nicolas and Ildefonse, 1996). *Black dashed line* marks the Moho boundary, which is underlain by a broad low velocity in the mantle underneath.

a matter of debate in the oceans (Kawakatsu et al., 2009; Karato, 2012).

Hydrothermal circulation

The temperature of the melt in the AML and AMC is ~1,200 °C, which is cooled by hydrothermal process as cold seawater enters the crust (thermal sink near the ridge axis), then rises back to the seafloor where it exists as hydrothermal vents (black smokers) with a water temperature up to 400 °C. Just below the ridge axis, the water circulates down to the AML, whereas it could potentially circulate down to Moho depths around the AMC (MacLennan et al., 2004). At one time it was believed that hydrothermal circulation mainly occurred across axis, but recent studies have shown that a significant amount of hydrothermal circulation takes place along the ridge axis (Fontaine and Wilcock, 2007; Tolstoy et al., 2008).

Comparing internal structures of the EPR and Oman ophiolite

With the exception of the seafloor itself, little of the internal geology of fast spreading ridges can be directly studied, in contrast to slow spreading ridges where kilometers of relief are exposed along deep transform faults and oceanic core complexes. Few exceptions are Hess Deep where the westward-propagating Galapagos Spreading Center cuts through young the EPR crust, exposing its crustal section. It has been drilled during the IODP program (Legs 147 in 1993 and 345 in 2012–2013). Another site is the Pito Deep, the tip of a propagator on the Easter microplate. So far only gabbros (in situ) have been recovered in these sites. This is why we briefly present some of the prominent geological features exposed in the field in the Oman ophiolite and show how they compare with the wealth of geophysical data discussed above (see also "Ophiolites" and "Oceanic Rifts" in this encyclopedia).

The Oman ophiolite is a piece of oceanic lithosphere formed 95 Ma ago that got detached and thrust on to the passive margin of Oman about 10 Ma later. It is best suited for a comparison with the EPR as it was formed at fast spreading to superfast (20 cm/year) spreading centers (Rioux et al., 2012). It is very large (500 km long and 60–80 km wide), compared with the well-studied 200 km by 50 km 8–10°N segment of the EPR (Figure 4). The Oman ophiolite has excellent exposures, is flat-lying, and has not been structurally dismembered. Its structural and petrologic study is remarkably complementary to that obtained by the detailed studies conducted around 9°N on the EPR. Two examples illustrate this complementarity. AMLs are well imaged at the EPR using seismic reflection techniques (see above) whereas in Oman, the existence of such a melt lens could not be imagined because AMLs are transient features of an active ridge, vanishing at a kilometer distance from the ridge axis. However, looking between the lid and the gabbro unit for evidences of their presence in Oman, it has been possible to identify their former existence along the ridge and to describe their past magmatic history, such as episodic filling by new melt or internal convection, obviously out of reach in active ridges. AMLs generate most of the gabbro unit by deposition of gabbroic mushes on their floor (Boudier and Nicolas, 2011). This is followed by subsidence of this mush from the floor to deeper in the crust (Nicolas et al., 2009). The typical layering of

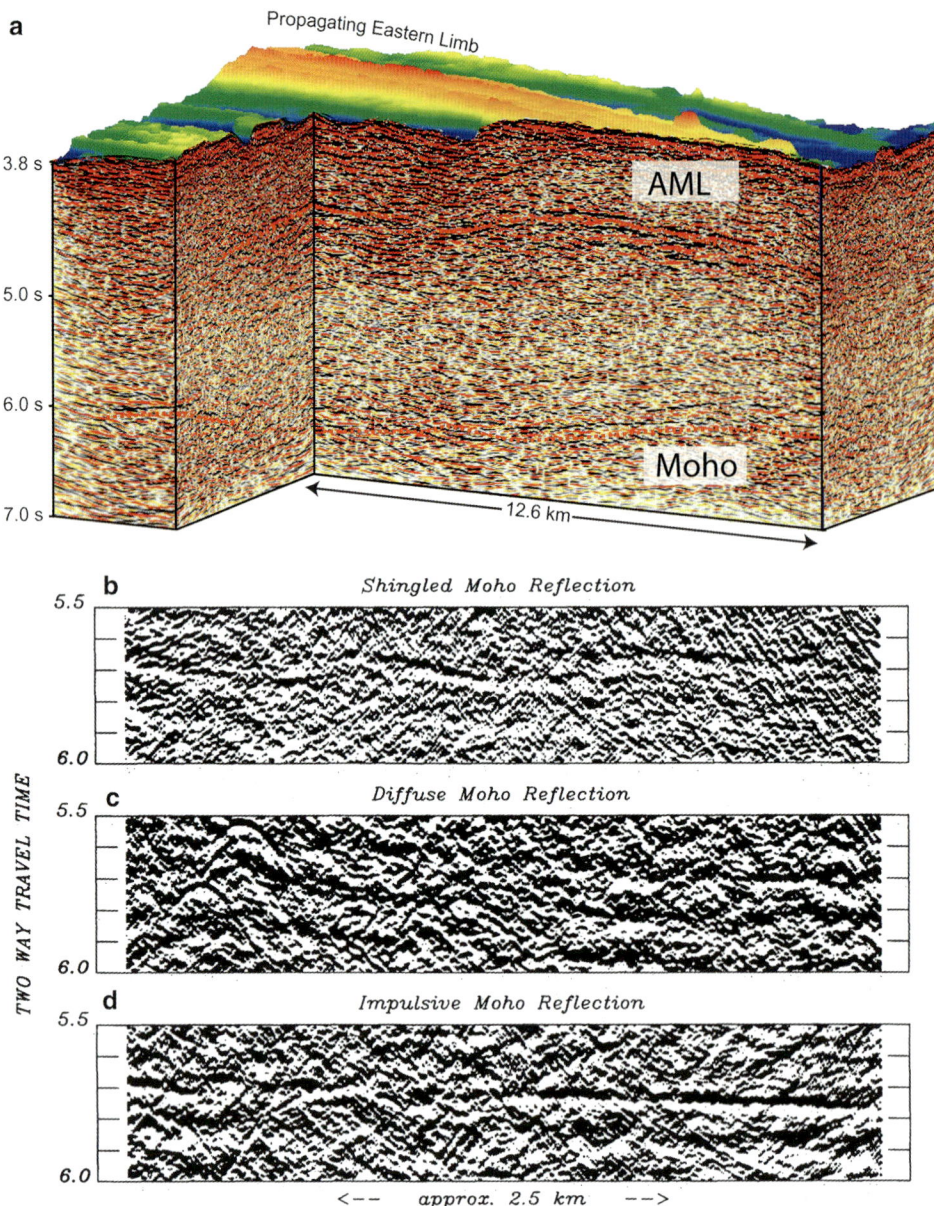

Oceanic Spreading Centers, Figure 7 (a) 3D seismic reflection image of Moho beneath the AML along the eastern limb of the overlapping spreading center at 9° N East Pacific Rise (Singh et al., 2006a), (b) seismic reflection image of shingled Moho, (c) diffuse Moho, and (d) impulsive Moho (Kent et al., 1994).

gabbros is dominantly inherited from the episodic filling and internal convection in AML, together with intrusion of gabbro and wehrlite sills. Layering is reinforced by large magmatic flowage (e.g., Figure 8).

Contrasting with the AML concept at the EPR, mantle diapirs were early discovered and mapped in the Oman ophiolite. The mantle diapirs are derived from hot, partially molten mantle rising beneath the ridge axis. They are ~10 km wide at the Moho level where the vertical mantle flow pattern is broken and diverges horizontally (Rabinowicz et al., 1984) (Figure 8c). Similar mantle upwellings have been interpreted to occur at the 9°N EPR from seismic refraction studies (Dunn and Toomey, 1997; Toomey et al., 2007). Beneath the Moho in the Oman ophiolite, there is a ~400 m thick Moho transition zone (MTZ) typically composed of a residual dunite impregnated by gabbroic melt. The MTZ at an active ridge should be largely molten with a viscosity accounting for the mantle flow rotation (Jousselin and Nicolas, 2000) (Figure 8c). It is this MTZ that seems to have been imaged by seafloor compliance studies at the EPR (Crawford and Webb, 2002). Based on field studies in Oman, at the ridge

Oceanic Spreading Centers, Figure 8 Typical images in the Oman ophiolite. (a) Lower gabbros with a flat-lying layering foliation. (b) Dynamic, magmatic foliation gabbro with black sills of wehrlites (bar = 10 cm). (c) The horizontal Moho. Below the light-colored layered gabbros, the geologists stand on the *brown*-colored mantle.

axis above the Moho, wehrlites, a mobile mixture of the melt-impregnated dunites derived from the MTZ, are injected mainly into the lower crust. When injected within the AMC, they form sills that get stretched by magmatic flow (Figure 8b). When they are injected outside the AMC, they constitute intrusions, commonly kilometers long (Figure 9). Interestingly, the tent shape of the magma chamber observed at the EPR by wide-angle seismic refraction (Detrick et al., 1987) and in Oman by geologists was discovered simultaneously (Nicolas et al., 1988). Later on, magma chamber models were further refined by Oman geologists (Mainprice, 1997) and marine geophysicists (Dunn and Toomey, 1997) with a good agreement on the overall shape and nature of the gabbroic mush that fills it.

There is also an agreement between the results from Oman and EPR regarding the hydrothermal alteration of the juvenile crust by seawater. The black smokers at the EPR are the surface manifestation of hydrothermal circulation at spreading ridges, whereas at depth the hydrothermal plumbing is known from drilling through the basalts and sheeted dikes in the upper crust to near the top of the melt lens (IODP leg 147, Hess Deep and leg 1256, EPR), showing seawater alteration at ~400 °C (Lowell et al., 2012). In the gabbro cores recovered from Hess Deep, evidence of a high temperature (~700 °C) alteration has been discovered (Manning et al., 1996). In Oman, where the 3D hydrothermal structure down to the Moho is fully exposed, this high temperature alteration (Manning et al., 2000) and a very high temperature (~1,000 °C) (Nicolas et al., 2003) metamorphism are related to a thermal boundary layer (TBL) plated along the walls of the magma chamber (Cathles, 1993) (Figure 9).

The transition between fast and slow spreading ridges

The transition from fast to slow spreading ridge occurs when the magmatic input controlling the spreading activity is no longer continuous and becomes time dependent. Typically, this is observed in slow spreading ridges with successive episodes of ridge accretion by new magmatic addition and by tectonic stretching of the lithosphere, in particular, creating spectacular oceanic core complexes (OCC). The distinction between fast and slow spreading ridges is around 4–5 cm/year (Phipps Morgan and Chen, 1993), a rate above which coincides with the appearance of a stable melt lens on top of the low-velocity zone. As noted above, this limit depends on the nature of the rising mantle. For instance, the Reykjanes ridge, which is fed by the nearby Iceland hot spot, has many attributes of a fast spreading ridge, such as a permanent crustal magma chamber, even though the spreading rate is ~2 cm/year (Sinha et al., 1997).

Slow spreading centers
Morphology

The morphology of slow spreading centers is characterized by an axial valley (median valley), which could be up to 15–25 km wide and ~500–1,000 m deep (Figure 10). The shallowest part of the ridge is generally at the center of the segment, possibly with a central volcano, and it is deepest at the segment ends (Figure 10). The lithosphere at slow ridge axes is quite variable but typically ~6 km thick (Figure 3b2). The median valley is bounded by active ridge-ward dipping faults. Commonly, there are many other smaller faults within the median valley. In contrast with fast spreading ridges, kilometers-deep transform faults, the extensions of which are called

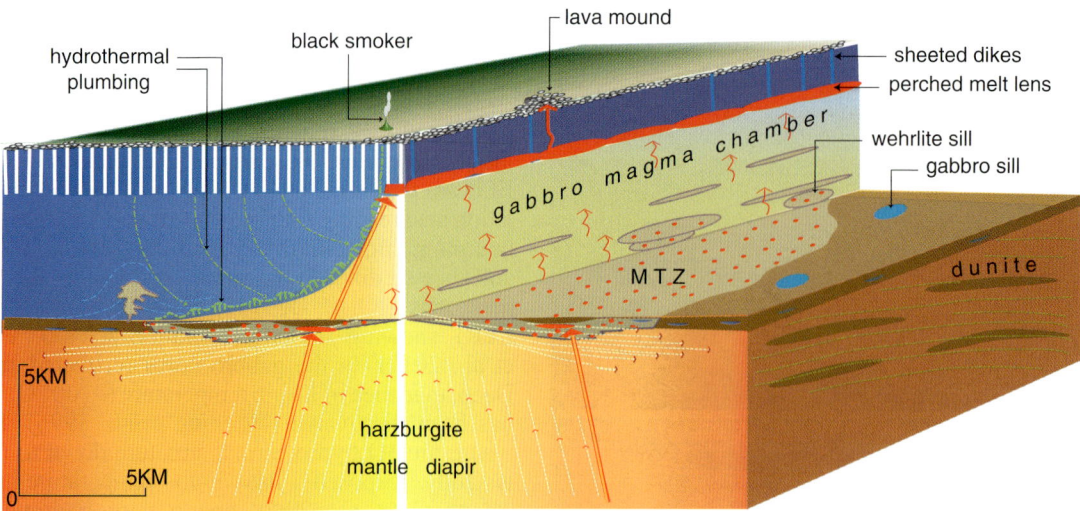

Oceanic Spreading Centers, Figure 9 A general structural model of accretion at a young ridge segment in Oman. The melt (*red*) rising through a mantle diapir accumulates in the MTZ, just beneath the Moho. In this mushy horizon of melt-impregnated dunites; the steep mantle flow rotates and diverges. Also in the MTZ, the melt chemically reacts and re-equilibrates with dunite to become MORB magma. Physically, it can be stored in the dunitic mush before an extrusion towards the AML, shorter than a few years (Boudier and Nicolas, 2011). It may also rise through the gabbro magma chamber, either by percolation or by wehrlite intrusions derived from the MTZ mush. On the *left* side of the figure, hydrothermal alteration is sketched (*green*). It refers to a nonconventional model (Cathles, 1993). In contrast with the conventional model based on progressive heating of the descending seawater flow, here low-temperature seawater reaches the Moho through fissures and fractures from a distance of ~5 km from the ridge axis: the approximate width limit of the magma chamber at the Moho level. Seawater upflow is heated by the return flow from a thin TBL that is plated against the magma chamber walls and its associated melt lens roof. The return seawater flow at 400 °C or above ascends through a low porosity channel at the edge of the TBL.

fracture zones, can expose sections of the entire crust. Between these dominant faults, segments can also be separated by non-transform offsets. Commonly located at the internal corner between ridge and fracture zones, long-lived detachment faults tend to generate oceanic core complexes (OCC), described below (Figure 13), although OCC have also been observed within segments (Escartin et al., 2008).

Structure

Because of faulting and rotation of blocks in OCC, the seafloor morphology of the crust formed at slow spreading ridges is very irregular and rugged. For this reason, only limited combined seismic reflection and refraction studies have been successful; the best results were obtained at the Lucky Strike segment of the Mid-Atlantic Ridge (Crawford et al., 2010). Layer 2A in this area is thicker (400–800 m) than that at fast spreading centers, which could be due to the accumulation of lava flow over a long time and slow spreading. The thick layer 2A could also result from a complex interaction of lava flow, faulting, and hydrothermal circulation. At the Lucky Strike segment, layer 2A does not seem to vary along the ridge axis, indicating that the thickness of layer 2A remains constant along the whole 70 km long segment, suggesting that volcanism plays an important role along the whole segment. The thickness of layer 2A decreases around the bounding faults, suggesting that the tectonic processes could decrease the thickness of volcanically constructed crust.

The imaging of layer 2B is rather difficult beneath slow spreading ridges because of the absence of AML in the crust and complex accretion process. However, a 3–4 km wide and 7 km long AML was recently imaged at the center of the Lucky Strike segment at ~3 km depth (Singh et al., 2006b) for the first time, suggesting that layer 2B there could be 2–2.5 km thick (Figure 11). There is a low velocity beneath the AML, which could be due to the partial melt in the lower crust, but the percentage of melt is much less than that observed at fast spreading centers (Dunn et al., 2000). The presence of AML also allows to characterize the lower crust, layer 3, which in the Lucky Strike area is about 3–4 km thick, leading to a crustal thickness of 7 km at the center of the segment (Seher et al., 2010). The crustal thickness at the segment ends smoothly decreases to 5 km, suggesting the crust at slow spreading centers are boudin shaped, i.e., thick in the center of segment and thin at segments end.

Seismic reflection images at the Lucky Strike segment also show faults that penetrate down to AML, which include not just the median valley bounding faults but many other faults within the axis (Figure 11). As expected, these faults are ridge-ward dipping. These faults could be responsible for hydrothermal circulation and may influence the positions of hydrothermal vent fields.

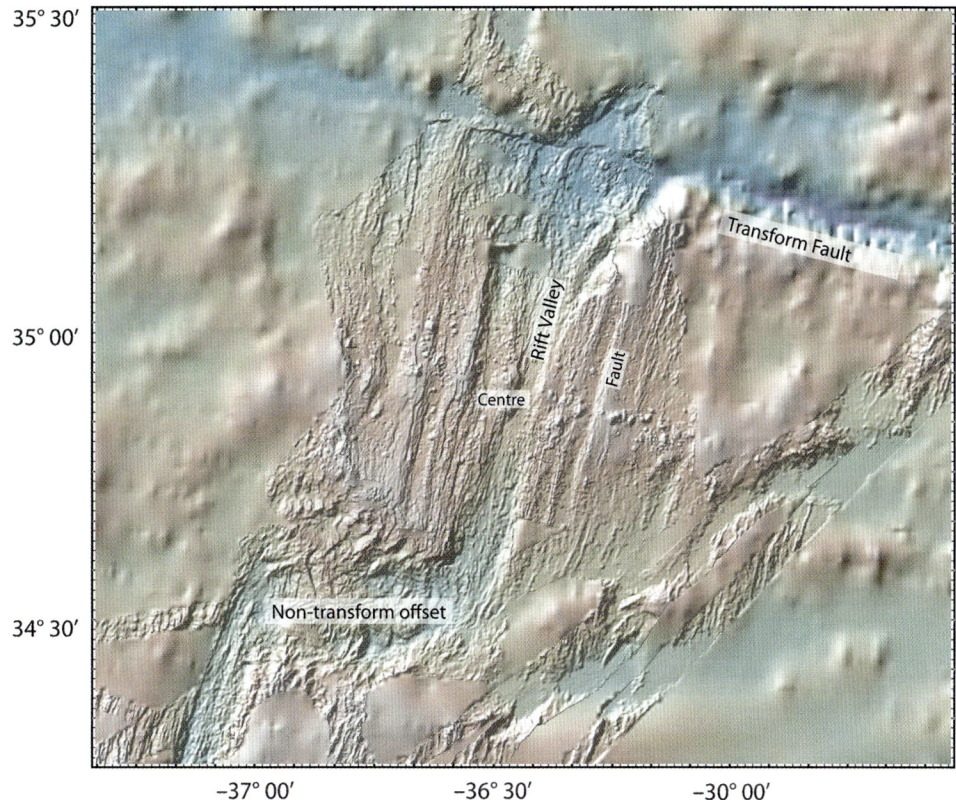

Oceanic Spreading Centers, Figure 10 Seafloor morphology of 35° N Mid-Atlantic Ridge, showing transform fault, non-transform offset, rift valley, and faults scarps (Data source: GeoMap App).

The crustal seismic structure at non-transform faults show some low velocities in the lower crust, which could be due to the serpentinization of ultramafic oceanic rocks that likely include parts of the mantle (Cannat, 1996; Canales et al., 2000). The crust beneath fracture zones is generally thin, 3–4.5 km, and shows absence of the layer 2A, indicating the absence of volcanism (Detrick et al., 1993).

Oceanic core complexes

The distinctive features of slow spreading ridges are due to their lower thermal budget as compared to fast spreading ridges. Because melt delivery is believed to be discontinuous in time and space, the continuous spreading imposed by plate tectonics leads to alternate episodes of magmatic and tectonic accretion. Magmatic accretion allows the temporary formation of a magma chamber, such as at Lucky Strike, with the presence of an AML. Tectonic accretion proceeds by the normal faulting responsible for the axial rifting and by the detachment faulting creating the spectacular OCCs (Tucholke et al., 1998). Mantle lithosphere dips about 30° away from the ridge axial plane (Figure 3b) and studies of ophiolites (Nicolas, 1989; Jousselin and Nicolas, 2000) suggest that these lithospheric slopes channel mantle flow parallel to the ridge axis between major transform faults. As a consequence, the MTZ would become a few kilometers thick below the ridge axis, compared to a few hundred meters at fast spreading ridges. However, it thins out at a short distance on each side of the ridge (Figure 13).

We briefly present the characteristic morphology of OCC from the Atlantic seafloor (Figure 12) and, tentatively, introduce (Figure 13) a model derived from the Mirdita ophiolite in Albania. It may complement the marine studies on OCCs that are mostly limited to seafloor observations. Introducing field observations on the deep structure of OCCs in this slow spreading ophiolite, we take into account the geophysical data obtained in OCCS from MAR (Nooner et al., 2003; Schroeder and John, 2004; DeMartin et al., 2007; Henig et al., 2012; Hansen et al., 2013).

Ultraslow spreading centers

Ultraslow spreading centers correspond to spreading rates of 1–2 cm/year, rates between those that characterize slow spreading ridges described above and those of oceanic rifts (see Oceanic Rifts). The Southwest Indian Ridge (SWIR) and the Gakkel Ridge in the Arctic have both been recently studied, and a few general features can be anticipated based on the mantle melting models presented in Figure 3. The fertile mantle rocks collected on seafloor of these two ridges are in the field of spinel lherzolites,

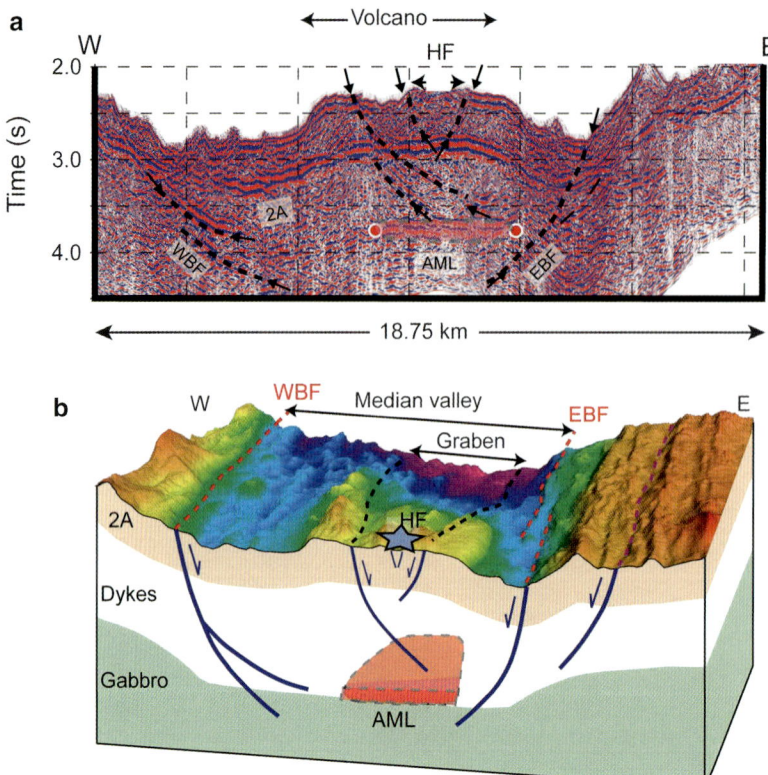

Oceanic Spreading Centers, Figure 11 (a) Seismic reflection image at the Lucky Strike volcano, Mid-Atlantic Ridge, and (b) schematic diagram showing the interpreted crustal structure (Singh et al., 2006b). *EBF* eastern bounding fault, *WBF* western bounding fault, *HF* hydrothermal field. 2A: Layer 2A. *AML* axial melt lens.

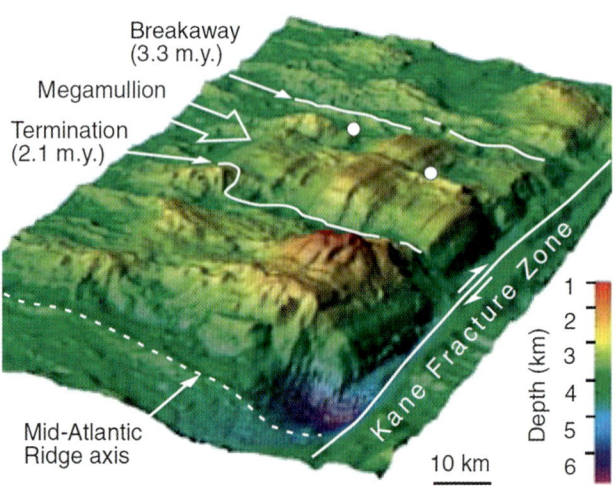

Oceanic Spreading Centers, Figure 12 Core complex at MAR, 23.5°N. The setting with the OCC at the internal corner between the ridge, looking southwest along the Kane fracture zone. The breakaway occurred along the former ridge site, where the detachment emerged and moved SW parallel to surface grooves (the megamullions). There are two successive detachments within the back of the breakaway normal fault, limiting the tectonic extension generated by these detachments and, in the front, the high-angle, outward-facing normal faults (*dots*) that are interrupted as detachments (Tucholke et al., 2008).

corresponding to a depth of ~30 km (Figure 3b3). Thus, the lithosphere thickness may attain this depth as a consequence of a much reduced melting as compared to faster ridges. These two ultraslow ridges share many similar features. The average seafloor depth is over 3 km and down to 6 km inside deepest rifts (Figure 14). Over a thin "crust," the seafloor is covered with basalts and mantle serpentinized peridotites, locally fresh in Gakkel Ridge. The SWIR rough topography seems to result from imbrication of detachment faults, lubricated by serpentinites, an expected consequence of spreading being dominantly tectonic rather than magmatic (Sauter et al., 2013). Along the western part of SWIR, the oblique spreading direction with respect to the rift alignment creates oblique and small-scale segmentation within the rift (Dick et al., 2003; Cannat et al., 2008). There are, however, sharp variations along axis, defining a large-scale segmentation (over 200 km) controlled by melting in the underlying mantle. In western parts of the SWIR, the rift rises to a depth of some 3 km and develop a thin, continuous mafic crust; the drilling has revealed a gabbro layer of ~2 km thick (Dick et al., 2003). In this region of the SWIR, the spreading may be influenced by the South African hot spot. Although spreading rate likely remained ultraslow over time, the excess heat and magmatism from the hot spot may explain both the topography and the accretion

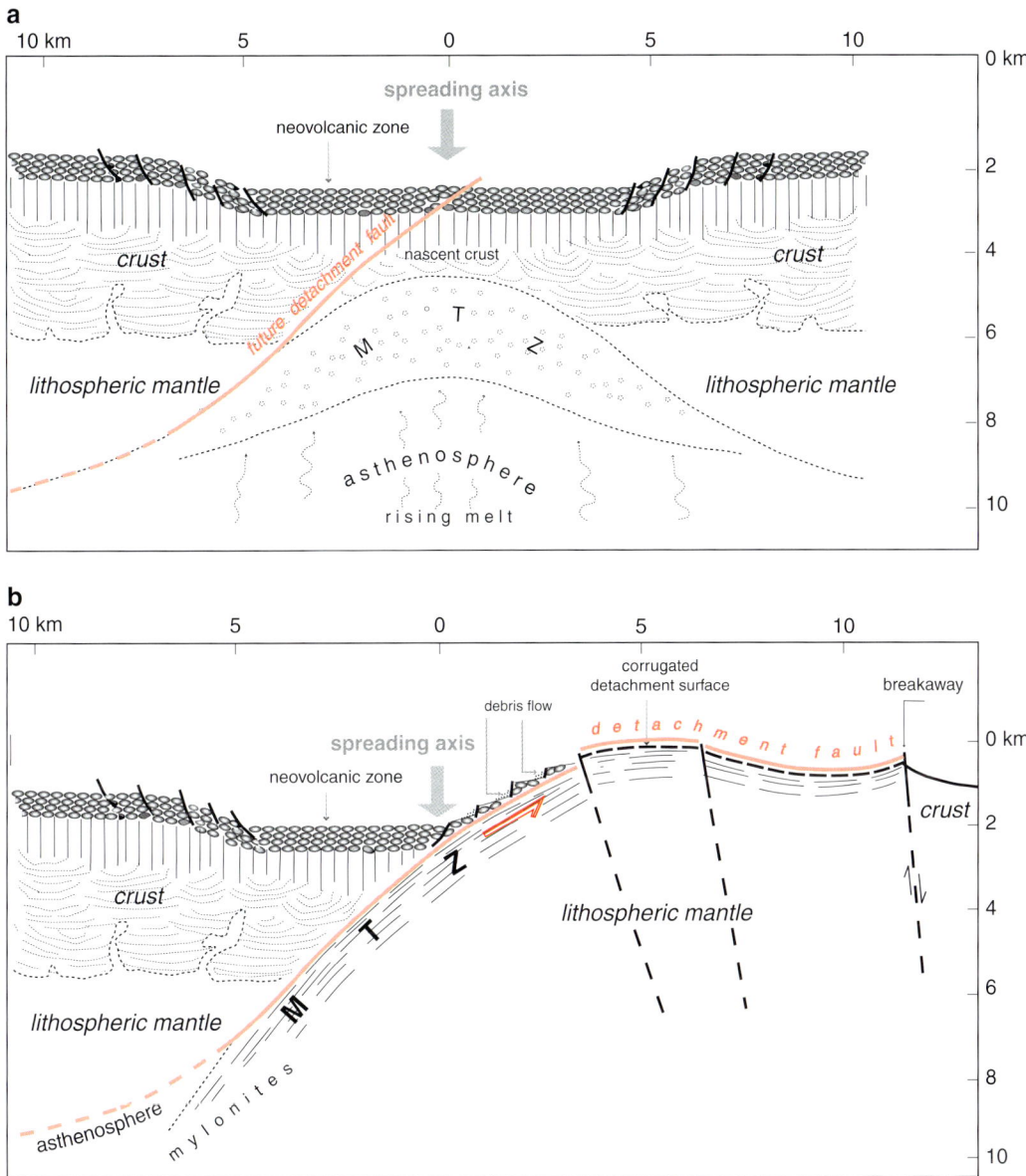

Oceanic Spreading Centers, Figure 13 Deep detachment model at ~10 km within the thick MTZ suggested by studies in the slow spreading Mirdita ophiolites (Nicolas et al., 1999) and that incorporates geophysical data from MAR. The lithosphere/asthenosphere shape has been modeled accordingly from Sleep and Rosendahl (1979) data.

of a mafic crust (Cannat et al., 2008). In the SWIR, the NS diverging plate motions are responsible for the EW trend of the rift shifts to SW-NE, presumably due to oblique asthenospheric channeling. It is only within the thinner and weaker rift floor that EW traces relating to the NS extension are observed, reflecting the lithosphere tectonic print (Cannat et al., 2008). Interestingly, below slow spreading ridges where the lithospheric walls penetrate deeper than 10 km, these walls seem to control somewhat the longitudinal upper asthenospheric flow.

In comparison to the SWIR, the Gakkel Ridge is more linear with few tectonic discontinuities, with spreading rates decreasing eastwards that are lower than those at the SWIR. From west to east, the topography deepens and magmatism progressively decreases, but it is controlled neither by lower spreading rates nor by lithospheric tectonics. Again, the source of these discrepancies lies in primary heterogeneities of the melting mantle, either due to temperatures and/or chemistry (variable levels of depletion/enrichment, as recently documented by Wanless

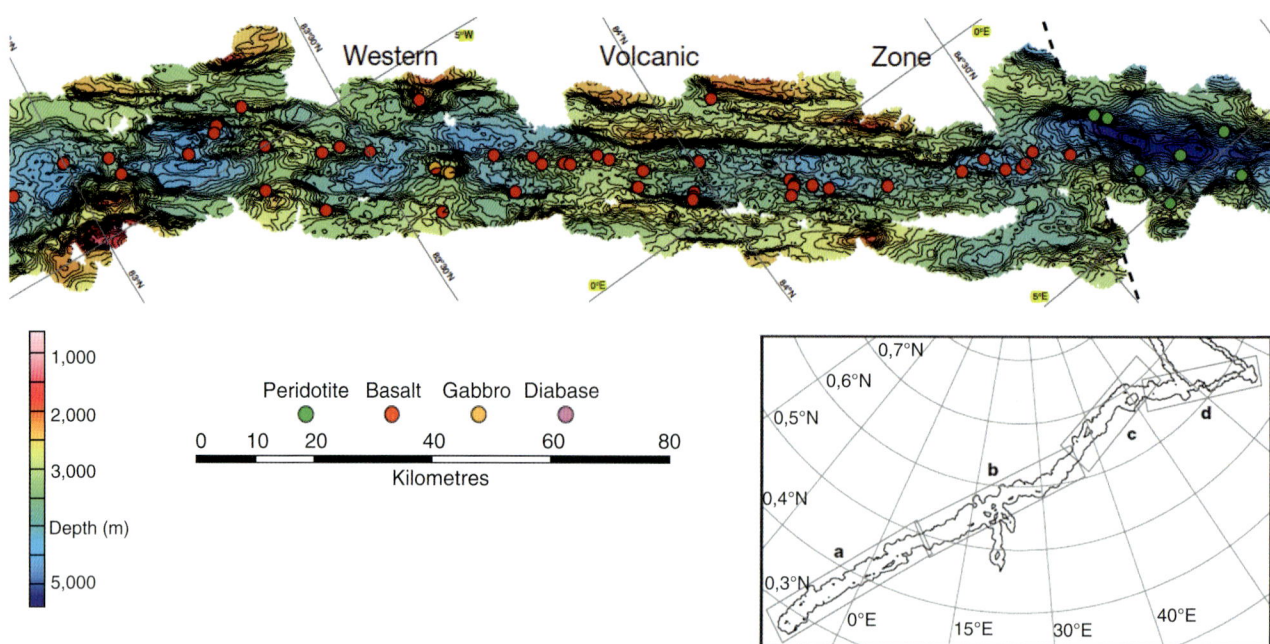

Oceanic Spreading Centers, Figure 14 The western segment of Gakkel Ridge (segment "a"), mainly in the volcanic zone (volcanoes, *red dots*), and to the right (segment "d"), the border of the central zone with rift floor at 6,000 m largely exposing the peridotite (Michael et al., 2003).

et al., 2014). Hydrothermal activity in Gakkel Ridge is unexpectedly abundant and may be a result of focusing of mantle melts feeding large volcanic centers, coupled with deep penetration of faults through the cold lithosphere (Michael et al., 2003).

Back-arc spreading centers

Apart from spreading at mid-ocean ridges, spreading can also occur in back-arc settings where the spreading is initiated due to the rollback of the subducted lithospheric slab beneath the active volcanic arc. Examples of such situations include the Lau Spreading Center in the Lau Basin behind the Tonga-Kermadec island arc and subduction zone (Turner et al., 1999; Jacobs et al., 2007), the East Scotia Sea spreading center behind the Scotia arc (Livermore et al., 1997), and the Andaman Sea spreading center (Kamesh Raju et al., 2004) behind Andaman-Nicobar subduction. Since they are related to the subduction systems, they are generally younger spreading centers, showing strong variability in terms of ridge morphology, petrology, and crustal structure, both in time and space. The Lau Basin spreading center is the most completely studied back-arc spreading center. It is intermediate in terms of spreading rate, ranging from 95 mm/year in the north to 40 mm/year in the south and is located 185–45 km from the Tonga-Kermadec volcanic arc. It is about 550 km long and is segmented by overlapping spreading centers, instead of transform faults. Because of this strong variation in spreading rate and the distance from the arc, crustal structure also varies rapidly along the ridge axis. The layer 2A thickness varies from 400 m in the north to 1 km in the south. AMLs are present in the north and south and lie at 1.5 km depth in the north to 2.8 km in the south. Similar to the 9°N OSC, a wide AML was imaged beneath, the Lau OSC (Collier and Sinha, 1990). The average layer 2A velocity varies from 3.2 to 2.5 km/s, whereas the average layer 2B velocity varies from 5 to 3 km/s; this large velocity variation is attributed to petrological changes from more mid-ocean ridge basalt-type to more silica-rich arc-type melt beneath the ridge axis (Taylor et al., 1996; Turner et al., 1999).

Summary

From fast or superfast to slow and ultraslow spreading rates, the respective roles of asthenosphere and lithosphere in the features of ocean ridges deserve some concluding comments. The faster the spreading rate, the thinner the lithosphere remains over wide areas and the more this thin lithosphere appears autonomous with respect to the underlying asthenosphere. This is illustrated by Toomey et al. (2007) who describe at an oblique alignment of the mantle individual diapirs compared to the ridge alignment along the 9°–10° N EPR and the associated orthogonal transform faults (Figure 4). It appears that this thin lithosphere is decoupled from the soft, underlying asthenosphere and responds by brittle fracturing to the large-scale plate tectonics. In contrast, the progressive thickening of lithospheres from slow to ultraslow spreading ridges is increasingly controlled by the uprising asthenosphere. This is visible in their morphology and their large-scale tectonics. There is some evidence from ophiolites (Nicolas et al., 1999) and slow spreading ridges, based on seismic anisotropy (Li and Detrick, 2003; Nowacki et al., 2012), suggesting that mantle flow is ridge parallel and not more ridge perpendicular. As discussed above, ultraslow ridge morphology and crustal composition respond to variations in mantle upwelling and associated melt delivery.

The combined study of ophiolites and geophysical data at oceanic ridges has allowed to comprehend the inner working of ocean spreading processes. The ophiolites allowed us to image deep accreted lithosphere and, indirectly, to investigate accretion processes, whereas the geophysics provide direct images of the accreting ridge systems. These observations showed that these accreting systems are different depending firstly on spreading rates and sometimes displaying significant variations along the same segment of a ridge axis. For example, along the most studied 9° N segment of the EPR, the width of the AML changes significantly and so does the melt-mush distribution along the ridge axis. Along slow spreading ridges, the crust is normally thickest at the center of a segment and thinnest at segment ends. The crustal accretion process at slow spreading centers is further complicated by the presence of ocean core complexes, which are not just present at segment ends, but could be present along a significant part of a segment (Escartin et al., 2008). The ultraslow spreading ridges, a special class of ridges, show extreme variations in seafloor morphology, crustal thickness, and geochemistry; more geophysical studies of ultraslow spreading ridges are required to fully appreciate the diversity of ridge processes on earth.

Acknowledgments

We would like to thank F. Boudier and M. Cannat for fruitful discussions and M.R. Perfit for a very comprehensive and thoughtful review.

Bibliography

Boudier, F., and Nicolas, A., 2011. Discovery of oceanic ridge axial melt lenses in the Oman ophiolite. *Earth and Planetary Science Letters*, **304**, 313–325.

Canales, J. P., Detrick, R. S., Lin, J., and Collines, J. A., 2000. Crustal and upper mantle seismic structure beneath the rift mountains and across nontransform offset at the Mid-Atlantic Ridge (35° N). *Journal of Geophysical Research*, **105**, 2699–2719.

Canales, J. P., et al., 2011. Network of off-axis melt bodies at the East Pacific Rise. *Nature Geoscience*, **5**, 279–283.

Cannat, M., 1996. How thick is the magmatic crust at slow spreading oceanic ridges? *Journal of Geophysical Research*, **101**, 2847–2857.

Cannat, M., Sauter, D., Bezos, A., Meyzen, C., Humler, E., and Le Rigoleur, M., 2008. Spreading rate, spreading obliquity, and melt supply at ultraslow spreading Southwest Indian Ridge. *Geochemistry, Geophysics, Geosystems*, **9**, Q04002.

Cathles, L. M., 1993. A capless 350 °C flow zone model to explain megaplumes, salinity variations and high-temperature veins in ridge axis hydrothermal systems. *Economic Geology*, **88**, 1977–1988.

Collier, J., and Singh, S. C., 1998. Poisson's ratio in young oceanic crust from two-ship wide-angle data using waveform inversion. *Journal of Geophysical Research*, **103**, 20981–20996.

Collier, J., and Sinha, M. C., 1990. Seismic images of a magma chamber beneath the Lau Basin back-arc spreading centre. *Nature*, **346**, 646–648.

Crawford, W. C., and Webb, S. C., 2002. Variations in the distribution of magma in the lower crust and at the Moho beneath the East Pacific Rise at 9°-10°N. *Earth and Planetary Science Letters*, **203**, 117–130.

Crawford, W. C., Singh, S. C., Seher, T., Combier, V., Dusunur, D., and Cannat, M., 2010. Crustal structure, magma chamber faults beneath the Lucky Strike hydrothermal fields. *AGU Geophysical Monograph Series*, **188**, 440.

DeMartin, B. J., Sohn, R. A., Canales, J. P., and Humphris, S. E., 2007. Kinematics and geometry of active detachment faulting beneath the Trans-Atlantic Geotraverse (TAG) hydrothermal field on the Mid-Atlantic Ridge. *Geology*, **35**(8), 711–714, doi:10.1130/G23718A.1.

Detrick, R. S., Buhl, P., Vera, E., Mutter, J., Orcutt, J., Madsen, J., and Brocher, T., 1987. Multi-channel seismic imaging of a crustal magma chamber along the East Pacific Rise. *Nature*, **326**, 35–41.

Detrick, R. S., White, R. S., and Purdy, G. M., 1993. Crustal structure of North Atlantic fracture zone. *Reviews of Geophysics*, **31**, 439–458.

Dick, H., Lin, J., and Schouten, H., 2003. An ultraslow spreading class of ocean ridge. *Nature*, **426**, 405–412.

Dunn, R. A., and Toomey, D. R., 1997. Seismological evidence for three-dimensional melt migration beneath the East Pacific Rise. *Nature*, **388**, 259–262.

Dunn, R. A., Toomey, D. R., and Solomon, S. C., 2000. Three-dimensional seismic structure and physical properties of the crust and shallow mantle beneath the East Pacific Rise 9°30′ N. *Journal of Geophysical Research*, **105**, 23537–23555.

Escartin, J., et al., 2008. Central role of detachment faults in accretion of slow-spreading oceanic lithosphere. *Nature*, **455**, 790–794

Fontaine, F. J., and Wilcock, W. S. D., 2007. Two-dimensional models of hydrothermal convection at high Rayleigh and Nusselt numbers: implications for mid-ocean ridges. *Geochemistry, Geophysics, Geosystems*, **8**, Q07010, doi:10.1029/2007GC001601.

Hansen, L. N., et al., 2013. Mylonitic deformation at the Kane oceanic core complex: implication for the rheological behavior of oceanic detachment faults. *Geochemistry, Geophysics, Geosystems*, **14**(8), doi:10.1002/ggge.20184.

Harding, A. J., et al., 1989. Structure of young oceanic crust at 13 N on the East Pacific Rise from expanding spread profiles. *Journal of Geophysical Research*, **94**, 12163–12196.

Harding, A. J., Kent, G. M., and Orcutt, J. A., 1993. A multichannel seismic investigation of upper crustal structure at 9° N at the East Pacific Rise: implications of crustal accretion. *Journal of Geophysical Research*, **98**, 13925–13944.

Henig, J. M., Blackman, D. K., Harding, A. J., Canales, J. P., and Kent, G. M., 2012. Downward continued multichannel seismic refraction analysis of Atlantis Massif oceanic core complex, 30°N, Mid-Atlantic Ridge. *Geochemistry, Geophysics, Geosystems*, **13**(1), doi:10:1029/2012GC004059

Jacobs, A. M., Harding, A. J., and Kent, G. M., 2007. Axial crustal structure of the Lau back-arc basin from velocity modeling of multichannel seismic data. *Earth and Planetary Science Letters*, **259**, 239–255.

Jousselin, D., and Nicolas, A., 2000. The Moho transition zone in the Oman ophiolite – relation with wehrlites in the crust and dunites in the mantle. *Marine Geophysical Researches*, **21**(3–4), 229–241.

Karato, S., 2012. On the origin of the asthenosphere. *Earth and Planetary Science Letters*, **321**, 95–103.

Kamesh Raju, K. A., Ramprasad, T., Rao, P. S., Ramalingeswara Rao, B., and Varghese, J., 2004, New insights into the tectonic evolution of the Andaman basin, northeast Indian Ocean. *Earth Planetary Science Letters*, **221**, 145–162.

Kawakatsu, H., Kumar, P., Takei, Y., Shinohara, M., Kanazawa, T., Araki, E., and Suyehiro, K., 2009. Seismic evidence for sharp lithosphere-asthenosphere boundaries of oceanic plates. *Science*, **324**, 499–502.

Kent, G. M., Harding, A. J., and Orcutt, J. A., 1993. Distribution of magma beneath the East Pacific Rise between the Clipperton transform and the 9° 17′ N Deval from forward modelling of common depth point data. *Journal of Geophysical Research*, **98**, 13945–13969.

Kent, G. M., Harding, A. J., Orcutt, J. A., Detrick, R. S., Mutter, J. C., and Buhl, P., 1994. Uniform accretion of oceanic crust south of the Garrett transform at 14° 15′ S on the East Pacific Rise. *Journal of Geophysical Research*, **99**, 9097–9116.

Kent, G. M., Singh, S. C., Harding, A. J., Sinha, M. C., Tong, V., Barton, P. J., Hobbs, R., White, R., Bazin, S., and Pye, J., 2000. Evidence from three-dimensional reflectivity images for enhanced melt supply beneath mid-ocean-ridge discontinuities. *Nature*, **406**, 614–618.

Li, A., and Detrick, R. S., 2003. Azimuthal anisotropy and phase velocity beneath Iceland: implication for plume-ridge interaction. *Earth and Planetary Science Letters*, **214**, 153–165.

Livermore, R., et al., 1997. Subduction influence on magma supply at the East Scotia Ridge. *Earth and Planetary Science Letters*, **123**, 255–268.

Lowell, R. P., Farough, A., Germanovich, L. N., Hebert, L. B., and Horne, R., 2012. A vent-field-scale model of the East Pacific Rise 9°50′N magma-hydrothermal system. *Oceanography*, **25**(1), 158–167, doi:10.5670/oceanog.2012.13.

Macdonald, K. C., Fox, P. J., Perram, L. J., Eisen, M. F., Haymon, R. M., Miller, S. P., Carbotte, S. M., Cormier, M. H., and Shor, A. N., 1988. A new view of the mid-ocean ridge from the behaviour of the ridge-axis discontinuities. *Nature*, **335**, 217–225.

MacLennan, J., Hulme, T., and Singh, S. C., 2004. Thermal model of oceanic crustal accretion: linking geophysical, geological and geochemical observations. *Geochemistry, Geophysics, Geosystems*, **5**, Q02F25.

MacLeod, C. J., Escartin, J., Banerji, D., Banks, G. J., Gleeson, M., Irving, D. H. B., Lilly, R. M., McCaig, A. M., Niu, Y., Allerton, S., and Smith, D. K., 2002. Direct geological evidence for oceanic detachment faulting: the Mid-Atlantic Ridge 15°45′N. *Geology*, **30**(10), 879–882.

Mainprice, D., 1997. Modelling the anisotropic seismic properties of partially molten rocks found at mid-ocean ridges. *Tectonophysics*, **279**, 161–179.

Manning, C. E., Weston, P. E., and Mahon, K. I., 1996. Rapid high-temperature metamorphism of East Pacific Rise gabbros from Hess Deep. *Earth and Planetary Science Letters*, **144**, 123–132.

Manning, G. E., MacLeod, C. J., and Weston, P. E., 2000. Lower-crustal cracking front at fast-spreading ridges: evidence from the East Pacific Rise and the Oman ophiolite. *Geological Society of America Special Papers*, **349**, 261–272.

Michael, P. J., et al., 2003. Magmatic and amagmatic seafloor generation at the ultraslow spreading Gakkel ridge, Arctic Ocean. *Nature*, **423**, 956–961.

Murase, T., and McBirney, A. R., 1973. Properties of some common igneous rocks and their melts at high temperature. *Geological Society of America Bulletin*, **84**, 3563–3592.

Nedimovic, M. R., Carbotte, S. M., Harding, A. J., Detrick, R. S., Canales, J. P., Diebold, J. B., Kent, G. M., Tischer, M., and Babcock, J. M., 2005. Frozen magma lenses below the oceanic crust. *Nature*, **436**, 1149–1152.

Nicolas, A., 1989. *Structures of Ophiolites and Dynamics of Oceanic Lithosphere*. Dordrecht: Kluwer Academic Publishers.

Nicolas, A., 1992. Les montagnes sous la mer. BRGM editions, Orléans. English translation: The Mid-Oceanic Ridges: Mountains Below Sea Level. 1995, Edition BRGM, Orlean.

Nicolas, A., and Ildefonse, B., 1996. Flow mechanism and viscosity in basaltic magma chambers. *Geophysical Research Letters*, **23**, 2013–2016.

Nicolas, A., Reuber, I., and Benn, K., 1988. A new magma chamber model based on structural studies in the Oman ophiolite. *Tectonophysics*, **151**, 87–105.

Nicolas, A., Boudier, F., and Meshi, A., 1999. Slow spreading accretion and mantle denudation in the Mirdita ophiolite (Albania). *Journal of Geophysical Research*, **104**(B7), 15155–15167.

Nicolas, A., Boudier, F., and Mainprice, D., 2003. High-temperature seawater circulation throughout crust of oceanic crust: a model derived from the Oman ophiolite. *Journal of Geophysical Research*, **108**(B8), 2371, doi:10.1029/2002JB002094.

Nicolas, A., Boudier, F., and France, L., 2009. Subsidence in magma chamber and the development of magmatic foliation in the Oman ophiolite gabbros. *Earth and Planetary Science Letters*, **284**, 76–87.

Nooner, S. L., et al., 2003. Constraints on crustal structure at the Mid-Atlantic Ridge from seafloor gravity measurements made at the Atlantis Massif. *Geophysical Research Letters*, **30**(8), 1446, doi:10.1029/2003GL017126.

Nowacki, A., Kendall, J. M., and Wookey, J., 2012. Mantle anisotropy beneath the Earth's mid-ocean ridges. *Earth and Planetary Science Letters*, **317–318**, 56–67, doi:10.1016/j.epsl.2011.11.044.

Phipps Morgan, J., and Chen, Y. J., 1993. Dependence of ridge-axis morphology on magma supply and spreading rate. *Nature*, **364**, 706–708.

Rabinowicz, M., Nicolas, A., and Vigneresse, J. L., 1984. A rolling mill effect in asthenosphere beneath oceanic spreading centers. *Earth and Planetary Science Letters*, **67**, 97–108.

Rioux, M., Bowring, S., Kelemen, P., Gordon, S., Miller, R., and Dudas, F., 2012. Tectonic development of the Samail ophiolite; high precision U-Zircon geochronology and Sm-Nd isotopic constraints on crustal growth and emplacement. *Journal of Geophysical Research*, **117**, B07201.

Sauter, D., Cannat, M., Rouméjon, S., Andreani, M., Birot, D., Bronner, A., Brunelli, D., Carlut, J., Delacour, A., Guyader, V., MacLeod, J., Manatschal, G., Mendel, V., Ménez, B., Pasini, V., Ruellan, E., and Searle, R., 2013. Continuous exhumation of mantle-derived rocks at the Southern Indian Ridge for 11 million years. *Nature Geoscience*, **6**, 314–320.

Schroeder, T., and John, B. E., 2004. Strain localization on an oceanic detachment fault system, Atlantis Massif, 30°N, Mid-Atlantic Ridge. *Geochemistry, Geophysics, Geosystems*, **5**(11), Q11007, doi:10.1029/2004GC000728.

Seher, T., Crawford, W., Singh, S. C., Cannat, M., Combier, V., Dusunur, D., and Canales, J.-P.,2010. Crustal velocity structure of the Lucky Strike segment of the Mid-Atlantic Ridge (37°N) from seismic refraction measurements. *J.G.R.*, **115**, B03103. doi:10.1029/2009.

Shipboard Scientific Party. 1999. Leg 179 Summary. In Pettigrew, T. L., Casey, J. F., and Miller, D. J. (eds.), *Proceeding of ODP, Initial Reports*. College Station: Ocean Drilling Program, Vol. 179, pp. 1–26.

Singh, S. C., Kent, G. M., Collier, J. S., Harding, A. J., and Orcutt, J. A., 1998. Melt to mush variations in crustal magma properties along the ridge crest at the southern East Pacific Rise. *Nature*, **394**, 874–878.

Singh, S. C., Collier, J. S., Kent, G. M., Harding, A. J., and Orcutt, J. A., 1999. Seismic evidence for a hydrothermal layer above the solid roof of axial magma chamber at the southern East Pacific Rise. *Geology*, **27**, 219–222.

Singh, S. C., Harding, A., Kent, G., Sinha, M. C., Combier, V., Hobbs, R., Barton, P., White, R., Tong, V., Pye, J., and Orcutt, J. A., 2006a. Seismic reflection images of Moho underlying melt sills at the East Pacific Rise. *Nature*, **442**, 287–290.

Singh, S. C., Crawford, W., Carton, H., Seher, T., Combier, V., Cannat, M., Canales, J., Dusunur, D., Escartin, J., and Miranda, M., 2006b. Discovery of a magma chamber and faults beneath a hydrothermal field at the Mid-Atlantic Ridge. *Nature*, **442**, 1029–1033.

Sinha, M. C., Navin, D., MacGregor, L., et al., 1997. Evidence for accumulated melt beneath the slow spreading Mid-Atlantic Ridge. *Philosophical Transactions of the Royal Society A*, **355**, 233–253.

Sleep, N. H., and Rosendhal, B. R., 1979. Topography and tectonics of Mid-Oceanic Ridge axes. *Journal of Geophysical Research*, **84**(B12), 6831–6839.

Taylor, B., Zellmer, K., Martinnez, F., and Goodliffe, A., 1996. Seafloor spreading in the Lau backarc basin. *Earth and Planetary Science Letters*, **144**, 35–40.

Tolstoy, M., et al., 2008. Seismic identification of along-axis hydrothermal flow on the East Pacific Rise. *Nature*, **451**, 181–185.

Toomey, D. R., Jousselin, D., Dunn, R., Wilcock, W., and Detrick, R., 2007. Skew of mantle upwelling beneath the East Pacific Rise governs segmentation. *Nature*, **446**, 409–414.

Toomey, D. R., Wilcock, W., Solomon, S and Orcutt, J.A., 1998. Mantle seismic structure beneath the MELT region of the East Pacific Rise from P and S wave tomography. *Science*, **280**, 1224–1227.

Tucholke, B. E., Lin, J., and Kleinrock, M. C., 1998. Megamullions and mullions structure defining oceanic metamorphic core complexes on the Mid-Atlantic Ridge. *Journal of Geophysical Research*, **103**, 9857–9866.

Tucholke, B. E., Behn, M. D., Buck, W. R., and Lin, J., 2008. Role of melt supply in oceanic detachment faulting and formation of megamullions. *Geology*, **36**, 455–458.

Turner, I. M., Peirce, C., and Sinha, M. C., 1999. Seismic imaging of the axial region of the Valu Fa Ridge, Lau Basin – the accretionary process of an intermediate back-arc spreading ridge. *Geophysical Journal International*, **138**, 495–519.

Wanless, V. D., Behn, M. D., Shaw, A. M., and Plank, T., 2014. Variations in melting dynamics and mantle compositions along the Eastern Volcanic Zone of the Gakkel Ridge: insights from olivine-hosted melt inclusions. *Contributions to Mineralogy and Petrology*, **167**, 1005.

Cross-references

Crustal Accretion
Layering of Oceanic Crust
Marginal Seas
Oceanic Rifts
Ophiolites
Plate Tectonics

OIL SPILL

Erich Gundlach
E-Tech International Inc., Boulder, CO, USA

Definition

An oil spill is the accidental release of liquid hydrocarbons to the environment as crude oil or a petroleum product.

Introduction

Experts in coastal geosciences play a key role in understanding the processes affecting a spill, measures appropriate to remove spilled oil, and, importantly, when to end cleanup activities and let nature continue the cleansing process. The major processes affecting oil on sedimentary shorelines are illustrated in Figure 1.

Oil composition and weathering

The spilled oil's composition and subsequent changes over time (weathering) greatly influence the interaction of oil in coastal environments. Oil is composed of a complex mixture of straight-chained and ringed/cyclic hydrocarbon compounds, with small amounts of N, S, O, and trace metals, depending on the source reservoir and petroleum processing (cracking). Spilled oil is commonly categorized into broad categories. Light oils have a low specific gravity (SG < 0.8) and low viscosity (0.5–2.0 cSt) and include gasoline/petrol, naphtha, and kerosene. Medium-grade oils are mostly crude oils and medium fuel oils (SG 0.80–0.95, 8–275 cSt), while heavy fuel oils comprise the heavy oil category (SG > 0.95 and viscosity >1,500 cSt).

Weathering refers to the biochemical processes affecting oil when released to the open environment and includes spreading, advection, evaporation, emulsification, dissolution, dispersion, photooxidation, sedimentation, and biodegradation (Reed et al., 1999). Of these, evaporation and emulsification are the primary affecting processes. Evaporation potentially reduces the quantity of remaining oil up to 40 % for light oils, whereas the generation of emulsified oil (a water-in-oil mixture commonly called "mousse") potentially doubles the amount of oily material. As oil weathers, viscosity, density, and adhesion (stickiness) to sediments and rock surfaces all increase. Biodegradation is a long-term process slowly reducing the amount of remaining oil. These major processes have been summarized for the Amoco Cadiz (Gundlach et al., 1983), *Exxon Valdez* (Wolfe et al., 1994), and Deepwater Horizon (FISG, 2010) oil spills.

Applying geosciences to oil spills

The impetus to understand the dynamic processes of spilled oil in coastal environments began in the 1970s as a result of an increasingly strong environmental movement, several major oil spills (e.g., Torrey Canyon, 1967; Santa Barbara, 1969), and offshore oil and gas

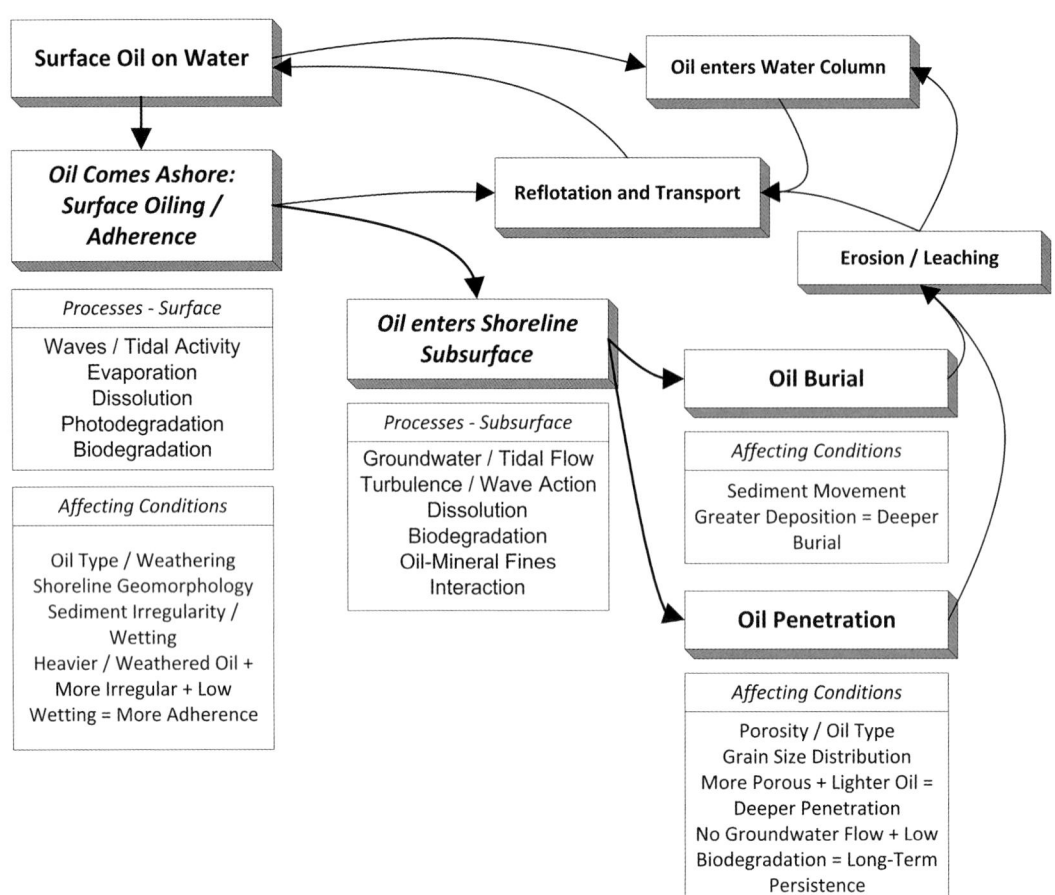

Oil Spill, Figure 1 Processes and conditions affecting spilled oil on sedimentary shorelines (modified from Page et al. 2013).

development. Field investigations in 1970 (*Arrow*), 1975 (*Metula*), 1976 (*Urquiola, Argo Merchant*), and 1978 (*Amoco Cadiz*, Ixtoc I) enabled geoscientists to begin understanding the complex interactions between spilled oil, shorelines, sediments, and coastal processes. Table 1 lists the major studies of oil/shoreline interactions and serves as references to the observations described in the following sections. *Exxon Valdez* studies have continued for more than 20 years and provide the longest detailed record of any oil spill to date (Wiens, 2013).

Spill interaction: sediment-dominated shorelines

On sedimentary shorelines, the extent of oil impact and persistence is determined by oil characteristics, changes in the beach profile in response to wave and tidal action, and grain size characteristics. Both depths of oil burial and penetration generally increase with grain size, with wide variability dependent on local conditions. Silt or fine-grained beaches show less burial/penetration and cobble/boulder shorelines the greatest. Sediments where pore spaces are filled with water, as on tidal flats or along lower portions of sedimentary shorelines, inhibit oil entry and generally remain unoiled.

As the spill progresses, sediment movement, wave action, and weathering tend to distribute the spill into ever smaller particles (tar balls), which are then exposed to further degradation by microorganisms. Stranded oil may also be reduced by adherence to very fine-grained material (mineral fines) (Bragg and Owens, 1995; Owens and Lee, 2003). Small quantities of weathered oil may become deeply incorporated into beach sediments, requiring several years for natural removal. Spilled oil may also interact with shoreline sediments and sink in the nearshore zone. Changes in wave energy or direction can cause a movement of this oil and sediments onto the beach and cause reoiling.

The *Exxon Valdez* is an ongoing case study where oil has deeply penetrated into coarse-grained environments and is still present at least 23 years later in well-known localities. The remaining oiled sediments reside in an area of restricted interstitial water flow and below the zone of active sediment movement. Therefore, physical as well as microbial processes are greatly reduced but not totally stopped.

Spill interaction: rock-dominated shorelines

Spilled oil tends to be held offshore by wave reflection on uniform vertical cliffs exposed to high wave action.

Oil Spill, Table 1 Major field and experimental studies of oil interaction with shoreline environments. Shoreline types: C coarse grained, R rocky, S sandy, T tidal flat, M marsh

Spill or study	Location	Shoreline type(s)	Years persistent[a]	Illustrative references
West Falmouth, 1969	Massachusetts, United States	M	35	Reddy et al. (2002), Peacock et al. (2005)
Arrow, 1970	Nova Scotia, Canada	C	35	Owens et al. (2008), Owens (2010)
Metula, 1974	Strait of Magellan, Chile	C, M, T	30	Gundlach et al. (1978b), Gundlach et al. (1982), Baker et al. (1993), Gundlach (1997), Owens and Sergy (2005)
Urquiola, 1976	Galicia, Spain	C, R, S		Gundlach et al. (1978a), Gundlach et al. (1978b)
Amoco Cadiz, 1978	Brittany, France	C, M, R, T	23	Gundlach et al. (1981), Oudot and Chaillan (2010)
BIOS, 1981	Baffin Island, Canada	C	20	Owens et al. (1987), Prince et al. (2002)
TROPICS, 1983	Panama	M	25	DeMicco et al. (2011)
Oiling experiments	Various locations	C, S		Little and Scales (1987), Harper and Harvey-Kelly (1994), Cheong (2004), Bernabeu et al. (2010)
Exxon Valdez, 1989	Alaska, United States	C	23	Owens et al. (2008), Hayes et al. (2010), Li and Boufadel (2011), Pope et al. (2010), Wiens (2013)
Gulf War spills, 1991	Kuwait and Saudi Arabia	S, T	12	Barth (2002), Hayes et al. (1993), Michel et al. (2005)
Erika, 1999	Brittany, France, and Spain	R, S		Kerambrun (2003)
Prestige, 2002	Galicia, Spain, and France	S	7	Bernabeu et al. (2006), González et al (2009)
Deepwater Horizon, 2010	Gulf of Mexico, United States	M, S		Hayworth et al. (2011), OSAT-2 (2011), Parham (2011), Wang and Roberts (2013)

[a]Persistence >6 years as of last publication

However, oil may settle in thick pools as the irregularity of the shoreline increases and in areas with lower wave energy. Light oils are rapidly removed by wave action. Heavy oils are the most persistent and commonly require extensive pressure washing to remove.

Spill interaction: vegetated shorelines

The generally quiescent nature of marsh and mangrove shorelines enables spilled oil to settle onto the silt-clay surfaces of the habitat. Oil may mix into soft sediment and/or penetrate by the infilling of animal burrows (Howard and Little, 1987). The loss of plants potentially causes shoreline erosion and habitat loss. Oil incorporated into these sediments, in the absence of wave action and reduced microbial degradation, has been shown to persist at least 35 years. The large-scale destruction of mangroves in the Ogoni area of Nigeria (UNEP, 2011) potentially shows the greatest loss of mangroves due to oil spillage to date.

Coastal geomorphology and oil spill sensitivity

Early studies identified the influence of shoreline characteristics in predicting spill persistence, difficulty of cleanup, and, because biota also vary by shoreline type, its potential biological effects. To enable use by nonscientists, a vulnerability index classification, later changed to environmental sensitivity index (ESI), was developed, which categorizes and ranks shoreline geomorphology according to sensitivity to spilled oil (Gundlach and Hayes, 1978). With the inclusion of ecological and social data (NOAA, 2002, 2012), ESI maps are now an international standard to provide summary information to governments and industry involved with spill response planning.

Geoscience advise during response operations

During major oil spills, geoscientists play a key advisory role in directing SCAT (Shoreline Cleanup Assessment Technique) teams. These teams provide site-specific data concerning the quantity of oil present (surface and subsurface), biological and social characteristics, and cleanup options and restrictions (Owens and Sergy, 2003). Importantly, SCAT teams include the representatives of the spiller, government, and landowner. First deployed during the *Exxon Valdez* spill, SCAT information enabled the long-term tracking of remaining oil on the shoreline (Taylor and Reimer, 2008; Wiens, 2013). SCAT presently has widespread international support by governments and industry and was used extensively during *Deepwater Horizon* (Santner et al., 2011).

Oil spill modeling

Advances in oil spill modeling occurred with the inclusion of oil-holding capacities (Gundlach, 1987) based on the sedimentary characteristics, the oil type, the depositional width of the shoreline, the tidal range, and the application of a removal rate parameter dependent on shoreline type (Reed et al., 1989; Howlett and Jayko, 1998; Etkin et al., 2007). The integration of coastal characteristics into oil spill models enables a more realistic representation of what happens when spilled oil reaches the shoreline. Some spill models simplify the entire process into a simple reflotation parameter (i.e., high reflotation from exposed rock surfaces, low reflotation from sheltered marshes and mangroves). The testing of various reflotation parameters for the same section of shoreline indicates their importance for spill modeling (Danchuk and Wilson, 2010).

Shore zone coastal mapping

Shore zone mapping uses low-altitude high-resolution photography and video surveying to classify the shoreline into standardized geomorphic and biological categories. It is used to characterize large sections of coastline (e.g., Alaska) with the benefit that images and data are able to be retrieved from any web-linked computer (Hamey et al., 2008; NOAA Fisheries, 2012). The access to these images is a valuable tool for both spill planning and response.

Summary and conclusions

The scientific understanding of oil spill interactions in the coastal and marine environments has progressed substantially from its initial beginnings in the mid-1970s. Because of their unique background in understanding these interactions and affecting processes, marine geoscientists play a key role in providing technical and scientific support during most major oil spills occurring today.

Bibliography

Baker, J. M., Guzman, L., Bartlett, P. D., Little, D. I., and Wilson, C. M., 1993. Long-term fate and effects of untreated thick oil deposits on salt marshes. In *Proceedings International Oil Spill Conference*, pp. 395–399. doi:10.7901/2169-3358-1993-1-395.

Barth, H.-J., 2002. *The 1991 Gulf War Oil Spill, its Ecological Effects and Recovery Rates of Intertidal Ecosystems at the Saudi Arabian Gulf coast – Results of a 10-year Monitoring Period*. Wissenschaftliche Arbeit Im Rahmen des Habilitationsverfahrens im Fach Geographie, 270 pp. http://www.jubail-wildlife-sanctuary.info/pdf/Barth2002.pdf. Accessed December 2012.

Bernabeu, A. M., Nuez de la Fuente, M., Rey, D., Rubio, B., Vilas, F., Medina, R., and González, M. E., 2006. Beach morphodynamics forcements in oiled shorelines: coupled physical and chemical processes during and after fuel burial. *Marine Pollution Bulletin*, **52**, 1156–1168, doi:10.1016/j.marpolbul.2006.01.013.

Bernabeu, A. M., Rey, D., Lago, A., and Villas, F., 2010. Simulating the influence of physicochemical parameters on subsurface oil on beaches: preliminary results. *Marine Pollution Bulletin*, **60**(8), 1170–1174, doi:10.1016/j.marpolbul.2010.04.001.

Bragg, J. R., and Owens, E. H., 1995. Shoreline cleansing by interactions between oil and fine mineral particles. In *Proceedings International Oil Spill Conference*, pp. 219–227. doi:10.7901/2169-3358-1995-1-219.

Cheong, C.-J., 2004. Penetration of dispersed and weathered oil and its ecological implication in model sandy beach. *Environmental Engineering Research*, **9**, 23–30.

Danchuk, S., and Wilson, C. S., 2010. Effects of shoreline sensitivity on oil spill trajectory modeling of the Lower Mississippi River. *Environmental Science Pollution Research*, **17**, 331–340, doi:10.1007/s11356-009-0159-8.

DeMicco, E., Schuler, P. A., Omer, T., and Baca, B., 2011. Net Environmental Benefit Analysis (NEBA) of dispersed oil on nearshore tropical ecosystems: TROPICS – the 25th year research visit. In *Proceedings International Oil Spill Conference*, 14 pp. doi:10.7901/2169-3358-2011-1-282.

Etkin, D., French-McCay, D., and Michel, J., 2007. *Review of the State-of-the-Art on Modeling Interactions between Spilled Oil and Shorelines for the Development of Algorithms for Oil Spill Risk Analysis Modeling*. OCS Study MMS 007–063, U.S. Department of Interior, 157 pp.

FISG, 2010. *Oil Budget Calculator, Deepwater Horizon*. Technical Documentation, November 2010. The Federal Interagency Solutions Group, 217 pp.

González, M., Medina, R., Bernabeu, A. M., and Novóa, X., 2009. Influence of beach morphodynamics in the deep burial of fuel in beaches. *Journal of Coastal Research*, **25**(4), 799–818.

Gundlach, E. R., 1987. Oil-holding capacities and removal coefficients for different shoreline types to computer simulate spills in coastal waters. In *Proceedings International Oil Spill Conference*, pp. 451–457. doi:10.7901/2169-3358-1987-1-451.

Gundlach, E. R., 1997. Comparative photographs of the Metula spill site, 21 years later. In *Proceedings International Oil Spill Conference*, pp. 1042–1044. doi:10.7901/2169-3358-1997-1-1042.

Gundlach, E. R., and Hayes, M., 1978. Classification of coastal environments in terms of potential vulnerability to oil spill damage. *Marine Technology Society Journal*, **12**(4), 18–27.

Gundlach, E. R., Ruby, C. H., Hayes, M. O., and Blount, A. E., 1978a. The Urquiola oil spill, La Coruña, Spain: impact and reaction on beaches and rocky coasts. *Environmental Geology*, **2**(3), 131–143.

Gundlach, E. R., Hayes, M. O., Ruby, C. H., Ward, L. G., Blount, A. E., Fischer, I. A., and Stein, R. J., 1978b. Some guidelines for oil spill control in coastal environments, based on field studies of four oil spills. In *Proceedings Symposium on Chemical Dispersants for the Control of Oil Spills, ASTM STP 659*. American Society for Testing and Materials, pp. 98–118. doi:10.1520/STP659-EB.

Gundlach, E. R., Berne, S., D'Ozouville, L., and Topinka, J. A., 1981. Shoreline oil two years after Amoco Cadiz, new complications from Tanio. In *Proceedings International Oil Spill Conference*, pp. 525–534. doi:10.7901/2169-3358-1981-1-525.

Gundlach, E. R., Domeracki, D. D., and Thebeau, L. C., 1982. Persistence of Metula oil in the Strait of Magellan, six and one-half years after the incident. *Journal Oil and Petrochemical Pollution*, **1**, 37–48.

Gundlach, E. R., Boehm, P. D., Marchand, M., Atlas, R. M., Ward, D. M., and Wolfe, D. A., 1983. The fate of Amoco Cadiz oil. *Science*, **221**, 122–129.

Hamey, J. N., Morris, M., and Harper, J. R., 2008. *ShoreZone Coastal Habitat Mapping Protocol for the Gulf of Alaska,*

146 pp. http://alaskafisheries.noaa.gov/shorezone/goa_protocol.pdf. Accessed March 2013.

Harper, J. R., and Harvey-Kelly, F., 1994. *Subsurface Oil Retention in Coarse Sediments Beaches.* Report EE-147, Environment Canada, 60 pp.

Hayes, M. O., Michel, J., Montello, T. M., Aurand, D. V., Al-Mansi, A. M., Al-Moaen, A. H., Sauer, T. C., and Thayer, G. W., 1993. Distribution and weathering of shoreline oil one year after the Gulf War oil spill. *Marine Pollution Bulletin*, **27**, 135–142.

Hayes, M. O., Michel, J., and Betenbaugh, D. V., 2010. The intermittently exposed, coarse-grained gravel beaches of Prince William Sound, Alaska: comparison with open-ocean gravel beaches. *Journal of Coastal Research*, **26**(1), 4–30.

Hayworth, T. P., Clement, T. P., and Valentine, J. F., 2011. Deepwater Horizon oil spill impacts on Alabama beaches. *Hydrology and Earth System Sciences*, **15**, 3639–3649.

Howard, S., and Little, D. I., 1987. Effect of infaunal burrow structure on oil penetration into sediments. In *Proceedings International Oil Spill Conference*, pp. 427–431. doi:10.7901/2169-3358-1987-1-427.

Howlett, E., and Jayko, K., 1998. COZOIL (Coastal Zone Oil Spill Model), model improvements and linkage to a graphical user interface. In *Proceedings Arctic Marine Oilspill Project (AMOP) Technical Seminar.* Environment Canada, 18 pp. http://www.asascience.com/about/publications/pdf/1998/cozoil.pdf. Accessed 3 March 2015.

Kerambrun, L., 2003. Erika oil spill: responding in difficult-to-access coves and on cliffs. In *Proceedings International Oil Spill Conference*, pp. 1085–1089. doi:10.7901/2169-3358-2003-1-1085.

Li, H., and Boufadel, M., 2011. A tracer study in an Alaskan gravel beach and its implications on the persistence of the Exxon Valdez oil. *Marine Pollution Bulletin*, **62**, 1261–1269.

Little, D. I., and Scales, D. L., 1987. The persistence of oil stranded on sediment shorelines. In *Proceedings International Oil Spill Conference*, pp. 433–438. doi:10.7901/2169-3358-1987-1-433.

Michel, J., Hayes, M. O., Getter, C. D., and Cotsapas, L., 2005. The Gulf War oil spill twelve years later: consequences of eco-terrorism. In *Proceedings International Oil Spill Conference.* American Petroleum Institute, 5 pp. doi:10.7901/2169-3358-2005-1-957.

NOAA, 2002. *Environmental Sensitivity Index Guidelines, Version 3.0.* NOAA Technical Memorandum NOS OR&R 11, U.S. National Oceanic and Atmospheric Administration, 100 p.

NOAA Fisheries, 2012. *Alaska ShoreZone Coastal Mapping and Imagery.* www.alaskafisheries.noaa.gov/shorezone/. Accessed March 2013.

NOAA, 2012. *Proceedings ESI Mapping Workshop*, May 1–3, 2012. http://response.restoration.noaa.gov/oil-and-chemical-spills/oil-spills/response-tools/esi-workshop-next-generation-esis.html. Acces-sed March 2013.

OSAT-2, 2011. *Summary Report for Fate and Effects of Remnant Oil in the Beach Environment.* Operational Advisory Team. Prepared for Federal On-Scene Coordinator, U.S. Coast Guard, 35 pp.

Oudot, J., and Chaillan, F., 2010. Pyrolysis of asphaltenes and biomarkers for the fingerprinting of the Amoco Cadiz oil spill after 23 years. *Comptes Rendus Chimie*, **13**, 548–552.

Owens, E. H., 2010. Shoreline response and long-term oil behavior studies following the 1970 Arrow spill in Chedabucto Bay, NS. In *Proceedings Arctic Marine Oilspill Project (AMOP) Technical Seminar.* Environment Canada, pp. 207–221.

Owens, E. H., and Lee, K., 2003. Interaction of oil and mineral fines on shorelines: review and assessment. *Marine Pollution Bulletin*, **47**(9–12), 397–405.

Owens, E. H., and Sergy, G. A., 2003. The development of the SCAT process for the assessment of oiled shorelines. *Marine Pollution Bulletin*, **47**, 415–422, http://dx.doi.org/10.1016/S0025-326X(03)00211-X.

Owens, E. H., and Sergy, G. A., 2005. Time series observations of marsh recovery and pavement persistence at three Metula spill sites after 30½ years. In *Proceedings Arctic Marine Oilspill Project (AMOP) Technical Seminar.* Environment Canada, pp. 463–472.

Owens, E. H., Harper, J. R., Robson, W., and Boehm, P. D., 1987. Fate and persistence of crude oil stranded on a sheltered beach. *Arctic*, **40**(1), 109–123.

Owens, E. H., Taylor, E., and Humphrey, B., 2008. The persistence and character of stranded oil on coarse-sediment beaches. *Marine Pollution Bulletin*, **56**, 14–26.

Page, D. S., Boehm, P. D., Brown, J. S., Gundlach, E. R., and Neff, J. M., 2013. Fate of oil on shorelines. Chapter 6. In Wiens, J. A. (ed.), *Oil in the Environment: Legacies and Lessons of the Exxon Valdez Oil Spill.* Cambridge: Cambridge University Press. ISBN 9781107027176.

Parham, P., 2011. Sedimentary evolution of Deepwater Horizon/Macondo oil on the gulfside beaches of Mississippi, Alabama and western Florida: observations of a SCAT team member. In *Proceedings South-Central Regional Geological Society of America*. 27 pp. http://gsa.confex.com/data/handout/gsa/2011SC/Paper_186995_handout_89_0.ppt. Accessed December 2012.

Peacock, E. E., Nelson, R. K., Solow, A. R., Warren, J. D., Baker, J. L., and Reddy, C. M., 2005. The West Falmouth oil spill: 100 kg of oil found to persist decades later. *Environmental Forensics*, **6**, 273–281.

Pope, G., Gordon, K., and Bragg, J., 2010. Using fundamental practices to explain field observations twenty-one years after the Exxon Valdez oil spill. In *Proceedings International Oil Spill Conference.* American Petroleum Institute, 13 pp. doi:10.7901/2169-3358-2011-1-214.

Prince, R. C., Owens, E. H., and Sergy, G. A., 2002. Weathering of an Arctic oil spill over 20 years: the BIOS experiment revisited. *Marine Pollution Bulletin*, **44**, 1236–1242.

Reddy, C. M., Eglinton, T., Hounshell, A., White, H. K., Xu, L., Gaines, R. B., and Frysinger, G. S., 2002. The West Falmouth oil spill after thirty years: the persistence of petroleum hydrocarbons in marsh sediments. *Environmental Science & Technology*, **36**, 4754–4760.

Reed, M., Gundlach, E. R., and Kana, T., 1989. A coastal zone oil spill model: development and sensitivity studies. *Oil and Chemical Pollution*, **5**, 419–449.

Reed, M., Johansen, O., Brankvik, P. J., Daling, P., Lewis, A., Fiocco, R., Mackay, D., and Prentki, R., 1999. Oil spill modeling towards the close of the 20th century: overview of the state of the art. *Spill Science & Technology Bulletin*, **5**, 3–16.

Santner, R., Cocklan-Vendl, M. C., Stong, B., Michel, J., Owens, E., and Taylor, E., 2011. The Deepwater Horizon, MC 252-Macondo shoreline cleanup assessment technique (SCAT) program. In *Proceedings International Oil Spill Conference.* American Petroleum Institute, 9 pp. doi:10.7901/2169-3358-2011-1-270.

Taylor, E., and Reimer, P. D., 2008. Oil persistence on beaches in Prince William Sound – a review of SCAT surveys conducted from 1989 to 2002. *Marine Pollution Bulletin*, **40**, 458–474.

UNEP, 2011. *Environmental Assessment of Ogoniland.* Nairobi: United Nations Environment Programme. 257 pp.

Wang, P., and Roberts, T. M., 2013. Distribution of surficial and buried oil contaminants across sandy beaches along NW Florida and Alabama coasts following the Deepwater Horizon oil spill in 2010. *Journal of Coastal Research*, **29**, 144–155, doi:10.2112/JCOASTRES-D-12-00198.1.

Wiens, J. A. (ed.), 2013. *Oil in the Environment: Legacies and Lessons of the Exxon Valdez Oil Spill*. Cambridge, NY: Cambridge University Press. ISBN 9781107027176.

Wolfe, D. A., Hameedi, M. J., Galt, J. A., Watabayashi, G., Short, J., O'Claire, C., Rice, S., Michel, J., Payne, J. R., Braddock, J., Hanna, S., and Sale, D., 1994. The fate of the oil spilled from the Exxon Valdez. *Environmental Science & Technology*, **28**(3), 560A–568A, doi:10.1021/es00062a712.

Cross-references

Beach Processes
Coasts
Energy Resources
Geohazards: Coastal Disasters
Mangrove Coasts

OPHIOLITES

Adolphe Nicolas
Geosciences Montpellier, University of Montpellier 2, Montpellier, France

Definition

Ophiolites represent fragments of oceanic lithosphere that have been detached from their ocean of origin and transported by plate tectonic processes onto a continent.

Historical development of the ophiolite concept

Ophiolite, the "snake stone" in Greek for its serpentine component appears for the first time with Brongniard in 1813. At the turn of the same century, Italian alpine geologists had fully recognized its deep marine origin and its main components: with, from sea bottom, radiolarian cherts and basaltic pillow lavas, over and associated to gabbros and serpentinites. The surge of plate tectonics in the 1960s, that was a direct result of oceanic floor mapping, fostered studies on ophiolites. Ten years earlier, De Roever (1957) and Brunn (1959) had equated with the so-called alpine ophiolites to the Atlantic oceanic ridge, and by that time, most European geologists knew that ophiolites derived from oceanic spreading centers. Interestingly, marine geophysicists remained for a long time reluctant to this connection. In the 1960s with only a few exceptions, American colleagues were not convinced by the comprehensive concept of "ophiolite." They had correctly identified "peridotite-gabbro" complexes as the basement of seafloor but were doubting any connection to the upper crust of ophiolites. Outstanding among American dissidents was Harry Hess, a geologist who was promoted admiral in the US Navy during the Second World War, launching the mapping of the oceans' seafloor that has been critical when he anticipated the seafloor spreading theory. During a famous field trip meeting in Oregon, European and American communities adopted the following manifesto on ophiolites, a modern concept that we revisit here.

Ophiolites refers to a distinctive assemblage of mafic to ultramafic rocks. It should not be used as a rock name or a lithologic unit in mapping. In a completely developed ophiolite, the rock types occur in the following sequence, starting from the bottom and working up:

- *Ultramafic complex*, consisting of variable proportions of harzburgite, lherzolite, and dunite, usually with a metamorphic and tectonic fabric (more or less serpentinized)
- *Gabbroic complex*, ordinarily with cumulus textures, commonly containing cumulus peridotites and pyroxenites and usually less deformed than the ultramafic complex
- *Mafic sheeted dike complex*
- *Mafic volcanic complex*, commonly pillowed
- Associated rock types include (1) an overlying sedimentary section typically including ribbon cherts, thin shelled interbeds, and minor limestones; (2) podiform bodies of chromite, generally associated with dunites; and (3) sodic felsic intrusive and extrusive rocks.

Faulted contacts between mappable units are common. Whole sections may be missing. An ophiolite may be incomplete, dismembered, or metamorphosed. Although ophiolite generally is interpreted to be oceanic crust and upper mantle, the use of the term should be independent of its supposed origin.

Importance of studies in ophiolites

Beyond the interest for mineral resources historically extracted from ophiolites (copper, chrome, PGE, dunites, ...), two scientific aspects should be mentioned. The structural, petrological, and geochemical features of ophiolites that have been implied in collision belts can be used to constrain their marine environment of origin, for instance, a mid-ocean ridge (MOR), a suprasubduction zone (SSZ), or any other marine environment. This concern appears in a vast number of studies on ophiolites using geochemical data, considered as discriminant for the different environments of origin. Initiated by Miyashiro in Cyprus (1973), this concept has been applied to the Oman ophiolite the conclusion that it was generated within a SSZ environment. Developed in the entry "Sea floor spreading," another aim in studying ophiolites deals with seafloor accretion processes operating at oceanic ridges where the 3D information obtained in ophiolites is important, complementing and interacting with the geophysical observations: for instance, on shape and dynamics of mantle diapirs and overlying magma chambers at fast-spreading ridges and on deep rooting of oceanic core complexes and kinematics of their emplacement at slow-spreading ridges.

Typology of ophiolites

The first attempt to classify ophiolites has been presented in 1985, independently by Ishiwatari and by Boudier and Nicolas. Both started from the idea to classify ophiolites on the degree of partial melting in the mantle of origin.

Ophiolites, Table 1 The three main ophiolite types reflecting decreasing mantle partial melting related to decreasing spreading rates. Specific references to the massifs of origin (their number in first line) are presented in Nicolas and Boudier, where a comparable typology based on information from oceanic environments is also presented

	Harzburgite ophiolite type (HOT)	Harzburgite lherzolite ophiolite type (LHOT)	Lherzolite ophiolite type (LOT)
Number of Massifs considered	3	12	4
Crustal section	Thickness: 4–6 km	Thickness: 2–3 km	Thickness : 0–1 km
Volcanics overlying sheeted dikes	Low-alumina tholeiite	High-alumina tholeiite	Alkali basalt
Sheeted dikes	Well expressed and steep	Steep to horizontal, poorly organized	Absent or poorly organized
Gabbros Exposure	Thick, continuous	Thin, continuous, or nestled	Absent, small bodies
Lithology	Foliated ol-gabbros locally gabbronorites	Dominant gabbronorites, ol-gabbros, ferrogabbros	Ol-gabbros
Structure	Lower, layered gabbros Upper, foliated gabbros Top, isotropic	Well to poorly layered-foliated gabbros, isotropic gabbros	Isotropic to foliated gabbros
Penetrative deformation	Very large, magmatic	Magmatic to plastic (flasergabbro)	
Seawater alteration	Variable, up to high-T	Variable	Variable, metamorphic
Wehrlite intrusions	Abundant	Present	Absent
Mantle section			
Moho transition zone (MTZ)	Dunite-gabbro, websterites 50–400 m	Locally thick (up to 2 km) dunite, interlayered with wehrlite, websterite	Dunite lenses
Shear zones	Uncommon, HT, vertical horizontal lineation	Common, mid-T to low-T, steep, steep lineation, oceanic core complex horizontal detachments	Very common, mid-T to low-T steep
Diabase occurrence	Uncommon	Common	Common dikes-sills
Chromite pods	Uncommon	Common to very large	Uncommon, absent
Serpentinization	Lizardite	Lizardite and antigorite	Antigorite

Ishiwatari mainly considered petrological and geochemical criteria in gabbros and basalts, whereas Boudier and Nicolas applied a structural and petrological approach, largely based on the mantle component of ophiolites. This mantle contribution to the fate of any ophiolite and its structural approach being essential, we wish to incorporate these wider criteria in Table 1.

In a rising mantle with an average mantle composition, mantle depletion increases with spreading rate of the ridge of origin. This approach does not account for ophiolites that depart much from the average norm. For instance, the present day Reykjanes Ridge is significantly hotter than other ridges because it is located close south to the Iceland hot spot.

Figure 1 emphasizes the contrast between fast-spreading (HOT) and slow-spreading (LHOT-LOT) ridges. The copious thermal budget in HOT supports the continuous activity at ~1,200 °C of a crustal magma chamber beneath the ridge axis (see "Oceanic Spreading Centers"). The asthenosphere/lithosphere upper limit (~1,100 °C) is the wall of this permanent magma chamber. At the Moho level, this limit flattens. As a consequence, the fast rising and melting mantle diapir rising below the ridge axis can diverge in nearly all directions. In contrast in LHOT, the thermal budget does not sustain a permanent magma chamber. The steepened limit between lithosphere and asthenosphere along the ridge axis channels mantle flow parallel to the ridge and generates just beneath Moho and only locally, a thickened MTZ compared to HOT. The LOT ophiolites ascribed to conditions of continental to oceanic rifts are structurally akin to LHOT with, as a consequence of a lower thermal budget, a steep limit between lithosphere and asthenosphere channeling mantle flow parallel to the ridge axis (see "Oceanic Rifts").

Modes of emplacement of ophiolites

Within the frame of plate tectonics, Moores (1982) and Coleman (1984) divided ophiolites into two groups. The *Tethyan ophiolites* are giant thrust sheets resting upon passive margins of continental and adjacent oceanic crust, as along the Oman margin of Arabia (Figure 2a). In contrast, the *Cordilleran ophiolites* are incorporated, often dismembered or incorporated in mélanges, into accretion terranes. This is typically achieved in California Cordillera where ophiolites are in a situation of forearc (Figure 2b). Clearly, the Western Pacific and its complex back-arc

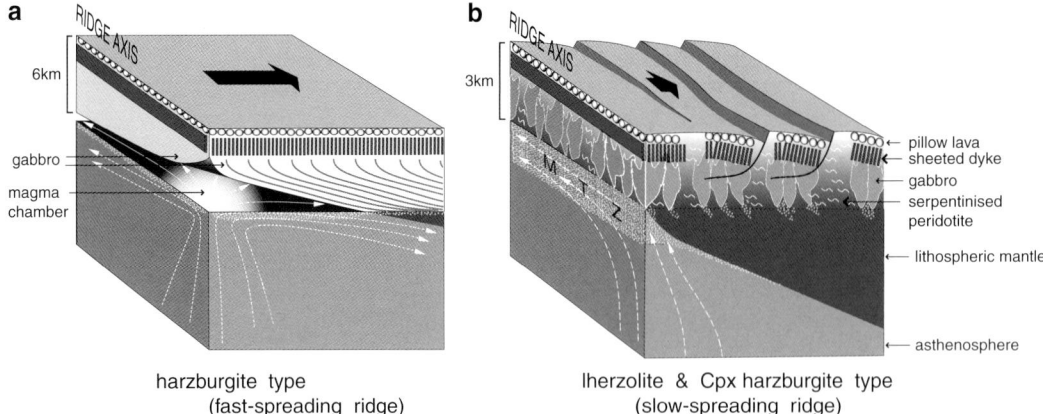

Ophiolites, Figure 1 Compared ridge structure and dynamics between (**a**) HOT and (**b**) LHOT-LOT deduced from studies in ophiolites. The scales are different. The fast-spreading ridge permanent magma chamber is transversally ~ 10 km wide, as wide as the underlying mantle diapir that it is covering. In slow-spreading ridges, upper mantle flow is channeled parallel to the ridge plane. Magma chamber are discontinuous with serpentinites screens. MTZ stands for mantle transition zone, a dunitic transition to the crust.

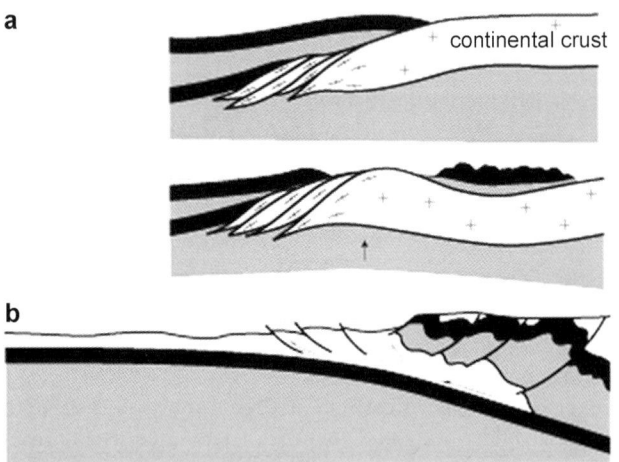

Ophiolites, Figure 2 The two main modes of ophiolites emplacement onto (**a**) passive and (**b**) active continent margins (oceanic crust, *black* and mantle *gray*). (**a**) Tethyan-type, oceanic thin lithospheric thrust. Following their detachment at the ridge of origin, ophiolites have been thrust on a passive continental margin before a crustal rebound, as recorded in Oman ophiolite (Boudier et al., 1988, Ishiwatari et al., 2005). (**b**) Cordilleran-type, accretion, and upheaval of oceanic sediments and of forearc oceanic crust in a subduction-related accretionary prism (Platt, 1986).

spreading over subduction zones has been a nursery for a number of ophiolites. These supra-subduction zone (SSZ) ophiolites are opposed to mid-oceanic ridge (MOR) ophiolites.

Tethyan ophiolites. This mode of emplacement, conceived as an MOR ophiolite being thrust onto a continental margin, could alternatively be detached from an SSZ back-arc ridge and thrust either on the arc itself or in the opposite direction. A compilation on 63 ophiolites from various environments throughout the world, with a fair knowledge on a number of them, has shown that over half of them have been emplaced as *Tethyan ophiolites* (Nicolas, 1989, p. 298). Some are still resting as giant thrust sheets, with a high-temperature metamorphic sole, upon passive margins, as in Oman (Glennie et al., 1974) or in Papua New Guinea (Finlayson et al., 1977) (Figure 3b). Some have been implied in collision belts and are identified thanks to their typical metamorphic sole.

Cordilleran ophiolites deriving from a forearc. These ophiolites are associated to margins that are still active or have been implied in collision belts with partial destruction during their emplacement. Before upheaval, forearc ophiolites (that have no a priori reason to be SSZ related) may have been entrained in the shallow subduction zone or, at the least, underthrust within the accretion prism to over 1 GPa (Ernst, 1981). Ophiolites may also derive from fragments scrapped off the subducting lithosphere. Whatever their origin, a number of cordilleran ophiolites have reached high-pressure metamorphic conditions before being incorporated into more superficial formations as in the Franciscan Mélange of California.

Thrusts with a high-temperature sole in Tethyan ophiolites

Few Tethyan ophiolites that still rest upon a passive margin have been saved from the destructive effects of collision. They are most favorable to retrieve the conditions of accretion at oceanic ridges. The flat structures of asthenospheric mantle flow (~1,200 °C) grade, within ~1 km of increasingly deformed peridotites, into a high-T metamorphic sole at the expense of the underlying oceanic crust and its sedimentary cover (Figure 3). At this contact, peridotites are mylonitic deforming at ~1,000 °C with, in the underlying crust, development of high-grade metamorphism. Overthrust basalts are transformed to amphibolites and radiolarian sediments to quartzites.

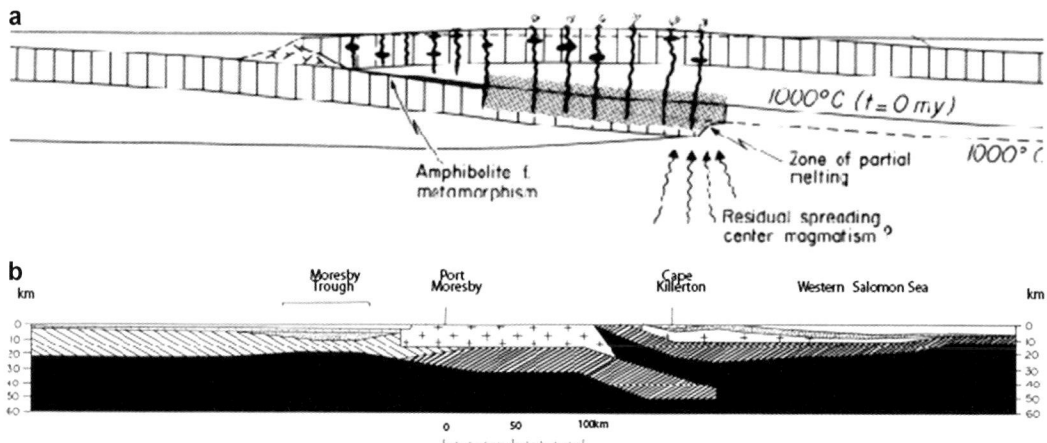

Ophiolites, Figure 3 Model of oceanic detachments and thrust of the rigid lithosphere slab. (a) Oman. The late volcanism (V2) has been contaminated with water released by the ironing effect of the hot thrust (Boudier et al., 1988; Ernewein et al., 1988; Ishiwatari et al., 2005). (b) Comparison with the geophysical modeling of oceanic thrusting of Papua New Guinea ophiolite (Finlayson et al., 1977).

With time and decreasing temperature, greenschist facies metamorphism migrates downward. This inverse metamorphism develops over a vertical distance of some 200 m. While moving away from the ridge, the ophiolite nappe shovels at its front and drags below its sole, marine sediments forming ophiolite mélanges. Commonly, this nappe rides over high-pressure mélanges, relating to some former active margin (review in Nicolas 1989).

This is best illustrated by the Oman ophiolite, where the oceanic thrust was ~15 km thick (Figure 3). The basal thrust plane remains parallel to the overlying Moho and was nearly horizontal. This leads to the model of an initially thin and flat lithospheric slab being detached and thrust over a sole of wet and heated oceanic crust. A nearly flat lithosphere points to Oman being derived from a fast to super-fast ridge, as confirmed by precise U-Pb dating (Rioux et al., 2012).

The "ironing effect" of the ophiolite nappe is only possible if the detachment occurred at very high temperatures near the ridge axis (Figure 3). This thermal constraint is confirmed by U-Pb geochronology on age difference between the last magmatic events related to ridge activity and melt products related to the metamorphic sole. In Oman, the age span is only ~300,000 years between accretion and early thrusting (Warren et al., 2005; Rioux et al., 2012). Such precise geochronological data open a new era in ophiolite studies.

MOR versus SSZ origin

For a long time, Oman has been dominantly considered, as most other ophiolites, as being originated in a back-arc environment. This conclusion is based on the geochemical discriminant diagrams. SSZ origin relies on hydrous fluids being implied in the ophiolite generation. The question is whether SSZ environment is compulsory for ophiolites? Recent contributions point to the importance of seawater contamination in the late products of the Oman ophiolite pile. Let us consider hydrous fluids at 1,000 °C reaching the magma chamber wall, or the secondary magmatism (wehrlites and plagiogranites) in the gabbro section (Koepke et al., 2009) or V2 (Ernewein et al., 1988) and boninitic lavas (Ishiwatari et al., 2005) related to fluids from the metamorphic sole (Figure 3). Large mantle shear zones, deep into the mantle section, have also been the path for the hydrous fluids that have generated high-Cr chromites (Boudier, 2014). This large data set suggests that MOR ridges exchange with seawater at all temperatures and renders the deeper SSZ implication, not compulsory.

Arc or back-arc setting is hardly compatible with regional geology in Oman as arc-related volcanics have never been described in the Hawasina basin that has been overridden by the ophiolite nappe. In terms of modeling, the recent reduction to a few hundreds thousand years the age span between accretion and thrusting of the ophiolite nappe from the ridge vicinity seems difficult to reconcile with the SSZ origin of the ridge. The thrust of Figure 3 has been assimilated to a "shallow subduction of hot oceanic lithosphere" by a few authors (Ishiwatari et al., 2005; Koepke et al., 2009). Here, we may be facing a semantic problem. If so, it would be sane to come back to original definitions (Coleman, 1971, 1977) who forged the term "obduction" to describe the thrusting of a hot Tethyan ophiolite nappe onto a continent margin, in contrast to a cold subduction, the one-way trip to the deep mantle.

Summary

The large variety of ophiolites is issued from most types of ocean spreading centers. Ophiolites are scattered onto continents, either on their passive margins or implied into collision belts. The latter are generally more deformed and the former less and more suited to study the processes that lead to the oceanic accretion. Because ophiolites display deep sections into lithosphere, they complement marine data mainly obtained from the seafloor. When implied into mountain belts, they are keys to reconstitute regional paleogeography.

Acknowledgments

Scientifically, this paper owes much to Françoise Boudier and, technically, to her help.

Bibliography

Anonymous, 1972. Penrose field conference on ophiolites. *Geotimes*, 17(12), 24–25.

Bosch, D., Jamais, M., Boudier, F., Nicolas, A., Dautria, J. M., and Agrinier, P., 2004. Deep and high-temperature hydrothermal circulation in the Oman ophiolite-petrological and isotopic evidence. *Journal of Petrology*, 45, 1181–1206.

Boudier, F., (2014). *Structural Control on Chromite Deposits in Ophiolite: The Oman Case*. Geological Society London Special Publication, Vol. 392, pp. 253–267.

Boudier, F., Ceulener, G., and Nicolas, A., 1988. Shear zones, thrusts and related magmatism in the Oman ophiolite: initiation of thrusting on an oceanic ridge. *Tectonophysics*, 151, 275–296.

Brongniard, A., 1813. Essai de classification des roches mélangées. *Journal des Mines Paris*, 199, 177–238.

Brunn, I. H., 1959. Contribution à l'étude du Pinde septentrional et d'une partie de la Macédoine occidentale. *Annales Géologiques des Pays Helléniques*, 7, 1–358.

Casey, J. F., and Dewey, J. F., 1984. *Initiation of Subduction Zones Along Transform and Accreting Plate Boundaries, Triple-Junction Evolution and Fore-Arc Spreading Centers-Implications for Ophiolitic Geology and Obduction*. London: Geological Society, Special Publications, Vol. 13, pp. 269–290.

Coleman, R. G., 1971. Plate tectonic emplacement of upper mantle peridotites along continental edges. *Journal of Geophysical Research*, 76, 1212–1222.

Coleman, R. G., 1977. *Ophiolites*. Berlin/Heidelberg/New York: Springer. pp. 229.

Coleman, R. G., 1984. The diversity of ophiolites. *Geologie en Mijnbouw*, 63, 141–150.

De Roever, W. P., 1957. Sind die alpinotypen Peridotitmassen vielleicht tektonisch verfrachtete bruchstücke der Peridotitschale? *Geologische Rundschau*, 46, 137–146.

Dick, H., Lin, J., and Schouten, H., 2003. An ultraslow spreading class of ocean ridge. *Nature*, 426, 405–412.

Dijkstra, A., Drury, M. R., and Vissers, R. L. M., 2001. Structural petrology of plagioclase peridotites in the West Othris mountains (Greece): melt impregnation in mantle lithosphere. *Journal of Petrology*, 42, 5–24.

Ernewein, M., Pflumio, C., and Whitechurch, H., 1988. The death of an accretion zone as evidenced by the magmatic history of the Sumail ophiolite (Oman). *Tectonophysics*, 151, 247–274.

Ernst, W. G., 1981. Petrotectonic setting of glaucophane schist belts and some implications for Taiwan. Geological Society of China Memoirs, Vol. 4, pp. 229–267.

Finlayson, D. M., Drummond, B. J., Collins, C. D. M., and Connely, J. B., 1977. Crustal structures in the region of Papuan ultramafic belt. *Physics of the Earth and Planetary Interiors*, 14, 13–20.

Glennie, K. W., Boeuf, M. G. A., Hugh Clark, M. W., Moody-Stuart, M., Pilaar, W. F. H., and Reinhardt, B. M., 1974. Geology of the Oman Mountains. *Mijnbouwkundig Genootschap, Deel*, 31, 1–423.

Ishiwatari, T., Fujisawa, S., Nagaishi, K., and Masuda, T., 2005. Trace element characteristics of the fluid liberated from amphibole-facies slab: inference from the metamorphic sole beneath the Oman ophiolite and implication for boninite genesis. *Earth and Planetary Science Letters*, 240, 355–377.

Ishiwatari, A., 1985. Alpine ophiolites: product of low-degree mantle melting in a Mesozoic transcurrent rift zone. *Earth and Planetary Science Letters*, 76, 93–108.

Koepke, J., Schoenborn, S., Oelze, M., Wittmann, H., Feig, S. T., Hellebrand, E., Boudier, F., Schoenberg, R., et al., 2009. Petrogenesis of crustal wehrlites in the Oman ophiolite: experiments and natural rocks. *Geochemistry Geophysics Geosystems*, 10, 1525–2027.

Le Roux, V., Bodinier, J. L., Tommasi, A., Alard, O., Dautria, J. M., Vauchez, A., and Riches, A. J. V., 2007. The Lherz spinel lherzolite: refertilized rather than pristine mantle. *Earth and Planetary Science Letters*, 259, 599–612.

Lippard, S. J., Shelton, A. W., and Gass, Y. G., 1986. *The ophiolite of northern Oman*. Oxford: Blackwell Scientific Publications. Geological Society of London, Memoir, Vol. 11. pp. 178.

Miyashiro, A., 1973. The Troodos ophiolite was probably formed in an island arc. *Earth and Planetary Science Letters*, 19, 249–281.

Moores, E. M., 1982. Origin and emplacement of ophiolites. *Reviews of Geophysics*, 20, 735–760.

Nicolas, A., 1989. Structures of ophiolites and dynamics of oceanic lithosphere. Kluwer Academic Publishers, Dordrecht, pp. 367.

Pearce, J. A., 1975. Basalt geochemistry used to investigate past tectonic environments. *Tectonophysics*, 25, 41–67.

Platt, J. P., 1986. Dynamics of orogenic wedges and the uplift of high-pressure metamorphic rocks. *Geological Society of America Bulletin*, 97, 1037–1053.

Rioux, M., Bowring, S., Kelemen, P., Gordon, S., Miller, R., and Dudas, F., 2012. Tectonic development of the Samail ophiolite; high precision U-Zircon geochronology and Sm-Nd isotopic constraints on crustal growth and emplacement. Journal of Geophysical Research, 1–17. doi:10.1029/2012JB009273

Warren, C. J., Parrish, R. R., Waters, D. J., and Searle, M. M., 2005. Dating the geologic history of Oman's Semail ophiolite: insights from U-Pb geochronology. *Contributions to Mineralogy and Petrology*, 150, 403–422.

Cross-references

Crustal Accretion
Layering of Oceanic Crust
Mid-Ocean Ridge Magmatism and Volcanism
Mohorovičić Discontinuity (Moho)
Oceanic Spreading Centers
Peridotites
Plate Tectonics

ORGANIC MATTER

Evgeny Romankevich[1,2] and Alexander Vetrov[1]
[1]P.P. Shirshov Institute of Oceanology, Russian Academy of Sciences, Moscow, Russia
[2]National Research Tomsk Polytechnic University, Thomsk, Russia

Definition

The organic matter is presented in the ocean as a basis of living organisms and their decomposition products in the particulate, dissolved, and colloid forms. Sequence of these forms in the synthesis–destruction system constitutes the biogeochemical cycling of matter in the ocean and the biosphere as a whole.

Forms of organic matter

Organic matter (OM) occurs in marine and oceanic water as dissolved compounds (DOM), colloids, and particulate OM. The colloidal and dissolved fractions can be separated from each other with methods such as ultrafiltration and ion exchange, but in practice, they are usually treated together. Hence, DOM includes the truly dissolved compounds (particles <1 nm) and colloids, including 1–10 nm and 10–450 nm structures. Large particles (>0.1–1.0 μm), which can be collected by filtration through a variety of filters, such as glass fiber Whatman GF/F (~0.6–0.7 μm pores), separators, and sediment traps, refer to particulate matter. OM of living beings producing and degrading the bulk of OM in the ocean represents only a small part of the total amount of OM. The three forms of OM in the World Ocean, i.e., living, particulate, and dissolved OM, make up the pyramid of mass 1: 13: 250 ($4 \cdot 10^{15}$, $50 \cdot 10^{15}$, and $1{,}000 \cdot 10^{15}$ g C_{org}), which reflects the dynamic equilibrium existing in the biogeochemical cycle. These forms are combined into an interconnected heterogeneous system that includes organic nitrogen, phosphorus, sulfur, and organometallic compounds of natural and anthropogenic origin. Since most of OM in the ocean is produced during photosynthesis, N- and P-containing compounds of OM are predominantly of natural origin, while among S-containing compounds, the fraction of anthropogenous substances increases.

Sedimentation of particulate matter leads to the fact that all sediments include OM: the maximum concentration is typical for muds of near-continental areas, especially for upwelling and high productive waters, and the minimum one is observed in coarse-grained deposits and red clay, common in areas where OM supply is scarce. Annually, $160–250 \cdot 10^{12}$ g C_{org} are buried in sediments of the World Ocean, including about 95 % of near-continental and 5 % of pelagic oceanic sediments (Hedges and Keil, 1995; Eglinton and Repeta, 2003; Romankevich and Vetrov, 2013).

Sources of organic matter

The main primary source of OM in the ocean is phytoplankton (Figure 1). Its average net production, including lifetime excretions, is ~$60 \cdot 10^{15}$ g C_{org} year^{-1}. Another source of OM is the production of ice algae (0.2 10^{15} g C_{org} annually) and phytobenthos (3 10^{15} g C_{org}) created by macrophytes and diatoms living in sediments, as well as a part of OM produced by chemolithotrophs in the rift zones of the ocean and cold seeps, widely developed in seas and on the periphery of all oceans. Assessment of the extent of bacterial chemosynthesis that proceeds through the oxidation of reduced compounds coming from the Earth's interior suggests its small contribution (<0.6 % of photosynthesis products) to OM balance (Romankevich, 1984).

To the autochthonous (marine) sources of OM more resistant to degradation, allochthonous (terrigenous) OM coming from the land is added (~$0.7 \cdot 10^{15}$ g C_{org} year^{-1}). The most important terrigenous source of OM in the ocean is the river runoff, which supplies ~$0.6 \cdot 10^{15}$ g C_{org} year^{-1} in the dissolved and particulate forms (~30 % and 70 %, respectively). There are also unevenly distributed organic compounds in the troposphere; its contribution to the ocean is estimated as <$0.1 \cdot 10^{15}$ g C_{org} year^{-1}. Finally, we should mention such minor, sometimes underestimated, sources of OM such as coastal erosion, thermoabrasion, and underground runoff giving in total ~$0.4 \cdot 10^{12}$ g C_{org} year^{-1}.

Living OM produced on land and in the ocean differs in ash contents (5 % and 44 %, respectively), proteins (5 and 62 %), carbohydrates (62 and 25 %), lipids (6 and 13 %), and lignin (27 % and traces). Marine OM contains about five times more nitrogen and sulfur, more trace elements, and less oxygen. The $\delta^{13}C$ value varies from −19 to −24 ‰ in the ocean and differs from the same parameter for land plants of C_3 type (from −27 to −28 ‰) and C_4 type (~ −12 ‰) of carbon fixation.

Dissolved organic matter

In the World Ocean, dissolved OM represents ~94 % of the total OM and its mass is estimated as $750–1{,}000 \cdot 10^{15}$ g C. This is roughly equal to the carbon amount in the atmosphere.

With the exception of viruses and small non-aggregated bacteria, DOM is mainly represented by nonliving OM, which is extremely small and cannot sink in water at appreciable rates. Seawater DOM affects light transmission in the ocean. Approximately 25–35 % of marine DOM occurs in molecules with masses of 1 kDa that correspond to nominal molecular sizes of 1 nm or more. Seawater DOM has numerous sources and sinks and a range of different reactions.

In the active surface layer, DOM concentrations vary over space and time within two orders of magnitude from 5–10 to 1,000 μM C (1 μM C = 0.012 mg Cl^{-1}). The main DOM pool is found in the deep ocean where concentrations vary little (mostly within 35–45 μM, Figure 2). General patterns of DOM distribution are expressed in reducing concentration from coastal areas with high productivity of plankton, including the areas of allochthonous supply from rivers to oligotrophic regions of the ocean, as well as decreasing concentration with depth in the pelagic areas. A typical depth profile of DOM concentration in the ocean shows that the presence of labile DOM is associated with only the upper part of the water column.

About 30 % of the DOM (sometimes up to 80 %) is represented by colloids scattering light (size >1 nm) and containing bioactive and bioavailable organic compounds. Submicron DOM components include liquid excretions of phytoplankton, phytobenthos, protozoa,

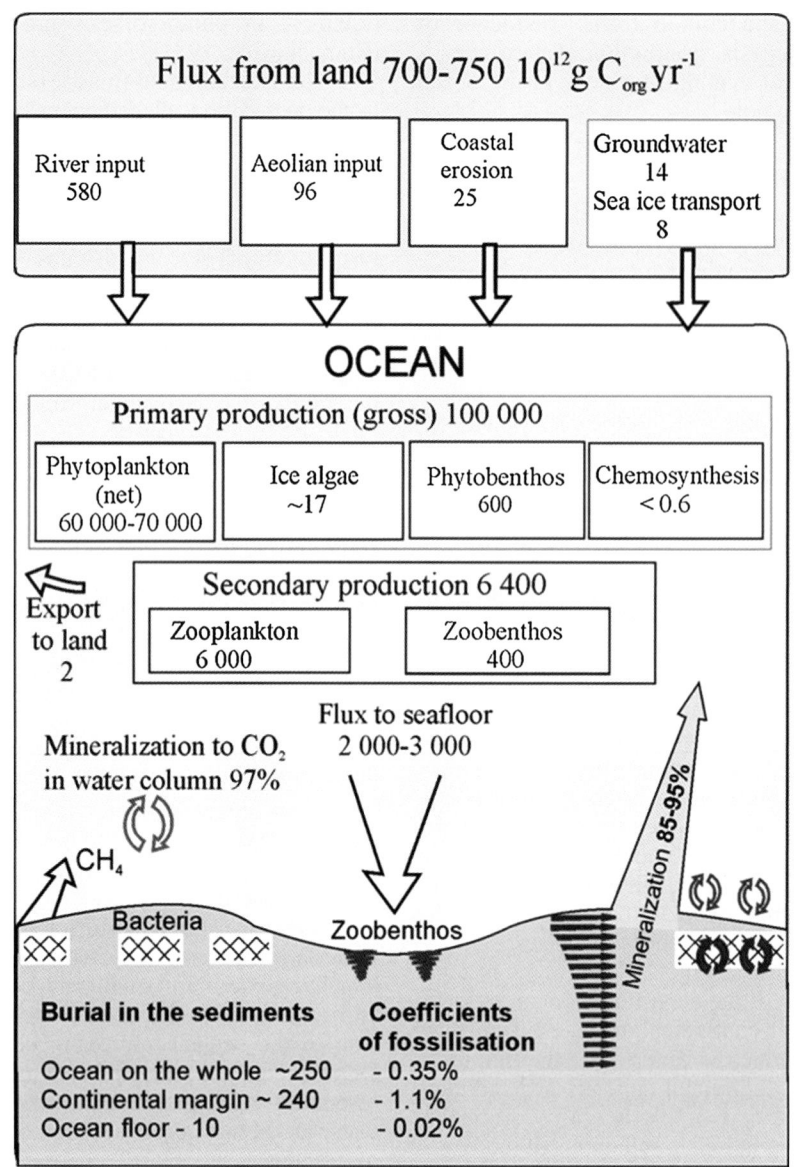

Organic Matter, Figure 1 Fluxes of organic carbon in the World Ocean, 10^{12} g year^{-1}.

and other animals, the products of bacterial decay of seston and fecal pellets, viral particles, and organic compounds that come with river runoff and from sediments. A small part of DOM is currently linked to the activities of people creating more and more xenobiotics. The loss of DOM is associated with photodegradation, sorption on sinking particles, and microbial utilization. Fungi, protozoa, and different mixotrophic organisms play a significant role in mineralization and transformation of OM, but these processes are not well understood and underestimated so far.

UV radiation leads to the mineralization of some DOM, its transformation, and formation of new organo-mineral compounds in the thin ocean surface and subsurface layers; this, in turn, affects the physicochemical, biological, and biochemical characteristics of the surface layer and gas exchange in the air–water system.

Humic substances play an important role in the transition of DOM in particulate OM (POM). Labile colloidal substance is a significant part of DOM in coastal waters. The conservative nature of terrigenous DOC and its direct correlation with salinity are violated in coastal conditions, which affects the behavior and transport distance of DOM offshore.

Marine DOM differs from river one by the lower content of unsaturated aromatic hydrocarbons, lignin phenols, a heavier isotope composition of carbon and nitrogen, lower C/N ratio, and much greater age according to $\Delta^{14}C$. The age of water at 2,000–3,000 m and 5,000–6,000 m depths is estimated as 2–3 and 5–6

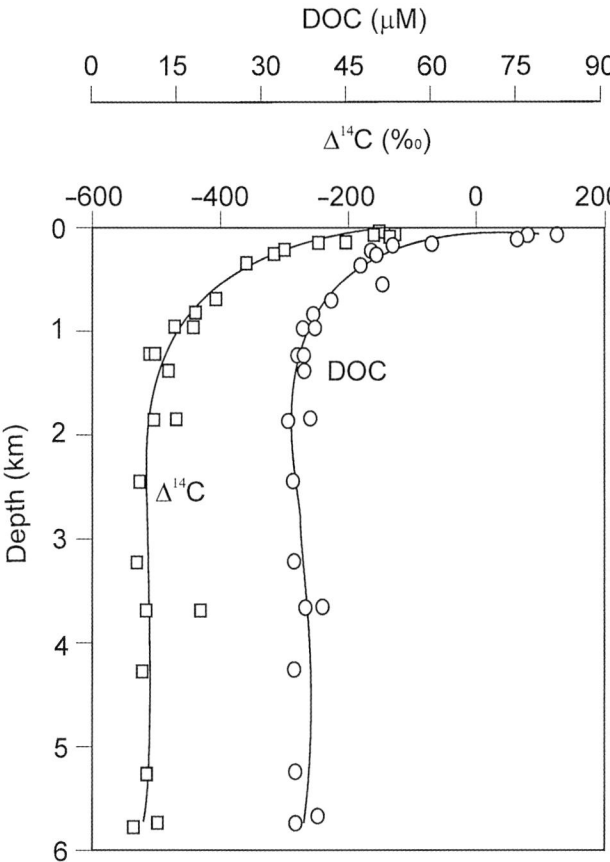

Organic Matter, Figure 2 Vertical profiles of concentration (*circles*) and $\Delta^{14}C$ (*squares*) versus water column depth for dissolved organic carbon in the temperate Pacific Ocean (after Druffel et al., 1989).

fractions, which increases in relative abundance with depth, is unknown (Hedges et al., 1997; Benner, 2002; Emerson et al., 2008).

Among amino acids, alanine, serine, and aspartic and glutamic acids are enriched in D-enantiomers (>10 mol%), as is typical of bacterial cell walls. As for the sugars, galactose dominates glucose in the DOM as opposed to living organisms. LMW compounds and some humic acids have the highest refractory in DOM composition (Hedges et al., 1997; Benner, 2002; Emerson et al., 2008).

The flux of DOM from sediment that is much richer than seawater DOM in the shelf and continental slope areas has not been studied yet and is a challenge for the future.

Particulate organic matter

Particulate OM is the second most common form ($50 \cdot 10^{15}$ g C_{org}) after dissolved OM and plays an important role in biogeochemical cycling of matter in the ocean. It has a great influence on the CO_2 content in the atmosphere binding CO_2 during photosynthesis and transporting organic compounds from the euphotic layer to the sediments. Regional, seasonal, and interannual variations of quantity and composition of the bulk of POM in the seas and oceans are mainly determined by the primary production of phytoplankton. This relationship may be corrupt or even absent in the nearshore zone, especially in the regions of a large supply of suspensions (rivers, technogenic waste release), due to high turbidity, light deficit, or unfavorable conditions for photosynthesis and living biota.

OM produced by phytoplankton is transformed by heterotrophs and reducers and precipitate relatively quickly (days, weeks, depending on their size, density, and depth of the ocean) in the form of aggregated particles, pellets, and residues of zooplankton. Dependence of POM mineralization on the depth is close to hyperbolic function, so only a few percent of POM reach the seabed in pelagic zone (Figure 3). OM of both phytoplankton and small zooplankton would have almost completely decayed and could not reach seabed but for the mechanism of pellet transport.

The most active POM destruction takes place in two zones of the ocean, namely, in the surface water (0–200 m) and on the seabed (benthic water, nepheloid layer, and bioturbated sediments). Changes in the elemental composition of POM in the water column reveal that the destruction of N- and P-containing organic compounds are, respectively, 40 and 70 % faster than mineralization of total OM. These processes continue in the sediments. The river/sea mixing zone plays an important role in sorption of microelements, sedimentation of various forms of OM, and its extent offshore. In this zone, a small part (5–15 %) of DOM can transfer into POM. Up to 40–>90 % of POM may precipitate in the estuarine zones and adjacent shelf (Lisitzin, 2004). However, over time,

1,000 years, respectively. Although marine DOM contains many organic molecules found in marine biota, the composition of the most part of DOM (up to 80–95 %) is not well studied due to hard selection and identification of trace quantity compounds. The high-molecular-weight (HMW) fraction of DOM (>1 kDa), which can be accounted for as chromatographically measured amino acids and sugars, is only 15 % in surface water and 6 % in the deep ocean, which is several times lower than in marine plankton (~80 %) and particulate OM (~30 %). Low-molecular-weight (LMW) fraction of DOM contains even less organic polymers and is more refractory.

Extremely low biochemical content of HMW DOM points toward a highly altered chemical structure. ^{15}N NMR analyses indicate that all the nitrogen is in amide structures that are almost certainly of biochemical origin. ^{13}C NMR spectra indicate large concentration of carbohydrates that decrease with depth and thus are relatively labile. These spectra indicate very little aromatic compounds, as is generally associated with humic acids. The mechanism of formation of the carbon-rich aliphatic

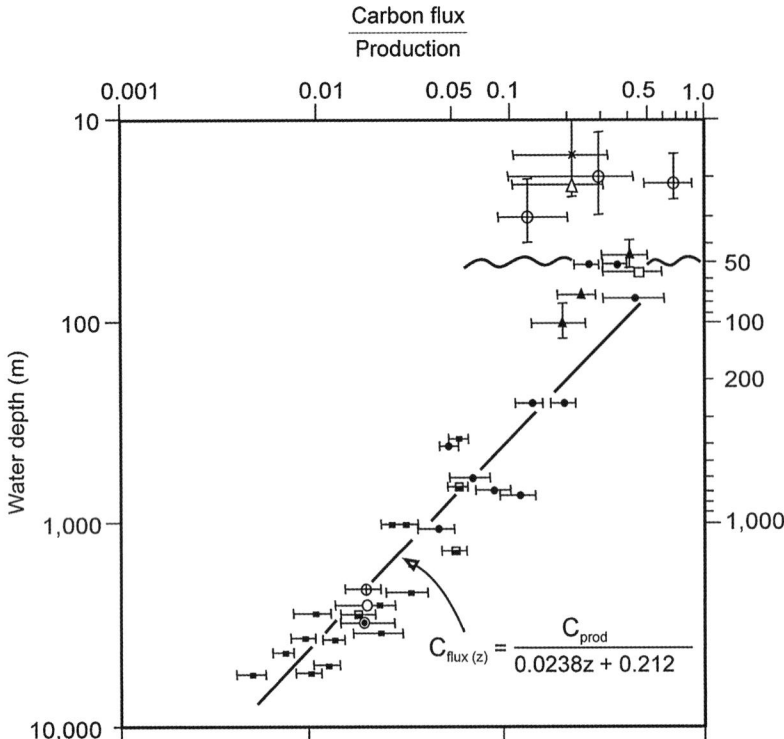

Organic Matter, Figure 3 Organic carbon fluxes with depth in the water column normalized to mean annual primary production rates at the sites of sediment trap deployment. The undulating line indicates the base of the euphotic zone (by Suess, 1980).

a part of settled particulate material and some part of surface sediments enriched by OM are transferred to the shelf depressions, lower part of the continental slope, and continental rise by currents, gravity, and tectonic movements. During mineralization, the balance between autochthonous and allochthonous matter changes considerably in favor of the latter as the more resistant one.

POM, which is mostly pellet material, is very different from phytoplankton and zooplankton as it has less protein, N, and P, more polysaccharides, aliphatic hydrocarbons, aromatic structures, and the presence of trace quantities of geopolymers (humic substances), which have different molecular and isotopic compositions. In the arctic seas, seasonal ice contains many of POM due to the sea algae (mainly diatoms and coccolitophorids) bloom. This substance contains a characteristic molecular and isotopic indicators of origin, namely, high content of the stable isotope ^{13}C up to $\delta^{13}C = -8$ ‰ compared with POM in open water (from -20 to -28 ‰). Increased level of ^{13}C is apparently caused by the low concentration of CO_2 in the pores of the sea ice (Thomas et al., 2010).

Organic matter of sediments

The particulate matter reaching the seabed and a thin surface layer of sediment (<1 cm) are enriched by OM. Resuspension, bioturbation, and microbial utilization deplete the OM content. At high accumulation rate, the zone of active OM degradation extends down to 2 m or more. Deeper enzymatic OM decay slows down. It depends a lot on the OM content and composition, sedimentation rate, and burial conditions. Most of the OM degrades in oxygen conditions. In anoxic conditions, the main degradation processes are sulfate reduction and methanogenesis, and rate of enzymatic OM degradation slows dramatically.

OM buried in sediments (Figure 4) contains a wide range of transformed compounds of marine and terrigenous genesis, gas hydrates of methane and other gases, as well as OM of interstitial water. Annually, approximately 0.35 % C_{org} of the primary production and OM supplied from land is buried in sediments of the World Ocean. During diagenesis and catagenesis, this OM forms kerogen, liquid and gaseous hydrocarbons, and other organic and nonvolatile compounds – prospective oil and gas resources.

OM of sediment contains different classes of organic compounds such as hydrocarbons, alcohols, amines, aldehydes, ketones, carbonic acids and their esters, amino acids, carbohydrates, amino sugars, peptides, and vitamins.

The basic change of OM at the stage of early diagenesis is confined in its humification, which represents a mixture of enzymatic depolymerization of biological molecules and the formation of new geopolymers, stable out of living

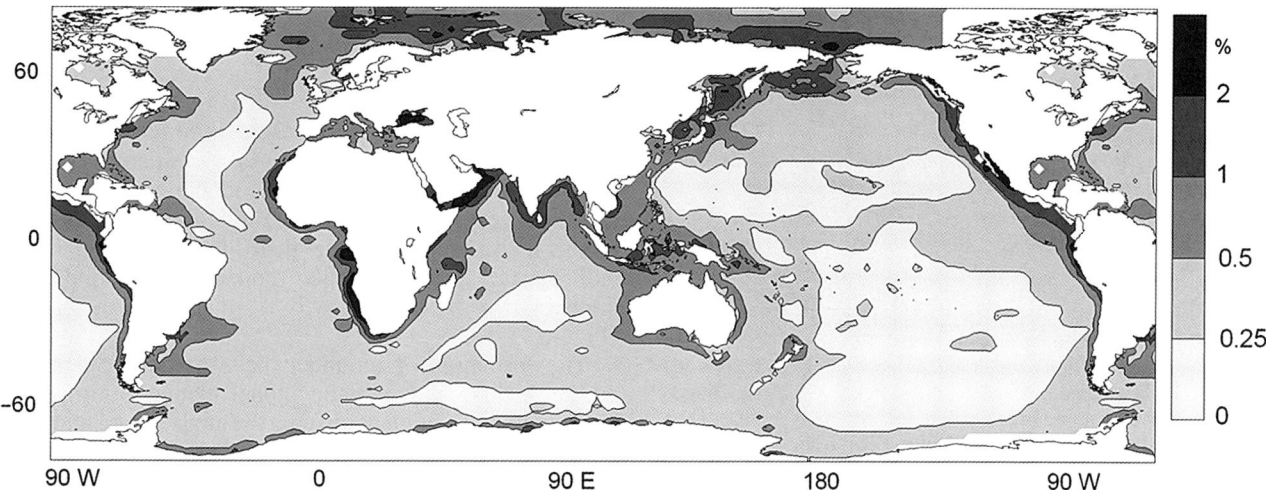

Organic Matter, Figure 4 Distribution map of total organic carbon in surface sediments, % of dry sediment.

being. A part of autochthonous humic acids (HA) is formed in melanoidin reaction and differs from allochthonous HA in increased content of nitrogen (2–7 %) and heavy metals. Autochthonous insoluble OM is genetically related to the HA, but has more C and a smaller content of N and P. HA and kerogen form the vast majority of OM of the ocean sediments (up to 95 %).

Conversion of lipids in both oxidative and reductive environments leads to the selective preservation of low-polar compounds, including hydrocarbons, fatty acids, and sometimes sterols. Alteration of the carbon isotopic composition of lipids is a good indicator of the degree of biogenic OM transformation. Hydrocarbon content in OM of recent marine sediments ranges from <0.02 % to 4 %. They contain cyclanes >50 %, alkanes <30 %, and arenes 10–15 %.

Among n-alkanes (or n paraffins) the most abundant hydrocarbons are C_{15}–C_{36}; among polysubstituded alkanes – isoprene, phytane and pristane; the isoparaffinic hydrocarbons are less common in marine sediments. Distribution of n-alkanes is a good indicator of changes in facies depositional environment, contribution of terrigenous components, and transformation of OM. Along with hydrocarbons, phenols of lignin are also undoubted indicators of the contribution of terrigenous material in sedimentation (Stein and Macdonald, 2004).

Aromatic hydrocarbons of oceanic sediments include monocyclic, uncondensed and condensed bicyclic, tricyclic, or polycyclic (PAH) structures. Benzene and its derivatives constitute 50–80 % of aromatic hydrocarbons, PAH – usually up to 1 %. Geochemical PAH background is formed in the sediments under the control of lithological, climatic, pyrogenic, and anthropogenic factors. The individual composition of PAH, along with n-alkanes and lignin phenols, is an informative indicator of OM transformation in the stage of sedimento-, dia-, and catagenesis.

Summary

The organic matter (OM) in the ocean is continuously created in the form of primary production of micro- and macroalgae and chemoautotrophic bacteria, simultaneously with processes of consumption, dying, and decay. The remains of dead organisms found in the form of a particulate matter are a source of dissolved OM. In addition, dissolved and suspended OM come to the ocean from the land as part of the river and groundwater flows, wave abrasion products, and aerosols. Dissolved OM contains proteins, amino acids, carbohydrates, pectin, fatty acids, and humus. Products of OM mineralization are involved in the next biogeochemical cycles. A small part of the OM, however, is buried in the sediments and forms the oil and gas fields during diagenesis and catagenesis. For the accumulation of OM, down-slope transport by turbidity currents/slumps and rapid burial are important processes for the preservation of OM in the sediments.

Bibliography

Benner, R., 2002. Chemical composition and reactivity. In Hansell, D. A., and Carlson, C. A. (eds.), *Biogeochemistry of Marine Dissolved Organic Matter*. New York: Elsevier Science, pp. 59–90.

Druffel, E. R., Williams, P. M., and Suzuki, Y., 1989. Concentrations and radiocarbon signatures of dissolved organic matter in the Pacific Ocean. *Geophysical Research Letters*, **16**, 991–994.

Eglinton, T. I., and Repeta, D. J., 2003. Organic matter in the contemporary ocean. In Holland, H. D., and Turekian, K. K. (eds.), *Treatise On Geochemistry*. Oxford: Elsevier-Pergamon, pp. 145–180.

Emerson, S., Hedges, J., and Whitehead, K., 2008. Marin organic geochemistry. In Hedges, J., and Emerson, S. (eds.), *Chemical Oceanography and the Marine Carbon Cycle*. Cambridge: Cambridge University Press, pp. 261–302.

Hedges, J. I., and Keil, R. G., 1995. Sedimentary organic matter preservation: an assessment and speculative synthesis. *Marine Chemistry*, **49**, 81–115.

Hedges, J. I., Keil, R., and Benner, R., 1997. What happens to terrestrial organic matter in the ocean? *Organic Geochemistry*, **27**, 195–212.

Lisitzin, A. P., 2004. Sediment fluxes, natural filtration, and sedimentary systems of a "living ocean". *Russian Geology and Geophysics*, **45**, 12–43.

Romankevich, E. A., 1984. *Biogeochemistry of Organic Matter in the Ocean*. Berlin: Springer, p. 334.

Romankevich, E. A., and Vetrov, A. A., 2013. Masses of carbon in the Earth's hydrosphere. *Geochemistry International*, **51**, 431–455.

Stein, R., and Macdonald, R. W. (eds.), 2004. *The Organic Carbon Cycle in the Arctic Ocean*. Berlin: Springer. 363 p.

Suess, E., 1980. Particulate organic carbon flux in the oceans – surface productivity and oxygen utilization. *Nature*, **288**, 260–263.

Thomas, D. N., Papadimitriou, S., and Michel, C., 2010. Biogeochemistry of sea ice. In Dieckmann, G. S., and Thomas, D. N. (eds.), *Sea Ice*. Oxford, Ames: Wiley-Blackwell, pp. 425–467.

Cross-references

Anoxic Oceans
Chemosynthetic Life
Cold Seeps
Deep-sea Sediments
Diatoms
Energy Resources
Estuary, Estuarine Hydrodynamics
Marine Gas Hydrates
Methane in Marine Sediments
Radiocarbon: Clock and Tracer
Sapropels
Turbidites

OROGENY

Bernhard Grasemann and Benjamin Huet
Department for Geodynamics and Sedimentology,
University of Vienna, Wien, Austria

Synonyms

Orogens: orogenic belts, mountain belts; Orogeny: mountain building, formation of orogens

Definition

Orogeny, from the Greek ὄρος (*óros*) "mountain" and γένεσις (*génesis*) "creation," refers to the geodynamic processes leading to deformation of the Earth lithosphere resulting in crustal thickening, surface uplift, and the formation of large-scale topography rising more or less abruptly from the surrounding level.

Introduction

Orogenic belts are large mountain belts forming usually at destructive plate boundaries, as part of an orogenic cycle. They record a number of complex geodynamic processes, often associated with subduction of an oceanic plate beneath another tectonic plate, which is eventually followed by collision between two continents or between a continent and an island arc. Contrary to the continental crust, which is thick (generally >30 km) and buoyant (density ~2,700 kg m^{-3}), the oceanic crust is thin (generally <10 km) and dense (density ~2,900 kg m^{-3}). It can be easily recycled into the mantle during subduction, due to its transformation to very dense eclogite (density >3,300 kg m^{-3}). Since subduction has been consuming oceanic lithosphere through the Earth history, almost no oceanic crust older than ~180 Ma (lower Jurassic) exists. Orogenic belts represent thus important archives for more than 95 % of the Earth history (Figure 1).

The formation of mountain belts has induced major environmental and climatic global changes during the Earth history, through modifying the atmospheric and oceanic circulations, the location of elevated terrains suitable for inland ice, the radiation balance, and the geochemical cycles. The present-day orogenic belts are associated with active tectonics and volcanism which manifest them by potentially deadly geological hazards (earthquakes, landslides, and volcanic eruptions). Understanding the formation and the destruction of orogens at both short and long term is, therefore, essential.

Although orogens are generally considered to form at destructive plate boundaries, it is important to note that deformation of the lithosphere and its resulting topography may develop in other geodynamic environments: along constructive plate boundaries (e.g., East African Rift), extensional orogens (e.g., Canadian Cordillera), along conservative plate boundaries (e.g., Dead Sea Transform fault system), and above hot spots (e.g., Hawaiian–Emperor Seamount Chain).

The orogenic cycle

The Canadian geologist John Tuzo Wilson (1908–1993) described the periodic opening and closing of oceans giving rise to the formation of orogens as the orogenic cycle (sometimes called *"Wilson Cycle"* or *"supercontinent cycle"*). Each stage of such a cycle combines a number of different geodynamic processes (Moores and Twiss, 1995). For example, deep tectonic processes controlling the formation and the destruction of an orogen are always associated with erosion and sedimentation, which act at the surface (England and Molnar, 1990; Avouac and Burov, 1996).

There are four characteristic stages in an orogenic cycle:

(i) *Continental rifting and opening of a new oceanic basin.* Thick shallow-water sediments accumulate along the margin of an oceanic basin. In the more central parts of the basin, deep-water sediments are deposited on top of basaltic flows and gabbroic intrusions forming at the mid-ocean ridge.

(ii) *Subduction of the oceanic lithosphere* initiates along a ridge-parallel fault, an oceanic fracture zone, a transform fault, or an ocean–continent transition. It may be associated with the obduction of oceanic

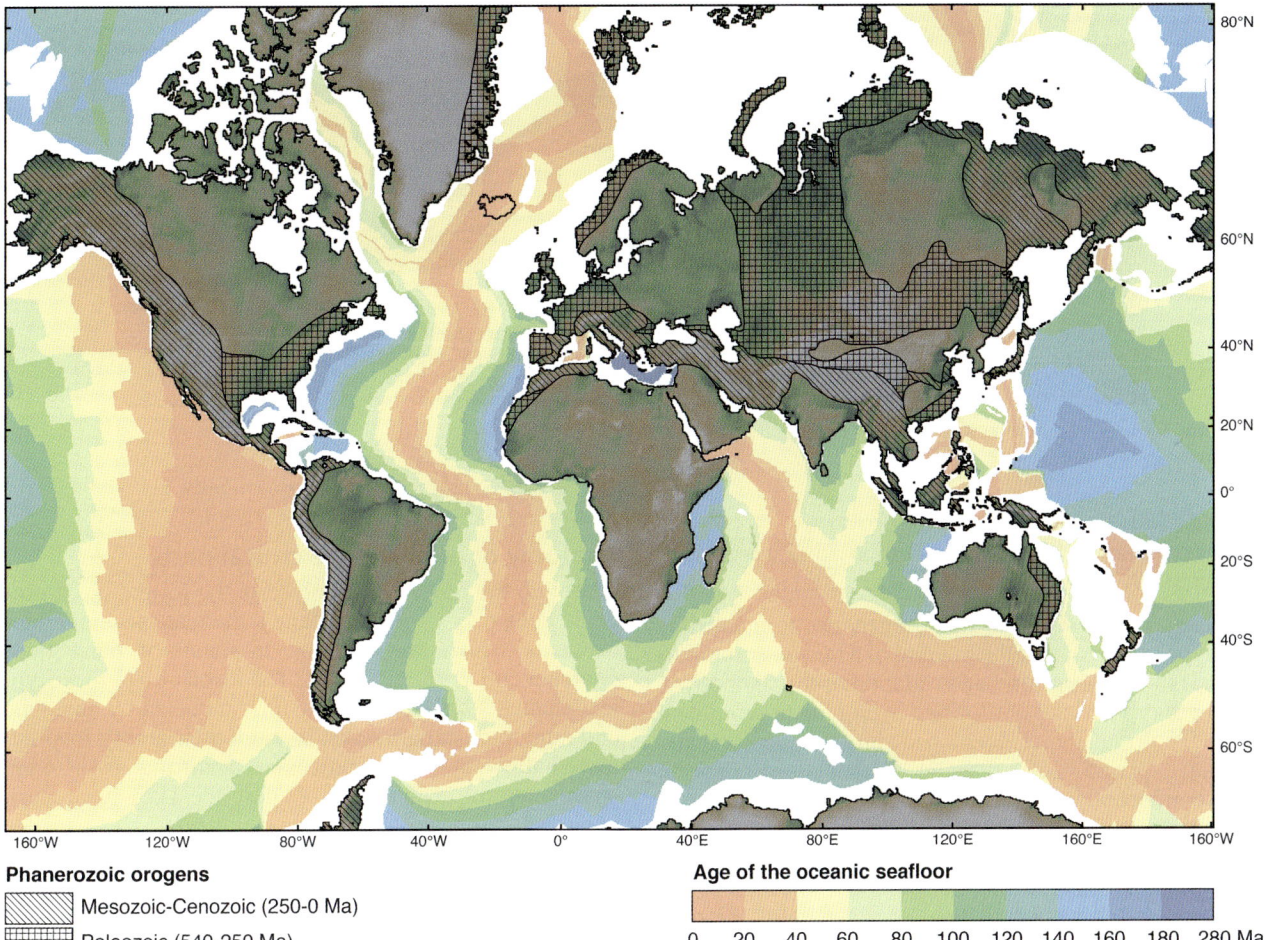

Orogeny, Figure 1 Global map of the Phanerozoic orogens. The Paleozoic and Mesozoic–Cenozoic orogens are distinguished by different hatches. Continental areas without pattern correspond to Precambrian domains (cratons and Pan-African orogen). The topography of the continents (SRTM30_PLUS, Becker et al., 2009) and the age of the oceanic seafloor (Müller et al., 2008) are also presented.

lithosphere onto the continental margin. Dehydration of the downgoing oceanic plate leads to melting of its overlying mantle wedge and induces the formation of a volcanic arc in the overriding plate.

(iii) *Continental subduction and collision of two continents* eventually occurs, as the continental margin at the other side of the previous oceanic basin arrives in the subduction zone. Ongoing shortening induces deformation, metamorphism, magmatism, and thickening of the crust. Surface uplift together with erosion of the rising orogenic belt results in deposition of synorogenic sediments. Convergence may be accommodated by lateral escape of crustal blocks along continent-scale strike-slip faults.

(iv) *Destruction of the orogen* is accompanied by lithospheric extension, crustal thinning, development of fault-bounded sedimentary basins, and intrusions of large magmatic bodies. Once the crust has reached a normal thickness, a new orogenic cycle can begin.

Although similar stages can be found in all Phanerozoic orogenic belts, such a cycle is a very broad simplification, which probably does not apply to all orogens in detail. In fact, it is very unlikely that rifted continental margins reassemble at the same location since in most continental collisions the two colliding continents were never previously close together. It must also be noted that orogeny and the orogenic cycle are here described in the framework of plate tectonics, the onset of which is debated (sometimes in the Precambrian, Eriksson et al., 2004). Other geodynamic processes than those described here have probably controlled formation of mountain ranges early in the Earth's history.

Types of orogens

Based on the different tectonic, metamorphic, and magmatic evolutions of orogenic belts, two main different types of orogens, which are in fact two different stages

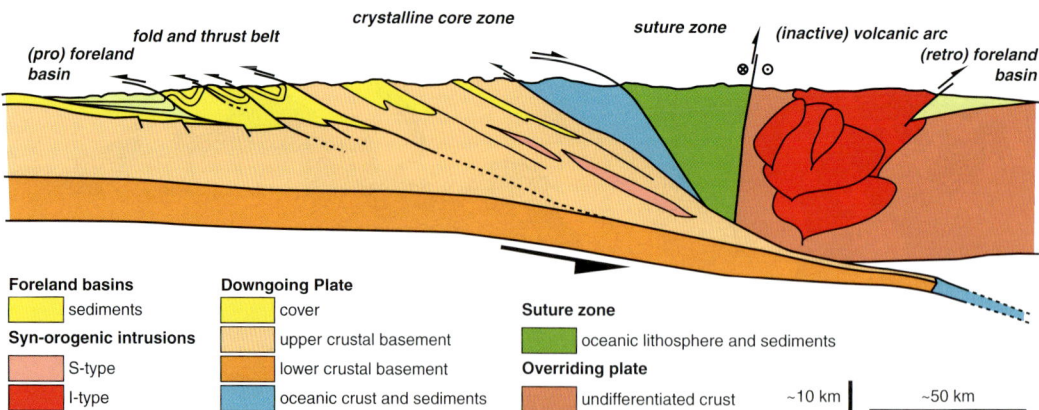

Orogeny, Figure 2 Section across a model composite orogenic belt.

of an orogenic cycle, have been proposed (Zwart, 1967; Pitcher 1979; Ernst, 2005; Frisch et al., 2011):

(i) *Cordilleran-type* or *arc-type* orogens are dominated by I-type (derived from the melting of a igneous protolith) andesitic volcanism and granodioritic batholiths. These orogens are not formed by continent–continent collision but rather developed during a geologically long period of oceanic subduction beneath a continental margin. Although these orogens may experience a protracted history including collisions (with seamounts, oceanic plateaus, or microcontinents), abyssal sedimentary rocks, a suture zone, and basement nappes are generally missing. Above the downgoing oceanic plate, huge sedimentary sequences form thrust sheets of an accretionary wedge. The distribution of metamorphic rocks in such orogens can be described as paired metamorphic belts parallels to the chain. Low pressure–high temperature metamorphism occurs along the magmatic arc, while high pressure–low temperature rocks are present as part of a mélange incorporated to the accretionary wedge.

(ii) *Alpine-type* orogens result from the collision of two continents after subduction of an oceanic plate. Such orogens form relatively narrow belts, which are dominated by thick nappe stacks involving basement of different metamorphic pressure–temperature evolution. The nappes experience considerable deformation and large-scale overthrusting in the order of 100 km or more. Ophiolitic material marks the suture zone between the initially separated continents, where the separating oceanic plate has been consumed. Typically, one or more tectonic units record high-pressure (blueschist and/or eclogite facies) or even ultrahigh-pressure metamorphism (>2.7 GPa). They are extruded up to the surface above a major thrust and below an apparent normal fault. Syn- and post-tectonic plutons are dominated by S-type granites (derived from the melting of a sedimentary protolith).

Anatomy of orogens: the critical parts

Although the processes involved in the formation of orogenic belts may vary significantly, there are several major elements which are common to many of these belts (Hatcher and Williams, 1986). The formation of these typical elements has been successfully modeled by analogue and numerical experiments, through their ability to simulate shallow and deep processes occurring at different space and time scales. Here we describe a composite orogenic belt, at the collision stage (Figure 2).

(i) *A foreland basin* forms parallel to the mountain front because the mass created by crustal thickening causes the lithosphere to bend (i.e., lithospheric flexure). The cross-sectional shape of the lithosphere can be described as an asymmetrical low close to the mountain front and a broad convex upwards deflection along the forebulge (Catuneau, 2004). The facies of the sediments deposited in foreland basins generally evolves during the life of the orogen, from flysch in the early stages to molasse in the late stages. The width and depth of the foreland basin is mainly controlled by the flexural rigidity of the underlying lithosphere and the size and thickness of the mountain belt. Basin fillings can reach up to 10 km thick sediment succession derived from the eroding mountains, thinning away from the mountain belt (Allen and Allen, 2005). Pro-foreland basins occur at the surface of the downgoing lithosphere (e.g., the Ganges Basin, south of the Himalaya). Retro-foreland basins occur on top of the overriding plate (e.g., Andean basins, east of the Andes). Foreland basin systems have an orogeny-parallel, structured internal geometry and comprise several distinct depozones that result from the flexural response to topographic loading of the overlying continent and subduction processes of the downgoing plate (DeCelles and Gilles, 1996).

Since the geometry of foreland basins is mainly influenced by fold-and-thrust belt migration, flexural subsidence, and sedimentation derived from the rising mountain belt, geodynamics models have been used to derive important parameters like the rheology of the lithosphere, thrusting rates, or erosion rates (DeCelles and DeCelles, 2001; Jordan and Watts, 2005; Pelletier, 2007). Numerical models of foreland basin successfully combined the effects of fold-and-thrust belt migration, flexural subsidence, and sedimentation including three-dimensional geometries and nonlinear crustal rheology (Garcia-Castellanos et al., 1997, Clevis et al., 2004).

(ii) *A fold-and-thrust belt* consists in intensely thrusted and folded unmetamorphosed or low-grade sediments, derived from the passive margins of the downgoing continent. In cross section, the fold-and-thrust belt has a wedge-shaped geometry with a sole fault (décollement) at its base. The deformation generally propagates outwards in a sequence so that the deformation near the core of the orogen and at high structural levels is older than the deformation near the foreland at the structural base of the thrust stack. The décollement characteristically contains weak rocks like shales or salt along which the deformation of the overlying wedge is decoupled from the underlying rocks, which frequently consists of crystalline basement rocks. This style of deformation where strongly shortened sedimentary rocks are detached from a basement is called thin-skinned tectonics. At the contrary, thick-skin tectonics describes a fold-and-thrust belt in which the basement is also deformed (typically by reactivation of normal faults as thrusts).

The shape of the wedge has been explained by the mechanics of a deformable sequence above a basal décollement, generally known as the critical Coulomb wedge theory (Davis et al., 1983). Critical parameters, which influence the deformation within the wedge, are the dip angle of the décollement, the basal traction along the décollement, the friction angle of the décollement, and the pore fluid pressure, which counteracts the overburden. Numerous physical and numerical models have tested the influence of various parameters on the development and geometry of fold-and-thrust belts like the décollement layer thickness, intermediate décollements within the sedimentary sequence and syn-shortening erosion (Costa and Vendeville, 2002; Ellis et al., 2004; Sherkati et al., 2006, Cruz et al., 2010, Fillon et al., 2013).

Since fold-and-thrust belts are estimated to contain 15 % of the undiscovered global recoverable hydrocarbons, numerous studies focused on the development of balanced cross-section techniques, which aimed to geometrically constrain the style of folding and thrusting and to kinematically model palinspastic reconstructions by unfolding and retro-deforming the stratigraphic sequence (Dahlstrom, 1969; Elliot, 1983; Groshong, 1994; Homza and Wallace, 1995). However, recent studies have demonstrated that the characteristics and the wavelength of folding in fold-and-thrust belts are strongly influenced by rheological parameters and mechanical isotropies of the stratigraphic pile (Yamato et al., 2011; Ruh et al., 2012). Therefore, future research will focus on mechanical reconstructions of fold-and-thrust belts (Simpson, 2009; Frehner et al., 2012; Vidal-Royo et al., 2012) including three-dimensional mechanical interaction of fold and fault growth (Schmid et al., 2008a; Grasemann and Schmalholz, 2012).

(iii) *A crystalline core zone* forms the central and highest part of a mountain belt, usually above the orogenic root, where the crust is the thickest. It mainly consists of metamorphic and magmatic rocks derived from the upper crust of the lower plate. Lower crustal rocks are rare (Jolivet et al., 2003). Units of oceanic material involved earlier in the subduction zone can also be found, overthrusting the continental units. The crystalline zone is mainly deformed by ductile flow, resulting in large shear zones and folded nappes recording complex multiple deformation events. Metamorphism reaching the amphibolite and eclogite facies is widespread. The metamorphic evolution is, however, very heterogeneous in time and space. Frequently the high-grade metamorphic nappes preserve an inverted metamorphic field gradient, showing an increase of the metamorphic pressure and temperature conditions towards structural higher levels. The presence of eclogites, some of which experienced ultrahigh-pressure metamorphism, suggests that at least parts of the core zone of mountain belts have been involved in continental subduction at mantle depth and later exhumed back to the surface (Chopin, 1984; Hacker and Peacock, 1995; Hacker and Liou, 1998; Jolivet et al., 2003).

Structural field observations demonstrate that the high-grade rocks are exhumed above high-strain shear zones with thrusting kinematic at the base and below normal shear zones at the top. Physical and numerical models together with geological data resulted in several extrusion wedge models accounting for more or less complex internal nappe stack structures (Grujic et al., 1996; Grasemann et al., 1999; Beaumont et al., 2001; Grujic et al., 2002; Gerya et al., 2008; Yamato et al., 2008), which have been applied to most Phanerozoic orogens (Himalayas, Alps, Hellenides, Dabie Shan, Ural, Hindu Kush, or Caledonides; Ernst, 2001; Law et al., 2006). Most of these models have in common that the extrusion of the high-grade rocks has to be at least partly balanced by surface erosion (Froitzheim et al., 2003; Beaumont et al., 2004; Schulmann et al., 2008) requiring huge sediment accumulations in the foreland of the orogen (White et al., 2002).

Synorogenic S-type granites result from partial melting of metasedimentary source rocks from the lower part of the continental crust that has been thickened by formation of a mountain root during collision. These plutons record a strong interaction and mechanical feedback with deformation in major shear zones (e.g., Schneider et al., 1999; Rosenberg, 2004).

(iv) *A suture zone belt* marks the original contact between two plates that have collided or accreted and is therefore of central interest in order to determine the timing and the nature of collision processes (Haynes and McQuillan, 1974; Klootwijk et al., 1984; Colchen et al., 1986; Agard et al., 2005; Anders et al., 2005; Schmid et al., 2008b; Jolivet and Brun, 2010). In orogen perpendicular cross sections, suture zones are generally steeply dipping zones, which are characterized by an asymmetry inherited from subduction separating block with a marked different geological history. A collisional suture marks the site of a subducted oceanic plate, which separated the collided continents. Therefore, the suture zone consists of various strongly deformed lithologies derived from the oceanic plate, exotic terrane, or accretionary wedges of the continents (tectonic mélange). The existence of ophiolitic remnants between continental blocks is the most direct evidence for the location of the suture zone.

Geodynamic models suggest that during the initial stage of collision, the suture zone forms a large shear band where most of the deformation is localized (Toussaint et al., 2004). This stage is followed by back thrusting around the suture zone and displacement of the initial crustal suture zone towards the overriding plate (Willet, 1994; Beaumont et al., 1996; Ellis, 1996). A zone of high topography, erosion, and exhumation grows above the area of active deformation. After further shortening the suture zone is no longer the major locus of deformation, but crustal scale thrusts dominate the surface deformation leading to the formation of another zone of surface uplift (Toussaint et al., 2004).

(v) *A volcanic arc* forms in the hanging wall of an oceanic subduction and is the result of melting of the mantle wedge induced by volatiles released from the downgoing plate (mainly water), which drastically lower the melting point of the mantle. Therefore, magma ascends above the subducting slab and forms an arc of volcanoes parallel to the subduction zone. This arc can form at the rim of a continental plate (continental arc, e.g., the Andes) or at the rim of an oceanic plate (island arc, e.g., the Lesser Antilles), depending if the oceanic crust subducts beneath a continent or another oceanic crust, respectively. The arcuate shape of volcanic arcs is the result of the curvature of the Earth's surface forcing the subducting plate to deform into a concave shape.

About a quarter of the magmatic rocks on Earth are produced above subduction zones to form volcanic arcs (Schminke, 2004). Water-bearing fluids from the subducted slab, which are mainly derived from destabilization of chlorite at approximately 50 km depth, rise into the overriding mantle wedge induces partial melting of and generates low-density, calc-alkaline magma (Stern, 2002). The magma buoyantly rises through the overlying crust because of its lower density. If the crust above the subduction zone is oceanic, the basaltic to andesitic magma is not significantly altered. Magmas, which intrude into the continental crust are altered by assimilation of crustal material and is enriched in SiO_2 and K_2O.

Recently, the geodynamic processes associated with subduction and volcanic arc growth have been studied by various numerical models including the subduction dynamics influenced by volatile release (Gerya et al., 2002; Rüpke et al., 2004), melting processes associated with the downgoing slab (Gerya and Yuen, 2003; Gerya et al., 2004), growth and mixing dynamics of magmas in mantle wedges (Gorczyk et al., 2007a; Gorczyk et al., 2007b), and the crustal growth of volcanic arcs (Nikolaeva et al., 2008). Some of these models include the physics of various geochemical–petrological–thermomechanical processes investigating slab retreat and bending, crust dehydration, fluid transport, mantle wedge melting, and melt extraction. The results of these numerical experiments show that the rate of plate retreat significantly influences the rate of crustal growth of the volcanic arc and the composition of newly formed crust because of the major contribution of a crustal component by hydrated partially molten upwellings rising from downgoing slabs (Gerya and Yuen, 2003; Nikolaeva et al., 2008).

Destruction of orogens

Continental collision leads to crustal thickening and building of high topography supported by a thick crustal root. The transition from the cold oceanic subduction regime to the warm collision regime weakens the orogenic root, inducing the formation of a continental plateau. The lower crust below the plateau can melt and flow out horizontally in response to crustal thickness variations (Royden, 1996; Rey et al., 2001). During ongoing convergence, lateral extrusion of crustal blocks can become the dominant mechanism accommodating crustal shortening, instead of crustal thickening (Tapponnier and Molnar, 1976; Ratschbacher et al., 1991). Depending on free or extensional boundary condition, crustal blocks can accommodate tens to hundreds of kilometers of lateral displacement along major strike-slip faults (Seyferth and Henk, 2004). When convergence between the two plates involved in the orogen stops, gravitational collapse occurs, inducing thinning of the crust and destruction of the orogenic belt (Vanderhaeghe, 2009). It may be associated, at depth, with slab rollback, slab detachment, delamination, or convective thinning of the lithospheric mantle (Houseman et al., 1981; Wortel and Spakman, 1992; Davis and von Blanckenburg, 1995) and reflected by generalized post-orogenic I- and S-type magmatic activity.

During gravitational collapse, the root of the orogen can be exhumed at the surface in metamorphic core complexes. These crustal scale structures are domes of high-grade rocks located in the hanging wall of low-angle normal fault systems or detachments (Whitney et al., 2013). The erosion products of the exhuming hanging wall are deposited in supradetachment basins. The formation of metamorphic core complexes is controlled by the rheological layering of the lithosphere at the onset of collapse (Buck, 1991). However, the exhumation of these dome-shaped structures below low-angle normal faults has long been considered as mechanically paradoxical. Thermomechanical modeling and field observations have shown that warm geotherms, intrusions (Tirel et al., 2008), partial melting (Rey et al., 2009), complex crustal layering inherited from the collision (Huet et al., 2011), and metamorphic reactions (Collettini and Holdsworth, 2004; Hürzeler and Abart, 2008; Grasemann and Tschegg, 2012) create the conditions for strain localization in low-angle normal faults and mechanically interacting high-angle normal faults which rotate to lower angles during ongoing extension. After gravitational collapse, continental crust of normal thickness and thermal regime is produced. Further extension then leads to continental rifting and the onset of a new orogenic cycle.

Outlook and conclusions

Much advance has been made in recent years in the understanding of the dynamics of orogenic belts in response to the large-scale motion of the lithospheric plates and numerous ideas, many of which developed from studies on the collision between India and Asia have led to a better understanding of the thermomechanical evolution of orogens (for a comprehensive overview see Stüwe, 2007; Turcotte and Schubert, 2014). Large, global geophysical datasets, including topography, gravity, geoid, and magnetic anomaly, together with new high-resolution geological datasets constraining the rheology of the lithosphere and the rates of geodynamic processes, increased our knowledge about the evolution and structure of orogens (Magni et al., 2013). Numerical thermomechanical models have demonstrated that the dynamics of orogens is strongly controlled by the temperature-dependent strength distribution within the lithosphere and the efficiency of erosion. Indeed, the recognition of a significant linkage of deep Earth dynamic processes with surface and/or near-surface geological processes is one of the important developments in Earth sciences over the past decade (Braun, 2010, Cloetingh and Willet, 2013).

Recently, it has been suggested that mantle dynamics has a major influence on building mountains. Two orogenic end-members have been distinguished, depending on the forces controlling the formation of the orogen (Faccenna et al., 2013). Slab pull dynamics is associated with slabs subducting in convective upper mantle, trench rollback, and the formation of mountains with moderate crustal thickness (e.g., the Mediterranean). Slab suction is associated with slabs penetrating through the phase transitions around 660 km, leading to the formation of mountains with thick crustal columns (e.g., the Himalayas).

In spite of this rapid increase of our knowledge level about orogeny, the effects of three-dimensionality in orogenic systems and their forming processes are still poorly understood. Recent three-dimensional mechanical models of folding (Schmalholz, 2008; Schmid et al., 2008a), faulting (Walsh et al., 2001; Imber et al., 2004; Seyferth and Henk, 2004; Soliva et al, 2008; Le Pourhiet et al., in review), rifting (Allken et al., 2012), crustal wedges (Braun and Yamato, 2010), metamorphic core complexes (Le Pourhiet et al., 2012), and mid-ocean ridges (Gerya, 2012) demonstrated the urgent need to use high-resolution three-dimensional numerical models in order to further develop our understanding of geodynamic processes in orogens.

Bibliography

Agard, P., Omrani, J., Jolivet, L., and Mouthereau, F., 2005. Convergence history across Zagros (Iran): constraints from collisional and earlier deformation. *International Journal of Earth Sciences*, **94**, 401–419.

Allen, P. A., and Allen, J. R., 2005. *Basin Analysis: Principles and Applications*, 2nd edn. Oxford: Wiley-Blackwell.

Allken, V., Huismans, R. S., and Thieulot, C., 2012. Factors controlling the mode of rift interaction in brittle-ductile coupled systems: a 3D numerical study. *Geochemistry, Geophysics, Geosystems*, **13**, Q05010, doi:10.1029/2012gc004077.

Anders, B., Reischmann, T., Poller, U., and Kostopoulos, D., 2005. Age and origin of granitic rocks of the eastern Vardar Zone, Greece: new constraints on the evolution of the Internal Hellenides. *Journal of the Geological Society of London*, **162**, 857–870.

Avouac, J.-P., and Burov, E. B., 1996. Erosion as a driving mechanism of intracontinental mountain growth. *Journal of Geophysical Research*, **110**, 17747–17769.

Beaumont, C., Ellis, S., Hamilton, J., and Fullsack, P., 1996. Mechanical model for subduction-collision tectonics of Alpine-type compressional orogens. *Geology*, **24**, 675–678.

Beaumont, C., Jamieson, R. A., Nguyen, M. H., and Lee, B., 2001. Himalayan tectonics explained by extrusion of a low-viscosity crustal channel coupled to focused surface denudation. *Nature*, **414**, 738.

Beaumont, C., Jamieson, R. A., Nguyen, M. H., and Medvedev, S., 2004. Crustal channel flows: 1. Numerical models with applications to the tectonics of the Himalayan-Tibetan orogen. *Journal of Geophysical Research*, **109**, B06406, doi:10.1029/2003JB002809.

Becker, J. J., Sandwell, D. T., Smith, W. H. F., Braud, J., Binder, B., Depner, J., Fabre, D., Factor, J., Ingalls, S., Kim, S.-H., Ladner, R., Marks, K., Nelson, S., Pharaoh, A., Trimmer, R., Von Rosenberg, J., Wallace, G., and Weatherall, P., 2009. Global bathymetry and elevation data at 30 arc seconds resolution: SRTM30_PLUS. *Marine Geodesy*, **32**(4), 355–371.

Braun, J., 2010. The many surface expressions of mantle dynamics. *Nature Geoscience*, **3**, 825–833.

Braun, J., and Yamato, P., 2010. Structural evolution of a three-dimensional, finite-width crustal wedge. *Tectonophysics*, **484**, 181–192.

Buck, W. R., 1991. Modes of continental lithospheric extension. *Journal of Geophysical Research*, **96**, 20,161–120,178.

Catuneanu, O., 2004. Retroarc foreland systems–evolution through time. *Journal of African Earth Sciences*, **38**, 225–242.

Chopin, C., 1984. Coesite and pure pyrope in high grade blueschists of the Western Alps: a first record and some consequences. *Contributions to Mineralogy and Petrology*, **86**, 107–118.

Clevis, Q., De Boer, P. L., and Nijman, W., 2004. Differentiating the effect of episodic tectonism and eustatic sea-level fluctuations in foreland basins filled by alluvial fans and axial deltaic systems: insights from a three-dimensional stratigraphic forward model. *Sedimentology*, **51**, 809–835.

Cloetingh, S., and Willett, S. D., 2013. Linking deep earth and surface processes. *Eos, Transactions American Geophysical Union*, **94**, 53–54.

Colchen, M., Mascle, G., and van Haver, T., 1986. Some aspects of collision tectonics in the Indus Suture Zone, Ladakh. In Coward, M. P., and Ries, A. C. (eds.), *Collision Tectonics*. London: The Geological Society of London, Special Publications, pp. 173–184.

Collettini, C., and Holdsworth, R. E., 2004. Fault zone weakening and character of slip along low-angle normal faults: insights from the Zuccale fault, Elba, Italy. *Journal of the Geological Society*, **161**, 1039–1051.

Costa, E., and Vendeville, B. C., 2002. Experimental insights on the geometry and kinematics of fold-and-thrust belts above weak, viscous evaporitic decollement. *Journal of Structural Geology*, **24**, 1729–1739.

Cruz, L., Malinski, J., Wilson, A., Take, W. A., and Hilley, G., 2010. Erosional control of the kinematics and geometry of fold-and-thrust belts imaged in a physical and numerical sandbox. *Journal of Geophysical Research*, **115**, B09404, doi:10.1029/2010jb007472.

Dahlstrom, C. D. A., 1969. Balanced cross sections. *Canadian Journal of Earth Sciences*, **6**, 743–757.

Davis, J. H., and von Blanckenburg, F., 1995. Slab breakoff: a model of lithosphere detachment and its test in the magmatism and deformation of collisional orogens. *Earth and Planetary Science Letters*, **129**, 85–102.

Davis, D., Suppe, J., and Dahlen, F. A., 1983. Mechanics of fold-and-thrust belts and accretionary wedges. *Journal of Geophysical Research*, **88**, 1153–1172.

DeCelles, P. G., and DeCelles, P. C., 2001. Rates of shortening, propagation, underthrusting, and flexural wave migration in continental orogenic systems. *Geology*, **29**, 135–138.

DeCelles, P. G., and Giles, K. A., 1996. Foreland basin systems. *Basin Research*, **8**, 105–123.

Elliott, D., 1983. The construction of balanced cross sections. *Journal of Structural Geology*, **5**, 101.

Ellis, S., 1996. Forces driving continental collision: reconciling indentation and mantle subduction tectonics. *Geology*, **24**, 699–702.

Ellis, S., Schreurs, G., and Panien, M., 2004. Comparisons between analogue and numerical models of thrust wedge development. *Journal of Structural Geology*, **26**, 1659–1675.

England, P. C., and Molnar, P., 1990. Surface uplift, uplift of rocks, and exhumation of rocks. *Geology*, **18**, 1173–1177.

Eriksson, P., Altermann, W., Nelson, D., Mueller, W., and Catuneanu, O., 2004. *The Precambrian Earth: Tempos and Events*. Amsterdam: Elsevier.

Ernst, G., 2001. Subduction, ultrahigh-pressure metamorphism, and regurgitation of buoyant crustal slices implications for arcs and continental growth. *Physics of the Earth and Planetary Interiors*, **127**, 253–275.

Ernst, W. G., 2005. Alpine and Pacific styles of Phanerozoic mountain building: subduction-zone petrogenesis of continental crust. *Terra Nova*, **17**, 165–188.

Faccenna, C., Becker, T. W., Conrad, C. P., and Husson, L., 2013. Mountain building and mantle dynamics. *Tectonics*, **32**, 80–93, doi:10.1029/2012TC003176.

Fillon, C., Huismans, R. S., and van der Beek, P., 2013. Syntectonic sedimentation effects on the growth of fold-and-thrust belts. *Geology*, **41**, 83–86.

Frehner, M., Reif, D., and Grasemann, B., 2012. Mechanical versus kinematic shortening reconstructions of the Zagros High Folded Zone (Kurdistan region of Iraq). *Tectonics*, **31**, TC3002, doi:10.1029/2011tc003010.

Frisch, W., Meschede, M., and Blakey, R. C., 2011. *Plate Tectonics – Continental Drift and Mountain Building*. Berlin, Heidelberg: Springer.

Froitzheim, N., Pleuger, J., Roller, S., and Nagel, T., 2003. Exhumation of high- and ultrahigh-pressure metamorphic rocks by slab extraction. *Geology*, **31**, 925–928.

Garcia-Castellanos, D., Fernàndez, M., and Torne, M., 1997. Numerical modeling of foreland basin formation: a program relating thrusting, flexure, sediment geometry and lithosphere rheology. *Computers & Geosciences*, **23**, 993–1003.

Gerya, T. V., 2012. Three-dimensional thermomechanical modeling of oceanic spreading initiation and évolution. *Physics of the Earth and Planetary Interiors*, **214**, 35–52.

Gerya, T. V., and Yuen, D. A., 2003. Rayleigh–Taylor instabilities from hydration and melting propel "cold plumes" at subduction zones. *Earth and Planetary Science Letters*, **212**, 47–62.

Gerya, T. V., Stöckhert, B., and Perchuk, A. L., 2002. Exhumation of high-pressure metamorphic rocks in a subduction channel: a numerical simulation. *Tectonics*, **21**, 1056, doi:10.1029/2002TC001406.

Gerya, T. V., Yuen, D. A., and Sevre, E. O. D., 2004. Dynamical causes for incipient magma chambers above slabs. *Geology*, **32**, 89–92.

Gerya, T. V., Perchuck, L. L., and Burg, J. P., 2008. Transient hot channels: perpetrating and regurgitating ultrahigh-pressure, high-temperature crust-mantle associations in collision belts. *Lithos*, **103**, 236–256.

Gorczyk, W., Gerya, T. V., Connolly, J. A. D., and Yuen, D. A., 2007a. Growth and mixing dynamics of mantle wedge plumes. *Geology*, **35**, 587–590.

Gorczyk, W., Willner, A. P., Gerya, T. V., Connolly, J. A. D., and Burg, J.-P., 2007b. Physical controls of magmatic productivity at Pacific-type convergent margins: numerical modelling. *Physics of the Earth and Planetary Interiors*, **163**, 209–232.

Grasemann, B., and Schmalholz, S. M., 2012. Lateral fold growth and fold linkage. *Geology*, **40**, 1039–1042.

Grasemann, B., and Tschegg, C., 2012. Localization of deformation triggered by chemo-mechanical feedback processes. *Geological Society of America Bulletin*, **124**, 737–745.

Grasemann, B., Fritz, H., and Vannay, J.-C., 1999. Quantitative kinematic flow analysis from the Main Central Thrust Zone (NW-Himalaya, India); implications for a decelerating strain path and the extrusion of orogenic wedges. *Journal of Structural Geology*, **21**, 837–853.

Groshong, R. H., Jr., 1994. Area balance, depth to detachment, and strain in extension. *Tectonics*, **13**, 1488–1497.

Grujic, D., Casey, M., Davidson, C., Hollister, L. S., Kündig, R., Pavlis, T., and Schmid, S., 1996. Ductile extrusion of the Higher Himalayan Crystalline in Bhutan: evidence from quartz microfabrics. *Tectonophysics*, **260**, 21–43.

Grujic, D., Hollister, L. S., and Parrish, R. R., 2002. Himalayan metamorphic sequence as an orogenic channel: insight from Bhutan. *Earth and Planetary Science Letters*, **198**, 177–191.

Hacker, B. R., and Liou, J. G., 1998. *When Continents Collide: Geodynamics and Geochemistry of Ultrahigh-Pressure Rocks*. Dordrecht: Kluwer Academic Publishers.

Hacker, B. R., and Peacock, S. M., 1995. Exhumation of ultrahigh-pressure metamorphic rocks: constraints from kinetic studies and thermal modeling. In Coleman, R. C., and Wang, X. (eds.),

Ultra-high Pressure Metamorphism. Cambridge: Cambridge University Press, pp. 159–181.

Hatcher, R. D., and Williams, R. T., 1986. Mechanical model for single thrust sheets part I: taxonomy of crystalline thrust sheets and their relationships to the mechanical behavior of orogenic belts. *Geological Society of America Bulletin*, **97**, 975–985.

Haynes, S. J., and McQuillan, H., 1974. Evolution of the Zagros Suture Zone, Southern Iran. *Geological Society of America Bulletin*, **85**, 739–744.

Homza, T. X., and Wallace, W. K., 1995. Geometric and kinematic models for detachment folds with fixed and variable detachment depths. *Journal of Structural Geology*, **17**, 575–588.

Houseman, G. A., McKenzie, D. P., and Molnar, P., 1981. Convective instability of a thickened boundary layer and its relevance for the thermal evolution of continental convergent belts. *Journal of Geophysical Research*, **86**, 6115–6132.

Huet, B., Le Pourhiet, L., Labrousse, L., Burov, E., and Jolivet, L., 2011. Post-orogenic extension and metamorphic core complexes in a heterogeneous crust: the role of crustal layering inherited from collision. Application to the Cyclades (Aegean domain). *Geophysical Journal International*, **184**, 611–625.

Hürzeler, J.-P., and Abart, R., 2008. Fluid flow and rock alteration along the Glarus thrust. *Swiss Journal of Geosciences*, **101**, 251–268.

Imber, J., Tuckwell, G. W., Childs, C., Walsh, J. J., Manzocchi, T., Heath, A. E., Bonson, C. G., and Strand, J., 2004. Three-dimensional distinct element modelling of relay growth and breaching along normal faults. *Journal of Structural Geology*, **26**, 1897–1911.

Jolivet, L., and Brun, J.-P., 2010. Cenozoic geodynamic evolution of the Aegean. *International Journal of Earth Sciences*, **99**, 109–138.

Jolivet, L., Faccenna, C., Goffé, B., Burov, E., and Agard, P., 2003. Subduction tectonics and exhumation of high-pressure metamorphic rocks in the Mediterranean orogens. *American Journal of Science*, **3003**, 353–409.

Jordan, T. A., and Watts, A. B., 2005. Gravity anomalies, flexure and the elastic thickness structure of the India–Eurasia collisional system. *Earth and Planetary Science Letters*, **236**, 732–750.

Klootwijk, C., Sharma, M. L., Gergan, J., Shah, S. K., and Tirkey, B., 1984. The Indus-Tsangpo Suture zone in Ladakh, Northwest Himalaya: further palaeomagnetic data and implications. *Tectonophysics*, **106**, 215–238.

Law, R., Searle, M. P., and Godin, L., 2006. *Channel Flow, Ductile Extrusion and Exhumation of Lower-Mid Crust in Continental Collision Zones*. London: Geological Society of London, Special Publication.

Le Pourhiet, L., Huet, B., and Traoré, M., in press (available online). Links between long-term and short-term rheology of the lithosphere: Insights from strike-slip fault modeling. Submitted to Tectonophysics.

Le Pourhiet, L., Huet, B., May, D. A., Labrousse, L., and Jolivet, L., 2012. Kinematic interpretation of the 3D shapes of metamorphic core complexes. *Geochemistry, Geophysics, Geosystems*, **13**, Q09002, doi:10.1029/2012gc004271.

Magni, V., Faccenna, C., van Hunen, J., and Funiciello, F., 2013. Delamination vs. break-off: the fate of continental collision. *Geophysical Research Letters*, **40**, 285–289, doi:10.1029/2012GL054404.

Moores, E. M., and Twiss, R. J., 1995. *Tectonics*. New York: W.H. Freeman and Company.

Müller, R. D., Sdrolias, M., Gaina, C., and Roest, W. R., 2008. Age, spreading rates and spreading symmetry of the world's ocean crust. *Geochemistry, Geophysics, Geosystems*, **9**, Q04006, doi:10.1029/2007GC001743.

Nikolaeva, K., Gerya, T. V., and Connolly, J. A. D., 2008. Numerical modelling of crustal growth in intraoceanic volcanic arcs. *Physics of the Earth and Planetary Interiors*, **171**, 336–356.

Pelletier, J. D., 2007. Erosion-rate determination from foreland basin geometry. *Geology*, **35**, 5–8.

Pitcher, W. S., 1979. The nature, ascent and emplacement of granitic magmas. *Journal of the Geological Society*, **136**, 627–662.

Ratschbacher, L., Merle, O., Davy, P., and Cobbold, P., 1991. Lateral extrusion in the Eastern Alps, Part 1: boundary conditions and experiments scaled for gravity. *Tectonics*, **10**, 245–256.

Rey, P., Vanderhaeghe, O., and Teyssier, C., 2001. Gravitational collapse of the continental crust: definition, regimes and modes. *Tectonophysics*, **342**, 435–449.

Rey, P. F., Teyssier, C., and Whitney, D. L., 2009. Extension rates, crustal melting, and core complex dynamics. *Geology*, **37**, 391–394.

Rosenberg, C. L., 2004. Shear zones and magma ascent: a model based on a review of the Tertiary magmatism in the Alps. *Tectonics*, **23**, 3002, doi:10.1029/2003TC001526.

Royden, L. H., 1996. Coupling and decoupling of crust and mantle in convergent orogens: implications for strain partitioning in the crust. *Journal Geophysical Research*, **101**(17), 679–617,692.

Ruh, J. B., Kaus, B. J. P., and Burg, J.-P., 2012. Numerical investigation of deformation mechanics in fold-and-thrust belts: influence of rheology of single and multiple décollements. *Tectonics*, **31**, TC3005, doi:10.1029/2011tc003047.

Rüpke, L. H., Morgan, J. P., Hort, M., and Connolly, J. A. D., 2004. Serpentine and the subduction zone water cycle. *Earth and Planetary Science Letters*, **223**, 17–34.

Schmalholz, S. M., 2008. 3D numerical modeling of forward folding and reverse unfolding of a viscous single-layer: implications for the formation of folds and fold patterns. *Tectonophysics*, **446**, 31–41.

Schmid, D. W., Dabrowski, M., and Krotkiewski, M., 2008a. Evolution of large amplitude 3D fold patterns: a FEM study. *Physics of the Earth and Planetary Interiors*, **171**, 400–408.

Schmid, S., Bernoulli, D., Fügenschuh, B., Matenco, L., Schefer, S., Schuster, R., Tischler, M., and Ustaszewski, K., 2008b. The Alpine-Carpathian-Dinaridic orogenic system: correlation and evolution of tectonic units. *Swiss Journal of Geosciences*, **101**, 139–183.

Schmincke, H. U., 2004. *Volcanism*. Berlin/Heidelberg/New York: Springer.

Schneider, D. A., Edwards, M. A., Kidd, W. S. F., Khan, A. M., Seeber, L., and Zeitler, P. K., 1999. Tectonics of Nanga Parbat, western Himalaya: synkinematic plutonism within the doubly vergent shear zones of a crustal-scale pop-up structure. *Geology*, **27**, 999–1002.

Schulmann, K., Lexa, O., Stipska, P., Racek, M., Tajcmanova, L., Konopasek, J., Edel, J. B., Peschler, A., and Lehmann, J., 2008. Vertical extrusion and horizontal channel flow of orogenic lower crust: key exhumation mechanisms in large hot orogens? *Journal of Metamorphic Geology*, **26**, 273–297.

Seyferth, M., and Henk, A., 2004. Syn-convergent exhumation and lateral extrusion in continental collision zones–insights from three-dimensional numerical models. *Tectonophysics*, **382**, 1–29.

Sherkati, S., Letouzey, J., and Frizon de Lamotte, D., 2006. Central Zagros fold-thrust belt (Iran): new insights from seismic data, field observation, and sandbox modeling. *Tectonics*, **25**, TC4007, doi:10.1029/2004tc001766.

Simpson, G. D. H., 2009. Mechanical modelling of folding versus faulting in brittle-ductile wedges. *Journal of Structural Geology*, **31**, 369–381.

Soliva, R., Benedicto, A., Schultz, R. A., Maerten, L., and Micarelli, L., 2008. Displacement and interaction of normal fault segments branched at depth: implications for fault growth and potential earthquake rupture size. *Journal of Structural Geology*, **30**, 1288–1299.

Stern, R. J., 2002. Subduction zones. *Reviews of Geophysics*, **40**, 1–38.

Stünitz, H., and Tullis, J., 2001. Weakening and strain localization produced by syn-deformational reaction of plagioclase. *International Journal of Earth Sciences*, **90**, 136–148.

Stüwe, K., 2007. *Geodynamics of the Lithosphere*, 2nd edn. Berlin: Springer, p. 493.

Tapponnier, P., and Molnar, P., 1976. Slip-line field theory and large-scale continental tectonics. *Nature*, **264**, 319–324.

Tirel, C., Brun, J.-P., and Burov, E., 2008. Dynamics and structural development of metamorphic core complexes. *Journal of Geophysical Research*, **113**, B04403, doi:10.1029/2005JB003694.

Toussaint, G., Burov, E., and Avouac, J. P., 2004. Tectonic evolution of a continental collision zone: a thermomechanical numerical model. *Tectonics*, **23**, TC6003, doi:10.1029/2003tc001604.

Turcotte, D. L., and Schubert, G., 2014. *Geodynamics*, 3rd edn. Cambridge: Cambridge University Press, p. 623.

Vanderhaeghe, O., 2009. Migmatites, granites and orogeny: flow modes of partially-molten rocks and magmas associated with melt/solid segregation in orogenic belts. *Tectonophysics*, **477**, 119–134.

Vidal-Royo, O., Cardozo, N., Muñoz, J. A., Hardy, S., and Maerten, L., 2012. Multiple mechanisms driving detachment folding as deduced from 3D reconstruction and geomechanical restoration: the Pico del Águila anticline (External Sierras, Southern Pyrenees). *Basin Research*, **24**, 295–313.

Walsh, J. J., Childs, C., Meyer, V., Manzocchi, T., Imber, J., Nicol, A., Tuckwell, G., Bailey, W. R., Bonson, C. G., Watterson, J., Nell, P. A., and Strand, J., 2001. Geometric controls on the evolution of normal fault systems. In Strachan, R. A., Magloughlin, J. F., and Knipe, R. J. (eds.), *The Nature and Significance of Fault Zone Weakening*. London: Geological Society Special Publication, pp. 157–170.

White, N. M., Pringle, M., Garzanti, E., Bickle, M., Najman, Y., Chapman, H., and Friend, P., 2002. Constraints on the exhumation and erosion of the High Himalayan Slab, NW India, from foreland basin deposits. *Earth and Planetary Science Letters*, **195**, 29–44.

Whitney, D. L., Teyssier, C., Rey, P., and Buck, W. R., 2013. Continental and oceanic core complexes. *Geological Society of America Bulletin*, **125**, 273–298.

Willett, S. D., and Beaumont, C., 1994. Subduction of Asian lithospheric mantle beneath Tibet inferred from models of continental collision. *Nature*, **369**, 642–645.

Wortel, M. J. R., and Spakman, W., 1992. Structure and dynamics of subducted lithosphere in the Mediterranean region. *Proceedings, Koninklijke Nederlandse Academic Wetenschappen*, **95**, 325–347.

Yamato, P., Burov, E., Agard, P., Le Pourhiet, L., and Jolivet, L., 2008. HP-UHP exhumation during slow continental subduction: self-consistent thermodynamically and thermomechanically coupled model with application to the Western Alps. *Earth and Planetary Science Letters*, **271**, 63–74.

Yamato, P., Kaus, B. J. P., Mouthereau, F., and Castelltort, S., 2011. Dynamic constraints on the crustal-scale rheology of the Zagros fold belt, Iran. *Geology*, **39**, 815–818.

Zwart, H. J., 1967. The duality of orogenic belts. *Geologie en Mijnbouw*, **46**, 283–309.

Cross-references

Accretionary Wedges
Active Continental Margins
Driving Forces: Slab Pull, Ridge Push
Earthquakes
High-pressure, Low-temperature Metamorphism
Hot Spots and Mantle Plumes
Intraoceanic Subduction Zone
Island Arc Volcanism, Volcanic Arcs
Lithosphere: Structure and Composition
Magmatism at Convergent Plate Boundaries
Marine Heat Flow
Metamorphic Core Complexes
Mid-ocean Ridge Magmatism and Volcanism
Mohorovičić Discontinuity (Moho)
Morphology Across Convergent Plate Boundaries
Oceanic Spreading Centers
Paired Metamorphic Belts
Serpentinization
Subduction
Transform Faults
Wilson Cycle

OXYGEN ISOTOPES

Mark Maslin[1] and Alexander J. Dickson[2]
[1]Department of Geography, University College London, London, UK
[2]Department of Earth Sciences, University of Oxford, Oxford, UK

Definition

Marine sediments provide long continuous records of past climate changes at intraannual, annual to centennial scale resolutions enabling insights into past changes within both oceanic and continental environments. Stable oxygen isotopes are a fundamental tool for paleoceanographers providing the means to reconstruct a range of variables including sea surface and bottom water temperature, sea surface salinity, sea-level, river discharge, and iceberg activity. Comprehensive introductions on oxygen isotopes and their physicochemical behavior/systematics are given in Craig and Gordon (1965), Garlick (1974), Hoefs (1997), Criss (1999), Rohling and Cooke (1999), Zeebe and Wolf-Gladrow (2001) and Pearson (2012).

Oxygen isotopes in marine archives

Oxygen isotope compositions can be measured in many forms of mineral, including calcite, aragonite, opal, and barite. The shells of carbonate producing organisms have most commonly been used for $\delta^{18}O$ analysis, and have underpinned many of the advances in stratigraphy and paleoceanographic reconstruction that have been made in recent decades. However, $\delta^{18}O$ in the other minerals has also been extremely important in allowing $\delta^{18}O$-based paleoclimate reconstructions to be extended into carbonate-poor regions of the global oceans and in complementing carbonate-derived data in regions where the different minerals coexist.

There are several major influences on the $\delta^{18}O$ compositions in marine minerals (Eq. 1). First is the $\delta^{18}O$ of the seawater in which minerals precipitate ($\Delta\delta^{18}O_w$). This is controlled by the globally significant impact of the storage and release of ^{16}O-enriched freshwater in continental ice sheets ($\Delta\delta_{GIV}$), and the local balance of evaporation verses freshwater inputs from precipitation, river, or iceberg discharge ($\Delta\delta_{local}$). Second is the water temperature at which the precipitate forms ($\Delta\delta_T$). A greater proportion of ^{16}O is incorporated into marine minerals at higher temperatures. The oxygen isotope composition of marine carbonates is also controlled by the carbonate ion concentration of the seawater from which they precipitate and by the impact of postdepositional dissolution under conditions of carbonate ion undersaturation. In the sections below, these influences are examined and examples given of how they can be used to reconstruct past climate on a range of timescales.

$$\Delta\delta^{18}O_M = \Delta\delta^{18}O_w + \Delta\delta_T \\ = \Delta\delta_{GIV} + \Delta\delta_{local} + \Delta\delta_T + \text{vital effects} \quad (1)$$

where:

$\Delta\delta^{18}O_M$ = oxygen isotopes of the mineral precipitated (e.g., calcite or opal)

$\Delta\delta^{18}O_w$ = oxygen isotope composition of the water which is a combined affect of variations in global ice volume ($\Delta\delta_{GIV}$) and local influences ($\Delta\delta_{local}$)

$\Delta\delta_T$ = water temperature

Vital effects = deviations from equilibrium due to the incorporation of oxygen into biogenetically produced calcite or opal. These effects include ontogeny, symbiont photosynthesis, respiration and life history (see Rohling and Cooke, 1999; Marchitto et al., 2014).

Global ice volume ($\Delta\delta_{GIV}$)

During particularly cold intervals of Earth history, ^{18}O-depleted precipitation is transferred from the oceans to continental ice sheets, where it is stored over timescales of 10^3–10^4 years. This transfer registers as a whole ocean shift in the oxygen isotope composition of seawater, given that the magnitude of the isotope perturbation is considerably longer than the ocean mixing time of ~2000 years. The oxygen isotope composition of marine minerals thus allow stratigraphic records from around the world to be correlated with a precision of a few thousand years (Figure 1).

Changes in seawater δ^{18}O have been calibrated to known variations in global sea-level (e.g., Chappell and Shackleton, 1986; Shackleton, 1987; Fairbanks, 1989; Bard et al., 1996; Schrag et al., 1996; Burns and Maslin, 1999; Shackleton, 2000). Using the best current estimate for the magnitude of shifts in both parameters over the past ~20,000 years (1‰ and 130 m) produces a relationship of 0.0077 ‰/m. However, it is known that the relationship between RSL and benthic foraminifera δ^{18}O is non-linear and is also different for both deglaciation and glaciation (see Figure 2). This is firstly because there is clear evidence that there were large differences between the average isotopic composition of each of the major ice sheets that existed during Quaternary glacial periods (Duplessy et al., 2002) with compositions ranging from −16‰ to −60‰ for Northern Hemisphere and Antarctic ice. The timing of the growth and melting of each ice sheet will consequently influence the temporal evolution of the RSL-$\Delta\delta_{GIV}$ relationship as illustrated in Figure 2. Secondly, the extent of isotopic fractionation during snowfall changes as ice sheets grow, typically increasing as the ice sheet matures through a combination of continentality and altitude effects.

Carbonate ion concentrations

pH can affect seawater δ^{18}O. In seawater, dissolved inorganic carbon (DIC) is comprised of the sum of several species:

$$\text{DIC} = \text{CO}_{2[aq]} + \text{H}_2\text{CO}_3 + \text{HCO}_3^- + \text{CO}_3^{2-}$$

At present day seawater pH, the relative abundance of these species is dominated by bicarbonate (HCO_3^-). At times of higher seawater pH, when the relative abundance of carbonate ions (CO_3^{2-}) is higher, the δ^{18}O of seawater decreases due to a lower equilibrium δ^{18}O composition in CO_3^{2-} compared to HCO_3^-. The magnitude of this effect has been quantified in culture experiments of the planktonic foraminifera in the range −0.0015 to −0.005‰/μmol/kg (Spero et al., 1997), which agrees closely with a gradient of −0.0024‰/μmol/kg calculated from inorganic precipitation theory (Zeebe, 1999). Quantifying the magnitude of the carbonate ion effect on long-term stratigraphic record of seawater δ^{18}O is, however, difficult due to problems reconstructing carbonate ion concentrations using proxy data.

Paleotemperature ($\Delta\delta_T$)

Temperature causes a change in the δ^{18}O of marine minerals due to inorganic kinetic fractionations during mineral formation. This effect has been quantified in several inorganic precipitation experiments (e.g., McCrea, 1950; Craig, 1965; O'Neil et al., 1969; Kim and O'Neil, 1997), examples of which include:

Epstein et al. (1953)

$$T = 16.5 - 4.3(\delta_c - \delta_w) + 0.14(\delta_c - \delta_w)^2 \quad (2)$$

Shackleton (1974) based on O'Neil et al. (1969)

$$T = 16.9 - 4.38(\delta_c - \delta_w) + 0.1(\delta_c - \delta_w)^2 \quad (3)$$

Hays and Grossman (1991) based on O'Neil et al. (1969)

$$T = 15.7 - 4.36(\delta_c - \delta_w) + 0.12(\delta_c - \delta_w)^2 \quad (4)$$

Shemesh et al. (2002)

$$T = 11.03 - 2.03\left(\delta^{18}O_{(diatom)} - \delta^{18}O_{(water)} - 40\right) \quad (5)$$

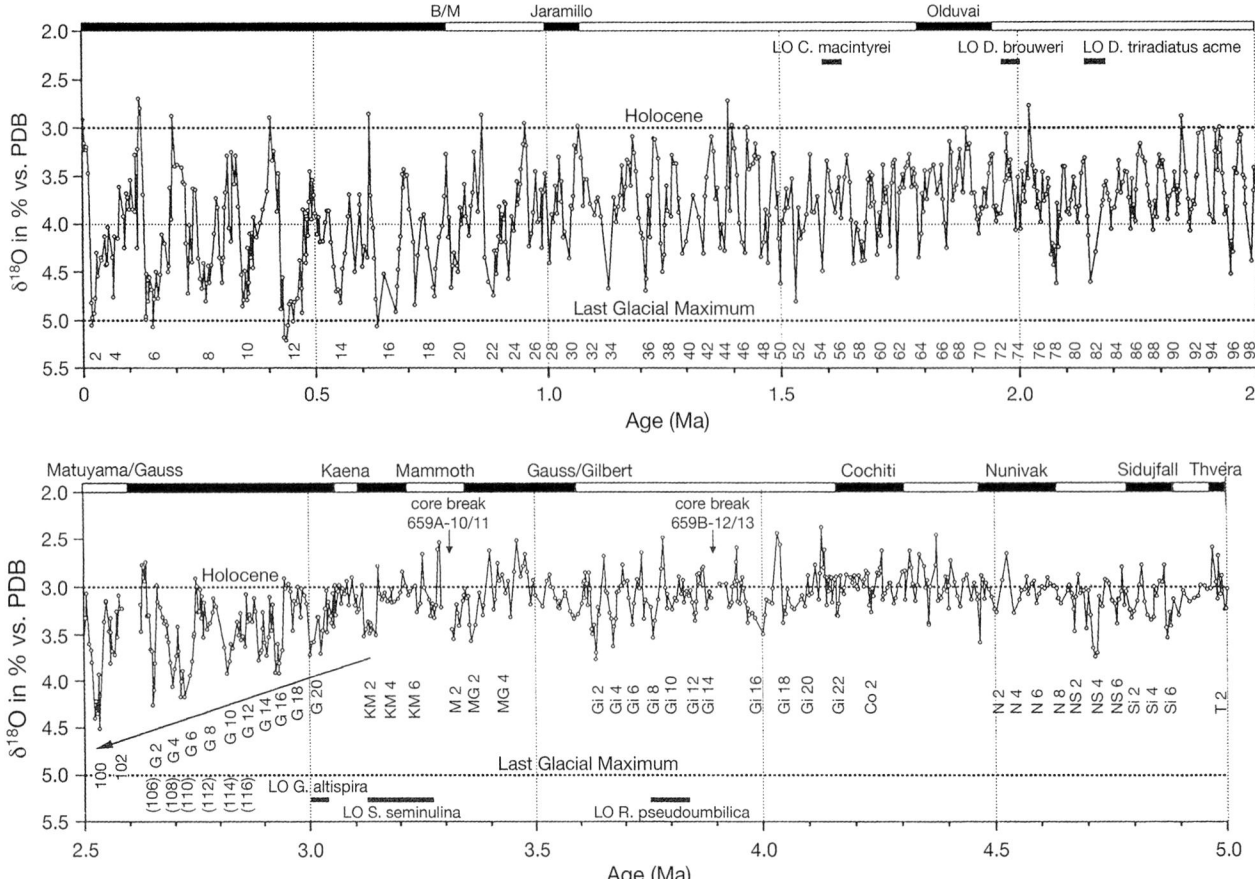

Oxygen Isotopes, Figure 1 Benthic foraminifera oxygen isotope record from ODP 659 in the tropical North East Atlantic with paleomagnetic reversal boundaries and marine oxygen isotope stages (MIOS) defined (adapted from Tiedemann et al., 1994).

When;

T = isotopic temperature estimate
$\delta_c = \delta^{18}O$ of the foraminifera calcite test
$\delta_w = \delta^{18}O$ of the surface waters,

Alternative equations have been proposed by Bemis et al. (1998), which take the form of linear relationships between (foraminifera) calcite $\delta^{18}O$ and temperature, e.g.,:

$$\Delta\delta_c = \Delta\delta_w - 0.23\Delta T \qquad (6)$$

Moreover different calibration equation have been developed for different species of planktonic and benthic foraminifera (Marchitto et al., 2014). Although these equations differ in their predictions of absolute temperature, they consistently predict a gradient of ~ −0.2 to −0.3‰/°C. An important pre-requisite to reconstructing absolute past temperatures from $\delta^{18}O$ is a knowledge of the seawater $\delta^{18}O$ (δ_w). Since δ_w has varied in the past, temperature reconstructions require that there should be little or no change in global ice volume and/or local conditions during the period of reconstruction or there should be independent estimates of both these influences.

Paleosalinity ($\Delta\delta_{local}$)

Calculation of surface water salinity or sea surface salinity (SSS) is possible in locations where the $\Delta\delta_{local}$ is primarily controlled either by the balance of evaporation verses precipitation (E-P) or the input of ^{18}O-depleted freshwater such as from melting icebergs or rivers. The local influence on the marine oxygen isotope record is calculated by subtracting the temperature influence and the global ice volume variations from the precipitate record:

$$\Delta\delta_{local} = \Delta\delta^{18}O_M - \Delta\delta_T - \Delta\delta_{GIV} \qquad (7)$$

Sea surface temperatures can be independently reconstructed in five ways; planktonic foraminifera assemblage data (e.g., Imbrie and Kipp, 1971; Pflaumann et al., 2003), ratios of n-alkenones with differing number of unsaturated bonds (U^K_{37}) (Prahl and Wakeham, 1987), the number of cyclic inclusions in archaea-derived

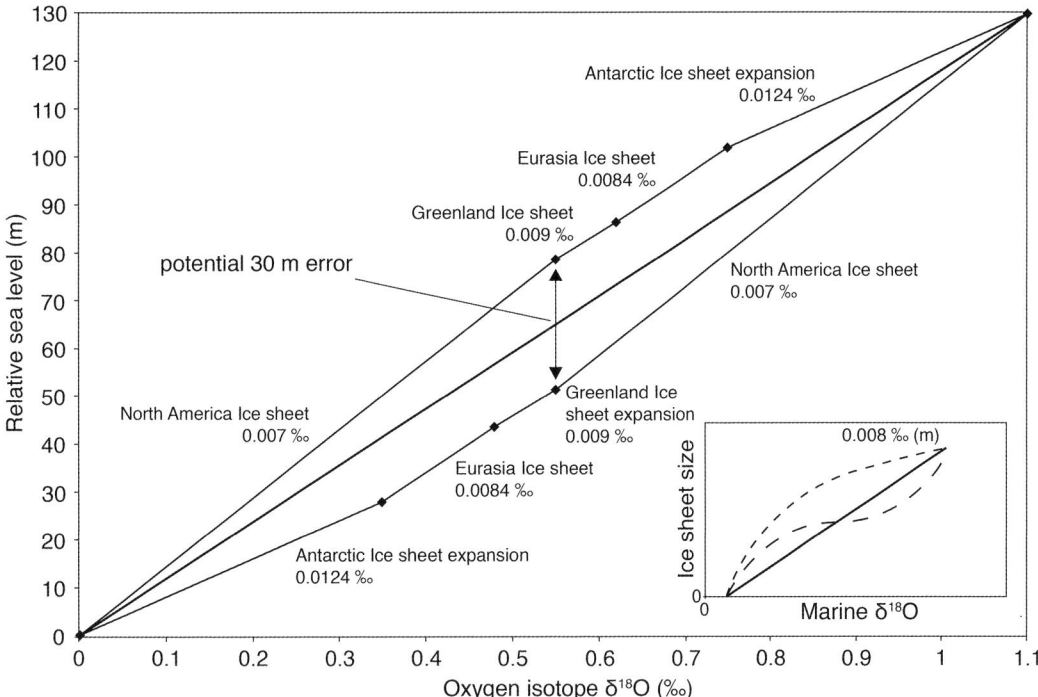

Oxygen Isotopes, Figure 2 Relationship between relative sea-level (*RSL*) and oceanic oxygen isotopes assuming the four major past ice sheets (Antarctic, Greenland, Eurasia, and North America) melted or grow in different successive orders. Note the different amount of fractionation that occurs within each ice sheet alters the RSL-δ^{18}O relationship and could create an error of up to 30 m RSL. *Insert* – The relationship between individual continental ice sheet size and oceanic oxygen isotopes may also not be linear.

biomarkers (TEX$_{86}$) (Schouten et al., 2002), clumped oxygen isotopes (Eiler, J.M., 2007, EPSL 3–4, p 309–327) and Mg/Ca ratios in planktonic foraminifera tests (Elderfield and Ganssen, 2000). These estimates of temperature can be converted to $\Delta\delta_T$ using Eqs 2, 3, 4, 5, and 6. The isotopic effect of changing global ice volume can be estimated by converting RSL or ESL to oxygen isotopic variation using a simple linear relationship of ~0.008–0.01 ‰ m^{-1}. Once $\Delta\delta_{local}$ has been calculated then the relationship between it and sea surface salinity can be estimated. This relationship depends on the oxygen isotope value of the end-member inputs i.e., freshwater (δ_{fresh}) and mean marine sea water (δ_{ocean}). These values vary with locality due to the input of meltwater or precipitation or river runoff compared with the influence of evaporation and ocean circulation. The oxygen isotope-salinity relationship can be defined as:

$$\Delta\delta_{local} = (\delta_{ocean} - \delta_{fresh}/S_{ocean} - S_{fresh})S_{local} + \delta_{fresh} \quad (8)$$

where salinity of freshwater (S_{fresh}) is equal to zero and can therefore be omitted and the equation solved for local Salinity:

$$S_{local} = (\delta_{fresh} - \Delta\delta_{local})(S_{ocean}/\delta_{fresh} - \delta_{ocean}) \quad (9)$$

Calculation of surface water salinity, however, is fraught with difficulties due to the propagation of the uncertainties associated with estimating each unknown parameter in Eq. 6 (Schmidt, 1999; Rohling, 2000). These include (i) uncertainties in the paleo-SST estimates, which are ± 1–$1.5\,^\circ$C for mid- to high-latitude sites and over at least $\pm 2\,^\circ$C in the tropics (see Elderfield and Ganssen, 2000; Pflaumann et al., 2003; Kucera et al., 2005), (ii) Estimation of the $\Delta\delta_{GIV}$ has been shown above to contain uncertainties, which are usually quoted as ± 0.001 ‰ $^\circ$C^{-1}, (iii) accurate determination of the mean ocean salinity (S$_{ocean}$) due to the concentration effect of changing global ice-volume effect. This can be calculated using Eq. 9, but suffers from the uncertainties in estimating past sea-levels, which can be on the order of centimeters to meters.

$$S_{ocean} = 34.74 \times RSL/(3900 - RSL) \quad (10)$$

(iv) The estimation of the oxygen isotope freshwater endmember (δ_{fresh}) also has inherent uncertainties. Selection of freshwater endmember compositions of ~ -20‰ or -40‰ can produce significantly different mixing gradients resulting in salinity estimates that vary greatly.

Not only are there large errors associated with each of the above assumptions but these errors are cumulative in the calculations (Rohling, 2000). Salinity reconstructions have been used to infer a large number of climatic parameters including: iceberg discharge (Maslin et al., 1995;

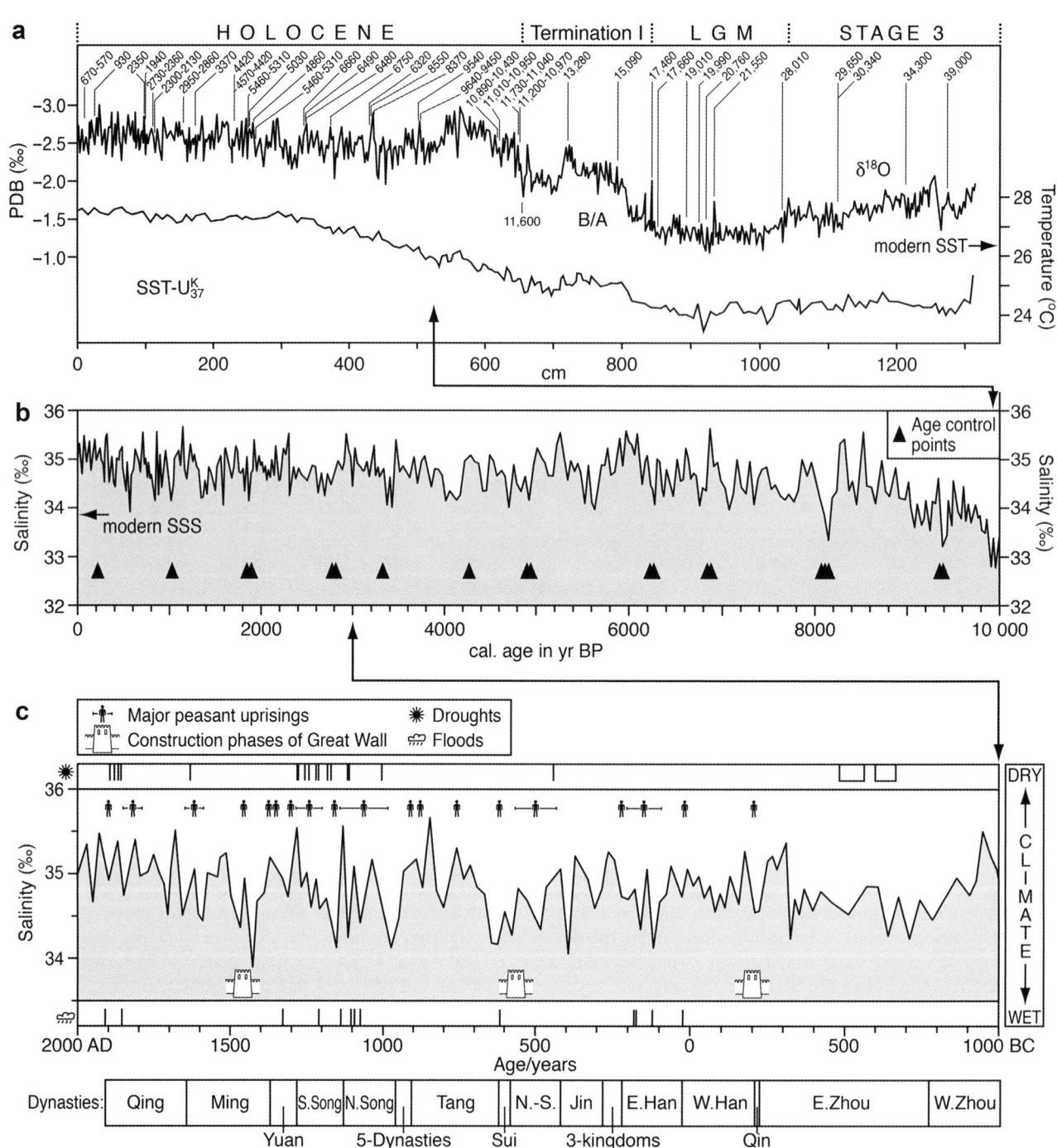

Oxygen Isotopes, Figure 3 South China Sea: Comparison of the planktonic foraminifera oxygen isotopes, sea surface temperature estimates and reconstructed sea surface salinity for the last 3000 years with documentary evidence from China of periods of droughts, floods, peasant uprisings, construction phases of the Great Wall and changes in the ruling Dynasties (redrawn from Wang et al., 1999).

Chapman and Maslin, 1999) river runoff (Maslin and Burns, 2000; Maslin et al., 2011), iceberg rafting (Maslin et al., 1995), surface ocean circulation and upwelling (Swann et al., 2006; Dickson et al., 2010; Martinez-Mendez et al., 2010), and monsoonal strength (Wang et al., 1999). An example is shown in Figure 3, where $\Delta\delta_{local}$ was isolated from planktonic foraminifera $\delta^{18}O$ data from the South China Sea to infer changes in the SE Asian monsoon over the last 3000 years (Wang et al., 1999).

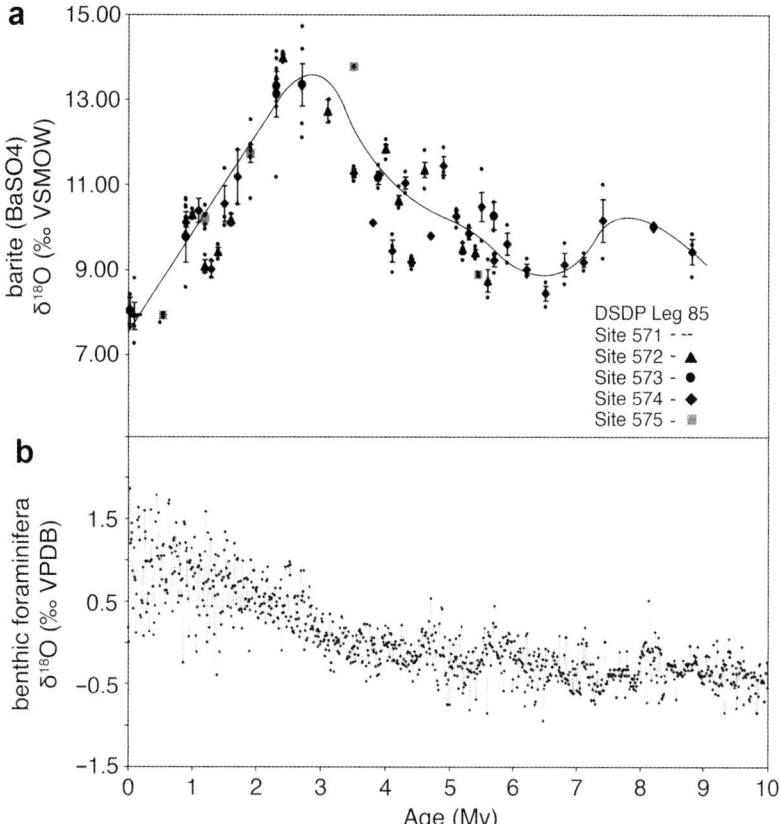

Oxygen Isotopes, Figure 4 Comparison of benthic foraminifera oxygen isotopes with barite oxygen isotopes for the last 8 Ma (adapted from Turchyn and Schrag, 2004).

Influence of diagenesis

Oxygen isotopes in marine minerals can become altered from their primary compositions following sediment burial. This effect arises due to the higher $\delta^{18}O$ composition of cold deep ocean water compared to warmer surface waters, where many planktonic (surface dwelling) organisms live. Following sedimentation on the seafloor, such planktonic organisms may partially dissolve due to carbonate or silicic acid undersaturation, or barite reduction (Schrag, 1999; Broecker, 2003), before recrystallizing in pore waters with a higher saturation state (Schrag et al., 2013). Diagenesis may in principal have a large effect on Paleoceanographic reconstructions, principally through biasing temperature reconstructions towards lower values.

Several studies had previously argued for relatively cool tropical surface ocean waters during the Mesozoic, a period of prolonged greenhouse warming (e.g., Shackleton and Boersma, 1981; Bralower et al., 1995). However, subsequent studies have utilized well-preserved planktonic foraminifera recovered from clay-rich lithologies that limit the impact of pore water recrystallization to demonstrate that cool temperatures were likely to have been the result of diagenesis (e.g., Wilson and Norris, 2001; Pearson et al., 2007; Pearson, 2012).

Sulfur cycle

Oxygen isotopes can also be used to examine the sulfur cycle (Turchyn and Schrag, 2004; Turchyn et al., 2009). The amount of sulfate in the ocean is controlled by three major processes: input from rivers, sulfate reduction and sulfide reoxidation on continental shelves and slopes, and burial of anhydrite and pyrite in ocean crusts. These controls on the biogeochemistry of the sulfur cycle can be monitored using oxygen isotopes as each source has a distinct isotopic signature. Natural riverine input has a $\delta^{18}O_{SO4}$ of -2 to $+7‰$ VSMOW which is a mixture of weathered evaporate deposits (11–13‰) and oxidative pyrite (-4 to $+2‰$). While sulfur reduction leaves the sulfate pool enriched by 8–10‰ (up to 25‰ enriched in sulfate-limited environments) and sulfide reoxidation on continental shelves ^{18}O-enriches sulfate by 8–17‰ compared with the $\delta^{18}O$ of seawater, depending on what metals are present. The final effect on $\delta^{18}O_{SO4}$ is direct sulfide oxidation in hydrothermal and some shallow sediment environments which produces sulfate with the same $\delta^{18}O$ as the seawater it was oxidized in. Turchyn and Schrag (2004) have modeled these various inputs to the $\delta^{18}O_{SO4}$ and suggest that the oxygen isotope shifts over the past 10 million years can be linked to sea-level

Oxygen Isotopes, Table 1 Environmental influences on marine oxygen isotopes

Environmental Factor	Higher	Lower
Temperature	$\delta^{18}O$ decrease	$\delta^{18}O$ increase
Global Ice Volume	$\delta^{18}O$ increase	$\delta^{18}O$ decrease
Salinity[a]	$\delta^{18}O$ increase	$\delta^{18}O$ decrease
Density[a]	$\delta^{18}O$ increase	$\delta^{18}O$ decrease

[a]an exception is sea ice formation which increases surface water salinity and density but lowers the surface water $\delta^{18}O$ (O'Neil 1968) as $\delta^{18}O$ always increases in the formation of ice from water. This affect though is minimal and is not believed to influence past climatic records

fluctuations during the Plio-Pleistocene glacial cycles, which influencing the burial and oxidation of pyrite on continental shelves (Figure 4).

Conclusions

Table 1 summarizes the relationship between marine oxygen isotopes and their main environmental controls, namely water temperature, ice volume, and salinity. This provides a guide to the causes of ^{18}O-depletion or ^{18}O-enrichment in ocean water, which form the basis of all Paleoreconstructions. In this article, we have demonstrated how oxygen isotopes records can provide stratigraphy as well as enabling the reconstruction of a range of environmental parameters such as global ice volume, ocean temperatures, sea-level, and the sulfur cycle. The advances in our understanding of past climates due to oxygen isotopes are fundamental, the most important of which are: confirmation of the theory of orbital forcing, variability of deep ocean circulation of a range of timescales from millions of years to century, variability of atmospheric weather systems such the El Nino Southern Oscillation, monsoonal and riverine systems, and discovery of large amplitude millennial variations in relative sea-level (Maslin and Swann, 2005).

Bibliography

Bard, E., Hamelin, B., Arnold, M., Montaggioni, L., Cabioch, G., Faure, G., and Rougerieet, F., 1996. Deglacial sea-level record from Tahiti corals and the timing of global meltwater discharges. *Nature*, **382**, 241–244.

Bemis, B. E., Spero, H., Bijma, J., and Lea, D. W., 1998. Reevaluation of the oxygen isotopic composition of planktonic foraminifera: experimental results and revised paleotemperature equations. *Paleoceanography*, **13**, 150–160.

Bralower, T. J., Zachos, J. C., Thomas, E., Parrow, M., Paull, C. K., Kelly, D. C., Pre-Moli Silva, I., Sliter, W. V., and Loh-Mann, K. C., 1995. Late Paleocene to Eocene paleoceanography of the equatorial Pacific Ocean: stable isotopes recorded at Ocean Drilling Program Site 865, Allison Guyot. *Paleoceanography*, **10**, 841–865.

Broecker, W. S., 2003. The oceanic $CaCO_3$ cycle. In Holland, H. D., and Turekian, K. K. (eds.), *Treatise on Geochemistry*. Amsterdam: Elsevier, Vol. 6, pp. 529–549.

Burns, S., and Maslin, M. A., 1999. Composition and circulation of bottom water in the western Atlantic Ocean during the last glacial, based on pore-water analyses from the Amazon Fan. *Geology*, **27**, 1011–1014.

Chapman, M., and Maslin, M. A., 1999. Low latitude forcing of meridional temperature and salinity gradients in the North Atlantic and the growth of glacial ice sheets. *Geology*, **27**, 875–878.

Chappell, J., and Shackleton, N. J., 1986. Oxygen isotopes and sea level. *Nature*, **324**, 137–140.

Craig, H., 1965. Measurement of oxygen isotope paleotemperatures. In Tongiorgi, E. (ed.), *Stable Isotopes in Oceanographic Studies and Paleotemperatures*. Spoleto: Consiglio Nazionale delle Ricerche, pp. 162–182.

Craig, H., and Gordon, L. I., 1965. Isotope oceanography: deuterium and oxygen 18 variations in the ocean and the marine atmosphere. In *Symposium on Marine Geochemistry*, University of Rhode Island Occasional Publications 3, USA, pp. 77–374.

Criss, R. E., 1999. *Principles of Stable Isotope Distribution*. New York: Oxford University Press, p. 254.

Dickson, A. J., Leng, M. J., Maslin, M. A., Sloane, H. J., Green, J., Bendle, J. A., McClymont, E. L., and Pancost, R. D., 2010. Atlantic overturning circulation and Agulhas leakage influences on southeast Atlantic upper ocean hydrography during marine isotope stage 11. *Paleoceanography*, **25**, PA3208, doi:10.1029/2009PA001830.

Duplessy, J.-C., Labeyrie, L., and Waelbroeck, C., 2002. Constraints on the ocean oxygen isotopic enrichment between the Last Glacial Maximum and the Holocene. *Quaternary Science Reviews*, **21**, 315–330.

Elderfield, H., and Ganssen, G., 2000. Past temperature and delta^{18}O of surface ocean waters inferred from foraminiferal Mg/Ca ratios. *Nature*, **405**, 442–445.

Epstein, S., Buchsbaum, R., Lowenstam, H. A., and Urey, H. C., 1953. Revised carbonate-water isotopic temperature scale. *Geological Society of America Bulletin*, **64**, 1315–1325.

Fairbanks, R. G., 1989. A 17,000 year glacio-eustatic sea level record: influence of glacial melting rates on the Younger Dryas event and deep-ocean circulation. *Nature*, **342**, 637–642.

Garlick, G. D., 1974. The stable isotopes of oxygen, carbon, hydrogen in the marine environment. In Goldberg, E. D. (ed.), *The Sea*. New York: Wiley, Vol. 5, pp. 393–425.

Hays, P. D., and Grossman, E. L., 1991. Oxygen isotopes in meteoric calcite cements as indicators of continental paleoclimate. *Geology*, **19**, 441–444.

Hoefs, J., 1997. *Stable Isotope Geochemistry*, 4th edn. Berlin: Springer.

Imbrie, J., and Kipp, N. G., 1971. A new micropaleontological method for quantitative paleoclimatology. In Turekian, K. K. (ed.), *Late Cenozoic Glacial Ages*. New Haven: Yale University Press, pp. 71–182.

Kim, S. T., and O'Neil, J. R., 1997. Equilibrium and nonequilibrium oxygen isotope effects in synthetic calcites. *Geochimica et Cosmochimica Acta*, **61**, 3461–3475.

Kucera, M., Weinel, M., Kiefer, T., Pflaumann, U., Hayes, A., Weinelt, M., Chen, M.-T., Mix, A. C., Barrows, T. T., Cortijo, E., Duprat, J., Juggins, S., and Waelbroeck, C., 2005. Reconstruction of sea-surface temperatures from assemblages of planktonic foraminifera: multi-technique approach based on geographically constrained calibration data sets and its application to glacial Atlantic and Pacific Oceans. *Quaternary Science Reviews*, **24**, 951–998.

Marchitto, T. M., Curry, W. B., Lynch-Stieglitz, J., Bryan, S. P., Cobb, K. M., and Lund, D. C., 2014. Improved oxygen isotope temperature calibrations for cosmopolitan benthic foraminifera. *Geochimica et Cosmochimica Acta*, **130**, 1–11.

Martinez-Mendez, G., Zahn, R., Hall, I. R., Peeters, F. J. C., Pena, L. D., Cacho, I., and Negre, C., 2010. Contrasting multiproxy reconstructions of surface ocean hydrography in the Agulhas Corridor and implications for the Agulhas Leakage dduring the last 345,000 years. *Paleoceanography*, **25**, PA4227, doi:10.1029/2009PA001879.

Maslin, M. A., and Burns, S. J., 2000. Reconstruction of the Amazon Basin effective moisture availability over the last 14,000 years. *Science*, **290**, 2285–2287.

Maslin, M. A., and Swann, G., 2005. Isotopes in marine sediments. In Leng, M. (ed.), *Isotopes in Palaeoenvironmnetal Research*. Dordrecht: Springer, pp. 227–290.

Maslin, M. A., Shackleton, N. J., and Pflaumann, U., 1995. Temperature, salinity and density changes in the Northeast Atlantic during the last 45,000 years: Heinrich events, deep water formation and climatic rebounds. *Paleoceanography*, **10**, 527–544.

Maslin, M. A., Ettwein, V. J., Wilson, K. E., Guilderson, T. P., Burns, S. J., and Leng, M. J., 2011. Dynamic boundary-monsoon intensity hypothesis: evidence from the deglacial Amazon River discharge record. *Quaternary Science Reviews*, **30**, 3823–3833.

McCrea, J. M., 1950. On the isotope chemistry of carbonates and a paleotemperature scale. *Journal of Chemical Physics*, **18**, 849–857.

O'Neil, J. R., Clayton, R. N., and Mayeda, T. K., 1969. Oxygen isotope fractionation on divalent metal carbonates. *Journal of Chemical Physics*, **51**, 5547–5558.

Pearson, P. N., 2012. Oxygen isotopes in foraminifera: overview and historical review. *Paleontological Society Papers*, **18**, 1–38.

Pearson, P. N., van Dongen, B. E., Nicholas, C. J., Pancost, R. D., Schouten, S., Singano, J. M., and Wade, B. S., 2007. Stable warm tropical climate through the Eocene epoch. *Geology*, **35**, 211–214.

Pflaumann, U., Sarnthein, M., Chapman, M., d'Abreu, L., Funnell, B., Huels, M., Kiefer, T., Maslin, M. A., Schulz, H., Swallow, J., van Kreveld, S., Vautravers, M., Vogelsang, E., and Weinelt, M., 2003. The Glacial North Atlantic: sea-surface conditions reconstructed by GLAMAP-2000. *Paleoceanography*, **18**, 1065, doi:10.1029/2002PA000774.

Prahl, F. G., and Wakeham, S. G., 1987. Calibration of unsaturation patterns in long-chain ketone compositions for palaeotemperature assessment. *Nature*, **330**, 367–369.

Rohling, E. J., 2000. Paleosalinity: confidence limits and future applications. *Marine Geology*, **163**, 1–11.

Rohling, E. J., and Cooke, S., 1999. Stable oxygen and carbon isotope ratios in foraminiferal carbonate. In Sen Gupta, B. K. (ed.), *Modern Foraminifera*. Dordrecht: Kluwer, pp. 239–258.

Schmidt, G. A., 1999. Error analysis of paleosalinity calculations. *Paleoceanography*, **14**, 422–429.

Schouten, S., Hopmans, E. C., Schefuß, E., and Sinninghe Damsté, J. S., 2002. Distributional variations in marine crenarchaeotal membrane lipids: a new organic proxy for reconstructing ancient sea water temperatures? *Earth and Planetary Science Letters*, **204**, 265–274.

Schrag, D. P., 1999. Effects of diagenesis on the isotopic record of late Paleogene tropical sea surface temperatures. *Chemical Geology*, **161**, 215–224.

Schrag, D. P., Hampt, G., and Murray, D. W., 1996. Pore fluid constraints on the temperature and oxygen isotopic composition of the glacial ocean. *Science*, **272**, 1930–1932.

Schrag, D. P., Higgins, J. A., Macdonald, F. A., and Johnston, D. T., 2013. Authigenic carbonate and the history of the global carbon cycle. *Science*, **339**, 540–543.

Shackleton, N. J., 1974. Attainment of isotopic equilibrium between ocean water and the benthonic foraminifera genus *Uvigerina*: isotopic changes in the ocean during the last glacial. *CNRS, Colloques Internationals*, **219**, 203–209.

Shackleton, N. J., 1987. Oxygen isotopes, ice volume and sea-level. *Quaternary Science Reviews*, **6**, 183–190.

Shackleton, N. J., 2000. The 100,000-year ice-age cycle identified and found to lag temperature, carbon dioxide, and orbital eccentricity. *Science*, **289**(5486), 1897–1902.

Shackleton, N., and Boersma, A., 1981. The climate of the Eocene ocean. *Journal of the Geological Society of London*, **138**, 153–157.

Shemesh, A., Hodell, D., Crosta, C., Kanfoush, S., Charles, C., and Guilderson, T., 2002. Sequence of events during the last deglaciation in Southern Ocean sediments and Antarctic ice cores. *Paleoceanography*, **17**, 1056, doi:10.1029/2000PA000599.

Spero, H. J., Bijma, J., Lea, D. W., and Bemis, B. E., 1997. Effect of seawater carbonate concentration on foraminiferal carbon and oxygen isotopes. *Nature*, **390**, 470–500.

Tiedemann, R., Sarnthein, M., and Shackleton, N. J., 1994. Astronomic timescale for the Pliocene Atlantic δ18O and dust flux records of ODP Site 659. *Paleoceanography*, **9**, 619–638.

Turchyn, A. V., and Schrag, D., 2004. Oxygen isotope constraints on the sulphur cycle over the last 10 million years. *Science*, **303**, 2004–2007.

Turchyn, A. V., Schrag, D. P., Coccioni, R., and Montanari, A., 2009. Stable isotope analysis of the Cretaceous sulfur cycle. *Earth and Planetary Science Letters*, doi:10.1016/j.epsl.2009.06.002.

Wang, L. J., Sarnthein, M., Erlenkeuser, H., Grootes, P., Grimalt, J., Pelejero, C., and Linck, G., 1999. Holocene variations in Asian monsoon moisture: a bidecadal sediment record from the South China Sea. *Geophysical Research Letters*, **26**, 2889–2892.

Wilson, P. A., and Norris, R. D., 2001. Warm tropical ocean surface and global anoxia during the mid-Cretaceous period. *Nature*, **412**, 425–429.

Zeebe, R. E., 1999. An explanation of the effect of seawater carbonate concentration on foraminiferal oxygen isotopes. *Geochimica et Cosmochimica Acta*, **63**, 2001–2007.

Zeebe, R. E., and Wolf-Gladrow, D. A., 2001. *CO_2 in Seawater: Equilibrium, Kinetics, Isotopes*. Amsterdam: Elsevier. Elsevier Oceanography Series, 65, p. 346.

Cross-references

Astronomical Frequencies in Paleoclimates
Carbon Isotopes
Foraminifers (Benthic)
Foraminifers (Planktonic)
Geologic Time Scale
Marine Microfossils
Mg/Ca Paleothermometry
Paleoceanographic Proxies
Paleoceanography

P

PAIRED METAMORPHIC BELTS

Wolfgang Frisch
Department of Geosciences, University of Tübingen, Tübingen, Germany

Definition

As first identified in Japan, pairs of elongate, parallel belts with significantly different metamorphic rocks are indicative of former subduction activity. One metamorphic belt consists of rocks formed by high-pressure metamorphism and the other consists of rocks of the Abukuma facies high-temperature metamorphic rocks. These two belts of metamorphic rocks, which were formed under extremely different pressure/temperature relations, occur in direct association and are called "paired metamorphic belts."

Orogenic processes involve uplift that raises portions of the metamorphic rocks toward the surface and into the zone of erosion. From the spatial arrangement of the two contrasting belts at the surface, it is possible to detect not only former subduction activity but also the polarity of the subduction zone (Figure 1). The high-pressure belt (subduction metamorphism) marks the area of the subduction zone, and the high-temperature belt (Abukuma-type metamorphism) marks the location of the magmatic arc. In Japan, the paired metamorphic belts indicate subduction beneath the Asian continent since the Permian epoch. Since that period, the subduction zone has shifted twice toward the present Pacific Ocean. The presently formed metamorphic rocks are still hidden at depth.

High-pressure or subduction metamorphism

Rocks in subduction zones are metamorphosed under conditions of high pressure at relatively low temperature, hence the terms high-pressure (properly high-pressure/low-temperature) metamorphism or subduction metamorphism (Figure 2). Such metamorphism results because cool rock material is rapidly subducted to great depths. Normally, at a depth of 50 km that corresponds to a pressure of 1.5 gigapascal, the temperature is approximately 800–900 °C. In contrast, oceanic crust that reaches the same depth in a subduction zone reaches temperatures between 200 to more than 500 °C. The lower value is valid for old, cool, rapidly subducting lithosphere, and the higher value is for young and warm, more slowly subducting lithosphere (Stern, 2002). The isotherms in the subduction zone are strongly declined downward because rocks are poor heat conductors and they need millions of years to adjust to the higher temperatures normally seen at depth (Figure 3).

Rocks that have been buried within subduction zones (to even more than 100 km) and subsequently exhumed during orogenic processes can be observed directly on outcrop. Metamorphic conditions in a subduction zone transform basalts, dolerites, and gabbros of the oceanic crust into glaucophane schists (blueschists) and eclogites (Figure 2). In addition, sedimentary rocks and splinters

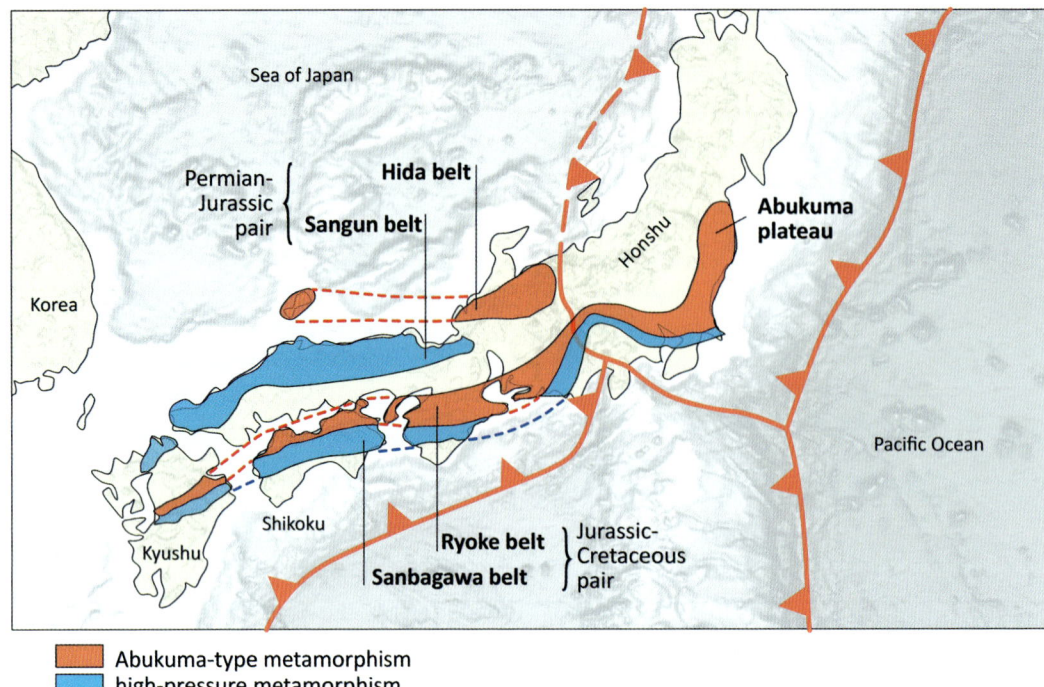

Paired Metamorphic Belts, Figure 1 Paired metamorphic belts in Japan. The different ages of the belts indicate that subduction lasted for more than 250 million years, each with subduction directed toward N or NW. The subduction zone shifted stepwise outward (southeastward) and is today offshore the Japanese coast in the Pacific (after Frisch et al., 2011).

of continental crust carried with the subducted material experience high-pressure metamorphism. Basalts and sediments contribute significantly to the dewatering in a subduction zone. Water released in the subduction zone causes intense changes in the mantle above and, from a certain depth, interacts with hot asthenospheric material and triggers melting processes which are responsible for magmatism above the subduction zone (Figure 3). The rising magmas cause increasing temperatures in the magmatic belt above a subduction zone, where Abukuma-type metamorphism occurs.

Abukuma-type metamorphism

The magmatic zones of island arcs and active continental margins are characterized by high geothermal gradients generated by the rising magmas. Whereas the temperature increases downward in the subduction zone and in the forearc area by less than 10 °C per kilometer, the geothermal gradient in the upper crust at the magmatic front jumps to 35–50 °C/km. Consequently, rocks of the crust below the volcanic belt experience completely different metamorphism than rocks in the subduction zone.

Metamorphism in the magmatic zone is characterized as low pressure/high temperature. This temperature-dominated metamorphism is named Abukuma-type metamorphism after the Abukuma Plateau in Japan (Figures 1 and 2) and produces characteristic mineral assemblages. Because of the high temperatures, the rocks are easily deformed. Due to the high geothermal gradient, the temperature increases to generate the highest grades of metamorphism, 650–900 °C, present at crustal depths. Rocks experience partial melting (anatexis) under these conditions, when water is present. Dry magmatic rocks and sedimentary rocks, which have already expelled their water, are not melted under these conditions; instead, granulites develop in the deep continental crust.

Abukuma-type metamorphism occurs below 7–10 km depth and reflects regional metamorphism. In contrast, intrusions that rise into shallow depths of only a few kilometers create a zone of contact along their margins, the metamorphic aureole. During this contact metamorphism (Figure 3), heat of the intrusive body reacts with the cold country rock and mobilizes aqueous fluids that trigger numerous mineral changes.

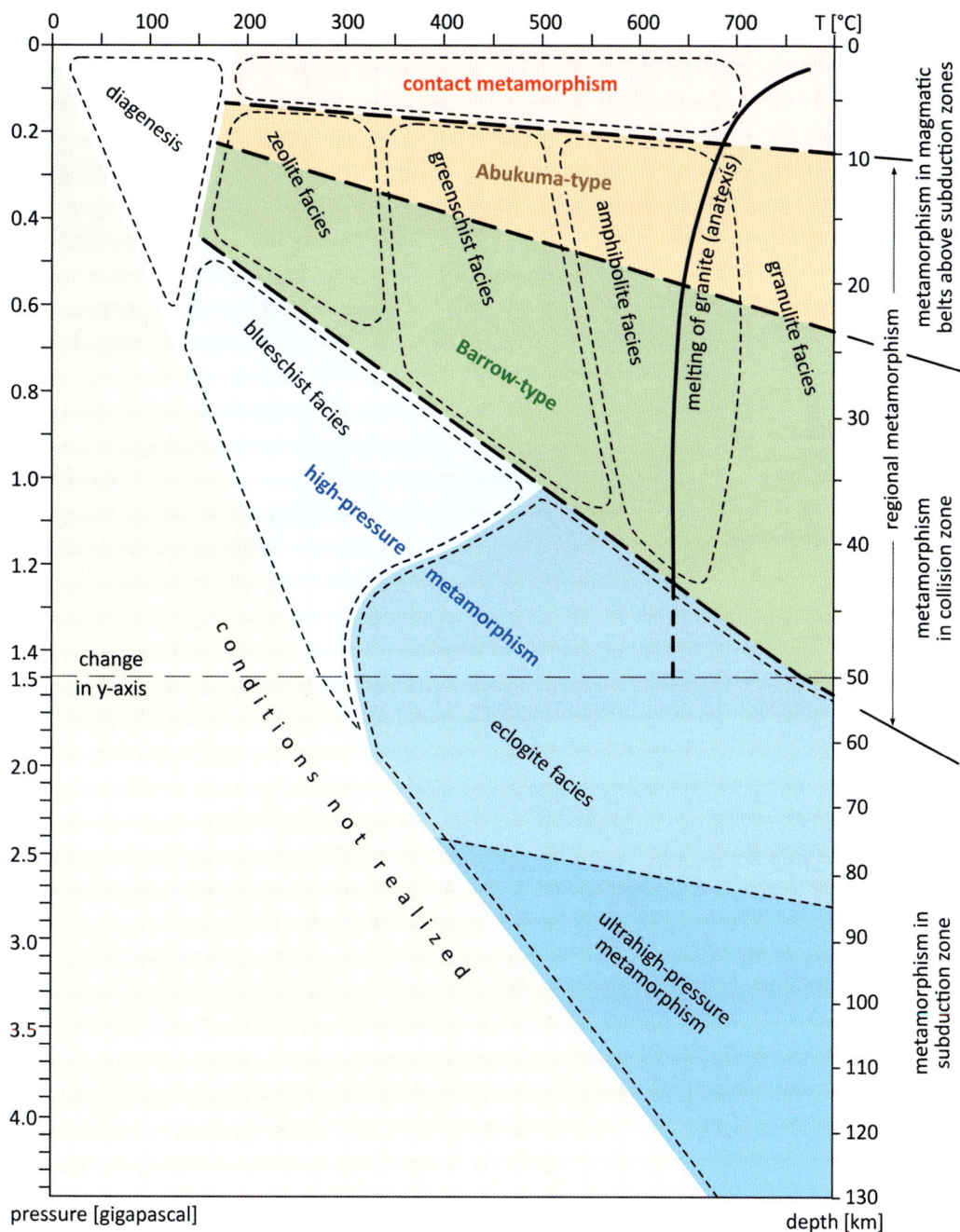

Paired Metamorphic Belts, Figure 2 Pressure-temperature (P-T) diagram showing different types of metamorphism and metamorphic facies. High-pressure metamorphism is typical for subduction zones, and Barrovian-type regional metamorphism typifies continent-continent collisions. Abukuma-type regional metamorphism and contact metamorphism occur in the magmatic belts above subduction zones (after Frisch et al., 2011).

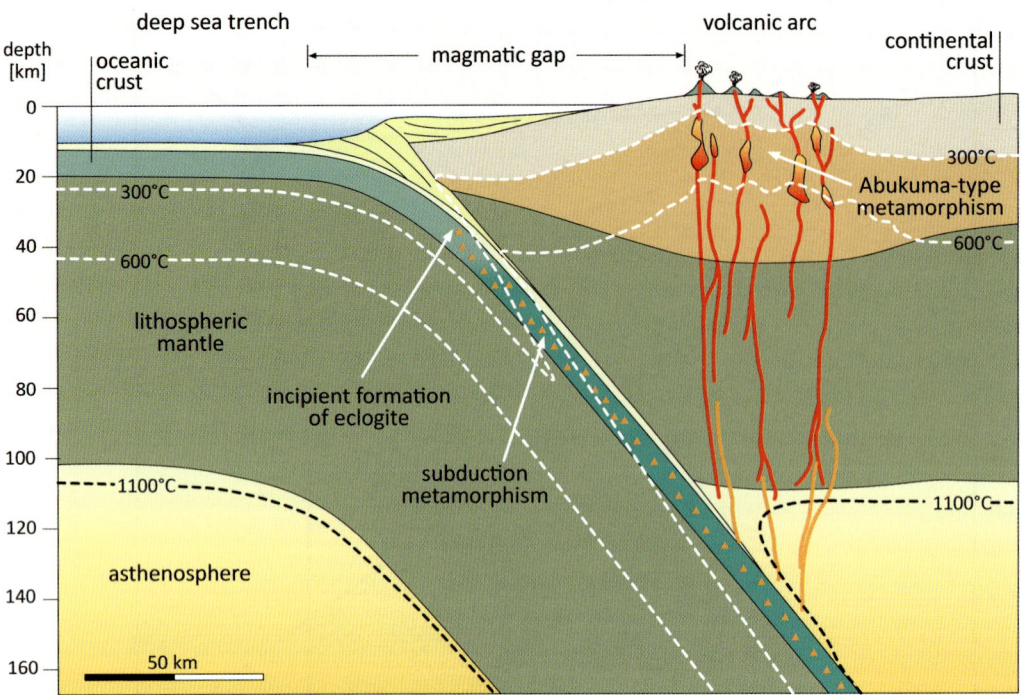

Paired Metamorphic Belts, Figure 3 Cross section through an active continental margin showing the location of subduction and Abukuma-type metamorphism. *Dashed lines* are isograds; numbers indicate temperatures in °C (Schubert et al., 1975).

Bibliography

Frisch, W., Meschede, M., and Blakey, R. C., 2011. *Plate Tectonics.* Heidelberg/Dordrecht/London/New York: Springer. 212 pp.

Schubert, G., Yuen, D. A., and Turcotte, D. L., 1975. Role of phase transitions in a dynamic mantle. *Geophysical Journal of the Royal Astronomical Society*, **42**, 705–735.

Stern, R. J., 2002. Subduction zones. *Reviews of Geophysics*, **40**, 3–1. to 3–38.

Cross-references

Magmatism at Convergent Plate Boundaries
Ophiolites
Subduction

PALEOCEANOGRAPHIC PROXIES

Gerold Wefer
MARUM-Center for Marine Environmental Sciences, University of Bremen, Bremen, Germany

Definition

Proxies stand in for direct measurements of past environmental conditions such as temperature, salinity, etc.

Introduction

Measured climate information such as temperature is only available back to the late nineteenth century. To extend the record further back in history, proxies have to be used. Proxies are physical and chemical parameters that can be transformed through a calibration process to real measured variables such as temperature. From the ratios of stable oxygen isotopes in the carbonate shells of marine organisms, e.g., past temperatures can be reconstructed if the isotopic composition of the seawater is known. The information is stored in sediments of the shelf areas, continental margins and abyssal plains, and in the skeletons of tropical or cold-water corals. Environmental information can be obtained from the quantity and composition of organic matter, carbonate, and opal shells and from the species compositions of plants and animals. An overview of the use of proxies in paleoceanography is given by Fischer and Wefer (1999).

New results on proxy research are presented at many conferences, particularly at the triennial International Conference on Paleoceanography (ICP). The 11th ICP took place in Barcelona (Spain) in 2013; the next will be held in Utrecht (Netherlands) in 2016.

What is a proxy

The definition of a proxy, according to Wefer et al. (1999):

Proxy, as used here and elsewhere, is short for proxy variable. The word "proxy" is commonly used to describe a stand-in, as in "proxy vote" or "fighting by proxy." By analogy, "proxy variables" in the parlance of paleoenvironmental reconstruction are measurable descriptors which stand in for desired (but unobservable) variables such as temperature, salinity, nutrient content, oxygen content, carbon dioxide concentration, wind speed, and productivity. These are the *target parameters*. The concept of a proxy, then, presumes the existence (in principle) of a target, that is, the proxied parameter of the system being reconstructed. Consequently, each proxy is associated with a rule or rules denoting how the transform from proxy to target is to be accomplished. Usually, this means that a transforming algorithm has to be established by calibration.

Source of proxies

Proxies are preserved in different types of sediments and corals. Depending on productivity and sediment transport by rivers and the atmosphere, the temporal resolution can vary from millennia in offshore and low-productive areas to centuries or decades on continental margins or in coastal areas. In oxygen-depleted areas, sediments can be deposited in annual layers, varves, caused by seasonal sedimentation patterns. Varves are only common in oxygen-free areas due to the lack of bioturbation, e.g., are formed in high-production areas or in anoxic basins on the continental slope. A particularly high temporal resolution is provided by skeletons of tropical corals, which even allow the reconstruction of annual cycles. Paleoclimate reconstructions are not restricted to the oceans. Often, they also include land-based phenomena such as tree rings, which permit global temperature reconstructions over the past 1500 years (Mann et al., 2009). Important information is also provided by loess deposits, especially in China (Guo et al., 2002).

What can be reconstructed?

Temperatures

Most often, temperature is reconstructed, because it is an important parameter and because temperatures can be determined quite easily in comparison to salinity or productivity and with relatively high accuracy. A few decades ago transfer functions were developed (Imbrie and Kipp, 1971). Planktonic foraminiferal communities were employed to calculate a temperature or temperature range (Figure 1). This method was the basis for the reconstruction of sea-surface temperatures during the Last Glacial Maximum (CLIMAP, Climate: Long range Investigation, Mapping, and Prediction, CLIMAP, 1981) (Figure 2).

Another proxy applies the ratios of stable oxygen isotopes ^{16}O and ^{18}O incorporated in the calcite shells, which are temperature dependent. If the isotopic composition of the surrounding water is known, it is possible to determine temperature or changes in temperature because the absolute temperatures are distorted by vital effects. In general, the oxygen-isotope ratios are used for stratigraphy, and other proxies, such as Mg/Ca ratios or alkenones, are used for temperatures.

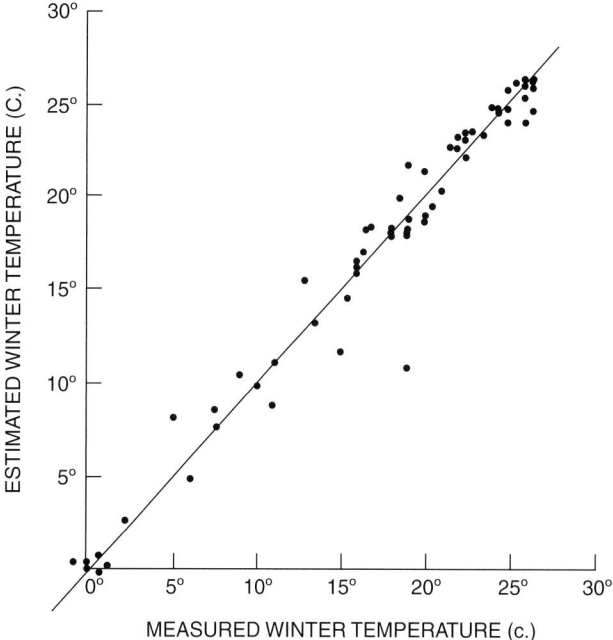

Paleoceanographic Proxies, Figure 1 Scatter diagram showing measured winter sea-surface temperatures versus those estimated from faunal assemblages in core-top samples using factor analysis and transfer function methods (From Imbrie and Kipp 1971).

Mg/Ca ratios in benthic and planktonic foraminifera and corals are frequently used for temperature reconstructions. At higher temperatures the amount of Mg replacing Ca in calcareous skeletons is greater. A number of equations have been specially developed for individual foraminiferal or coral species. Several effects must be considered in this method including impurities, vital effects, partial solution, or postmortem mineral formation.

Another proxy for temperatures employs Sr/Ca ratios. Corals incorporate Sr in a complicated way in their aragonite skeletons, but in relation to water temperature (Cohen et al., 2002). If the skeletons are well preserved, temperatures can be reconstructed with high temporal resolution, even up to seasonal cycles. One example is the reconstruction of tropical Atlantic temperature seasonality at the end of the last interglacial, about 117 kyr ago (Felis et al., 2015).

For temperature reconstructions, lipid biomarkers such as alkenones (UK_{37}) or isoprenoid glycerol dialkyl glycerol tetraethers (GDGTs or TEX_{86}) are also frequently used. Alkenones are long-chain organic molecules, which are produced by coccolithophores in surface waters. These chains are formed depending on the ambient temperature, and different calibration curves are used. GDGTs are membrane lipids of marine planktonic Thaumarchaeota. Not only is it important to understand the calibration relationship between the contents of these lipids and temperatures but also how these lipids are transported to the deep

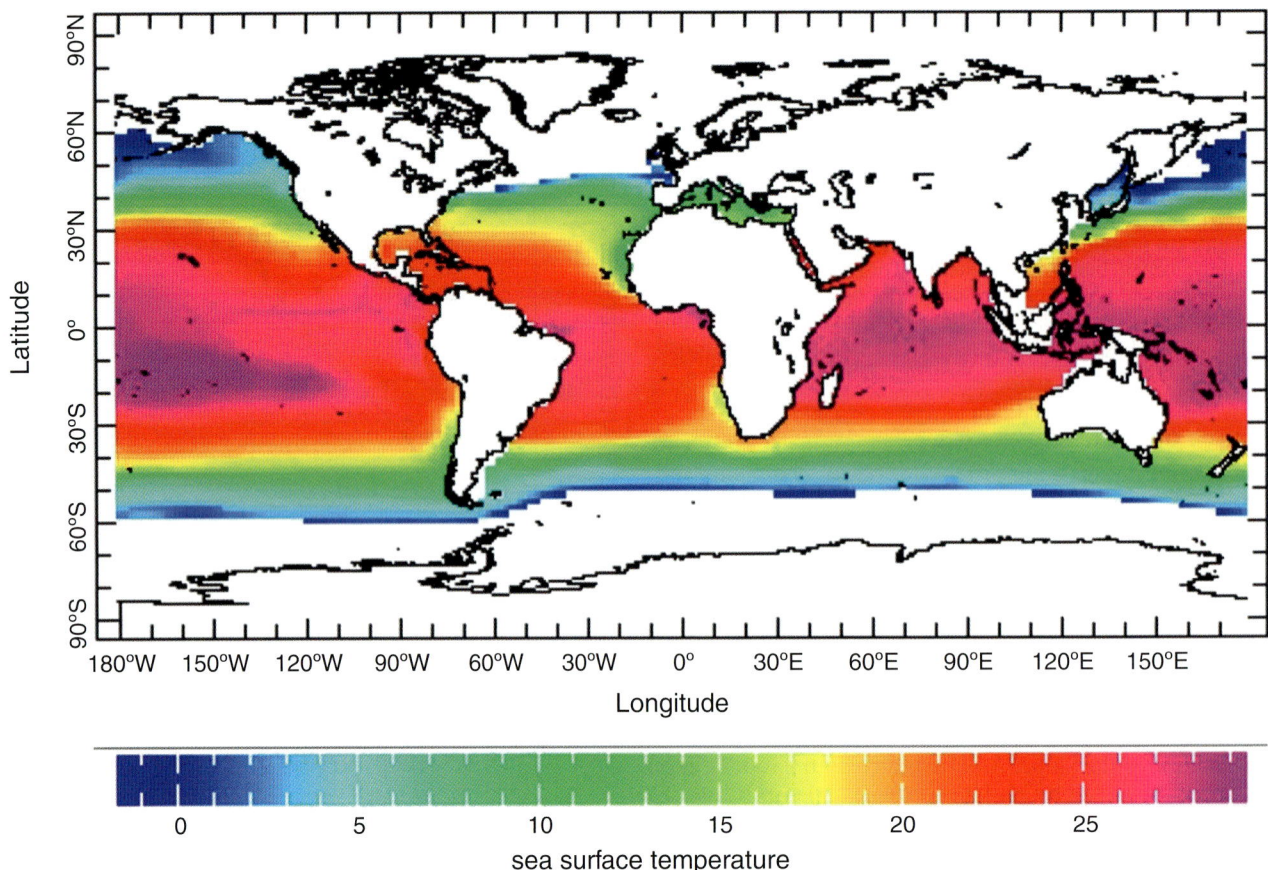

Paleoceanographic Proxies, Figure 2 Sea-surface temperatures during the Last Glacial Maximum (February) from Data Library CLIMAP LGM (http://iridl.ldeo.columbia.edu/SOURCES/.CLIMAP/.LGM/.feb/.sst/figviewer.html?map.url=X+Y+fig−+colors+coasts+lakes+−fig).

sea and how they can be changed, both on their way and at the seabed (see, e.g., Mollenhauer et al., 2015).

Salinity

Salinity is much more difficult to determine than temperatures. One possibility is to use stable oxygen isotopes, when temperatures are determined by other methods, e.g., Mg/Ca ratios or alkenones.

Productivity

In the reconstruction of productivity, it is necessary to distinguish between flux proxies (export of biogenic material formed in the surface water) and nutrient proxies (containing information on nutrients such as phosphate and nitrogen).

General information on the productivity of surface waters is provided by the accumulation of organic carbon. However, this information is qualitative, since the settling of organic matter to the seafloor does not exactly correspond to primary production. The fraction of the primary production depends on the production system or on the water depth. Also, laterally introduced organic matter influences the results. Another method was developed by Herguera and Berger (1994). They used the abundance of benthic foraminifera in the sediment as a measure of productivity, because their frequency is related to the food supply, the amount of organic matter deposited on the seafloor.

The ratios of stable carbon and nitrogen isotopes are nutrient proxies, expressed as deviations from a standard $\delta^{13}C$ or $\delta^{15}N$. A measure for fertility is the difference in $\delta^{13}C$ between foraminifera living in different areas in the ocean, e.g., shallow-living versus deep-living planktonic foraminifera or planktonic versus benthic foraminifera. Reconstructions depend on a variety of background information, e.g., on exchange with the atmosphere or calcification depth of planktonic foraminifera, or whether benthic foraminifera live on the surface or in the sediment.

Nitrogen isotopes, the ratio of $^{14}N/^{15}N$, are another tool. During the uptake of nitrogen by phytoplankton, isotope fractionation takes place with a preferential uptake of ^{14}N, so ^{15}N becomes enriched in the seawater. The organic matter that is exported from the photic zone is enriched in ^{14}N, which will subsequently be set free in the deep ocean or on the seafloor by oxidation of the organic matter. For a review of the marine nitrogen cycle, see Altabet (2006).

Circulation of the ocean

Several proxies are available to decipher the history of ocean circulation. The ratio of the stable carbon isotopes ($^{12}C/^{13}C$) in benthic foraminifera is a measure of the age of the bottom water. During descent of oxygenated surface water to the deep sea, organic matter enriched with the stable isotope ^{12}C is degraded. Through this process, water in the deep ocean becomes more enriched in ^{12}C, which is recorded in the shells of benthic foraminifera. The more organic matter that was decomposed using oxygen, the more ^{12}C is present in the shells, and the isotopic ratio is lighter. A similar mechanism operates for the cadmium/calcium isotope ratios. The Cd/Ca isotope ratios and the phosphate content of the water show a linear relationship. With the decomposition of organic matter, cadmium and phosphate are released, and therefore, the Cd/Ca ratios in the shells of benthic foraminifera provide information on the phosphate content, which is related to the age of the bottom water and ventilation of the ocean in the past.

A relatively new proxy is the isotopic ratio of protactinium ^{231}Pa to thorium ^{230}Th. Both isotopes are formed by the radioactive decay of uranium dissolved in the water column, and ^{231}Pa has a longer residence time in the water column due to its greater half-life. Changes in the $^{231}Pa/^{230}Th$ ratio are related to the intensity of the thermohaline circulation. Another proxy is the ratio of the neodymium isotopes ^{143}Nd and ^{144}Nd. This ratio characterizes specific bodies of water. The neodymium isotopes are incorporated into manganese crusts and nodules, and, because of the low growth rates of the nodules, the temporal resolution is low, but the records can go far back into the Earth's history. There are also a number of sedimentological proxies, e.g., the proportion of sortable silt (10–63 μ) of deep-sea sediments, which provides information about bottom currents.

Not only currents or water masses are reconstructed; diatoms and biomarkers also allow the reconstruction of past sea-ice boundaries (Gersonde et al., 2014). Promising new biomarkers include sea-ice diatom-specific highly branched isoprenoids with 25 carbon isotopes. A review is given by Stein et al. (2012) describing efficient kinds of proxies that can be used to reconstruct sea-ice cover and temperatures in polar environments.

Sea-level

One specific issue is the past sea-level stand in relation to climate. There are many regional sea-level reconstructions that need to be related to the eustatic sea-level. The problem is the dynamics of ice sheets. During the melting of ice shelves and the associated rise of the land, the ice body is stabilized. For the correction of sea-level reconstructions, it is important to know where the ice was deposited or melted. For an example of a sea-level reconstruction that takes into account the melting of the Greenland Ice Sheet and West Antarctic Ice Sheet, see Raymo and Mitrovica (2012). While for the present sea-level rise of between 1.5 and 2 mm/year thermal expansion plays a major role, ice dynamics are dominant over longer periods.

CO_2 content of the atmosphere

Boron isotope ratios ($^{11}B/^{10}B$) can be used to reconstruct the past CO_2 content of the atmosphere. The boron isotope ratio depends on the acidity, pH, and alkalinity of the water, which are determined by the CO_2 concentration of the atmosphere and the concentration of bicarbonate ions in the ocean. The boron isotope ratios are incorporated into the shells of planktonic foraminifera and are a proxy for CO_2 levels during earlier times. The theory is understood (e.g., Hönisch and Hemming, 2004), but the measurements are complicated, and hence the data contain large uncertainties.

What role the ocean plays in the CO_2 content of the atmosphere depends on the physical and biological pumps. An important location for studying the effectiveness of the biological pump is the Southern Ocean, as there are unused nutrients available. Therefore, many investigations of the effectiveness of the biological pump during past times are carried out with respect to the position of fronts, sea ice, and dust flux (e.g., Ziegler et al., 2013).

Land vegetation and rainfall

The pollen of land plants is very resistant to degradation. It is transported by rivers or through the atmosphere into the ocean, settles to the seafloor, and is embedded in the sediment. With pollen communities, the history of vegetation and precipitation patterns of land masses can be reconstructed when the transport routes are known. Recently, pollen distributions have been combined with organic proxies to reconstruct vegetation history and precipitation patterns using the signatures of plant-wax-specific hydrogen (δD) and carbon isotopes ($\delta^{13}C$) (Kuechler et al., 2013) (Figure 3).

Stratigraphy

A prerequisite for climate reconstructions is a sound chronological classification of the sedimentary record. Depending on the time period, different methods are available. Approximately the last 60,000 years can be dated with the radioactive isotope ^{14}C, which has a half-life of 5730 years. New possibilities of dating were created by the development of accelerator mass spectrometry (AMS) because significantly smaller sample sizes were sufficient for the measurements. In contrast to 14C, which has a relatively long half-life of 5730 years and thus does not allow high temporal resolution, 210Pb produced in the atmosphere and deposited in sediments over the past 100–150 years can be dated. This dating method is used in marine and lacustrine sediments, peat and ice deposits.

Since the ^{14}C content of the atmosphere is subject to fluctuations caused by the cycles of sunspots and fluctuations of the Earth's magnetic field, calibration curves are needed based on dendrochronology and other

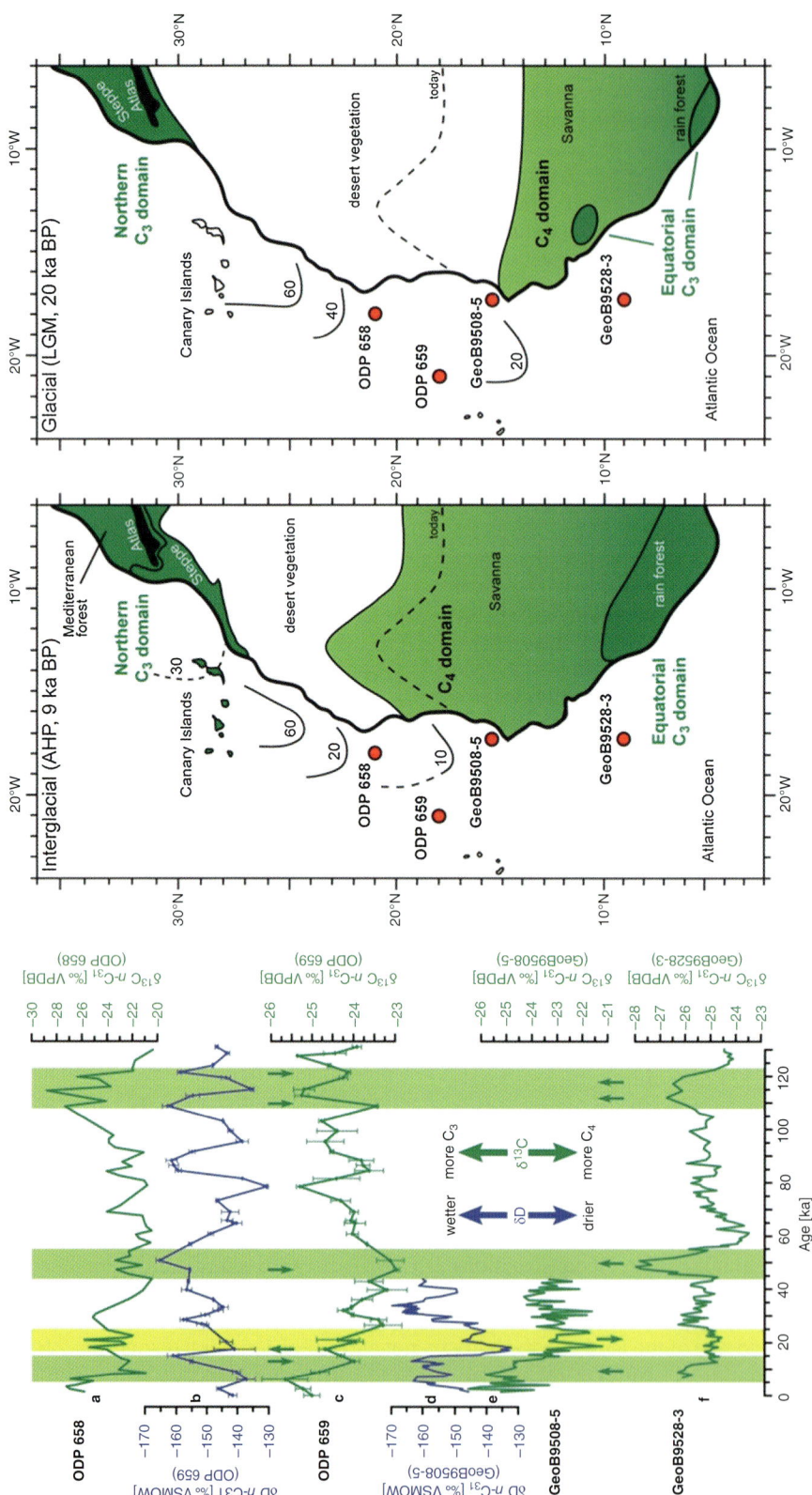

Paleoceanographic Proxies, Figure 3 Distribution of C3 and C4 plants in North Africa during the interglacial (African Humid period, AHP, ca. 9 ky BP) and glacial (Last Glacial Maximum, LGM, ca. 20 ky) (from Kuechler et al., 2013).

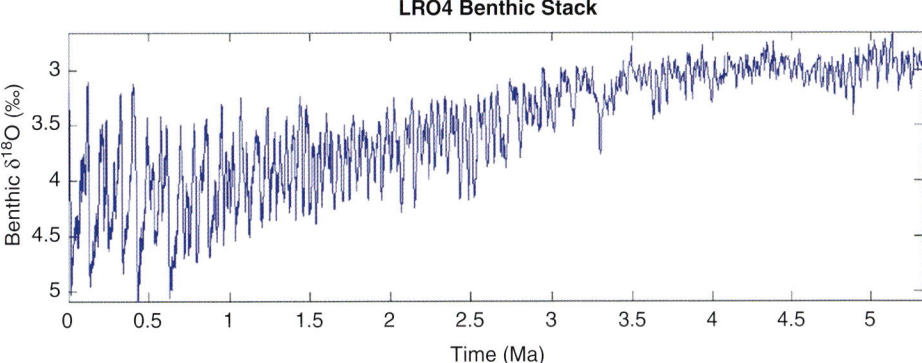

Paleoceanographic Proxies, Figure 4 LR04 stack of Lisiecki and Raymo (2005), using 57 published benthic $\delta^{18}O$ records and orbital tuning to develop an age model for the last 5.3 my.

information. The knowledge of the age of the water at the time of the incorporation of ^{14}C into the carbonate shells of planktonic and benthic organisms is important (reservoir effect). In the analysis of organic matter, a possible admixture of older material (without ^{14}C) must be considered. When dating young sediments, the influence of industrialization and the atomic bomb effect has to be taken into account. The combustion of fossil fuels releases ^{14}C-free carbon into the atmosphere (Suess effect), and with the atomic bomb tests (1945–1993), additional ^{14}C was released.

For older deposits, different methods of stratigraphy are used, e.g., biostratigraphy. The first appearances and extinctions of fossils provide time markers. Also widely used is magnetostratigraphy, which can divide the younger geological history on the basis of magnetic field reversals.

The backbone of paleoceanography is oxygen-isotope stratigraphy. The ratio of the stable oxygen isotopes ^{16}O and ^{18}O, which is incorporated into the calcite shells, depends on the isotope ratio of the water and the temperature, documenting climate development, and in particular the growth and dismantling of ice sheets. Since they are related to the Earth's orbital parameters, time intervals can be defined based on these cycles. A frequently used time scale based on $\delta^{18}O$ values of benthic foraminifera was published by Lisiecki und Raymo (2005) (Figure 4). Lisiecki und Raymo used 57 globally distributed records, recording global ice volume and deep-sea temperatures. Orbital tuning was used to establish a time frame of 5.3 my (LR04 stack).

Conclusions

The development of proxies is an ongoing process and always closely linked to the reconstruction of past environmental conditions. In recent decades, great progress has been made in the accuracy of the reconstructions, for instance in reconstructing temperatures with multiple proxies. A great potential lies in the field of biomarkers. They offer, for example, information about the CO_2 system of the ocean and precipitation patterns on the continent.

The further you go back in Earth's history, the less accurate are the reconstructions. Often, however, it is not the proxy, but the time frame of the deposits. Through the use of orbital forcing during the Neogene, supported by a biostratigraphic framework, the chronological time frame has improved substantially (e.g., Röhl et al., 2000). The use of nondestructive measurements of sediment characteristics such as sediment color or magnetic susceptibility has been very beneficial for the production of high-resolution records.

Bibliography

Altabet, M. A., 2006. Isotopic tracers of the marine nitrogen cycle: present and past. *The Handbook of Environmental Chemistry*, **2N**, 251–293, doi:10.1007/698_2_008.

CLIMAP, 1981. Seasonal reconstructions of the Earth's surface at the last glacial maximum in Map Series, Technical Report MC-36. Boulder: Geological Society of America.

Cohen, A. L., Owens, K. E., Layne, G. D., and Shimizu, N., 2002. The effect of algal symbiosis on the accuracy of Sr/Ca paleotemperatures from coral. *Science*, **296**(5566), 331–333.

Felis, T., Giry, C., Scholz, D., Lohmann, G., Pfeiffer, M., Pätzold, J., Kölling, M., and Scheffers, S. R., 2015. Tropical Atlantic temperature seasonality at the end of the last interglacial. *Nature Communications*, **6**, 6159, doi:10.1038/ncomms7159.

Fischer, G., and Wefer, G. (eds.), 1999. *Use of Proxies in Paleoceanography: Examples from the South Atlantic*. Berlin/Heidelberg: Springer, p. 735.

Gersonde, R., de Vernal, A., and Wolff, E. W., 2014. Past sea ice reconstruction – proxy data and modeling. *PAGES Magazine*, **22**(2), 97.

Guo, Z. T., Ruddiman, W. F., Hao, Q. Z., Wu, H. B., Qiao, Y. S., Zhu, R. X., Peng, S. Z., Wei, J. J., Yuan, B. Y., and Liu, T. S., 2002. Onset of Asian desertification by 22 Myr ago inferred from loess deposits in China. *Nature*, **416**, 159–163.

Heguera, J. C., and Berger, W. H., 1994. Glacial to postglacial drop in productivity in the western equatorial Pacific: mixing rate vs. nutrient concentrations. *Geology*, **22**, 629–632.

Hönisch, B., and Hemming, N. G., 2004. Ground-truthing the boron isotope-paleo-pH proxy in planktonic foraminifera shells: partial dissolution and shell size effects. *Paleoceanography*, **19**(4). doi: 10.1029/2004PA001026

Imbrie, J., and Kipp, N. G., 1971. A new micropaleontological method for quantitative paleoclimatology: application to a Late Pleistocene Caribbean core. In Turekian, K. (ed.), *The Late Cenozoic Glacial Ages*. New Haven: Yale University Press, pp. 71–181.

Kuechler, R. R., Schefuß, E., Beckmann, B., Dupont, L. M., and Wefer, G., 2013. NW African hydrology and vegetation during the last glacial cycle reflected in plant-wax-specific hydrogen and carbon isotopes. *Quaternary Science Reviews*, **82**, 56–67, doi:10.1016/j.quascirev.2013.10.013.

Lisiecki, L. E., and Raymo, M. E., 2005. A Pliocene-Pleistocene stack of 57 globally distributed benthic $\delta^{18}O$ records. *Paleoceanography*, **20**, PA1003, doi:10.1029/2004PA001071.

Mann, M. E., Zhan, Z., Rutherford, S., Bradley, R. S., Hughes, M. K., Shindell, D., Ammann, C., Faluvegl, G., and Ni, F., 2009. Global signatures and dynamical origins of the Little Ice Age and Medieval Climate Anomaly. *Science*, **326**, 1256–1260, doi:10.1126/science.1177303.

Mollenhauer, G., Basse, A., Kim, J. H., Sinninghe Damsté, J. S., and Fischer, G., 2015. A four-year record of – and TEX86-derived sea surface temperature estimates from sinking particles in the filamentous upwelling region off Cape Blanc, Mauritania. *Deep Sea Research I*, **97**, 67–79.

Raymo, M. E., and Mitrovica, J. X., 2012. Collapse of polar ice sheets during the state 11 interglacial. *Nature*, **483**, 453–456, doi:10.1038/nature10891.

Röhl, U., Bralower, T. J., Norris, R. D., and Wefer, G., 2000. New chronology for the late Paleocene thermal maximum and its environmental implications. *Geology*, **28**(10), 927–930.

Stein, R., Fahl, K., and Müller, J., 2012. Proxy reconstruction of Arctic Ocean sea ice history – From IRD to IP25. *Polarforschung*, **82**, 37–71.

Wefer, G., Berger, W. H., Fischer, G., and Bijma, J., 1999. Clues to ocean history – a brief overview of proxies. In Fischer, G., and Wefer, G. (eds.), *Use of Proxies in Paleoceanography: Examples from the South Atlantic*. Berlin/Heidelberg: Springer, pp. 1–68.

Ziegler, M., Diz, P., Hall, I. R., and Zahn, R., 2013. Millennial-scale Agulhas Current variability and its implications for salt-leakage through the Indian–Atlantic Ocean Gateway. *Earth and Planetary Science Letters*, **383**, 101–112.

Cross-references

Biogenic Barium
Marine Microfossils
Mg/Ca Paleothermometry
Nitrogen Isotopes in the Sea
Paleoceanography
Paleoproductivity
Radiocarbon: Clock and Tracer
Radiogenic Tracers

PALEOCEANOGRAPHY

Jörn Thiede
Institute of Earth Sciences, Saint Petersburg State University, St. Petersburg, Russia
AdW Mainz c/o GEOMAR, Kiel, Germany

Synonyms

History of ocean basins; Paleo-Oceanography; Paleooceanography; Paleoceanology

Definition

Paleoceanography is the science of the history of the world ocean and its subbasins, of their physiography, benthic and planktonic biota, water masses and their properties, circulation patterns of surface, intermediate, and bottom water masses. Quantitative paleoceanography can in essence only be done from pelagic and hemipelagic sediments of the Mesozoic and Cenozoic ocean, basically from the age of the oldest undisturbed ocean crust to modern times. It is closely linked to the time of collecting ocean-wide geophysical data (marine seismics, magnetics), of deep-sea drilling and our increasing ability to collect long sediment cores, providing a global coverage and long, undisturbed time series of paleoceanographic data. It is henceforth a young subdiscipline of the marine geosciences.

Paleoceanography from infancy to maturity

When Schopf (1980, see also Kennett, 1982) wrote the first textbook about "paleoceanography," important steps toward development of this new subdiscipline of marine geosciences had already been taken, and the term had been introduced in earlier publications (Van Andel et al., 1975). During the late 1950s and early 1960s one had learned the techniques to take long and undisturbed sediment cores and to drill through/sample ocean sediment sequences, to date these with great precision, to interpret their contents of microfossils and their chemical (isotopic) composition (Emiliani, 1992) for establishing the properties of past ocean surface and bottom water masses. However, with his devotion to time spans older than Mesozoic Schopf (1980) missed one important aspect of modern paleoceanography, namely its importance of climatic and paleoclimatic modeling efforts for our attempts to deduce possible future climate scenarios from past analogies.

Now, some 40 years later, "Paleoceanography" has developed into a mature discipline of the marine geosciences, which is pursued at many universities and research institutions with great sophistication, as can be demonstrated by many entries in this Springer Encyclopedia as well as in Gornitz (2009). The examples selected for this entry concern the Late Mesozoic and Cenozoic Oceans whose physiographic evolution can be deduced from the through-time-increasing areas of undisturbed and undestroyed ocean crust (cf. Müller, this volume). The salinity of the marine water masses also demonstrates that we are dealing with a compartment of the environment, like the atmosphere of truly global dimensions (and contrary to terrestrial, limnic, or ice core data which record regional or local phenomena). After decades of sediment coring and drilling in most of the ocean basins, the modern development of powerful research ice breakers capable to enter the central ice-covered Arctic Ocean (Fütterer and Shipboard Scientific Party, 1992; Moran et al., 2006) has opened up the last virtually unexplored deep-sea basin for paleoceanography. Here an ecclectic and subjecive overview of some of the *most important discoveries* of the past decades will be presented.

Paleophysiography: When Alfred Wegener hypothesized about the "origin of continents and ocean" between 1912 and 1929 he had to do so without sufficient information about the nature of the ocean floor; despite this deficiency he succeeded to interpret correctly many of the paleogeographic changes of the distribution of land and sea through time. In modern times (cf. Müller et al., this volume) the information deduced from the age distribution of sea-floor spreading magnetic anomalies and of our understanding of the subsidence of ocean crust after its formation due to its thermal history (Sclater et al., this volume) allows us to draw a detailed picture of the paleophysiography of the location, shape including depths, size, and volume of the ocean basins and their evolution through time, at least since mid-Mesozoic times (cf. Müller et al., this volume). Basically it is the evolution of planet Earth from a geography with one large continental land mass versus one contiguous ocean basin at the end of the Paleozoic to the modern one with numerous "small" continents and an ocean basin fragmented into many subbasins.

Plate tectonics and climate: The major changes of the physiography of ocean basins, in particular tectonically driven changes of the seaways between ocean basins, were clear since Wegener's (1912–2005) publications. A modern example is described in Haug and Tiedemann (1998), who reasoned that the closure of the Isthmus of Panama led to environmental changes resulting in the deteriorating livelihood of the late Holocene Central American populations. However, more recently, a much closer (and hitherto) poorly understood correlation between volcanic/tectonic processes at the mid-ocean ridges and the evolution of climate seems to emerge. Van der Meer et al., 2014 have determined plate tectonic controls on atmospheric carbon dioxide levels in the atmosphere since the Triassic and speculate about an intricate interrelationship between subduction rates and carbon dioxide levels in the atmosphere, controlled by processes over time scales of 10 to 100 million years. However, recent comparisons of volcanic activity on Iceland and the observations of the similarity of the fine grain of the morphology of volcanic structures of the mid-ocean ridge regions with Late Cenozoic climate variability seem to suggest that plate tectonic and climate variability are much more closely linked and over much shorter time scales than presumed hitherto.

Plate stratigraphy: The mode of the general subsidence of the ocean floor and the mechanisms of pelagic sediment formation generate a characteristic stratigraphic pattern of sediments on top of the ocean crust, as demonstrated in many cores of the various deep-sea drilling projects (cf. Suyehiro, this volume, and discussed by Kennett, 1982 and Van Andel et al., 1975). The global patterns of pelagic sediment formation (cf. Lyle, this volume) and in particular the water depth-dependent aragonite and calcite differential dissolution as well as the existence of the calcite lysocline and Calcite Compensation Depth (CCD, cf. Berger, this volume) result in the preservation of the ubiquitous calcareous deep-sea deposits only in well-defined regions and depth intervals (Van Andel et al., 1977; Ehrmann and Thiede, 1985). Young ocean crust lies usually above the CCD giving place to the formation of calcareous oozes if not biogenic siliceous oozes are formed under highly productive ocean surface waters. In high latitudes the input of terrigeneous fines dominates ocean sediment formation. Old ocean crust which had sunk below the CCD in temperate to tropical regions is mostly covered by red deep-sea clays.

Marine microfossils and their environmental record: Marine microfossils have been producing the first proxies for paleoceanographic and paleoclimatic records deduced from ocean sediment cores. This was first realized by Schott (1935, in Correns and Schott, 1935/37), who mapped out planktonic foraminiferal distributions in sediment cores from the South Atlantic and who was able to argue that these cores covered the transition from the last glacial to the much warmer Holocene. In the meantime it has been recognized that all planktonic microfossils reveal distribution patterns according to temperature zonations in the ocean and that they are superb indicators which can be exploited by using modern analog techniques (Hale and Pflaumann, 1999), a technique which has been developed by Imbrie and Kipp (1971) during the early stages of the CLIMAP project (Cline and Hays, 1976). In the meantime this method has been applied to many other examples, cf. Pflaumann et al. (2003).

Paleotemperature, paleosalinity, ice in the ocean: When reconstructing the hydrography of former oceans, one needs to quantify its basic hydrographic properties. This is relatively easy but sometimes ambiguous (Stärz et al., 2012) for ice in the oceans, because it leaves traces in the distribution of ice-rafted sediment components and of biotic remains of organisms (diatoms) adapted to the life in and under the sea ice (Thiede et al., 2011; Stickley et al., 2009). The relatively recent drilling on Lomosow Ridge close to the North Pole (Moran et al., 2006) suggests that the central Arctic Ocean at least was seasonally sea-ice covered in the middle Eocene, milions of years before the assumed initial glaciation of the northern hemisphere. Paleotemperatures and –salinities can be quantitatively established by a wide variety of biotic and geochemical methods as summarized by Fischer and Wefer, 1999.

Fractionation of Oxygen isotopes in relation to global ice volume: When Emiliani studied the O-Isotpe ratios in forminiferal shells from Upper Quaternary deep sea sediments (Emiliani, 1955), he considered the O-isotopic variations to be mainly due to temperature changes between Glacials and Interglacials. However, even though benthic foraminifers showed similar variations with a clear offset, it would not have been possible to invoke the same temperature variations as their cause. And it was Shackleton (1987) who linked the O-isotope fractionation observed in the foraminiferal records to the late Quaternary climate variability (ice volume, sea-level) assuming that the major effects of the isotope fractionation were mainly caused by the evaporation of large quantities of water from the ocean to be bound in the glacial ice sheets during Glacials, both on the northern and the southern hemisphere, respectively released during Interglacials.

Paleoceanographic proxies (or proxy variables): There are numerous proxies which have been developed over the past decades which allow describing paleoceanographic properties of past oceans (Fischer and Wefer, 1999). Beside biotic proxies such as pelagic microfossils there are numerous geochemical and oganochemical tools to address variations of the surface water and bottom properties and circulation (including temperature and salinity) and water mass stratification. Biogenic Barium can be used to describe paleoproductivity and nutrient distributions. A variety of radiogenic isotopes can be used as tracers for bioproductivity and particle fluxes in the oceans. The carbon isotope ratios of alkenones help to assess past carbon dioxide distributions (Paleoceanographic Proxies).

Grain size and current strength: McCave et al. (1995) used sophisticated grain size analyses of relatively fine-grained terrigenous sediments, in their attempts to reconstruct the vigor of bottom water currents in the North Atlantic during the transition from the Last Glacial Maximum (LGM) to the Holocene. They used well-dated sediment cores from the Rockall area and were able to prove the stratification and strength of the bottom currents during a time of an intense reorganization of the North Atlantic current systems. This is but one example of numerous in the world ocean. Similar attempts have been made to describe the flow of the Antarctic Bottom Water in the Southern Ocean.

Global Conveyor Belt: Ocean surface waters are warm in the tropics but cold close to the poles. But it was known since the Russian explorer Krusenstern's circumglobal expedition almost 200 years ago that the deep oceans were cold all over. With the first systematic hydrographic investigations such as carried out during the German "Meteor" expedition to the South Atlantic Ocean (Wüst, 1936) it was recognized that cold waters which had gotten their most important hydrologic properties in the polar regions filled many deep-sea basins. They were dense enough to sink as part of the thermohaline circulation into the oceanic deep-sea basins where their flow was controlled by the morphology of the gateways. Minor differences in their densities controlled the mode of their layering in the Atlantic Ocean with water of Arctic in many areas below those of Antarctic origin. Broecker (1991), after having determined by means of radiocarbon dating and thus the last time when these waters were able to exchange carbon dioxide with the atmosphere, established the young "ages" of bottom waters in the deep Atlantic Ocean, contrasting the old ones in the deep North Pacific Ocean. Based on this pattern and combining it with the large-scale circulation system of ocean surface waters he coined the term "great ocean conveyor belt." How this ocean circulation system has been changing going back in time has been subject of controversial debates ever since, because there is no question that ocean deep-water masses were much warmer during the earliest than during the late Cenozoic (Zachos et al., 1993).

Atmospheric circulation: The atmospheric circulation and its past variations leave many traces in deep-sea sediments. Dust plumes originating from desert systems like the Sahara in North Africa are exported across the African coast. They document through increased accumulation rates (Thiede et al., 1982) and grain sizes (Sarnthein and Koopmann, 1980) during glacial intervals an intensification of the glacial atmospheric circulation. Similar messages can be deduced from the dust records in ice cores (Svensson et al., 2000) or South Pacific dust records (Thiede, 1979). It can be concluded that the glacial atmospheric circulation not only changed its patterns but also increased in intensity due to the steeper temperature gradients between low and high latitudes. Conclusions about former atmospheric circulation patterns can also be reached by looking at the distributions of pollen, spores, and other biotic remains of terrestrial origin preserved in deep-sea sediments as well as the distribution of organic matter of terrestrial origin (Fischer and Wefer, 1999).

Time slices versus time series: The progress in dating techniques of pelagic sediments was a prerequisite for establishing reconstructions of the paleoceanographic properties of certain regions both in terms of a well-defined time slice as well as time series. One of the best examples for the former is the reconstruction of the surface properties of planet Earth during the Last Glacial Maximum produced by CLIMAP (CLIMAP Project Members, 1976). The first time series which allowed proving Milankovitch frequency controlled time series from the ocean was published by Hays et al. (1976), but since then numerous time series of paleoceanographic data sets have been documented in the results of the deep-sea drilling projects.

Milankovitch Frequencies: When Köppen and Wegener (1924) published their monograph "The Climate of the Geological Past," the time scales of the late Cenozoic alterations between Glacials and Interglacials were only vaguely known. However, they had established a close cooperation with Milutin Milankovitch. They were convinced that the geometry of the Earth´s orbit around the sun with its regular changes of eccentricity, obliquity, and precession are controlling the insolation (amount of incoming solar radiation) and hence are important climate boundary conditions. Milankovitch published (Milankovitch, 1941) the detailed data of the Milankovitch frequencies only decades later. But it was one of the major successes of the CLIMAP project (Hays et al., 1976) that it found two sediment cores from the southernmost Indian Ocean which could be dated precisely enough and which reflected the paleoceanographic responses to the Late Quaternary glacial-interglacial changes. Thus Köppen and Wegener`s (1924) claim could be substantiated by well-dated time series derived from the very best global climate archive. In the meantime the

Milankovitch frequencies have been documented also in numerous other historic records, such as those derived from ice cores in Antarctica and long lake sediment cores. Milankovitch frequencies can be extrapolated into the future, which raised the possibility of the predictability of future climate change (Thiede and Tiedemann, 1998). The paper of Berger (2012) represents a modern perspective of the Milankovitch frequencies (cf. also Berger, this volume).

Climate Modeling: Paleoceanograhic studies provide indispensable data for paleoceanographic and paleoclimatological modeling (cf. Sündermann, this volume), both as boundary conditions to run the models as well as for their validation. The first true global data set which could be used as input parameters for modeling glacial climates (Herterich et al., 1999) has been produced by CLIMAP (Climate: Long-range Investigation and Prediction), a large paleoclimatic research project in the USA (CLIMAP Project Members, 1976). Since that time available models (GCMs) developed by various institutions (NCAR, Boulder/US; MPI Meterology in Hamburg; Hadley Center – Met Office/UK, to name a few) have all delivered their own modeling approaches aiming to model various compartments of the global environment (e.g., like the ECHAM model series in Hamburg). For a more detailed discussion see Paleoclimte Modelling Intercomparison Project (http://pmip.lsce.ipsl.fr/). Many of these efforts of generating past oceanographic and climatic scenarios are trying to understand through the reconstruction of the past, how the future oceans might behave, e.g., with a seasonally ice-free Arctic Ocean which would affect the entire northern hemisphere.

Summary

Paleoceanography is fed by data sets from the only truly global climate archive, namely ocean sediments. It has developed from modest beginnings during the early 1960s (prior to the onset of deep-sea drilling) to a highly sophisticated discipline of the marine geosciences. And it provides quantitative reconstructions of former global climates as boundary conditions for modeling climates with no modern analogue.

Bibliography

Berger, A., 2012. A brief history of the astronomical theories of paleoclimates. In Berger, A., et al. (eds.), *Climate Change*. Vienna: Springer, pp. 107–129, doi:10.1007/978-3-7091-0973-1_8.

Broecker, W. S., 1991. The great ocean conveyor. *Oceanography*, **4**, 79–89.

CLIMAP Project Members, 1976. The Surface of the Ice-Age Earth. *Science*, **191**(4232), 1131–1137.

Cline, R. M., and Hays, J. D. (eds.), 1976. *Investigations of Late Quaternary Paleoceanography and Paleoclimatology*. Boulder, CO: Geological Society of America. Geological Society of America memoirs, **145**, 464 pp.

Correns, C. W., and W. Schott, 1935/1937. Die Sediments des Äquatorialen Atlantischen Ozeans. 1. Lfg.: A. Die Verfahren der Gewinnung und Untersuchung der Sediemtns (Correns). B. Die Foraminiferen in dem äquatorialen Teil des Atlantischen Ozeans (Schott).- 2. Lfg.: C. Zusammenstellung der Untersuchungsergebnisse nach Stationen geordnet. D. Die Auswertung der Ergebnisse.-Wiss. Ergebn. der dt. Atlantischen Expedition auf dem Forschungs- und Vermessungsschiff "Meteor",Wiss. Ergebn., Vol. 3.3., Berlin: de Gruyter, 298 p.

Ehrmann, W., and Thiede, J., 1985. History of Mesozoic and Cenozoic sediment fluxes to the North Atlantic Ocean. *Contributions to Sedimentology*, **15**, 109 pp.

Emiliani, C., 1955. Pleistocene temperatures. *Journal of Geology*, **63**, 538–578.

Emiliani, C., 1992. *Planet Earth: Cosmology, Geology, and the Evolution of Life and the Environment*. Cambridge: Cambridge University Press. ISBN 0-521-40949-7.

Fischer, G., and Wefer, G. (eds.), 1999. *Use of Proxies in Paleoceanography. Examples from the South Atlantic*. Heidelberg: Springer, 735 pp, figs, tables.

Fütterer, D., and Shipboard Scientific Party, 1992. *Arctic '91: the expedition ARK-VIII/3 of RV POLARSTERN in 1991*. Bremerhaven: Alfred-Wegener-Institute, Report of Polar Research, Vol. **107**, 267 pp.

Gornitz, V., 2009. *Encyclopedia of Paleoclimatology and Ancient Environments*. Springer Heidelberg.

Hale, W., and Pflaumann, U., 1999. Sea-surface temperature estimations using the modern analog technique with Foraminiferal assemblages from Western Atlantic Quaternary sediments. In Fischer, G., and Wefer, G. (eds.), *Use of Proxies in Paleoceanography. Examples from the South Atlantic*. Heidelberg: Springer, pp. 69–90, 10 figs.

Haug, G. H., and Tiedemann, R., 1997. Effect of the formation of the Isthmus of Panama on Atlantic Ocean thermohaline circulation. *Nature*, **393**, 673–676, doi:10.1038/31447. 18 June 1998.

Hays, J. D., Imbrie, J., and Shackleton, N. J., 1976. Variations in the Earth's orbit: pacemaker of the Ice Ages. *Science*, **194**(4270), 1121–1132.

Herterich, K., et al., 1999. Reconstructing and modelling the last glacial maximum beyond CLIMA. In Fischer, G., and Wefer, G. (eds.), *Use of Proxies in Paleoceanography. Examples from the South Atlantic*. Heidelberg: Springer, pp. 687–714, 15 figs.

Imbrie, J., and Kipp, N., 1971. A new micropaleontological method for quantitative paleoclimatology. In Turekian, K. K. (ed.), *The Late Cenozoic glacial ages*. New Haven: Yale University Press, pp. 71–181, figs.

Kennett, J., 1982. *Marine Geology*. Englewood Cliffs: Prentice-Hall, 813 pp, figs, tables.

Köppen, W., and Wegener, A., 1924. *Die Klimate der geologischen Vorzeit*. Berlin: Publ. Gebr. Borntraeger, 255 pp., figs, 1 plate (republished by Borntraeger Science Publ. 2015 together with a translation into English).

McCave, I. N., Manighetti, B., and Beveridge, N. A. S., 1995. Circulation in the glacial North Atlantic inferred from grain-size measurements. *Nature*, **374**, 149–152.

Milankovitch, M., 1941. *Kanon der Erdbestrahlung und seine Anwendung auf das Eiszeitenproblem*. Belgrad: Royal Serbian Sciences, Special publication 132, Natural Sciences and Mathematics, Vol. **33**, 633 pp.

Moran, K., Backman, J., and ASEX Scientific Party, 2006. The Cenozoic palaeoenvironment of the Arctic Ocean. *Nature*, **441**, 601–605, doi:10.1038/nature04800.

Pflaumann, U., Sarnthein, M., and Chapman, M. R., 2003. Glacial North Atlantic: sea-surface conditions reconstructed by GLAMAP 2000. *Paleoceanography*, **18**(3), 1065, doi:10.1029/2002PA000774.

Sarnthein, M., and Koopmann, B., 1980. Late Quaternary deep-sea record in Northwest African dust supply and wind circulation. *Paleoecology of Africa and Surrounding Islands*, **9**, 239–253, Rotterdam: A. A. Balkema.

Schopf, T. J. M., 1980. *Paleoceanography*. Cambridge, MA: Harvard University Press, 341 pp, figs, tables.

Shackleton, N. J., 1987. Oxygen isotopes, ice volume and sea-level. *Quaternary Science Reviews*, **6**, 183–190.

Stärz, M., Gong, X., Stein, R., Darby, D. A., Kauker, F., and Lohmann, G., 2012. Glacial shortcut of Arctic sea-ice transport. *Earth Plant Science Letters*, **357–358**, 257–267.

Stickley, C. E., St. John, K., and Koç, N., 2009. Evidence for middle Eocene Arctic sea ice from diatoms and ice-rafted debris. *Nature*, **460**, 376–379, doi:10.1038/nature 08163, 16 July 2009.

Svensson, A., Bicaye, P., and Grousset, F. E., 2000. Characterization of late glacial continental dust in the Greenland Ice Core Project ice core. *Journal of Geophysical Research*, **105**(D4), 4637–4656.

Thiede, J., 1979. Wind regimes over the late Quaternary Southwest Pacific Ocean. *Geology*, **7**, 259–262.

Thiede, J., Jessen, C., and Kuijpers, A., 2011. Millions of years of Greenland Ice Sheet history recorded in ocean sediments. *Polarforschung*, **80**(3), 141–159.

Thiede, J., Suess, E., and Müller, P., 1982. Late Quaternary fluxes of major sediment components to the seafloor along the northwest African continental slope. In von Rad, U., et al. (eds.), *Geology of the Northwest African Continental Margin*. Heidelberg: Springer, pp. 605–631.

Thiede, J., and Tiedemann, R., 1998. Die Alternative: NatürlicheKlimaveränderungen – Umkippen zu einer neuen Kaltzeit? In LOZAN, J. L., Graßl, H., and Hupfer, P. (eds.), *Das Klima des 21*. Hamburg: Jahrhunderts, pp. 190–196.

Van Andel, T. H., Heath, G. R., and Moore, T. C., 1975. *Cenozoic History and Paleoceanography of the Central Equatorial Pacific Ocean*. Boulder: Geological Society of America, Geological Society of America memoirs, Vol. **143**, 134 pp., 68 figs.

Van Andel, T. H., Thiede, J., Sclater, J. G., and Hay, W. W., 1977. Depositional history of the South Atlantic Ocean during the last 125 million years. *Journal of Geology*, **85**(6), 651–698.

Van der Meer, D. G., et al., 2014. Plate tectonic controls on atmospheric CO_2 levels since the Triassic. *Proceedings of the National Academy of Sciences of the United States of America*, **111**, 4380–4385.

Wegener, A., 1915/1929/2005. *Die Entstehung der Kontinente und Ozeane*. Reprint of the 1st ed. with handwritten comments of him and of the 4th completely revised ed., Berlin: Borntraeger, 481 pp.

Wüst, G., 1936. Schichtung und Zirkulation des Atlantischen Ozeans.-Wiss. Ergebn. Dt. Atlant. Expedition VFS Meteor 1925–1927: 6(1), Berlin, 1–288.

Zachos, J. C., Lohmann, K. C., Walker, J. C. G., and Wise, S. W., 1993. Abrupt climate change and transient climates in the Paleogene: a marine perspective. *Journal of Geology*, **101**, 193–215.

Cross-references

Anoxic Oceans
Astronomical Frequencies in Paleoclimates
Carbon Isotopes
Geologic Time Scale
Marine Microfossils
Modelling Past Oceans
Oxygen Isotopes
Paleoceanographic Proxies
Paleophysiography of Ocean Basins
Paleoproductivity
Plate Tectonics
Regional Marine Geology
Sequence Stratigraphy

PALEOMAGNETISM AND JURASSIC PALEOGEOGRAPHY

María Paula Iglesia Llanos
IGEBA, National Scientific and Research Council, University of Buenos Aires, Buenos Aires, Argentina

Introduction

From the Late Carboniferous until the Middle Jurassic, continents were assembled in a quasi-rigid supercontinent called Pangaea, which occupied most of a hemisphere, while the rest of the Earth's was part of a large ocean called Panthalassa. For decades, authors interpreted that South America remained stationary in similar present-day latitudes during most of the Mesozoic and the end of the Paleozoic, because paleomagnetic poles (PP) clustered close to the geographical pole (e.g., Valencio et al., 1983; Oviedo and Vilas, 1984; Beck, 1999). More recent data obtained in Patagonia, Argentina, show that Early Jurassic PP actually fall far away from the geographic pole (Iglesia Llanos et al., 2006).

Using these and other selected poles from South America, a substantially different and more refined Jurassic apparent polar wander (APW) path was constructed (Iglesia Llanos et al., 2006; Iglesia Llanos and Prezzi, 2013). APW paths depict the past positions of the Earth's spin axis (= averaged paleomagnetic pole) with respect to a certain lithospheric plate, hence the name "apparent" (Creer et al., 1954). The path is constructed as a sequence of paleomagnetic poles tracking away from the geographic pole with increasing age. Tracks represent rotations about Euler poles, whereas cusps symbolize times of reorganization of the lithospheric plate boundaries and resulting driving forces (Cox and Hart, 1986).

In this study, Jurassic APW paths of Eurasia, North America, and Africa were built and compared to those obtained in South America. PP were selected mostly from the Global Paleomagnetic Database v.4.6 2005 (http://www.tsrc.uwa.edu.au/data_bases) and from other recent studies. Such PP are considered to be representative of stable cratons (Torsvik et al., 2008; Kent and Irving, 2010).

APW paths from Eurasia, North America, and Africa were rotated to South American present-day coordinates, using classical paleoreconstructions available in the literature for Jurassic Pangaea. Among all kinematic models, those which provided the best clustering of paleopoles were taken into account for this study. Thus, Africa was rotated to South American coordinates using the paleoreconstructions proposed by Lawver and Scotese (1987). Eurasia was translated first to North America using the paleoreconstruction of Frei and Cox (1987), then to Africa, and subsequently to South America using Lawver and Scotese (1987). North America, on the other hand, was first moved to Africa with the paleoreconstruction of Klitgord and Schouten (1986) and then to South America using the parameters of Lawver and Scotese (1987). The resultant paths proved to be almost identical in terms of shape and chronology. Subsequently, by averaging nonoverlapping poles with 5–10 Myr windows, a master path was derived for Pangaea that could be used in any continent in the Jurassic (Figure 1). The resultant curve shows, when observed from South America, a W-E track between 215 and 197 Ma (Late Triassic-Earliest Jurassic), followed by a SE-NW track between 197 and 185 Ma (Early Jurassic). The intersection of these two tracks defines a cusp at ∼197 Ma. After 185 Ma until 170 Ma, a smaller track and consecutive standstill that might have lasted until the Late Jurassic are observed. The 197–185 Ma (Early Jurassic) track reveals a c. 50° shift in the pole positions and a minimum angular change of poles of approximately $4° \text{ Myr}^{-1}$.

Discussion

There are a few phenomena that could have caused such shift in the pole positions: either the geomagnetic field was not dipolar at that time or some geodynamic event took place. With regard to the first, it is well accepted that the Earth's magnetic field behaved essentially as a geocentric dipole (e.g., McElhinny and Brock, 1975; Evans, 1976; Livermore et al., 1983, 1984). This means that the c. 40° polar shift in the APW path cannot be attributed to nondipole components. The cause of the polar shift is rather related to some kind of geodynamic event, such as lithospheric motion or true polar wander (TPW). In order to appraise lithospheric motion, the South American plate constitutes an illustrative case. Paleomagnetic data indicate that, during the Late Triassic and lowermost Jurassic, the continent rotated 50° CW (Figure 1). Had such rotation resulted from sole lithospheric motion, South America would have been forced to override the trench as well as the Pacific plate, thus causing major compression in the overriding plate. Yet, the formation of the Andes comprised two main deformation stages, the first (∼ 200 to 90 Ma) characterized by comprehensive extension in the overriding plate, during which major rift systems like in Peru, northern Chile (e.g., Cobbing et al., 1981; Åberg et al., 1984; Mpodozis and Allmendinger, 1993), and Argentina (e.g., Giambiagi et al., 2008) are formed. The last stage on the other hand occurred in the mid-Cretaceous and characterized a major compressional regime (Somoza and Zaffarana, 2008).

Therefore, the polar shift might have been resulted from the other geodynamical event which is TPW, defined as the drift of the spin axis in relation with the solid Earth over geological time. TPW is determined from hot spots, which are assumed to form an array of fixed points in the mantle-mantle reference frame. Defined as it is, TPW makes all continents and mantle plumes rotate around an Euler pole located at the Equator (Goldreich and Toomre, 1969; Jurdy and van der Voo, 1975). TPW is a global type of event; therefore, it should be recognizable in all APW paths from that time. The similarity in all four curves (Figure 1) is hence the first hint of TPW. Despite that fixity of hot spots is still hot matter of debate, many authors such as Besse and Courtillot (2002) use this uniquely available tool to determine TPW in the Jurassic.

To assess the occurrence of TPW, the most straightforward way is paleolatitudes plots. For this study, two South American localities were analyzed (Figure 2). To do this, continents were first rotated to South American present-day coordinates and rotated once again to position them with respect to the mantle using Morgan's (1983) grid of hot spots. Two sets of paleolatitudes were thus calculated for each locality, one derived from the master APW path and the other from the array of hot spots (Figure 2). Had no TPW existed, both paleolatitudinal sets should coincide. But data show a significant divergence of the curves after 200 Ma, which could be interpreted as caused by TPW. The curves show that whereas no evident latitudinal change occurred from the hotspots reference frame, the South American continent was actually moving northward. This is solid evidence of the occurrence of a clear TPW event in the Early Jurassic.

In addition, the TPW path constructed after removing plate motions in the APW path reveals that at ∼197 Ma, poles displaced approximately 45° over a 12-Myr period or a minimum angular change of $4° \text{ Myr}^{-1}$ (0.4 Myr^{-1}). This magnitude is consistent with the values predicted by Prévot et al. (2000) of $1–5° \text{ Myr}^{-1}$, or the $3°–10° \text{ Myr}^{-1}$ proposed by Sager and Koppers (2000) for the Pacific plate. Subsequently by the end of the Early Jurassic, the path turned back on itself.

Paleomagnetism can provide paleolatitudes and orientations of the continents, but because the time-averaged geomagnetic field is symmetric about the rotation axis, the paleomagnetic method alone cannot assess paleolongitudes. There are other methods to calculate paleolongitudes such as hot spots. Thus, if the motion of, e.g., South America, can be established in relation with the grid of hot spots, all other plates can be put together within this frame after applying the corresponding kinematic model. Therefore, by combining relative motions of the plates, their motions relative to the hot spots, and paleomagnetic data which indicate the ancient position of the geomagnetic pole (paleopole) for a specific locality and time, "absolute" paleoreconstructions can be achieved.

Paleogeographic reconstructions thus obtained reveal (Figure 3) that during the Late Triassic-Sinemurian (∼210 to 200 Ma), Pangaea would have been located at its southernmost position with continents rotated CCW with respect to present-day orientation. Subsequently, the supercontinent moved northward while rotated approximately 50° CW at higher speeds (c. $4° \text{ Myr}^{-1}$). By the Pliensbachian (∼185 Ma), the supercontinent achieved its northernmost latitudes. In the end of the Early Jurassic, Pangaea moved to the south again until the Middle Jurassic when it positioned similar present-day latitudes.

Since latitude affects climate, it is possible to correlate paleolatitudes derived from the paleomagnetic method with paleoclimatic data derived from fossils having a

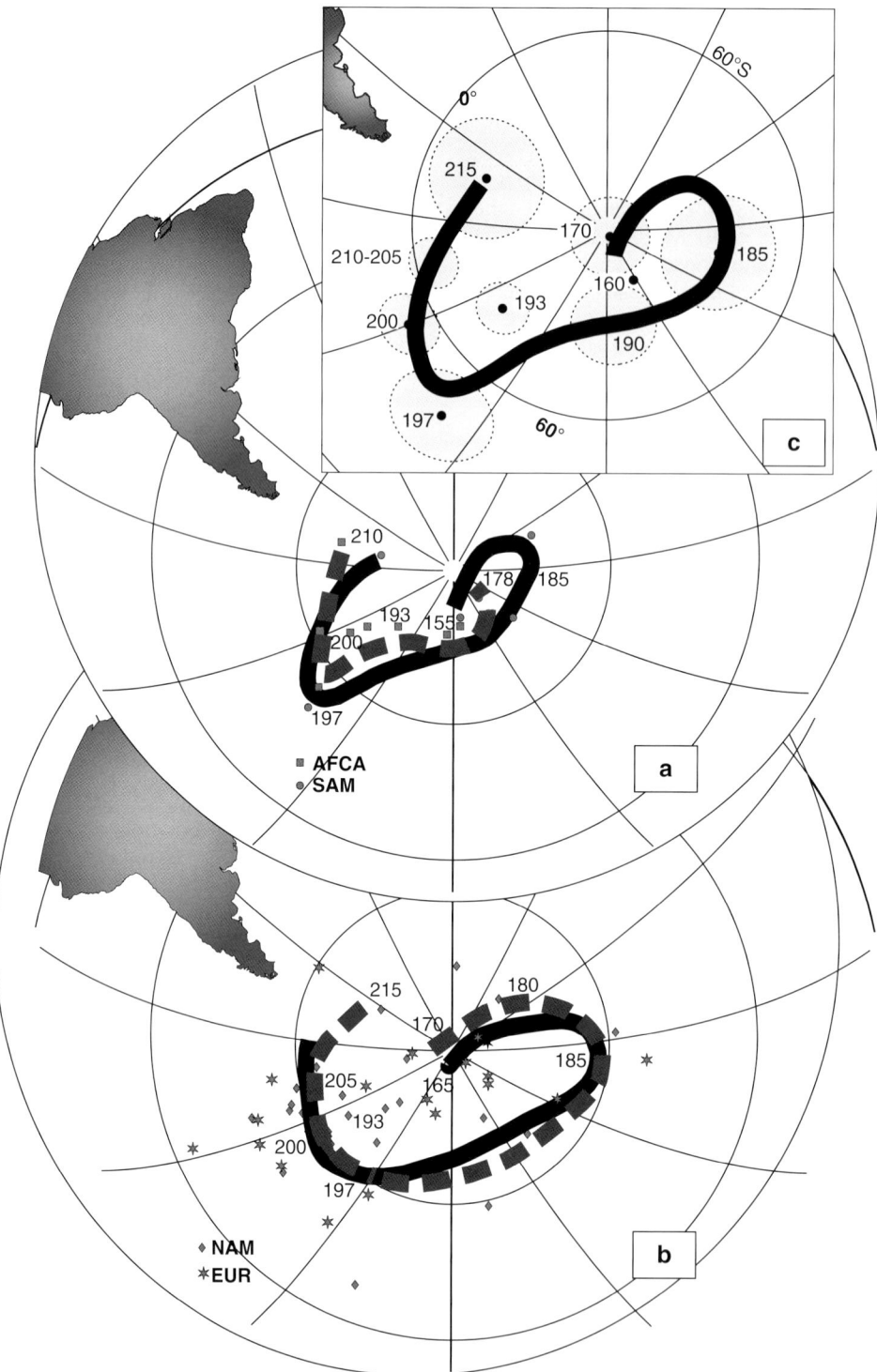

Paleomagnetism and Jurassic Paleogeography, Figure 1 Jurassic APW paths with mean ages for (**a**) Africa and South America, (**b**) North America and Eurasia. Africa, North America, and Eurasia have been rotated to South American present-day coordinates (**c**) Master APW path for Pangaea with 95 % confidence interval.

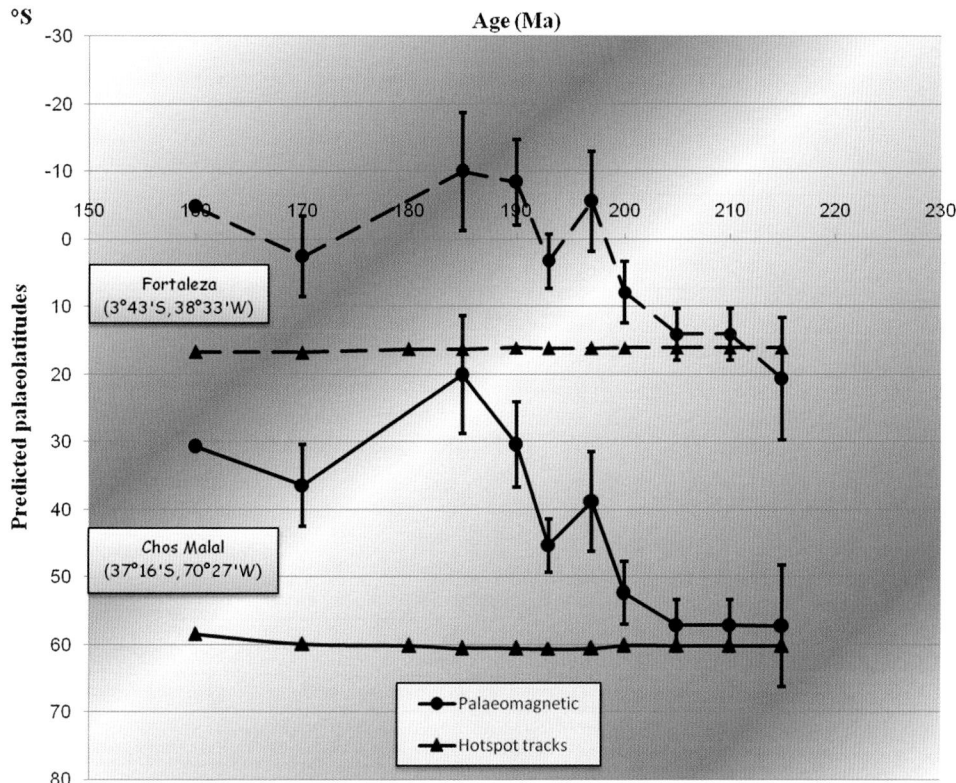

Paleomagnetism and Jurassic Paleogeography, Figure 2 Paleolatitudes plot support the occurrence of TPW after 200 Ma. Two paleolatitude sets are shown for two South American localities, one from Brazil and the other from Argentina, the *first* representing paleolatitudes calculated from the master APW path paleomagnetic reference frame and the *second* set from Morgan's (1983) grid of HS (mantle reference frame). If no TPW occurred, both *curves* should be coincident.

geographic distribution limited to certain climatic regions. In this study, paleobiogeographic data were analyzed in order to confront paleolatitudes with paleoclimatic proxies (Iglesia Llanos and Prezzi, 2013).

When taking into account marine invertebrates, it is known that bivalves are very sensitive to water temperatures. In the Southern Hemisphere (Figure 3) during the Hettangian-Sinemurian (∼200 Ma), the boundary between the South Pacific high latitudes and Tethyan low latitudes realms was located in northern Chile (Damborenea, 2001, 2002). This coincides with the time that the supercontinent was located at its southernmost position (Iglesia Llanos and Prezzi, 2013). By the end of the Sinemurian and throughout the Early Jurassic, the same boundary shifted to the south until the Toarcian, high-latitude bivalves became restricted to southernmost South America. At this time, the continent was located far off to the north (Figure 3). The Pliensbachian marked the first expansion of colonial corals (warm water temperatures) in west-central Argentina. Meanwhile in the Northern Hemisphere, average water temperatures indicated a major drop – from 18 °C to 14 °C – during the Pliensbachian–Toarcian boundary (Rosales et al., 2004). This drop was also recorded in the Arctic and was the likely cause of the significant turnover registered in Boreal marine invertebrates during this time. Moreover, the sub-Tethyan and Low Boreal bivalves that dwelt northward of 68° N present-day coordinates in the terminal Pliensbachian migrated southward to the latitude 55°N at the beginning of the Toarcian and returned to North Siberia in the Late Toarcian (Zakharov et al., 2006). Likewise, ostracods experienced an extinction episode during the Late Pliensbachian and Early Toarcian (Arias, 2009). On the other hand, ammonites of the Tethyan Realm kept moving to the north until the Late Pliensbachian, when the Boreal Realm commenced to shift southward replacing the Mediterranean fauna. Subsequently throughout the Late Pliensbachian and Earliest Toarcian, ammonites were affected by a major extinction which caused particularly the Tethyan species to disappear and be replaced in the Early Toarcian (Macchioni and Cecca, 2002), when the continent was located at its northernmost position (Figure 3).

The paleoflora, vertebrates, and geological records from both the Southern and Northern Hemispheres also are consistent with the paleolatitudes determined from the paleomagnetic record (Iglesia Llanos and Prezzi, 2013).

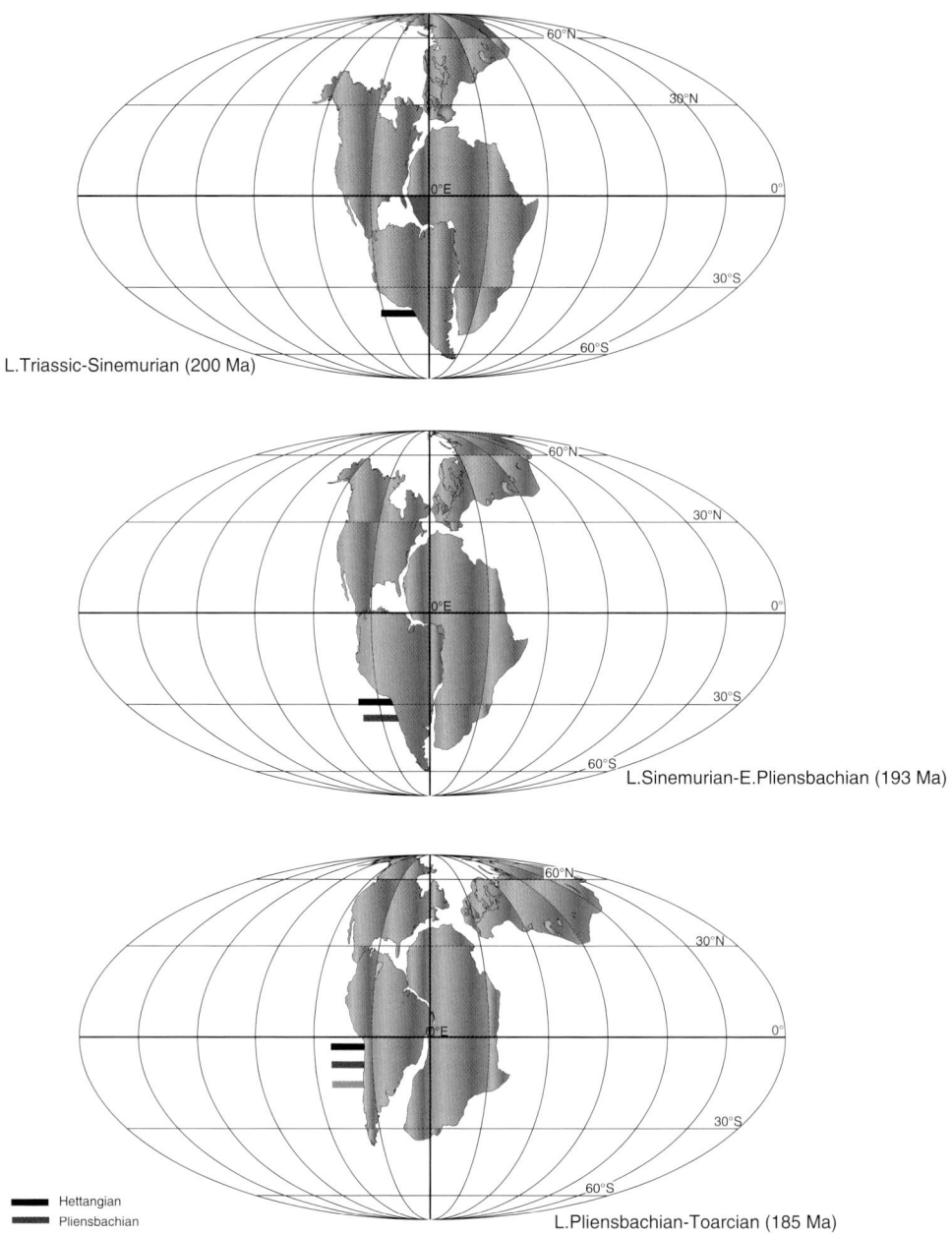

Paleomagnetism and Jurassic Paleogeography, Figure 3 "Absolute" paleogeographical reconstructions of Pangaea for the Early Jurassic. During the Late Triassic-Early Sinemurian, Pangaea was placed at its southernmost position. At this time, the boundary – solid thick line between the South Pacific (high latitudes) and Tethyan (low latitudes) realms of bivalves – was located in northern Chile (Damborenea, 2001, 2002). Subsequently and throughout the Early Jurassic, Pangaea moved northward and rotated CW, until the Late Pliensbachian-Early Toarcian when it reached the northernmost position. During this time, the boundary kept shifting toward the south until in the Early Toarcian; high-latitude bivalves became restricted to the southern extreme of the continent.

Conclusions

A master APW path for the Jurassic Pangaea was constructed using selected Eurasian, North American, African, and South American high-quality poles. The resultant path reveals a distinctive cusp at ~197 Ma and a subsequent standstill which seemed to persist until the Late Jurassic. The 197–185 Ma Early Jurassic track reveals a c. 50° shift in the pole positions during this time and a minimum angular change of poles of approximately $4°$ Myr^{-1}. This path clearly shows that Pangaea would have been subjected to a 50° CW rotation between the Sinemurian (~197 Ma) and Pliensbachian (~185 Ma), while it moved to the north. The motion of the supercontinent is interpreted to be primarily caused by TPW defined as the drift of the spin axis with respect to entire solid Earth, rather than other types of phenomenon such as lithospheric motion or an artifact produced by nondipole components. The c. 50° rotation is interpreted to result from the motion of the lithosphere AND the mantle. Estimated paleolatitudes are in good agreement with paleoclimatic proxies derived from fossil and geological records having a geographic distribution limited to certain climatic regions.

Bibliography

Åberg, G., Aguirre, L., Levi, B., and Nyström, J. O., 1984. Spreading-subsidence and 382 generation of ensialic marginal basins: an example from the early Cretaceous of central Chile. In Vokelaar, B. P., Howells, M. F. (eds.), *Marginal Basins*. Geological Society of London, Special Publication, 16, pp. 185–193.

Arias, C., 2009. Extinction pattern of marine Ostracoda across the Pliensbachian-Toarcian boundary in the Cordillera Ibérica, NE Spain: causes and consequences. *Geobios*, **42**, 1–15.

Beck, M. E., Jr., 1999. Jurassic and Cretaceous apparent polar wander relative to South America: some tectonic implications. *Journal of Geophysical Research*, **104**, 5063–5067.

Besse, J., and Courtillot, V., 2002. Apparent and true polar wander and the geometry of the geomagnetic field over the last 200 Myr. *Journal of Geophysical Research*, **107**, doi:10.1029/2000JB000050.

Cobbing, E. J., Pitcher, W. S., Wilson, J. J., Baldock, J. W., Taylor, W. P., McCourt, W., and Snelling, N. J., 1981. The geology of the western cordillera of northern Perú. *Institute of Geological Sciences, Overseas Memoir*, **5**, 1–143.

Cox, A., and Hart, R. B., 1986. *Plate Tectonics: How It Works*. Blackwell Scientific Publications, Palo Alto, Calif., 392 pp.

Creer, K. M., Irving, E., and Runcorn, S. K., 1954. The direction of the geomagnetic field in remote epochs in Great Britain. *Journal of Geomagnetism and Geoelectricity*, **6**, 163–168.

Damborenea, S. E., 2001. Unidades paleobiogeográficas marinas jurásicas basadas sobre moluscos bivalvos: una visión desde el Hemisferio Sur. *Anales Academia Nacional de Ciencias Exactas, Físicas y Naturales*, **53**, 141–160.

Damborenea, S. E., 2002. Jurassic evolution of Southern Hemisphere marine palaeobiogeographic units based on benthonic bivalves. *Geobios*, **35**, 51–71.

Evans, M. E., 1976. Test of the dipolar nature of the geomagnetic field throughout Phanerozoic time. *Nature*, **262**, 676–677.

Frei, L. S., and Cox, A., 1987. Relative displacement between Eurasia and North America prior to the formation of oceanic crust in the North Atlantic. *Tectonophysics*, **142**, 111–136.

Giambiagi, L., Bechis, F., García, V., and Clark, A., 2008. Temporal and spatial relationships of thin-skinned deformation: a case study from the Malargüe fold and thrust belt. *Southern Central Andes Tectonophysics*, doi:10.1016/j.tecto.2007.11.069.

Goldreich, P., and Toomre, A., 1969. Some remarks on polar wandering. *Journal of Geophyical Research*, **74**, 2555–2567.

Iglesia Llanos, M. P., and Prezzi, C. B., 2013. The role of true polar wander on the Jurassic palaeoclimate. *International Journal of Earth Sciences*, **102**, 745–759.

Jurdy, D. M., and van der Voo, R., 1975. True polar wander since the Early Cretaceous. *Science*, **187**, 1193–1196.

Kent, D.V., and Irving, E., 2010. Influence of inclination error in sedimentary rocks on the Triassic and Jurassic apparent pole wander path for North America and implications for Cordilleran tectonics. *Journal of Geophysical Research*, **115**, doi:10.1029/2009JB007205.

Klitgord, K. D., and Schouten, H., 1986. Plate kinematics of the central Atlantic. In Vogt, P. R., and Tulchoke, B. E. (eds.), *The Geology of North America, vol. M, The Western North Atlantic Region*. New York: Geological Society of America, pp. 351–378.

Lawver, L., and Scotese, C. R., 1987. A revised reconstruction of Gondwanaland. In McKenzie, G. D. (ed.), *Gondwana Six: Structure, Tectonics and Geophysics*. Washington, DC: American Geophysical Union Monograph, Vol. 40, pp. 17–23.

Livermore, R. A., Vine, F. J., and Smith, A. G., 1983. Plate motions and the geomagnetic field – I. Quaternary and late tertiary. *Geophysical Journal of the Royal Astronomical Society*, **73**, 153–171.

Livermore, R. A., Vine, F. J., and Smith, A. G., 1984. Plate motions and the geomagnetic field – II. Jurassic to tertiary. *Geophysical Journal of the Royal Astronomical Society*, **79**, 939–961.

Llanos, I., Riccardi, A. C., and Singer, S. E., 2006. Palaeomagnetic study of lower Jurassic marine strata from the Neuquén Basin, Argentina: a new Jurassic apparent polar wander path for South America. *Earth Planetary Science Letters*, **252**, 379–397.

Macchioni, F., and Cecca, F., 2002. Biodiversity and biogeography of middle-late liassic ammonoids: implications for the Early Toarcian mass extinction. *Geobios*, **35**, 165–175.

McElhinny, M. W., and Brock, A., 1975. A new palaeomagnetic result from East Africa and estimates of the Mesozoic palaeoradius. *Earth Planetary Science Letters*, **27**, 321–328.

Morgan, W. J., 1983. Hotspot tracks and the early rifting of the Atlantic. *Tectonophysics*, **94**, 123–139.

Mpodozis, C., and Allmendinger, R. W., 1993. Extensional tectonics, Cretaceous Andes, northern Chile (27°S). *Geological Society American Bulletin*, **105**, 1462–1477.

Oviedo, E., and Vilas, J. F., 1984. Movimientos recurrentes en el Permo-Triásico entre el Gondwana Occidental y el Oriental. In *Actas 9° Congreso Geológico Argentino*, Vol. 3, pp. 97–114.

Prévot, M., Mattern, E., Camps, P., and Daignières, M., 2000. Evidence for a 20° tilting of the Earth's rotation axis 110 million years ago. *Earth Planetary Science Letters*, **179**, 517–528.

Rosales, I., Quesada, S., and Robles, S., 2004. Paleotemperature variations of Early Jurassic seawater recorded in geochemical trends of belemnites from the Basque – Cantabrian basin, northern Spain. *Palaeogeography Palaeoclimatology Palaeoecology*, **203**, 253–275.

Sager, W. W., and Koppers, A. A. P., 2000. Late Cretaceous polar wander of the pacific plate: evidence of a rapid true polar wander event. *Science*, **287**, 455–459.

Somoza, R., and Zaffarana, C. B., 2008. Mid-Cretaceous polar standstill of South America, motion of the Atlantic hotspots and the birth of the Andean cordillera. *Earth Planetary Science Letters*, **271**, 267–277.

Torsvik, T. H., Dietmar Müller, R., Van der Voo, R., Steinberger, B., and Gaina, C., 2008. Global plate motion frames: toward a unified mode. *Reviews of Geophysics*, **46**, RG3004. 1–44.

Valencio, D. A., Vilas, J. F., and Pacca, I. G., 1983. The significance of the palaeomagnetism of Jurassic-Cretaceous rocks from South America: predrift movements, hairpins and Magnetostratigraphy. *Geophysical Journal of the Royal Astronomical Society*, **73**, 135–151.

Zakharov, V. A., Shurygin, B. N., Il'ina, V. I., and Nikitenko, B. L., 2006. Pliensbachian–Toarcian biotic turnover in North Siberia and the Arctic region. *Stratigraphy and Geological Correlation*, **14**, 399–417.

Cross-references

Paleophysiography of Ocean Basins
Paleomagmatism and Jurassic Paleography
Plate Tectonics

PALEOPHYSIOGRAPHY OF OCEAN BASINS

R. Dietmar Müller and Maria Seton
School of Geosciences, The University of Sydney,
Sydney, NSW, Australia

Synonyms
Paleogeography of Ocean Basins

Definition
Plate motions and the history of plate boundary geometries through time are the primary drivers for the large-scale paleophysiography of the ocean basins. These in turn determine the history of seafloor spreading and subduction, driving the time dependence of the age-area distribution of ocean floor. The depth of the ocean floor and volume of the ocean basins are primarily dependent on its age. Reconstructions of the age and depth distribution of the ocean floor combined with estimates of sediment thickness through time and the reconstruction of oceanic plateaus yield broad-scale paleophysiographic maps of the ocean basins.

Introduction
The paleophysiography of the ocean basins relies on an understanding of the current physiography of the oceans and the processes governing its development through geological time. The most fundamental parameter driving the depth distribution of ocean basins is the age of the oceanic lithosphere. The recognition that age and depth are linked via the cooling of oceanic lithosphere as newly formed crust moves away from mid-ocean ridges provides a framework for mapping ocean basin depth through time. Constructing a present-day physiographic map of the oceans is relatively straightforward due to the preservation of oceanic lithosphere. However, to generate complete oceanic physiographic maps in the past involves recreating ocean crust that has since been subducted. This becomes progressively more difficult back through time.

History
Turcotte and Oxburgh (1967) were the first to suggest that the subsidence of the seafloor away from mid-ocean ridges can be explained by the cooling of the lithosphere as a thermal boundary layer. Ten years later, a major analysis of the relationship between depth and age of the ocean floor (Parsons and Sclater, 1977) showed that the cooling of a simple rigid boundary layer can explain the depth–age relationship of crust younger than 80 Ma of age. For older ocean floor, the depth was shown to increase asymptotically to a constant value, i.e., to "flatten." The flattening of the depth implied the addition of extra heat under the older ocean floor. Parsons and Sclater (1977) showed that a plate of constant thickness gave a good approximation to this concept and gave birth to so-called plate models describing oceanic depth–age relationships. The severe computational restrictions during the 1970s compared to today's standards meant that the observational parameter space could not be efficiently explored for the best-fit solution. A formal best-fit solution was first found by Stein and Stein (1992) via their widely used GDH1 model (global depth and heat flow #1). More recently, Crosby and McKenzie (2009) suggested a revised age-depth curve by adding a damped sinusoidal perturbation to the flattening of seafloor older than 80 Ma. This approach is based on the observed sinusoidal shallowing of the reference depth–age curve for the North Pacific and to a lesser extent the North Atlantic between the ages of 80 and 130 Ma. They argued that this sinusoidal shallowing resembles the results of early numerical models where a surface boundary layer cools by conduction and then becomes unstable once its local Rayleigh number exceeds a critical value. The growing instability then suddenly increases as the base of the lithospheric boundary layer falls off and is replaced by hotter asthenosphere from below. The new material then cools again by conduction, until it in turn becomes unstable, resulting in a series of decaying oscillations about an asymptotic steady-state value. As well as these "thermal" models, there exist other age-depth models, which are categorized as either "dynamic" or "chemical" models. In dynamic models, increasing pressure in the asthenosphere prevents the seafloor from subsiding (Phipps Morgan and Smith, 1992). Chemical models take into account the compositional and rheological stratification on small-scale convection (Afonso et al., 2008) or consider the lateral flow and spreading of depleted, buoyant mantle residuum left over from the melt extraction process at mid-ocean ridges (Phipps Morgan et al., 1995). Despite the ongoing debate about the details of the depth–age relationship of relatively old ocean floor, models such as GDH1 are a good approximation to construct paleophysiographic ocean basin models. Other uncertainties such as the effect of large-scale mantle convection on oceanic depth (Spasojevic and Gurnis, 2012) and uncertainties in the elevation of oceanic plateaus and sediment thickness through time (Müller et al., 2008b) potentially play a large role in reconstructing the paleo-depth of ocean basins.

Reconstructing mid-ocean ridges and flanks
The reconstruction of the ocean basins relies on a global plate motion model to establish the location and geometry of mid-ocean ridges through time. This is achieved through marine magnetic anomaly identifications, geological information such as paleomagnetic data from terranes and microcontinents (especially in the Tethys Ocean), mid-ocean ridge subduction events, and the rules of plate tectonics (Cox and Hart, 1986). Uncertainties in reconstructing mid-ocean ridges and flanks grow progressively as you travel back through time as more and more mid-ocean ridge flanks are subducted and, hence, need to be recreated. Even though the details of now subducted mid-ocean ridge geometries cannot be known, reasonable

and conservative estimates can be made for their geometry, orientation, and location based on available geophysical and geological data while adhering to the rules of plate tectonics. Pre-Jurassic ocean basin physiographies have never been explicitly reconstructed to date.

Preserved magnetic lineations in the Pacific Ocean provide unequivocal evidence that a vast mid-ocean ridge system existed in the Pacific Ocean in the mid/late Cretaceous, significantly longer than today's ridge system, and much of which is now subducted – this was first recognized by Larson and Chase (1972). Subsequent detailed mapping and compilation of magnetic M-sequence anomalies in the northwestern Pacific Ocean (Nakanishi et al., 1992a) revealed the complete Mesozoic magnetic anomaly lineation pattern in this area and the origin of the Pacific plate as a triangularly shaped microplate at a triple junction between the Izanagi, Farallon, and Phoenix plates in the Panthalassa Ocean basin, the predecessor of the Pacific Ocean. This work forms the basis of Pacific Ocean reconstructions. In the southwest Pacific, the reconstructions from Seton et al. (2012) implement the opening of ocean basins between fragments of the Ontong Java–Manihiki–Hikurangi large igneous provinces (LIPs) (Viso et al., 2005; Taylor, 2006), a major tectonic event that was missing from previous reconstructions of the Cretaceous Pacific Ocean (Müller et al., 2008b). The Tethys Ocean, separating Laurasia from Gondwanaland, is reconstructed by accounting for the migration history of a series of Gondwanaland ribbon continents that were rifted off the northern margin of Gondwanaland and accreted to southern Eurasia in several stages. By using combined evidence from preserved magnetic lineations in the present-day Indian Ocean and geological data from the terranes that accreted to southern Eurasia and the rules of plate tectonics (Cox and Hart, 1986), it is possible to constrain the overall geometries of the Tethyan mid-ocean ridge system. The closure of the Mongol–Okhotsk Ocean between 200 and 150 Ma is modeled after van der Voo et al. (1999).

The overall physiography of the global oceans since the Mesozoic follows an alternating pattern mid-ocean ridge length increase and decrease as well as changes in the average age of the oceanic lithosphere. A number of old mid-ocean ridge flanks in Panthalassa and the Tethys oceans were gradually being destroyed between 200 and 150 Ma, while new mid-ocean ridge systems were initiated in a stepwise fashion between the late Jurassic and early Cretaceous, leading to a substantial younging and shallowing of the ocean basins from the Jurassic to the Cretaceous period. After the Cretaceous, the length of the mid-ocean ridge system decreased and was accompanied by an aging and deepening of the ocean basins toward the present day.

To produce a model of the age-area distribution of the ocean basins since the Mesozoic, we take the global set of plate rotations, reconstructed plate boundaries and seafloor magnetic anomalies, and fracture zones imbedded in the global plate model of Seton et al. (2012) and construct a set of seafloor spreading isochrons, equivalent to paleo-mid-ocean ridge geometries and locations. Seafloor isochrons are then gridded (Müller et al., 2008a) to produce smooth, continuous models for the age-area distribution of ocean floor through time (see "Plate Motion"). The resultant age grid becomes the primary input to producing models of the depth of the ocean basins through time, based on a selected depth–age relationship. To convert age to depth, we use the GDH1 model (Stein and Stein, 1992). Other recent depth–age models, e.g., Crosby and McKenzie (2009), result in an extremely similar mapping of age to depth (see Müller et al. (2008a) for details) and would make little difference for the reconstruction of ocean basin depths.

Oceanic plateaus and sediment cover

Two additional factors play a significant role in controlling ocean basin depth through time, namely, the generation of oceanic large igneous provinces (LIPs) and oceanic sedimentation. To account for the effect of LIPs on oceanic depth, we take a subset of the Coffin et al. (2006) dataset to include LIPs that were generated by a transient plume head as we are most interested in capturing the voluminous volcanism rather than the small scale. We use Schubert and Sandwell's (1989) method to determine the average elevation of major oceanic plateaus relative to the surrounding seafloor, based on the difference between the modal depths within two polygons, one outlining the perimeter of a given oceanic plateau and another larger polygon including surrounding ocean floor (Figure 1). Individual plateaus are included in basement depth grids at their emplacement time based on a compilation of eruption ages (Müller et al., 2008b). Where available, the paleo-depth estimates of LIPs were cross-referenced with published data (e.g., back-stripping from well data or from seismic data).

As the ocean floor ages, the thickness of its sediment cover increases, but abyssal sediment thickness is also latitude dependent, as illustrated in a polynomial surface fit of global sediment thickness as a function of oceanic crustal age and latitude (Figure 2). The observed latitudinal sediment thickness variation primarily reflects differences in productivity between zones of tropical upwelling and an increase in sediment thickness toward high latitudes, reflecting terrigenous runoff. The simple relationship provides a good first approximation for reconstructing an estimate for regional oceanic sediment thickness through time. We isostatically correct for sediment loading based on Sykes (1996).

Paleophysiography of Mesozoic and Cenozoic ocean basins

Overview

Broad-scale paleo-bathymetries of the ocean basins can be constructed by combining oceanic basement depths derived from a depth–age relationship with oceanic plateaus and sediment thickness estimates (Figure 3). The

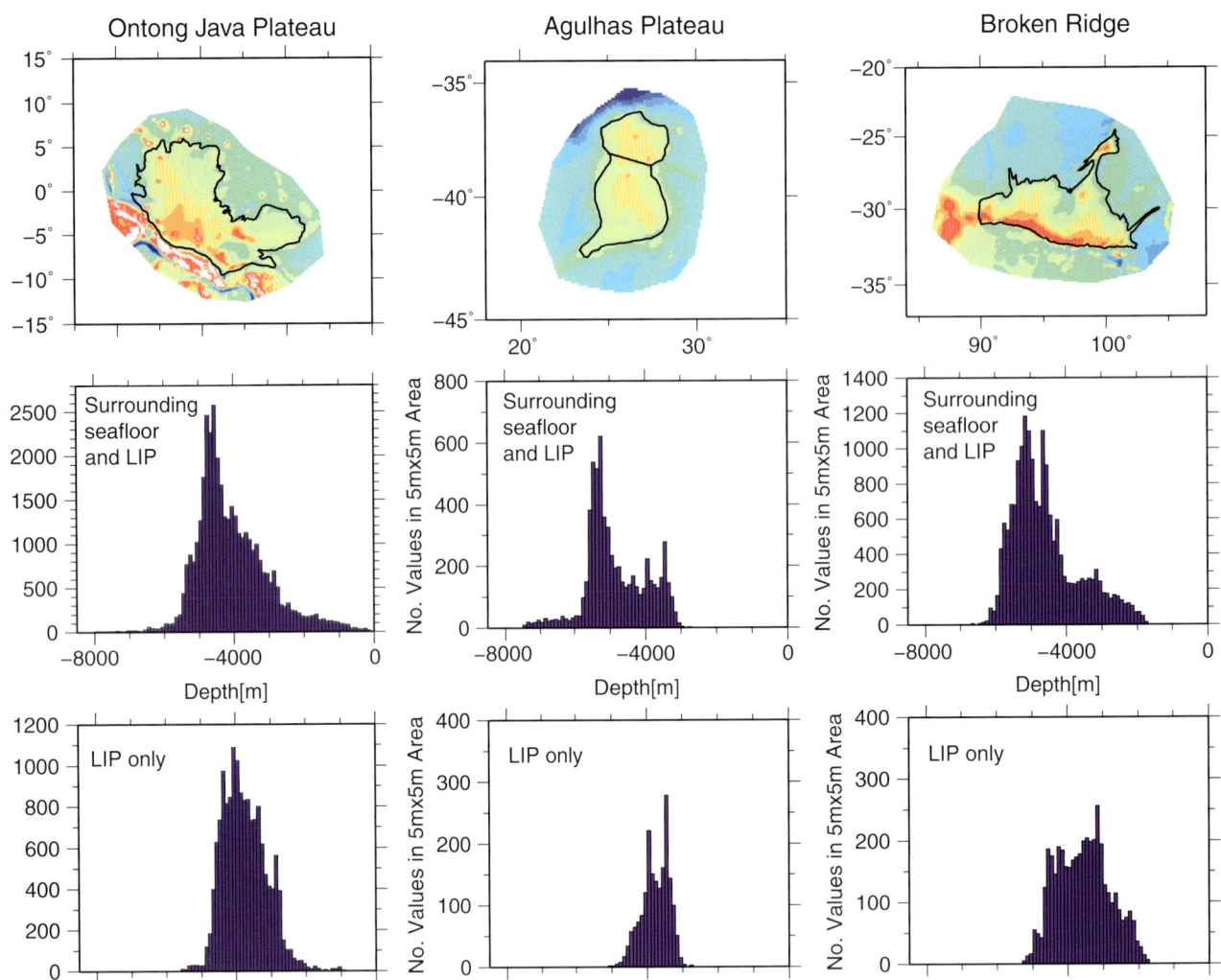

Paleophysiography of Ocean Basins, Figure 1 Accounting for the eruption and subsidence of LIPs through time is an important aspect of paleo-depth models of the ocean floor. We show three examples of LIPs (Ontong Java and Agulhas Plateaus and Broken ridge) and the method applied to predict LIP depth. *Top panel* shows present-day bathymetry from ETOPO with *black outline* indicating the outline of the LIP. *Middle panel* shows the depths of the seafloor surrounding the LIP as well as the LIP itself per 5 × 5 m area. *Bottom panel* shows depths for the LIP only.

paleo-bathymetries presented here, spanning the period from the early Jurassic (200 Ma) to present day, focus on the physiogeography of the main ocean basins, i.e., mid-ocean ridges, their flanks, and abyssal plains, but exclude detailed depth estimates of passive continental margins and submerged continental plateaus and microcontinents. These depths in these regions are difficult to derive as they require a global compilation of well and seismic data to reconstruct the regional tectonic subsidence and sedimentary history, in areas where data coverage is generally sparse.

200 Ma

At 200 Ma, the supercontinent Pangaea is still intact and is surrounded by the Panthalassa "super-ocean" and the Tethys ocean basin (Figure 3a). Panthalassa primarily consists of ocean floor that formed by seafloor spreading between the Izanagi, Farallon, and Phoenix plates. The Cache Creek Ocean exists along the northeastern margin of Panthalassa, adjacent to the North American continent. This smaller ocean basin owes its name to the Cache Creek Terrane, a mid-Paleozoic to mid-Jurassic oceanic terrane with exotic Permian Tethyan faunas in limestone blocks and long-lived island edifices (Nelson and Mihalynuk, 1993; Mihalynuk et al., 1994). At this time, the Tethys Ocean is divided into the northern paleo-Tethys, a narrow abyssal seaway, which is a remnant of a once vast ocean basin whose mid-ocean ridge was previously subducted, and the southern meso-Tethys, which formed by means of a number of continental blocks and slivers rifting off northern Gondwanaland, separating the paleo- from the meso-Tethys by a continental barrier

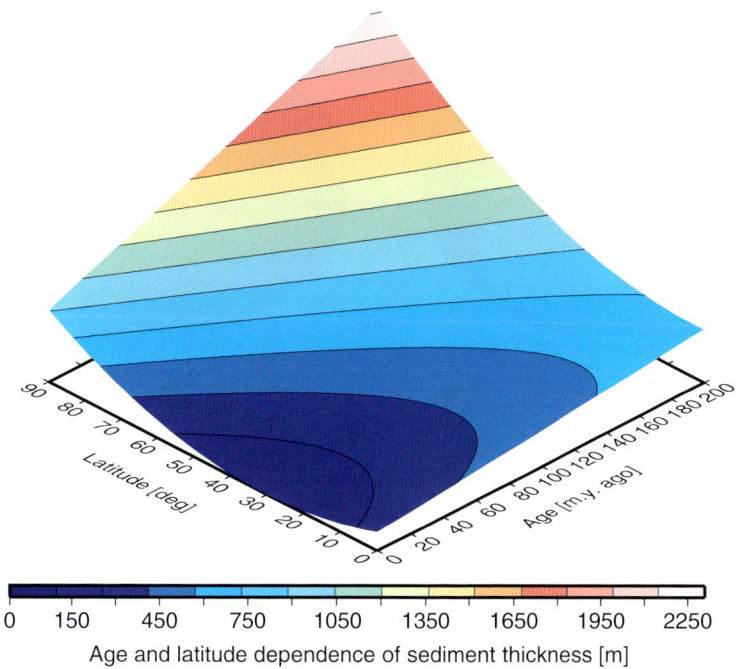

Paleophysiography of Ocean Basins, Figure 2 Sediment thickness as function of latitude and crustal age. The observed latitudinal sediment thickness variation primarily reflects differences in productivity between zones of tropical upwelling and an increase in sediment thickness toward high latitudes, reflecting terrigenous runoff. Regional marine sediment thickness also depends on many other factors, including depth of carbonate compensation depth and regional changes in terrigenous flux. The above relationship provides a good approximation for our aim to reconstruct an estimate for mean oceanic sediment thickness through time.

referred to as the Cimmerian terrane. The detailed configuration of the Cimmerian terrane is unknown. In addition, there is evidence for the existence of the Mongol–Okhotsk Basin in central Asia (Figure 3a) from a series of remnant island arc volcanics and ophiolites adjacent to the suture zone as well as a large volume of seismically fast material in the lower mantle underlying Siberia, imaged in seismic tomography, and representing subducted Mongol–Okhotsk Ocean slabs (Van der Voo et al., 1999). The initiation of breakup of Pangaea is focused in the central North Atlantic, where seafloor spreading may have initiated as early as 200 Ma (Labails et al., 2010).

180 Ma

At 180 Ma the configuration of the global oceans becomes simplified as they are nearly entirely dominated by Panthalassa and the meso-Tethys, with all smaller ocean basins reduced to narrow remnant strips of deep ocean crust. The closure of the Cache Creek Ocean is tightly constrained to around 172–174 Ma (Colpron et al., 2007) and references therein. The closure of the paleo-Tethys and accretion of the Cimmerian terrane occurred along the southern Laurasian margin at 170 Ma in this model (Figure 3b, c). This time also sees the birth of the Pacific plate, visible as a triangular area in Figure 3b in the center of Panthalassa. The "Pacific triangle" is an area of the western Pacific where three Mesozoic magnetic lineation sets (Japanese, Hawaiian, and Phoenix lineations) intersect, recording the birth of the Pacific plate at a triple junction from three parent plates: the Farallon, Izanagi, and Phoenix plates. The northwestern (Japanese) lineations represent spreading between the Pacific and Izanagi plates and young toward the west–northwest, the easternmost (Hawaiian) lineations represent spreading between the Pacific and Farallon plates and young toward the east, and the southernmost (Phoenix) lineations represent spreading between the Pacific and Phoenix plates and young toward the south (Atwater, 1970; Nakanishi et al., 1992b). These three plates existed prior to the establishment of the Pacific plate in a simple ridge–ridge–ridge configuration (Figure 3a).

160 Ma

Rapid growth of the Pacific plate is recorded during this time with a gradual increase in the spreading rate. Seafloor spreading is well underway in the central North Atlantic, signifying the breakup of Pangaea into Laurasia and Gondwanaland (Figure 3c). The earliest onset of breakup between North America and Africa has recently been redated at around 200 Ma (Labails et al., 2010) coincident with ultraslow seafloor spreading but with an increase in spreading rate from 170 Ma. Further south, the deformation within South America associated with the breakup of the South Atlantic is believed to have initiated at around

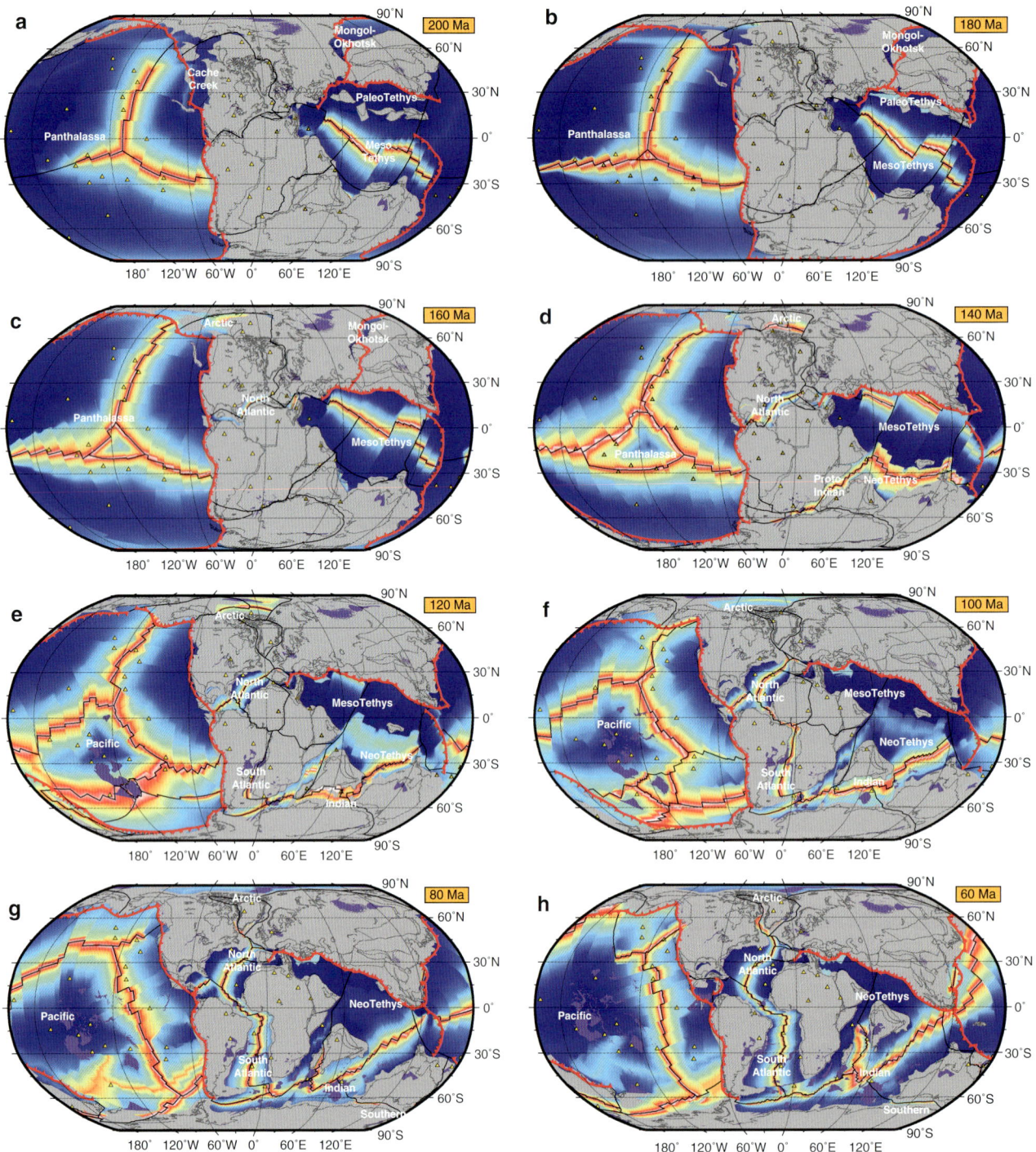

Paleophysiography of Ocean Basins, Figure 3 (continued)

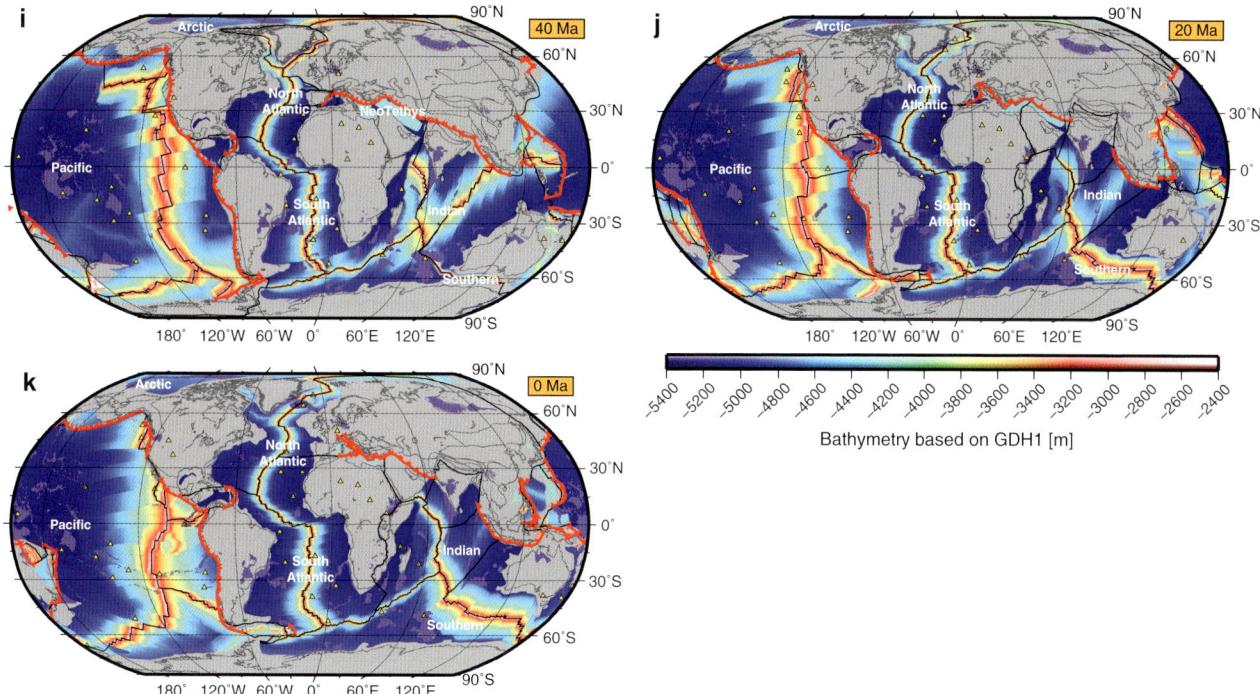

Paleophysiography of Ocean Basins, Figure 3 Physiogeography of ocean basins through time from 200 Ma to present in 20 Ma intervals. Basemap shows predicted oceanic depth, *gray* denote non-oceanic areas. *Blue* (i.e., *deep*) areas mark the abyssal plains. *Red* (*shallow*) areas mark the mid-ocean ridge system. *Black lines* show plate boundaries and *yellow triangles* are present-day hotspot locations.

150 Ma (Torsvik et al., 2009). The Mongol–Okhotsk Ocean is almost entirely closed, with complete closure dated at around 150 Ma based on the cessation of compression in the suture zone region (Zorin, 1999).

140 Ma

This reconstruction illustrates the initial formation of the proto-Caribbean between North and South America as a consequence of the divergence between Laurasia and Gondwanaland. In the Arctic Ocean, the Canada Basin started forming via counterclockwise rotation of the North Slope of Alaska away from the northern Canadian margin (Halgedahl and Jarrard, 1987), with seafloor spreading starting at 142 Ma (Alvey et al., 2008). The Canada Basin spreading ridge is connected with the North Atlantic rift zone, which in turn connects to the northern Tethyan subduction zone and the central North Atlantic spreading ridge (Figure 3d). The neo-Tethys started forming prior to 140 Ma, best documented by the separation of Argoland from the northwest shelf of Australia around 155 Ma (Gibbons et al., 2012). The neo-Tethyan mid-ocean ridge likely wrapped around northern India and is assumed to have connected with seafloor spreading in the Somali Basin (Figure 3d). This incipient spreading system continued south into the Mozambique Basin, the Riser-Larson, and the Weddell seas, marking the fragmentation of Gondwanaland. The southwest Panthalassic margin, along eastern Australia, involved the opening of the South Loyalty Basin, due to rollback of the southwest Panthalassic subduction zone from 140 Ma.

120 Ma

This time marks a significant increase in seafloor spreading rates in Panthalassa corresponding to the mid-Cretaceous seafloor spreading pulse, with spreading between the Pacific, Farallon, Izanagi, and Phoenix plates. The breakup of the Phoenix plate occurred due to the eruption of a suite of LIPs, most notably the Ontong Java, Manihiki, and Hikurangi Plateaus, which originally formed one giant oceanic plateau (Taylor, 2006) (Figure 3e). The eruption of this mega-LIP led directly to the formation of the Hikurangi, Manihiki, Chasca, and Catequil plates (Figure 3e). Additional two triple junctions were active in the region leading to the breakup of the eastern Manihiki Plateau and the development of the Tongareva triple junction. The South Loyalty Basin off northeast Australia was actively opening until 120 Ma, when it was halted by a major change in the plate configurations in the SW Panthalassic Ocean. After a landward ridge jump of the neo-Tethys ridge at 135 Ma, the mid-ocean ridge propagated southward to open the Gascoyne, Cuvier, and Perth Abyssal Plains between India and Australia (Gibbons et al., 2012). The West Australian spreading ridge system joined with the Enderby

Basin spreading ridge, separating Antarctica from the Elan Bank/India, to the west and to the rift between Australia and Antarctica to the east. Spreading extended southward along the South Atlantic ridge with a northward propagation leading to seafloor spreading in the "central" segment by 120 Ma and in the "equatorial" segment by 110 Ma. Breakup between Iberia and North America occurred around 110 Ma. The North Atlantic rift zones connected with the spreading center in the Canada Basin until about 118 Ma, marking the cessation of seafloor spreading in the Canada Basin. Spreading terminated when the rotation of North Slope Alaska ceased, coincident with a change in the southern North Slope margin from largely strike-slip to convergence due to a change in spreading direction in Panthalassa.

100 Ma

The mid and South Atlantic ridges were well established from 100 Ma and spread along the central North Atlantic ridge and continued into the proto-Caribbean Sea until this time (Figure 3f). Evidence from the seafloor fabric for a global plate motion change is abundant between 105 and 100 Ma, with either fracture zone bends or terminations preserved in all ocean basins (Matthews et al., 2012). For example, a change in spreading direction is recorded in the Mendocino, Molokai, and Clarion fracture zones (associated with Pacific–Farallon spreading), which is dated to 103–100 Ma (Seton et al., 2012) coincident with an observed bend in the hotspot trails on the Pacific plate, suggesting a plate reorganization at this time. A major change in spreading direction is also recorded in fracture zone trends in the Indian Ocean around 100 Ma (Gibbons et al., 2012). Spreading between India and Australia subsequently became dominantly N–S directed, establishing the Wharton Basin. The West Australian mid-ocean ridge system formed a triple junction with the Australian–Antarctic ridge around 100 Ma (initiation of ultraslow seafloor spreading) and spreading between India and Antarctica north of the Elan Bank microcontinent (Figure 3f). The Indian–Antarctic ridge (or southeast Indian ridge) connected with the African–Antarctic ridge (or southwest Indian ridge) from 100 Ma.

80 Ma

The Pacific was dominated by the breakup of the Farallon plate into the Kula plate with seafloor spreading initiating at 79 Ma along the E–W trending Kula–Pacific ridge and the NE–SW trending Kula–Farallon ridge (Figure 3g). The Kula–Farallon ridge follows the location of the Yellowstone hotspot and intersects the North American margin in Washington/British Columbia before migrating northward along the margin. At 86 Ma, the Hikurangi Plateau docked with the Chatham Rise triggering a cessation in spreading associated with the Ontong Java, Manihiki, and Hikurangi Plateaus. After the cessation of spreading along these ridge axes, the locus of extension jumped southward between Antarctica and the Chatham Rise, establishing the Pacific–Antarctic spreading ridge (Figure 3g). To the east, the Pacific–Farallon ridge extended to the south connecting with the Pacific–Antarctic ridge at the Pacific–Antarctic–Farallon triple junction. After the cessation of the spreading centers associated with the Pacific mega-LIP formation and breakup, the Pacific plate became the dominant plate in Panthalassa, and it is at this time that Panthalassa morphs into the Pacific Ocean.

In the Caribbean, spreading in the proto-Caribbean Sea ceased at 80 Ma, whereas the Caribbean Arc subduction zone continued its northeastward rollback. The rollback of this subduction zone along the Caribbean Arc led to the consumption of the actively spreading proto-Caribbean ocean floor and encroachment of the Farallon plate into the Caribbean domain (Figure 3f, g). The continued rollback of the Caribbean Arc subduction zone led to the formation of the Yucatan Basin as a back arc in the late Cretaceous with cessation occurring at 70 Ma when the Caribbean Arc accreted to the Bahaman Platform. The accretion led to a jump in the locus of subduction westward along the newly developed Panama–Costa Rica to accommodate the continued eastward motion of the Farallon plate, trapping Farallon oceanic lithosphere onto the Caribbean plate.

Rifting between India and Madagascar in the Mascarene Basin initiated at 87 Ma (Gibbons et al., 2013). The southwest Indian ridge connected with spreading in the Malvinas plate in the southernmost Atlantic at 83.5 Ma and the American–Antarctic ridge (established after the cessation of spreading in the Weddell Sea) (Figure 3g). The mid-Atlantic ridge propagated northward into the North Atlantic with the initiation of seafloor spreading in the Labrador Sea (between North America and Greenland) and between the Rockall Plateau and Greenland at 79 Ma. Spreading propagated from the Labrador Sea to Baffin Bay by 63 Ma across the Davis Straits via left-lateral transform faults and connected to the Arctic via the Nares Strait.

60 Ma

In the Pacific, the Pacific–Izanagi ridge started to subduct under the East Asian margin between 55 and 50 Ma, signaling the death of the Izanagi plate coincident with a dramatic change in spreading direction from N–S to NW–SE between Kula–Pacific spreading. The Kula–Pacific ridge connected with the Pacific–Farallon ridge and Kula–Farallon ridge from 60 to 55 Ma (Figure 3h). After 55 Ma, the eastern Pacific was dominated by the rupture of the Farallon plate close to the Pioneer fracture zone, forming the Vancouver plate. The breakup resulted in minor relative motion along the Pioneer fracture zone. Further south, spreading continued along the Pacific–Farallon, Pacific–Antarctic, Farallon–Antarctic, and Pacific–Aluk ridges (Figure 3h). The fracture zones associated with the Pacific–Antarctic ridge close to the Campbell Plateau record a change in spreading direction at

55 Ma, coincident with other events that occurred in the Pacific at this time.

In the southern Pacific, spreading continued along the Pacific–Antarctic ridge, extending eastward to connect with the Pacific–Farallon and Farallon–Antarctic spreading ridges. At 67 Ma, a change in spreading direction is recorded in the fracture zones of the South Pacific initiating the opening of the Aluk plate in the South Pacific. In the Indian Ocean, spreading was occurring along the Wharton ridge, SEIR, SWIR, and in the Mascarene Basin (Figure 3h). Spreading in the Mascarene Basin ceased at 64 Ma jumping northward, isolating the Seychelles microcontinent, and initiating spreading between India and the Seychelles along the Carlsberg ridge. The SWIR connected with spreading in the Malvinas plate until 66 Ma. After this, the SWIR connected directly with the American–Antarctic and South Atlantic ridge. Seafloor spreading propagated into the Eurasia–Greenland margin along the Reykjanes ridge by 58 Ma, forming a triple junction between North America, Greenland, and Eurasia (Figure 3h). The Jan Mayen microcontinent rifted off the margin forming the fan-shaped Norway Basin along the Aegir ridge. The Aegir ridge connected to the Mohns ridge to the north and Reykjanes ridge to the south via a series of transform faults. Spreading in the Eurasian Basin to the north initiated around 55 Ma along the Gakkel/Nansen ridge.

40 Ma

In the Pacific, spreading between the Kula–Pacific and Kula–Farallon ceased at 42 Ma, leading to the Pacific plate consisting of the Pacific, Vancouver, Farallon, Aluk, and Antarctic plates (Figure 3i). The intersection of the Murray transform fault with the North American subduction zone around 30 Ma led to the establishment of the San Andreas Fault and corresponds to the establishment of the Juan De Fuca plate at the expense of the Vancouver plate.

In the western Pacific, spreading in the proto-South China Sea ceased at 50 Ma coincident with the clockwise rotation of the neighboring Philippine Sea plate. The dramatic change in motion of the Philippine Sea plate reorganized the plate boundaries in the area leading to the establishment of a subduction zone between Palawan and the proto-South China Sea and the subduction of the proto-South China Sea after 50 Ma (Figure 3i). Further south, spreading initiated in the North Loyalty Basin behind the proto-Tonga–Kermadec Trench. Spreading in the West Philippine Basin ceased at 38 Ma, whereas spreading continued in the Celebes Sea. The formation of the Caroline Sea occurred behind a rapidly southward migrating subduction zone. By 30 Ma, spreading initiated in the Shikoku and Parece Vela Basins behind the west-dipping Izu–Bonin–Mariana Arc. Spreading terminated in the Celebes Sea. In the SW Pacific, spreading initiated in the Solomon Sea at 40 Ma and in the South Fiji Basin at 35 Ma.

The Indian Ocean was dominated by a series of mid-ocean ridges such as the Wharton ridge, SEIR, SWIR, and Carlsberg ridge (Figure 3i). Prior to 55 Ma, subduction was occurring along the Tethyan subduction zone, consuming crust that formed during meso- and neo-Tethys spreading. At 55 Ma, the northern tip of Greater India marks the start of collision between India and Eurasia and the uplift of the Himalayas. Closure of the Tethys Ocean in this area occurred by about 43 Ma. Full closure of the neo-Tethys between India and Eurasia also corresponds to the cessation of spreading in the Wharton Basin, which describes Australia–India motion.

20 Ma to present day

In the Pacific, spreading continued along the Pacific–Juan De Fuca, Pacific–Nazca, Pacific–Cocos, Cocos–Nazca, Pacific–Antarctic, and Nazca–Antarctic ridges (Figure 3j, k). The Bauer microplate formed along the East Pacific Rise at 17 Ma and continued until 6 Ma. The locus of spreading then jumped back to the East Pacific Rise (between the Pacific and Nazca plates). The East Pacific Rise is the fastest spreading ridge system (excluding back-arc opening) and currently encompasses microplate formation at the Easter, Juan Fernandez, and Galapagos plates. Currently, the Juan De Fuca plate is limited at its southern end by the Mendocino Fracture Zone and is subducting slowly along the Cascadia subduction zone.

The western Pacific is dominated by the opening of a series of back-arc basins due to the rollback of the subduction hinge of the Tonga–Kermadec and Izu–Bonin–Mariana trenches (Figure 3j, k). Spreading in the Shikoku and Parece Vela Basins and South China Sea ceased at 15 Ma. By 9 Ma, spreading initiated in the Mariana Trough. We model complete closure of the proto-South China Sea at around 10 Ma behind a subduction zone located along Palawan and the North Borneo/Kalimantan margin. In the SW Pacific, spreading in the Lau Basin initiated by 7 Ma with back-arc extension occurring in the Havre Trough.

A further rupture of the Farallon plate occurred at 23 Ma leading to the establishment of the Cocos and Nazca plates and initiation of the East Pacific Rise, Galapagos Spreading Center, and Chile ridge. Cessation of spreading in the South Fiji Basin occurred at 25 Ma.

In the Indian Ocean, spreading continued along the SWIR, SEIR, CIR, and Carlsberg ridge (Figure 3j, k). Extension along the East Africa rifts was established at 30 Ma leading to the breakup of Africa into Somalia plate. Rifting along the Sheba ridge separating Arabia from Africa/Somalia initiated at 30 Ma. A zone of diffuse deformation occurring in the middle of the Indo-Australian plate led to the development of the Capricorn plate in the central-east Indian Ocean at 20 Ma. Further west, spreading is initiated along the Sheba ridge at 20 Ma. The Sheba ridge propagated into the Red Sea at 15 Ma.

Spreading in the South and North Atlantic continued unabated, with the westward motion of the North

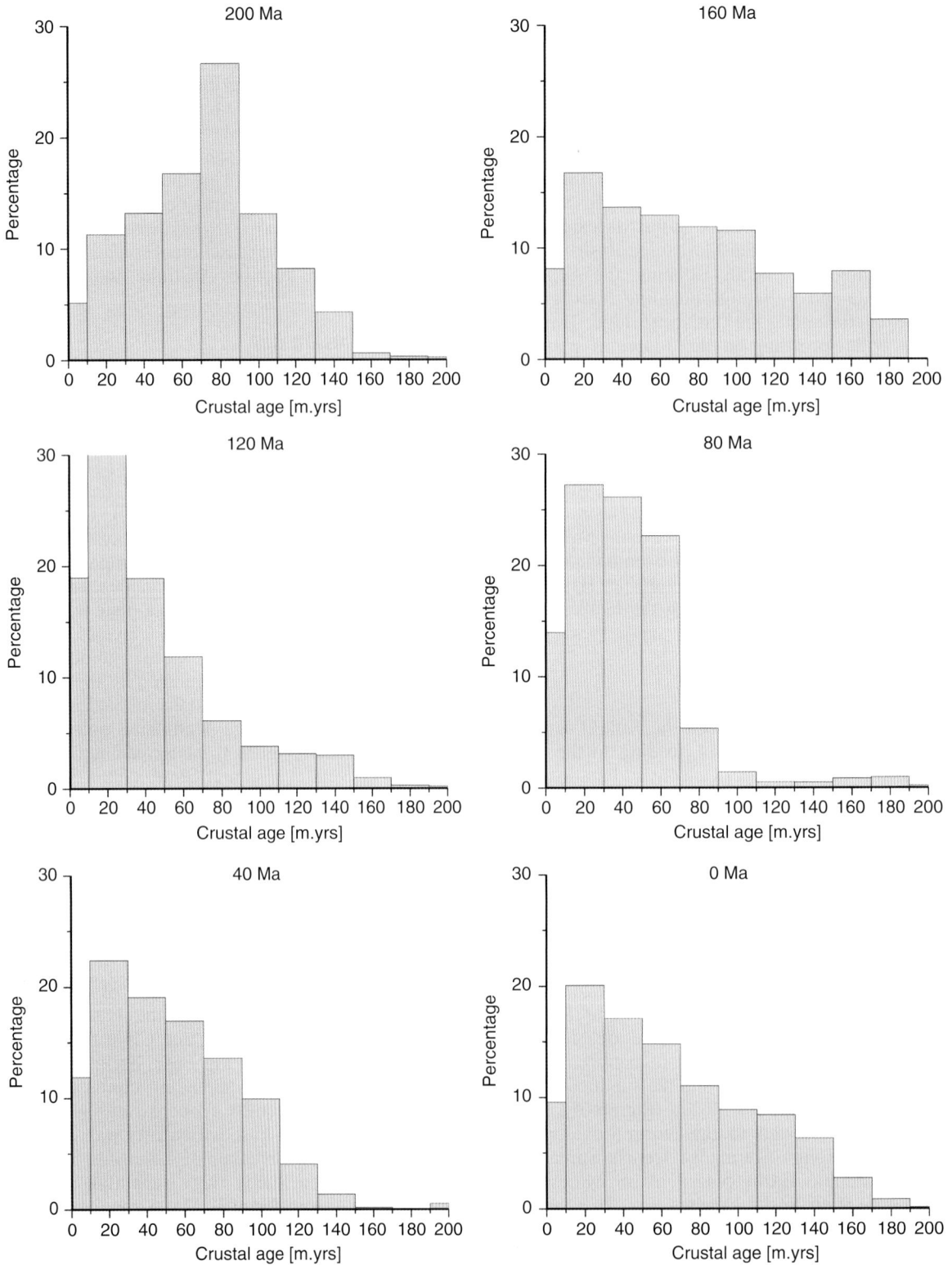

Paleophysiography of Ocean Basins, Figure 4 Age-area histograms of ocean basins from 200 Ma to the present day in 40 Ma intervals. The histograms reflect changes in the average age of the ocean basins through time, from a generally older ocean floor at the time of Pangaea breakup to generally young seafloor in the Cretaceous corresponding to the Cretaceous seafloor spreading pulse, toward an aging and deepening of the ocean basins toward the present day.

American plate relative to the slow moving Caribbean plate causing the opening of the Cayman Trough (Figure 3j, k). Subduction of Atlantic Ocean crust is observed along the Lesser Antilles Trench.

Ocean gateways

The present-day oceanic realm is divided into numerous large ocean basins, the Pacific being the largest. Each of these ocean basins is surrounded by continental areas and connected to neighboring oceans via narrow, shallow, and diffuse connections, called oceanic gateways. These oceanic gateways are focal points for the large-scale exchange of heat, water, salinity, and genes between the oceans and play a major role in influencing global climate, ocean circulation, and species diversity. The most well-known examples of ocean gateways include the Tasman Gateway, Drake Passage, Fram Strait, Panama Isthmus, Greenland–Scotland ridge, Tethyan Seaway, and Indonesian Throughflow. Their shape, size, and depth change over geological timescales and are largely controlled by large-scale lateral and vertical motions of the many blocks and basins that form during protracted periods of continental breakup and extension (or collision, uplift, and shortening during oceanic gateway closure). Global ocean basin reconstructions, such as those presented in Figure 3, provide a broad basis for assessing the opening and closure of deep-water ocean gateways through time and thus how plate tectonic movements influence global climate and oceanography. The physiographic reconstructions presented here highlight the timing of separation or amalgamation of continental blocks, assuming these areas to be rigid features. However, we know that extensive deformation occurs over protracted periods at plate boundaries. Therefore, the detailed evolution of many gateways such as the Drake Passage, the Tasman–Antarctic Passage, and the Panama gateway is dependent on understanding the subsidence history of submerged fragments of continental crust or volcanic arc fragments (Lawver et al., 2011). These are not included in the reconstructions presented in Figure 3.

Ocean basin physiography and the age-area distribution of ocean floor through time

Ocean basin reconstructions from 200 Ma to the present imply that old mid-ocean ridge flanks in the proto-Pacific and the Tethys oceans are gradually destroyed between 200 and 150 Myr ago, while new mid-ocean ridge systems are initiated in a stepwise fashion between the late Jurassic and early Cretaceous. This cycle leads to a substantial younging of the ocean basins from the Jurassic to the Cretaceous period, followed by a gradually aging of the oceanic lithosphere from the Cretaceous seafloor spreading pulse to the present day (Figure 4). The age-area distribution of ocean crust from 200 Ma to the present day implies a mid/late Cretaceous sea-level high linked directly to the transition from supercontinent stability to initial and mature dispersal, first creating many new mid-ocean ridges and flanks, which later experience stepwise subduction (Müller et al., 2008b). These cycles can also be linked with long-term fluctuations in seawater chemistry, leading to alternations between so-called aragonite and calcite seas (Müller et al., 2013).

Summary and conclusions

Complete reconstructions of the physiogeography of the ocean basins, starting with the breakup of Pangaea, have only been achieved recently for the first time, reflecting the difficulties involved in recreating now vanished ocean basins and assimilating many types of geological and geophysical observations into complete models through time. Although these ocean physiogeographic reconstructions are limited as they only account for broad-scale processes, they do provide an essential framework for creating boundary conditions for understanding ocean circulation, sediment accumulation, and ocean chemistry through time, as well as paleoclimate and long-term sea-level fluctuations. Further studies on refining global plate motions within a deforming plate framework as well as accounting for vertical motions associated with areas of continental breakup and collision will lead to more accurate reconstructions of ocean basins.

Bibliography

Afonso, J., Zlotnik, S., and Fernández, M., 2008. Effects of compositional and rheological stratifications on small-scale convection under the oceans: Implications for the thickness of oceanic lithosphere and seafloor flattening. *Geophysical Research Letters*, **35**, L20308.

Alvey, A., Gaina, C., Kusznir, N., and Torsvik, T., 2008. Integrated crustal thickness mapping and plate reconstructions for the high Arctic. *Earth and Planetary Science Letters*, **274**(3–4), 310–321.

Atwater, T., 1970. Implications of plate tectonics for the Cenozoic tectonic evolution of western North America. *Geological Society of America Bulletin*, **81**, 3513–3536.

Coffin, M. F., Duncan, R. A., Eldholm, O., Fitton, J. G., Frey, F. A., Larsen, H. C., Mahoney, J. J., Saunders, A. D., Schlich, R., and Wallace, P. J., 2006. Large igneous provinces and scientific ocean drilling: status quo and a look ahead. *Oceanography*, **19**(4), 150–160.

Colpron, M., Nelson, J. A. L., and Murphy, D. C., 2007. Northern Cordilleran terranes and their interactions through time. *GSA Today*, **17**(4), 4–10.

Cox, A., and Hart, B. R., 1986. *Plate Tectonics: How It Works*. Boston: Blackwell Science Inc. 400 pp.

Crosby, A. G., and McKenzie, D., 2009. An analysis of young ocean depth, gravity and global residual topography. *Geophysical Journal International*, **178**(3), 1198–1219.

Gibbons, A. D., Barckhausen, U., van den Bogaard, P., Hoernle, K., Werner, R., Whittaker, J. M., and Müller, R. D., 2012. Constraining the Jurassic extent of Greater India: tectonic evolution of the West Australian margin. *Geochemistry, Geophysics, Geosystems*, **13**, Q05W13.

Gibbons, A. D., Whittaker, J. M., and Dietmar Müller, R., 2013. The breakup of East Gondwana: assimilating constraints from Cretaceous ocean basins around India into a best-fit tectonic model. *Journal of Geophysical Research*, **118**, 808–822.

Halgedahl, S., and Jarrard, R., 1987. Paleomagnetism of the Kuparuk River formation from oriented drill core: evidence for rotation of the North Slope block. In Tailleur, I. L., and Weimer, P. (eds.), *Alaskan North Slope Geology*. Los Angeles: Society of Economic Paleontologists and Mineralogists, Pacific Section, pp. 581–617.

Labails, C., Olivet, J., Aslanian, D., and Roest, W., 2010. An alternative early opening scenario for the Central Atlantic Ocean. *Earth and Planetary Science Letters*, **297**, 355–368.

Larson, R. L., and Chase, C. G., 1972. Late Mesozoic evolution of the western Pacific. *Geological Society of America Bulletin*, **83**, 3627–3644.

Lawver, L. A., Gahagan, L. M., and Dalziel, I. W. D., 2011. A different look at gateways: Drake Passage and Australia/Antarctica. In Anderson, J. B., and Wellner, J. S. (eds.), *Tectonic, Climatic, and Cryospheric Evolution of the Antarctic Peninsula*. Washington, DC: AGU, Vol. 63, pp. 5–33.

Matthews, K. J., Seton, M., and Müller, R. D., 2012. A global-scale plate reorganization event at 105–100 Ma. *Earth and Planetary Science Letters*, **355**, 283–298.

Mihalynuk, M. G., Nelson, J. A., and Diakow, L. J., 1994. Cache Creek terrane entrapment: oroclinal paradox within the Canadian Cordillera. *Tectonics*, **13**(3), 575–595.

Müller, R. D., Sdrolias, M., Gaina, C., and Roest, W. R., 2008a. Age, spreading rates, and spreading asymmetry of the world's ocean crust. *Geochemistry, Geophysics, Geosystems*, **9**, Q04006, doi:10.1029/2007GC001743.

Müller, R. D., Sdrolias, M., Gaina, C., Steinberger, B., and Heine, C., 2008b. Long-term sea level fluctuations driven by ocean basin dynamics. *Science*, **319**(5868), 1357–1362.

Müller, R. D., Dutkiewicz, A., Seton, M., and Gaina, C., 2013. Seawater chemistry driven by supercontinent assembly, break-up and dispersal. *Geology*, **41**, 907–910.

Nakanishi, M., Tamaki, K., and Kobayashi, K., 1992a. Magnetic anomaly lineations from Late Jurassic to Early Cretaceous in the west central Pacific Ocean. *Geophysical Journal International*, **109**(3), 701–719.

Nakanishi, M., Tamaki, K., and Kobayashi, K., 1992b. A new Mesozoic isochron chart of the northwestern Pacific Ocean: paleomagnetic and tectonic implications. *Geophysical Research Letters*, **19**(7), 693–696.

Nelson, J. A., and Mihalynuk, M., 1993. Cache Creek ocean: closure or enclosure? *Geology*, **21**(2), 173.

Parsons, B., and Sclater, J. G., 1977. An analysis of the variation of ocean floor bathymetry and heat flow with age. *Journal of Geophysical Research*, **82**(5), 803–827.

Phipps Morgan, J., and Smith, W. H. F., 1992. Flattening of the sea-floor depth-age curve as a response to asthenospheric flow. *Nature*, **359**(6395), 524–527.

Phipps Morgan, J., Morgan, W. J., and Price, E., 1995. Hotspot melting generates both hotspot volcanism and a hotspot swell? *Journal of Geophysical Research*, **100**(B5), 8045–8062.

Schubert, G., and Sandwell, D., 1989. Crustal volumes of the continents and of oceanic and continental submarine plateaus. *Earth and Planetary Science Letters*, **92**, 234–246.

Seton, M., Müller, R. D., Zahirovic, S., Gaina, C., Torsvik, T., Shephard, G. E., Talsma, A. S., Gurnis, M., Turner, M., Maus, S., and Chandler, M. T., 2012. Global continental and ocean basin reconstructions since 200 Ma. *Earth Science Reviews*, **113**, 212–270.

Spasojevic, S., and Gurnis, M., 2012. Sea level and vertical motion of continents from dynamic earth models since the Late Cretaceous. *AAPG Bulletin*, **96**(11), 2037–2064.

Stein, C., and Stein, S., 1992. A model for the global variation in oceanic depth and heat flow with lithospheric age. *Nature*, **359**(6391), 123–129.

Sykes, T. J., 1996. A correction for sediment load upon the ocean floor: uniform versus varying sediment density estimations – implications for isostatic correction. *Marine Geology*, **133**(1), 35–49.

Taylor, B., 2006. The single largest oceanic plateau: Ontong Java-Manihiki-Hikurangi. *Earth and Planetary Science Letters*, **241** (3–4), 372–380.

Torsvik, T. H., Rousse, S., Labails, C., and Smethurst, M. A., 2009. A new scheme for the opening of the south atlantic ocean and the dissection of an aptian salt basin. *Geophysical Journal International*, **177**(3), 1315–1333.

Turcotte, D. L., and Oxburgh, E. R., 1967. Finite amplitude convection cells and continental drift. *Journal of Fluid Mechanics*, **28**, 29–42.

Van der Voo, R., Spakman, W., and Bijwaard, H., 1999. Mesozoic subducted slabs under Siberia. *Nature*, **397**(6716), 246–249.

Viso, R. F., Larson, R. L., and Pockalny, R. A., 2005. Tectonic evolution of the Pacific-Phoenix-Farallon triple junction in the South Pacific Ocean. *Earth and Planetary Science Letters*, **233** (1–2), 179.

Zorin, Y. A., 1999. Geodynamics of the western part of the Mongolia-Okhotsk collisional belt, Trans-Baikal region (Russia) and Mongolia. *Tectonophysics*, **306**(1), 33–56.

Cross-references

Deep-sea Sediments
Energy Resources
Marginal Seas
Oceanic Spreading Centers
Paleoceanographic Proxies
Paleoceanography
Regional Marine Geology
Sea-Level

PALEOPRODUCTIVITY

Paul Loubere
Department of Anthropology, Northern Illinois University, DeKalb, IL, USA

Definition

Biological productivity involves the synthesis of biogenic materials (skeletal and nonskeletal). Paleoproductivity refers to the record of synthesis in the past. Generally, the term is applied to production that occurred prior to times when direct measurements are available.

Introduction

Production can be primary, when organic matter is synthesized from inorganic antecedents (e.g., plant synthesis based on chemicals and sunshine), or secondary, when it is based on previously made organic matter. Most analysis of paleoproductivity aims to reconstruct past changes in primary production.

The interest in paleoproductivity stems mostly from research into three different issues in paleo-oceanic/paleoclimate studies. Ocean-basin-scale patterns of production reflect the location and extent of past/current systems. Also, regional changes provide information about variation in exchange between deep and surface ocean

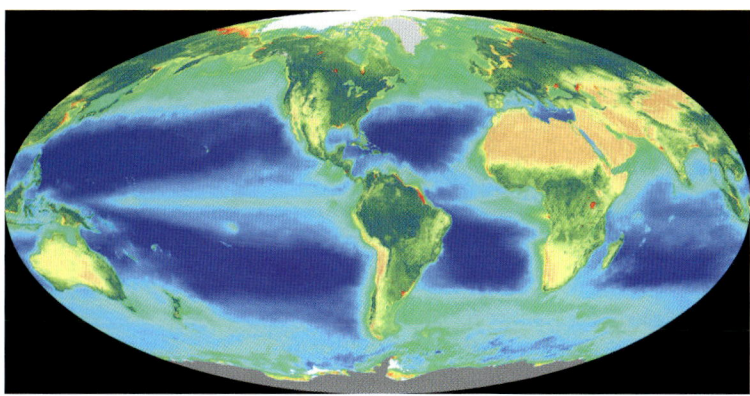

Paleoproductivity, Figure 1 Distribution of marine algae in the surface ocean, revealed by chlorophyll concentrations (*darker green* = more chlorophyll) calculated from satellite observations (source: http://oceancolor.gsfc.nasa.gov/SeaWiFS/). Higher algal concentrations are associated with upwelling of deeper, nutrient-rich waters or with enhanced vertical mixing of the water column.

waters. Also, decay of settling organic material alters the chemical properties of the subsurface ocean (e.g., oxygen concentration). Finally, changes in both quantities and systems of production could influence global climate through regulation of atmospheric carbon dioxide content (Loubere, 2012; Sigman and Hain, 2012).

What controls marine production?

Biological production depends on the availability of sunlight and required macronutrients (classically: carbon, nitrogen, phosphorus) and micronutrients (with iron receiving most attention, Sarmiento and Gruber, 2006; Boyd et al., 2007). The intensity of sunlight useable by phytoplankton (marine algae, the dominant marine primary producers) depends on latitude, water depth, water clarity, and water column stability. Latitude controls the day length and the angle of the sun on the sea surface. Water depth is important because light intensity drops exponentially with depth and penetration is generally negligible below 50–100 m. Because of this, any production over most of the ocean will depend on populations of floating algae. A stable water column will be resistant to vertical mixing, which means that plankton can remain close to the sunlit surface most of the time. Near shore, an unstable water column and vertical mixing can lead to increased turbidity from sediments, reducing light.

Essential nutrients are concentrated at depth in the oceans, so resupply for upper ocean plankton consumption depends on processes which bring deep water to the surface (Figure 1; upwelling, deep vertical mixing; Sarmiento and Gruber, 2006). Some nutrients are also delivered by rivers and prevailing winds (as dust). Supply depends on continental geography and climatology.

New production: a key global environmental variable

Algal primary production in the oceans is based on locally recycled nutrients and on "new" nutrients supplied from the deeper ocean reservoir (Sarmiento and Gruber, 2006). In most of the ocean, the majority of production is maintained by local recycling of nutrients in the near-surface water column. Since this recycling is not 100 % efficient, however, there is transfer from the shallow to the deep ocean. Consequently, replenishment of the surface, via upwelling and mixing, is necessary for biological production to continue (Palter et al., 1991).

The supply of biological materials to the seafloor and geological record depends on the production engendered from the "new" nutrients reaching the upper ocean. This is different from the "total" (recycled + new) production which biologists routinely measure. In this way, analysis of paleoproductivity provides insight to the rates of exchange of waters (and their load of nutrients) between ocean depths and the surface. This vertical cycling of chemicals, since it includes carbon dioxide, is important to the global climate (Sigman and Hain, 2012).

The efficiency of biological recycling in the upper ocean's productive zone is not constant. Efficiency drops as production increases (Wassman, 1998; Antia et al., 2001). Variation of recycling efficiency and associated change in plankton communities can also affect the distribution of CO_2 in the oceans and global climate (Matsumoto et al., 2002).

Making a record of past changes in productivity

In principle, reconstruction of paleoproductivity should be simple, requiring only measurement of accumulation rates of sedimentary biogenic material. Sadly, this is complicated by a number of factors.

The materials incorporated into the ocean floor are those which were produced in the upper ocean, survived recycling in the upper water column while sinking, and then survived the passage to greater depth and the recycling processes at the seabed. Observations with sediment traps in the ocean have demonstrated that attenuation of sinking biological particles is strong, especially in

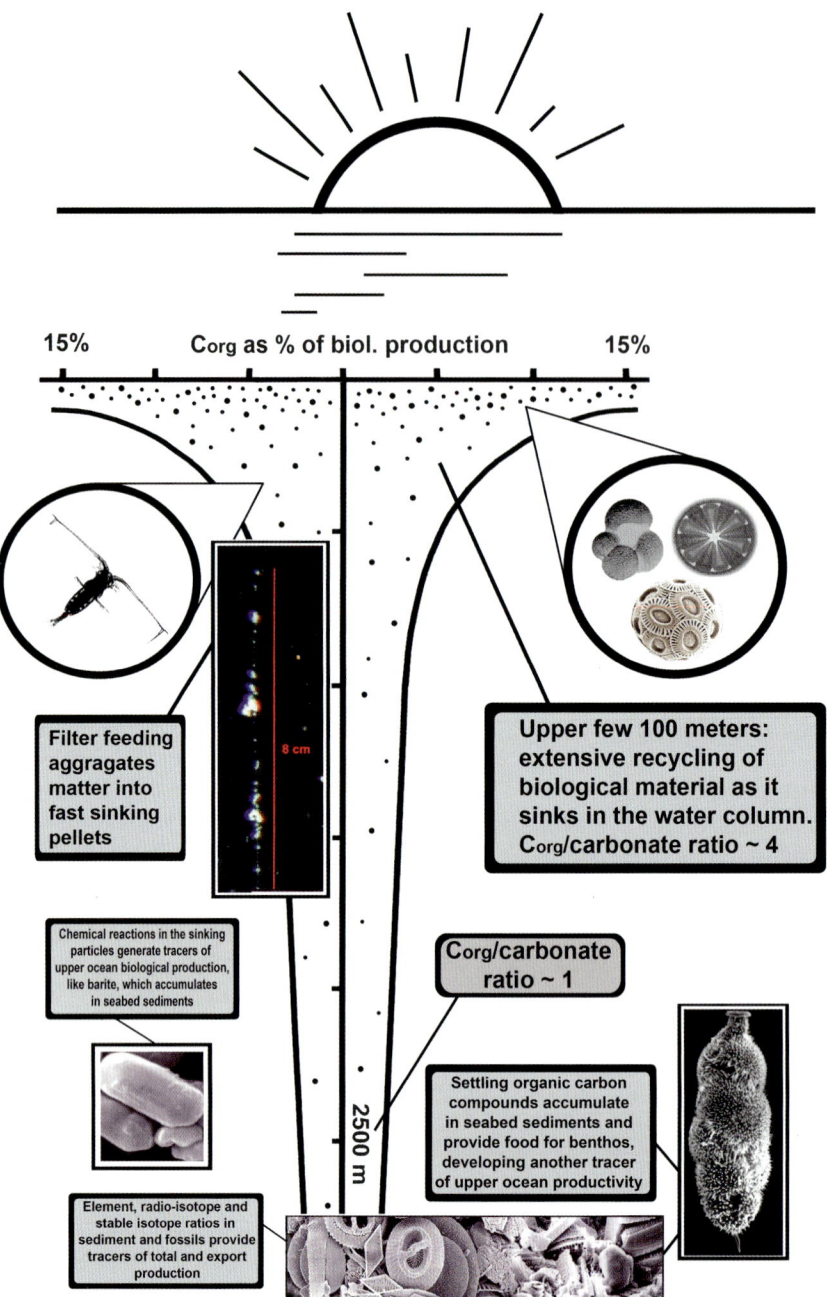

Paleoproductivity, Figure 2 Transfer of materials from the productive upper ocean to the deep sea. Plot of organic carbon flux with water depth as percent of total primary production (after Antia et al., 2001).

the upper few hundred meters of the water column (Figure 2; Antia et al., 2001; Honjo et al., 2008). There is an exponential decrease in flux with depth (flux = weight of material, per volume of water or area of seabed, per unit of time). Additionally, different components of the sinking material degrade at different rates, so the composition of the sinking material changes with depth. Particle incorporation into larger, fast sinking pellets by filter feeders in the water column is important to material delivery to the seabed.

The changes in the flux of biological material through the water column are strongest for labile (high food quality) organic matter and for mineral matter which is undersaturated or reactive to the conditions in the water column or pellet environment. The latter includes, for example, silicate (as in the frustules of diatoms). By contrast, calcite shell or bone material is far more resistant until carried to undersaturated conditions usually at considerable depth in the ocean (Dittert et al., 1999; Berger and Wefer, 2008) or within seabed sediments.

Interestingly, as organic matter undergoes diagenesis in sinking particles, reactions occur which add new material to the debris. The reactions lead to precipitation of new minerals, like barite. It appears that the amount of biogenic barite created is related to the quantity of reactive organic matter settling to depth (Paytan and Griffith, 2007). The diagenetic products accumulate at the seabed and can provide a record of biological organic matter production.

Diagenesis (recycling, consumption) continues at the seafloor, which is actually a highly reactive interface where biological activity and chemical reactions alter what arrives. Benthic fauna feed on reactive organic compounds and biotic activity combined with dissolution reactions destroy skeletal materials (Ragueneau et al., 2000; Zeebe and Wolf-Gladrow, 2001; Sarmiento and Gruber, 2006).

Tracers of past biological production

Historically, tracking paleoproduction was attempted first by measuring contents of organic carbon and carbonate in deep-sea sediments. In the equatorial Pacific, for example, there are distinct cycles of carbonate percent in Pleistocene-age sediments (Berger, 1992). Percent organic carbon content was also observed to vary regionally and in time, in ways that related to patterns of biological production in the ocean. However, the analysis of the carbonate cycles was confounded by argument about whether they were due to changes in production or preservation (Berger, 1992). The cycles could alternatively reflect variation in biological production in the upper ocean or changes in the properties of deep ocean waters (changes in degree of corrosiveness to carbonate) related to climate shifts.

The problem with contents is that, for each biogenic material, it depends on the supply and preservation of *all* the components going in the sediments, so interpreting content data is complicated. Because of this, measuring accumulation rates (weight of material accumulated per cm^2 of seabed per 1,000 years) offers a better avenue to reconstruction paleoproduction.

Accumulation rates of biogenic components have been used extensively to infer changes in productivity in times past. However, these rates are still subject to the problem of varying preservation (what accumulates is a function of supply minus diagenetic loss). Also, the simple use of accumulation rates for inference of supply changes assumes that delivery of materials from the surface ocean to the seabed is completely vertical, with no lateral transport or redeposition.

Examination of accumulation rates requires correction for changes in preservation and for possibly variable water column or seabed transportation. These issues remain problematic. Methods have been developed to quantify changes in carbonate preservation (Mekik et al., 2002), and methods based on thorium-230 concentrations in deep-sea sediments have been derived for measuring lateral sediment transport (Francois et al., 2004). Application of the method has produced some notable disagreements over the Pleistocene cycles of carbonate and organic carbon accumulation in the eastern equatorial Pacific (Loubere and Richaud, 2006).

Tracers not dependent on accumulation rate calculations offer escape from the complications described above. Relative abundances (percentages) of benthic (bottom dwelling) foraminifera (shell-producing microorganism) species in deep-sea sediments have been found to reflect flux of labile organic carbon reaching the seafloor. This is because the carbon flux is food to the benthic community whose ecology is structured depending on how large that flux is. Composition of benthic fossil assemblages has been empirically calibrated to both average annual organic carbon production and seasonality of that production for the open ocean (Loubere and Fariduddin, 1999a, b).

Tracers of paleoproductivity are also available in biogenic components which are thought to be highly resistant to alteration in the water column and seabed or whose characteristics are set in the upper ocean. These include barite (biogenic barite produced as explained above and a tracer of reactive organic carbon flux from the upper ocean), alkenones (lipids) whose concentration has been taken to reflect productivity of the *Prymnesiophyceae* algae (including coccolithophorids and *Phaeocystis*) (e.g., Sachs and Anderson, 2005), elemental ratios in pelagic sediments (Anderson and Winckler, 2005), and elemental or isotope ratios in planktonic material (e.g., Stoll et al., 2007; see Fischer and Wefer, 1999, pp. 315–468).

Has oceanic productivity varied in the past?

In principle, we would expect that global and regional marine productivity would vary in response to changes in nutrient concentrations in seawater and to the structure of the upper ocean water column (degree of stability influences availability of sunlight). In short to intermediate geological timescales, nutrient concentrations will depend on the whole ocean exchange between deep and surface waters and on regional patterns of upwelling and vertical mixing (Sarmiento and Gruber, 2006). Changes in ocean circulation should be linked to global climate change. On the longer geological timescale, ocean circulation will change in response to shifting continents, and nutrient concentrations might change in the global ocean due to variations in the ultimate supply, the erosion of continents (the key element phosphorus is an example; Filippelli, 2002).

On the grandest scale, strong decreases in marine biological productivity have been inferred for major extinction events in global history (e.g., Coxall et al., 2006). The exact causes are unclear as yet, but impacts on both planktonic community composition and changes in ocean circulation and nutrient supply likely all played a role.

Closer to the present, large changes in marine productivity have been inferred from the Neogene to Pleistocene glacial-interglacial cycles (e.g., Schmiedl and Mackensen, 1997; Diester-Haass et al., 2005; Loubere et al., 2011; Murray et al., 2012). These have been ascribed to changes in exchange between the deep and upper ocean (primarily in the Southern Ocean), change in strength of winds that induce upwelling in tropical regions, or fluctuating supply of micronutrients like iron. There is currently disagreement on the relative importance of the local, wind-related, and more distant, ocean circulation-related factors in productivity change.

Variation in exchange between the deep and surface ocean on the glacial-interglacial timescale provides a mechanism for explaining why atmospheric carbon dioxide concentrations have changed in step with climate change (and, indeed, helped cause it; Toggweiler, 1999). Another possible explanation is change in the relative fluxes of carbonate and non-carbonate biological material produced in the upper ocean and delivered to the deep sea (Matsumoto et al., 2002). A shift toward a higher ratio of organic carbon to carbonate in the sinking biogenic matter of the oceans could provide lower atmospheric carbon dioxide concentrations during glacial intervals. This requires a change in the relative production of different materials by the ocean's planktonic organisms. There is some evidence that this happened (Richaud et al., 2007).

Summary

Biological production in the oceans generates a complex mix of materials, much of which is recycled near the ocean's surface, but some of which sinks into the deep sea and influences global climate, as well as generates a record of past production. Tracking the history of the major components of production has only been partly accomplished due to complications with interpretation of that record. Nevertheless, the evidence, at least qualitatively, shows clearly that oceanic biological production has been variable in the past in ways linked to global climate and oceanographic conditions.

Bibliography

Anderson, R., and Winckler, G., 2005. Problems with paleoproductivity proxies. *Paleoceanography*, **20**(3), PA 3012, doi:10.1029/2004PA001107.

Antia, A., Koeve, W., Fischer, G., Blanz, T., Schulz-Bull, D., et al., 2001. Basin-wide particulate carbon flux in the Atlantic Ocean: regional export patterns and potential for atmospheric CO_2 sequestration. *Global Biogeochemical Cycles*, **15**(4), 845–862.

Berger, W., 1992. Pacific carbonate cycles revisited: arguments for and against productivity control. In Ishizaki, K., and Saito, T. (eds.), *Centenary of Japanese Micropaleontology*. Tokyo: Terra Science, pp. 15–25.

Berger, W. H., and Wefer, G., 2008. Marine biogenic sediments. In Gornitz, V. (ed.), *Encyclopedia of Paleoclimatology and Ancient Environments*. Amsterdam: Springer, pp. 525–533.

Boyd, P. W., et al., 2007. Mesoscale iron enrichment experiments 1993–2005: synthesis and future directions. *Science*, **315**, 612–617, doi:10.1126/science.1131669.

Coxall, H., D'Hondt, S., and Zachos, J., 2006. Pelagic evolution and environmental recovery after the cretaceous-paleogene mass extinction. *Geology*, **34**, 297–300.

Diester-Haass, L., Billups, K., and Emeis, K., 2005. In search of the late miocene early-pliocene "biogenic bloom" in the Atlantic Ocean (Ocean Drilling Program Sites 982, 925 and 1088). *Paleoceanography*, **20**, PA4001.

Dittert, N., Baumann, K., Bickert, T., Henrich, R., Huber, R., Kinkel, H., and Meggers, H., 1999. Carbonate dissolution in the deep-sea: methods, quantification and paleoceanographic application. In Fischer, G., and Wefer, G. (eds.), *Use of Proxies in Paleoceanography*. Berlin: Springer, pp. 255–284.

Filippelli, G., 2002. The global phosphorus cycle. *Reviews in Mineralogy and Geochemistry*, **48**, 391–425.

Fischer, G., and Wefer, G. (eds.), 1999. *Paleoproductivity and Nutrients, Use of Proxies in Paleoceanography*. Berlin: Springer, pp. 315–468.

Francois, R., Frank, M., Rutgers van der Loeff, M., and Bacon, M., 2004. 230Th normalization: an essential tool for interpreting sedimentary fluxes during the Late Quaternary. *Paleoceanography*, **19**, PA1018, doi:10.1029/2003PA000939.

Honjo, S., Manganini, S., Krushfield, R., and Francois, R., 2008. Particulate organic carbon fluxes to the ocean interior and factors controlling the biological pump: a synthesis of global sediment trap programs since 1983. *Progress in Oceanography*, **76**, 217–285.

Loubere, P., 2012. The global climate system, nature education knowledge. **3**(5), 2. www.nature.com/scitable/knowledge/library/the-global-climate-system-74649049

Loubere, P., and Fariduddin, M., 1999a. Quantitative estimation of global patterns of surface ocean biological productivity and its seasonal variation on time scales from centuries to millennia. *Global Biogeochemical Cycles*, **13**, 115–133.

Loubere, P., and Fariduddin, M., 1999b. Benthic foraminifera and the flux of organic carbon to the seabed. In Sen Gupta, B. (ed.), *Modern Foraminifera*. Amsterdam: Kluwer Press, pp. 181–200.

Loubere, P., and Richaud, M., 2006. Some reconciliation of glacial-interglacial calcite flux reconstructions for the Eastern equatorial Pacific. *Geochemistry, Geophysics, Geosystems*, **8**, Q03008, doi:10.1029/2006GC001367.

Loubere, P., Fariduddin, M., and Richaud, M., 2011. Glacial marine nutrient and carbon redistribution: evidence from the tropical ocean. *Geochemistry, Geophysics, Geosystems*, **12**(8), Q08013, doi:10.1029/2011GC003546.

Matsumoto, M., Sarmiento, J., and Brzezinski, M., 2002. Silica acid leakage from the southern ocean: a possible explanation for glacial atmospheric CO_2. *Global Biogeochemical Cycles*, **5**, 1–23, doi:10.1029/2001GB001442.

Mekik, F., Loubere, P., and Archer, D., 2002. Organic carbon flux and organic carbon to calcite flux ratio recorded in deep sea carbonates: demonstration and a new proxy. *Global Biogeochemical Cycles*, **16**(3), 1052.

Murray, R., Leinen, M., and Knowlton, C., 2012. Links between iron input and opal deposition in the Pleistocene equatorial Pacific Ocean. *Nature Geoscience*, **5**, 270–274, doi:10.1038/NGEO1422.

Open University, 2001. *Ocean Circulation*, 2nd edn. Oxford: Butterworth-Heinemann, p. 281.

Palter, J. B., et al., 1991. Fueling export production: nutrient return pathways from the deep ocean and their dependence on the meridional overturning circulation. *Biogeosciences*, **7**, 3549–3568, doi:10.5194/bg-7-3549-2010.

Paytan, A., and Griffith, E., 2007. Marine barite: recorder of variations in ocean export productivity. *Deep-Sea Research Part II*, **54**, 687–705.

Ragueneau, O., Treguer, P., Leynaert, A., Anderson, R., and Brzezinski, M. A., 2000. A review of the Si cycle in the modern ocean: recent progress and missing gaps in the application of biogenic opal as a paleoproductivity proxy. *Global and Planetary Change*, **26**, 317–365.

Richaud, M., Loubere, P., Pichat, S., and Francois, R., 2007. Changes in opal flux and the rain ratio during the last 50,000 years in the equatorial Pacific. *Deep-Sea Research Part II*, **54**, 762–771.

Sachs, J., and Anderson, R., 2005. Increased productivity in the subantarctic ocean during Heinrich events. *Nature*, **434**, 1118–1121.

Sarmiento, J., and Gruber, N., 2006. *Ocean Biogeochemical Dynamics*. Princeton: Princeton University Press, p. 526.

Schmiedl, G., and Mackensen, A., 1997. Late quaternary paleoproductivity and deep water circulation in the eastern South Atlantic Ocean: evidence from benthic Foraminifera. *Palaeogeography Palaeoclimatology Palaeoecology*, **130**, 43–80.

Sigman, D. M., and Hain, M. P., 2012. The biological productivity of the Ocean: section 1. *Nature Education Knowledge*, **3**(10), 21. www.nature.com/scitable/knowledge/library/the-biological-productivity-of-the-ocean-section-70631104

Stoll, H., Ziveri, P., Shimizu, N., Conte, M., and Theroux, S., 2007. Relationship between coccolith Sr/Ca ratios and coccolithophore production and export in the Arabian Sea and Sargasso Sea. *Deep-Sea Research Part II*, **54**, 581–600.

Toggweiler, J., 1999. Variation of atmospheric CO_2 by ventilation of the ocean's deepest water. *Paleoceanography*, **14**, 571–588, doi:10.1029/1999PA900033.

Wassman, P., 1998. Retention versus export food chains: processes controlling sinking loss from marine pelagic systems. *Hydrobiologia*, **363**, 29–57.

Zeebe, R., and Wolf-Gladrow, D., 2001. CO_2 *in Seawater: Equilibrium, Kinetics, Isotopes*. Amsterdam: Elsevier. Oceanography Series, Vol. 65, p. 346.

Cross-references

Biogenic Barium
Carbonate Dissolution
Calcite Compensation Depth (CCD)
Deep-sea Sediments
Diatoms
Dust in the Ocean
Export Production
Foraminifers (Benthic)
General ocean circulation - its signals in the sediments
Marine Microfossils
Organic Matter
Paleoceanographic Proxies
Silica

PALYNOLOGY (POLLEN, SPORES, ETC.)

Anne de Vernal
Centre GEOTOP, University of Québec in Montréal, Montréal, QC, Canada

Palynology sensu stricto and sensu lato

The word "palynology," which is the study of pollen, comes from the Greek word palunein (to sprinkle) in reference to the pollen that is sprinkled as dust during the blooming season. Palynology became a scientific discipline in the early 1990s after the pioneer works of Lennart von Post, a Swedish geologist who produced the first pollen diagrams based on the identification and count of pollen grains from peat deposits, to reconstruct the postglacial vegetation of Western Europe (cf. Manten, 1967). Since then, palynology has been widely used to reconstruct the history of vegetation through time and past climate/environmental conditions. In botany and ecology, palynology is often associated with the study of pollen from flowering plants. In geosciences at large, palynology is used in a much broader sense because palynological slides prepared from sediment for the observation of pollen grains in optical microscopy contain many other biogenic remains. By extension, palynology thus corresponds to the study of all microfossils composed of highly resistant organic matter, also called sporopollenin, which is recovered after chemical digestion of carbonate and silica particles with hydrochloric and hydrofluoric acid treatments, respectively.

In marine sediments, the acid-resistant organic-walled microfossils or palynomorphs may include remains from the planktonic organisms such as cysts of dinoflagellates (or dinocysts), phycoma of prasinophytes, acritarchs, tasmanids, cysts and lorica of tintinnids, copepod eggs, etc. (e.g., Jansonius and McGregor, 1996). They may also include organic linings of benthic foraminifers (de Vernal et al., 1992) and allochthonous remains from freshwater environments such as Chlorococcales (e.g., *Pediastrum* spp.), in addition to pollen and spores from the terrestrial vegetation (e.g., Traverse, 2007). Here, we focus on pollen and spores that are particularly useful in marine geosciences as they permit land-sea correlations in addition to the reconstruction of past vegetation and climate in terrestrial areas from where pollen and spore assemblages originate.

Sediment sample processing for pollen analyses

The preparation of sediment samples for palynological analyses consists in mechanical and chemical procedures to concentrate organic-walled microfossils (e.g., Traverse, 2007; de Vernal et al., 2010). Mechanical procedures include wet sieving for eliminating small silt size and sand fractions and/or dense liquid separation. Chemical procedures consist in hydrochloric acid (HCl) and hydrofluoric acid (HF) maceration to dissolve carbonate and silica particles. They may also include oxidation with potassium hydroxide (KOH) or acetolysis, which is a treatment with acetic anhydride $((CH_3CO)_2O)$ and sulfuric acid (H_2SO_4) to eliminate labile organic matter. While oxidation techniques are often used for pollen and spore studies from lake sediment or peat, they are avoided in marine palynology as they can alter the preservation of some marine palynomorphs such as dinoflagellate cysts. Moreover, the palynology of marine sediment rarely requires oxidation as the organic carbon content is generally low especially in deep-sea settings due to its degradation in the water column and because of diagenetic processes.

Pollen and spores are usually observed, identified, and counted with optical microscope at $\times 400$ to $\times 1,000$

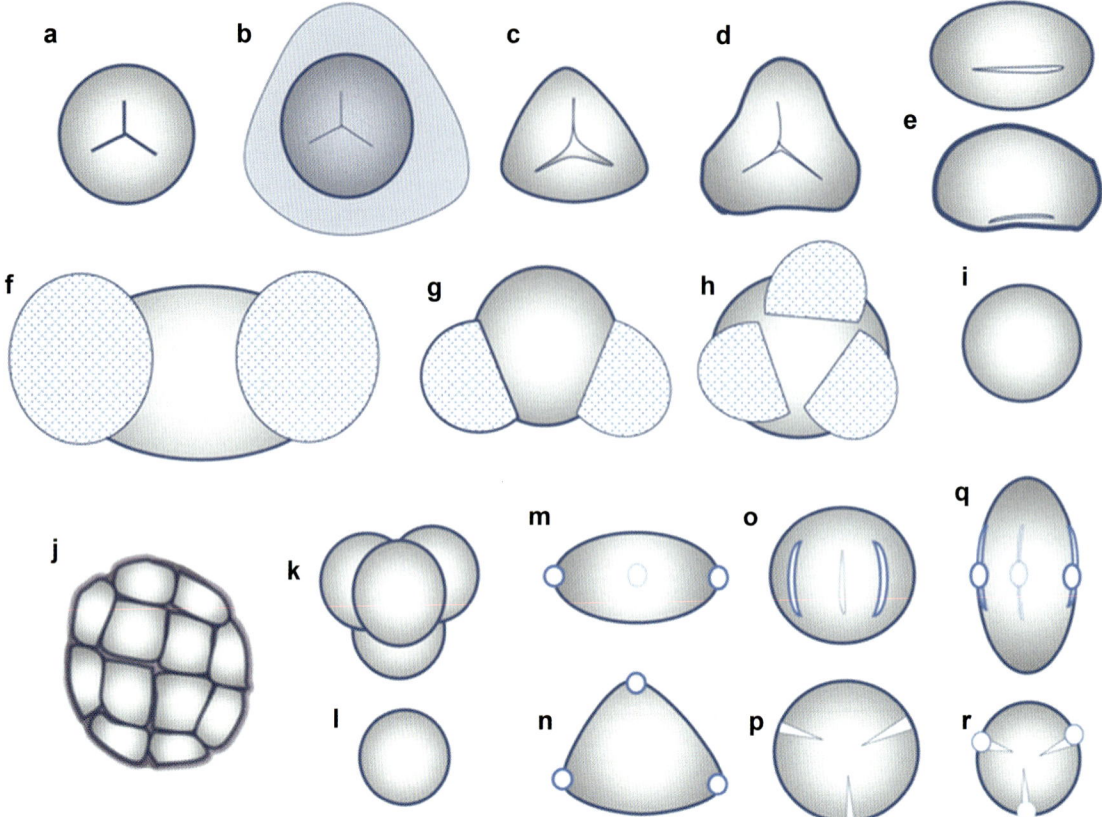

Palynology (Pollen, Spores, etc.), Figure 1 Schematic illustration of the main morphological types of spores and pollen grains. (**a–e**) Spores from seedless plants: trilete spores with circular shape (**a**) and a perine (**b**), trilete spores with subcircular and triangular shape (**c, d**), monolete spores (**e**). (**f–i**) Pollen grains of gymnosperms: bisaccate pollen of conifers in proximal (**f**) and lateral (**g**) views, polysaccate pollen of fossil gymnosperm (**h**), and inaperturate pollen without saccus (**i**). (**j–r**) Angiosperm pollen grains: polyade (**j**), tetrad (**k**), and single pollen grains with no aperture (inaperturate (**l**)), with pores (**m, n**), colpi (**o, p**), and colpi and pores (**q, r**). The figures **m**, **o** and **q** show oblate, spherical, and prolate pollen grains in equatorial view, and the figures **n**, **p** and **r** show the same pollen types in polar view. The schemes of figures **m–r** correspond to pollen type with three apertures: triporate, tricolpate, or tricolporates. Any combination with 1 to >10 apertures in equatorial or evenly distributed position is possible and serves in identification key. Beyond the geometry of apertures in pollen grains, the structure of the exine and ornamentation are important criteria for taxa identification.

magnification. Ideally, a minimum of 200 or 300 pollen grains are identified and counted for calculation of concentration (number of pollen grains per dry weight or volume) and percentages of taxa. The resulting spectrum (pollen representation in a given sample) is then combined with others to prepare a diagram of pollen assemblages against depth or age, which provides a picture of pollen input changes through time. Concentration can be estimated from the analyses of an aliquot of the sample from the marker-grain method, which consists in adding a known number of exotic grains to the sample (e.g., Matthews, 1969; de Vernal et al., 2010).

Systematics, biology, and morphology of spores and pollen grains

Spores and pollen grains directly relate to the reproductive stage in the life cycle of land plants also designated as embryophytes. Hence, the systematics of pollen grains and spores follow that of the plant kingdom, with taxa designated after the botanical name of their corresponding plants.

Spores are produced by seedless plants, including mosses (bryophytes), club mosses (lycophytes), and ferns (pteridophytes). They develop in capsule or sporangium on the sporophyte and are adapted for dispersal and survival over extended periods. Spore germination produces gametophytes, which in turn produce the sperms and eggs merging into a zygote to form a new sporophyte. Spore size ranges from 5 to 150 µm. The content of the spore is protected by a membrane composed of sporopollenin, which can fossilize. The membrane or sporoderm is usually characterized by a simple shape (round, triangular, ellipsoid, or bean shaped) and by an aperture, which can be simple (monolete slot) or triaxial (trilete slot) (see Figures 1a–e and 2a–f). The sporoderm can be

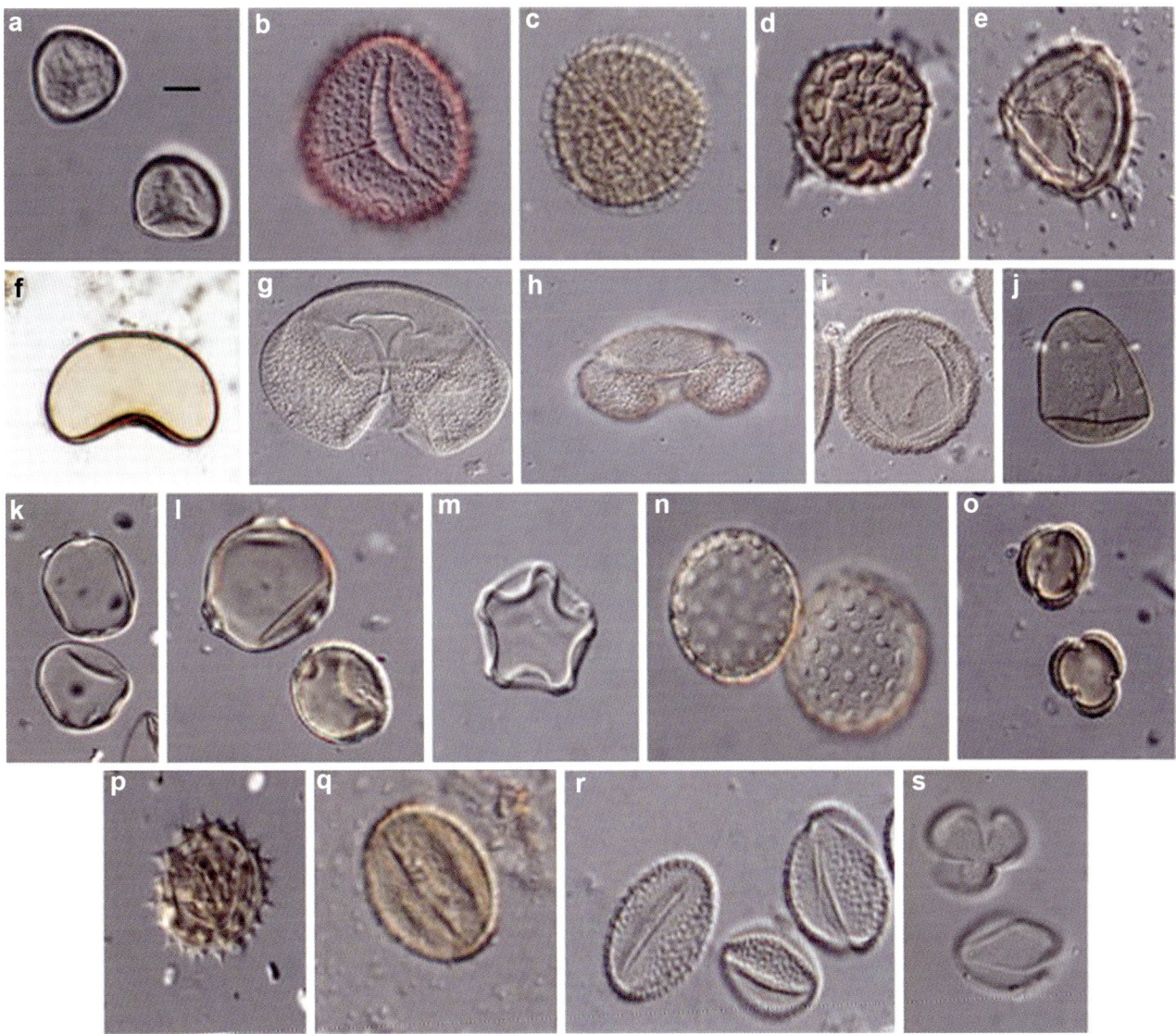

Palynology (Pollen, Spores, etc.), Figure 2 Photographs of common spore and pollen taxa. The magnification is the same for all photographs and the scale shown in **a** is 10 μm. (**a–f**) Photographs of spores from seedless plants: trilete spores of *Sphagnum*, which belongs to bryophyte (**a**); trilete spores of the pteridophyte *Osmunda* (**b**); trilete spores of the lycopodiophytes *Lycopodium* (**c, d**) and *Selaginella* (**e**); monolete spore of undetermined pteridophyte (**f**). Photographs of pollen grains from conifers: bisaccate pollen grain of *Picea* (**g**) and *Pinus* (**h**) and vesiculate pollen of *Tsuga* (**i**). (**j–s**) Photographs of pollen grains from angiosperms: Cyperaceae (**j**), Poaceae (**k**), *Betula* (**l**), *Alnus* (**m**), Chenopodiaceae (**n**), *Artemisia* (**o**), Tubiflorae (**p**), *Quercus* (**q**), *Salix* (**r**), and Rosaceae (**s**).

characterized by ornamentation (reticulation, ridges, spines, etc.), and some taxa may have an outer wall called perine.

Pollen grains are produced by seed plants or spermatophytes, which involve reproduction on the sporophyte instead of reproduction with alternation of generations (sporophyte and gametophyte) as for seedless plants. Spermatophytes include the flowering plants or angiosperms (magnoliophytes) and the gymnosperms (pinophytes or conifers, ginkgophytes, gnetophytes, cycadophytes). The pollen grains contain the male reproductive body, which is protected by the exine, an envelope composed of sporopollenin. Pollen is formed in the anther of flowers or in the male cone of gymnosperms. When mature, pollen grains are blown away by wind or carried by animal vectors to fertilize either the ovule in the pistil of flowers or the female cones of gymnosperms. The fertilization of the ovule also called pollination can be referred to as anemophilous or entomophilous in the case of pollination from wind or insects, respectively. In general, anemogamous plants produce more pollen grains than entomophilous plants.

Pollen grains of gymnosperms are generally distinguished on the basis of their shape, subspherical with or without vesicles or sacci (see Figures 1f–i and 2g–i). Conifer pollen grains are often characterized by a pair of sacci and are designated as bisaccate. The size of gymnosperm pollen is variable, ranging from about 10 μm in simple forms to about 200 μm for large bisaccate taxa.

The pollen grains of angiosperms have a great morphological diversity (see Figures 1j–r and 2j–s). They are identified on the basis of their geometry (spherical, oblate, prolate), the type of aperture (circular (pore) or elongate (colpus)), the number and distribution of apertures, the ornamentation (smooth, spiny, reticulate, striated), and structure of the exine or pollen grain outer membrane. Their size is usually comprised between 5 and 50 μm.

In contrary to many other microfossil groups, pollen grains and spores are characterized by homogenous morphology within their taxonomical categories, at family, genus, or species level. Hence, keys based on the abovementioned characters have been developed for the identification of pollen and spores (e.g., Faegri and Iversen, 1989; Moore et al., 1991; Reille, 1992; Kapp et al., 2000).

One particularity of pollen and spores is the chemical compound of the sporopollenin that forms the exine and sporoderm, respectively, and which consists in extremely resistant biological polymers (e.g., Traverse, 2007). The chemical composition of sporopollenin, however, is very complex and differs according to the taxonomical affinities of pollen and spores (e.g., Hemsley et al., 1993; Hemsley and Poole, 2004).

Evolution of land plants and pollen and spore stratigraphy

The evolution of land plants through the geological history has been mostly documented from microfossils, including pollen and spores. The land plants or embryophytes developed from green algae ancestors in the Paleozoic. Their first occurrence might be inferred from spores recovered in mid-Ordovician deposits dated of ca. 470–460 Ma (e.g., Steemans et al., 2009), and the diversification of vascular plants occurred in the Silurian about 420 Ma ago (e.g., Kenrick and Crane, 1997; Kenrick et al., 2012). Seedless plants remained dominant until about 320–300 Ma ago when the first spermatophytes developed. Seed plants were first represented by the gymnosperms, which diversified during the Carboniferous and Permian with the successive development of cycadophytes, ginkgophytes, gnetophytes, and pinophytes (e.g., Magallón et al., 2013). The occurrence of angiosperms or flowering plants, which are characterized not only by a reproduction from the sporophyte but also by flowers and protected seeds, took place much later but with a spectacular diversification of taxa during the Jurassic and early Cretaceous, from about 200 to 130 Ma (e.g., Magallón et al., 2013).

From a stratigraphical point of view, spores date from the Ordovician, whereas inaperturate and saccate pollen grains date from the Carboniferous. Pollen grains of angiosperms bearing pores and colpi range mostly from the Cretaceous.

Because the sporopollenin is highly resistant to diagenetic alteration, except under highly oxidizing conditions, all sedimentary rocks dating from the Ordovician to recent potentially contain fossil pollen and spores. Hence, reworked Paleozoic or Mesozoic pollen and spores are frequently recovered in late Cenozoic sediments. The distinction of "fresh" versus old reworked pollen and spores can be made from taxa identification and from the state of preservation as aged pollen grains and spores have lost the ultrastructure and fluorescent properties of their exine or sporoderm through diagenetic processes.

Pollen and spore transport into marine sediments

Pollen and spores are routinely recovered in marine sediments from coastal areas and continental margins. They are necessarily allochthonous, and their occurrence depends upon production in the source area in addition to transport mechanisms through air or water. Pollen and spore deposition decreases rapidly with the distance from source areas. Pollen grains produced at high elevation above ground, from the tree canopy, for example, are more prone to be carried by wind than pollen produced from the herb strata closer to the ground. Moreover, pollen grains from anemophilous plants have attributes fostering transport by wind. in particular, the pinophytes characterized by saccate morphology. In addition to atmospheric transport, hydrodynamic transport through runoff, rivers, and marine currents has to be taken into account.

The long-distance atmospheric transport of pollen results in a nearly asymptotic decrease of pollen and spore concentration with the distance of the source area (e.g., Rochon and de Vernal, 1994). It also results in important modifications of taxa percentages in assemblages, because of their selective transport by wind. For example, as shown from the distribution of pollen in surface sediment samples, there is an offshore overrepresentation of some tree taxa and especially of *Pinus* pollen grains that have particularly efficient aerial dispersion properties (e.g., Heusser and Balsam, 1977; Mudie, 1982; Sun et al., 1999).

The transport of pollen grains and spores with water is less selective than that in the atmosphere. In estuarine environment, deltas, and epicontinental seas, pollen grains and spores can be very abundant reaching concentration of $10^5 \, g^{-1}$, which are comparable to those of lakes (e.g., de Vernal and Giroux, 1991; Marret et al., 2001; Traverse, 2007). Detailed studies from neritic settings along southwest European margins, where relatively humid climate prevails, have shown that most pollen and spores recovered in marine sediments are intimately related to fluvial inputs from adjacent lands, allowing direct land-sea correlations (e.g., Sánchez Goñi et al., 1999; see Figure 3a). In a different context, the distribution pattern of pollen grains off northwest Africa permitted to identify aeolian transport through trade winds, in addition to input by rivers,

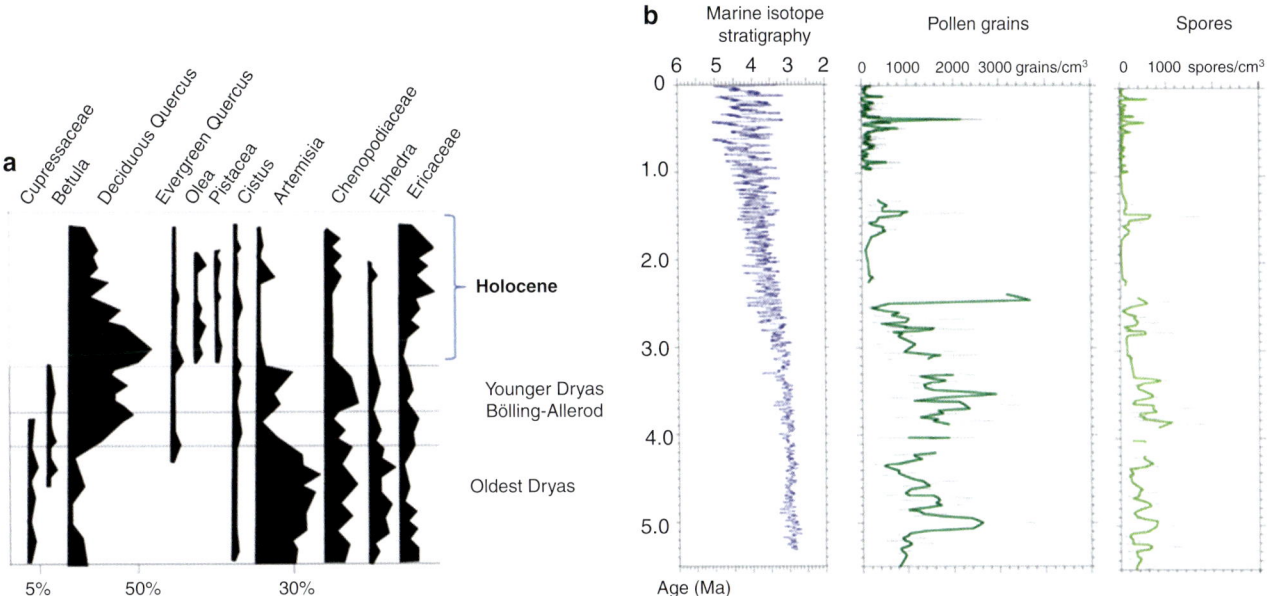

Palynology (Pollen, Spores, etc.), Figure 3 Examples of pollen diagrams from marine sediment cores. (a) Summary pollen diagram (percentages of the main taxa) of the Late and Postglacial interval (last 15,000 years) from core SU8118 (37°46′N, 10°11′W; 3,135 m water depth) off the Iberian margin (Redrafted from Sánchez Goñi et al., 2000). High percentages of *Artemisia* during the Late Glacial indicate input from open tundra or steppic-like vegetation. The overall pollen assemblages permit direct correlation with the land-based climatostratigraphy (cf. also Sánchez Goñi et al., 1999). (b) Isotope stratigraphy from Lisiecki and Raymo (2005) and diagram of pollen and spore concentrations in Pliocene and Quaternary sediments at ODP Site 646 (58°12.56 N, 48°22.15 W; water depth 3,460 m) off southwest Greenland (cf. de Vernal and Mudie, 1989). Relatively high concentrations of pollen and spores are recorded in Pliocene sediments and some interglacial stages of the Quaternary, indicating inputs from dense vegetation on the Greenland (cf. also de Vernal and Hillaire-Marcel, 2008). Pollen assemblages are dominated by gymnosperm taxa (mostly *Pinus* and *Picea*), which are selectively transported over long distance, in addition to angiosperm taxa such as *Betula* and *Alnus*. Among spores, *Lycopodium* (lycopodiophytes) and *Sphagnum* (bryophyte) dominate.

which together relate to vegetation belt displacement on adjacent continents in response to climatic changes (e.g., Hooghiemstra et al., 2006).

The use of pollen and spores in marine geosciences

Pollen and spores are generally well preserved in sediments that contain a fine silt or clay-sized matrix. While oxidation can be an issue in some continental deposits, marine sediments are usually suitable for a good conservation of pollen grains and spores. Because pollen and spores originate from land vegetation, their occurrence relates to biogenic inputs from the continent. Hence, the ratio of pollen and spore versus marine palynomorphs such as dinocysts has been used to document the origin of organic matter, allochthonous versus pelagic, as the $\delta^{13}C$-values or C/N ratios, which also serve to identify organic input from land (e.g., Thibodeau et al., 2006; de Vernal, 2009). Since the sporopollenin of palynomorphs is less affected by early diagenetic processes than other organic compounds, the use of pollen/dinocyst ratio is particularly suitable to document the source of biogenic matter in sedimentary sequences. Moreover, the nature of pollen and spores recovered in sediment may give a clue on erosional activity, notably from the occurrence of spores and pollen reworked from old sedimentary rocks (e.g., de Vernal and Mudie, 1989).

In marine geosciences, pollen and spores from sediment cores have been particularly useful to document the vegetation and climate over terrestrial areas adjacent to study sites and to establish land-sea correlations. Because such sedimentary sequences are more continuous than those in terrestrial setting, pollen and spore records from marine sediment cores revealed much useful in paleoclimatology. Examples can be given from Pleistocene and Holocene records that document large amplitude changes of climate over Western Africa (e.g., Dupont and Agwu, 1992), western North America (e.g., Heusser and Shackleton, 1979), New Zealand (Heusser and Van de Geer, 1994), or Western Europe (e.g., Sánchez Goñi et al., 2008; see Figure 3a). Other examples include records documenting variation in vegetation and climate in areas now occupied by ice sheet, such as Greenland (de Vernal and Mudie, 1989; de Vernal and Hillaire-Marcel, 2008; see Figure 3b) or Antarctica (Warny et al., 2009).

Advantages and disadvantages of palynological approaches in marine geology and paleoclimatology

Marine palynology is an important discipline in the field of marine geosciences with applications in biostratigraphy, paleoclimatology, and paleogeography. There are three main advantages of palynological approaches among other micropaleontological methods for the study of marine sediment. First, organic-walled microfossils such as pollen and spores are well preserved in sediments, whereas biogenic minerals such as opal silica and calcium carbonate are susceptible to dissolution leading to poor representation of the original diatom, foraminifer, or coccolith populations in the sediments. Second, organic-walled microfossils include representatives from several ecological settings (marine, terrestrial, lacustrine, pelagic, benthic) and all latitudes, from poles to equator, thus allowing assessment on the depositional environments, biogenic productivity, and biofacies, in addition to provide biostratigraphic markers. Third, among palynomorphs, pollen and spores occupy an important place for paleoclimatological studies as they permit to document the vegetation and climate. Hence, in marine environments, the advantage of their study is for establishing direct correlation between the continental and oceanic stratigraphy. Because terrestrial sedimentary sequences are usually discontinuous unlike marine series, pollen and spores offer a unique tool for direct land-sea correlation.

Palynological approaches, however, are moderately useful in distal offshore ocean settings characterized by oligotrophic conditions because of low concentration of palynomorphs, especially pollen and spores, making it difficult to do statistically representative counts. Another difficulty of the palynological approaches is the post-sedimentary reworking of palynomorphs that can be preserved despite erosion and subsequently transported with bottom current and redeposited. Hence, it is sometime difficult to distinguish reworked palynomorphs from others. Finally, one important limitation, which concerns mostly pollen and spores, is the fact that they necessarily relate to long-distance transport through winds and water resulting in distortion in the proportion of taxa and making it difficult to identify the precise source area.

Summary

Marine palynology is an extremely useful discipline as it may provide information on biostratigraphy in addition to environmental and climatic conditions. In particular, pollen grains and spores may give clues on terrestrial vegetation and climate, whereas the respective proportion of terrestrial and marine palynomorphs may provide indication on the paleogeography. The interpretation of palynological records from marine sediments is not unequivocal, and the transcription of pollen assemblages in terms of vegetation is not straightforward as it requires taking into account specific properties of pollen grains and spores: the species morphology and size as well as taphonomic processes such as dispersion and sedimentation. Nevertheless, marine palynology offers one of the best means for direct land-sea correlations and linkage of paleoceanography with continental paleoclimatology.

Bibliography

de Vernal, A., 2009. Marine palynology and its use for studying nearshore environments. *IOP Conference Series: Earth and Environmental Science*, **5**, 012002.

de Vernal, A., and Giroux, L., 1991. Distribution of organic walled microfossils in recent sediments from the Estuary and Gulf of St. Lawrence: some aspects of the organic matter fluxes. *Canadian Journal of Fisheries and Aquatic Sciences*, **113**, 189–199.

de Vernal, A., and Hillaire-Marcel, C., 2008. Natural variability of Greenland climate, vegetation and ice volume during the last million years. *Science*, **320**, 1622–1625.

de Vernal, A., and Mudie, P. J., 1989. Pliocene to recent palynostratigraphy at ODP sites 646 and 647, eastern and southern Labrador Sea. *Proceedings of the Ocean Drilling Program*, **105B**, 401–422.

de Vernal, A., Bilodeau, G., Hillaire-Marcel, C., and Kassou, N., 1992. Quantitative assessment of carbonate dissolution in marine sediments from foraminifer linings vs. shell ratios: example from Davis Strait, NW North Atlantic. *Geology*, **20**, 527–530.

de Vernal, A., Bilodeau, G., and Henry, M., 2010. *Micropaleontological Preparation Techniques and Analyses*. Cahier du Geotop n°3. http://www.geotop.ca/en/publications/cahiers-de-laboratoire-et-protocoles.html.

Dupont, L., and Agwu, C. O. C., 1992. Latitudinal shifts of forest and savanna in NW Africa during the Brunhes chron: further marine palynological results from site M 16415 (9°19′W). *Vegetation History and Archaeobotany*, **1**, 163–175.

Faegri, K., and Iversen, J., 1989. *Textbook of Pollen Analysis*. Chichester: Wiley.

Hemsley, A. R., and Poole, I., 2004. *The Evolution of Plant Physiology*. London: Elsevier Academic Press.

Hemsley, A. R., Barrie, P. J., Chaloner, W. G., and Scott, A. C., 1993. The composition of sporopollenin and its use in living and fossil plant systematics. *Grana*, **32**(Suppl. 1), 2–11.

Heusser, L. E., and Balsam, W. L., 1977. Pollen distribution in the northeast Pacific Ocean. *Quaternary Research*, **7**, 45–62.

Heusser, L. E., and Shackleton, N. J., 1979. Direct marine-continental correlation: 150,000-year oxygen isotope-pollen record from the North Pacific. *Science*, **204**, 837–839.

Heusser, L. E., and Van de Geer, G., 1994. Direct correlation of terrestrial and marine paleoclimatic records from four glacial-interglacial cycles – DSDP site 594 Southwest Pacific. *Quaternary Science Reviews*, **13**, 273–282.

Hooghiemstra, H., Lézine, A.-M., Leroy, S. A. M., Dupont, L., and Marret, F., 2006. Late Quaternary palynology in marine sediments: a synthesis of the understanding of pollen distribution patterns in the NW African setting. *Quaternary International*, **148**, 29–44.

Jansonius, J., and McGregor, D. C. (eds.), 1996. *Palynology: Principles and Applications*. College Station: American Association of Stratigraphic Palynology.

Kapp, R. O., Davis, O. K., and King, J. E., 2000. *Guide to Pollen and Spores*, 2nd edn. College Station: American Association of Stratigraphic Palynology.

Kenrick, P., and Crane, P. R., 1997. *The Origin and Early Diversification of Land Plants: A Cladistic Study*. Washington, DC: Smithsonian Institution Press.

Kenrick, P., Wellman, C. H., Schneider, H., and Edgecombe, G. D., 2012. A timeline for terrestrialization: consequences for the carbon cycle in the Palaeozoic. *Philosophical Transactions of the Royal Society, B: Biological Sciences*, **367**, 519–536.

Lisiecki, L. E., and Raymo, M. E., 2005. A Pliocene-Pleistocene stack of 57 globally distributed benthic $\delta^{18}O$ records. *Paleoceanography*, **20**, PA1003, doi:10.1029/2004PA001071.

Magallón, S., Hilu, K. W., and Quant, D., 2013. Land plant evolution time lines: gene effects are secondary to fossil constraints in relaxed clock estimation of age and substitution rates. *American Journal of Botany*, **100**, 556–573.

Manten, A. A., 1967. Lennart Von Post and the foundation of modern palynology. *Review of Palaeobotany and Palynology*, **1**, 11–22.

Marret, F., Scourse, J. D., Versteegh, G., Jansen, J. H. F., and Schneider, R., 2001. Integrated marine and terrestrial evidence for abrupt Congo River palaeodischarge fluctuations during the last deglaciation. *Journal of Quaternary Science*, **16**, 761–766.

Matthews, J., 1969. The assessment of a method for the determination of absolute pollen frequency. *The New Phytologist*, **68**, 161–166.

Moore, P. D., Webb, J. A., and Collinson, M. E., 1991. *Pollen Analysis*, 2nd edn. Oxford: Blackwell Scientific Publications.

Mudie, P. J., 1982. Pollen distribution in recent marine sediments, eastern Canada. *Canadian Journal of Earth Sciences*, **19**, 729–747.

Reille, M., 1992. *Pollen et Spores d'Europe et d'Afrique du Nord*. Marseille: Laboratoire de Botanique historique et Palynologie.

Rochon, A., and de Vernal, A., 1994. Palynomorph distribution in recent sediments from the Labrador Sea. *Canadian Journal of Earth Sciences*, **31**, 115–127.

Sánchez Goñi, M. F., Eynaud, F., Turon, J.-L., and Shackleton, N. J., 1999. High resolution palynological record off the Iberian margin: direct land-sea correlation for the Last Interglacial complex. *Earth and Planetary Science Letters*, **171**, 123–137.

Sánchez Goñi, M. F., Turon, J.-L., Eynaud, F., Shackleton, N. J., and Cayre, O., 2000. Direct land/sea correlation of the Eemian, and its comparison with the Holocene: a high-resolution palynological record off the Iberian margin. *Netherlands Journal of Geosciences*, **79**, 345–354.

Sánchez Goñi, M. F., Landais, A., Fletcher, W. J., Naughton, F., Desprat, S., and Duprat, J., 2008. Contrasting impacts of Dansgaard-Oeschger events over a western European latitudinal transect modulated by orbital parameters. *Quaternary Science Reviews*, **27**, 1136–1151.

Steemans, P., Le Hérissé, A., Melvin, J., Miller, M. A., Paris, F., Verniers, J., and Wellman, C. H., 2009. Origin and radiation of the earliest vascular land plants. *Science*, **324**, 353.

Sun, X., Li, X., and Beug, H.-J., 1999. Pollen distribution in hemipelagic surface sediments of the South China Sea and its relation to modern vegetation distribution. *Marine Geology*, **156**, 211–226.

Thibodeau, B., de Vernal, A., and Mucci, A., 2006. Enhanced primary productivity, organic carbon fluxes and the development of hypoxic bottom waters in the Lower St. Lawrence Estuary, Eastern Canada: micropaleontological and geochemical evidence. *Marine Geology*, **231**, 37–50.

Traverse, A., 2007. *Paleopalynology*, 2nd edn. Dordrecht: Springer.

Warny, S., Askin, R. A., Hannah, M. J., Mohr, B. A. R., Raine, J. I., Harwood, D. M., Florindo, F., The, S. M. S., and Team, S., 2009. Palynomorphs from a sediment core reveal a sudden remarkably warm Antarctica during the middle Miocene. *Geology*, **37**, 955–958.

Cross-references

Deep-sea Sediments
Dinoflagellates
Marine Microfossils
Paleoceanographic Proxies

PASSIVE PLATE MARGINS

Paul Mann
Department of Earth and Atmospheric Sciences,
University of Houston, Houston, TX, USA

Synonyms

Trailing edge margin, rifted margin, divergent margin, Atlantic-type margin, geosyncline (pre-plate tectonic term), pull-apart margin (used by some in the oil and gas community for non-strike-slip, Atlantic-type passive margins)

Definition

Passive margins are the most common type of crustal boundary on Earth with present-day passive margins having an aggregate length of 105,000 km which is greater than either spreading ridges (65,000 km) or convergent plate boundaries (53,000 km) (Bradley, 2008; Figure 1). In general, passive margins are not zones of crustal tectonic activity and therefore lack associated earthquake activity (Figure 1).

Passive margins are one of the main stages in the Wilson cycle of ocean opening and their ages, and ages of deformation are key indicators of the timing of ocean openings and closings through geologic time. Passive margins are known to contain a major part of the Earth's hydrocarbon reserves. In their compilation of giant oil fields of the world, Mann et al. (2003) identified passive margins as containing 35 % of the world's giant oil fields discovered up to 1999.

Passive margins are sedimentary wedges that overlie an inactive and subsiding weld between rifted continental crust and newly formed and younger oceanic crust. When rifting continental margins overlain by passive margins face each other, they are known collectively as "conjugate margins" as shown in the Black Sea example in Figure 2.

Rifted margins underlie passive margins

Passive margins are thick, sedimentary wedges overlying rifted, continental margins. The fundamental crustal weld or boundary for rifted continental margins is also called a "continent-ocean boundary"(COB) and can form narrow or broad features – as defined and imaged using deeply penetrating geophysical methods as seen in Figures 2, 3, 4, and 5. The stretched continental crust or "transitional crust" is generally characterized by seaward-dipping normal faults, a prominent "breakup" unconformity that records the end of the syn-rift phase, and an overlying, post-rift sag basin formed by slower and gradually dissipating thermal subsidence following rifting. Faulting deforming the passive margin section that postdates the breakup unconformity is commonly associated with gravitational sliding, rather than tectonically driven crustal deformation as observed on the western end of the Black Sea line as shown in Figure 2b.

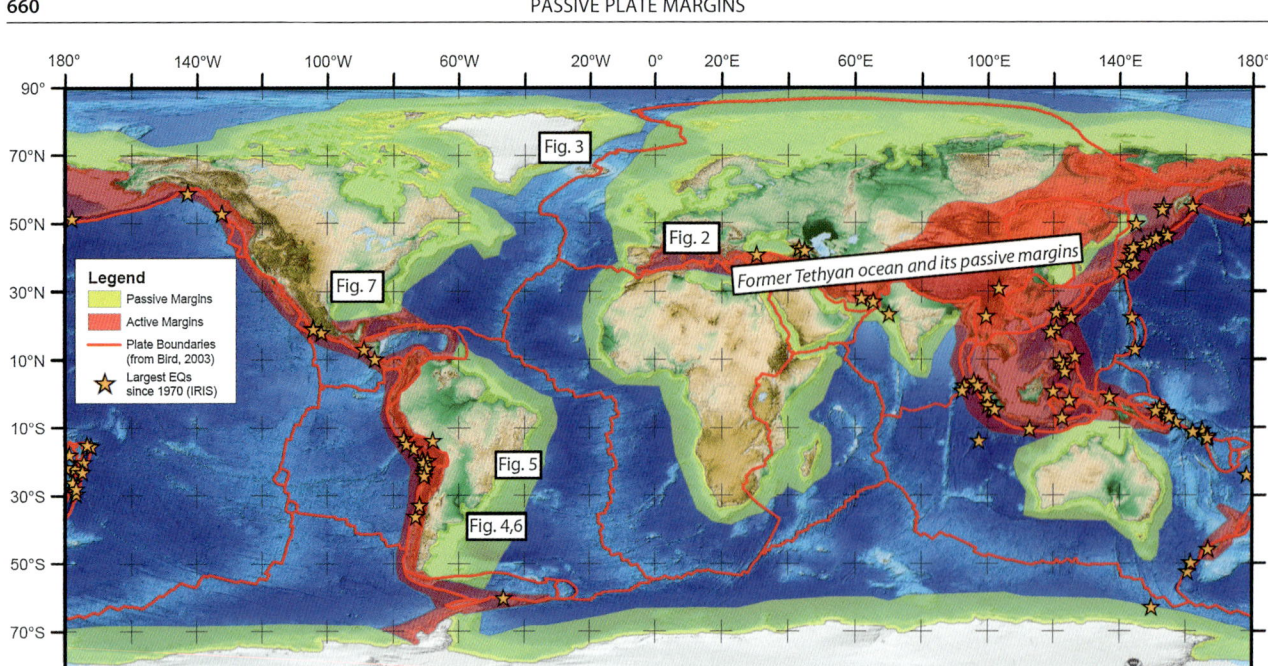

Passive Plate Margins, Figure 1 Map of passive (*green*) and active (*red*) margins of the world modified from passive margin compilation by Bradley (2008). Stars show largest earthquakes since 1970 that are confined to the active plate margins shown with red shading. Locations of data shown in Figures 2, 3, 4, 5, 6, and 7 are indicated.

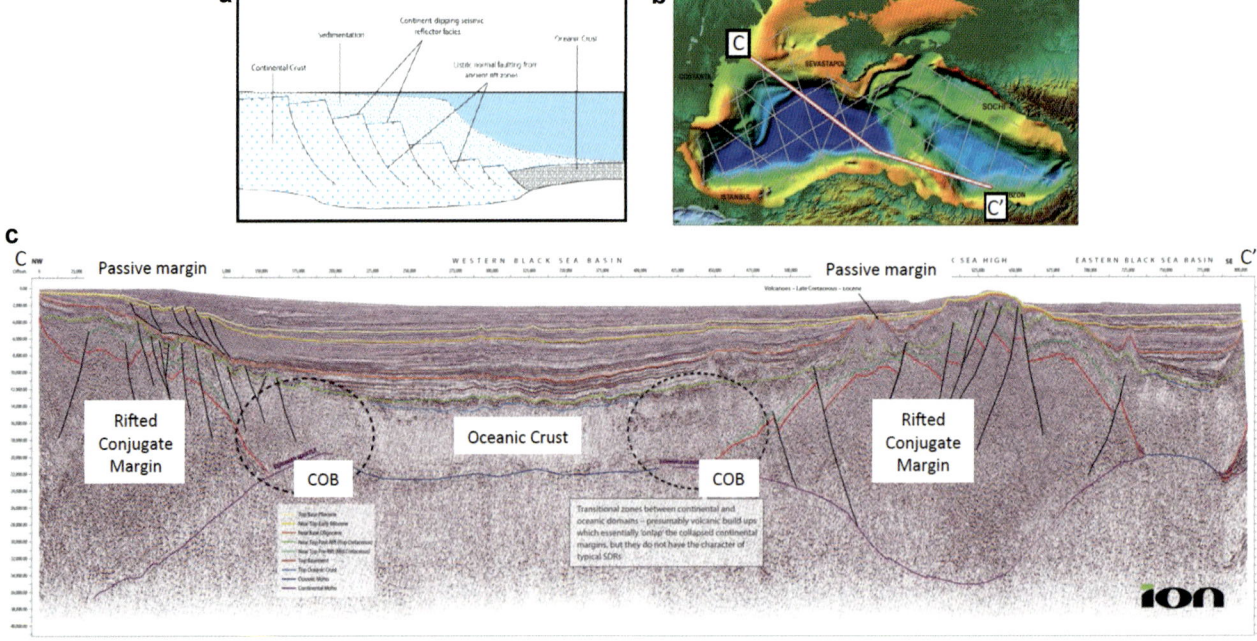

Passive Plate Margins, Figure 2 (**a**) Schematic section of a rifted continental margin overlain by a wedge of passive margin sedimentation extending seaward and tapering onto oceanic crust from https://en.wikipedia.org/wiki/Passivemargin. (**b**) Map of the Black Sea showing the location of the long seismic line shown in **c**. (**c**) Deep-penetration seismic reflection image of conjugate, continental rifted margins of the Black Sea modified from GEO ExPro (2013) that illustrate the main characteristics of passive margins and their relation to their underlying rifted margins. Note complexities in the conjugate passive margins including downslope slumping to the west and eruption of late Cretaceous-Eocene volcanoes to the east.

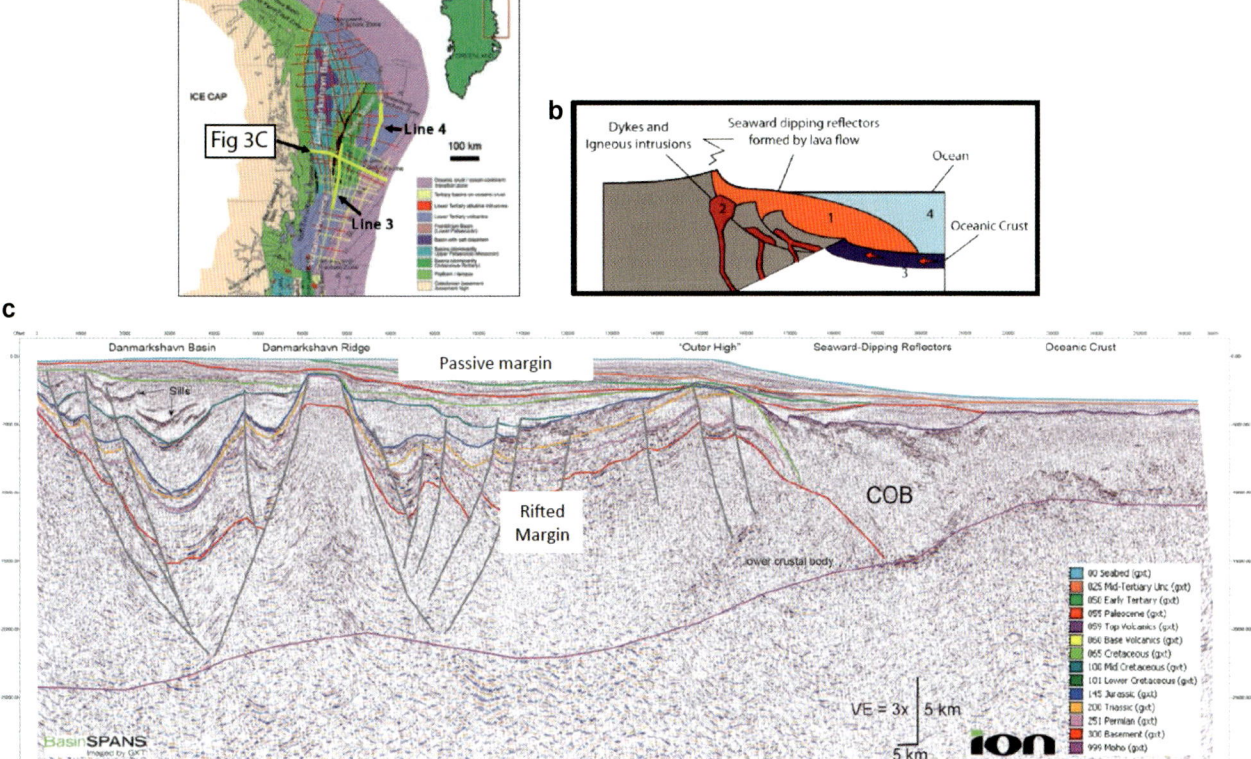

Passive Plate Margins, Figure 3 (**a**) Major tectonic elements of northeast Greenland from GEO ExPro (2010) showing existing, deep-penetration seismic data in yellow and planned lines in *red*. Location of line shown in **c** is indicated. (**b**) Schematic section of a volcanic passive margin with volcanic and intrusive rocks in *red* from https://en.wikipedia.org/wiki/Passive_margin. Note that layered, volcanic flows, or seaward-dipping reflectors, prograded onto oceanic crust in the area of the continent-ocean boundary. (**c**) Deep-penetration seismic profile (vertical scale is 30 km) showing the volcanic passive margin of northeast Greenland. The Danmarkshavn and Thetis basins are continental rift basins overlain by a relatively thin, passive margin section. The volcanic nature of the passive margin is indicated by the presence of sills in the Danmarkshavn rift, the presence of SDRs near the continent-ocean boundary (*COB*), and the nearby presence of a "lower crustal body" likely composed of high-density, intrusive rocks.

Tectonic origins of passive margins

The subsidence of passive margins is driven by thermal subsidence of the lithosphere following rifting and seafloor spreading and for that reason forms the thickest sedimentary environments on Earth with up to 18 km of sedimentary rock, especially in the presence of deltas fed by river systems draining large continents. Due to their great thickness and wide extent, sedimentary rocks originally deposited along passive margins are commonly preserved in deformed, on-land mountain belts.

Passive margins occur as conjugate margins that originated as contiguous areas of continental crust that become separated by oceanic basins of varying widths and can be reconstructed by removing the area of oceanic crust and subtracting the width of extended continental crust deformed during the early rift period. Well-known examples of passive margins include the conjugate margins of the Arctic Ocean, most of the margins of Africa, Greenland, India, Australia, and the South Atlantic Ocean as shown on Figures 2, 3, and 4.

Passive margins are distinct from active margins that are characterized by strike-slip faulting, collisional faulting, and subduction-related faulting along major plate boundaries (Lallemand, this volume). Active margins exhibit plate-driven, active subsidence and uplift, and the occurrence of earthquakes in linear belts along the plate boundary (Figure 1). In addition to thermal subsidence, passive margins can locally exhibit active, gravity-driven deformation that includes both normal faulting in the higher elevation and updip areas and thrust faulting in the lower elevation and downdip areas. Downdip areas of thrust faulting are called "passive margin thrust belts" and are a major object of deepwater oil exploration (Rowan et al., 2004; Figures 6 and 7).

Passive margins in cross section

The direction of early rifting and continental separation fundamentally affects the cross-sectional profile of passive margins. Transform passive margins can form above former transform plate boundaries characterized by

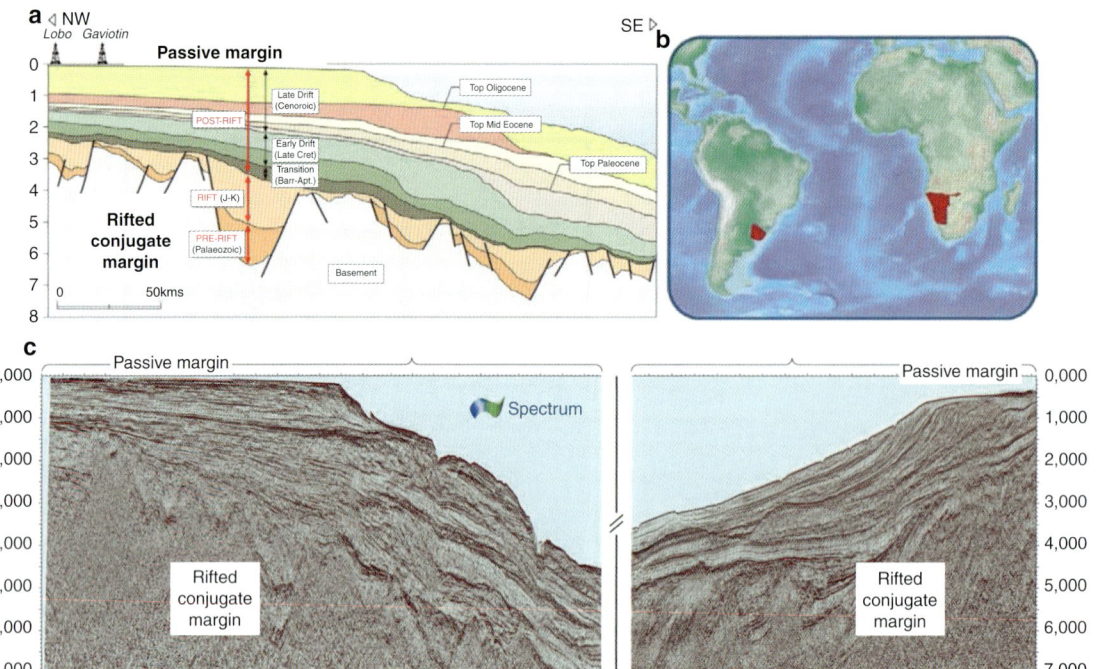

Passive Plate Margins, Figure 4 Conjugate passive margins overlying rifted continental margins of Uruguay in South America and Namibia in West Africa now widely separated by oceanic crust of the South Atlantic Ocean modified from GEO ExPro (2012). (**a**) Section showing underlying rifts overlain by passive margin of Uruguay. (**b**) Locations of conjugate margins in Uruguay and Namibia. (**c**) Seismic sections showing conjugate rift and passive margins.

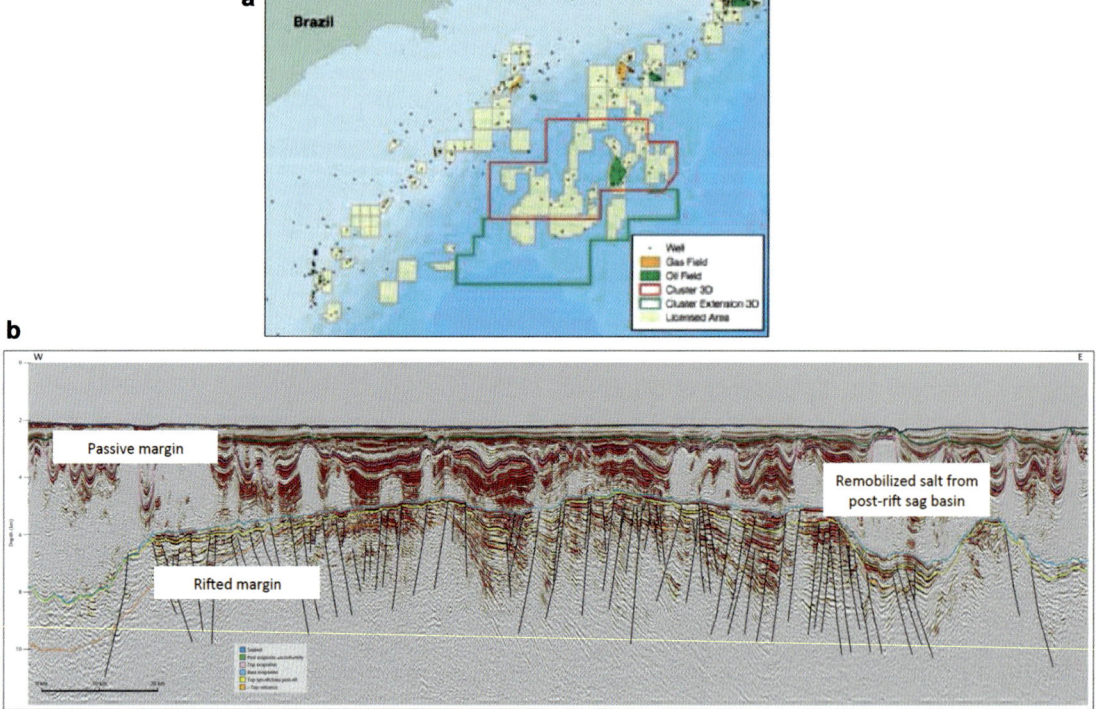

Passive Plate Margins, Figure 5 (**a**) Location of the Santos basin of offshore Brazil from GEO ExPro (2012). Seismic line in **b** is from within the area of the *red box*. (**b**) Sag basins of the passive margin modified from GEO ExPro (2012). Salt mobilization mainly occurs by sediment loading onto the salt layer, rather than downslope gravitational sliding.

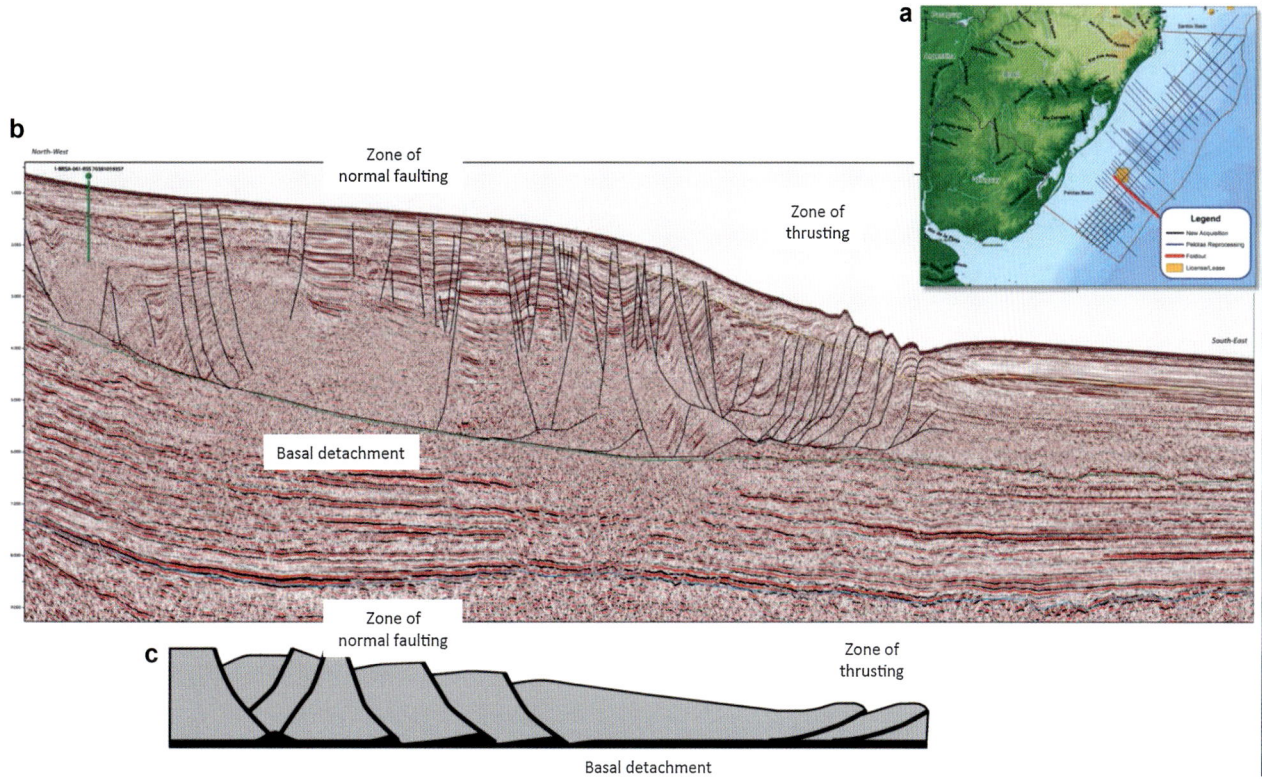

Passive Plate Margins, Figure 6 (a) Map of the southeastern coast of Brazil showing location of the deep-penetration seismic lines of the Pelotas basin; the heavy *red line* is the line shown in **b**. (b) Deep-penetration seismic line of the Pelotas passive margin fold belt modified from GEO ExPro (2013). The updip zone of normal faulting gives way to a downdip zone of folding and thrusting. (c) Schematic cross section of a passive margin fold belt from Rowan et al. (2004) illustrating updip normal faulting and downdip shortening as seen in b.

transtensional faults, pull-apart basins, and narrow, marginal continental ridges parallel to the transform fault as observed in the equatorial region of the South Atlantic Ocean. Passive margins formed above rifts accommodating orthogonal extension are generally much wider that transform passive margins as can be observed for the passive margins bounding the South Atlantic (Figure 4).

Volcanic versus nonvolcanic margins

This distinction is determined from the amount of volcanism that accompanies the rifting process. Rift-related volcanism can occur in large piles of "seaward-dipping reflectors" centered near the continent-ocean boundary as seen in the Greenland passive margin (Figure 3). Nonvolcanic boundaries lack widespread and voluminous volcanic eruptions and are generally characterized by slower rates of extension and a higher degree of crustal stretching.

Passive margin morphology and sedimentation

In profile, wedge-shaped passive margins share common morphologies including a continental shelf, a continental slope, a continental rise, and an abyssal plain, although these can vary in width, largely controlled by the extensional or strike-slip origin of the margin (Figure 4). Wedge-shaped geometries as seen on Figure 4 result from the seaward progradation of shelf facies into deeper water.

Accommodation space for continued sedimentation is driven by continued thermal subsidence of the underlying and rifted continental crust and lithosphere. Sea-level lowstands imprint passive margins with canyons incised into the outer continental rise and slope. Semicircular protuberances on passive margins generally mark the sites of either active river deltas, including the modern deltas of the Mississippi, Nile, and Amazon deltas, or ancient deltas no longer connected to an active river system. Deltaic settings on passive margins are excellent areas for the concentration of hydrocarbon resources due to the merger of high-quality, "big river" sandstone reservoirs, maturation of source rocks due to the enhanced burial by the delta, and structural traps related to gravitationally induced stabilities of the delta front. Examples of productive, large deltas include the Mississippi, Nile, and Niger.

Sag basins as transitions between rifting and passive margins

Sag basins above rifts and below passive margins represent structural and stratigraphic transitions between the rift

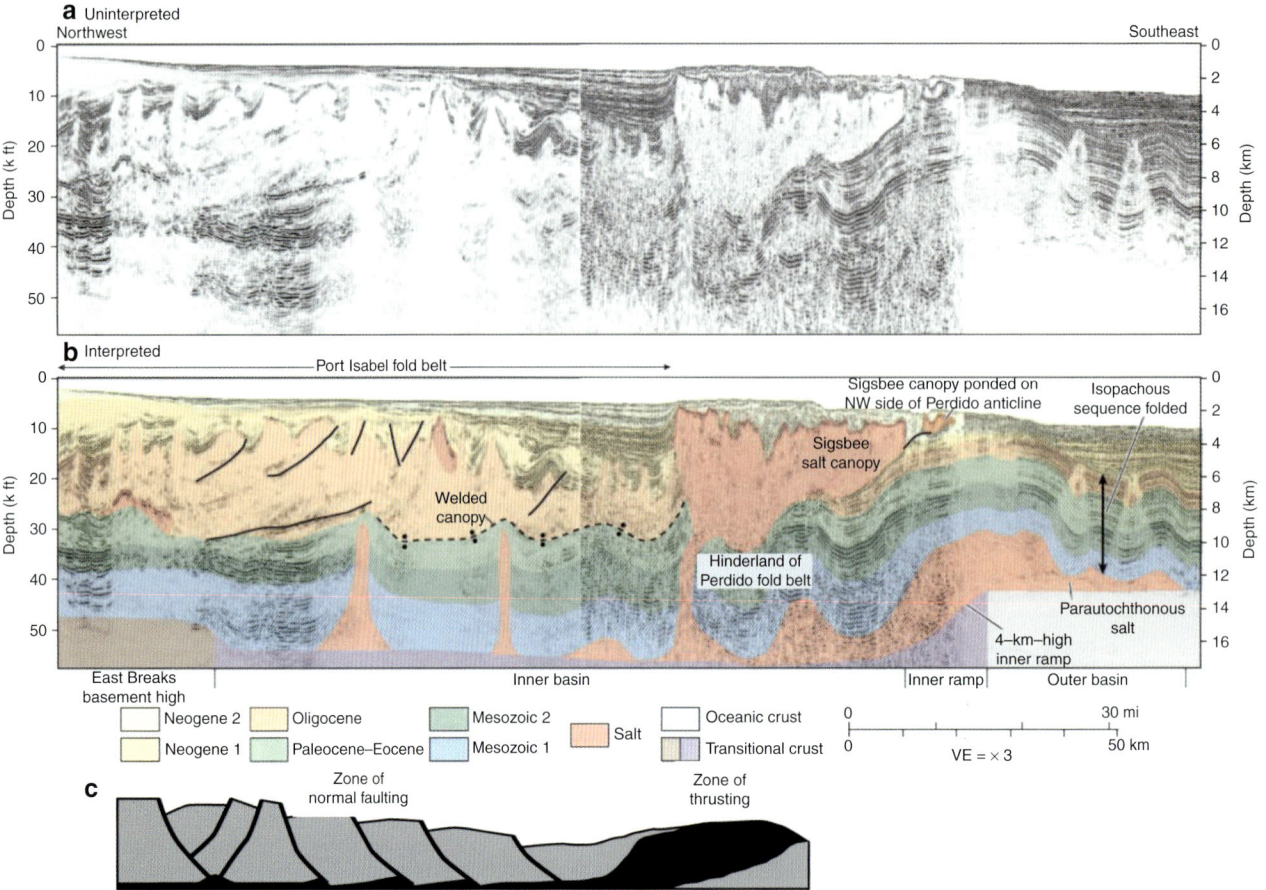

Passive Plate Margins, Figure 7 (a) Uninterpreted seismic line through the Port Isabel passive margin fold belt of the northern Gulf of Mexico from Hudec et al. (2013). (b) Interpreted seismic line from **a** showing a large fold forming above the 4-km-high step between continental crust to the northwest and oceanic crust of the southeast. (c) Schematic cross section of a passive margin fold belt from Rowan et al. (2004) showing how the extrusion of a large salt mass in the lower contractional zone.

sequence below and the unfaulted, passive margin sequence above (Figure 5). Sag basins in areas like Brazil are less deformed by gravitational movements of passive margin fold belts and more affected by salt tectonics activated by sedimentary loading (Figure 5). Sag basins have been shown in recent years to be excellent habitats for petroleum due to their more moderate faulting than the underlying rift basin, their finer-grained facies – including high-quality, lacustrine source rocks and carbonate-hosted reservoir rocks – and regional sealing by overlying evaporate deposits.

Definition and significance of passive margin fold belts

Passive margin fold belts are broad zones of deformed passive margin deepwater areas that have resulted from the horizontal translation of the post-rift cover driven by gravitational failure of the margin (Rowan et al., 2004). Margin failure – occurring on a large scale and spanning zones that are hundreds of kilometers wide – is driven by a combination of gravity gliding above a basinward-dipping detachment and gravity spreading of a sedimentary wedge with a seaward-dipping bathymetric surface. For example, the Pelotas passive margin of Brazil exhibits an updip zone of normal faults that generally dips seaward and soles onto a basal, basinward-dipping detachment (Figure 6). The upper surface of the passive margin also forms a seaward-dipping bathymetric surface. Sliding occurs on a mobile salt layer. The base of the zone is characterized by a zone of thrusting that is more narrow than the upslope zone of normal faulting (Figure 6). These thrusts also sole onto the same basinward-dipping detachment as the normal faults upslope. The material between the thrusts is folded and forms large folds at the level of the seafloor.

The Port Isabel passive margin fold belt in the northern Gulf of Mexico is wider and more complex than the Pelotas example. In this case, the downslope transport of material has occurred on two different salt detachments with the upper plane largely evacuated of salt and marked by a weld (Figure 7). A 4-km-high inner ramp marks the COB (oceanic side to the south in the Gulf of Mexico is higher; stretched continental crust to the north is lower) (Hudec et al., 2013). This preexisting crustal ramp causes the salt at both levels to ramp upward to form a large anticlinal structure in the contractional part of the passive margin fold belt that contrasts to the smaller and more distributed folds in the Pelotas example shown in Figure 4. Both the more distributed folds and single larger fold localized along the crustal ramp provide excellent structural traps for hydrocarbons in these deepwater, passive margin settings.

Bibliography

Bradley, D., 2008. Passive margins through earth history. *Earth-Science Reviews*, **91**, 1–26.
GEO ExPro, 2010. The northeast Greenland continental margin. http://assets.geoexpro.com/uploads/e4172077-a931-4725-955a-fafa7c856af4/GEO_ExPro_v7i6_Full.pdf
GEO ExPro, 2012. Perfect Atlantic analogues. http://www.geoexpro.com/articles/2012/06/perfect-atlantic-analogues
GEO ExPro, 2013. Mysteries of the Black Sea revealed. http://assets.geoexpro.com/uploads/fb83ded7-634e-4a03-92b4-5e213a94e9e0/GEO_ExPro_v10i5_Full.pdf
GEO ExPro, 2013. The Pelotas basin oil province revealed. http://www.geoexpro.com/articles/2013/12/the-pelotas-basin-oil-province-revealed
Hudec, M., Jackson, M., and Peel, F., 2013. Influence of deep Louann structure on the evolution of the northern Gulf of Mexico basin. *AAPG Bulletin*, **97**, 1711–1735.
Mann, P., Gahagan, L., and Gordon, M., 2003. Tectonic setting of the world's giant oil and gas fields. In Halbouty, M. T. (ed.), *Giant Oil and Gas Fields of the Decade, 1990–1999*. Tulsa: American Association of Petroleum Geologists. AAPG Memoir, Vol. 78, pp. 15–105.
Rowan, M., Peel, F., and Vendeville, B., 2004. Gravity-driven fold belts on passive margins. In Clay, K. R. (ed.), *Thrust Tectonics and Hydrocarbon Systems*. AAPG Memoir, Vol. 82, pp. 157–182.

Cross-references

Marine Sedimentary Basins
Oceanic Rifts
Plate Tectonics
Regional Marine Geology
Sequence Stratigraphy

PERIDOTITES

Eric Hellebrand
Department of Geology & Geophysics, SOEST, University of Hawaii, Honolulu, HI, USA

Definition

Peridotites are the main lithology of the Earth's upper mantle. According to the IUGS classification (Le Bas and Streckeisen, 1991), they are defined as a coarse-grained ultramafic rock type, containing more than 40 % olivine. Other minerals that coexist with olivine are low-Ca orthopyroxene (opx), high-Ca clinopyroxene (cpx), and chromian spinel (always present in very low modal abundance). In oceanic mantle rocks, plagioclase can also occur as a minor phase in about 20 % of all collected samples, attesting to a low-pressure (<0.8 GPa) reequilibration predicted by laboratory experiments. At high pressures (>2.5 GPa), spinel peridotites transform into garnet peridotites, which have not been found on the ocean floor, but which are occasionally brought to the Earth's surface as xenoliths in Ocean Island Basalts.

Peridotite subgroups

Peridotites are subgrouped into:

- Dunites (>90 % olivine)
- Harzburgites (<5 % cpx, >5 % opx, 40–90 % olivine)
- Wehrlites (<5 % opx, >5 % cpx, 40–90 % olivine)
- Lherzolites (>5 % opx, >5 % cpx, 40–90 % olivine)

Abyssal peridotite

Peridotites exposed on the ocean floor are called "abyssal peridotites." Hydrothermal interaction with seawater at or near the seafloor transformed most of the mantle minerals into serpentine. Despite the seawater overprint, original minerals can still be identified, and mineral proportions can be reconstructed by point counting. Most abyssal peridotites are harzburgites and lherzolites. Dunites are rare, wehrlites are exceptionally rare. Pyroxenites, which are also coarse-grained ultramafic mantle rocks, mainly occur as rare cm-dm sized veins inside peridotite (Dick et al., 1984).

Melting processes

The MORB source, depleted MORB mantle (DMM), is believed to be dominantly composed of fertile lherzolite. During adiabatic decompression melting, basaltic components are incrementally removed from the peridotite. With increasing degrees of melting, the modal abundance of cpx decreases until all cpx is exhausted, producing a refractory harzburgite. Higher degrees of melting, which would turn harzburgites into dunites, are generally not found at mid-ocean ridges. Instead, dunites are interpreted as former melt transport channels, an inference largely built on observations in ophiolites. At subduction zones, where flux melting in the presence of slab-released fluids reduces the melting temperature of peridotite, melting can continue beyond cpx-out, leading to more refractory residues.

Bibliography

Dick, H. J. B., Fisher, R. L., and Bryan, W. B., 1984. Mineralogical variation of the uppermost mantle along mid-ocean ridges. *Earth and Planetary Science Letters*, **69**, 88–106.
Le Bas, M. J., and Streckeisen, A. L., 1991. IUGS systematics of igneous rocks. *Journal of the Geological Society, London*, **148**, 825–833.

Cross-references

Cumulates
Depleted Mantle
Gabbro
Intraplate Magmatism
Magmatism at Convergent Plate Boundaries
Mid-ocean Ridge Magmatism and Volcanism
Ophiolites

PHOSPHORITES

Hermann Kudrass
MARUM-Center for Marine Environmental Sciences,
University of Bremen, Bremen, Germany

Definition

Phosphorite is a sedimentary component consisting of a significant portion of phosphate minerals, mostly francolite of the apatite group. Some large deposits with a high concentration of phosphorite possibly have an economic value.

Phosphorite types and regional distribution

Phosphorite mainly consists of microcrystalline francolite, a calcium fluorapatite, which is chemically and mechanically stable in most marine sedimentary environments. Phosphorite occurs in various forms and shapes as pellets, ooids, peloids, nodules, laminae, crusts, and friable nodules; as phosphatized shells, tests, whale bones, and coprolites; as phosphatized limestone clasts, chalk surfaces, and hardgrounds, as well as molds of foraminiferal tests and mollusk shells; and as a fill of cracks in manganese crusts. This variety reflects different types of authigenic precipitation, diagenetic replacement, and repeated mechanical reworking (Bentor, 1980; Föllmi, 1996). Fine-grained phosphorites and friable phosphatic nodules are often dispersed in diatomaceous ooze, while the coarser phosphoritic components are typical for sedimentary hiatus and often associated with glauconitic sand.

Most of the phosphorite-bearing sediments occur in the condensed sediment sequences of the western tropical and subtropical continental margins, which have a long history of upwelling (Figure 1). The predominant portion of this phosphorite is relict deposit from the Miocene. Isolated phosphorite occurrences on seamounts of the western Pacific Ocean are mostly early tertiary phosphorite-bearing manganese crusts or fossil guano deposits.

Authigenic precipitation and diagenetic phosphatization

Modern precipitation of phosphorite is confined to the upwelling areas off Baja California/Mexico (Jahnke et al., 1983), Peru (Burnett et al., 2000), Chile (Schulz and Schulz, 2005), Namibia (Baturin 2000; Goldhammer et al., 2010), and the margins of the Arabian Sea (Schenau et al., 2000; Figure 1). In these fertile areas the high flux of marine organic detritus with a C:P ratio of 106–117:1 from the productive zone at the surface to the sea floor is the main sink of phosphorus in the oceans (Benitez-Nelson, 2000).

The decay of the highly reactive organic detritus creates suboxic to anoxic conditions in the water masses on the shelf and upper continent slope. In this environment abundant colonies of the sulfur bacteria like the genera *Beggiatoa*, *Thioploca*, and *Thiomargarita* flourish in uppermost 10-cm layer of the Corg-rich sediments. These bacteria feed on the rich organic material by oxidizing hydrogen sulfide from the mud with nitrate from the sea water. In their cells they accumulate polyphosphate as an additional energy supply, which can be rapidly released during oxic/anoxic changes of the bottom water (Schulz and Schulz, 2005; Goldhammer et al., 2010). Consequently the phosphate concentration in the pore water becomes high enough to precipitate as Ca phosphate (Schenau et al., 2000), to phosphatize carbonate components, or to precipitate with calcium and fluorine as francolite. The authigenic francolite or its precursor phases preferentially aggregate on existing phosphorite grains forming laminae or ooids (Arning et al., 2009). In addition, but with much smaller contributions, dissolved phosphorus near the surface of the upwelling mud can be released by the comparatively slow decay of organic material, by the desorption from oxic/hydroxic iron phases (Froelich et al., 1988), and the direct input from organic phosphates from fish and cetacean teeth and bones (Suess, 1981).

Formation of phosphorite mineral deposits

Deposition of marine organic matter in coastal upwelling systems is the main sink for phosphorus in the oceans (Benitez-Nelson, 2000), and most phosphorite deposits presumably originated in this oceanographic setting. As the content of phosphorus in the marine organic matter is close to 1 %, the accumulation of a large phosphorite deposit requires many cycles of mud deposition and phosphorus sequestration. The recent diatomaceous ooze of the Namibia upwelling system, which is accumulating with a rate of 0.5 cm year^{-1}, contains about 6 µmol cm^{-3} phosphorus in the marine organic matter (Goldhammer et al., 2010). The entire organic-bound phosphorus contained in 30-m-thick Namibian mud would produce a condensed layer of 0.6 g phosphorus cm^{-2} accumulated in 6,000 years. However, as only a small portion of the phosphorus of the organic matter is finally transferred into phosphorite (Föllmi, 1996; Schenau et al., 2000) and as the bacterial-mediated phosphorite precipitation is concentrated in the uppermost 10 cm of sediment, the upwelling sediments had to be frequently removed for concentrating the phosphoritic grains.

Thus upwelling, the associated phosphorite precipitation and episodic reworking must have continued over much longer periods to produce a sizable phosphorite

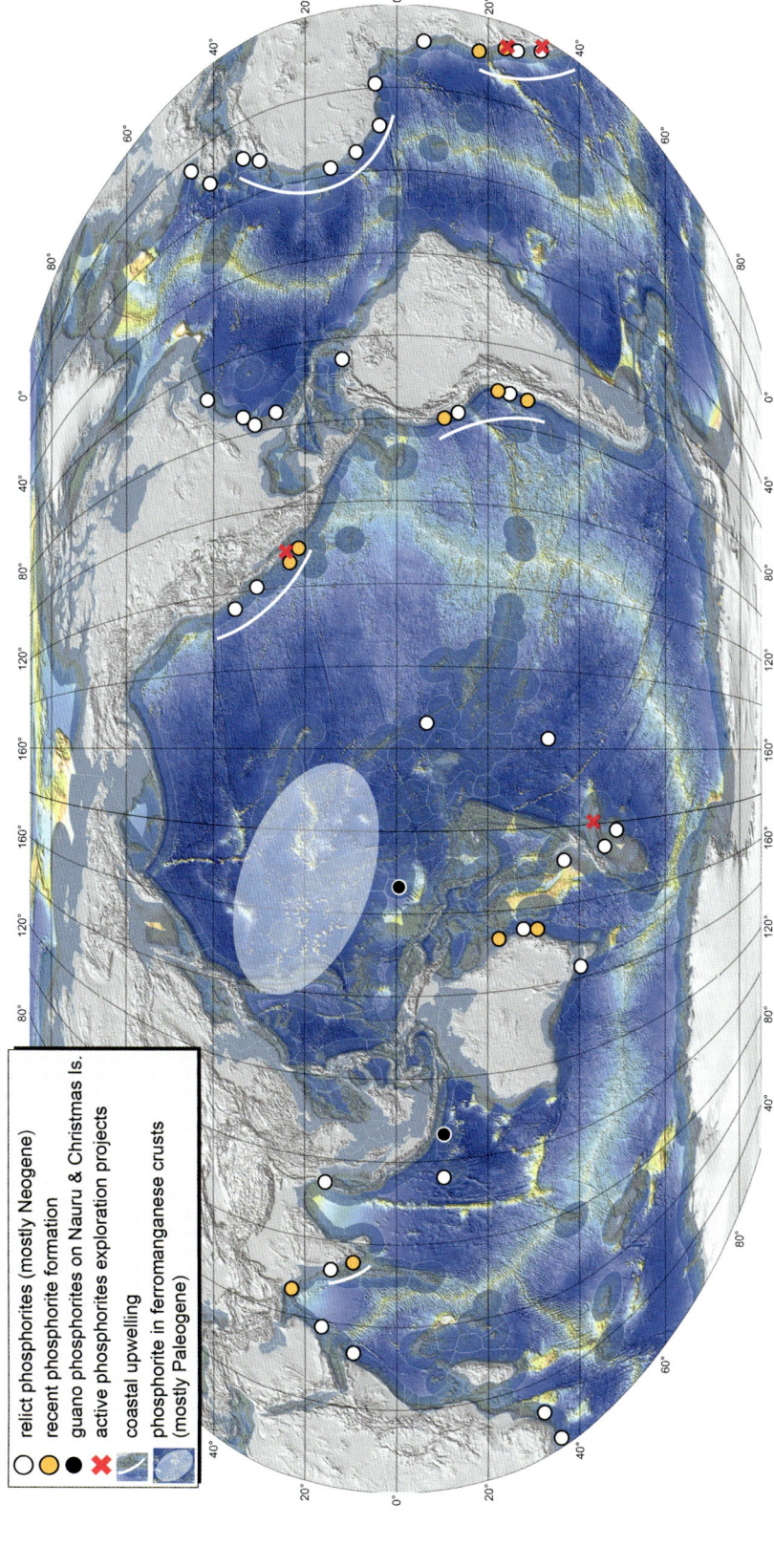

Phosphorites, Figure 1 Distribution of upwelling areas and occurrences of relict and recent phosphorites (Modified from Föllmi 1996). Ongoing offshore exploration projects are indicated.

deposit. Climatic cycles like the presently active El Niño/ La Niña changes of the upwelling areas off Peru and Chile with their drastically changing current regimes offer a suitable model for reworking with a decadal frequency. Sea-level changes with a much smaller frequency are further significant agents to enrich the dispersed phosphorite aggregates in lag deposits (Föllmi, 1996). Thus, the larger phosphorite deposits off South Africa (Compton et al., 2002), Baja California/Mexico (D'Anglejan 1967), and on the Chatham Rise/New Zealand (Kudrass and von Rad, 1984), which are presently explored by various industrial groups, have a complex accumulation history, which probably extends over several hundred thousand years (Arning et al., 2009; Burnett et al., 2000).

Summary

Phosphorite deposits are confined to upwelling at continental margins, where a high flux of organic material supplies phosphorus to the sea floor. In this environment anaerobic sulfur bacteria concentrate and release phosphorus, which then precipitates as francolite. The formation of a phosphorite deposit requires a long-term delicate balance of mud sedimentation and episodic reworking, and consequently phosphorite deposits only accumulate at continental margins with a very slow average sedimentation rate.

Bibliography

Arning, E. T., Lückge, A., Breuer, C., Gussone, N., Birgel, D., and Peckmann, J., 2009. Genesis of phosphorite crusts off Peru. *Marine Geology*, doi:10.1016/j.margeo.2009.03.006.

Baturin, G. N., 2000. Formation and evolution of phosphorite grains and nodules on the Namibian shelf, from recent to Pleistocene. In Glenn, C. R., Prevot-Lucas, L., and Lucas, J. (eds.), *Marine Authigenesis: From Global to Microbial. Society of Economic Paleontologists and Mineralogists, Special Publication*. Tulsa Oklahoma U S A, Vol. 66, pp. 185–199.

Benitez-Nelson, C. R., 2000. The biogeochemical cycling of phosphorus in marine systems. *Earth-Science Reviews*, **51**, 109–135.

Bentor, Y. K., 1980. (ed.), *Marine Phosphorites – Geochemistry, Occurrence, Genesis. Society of Economic Paleontologists and Mineralogists, Special Publication*. Tulsa Oklahoma U S A, Vol. 29, pp. 1–249.

Burnett, W. C., Glenn, C. R., Yeh, C. C., Schultz, M., Chanton, J., and Kashgarian, M., 2000. U-series, ^{14}C, and stable isotope studies of recent phosphatic "protocrusts" from the Peru margin. In Glenn C. R., Prevot-Lucas L., and Lucas J., (eds.), *Marine Authigenesis: From Global to Microbial. Society of Economic Paleontologists and Mineralogists, Special Publication*. Tulsa Oklahoma U S A, Vol. 66, pp. 163–183.

Compton, J. S., Mulabasina, J., and McMillan, I. K., 2002. Origin and age of phosphorite from the Last Glacial Maximum to Holocene transgressive succession off the Orange River, South Africa. *Marine Geology*, **186**, 243–261.

D'Anglejan, B. F., 1967. Origin of marine phosphorites off Baja California, Mexico. *Marine Geology*, **5**, 15–44.

Föllmi, B., 1996. The phosphorus cycle, phosphogenesis and marine phosphate-rich deposits. *Earth-Science Review*, **40**, 55–124.

Froelich, P. N., Arthur, M. A., Burnett, W. C., Deakin, N., Hensley, V., Jahnke, R., Kaul, L., Kim, K.-H., Roe, K., Soutar, A., and Vathakanon, C., 1988. Early diagenesis of organic matter in Peru continental margin sediments: phosphorite precipitation. *Marine Geology*, **80**, 309–343.

Goldhammer, T., Brüchert, V., Ferdelmann, T. G., and Zabel, M., 2010. Microbial sequestration of phosphorus in anoxic upwelling sediments. *Nature Geoscience*, **3**, 557–561, doi:10.1038/ngeo913.

Jahnke, R. A., Emerson, S. R., Roe, K. K., and Burnett, W. C., 1983. The present day formation of apatite in the Mexican continental margin sediments. *Geochimica Cosmochimica Acta*, **47**, 259–266.

Kudrass, H. R., and von Rad, U., 1984. Geology and some mining aspects of the Chatham Rise phosphorite: a synthesis of the SONNE-17-results. *Geologisches Jahrbuch*, **D65**, 23–252.

Schenau, S. J., Slomp, C. P., and De Lange, G. J., 2000. Phosphogenesis and active formation in sediments from the Arabian Sea oxygen minimum zone. *Marine Geology*, **169**, 1–20.

Schulz, H. N., and Schulz, H. D., 2005. Large sulphur bacteria and the formation of phosphorite. *Science*, **307**, 416–418.

Suess, E., 1981. Phosphate regeneration from sediments of the Peru continental margin by dissolution of fish debris. *Geochimica Cosmochimica Acta*, **45**, 577–588.

Cross-references

Anoxic Oceans
Deep Biosphere
Marine Mineral Resources
Upwelling

PLACER DEPOSITS

Hermann Kudrass
MARUM-Center for Marine Environmental Sciences, University of Bremen, Bremen, Germany

Placer deposits are accumulations of dense "heavy" minerals ($\rho > 2.85$ g cm^{-3}), which are enriched in comparison to the common "light" minerals ($\rho < 2.65$ g cm^{-3}) like quartz and feldspar. The placer minerals must be mechanically and chemically stable to survive weathering and transport in rivers and along the beach. Most of the placer minerals have a grain size of fine and medium sand (63–250 μm), and they can only be enriched to form commercially interesting deposits under special sorting conditions at the beach. In the highly turbulent flow of the waves entering the surf zone, light and heavy minerals are jointly transported according to the principles of hydraulic equivalent diameters (Figure 1a, Rittenhouse, 1943), i.e., grains of larger light minerals are suspended and deposited together with smaller heavy minerals with the same settling velocity (Figure 1a, b). A strong vertical gradient develops in the laminar flow at the bottom-water interface of the water masses flowing back from the beach. Thus, larger grains of light minerals are faster transported, and the smaller heavy minerals are enriched as a lag deposit (Figure 1c). With a continuing supply of unsorted sediment, this panning and film sizing processes can form finely laminated heavy-mineral deposits (Figure 1d).

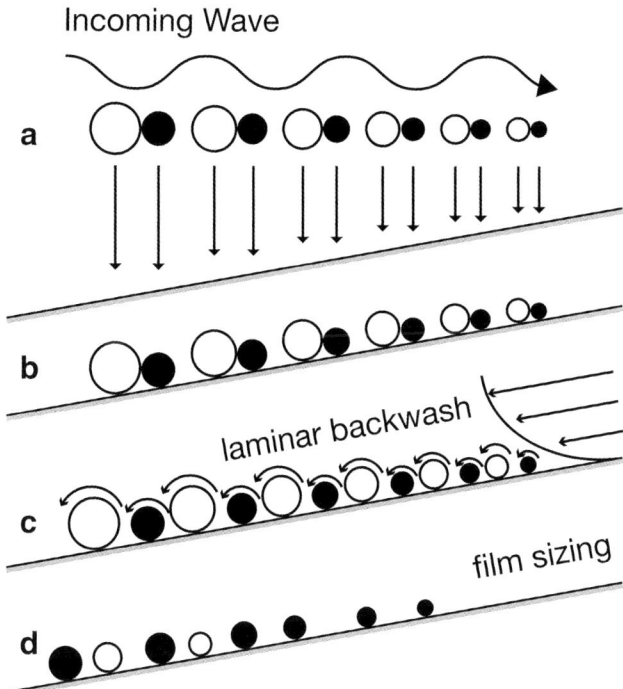

Placer Deposits, Figure 1 Transport and enrichment of light (*circles*) and heavy minerals (*solid*) in the surf zone (modified from Seibold and Berger, 1993).

Rutile (TiO_2), ilmenite ($FeTiO_3$), magnetite (Fe_3O_4), zircon ($ZrSiO_4$), monazite ($(Ce,La,Th)PO_4$), and garnet ($Fe_3Al_2(SiO_4)_3$) are the commercially most important minerals in these beach placer deposits. In most shelf areas, the placer deposits formed during glacial periods of lower sea-level were not preserved, as the bulldozing effect of the transgrading sea completely destroyed the former deposits or partially moved them landwards (Kudrass, 1987). The Holocene transgression disseminated a rich placer deposit with ilmenite, rutile, and zircon on the outer shelf off the Zambezi mouth/Mozambique (Beierdorf et al., 1980). The magnetite-bearing shelf sediments off the North Island/New Zealand are another potential offshore placer deposit.

The more dense heavy minerals like cassiterite (SnO_2, $\rho = 7$ g cm^{-3}) and gold (Au, $\rho > 15$ g cm^{-3}) usually are coarser grained, and consequently these minerals are transported as bed load in the river beds. Their offshore deposits as the cassiterite deposits of Sundaland between Indonesia and Malaysia are exploited from drowned river valleys, which originated during glacial periods of lower sea-level. Diamonds (C, $\rho = 3.5$ g cm^{-3}), the most valuable heavy mineral, are presently mined on the shelf off South Africa and Namibia from beach deposits and the coarse sediments directly overlying the rocky bedrock (Garnett, 2000a). Moraines of the high latitudes may locally contain rich gold placers in Alaska (Garnett, 2000b) and offshore New Zealand.

Bibliography

Beierdorf, H., Kudrass, H. R., and von Stackelberg, U., 1980. Placer deposits of ilmenite and zircon on the Zambezi shelf. *Geologisches Jahrbuch*, **D-36**, 5–85.

Garnett, R. H. T., 2000a. Marine placer diamonds, with particular reference to southern Africa. In Cronan, D. S. (ed.), *Handbook for Marine Mineral Deposits*. Boca Raton: CRC Press, pp. 102–141.

Garnett, R. H. T., 2000b. Marine placer gold, with particular reference to Nome, Alaska. In Cronan, D. S. (ed.), *Handbook of Marine Mineral Deposits*. Boca Raton: CRC Press, pp. 6–101.

Kudrass, H. R., 1987. Sedimentary models to estimate the heavy-mineral potential of shelf sediments. In Teleki, P. G. (ed.), *Marine Minerals*. Dordrecht: Reidel Publishing Company, pp. 39–59.

Rittenhouse, G., 1943. Transportation and deposition of heavy minerals. *Geological Society of America Bulletin*, **54**, 1725–1780.

Seibold, E., and Berger, W. H., 1993. *The Sea Floor, An Introduction to Marine Geology*. Berlin: Springer.

Cross-references

Beach
Beach Processes
Marine Mineral Resources

PLATE MOTION

R. Dietmar Müller and Maria Seton
School of Geosciences, The University of Sydney, Sydney, NSW, Australia

Definition

Plate motion can be relative or absolute. Relative plate motion describes the motion of one tectonic plate relative to another. Absolute plate motion describes the motion of one plate relative to a fixed reference system. Plate motion can be described by a pole of rotation and an angular velocity about this pole.

Introduction

A tectonic plate is defined as a portion of the outer shell of the Earth that moves coherently as a rigid body without any significant internal deformation over geological timescales. The Earth's surface is composed of a mosaic of rigid plates that move relative to one another over hotter, more mobile mantle material. Plates interact at plate boundaries, which are dynamically evolving and continuous features.

History

Alfred Wegner's idea of "continental drift" (Wegener, 1915) to explain the geometrical, geological, environmental, and paleontological similarities between now distant continents lacked a physically plausible mechanism to explain the vast distances travelled by the continents. The interpretation of ocean floor data during a rapid increase in seafloor mapping after World War II led Hess

(1962) and Dietz (1961) to propose the concept for seafloor spreading. They suggested that new seafloor is created at mid-ocean ridges, where a cold, strong surface boundary layer diverges, dividing tectonic plates. The new seafloor then spreads away from the mid-ocean ridge as it ages and is eventually subducted at deep-sea trenches, where it detaches from the surface and is recycled back into the Earth's convecting mantle. Wilson (1965) developed the concept of plates and transform faults, suggesting that the active mobile belts on the surface of the Earth are not isolated but continuous and that these mobile belts, marked by active seismicity, separate the Earth into a rigid set of plates. These active mobile belts consist of ridges where plates are created, trenches where plates are destroyed, and transform faults, which connect the other two belts to each other.

Euler's theorem

All tectonic plates can be viewed as rigid caps on the surface of a sphere. The motion of a plate can be described by a rotation about a virtual axis that passes through the center of the sphere (Euler's theorem). In terms of the Earth, this implies that a single angular velocity vector originating at the center of the globe can describe the motion of a plate. The most widespread parametrization of such a vector is using latitude and longitude to describe the location where the rotation axis intersects the surface of the Earth and a rotation rate that corresponds to the magnitude of the angular velocity (in degrees per million years or microradians per year). The latitude and longitude of the angular velocity vector constitute the so-called Euler pole (Figure 1).

Plate tectonic hypothesis

The formal hypothesis of plate tectonics states that the Earth is composed of an interlocking internally rigid set of plates in constant motion. These plates are rigid except at plate boundaries, which are lines between adjoining plates. The relative motion between plates gives rise to earthquakes; these earthquakes in turn define the plate boundaries. There are three types of plate boundaries: (1) divergent boundaries, where new crust is generated as the plates diverge; (2) convergent boundaries, where crust is destroyed as one plate dives beneath another; and (3) transform boundaries, where crust is neither produced nor destroyed as the plates slide horizontally past each other. Not all plate boundaries are as simple as the main types discussed above. In some regions, the boundaries are not well defined because the deformation there extends over a broad belt, a plate-boundary zone (Gordon, 2000). These plate-boundary zones tend to have complicated geological structures and earthquake patterns as they involve at least two large plates and one or more microplates caught up between them.

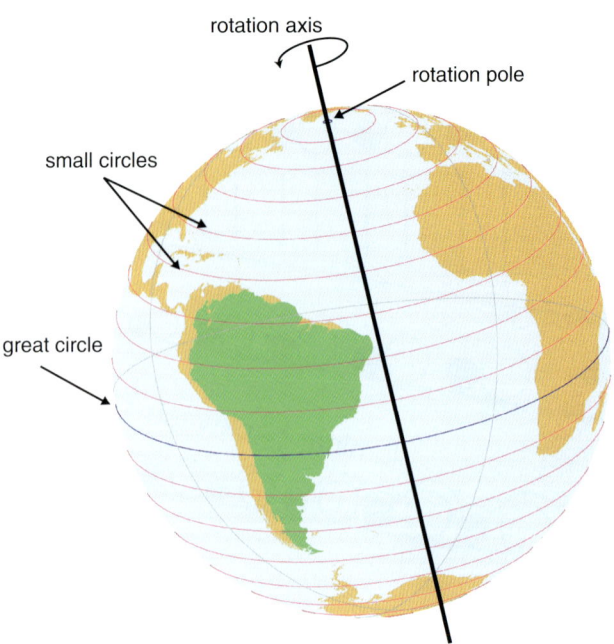

Plate Motion, Figure 1 Euler's theorem describes the motion of a tectonic plate by a rotation about a virtual axis that passes through the center of the sphere. The Euler poles are the intersections between the rotation axis and the surface of the sphere. *Small circles* describe the angular motion of a given plate. The set of *small* and *great circles* describing the rotation pole coordinate system is equivalent to a tilted (rotated) version of the familiar geographic coordinate system of the Earth represented by parallels and meridians.

Relative plate motions

Present-day plate motions

Relative plate motions at the present are represented by so-called instantaneous rotations described by angular velocity vectors and follow standard vector algebra, i.e., the angular velocity vectors $_A\mathbf{w}_B$ and $_B\mathbf{w}_C$ describing the rotations of plate B relative to plate A and plate C relative to plate B can simply be added vectorially to obtain $_A\mathbf{w}_C$, the rotation of plate C relative to plate A. Models of present-day plate motions are often called "geologically current" plate models and can be measured using satellite technology (e.g., GPS measurements (Argus and Heflin, 1995) or space geodesy (Argus et al., 2010)) and/or a number of other observations:

1. The orientation of active transform faults between two plates can be used to compute their direction of relative motion. The relative motion of two plates sharing a mid-ocean ridge is assumed to be parallel to the transform faults and in turn assumed to follow small circles. Based on two or more transform faults between a plate pair, the intersection of the great circles, which are perpendicular to the small circles paralleling transform faults, approximates the position of the rotation pole (e.g., Morgan, 1968).

2. The spreading rates along a mid-ocean ridge as determined from magnetic anomaly patterns (e.g., Müller et al. (2008)) can be used to compute a rotation pole, since the spreading rate varies as the sine of the colatitude (i.e., angular distance) from the rotation pole.
3. Fault plane solutions (focal mechanisms) of earthquakes at plate boundaries can be utilized to compute the direction of relative motion between plates. Only the location of the pole, but not the spreading rate, can be determined this way. A global model for current plate motions based on data of types (1), (2), and (3) has been constructed and reviewed by DeMets et al. (2010), utilizing data which cover 3.2 million years of plate motion.
4. Geological markers are used along plate boundaries on land, in particular along strike-slip faults, to determine geologically recent local relative motion. Markers used include streambed channels, roads, and field boundaries that have been offset by strike-slip motion.
5. A satellite method called very-long-baseline interferometry (VLBI) uses quasars as signal source and terrestrial radio telescopes as receivers. The difference in distance between two telescopes is measured over a period of years.
6. Satellites have made it possible to measure present-day plate motions in real time at many more stations than possible using the VLBI technique, which depends on permanent radio telescopes as the receivers. Satellite laser ranging techniques using the Global Positioning System (GPS) have been used very successfully recently to measure plate motions all over the world, especially in areas that are subject to earthquake hazards. A current plate motion model based on a few decades of GPS data was constructed by Kogan and Steblov (2008).
7. Coupled geodynamic plate models, which model plate boundary locations and mantle density heterogeneity, have been used to predict present plate motions (e.g., (Stadler et al., 2010)). These predicted motions are entirely model driven and sensitive to the mantle properties used but can be compared to present-day observations.

Plate motions through geological time

The reconstruction of plate motions through geological "deep" time requires the use of finite rotations whose manipulation is considerably more complex than those used for current plate motion (Cox and Hart, 1986). Relative plate motions for plate pairs which have preserved ocean floor generated by seafloor spreading at a common mid-ocean ridge can be reconstructed by fitting marine magnetic anomalies and fracture zones from conjugate ridge flanks. The Earth's magnetic field experiences reversals over geological time, leading to linear bands of ocean crust magnetized during a normal or reversed state of the Earth's magnetic field as seafloor spreading progresses. Thus, the ocean floor acts like a giant tape recorder and the magnetic anomalies caused by the alternating bands of normal and reversely polarized crust on the ocean floor can be fitted back together with the additional aid of fracture zones (Matthews et al., 2011), which are passive traces left on the seafloor by the transform fault offsets at mid-ocean ridges. A qualitative method for fitting such data visually ("visual fitting method") is provided by the community GPlates software (Boyden et al., 2011), and a quantitative method including the computation of uncertainties ("Hellinger method") has been developed by Hellinger (1981) and Chang (1988).

When only one flank of a spreading system is preserved due to partial subduction of an ocean basin, the computation of rotations describing plate motion is more complicated and relies on a "half-stage rotation" methodology described by Stock and Molnar (1988). This method involves computing a stage rotation between adjacent seafloor spreading isochrons on one flank and doubling the angle (assuming spreading symmetry) to obtain a "full-stage rotation." In instances where crust from both ridge flanks has been subducted, we rely on the onshore geological record (e.g., mapping of major sutures, terrane boundaries, and active and ancient magmatic arcs) to help define the locations of paleo-plate boundaries and use inferences from younger, preserved crust to estimate earlier spreading directions and rates. Where continental terranes have crossed ocean basins, a combination of paleomagnetic and geological data can be used to reconstruct terrane migration (Stampfli and Borel, 2002) and the now subducted ocean basins (Seton et al., 2012).

Absolute plate motions

Absolute plate motions represent the motion of individual plates relative to a reference system regarded as fixed, such as the Earth's mesosphere or the center of the Earth, in accordance with the forces that drive the plates. Relative plate motions are merely consequences of absolute plate motions, yet absolute plate motions are much more difficult to reconstruct. The main methodologies to constrain absolute plate motions are based on paleomagnetic data and volcanic hot spot tracks (Torsvik et al., 2008). Paleomagnetic data can be used to determine the orientation and latitude of a plate through time. However, since the Earth's magnetic dipole field is radially symmetric, paleo-longitudinal information cannot be deduced from paleomagnetic data alone. Seamount chains with a linear age progression (i.e., hot spot tracks) can be used to restore plates to their paleopositions with the assumption that hot spots are either fixed or nearly fixed relative to each other ("fixed hot spot hypothesis" (Morgan, 1971)). In considering seamounts and island chains as markers of past plate motion over stationary hot spots, Morgan (1971, 1972) suggested that such trails should form sets of co-polar small circles on the Earth's surface as a consequence of Euler's theorem for plate motions on a sphere. Wessel and Kroenke (1997) refined this method by introducing

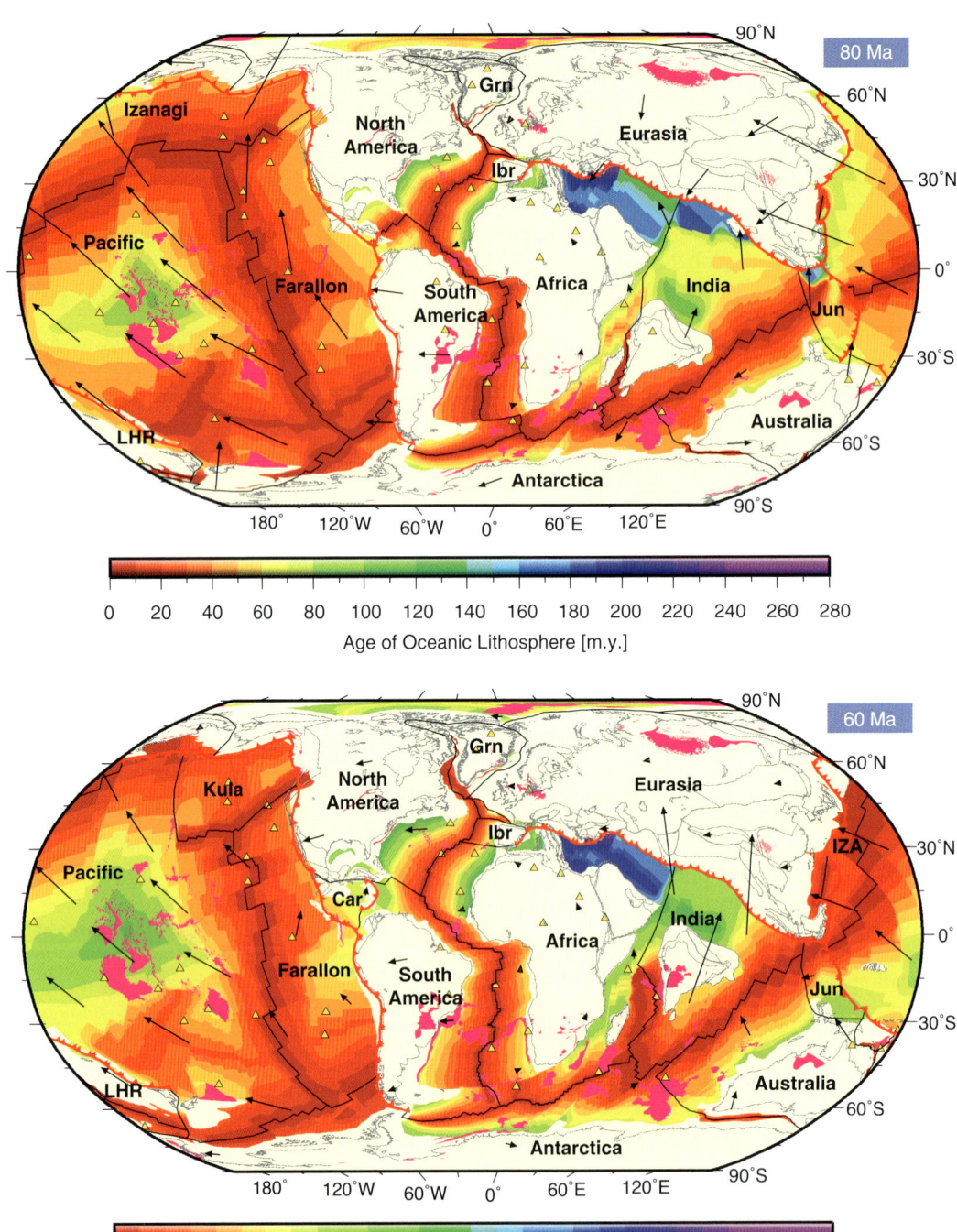

Plate Motion, Figure 2 (continued)

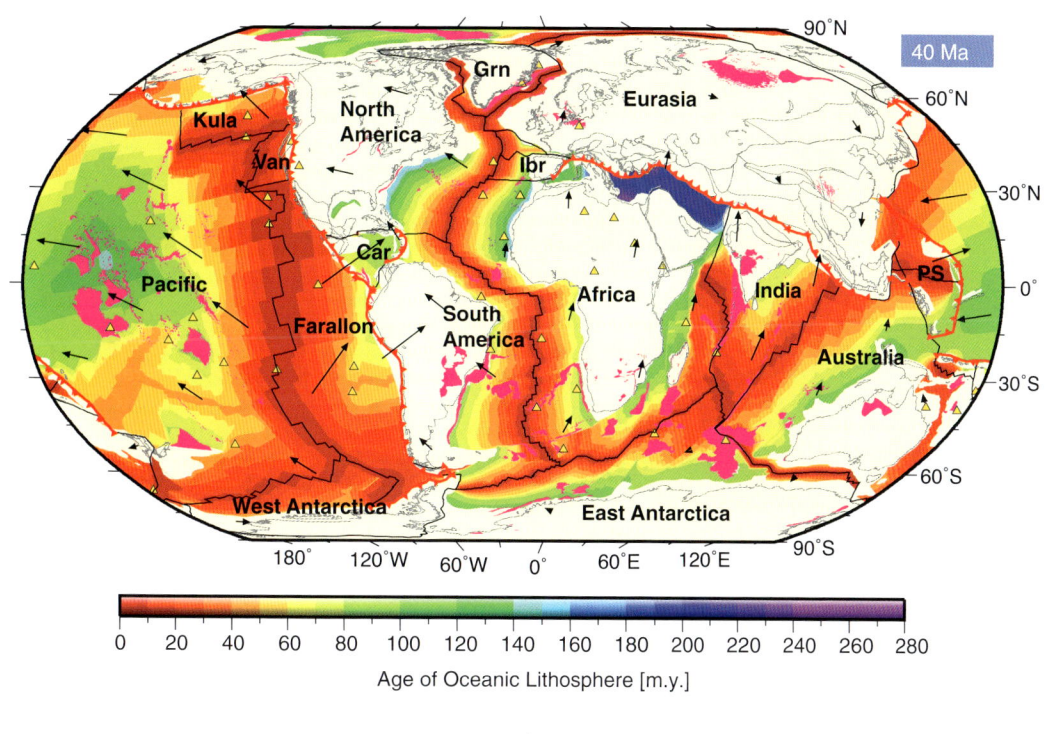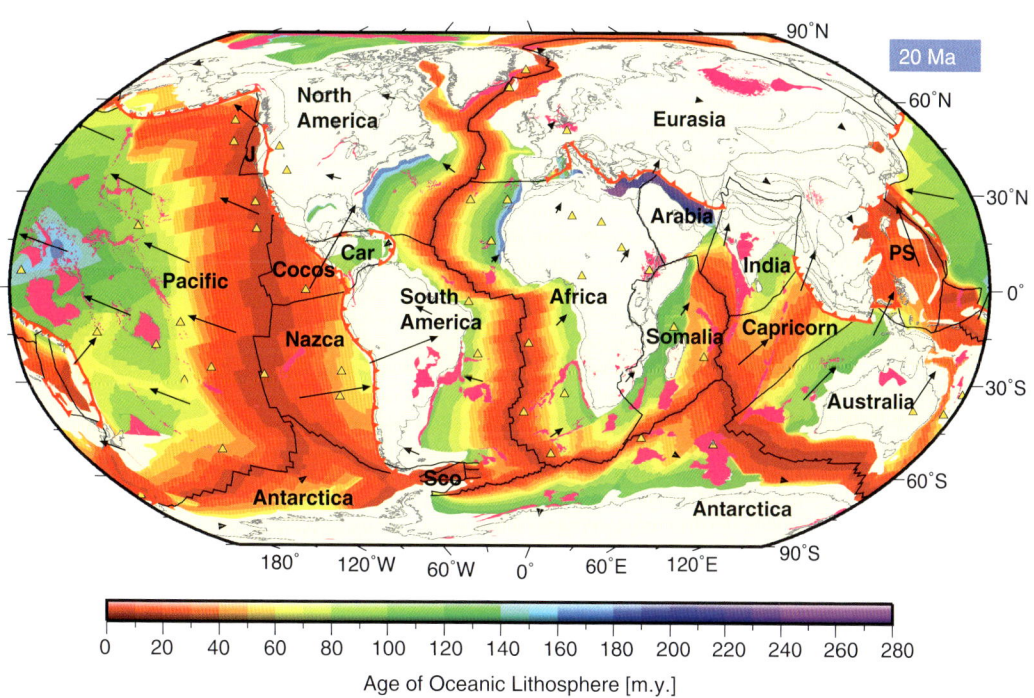

Plate Motion, Figure 2 (continued)

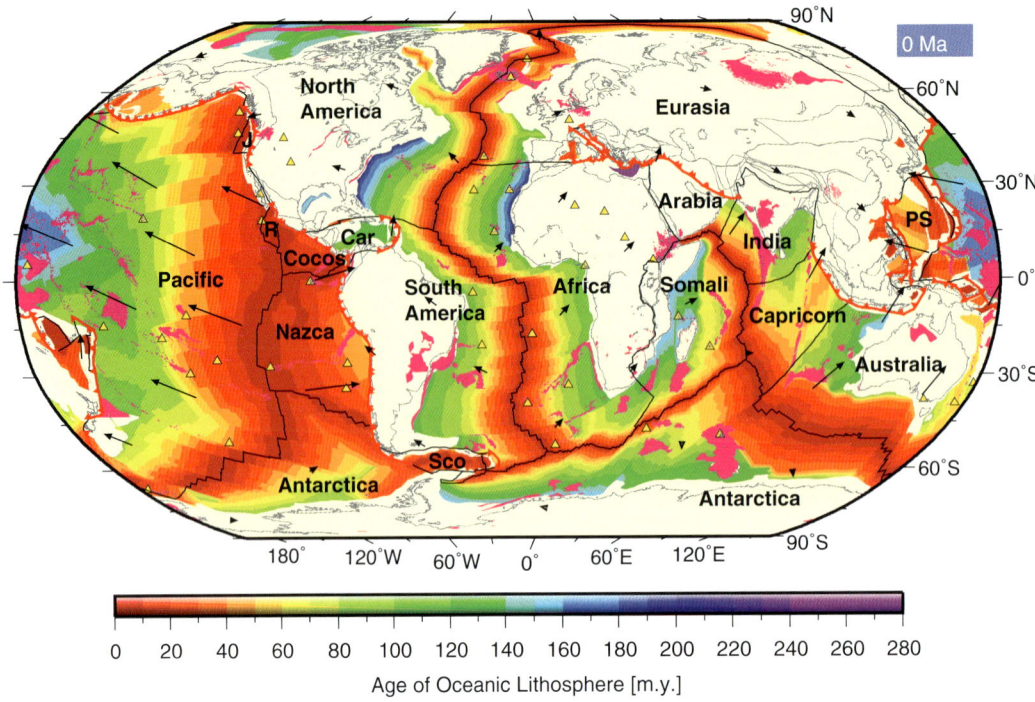

Plate Motion, Figure 2 Global plate reconstructions from 200 Ma to the present day in 20 million year time intervals. Base map shows the age-area distribution of oceanic lithosphere. *Red lines* denote subduction zones and *black lines* denote mid-ocean ridges and transform faults. *Pink polygons* indicate products of plume-related excessive volcanism (e.g., LIPs and volcanic plateaus). *Yellow triangles* indicate present-day hot spot locations. Absolute plate velocity vectors are denoted as *black arrows*. Major plates are labeled. Abbreviations include *Car* caribbean, *Col* colorado, *Flk* falkland, *Grn* greenland, *Ibr* iberia, *J* juan de fuca, *Jun* junction, *LHR* lord howe rise, *Man* manihiki, *Pat* patagonia, *PS* philippine sea, *R* rivera, *Sco* scotia sea, *Van* vancouver.

the "hot-spotting" technique which constructs seafloor flow lines backward in time using a combination of seamounts with and without age constraints. Another class of models combining plate motion and mantle convection (Steinberger and O'Connell, 1998) is designed to move away from the fixed hot spot hypothesis and consider the deflection of plumes through time in a convecting mantle. Absolute plate motion models in which seamount chains with age progression are assumed to be due to the interaction of plates relative to non-stationary hot spots are termed moving hot spot models (Doubrovine et al., 2012; O'Neill et al., 2005). Moving hot spot models only work well for the last 70–100 million years at most and thus require a changeover to an alternative reference frame (e.g., paleomagnetic-based) for earlier times. These models, which combine reference frames using two techniques, are termed "hybrid models." Another complication for constructing absolute plate models is the process of true polar wander, the wholesale rotation of the Earth relative to its spin axis (Torsvik et al., 2002). True polar wander is believed to occur mainly in response to the changing mass distribution in the Earth's convecting mantle due to the time dependence of subduction (Steinberger and Torsvik, 2010). When true polar wander is much faster than the average speed of the tectonic plates, it is expressed in all plates on the same hemisphere exhibiting the same sense of rotation. This method can be used to construct a true polar wander corrected absolute plate motion model (Steinberger and Torsvik, 2008).

Global plate motion models through time

The most up-to-date global plate motion model is that by Seton et al. (2012). It combines relative and absolute plate motions in a global plate model, including the reconstruction of the global network of plate boundaries and the plates themselves through time. It uses a hybrid reference frame, which merges a moving Indian/Atlantic hot spot reference frame (O'Neill et al., 2003) back to 100 Ma (from "Megannum," meaning millions of years before present) with a paleomagnetically derived true polar wander corrected reference frame (Steinberger and Torsvik, 2008) back to 200 Ma. This reference frame links to the global plate circuit through Africa, as Africa has been surrounded by mid-ocean ridges for at least the last 170 million years and has moved relatively little since the breakup of the Pangaea supercontinent. The model (Figure 2) is freely available on the internet in a GPlates-compatible (Williams et al., 2012) format.

Pre-200 Ma plate motions are based entirely on continental paleomagnetic data due to the absence of preserved seafloor spreading histories. Although paleomagnetic data on continents do not provide paleo-longitudes, the relative plate motions and tectonic unity of two continental blocks can be inferred from commonalities in the apparent polar wander (APW) paths (Van der Voo, 1990). If two or more continents share a similar APW path for a time period, then it can be inferred that these continents were joined for these times. Similarly, tectonic affinities can be deduced from the continuity of orogenic belts, sedimentary basins, volcanic provinces, biofacies, and other large-scale features across presently isolated continents (Wegener, 1915). A community plate motion model covering most of the Phanerozoic Eon, i.e., the last 550 million years, has been published by Wright et al. (2013). It builds on the models of Golonka (2007) and Scotese (2004) and provides examples for the use of paleobiology data as additional constraints for absolute and relative plate motions.

Summary and conclusions

The main outstanding challenge for understanding plate motion through time is to further develop models that treat the tectonic plates and the convecting mantle as a coupled system, as well as extending plate models further in time that include plate boundary geometries and locations as well as reconstructed ocean basins. This is essential for understanding absolute plate motions in terms of plate driving forces through time and to evaluate to what extent mantle plumes and the surface hot spots they cause may have been moving relative to each other. In order to advance our understanding of the coupling and feedbacks between deep earth processes and plate kinematics, new software tools and workflows need to be established in which observations, plate kinematics through time, geodynamic modeling, and model/data visualization are seamlessly linked, based on open standards and open-source tools. The burgeoning fields of simulation and modeling and "big data" analysis are likely to enable key advances in this area.

Bibliography

Argus, D. F., and Heflin, M. B., 1995. Plate motion and crustal deformation estimated with geodetic data from the Global Positioning System. *Geophysical Research Letters*, **22**(15), 1973–1976.

Argus, D. F., Gordon, R. G., Heflin, M. B., Ma, C., Eanes, R. J., Willis, P., Peltier, W. R., and Owen, S. E., 2010. The angular velocities of the plates and the velocity of Earth's centre from space geodesy. *Geophysical Journal International*, **180**(3), 913–960.

Boyden, J. A., Müller, R. D., Gurnis, M., Torsvik, T. H., Clark, J. A., Turner, M., Ivey-Law, H., Watson, R. J., and Cannon, J. S., 2011. Next-generation plate-tectonic reconstructions using GPlates. In Randy Keller, G., Randy Keller, G., and Baru, C. (eds.), *Geoinformatics: Cyberinfrastructure for the Solid Earth Sciences*. Cambridge: Cambridge University Press, pp. 95–114.

Chang, T., 1988. Estimating the relative rotation of two tectonic plates from boundary crossings. *Journal of the American Statistical Association*, **83**(404), 1178–1183.

Cox, A., and Hart, R. B., 1986. *Plate Tectonics: How it Works*. Oxford: Blackwell. 392 p.

DeMets, C., Gordon, R. G., and Argus, D. F., 2010. Geologically current plate motions. *Geophysical Journal International*, **181**, 1–80.

Dietz, R. S., 1961. Evolution by spreading of the sea floor. *Nature*, **190**, 854–857.

Doubrovine, P. V., Steinberger, B., and Torsvik, T. H., 2012. Absolute plate motions in a reference frame defined by moving hot spots in the Pacific, Atlantic, and Indian oceans. *Journal of Geophysical Research*, **117**(B9), B09101.

Golonka, J., 2007. Late triassic and early jurassic palaeogeography of the world. *Palaeogeography Palaeoclimatology Palaeoecology*, **244**(1–4), 297–307.

Gordon, R. C., 2000. Diffuse oceanic plate boundaries: strain rates, vertically averaged rheology, and comparisons with narrow plate boundaries and stable plate interiors. In Richards, M. A., Gordon, R. C., and van der Hilst, R. D. (eds.), *The History and Dynamics of Global Plate Motions*. Washington, DC: American Geophysical Union, Vol. 121, pp. 145–159.

Hellinger, S. J., 1981. The uncertainties of finite rotations in plate tectonics. *Journal of Geophysical Research*, **86**(B10), 9312–9318.

Hess, H. H., 1962. History of Ocean Basins: Geological Society of America Bulletin. In *Petrologic Studies: A Volume to Honour A.F. Buddington*. Boulder, CO: Geological Society of America, pp. 559–620.

Kogan, M. G., and Steblov, G. M., 2008. Current global plate kinematics from GPS (1995–2007) with the plate-consistent reference frame. *Journal of Geophysical Research*, **113**(B4), B04416.

Matthews, K. J., Müller, R. D., Wessel, P., and Whittaker, J. M., 2011. The tectonic fabric of the ocean basins. *Journal of Geophysical Research*, **116**, B12109, 1–28.

Morgan, W. J., 1968. Rises, trenches, great faults and crustal blocks. *Journal of Geophysical Research*, **73**, 1959–1982.

Morgan, W. J., 1971. Convection plumes in the lower mantle. *Nature*, **230**, 43–44.

Morgan, W. J. 1972. *Plate Motions and Deep Mantle Convection*. Geological Society of America Memoirs. Boulder, CO: Geological Society of America, Vol. 132, pp. 7–22.

Müller, R. D., Sdrolias, M., Gaina, C., and Roest, W. R., 2008. Age, spreading rates, and spreading asymmetry of the world's ocean crust. *Geochemistry, Geophysics, Geosystems*, **9**(Q04006), 18–36, doi:10.1029/2007GC001743

O'Neill, C., Muller, D., and Steinberger, B., 2003. Geodynamic implications of moving Indian Ocean hotspots. *Earth and Planetary Science Letters*, **215**(1–2), 151–168.

O'Neill, C., Müller, D., and Steinberger, B., 2005. On the uncertainties in hot spot reconstructions and the significance of moving hot spot reference frames. *Geochemistry, Geophysics, Geosystems*. **6**(4), Q04003.

Scotese, C. R., 2004. A continental drift flipbook. *Journal of Geology*, **112**(6), 729–741.

Seton, M., Müller, R. D., Zahirovic, S., Gaina, C., Torsvik, T., Shephard, G. E., Talsma, A. S., Gurnis, M., Turner, M., Maus, S., and Chandler, M. T., 2012. Global continental and ocean basin reconstructions since 200 Ma. *Earth Science Reviews*, **113**, 212–270.

Stadler, G., Gurnis, M., Burstedde, C., Wilcox, L. C., Alisic, L., and Ghattas, O., 2010. The dynamics of plate tectonics and mantle flow: from local to global scales. *Science*, **329**, 1033–1038.

Stampfli, G. M., and Borel, G. D., 2002. A plate tectonic model for the Paleozoic and Mesozoic constrained by dynamic plate boundaries and restored synthetic oceanic isochrons. *Earth and Planetary Science Letters*, **196**(1–2), 17–33.

Steinberger, B., and O'Connell, R. J., 1998. Advection of plumes in mantle flow: Implications for hot spot motion, mantle viscosity and plume distributions. *Geophysical Journal International*, **132**, 412–434.

Steinberger, B., and Torsvik, T. H., 2008. Absolute plate motions and true polar wander in the absence of hotspot tracks. *Nature*, **452**(7187), 620–623.

Steinberger, B., and Torsvik, T., 2010. Toward an explanation for the present and past locations of the poles. *Geochemistry, Geophysics, Geosystems*, **11**(6).

Stock, J., and Molnar, P., 1988. Uncertainties and implications of the late cretaceous and tertiary position of North America relative to the Farallon, Kula, and Pacific plate. *Tectonics*, **7**(6), 1339–1384.

Torsvik, T. H., Van der Voo, R., and Redfield, T. F., 2002. Relative hotspot motions versus true polar wander. *Earth and Planetary Science Letters*, **202**(2), 185–200.

Torsvik, T., Müller, R. D., Van der Voo, R., Steinberger, B., and Gaina, C., 2008. Global plate motion frames: toward a unified model. *Reviews of Geophysics*, **46**(RG3004), doi:10.1029/2007RG000227

Van der Voo, R., 1990. The reliability of paleomagnetic data. *Tectonophysics*, **184**(1), 1–9.

Wegener, A., 1915. The Origin of Continents and Oceans. New York, NY: Courier Dover Publications.

Wessel, P., and Kroenke, L. W., 1997. A geometric technique for relocating hotspots and refining absolute plate motions. *Nature*, **387**(6631), 365–369.

Williams, S. E., Müller, R. D., Landgrebe, T. C. W., and Whittaker, J. M., 2012. An open-source software environment for visualizing and refining plate tectonic reconstructions using high-resolution geological and geophysical data sets. *GSA Today*, **22**(4/5), 4–9.

Wilson, J. T., 1965. A new class of faults and their bearing on continental drift. *Nature*, **207**, 343–347.

Wright, N., Zahirovic, S., Müller, R. D., and Seton, M., 2013. Towards community-driven paleogeographic reconstructions: integrating open-access paleogeographic and paleobiology data with plate tectonics. *Biogeosciences*, **10**(3), 1529–1541.

Cross-references

Hot Spots and Mantle Plumes
Lithosphere: Structure and Composition
Oceanic Spreading Centers
Plate Tectonics
Regional Marine Geology
Subduction
Transform Faults

PLATE TECTONICS

Martin Meschede
Institute of Geography and Geology, Ernst-Moritz-Arndt University, Greifswald, Germany

The Earth is a dynamic planet which continuously changes its surficial shape since it came into existence about 4.5 billion years ago. All the changes are caused by convection processes in the Earth's mantle which formed by segregation of different spheres in early times of the Earth. The processes of plate tectonics as we know it today started about 2.5–3 billion years ago and operated since that time (Stern, 2005, 2008; Hamilton, 2007). However, nothing of the Earth's surface is of constancy even if it appears so in the form of several-billion-year-old rocks. The entire surface of the Earth is subject to a permanent reconstruction. Heat either inherited from the early times of the Earth and stored in its interior or newly formed by radioactive decay of unstable elements is constantly released. The heat conductivity of rocks, however, is poor, and a rigid and stiff outer crust formed over time which prevents the direct heat loss. Therefore, convection currents in the more plastically behaving mantle are needed to maintain the heat output (Figure 1). These currents are responsible for the upward transportation of hot rock material, which is a little bit lighter than cold material of the same composition, to the surface. In return cold rock material is being transported down into deeper regions of the Earth. The result is a kind of conveyor system to provide the thermal output that is transporting the rigid plates along at the surface.

Heat emission occurs at zones of weakness. The most important are the mid-oceanic spreading centers that circle the globe for more than 67,000 km in length (Bird, 2003). Most of the volcanic activity on Earth occurs at these divergent plate boundaries (Perfit and Davidson, 2000). On the other hand, cold rock material is being transported into the Earth in subduction zones at convergent plate boundaries. There exist two fundamentally different types of rigid and stiff lithospheric plates (Greek *lithos*: stiff): oceanic and continental lithosphere. The differences of these crustal types have already been recognized by Alfred Wegener (1880–1930), a German meteorologist and pioneer of plate tectonics who first formulated the continental drift theory (Wegener 1912, 1915). Wegener recognized the different mean heights of continents and oceans. He could, however, not yet explain the reason for the plate movements, which is why he encountered strong resistance against his theory. Despite his conclusive explanation for, e.g., identical fossils on both sides of the Atlantic Ocean or traces of glaciation in the Sahara desert in Africa, it needed another 50 years to establish the actual plate tectonic theory. Today, this theory is generally accepted and discussed in a number of textbooks (e.g., Frisch et al., 2011).

Movements resulting from convection currents leave their traces on the Earth's surface. Normally, these movements occur very slowly with only a few millimeters or centimeters per year, and we are not able to perceive it directly. However, we recognize the dynamics of the Earth in cases of large earthquakes or spectacular volcanic eruptions. This happens when stress has been accumulated in the rigid and stiff lithosphere that is suddenly released or when a magma chamber erupts, which has been filled and pressurized until it loses its resistibility.

The stiff lithospheric plates are very different in size (Bird, 2003). Their thickness varies in general between 70 and 150 km, and they are thicker underneath the continents as compared to the oceans. In some cases below large mountain chains, they may achieve a thickness of

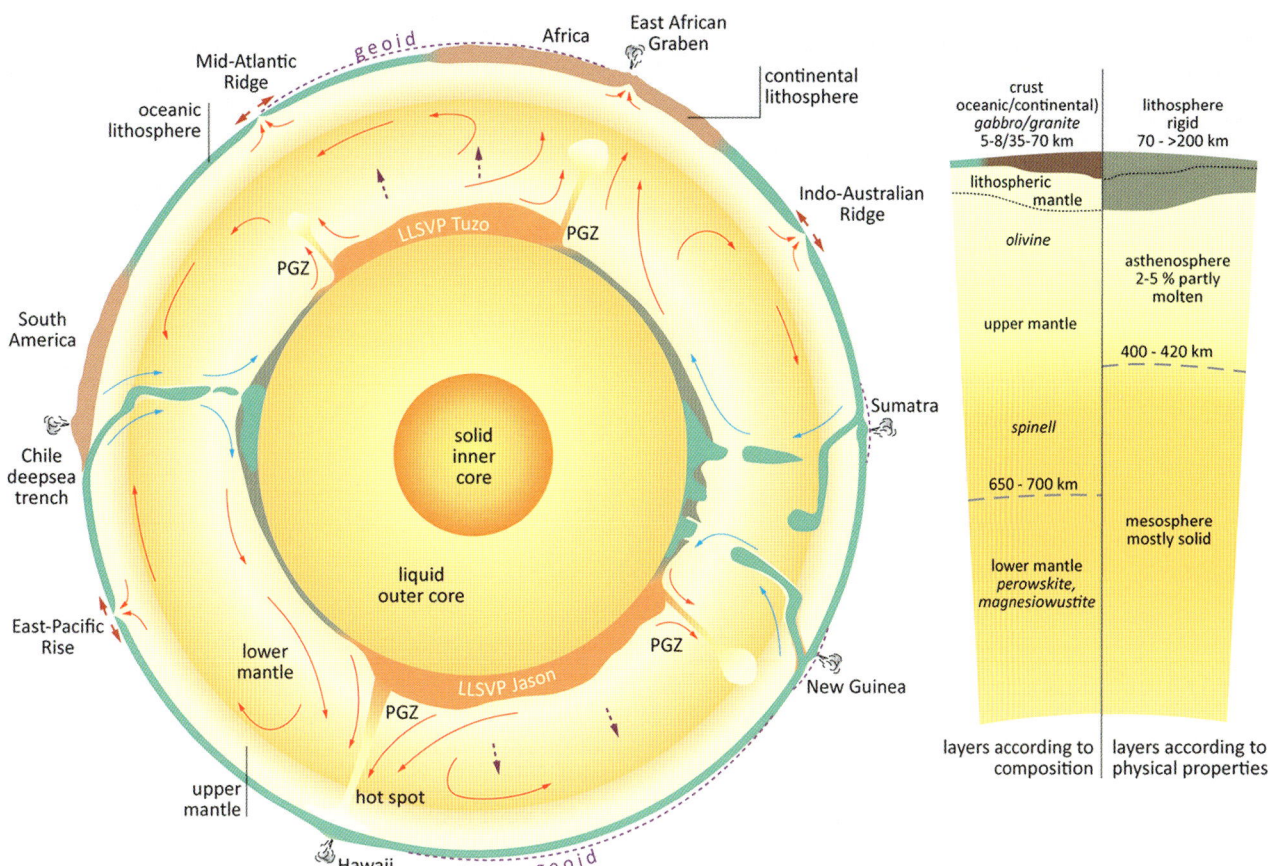

Plate Tectonics, Figure 1 *Left*: Convection currents in the Earth's interior. The thickness of oceanic and continental lithosphere is exaggerated to clarify the differences between the two crustal types: if drawn with true proportions, the lithosphere could be less than 1 mm thick (see also Fig. 2). The viewing direction of the cross section through the Earth's interior is from the South Pole (Modified and completed according to Condie (1997), Schubert et al. (2001), Tackley (2008), and Torsvik et al. (2014)). *LLSVP* large low shear-wave velocity provinces beneath Africa (Tuzo) and the Pacific (Jason), *PGZ* plume generation zones; *thin arrows* above Tuzo and Jason indicate the rise of lighter material resulting in a deflection of the geoid at the surface (*dashed line*). *Right*: comparison of the mantle and lithosphere layers according to composition (mineralogy, petrology) and physical properties (geophysical characteristics) (modified after Stüwe 2000).

more than 200 km. Lithospheric plates are composed of the Earth's crust, either of oceanic or continental type, and the lithospheric mantle (Figure 1). Oceanic crust has an average thickness of 5–8 km and is mainly composed of gabbroic and basaltic rock material. The mean density of oceanic crust is about 3.0 g/cm^3. Continental crust on the other hand has an average thickness of 30–35 km and may reach more than 70 km beneath young mountain ranges. In its upper part, it comprises of less dense rock material (e.g., granites, granitoids, sedimentary rocks) and has a mean density of 2.6–2.8 g/cm^3.

The lithospheric mantle and the asthenosphere differ in their physical properties but not in their average composition (Figure 1). In contrast to the solid lithospheric mantle, the asthenosphere is to a small degree (about 1–5 %) molten causing its ability for plastic flow. However, the molten part encompasses only a few percent of the total rock material. The velocity of movement of the lithospheric plates ranges in the order of centimeters per year. This may add to several thousand kilometers in geological periods. To illustrate the dimension of movement, let's have a look on a lithospheric plate slowly moving with 1 cm/year. With this velocity, it moves 10 km in 1 million years and sums up to 1,000 km in 100 million years. Recent plate movement velocities range from less than 1 cm/year to more than 15 cm/year at the central East Pacific Rise (i.e., 7.5 cm/year to each side). In the past, even faster velocities of more than 20 cm/year have been detected (e.g., Kumar et al., 2007). Therefore, it is obvious that the distribution of continents and oceans on the surface of the Earth is only a snapshot in the Earth's history and has consistently changed since the onset of the currently active plate tectonic processes.

Lithospheric plates move with different velocities and different directions and interact with other plates. If plates move apart, new oceanic lithosphere is formed in between. These plate boundaries are called constructive or divergent boundaries (Figure 2), and they are characterized by

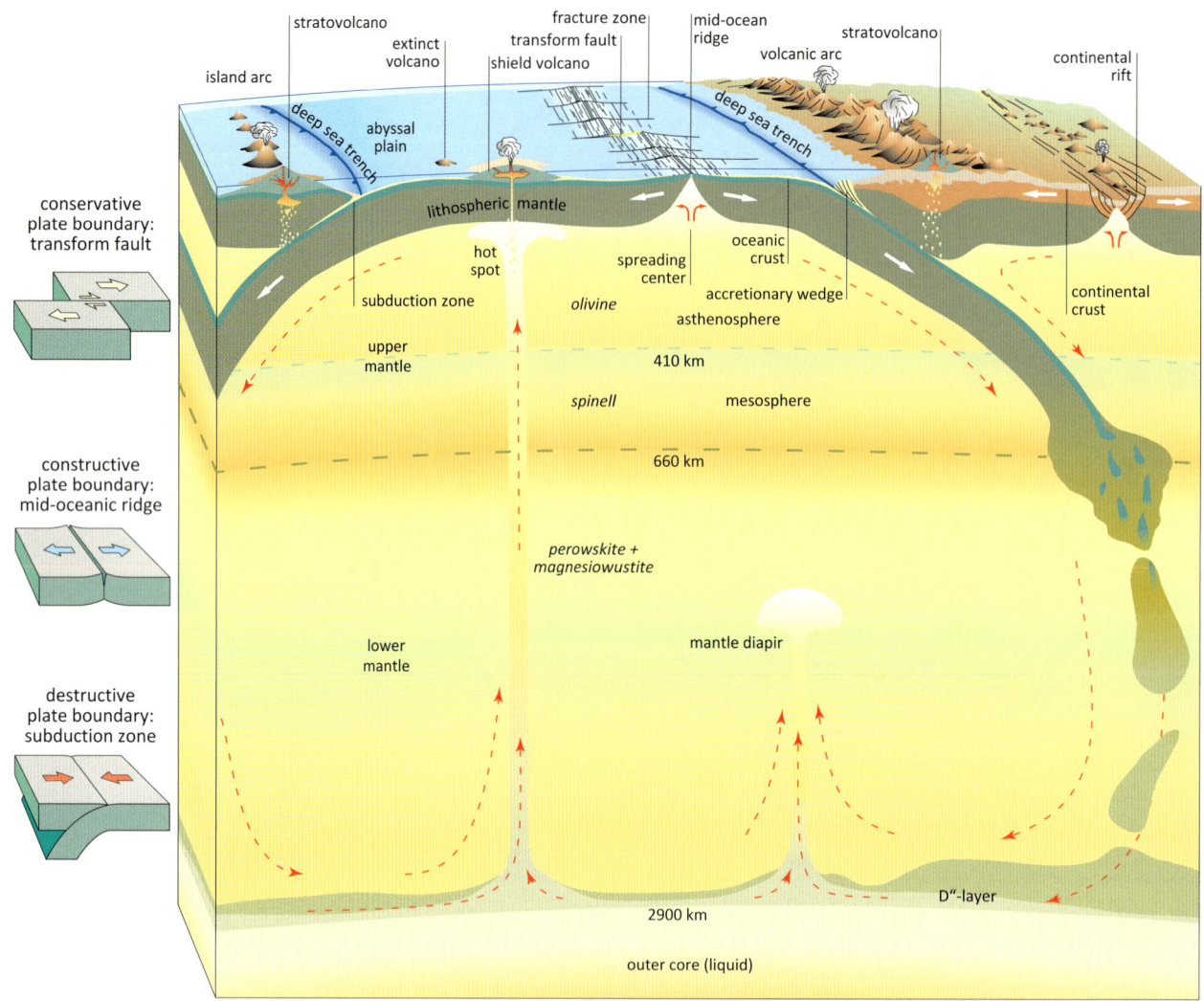

Plate Tectonics, Figure 2 Schematic overview of plate tectonic processes at the surface and in the Earth's crust and mantle.

a mid-oceanic spreading center. On the other hand, plates collide and subduction zones form when one lithospheric plate descents beneath the other and is returned back into the mantle. These plate boundaries are called destructive or convergent plate margins. Frequently, a deep trench evolved along the plate boundary, and the deepest point on Earth, the Challenger Deep with a depth of 10,994 ($\pm$40) m, was determined in the typical deep-sea trench of the Mariana arc. In many cases accretionary wedges are developed at the surface along convergent plate boundaries where sedimentary material from the subducting plate is offscraped and thrust as nappes onto the overriding plate. About half of the convergent plate boundaries are characterized by such accretionary wedges, whereas the other half is dominated by subduction erosion or at least non-accretion. A third type of plate boundaries exists where two plates slide along each other and lithospheric material is neither formed nor destroyed.

This plate boundary is characterized by transform faults and is called a conservative boundary.

Continent-ocean boundaries are subdivided into convergent and passive plate margins. In the case of a subduction zone along the continent, e.g., the west coast of South America, it is called an active continental margin. If the continental lithosphere is connected to the oceanic lithosphere without a plate boundary, it is called a passive continental margin. Active continental margins are mostly characterized by a high mountain range formed mainly by magmatic activity. Passive continental margins, on the other hand, frequently have a low coastal mountain range as a remnant of the former rifting stage. They are characterized by a wide shelf area where strong subsidence caused by dilatational thinning of the continental crust occurs and thick layers of sediments were accumulated.

Oceanic crust is constantly being formed at mid-oceanic spreading centers of constructive plate boundaries. The mid-oceanic spreading centers are in the crest of long submarine ridges in the middle of an ocean, which are in part more than 3,000 m higher than the abyssal plains at its flanks. For instance, the water depth at the Mid-Atlantic Ridge is in average about 2,500 m below sea-level; the neighboring abyssal plains, on the other hand, are more than 5,000 m deep. The spreading centers are the most active volcanic chains on Earth. More than three-fourths of the global annual volcanic production happens on these ridges (Schmincke, 2004). However, since these volcanoes are nearly entirely hidden under water, we generally do not notice their activity. In contrast, subduction zones are characterized by a large number of superficially visible volcanoes arranged at the surface in long chains. The major part of about 550 historically active volcanoes on Earth is located above subduction zones.

The composition of continental crust is much more complicated and variable than oceanic crust. Oceanic crust, mainly made up of gabbroic and basaltic rocks, is normally transported back into the Earth's mantle where it becomes recycled. Based on high-pressure metamorphic processes, basalt and gabbro are turned into eclogite when they enter a depth of more than 35–40 km. Eclogite is with an average density of 3.4 g/cm^3 heavier than peridotite of the lithospheric mantle with an average density of 3.3 g/cm^3 and is thus able to pull down the subducting slab into the asthenosphere. Continental crust, in contrast, remains at the surface since it is too light to become subducted. The modern continental crust is thus a product of several billions of years of relocations, deformations, and transformations such as metamorphism. The rocks of the continental crust conserve the entire history of the Earth, however, often in very concealed or even unrecognizable conditions. On the other hand, the oldest oceanic crust preserved in its original context has an age of only about 195 million years (Muller et al., 2008). Captured oceanic crust of the Neotethys Ocean in the eastern Mediterranean, which is not yet added to the continental lithosphere by obduction, was formed in the Triassic. All other, older remnants of oceanic crust are part of ophiolites that were added to the continental lithosphere by accretionary processes along convergent plate margins. Therefore, rocks of the continental crust are partly more than 20 times older than the present oceanic lithosphere. This indicates that the oceanic crust may have been completely recycled more than 20 times through the Earth's history.

Subduction zones are characterized by strong earthquakes and volcanic activity. The collision of lithospheric plates leads to the accumulation of considerable stresses that often result in devastating earthquakes such as the earthquakes at Sumatra (2004), Japan (2011), and Chile (2010, 2014). In these zones the oceanic lithosphere is subducted into a depth where it becomes metamorphosed and molten. Parts of the downgoing slab may be transported down to the mantle-core boundary where it is incorporated in the so-called D" layer (named by Bullen 1950; Figure 2). Caused by thermal convection and heat transfer from the core to the mantle, the material of the D" layer is heated. This results in the development of mantle diapirs fed from the D" layer (Peltier, 2007). Thus, after hundred millions of years, remnants of subducted material may be transported back to the surface in mantle diapirs where hot mantle material rises upwards forming hot spots on the surface (Hofmann and White, 1982).

Sediments are deposited on the oceanic lithospheric plates during their drift from the mid-oceanic spreading center to the subduction zone. Sediments, however, are too light to be transported downward into the asthenosphere. Therefore, they are scraped off at the plate boundary and accumulated in an accretionary wedge or transported down to a relatively shallow depth of 110–120 km. In this depth the light rocks are molten and incorporated into magmatic melts which rise to the surface and may feed volcanoes or crystallize in plutonic bodies of the magmatic arc. The magmatic rocks are mostly of andesitic/dioritic to rhyolitic/granitic composition.

Volcanic eruptions far away from plate boundaries are mostly caused by hot spot activities. At these locations mantle diapirs with hot asthenospheric material rising from deep areas in the mantle form plume heads beneath the rigid lithosphere that feed magma chambers in the continental or oceanic crust (Figure 2) and cause upward bulges in the lithosphere. In addition to the mid-oceanic spreading centers, these mantle diapirs also play an important role in the thermal budget of the Earth. Mantle diapirs may feed volcanic eruptions of gigantic dimensions and global catastrophic consequences, e.g., Yellowstone, about 640,000 years b.p., or Toba, about 74,000 years b.p. (Sparks et al., 2005). These types of volcanoes are called supervolcanoes, which are not expressed by volcanic edifices on the surface but by large calderas of several tens of kilometers in diameter (de Silva, 2008).

Plate tectonic processes repeat in long-lasting cycles, called Wilson cycle, named after J. Tuzo Wilson (1908–1993). The Wilson cycle starts with the continental breakup and the development of a rift system between two newly diverging plates. Rising and partly molten asthenosphere forms new oceanic lithosphere in the middle of an evolving ocean. This ocean may evolve from an elongate narrow basin to a wide ocean of several-thousand-kilometer width. The oceanic lithosphere is being subducted and transported back into the mantle after its drift on the surface, which lasts in maximum about 200 million years. The collision of two continental plates leads to the orogenesis of a mountain chain. After the gradation of this mountain chain, the Wilson cycle may start again with the breakup of the continental lithosphere.

Bibliography

Bird, P., 2003. An updated digital model of plate boundaries. *Geochemistry, Geophysics, Geosystems*, **4**, doi:10.1029/2001GC000252.

Bullen, K. E., 1950. An earth model based on a compressibility-pressure hypothesis. *Monthly Notices, Royal Astronomical Society: Geophysics*, **6**(Supplement), 50–59.

Condie, K., 1997. *Plate Tectonics and Crustal Evolution*, 4th edn. Oxford: Butterworth-Heinemann; 282pp.

De Silva, S., 2008. Arc magmatism, calderas, and supervolcanoes. *Geology*, **36**, 671–672.

Frisch, W., Meschede, M., and Blakey, R. C., 2011. *Plate Tectonics*. Heidelberg/Dordrecht/London/New York: Springer; 212pp.

Hamilton, W. B., 2007. Earth's first two billion years – the era of internally mobile crust. In Hatcher, R. D., Jr., Carlson, M. P., McBride, J. H., and Martínez Catalán, J. R. (eds.), *4-D Framework of Continental Crust*. Boulder: Geological Society of America. Geological Society of America, Memoir, 200, pp. 233–296.

Hofmann, A. W., and White, W. M., 1982. Mantle plumes from ancient oceanic crust. *Earth and Planetary Science Letters*, **57**, 421–436.

Kumar, P., Yuan, X., Kumar, M. R., Kind, R., Li, X., and Chadha, R. K., 2007. The rapid drift of the Indian tectonic plate. *Nature*, **449**, 894–897.

Müller, R. D., Sdrolias, M., Gaina, C., and Roest, W. R., 2008. Age, spreading rates and spreading symmetry of the world's ocean crust. *Geochemistry, Geophysics, Geosystems*, **9**, Q04006, doi:10.1029/2007GC001743.

Peltier, W. R., 2007. *Mantle Dynamics and the D" Layer. Impacts of the Post-perovskite Phase*. Washington, DC: AGU books. AGU Geodynamics Series Monograph, pp. 217–227.

Perfit, M. R., and Davidson, J. P., 2000. Plate tectonics and volcanism. In Sigurdsson, H., et al. (eds.), *Encyclopedia of Volcanoes*. San Diego: Academic Press, pp. 89–113.

Schmincke, H.-U., 2004. *Volcanism*. Heidelberg/Dordrecht/London/New York: Springer; 324pp.

Schubert, G., Turcotte, D. L., and Olson, P., 2001. *Mantle Convection in the Earth and Planets*. Cambridge: Cambridge University Press; 941pp.

Sparks, S., Self, S., Grattan, J., Oppenheimer, C., Pyle, D., and Rymer, H., 2005. *Super-Eruptions: Global Effects and Future Threats: Report of a Geological Society of London Working Group*. London: Geological Society; 25pp.

Stern, R. J., 2005. Evidence from ophiolites, blueschists, and ultrahigh pressure metamorphic terranes that the modern episode of subduction tectonics began in Neoproterozoic time. *Geology*, **33**, 557–560.

Stern, R. J., 2008. Modern-style plate tectonics began in Neoproterozoic time: an alternative interpretation of Earth's tectonic history. In Condie, K., and Pease, V. (eds.), *When did Plate Tectonics Begin?* Boulder: Geological Society of America. Geological Society of America, Special Papers, 440, pp. 265–280.

Stüwe, K., 2000. *Einführung in die Geodynamik der Lithosphäre – Quantitative Behandlung geowissenschaftlicher Probleme*. Berlin: Springer; 405pp.

Tackley, P. J., 2008. Geodynamics: layer cake or plum pudding? *Nature Geoscience*, **1**, 157–158.

Torsvik, T. H., van der Voo, R., Doubrovine, P., Burke, K., Steinberger, B., Ashwal, L.D., Tronnes, R., Webb, S. J., and Bull, A. L., 2014. Deep mantle structure as a reference frame for movements in and on the Earth. *Proceedings of the National Academy of Sciences of the United States of America (PNAS)*, **111**, 8735–8740.

Wegener, A., 1912. Die Entstehung der Kontinente. *Geologische Rundschau*, **3**, 276–292.

Wegener, A., 1915. *Die Entstehung der Kontinente und Ozeane*. Braunschweig: Vieweg; 94pp.

Cross-references

Abyssal Plains
Active Continental Margins
Driving Forces: Slab Pull, Ridge Push
Earthquakes
Gabbro
Hot Spots and Mantle Plumes
Intraplate Magmatism
Island Arc Volcanism, Volcanic Arcs
Layering of Oceanic Crust
Magmatism at Convergent Plate Boundaries
Mid-ocean Ridge Magmatism and Volcanism
Oceanic Rifts
Oceanic Spreading Centers
Ophiolites
Passive Plate Margins
Peridotites
Plate Motion
Regional Marine Geology
Shelf
Subduction
Subduction Erosion
Subsidence of Oceanic Crust
Tethys in Marine Geosciences
Transform Faults
Wilson Cycle

PTEROPODS

Annelies C. Pierrot-Bults[1,2] and Katja T. C. A. Peijnenburg[1,2]
[1]Institute for Biodiversity and Ecosystem Dynamics, University of Amsterdam, Amsterdam, The Netherlands
[2]Naturalis Biodiversity Center, Leiden, The Netherlands

Definition

Pteropods were first described as a group in 1804 by Georges Cuvier. The name derives from the fact that the molluscan foot (−poda) has been modified to form paired swimming wings (ptero-). There are two different orders, (1) the Thecosomata, also known as "sea butterflies" with suborders Euthecosomata (shelled and with a good fossil record) and Pseudothecosomata (shelled and shell-less), and (2) the Gymnosomata (shell-less as adults), commonly known as "sea angels" (Table 1). Recent molecular studies (Klussmann-Kolb and Dinapoli, 2006; Jennings et al., 2010) provided convincing evidence that the two orders are sister groups, hence the superorder Pteropoda is now considered a taxonomic entity.

Introduction

Pteropods are found throughout the world's oceans and are holoplanktonic, meaning that they spend their entire life cycle in the open water column. They represent an extraordinary diversity in morphological forms (Figure 1) and can be numerically and functionally important components of marine food webs (Lalli and Gilmer,

Pteropods, Table 1 Hierarchical subdivisions of the Pteropoda including the orders Thecosomata and Gymnosomata. Genus names and species numbers are taken from the World Register of Marine Species (WORMS). Pending further revisions we have taken a conservative approach and have not included recent revisions including superfamilies and subgenera

Phylum Mollusca

Class Gastropoda
Subclass Heterobranchia
Order Thecosomata
    Suborder Euthecosomata
        Family Cavoliniidae
            Genus   *Cavolinia* 6 spp.
                        Diacavolinia 19 spp.
                        Diacria 8 spp.
                        Clio 9 spp.
                        Creseis 5 spp.
                        Hyalocylis 1 sp.
                        Styliola 1 sp.
                        Cuvierina 5 spp.
                        ? *Vaginella*
        Family Limacinidae
            Genus   *Heliconoides* 1 sp.
                        Limacina 6 spp.
                        Thielea 1 sp.
    Suborder Pseudothecosomata
        Family Cymbuliidae
            Genus   *Cymbulia* 4 spp.
                        Corolla 6 spp.
                        Gleba 1 sp.
        Family Desmopteridae
            Genus   *Desmopterus* 4 spp.
        Family Peraclidae
            Genus   *Peracle* 8 spp.

Order Gymnosomata
    Family Clionidae
        Genus   *Clione* 1 sp.
                    Fowlerina 2 spp.
                    Paedoclione 1 sp.
                    Paraclione 2 spp.
                    Thalassopteris 1 sp.
                    Thliptodon 3 spp.
                    Cephalobrachia 2 spp.
                    Massya 1 sp.
    Family Cliopsidae
        Genus   *Cliopsis* 2 spp.
                    Pruvotella 1 sp.
    Family Notobranchaeidae
        Genus   *Notobranchaea* 8 spp.
    Family Pneumodermatidae
        Genus   *Abranchaea*
                    Platybrachium 1 sp.
                    Pneumoderma 3 spp.
                    Pneumodermopsis 9 spp.
                    Schizobrachium 1 sp.
                    Spongiobranchaea 2 spp.
    Family Hydromylidae
        Genus   *Hydromyles* 1 sp.
    Family Laginiopsidae
        Genus   *Laginopsis* 1 sp.

1989). Shelled pteropods affect the ocean carbon cycle by producing aragonite shells that can accelerate the export of organic matter from the surface into the deep ocean. Because of their delicate aragonite shells, pteropods have been identified as extremely vulnerable to the effects of ocean acidification (e.g., Feely et al., 2004). Hence, they have become the subject of intense recent research in this area. About 40 gymnosomatous and ~100 thecosomatous pteropod species are currently recognized. The fossil record almost exclusively concerns members of the Euthecosomata so the focus in this short overview will be put there. About 200 fossil euthecosome species are currently described.

Thecosomata

Suborders Euthecosomata and Pseudothecosomata

For identification of Thecosomata, mainly the form and dimensions of the shells are used. Figure 2 shows the main features of the soft body and the shell. The paired wings are homologous to the gastropod foot and serve a dual function for swimming and feeding. The slit-like mouth between the lips opens into the radula sac where the radula and salivary glands are found. The radula can also be a distinguishing character for species. The shells may be sinistrally coiled as in Limacinidae, or uncoiled, conical or bilaterally symmetrical as in Cavoliniidae (Figures 2 and 3). The small and fragile shells can be found along the beach but they may be easily overlooked, and only fragments are usually found. The shells of both Limacinidae and Cavoliniidae are constructed of aragonite, a less stable form of calcium carbonate than calcite, but their microstructures are very different. The Limacinidae possess a crossed-lamellar shell microstructure. By contrast, the Cavoliniidae have a helical microstructure that is unlike that of any other molluscan group (Bé et al., 1972; Bé and Gilmer, 1977). The poorly known Pseudothecosomata belong to the Thecosomata but two of the three families do not have a shell. Only in the genus *Peracle* is a shell present in the adult stage.

Gymnosomata

The Gymnosomata include some 50 described species though they remain relatively little known. Gymnosomes are shelled only in an early life stage, the protoconch shell is formed in the egg and shed after hatching. They usually do not form swarms and do not contribute much to the zooplankton biomass except for some species in the polar seas like *Clione limacina*. Gymnosomes have very different anatomical features compared to thecosomes. Besides

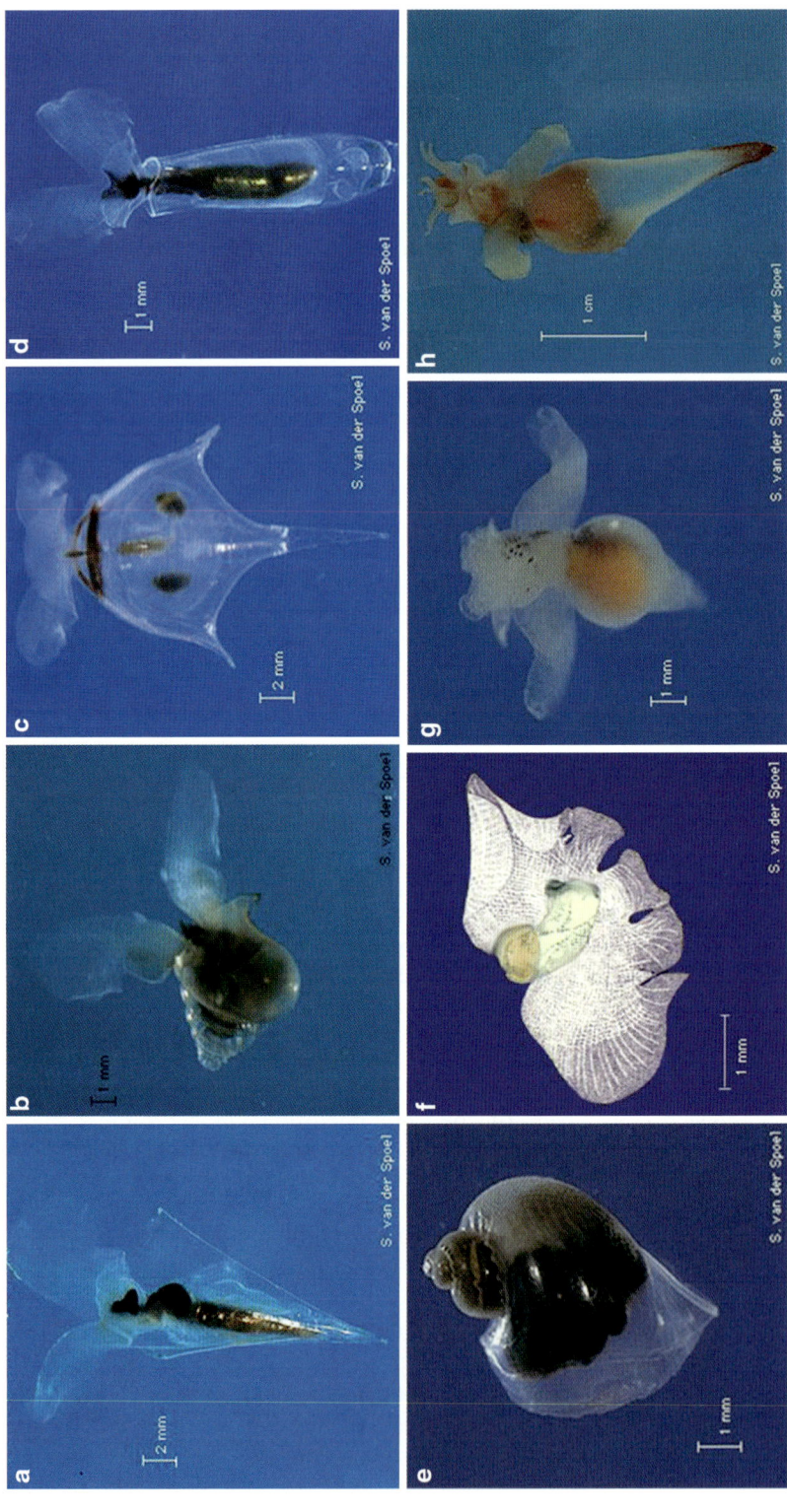

Pteropods, Figure 1 (a) *Clio pyramidata* (Euthecosomata) (b) *Limacina retroversa* (Euthecosomata) (c) *Diacria major* with protoconch (Euthecosomata) (d) *Cuvierina atlantica* (e) *Peracle bispinosa* (f) *Desmopterus pacificus* (g) *Pneumodermopsis macrochira* (Gymnosomata) (h) *Clione limacina* (Gymnosomata) (Permission of ETI).

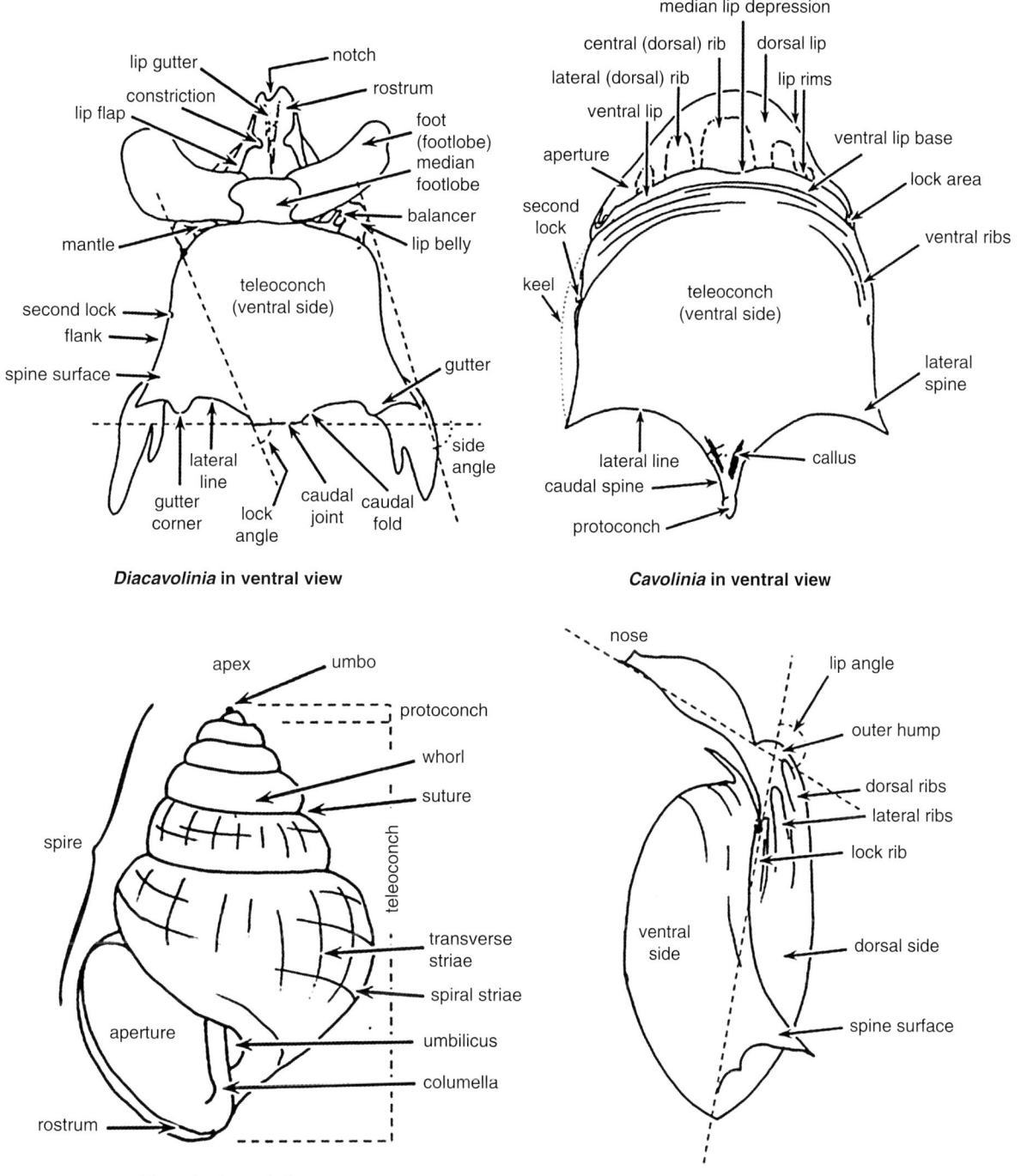

Pteropods, Figure 2 Diagrams of Diacavolinia, Cavolinia, Limacina, and Diacavolinia in different orientations (from Van der Spoel and Dadon, 1999).

the shell also the mantle and mantle cavity are absent. The anatomy of the wings and footlobes differs, and the head and mouth are found anterior to the wings. The digestive tract is shorter than in the Thecosomata so that the caudal part of the body seems empty. Van der Spoel and Diester-Haass (1976) found the first fossil gymnosomatous protoconchae in the East Atlantic and several more gymnosomatous larval shells have since been found, though none of them could be reliably identified to species.

Pteropods, Figure 3 (a) Shell of *Cavolinia teschi* (Euthecosomata) (b) Shell of *Clio cuspidata* (Euthecosomata) (c) Shell of *Diacria schmidti* (Euthecosomata) (d) Shell of Limacina bulimoides (e) Shell of *Creseis virgula* (Euthecosomata) (f) Shell of Peracle reticulata (Pseudothecosomata) (Permission of ETI).

Distribution

Pteropods are widely distributed and abundant in all oceans, and the distribution patterns of individual species are generally closely related to the hydrographic conditions. As in other holozooplankton typical horizontal distribution patterns are (1) broad in the tropics and subtropics in all three oceans, (2) more narrowly distributed in the tropics usually restricted to the Indo-Pacific, (3) distributed in northern and southern transition or temperate zones, and (4) cold-water species, including bipolar distributions. Most pteropod species occur in warm water. A striking pattern is the high species diversity in tropical and subtropical oceans, which contrasts with the low species diversity in cold water regions. Angel (1997) stated "species richness decreases and dominance increases as organic inputs become more seasonally pulsed." The lowest biomass and the highest species richness is found in the permanently stratified subtropical gyres, with more than 80 taxa of Thecosomata and Gymnosomata. Even though poor in species richness, pteropods are most abundant in Subarctic and Subantarctic waters and sometimes dominant components of the plankton. Here they are the main contributors to vertical particle flux and also an important food source for whales and large fish such as cod. Swarms were found of *Limacina retroversa* with densities of >13,600 ind. per 1,000 m^3 in Norwegian waters (Bathmann et al., 1991) and of *L. helicina* with densities of >10,000 ind. per 1,000 m^3 south of the Antarctic Convergence near 65°S (Chen, 1968b).

Species with large latitudinal ranges or occurring in different ocean basins tend to reflect the influences of environment as phenotypic variation in e.g., shell shape, size, and color. Many species are thus represented by one or more subspecific taxa (Van der Spoel, 1967; Van der Spoel et al., 1997). It has long been thought that isolation is rare in the marine pelagic realm where all water masses eventually mix and isolating barriers are essentially lacking resulting in slowly evolving species. However, upon reviewing all primary population genetic studies of open ocean zooplankton, Peijnenburg and Goetze (2013) find that genetic isolation can be achieved at the scale of gyre systems in open ocean habitats (100 s to 1,000s of km) and suggest that most marine zooplankton are well poised for evolutionary change. Molecular work on pteropods has shown that some circumglobal species or subspecific taxa are supported by genetic evidence, whereas others are not (e.g., Jennings et al., 2010; Gasca and Janssen, 2014; Maas et al., 2013; Burridge et al. 2015), suggesting the need for further work and taxonomic revisions. One example is the euthecosome *Limacina helicina* with a cold-water bipolar distribution pattern. Based on morphological evidence, this species is currently recognized as a species complex comprising two subspecies and at least five "formae." Molecular results, however, convincingly demonstrated that the Arctic *L. helicina helicina* and the Antarctic *L.h. antarctica* differ at least at the species level (Hunt et al., 2010).

Although pteropods occur at all depth layers, most species and higher abundances are found in the epipelagic. Species richness and abundances are lower in the meso- and bathypelagic as in many other pelagic groups. For example, of the 92 species found in the South Atlantic 50 are epipelagic (0–200 m), 29 occur both in the epipelagic and mesopelagic (200–1,000 m), four occur only in the mesopelagic, five both in the meso- and bathypelagic (>1,000 m) and four are only found in the bathypelagic (Van der Spoel and Dadon, 1999). Typical mesopelagic species are *Clio recurva*, *Peracle* spp. The euthecosomes *Thielea helicoides*, *Clio chaptalii*, and *Clio andreae*, and the gymnosomes *Massya* sp. and *Schizobrachium* sp. are bathypelagic.

Although there are no species restricted to the neritic environment several Creseis and Diacavolinia taxa may occur in very high abundances in coastal areas.

Ecology

Pteropods are protandric hermaphrodites, meaning that the same individuals first have a male stage and later a female stage. It is thought that cross fertilization is the rule, although self fertilization may also be possible. The sperm is stored in an accessory sac of the gonad and oocytes are fertilized internally by the stored sperm. The fertilized eggs are usually released into the water as floating egg masses in the form of gelatinous ribbons or spheres. A veliger larva hatches, metamorphoses, and develops into a juvenile stage. In the Thecosomata, the juvenile produces the protoconch II after the embryonic stage and then the soft parts grow regularly to maturity, when the teleoconch shell is formed. The euthecosome *Heliconoides inflata* has an unusual development with brood protection. Adults of this species brood their embryos and early veligers in their mantle cavity. Some deep-sea species such as *Thielea helicoides* and *Clio chaptalii* are viviparous and do not have free swimming larval stages. Like in other plankton groups growth rate and generation time is dependent on seasonality and temperature. For most species, there is probably one generation per year.

Most Thecosomata are herbivorous or omnivorous mucus feeders. They catch their food by making a mucous web that is much larger than the organism itself. The food consists of particulate matter as well as small organisms such as dinoflagellates, coccolithophorids, radiolaria, tintinnids, foraminifera, and diatoms, the particles are wrapped in mucus, swallowed, and ground by the radula. Gannefors et al. (2005) found a change in diet from diatom-dominated food items in spring to dinoflagellates in summer/autumn and also evidence of ingestion of copepods in *Limacina helicina*. Herbivorous euthecosomes often show swarming behavior resulting in very patchy distributions and varying effects on the local ecosystem. Some bathypelagic euthecosomes, e.g., *Thielea helicoides*, are carnivorous and usually occur as solitary individuals. Gymnosomata are active carnivores and usually occur as solitary individuals as well. They are often specialized feeders preying upon certain species of Thecosomata. An extreme example being that of *Clione limacina* feeding exclusively on *Limacina helicina* in the North Atlantic and on *L. helicina* and *L. retroversa* in the South Atlantic. Most other species are less specialized in their diet but all feed either exclusively or mainly on a few Thecosomata species (Conover and Lalli, 1974).

Many pteropods undergo more or less pronounced diel vertical migrations, feeding at night near the surface and migrating to greater depths during the day. Most species are influenced by specific hydrographic conditions so that day and night levels differ by place and season. Migration ranges vary in different species, for example, *Diacavolinia* species migrate daily only over some 10 m, whereas *Clio pyramidata* migrates over hundreds, even up to 1,500 m (van der Spoel, 1973). *Clio andreae* and *Thielea helicoides* undergo ontogenetic vertical migration, juveniles occurring higher in the water column than the adults.

Bednaršek et al. (2012) found that shelled pteropods constitute a mean global carbonate biomass of ~23 mg $CaCO_3$ m^{-3} (based on non-zero records). Total biomass values were lowest in the equatorial regions and equally high at both poles. Pteropods were found at least to depths of 1,000 m, with the highest biomass values located in the surface layer (0–10 m) and gradually decreasing with depth, with values in excess of 100 mg carbon (C) m^{-3} only found above 200 m depth. By extrapolating regional biomass to a global scale, they established global pteropod biomass to add up to 500 Tg C.

Geology

Pteropods are geologically a rather recent group only present since the Caenozoic. The oldest pteropod species, *Heliconoides mercinensis* is known from Late Paleocene deposits in Europe and the United States (e.g., Janssen and Peijnenburg, 2014). Pteropods probably originated from an Atlantic center of speciation (van der Spoel and Heyman, 1983). This hypothesis is based on the distribution patterns of some groups, such as the mesopelagic genus *Peracle*, whose species are all found in the Northeastern Atlantic off Dakar. In bathypelagic species, like *Thielea helicoides*, climatic influences on distribution patterns are probably negligible. Restriction of this bathypelagic species to the Atlantic Ocean and its deep water outflow may indicate that it originated in the oldest (late Cretaceous) deep-sea basin, the North Atlantic Ocean.

Conclusions

Shelled pteropods can be important in paleoclimatic, paleoceanographic, and paleoecologic studies (Herman and Rosenberg, 1969; Diester-Haass, 1972; Janssen and Peijnenburg, 2014; Wall-Palmer et al. 2012). The fact that many groups have extensive horizontal distribution patterns and show a rapid morphological evolution makes them ideal candidates as index fossils for biostratigraphical purposes. However, aragonite is highly undersaturated in the deep sea and dissolves away before it can be buried. Consequently, sediments containing >30 % pteropod shells ("pteropod ooze") are found predominantly in the tropics and sub-tropics, at <2,800 m depth in the Atlantic and <500 m in the Pacific, and none are present in the Southern Ocean (Bednaršek et al., 2012). Overall, only 2.4 % of the Atlantic Ocean sea floor comprises pteropod ooze. Nevertheless, one major advantage of using pteropods for biostratigraphy is that their fragile shells do not survive extensive transportation and are rarely reworked from one sedimentary unit into another (Janssen and Peijnenburg, 2014). Biostratigraphic applications were made by Chen (1968a) who correlated the Pleistocene pteropod zones with planktonic foraminifera zones, oxygen isotope analyzes and ^{14}C data. Gürs and Janssen (2004) refined the pteropod zonation for the later Miocene in the North Sea Basin and Cahuzac and Janssen (2010) correlated time intervals in the Aquitaine Basin with the zonation scheme for the North Sea Basin.

Bibliography

Angel, M. V., 1997. Pelagic biodiversity. In Ormond, R. F. G., Gage, J. D., and Angel, M. V. (eds.), *Marine Biodiversity. Patterns and Processes*. New York: Cambridge University Press, pp. 35–68.

Bathmann, U., Noji, T. T., and von Bondungen, B., 1991. Sedimentation of pteropods in the Norwegian Sea in autumn. *Deep Sea Research A*, 38, 1341–1360.

Bé, A. W. H., and Gilmer, R. W., 1977. A zoogeographic and taxonomic review of Euthecosomatous Pteropoda. In Ramsay, A. T. S. (ed.), *Oceanic Micropalaeontology*. London: Academic, pp. 733–800.

Bé, W. H., MacClintock, C., and Currie, D. C., 1972. Helical shell structure and growth of the pteropod Cuvierina columnella (Rang) (Mollusca, Gastropoda). *Biomineral Research Reports*, 4, 47–79.

Bednaršek, N., Možina, J., Vogt, M., O'Brien, C., and Tarling, G. A., 2012. The global distribution of pteropods and their contribution to carbonate and carbon biomass in the modern ocean. *Earth System Science Data*, 4, 167–186.

Burridge, A. K., Goetze, E., Raes, N., Huisman, J., and Peijnenburg, K. T. C. A., (2015). Global biogeography and evolution of *Cuvierina* pteropods. *BMC Evolutionary Biology* 15:39. doi:10.1186/s12862-015-0310-8.

Cahuzac, B., and Janssen, A. W., 2010. Eocene to miocene holoplanktonic mollusca (gastropoda) of the Aquitaine basin, southwest France. *Scripta Geologica*, 141, 1–193.

Chen, C., 1968a. The distribution of thecosomatous pteropods in relation to the Antarctic convergence. *Antarctic Journal of the United States*, 3, 155–157.

Chen, C., 1968b. Pleistocene pteropods in pelagic sediments. *Nature*, 210(5159), 1145–1149.

Conover, R. J., and Lalli, C. M., 1974. Feeding and growth in *Clione limacina* (Phipps), a pteropod mollusc, 2. Assimilation, metabolism, and growth efficiency. *Journal of Experimental Marine Biology and Ecology*, 16, 131–154.

Cuvier, G., 1804. Mémoire concernant l'animal de l'*Hyale*, un nouveau genre de mollusques nus, intermédiaire entre l'*Hyale* et le *Clio* et l'établissement d'un nouvel ordre dans la classe des mollusques. *Annales du Muséum National d'Histoire naturelle Paris*, 4, 223–234.

Diester-Haass, L., 1972. Late Pleistocene and Holocene sedimentation in the central and eastern Persian Gulf. *"Meteor" Forschungsergebnisse (C)*, 8, 37–83.

Feely, R. A., Sabine, C. L., Lee, K., Berelson, W., Kleypas, J., Fabry, V. J., and Millero, F. J., 2004. Impact of anthropogenic CO_2 on the $CaCO_3$ system in the oceans. *Science*, **305**, 362–366.

Gannefors, C., Marco, B., Gerhard, K., Martin, G., Ketil, E., Bjørn, G., Haakon, H., and Falk-Petersen, S., 2005. The Arctic sea butterfly *Limacina helicina*: lipids and life strategy. *Marine Biology*, **147**, 169–177.

Gasca, R., and Janssen, A. W., 2014. Taxonomic review, molecular data and key to the species of Creseidae from the Atlantic Ocean. *Journal of Molluscan Studies*, **80**, 35–42.

Gürs, K., and Janssen, A. W., 2004. Sea-level related molluscan plankton events (Gastropoda, Euthecosomata) during Rupelian (Early Oligocene) of the North Sea Basin. *Netherlands Journal of Geoscience*, **83**, 199–208.

Herman, Y., and Rosenberg, P. E., 1969. Pteropods as bathymetric indicators. *Marine Geology*, **7**, 169–173.

Hunt, B., Strugnell, J., Bednarsek, N., Linse, K., Nelson, R. J., Pakhomov, E., Seibel, B., Steinke, D., and Würzberg, L., 2010. Poles apart: the "Bipolar" Pteropod species *Limacina helicina* is genetically distinct between the Arctic and Antarctic Oceans. *Plos One*, **5**(3), e983.

Janssen, A. W., and Peijnenburg, K. T. C. A., 2014. Holoplanktonic Mollusca: development in the Mediterranean Basin during the last 30 million years and their future. In Goffredo, S., and Dubinsky, Z. (eds.), *The Mediterranean Sea: Its History and Present Challenges*. Dordrecht: Springer, pp. 341–362.

Jennings, R. M., Bucklin, A., Ossenbrügger, H., and Hopcroft, R. R., 2010. Species diversity of planktonic gastropods (Pteropoda and Heteropoda) from six ocean regions based on DNA barcode analysis. *Deep-Sea Research II*, **57**, 2199–2210.

Klussmann-Kolb, A., and Dinapoli, A., 2006. Systematic position of the pelagic Thecosomata and Gymnosomata within Opisthobranchia (Mollusca, Gastropoda)- revival of the Pteropoda. *Journal of Zoological Systematics and Evolutionary Research*, **44**, 118–129.

Lalli, C. M., and Gilmer, R. W., 1989. *Pelagic Snails. The Biology of Holoplanktonic Gastropod Mollusks*. Stanford: Stanford University Press, pp. 1–259.

Maas, A. E., Blanco-Bercial, L., and Lawson, G. L., 2013. Reexamination of the species assignment of *Diacavolinia* pteropods using DNA barcoding. *PLoS One*, **8**(1), e53889, doi:10.1371/journal.pone.0053889.

Peijnenburg, K. T. C. A., and Goetze, E., 2013. High evolutionary potential of marine zooplankton. *Ecology and Evolution*, **3**(8), 2765–2781, doi:10.1002/ece3.644.

Van der Spoel, S., 1967. *Euthecosomata: A Group with Remarkable Developmental Stages (Gastropoda, Pteropoda)*. Gorinchem: J. Noorduijn en Zoon N.V., p. 375.

Van der Spoel, S., 1973. Growth, reproduction and vertical migration in *Clio pyramidata* Linnaeus, 1767 forma *lanceolata* (Lesueur, 1813) with notes on some other Cavoliniidae (Mollsuca, Pteropoda). *Beaufortia*, **21**, 117–134.

Van der Spoel, S., and Dadon, J., 1999. Pteropoda. In Boltovskoy, D. (ed.), *South Atlantic Zooplankton*. Leiden: Backhuys Publishers, Vol. 1, pp. 868–1706.

Van der Spoel, S., and Diester-Haas, L., 1976. First records of fossil gymnosomatous protoconchae (Pteropoda, Gastropoda). *Bulletin Zoölogisch Museum Universiteit van Amsterdam*, **5**, 85–88.

Van der Spoel, S., and Heyman, R. P., 1983. *A Comparative Atlas of Zooplankton. Biological Patterns in the Oceans*. Utrecht/Berlin: Bunge/Springer Verlag, pp. 1–186.

Van der Spoel, S., Newman, L. J., and Estep, K. W., 1997. *Pelagic Molluscs of the World*. http://species-identification.org/.

Wall-Palmer, D., Hart M. B., Smart C. W., Sparks R. S. J., Le Friant, A., Boudon, G., Deplus, C., and Komorowski J. C., 2012. Pteropods from the Caribbean sea: variations in calcification as an indicator of past ocean carbonate saturation. *Biogeosciences* 9, 309–315.

Cross-references

Calcite Compensation Depth (CCD)
Carbonate Dissolution
Marine Microfossils

PULL-APART BASINS

Alper Gürbüz
Department of Geological Engineering, University of Ankara, Tandoğan, Ankara, Turkey
Department of Geological Engineering, University of Niğde, Niğde, Turkey

Synonyms

Dilatational fault jog; Extensional duplex; Releasing bend; Rhomb graben; Rhombochasm; Stepover basin; Tensile bridge; Wrench graben

Definition

Pull-apart basins are structural depressions that are localized on the geometric irregularities along strike-slip faults where the master fault is overstepping or bending.

Introduction

Strike-slip faults form linear and continuous fault systems, but they are typically segmented, resulting in localized regime changes across a variety of discontinuities or steps (e.g., Dooley and Schreurs, 2012). Pull-apart basins have been recognized originally along major strike-slip faults throughout the world where the segmented faults cause extension/transtension. Quennell (1956), who was the first to recognize a pull-apart basin without using this term, proposed that the Dead Sea is a void in the crust caused by the overlapping segments of the Dead Sea fault zone. For the same purpose, the first use of the term "pull-apart basin" by Burchfiel and Stewart (1966) was chosen for the interpretation of the Death Valley basin. The term pull-apart basin is a well-used and well-understood term for strike-slip basins (Mann et al., 1983) both in marine and terrestrial environments.

Model for pull-apart basin development

As indicated above, there are several synonyms of pull-apart basins (see Mann, 2007 for references) that reproduced from the identified tectonic, structural, geometric, and geomorphic features of pull-apart basins in nature. Tectonically, the basins are located along strike-slip faults and transform zones; structurally, they form when strike-slip faults bend or overstep; geometrically, while the overstepping master strike-slip faults are parallel or subparallel, the secondary basin-bounding faults perpendicular or diagonal to the main faults give the basin its characteristic shape; and geomorphologically, they are sharp-bounded and deep depressions. All these characteristics of pull-apart basins in nature help us to draw an ideal model for pull-apart basin that represents its mechanism of formation (Figure 1).

Detailed properties of natural pull-apart basins have been obtained through using experimental models. Although experimental laboratory studies have been used to investigate strike-slip faults since approximately the first quarter of twentieth century (e.g., Cloos, 1928), they have been used to simulate the mechanism and geometries of pull-apart basin formation in the last quarter. Because these basins differ considerably from simple strike-slip mechanism, their development processes share properties with both strike-slip and extensional settings (Rahe et al., 1998). Physical and/or numerical model studies have been a profitable approach, producing good geometric matches with natural examples of pull-apart basins (e.g., Rodgers, 1980; Gölke et al., 1994; McClay and Dooley, 1995; Katzman et al., 1995; Bertoluzza and Perotti, 1997; Dooley and McClay, 1997; Rahe et al., 1998; Basile and Brun 1999; Sims et al., 1999; Atmaoui et al. 2006; Wu et al., 2009; Mitra and Paul, 2011; Dooley and Schreurs, 2012). While the physical (analogue) modeling experiments used a variety of apparatus and different materials (e.g., sand, clay, and rock), the numerical experiments use a wide range of parameters and quantitative values. Thus, the results are helpful to understand the mechanical and geometrical aspects of natural large-scaled cases under different conditions.

Mechanical aspects

As detailed above, the basic mechanism of pull-apart basin development can be defined simply as local extension between two overstepping strike-slip faults (e.g., Aydın and Nur, 1982; Mann et al., 1983, Figure 1). Distributed strike-slip and Riedel shear mechanisms are two alternative mechanisms that have been proposed for pull-apart basin formation (e.g., Atmaoui et al., 2006; Dooley and Schreurs, 2012). Distributed shear deformation is developed in weak layers, such as evaporates or shales, above a broad strike-slip fault zone in the basement, and pull-apart basins can develop at releasing steps during fault interaction, coalescence, and linkage (An and Sammis, 1996). The Riedel shear mechanism causes development of pull-apart basins along Riedel faults connected by segments of strike-slip faults in the basement (Dewey, 1978; Hagglauer-Ruppel, 1991).

Independent of the formation mechanism, the extension which develops within pull-apart basins depends on the local crustal rheology. While the displacement along the master strike-slip faults links to each other by normal faults in a brittle crust, in a ductile crust it accommodates by other mechanisms (e.g., continuous extension inside the basin; Reches, 1987). Another necessary condition for a pull-apart basin formation is a mechanical detachment between the brittle part of the crust and rheologically strong uppermost mantle (Petrunin and Sobolev, 2006). The strength of décollement zones of pull-apart basins, which are observed or postulated to form within ductile or viscous zones at depth, also affects structural development in most tectonic regimes (Molnar and Lyon-Caen, 1988) and likely exhibits some control on the development process (Sims et al., 1999).

The elastic stress field in the surroundings of a fault step was firstly analyzed by Rodgers (1980) and Segall and Pollard (1980) with two different approaches. However, their conclusions were consistent with each other: while the mean shear stress decreases inside the pull-apart basin, the maximum stress increases at fault ends with a slight larger increase outside the basin boundaries (Reches, 1987). The theoretical models of these authors provide clues to the orientation of the different faults that can form inside a pull-apart basin (Bertoluzza and Perotti, 1997).

The development of new strike-slip faults inside the basins which are diagonal to master faults, has defined as the extinction process of pull-apart basins first by Zhang et al. (1989). Similar tendencies and real-time observations have been made by different researchers in experimental models and earthquakes (e.g., McClay and Dooley, 1995; Sims et al., 1999; Gürbüz and Gürer, 2008). The pull-apart basins are structurally defined by three evolutionary stages: incipient, early, and mature stages by comparison with physical modeling experiments (e.g., Rahe et al., 1998). Dramatically this type of basins tends to link the master faults in the last stage (Rahe et al., 1998; Sims et al., 1999). While the development of through-going cross-basin faults in the mature stage indicates the decreasing activity along major boundary faults, their formations increase the probability for large-magnitude earthquakes. The best examples of such process and related destructive earthquakes have been seen in China and Turkey in the last century (e.g., Zhang et al., 1989; Gürbüz and Gürer, 2009).

Geometrical aspects

The shape, fault system, and sedimentation features of a pull-apart basin depend upon the geometry associated

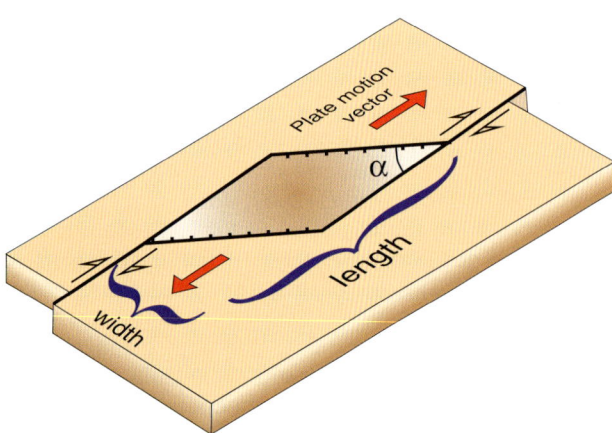

Pull-apart Basins, Figure 1 Formation and geometric model of pull-apart basin.

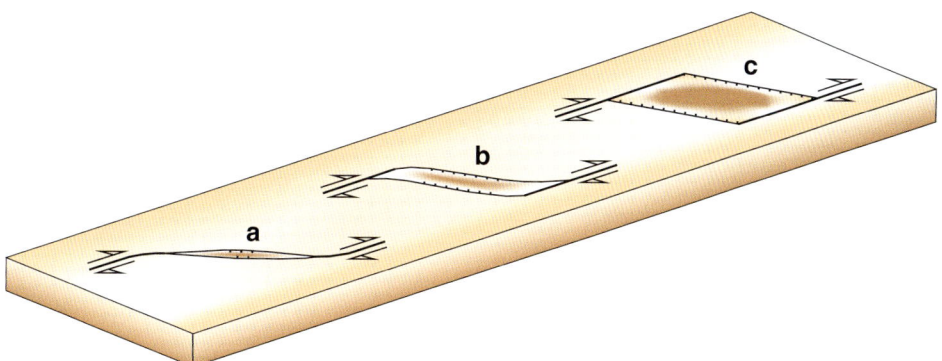

Pull-apart Basins, Figure 2 Typical shapes of pull-apart basins: (a) spindle-shaped, (b) lazy Z-shaped, (c) rhomboidal (modified from Mann, 2007).

with the step in the master strike-slip fault system (Carton et al., 2007). Likewise, the geometry of a pull-apart basin is controlled by the amount of master fault overlap, separation, and displacement, and the size of the basin is varied and well defined by these parameters (Rodgers, 1980). Therefore to understand their anatomies and evolutions, we need to define some geometric parameters because many uncertainties complicate their interpretations. As demonstrated by some authors, shape, scale, and angular characteristics of pull-apart basins are useful to discuss their initiation and structural evolutions (e.g., Aydın and Nur, 1982, 1985; Bahat, 1983; Mann, 2007; Gürbüz, 2010).

Pull-apart basin structures are usually rhomb-shaped depressions, but they vary from the lazy Z- or S-shaped to spindle- or almond-shaped to extreme-shaped examples in nature (e.g., Mann, 2007, Figure 2). Mann et al. (1983) proposed that the rhomboidal form of pull-apart basins results from the lengthening of an S- or Z-shaped basin with increased master fault overlap. Regardless of offset geometry, they evolve progressively from narrow grabens bounded by oblique-slip-linked faults to wider rhombic basins (Dooley and McClay, 1997). Within these shapes, they can be asymmetric, symmetric, or hybrid forms depending on the activity and connectivity of basin-bounding faults (e.g., Rahe et al., 1998).

Within angular meaning, the results of substantial experimental studies and natural pull-apart basins from all over the world represent the acute angles (α angle in Figure 1) between the oblique bounding faults and the master strike-slip faults of pull-apart basins are generally clustered at 30°–35° (e.g., Bahat, 1983; Gürbüz, 2010). Otherwise, the boundary condition (i.e., pure strike-slip, transtension, transpression) is the major parameter determining the acute angle thereby the shape of the basin (Bertoluzza and Perotti, 1997).

Pull-apart basins have a specific scale ratio. Aydın and Nur (1982) showed a well-defined linear correlation between the length and width of pull-apart basins with a value of 3:1 through analyzing of over 60 natural examples and proposed that the value may vary widely, depending on the structural, physiographic, or active dimensions of the basin. In addition, experimental models show similar results with an approximately constant length versus width ratio (2.2:1–3.8:1; Basile and Brun, 1999). In the same vein, a step further, a three-dimensional relationship between the length, width, and depth of pull-apart basins has suggested by Gürbüz (2010) through using the natural examples.

Pull-apart basins represent the same mechanical and geometrical features both in marine and terrestrial environments (except extreme types which can lengthen very much due to spreading ridges; see Mann, 2007 for details), and sometimes they represent a complex basin system scene. Marmara Sea of Turkey is located on the North Anatolian fault zone which is one of the world's famous strike-slip faults with its length of ~1,200 km (e.g., Şengör et al., 2005), and its internal structure includes such a basin system (Figure 3). Each of the basins has developed by the strike-slip tectonics of the North Anatolian fault zone (e.g., Okay et al., 2000; Armijo et al., 2002) and represents similar geometric features while varying in shapes, sizes, and also depths (e.g., Gürbüz, 2010).

Sedimentary aspects

Sedimentation features of pull-apart basins are generally related to kinematic, geometric, geomorphic, and crustal properties. Strike-slip faults that bound these basins have a large component of normal slip. Thus, while the basin inside is represented by a deep depression, the basin margins surrounded by steeper uplifted and deeply eroded highlands.

In continental pull-apart basins, along the strike-slip margins of the basin, coarser sediments in the form of talus breccia and debris-flow-dominated alluvial fans grade into fine-grained sediments toward the center of the basin (Crowell, 1974; Hempton and Dunne, 1984). As a result of such geomorphic and geometric characteristics, higher sedimentation rates, greater thicknesses in smaller areas,

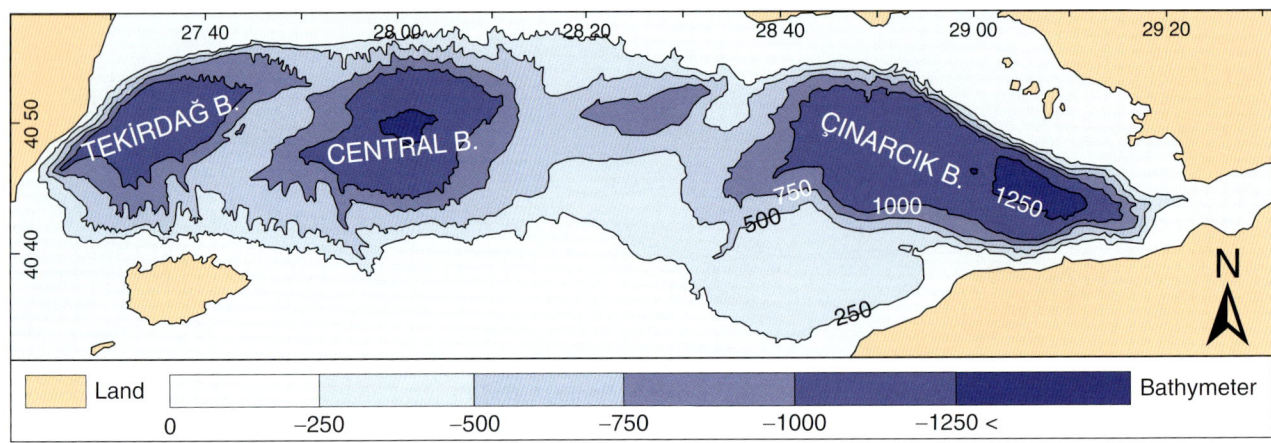

Pull-apart Basins, Figure 3 Bathymetric map of the Marmara Sea pull-apart basin system.

displaced fan-source relationships and skewed fans, and depocenter migrations differentiate pull-apart basins from other extensional depressions (Hempton and Dunne, 1984). In marine environments, similar sedimentary patterns are valid for pull-apart basins by coarser sediments along basin margins transported through submarine canyons instead of rivers and submarine landslides along fault scarps and fine sediments toward the deepest parts (e.g., Beck et al., 2007; Görür and Çağatay, 2010). Basins far away from efficient sediment sources may include pelagic sediments and deposits of gravity mass movements (Einsele, 2000).

Strike-slip basins have high amounts of sediment thicknesses in spite of their small sizes and short life spans, typically in existence in 3–10 My (Woodcock, 2004). This would be related to their rapid cooling character within thermal meaning which gives the way to rapid subsidence (e.g., Pitman and Andrews, 1985; Mann, 2007). The major parameter that controls the basin-fill thickness during this process is the thickness of the brittle layer (Petrunin and Sobolev, 2006).

Summary

Pull-apart basins are located along strike-slip faults and transform zones and are developed when the master fault bends or oversteps extensionally. While the master faults are parallel or subparallel, the secondary basin forms with faults perpendicular or diagonal to the master faults giving the basin its characteristic shape. The extension which develops within the pull-apart basins depends on the local crustal rheology. Stress fields suggest that the mean shear stress decreases inside the basin and the maximum stress increases at fault ends. The anatomies and evolutions of pull-apart basins are closely related to their geometric aspects. Their shapes, scales, and angular characteristics are useful to discuss their structures. Pull-apart basins are generally shown as rhomb-shaped depressions. But the shape of the basin varies widely, from spindle to lazy Z or S to other extreme examples. In an ideal rhomb-shaped pull-apart basin, while the acute angle between the master faults and basin-bounding diagonal faults is generally a value between 30° and 35°, the 2D scale characteristics of pull-apart basins represent a mean aspect ratio of 3:1, and 3D characteristics represent a quantitative relationship between length, width, and depth. These values are suggested for basins shown in ideal geometries, and they can vary in a wide range due to their spatial and temporal positions in an evolving environment. The changing and intersecting tectonic regimes and local crustal, lithological, or climatic properties can also influence basin geometry. In contrast to their small-sized and short-lived characters, pull-apart basins have great thicknesses of basin-fill deposits. Higher sedimentation rates, displaced fan-source relationships and skewed fans, and migrated depocenters are some sedimentary aspects of pull-apart basins that differentiated this type of basin from the other tectonic depressions.

Bibliography

An, L.-J., and Sammis, C. G., 1996. Development of strike-slip faults: shear experiments in granular materials and clay using a new technique. *Journal of Structural Geology*, **18**, 1061–1077.

Armijo, R., Meyer, B., Navarro, S., King, G., and Barka, A., 2002. Asymmetric slip partitioning in the Sea of Marmara pull-apart: a clue to propagation processes of the North Anatolian Fault? *Terra Nova*, **14**, 80–86.

Atmaoui, N., Kukowski, N., Stockhert, B., and Konig, D., 2006. Initiation and development of pull-apart basins with. Riedel shear mechanism: Insights from scaled clay experiments. *International Journal of Earth Sciences*, **95**(2), 225–238.

Aydın, A., and Nur, A., 1982. Evolution of pull-apart basins and their scale independence. *Tectonics*, **1**, 91–105.

Aydın, A. A., and Nur, A., 1985. The types and roles of stepovers in strike-slip tectonics. In Biddle, K. T., and Christie-Blick, N. (eds.), *Strike-slip deformation, basin formation, and sedimentation*. SEPM Special Publication, 37, pp. 35–45.

Bahat, D., 1983. New aspects of rhomb structures. *Journal of Structural Geology*, **5**, 591–601.

Basile, C., and Brun, J. P., 1999. Transtensional faulting patterns ranging from pull-apart basins to transform continental margins.

An experimental investigation. *Journal of Structural Geology*, **21**, 23–37.

Beck, C., Mercier de Lapinay, B., Schneider, J. L., Cremer, M., Çağatay, N., Wendenbaum, E., Boutareaoud, S., Menot, G., Schmidt, S., Webe, O., Eriş, K., Armijo, R., Meyer, B., Pondard, N., Gutcher, M. A., Turon, J. L., Labeyrie, L., Cortijo, E., Gallet, Y., Bouquerel, H., Görür, N., Geravis, A., Castera, M. H., Londeix, L., de Resseguier, A., and Jaouen, A., 2007. Late Quaternary co-seismic sedimentation in the Sea of Marmara's deep basins. *Sedimentary Geology*, **199**, 65–89.

Bertoluzza, L., and Perotti, C. R., 1997. A finite-element model of the stress field in strike-slip basins: implications for the Permian tectonics of the southern Alps (Italy). *Tectonophysics*, **280**, 185–197.

Burchfiel, B. C., and Stewart, J. H., 1966. "Pull-apart" origin of the central segment of Death Valley, California. *Geological Society of America Bulletin*, **77**, 439–442.

Carton, H., Singh, S. C., Hirn, A., Bazin, S., de Voogd, B., Vigner, A., Ricolleau, A., Çetin, S., Ocakoğlu, N., Karakoç, F., and Sevilgen, V., 2007. Seismic imaging of the three-dimensional architecture of the Çınarcık Basin along the North Anatolian Fault. *Journal of Geophysical Research*, **180**, B06101.

Cloos, H., 1928. Experimente zur inneren Tektonik. *Cetralblatt fur Mineralogie*, **5**, 609–621.

Crowell, J. C., 1974. Sedimentation along the San Andreas fault, California. In Dott, R. H., and Shaver, R. H. (eds.), *Modern and Ancient Geosynclinal Sedimentation*. Tulsa, OK: Society of Economic Paleontologists and Mineralogists. Society of Economic Paleontologists and Mineralogists (SEPM) Special Publication, 19, pp. 292–303.

Dewey, J. F., 1978. Origin of long transform-short ridge systems. *Geological Society of America*, **10**, 388.

Dooley, T., and McClay, K., 1997. Analog modeling of pull-apart basins. *American Association of Petroleum Geologists Bulletin*, **81**, 1804–1826.

Dooley, T. P., and Schreurs, G., 2012. Analogue modelling of intraplate strike-slip tectonics: a review and new experimental results. *Tectonophysics*, **574–575**, 1–71.

Einsele, G., 2000. *Sedimentary Basins: Evolution. Facies, and Sediment Budget*, 2nd edn. Heidelberg: Springer.

Gölke, M., Cloetingh, S., and Fuchs, K., 1994. Finite-element modeling of pull-apart basin formation. *Tectonophysics*, **240**, 45–57.

Görür, N., and Çağatay, N., 2010. Geohazards rooted from the northern margin of the Sea of Marmara since the late Pleistocene: a review of recent results. *Natural Hazards*, **54**, 583–603.

Gürbüz, A., 2010. Geometric characteristics of pull-apart basins. *Lithosphere*, **2**, 199–206.

Gürbüz, A., and Gürer, Ö. F., 2008. Tectonic Geomorphology of the North Anatolian Fault Zone in the Lake Sapanca Basin (Eastern Marmara Region, Turkey). *Geosciences Journal*, **12**, 215–225.

Gürbüz, A., and Gürer, Ö. F., 2009. Middle Pleistocene extinction process of pull-apart basins along the North Anatolian fault zone. *Physics of the Earth and Planetary Interiors*, **173**(1–2), 177–180.

Hagglauer-Ruppel, B., 1991. *Kinematik und Begleitstrukturen von Scherzonen: Experimente und Beispiele Mitteleuropas (mit besonderer Berücksichtigung des Osning-Lineamentes)*. Unpublished PhD thesis, Germany, Ruhr-University of Bochum.

Hempton, M. R., and Dunne, L. A., 1984. Sedimentation in pull-apart basins. Active example in eastern Turkey. *The Journal of Geology*, **92**, 513–530.

Katzman, R., ten Brink, U. S., and Lin, J., 1995. 3-D modelling of pull-apart basins: implications for the tectonics of the Dead Sea Basin. *Journal of Geophysical Research*, **100**, 6295–6312.

Mann, P., 2007. Global catalogue, classification and tectonic origins of restraining and releasing bends on active and ancient strike-slip fault systems. In Cunningham, W. D., and Mann, P. (eds.), *Tectonics of Strike-Slip Restraining and Releasing Bends*. London: Geological Society of London. Special Publication, 290, pp. 13–142.

Mann, P., Hempton, M. R., Bradley, D. C., and Burke, K., 1983. Development of pull-apart basins. *The Journal of Geology*, **91**, 529–554.

McClay, K., and Dooley, T., 1995. Analogue models of pull-apart basins. *Geology*, **23**, 711–714.

Mitra, S., and Paul, D., 2011. Structural geometry and evolution of releasing and restraining bends: Insights from laser-scanned experimental models. *AAPG Bulletin*, **95**, 1147–1180.

Molnar, P., and Lyon-Caen, H., 1988. Some simple physical aspects of the support, structure, and evolution of mountain belts. In Clark, S. P., Jr., Burchfiel, B. C., and Suppe, J. (eds.), *Processes in Continental Lithospheric Deformation*. Boulder: Geological Society of America. Geological Society of America, Special Paper 218.

Okay, A. İ., Kaşlılar-Özcan, A., İmren, C., Boztepe-Güney, A., Demirbağ, E., and Kuşçu, İ., 2000. Active faults and evolving strike-slip basins in the Marmara Sea, northwest Turkey: a multichannel seismic reflection study. *Tectonophysics*, **321**, 189–218.

Petrunin, A., and Sobolev, S. V., 2006. What controls thickness of sediments and lithospheric deformation at a pull-apart basin. *Geology*, **34**(5), 389–392.

Pitman, W. C., and Andrews, J. A., 1985. Subsidence and thermal history of small pull-apart basins. In Biddle, K. T., and Christie-Blick, N. (eds.), *Strike-Slip Deformation, Basin Formation and Sedimentation*. Tulsa, OK: Society of Economic Paleontologists and Mineralogists. SEPM Special Publications, 37, pp. 45–49.

Quennell, A. M., 1956. Tectonics of the Dead Sea rift. In *Proceedings, Congreso Geologico Internacional, 20th*, Asociacion de Servicios Geologicos Africanos, Mexico City, pp. 385–405.

Rahe, B., Ferrill, D. A., and Morris, A. P., 1998. Physical analog modeling of pull-apart basin evolution. *Tectonophysics*, **285**, 21–40.

Reches, Z., 1987. Mechanical aspects of pull-apart basins and push-up swells with application to the Dead Sea transform. *Tectonophysics*, **141**, 75–88.

Rodgers, D. A., 1980. Analysis of pull-apart basin development produced by en-echelon strike-slip faults, In: Balance, P. F., and Reading, H. G., (eds.), *Sedimentation in Oblique-Slip Mobile Zones*. International Association of Sedimentologists Special Publication, 4, pp. 27–41.

Segall, P., and Pollard, D. O., 1980. Mechanics of discontinuous faults. *Journal of Geophysical Research*, **85**, 4337–4350.

Şengör, A. M. C., Tüysüz, O., İmren, C., Sakınç, M., Eyidoğan, H., Görür, N., Le Pichon, X., and Rangin, C., 2005. The north anatolian fault: a new look. *Annual Review of Earth and Planetary Sciences*, **33**, 1–75.

Sims, D., Ferrill, D. A., and Stamatakos, J. A., 1999. Role of ductile décollement in the development of pull-apart basins. Experimental and natural examples. *Journal of Structural Geology*, **21**, 533–554.

Woodcock, N. H., 2004. Life span and fate of basins. *Geology*, **32**, 685–688.

Wu, J. E., McClay, K., Whitehouse, P., and Dooley, T., 2009. 4D analogue modelling of transtensional pull-apart basins. *Marine and Petroleum Geology*, **26**, 1608–1623.

Zhang, P.-Z., Burchfiel, B. C., Chen, S., and Deng, Q., 1989. Extinction of pull-apart basins. *Geology*, **17**, 814–817.

Cross-references

Marine Sedimentary Basins
Plate Motion
Push-up Blocks
Transform Faults

PUSH-UP BLOCKS

Alper Gürbüz
Department of Geological Engineering, University of Ankara, Tandoğan, Ankara, Turkey
Department of Geological Engineering, University of Niğde, Niğde, Turkey

Synonyms

Antidilational fault jog; Compressional bridge; Compressional stepover; Contractional duplex; Pop-ups; Pressure ridge; Push-up swell; Restraining bend; Restraining stepover; Rhomb horst; Structural knot; Transpressional uplift

Definition

Push-up blocks are topographic uplifts that localize on the geometric irregularities along strike-slip faults where fault segments form a compressive stepover or bend.

Introduction

Push-up blocks are an integral part of intraplate and interplate strike-slip fault zones (e.g., Sylvester and Smith, 1976; Christie-Blick and Biddle, 1985; Sylvester, 1988) and are very important structures for hydrocarbon accumulations. They represent structural traps as *en échelon* anticlines in places combined with stratigraphic traps (Harding, 1974, 1990; McClay and Bonora, 2001). Many large strike-slip fault zones commonly have large-scale push-up blocks associated with overstepping and bending both in marine and terrestrial environments (e.g., Angelier et al., 2004; Wakabayashi et al., 2004; Mann, 2007; Dooley and Schreurs, 2012). Probably, Trevisan (1939) was the first to define the push-up blocks, without using the term, to explain the development of the Brenta Mountains in the Lombardy Alps as the displacement along two wrench faults, which stepped with respect to each other and generated local folding and thrusting (Reches, 1987). The first use of the term "push-up block" by Mann et al. (1983) was chosen for describing compressional uplifts along strike-slip fault discontinuities.

Model for push-up block development

Whereas pull-apart basins reflect localized extension along a strike-slip fault, push-up blocks reflect local contraction. Basins develop where a right-lateral sense of slip steps to the right and left-lateral to the left, but push-up blocks develop when the sense of slip along the master faults falls afoul to stepped sides (Figure 1). Push-up blocks typically form uplifts, commonly with folds and reverse and/or thrust faults. In plan view they are broadly in form, whereas in cross section they are commonly bounded convex-up faults forming positive flower or palm tree structures (e.g., Sylvester, 1988; Dooley and Schreurs, 2012).

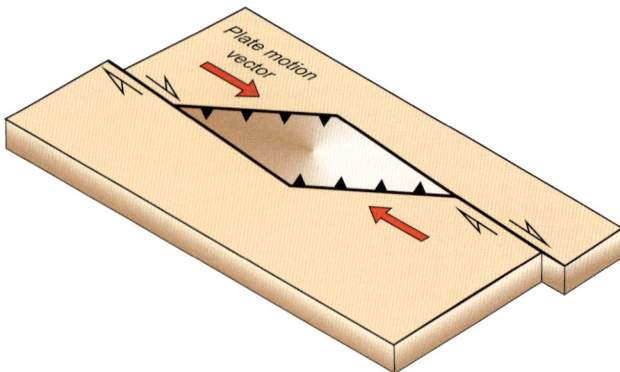

Push-up Blocks, Figure 1 Formation model of push-up block along a right-lateral fault.

Mechanical aspects

Push-up blocks combine elements of strike-slip and compressional regimes. Thus, they share the properties of both simple shear and pure shear. In these blocks, exposure of older rocks is produced as the result of the oblique intersection of the fault with the direction of the plate or block motion (Mann, 2007).

The kinematic mechanism for the formation and extinction processes of push-up blocks is related to the "switchyard behavior" of the main strike-slip fault (Brown et al., 1991). These uplifted regions develop if the master fault oversteps or bends displacement accommodation provided by oblique-slip flanking fault which rejoins the master fault on another lateral ramp further along the strike (Swanson, 1989). In case the accommodated displacement bypasses the flanking faults during progressive offset of the master fault, push-up blocks become extinct (e.g., Mann, 2007). Experimental models also show cutoff faults that transect the uplifted zone (e.g., McClay and Bonora, 2001), particularly where the model does not utilize a set of basement plates, in order to generate the restraining stepover (e.g., Dooley and Schreurs, 2012).

Like several other features, the stress field in and around a push-up block represents contrasting characteristics to pull-apart basins: the mean stress decreases outside the step, whereas the maximum shear stresses increase inside the block (Segall and Pollard, 1980). Thus, the internal structure of push-up blocks is highly variable both in space and time, owing to this stress field and heterogeneous internal lithology and crustal rheology.

Several cases of block rotations have been identified in push-up blocks both in nature and analogue models (e.g., Ron et al., 1984; Sarıbudak et al., 1990; McClay and Bonora, 2001). The paleomagnetic recognition of significant vertical axis rotations in the field has led to block models which propose that large rigid blocks experience greatest rotations in the restraining bend areas (e.g., Luyendyk, 1991; Dickinson, 1996). According to physical modeling studies, the upper surfaces of the push-ups showed rotation with a maximum 16° after 10 cm displacement for the 150° overlapping stepovers (McClay and Bonora, 2001).

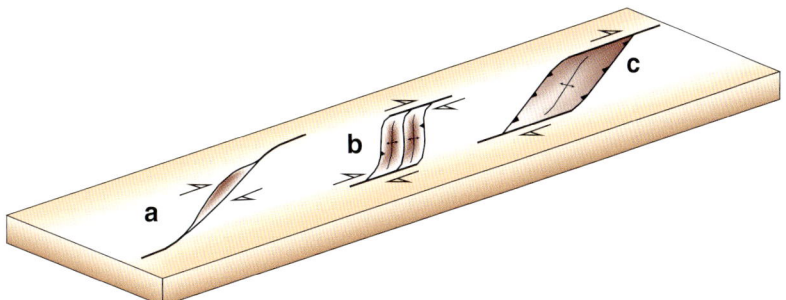

Push-up Blocks, Figure 2 Types of push-up structures: (a) transpressional uplift, (b) sharp restraining bend, and (c) gentle restraining bend (modified from Mann, 2007).

Geometrical aspects

Push-up blocks represent various shapes, like the planforms of pull-apart basins, but commonly in lozenge-shaped types (e.g., Dooley and Schreurs, 2012, Figure 1). Although there are some published works that studied natural examples, fundamental data about the geometric features of push-up blocks are based on physical modeling studies. According to these researches, their geometries were strongly controlled by the geometry of stepover (i.e., underlapping and overlapping), the width of the stepover in the rigid basement beneath the experiment material (e.g., sandpack), and the thickness of this material (McClay and Bonora, 2001; Mitra and Paul, 2011, Dooley and Schreurs, 2012). Related to increase in the amount of stepover, push-up blocks become wider (e.g., McClay and Bonora, 2001) like the pull-apart basins (e.g., Hempton and Dunne, 1984; Gürbüz, 2010).

Geomorphic aspects

Push-up blocks are topographic uplifts that sometimes reach several kilometers in elevation and are areas of exposed older and crustal rocks (Mann, 2007). Therefore, geomorphic features are more important for their understanding. On the other hand, well-described natural examples of push-up blocks are uncommon due to the complex three-dimensional geometries of the fault systems, and also due to their easy erodible structures as topographic uplifts (e.g., McClay and Bonora, 2001).

Mann (2007) has classified the restraining bends into three types: transpressional uplifts, sharp restraining bends, and gentle restraining bends (Figure 2). Transpressional uplifts are adjacent to straight segments of strike-slip faults which are directed oblique to plate or block movement. Sharp restraining bends represent the push-ups which form at fault stepovers and produce rhomboidal uplifts. These two structures are less common than the last gentle one. Gentle restraining bends produce elongate to domal uplifts. The inner parts consist of exposed older and deeper crustal igneous and metamorphic rocks, and their boundaries are surrounded by a radiating pattern of alluvial fans and fan deltas (Mann and Gordon, 1996).

The topographic lineaments related to the master fault traces inside the push-up block are sometimes bedimmed by the crinkled topography of the core of block but are well denoted by the valley of the master faults on either side of the block (Mann, 2007).

Summary

Large strike-slip fault zones are fundamentally segmented (e.g., the North Anatolian fault zone, Herece and Akay, 2003, Şengör et al., 2005; San Andreas fault zone, Sylvester, 1988, Powell et al., 1993) and have contractional structures (i.e., folds and reverse/thrust faults) formed in regions of overstepping and bending in the fault zone. Push-up blocks are located along strike-slip faults and transform zones and are developed between steps or long bends which produce localized contraction. They are very similar to pull-apart basins within the tectonic origin and plan view geometries, but they generate positive topography along the fault zone. Unlike the pull-apart basins, push-up blocks develop when the sense of step, or bending, of the master faults is opposite to the sense of slip on the fault zone, i.e., restraining stepovers in left-lateral faults step to the right, and in right-lateral systems, they step to the left. Push-up blocks share properties of both strike-slip and compressional regimes. The mean shear stress decreases outside the step, whereas the maximum stresses increase inside the block. They have been classified by Mann (2007) as transpressional uplifts, sharp and gentle restraining bends according to their morphologies and setting. The internal structure of push-up blocks is highly variable both in space and time owing to stress field changes and heterogeneous internal lithology and crustal rheology. Although push-up blocks are the opposite reflections of pull-apart basins, there are more synonyms for these uplifted areas as demonstrated under the title. However, pull-apart basins are far more numerous than push-up blocks in nature (e.g., Aydın and Nur, 1982; Bahat, 1983). Push-up blocks are not as well understood and well-studied as pull-apart basins, probably because of the complexity of fault geometries, and because they are regions of uplift, which once formed rapidly become eroded (e.g., McClay and Bonora, 2001).

Therefore, more surface and subsurface methods (i.e., seismic imaging) should be operated for a better understanding of this tectonic structure.

Bibliography

Angelier, J., Bergerat, F., Bellou, M., and Homberg, C., 2004. Co-seismic strike-slip fault displacement determined from push-up structures: the Selsund Fault case, South Iceland. *Journal of Structural Geology*, **26**, 709–724.

Aydın, A., and Nur, A., 1982. Evolution of pull-apart basins and their scale independence. *Tectonics*, **1**, 91–105.

Bahat, D., 1983. New aspects of rhomb structures. *Journal of Structural Geology*, **5**, 591–601.

Brown, N. N., Fuller, M. D., and Sibson, R. H., 1991. Paleomagnetism of the Ocotillo Badlands, southern California, and implications for slip transfer through an antidilational fault jog. *Earth and Planetary Science Letters*, **102**, 277–288.

Christie-Blick, N., and Biddle, K. T., 1985. Deformation and basin formation along strike-slip faults. In Biddle, K. T., and Christie-Blick, N. (eds.), *Strike-Slip Deformation, Basin Formation, and Sedimentation*. Tulsa (Oklahoma): SEPM (Society of Economic Paleontologists and Mineralogists). SEPM Special Publication, No 37, pp. 1–35.

Dickinson, W. R., 1996. Kinematics of transrotational tectonism in the California tranverse ranges and its contribution to cumulative slip along the san andreas transform fault system. *Geological Society of America Special Papers*, **305**, 1–14.

Dooley, T. P., and Schreurs, G., 2012. Analogue modelling of intraplate strike-slip tectonics: a review and new experimental results. *Tectonophysics*, **574–575**, 1–71.

Gürbüz, A., 2010. Geometric characteristics of pull-apart basins. *Lithosphere*, **2**, 199–206.

Harding, T. P., 1974. Petroleum traps associated with wrench faults. *AAPG Bulletin*, **58**, 1290–1304.

Harding, T. P., 1990. Identification of wrench faults using subsurface structural data: criteria and pitfalls. *AAPG Bulletin*, **74**, 1590–1609.

Hempton, M. R., and Dunne, L. A., 1984. Sedimentation in pull-apart basins. Active example in eastern Turkey. *The Journal of Geology*, **92**, 513–530.

Herece. E., and Akay, E., 2003. Atlas of North Anatolian Fault (NAF). *General Directorate of Mineral Research and Exploration*. Special Publication series-2, Ankara, 61 p and 13 appendices as separate maps.

Luyendyk, B. P., 1991. A model for Neogene crustal rotations, transtension, and transpression in Southern California. *Geological Society of America Bulletin*, **103**, 1528–1536.

Mann, P., 2007. Global catalogue, classification and tectonic origins of restraining- and releasing bends on active and ancient strike-slip fault systems. In Cunningham, W. D., and Mann, P. (eds.), *Tectonics of Strike-Slip Restraining and Releasing Bends*. London: Geological Society. Special Publication 290, pp. 13–142.

Mann, P., and Gordon, M. B., 1996. Tectonic uplift and exhumation of blueschist belts along transpressional strike-slip fault zones. In Bebout, G. E., Scholl, D. W., Kirby, S. H., and Platt, J. P. (eds.), *Subduction Top to Bottom*. Washington, DC: AGU. American Geophysical Union, Geophysical Monograph Series, 96, pp. 143–154.

Mann, P., Hempton, M. R., Bradley, D. C., and Burke, K., 1983. Development of pull-apart basins. *The Journal of Geology*, **91**, 529–554.

McClay, K., and Bonora, M., 2001. Analog models of restraining stepovers in strike-slip fault systems. *AAPG Bulletin*, **85**, 233–260.

Mitra, S., and Paul, D., 2011. Structural geometry and evolution of releasing and restraining bends: insights from laser-scanned experimental models. *AAPG Bulletin*, **95**, 1147–1180.

Powell, R. E., Weldon, R. J., and Matti, J. C., 1993. *The San Andreas fault system: displacement, palinspastic reconstruction and geologic evolution*. Boulder, CO: Geological Society of America. Geological Society of America Memoir, 178, 332 p.

Reches, Z., 1987. Mechanical aspects of pull-apart basins and push-up swells with application to the Dead Sea transform. *Tectonophysics*, **141**, 75–88.

Ron, H., Freund, R., Garfunkel, Z., and Nur, A., 1984. Block rotation by strike-slip faulting; structural and paleomagnetic evidence. *Journal of Geophysical Research B*, **89**, 6256–6270.

Sarıbudak, M., Sanver, M., Şengör, A. M. C., and Görür, N., 1990. Paleomagnetic evidence for substantial rotation of the Almacik flake within the North Anatolian Fault zone, NW Turkey. *Geophysical Journal International*, **102**, 563–568.

Segall, P., and Pollard, D. O., 1980. Mechanics of discontinuous faults. *Journal of Geophysical Research*, **85**, 4337–4350.

Şengör, A. M. C., Tüysüz, O., İmren, C., Sakınç, M., Eyidoğan, H., Görür, N., Le Pichon, X., and Rangin, C., 2005. The North Anatolian fault: a new look. *Annual Review of Earth and Planetary Sciences*, **33**, 1–75.

Swanson, M. T., 1989. Sidewall ripouts in strike-slip faults. *Journal of Structural Geology*, **11**, 933–948.

Sylvester, A. G., 1988. Strike-slip faults. *Geological Society of America Bulletin*, **100**, 1666–1703.

Sylvester, A. G., and Smith, R. R., 1976. Tectonic transpression and basement-controlled deformation in the San Andreas fault zone, Salton trough, California. *AAPG Bulletin*, **60**, 74–96.

Trevisan, L., 1939. Il Gruppo di Brenta. *Memorie degli Istituti di Geologia e Mineralogia dell'Università di Padova*, **13**, 1–128.

Wakabayashi, J., Hengesh, J. V., and Sawyer, T. L., 2004. Four-dimensional transform fault processes: progressive evolution of step-overs and bends. *Tectonophysics*, **392**, 279–301.

Cross-references

Plate Motion
Pull-apart Basins
Transform Faults

R

RADIOCARBON: CLOCK AND TRACER

Pieter M. Grootes
Institute for Ecosystem Research, Christian-Albrechts-University Kiel, Kiel, Germany

Synonyms
Dating; Isotopes; Tracer

Introduction
Carbon
Carbon (chemical symbol C) is an element with six protons in a nucleus circled by six electrons. In addition, the nucleus may contain six, seven, or eight neutrons, which leads to three forms of carbon having 12, 13, and 14 mass units (^{12}C, ^{13}C, ^{14}C) with a natural relative abundance of 98.9 %, 1.1 %, and ca. 10^{-10} %, respectively. These three natural forms of carbon, called isotopes, have the same chemical properties but slightly different, mass-dependent, physical properties. The isotope carbon-14 (^{14}C) is also called *radiocarbon*, because it is radioactive, showing beta decay. Carbon is a key building block for life on earth, and its isotopes can be used to study physiological and environmental processes.

Production, dispersion, and decay
Radiocarbon is produced in the upper atmosphere, near the boundary between the stratosphere and troposphere (9–15 km altitude). There a neutron (n), produced by incoming cosmic radiation and slowed to thermal energies by collisions, enters the nucleus of an atmospheric nitrogen-14 and shoots out a proton (p) ($^{14}_{7}N + n \rightarrow {}^{14}_{6}C + p$). This ^{14}C is instable and decays back to nitrogen-14, when a neutron turns into a proton, while emitting an electron (e^-, beta particle) and an antineutrino ($\bar{\nu}$) ($^{14}_{6}C \rightarrow {}^{14}_{7}N + e^- + \bar{\nu}$). By consensus, the value used for the half-life of this decay is 5,730 ± 40 years (Godwin, 1962; Roberts and Southon, 2007), which means half of the ^{14}C atoms decay in this time.

The newly formed radiocarbon is oxidized to $^{14}CO_2$, mixed globally with CO_2 in the atmosphere, incorporated into organic compounds by photosynthesis, and dissolved as bicarbonate in rivers, lakes, oceans, and groundwater. Carbon exchange between the different global carbon reservoirs (atmosphere, biosphere, soils, sediments, and ocean (surface, intermediate, deep)) maintains a flow of radiocarbon from its atmospheric production site to its decay in the various reservoirs and, thereby, offers the possibility to study the dynamics of the global carbon cycle.

Measurement
The potential of radiocarbon as a research tool was recognized and developed by Libby in the 1940s (Libby, 1965). He determined the ^{14}C concentration of a sample by comparing its rate of ^{14}C decay with that of a modern standard and a ^{14}C-free background sample, measured under identical conditions. Libby initially used Geiger-Müller counters with solid carbon to count the number of ^{14}C decays per minute; later proportional gas counters and liquid scintillation counters were used. Direct mass spectrometric measurement of the relative abundance of ^{14}C was prevented by its very low concentration (one ^{14}C atom per 10^{12} atoms ^{12}C) and the abundance of ions of ^{14}N and isobaric molecular fragments like $^{12}CH_2$ and ^{13}CH in the ion beam. Application of nuclear physics technology in the 1970s made direct measurement of ^{14}C via accelerator mass spectrometry (AMS) feasible (Tunis et al., 1998). Crucial were the initial use of negative ions, which eliminates the isobaric ^{14}N, and the stripping-off of electrons after initial acceleration to produce positive ions, which destroys isobaric molecular fragments, plus the ability to identify and count single ions. Typical AMS ^{14}C count rates range from several hundreds to 0.1 particles per second for samples from modern to over 50,000

years old. For a modern sample, the usual 0.3 % standard deviation (equivalent to ± 25 years) of ^{14}C counting statistics can be obtained in less than half an hour, compared with several days needed in ^{14}C decay counting. The main advantage of AMS is, however, the direct detection of all ^{14}C atoms in the sample, instead of only those that decay during the measurement (one in a million in a typical 3-day measurement). Thus, despite a low efficiency in the production of negative ions in the ion source (only a few percent), AMS has reduced the sample size routinely required for a radiocarbon measurement from grams to milligrams of carbon and allows special measurements down to the 10 μg level. This makes it possible to select for analysis organic macrofossils or specific organic compounds closely associated with the event to be studied. In ocean studies, the size of water samples can be reduced by a factor of 1,000 or more, and, instead of bulk carbonate, monospecific planktic foram samples can be used to date sediments. Dissolved organic carbon (DOC) can complement DIC studies of the water column; benthic forams and pore waters can reveal carbon cycling at the ocean-sediment interface and in the sediment.

Clock: radiocarbon dating
The clock
Radiocarbon decay provides a clock that starts running the moment exchange of carbon with the environment, in particular with the atmosphere, stops and thus tells us how long ago this happened. Through metabolic processes and gas exchange, the radiocarbon concentration in living organisms and surface waters is kept in equilibrium with the environment. Upon death and when waters lose contact with the atmosphere, the radiocarbon uptake stops, and radioactive decay results in a ^{14}C concentration decreasing with time. This is the "radiocarbon clock." Measurement of the remaining ^{14}C concentration allows the calculation of the time elapsed since "death" with a precision of a few decades, provided several assumptions are valid: (i) after death no carbon was lost from, or added to, the sample (closed system assumption), (ii) the initial ^{14}C concentration is known, and (iii) the decay rate is constant and known. Of these three assumptions, the last one is valid. Extensive chemical cleaning to remove contaminants is usually needed to satisfy the first.

Initial concentration: tree-ring calibration
The original assumption (ii) that the atmospheric ^{14}C concentration was constant, everywhere on earth and over the full range of the radiocarbon method (ca. 50,000 years), was proven invalid early on (Suess, 1955; de Vries, 1958). This resulted in much research aimed at establishing a high-resolution, high-precision record of atmospheric ^{14}C concentrations in the past with which a measured ^{14}C age can be translated in a "real or calendar" age. Tree-ring research over the past 50-plus years has resulted in a dendrochronological ^{14}C calibration record going back about 14,000 years (Reimer et al., 2013). This record reveals atmospheric ^{14}C concentrations of 20 % or more above the standard atmosphere during the last part of the glacial-to-interglacial transition (14–11.7 ka), implying ^{14}C ages too young by 1,460 years or more. The Holocene has initially a varying ^{14}C excess of 7–11 % that decreases gradually over the last 5,000 years and reaches a minimum of a 2 % deficit (^{14}C age 160 years too old) around 1,400 years ago (Stuiver and Braziunas, 1993; Reimer et al., 2013). This long-term trend probably reflects changes in ^{14}C production related to the strength of the earth's magnetic field. Climate-related redistribution of carbon between ocean and atmosphere may contribute ^{14}C changes on centennial and longer time scales. Variability on time scales of decades to centuries is attributed to the modulation of the atmospheric ^{14}C production by variable magnetic shielding by the sun (Stuiver and Quay, 1980), while solar energetic particle (SEP) events may cause sporadic jumps in atmospheric ^{14}C concentration (Miyake et al., 2012; Usoskin et al., 2013).

Initial concentration: carbonates
Carbonate fixed in foraminifera in high-quality ocean-sediment cores, in corals, and in speleothems has been used to determine (local) initial ^{14}C concentrations in the surface ocean and in cave waters, going back over 50,000 years. Under the assumption of a constant offset in ^{14}C concentration between the local surface ocean and the atmosphere and between cave waters and the atmosphere, respectively, these data have also been used as an approximation of atmospheric ^{14}C concentrations beyond the range of tree-ring calibration (Reimer et al., 2013). The Japanese Lake Suigetsu has provided the first continuous atmospheric ^{14}C record, based on terrestrial macrofossils in a set of laminated lake-sediment cores, going back beyond 14,000 to more than 50,000 years ago (Bronk Ramsey et al., 2012). Although construction of a reliable absolute Suigetsu time scale based on annual layer counting is difficult and still needs further work, comparison of the atmospheric Suigetsu ^{14}C record with the carbonate-based records indicates that the assumption of a constant offset, used so far to convert the carbonate-derived ^{14}C data to "atmospheric" values, is violated to varying degrees. This is a problem for the construction of a reliable atmospheric ^{14}C calibration record beyond 14,000 years. Yet, it offers the perspective that once an atmospheric ^{14}C calibration record has been established, the ^{14}C record of ocean-sediment cores, with an age-depth relationship determined independently, may be used to determine local ocean-atmosphere gas exchange and local oceanic mixing in the past.

Tracer: exploring the carbon cycle
Mixing and reservoir age
The use of radiocarbon as a clock requires that the ^{14}C concentration of a sample has a simple, known relation to that of the atmosphere (extended assumption ii). For carbonates this is generally not the case. The dissolved

inorganic carbon (DIC), mostly bicarbonate HCO_3^-, in the waters from which carbonates precipitate, may derive in varying proportions from sources with different ^{14}C concentration (e.g., the atmosphere, young and old organic material, and carbonates). The difference in ^{14}C concentration between DIC and contemporaneous atmospheric CO_2 is frequently expressed as a "reservoir age," defined as the time it would take the atmospheric ^{14}C concentration to reach the concentration of the reservoir DIC by decay. This reservoir age, although it may reflect mixing instead of the age of the water, can be used to trace DIC origins and through them large-scale oceanic transport and mixing processes.

^{14}C in oceanic circulation

Early water samples from the Atlantic and Pacific Oceans (Fonselius and Ostlund, 1959; Broecker et al., 1959; Bien et al., 1960) showed significant differences in oceanic DIC ^{14}C concentrations. Large-scale systematic sampling of the Atlantic, Pacific, and Indian Oceans for ^{14}C was first carried out during the Geochemical Ocean Sections Study (GEOSECS) of the 1970s. This revealed a systematic apparent aging of abyssal waters by up to 1,680 years, when going from the northern North Atlantic, via the circumpolar waters south of 50°S, to the north of the Indian and Pacific Oceans (Stuiver et al., 1983). This provided a time frame for the deep limb of the thermohaline circulation (Stommel, 1961; Rahmstorf, 2006), which as "the great ocean conveyor" (Broecker, 1991) plays a major role in global climate variability. The measure of time, introduced by the decay of ^{14}C in the ocean, gave a better understanding of large-scale oceanic transport and mixing processes, documented by conservative tracers such as salinity and nutrients. It indicated that deep waters (>1,500 m) of the entire world ocean appear to be replaced on average every 500 years, with a replacement time for deep waters of the Pacific, Indian, and Atlantic Ocean of approximately 510, 250, and 275 years, respectively (Stuiver et al., 1983). Penetration of atmospheric ^{14}C excess, created by atmospheric testing of nuclear weapons, into the ocean provided a sensitive tracer for atmosphere-ocean CO_2 exchange and oceanic mixing (Rafter and O'Brien, 1970; Ostlund, 1983; Quay et al., 1983; Broecker et al., 1985; Nydal and Gislefoss, 1996).

Modern ocean and climate: latest radiocarbon survey

The oceans have been identified as a key element in the global climate system because of their capacity to transport heat and freshwater and to sequester from (and release to) the atmosphere major amounts of carbon. Thus, the World Ocean Circulation Experiment (WOCE) was performed as part of the World Climate Research Program (WCRP) to provide coherent global ocean observations of unprecedented extent and quality between 1990 and 1998. WOCE included ^{14}C as a tracer for carbon, complemented by CFCs and tritium as tracers of anthropogenic influences. Initial ^{14}C analyses were via β-decay counting like in GEOSECS (Stuiver et al., 1996), because AMS was not ready yet to meet the stringent quality requirements of WOCE. Thousandfold reduced sample requirements for AMS allowed detailed sampling of ocean profiles and produced large numbers of samples. These were analyzed in the newly created NOSAMS laboratory at Woods Hole (Key et al., 1996; McNichol et al., 2000). In GLODAP (Global Ocean Data Analysis Project), the data gained under GEOSECS, WOCE, and further ocean research programs were integrated and processed into a "global ocean carbon climatology," providing a detailed picture of the state of the current ocean for use in ocean and climate modeling (Key et al., 2004).

Ocean, carbon, and climate of the past

Oceanic sediments were radiocarbon dated on carbonate right from the start. A preindustrial ocean reservoir age of 400 years for the Northern Hemisphere (360 years in the Southern Hemisphere), based on limited data measured on pre-bomb mollusks (Stuiver and Braziunas, 1993), was used later to correct for the oceanic DIC offset from the atmospheric ^{14}C concentration. These reservoir ages were assumed to be constant. GLODAP shows, however, large spatial variability for the modern surface ocean as well as for the deep ocean (Matsumoto, 2007).

The advent of AMS in the late 1970s allowed the selection of monospecific planktic foram samples for ^{14}C dating, which significantly improved the quality and interpretability of the results. Over the past years, comparisons of planktic foram ages in sediment cores with independent age information increasingly showed that local reservoir ages of the surface ocean have varied over time (Ingram and Southon, 1996; Voelker et al., 1998; Reimer and Reimer, 2001; Thornalley et al., 2011; Bard et al., 2013). Together with variations in the difference in ^{14}C concentration between paired benthic and planktic foram samples, this provides snapshots of the carbon cycle under different climatic and deep-ocean conditions in the past. These offer the potential to further develop and test models of the interaction between climate, the ocean, and the carbon cycle under different boundary conditions. A new method of age control in ocean-sediment cores, complementary to the classical volcanic event markers and isotopic correlations, may allow a significant systematic increase in available data. The so-called ^{14}C plateau tuning (Sarnthein et al., 2007, recalibrated to the Lake Suigetsu record by Sarnthein et al., 2015), uses the close connection between ^{14}C concentrations in the surface ocean and the atmosphere that has been the basis for the use of oceanic data in the construction of the atmospheric ^{14}C calibration beyond the realm of tree-ring calibration. High-precision, high-resolution dating of sediment cores with high sedimentation rates, made feasible by AMS, produces irregular age-depth curves with a series of plateaus that can be correlated with the distinct suite of plateaus in the atmospheric ^{14}C calibration curve. Although local oceanographic and/or sedimentological changes may deform and mask some of the plateaus, the use of a suite of seven or more plateaus usually allows the correlation of several and thus provides both a calibrated age of the plateau boundaries and a ^{14}C offset for the locality at that time. Provided an apparent global linear relationship of ~1.22 μmol DIC/kg water per −1‰ ^{14}C between ^{14}C offset and concentration of modern DIC in the deep ocean (>2,000 m) also is valid in the past (Sarnthein et al., 2013), the paleo-^{14}C offsets offer a unique opportunity to reconstruct past changes in ocean circulation and the ocean carbon cycle over glacial-to-deglacial times. Although many of the available ocean climate-carbon models have not been tuned to these data yet, first positive results using this information have been obtained. It may thus be possible to further exploit the power of ^{14}C as a paleo-oceanographic tracer and dating tool for long-distance correlation and understanding of oceanographic and climatic changes in the coming years.

Summary

Radiocarbon in the ocean provides both a dating tool and a tracer for the carbon cycle. Because oceanic mixing processes on time scales of centuries to millennia are not short compared to the half-life of ^{14}C, they can be studied via the differences in ^{14}C concentration caused by ^{14}C decay during this mixing. Detailed sampling, made possible by AMS, has produced a ^{14}C database that contributes significantly to our understanding of the modern ocean and its role in the global climate. Quantifying ^{14}C deficiencies in key ocean-sediment cores from the LGM up to the present may be expected to provide further important insights.

Bibliography

Bard, E., Ménot, G., Rostek, F., Licari, L., Böning, P., Edwards, R. L., Cheng, H., Wang, Y.-J., and Heaton, T. J., 2013. Radiocarbon calibration/comparison records based on marine sediments from the Pakistan and Iberian margins. *Radiocarbon*, **55**(4), 1999–2019.

Bien, G. S., Rakestraw, N. W., and Suess, H. E., 1960. Radiocarbon concentration in Pacific Ocean water. *Tellus*, **12**, 436–443.

Broecker, W. S., 1991. The great ocean conveyor. *Oceanography*, **4**(2), 49–89.

Broecker, W. S., Tucek, C. S., and Olson, E. A., 1959. Radio-carbon analysis of oceanic CO_2. *International Journal of Applied Radiation and Isotopes*, **7**, 2903–2931.

Broecker, W. S., Peng, T. S., Ostlund, G., and Stuiver, M., 1985. The distribution of bomb radiocarbon in the ocean. *Journal of Geophysical Research*, **90**(C4), 6953–6970.

Bronk Ramsey, C., Staff, R. A., Bryant, C. L., Brock, F., Kitagawa, H., van der Plicht, J., Schlolaut, G., Marshall, M. H., Brauer, A., Lamb, H. F., Payne, R. L., Tarasov, P. E., Haraguchi, T., Gotanda, K., Yonenobu, H., Yokoyama, Y., Tada, R., and Nakagawa, T., 2012. A complete terrestrial radiocarbon record for 11.2 to 52.8 kyr B.P. *Science*, **338**(6105), 370–374.

De Vries, H., 1958. Variation in concentration of radiocarbon with time and location on earth. *Proceedings of the Koninklijke Nederlandse Akademie Van Wetenschappen Series B*, **61**, 94–102.

Fonselius, S., and Ostlund, H. G., 1959. Natural radiocarbon measurements on surface water from the North Atlantic and the Arctic Sea. *Tellus*, **11**, 77–82.

Godwin, H., 1962. Half-life of radiocarbon. *Nature*, **195**, 984.

Ingram, B. L., and Southon, J. R., 1996. Reservoir ages in Eastern Pacific coastal and estuarine waters. *Radiocarbon*, **38**(3), 573–582.

Key, R. M., Quay, P. D., Jones, G. A., McNichol, A. P., Von Reden, K. F., and Schneider, R. J., 1996. WOCE AMS radiocarbon I: Pacific Ocean results (P6, P16 and P17). *Radiocarbon*, **38**(3), 425–518.

Key, R. M., Kozyr, A., Sabine, C. L., Lee, K., Wanninkhof, R., Bullister, J., Feely, R. A., Millero, F., Mordy, C., and Peng, T.-H., 2004. A global ocean carbon climatology: results from Global Data Analysis Project (GLODAP). *Global Biogeochemical Cycles*, **18**, GB4031.

Libby, W. F., 1965. *Radiocarbon Dating*. Chicago, IL: University of Chicago Press.

Matsumoto, K., 2007. Radiocarbon-based circulation age of the world oceans. *Journal of Geophysical Research*, **112**, C09004, doi:10.1029/2007JC0040952007.

McNichol, A. P., Schneider, R. J., Von Reden, K. F., Gagnon, A. R., Elder, K. L., Key, R. M., and Quay, P. D., 2000. Ten years after – the WOCE AMS radiocarbon program. *Nuclear Instruments and Methods in Physics Research, Section B: Beam Interactions with Materials and Atoms*, **172**(1–4), 479–484.

Miyake, F., Nagaya, K., Masuda, K., and Nakamura, T., 2012. A signature of cosmic-ray increase in AD 774–775 from tree rings in Japan. *Nature*, **486**, 240–242.

Nydal, R., and Gislefoss, J. S., 1996. Further application of bomb ^{14}C as a tracer in the atmosphere and ocean. *Radiocarbon*, **38**(3), 389–406.

Ostlund, H. G., 1983. *TTO North Atlantic Studies, Tritium and Radiocarbon, Data Rel*. Miami: Tritium Laboratory University of Miami, pp. 83–85.

Quay, P. D., Stuiver, M., and Broecker, W. S., 1983. Upwelling rates for the equatorial Pacific Ocean derived from bomb ^{14}C distribution. *Journal of Marine Research*, **41**, 769–792.

Rafter, T. A., and O'Brien, B. J., 1970. Exchange rates between the atmosphere and the ocean as shown by recent C-14 measurements in the South Pacific. In Olsson, I. U. (ed.), *Nobel Symposium 12, Radiocarbon Variations and Absolute Chronology*. New York: Wiley, pp. 355–377.

Rahmstorf, S., 2006. Thermohaline ocean circulation. In Elias, S. A. (ed.), *Encyclopedia of Quaternary Sciences*. Amsterdam: Elsevier, pp. 1–10.

Reimer, P. J., and Reimer, R. W., 2001. A marine reservoir correction database and on-line interface. *Radiocarbon*, **43**(2A), 461–463.

Reimer, P. J., Bard, E., Bayliss, A., Warren Beck, J., Blackwell, P. G., Bronk Ramsey, C., Buck, C. E., Hai, C., Edwards, R. L., Friedrich, M., Grootes, P. M., Guilderson, T. P., Haflidason, H., Hajdas, I., Hatté, C., Heaton, T. J., Hoffmann, D. L., Hogg, A. G., Hughen, K. A., Kaiser, K. F., Kromer, B., Manning, S. W., Niu, M., Reimer, R. W., Richards, D. A., Scott, E. M., Southon, J. R., Staff, R. A., Turney, C. S. M., and van der Plicht, J., 2013. IntCal13 and Marine13 radiocarbon age calibration curves 0–50,000 years cal BP. *Radiocarbon*, **55**, 1869–1887.

Roberts, M. L., and Southon, J. R., 2007. A preliminary determination of the absolute $^{14}C/^{12}C$ ratio of OX-I. *Radiocarbon*, **49**(2), 441–445.

Sarnthein, M., Grootes, P. M., Kennett, J. P., and Nadeau, M.-J., 2007. ^{14}C reservoir ages show deglacial changes in ocean currents and carbon cycle. In Schmittner, A., Chiang, J. C. H., and Hemming, S. R. (eds.), *Ocean Circulation: Mechanisms and Impacts – Past and Future Changes of Meridional Overturning*. Washington, DC: American Geophysical Union. AGU geophysics. Monographs, Vol. 173, pp. 175–196, doi:10.1029/173GM13.

Sarnthein, M., Schneider, B., and Grootes, P. M., 2013. Peak glacial ^{14}C ventilation ages suggest major draw-down of carbon into the abyssal ocean. *Climate of the Past*, **9**, 2595–2614, doi:10.5194/cp-9-2595-2013.

Sarnthein, M., Balmer, S., Grootes, P. M., and Mudelsee, M., 2015. Planktic and benthic ^{14}C reservoir ages for three ocean basins, calibrated by a suite of ^{14}C plateaus in the glacial-to-deglacial Suigetsu atmospheric ^{14}C record. *Radiocarbon*, **57**(1), 129–151.

Stommel, H., 1961. Thermohaline convection with two stable regimes of flow. *Tellus*, **13**, 224–230.

Stuiver, M., and Braziunas, T. F., 1993. Modeling atmospheric ^{14}C influences and ^{14}C ages of marine samples to 10,000 BC. *Radiocarbon*, **35**(1), 137–189.

Stuiver, M., and Quay, P. D., 1980. Changes in atmospheric carbon-14 attributed to a variable sun. *Science*, **207**, 11–19, doi:10.1126/science.207.4426.11.

Stuiver, M., Quay, P. D., and Östlund, H. G., 1983. Abyssal water carbon-14 distribution and the age of the world oceans. *Science*, **219**, 849–851.

Stuiver, M., Ostlund, H. H., Key, R. M., and Reimer, P. J., 1996. Large-volume WOCE radiocarbon sampling in the Pacific Ocean. *Radiocarbon*, **38**(3), 519–561.

Suess, H. E., 1955. Radiocarbon concentration in modern wood. *Science*, **122**, 415–417.

Thornalley, D. J. R., Barker, S., Broecker, W. S., Elderfield, H., and McCave, I. N., 2011. The deglacial evolution of the North Atlantic deep convection. *Science*, **331**, 202–205.

Tunis, C., Bird, J. R., Fink, D., and Herzog, G. F., 1998. *Accelerator Mass Spectrometry. Ultrasensitive Analysis for Global Science*. Boca Raton: LLC, CRC Press, p. 371.

Usoskin, G., Kromer, B., Ludlow, F., Beer, J., Friedrich, M., Kovaltsov, G. A., Solanki, S. K., and Wacker, L., 2013. The AD775 cosmic event revisited: the Sun is to blame. *Astronomy and Astrophysics*, **552**, L3, doi:10.1051/0004-6361/201321080.

Voelker, A. H. L., Sarnthein, M., Grootes, P. M., Erlenkeuser, H., Laj, C., Mazaud, A., Nadeau, M.-J., and Schleicher, M., 1998. Correlation of marine ^{14}C ages from the Nordic Seas with the GISP2 isotope record: implications for ^{14}C calibration beyond 25 ka BP. *Radiocarbon*, **40**, 517–534.

Cross-references

Carbonate Dissolution
Carbon Isotopes
Currents
Foraminifers (Benthic)
Foraminifers (Planktonic)
General ocean circulation - its signals in the sediments
Laminated Sediments
Modelling Past Oceans
Ocean Drilling
Oxygen Isotopes
Paleoceanographic Proxies
Paleoceanography
Radiogenic Tracers
Technology in Marine Geosciences

RADIOGENIC TRACERS

Anton Eisenhauer
GEOMAR Helmholtz Centre for Ocean Research, Kiel, Germany

Synonyms

Dating; Isotopes; Stratigraphy; Tracers

Isotopes can be characterized to be stable, cosmogenic, radioactive, or radiogenic. Stable isotopes do not change their relative abundance in the environment as long as no isotopes are added or extracted from the system. Cosmogenic isotopes are produced by high energetic cosmic rays in the upper layers of the atmosphere and are usually radioactive also. Radioactive isotopes change their abundance in the environment according to their half-life ($T_{1/2}$), whereas radiogenic isotopes are not radioactive itself but change their abundance according to the half-life of their mother isotope. There are several important radiogenic systems known, and the most important are ^{238}U/^{206}Pb, ^{190}Pt/^{186}Os, ^{147}Sm/^{143}Nd, ^{87}Rb/^{87}Sr, ^{187}Re/^{187}Os, ^{176}Lu/^{176}Hf, ^{232}Th/^{208}Pb, ^{40}K/^{40}Ar, ^{40}K/^{40}Ca, ^{235}U/^{207}Pb, ^{129}I/^{129}Xe, ^{10}Be/^{10}B, ^{26}Al/^{26}Mg, ^{36}Cl/^{36}Ar, ^{14}C/^{14}N, etc. For the geological sciences, the most important radiogenic isotope systems are the radiogenic systems of three natural decay chains ^{238}U/^{208}Pb ($T_{1/2}$:4.468 Gyr; Gyr = 1 billion years), ^{235}U/^{207}Pb ($T_{1/2}$:0.7038 Gyr), and ^{232}Th/^{208}Pb ($T_{1/2}$:14.01 Gyr) as well as ^{87}Rb/^{87}Sr ($T_{1/2}$:49.44 Gyr) and ^{147}Sm/^{143}Nd (106 Gyr), respectively. For practical reasons and their application, the radiogenic isotopes are usually normalized to a stable isotope of the same element, for example, ^{208}Pb/^{204}Pb, ^{207}Pb/^{204}Pb, ^{206}Pb/^{204}Pb, ^{87}Sr/^{86}Sr, and ^{143}Nd/^{144}Nd. The latter "radiogenic isotope ratios" are considered tracers because of their radiogenic characteristic; they can be applied for absolute (isochron age dating) and stratigraphic age dating as well as for provenance studies. Provenance studies are possible because the radiogenic isotope ratios in minerals and mineral-forming rocks are characterized by both the time interval of the mineral since mineral formation and the abundance of the radiogenic mother nuclide.

Concerning age dating besides the ^{87}Rb/^{87}Sr and ^{147}Sm/^{143}Nd systems, the U/Pb dating system is the most important one, because its great advantage over other absolute dating methods is that any sample provides two clocks, one based on ^{235}U/^{207}Pb with a $T_{1/2}$ of ~0.7 Gyr and one based on ^{238}U/^{206}Pb with a $T_{1/2}$ of ~4.5 Gyr, both providing independent verification of the calculated ages that allows accurate determination of the age of the sample even if the "closed-system behavior" of the sample is violated. The data are plotted in a "concordia diagram," where the samples plot along an isochron, line of equal ages, which intersects the concordia curve at the age of the sample (Faure, 1986; Dickin, 1995).

Radiogenic isotope stratigraphy relies on the recognized variations of certain radiogenic isotope ratios (^{87}Sr/^{86}Sr, ^{88}Sr/^{86}Sr, ^{44}Ca/^{40}Ca, etc.) over time in seawater and their recording in marine inorganic and biogenic carbonates. Records of radiogenic isotopes in marine carbonates are unique time series and documents (e.g., Veizer, 1989) of the chemical history of seawater since its formation until today as well as of the tectonic, biological, and climatic history of the embedded continental land masses.

Bibliography

Dickin, A. P., 1995. *Radiogenic Isotope Geology*. Cambridge: Cambridge University Press.
Faure, G., 1986. *Principles of Isotope Geology*. Cambridge: Cambridge University Press.
Veizer, J., 1989. Strontium isotopes in seawater through time. *The Annual Review of Earth and Planetary Sciences*, **17**, 141–167.

Cross-references

Paleoceanographic Proxies
Radiocarbon: Clock and Tracer

RADIOLARIANS

Kjell R. Bjørklund
Museum of Natural History, University of Oslo, Blindern, Oslo, Norway

Definition

Radiolaria are single-celled marine eukaryotes, also some colonial forms, existing from the Cambrian (ca. 530 Ma) to recent. Thus, radiolarians are one of the longest ranging groups of fossil microorganisms. The founders of radiolarian taxonomy were two German scientists, C.G. Ehrenberg (1795–1876) and E. Haeckel (1834–1919). Ehrenberg described more than 70 genera and 500 species, while Haeckel contributed more than 700 genera and 4,000 species. Haeckel's (1887) system is the basis of some modern taxonomic systems. One proposed by Riedel (1967) is the most commonly used: *kingdom* Protista, *phylum* Sarcomastigophora, *subphylum* Sarcodina, *class* Actinopoda, *subclass* Radiolaria, *superorder* Polycystina, and *orders* Spumellaria (SiO_2), Nassellaria (SiO_2), and Phaeodaria (SiO_2 + organic compound). In this scheme, Acantharia ($SrSO_4$) and Heliozoa (SiO_2) also are assigned to the class Actinopoda but in separate subclasses. This system is mainly based on skeletal material and general morphology, as well as internal cellular structures. New phylogenomic studies show that this Haeckelian system is artificial (not representing evolutionary relationships), and new systems based on molecular genetic data establish more natural relationships among the six supergroups of eukaryotes as proposed by Cavalier-Smith (2002, 2003).

The first description of living colonial forms, such as *Sphaerozoum fuscum*, was made by Meyen (1834). Since then, we have recorded about 2,500 high-rank taxa (genera/subgenera) and 15,000 species, of which, 800–1,000 are currently living in the oceans (Suzuki and Aita, 2011).

Taxonomic position

Today, molecular work on radiolarians is an active field of research, and the taxonomical scheme is constantly adjusted as new data are made available. Cavalier-Smith (2003) suggested the following system: kingdom Protozoa, subkingdom Biciliata, phylum Retaria, and subphylum Radiozoa, with classes Acantharea, Polycystinea, and Sticholonchea, a system that has not stabilized and is under development. The Phaeodaria are no longer included in the general category of "Radiolaria" (previously in the phylum Retaria), but placed in phylum Cercozoa. The Radiolaria 18S rDNA phylogeny trees presented in recent publications are rather similar to each other and suggest a polyphyletic status for the polycystines but are limited by using only the 18S gene. By contrast, Krabberød et al. (2011), using both 18S and 28S rDNA, show that Spumellaria group with Nassellaria, thus suggesting they form a natural phylogenetic group (Figure 1). Their data provide a more convincing monophyletic status for the polycystines (Spumellaria together with Nassellaria), as well as a close relation between Acantharea and Sticholonchea, a group given the name Spasmaria (Cavalier-Smith, 1993). In contrast, when only 18S rDNA is used, Acantharea branch with Spumellaria. These results show that combining the 18S and 28S rDNA genes may greatly improve the resolution of the radiolarian phylogeny.

Organization of the cell

The *central capsule* is the inner part of the cell, made up of endoplasm and some organelles like the nucleus and the axoplast. The shape of the siliceous skeleton depends on the organization of the central capsule, and an individual that is reduced to this stage can regenerate all missing parts. Its shape varies from species to species and increases with age. The soft-bodied central capsule is not fossilized, but Dumitrica (1999) observed some very rare, well-preserved, Mesozoic capsular membranes. The *capsular membrane* (made of chitinous or pseudochitinous material) is perforated by small fusules and fissures, through which axopodial filaments extend from the axoplast to the outer parts of the shell. In Spumellaria, the fusules and fissures are evenly distributed over the surface of the central capsule, while in Nassellaria, they are located at one end on the capsular membrane.

The *endoplasm*, on the inside of the capsular membrane, contains one or several *nuclei* (latter typically during reproduction) in addition to *mitochondria* (cell's supply of energy, adenosine triphosphate (ATP)), the *Golgi* apparatus (processing of proteins for secretion), and some *vacuoles* (lipid droplets) and the *axoplast* (area where the microtubules of the slender and stiffened *axopodia* originate). Often, there is a main axopod called *axoflagellum* (Figure 2).

The *ectoplasm*, outside of the capsular membrane, is a complex body, appears spongy, and can retract if disturbed by external objects. Ectoplasm is divided in three parts: (a) an inner thin layer, *sarcomatrix*, a pigmented and granular cytoplasm; (b) a middle part, *calymma*, a sticky, gelantinous cytoplasm with numerous *alveoles* (filled

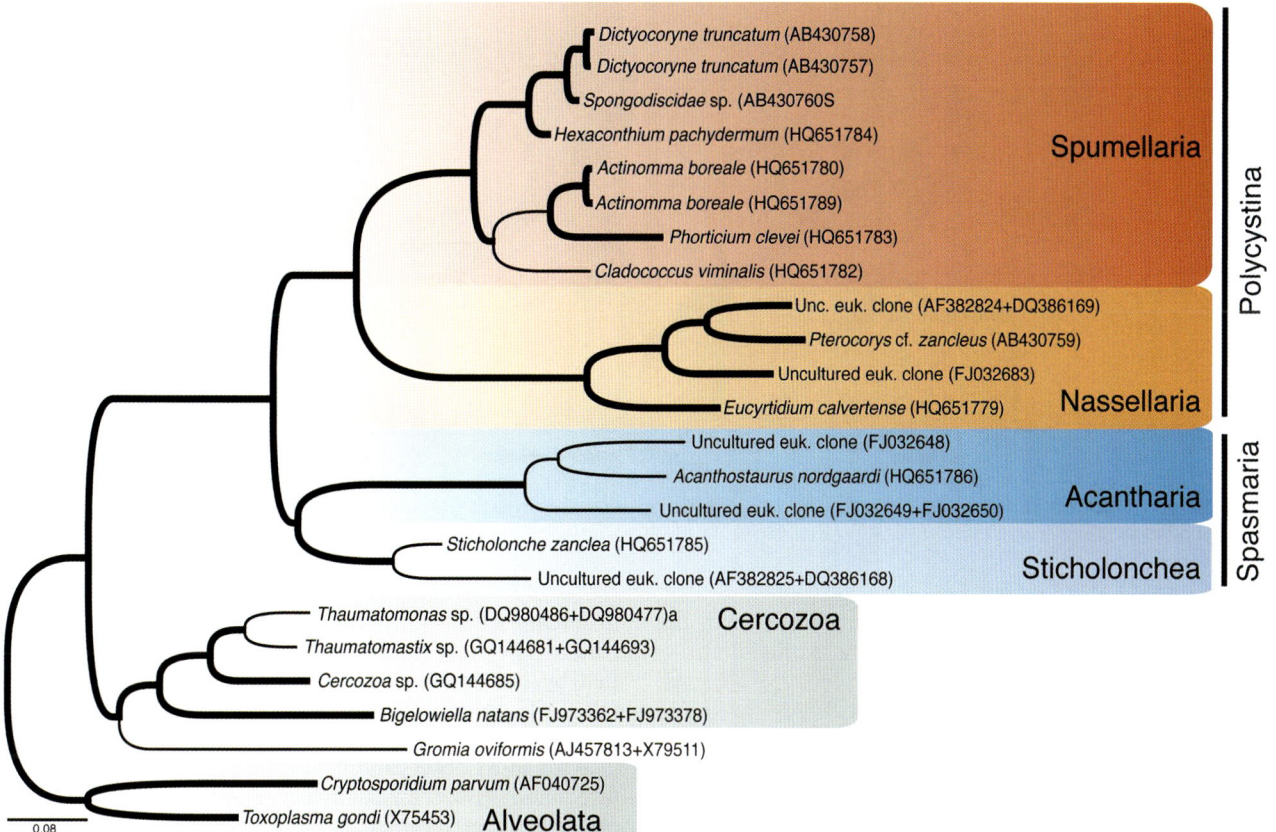

Radiolarians, Figure 1 A phylogenetic tree of Radiolaria using the 18S and 28S rDNA genes. *Thick* branches are well supported with maximum likelihood bootstrap value higher than 95 and Bayesian posterior probability higher than 0.95 (modified from Krabberød et al., 2011).

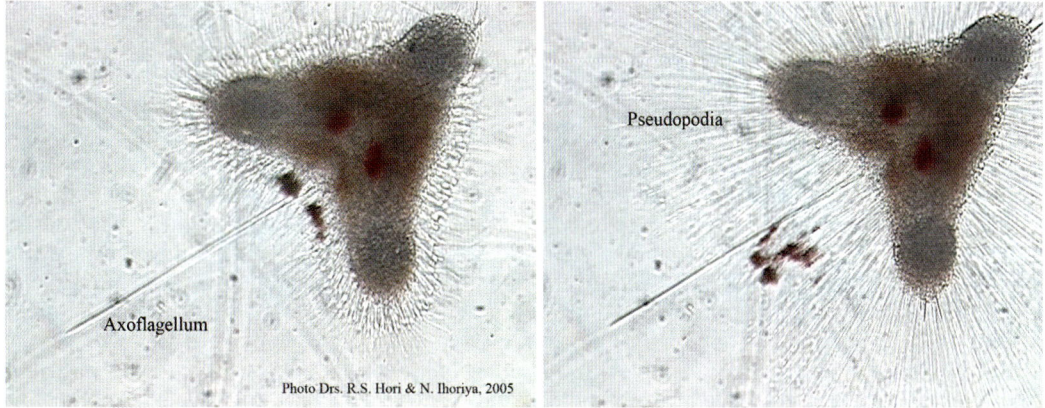

Radiolarians, Figure 2 *Dictyocoryne truncatum* with a long and well-extended axoflagellum and with short and long pseudopodia, *left* and *right* images, respectively. The two *red* areas are part of the large U-shaped central capsule (photo by Drs. R.S. Hori and N. Ihoriya, 2005).

with liquid or gas, looks like a soap-bubble-like frothy layer, especially in Spumellaria) and digestive vacuoles and often colored by the presence of symbionts (*zooxanthellae*, often dinoflagellates); and (c) the outer part, the *sarcodictyum* (a thin network of protoplasm on the surface of the calymma of a radiolarian).

The *pseudopodia* are mobile elongated extensions of the cytoplasm extending outside the cell. This includes

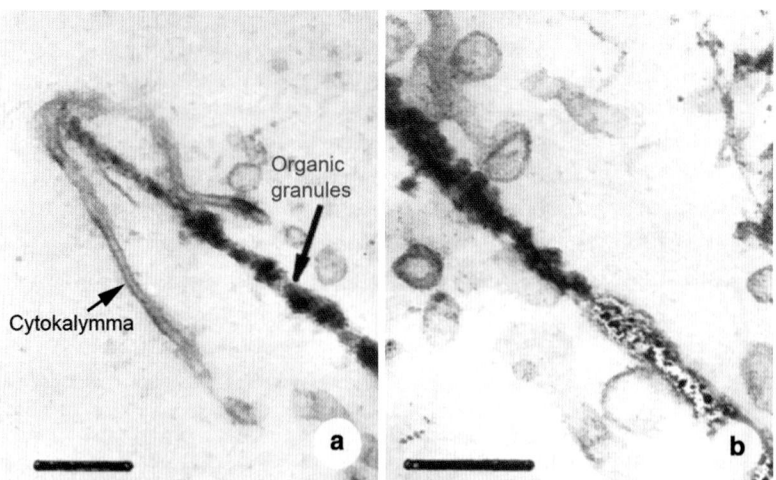

Radiolarians, Figure 3 Ultrathin sections of early shell deposition showing organic granules (a) and preceding silica deposition (b). Sample B has been HF treated demonstrating the resistance of these particles. Scale = 250 nm (after Anderson and Swanberg, 1981).

the filopodia (slender, tapered, unstiffened) and the axopodia (stiffened), sometimes also the axoflagellum.

Organization of the skeleton

For the micropaleontologists, the skeleton is the most interesting part of a radiolarian cell. In the following, only the Polycystina with a pure silica skeleton ($SiO_2 + nH_2O$) will be discussed, as the Acantharia and Phaeodaria do not produce skeletons that are fossilized (see above). Paleontologists study the skeletons in soft sediment and hard rock samples and thereby have limited information on the growth stages of the different species and specimens. However, this can be done when living specimens are taken from the plankton and studied in the laboratory. In culture experiments, juvenile radiolarians have been kept alive until gamete formation (Anderson, 1983; Kimoto et al., 2011), but nobody has so far succeeded rearing radiolarians through successive generations.

In general terms, the skeleton in all radiolarians is extracapsular, only in adult Spumellaria and Entactinaria can the innermost shells be found in the capsular cytoplasm (De Wever et al., 2001). The radiolarian skeleton is enclosed in a sheath of cytoplasm that protects the surface against immediate dissolution by seawater and also provides a strict physiological control over the rate of deposition and the ultimate shape of the skeleton.

The skeletal morphology is also affected by environmental conditions but even more dependent on the physiology of the cell (Matsuoka, 1992). Immediately before the silica deposition starts (Figure 3a), a granular matrix was observed that is resistant to HF dissolution (Figure 3b) showing that the granules are not silica (Anderson and Swanberg, 1981).

They concluded that the densities of these siliceous granules dictated the porosity of the shell, an important aspect with respect to fossilization of the skeleton, different for each species, depending on its cellular organization. The silica, absorbed from the seawater, is deposited by intermittent episodes, based on observations of rates of growth of Radiolaria in laboratory culture (Figure 4).

Also SEM observations of concentric layering in fractured and dissolved segments of radiolarian skeletons obtained from sediments (Figure 5) document the stepwise growth. Brandt (1881) believed that a still not identified organic substance is connected with the silica deposition. He noticed that some shells of *Collosphaera huxleyi* consisted only of albuminous substances and had the size and shape of the future latticed shell. Brandt never came to a final conclusion whether the organic substance in the *C. huxleyi* shells is, completely or only partially, subject to the silicification process. This has so far not been clarified. However, Anderson (1983) used the term "cytokalymma" and defined it as a cytoplasmic sheath that encloses, molds, and deposits the skeleton. This is in remarkable good harmony with what was proposed by Brandt (1881). It is also important to note that within the cytokalymma, silica deposition at the growing edge of the skeleton is deposited on apparently organic granules that form the initial framework for the growing skeleton (Anderson and Swanberg, 1981). The spumellarian shells have a centrifugal, concentric, growth pattern, meaning that the innermost shell, the *microsphere*, develops first, and outer shells are constructed sequentially outward. Therefore, the innermost structures are more conservative and are of more importance in taxonomic work than structures on the outermost (cortical) shells. The radial spines grow first from the microsphere, and their number and position are fixed by the position of the pores (*fusules*) on the capsular membrane, so is also the number of spines. Spumellarian skeletons have different structures and shapes. In spherical forms, all three axes are of equal length. A shortened z-axis gives a discoidal shape, and

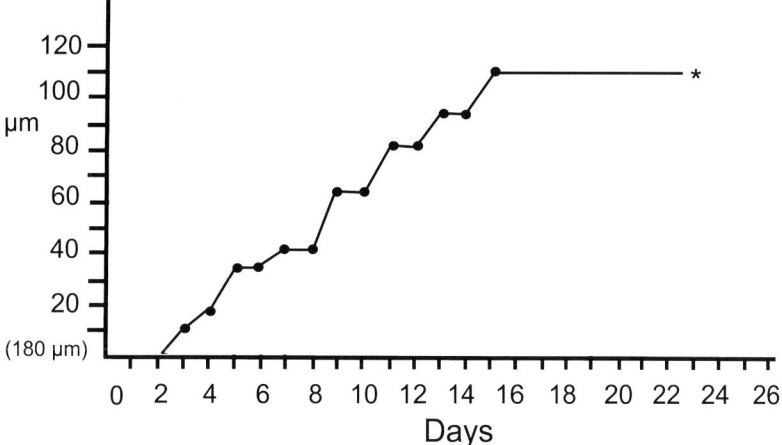

Radiolarians, Figure 4 Stepwise growth curves of individual *Spongaster tetras tetras* grown in laboratory culture (Anderson et al., 1989).

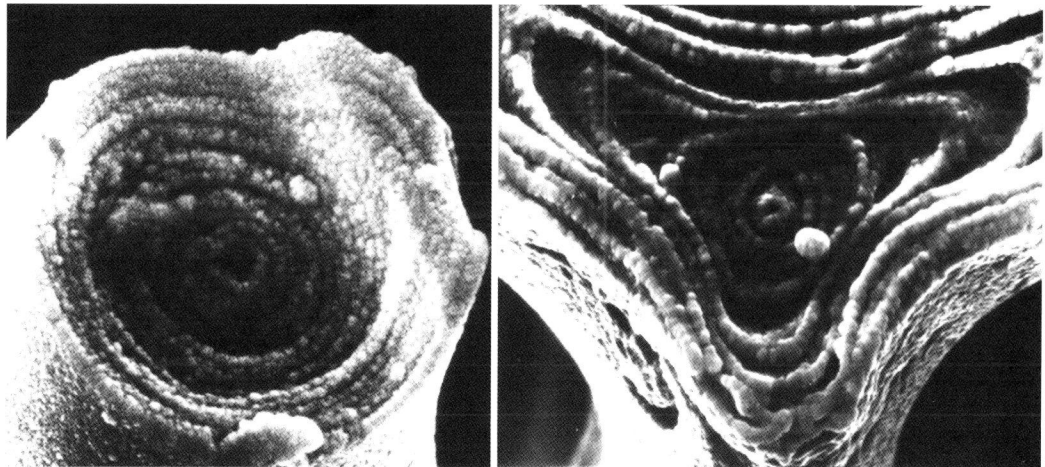

Radiolarians, Figure 5 Growth rings: *left, Rhizosphaera* sp. from Upper Eocene, the Norwegian Sea. *Right, Gorgospyris perplexus* from Upper Miocene, California (photo by K.R. Bjørklund and R.M. Goll, 1982).

when the other axes are more extended, the shape becomes ellipsoidal. If all three axes are of different length, the skeletons will have a system of three successively larger elliptical girdles distributed in three mutually perpendicular planes. A general morphologic outline of a spherical spumellarian shell includes the radial spines, fine byspines, and thickened bars connecting concentric shells (Figure 6).

For the nassellarians, one character is their *bilateral symmetry* situated around the first skeletal element, the *median bar (MB)*. From one end of this structure, an *apical spine (A)* points upward, while one *dorsal spine (D)* and two *secondary lateral spines (l_l and l_r)* point downward. From the other end of the MB, an unpaired *ventral spine (V)* (also called vertical) points upward, while a pair of *primary lateral spines (L_L and L_R)* points downward. These principal bar and spine structures are the first skeletal elements to be seen in juvenile forms, Figure 7.

The next skeletal element to develop is the first segment, the *cephalis*, then the *thorax*, and finally the *abdomen* (one or more segments). At the junction between two segments, there may appear heavily silicified internal rings, called the *collar ring* (or stricture) and the *lumbar ring*, separating the cephalis-thorax-abdomen, respectively. In addition to the arrangement of pores (longitudinally, transversally, or obliquely), the presence of spines and their different shapes on different segments are of taxonomic importance. See De Wever et al. (2001) for further skeletal characters and taxonomical notes.

Longevity

Of the estimated 800–1,000 living radiolarian species (Suzuki and Aita, 2011), there are no data of the exact *longevity* of any species. Different estimates have been made for cultured juvenile forms and until they made

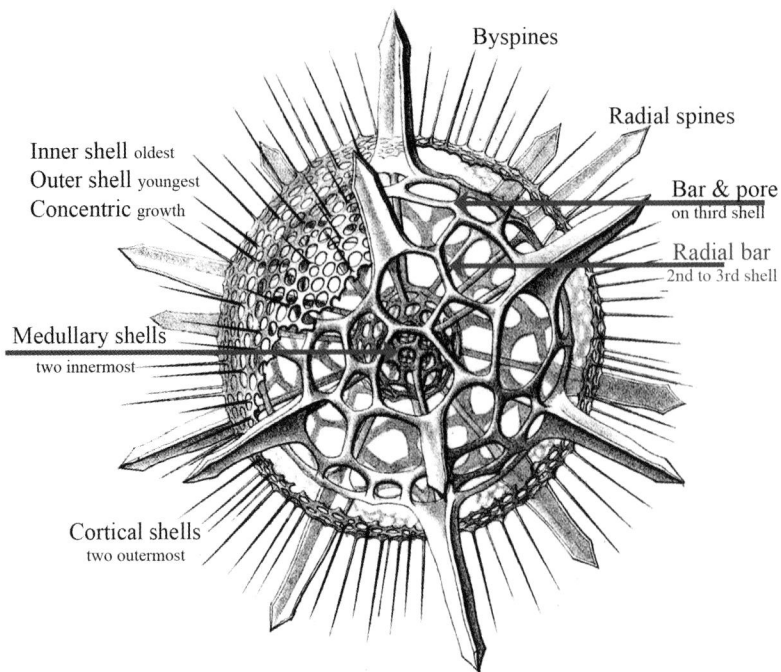

Radiolarians, Figure 6 Line drawing modified from Haeckel (1887, pl. 30 Figure 1), the most important structures are named.

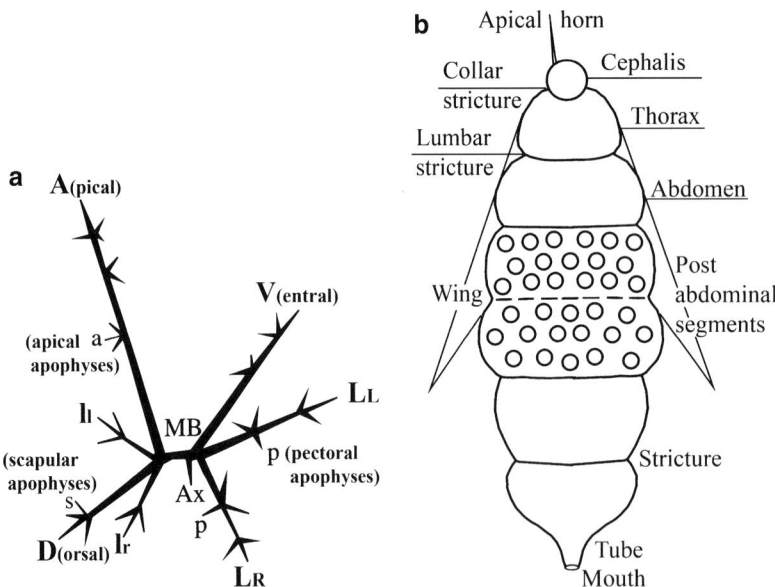

Radiolarians, Figure 7 (a) Principal internal spines in a nassellarian skeleton. See text for explanation of legend. (b) External skeletal element in a typical multichambered nassellarian (modified after De Wever et al., 2001).

swarmer cells. The longest survival observed in the laboratory was 75 days (Matsuoka, pers. com., 2009), but other observations indicate much shorter periods, longer than 3 weeks. Other experiments by estimating the turnover time based on the study of sediment trap samples gave a shorter longevity, 0.44–37 days in the California Current (Casey et al., 1971) and 16–42 days in the North Pacific (Takahashi, 1983).

Reproduction

The reproduction within radiolarians is still not fully understood. Stages with swarmer production have been observed by Anderson (1976, 1978), a probable sign for sexual reproduction. Similar findings were done by Kimoto et al. (2011) in *Cypassis irregularis* Nigrini (Spumellaria). These swarmers, in the range of <3 μm, gave 18S rDNA sequences that fell within those of typical radiolarian families. This may explain that the deep-water radiolarian 18S rDNA sequences found in the picoeukaryote size fraction from deep water, by several authors (Not et al., 2009; Lovejoy and Potvin, 2011), are gametes or swarmers released for reproduction.

Binary fission or asexual reproduction was demonstrated already by Brandt (1902) and confirmed by Hollande and Enjumet (1953). Vegetative reproduction is probably an important aspect as Itaki and Bjørklund (2007) reported on conjoined skeletons, in several combinations, as two sets of medullar shells are incorporated within one cortical shell. This is also observed within and between different species and genera, indicating that the present taxonomy based on morphology alone needs to be modified and combined with molecular support. Finally, hybridization was discussed by Goll (1976) and Lazarus (1983). Goll convincingly showed that the geographic distributions of his inferred hybrid taxa correspond to regions of geographic overlap in the inferred parental species and to oceanic regions of upwelling. Some of the old radiolarian reports (Hertwig, 1879; Brandt, 1902; Enriques, 1919, to mention a few) described different types of spores which were associated with reproduction. However, these have by later workers been demonstrated to be motile stages of endosymbiotic dinoflagellates.

Nutrition

Radiolarians feed on a variety of prey and can be divided into herbivorous, carnivorous, and omnivorous. Anderson et al. (1984) demonstrated that some of the spinose species easily can trap and digest zooplankton prey. As for the non-spinose species, algae are more common than animal prey. Other species may be more specialized with preferential prey (Anderson, 1978). The prey is trapped by the numerous pseudopodia. When prey has been made immobile, it is encapsulated in tick membranous vacuoles where digestion takes place. A practical upper size of the prey can be set at about 1 cm, but prey as small as bacteria are caught. Furthermore, some species, especially those living in the photic zone, may be the host for **symbionts**, most frequently these are dinoflagellates. Colonial spumellarians have the highest concentration of symbionts, and Swanberg and Harbison (1980) described *Collozoum longiforme*, up to 3 m long, with 14–50 algae per central capsule. Swanberg and Anderson (1981) described *C. caudatum*, up to 2 m long, with 58 (40–70) central capsules per cm, and 311 (250–350) algae per central capsule, indicating that a 2 m long colony would have a number of about 3.6×10^6 algae. Swanberg and Harbison (1980) found that colonial radiolarians can produce 14,000–41,000 ng C/colony/h. Similarly, Anderson et al. (1989) found that some large solitary radiolarians could produce from 35,000 to 110,000 ng C/individual/h. These values are directly linked to the large size of the species investigated and the high number of symbiotic algae.

Migration

Radiolarians are planktonic organisms passively drifting around, carried by the different currents in the world ocean. Therefore, their lateral motion is not controlled by the organism itself. **Lateral transportation** is well known (see below), as well as their **vertical migration** has been observed. Their buoyancy is explained by the vacuolar cytoplasm and its content of oil droplets, important for their position in the water column, as well as consuming (eating) or shedding their oil droplets and symbionts. The volume of an individual radiolarian cell can be controlled by extending and retracting its pseudopods. By extension, the volume increases, so does the buoyancy, and the specimen is lifted in the water column. Similarly, the symbionts will be exposed for the diurnal photosynthetic cycle. During daytime, CO_2 is consumed due to the photosynthetic activity in the cell, causing a downward migration, while during the nocturnal phase, CO_2 is accumulated in the cell. This increases the uplift resulting in an upward migration. Anderson (1983) indicated a diurnal migration in the range of 200–350 m up and down. In the acantharians, Schewiakoff (1926) described slender contractile myonemes in the pericapsular cytoplasm that can contract (Figure 8, left) and relax (Figure 8, right) and thus change the volume of the cell, causing it to rise or sink. The two images are one and the same specimen, demonstrating the effect of the myonemes.

Luminescence

This phenomenon has been reported to take place in both solitary and colonial radiolarians. It seems to take place in the gelatinous extracapsulum (Anderson, 1983), but at present, it is not known whether it is the dinoflagellate symbionts or the radiolarian cytoplasm that is the source of this luminescence. However, Herring (1979) reported that the rhythmic discharges within the unicellular *Thallasicolla* sp. are clearly very different from that of dinoflagellates both in its physiology and its chemistry.

Diversity

It has been estimated (Suzuki and Aita, 2011) that at present we have about 800 living polycystine species. Some species are cosmopolitan, but the majority are confined to particular water masses, where they are most adapted. The fauna is further restricted to zoogeographical provinces (tropical, subtropical, transitional, subpolar, and Antarctic zones) (Johnson and Nigrini, 1980) as well as in vertical or environmental zones (surface, subsurface, moderately deep, bathypelagic, and abyssal (Reshetnyak, 1955)). The highest number of species is found in the tropics diminishing toward the poles, being most abundant in the upper ca 200 m, with 5,000–15,000 individual/m^3 (Petrushevskaya, 1971), but abundances of up to 394,000 individuals/m^3 have been reported (Boltovskoy et al., 2003) of a monospecific assemblage of *Lophophaena rioplatensis* of Argentina.

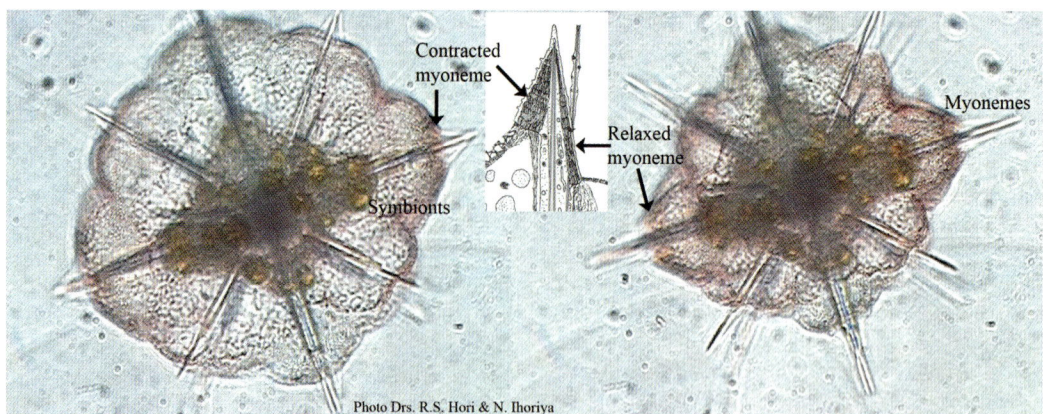

Radiolarians, Figure 8 An acantharian cell with perispicular cytoplasm that includes myonemes. Same cell with contracted (*left* image) and relaxed (*right* image) myonemes (photo by Drs. R.S. Hori and N. Ihoriya 2005).

Radiolarians were earlier considered to be oceanic indicators (Casey et al., 1970; Casey, 1971) and only adventitiously found in coastal waters (Casey, 1971). However, Swanberg and Bjørklund (1986) demonstrated that typical abundances of radiolarians greater than $2000/m^3$ are present in landlocked Norwegian fjords, with a maximum of $9800/m^3$ in Finnabotn.

Currents can transport radiolarians and carry them far from their home areas, where they may occur abundantly in alien oceans, or from one ocean to another where they may have variable reproduction success. Recently, it was documented that a warmwater spumellarian *Didymocyrtis tetrathalamus* was observed in plankton off northern Svalbard, together with several other warmwater (tropical) spumellarian and nassellarian species (Bjørklund et al., 2012). Assuming these radiolarians originated in the Gulf of Mexico, where an estimated 450–500 species occur, most of this fauna will not survive during advection from the tropical-subtropical region to the Arctic. Based on an advection time for the warm water of 5–7 years from the Florida Strait to the Fram Strait and with a radiolarian longevity of about 30 days, this warmwater assemblage would require 60–84 generations to complete this traverse. This is an example of unexpected warmwater fauna in a cold water realm. Keeping this in mind, there have been similar events in the past, thus obscuring paleoecological interpretations. For example, an almost 100 % tropical planktonic foraminifera associations have been reported in the Laptev Sea (Matul et al., 2007).

Preservation in sediments

The ocean is undersaturated with respect to dissolved SiO_2. Diatoms and radiolarians, the two most important contributors of suspended silica, are most common in the photic zone, from where the silica for the skeletons is extracted. This results in a strong dissolution of the skeletons, and some loss of siliceous microfossils occurs initially in the upper part of the water column. Lizitsin (1972) indicates that only 1–10 % of siliceous skeletons produced in the upper 200 m are eventually deposited in the bottom sediments. Takahashi (1991) reported that radiolarian skeletons collected in shallow sediment traps have porous surface layers due to dissolution. Similarly, I have observed completely hollow skeletons of *Lithomitra lineata* in surface sediments in the Norwegian Sea. Johnson (1974) reported on selective dissolution of siliceous skeletons with increasing detrital input in the eastern equatorial Pacific. Abundance and preservation patterns of radiolarian skeletons in the surface sediments of the northern and southern Atlantic (Goll and Bjørklund, 1971, 1974, respectively) indicated that the preservation of their skeletons is closely related to their relative abundances. Radiolarian ooze in the sediments (>30 % radiolarian SiO_2) is located in the equatorial regions of the Pacific and Indian Oceans, which is a direct result of surface water production as well as $CaCO_3$ dissolution at great depths.

The living radiolarian surface water associations will certainly be different from the fauna associations observed for their skeletons in the bottom sediments, due to dissolution, as exemplified by Boltovskoy et al. (1993). In the east equatorial Atlantic, they compared radiolarian faunas in sediment traps and surface sediments. They convincingly showed that the spumellarian shells occurred with 35 % in the sediments while only 19 % in the water column. The nassellarians showed the opposite, being more abundant in the water column (80 %) than in the bottom sediments (64 %).

A rather interesting phenomenon is found in the anoxic and superhaline basins in the Mediterranean Sea (Bjørklund and De Ruiter 1987), Red Sea, and Gulf of Mexico. In the Orca Basin, the salinities are >260 psu, due to seepage from salt domes, and with measured silica levels of up to 235 μM, while it is 25 μM in normal ocean water. In these basins, siliceous skeletons are found in high abundance and with perfect preservation of fine skeletal structures.

Zoogeography through time

Radiolarians were one of the main groups to be studied in the 1970s and 1980s during the CLIMAP project (Climate: Long-range Investigation, Mapping, and Prediction). Its main objective was to study and map the climatic situation during the last glacial maximum (LGM) versus the present-day conditions. In the north Atlantic, one of the main results was that the polar front had moved down from the present position between Iceland and Greenland to about 45 °N, almost east-west from Newfoundland to the northern tip of the Iberian Peninsula. This can be illustrated by the arctic radiolarian *Amphimelissa setosa*, at present making up ca. 70 % of the cold water fauna on the Iceland Plateau. However, traces occurred about 5 °N of the Azores during LGM, indicating much colder water. Similarly, the apparently deep and cold water radiolarian *Cycladophora davisiana* reached values in excess of 20 % (Morley, 1983).

The onset of the Deep Sea Drilling Project (DSDP) in the late 1960s, and continuing under different names, has cored the ocean bottom sediments and produced a tremendous sediment core archive for scientific use, especially

a

Series/ Subseries	Leg 104 Holes					Radiolaria Zone	Boundary Criteria	Interpreted age (in Ma)
	642B	642C	642D	643A	644A			
Pleist.	–	–	–	–	Top to 30-3, 25-27cm 0-229.4mbsf	Cyclaclophora davisiana davisiana Taxon range zone	First occurrence of *Cycladophora davisiana davisiana*	2.6
Pliocene	–	–	–	–	31-3, 25-27cm to 34-cc 236.9-252.8mbsf	*Spongaster ? tetras* Taxon range zone	First occurrence of *Spongaster ? tetras*	2.9-3.1
	9-1, 75-77cm to 10-1, 75-77cm 67.2-76.7mbsf	9-cc to 11-3, 70-72cm 63.5-76.7mbsf	–	–	–	*Pseudodictyophimus gracilipes tetracanthus* Partial range zone	First occurrence of *P. gracilipes tetracanthus*	4.0
	10-2, 25-27cm to 11-6, 25-27cm 77.7-93.2mbsf	11-4, 70-72cm to 12-cc 78.2-92.0mbsf	–	–	–	*Antarctissa whitei* Interval zone	Last occurrence of *L. cricus*	5.2
upper Miocene	11-7, 25-27cm to 12-cc 94.7-104.2mbsf	13-1, 70-72cm to 14-2, 70-72cm 92.7-103.7mbsf	–	–	–	*Liriospyris cricus* Taxon range zone	First occurrence of *L. cricus*	5.5
	13-1, 24-26cm to 15-1, 27-29cm 104.4-123.8mbsf	14-3, 70-72cm to 15-6, 25-27cm 105.2-118.8mbsf	–	8-1, 103-105cm to 8-5, 103-105cm 63.3-69.3mbsf	–	*Tessarastrum thiedei* Partial range zone	First occurrence of *Tessarastrum thiedei*	6.2
	–	15-7, 25-27cm to 15-cc 120.3-120.5mbsf	–	–	–	*Larcospira bulbosa* Partial range zone	First occurrence of *Larcospira bulbosa*	6.3-6.9
	15-2, 27-29cm 1253.3mbsf	16-1, 25-27cm to 16-3, 25-27cm 120.8-123.8mbsf	–	8-6, 103-105cm to 9-cc 70.8-81.3mbsf	–	*Hexalonche esmarki* Partial range zone	First occurrence of *Hexalonche esmarki*	7.7-7.9
	15-3, 27-29cm to 16-2, 31-33cm 126.8-129.9mbsf	16-4, 25-27cm to 16-5, 25-27cm 125.3-126.8mbsf	–	10-1, 105-107cm to 10-5, 105-107cm 82.4-88.4mbsf	–	*Spongurus cauleti* Partial range zone	First occurrence of *Spongurus cauleti*	8.1
	16-3, 31-33cm to 16-7, 31-33cm 131.4-137.4mbsf	16-6, 25-27cm to 17-2, 25-27cm 128.3-131.8 mbsf	–	10-6, 105-107cm to 11-1, 106-108cm 89.9-91.9mbsf	–	*Corythospyris reuschi* Interval zone	Last consistent occurrence of *E. fridtjofnanseni*	8.3-8.6
	16-cc to 18-4, 24-26cm 138.0-152.5mbsf	17-3, 25-27cm to 19-2, 25-27cm 133.3-150.8mbsf	–	11-2, 106-108cm to 11-cc 93.4-100.3mbsf	–	*E. fridtjofnanseni/C.reuschi* Concurrent range zone	First occurrence of *Corythospyris reuschi*	9.5
	18-5, 24-26cm to 19-5, 24-26cm 154.0-163.9mbsf	19-3, 25-27cm to 20-2, 25-27cm 152.3-160.3mbsf	–	–	–	*Eucoronis fridtjofnanseni* Partial range zone	First occurrence of *E. fridtjofnanseni*	~13.5-13.6
middle Miocene	–	–	–	12-1, 105-107cm to 12-2, 105-107cm 101.4-102.9mbsf	–	*Clathrospyris vogti* Taxon range zone	First occurrence of *Clathrospyris vogti*	~13.7-13.8
	19-6, 24-26cm to 19-cc 165.4-167.2mbsf	20-3, 25-27cm to 20-6, 25-27cm 161.8-166.3mbsf	–	–	–	Interzone B	Last consistent occurrence of *P. horrida*	~14.1
	20-1, 24-26cm to 21.5, 24-26cm 167.6-183.3mbsf	20-7, 25-27cm to 22-3, 25-27cm 167.8-180.8mbsf	–	12-3, 105-107cm to 13-3, 105-107cm 104.4-113.8mbsf	–	*Pseudodictyophimus horrida* Taxon range zone	First occurrence of *P. horrida*	~14.5

Radiolarians, Figure 9 (continued)

b

Series/Subseries	Leg 104 Holes					Radiolaria Zone	Boundary Criteria	Interpreted age (in Ma)
	642B	642C	642D	643A	644A			
middle Miocene	21-6, 24-26cm to 22-1, 75-77cm 184.8-187.0mbsf	22-4, 25-27cm to 23-2, 24-26cm 182.3-186.0mbsf	–	–	–	Actinomma plasticum Interval zone	Last consistent occurrence of C. ampullacea	~14.7
	22-2, 75-77cm to 22-3, 75-77cm 188.5-190.0mbsf	23-3, 24-26cm to 23-cc 187.5-192.4mbsf	2-1, 25-27cm 190.2mbsf	–	–	C. Kladaros Subzone b. (Cyrtocapsella ampullacea) Taxon range subzone	First occurrence of C. ampullacea	~14.8
	22-4, 75-77cm to 22-5, 75-77cm 191.5-193.0mbsf	24-1, 25-27cm to 24-2, 25-27cm 192.7-194.2mbsf	2-2, 25-27cm to 2-4, 25-27cm 191.7-194.7mbsf	13-4, 104-106cm to 13-cc 115.4-119.3mbsf	–	C. kladaros Subzone a. (Cyrtocapsella kladaros) Partial range subzone	First occurrence of Cyrtocapsella kladaros	~14.9
	22-6, 75-77cm to 22-cc 194.5-196.0mbsf	24-3, 25-27cm to 24-4, 25-27cm 195.7-197.2mbsf	2-6, 25-27cm 197.7mbsf	14-1, 109-111cm to 15-4, 25-27cm 120.4-133.6mbsf	–	Ceratocyrtis broeggeri Interval zone	First occurrence of Ceratocyrtis stoermeri	~15.5
	23-1, 75-77cm to 23-3, 75-77cm 196.8-199.8mbsf	24-5, 25-27cm to 24-cc 198.7-199.6mbsf	2-cc to 3-2, 25-27cm 199.7-201.4mbsf	–	–	Cyrtocapsella eldholmi Partial range zone	First occurrence of Cyrtocapsella eldholmi	~15.7
	–	–	–	15-5, 105-107cm to 17-cc 135.9-148.5mbsf	–	Ceratocyrtis manumi Partial range zone	First occurrence of Ceratocyrtis manumi	~16.0-16.2
	–	–	–	18-1, 105-107cm to 20-2, 105-107cm 158.4-178.9mbsf	–	Cycladophora davisiana cornutoides Partial range zone	First occurrence of C. davisiana cornutoides	~16.7
	–	–	–	20-3, 105-107cm to 20-5, 105-107cm 180.4-183.4mbsf	–	Spongotrochus vitabilis Partial range zone	First occurrence of Spongotrochus vitabilis	~16.8
	–	–	–	20-6, 105-107cm to 20-cc 184.9-185.8mbsf	–	Interzone A	Last consistent occurrence of P. amundseni	~16.9
lower Miocene	23-4, 75-77cm to 23-cc 201.3-205.6mbsf	–	3-3, 25-27cm to 3-cc 202.9-207.5mbsf	21-1, 35-37cm to 23-4, 105-107cm 186.2-210.7mbsf	–	Pseudodictyophimus amundseni Taxon range zone	First occurrence of P. amundseni	~17.5
	24-1, 75-77cm to 24-5, 75-77cm 206.4-212.4mbsf	–	4-1, 24-26cm to 4-6, 24-26cm 209.4-217.0mbsf	23-5, 105-107cm to 25-3, 105-107cm 212.2-228.8mbsf	–	Corythospyris jubata jubata severdrupi Partial range zone	First occurrence of C. jubata sverdrupi	~17.8
	24-cc to 25-cc 213.0-221.1mbsf	–	4-cc to 5-cc 218.9-228.5mbsf	25-4, 105-107cm to 27-cc 230.3-254.0mbsf	–	Clathrospyris sandellae Partial range zone	First occurrence of Clathrospyris sandellae	~18.5
	–	–	6-1, 30-32cm to 8-3, 25-27cm 228.8-251.5mbsf	28-1, 105-107cm to 30-cc 255.2-274.2mbsf	–	Gondwanaria japonica kiaeri Partial range zone	First occurrence of G. japonica kiaeri	~20.0-20.5
	–	–	8-4, 25-27cm to 9-cc 253.0-267.1mbsf	–	–	Eucyrtidium saccoi Partial range zone	First occurrence of E. saccoi	~21.2
	–	–	10-1, 25-27cm to 11-1, 65-67cm 267.4-277.5mbsf	–	–	Actinomma henningsmoeni Partial range zone	First occurrence of A. henningsmoeni	~21.7

Radiolarians, Figure 9 An example of a detailed Quaternary – middle Neogene (0–14.5 Ma) Norwegian Sea radiolarian biostratigraphy (IODP Leg 104). Many endemic species used as zonal markers allow this scheme only to be used as a local to regional biostratigraphic tool. Early Neogene (14.5–21.7 Ma) Norwegian Sea radiolarian biostratigraphy (IODP Leg 104).

for biostratigraphy, paleoceanography, and paleoclimatology. The Integrated Ocean Drilling Program (IODP) expedition 306 to the north Atlantic cored a continuous sediment column spanning the last ca. 3 Ma at site U1314. At this site, it was nicely demonstrated that the northern hemisphere glaciation started at about 2.7 Ma and that the radiolarian fauna fluctuated between a glacial, transitional, and interglacial assemblage corresponding to the 41,000 year Milankovitch obliquity cycle. The fauna changes became even more pronounced when the glacial-interglacial cycles shifted from a recurrence of 41,000–100,000 years at about one million years ago.

When an ocean site over time receives alternating warm, transitional, and cold water radiolarian fauna, this indicates that the different fauna provinces are migrating from north to south during glacial periods (an expanding arctic fauna province that shrinks the other fauna provinces) and opposite during interglacial periods (a shrinking arctic fauna province that expands the other fauna provinces).

At a longer time scale, such as during ocean basin development, for example, opening of the Norwegian Sea, different episodes in the tectonic evolution (opening and closing of water gateways) are depicted in the radiolarian biostratigraphy. Biostratigraphy is important for correlation between sites. It is not possible to compose one radiolarian biostratigraphic scheme applicable for the entire world, which would have been possible if all species had been cosmopolitan. Due to the phenomenon of endemic species, and particular species dwelling in zoogeographical provinces, only the tropical provinces of the Pacific, Indian, and Atlantic Oceans can be worked out with a common biostratigraphic framework. High-latitude faunas in the Atlantic are different from those in the Pacific, which again are different than in high southern hemisphere latitudes. In other words, the radiolarian biostratigraphic schemes are only of regional use and very dependent upon the radiolarian preservation. An example of a rather detailed local Neogene-Quaternary stratigraphy established by Goll and Bjørklund (1989) is shown in Figure 9.

Bibliography

Anderson, O. R., 1976. Fine structure of a collodarian radiolarian (*Sphaerozoum punctatum* Müller 1858) and cytoplasmic changes during reproduction. *Marine Micropaleontology*, **1**(4), 287–297.

Anderson, O. R., 1978. Light and electron microscopic observations of feeding behaviour, nutrition, and reproduction in laboratory cultures of *Thalassicolla nucleata*. *Tissue and Cell*, **10**(3), 401–412.

Anderson, O. R., 1983. *Radiolaria*. New York: Springer, pp. 1–355.

Anderson, O. R., and Swanberg, N. R., 1981. Skeletal morphogenesis in some living collosphaerid radiolaria. *Marine Micropaleontology*, **6**, 385–396.

Anderson, O. R., Swanberg, N. R., and Bennett, P., 1984. An estimate of predation rate and relative preference for algal versus crustacean prey by a spongiose skeletal radiolarian. *Marine Biology*, **78**(2), 205–207.

Anderson, O. R., Bennett, P., and Bryan, M., 1989. Experimental and observational studies of radiolarian physiological ecology: 1. Growth, abundance and opal productivity of the spongiose radiolarian *Spongaster tetras tetras*. *Marine Micropaleontology*, **14**(4), 257–266.

Bjørklund, K. R., and De Ruiter, R., 1987. Radiolarian preservation in Eastern Mediterranean anoxic sediments. *Marine Geology*, **75**, 271–281.

Bjørklund, K. R., Kruglikova, S. B., and Anderson, O. R., 2012. Modern incursions of tropical radiolaria into the Arctic Ocean. *Journal of Micropalaeontology*, **31**, 139–158.

Boltovskoy, D., Alder, V. A., and Abelmann, A., 1993. Radiolarian sedimentary imprint in Atlantic equatorial sediments: composition with the yearly flux at 853 m. *Marine Micropaleontology*, **23**, 1–12.

Boltovskoy, D., Kogan, M., Alder, V. A., and Mianzan, H., 2003. First record of a brachish radiolarian (Polycystina): *Lophophaena rioplatensis* n. sp. in the Río de la Plata estuary. *Journal of Plankton Research*, **25**(12), 1551–1559.

Brandt, K., 1881. *Untersuchungen an Radiolarian*. Königliche Preussische Akademie der Wissenschaften zu Berlin, Monatsbericht, pp. 388–404.

Brandt, K., 1902. Beitrage zur Kenntnis der Colliden. *Archiv für Protistenkunde*, **1**(1), 59–88.

Casey, R., 1971. Radiolarians as indicators of past and present water masses. In Funnell, R. M., and Riedel, W. R. (eds.), *Micropaleontology of the Oceans*. Cambridge: Cambridge University Press, pp. 331–341.

Casey, R., Partridge, T. M., and Sloan, J. R., 1970. Radiolarian life spans, mortality rates, and seasonality gained from recent sediment and plankton samples. In *Proceedings IInd Planktonic Conference, Tecnoscienzia, Roma*, Vol. 1, pp. 159–165.

Casey, R. E., Partridge, T. M., and Sloan, J. R., 1971. Radiolarian life spans, mortality rates and seasonality gained from sediment and plankton samples. In Farinacci, A. (ed.), *Proceedings of the II Planktonic Conference, Roma 1970*. Roma: Edizioni Tecnoscienza, pp. 159–165.

Cavalier-Smith, T., 1993. Kingdom protozoa and its 18 phyla. *Microbiology and Molecular Biology Reviews*, **57**, 953–994.

Cavalier-Smith, T., 2002. The phagotrophic origin of eukaryotes and phylogenetic classification of Protozoa. *International Journal of Systematic and Evolutionary Microbiology*, **52**, 297–354.

Cavalier-Smith, T., 2003. Protist phylogeny and the high-level classification of Protozoa. *European Journal of Protistology*, **39**, 338–348.

De Wever, P., Dumitrica, P., Caulet, J. P., Nigrini, C., and Caridroit, M., 2001. *Radiolarians in the Sedimentary Record*. Yverdon: Gordon and Breach Science Publishers, pp. 1–533.

Dumitrica, P., 1999. On the presence of central capsular membranes of Radiolaria in fossil state. *Revista Española de Micropaleontología*, **31**(2), 155–183.

Enriques, P., 1919. Ricerche sui Radiolari Coloniali. R Comitato Talassografico Italiano. *Memoria*, **71**(1), 1–177.

Goll, R. M., 1976. Morphological intergradation between modern populations of lophospyris and phormospyris (Trissocyclidae, radiolaria). *Micropaleontology*, **22**(4), 379–418.

Goll, R. M., and Bjørklund, K. R., 1971. Radiolaria in surface sediments of the North Atlantic Ocean. *Micropaleontology*, **17**(4), 434–454.

Goll, R. M., and Bjørklund, K. R., 1974. Radiolaria in surface sediments of the North Atlantic Ocean. *Micropaleontology*, **20**(1), 38–75.

Goll, R. M., and Bjørklund, K. R., 1989. A new radiolarian biostigraphy for the Neogene of the Norwegian Sea: ODP Leg 104. *Proceedings of the Ocean Drilling Program Scientific Results*, **104**(35), 697–738.

Haeckel, E., 1887. Report on the radiolaria collected by H.M.S. Challenger during the years 1873–1876. *Representative Science Research Voyage H.M.S. Challenger, Zoology*, **18**, 1–1803.

Herring, P. J., 1979. Some features of the bioluminescence of the radiolarian *Thalassicolla* sp. *Marine Biology*, **53**, 213–216.

Hertwig, R., 1879. *Der Organismus der Radiolarien*. Jena: Verlag von Gustav Fischer, pp. 1–149.

Hollande, A., and Enjumet, M., 1953. Contribution à l'étude biologique des Sphaerocollidés (Radiolaires Collodaires et Radiolaires polycyttaires et leurs paracites) 1 – Thalassicollidae, Physematidae, Thalassophysidae (Biological study of the Sphaerocollidae (Radiolaria Collodaria and Polycyttaria with their parasites) 1 – Thalassicollidae, Physematidae, Thalassophysidae). *Annales des Sciences Naturelles, Séries 11*, **15**, 99–183.

Itaki, T., and Bjørklund, K. R., 2007. Conjoined radiolarian skeletons (Actinommidae) from the Japan Sea sediments. *Micropaleontology*, **53**(5), 371–389.

Johnson, T. C., 1974. The dissolution of siliceous microfossils in surface sediments of the eastern tropical Pacific. *Deep-Sea Research*, **21**, 851–864.

Johnson, D. A., and Nigrini, C., 1980. Radiolarian biogeography in surface sediments of the western Indian Ocean. *Marine Micropaleontology*, **5**(2), 111–152.

Kimoto, K., Yuasa, T., and Takahashi, O., 2011. Molecular identification of reproductive cells released from *Cypassis irregularis* Nigrini (Radiolaria). *Environmental Microbiology Reports*, **3**(1), 86–90.

Krabberød, A. K., Bråte, J., Dolven, J. K., Ose, R. F., Klavenes, D., Kristensen, T., Bjørklund, K. R., and Shalchian-Tabrizi, K., 2011. Radiolaria divided into polycystina and spasmaria in combined 18S and 28S rDNA phylogeny. *PLoS One*, **6**(8), 1–10, doi:10.1371/journal.pone.0023526.

Lazarus, D., 1983. Speciation in pelagic protista and its study in planktonic microfossil record: a review. *Paleobiology*, **9**(4), 327–340.

Lizitsin, A. P., 1972. *Sedimentation in the World Ocean. Society of Economic Paleontologists and Mineralogists*. Menasha: Georg Banta Company. Special publication number 17, 218 pp.

Lovejoy, C., and Potvin, M., 2011. Microbial eukaryotic distribution in a dynamic Beaufort Sea and the Arctic Ocean. *Journal of Plankton Research*, **33**(3), 431–444.

Matsuoka, A., 1992. Skeletal growth of a spongiose radiolarian *Dictyocoryne truncatum* in laboratory culture. *Marine Micropaleontology*, **19**(4), 287–297.

Matul, A. G., Khusid, T. A., Mukhina, V. V., Chekhovskaya, M. P., and Safarova, S. A., 2007. Recent and late holocene environments on the southeastern shelf of the Laptev Sea as inferred from microfossil data. *Oceanology*, **47**, 80–90.

Meyen, F. J. F., 1834. Beiträge zur Zoologie: gesammelt auf einer Reise um die Erde Fünfte Abhandlung von F.J.F. Meyen. Über das Leuchten des Meeres und Beschreibung einiger Polypen und anderer niederer Thiere. *Nova Acta Physico-Medica. Academiae Caesariae Leopoldino-Carolinae. Naturae Curiosorum (Verhandlungen der Kaiserlichen Leopoldinisch-Carolinischen Akademie der Naturforscher)*, **16**(Suppl 1), 125–216.

Morley, J., 1983. Identification of density-stratified waters in the Late-Pleistocene North Atlantic: a faunal derivation. *Quaternary Research*, **20**, 374–386.

Not, F., del Campo, J., Balague, V., de Vargas, C., and Massana, R., 2009. New insights into the diversity of marine picoeukaryotes. *PLoS One*, **4**, eE7143.

Petrushevskaya, M. G., 1971. *Radiolyari Nassellaria v planktone Mirovogo Okeana. Issledovaniya Fauny Morei*. Leningrad: Nauka, Vol. 9. 294 pp (in Russian).

Reshetnyak, V. V., 1955. Vertikalnoe raspredelenie radiolaryarii Kurilo-Kamchtskoi vpadiny. *Trudy Zool Akad Nauk U.S.S.R*, **21**, 94–101 (in Russian).

Riedel, W. R., 1967. Subclass Radiolaria. In Harland, W. B. (ed.), *The Fossil Record. A Symposium with Documentation*. London: Geological Society of London, pp. 291–298.

Schewiakoff, W., 1926. Die Acantharia des Golfes von Napoli. Fauna e Flora del Golfo di Napoli. *Monografia*, **37**, 1–755. Friedland e Sohn, Berlin.

Suzuki, N., and Aita, Y., 2011. Radiolaria: achievements and unresolved issues: taxonomy and cytology. *Plankton & Benthos Research*, **6**(2), 69–91.

Swanberg, N. R., and Anderson, O. R., 1981. Collozoum caudatum sp. nov.: a giant colonial radiolarian from equatorial and Gulf Stream waters. *Deep-Sea Research Part A*, **28**(9A), 1033–1047.

Swanberg, N. R., and Bjørklund, K. R., 1986. The radiolarian fauna of western Norwegian fjords: patterns of abundance in the plankton. *Marine Micropaleontology*, **11**, 231–241.

Swanberg, N. R., and Harbison, G. R., 1980. Distribution, ecology of a new species of *Collozoum* in the tropical Atlantic. *Deep-Sea Research Part A*, **27**(9A), 715–732.

Takahashi, K., 1983. Radiolaria, sinking population, standing stock, and production rate. *Marine Micropaleontology*, **8**(3), 171–181.

Takahashi, K., 1991. *Radiolaria: Flux, Ecology, and Taxonomy in the Pacific and Atlantic*. Woods Hole: Woods Hole Oceanographic Institution Ocean. Ocean Biocoenosis Series 3, pp. 1–303.

Cross-references

Biochronology, Biostratigraphy
Deep-sea Sediments
Fjords
Geologic Time Scale
Marine Microfossils
Radiolarians
Silica

REEF COASTS

Charles W. Finkl
Department of Geosciences, Florida Atlantic University, Boca Raton, FL, USA
The Coastal Education and Research Foundation, Coconut Creek, FL, USA

Synonyms

Atoll; Bank; Barrier reef; Carbonate mound; Cay; Coral island; Coral reef; Fringing reef; Key; Organic reef; Rock barrier; Sandbar; Shoal; Skerry

Definition

In nautical terms, a *reef* is a rock, sandbar, shoal, or other topographic feature lying at 10 fathoms (20 m) or less beneath the surface of the water (Jackson 1997) and is so recognized because these features form an obstruction to navigation. From a geological point of view, a reef may be a mass or ridge of rock (i.e., hogback ridge, whaleback, dike, or sill) that lies below, at, or somewhat above the surface of the sea or lake.

A more commonly recognized form of reef is the contemporary *coral reef* (e.g., Done 2011), which may constitute the shore with no other land (atoll), may touch a terrestrial shore (fringing reef), or may occur some distance offshore with an adjacent landward lagoon on the continental shelf (barrier reef). Coral reefs are thus commonly defined as: "A mound or ridge of living coral, coral skeletons, and calcium carbonate deposits from reef biota such as calcareous algae, mollusks, and protozoans," composed mainly of coral and other organic matter of which parts have solidified into successive layers of limestone (e.g., Done 2011; Jackson 1997). Coral reefs that are now separated from the shore by sea-level change and/or tectonic movements (i.e., uplifted or drowned) are referred to as bioherms or biostromes.

Artificial reefs such as shipwrecks or debris piles are sometimes created to enhance physical complexity on generally featureless sand bottoms in order to attract a diverse assemblage of organisms, especially fish.

A reef coast thus may be regarded as a shore that is dominated by underwater or shallowly surfacing rock ridges, sandbars, coral mounds, ridges, or islets,, including short coastal segments that are dominated by the placement of artificial reefs.

Etymology and usage

According to Fox (2005) and Gove (1971), for example, the word reef seems to be derived from the Old Norse word *rif* (ridge), which literally means rib (but by the fourteenth century meant "horizontal section of sail"), and hence "reef of a sail," the connotation eventually referring to the shape or morphology of this coastal feature. Other derivations probably stem from Middle English *rif*, from Proto-Germanic *rebjan* ("rib, reef"), from Proto-Indo-European *rebh-* ("arch, ceiling, cover") and its cognate with Dutch *rif* ("reef"), Low German *riff, ref*, and *reff* ("reef"), German *riff* ("reef, ledge"), Danish *reb*, Swedish *ref*, and Old English *ribb* ("rib"). The various origins show similarity of intent to convey the same message that was subsequently used to describe these coastal marine features in the scientific literature. When the hypernym reef is used to modify a noun as in the case of *reef coast*, it refers to a special kind of coast (q.v.) that is dominated by non-coral and coral reefs.

Rock reefs and non-biogenic reefs

Rock reefs result from abiotic processes, such as the deposition of sand, wave erosion planning down rock outcrops, glaciation of bedrock and surficial covers to form glacial residuals (i.e., moraines, kames, eskers, drumlins, whalebacks [rock drumlins, tadpole rocks, streamlined hills], roche moutonnée ["sheepback"] rock), drowning of weathering residuals (i.e., tors, corestones ["woolsacks"]), and other natural processes that result in exposure and drowning of rock or geochemically indurated materials such as laterite or silcrete. Modern non-biogenic reefs thrive in a wide range of environments extending from sponge reefs in the Arctic to chemosynthetic algae and *Halimeda* reefs found near methane seeping faults at depths of 600 m (Wood 1999).

Coral reefs

The best-known reefs, however, are the coral reefs of clear tropical and subtropical waters where they occur adjacent to land or around submerged volcanic peaks, as so clearly described long ago for the Pacific Ocean, for example, by Agassiz (1898), Dana (1872), and Darwin (1842). These reefs were developed through biotic processes dominated by communities of corals and calcareous (coralline) algae.

Although rock reefs may occur along all coasts, coral reefs exhibit restricted geographic distribution due to ecological requirements of the coral ecosystem. Figure 1 shows the global distribution of major coral reef systems in tropical and subtropical regions. Shallow-water coral reefs do not occur just anywhere in this geographical zone but are concentrated in specific areas indicated by the red dots on the maps. Reefs thrive in these areas where ecological conditions are favorable for their growth.

Coral reefs are found in circumtropical shallow tropical waters (mainly in regions where the water temperature is always above 18 °C in winter) along the shores of islands and continents (Figure 1). Coral reef coasts are found in the Indo-Pacific province, which is traditionally subdivided into eastern (Red Sea, the western part of the Indian Ocean, and the eastern part of the Indian Ocean) and western (central part of the tropical Pacific, Malaysia, the Philippines, and Indonesia) subregions, and the

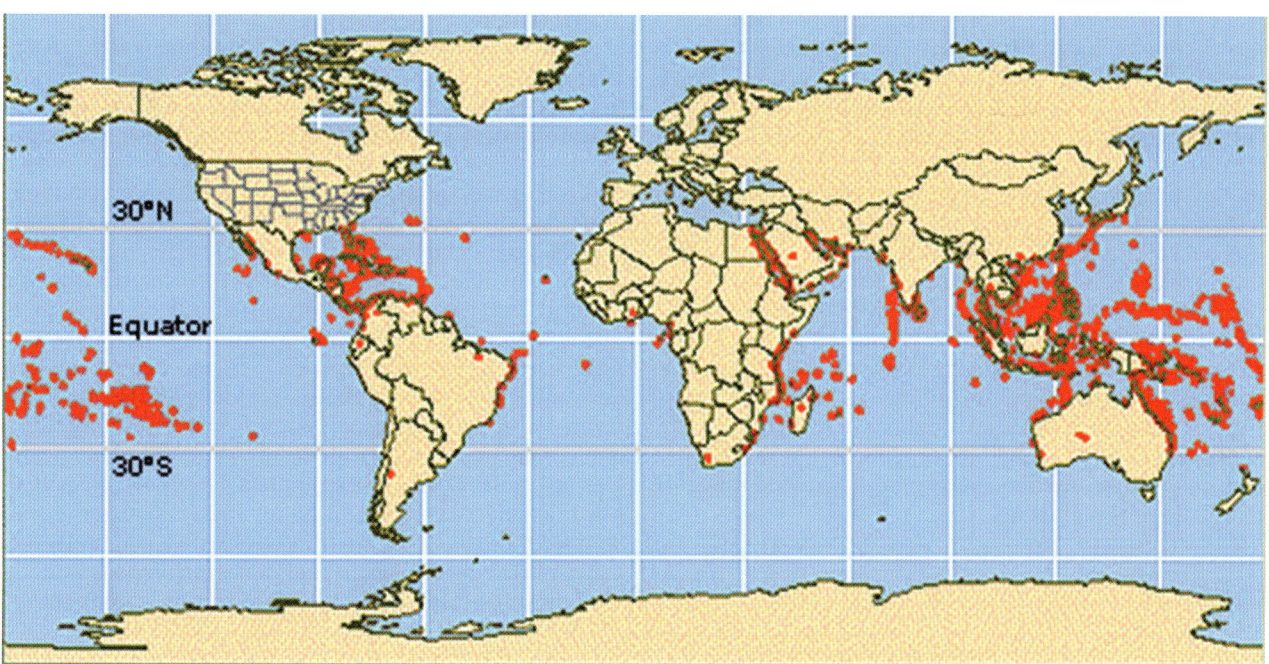

Reef Coasts, Figure 1 Global distribution of coral reefs showing concentrations in tropical and subtropical regions (source: NOAA Coral Reef Information System: http://coris.noaa.gov/about/what_are/welcome.html).

Atlantic (Bermudas, Caribbean Sea area, coasts of northern Brazil and West Africa) province (i.e., Woodroffe 2002). The Indo-Pacific province contains about 500 species of corals where the dominant genus of branching corals (*Acropora*) consists of more than 150 species, compared to only two species prominent in this genus in the Atlantic province (i.e., Wood 1999; Woodroffe 2002).

Shallow-water reefs require a minimum temperature of 18 °C and therefore occur less frequently on the eastern Atlantic and Pacific coasts than on the warmer western coasts. The reef substrate is mainly composed of calcium carbonate from living and dead scleractinian corals. Corals live in very nutrient-poor waters and have certain zones of tolerance to water temperature, salinity, UV radiation, opacity, and nutrient quantities (Wood 1999). In spite of these ecological requirements, coral communities can thrive under the proper conditions to form large features in the sea and adjacent to land masses.

Australia has the most extensive and pristine coral reef systems in the world where the northeast coast is dominated by the Great Barrier Reef (GBR). Ningaloo Reef, along the northwest coast of Western Australia, is the largest fringing reef in the world (Short and Woodroffe 2009). The GBR, a continental-shelf reef system that can be seen from the outer space, is the world's most extensive coral reef system that stretches 2,300 km over an area of approximately 344,400 km^2 making it the largest biologically built structure in the world (Short and Woodroffe 2009; Woodroffe 2002). The GBR contains more than 3,600 individual reefs and 900 islands. More than 750 fringing reefs occur along the mainland coast and on "high" islands where bedrock outcrops. Other notable coral coasts in the world include the New Caledonian reefs, the Mayotte reefs, the Belizean reef system, and the Florida Reef Tract.

All coral reefs have distinctive horizontal and vertical zones created by differences in depth, wave action, current movement, light, temperature, and sediment along different parts of the reef (Woodroffe 2002). Although the exact characteristics of each zone may vary slightly depending on the location and type of reef, all shallow-water coral reefs have a forereef, or seaward slope, and a backreef. The backreef is the part of the reef closest to shore, while the forereef is farther out to sea. In addition to these distinctive zones, coral reefs are grouped together on the basis of dominant morphological characteristics and classified on the basis of variable properties (i.e., Stoddart 1969a). Although there are different classification systems, several basic types are commonly recognized, viz., atolls, fringing reefs, and barrier reefs. All of these types of coral reef are subdivided to produce a wide range of morphological reef coasts.

Classification

There are many different types of coral reefs and their subdivision can become quite complicated. Shallow-water coral reefs vary in form, plan, and origin in an almost infinite number of ways (Nunn 1994; Parnell 2011); it is also difficult to get scientists to agree on any one particular schema because of the complexity involved. Even though it is impossible to identify all types of coral reef coasts, general impressions may be obtained from the following brief summaries that are based on seminal and much earlier works that are still relevant today. Darwin's (1838, 1842) original descriptions of the basic reef forms (i.e., fringing, barrier, and atoll reefs) are still in common use today. Although the original reef coast terms are somewhat modified and subdivided, they still form the basis of modern understanding. Working in the Australian Indo-Pacific region, for example, Fairbridge (1950) originally focused on the Great Barrier Reef (GBR) but later extended his system to include the whole of the Australian region (Fairbridge 1967), detailing reefs on the west coast continental platform as well as those occurring along the northern margins of the continent in the Arafura Sea and Torres Strait. This basic classification of fringing, barrier, atoll, and platform reefs, along with five varieties of coral island, was expanded by Maxwell (1968) and later more completely in examples from Hopley (1982, 2011), Hopley et al. (2007), Wood (1999), and Woodroffe (2002).

Atoll

An atoll is a ring (annular) reef that surrounds a shallow lagoon (Darwin 1842), the definition being based on the only forms that Darwin visited in the Cocos (Keeling) Islands in the eastern Indian Ocean. The term "atoll" is derived from the Maldivian word, *atolu* (Woodroffe and Biribo 2011). Most atolls occur in the Pacific Ocean as well as the central Indian Ocean, but there are only a few in the Caribbean where morphological descriptions and actual number vary between 10 and 27 according to the authority cited (cf. Stoddart 1965; Milliman 1973; Riegl and Dodge 2008). The three Belizean atoll reefs are formed on submarine ridges, Turneffe and Glover's on one ridge and Lighthouse (Figure 2) on another to the east. Atolls were classified by Ladd (1977) into ocean atolls and shelf atolls, but a more complete description is provided by Woodroffe and Biribo (2011). The reef rim may be continuous or broken by a few channels as occur in the central Pacific, or many shallow passes (*hoa*) as in French Polynesia, or deep passages as in the Maldives (McLean 2011). Fairbridge (1950) reports that there is no marked difference in appearance between an "annular reef" (one that is perfectly enclosed) rising from an oceanic cone or platform and one rising from the continental shelf. Examples of some morphologically perfect atolls occur on Australia's North West Shelf in the Indian Ocean and on the Sahul Shelf where Seringapatam Reef is a *perfectly enclosed reef* with a central lagoon approximately 36 m deep. Boulders typically occur along reef crests and small shingle often forms "hammerhead" spits. Wide reef flats may accommodate a radial zone that is produced by algae, boulders, and sediment. A "sanded zone" often breaks up into coral patches toward the lagoon.

Reef Coasts, Figure 2 This fringing reef in Belize is shore attached, growing seaward in an irregular manner where it can find suitable seafloor substrate. The seaward margin of the coral reef drops off precipitously to deeper water, whereas the landward margin merges with the carbonate rock of the island on Lighthouse Reef Atoll, where this fringing reef is part of the larger atoll development (photo: Istockphoto.com).

Some atoll-like reefs have a facing edge of stronger established coral reef and a more weakly developed back edge, due to prevailing ocean currents and dominant weather. The well-developed edge faces the direction of high oceanic energy where corals access nutrients that provide greater growth potential (i.e., Nunn 1994; Short and Woodroffe 2009).

Microatolls are individual subcircular colonial corals, usually of the genus *Porites*, that occur on reef flats (i.e., Nunn 1994). These reefs expand by growth of their outer edges, but they are very sensitive to emergence as described by Woodroffe and McLean (1990).

Fringing reef

Fringing reefs develop close to the shore and may or may not be attached (approach or touch the foreshore) to the land, which may be mainland or an island (cf. Figure 2), as seen in Figure 3 for a reef system on the seaward margin of Long Island in the Bahamas. They are linear reefs that grow along shelving coastlines and across embayments (Done 2011), as shown in the center of Figure 3, with a backreef area that may be shallowly submerged (Smithers 2011). These relatively simple structures can be geomorphologically subdivided into three main zones: forereef, reef crest, and backreef (e.g., Kennedy and Woodroffe, 2002). The width of a fringing reef is an indication of its age and the kind of substrate available as a foundation for reef growth. Symmetries of reefs and islands seem to be controlled by wave direction where the reefs may be markedly asymmetrical around islands that are affected by waves from a constant direction (Parnell 2011). Some examples of well-known fringing reefs include Hanauma Reef on the island of Oahu in Hawaii, along the coast of Mahé in the Seychelles, Kolombangara and Gizo Islands in the Solomon Islands (Stoddart 1969b), the Tuamotu Archipelago (Montaggioni et al. 1985), eastern coast of Belay (Palau in the northwest Pacific) (Easton and Ku 1980), and off many islands in the GBR (i.e., the Whitsundays, Lizard Island, Fantome, Orpheus, Hinchinbrook Island, Cape Tribulation) (Short and Woodroffe 2009).

Fringing reefs are an important component of the GBR with more than 750 being identified and an additional 200 incipient fringing reefs that are shore attached or nearshore but not yet reaching the sea surface. Fringing reefs along mainlands tend to be the poorest of reef types due to rigorous conditions near the shore, excessive sedimentation, dilution of seawater by fresh water, and acidic waters running off the land (Fairbridge 1950). More vigorous fringing reefs on the other hand often encircle offshore continental islands where growth conditions are more favorable. Hopley, Smithers, and Parnell (2011) have additionally shown that fringing reefs of the GBR have experienced several critical growth phases since the mid-Holocene where there is apparent synchronicity of growth-and-quiescent phases over wide geographical areas that influenced by climatic cycles and sea-level change. But, in general, episodic living coral growth is confined to the outer reef edge, where there may be a *Lithothamnion* ridge, whereas behind it there is dead reef,

Reef Coasts, Figure 3 Example of Bahamian coral reef complex. In the photo foreground are deeper-water (>20 m depth) reefs on the edge of the Bahamian carbonate platform fronting the open Atlantic Ocean. Fringing reefs are shown along the photo center as shore-attached coral reefs that front beach-dune systems and low calcarenite cliffs but also which cross small embayments. Barrier reefs lie farther offshore as shelf-edge reefs. Both barrier and fringing reefs contain reef gaps filled with carbonate sands (photo: C. W. Finkl).

about a meter below low-water springs, lightly sand covered, and marked by the radial pattern of seaweed (Fairbridge 1950). Lord Howe Island has the southernmost fringing reef in the Pacific Ocean (Short and Woodroffe 2009).

Barrier (ribbon) reef

A barrier reef forms a calcareous barrier along a segment of mainland or around an island resulting in a lagoon between the shore and the reef. They are also often referred to as platform, continental shelf, or shelf-edge reefs such that dominate the GBR (e.g., Done 2011; Fairbridge 1950; Hopley 1982; Short and Woodroffe 2009) or the Florida Reef Tract (Florida Keys reef tract plus northern extensions), which stretches 125 km along the southeast coast of the Florida Peninsula (Banks et al. 2008). The Belizean section of the Mesoamerican Barrier Reef System (Mexican Yucatan, Belize, Guatemala, and Honduras) is about 200 km long (Andréfouët and Cabioch 2011), but the entire system stretches for about 1,000 km. The example shown in Figure 4 is part of the barrier reef around Roatan (Honduras) in the Caribbean. Platform reefs colonize topographic highs that provide suitable foundations. Barrier reefs are always strongly asymmetric in plan view and cross section, being steep-to on the ocean side and grading toward the land with a sediment wedge dotted by small reef patches, pinnacles, and coral heads (i.e., Fairbridge 1967; Nunn 1994; Short and Woodroffe 2009). Figure 5 shows a Bahamian example of a complex barrier reef system composed of a seaward shelf-edge barrier with landward complexes of patch reefs and aggregated reefs. Incipient barriers may include small fringing reefs around islands within the main lagoon. Discontinuous reefs may be genetically linked to the barrier type, but from a morphological point of view, they are only patch or platform reefs, as described by Fairbridge and Teichert (1948).

Ribbon-shaped reefs, also referred to as "linear reefs" by Jukes (1847), occur as festoons along the outer edges of continental shelves. The ribbons are separated from one another by narrow passages, some up to a kilometer or more in width, called reef gaps (Finkl 2004; Finkl and Andrews 2008; Lidz et al. 2006) as found in the Florida Reef Tract. Typically there is an outer, parapet-like fringe of living corals that are furrowed by wedge-shaped channels or chutes, also referred to as spur and groove topography. Grooves are commonly floored with coral rubble or carbonate sand and may function as sediment conduits from backreef or inter-reefal areas to deeper water (Finkl 2004).

In the Australian region, for example, the reef crest rises gently from here, often including a *Lithothamnion* rim of calcareous algae about 1–2 m above low-water springs. Landward of the crest there is debris containing broken corals, boulders, shells, and sand, all cemented together. The so-called trickle zone or radial zone occurs on wider parts of the backreef where sediment debris and algae form a definite pattern normal to the reef edge. A shallow "inner moat" occurs inside of the rim, partly filled with clumps of living coral, boulders, and sediment. From here the floor deepens gradually and tends to be covered with white coral sand of the "sanded zone." Farther out in deeper waters of the lagoon, only a few larger coral heads reach the surface. This typical morphological sequence was reported early on by Agassiz (1898).

Reef Coasts, Figure 4 Barrier reef on Roatan (Bay Islands), Honduras, as an example of the Caribbean-type coral barrier. The Bonacca Ridge is the second longest barrier reef system (Great Western Barrier Reef) in the world (after the Great Barrier Reef in Australia) and extends over 400 km from Roatan to Belize. The deeper dark blue water lies seaward of the barrier, whereas the light green-colored water occurs in the lagoon between the reef and shore. Roatan lies on the edge of the Cayman Trench that provides clear water from the depths where the trench plunges thousands of meters off the west end of the island (photo: C. W. Finkl).

Reef Coasts, Figure 5 Example of a barrier-type coral reef system along the ocean side of Long Island, Bahamas, showing the forereef spur and groove topography in the lower right-hand corner, numerous patch reefs on the reef flat, and a lagoon between the reefs and shore. The reef edge drops off into water that is hundreds of meters deep (photo: C. W. Finkl).

Patch reefs

These are young patches of coral, roughly equant in plan view, that grow up off the bottom of the sea to form an isolated small platform in a lagoon or behind a barrier reef or atoll rim (Hopley 1982). These comparatively small reefs (cf. Figure 5), which often reach the surface and spread outward forming a pool in the center, are commonly surrounded by a sand halo or sea grass (i.e., Finkl and Vollmer 2011). An example of non-lagoonal patch reef coast occurs off northern St. Croix (Caribbean Sea) where small reefs (up to 20 m across) rise out of 10–15 m of water. They are locally referred to as "haystacks" (KellerLynn 2011) because of their broken and piled-up coral branches. Fairbridge (1950, 1967) applied the term

Reef Coasts, Figure 6 Construction debris used to build an artificial reef offshore southeast Florida, USA. This man-made reef structure was constructed by placing steel rails and circular concrete culverts within a specific zoned area on a sandy seafloor landward of the Florida Reef Tract. Over time, marine organisms, such as sponges and scleractinian coral colonies, utilized the artificial reef substrates to adhere and grow, just as they would on natural reefs. Colonization along the steel rail transecting the photo center shows a wide range of biological marine life upon its surface (photo: Christopher Makowski).

"patch reef" (also referred to as shelf, bank, table, or hummock reef) to the smaller reefs (<2 km long) and "platform reef" to larger ones (>2 km long), but there are many other definitions of the types of reef coral formations that constitute a patch reef (e.g., Hubbard 1997). Many patch reefs occur in the GBR lagoon, in the Torres Straits almost blocking the channel between Australia and Papua New Guinea, and scattered along the northern edge of the Sahul Shelf.

Coral reefs with islands

Many different types of islands are associated with coral reefs. Initial studies of reef islands were made in the nineteenth century in Southeast Asia and the Australian GBR and later in the Pacific and Indian Oceans and in the Caribbean Sea (Steers 1937; Smithers and Hopley 2011).

The islands have traditionally been differentiated as "high" islands composed of continental rocks (with associated coral reefs) (e.g., Steers 1929) and "low" islands (coral cays) composed of biogenic carbonate sediments produced by reef organisms. There is a great diversity of forms of coral cays and numerous classifications have been developed from investigations of the GBR, the largest shallow-water coral reef province in the world (combined with Torres Strait) that contains about 350 coral cays (Smithers and Hopley 2011). Most coral cays occur on reef flats (many occurring on small reefs <1 km^2) at or very close to sea-level as, for example, on Pacific atolls and on the Cocos (Keeling) Islands where less than one-third of the island surface is more than 2 m above sea-level (i.e., Woodroffe and Biribo 2011), but some may develop on lagoonal sediments (Kench et al. 2005).

By way of an example from one approach, Fairbridge (1950, 1967) grouped reef islands into (1) simple sand cays (islands that are commonly unstable and which migrate seasonally, found in all areas except on fringing reefs), (2) vegetated sand cays (islands that are moderately stabilized with beach rock, widely distributed but generally lacking from the outer reefs), (3) shingle cays (moderately stabilized islands that are widely distributed but found mostly on smaller, more exposed reefs), (4) sand cays with shingle ramparts (unvegetated to vegetated islands, including mangrove swamp), and (5) emerged reef islands (reef islands with or without a fringe of recent sand or shingle beach ridges or ramparts). More comprehensive summaries of these complicated coral reef coast features may be found in examples from Barnes and Hughes (1999), Hopley (1982, 2011), Ladd (1977), Maxwell (1968), Stoddart and Steers (1977), and Wood (1999), among others.

Artificial reefs

An artificial reef is an underwater structure that is positioned to promote marine life in areas with a generally featureless bottom, control erosion, block ship passage, or improve surfing. The purpose of fish attracting devices (e.g., concrete blocks, streamers, scuttled ships, and oil-drilling rigs) is to enhance fish abundance, species richness, and biomass (Sherman et al. 2002). Figure 6 is an example of construction debris that was placed on a featureless sandy seafloor to provide habitat for marine life off the coast of southeast Florida along the leeside of the Florida Reef Tract. These types of reefs contain objects of natural or human origin (i.e., rocks, ships, railcars, tanks, sculptures) deployed on the seafloor to favorably influence physical, biological, or socioeconomic processes related to living marine resources. Initial attempts to construct an artificial surfing reef were made in El Segundo (near Los Angeles, in California) and at Mosman Beach (Perth, Western Australia). Many coastal segments around the world now host various types of artificial reefs.

Conclusions

The term *reef coast* refers to a complex of coastal types that range from rock reef coast to coral reef coast to artificial reef coast. Of these types, shallow-water coral reefs are the most well known, most widely studied, and most used for recreation and tourism. Rock reefs occur along most shores of coastal oceans, whereas coral reefs are restricted to ecologically favorable sites in tropical and subtropical regions. Artificial reef coasts are most common in warmer waters where they attract sports fishermen, snorkelers, and scuba divers.

Bibliography

Agassiz, A., 1898. A visit to the Great Barrier Reef of Australia in the streamer "Croydon". *Bulletin of the Museum of Comparative Zoology Harvard*, **28**, 95–148.

Andréfouët, S., and Cabioch, G., 2011. Barrier reef (ribbon reef). In Hopley, D. (ed.), *Encyclopedia of Modern Coral Reefs*. Dordrecht: Springer, pp. 102–107.

Banks, K. W., Riegel, B. M., Richards, V. P., Walker, B. K., Helmle, K. P., Joprdan, L. K. B., Phipps, J., Shivji, M. S., Spieler, R. E., and Dodge, R. E., 2008. The reef tract of continental southeast Florida (Miami-Dade, Broward and Palm Beach counties, USA). In Riegal, B. M., and Dodge, R. E. (eds.), *Coral Reefs of the USA*. Dordrecht: Springer, pp. 175–220.

Barnes, R. S. K., and Hughes, R. N., 1999. *An Introduction to Marine Ecology*. Oxford, UK: Blackwell Science, p. 296p.

Dana, J. D., 1872. *Corals and Coral Islands*. New York: Dodd and Mean, p. 406p.

Darwin, C., 1838. On certain areas of elevation and subsidence in the Pacific and Indian Oceans as deduced from the study of coral formations. *Proceedings of the Geological Society of London*, **2**, 552–554.

Darwin, C., 1842. *The Structure and Distribution of Coral Reefs*. London: Smith Elder, p. 214p.

Done, T., 2011. Coral reef, definition. In Hopley, D. (ed.), *Encyclopedia of Modern Coral Reefs*. Dordrecht: Springer, pp. 261–267.

Easton, W. H., and Ku, T. L., 1980. Holocene sea-level changes in Palau, western Caroline Islands. *Quaternary Research*, **14**, 199–209.

Fairbridge, R. W., 1950. Recent and Pleistocene coral reefs of Australia. *Journal of Geology*, **58**, 330–401.

Fairbridge, R. W., 1967. Coral reefs of the Australian region. In Jenings, J. N., and Mabbutt, J. A. (eds.), *Landform Studies from Australia and New Guinea*. Canberra: Australian National University Press, pp. 386–417.

Fairbridge, R. W., and Teichert, C., 1948. The low isles of the Great Barrier Reef: a new analysis. *Geographical Journal*, **3**, 67–88.

Finkl, C. W., 2004. Leaky valves in littoral sediment budgets: Loss of nearshore sand to deep offshore zones via chutes in barrier reef systems, southeast coast of Florida, USA. *Journal of Coastal Research*, **20**(2), 605–611.

Finkl, C. W., and Andrews, J. L., 2008. Shelf geomorphology along the southeast Florida Atlantic continental platform: Barrier coral reefs, nearshore bedrock, and morphosedimentary features. *Journal of Coastal Research*, **24**(4), 823–849.

Finkl, C. W., and Vollmer, H., 2011. Interpretation of bottom types from IKONOS satellite images of the southern Key West National Wildlife Refuge, Florida, USA. *Journal of Coastal Research*, Special Issue No. 64 (*Proceedings of the 11th International Coastal Symposium, Szczecin, Poland*), pp. 731–735.

Fox, W. T., 2005. Reefs, non-coral. In Schwartz, M. L. (ed.), *The Encyclopedia of Coastal Science*. Dordrecht: Springer, p. 795.

Gove, P. B. (ed.), 1971. *Webster's Third International Dictionary of English Language Unabridged*. Springfield, MA: G. & C. Merriam Company, 2662 p. (Updated 2002 version on the Internet).

Hopley, D., 1982. *Geomorphology of the Great Barrier Reef: Quaternary Development of Coral Reefs*. New York: Wiley Interscience, p. 453p.

Hopley, D. (ed.), 2011. *Encyclopedia of Modern Coral Reefs*. Dordrecht: Springer, p. 1205p.

Hopley, D., Smithers, S. G., and Parnell, K. E., 2007. *Geomorphology of the Great Barrier Reef: Development, Diversity and Change*. Cambridge: Cambridge University Press, p. 532p.

Hubbard, D. K., 1997. Reefs as dynamic systems. In Birkeland, C. (ed.), *Life and Death of Coral Reefs*. New York: Chapman & Hall, pp. 43–67.

Jackson, J. A., 1997. *Glossary of Geology*. Alexandria, VA: American Geological Institute, p. 769p.

Jukes, J. B., 1847. Narrative of the Surveying Voyage of H.M.S. Fly... During the Years 1842–1846. London: T. and W. Boone, Vol. I, p. 423.

KellerLynn, K., 2011. *Buck Island Reef National Monument: Geologic Resources Inventory Report*. Fort Collins, CO: National Park Service, National Resource Report NPS/NRSS/GRD/NRR-2011/462.

Kench, P. S., McLean, R. F., and Nichol, S. L., 2005. New model of reef-islands formation: Maldives. *Geology*, **33**, 145–148.

Kennedy, D. M., and Woodroffe, C. D., 2002. Fringing reef growth and morphology: a review. *Earth-Science Reviews*, **57**(3–4), 255–277.

Ladd, H. S., 1977. Types of coral reefs and their distribution. In Jones, O. A., and Endean, R. (eds.), *Biology and Geology of Coral Reefs, Vol. 4, Geology 2*. New York: Academic, pp. 1–19.

Lidz, B. H., Reich, C. D., Peterson, R. L., and Shinn, E. A., 2006. New maps, new information: coral reefs of the Florida Keys. *Journal of Coastal Research*, **22**, 260–282.

Maxwell, W. G. H., 1968. *Atlas of the Great Barrier Reef*. Amsterdam: Elsevier, p. 258p.

McLean, R., 2011. Atoll islands (motu). In Hopley, D. (ed.), *Encyclopedia of Modern Coral Reefs*. Dordrecht: Springer, pp. 47–51.

Milliman, J. D., 1973. Caribbean coral reefs. In Jones, O. A., and Endean, R. (eds.), *Biology and Geology of Coral Reefs, Vol. 1, Geology 1*. New York: Academic, pp. 1–50.

Montaggioni, L. F., Richard, G., Bourrouilh-Le Jan, F., Gabrié, C., HJumbert, L., Montforte, M., Naim, O., Payri, C., and Salvat, B., 985. Geology and marine biology of Makatea, an uplifted atoll, Tuamotu archipelago, central Pacific Ocean. *Journal of Coastal Research*, **1**, 165–171.

Nunn, P. D., 1994. *Oceanic Islands*. Oxford, UK: Blackwell, p. 413p.

Parnell, K. E., 2011. Fringing reef circulation. In Hopley, D. (ed.), *Encyclopedia of Modern Coral Reefs*. Dordrecht: Springer, pp. 427–430.

Riegl, B. M., and Dodge, R. E. (eds.), 2008. *Coral Reefs of the USA*. Dordrecht: Springer, p. 803p.

Sherman, R. L., Gilliam, D. S., and Spieler, R. E., 2002. Artificial reef design: void space, complexity, and attractants. *ICES Journal of Marine Science*, **59**, S196–S200.

Short, A. D., and Woodroffe, C. D., 2009. *The Coast of Australia*. Cambridge: Cambridge University Press. 288p.

Smithers, S., 2011. Fringing reefs. In Hopley, D. (ed.), *Encyclopedia of Modern Coral Reefs*. Dordrecht: Springer, pp. 430–446.

Smithers, S., and Hopley, D., 2011. Coral cay classification and evolution. In Hopley, D. (ed.), *Encyclopedia of Modern Coral Reefs*. Dordrecht: Springer, pp. 237–254.

Steers, J. A., 1929. The Queensland coast and the Great Barrier Reef. *Geographical Journal*, **74**(232–257), 341–370.

Steers, J. A., 1937. The coral islands and associated features of the Great Barrier Reef. *Geographical Journal*, **89**(1–28), 119–146.

Stoddart, D. R., 1965. The shape of atolls. *Marine Geology*, **3**, 369–383.

Stoddart, D. R., 1969a. Ecology and morphology of recent coral reefs. *Biological Reviews*, **44**, 433–498.

Stoddart, D. R., 1969b. Geomorphology of the Solomon Islands coral reefs. *Philosophical Transactions of the Royal Society of London, B*, **255**, 355–382.

Stoddart, D. R., and Steers, J. A., 1977. The nature and origin of coral reef islands. In Jones, O. A., and Endean, R. (eds.), *Biology and Geology of Coral Reefs, Volume IV, Geology 2*. London: Academic, pp. 59–105.

Wood, R. A., 1999. *Reef Evolution*. Oxford: Oxford University Press, p. 414p.

Woodroffe, C. D., 2002. *Coasts: Form, Process and Evolution*. Cambridge: Cambridge University Press, p. 623.

Woodroffe, C. D., and Biribo, N., 2011. Atolls. In Hopley, D. (ed.), *Encyclopedia of Modern Coral Reefs*. Dordrecht: Springer, pp. 51–71.

Woodroffe, C. D., and McLean, R., 1990. Microatolls and recent sea level change on atolls. *Nature*, **344**, 531–534.

Cross-references

Carbonate Factories
Coasts
Eustasy
Guyot, Atoll
Lagoons
Reefs (Biogenic)
Waves

REEFS (BIOGENIC)

Christian Dullo
GEOMAR Helmholtz Centre for Ocean Research, Kiel, Germany

Definition

Reefs are submarine, three-dimensional elevations on the shelf or on the continental slope formed by the activity of sessile organisms.

Introduction

Reefs are in situ organic deposits which exhibit different sizes from few cubic meters to several hundreds of kilometers length and even several hundreds of meters in thickness (Spalding et al., 2001). Modern reefs are formed predominantly by stony corals which are the equivalent of the taxonomic order Scleractinia. Corals and other calcifying organisms produce hard skeletons, which lead to the accumulation of biogenic carbonates as a result of individual growth, bioerosion, sedimentation, and cementation due to wave energy during several hundreds of years. On millennial and much longer timescales, sea-level changes are the major driving force of reef growth. Today coral reefs cover more than 284,000 km^2 (Spalding et al., 2001), and they are the largest marine structures on earth formed by biota, having a long geological record and evolution (Wood, 1999). At present, coral reefs suffer from overfishing, climate change (Wilkinson, 2004), and ocean acidification, which impacts on calcification and reef stability (Andersson and Gledhill, 2013). The classical coral reef system thrives in subtropical to tropical, warm, and sunlit waters, thereby forming wave-resistant structures. Stony corals do also exist in deep, dysphotic to aphotic depths where water temperatures are much lower where they can form three-dimensional structures of substantial size. These cold-water coral systems are referred to as coral mounds and mud mounds (Roberts et al., 2009).

Tropical reefs

Tropical shallow-water reefs are built by zooxanthellate scleractinian corals, which live in symbiosis with photosynthetic dinoflagellates. The lower temperature limit of tropical coral reefs is 20 °C, which corresponds to the tropical realm between 30°N and 30°S (Spalding et al., 2001). The most northern coral reefs occur in the Gulf of Aqaba (29°30′ N) and the southernmost ones around Lord Howe Island (31°30′ S). The photosynthetic activity of the symbionts further limits their depth range between the base of the photic zone and the surf zone of the intertidal environment. Since wave action influences such setting, shallow-water reefs exhibit a distinct zonal pattern of biota and sedimentary facies. The beach and shore facies is under the influence of the tides and shows accumulations of skeletal rubble and sand. Sea grass, sea urchins and other echinoderms, several calcified and noncalcified algae, and few soft corals may settle the lagoon. The back reef exhibits isolated scleractinians; some may form microatolls. The reef crest is a small and elongated area of intensive scleractinian coral growth where the waves brake and which may be subaerially exposed during nip tides. Together with the fore reef, the reef crest is characterized by the highest biodiversity of reef-building and reef-dwelling organisms, therefore representing the center of biogenic reef accretion (Figure 1). The fore-reef area, including the slope with the reef talus, faces the open ocean. Due to sea-level changes in course of the last glacials, the fore-reef slope may be structured in terraces witnessing ancient sea-level positions.

Reefs (Biogenic), Figure 1 View from the reef crest down to shallow fore reef. Fringing reef, Gulf of Aqaba, Red Sea Saudi Arabia (photo: C. Dullo).

Major reef types

Tropical coral reefs can be differentiated with respect to their setting and formation (Spalding et al., 2001). *Fringing reefs* follow parallel to the coastline and exhibit elongated depositional environments of up to several tenth of kilometers length but normally less than 100 m of width. They start to grow close to the coast as possible and migrate toward the open sea during time. Their lateral extension is controlled by the inclination of the underlying slope. Some older fringing reefs have lagoons developed. The highest variation of fringing reefs is seen in the Red Sea.

In contrast, *barrier reefs*, among which the Great Barrier Reef off Australia is one of the most prominent reefs of today, have very large lagoons of several kilometers in width and may cover several thousands of square kilometers. The reef front and crest has not migrated from the shore in offshore direction but started to grow on a preexisting relief and remained where it is. Major driving forces are sea-level changes and subsidence; therefore, barrier reefs have a much longer history than fringing reefs.

Atolls are mid-ocean reefs that enclose a central lagoon. Atolls started their evolution as a fringing reef on an oceanic volcano. Continued slow subsidence of the volcanic edifice forces the reef-building organisms to form barrier reefs and finally atolls, with or without a central island in the center. Due to the biogenic activity and in combination with sea-level variations, some of the atoll reef crests turn into "land." The most prominent atolls are found in Polynesia or in the Maldives.

Platform reefs grow on the open shelf, where the shallow bathymetry of the sea floor promotes reef growth. They expand in all directions and reach dimensions of several square kilometers. Due to coral growth and carbonate accumulation, some parts of the platform reefs may turn into land, forming small islands, which can be fringed by reefs themselves. Prominent representatives are found in the southern Red Sea and in the southern part of the Great Barrier Reef. A very old and extremely expanded platform reef is the Bahama platform.

Coral mounds and mud mounds

All non-zooxanthellate stony corals may live in much greater water depth under much cooler conditions (Roberts et al., 2009). Most of them are solitary corals, but few species form real reef structures. At present, the largest cold-water reef is the chain of biogenic structures of Europe, ranging from the Iberian Peninsula up to northern Norway (Figure 2). Distinct cold-water reef structures of living corals are up to 50 m high and up to 2 km long. However, some of these reefs have a long geological history, like their shallow-water counterparts. Such structures cover much larger areas of several square kilometers and form dome or mound-shaped elevations of more than 350 m above the surrounding seafloor. In contrast to shallow-water tropical reefs, they generally do not show any strong unidirectional zonation pattern of different depositional environments but concentrically shaped facies patterns. Coral mounds and mud mounds occur in different water depths ranging from 100 m down to 1,000 m. The habitat-forming corals act primarily as

Reefs (Biogenic), Figure 2 Living cold-water coral reef in the Stjernsund, Northern Norway, seen from the submersible JAGO (photo: GEOMAR).

sediment baffler rather than as frame builders like in the shallow water. Moreover, the nutrient level in these "mounds" is much higher than in the oligotrophic environments of tropical shallow-water reefs. That is why their occurrence is ascribed to the DDD environment (dark, deep, and dirty = lots of nutrients, Wilson, 1975).

Reef growth

The accumulation of carbonate and reef growth is controlled by four major processes: carbonate accretion, carbonate destruction, sedimentation, and cementation (Flügel, 2004). *Carbonate accretion* consists of the formation of a primary biogenic fabric being constructed by colonial and sessile organisms such as sponges, hydrozoans or scleractinian corals, and algae. This primary fabric is subsequently stabilized by secondary frame builders due to incrustations by bryozoans, benthic foraminifera, serpulids, sponges, and calcareous algae. Sediment fixation and stabilization play important roles as baffler and binder such as in sea grass meadows, crinoids, and cyanobacteria. *Carbonate destruction* is controlled by the physical processes of wave action and currents, frequently intensified by storms. This mechanism is accompanied by bioerosion which makes the primary and secondary fabrics prone to storm damage. Bioerosion takes place by boring, rasping, biting, and chemical leaching by algae, cyanobacteria, sponges, worms, mollusks, echinoids, and fish. Although destruction seems to have a negative impact on reef growth, it is essential and an important process to produce carbonate particles which fill the interstices between the biogenic fabrics. Therefore, *sedimentation* leads finally to the formation of a reef rock. In addition to particles derived from bioerosion, several other calcareous reef biotas contribute to the volume of carbonate by disintegration after death, like the green algae *Halimeda*. *Cementation* is the final stabilization by marine carbonate cements, which precipitate in intraparticle and interparticle porosities.

Reefs in the geological past

Both reef types are primarily made up of carbonate and carbonate-secreting biotas, among which corals, sponges, and algae are very prominent. Therefore, reef growth depends on calcification rates of reef biota (Dullo, 2005). During earth history, reefs played an essential role since almost all carbonates originated from biogenic activity (Wood, 1999). Most parts of the Northern Calcareous Alps, the Dolomites, and other mountain ranges containing thick carbonate sequences are formed by reefs and associated sediments. Due to the evolution of organisms, also the ecosystem reef exhibits dramatic evolutionary changes during earth history. The earliest reef assemblages were formed by stromatolites, a unique ecosystem consisting primarily of cyanobacteria and other microbes, which fixed and trapped sediment. During the Cambrian spongelike organisms, the archaeocyathids took over and built larger carbonate bodies. Paleozoic reefs were formed by tabulate "corals" and stromatopores, which belong to the large group of sponges. During the Devonian, rugose corals were essential until the end of the Permian. During the Carboniferous and mainly during the Permian, phylloid algae were important reef builders. Although many of the Paleozoic reefs look more like the modern cold-water reef bodies being either coral mounds or mud mounds, some Devonian (Canning Basin: Figure 3) or Permian reefs (Guadeloupe Mountains) exhibit already a modern type of facies arrangement. The onset of reef growth after the Permo-Triassic extinction event, which had led to a complete decline in reef formation, was characterized by the first occurrence of scleractinian corals. Since then, they are the major frame builders, except few interruptions during the Upper Cretaceous. Together with calcareous sponges, they formed large carbonate bodies in all Alpine mountain belts. Siliceous sponges acted during the Upper Jurassic as reef builders in some settings, while the Upper *Cretaceous* witnessed giant reefs formed by sessile mollusks, so-called rudists.

Reefs as reservoir rocks

Reef building irrespective of shallow-water reef or cold-water reef settings results in the accumulation of carbonate which exhibits distinct elevations with respect to the surrounding beds. Such structures are called bioherms in contrast to biostromes, which consist of calcified reef biota as well; however, biostromes are not elevated structures but rather thick non-stratified limestone beds. In both cases, the resulting rock is frequently not entirely cemented and filled with sediment. Therefore, a primary porosity may remain. In addition, reef biotas are predominantly formed by metastable carbonate minerals which are affected by diagenesis subsequently to their formation, resulting in

Reefs (Biogenic), Figure 3 The classic face of the Devonian reef outcropping in Windjana Gorge, Canning Basin, Australia. The inclined fore-reef beds, the massive reef limestone, and the well-bedded lagoonal strata are well exposed (photo: C. Dullo).

secondary porosity. Furthermore, dolomitization as a diagenetic process may also increase porosity. This makes carbonates important reservoir rocks for hydrocarbon accumulations at distinct times during earth history (Ahr, 2011). The majority of giant oil fields hosted in carbonates are of Permian, Jurassic, predominantly of Cretaceous, and Cenozoic ages. Fore-slope reef talus, reef debris, and reef crest sediments as well as in part lagoonal sediments are the preferred facies types to develop larger quantities of porosity to accumulate hydrocarbons. The most giant hydrocarbon field occurs in the Cretaceous rudist reefs of the Arabian Peninsula. Apart of hydrocarbons, lead-zinc accumulations are also frequently linked to carbonates and associated reef limestones which have been exposed subaerially and became karstified prior to the impregnation of lead and/or zinc.

Bibliography

Ahr, W. M., 2011. *Geology of Carbonate Reservoirs*. Hoboken: Wiley.
Andersson, A. J., and Gledhill, D., 2013. Ocean acidification and coral reefs: effects on breakdown, dissolution, and net ecosystem calcification. *Annual Review of Marine Science*, **5**, 321–348.
Dullo, W. C., 2005. Coral growth and reef growth: a brief review. *Facies*, **51**, 33–48.
Flügel, E., 2004. *Microfacies of Carbonate Rocks: Analysis, Interpretation and Application*. New York: Springer.
Roberts, J. M., Wheeler, A. J., Freiwald, A., and Cairns, S. D., 2009. *Cold-Water Corals – The Biology and Geology of Deep-Sea Coral Habitats*. Cambridge, UK: Cambridge University Press.
Spalding, M. D., Green, E. P., and Ravilious, C., 2001. *World Atlas of Coral Reefs*. Berkeley: University of California Press.
Wilkinson, C., 2004. *Status of Coral Reefs of the World*. Townsville: Australian Institute of Marine Science.
Wilson, J. L., 1975. *Carbonate Facies in Geologic History*. New York: Springer.
Wood, R., 1999. *Reef Evolution*. Oxford: Oxford University Press.

Cross-references

Carbonate Factories
Coasts
Eustasy
Guyot, Atoll
Lagoons
Reef Coasts

REFLECTION/REFRACTION SEISMOLOGY

Christian Hübscher[1] and Karsten Gohl[2]
[1]Institute of Geophysics, Center for Earth System Research and Sustainability, University of Hamburg, Hamburg, Germany
[2]Alfred Wegener Institute, Helmholtz-Centre for Polar and Marine Research, Bremerhaven, Germany

Synonyms

Controlled-source seismology; Multichannel reflection seismics; Wide-angle reflection/refraction profiling (WARRP)

Definition

Methods to image and to physically parameterize the subsurface geologic structures and conditions by means of artificially generated shock waves that travel through the subsurface and return to the surface.

Overview

Controlled-source seismology comprises a variety of geophysical methods to image and to physically parameterize the subsurface geologic structures and conditions. All these methods base upon the principle that artificially generated shock waves travel through the subsurface and that the returned signal can be analyzed regarding the properties of the subsurface. During the last decades the discrepancies between marine seismic equipment as used by the academic marine research community and that used by the hydrocarbon (HC) exploration industry increased significantly, caused by different research targets and the simple fact that gear commonly used for HC exploration is much too costly. In this entry we will focus on those systems which are typical for academic marine research.

In principle, seismic data acquisition requires an energy source, a receiver, and a recording system. The two most important seismic methods are reflection and refraction seismology (Figure 1).

Reflection seismologists deal mainly with steep angle reflections, which means that the source to receiver distance is small compared to the target depth. This method utilizes the fact that a small part of the down-going energy is reflected on geological layer boundaries. The main fraction of the energy is transmitted and travels deeper where reflections occur at the next layer boundary and so on. This method results in a good vertical and horizontal structural resolution of the subsurface. Earth scientists benefit from the technical and methodical developments of the exploration industry which uses reflection seismology for the detection of oil and gas in depths of up to several kilometers below the seafloor.

In refraction seismology, seismic waves are recorded that propagate along layer boundaries or as arcuate "diving waves" mainly subhorizontally. This method is either used in engineering geology for near-surface investigations or (the other extreme) to analyze deep crustal structures, the crust–mantle boundary and the upper mantle. Wide-angle reflections recorded at large distances between source and receivers are part of this data analysis scheme. Geophysicists often use the abbreviation WARRP (wide-angle reflection/refraction profiling) for these techniques. The main advantage of WARRP is that the inversion procedure directly results in crustal depth sections; its major disadvantage is the relatively low structural resolution.

History of refraction and reflection seismology

One of the founders of the seismic refraction method was German scientist Ludger Mintrop (1880–1956) who received a patent in 1917 for a so-called portable field seismograph and a method to locate artificial shock sources. In fact, he used this method in World War I to locate the position of Allied heavy artillery pieces. After the war he reversed this method: He released an explosive blast in a known distance to the seismograph and used the underground travel time of the shock wave to calculate the depth of geological features. In 1919 the seismic refraction method was successfully used to constrain the location of a salt dome in northern Germany.

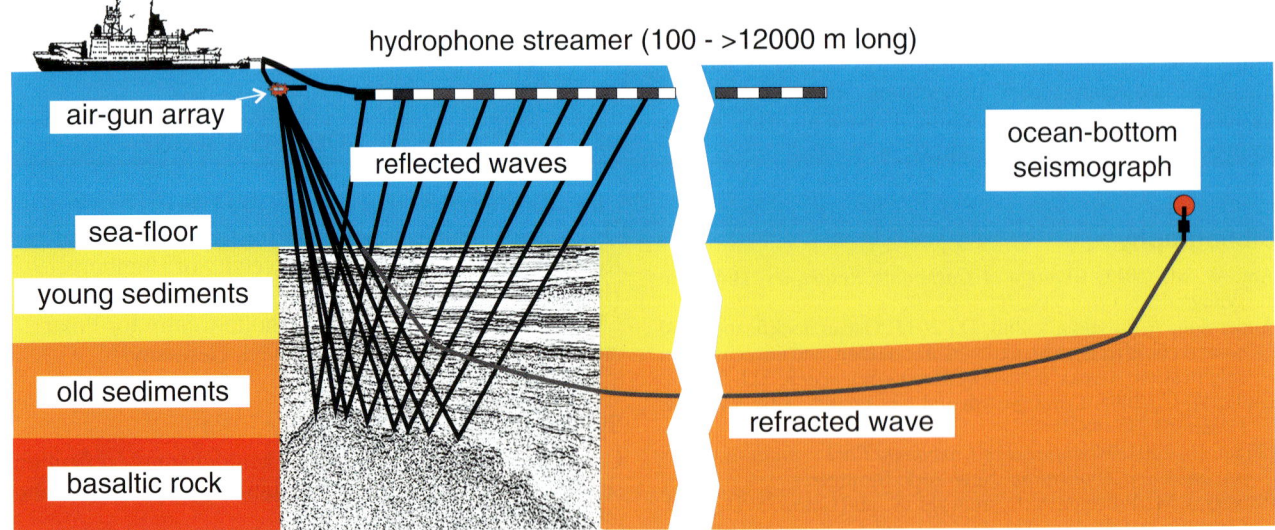

Reflection/Refraction Seismology, Figure 1 Basic principles of reflection and refraction seismology. Reflection seismologists deal mainly with steep angle reflections, which means that the source to receiver distance is small compared to the target depth. Refracted waves propagate along layer boundaries or as arcuate "diving waves" mainly horizontally. This method is either used in engineering geology for near-surface investigations or (the other extreme) to analyze deep crustal structures, the earth crust–mantle boundary and the upper mantle.

Marine reflection seismology evolved from work by Reginald Fessenden who developed a sonic sounder to find icebergs (after the sinking of the *Titanic*) (Roden, 2005). The first reported usage of seismic reflections for subsurface studies was in 1921 when a small team of geophysicist and geologists performed a historical experiment in Oklahoma (Dragoset, 2005). The team analyzed the return time of reflected waves generated by small dynamite charges and detected the boundary between two subsurface geological layers. The success of this so-called Vines Branch experiment stimulated a boom in seismic exploration when the oil price rose in 1929.

During the first decades each subsurface point was imaged by a single set of source and receiver. Multichannel seismics became more common in the 1950s. This method allowed for significant noise reduction and for computing of layer velocities, enabling the conversion of travel times to depth. The advent of the digital technology in the 1960s represented a giant step forward because digital post-processing opened a wide variety of possibility to enhance seismic data and image quality. Several technical developments in the 1980s such as plotters for wiggle trace display or color maps as well as methodical progress including seismic stratigraphy or seismic attribute analysis became available for earth scientists and stimulated academic frontier research.

The hydrocarbon industry benefited largely from the introduction of the 3D seismic method which enhanced understanding of petroleum reservoirs and reduced exploration risks. However, the earth science community has had quite limited access to these data, also because 3D seismic experiments for nonprofit-oriented research are generally too costly.

Seismic energy sources

The theoretical optimum seismic source produces a Dirac delta function like signal (called wavelet) which is infinitesimal short in time by having a maximum spectral bandwidth. In the real world, a "good" seismic source creates a high energy and temporally short signal with repeatable characteristics in terms of wavelet signature and frequency content.

Marine seismic sources are subdivided into exploders and imploders. Exploders create the wavelet by an expanding volume (solid material or gas/air). Imploders generate the signal by the collapse of a volume, created for instance by cavitation. We can further distinguish between chemical, mechanical, electrical, and pneumatic/hydraulic sources (Parkes and Hatton, 1986). Chemical sources comprise various explosives. Explosives can create very strong signals; however, the shot interval is relatively low and the signal characteristics are reproducible to a limited amount due to surface waves and explosion depth. Mechanical sources, such as a boomer or piezoelectric transducers, rely on physical vibration to transmit seismic waves into the water. Sparkers are electrical sources that discharge large electrical charges into the water. The steam bubble that vaporizes along the spark collapses, thus producing the desired signal. Pneumatic/hydraulic sources dispel gas (usually air) or fluid (usually water) under high pressure into the water. For instance, a water-ejecting water gun produces a cavitation and its subsequent collapse generates a seismic wave-field. The water is accelerated in the water gun by a piston, whereas the driving pressure consists usually of compressed air.

The pneumatic so-called air gun is nowadays the most common source in marine seismics (Dragoset, 1990). Compressed air of usually 140–210 bar is released into the water every few seconds. The rapid expansion creates the highly reproducible primary signal (Figure 2a). The reflection at the sea surface creates a similar strong but phase-reversed signal, the so-called ghost. Signal strength is described either as the peak amplitude pk (level A in Figure 2a) or as pk-pk amplitude which includes the amplitude of the ghost arrival (A–B in Figure 2a), both traditionally expressed in bar-meter (bar-m). If the peak amplitude is 1.5 bar-m as in Figure 2a, a hydrophone located in a distance of 1 m to the source would measure peak amplitude of 1.5 bar, in 3 m distance 0.5 bar. A second important source characteristic is the spectral bandwidth (Figure 2b). Amplitude spectra are usually shown on a logarithmic scale. A reasonable parameter for the bandwidth is the range between the $-3db$ (71 % of maximum) or $-6db$ (50 % of maximum) points. The frequency notch in the spectra results from the destructive interference between the primary and the surface (ghost) reflection. It is a function of the source towing depth and can be easily calculated by $f_{Notch} = V_{H_2O}/(2 * \text{towing depth})$. V_{H_2O} is the seismic wave speed in water which amounts to ca. 1,500 m/s.

The bubble signal is an undesired effect which results from an oscillation of the air bubble. Consequently, an important characteristic of a single air gun or an array of air guns is the peak to bubble (amplitude) ratio (PBR) (Figure 2a). There are several possibilities to enhance the PBR. As the bubble period depends on the same parameters as the primary signal, the bubble can be suppressed by simultaneously firing several air guns with different bubble frequencies. The PBR can be further enhanced if the separation distance between the air guns is small enough that the air bubbles interact (Strandenes and Vaage, 1992), which is the so-called cluster effect. Many modern air gun arrays consist of several pairs of clustered air guns. A high-end system frequently used by geophysicists is the so-called GI-Gun (by Sercel), which consists of two air gun chambers built into a single housing. The first chamber, called generator, releases highly compressed air. The air bubble expands, and just before it starts to oscillate, an additional volume is injected by the second air gun chamber (the injector), thus stabilizing the air volume in order to prevent oscillations.

Amplitude and spectral bandwidth (Figure 2b) can be adjusted by air pressure, air chamber volume, towing depth, and geometry of discharge ports. Air chamber

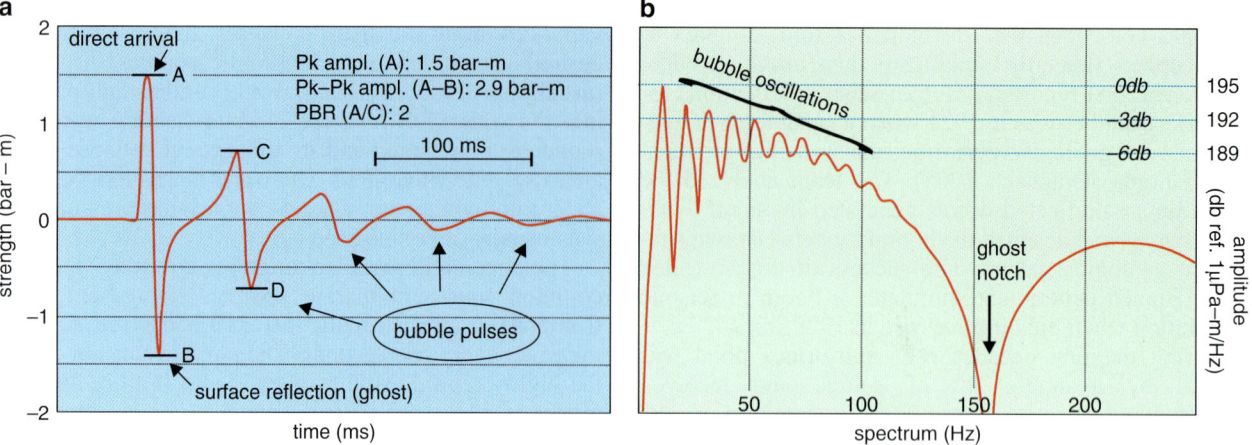

Reflection/Refraction Seismology, Figure 2 (**a**) Definitions of time-domain air gun specifications (after Dragoset, 1990, 2005). The signal is composed of the primary signal, the phase-reversed sea surface reflection (ghost), and the bubble signal, caused by oscillations of the air bulb. Signal strength is measured in bar-m. Values are either given for the strength of the peak amplitude of the primary signal (pk amplitude. Marked by an "A") or the ghost amplitude marked by a "B" is included (pk-pk amplitude). (**b**) The spectral bandwidth is usually shown on a logarithmic scale. A reasonable parameter to characterize the bandwidth is the range between the −3db (71 % of maximum; ca. 10–52 Hz) and −6db (50 % of maximum; 8–80 Hz) points. The bubble oscillations are visible in the low-frequency domain. The ghost signal forms a notch at a frequency depending on the source depth due to destructive interferences.

volumes may vary between 32 l to less than 1 l. Large air chambers are used for low-frequency, high-amplitude sources which are desirable for deep crustal refraction and wide-angle reflection seismics. Small air guns of up to a few liters chamber volumes are used for high-resolution shallow reflection seismics.

Data acquisition and processing: reflection seismology

A seismic recording system comprises a sensor or receiver that converts the seismic wave into an electrical signal, an analog–digital converter and a recorder. The most common sensors in marine seismics are hydrophones, which are piezoelectric elements that produce an electric potential difference caused by the pressure pulses of the seismic wave. Usually groups of hydrophones are used to eliminate translational accelerations and to enlarge the signal-to-noise ratio by signal summation (Telford et al., 1990).

Seismic streamers

The standard device for marine reflection seismics is the streamer (also towed array, seismic cable) which comprises hydrophone groups (so-called channels) or single hydrophones (Figure 1). There is a huge variety in the length and number of channels. A streamer geometry quite common for systems operated by academic research institutes comprises a lead-in cable (also tow-lead) which is a heavy, steel mash protected cable with negative buoyancy which connects the streamer with the streamer winch (Figure 3). The next part of the streamer is the elastic stretch section (also passive or compliant section) which mechanically decouples irregular ship accelerations from the active sections, consisting of the hydrophone channels. Another stretch section decouples the streamer from the tail buoy.

The length of the individual sections may vary. Typical values for systems operated by academic institutions are front stretch 50–200 m, active length 300–3,000 m, and end stretch 50–100 m. Depth levelers, so-called birds, are connected every 100–200 m.

Typical hydrophone group distances are integer multiples of 6.25 m. This number results from times when no satellite positioning was available. A good trade-off between water current-induced noise and progress is a ship speed of about 5 kn (5 nautical miles per hour). This speed corresponds to 2.5 m/s. A shot interval of 5 s results in 12.5 m shot distance, an interval of 10 s in 25 m. With those figures the geometry of the multichannel data could be easily calculated (see below). The number of hydrophones within and the total length of a hydrophone group control its spatial response characteristic (Hübscher and Spieß, 1997).

The towing depth of the streamer should be a quarter of the wavelength ($\lambda/4$) of the expected seismic signal in order to prevent destructive interferences within the frequency range of the source. Buoyancy of a streamer is controlled both by the average density of the stretch and active sections and by cable levelers. For several decades, the sections have been filled by kerosene and later by Isopar. These fluids have a density less than that of seawater. The volume of the oil is calculated in a way that the average density of the section containing hydrophones and towing and data cables is close to that of seawater to yield neutral buoyancy. So-called solid streamers filled with polyurethane and similar solid materials have

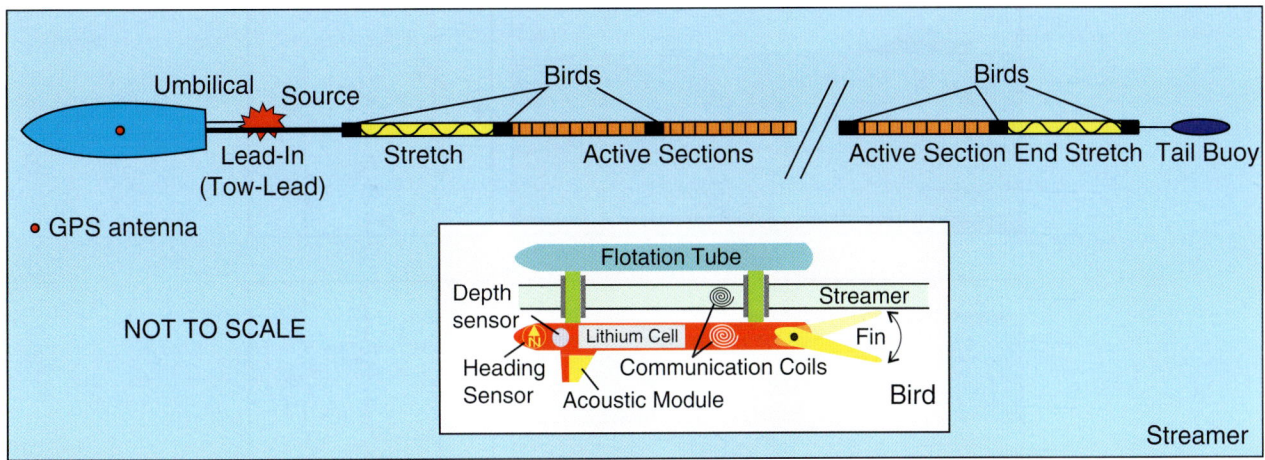

Reflection/Refraction Seismology, Figure 3 A marine seismic streamer typically used for academic 2D reflection seismic surveys comprises a lead-in cable, a stretch section that mechanically decouples ship movements from the active sections, an end stretch, and a tail buoy. The streamer is depth controlled by cable levelers also called birds (insert).

replaced more and more the conventional oil-filled ones, because they have proven to be more robust. Birds allow for an active control of the streamer depth (insert Figure 3). The bird is attached to the streamer by bird collars and can rotate freely around. Depth sensors determine the actual depth. Inductive communication coils within birds and streamer allow data transmitting to the shipboard controller system where depth values are averaged depending on the swell. If depth correction is necessary, the controller sends a signal to the birds, which turns the fins accordingly, and the streamer segment is dragged up or down. Heading sensors (fluxgate magnetometers or mechanical compass) allow for geometric corrections if the streamer drifts off the track by currents. On the opposite side of the bird, a flotation tube is also attached to the collars. The flotation tube eliminates the negative buoyancy of the bird and keeps the bird always beneath the streamer and, consequently, the fins parallel to the sea surface. For 3D applications the precise position of each streamer segment with respect to the air guns is determined by acoustic triangulation. This requires acoustic senders (so-called pingers), e.g., close to the air guns, and acoustic modules (Figure 3) in the birds which receive the pings and send travel times to a control PC.

Seismographs

The term "analog acquisition" refers to a receiver–recorder system where the electrical signal is digitized by a seismic recording unit, which is a data recorder specially designed for seismic data recording, located in the vessel. The seismic recorder fulfills several tasks. Prior to analog–digital conversion, data is usually preamplified and low-pass filtered in order to avoid spectral aliasing. Since the suppression of frequencies becomes significant at about 60–70 % of the Nyquist frequency, the A/D conversion frequency should be about three times the maximum signal frequency. High-pass filtering can be wise to apply in order to suppress high-amplitude ship propeller noise. After A/D conversion, data are normally formatted to standard formats developed by the Society of Exploration Geophysicists (SEG), e.g., the SEG-D or SEG-Y formats. Analog data transfer via cables results in some loss in data quality, e.g., due to the antenna effects. This disadvantage is overcome by digital systems which digitize the electric signal representing the incoming seismic wave inside the streamer. More expensive digital receivers are standard in exploration geophysics, while the scientific community still operates many analog systems. The disadvantage is less significant for the often near-surface applications.

Acquisition schemes

The classical acquisition scheme for academic marine reflection seismic surveys is the 2D geometry. The first dimension is the 1D acquisition geometry and the second one the travel time of the seismic waves. This acquisition scheme is designed for common midpoint (CMP)-based processing (Figure 4). If a shot is released, the seismic signal travels from the shot point down to the seafloor. Assuming a horizontal seafloor, the reflection points have a distance half of the hydrophone group spacing $\Delta H/2$ (Figure 4). Let us focus on the reflection point of the ray traveling from the shot point down to the seafloor and up to the nearest hydrophone group (Figure 4a). If the shot spacing is also $\Delta H/2$, the same reflection point on the seafloor is covered by the ray between the second shot point and the second channel (Figure 4b), but the travel time is longer and so is the travel time of the wavelet between shot 3 and channel 3 (Figure 4c). The effect that travel time increases with offset is called normal move-out (NMO). During data processing, the individual shot records are sorted to CMP gathers which comprise all those records which have the common midpoint between shot and

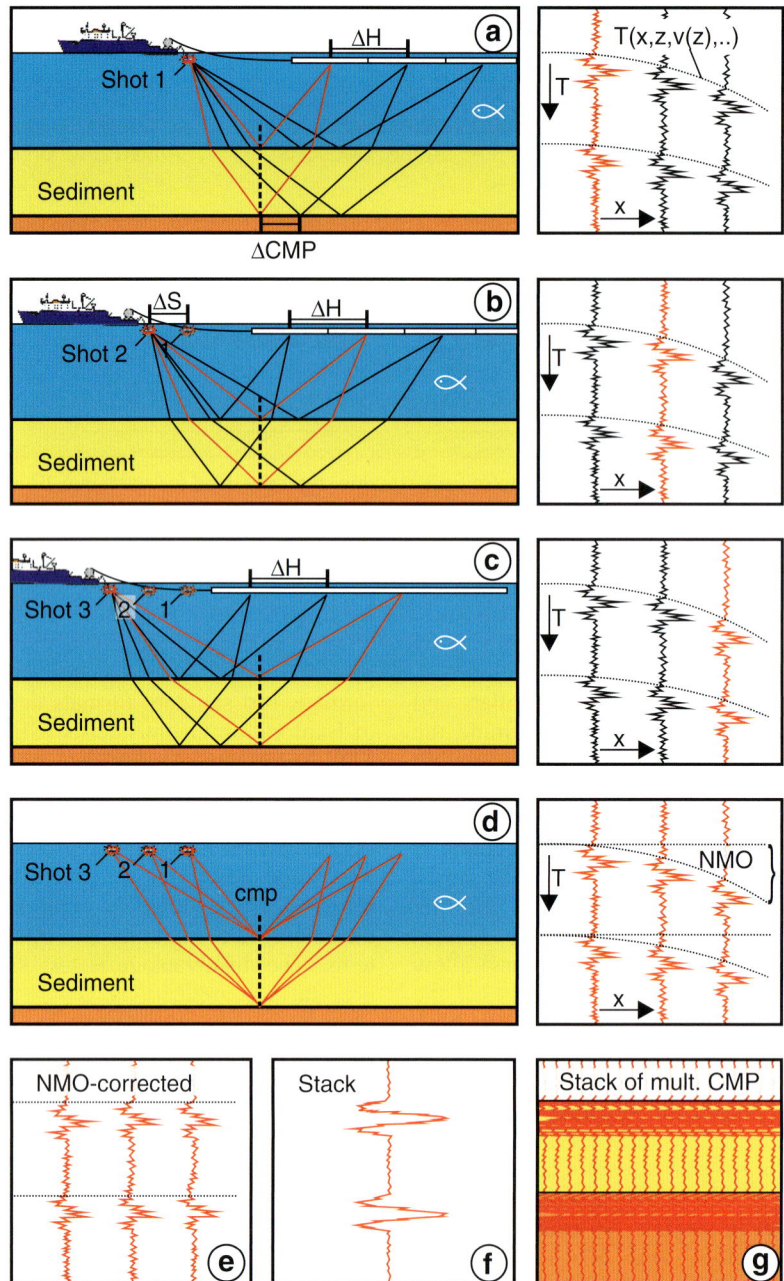

Reflection/Refraction Seismology, Figure 4 CMP-based 2D multichannel seismic acquisition scheme (after Mutter, 1987). See text for detailed description.

receiver coordinates (Figure 4d). Analysis of the offset-dependent travel time discrepancies allow – if offset is not much less than reflector depth – calculating interval velocities and reflector depth. During data processing (see below), recordings of each CMP gather (Figure 4e) are NMO corrected and stacked (summed to one single trace) which results in a constructive superposition of primary information (reflections) and destructive superposition of (incoherent) noise (Figure 4f). Displaying of all CMP traces along a profile (Figure 4g) gives an image of subsurface strata.

In the exploration industry, 2.5D, 3D, and 4D data acquisition schemes are more common. Sequential acquisition of parallel and fairly narrow-spaced 2D lines results in a 3D subsurface model. However, this is called 2.5D seismics because azimuth-dependent factors are not considered. In a strict sense, the term 3D seismics refers to acquisition geometry only if seismic waves are recorded by a 2D pattern of receivers; thus, the seismic waves are measured with different azimuth angles. Repeated measurements within the same area with a 3D geometry are called time-lapse monitoring or 4D seismics, enabling the interpreter concluding on subseafloor fluid dynamics. Recently a highly mobile 3D seismic system has been introduced which allows imaging the upper 1–2 km with high lateral and vertical resolution. A bundle of 12 short streamers containing 8–16 channels is towed behind the survey vessel (e.g., Berndt et al., 2012).

Data processing

Seismic data processing comprises a vast of different methods for the analysis of recorded seismic signals to reduce or eliminate unwanted components (noise), to create an image of the subsurface to enable geological interpretation, and eventually to obtain an estimate of the distribution of physical material properties in the subsurface (inversion) (Chowdhury, 2011). Processing is a science by its own and it is impossible to give a complete overview here (for a comprehensive text book, see Yilmaz, 2008).

One of the primary goals is the enhancement of the seismic *signal* and the suppression of *noise*. The term signal refers commonly to primary reflections, which is seismic energy reflected only once during its path from source to receiver. Everything else can be considered as noise; it comprises random and coherent noise (Kumar and Ahmed, 2011). Random noise does not correlate with neighboring channels. Coherent noise includes multiple reflected energy, side wipes (out-of-plane reflections), and energy from previous shots or processing artifacts. Ambient noise means noise resulting from other sources such as waves, swell, ship propeller, or marine animals. Noise suppression is easy if its frequency lies outside the signal frequency range; otherwise more sophisticated methods have to be applied.

A fundamental parameter which is crucial to know for several processing steps is the distribution of the seismic wave speed, called "velocity." For example, the velocity controls the propagation of the seismic wave, its reflection, refraction, and diffraction as well as the geometrical spreading of the wave front and the related amplitude loss (spherical divergence). Vice versa, seismic velocities are crucial to correct for amplitude losses, travel time differences within a CMP gather, and dislocated reflections and to bring the seismic energy smeared along diffraction hyperbola into focus (called migration). The conversion from a seismic time section to a depth section can be only correct if the velocity distribution is well known. Further, velocity functions can be used to remove multiple reflections. Basically, the velocity distribution can be calculated from the move-out in CMP gathers. The bigger the move-out, the more accurate are the derived velocities. If the reflector depth is much higher than the maximum offset (streamer length), the move-out becomes neglectable and no reliable velocity calculation is possible. For example, if a 600 m streamer is used – a quite typical value for streamers operated by academic institutions – accurate velocities beneath depths of 1,000 m can hardly be derived. The necessary computation of velocity is a relatively easy task if the subsurface is a horizontal layer cake, but geological formations are often not formed in this simple matter. It becomes more difficult if strata are tilted, deformed, disrupted, and anisotropic, all resulting in lateral velocity changes.

The supreme discipline in seismic data processing is seismic migration (Yilmaz, 2008; Gray, 2011). It comprises a set of techniques for transforming reflection seismic data into an image of reflecting boundaries in the subsurface. It brings the seismic energy distributed along a diffraction hyperbola into focus and corrects geometric effects caused by dipping reflectors and velocity effects. If the velocity model is well known, pre-stack migration gives best results; however, it is quite costly in terms of CPU time. Migration can result either in time or depth sections. Pre-stack depth migration (PSDM) is the ultimate goal of seismic data processing.

A basic processing sequence for marine seismic data includes the following steps:

- Preprocessing: Set amplitudes of bad traces to zero, correct misfires, assign geometry (CMP numbers) to traces. Suppress noise outside signal frequencies by band-pass filtering.
- Pre-stack processing I: Mute direct wave; balance energy between individual channels.
- Velocity analysis: Sort shot gathers to CMP gathers; calculate velocity model.
- Pre-stack processing II (based on velocity model): Correct amplitudes for spherical spreading, absorption, and energy partitioning at interfaces; suppress multiples with the help of velocity model.
- Pre-stack time or depth migration (optional).
- Stacking: Flatten reflection events by NMO correction using velocity model within CMP gather; stack all traces.
- Post-stack time or depth migration (optional).

Data acquisition: refraction seismology

Modern ocean-bottom seismometers (OBS) or hydrophones (OBH) are autonomous receiver and recording systems which are deployed on the seafloor where they remain for days or even several months and where they record seismic profiles during this time. The decoupling between seismic source and receiver allows for great distances which is necessary to receive refracted waves from the deep crust and upper mantle. These systems consist of a data logger, batteries, a three-component seismometer and/or a hydrophone, a radio beacon, a xenon flashlight, and a flag. These components are either mounted on a steel frame, which is kept afloat by syntactic foam bodies, or are installed within and on top of a floatable glass sphere. Either system type is connected to an anchor frame via an acoustic releaser. On hydroacoustic signaling, the releaser opens a hook to release the OBS/OBH system from its anchor and the system ascends to the sea surface. Recording parameters such as sample rate (e.g., 200 Hz) and gain are set before deployment, and data are retrieved from the data logger after recovery. Most commonly the data logger is programmed such that it records continuously.

Data processing and signal enhancement

Data processing of standard seismic refraction records is – compared to that of the seismic reflection technique – relatively straightforward. As a major component of the record analysis and modeling is focused on extracting seismic velocities from the relationship of seismic travel times to source-receiver offsets, the accurate determination of source and receiver coordinates is of great importance. While this is a trivial problem to be solved in land surveys, the exact location of an ocean-bottom seismometer, which may have drifted on its sinking path to the seafloor, can in most cases only be estimated by using the direct acoustic water wave phase in the record. After this so-called re-localization process of each OBS, the remaining data processing includes mainly band-pass filtering for enhancement of the signals which have their largest amplitudes between 5 and 12 Hz in deep crustal seismics. Deconvolution filters can be applied in some instances to improve the identification of the time onset of particular refraction phases. Some workers use coherence filters in order to enhance seismic phases of a particular dip range or to suppress the so-called wraparound, which is the direct acoustic wave-field through the water emitted from the previous shot. As this direct wave travel with only 1.5 km/s water velocity, it is still recorded at large offsets and is sometimes superposed on reflected or refracted phases from the deep crust or upper mantle. Most commonly, seismic refraction records are displayed with travel times reduced (Figure 5) in order to include all refraction and wide-angle reflection phases of interest into a reasonably sized display window and to provide the viewer with a visual impression of the velocities of the main refraction phases. For instance, a record displayed with travel time T reduced with a velocity V_{red} of 8.0 km/s ($T_{red} = T - X/V_{red}$, where X is offset) shows refractions with an apparent P-wave velocity of 8.0 km/s as horizontal phases.

The processing flow is similar for the four components (hydrophone h, vertical z, horizontal x, horizontal y) of an OBS record or the three components of a land instrument record. When searching for S-waves or converted wave phases (P-to-S or S-to-P), records are displayed with the corresponding main reduction velocity applied to travel time.

Modeling, inversion, and tomography

Various modeling strategies are exercised to derive physical properties of the subsurface, the crust and/or the upper mantle from the seismic refraction, and wide-angle

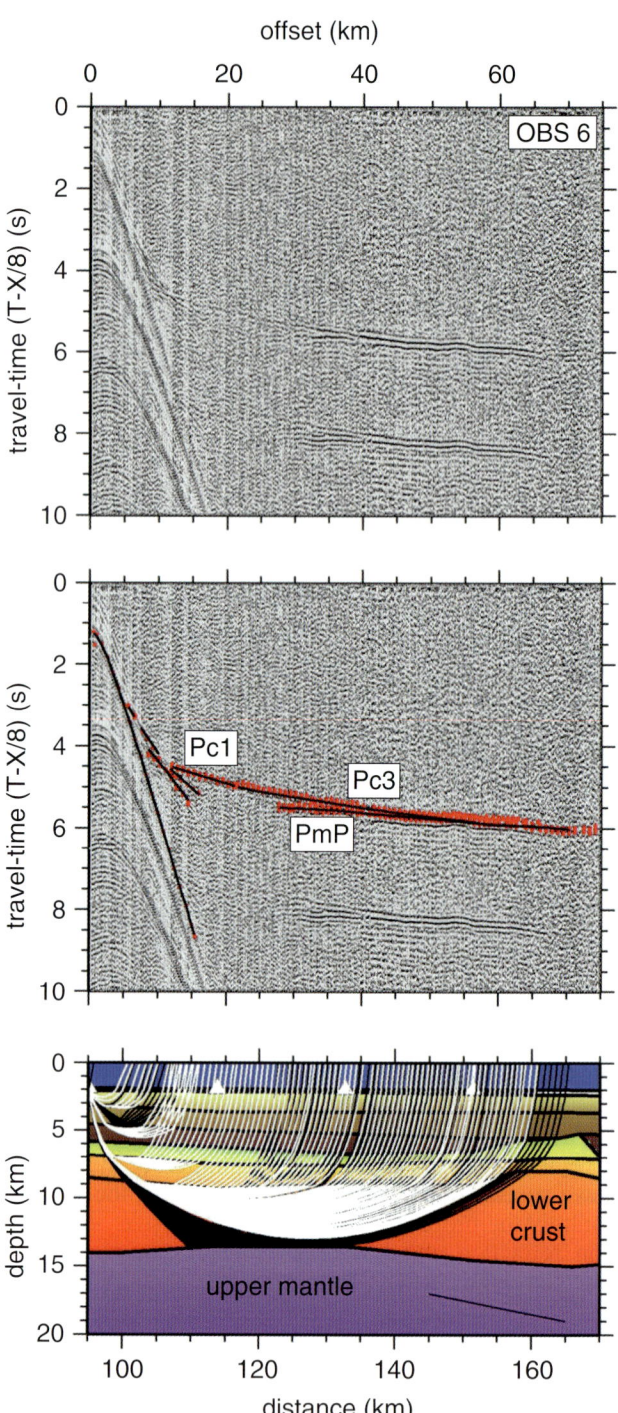

Reflection/Refraction Seismology, Figure 5 Example of a seismic refraction OBS record with travel times reduced with 8 km/s (*top*), picked and modeled refraction phases (*middle*), and a crustal cross-section model with traced rays corresponding to the picked refraction phases (*bottom*) (from Suckro et al., 2012). Pc1 and Pc3 are refraction phases from the middle to lower crust, while PmP denotes a wide-angle reflection from the crust–mantle boundary (Moho discontinuity).

reflection records. Deriving the seismic velocity–depth distribution along a surveyed 2D profile or in a 3D area is the most widely desired aim for modeling because of the strength of seismic refraction technique to provide firsthand information on P- and S-wave velocities of the layers penetrated. While basic geometric methods such as the plus-minus method, the intercept-time method, the delay-time concept, and reciprocal methods (e.g., Palmer, 1986) are sufficient to represent simple 2D subsurface geometries, even with irregular interfaces, as often approximated for the shallow subsurface, deep crustal complex layer structures and heterogeneous velocity–depth distributions can only be modeled with more sophisticated numerical methods using a ray-geometric or elastic wave-field inversion approaches. A short overview of commonly used modeling schemes is given here.

Forward ray-tracing and travel time inversion: Based on acoustic wave propagation laws, software for ray-tracing in complex multilayered media was first developed in the 1970s (e.g., Cerveny et al., 1977). The method of ray-tracing is performed by an efficient numerical solution of the 2D ray-tracing equations coupled with an automatic determination of ray takeoff angles. The model response is iteratively optimized by approaching a best fit to the observed travel times. This ray-tracing method was further optimized by Zelt and and Smith (1992) who implemented a travel time inversion by a best-fit approximation with a damped least-square algorithm. Due to its nonlinear nature, the inversion process is controlled by strict boundary conditions on travel time picking uncertainties as well as model parameterization and resolution bounds. Due to its efficiency, the open-source software RAYINVR, originally developed by Zelt and and Smith (1992), is still the most widely used 2D ray-tracing scheme in academic research to date (Figure 5), although different modifications exist, mainly improving model parameterization and visualization and adding more features (Zelt, 1999). The software was further developed to allow modeling in 3D subsurface media (Zelt and and Barton, 1998). Ray-tracing can also be used to calculate synthetic amplitudes that can be compared with the observed ones. However, this has limitations as it does not take into account the full elastic properties of the subsurface layers and interfaces.

Travel time tomography: A 2D and 3D travel time inversion by using an iterative first-arrival seismic tomographic (FAST) algorithm, in which a regularized inversion routine is computed rapidly over a regular grid using finite-difference extrapolation (Vidale, 1990), was developed by Zelt and and Barton (1998). This routine was originally developed for mid- and deep crustal seismic refraction tomography (e.g., Zelt et al., 2004; Stankiewicz et al., 2008) but has been utilized for shallow subsurface imaging by appropriately adjusting model parameters (Deen and and Gohl, 2002). As required for a linearized inversion, an initial model has to be implemented first. From this initial velocity–depth model, a series of iterations improves the model with respect to the picked travel times. Each subsequent iteration of the model is developed using both a finite-difference forward calculation of travel times and ray-paths and a regularized inversion incorporating a combination of smallest, flattest, and smoothest perturbation constraints, the weights of each being allowed to vary with depth. An advanced tomographic inversion based on a Monte Carlo scheme with an implementation of automated model regularization and a polynomial expansion of the probability function has been developed by Korenaga and and Sager (2012).

Wave-field inversion: One of the first widely used wave-field inversion techniques was developed by Pratt et al. (1996) and is based on finite-difference modeling of the wave equation, thus allowing very general wave types to be incorporated and enhancing the resolution when compared to travel time methods. Their method operates in the frequency domain and allows modeling and inverting velocity structure as well as inelastic attenuation factors. The inversion can be iterated to improve the data fit and to take account of some nonlinearity. Some of the nonlinearity can be dealt with by using starting models that have been developed with standard forward ray-tracing. Despite advances in computational speed, full wave-field inversion techniques that incorporate (visco-)elastic and anisotropic wave equations on full crustal model scales still remain in a small user niche as parameterization is cumbersome and computational requirements are not always met. However, it is a matter of time until this method will be taken to the next level of being accepted for more regular use in the academic research and exploration communities.

Seismic interpretation

As seismic methods can be considered as a way of remote sensing into the earth, all interpretations can be significantly corroborated by ground truthing, meaning to calibrate the interpretation by drilling results, which includes both coring and well logging. The understanding of large-scale crustal structures can be significantly supported by modeling and inversion of gravity and magnetic data.

The interpretation of reflection seismic data consists of two major steps. During the first one data are described by an association-free nomenclature in order to prevent a biased point of view. For example, "disrupted horizons" is descriptive; the denotation "normal fault" is a possible interpretation of this pattern. From that objective description of the data, conclusions are drawn regarding tectonic and sedimentary processes.

The seismic interpreter should start with the evaluation of recording and processing parameters. From this he/she will get a first idea about the reliability of his observations. The description of reflection seismic images should follow the following scheme:

1. Identification and mapping of unconformities and correlated conformities

2. Determination of termination style (e.g., onlap, toplap, downlap, offlap, etc.)
3. Classification of internal reflection pattern (e.g., parallel, wavy, divergent, etc.)
4. Recognition of clinoforms, structural highs and lows
5. Categorization of seismic attributes (interval velocity, instantaneous frequency, etc.)

From the yet described data, the interpreter derives the processes which led to the observed features. Both the inductive and deductive approaches are used. The consensus about earth processes is used for deductive reasoning, which means propositions generally accepted by the scientific community are used for explaining specific observations. New general propositions about earth processes evolve from inductive reasoning by arguing that a specific example is representative for other cases. A "good" interpretation explains a maximum of observations and makes minimum assumptions about unknowns.

In all cases the interpreter should have a sound knowledge about both geology and geophysics. Since a quantitative failure discussion is often not possible, the interpreter needs a sound knowledge and experience to distinguish between imaging or numerical artifacts and geological features.

The primary information content of seismic refraction data consists of P-wave and – if properly recorded – S-wave velocities of the penetrated subsurface layers. As seismic velocities translate to ranges of sediment and rock types, velocity–depth distribution models are used to help interpret shallow subsurface to deep crustal 2D cross-sections or 3D depth-slices, ideally in combination with complimentary seismic reflection and/or potential field (gravity, magnetic) data. Interfaces of first-order impedance discontinuities or layers with strong vertical velocity gradients with respect to wavelength are well recognizable in travel time phases and can be meaningfully modeled. Amplitude analyses of either refracted or wide-angle reflected wave-fields, using ray-geometric or wave-field inversion techniques, are useful for characterizing layer interface properties such as thin lamination or intercalation. For instance, such analyses have revealed a complex formation and composition of the crust–mantle boundary in some regions (e.g., Levander et al., 2005).

Contribution to earth science

Classical applications of the reflection and seismic refraction method to earth science are seismic stratigraphy and structural geology. Both approaches are needed to unravel the dynamics of continental and oceanic basins. Structural geology deals with the imaging of 3D distribution of rock units; its interpretation aims on the reconstruction of their deformational histories.

Seismic refraction surveying has largely contributed to the understanding of the architecture and composition of the earth's crust and uppermost mantle. Crustal thicknesses of the ocean and continental lithosphere have been determined largely due to the frequently encountered strong impedance contrasts at the crust–mantle boundary.

Intra-crustal reflection interfaces are deciphered from analyses of amplitude characteristics in wide-angle reflection and refraction records. Being a powerful tool in the studies of continent–ocean boundaries, refraction surveying and the resulting velocity–depth models have revealed that these boundaries are often transitional zones of various widths in continental margins of both volcanic and non-volcanic types (e.g., Mjelde et al., 2005; Voss and and Jokat, 2007; Suckro et al., 2012). While the normal-incidence seismic reflection technique largely fails to properly image sub-salt and sub-basalt sedimentary formations and structures, the seismic refraction technique is used as a complimentary tool to provide velocity information from such "hidden" zones.

Sequence stratigraphy is a methodology that provides a framework for the elements of any depositional setting, facilitating paleogeographic reconstructions and the prediction of facies and lithologies away from control points (Catuneanu et al., 2011). Seismic stratigraphy is the approach to conclude on sequence stratigraphy by seismic interpretation. A special discipline of seismic stratigraphy is the analysis of depositional geometries in terms of oceanic currents, which may influence deposition processes forming so-called contourites or contourite drifts (Nielsen et al., 2008). Geophysicist therefore may contribute to paleoceanographic and related paleoclimate studies.

The contribution of the reflection seismic method to chemical processes and the migration of fluids and volatiles, both producing particular characteristics in seismic images, were acknowledged in the 1990s, stimulating research cooperation with biologists and chemists. Further, the transition of free gas to overlying gas hydrate, mainly consisting of methane trapped inside cages of hydrogen-bonded water molecules, gives a strong reflection crosscutting depositional strata, which is easy to detect in seismic data.

The assessment of geohazard potential caused by landslides including their tsunamigenic potential is a fast-developing research field. Owing to the specific internal and external deformation pattern, large submarine landslides are easy to detect in seismic data. The imaging of the internal geometry of volcanoes allows for conclusions on acting volcanism processes and recurrence rates.

The seismic refraction technique is successfully employed in geotechnical and engineering surveys to determine the depth to bedrock. Applications range from the detection of fracture zones in hard rocks in connection with groundwater prospecting, to delineating unconsolidated strata in civil engineering (road and tunnel constructions), to hazardous waste disposal projects.

Conclusions

Seismic reflection and refraction methods are two of the most commonly used geophysical methods in exploring the earth's subsurface for both hydrocarbon exploration and academic research of the buildup of the crust and its sedimentary cover. Both techniques have provided data to constrain the development of the earth's crust from the sedimentary basins down to the transition to the uppermost mantle of the continents and oceans. Their relatively high structural resolution capacity is one of their greatest advantages. Future developments will see further improvement of 3D imaging and 4D monitoring capabilities from observing resource reservoirs and hydrological systems to geotectonic hazards in regions at risk of earthquake and tsunami.

Bibliography

Berndt, C., Costa, S., Canals, M., Camerlenghi, A., de Mol, B., and Saunders, M., 2012. Rpepeated slope failure linked to fluid migration: the Ana submarine landslide complex, Eivissa Channel, Western Mediterranean. *Earth and Planetary Science Letters*, **319–230**, 65–74.

Catuneanu, O., Galloway, W. E., Kendall, C. G. S. C., Miall, A. D., Posamentier, H. W., Straser, A., and Tucker, M. E., 2011. Sequence stratigraphy: methodology and nomenclature. *Newsletter on Stratigraphy*, **44**(3), 173–245.

Cerveny, V., Molotkov, I., and Psencik, I., 1977. *Ray Method in Seismology*. Prague, Czechoslovakia: University of Karlova.

Chowdhury, K. R., 2011. Seismic data acquisition and processing. In Gupta, H. K. (ed.), *Encyclopedia of Solid Earth Geophysics*. Dordrecht: Springer, pp. 1081–1097.

Deen, T. J., and Gohl, K., 2002. 3-D tomographic seismic inversion of a paleochannel system in central New South Wales, Australia. *Geophysics*, **67**, 1364–1371.

Dragoset, W. H., 1990. Air-gun array specs: a tutorial. *The Leading Edge*, **9**(1), 24–32.

Dragoset, W., 2005. A historical reflection on reflections. *The Leading Edge*, **24**, 47–71.

Gray, S. H., 2011. Seismic, migration. In Gupta, H. K. (ed.), *Encyclopedia of Solid Earth Geophysics*. Dordrecht: Springer, pp. 1236–1244.

Hübscher, C., and Spieß, V., 1997. An integrated marine seismic approach for time efficient high-resolution surveys. In *Proceedings of the Offshore Technology Conference*, Vol. 1, pp. 387–391.

Korenaga, J., and Sager, W. W., 2012. Seismic tomography of Shatsky Rise by adaptive importance sampling. *Journal of Geophysical Research*, **117**, B08102, doi:10.1029/2012JB009248.

Kumar, D., and Ahmed, I., 2011. Seismic Noise. In: Gupta H. K. (ed.) *Encyclopedia of Solid Earth Geophysics*. Dordrecht: Springer, pp. 1157–1161.

Levander, A., Zelt, C. A., and Magnani, M. B., 2005. Crust and upper mantle velocity structure of the Southern Rocky Mountains from the Jemez Lineament to the Cheyenne belt. In Karlstrom, K. E., and Keller, G. R. (eds.), *The Rocky Mountain Region: An Evolving Lithosphere*. Washington, DC: American Geophysical Union. Geophysical Monograph Series, Vol. 154, pp. 293–308.

Mjelde, R., Raum, T., Myhren, B., Shimamura, H., Murai, Y., Takanami, T., Karpuz, R., and Naess, U., 2005. NE Atlantic, derived from densely spaced ocean bottom seismometer data. *Journal of Geophysical Research*, **110**, B05101, doi:10.1029/2004JB003026.

Mutter, J. C., 1987. *Seismische Bilder von Plattengrenzen*. Dynamik der Erde. Spektrum der Wissenschaft Verlagsgesellschaft, Dordrecht: pp. 132–143.

Nielsen, T., Knutz, P. C., and Kuijpers, A., 2008. Seismic expression of contourite depositional systems. In Rebesco, M., and Camerlenghi, A. (eds.), *Contourites*. Amsterdam: Oxford. Developments in Sedimentology, Vol. 60, pp. 301–321.

Palmer, D., 1986. Refraction seismics – the lateral resolution of structure and seismic velocity. In Helbig, K., and Treitel, S. (eds.), *Handbook of Geophysical Exploration, Section I. Seismic Exploration*. London/Amsterdam: Geophysical Press.

Parkes, G., and Hatton, L., 1986. *The Marine Seismic Source*. Dordrecht: Kluwer Academic Press, p. 111.

Pratt, R. G., Song, Z.-M., Williamson, P., and Warner, M., 1996. Two- dimensional velocity models from wide-angle seismic data by wavefield inversion. *Geophysical Journal International*, **124**, 323–340.

Roden, R., 2005. The evolution of the interpreter's toolkit – past, present, and future. *The Leading Edge*, **24**, 72–78.

Stankiewicz, J., Parsiegla, N., Ryberg, T., Gohl, K., Weckmann, U., Trumbull, R., and Weber, M., 2008. Crustal structure of the southern margin of the African continent: results from geophysical experiments. *Journal of Geophysical Research*, **113**, B10313, doi:10.1029/2008JB005612.

Strandenes, S., and Vaage, S., 1992. Signatures from clustered airguns. *First Break*, **10**(8), 305–311.

Suckro, S. K., Gohl, K., Funck, T., Heyde, I., Ehrhardt, A., Schreckenberger, B., Gerlings, J., Damm, V., and Jokat, W., 2012. The crustal structure of southern Baffin Bay: implications from a seismic refraction experiment. *Geophysical Journal International*, **190**, 37–58, doi:10.1111/j.1365-246X.2012.05477.x.

Telford, W. W. M., Geldart, L. P., and Sheriff, R. E., 1990. *Applied Geophysics*. New York: Cambridge University Press.

Vidale, J. E., 1990. Finite-difference calculation of travel times in three dimensions. *Geophysics*, **55**, 521–526.

Voss, M., and Jokat, W., 2007. Continent–ocean transition and voluminous magmatic underplating derived from P-wave velocity modelling of the East Greenland continental margin. *Geophysical Journal International*, **170**, 580–604.

Yilmaz, Ö., 2008. *Seismic Data Analysis: Processing, Inversion, and Interpretation of Seismic Data*. Tulsa: SEG, Vol. I and II.

Zelt, C. A., 1999. Modelling strategies and model assessment for wide-angle seismic traveltime data. *Geophysical Journal International*, **139**, 183–204.

Zelt, C. A., and Barton, P. J., 1998. Three-dimensional seismic refraction tomography: a comparison of two methods applied to data from the Faeroe Basin. *Journal of Geophysical Research*, **103**, 7187–7210.

Zelt, C. A., and Smith, R. B., 1992. Seismic traveltime inversion for 2-D crustal velocity structure. *Geophysical Journal International*, **108**, 16–34.

Zelt, B. C., Taylor, B., Weiss, J. R., Goodliffe, A. M., Sachpazi, M., and Hirn, A., 2004. Streamer tomography velocity models for the Gulf of Corinth and Gulf of Itea, Greece. *Geophysical Journal International*, **159**, 333–346, doi:10.1111/j.1365-246X.2004.02388.x.

Cross-references

Sequence Stratigraphy
Technology in Marine Geosciences

REGIONAL MARINE GEOLOGY

Martin Meschede

Institute of Geography and Geology, Ernst-Moritz-Arndt University, Greifswald, Germany

The oceans and their marginal seas cover approximately 71 % of the Earth's surface. They contain about 97 % of the total water of the Earth and they have an average depth

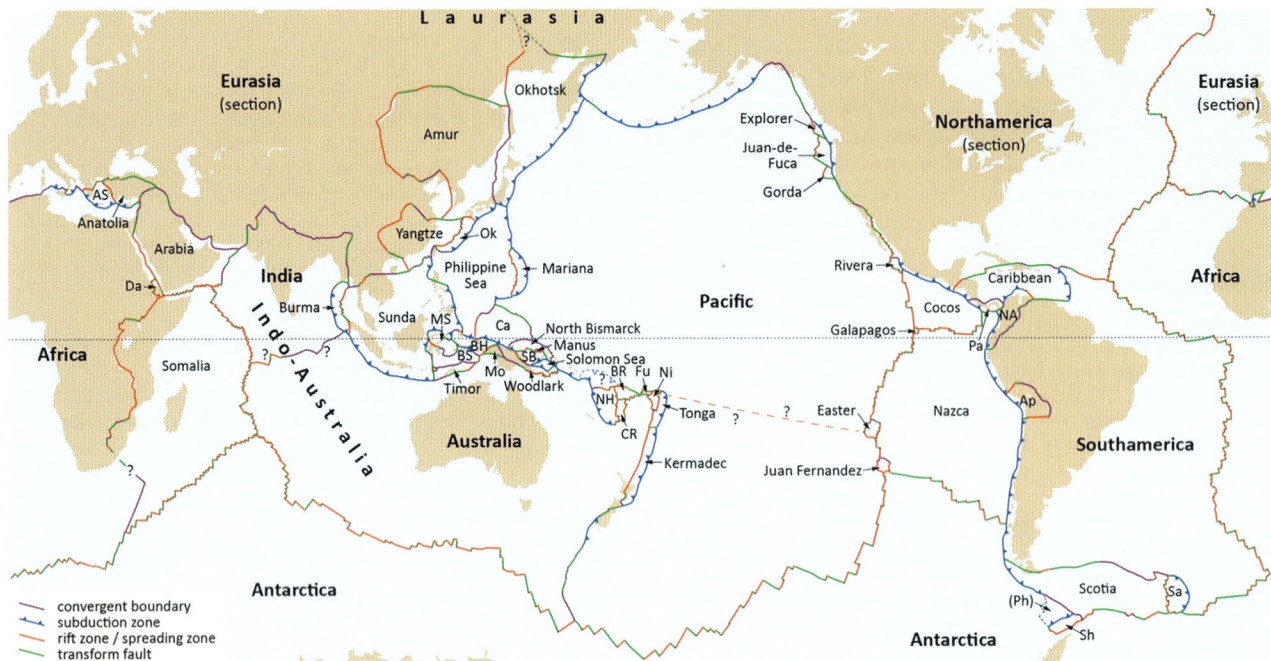

Regional Marine Geology, Figure 1 Lithospheric plates on Earth: *AS* aegean sea, *Ap* altiplano, *BH* birds head, *BR* balmoral reef, *BS* banda sea, *Ca* carolina, *CR* conway reef, *Da* danakil, *Fu* futuna, *Ke* kermadec, *Mn* manus, *Mo* maoke, *MS* molucca sea, *NA* north andes, *NH* new hebrides-fiji, *Ni* niuafo'ou, *Ok* okinawa, *Pa* panama, *(Ph)* former phoenix (now merged with Antarctica), *SB* south bismarck, *Sa* sandwich, *Sh* shetland, *To* tonga. Other names are indicated in the figure (based on Bird (2003)).

of approximately 3,680 m. The largest ocean is the Pacific Ocean with more than 181.3 Mio. km^2 corresponding to 35 % of the Earth's surface. The Atlantic Ocean encompasses almost 89.8 Mio. km^2, corresponding to about 18 % of the Earth's surface. The Indian Ocean (74.9 Mio. km^2), Southern (Antarctic) Ocean (20.3 Mio. km^2), and Arctic Ocean (14.1 Mio. km^2) are smaller and cover about 14.7 %, 4.0 %, and 2.7 % of the Earth's surface. The deepest known structure in the oceans is the Mariana Trench in the western Pacific Ocean near the Northern Mariana Islands. The deepest point of 10,920 m ($\pm$10 m) was named "Challenger Deep" after the British survey ship HMS Challenger that made the first recordings of this depth (expedition 1872–1876).

The geographical subdivision of the Earth into oceans and continents does, however, not match the subdivision of oceanic and continental plates in terms of plate tectonics (Figure 1). In the following the regional distribution of plates is described with special emphasis on the oceanic areas.

The lithosphere of the Earth is subdivided into a number of lithospheric plates. All plates are separated from the others by plate boundaries, which are defined by differential movement of the two plates on both sides of the boundary. Three types of plate boundaries are subdivided (Figure 2): divergent or constructive (spreading center, rift zone), convergent or destructive (subduction zone, continental collisional), and conservative (transform fault). New lithosphere is formed by seafloor spreading at divergent plate boundaries. At convergent boundaries two plates move against each other forming either a subduction zone where oceanic lithosphere descents into the asthenophere or a collisional orogen where two continental plates collide. Transform boundaries occur at locations where two plates slide past each other along transform faults without forming of destructing lithospheric material. All plate boundaries are characterized by seismic activity.

The seven largest plates are the African, Antarctic, Eurasian, North American, South American, Australian, and Pacific plates (Figure 1), and there is no question about smaller plates like the Caribbean, Cocos, Nazca, Juan-de-Fuca, Scotia, Arabian, Philippine Sea, or the Sunda Plate. So-called microplates like the Easter, Explorer, Juan Fernandez, Gorda, Caroline, Woodlark, Sandwich, and several other microplates are also accepted as plate tectonic entities. However, a number of small plates and microplates exist, which are not clearly defined in every case, e.g., Somalia, Danakil, Panama, Altiplano, and several others. Bird (2003) counts a total number of 52 plates, in Figure 1 a total number of 55 plates are shown.

The North American and Eurasian plates are generally taken as two separate plates. The plate boundary between these two plates in the area of eastern Siberia is, however, not clearly defined. The mid-oceanic spreading center in the North Atlantic separating the North American and Eurasian plates continues into the spreading zone of the Gakkel ridge. The spreading velocity at the Gakkel ridge

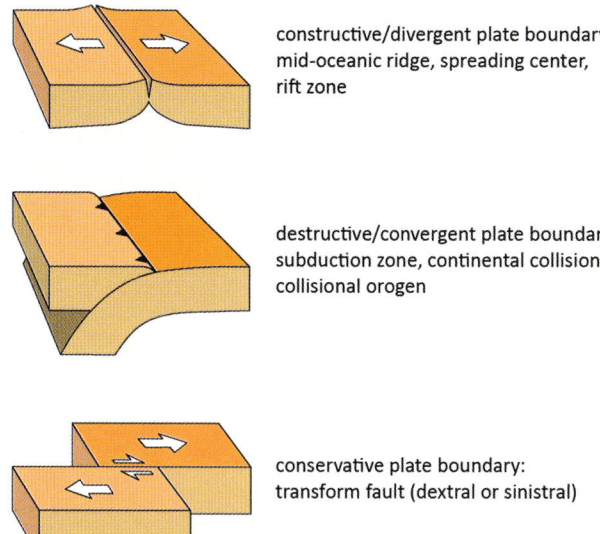

Regional Marine Geology, Figure 2 Three different types of plate boundaries (after Frisch et al., 2011).

decreases from 1.24 cm/year in the area of Northern Greenland to 0.73 cm/year at the Siberian shelf. The spreading center continues toward south into a rift zone in the Laptev Sea and follows the valley of the Lena River in Northern Siberia. A distinct plate boundary is missing further to the south. On the other side, a convergent plate boundary exists west of Japan running through the eastern part of the Japan Sea. It separates the Amur and Okhotsk plates (Figure 1), which in other plate distribution maps belong to the Eurasian Plate. The northern continuation of the convergent plate boundary is, however, unclear and there is no direct connection to the rift valley of the Lena River. It is, therefore, suggested that the North American and Eurasian plates are still connected in their central part in Northeastern Siberia. The two plates are thus defined as the North American and Eurasian sections of the large Laurasian Plate.

The Indo-Australian Plate is marked by two large landmasses, India and Australia. An ongoing debate exists whether a plate boundary separates the Australian from the Indian Plate. The suggested mostly convergent plate boundary (Bird, 2003) is located west of Sumatra and south of the Maldives islands (Figure 1). Many other authors, however, do not subdivide an Indian and Australian plate. The formation of a new plate boundary may be indicated by strong earthquakes occurring in the equatorial area of the Indo-Australian Plate (Wiens et al., 1986; Delescluse et al., 2012; Shen, 2012). The earthquakes are interpreted to indicate the ongoing process of plate boundary formation. At the eastern plate boundary of the Australian Plate a flip in the subduction polarity can be observed in the area of New Zealand. The North Island of New Zealand is characterized by the subduction of the Pacific Plate beneath the small Kermadec Plate, which is separated by a back-arc spreading center from the Australian Plate. The morphological expression of the subduction zone is the Hikurangi Trough. The subduction zone continues on the South Island into the New Zealand Alpine Transform Fault. Further to the southwest in the Tasman Sea the Australian Plate is subducted beneath the Pacific Plate thus having an opposite subduction polarity.

The Somalia Plate is separated from the African Plate by the Central African rift system. The rift system was initiated about 25 Myrs ago (Roberts et al., 2012). It is characterized by strongly thinned continental lithosphere beneath the graben axis and a deep graben structure on top of the rising asthenosphere that produces a typical rift volcanism. The volcanoes of Central Africa, e.g., the Kilimanjaro, the Nyiragongo, or the Ol Doinyo Lengai illustrate the ongoing rifting process. The southern boundary of this plate, however, is not clearly defined and still under discussion (Calais et al., 2006; Stamps et al., 2008).

The most complicated area in terms of plate tectonic boundaries is the Indonesian archipelago. A large number of small plates are acting together forming a number of special situations. Up to about 20 different small plates, microplates, or terranes are subdivided, all of them separated by one of the three types of plate boundaries (Figure 2). A well investigated situation exists in the Woodlark Basin where an asymmetric rift led to the denudation of mantle material in the direct neighborhood of an oceanic spreading ridge (e.g., Taylor et al., 1994). Other plate boundaries are apparently just in the stage of formation as there are indications that a morphological trench between the Tuvalu and Solomon islands represents the initial stage of a forming subduction zone (Okal et al., 1986).

The Tonga-Kermadec subduction zone marks the western boundary of the Pacific Plate. It is a classical situation of an oceanic convergent plate boundary with the formation of a back-arc spreading center that defines the small Kermadec and Tonga plates located between the Pacific and Australian plates. A similar situation exists with the Mariana Plate between the Pacific Plate subducted beneath the Mariana Plate and the Philippine Sea Plate west of the back-arc spreading center of Mariana. Other situations with back-arc spreading centers forming small independent plates exist in the New Hebrides-Fiji Plate separated by a back-arc spreading center from the Conway Reef Plate, the Okinawa Plate separated from the Yangtze Plate, and the Sandwich Plate separated from the Scotia Plate (Figure 1).

The Pacific Ocean is with approximately 181.3 Mio. km^2, the largest oceanic basin of the Earth. However, only a little less than two thirds of the ocean belongs to the Pacific Plate which is the largest lithospheric plate on Earth covering an area of about 108 Mio. km^2. It is mainly made up of oceanic crust except some small areas in its surroundings, e.g., the area of Baja California in northwestern Mexico or part of the Southern Island of New Zealand. It is bordered by the East Pacific Rise in the east,

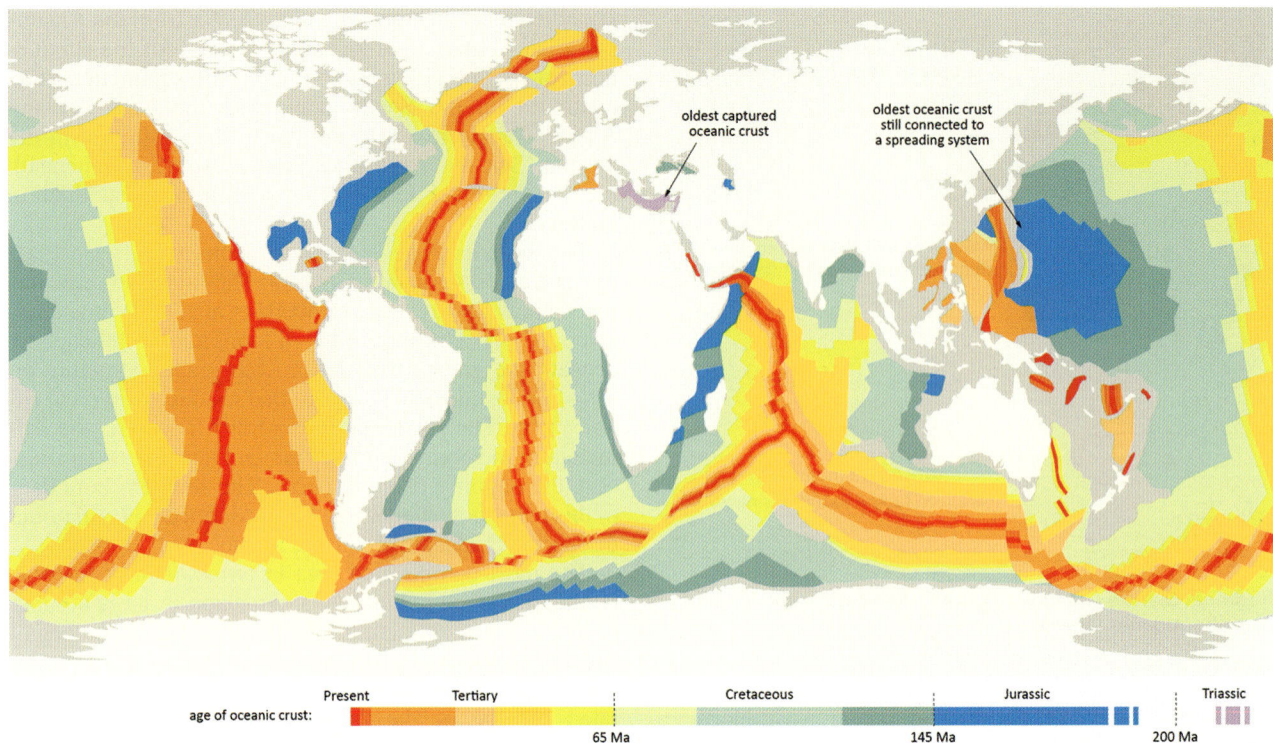

Regional Marine Geology, Figure 3 Distribution of ages of oceanic lithosphere (based on Müller et al. (2008) and Frisch et al. (2011)).

a mid-oceanic rift system of about 9,000 km length. The East Pacific Rise continues toward north into the transform/rift system of the Gulf of California, the San Andreas transform fault, and is followed by the rift system between the Pacific and Gorda/Juan-de-Fuca/Explorer plates and further to the north by the Queen-Charlotte transform fault. This transform fault ends in a series of destructive plate margins: Aleutian trench, Kurile-Kamchatka trench, and the so called IBM system consisting of the Izu, Bonin, and Mariana trenches. The southernmost part of the East Pacific Rise is the rift zone between the Pacific and Antarctic plates. The western part of the Pacific Plate encompasses the oldest oceanic crust on Earth with more than 185 Ma (Figure 3; Müller et al., 2008) that is still in direct connection to the mid-oceanic ridge where it was formed. A discussion exists whether the Pacific Plate may split up in the near future into two parts along a break-up zone between the Tonga Plate in the west and the Easter Plate in the east (Figure 1). This zone is marked by an abnormal concentration of intraplate volcanism (Clouard and Gerbault, 2008), which may represent the first indications of a future plate reorganization that will break the Pacific Plate into a southern and a northern plate.

The Pacific Plate is bordered in the east by a number of differently sized plates (from north to south): Explorer, Juan-de-Fuca, Gorda, Rivera, Cocos, Galapagos, Nazca, Easter, Juan Fernandez, and Antarctic plates. With the exception of the Antarctic plate, all these plates are entirely composed of oceanic crust and are partly remnants of the former Farallon Plate, which has been one large single oceanic plate formed at the East-Pacific Rise spreading center formed in the Mesozoic and early Cenozoic until it was nearly completely subducted beneath the American continents. It was split up into several parts by the subduction of the mid-oceanic rift system about 30 Ma ago. 25 Ma ago the three-phase Cocos-Nazca spreading system came into existence (Meschede et al., 1998; Meschede and Barckhausen, 2001) that separated the largest remnant of the Farallon Plate into the Cocos and Nazca plates. The Cocos Plate is subducted beneath the North American and Caribbean plates. The easternmost part is subducted beneath the Panama Plate whose boundaries with the Caribbean Plate, however, are not clearly defined in all places. The Nazca Plate is subducted beneath the South American Plate that includes the North Andes and Altiplano plates defined by Bird (2003).

With exceptions in the eastern Caribbean and eastern Sandwich plates, the Atlantic Ocean is nearly entirely bordered by passive plate margins without plate boundaries. The oceanic crust of the northern part of the Atlantic Ocean belongs to the North American and Eurasian plate sections of the Laurasian Plate. The southern part belongs to the South American, Scotia/Sandwich, African, and Antarctic plates. There is no separate Atlantic Plate as it exists in the Pacific. In the middle of the Atlantic Ocean, the Mid-Atlantic rift zone separates the American plates in the west from the European and African plates in the east. Subduction of oceanic crust of the Atlantic Ocean occurs currently along the eastern boundaries of the Caribbean and Sandwich plates in the north and south of South America, respectively. According to seismic data and earthquake observations (Gutscher et al., 2002; Gutscher, 2004; Spakman and Wortel, 2004), a subduction zone in the process of formation is interpreted beneath the Betic cordillera in southern Spain and the Atlas mountains in Morocco. The oldest part of the Atlantic oceanic crust is located in the Central Atlantic along the coasts of the eastern United States and western Africa (Figure 3). With an age of about 175 Ma, it is only slightly younger than the oldest oceanic crust of the Pacific Plate.

The Indian Ocean is the third largest of the world's oceans. In its northern part the Carlsberg Ridge as part of the Central Indian Ridge separates the African Plate from the Indo-Australian Plate (Figure 1). The Carlsberg Ridge ends at the Rodrigues Triple Junction. The mid-oceanic ridge running from this triple junction towards southwest separates the African and Antarctic plates whereas the other branch running towards southeast separates the Indo-Australian and Antarctic plates. An ongoing discussion exists about a plate boundary separating the northern Indian part of the Indo-Australian Plate from its southern Australian part (see above). The oldest oceanic crust of the Indian Ocean is located in the western part off the coast of Somalia, Kenya, and Mozambique (Figure 3).

Bibliography

Bird, R., 2003. An updated digital model of plate boundaries. *Geochemistry, Geophysics, Geosystems*, **4**, 1027, doi:10.1029/2001GC000252.

Calais, E., Hartnady, C., Ebinger, C., and Nocquet, J. M., 2006. Kinematics of the East African Rift from GPS and earthquake slip vector data. In Yirgu, G., Ebinger, C. J., and Maguire, P. K. H. (eds.), *Structure and Evolution of the Rift Systems Within the Afar Volcanic Province, Northeast Africa*. London: Geological Society. Geological Society special publication, Vol. 259, pp. 9–22.

Clouard, V., and Gerbault, M., 2008. Break-up spots: could the Pacific open as a consequence of plate kinematics? *Earth and Planetary Science Letters*, **265**, 195–208.

Delescluse, M., Chamot-Rooke, N., Cattin, R., Fleitout, L., Trubienko, O., and Vigny, C., 2012. April 2012 intra-oceanic seismicity off Sumatra boosted by the Banda-Aceh megathrust. *Nature*, **490**, 240–244.

Frisch, W., Meschede, M., and Blakey, R., 2011. *Plate Tectonics, Continental Drift and Mountain Building*. Berlin/Heidelberg: Springer. 212 p.

Gutscher, M. A., 2004. What caused the great Lisbon earthquake? *Science*, **305**, 1247.

Gutscher, M. A., Malod, J., Rehault, J. P., Contrucci, I., Klingelhoefer, F., Mendes-Victor, L., and Spakman, W., 2002. Evidence for active subduction beneath Gibraltar. *Geology*, **30**, 1071–1074.

Meschede, M., and Barckhausen, U., 2001. The relationship of the Cocos and Carnegie ridges – age constraints from palinspastic reconstructions. *International Journal of Earth Sciences*, **90**(2), 386–392.

Meschede, M., Barckhausen, U., and Worm, H.-U., 1998. Extinct spreading on the Cocos Ridge. *Terra Nova*, **10**, 211–216.

Müller, R. D., Sdrolias, M., Gaina, C., and Roest, W. R., 2008. Age, spreading rates and spreading symmetry of the world's ocean crust. *Geochemistry, Geophysics, Geosystems*, **9**, Q04006, doi:10.1029/2007GC001743.

Okal, E. A., Woods, D. F., and Lay, T., 1986. Intraplate deformation in the Samoa-Gilbert-Ralik area: a prelude to a change of plate boundaries in the southwest Pacific? *Tectonophysics*, **132**, 69–77.

Roberts, E. M., Stevens, N. J., O'Connor, P. M., Dirks, P. H. G. M., Gottfried, M. D., Clyde, W. C., Armstrong, R. A., Kemp, A. I. S., and Hemming, S., 2012. Initiation of the western branch of the East African Rift coeval with the eastern branch. *Nature Geoscience*, **5**, 289–294, doi:10.1038/ngeo1432.

Shen, H., 2012. Unusual Indian Ocean earthquakes hint at tectonic breakup. *Nature*, doi:10.1038/nature.2012.11487.

Spakman, W., and Wortel, R., 2004. A tomographic view on western Mediterranean geodynamics. In Cavazza, W., Roure, F., Spakman, W., Stampfli, G. M., and Ziegler, P. (eds.), *The TRANSMED Atlas, The Mediterranean Region from Crust to Mantle*. Berlin: Springer, pp. 31–52.

Stamps, D. S., Calais, E., Saria, E., Hartnady, C., Nocquet, J. M., Ebinger, C. J., and Fernandes, R. M., 2008. A kinematic model for the East African Rift. *Geophysical Research Letters*, **35**, L05304, doi:10.1029/2007GL032781.

Taylor, B., Goodliffe, A., Martinez, F., and Hey, R., 1994. Continental rifting and initial sea-floor spreading in the Woodlark basin. *Nature*, **374**, 534–537.

Wiens, D. A., Stein, S., Demets, C., Gordon, R. G., and Stein, C., 1986. Plate tectonic models for Indian Ocean "intraplate" deformation. *Tectonophysics*, **132**, 37–48.

Cross-references

Active Continental Margins
Marginal Seas
Mid-ocean Ridge Magmatism and Volcanism
Morphology Across Convergent Plate Boundaries
Oceanic Rifts
Plate Motion
Plate Tectonics
Subduction
Terranes
Transform Faults
Triple Junctions

RELATIVE SEA-LEVEL (RSL) CYCLE

Jan Harff
Institute of Marine and Coastal Sciences, University of Szczecin, Szczecin, Poland

Definition

Relative sea-level (RSL) is the position of sea surface relative to the base level. Its cyclic change between transgression and regression takes into account two main components: eustatic change and vertical movement (tectonic/isostatic) of the sea floor (modified from Posamentier and Allen, 1999).

Introduction

One of the main achievements in stratigraphy in the beginning of the twentieth century was to use sedimentological fingerprints of global transgressions and regressions for a general division of the Phanerozoic earth history. Suess (1906) developed the concept of eustasy to link physical stratigraphy of different regions with the frame of global transgressions and regressions. This concept was refined by von Bubnoff (1954) who subdivided the geological history during the Phanerozoic on the global scale into six megacycles, each consisting of two transgression phases followed by an inundation, differentiation, regression, and emersion phase. The assumption of a decreasing period of the cycle from 175 Mio yrs ("pre-Caledonian cycle") to 20 Mio yrs for the latest (still ongoing) cycle led to the "time-snail" model (von Bubnoff, 1954) describing the tectonically driven history of the earth.

Sloss' (1963) and Vail et al.'s (1977) pioneering studies of global and regional relative sea-level change and its stratigraphic response on passive continental margins are used in geology today to provide a spatial–temporal framework for genetic and environmental studies of sedimentary rocks. Vail et al.'s work was based on the interpretation of onshore seismic reflection profiles at Exxon Production Research Company (EPR) to identify sea-level changes using unconformities in the stratigraphic record. The Ocean Drilling Program (ODP) provided the tools to link the onshore records with offshore sediment records starting with the New Jersey sea-level transect. Based on oxygen isotope ratios ($\delta^{18}O$), the relative sea-level change has been reconstructed from the Mesozoic to the Cenozoic (Haq et al., 1987). High-resolution sea-level records for the Quaternary are based on interpretation of isotope signals from corals and the Late Pleistocene and Holocene geodetic position of radiocarbon-dated paleo-coastlines (Fairbanks, 1989; Lambeck and Chappell, 2001).

Factors influencing relative sea-level cycles

It is assumed that the change of relative sea-level can be split into different components:

$$\Delta S = \Delta E + \Delta V + \sum_i \Delta E_i. \quad (1)$$

Here ΔS stands for the relative sea-level change, ΔE marks eustatic change, ΔV is the vertical crustal movement (tectonically or isostatically induced), and ΔE_i stands for other effects of varying importance, such as compaction, local tectonics, gravitational forces, and others. Global eustatic changes are controlled by the volume of water filling the ocean basins or by changes in the volume of these basins. According to Allen and Allen (1990), these changes are due to differentiation of lithospheric material as a result of plate tectonic processes, changes in the basin volume because of accumulation of sediments, and changes in the volumetric capacity of ocean basins because of volume changes in the mid-ocean ridge systems. Here also the subduction of oceanic crust and the resulting shrinking of marine basin volume have to be taken into account (ocean closure). Cloetingh and Haq (2015) give a comprehensive view of the interplay between geodynamic processes and sea-level change. A very important factor is the change of water volume due to climate-related terrestrial storage of water in the form of continental ice sheets. To these can be added changes in volume due to storage and release of groundwater (Hay and Leslie, 1990).

The isostatic/tectonic vertical crustal movement is specific for different regions and superimposed on the global eustatic changes. Basin formation, changes in subsidence rates due to internal mantle dynamics or external loading, folding, faulting, magmatism, and diapirism determine the regional vertical displacement of the crust.

Eustatic and tectonic/isostatic cycles are superposed and on passive continental margins result in cyclic successions of transgressions and regressions. On continental margins this process produces time-equivalent sediment packages bounded by correlative physical surfaces (sedimentary sequences and parasequences). Based on the sedimentary correlation of the relative sea-level changes, Vail et al. (1991) ordered the sea-level cycles by their periodicity (duration) into six classes (see Table 1).

According to Allen and Allen (1990), long-term changes in the volume of mid-ocean ridges resulting from changes in spreading rates are the cause of first-order cycles. Second-order cycles ("supercycles" after Vail et al., 1977) have been globally traced by Vail et al. (1977) based on onlapping of the strata of a sequence in relation to its lower boundary. Figure 1 compares global (second-order) cycles of Vail et al. (1977) and the age of rift–drift transitions on different continental margins after Watts (1982). The latter author suggests a close relation between the rift–drift transition during the continental breakup of the supercontinent Pangaea and the initiation of coastal onlap.

Other possible causes of the second-order relative sea-level changes are plate boundary reorganizations and glacial/deglacial cycles (Allen and Allen, 1990).

Relative Sea-Level (RSL) Cycle, Table 1 Duration (period) of each order of RSL cycles (Vail et al., 1991)

Order	Duration (period)
1	50 + Ma
2	3–50 Ma
3	0.5–3 Ma
4	0.08–0.5 Ma
5	0.03–0.08 Ma
6	0.01–0.03 Ma

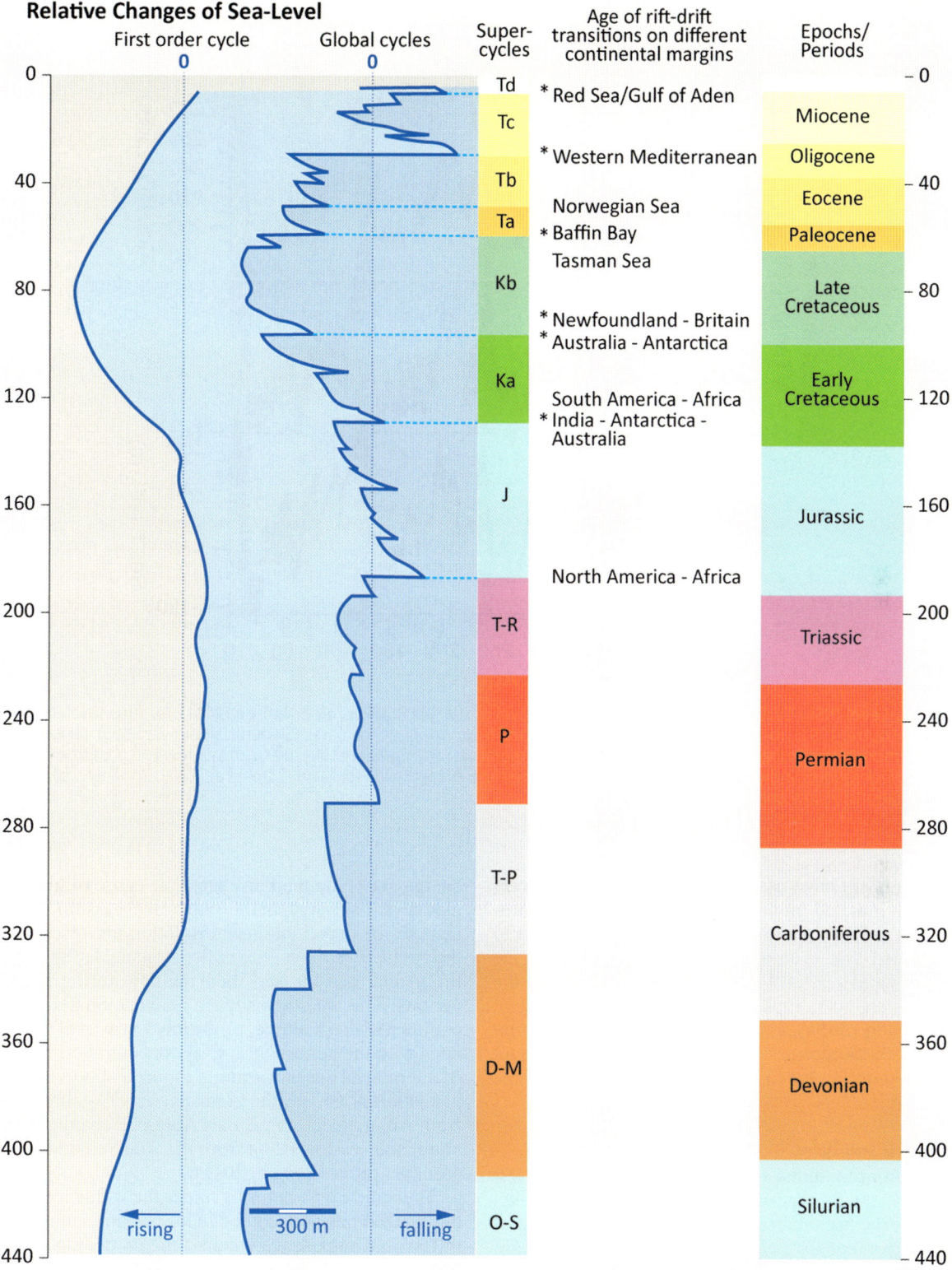

Relative Sea-Level (RSL) Cycle, Figure 1 Comparison of global (2nd-order) relative sea-level cycles after Vail et al. (1977) and the age of the rift–drift transition at different continental margins (modified from Allen and Allen, 1990; Watts, 1982).

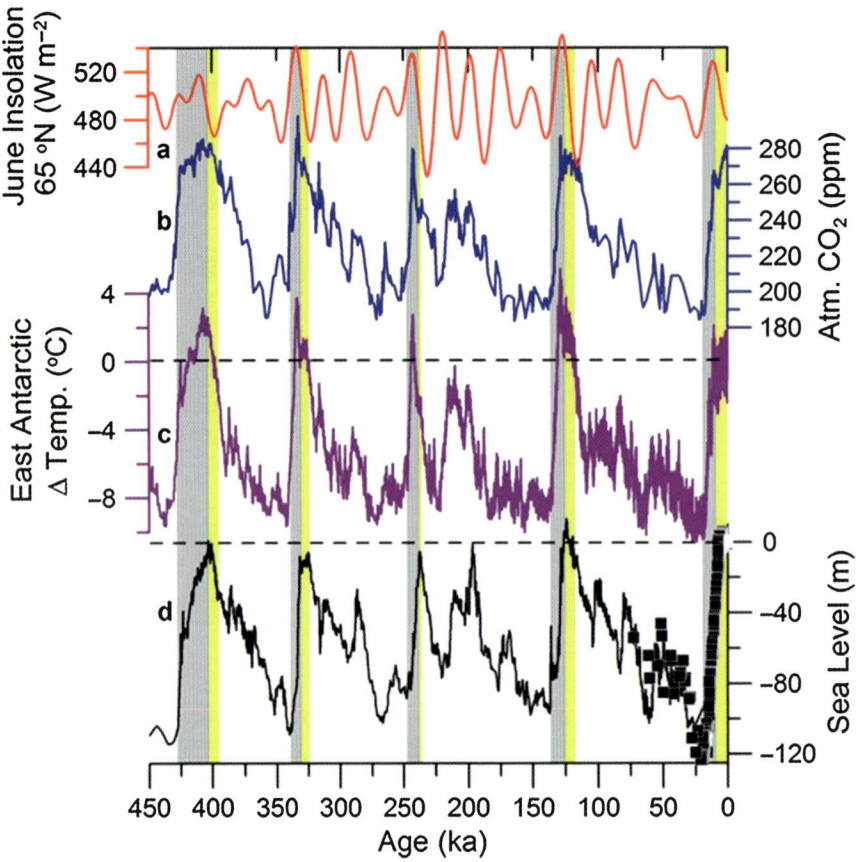

Relative Sea-Level (RSL) Cycle, Figure 2 Climate and sea-level of the last 450 kyrs after Carlson (2011): (**a**) June insolation at 65° N (Berger and Loutre, 1991), (**b**) atmospheric CO_2 concentrations from Antarctic ice cores (Siegenthaler et al., 2005), (**c**) East Antarctic change in temperature (*dashed line* denotes present-day temperature) after Jouzel et al. (2007). (**d**) Sea-level records from the Red Sea (*black line*): *black squares*, individual sea-level estimates; *dashed line*, present-day sea level (Clark et al., 2009; Rohling et al., 2009); *gray bars*, deglaciations; *yellow bars*, interglaciations.

Haq et al. (1987) have used the concept of second-order tectonically initiated global sea-level change for a chronology of the Earth's history since the Triassic. It has to be mentioned that even after 50 years of research, the causes for rapid sea-level shifts, in particular during the Cretaceous, are not completely understood. Due to recent studies of Haq (2014), a fourth-order (400 kyr) cyclicity related to orbital eccentricity seems to play an important role not only during the Cretaceous but also for the whole Phanerozoic.

Glacial and postglacial relative sea-level cycles

Cycles of third-to sixth-order can have different causes. Whereas tectonic processes affect relative sea-level changes on the regional scale, global sea-level cycles of third-to sixth-order are mainly caused by climatically controlled changes in the marine water volume (Vail et al., 1991). The history of climate change is recorded in high resolution for the Quaternary. Milankowitch (1930) calculated the cyclic variations of the earth's orbital parameters proposing that the changes in insolation and consequently energy supply to the Earth were the cause of glacial/interglacial shifts. Berger and Loutre (1991) assume that cyclic changes in boreal summer insolation are responsible for the cyclicity in the buildup of continental ice sheets during the glacial periods and their melting during interglacial periods. This interdependence between boreal insolation, continental ice volume, and global sea-level also correlates with the atmospheric CO_2 concentration as an effect of interglacial ocean warming. Figure 2 shows the summer insolation at 65° N, the atmospheric CO_2 concentrations from Antarctic ice cores, East Antarctic change in temperature, and sea-level data from the Red Sea during the last 450 kyrs (after Carlson, 2011).

The solid Earth deforms under the influence of changing loads on its surface; the loads of continental ice cause differential subsidence whereas the unloading by melting during the interglacial results in uplift of the crust. This means that the ice/water volume change has quite different effects on the changes in relative sea-level observed in glaciated and non-glaciated parts of the world. Different models have been elaborated to describe these processes

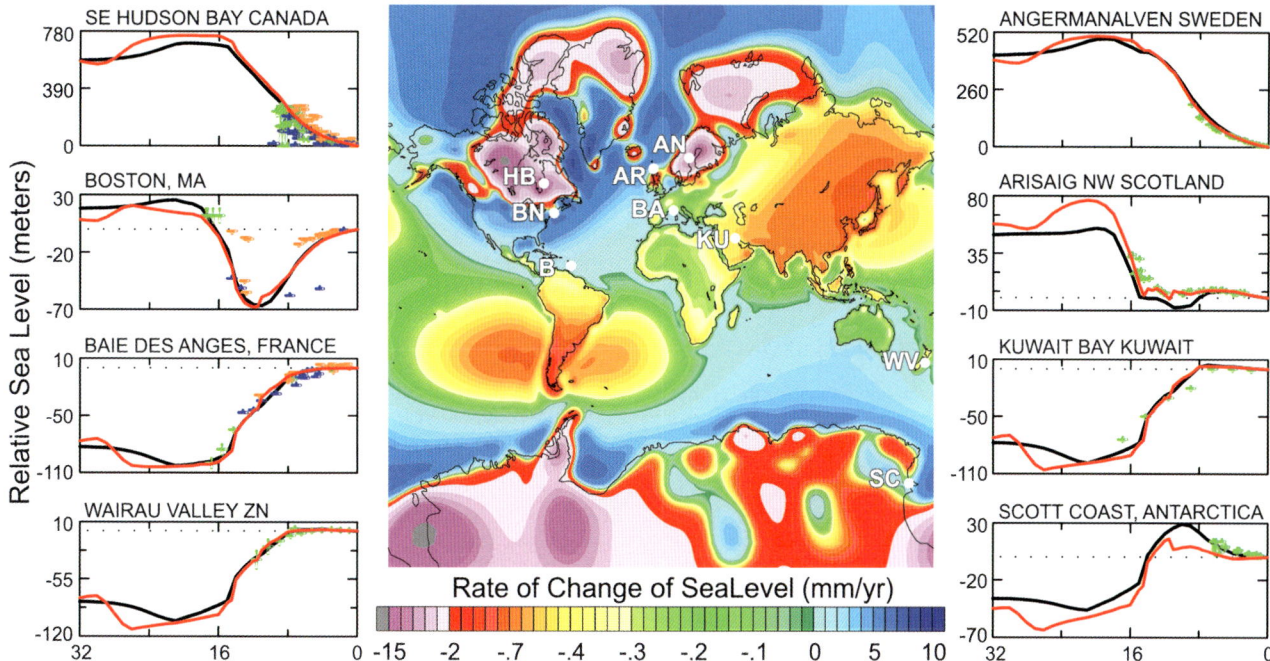

Relative Sea-Level (RSL) Cycle, Figure 3 Intercomparisons between observed and predicted relative sea-level histories at eight different geographical locations using two different versions of the ICE-5G(VM2) model of the deglaciation process: versions 1.1 (LGM at 21 ka BP, *black curve*) and 1.2 (LGM at 26 ka BP, *red curve*) (after Peltier, 2007). The positions of the eight sites are shown in the central part of the figure superimposed upon a Mercator projection of the present-day predicted RSL rate using Peltier's (1998) SLE.

for the last glacial cycle (Lambeck and Chappell, 2001; Peltier, 2007).

Buildup and decay of the continental ice sheets cause glacio-isostatic adjustments (GIA) of the Earth's crust, and redistribution of meltwaters filling marine and freshwater basins causes hydro-isostatic processes. Peltier (1998) has developed a sea-level equation (SLE) that describes the complex interactions between melting ice sheets, crustal deformations, and sea-level changes.

Applying this SLE Peltier (2007) has used the ice load history model ICE-5G(VM2) (Peltier, 2004) to predict the relative sea-level curves for eight different geographical locations in high and low latitudes from 32 ka BP to present day including the Last Glacial Maximum (LGM) and deglaciation period. In Figure 3 the relative sea-level curves for six locations are depicted for two different assumptions: the occurrence of the LGM, i.e., the time of which the greatest concentration of land ice existed, at 21 ka BP (black curve) which was conventionally assumed, and 26 ka BP (red curve) which corresponds to more recent assumptions about the timing of the maximum ice sheet extent during the LGM. According to Clark et al. (2009) nearly most of the continental ice sheets had reached their maximum extents around 26.5 ka BP. The comparison with geologically inferred relative sea-level data (marked in the RSL plots of Figure 3) shows that the model reasonably reconstructs the sea-level history for all selected sites despite different assumptions about the LGM maximum. The differences in the vertical crustal deformation are evident in comparing the sea-level curves from the formerly glaciated high-latitude regions and those at low latitudes not influenced by continental ice sheets. Formerly glaciated regions show the relative sea-level fall in response to glacio-isostatic uplift beginning about 16 ka BP whereas the neotectonically relative stable low-latitude sites show the continuous sea-level rise due to the water volume effect of the deglaciation between 16 and 6 ka BP. In a global atlas Pirazzoli (1991, 1996) gives an overview of the empirical RSL data for the Holocene showing the regional differences caused by tectonic and GIA effects.

For future projection of relative sea-level change, in particular in its regional effect on coastline migration, besides the already mentioned factors, gravitational forces (geoid effects) have to be taken into account. So postglacial GIA (for instance, of Scandinavia, see Richter et al., 2012) and climatically controlled changes of ice sheet mass (for instance, of Greenland, see Kopp et al., 2010) cause considerable deformation of the adjacent sea surface.

Summary

The relative sea-level changes periodically between regression and transgression of the sea and acts on different – hierarchically superimposed – temporal and spatial scales. Eustasy and tectonic (isostatic) vertical crustal movements are the main components of relative sea-level records reconstructed for the Phanerozoic by indirect geological methods. The younger geological history, in particular the glacio-isostatic adjustment's influence on coastline change, can be reconstructed by models describing the effect of loading and unloading of continental ice shields. The empirical data have to be analyzed very carefully in order to separate vertical crustal movements from the eustatic (climatically controlled) components.

Bibliography

Allen, P. A., and Allen, J. R., 1990. *Basin Analysis – Principles and Application*. Oxford: Blackwell. 451 p.g.

Berger, A., and Loutre, M. F., 1991. Insolation values for the climate of the last 10 million years. *Quaternary Science Reviews*, **10**, 297–317.

Carlson, A. E., 2011. Ice sheets and sea level in earth's past. *Nature Education Knowledge* 3(5), 3 (http://www.nature.com/scitable/knowledge/library/ice-sheets-and-sea-level-in-earth-24148940)

Clark, P. U., Dyke, A. S., Shakun, J. D., Carlson, A. E., Clark, J., Wohlfahrt, B., Mitrovica, J. X., Hostetler, S. W., and McCabe, A. M., 2009. The last glacial maximum. *Science*, **325**, 710–714.

Cloetingh, S., and Haq, B. U., 2015. Inherited landscapes and sea level change. *Science*, **347**, 1258375-1–1258375-10.

Fairbanks, R. G., 1989. A 17,000 years glacio-eustatic sea level record: influence of glacial melting rates on the Younger Dryas event and deep ocean circulation. *Nature*, **342**, 637–642.

Haq, B. U., 2014. Cretaceous eustasy revisited. *Global and Planetary Change*, **113**, 44–58.

Haq, B. U., Hardenbol, J., and Vail, P. R., 1987. The chronology of fluctuating sea levels since the Triassic. *Science*, **235**, 1156–1167.

Hay, W. W., and Leslie, M. A., 1990. Could possible changes in global groundwater reservoir cause eustatic sea-level fluctuations? In Revelle, R. (ed.), *(Panel Chairman), Sea- Level Change*. Washington, DC: National Academy Press, pp. 161–170.

Jouzel, J., Masson-Delmotte, V., Cattani. O., Dreyfus, G., Falourd, S., Hoffmann, G., Minster, B., Nouet, J., Barnola, J.M., Chappellaz, J., Fischer, H., Gallet, J.C., Johnsen, S., Leuenberger, M., Loulergue, L., Luethi, D., Oerter, H., Parrenin, F., Raisbeck, G., Raynaud, D., Schilt, A., Schwander, J., Selmo, E., Souchez, R., Spahni, R., Stauffer, B., Steffensen, J.P., Stenni, B., Stocker, T.F., Tison, J.L-. Werner, M., Wolff, E.W., 2007. Orbital and millennial Antarctic climate variability over the past 800,000 years. Science, 317, 793–796.

Kopp, R. E., Mitrovica, J. X., Griffies, S. M., Yin, J., Hay, C. C., and Stouffer, R. J., 2010. The impact of Greenland melt on local sea levels: a partially coupled analysis of dynamic and static equilibrium effects in idealized water-hosing experiments. *Climatic Change*, **103**, 619–625.

Lambeck, K., and Chappell, J., 2001. Sea level change through the last glacial cycle. *Science*, **292**, 679–686.

Milankowitch, M., 1930. Mathematische Klimalehre und Astronomische Theorie der Klimaschwankungen. In Koppen, I. W., and Geiger, R. (eds.), *Handbuch der Klimatologie*. Berlin: Gebrüder Borntraeger.

Peltier, W. R., 1998. Postglacial variations in the level of the sea: implications for climate dynamics and solid-earth geophysics. *Reviews of Geophysics*, **36**, 603–689.

Peltier, W. R., 2004. Global glacial isostasy and the surface of the ice-age earth: the ICE-5G(VM2) model and GRACE. *Annual Review of Earth and Planetary Sciences*, **32**, 111–149.

Peltier, W. R., 2007. Postglacial coastal evolution: ice-ocean-solid earth interactions in a period of rapid climate change. In Harff, J., Hay, W. W., and Tetzlaff, D. (eds.), *Coastline Change – Interrelation of Climate and Geological Processes*. The Geological Society of America, Special Paper, Vol. 426, pp. 5–28.

Pirazzoli, P. A., 1991. *World Atlas of Holocene Sea Level Changes*. Amsterdam: Elsevier. Oceanography Series, Vol. 58. 300 p.

Pirazzoli, P. A., 1996. *Sea Level Changes: The Last 20,000 Years*. New York: Wiley. 211 p.

Posamentier, H. W., and Allen, G. P., 1999. Siliciclastic: concepts and applications. *SEPM Concepts in Sedimentology and Paleontology:*, **7**(210), 19.

Raisbeck, G., Raynaud, D., Schilt, A., Schwander, J., Selmo, E., Souchez, R., Spahni, R., Stauffer, B., Steffensen, J. P., Stenni, B., Stocker, T. F., Tison, J. L., Werner, M., and Wolff, E. W., 2007. Orbital and millennial Antarctic climate variability over the past 800,000 years. *Science*, **317**, 793–796.

Richter, A., Groh, A., and Dietrich, R., 2012. Geodetic observation of sea-level change and crustal deformation in the Baltic Sea region. *Physics and Chemistry of the Earth*, **53–54**, 43–53.

Rohling, E. J., Grant, K., Hemleben, C., Siddall, M., Hoogakker, B. A. A., Bolshaw, M., and Kucera, M., 2008. High rates of sea-level rise during the last interglacial period. *Nature Geoscience*, **1**, 38–42.

Rohling, E. J., Grant, K., Bolshaw, M., Roberts, A. P., Siddall, M., Hemleben, C., Kucera,M., 2009: Antarctic temperature and global sea level closely coupled over the past five glacial cycles. Nature Geoscience, **2**, 500–504.

Siegenthaler, U., Stocker, T. F., Monnin, E., Lüthi, D., Schwander, J., Stauffer, B., Raynaud, D., Barnola, J.-M., Fischer, H., Masson-Delmotte, V., and Jouzel, J., 2005. Stable carbon cycle–climate relationship during the Late Pleistocene. *Science*, **25**(310), 1313–1317, doi:10.1126/science.1120130.

Sloss, L. L., 1963, Sequences in the cratonic interior of North America: Geological Society of America Bulletin, v. 74, p. 93–114.

Suess, E., 1906. *Das Antlitz der Erde*. Wien/Leipzig: F. Tempsky/G. Freytag. 789 p.

Vail, P. R., Mitchum, R. M., Jr., and Thompson, S., 1977. Relative changes of sea level from coastal onlap. In Payton, C. E. (ed.), *Seismic Stratigraphy – Application to Hydrocarbon Exploration*. American Association Petroleum Geologists Memoir 26, pp. 63–82.

Vail, P. R., Audemard, F., Bowman, S. A., Eisner, P. N., and Perez-Cruz, C., 1991. The stratigraphic signatures of tectonics, eustasy and sedimentology – an overview. In Einsele, G., Ricken, W., and Seilacher, A. (eds.), *Cycles and Events in Stratigraphy*. Berlin: Springer, pp. 617–659.

von Bubnoff, S., 1954. *Grundprobleme der Geologie*. Berlin: Akademie Verlag. 234 p.

Watts, A. B., 1982. Tectonic subsidence, flexure and global change in sea level. *Nature*, **297**, 49–474.

Wilgus, C. K., Hastings, B. S., Kendall, C. G., Posamentier, H. W., Ross, C. A., and VanWagoner, J. C. (eds.), 1988. *Sea Level Changes: An Integrated Approach*. Society of Economic Paleontologists and Mineralogists, Special Publication, Vol. 42. 407 p.

Cross-references

Marine Regression
Marine Transgression
Sea-Level
Sedimentary Sequence
Sequence Stratigraphy

SALT DIAPIRISM IN THE OCEANS AND CONTINENTAL MARGINS

Sergey S. Drachev
ExxonMobil House, ExxonMobil International Ltd.,
Leatherhead, UK

Definition

Salt diapir is a multiform-shaped body of crystalline aggregate of evaporite minerals (mainly halite) that has discordant contacts with the encompassing sedimentary strata and formed due to salt movement. The process of SD formation, often referred to as diapirism, is accompanied by deformations significant in scale and magnitude of both overburden rocks and salt itself. It is an integral part of the phenomenon known as salt tectonics, or halokinesis. The word "diapir" comes from the Greek verb diaperein ($διαπερείν$) – "to pierce."

Introduction

Historically the salt diapirs have for long been known in the northern German Plain, around the northern Caspian Sea, in western Iran, coastal areas of the Mediterranean Sea and the Red Sea, East Texas, Zagros, and many other areas on all continents except for Antarctica (Figure 1). In 1901, the Spindletop Dome discovery in Beaumont (Texas) revealed an important role of salt in formation of the hydrocarbon accumulations, which has later been confirmed by discovery of the world's largest hydrocarbon provinces in salt-bearing sedimentary basins (e.g., Gulf of Mexico, North Sea, Campos and Santos basins, Lower Congo Basin, Zagros Fold Belt). Since then the prime interest in salt tectonics comes from the petroleum industry.

Post-WWII expansion of petroleum exploration offshore and development of marine geophysical methods, especially multichannel seismic reflection technique, led to discovery of the salt diapirs on many continental margins of Atlantic and Indian oceans (Figure 1). In the recent times with the intensification of deep-water petroleum exploration the salt diapirs have been revealed at the water depth of 3000 m and greater.

As many deep-water salt-bearing basins have been proven to contain tremendous volumes of hydrocarbons, the structural styles produced by the salt tectonics are well studied and characterized. Among those, the Gulf of Mexico, Brazilian, and Angolan Atlantic continental margins (Figure 1) are the best explored and represent a unique natural laboratory for studying formation of the salt diapirs. Significant progress in imaging complex salt-related deformations has been achieved due to recent developments in the 3D multichannel seismic reflection method and in the seismic data processing with implementation of new migration algorithms such as reverse-time migrations (Leveille et al., 2011; Davison et al., 2013). A considerable effort has also been focused on development of various numerical (Poliakov et al., 1996; Chemia et al., 2007; Fuchs et al., 2011) and analog (Vendeville et al., 1995; Rowan and Vendeville, 2006; Fort and Brun, 2012; Warsitzka et al., 2013) models, which represent inexpensive and powerful tools for studying the salt tectonics in various structural and geodynamic settings.

Classification of salt diapirs

There are several ways to characterize the salt diapirs. The most common classifications are based on morphological characteristics of the features and on inferred mechanisms of their formation.

Jackson and Talbot (1991) characterized the salt diapirs as masses of salt that have flowed ductilely and appear to have discordantly pierced or intruded the overburden. Diapirs begin as anticlinal or domal uplifts and evolve into

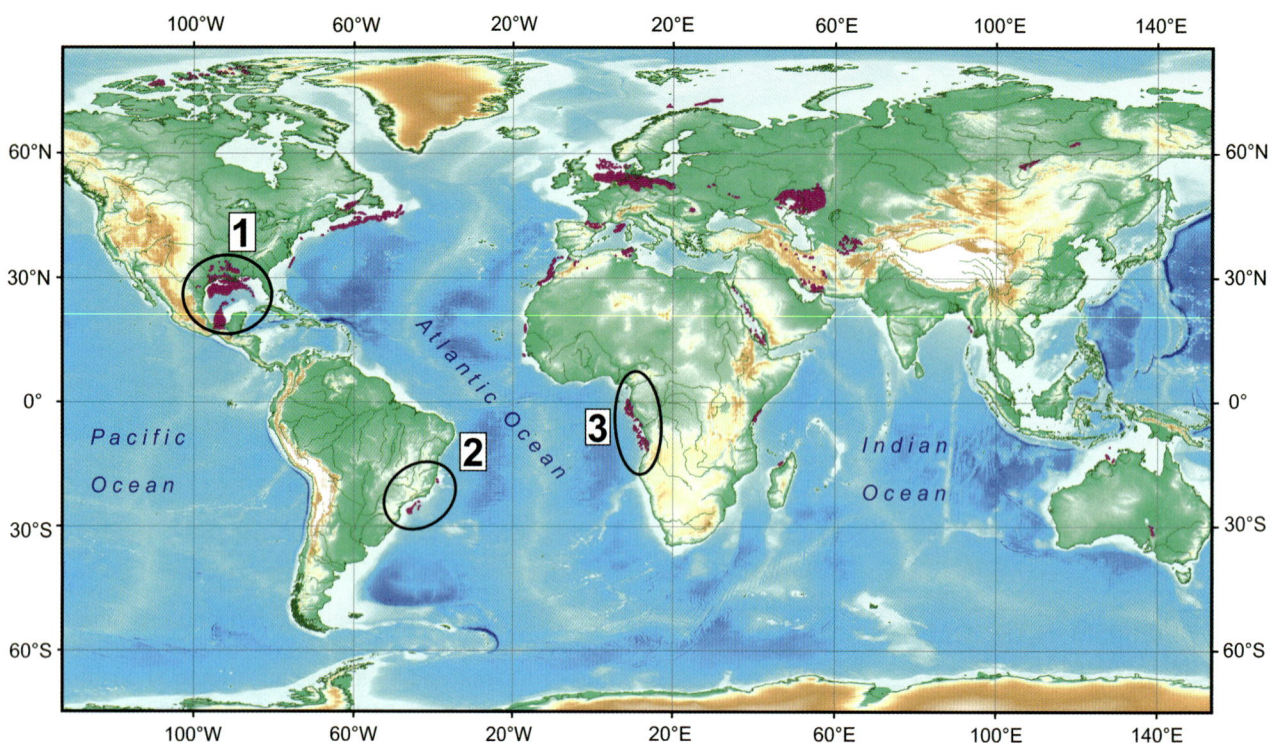

Salt Diapirism in the Oceans and Continental Margins, Figure 1 Global occurrence of salt-bearing sedimentary basins. Salt diapirs are shown in magenta. Black ellipses outline three deep-water salt tectonics regions referred to in text: *1*, Gulf of Mexico; *2*, Brazilian Atlantic margin; *3*, West African Atlantic margin.

pillows, domes, and mushroom-shaped diapirs, columns, canopies, bulbs, tongues, walls, and more complex features (Figure 2).

Salt density can be higher or equivalent to the overburden density during commencement of salt deformation (Hudec et al., 2009), and in order for a salt diapir to be emplaced into its overburden any rock occupying that space must be removed or displaced (Hudec and Jackson, 2007). Vendeville and Jackson (1992) demonstrated that even when a stiff overburden buries a less dense rock salt, this may never lead to formation of a diapir unless the overburden is thinned and weakened by a tectonic extension. Therefore, the overburden has to be weakened by tectonic faulting until salt starts to rise up (Vendeville and Jackson, 1992; Jackson and Vendeville, 1994). This concept has been utilized by numerous analog models (Nalpas and Brun, 1993; Withjack and Callaway, 2000; Le Calvez and Vendeville, 2002) and successfully applied to natural structures interpreted on seismic lines (Seni and Jackson, 1983; Koyi et al., 1993).

Dynamically, the salt diapirs can be reactive, active, passive, or dormant (Yin and Groshong, 2006). Figure 3 illustrates different stages of the diapir formation. The diapirism can be triggered when the overburden rocks are subjected to extension, compression, uplift, and erosion (reactive diapirism). Very often salt rises in the footwalls of normal faults where the sedimentation rate remains low (Vendeville and Jackson, 1992; Quirk and Pilcher, 2012). The faulted overburden lets a reactive diapir move up (Figure 3a). This gives way to an active diapirism (b). Once salt reaches the surface, it can continue to rise by passive diapirism (c), in which the diapir grows as sediments accumulate around it. A rapidly rising passive diapir may spread over the sediment surface to form an allochthonous salt sheet (d).

According to Koyi (1998), the salt diapir geometry may be the result of different combinations of six parameters, namely, the rates of sediment accumulation, rate of salt supply, and rates of extension/erosion/dissolution/shortening at different times.

Driving forces

The initiation of salt diapirs can be realized by several driving forces and/or their combinations including buoyancy, tectonic stresses, gravity-driven gliding or differential loading (Ge et al., 1997; Hudec and Jackson, 2007). The latter two are being discussed as primary forces causing salt movement at the continental margins (Brun and Fort, 2011, 2012; Rowan et al., 2012).

According to Brun and Fort (2011), available numerical and physical models of salt tectonics at the passive continental margins fall into two end-member categories: (i) dominant gravity gliding primarily driven by margin tilt and (ii) pure gravity spreading driven only by sedimentary loading (Figure 4).

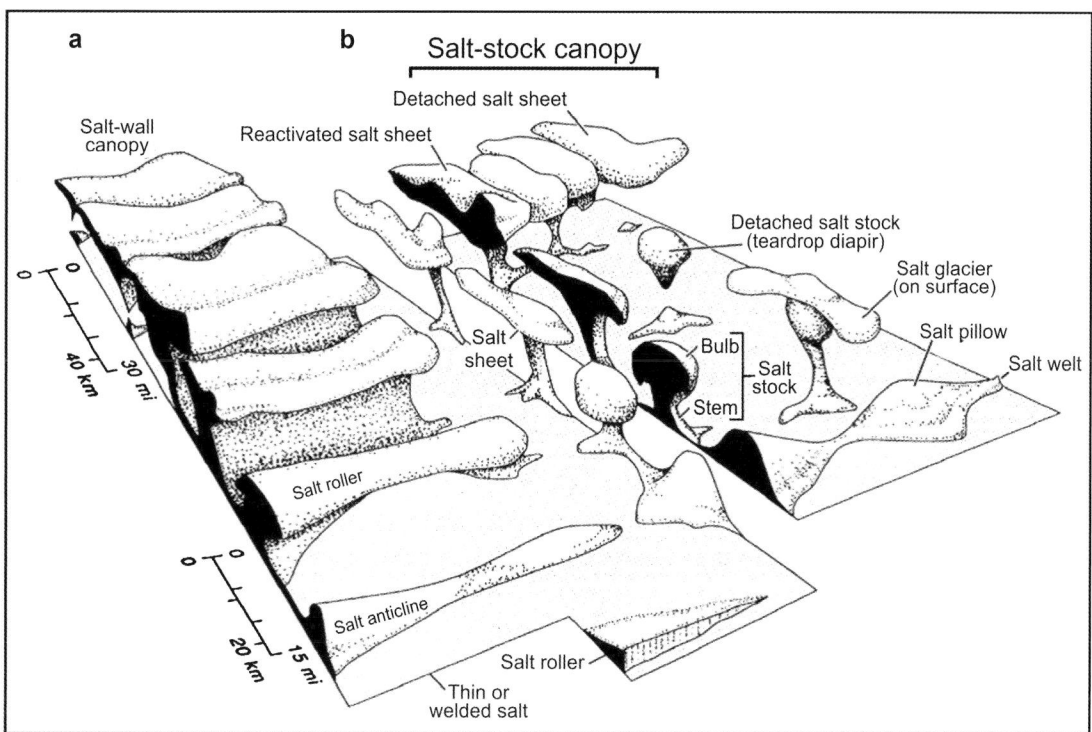

Salt Diapirism in the Oceans and Continental Margins, Figure 2 Main types of salt structures that develop from linear sources (**a**) and point sources (**b**) (modified from Hudec and Jackson 2007).

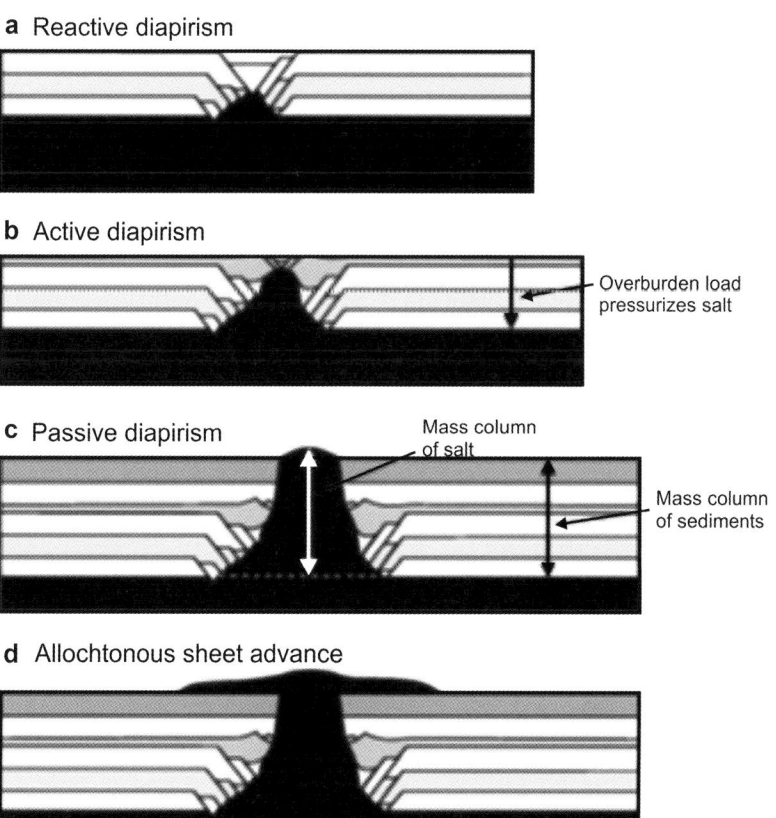

Salt Diapirism in the Oceans and Continental Margins, Figure 3 Progressive development of a diapir from reactive to passive stage. *Black* represents the salt layer and *white-gray* layers sedimentary overburden (modified from Hudec and Jackson 2007).

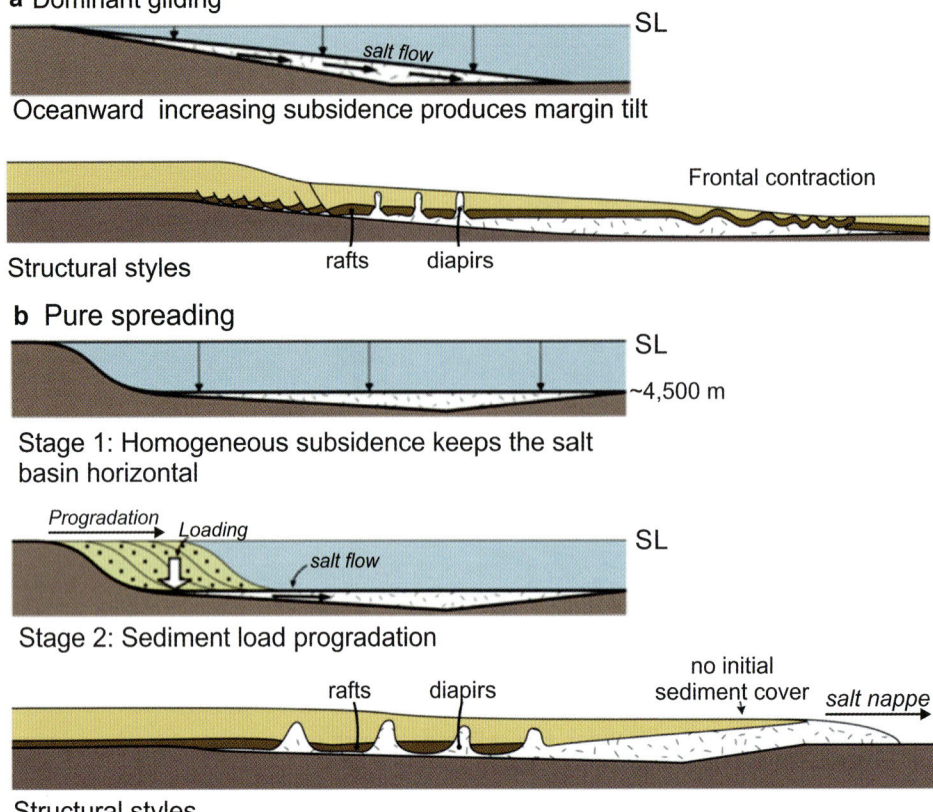

Salt Diapirism in the Oceans and Continental Margins, Figure 4 Two models of salt tectonics: Dominant gliding (a) and pure spreading (b) (modified from Brun and Fort 2011).

Dominant gliding models: salt tectonics primarily driven by margin tilt

The dominant gliding concept that drives a two-layer salt-sediment system down an inclined basal surface has mostly been inspired by studies carried out on the South Atlantic Western African (Duval et al., 1992; Tari et al., 2003) and Brazilian (Cobbold and Szatmari, 1991; Demercian et al., 1993) continental margins and in the North Sea (Nalpas and Brun, 1993; Stewart and Coward, 1995). Different types of laboratory experiments (Cobbold and Szatmari, 1991; Mauduit et al., 1997; Fort et al., 2004; Brun and Mauduit, 2009) or numerical modelling (Ings et al., 2004; Albertz et al., 2010; Quirk et al., 2012) have been carried out to characterize the structures produced by this mechanism.

The dominant gliding model (Figure 3a) utilizes the updip extension represented by a broad rafted domain, which is balanced by the downdip contraction in the form of salt tongues, sheets, canopies, and the progressive inflation of a massive salt domain at the basinward edge of the salt basin. Downdip migration of extension and updip migration of contraction lead to a contractional inversion of previously extensional structures and produce the squeezed diapirs (Brun and Fort, 2011). The efficiency of this gravity sliding across the whole margin is due to the more or less uniform original distribution of salt in the postrift succession forming a continuous detachment level (Tari et al., 2003).

Pure spreading models: salt tectonics only driven by differential sedimentary loading

Apart from the buoyancy, basal slope, and tectonic forces, diapirism can also be driven by a pressure gradient caused by differential loading induced by thickness or density variation in the overburden (Ge et al., 1997; Gaullier and Vendeville, 2005; Vendeville, 2005; Hudec and Jackson, 2007). This concept rose from the Gulf of Mexico studies (Rowan, 1995), where the salt is very thick and sedimentation rates rather high. Different types of laboratory experiments (Ge et al., 1997; Gaullier and Vendeville, 2005; Vendeville, 2005; Krézsek et al., 2007) and numerical modeling (Cohen and Hardy, 1996; Gemmer et al., 2004; Ings et al., 2004; Ings and Shimeld, 2006; Gradmann et al., 2009) were carried out to characterize the salt tectonic systems driven by sedimentary wedge progradation on top of salt either covered or not covered by sediments at the distal oceanward flank of the basin.

The pure spreading models require (Brun and Fort, 2011): (a) thick sedimentary overburden and significant water depths (c. 4,000–4,500 m), (b) high sediment density, (c) high fluid pore pressure in the sediments, and (d) a seaward free boundary of the salt basin. In pure spreading, extension is located inside and contraction at the tip of the prograding sedimentary wedge (Figure 3b). Both extension and contraction migrate seaward with the sediment progradation. Migration of the deformation can create an extensional inversion of previously contractional structures (Brun and Fort, 2011).

The growth of an individual salt diapir is considered in many models and interpretations as a product of upbuilding due to the classical Rayleigh–Taylor instability (Schmeling, 1987). However, it has for long been clear that many salt diapirs formed already during sedimentation by downbuilding in a brittle overburden when the salt diapir top remains at the surface all the time during its formation (Barton, 1933; Parker and McDowell, 1955; Seni and Jackson, 1983; Vendeville and Jackson, 1992; Koyi, 1998).

Bibliography

Albertz, M., Beaumont, C., Shimeld, J. W., Ings, S. J., and Gradmann, S., 2010. An investigation of salt tectonics structural styles in the Scotian Basin, offshore Atlantic Canada: paper 1, comparison of observations with geometrically simple numerical models. *Tectonics*, **29**, TC4017, doi:10.1029/2009TC002539.

Barton, D. C., 1933. Mechanics of formation of salt domes with special reference to Gulf Coast salt domes of Texas and Louisiana. *American Association of Petroleum Geologists Bulletin*, **17**, 1025–1083.

Brun, J.-P., and Fort, X., 2011. Salt tectonics at passive margins: geology versus models. *Marine and Petroleum Geology*, **28**, 1123–1145.

Brun, J.-P., and Fort, X., 2012. Salt tectonics at passive margins: geology versus models – response. *Marine and Petroleum Geology*, **37**, 195–208.

Brun, J.-P., and Mauduit, T., 2009. Salt rollers: structure and kinematics from analogue modelling. *Marine and Petroleum Geology*, **26**, 249–258.

Chemia, Z., Koyi, H., and Schmeling, H., 2007. Numerical modelling of rise and fall of a dense layer in salt diapirs. *Geophysical Journal International*, **172**, 798–816.

Cobbold, P. R., and Szatmari, P., 1991. Radial gravitational gliding on passive margins. *Tectonophysics*, **188**, 249–289.

Cohen, H. A., and Hardy, S., 1996. Numerical modelling of stratal architectures resulting from differential loading of a mobile substrate. In Alsop, G. I., Blundell, D. J., and Davison, I. (eds.), *Salt Tectonics*. London: Geological Society, Special publication, 100, pp. 265–273.

Davison, I., Jones, I., and Waltham, D., 2013. Seismic imaging of salt diapirs: problems and pitfalls. In *Proceedings of 13th International Congress of the Brazilian Geophysical Society and EXPOGEF*, pp. 1332–1336.

Demercian, L. S., Szatmari, P., and Cobbold, P. R., 1993. Style and pattern of salt diapirs due to thin-skinned gravitational gliding, Campos and Santos basins, offshore Brazil. *Tectonophysics*, **228**, 393–433.

Duval, B., Cramez, C., and Jackson, M. P. A., 1992. Raft tectonics in the Kwanza Basin, Angola. *Marine and Petroleum Geology*, **9**, 389–404.

Gaullier, V., and Vendeville, B. C., 2005. Salt tectonics driven by sediment progradation. Part II: radial spreading of sedimentary lobes prograding above salt. *American Association of Petroleum Geologists Bulletin*, **89**, 1081–1089.

Ge, H., Jackson, M. P. A., and Vendeville, B. C., 1997. Kinematics and dynamics of salt tectonics driven by progradation. *American Association of Petroleum Geologist Bulletin*, **81**, 398–423.

Gemmer, L., Ings, S. J., Medvedev, S., and Beaumont, C., 2004. Salt tectonics driven by differential sediment loading: stability analysis and finite element experiments. *Basin Research*, **16**, 199–219.

Gradmann, S., Beaumont, C., and Albertz, M., 2009. Factors controlling the evolution of the Perdido Fold Belt, northwestern Gulf of Mexico, determined from numerical models. *Tectonics*, **28**, 1–28.

Fort, X., Brun, J.-P., and Chauvel, F., 2004. Salt tectonics on the Angolan margin, synsedimentary deformation processes. *American Association of Petroleum Geologists Bulletin*, **88**, 1523–1544.

Fort, X., Brun, J.-P., 2012. Kinematics of regional salt flow in the northern Gulf of Mexico. In Alsop, G. I., Archer, S. G., Hartley, A. J., Grant, N. T., and Hodgkinson, R. (eds.), *Salt Tectonics, Sediments and Prospectivity*. London: Geological Society, Special publications, 363, pp. 265–287.

Fuchs, L., Schmeling, H., and Koyi, H., 2011. Numerical models of salt diapir formation by down-building: the role of sedimentation rate, viscosity contrast, initial amplitude, and wavelength. *Geophysical Journal International*, **186**, 390–400.

Hudec, M. R., Jackson, M. P. A., and Schultz-Ela, D. D., 2009. The paradox of mini-basin subsidence into salt: clues to the evolution of crustal basins. *Geological Society of America Bulletin*, **121**, 201–221.

Hudec, M. R., and Jackson, M. P. A., 2007. Terra infirma: understanding salt tectonics. *Earth-Science Reviews*, **82**, 1–27.

Ings, S., Beaumont, C., and Gemmer, L., 2004. Numerical modeling of salt tectonics on passive continental margins: preliminary assessment of the effects of sediment loading, buoyancy, margin tilt, and isostasy. In Post, P.J., Olson, D.L., Lyons, K.T., Palmes, S.L., Harrison, P.F., and Rosen, N.C. (eds.), Salt Sediment Interactions and Hydrocarbon Prospectivity: Concepts, Applications, and Case Studies for the 21st Century. 24th Annual Gulf Coast Section SEPM Foundation Bob F. Perkins Research Conference, Houston, Texas, USA. Proceedings, pp. 36–68.

Ings, S. J., and Shimeld, J. W., 2006. A new conceptual model for the structural evolution of a regional salt detachment on the northeast Scotian margin, offshore Eastern Canada. *American Association of Petroleum Geologists Bulletin*, **90**, 1407–1423.

Jackson, M. P. A., and Talbot, C. J., 1991. *A Glossary of Salt Tectonics*. Bureau of Economic Geology, University of Texas at Austin, Geological Circular 91-4.

Jackson, M. P. A., and Vendeville, B. C., 1994. Regional extension as a geological trigger for diapirism. *Geological Society of America Bulletin*, **106**, 57–73.

Koyi, H. A., Jenyon, M. K., and Petersen, K., 1993. The effect of basement faulting on diapirism. *Journal of Petroleum Geology*, **16**(3), 285–312.

Koyi, H., 1998. The shaping of salt diapirs. *Journal of Structural Geology*, **20**, 321–338.

Krézsek, C., Adam, J., and Grujic, D., 2007. Mechanics of fault and expulsion rollover systems developed on passive margins detached on salt: insights from analogue modelling and optical strain monitoring. In Jolley, S. J., Barr, D., Walsh, J. J., and Knipe, R. J. (eds.), *Structurally Complex Reservoirs*. London: Geological Society, Special publications, 292, pp. 103–121.

Le Calvez, J. H., and Vendeville, B. C., 2002. Physical modeling of normal faults and graben relays above salt: a qualitative and quantitative analysis. *Gulf Coast Association of Geological Societies Transactions*, **52**, 599–606.

Leveille, J. P., Jones, I. F., Zhou, Z., Wang, B., and Liu, F., 2011. Subsalt imaging for exploration, production, and development: a review. *Geophysics*, **76**, WB3–WB20.

Mauduit, T., Guérin, G., Brun, J.-P., and Lecanu, H., 1997. Raft tectonics: the effects of basal slope value and sedimentation rate on progressive extension. *Journal of Structural Geology*, **19**, 1219–1230.

Nalpas, T., and Brun, J. P., 1993. Salt flow and diapirism related to extension at crustal scale. *Tectonophysics*, **228**, 349–362.

Parker, T. J., and Mcdowell, A. N., 1955. Model studies of salt-diapir tectonics. *American Association of Petroleum Geologists Bulletin*, **39**, 2384–2470.

Poliakov, A. N. B., Podladchikov, Yu., Dawson, E. C, and Talbot, C. J., 1996. Salt diapirism with simultaneous brittle faulting and viscous flow. In Alsop, G. I., Blundell, D. J., and Davison. I. (eds.), *Salt Tectonics*. London: Geological Society, Special publications, 100, pp. 291–302.

Quirk, D. G., and Pilcher, R. S., 2012. Flip-flop salt tectonics. In *Salt Tectonics, Sediments and Prospectivity*. London: Geological Society, Special publications, 363, pp. 245–264.

Quirk, D.G., Hsu, D., Bissada, M., Ambirk, D., Emily Ferguson, E., Kendrick, T., Chigozie, T., Nwokeafor, C., Seidler, L., and Nielsen, M., 2012. Extensional salt tectonics on passive margins: examples from Santos, Campos and Kwanza basins. In *Salt Tectonics, Sediments and Prospectivity*. London: Geological Society, Special publications, 363, pp. 207–244.

Rowan, M. G., 1995. Structural styles and evolution of allochthonous salt, central Louisiana outer shelf and upper slope. In Jackson, M. P. A., Roberts, D. G., and Snelson, S. (eds.), *Salt Tectonics, A Global Perspective*. American Association of Petroleum Geologists Memoir 65, pp. 199–228.

Rowan, M. G., and Vendeville, B. C., 2006. Foldbelts with early salt withdrawal and diapirism: physical model and examples from the northern Gulf of Mexico and the Flinders Ranges, Australia. *Marine and Petroleum Geology*, **23**, 871–891.

Rowan, M. G., Peel, F. J., Vendeville, B. C., and Gaullier, V., 2012. Salt tectonics at passive margins: geology versus models – discussion. *Marine and Petroleum Geology*, **37**, 184–194.

Schmeling, H., 1987. On the relation between initial conditions and late stages of Rayleigh-Taylor instabilities. *Tectonophysics*, **133**, 65–80.

Seni, S. J., and Jackson, M. P. A., 1983. Evolution of salt structures, east Texas diapir province. Part 2: patterns and rates of halokinesis. *American Association of Petroleum Geologists Bulletin*, **67**, 1245–1274.

Stewart, S. A., and Coward, M. P., 1995. Synthesis of salt tectonics in the southern North Sea, UK. *Marine and Petroleum Geology*, **12**, 457–475.

Tari, G., Molnar, J., and Ashton, P., 2003. Examples of salt tectonics from West Africa: a comparative approach. In Arthur, T. J., Macgregor, D. S., and Cameron, N. R. (eds.), *Petroleum Geology of Africa: New Themes and Developing Technologies*. London: Geological Society, Special publications, 207, pp. 85–104.

Vendeville, B. C., 2005. Salt tectonics driven by sediment progradation: part I – mechanics and kinematics. *American Association of Petroleum Geologists Bulletin*, **89**, 1071–1079.

Vendeville, B. C., and Jackson, M. P. A., 1992. The rise of diapirs during thin-skinned extension. *Marine and Petroleum Geology*, **9**, 331–353.

Vendeville, B. C., Ge, H., and Jackson, M. P. A., 1995. Scale models of salt tectonics during basement-involved extension. *Petroleum Geoscience*, **1**, 179–183.

Warsitzka, M., Kley, J., and Kukowski, N., 2013. Salt diapirism driven by differential loading – some insights from analogue modelling. *Tectonophysics*, **591**, 83–97.

Withjack, M.O., Callaway, J.S., 2000. Active normal faulting beneath a salt layer-an experimental study of deformation in the cover sequence: American Association of Petroleum Geologists. *Bulletin* **84**, 627–651.

Yin, H., and Groshong, R. H., Jr., 2006. Balancing and restoration of piercement structures: geologic insights from 3D kinematic models. *Journal of Structural Geology*, **28**, 99–114.

Cross-references

Marine Evaporites
Marine Sedimentary Basins

SAPROPELS

Rüdiger Stein
Alfred Wegener Institute, Helmholtz Centre for Polar and Marine Research (AWI), Bremerhaven, Germany

The term "sapropel," originating from the Greek *sapros* (rotten) and *pelos* (soil) and already introduced by Potonié (1904) as international term for the German word *faulschlamm*, is used in marine geosciences to describe dark-colored, unconsolidated, fine-grained, and often laminated sediments that are enriched in organic carbon. Following the first more quantitative definition proposed by Kidd et al. (1978), the organic carbon content of common sapropels is greater than 2 wt%.

Sapropels seem to be a characteristic deposit in a variety of settings, including semi-isolated basins with restricted bottom circulation (e.g., Mediterranean Sea, Japan Sea, Santa Barbara Basin) and portions of continental margins that lie within the mid-water oxygen minimum zone and below upwelling zones. The late Pleistocene to Holocene Mediterranean sapropels, however, are probably the best studied ones. After the first recovery of these sapropels during the Swedish Deep-Sea Expedition in 1947 (Kullenberg, 1952), they have been sampled in numerous conventional gravity cores as well as drill cores (recovered during the Deep Sea Drilling Project – DSDP – and the Ocean Drilling Program – ODP) over the last decades, and several different hypotheses on their origin and significance have been published since then (e.g., Ryan and Cita, 1977; Rohling and Hilgen, 1991; Cramp and O'Sullivan, 1999; Emeis and Weissert, 2009). During ODP Expedition 160, a complete Pliocene to Holocene sequence containing more than 80 discrete sapropels was recovered in the Eastern Mediterranean (Emeis and Weissert, 2009), though the earliest reported occurrence is from the Middle Miocene (Kidd et al., 1978).

The strong correlation between sapropel formation and minima in the precession index, already recognized by Rossignol-Strick (1985), suggests an orbital forcing control. That means, an intensified summer monsoon may have resulted in increased freshwater/river discharge in the eastern Mediterranean. At that time periods, the freshening (and warming) of surface waters enhanced stratification and inhibited or weakened deepwater formation and oxygen advection, and – finally – led to dysoxic to anoxic bottom waters that allowed the formation and preservation of sapropels (Rohling and Hilgen, 1991; Emeis and Weissert, 2009). In addition, increased river discharge may also have caused enhanced primary production due to river-borne nutrient supply as well as increased input of terrigenous organic matter, both positive feedback mechanisms for the formation of organic carbon rich sediments. For further detailed discussion of different hypotheses of sapropel formation, the reader is referred to the reviews by Rohling and Hilgen (1991) and Emeis and Weissert (2009).

Bibliography

Cramp, A., and O'Sullivan, G., 1999. Neogene sapropels in the Mediterranean: a review. *Marine Geology*, **153**, 11–28.

Emeis, K.-C., and Weissert, H., 2009. Tethyan-Mediterranean organic carbon rich sediments from Mesozoic black shales to sapropels. *Sedimentology*, **56**, 247–266.

Kidd, R. B., Cita, M. B., and Ryan, W. B. F., 1978. Stratigraphy of eastern Mediterranean sapropel sequences recovered during Leg 42A and their paleoenvironmental significance. *Initial Rep DSDP*, **42A**, 421–443.

Kullenberg, B., 1952. On the salinity of water contained in marine sediments. *Meddelanden fran Oceanografiska Institutet i Göteborg*, **21**, 1–38.

Potonié, H., 1904. Über Faulschlamm-(Sapropel)-Gesteine. Sitz. Gesell. Nat. V. Berlin, pp. 243–245.

Rohling, E. J., and Hilgen, F. J., 1991. The eastern Mediterranean climate at times of sapropel formation: a review. *Geologie en Mijnbow*, **70**, 253–264.

Rossignol-Strick, M., 1985. Mediterranean Quaternary sapropels: an immediate response of the African monsoon to variation of insolation. *Palaeogeography, Palaeoclimatology, Palaeoecololgy*, **49**, 237–265.

Ryan, W. B. F., and Cita, M. B., 1977. Ignorance concerning episodes of ocean-wide stagnation. *Marine Geology*, **23**, 193–215.

Cross-references

Anoxic Oceans
Laminated Sediments

SCLEROCHRONOLOGY

Christian Dullo
GEOMAR Helmholtz Centre for Ocean Research, Kiel, Germany

Sclerochronology is the record of different periodicities expressed as chemical and physical variations in mineralized endo- or exoskeletons of living, fossil, and even extinct aquatic organisms.

Sclerochronological periodicities are documented in growth increments of different shapes and sizes, depending on the time span they represent (Schöne and Surge, 2005). Such increments may portray days, lunar cycles, months, or years. Their individual width and pattern reflect either specific taxonomically related and therefore biologically controlled signals or environmental conditions in which the organism grew. These skeletal chronologies may comprise individual lifetimes of organisms ranging from several decades to several hundreds of years. Prominent organisms are scleractinian corals, calcified sponges, mollusk shells, or otoliths from fish. Along with the record of growth increments, the chemical composition of the mineralized skeleton based on elemental ratios such as Sr/Ca and Mg/Ca or the isotopic ratios of oxygen or carbon provides valuable information for the proxy-related reconstruction of paleoenvironmental, paleoecological, and paleoclimatological conditions as well as for the evaluation of pollution and global change. Since direct measurements of marine physical and chemical parameters are very limited in time, such sclerochronological data provide information both on the lifetime and history of the organism and on the dynamics of the environment for longtime series. This sclerochronological approach is the only way to study past climate dynamics on oceanographic and meteorological scales for several 100 years and to constrain and validate models for prediction. The term sclerochronology was first coined by Buddemeier et al. (1974) in order to present the growth bands of coral revealed by radiographic studies. Sclerochronological patterns are the aquatic equivalent to tree rings, known as dendrochronology from terrestrial settings. Like in dendrochronology, composed time series of skeletons of different ages enlarge the chronology further back in time, which is currently under development for mollusks.

Bibliography

Buddemeier, R. W., Maragos, J. E., and Knutson, D. W., 1974. Radiographic studies of reef coral exoskeletons: rates and patterns of coral growth. *Journal of Experimental Marine Biology and Ecology*, **14**, 179–199.

Schöne, B. R., and Surge, D., 2005. Looking back over skeletal diaries – high-resolution environmental reconstructions from accretionary hard parts of aquatic organisms. *Palaeogeography, Palaeoclimatology, Palaeoecology*, **228**, 1–3.

Cross-references

Reefs (Biogenic)

SEA WALLS/REVETMENTS

Louise Wallendorf
Hydromechanics Laboratory, United States Naval Academy, Annapolis, MD, USA

Seawalls are vertical walls built to delineate the border between sea and land, in an area where the upland contains infrastructure that requires protection from storm surge and wave overtopping during an extreme storm event. Seawalls are often constructed in regions that have experienced high erosion and have limited or no beachfront, unless with foresight they are embedded in a natural dune on a beach and only become exposed during an extreme storm; one of the most well-known seawalls is the Galveston Seawall constructed after the hurricane of 1900 and 1915 and chronicled in detail (Larson, 1999). Seawalls are typically constructed from concrete, with deep heavy foundations often tied to the backshore to prevent overturning of the structure and caps to prevent erosion of the backfill. Critical design parameters during extreme events are the crest elevation and the depth of the wall below the existing sea bottom. The wall structure needs to withstand the weight of the soil backfill and cap and the hydraulic forces on the front and back of the wall; a means of draining water from rainfall and flooding from behind the wall is necessary. In contrast, a bulkhead is a vertical wall constructed of timber and pilings or sheet pile and designed to retain soil at the shoreline without accounting for flooding; its strength is derived solely from the soil backfill. Seawall construction on a shoreline with existing beach is a controversial topic with the public, who feels they cause excessive beach erosion and/or changes in bottom topography which alter swimming and surfing wave conditions. A thorough investigation of the effects of seawalls on the beach (Krauss, 1988) provides recommendations for their use.

Revetments are sloped surface layers designed to protect the shoreline from erosion from waves and currents. They can be as simple as grass on packed clay, as is in river levees, or as more complex armored layers of rock or interlocking man-made elements on top of a soil slope. The revetment design evolved from dike construction in the Netherlands and is discussed in detail (Pilarczyk, 1990). Slopes are selected to prevent sliding of the soil and armor material, with protection at the toe to prevent undercutting. Heights are chosen to prevent flooding in the lee.

Geotextiles are used in the design of both revetments and seawalls to (a) stabilize an upland bank/bluff, (b) retain the soil behind the structure, and (c) protect the toe and prevent the erosion of soil back to the sea. Care needs to be taken at the transition areas at the ends of a revetment/seawall to prevent flanking erosion, defined as leakage of fill at the ends, or discontinuities in the shoreline erosion rates. A smooth vertical front face of a seawall is a perfect wave reflector. Curving or roughening the front surface helps attenuate some of the wave energy.

Bibliography

Krauss, N. C. (ed.), 1988. *The Effects of Seawalls on the Beach.* Journal of Coastal Research, Special Issue #4. Charlottesville: CERF.
Larson, E., 1999. *Isaac's Storm: A Man, A Time and the Deadliest Storm in History.* New York: Crown.
Pilarczyk, K. W., 1990. Design of seawalls and dikes – including overview of revetments. In *Coastal Protection.* Rotterdam: Balkeema, pp. 197–288.

Cross-references

Coastal Engineering

SEA-LEVEL

Reinhard Dietrich
Institute of Planetary Geodesy, Technical University of Dresden, Dresden, Germany

Definition

The surface which separates the global ocean and the atmosphere is called the sea surface. A measure of this boundary surface is called a sea-level measurement. There are several definitions for quantities related to sea-level (Figure 1). The distance between a reference ellipsoid and the instantaneous sea-level is called an instantaneous sea-surface height. The distance between the reference ellipsoid and a mean sea-level is called mean sea-surface height. The geoid can be defined as the equipotential surface of the Earth's gravity field which fits best to the global mean sea surface at a given epoch. The separation between geoid and mean sea surface is called mean sea-surface topography, whereas the distance from the geoid to the instantaneous sea-level is called absolute dynamic topography. The sea-level with respect to the solid Earth is called relative sea-level.

Observation of the sea-level

Geological records

There exist a variety of geological indications of former (relative) sea-level heights. For a good review, see Church et al. (2010). Techniques include the use of salt marsh data (e.g., Gehrels and Woodworth, 2013). Also corals can help to reconstruct sea-level curves at specific sites (e.g., Yokoyama et al., 2006). Deposits of marine shells or other species which are located above the present sea-level (Pluet and Pirazzoli, 1991; Weidick, 1996a) can provide limits on former sea levels heights. Also archeological sites below or above the present sea-level can be used to set constraints on historical sea-level heights (Weidick, 1996b; Antonioli et al., 2007; Lübke et al., 2011).

Instrumental period

The relative sea-level can be observed with tide gauges. Any changes of the sea-level are in this case related to the tide gauge benchmark. If a parallel set of measurements of the vertical movement of the tide gauge benchmark with respect to the ellipsoid is obtained, e.g., by GPS (Figure 2), relative sea-level changes can be converted into absolute sea-level changes with respect to the ellipsoid.

In the nineteenth century, the first recording tide gauges were developed which were installed especially in harbor sites for local navigation purposes (Figure 3).

For the open sea, there exist pressure gauges which are moored on the sea bottom (Figure 4). Here, the pressure caused by the water column above the gauge is recorded. Since the air column (air pressure) above the sea surface also affects the recorded pressure, it has to be known in

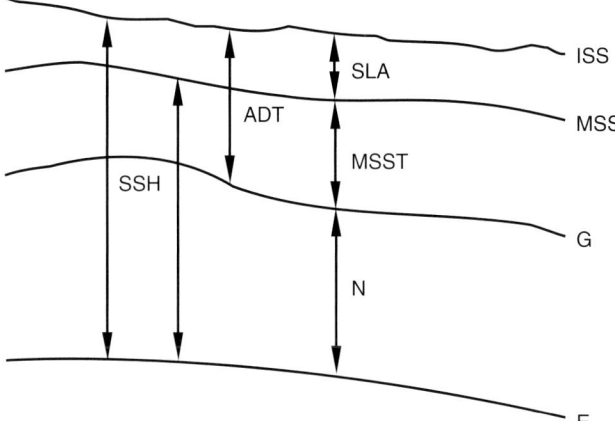

Sea-Level, Figure 1 Sea-level and related quantities (*E* ellipsoid, *G* geoid, *MSS* mean sea surface, *ISS* instantaneous sea surface, *SSH* sea-surface height, *N* geoid height, *ADT* absolute dynamic topography, *MSST* mean sea-surface topography, *SLA* sea-level anomaly).

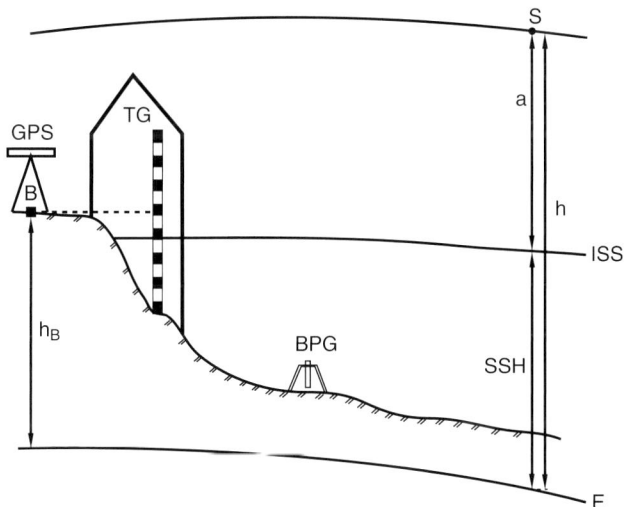

Sea-Level, Figure 2 Observation of sea-level with tide gauge (*TG*, *B* tide gauge benchmark, *GPS* provides the ellipsoidal height h of B), bottom pressure gauge (*BPG*) and altimeter satellite S (the known ellipsoidal height h of S and the altimeter measurement a allow the determination of SSH).

Sea-Level, Figure 3 Tide gauge in a harbor site with GPS setup at a tripod (*left*).

order to reduce the obtained pressure record. In order to convert the (reduced) pressure into the height of the water column above the gauge, the density of the water column (i.e., its temperature and salinity) and the local acceleration due to gravity also have to be known.

For several decades now, the sea-level is also observed by satellite altimetry. In this case the travel time of a microwave or a laser pulse emitted by the satellite and reflected by the ocean surface back to the satellite is measured. With known satellite position with respect to the reference ellipsoid, the range measurement is used to calculate the instantaneous sea-surface height (see, e.g., Bosch et al., 2014).

The variability of sea-level in time and space

The sea-level is never on rest. There exists a variety of effects which cause an instantaneous change of the sea-surface height. On short time scales of seconds to minutes, there are waves (see "Waves"); on minutes to hours, there may be tsunamis (see "Tsunamis"), storm surges, and seiches; and from hours to days, there are the tides. Furthermore, a variation of air pressure will cause a change of the sea-surface height (inverse barometer effect – IB). If these short-term variations of the sea-surface height are averaged, one gets the mean sea-surface height.

The mean sea-surface height also changes on longer time scales. There are effects due to ocean dynamics (weeks to decades), thermal expansion (months and longer), the glacial eustacy (see "Eustasy") (months to millennia) and glacial isostasy (see "Glacio(hydro)-isostatic Adjustment") (years to tens of millennia), and finally also tectonic processes that may be either fast (earthquakes) or slow (time scales of millennia to millions of years). For more details of many of these processes, see Pugh and Woodworth (2014).

Tides

One important component of sea-level variability concerns the ocean tides. The tides are caused by the gravitational forces of the moon and the sun. The tidal potential can be separated into semidiurnal, diurnal, and long-period tides (Figure 5). The tidal variations are not uniform around the globe. The distribution of land masses and the bathymetry of the ocean floor are the reason for large variations of tidal amplitudes and phases, with maximum amplitudes up to 8…10 m, e.g., in the Bay of Fundy and the Bristol Channel. Tidal currents play an important role in sediment dynamics (see "Sediment Dynamics").

Seiches

Seiches are oscillations of the sea-level in semi-enclosed basins, bays and harbors. The period depends on the size and the depth of the water body. In the Baltic Sea, for example, the largest periods have been determined to be 39.4, 22.5, 17.9, and 12.9 h (Lisitzin, 1974). Seiches are resonance phenomena, and therefore even a small forcing caused, for example, by sudden changes of wind conditions may result in large amplitudes.

Storm surges

Specific meteorological conditions, especially strong and rapidly changing wind fields in shallow water areas, play a key role in generating storm surges. Usually catastrophic floodings are the result of an overlay of wind-driven water mass changes with high tides and/or seiches (Figure 6).

Tsunamis

Tsunamis occur whenever large volumes of water are displaced and equilibrium is restored by the propagation of long waves across the ocean. The most devastating tsunamis are caused by undersea earthquakes with the resulting waves traveling across the ocean at the same speed as the tide, often impacting on distant shorelines. However, they can also originate from geological

Sea-Level, Figure 4 Bottom pressure gauge with mooring frame in operation.

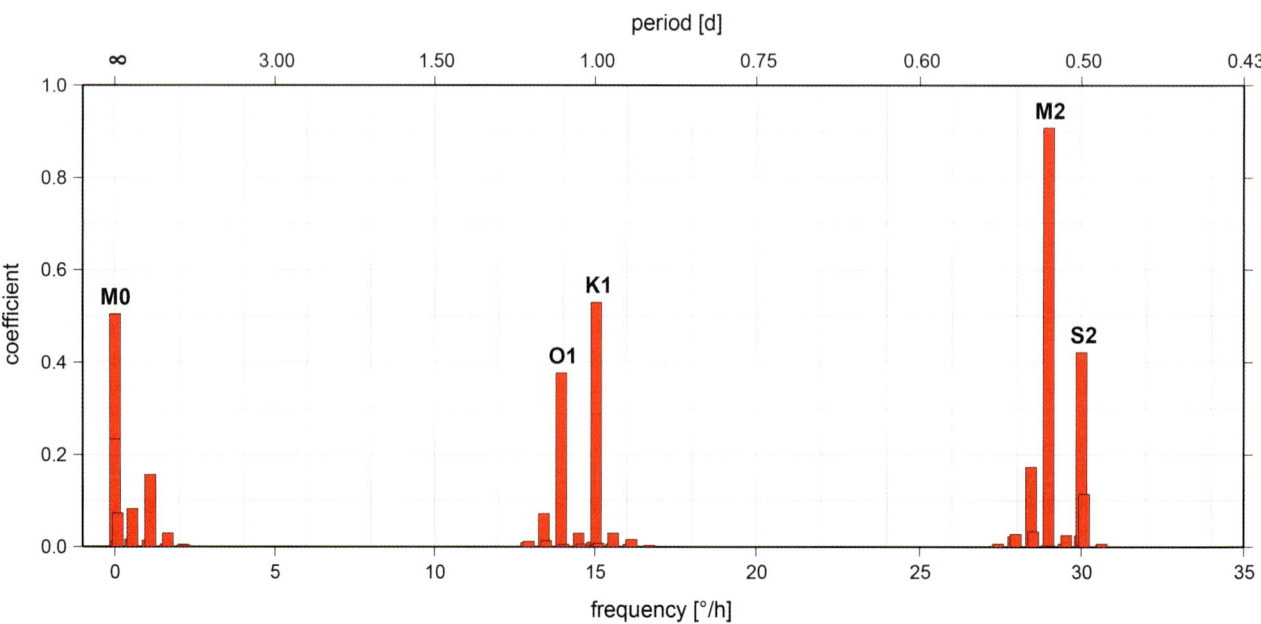

Sea-Level, Figure 5 Spectrum of tides with semidiurnal (*right*), diurnal (*center*), and long-period (*left*) tides.

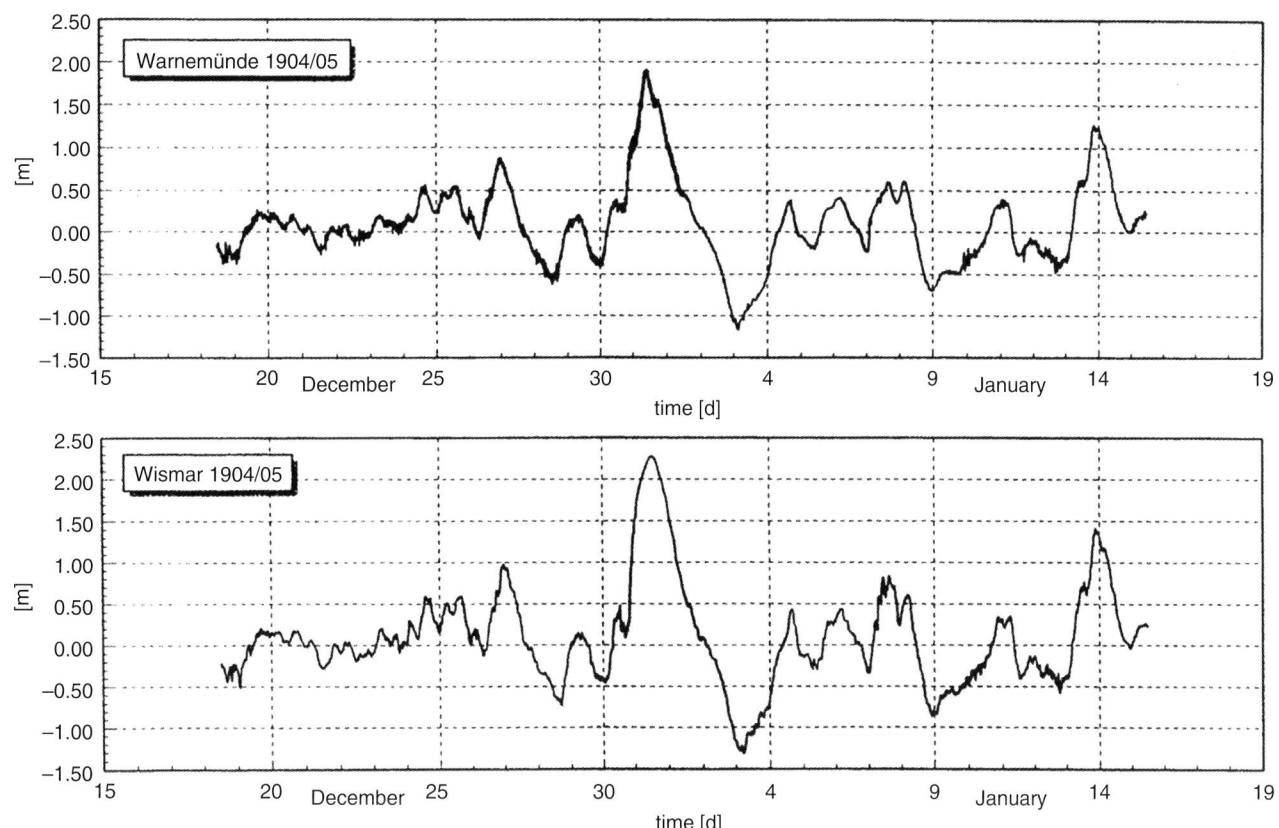

Sea-Level, Figure 6 Historical tide gauge record of a storm surge at the southern coast of the Baltic Sea in 1904/1905.

processes such as terrestrial and submarine landslides and volcanic eruptions and from asteroid impacts. They can also be caused by man-made explosions. Meteotsunamis are tsunami-like waves caused by abrupt changes in meteorological conditions that are amplified at the coast by local seiching (see also "Tsunamis").

Ocean dynamics

The deviation of the sea-level from the geoid (the dynamic topography) is a result of ocean dynamics. It can be obtained by oceanographic modeling or by subtracting a precise geoid from instantaneous sea-level heights observed by satellite altimetry (Bosch et al., 2012). The mean value of this deviation is in the order of 1–2 m. Specific spatial patterns (e.g., the Gulf Stream, the Kuroshio, or the Antarctic Circumpolar Current) are clearly visible. Furthermore, also phenomena with time scales of weeks to years like eddies or the El Niño Southern Oscillation (ENSO) may cause sea-level variations in the order of several decimeters (Figure 7).

Information on long-term trends, present-day sea-level rise, and future projections

For large time scales (millennia and more), there exists an excellent compilation on relative sea-level data from geological measurements (Pluet and Pirazzoli 1991). For the instrumental period, many data are collected by the Permanent Service for Mean Sea Level (PSMSL). Here tide gauge records with revised local reference (RLR) are especially valuable (Figure 8). RLR means that the observed sea-level is related to a fixed tide gauge benchmark. Sea-level changes observed with satellite altimetry within the last decades are summarized, e.g., by Bosch et al. (2014). Recent sea-level trends and projections up to the year 2100 are discussed in Church et al. (2010) and IPCC (2013).

Summary and conclusions

The sea-level is never on rest. Its dynamics affects many marine geoprocesses in time scales from seconds (waves) up to millennia (glacial isostasy). Of special socioeconomic importance are sudden events like storm surges and tsunamis as well as the observed present-day mean sea-level rise.

The observation of the sea-level on ground (tide gauges) and from space (altimetry) is of special importance for an improved understanding of all processes connected with the sea-level and for a more reliable prediction of its future changes.

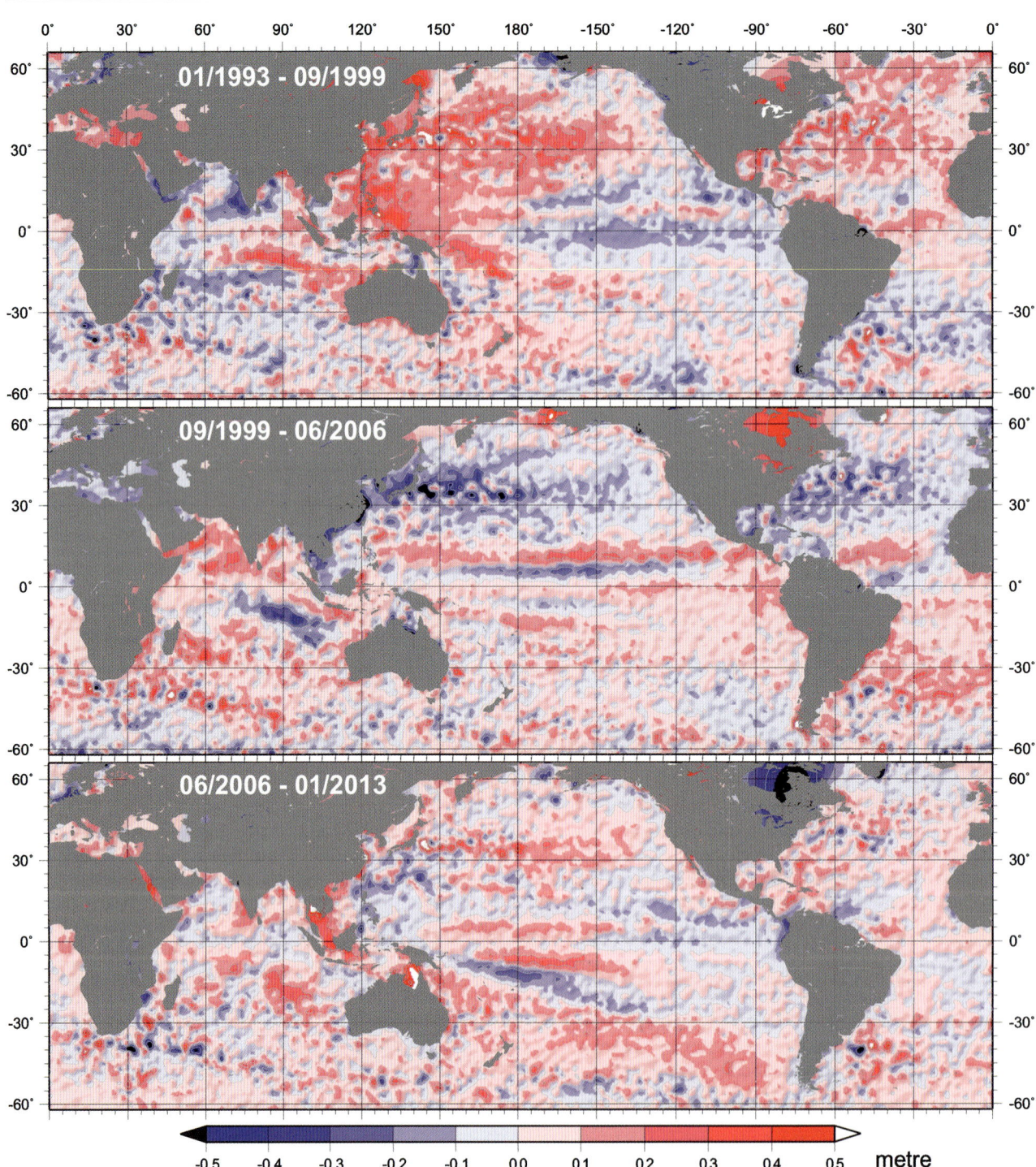

Sea-Level, Figure 7 Regional sea-level change [m] for three equally long sections of the two-decade period from 01/1993 to 01/2013 as derived by satellite altimetry. Each panel exhibits different pattern with changes of up to ± (0.2–0.4) m. For 6.67 years this is more than 10 times the mean rate of global sea-level rise of some 3 mm/year (courtesy Wolfgang Bosch, München).

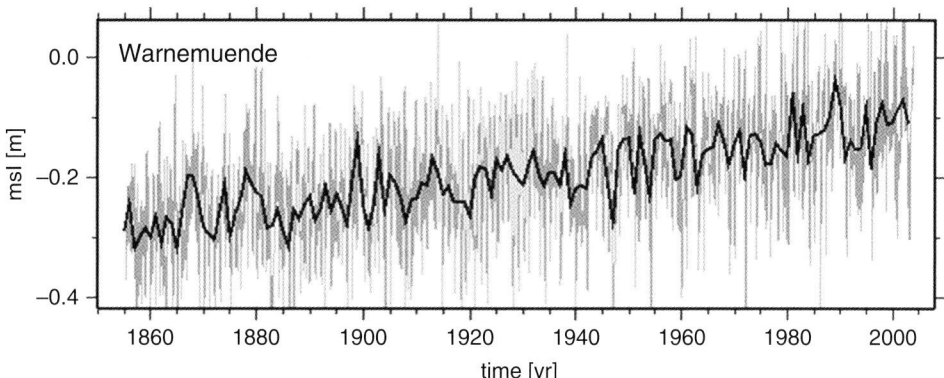

Sea-Level, Figure 8 Sea-level record at the southern coast of the Baltic Sea in Warnemünde (*gray*, monthly means; *black*, annual means). The record is based on a Revised Local Reference (*RLR*).

Bibliography

Antonioli, F., Anzidei, M., Lambeck, K., et al., 2007. Sea-level change during the Holocene in Sardinia and in the northeastern Adriatic (central Mediterranean Sea) from archaeological and geomorphological data. *Quaternary Science Reviews*, **26** (19–21), 2463–2486.

Bosch, W., Savcenko, R., Dettmering, D., Schwatke, C., 2012. A - two-decade time series of eddy-resolving dynamic ocean topography (iDOT). In Ouwehand, L. (ed.), *Proceedings of "20 Years of Progress in Radar Altimetry"*, Venice, Italy, ESA SP-710 (CD-ROM), ISBN 978-92-9221-274-2, ESA/ESTEC, 2013.

Bosch, W., Dettmering, D., and Schwatke, C., 2014. Multi-mission cross-calibration of satellite altimeters: constructing a long-term data record for global and regional sea level change studies. *Remote Sensing*, **6**(3), 2255–2281, doi:10.3390/rs6032255.

Church, J., Aarup, T., Woodworth, P., et al., 2010. Sea-level rise and variability: synthesis and outlook for the future. In Church, J., Woodworth, P., Aarup, T., and Wilson, W. S. (eds.), *Understanding Sea-Level Rise and Variability*. Chichester: Wiley-Blackwell, pp. 402–419.

Emery, K. O., and Aubrey, D. G., 1991. *Sea Levels, Land Levels, and Tide Gauges*. New York/Berlin/Heidelberg/London/Paris/Tokyo/Hong Kong/Barcelona: Springer.

Gehrels, W. R., and Woodworth, P. L., 2013. When did modern rates of sea-level rise start? *Global and Planetary Change*, **100**, 263–277, doi:10.1016/j.gloplacha.2012.10.020.

IPCC, 2013. Summary for policymakers. In Stocker, T. F., Qin, D., Plattner, G.-K., Tignor, M., Allen, S. K., Boschung, J., Nauels, A., Xia, Y., Bex, V., and Midgley, P. M. (eds.), *Climate Change 2013: The Physical Science Basis. Contribution of Working Group I to the Fifth Assessment Report of the Intergovernmental Panel on Climate Change*. Cambridge, UK/New York: Cambridge University Press.

Lisitzin, E., 1974. *Sea Level Changes*. Amsterdam: Elsevier.

Lübke, H., Schmölcke, U., and Tauber, F., 2011. Mesolithic hunter-fishers in a changing world: a case study of submerged sites on the Jäckelberg, Wismar Bay, northeastern Germany. In Benjamin, J., Bonsall, C., Pickard, C., and Fischer, A. (eds.), *Submerged Prehistory*. Oxford: Oxbow Books, pp. 21–37.

Pluet, J., and Pirazzoli, P. A., 1991. *World Atlas of Holocene Sea-Level Changes*. Amsterdam: Elsevier. Elsevier Oceanographic Series. Vol. 58

PSMSL. http://www.psmsl.org/data/obtaining/

Pugh, D. T., and Woodworth, P. L., 2014. *Sea-Level Science: Understanding Tides, Surges, Tsunamis and Mean Sea-Level Changes*. Cambridge: Cambridge University Press. 408pp. ISBN ISBN 9781107028197.

Rasch, M., and Jensen, J. F., 1997. Ancient Eskimo dwelling sites and Holocene relative sea-level changes in southern Disko Bugt, central West Greenland. *Polar Research*, **16**(2), 101–115.

Weidick, A., 1996a. Late Holocene and historical changes of glacier cover and related relative sea level in Greenland. *Zeitschrift für Gletscherkunde und Glazialgeologie*, **32**, 217–224.

Weidick, A., 1996b. Neoglacial changes of ice cover and sea level in Greenland – a classical enigma. In Grønnow, B. (ed.), *The Paleo-Eskimo Cultures of Greenland*. Copenhagen: Danish Polar Center, pp. 257–270.

Yokoyama, Y., Purcell, A., Marshall, J., et al., 2006. Sea-level during the early deglaciation period in the Great Barrier Reef, Australia. *Global and Planetary Change*, **53**, 147–153.

Cross-references

Coasts
El Niño (Southern Oscillation)
Eustasy
Geohazards: Coastal Disasters
Glacio(hydro)-isostatic Adjustment
Gravity Field
Marine Regression
Marine Transgression
Relative Sea-level (RSL) Cycle
Sediment Dynamics
Tidal Depositional Systems
Tsunamis
Underwater Archaeology
Waves

SEAMOUNTS

David M. Buchs[1], Kaj Hoernle[2] and Ingo Grevemeyer[2]
[1]School of Earth and Ocean Sciences, Cardiff University, Cardiff, UK
[2]GEOMAR Helmholtz Centre for Ocean Research, Kiel, Germany

Definition

Seamounts are literally mountains rising from the seafloor. More specifically, they are "any geographically isolated topographic feature on the seafloor taller than 100 m, including ones whose summit regions may temporarily emerge above sea-level, but not including features that are located on continental shelves or that are part of other major landmasses" (Staudigel et al., 2010). The term "guyot" can be used for seamounts having a truncated cone shape with a flat summit produced by erosion at sea-level (Hess, 1946), development of carbonate reefs (e.g., Flood, 1999), or partial collapse due to caldera formation (e.g., Batiza et al., 1984). Seamounts <1,000 m tall are sometimes referred to as "knolls" (e.g., Hirano et al., 2008). "Petit spots" are a newly discovered subset of sea knolls confined to the bulge of subducting oceanic plates of oceanic plates seaward of deep-sea trenches (Hirano et al., 2006).

Charting, abundance, and distribution

Seamounts form one of the most common bathymetric features on the seafloor. Two techniques, both with advantages and disadvantages, have been used to chart them. The first technique is satellite altimetry that measures the height between a satellite and the instantaneous sea surface. This distance can be used after correction for oceanographic effects on the sea surface to derive the geoid undulation, which (with assumptions) can be used to derive gravity anomalies. These are then converted (with assumptions) to seafloor bathymetry (Wessel, 2001). This technique permits full coverage of the ocean floor in the $\pm 72°$ latitude range. However, oceanographic noise at the sea surface limits the identification of seamounts smaller than ~1.5 km in most areas of the ocean (Wessel et al., 2010). The second type of technique uses ship SONAR (e.g., multibeam echo-sounding systems) or towed/autonomous underwater vehicles (AUVs), allowing topographic mapping at high resolution of the ocean floor. This technique permits recognition of seamounts of virtually any size provided that they are not entirely covered by sediment. However, the spatial coverage of the multi-beam swath is restricted to a narrow stripe beneath the vehicle conducting the measurements, with the width of the stripe increasing with water depth between the vehicle and the seafloor. As a consequence, <10 % of the global seafloor has been charted using this method.

Size–frequency relationship of seamounts based on satellite and ship track data can be combined to estimate the total number of seamounts (Hillier and Watts, 2007). It is therefore speculated that there are 3–80 million seamounts greater than 100 m in height on the seafloor. The total number and global distribution of smaller seamounts, however, remains poorly constrained due to relatively limited coverage of the seafloor by echo sounding. Around 90 % of the seamounts, a fraction primarily including the smallest, <100 m-tall edifices, remain to be charted (Wessel et al., 2010). Only a very limited number of seamounts are currently included in open-source databases like the Seamount Catalog (http://earthref.org/SC/) or the General Bathymetric Chart of the Oceans (GEBCO, http://www.gebco.net/). There could be more than 50,000 seamounts taller than 1 km, representing only a minor fraction of the total seamount number (Figure 1). Many of these larger seamounts are distributed along clusters or lines that can reach several thousands of kilometers in length (e.g., the Hawaiian–Emperor seamount chain in the NW Pacific). These seamount/island chains provide important insights into plate motions, the formation of oceanic intraplate volcanoes, and the dynamics of the oceanic lithosphere and underlying mantle.

Origins

Most seamounts are formed by igneous activity close to mid-ocean ridges, island arcs, or in mid-plate settings (Koppers and Watts, 2010). Due to percolation of fluid-rich material through the upper plate, small mud or serpentine seamounts (volcanoes) can form between the volcanic arc and trench in intra-oceanic subduction zones (Kopf, 2002). Blocks of continental crust, stranded during the opening of ocean basins, can form nonvolcanic seamounts (e.g., Stuttgart seamount around 1,000 km east of New Zealand, Mortimer et al., 2006).

Seamounts of igneous origin formed at mid-ocean ridges are called *on-axis* seamounts. Those formed within a few hundreds of kilometers of mid-ocean ridges on relatively young lithosphere, but not directly at the ridge axis, are commonly referred to as *near-ridge* or *off-axis* seamounts. Estimates suggest that more than 50 % of seamounts are emplaced on oceanic crust younger than 20 Ma (Hillier, 2007), and, thus, on-/near-/off-axis seamounts probably compose the largest part of the seamount population. Even large seamount provinces, such as the Christmas Island seamount province in the northeast Indian Ocean (e.g., Hoernle et al., 2011) or the earlier Mid-Pacific Mountains in the central western Pacific Ocean (e.g., Hillier, 2007), are believed to have been formed in a near spreading center environment. Their emplacement is intrinsically related to upwelling beneath spreading centers and the formation of the oceanic crust (entries "Lithosphere: Structure and Composition" and "Mid-ocean Ridge Magmatism and Volcanism").

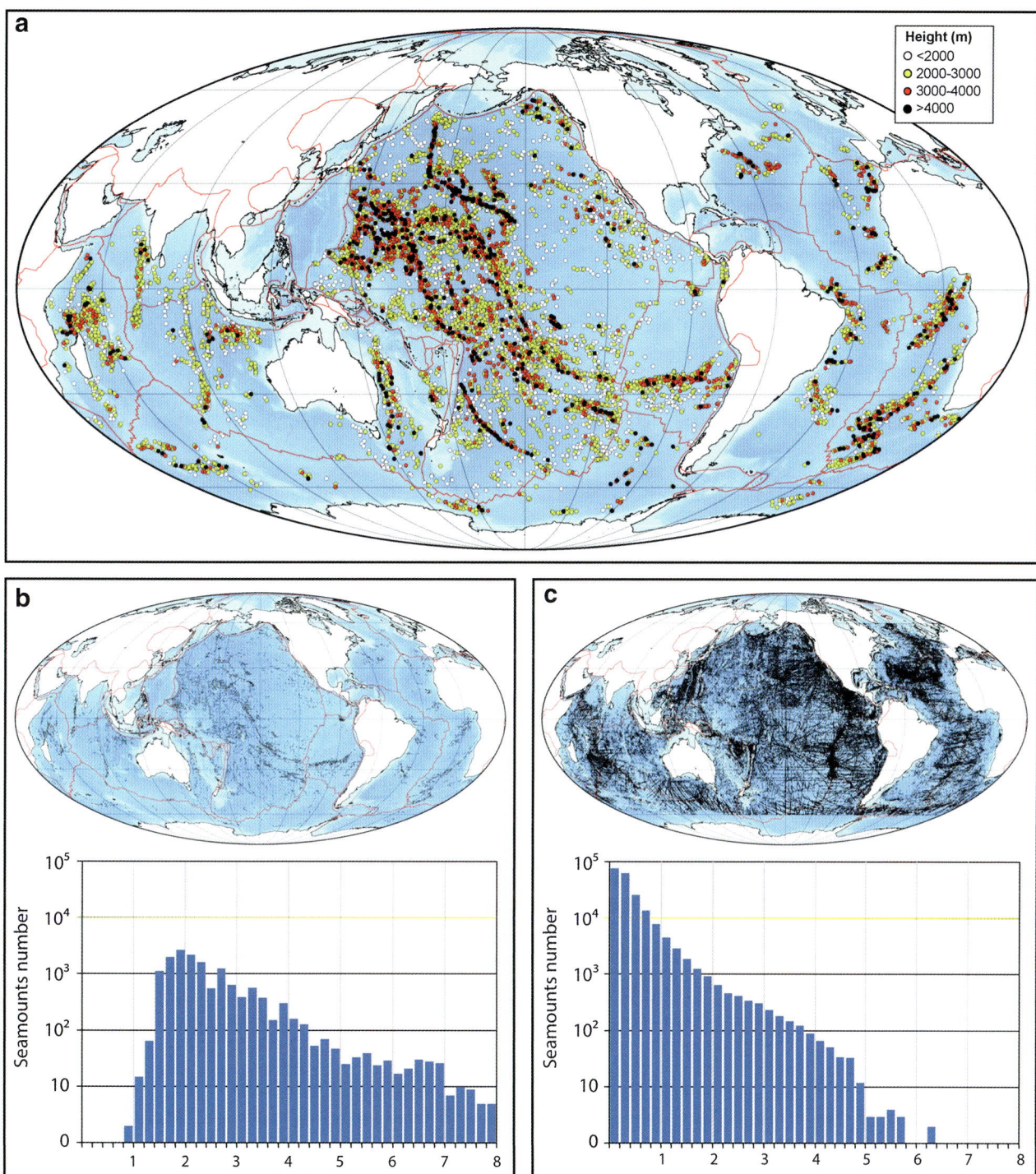

Seamounts, Figure 1 Global distribution of seamounts. (**a–b**) Large seamounts identified by satellite altimetry in the ±72° latitude range (n = 14′675) (Wessel, 2001). It should be noted that island volcanoes have also been included in as "seamounts," extending to heights >8 km. (**c**) Seamounts identified by ship echo sounding over an area representing <10 % of the total ocean floor surface (n = 201′055) (Hillier and Watts, 2007). This histogram illustrates that the largest number of seamounts has sizes that cannot be detected with satellite altimetry. Only submarine seamounts are included, explaining why there are no seamounts with elevations of >6.5 km.

Seamounts formed in mid-plate settings are called "oceanic intraplate volcanoes." Intraplate volcanoes that are formed near spreading centers, i.e., on young and thus thin lithosphere, consist primarily of tholeiitic magmas (e.g., Niu et al., 2002). Most intraplate seamounts, however, are emplaced on older and therefore thicker lithosphere. These volcanoes comprise abundant alkaline igneous rocks that result from lower degrees of partial melting at greater depth compared to the production of tholeiitic volcanism (entry "Intraplate Magmatism"). The formation of intraplate seamounts can be due to local melting anomalies or "hotspots" that underlie the lithosphere (entry "Hot Spots and Mantle Plumes"), plate cracking allowing preexisting melt in the upper mantle to rise to the surface (e.g., Natland and Winterer, 2005; Valentine and Hirano, 2010), or localized mantle upwelling in the shallow asthenosphere (e.g., Geldmacher et al., 2008). Hotspot volcanism is commonly associated with the presence of mantle plumes that bring hot, and thus buoyant, fertile mantle from a thermal boundary layer, such as the core–mantle or upper–lower mantle boundary, to the base of the lithosphere where the mantle melts. As the lithosphere moves over the plume, volcanism forms an age-progressive island/seamount chain (Morgan, 1972). The most prominent and thus well-known example of hotspot/plume volcanism is the Hawaiian–Emperor island/seamount chain, where seamounts become progressively older with increasing distance from the active volcanoes forming the Hawaiian Islands (Clague and Dalrymple, 1987). Seamounts can also form along cracks resulting from stress in the oceanic lithosphere, e.g., as a result of plate cooling, plate bending outboard of subduction zones, collision of continental blocks, or abrupt changes in plate motions (Natland and Winterer, 2005; Valentine and Hirano, 2010). Petit-spot volcanoes near Japan and Pliocene volcanism on Christmas Island near Indonesia, for example, formed outboard of subduction zones due to plate bending (Hirano et al., 2006; Hoernle et al., 2011). Plate cracking formed by extension can not only facilitate the rise of melts through the lithosphere but also can trigger asthenospheric upwelling causing decompression melting (O'Connor et al. 2015). Because seamounts are generally formed by lower degrees of partial melting and over a larger range of melting than normal mid-ocean ridge basalts (MORB), their igneous rocks exihibit a much larger spectrum of chemical compositions than MORB, since they preferentially sample fertile anomalies in the asthenospheric mantle (e.g., Niu et al. 2002; Hoernle et al. 2011). Seamount volcanism has played an important role in our understanding of mantle dynamics, chemical heterogeneity of the mantle, and magmatism in general (entries "Intraplate Magmatism" and "Hot Spots and Mantle Plumes").

Formation and evolution

Seamounts are formed by eruptive and intrusive products with an estimated ratio of approximately 70:30 (Staudigel and Clague, 2010; Flinders et al., 2013). Eruptive products broadly include pillow lavas that reflect low to medium effusive rates in submarine environments, massive sheet flows formed at higher effusive rates, and volcaniclastics formed through disaggregation of pillows or sheet flows on steep slopes or through explosive volcanism. Intrusive products generally consist of dikes, sills, and gabbroic to ultramafic rocks. The growth of a seamount is controlled by successive stacking of eruptive products and inflation due to emplacement of intrusives. Explosive volcanism occurs when the magma interacts explosively with water or when the magma rapidly outgasses in response to a decrease of pressure. Degassing of CO_2 can occur at great pressure and can lead to explosive volcanism in the deepest parts of the ocean. Overall, the style of submarine eruptions reflects variation in the composition, physical properties, and supply rate of magma, as well as the tectonic setting of formation of the volcanic edifice (Rubin et al., 2012). Volcanoes several kilometers tall can form over hundreds of thousands of years (Clague and Dalrymple, 1987), whereas smaller seamounts may form in days to months.

Large seamounts follow a particular pattern of evolution primarily controlled by the relative intensity of magma supply and erosional processes through time. This pattern of evolution can be subdivided into six main phases (Staudigel and Clague, 2010):

1. A small seamount forms during incipient volcanism after a possible phase of magma interaction with seafloor sediments. The magmatic plumbing system of the volcano is restricted within the oceanic crust. Commonly, the magma supply wanes quickly and the seamount remains small.
2. More rarely, the seamount will experience continued or increased volcanic activity that results in formation of a much larger volcano. The magmatic plumbing system migrates upward within the volcanic structure.
3. When the summit of the seamount approaches sea-level, explosive volcanism becomes more abundant. The amount of clastic deposits emplaced on the volcano slopes increases. The seamount emerges and forms an island when the eruption rate is high enough to overcome erosion by sea-surface waves, mass wasting, eustatic changes, and subsidence due to isostatic adjustment of the lithosphere under the fast-growing seamount.
4. The seamount goes through a subaerial, island phase with continued volcanism. Incipient carbonate reefs can develop along the shorelines.
5. The main magmatic phase ceases and the island volcano stops growing and starts to erode. Slope failure can lead to removal of major parts of the volcanic edifice with the potential for tsunami formation; such events may occur throughout the evolution of the seamount (Masson et al., 2006). The volcano can go through a period of quiescence for several million years until small volumes of lava are erupted during short periods of volcanic rejuvenation. Continued erosion eventually leads to formation of a flat-topped to

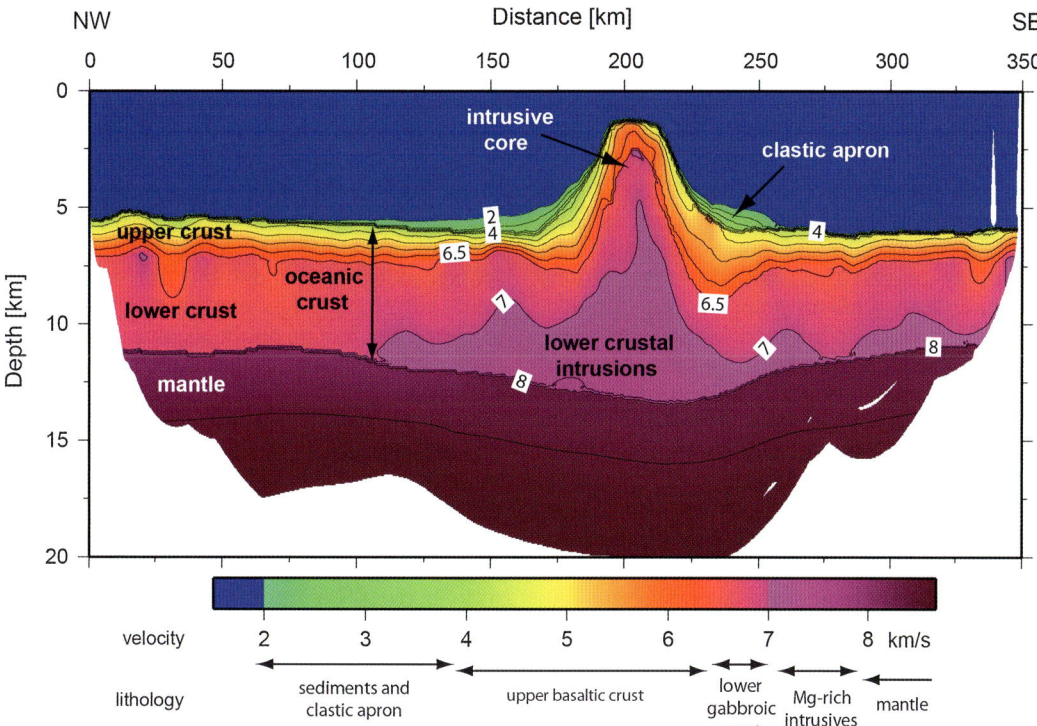

Seamounts, Figure 2 Seismic velocity structure of the Louisville seamount chain (data from Contreras-Reyes et al., 2010).

slightly dome-shaped (if subsidence occurs during erosion) seamount with a plateau just beneath wave base. As the volcano is cut off, i.e., moved away, from its heat/magma source, lithospheric cooling leads to subsidence of the "guyot" seamount to depths up to thousands of meters below sea-level. Under appropriate environmental conditions, carbonate reefs can develop on top of the volcano and form an atoll, while the top is still in shallow water. The atoll becomes submerged when the growth of the reef can no longer compensate for seamount subsidence and/or sea-level rise. The location at which an atoll drowns is sometimes referred to as the "Darwin Point" (e.g., Flood, 2001).

6. Most seamounts are ultimately transported to subduction zones, where flexure of the downgoing plate and the formation of bending-related faults can cause partial collapse of the volcanic edifice. During subduction, the seamount and related collapse products can be accreted (transferred to) and preserved in the accretionary wedge of a subduction zone (Hoernle et al., 2002; Portnyagin et al., 2008; Buchs et al., 2011). If not accreted, seamounts can contribute to the petrogenesis of arc magmas (e.g., in Central America, Hoernle et al., 2008).

Architecture and structure

Crustal structure derived from seismic surveys reveals insights into the melt generation and the internal architecture of volcanic seamounts. The architecture of a seamount can be revealed using a structural comparison between an intraplate seamount and normal oceanic crust, using the correlation between P-wave velocity structure of seamounts and typical oceanic crust. Since non-explosive extrusive processes prevail during the deepwater stage, dominantly producing pillow basalts (~75 %) (Staudigel and Clague, 2010), the seismic velocity structure of small seamounts of <1,000 m roughly mimic thickened oceanic crust. However, tall edifices show average P-wave velocities, which are significantly lower than those generally found in oceanic crust. The low velocities found in the uppermost crust beneath the summits and the clastic aprons of tall seamounts (Figure 2) are indicative of low densities and high porosities, consistent with shoaling sequences composed dominantly (~70 %) of clastic deposits (Staudigel and Clague, 2010). Accumulation of large volumes of clastic material in the summit region and upper flanks of a seamount volcano can generate debris flows that contribute to the clastic apron around the volcano and can extend into the abyssal plains beyond the apron. Some of the individual debris avalanches are 100–200 km long and 100–500 km^3 in volume (e.g., Watts and Masson, 1995). Apron sediments may extend out to several hundreds of kilometers from a seamount or ocean island (e.g., Wolfe et al., 1994). On the summit and flanks, clastic sediments and rocks show low seismic velocities of 2.8–3.2 km/s (Weigel and Grevemeyer, 1999; Contreras-Reyes et al., 2010), but can also reach velocities of 4.0–5.0 km/s in the deep portions of the moat around the volcano (Wolfe et al., 1994; Weigel and Grevemeyer, 1999).

Below the summit, the transition to rocks with velocities of >6.5 km/s occurs generally at 2–4-km depth (Weigel and Grevemeyer, 1999; Contreras-Reyes et al., 2010), indicating the presence of gabbroic rocks and hence a central conduit or a plutonic suite (Figure 2). Seismic velocities exceeding 7.0–7.2 km/s in the lower crust beneath seamounts suggest that significant proportions of magma feeding seamounts are trapped, i.e., intruded, near the base of the crust (White and McKenzie, 1989; Holbrook, 1995). The high-magnesium basalts generated from abnormally hot mantle typically exhibit high densities and hence velocities when they crystallize (White and McKenzie, 1989; Keleman and Holbrook, 1995). The base of the crust forms a marked density contrast, so it is likely that melts generated at depth will accumulate near the base of the crust, where they either underplate the crust or intrude it as sills. High-velocity, lower-crustal rocks are an integral part of large seamounts, including the Hawaiian chain (Watts et al., 1985), Ninetyeast Ridge (Grevemeyer et al., 2001), and Louisville seamounts (Contreras-Reyes et al., 2010).

Implications for topographic anomalies on the seafloor

The morphology of seamounts in the oceans controls several processes of global significance.

Seamounts play an important role in influencing local and global oceanic currents by redistributing the energy from surface tides, blocking water flow and creating eddies, and generating hydrothermally induced plumes in the water column (Lavelle and Mohn, 2010). For example, it was suggested that the New England Seamount chain in the Northern Atlantic can induce gyres with a stabilizing effect on the dynamics of the Gulf Stream (Ezer, 1994).

Large subducting seamounts can substantially modify the geometry of the plate interface at convergent margins, causing indentations in margins. They can locally "jam" subduction zones, trigger earthquakes or modify regional patterns of seismicity (Watts et al., 2010). They may also influence the patterns of fluid release in the subduction zone by creating preferential circulation paths within the deforming upper plate.

Seamounts represent areas of the ocean floor where the volcanic basement outcrops above the relatively impermeable layer of pelagic sediments on the seafloor. They therefore serve as preferential areas of recharge and discharge for hydrothermal fluids and permit extended exchange of heat and solutes between the ocean and the igneous crust (Fisher and Wheat, 2010; entry "Hydrothermalism").

Seamounts host a large variety of mineral deposits, including hydrogenous ferromanganese crusts (Fe–Mn crusts) or nodules, hydrothermal iron and manganese oxides, hydrothermal sulfide, sulfate and sulfur deposits, and phosphorite deposits (Hein et al., 2010). Formation of these mineral deposits is related to hydrothermal, diagenetic, or precipitation processes that take advantage of magmatic heat. As the supply of some critical metals on land becomes more challenging (mining moves to greater depths and/or to more remote locations) or inaccessible for political or environmental reasons, these resources become more expensive, and new techniques to explore and mine the seafloor are currently being developed. The exploitation of mineral resources at seamounts will no doubt become economically advantageous at some stage in the future (entry "Marine Mineral Resources").

Volcanic islands and atolls constitute particular habitats that can remain isolated from major landmasses for tens of millions of years. This isolation fosters development of endemic taxa, which stimulated the postulation of the theory of natural selection in the nineteenth century. Similarly, seamounts (or submerged parts of oceanic islands) form barriers for ocean currents, causing upwelling of deeper nutrient-laden waters that supply sustenance for a wide range of marine organisms. Larger fauna, e.g., fish, feeds on the smaller organisms, and thus seamounts serve as oases for life in the ocean (entry "Deep Biosphere"; Clark et al., 2010). Interestingly, the location of seamounts can often be inferred from the presence of large numbers of seabirds feeding off fish above the seamount. Understanding the exact nature of biodiversity and fauna connectivity on seamounts, however, is still limited due to spatially and temporally restricted sampling. Fishing and trawling has caused substantial damage to some seamount habitats, and monitoring is needed to permit sustainable development of anthropogenic activities at seamounts.

Lithospheric flexure beneath seamounts

Seamounts define loads that deform the oceanic lithosphere on which they sit. It has been shown that seamounts emplaced on young oceanic crust are characterized by local Airy-type isostasy (Watts, 2001), similar to icebergs floating in the ocean. With increasing age, however, the oceanic lithosphere becomes stronger, and rather than forming a local rootlike structure, it deforms elastically by flexure, like an elastic plate. The Dutch geophysicist and geodesist Felix Vening Meinesz introduced the effective elastic thickness (Te), to describe the deformation of the lithosphere. Thus, loads emplaced on older lithosphere are compensated regionally (see Figure 2), causing prominent flexural moats and bulges along major seamounts chains, like the Hawaiian–Emperor seamounts (Watts, 1978). Oceanic settings provide Te values ranging from zero (Airy-type isostasy) to 40 km. Since the oceanic crust is approximately 6 km thick, it is clear that not only the crust but also the lithospheric mantle is involved in supporting the load of seamounts (Watts, 2001).

Seamounts and true polar wander

Paleomagnetic data from seamounts support a mobile Earth and hence plate tectonics. Assuming that the Earth's magnetic field is aligned with the spin axis, paleomagnetic data from volcanic and sedimentary rocks allow the reconstruction of the location of the magnetic pole. In the past 5 Myr, the geomagnetic field was axial geocentric dipolar within confidence limits of ~3° (Merrill and McElhinny, 1983). However, in older geological time intervals, the geomagnetic pole from a given region seems to move away from the spin axis (Creer et al., 1954) and is called apparent polar wander (APW). APW paths are composed of two components: plate motion and polar wander. Polar wander is the shift of the entire Earth with respect to the rotation axis. On an Earth where plates are constantly moving, it is difficult to define a reference frame. Therefore, true polar wander (TPW) has been introduced as a definition of polar wander, where paleomagnetic poles are viewed in a fixed hotspot reference frame (e.g., Besse and Courtillot, 1991). However, measurement of the paleomagnetism of samples from the Hawaiian–Emperor seamount chain, i.e., volcanic track formed above the Hawaiian hotspot, provided evidence for a significant southward motion of the Hawaiian mantle plume from 81 to 47 Myr ago. This motion was a dominant factor causing the trend and morphology of the Hawaiian–Emperor chain and the famous bend, i.e., change in orientation, of the Hawaiian hotspot at ~47 Ma (Tarduno et al., 2003; O'Connor et al., 2015). Tarduno (2007) concluded that "an important corollary of this finding is that true polar wander (i.e., polar wander calculated in a fixed hotspot reference frame) is illusory; the solid Earth has been relatively stable with respect to the spin axis for at least the past 130 million years. The paleomagnetic data from the Hawaiian–Emperor tracks, and other oceanic sites, indicate that the geometry of hotspot tracks contains a rich history of both mantle flow and plate processes."

Summary and conclusions

Millions of seamounts on the seafloor have a large, multidisciplinary significance. Seamounts can provide constraints on the tectonic evolution of the seafloor and serve as a window into mantle compositional heterogeneity and dynamics. They can cause significant hazards for coastal populations by triggering earthquakes during subduction or tsunamis during major flank collapse. Seamounts can host extensive hydrothermal activity and mineral resources of increasing significance to a planet with an exploding population growth. They are morphologic oases in the deep sea serving as stepping stones for marine biota. Clearly, there is a strong need for additional seamount research, and it is expected that forthcoming decades will see significant advances in our understanding of one of the most common but least studied features on Earth.

Acknowledgment

We thank Anthony Watts for his constructive and insightful review.

Bibliography

Batiza, R., Fornari, D. J., Vanko, D. A., and Lonsdale, P., 1984. Craters, calderas, and hyaloclastites on young Pacific seamounts. *Journal of Geophysical Research, Solid Earth*, **89**, 8371–8390.

Besse, J., and Courtillot, V., 1991. Revised and synthetic apparent polar wander paths of the African, Eurasian, North-American and Indian plates, and true polar wander since 200 Ma. *Journal of Geophysical Research*, **96**, 4029–4050.

Buchs, D. M., Arculus, R. J., Baumgartner, P. O., and Ulianov, A., 2011. Oceanic intraplate volcanoes exposed: example from seamounts accreted in Panama. *Geology*, **39**, 335–338.

Clague, D. A., and Dalrymple, G. B., 1987. The Hawaiian-Emperor volcanic chain, Part I, Geologic evolution. In Wright, T. L., Stauffer, P. H., and Decker, R. W. (eds.), *Volcanism in Hawaii*. Washington: U.S. Government Printing Office, Washington: US Geological Survey Professional Paper 1350. pp. 5–54.

Clark, M. R., Rowden, A. A., Schlacher, T., Williams, A., Consalvey, M., Stocks, K. I., Rogers, A. D., O'Hara, T. D., White, M., Shank, T. M., and Hall-Spencer, J. M., 2010. The ecology of seamounts: structure, function, and human impacts. *Annual Review of Marine Science*, **2**, 253–278.

Contreras-Reyes, E., Grevemeyer, I., Watts, A. B., Planert, L., Flueh, E. R., and Peirce, C., 2010. Crustal intrusion beneath the Louisville hotspot track. *Earth and Planetary Science Letters*, **289**(3–4), 323–333.

Creer, K. M., Irving, E., and Runcorn, S. K., 1954. The direction of the geomagnetic field in remote epochs in Great Britain. *Journal of Geomagnetism and Geoelectricity*, **6**, 163–168.

Ezer, T., 1994. On the interaction between the Gulf stream and the New England seamount chain. *Journal of Physical Oceanography*, **24**, 191–204.

Fisher, A. T., and Wheat, C. G., 2010. Seamounts as conduits for massive fluid, heat, and solute fluxes on ridge flank. *Oceanography*, **23**(1), 74–87.

Flinders, A. F., Ito, G., Garcia, M. O., Sinton, J. M., Kauahikaua, J., and Taylor, B., 2013. Intrusive dike complexes, cumulate cores, and the extrusive growth of Hawaiian volcanoes. *Geophysical Research Letters*, **40**, 3367–3373.

Flood, P., 1999. Development of northwest Pacific guyots: general results from ocean drilling program legs 143 and 144. *Island Arc*, **8**, 92–98.

Flood, P. G., 2001. The "Darwin Point" of Pacific Ocean atolls and guyots: a reappraisal. *Palaeogeography, Palaeoclimatology, Palaeoecology*, **175**, 147–152.

Geldmacher, J., Hoernle, K., Klügel, A., van den Bogaard, P., and Bindeman, I., 2008. Geochemistry of a new enriched mantle type locality in the northern hemisphere: implications for the origin of the EM-I source. *Earth and Planetary Science Letters*, **265**, 167–182.

Grevemeyer, I., Flueh, E. R., Reichert, C., Bialas, J., Klaeschen, D., and Kopp, C., 2001. Crustal architecture and deep structure of the Ninetyeast Ridge hotspot trail from active-source ocean bottom seismology. *Geophysical Journal International*, **144**, 414–431.

Hein, J. R., Conrad, T. A., and Staudigel, H., 2010. Seamount mineral deposits. *Oceanography*, **23**(1), 184–189.

Hess, H. H., 1946. Drowned ancient islands of the Pacific Basin. *American Journal of Science*, **244**, 772–791.

Hillier, J. K., 2007. Pacific seamount volcanism in space and time. *Geophysical Journal International*, **168**, 877–889.

Hillier, J. K., and Watts, A. B., 2007. Global distribution of seamounts from ship-track bathymetry data. *Geophysical Research Letters*, **34**, L13304.

Hirano, N., Takahashi, E., Yamamoto, J., Abe, N., Ingle, S. P., Kaneoka, I., Hirata, T., Kimura, J. I., Ishii, T., Ogawa, Y., Machida, S., and Suyehiro, K., 2006. Volcanism in response to plate flexure. *Science*, **313**, 1426–1428.

Hirano, N., Koppers, A. A. P., Takahashi, A., Fujiwara, T., and Nakanishi, M., 2008. Seamounts, knolls and petit-spot monogenetic volcanoes on the subducting Pacific Plate. *Basin Research*, **20**, 543–553.

Hoernle, K. A., Bogaard, P. V. D., Werner, R., Lissinna, B., Hauff, F., Alvarado, G., and Garbe-Schönberg, D., 2002. The missing history (16–71 Ma) of the Galápagos hotspot: implications for the tectonic and biological evolution of the Americas. *Geology*, **30**, 795–798.

Hoernle, K., Abt, D. L., Fischer, K. M., Nichols, H., Hauff, F., Abers, G., van den Bogaard, P., Heydolph, K., Alvarado, G., Protti, J. M., and Strauch, W., 2008. Arc-parallel flow in the mantle wedge beneath Costa Rica and Nicaragua. *Nature*, **451**, 1094–1097.

Hoernle, K., Hauff, F., Werner, R., van den Bogaard, P., Gibbons, A. D., Conrad, S., and Muller, R. D., 2011. Origin of Indian Ocean seamount province by shallow recycling of continental lithosphere. *Nature Geoscience*, **4**(12), 883–887.

Holbrook, W. S., 1995. Underplating over hotspots. *Nature*, **373**, 559.

Keleman, P. B., and Holbrook, W. S., 1995. Origin of thick, high-velocity igneous crust along the U.S. East Coast margin. *Journal of Geophysical Research*, **100**, 10077–10094.

Kopf, A. J., 2002. Significance of mud volcanism. *Reviews of Geophysics*, **40**, 1005.

Koppers, A. A. P., and Watts, A. B., 2010. Intraplate seamounts as a window into deep earth processes. *Oceanography*, **23**(1), 42–57.

Lavelle, J. W., and Mohn, C., 2010. Motion, commotion, and biophysical connections at deep ocean seamounts. *Oceanography*, **23**(1), 90–103.

Masson, D. G., Harbitz, C. B., Wynn, R. B., Pedersen, G., and Løvholt, F., 2006. Submarine landslides: processes, triggers and hazard prediction. *Philosophical Transactions of the Royal Society A: Mathematical, Physical and Engineering Sciences*, **364**, 2009–2039.

Merrill, R. T., and McElhinny, M. W., 1983. *The Earth's Magnetic Field: Its History, Origin and Planetary Perspective*. San Diego: Academic Press. pp. 401.

Morgan, W. J., 1972. Deep mantle convection plumes and plate motions. *AAPG Bulletin*, **56**, 203–213.

Mortimer, N., Hoernle, K., Hauff, F., Palin, J. M., Dunlop, W. J., Werner, R., and Faure, K., 2006. New constraints on the age and evolution of the Wishbone Ridge, southwest Pacific Cretaceous microplates, and Zealandia-West Antarctica breakup. *Geology*, **34**(3), 185–188.

Natland, J.H., and Winterer, E.L., 2005. Fissure control on volcanic action in the Pacific. In: Foulger, G.R., Natland, J.H., Presnall, D.C., and Anderson, D.L. (eds.), *Plates, Plumes and Paradigms*. Boulder, CO: Geological Society of America, Special volume 388, pp. 687–710.

Niu, Y., Regelous, M., Wendt, I. J., Batiza, R., and O'Hara, M. J., 2002. Geochemistry of near-EPR seamounts: importance of source vs. process and the origin of enriched mantle component. *Earth and Planetary Science Letters*, **199**, 327–345.

O'Connor, J., Hoernle, K., Butterworth, N., Müller, D., Hauff, F., Sandwell, D., Phipps Morgan, J., Jokat, W., Wijbrans, J., and Stoffers, P., (2015) Deformation-related volcanism in the Pacific Ocean linked to the Hawaiian–Emperor bend. *Nature Geoscience*, **8**, 393–397. DOI: 10.1038/NGEO2416.

Portnyagin, M., Savelyev, D., Hoernle, K., Hauff, F., and Garbe-Schönberg, D., 2008. Mid-Cretaceous Hawaiian tholeiites preserved in Kamchatka. *Geology*, **36**(11), 903–906.

Rubin, K. H., Soule, S. A., Chadwick, W. W., Fornari, D. J., Clague, D. A., Embley, R. W., Baker, E. T., Perfit, M. R., Caress, D. W., and Dziak, R. P., 2012. Volcanic eruptions in the deep sea. *Oceanography*, **25**(1), 142–157.

Staudigel, H., and Clague, D. A., 2010. The geological history of deep-sea volcanoes. *Oceanography*, **23**(1), 58–71.

Staudigel, H., and Schmincke, H.-U., 1984. The Pliocene seamount series of La Palma, Canary Islands. *Journal of Geophysical Research*, **89**, 11195–11215.

Staudigel, H., Koppers, A. A. P., Lavelle, J. W., Pitcher, T. J. P., and Shank, T. M., 2010. Defining the word "seamount". *Oceanography*, **23**(1), 20–21.

Tarduno, J. A., 2007. On the motion of Hawaii and other mantle plumes. *Chemical Geology*, **241**, 234–247.

Tarduno, J. A., Duncan, R. A., Scholl, D. W., Cottrell, R. D., Steinberger, B., Thordarson, T., Kerr, B. C., Neal, C. R., Frey, F. A., Torii, M., and Carvallo, C., 2003. The Emperor seamounts: southward motion of the Hawaiian hotspot plume in Earth's mantle. *Science*, **301**, 1064–1069.

Valentine, G. A., and Hirano, N., 2010. Mechanisms of low-flux intraplate volcanic fields–Basin and Range (North America) and northwest Pacific Ocean. *Geology*, **38**, 55–58.

Watts, A. B., 1978. An analysis of isostasy in the world's oceans: 1. Hawaiian-Emperor seamount chain. *Journal of Geophysical Research*, **83**, 5989–6004.

Watts, A. B., 2001. *Isostasy and Flexure of the Lithosphere*. Cambridge: Cambridge University Press, p. 458.

Watts, A. B., and Masson, D. G., 1995. A giant landslide on the north flank of Tenerife, Canary Islands. *Journal of Geophysical Research*, **100**, 24487–24498.

Watts, A. B., ten Brink, U. S., Buhl, P., and Brocher, T. M., 1985. A multichannel seismic study of the lithospheric flexure across the Hawaiian-Emperor seamount chain. *Nature*, **315**, 105–111.

Watts, A. B., Koppers, A. A. P., and Robinson, R. T., 2010. Seamount subduction and earthquakes. *Oceanography*, **23**(1), 166–173.

Wessel, P., 2001. Global distribution of seamounts inferred from gridded Geosat/ERS-1 altimetry. *Journal of Geophysical Research*, **106**, 19421.

Wessel, P., Sandwell, D. T., and Kim, S.-S., 2010. The global seamount census. *Oceanography*, **23**(1), 24–33.

White, R. S., and McKenzie, D. P., 1989. The generation of volcanic continental margins and flood basalts. *Journal of Geophysical Research*, **94**, 7685–7729.

White, R. S., MCKenzie, D. P., and O'Nions, R. K., 1992. Oceanic crustal thickness from seismic measurements and rare earth element inversions. *Journal of Geophysical Research*, **97**, 19683–19715.

Wolfe, C. J., McNutt, M. K., and Detrick, R. S., 1994. The Marquesas archipelagic apron: seismic stratigraphy and implications for volcano growth, mass wasting, and crustal underplating. *Journal of Geophysical Research*, **99**, 13591–13608.

Weigel, W., and Grevemeyer, I., 1999. The Great Meteor seamount: seismic structure of a submerged intraplate volcano. In Charvis, P., and Danobeitia, J. J. (eds.), *Hotspot and Oceanic Crust Interaction*. Journal of geodynamics, Vol. 28, pp. 27–40.

Cross-references

Deep Biosphere
Hot Spots and Mantle Plumes
Hydrothermalism
Marine Mineral Resources
Oceanic Spreading Centers

SEDIMENT DYNAMICS

Wenyan Zhang
MARUM-Center for Marine Environmental Sciences, University of Bremen, Bremen, Germany

Synonyms
Sedimentary processes

Definition
Sediment dynamics refers to the motion of sediment particles during their formation, transport, and settling processes.

Introduction
Sediment dynamics is an interdisciplinary research field integrating the knowledge of physical oceanography, geology, biology, and chemistry. Since centuries ago, it has been a subject of study for geologists and engineers. Many different theories and formulas have been developed to describe the motion of sediment particles during their formation, transport, and settling. Some of the approaches have been applied widely to solve engineering and environmental problems, and some still remain on a theoretical level. Since most of the existing formulas describing sediment dynamics are derived empirically, results obtained from different approaches often differ drastically from each other even though they are based on the same initial conditions (Yang, 1996). Gaps are always shown between theoretical results and field observations. The difficulty in matching theoretical results to sediment dynamics is due to the vulnerability of sediment particle mobility to many environmental factors and processes occurring at different temporal and spatial scales. Although the history of study on sediment dynamics is long, some of the basic concepts, driving mechanisms, and interrelationships among them have become clear to scientists only in recent decades. The complexity of sediment dynamics makes it among the most complicated and least understood subjects in earth sciences.

Rather than gaining more insight into the complexity of sediment dynamics, this entry aims at providing a summary of the most basic concepts and driving mechanisms involved in the motion of sediment particles in coastal and marine environment.

Origin of sediment
Any solid fragment of organic or inorganic material can be termed sediment. Marine sediments originate from a variety of sources. The origin of a sediment particle determines its dynamics to a large extent. According to the ways a sediment particle is formed, sediments can be classified into two broad categories: granular sediments and chemical sediments. Granular sediments result from the fragmentation of inorganic or organic materials. Fragmentation of the parent material may be caused by mechanical (e.g., weathering and erosion of rocks), biological (e.g., shell secretion and cell wall mineralization), or volcanic processes (ejection). According to these production processes, granular sediments can be further categorized into three major groups: lithogenic, biogenic, and volcanogenic sediments. Chemical sediments are formed from chemical reactions of dissolved compounds in seawater. They include the precipitation of minerals in seawater (e.g., manganese and phosphorus nodules), evaporites (salts), metalliferous compounds produced by hydrothermal vents, as well as fragments of limestone and limestone-like rocks (e.g., chert, chalk, and dolomite) that come from compacted, buried, and mineralized remains of marine organisms. Chemical sediments are also called authigenic or hydrogenic sediments.

Sediments can also be classified by particle grain size. The three major categories are gravel (>2 mm), sand ($0.063-2$ mm) and mud (<0.063 mm). They can be further classified into subgroups according to a refinement of the grain-size distribution. Gravel, sand, and silt (a subgroup of mud with diameter > 0.002 mm) are granular sediments originating mainly from weathering or erosion of rocks. Their dynamics can be studied by tracing the motions of individual particles driven by external forces. This marks their basic distinction from another subgroup of mud, i.e., clay that refers to the finest-grained sediments with diameter <0.002 mm and characterized by cohesion. Clay includes many combinations of one or more clay minerals with traces of metal oxides and organic matter. It can be produced by chemical weathering of rocks, biogenic processes, and chemical reactions of dissolved compounds in seawater. The cohesive force inherent in clay sediment enables the formation of bonds between particles and eventually leads to aggregation of tens of thousands of fine particles. Due to the cohesive force between particles, dynamics of clay sediments is quite different from that of non-cohesive sediments. As in natural environments, clay is normally mixed with silt; thus, they are both considered as cohesive sediment in engineering practice, despite the fact that the cohesive properties of silt are primarily due to the existence of clay. Gravel and sand sediments are termed non-cohesive sediments.

Sediment transport
Transport of sediments is triggered mainly by currents in marine environment, by waves in the near shore and by wind in subaerial coastal environment. When the near-bed fluid velocity increases to a certain level, some sediment particles start to move intermittently by sliding and rolling along the seabed. With a further increase of fluid velocity, more sediment particles are involved in migration and the bed form tends to become uneven due to different movement speeds and actions of different-sized particles, forming small perturbations on the seabed. These perturbations increase the roughness of the bed and induce an enhanced fluid friction, which further enrolls more sediment particles into the movement as a

result (e.g., Raudkivi, 1990; Allen, 1997). Smaller particles may roll upward along the upstream slope of these perturbations and jump out from the crest into the fluid flow, while larger particles roll to the downstream side of the perturbations. Some particles entrained by the flow may travel for a short distance before they fall down to the seabed again (i.e., saltation), while others may stay in suspension for a longer period, depending upon their physical (e.g., density, grain size, and shape) and biochemical properties (e.g., interaction with organic matter and ions). In general, three types of movement occur: rolling, saltation, and suspension. Gravels and sands are usually transported in the form of rolling and saltation in marine environment, while very fine sands, silts, and clays can be easily entrained by fluid flow and are normally transported as suspended matter. Cohesive sediments may interact with each other, flocculate or de-flocculate during the collision. During the movement, some particles may hit others and transfer their energy and momentum, increasing the complexity of the entire system.

Even in the simplest steady flow in a straight channel, sediment movements reveal a high degree of complexity. For non-cohesive sediments, a reasonable way to quantitatively describe their dynamics is to trace the motion of individual particles under different boundary conditions and via the multitude of collisions. With a development of computational devices, it is possible now to examine the validity of existing theories on granular sediment dynamics by the use of particle-based modeling (e.g., Herrmann et al., 2007; Radjaï and Dubois, 2011). The dynamics of cohesive sediments is more complicated and least understood. The major uncertainty of their movement comes from the cohesion between particles, which allows specific types of interactions (flocculation and de-flocculation) among sediment particles. Brownian motion, turbulence, and differential settling are recognized as the three major mechanisms to cause collision within a suspension under laboratory conditions (Krone, 1972; Winterwerp, 1999). In natural estuarine and coastal marine environments, studies (e.g., Stolzenbach and Elimelich, 1994) show that the contribution of Brownian motion and differential settling to sediment flocculation is minor, and turbulence plays the major role in controlling the interaction among cohesive sediments. Only in the case of high sediment concentration (e.g., on the order of magnitude of 10 kg/m^3 and higher), the effect of differential settling becomes substantial. Other parameters that affect the sediment flocculation in marine environment are salinity, the existence of ions, temperature, organic matter, temperature, and local suspended matter concentration (e.g., McAnally and Mehta, 2001; Winterwerp and Van Kesteren, 2004).

Sediment settling and recycling

Sediments settle down on the seabed when the fluid velocity drops below a certain level. Grain size is an important factor in determining the settling velocity of granular sediment particles. Large grains may settle quickly, while small, unflocculated grains like silts and clays may stay in the water column for a much longer time. Stokes' law is normally used to calculate the settling velocity of fine grain-size (<0.125 mm) particles. It is originally derived based on the assumptions of (1) small Reynolds number (i.e., laminar flow), and (2) spherical particles consist of homogeneous material and (3) none particle interactions. This hinders its application to particles with irregular shapes in turbulent flows. In natural estuarine environments, direct observations show that micro-flocs (<0.125 mm) are almost spherical, with width to length ratios normally in the range 0.8–1.5. Within this range, Stokes' law can be considered to apply (Gibbs, 1985). Macro-flocs (>0.125 mm) often show more complicated outlines and have variations in transparency that suggests a complex composition made up of smaller denser micro-flocs. Some researchers proposed modified versions of Stokes' law to take into account the shape irregularity of large particles (e.g., Lerman et al., 1974; Sternberg et al., 1999), while some other researchers tried to separate the numerical description of settling velocity of large particles from small particles (e.g., Van Rijn, 1984; Baugh and Manning, 2007). Grain size is a poor predictor for settling velocity of carbonate sediments, and some researchers built up relationships between fluid threshold velocity and settling velocity to describe the resuspension and settling of bioclastic sediments (e.g., Kench, 1997; Paphitis et al., 2002).

The sinking process of sediment particles is complicated due to the influence of turbulence and interactions among particles. Higher sediment concentration in the ambient water increases the possibility of particle interactions. This facilitates the formation of macro-flocs from micro-flocs and thus increases the sediment settling velocity. A certain level of turbulence is also proven to facilitate flocculation, although a strong turbulent shear will act to destroy the bonds between particles and induce a de-flocculation. The transition from positive to negative impact on particle flocculation by turbulence is often ambiguous, depending on not only the mineralogy and biological properties of the particles but also the local water environmental factors, e.g., sediment concentration, temperature, and salinity. The relationship between turbulence and sediment movement is not one-way. Sediment can also impose significant effects on the turbulence when its concentration is high enough. For example, close to the seabed often a concentrated benthic suspension is formed where suspended sediments interact strongly with turbulent flow through buoyant effects, and this may eventually lead to the formation of a fluid-mud vertical structure with a lutocline developed between the relatively clear water on top and the fluid-mud underneath. Sediment-induced density gradient (e.g., at the lutocline) as well as thermohaline pycnocline acts to damp turbulence and allows deposition of fine grain-size sediments.

Different from a direct resting of non-cohesive sediment particles on a rigid seabed, settling of cohesive sediment particles goes through two processes. Initially the freshly deposited sediments form a loose layer called fluffy layer with a thickness of several centimeters or even develop a fluid mud with a thickness of some tens of centimeters, depending on the total amount of sediment that settled within a certain time interval. In the latter case, the entire suspended sediment layer may act as a non-Newtonian fluid (Williams and Williams, 1989). If these sediments are undisturbed, they will gradually consolidate by a depletion of the interstitial water. Primary consolidation is caused by the self-weight of sediment, as well as the deposition of additional materials. It normally has a period of one or two days. Secondary consolidation is caused by the plastic deformation of the bed under a constant overburden. It begins during the primary consolidation and may last for weeks or months (Yang, 2006).

Once the fluid velocity increases above a certain level, sediment particles resting on the seabed surface come into recycling again. The most widely used semiempirical criteria to determine the thresholds for sediment motion are Shields (Shields, 1936) and Hjulström-Sundborg (Hjulström, 1939; Sundborg, 1956) diagrams. The threshold of fluid velocity for erosion of a sediment particle is normally larger than that for deposition. The difference between the relevant threshold velocities can play an important role in the landward transport of sediments through two processes called settling and scour lag (e.g., Postma, 1954). Non-cohesive sediments require larger threshold velocity for erosion than cohesive sediments that are resting loosely on the seabed surface. However, the threshold velocity for erosion of cohesive sediments increases drastically when they are consolidated. Macroflocs are likely to break into micro-flocs once they are resuspended by strong shear stress. A mixture of different-sized sediments on the seabed may increase the threshold velocity for erosion (Wilcock and Crowe, 2003). This is because that fine particles fill the interspaces among large particles, increasing their resistance to pore water erosion. Also large, heavier particles may help to increase the rate of consolidation of fine particles (e.g., mud).

Biological impacts may significantly influence the sediment settling and recycling processes. Their extension still remains poorly understood. Benthic organisms are able to entrain, disperse, and sort sediments (e.g., Scoffin, 1987). In particular, bioturbation by benthic fauna has been found to counteract the consolidation process by loosening the surface bed structure. Contrariwise, biostabilization by bacteria and benthic diatoms helps to increase the cohesion of the seabed through secretion of extracellular polymetric substances. Macrophytes also help to increase the seabed resistance to erosion through their root system.

Summary

Sediment dynamics in coastal and marine environment is an extremely complicated and not fully understood process. Movement of sediment particles is subject to the influence of many environmental factors and processes occurring at different temporal and spatial scales. A comprehensive understanding of sediment dynamics requires systematic analysis of the joint contributions of geophysical, biological, and chemical processes and their coupling.

Bibliography

Allen, P. A., 1997. *Earth Surface Processes*. Oxford: Blackwell Science.

Baugh, J. V., and Manning, A. J., 2007. An assessment of a new settling velocity parameterisation for cohesive sediment transport modelling. *Continental Shelf Research*, **27**(13), 1835–1855.

Gibbs, R. J., 1985. Estuarine flocs: their size settling velocity and density. *Journal of Geophysical Research*, **90**(C2), 3249–3251.

Herrmann, H. J., Andrade, J. S., Araujo, A. D., and Almeida, M. P., 2007. Particles in fluids. *European Physical Journal-Special Topics*, **143**, 181–189.

Hjulström, F., 1939. Transportation of debris by moving water. In Trask, P. D. (ed.), *Recent Marine Sediments; a Symposium: Tulsa*. Oklahoma: American Association of Petroleum Geologists, pp. 5–31.

Kench, P. S., 1997. Contemporary sedimentation in the Cocos (Keeling) Islands, Indian Ocean: interpretation using settling velocity analysis. *Sedimentary Geology*, **114**, 109–130.

Krone, R. B., 1972. *A Field Study of Flocculation as a Factor in Estuarial Shoaling Processes*. Vicksburg, MS: U.S. Army Corps of Engineers. Technical Bulletin 19, pp. 32–62.

Lerman, A., Lal, D., and Dacey, M. F., 1974. Stokes' settling and chemical reactivity of suspended particles in natural waters. In Gibbs, R. J. (ed.), *Suspended Solids in Water*. New York: Plenum Press, pp. 17–47.

McAnally, W. H., and Mehta, A. J., 2001. Collisional aggregation of fine estuarial sediment. In McAnally, W. H., and Mehta, A. J. (eds.), *Coastal and Estuarine Sediment. Proceedings in Marine Science*. Amsterdam: Elsevier, Vol. 3, pp. 19–39.

Paphitis, D., Collins, M.B., Nash, L.A., and Wallbridge, S., 2002. Settling velocities and entrainment thresholds of biogenic sands (shell fragments) under unidirectional flow. *Sedimentology*, **49**, 211–225.

Postma, H., 1954. Hydrography of the Dutch Wadden Sea. A study of the relations between water movement, the transport of suspended materials and the production of organic matter. *Archives Néerlandaises de Zoologie*, **10**, 405–511.

Radjaï, F., and Dubois, F., 2011. *Discrete-Element Modeling of Granular Materials*. New York: John Wiley & Sons.

Raudkivi, A. J., 1990. *Loose Boundary Hydraulics*, 3rd edn. Oxford: Pergamon Press.

Scoffin, T. P., 1987. *An Introduction to Carbonate Sediments and Rocks*. Glasgow: Blackie, p. 274.

Shields, A., 1936. *Anwendung der Ahnlichkeitsmechanik und der Turbulenzforschung auf die Geschiebebewegung. Mitteilung der preussischen Versuchsanstalt fur Wasserbau und Schiffbau*. Berlin: Triltsch & Huther, Vol. 26.

Sternberg, R. W., Berhane, I., and Ogston, A. S., 1999. Measurement of size and settling velocity of suspended aggregates on the northern California continental shelf. *Marine Geology*, **154**, 43–53.

Stolzenbach, K. D., and Elimelich, M., 1994. The effect of density on collisions between sinking particles: implications for particle aggregation in the ocean. *Journal of Deep Sea Research I*, **41**(3), 469–483.

Sundborg, A., 1956. The River Klarålven: Chapter 2. The morphological activity of flowing water erosion of the stream bed. *Geografiska Annaler*, **38**, 165–221.

van Rijn, L. C., 1984. Sediment transport. Part II: suspended load transport. *Journal of Hydraulic Engineering*, **110**(11), 1613–1641.

Wilcock, P., and Crowe, J., 2003. Surface-based transport model for mixed-size sediment. *Journal of Hydraulic Engineering*, **129**(2), 120–128.

Williams, D. J. A., and Williams, P. R., 1989. Rheology of concentrated cohesive sediments. *Journal of Coastal Research*, **5**, 165–173.

Winterwerp, J. C., 1999. *On the Dynamics of High-Concentrated Mud Suspensions*. Delft: Judels Brinkman & Ammerlaan, p. 172.

Winterwerp, J. C., and Van Kesteren, W. G. M., 2004. *Introduction to the Physics of Cohesive Sediment in the Marine Environment. Developments in Sedimentology 56*. Amsterdam: Elsevier.

Yang, C. T., 1996. *Sediment Transport, Theory and Practice*. New York: McGraw-Hill, pp. 211–266.

Yang, C. T., 2006. *Erosion and Sedimentation Manual*. Denver, CO: U.S. Department of the Interior Bureau of Reclamation, Technical Service Center Sedimentation and River Hydraulics Group.

Cross-references

Beach Processes
Currents
Sediment Transport Models
Placer Deposits
Waves

SEDIMENT TRANSPORT MODELS

Wenyan Zhang
MARUM-Center for Marine Environmental Sciences, University of Bremen, Bremen, Germany

Synonyms

Morphodynamic models

Definition

Sediment transport models refer to numerical models that describe mobilization, migration, and settling of sediment in fluids (e.g., water and air).

Introduction

Sediment transport is a dynamic process occurring persistently in the earth system wherever there is moving water or air. Its direct consequence is a gradual change of the earth surface landform, which may significantly affect the habitats not only for human but also for large ecosystems. In shallow water, sediment transport is caused mainly by a combined action of surface gravity waves and currents, while in deep sea it is mainly controlled by dense water circulations and internal waves. Benefited from a continuous development of computational facilities, sediment transport modeling has become a popular tool for addressing many environmental and engineering problems.

Sediment transport in natural waters can be divided into two major modes, namely bed-load and suspended-load transport (Figure 1). Bed-load transport refers to the migration of sediment grains within a thin layer just above the seabed. It includes rolling, sliding, and saltation of grains. Though some literature consider saltation as a form of suspended-load transport, it seems more proper to treat it as bed-load transport in numerical models. This is not only because of a high settling velocity of saltating particles which is difficult to resolve numerically but also due to a strong interaction between saltating particles and bed surface. Suspended-load transport refers to the entrainment and transport of sediment by a fluid flow in which sediment particles sink slowly and are carried over a long distance before settling. There is no clear boundary between these two modes because the range of relocation of a sediment particle greatly depends on the ambient water motions.

Although our understanding and numerical modeling of waves and currents have improved greatly in recent decades, sediment transport by water motions remains poorly understood. The difficulty in matching theoretical results to observations is due to the sensitivity of sediment particle mobility to many environmental factors and processes at different temporal and spatial scales. By now there exist no mathematical formulas which can accurately predict the motion of a large number of sediment particles in natural environment. Most existing formulas are derived empirically and their applicability is restricted to a narrow range of environmental settings. Results obtained from different models often differ drastically from each other even though they are based on the same initial conditions (Yang, 1996).

Three major obstacles hinder precise modeling of sediment transport in marine environment. The first is the genetic diversity of sediment particles. Their different physical and biochemical properties lead to various dynamic responses to ambient water motions. For example, grains of granular, noncohesive sediment like sands and gravels normally migrate as individual particles, possibly interacting with each other by direct collisions, through which momentum and energy is transferred. Fine sediment particles with a diameter <0.063 mm show quite different behaviors due to the existence of a cohesive force. They rarely exist as individual particles but rather tend to aggregate together, forming much larger structures which may consist of tens of thousands of particles. During the transport, these flocculated structures may further attract more fine particles or break down to smaller ones, depending on a bunch of parameters (e.g., turbulence strength, particle concentration, ion concentration, temperature, organic matters) in the ambient water. Such dynamics are still poorly understood.

Another major obstacle is the stochastic nature of sediment particle movement in fluids. A primary source of randomness of particle movements is turbulence, which remains still one of the most challenging problems in fluid dynamics. Another important source contributing to this

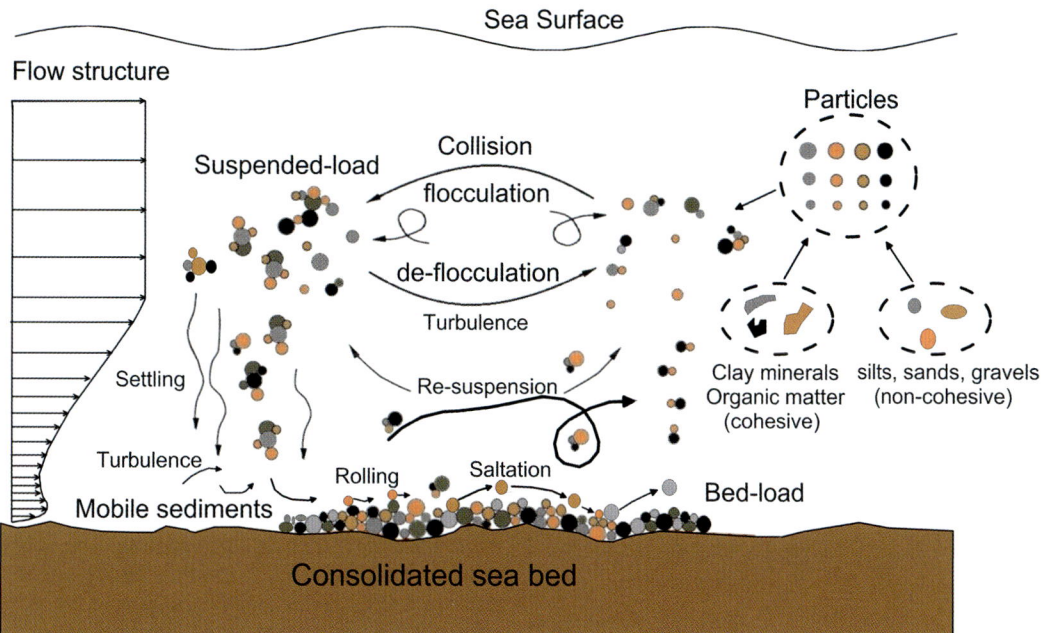

Sediment Transport Models, Figure 1 Sketch of sediment transport in open water (modified from Wang and Andutta, 2013). Note that the vertical flow structure is not scaled.

uncertainty is the fundamental ambiguity of pickup process of particles from the seabed. It is commonly recognized that sediment particles only start to move when the moments of the driving forces (exerted by the ambient water motion) exceed those of the stabilizing forces. Although the physics behind this phenomenon is clear, mathematical description of the pickup threshold is still ambiguous. Most sediment transport models treat the pickup threshold by the use of a critical shear stress following Shields (1936). Some other studies (e.g., Einstein, 1942; van Rijn, 1984; Nelson et al., 1995) suggest that such critical shear stress is not a deterministic value but rather lies in a range of values characterized by certain probability distributions (e.g., a Gaussian or Gamma distribution). The usual presence of a mixture of different-sized sediment particles on the seabed further complicates the problem to define a reliable pickup threshold.

The third major obstacle is the presence of greatly different scales in the process of sediment transport. Coastal and ocean sediment transport models are normally developed at a scale (both spatially and temporally) that is much larger than the one on which turbulence, sediment particle–particle interactions and particle–fluid interactions occur. These small-scale processes have to be either parameterized by subgrid modeling techniques or simplified by empirical formulae. In either case errors are inevitably induced and different formulations may produce significant differences in the results (Amoudry and Souza, 2011). Despite great efforts in recent decades to reduce errors of medium to large scale sediment transport models, such as data assimilation (e.g., van Dongeren et al., 2008; Lahoz et al., 2010) and improvement of numerical schemes (e.g., Spotz and Carey, 2001; Roelvink, 2006; Zhang et al., 2012), even the most advanced model to date can only predict sediment transport within an accuracy of $\pm 50\ \%$ at best and higher uncertainties are common (Amoudry and Souza, 2011).

According to the strategy in treating the uncertainty of sediment motion in fluids, existing sediment transport models can be divided into deterministic and probabilistic models. Deterministic models treat sediment transport in a time- and space-averaged manner. Usually the entire system is divided into smaller (grid) cells and a single value for each physical parameter is evaluated as representing its properties in the entire cell at each time step. Probabilistic models specify a possible range of the parameters and provide a collection of values characterized by different possibilities. Microscale (particle-based) sediment transport models such as discrete element models are not discussed here as they are not applied to open environments due to an intrinsic restriction of computational capacity and memory space. Nevertheless, outcomes of microscale particle-based models that are further validated by laboratory observations may help to improve numerical models for sediment transport in natural environments.

Deterministic models

Although transport of individual sediment particle is often seemingly stochastic, it does not hinder a development of numerical models that describe the sediment transport in a deterministic manner (Hardy, 2013). Deterministic models are mostly built on process-based descriptions of sediment transport and they are able to provide detailed insights of the physics of transport, which is often not possible in probabilistic models. Quite often the deterministic models show good performance on capturing medium to large spatial scale (of tens of meters to tens of kilometers) and short-term (of days to months) sediment transport phenomena.

Deterministic sediment transport models usually treat sediment as a continuum rather than individual particles. This assertion makes it possible to represent the presence of sediment in a spatial unit by a concentration. In this case suspended sediment transport in each fluid cell is calculated by either a presumption of a Rouse profile (Rouse, 1937) or by solving a mass balance equation formulated in a fully three dimensional Eulerian framework:

$$\frac{\partial C}{\partial t} + \frac{\partial uC}{\partial x} + \frac{\partial vC}{\partial y} + \frac{\partial (w - W_s)C}{\partial z}$$
$$= \frac{\partial}{\partial x}\left(A_h \frac{\partial C}{\partial x}\right) + \frac{\partial}{\partial y}\left(A_h \frac{\partial C}{\partial y}\right)$$
$$+ \frac{\partial}{\partial z}\left(K_h \frac{\partial C}{\partial z}\right) + S, \quad (1)$$

where C is the suspended sediment concentration, Ws is the setting velocity, A_h is the horizontal eddy diffusivity, K_h is the vertical eddy diffusivity, and S represents the sediment source/sink term at the boundary. With a set of appropriate initial and boundary conditions the models produce values for sediment distribution at each time step.

Because sediment is implicitly assumed to be homogeneously distributed over each computational cell, deterministic models are not able to directly resolve near-bed sediment transport processes such as bed-load transport that is largely influenced by particle–particle interactions in a thin near-bed region of high sediment concentration. In order to approximate bed-load transport and sediment flux at the seabed, deterministic models usually relate these transport rates to the bottom shear stress. For example, the bed-load transport rate (Q) is usually represented as

$$\frac{Q}{\sqrt{(\rho_s - \rho)gD_{50}/\rho}} = m\theta^n(\theta - \theta_{cr})^p, \quad (2)$$

where ρ_s and ρ are sediment and fluid density, respectively; m, n, p are empirical parameters; θ is the nondimensional bed shear stress (i.e., Shields parameter); and θ_{cr} is the critical Shields parameter for initiation of sediment particle motion. A representative particle grain size D_{50} is often adopted to derive the critical values for initiation of the motion.

Deterministic models normally classify sediment into two types, namely noncohesive and cohesive, and calculate the transport of these two types of sediment independently. Each sediment type may be further divided into several groups represented by different D_{50} values. Though some models are able to replicate the interaction among the cohesive sediments (flocculation and deflocculation), there seems to be a lack of model which treats the interaction between cohesive and noncohesive sediment explicitly.

By the use of mean or representative parameters, elaborate deterministic models are able to capture medium to large scale sediment transport phenomena in an efficient way. This facilitates their development for both engineering and research purposes. However, much progress is still needed to improve the accuracy of deterministic models, e.g., on the choice of parameters. As deterministic models are very sensitive to initial and boundary conditions, any perturbation may lead to a significant difference in the result. Thus a proper way to use deterministic models involves a sensitivity study which allows us to derive the "what if" scenarios for an optimum solution of the problem.

Probabilistic models

Probabilistic models emphasize the uncertainty of sediment transport in their formulation. The scope of uncertainty encompasses many aspects, from basic measurements to our inability to derive the true underlying relationship that governs the complex phenomena observed in the natural environments. This imposes great challenges to researchers to construct a numerical model to properly represent such uncertainties. Many existing probabilistic models deal with only a certain part of the uncertainties which are believed to be most crucial to sediment transport at the scales of interest.

The first probabilistic model to describe sediment transport was introduced by Einstein (1942). It was stated that the threshold of a sediment particle motion is when the ratio of the hydrodynamic force to the submerged weight force acting on the particle exceeds a certain limit that is determined by the particle geometry and the particle Reynolds number. The submerged weight of a particle is a constant for a given diameter while the hydrodynamic lift force varies depending on the near-bed turbulent flow and the relative position of the particles. Hence, the pickup probability for a bed particle can be defined as that of the instantaneous lift force being greater than the submerged weight force of the particle. With Einstein's approach, a pickup probability may be used to derive the total load transport function. This inspired a series of studies into probabilistic transport such as Kalinske (1947), Engelund and Fredsoe (1976), and Fredsøe and Deigaard (1992). In addition to Einstein's type transport theories, new probabilistic methods such as Bayesian statistical models have been proposed in recent years (e.g., Schmelter et al., 2011) for a better estimation of credible intervals for the critical shear stresses (for incipient motion and deposition) and their variances.

Another well-known type of probabilistic transport models uses the Lagrangian framework. Such models are widely used to trace the suspended transport of individual sediment particles in coastal and marine environment. Lagrangian models may include both advective and diffusive components of the motion of particles. Advection is the transport associated with the numerically simulated flow properties in grid cells while diffusion is associated with subgrid-scaled motions that are not resolved by the underlying flow model. The flow field is usually provided by an Eulerian model, and the random motions can be calculated by a random walk scheme (e.g., Hunter et al., 1993) or a trajectory path scheme (e.g., Döös et al., 2013).

Cellular Automata (CA) models, which include Lattice Gas and Lattice Boltzmann models, are a relatively new type of probabilistic transport models (Fonstad, 2013). Common characteristics for CA type models include: (1) The model area is divided into a set of connected cells (i.e., the lattice), (2) A limited number of state variables are defined at each cell, (3) The change of cell state is defined by a limited set of rules which act only on the local cell and its limited neighbors, and (4) The rules are applied at every time step of simulation. Although based on relatively simple rules that govern the evolution of the cell states, CA models have amazing capabilities in reproducing almost all types of transport patterns with a high computational efficiency (e.g., Werner, 1995; Pilotti and Menduni, 1997; Ladd and Verberg, 2001).

Summary

When targeting the numerical solution of a specific sediment transport problem, the choice of a model has to consider several aspects, e.g., the nature and complexity of the problem, the capabilities of the model to simulate the problem adequately, data availability for model calibration and verification, and overall available time and budget for solving the problem (Papanicolaou et al., 2008). One should be cautious of the advantages as well as disadvantages of a model when applying it to a specific problem. Error estimation is always necessary to define credible intervals of the simulation results. A hybrid coupling of deterministic and probabilistic models seems to provide the optimum means to study coastal and marine sediment transport phenomena but much effort is still needed to develop such models.

Bibliography

Amoudry, L. O., and Souza, A. J., 2011. Deterministic coastal morphological and sediment transport modeling: a review and discussion. *Review of Geophysics*, **49**, RG2002, doi:10.1029/2010RG000341.

Döös, K., Kjellsson, J., and Jönsson, B., 2013. TRACMASS – A lagrangian trajectory model. In Soomere, T., and Quak, E. (eds.), *Preventive Methods for Coastal Protection: Towards the Use of Ocean Dynamics for Pollution Control*. Heidelberg: Springer, pp. 225–249.

Einstein, H. A., 1942. Formula for the transportation of bed load. *Transportation ASCE*, **107**, 561–597.

Engelund, F., and Fredsoe, J., 1976. A sediment transport model for straight alluvial channels. *Nordic Hydrology*, **125**(5), 293–306.

Fonstad, M. A., 2013. Cellular automata in geomorphology. In Shroder, J. (Editor in chief), Baas, A. C. W. (ed.), *Treatise on Geomorphology*. San Diego, CA: Academic Press, Vol. 2, Quantitative Modeling of Geomorphology, pp. 117–134.

Fredsøe, J., and Deigaard, R., 1992. *Mechanics of Coastal Sediment Transport*. World Scientific, Singapore, pp. 369.

Hardy, R. J., 2013. Process-based sediment transport modeling. In Shroder, J. (Editor in chief), Baas, A. C. W. (ed.), *Treatise on Geomorphology*. San Diego, CA: Academic Press, Vol. 2, Quantitative Modeling of Geomorphology, pp. 147–159.

Herrmann, H. J., Andrade, J. S., Araujo, A. D., and Almeida, M. P., 2007. Particles in fluids. *European Physical Journal*, **143**, 181–189 -Special Topics.

Hunter, J. R., Crais, P. D., and Phillips, H. E., 1993. On the use of random walk models with spatially variable diffusivity. *Journal of Computational Physics*, **106**, 366–376.

Kalinske, A., 1947. Movement of sediment as bed load in rivers. *American Geophysical Union*, **28**(4), 615–620.

Ladd, A. J. C., and Verberg, R., 2001. Lattice-Boltzmann simulations of particle fluid suspensions. *Journal of Statistical Physics*, **104**, 1191–1251.

Lahoz, W., Khattatov, B., and Ménard, R., 2010. *Data Assimilation: Making Sense of Observations*. Springer-Verlag, Berlin.

Nelson, J. M., Shreve, R. L., McLean, S. R., and Drake, T. G., 1995. Role of near-bed turbulence structure in bed load transport and bed form mechanics. *Water Resources Research*, **31**(8), 2071–2086.

Papanicolaou, A., Elhakeem, M., Krallis, G., and Edinger, S. J., 2008. Sediment transport modeling review – current and future developments. *Journal of Hydraulic Engineering*, **134**(1), 1–14.

Pilotti, M., and Menduni, G., 1997. Application of lattice gas techniques to the study of sediment erosion and transport caused by laminar sheetflow. *Earth Surface Processes and Landforms*, **22**, 885–893.

Roelvink, J. A., 2006. Coastal morphodynamic evolution techniques. *Coastal Engineering*, **53**, 277–287.

Rouse, H. 1937. *Nomogram for the Settling Velocity of Spheres*. Division of Geology and Geography, Exhibit D of the Report of the Commission on Sedimentation, 1936–37, Washington, D.C.: National Research Council, pp. 57–64.

Schmelter, M. L., Hooten, M. B., and Stevens, D. K., 2011. Bayesian sediment transport model for unisize bed load. *Water Resources Research*, **47**, W11514, doi:10.1029/2011WR010754.

Shields, A., 1936. Anwendung der Ahnlichkeitsmechanik und Turbulenz forschung auf die Geschiebebewegung, Mitt. Preuss. Vers. Wasserbau Schiffbau, 26, 5–24.

Spotz, W., and Carey, G. F., 2001. Extension of high-order compact schemes to time-dependent problems. *Numerical Methods for Partial Differential Equations*, **17**, 657–672.

van Dongeren, A., Plant, N., Cohen, A., Roelvink, D., Haller, M. C., and Catalan, P., 2008. Beach wizard: nearshore bathymetry estimation through assimilation of model computations and remote observations. *Coastal Engineering*, **55**, 1016–1027.

van Rijn, L. C., 1984. Sediment pick-up function. *Journal of Hydraulic Engineering ASCE*, **110**(10), 1494–1502.

Wang, X. H., and Andutta, F. P., 2013. Sediment transport dynamics in ports, estuaries and other coastal environments. In Manning, A. J. (ed.), *Sediment Transport Processes and Their Modelling Application*. InTech, Croatia, pp. 3–35.

Werner, B. T., 1995. Eolian dunes: computer simulation and attractor interpretation. *Geology*, **23**, 1107–1110.

Yang, C. T., 1996. *Sediment Transport, Theory and Practice*. New York: McGraw-Hill, pp. 211–266.

Zhang, W. Y., Schneider, R., and Harff, J., 2012. A multi-scale hybrid long-term morphodynamic model for wave-dominated coasts. *Geomorphology*, **149–150**, 49–61.

Cross-references

Beach Processes
Coasts
Currents
Placer Deposits
Sediment Dynamics

SEDIMENTARY SEQUENCE

Christopher George St. Clement Kendall
Earth and Ocean Sciences, University of South Carolina, Columbia, SC, USA

Introduction

Sequence stratigraphy

The methodology of sequence stratigraphy uses a framework of genetically related sedimentary geometries and their bounding surfaces to determine the depositional setting of the sedimentary section of the earth's crust. This framework is built from the interpretation of seismic, well data, adjacent outcrops and knowledge of regional geology tied to an understanding of sedimentology, and modern and ancient depositional systems (Figure 1). This template can then be extrapolated to predict sediment character in the nearby subsurface and outcrops.

NB: *Visit the SEPM STRATA website (http://www.sepmstrata.org/) for explanations of the sequence stratigraphic terminology used here. Terminology on the sites is displayed on pop-up boxes containing information that clarify the understanding and use of this discipline of sequence stratigraphy.*

Sequence stratigraphic units

The sequence stratigraphic framework of erosional and depositional surfaces subdivides the stratigraphic section into gross sediment geometric units. In order of increasingly higher frequency, these sequence stratigraphic packages are:

- Sequences
- Systems tracts
- Parasequences and/or cycles

These geometric end members have similar sediment strata and are defined by their lithology, sedimentary structures, fauna, and bounding surfaces. These evolved in response to changes in accommodation (the space available to be filled) created by changes in eustasy (worldwide sea-level variation), tectonics, and sediment fill (Jervey, 1988). The characterization of these packages is the key to interpreting the depositional setting of the sedimentary sections that contain them. Once the depositional systems are better understood and described in terms of the above hierarchy, their lithofacies, and elements, better stratigraphic interpretations and predictions result (Figure 2).

Sedimentary sequence

Sedimentary sequences are the fundamental low-frequency stratal units of sequence stratigraphy (Catuneanu et al., 2011). A "sequence," as originally defined by Sloss et al. (1949) and Sloss (1963), is an unconformity-bounded stratigraphic unit. Mitchum Jr. (1977) modified this to "a relatively conformable succession of genetically related strata bounded by unconformities or their correlative conformities. Through the 1980s and 1990s, these definitions (Catuneanu et al., 2011) evolved to:

- *Depositional sequences*: bounded by subaerial unconformities and their marine correlative conformities (Vail, 1987; Posamentier et al., 1988; Van Wagoner et al., 1988, 1990; Vail et al., 1991; Hunt and Tucker, 1992)
- *Genetic stratigraphic sequences*: bounded by maximum flooding surfaces (Galloway, 1989)
- *Transgressive-regressive (T-R) sequences or T-R cycles*: bounded by maximum regressive surfaces (Johnson and Murphy, 1984; Johnson et al., 1985), subsequently redefined as a unit bounded by composite surfaces including subaerial unconformities and the marine portion of the maximum regressive surface (Embry and Johannessen, 1992)

These depositional sequences represent a relatively conformable succession of genetically related sedimentary strata generated by cyclic changes in accommodation. Enclosed by similar stratigraphic surfaces with the same general origin through geologic time, sequence fill is the result of the same geological events and processes that generate and occlude accommodation. The cyclic character of a sequence may be symmetrical or asymmetrical. The main cause of the cyclic character of sequences is commonly ascribed to eustasy since a sequence accumulates between eustatic-fall inflection points (Posamentier et al., 1988) forming successions of genetically linked deposition systems (systems tracts).

Sequences start and end with a surface generated by the same kind of event, including the onset of a relative sea-level fall (sequence boundary – SB), the end of relative sea-level fall (sequence boundary – SB), the end of regression (regressive surface of marine erosion – RSME), the start of a transgression (transgressive surface of erosion –TS), or the end of transgression (maximum flooding surface – MFS). These erosional and depositional surfaces mark changes in depositional regime that abrupt thresholds in depositional processes cause. The boundaries of sequences contrast with systems tract and parasequence boundaries, matching different "events" within a relative sea-level cycle and enabling the distinction of sequences from component systems tracts and parasequences.

Sedimentary systems tracts

Systems tracts are subdivisions of a sequence. Generated by "contemporaneous depositional systems" (Brown and Fisher, 1977), their fill of accommodation is associated with specific phases of the relative sea-level cycle (Posamentier et al., 1988; Hunt and Tucker, 1992; Posamentier and Allen, 1999; Catuneanu, 2006; Catuneanu et al., 2009; Catuneanu et al., 2011). Systems tracts commonly include falling stage systems tract

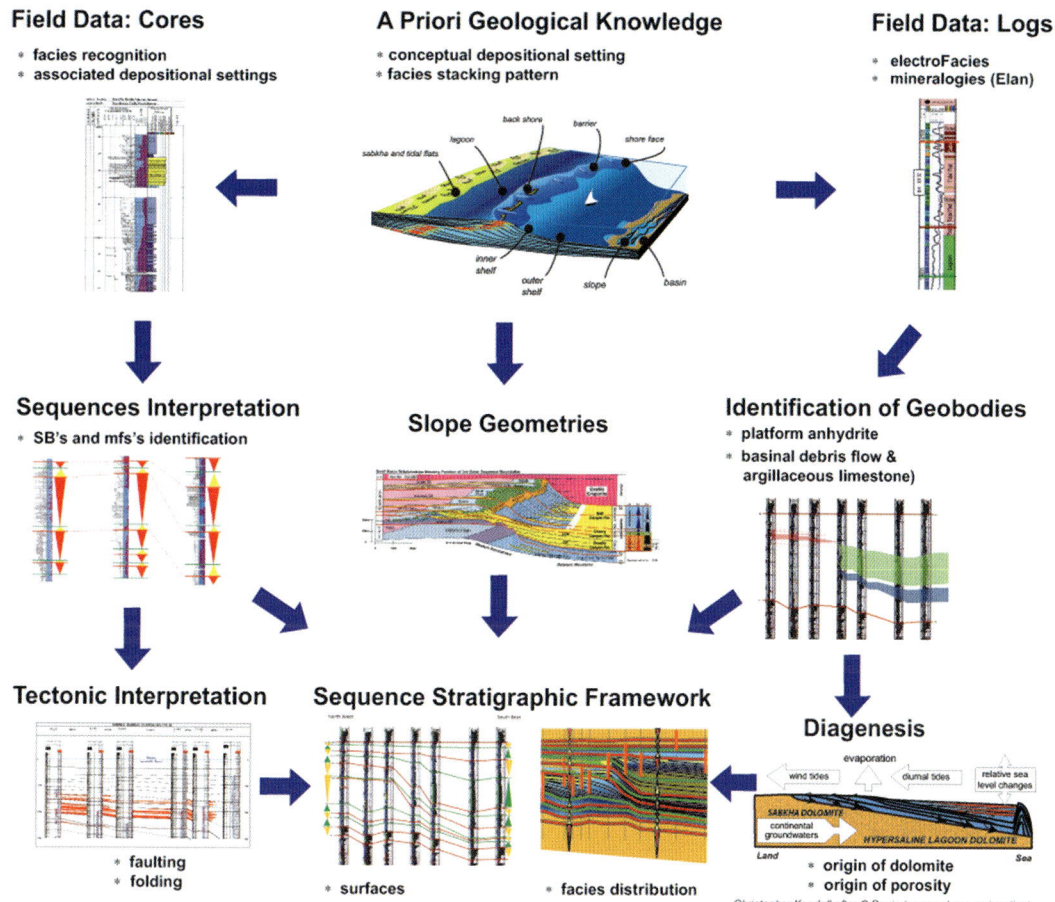

Sedimentary Sequence, Figure 1 Linkage between "a priori" geological knowledge, subsurface, and outcrop data with a sequence stratigraphic framework to predict facies distribution.

(FSST), lowstand systems tract (LST), transgressive systems tract (TST), and highstand systems tract (HST) (Figure 2). An exception to this is the regressive systems tract (RST) of Embry (1995), interpreted as regressive fill of accommodation generated independently of a specific phase of sea-level.

The internal architecture of systems tracts includes parasequence sets and/or high-frequency cycles driven by orbital forcing of eustasy or higher-frequency tectonic cycles, though the latter are usually assumed to be low-frequency events. Systems tracts relationships to the phase of the sea-level cycle are interpreted on the basis of stratal stacking patterns, position within the sequence, and types of bounding surface (Van Wagoner et al., 1987, 1988, 1990; Posamentier et al., 1988; Van Wagoner, 1995; Posamentier and Allen, 1999). Tracts consist of a relatively conformable succession of genetically related strata bounded by conformable or unconformable sequence stratigraphic surfaces. Systems tracts may be linked to particular types of shoreline trajectory, be this a forced regression, normal regression, or transgression and, defined from stratal stacking patterns (Figures 2 and 3), where a genetic link to coeval shorelines cannot be determined, may be shoreline independent (Catuneanu et al., 2011).

Parasequence

Though the term may be applied to carbonates, it is more common for siliciclastics. Parasequences are a relatively conformable succession of genetically related beds or bedsets bounded by marine flooding surfaces or their correlative surfaces (Van Wagoner, 1985). Patterns of stacked parasequences, parasequence sets, are used in conjunction with boundaries and their position within a sequence to define systems tracts (Van Wagoner et al., 1988) and their relationship to eustasy (Figure 3).

Parasequences are commonly characterized by a cycle of sediment that either coarsens or fines upward and the subdividing flooding surfaces are usually identified by abrupt and correlatable changes of the grain size of the sediments on either side of that flooding surface. This

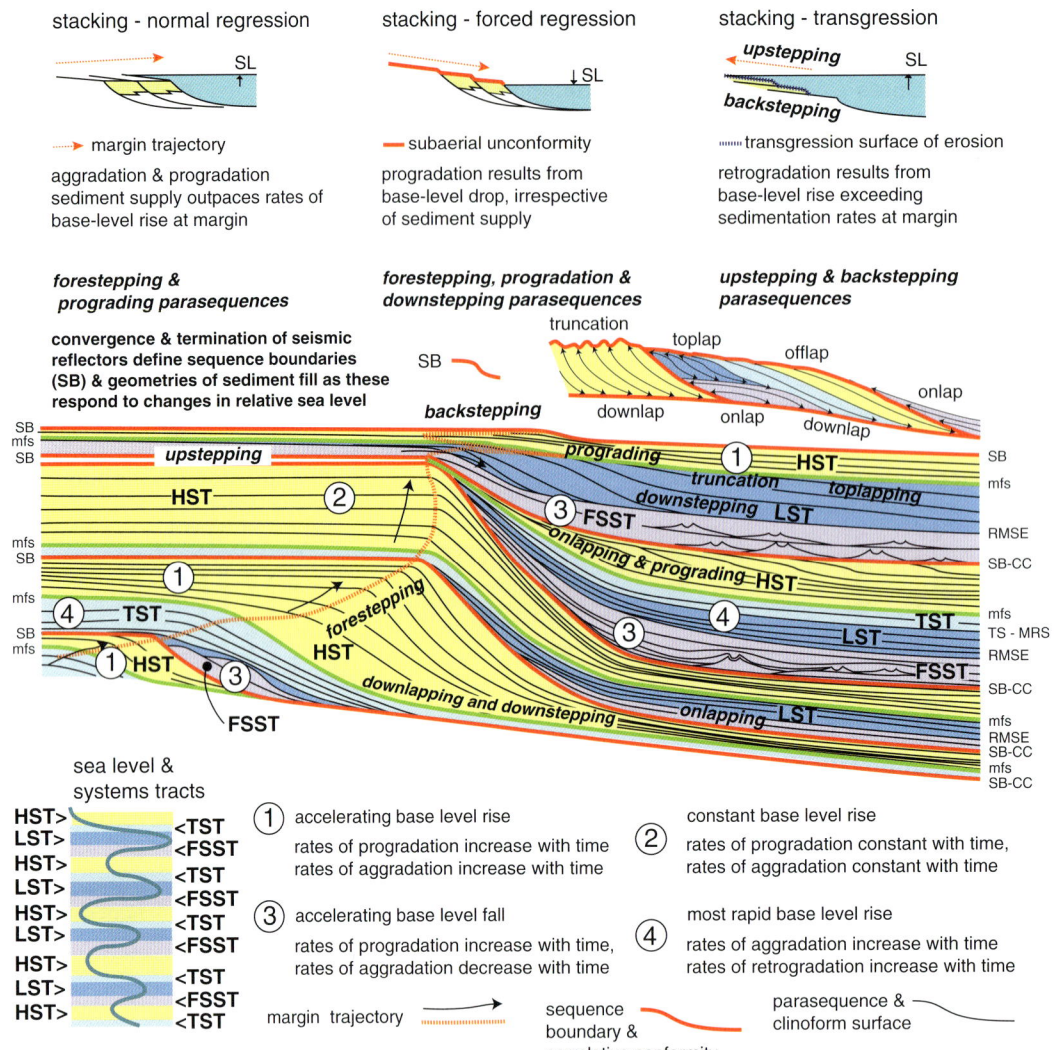

Sedimentary Sequence, Figure 2 Sequences defined by sequence boundaries (*SBs*); systems tracts defined by geometry of stacking patterns of parasequences and phase of sea-level curve, namely: aggrading, thinning, and prograding parasequences, highstand systems tract (*HST*); downstepping and prograding parasequences, forced by falling stage systems tract (*FSST*); aggrading, thickening, and prograding parasequences, lowstand systems tract (*LST*); and aggrading, thickening, and retrograding parasequences, transgressive systems tract (*TST*). Right-hand side of the diagram illustrates a deeper-water setting, with submarine fans just below wave base form forced regression of the FSST. The base of the FSST is an erosional surface, or sequence boundary that extends seaward and deeper to the correlative conformity (*CC*) sensu Posamentier and Allen (1999). The surface capping the FSST is a regressive surface of marine erosion (*RSME*) (Plint, 1988) within the depth of wave-cut scour related to a shallow-water setting, too shallow for a deeper-water CC sensu Hunt and Tucker (1992).

Step 1 cross section has an "accelerating base level rise," at the base of HST. This is followed by a decelerating base level rise and approaching a zero rate at the end of rise. The initial accelerating base level rise marks the onset of transgression. When the base level is falling, whether accelerating or decelerating, during the entire falling stage, the unit is an FSST.

change in grain size is often caused by the abrupt changes in energy associated with the waves or currents of the sea transgressing across the sediment interface and can be identified in well logs, outcrop, and seismic. In carbonate settings, a parasequence corresponds to a succession of facies commonly containing lag deposit or thin deepening interval followed by a thicker shallowing-upward part, as, for example, in peritidal cycles.

In contrast to sequences and systems tracts, which may extend across an entire sedimentary basin from fluvial into the deep-water setting, parasequences are geographically restricted to the coastal to shallow-water areas where marine flooding surfaces may form (Posamentier and Allen, 1999). In the case of carbonate settings, peritidal cycles can in some cases be correlated into slope and basinal facies (e.g., Tinker, 1998; Chen and Tucker, 2003).

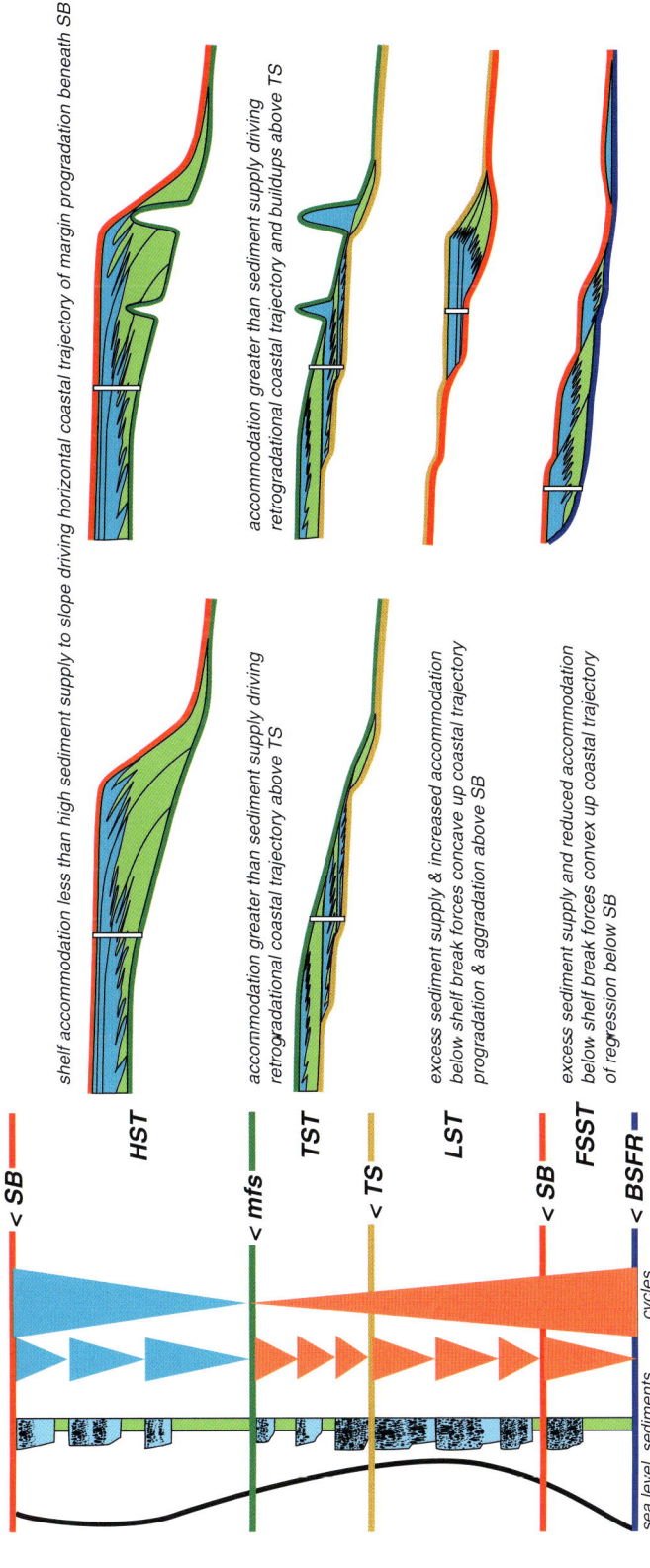

Sedimentary Sequence, Figure 3 Systems tracts and parasequence geometries of a composite sequence stratigraphic framework for a carbonate platform with (*right-hand side*) and without (*left-hand side*) buildups. In the falling stage system tract (*FSST*), initially a forced regression causes the parasequences of the carbonate margin to prograde and downstep seaward in response to lost accommodation of a relative sea-level fall. The three downstepping parasequence wedges build out over the basal surface of forced regression (*BSFR*) and terminate seaward in a basin-floor fan. Above the fan is a correlative conformity, while up-dip a subaerial unconformity forms a sequence boundary (SB) over the downstepping parasequence wedges of the forced regression. The lowstand systems tract (*LST*) overlies the unconformity, and the trajectory of shoreline parasequences rises seaward. There follow three onlapping retrograding carbonate parasequences of the transgressive systems tract (*TST*) overlying a transgressive surface (*TS*) topped by a maximum flooding surface (*MFS*). A highstand systems tract (*HST*) carbonate margin follows expressed as three shoaling-upward parasequences which lie below the next depositional sequence boundary (*SB*). The designation of sequence boundary in this diagram follows Hunt and Tucker (1992), while Posamentier and Allen (1999) choose the BSFR as the sequence boundary. The vertical section on the left-hand side is a schematic column of the stacked sediments that accumulated within each systems tract. Approximate position of sedimentary column indicated by white strips on sections, no one locality provides a complete section through all the systems tracts and parasequences, the FSST is represented by a section that is situated and extending basinward, and the LST has a section even farther basinward, and the section for the TST and HST parasequences is better expressed further onto platform.

Summary and conclusions

Sequence stratigraphy uses a framework of genetically related sedimentary geometries, namely:

- Sequences
- Systems tracts
- Parasequences and/or cycles

and their bounding surfaces:

- Sequence boundary – SB
- Regressive surface of marine erosion – RSME
- Transgressive surface of erosion – TS
- Maximum flooding surface – MFS

These are used to determine the depositional setting of the sedimentary section.

- The framework is built using the interpretation of seismic, well data, adjacent outcrops and knowledge of regional geology tied to an understanding of sedimentology, and modern and ancient depositional systems
- The sequence stratigraphic framework is extrapolated to predict sediment character in the subsurface and outcrops.

Bibliography

Brown, L. F., Jr., and Fisher, W. L., 1977. Seismic stratigraphicinterpretation of depositional systems: examples fromBrazilian rift and pull apart basins. In Payton, C. E. (ed.), *Seismic Stratigraphy – Applications to Hydrocarbon Exploration*. American Association of Petroleum Geologists Memoir, Vol. 26, pp. 213–248.

Catuneanu, O., 2002. Sequence stratigraphy of clastic systems: concepts, merits, and pitfalls. *Journal of African Earth Sciences*, **35**, 1–43.

Catuneanu, O., 2006. *Principles of Sequence Stratigraphy*. Amsterdam: Elsevier, p. 375.

Catuneanu, O., Galloway, W. E., Kendall, C. G. St. C., Miall, A. D., Posamentier, H. W., Strasser, A., and Tucker, M. E., 2011. Sequence stratigraphy: methodology and nomenclature. *Newsletters on Stratigraphy, Stuttgart*, **44/3**, 173–245.

Chen, D., and Tucker, M. E., 2003. The Frasnian-Famennian mass extinction: insights from high-resolution sequence stratigraphy and cyclostratigraphy in South China. *Palaeogeography, Palaeoclimatology, Palaeoecology*, **193**, 87–111.

Embry, A. F., 1995. Sequence boundaries and sequence hierarchies: problems and proposals. In *Sequence Stratigraphy on the Northwest European Margin, Proceedings of the Norwegian Petroleum Society Conference*. Norwegian Petroleum Society Special Publications 5, pp. 1–11.

Embry, A. F., and Johannessen, E. P., 1992. T-R sequence stratigraphy, facies analysis and reservoir distribution in the uppermost Triassic-Lower Jurassic succession, western Sverdrup Basin, Arctic Canada. In Vorren, T. O., Berg – sager, E., Dahl-Stamnes, O. A., Holter, E., Johansen, B., Lie, E., and Lund, T. B. (eds.), *Arctic Geology and Petroleum Potential*. Norwegian Petroleum Society (NPF), Special Publication, Vol. 2, pp. 121–146.

Galloway, W. E., 1989. Genetic stratigraphic sequences in basin analysis I. Architecture and genesis of flooding surface bounded depositional units. *American Association of Petroleum Geologists Bulletin*, **73**, 125–142.

Hunt, D., and Tucker, M. E., 1992. Stranded parasequences and the forced regressive wedge systems tract: deposition during base-level fall. *Sedimentary Geology*, **81**, 1–9.

Jervey, M. T., 1988. Quantitative geological modeling of siliciclastic rock sequences and their seismic expression. In Wilgus, C. K., Hasting, B. S., Kendall, C., Posamentier, H. W., Ross, C. A., and Van Wagoner, J. C. (eds.), *Sea-level changes: an integrated approach*. Tulsa, OK: Society of Economic Paleontologists and Mineralogists. Special Publication No. 42, pp. 47–69.

Johnson, J. G., and Murphy, M. A., 1984. Time-rock model for Siluro-Devonian continental shelf, western United States. *Geological Society of America Bulletin*, **95**, 1349–1359.

Johnson, J. G., Klapper, G., and Sandberg, C. A., 1985. Devonian eustatic fluctuations in Euramerica. *Geological Society of America Bulletin*, **96**, 567–587.

Mitchum, Jr., R. M., 1977. *Seismic Stratigraphy and Global Changes of Sea Level: Part 11. Glossary of Terms used in Seismic Stratigraphy: Section 2. Application of Seismic Reflection Configuration to Stratigraphic Interpretation*, Memoir No. 26, pp. 205–212.

Plint, A. G., 1988. Sharp-based shoreface sequences and "offshore bars" in the Cardium Formation of Alberta; their relationship to relative changes in sea level. In Wilgus, C. K., Hastings, B. S., Kendall, C. G. S. C., Posamentier, H. W., Ross, C. A., and Van Wagoner, J. C. (eds.), *Sea Level Changes – An Integrated Approach*. Tulsa, OK: Society of Economic Paleontologists and Mineralogists. SEPM Special Publication No. 42, pp. 357–370.

Posamentier, H. W., and Allen, G. P., 1999. *Siliciclastic sequence stratigraphy: concepts and applications*. Tulsa, OK: SEPM Concepts in Sedimentology and Paleontology, Vol. 7, p. 210.

Posamentier, H. W., Jervey, M., and Vail, P. R., 1988. Eustatic controls on clastic deposition. I. Conceptual framework. In Wilgus, C. K., Hastings, B. S., Kendall, C. G., Posamentier, H. W., Ross, C. A., and Van Wagoner, J. C. (eds.), *Sea Level Changes—An Integrated Approach*. SEPM Special Publication 42, pp. 110–124.

Sloss, L. L., 1963. Sequences in the cratonic interior of North America. *Geological Society of America Bulletin*, **74**, 93–114.

Sloss, L. L., Krumbein, W. C., and Dapples, E. C., 1949. Integrated facies analysis. In Longwell, C. R. (ed.), *Sedimentary Facies in Geologic History*. New York: Geological Society of America Memoir, Vol. 39, pp. 91–124.

Tinker, S. W., 1998. Shelf-To-basin Facies Distributions and sequence stratigraphy of a Steep-Rimmed carbonate Margin: Capitan depositional system McKittrick Canyon, New Mexico and Texas. *Journal of Sedimentary Research*, **68**, 1146–1174.

Vail, P. R., 1987. Seismic stratigraphy interpretation procedure. In Bally, A. W. (ed.), *Atlas of Seismic Stratigraphy*. Tulsa, OK: American Association of Petroleum Geologists Studies in Geology, Vol. 27, pp. 1–10.

Vail, P. R., Audemard, F., Bowman, S. A., Eisner, P. N., and Perez-Cruz, C., 1991. The stratigraphic signatures oftectonics, eustasy and sedimentology – an overview. In Einsele, G., Ricken, W., and Seilacher, A. (eds.), *Cycles and Events in Stratigraphy*. Berlin: Springer–Verlag, pp. 617–659.

Van Wagoner, J. C., 1995. Sequence Stratigraphy and Marine to Nonmarine Facies Architecture of Foreland Basin Strata, Book Cliffs, Utah, U.S.A. In Van Wagoner, J. C., and Bertram, G. T. (eds.), *Sequence Stratigraphy of Foreland Basin Deposits*. Tulsa, OK: American Association Petroleum Geologists Memoir, Vol. 64, pp. 137–223.

Van Wagoner, J. C., Mitchum, R. M., Posamentier, H. W., and Vail, P. R., 1987. An overview of sequence stratigraphy and key definitions. In Bally, A. W. (ed.), *Atlas of Seismic Stratigraphy, volume 1*. Tulsa, OK: AAPG. Studies in Geology 27, Vol. 1, pp. 11–14.

Van Wagoner, J. C., Posamentier, H. W., Mitchum, R. M., Vail, P. R., Sarg, J. F., Loutit, T. S., and Hardenbol, J., 1988.

An overview of sequence stratigraphy and key definitions. In Wilgus, C. K., Hastings, B. S., Kendall, C. G. S. C., Posamentier, H. W., Ross, C. A., and Van Wagoner, J. C. (eds.), *Sea Level Changes – An Integrated Approach*. Tulsa, OK: SEPM. Special Publication 42, pp. 39–45.

Van Wagoner, J. C., Mitchum Jr., R. M., Campion, K. M., and Rahmanian, V. D., 1990. *Siliciclastic Sequence Stratigraphy in Well Logs, Core, and Outcrops: Concepts for High Resolution Correlation of Time and Facies*. American Association of Petroleum Geologists Methods in Exploration Series, Vol. 7, 55 pp.

Cross-references

Convergence Texture of Seismic Reflectors
Lithostratigraphy
Reflection/Refraction Seismology
Relative Sea-level (RSL) Cycle
Sequence Stratigraphy

SEISMOGENIC ZONE

Eli Silver
Earth and Planetary Sciences Department, University of California, Santa Cruz, CA, USA

The seismogenic zone is the depth interval of unstable (stick-slip) behavior where earthquakes can nucleate and slip coseismically (Scholz, 2002; Marone and Richardson, 2010). Earthquake nucleation usually occurs in the range of a few kilometers to 40 km depth. Variations are related to lithologic differences along the fault interface (Sibson, 1986), cementation, mineral dehydration, and diagenesis (Moore et al., 2007); pore fluid pressures and stress state (Saffer and Tobin, 2011); the structure of the lower plate interface (Heuret et al., 2011); the depth of the crust-mantle boundary beneath the upper plate; and the thermal regime (Oleskevich et al., 1999).

Updip and downdip limits are determined by locating aftershock distributions of earthquakes along the fault zone and by geodetically determining the zone over which locking occurs on the interface (Moore et al., 2007). The zone of coseismic slip can exceed that of nucleation alone. Tsunami modeling can constrain the slip distribution during large earthquakes (Yamazaki et al., 2011) and provides one measure of the full coseismic displacement.

The physical characteristics of the seismogenic zone are generally inferred from geophysical observations, such as seismic velocities, density, and electrical conductivity. Direct measurements have included drilling through the San Andreas Fault (Zoback et al., 2010) and through splay faults of the Nankai Trough subduction zone (Tobin and Kinoshita, 2006), which have provided material samples and allow long-term monitoring of temperature, fluid pressure, and fluid chemistry that may change during seismic events.

Slow earthquakes, including tsunami earthquakes (Kanamori and Kikuchi, 1993), and episodic tremor are often generated in the transition between the seismogenic zone where high-velocity slip earthquakes are generated and the region of conditionally stable or stable slip behavior above and below this zone (Schwartz and Rokosky, 2007). Slow slip is generally determined by geodetic observations, while episodic tremor is seen on seismographs.

Bibliography

Heuret, A., Lallemand, S., Funiciello, F., Piromallo, C., and Faccenna, C., 2011. Physical characteristics of subduction interface type seismogenic zones revisited. *Geochemistry, Geophysics, Geosystems*, **12**. doi:10.1029/2010GC003230

Kanamori, H., and Kikuchi, M., 1993. The 1992 Nicaragua earthquake: a slow tsunami earthquake associated with subducted sediments. *Nature*, **361**, 714–716.

Marone, C., and Richardson, E., 2010. Learning to read fault-slip behavior from fault-zone structure. *Geology*, **38**, 767–768.

Moore, C., Rowe, C., and Meneghini, F., 2007. How accretionary prisms elucidate seismogenesis in subduction zones. In Dixon, T., and Moore, C. (eds.), *The Seismogenic Zone of Subduction Thrusts*. New York: Columbia University Press, pp. 288–315.

Oleskevich, D. A., Hyndman, R. D., and Wang, K., 1999. The updip and downdip limits to great subduction earthquakes: thermal and structural models of Cascadia, south Alaska, SW Japan, and Chile. *Journal of Geophysical Research*, **104**, 14965–14991.

Saffer, D. M., and Tobin, H. J., 2011. Hydrogeology and mechanics of subduction zone forearcs: fluid flow and pore pressure. *Annual Reviews of Earth and Planetary Sciences*, **39**, 157–186.

Scholz, C. H., 2002. *The Mechanics of Earthquakes and Faulting*. Cambridge: Cambridge University Press.

Schwartz, S.Y., and Rokosky, J.M., (2007). Slow-slip events and seismic tremor at circum-Pacific subduction zones. *Reviews of Geophysics*, **45**. DOI: 10.1029/2006RG000208.

Sibson, R., 1986. Earthquakes and rock deformation in crustal fault zones. *Annual Reviews of Earth and Planetary Sciences*, **14**, 149–175.

Tobin, H., and Kinoshita, M., 2006. Nantroseize: the IODP Nankai Trough seismogenic zone experiment. *Scientific Drilling*, **2**. doi:10.2204/iodp.sd.2.06.2006

Yamazaki, Y., Lay, T., Cheung, K.F., Yue, H., and Kanamori, H., 2011. Modeling near-field tsunami observations to improve finite-fault slip models for the 11 March 2011 Tohoku earthquake. *Geophysical Research Letters*, **38**. doi:10.1029/2011GL049130

Zoback, M., Hickman, S., and Ellsworth, W., 2010. Scientific drilling into the San Andreas Fault zone. *EOS*, **91**, 197–204.

Cross-references

Earthquakes
Subduction
Tsunamis

SEQUENCE STRATIGRAPHY

Christopher George St. Clement Kendall
Earth and Ocean Sciences, University of South Carolina, Columbia, SC, USA

Definition

A methodology of stratigraphic interpretation that uses a framework of genetically related stratigraphic surfaces,

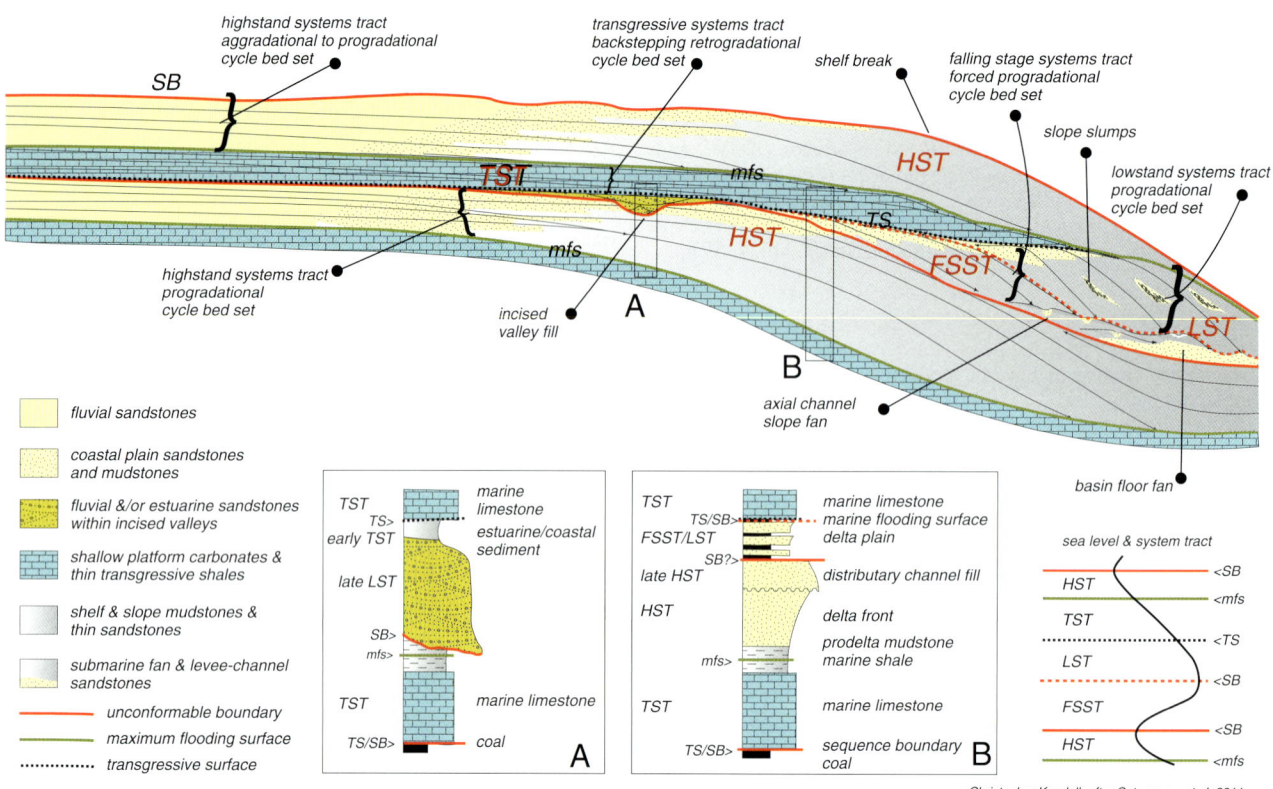

Sequence Stratigraphy, Figure 1 Sequence stratigraphic model and cycle types for a mixed carbonate-clastic succession based on the Yoredale cycles of the mid-Carboniferous of northern England. The schematic logs show (A) a cycle with an incised-valley fill sandstone cutting into shales that overlie marine carbonate and (B) coarsening-up deltaic clastics that have onlapped the marine carbonate.

Introduction

Stratigraphy is the science of the layered character of rocks, be these sedimentary, volcanic, metamorphic, or igneous rocks. Sequence stratigraphy, a branch of sedimentary stratigraphy, is a methodology that uses the order sedimentary strata accumulated within a framework of major depositional and erosional surfaces to interpret the depositional setting of clastic and carbonate sediments from continental, marginal marine, basin margins, and downslope settings. The surfaces that bound and subdivide the strata are often interpreted to be generated during changes in relative sea-level causing associated deposition and erosion. The resulting template of surfaces and the lithologic facies, sedimentary structures, and fauna they bound are used to interpret the depositional setting and predict the heterogeneity, extent, and character of the lithofacies.

Sequence stratigraphy is based on the premise in that "the present is the key to the past" (Lyell, 1830, 1832, 1833). It supports the contention that the sedimentary record of the earth's crust is the product of uniform and common physical processes that interacted with sediments as they accumulated. The section that follows is based on Catuneanu et al. (2011) and a summary of a collective understanding of sequence stratigraphy that is available on the website SEPM STRATA.

A major problem with sequence stratigraphy is that the definition, terminology, and interpretation of the surfaces of sequence stratigraphy can be complex and sometimes contentious. **NB**: *Visit the SEPM STRATA website* (http://www.sepmstrata.org/) *for explanations of the sequence stratigraphic terminology used here, which is linked to pop-up boxes containing information that clarify the understanding and use of this discipline of stratigraphy.*

Bounding surfaces of sequence stratigraphy

Evidence that the relative position of the sea has varied through geologic time is recorded in the Earth's marine sedimentary strata from the Precambrian through present by the major bounding and subdividing surfaces of the sedimentary section (Figure 1). These surfaces are commonly generated by the changing position of relative sea-level when the volume of water in the oceans varies and/or ocean relief changes. The bounding surfaces are the key to sequence stratigraphy.

A cycle of the rise and fall of relative sea-level will likely produce a corresponding response in the sedimentary section and marked changes in the flow regime that form subdividing sequence stratigraphic surfaces. For instance, when eustasy, worldwide sea-level, rises or the local crustal substrate subsides, a relative rise in sea-level occurs. The resulting transgression floods the shore and a nearshore transgressive surface (TS) forms. When the rate of sea-level rise reaches its most rapid increase, sediment accumulation seaward of the shore is reduced. Meanwhile pelagic and benthic fossils and organic matter continue to accumulate in the open sea. As a result condensed accumulations of fossils are found on the sediment surface or within a thin sedimentary zone, and, coincidentally, the organics sequestered radioactive elements producing a strong radioactive signal on the gamma logs. This zone and/or surface is known as the maximum flooding surface (mfs). Following the highest position of the sea, a drop in sea-level may cause the shore and the nearshore to be eroded, forming an unconformity or sequence boundary (SB).

Sequence stratigraphic interpretation identifies the subdividing surfaces of TS, mfs, and SB described above. These surfaces envelope the discrete sediment geometries of the sedimentary section. Interpretation involves conceptually reversing the order of deposition and back-stripping the geometries from oldest to youngest. They are then reassembled in the order of accumulation, using as a template the subdividing surfaces, lithofacies geometry, and fauna to interpret the evolving character of depositional setting. The reassembly tracks the evolution of the sedimentary system, its hydrodynamic setting, and accommodation.

The back-stripping analysis is aided by the subdivision of the sequence stratigraphic section on the basis of major depositional and erosional surfaces alluded to above. A variety of elements are subdivided by the surfaces. Their hierarchy from low frequency to high frequency includes:

- Sequences: fundamental stratal units of sequence stratigraphy (Catuneanu et al., 2011) composed of relatively conformable successions of genetically related strata generated by cycles of change in accommodation (the space available to be filled) (Jervey, 1988), and/or sediment supply are bounded by similar stratigraphic surfaces which have same origin through geologic time.
- Systems tracts: subdivisions of a sequence linked to "contemporaneous depositional systems" (Brown and Fisher, 1977) and sea-level position. Include highstand systems tract, falling stage systems tract, lowstand systems tract, and transgressive systems tract.
- Parasequences and/or cycles: relatively conformable successions of genetically related beds or bed sets (within a parasequence set) bounded by marine flooding surfaces or their correlative surfaces (Van Wagoner, 1995). Stacking of parasequence sets is used in conjunction with boundaries and position within a sequence to define systems tracts (Van Wagoner et al., 1988) and their relationship to eustacy.

As sediments are reassembled, the genetic character of the sequences, systems tracts, and parasequences is determined as products of changes in accommodation and the regional setting. A limit to this analytical strategy is often the extent of the understanding of the inferred depositional setting.

Sequence stratigraphy of depositional systems

Understanding how modern and ancient depositional systems are tied to sequence stratigraphy and the processes that formed them aids the interpretation of depositional systems. The sequence stratigraphic framework is used to analyze and explain how sedimentary rocks acquire their layered character, lithology, texture, faunal associations, and other properties. These properties explain how the mechanisms of sediment accumulation, erosion, and interrelated processes produced the current configuration of the rocks. This strategy leads to new questions and more realistic interpretations and enhances predictions of lithofacies heterogeneity. These systems include:

- Clastic systems
  - Marine: barrier island coasts; deltaic systems; deepwater fans; deepwater basins
  - Continental: glacial; eolian; alluvial fans; braided streams; coarse and fine-grained fluvial systems; lacustrian
- Carbonate systems
  - Inner carbonate shelf; outer carbonate shelf and margins; deepwater carbonates

Both respond to changes in base level and are subdivided by similar surfaces, the major difference being carbonate accumulation tending to be "in situ production" while clastics are transported to their depositional setting. Rates of carbonate production are linked to photosynthesis and depth dependent with rates greatest close to the air/sea interface, so carbonate facies and their fabrics are often clear indicators of sea-level position. Additionally organisms have the capacity to produce and accumulate sediments above the hydrodynamic thresholds associated with clastic systems generating the ecological accommodation of Pomar (2001).

Niels Stensen and Johannes Walther

The interpretation is better, and predictions of local and regional stratigraphy are more accurate when the sequence stratigraphic framework is integrated with:

- Steno's laws of sediment accumulation (Dott and Batten, 1976)
- Walther's law (Middleton, 1973) of the vertical and lateral equivalence of sediments

Steno established that younger rock sedimentary layers overlie in order older layers (principle of superposition),

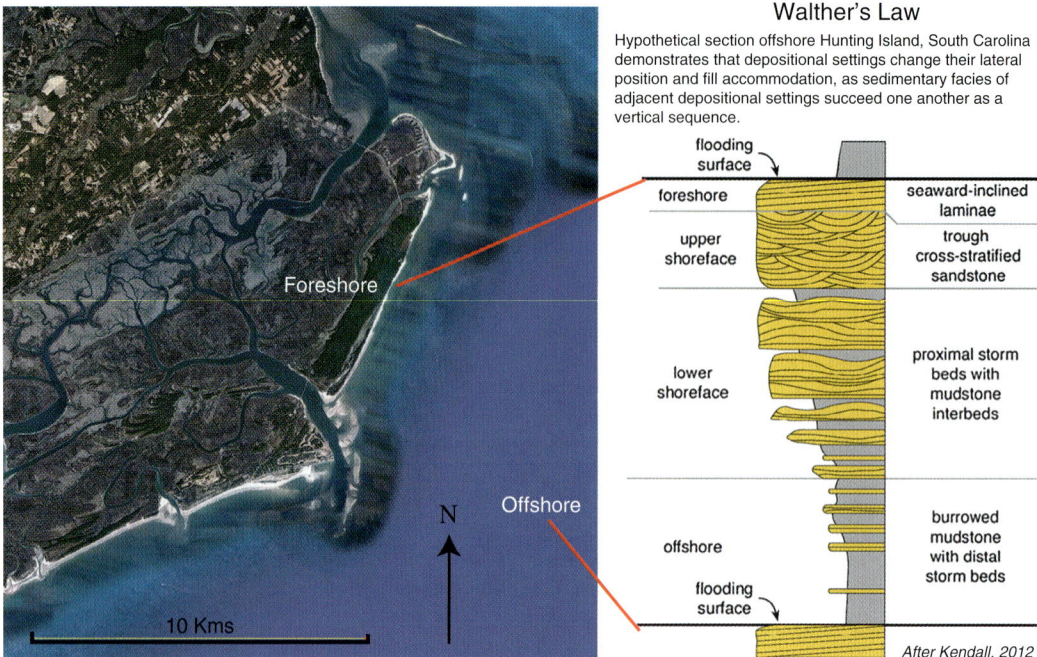

Sequence Stratigraphy, Figure 2 Hypothetical section offshore Hunting Island, South Carolina, demonstrating that, as Steno (Dott and Batten, 1976) indicated, sediments accumulate vertically from oldest to youngest and extend laterally while, as Walther (Middleton, 1973) indicated, depositional settings change their lateral position and fill accommodation, so that sedimentary facies of adjacent depositional settings succeed one another as a vertical sequence.

sediments fill over basal irregular surfaces enclosed above by a smooth surface (principle of original horizontality), and sediment layers have continuity to where they pinch out, or are prevented by a barrier to spread further, or are disrupted by folding and/or faulting (principle of lateral continuity). Walther recognized that as sediments fill the accommodation of their depositional settings, those laying laterally side by side change their lateral position as the sedimentary facies of adjacent depositional settings succeed one another as a vertical sequence (Figure 2).

Stacking patterns and geometries

Sequence stratigraphy methodology establishes the depositional setting of a sedimentary section by analyzing the geometric arrangement of sedimentary fill or stacking patterns. These patterns are represented by the vertical succession of sedimentary facies geometries and their enveloping surfaces. The geometries and stacking patterns of uncemented carbonates and clastics are similar, as both respond to changes in base level in the same way, both can be subdivided by similar surfaces and both respond to wave and current movement similarly.

However major differences in the sequence stratigraphy of the sediments exist. All clastics are transported to their depositional resting place while carbonates are produced and accumulate "in situ." Rates of carbonate production, and so accumulation, are linked to photosynthesis and are depth dependent. Their rates of accumulation are greatest close to the air/sea interface. Thus carbonate facies and their fabrics are often used as indicators of sea-level position. Similarly siliciclastic coastal sediments, especially tidal sediments, can also be used as sea-level indicators. Additionally rates of carbonate accumulation often have a biochemical and physicochemical origin influenced by the chemistry of the water from which they are precipitated. Stacking patterns of both sediments are expressed by geometric bodies that may be (Figure 3):

- Unconfined by topography
- Confined within eroded topography

Stacking patterns for both clastics and carbonates that are the product of physical accommodation vary between:

- Unconfined sheets that:
  - Prograde (step seaward)
  - Retrograde (step landward)
  - Aggrade (built vertically)
- Unconfined prograding carbonate sheets that are the product of physical accommodation are further subdivided below into:
  - Low-angle ramps of fine shallow-water carbonate that in deeper water pass to gravels
  - Homoclinal ramps of fine shallow-water carbonate
  - Distally steepened ramps of shallow-water grain-dominated carbonate

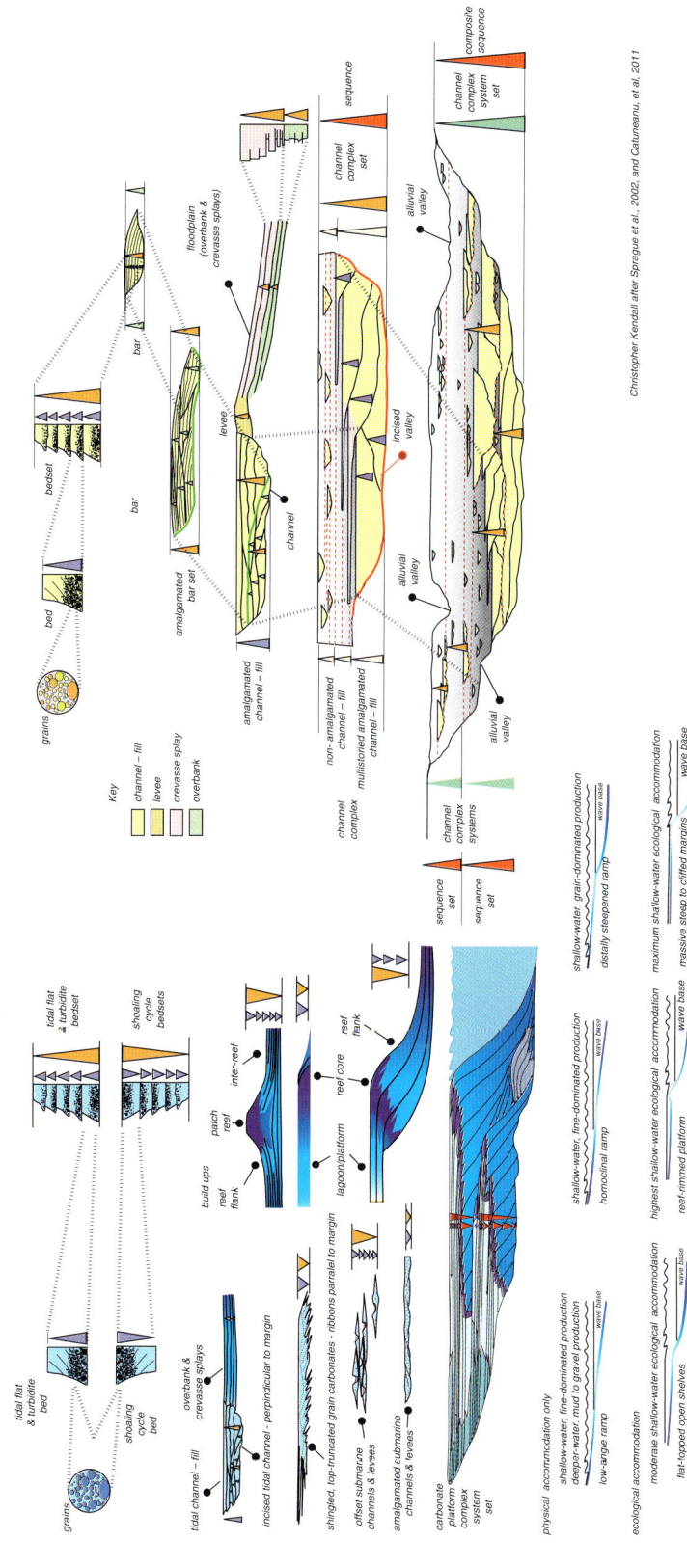

Sequence Stratigraphy, Figure 3 Hierarchical differences between carbonate and clastic architectural elements. Comparison of grains to cycles and bed sets to channels to reefs, carbonate platform, and channel complex system sets that respond to physical and/or ecological accommodation and are used to define depositional sequence boundaries, transgressive surfaces (TS), and maximum flooding surfaces (mfs).

- Unconfined carbonate platform sheet geometries influenced by organisms or ecological accommodation form
  - Flat-topped open shelves with moderate shallow-water ecological accommodation
  - Reef-rimmed platform with highest shallow-water ecological accommodation
  - Massive steep to cliffed margins with maximum shallow-water ecological accommodation

Confined bodies represented by fill of incised topography include:

- Subaerial incised valleys
- Submarine incised valleys

Channel fill and stacking of confining valleys and unconfined lobes and sheets may be expressed as:

- Organized bodies
  - Randomly organized bodies
  - Multistoried
  - Amalgamated

Other stratigraphic tools utilized with sequence stratigraphy

Prediction and interpretation improves not only when sequence stratigraphy is coupled to the laws of Steno and Walther but when tied to indicators of deposition and time. Indicators of depositional setting include:

- Ichnofacies and fossils
- Sedimentary structures
- Volcanics
- Storm layers or tempestites
- Sequence stratigraphic boundaries

Chronostratigraphic markers include:

- Fossils
- Magnetostratigraphic
- Radioactive markers or gamma ray log signal markers
- Radiometric markers

Terminology

Though the linkage between the sequence stratigraphy and the other subdisciplines of stratigraphy can be "fuzzy," these links are important to prediction and interpretation. The links are strengthened when sequence stratigraphic terminology carries connotations related to the interpretation of the surfaces used to interpret the stratigraphic section but also a consideration of sedimentology and chronostratigraphy. How the terminology is defined and used and/or fits preconceived classifications is tied to the character of the data and stratigraphic techniques used. In the end it is up to the users to consider their data and the goals of their interpretations. They should be able to explain their choice of terms and then make their interpretation!

Use of the "over simplification" of time as it relates to sequence stratigraphy

The sedimentary layering of a stratigraphic section has a vast array of dimensional hierarchies. These range from units millimeters thick that may have accumulated over seconds to thousands of meters thick and collected over millions of years. As much of the literature related to surfaces and the layers they enclose indicate, whatever the dimension of a layer is and the time involved in its deposition, each layer is bounded by surfaces that transgress time (Wheeler, 1958; Middleton, 1973; Vail et al., 1977; Galloway, 1989; Catuneanu et al., 1998; Schwarzacher, 2000; Catuneanu, 2002; Embry, 2002). All sedimentary sections are subdivided by these diachronous surfaces. If most sedimentary sections are cut by diachronous surfaces, then their depositional setting cannot be interpreted using Walther's law, since this would be contravened. However most interpreters accept that the layered units bounded by these surfaces formed at different times, and believe that the subdividing surfaces are of a higher order frequency than the time envelope of the section being considered. In other words, the setting and problem are oversimplified and the surfaces are ignored and it is assumed that the sediments accumulated continuously. So despite the surfaces and the sedimentary layers transgressing time for the sake of interpretation, these are assumed to have filled continuously.

Thus it should be recognized that in sedimentary interpretation the application of Steno's principles and Walther's law assumes the simplification that the sediments packaged by surfaces accumulated within length of time of the accumulation. These simplifications do not contravene logic (which is literally fuzzy) but aid in the interpretation of the sedimentary section. For a more complete and thorough discussion of this topic, one should read Catuneanu et al. (2012).

Summary and conclusions

The flow of sequence stratigraphic interpretations of sedimentary sections involves:

- Data sources
  - 2-D and 3-D seismic sections
  - Well log data
  - Outcrops
- Subdivision of sections into sequences, parasequences, and/or their associated systems tracts
- Identification and correlation of the following surfaces
  - Erosion and nondeposition (sequence boundaries [SB])
  - Transgressive surfaces [TS]
  - Maximum flooding surfaces [mfs]
- These surfaces provide:
  - A relative time framework for the sedimentary succession
  - The interrelationship of the depositional settings and their lateral correlation
  - A compartmentalization of hydrocarbon reservoirs

- Results include:
  - Depositional setting is determined.
  - Extent of their lithofacies is characterized.
  - Predict the above into unknown areas, particularly when associated with hydrocarbon reservoirs and aquifers.

In summary this section explains how "sequence stratigraphy" can be used to study sedimentary rock relationships within a time-stratigraphic framework of repetitive, genetically related strata bounded by surfaces of erosion or nondeposition or their correlative conformity (Posamentier et al., 1988; Van Wagoner et al., 1988).

Bibliography

Brown, L. F., Jr., and Fisher, W. L., 1977. Seismic stratigraphic interpretation of depositional systems: examples from Brazilian rift and pull apart basins. In Payton, C. E. (ed.), *Seismic Stratigraphy – Applications to Hydrocarbon Exploration*. Tulsa: American Association of Petroleum Geologists. American Association of Petroleum Geologists Memoir, Vol. 26, pp. 213–248.

Catuneanu, O., 2002. Sequence stratigraphy of clastic systems: concepts, merits, and pitfalls. *Journal of African Earth Sciences*, **35**, 1–43.

Catuneanu, O., 2006. *Principles of Sequence Stratigraphy*. Elsevier, p. 375.

Catuneanu, O., Galloway, W. E., Christopher, G. S., Kendall, C., Miall, A. D., Posamentier, H. W., Strasser, A., and Tucker, M. E., 2011. Sequence stratigraphy: methodology and nomenclature. *Stuttgart, Newsletters on Stratigraphy*, **44**(3), 173–245.

Dott, R. H., and Batten, R. L., 1976. *Evolution of the Earth*, 2nd edn. New York: McGraw-Hill. 504 p. ISBN 0070176191.

Galloway, W. E., 1989. Genetic stratigraphic sequences in basin analysis. I. Architecture and genesis of flooding surface bounded depositional units. *American Association of Petroleum Geologists Bulletin*, **73**, 125–142.

Jervey, M. T., 1988. Quantitative geological modeling of siliciclastic rock sequences and their seismic expression. In Wilgus, C. K., Hasting, B. S., Kendall, C. G. S. C., Posamentier, H. W., Ross, C. A., and Van Wagoner, J. C. (eds.), *Sea-Level Changes: An Integrated Approach*. Tulsa: Society of Economic Paleontologists and Mineralogists. Special Publication No. 42, pp. 47–69.

Kendall, C. G, St. C., 2012. SEPM STRATA. Website. http://www.sepmstrata.org/

Lyell, C., 1830. *Principles of Geology*, 1st edn. London: John Murray, Vol. 1.

Lyell, C., 1832. *Principles of Geology*, 1st edn. London: John Murray, Vol. 2.

Lyell, C., 1833. *Principles of Geology*, 1st edn. London: John Murray, Vol. 3.

Middleton, G. V., 1973. Johannes Walther's law of the correlation of facies. *Geological Society of America Bulletin*, **84**, 979–988.

Pomar, L., 2001. Ecological control of sedimentary accommodation: evolution from a carbonate ramp to rimmed shelf, Upper Miocene, Balearic Islands. *Palaeogeography, Palaeoclimatology, Palaeoecology*, **175**, 249–272.

Posamentier, H. W., Jervey, M. T., and Vail, P. R., 1988. Eustatic controls on clastic deposition. I. Conceptual framework. In Wilgus, C. K., Hastings, B. S., Kendall, C. G. St. C., Posamentier, H. W., Ross, C. A., and Van Wagoner, J. C. (Eds.), *Sea Level Changes – An Integrated Approach. SEPM Special Publication No 42*, pp. 110–124.

Schwarzacher, A., 2000. Repetition and cycles in stratigraphy. *Earth-Science Reviews*, **50**, 51–75.

Sprague, A. R., Patterson, P. E., Hill, R. E., Jones, C. R., Campion, K. M., Van Wagoner, J. C., Sullivan, M. D., Larue, D. K., Feldman, H. R., Demko, T. M., Wellner, R.W., Geslin, J. K., 2002. *The Physical Stratigraphy of Fluvial Strata: A Hierarchical Approach to the Analysis of Genetically Related Stratigraphic Elements for Improved Reservoir Prediction*. (Abs) AAPG Annual Meeting, Official Program, p. A167.

Vail, P. R., Mitchum, R. M., Jr., and Thompson, S., III, 1977. Seismic stratigraphy and global changes of sea level, part four: global cycles of relative changes of sea level. *American Association of Petroleum Geologists Memoir*, **26**, 83–98.

Van Wagoner, J. C., 1995. Sequence stratigraphy and marine to nonmarine facies architecture of Foreland Basin Strata, Book Cliffs, Utah, U.S.A. In Van Wagoner, J. C., and Bertram, G. T. (Eds.), *Sequence Stratigraphy of Foreland Basin Deposits. American Association Petroleum Geologists Memoir*. Vol. 64, pp. 137–223.

Van Wagoner, J. C., Posamentier, H. W., Mitchum, R. M., Vail, P. R., Sarg, J. F., Loutit, T. S., and Hardenbol, J., 1988. An overview of sequence stratigraphy and key definitions. In Wilgus, C. K., Hastings, B. S., Kendall, C. G. St. C., Posamentier, H. W., Ross, C. A., and Van Wagoner, J. C. (Eds.), *Sea Level Changes – An Integrated Approach. SEPM Special Publication*. Vol. 42, pp. 39–45.

Wheeler, H. E., 1958. Time stratigraphy. *American Association of Petroleum Geologists, Bulletin*, **42**, 1047–1063.

Cross-references

Convergence Texture of Seismic Reflectors
Lithostratigraphy
Relative Sea-level (RSL) Cycle
Sedimentary Sequence

SERPENTINIZATION

Niels Jöns[1] and Wolfgang Bach[2,3]
[1]Department of Geology, Mineralogy and Geophysics, Ruhr-University Bochum, Bochum, Germany
[2]Department of Geosciences, University of Bremen, Bremen, Germany
[3]MARUM-Center for Marine Environmental Sciences, University of Bremen, Bremen, Germany

Synonyms

Abyssal peridotite; Abyssal serpentinite; Bastite; Mesh texture; Serpentine minerals; Serpentinite

Definition

"Serpentinization" is the process leading to the formation of serpentinites, metamorphic rocks derived from hydrous alteration of olivine-rich ultramafic rocks and consisting mainly of serpentine group minerals.

Introduction

Serpentinization denotes the hydrous alteration of olivine-rich ultramafic rocks, which can occur everywhere, where

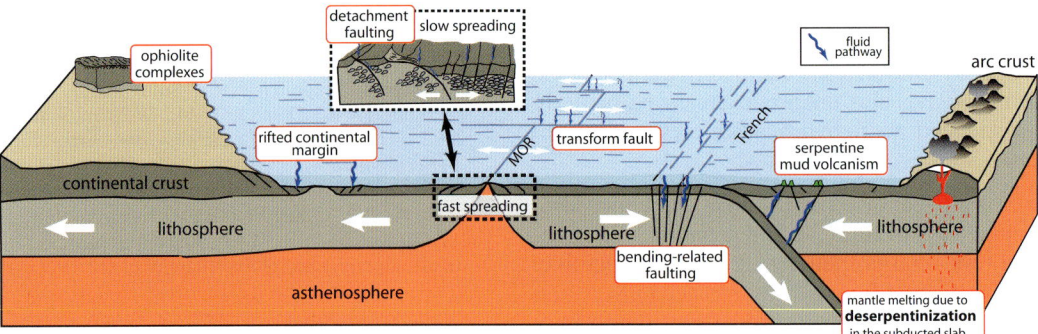

Serpentinization, Figure 1 Schematic cross section through the oceanic lithosphere showing the main locations where serpentinization takes place in the marine realm: at detachment faults and transform faults at mid-oceanic ridges, at rifted continental margins, due to seawater ingress on bending-related faults near the subduction trench and above the subduction zone causing serpentine mud volcanism. Ophiolite complexes are parts of the oceanic lithosphere obducted onto continents and often featuring serpentinization of the lithospheric mantle.

these rocks and aqueous solutions react under suitable pressure and temperature conditions. In the oceanic realm, seawater-derived fluids are abundant; however, ultramafic rocks (peridotites) are typically found in the Earth's upper mantle. Serpentinization will only take place where these rocks get in contact with seawater fluids, either by deep fluid infiltration in fracture zones and bend faults or by uplift and exposure of peridotites at or near the seafloor. During serpentinization, the primary minerals contained in a peridotite (i.e., mainly olivine + orthopyroxene + clinopyroxene) are altered to form mostly serpentine group minerals (e.g., lizardite, chrysotile). The stability of these OH-bearing phyllosilicates ($T < 400\ °C$; at low pressures) sets the limit for the occurrence of serpentinization in nature to low-temperature conditions (i.e., $T < 400\ °C$; at low pressures). If all boundary conditions (P-T-X) are suitable, serpentinization causes dramatic changes of the physical and chemical properties of peridotites, i.e., transformation of a dense ($\rho = 3.3$ g cm^{-3}), anhydrous, and weakly magnetic rock into a less dense ($\rho = 2.4–2.7$ g cm^{-3}), hydrous (>12 wt% H_2O), and strongly magnetic rock. In consequence, these changes influence the rheological behavior of the oceanic lithosphere, and the large amount of water contained in serpentinites makes them an important aspect of the global water cycle. Furthermore, hydrogen produced during serpentinization plays an important role for life in the deep sea. These various aspects underline that serpentinization is a critical process that acts at the interface between the lithosphere, hydrosphere, and biosphere.

Geological setting of serpentinization in the marine environment

In marine environments, serpentinization occurs where fractures or faults allow for seawater ingress into the oceanic crust to reach ultramafic rocks (Figure 1). Oceanic fracture zones, transform faults, and detachment faults facilitate serpentinization at constructive plate boundaries. At convergent plate boundaries, serpentinization is associated with bending-related faults as well as in the supra-subduction mantle wedge. Finally, serpentinites are found at nonvolcanic continental rifted margins, where peridotites have been emplaced during continental rifting in the earliest oceanic crust or stretched continental crust. Many insights into serpentinization processes are derived from outcrops in ophiolite complexes, where serpentinized rocks are mantle peridotites and dunites as well as cumulate dunite in the lowermost ocean crust.

Mid-oceanic ridge spreading centers

Intermediate- and fast-spreading centers are magmatically robust and feature a layer-cake crustal architecture, consisting of a 6–7 km thick succession of gabbros, sheeted dikes, and lavas. This mafic crust does not serpentinize and protects the underlying ultramafic rocks from reaction with seawater. With the exception of sparse fracture zones and in the Hess Deep, a tectonic window at the intersection of the East Pacific Rise and the Cocos-Nazca spreading center, serpentinization is uncommon where spreading rates exceed 50 mm year^{-1} full rate.

At slower spreading rates, however, lithospheric extension is largely accommodated by tectonic processes and magmatism may be episodic. Depending on the magma budget and lithospheric thickness of a slow-spreading center, the extension may be dispersed across many normal faults or it is focused on low-angle detachment faults. The latter commonly lead to the formation of oceanic core complexes, which expose mantle rocks at the seafloor. In recent years more and more exposures of harzburgites and dunites at the seafloor have been found (e.g., Smith et al., 2006), which led Cannat et al. (2010) to suggest that up to 25 % of slow-spreading seafloor consist of ultramafic rocks. Slow-spreading ridges are dissected by numerous large-offset transform faults, which are an important, albeit understudied, setting where serpentinization is common, because the strike slip faults extend into the lithospheric mantle and tectonic movement facilitates seawater ingress. A prominent example is the Vema fracture zone in the Atlantic Ocean, where flexural uplift of the oceanic plate in a transverse regime generates suitable conditions for deep fluid ingress and serpentinization of mantle rocks (Bonatti et al., 1971; Melson and Thompson, 1971; Bonatti et al., 2005). This strong segmentation is missing along some ultraslow-spreading ridges (full rates <18 mm year^{-1}), where long stretches of amagmatic spreading with abundant serpentinite at the seafloor are separated by large volcanic centers (Dick et al., 2003).

Both slow- and ultraslow-spreading centers feature hydrothermal activity, which is hosted in basaltic or ultramafic rocks. Some of these systems are clearly related to magmatic activity, but abundant hydrothermal activity has also been documented for the amagmatic and ultraslow-spreading SW Indian Ridge 10–16°E (Bach et al., 2002), the Gakkel Ridge in the Arctic Ocean (Edmonds et al., 2003), and the Cayman Trough (German et al., 2010).

While the structural controls on venting along ultraslow-spreading ridges are to be revealed, there appears to be a relation between detachment faulting and hydrothermal vent distribution in slow-spreading segments. McCaig et al. (2007) proposed that fluids emanating from black smokers in hydrothermal systems are focused along the low-angle detachment faults, allowing for occurrence of hydrothermal vent fields far away from magmatically active zones.

Convergent plate boundaries

At convergent plate boundaries, serpentinization and deserpentinization (i.e., breakdown of serpentine and brucite to olivine) processes are intimately linked. Serpentinization occurs at the trench where the incoming lithospheric plate bends down prior to subduction (Figure 1). This flexure causes the development of bend faults, which can extend down to 20 km off Costa Rica (Ranero et al., 2003; Grevemeyer et al., 2005). These faults provide pathways for seawater ingress and serpentinization of the mantle. Although direct sampling of this system is currently not possible, geophysical measurements provide insights into hydration of mantle rocks in the subduction trench area. Furthermore, serpentinized peridotites have been recovered from several non-accretionary convergent margins (e.g., the Puerto Rico trench and the Tonga trench). Serpentinites in these areas may have been exposed due to deep faulting during initiation of the subduction zone, or they represent sheared-off and exhumed portions of the subduction channel. When the hydrated slab experiences increased pressure and temperature conditions in the subduction zone, partial deserpentinization takes place. At higher temperatures, the fluids released from the downgoing slab can trigger mantle wedge melting and thus arc volcanism; at lower temperatures serpentinization takes place in the mantle wedge (e.g., Fryer, 2002). Subduction of serpentinites is critical in the Earth's water cycles, as it keeps the mantle viscosity low and sustains mantle convection (Rüpke et al., 2004).

Insights into serpentinization in subduction zone settings have been gained by studies of serpentinites recovered from mud volcanoes in forearc regions (e.g., the Izu Bonin/Mariana arc system). There, the involvement of slab-derived fluids in serpentinization is reflected in the stable isotope composition of serpentinite clasts (Alt and Shanks, 2006) and the composition of fluids issuing from the mud volcanoes (Mottl et al., 2004).

Nonvolcanic continental rifted margins

The occurrence of ultramafic rocks in passive margin settings is explained with tectonic movements during continental breakup and subsequent development of an oceanic spreading center. Serpentinites occurring at continental rifted margins generally represent parts of the subcontinental lithospheric mantle that have been emplaced due to low-angle detachment faulting (e.g., Lavier and Manatschal, 2006). Whether emplacement of serpentinites as diapiric bodies can occur in this setting is under debate. In the case of the Iberian abyssal plane, it has been observed that shearing and uplift occurred prior to serpentinization. Sediment pore waters were involved in the serpentinization reactions (Sharp and Barnes, 2004); the fully serpentinized rocks show unusual hydrogarnet- and diaspore-bearing assemblages (Beard and Hopkinson, 2000). Sulfur isotope studies of serpentinites from the Iberian abyssal plane reveal that sulfate-reducing bacteria formed some of the pyrite found in veins, indicating the presence of a deep serpentinite-hosted biosphere (Alt and Shanks, 1998).

Ophiolite complexes

Ophiolites are remnants of oceanic lithosphere that are obducted onto continental margins. Outcrops in ophiolite complexes have contributed to our understanding of the architecture of the oceanic lithosphere. Prominent examples are the Semail ophiolite in Oman, the Troodos massif in Cyprus, and the Josephine ophiolite in the USA. Although serpentinization is commonly observed in the exposed lithospheric mantle, the process of serpentinization did not necessarily take place under in situ conditions. Serpentinization might well have occurred during or after obduction, possibly involving meteoric waters under comparatively low temperature (e.g., Neal and Stanger, 1985; Harper et al., 1988). Therefore, examination of ophiolites cannot fully substitute observations made in in situ oceanic lithosphere.

Mineralogical and geochemical changes during serpentinization

The interaction of seawater and peridotite, which consists mainly of olivine of forsterite-rich composition, can be described by the following simplified exothermic mineral reaction:

$$Mg_{1.8}Fe_{0.2}SiO_4 + 1.37\,H_2O = 0.5\,Mg_3Si_2O_5(OH)_4 + 0.3\,Mg(OH)_2 + 0.067\,Fe_3O_4 + 0.067\,H_2 \quad (1)$$
$$\text{Olivine} + H_2O = \text{Serpentine} + \text{Brucite} + \text{Magnetite} + H_2$$

When rocks are undeformed, olivine is pseudomorphously replaced by the secondary mineral assemblage, which builds up an hourglass or mesh texture. Such a mesh is bordered by magnetite inwards followed by serpentine (mostly lizardite) and eventually a relict of olivine in the mesh center (see Figure 2a, b). Brucite is not always present, but may form at the interface between olivine and serpentine or is intergrown with serpentine. Primary pyroxenes of the ultramafic precursor rock also undergo serpentinization:

$$3\,Mg_2Si_2O_6 + 3\,H_2O = Mg_3Si_2O_5(OH)_4 + Mg_3Si_4O_{10}(OH)_2 \quad (2)$$
$$\text{Orthopyroxene} + H_2O = \text{Serpentine} + \text{Talc (simplified reaction in an Fe–free system)}.$$

Pseudomorphic replacement textures of pyroxenes (in rare cases also of amphibole or phlogopite) are referred to as bastite (Figure 2c, d). Bastites typically lack magnetite, but often also do not contain talc. This indicates that Mg and Si can be exchanged across local domains. Silica may react with brucite in adjacent mesh textures to form additional serpentine, explaining why serpentinized harzburgites rarely contain talc.

Apart from the aforementioned replacement textures, serpentine can also form veins of variable thickness (μm to cm) crosscutting serpentinites (Figure 2e, f). They consist of fine-fibrous serpentine and have been interpreted as evidence for volume expansion related to serpentinization (e.g., Andreani et al., 2007).

Unlike in the simplified mineral reactions given above, the product phases serpentine and brucite are typically not Mg end-members; they contain some iron, which results in a smaller amount of magnetite being formed. Magnetite contains both divalent and trivalent iron, and the amount of primary Fe^{+2} that is oxidized to secondary Fe^{+3} in magnetite determines the amount of hydrogen produced. With decreasing serpentinization temperature successively more Fe-rich brucite and serpentine are formed and the amount of magnetite decreases. However, ferric iron in serpentine can also cause the release of hydrogen, although the yields are lower than in magnetite-pulled hydrogen production (Klein et al., 2009). Recent field, experimental, and spectroscopic work has provided evidence for a lack of magnetite and the existence of serpentine rich in ferric iron (Evans et al., 2009; Andreani et al., 2013).

Geochemical reaction path models have been employed successfully in examining the reaction sequence during serpentinization in the course of decreasing temperatures and water-to-rock ratios (e.g., Klein et al., 2009; McCollom and Bach, 2009). The isobaric (1 kbar) model presented in Figure 3 predicts that olivine is stable (bar minor reaction to talc and magnetite) at $T > 430\,°C$, where ortho- and clinopyroxene are expected to be completely replaced by tremolite, talc, and magnetite. Serpentine appears just below 430 °C, initially from breakdown of olivine, talc, and magnetite. Below 415 °C, olivine reacts with dissolved silica to produce serpentine and magnetite. Tremolite gives way to secondary clinopyroxene and serpentine at $T < 405\,°C$. Below 350 °C, brucite is stable, and olivine is predicted to be consumed by serpentine and brucite. The strongly decreasing silica activities associated with the olivine-out reaction stabilize magnetite (Frost and Beard, 2007) and generate driving force for hydrogen production (Klein et al., 2009).

Below about 290 °C, grossular-rich hydrogarnet is predicted to replace secondary clinopyroxene. In the course of continued cooling and intensified water flux, magnetite should decrease in abundance at the expense of increasing Fe contents in the hydrogarnet and brucite. Below 100 °C, magnetite is no longer stable in serpentinites. The thermodynamically stable assemblages at the lower-temperature end of the reaction path do form occasionally, but clinopyroxene and magnetite are often preserved due to kinetic inhibition of the reactions. The metastable diopside-brucite-serpentine assemblage accounts for the exceptionally high pH and high Ca concentrations in many serpentinization fluids.

Spinel rich in chromite ($FeCr_2O_4$) is an accessory primary mineral commonly contained in oceanic serpentinites. Under typical P-T-fluid conditions of serpentinization, chromite is comparatively unreactive and therefore preserved even in completely serpentinized peridotites. Furthermore, Fe-Ni-Co sulfide minerals as well as native metals and alloys are commonly found, e.g., pentlandite ($[Fe, Ni]_9S_8$), awaruite (Ni_3Fe), Heazlewoodite (Ni_3S_2), and wairauite (CoFe). Their

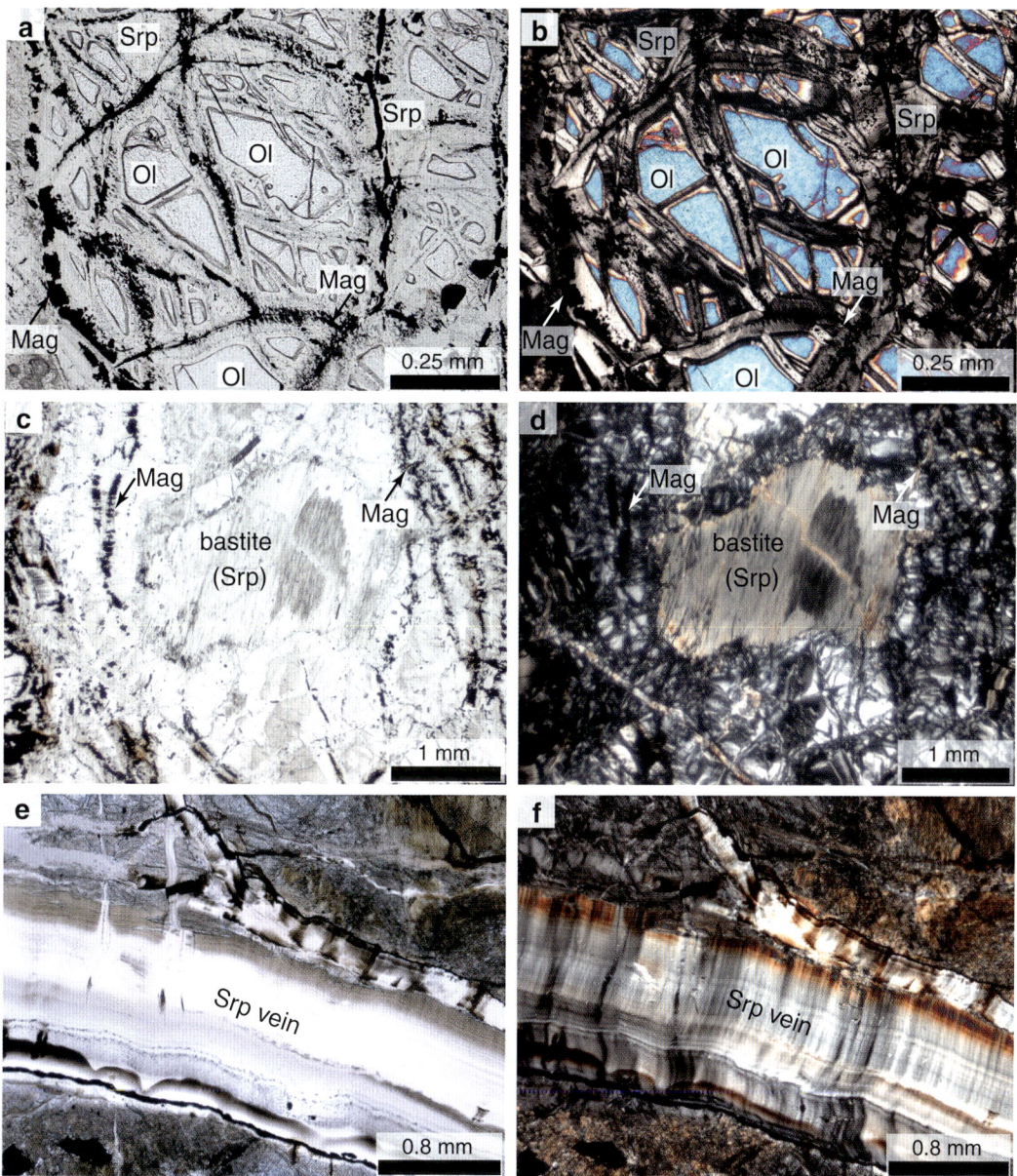

Serpentinization, Figure 2 Microtextures of serpentinites. (**a, b**) Partially serpentinized harzburgite showing olivine relicts in centers of a mesh texture. Olivine breaks down to form serpentine and magnetite. (**c, d**) Serpentine as breakdown product of former orthopyroxene, forming a bastite texture. The bastite is almost magnetite free. (**e, f**) Serpentine forming a vein crosscutting a serpentinized harzburgite. A vein-parallel zonation points to multiple crack openings and polystage serpentine growth. Figures (**a, c,** and **e**) taken under plane-polarized light; (**b, d,** and **f**) under cross-polarized light.

occurrence is a result of the strongly reducing nature of the serpentinization fluids (Frost, 1985; Klein and Bach, 2009). Completely serpentinized rocks have exhausted their reducing capacity and may contain polysulfides (pyrite, FeS_2; vaesite, NiS_2; cattierite, CoS_2) and thiospinels (polydymite, Ni_3S_4; violarite, $FeNi_2S_4$), indicative of higher oxygen and sulfur fugacities. Other minerals that are occasionally reported are iowaite ($Mg_4Fe[OH]_8OCl \times 2$–$4\ H_2O$), calcite, aragonite, and dolomite. These minerals, however, require addition of chemical constituents such as Ca, Cl, or CO_2, whereas serpentinization sensu stricto is defined as a pure hydration process.

The major chemical change due to serpentinization is the addition of water. The whole-rock geochemical composition of abyssal serpentinites therefore has a strong control on the abundance of secondary phases in serpentinites. The composition of mantle peridotites is

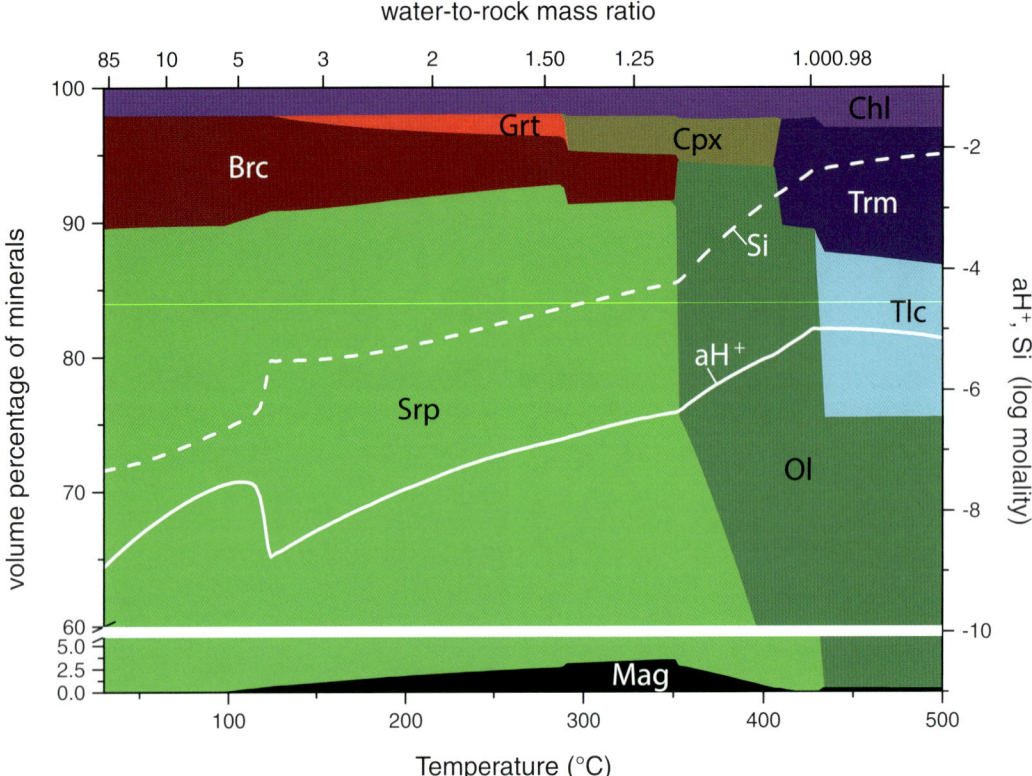

Serpentinization, Figure 3 Predicted modal abundances of a mantle peridotite (80 % olivine, 15 % orthopyroxene, 5 % clinopyroxene; see Klein et al., 2009 for details) that underwent reaction with seawater at a pressure of 1 kbar. The influx of seawater will cool the system, and hence temperature and water-to-rock rations are inversely correlated in the model. Note that the reaction of olivine and water to serpentine and brucite causes a marked decrease in silica activity.

affected by the degree of mantle melting and melt extraction as well as melt-refertilization (Niu, 2004). The consequences of modal variability of abyssal ultramafics and the addition of melts have been investigated by Klein et al. (2013) and Jöns et al. (2009).

Serpentinites exposed at the seafloor are not stable and weathering reactions will dissolve brucite and turn relict olivine into clay and iron-oxyhydroxide (iddingsite). Serpentine is often preserved metastably, but may be partially replaced by talc and calcite in heavily weathered serpentinites.

Related rock types

The thick lithospheric mantle underneath slow-spreading ridges features large gabbroic bodies and dikes of variable thickness. Soapstones (monomineral talc rocks, also known as steatites) form when seawater-derived fluids equilibrated with gabbro encounter serpentinite. In contrast, rodingites (Ca-metasomatized gabbros) form when serpentinization fluids react with gabbroic rocks. These metasomatic rocks are common in abyssal serpentinite massifs, and the abundance of talc at the interface between large gabbroic bodies and hosting mantle rock constitutes a discrete zone of mechanical weakness that may play a role in the tectonic exhumation of oceanic core complexes (Ildefonse et al., 2007).

Serpentinization, hydrothermal vents, and life

Ultramafic-hosted hydrothermal systems in the deep sea are below the zone where photosynthesis provides energy to make life possible. Instead, chemosynthetic microbial organisms gain metabolic energy from hydrogen contained in hydrothermal fluids (cp. Reaction 1) and use this energy to produce biomass. Consumption of the biomass by higher organisms is the basis for establishment of complex food webs that are independent of photosynthesis. The presence of hydrogen also promotes abiotic formation of methane and other hydrocarbons, and it has been speculated that the early development of life on the Earth took place under similar conditions.

Known deep-sea hydrothermal vents hosted in serpentinite comprise high-temperature (350–370 °C) systems such as Logatchev (MAR 14°45′N) or Rainbow (MAR 36°14′N), and the low-temperature (90–110 °C) Lost City Hydrothermal Field (MAR 30°N). Hydrogen and methane provide most of the energy for microbial metabolism in all these systems, with reduced metals being another energy source in the acidic hot fluids (Amend et al., 2011).

The less hot fluids from Lost City are alkaline and low in metal concentrations (Kelley et al., 2001). The Lost City system features spectacular carbonate-brucite chimneys that are up to 60 m tall. Lost City is situated on top of an oceanic core complex, and the hosting rocks reveal a multistage, coupled tectonic and metamorphic history (Boschi et al., 2006). Geochemical and geochronological studies show that the systems have been active for >30,000 years (Früh-Green et al., 2003). The seawater-like isotopic compositions indicate that the carbonates form in response to venting of the highly alkaline fluids (pH 11).

The serpentine mud volcano vents in the Mariana forearc feature even more extreme compositions with pH ranging up to 12.5 (Mottl et al., 2004). Serpentinization fluids are strongly enriched in H_2, and dissolved CO_2 is no longer thermodynamically stable. Reduction of CO_2 leads to the abiotic generation of alkanes and organic acids (Proskurowski et al., 2008; Lang et al., 2010).

The organisms that inhabit these serpentinization systems are specifically adapted to the energy and carbon sources available (Takai et al., 2005; Perner et al., 2007; Petersen et al., 2011). Because of the unique capacity for reducing CO_2 and the highly alkaline nature of the vent fluids, serpentinization systems are believed by some to resemble hydrothermal systems in which life may have developed on Earth and perhaps other planetary bodies (Martin et al., 2008).

Societal relevance of serpentinites

More than half of the world's Ni resources are bound in Ni-bearing regoliths/laterites, which originate from intensive weathering of serpentinites under tropical humid conditions. While the precursor rocks – olivine-rich ultramafic rocks – contain less than 0.5 wt% Ni, enrichment to >5 wt% Ni is observed during regolith formation and lateritization. Nickel is hosted in either oxide minerals (limonitic ore) or hydrous Mg silicates (e.g., népouite, kerolite, garnierite) or clay silicates (nontronite, saponite; for details, see Butt and Cluzel, 2013). Additionally, these ores are important cobalt deposits. Furthermore, starting in the nineteenth century, serpentinites were intensively mined for asbestos, thin fibrous forms of serpentine and amphibole. Asbestos was used as electrical and thermal insulators and building material until the late 1980s. Already in the 1930s, asbestos-related diseases were reported, and nowadays, the use of asbestos is banned due to serious health concerns (causing mesothelioma and asbestosis). Talc and magnesite are other mining products from metasomatized portions of serpentinite bodies on land.

The replacement of serpentine and olivine by carbonate is a process predicted to take place at low temperatures and low CO_2 contents of the interacting fluid (e.g., Klein and Garrido, 2011). Because the carbonation reactions are expected to proceed rapidly at moderate temperature and create permeability, ultramafic rocks have been proposed as possible site for CO_2 storage (Kelemen and Matter, 2008).

Summary and conclusions

Serpentinization is a common process in marine environments and is of both environmental and societal relevance. In abyssal peridotites, olivine is replaced by lizardite or chrysotile, which might be intergrown with brucite. Magnetite and hydrogen are important by-products of serpentinization. The mineralogical transformations during serpentinization cause dramatic changes in the mantle rock physical properties, which have important implication for tectonic processes. The chemical exchange with seawater makes serpentinization an important part of the global cycles of water and solutes. Serpentinization-related hydrothermal systems in the deep sea harbor abundant microbial life, which is intimately linked with serpentinization reactions. Some major ore deposits nowadays mined on land trace back to serpentinization processes in the marine environment, and there is potential future relevance of peridotites and serpentinites for sequestering CO_2. In summary, serpentinization occurs at the interface between the lithosphere, hydrosphere, and biosphere and is demonstrably one of the most important hydration processes on Earth.

Bibliography

Alt, J. C., and Shanks, W. C., 1998. Sulfur in serpentinized oceanic peridotites: serpentinization processes and microbial sulfate reduction. *Journal of Geophysical Research*, **103**, 9917–9929.

Alt, J. C., and Shanks, W. C., 2006. Stable isotope compositions of serpentinite seamounts in the Mariana forearc: serpentinization processes, fluid sources and sulfur metasomatism. *Earth and Planetary Science Letters*, **242**, 272–287.

Amend, J. P., McCollom, T. M., Hentscher, M., and Bach, W., 2011. Catabolic and anabolic energy for chemolithoautotrophs in deep-sea hydrothermal systems hosted in different rock types. *Geochimica et Cosmochimica Acta*, **75**, 5736–5748.

Andreani, M., Mével, C., Boullier, A.-M., and Escartín, J., 2007. Dynamic control on serpentine crystallization in veins: constraints on hydration processes in oceanic peridotites. *Geochemistry, Geophysics, Geosystems*, **8**, Q02012.

Andreani, M., Muñoz, M., Marcaillou, C., and Delacour, A., 2013. μXANES study of iron redox state in serpentine during oceanic serpentinization. *Lithos*, doi:10.1016/j.lithos.2013.04.008.

Bach, W., Banerjee, N. R., Dick, H. J. B., and Baker, E. T., 2002. Discovery of ancient and active hydrothermal systems along the ultra-slow spreading Southwest Indian Ridge 10–16°E. *Geochemistry, Geophysics, Geosystems*, **3**, doi:10.1029/2001GC00027.

Beard, J. S., and Hopkinson, L., 2000. A fossil, serpentinization-related hydrothermal vent, Ocean Drilling Program Leg 173, Site 1068 (Iberia Abyssal Plain), some aspects of mineral and fluid chemistry. *Journal of Geophysical Research*, **105**, 16527–16539.

Bonatti, E., Honnorez, J., and Ferrara, G., 1971. Ultramafic rocks: Peridotite-Gabbro-Basalt complex from the equatorial mid-Atlantic ridge. *Philosophical Transactions of the Royal Society of London. Series A. Mathematical and Physical Sciences*, **268**, 385–402.

Bonatti, E., Brunelli, D., Buck, W. R., Cipriani, A., Fabretti, P., Ferrante, V., Gasperini, L., and Ligi, M., 2005. Flexural uplift of a lithospheric slab near the Vema transform (Central Atlantic), timing and mechanisms. *Earth and Planetary Science Letters*, **240**, 642–655.

Boschi, C., Früh-Green, G. L., Delacour, A., Karson, J. A., and Kelley, D. S., 2006. Mass transfer and fluid flow during detachment faulting and development of an oceanic core complex, Atlantis Massif (MAR 30°N). *Geochemistry, Geophysics, Geosystems*, **7**, Q01004.

Butt, C. R. M., and Cluzel, D., 2013. Nickel laterite ore deposits: weathered serpentinites. *Elements*, **9**, 123–128.

Cannat, M., Fontaine, F., and Escartín, J., 2010. Serpentinization and associated hydrogen and methane fluxes at slow spreading ridges. In *Diversity of hydrothermal systems on slow spreading ocean ridges*. Washington, DC: AGU, pp. 241–264.

Dick, H. J. B., Lin, J., and Schouten, H., 2003. An ultraslow-spreading class of ocean ridge. *Nature*, **426**, 405–412.

Edmonds, H. N., Michael, P. J., Baker, E. T., Connelly, D. P., Snow, J. E., Langmuir, C. H., Dick, H. J. B., Mühe, R., German, C. R., and Graham, D. W., 2003. Discovery of abundant hydrothermal venting on the ultraslow-spreading Gakkel ridge in the Arctic Ocean. *Nature*, **421**, 252–256.

Evans, B. W., Kuehner, S. M., and Chopelas, A., 2009. Magnetite-free, yellow lizardite serpentinization of olivine websterite, Canyon Mountain complex, N.E. Oregon. *American Mineralogist*, **94**, 1731–1744.

Frost, B. R., 1985. On the stability of sulfides, oxides, and native metals in serpentinite. *Journal of Petrology*, **26**, 31–63.

Frost, B. R., and Beard, J. S., 2007. On silica activity and serpentinization. *Journal of Petrology*, **48**, 1351–1368.

Früh-Green, G. L., Kelley, D. S., Bernasconi, S. M., Karson, J. A., Ludwig, K. A., Butterfield, D. A., Boschi, C., and Proskurowski, G., 2003. 30,000 years of hydrothermal activity at the Lost City vent field. *Science*, **301**, 495–498.

Fryer, P., 2002. Recent studies of serpentinite occurrences in the oceans: mantle-ocean interactions in the plate tectonic cycle. *Chemie der Erde – Gechemistry*, **62**, 257–302.

German, C. R., Bowen, A., Coleman, M. L., Honig, D. L., Huber, J. A., Jakuba, M. V., Kinsey, J. C., Kurz, M. D., Leroy, S., McDermott, J. M., de Lépinay, B. M., Nakamura, K., Seewald, J. S., Smith, J. L., Sylva, S. P., Van Dover, C. L., Whitcomb, L. L., and Yoerger, D. R., 2010. Diverse styles of submarine venting on the ultraslow spreading Mid-Cayman Rise. *Proceedings of the National Academy of Sciences*, **107**, 14020–14025.

Grevemeyer, I., Kaul, N., Diaz-Naveas, J. L., Villinger, H. W., Ranero, C. R., and Reichert, C., 2005. Heat flow and bending-related faulting at subduction trenches: case studies offshore of Nicaragua and Central Chile. *Earth and Planetary Science Letters*, **236**, 238–248.

Harper, G. D., Bowman, J. R., Kuhns, R 1988. A field, chemical, and stable isotope study of subseafloor metamorphism of the Josephine ophiolite, California-Oregon. *J Geophys Res*, **93**, 4625–4656.

Ildefonse, B., Blackman, D. K., John, B. E., Ohara, Y., Miller, D. J., MacLeod, C. J., and Integrated Ocean Drilling Program Expeditions 304/305 Science Party, 2007. Oceanic core complexes and crustal accretion at slow-spreading ridges. *Geology*, **35**, 623–626.

Jöns, N., Bach, W., and Schroeder, T., 2009. Formation and alteration of plagiogranites in an ultramafic-hosted detachment fault at the Mid-Atlantic Ridge (ODP Leg 209). *Contributions to Mineralogy and Petrology*, **157**, 625–639.

Kelemen, P. B., and Matter, J. M., 2008. In situ carbonation of peridotite for CO_2 storage. *Proceedings of the National Academy of Sciences*, **105**, 17295–17300.

Kelley, D. S., Karson, J. A., Blackman, D. K., Früh-Green, G. L., Butterfield, D. A., Lilley, M. D., Olson, E. J., Schrenk, M. O., Roe, K. K., Lebon, G. T., Rivizzigno, P., and AT3-60 Shipboard Party, 2001. An off-axis hydrothermal vent field near the Mid-Atlantic Ridge at 30°N. *Nature*, **412**, 145–149.

Klein, F., and Bach, W., 2009. Fe-Ni-Co-O-S phase relations in peridotite-seawater interactions. *Journal of Petrology*, **50**, 37–59.

Klein, F., and Garrido, C. J., 2011. Thermodynamic constraints on mineral carbonation of serpentinized peridotite. *Lithos*, **126**, 147–160.

Klein, F., Bach, W., Jöns, N., McCollom, T. M., Moskowitz, B., and Berquó, T., 2009. Iron partitioning and hydrogen generation during serpentinization of abyssal peridotites from 15°N on the Mid-Atlantic Ridge. *Geochimica et Cosmochimica Acta*, **73**, 6868–6893.

Klein, F., Bach, W., and McCollom, T. M., 2013. Compositional controls on hydrogen generation during serpentinization of ultramafic rocks. *Lithos*, doi:10.1016/j.lithos.2013.03.008.

Lang, S. Q., Butterfield, D. A., Schulte, M., Kelley, D. S., and Lilley, M. D., 2010. Elevated concentrations of formate, acetate and dissolved organic carbon found at the Lost City hydrothermal field. *Geochimica et Cosmochimica Acta*, **74**, 941–952.

Lavier, L. L., Manatschal, G., 2006. A mechanism to thin the continental lithosphere at magma-poor margins. *Nature*, **440**, 324–328.

Martin, W., Baross, J. A., Kelley, D. S., and Russell, M. J., 2008. Hydrothermal vents and the origin of life. *Nature Review Microbiology*, **6**, 805–814.

McCaig, A. M., Cliff, R. A., Escartín, J., Fallick, A. E., and MacLeod, C. J., 2007. Oceanic detachment faults focus very large volumes of black smoker fluids. *Geology*, **35**, 935–938.

McCollom, T. M., and Bach, W., 2009. Thermodynamic constraints on hydrogen generation during serpentinization of ultramafic rocks. *Geochimica et Cosmochimica Acta*, **73**, 856–875.

Melson, W. G., and Thompson, G., 1971. Petrology of a transform fault zone and adjacent ridge segments. *Philosophical Transactions of the Royal Society of London. Series A. Mathematical and Physical Sciences*, **268**, 423–441.

Mottl, M. J., Wheat, C. G., Fryer, P., Gharib, J., and Martin, J. B., 2004. Chemistry of springs across the Mariana forearc shows progressive devolatilization of the subducting plate. *Geochimica et Cosmochimica Acta*, **68**, 4915–4933.

Neal, C., and Stanger, G., 1985. Past and present serpentinisation of ultramafic rocks; An example from the Semail Ophiolite Nappe of Northern Oman. In Drever, J. I. (ed.), *Chemistry of weathering*. Dordrecht/Boston: Riedel Publishing Company, pp. 249–276.

Niu, Y., 2004. Bulk-rock major and trace element compositions of abyssal peridotites: implications for mantle melting, melt extraction and post melting processes beneath Mid-Ocean ridges. *Journal of Petrology*, **45**, 2423–2458.

Perner, M., Kuever, J., Seifert, R., Pape, T., Koschinsky, A., Schmidt, K., Strauss, H., and Imhoff, J. F., 2007. The influence of ultramafic rocks on microbial communities at the Logatchev hydrothermal field, located 15°N on the Mid-Atlantic Ridge. *FEMS Microbiology Ecology*, **61**, 97–109.

Petersen, J. M., Zielinski, F. U., Pape, T., Seifert, R., Moraru, C., Amann, R., Hourdez, S., Girguis, P. R., Wankel, S. D., Barbe, V., Pelletier, E., Fink, D., Borowski, C., Bach, W., and Dubilier, N., 2011. Hydrogen is an energy source for hydrothermal vent symbioses. *Nature*, **476**, 176–180.

Proskurowski, G., Lilley, M. D., Seewald, J. S., Früh-Green, G. L., Olson, E. J., Lupton, J. E., Sylva, S. P., and Kelley, D. S., 2008. Abiogenic hydrocarbon production at Lost City hydrothermal field. *Science*, **319**, 604–607.

Ranero, C. R., Morgan, J. P., McIntosh, K., and Reichert, C., 2003. Bending-related faulting and mantle serpentinization at the Middle America trench. *Nature*, **425**, 367–373.

Rüpke, L. H., Phipps Morgan, J., Hort, M., and Connolly, J. A. D., 2004. Serpentine and the subduction zone water cycle. *Earth and Planetary Science Letters*, **223**, 17–34.

Sharp, Z. D., and Barnes, J. D., 2004. Water-soluble chlorides in massive seafloor serpentinites: a source of chloride in subduction zones. *Earth and Planetary Science Letters*, **226**, 243–256.

Smith, D. K., Cann, J. R., and Escartín, J., 2006. Widespread active detachment faulting and core complex formation near 13°N on the Mid-Atlantic Ridge. *Nature*, **442**, 440–443.

Takai, K., Moyer, C. L., Miyazaki, M., Nogi, Y., Hirayama, H., Nealson, K. H., and Horikoshi, K., 2005. *Marinobacter alkaliphilus* sp. nov., a novel alkaliphilic bacterium isolated from subseafloor alkaline serpentine mud from Ocean Drilling Program Site 1200 at South Chamorro Seamount, Mariana Forearc. *Extremophiles*, **9**, 17–27.

Cross-references

Chemosynthetic Life
Hydrothermal Plumes
Hydrothermal Vent Fluids (Seafloor)
Hydrothermalism
Metamorphic Core Complexes
Oceanic Spreading Centers
Ophiolites
Peridotites
Subduction
Transform Faults

SHELF

William W. Hay
Department of Geological Sciences, University of Colorado at Boulder, Estes Park, CO, USA

Synonyms

Continental shelf; island shelf

Definition

In marine science, a "shelf" is a shallow underwater extension of a land area, either a continent "continental shelf" or large island "island shelf."

The first use of the term "continental shelf" appears to have been by Hugh Robert Mill in his 1888 publication titled "Sea-temperatures on the continental shelf": "The term "continental shelf" is applied to the shallow portion of the continental slope, lying within the 100-fathom line, which is usually terminated seawards by a very abrupt descent to abysmal soundings." This definition has generally been followed but restated in various forms:

1. The relatively shallow belt of sea-bottom bordering a continental mass, the outer edge of which sinks rapidly to the deep ocean-floor (*Oxford English Dictionary*)
2. The extended perimeter of each continent and associated coastal plain (Wikipedia)
3. A submerged border of a continent that slopes gradually and extends to a point of steeper descent to the ocean bottom (Free Online Dictionary)
4. A shallow submarine plain of varying width forming a border to a continent and typically ending in a comparatively steep slope to the deep ocean floor (Merriam-Webster)
5. The edge of a continent that lies under the ocean (National Geographic)
6. The sea bed surrounding a continent at depths of up to about 200 m (100 fathoms), at the edge of which the continental slope drops steeply to the ocean floor (*Collins English Dictionary*)

The United Nations' Law of the Sea Treaty Article 76 gives a legal definition of the "continental shelf" in very different terms, largely unrelated to its geomorphological-physiographic definition (http://www.un.org/Depts/los/convention_agreements/texts/unclos/part6.htm).

History

Knowledge of the configuration of the ocean floor was very limited before the middle of the nineteenth century. Depth soundings were made by "lead line" with a heavy weight at the end of a rope or wire, sensing the change when the weight touched the bottom. However, at water depths greater than a few hundred fathoms, it became impossible to determine when the weight had reached the bottom. It was the laying of the first telegraph cables across the Atlantic that provided insight into the nature of the boundary between the continent and deep ocean.

Because of the time and difficulty of making soundings by lead line, progress in mapping the seafloor was slow until the development of SONAR (*SO*und *N*avigation *A*nd *R*anging). The idea of using sound arose as a means of detecting icebergs after the sinking of the Titanic in 1912. It was rapidly developed as a means of detecting submarines during World War II, but it incidentally provided information on depths of the seafloor and detected the presence of the flat-topped submarine mountains known as "*guyots*" in the Pacific, one of the first clues to the mobility of the seafloor. After the war, SONAR became a means of mapping the ocean floor, at first for Cold-War military purposes (Doel et al., 2006). Although the developing knowledge of the detailed bathymetry, primarily by the United States and Soviet Union, was secret, Bruce Heezen and Marie Tharp, at Lamont Geological Observatory in the United States, devised a method of producing a physiographic depiction the data that was allowed to be published. The first detailed maps showing the edges of the continental shelves on both sides of the North Atlantic as marked by the break in slope were made by Heezen and Tharp and were published in Elmendorf and Heezen (1957) and Heezen et al. (1959). Although known best for its graphic depiction of the ocean floor, the map showed that the edges of the continental shelves, marked by the shelf breaks on the US North American and African margins, fitted together much better than the shorelines used by Alfred Wegener in developing his theory of continental drift. At the time Heezen favored the

Shelf, Table 1 Areas of shelves to the break in slope along the margins of oceans bordering continents, on fragments of continental crust, and in marginal seas. The areas of major epicontinental seas, representing flooding onto the continental blocks, are also shown (After data from Hay and Southam 1977)

Continental margin shelves	Area (10^3 km^2)	Island margin shelves	Area (10^3 km^2)	Marginal sea shelves	Area (10^3 km^2)	Epicontinental seas	Area (10^3 km^2)
Arctic	4,020	Iceland	101	Baffin Bay	375	Canadian Archipelago	495
				Gulf of Mexico	296	Hudson Bay	1,082
Western North Atlantic	1,335			Caribbean Sea	362	Baltic Sea	378
Eastern North Atlantic	1,763			Mediterranean Sea	567	Persian Gulf	267
Western South Atlantic	1,299			Black Sea	148	Oman/Red Sea	248
Eastern South Atlantic	333			Bering Sea	1,031	Arafura Sea	650
Western Indian Ocean	576			Aleutians	116	Gulf of Carpentaria	762
Madagascar	132			Sea of Okhotsk	700	Yellow Sea	380
Kerguelen Plateau	127			Sea of Japan	163	East China Sea	681
Eastern Indian Ocean	6,011					South China Sea	1,235
Western North Pacific	90					Gulf of Thailand	320
Western South Pacific	94						
Australian Pacific	288						
New Zealand	277						
Eastern North Pacific	792						
Eastern South Pacific	644						
Antarctic	376						
Totals	18,258				3,758		6,398

idea of an expanding Earth, and this was seen as supporting evidence.

Detailed publicly available mapping of the continental shelves varies from region to region because of coastal states proprietary economic and military interests. The GEBCO (General Bathymetric Chart of the Ocean) sheets generally show only a 100 m contour in the shelf regions, but in a few areas, shallower contours are also shown. Much more detailed information is available in the ETOPO5 global elevation model data set (US NOAA National Geophysical Data Center [NGDC] Marine Geology & Geophysics Division), and Google Earth includes several large data sets of seafloor depths. There are many local and regional studies of the continental shelves, but the only comprehensive basin-wide account is that of Emery and Uchupi (1984) in "The Geology of the Atlantic Ocean."

The geologic basis for the continental and island shelves

Continental shelves are, by definition, subsea extensions of the peripheries of the seven continents. Geologically, they represent the regions over continental crust thinned by erosion during uplift early in the process of rifting to form passive margins as the continents separated. The average width of the thinned crust is of the order of 130 km; the average width of the subsea portion, the continental shelf, is 70 km. As subsidence has occurred with drifting, they have accumulated thickness of several kilometers of sediment; in the Atlantic margins, these are typically 4–8 km thick. Oceanward slope of the shelf is of the order of 1–2 m/km, but some areas may be flat or have local relief of up to 18 m reflecting sand waves, megaripples, sand ribbons, furrows, and scour. Continental shelves are much narrower along active margins, such as those around the Pacific, and typically have only thin accumulations of sediment. However, back-arc seas along active margins, such as the East and South China Seas, may have broad shelves and thick accumulations of sediment. Also, although not continents, the shelves of large islands, such as Iceland, New Zealand, and Kerguelen, and carbonate banks, such as the Bahamas, are often included in discussions of continental shelves, sometimes as "island shelves." In some areas, continental shelves merge inland with low-lying areas of the interior of the continental block that are flooded as shallow marginal seas: the Baltic Sea, Hudson Bay, the North Sea/English Channel, Arafura Sea/Gulf of Carpentaria, Gulf of Thailand, etc. The areas of continental shelves, marginal seas, and carbonate banks are indicated in Table 1.

Along passive margins, the regions over thinned crust extend inland as coastal plains. At present these are generally less wide than the submerged shelf, but during glacial maxima the coastal plains extended to the shelf edge in many areas. The shoreline is an almost arbitrary divide between the coastal plain and continental shelf and is typically in motion, moving inland or seaward in response to ongoing isostatic adjustment, wave action, sea-level rise or fall, the detrital sediment supply from rivers, and carbonate deposition. Similarly, the depth of the oceanward limit of the continental shelf, the shelf break, varies according to the balance between sediment supply and subsea erosion by waves and currents. Figure 1 shows the varying widths and depths of the shelf break along the classic passive margins of the Atlantic. There is great variability on both sides of the Atlantic, but as a gross generalization, the shelves are wider and the shelf breaks deeper in high latitudes and narrower and shallower in low latitudes. This is due to the state of postglacial isostatic readjustment, sediment supply, and carbonate content of the sediments. Although there may be local pockets of carbonate shell debris at high latitudes, the carbonate content of the sediment generally increases from minimal amounts (isolated shells) at high latitudes to >30 % about 35°N and S and may approach 100 % between 20° N and S. At lower latitudes most of the carbonate is aragonite, produced by algae. It is converted to calcite upon exposure to fresh water during sea-level falls and becomes cemented into a solid layer unaffected by waves and current.

Figure 1 also shows that the original choice of 100 fathoms (~200 m), a typical depth of the shelf break off Scotland included in many definitions, is abnormally deep. The average shelf break depth in the Atlantic is 122 m (Emery and Uchupi, 1984). The global average for the shelf break is variously given as 133 m (*Encyclopedia Britannica online*) and 167 m (Brown et al., 1989).

The shelves are narrowest where rivers supply sediment and build up deltas which have surfaces above sea-level and where they are the margins of carbonate banks now land (e.g., South Florida, Yucatan). They are widest where estuaries and bays extend inland. Similarly, the shelf break is shallowest off large rivers and deepest where there is no nearby sediment source.

Antarctic shelves are different from those of the rest of the world. The shelf breaks are very deep, 400–800 m, and the slope is downward toward the ice-covered continent.

What controls the configuration of the continental shelf?

The examples cited here are from the Atlantic, the most classic passive margin region, described in detail by Emery and Uchupi (1984):

1. Original width of thinning of continental crust. The pull-apart separations between the east coast of the United States and northwest Africa and between South America south of the eastern tip of Brazil and southern Africa south of the Niger Delta occurred about 180 and 130 Ma, respectively. The separation between the northeast coast of South America east of the Amazon and the Ivory Coast of Africa occurred along a transform fault about 130 Ma, with much less thinning of the adjacent continental blocks. All of these are now "old" slowly subsiding margins. Shelves along the pull-apart margins are typically 100 km wide; those along the transform margin are much narrower, only about 25 km wide.
2. Detrital sediment supply. Rivers act as point sources for sediment which can then be redistributed by wave and current action. Globally, the total sediment supply by rivers is larger than can be accommodated in the space available on shelves, but it is very unevenly distributed. Some areas are oversupplied, others sediment starved. Where the local sediment supply is very large, as in the case of the Mississippi, Niger, or Ganges/Brahmaputra complex, the river's delta may extend the shelf edge. However, if there are strong coastal currents, as in the case of the Amazon, the water with its sediment load can flow out into the ocean as a plume and deposit its load over broad areas in the deep sea.
3. Carbonate deposition. Where there is no detrital sediment supply, as in southern Florida and Yucatan, the continental margin resembles a carbonate bank. Carbonate banks accumulate sediment until they are typically only a few meters below sea-level highstands. Because the sea-level highstand of the last interglacial was about 10 m higher than today, some of the carbonate areas are exposed as land today with only a narrow flooded fringe, a very shallow (20 m) shelf edge break, and a very steep continental slope.
4. Wave action. Where the water depth is less than half the wavelength, wave motion affects the bottom sediment and can cause detrital particles to be brought into suspension. Below the littoral zone between the shore and depths of 10–20 m where waves continuously perturb the bottom sediment, it is storm waves that periodically cause sediment to be set in suspension. Once suspended, the sediment can be carried by tidal or other currents on the shelf and redeposited. However, although not yet well documented, it is possible that the depth of the edge of detrital shelves is controlled by the persistent swell coming from distant sources. Swell typically has wavelengths of 300–600 m and so can set sediment at depths of 150–300 m into suspension. The most active region generating swell is the Circum-Antarctic, where the constant westerly winds force a circum-global current unobstructed by land and the fetch is so long that huge waves can develop. The swell generated in this region extends throughout the Pacific Basin, but in the Atlantic is restricted by the S-shaped configuration of the basin. The Antarctic shelf margin is covered by cobbles and boulders rather than sand and clay.
5. Currents. Shelf seas are characterized by persistent longshore currents and short-term tidal currents. Tidal currents onto and off the shelves are not symmetrical either in terms of velocity or location. They can move the sediment suspended by storm waves and cause it to redeposit far from its original site. Similarly, shelf currents can redistribute sediment.

One of the peculiarities of some shelf seas, but particularly the Middle Atlantic Bight, the eastern Bering Sea, Norwegian Sea, Celtic Sea, and along the margin of Southern Australia, is that the shelf waters are separated from those of the open ocean by a frontal system tied to the shelf break (e.g., Gawarkiewicz and Chapman, 1992; Condie, 1993). This separates the muddier "brown" waters of the shelf from the blue waters of the

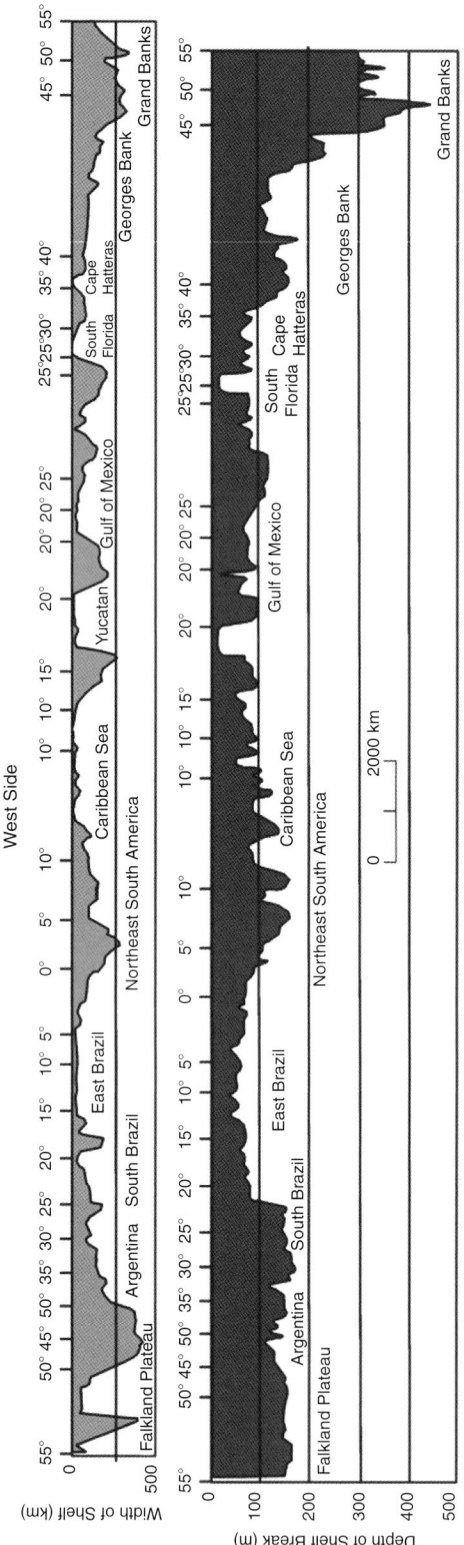
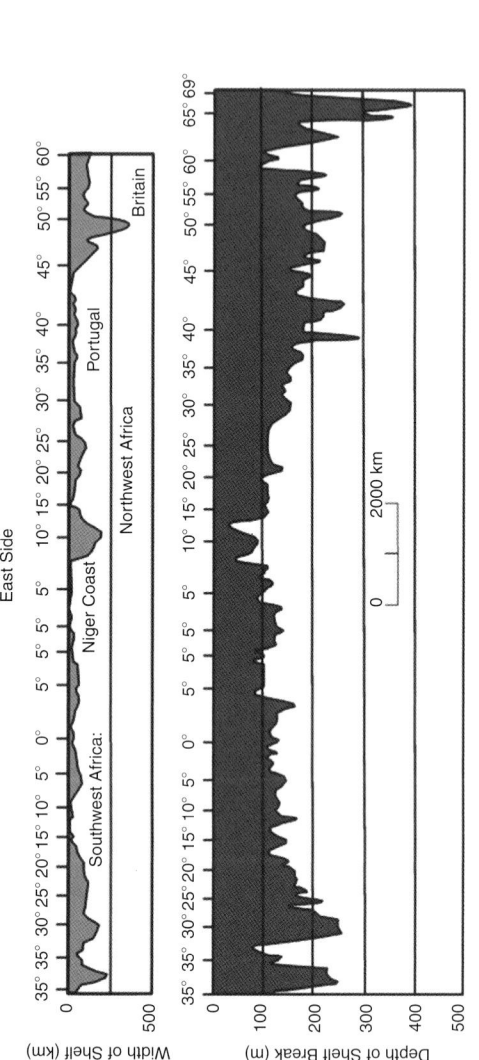

Shelf, Figure 1 Width and depth to the shelf break of shelves along the margins of the Atlantic Ocean (after Emery and Uchupi, 1984).

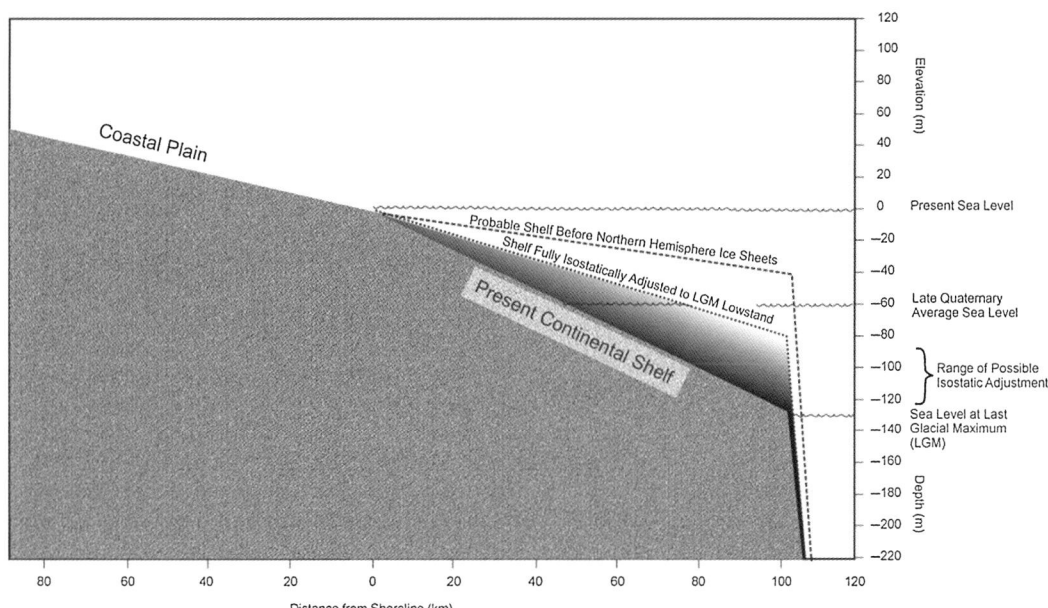

Shelf, Figure 2 The present continental shelf in the context of the Quaternary rises and falls of sea-level. In lower latitudes the modern shelf is almost completely isostatically adjusted to the deglacial flooding, but at higher latitudes near the centers of glaciation isostatic adjustment continues. The rapidity of sea-level change is such that the shelves are always out of adjustment during rises and falls (after Hay and Southam, 1977).

open ocean and inhibits transport of suspended sediment beyond the shelf break.

6. Sea-level. High sea-level stands like that of the modern interglacial are characteristic on less than 10 % of the past few million years. Similarly the extreme lowstands are characteristic of only relatively brief periods of time. Water depths change not only because the global sea-level rises and falls, but in response to isostatic adjustment due to water and ice loading and unloading. The most rapid sea-level changes are in excess of 45 mm/year over times of several thousands of years, whereas isostatic response requires many thousands of years. Hay and Southam (1977) analyzed the effects of water and sediment loading and unloading during glacials and interglacials and concluded that the present shelf slope is adjusted to sea-level lowstands. During those Quaternary lowstands, sediment would have been eroded off the shelves and deposited in the deep sea. As sea-level has risen since the last glacial, the shelf has been depressed by the load of water. However, both the highstands and lowstands are brief; sea-level is usually moving upward or downward with isostatic adjustment following. The average sea-level over the past half million years was about −60 m relative to that of today.

The effects are summarized in Figure 2.

Hay and Southam (1977) noted that when the glaciations began and sea-level fell, the shelves would be eroded and reach a new equilibrium state. They concluded that the detrital sediment supply from rivers during the brief highstands could not rebuild the shelves to their original height and that the present configuration of the shelves is in equilibrium with sea-level lowstands. Their estimate of the original surface, prior to the onset of the Northern Hemisphere glaciations, is shown in Figure 2. The slope on the shelf would be slightly less than that of the adjacent coastal plain. This would reflect the lesser angle of repose of sediment underwater compared to that of drier sediment above water.

Summary

A "shelf" is a shallow underwater extension of a land area, either a continent "continental shelf" or large island "island shelf." Most shelves represent the heavily sedimented regions over continental crust thinned by erosion during uplift early in the process of rifting to form passive margins as the continents broke up. The configuration of the shelf is determined by subsidence rate, sediment supply, waves and currents, and rises and falls of sea-level.

Bibliography

Brown, J., Colling, A., Park, D., Phillips, J., Rothery, D., and Wright, J., 1989. *Waves, Tides, and Shallow Water Processes*. Milton Keynes: The Open University. 187 pp.

Condie, S., 1993. Formation and stability of shelf break fronts. *Journal of Geophysical Research*, **98**(C7), 12405–12416.

Doel, R. E., Levin, T. J., and Marker, M. K., 2006. Extending modern cartography to the ocean depths: military patronage, Cold War priorities, and the Heezen-Tharp mapping project, 1952–1959. *Journal of Historical Geography*, **32**, 605–626.

Elmendorf, C. H., and Heezen, B. C., 1957. Oceanographic information for engineering submarine cable systems. *The Bell System Technical Journal*, **36**(5), 1047–1093.

Emery, K. O., and Uchupi, E., 1984. *The Geology of the Atlantic Ocean*. New York: Springer. 1050 pp. + Map set.

Gawarkiewicz, G., and Chapman, D. C., 1992. The role of stratification in the formation and maintenance of shelf-break fronts. *Journal of Physical Oceanography*, **22**, 753–772.

Hay, W. W., and Southam, J. R., 1977. Modulation of marine sedimentation by the continental shelves. In Andersen, N. R., and Malahoff, A. L. (eds.), *The Fate of Fossil Fuel CO_2 in the Oceans*. New York: Plenum Press, pp. 569–604.

Heezen, B.C., Tharp, M., and Ewing, M., 1959. *The floors of the oceans: I. The North Atlantic*. Geological Society of America Special Paper 65.

Mill, H. R., 1888. Sea-temperatures on the continental shelf. *Scottish Geographical Magazine*, **4**(10), 544–549.

Cross-references

Continental Rise
Continental Slope
Ocean Margin Systems

SILICA

Paul Treguer
European Institute for Marine Studies (IUEM), University of Bretagne Occidentale (UBO), Plouzané, France

Silicon is the seventh most abundant element in the universe and the second most abundant element in the Earth's crust. It is a key nutrient element in the ocean, required for the growth of diatoms and some sponges and utilized by radiolarians, silicoflagellates, several species of choanoflagellates, and potentially picocyanobacteria. The world ocean silica cycle has been recently updated by Tréguer and De La Rocha (2013) who described and quantified source and sink fluxes.

Four pathways serve as external sources of silicic acid (also called dissolved silica, DSi) to the ocean, all of which ultimately derive from the weathering of Earth's crust. DSi from the chemical weathering of continental rocks is discharged into the coastal zone by rivers and groundwater. Rivers also transport significant quantities of particulate amorphous silica that may dissolve, as may dust deposited on the ocean's surface. Lastly, terrigenous silicates in sediments of continental margins may dissolve, and submarine basalts react with high- and low-temperature hydrothermal fluids, releasing DSi. The total net input of DSi to the ocean is 9.4 ± 4.7 T $(=10^{12})$ mol Si year^{-1}.

The permanent removal of silicon from the water column of the ocean occurs mostly via the burial of biogenic silica in sediments. Abiotic precipitation of amorphous silica can occur in hydrothermal vent plumes, and authigenic aluminosilicate formation may occur in sediments, but these two processes support minor export fluxes compared with biogenic silica burial. Siliceous sponges might also act as a significant silica sink, at least regionally. The total net output of biogenic silica to the ocean is 9.9 ± 7.3 T mol Si year^{-1}.

The overall residence time for silicon in the ocean (τ_G) is equal to the total amount of DSi in the ocean (97,000 T mol Si) divided by the net input (or output) flux. The updated budget suggests a mean residence time of approximately 10,000 years. The residence time relative to biological uptake of DSi (τ_B) from surface waters is calculated by dividing the total DSi content of the world ocean by gross silica production by diatoms (240 T mol Si year^{-1}) and sponges (3.7 T mol Si year^{-1}): τ_B is approximately 400 years. The difference between τ_B and τ_G implies that silicon delivered to the ocean passes through the biological uptake and dissolution cycle an average of 25 times $[(240 + 3.7)/9.9]$ before being removed to the seabed.

Bibliography

Tréguer, P. J., and De La Rocha, C. L., 2013. The world ocean silica cycle. *Annual Review of Marine Science*, **5**, 5.1–5.25, doi:10.1146/annurev-marine-121211-172346.

Cross-references

Deep-sea Sediments
Diatoms
Marine Microfossils
Radiolarians

SOURCE ROCKS, RESERVOIRS

Rüdiger Stein
Alfred Wegener Institute, Helmholtz Centre for Polar and Marine Research (AWI), Bremerhaven, Germany

Definition

A *source rock* is any type of rock that may generate hydrocarbons in the future, depending on the amount of organic matter present and adequate temperature and pressure conditions at a specific burial depth as well as sufficient time to reach the needed maturity. A good source rock should have a total organic carbon (TOC) content of larger than 1 %. Most prominent source rocks are sapropels and black shales but also TOC-rich carbonate rocks.

Reservoir rocks are rocks that have sufficient porosity and permeability to store and transmit fluids and gas. Typical reservoir rocks are sandstones and carbonate rocks.

"Source rocks" and "reservoirs" are major components of the *Petroleum System*, consisting of a mature source rock, migration pathway, reservoir rock, trap, and seal (e.g., Tissot and Welte, 1984). For the accumulation and preservation of hydrocarbons, the appropriate relative timing of the formation of these elements is of major relevance.

Source rocks (or organic carbon-rich sediments in general) can be formed under very different environments, including oceans, lakes, deltas, and swamps. In marine environments, organic carbon-rich sediments are especially formed in anoxic basins (such as the modern Black Sea characterized by increased preservation rate due to oxygen deficiency), upwelling areas (characterized by high primary production causing increased marine organic carbon flux), and continental margins with high fluvial input (characterized by increased flux of terrigenous organic matter and high sedimentation rates causing rapid burial and increased preservation). Anoxic environments as well as high-productivity environments allow the accumulation and preservation of a large amount of marine (algae type) of organic matter (kerogen type I and II), resulting in the formation of oil-prone source rocks. Under oxic conditions and in environments with high fluxes of terrigenous organic matter, on the other hand, kerogen type III (originated from higher plants) is predominant, resulting in the formation of gas-prone source rocks.

In Earth's history, the most prominent occurrence of good to excellent source rocks was during Jurassic and Cretaceous times, i.e., during a period with widespread occurrence of regional to global oceanic anoxia and black-shale formation (e.g., Schlanger and Jenkyns, 1976; Stein et al., 1986; Emeis and Weissert, 2009; Jenkyns, 2010; and references therein). More than 50 % of the known petroleum was probably generated from these (dominantly oil-prone) source rocks (Klemme and Ulmishek, 1991; Emeis and Weissert, 2009). The huge deltaic systems such as the Gulf Coast, Niger, and Mackenzie deltas characterized by an enormous supply and accumulation of sediment (including terrigenous – kerogen type III – organic matter) during the Tertiary are further important origins of (predominantly gas-prone) source rocks (e.g., Tissot and Welte, 1984).

Reservoir rocks are rocks that have sufficient porosity and permeability to store and transmit fluids and gas. Typical reservoir rocks are sandstones and carbonate rocks.

Bibliography

Emeis, K.-C., and Weissert, H., 2009. Tethyan-Mediterranean organic carbon-rich sediments from Mesozoic black shales to sapropels. *Sedimentology*, **56**, 247–266.

Jenkyns, H. C., 2010. Geochemistry of oceanic anoxic events. *Geochemistry, Geophysics, Geosystems*, **11**, Q03004, doi:10.1029/2009GC002788.

Klemme, H. D., and Ulmishek, G. F., 1991. Effective petroleum source rocks of the world: stratigraphic distribution and controlling depositional factors. *AAPG Bulletin*, **75**, 1809–1851.

Schlanger, S. O., and Jenkyns, H. C., 1976. Cretaceous oceanic anoxic events: causes and consequences. *Geologie en Mijnbouw*, **55**, 179–184.

Stein, R., Rullkötter, J., and Welte, D. H., 1986. Accumulation of organic-carbon-rich sediments in the Late Jurassic and Cretaceous Atlantic Ocean – a synthesis. *Chemical Geology*, **56**, 1–32.

Tissot, B. P., and Welte, D. H., 1984. *Petroleum Formation and Occurrence*. Heidelberg: Springer. 699 pp.

Cross-references

Energy Resources
Organic Matter
Sapropels

SUBDUCTION

Serge Lallemand
Geosciences Montpellier, University of Montpellier, Montpellier, France

Synonyms

Convergent plate boundary; Destructive plate boundary; Plate consumption

Definition

At a convergent plate boundary, one of the plates necessarily passes beneath the other. This process, first termed *subduction* by A. Amstutz in 1951 when discussing the evolution of alpine structures, allows nearly all oceanic lithosphere to be recycled into the mantle. By contrast, only the lower portion of the continental lithosphere (the lithospheric mantle) may, under certain conditions, be transported into the deep mantle. Since the continental crust resists subduction at depths beyond about 100 km due to buoyancy forces, this type of convergence is typically called *collision* and forms an "orogenic" or a "mountain" belt of which the Alps or the Himalayas are classical examples. Subduction zones can generally be traced at the surface of the Earth because they are associated with large earthquakes, active volcanoes, and sometimes mountain building.

Classification

There are two main types of subduction (Figure 1), depending on the nature of the downgoing plate (hereafter called *slab*):

- Oceanic subduction (often just called subduction) when an oceanic plate subducts
- Continental subduction (often just called collision) when a continental plate subducts

Oceanic subduction can last tens or hundreds of millions of years without a significant change in the topography and tectonics in the upper plate, whereas continental subduction, which typically occurs after a period of oceanic subduction, rapidly causes either a plate reorganization or mountain building. Below, we will focus primarily on oceanic subduction processes (see "Orogeny" entry for details on continental subduction).

Each of these subduction types can be divided in two "subtypes" depending on the nature of the overriding plate. The most common situation at a convergent plate boundary is the subduction of an oceanic plate beneath

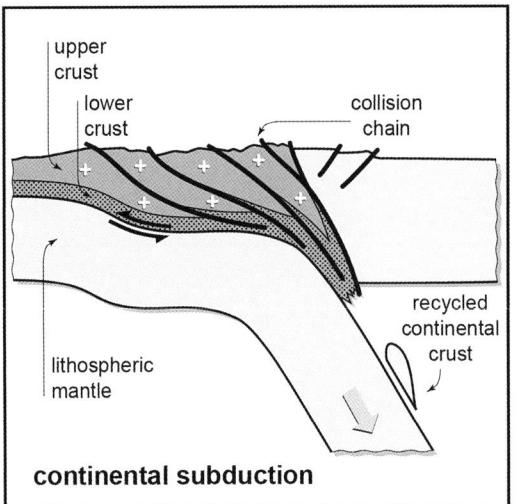 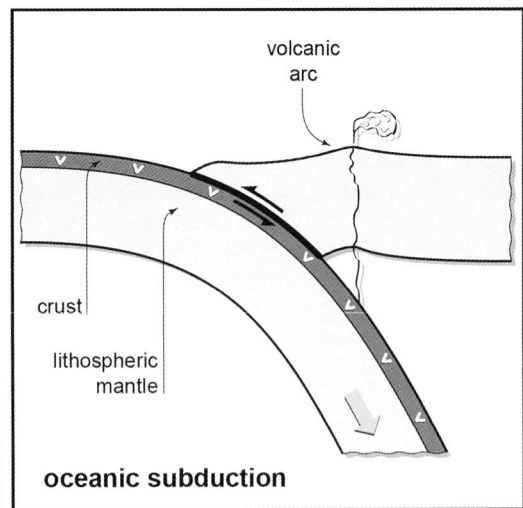

Subduction, Figure 1 Two types of subduction mainly depending on the nature of the downgoing plate: oceanic subduction, 82 %, and continental subduction, 18 %, modified after Lallemand (1999).

a continental plate (~67 %, 45,000 km, see entry "Active Continental Margins"). Among the remaining 33 %, the converging continents contribute for about half, i.e., 17 % (11,200 km), of the total, intra-oceanic subduction zones for about 15 % (10,000 km) and continents subducting beneath oceanic plates for only 1 % (600 km) (Lallemand et al., 2005a).

Many other criteria or combinations of criteria have been proposed to describe the wide variety of subduction zones (Jarrard, 1986). The most commonly used, first introduced by Uyeda in 1984, is the opposition between the Chilean-type and the Mariana-type subductions. The Chilean-type is characterized by a young subducting plate with a shallow slab dip, a high degree of interplate coupling, inducing high stresses and producing large earthquakes, mountains, and thus terrigenous sediment supply feeding an accretionary wedge. On the contrary, the Mariana type is characterized by an old subducting plate with a steep slab dip; a low degree of interplate coupling, producing low stresses; and only moderate seismicity, the presence of back-arc extension, and a low topography unable to feed the trench. Despite the correct description of these two end-member subduction zones, Heuret and Lallemand (2005), based on revised global datasets, have shown that some of these associations of parameters are not statistically verified. Those which are statistically verified in modern subduction zones are the combination of upper plate compressional stress (typically Chile), shallow-dipping slabs (less than 50°), and upper plate seaward absolute motion, or, reversely, upper plate extensional stress (typically Mariana), steeply dipping slabs (more than 50°), and upper plate landward absolute motion (Lallemand et al., 2005b). The other characteristics such as plate age, seismicity, or the presence of an accretionary wedge do not show simple correlations.

Distribution on Earth

Ninety percent of the oceanic subduction zones are concentrated around the Pacific Ocean and in Southeast Asia (48,700 km). The rest are distributed in the Atlantic Ocean (5,000 km) and the Mediterranean Sea (1,350 km). This asymmetric distribution at Earth's surface results from the breakup of the Pangaea in Late Paleozoic and Early Mesozoic time accompanied by the rifting of the Atlantic and Indian oceans and the closure of the Tethys and Panthalassa oceans. Because the most rapid subduction rates are presently observed in the Western Pacific and Southeast Asia, Xavier Le Pichon in 1990 (lessons at Collège de France) compared this relatively small region with a huge pit centered on the Philippine Sea where oceanic plates are recycled into the mantle. Subduction rates generally range between 1 cm/year off Sicily, for example, and 10 cm/year off Japan, except in Southwest Pacific where it exceptionally reaches 24 cm/year offshore the northernmost Tonga archipelago.

Distinctive characteristics of subduction zones

The plane forming the interface between converging plates has distinctive properties compared with a "classical" active fault. It accommodates the underthrusting of one plate beneath another. Its dip angle may vary between 10° and 35° for the shallow part which is seismogenic (Heuret et al., 2011), and it can reach 60° in its deepest part. The subduction interface outlines the contact surface between the plates, i.e., from the trench down to the limit with the convecting mantle. The main differences with a "classical" active fault are:

- A great size in both downdip width (typically between one and two hundreds of km) and lateral extent (length between a few hundreds and a few thousands of km).

- A lower thermal gradient along the interface because of the continuous advection of low-temperature rocks (top of subducting plate). The thermal state depends on the age of the incoming plate and the rate of subduction (Stein and Stein, 1996).
- Fluid overpressure along the interface caused by the progressive dehydration of the subducting plate. Fluid pressure at a given depth depends on the available input from the subducting slab, temperature, and pressure that control phase changes and permeability of the rocks of the hanging wall (see review by Saffer and Tobin, 2011).

The great size of the fault and the low temperature increases the seismic potential. This is the reason why all giant earthquakes occur at subduction zones. When the subducting plate is old and the subduction rate is fast (e.g., below Japan), the subducting plate is characterized by a cloud of intraslab earthquakes termed the "Wadati-Benioff" or "Benioff" zone (Wadati, 1928; Benioff, 1949, see entry "Wadati-Benioff-Zone") down to 660 km, which is the depth of the discontinuity between the upper and the lower mantle.

The last distinctive characteristic is the volcanic arc that generally outlines the oceanic subduction zones. Fluids stored in the subducting plate are released at various depths from the first kilometers of subduction near the trench (pore fluids) down to about 150 km (hydrous minerals) depending on the thermal state of the slab. At depths between 80 and 150 km, slab dehydration is responsible for the metasomatism of the overlying mantle wedge. A small fraction of the asthenospheric wedge undergoes partial melting and migrates upward. Volcanoes thus develop at a given distance depending on the slab dip because melts migrate vertically from the slab.

Slab shape and dynamics

At a large scale, oceanic plates behave elastically when they are forced to bend in subduction zones. We often observe a small bulge seaward of the trench before the plate sinks into the mantle. The trench is the deepest surface expression of the oceanic plate. Its depth depends on the age of the subducting plate and the amount of sediment infill. The deepest trenches are the Mariana Trough (10,938 m revised depth by Fujioka et al., 2002) and the Philippine trench (10,100 m after Lallemand et al., 1998). It is interesting to note that the deepest trenches are located right in the area of maximum downwelling of oceanic lithosphere on Earth as computed by Lithgow-Bertelloni and Richards (1998). Some authors argue that the elastic properties of the sinking lithosphere help the slab to unbend below the arc so that it recovers a flat profile. In detail, other forces may also contribute in the unbending of the plate such as the suction force exerted by the overriding plate and mantle wedge. Slabs may exhibit various shapes such as flat and horizontal in their upper section (Peru or Central Chile), shallow dipping (Japan or Kurile), intermediate dipping (Tonga or Kermadec), steeply dipping (Izu-Bonin), vertically dipping (Mariana), or even overturned (Luzon or New Britain).

In modern subduction zones, active compression in the overriding plate is always observed for shallow-dipping slabs (<50°), whereas active back-arc extension is always observed for steeply dipping slabs (>50°). Lallemand et al. (2005b, 2008) have shown that the slab shape is not simply linked with the age of the subducting plate (the common assumption is that an old plate is thick and dense and thus prone to sink steeply or even vertically in the mantle) but rather with the combination of absolute velocities between converging plates and the regional mantle dynamics. Moreover, the upper plate strain appears to directly result from this kinematic interaction.

Main driving and resistive forces in subduction zones

Today, the main driving mechanism for plate motion in general (see "Driving Forces: Slab Pull, Ridge Push"), and for subduction in particular, is the gravitational sinking of the slab (Turcotte and Schubert, 1982) which is usually slightly denser (about 1 %) than the surrounding mantle because of its lower temperature (thermal contraction). This force is called *slab pull* (Figure 2). The viscous mantle opposes a resistance to plate motion from the ridge to the trench called *mantle drag*. The upper portion of the asthenosphere is known to have a low viscosity so that the *mantle drag* should be moderate. The viscous shearing increases considerably as the plate sinks into the mantle because it exerts on both sides – top and bottom – of the descending slab and because the viscosity is pressure dependent (Karato and Wu, 1993). It probably increases by at least one order of magnitude between the top and the bottom of the upper mantle. This resistance to penetration drops again when the slab reaches the discontinuity between the upper and the lower mantle near 660 km as a result of phase changes in minerals. The mantle opposes another resistance called *anchoring*, which accounts for mantle reaction to slab facewise translation (Scholtz and Campos, 1995; Lallemand et al., 2008). Since it is a viscous pressure force, it will depend on the velocity of slab rollback or advance associated with trench migration and increase with depth. The slab also resists to bend and unbend when passing the trench. This resisting force called *bending moment* increases with the stiffness of the plate and thus its age. It is inversely proportional to the radius of curvature of the slab (Buffett and Rowley, 2006; Wu et al., 2008).

Among these main internal forces, two are always resisting: the *mantle resistance to penetration*, by definition, which is effective as soon as a slab dives into the mantle, and the *bending moment*, also called *bending resistance*, as the plate bends and unbends. All the other forces can shift from driving to resisting depending on the geodynamic context. Let us first consider *slab pull* whose name comes, by definition, from the capacity of

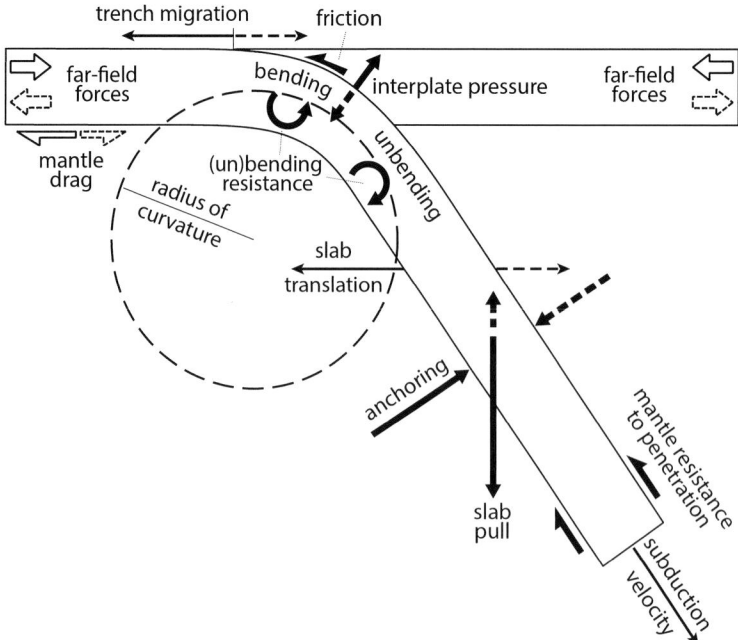

Subduction, Figure 2 Description of driving and resistive forces in a subduction zone modified after Lallemand et al. (2008). Many forces can point in two opposite directions depending on the geodynamic context. *Full lines* are the most frequent directions, and *dotted lines* are less frequent directions but still observed. Far-field forces commonly originate from ridge-push, slab pull, or collision.

the slab to sink into the mantle because of its negative buoyancy. Most of the time, it is considered as a driving force, but in about 10 % of cases, we observe "flat" slabs, i.e., subducting slabs sliding horizontally beneath the upper plate over hundreds of kilometers before sinking. Gutscher et al. (2000) attributed this flatness to a positive buoyancy of the oceanic plate and a delay in the basalt to eclogite transition due to the cool thermal structure of the two overlapping lithospheres. The best examples are the Nankai or Cascadia subduction zone where the subducting crust is very young, or the Peru and Central Chile segments along the Andean margin where buoyant ridges or plateaus are subducting. The *mantle drag* generally offers resistance to plate motion near subduction zones except when mantle convective cells, produced by another mechanism rather than the subduction, drive the plate down. This could be the case when a spreading ridge subducts like in South Chile or in Cascadia. The *anchoring force* always offers resistance to facewise translation of the slab but not much to its downward penetration.

There are also second-order external forces originating from the far field (Figure 2). Their magnitude and direction depend on the geodynamic context. The so-called ridge-push results from the gravitational collapse or subsidence of the oceanic plates as they drift away from the ridge. Fowler (1990) estimates that, on average, the *slab pull* force is one order of magnitude greater than the *ridge-push* force. If the active ridge is located seaward of the trench, the *ridge-push* force will push the plate toward the trench, but if the spreading ridge is located landward of the trench, it will push the upper plate onto the subducting plate (like in South America with the East Pacific Rise on one side and the mid-Atlantic Ridge on the other). Far-field slab pull can also provide an additional force acting in the subduction zone. For example, the Izu-Bonin-Mariana arc overrides the subducting Pacific plate but is carried by the Philippine Sea plate which is dragged northwestward as a consequence of the slab pull acting along the Ryukyu subduction zone (Pacanovsky et al., 1999). In this case, the upper plate is pulled away from the trench. Another source of external stress is the effect of a collision like the Ontong Java Plateau with the Solomon trench that impacts the stress field over a huge area from New Guinea to New Hebrides (Mann and Taira, 2004).

Subduction interface

The shallowest part of the subduction interface often corresponds to an aseismic *décollement zone* characterized by overpressures resulting from the pore fluids release as lithostatic pressure increases (Saffer and Tobin, 2011). At rather low temperatures between 100–150° and 350–450° (Oleskevich et al., 1999; Hyndman, 2007), the plate interface becomes frictional and thus seismogenic (see "Seismogenic Zone" and Dixon and Moore, 2007). All "subduction earthquakes" nucleate in this region. The *seismogenic zone* generally extends from a depth of 11 ± 4 to 51 ± 9 km. It is about 112 ± 40 km wide and dips

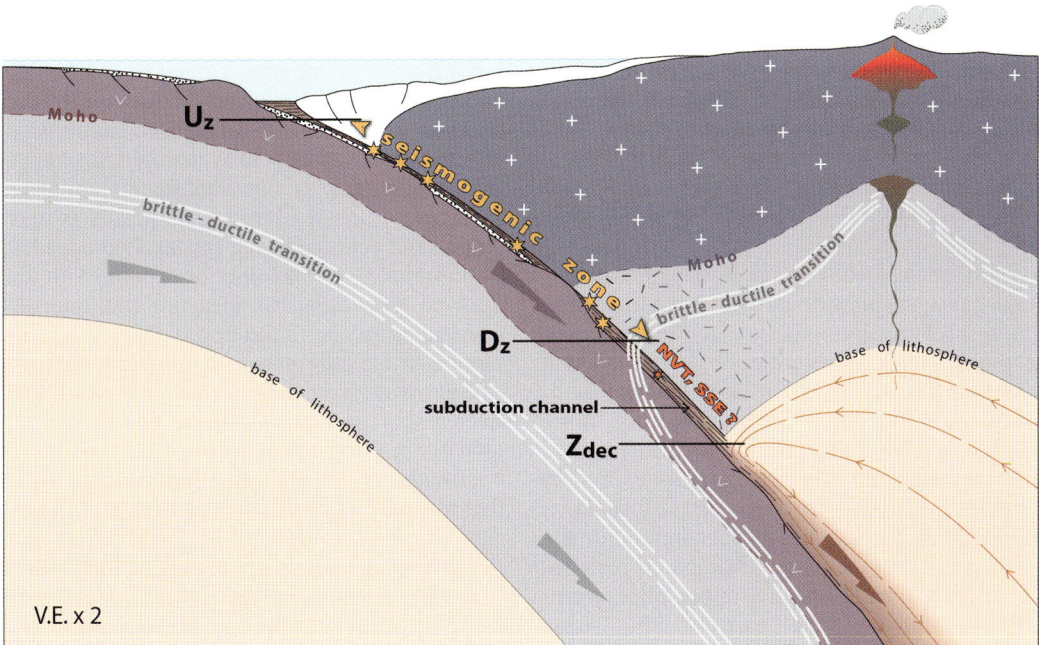

Subduction, Figure 3 Schematic illustration of the subduction interface in an oceanic subduction zone. U_z is the upper limit and D_z the downdip limit of the seismogenic zone. Z_{dec} is the decoupling depth. Below that depth, the convective mantle is dragged down (coupled) with the subducting plate. *NVT* nonvolcanic tremors and *SSE* slow-slip events are observed in some subduction zones at depths between D_z and Z_{dec}.

$23 \pm 8°$ (Heuret et al., 2011). At temperatures higher than 350–450°, the slip along the interface is either continuous (creep) or characterized by slow-slip events (SSE) associated with nonvolcanic tremors (NVT) (Rogers and Draggert, 2003). The downdip limit of the subduction interface is marked by the coupling between the descending plate and the overlying mantle (Figure 3). Furukawa (1993) called the depth below which the plates are uncoupled on the long term Z_{dec}. Downdip of this depth, a significant thickness of mantle is dragged with the slab, called "viscous blanket" by Kincaid and Sacks (1997). The thickness depends on the temperature contrast between the slab and the mantle wedge. Beyond Z_{dec}, there is no more plate interface, the slab being overlain by the convective mantle wedge.

There is a debate on the nature of the plate interface. Is it an ~1–20 m thin "décollement" zone like those drilled in Nankai (Bangs et al., 1996) or an ~100–1,000 m thick channel across which the shear distributes like in the exhumed Franciscan complex (Cloos and Shreve, 1988)? The discovery of erosional processes at the hanging wall of the subduction interface in Japan, Middle America, and Izu-Bonin-Mariana trenches since the early 1980s (von Huene et al., 1980; von Huene and Lallemand, 1990; von Huene and Scholl, 1991; Lallemand et al., 1992; Meschede et al., 1999) has been decisive in promoting the concept of *subduction channel*. It has been later confirmed, thanks to the increasing resolution of seismic images as in the Ecuadorian margin (Sage et al., 2006; Collot et al., 2008) and the reinterpretation of some outcrops of exhumed subduction interface like in the Shimanto belt (Kitamura et al., 2005) or in the Apennines (Vannucchi et al., 2008). Several processes of upward migration of the roof décollement or downward migration of the basal décollement delimiting the subduction channel have been discussed by Vannucchi et al. (2012) and are not yet completely understood. Hydrofracturing of the hanging wall rocks under increasing fluid pressure coming from the dewatered downgoing plate (von Huene et al., 2004) very probably contributes to the upward migration of the deformation front as well as the subduction of oceanic reliefs such as seamounts and ridges (Dominguez et al., 1998, 2000).

At the scale of the plate interface, two forces govern local tectonics and hazards: friction and pressure. The net pressure or normal stress exerted along the subduction interface combines the lithostatic pressure, increasing with depth, and the non-isostatic "dynamic" pressure, which accounts for the slab pull, the slab-mantle interactions, and the far-field "tectonic" forces. The idea, developed by Shemenda (1992) in laboratory analog models, where far-field forces and mantle drag and anchoring are neglected, is that a dense (old) and long subducting plate tends to pull the upper plate down producing a subsidence of the forearc, whereas a light (young) or a short subducting plate tends to push the upper plate up producing uplift of the forearc. In nature, not only the age and length of the subducting plate but mantle and far-field forces also contribute to the normal stress along the interface. The second force is frictional shear, which resists subduction. The stress is released intermittently as it overcomes the yield strength of the fault, through repeated earthquakes and subsequent fault ruptures.

Seismicity with special emphasis on mega-thrust earthquakes

Subduction zones are the locus of most earthquakes occurring on Earth because the plate boundaries have to accommodate the largest relative motions and subsequent strain, because the plate interface is continuously cooled and because the slab still remains cool at large depths. "Subduction earthquakes" are thrust events occurring along the seismogenic plate interface. They represent most of the seismic energy released at subduction zones and in the world (Scholz, 2002). There are also intraslab shallow, intermediate, and deep earthquakes that accommodate the high level of stress during the plate bending and penetration into the viscous mantle and the upper plate earthquakes that accommodate the part of the stress transmitted through the plate interface. See entries "Earthquakes", "Active Continental Margins" and in particular Figure 2 of "Active Continental Margins."

One specificity of the subduction zones is that they can generate giant earthquakes with magnitudes ≥ 8.5, most of them being tsunamigenic like the 2004 M9.2 event off Sumatra (Indonesia) or the 2011 M9.0 event off Tohoku (Japan). Back to the 1900–2012 period, Figure 4 shows the rupture areas associated with the $M \geq 8.0$ earthquakes and the segmentation of the seismogenic zones mainly based on past earthquake ruptures. It is easy to see that they don't distribute equally along the subduction zones. Based on available global datasets, Heuret et al. (2011, 2012) have shown that giant earthquakes preferentially occurred near slab edges, where the upper plate was continental, the back-arc strain neutral to compressive, and the trench fill larger than 1 km. A possible explanation would be that the rupture propagation could be facilitated if the plate interface is smooth (Ruff, 1989) and the normal stress on the interface large enough to accumulate strain but not too large to allow rupture propagation. The continental nature of the upper plate could be preferred because it is associated with wider (downdip) seismogenic zones. The dependence of earthquake magnitude with the subduction velocity and plate age, first proposed by Ruff and Kanamori (1980, 1983), is not statistically verified using revised databases. However, none of the Sumatra and Tohoku earthquakes were expected to occur based on our knowledge at the time they happened, because the recurrence time of such huge events is often underestimated and because we generally do not have access to paleoearthquake archives (McCaffrey, 1997, 2008; Lay and Kanamori, 2011).

The degree of coupling between the converging plates may be used to estimate the seismic hazard at a given subduction zone. For many years, authors calculated the *seismic coupling coefficient* as the ratio between the observed seismic moment release rate and the rate calculated from plate tectonic velocities (Brune, 1968; Davies and Brune, 1971). Scholz and Campos (1995) have shown that this coefficient should be used cautiously because the seismic slip was generally computed over a short time period (typically a hundred years) compared with the recurrence time for large earthquakes (sometimes up to a thousand years). During the last two decades, the advent of monitoring of subduction zones with GPS stations allowed geodesists to measure the interseismic strain accumulation rate in the upper plate, which gives a better picture of the interplate coupling than the earlier measure of energy release rate by summing the earthquake moment over a short period. By doing this, Scholz and Campos (2012) have revised their estimate of the seismic coupling coefficient in many subduction settings. They concluded that this new approach agrees well with the revised "classical way" of measuring the coefficient accounting for major earthquakes older than a century when available. It is also more satisfactory for the balance of forces acting along the plate interface like the normal stress and the friction making assumptions on their respective origins (see above). They confirm the observation of Chemenda et al. (2000) that high coupling, and thus high friction, is observed in regions of high normal stress. The limitation of the method, consisting of measuring the interseismic elastic strain, is that most of the time the coupled area in a subduction zone is located offshore, whereas GPS receivers are deployed onshore. The resolution of the models should be improved in the future when a significant number of GPS stations can be installed offshore like off Japan. Tremendous developments must be further done to precisely map the complexity of the coupling distribution along the subduction interface because large lateral variations are predicted by authors (Ruff, 1992; Bilek, 2007; Scholz and Campos, 2012).

Upper plate tectonics

The upper plate is the visible part of a subduction zone, also called *active margin*. The tectonic regime of the active margin depends on interactions between the converging plates at the level of the trench, the plate interface, and even the whole overriding plate.

At the trench, sediment carried by the subducting oceanic plate may be offscraped and incorporated at the front of the overriding plate. This process, called *frontal accretion*, contributes to the growth of an *accretionary wedge* (see entry "Accretionary Wedges" for more details or Moore and Silver, 1987). Von Huene and Scholl (1991) have shown that half of the growth of an accretionary prism can be accounted for by subcrustal or basal accretion, also called *underplating*. They also estimated that only half of the modern convergent margins are accretionary. The other half undergoes what they called *tectonic erosion*. This process has been definitely demonstrated in the late 1970s (e.g., Scholl et al., 1980) after

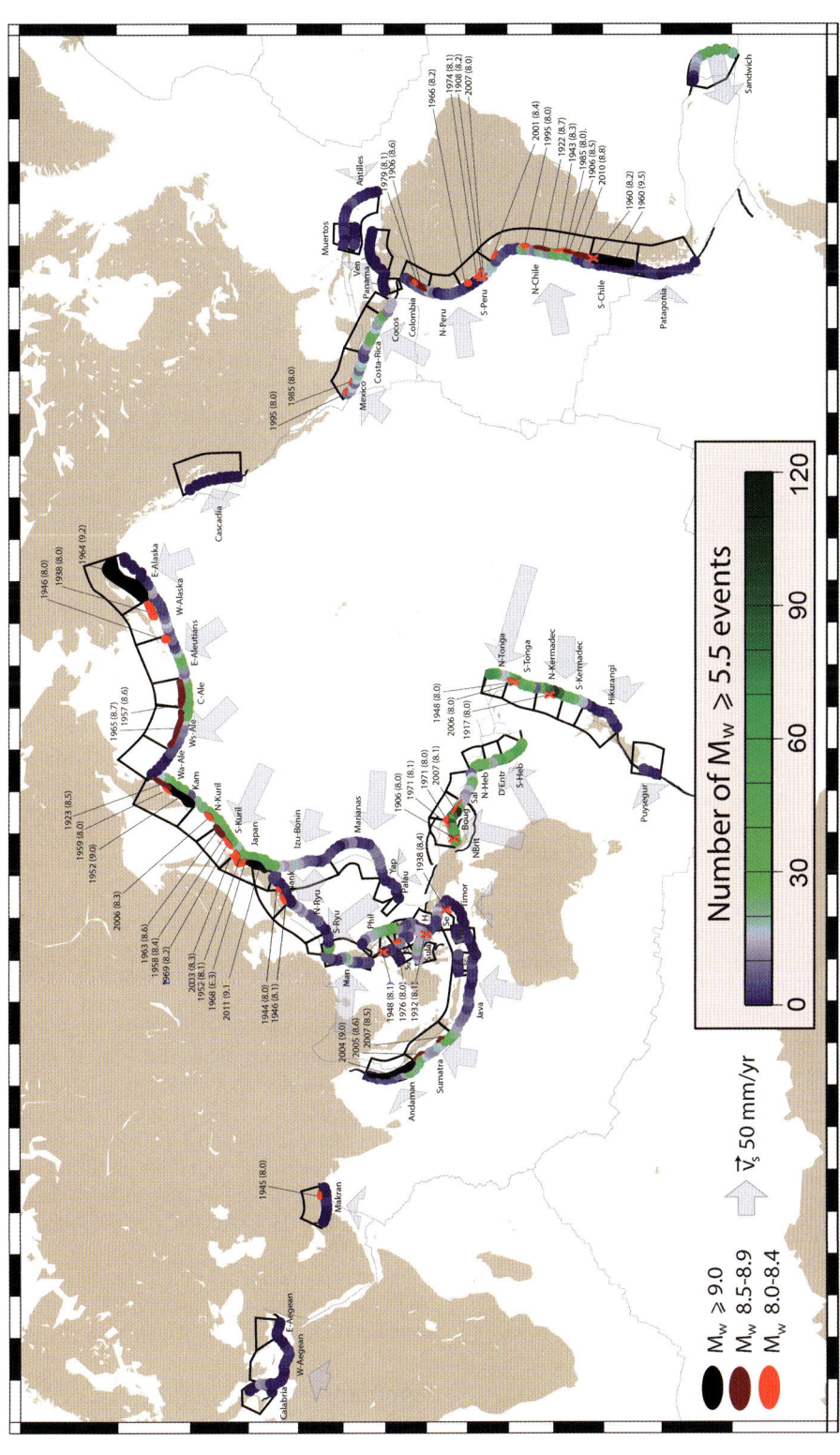

Subduction, Figure 4 Map of the subduction interface seismic rate, trench segments, and major rupture areas of the $M_w \geq 8.0$ 1900–2012 events updated from Heuret et al. (2011). *Pale blue arrows* indicate subduction relative velocities after Heuret and Lallemand (2005).

observations from both deep-sea drilling and reflection seismic imaging, which the Japan margin underwent a Miocene subsidence of at least 2 km. Such large subsidence increasing seaward, together with the absence of an accretionary prism, erosional features at margin's front, and simultaneous retreat of the volcanic arc, led the authors to propose a model in which part of the margin was removed both frontally and subcrustally during subduction. Since 1980, other margins appeared to also undergo tectonic erosion like in mid-America, Peru, Tonga, Mariana, or Izu-Bonin (von Huene and Lallemand, 1990; Lallemand, 1995; Meschede et al., 1999; Clift and Vannucchi, 2004). See entry "Subduction Erosion" for more details.

The convergent margins deform in response to the stress transmitted across the plate interface but also in response either to frontal or basal accretion or frontal or basal tectonic erosion. The tectonic regime is often compressive in the frontal part of the overriding plate, but it is also quite common to observe normal faults dipping either landward or seaward in active convergent margins, especially when they are subject to tectonic erosion. These faults reflect either a trenchward collapse as the margin progressively steepens (Aubouin et al., 1984), or a buckling as underplating occurs (Lallemand et al., 1994). Another mechanism, coseismically induced, has been recently suggested. Indeed, a seaward dipping normal fault has been activated at the front of the Japan margin during the 2011 Tohoku mega-earthquake. The explanation provided by McKenzie and Jackson (2012) would be that both the release of gravitational potential energy and the elastic strain account for this coseismic behavior.

Trench-parallel strike-slip faults may also be observed in the forearc in case of oblique convergence like off Sumatra (Mentawai Fault) or the southern Ryukyus (Yaeyama Fault). These transcurrent subvertical faults accommodate part of the lateral component of the oblique convergence, whereas the slip along the frontal part of the subduction thrust is closer to trench normal (e.g., McCaffrey, 1992). This mechanism is verified by the direction of slip vectors of subduction earthquakes that are often deflected toward trench normal when the convergence is oblique. This process called *slip partitioning* has been observed not only at the scale of the forearc but also at the scale of tectonic plates. Indeed, several oblique subduction systems develop transcurrent faults. Chemenda et al. (2000), for example, have investigated using physical models the effect of the pressure and the friction along the plate interface on slip partitioning in a context of oblique convergence. They conclude that slip partitioning can occur in the models only when interplate friction is high and when the overriding plate contains a weak zone. As a matter of fact, lithospheric transcurrent faults develop either at the toe of the rigid backstop (rear of accretionary wedge like off East Taiwan), along the volcanic arc (Sumatra, the Philippines, South Kuriles), at the back-arc rift axis, or spreading center (Andaman Sea, Le Havre Trough).

Ultimately, the global strain rate map (Kreemer et al., 2003) indicates that most subduction systems deform far from the plate boundaries, or one may consider that the plate boundaries are wider than a single fault. They can potentially affect a strip several hundreds of kilometers large at upper plate edge. In addition to interseismic elastic deformation, there are many areas where permanent (plastic) deformation localizes. We have seen above that, in some oblique geodynamic settings, transcurrent faulting may occur, but even more *shortening*, *rifting*, or *spreading* can also occur. Significant shortening (>1 cm/year) is observed, for example, along the eastern margin of the Japan Sea, the southern margin of the Okhotsk Sea, or the eastern cordillera of the Andes. Opposingly, rifting or spreading develops in the Sumisu rift and the Mariana Trough (Izu-Bonin-Mariana arc), the Lau Basin and Le Havre Trough (Tonga-Kermadec arc), the Andaman Sea (Andaman arc), the Central Kamchatka graben (North Kurile arc), the Manus Basin (New Britain arc), the North and South Fiji basins (New Hebrides arc), the Scotia Sea (South Sandwich arc), the Tyrrhenian Sea (Calabria arc), or the Aegean Sea (Hellenic arc). Several mechanisms have been invoked to account for this upper plate deformation playing with interplate stress resulting from plate kinematics (Dewey, 1980; Lallemand et al., 2008) and/or mantle wedge dynamics (Sleep and Toksöz, 1971; Lagabrielle et al., 1997; Faccenna et al., 2010).

Magmatism at subduction zones

Volcanic arcs are the second worldwide magmatic contributor to crustal growth on Earth after the oceanic spreading centers (nearly 1 km^3/year according to Crisp, 1984). Most subduction zones are characterized by an active *volcanic arc*. Sometimes, the volcanic activity ceases for a few million years and resumes again. The reason for the cessation of volcanism is commonly either the collision with a continent or the geometry of the slab which is incompatible with the presence of a mantle wedge. This is the case, for example, with flat slab segments in the Andes (Gutscher et al., 2000). Arc magmatism results from the metasomatism and partial melting of rocks of the mantle wedge that are hydrated by the water extracted from hydrous minerals in the subducting plate at depths ranging between 80 and 150 km. Further details on arc magmatism can be found in Tatsumi (1989) or Tatsumi and Kosigo (2003). See also entries "Magmatism at Convergent Plate Boundaries", and "Island Arc Volcanism, Volcanic Arcs".

Access to a database on kinematic, geometric, seismologic, and structural parameters of oceanic subduction zones can be found via url: *submap.fr*.

Summary

Subduction is a lithospheric process which occurs at convergent plate boundaries. One plate underthrusts another and sinks into the Earth mantle. The subduction mainly involves a subducting oceanic plate, but sometimes, a continental plate enters into the subduction zone. In this last case, a collisional chain develops. This process generates a high level of seismicity as the rocks deform in the vicinity of the plate boundary. All mega-earthquakes occur in subduction zones. Oceanic subduction creates favorable conditions for arc magmatism.

Bibliography

Amstutz, A., 1951. Sur l'évolution des structures alpines. *Archives Science*, **4**(5), 323–329.

Aubouin, J., Bourgois, J., and Azéma, J., 1984. A new type of active margin: the convergent extensional margin, as exemplified by the Middle America Trench off Guatemala. *Earth Planet Science Letters*, **67**, 211–218.

Bangs, N. L., Shipley, T. H., and Moore, G. F., 1996. Elevated fluid pressure and fault zone dilation inferred from seismic models of the northern Barbados Ridge décollement. *Journal of Geophysical Research*, **101**(B1), 627–642.

Benioff, H., 1949. Seismic evidence for the fault origin of oceanic deeps. *Geological Society of America Bulletin*, **60**, 1837–1856.

Bilek, S. L., 2007. Influence of subducting topography on earthquake rupture. In Dixon, T. H., and Moore, J. C. (eds.), *The Seismogenic Zone of Subduction Thrust Faults*. New York: Columbia University Press, pp. 123–146.

Brune, J., 1968. Seismic moment, seismicity, and rate of slip along major fault zones. *Journal of Geophysical Research*, **73**, 777–784.

Buffett, B. A., and Rowley, D. B., 2006. Plate bending at subduction zones: consequences for the direction of plate motions. *Earth Planet Science Letters*, **245**, 359–364, doi:10.1016/j.epsl.2006.03.011.

Chemenda, A., Lallemand, S., and Bokun, A., 2000. Strain partitioning and interplate friction in oblique subduction zones: constraints provided by physical modeling. *Journal of Geophysical Research*, **105**(3), 5567–5582.

Clift, P., and Vannucchi, P., 2004. Controls on tectonic accretion versus erosion in subduction zones: implications for the origin and recycling of the continental crust. *Reviews of Geophysics*, **42**, RG2001, doi:10.1029/2003RG000127.

Cloos, M., and Shreve, R., 1988. Subduction channel model of prism accretion, melange formation, sediment subduction, and subduction erosion at convergent plate margins: 2. Implications and discussion. *PAGEOPH*, **128**(3/4), 501–545.

Collot, J.-Y., Agudelo, W., Ribodetti, A., and Marcaillou, B., 2008. Origin of a crustal splay fault and its relation to the seismogenic zone and underplating at the erosional north Ecuador–south Colombia oceanic margin. *Journal of Geophysical Research*, **113**, B12102, doi:10.1029/2008JB005691.

Crisp, J. A., 1984. Rates of magma emplacement and volcanic output. *Journal of Volcanology and Geothermal Research*, **20**, 177–211.

Davies, G., and Brune, J. N., 1971. Global plate motion rates from seismicity data. *Nature*, **229**, 101–107.

Dewey, J. F., 1980. Episodicity, sequence, and style at convergent plate boundaries. In Strangway, D. W. (ed.), *The continental crust and its mineral deposits*, Geological Association of Canada Special paper, Vol. 20, pp. 553–573.

Dixon, T. H., and Moore, J. C. (eds.), 2007. *The Seismogenic Zone of Subduction Thrust Faults*. New York: Columbia University Press. 680 pp.

Dominguez, S., Lallemand, S. E., Malavieille, J., and von Huene, R., 1998. Upper plate deformation associated with seamount subduction. *Tectonophysics*, **293**, 207–224.

Dominguez, S., Malavieille, J., and Lallemand, S. E., 2000. Deformation of accretionary wedges in response to seamount subduction – insights from sandbox experiments. *Tectonics*, **19**(1), 182–196.

Faccenna, C., Becker, T. W., Lallemand, S., Lagabrielle, Y., Funiciello, F., and Piromallo, C., 2010. Subduction-triggered magmatic pulses : a new class of plumes? *Earth Planet Science Letter*, **299**, 54–68, doi:10.1016/j.epsl.2010.08.012.

Fowler, C. M. R., 1990. *The Solid Earth: An Introduction to Global Geophysics*. New York: Cambridge University Press. 472 pp.

Fujioka, K., Okino, K., Kanamatsu, T., and Ohara, Y., 2002. Morphology and origin of the challenger deep in the southern Mariana trench. *Geophysical Research Letters*, **29**(10), 1372.

Furukawa, Y., 1993. Depth of the decoupling plate interface and thermal structure under arcs. *Journal of Geophysical Research*, **98**, 20005–20013.

Gutscher, M.-A., Spakman, W., Bijwaard, H., and Engdahl, E. R., 2000. Geodynamics of flat subduction: seismicity and tomographic constraints from the Andean margin. *Tectonics*, **19**(5), 814–833.

Heuret, A., and Lallemand, S., 2005. Plate motions, slab dynamics and back-arc deformation. *Physics of the Earth and Planetary Interiors*, **149**, 31–51.

Heuret, A., Lallemand, S., Funiciello, F., Piromallo, C., and Faccenna, C., 2011. Physical characteristics of subduction interface type seismogenic zones revisited. *Geochemistry, Geophysics, Geosystems*, **12**, Q01004, doi:10.1029/2010GC003230.

Heuret, A., Conrad, C. P., Funiciello, F., Lallemand, S., and Sandri, L., 2012. Relation between subduction megathrust earthquakes, trench sediment thickness and upper plate strain. *Geophysical Research Letters*, **39**, L05304, doi:10.1029/2011GL050712.

Hyndman, R. D., 2007. The seismogenic zone of subduction thrust faults: What we know and don't know. In Dixon, T. H., and Moore, J. C. (eds.), *The Seismogenic Zone of Subduction Thrust Faults*. New York: Columbia University Press, pp. 15–40.

Jarrard, R. D., 1986. Relations among subduction parameters. *Reviews of Geophysics*, **24**, 217–284.

Karato, S., and Wu, P., 1993. Rheology of the upper mantle: a synthesis. *Science*, **260**, 771–778.

Kincaid, C., and Sacks, I., 1997. Thermal and dynamical evolution of the upper mantle in subduction zones. *Journal of Geophysical Research*, **102**, 12295–12315.

Kitamura, Y., Sato, K., Ikesawa, E., Ikehara-Ohmori, K., Kimura, G., Kondo, H., Ujiie, K., Tiemi Onishi, C., Kawabata, K., Hashimoto, Y., Mukoyoshi, H., and Masago, H., 2005. Mélange and its seismogenic roof décollement: a plate boundary fault rock in the subduction zone—an example from the Shimanto Belt, Japan. *Tectonics*, **24**, TC5012, doi:10.1029/2004TC001635.

Kreemer, C., Holt, W. E., and Haines, A. J., 2003. An integrated global model of present-day plate motions and plate boundary deformation. *Geophysical Journal International*, **154**, 8–34.

Lagabrielle, Y., Goslin, J., Martin, H., Thirot, J. L., and Auzende, J. M., 1997. Multiple active spreading centers in the hot North Fiji Basin (Southwest Pacific): a possible model for Archean sea floor dynamics? *Earth and Planetary Science Letters*, **149**, 1–13.

Lallemand, S. E., Schnurle, P., and Manoussis, S., 1992. Reconstruction of subduction zone paleogeometries and quantification of upper plate material losses caused by tectonic erosion. *Journal of Geophysical Research*, **97**(B1), 217–240.

Lallemand, S. E., Schnurle, P., and Malavieille, J., 1994. Coulomb theory applied to accretionary and non-accretionary wedges – possible causes for tectonic erosion and/or frontal accretion. *Journal of Geophysical Research*, **99**(B6), 12033–12055.

Lallemand, S. E., 1995. High rates of arc consumption by subduction processes: some consequences. *Geology*, **23**(6), 551–554.

Lallemand, S. E., Popoff, M., Cadet, J.-P., Deffontaines, B., Bader, A.-G., Pubellier, M., and Rangin, C., 1998. Genetic relations between the central & southern Philippine Trench and the

Philippine Trench. *Journal of Geophysical Research*, **103**(B1), 933–950.

Lallemand, S., 1999. *La subduction océanique*. Amsterdam: Gordon and Breach Science Publishers. 208 pp.

Lallemand, S., Huchon, P., Jolivet, L., and Prouteau, G., 2005a. *Convergence lithosphérique*. Paris: Vuibert. 182 pp.

Lallemand, S., Heuret, A., and Boutelier, D., 2005b. On the relationships between slab dip, back-arc stress, upper plate absolute motion and crustal nature in subduction zones. *Geochemistry, Geophysics, Geosystem*, **6**, 1, doi:10.1029/2005GC000917.

Lallemand, S., Heuret, A., Faccenna, C., and Funiciello, F., 2008. Subduction dynamics revealed by trench migration. *Tectonics*, **27**, TC3014, doi:10.1029/2007TC002212.

Lay, T., and Kanamori, H., 2011. Insights from the great 2011 Japan earthquake. *Physics Today*, **64**, 33–39.

Lithgow-Bertelloni, C., and Richards, M. A., 1998. The dynamics of Cenozoic and Mesozoic plate motions. *Reviews of Geophysics*, **36**(1), 27–78.

Mann, P., and Taira, P., 2004. Global tectonic significance of the Solomon Islands and Ontong Java Plateau convergent zone. *Tectonophysics*, **389**, 137–190.

McCaffrey, R., 1992. Oblique plate convergence, slip vectors, and forearc deformation. *Journal of Geophysical Research*, **97**, 8905–8915.

McCaffrey, R., 1997. Statistical significance of the seismic coupling coefficient. *Seismological Society of America Bulletin*, **87**, 1069–1073.

McCaffrey, R., 2008. Global frequency of magnitude 9 earthquakes. *Geology*, **36**(3), 263–266, doi:10.1130/G24402A.

McKenzie, D., and Jackson, J., 2012. Tsunami earthquake generation by the release of gravitational potential energy. *Earth and Planetary Science Letters*, **345–348**, 1–8.

Meschede, M., Zweigel, P., and Kiefer, E., 1999. Subsidence and extension at a convergent plate margin: evidence for subduction erosion off Costa Rica. *Terran Nova*, **11**(2/3), 112–117.

Moore, J. C., and Silver, E. A., 1987. Continental margin tectonics: Submarine accretionary prisms. *Reviews of Geophysics*, **25**(6), 1305–1312.

Oleskevich, D. A., Hyndman, R. D., and Wang, K., 1999. The updip and downdip limits to great subduction earthquakes: thermal and structural models of Cascadia, south Alaska, SW Japan, and Chile. *Journal of Geophysical Research*, **104**, 14,965–14,991.

Pacanovsky, K. M., Davis, D. M., Richardson, R. M., and Coblentz, D. D., 1999. Intraplate stresses and plate-driving forces in the Philippine sea plate. *Journal of Geophysical Research*, **104** (B1), 1095–1110.

Rogers, G. C., and Dragert, H., 2003. Episodic tremor and slip on the Cascadia subduction zone: the chatter of silent slip. *Science*, **300**(5627), 1942–1943.

Ruff, L., and Kanamori, H., 1980. Seismicity and the subduction process. *Physics of the Earth and Planetary Intriors*, **23**, 240–252.

Ruff, L., and Kanamori, H., 1983. Seismic coupling and uncoupling at subduction zones. *Tectonophysics*, **99**, 99–117.

Ruff, L., 1989. Do trench sediment affect great earthquake occurrence in subduction zones? *PAGEOPH*, **129**, 263–282.

Ruff, L. R., 1992. Asperity distributions and large earthquake occurrence in subduction zones. *Tectonophysics*, **211**, 61–83.

Saffer, D. M., and Tobin, H. J., 2011. Hydrogeology and mechanics of subduction zone forearcs: fluid flow and pore pressure. *Annual Review of Earth and Planetary Sciences*, **39**, 157–186.

Sage, F., Collot, J.-Y., and Ranero, C. R., 2006. Interplate patchiness and subduction-erosion mechanisms: evidence from depth migrated seismic images at the central Ecuador convergent margin. *Geology*, **34**(12), 997–1000, doi:10.1130/G22790A.1.

Scholl, D. W., von Huene, R., Vallier, T. L., and Howell, D. G., 1980. Sedimentary masses and concepts about tectonic processes at underthrust ocean margins. *Geology*, **8**, 564–568.

Scholz, C. H., and Campos, J., 1995. On the mechanism of seismic decoupling and back-arc spreading in subduction zones. *Journal of Geophysical Research*, **100**, 22,103–22,115.

Scholz, C. H., 2002. *The Mechanics of Earthquakes and Faulting*, 2nd edn. Cambridge: Cambridge University Press. 471 pp.

Scholz, C. H., and Campos, J., 2012. The seismic coupling of subduction zones revisited. *Journal of Geophysical Research*, **117**, B05310, doi:10.1029/2011JB009003.

Shemenda, A., 1992. *Subduction: Insights from Physical Modeling*. Netherlands: Kluwer Academic Publishers. Modern approaches in Geophysics. 215 pp.

Sleep, N., and Toksöz, M. N., 1971. Evolution of marginal basins. *Nature*, **33**, 548–550.

Stein, S., and Stein, C. A., 1996. Thermo-mechanical evolution of oceanic lithosphere: Implications for the subduction process and deep earthquakes. In Bebout, G. E., Scholl, D. W., Kirby, S. H., and Platt, J. (eds.), *Subduction: Top to Bottom*. Washington, DC: AGU. Geophysical monograph 96.

Tatsumi, Y., 1989. Migration of fluid phases and genesis of basalt magmas in subduction zones. *Journal of Geophysical Research*, **94**(B4), 4697–4707.

Turcotte, D. L., and Schubert, G., 1982. *Geodynamics: Applications of Continuum Physics to Geological Problems*. New York: Wiley. 450 pp.

Tatsumi, Y., and Kogiso, T., 2003. The subduction factory: Its role in the evolution of the Earth's crust and mantle. In Larter R. D., and Leat P. T.(eds.), Geological Society of London, Special Publication, pp. 55–80.

Uyeda, S., 1984. Subduction zones: their diversity, mechanisms and human impacts. *GeoJournal*, **8**(4), 381–406.

Vannucchi, P., Remitti, F. and Bettelli, G., 2008. Geological record of fluid flow and seismogenesis along an erosive subducting plate boundary. *Nature*, **451**, doi:10.1038/nature06486.

Vannucchi, P., Sage, F., Morgan, J. P., Remitti, F., and Collot, J.-Y., 2012. Toward a dynamic concept of the subduction channel at erosive convergent margins with impla-ications for interplate material transfer. *Geochemistry, Geophysics, Geosystems*, **13**, 1, doi:10.1029/2011GC003846.

von Huene, R., Langseth, M., Nasu, N., 1980. Summary, Japan trench transect. Leg 56–57. In *Scientific Party*, Init. Report DSDP, 56–57, part 1. Washington: U.S. Govt. Printing Office, pp. 473–488.

von Huene, R., and Lallemand, S., 1990. Tectonic erosion along convergent margins. *Geological Society of America Bulletin*, **102**, 704–720.

von Huene, R., and Scholl, D. W., 1991. Observations at convergent margins concerning sediment subduction, subduction erosion, and the growth of continental crust. *Reviews of Geophysics*, **29**(3), 279–316.

von Huene, R., Ranero, C. R., and Vannucchi, P., 2004. Generic model of subduction erosion. *Geology*, **32**(10), 913–916.

Wadati, K., 1928. Shallow and deep earthquakes. *Geophysical Magazine*, **1**, 162–202.

Wu, B., Conrad, C., Heuret, A., Lithgow-Bertelloni, C., and Lallemand, S., 2008. Reconciling strong slab pull and weak plate bending: the plate motion constraint on the strength of mantle slabs. *Earth and Planetary Science Letters*, **272**, 412–421, doi:10.1016/j.epsl.2008.05.009.

Cross-references

Accretionary Wedges
Active Continental Margins
Crustal Accretion

Driving Forces: Slab Pull, Ridge Push
Earthquakes
Intraoceanic Subduction Zone
Island Arc Volcanism, Volcanic Arcs
Lithosphere: Structure and Composition
Magmatism at Convergent Plate Boundaries
Marine Evaporites
Morphology Across Convergent Plate Boundaries
Ocean Margin Systems
Orogeny
Seamounts
Seismogenic Zone
Subduction Erosion
Wadati-Benioff-Zone
Wilson Cycle

SUBDUCTION EROSION

Martin Meschede
Institute of Geography and Geology, Ernst-Moritz-Arndt University, Greifswald, Germany

Definition

Subduction erosion defines a process in a subduction zone where material of the overriding upper plate is scraped off from its bottom by the downgoing lower plate. The offscraped material is transported downward within a subduction mélange into deeper parts of the subduction zone where it is added to the upper plate by melting and comprehension into magmatic melts of the volcanic arc.

Occurrences

About half of the convergent plate boundaries on Earth are dominated by the process of subduction erosion or at least non-accretion instead of accretion in an accretionary wedge (von Huene and Scholl 1991), which is the opposite process characterized by the formation of sedimentary accretionary wedges (Figure 1). Examples of such accretionary wedges are represented in SW Japan, Sumatra, large portions of the Gulf of Oman (Makran subduction zone), in the Lesser Antilles, and in smaller areas of western North America. During subduction erosion, rock material is scraped off from the bottom of the upper plate and transported downward with the subducting plate. Such basal erosion of the upper plate is known, e.g., from the Marianas, the Tonga Islands, Costa Rica, and Chile. In these cases, an accretionary wedge has not evolved or persisted only temporary before also being subducted. The identification of subduction erosion is thus difficult, since material to record the process does not evolve. Indications, therefore, have to be derived indirectly from other evidences.

Tectonic erosion in a subduction zone

It is assumed that at many modern convergent plate boundaries, accretion occurs at the tip of the upper plate and transport of at least parts of the incoming sedimentary material farther down into the subduction zone. Subduction erosion, on the other hand, transfers a considerable amount of material scraped off from the upper plate to greater depths. In contrast to accretion which leaves a geologic record, subduction erosion produces a hiatus making the process of erosion generally hypothetical (Von Huene, 1986). It is, therefore, necessary to determine secondary effects from subduction erosion such as subsidence of the convergent margin, extensional structures, retreat of the volcanic arc, or the content of some particular radiogenic isotopes.

Subducting pelagic sediments of the incoming plate as well as slope sediments from the convergent margin and volcanic ash layers from the volcanic arc volcanism nearby, oceanic crust of the incoming plate, and tectonically eroded material from the margin wedge of the overriding plate are mixed in a subduction mélange which is transported down into deeper parts of the subduction zone. A well-exposed example of such a subduction mélange can be studied at the Osa Peninsula in Costa Rica. This mélange zone which was already subducted to a depth of more than 10 km has been exhumed by the indentation of the Cocos Ridge since about 4 Myrs (Meschede et al., 1999a) for which we can study it today on land. The mélange is composed of highly sheared basaltic rocks and radiolarian cherts which belong to the Cretaceous Nicoya ophiolite complex occurring at the Central American land bridge, Cenozoic pelagic sediments and carbonatic rocks from seamounts and submarine ridges from the East Pacific basin, and probably of younger basaltic material from the subducting Cocos Plate. The subducted sedimentary material, however, is too light to be recycled into the mantle; it is instead incorporated into the melts feeding the volcanic arc volcanism above the subduction zone. Morris et al. (1990) realized that the occurrence of boron and the cosmogenic isotope beryllium-10 which has a half-life of about 1.5 Myrs indicates the subduction of sediments and their inclusion into melting processes above the subduction zone. ^{10}Be is constantly produced in the atmosphere by sun radiation and subsequently deposited in sediments. However, detection of ^{10}Be in recently formed volcanic rocks above subduction zones without an accretionary wedge indicates subduction of sediments and incorporation into magmatic melts (e.g., Morris et al., 2002). It is assumed that sediments containing a certain amount of ^{10}Be are involved in the melting process above the subduction zone after they were transported downward into deeper parts of the subduction zone. This process has to be performed within a short time span because after about 2–2.5 Myrs, a detection of ^{10}Be would not be possible anymore. The occurrence of ^{10}Be in recent volcanic rocks is thus taken as a reference for subduction of even the youngest sediments going into the subduction zone.

Snyder and Fehn (2002) determined an unusual ^{129}J/J ratio in young basaltic rocks from the Costa Rican volcanic arc. The ^{129}J/J ratio indicates minimum ages of iodine in fluids and is used to determine the age of the

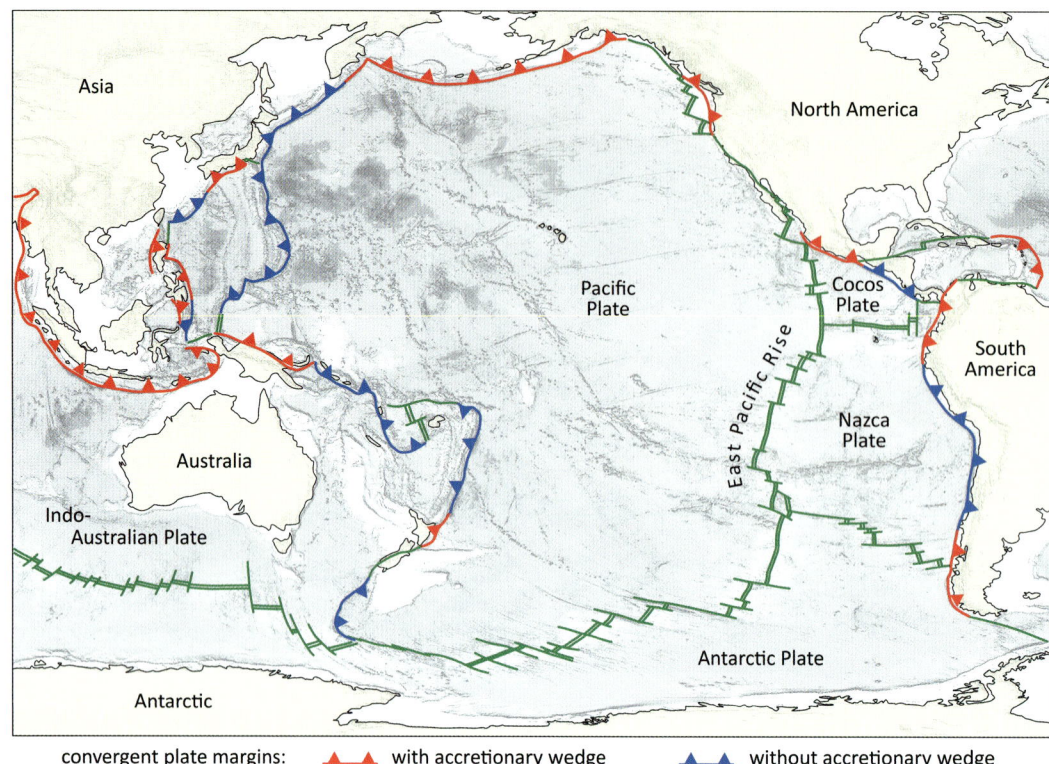

Subduction Erosion, Figure 1 Map of Circum-Pacific showing location of subduction zones with and without accretionary wedges (after von Huene and Scholl 1991; modified from Frisch et al. 2011).

subducted material. Some of the results, however, are too old to have originated from subducted sediments whose age is well determined. They interpret their results as probably caused by subduction erosion where old sedimentary rocks of the Cretaceous Nicoya ophiolite complex as part of the upper plate are scraped off by the subducting plate and transported down into the subduction zone together with younger sedimentary material. The material mix has been incorporated into the melt above the subduction zone, thus adding the old inherited ^{129}J/J ratios to the ^{129}J/J ratios of younger subducted sediments. Because of this mixing, the ^{129}J/J ratio of the calculated minimum ages becomes too old for being related only to the subducted sediments. The modification of this ratio may thus be taken as an indication for subduction erosion in the Costa Rica convergent margin.

In places where subduction erosion occurs, the outer ridge above a subduction zone as it is evolved in case of an accretionary wedge (e.g., the island of Barbados in the Lesser Antilles or the Mentawai Ridge in front of Sumatra) is not a significant feature because sedimentary material does not accumulate as in an accretionary wedge. The forearc basin commonly grades directly into the trench with little or no topographic rise between the two.

Subduction erosion is particularly effective when the lower plate has a roughly textured surface and is only covered by a thin sedimentary layer (Figure 2). As the lower plate enters the subduction zone and is bent, the upper part of the plate is extended; a series of trench-parallel horst and graben structures are subsequently formed. The roughness caused by these structures acts like a grater on the basal part of the upper plate (basal erosion, Figure 2), much like a cheese grater removes cheese from the base of the cheese block. Material that is scraped off the upper plate enters the subduction zone where it is transported downward into deeper parts of the subduction zone. This process is currently happening in the Mariana Arc.

Subduction of seamounts

Seamounts on the subducting plate create a somewhat different kind of subduction erosion. Single seamounts scrape off material from the frontal tip of the accretionary wedge along the upper plate and carry it into the subduction zone (frontal erosion). A superb example of this setting occurs off the Pacific coast of Costa Rica where seamounts currently buried by the tip of the upper plate leave obvious bulges followed by indentations in the slope of the upper plate. Subsequently, sediment gravity slides may fill up or obscure these indentations (Figure 3).

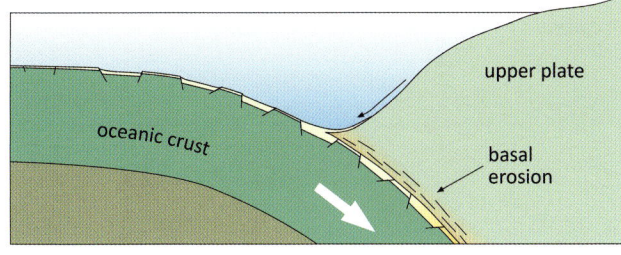
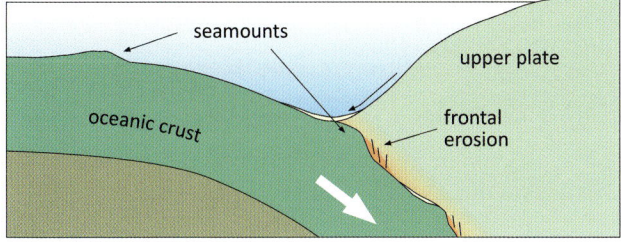

Subduction Erosion, Figure 2 Subduction erosion *upper*: at the base of the upper plate caused by subducted horsts and grabens; *lower*: at the front of the upper plate caused by subducted seamounts (modified from Frisch et al. 2011).

Migration of the subduction zone

Basal erosion on the upper plate causes the plate boundary to migrate toward the arc, which itself is moving backward in response to the migrating subduction zone. This process occurred along Costa Rica and in parts of the Andes between the Miocene and present (Figure 4). A tectonic mélange, consisting of parts of the scraped off upper plate, has evolved along the decollement zone. The mélange is transported into the depths of the subduction zone along with the plate being subducted.

Controlling factors

Subduction erosion can occur where the subduction zone angle is relatively low and there is a strong coupling between the two plates (Chile type) and where the subduction angle is high and subduction rollback occurs (Mariana type). Examples for the first case include Central America (see below) and parts of South America, and for the second case the Marianas, Tonga, and NE Japan. In Chile-type subduction, subduction erosion occurs because of the strong contact pressure caused by strong plate coupling. In Mariana-type subduction, erosion is caused by

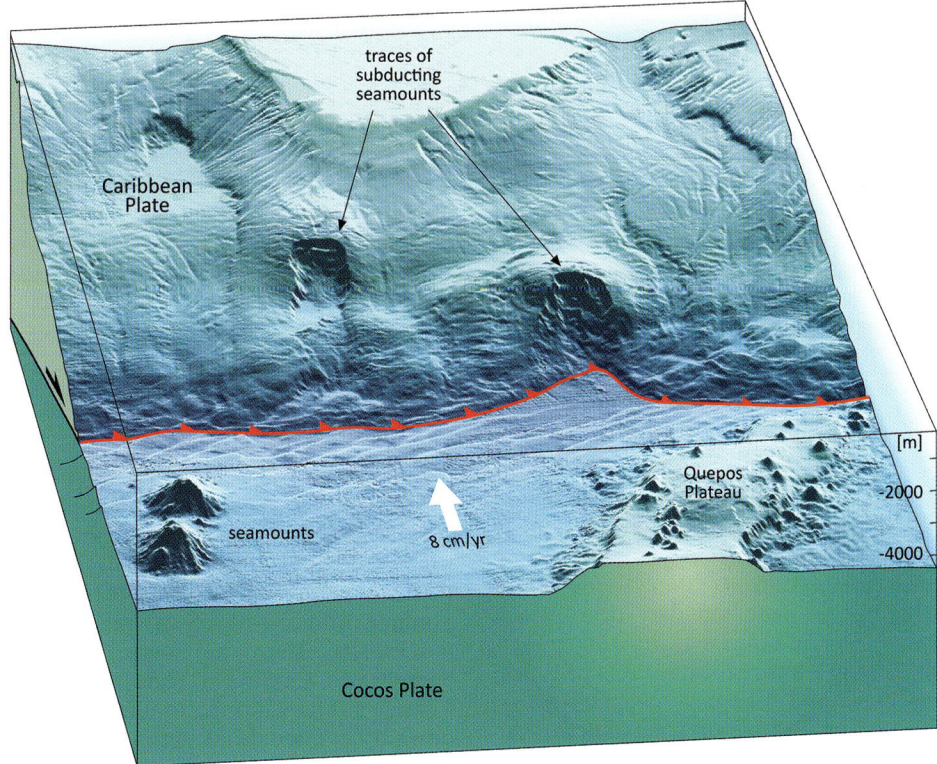

Subduction Erosion, Figure 3 Subduction of seamounts at the plate margin off Costa Rica (von Huene et al., 2000). The indentation of the seamount into the subduction zone leaves deep furrows in the upper plate (Figure after bathymetric data of the research vessel Sonne, made by Wilhelm Weinrebe, GEOMAR Kiel, Germany; modified from Frisch et al., 2011).

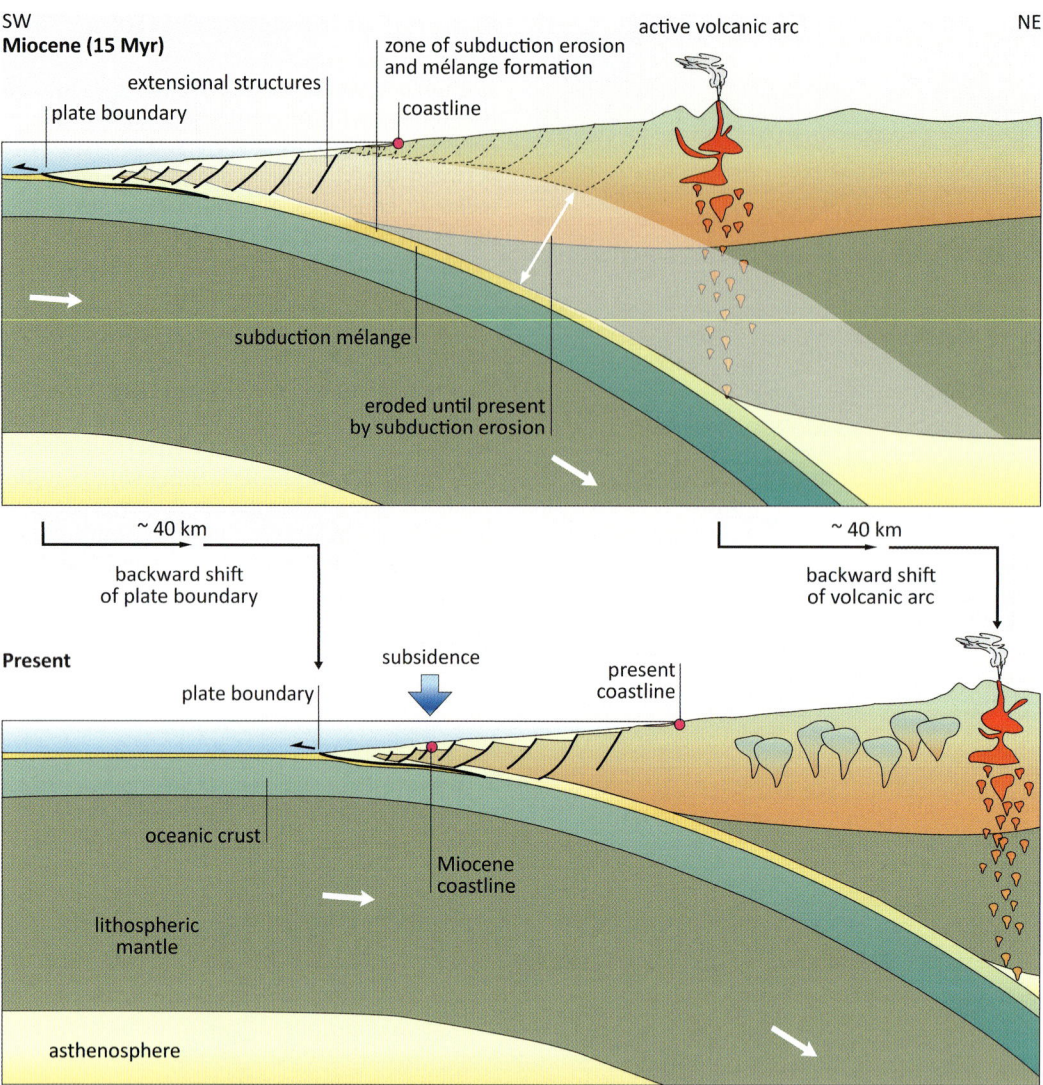

Subduction Erosion, Figure 4 Subduction erosion at the plate margin off Costa Rica where the Cocos Plate subducts beneath the Caribbean Plate (Meschede et al., 1999). Backward shift (toward the back-arc) of the volcanic arc is caused by subduction erosion of the upper plate since the Miocene (modified from Frisch et al., 2011).

the rough, sawtooth-like morphology on the subducting plate.

Pore fluids have important consequences in subduction zones and on subduction erosion. Pore fluids originate both from dewatering of hydrous minerals and from expulsion of water from sediments. Pore fluids in the high-stress zone along the plate-boundary decollement actually increase mobility and slippage in this zone by acting as a lubricant. During explosive expulsions of fluids, framework grains break and lose their connectiveness, and the rocks become ground by shearing. Such rocks are called cataclasites. Cataclasis (Greek: strong fracturing) is a common process associated with subduction erosion. Fluids that escape upward through the upper plate may generate mud volcanoes.

Record of subduction erosion

Subduction erosion can be proven only indirectly as the process occurs at depth and the eroded rocks become subducted to even greater depth. In contrast, accretionary wedges can be directly studied. The Central American subduction zone adjacent to Costa Rica is a well-studied example of an active plate margin being the site of subduction erosion. Here, the Cocos Plate is being subducted beneath the Caribbean Plate with a velocity of about 9 cm/year. The volcanic front has shifted 40 km backward (toward the Caribbean) during the last 15 Ma. This shift more or less corresponds to the amount of shortening formed by subduction erosion (Meschede et al., 1999a; Figure 4).

As subduction erosion thins the upper plate, the outer part of the wedge structure subsides. Seismic profiles along erosive convergent margins indicate the existence of tilted blocks and normal faults (Meschede et al., 1999b; Ranero and von Huene, 2000). Sediments deposited in the forearc area can be used to document this subsidence. For instance, benthic foraminifera, single-celled organisms which live within the sediment, indicate the approximate sedimentation depth based on their shell shape and design. If foraminifera and other sedimentological features in layers of decreasing age at a convergent margin indicate increasing water depth, subduction erosion is a likely explanation. At a drill site of the Ocean Drilling Program offshore Costa Rica (ODP Leg 170, Costa Rica convergent margin), shallow-water sediments occur at the deepest part of a hole, whereas near the top of the well core, deepwater sediments occur (Meschede et al., 2002). Accretionary sequences would produce the opposite pattern because the gradual buildup of the accretionary wedge causes a shallowing-upward sedimentation pattern over the wedge.

Bibliography

Frisch, W., Meschede, M., and Blakey, R. C., 2011. *Plate Tectonics*. Heidelberg/Dordrecht/London/New York: Springer, p. 212.

Meschede, M., Zweigel, P., Frisch, W., and Völker, D., 1999a. Mélange formation by subduction erosion: the case of the Osa mélange in southern Costa Rica. *Terra Nova*, **11**, 141–148.

Meschede, M., Zweigel, P., and Kiefer, E., 1999b. Subsidence at a convergent plate boundary: evidence for tectonic erosion off Costa Rica. *Terra Nova*, **11**, 112–117.

Meschede, M., Schmiedl, G., Weiss, R., and Hemleben, C., 2002. Benthic foraminiferal distribution and sedimentary structures suggest tectonic erosion at the Costa Rica convergent plate margin. *Terra Nova*, **14**, 1–12.

Morris, J., Leeman, W. P., and Tera, F., 1990. The subducted component in island arc lavas: constraints from Be isotopes and B–Be systematics. *Nature*, **344**, 31–36.

Morris, J., Valentine, R., and Harrison, T., 2002. ^{10}Be imaging of sediment accretion and subduction along the northeast Japan and Costa Rica convergent margins. *Geology*, **30**, 59–62.

Ranero, C., and von Huene, R., 2000. Subduction erosion along the Middle America convergent margin. *Nature*, **344**, 31–36.

Snyder, G. T., and Fehn, U., 2002. The origin of iodine in volcanic fluids: ^{129}I results from the Central American Volcanic Arc. *Geochimica et Cosmochimica Acta*, **66**, 3827–3838.

von Huene, R., 1986. To accrete or not accrete, that is the question. *Geologische Rundschau*, **75**, 1–15.

von Huene, R., and Scholl, D. W., 1991. Observations at convergent margins concerning sediment subduction, subduction erosion, and the growth of continental crust. *Review of Geophysics*, **29**, 279–316.

von Huene, R., Ranero, C. R., and Weinrebe, W., 2000. Quaternary convergent margin tectonics off Costa Rica, segmentation of the Cocos Plate, and Central American volcanism. *Tectonics*, **19**, 314–334.

Cross-references

Accretionary Wedges
Foraminifers (Benthic)
Mud Volcano
Ocean Drilling
Seamounts
Subduction
Subsidence of Oceanic Crust

SUBMARINE CANYONS

William W. Hay
Department of Geological Sciences, University of Colorado at Boulder, Estes Park, CO, USA

Definition

A submarine canyon is a valley cut into the seafloor of the continental shelf, forming a notch in the shelf break and continuing as a steep-sided valley down the continental slope.

History

The discovery of submarine canyons can be traced back to the late nineteenth century. James Dana reported on soundings revealing the Hudson submarine river channel in 1890.

The distribution of submarine canyons

Small submarine canyons are common along active continental margins. They are more widely separated along passive margins. Some submarine canyons have their heads on the continental shelf and are directly associated with large rivers. Others have their head near the break in the slope and have no obvious association with existing rivers.

Canyons cut into continental slopes can be traced down to depths of 2 km or more below sea-level. They can continue as submarine channels on the continental rise extending for hundreds of kilometers.

Origin of submarine canyons

The origin of submarine canyons was proposed to be from flows of dense turbid waters (later "turbidity currents") by R. A. Daly (1936). As isostatic response to sea-level change associated with glaciation and deglaciation became better understood, it became evident that rivers might have cut the canyons into the shelf break at sea-level lowstands. However, the existence of turbidity currents capable of eroding the continental slope was considered hypothetical until the investigations of the effects of the Grand Banks earthquake of 1929 on the seafloor of Nova Scotia.

Bibliography

Daly, R. A., 1936. Origin of submarine "canyons". *American Journal of Science*, **31**, 401–420.

Dana, J. D., 1890. Long Island sound in the quaternary era, with observation on the Hudson submarine river channel. *American Journal of Science*, **40**, 425–437.

Emery, K. O., and Uchupi, E., 1984. *The Geology of the Atlantic Ocean*. New York: Springer, 1050 pp. + Map set.

Heezen, B. C., and Ewing, M., 1952. Turbidity currents and submarine slumps and the 1929 Grand Banks earthquake. *American Journal of Science*, **250**, 849–873.

SUBMARINE LAVA TYPES

Michael R. Perfit[1] and Samuel Adam Soule[2]
[1]Department of Geological Sciences, University of Florida, Gainesville, FL, USA
[2]Woods Hole Oceanographic Institution, Woods Hole, MA, USA

Introduction

Variations: Pillow lava, lobate flow, sheet flow

Most of the Earth's crust surface is comprised of lava – the solid igneous rock created by the extrusion, cooling, and crystallization of magma generated within the Earth's mantle or by partially melting preexisting crust. Lava is the nonexplosive extrusive product of eruptions from volcanoes and rifts in a variety of tectonic environments. Volcanism in the oceans occurs in a number of different tectonic settings including divergent plate boundaries (mid-ocean ridges), convergent plate boundaries (oceanic island arcs and back-arcs), and intraplate settings (hot spot-generated ocean islands and oceanic plateaus). These settings produce eruptive products that are geochemically and physically distinct. At arcs and back-arcs, chemically evolved ($SiO_2 > 52$ wt%) and volatile-rich ($H_2O > 2$ wt%) magmas can lead to explosive eruptions, although effusive eruptions including flows and domes are common. At ocean islands and oceanic plateaus, magmas contain less SiO_2 and are characteristically more mafic (i.e., contain more MgO and FeO) and less volatile rich ($H_2O < 1$ wt%). Consequently, quiescent effusive eruptions are dominant though some explosive eruptions do occur. At ocean islands, eruptions are typically focused at volcanoes leading to the construction of edifices that may eventually breach the sea surface. At oceanic plateaus volumetrically significant outpourings of lava produce thick, aerially extensive platforms. Outside of large oceanic plateaus, mid-ocean ridges produce the majority of erupted lava in the oceans, representing as much as two-thirds of the Earth's annual volcanic output. Mid-ocean ridge magmas are broadly mafic ($SiO_2 \sim$ 50 wt%, MgO $\sim$ 8 wt%) and volatile poor ($H_2O \sim$ 0.25 wt%) and define a geochemical class appropriately named mid-ocean ridge basalt (MORB) that is thought to represent partial melting of an incompatible element-depleted mantle source. Most lava erupted at mid-ocean ridges (MOR) forms lava flows that have morphologic and textural characteristics distinct from those at subaerial volcanoes due to differences in their physical environment of eruption and chemical characteristics. Submarine eruptions are rarely observed because of their great depths and remote locations. However, recent high-resolution mapping and sampling of MOR spreading centers and seamounts together with studies of historic eruptions have provided important information about the various types of lavas that exist in the oceans and the mechanisms that control their emplacement.

Lava flows on mid-ocean ridges

A great majority of lavas erupted on MORs are basaltic in composition and distinctive enough from basalts erupted in other environments that they are known as ocean floor basalts or mid-ocean ridge basalts (MORB). They are characterized by having relatively low silica contents (<52 wt%), high magnesium and iron concentrations, and very low alkali and volatile contents (most notably H_2O and CO_2). In some ridge environments such as overlapping spreading centers and ridge-transform intersections, a greater proportion of cooler, more chemically evolved varieties such as ferroandesites, dacites, and rhyodacites that have higher silica and alkalis and lower magnesium and calcium along with higher concentrations of volatiles have been recovered (e.g., Wanless et al., 2010). MORB exhibit a spectrum of lava flow morphologies (Figure 1) that range from bulbous, pillow shapes to flatter oblate forms known as "lobates" and even to flat, ropy, and folded types of sheets (Ballard et al., 1975; Lonsdale, 1977; Francheteau and Ballard, 1983; Bonatti and Harrison, 1988; Embley et al., 1990; Embley and Chadwick, 1994; Perfit and Chadwick, 1998).

Significant differences in lava morphology, axial structure, and frequency and distribution of magmatic activity along MOR vary with the rate of spreading – a reflection of magma supply rate at ridge crests (Bonatti and Harrison, 1988; Perfit and Chadwick, 1998; Sinton et al., 2002). Slowly diverging plate boundaries, which have relatively lower magma supply rates and volcanic outputs, are dominated by pillow lava, which tends to construct hummocks (<50 m high, <500 m diameter), hummocky ridges (1–2 km long), and small circular seamounts (10–100s m high and 100s–1,000s m in diameter). These constructs commonly coalesce to form axial volcanic ridges (AVR) along the valley floor of slow-spread axial rift zones (Smith and Cann, 1993; Searle et al., 2010; Yeo et al., 2012). At fast-spreading MORs where magma supply is greater, volcanic output is dominated by lobate flows and sheet flows. Sheet flows are further subdivided based on their surface texture as flat, ropy/folded, and jumbled/hackly (e.g., Engels et al., 2003; Soule et al., 2005; Garry et al., 2006; Fundis et al., 2010). Intermediate-spreading rate ridges have mixes of lava types due to periods of low magmatic output punctuated by more voluminous magmatism (Smith et al., 1994; Stakes et al., 2006) or variable magmatic budgets due to differences in along-axis melt supply (Colman et al., 2012). It has also been documented on the fast-spreading

Submarine Lava Types, Figure 1 MORB exhibit a wide variety of lava flow morphologies: (**a**) Pillow lava with characteristic bread crust surface texture. Edges are decorated with small, glassy, finger-like protuberances. (**b**) Partial vertical exposure of sheet flow exhibiting very flat upper, lineated surface overlying another sheet flow and lobate flow. (**c**) Young, glassy lobate flow of 1992 East Pacific Rise flow in contact with older pillow lava on right side. (**d**) Folded to jumbled sheet flow. Scale across bottom of image (**b**) is ~6 m; others are all ~3 m.

East Pacific Rise (EPR) that the relative proportion of flow types correlates with segmentation of the melt lens such that the abundance of pillows increases at segment ends where magma supply is thought to be lower (White et al., 2002).

High hydrostatic pressure and low volatile contents in MORB limit explosive eruptions that would produce submarine pyroclastic deposits. However, accumulations of gasses in sub-seafloor magma chambers are considered a mechanism that can drive mildly explosive eruptions that produce volcaniclastic ash deposits and delicate bubble wall and curved fragments of glass know as "limu o Pele" in the deep sea (Clague et al., 2003; Sohn et al., 2008; Clague et al., 2009; Barreyre et al., 2011; Pontbriand et al., 2012). Explosive eruptions appear to be more common at ocean islands (i.e., hotspots and in submarine arcs (e.g., subduction settings)) likely due to higher initial volatile contents and, in some cases, higher magma viscosity (e.g., Wright et al., 2006; Clague et al., 2011; Helo et al., 2011; Resing et al., 2011).

Lava morphology types

The primary differences between submarine lava flow morphologies (pillows, lobates, and sheets) are the size of individual flow units and their interconnectivity. Pillow flows extrude as ~1-m-scale individual spherical or cylindrical tubes of lava that develop a glassy crust on all sides, generally inhibiting coalescence with neighboring pillows (Moore, 1975). Pillow flows and mounds are produced by the piling up of individual pillow lava lobes whereby the newest pillows are erupted from the top of the stack and flow outward a limited distance before freezing, resulting in the formation of steep-sided mounds, ridges, and haystacks (Figure 2a–c) (Chadwick and Embley, 1994; Perfit and Chadwick, 1998; Embley et al., 2000). Although, by necessity, some continuous pathways must exist within a pillow flow to deliver lava from the seafloor vent to the top or sides of an actively forming mount, the high relief of pillow flows and absence of collapse features that span multiple pillows (although individual pillows may collapse and drain) suggest limited hydraulic connectivity.

Submarine Lava Types, Figure 2 Pillow lavas forming steep-sided mounds, ridges, and haystacks. (**a**) Upper part of the "new mounds" pillow mounds on northern cleft segment (Juan de Fuca Ridge). (**b**) Haystack off axis on southern cleft segment showing tubular forms of pillow lavas. (**c**) Pillow mound with well-formed pillow and elongate tubes formed from breakouts from pillows formed as the mound grew. (**d**) Large pillows and tubes formed from high-silica lavas. The pillows are > 2 m in diameter and have well-defined surficial bread crust textures. Field of view across bottom ~6 m in each photo.

In rare MOR environments where high-silica lavas have erupted, unusually large (>2 m diameter) pillows or glassy angular blocks are observed. These pillows may exhibit flow banding, have well-defined surficial bread crust textures (Figure 2d), and are commonly highly vesicular (Stakes et al., 2006).

Lobate flows are comprised of individual flow units that are similar to pillows but larger in width, several to >10 m. The lobes develop a quenched crust on their top and sides but are emplaced rapidly enough they can coalesce into one interconnected flow (Figure 3a). Lobate flows commonly inflate (similar to terrestrial inflated pahoehoe sheet flows (Hon et al., 1994)) producing relatively smooth and flat upper crusts that may later collapse if the molten interior of the flow subsides or trapped vapor from heated seawater condenses beneath them (Perfit et al., 2003). Flows that become ponded and later collapse are sometimes referred to as "lava lakes" but lack a direct connection to a crustal magma supply and are more akin to lava ponds described in subaerial basaltic volcanic terranes (e.g., Wilson and Parfitt, 1993). Inflated lobate flows and lava lakes commonly display "lava pillars" in their collapsed interiors (Figures 3b and 6b; Ballard et al., 1979; Francheteau et al., 1979; Gregg and Chadwick, 1996; Chadwick et al., 2013).

Sheet flows erupt with an upper surface of uniform height, but they are typically modified into folded, lineated, jumbled, or hackly morphologies (Figure 4a–c) that reflect a variety of processes that may include flow through restrictions, flow along channel margins, and repeated inflation/deflation episodes. Sheet flows are typically laterally extensive but can also be confined to axial valleys or the axial summit trough (AST). In some locations on fast-spread MOR, sheet flows extend as far as ~3 km from the AST as a result of rapid, high-volume flow that forms lava channels (Sinton et al., 2002; Soule et al.,

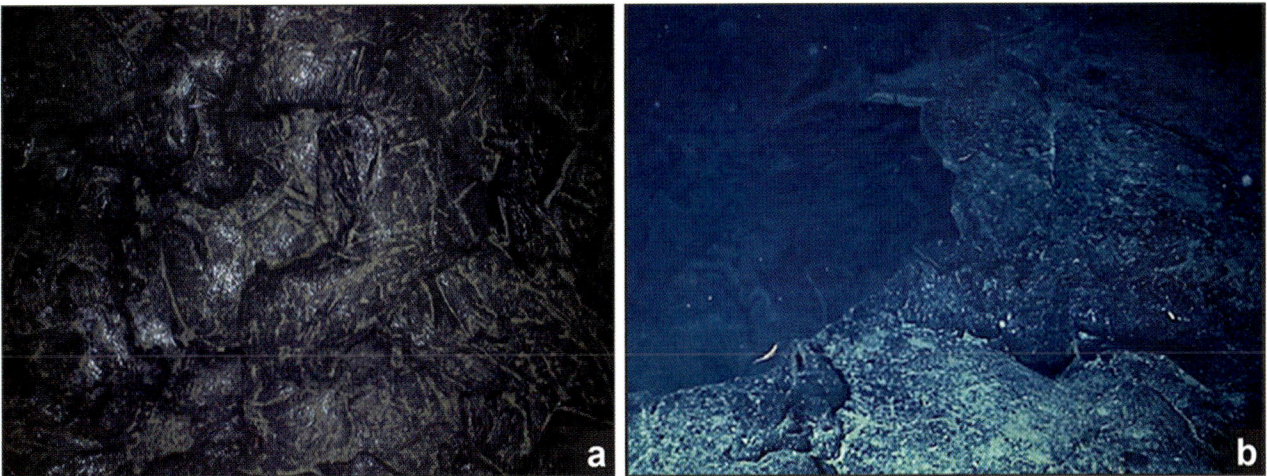

Submarine Lava Types, Figure 3 Lobate flows. (**a**) Down-looking image of glassy lobate flow erupted on the East Pacific Rise in 2005–2006. The flow was emplaced rapidly enough that individual lobes coalesce into one interconnected flow. (**b**) Inflated lobate lava showing characteristic collapse features with thin, smooth upper crust, with talus and a "lava pillar" in the evacuated interior. Scale across bottom of image (**b**) is ~3 m.

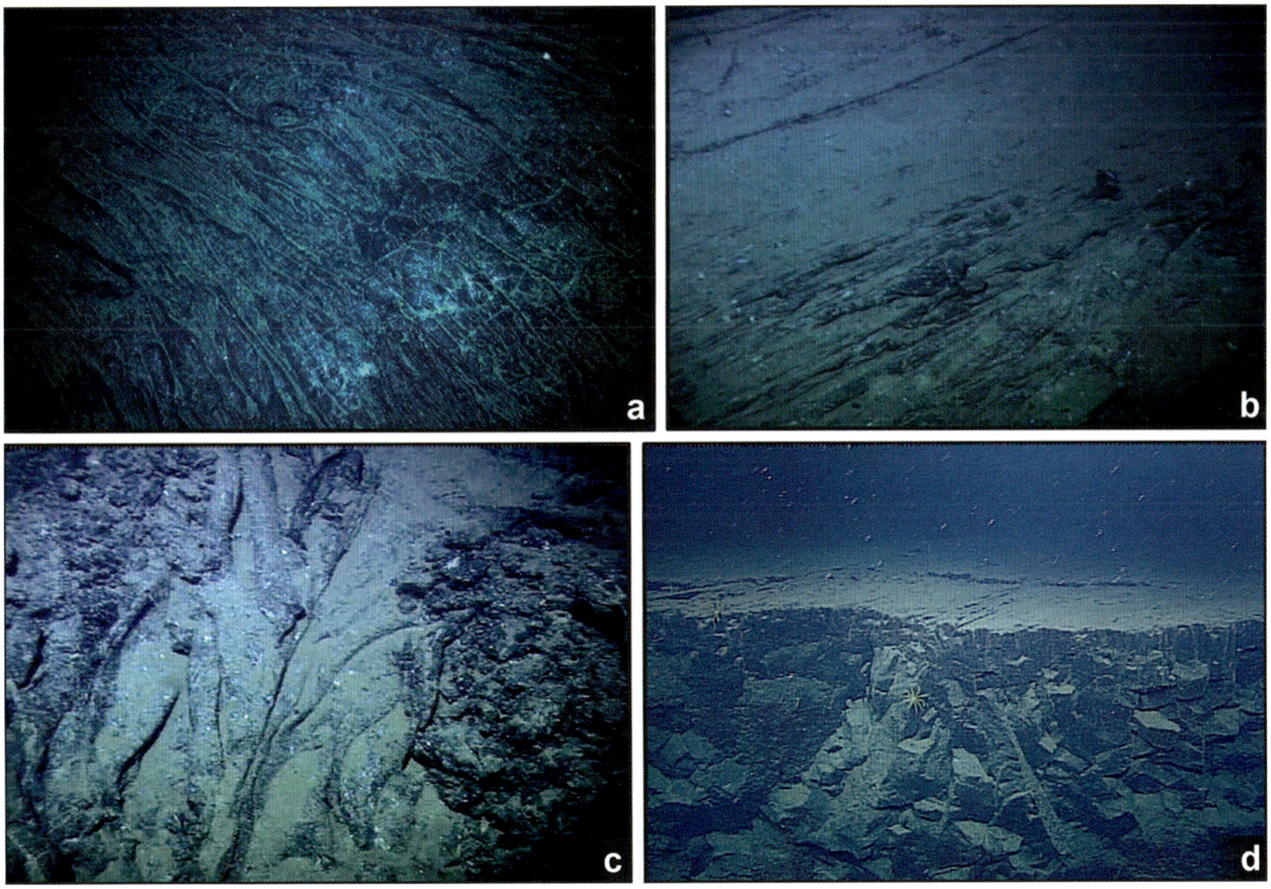

Submarine Lava Types, Figure 4 Various morphologic types of sheet flows. (**a**) Young-looking, glassy, lineated sheet flow (down-looking picture). (**b**) Lineated sheet flow with thin sediment cover showing some folding at edge of flow. (**c**) Lightly sedimented folded sheet flow transitioning to hackly flow on margins of the flow. (**d**) Vertical exposure of sheet flow. Thicknesses can range from a few tens of centimeters to a few meters and likely transition into what are known as "massive flows" (>3 m thick) that may have formed as lava ponds. Horizontal scale in images (**d**) ~6 m and ~2–3 m in the others.

2005, 2007). Vertical sections of oceanic crust exposed along fault scarps (e.g., Karson et al. 2002a, Karson et al. 2002b; Stakes et al., 2006) or revealed by drilling during the Integrated Ocean Drilling Program (IODP) (e.g., Tominaga and Umino, 2010) show that some sheet flows can attain thicknesses of a few meters and likely transition into what are known as "massive flows" (>3 m thick) that may have formed as lava ponds (Figure 4d). A massive flow known to be at least 75 m thick drilled at IODP site 1,256 is proposed to represent a flow that originated at the axis and ponded in a depression more than 3 km off axis (Tominaga et al., 2009).

Variations in lava flow type are observed over short distances along mid-ocean ridges. On most MOR, lavas extrude along linear eruptive fissures commonly found within the axial summit trough and axial valley that can be a few to tens of kilometers long. The eruptions can be sourced from melt lenses directly below the fissures (Carbotte et al., 2013) or from magmatic dikes that travel along the length of ridge axes (Dziak et al., 1995; Delaney et al., 1998). At fast-spreading ridges, it has been noted that lava at the ends of magmatic segments produces markedly more pillows than at the middle of the segments (White et al., 2000). A similar increase in pillows has been observed at the ends of eruptive fissures, which commonly coincide with magmatic segment boundaries (Sinton et al., 2002; Fundis et al., 2010). At slow- and intermediate-spreading rate ridges, lavas erupted near the segment center where ascending melts are focused within the shallow crust have greater abundance of lobate and sheet flows relative to pillows than toward the segment ends, where melts are transported through lateral diking to eruption sites (Cann and Smith, 2005; Behn et al., 2006). At the intermediate-spreading rate Galapagos Spreading Center, melt supply systematically increases along axis toward the intersection with the Galapagos hot spot, and the relative abundances of lobate and sheet flows to pillow flows are positively correlated with inferred melt supply (e.g., Behn et al., 2004; White et al., 2008; Colman et al., 2012).

Controls on morphology

Early studies suggested a variety of intrinsic and extrinsic parameters may influence the development of distinct flow morphologies including lava rheology (i.e., composition, temperature, and crystallinity that all control viscosity), preexisting slope and terrain, and eruption rates (e.g., Bonatti and Harrison, 1988). Included in these studies are laboratory analogue experiments involving the extrusion of molten wax into a cold liquid under various physical conditions that have greatly increased our understanding of the significance of submarine flow morphologies (Fink and Griffiths, 1992; Griffiths and Fink, 1992; Gregg and Fink, 2000; Garry et al., 2006). These experiments have clearly shown that submarine flow morphology is a function of flow rate as well as cooling rate and lava viscosity.

Detailed investigations of historic eruptions suggest that preexisting slope and terrain do not significantly affect the morphology of the flows on mid-ocean ridges where the slopes are low (<5°, e.g., Kurras et al., 2000), although other studies have found that correlations are present where slopes are higher (Gregg and Fink, 2000). MORB viscosity, largely controlled by lava composition and eruption temperature, does not vary significantly along the global MOR (Rubin and Sinton, 2007). In addition, petrologic and geochemical analyses of lava samples indicate that lava rheology remains relatively constant throughout the course of an eruption (e.g., Soule et al., 2012). Cooling rate, controlled by the temperature difference between seawater and lava, is likewise similar for nearly all MOR eruptions. As a result, flow morphology can be thought of as a proxy for the rate of lava extrusion or magma supply rate. Pillow lavas form at the slowest flow rates, lobate flows at intermediate rates, and sheet flows at the highest flow rates.

Intra-flow variations

Variations in flow rate during emplacement can lead to the formation of multiple morphologies within a single submarine lava flow. Transitions between different surface morphologies can be gradational due to changing flow conditions or abrupt due to superposition of separate flow units (e.g., Cann and Smith, 2005; Soule et al., 2005; Fundis et al., 2010; Chadwick et al., 2013). Gradational transitions are observed between sheet and lobate flows and lobate and pillow flows but not between pillow and sheet flows (Figure 5). Transitions between sheet flows of different varieties are common. For example, it is common for a smooth sheet flow to transition into a folded flow – with a surface characterized by small folds – as the partially cooled and more brittle surface deforms due to continued flow of the underlying molten lava. Folded flows commonly transition into jumbled or hackly flows as the folded surface continues to deform, ultimately breaking the glassy surface crust. The gradation from sheet to hackly flows typically occurs over distances of 1–10 m (Fundis et al., 2010) and tends to occur on the margins of lava channels where there are significantly higher rates of shear (Soule et al., 2005). Transitions from lobate to pillow lavas occur over short distances (<1 m) and are characterized by pillows branching out of well-defined lobate flows. Lobate flows near these transitions have more surface decoration and have distinct lobes that are minimally coalesced.

Overall, the general trend in the distribution of flow types normal to the ridge axis is that lava morphologies associated with higher flow rates (i.e., sheet, hackly, and flat-lying lobate flows) occur near the eruptive source and transition into morphologies that are typically associated with lower flow rates toward the termini of the flow. Collapse pits and collapsed lava lakes and ponds are common features of submarine lava flows at fast- and superfast-spreading MORs that form in response to flow

Submarine Lava Types, Figure 5 Contact relations between different flow types. (**a**) Terminus of a well-decorated pillow flow on an older slightly folded sheet flow. (**b**) Pillow mound, possibly formed at an eruptive fissure to the *upper left*, overlying a sheet flow. Horizontal scale in (**a**) is ~3 m and in (**b**) is ~5 m.

inflation followed by post-eruption drain back (Fornari et al., 1998; Perfit and Chadwick, 1998; Sinton et al., 2002). Collapse pits are typically concentrated near eruptive vents and decrease in abundance with distance from eruptive centers (Engels et al., 2003; Fundis et al., 2010; Figure 6a). Larger collapse areas form where lava accumulates in preexisting depressions in the terrain (e.g., fault bounded, between inflated flows, or in preexisting AST) and drains before solidifying causing the upper crust to founder, leaving a talus-strewn floor and lava pillars that commonly support archways in the upper surface (Gregg and Chadwick, 1996; Gregg et al., 1996; Figure 6b). These pillars form where vapor phase and hydrothermal fluids flow through and exit the ponded flows and provide structural buttresses after roof collapse (Engels et al., 2003). Features such as lava drips and cusps on the undersides of lava crusts also indicate that vapor formed by lava-seawater interactions is present within lobate flows, allowing lava drips to form in their vapor-supported cavities and ultimately can result in the development of collapse pits as the vapor cools and condenses (Chadwick, 2003; Perfit et al., 2003; Soule et al., 2006, 2012; Figure 6c).

Lavas in subduction zone environments

Discoveries and studies of deep-sea eruptions in subduction zone settings (volcanic arcs and back-arc basins) provide another perspective on submarine lava types because the compositions and mode of eruption can be quite different from those at MOR (Chadwick et al. 2005; Embley et al., 2006; Chadwick et al. 2008; Deardorff et al., 2011). In these settings, eruptions can occur along linear fissures, in the case of back-arc spreading centers, and commonly form localized, long-lived vents that result in the development of seamounts in submarine arc environments. Although much of the volcanism along the Lau Basin back-arc spreading centers in the southwest Pacific is quite similar to that along MOR, active submarine eruptions in the northeast Lau Basin, associated with the Tonga-Kermadec arc, were dominated by dramatic pyroclastic and active effusive pillow lava eruptions (Clague et al., 2011; Resing et al., 2011; Rubin et al., 2012). Volcanic activity at the submarine volcano NW Rota-1 in the Mariana arc is characterized by the repeated buildup and collapse of a cinder cone at the vent. Both eruptions are considered to be Strombolian-like (i.e., driven by magmatic gas rising through the eruptive conduit) producing mildly explosive eruptions that generate volcaniclastic material including scoria. Detailed studies and real-time observations indicate that eruptions in subduction zone environments are generally more explosive (generating large volumes of volcaniclastics) and produce more continuous, low-effusion-rate eruptions that construct isolated composite cones with unstable slopes (Rubin et al., 2012). In large part these characteristics are due to the inherently higher volatile content of the magmas and their higher viscosity due to greater silica content and crystallinity.

Conclusions

Submarine volcanism on MORs is dominated by quiescent eruptions due to high hydrostatic pressure and low volatile contents. Submarine lavas differ from their subaerial counterparts in that their outer surfaces cool rapidly in the cold seawater that surrounds them. With variations in eruption rate, three morphologic types are commonly produced including pillows, lobates, and sheets. Eruptions in subduction zone environments can be explosive due to the more viscous and volatile-rich nature of arc lavas, and consequently pillows and volcaniclastics are the dominant effusive types of deposits. However, the discovery of widespread volcaniclastic deposits along some MORs (Gakkel Ridge – Sohn et al., 2008; EPR and Juan de Fuca Ridge – Clague et al., 2009) suggests that there are exceptions to this model.

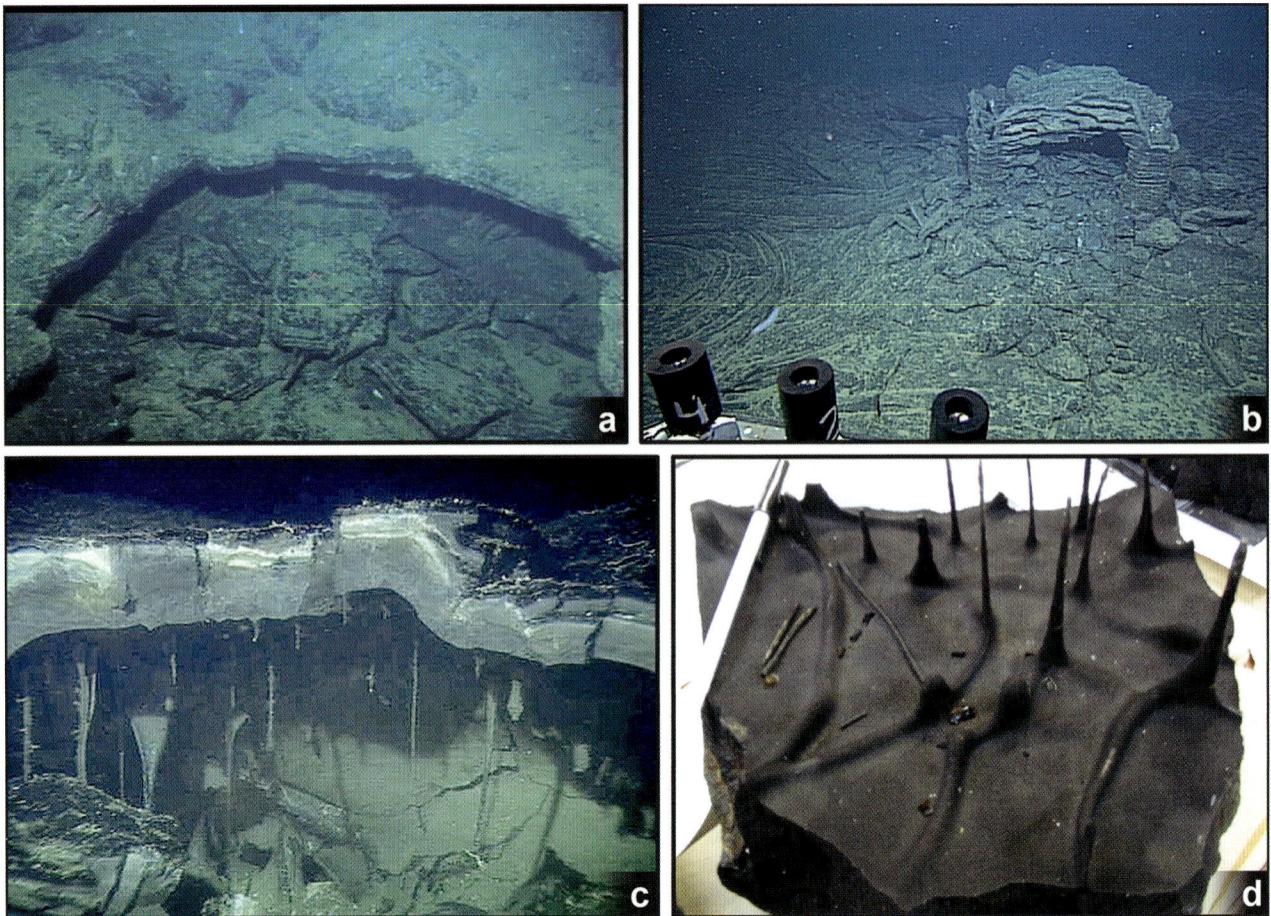

Submarine Lava Types, Figure 6 Submarine features associated with lava collapse and possible interaction of lava flows with a seawater vapor phase. (**a**) Collapsed lobate flow showing pieces of thin roof crust lying in collapse pit. These features are typically concentrated near eruptive vents and axial summit troughs (*AST*) and decrease with distance from the axis. Horizontal scale ~2 m (**b**) Portion of collapsed lava lake with remnant lava pillars and archway surrounded by talus formed from collapsed roof. To the *left* is a largely undisturbed sheet flow with lava whorl. Horizontal scale ~10 m. (**c**) Closeup view of lava drips on underside of lobate roof crust in place on the edge of northern East Pacific Rise AST. Vapor that filled the cavity had to be hot enough to allow molten lava to drip from the upper crust (see Perfit et al., 2003). (**d**) Underside of lobate roof crust showing delicate lava drips and rills formed in a vapor-supported cavity as shown in image (**c**). Pen in *upper left* of image for scale.

The observation that different flow morphologies are dominant at different spreading rates leads to the conclusion that fast-spreading ridges typically have high-effusion-rate eruptions, whereas eruptions at slow-spreading ridges have lower effusion rates. This suggests that dikes may be intruded at higher magma pressures at fast-spreading ridges than at slow-spreading ridges and is consistent with the concept that magmatism dominates over tectonism at fast ridges. At intermediate-spreading rate ridges like the Juan de Fuca Ridge, where the ridge may go through cycles dominated by tectonism or magmatism, the eruption style may vary with time.

In summary, magma supply related to overall spreading rate exerts a major control on eruption rates and the consequent types of lava morphologies that form. It should be kept in mind, however, that very few ridges have been sufficiently well mapped or eruptive events documented in the detail in order to fully understand the volcanic processes occurring in the deep sea.

Bibliography

Ballard, R. D., Bryan, W. B., Heirtzler, J. R., Keller, G. R., Moore, J. G., and van Andel, T. H., 1975. Manned submersible observations in the famous area. *Science*, **190**, 103–108.

Ballard, R. D., Holcomb, R. T., and van Andel, T. H., 1979. The Galapagos Rift at 86°W: sheet flows, collapse pits, and lava lakes of the rift valley. *Journal of Geophysical Research*, **84** (B10), 5407–5422.

Barreyre, T., Soule, S. A., and Sohn, R. A., 2011. Dispersal of volcaniclasts during deep sea eruptions: settling velocities and entrainment in buoyant seawater plumes. *Journal of Volcanology and Geothermal Research*, **205**(3), 84–93. ISSN 0377-0273. 10.1016/j.jvolgeores.2011.05.006.

Behn, M. D., Sinton, J. M., and Detrick, R. S., 2004. Effect of the Galapagos hotspot on seafloor volcanism along the Galapagos

Spreading Center (90.9–97.6 W). *Earth and Planetary Science Letters*, **217**(3), 331–347.

Behn, M., Buck, W., and Sacks, I., 2006. Topographic controls on dike injection in volcanic rift zones. *Earth and Planetary Science Letters*, **246**(3–4), 188–196, doi:10.1016/j.epsl.2006.04.005.

Bonatti, E., and Harrison, C. G. A., 1988. Eruption styles of basalt in oceanic spreading ridges and seamounts: effect of magma temperature and viscosity. *Journal of Geophysical Research*, **93**(B4), 2967–2980.

Cann, J. R., and Smith, D. K., 2005. Evolution of volcanism and faulting in a segment of the Mid-Atlantic Ridge at 25° N. *Geochemistry, Geophysics, Geosystems*, **6**(9). doi:10.1029/2005GC000954.

Carbotte, S. M., Marjanović, M., Carton, H., Mutter, J. C., Canales, J. P., Nedimović, M. R., and Perfit, M. R., 2013. Fine-scale segmentation of the crustal magma reservoir beneath the East Pacific Rise. *Nature Geoscience*, **6**(10), 866–870.

Chadwick, W. W., Jr., 2003. Quantitative constraints on the growth of submarine lava pillars from a monitoring instrument that was caught in a lava flow. *Journal of Geophysical Research*, **108**(B11), 2534, doi:10.1029/2003JB002422.

Chadwick, W. W., Jr., and Embley, R. W., 1994. Lava flows from a mid-1980s submarine eruption on the cleft segment, Juan de Fuca Ridge. *Journal of Geophysical Research*, **99**(B3), 4761–4776.

Chadwick, W. W., Jr., Embley, R. W., Johnson, P. D., Merle, S. G., Ristau, S., and Bobbitt, A., 2005. The submarine flanks of Anatahan Volcano, commonwealth of the Northern Mariana Islands. *Journal of Volcanology and Geothermal Research*, **146**(1), 8–25.

Chadwick, W. W., Jr., Cashman, K. V., Embley, R. W., Matsumoto, H., Dziak, R. P., de Ronde, C. E. J., Lau, T.-K., Deardorff, N., and Merle, S. G., 2008. Direct video and hydrophone observations of submarine explosive eruptions at NW Rota-1 Volcano, Mariana Arc. *Journal of. Geophysical. Research.* **113**, B08S10, doi:10.1029/2007JB005215.

Chadwick, W. W., Jr., Clague, D. A., Embley, R. W., Perfit, M. R., Butterfield, D. A., Caress, D. W., and Bobbitt, A. M., 2013. The 1998 eruption of Axial Seamount: new insights on submarine lava flow emplacement from high-resolution mapping. *Geochemistry, Geophysics, Geosystems*, **14**(10), 3939–3968.

Clague, D. A., Davis, A. S., and Dixon, J. E., 2003. Submarine strombolian eruptions along the Gorda mid-ocean ridge. In *Explosive Subaqueous Volcanism*. Washington, DC: American Geophysical Union. Monograph, Vol. 140, pp. 11–128.

Clague, D. A., Paduan, J. B., and Davis, A. S., 2009. Widespread strombolian eruptions of midocean ridge basalt. *Journal of Volcanology and Geothermal Research*, **180**, 171–188, doi:10.1016/j.jvolgeores.2008.08.007.

Clague, D. A., Paduan, J. B., Caress, D., Thomas, H., and Chadwick, W. W., Jr., 2011. Volcanic morphology of West Mata Volcano, NE Lau Basin, based on high-resolution bathymetry and depth changes. *Geochemistry, Geophysics, Geosystems*, **12**, Q0AF03, doi:10.1029/2011GC003791.

Colman, A., Sinton, J. M., White, S. M., McClinton, J. T., Bowles, J. A., Rubin, K. H., Behn, M. D., Cushman, B., Eason, D. E., Gregg, T. K. P., Gronvold, K., Hidalgo, S., Howell, J., Neill, O., and Russo, C., 2012. Effects of variable magma supply on mid-ocean ridge eruptions: constraints from mapped lava flows along the Galapagos Spreading Center. *Geochemistry, Geophysics, Geosystems*, **13**(8). doi:10.1029/2012GC004163.

Deardorff, N. D., Cashman, K. V., and Chadwick, W. W., Jr., 2011. Observations of eruptive plumes and pyroclastic deposits from submarine explosive eruptions at NW Rota-1, Mariana Arc. *Journal of Volcanology and Geothermal Research*, **202**(1–2), 47–59, doi:10.1016/j.jvolgeores.2011.01.003.

Delaney, J. R., Kelley, D. S., Lilley, M. D., Butterfield, D. A., Baross, J. A., Embley, R. W., and Summit, M., 1998. The quantum event of oceanic crustal accretion: impacts of diking at mid-ocean ridges. *Science*, **281**, 222–230.

Dziak, R. P., Fox, C. G., and Schreiner, A. E., 1995. The June-July 1993 seismo-acoustic event at CoAxial segment, Juan de Fuca Ridge: evidence for a lateral dike injection. *Geophysical Research Letters*, **22**, 135–138.

Embley, R. W., and Chadwick, W. W., Jr., 1994. Volcanic and hydrothermal processes associated with a recent phase of seafloor spreading at the northern cleft segment: Juan de Fuca Ridge. *Journal of Geophysical Research*, **99**(B3), 4741–4760, doi:10.1029/93JB02038.

Embley, R. W., Murphy, K. M., and Fox, C. G., 1990. High resolution studies of the summit of Axial Volcano. *Journal of Geophysical Research*, **95**, 12,785–712,812.

Embley, R. W., Chadwick, W. W., Perfit, M. R., Smith, M. C., and Delaney, J. R., 2000. Recent eruptions on the CoAxial segment of the Juan de Fuca Ridge: implications for mid-ocean ridge accretion processes. *Journal of Geophysical Research*, **105**, 16501–16525.

Embley, R. W., Chadwick, W. W., Jr., Baker, E. T., Butterfield, D. A., Resing, J. A., de Ronde, C. E. J., Tunnicliffe, V., Lupton, J. E., Juniper, S. K., Rubin, K. H., et al., 2006. Long-term eruptive activity at a submarine arc volcano. *Nature*, **441**, 494–497.

Engels, J. L., Edwards, M. H., Fornari, D. J., Perfit, M. R., and Cann, J. R., 2003. A new model for submarine collapse formation. *Geochemistry, Geophysics, Geosystems*, **4**(9), 1077, doi:10.1029/2002GC000483.

Fink, J. H., and Griffiths, R. W., 1992. A laboratory analog study of the surface morphology of lava flows extruded from point and line sources. *Journal of Volcanology and Geothermal Research*, **54**, 19–32.

Fornari, D. J., Haymon, R. M., Perfit, M. R., Gregg, T. K. P., and Edwards, M. H., 1998. Axial summit trough of the East Pacific Rise 9°–10°N: geological characteristics and evolution of the axial zone on fast spreading mid-ocean ridges. *Journal of Geophysical Research*, **103**(B5), 9,827–9,855.

Francheteau, J., and Ballard, R. D., 1983. The East Pacific Rise near 21°N, 13°N and 20°S: inferences for along-strike variability of axial processes of the mid-ocean ridge. *Earth and Planetary Science Letters*, **64**, 93–116.

Francheteau, J., Juteau, T., and Rangan, C., 1979. Basaltic pillars in collapsed lava-pools on the deep ocean-floor. *Nature*, **281**, 209–211.

Fundis, A., Soule, S. A., Fornari, D. J., and Perfit, M. R., 2010. Paving the seafloor: volcanic emplacement processes during the 2005-06 eruption at the fast-spreading East Pacific Rise, 9°50′N. *Geochemistry, Geophysics, Geosystems*, doi:10.1029/2010GC00305.

Garry, W. B., Gregg, T. K. P., Soule, S. A., and Fornari, D. J., 2006. Formation of submarine lava channel textures: insights from laboratory simulations. *Journal of Geophysical Research*, **111**, B03104, doi:10.1029/2005JB003796.

Gregg, T. K. P., and Chadwick, W. W., Jr., 1996. Submarine lava-flow inflation: a model for the formation of lava pillars. *Geology*, **24**, 981–984.

Gregg, T. K. P., and Fink, J. H., 1995. Quantification of submarine lava-flow morphology through analog experiments. *Geology*, **23**, 73–76.

Gregg, T. K. P., and Fink, J. H., 2000. A laboratory investigation into the effects of slope on lava flow morphology. *Journal of Volcanology and Geothermal Research*, **96**, 145–159.

Gregg, T. K. P., Fornari, D. J., Perfit, M. R., Haymon, R. M., and Fink, J. H., 1996. Rapid emplacement of a mid-ocean ridge lava flow on the East Pacific Rise at 9°46′–51′N. *Earth and Planetary Science Letters*, **144**, E1–E7.

Griffiths, R. W., and Fink, J. H., 1992. The morphology of lava flows in planetary environments: predictions from analog experiments. *Journal of Geophysical Research*, **97**(B13), 19,739–19,748.

Helo, C., Longpre, M.-A., Shimizu, N., Clague, D. A., and Stix, J., 2011. Explosive eruptions at mid-ocean ridges driven by CO_2-rich magmas. *Nature Geoscience*, **4**, 260–263, doi:10.1038/NGEO1104.

Hon, K., Kauahikaua, J., Denlinger, R., and Mackay, K., 1994. Emplacement and inflation of pahoehoe sheet flows: observations and measurements of active lava flows on Kilauea volcano, Hawaii. *Geological Society of America Bulletin*, **106**, 351–370.

Karson, J. A., Klein, E. M., Hurst, S. D., Lee, C. E., Rivizzigno, P. A., Curewitz, D., Morris, A. R., and Party, H. D. S., 2002a. Structure of uppermost fast-spread oceanic crust exposed at the Hess Deep Rift: implications for subaxial processes at the East Pacific Rise. *Geochemistry, Geophysics, Geosystems*, **3**, 2001GC000155.

Karson, J. A., Tivey, M. A., and Delaney, J. R., 2002b. Internal structure of uppermost oceanic crust along the western Blanco Transform Scarp: implications for subaxial accretion and deformation at the Juan de Fuca Ridge. *Journal of Geophysical Research*, **107**, 2000JB000007.

Kurras, G. J., Fornari, D. J., Edwards, M. H., Perfit, M. R., and Smith, M. C., 2000. Volcanic morphology of the East Pacific Rise Crest $9°49'-52'$: implications for volcanic emplacement processes at fast-spreading mid-ocean ridges. *Marine Geophysical Research*, **21**, 23–41, doi:10.1023/A:1004792202764.

Lonsdale, P., 1977. Abyssal pahoehoe with lava coils at the Galapagos rift. *Geology*, **5**, 147–152.

Moore, J. G., 1975. Mechanism of formation of pillow lava. *American Journal of Science*, **63**, 269–277.

Perfit, M. R., and Chadwick, W. W., Jr., 1998. Magmatism at mid-ocean ridges: constraints from volcanological and geochemical investigations. In Buck, W. R. (ed.), *Faulting and Magmatism at Mid-Ocean Ridges*. Washington, DC: AGU. Geophysical Monograph Series, Vol. 106, pp. 59–116.

Perfit, M. R., Cann, J. R., Fornari, D. J., Engels, J., Smith, D. K., Ridley, W. I., and Edwards, M. H., 2003. Interaction of seawater and lava during submarine eruptions at mid-ocean ridges. *Nature*, **426**, 62–65.

Pontbriand, C. W., Soule, S. A., Sohn, R. A., Humphris, S. E., Kunz, C., Singh, H., and Shank, T., 2012. Effusive and explosive volcanism on the ultraslow-spreading Gakkel Ridge, 85° E. *Geochemistry, Geophysics, Geosystems*, **13**(10). doi:10.1029/2012GC004187.

Resing, J., Rubin, K. H., Embley, R., Lupton, J., Baker, E. T., Dziak, R., Baumberger, T., Lilley, M., Huber, J., Shank T., et al., 2011. Active submarine eruption of boninite in the northeastern Lau Basin. *Nature Geoscience*, **4**, 799–806. doi:10.1038/NGEO1275.

Rubin, K. H., and Sinton, J. M., 2007. Inferences on mid-ocean ridge thermal and magmatic structure from MORB compositions. *Earth and Planetary Science Letters*, **260**(1), 257–276.

Rubin, K. H., Soule, S. A., Chadwick, W. W., Jr., Fornari, D. J., Clague, D. A., Embley, R. W., Baker, E. T., Perfit, M. R., Caress, D. W., and Dziak, R. P., 2012. Volcanic eruptions in the deep sea. *Oceanography*, **25**(1), 142–157.

Searle, R. C., Murton, B. J., Achenbach, K., LeBas, T., Tivey, M., Yeo, I., Cormier, M. H., Carlut, J., Ferreira, P., Mallows, C., et al., 2010. Structure and development of an axial volcanic ridge: Mid-Atlantic Ridge, 45°N. *Earth and Planetary Science Letters*, **299**, 228–241, doi:10.1016/j.epsl.2010.09.003.

Sinton, J., Bergmanis, E., Rubin, K. H., Batiza, R., Gregg, T. K. P., Gronvold, K., Macdonald, K. C., and White, S. M., 2002. Volcanic eruptions on mid-ocean ridges: new evidence from the superfast spreading East Pacific Rise, $17°-19°S$. *Journal of Geophysical Research*, **107**(B6), 2115, doi:10.1029/2000JB000090.

Smith, D. K., and Cann, J. R., 1993. Building the crust at the Mid-Atlantic Ridge. *Nature*, **365**, 707–715.

Smith, M. C., Perfit, M. R., and Jonasson, I. R., 1994. Spatial and temporal variations in the geochemistry of lavas from the S. Juan de Fuca Ridge: implications for petrogenesis. *Journal of Geophysical Research*, **99**, 4787–4812.

Smith, D. K., Cann, J. R., Dougherty, M. E., Lin, J., Spencer, S., MacLeod, C., and Robertson, W., 1995. Mid-Atlantic Ridge volcanism from deep-towed side-scan sonar images, 25–29 N. *Journal of Volcanology and Geothermal Research*, **67**(4), 233–262.

Sohn, R. A., Willis, C., Humphris, S. E., Shank, T. M., Singh, H., Edmonds, H. N., Kunz, C., Hedman, U., Helmke, E., Jakuba, M., et al., 2008. Explosive volcanism on the ultraslow-spreading Gakkel ridge, Arctic Ocean. *Nature*, **453**, 1236–1238.

Soule, S. A., Fornari, D. J., Perfit, M. R., Tivey, M. A., Ridley, W. I., and Schouten, H., 2005. Channelized lava flows at the East Pacific Rise crest $9°-10°N$: the importance of off-axis lava transport in developing the architecture of young oceanic crust. *Geochemistry, Geophysics, Geosystems*, **6**, Q08005, doi:10.1029/2005GC000912.

Soule, S. A., Fornari, D. J., Perfit, M. R., Ridley, W. I., Reed, M. H., and Cann, J. R., 2006. Incorporation of seawater into mid-ocean ridge lava flows during emplacement. *Earth and Planetary Science Letters*, **252**, 289–307.

Soule, S. A., Fornari, D. J., Perfit, M. R., and Rubin, K. H., 2007. New insights into mid-ocean ridge volcanic processes from the 2005–2006 eruption of the East Pacific Rise, $9°46'N-9°56'N$. *Geology*, **35**, 1079–1082.

Soule, S. A., Nakata, D. S., Fornari, D. J., Fundis, A. T., Perfit, M. R., and Kurz, M. D., 2012. CO_2 variability in mid-ocean ridge basalts from syn-emplacement degassing: constraints on eruption dynamics. *Earth and Planetary Science Letters*, **327**, 39–49.

Stakes, D. S., Perfit, M. R., Tivey, M. A., Caress, D., Ramirez, T. M., and Maher, N., 2006. The cleft revealed: geologic, magnetic and morphologic evidence for construction of upper oceanic crust along the southern Juan de Fuca Ridge. *Geochemistry, Geophysics, Geosystems*, **7**, Q04003, doi:10.1029/2005GC001038.

Staudigel, H., and Clague, D. A., 2010. The geological history of deep-sea volcanoes. *Oceanography*, **23**, 59–71.

Tominaga, M., and Umino, S., 2010. Lava deposition history in ODP Hole 1256D: insights from log-based volcanostratigraphy. *Geochemistry, Geophysics, Geosystems*, **11**(5). doi:10.1029/2009GC002933.

Tominaga, M., Teagle, D. A., Alt, J. C., and Umino, S., 2009. Determination of the volcanostratigraphy of oceanic crust formed at superfast spreading ridge: electrofacies analyses of ODP/IODP Hole 1256D. *Geochemistry, Geophysics, Geosystems*, **10**(1). doi:10.1029/2008GC002143.

Wanless, V. D., Perfit, M. R., Ridley, W. I., and Klein, E. M., 2010. Dacite petrogenesis on mid-ocean ridges: evidence for crustal melting and assimilation. *Journal of Petrology*, **51**, 2377–2410.

White, S. M., Macdonald, K. C., and Haymon, R. M., 2000. Basaltic lava domes, lava lakes, and volcanic segmentation on the southern East Pacific Rise. *Journal of Geophysical Research*, **105** (B10), 23519, doi:10.1029/2000JB900248.

White, S. M., Haymon, R. M., Fornari, D. J., Perfit, M. R., and Macdonald, K. C., 2002. Correlation between volcanic and tectonic segmentation of fast-spreading ridges: evidence from volcanic structures and lava flow morphology on the East Pacific Rise at $9°-10°N$. *Journal of Geophysical Research*, **107**(B8), 2173, doi:10.1029/2001JB000571.

White, S. M., Haymon, R. M., and Carbotte, S., 2006. A new view of ridge segmentation and near-axis volcanism at the East Pacific Rise, 8°–12°N, from EM300 multibeam bathymetry. *Geochemistry, Geophysics, Geosystems*, **7**, Q12O05, doi:10.1029/2006GC001407.

White, S. M., Meyer, J. D., Haymon, R. M., Macdonald, K. C., Baker, E. T., and Resing, J. A., 2008. High–resolution surveys along the hot spot–affected Galápagos Spreading Center: 2. Influence of magma supply on volcanic morphology, *Geochemistry, Geophysics, Geosystems*, **9**(9).

Wilson, L., and Parfitt, E. A., 1993. The formation of perched lava ponds on basaltic volcanoes: the influence of flow geometry on cooling-limited lava flow lengths. *Journal of Volcanology and Geothermal Research*, **56**(1), 113–123.

Wright, I. C., Worthington, T. J., and Gamble, J. A., 2006. New multibeam mapping and geochemistry of the 30–35 S sector, and overview, of southern Kermadec arc volcanism. *Journal of Volcanology and Geothermal Research*, **149**(3), 263–296.

Yeo, I., Searle, R. C., Achenbach, K. L., Le Bas, T. P., and Murton, B. J., 2012. Eruptive hummocks: building blocks of the upper ocean. *Geology*, **40**, 91–94, doi:10.1130/G31892.

Cross-references

Axial Summit Troughs
Explosive Volcanism in the Deep Sea
Intraplate Magmatism
Island Arc Volcanism, Volcanic Arcs
Marginal Seas
Mid-ocean Ridge Magmatism and Volcanism
Ocean Drilling
Oceanic Rifts
Oceanic Spreading Centers

SUBMARINE SLIDES

Roland von Huene
GEOMAR Helmholtz Centre for Ocean Research,
Kiel, Germany

Definition

Failure of submarine slopes results in underwater landslides, a mass movement of sediment and rock that spills down ocean margin slopes (Hampton and Locat, 1996). Visual observations during submarine mass movement are few, and therefore landslides are commonly recognized by the remnant morphology they leave behind. These remnant slope failure features have been observed along the world's continental margins, on the slopes of volcanic islands, and volcanic island arcs. Initiation of landslides commonly results from dynamic events, generally earthquakes. Landslides are most frequent along margins where converging lithospheric plates form subduction zones and the shear between the upper and lower plates create large earthquakes, however, some large landslides occur without tectonic or seismic triggers. Sedimentary processes have caused failure from loading or unloading. Submarine landslides in the spectrum from fast to slow can be catastrophic, incremental, or creeping. Large catastrophic slides have created unexpected tsunamis that inundate adjacent shores within minutes of an earthquake, making it difficult to mitigate hazards in the affected area.

History

An awareness of how wide spread submarine landslides have been grew through the means to image the seafloor with acoustic instrument systems. Shallow water slides were reported before WWII; however, recognition became more frequent when oceanographic science received increased attention because of its military importance. Convincing evidence of slides was reported as the *precision depth recorder* of the early 1950s imaged profiles of the seafloor rather than spot soundings that controlled contours, as was done with "lead lines" and later with acoustic spot sounding. Perhaps the greatest advance was made with *multibeam bathymetric imaging systems* that map a swath of the seafloor and provide greater resolution than a series of profiles. Satellite navigation with GPS allowed individual multibeam bathymetric swaths to be joined precisely, and the improved resolution revealed many landslide features. Numerous features are inferred to have generated local tsunamis because the morphological character of remnant landslide scars can commonly show whether the slide was sufficiently large and perhaps catastrophic rather than a small translational debris flow. Indeed, the mounting evidence of submarine landslides has reinforced assertions that they were responsible for unexpected tsunamis. As improved resolution of bathymetric imaging became more widespread, it showed that submarine landslides are as common as those on land and can occur on very gentle slopes, sometimes as little as 2°.

Environments of submarine landslides

Slope failure occurs in all ocean depths, from coastal waters to the shelf edge and then down slope to tectonically active features of deep trenches. Along steep shores, landslides that begin above sea-level can travel great distances under water. In 1958, a M = 8.3 earthquake triggered a subaerial rock slide that plunged catastrophically into Lituya Bay, Alaska, generating a wave that ran 1,720 ft (525 m) up the opposite shore (i.e., Geist, 2002). Shallow water failure of delta fronts at the heads of fjords in Norway and Alaska have traveled tens of kilometers and produced damaging tsunamis. For example, during the 1964 Alaska earthquake, the front of the glacial delta on which the town of Valdez was constructed collapsed in a catastrophic 4,000 ft long by 600 ft wide landslide that took with it the harbor and docking area where 32 persons were unloading a freighter (Coulter and Migliaccio, 1966). The slump generated a 30–40 ft tsunami before the shaking of the earthquake subsided carrying the freighter onto land where it nearly capsized before the backwash refloated the ship and returned it to the fjord. Deltaic landslides not associated with an earthquake are

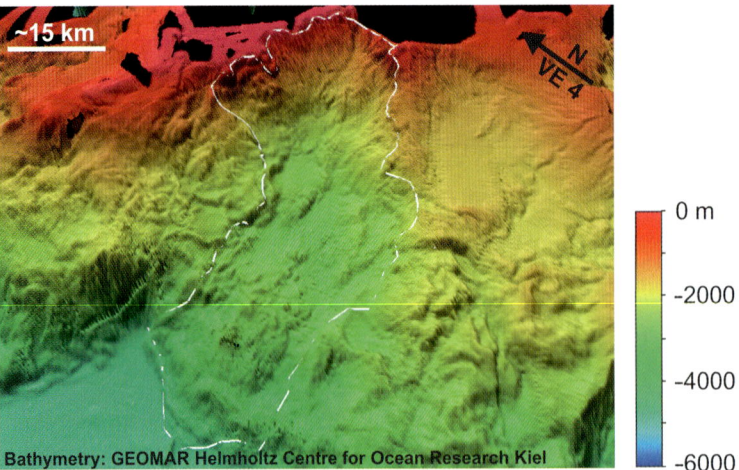

Submarine Slides, Figure 1 A large remnant slide scar on the Chilean margin south of Concepcion extends from the edge of the shelf to the trench (outlined with *white line*). It was probably formed by multiple slides some if which did not reach the trench (from Geersen et al., 2011).

illustrated by the large landslides of the submerged Mississippi River delta (Chaytor et al., 2010).

On continental shelves, landslides begin in local steep topography and small slide scars are observed on the sides of canyons. At the edge of continental shelves where slopes steepen, numerous landslides occur along both convergent and nonconvergent margins. Notable along Atlantic nonconvergent margins are the Storegga slide off Norway (Bryn et al., 2005) and the 1929 Grand Banks slide off North America (Mosher and Piper, 2007). Twenty large slide scars tens of kilometers wide have been identified along the US Atlantic margin (Chaytor et al., 2007). Along tectonically active Caribbean island margins, several slide scars have been mapped with multibeam systems as for instance off Puerto Rico.

The convergent margins of the Pacific have been extensively studied in various places. Along Central America where many smaller slides are documented as well as the 55 km wide rotational slump imaged off the Nicoya Peninsula (Harders et al., 2011). The continental shelf edges off Nicaragua and Costa Rica have 10 large landslide scars some of which may have been caused by subducting relief.

Multibeam mapping of the Chilean margin has revealed multiple huge slides (Geersen et al., 2011) (Figure 1).

Several occur offshore Central Chile and are probably triggered by large earthquakes.

In the north Pacific at the end of the Alaska Peninsula, a local tsunami that ran 42 m high up the rugged terrain destroyed the Scotch Cap lighthouse complex (Fryer and Tryon, 2005; Rathburn et al., 2009). The principal slide mass may have been a 15 km wide block detached from the upper slope that came to rest 10 km from its source on the mid-slope terrace (Miller et al., 2014). Upper slope slides along the US Pacific margin have been observed, such as the one off the Palos Verdes coast of Southern California. An example on a western Pacific margin is the 1998 landslide off Papua New Guinea that created a tsunami wave locally 15 m high and resulted in more than 2,200 fatalities (Tappin et al., 2001; Geist, 2000; Sweet and Silver, 2003). This tsunami showed the potential for small earthquakes to trigger large tsunamis, if they trigger undersea landslides.

In multibeam imaged morphology of the lower trench slopes of convergent margins, rapid tectonic thickening that produces steep slopes results in considerable small to modest slope failure. Such slope failure is commonly at the limits of resolution of surface ship surveys. The great water depths and long acoustic travel path reduces the resolution because high frequency acoustic waves are attenuated. High resolution is afforded by bottom towed instruments that are deployed near the seafloor. Slope failure at small to modest scales is created as contractile deformation thickens trench sediment and pushes it upward in folds to its critical slope limit. However, much larger landslide features are produced by subducting topographic relief on the lower plate (Harders et al., 2011). Not only are the horst and graben created when the lower plate bends into the trench axis, but numerous seamounts and ridges formed by ocean plate processes produce a disruptive subducting morphology (see Figure 2 below).

Notable are the trails of remnant slope failure left by subducting seamounts as they plow into the lower slope and destabilize it. As relief tunnels into the slope and beneath the seafloor, the dome uplifted above it has steep slopes and causes debris avalanches that flow toward the trench. Theoretical studies indicate that the volume of debris in such an avalanche can be sufficiently large to cause a local tsunami. Unstable kilometer wide slope failure scars and translational debris streams are well imaged off central Costa Rica. Also imaged there and off Peru are rotational slumps 50 km and more wide. If these slumps

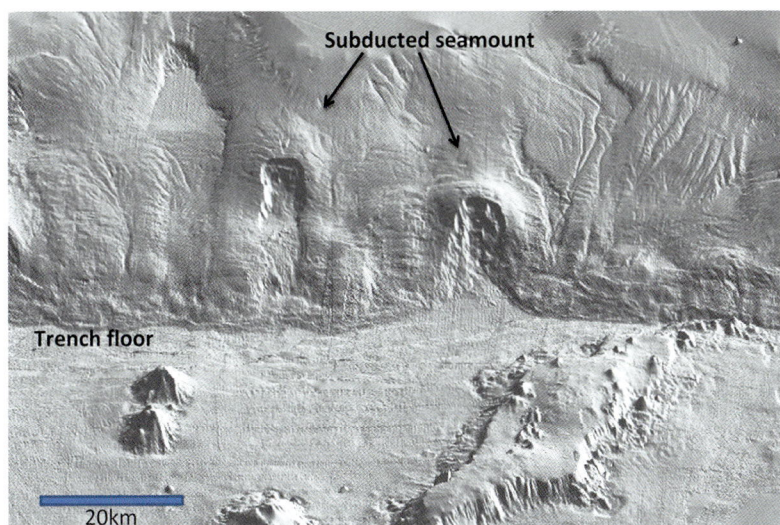

Submarine Slides, Figure 2 Submarine slide landslides in the trailing wake of subducting seamounts in the Middle America Trench offshore Costa Rica. Steepening of the seafloor over the seamounts causes translational slides that disintegrate into debris flows and turbidity currents.

failed catastrophically, even a 3–4 km^3 volume of slope sediment would displace sufficient water to have created large tsunamis as was shown off Papua New Guinea (Sweet and Silver, 2003).

Causes of slope instability and failure

In some instances the origins of a remnant landslide scar seem obvious, whereas in other instances they are puzzling. More than one factor is commonly involved and one type of failure transforms to another. For instance, rotational slumps or detached blocks can disintegrate into debris flows and then turbidity currents (i.e., Hampton and Locat, 1996). Large scale tectonic addition by accretion and removal by the erosion of material along continental margins are basic causes of the slope instability. These tectonic processes are a precondition to submarine slumps and landslides. Plate convergence can result in accretion of trench sediment thereby steepening the slope by thickening the upper plate. Conversely, convergence can remove material from the base of the upper plate thereby thinning it and causing subsidence. Slope steepening can also be caused by scouring of the seafloor from currents.

Over steepened slopes prone to failure during local earthquakes are a key factor in initiating landslides. During earthquake shaking, stresses increase and the frictional strength along fracture interfaces decreases as pore water pressures fluctuate instantaneously. A slower rise in pore water pressure can also result from accumulation of sediment in deltas, and if sediment permeability and fluid flow is restricted by overlying rock or sediment, the pressures can elevate to a level where the superimposed mass will almost float on a water rich layer. Weak geologic layers certainly contribute to landslides but few observations have confirmed the mechanics that contribute to their failure. Studies have indicated that gas hydrates (an ice like state of gas and water under pressure) might be at the base of some slides. When gas hydrate dissociates, it forms gas charged water as temperatures rise or pressure decreases. If drainage of this fluid is inhibited, the fluid rich layer may reach a level at which slopes become unstable and fail without being triggered by an earthquake.

At non-convergent Atlantic type margins, slides have broken loose at low slope angles (2° and less) and become unstable because of sediment accumulation (Hühnerbach and Masson, 2004; Chaytor et al., 2009, 2010). With increasing sediment thickness, strata of different lithologies change physical properties at different rates as the sediment load increases. The physical property contrasts of strong and weak horizons can result in slope failure under loading by storms or changes in morphology.

At volcanic islands and island arcs, submarine landslides often result from collapse of volcano segments as the volcanos grow with the addition of lava and phyoclastic flows (Masson et al., 2002; Coombs et al., 2007). The large and extensive landslides of the Hawaiian Islands were recognized once high resolution bathymetric data were acquired. Most have occurred within the past four million years and the youngest is thought to be only 100,000 years old. Large tsunami waves carried rocks and sediments as high as 1,000 ft (30 m) above sea-level. Such landslides have the potential for enormous loss of life and property.

Hazards

Landslides that begin near sea-level have destroyed port facilities such as the catastrophic collapse of Port Valdez in 1964 (Coulter and Migliaccio, 1966). They have ripped apart submarine cables in the Grand Banks slide of 1929 almost 600 km from the head of the slide (Mosher and Piper, 2007). They have destroyed seafloor infrastructure associated with offshore oil production. However, little is known of recurrence rates, trigger mechanisms, pre- and post-failure geotechnical conditions, and the role of landslide processes in deep water sediment. Increasing development of seafloor-based offshore platforms, telecommunication and energy transport facilities, and coastal infrastructures asks for better knowledge of seafloor stability conditions, submarine landslide processes and tsunamis.

A potential frequency of landslides has been addressed in theoretical studies. The locations and extent of morphological landslide scars has been documented locally but the observational data for a global assessment is not yet sufficiently complete to produce meaningful constraints. Elementary quantitative models of landslides that produce tsunamis have been formulated and tested against natural examples (Ward, 2001). Another approach is with probabilistic analysis (Geist and Parsons, 2009) but again the observational base requires more data. Paleoseismic studies in coastal areas have documented histories of inundation preserved in sediment sequences but whether the cause is from transoceanic sources or local events is difficult to determine without a broad regional data base.

Summary

Areas of the seafloor mapped with multibeam sounding instruments show morphological evidence of considerable mass movement on seafloor slopes from the shore to the deep trenches. Slides and slumps are especially prevalent where large earthquakes occur. Submarine landslides have damaged coastal infrastructure and are costly, they can strike without warning and have been deadly. Key factors in determining whether a landslide will be tsunamigenic are slide volume and the acceleration of a detached mass. Modern oceanographic instrumentation can resolve remnant features in areas of potential future slides and outline areas where past slides indicate dangerously unstable slopes. Monitoring critical areas with bottom instrumentation is a means to show where future slope failure can be anticipated. But the distribution and magnitude of tsunamigenic landslides has yet to be fully appreciated. To raise the proper degree of awareness, the hazards need to be assessed quantitatively which requires more data on the physical properties related to slope failure. Long term assessments of all natural hazards should be part of an area's infrastructure.

Landslides are common on inclined areas of the seafloor, particularly in environments where weak geologic materials such as rapidly deposited fine-grained sediment or fractured rock are subjected to strong environmental stresses. Such stresses are inflicted by earthquakes, large storm waves, and high internal pore pressures. Submarine landslides can involve huge amounts of material and can move great distances: slide volumes as large as 20,000 km^3 and runout distances in excess of 140 km have been reported. They occur at locations where the downslope component of stress exceeds the resisting stress, causing movement along one or several concave to planar rupture surfaces. Some recent slides that originated near shore were conspicuous by their direct impact on human life and activities. Most known slides, however, occurred far from land in prehistoric time and were discovered by noting distinct to subtle characteristics, such as headwall scarps and displaced sediment or rock masses, on acoustic-reflection profiles and side-scan sonar images. Submarine landslides can be analyzed using the same mechanics principles as are used for occurrences on land. However, some loading mechanisms are unique, for example, storm waves. There is a potential for landslides to transform into sediment flows that can travel exceedingly long distances and this is related to the density of the slope material and the amount of shear strength that is lost during catastrophic slope failure.

Bibliography

Bryn, P., Berg, K., Forsberg, C. F., Solheim, A., and Kvalstad, T. J., 2005. Explaining the Storegga Slide. *Marine and Petroleum Geology*, **22**, 11–19.

Chaytor, J., Twichell, D., ten Brink, U., and Buczkowski, B., 2007. Revisiting submarine mass movements along the U.S. Atlantic continental margin; implications for tsunami hazards. In Lykousis, V., Sakellariou, D., and Locat, J. (eds.), *Submarine Mass Movements and Their Consequences*. Santorini: Springer.

Chaytor, J., ten Brink, U., Solow, A., and Andrews, B., 2009. Size distribution of submarine landslides along the U.S. Atlantic margin. *Marine Geology*, **264**(1–2) (also available at http://dx.doi.org/10.1016/j.margeo.2008.08.007).

Chaytor, J., Twichell, D., Lynett, P., and Geist, E., 2010. Distribution and tsunamigenic potential of submarine landslides in the Gulf of Mexico. In Mosher, D. C., Shipp, R. C., Moscardelli, L., Chaytor, J.D., Baxter, C.D.P., Lee, H.J., and Urgeles, R. (eds.), *Submarine Mass Movements and Their Consequences, 4th International Symposium* (also available at http://dx.doi.org/10.1007/978-90-481-3071-9_60).

Coombs, M., White, S. M., and Scholl, D., 2007. Massive edifice failure at Aleutian arc volcanoes. *Earth and Planetary Science Letters*, **256**, 403–418.

Coulter, H. W., and Migliaccio R. R., 1966. *Effects of the earthquake of March 27, 1964 at Valdez Alaska*. Washington, DC: US Government Printing Office. United States Geological Survey Professioal Paper 542C.

Fryer, G. J., and Tryon, M. D., 2005. Great earthquakes, gigantic landslides, and the continuing enigma of the April Fool's tsunami of 1946. *EOS transactions AGU*, 86(52). T11A-0355.

Geersen, J., Voelker, D., Behrmann, J. H., Reichert, C., and Krastel, S., 2011. Pleistocene giant slope failures offshore Arauco Peninsula, Southern Chile. *Journal of the Geological Society of London*, **168**, 1237–1248, doi:10.1144/0016-76492011-027.

Geist, E. L., 2000. Origin of the 17 July 1998 Papua New Guinea tsunami: earthquake or landslide? *Seismological Research Letters*, **71**(3), 344–351.

Geist, E. L., 2002. Complex earthquake rupture and local tsunamis. *Journal of Geophysical Research*, **107**, ESE 2–1–ESE 2–16. doi:10.1029/2000JB000139. [Download PDF (399 K)].

Geist, E. L., and Parsons, T., 2009. Assessment of source probabilities for potential tsunamis affecting the U.S. Atlantic coast. *Marine Geology*, **264**, 98–108.

Hampton, M., and Locat, J., 1996. Submarine landslides. *Reviews of Geophysics*, **34**, 33–59.

Harders, R., Ranero, C. R., Weinrebe, W., and Behrmann, J. H., 2011. Submarine slope failures along the convergent continental margin of the Middle America Trench. *Geochemistry, Geophysics, Geosystems*, **12**(6), Q05S32, doi:10.1029/2010GC00340.

Hühnerbach, V., and Masson, D. G., 2004. Landslides in the North Atalntic and its adjacent seas: an analysis of their morphology, setting and vehaviour. *Marine Geology*, **213**, 343–362.

Locat, J, Lee, H. J, Schwab, W. C., and Twichell, D. C, 1996. Analysis of the mobility of far reaching debris flows on the Mississippi Fan, Gulf of Mexico. In Senneset, K. (ed.), *Landslides 7th International Symposium, Proceedings*. Trondheim, Vol. 1.

Masson, D. G., Watts, A. B., Gee, M. J. R., Urgeles, R., Mitchell, N. C., Lee Bes, T. P., and Canals, M., 2002. Slope failures of the western Canary Islands. *Earth Science Reviews*, **57**, 1–35.

Miller, J. J., von Huene, R., and Ryan, H., 2014. *The Unimak tsunami earthquake area: revised tectonic structure in reprocessed seismic images and a suspect near field seismic source*. U.S. Geological Survey open file report 140002.

Mosher, D. C., and Piper, D. J. W., 2007. Analysis of multibeam seafloor imagery of the Laurentian Fan and the 1929 Grand banks landslide. In Lykousis, V., Sakellariou, D., and Locat, J. (eds.), *Submarine Mass Movements and Their Consequences*. Santorini: Springer, pp. 77–88.

Rathburn, A. E., Levin, L. A., Tryon, M., Gieskes, J. M., Martin, J. B., Perez, M. E., Fodrie, F. J., Neira, C., Fryer, G. J., Mendoza, G., McMillan, P. A., Kluesner, J., Adamic, J., and Zeibis, W., 2009. Geological and biological heterogeneity of the Aleutian margin (1965-4822 m). *Progress in Oceanography*, **80**(2009), 22–50.

Sweet, S., and Silver, E. A., 2003. Tectonics and slumping in the source region of the 1998 Papua New Guinea tsunami from seismic reflection images. *Pure and Applied Geophysics*, **160**, 1945–1968, doi:10.1007/s00024-003-2415-z. 0033–4553/03/111945–24.

Tappin, D. R., Watts, P., McMurty, G. M., Lafoy, Y., and Matsumoto, T., 2001. The Sissano Papua New Guinea tsunami of July 1998 – offshore evidence on the source mechanism. *Marine Geology*, **175**, 1–24.

Ward, S., 2001. Landslide tsunami. *Journal of Geophysical Research-Solid Earth*, doi:10.1029/2000JB900450.

Cross-references

Earthquakes
Geohazards: Coastal Disasters
Mass Wasting
Tsunamis
Turbidites

SUBSIDENCE OF OCEANIC CRUST

David Voelker
MARUM-Center for Marine Environmental Sciences, University of Bremen, Bremen, Germany

Definition

The average depth of the deep-sea floor relative to the sea surface is a function of the age of the oceanic *crust*. (This is by definition the floor of the oceans that is underlain by normally thick (5–8 km) basaltic ocean crust under normal tectonic loads. Consequently, the age-depth relation does not apply to (a) those parts of the oceans that are underlain by continental crust such as continental shelves, some epeiric seas, or continental fragments, (b) convergent plate boundaries where tectonic loading by subduction leads to deep-sea trenches that exceed the "normal" average depth of the seafloor, and (c) regions where the seafloor was formed by flood basalt events (Large Igneous Provinces) and the ocean crust is thickened.) Newly formed oceanic crust is found at water depths around 2,500 m globally, whereas its depth increases by 1,000 m during the first 10 million years after its formation. This deepening with time is called (thermal) subsidence. Subsidence slows down with time, e.g., it needs ~26 million years more for the seafloor to sink another 1,000 m.

Morphological expression of subsidence

The most obvious demonstration of the age-depth relation of the seafloor is given by *mid-ocean ridges*. Mid-ocean ridges are the places where new ocean floor forms. They stand out morphologically as ridges because of the relative low density of young ocean floor. The average elevation of the crest of the Mid-Atlantic Ridge is rather uniform along all of its length (~10,000 km), and a similar crest elevation is valid for other mid-ocean ridges. Away from the ridge axis, the average depth of the seafloor increases to both sides, resulting in the striking mirror symmetry of mid-ocean ridges. As the age increase of the oceanic crust with distance from the spreading axis is directly related to the spreading rate, fast-spreading ridges are much wider than slow-spreading ridges (Figure 1 cross section across the Mid-Atlantic and Pacific Ridge).

At oceanic fracture zones, where ocean floor segments of different crustal age are aligned by displacement by transform faults, we observe steps (fault escarpments) with a vertical displacement of up to 1,000 m (Figure 1).

The age-depth relation can be expressed as

$$Z = Z_0 - k\sqrt{t}$$

Z: depth of the seafloor of crustal age t.

$Z0$: depth of the seafloor at the spreading axis or zero-age depth (typically 2,500 m).

k: *factor of proportionality or subsidence rate*; this factor was proposed to be ca. $350 \text{ m Ma}^{-0.5}$ by Parsons and Sclater (1977), Stein and Stein (1992), and others. A newer statistical analysis gives a lower value of $320 \text{ m Ma}^{-0.5}$ as global average (Korenaga and Korenaga, 2008). Although this global average holds for many regions of the world, there are significant regional variations (Marty and Cazenave, 1989):

t: crustal age in Ma

This formula and its graphical expression are known as "Sclater curve" and were originally published in 1977 (Parsons and Sclater, 1977). It describes the age-depth relation reasonably well over 70–80 My of crustal age.

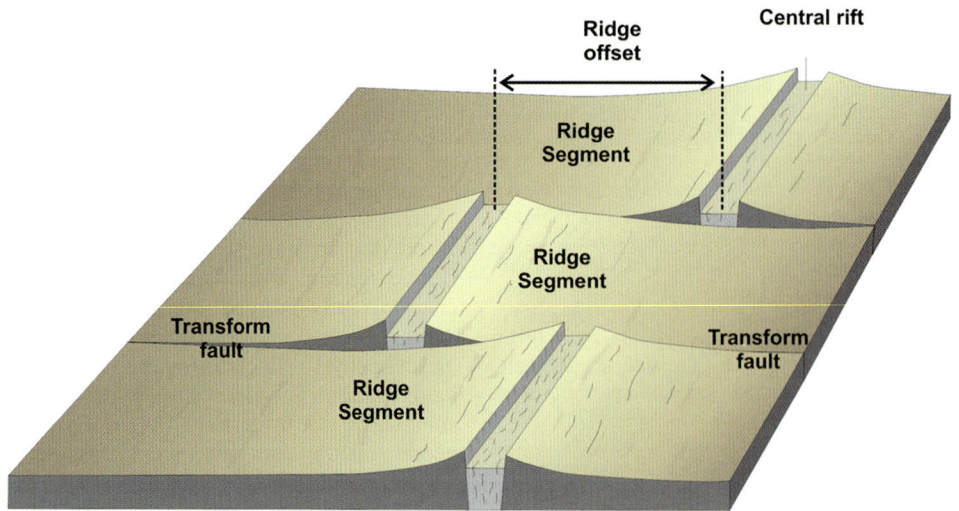

Subsidence of Oceanic Crust, Figure 1 Transform faults forming vertical offsets at a mid-ocean ridge.

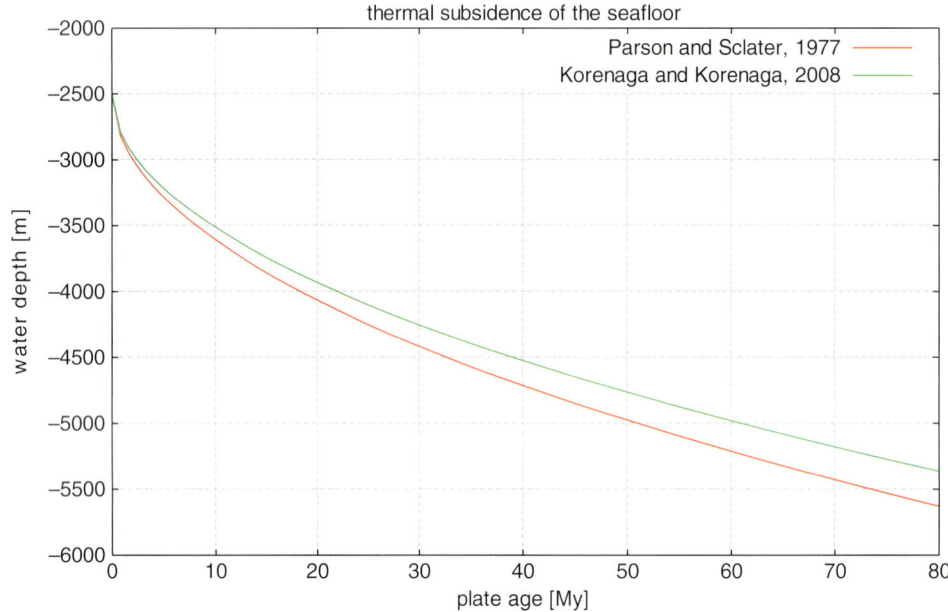

Subsidence of Oceanic Crust, Figure 2 Graphical representation of the Sclater relation between crustal age and water depth.

For older crust, the subsidence is lower in reality, an effect known as "seafloor flattening" Figure 2.

Cause of thermal subsidence of the seafloor

The relative height of the seafloor is determined by the density and thickness of the oceanic lithosphere. As the chemical composition and thickness of the oceanic crust that is formed globally at ocean spreading centers show a close similarity on a global scale (the group of *mid-ocean ridge basalts or MORBs*), the density variations of the oceanic lithosphere are closely correlated to its thermal structure.

Ridges are elevated above the surrounding seafloor as they consist of rock that is hotter and less dense than the surrounding older seafloor. As the seafloor cools, it contracts and gets more dense. The hot suboceanic mantle beneath mid-ocean ridges is gradually cooled from above by the seawater (which has a mean temperature of ~3 °C below ~500 m water depth) as it moves away from the spreading centers by plate motion. As *oceanic lithosphere* is constantly recycled and fed back to the mantle at *subduction zones*, its age rarely exceeds 150 Ma and nowhere 200 Ma. Given the very low thermal diffusivity of mantle rocks of ~10^{-6} m^{2}s^{-1} and over this time span, this cooling only affects the uppermost 100 km of the oceanic crust and mantle.

This thermal evolution can be simulated by a so-called half-space cooling model (semi-infinite half-space). This kind of models describes the thermal evolution by heat diffusion of a body of infinite extension in one direction (ideally infinite in one direction; in reality such a model can be applied to a body of dimensions that is large compared to the scale that is affected by cooling. For the Earth model, a half-space cooling model can be applied as the surface cooling of the oceanic mantle affects the uppermost 100 km of the mantle at most, which is small compared to the depth extent of the Earth's mantle of ~3,000 km (the oceanic mantle) that is in contact with another body of constant temperature, the ocean). A textbook mathematical description is given, e.g., by Turcotte and Schubert (1982) and by Stüwe (2000). A half-space cooling model of the oceanic mantle can predict the temperature-depth relation for ocean floor of a given age as well as the amount of heat loss per surface unit of the seafloor. This latter value can be measured instrumentally as heat flow and a global set of seafloor heat flow measurements were successfully used to constrain the values that were predicted by half-space models (Sclater et al., 1980; Stein and Stein, 1992).

Seamounts and atolls

One of the most striking consequences of thermal subsidence is the fate of volcanic islands that form at *volcanic hotspots* and slowly sink along with the aging seafloor until they are abraded by the breakwater to form flat-topped edifices, called *guyots*. As subsidence continues, those guyots are found as flat-topped seamounts that rise from the seafloor but have their summits at water depths between some hundreds or thousands of meters with their flat tops still documenting wave abrasion. If the oceanic plate moves in relation to the volcanic hotspot, linear chains of slowly subsiding seamounts like the 6,100 km long *Hawaiian-Emperor seamount chain* form.

In tropical seas, coral reefs form on top of subsiding guyots. If coral reef growth can keep up with the rate of subsidence, then the top of the edifice may never sink below the *photic zone*. In this case a thick carbonate cap can develop on top of the volcanic basal edifice. Reef structures on top of seamounts form *atolls*. This explanation for the evolution of atolls was firstly delineated by Charles Darwin and solved the apparent enigma that coral reefs that need light for their growth rise from lightless depths of the ocean to the sea surface.

Summary

The water depth of the ocean floor in the open ocean (away from the continents) shows a clear relation to the time at which the seafloor was originally formed. Young seafloor that is still close to its place of formation forms chains of submarine mountains, the mid-ocean ridges. When moving away from the axes of the mid-ocean ridges, the seafloor sinks (subsides) with a rate that is related to the square root of the plate age. The reason for subsidence is the progressive cooling of the crust, resulting in contraction and an increase in density of the material, which makes the oceanic lithosphere sink.

Bibliography

Korenaga, T., and Korenaga, J., 2008. Subsidence of normal oceanic lithosphere, apparent thermal expansivity, and seafloor flattening. *Earth and Planetary Science Letters*, **268**, 41–51.

Marty, J. C., and Cazenave, A., 1989. Regional variations in subsidence rate of oceanic plates: a global analysis. *Earth and Planetary Science Letters*, **94**, 301–315.

Parsons, B. S., and Sclater, J. C., 1977. An analysis of the variation of ocean floor bathymetry with age. *Journal of Geophysical Research*, **82**, 803–827.

Sclater, J. G., Jaupart, C., and Galson, D., 1980. The heat flow through oceanic and continental crust and the heat loss of the earth. *Reviews of Geophysics and Space Physics*, **18**, 269–311.

Stein, C., and Stein, S., 1992. A model for the global variation in oceanic depth and heat flow with lithospheric age. *Nature*, **359**, 123–129.

Stüwe, K., 2000. *Geodynamik der Lithosphäre*. Berlin: Springer. 405 p.

Turcotte, D., and Schubert, G., 1982. *Geodynamics – applications of continuum physics to geological problems*. Hoboken: Wiley. 454 pp.

Cross-references

Guyot, Atoll
Geohazards: Coastal Disasters
Lithosphere: Structure and Composition
Marine Heat Flow
Oceanic Spreading Centers
Transform Faults

T

TECHNOLOGY IN MARINE GEOSCIENCES

Sven Petersen
GEOMAR Helmholtz Centre for Ocean Research,
Kiel, Germany

Definition

Technology within this chapter defines the instrumentation and technologies used mainly to observe and sample the seafloor and subseafloor for geoscientific purposes. Equipment used specifically to observe or sample for other scientific disciplines may not be listed.

Introduction

Advances in our understanding of ocean floor processes are closely linked to advances in the technology that is being used. We came a long way from measuring ocean depths using simple lead lines to complex (and expensive) submersibles, remotely operated or autonomous vehicles, or tectonic plate-wide cabled observatories (Kelley et al., 2014; Favali et al., 2015). This is largely related to our needs to better understand marine processes and therefore to investigate over longer periods of time and with higher precision and resolution. Also, seabed sampling has changed dramatically from the first sampling programs performed, for instance, during the famous voyage of the VMS Challenger in 1872–1876 to the drilling vessel Chikyu used today by the Integrated Ocean Drilling Program (IODP). The needs to accurately and precisely take samples from the seabed and the data volume collected from the seabed both increased substantially. At the same time working at sea and at the seabed is increasingly logistically challenging due to types of instruments now being used. This new technology in marine geosciences profited over the last couple of years from the advances in technology made by the oil and gas industry associated with their move to deep ocean off-shore fields.

Here we present a list and short description of seabed sampling techniques and instruments used to observe and measure processes relevant for geosciences. This list is by no means exhaustive and complete but is intended to show the main technologies that are being used. Overall, sampling quality and precision is becoming more and more important also because of a changing vision of the ocean and the necessity to minimize our impact on the seabed during scientific sampling (e.g., InterRidge code of conduct, Devey et al., 2007; IMMS, 2011).

Research vessels

Arguably the most important instrument for any seagoing activity is a ship. These research vessels are usually designed, built, and equipped to carry out research at sea with specific winches and cranes to allow scientists to deploy and recover their instruments from various depths. The size of the research vessel often correlates with the working area. Smaller ships tend to stay in coastal environments while larger vessels are working the high seas. Many countries run several research vessels, and even a basic description of their capabilities and differences is beyond the scope of this paper. The interested reader is referred to publicly accessible lists of research vessels (research vessels by country) and live-feeds of ship tracks that are available online under http://www.sailwx.info/shiptrack/researchships.phtml.

Research vessels are commonly built to allow various scientific instruments to be used on a single vessel. Modern ships have increased deck space and storage capacity to allow better use of containerized equipment. Also, the multidisciplinary approach of modern science and the increasing use of high-tech instrumentation with their support crew require more lab space on board as well as additional berths. Certain tasks may, however, still require a dedicated vessel. Due to the demanding nature of working in polar regions some research vessels have to be icebreakers. Others use a moon pool to allow better launch and recovery of instrumentation. Additionally, some dedicated vessels support larger infrastructure such as submersibles, large autonomous vehicles, or special seismic systems. Some historical background on research vessels can be found in Wuest (1964) and Thiede and Harff (2015, Marine Geosciences: a Short, Ecclectic and Weighted Historic Account).

Seafloor observation methods

Depth measurements and seafloor topography

One, if not the most, important parameter to understand geologic processes in the oceans is water depth and its variability (topography or bathymetry). Early direct measurements included ropes with rocks or lead weights; however, this is only practical in shallow waters. To fully map the deep oceans other indirect methods have to be used. Since radar, used on land to provide the most accurate topographic maps down to centimeter resolution, cannot penetrate water we need other means to map the ocean floor. Conventional seafloor mapping is using a sound signal that is transmitted from the ship downward to the ship's nadir, and the reflection of the signal is received by a ship-based receiver. Using knowledge about the travel time of sound in seawater and being able to accurately measure time the water depth can be deduced using such acoustic echo sounders. While such single-beam echo sounders were already used in the early to mid-twentieth century, the development of multibeam echo sounding (or swath echo sounding) in the 1960s resulted in a revolution of our capabilities to map the ocean floor topography. These systems use electronic beamforming to create a fan-shaped array of narrow beams (currently several hundred beams) that can map a wide swath on each side of the ship's track reaching out several kilometers (up to 20 times the water depth) resulting in a dramatic increase in coverage compared to single-beam echo sounders. These beams are also narrower than single-beam echo sounders and can obtain a higher resolution (commonly 1 % of the water depth) than single-beam systems. At the same time the characteristic and strength of the returned signal (backscatter) can help to characterize the seabed (grain size, lithology, relative sediment thickness, relative age, roughness). These more accurate multibeam mapping systems came into use on ships in the 1980s, but in the deep ocean, these were deployed primarily along mid-ocean ridges, and mapping the deep ocean basins is still limited. The reason for this is mainly that seafloor mapping is difficult, expensive, and slow. Global seafloor mapping programs such as the "General Bathymetric Chart of the Oceans" (Schenke, General Bathymetric Chart of the Oceans (GEBCO)), the "International Bathymetric Chart of the Arctic Ocean" (Jakobsson, International Bathymetric Chart of the Arctic Ocean (IBCAO)), and the visualization tool GeoMapApp from Lamont-Doherty Earth Observatory (www.geomapapp.org) are combining published data from many sources in order to establish minimum standards for data processing and to provide better access to the data. It has been indicated by Becker et al. (2009) that a systematic mapping of the deep oceans by ships would take more than 120 ship-years of survey time while surveying the shallower parts would take even longer (750 ship years) due to the reduced swath width at shallower water depth.

In order to provide a map of the global ocean sparse ship-based measurements are combined with depth estimates derived from other sources (Smith and Sandwell, 1994, 1997). Currently, the best global data sets for ocean topography come from indirect measurements combining available depth soundings from ships with satellite altimetry measuring the ocean surface height and deriving high-resolution marine gravity information (Ryan et al., 2009; Sandwell and Smith, 1997; Sandwell et al., 2014). Satellite altimetry uses pulse-limited radar to measure height of the satellite over the sea surface with high precision. The difference between this height measurement and an independent value derived from precise tracking of the satellites provides a measure for the sea surface height, which, in turn, is related to the geoid. It is possible to use this satellite-derived data because it has been shown that gravity anomalies are highly correlated with seafloor topography and can be used to recover topography (Dixon et al., 1983). The most recent update of the global data was published in late 2014 (http://topex.ucsd.edu/marine_topo/). The resolution of the topographic data derived from this processing has improved over the years, and the global grid cell size is now close to 1 km. Still, today the topography of the Moon, Mars, and Venus is mapped with higher resolution than most areas of the ocean floor (e.g., Tanaka et al., 2014). The current global ocean floor topographic map shows unprecedented detail and is great for recognizing large-scale (global or regional) morphological features (Harris et al., 2014) and understanding processes such as the evolution of oceanic basins, spreading centers, or microplates over time (Sandwell et al., 2014), it is, however, not good enough for local volcanological processes, landslides, or mapping of small-scale variability. Here we rely on the ship-based multibeam bathymetry with a resolution of several tens to 100 m recognizing our needs to go even closer to the seafloor to improve resolution. Over the last couple of years multibeam echosounding systems have been reduced in size so that they can now routinely be used on remotely operated vehicles (ROVs) and autonomous underwater vehicles (AUVs) providing topographic data with a resolution down to a few centimeters, depending on the altitude at which these surveys are flown (Caress et al., 2012; Yoshikawa et al., 2012; Jamieson et al., 2014; Clague et al., 2014). An example of the difference in resolution from satellite-derived topography and ship-based and high-resolution AUV-based bathymetry is provided in Figure 1.

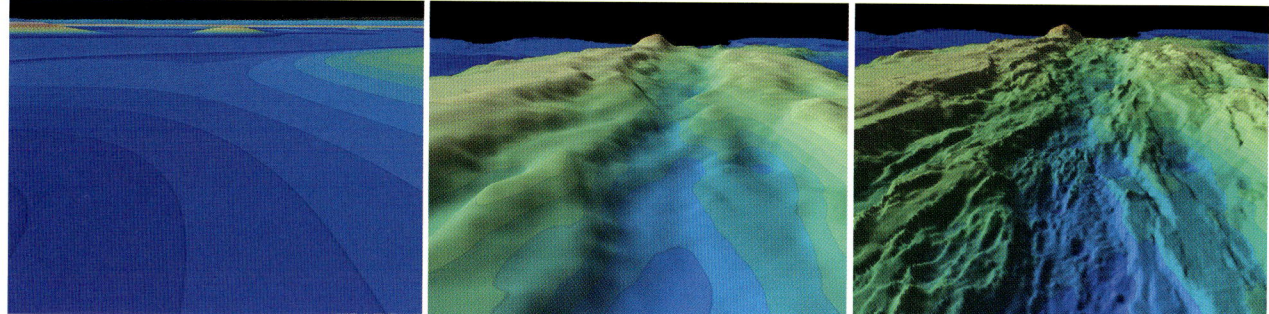

Technology in Marine Geosciences, Figure 1 Example of resolution increase from satellite-altimetry-derived topography (*left*, 3 km per pixel) to ship-based multibeam echosounding (*middle*; 50 m per pixel) to AUV-based high-resolution data (*right*; 1 m per pixel). Example for the Endeavour Segment on the Juan de Fuca Ridge. Data sources: Sandwell and Smith, 1997 for satellite data; ship-based data recorded during cruise TN146 on RV Thomas G. Thomspson; source GeoMapApp; Ryan et al., 2009; AUV-based bathymetry recorded by MBARI's AUV "D.Allan B." in 2008. Width of the rift valley in each image is approximately 1.5 km (image modified from Jamieson et al. (2014)).

Sidescan sonar and synthetic aperture sonar

The method of choice for imaging large areas of the seafloor in high resolution is sidescan sonar imaging. These sensors transmit one beam on each side (broad in the vertical plane and narrow in the horizontal plane) and cover a large portion of the seabed away from the surveying vessel or platform (Blondel, 2010). A coverage ranging from a few tens of meters to 60 km or more can be obtained depending on the altitude of the sensor. Sidescan sonars can, depending on the frequencies, configuration, and altitude, achieve resolutions down to centimeter scale, by far exceeding those of multibeam echosounding systems flown at similar altitude. The most commonly used sidescan sonars are deep-towed systems for regional mapping (Figure 2a), but sidescan sonar sensors are now also routinely operated on autonomous underwater vehicles for high-resolution surveys at the seafloor. A recent development is the use of synthetic aperture sonar in conjunction with AUVs that provides even better along-track resolution and has shown its potential for seabed resource exploration purposes (Hansen et al., 2015; Denny et al., 2015).

Seismology

The use of seismology in marine geosciences is threefold: (1) usually on-shore global networks record distant arrivals from larger earthquakes that are being used to delineate active plate boundaries and to understand the nature of plate motion and large-scale faulting; (2) passive marine seismology using local networks of seismic recorders at the seafloor (ocean-bottom seismometers, OBS) to detect natural earthquakes, and (3) controlled-source experiments that use pulses of compressed air generated by air guns on the ocean surface (Prodehl and Mooney, 2014).

The latter two are being used to investigate the local structures and processes within the Earth's crust and mantle. These include the supply of magma to form the oceanic crust, faulting, convective and hydrothermal processes, the fate of subducting slabs, the formation of seamounts and hot spots, as well as processes effecting continental margins such as submarine landslides that may generate coastal hazards, or the distribution and fluid flow associated with gas hydrates and mud volcanoes.

Passive seismology uses a local network of ocean bottom seismometers that is deployed for prolonged time (up to 6 months or 1 year) to record natural earthquake swarms or tremor to provide information related to the source mechanisms and source area at depth. Ocean-bottom seismometers (Figure 2b) are, in principle, a carrier platform for a seismic sensor and the recording unit consisting of pressure-tight cylinders hosting the sensors, electronics and batteries in addition to an anchor weight that can be released acoustically, the floats for buoyancy, as well as equipment to ease detection and recovery when afloat such as lights, a flag, and the recovery rope.

Controlled-source or active seismology is aimed at obtaining higher-resolution images of the upper layers of the earth's crust than can be obtained by passive seismology. It is based on the principle that an artificially generated sound wave travels through the subsurface and returns to the surface where it can be analyzed. Two controlled-source seismic methods are being used: reflection and refraction seismology. The classical acquisition method in marine reflection seismology is a 2D geometry where a single seismic streamer is towed behind the vessel recording the reflected waves. New systems use a bundle of several short streamers that is towed allowing for a 3D geometry. In refraction seismology air guns are also being used as transmitters, but the receiving units (OBS) that are similar to those used for passive seismology are stationary at the seafloor. A detailed account of reflection and refraction seismology and their instrumentation used is given in Hübscher and Gohl (Reflection/Refraction Seismology) and (Pecher et al., 2011).

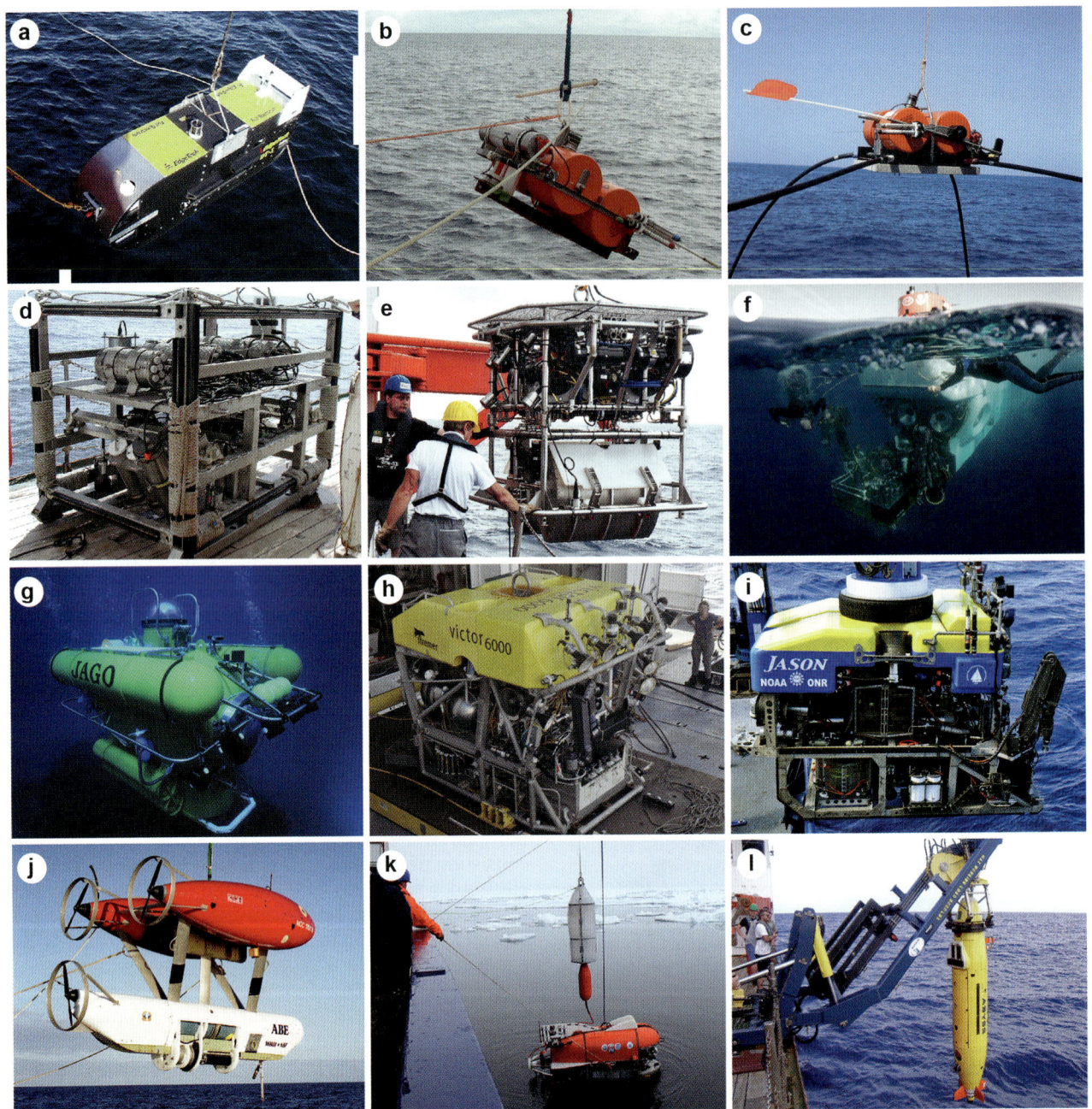

Technology in Marine Geosciences, Figure 2 Sensor platforms used in marine geosciences: (**a**) deep-towed sidescan sonar (Photo J. Bialas), (**b**) ocean-bottom seismometer (OBS; photo U. Büchele), (**c**) electromagnetic ocean-bottom receiver unit (OBEM, note the long antennas; photo H. Grant), (**d**) towed camera sled, (**e**) the instrument platform "Hybis" (photo C. Berndt), (**f**) submersible "Alvin" prepared for diving (photo C. Linder, WHOI) (**g**) submersible "Jago" (source Geomar), (**h**) ROV "Victor" from Ifremer, (**i**) ROV "Jason-2" (photo J. McDermott; WHOI), (**j**) AUV "Abe" (photo J. Hayes, WHOI), (**k**) the under-ice-going AUV "Nereid" also from WHOI, and (**l**) the torpedo-shaped 6,000 m-rated AUV "Abyss" from Geomar.

Magnetics

As new oceanic crust is formed at the mid-ocean ridges, it cools and acquires a record of past geomagnetic field variations forming a pattern of normal and reverse polarized crust that are readily identifiable in the magnetic data measured at the sea surface (Gee and Kent, 2007). These patterns of the ocean's magnetic field were fundamental in developing the seafloor spreading and plate tectonic theories (Vine and Matthew 1963; Heirtzler et al., 1968). Marine magnetic measurements are important for providing a better understanding of the global magnetic field and polarity changes (Gee and Kent, 2007; Tominaga et al., 2008), as global bathymetric data affording to combine modern and old measurements into a single global model are important (Quesnel et al., 2009). Presently a large range of instruments exists that measure from towed platforms thereby avoiding the ship's electric and magnetic field and, more recently, from submersibles, deep-towed vehicles, ROVs, and AUVs (Tivey and Dymont, 2010). These high-resolution surveys allow for a better understanding of local magmatic activity (Shah et al., 2003) and for exploration of seafloor massive sulfides due to changes in the magnetization caused by ascending hydrothermal fluids (Tivey and Johnson, 2002; Tivey and Dymont, 2010).

Electromagnetics (EM)

Similar to seismology marine electromagnetics can be divided into passive (magnetotelluric) or active methods (controlled-source electromagnetics, CSEM). Electromagnetic surveying is used to investigate the distribution of electrical properties, mostly electrical conductivity, of geologic formations in the subseafloor. Marine magnetotelluric EM measures the natural variations of the Earth's magnetic field, which induce electromagnetic fields in the seabed. Controlled-source EM methods measure the electromagnetic fields associated with artificially induced subsurface electrical currents created by magnetic coils or grounded electric wires (Constable, 2013).

For a CSEM survey, an electric dipole transmitter is towed from a ship and passes a time-varying current between electrodes that are spaced apart by up to a few hundred meters, producing an electromagnetic field that diffuses through the conductive surrounding, i.e., the ocean and the seabed, as well as through the air in shallow-water applications. The transmitter is usually towed close to the seafloor to maximize the energy that couples to the seabed. The diffusion of the electromagnetic field away from the transmitter depends on the conductivity of seawater and the seafloor. The resulting attenuation and phase shift of the transmitted electromagnetic field is recorded by an array of receivers. These receivers can be stationary at the seafloor (Figure 2c) or can be towed together with the transmitter (Schwalenberg et al., 2010; Anderson and Mattson, 2010). Details of the varying methods currently in use are given in Edwards (2005) and Constable (2013). The replacement of saline pore fluids in sediments by hydrocarbons and the resulting formation of resistive layers in the subseafloor make controlled-source EM a prime choice in oil and gas exploration and gas hydrate research (Constable, 2010) and the investigation of mud volcanoes (Hölz et al., 2015). Recently, controlled-source EM methods have shown their potential to identify layers with higher conductivity in the subseafloor (Swidinsky et al., 2012) extending marine mineral exploration of seafloor massive sulfides using CSEM (Kowalczyk, 2008) to the subseafloor.

Gravity

We have seen above that gravity data can be important in marine geosciences even from space. However, marine gravity data is also measured from ships and other platforms and is important in recognizing crustal structure and crustal thickness variations because they reflect variations in subsurface mass and density (Watts and Talwani, 1974). Measurements of the gravitational field at sea are compared to the theoretical gravity given by a reference Earth model for the same location and presented as the free-air gravity anomaly (Barckhausen and Heyde, Gravity Field). Variations in the sea surface have to be taken into account and result in the Bouguer anomaly, which is assumed to reflect variations in the density distribution below the seafloor. Considering effects of water/crust, crust/mantle interfaces, as well as temperature, density variations in the underlying mantle are used to calculate the residual mantle Bouguer anomaly that can be interpreted with respect to crustal thickness and structure (Lin et al., 1990, Blackman et al., 2008).

Heat flow

Heat flow measurements in sedimented areas of the ocean crust were instrumental in understanding the processes of seafloor spreading and plate tectonics. Soon after recognizing that seafloor spreading takes place at mid-ocean ridges heat flow studies indicated that conduction of heat through the lithosphere cannot be the only process governing energy exchange between solid earth and the oceans and that convective processes must also occur (Lister 1972; Davis and Lister, 1977; Sleep and Wolery, 1978; Stein and Stein, 1992). These studies came just before hydrothermal activity was actually found near the Galapagos Islands in 1977 (Corliss et al., 1979) and on the East Pacific Rise in 1979 (Spiess et al., 1980). Later heat flow studies have been used to constrain numerical models of crustal fluid flow at ridge flanks and in off-axis sedimented regions where basement outcrops and seamounts have been shown to be important for lithospheric cooling (Fisher et al., 2003) indicating that this off-axis heat flux can be comparable to that of black smoker vent fields (Hutnak et al., 2008). A detailed account of marine heat flow is given in Sclater et al. (Marine Heat Flow).

For heat flow studies temperature gradients are measured by an array of thermistors in a narrow tube that is inserted into the sediment by a corer. Thermal conductivity is obtained by heating a wire for short times and measuring the temperature decay as the heat dissipates through the sediment. Smaller versions of such instruments are also used from submersibles (Johnson et al., 1993). Recent developments using thermal blankets also allow for high-resolution measurement in nonsedimented areas such as volcanic ridges and within hydrothermal vent fields (Johnson et al., 2010).

Sediment physics

The in situ physical properties of sediments are important for marine geoscience related to coastal hazards but also for constructional work in sedimented environments off shore. Conventional and *free-fall cone penetrometers* (CPTs) have been developed that measure cone resistance, sleeve friction, pore pressure, tilt, and temperature (Lunne et al., 1997; Stegmann et al., 2006). These parameters are used to determine shear strength and excess pore pressure that are important trigger mechanisms for sediment movement and slope stability.

Visual imaging

Visual imaging of the seafloor is a fundamental method in geology. Today a number different systems are in use ranging from simple deep-towed camera systems (Figure 2d) that only use photography and video to image the seafloor to multipurpose towed platforms that can also additionally run other sensors (Figure 2e) to even more complex systems such as manned submersibles, remotely operated vehicles (ROVs), and autonomous underwater vehicles (AUVs). Examples of these kinds of instrumentation are provided in Figure 2f–l.

Today, high-definition cameras are standard on many submersibles and ROVs allowing better control on positioning of the instrument and selection of images. Numerous submersibles (or human occupied vehicles, HOVs) are run by research institutions and include "Alvin" from Woods Hole Oceanographic Institution (WHOI), which started its life 50 years ago but has been refurbished a couple of times (Figure 2f; Humphris et al., 2014). France, Japan, Russia, and China currently also run research submersibles for deep ocean research with Chinas "Jiaolong" being able to reach 7,000 m water depth. The French "Nautile," Japan's "Shinkai6500," and the Russian "Mir" submersibles all have depth ratings beyond 6,000 m. Manned submersibles are usually occupied by three persons and can take one to two scientists to the seafloor. They stay at the bottom for several hours depending on water depth and battery capacity and enable traversing speeds of 1 knot (0.5 m/s). Submersibles have lighting, multiple cameras, and manipulator arms for imaging and working at the seabed. Additionally, they can take various scientific payloads of up to 100 kg or so to the seafloor. Most submersibles need a dedicated ship due to the logistical requirements for deployment and recovery and a large number of crew for support. An exception to this rule is Germany's manned submersible "Jago," which is containerized and can be deployed from smaller vessels but has a depth limit of 400 m (Figure 2g).

Over the last decades manned submersibles have been replaced to some extent by deep-sea remotely operated vehicles that were developed based on existing towed platforms that carried instruments to the seafloor (Figure 2h). They have a couple of advantages over manned submersibles including, in theory, unlimited power supply through the umbilical providing longer endurance. In theory ROVs can stay at the bottom for several days with pilots and scientists being exchanged at regular intervals. ROVs are also smaller, cheaper to run, commonly use less personnel, and provide no risk to humans working at the seafloor.

The ROVs are attached to the vessel by a cable providing energy and allowing data transmission including live video (Figure 2i). The ROVs are not towed but use their thrusters to move at similar speeds as submersibles despite the fact that the ROV and the vessel have to be moved in parallel. The ROV pilots are maneuvering from a control container on the vessel guided by scientists. Instrumentation and payload for work-class ROVs is similar to that of the manned submersibles. Navigation for both types of instruments is by Ultra Short Base Line (USBL) acoustic communication and dead reckoning.

Both manned submersibles and ROVs move slowly at the seafloor, and in order to map larger areas of the seafloor faster and more efficiently autonomous underwater vehicles (AUVs) have been developed that are not attached to the vessel, drive self-propelled, and use batteries to power the instrument. They commonly follow a predefined track and use several sensors (such as sidescan sonar, multibeam echosounding, subbottom profilers, magnetometers, temperature sensors, chemical sensors, or cameras) to investigate the seafloor or the water column (Wynn et al., 2014). AUVs are able to navigate using either acoustic beacons on the seafloor (so-called Long Base Line (LBL) navigation) or a combination of USBL acoustic communication and inertial navigation (based on dead reckoning using a combination of depth sensors, inertial sensors, and Doppler velocity sensors).

Most AUVs are torpedo shaped and can reach a speed of several knots allowing larger areas to be covered with high resolution. Others have a more complex structure and move more slowly providing them with the capability to move or hover over complex terrain at lower altitude. One of the earliest multisensor AUVs for the deep sea was "Abe" run by Woods Hole Oceanographic Institution that was commissioned in 1994 and was the prime AUV of its time before being lost at sea in 2010 after more than 200 scientific dives. It was also famous for its V-shape structure resembling "Starship Enterprise" (Figure 2j). Numerous AUVs are now being used around the globe; however, few systems are able to work at great depths of 5,000 m or more or are working under ice cover such as the most recent addition to WHOI's fleet, the "Nereid" (Figure 2k). The usefulness of such deep-diving AUVs outside of its science usage was demonstrated to the public by the use of three REMUS6000 AUVs during the search of the lost Air France flight AF447 in 2010 and 2011, when these systems found the wreck site (Figure 2l).

One of the biggest advantages of AUVs is that once the AUV is following its predefined track the research vessel is free for other research even at a greater distance to the AUV survey area maximizing ship time use.

Seafloor sampling tools

Simple geologic sampling tools can be divided into three groups (Sterk and Stein, 2015): (1) towed systems that are dragged along the seafloor, (2) cable-linked spot-sampling systems that take a sample from a well-defined location (box corer, grab systems), and (3) devices that can take samples from the subseafloor. These include simple devices such as gravity corers, lander-type seabed drills, as well as complex drilling operations from surface vessels.

Towed sampling instruments

One of the oldest and most reliable sampling tools are dredges (Figure 3a). These box, bucket, or chain bag dredges are deployed on a steel cable and towed behind the ship for several hundreds of meters or even longer. These simple and cost-effective instruments collect samples over the entire track length, and since they commonly lack a navigation device, the location of the samples from the seabed is only crudely known. Dredges, however, have the advantage that they can be deployed at severe weather conditions, when most sophisticated instruments are long out of business.

Spot-sampling instruments

For surface sediment sampling *box corers* or *multicorers* are commonly used (Figure 3b). Box corers are deployed from a steel wire and use a heavy weight to press a rectangular or square steel box into the substrate. Lifting the box corer first moves a sliding plate across the base of the box enclosing the sediment before lifting the instrument from the seabed. Penetration into the sediment is several tens of centimeters and the area of the seafloor that is sampled is in the order of $0.5\ m^2$. Box corers cannot be used for sampling consolidated hard rocks. A similar tool for sampling in soft sediment is a multicorer that uses several plastic tubes that are pushed into the sediment by a weight (Figure 3c). The multicorer is also deployed from a wire. Penetration is again in the order of several tens of centimeters; however, the area that is being sampled is much smaller. Multicorers are often used to collect multiple samples from a given area for investigations of the sediment-water interface. For better control and documentation of the actual sampling site camera systems can be used to take either photographs or video information.

A suite of grab samplers that use opposing jaws is used for sediment sampling (simple bucket grabs) or for sampling of indurated material, hard crusts, and rocks (powered TV-guided grab systems). The latter were originally developed for sampling of marine minerals from the seabed in the 1980s and use opposing jaws that close hydraulically and thereby breaking off even massive sulfides and basalt (Figure 3d). These systems have the advantage of being able to sample larger, more representative pieces of the seabed than sophisticated manipulator arms such as those used by submersibles and robotic systems (see above). Such TV-guided grab systems can sample close to $1\ m^2$ of seabed up to 50 cm deep. The live camera feed allows for a controlled sampling and even resampling (after opening the jaws again) in case the sample size or nature was not sufficient.

Sediment traps are sampling devices in the water column used to measure the quantity of sinking particles over prolonged periods of time (often several months) and to collect them for shore-based analyses. In sedimentology they allow the characterization of the sedimentation processes, their rates, and the fluxes of chemical components to the seafloor. Sediment traps usually consist of an upward-facing funnel that collects sinking particles into a trap. Rotating sampling devices below the funnel commonly allow for discrete subsampling using predefined time schedules thereby recording changes over time.

Subseafloor sampling instruments

In order to sample the upper few meters of sediments a number of corer devices can be used. These include simple gravity corers and piston corers as well as more advanced vibro corers. *Gravity corers* are the simplest form of coring device and consist of a steel pipe, typically with several tens to hundreds of kilograms of weight added to the top, a plastic liner for easier sample retrieval, and a core catcher to prevent sample loss (Figure 3e). The gravity corer is lowered close to the seafloor on a wire and then allowed to freely fall into the sediment. The length of the core barrel and the weight used can vary due to the sampling needs and/or the stiffness of the sediment preventing easy penetration. For longer sediment cores *piston corers* are used that are similar to gravity corers but use a small piston that helps to draw the core into the barrel more effectively. This piston is set at the bottom entrance to the barrel and moves up the barrel as it penetrates the seabed. The longest core recovered using such a system was collected aboard the French research vessel R/V Marion Dufresne and collected 64.4 m of sediment from a single deployment (Sterk and Stein, 2015). In gas hydrate research autoclave piston cores have been developed to retrieve the sediments under in situ hydrostatic pressure (Abegg et al., 2008).

Vibro corers consist in general of a tripod-supported frame containing a vibrating head that is driven by hydraulic, pneumatic, mechanical, or electrical power and which uses vibration and gravity to drive a core barrel into the seabed. The entire vibro corer is launched from the surface vessel via a cable. After sampling the vibration is turned off and the core barrel is withdrawn back into the housing followed by retrieval of the instrument using a winch and cable.

For hard rock sampling a different toolset is necessary. Here, *remotely controlled robotic seabed drills* are being used that are lowered to the seafloor via a winch and cable (McGinnis, 2009). The drill string is handled via robotic machinery at the seafloor and remotely controlled by an operator on the vessel. This isolates the drilling process from vessel motion and subsequently enables critical drilling parameters to be precisely controlled. Research institutes currently run a number of these systems globally, but commercial companies are entering this market. Early versions used single-pass drilling (Figure 3f) while modern systems use

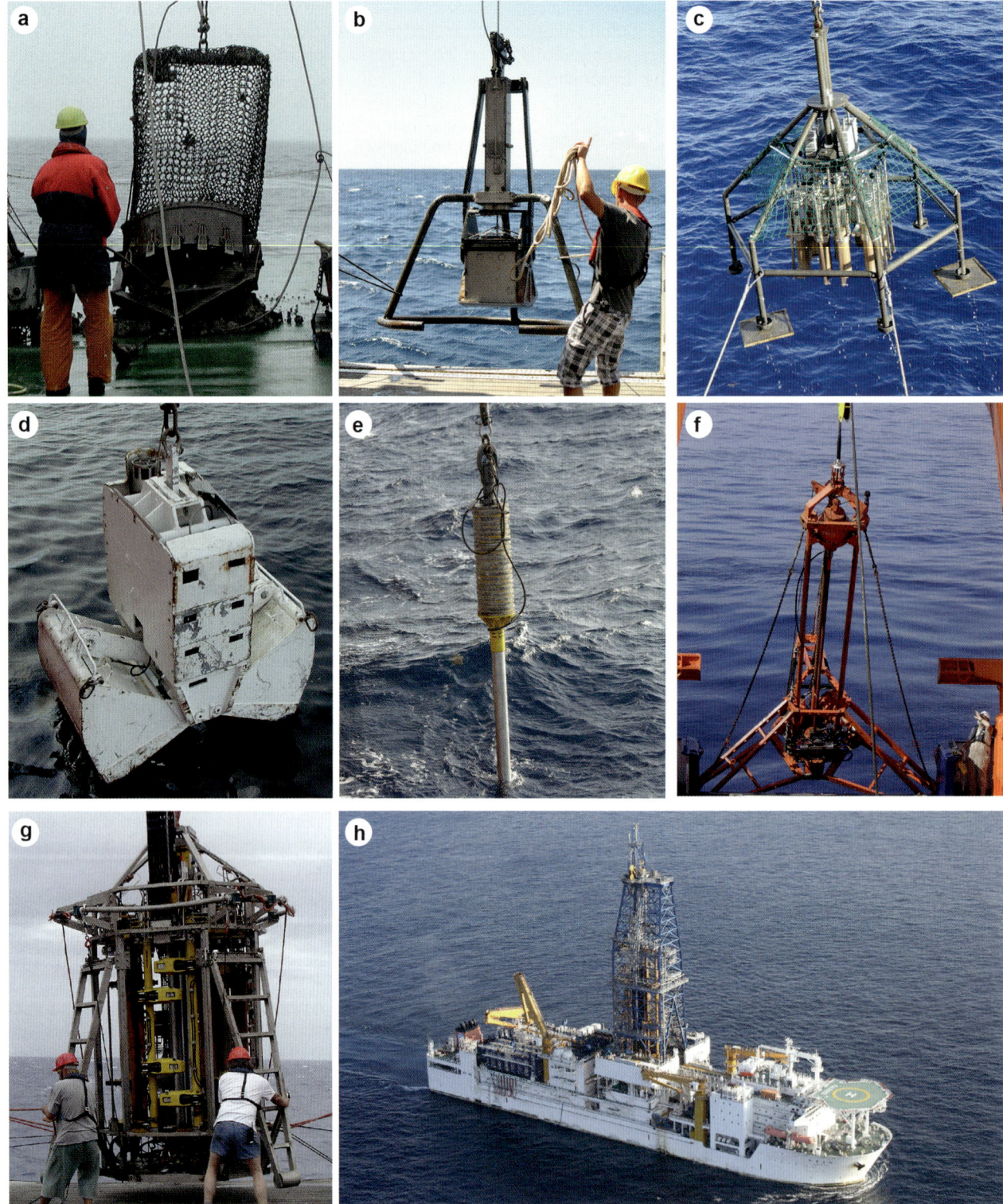

Technology in Marine Geosciences, Figure 3 Sampling gear used in marine geosciences: (**a**) chain bag dredge, (**b**) box corer (photo P. Ribeiro), (**c**) multicorer (photo T. Walter), (**d**) TV-guided grab, (**e**) gravity corer (photo T. Walter), (**f**) vibro corer Rockdrill-1 (source British Geological Survey), (**g**) Rockdrill-2 (source British Geological Survey), and (**h**) Drillship Chikyu (source JAMSTEC).

built-in rod racks or carousels (Figure 3g) to store additional drill pipe, casing, and core barrels (Freudenthal and Wefer, 2007). Maximum penetration depths of these drill rigs are currently around 200 m and are determined by the number of rods and core barrels that can be stored. Most drill rigs, however, penetrate only a few tens of meters into the seabed. While this is significant it is not comparable to the penetration depths that can be achieved by *vessel-mounted drill systems* such as the drill ships "Joides Resolution" or "Chikyu," the latter having drilled to 2,466 m below the seafloor in 1,180 m water depth (Figure 3h; Suyehiro, Ocean Drilling). Seabed drilling is, in general, cheaper than the use of dedicated drilling vessels and since the systems are mobile can be done on ships of opportunity. A detailed account of ship-based ocean drilling programs is provided by Suyehiro (Ocean Drilling).

Summary

This chapter describes the most commonly used technologies and instruments for observing and sampling in the field of marine geosciences and refers to a number of special papers within this encyclopedia and beyond. This chapter is meant as an introduction showing the variability and broad scale of intruments being used. Since technology is a vast and rapidly changing field, numerous specific instruments used throughout marine geoscience are undouptfully missing in this compilation.

Bibliography

Abegg, F., Hohnberg, H. J., Pape, T., Bohrmann, G., and Freitag, J., 2008. Development and application of pressure-core-sampling systems for the investigation of gas- and gas-hydrate-bearing sediments. *Deep-Sea Research I: Oceanographic Research Papers*, 55, 1590–1599.

Anderson, C., and Mattson, J., 2010. An integrated approach to marine electromagnetic surveying using a towed streamer and source. *First Break*, 28, 71–75.

Becker, J. J., Sandwell, D. T., Smith, W. H. F., Braud, J., Binder, B., Depner, J., Fabre, D., Factor, J., Ingalls, S., Kim, S.-H., Ladner, R., Marks, K., Nelson, S., Pharaoh, A., Trimmer, R., Von Rosenberg, J., Wallace, G., and Weatherall, P., 2009. Global bathymetry and elevation data at 30 arc seconds resolution: SRTM30_PLUS. *Marine Geodesy*, 32, 355–371.

Blackman, D. K., Karner, G. D., and Searle, R., 2008. Three-dimensional structure of oceanic core complexes: effects on gravity signature and ridge flank morphology, Mid-Atlantic Ridge 30°N. *Geochemistry, Geophysics, Geosystems*, 9, Q06007, doi:10.1029/2008GC001951.

Blondel, P., 2010. *The Handbook of Sidescan Sonar*. Berlin/Heidelberg: Springer, 316 pp.

Caress, D. W., Clague, D. A., Paduan, J. B., Martin, J. F., Dreyer, B. M., Chadwick, W. W., Jr., Denny, A., and Kelley, D. S., 2012. Repeat bathymetric surveys at 1-metre resolution of lava flows erupted at Axial Seamount in April 2011. *Nature Geoscience*, 5, 483–488.

Clague, D. A., Dreyer, B. M., Paduan, J. B., Martin, J. F., Caress, D. W., Gill, J. B., Kelley, D. S., Thomas, H., Portner, R. A., Delaney, J. R., Guilderson, T. P., and McGann, M. L., 2014. Eruptive and tectonic history of the Endeavour Segment, Juan de Fuca Ridge, based on AUV mapping data and lava flow ages. *Geochemistry, Geophysics, Geosystems*, 15, 3364–3391.

Constable, S., 2010. Ten years of marine CSEM for hydrocarbon exploration. *Geophysics*, 75, 75A67–75A81.

Constable, S., 2013. Review paper: instrumentation for marine magnetotelluric and controlled source electromagnetic sounding. *Geophysical Prospecting*, 61, 505–532.

Corliss, J. B., Dymond, J., Gordon, L. I., Edmond, J. M., Von Herzen, R. P., Ballard, R. D., Green, K., Williams, D., Bainbridge, A., Crane, K., and van Andel, T. H., 1979. Submarine thermal springs on the Galapagos rift. *Science*, 203, 1073–1083.

Davis, E. E., and Lister, C. R. B., 1977. Heat flow measured over the Juan de Fuca Ridge : evidence for widespread, hydrothermal circulation in a highly heat transportive crust. *Journal of Geophysical Research*, 82, 4845–4860.

Denny, A. R., Saebo, T. O., Hansen, R. E., and Pedersen, R. B., 2015. The use of synthetic aperture sonar to survey seafloor massive sulfide deposits. *The Journal of Ocean Technology*, 10, 37–53.

Devey, C. W., Fisher, C. R., and Scott, S., 2007. Responsible science at hydrothermal vents. *Oceanography*, 20, 162–171.

Dixon, T. H., Naraghi, M., McNutt, M. K., and Smith, S. M., 1983. Bathymetric prediction from Seasat altimeter data. *Journal of Geophysical Research*, 88, 1563–1571.

Edwards, N., 2005. Marine controlled source electromagnetics: principles, methodologies, future commercial applications. *Surveys in Geophysics*, 26, 675–700.

Favali, P., Beranzoli, L., and de Santis, A., 2015. *Seafloor Observatories. A New Vision of the Earth from the Abyss*. Berlin/Heidelberg: Springer, 676 pp.

Fisher, A. T., Davis, E. E., Hutnak, M., Spiess, V., Zühlsdorff, L., Cherkaoui, A., Christiansen, L., Edwards, K., Macdonald, R., Villinger, H., Mottl, M. J., Wheat, C. G., and Becker, K., 2003. Hydrothermal recharge and discharge across 50 km guided by seamounts on a young ridge flank. *Nature*, 421, 618–621.

Freudenthal, T., and Wefer, G., 2007. Scientific drilling with the sea floor drill rig MEBO. *Scientific Drilling*, 5, doi: 10.2204/iodp.sd.5.11.2007.

Gee, J. S., and Kent, D. V., 2007. Source of oceanic magnetic anomalies and the geomagnetic polarity timescale. *Treatise in Geophysics*, 5, 455–507.

Hansen, R. E., Callow, H. J., Sabo, T. O., and Synnes, S. A. V., 2015. Challenges in seafloor imaging and mapping with synthetic aperture sonar. *IEEE Transactions on Geoscience and Remote Sensing*, 49, 3677–3687.

Harris, P.T., Macmillan-Lawler, M., Rupp, J., Baker, E.K., 2014. Geomorphology of the oceans. Marine Geology 352, 4–24. doi:10.1016/j.margeo.2014.01.011.

Heirtzler, J. R., Dickson, G. O., Herron, E. M., Pitman, W. C., and Le Pichon, X., 1968. Marine magnetic anomalies, geomagnetic field reversals, and motions of the ocean floor and continents. *Journal of Geophysical Research*, 73, 2119–2136.

Hölz, S., Swidinsky, A., Sommer, M., Jegen, M., and Bialas, J., 2015. The use of rotational invariants for the interpretation of marine CSEM data with a case study from the North Alex Mud Volcano, West Nile Delta. *Geophysical Journal International*, 201, 224–245.

Humphris, S. E., German, C. R., and Hickey, J. P., 2014. Fifty years of deep ocean exploration with the DSV Alvin. *Eos, Transactions American Geophysical Union*, 95, 181–182.

Hutnak, M., Fisher, A. T., Harris, R., Stein, C., Wang, K., Spinelli, G., Schindler, M., Villinger, H., and Silver, E., 2008. Large heat and fluid fluxes driven through mid-plate outcrops on ocean crust. *Nature Geoscience*, 1, 611–614.

IMMS, 2011. International Marine Minerals Society Code for Environmental Management of Marine Mining (www.immsoc.org/IMMS_code.htm).

Jamieson, J. W., Clague, D. A., and Hannington, M. D., 2014. Hydrothermal sulfide accumulation along the Endeavour Segment, Juan de Fuca Ridge. *Earth and Planetary Science Letters*, **395**, 136–148.

Johnson, H. P., Becker, K., and Von Herzen, R. P., 1993. Near-axis heat flow measurements on the northern Juan de Fuca Ridge: implications for fluid circulation in oceanic crust. *Geophysical Research Letters*, **20**(17), 1875–1878.

Johnson, H. P., Tivey, M. A., Bjorklund, T. A., and Salmi, M. S., 2010. Hydrothermal circulation within the Endeavour Segment, Juan de Fuca Ridge. *Geochemistry, Geophysics, Geosystems*, **11**, doi:10.1029/2009GC002957.

Kelley, D. S., Delaney, J. R., and Juniper, S. K., 2014. Establishing a new era of submarine volcanic observatories: Cabling Axial Seamount and the Endeavour Segment of the Juan de Fuca Ridge. *Marine Geology*, **352**, 426–450.

Key, K., 2011. Marine electromagnetic studies of seafloor resources and tectonics. *Surveys in Geophysics*, **33**, 135–167.

Kowalczyk, P., 2008. Geophysical prelude to first exploitation of submarine massive sulphides. *First Break*, **26**, 99–106.

Lin, J., Purdy, G. M., Schouten, H., Sempere, J. C., and Zervas, C., 1990. Evidence from gravity data for focused magmatic accretion along the Mid-Atlantic Ridge. *Nature*, **344**, 627–632.

Lister, C. R. B., 1972. On the thermal balance of a mid-ocean ridge. *Geophysical Journal of the Royal Astronomical Society*, **26**, 515–535.

Lunne, T., Robertson, P. K., and Powell, J. J. M., 1997. *Cone Penetrating Testing. Geotechnical Practice.* Spon Press. New York: Blackie Academic/Routledge Publishing, 312 pp.

McGinnis, T., 2009. Seafloor drilling. In Bar-Cohen, Y., and Zacny, K. (eds.), *Drilling in Extreme Environments: Penetration and Sampling on Earth and Other Planets*. Weinheim: Wiley, pp. 309–345.

Pecher, I. A., Bialas, J., and Flueh, E. R., 2011. Ocean bottom seismics. In Gupta, H. K. (ed.), *Encyclopedia of Solid Earth Geophysics*. Heidelberg: Springer. Encyclopedia of Earth Science Series, Vol. 1–2, pp. 901–918.

Prodehl, C., and Mooney, W. D., 2014. Exploring the Earths Crust: History and Results of Controlled-source Seismology. *Geological Society of America Memoir*, **208**, 764 pp.

Quesnel, Y., Catalan, M., and Ishihara, T., 2009. A new global marine magnetic anomaly data set. *Journal of Geophysical Research*, **114**, B04106, doi:10.1029/2008JB006144.

Ryan, W. B. F., Carbotte, S. M., Coplan, J. O., O'Hara, S., Melkonian, A., Arko, R., Weissel, R. A., Ferrini, V., Goodwillie, A., Nitsche, F., Bonczkowski, J., and Zemsky, R., 2009. Global multi-resolution topography synthesis. *Geochemistry, Geophysics, Geosystems*, **10**, Q03014, doi:10.1029/2008GC002332.

Sandwell, D. T., and Smith, W. H. F., 1997. Marine gravity anomaly from Geosat und ERS1 satellite altimetry. *Journal of Geophysical Research*, **102**, 10039–10054.

Sandwell, D. T., Muller, R. D., Smith, W. H. F., Garcia, E., and Francis, R., 2014. New global marine gravity model from CryoSat-2 and Jason-1 reveals buried tectonic structure. *Science*, **346**, 65–67, doi:10.1126/science.1258213.

Schwalenberg, K., Haeckel, M., Poort, J., and Jegen, M., 2010. Evaluation of gas hydrate deposits in an active seep area using marine controlled source electromagnetics: results from Opouawe Bank, Hikurangi Margin, New Zealand. *Marine Geology*, **272**, 79–88.

Shah, A. K., Cormier, M.-H., Ryan, W. B. F., et al., 2003. Episodic dike swarms inferred from near-bottom magnetic anomaly maps at the southern East Pacific Rise. *Journal of Geophysical Research*, **108**(B2), 2097, doi:10.1029/2001JB000564.

Sleep, N. H., and Wolery, T. J., 1978. Egress of hot water from mid-ocean ridge hydrothermal systems : some thermal constraints. *Journal of Geophysical Research*, **83**, 5913–5922.

Smith, W. H. F., and Sandwell, D. T., 1994. Bathymetric prediction from dense satellite altimetry and sparse shipboard bathymetry. *Journal of Geophysical Research*, **99**, 21803–21824.

Smith, W. H. F., and Sandwell, D. T., 1997. Global sea floor topography from satellite altimetry and ship depth soundings. *Science*, **277**, 1956–1962.

Spiess, F. N., Macdonald, K. C., Atwater, T., Ballard, R., Carranza, A., Cordoba, D., Cox, C., Diaz Garcia, V. M., Francheteau, J., Guerrero, J., Hawkins, J., Haymon, R., Hessler, R., Juteau, T., Kastner, M., Larson, R., Luyendyk, B., Macdougall, J. D., Miller, S., Normark, W., Orcutt, J., and Rangin, C., 1980. East Pacific Rise: hot springs and geophysical experiments. *Science*, **207**, 1421–1433.

Stegmann, S., Mörz, T., and Kopf, A., 2006. Initial results of a new free fall-cone penetrometer (FF-CPT) for geotechnical in situ characterisation of soft marine sediments. *Norwegian Journal of Geology*, **86**, 199–208.

Stein, C. A., and Stein, S., 1992. A model for the global variation in oceanic depth and heat flow with lithospheric age. *Nature*, **359**, 123–129.

Sterk, R., and Stein, J. K., 2015. Seabed mineral deposits – an overview of sampling techniques and future developments, In: *Paper Presented at the Deep-Sea Mining Summit*, Aberdeen, pp. 1–29.

Swidinsky, A., Hölz, S., and Jegen, M., 2012. On mapping seafloor mineral deposits with central loop transient electromagnetics. *Geophysics*, **77**, E171–E184.

Tanaka, K. L., Robbins, S. J., Fortezzo, C. M., Skinner, J. A., and Hare, T. M., 2014. The digital global geologic map of Mars: chronostratigraphic ages, topographic and crater morphologic characteristics, and updated resurfacing history. *Planetary and Space Science*, **95**, 11–24.

Tivey, M. A., and Dymont, J., 2010. The magnetic signature of hydrothermal systems in slow spreading environments. In Rona, P. A. (ed.), *Diversity of Hydrothermal Systems on Slow Spreading Ocean Ridges*. Washington, DC: American Geophysical Union. Geophysical Monograph Series, Vol. 188, pp. 43–66.

Tivey, M. A., and Johnson, H. P., 2002. Crustal magnetization reveals subsurface structure of Juan de Fuca Ridge hydrothermal vent fields. *Geology*, **30**, 979–982.

Tominaga, M., Sager, W. W., Tivey, M. A., and Lee, S.-M., 2008. Deep-tow magnetic anomaly study of the Pacific Jurassic Quiet Zone and implications for the geomagnetic polarity reversal timescale and geomagnetic field behavior. *Journal of Geophysical Research*, **113**, B07110, doi:10.1029/2007JB005527.

Vine, F. J., and Matthew, D. H., 1963. Magnetic anomalies over oceanic ridges. *Nature*, **199**, 947–949.

Watts, A. B., and Talwani, M., 1974. Gravity anomalies seaward of deep-sea trenches and their tectonic implications. *Geophysical Journal of the Royal Astronomical Society*, **36**, 57–90.

Wuest, G., 1964. The major deep-sea expeditions and research vessels 1873–1960. *Progress in Oceanography*, **2**, 3–52.

Wynn, R. B., Huvenne, V. A. I., Le Bas, T. P., Murton, B. J., Connelly, D. P., Bett, B. J., Ruhl, H. A., Morris, K. J., Peakall, J., Parsons, D. R., Sumner, E. J., Darby, S. E., Dorrell, R. M., and Hunt, J. E., 2014. Autonomous Underwater Vehicles (AUVs): their past, present and future contributions to the advancement of marine geoscience. *Marine Geology*, **352**, 451–468.

Yoshikawa, S., Okino, K., and Asada, M., 2012. Geomorphological variations at hydrothermal sites in the southern Mariana Trough: relationship between hydrothermal activity and topographic characteristics. *Marine Geology*, **303–306**, 172–182.

Cross-references

Bottom Simulating Seismic Reflectors (BSR)
Continental Slope

General Bathymetric Chart of the Oceans (GEBCO)
Geohazards: Coastal Disasters
Gravity Field
Hot Spots and Mantle Plumes
International Bathymetric Chart of the Arctic Ocean (IBCAO)
Marine Gas Hydrates
Marine Geosciences: a Short, Ecclectic and Weighted Historic Account
Marine Heat Flow
Marine Mineral Resources
Mid-ocean Ridge Magmatism and Volcanism
Ocean Drilling
Oceanic Spreading Centers
Plate Motion
Plate Tectonics
Reflection/Refraction Seismology
Submarine Slides
Volcanogenic Massive Sulfides

TERRANES

Maurice Colpron[1] and JoAnne Nelson[2]
[1]Yukon Geological Survey, Whitehorse, Yukon, Canada
[2]British Columbia Geological Survey, Victoria, BC, Canada

Synonyms

Allochthonous terrane; Exotic terrane; Suspect terrane; Tectonostratigraphic terrane

Definition

Rock assemblages of regional extent within an orogenic belt that show internal geological consistency and that differ significantly from rock assemblages in adjacent terranes (Nelson et al., 2013).

Introduction

The terrane concept was first applied to the tectonic analysis of the North American Cordillera, one of the world's classic accretionary orogens (Figure 1; Coney et al., 1980; Jones et al., 1983). Terranes were introduced as a means to describe the collage of disparate elements that makes up the Cordilleran orogen (e.g., Helwig, 1974). In the original definition, terranes were considered "suspect" because little was known about their paleogeographic affinities and because most terranes were bounded by faults, thus raising questions about relationships between adjacent terranes. The terrane concept continues to guide tectonic analysis of the North American Cordillera, although new studies have yielded more information about paleogeographic origins and the relationships among many terranes, leading to refinement and evolution of the concept (e.g., Colpron et al., 2007).

Evolution of the terrane concept

Since the original inception of the terrane concept, detailed studies of Cordilleran terranes have repeatedly documented stratigraphic or intrusive links between some of the major terranes (e.g., Klepacki and Wheeler, 1985; Gardner et al., 1988; Colpron and Price, 1995; Colpron et al., 2007). The application of geodynamic interpretations to Cordilleran terranes has also revealed genetic links between paired arcs and accretionary complexes (Nokleberg et al., 2005). In many cases, terrane-bounding faults are now clearly identified as younger, post-accretionary structures that simply reshuffled previously accreted terranes. In places, these younger structures coincide with, or occur near, facies transitions between adjacent geodynamic elements. A useful current definition needs to acknowledge the enduring importance of terranes as recognizable geologic entities while accepting that they may have developed as adjacent, locally linked tectonic elements.

Continuing investigations of the North American Cordillera have also led to a greater understanding of the paleogeographic origins and tectonic evolution of many terranes, such that the "suspect" prefix is no longer appropriate. A growing body of faunal, isotopic, geochemical, paleomagnetic, and geochronological evidence now provides the basis for increasingly more refined paleogeographic and tectonic models of the northern Cordillera in Canada and Alaska (Monger and Price, 2002; Colpron and Nelson, 2009; Nelson et al., 2013, and references therein). Based on the region of origin, sets of related terranes in the northern Cordillera can be grouped into paleogeographic realms in a similar fashion to those identified in the Appalachian orogen (e.g., van Staal, 2007; Figure 2): (1) the Laurentian realm, which includes the autochthon and parautochthon of ancestral North America; (2) the peri-Laurentian realm of marginal arc, basin and pericratonic terranes, and forearc accretionary complexes that evolved in proximity to the western (present coordinates) continent margin of ancestral North America; (3) the Arctic-northeastern Pacific realm, including pre-Devonian terranes of Baltican, Caledonian, and/or Siberian affinity that originated in the present circum-Arctic region and then evolved at high latitudes in the ancestral northeastern Pacific basin, along with younger associated arc and accretionary terranes; and (4) the Coastal realm of later Mesozoic and Cenozoic accretionary complexes that originated near or on the eastern side of the Pacific basin and are accreted (or accreting) along the present North American-Pacific plate boundary (Figure 2). It emerges from this analysis that only terranes of the Arctic-northeastern Pacific realm are truly exotic to western North America.

Applications to other orogens

After its introduction in the North American Cordillera, the terrane concept was variably applied to other orogens around the world (e.g., Williams and Hatcher, 1983; Howell, 1989; Audley-Charles et al., 1990; Bluck, 1990) and also debated (Hamilton, 1990; Sengör and Dewey, 1990). Terranes, as used in the North American Cordillera,

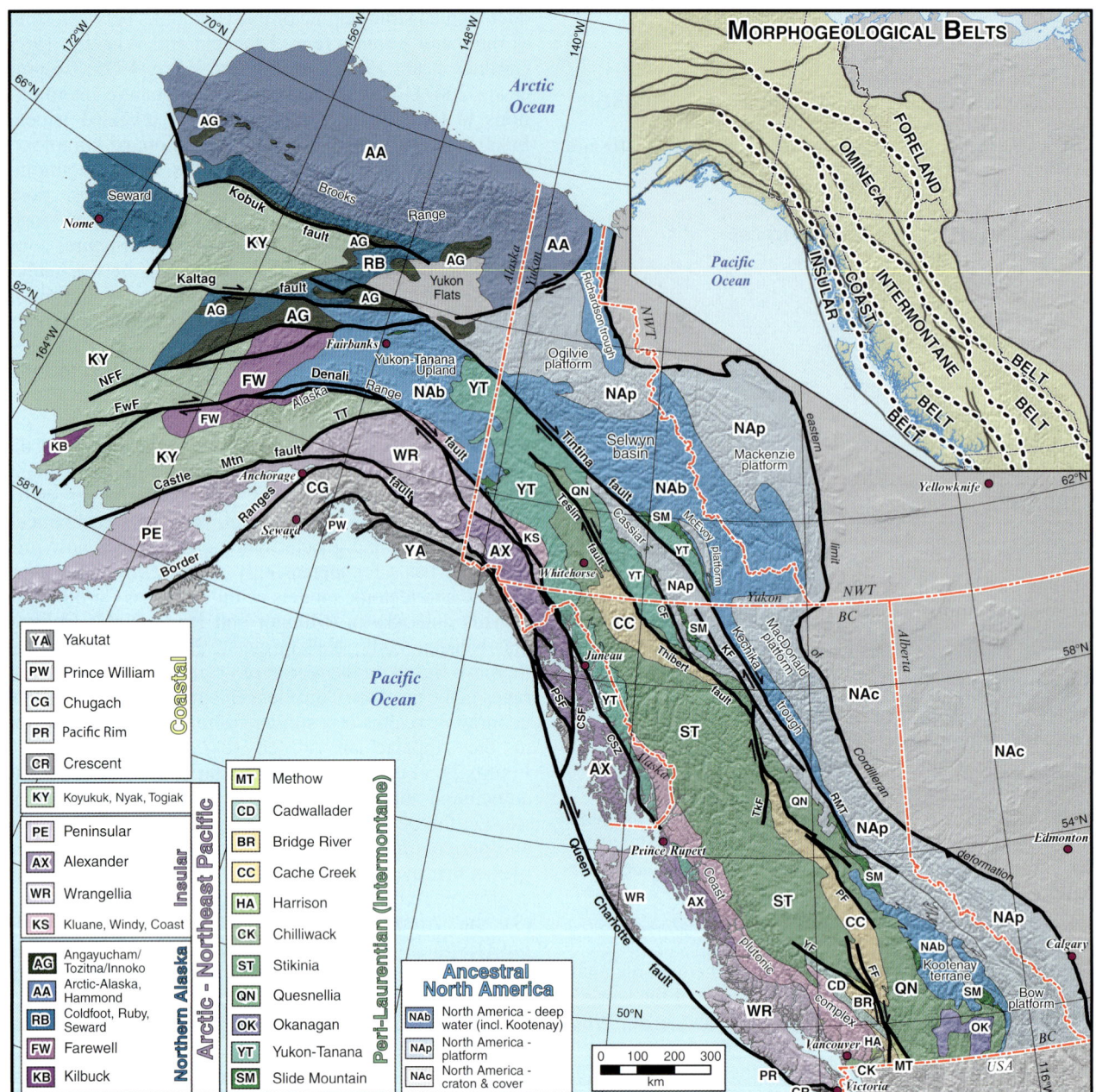

Terranes, Figure 1 Terranes of the Canadian-Alaskan Cordillera (after Colpron and Nelson, 2011). Terranes are grouped in the legend according to paleogeographic affinities shown in Figure 2. Inset shows morphogeological belts of the northern Cordillera after Gabrielse et al. (1991). Fault abbreviations: *CF* Cassiar fault, *CSF* Chatham Strait fault, *FF* Fraser fault, *FwF* Farewell fault, *KF* Kechika fault, *NFF* Nixon Fork-Iditarod fault, *PF* Pinchi fault, *PSF* Peril Strait fault, *RMT* northern Rocky Mountain trench, *TkF* Takla-Finlay-Ingenika fault system, *TT* Talkeetna thrust, *YF* Yalakom fault, *CSZ* Coast shear zone.

may be in part equivalent to "allochthons" or "nappes" as used in other orogens, particularly where the geologic record of an allochthon (or nappe) differs from adjacent allochthons. However, the use of allochthon or nappe to describe individual thrust imbrications of a single tectonic element does not correspond to terranes as defined in this article.

Summary

Terranes are regions within an accretionary orogen that are characterized by rock assemblages of internal geological consistency that differ from assemblages contained in adjacent terranes. The terrane concept has been, and continues to be, an important tool in the tectonic analysis of the North American Cordillera and other orogens

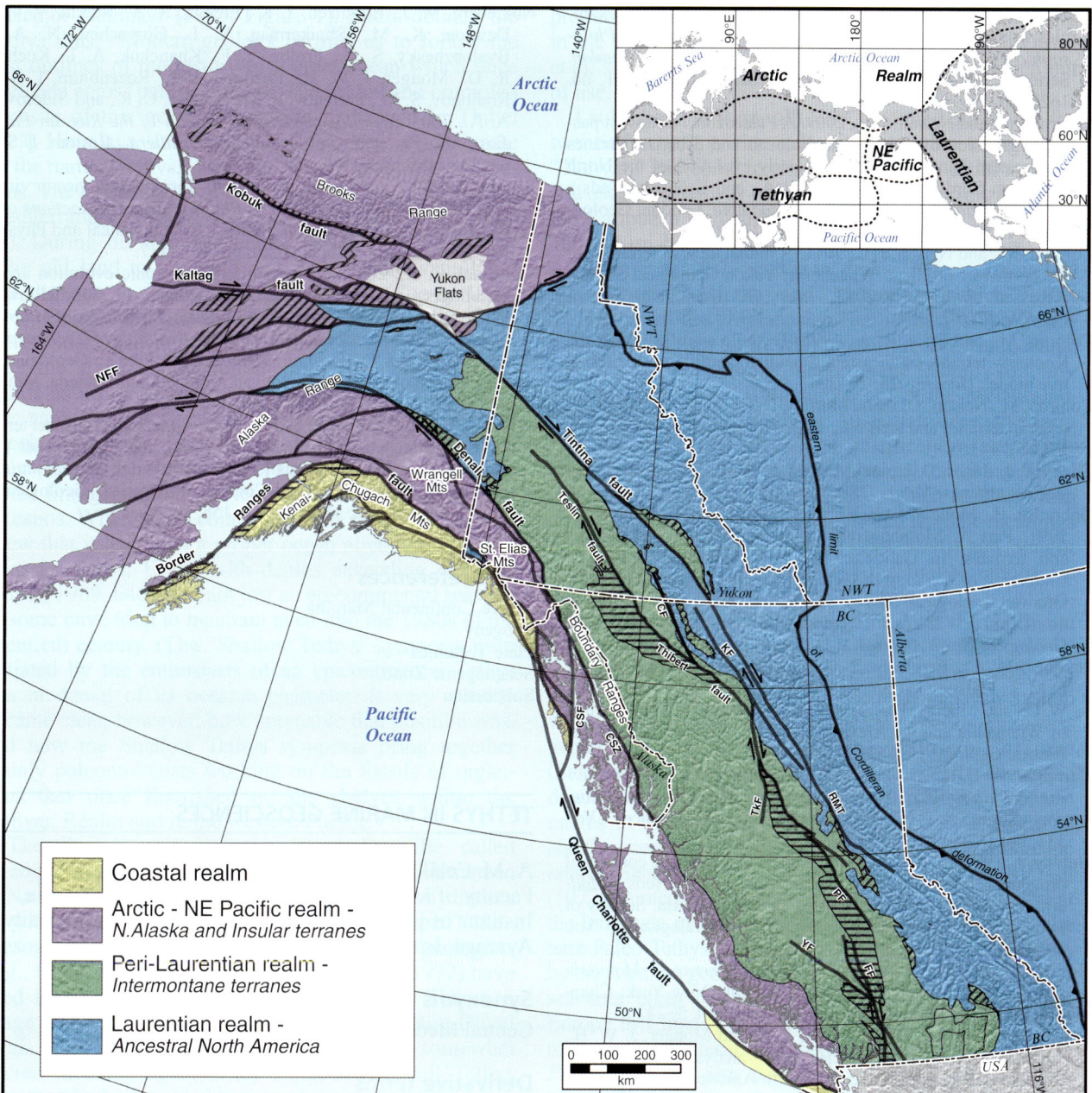

Terranes, Figure 2 Cordilleran terranes grouped by paleogeographic affinities. Paleogeographic affinity is assigned according to region of origin in Paleozoic time, except for the Coastal terranes, which originated along the late Mesozoic-Cenozoic eastern Pacific plate margin. Inset shows the main distribution of Paleozoic paleogeographic realms in the circum-Pacific region. Diagonal hatching indicates oceanic terranes in the peri-Laurentian and Arctic realms; horizontal hatching indicates accretionary complex containing elements of Tethyan affinity (e.g., Cache Creek terrane). Fault abbreviations: *CF* Cassiar fault, *CSF* Chatham Strait fault, *FF* Fraser fault, *KF* Kechika fault, *NFF* Nixon Fork-Iditarod fault, *PF* Pinchi fault, *RMT* northern Rocky Mountain trench, *TkF* Takla-Finlay-Ingenika fault system, *YF* Yalakom fault, *CSZ* Coast shear zone.

around the world. In the North American Cordillera, the terrane concept has evolved since the original inception to recognize that not all terranes are necessarily fault bounded and to integrate evolving geodynamic interpretations into the tectonic analysis of the orogen (Colpron et al., 2007).

Bibliography

Audley-Charles, M. G., Harris, R. A., and Clift, P. D., 1990. Allochthonous terranes of the southwest Pacific and Indonesia. In *Allochthonous Terranes: Philosophical Transactions of the Royal Society of London*. Series A, Mathematical and Physical Sciences, Vol. 331, no. 1620, pp. 571–587.

collisions along the Hercynian chains in Europe, the "Great Appalachides" comprising the Appalachians, the Ouachitas, the Glass Mountains, and the Huastecan Belt in Canada, the United States, and northern Mexico (Stille, 1940) and the Mauritanides in northwest Africa and the Uralides between the Russian Platform and the Altaids in Asia. Before the collisional events of the Hercynides, the Uralides, and the Altaids, it is meaningless to speak of a Tethys, because the ocean that separated Gondwana-Land from the northern continents did not as yet have a Laurasian counterpart in the north. That counterpart originated only after the closure of the Pleonic Ocean that created the Uralides (McKerrow and Ziegler, 1972). The Paleo-Tethys thus formed by gathering the continents around its frame forming the Pangaea as opposed to the Neo-Tethys that later formed by rifting.

Some of the more recent general publications on the Tethyan Realm are the following. They contain good bibliographies that would lead the interested reader to more specialized publications: Şengör (1990b: this paper is almost a textbook of orogenic structures found in the entire Tethysides), Buffetaut et al. (2009), Crasquin-Soleau and Barrier (1998a, 1998b, 2000), Dercourt et al. (1993, 2000), Hall (2002, 2012), Hall et al. (2011), Metcalfe (1999), Nairn et al. (1996), Roeder (2009), Roure (1994), Şengör (1990b, 1998: this long paper in particular contains a very full bibliography), Şengör and Natal'in (1996), Şengör and Atayman (2009), Sinha et al. (1997), Stampfli et al. (1991, 1998, 2001), Stampfli and Borel (2002), Veevers (2000, 2001, 2004), Yin et al. (2000), Ziegler and Horvath (1996), and Ziegler et al. (2001); for a critique of some of the Tethyan tectonic models in the eastern part of the European Alpides, see Zachner and Lupu (1999). Sonnenfeld (1981) is a book on benchmark papers on the Tethys, but it is thoroughly inadequate and even misleading. An outstanding textbook intended for undergraduates by a master paleoceanographer and sedimentologist, as well as an Alpine geologist, is Weissert and Stössel (2010), in which the evolution of the Neo-Tethys is narrated on classic Alpine examples. Barrier and Francou (1996) and Stow (2010) are two recent popular books on the Tethys. For the most reliable reconstructions of the entire Tethyan Realm through time, go to http://www.scotese.com/ on the web owned by Christopher Scotese, where continuously updated reconstructions through the Phanerozoic are available for free.

Figure 3a shows a schematic tectonic map of the Pangaea during the time when Paleo-Tethys existed, and Figure 3b shows its geomorphology. Figure 3b also shows a hypothetical display of the internal bathymetry of the Paleo-Tethys based on the distribution of the plate boundaries within it that are compatible with the tectonics of its margins (see Şengör and Atayman, 2009, for the details).

Pangaea A versus Pangaea B and the paleo-Tethys

Our view of the geometry of the Paleo-Tethys is dependent on what we think Pangaea looked like. After the publication of the famous "Bullard fit" of the continents around the Atlantic Ocean (Bullard et al., 1965), it was assumed for more than 20 years that a geometry similar to it had been maintained from the late Carboniferous Hercynide collisions until the early Jurassic opening of the Central Atlantic. That this may not have been so was first recognized by Irving in 1977 when he noticed a major discrepancy between the paleomagnetically determined positions of Gondwana-Land and Laurasia. He pointed out that the paleomagnetic data could only be reconciled with the reconstruction of the Pangaea if Gondwana-Land is shifted some 3,000 km eastward with respect to Laurasia, bringing northern South America against the Appalachian margin of North America. The resulting geometry he named Pangaea B (Irving, 1977). As Van der Voo and French had already improved on the Bullard et al. (1965) fit by making a tighter reconstruction in the area of the Gulf of Mexico between the cratonic edges of northern South America and southern North America, Morel and Irving (1981) decided to call the Pangaea of Van der Voo and French Pangaea A2 and the Bullard et al. Pangaea Pangaea A1 (Figure 4).

In 1981, Smith et al. complained that Irving's solution still did not honour the paleomagnetic data completely, and they instead proposed a more radical Pangaea reconstruction which they called Pangaea C. This reconstruction shoved northern South America all the way under the belly of Central Europe and all but eliminated the Paleo-Tethys, but created what might be called an "Anti-Tethys" between eastern North America and western South America (Figure 4d). Because Pangaea C violates so many geological relationships, it has so far found no adherents except among those who proposed it. Even they later abandoned it (Livermore et al., 1986).

It was later recognized that Irving's original timing of the Pangaea B to Pangaea A2 transition, from the latest Carboniferous to the Jurassic, was wrong, and Irving himself modified his original suggestion by indicating that an early Permian (~280 Ma ago) Pangaea B had already become Pangaea A2 in the late Permian (~250 Ma ago). Geological work of the last two decades has discovered more and more Permian dextral strike-slip systems in Central and Southern Europe with a cumulative offset of some 2,500 km (see Şengör, 2013, for a discussion and references) providing strong support for Irving's view. In addition, data on Permian quadrupeds make it likely that they migrated from northern Laurasia to southern Gondwana-Land via western Pangaea that must have had a geometry closer to that of Pangaea B than to that of Pangaea A2 (Cisneros et al., 2012). Figure 5 shows a probable Pangaea geometry that must have existed during the early to medial Permian (i.e., Pangaea B), whereas Figure 3 displays a late Permian Pangaea (Pangaea A2).

The demise of the paleo-Tethys and the opening of the neo-Tethys

Figure 3a shows a continuous rift system between the future Apulian Shelf and Southeast Asia appearing as a

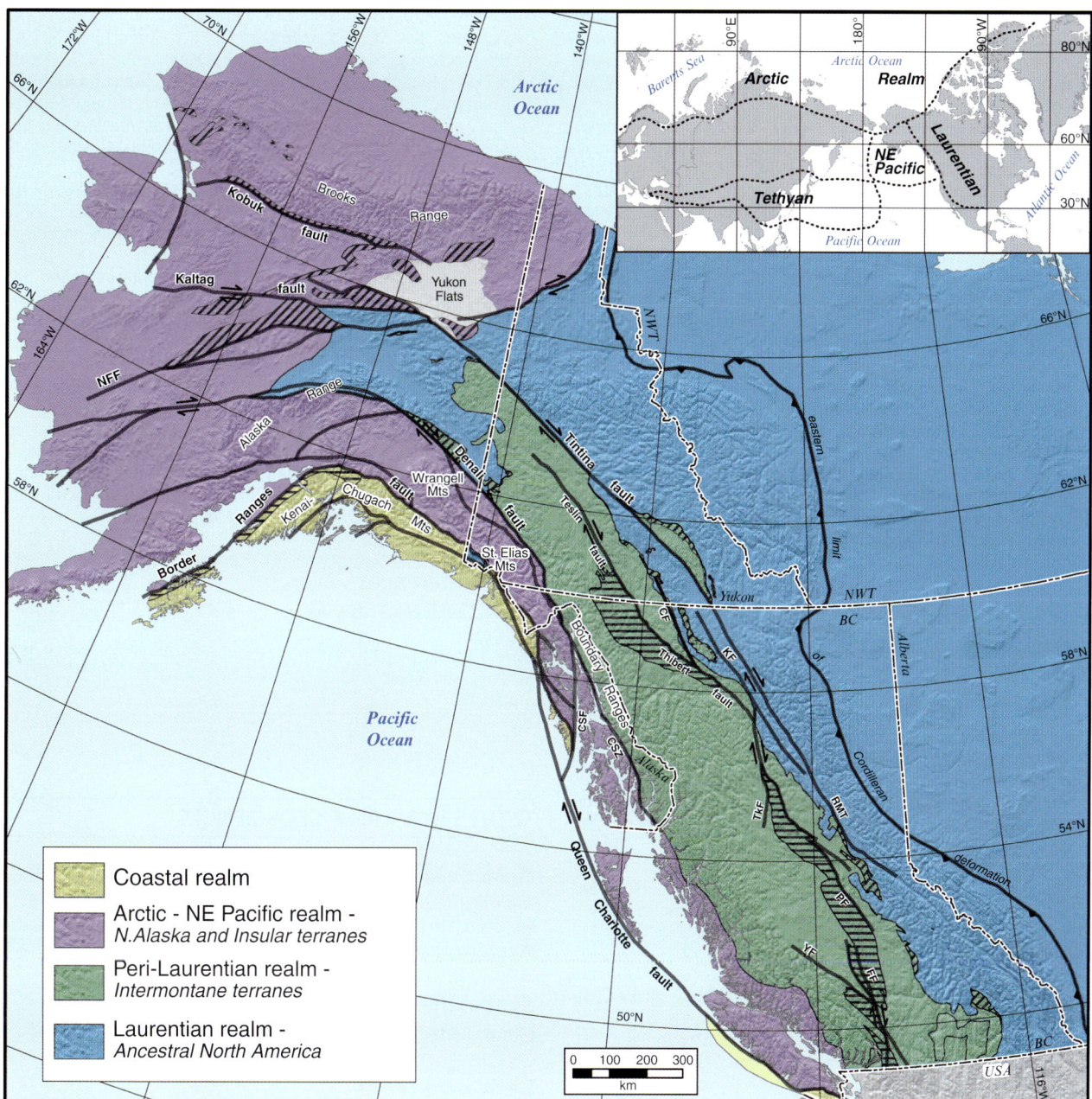

Terranes, Figure 2 Cordilleran terranes grouped by paleogeographic affinities. Paleogeographic affinity is assigned according to region of origin in Paleozoic time, except for the Coastal terranes, which originated along the late Mesozoic-Cenozoic eastern Pacific plate margin. Inset shows the main distribution of Paleozoic paleogeographic realms in the circum-Pacific region. Diagonal hatching indicates oceanic terranes in the peri-Laurentian and Arctic realms; horizontal hatching indicates accretionary complex containing elements of Tethyan affinity (e.g., Cache Creek terrane). Fault abbreviations: *CF* Cassiar fault, *CSF* Chatham Strait fault, *FF* Fraser fault, *KF* Kechika fault, *NFF* Nixon Fork-Iditarod fault, *PF* Pinchi fault, *RMT* northern Rocky Mountain trench, *TkF* Takla-Finlay-Ingenika fault system, *YF* Yalakom fault, *CSZ* Coast shear zone.

around the world. In the North American Cordillera, the terrane concept has evolved since the original inception to recognize that not all terranes are necessarily fault bounded and to integrate evolving geodynamic interpretations into the tectonic analysis of the orogen (Colpron et al., 2007).

Bibliography

Audley-Charles, M. G., Harris, R. A., and Clift, P. D., 1990. Allochthonous terranes of the southwest Pacific and Indonesia. In *Allochthonous Terranes: Philosophical Transactions of the Royal Society of London*. Series A, Mathematical and Physical Sciences, Vol. 331, no. 1620, pp. 571–587.

Bluck, B. J., 1990. Terrane provenance and amalgamation: examples from the Caledonides. In *Allochthonous Terranes: Philosophical Transactions of the Royal Society of London.* Series A, Mathematical and Physical Sciences, Vol. 331, no. 1620, pp. 599–609.

Colpron, M., and Nelson, J. L., 2009. A Palaeozoic Northwest passage: incursion of Caledonian, Baltican and Siberian terranes into eastern Panthalassa, and the early evolution of the North American Cordillera. In Cawood, P. A., and Kröner, A. (eds.), *Earth accretionary systems in space and time.* London: Geological Society of London, Vol. 318, pp. 273–307.

Colpron, M., and Nelson, J. L., 2011. A digital atlas of terranes for the northern Cordillera. Whitehorse: Yukon Geological Survey. Accessed on 15 May 2013 *(also, BC Geological Survey, GeoFile 2011–11).*

Colpron, M., and Price, R. A., 1995. Tectonic significance of the Kootenay terrane, southeastern Canadian Cordillera: an alternative model. *Geology,* **23**, 25–28.

Colpron, M., Nelson, J. L., and Murphy, D. C., 2007. Northern Cordilleran terranes and their interactions through time. *GSA Today,* **17**(4/5), 4–10.

Coney, P. J., Jones, D. L., and Monger, J. W. H., 1980. Cordilleran suspect terranes. *Nature,* **288**, 329–333.

Gabrielse, H., Monger, J. W. H., Wheeler, J. O., and Yorath, C. J., 1991. Part A. Morphogeological belts, tectonic assemblages and terranes. In Gabrielse, H., and Yorath, C. J., (eds.), *Chapter 2 of Geology of the Cordilleran Orogen in Canada.* Ottawa: Geological Survey of Canada, Geology of Canada, no. 4, pp. 15–28. *(Also Geological Society of America, The Geology of North America, v. G-2).*

Gardner, M. C., Bergman, S. C., Cushing, G. W., MacKevett, E. M. J., Plafker, G., Campbell, R. B., Dodds, C. J., McClelland, W. C., and Mueller, P. A., 1988. Pennsylvanian pluton stitching of Wrangellia and the Alexander terrane, Wrangell Mountains, Alaska. *Geology,* **16**, 967–971.

Hamilton, W. B., 1990. On terrane analysis. In *Allochthonous Terranes: Philosophical Transactions of the Royal Society of London.* Series A, Mathematical and Physical Sciences, Vol. 331, no. 1620, pp. 511–522.

Helwig, J., 1974. Eugeosynclinal basement and a collage concept of orogenic belts. In Dott, R. H., Jr., and Shaver, R. H., (eds.), *Modern and Ancient Geosynclinal Sedimentation.* Tulsa: Society of Economic Paleontologists and Mineralogists, Vol. 19, pp. 359–380

Howell, D. G., 1989. *Tectonics of Suspect Terranes: Mountain Building and Continental Growth.* London, New York: Chapman and Hall. Topics in the earth sciences, Vol. 3. 232 p.

Jones, D. L., Howell, D. G., Coney, P. J., and Monger, J. W. H., 1983. Recognition, character and analysis of tectonostratigraphic terranes in western North America. In Hashimoto, M., and Uyeda, S. (eds.), *Accretion Tectonics in the Circum-Pacific Region.* Tokyo: Terra, pp. 21–35.

Klepacki, D. W., and Wheeler, J. O., 1985. Stratigraphic and structural relations of the Milford, Kaslo and Slocan Groups, Goat Range, Lardeau and Nelson map areas, British Columbia. In: *Current Research, Part A: Geological Survey of Canada.* Paper 85-1A, pp. 277–286.

Monger, J. W. H., and Price, R. A., 2002. The Canadian Cordillera: geology and tectonic evolution. *Canadian Society of Exploration Geophysicists Recorder,* **27**, 17–36.

Nelson, J. L., Colpron, M., and Israel, S., 2013. The Cordillera of British Columbia, Yukon, and Alaska: tectonics and metallogeny. In Colpron, M., Bissig, T., Rusk, B. G., and Thompson, J. F. H. (eds.), *Tectonics, Metallogeny and Discovery: The North American Cordillera and Similar Accretionary Settings.* Littleton: Society of Economic Geologists. Special Publication, Vol. 17, pp. 53–103.

Nokleberg, W. J., Bundtzen, T. K., Eremin, R. A., Ratkin, V. V., Dawson, K. M., Shpikerman, V. I., Goryachev, N. A., Byalobzhesky, S. G., Frolov, Y. F., Khanchuk, A. I., Koch, R. D., Monger, J. W. H., Pozdeev, A. I., Rozenblum, I. S., Rodionov, S. M., Parfenov, L. M., Scotese, C. R., and Sidorov, A. A., 2005. *Metallogenesis and tectonics of the Russian Far East, Alaska, and the Canadian Cordillera.* Reston: U.S. Geological Survey, Vol. 1697. 397 p.

Şengör, A. M. C., and Dewey, J. F., 1990. Terranology: vice or virtue? In *Allochthonous Terranes: Philosophical Transactions of the Royal Society of London.* Series A, Mathematical and Physical Sciences, Vol. 331, no.1620, pp. 457–477.

van Staal, C. R., 2007. Pre-carboniferous tectonic evolution and metallogeny of the Canadian Appalachians. In Goodfelllow, W. D. (ed.), *Mineral Deposits of Canada: A Synthesis of Major Deposit Types, District Metallogeny, the Evolution of Geological Provinces and Exploration Methods.* St. John's: Geological Association of Canada. Geological Association of Canada, Mineral Deposits Division, Vol. 5, pp. 793–818.

Williams, H., and Hatcher, R. D., Jr., 1983. Appalachian suspect terranes. In Hatcher, R. D., Jr., Williams, H., and Zietz, I. (eds.), *Contributions to the tectonics and geophysics of mountain chains.* Boulder: Geological Society of America. Geological Society of America, Memoir, Vol. 158, pp. 33–53.

Cross-references

Active Continental Margins
Orogeny
Plate Tectonics
Seismogenic Zone
Subduction

TETHYS IN MARINE GEOSCIENCES

A. M. Celâl Şengör
Faculty of Mines, Department of Geology and Eurasia, Institute of Earth Sciences, Istanbul Technical University, Ayazaga, Istanbul, Turkey

Synonyms

Central Mediterranean; Mesogea

Derivative terms

Antitethys, Eotethys, Eoparatethys, Greater Tethys, Hemitethys, Meso-Tethys, Mesoparatethys, Neo-Tethys, Neoparatethys, Paleo-Tethys, Paleotethys, Paleothetis, Paratethys, Protoparatethys, Proto-Tethys

Definition

Tethys is the name given by Eduard Suess to an equatorial ocean that extended from the westernmost Mediterranean region of today to eastern Asia during the time interval from the late Paleozoic to the early Cainozoic, the demise of which gave rise to the present-day Alpine-Himalayan mountain ranges. This seaway had been identified earlier, but for the Jurassic time only and called the "Central Mediterranean" by Suess' son-in-law, the great paleontologist,

Melchior Neumayr (1885; Figure 1). Suess defined the Tethys thus: "Modern geology permits us to follow the first outlines of the history of a great ocean which once stretched across part of Eurasia. The folded and crumpled deposits of this ocean stand forth to heaven in Thibet [sic!], Himalaya, and the Alps. This ocean we designate by the name "Tethys," after the sister and consort of Oceanus. The latest successor of the Tethyan sea is the present Mediterranean" (Suess, 1893, 183; see also Suess, 1901, 25). During his farewell lecture in 1901, Suess added: "The old land to its south was called Gondwana-Land and the one to its north Angara-Land" (Suess, 1902, S. 4). Despite the fact that the theoretical framework in which Suess had defined the Tethys is now abandoned (i.e., the fixist contraction theory), his definition still stands as a precise description of what the Tethys was. The name itself derives, as Suess says, from the ancient Greek mythology, from Τηθύς (Tethys), a daughter of Gaia (the earth) and Uranos (the sky) and as such one of the first generation of Titans and sister and consort of Okeanos. When Suess coined the term Tethys, he already knew that it had been a proper ocean already during the Triassic (Suess, 1895), with depths exceeding 4,000 m (Suess, 1909, 646), and not just an epicontinental seaway, as some have tried to maintain even into the 1980s of the twentieth century. (The "Shallow Tethys" symposia were initiated by the enthusiasts of an epicontinental Tethys Sea in denial of its oceanic character. It very quickly became clear, however, how untenable that position was, and now the Shallow Tethys symposia bring together mainly paleontologists working on the fossils of organisms that once flourished on the shelves within the Tethyan Realm and in the western Pacific.)

The French paleontologist Henri Douvillé called *Mésogée* "a particular phase of the Central Mediterranean of Neumayr or the Tethys of Suess; it is specifically the sea in which the Rudists have lived and developed" (i.e., Mesogea; Douvillé, 1900, 223). Later French geologists (e.g., Haug, 1908–1911, 661; Aubouin et al., 1977) have used Douvillé's term as an equivalent of the Tethys, a usage Douvillé clearly did not envisage. Biju-Duval et al. (1977) resuscitated this term, but in a somewhat altered meaning again. They suggested to call Tethys "the ocean which existed between Europe and Africa during the early Mesozoic" and to call Mesogea "the fairly broad ocean which was created during Mesozoic." Thus, Biju-Duval et al. (1977) and Biju-Duval and Dercourt (1980) would call the Eastern Mediterranean and the now-closed Bitlis/Zagros ocean (the "southern branch of the Neo-Tethys" of Şengör, 1979) Mesogea and the oceans, the remnants of which are now found in the Carpathian, Dinaride, Hellenide, and the north Turkish orogens (the "northern branch of the Neo-Tethys" of Şengör, 1979) they would call Tethys. Following them Boccaletti (1977) employed a similar terminology. This usage is confusing, because all of the oceans mentioned began opening largely during the Mesozoic; only what Biju-Duval et al. (1977) then singled out as later Mesozoic products have now turned out to have started their rifting in the Permian (e.g., Stampfli et al., 1991). Biju-Duval et al.'s (1977) suggested terminology has since fallen out of use.

Laubscher and Bernoulli (1977) pointed out that in the twentieth century, the paleobiogeographic designation "Tethyan" had come to mean simply "equatorial marine" and used for a seaway of Mesozoic and Cenozoic age from the Caribbean (which Stille, 1940, had called "Antillian Tethys" on his p. 153 and "Antitethys" on his p. 581; Aubouin et al., 1977, called it "*Téthys de la reconquête*") to northern Australia and Melanesia (which Stille, 1958a, called the "Melanesian Hemitethys"). The Caribbean had nothing to do with any of the Tethyan oceans, and it is therefore simply inappropriate to apply to it any Tethys-based terms. By contrast, not in the Melanesian, but in parts of the northern Australian area, the term "Hemitethys" does have some justification (see Görür and Şengör, 1992). For a controversy on the original meaning of Tethys in Suess' usage, which resulted from a confusion of the paleogeographic and tectonic meanings of the term Tethys, see Tozer (1989, 1990) and Şengör (1990a). Stille (1958b, 153) was the first to emphasize the double meaning of the term Tethys, i.e., a paleogeographic and a tectonic one, and pointed out the confusion their mixing usually causes.

In the 1970s of the twentieth century, it was recognized that the mountain ranges of the Alpine-Himalayan System were the products of more than one major ocean. The Swiss geologist Jovan Stöcklin, working in Iran at the time (Stöcklin, 1974), called the ocean, the early Mesozoic demise of which had given rise to a part of the mountain ranges in northern Iran, "Paleo-Tethys" (i.e., *old* Tethys) and the ocean which generated the late Mesozoic-Cenozoic ranges (Şengör 1979) "Neo-Tethys" (i.e., *new* Tethys). Hsü (1977) pointed out that the remnants of the Palaeo-Tethys at the longitude of Turkey be looked for in the Crimea. The term Paleo-Tethys had been first introduced, as *Paläothetis*, by the Austrian geologist Franz Kahler (1939) for a shallow sea area north of the then-known Tethys from Eastern Europe to China. Kahler's teacher Franz Heritsch (1940) took over this term and corrected the orthography to *Paläotethys* but also used it in a tectonic sense, for a geosyncline.

The German geologist Hans Stille also adopted the term Paleo-Tethys but defined it as the Tethys of the Variscan era (1949, 1951). Later, he changed his mind and used the term Paleo-Tethys for a hypothetical Tethyan geosyncline of the "Caledonian Era" and introduced the terms Eotethys or Greater Tethys (*Urtethys*) for the Tethyan geosyncline of the Assyntic era and "Mesotethys" and "Neotethys" for the Tethyan geosynclines of the "Variscan" and the "Neoidic" (i.e., Mesozoic + Cenozoic) eras, respectively (Stille, 1958b). All of Stille's terms are now obsolete as his tectonic world picture in which they were defined. To avoid all confusion, it is best to spell the terms Paleo-Tethys and Neo-Tethys with hyphens, which neither Stille nor his predecessors did.

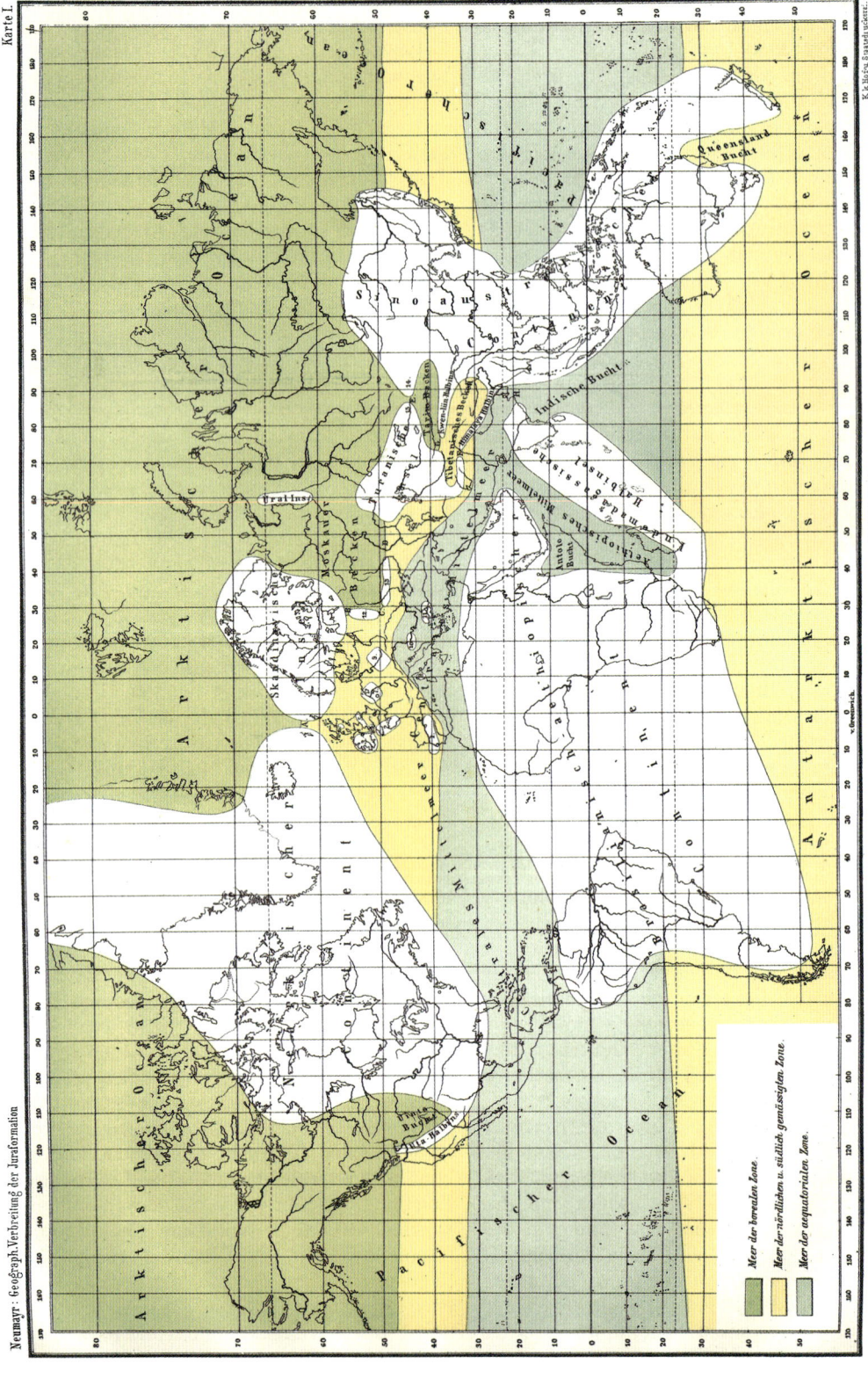

Tethys in Marine Geosciences, Figure 1 The map of the Central Mediterranean during the Jurassic by Melchior Neumayr (1885). Suess showed within 3 years of the publication of this map that the same ocean minus its Central American parts already existed in the Triassic and that it survived into the Eocene.

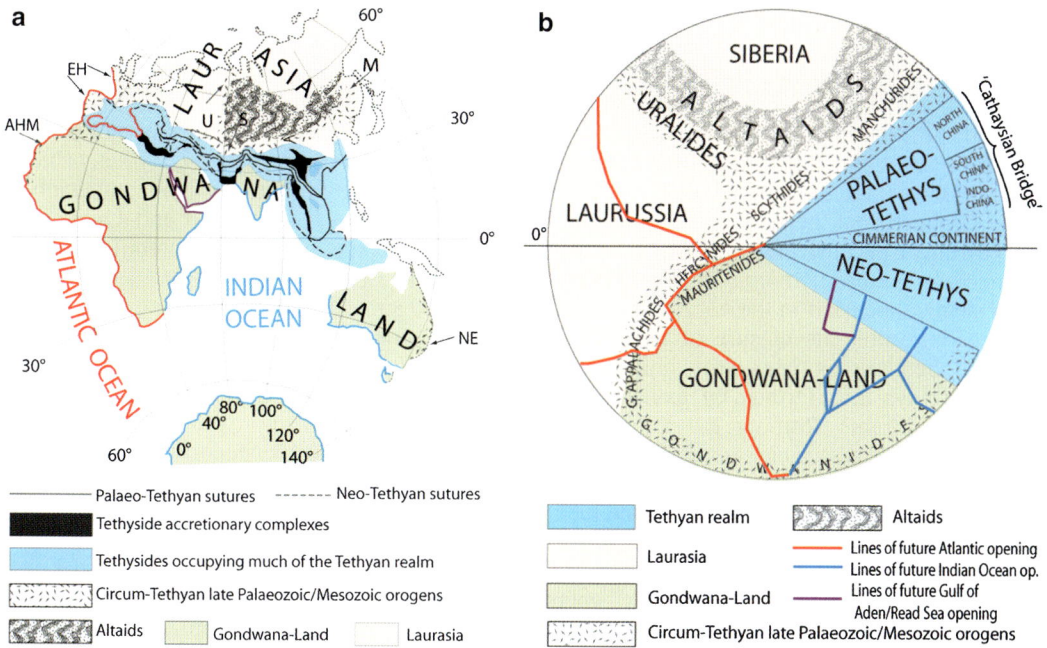

Tethys in Marine Geosciences, Figure 2 (a) Some of the late Paleozoic to the present tectonic elements of the Old World showing the position of the Tethyan Realm within it. The rifted continental margins of the two different oceans disrupting the Old World are shown in the colors corresponding to the color in which their names are written (except the Red Sea and the Gulf of Aden, whose names could not be written because of limited space). The colours and patterns in this figure correspond to those in **b**. Key to abbreviations: *AHM* African Hercynides including Mauritanides, *EH* European Hercynides, *M* Manchurides, *NE* New England Fold Belt, *S* Scythides, *U* Uralides. (**b**) A schematic illustration of the main tectonic elements of the world during the late Permian including the Tethyan Realm. The colours and patterns in this figure correspond to those in **a**.

Proto-Tethys is a term used by some geologists to refer to a Tethyan ocean before the Tethys (e.g., Flügel, 1972), but this is a contradiction in terms, because there was no space that resembled any of the Tethyan oceans before the Tethys. The term Tethys today has a meaning only with respect to Suess' definition and with respect to Pangaea, in which it was located. When no Pangaea existed, it is meaningless to talk about a "Tethys."

It was later recognized that an entire late Paleozoic-early Mesozoic mountain range lay hidden in the northern part of the Alpine-Himalayan ranges, including the Turkish-Iranian and the Tibetan high plateaux, between the Black Sea and Southeast Asia, which had been created by the subductive removal of an ocean between the early Carboniferous and medial Jurassic; some satellite oceans of this major ocean (marginal basins) survived until the early Cretaceous. Şengör (1979) recognized that this ocean corresponded to what Stöcklin (1974, 1977) and Hsü (1977) had called the paleo-Tethys. Şengör called the resulting orogenic system the "Cimmerides" from the Cimmerian Mountains comprising the Yayla Range in the Crimea, the tiny Serpent Island off the Danube Delta (at 45°15′N and 30°12′E; Острів Зміїний in Ukrainian and *Insula şerpilor* in Romanian), and the North Dobruja (Suess, 1909, 22). The name "Alpides" he confined to the products of the Neo-Tethys. Both the Cimmerides and the Alpides comprised, in Şengör's definition, the "Tethysides." A long continental strip that rifted off from northern Gondwanaland creating the Neo-Tethys in its wake and closing the Paleo-Tethys in its front, Şengör (1979) called the "Cimmerian Continent." Some later authors, either ignorant of or inattentive to the tradition of geographical nomenclature, have called this major continental strip "Cimmeria," but I would recommend using the term Cimmerian Continent, because Cimmeria, a geographical term in a form normally used for countries (e.g., Mongolia) or geographical regions (e.g., Scythia), gives no information about the geological nature of the object.

Şengör called the entire group comprising the *Paleo-Tethys + the Cimmerian Continent + the Neo-Tethys and their continental margins* the "Tethyan Realm," also called the "Tethyan Domain." Figure 2a shows where in the Old World the remnants of the Tethyan Realm are now located, and Figure 2b is a schematic representation of the Tethyan Realm to give an idea of its internal organization sometime in the early Triassic.

Present usage of the Tethys

Today the term Tethys is used in the sense of the Tethyan Realm as defined above and shown in Figure 2b. Within the Tethyan Realm, the Paleo-Tethys came into existence by the formation of the Pangaea following the Hercynide collisions along the Hercynian chains in Europe, the "Great Appalachides" comprising the Appalachians, the Ouachitas, the Glass Mountains, and the Huastecan Belt in Canada, the United States, and northern Mexico (Stille, 1940) and the Mauritanides in northwest Africa and the Uralides between the Russian Platform and the Altaids in Asia. Before the collisional events of the Hercynides, the Uralides, and the Altaids, it is meaningless to speak of a Tethys, because the ocean that separated Gondwana-Land from the northern continents did not as yet have a Laurasian counterpart in the north. That counterpart originated only after the closure of the Pleonic Ocean that created the Uralides (McKerrow and Ziegler, 1972). The Paleo-Tethys thus formed by gathering the continents around its frame forming the Pangaea as opposed to the Neo-Tethys that later formed by rifting.

Some of the more recent general publications on the Tethyan Realm are the following. They contain good bibliographies that would lead the interested reader to more specialized publications: Şengör (1990b: this paper is almost a textbook of orogenic structures found in the entire Tethysides), Buffetaut et al. (2009), Crasquin-Soleau and Barrier (1998a, 1998b, 2000), Dercourt et al. (1993, 2000), Hall (2002, 2012), Hall et al. (2011), Metcalfe (1999), Nairn et al. (1996), Roeder (2009), Roure (1994), Şengör (1990b, 1998: this long paper in particular contains a very full bibliography), Şengör and Natal'in (1996), Şengör and Atayman (2009), Sinha et al. (1997), Stampfli et al. (1991, 1998, 2001), Stampfli and Borel (2002), Veevers (2000, 2001, 2004), Yin et al. (2000), Ziegler and Horvath (1996), and Ziegler et al. (2001); for a critique of some of the Tethyan tectonic models in the eastern part of the European Alpides, see Zachner and Lupu (1999). Sonnenfeld (1981) is a book on benchmark papers on the Tethys, but it is thoroughly inadequate and even misleading. An outstanding textbook intended for undergraduates by a master paleoceanographer and sedimentologist, as well as an Alpine geologist, is Weissert and Stössel (2010), in which the evolution of the Neo-Tethys is narrated on classic Alpine examples. Barrier and Francou (1996) and Stow (2010) are two recent popular books on the Tethys. For the most reliable reconstructions of the entire Tethyan Realm through time, go to http://www.scotese.com/ on the web owned by Christopher Scotese, where continuously updated reconstructions through the Phanerozoic are available for free.

Figure 3a shows a schematic tectonic map of the Pangaea during the time when Paleo-Tethys existed, and Figure 3b shows its geomorphology. Figure 3b also shows a hypothetical display of the internal bathymetry of the Paleo-Tethys based on the distribution of the plate boundaries within it that are compatible with the tectonics of its margins (see Şengör and Atayman, 2009, for the details).

Pangaea A versus Pangaea B and the paleo-Tethys

Our view of the geometry of the Paleo-Tethys is dependent on what we think Pangaea looked like. After the publication of the famous "Bullard fit" of the continents around the Atlantic Ocean (Bullard et al., 1965), it was assumed for more than 20 years that a geometry similar to it had been maintained from the late Carboniferous Hercynide collisions until the early Jurassic opening of the Central Atlantic. That this may not have been so was first recognized by Irving in 1977 when he noticed a major discrepancy between the paleomagnetically determined positions of Gondwana-Land and Laurasia. He pointed out that the paleomagnetic data could only be reconciled with the reconstruction of the Pangaea if Gondwana-Land is shifted some 3,000 km eastward with respect to Laurasia, bringing northern South America against the Appalachian margin of North America. The resulting geometry he named Pangaea B (Irving, 1977). As Van der Voo and French had already improved on the Bullard et al. (1965) fit by making a tighter reconstruction in the area of the Gulf of Mexico between the cratonic edges of northern South America and southern North America, Morel and Irving (1981) decided to call the Pangaea of Van der Voo and French Pangaea A2 and the Bullard et al. Pangaea Pangaea A1 (Figure 4).

In 1981, Smith et al. complained that Irving's solution still did not honour the paleomagnetic data completely, and they instead proposed a more radical Pangaea reconstruction which they called Pangaea C. This reconstruction shoved northern South America all the way under the belly of Central Europe and all but eliminated the Paleo-Tethys, but created what might be called an "Anti-Tethys" between eastern North America and western South America (Figure 4d). Because Pangaea C violates so many geological relationships, it has so far found no adherents except among those who proposed it. Even they later abandoned it (Livermore et al., 1986).

It was later recognized that Irving's original timing of the Pangaea B to Pangaea A2 transition, from the latest Carboniferous to the Jurassic, was wrong, and Irving himself modified his original suggestion by indicating that an early Permian ($\sim$280 Ma ago) Pangaea B had already become Pangaea A2 in the late Permian ($\sim$250 Ma ago). Geological work of the last two decades has discovered more and more Permian dextral strike-slip systems in Central and Southern Europe with a cumulative offset of some 2,500 km (see Şengör, 2013, for a discussion and references) providing strong support for Irving's view. In addition, data on Permian quadrupeds make it likely that they migrated from northern Laurasia to southern Gondwana-Land via western Pangaea that must have had a geometry closer to that of Pangaea B than to that of Pangaea A2 (Cisneros et al., 2012). Figure 5 shows a probable Pangaea geometry that must have existed during the early to medial Permian (i.e., Pangaea B), whereas Figure 3 displays a late Permian Pangaea (Pangaea A2).

The demise of the paleo-Tethys and the opening of the neo-Tethys

Figure 3a shows a continuous rift system between the future Apulian Shelf and Southeast Asia appearing as a

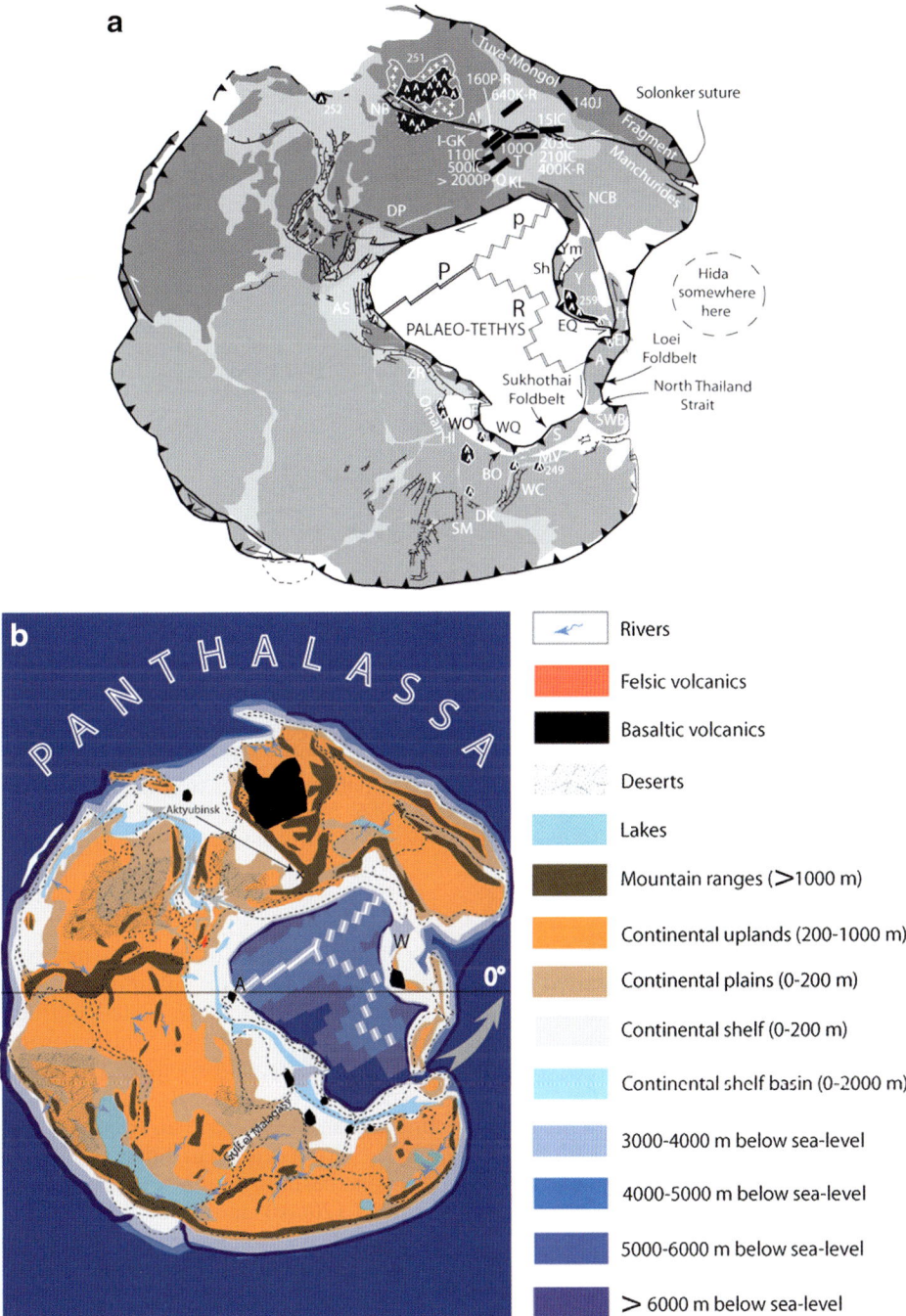

Tethys in Marine Geosciences, Figure 3 (a) Latest Permian tectonics of Pangaea and the Paleo-Tethys. Key to lettering: *A* Annamia, *Al* Altay Mountains, *AS* Apulian Shelf, *BO* Banggong Co-Nu Jiang Ocean, *DK* Damodar-Koel Valley rift, *DP* Port of Dobruja, *El* Emei Shan basalts on Annamia, *EQ* eastern Qangtang, *F* Farah Block, *H* Huanan Block, *Hl* Helmand Block, *I-GK* Irtysh-Gornostaev keirogen, *K* Kurduvadi rift, *KL* Kunlun, *MV* Mount Victoria Land block, *NB* Narym Basin, *NCB* North China block, *S* Sibumasu, *Sh* Shaluli Shan ensimatic island arc, *SM* Son-Mahanadi rift, *SWB* south-west Borneo, *T* Tarim Block, *WC* western coastlands taphrogen, *WQ* western Qangtang, *Y* Yangtze Block, *Ym* Yajiang marginal basin, *ZR* future Zagros Mountains. Lone figures are isotopic ages (in Ma) of extensional and/or plume-related igneous rocks, mostly traps. The figures with letters after them represent observed amounts (in km) and times of crustal shortening across the bars associated with them: >2000P-Q: Permian to Quaternary; 160P-R: Permian to Recent; 140 J: Jurassic; 640 K-R: Cretaceous to Recent; 400 K-R: Cretaceous to Recent; 203C: Cenozoic; 203C: Cenozoic; 500lC: late Cenozoic; 110lC: late Cenozoic; 15lC: late Cenozoic; 210lC: late Cenozoic; 100Q: late Quaternary. From Şengör and Atayman (2009). (b) The geomorphology of the Permian world including the floor of the Paleo-Tethys. No morphological elements are shown within the Panthalassa except the inferred circum-Panthalassan trench, which, in fact, may not have existed as a topographic feature everywhere. A is where the Aksu Flysch in northern Turkey is seen; W is Wolong in western Sichuan (ancient Xikang) (from Şengör and Atayman (2009)).

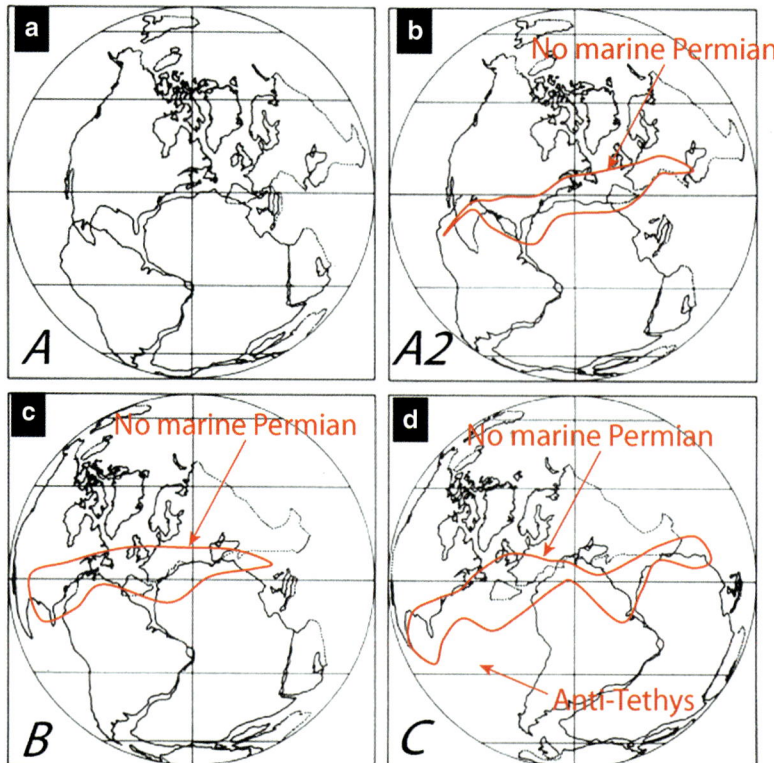

Tethys in Marine Geosciences, Figure 4 The four Pangaeas proposed by different authors to account for the large-scale morphological and paleomagnetic data (modified from Livermore et al., 1986). Areas where there are no marine Permian deposits are indicated to show how compatible each reconstruction is with some of the decisive geological data.

long and narrow continental shelf basin in Figure 3b. Along this basin system rifting had already started in the Tournaisian in Southeast Asia (in Thailand) and was underway in the Permian all the way into the Apulian Shelf. However, rifting did not progress regularly from Southeast Asia to future Europe as shown by the fact that continental margin stretching was going on both in the Himalaya and in the Zagros, while an ocean had already opened in Oman. This was largely because an earlier rift separating northern and southern Tibet and northern and southern Afghanistan had turned south along the Sistan suture in Iran and joined a newly developing rift along the Himalaya and the Zagros in what is today Oman (Şengör and Natal'in, 1996). In any event, by the early Triassic, ocean opening had commenced along this southern rift as far west as the Carpathians. It did not propagate westward except to deepen the already existing rift basins on the Apulian Shelf including the southern Alps and the Northern Calcareous Alps, the Transdanubian, and the Bükk Mountains in Hungary (the so-called "ALCAPA" unit). Only with the early opening of the Central Atlantic Ocean did rift branches enter the future Betic-Rif-Tellian Atlas-Apennine-Alpine region and tore open an oceanic corridor that united with the earlier Triassic rifts and rifted to their north new oceanic corridors, for example, along the Pennine Ocean in the Alps and in northern Turkey along the Intra-Pontide suture.

The rifts that had separated northern Tibet, central Pamirs, and northern Afghanistan from southern Tibet, southern Pamirs, and southern Afghanistan opened what is known as the Wašer/Rushan-Pshart/Banggong Co-Nu Jing Ocean in the Permian (Şengör, 1984; Şengör et al., 1988). That ocean had branches into the main Paleo-Tethyan basin through the central Pamirs (Gaetani, undated (2006)) and along the future Chasa suture in northern Tibet (Şengör et al., 1988; Şengör and Natal'in, 1996). The continental fragments thus created must not have moved independently, however, as they all collided with a long continental strip consisting of southern Tibet, southern Pamirs, and southern Afghanistan before the Aptian/Albian as limestones of that age cover their suture from Iran to Burma. That is why Şengör (1984), Şengör et al. (1988), and Şengör and Natal'in (1996) considered the Wašer/Rushan-Pshart/Banggong Co-Nu Jing Ocean a marginal basin complex of the Paleo-Tethys (which Belov, 1980, called the Meso-Tethys following an earlier established Russian tradition; Shvolman, 1978, and Burtman, 1994, also used the term Meso-Tethys for Tethyan oceans that closed during the Mesozoic, but as Burtman himself admitted, his Meso-Tethyan oceans also included some that closed during the Cenozoic, blurring the precision of the terminology; that is why I recommend Meso-Tethys not be used).

Şengör (1979) called the entire continental strip that rifted from northwestern Gondwana-Land the Cimmerian Continent. Thus, the Wašer/Rushan-Pshart/Banggong Co-Nu Jing Ocean was an "intra-Cimmerian-Continent ocean" (Şengör, 1986, 1998). The northern continental fragments were active frontal arcs above a southwest-dipping major subduction zone and the southern continental fragments "remnant arcs."

The frontal arcs of the Cimmerian Continent collided with Laurasia during the late Triassic from Southeast Asia to Iran, while farther west in Turkey, the final collision was delayed until the early Jurassic. Some of the frontal arcs of the Cimmerian Continent consisted entirely of subduction-accretion material of earlier arcs onto which the magmatic arc fronts had migrated. This has misled some into thinking that in such places the Cimmerian Continent did not exist (Topuz et al., 2012).

The evolution and end of the neo-Tethys

After the late Triassic-early Jurassic closure of the main Paleo-Tethyan basin, its only remnant was represented by the Wašer/Rushan-Pshart/Banggong Co-Nu Jing Ocean until the early Cretaceous. After the early Cretaceous, the Neo-Tethys became the sole occupier of the Tethyan Realm. It was at this time that a cooling global climate is thought to have caused density stratification and black shale deposition (Weissert et al., 1979) in restricted basins along the Neo-Tethyan north margin (Görür, 1991) as well as in the juvenile Atlantic Ocean (Weissert, 1981).

A major event in the Neo-Tethys was the generation of at least two major subduction zones from present-day Turkey in the east to the Himalaya and possibly beyond all the way into northwestern Sumatra in the east. The northerly subduction zone may have become already active in places such as Iran during the medial to late Jurassic, but it was during the Aptian-Albian interval that a united subduction zone extended from the eastern Alps all the way to Burma. This subduction zone constructed the major continental margin arcs of the southern margin of Eurasia during the Cretaceous to Eocene interval. The other subduction zone must have originated within the Neo-Tethys, much closer to the northern Gondwanian margin, because it soon collided with its continental margin between eastern Libya and Oman and also possibly in the Himalaya, causing gentle orogenic deformation where no ophiolite obduction occurred between Libya and Lebanon and spectacular ophiolite nappes and associated intense metamorphism and deformation from northeastern Syria (Baer-Bassit region) to Oman during the Coniacian to Campanian interval (in some places lasting into the early Maastrichtian). Searle (1986), Corfield and Searle (2000), and Corfield et al. (2005) suggested that the ophiolite obduction in the Himalaya was most likely also of late Cretaceous age. Jagoutz and Royden (submitted) in fact argued that there must have been two parallel subduction zones active within the Neo-Tethys to account for the very high relative rate of motion of India with respect to Asia during the Cretaceous to Eocene interval. Beyond the Himalaya the two subduction zones probably are now represented by the late Cretaceous Woyla suture in northwestern Sumatra and the presently active subduction zone extending from Burma to the Banda Sea (Görür and Şengör, 1992).

Of the Gondwanian fragments to the south of the Neo-Tethys, the Apulian Platform was probably the first to collide with Eurasia, which happened only in the easternmost Alps in the late Cretaceous. India, after having left the Madagascar margin at the beginning of the Cainozoic, ran into Asia sometime during the medial Eocene.

Between the Alps and Afghanistan, the geometry of the Tethyan collisions was much more complicated, because of the presence of such smaller ensialic arcs in the Neo-Tethys as the Kırşehir in central Turkey and the extreme irregularity of the colliding margins, caused both by the original rifting that created them and by later margin subparallel strike-slip faulting. However, all oceans between the Alps and Iran were closed by the Oligocene, except the Eastern Mediterranean and its continuation into southeastern Turkey and the Zagros. In the latter two places, the Neo-Tethys finally closed between 30 and 10 ma ago (McQuarrie et al., 2003; Şengör et al., 2008). This was coeval with the final collisions in the Rif and Betic Cordilleras at the western extremity of the Mediterranean (Şengör, 1993) finally terminating the circum-global equatorial current and probably triggering the growth of an ice cap in Antarctica (Coxall et al., 2005). Jovane et al. (2007) ascribed the global cooling to an Eocene-Oligocene closure of the Arabia/Eurasia collision. This is probably true because in Eastern Turkey the toe of the East Anatolian Accretionary Complex had already collided with the combined Bitlis/Arabia margin (Şengör et al., 2008) and began blocking deep oceanic circulation, although the full tectonic effects of the collision were not felt until the Miocene.

The Eastern Mediterranean and the Arabian Sea are the only surviving remnants of the Neo-Tethys in our days.

The paratethys

Eduard Suess had already recognized in 1866 that a peculiar sea basin containing brackish waters had extended during the Miocene from Vienna to almost the Aral Sea via the Styrian Basin, the Pannonian Basin, the Euxine Sea, and the Caspian Sea. It was later recognized that this sea had been a dependent of the Neo-Tethys cut off from it by the Alpide orogeny. This isolation helped the formation of a species-poor but individual-rich marine fauna. The Serbian geologist V. D. Laskarev called it Paratethys in 1924, which is a very apt designation, and it is consequently still in use. Today it is known that the Paratethys

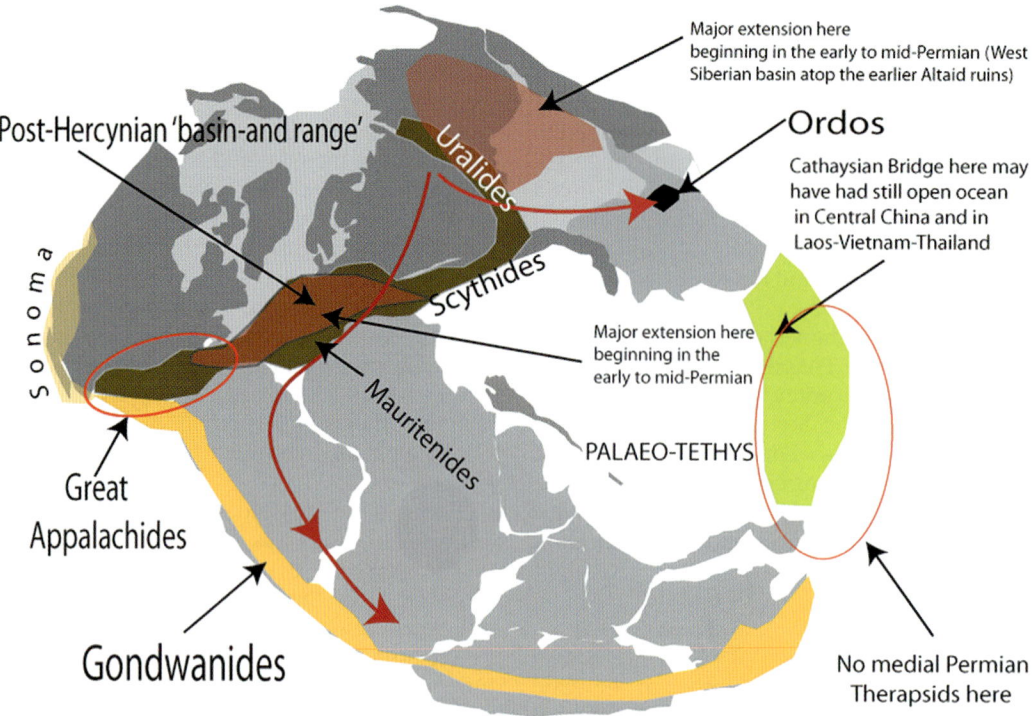

Tethys in Marine Geosciences, Figure 5 A Pangaea B reconstruction modified from Cisneros et al. (2012). *Red arrows* show the migration path of some of the therapsids from Russia into southern Africa and northern China.

existed from the latest Eocene to the Quaternary ice ages. The present-day Black Sea, Caspian Sea, and Aral Sea are its last remnants. The early, medial, and late phases of the Paratethys have recently been divided (Senes and Marinescu, 1974; Rusu, 1988; Olteanu and Jipa, 2006) into Protoparatethys (early to medial Oligocene, i.e., to late Kiscellian), Eoparatethys (latest Kiscellian to Otnangian), Mesoparatethys (Karpathian-Badenian), and Neoparatethys (Sarmatian-Recent). For two most informative reviews of the geology and evolution of the Paratethys, see Steininger and Wessely (2000) and Popov et al. (2004).

Paleo-Tethys and the Permian extinctions

Şengör and Atayman (2009) pointed out that the Paleo-Tethys had become completely continent locked during the late Permian, because a Cathaysian Bridge cut it off from the Panthalassa during the late Permian (see Figures. 2b and 5). They called such a situation, i.e., the presence of an ocean on earth completely cut off from other oceans, a *Ptolemaic condition*. They documented that the waters of the Paleo-Tethys had started becoming anoxic already during the Carboniferous in the deep subduction trenches, and this anoxia gradually invaded the entire ocean, finally flooding its shelves during the latest Permian. They further noticed that almost the entire Capitanian and Changhsingian extinctions were confined to areas that were either within the Paleo-Tethys or in a 2,000-km-wide halo around it and in regions that were receiving the poisonous waters of the Paleo-Tethys. It was also noteworthy that extinctions began in the deep basins of the ocean and gradually spread to its shallower parts. Outside such areas, as seen in many Panthalassan sections in Japan, Klamath Mountains, and New Zealand, there was no anoxia at these times and no extinction. They suggested that this state of affairs may be a result of the eruption of poisonous gases from the Paleo-Tethys that not only killed its inhabitants but also the air-breathing organisms (including the insects!) around it.

Ptolemaic conditions on earth invite marine scientists also to consider the effects of oceans cut off from the major oceans on the evolution of the earth system.

Conclusions

The name Tethys designates a mainly Mesozoic equatorial oceanic area that extended from the Gibraltar to Southeast Asia. In reality, there were two Tethyses: one that literally "popped into existence" when Laurasia and Gondwana-Land collided in the late Paleozoic along the Hercynides and was destroyed by the early Cretaceous when the last remnants of a continental strip, called the Cimmerian Continent, collided with Laurasia. That ocean is known under the designation Paleo-Tethys. The Cimmerian Continent rifted off from the northern margin of Gondwana-Land mostly during the Permo-Triassic opening behind it the Neo-Tethys. The combined areas of the Paleo-Tethys, the Cimmerian Continent, and the Neo-Tethys constitute, together with their continental margins, the Tethyan Realm. A double orogenic system resulted from the destruction of the Tethyan Realm: the products of the closure of the Paleo-Tethys are the Cimmerides, and the products of the closure of major parts of the Neo-Tethys are called the Alpides. The Eastern Mediterranean and the Arabian Sea are the remnants of the Neo-Tethys, to the north of which the Alpide evolution today forms Andean-type orogenic belts forming parts of the Alpides.

The Tethyan Realm was important both paleobiogeographically and paleoceanographically: It constituted a paleobiogeographic realm that went beyond its borders all the way into the future Central American area. The opening of an equatorial oceanic corridor from Central America to Southeast Asia led to the formation of an equatorial current system that greatly influenced world climate. When the Paleo-Tethys became an entirely closed ocean, it developed a major anoxic water body that probably triggered the end-Permian extinctions. A major water body cut off from it during the latest Eocene by Alpide collisions formed a closed basin. It is known as Paratethys whose deposits range from the Vienna Basin to the Aral Sea.

Bibliography

Aubouin, J., Blanchet, R., Stephan, J.-F., and Tardy, M., 1977. Téthys (Mésogée) et Atlantique: données de la géologie. *Comptes Rendus hébdomadaires de l'Académie des Sciences (Paris)*, **285**, 1025–1028.

Barrier, P., and Francou, C., 1996. *Téthys à la Recherche d'un Océan Disparu – The Search for a Lost Ocean*. Pau: Elf Aquitaine, 135 pp.

Belov, A. A., 1981. *Tektonicheskoe Razvitie Alpinskoi Skladchatoi Oblasti v Paleozoe*. Moskva: Nauka, 211 pp.

Biju-Duval, B., and Dercourt, J., 1980. Les bassins de la Méditerranée orientale représentant-ils les restes d'un domaine océanique, la Mesogée, ouvert au Mésozoique et distinct de la Téthys? *Bulletin de la Societe Geologique de France, sér. 7*, **22**, 43–60.

Biju-Duval, B., Dercourt, J., and Le Pichon, X., 1977. From the tethys ocean to the mediterranean seas: a plate tectonic model of the evolution of the western alpine system. In Biju-Duval, B., and Montadert, J. (eds.), *Structural History of the Mediterranean Basins*. Paris: Editions Technip, pp. 143–164.

Boccaletti, M., 1977. Mesogea, Mesoparatethys, Mediterranean and Paratethys: their possible relations with the Tethys ocean development. *Ofioliti*, **4**, 83–96.

Buffetaut, E., Cuny, G., Le Loeuff, J., and Suteethorn, V., (eds.) 2009. *Late Palaeozoic and Mesozoic Ecosystems in SE Asia*. The Geological Society (London). Special Publications, 315, vi + 306 pp.

Bullard E., Everett J. E., and Smith A. G., 1965. The fit of the continents around the Atlantic. *Philosophical Transactions of the Royal Society of London, Series A, Mathematical and Physical Sciences*, **258**(1088), 41–51, A Symposium on Continental Drift.

Burtman, V. S., 1994. Meso-Tethyan oceanic sutures and their deformation. *Tectonophysics*, **234**, 305–327.

Cisneros, J. C., Abdala, F., Atayman-Güven, S., Rubidge, B. S., Şengör, A. M. C., and Schultz, C. L., 2012. Carnivorous dinocephalian from the Middle Permian of Brazil and tetrapod dispersal in Pangaea. *Proceedings of the National Academy of Sciences of the United States of America*, **109**, 1584–1588.

Corfield, R. I., and Searle, M. P., 2000. Crustal shortening estimates across the north Indian continental margin, Ladakh, NW India. In Khan, M. A., Treloar, P. J., Searle, M. P., and Jan, M. Q., (eds.), *Tectonics of the Nanga Parbat Syntaxis and the Western Himalaya*. London: Geological Society. Special Publications, 170, pp. 395–410.

Corfield, R. I., Watts, A. B., and Searle, M. P., 2005. Subsidence history of the north Indian continental margin, Zanskar–Ladakh Himalaya, NW India. *Journal of the Geological Society (London)*, **162**, 135–146.

Coxall, H. C., Wilson, P. A., Palike, H., Backman, J., and Lear, C. H., 2005. Rapid stepwise onset of Antarctic glaciation and deeper calcite compensation in the Pacific Ocean. *Nature*, **433**, 53–57.

Crasquin-Soleau, S., and Barrier, E. (eds.), 1998a. *Peri-Tethys Memoir 3: Stratigraphy and Evolution of Peri-Tethyan Platforms*. Paris: Mémoires du Muséum National d'Histoire Naturelle, Vol. 177, 264 pp.

Crasquin-Soleau, S., and Barrier, E. (eds.), 1998b. *Peri-Tethys Memoir 4: Epicratonic Basins of Peri-Tethyan Platforms*. Paris: Mémoires du Muséum National d'Histoire Naturelle, Vol. 179, 294 pp.

Crasquin-Soleau, S., and Barrier, E. (eds.), 2000. *Peri-Tethys Memoir 5: New data on Peri-Tethyan Sedimentary Basins*. Paris: Mémoires du Muséum National d'Histoire Naturelle, Vol. 182, 266 pp.

Dercourt, J., Ricou, L. E., and Vrielynck, B., 1993. *Atlas Tethys Paleoenvironmental Maps*. Paris: Gauthier-Villars. 14 maps + 1 Table + Explanatory Notes, 307 pp.

Douvillé, H., 1900. Sur la distribution géographique des rudistes, des orbitolines et des orbitoïdes. *Bulletin de la Societe Geologique de France, sér. 3*, **28**, 222–235.

Flügel, H. W., 1972. Zur Entwicklung der "Prototethys" im Paläozoikum Vorderasiens. *Neues Jahrbuch für Geologie und Paläontologie Monatshefte*, **10**, 602–610.

Gaetani, M., undated (2006). *Geologia del Nord Karakorum – Stato dell'Arte*. [Milano], 24 pp.

Görür, N., 1991. Aptian-Albian palaeogeography of the Tethyan domain. *Palaeogeography Palaeoclimatology Palaeoecology*, **87**, 267–288.

Görür, N., and Şengör, A. M. C., 1992. Paleogeography and tectonic evolution the eastern Tethysides: Implications for the northwest Australian margin breakup history. In von Rad, U., and Haq, B., et al. (eds.), *Proceedings of the Ocean Drilling Program, Scientific Results*, Vol. 122, pp. 83–106.

Hall, R., 2002. Cenozoic geological and plate tectonic evolution of SE Asia and the SW Pacific: computer-based reconstructions, model and animations. *Journal of Asian Earth Sciences*, **20**, 353–431. + 1 CD-ROM.

Hall, R., 2012. Late Jurassic–Cenozoic reconstructions of the Indonesian region and the Indian Ocean. *Tectonophysics*, **570–571**, 1–41.

Hall, R., Cottam, M. A., and Wilson, M. E. J., (eds.) 2011. *The SE Asian Gateway: History and Tectonics of the Australia-Asia Collision*. London: Geological Society. Special Publication, 355, 381 pp.

Haug, É., 1908–1911. *Traité de Géologie, v. 2 [in 3 volumes] (Les Périodes Géologiques)*. Paris: Librairie Armand Colin, pp. 539–2024. + 64 plates.

Heritsch, F., 1940. Das Mittelmeer und die Krustenbewegungen des Perm. *Wissenschaftliches Jahrbuch der Universität Graz*, **1**, 305–338.

Hsü, K. J., 1977. Tectonic evolution of the Mediterranean Basins. In Nairn, A. E. M., Kanes, W. H., and Stehli, F. G. (eds.), *Ocean Basins and Margins, v. 4A, The Eastern Mediterranean*. New York: Plenum, pp. 29–75.

Irving, E., 1977. Drift of the major continental blocks since the Devonian. *Nature*, **270**, 304–309.

Irving, E., 2004. The case for Pangea B, and the intra-Pangean megashear. In Channell, J. E. T., Kent, D. V., Lowrie, W., and Meert, J. G. (eds.), *Timescales of the Paleomagnetic Field (Opdyke volume)*. Washington, DC: American Geophysical Union. Geophysical Monograph Series, Vol. 145, pp. 13–27.

Jovane, L., Sprovieri, M., Florindo, F., Acton, G., Coccioni, R., DallAntonia, B., and Dinarès-Turrell, J., 2007. Eocene- Oligocene paleoceanographic changes in the Stratotype section, Massignano, Italy: clues from rock magnetism and stable isotopes. *Journal of Geophysical Research*, **112**, B11101, doi:10.1029/2007JB004963.

Kahler, F., 1939. Verbreitung und Lebensdauer der Fusuliniden-Gattungen *Pseudoschwagerina* and *Paraschwagerina* und deren Bedeutung für die Grenze Karbon/Perm. *Senckenbergiana*, **21**, 169–215.

Laskarev, V., 1924. Sur les equivalents du Sarmatien supérieur en Serbie. In Vujevic, P. (ed.), *Receuil des Travaux offert à M. Jovan Cvijič par ses Amis et Collaborateurs*. Beograd: Drzhavna Shtamparija, pp. 73–85.

Laubscher, H., and Dernoulli, D., 1977. Mediterranean and Tethys. In Nairn, A. E. M., Kanes, W. H., and Stehli, F. G. (eds.), *Ocean Basins and Margins, v. 4A, The Eastern Mediterranean*. New York: Plenum, pp. 1–28.

Livermore, R. A., Smith, A. G., and Vine, F. J., 1986. Late Palaeozoic to early Mesozoic evolution of Pangaea. *Nature*, **322**, 162–165.

McKerrow, W. S., and Ziegler, A. M., 1972. Palaeozoic oceans. *Nature*, **240**, 92–94.

McQuarrie, N., Stock, J. M., Verdel, C., and Wernicke, B. P., 2003. Cenozoic evolution of Neotethys and implications for the causes of plate motions: Geophysical research letters, v. 30, DOI: 10.1029/2003GL017992.

Metcalfe, I., 1999. *The Ancient Tethys Oceans of Asia: How Many? How Old? How Deep? How Wide?* UNEAC Asia Papers 1, pp. 1–9, + 7 figures. (Electronic only, see at http://www.une.edu.au/asiacentre/PDF/Metcalfe.pdf)

Morel, P., and Irving, E., 1981. Paleomagnetism and the evolution of Pangea. *Journal of Geophysical Research*, **86**, 1858–1872.

Nairn, A. E. M., Ricou, L.-E., Vrielynck, B., and Dercourt, J. (eds.), 1996. *The Ocean Basins and Margins, v. 8, The Tethys Ocean*. New York: Plenum Press, XXII + 530 pp.

Neumayr, M., 1885. *Die geographische Verbreitung der Juraformation*. Denkschriften der kaiserlichen Akademie der Wissenschaften (Wien), mathematisch-naturwissenscaftliche Classe, Vol. 50, pp. 57–86.

Olteanu, R., and Jipa, D., 2006. Dacian Basin environmental evolution during the Upper Neogene within the Paratethys domain. Geo-Eco-Marina 12/2006 Coastal Zones and Deltas, pp. 91–105.

Popov, S. V., Rögl, F., Rozanov, A. Y., Steininger, F. F., Shcherba, I. G., and Kovac, M. (eds.), 2004. *Lithological-Paleogeographic Maps of Paratethys. 10 Maps Late Eocene to Pliocene*. Stuttgart: E. Schweizerbart'sche Verlagsbuchhandlung. Courier Forschungsinstitut Senckenberg, 250, 46 pp.

Ricou, L.-E., 1994. Tethys reconstructed: plates, continental fragments and their boundaries since 260 Ma from Central America to South-eastern Asia. *Geodinamica Acta*, **7**, 169–218.

Roeder, D., 2009. *American and Tethyan Fold-Thrust Belts: Beiträge zur Reginalen Geologie der Erde*. Berlin: Gebrüder Borntraeger, Vol. 31, VII + 168 pp.

Roure, F., (ed.) 1994. Peri-Tethyan platforms. In *Proceedings of the IFP/Peri-Tethys Research Conference Held in Arles*, France, March 23–25 1993, Paris: Éditions Technip, I-XVI + 275 pp.

Rusu, A., 1988. Oligocene events in Transylvania (Romania) and the first separation of the Paratethys. *Dari de Seama ale Sedintelor – Institutul de Geologie si Geofizica*, **72–73**, 207–223.

Searle, M. P., 1986. Structural evolution and sequence of thrusting in the High Himalayan, Tibetan–Tethys and Indus suture zones of Zanskar and Ladakh, Western Himalaya. *Journal of Structural Geology*, **8**, 923–936.

Senes, J., and Marinescu, F., 1974. Cartes paléogéographiques de la Paratéthys centrale. *Mémoires du BRGM*, **78**, 785–792.

Şengör, A. M. C., 1979. Mid-Mesozoic closure of Permo-Triassic tethys and its implications. *Nature*, **279**, 590–593.

Şengör, A. M. C., 1986. The dual nature of the Alpine-Himalayan system, progress, problems, and prospects. *Tectonophysics*, **127**, 177–195.

Şengör, A. M. C., 1990a. Tethys, Thethys, or Thetys? What, where, and when was it? Comment. *Geology*, **18**, 575.

Şengör, A. M. C., 1990b. Plate tectonics and orogenic research after 25 years: a Tethyan perspective. *Earth-Science Reviews*, **27**, 1–201.

Şengör, A. M. C., 1993. Some current problems on the tectonic evolution of the Mediterranean during the Cainozoic. In Boschi, E., Mantovani, E., and Morelli, A. (eds.), *Recent Evolution and Seismicity of the Mediterranean Region*. Dordrecht: Kluwer Academic Publishers. NATO ASI Series, Series C: Mathematical and Physical Sciences, Vol. 402, pp. 1–51.

Şengör, A. M. C., 1998. Die Tethys: vor hundert Jahren und heute. *Mitteilungen der Österreichischen Geologischen Gesellschaft*, **89**, 5–176.

Şengör, A. M. C., 2013. The Pyrenean Hercynian Keirogen and the Cantabrian Orocline as genetically coupled structures. *Journal of Geodynamics*, **65**, 3–21 (Koçyiğit volume).

Şengör, A. M. C., 1984. *The Cimmeride Orogenic System and the Tectonics of Eurasia*. Geological Society of America Special Paper 195, xi + 82 pp.

Şengör, A. M. C., and Atayman, S., 2009. The Permian Extinction and the Tethys: An Exercise in Global Geology: *Geological Society of America Special Paper* 448, x+96 pp.

Şengör, A. M. C., and Hsü, K. J., 1984. The Cimmerides of Eastern Asia: history of the eastern end of Palaeo-Tethys. In Buffetaut, E., Jaeger, J.-J., and Rage, J.-C. (eds.), *Paléogéographie de l'Inde, du Tibet et du Sud-Est Asiatique: Confrontations des Données Paléontologiques aves les Modèles Géodynamiques*. Paris: Mémoires de la Société Géologique de France. Nouvelle série No. 147, pp. 139–167.

Şengör, A. M. C., and Natal'in, B. A., 1996. Palaeotectonics of Asia: fragments of a synthesis. In Yin, A., and Harrison, M. (eds.), *The Tectonic Evolution of Asia, Rubey Colloquium*. Cambridge: Cambridge University Press, pp. 486–640.

Şengör, A. M. C., Yılmaz, Y., and Sungurlu, O., 1984. Tectonics of the Mediterranean Cimmerides: nature and evolution of the western termination of Palaeo-Tethys. In Dixon, J. E., and Robertson, A. H. F. (eds.), *Geological Evolution of the Eastern Mediterranean*. London: Geological Society. Special Publication 17, pp. 77–112.

Şengör, A. M. C., Altıner, D., Cin, A., Ustaömer, T., and ve Hsü, K. J., 1988. Origin and assembly of the Tethyside orogenic collage at the expense of Gondwana-Land. In Audley-Charles, M. G., and Hallam, A., (eds.), *Gondwana and Tethys*. London: Geological Society of London. Special Publication 37, pp. 119–181.

Şengör, A. M. C., Cin, A., Rowley, D. B., and Nie, S. Y., 1991. Magmatic evolution of the Tethysides: a guide to reconstruction of collage history. *Palaeogeography Palaeoclimatology Palaeoecology*, **87**, 411–440.

Şengör, A. M. C., Cin, A., Rowley, D. B., and Nie, S. Y., 1993. Space-time patterns of magmatism along the Tethysides: a preliminary study. *Journal of Geology*, **101**, 51–84.

Shvolman, V. A., 1978. Relics of Mesotethys in the Pamirs. *Himalayan Geology*, **8**, 369–378.

Sinha, A. K., Sassi, F. P., and Papanikoláou, D. (eds.), 1997. *Geodynamic Domains in the Alpine Geodynamic Domains in the Alpine-Himalayan Tethys: (A publication of the IGCP Project 276)*. Rotterdam: A.A. Balkema, I-XV + 441 pp.

Smith, A. G., Hurley, A. M., and Briden, J. C., 1981. *Phanerozoic Paleocontinental World Maps: Cambridge Earth Science Series.* Cambridge: Cambridge University Press, 102 pp.

Sonnenfeld, P. (ed.), 1981. *Tethys – The Ancestral Mediterranean: Benchmark Papers in Geology.* Stroudsburg: Hutchinson Ross, Vol. 53, XIII + 331 pp.

Stampfli, G. M., and Borel, G. D., 2002. A plate tectonic model for the Paleozoic and Mesozoic constrained by dynamic plate boundaries and restored synthetic oceanic isochrons. *Earth and Planetary Science Letters*, **196**, 17–33.

Stampfli, G., Marcoux, J., and Baud, A., 1991. Tethyan margins in space and time. *Palaeogeography Palaeoclimatology Palaeoecology*, **87**, 373–409.

Stampfli, G., Mosar, J., de Bono, A., and Vavasis, I., 1998. Late Paleozoic, early Mesozoic plate tectonics of the western Tethys. *Bulletin of the Geological Society of Greece*, **32**, 113–120.

Stampfli, G. M., Mosar, J., Favre, P., Pillevuit, A., and Vannay, J.-C., 2001. Permo-Mesozoic evolution of the western Tethys realm: the Neo-Tethys East Mediterranean basin connection. In Ziegler, P. A., Cavazza, W., Robertson, A. H. F., and Crasquin-Soleau, S. (eds.), *Peri-Tethys Rift/Wrench Basin and Passive Margins.* Paris: Mémoirs du Muséum National d'Histoire Naturelle, Vol. 186, pp. 51–108.

Steininger, F. F., and Wessely, G., 2000. From the Tethyan ocean to the Paratethys sea: Oligocene to neogene stratigraphy, paleogeography and paleobiogeography of the circum-Mediterranean region and the Oligocene to Neogene basin evolution in Austria. *Mitteilungen der Österreichischen Geologischen Gesellschaft*, **92**, 95–116.

Stille, H., 1940. *Einführung in den Bau Amerikas.* Berlin: Gebrüder Borntraeger, XX + 717 pp.

Stille, H., 1949. Die jungalkonkische Regeneration im Raume Amerikas: Abhandlungen der Deutschen Akademie der Wissenschaften zu Berlin. *Mathematisch-naturwissenschaftliche Klasse*, **1948** (3), 38 pp.

Stille, H., 1951. *Das Mitteleuropäische Varizische Grundgebirge im Bilde des Gesamteuropäischen.* Hannover: Bundesanstalt für Geowissenschaften und Rohstoffe. Beihefte zum Geologischen Jahrbuch, No. 2, 138 pp.

Stille, H., 1958a. Einiges über die Weltozeane und ihre Umrahmungsräume. *Geologie*, **7**, 284–306 (von Bubnoff volume).

Stille, H., 1958b, *Die Assyntische Tektonik im Geologischen Erdbild.* Beihefte zum Geologischen Jahrbuch, No. 22, 255 pp.

Stöcklin, J., 1974. Possible ancient continental margins in Iran. In Burk, C. A., and Drake, C. A. (eds.), *The Geology of Continental Margins.* Berlin: Springer, pp. 873–887.

Stöcklin, J., 1977. Structural correlation of the Alpine ranges between Iran and Central Asia. In *Livre à la Mémoire de Albert F. De Lapparent.* Paris: Société Géologique de France. Mémoire hors-série no. 8, pp. 333–353.

Stow, D., 2010. *Vanished Ocean – How Tethys Reshaped the World.* Oxford: Oxford University Press, I-XII; 300 p.

Suess, E., 1893. Are great ocean depths permanent? *Natural Science*, **2**, 180–187.

Suess, E., 1895. Note sur l'histoire des océans. *Comptes Rendus hébdomadaires de l'Académie des Sciences (Paris)*, **121**, 1113–1116.

Suess, E., 1901. *Das Antlitz der Erde, v. III/1 (Dritter Band. Erste Hälfte).* Prag and Wien/Leipzig: F. Tempsky/G. Freytag, IV + 508 pp.

Suess, E., 1902. Abschieds-Vorlesung des Professors Eduard Suess bei seinem Rücktritte vom Lehramte gehalten am 13. Juli 1901 im Geologischen Hörsaale der Wiener Universität: Beiträge zur Paläontologie und Geologie Österreich-Ungarns und des Orients, Bd. 14, Heft 1, pp. 1–8.

Suess, E., 1909. *Das Antlitz der Erde, v. III/2 (Dritter Band. Zweite Hälfte. Schluß des Gesamtwerkes).* Prag and Wien/Leipzig: F. Tempsky/G. Freytag, IV+789 pp.

Topuz, G., Göçmengil, G., Rolland, Y., Çelik, Ö. F., Zack, T., and Schmitt, A. K., 2012. Jurassic accretionary complex and ophiolite from northeast Turkey: no evidence for the Cimmerian continental ribbon. *Geology*, **41**, 255–258.

Tozer, E. T., 1989. Tethys, Thethys, or Thetys? What, where, and when was it? *Geology*, **17**, 882–884.

Tozer, E. T., 1990. Tethys, Thethys, or Thetys? What, where, and when was it? Reply. *Geology*, **18**, 575–576.

Van der Voo, R., and And French, R. B., 1974. Apparent polar wander for the Atlantic-bordering continents: late Carboniferous to Eocene. *Earth Science Reviews*, **10**, 99–119.

Veevers, J. J. (ed.), 2000. *Billion-Year Earth History of Australia and Neighbours in Gondwanaland.* Sydney: Gemoc, xi + 388 pp.

Veevers, J. J., 2001. *Atlas of Billion-Year Earth History of Australia and Neighbours in Gondwanaland.* Sydney: Gemoc Press, iv + 76 pp.

Veevers, J. J., 2004. Gondwanaland from 650–500 Ma assembly through 320 Ma merger in Pangea to 185–100 ma breakup: supercontinental tectonics via stratigraphy and radiometric dating. *Earth-Science Reviews*, **68**, 1–132.

Weissert, H., 1981. The environment of deposition of black shales in the early Cretaceous: an ongoing controversy. *Society of Economic Paleontologists and Mineralogists, Special Publication*, **32**, 547–560.

Weissert, H., and Stössel, I., 2010. *Der Ozean im Gebirge: Eine Geologische Zeitreise Durch die Schweiz*, 2nd edn. Zürich: vdf, 185 p.

Weissert, H., McKenzie, J., and Hochuli, P., 1979. Cyclic anoxic events in the Early Cretaceous Tethys Ocean. *Geology*, **7**, 147–151.

Yin, H. F., Dickins, J. M., Shi, G. R., and Tong, J. (eds.), 2000. *Permian-Triassic Evolution of Tethys and Western Circum-Pacific.* Amsterdam: Elsevier, xix + 392 pp.

Zachner, W., and Lupu, M., 1999. Pitfalls on the race for an ultimate Tethys model. *Journal of International Earth Sciences (Geologische Rundschau)*, **88**, 111–115.

Ziegler, P. A., and Horvath, F., 1996. *Peri-Tethys Memoirs, 2: Structure and Prospects of Alpine Basins and Forelands.* Paris: Mémoires du Muséum National d'Histoire Naturelle, Vol. 170, 552 pp.

Ziegler, P. A., Cavazza, W., Robertson, A. H. F., and Crasquin-Soleau, S. (eds.), 2001. *Peri-Tethys Memoir 6: Peri-Tethyan Rift/Wrench Basins and Passive Margins.* Paris: Mémoires du Muséum National d'Histoire Naturelle, Vol. 186, 765 pp.

Cross-references

Paleophysiography of Ocean Basins
Plate Tectonics
Wilson Cycle

TIDAL DEPOSITIONAL SYSTEMS

Achim Wehrmann
Marine Research Department, Senckenberg am Meer, Wilhelmshaven, Germany

Synonyms

(Inter)tidal depositional environments; Intertidal deposits; Peritidal deposits; Peritidal environments; Tidal

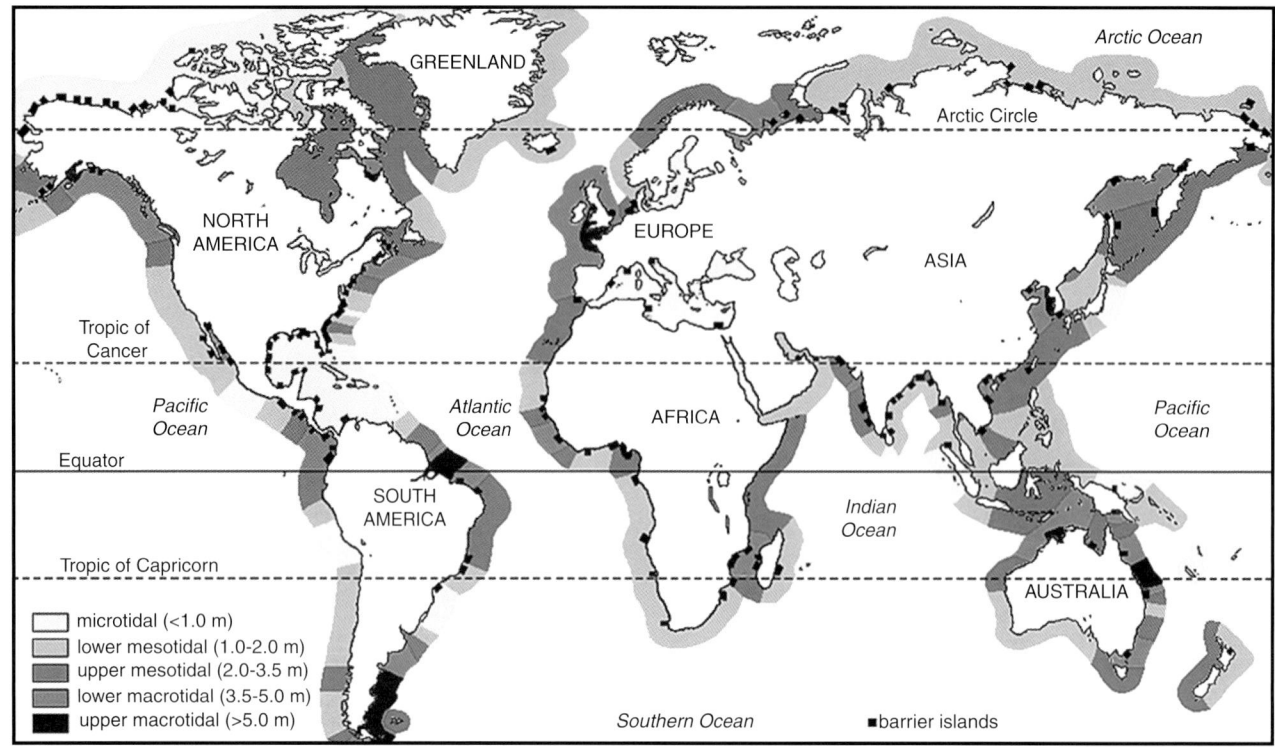

Tidal Depositional Systems, Figure 1 Global distribution of tidal ranges following the classification scheme of Hayes (1979) (from Flemming (2012)).

sedimentary environments; Tidal sedimentary facies; Tidalites; Tide-dominated environments; Tide-influenced deposits

Definition

Tidal depositional systems are sedimentary environments which are mainly influenced and controlled by tides and tide-related processes. Their occurrence is limited to coasts and shallow shelves having a pronounced tidal range and, in relation to that, moderate wave action. In their vertical extension, tidal depositional systems are not restricted to the intertidal or eulittoral part of the coastal zone which is defined by the tidal range, i.e., the vertical distance between the mean high water level and the mean low water level. They can also be found in supratidal (above mean high water level) as well as in shallow subtidal (below mean low water level) environments. In contrast, deepwater sedimentary environments can also be influenced by tides (for review see Dykstra, 2012) but are not tidal depositional systems sensu stricto, as tides are an ocean-wide phenomenon. Tidal depositional systems are characterized by steep environmental gradients. Besides tide-influenced physical characteristics, numerous other processes like biological activity can play an important role in the formation of tidal sedimentary facies. Tidal depositional systems can both be dominated by siliciclastic and/or biogenic (carbonate or organic-rich) sediments.

Introduction

The majority of tidal depositional systems are linked to coastal environments. Their morphodynamic and sedimentary processes are therefore not only controlled by tides but also by waves. Whether a coast is classified as tide dominated, mixed energy, or wave dominated depends mainly on the ratio of the regional tidal range and wave energy. In general, tidal depositional systems are most frequent where the tidal regime has a meso- to macrotidal character.

The classification of coastal environments with respect to the tidal range is based on Davis (1964, 1980) and Hayes (1979). In contrast to the relatively simple scheme of Davis, who defined only three categories (<2 m microtidal; 2–4 m mesotidal; >4 m macrotidal), the classification scheme of Hayes shows a more detailed subdivision, which takes specific process-related morphological units (e.g., occurrence of barrier islands) into account (Figure 1). Besides the astronomical forces, which regionally can result in extreme differences in neap and spring tides, wind stress in combination with specific shelf topography and shape of the coast can significantly modify the tidal water levels.

Tidal inlets and tidal deltas

Tidal inlets and tidal deltas are typical morphological units of barrier coasts resulting from the tide-controlled

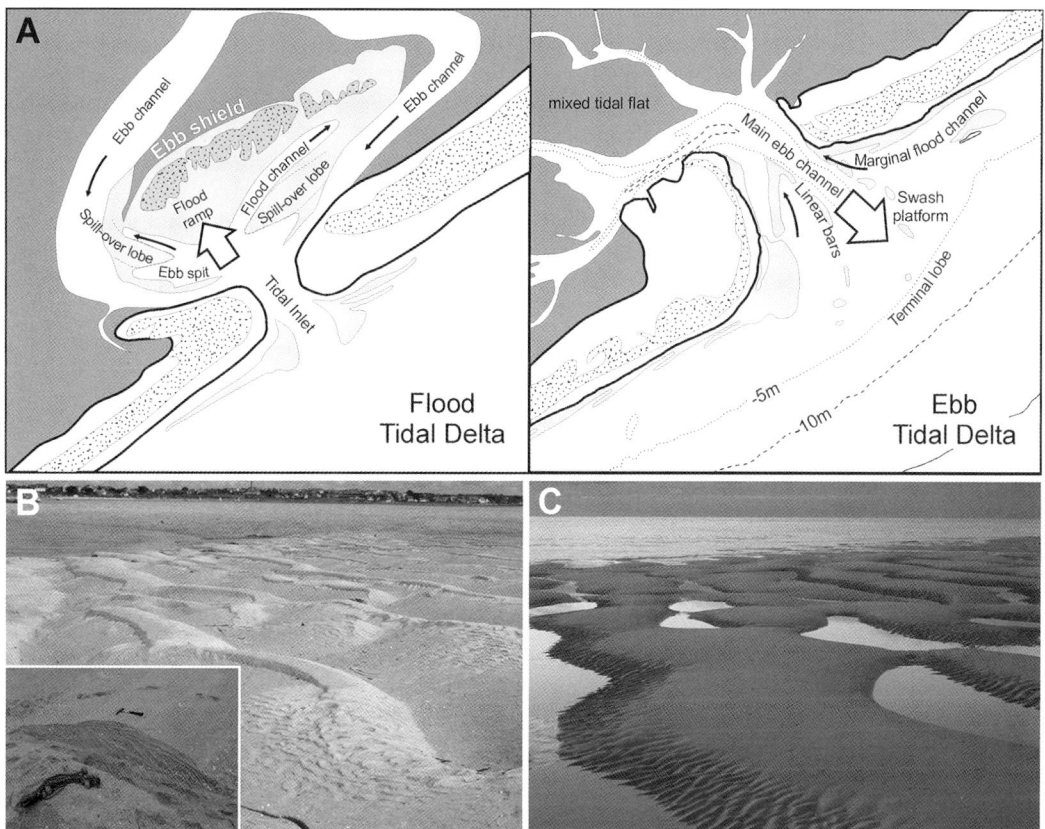

Tidal Depositional Systems, Figure 2 (a) Ebb–tidal delta and flood–tidal delta model (modified after Hayes (1979)) with its morphological units. (b) 3D compound dunes in a flood–tidal delta of a macrotidal lagoon (Northern Brittany, France, max. tidal range ~9.3 m). Dunes were formed during flood (currents from *left* to *right*) and modified on top during ebb phase, separated by a reactivation surface. Bedforms become inactive during neap tides whose tidal range is half of those of spring tides. (c) Dunes of a marginal flood channel from a mesotidal (tidal range 2.7 m) ebb–tidal delta (East Frisian barrier islands, Wadden Sea). Flood currents from *left* to *right*.

exchange of water between the open ocean and the back-barrier tidal basins (see "Wadden Sea," this volume). They are highly dynamic sedimentary systems showing a complex interaction of tide-dominated processes and wave-dominated longshore sediment transport. Also, the tidal prism, i.e., the water volume (as a function of average surface area of the tidal basin and the tidal range) passing the tidal inlet twice during each tidal cycle, and the sediment supply strongly control size, geometry, and morphodynamics of the tidal inlet–tidal delta complex (Hayes, 1979; Davis and Hayes, 1984). According to Fitz-Gerald et al. (2012), the formation of tidal inlet-associated flood–tidal delta and/or ebb–tidal delta results from different processes. Flood–tidal deltas initially originate from storm events, whereas ebb–tidal deltas are mainly influenced by longshore processes and the interaction of tidal and wave energy.

Flood–tidal deltas are characterized by their typical horseshoe-like shape and are situated in the sheltered back-barrier area. Internal morphological units are (i) flood ramp, (ii) flood channels, (iii) ebb shield, (iv) ebb spits, and (v) spillover lobes (FitzGerald, 2005) (Figure 2a). Flood–tidal deltas are best developed under moderate mesotidal conditions. Sediment transport to and from the flood–tidal delta is mainly controlled by the asymmetry in current velocity during tidal cycle (Boothroyd and Hubbard, 1975). At flood phase, highest velocities occur near high water when the flood–tidal delta is completely submerged. This results in a sand transport up the flood ramp (Figure 2b), through the flood channels, onto the ebb shield (Davis and FitzGerald, 2004). During ebb phase, highest velocities of tidal currents are reached at mid- to low tide, when the ebb shield of the flood–tidal delta is already emerged. As a consequence, the ebb flow is split and has to pass the delta laterally, eroding sand from the outer ebb shield and transporting it along the ebb spits. Sand reaching the tidal inlet will then become again part of the above-described transport gyre.

Ebb–tidal deltas are located at the offshore side of the tidal inlet and are therefore exposed to wave action. They display a complex pattern of channels and shoals.

Characteristic morphological units are (i) main ebb channel, (ii) terminal lobe, (iii) swash platform, (iv) channel margin linear bars, (v) swash bars, and (vi) marginal flood channels.

The morphology of ebb–tidal deltas is mainly controlled by the hydrodynamic energy of the tidal forces on the one hand, and the combination of wave action and longshore transport on the other hand. A tide-dominated inlet is characterized by an elongate ebb–tidal delta and a main ebb channel bordered by linear bars extending relatively far offshore. Under mixed energy conditions, the ebb current forms a well-developed main ebb channel. In relation to its narrow inlet, the corresponding swash platform has a broad shape widely overlapping the inlet shoreline. Contrastingly, wave-dominated tidal inlet/ebb–tidal delta complexes are relatively small. Their terminal lobe and/or swash bars just form a small arc marking the periphery of the delta which commonly is entirely subtidal (Davis and FitzGerald, 2004).

Sediment transport in the inlet–ebb delta system follows characteristic pathways. During ebb phase, the water masses of the tidal prism have to pass the narrow inlet in a seaward direction. Once passed, the ebb flow expands laterally which results in a marked decrease in transport energy. This led to a deposition of sediment on the terminal lobe and the formation of ebb-oriented sand waves and swash bars. In the very initial flood phase, water first flows into the tidal basin through the marginal flood channels, while due to momentum, in the main channels still ebb currents prevail. Bedforms in the marginal flood channels (Figure 2c) are therefore flood oriented (landward). Also, the breaking waves generate a landward-oriented flow across the swash platform.

The ebb–tidal deltas in general function as sand depocenters. According to Walton and Adams (1976), the volume of the ebb–tidal delta is a function of the tidal prism. By a given tidal prism, the sediment volume of the ebb–tidal delta is smaller under high wave-energy conditions than under low wave-energy conditions.

Tidal flats

Intertidal flats as morphological units frequently occur in numerous coastal depositional environments (back-barrier areas, embayments, estuaries, deltas, lagoons, and coastal plains). Their formation and depositional processes are mainly controlled by the hydrodynamic energy of tides and waves, sediment supply, climate, and biological processes. Tidal flats are composed of unconsolidated sediments and have a typically gentle sloping surface. Especially when sheltered, the intertidal sand and mud banks are dissected by a network of tidal channels and creeks. Open-coast tidal flats are predominantly related to the large tide-dominated estuaries or river delta complexes and their associated chenier plains (Galloway, 1976; Lee et al., 1994; Allen and Duffy, 1998; Fan et al., 2006). These tidal flats are often composed of mud originating from flocculated suspended material or from resuspended delta deposits distributed by longshore transport processes. At coasts where river-derived suspension load is missing, open tidal flats can be composed of sandy material. Sheltered tidal flats are mostly related to barrier islands, sand spits, and bars. According to their position on the landward side of these structural coastal elements, they are termed back-barrier tidal flats (see "Wadden Sea," this volume). As barrier islands and sand spits are restricted to micro- and mesotidal coastal environments, the respective tidal flats also develop under low to moderate hydrodynamic conditions. Principally, the continuous decrease in energy from the open sea toward the coastline is still visible in an accordant grain-size trend, from sandy toward muddy sediments. This is also true for open tidal flats. Most prominent physical sedimentary structures are tidal current- and wave-generated ripples of variable size and shape (2D to 3D small-scale ripples, mega ripples, dunes, linguoid current ripples, ladder-back ripples), current parallel alignments, even fine sand to mud sheets, scour structures, water-level marks, and runoff features. Depending on grain-size composition, these surface structures are internally visible as different kinds of ripple cross-bedding, flaser and lenticular bedding, as well as even lamination. Additionally, a high number of surficial and internal sedimentary structures are related to the activity of endo- to epibenthic organisms, including burrows, tubes, traces of locomotion, feeding, and resting (Figure 3a). Biostromal to reef-like structures are formed by microbial mats, polychaetes, and bivalves (e.g., mussel beds, oyster reefs; Figure 3b). Shell pavements reflect autochthonous erosion, concentration, and accumulation of endobenthic bivalve communities or epibenthic mussel beds. In temperate climates (e.g., Georgia Coast), tidal flats are regionally intensely vegetated.

In the main tidal channels, the water level corresponds to the tidal curve at any time as their channel floor lies below the low water level. In the topographic higher portions of the intertidal flats, a dendritic drainage network of subordinate creeks is developed in which water flow is gravity controlled after tidal flats fell dry during ebb phase. In the creeks the water level is inclined, resulting in ebb-directed flow also in the initial flood phase.

Most prominent bedforms in larger channels are subaqueous dunes whose asymmetry indicates predominant formation by tide-related currents. Reverse current direction during flood will modify just the surface of ebb-oriented subaqueous dunes documented by smaller superimposed ripples on top and vice versa. In smaller channels and creeks, meandering causes lateral erosion of tidal flat deposits, as well as point-bar sedimentation (Figure 3c). Lateral shift of channel beds is indicated by shell-lag deposits (partly with mud clasts) and longitudinal cross-bedding (Reineck and Singh, 1980; also termed inclined heterolithic stratification (IHS) Thomas et al., 1987).

Coastal wetlands: salt marshes and mangroves

The intertidal areas can be vegetated by dense stocks of vascular plants in parts where the input of hydrodynamic

Tidal Depositional Systems, Figure 3 Tidal flats of the central Wadden Sea. (**a**) Sandy to mixed tidal flats intensely bioturbated by highly abundant polychaetes. (**b**) Dense oyster reefs forming extended biogenic structures in the tidal flats. (**c**) In the topographic higher portions of the intertidal flats, a dendritic drainage network of subordinate creeks is developed. Note semi-consolidated mud beds of former point-bar sedimentation with inclined heterolithic stratification.

Tidal Depositional Systems, Figure 4 (**a**) Salt marshes at the transition from the uppermost intertidal to the lower supratidal environment (Jade Bay, Wadden Sea) under lower macrotidal regime (tidal range 4.1 m). The low hydrodynamic energy and extreme environmental conditions favor the establishment of microbial mats. Both microbial mats and the vegetation of the salt marsh baffle, trap, and stabilize fine-grained sediment particles. (**b**) Dense standing mangrove stocks significantly reduce hydrodynamic energy by their closely spaced root system which favors settlement and accumulation of suspended fine-grained sediment particles. Additionally, mangroves provide ecosystem services as they protect coasts against storm surges.

energy is low. These conditions generally occur in estuaries, in back-barrier areas, and in higher elevated areas close to the high water level. Two different kinds of ecosystems can be distinguished: the grasslike salt marshes and the mangrove forests covered by woody shrubs and trees. Their global distribution is controlled by climatic factors, of which air temperature is the most limiting, as mangroves do not tolerate freezing. Therefore, salt marshes typically occur in mid- to higher latitudes, whereas mangroves fringe parts of tropical coasts. As a general characteristic of both systems, they mark the transitional zone from the marine toward the terrestrial influenced coastal environment and have therefore the potential to be used as sea-level index points. Additionally, they both strongly act as sediment sinks as their vegetation traps and stabilizes fine-grained sediments. They are the most productive tidal ecosystems and provide ecosystem services in terms of coastal protection.

Salt marshes start to grow above neap high water level where disturbance by wave action is low, allowing the deposition of fine-grained organic-rich sediments. Zonation of the salt marsh vegetation displays the tolerance or adaptation of the plants to salt water as a function of the topographic level with respect to the tidal range and the frequency of flooding. The lowest part of the salt marsh, the silting-up zone, is typically characterized by the formation of multilayered microbial mats stabilizing and binding the barren sediment surface in front of the densely vegetated salt marsh (Figure 4a).

Sedimentation in the salt marsh environment is restricted to high water conditions (spring tides, extraordinary wind-stressed high waters, storm surges) when the salt marsh is flooded by suspension-loaded waters. The dense vegetation favors baffling and deposition of fine-grained sediment particles by reducing water motion to a minimum. Once settled, the roots of the vegetation cover will stabilize the muddy sediments. During storm surges coarse-grained sediments will be transported into the salt marshes by bed load processes. Most obvious are distinct shell-rich sand layers intercalating the muddy sediments. Layers of plant debris also indicate storm events. In this context, salt marshes situated on the sheltered side of back-barrier islands can completely be covered by thick layers of sandy wash-over fan deposits. As most of the salt marsh vegetation is resistant against burial, the marsh vegetation will recover through the storm layers within months. Besides well-bedded laminated muddy sediments and intercalated coarse-grained storm deposits, distinct bio- and phytoturbation is common in salt marsh sequences.

Mangroves cover a wider range within the tidal environment. Nevertheless, they are also restricted to sheltered parts of the coast and to estuaries and creeks where energy input is low. Similar to the salt marshes, they show a zonation depending on their vertical position within the tidal range. The lowest level is covered by the so-called red mangrove *Rhizophora mangle* which passes into the black mangrove of *Avicennia sp.* in the higher intertidal. At high tide levels the mangroves are built up by *Laguncularia racemosa*, the white mangrove. Dense standing stocks especially of the genus *Rhizophora* have a strong impact on sedimentary processes. Their closely spaced root system significantly reduces hydrodynamic energy which favors settlement and accumulation of suspended fine-grained sediment particles (Figure 4b). Additionally, the roots represent a substrate colonized by numerous benthic organisms like oysters which also produce a high amount of fine-grained aggregates (feces and pseudofeces) by their filter-feeding activity. The muddy sediment in between the root system is intensely burrowed by decapod crabs. Due to the resistant root systems, the mangroves display an effective ecosystem protecting the low-energy coasts against physical disturbance of tropical storms and hurricanes by functioning as wave breakers.

Tide-dominated river deltas

In the ternary diagram of river delta classification (Galloway, 1975), which is based on the relative influence of the factors of sediment supply (river), wave energy, and tidal energy, numerous larger river deltas are classified to be tide dominated. Tide-dominated river deltas are characterized by predominantly high-energy sediment transport and a complex hydrodynamic. Due to cyclic fluctuation in tidal energy and seasonal changes in river discharge and sediment load, the sedimentary sequences of tide-dominated river deltas show a vivid intercalation of sandy to muddy beds representing coarsening-upward as well as fining-upward trends in facies association. The most characteristic feature of a tide-dominated river delta is the formation of a clinothem, i.e., an internally sigmoidal-shaped sequence of the topset-foreset-bottomset, indicating deposition and a progradational trend. Typically, they occur in open coastal settings with meso- to macrotidal conditions and active orogens in the hinterland which, by their enormous sediment supply, allows delta formation under high-energy conditions. Modern examples are Fly, Ganges-Brahmaputra, Indus, Amazon, Irrawaddy, Yangtze, Mahakam, and Mekong. Principally, the sedimentary environments of a tide-dominated river delta can be assigned to the subaerial (upper and lower delta plain) and subaqueous portion (delta front platform, delta front slope, and prodelta) of the compound clinoform (Figure 5; Hori and Saito, 2007). In this context, the boundary of the lower to the upper delta plain marks the limit of tidal influence and other marine processes. Additionally, the lower delta plain and the delta front platform with their respective prograding clinoforms are separated by a broad zone of limited sediment accumulation (Swenson et al., 2005). The lower delta plain is characterized by extended muddy tidal flats and salt marshes and/or mangroves pervaded by a network of tidal channels and tidal creeks. Flaser to lenticular bedding, wavy lamination, and longitudinal cross-bedding related to point-bar sedimentation of laterally migrating channels are typical sedimentary structures of this depositional environment. Rootlets and plant debris-rich sediments indicate the transition to the vegetated delta plain. Large channel-mouth bars, extending from the shallow subtidal into the supratidal, mark the active distributaries of the rivermouth. When accreting laterally and vertically, these bars can be transformed to tidal flats and become part of the delta plain when vegetated (Allison et al., 2003). On tide-dominated river deltas which are also affected by significantly high wave action at their shoreline, sandy beach ridges (partly with eolian dunes) and intertidal longshore bars are frequently developed. At the transition of the outer delta plain platform to the delta front slope hydrodynamic energy increases resulting in coarsening of the sediments. Typical sedimentary structures in this zone are fine- to medium-scale bedding with wave ripples, hummocky and trough cross-stratification, subaquatic dunes, and frequent sharp-based erosional contacts formed by storm-wave scour (Kuehl et al., 1997; Goodbred and Saito, 2012). Toward the deeper part of the delta front slope, the sediment often shows coarsening-upward succession of sands and laminated to bioturbated muds. Internally, beds frequently have a turbidity character with normal grading and sharp basal contacts. The prodelta is dominated by highly bioturbated muddy sediments. Distinct signatures from tidal processes in this part of the delta are strongly overprinted by wave action, storm surges, and bioturbation.

Tide-dominated estuaries

Tide-dominated estuaries are characterized by their funnel-shaped geometry reflecting decrease in tidal flux

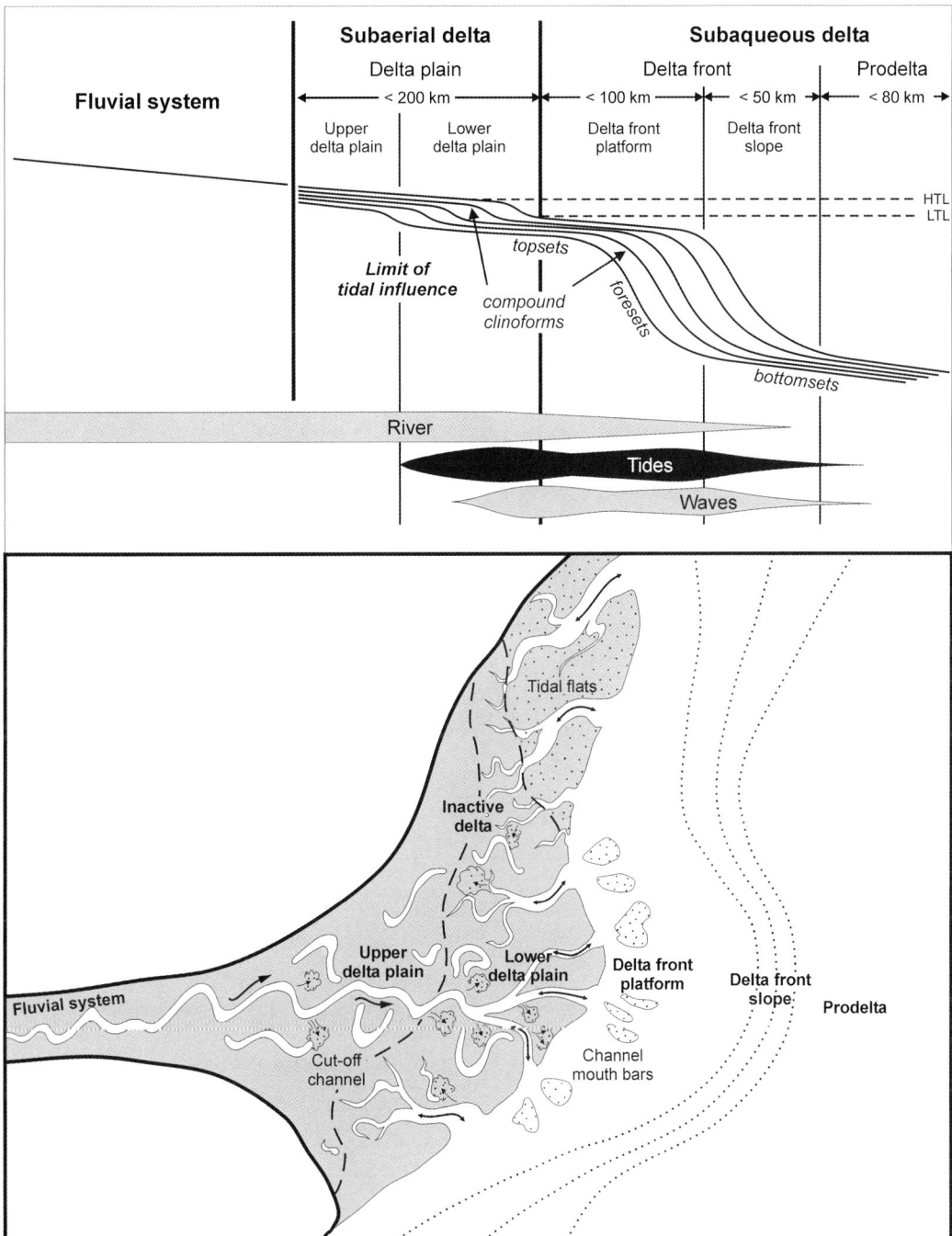

Tidal Depositional Systems, Figure 5 Major morphologic units of tide-dominated river delta systems (modified from Hori and Saito (2007) and Wehrmann et al. (2010)).

upstream. The estuary can principally be divided in a broader outer estuary and a narrow inner estuary (Figure 6a, Dalrymple et al., 1990; Dalrymple and Choi, 2007).

The outer estuary is characterized by a complex pattern of elongated, current parallel tidal bars and/or sand flats which are separated by several larger ebb- and flood-dominated channels. Internally, the linear bars are dissected by minor oblique channels ("swatchways" sensu Robinson, 1960). The sandbar–swatchway–channel system is characterized by a high morphodynamic but becomes simpler organized toward the inner limit of the outer estuary. Typical sedimentary structures are subaqueous dunes whose dimension is mainly controlled by water

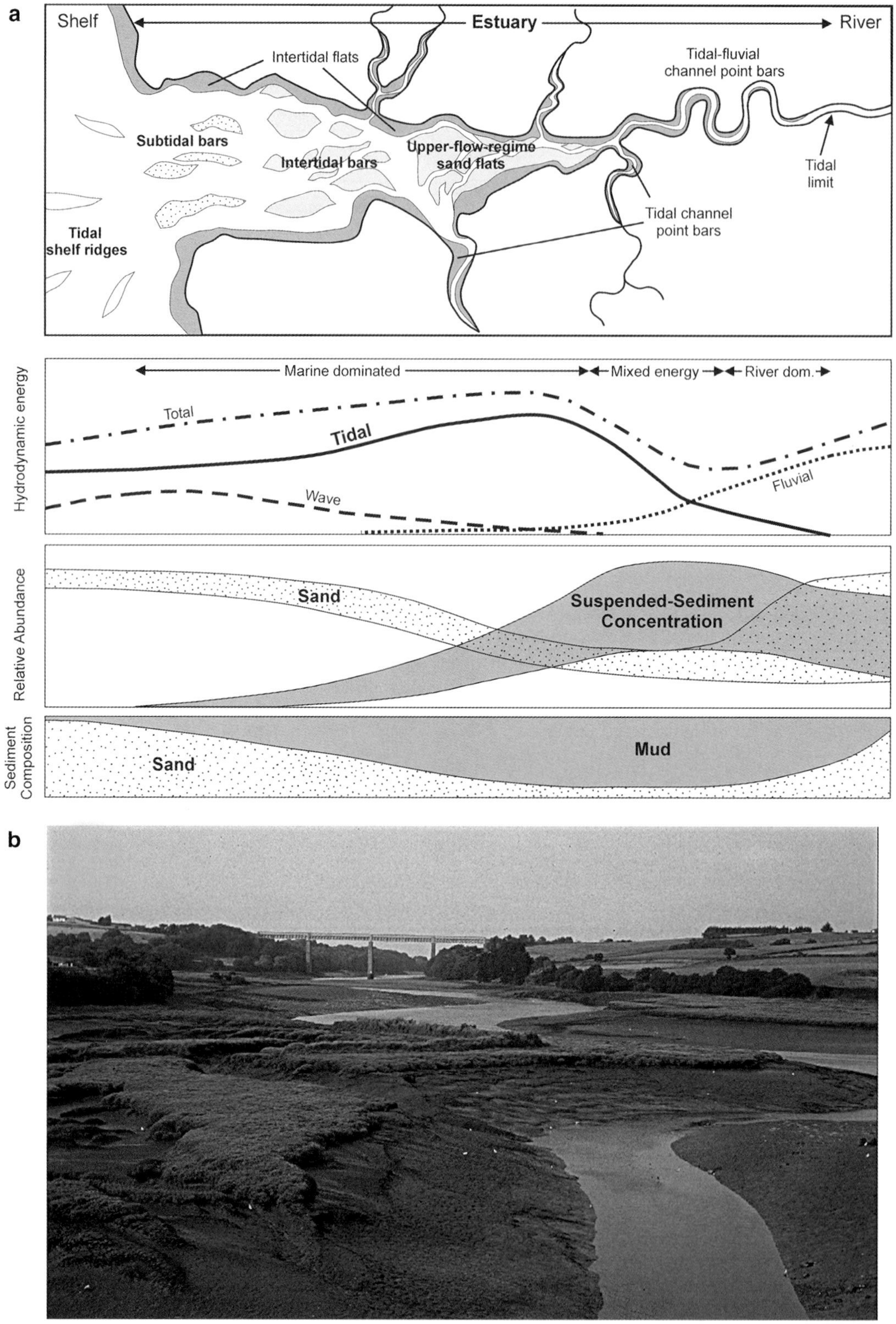

Tidal Depositional Systems, Figure 6 (a) Schematic model of a tide-dominated estuary and the variation of parameters and processes (modified from Dalrymple and Choi (2007)). (b) Macrotidal estuary of the Penzé River (Northern Brittany, France). The maximum tidal range during spring tide is 9.3 m. The channel point bars of the upper estuary are bordered by muddy tidal flats and salt marshes which are internally dissected by a network of tidal gullies. Note slumping structures by collapse of the semi-consolidated mud deposits.

depth, grain-size composition, current velocity, and sediment supply. Dunes can be modified surficially by smaller dunes and ripples when direction of tidal current reverses during tidal cycle, forming compound dunes. Internally, compound dunes mainly show uni- to bipolar cross-bedding frequently separated by a set of reactivation surfaces (Dalrymple and Rhodes, 1995). Bioturbation in this high dynamic sedimentary environment is rare. Laterally, the outer estuary fades into wave-generated beaches and shorefaces as waves are able to enter the estuary at high tide when tidal currents are low and most parts of the estuary are submerged. Where sand flats are present, ripple cross-bedding and parallel lamination (upper plane bed) is most obvious. Longitudinal cross-bedding (or inclined heterolithic stratification (IHS)) indicates point-bar sedimentation in meandering channels.

The inner estuary displays the transition from tide-controlled sedimentary system to the fluvial part of the river–estuary complex. The inner estuary is marked by a main ebb channel partly flanked by flood channels separated from each other by elongate tidal bars. In the channels, subaqueous dunes, internally cross-bedded, are common. The troughs between those dunes can be covered by fluid mud (Schrottke et al., 2006) forming an interbedding with the coarser dune material. Where noticeable input of coarse river-derived sediment is missing, thick fluid mud layers are the prevailing channel bottom deposits. The channels are bordered by muddy tidal flats and salt marshes (or mangroves) which narrow upstream. The mudflats and salt marshes are internally also dissected by a network of tidal gullies, principally oriented perpendicular to the main channel (Figure 6b). Lateral migration of the gullies leads to the formation of mud pebble layers by erosion and collapse of the semi-consolidated mud deposits from the steep outer bank. Bioturbation in the mud flat deposits and associated facies is common to abundant.

Carbonate tidal systems

Modern tide-controlled carbonate depositional environments are globally less widely distributed than their ancient analogs. Today, they are restricted to very distinct regions in the subtropics and tropics (Bahamian Archipelago, Arabian Gulf, Western Australia), whereas modern carbonate factories from the subtidal down to the abyssal occur in all climates up to the polar regions. Looking on the modern examples, carbonate tidal depositional systems are limited to microtidal and lower mesotidal conditions where, however, the influence of local wind effects can significantly exceed the astronomic tidal range. Similar to their siliciclastic counterparts, most of the sediments have an allochthonous origin imported from nearshore subtidal sources. Terrigenous input by eolian sediment transport or episodic sheet floods is typical for arid shorelines. The most important processes in sediment transport of carbonate tidal settings are related to high hydrodynamic energy events (i.e., storm surges) which transport resuspended sediment from shallow subtidal environments onto the adjacent tidal flats. In supratidal settings microbial mats frequently occur.

In the microtidal *Bahamian Archipelago*, subtidal conditions occur in nearshore regions and tidal channels. The sediment surface is sparsely covered with sea grass and calcareous algae. The sediment consists of gray peloidal mud, with scattered skeletal fragments of benthic foraminifera and mollusks, and is extensively bioturbated by thalassinid shrimps. On the flat-topped shallow subtidal platforms, oolitic sand shoal complexes occur predominantly at its margins where tidal currents can reach velocities of more than 2 m/s (Reeder and Rankey, 2009). Bedforms of oolitic sands are of various sizes, ranging from small-scale current ripples to large subaqueous dunes. Due to reverse current direction, most of the active bedforms internally show reactivation surfaces and bidirectional cross-bedding leading to composite dunes. Where shoal sands are not affected by sediment transport for some time, they are stabilized by sea grass, (calcareous) algae, and microbial mats. The tidal flats are affected by semidiurnal tides as well as by wind-driven water-level fluctuations, both onshore and offshore. The intertidal is characterized by red and black mangroves and microbial mat marshes. In the lower intertidal zone, sedimentary structures are almost completely destroyed by bioturbation of polychaetes, shrimps, and rhizoturbation. Decreased bioturbation in the upper intertidal and sediment stabilization by microbial mats result in preservation of the internal sedimentary fabric (lamination, fenestrae, desiccation cracks, intraclasts). Typical morphological units of the supratidal environment are levees, beach ridges, and plains bordering widespread tidal flats. Flooding of this area is limited to major storms resulting in pronounced cm-scale lamination of muddy sediments. Vegetation is dominated by *Scytonema* pincushions and sparse black mangrove. Burrows of the fiddler crab *Uca* are most common at flanks of the levees.

Sabkhas are inter- to supratidal flats which typically occur under semiarid to arid climatic conditions with strong evaporation. Modern sabkha environments can be found along the coasts of the Arabian Peninsula and Africa (e.g., Tunisia, Namibia; Figure 7), but only the sabkhas of the Arabian Peninsula are mainly calcareous. In the subtidal zone, gray muddy peloidal sediments (with up to 30 % dolomite) to coarse bioclastic (gastropods and ooids) sands are deposited, originating from near offshore. Longshore drift form hook-shaped sand spits led to the formation of sheltered lagoons and intertidal flats which are randomly vegetated by black mangrove. The intertidal flats are bordered landward by extended supratidal flats (sabkhas sensu stricto) which are only flooded by extremely high water levels.

Typical sedimentary structures are muddy microbial laminated sediments with fenestrae, intercalated by bioclastic storm lag deposits, anhydrite, halite, and terrestrial-derived eolian quartz sand (Wehrmann et al., 2011). Dolomitization in the supratidal flats results from hypersaline brines due to intense evaporation.

Tidal Depositional Systems, Figure 7 Supratidal sabkha at the arid Namibian coast (lower mesotidal). The sediment surface is covered by microbial mats showing a wavy–crinkly texture. Groundwater level (brackish to hypersaline) is around 40 cm below sediment surface.

Summary

Tidal depositional systems are widely distributed along coastal environments. The character of their sediments, both siliciclastic and biogenic, reflects a complex interaction of numerous physical and biotic factors of which tidal forces are the most prominent. Ranging up from the nearshore subtidal into the supratidal, they represent the direct transition from marine to terrestrial environments and, by tide-dominated deltas and estuaries, also the transition to fluvial controlled processes. They are typically characterized by steep environmental gradients. As tidal depositional systems are highly sensitive to sea-level changes, they are strongly affected by climate change. It has to be noted that tidal signatures can also be found in deeper water settings as tides, in general, are an ocean-wide phenomenon.

Bibliography

Allen, J.R.L., and Duffy, M. J., 1998. Medium-term sedimentation on high intertidal mudflats and salt marshes in the Severn Estuary, SW Britain: the role of wind and tide. *Marine Geology*, **150**, 1–27.

Allison, M. A., Khan, S. R., Goodbred, S. L., and Kuehl, S. A., 2003. Stratigraphic evolution of the late Holocene Ganges-Brahmaputra lower delta plain. *Sedimentary Geology*, **155**, 317–342.

Boothroyd, J. C., and Hubbard, D. K., 1975. Genesis of bedforms in mesotidal estuaries. In Cronin, J. E. (ed.), *Estuarine Research Geology and Engineering*. New York: Academic Press, Vol. 2, pp. 217–234.

Dalrymple, R. W., Knight, R. J., Zaitlin, B. A., and Middleton, G. V., 1990. Dynamics and facies model of a macrotidal sand bar complex. *Sedimentology*, **35**, 577–612.

Dalrymple, R. W., and Rhodes, R. N., 1995. Estuarine dunes and bar-forms. In Perillo, G. M. (ed.), *Geomorphology and Sedimentology of Estuaries*. Amsterdam: Elsevier. Developments in Sedimentology, Vol. 53, pp. 359–422.

Dalrymple, R. W., and Choi, K. S., 2007. Morphologic and facies trends through the fluvial-marine transition in tide-dominated depositional systems: a systematic framework for environmental and sequence-stratigraphic interpretation. *Earth Science Review*, **81**, 135–174.

Davis, J. L., 1964. A morphogenetic approach to world shorelines. *Zeitschrift Geomorphologie*, **8**, 27–42.

Davis, J. L., 1980. *Geographical Variation in Coastal Development*. London: Longman. Geomorphology Texts, Vol. 4.

Davis, R. A., and Hayes, M. O., 1984. What is a wave-dominated coast? *Marine Geology*, **60**, 313–329.

Davis, R. A. and FitzGerald, D. M., 2004. *Beaches and Coasts*. Oxford: Blackwell, p 419.

Dykstra, M., 2012. Deep-water tidal sedimentology. In Davis, R. A., and Dalrymple, R. W. (eds.), *Principles of Tidal Sedimentology*. Dordrecht: Springer, pp. 371–395.

Fan, D., Guo, Y., Wang, P., and Shi, J. Z., 2006. Cross-shore variations in morphodynamic processes of an open-coast mudflat in the Changjiang Delta, China: With an emphasis on storm impacts. *Continental Shelf Research*, **26**, 517–538.

Flemming, B. W., 2012. Siliciclastic Back-barrier tidal flats. In Davis, R. A., and Dalrymple, R. W. (eds.), *Principles of Tidal Sedimentology*. Dordrecht: Springer, pp. 231–267.

FitzGerald, D., 2005. Tidal inlets. In Schwartz, M. (ed.), *Springer Encyclopedia of Coastal Science*. Berlin: Springer, pp. 958–965.

FitzGerald, D., Buynevich, I., and Hein, C., 2012. Morphodynamics and facies architecture of tidal inlets and tidal deltas. In Davis, R. A., and Dalrymple, R. W. (eds.), *Principles of Tidal Sedimentology*. Dordrecht: Springer, pp. 301–333.

Galloway, W. E., 1975. Process framework for describing the morphologic and stratigraphic evolution of deltaic depositional systems. In Broussard, M. L. (ed.), *Deltas. Models for Exploration*. Houston: Houston Geological Society, pp. 87–98.

Galloway, W. E., 1976. Sediment and stratigraphic framework of the Copper River fan delta. *Journal of Sedimentary Petrology*, **46**, 726–737.

Goodbred, S. L., and Saito, Y., 2012. Tide-dominated deltas. In Davis, R. A., and Dalrymple, R. W. (eds.), *Principles of Tidal Sedimentology*. Dordrecht: Springer, pp. 129–149.

Hayes, M. O., 1979. Barrier island morphology as a function of tidal and wave regime. In Leatherman, S. P. (ed.), *Barrier Islands*. New York: Academic Press, pp. 1–27.

Hori, K., and Saito, Y., 2007. Classification, architecture and evolution of large-river deltas. In Gupta, A. (ed.), *Large Rivers: Geomorphology and Management*. Chichester: Wiley, pp. 75–96.

Kuehl, S. A., Levy, B. M., Moore, W. S., and Allison, M. A., 1997. Subaqueous delta of the Ganges-Brahmaputra river system. *Marine Geology*, **144**, 81–96.

Lee, H. J., Chun, S. S., Chang, J. H., and Han, S. J., 1994. Landward migration of isolated shelly sand ridges (chenier) on the macrotidal flats of Gomso Bay, west coast of Korea: controls of storms and thyphoon. *Journal of Sedimentary Research*, **64**, 886–893.

Reeder, S. L., and Rankey, E. C., 2009. Controls on morphology and sedimentology of carbonate tidal deltas, Abacos, Bahamas. *Marine Geology*, **267**, 141–155.

Reineck, H.-E., and Singh, I. B., 1980. Depositional sedimentary environments. Berlin, Heidelberg, New York: Springer, p 551.

Robinson, A. H. W., 1960. Ebb-flood channel systems in sandy bays and estuaries. *Geography*, **45**, 183–199.

Schrottke, K., Becker, M., Bartholomä, A., Flemming, B. W., and Hebbeln, D., 2006. Fluid mud dynamics in the Weser estuary turbidity zone tracked by high-resolution side-scan sonar and parametric sub-bottom profiler. *Geo-Marine Letters*, **26**, 185–198.

Swenson, J. B., Paola, C., Pratson, L., Voller, V. R., and Murray, A. B., 2005. Fluvial and marine controls on combined subaerial and subaqueous delta progradation. Morphodynamic modeling of compound-clinoform development. *Journal of Geophysical Research*, **110**, 1–16.

Thomas, R. G., Smith, D. G., Wood, J. M., Visser, J., Calverley-Range, E. A., and Koster, E. H., 1987. Inclined heterolithic stratification – terminology, description, interpretation and significance. *Sedimentary Geology*, **53**, 123–179.

Walton, T.L., and Adams, W.D., 1976. Capacity of inlet outer bars to store sand. In *Proceedings of the 15th Coastal Engineering Conference*,Honolulu: ASCE, pp. 1919–1937.

Wehrmann, A., Wilde, V., Schindler, E., Brocke, R., and Schultka, S., 2010. High-resolution facies analysis of a Lower Devonian deltaic marine-terrestrial transition (Nellenköpfchen Formation, Rheinisches Schiefergebirge, Germany): implications for small scale fluctuations of coastal environments. Neues Jahrbuch Geologie Paläontologie Abhandlungen, 256, 317–334.

Wehrmann, A., Gerdes, G., and Höfling, R., 2011. Microbial mats in Lower Triassic siliciclastic playa environment (Middle Buntsandstein, North Sea). In Chafetz, H., and Noffke, N. (eds.), *Microbial Mats in Siliciclastic Sediments*. Tulsa: SEPM Special Publication, 101, pp. 177–190.

Cross-references

Beach Processes
Coasts
Integrated Coastal Zone Management
Ocean Margin Systems
Sea-Level
Shelf
Wadden Sea

TRANSFORM FAULTS

A. M. Celâl Şengör
Faculty of Mines, Department of Geology and Eurasia, Institute of Earth Sciences, Istanbul Technical University, Ayazaga, Istanbul, Turkey

Synonyms
Conservative plate boundary

Definition
A *transform fault* is a plate boundary along which plate motion is parallel with the strike of the boundary. Along such a boundary, ideally, crust is neither generated nor destroyed, and that is why they are also called *conservative plate boundaries*. In real life, the thermal and mechanical properties of the crust and upper mantle and the time-averaged behaviour of the spreading centre and subduction zones in the oceans impose a finite width on transform faults in which deformation is complex, forming a fault zone rather than a single clean fault. Large, active, continental transform faults, such as the San Andreas Fault system in California, the North Anatolian Fault system in northern Turkey, the Alpine Fault in New Zealand, and the Altyn Tagh Fault in northern Tibetan Plateau, constitute veritable *keirogens*. In this entry, the emphasis is on the oceanic transform faults, in keeping with the theme of the volume.

Introduction

The concept of transform faults arose from the observation that many mobile belts taking the form of mountains, mid-ocean ridges, or major strike-slip faults and the seismic activity associated with them often appear to end abruptly at their strike terminations. Lotze (1937), Ketin (1948), and Wilson (1954) offered kinematic solutions to this problem which were revived by Wilson (1965) later. Wilson pointed out that large mobile belts do not end, but that they are connected into a continuous *network of mobile belts* about the earth dividing the surface of the planet into several rigid *plates* in motion with respect to one another (Wilson, 1965). Wilson further pointed out that "Any feature at its apparent termination may be transformed into another feature of one of the other two types (Figure 1). For example, a fault may be transformed into a mid-ocean ridge as illustrated in Figure 1a. At the point of transformation, the horizontal shear ends abruptly by being changed into an expanding tensional motion across the ridge or rift with a change in seismicity" (Wilson, 1965, 343). As is clear, this sentence is also the birth cry of plate tectonics.

Wilson called the terminations where a feature is transformed into another one *transforms*. Thus, a plate boundary of any one type *must* extend between two transforms (*not* transform faults, although they may indeed extend between two transform faults, to which they must attach at transforms).

Wilson defined transform faults as follows: "Faults in which the displacement suddenly stops or changes form and direction are not true transcurrent faults. It is proposed that a separate class of horizontal shear faults exists which terminate abruptly at both ends, but which nevertheless may show great displacements. Each may be thought of as a pair of half-shears joined end to end. Any combination of pairs of the three dextral half shears may be joined giving rise to the six types illustrated in Figure 2 (herein Figure 1). Another six sinistral forms can also exist. The name transform fault is proposed for the class, and members may be described in terms of the features which they connect (for example, dextral transform fault, ridge-convex arc type)" (Wilson, 1965, 343).

When Wilson pointed out that transform faults should be considered as a separate class of faults, he had Anderson's theory of faulting (Anderson, 1951) in mind, in which the three kinds of faults, namely, normal dip slip, reverse dip slip, and wrench, have been considered to have formed in a medium where the medium is continuous and conserved during faulting. By contrast, transform faults can only form if at both their ends there is material consumption or generation (Wilson, 1965, 344), in other words, if the Earth's outer rocky shell, i.e., the lithosphere, is divided into independently moving spherical caps

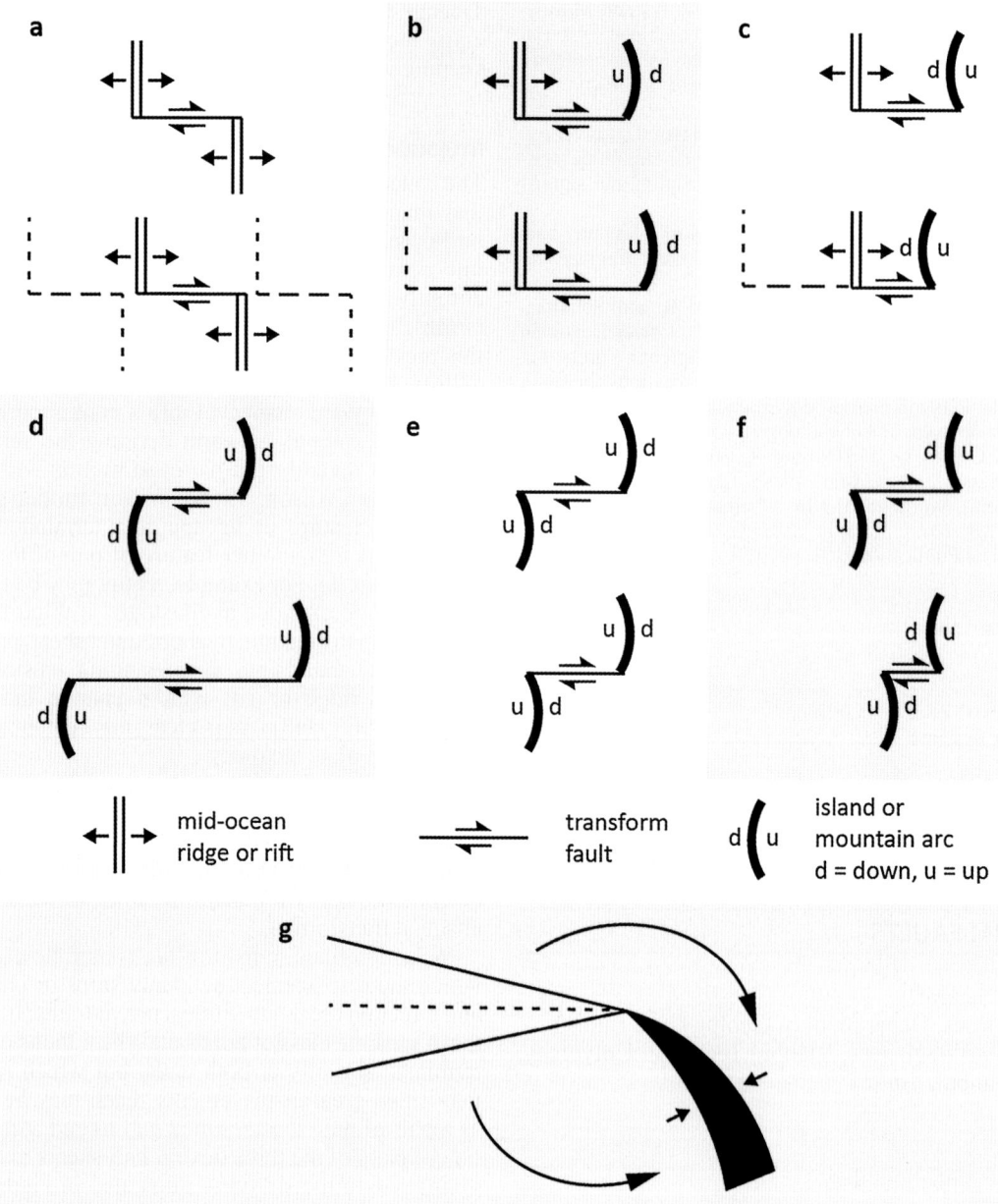

Transform Faults, Figure 1 Kinds and behaviours of transform faults after Wilson (1965, Figs. 3 and 4). (**a**) Ridge-ridge, (**b**) ridge-arc facing away from the ridge (later this type was called ridge-trench facing away from the ridge), (**c**) ridge-arc facing toward the ridge (ridge-trench facing toward the ridge), (**d**) arc-arc with arcs facing away from one another (trench-trench with trenches facing away from one another), (**e**) arc-arc with arcs facing the same way (trench-trench with trenches facing the same way), and (**f**) arc-arc with arcs facing toward each other (trench-trench with trenches facing toward one another). Note that the ridge-ridge transform faults and those between similarly facing ridges do not change their length during finite plate evolution (**a** and **e**). Those connecting ridges to trenches facing away from the ridge (**b**) get longer by ½ of the spreading velocity. Those connecting ridges to trenches facing toward the ridge (**c**) become shorter by ½ the spreading velocity. Transform faults connecting two ridges facing away from each other (**d**) become longer with the same velocity as the spreading, and those connecting two facing trenches become shorter with the full spreading velocity. (**g**) Transform fault at which extension changes to shortening.

(Teichert, 1979) that can be destroyed and generated along their boundaries.

Wilson further pointed out that transform faults display habits of growth and diminution different from faults described by Anderson (1951), and he believed that therein lay the great significance of their recognition: They could only occur on a planet where there is considerable horizontal displacement coupled with material generation

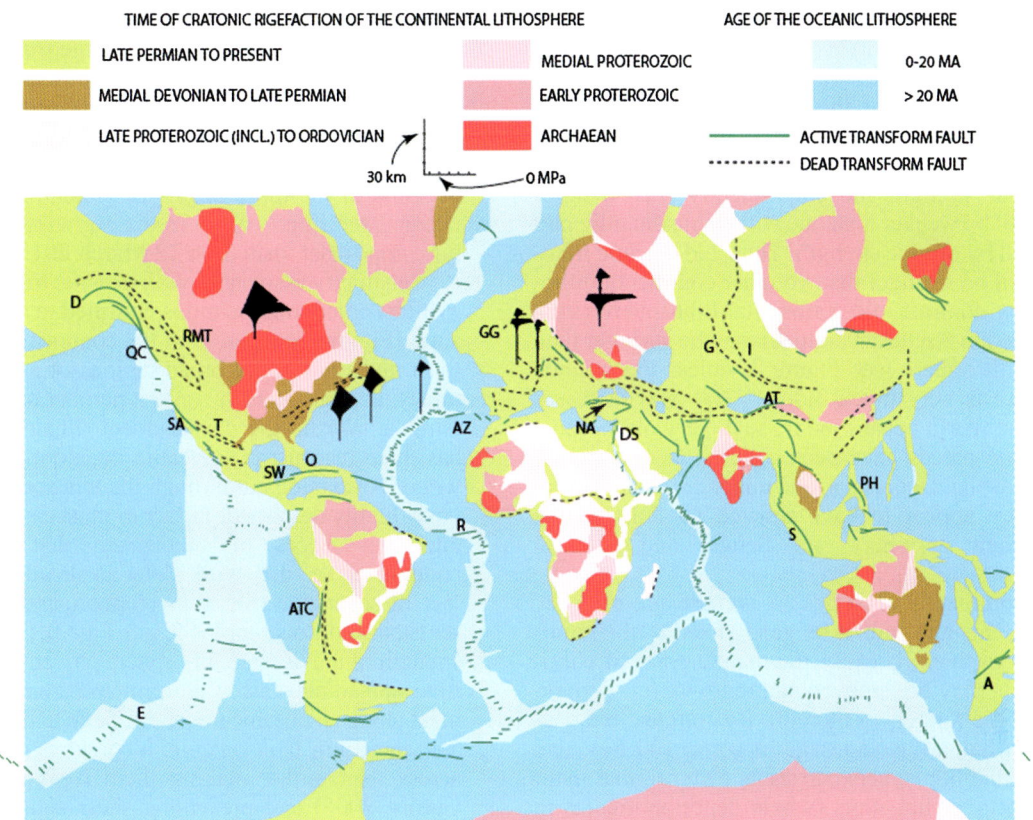

Transform Faults, Figure 2 Map showing the active transform faults of the world and some of the dead ones of the Paleozoic and Mesozoic ages. Notice that all transform faults shown in this figure are located in oceanic lithosphere younger than 20 Ma (with a few exceptions as in the Indian Ocean) or in continental lithosphere that was last heated and deformed (and thus began its rigefaction) after the Permian. The *black* tree-shaped figures are strength profiles (for scale, see the small graph showing strength in 1,000 Mpa and steps and depth in 30 km steps) with compressional strength to the *right* and tensional strength to the *left*. Note the immense strengths of the Precambrian cratons. That is why no post-Paleozoic transform fault was able to cut through them (unless they are very narrow as in the Gulf of Guinea/Northeast Brazilian conjugate margins). The strength profiles have been copied from Şengör (1999) with the tensional strength of the Canadian Shield simply extrapolated from the Baltic Shield, because for it no tensional calculations are available.

and destruction at the surface. Figure 1 shows also the six types of behaviour of transform faults in terms of their changes in size; some become longer; others become shorter and eventually disappear; yet others first shorten and then lengthen (for an analysis of the geological implications of these length changes, see Dewey, 1976). In an epoch-making paper, Sykes (1967) showed that what Wilson identified as transform faults indeed showed senses of motion as he had predicted as opposed to what the earlier geological wisdom would have proposed. In the same year, McKenzie and Parker (1967) pointed out that the fault systems in the northwest Pacific Ocean bore out fully the predictions of Wilson's theory of plate tectonics, of which transform faults were a critical component, when tested rigorously by earthquake slip vectors, fault orientations as evaluated within the bounds of Euler's theorem of the motion of a point on the surface of a sphere and as also analyzed in greater detail in an independently conceived presentation by Morgan (1968).

Transform faults as large-scale geological structures

General. Transform faults are strike-slip faults, but they commonly appear not as single faults. Instead, they are commonly broad shear zones in which there are a large number of parallel to subparallel faults that may all be active simultaneously or at different times. The geological behaviour of transform faults depends on four main factors: (1) the nature of the lithosphere they cut and displace, (2) their length and how stable that length is, (3) the behaviour of the plate boundaries they connect (for instance, whether they move in an environment where rotation poles governing their motion are stable for significant periods of time or whether they shift rapidly), and (4) the manner in which they form.

1. *The nature of the lithosphere in which they move has three main elements that are critical for their geological behaviour*: (a) the composition of the lithosphere,

(b) the presence and nature of preexisting structures in the lithosphere, and (c) its thermal structure. All of these, together with the strain rate, determine the deformational behaviour of the lithosphere. If the lithosphere is crust dominated, i.e., if its crust is thicker than its mantle part, its behaviour will be dominated by feldspar strength and consequently such a lithosphere will be weaker than one with a thick mantle part dominated by olivine strength. Thus, old oceanic lithosphere will be stronger than younger continental lithosphere. By contrast, old continental lithosphere (such as observed in Archaean cratons) may be stronger than oceanic lithosphere (see Figure 2; unless the oceanic lithosphere is very much older (>200 Ma) than any seen in the world oceans today; such very old oceanic lithospheric pieces have been claimed to exist under the Pre-Caspian Depression and under the Tarim Basin, where they appear to have resisted all deformation around their periphery: see Şengör and Natal'in (1996)).

The continental crust has an average age of 2 Ga. However, this age cannot be generalized to the continental lithosphere, because the mantle parts of continental rafts may be removed and regrown (cf. Şengör, 2001 and the references therein). By contrast, the oceanic crust has an average age of some 100 Ma, and most oceanic lithosphere carries its own original mantle, except over hot spots (where depth anomalies in the oceans exist; see Şengör (2001) and the references therein). For a continent and an ocean with similar thickness mantle lithospheres, the continents contain a much richer assortment of structures. This renders them weaker.

Young orogenic belts for long have been assumed not to have a thick lithosphere because it was believed that thickened lithosphere would become buoyantly unstable and sink into the asthenosphere, leaving the overlying orogenic belt devoid of a mantle lithosphere (Houseman et al., 1981). However, recent observations have shown that the underthrusting forelands tend to keep their lithospheric keels (Priestley and McKenzie, 2006), but that long-lived continental-margin arc orogens, such as the Andes, necessarily have no mantle lithosphere under them (Şengör, 1990). Because of active magmatism along their axes, arc cores are hot, and without underlying mantle lithosphere, they are among the weakest places in the continents. By contrast, old cratons are extremely strong and are essentially undeformable unless by mantle plumes that may thermally erode their lithospheric armor (Şengör, 1999). Young oceanic lithosphere is also weak and deforms easily.

From the above, one would expect large active transform faults to exist in the younger parts of the oceans and in and near young orogenic belts (see Sandwell, 1986). This is by and large true with exceptions existing only in areas of old continental crust that somehow had lost its equally old lithosphere as in wide areas within the Eurasian foldbelts or under the Pan-African terrain of Afro-Arabia (Figure 2; for a list of some of the more prominent fracture zones in the oceans, see http://en.wikipedia.org/wiki/List_of_fracture_zones last visited on 22 March 2013).

Seismicity associated with oceanic transform faults exhibits epicentral locations confined to a narrow fault zone (Figure 3), whereas in the continents, large transform faults have very broad regions of associated seismicity, because in such zones a number of parallel fault families take up the displacement (Figure 3b, c). This has three reasons: (1) Oceanic transform faults almost without exception cut through the lithosphere not previously highly deformed; in fact in most cases, they separate two pieces of lithosphere that were never continuous, i.e., they work along an already existing surface of separation (for one exception among a number, see Stein and Cochran, 1985, esp. Fig. 2); (2) the superior strength of the oceanic lithosphere prevents deformation from spreading outside the principal zone of displacement along an oceanic transform fault; and (3) most oceanic transform faults acquire a component of extension across their strike, making them weak (cf. Hall and Gurnis, 2005) and preventing shear strain from being exported outside the zone of principal displacement.

2. *The length of transform faults and the stability of that length* can be very variable: Some are very short (on the order of 10 km; there are even shorter ones forming links in a continuous spectrum beginning with intra-rift transfer faults (Karson, 1990) through zero-offset transform faults (Schouten and White, 1980) and ending with transform faults with significant transform separation) connecting two nearby plate boundaries, such as the ones seen offsetting the Mid-Atlantic Ridge in the north and the south Atlantic Oceans (Figure 2). Among the ridge-ridge transform faults, the longest (such as the Romanche or the Eltanin: see Figure 2 for locations) hardly exceed 600 km in length. One factor controlling the length of ridge-ridge transform faults may be large, transform-parallel tensile stresses arising from the thermal contraction of the oceanic lithosphere as it cools (Sandwell, 1986; Haxby and Parmentiar, 1988). The transform faults connecting two subduction zones (such as the Swan and Oriente pair in the northern Caribbean and the complex system bounding the Caribbean Plate to the south, of which the El Pilar Fault is the longest, or the Alpine Fault in New Zealand or the Philippine Fault in eastern Asia; see Figure 2 for locations) reach lengths between 1,000 and 2,000 km.

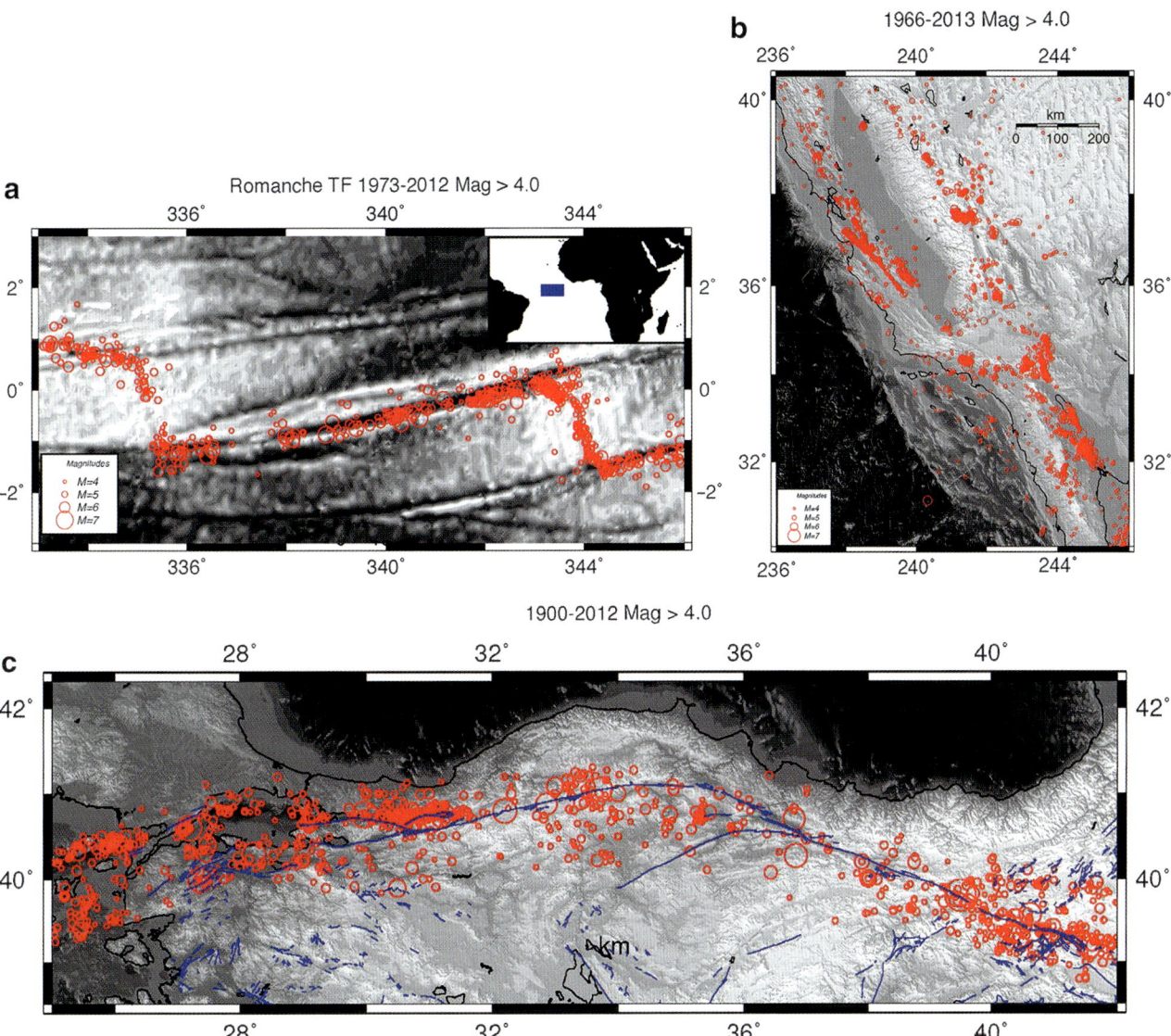

Transform Faults, Figure 3 Seismicity associated with transform faults. (**a**) Romanche transform fault, one of the longest ridge-ridge-type oceanic transform faults. The seismicity is that of the 1973–2012 interval showing Mw > 4 earthquakes from the National Earthquake Information Center (NEIC) catalog. Bathymetry is taken from GEBCO 30 arc second dataset. Note the extremely narrow seismic zone coincident with the bathymetric expression of the fault zone. (**b**) San Andreas Keirogen. Seismicity of the San Andreas Fault (SAF) system between 1966 and 2013 with Mw > 4. Seismicity of the SAF from the Northern California Earthquake Catalog and National Earthquake Information Center (NEIC) catalog. Bathymetry and topography are taken from GEBCO 30 arc second dataset. The map has the same scale as the map in Figure 6a. Note that the seismicity here is much more diffuse than that in the case of the oceanic Romanche transform fault. However, the diffuseness of the seismicity is largely a function of the keirogenic character of the transform fault system, i.e., of the fact that here we see more than one strand of strike-slip faults that are now active. Along each fault, the scatter of the epicenters is not incomparable to that along the Romanche. The seismicity gets scattered only in the Transverse Ranges (marked by TR) where there are substantial crustal shortening and disruption along many faults (cf. Dickinson, 1996). This is in marked contrast to the North Anatolian Keirogen illustrated in Figure 6c and is commonly attributed to the main strand of the San Andreas' being a weak fault. (**c**) Anatolian Keirogen. The seismicity is that of the 1900–2012 interval with Mw > 4 from the KOERI catalog. Bathymetry and topography are taken from GEBCO 30 arc second dataset. The map has the same scale as the map in Figure 6a, b. Note that although the North Anatolian Fault has only one main strand along much of its length, the seismicity along the entire keirogen is extremely scattered. Although the North Anatolian Fault also moves in a serpentinite-rich milieu (cf. Şengör et al., 2005), it does so in a much more complexly structured suture region, as opposed to the uncollided accretionary complex milieu of the San Andreas.

Others in more complex settings also can be very long (to select from among the active ones: San Andreas, Queen Charlotte, Atacama, Sumatra and North Anatolian; see Figure 2 for locations), being on the order of 1,000–1,500 km. Some of the largest fossil ones have been reported to have lengths of some 2,000–2,500 km (e.g., the Irtysh-Gornostaev, or the Silk Road Keirogens: for locations, see Figure 2).

Because material at the ends of transform faults is not conserved and the faults can change their lengths very considerably during their evolution (as Wilson already emphasized in his seminal 1965 paper), it is meaningless to talk about "displacements" along them (Figure 4). One can naturally say that the displacement along a transform fault equals the amount of relative plate motion that has taken place along it, but in many cases, this may not be measurable by any observations one can make along the fault.

3. *The behaviour of the plate boundary*, of which a transform fault forms a part, determines other aspects of the geometry and the kinematics of a transform fault than just its length and how much plate passage may occur along it. Almost all plate boundaries are unstable if given enough time. They change location, slip, and shape, because in any system, where more than two plates interact, at least one of the poles of rotation must shift in the framework of the plate boundary, the evolution of which it governs (Dewey, 1975). Transform faults along a boundary whose pole is shifting must change slip and character or give rise to entirely new plate boundary configurations. A discussion of this question will also illustrate how new transform faults may arise.

Consider the simple plate boundary configuration depicted in Figure 5a. A single plate boundary between plates *A* and *B* consists of three ridge segments and two transform faults. The pole AB_1 governs the motion of *B* with respect to *A*, here held fixed arbitrarily. In Figure 5b, the pole of rotation has moved to a new place shown with AB_2. The plate boundary configuration shown in Figure 5a can no longer satisfy the requirements of the new rotation as indicated in Figure 5b. A new plate boundary configuration is necessary. Figure 5c shows what shape this new boundary most likely will take. The reason for the shape shown in Figure 5c is that the spreading centers try to keep a time-averaged symmetric position between separating plates. They must tear through the weakest part of the lithosphere which is indicated with the brick-coloured strip in Figure 5b. In the new configuration, the spreading centers ideally must keep their orientation along the great circles ("meridians" of the new pole) going through the new pole of rotation, and the transform faults must lie along the small circles ("parallels" of the new pole) concentric around the pole (for possible origins of this orthogonal geometry, involving a minimum work argument for the ridges and the anisotropy of the accreting oceanic crust, see Freund and Merzer (1976) and the references 2 through 5 in that paper, but Sandwell (1986) has shown that even if it were not for these factors, the transform faults would still exist as an inherent part of the spreading process). The resultant new plate boundary and plate configuration will be that shown in Figure 5d. Here, one additional factor determining transform fault spacing may be due to thermal bending moments that concentrate bending stresses at a distance from the fracture zone determined by the degree of coupling across the fracture zone and the flexural length of the plate, thus providing a natural length scale controlling the spacing of transforms (Haxby and Parmentier, 1988). In Figure 6c and d, instead of the two original long transform faults, numerous shorter transform faults originated, reminding one of the geometry along the Mid-Atlantic Ridge. In a recent study, Gerya (2010) argued, on the basis of numerical modeling, that asymmetric spreading (see Schouten et al., 1980; Müller et al., 2008, for observations; and Katz, 2010, for possible causes) on alternate spreading segments of a previously straight ridge segment may impose on it an undulatory structure, which, in a few million years, would generate transform faults to avoid further asymmetric spreading. This model implies continuous change of the length of ridge-ridge transform faults with time, and it is not mutually exclusive with the pole shift-generated ridge jumps as shown above.

In Figure 5, had the pole shifted toward the right, rather than to the left, obliquely consuming subduction zones would have formed along the former transform faults. This seems a common process (Dewey, 1975; Karig, 1982; Casey and Dewey, 1984, esp. Fig. 7; Dewey and Casey, 2011). Because subduction zones can easily consume the oceanic lithosphere obliquely, there would have been no great necessity to slice the new subduction zones with numerous transform faults, as it happened to the new ridges.

Let us now consider the geometry shown in Figure 6a. This too shows a simple plate boundary between plates *A* and *B*, whereby *B* rotates around the pole AB_1 along two spreading ridges separated by a long transform fault. In Figure 6b, the pole is shown to have shifted to *CD* by 90°. The change here is so drastic that entirely new plate boundaries would need to form. First of all, let us consider the situation along the transform fault: To understand the three-dimensional geometry along the fault and along its non-transform fracture zone extensions (shown by dashed green lines in Figure 6b), look at Figure 7 showing two ridge segments separated by a dextral

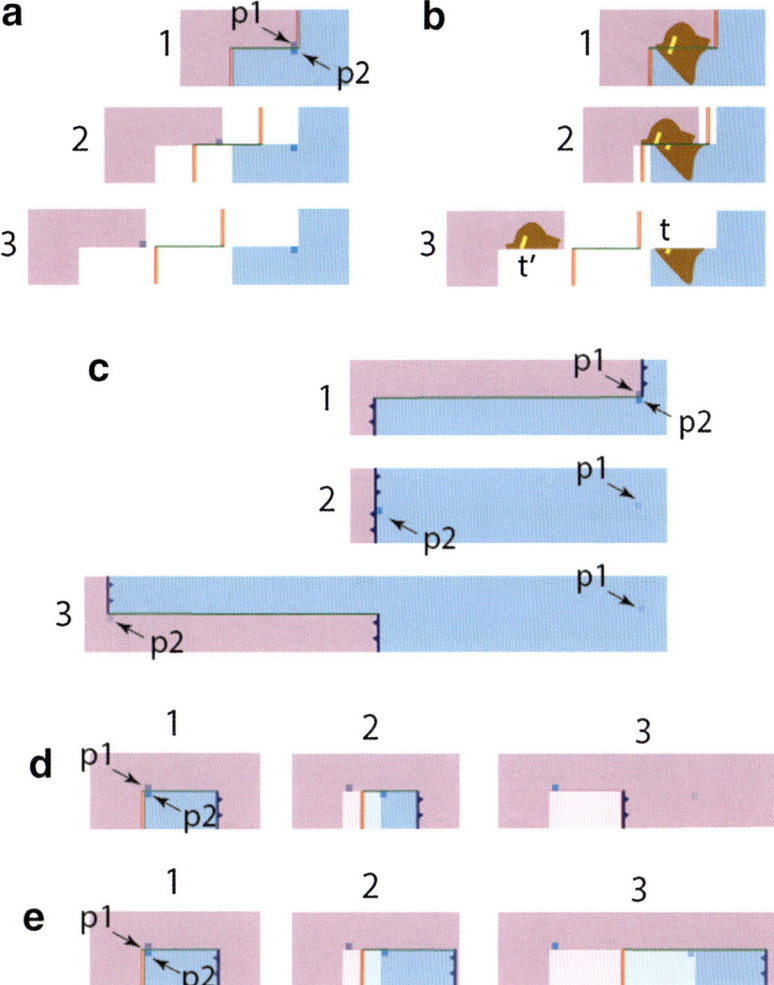

Transform Faults, Figure 4 "Difficulty of defining 'displacement'" along different kinds of transform faults: (**a**) In a ridge-ridge-type transform fault, the points *p1*, and *p2* are contiguous, but not originally continuous (**a**1). Therefore, they cannot measure the "displacement" along the transform fault, but their separation in time is equal to the plate displacement that at any one time takes place along the transform fault (**a**2 and **a**3). But that displacement cannot be said to be the displacement along the transform fault, because once *p1* and *p2* clear the fault (**a**3), they are no longer on it and therefore no longer undergo any strike-slip displacement. Their displacement cannot be converted to shear strain along the fault. But if a cylinder with a vertical axis fixed with respect to the two ridges and with surface touching the fault mirrors is assumed to be present along the fault, it will rotate (anticlockwise in a left-lateral fault as shown in Figure 4a and clockwise in a right-lateral fault) as long as the fault moves and will record the total separation of *p1* and *p2* in time. (**b**) If a transform fault cuts through a continent (as in the case of the San Andreas, North Anatolian, and Alpine Keirogens or of the conjugate margins of the northern margin of the Gulf of Guinea and Northeastern Brazil: the *brown* object in **b**1), the displacement of any originally continuous feature cut by the transform fault (for instance, the *yellow* band in **b**1) system will give the real displacement along the fault (**b**2), until the separated segments of the originally continuous feature clear the fault (**b**3).
(**c**) The three maps shown here illustrate a case where a transform fault first becomes shorter (**c**1), disappears for an instant (**c**2), and then grows along the same strike, but with a reversed sense of motion. Although "the fault" (how reasonable is it here to talk about "one fault"?) reverses its motion, the displacement of the points *p1* and *p2* will continue uninterrupted, in the sense that they will get farther and farther apart from one another, but not along the same fault! In fact, only *p2* will stay visible until the fault length reduces to zero, and then it too be subducted. (**d**) In **d**1, the two points *p1* and *p2* are contiguous, as in Fig. **a**1, but not continuous. In **d**2, the displacement of the points *p1* and *p2* is exactly twice as large as the original length of the fault, but the fault has already lost ¼ of its length. It will eventually totally disappear and we can have no means of defining a meaningful displacement along it. (**e**) In this case, both of the points *p1* and *p2* will remain visible during the evolution of the transform fault that here grows with a rate of ½ the spreading rate. But this is not the rate of displacement, which cannot be defined as the point *p1* is placed in an intraplate position an instant after the situation depicted in **e**1; i. e., it no longer faces a fault along which to be displaced.

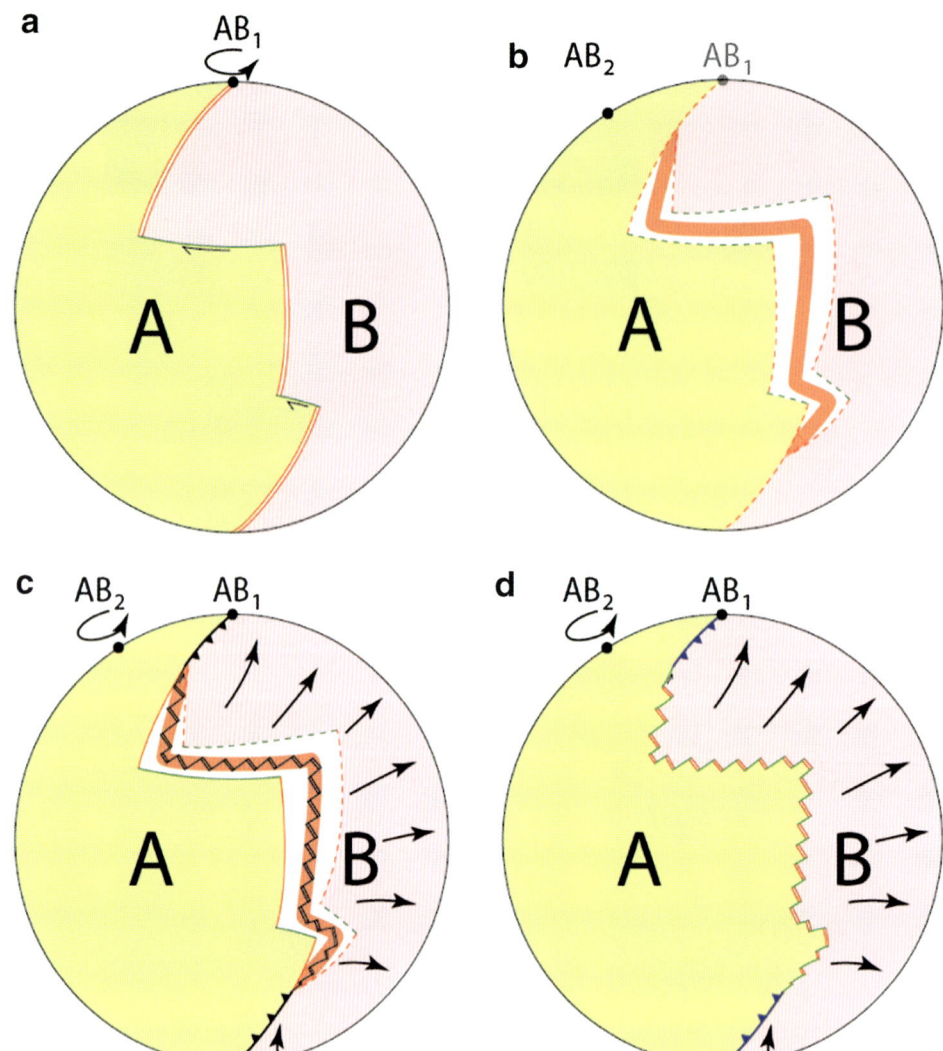

Transform Faults, Figure 5 (a) A simple two-plate geometry depicted on a stereographic lower-hemisphere projection using a Wulff net. *Red double lines* show spreading centers and single green lines are transform faults. *AB1* is the rotation pole of plate *B* with respect to plate *A*. (b) The pole of rotation of the plate *B* with respect to plate *A* has moved to its new location *AB2*. The brick-coloured stripe shows the hottest and thinnest lithosphere. The new plate boundary will try to localize itself along that area. (c) The geometry of the new plate boundary between the plates *A* and *B*. *Black lines* are subduction zones with teeth on the overriding plate. Note the numerous new, short transform faults that originated along the new spreading center. (d) The new plate boundary between plates *A* and *B*.

transform fault. The transform meets the ridges at the transforms T1 and T2. The two plates subside as they spread away from the ridges following the √age relationship. As a consequence, the left-hand ridge stands higher than the crust created at the right-hand ridge at T1. There is a crossover point, when both segment runs (Tucholke and Lin, 1994) are at the same elevation on both sides of the transform fault (cf. Figures. 6b and 7). When the entire transform fault/fracture zone system is placed under shortening across their strike, the ridge T1 will tend to overthrust the green crustal segment in front of it toward the viewer, whereas the ridge T2 will try to thrust the black crustal segment away from the viewer. As a consequence, a new subduction boundary may come into existence along the former fracture zone with opposite vergences on both sides of the crossover point. As the two subduction segments recede from one another, a completely new and continuously lengthening transform fault will connect them. In such a drastic change, the former spreading centers between the plates *A* and *B* may cease their activity, and the two entirely new plates *C* and *D* may come into existence, carrying the former spreading centers as fossils in them. Such an event often heralds the obduction of large ophiolite nappes (Casey and Dewey, 1984; Dewey and Casey, 2011).

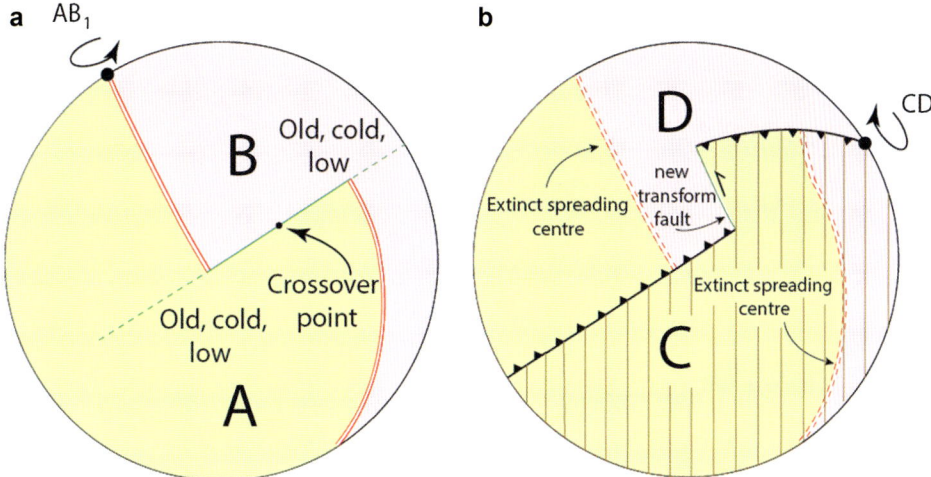

Transform Faults, Figure 6 (a) A simple plate boundary between plates A and B. B is rotating with respect to A around the pole AB_1 in the sense shown. Symbols as in Figure 5. (b): The rotation pole has moved to CD. As a result of this drastic shift, two new plates have come into existence, C, and D. Their boundary is a subduction zone offset by a new transform fault that formed at the point of the crossover shown in (a).

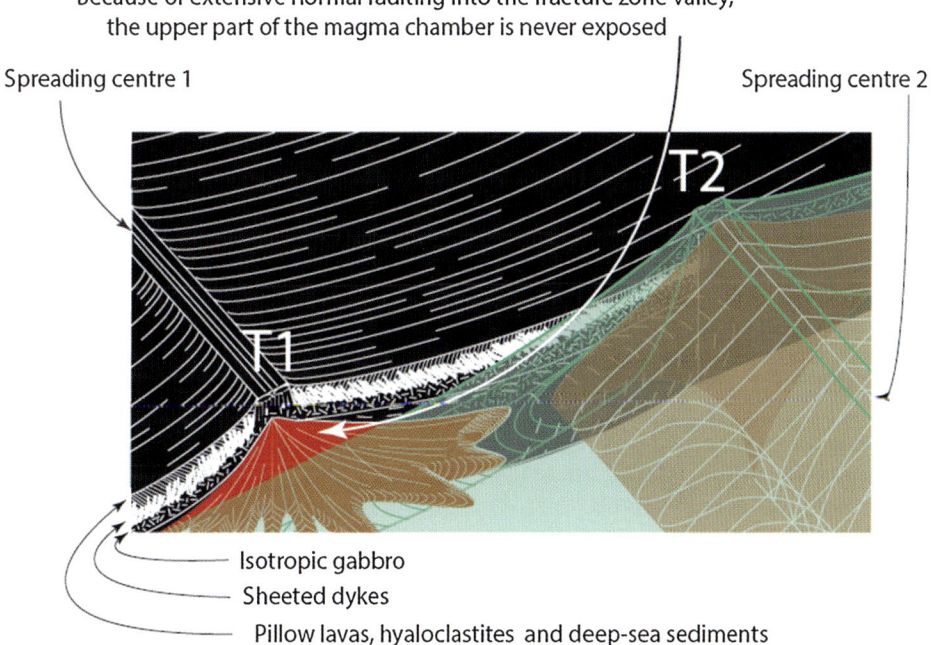

Transform Faults, Figure 7 Basic anatomy of an oceanic fracture zone including a transform fault extending between the transforms T1 and T2 at which the transform fault is attached to two ridges. The *green* segment is shown transparent so as to expose the interior geometry. *Red* are the magma chambers, here shown to be continuous as in fast spreading centers (ideally). At T1, the magma chamber is shown to send dykes into the neighboring cold and low segment causing their metamorphism from granulite to greenschist facies depending on depth and proximity to the magma chamber. The upper part of the magma chamber is never exposed because of extensive normal faulting into the fracture zone valley (here not shown so as to emphasize the ideal form of the magma chamber). The same will happen in the *black* segment at T2, but is not seen here because the *black* segment was drawn opaque.

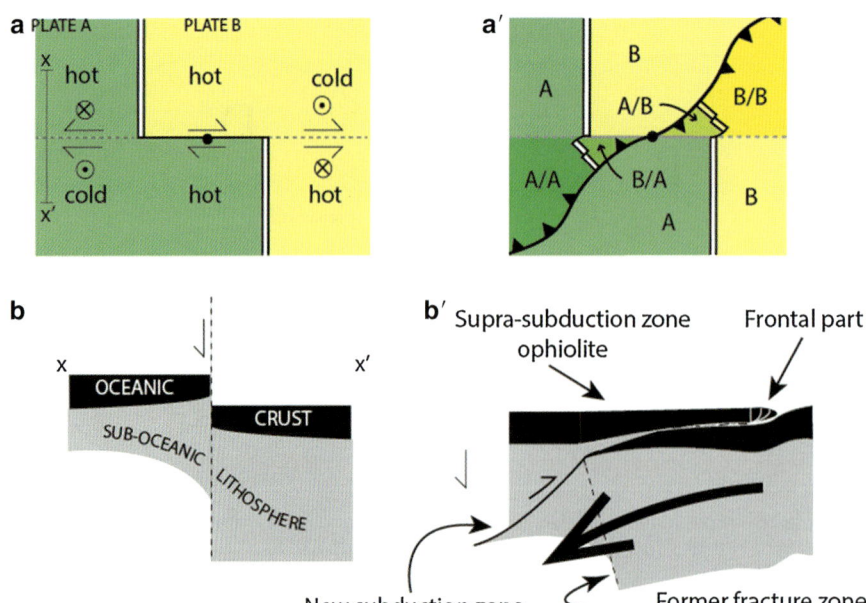

Transform Faults, Figure 8 Maps (**a** and **a'**) and cross sections (**b** and **b'**) showing what might happen when a fracture zone is put into shortening across its strike. The *black* circle in the middle of the transform fault is the crossover point. To the *left* of the crossover point, the upper segment will thrust over the lower one, because the lower segment there is topographically lower. By contrast, the lower spreading segment to the *right* of the crossover point will thrust over the upper one, because the lower segment here is the topographically higher one. This may induce a rotation around the vertical axis of the entire newly formed subduction zone. Such a development, if near a continental margin, may lead to the obduction of giant ophiolite nappes.

However, a tear along the crossover point may not materialize immediately. In such a case, the situation shown in Figure 8 may occur: The subduction zone, which switches its vergence at where the old crossover was, may begin rotating about it in the manner shown in Figure 8a'. The large anticlockwise rotation of Cyprus (50–60°) may have happened because of a geometry similar to the one shown in Figure 8 (cf. Morris et al., 2006).

4. *The manner in which transform faults can form* is discussed above, but the scenarios developed there are not the only ones for the origin of transform faults. They may form intracontinentally in diverse ways. In one of the classic papers relating plate tectonics to continental geology, Atwater (1970) showed how the San Andreas keirogenic system originated when the strike-slip motion between the Pacific and the North American Plates was carried into the continent, creating a broad keirogen with many strike-slip fault segments. Wilson showed that transform faults may be inherited from structures that formed during continental separation (Wilson, 1965, Fig. 6). Figure 9a shows the continent *P* with a number of plume-generated triple junctions of rifts (tj), similar to the present-day Africa (Burke and Dewey, 1973). In Figure 9b, it is shown that the continent *P* was split along the new spreading center *BB*. Notice that the transform fault *t* was inherited from the transform fault margins (for the geology of the sheared margins, see Scrutton, 1979, 1982) connecting two former triple junctions, the now inactive rifts of which have been left on the continental margins as aulacogens (*A*). Sage et al. (2000) found along the Côte d'Ivoire-Ghana transform fault margin that the thermal exchange between the continental margin and the adjacent spreading center was lower than predicted by available geological (e.g., Scrutton, 1979, 303) and thermal models. They ascribed this to the effect of the transform fault and compared it with diminished heat-related features (thinner oceanic crust, reduced magmatism, deeper bathymetry) at the end of spreading segments where a spreading ridge abuts against a transform fault. Especially, the thinner crust would here facilitate the onset of subduction at the ocean/continent boundary, because it reduces the buoyancy of the oceanic lithosphere. In fact, initiation of subduction at continental margins is difficult, especially along the margins of large and old oceans (Cloetingh et al., 1984), and thus it is probably most common along transform fault margins. The short transform faults along the spreading center *BB* in Figure 9b formed shortly after continental separation to avoid significant oblique spreading along the mid-ocean ridge.

Old oceanic lithosphere may also be torn (although not easily) along a new spreading center, which would have its own transform faults (e.g., the Sheba Ridge in the Indian Ocean; Stein and Cochran, 1985, esp. Fig. 2). In that case, formerly continuous markers along

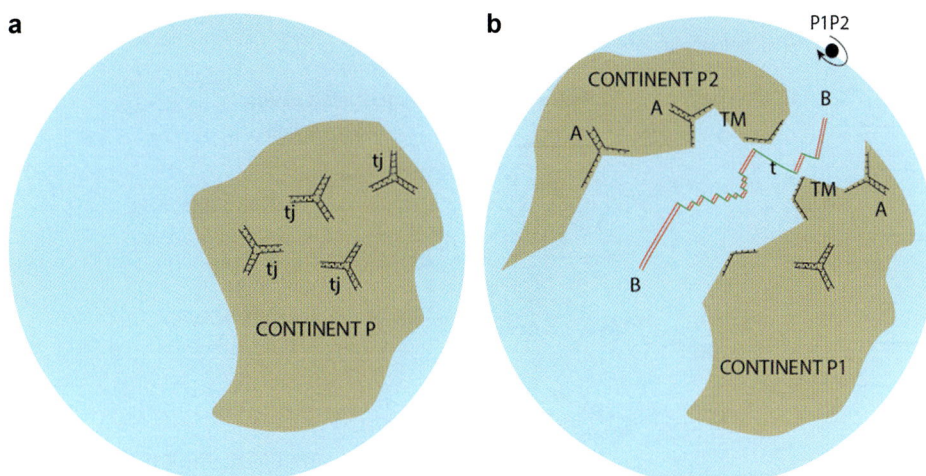

Transform Faults, Figure. 9 (a) Continent P (drawn on a Wulff net, lower-hemisphere projection) with a number of plume-generated rift triple junctions (*tj*) forming rift stars. (b) Continent P is now in two pieces (*P1* and *P2*) rifted along the new spreading center *BB*. Of the old rift stars, the now inactive rift arms are left as aulacogens. Note that the transform fault *t* was inherited from an intracontinental transform fault, whose continental traces are now seen at the conjugate transform fault margins (*TM*).

an oceanic transform fault may be found. Another way of tearing an old lithosphere is shown in Figures. 6a and b. As a transform-fault-bounded subduction zone advances onto a lithospheric plate, the downgoing plate must tear, as the part of it outside the subduction front must stay at the surface, whereas the rest must be subducted. This "edge effect" (DeLong et al., 1975, 734) along what Govers and Wortel (2005, 505) called a "Subduction-Transform Edge Propagator," or, in short, STEP, creates complex structural (Şengör, 1983), magmatic (DeLong et al., 1975), and large-scale tectonic (Govers and Wortel, 2005; özbakır et al., 2013) effects. Baes et al. (2011) pointed out that the initiation of subduction along such STEP faults, especially when they occur adjacent to the continents, is facilitated and combined with the thermally controlled weakness reported by Sage et al. (2000), such areas may be preferred locations for new subduction zones to come into existence.

Transform faults that form in the manner of *t*, i.e., by tearing a preexisting piece of lithosphere, do so by following the same steps as all strike-slip faults. These steps are illustrated in Figure 10 (see Tchalenko, 1970 and the synthesis in Şengör, 1995; also see Ahlgren, 2001; Katz et al, 2004). As shearing begins, the associated strain is initially taken up along a broad zone. If this zone contains a horizontally layered structure, folding will be the first response to shearing, and folds (and other sorts of structures of shortening such as thrust faults and/or horizontal stylolites) will form at angles at or close to 135° (measured clockwise from the shear, i.e., the Y plane). Tension gashes at 45° to the shear plane may form earlier or later. When the peak strength stage (Tchalenko, 1970) is reached, first the anti-Riedel (R') and then the Riedel shears (R) begin forming. In the early post-peak strength stage, the anti-Riedels generally rotate and lock, but the Riedels continue to evolve. They develop "flatter" tips which, in a later stage, join the P shears. In some cases, the X shears also form at this stage. Finally, the R and P shears form a throughgoing fault (generally referred to as the Y shear). The important thing in this evolution is that numerous extensional structures other than just tension gashes may form by the "opening" of the other shear structures giving rise to very complicated geometries of extension. Some of these extensional nodes may localize vulcanicity.

When the transform fault clears the continental margins, which it had originally formed, all such shear-related activities outside the surface of displacement must ideally cease, because the fault in the oceanic lithosphere is born as a fault without needing to tear a previously continuous medium. Indeed, in most oceans, the ridge-ridge transform faults have almost no shear-related structures outside the narrow fault zone. However, this is not universally the case and we must look at the anatomy of an oceanic fracture zone in some detail to see what goes on within it.

The finer geology of oceanic fracture zones

Kohlstedt et al. (1995) and Bercovici (2003) noted that the lower surface of the *oceanic mechanical lithosphere* corresponds with about 600–700 °C and, when about 60 Ma old, with a depth of about 40–50 km. This means that below that depth range, deformation is wholly ductile. The zone of ductile deformation is shallower in younger oceanic lithosphere (ideally at 0 km at the spreading centers, but dyke formation shows that even there the brittle deformation claims a layer whose thickness may vary from a few hundreds of meters to 1.5 km) and deeper in older. In fact, Mckenzie et al. (2005) claimed that the depth of brittle deformation in both oceanic and

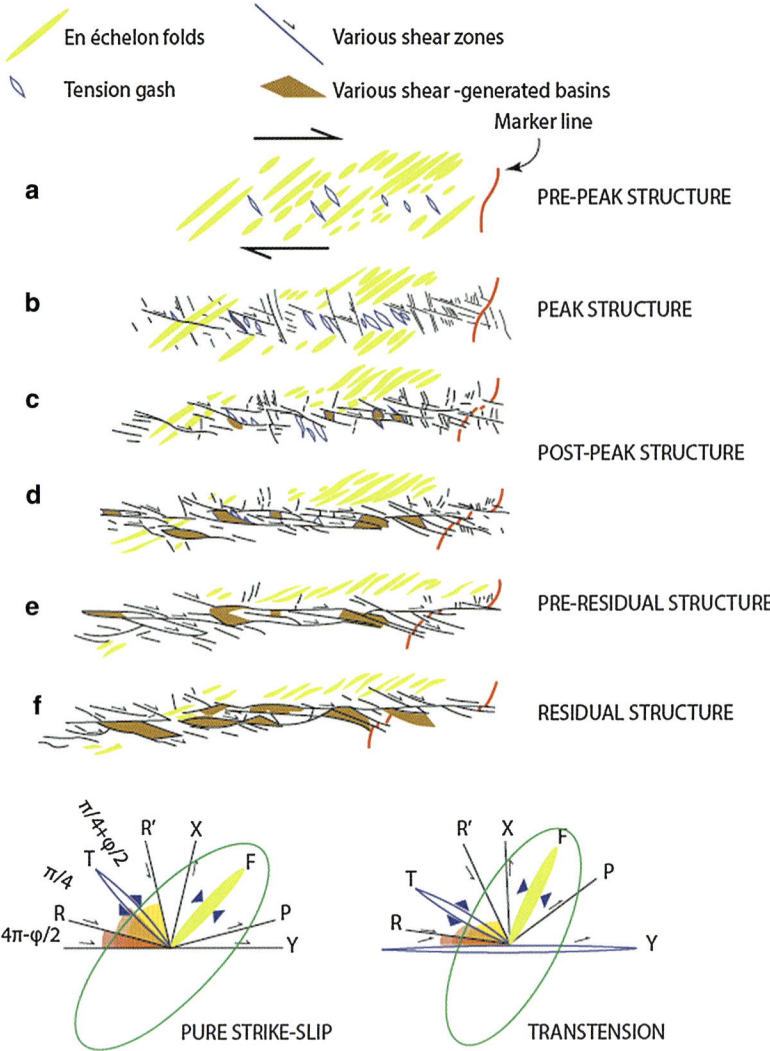

Transform Faults, Figure 10 Evolution of a shear zone (modified from Şengör, 1995, Fig. 2.27). The *brown* areas may be basins, or veins or intrusion receptacles depending on where and what scale the shear zone is active. In transtension, notice that all extensional structures become "flatter" and all shortening structures become "steeper" with respect to the Y shear. However, this is only an instantaneous picture, because as the shear zone evolves, all structures will rotate and strains will increase. Key to lettering: *F* is fold axial trace, *P* is P shear, *R* is a Riedel shear, *R'* is an anti-Riedel shear, *T* is a tension gash, and *X* is an X shear. φ is the angle of internal friction of the material being sheared.

continental lithospheres is controlled by the depth of the 600 °C isotherm, although Géli and Sclater (2008) later pointed out that McKenzie et al.'s conclusion was largely, but not entirely, correct, because they found earthquakes deeper than the 600 °C isotherm in the oceans, especially near fracture zones. They noted that "the strong tendency for the intra-plate earthquakes in the Indian Ocean to be located on or near fracture zones suggests that these processes are most important at fracture zones. Hydrothermal circulation, for instance, could be one such process. Seawater circulation likely results in lowering isotherms immediately below faults, thus increasing the maximum depth of brittle failure and explaining the unexpectedly deep occurrences of earthquakes. This provides indirect evidence that seawater circulates deeply along faults off axis in young lithosphere, a process which so far has been documented only at the crest of spreading centers" (Géli and Sclater, 2008; see also Sage et al., 2000). Kusznir and Cooper (2011) made estimates of the density and depth extent of serpentinized mantle at the Iberia-Newfoundland and Nova Scotia margins by comparing seismic refraction and gravity inversion Moho depths for these margins. They argue that "the proposed serpentinisation-depth relationships and gravity inversion suggest that there is little mantle serpentinisation deeper than 5 km, and in which case the average density of serpentinised mantle is 3,000 and 3,100 kg m^{-3}. We predict that where serpentinised mantle penetrates deeper

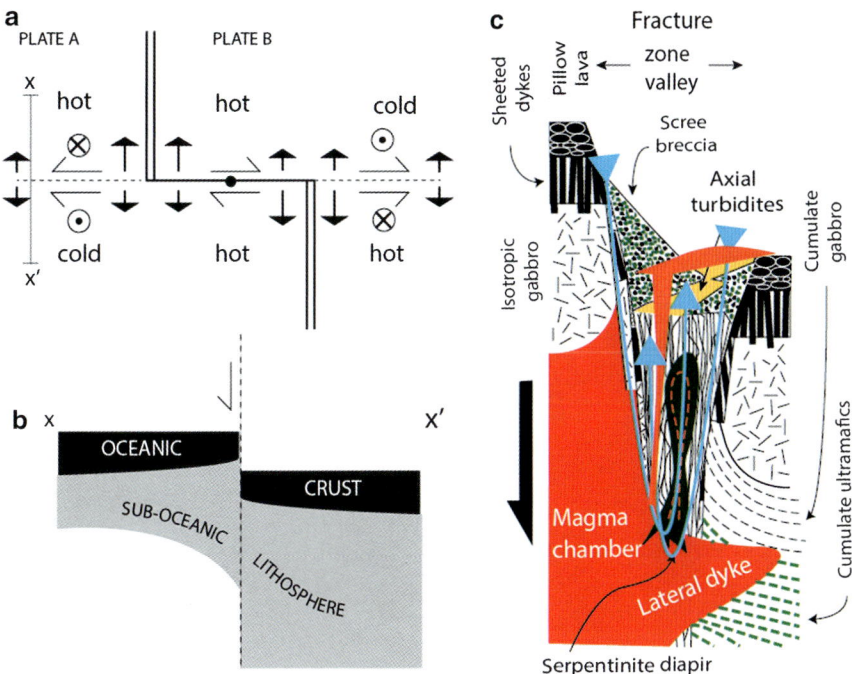

Transform Faults, Figure 11 (**a**): Map view of a fracture zone including a transform fault segment. Broad-headed *arrows* show extension across the fracture zone due to thermal contraction. The lengths of the *arrows* are indicative of the intensity of extension. (**b**): Cross section along x-x' (in **a**). (**c**) Schematized geological cross section across a fracture zone (only the crustal part is shown except the cumulate ultramafics). Blue arrows schematically represent the hydrothermal circulation.

than ~2.5 km then its average density exceeds oceanic and continental crustal basement density. This work also suggests that it is unlikely that significant serpentinisation occurs under thinned continent crust at the OCT [ocean-continent transition]." Most Atlantic-type continental margins are under margin-perpendicular compression as revealed by in situ instantaneous strain observations (Zoback, 1992) including significant earthquakes exceeding a magnitude of 7 (Wolin et al., 2012). It is reasonable to assume, in regions where discontinuities in the oceanic lithosphere are being compressed across their planes, that seawater would not circulate to any significant extent, and this would explain the little serpentinization in such environments as noted by Kusznir and Cooper (2011). By contrast, oceanic fracture zones are places where there is thermal-contraction-driven extension across the transform faults and their fracture zone extensions (Collette, 1974; Turcotte, 1974; Sandwell, 1986; Kastens, 1987; Kumar and Gordon, 2009 and further literature therein). Morphologically, fracture zone valleys may be as impressive as some rift valleys (see esp. Franchateau et al., 1976, Fig. 7; and Fox et al., 1980, Fig. 3b), and fracture zone tectonism may very significantly disrupt the coherence of the oceanic lithosphere thousands of kilometers away from spreading centers.

Figure 11 illustrates some of the salient cross-sectional elements of the geology of the fracture zones (for details, see *general*: DeLong et al., 1977, 1979; *in the oceans*: Franchateau et al., 1976; Bonatti, 1978; Bonatti and Hamlyn, 1978; Bonatti et al., 1979; Fox et al., 1980; Hamlyn and Bonatti, 1980; Schouten and White, 1980; Schouten et al., 1980; Karson and Dick, 1983; Searle, 1983; Bonatti and Crane, 1984; CYAGOR II Group, 1984; Fox and Gallo, 1984; Honnorez et al., 1984; Garfunkel, 1986; Loudenr et al., 1986; Sandwell, 1986; Kastens, 1987; Rutter and Brodie, 1987; Allerton, 1989; Smoot, 1989; Tamsett and Searle, 1990; Müller and Roest, 1992; Kastens et al., 1998; Okal and Langenhorst, 2000; Sage et al., 2000; Beutel and Okal, 2003; Behn et al., 2007; Sørensen et al., 2007; *in ophiolites, general*: Nicolas, 1989, Chap. 5).

Figure 11a illustrates the map view of an oceanic fracture zone including a transform fault segment between plates *A* and *B*. The black circle in the middle of the transform fault shows the crossover point as explained above (see also Figures. 8 and 9). Beyond the transform fault, the fracture zone displays a complex displacement history: The most obvious motion along the non-transform fault extension of a fracture zone is that it "unfaults itself" in that the high side moves downward with respect to the low side because of faster cooling. At the same time, contraction down the thermal gradient will impose a strike-slip component opposite to that of the main transform fault. Finally, thermal contraction will also cause extension across the fracture zone along its entire length, including the transform segment. Because of this, the cross-sectional morphology of fracture zone valleys greatly resembles that of rift valleys (e.g., Fox et al.,

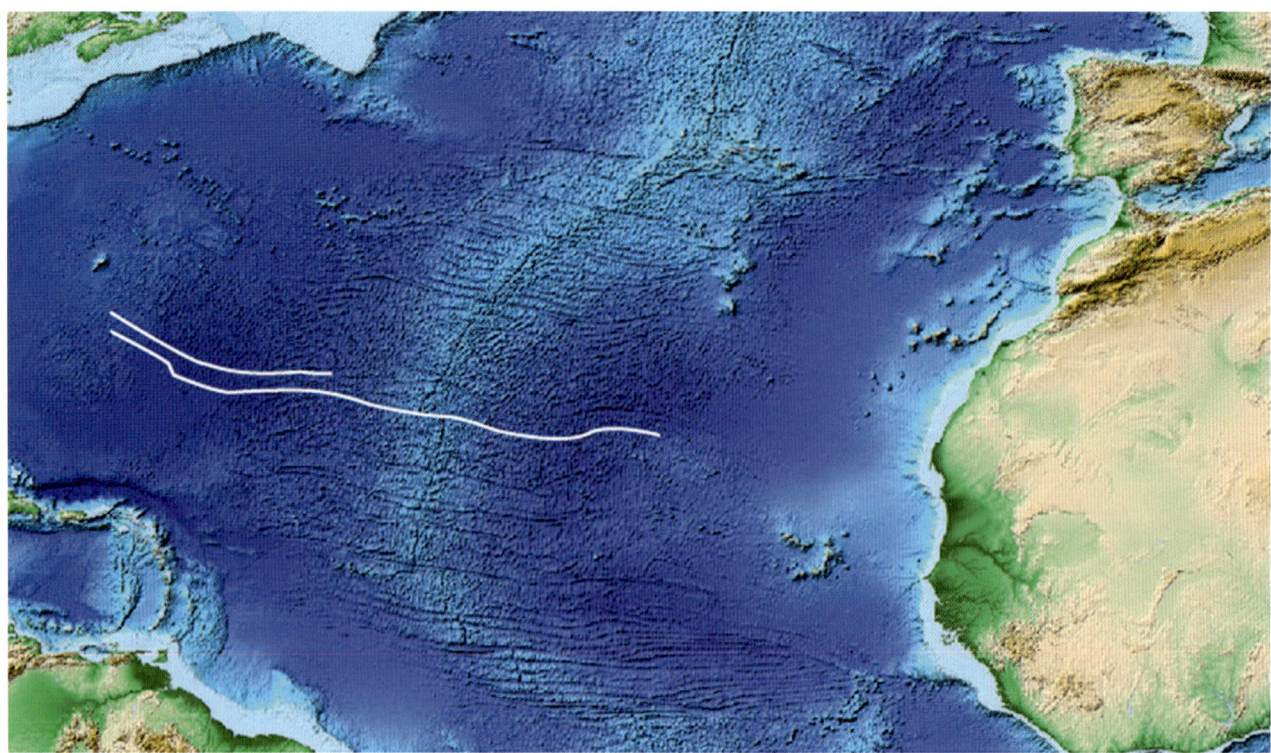

Transform Faults, Figure 12 Undulatory fracture zone strikes in the Central Atlantic.

1980, Fig. 3b; Tucholke and Lin, 1994, plate 3). The morphology of the fracture zone valleys differs from the morphology of rift valleys only in their perfect continuity and near perfect linearity (see esp. Tucholke and Lin's plate 3) and by the steepness of their sides. This linearity can only be disturbed, if there were one or more pole changes in the history of a transform fault. Such "ondulations" along the strike of many of the long fracture zones in the world's oceans are due to pole changes governing the spreading (cf. Figure 12 for some of the Central Atlantic examples).

The fracture valley slopes are steep ($>10°$) and both mass wasting and normal faulting are ubiquitous along them (Franchateau et al., 1976; DeLong et al., 1979; Fox et al., 1980). Extension, combined with strike-slip movement, induces a very complex *transtensional* tectonics on the fracture zone. What makes the whole structural picture so complex is that the strike-slip sense along the fracture zone valley reverses past the transforms. So every bit of fracture zone valley rock beyond the transforms will show the effects of two different episodes of transtensional strain (Dewey, 2002) in which the strike-slip components are opposite (but not necessarily of equal magnitude) of each other. No one observational report has yet reported all the expected complexities of the structural picture from any single fracture zone, but, combined, the available observations certainly indicate that what is expected (see esp. DeLong et al., 1979; Dewey, 2002) indeed happens.

The extension across the fracture zone will not only generate structures appropriate for a transtensional regime (which means strong constrictional strain in the more ductile lower parts: Dewey, 2002) but also magmatism (more alkalic as one goes away from the spreading centre), hydrothermal fields, and serpentinite diapirism. Water circulation down into the oceanic layer 3 (i.e., the magma chamber) will not only cause extensive mineralization in hydrothermal fields with varied and rich mineralization (e.g., Gibbs et al., 1993; Hein et al., 1999; Mazarovich et al., 2001) but also high-grade wet metamorphism creating amphibolite gneisses (Honnorez et al., 1984). As the segment run of the lower side of the fracture zone clears the transform fault, it is abutted against the ridge on the other segment. Here, the magma chamber of the high side, if present, will cause injection of mafic dykes into the plutonic foundation of the oceanic crust and the upper mantle. These dykes, depending on the length of the transform fault, can be significantly younger than the plutonic rocks they intrude (cf. Casey et al., 1981).

The steepness of the scarps will also cause the generation of scree breccias, the clasts of which may represent the entire suite of the oceanic crust and even upper mantle. As such, scree breccias accumulate into fans at the base of the slopes; they deform because of the tectonism of the fracture zone, commonly overlie hydrothermal vents, and become rebrecciated. Thus, rich, multicoloured calcite deposits cement them, giving rise to ophicalcites that are in places covered by radiolarites, in other places by distal turbidites, or hemipelagic deposits including deep-sea muds (DeLong et al., 1979; Bernoulli and Weissert,

1985), and in yet others they may be subaerial, possibly forming atop risen fracture zone rocks (Folk and McBride, 1976). Such breccias are often cut by dykes and show a polyphase origin indicated by a number of phases of fragmentation and different generations of cement fills. Ophicalcites form, under different names (e.g., *verd antique*, *verde aver*, *verde tinos*), one of the most popular ornamental stones in the Mediterranean region since antiquity.

Vertical tectonics along fracture zones, especially along their transform fault segments, is considerable (Bonatti, 1978), and atop risen segments, not only subaerial deposits but also localized shallow-water carbonate deposits may form. In the geological record, especially in ophiolitic mélange complexes, such shallow-water carbonate deposits may be confused with former atoll caps or fragments of oceanic plateaux.

Oceanic fracture zones in the geological record

The occurrence of proper oceanic fracture zones outside the present-day oceans is so far confined to ophiolite complexes. Some very young transfer faults in subaerial oceanic spreading centres, similar to those reported by Karson (1990) from the submarine ridges, have been mapped in Iceland (Einarsson, 1967; Sigmundsson, 2006), where even probably shear-related folding has been observed (Sigurdsson, 1967), and in Afar (Tapponnier et al., 1990; Abbate et al., 1995), but these places sit atop large, active mantle plumes. Nevertheless, they would give at least a first-order idea of what the oceanic fracture zones might look like at outcrop. None has been recognized in regions hypothesized to be underlain by trapped very old oceanic lithosphere such as the North Caspian Depression and the Tarim Basin. As yet, there is no modern synthesis of oceanic transform faults and fracture zones combining all the data from the ophiolites, subaerial spreading centres, and the oceans.

Bibliography

Abbate, E., Passerini, P., and Zan, L., 1995. Strike-slip faults in a rift area: a transect in the Afar triangle, East Africa. *Tectonophysics*, **241**, 67–97.

Ahlgren, S. G., 2001. The nucleation and evolution of Riedel shear zones as deformation bands in porous sandstone. *Journal of Structural Geology*, **23**, 1203–1214.

Allerton, S., 1989. Distortions, rotations, and crustal thinning at ridge-transform intersections. *Nature*, **340**, 626–632.

Anderson, E. M., 1951. *The Dynamics of Faulting and Dyke Formation with Applications to Britain*, 2nd revised edn. Edinburgh: Oliver and Boyd, x + 206 pp.

Atwater, T., 1970. Implications of plate tectonics for the Cenozoic tectonic evolution of western North America. *Geological Society of America Bulletin*, **81**, 3513–3536.

Baes, M., Govers, R., and Wortel, R., 2011. Subduction initiation along the inherited weakness zone at the edge of a slab: insights from numerical models. *Geophysical Journal International*, **184**, 991–1008.

Behn, M. D., Boettcher, M. S., and Hirth, G., 2007. Thermal structure of oceanic transform faults. *Geology*, **35**, 307–310.

Bercovici, D., 2003. The generation of plate tectonics from mantle convection. *Earth and Planetary Science Letters*, **205**, 107–121.

Bernoulli, D., and Weissert, H., 1985. Sedimentary fabrics in Alpine ophicalcites, South Pennine Arosa zone, Switzerland. *Geology*, **13**, 755–758.

Beutel, E. K., and Okal, E. M., 2003. Strength asperities along oceanic transform faults: a model for the origin of extensional earthquakes on the Eltanin transform system. *Earth and Planetary Science Letters*, **216**, 27–41.

Bonatti, E., 1978. Vertical tectonism in oceanic fracture zones. *Earth and Planetary Science Letters*, **37**, 249–251.

Bonatti, E., and Crane, K., 1984. Oceanic fracture zones. *Scientific American*, **250**(5), 36–47.

Bonatti, E., and Hamlyn, P. R., 1978. Mantle uplifted block in the western Indian Ocean. *Science*, **201**, 249–251.

Bonatti, E., Chermak, A., and Honnorez, J., 1979. Tectonic and igneous emplacement of crust in oceanic transform zones. In Talwani, M., Harrison, C. G., and Hayes, D. E. (eds.), *Deep Drilling Results in the Atlantic Ocean: Ocean Crust*. Washington: American Geophysical Union. Maurice Ewing Series, Vol. 2, pp. 239–248.

Burke, K. C. A., and Dewey, J. F., 1973. Plume-generated triple junctions: key indicators in applying plate tectonics to old rocks. *Journal of Geology*, **81**, 406–433.

Casey, J. F., and Dewey, J. F., 1984. Initiation of subduction zones along transform and accreting plate boundaries, triple junction evolution, and forearc spreading centres – implications for ophiolite geology and obduction. In Gass, I. G., Lippard, S. J., and Shelton, A. W. (eds.), *Ophiolites and Oceanic Lithosphere*, London: Geological Society Special Publication 13, pp. 269–290.

Casey, J. F., Dewey, J. F., Fox, P. J., Karson, J. A., and Rosenkrantz, E., 1981. Heterogeneous nature of oceanic crust and upper mantle: a perspective from the Bay of Islands ophiolite complex. In Emiliani, C. (ed.), *The Oceanic Lithosphere, The Sea*. New York: John Wiley & Sons, Vol. 7, pp. 305–338.

Cloetingh, S. A. P. L., Wortel, M. J. R., and Vlaar, N. J., 1984. Passive margin evolution, initiation of subduction and the Wilson Cycle. *Tectonophysics*, **109**, 147–163.

Collette, B. J., 1974. Thermal contraction joints in a spreading seafloor as origin of fracture zones. *Nature*, **251**, 299–300.

CYAGOR II Group, 1984. Intraoceanic tectonism on the Gorringe Bank: observations by submersible. In Gass, I. G., Lippard, S. J., and Shelton, A. W. (eds.), *Ophiolites and Oceanic Lithosphere*, London: Geological Society Special Publication 13, pp. 113–130.

DeLong, S. E., Hodges, F. N., and Arculus, R. J., 1975. Ultramafic and mafic inclusions, Kanaga Island, Alaska, and the occurrence of alkaline rocks in island arcs. *Journal of Geology*, **83**, 721–736.

DeLong, S. E., Dewey, J. F., and Fox, P. J., 1977. Displacement history of oceanic fracture zones. *Geology*, **5**, 199–202.

DeLong, S. E., Dewey, J. F., and Fox, P. J., 1979. Topographic and geologic evolution of fracture zones. *Journal of the Geological Society of London*, **136**, 303–310.

Dewey, J. F., 1975. Finite plate evolution: some implications for the evolution of rock masses at plate margins. *American Journal of Science*, **275-A**(John Rodgers volume), 260–284.

Dewey, J. F., 1976. Ancient plate margins: some observations. *Tectonophysics*, **33**, 379–385.

Dewey, J. F., 2002. Transtension in arcs and orogens. *International Geology Review*, **44**, 402–438.

Dewey, J. F., and Casey, J. F., 2011. The origin of obducted large-slab ophiolite complexes. In Brown, D., and Ryan, P. D. (eds.), *Arc Continent Collision*. Heidelberg: Springer. Frontiers in Earth Sciences, pp. 431–444.

Dickinson, W. R., 1996. *Kinematics of transrotational tectonism in the California transverse ranges and its contribution to cumulative slip along the San Andreas transform fault system*, Boulder Colorado: Geological Society of America Special Paper 305, iv + 46 pp.

Einarsson, T., 1967. The Icelandic fracture system and the inferred crustal stress field. In Björnsson, S. (ed.), *Iceland and Mid-Ocean Ridges, Visindafélag. Íslendinga (Societas Scientiarum Islandica)*. Reykjavik: Prentsmidjan Leiftur, pp. 128–141.

Folk, R. L., and McBride, E. F., 1976. Possible pedogenic origin of Ligurian ophicalcite: a Mesozoic calichified serpentinite. *Geology*, **4**, 327–332.

Fox, P. J., and Gallo, D. G., 1984. A tectonic model for ridge-transform-ridge plate boundaries: implications for the structure of oceanic lithosphere. *Tectonophysics*, **104**, 205–242.

Fox, P. J., Detrick, R. S., and Purdy, G. M., 1980. Evidence for crustal thinning near fracture zones: implications for ophiolites. In *Proceedings International Ophiolite Symposium Cyprus 1979*. Nicosia: Cyprus Geological Survey Department, pp. 161–168.

Franchateau, J., Choukroune, P., Hekinian, R., Le Pichon, X., and Needham, H. D., 1976. Oceanic fracture zones do not provide deep sections in the crust. *Canadian Journal of Earth Sciences*, **13**, 1223–1235.

Freund, R., and Merzer, A. M., 1976. Anisotropic origin of transform faults. *Science*, **192**, 137–138.

Garfunkel, Z., 1986. Review of oceanic transform activity and development. *Journal of the Geological Society of London*, **143**, 775–784.

Géli, L., and Sclater, J., 2008. On the depth of oceanic earthquakes: brief comments on "The thermal structure of oceanic and continental lithosphere" by McKenzie, D., Jackson, J. and Priestley, K. Earth Plan. Sci. Let., 233, [2005], 337–349. *Earth and Planetary Science Letters*, **265**, 769–775, http://dx.doi.org/10.1016/j.epsl.2007.08.029

Gerya, T., 2010. Dynamical instability produces transform faults at mid-ocean ridges. *Science*, **329**, 1047–1050.

Gibbs, A. E., Hein, J. R., Lewis, S. D., and McCulloch, D. S., 1993. Hydrothermal palygorskite and ferromanganese mineralization at a central California margin fracture zone. *Marine Geology*, **115**, 47–65.

Govers, R., and Wortel, M. J. R., 2005. Lithosphere tearing at STEP faults: response to edges of subduction zones. *Earth and Planetary Science Letters*, **236**, 505–523.

Hall, C., and Gurnis, M., 2005. Strength of fracture zones from their bathymetric and gravitational evolution. *Journal of Geophysical Research*, **110**, B01402, doi:10.1029/2004JB003312.

Hamlyn, P. R., and Bonatti, E., 1980. Petrology of mantle-derived ultramafics from the Owen Fracture Zone, northwest Indian Ocean: implications for the nature of the oceanic upper mantle. *Earth and Planetary Science Letters*, **48**, 65–79.

Haxby, W. F., and Parmentiar, E. M., 1988. Thermal contraction and the state of stress in the oceanic lithosphere. *Journal of Geophysical Research*, **93**, 6419–6429.

Hein, J. R., Koski, R. A., Embley, R. W., Reid, J., and Chang, S.-W., 1999. Diffuse-flow hydrothermal field in an oceanic fracture zone setting, Northeast Pacific: deposit composition. *Exploration and Mining Geology*, **8**, 299–322.

Honnorez, J., Mével, C., and Montigny, R., 1984. Occurrence and significance of gneissic amphibolites in the Vema fracture zone, equatorial Mid-Atlantic Ridge. In Gass, I. G., Lippard, S. J., and Shelton, A. W. (eds.), *Ophiolites and Oceanic Lithosphere*, London: Geological Society Special Publication 13, pp. 121–130.

Houseman, G., McKenzie, D., and Molnar, P., 1981. Convective instability of a thickened boundary layer and its relevance for the thermal evolution of continental convergent belts. *Journal of Geophysical Research*, **86**, 6115–6132.

Karig, D. E., 1982. Initiation of subduction zones: implications for arc evolution and ophiolite emplacement. In Leggett, J. K. (ed.), *Trench-Forearc Geology: Sedimentation and Tectonics on Modern and Ancient Plate Margins*, London: Geological Society Special Publication 10, pp. 563–576.

Karson, J. A., 1990. Accommodation zones and transfer faults: integral components of Mid-Atlantic extensional systems. In Peters, T., Nicolas, A., and Coleman, R. G. (eds.), *Ophiolite Genesis and Evolution of the Oceanic Lithosphere – Proceedings of the Ophiolite Conference, Held in Muscat, Oman, 7–18 January 1990*. Dordrecht: Kluwer Academic Publishers, pp. 21–37.

Karson, J. A., and Dick, H. J. B., 1983. Tectonics of ridge-transform intersections at the Kane fracture zone. *Marine Geophysical Researches*, **6**, 51–98.

Kastens, K., 1987. A compendium of causes and effects of processes at transform faults and fracture zones. *Reviews of Geophysics*, **25**, 1554–1562.

Kastens, K., Bonatti, E., Caress, D., Carrara, G., Dauteuil, O., Frueh-Green, G., Ligi, M., and Tartarotti, P., 1998. The Vema transverse ridge (central Atlantic). *Marine Geophysical Researches*, **20**, 533–556.

Katz, R. F., 2010. Porosity-driven convection and asymmetry beneath mid-ocean ridges. *Geochemistry Geophysics Geosystems G^3*, **11**, doi:10.1029/2010GC003282.

Katz, Y., Weinberger, R., and Aydın, A., 2004. Geometry and kinematic evolution of Riedel shear structures, Capitol Reef National Park, Utah. *Journal of Structural Geology*, **26**, 491–501.

Ketin, İ., 1948. Über die tektonisch-mechanischen Folgerungen aus den grossen anatolischen Erdbeben des letzten Dezenniums. *Geologische Rundschau*, **36**, 77–83.

Kohlstedt, D., Evans, B., and Mackwell, S., 1995. Strength of the lithosphere: constraints imposed by laboratory experiments. *Journal of Geophysical Research*, **100**, 17587–17602.

Kumar, R. R., and Gordon, R. G., 2009. Horizontal thermal contraction of oceanic lithosphere: the ultimate limit to the rigid plate approximation. *Journal of Geophysical Research*, **114**, B01403, doi:10.1029/2007JB005473.

Kusznir, N. J., and Cooper, C., 2011. The depth distribution of mantle serpentinization at magma poor rifted margins: geophysical evidence from the Iberian, Newfoundland and Nova Scotia margins. *American Geophysical Union, Fall Meeting, Abstracts*, Abstract #T23A-2374.

Lotze, F., 1937. Zur Methodik der Forschungen über saxonische Tektonik. *Geotektonische Forschungen*, **1**, 6–27.

Loudenr, K. E., White, R. S., Potts, C. G., and Forsyth, D. W., 1986. Structure and seismotectonics of the Vema Fracture Zone, Atlantic Ocean. *Journal of the Geological Society (London)*, **143**, 795–805.

Mazarovich, A. O., Simonov, V. A., Peive, A. A., Kovyazin, S. V., Tret'yakov, G. A., Raznitsin, Y. N., Savel'eva, G. N., Skolotnev, S. G., Sokolov, S. Y., and Turko, N. N., 2001. Hydrothermal mineralization in the Sierra Leone Fracture Zone (Central Atlantic). *Lithology and Mineral Resources*, **36**(5), 460–466.

McKenzie, D., and Parker, R., 1967. The North Pacific: an example of tectonics on a sphere. *Nature*, **216**, 1276–1280.

McKenzie, D., Jackson, J., and Priestley, K., 2005. The thermal structure of oceanic and continental lithosphere. *Earth and Planetary Science Letters*, **233**, 337–349.

Morgan, W. P., 1968. Rises, trenches, great faults, and crustal blocks. *Journal of Geophysical Research*, **73**, 1959–1982.

Morris, A., Andereson, M. W., Inwood, J., and Robertson, A. H. F., 2006. Palaeomegnetic insights into the evolution of Neotethyan oceanic crust in the eastern Mediterranean. In Robertson, A. H. F., and Mountrakis, D. (eds.), *Tectonic Development of the Eastern Mediterranean Region*, London: Geological Society (London) Special Publication 260, pp. 351–372.

Müller, R. D., and Roest, W. R., 1992. Fracture zones in the North Atlantic from combined Geosat and Seasat data. *Journal of Geophysical Research*, **97**, 3337–3350.

Müller, R. D., Sdrolias, M., Gaina, C., and Roest, W. R., 2008. Age, spreading rates, and spreading asymmetry of the world's ocean crust. *Geochemistry Geophysics Geosystems* G^3, **9**, doi:10.1029/2007GC001743.

Nicolas, A., 1989. *Structures of Ophiolites and Dynamics of Oceanic Lithosphere*. Dordrecht: Kluwer Academic Publishers. Petrology and Structural Geology, Vol. 4, xiii+367 pp.

Ohnenstetter, M., Bechon, F., and Ohnenstetter, D., 1990. Geochemistry and mineralogy of lavas from the Arakapas Fault Belt, Cyprus: consequences for magma chamber evolution. *Mineralogy and Petrology*, **41**, 105–124.

Okal, E. A., and Langenhorst, A. R., 2000. Seismic properties of the Eltanin Transform System, South Pacific. *Physics of the Earth and Planetary Interiors*, **119**, 185–208.

Özbakır, A. D., Şengör, A. M. C., Wortel, M.J.R, Gover, R., 2013. The Pliny-Strabo trench region: A large scale shear zone resulting from slab tearing: *Earth and Planetary Science Letters*, **375**, pp. 188–195

Priestley, K., and McKenzie, D., 2006. The thermal structure of the lithosphere from shear wave velocities. *Earth and Planetary Science Letters*, **244**, 285–301.

Rutter, E. H., and Brodie, K. H., 1987. On the mechanical properties of oceanic transform faults. *Annales Tectonicae*, **1**, 87–96.

Sage, F., Basile, C., Mascle, J., Pontoise, B., and Whitmarsh, R. B., 2000. Crustal structure of the continent-ocean transition off the Côte d'Ivoire-Ghana transform margin: implications for thermal exchanges across the palaeotransform boundary. *Geophysical Journal International*, **143**, 662–678.

Sandwell, D. T., 1986. Thermal stress and the spacings of transform faults. *Journal of Geophysical Research*, **91**, 6405–6417.

Schouten, H., and White, R. S., 1980. Zero-offset fracture zones. *Geology*, **8**, 175–179.

Schouten, H., Karson, J. A., and Dick, H., 1980. Geometry of transform zones. *Nature*, **288**, 470–473.

Scrutton, R. A., 1979. On sheared passive continental margins. *Tectonophysics*, **59**, 293–305 (reprinted in Keen, C. E. (ed.), *Crustal Properties Across Passive Margin*. Developments in Geotectonics 15. Amsterdam: Elsevier).

Scrutton, R. A., 1982. Crustal structure and development of sheared passive continental margins. In Scrutton, R. A. (ed.), *Dynamics of Passive Margins*. Washington, DC/Boulder: American Geophysical Union/Geological Society of America. Geodynamics Series, Vol. 6, pp. 133–140.

Searle, R. C., 1983. Multiple, closely spaced faults in fast-slipping fracture zones. *Geology*, **11**, 607–610.

Şengör, A. M. C., 1983. Transform faylar – Genel. In Canıtez, N. (ed.), *Levha Tektoniği*. İstanbul: İTÜ Maden Fakültesi/Ofset Baskı Atölyesi, pp. 547–569.

Şengör, A. M. C., 1990. Plate tectonics and orogenic research after 25 years: a Tethyan perspective. *Earth Science Reviews*, **27**, 1–201.

Şengör, A. M. C., 1995. Sedimentation and tectonics of fossil rifts. In Busby, C. J., and Ingersoll, R. V. (eds.), *Tectonics of Sedimentary Basins*. Oxford: Blackwell, pp. 53–117.

Şengör, A. M. C., 1999. Continental interiors and cratons: any relation? *Tectonophysics*, **305**, 1–42.

Şengör, A. M. C., 2001. Elevation as indicator of mantle plume activity. In Ernst, R., and Buchan, K. (eds.), Mantle Plumes: Their Identification Through Time; Colorado: Geological Society of America Special Paper 352, pp. 183–225.

Şengör, A. M. C., and Natal'in, B. A., 1996. Palaeotectonics of Asia: fragments of a synthesis. In Yin, A., and Harrison, M. (eds.), *The Tectonic Evolution of Asia, Rubey Colloquium*. Cambridge: Cambridge University Press, pp. 486–640.

Şengör, A. M. C., Tüysüz, O., İmren, C., Sakınç, M., Eyidoğan, H., Görür, N., Le Pichon, X. and Rangin, C., 2005. The North Anatolian Fault: A new look: *Annual Review of Earth and Planetary Sciences*, **33**, pp. 37–112.

Sigmundsson, F., 2006. *Iceland Geodynamics – Crustal Deformation and Divergent Plate Tectonics*. Heidelberg/Chichester: Springer/Praxis, xxiv+209 pp. +plates in the back.

Sigurdsson, H., 1967. Dykes, fractures and folds in the basalt plateau of Western Iceland: Einarsson, T., 1967, The Icelandic fracture system and the inferred crustal stress field. In Björnsson, S. (ed.), *Iceland and Mid-Ocean Ridges, Vísindafélag. Íslendinga (Societas Scientiarum Islandica)*. Reykjavik: Prentsmiðjan Leiftur, pp. 162–169.

Smoot, N. C., 1989. North Atlantic fracture-zone distribution and patterns shown by multibeam sonar. *Geology*, **17**, 1119–1122.

Sørensen, M. B., Ottemöller, L., Havzkov, J., Atakan, K., Hellevang, B., and Pedersen, R. B., 2007. Tectonic processes in the Jan Mayen Fracture Zone based on earthquake occurrence and bathymetry. *Bulletin of the Seismological Society of America*, **97**, 772–779.

Stein, C. A., and Cochran, J. R., 1985. The transition between the Sheba Ridge and Owen Basin: rifting of old oceanic lithosphere. *Geophysical Journal of the Royal Astronomical Society*, **81**, 47–74.

Sykes, L., 1967. Mechanism of earthquakes and nature of faulting on the mid-oceanic ridges. *Journal of Geophysical Research*, **72**, 2131–2153.

Tamsett, D., and Searle, R., 1990. Structure of the Alula-Fartak Fracture Zone, Gulf of Aden. *Journal of Geophysical Research*, **95**, 1239–1254.

Tapponnier, P., Armijo, R., Manighetti, I., and Courtillot, V., 1990. Bookshelf faulting and horizontal block rotations between overlapping rifts in southern Afar. *Geophysical Research Letters*, **17**, 1–4.

Tchalenko, J. S., 1970. Similarities between shear zones of different magnitudes. *Geological Society of America Bulletin*, **81**, 1625–1640.

Teichert, C., 1979. Spherical cap tectonics. *Geotimes*, October issue.

Tucholke, B. E., and Lin, J., 1994. A geological model for the structure of ridge segments in a slow spreading ocean crust. *Journal of Geophysical Research*, **99**, 11937–11958.

Turcotte, D. L., 1974. Are transform faults thermal contraction cracks? *Journal of Geophysical Research*, **79**, 2573–2577.

Wilson, J. T., 1954. The development and structure of the crust. In Kuiper, G. P. (ed.), *The Earth as a Planet: The Solar System*. Chicago: The University of Chicago Press, Vol. II, pp. 138–214.

Wilson, J. T., 1965. A new class of faults and their bearing on continental drift. *Nature*, **207**, 343–347.

Wolin, E., Stein, S., Pazzaglia, F., Meltzer, A., Kafka, A., and Berti, C., 2012. Mineral, Virginia, earthquake illustrates seismicity of a passive-aggressive margin. *Geophysical Research Letters*, **39**, doi:10.1029/2011GL050310.

Zoback, M. L., 1992. First and second-order patterns of stress in the lithosphere: the World Stress Map Project. *Journal of Geophysical Research*, **97**, 11703–11728 + coloured foldout map.

Some Complementary Web Sites for School and Elementary University Levels

http://plateboundary.rice.edu/. Last visited on 22 March 2013.

http://web.viu.ca/earle/transform-model/. Last visited on 22 March 2013.

http://www.bioygeo.info/Animaciones/TransformFaultsV2.swf. Last visited on 22 March 2013.

http://www.earthds.info/pdfs/EDS_20.PDF. Last visited on 22 March 2013.

http://www.earthlearningidea.com/PDF/84_Transform_faults.pdf. Last visited on 22 March 2013.
http://www.open.edu/openlearn/science-maths-technology/science/geology/plate-tectonics/content-section-3.8. Last visited on 22 March 2013.
http://www.wwnorton.com/college/geo/oceansci/animations.asp#ch4. Last visited on 22 March 2013.

Advanced Undergraduate and Postgraduate Levels

http://geoscience.wisc.edu/~chuck/MORVEL/trf_flts.html. Last visited on 22 March 2013.

Cross-references

Oceanic Spreading Centers
Plate Tectonics
Pull-apart Basins
Push-up Blocks
Regional Marine Geology
Triple Junctions

TRIPLE JUNCTIONS

A. M. Celâl Şengör
Faculty of Mines, Department of Geology and Eurasia, Institute of Earth Sciences, Istanbul Technical University, Ayazaga, Istanbul, Turkey

Synonyms

Triple junctures; Triple plate junctures

Definition

A *triple junction* is ideally a *point* at which boundaries of three lithospheric plates come together. At such a point, geological processes belonging to one plate boundary abruptly give place to others belonging to the other two boundaries. In real life, thermal and mechanical properties of the crust and upper mantle and the time-averaged behavior of the spreading centers and subduction zones in the oceans impose a *finite volume* on triple junctions in which deformation and magmatism are complex. Triple junctions may be stable or unstable, i.e., they may exist for finite amounts of time or only for an instant, depending on the geometry of the boundaries meeting at the triple junction and the motion velocities across them. Triple junctions also behave differently in the oceanic and in the continental lithosphere. In this entry, the emphasis is on oceanic triple junctions, in keeping with the theme of the volume, but at least two important types of triple junction involving continental plates are discussed, because they may be important for some oceanic settings also.

Introduction

The concept of triple junctions became prominent with the rise of plate tectonics, but before that, geologists dealing with areas characterized by the motion of rigid to semi-rigid blocks with respect to one another did come across triple junctions of boundaries of increased displacement and strain, and they recognized their geological significance (e.g., Lotze, 1937). After the recognition of the plate structure of the lithosphere and the three types of plate boundaries by Wilson (1965), the subject of triple junctions acquired a rigorously quantitative aspect (e.g., McKenzie and Parker, 1967; Morgan, 1968), and in 1969, McKenzie and Morgan presented an analysis of 16 types of triple junctions that they claimed to represent all possible triple junctions. Figure 1 shows these 16 types of triple junctions together with their velocity triangles (see below). In this entry, Figure spelled with a capital F refers to this entry's figures, whereas if spelled with a lowercase f, it refers to a figure in the literature not here reproduced.

Stability of triple junctions

Some triple junctions can exist during a finite lapse of time without changing their geometries and others cannot. Because plates are rigid and under ordinary circumstances holes are not allowed to open at triple junctions, the sum of the velocities of the plates with respect to one another around a triple junction must add up to zero. As the plates are required to be rigid caps moving on a spherical surface, McKenzie and Parker (1974) showed that the velocity v of a plate at any point on its surface can be expressed as

$$v = \Omega \times r \tag{1}$$

where Ω represents the plate's angular velocity in a nonrotating reference frame and r is the spherical coordinate of the point, with the centre of the sphere as the origin for both r and Ω.

The relative velocity v_{rel} between plates A and B can thus be expressed as:

$$v_{\text{rel}} = (\Omega_A - \Omega_B) \times r \tag{2}$$

McKenzie and Parker (1974) pointed out that as only relative motion between plates can be observed with certainty, it is convenient to consider only relative angular velocities given by:

$$_A\omega_B = \Omega_A - \Omega_B \tag{3}$$

The piercing point at the surface of the sphere of $_A\omega_B$ is known as the *pole of relative motion* between the plates A and B. McKenzie and Parker (1967) and McKenzie and Morgan (1969) pointed out that for a stable triple junction amidst plates A, B, and C, the following condition must be satisfied:

$$_A\omega_B + {_B\omega_C} + {_C\omega_A} = 0 \tag{4}$$

This equation allows one to determine the velocity along a plate boundary at a triple junction where the velocities on the other two boundaries are known. This can be of great help in historical geology in reconstructing past plate geometries.

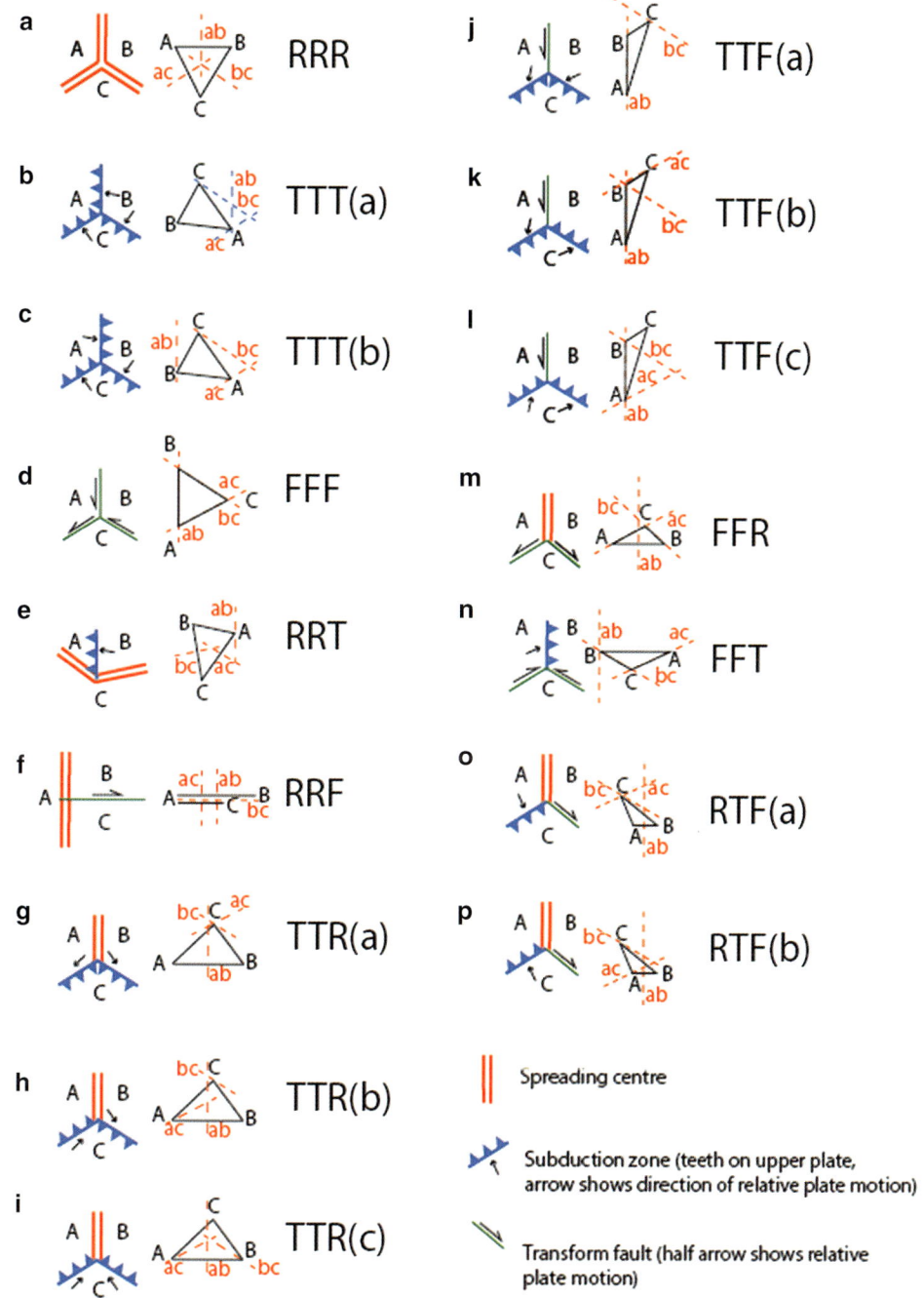

Triple Junctions, Figure 1 The geometry and stability of all possible triple junctions according to McKenzie and Morgan (1969, redrawn from their Figure 3. (**a**) *RRR* (Ridge-Ridge-Ridge) junction: stable in all orientations, as long as the three arms are spread more than 180° (York's condition: York, 1973a). (**b**) *TTT* (Trench-Trench-Trench) junction: stable if *ab*, *ac* form a straight line or if *bc* is parallel with the slip vector *CA*. (**c**) Another *TTT* junction: stable if the complicated general condition for *ab*, *bc*, and *ac* to meet at a point is satisfied. (**d**) *FFF* (Fault-Fault-Fault) junction: always unstable. (**e**) *RRT* junction: for stability, *ab* must go through the centroid. (**f**) *RRF* junction: always unstable except where the two ridges make a right angle with one another (York, 1973b; see Figure 5c below). (**g**) *TTR* junction: stable if *ab* goes through *C* or if *ac*, *bc* form a straight line. (**h**) Another *TTR* junction: can be stable if the *ab* boundary can be orientated so as *ab* to go through the centroid. (**i**) Another *TTR* junction: stable if the angles between *ab* and *ac*, *bc*, respectively, are equal or if *ac*, *bc* form a straight line. (**j**) TTF junction: stable if *ac*, *bc* form a straight line or if *C* lies on *ab*. (**k**) Another *TTF* junction: stable if *bc*, *ab* form a straight line or if *ac* goes through *B*. (**l**) Another *TTF* junction: stable if *ab*, *ac* form a straight line or if *ab*, *bc* do so. (**m**) *FFR* junction: stable if *C* lies on *ab* or if *ac*, *bc* form a straight line. (**n**) *FFT* junction: stable if *C* lies on *ab* or if *ac*, *bc* form a straight line. (**o**) *RTF* junction: stable if *ab* goes through *C* or if *ac*, *bc* form a straight line. (**p**) Another *RTF* junction: stable if *ac*, *ab* cross on *bc*.

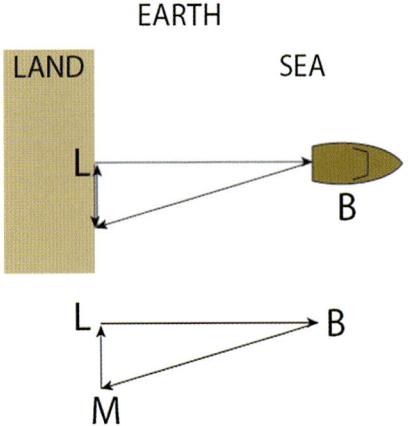

Triple Junctions, Figure 2 A simple visualization of how to compute relative velocities of three rigid objects with respect to one another. See text for explanation.

McKenzie and Morgan (1969) discussed the stability conditions of triple junctions using the *velocity space*. To do that, a triangle is formed by using the vectors of relative velocities of the plates meeting at the triple junction. Figure 2 illustrates how the velocity vectors are treated in drawing the necessary triangle: The problem is to determine the relative velocities of three rigid entities with respect to one another around a single point (Figure 2). Imagine a boat B leaving the shore of land L and speeding away along the vector LB on the surface of the sea (since both land and the sea are here assumed fixed with respect to the earth, we can use both for land and the sea the same label L). A man M begins walking along the shore along the vector LM at the instant the boat leaves the shore along the vector LB. What is the relative velocity of the man (M) with respect to the boat (B)? In other words, what is the orientation and the length of the vector MB?

To obtain that we must add the velocities of the man and the boat. However, here is a fine point: the boat moves with respect to land along the vector LB. The velocity of the man also with respect to the land is ML. But we need to fix our reference point *not* to the land but to the boat (or to the man) to obtain the required answer. So we keep M where he is with respect to the origin of the LB vector and move the land L. Hence we add not LB to LM but to LB to ML. But ML is only −LM. So our addition becomes: LB + (−LM) = BM. BM is the velocity of separation of B with respect to M. We could also say that MB is the separation velocity of M with respect to B. In all discussions on plate kinematics, it is of the utmost importance to make sure what reference frame is being adhered to.

To see whether a junction is stable, it is important to find a single point from the viewpoint of which the triple junction does not move (see also Cronin, 1992, but some of the triple junctions he lists simply cannot exist, so there is little point in listing them). Let us take the simple case of the three ridges with inter-ridge separation of 120° meeting at a triple junction. This is the situation depicted in Figure 1a left side, and the corresponding velocity triangle is shown on the right side. To seek the singularity from the vantage point of which the triple junction does not move, consider that AB represents the relative velocity of the plate B with respect to A. So any point on B moves with respect to the ridge separating the plate A from plate B with a velocity of AB/2. This corresponds to the midpoint of the AB vector. If we now fix a reference frame to the perpendicular bisector of AB, this will be equivalent to hovering over the spreading ridge and moving with it with respect to the reference frames fixed to the plates it separates. If our reference frame is fixed to the bisector, a line, the frame can move up and down the line but not across it. Now if we do it for all the ridges, our bisectors meet at a single point J, called the centroid (McKenzie and Morgan, 1969). If the centroid of any velocity triangle around a triple junction is a point, then that triple junction must be stable, because there is a single point from the vantage point of which that triple junction does not change its shape.

Although McKenzie and Morgan (1969) believed that all orientations of a ridge-ridge-ridge (abbreviated RRR) triple junction are stable, York (1973a) showed that unless the spread of the ridges is greater than 180° around the triple junction, the RRR junctions would be unstable and illustrated this with the example shown in Figure 3. Figure 3a shows an RRR triple junction that is unstable (cf. Figure 3c), if spreading is assumed to be bilaterally symmetric about the ridge axes. Figure 3b shows that if the junction is allowed to evolve, the ridge AB must develop a strong coaxial asymmetry of spreading, i.e., it must accrete more oceanic lithosphere to one of its sides than to the other. Although coaxial asymmetry is known from active spreading centres, it is always very short lived, and all ridges maintain bilaterally symmetric spreading for long periods of time even if this means to jump into the faster-spreading side several times (e.g., Stein et al., 1977; Müller et al., 2008). York (1973a) pointed out that a geometry similar to the Figure 3c, d, illustrating *ridge abandonment*, probably was responsible for the now inactive ridges seen within the Nazca Plate in the southeastern Pacific Ocean. York (1973b; Figure 1) also showed that an RRF junction, which brings together two ridges making an angle of 90° between them and a transform fault, is also stable, thus increasing the number of the triple junction configurations that can be stable under certain specific conditions from 14 to 15.

After the establishment of the stability conditions of all the 16 possible types of triple junctions using velocity triangles, McKenzie and Parker (1974) asked the question whether a similar thing may be done for accelerations instead of velocities. In other words, whether acceleration triangles for triple junctions may be drawn to render useful information about the stability conditions of triple junctions. They pointed out that since, in general, the relative accelerations between plates will not be related to relative velocities, junctions that are stable in v space may be unstable in dv/dt space. Thus, unless specific conditions

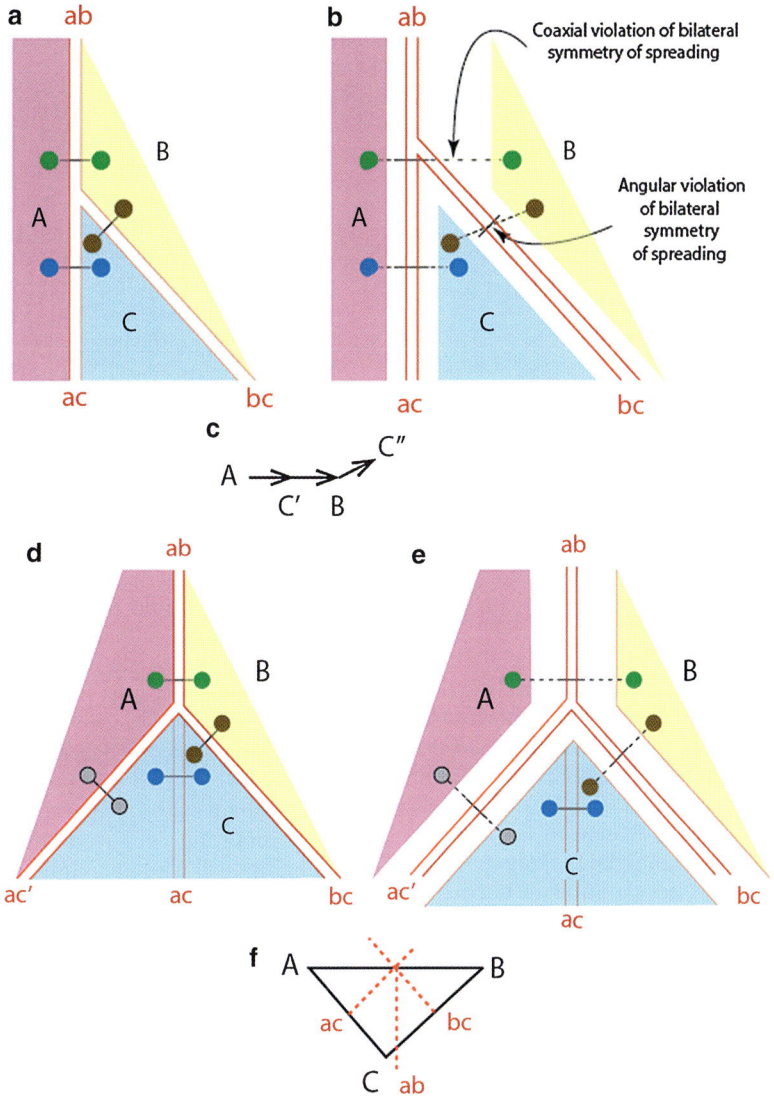

Triple Junctions, Figure 3 (a) York's (1973a) unstable RRR junction. (b) The same junction after a finite amount of time has elapsed. Note the great amount of coaxial asymmetric spreading imposed on the spreading centre *ab*. (c) No velocity triangle can be closed. (d) The ridge *ac* is extinguished and a new ridge *ac'* has formed cutting across the former *plate A* and adding the part of it between the new ridge *ac'* and *ac* to *plate C*. (e) The new triple junction geometry is stable. The *coloured circles* are markers showing relative displacement between pairs of plates. (f) The velocity *triangle* in this case can be closed and indicates a stable triple junction. The symbolism as in Figure 1.

are imposed on $d\omega/dt$, the changes in v space will cause a stable junction to become unstable.

One remarkable result of their study is that if all of the $dv/dt \neq 0$, the paths of the triple junction with respect to the plates are all curved. They indicated that if $_A v_T$ represents the relative motion between the triple junction T and plate A and if $d_A v_T/dt$ is the corresponding acceleration, the important component of the latter is at right angles to $_A v_T$ and McKenzie and Parker (1974) denote it as $(d_A v_T/dt)N$. They point out that the component of $d_A v_T/dt$ parallel with the ridge causes changes in the rate of spreading. The $(d_A v_T/dt)N$ produces a curved path of the triple junction T with respect to the plate A with curvature κ and radius of curvature ρ (Figure 4). The former is obtained by employing the following equation:

$$\kappa = 1/\rho = |(d_A v_T/dt)N| / |_A v_T|^2 \qquad (5)$$

κ will change with time, unless $d_A v_T/dt \times {_A v_T} = 0$. One may use the curvature of the ridge axis to obtain the component of the acceleration normal to the vector velocity between the triple junction and the ridge. This is not dependent on whether the spreading is orthogonal to the ridge or not. McKenzie and Parker (1974) further noted that using maps such as the one shown in Figure 4, one

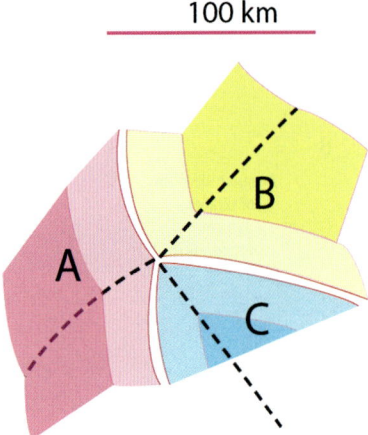

Triple Junctions, Figure 4 Map of an RRR triple junction among the plates *A*, *B*, and *C* in a plane tangent to the earth at the junction when all of the ***dv/dt*** ≠ 0. Note that in this case the three ridges curve. The symbolism as in Figure 1. The *black dashed line* is the trace of the triple junction on each plate.

could draw an acceleration triangle. All one needs to do is to trace the track of the triple junction T with respect to the three plates and measure the curvature of the ridges. Acton et al. (1988) interpreted such curved magnetic anomalies along the Galapagos Ridge between the Nazca and the Cocos plates as resulting from a change in the velocity (rate and/or direction) of the spreading along a ridge. Since in a rigid plate model, acceleration is needed to change velocity, they might have applied the ideas of McKenzie and Parker (1974), but they do not cite them. McKenzie and Parker (1974) themselves have used the curved anomalies of the Galapagos Ridge, but they were unable to fit the observations to their model because of the insufficiency of the precision of the data at the time. Data such as those presented by Acton et al. (1988) may be used to obtain the accelerations responsible for the changes in spreading velocities. In such situations, one can see how some stable triple junctions may be pushed into unstable geometries during finite plate evolution and might obtain an idea of the driving forces of the plates.

Triple junctions and ridge propagation

Ridge propagation has long been popular since the rise of plate tectonics, because continental splitting has been ascribed to various forms of it from centres of extension caused by diverse mechanisms (e.g., Burke and Dewey, 1973; Bhattacharji and Koide, 1975; McKenzie and Weiss, 1975; Courtillot, 1982), but even before, much had been written on various forms of rift propagation in multifarious contexts. Also, in oceanic areas, in certain cases, one of the two spreading centres joined by a transform fault propagates across the transform fault coevally with the recession of the other, creating complex magnetic anomaly patterns on the ocean floor (Hey, 1977; Atwater, 1981; McKenzie, 1982). Such propagation processes have various causes:

1. In continents, three-armed rift stars (called rrr junctions formed of intracontinental rifts as opposed to RRR junctions formed of spreading centers: see Burke and Dewey, 1973; Figure 2; Şengör, 1995, 2001; Şengör and Natal'in, 2001) are sometimes connected to one another by the propagation one or more of their arms, and they thus split a continent. Şengör (1995) and Şengör and Natal'in (2001) discussed diverse examples of such connections. The nature of the mechanism of propagation may be gravitational (Şengör, 2001) or any other cause that stretches the lithosphere (Şengör, 1995, 2001; Şengör and Natal'in, 2001) or fluid injection from below (Bhattacharji and Koide, 1975; Casey et al., 2006). Courtillot (1982) discussed a possible geometry of such propagating tips in continental and oceanic areas; although he called the process he described *propagation*, McKenzie (1986) thought *localization* would be a more appropriate appellation. It is however clear that McKenzie's localization is what makes propagation possible in Courtillot's scheme.

2. In the oceans, triple junction evolution leads to rift propagation or *recession* that some authors also term *extinction* or *death*. Recession leads to the formation of *fossil spreading centres* (e.g., Hey, 1977). Zonenshain et al. (1980, esp. Figure 6) first had the idea of placing the velocity triangle directly on top of the geographical triple junction and showed how this geometric construction allows to see at a glance which ridges must propagate and which must recede.

Patriat and Courtillot (1984) followed up Zonenshain et al.'s (1980) idea and pointed out that if the geographical triple junction lies within its own velocity triangle, then all ridges meeting at it propagate toward the interior of the velocity triangle. Figure 5 illustrates the simplest case of an RRR junction with legs spread at 120°. Figure 5a shows the triple junction amidst plates A, B, and C and its velocity triangle as drawn by McKenzie and Morgan (1969; see Figure 1). Figure 5a' is how Patriat and Courtillot draw it showing that all three ridges, ab, ac, and bc, must propagate. In Figure 5b, the triple junction among the plates A, B, and C is a right triangle. In this case, the ridges *ac* and *bc* propagate, but the ridge ab neither propagates nor recedes. This is an important property of right velocity triangles among three spreading centres. Figure 5c shows that even if we change the triple junction into an RRF junction, it remains stable and again the two previous ridges lengthen. However, in this case the transform fault connecting the ridge *ab* to the triple junction also lengthens without destroying the stability of the triple junction. In the FFR junction depicted in Figure 5d, both ridges *ac* and *ab* neither shorten nor lengthen, but the bc ridge propagates. Although the triple junction shown in Figure 5d has the same isosceles velocity triangle as the triple junction shown in Figure 5e, in 5e all ridges

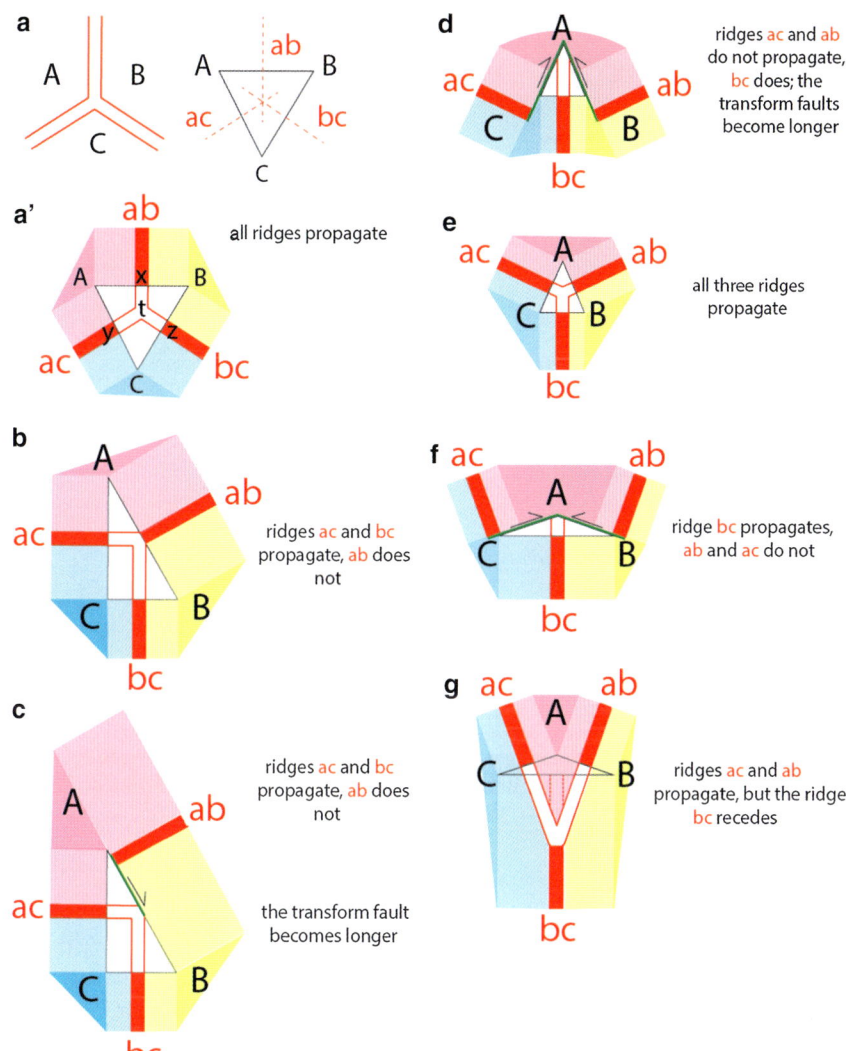

Triple Junctions, Figure 5 (**a**) An ordinary RRR junction and its equilateral velocity triangle. (**a′**) This figure, after Patriat and Courtillot (1984), shows that in this geometry, all ridges must propagate. The white areas in the ridges show the sections that formed as a result of propagation. (**b**) Another RRR junction, this time with a right velocity triangle. Note that in this case only, two ridges propagate and one does not change its length. (**c**) If the third ridge in Figure 5b is connected to the junction by a transform fault, again two ridges propagate as before, but in this case, the transform fault becomes longer. (**d**) In this configuration of an FFR junction, only the *bc* ridge propagates. The two transform faults also become longer. (**e**) The same isosceles velocity triangle with three ridges meeting at its centroid. In this case, all three ridges propagate similar to the case in Figure 5a′. (**f**) The same situation as in Figure 5d but with an obtuse isosceles velocity triangle. (**f**) The same velocity triangle as in Figure 5f, but here the RRR junction lies outside it. As a consequence, the ridges *ac* and *bc* propagate while *bc* recedes. Symbolism as in Figure 1.

propagate as already discussed in the case of the junction depicted in Figure 5a′, where the special case of an equilateral velocity triangle is shown.

In the RRR junction shown in Figure 5f, only the *bc* ridge propagates. Here the triple junction is coincident with the top of its obtuse isosceles velocity triangle. In Figure 5g, the triangle lies outside the velocity triangle. Here ridges *ac* and *bc* propagate, but the ridge *bc* recedes. Where a triple junction involving only ridges and transform faults lies outside its velocity triangle, at least one ridge recedes.

Patriat and Courtillot (1984) discussed some tectonic effects of the propagation and recession of ridges and lengthening and shortening of transform faults around triple junctions. As transform faults lengthen, they thought that friction between plates will increase, and the evolution of the triple junction will consume more energy. This is not necessarily true, because as they lengthen, the transform fault walls also cool and much lubricating material (serpentinite diapirs, magma) invades the thus opened space facilitating slippage (see "Transform Faults"). They also pointed out that triple junction evolution would

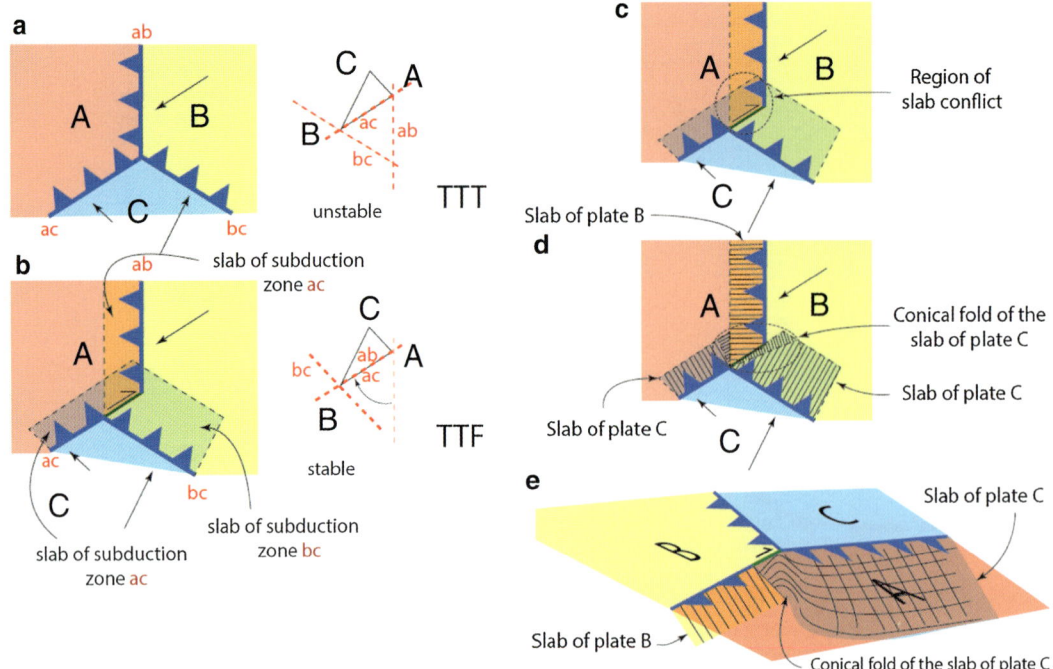

Triple Junctions, Figure 6 (**a**) A TTT junction with its velocity triangle. (**b**) The same triple junction after a finite amount of time has elapsed. Note that the slabs interfere with one another at depth, if they are assumed to be flat, which is unrealistic. (**c**) The *dotted circle* shows the area of slab conflict. (**d**) The *dotted ellipse* shows the volume in which a conical fold must be formed by the slab attached to *plate C*. The slab of *plate B* accommodates itself into that slab. (**e**) 3-D view of the conical fold of the slab of *plate C* and how the slab *of plate B* descends through the opening created by the conical fold of the slab of *plate C*.

influence magma chamber connectivity, which must be true. Some triple junctions generate much magma forming oceanic plateaux atop them (e.g., the Mid-Pacific Mountains).

Triple junctions and the third spatial dimension

Most studies on triple junctions have dealt with their 2-D aspects plus other dimensions like velocity and acceleration spaces. Additional complications in tectonics arise when the third spatial dimension is considered. Figure 6a shows a TTT triple junction amidst plates A, B, and C. The junction connects the subduction zones *ab*, *ac*, and *bc*. Figure 6a shows the initial condition and Figure 6b the situation after the lapse of a finite interval of time. Note that the two subduction zone slabs *ac* and *bc* are attached to the same plate C (Figure 6d). As they bend down to descend into the mantle, a space problem will arise at depth. Moreover, around the triple junction, the descent of the plate B will cause further interference problems among the slabs of the subduction zones *ab*, *ac*, and *bc* (Figure 6c). Interestingly, in this case, the slabs of the plate C will form a conical fold with an axis pointing upward to the triple juction (Figure 6d). The slab of the plate B would have to accomodate itself into the core of this fold as shown in Figure 6e so as not to cause a slab conflict at depth. This geometry might have important implications for the magmatism and the metamorphism of the slab of the plate B, because that slab may receive volatiles liberated from the slab of the subduction zone *bc* that would help further melt generation. The extensive Neogene to Quaternary volcanics of the Fossa Magna in Japan may well be a result of such a situation. Such significant tectonic effects are to be expected in all triple junctions where more than one subduction zone is involved.

Triple junctions and lithospheric "holes"

When triple junctions occur in non-subductable lithospheric material, they lead to compatibility problems and opening of holes in the lithosphere that express themselves geologically as complex rift basins, which Şengör et al. (1985) called "incompatibility basins."

Consider the triple junction among plates A, B, and C depicted in Figure 7a, which is assumed to be placed entirely in non-subductable continental crust. Its features actually closely correspond to the Karlıova triple junction in eastern Turkey (Şengör, 1979; Şengör et al., 1985), and that is why such triple junctions are herein termed *Karlıova-type*. The boundaries between C and the other two plates are strike-slip faults in Figure 6a. The boundary between A and B is a convergent but a nonconsuming type. Here convergence is turned into convergent strain as depicted by the deformation of the circle placed on it in Figure 7a that turns into an ellipse in Figure 7b.

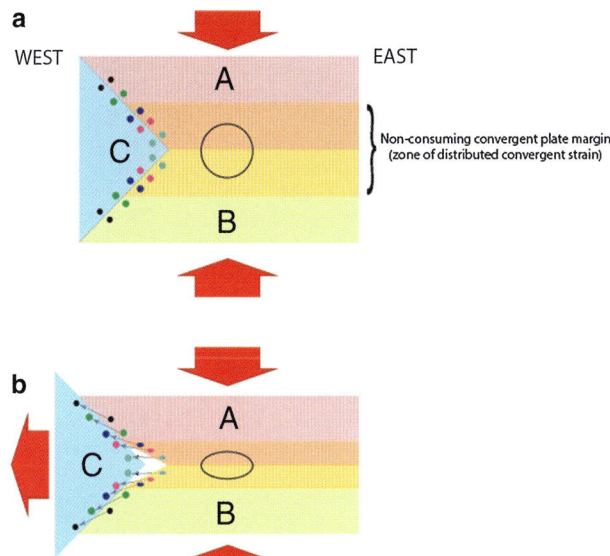

Triple Junctions, Figure 7 A Karlıova-type triple junction. The *coloured circles* are markers showing displacement across plate boundaries.

In Figure 7b, the plate boundary zone between A and B is shortened by 50 %, while plate C remains entirely rigid. Note that a boomerang-shaped hole opens behind the plate C as it is being expelled from the converging jaws of the plates A and B. Within this basin the extension direction changes from E to W in the center of the basin to progressively more northwesterly and southwesterly directions as shown by small arrows in Figure 7b. Correspondingly, the normal faults generated within the basin are expected to have curved map views displaying complex strains.

Another type of incompatibility basin arises at what is herein called the *Maraş-type* triple junctions, after the most prominent and best-studied example around the town of Kahramanmaraş in southern Turkey. Such a junction is shown in Figure 8.

Figure 8a shows the initial situation, where at least plates A and B consist of non-subductable continental material. B and C converge toward A, and this convergence is accommodated by left-lateral strike-slip faulting between A and B. C converges at a slower rate than B, and as a consequence, a hole opens up north of it. In the real case of the Kahramanmaraş triple junction, this basin is the Hatay-Adana-Cilicia basin complex (Şengör et al., 1985). The basin is elongated ENE–WSW, but its direction of opening is almost E–W. By contrast, it shortens in a N–S direction. These directions naturally change as the basin evolves, but here we see all the complexities of a transtensional basin as elaborated by Dewey (2002).

Triple junctions in non-subductable material lead to very great complexities and complex geological histories, but space here fails to elaborate on all.

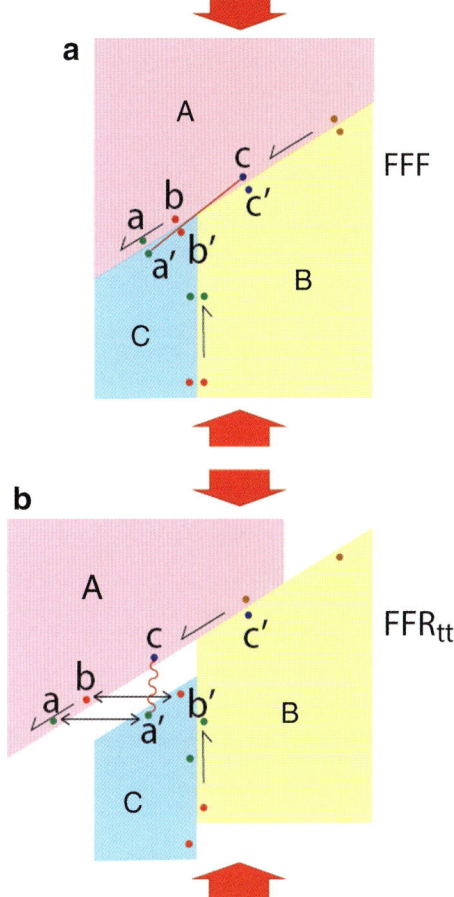

Triple Junctions, Figure 8 A Maraş-type triple junction. The *coloured circles* are markers showing displacement across plate boundaries. The *double-headed arrows* show cords connecting some markers to indicate shortening or extension between the markers thus connected.

Conclusions

Triple junctions are critically important elements in the kinematics of plate tectonics. In real life, they give rise to a very complex geological history wherever they occur, where geological events may abruptly overprint each other. McKenzie and Morgan (1969) said that only stable triple junctions allow continuous plate evolution implying that in plate reconstructions, one should allow only such junctions. Dan McKenzie once said that drawing an unstable triple junction would be like drawing a pin on its head. It can only stay at that unstable position for an instant before falling onto the floor to assume a stable position. This is theoretically true, but not in practice. The triple junction among the Japan, the Nankai, and the Izu-Bonin trenches is an unstable FFF junction, but it has been there for the last 60 Ma at least, albeit with small adjustments to the plate velocities around it. Therefore, unstable triple junctions also have their justification in reconstruction of past tectonics of any given area if the data demand them.

However, any reconstruction that does not honour the kinematic and geological requirements of the triple junctions it gives rise to is wrong from the outset.

Bibliography

Acton, G., Stein, S., and Engeln, J. F., 1988. Formation of curved seafloor fabric by changes in rift propagation velocity and spreading rate: application to the 95.5° W Galapagos propagator. *Journal of Geophysical Research*, **93**, 11845–11861.

Atwater, T., 1981. Propagating rifts in seafloor spreading patterns. *Nature*, **290**, 185–186.

Bhattacharji, S., and Koide, H., 1975. Mechanistic model for triple junction fracture geometry. *Nature*, **255**, 21–24.

Burke, K., and Dewey, J., 1973. Plume-generated triple junctions: key indicators in applying plate tectonics to old rocks. *Journal of Geology*, **81**, 406–433.

Casey, M., Ebinger, C., Keir, D., Gloaguen, R., and Mohamed, F., 2006. Strain accommodation in transitional rifts: extension by magma intrusion and faulting in Ethiopian rift magmatic segments. In Yirgou, G., Ebinger, C. J., and Maguire, P. K. H. (eds.), *The Afar Volcanic Province Within the East African Rift System, Geological Society Special Publication*. Geological Society (London) in London Vol. 259, pp. 143–163.

Courtillot, V., 1982. Propagating rifts and continental breakup. *Tectonics*, **1**, 239–250.

Cronin, V. S., 1992. Types and kinematic stability of triple junctions. *Tectonophysics*, **207**, 287–301.

Dewey, J. F., 2002. Transtension in arcs and orogens. *International Geology Review*, **44**, 402–438.

Hey, R., 1977. A new class of "pseudofaults" and their bearing on plate teconics: a propagating rift model. *Earth and Planetary Science Letters*, **37**, 321–325.

Lotze, F., 1937. Zur Methodik der Forschungen über saxonische Tektonik. *Geotektonische Forschungen*, **1**, 6–27.

McKenzie, D. P., 1982. The evolution of propagating rifts. *Nature*, **300**, 740–741.

McKenzie, D. P., 1986. The geometry of propagating rifts. *Earth and Planetary Science Letters*, **77**, 176–186.

McKenzie, D., and Morgan, W. J., 1969. Evolution of triple junctions. *Nature*, **224**, 125–133.

McKenzie, D., and Parker, R. L., 1967. The North Pacific: an example of tectonics on a sphere. *Nature*, **216**, 1276–1280.

McKenzie, D., and Parker, R. L., 1974. Plate tectonics in ω space. *Earth and Planetary Science Letters*, **22**, 285–293.

McKenzie, D. P., and Weiss, N., 1975. Speculations on the thermal and tectonic history of the Earth. *Geophysical Journal of the Royal Astronomical Society*, **42**, 131–174.

Morgan, W. J., 1968. Rises, trenches, great faults and crustal blocks. *Journal of Geophysical Research*, **73**, 2131–2153.

Müller, R. D., Sdrolias, M., Gaina, C., and Roest, W. R., 2008. Age, spreading rates, and spreading asymmetry of the world's ocean crust. *Geochemistry Geophysics Geosystems G^3*, **9**, doi:10.1029/2007GC001743.

Patriat, P., and Courtillot, V., 1984. On the stability of triple junctions and its relation to episodicity in spreading. *Tectonics*, **3**, 317–332.

Şengör, A. M. C., 1979. The North Anatolian Transform Fault: its age, offset and tectonic significance. *Journal of the Geological Society of London*, **136**, 269–282.

Şengör, A. M. C., 1995. Sedimentation and tectonics of fossil rifts. In Busby, C. J., and Ingersoll, R. V. (eds.), *Tectonics of Sedimentary Basins*. Oxford: Blackwell, pp. 53–117.

Sengor, A. M. C., 2011. Continental rifts. In Gupta, H. (ed.), *Encyclopedia of Solid Earth Sciences*. Dordrecht: Springer, Vol. 1, pp. 41–55.

Şengör, A. M. C., 2001. Elevation as indicator of mantle plume activity. In Ernst, R., and Buchan, K. (eds.), *Geological Society of America Special Paper*. Geological Society of America, Inc in Boulder, Colorado, Vol. 352, pp. 183–225.

Şengör, A. M. C., and Natal'in, B. A., 2001, Rifts of the world. In Ernst, R., and Buchan, K. (eds.), *Geological Society of America Special Paper*. Geological Society of America, Inc in Boulder, Colorado, Vol. 352, pp. 389–482.

Şengör, A. M. C., Görür N., and Şaroğlu, F., 1985. Strike-slip faulting and related basin formation in zones of tectonic escape: Turkey as a case study. In Biddle, K. T., and Christie-Blick, N. (eds.), *Strike-slip Deformation, Basin Formation, and Sedimentation, Society of Economic Paleontologists and Mineralogists Special Publication* (in honor of J.C. Crowell). The Society of Economic Paleontologists and Mineralogists in Tulsa, Oklahoma Vol. 37, pp. 227–264.

Stein, S., Melosh, H. J., and Minster, J. B., 1977. Ridge migration and asymmetric sea-floor spreading. *Earth and Planetary Science Letters*, **36**, 51–62.

Wilson, J. T., 1965, A new class of faults and their bearing on continental drift: Nature, Vol. 207, pp. 343–347.

York, D., 1973a. Evolution of triple junctions. *Canadian Journal of Earth Sciences*, **12**, 516–519.

York, D., 1973b. Evolution of triple junctions. *Nature*, **244**, 341–342.

Zonenshain, L. P., Kogan, L. I., Savostin, L. A., Golmstock, A. Y., and Gorodnitskii, A. M., 1980. Tectonics, crustal structure and evolution of the Galapagos triple junction. *Marine Geology*, **37**, 209–230.

Cross-references

Oceanic Spreading Centers
Plate Tectonics
Regional Marine Geology
Subduction
Transform Faults

TSUNAMIS

Jose C. Borrero
eCoast Ltd., Marine Consulting and Research, Raglan, New Zealand

Definition

A *tsunami* (plural, tsunami or tsunamis) is a series of water waves caused by an impulsive disturbance to a body of water. Tsunamis are most commonly caused in the ocean through the deformation of the seafloor during an earthquake or as a result of a submarine landslide. Additional source mechanisms include volcanic eruptions, subaerial landslides, asteroids falling into a body of water, or meteorological forcing (also referred to as "meteo-tsunami"). The word tsunami is Japanese (Kanji: 津波) and translates literally to "harbor wave," due to the fact that this phenomenon was frequently observed in bays and harbors in Japan. Tsunamis are also referred to as "tidal waves" and "seismic sea waves."

Tsunami generation

Tsunamis can generally be categorized as "near field" or "far field" depending on where the majority of its effects occur. A far-field tsunami (also "tele-tsunami" or "trans-oceanic tsunami") is one which causes damage or serious effects at great distances from the source region while the effects of a near-field or "local" tsunami are constrained roughly to the tsunami source region. This term is somewhat ambiguous in that tsunamis with appreciable effects in the far field also produce significant effects in the near field and thus have both near- and far-field components.

Tectonically generated tsunami

Most destructive tsunamis have been caused by thrust earthquakes along subduction zones such as those existing around the Pacific Rim and the Sunda Arc south and west of Indonesia. Notable recent examples include the Pacific tsunamis of 1960 (Chile), 1964 (Alaska), and 2011 (Japan) and the 2004 Indian Ocean tsunami (north Sumatra). Another common tectonic source mechanism is that of the outer-rise earthquake characterized by normal faulting and subsequent large-scale subsidence in the oceanic tectonic plate seaward of a subduction zone trench; the 1933 Sanriku tsunami and 2009 Samoa (South Pacific) tsunami are examples of this type of generation mechanism. Tsunamis have also been caused through strike-slip faulting where one side of the fault is displaced laterally relative to the other, and there is very little vertical deformation of the seabed. However, these types of tsunami may also be associated with additional source mechanisms such as submarine slumps or landslides which contribute to the overall strength of the tsunami and its impact.

Among the subduction zone thrust-type earthquakes exists a special class of events where the resulting tsunami is much larger than would be expected based on the earthquake magnitude alone. Such events have been termed "tsunami earthquakes" (Kanamori, 1972) and are generally characterized by earthquake rupture occurring on the shallower portion of the subduction zone plate interface, a slower rupture velocity and rate of seismic energy release, and a larger sea floor deformation relative to other earthquakes of similar magnitude. Each of these components contributes to an overall larger tsunami, particularly in the near field. Examples of this type of event include Sanriku, Japan, 1896; Nicaragua, 1992; East Java, Indonesia, 1994; West Java Indonesia, 2005; and the Mentawai Islands, Indonesia, 2010.

Landslide-generated tsunami

Tsunami can also be generated by slope failures – commonly referred to as landslides, slumps, slips, or debris flows. Tsunamigenic slope failures can occur either above or below water (i.e., "subaerial" or "submarine"). Tsunami waves are generated as water is displaced; in the case of a subaerial event, as the material crashes on to the water surface or when submarine, as water above the slide is pulled downward in to the void left as the slide material slides downslope. Tsunamigenesis by slope failure is an active and ongoing area of scientific research. The key parameters controlling the height of the initial wave are the slope of the sea bed, the amount of material displaced, the thickness of the slide, the depth at which the slide occurs, the speed at which it fails, and the distance over which the slide material translates (Bardet et al., 2003). Also critical to the study of landslide-generated tsunami is the ability to define the recurrence intervals of such events. This is particularly true for the large-scale flank collapse of volcanic islands, a rare but potentially catastrophic tsunami generator (Ward, 2002).

Perhaps the most famous and extreme example of a landslide-generated tsunami is that which occurred in Lituya Bay in southeastern Alaska on July 9, 1958. In this event, an on-land earthquake along the Fairweather Fault induced a subaerial landslide that displaced some 30 million cubic meters of material which collapsed into the head of Lituya Bay displacing a similar amount of water. The subsequent wave reached a height of 524 m above sea-level proximal to the landslide with wave heights tapering to ~10 m at the mouth of the bay (Miller, 1960). A submarine landslide is believed to be the primary cause of the 1998 Papua New Guinea tsunami which killed more than 2100 people (Synolakis et al., 2002).

Volcanic and other source mechanisms

Tsunamis can occur as the result of a volcanic eruption if the eruptions causes a caldera collapse and results in the displacement of a large volume of water. Additionally, volcanic eruptions and associated earthquakes can produce landslides or other subaerial mass movements which then fall into a body of water causing a wave. The 1883 eruption of Krakatoa in the Sunda Strait between the Islands of Sumatra and Java is the most significant contemporary example of this phenomenon (Winchester, 2005).

Other causes of tsunami or tsunami-like waves include meteorological forcing, i.e., "meteo-tsunami," most commonly through the translation of an area of low atmospheric pressure across a body of water. Since low atmospheric pressure causes a displacement of the water surface, if the translation speed of the pressure anomaly matches the local phase speed of the induced water wave, resonant coupling results in an exponential increase in the water wave height (see Dean and Dalrymple, 1991, p. xx). An example of this effect includes the tsunami-like wave that hit Daytona Beach, Florida, on the night of July 3, 1992 (Sallenger et al., 1995). A final (perhaps literally!) tsunami generation mechanism is that caused by an asteroid impact upon an ocean basin. While no such event is known to have occurred in human history, attempts have been made to quantify the uncertainty of such an occurrence (Chapman and Morrison 1994) and the size of the waves it might generate (Ward and Asphaug, 2000).

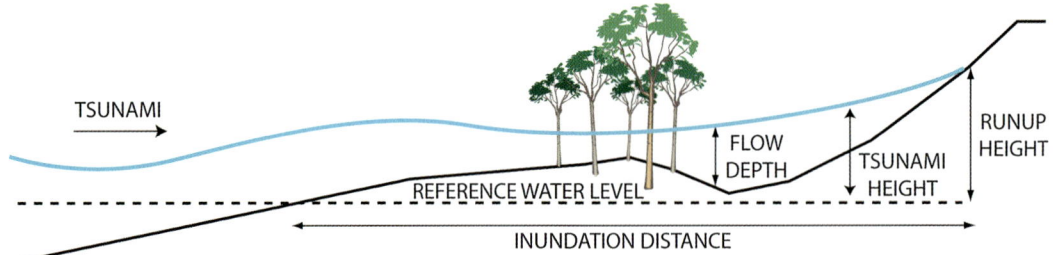

Tsunamis, Figure 1 Definition sketch for tsunami inundation and run-up.

Tsunami propagation, inundation, and run-up

Due to the fact that the wavelength (the distance between successive peaks in a series of waves) of most tectonically generated tsunami is usually much greater than the average depth of the ocean basins, they propagate as shallow water waves. An important consequence of this is that the speed of propagation is governed only by the depth of the water through the relationship:

$$c = (gd)^{1/2}$$

where c is the wave speed, g the acceleration due to gravity (9.8 m/s^2), and d the depth of the water. Taking d as 5,000 m (the average depth of an oceanic basin) yields a tsunami propagation speed of 221 m/s (797 km/h, 495 mi/h). This simple calculation gives rise to the popular maxim that "tsunamis travel at the same speed as a jet airliner." However, as the water depth decreases, so too does the propagation speed, and as a tsunami wave front approaches dry land, it has slowed down considerably relative to its propagation speed in the deep ocean.

It is this slowing of the wave front in shallow water that ultimately leads to the destructive inundation of tsunami waves. Due to their very long wavelengths, the speed of the leading edge of a tsunami wave will approach zero as following sections of the wave are still traveling at full speed. This results in the "train wreck" effect where the advancing wave piles up behind the slower moving wave front. Due to conservation of mass, the water contained in the wave train must increase in height ultimately flooding a coastal area. Depending on the specifics of the tsunami itself and the receiving environment, this flooding can manifest as a rapidly rising water level, similar to a tide (hence the oft-misused moniker "tidal wave"), or it can result in violently breaking waves and destructive high-speed inundation. The terms for tsunami inundation and run-up are defined in Figure 1.

Recent advances in tsunami science

Since the early 1990s there have been significant advances in many aspects of tsunami science and hazard mitigation. This is partially attributable to two tsunamis occurring in 1992. The first, caused by the April 25 Cape Mendocino earthquake at the southern end of the Cascadia subduction zone (CSZ) in Northern California, generated only a small, non-damaging tsunami (~1 m) that affected the coast immediately after. This event was important in that it served as a reminder of the potential for locally generated, near-field tsunamis occurring along the US West Coast and as a reminder of the seismicity of the CSZ, recently identified as a source of very large earthquakes and subsequent tsunamis (Atwater, 1987). This was followed on September 2, 1992 by a M ~7.2 earthquake off the Pacific Coast of Nicaragua. Although the earthquake was relatively small (roughly the same surface wave magnitude of the Cape Mendocino event), it produced a highly localized tsunami with wave heights of up to ~10 m and killed more than 100 people. It was following the Nicaragua tsunami that the modern era of systematic post-tsunami reconnaissance surveys began. This trend has continued since then and has led to a more nuanced and complete understanding of tsunami effects in both the near and far field (Synolakis and Okal, 2005). Around the same time, significant advances in computer hardware made it possible for the first time to begin to accurately simulate the details of tsunami propagation and inundation, and the mid-1990s saw the development of several numerical models designed specifically for this task. In the USA, efforts toward tsunami hazard mitigation also increased in this time frame. In 1997 the US Congress created the National Tsunami Hazard Mitigation Program (NTHMP) which directed the National Oceanic and Atmospheric Administration (NOAA) to develop and lead a program aimed at improving tsunami preparedness in the USA through hazard assessment, warning guidance, and mitigation (Bernard, 1998).

Tsunami deposits and paleo-tsunami studies

The study of tsunami sedimentology has advanced considerably in the past 25 years. Since Atwater's (1987) association between the depositional sand layers on the Pacific Coast of Washington State and the occurrence of great earthquakes and associated tsunamis, more and more researchers have integrated sedimentological analyses into post-tsunami field survey efforts. The resulting database has expanded the understanding of the depositional mechanisms responsible for tsunami deposits and assisted in the identification of tsunami deposits in other parts of the world. Recognizing and identifying tsunami deposits is an important part of paleo-seismological studies since deposits can be used to determine the magnitude and extent of prehistoric tsunamis or to verify the occurrence of such an event. In the early 2000s deposits resulting from an earthquake and tsunami in 869 AD were identified on the Sendai plain in Japan (Minoura et al., 2001) and suggested that this area was susceptible to large-scale tsunami inundation. This unfortunate reality was confirmed 10 years later as the 2011 Tohoku earthquake created a tsunami, which inundated the same area.

Tsunami in ports and harbors

The tsunamis of 2004 (Sumatra), 2010 (Chile), and 2011 (Japan) affected ports and maritime infrastructure throughout the Indian and Pacific Oceans. In the 40+ years since the last major transoceanic tsunami (the 1960 Chile and 1964 Alaska events), there has been a significant increase in shipping traffic, commerce, and related port development which has drastically increased the risk from tsunami exposure to the maritime sector. The tsunami events mentioned above were notable in that the damaging effects of tsunami-induced currents and the extended duration of tsunami activity were widely reported and quantified in terms of their effect on ports (Lynett et al., 2012; Borrero and Greer, 2012). These events also illustrated how tsunami-induced currents can be damaging and economically disruptive without causing any overland flooding or inundation. This fact suggests that previous hazard mitigation efforts which focused primarily on inundation extent are not sufficient for application to ports and harbors and the activities conducted therein.

Tsunami warning systems

Tsunami warning systems (TWS) are designed to detect tsunamis and issue appropriate warnings. The current TWS are administered through a system of intergovernmental agreements and treaties defining areas of responsibility and operational protocols. In the Pacific Ocean, the Pacific Tsunami Warning Center (PTWC), located in Hawaii, monitors global seismic activity and provides warning for the nations of the Pacific Rim. After the 2004 Indian Ocean tsunami, additional warning systems have been established for the Indian Ocean. Tsunami warning systems are under development in the North Atlantic/Mediterranean and Caribbean Seas.

Existing TWS are comprised of two components: (1) networks of sensors for detecting earthquakes and the tsunami itself and (2) communications infrastructure for the dissemination of the warning messages. Earthquakes detected on seismometers located around the world are used to initiate tsunami watches and warnings, while ocean-based sensors such as the DART® (Deep-ocean Assessment and Reporting of Tsunamis) buoys can directly verify the existence and size of a tsunami (Bernard and Meinig, 2011). The DART tsunameters were first deployed in 1995 and only became operational since the early 2000s. Prior to this, tsunami warnings were based only on the characteristics of earthquakes and monitoring coastal tide gauge networks. Warning messages are disseminated to the public through a variety of communications channels including radio, fax, telex, SMS, and e-mail. Dedicated communication links have been established for first responders and emergency managers.

Tsunamis, Table 1 Tsunami events worldwide since 1992

Date	Location
April 25, 1992	Cape Mendocino, California
September 2, 1992	El Popoyo, Nicaragua
December 12, 1992	Flores Island, Indonesia
July 12, 1993	Okushiri Island, Hokkaido, Japan
June 2, 1994	East Java, Indonesia
October 9, 1994	Shikotan, Kuril Islands, Russia
November 14, 1994	Mindoro, Philippines
October 9, 1995	Manzanillo, Mexico
February 17, 1996	Biak, Indonesia
February 21, 1996	Chimbote, Peru
July 17, 1998	Sissano, Papua New Guinea
August 17, 1999	Izmit, Turkey
November 26, 1999	Fatu Hiva, Marquesas Islands
November 26, 1999	Vanuatu
June 23, 2001	Camaná, Peru
September 8, 2002	Wewak, Papua New Guinea
December 26, 2004	North Sumatra, Indonesia (great Indian Ocean tsunami)
March 28, 2005	Nias–Simeulue, Indonesia
July 17, 2006	West Java, Indonesia
November 15, 2006	Kuril Islands, Russia
April 1, 2007	Ghizo, Solomon Islands
August 15, 2007	Southern Peru
September 12, 2007	Bengkulu, West Sumatra, Indonesia
September 29, 2009	Samoa (South Pacific tsunami)
February 27, 2010	Maule, Chile; Mentawai Islands, Indonesia
October 25, 2010	Mentawai Islands, Sumatra, Indonesia
March 11, 2011	Tohoku, Japan tsunami
August 27, 2012	Jiquilisco, El Salvador
February 6, 2013	Nendo, Solomon Islands

Summary

While significant advances have been made in tsunami science and hazard mitigation over recent decades, the work is never done. Tsunamis are not as rare an occurrence globally as we may think; in the past 20 years, there have been at least 29 tsunami events responsible for more than 300,000 deaths worldwide (Table 1). However, on regional or local scales, tsunamis tend to occur only once a generation or less. Furthermore, as Satake and Atwater (2007) recognized, the "great" tsunamis in human history do not conform to expected geological rules of thumb. This unfortunate reality was illustrated again in 2011 with the great Tohoku earthquake and tsunami, which, like the 2004 Indian Ocean event, was not expected along that plate boundary based on historical precedent. If we accept for a moment that hard science might not provide the answers we need in terms of predicting future events, perhaps for the tsunami problem we should not underestimate the importance of enhancing and expanding efforts in public education to maintain awareness of tsunami hazards over generational timescales.

Bibliography

Atwater, B. F., 1987. Evidence for great holocene earthquakes along the outer coast of Washington State. *Science*, **236**(942), 944.

Bardet, J.-P., Synolakis, C.E., Davies, H.L., Imamura, F., and Okal, E. A. (eds.), 2003. Landslide tsunamis: recent findings and research directions, *Pure and Applied Geophysics*, Topical Volume, ISBN:978-3-7643-6033-7.

Bernard, E. N., 1998. Program aims to reduce impact of tsunamis on Pacific states. *Eos, Transactions of the American Geophysical Union*, **79**(22), 258, 262–263, doi:10.1029/98EO00191.

Bernard, E., and Meinig, C., 2011. History and future of deep-ocean tsunami measurements. In *Proceedings of Oceans' 11*. MTS/IEEE, No. 6106894, September 19–22, 2011, p. 7.

Borrero, J. C., and Greer, S. D., 2012. Comparison of the 2010 Chile and 2010 Japan tsunamis in the Far-field. *Pure and Applied Geophysics*, doi:10.1007/s00024-012-0559-4.

Chapman, C. R., and Morrison, D., 1994. Impacts on the Earth by asteroids and comets: assessing the hazard. *Nature*, **367**, 33–40.

Dean, R. G., and Dalrymple, R. A., 1991. *Water Wave Mechanics for Scientists and Engineers*. Singapore: World Scientific.

Kanamori, H., 1972. Mechanism of Tsunami Earthquakes. *Physics of the Earth and Planetary Interiors*, **6**, 346–359.

Lynett, P., Borrero, J., Weiss, R., Son, S., Greer, D., and Renteria, W., 2012. Observations and Modeling of Tsunami-Induced Currents in Ports and Harbors. *Earth and Planetary Science Letters*, **327–328**, 68–74.

Miller, D.J., 1960. *Giant Waves in Lituya Bay Alaska*. U.S. Geological Survey Professional Paper 354-c.

Minoura, K., Imamura, F., Sugawara, D., Kono, Y., and Iwashita, T., 2001. The 869 Jogan tsunami deposit and recurrence interval of large-scale tsunami on the Pacific coast of northeast Japan. *Journal of Natural Disaster Science*, **23**(2), 83–88.

Sallenger, A. H., Jr., List, J. H., Gelfenbaum, G., Stumpf, R. P., and Hansen, M., 1995. Large wave at Daytona Beach, Florida, explained as a squall-line surge. *Journal of Coastal Research*, **11**, 1383–1388.

Satake, K., and Atwater, B., 2007. Long-term perspectives on giant earthquakes and tsunamis at subduction zones. *Annual Reviews of Earth and Planetary Science*, **35**, 349–374.

Synolakis, C. E., and Okal, E. A., 2005. 1992–2002: perspective on a decade of post-tsunami surveys. In Satake, K. (ed.), *Tsunamis: case studies and recent developments*. Dordrecht/New York: Springer. Advances in Natural and Technological Hazards Research, Vol. 23, pp. 1–30.

Synolakis, C.E., Bardet, J.-P., Borrero, J.C., Davies, H.L., Okal, E. A., Silver, E., Sweet, S., and Tappin, D.R., 2002. The slump origin of the 1998 Papua New Guinea tsunami. In *Proceedings of the Royal Society*, London, Ser. A, Vol. 458, pp. 763–789.

Ward, S. N., 2002. Slip-sliding away. *Nature*, **415**, 973–974.

Ward, S. N., and Asphaug, E., 2000. Asteroid impact tsunami: a probabilistic hazard assessment. *Icarus*, **145**, 64–78.

Winchester, S., 2005. *Krakatoa: The Day the World Exploded: August 27, 1883*. New York: Harper Collins.

Cross-references

Beach Processes
Earthquakes
Geohazards: Coastal Disasters
Intraoceanic Subduction Zone
Marine Impacts and Their Consequences
Waves

TURBIDITES

Thierry Mulder[1] and Heiko Hüneke[2]
[1]University of Bordeaux, Talence, France
[2]Institute of Geography and Geology, Ernst Moritz Arndt University, Greifswald, Germany

Synonyms

Turbidity-current deposit; turbidity-flow deposit (Bouma sequence)

Definition

A turbidite is a sedimentary bed deposited by a turbidity current or turbidity flow. It is composed of layered particles that grade upward from coarser to finer sizes and ideally display a (complete or incomplete) Bouma sequence (Bouma, 1962). Mud-dominated turbidites (fine-grained turbidites) may show the sequence detailed by Stow and Shanmugam (1980). The etymology of the term "turbidity current" relates to a turbid flow, which means a flow containing suspended particles.

In a much looser sense, the term turbidite is also used for deposits of other types of density flows that develop a non-cohesive (frictional/granular) behavior. These include deposits of hyperconcentrated flows and concentrated flows displaying divisions of the Lowe sequence (Lowe, 1982). Sensu stricto, a turbidite is the deposit resulting from a turbulent flow.

Deep-sea turbidite systems along ocean margins and, in particular, submarine fans in front of major river mouths host thick successions of turbidites as a major integral component. In addition, individual turbidites cover abyssal plains beyond the limits of turbidite systems or may occur in more shallow marine basins and even continental lakes, wherever gravity processes transport suspended sediment sustained mainly by fluid turbulence (Middleton and Hampton, 1973).

Turbidity currents and related density flows

There are various perspectives appropriate to classify submarine gravity processes as summarized by Mulder (2011). The predominant particle-support mechanism is the most commonly used criterion to classify gravity processes in present-day deep-sea environments (Lowe, 1979; Middleton and Hampton, 1973; Nardin et al., 1979). Four types of particle-support mechanism are usually distinguished: matrix cohesive strength, grain-to-grain collisional interactions, fluid support (excess pore fluid pressure, upward escape of pore water), and turbulent suspension.

The flow processes are best classified employing the rheology (mechanical behavior) of the density flow (Dott, 1963; Mulder and Cochonat, 1996; Shanmugam, 2000). This type of classification is also useful to choose an appropriate numerical model.

Classification relying on flow concentration includes characteristics that are derived from the observation of deposits (Mulder and Alexander, 2001). It principally takes into account the flow transformation with space and time and its progressive dilution. Because of fluid entrainment, flows may transform from hyperconcentrated to concentrated flow and finally to a turbidity flow.

Finally, much of our knowledge of turbidity-current deposition is inferred from extensive studies of their deposits (Mutti and Ricci Lucchi, 1975; Pickering et al., 1989). This approach is extensively used in research on ancient environments and in the oil industry. It is based on the classification of sedimentary facies and its change along the pathway of the flow.

A turbidity current represents a "particulate density flow." Widely accepted definitions of turbidity currents (Lowe, 1982; Middleton and Hampton, 1973; Sanders, 1965) state that they are sediment gravity flows in which the sediment is supported mainly by the upward component of fluid turbulence. Within such turbulent flows, additional particle-support mechanisms may act near the bed (turbidity current sensu lato). Mulder and Alexander (2001) used the Bagnold limit for full turbulent particle support as a threshold value to define turbidity currents sensu stricto (9 % sediment concentration by volume; Bagnold, 1962).

Based on flow duration and uniformity (Lüthi, 1980), Mulder and Alexander (2001) differentiate between (1) instantaneous surges (lasting over a period of hours to days), (2) longer-duration surge-like flows, and (3) quasi-steady currents (persisting over a period of weeks to months). The distinction between continuously moving currents and shorter-duration flows is important for understanding the nature of its deposits.

Surges are very short-lived flow events that have the form of a traveling, isolated flow head. Surge-like flows are also short lived but comprise a short flow body following the head. Probably, such turbidity flows most often result from flow differentiation through dilution (water entrainment), erosion, and acceleration from a density-flow type with higher sediment concentration (e.g., debris flows, concentrated density flows) created by failures on continental slopes and canyon walls. Quasi-steady turbidity currents typically form where a river discharges into a basin with the bulk density of the effluent (sediment and water) being greater than that of the receiving ambient water, and thus flow continues from the river onto the floor of the receiving shelf and into the deep sea (Mulder and Syvitski, 1995). Such (suspended-load) hyperpycnal turbidity currents are termed quasi-steady, as the flow is supplied by prolonged river flow. These flood-related flows frequently occur in rapidly subsiding basins where steep slopes are present (Mutti et al., 1996).

The flow steadiness controls the organization of the resulting deposits (Kneller and Branney, 1995). A surge-like turbidity flow tends to produce classical Bouma sequences, whose detailed characteristics at any one site depend on factors such as flow size, sediment type, and proximity to source. In contrast, a quasi-steady turbidity current, generated by hyperpycnal river effluent, can deposit a negatively graded division capped by positively graded divisions (because of waxing and waning conditions, respectively) and may also include a thick division of uniform character (resulting from prolonged periods of near-steady conditions (Kneller and Branney, 1995).

Any gravity flow may progressively change its character both with distance from the source of the flow and with time at any one point (Fisher, 1983; Shanmugam, 2000). Typically, a slump-generated hyperconcentrated density flow or debris flow is transformed along the transport path into concentrated density flow and finally into turbulent flow. These changed are controlled by parameters related to the flow behavior (internal parameters) such as the rate of fluid entrainment and fine-particle elutriation and by parameters related to the depositional environment such as gradient of the continental slope, lateral confinement by channel levees, and bed roughness.

Flow transformation from a homogeneous sediment failure generates rapidly stratified flows with a strong vertical velocity gradient. During most of its pathway, such a gravity flow remains bipartite, in which the basal part is hyperconcentrated with a laminar regime (or concentrated) and the upper part is more dilute where turbulence progressively develops (Sanders, 1965). A fining-upward vertical grain-size distribution progressively forms only in the most diluted uppermost part of the flow, while the basal part stays ungraded or displays a coarsening-upward grain-size distribution. Successive sediment deposition by such a bipartite flow typically produces a Lowe sequence or parts of it (Lowe, 1982; Figure 1).

Deep-sea turbidite systems

Turbidites are the most common and archetypal deposits forming deep-marine clastic systems at the base of a slope along continental margins – together with related gravity-flow deposits and hemipelagic and pelagic sediments. Reading and Richards (1994) provided a "matrix-shaped" environmental classification of such turbidite systems based essentially on the feeder system and grain size. The nature of the supply system (point-source submarine fan, multiple-source ramp, or linear-source slope apron) regulates the feeder channel stability, the organization of the depositional sequences, and the length/width ratio of the depositional system. Grain size and volume of the sediment supplied (mud rich, mud/sand rich, sand rich, or gravel rich) influence on the slope gradient, the persistence and frequency of flows, the pattern of morphological elements, and the tendency for channels to migrate. Altogether, 12 classes are recognized in depositional systems of deepwater basin margins (Reading and Richards, 1994). The character of these systems reflects a complex interplay between a range of autocyclic and allocyclic controls including sea-level fluctuations, basinal tectonics, and the rate, type, and nature of sediment supply.

Mutti (1985) proposed the efficiency concept, to differentiate between low-efficiency and high-efficiency fan

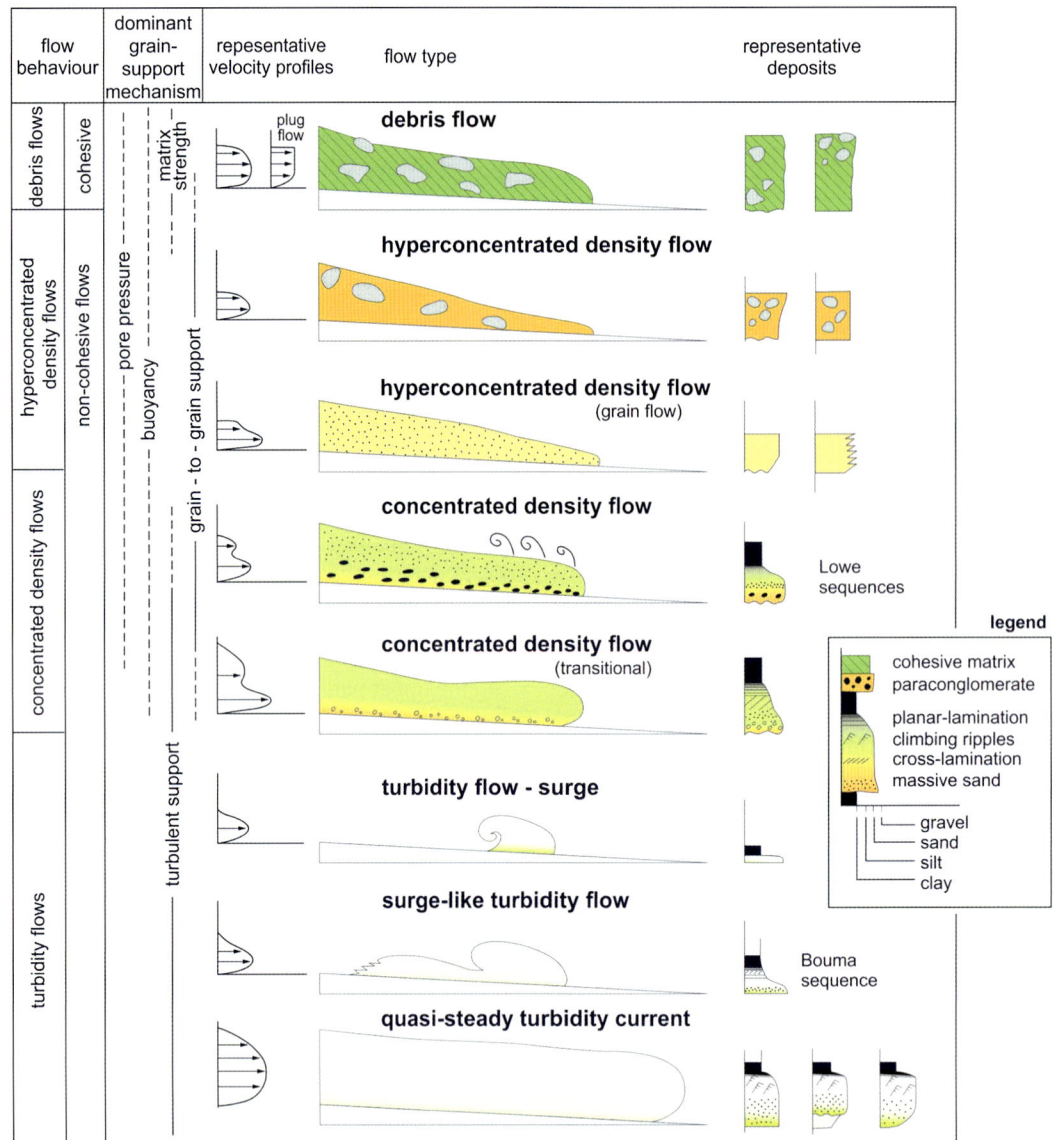

Turbidites, Figure 1 Diagram illustrating the physical character of subaqueous sedimentary density flows and their deposits (modified from Mulder and Alexander, 2001). Colors indicate sediment composition: *green,* diamictic sediment composition with high proportions of cohesive material (mud); *orange,* sediments with low amounts of cohesive particles; *pastel orange,* predominantly sand-sized sediment composition; *yellow to light yellow,* predominantly sand- to silt-sized sediment composition, displaying vertical trends in grain size and sorting during flow (including variable amounts of cohesive particles) and within the deposit.

systems. High-efficiency fan systems (mud-rich systems) are those where gravity flows are capable of transporting sand over long distances. Low-efficiency fan systems (sand-rich systems), in contrast, are those where gravity flows are not capable of transporting sand over long distances.

Architectural elements of deep-sea turbidite systems

Turbidite systems are best described and interpreted in terms of its architectural elements (Miall, 1985; Pickering et al., 1995; Stow and Mayall, 2000). Firstly defined to describe river systems, architectural elements are the elementary component parts of clastic depositional system which can be delineated by a set of hierarchically grouped bounding surfaces and are classified by its facies (sedimentary, acoustic, and seismic, respectively) and its architectural geometry (3D geometry and orientation). Once these architectural elements are identified, their character and interrelation allow the interpretation of sedimentary processes that formed them. It may help unravel complex depositional histories and understand the growth stages within turbidite systems (Clark and Pickering, 1996).

Turbidites, Figure 2 Turbidite beds of the Carboniferous age at Crackington Haven, Cornwall, United Kingdom. (**a**) Individual beds have sharp basal contacts, are tabular, and can be traced for long distances. Note that the succession is tectonically overturned. Younging direction is to the lower right. (**b**) A complete Bouma sequence is preserved within some of the positively graded sand-mud couplets.

Architectural elements are commonly termed the large-scale building blocks of a "do-it-yourself" depositional system. Both erosional and depositional processes control their formation. Architectural elements of turbidite system typically include canyons, channels and gullies, channel levees and overbank deposits, and various types of lobes, CLTZ (channel-levee transition zone; Wynn et al., 2002), mass-transport complexes, or erosional slide and slump scours (see Mulder, 2011; Stow and Mayall, 2000).

The concept of (architectural) elements in turbidite systems was introduced by Mutti and Normark (1987). These authors established a spatial and temporal hierarchy of five nested orders (turbidite complex, turbidite system, turbidite stage, turbidite substage, and turbidite bed) that allowed to connect between observations at ancient systems in outcrop scale (analysis of sedimentary facies, facies assemblages, and high-resolution stratigraphic correlation) and observations at recent systems by means of geophysical investigations (seismic and acoustic data).

Most turbidite systems in modern oceans are point-source submarine fans fed by a major delta and controlled by mud-dominated supply (Reading and Richards, 1994) although during a sea-level highstand, a substantial number of turbidite systems are fed by coastal drift. Their range extends from 250 km for the Astoria fan, through larger ones like the Mississippi fan (540 km), the Zaire (>800 km), and the Indus fan (1,500 km), to 2800 km for the giant Bengal fan. These turbidite systems display a similar organization into three parts: (1) a canyon, (2) an erosional and then depositional channel bordered with levees becoming progressively thinner and finer grained, and (3) a distal lobe. The Amazon turbidite fan, which has a length of more than 700 km, is a well-documented example (Flood and Damuth, 1987; Lopez, 2001; Figure 2).

Conclusions

The present review shows that turbidite systems cannot be described simply by a few number of parameters, and it is illusory to classify them according to a few number end-member models. However, most of the data, ideas, and concepts come from a system fed by a siliciclastic source.

Morphology, geometry, architecture, nature of deposits, and transport sedimentary process depend on a large number of parameters acting at different space and time scale: margin and basin type, source of sediment (size and elevation of the drainage basin on continent, nature of its substratum, type of climate), mean slope of the submarine basin, subsidence rate, exhumation, or isostasy rate on the continent. At short time scale, other parameters may be of importance to understand the triggering and deposition of individual gravity-flow beds: structural context and neotectonics (earthquake frequency and magnitude, seepage of fluids), eustatic changes (potential sediment storage on the continent, shelf processes, local bathymetry changes), and climatic changes (erosion, dissolution process and CCD (carbonate compensation depth) variations, vegetal cover on the continent, flood rate and magnitude).

Some of these processes are cyclic (eustatic changes, climate); some are not (earthquakes). A global understanding necessitates a complete reconstruction of paleoenvironment and efficient tools to obtain a proper stratigraphic frame in deep-sea clastic systems.

Bibliography

Bagnold, R. A., 1962. Auto-suspension of transported sediment: turbidity currents. *Proceedings. Royal Society of London*, **A265**, 315–319.

Bouma, A. H., 1962. *Sedimentology of Some Flysch deposits. A Graphic Approach to Facies Interpretation*. Amsterdam: Elsevier.

Clark, J. D., and Pickering, K. T., 1996. *Submarine Channels: Processes and Architecture*. London: Vallis Press.

Dott, R. H., 1963. Dynamics of subaqueous gravity depositional processes. *American Association of Petroleum Geologists Bulletin*, **47**, 104–128.

Fisher, R. V., 1983. Flow transformations in sediment gravity flows. *Geology*, **11**, 273–274.

Flood, R. D., and Damuth, J. E., 1987. Quantitative characteristics of sinuous distributary channels on the Amazon deep-sea fan. *Geological Society of America Bulletin*, **98**, 728–738.

Kneller, B. C., and Branney, M. J., 1995. Sustained high-density turbidity currents and the deposition of thick massive beds. *Sedimentology*, **42**, 607–616.

Lopez, M., 2001. Architecture and depositional pattern of the Quaternary deep-sea fan of the Amazon. *Marine and Petroleum Geology*, **18**(4), 479–486.

Lowe, D. R., 1979. Sediment gravity flows: their classification and some problems of application to natural flows and deposits. In Doyle, N. J., and Pilkey, O. H. (eds.), *Geology of Continental Slopes*. Tulsa: Society of Economic Paleontologists and Mineralogists. SEPM Special Publication, Vol. 27, pp. 75–82.

Lowe, D. R., 1982. Sediment gravity flows: II. Depositional models with special reference to the deposits of high-density turbidity currents. *Journal of Sedimentary Petrology*, **52**, 279–297.

Lüthi, S., 1980. Some new aspects of two-dimensional turbidity currents. *Sedimentology*, **28**, 97–105.

Miall, A. D., 1985. Architectural-element analysis: a new method of facies analysis applied to fluvial deposits. *Earth-Science Reviews*, **22**, 261–308.

Middleton, G.V., and Hampton, M.A., 1973. Sediment gravity flows: mechanics of flow and deposition. In Middleton, G.V., and Bouma, A.H. (eds.), *Turbidity and Deep Water Sedimentation*. SEPM, Pacific Section, Short Course Lecture Notes, Anaheim, pp. 1–38.

Mulder, T., 2011. Gravity processes on continental slope, rise and abyssal plains. In Hüneke, H., and Mulder, T. (eds.), *Deep-Sea Sediments*. Amsterdam: Elsevier, pp. 25–148.

Mulder, T., and Alexander, J., 2001. The physical character of sedimentary density currents and their deposits. *Sedimentology*, **48**, 269–299.

Mulder, T., and Cochonat, P., 1996. Classification of offshore mass movements. *Journal of Sedimentary Research*, **66**, 43–57.

Mulder, T., and Syvitski, J. P. M., 1995. Turbidity currents generated at river mouths during exceptional discharges to the world oceans. *Journal of Geology*, **103**, 285–299.

Mutti, E., 1985. Turbidite systems and their relations to depositional sequences. In Zuffa, G. (ed.), *Provenance of Arenites*. Dordrecht: D. Reidel Publishing Co, pp. 65–93.

Mutti, E., and Normark, W. R., 1987. Comparing examples of modern and ancient turbidite systems: problems and concepts. In Legget, J. K., and Zuffa, G. G. (eds.), *Marine Clastic Sedimentology*. London: Graham and Trotman, pp. 1–38.

Mutti, E., and Ricci Lucchi, F., 1975. Turbidite facies and facies association. In Mutti, E., Parea, G.C., Ricci Lucchi, F., Sagri, M., Zanzucchi, G., Ghibaudo, G., and Laccarino, S. (eds.), *Examples of Turbidite Facies and Facies Association from Selected Formations of the Northern Apennines*, A-11 of Fieldtrip Guidebook: Nice, France, Ninth International Congress Sedimentology, pp. 21–36.

Mutti, E., Davoli, G., Tinterri, R., and Zavala, C., 1996. The importance of ancient fluviodeltaic systems dominated by catastrophic flooding in tectonically active basins. *Sciences Géologiques Mémoires (Strasbourg)*, **48**, 233–291.

Nardin, T. R., Hein, F. J., Gorsline, D. S., and Edwards, B. D., 1979. A review of mass movement processes, sediment and acoustic characteristics, and contrasts in slope and base-of-slope systems versus canyon-fan-basin floor systems. In Doyle, L. J., and Pilkey, O. H. (eds.), *Geology of Continental Slopes*. Tulsa: SEPM. Special Publication, Vol. 27, pp. 61–73.

Pickering, K. T., Hiscott, R. N., and Hein, F. J., 1989. *Deep Marine Environments: Clastic Sedimentation and Tectonics*. London: Unwin Hyman.

Pickering, K. T., Clark, J. D., Smith, R. D. A., Hiscott, R. N., Ricci Lucchi, F., and Kenyon, N. H., 1995. Architectural element analysis of turbidite systems, and selected topical problems for sand-prone deep-water systems. In Pickering, K. T., Hiscott, R. N., Kenyon, N. H., Lucchi, R., and Smith, R. D. A. (eds.), *Atlas of Deep-Water Environments; Architectural Style in Turbidite Systems*. London: Chapman and Hall, pp. 1–10.

Reading, H. G., and Richard, M. T., 1994. The classification of deep-water siliciclastic depositional systems by grain size and feeder systems. *American Association of Petroleum Geologists Bulletin*, **78**, 792–822.

Sanders, J. E., 1965. Primary sedimentary structures formed by turbidity currents and related resedimentation mechanisms. In Middleton, G. V. (ed.), *Primary Sedimentary Structures and their Hydrodynamic Interpretation*. Tulsa: SEPM. Special Publication, Vol. 12, pp. 192–219.

Shanmugam, G., 2000. 50 years of the turbidite paradigm, (1950s–1990s): deep-water processes and facies models – a critical perspective. *Marine and Petroleum Geology*, **17**, 285–342.

Stow, D. A. V., and Mayall, M., 2000. Deep-water sedimentary systems: new models for the 21st century. *Marine and Petroleum Geology*, **17**, 125–135.

Stow, D. A. V., and Shanmugam, G., 1980. Sequence of structures in fine-grained turbidites: comparison of recent deep-sea and ancient flysch sediments. *Sedimentary Geology*, **25**, 23–42.

Wynn, R. B., Kenyon, N. H., Masson, D. G., Stow, D. A. V., and Weaver, P. P. E., 2002. Characterization and recognition of deep-water channel-lobe transition zones. *American Association of Petroleum Geologists Bulletin*, **86**, 1441–1446.

Cross-references

Bouma Sequence
Continental Rise
Continental Slope
Deep-sea Fans
Deep-sea Sediments
Submarine Canyons

U

UNDERWATER ARCHAEOLOGY

Geoffrey N. Bailey
Department of Archaeology, University of York, York, UK

Synonyms

Continental shelf archaeology; maritime archaeology; nautical archaeology; shipwreck archaeology; submerged landscape archaeology

Definition

Underwater archaeology is the systematic recovery, study, and interpretation of the material remains of human activity that are now located below present sea-level on or under the seabed or below water in inland water bodies such as lakes and rivers.

Introduction

Underwater archaeology covers a wide range of objectives and interests, which fall into two main categories. The first, and the one most commonly recognized, is the study of shipwrecks and other remains of maritime activity, such as harbor walls and port facilities partially or wholly submerged by minor changes of relative sea-level. Here the focus is on recent millennia, from the expansion of seafaring activity in the Bronze Age as much as 5,000 years ago up to the modern era, and on maritime history, particularly the development of seafaring technology and its social and political consequences in terms of trade, mobility, and warfare. "Nautical archaeology" emphasizes the technological aspect of seafaring, and "maritime archaeology" encompasses the social and political dimension (Muckelroy, 1978; Gould, 2000).

The second category comprises the study of the terrestrial landscapes that were occupied by human populations during periods of lower sea-level associated with the glacial periods of the Pleistocene era and which have been drowned by sea-level rise since the last glacial maximum and is variously labeled as submerged landscape archaeology, submerged prehistory, or continental shelf archaeology (Benjamin et al., 2011; Evans et al., 2014). Since modern sea-level was finally established about 6,000 years ago, the focus here is on the earlier time ranges of prehistory; on the reconstruction of the now-submerged landscape – its topography, paleoenvironmental setting, and evidence of prehistoric settlement – and on the social, economic, demographic, and evolutionary consequences of changes in sea-level and paleogeography.

Both categories are concerned with submerged material; the main difference is that the former concentrates on maritime and nautical developments in recent millennia, the latter on the terrestrial features of a more or less extensive and now-submerged lowland landscape that existed during the major glacial episodes of the Pleistocene epoch.

Nevertheless, there is some overlap. For example, the history of seafaring and maritime activity certainly extends much further back than 6,000 years ago. This is demonstrated by evidence for sea crossings at least 12,000 years ago in the Mediterranean and even earlier in the Pacific, the earliest currently accepted example being human entry into Australia and New Guinea at about 50,000 years ago (Anderson et al., 2010). Actual finds of boats from these earlier periods are rare, but an increasing number of wooden dugout canoes have been recovered from prehistoric waterlogged or submerged sediments in Europe, the earliest dated at about 10,000 years (Crumlin-Pedersen, 2010; Cunliffe, 2010).

Conversely, coastal settlements and terrestrial landscape features younger than 6,000 years have been drowned in some parts of the world because of land submergence resulting from recent tectonic or isostatic movements.

Both types of underwater investigation also share a common interest in the technology and logistics of underwater exploration and in stressing the different types and qualities of information that can be obtained from underwater archaeological data compared to data on land. Both of course may involve the study of material on dry land, and practitioners in both fields usually emphasize the desirability of treating underwater work as an extension of archaeological research on land, the two types of activity differing only in their technical requirements.

Geoscientific research has an important role to play in all types of archaeological underwater research, since it is critical to understanding the different processes by which artifacts are variously buried or exposed by marine currents and movements of sediment and the likelihood of long-term preservation or decay. In addition, the exploration of submerged landscapes, and the improved understanding of relative sea-level change that can come from combined archaeological and geological research, opens up areas of mutual scientific interest and collaboration between marine geoscience and archaeology, and this is the main emphasis in what follows.

Underwater Archaeology, Figure 1 A dugout canoe recovered from the Danish Mesolithic site of Tybrind Vig. Photograph courtesy of Søren Andersen and the Moesgaard Museum.

History

Technological developments have been a key factor. The invention of the aqua lung or SCUBA (Self-Contained Underwater Breathing Apparatus) after the Second World War opened up an era of underwater exploration for thousands of individuals and enthusiasts, and systematic exploration of historical shipwrecks and submerged harbor installations soon followed from the 1950s onwards with the work of pioneers such as George Bass, Honor Frost, and Nicholas Flemming. Military and industrial operations as well as archaeological and scientific questions have stimulated many new technologies, resulting in a modern array of vessels, sonar devices, underwater cameras and vehicles, and technical diving with mixed gases (Bass 1966; Blot, 1996; Delgado, 1998; Ruppé and Barstad, 2002).

Submerged forests with prehistoric archaeological finds were known about in the nineteenth century (Dawkins, 1870), and geologists studying sea-level change in the 1960s recognized the potential archaeological importance of the continental shelf (Emery and Edwards, 1966). Archaeological interest, however, remained limited and intermittent through the later twentieth century. In part this was due to technical and financial obstacles, but more potent inhibitions were the difficulty of locating and identifying target material comprising diffuse scatters of stone tools or potsherds barely distinguishable from the natural background on the seabed and a prevalent belief – still quite widely held – that archaeological material has either not survived, cannot be recovered, or cannot add significant new knowledge.

Pioneering work took place in the 1970s and early 1980s (Gagliano et al., 1982; Masters and Flemming, 1983; Andersen, 1985; Fischer, 1995a), but there was little follow-up. Finds of prehistoric settlements that demonstrated the unusually good quality of preservation in underwater sediments, such as Tybrind Vig, a 6,000-year old site in Denmark with remains of wooden canoes, paddles, and fish weirs (Andersen, 1985, 2011; Figures 1 and 2), and Atlit Yam off the coast of Israel, a 9,000-year old settlement with evidence of farming, fishing, seafaring, and the construction of stone-lined wells (Galili et al., 1993), were regarded as exceptions rather than the norm. Only within the past decade has there been a noticeable acceleration of research, spurred by a growing recognition of the likely importance of coastal regions and marine resources in the earlier stages of human evolution (Erlandson, 2001; Bailey and Milner, 2002; Bailey et al., 2008); by a steady increase in the number of reported underwater finds and a growing realization of how much significant evidence may now be awaiting discovery (Flemming, 1998; Bailey and Flemming, 2008); by the intensification of industrial and commercial activities on the sea floor, which have provided both new opportunities for acquiring archaeological and geologically relevant data, and also an increasing demand for government regulation and management of the underwater heritage, enshrined in the UNESCO convention of 2001 (Flemming, 2004; Gaffney et al. 2007, 2009; Evans et al., 2014); and by the development of new

Underwater Archaeology, Figure 2 A decorated wooden paddle from Tybrind Vig. Scale in centimetres. Photograph by Derrick Butler, courtesy of Claus Skriver.

interdisciplinary and international collaborations and sources of funding (Bailey, Sakellariou et al. 2012).

The archaeological importance of sea-level change and submerged landscapes

The general pattern of changes in global ocean volume with changes in glaciation during the Quaternary period is well known. One well-recognized consequence of low sea-levels is the creation of land bridges that facilitated population movement between continental land masses during the Pleistocene. A more significant consequence of sea-level change is the fact that, for most of human existence on this planet, sea-levels have been lower than present, by depths ranging between −20 m and −120 m for the greater part of the glacial cycle and a modal value of −40 to −60 m. This means that Pleistocene shorelines and extensive coastal regions representing prime territory for human occupation are now mostly submerged and that the apparent explosion of evidence for coastal settlements and maritime activities visible worldwide after about 6,000 years ago, when global sea-level rise at the end of the last glaciation ceased, has more to do with the likely loss of earlier evidence than with some global "revolution" involving social and economic intensification.

These now-submerged coastal regions were also probably the focus of more attractive terrestrial conditions of ecology, food resources, water supplies and local climate, and higher population densities than their adjacent hinterlands and were likely also major pathways of population movement and cultural interchange. It follows that we are missing some of the most important evidence for the early transformations of human biological and social evolution over the past 2 million years, including evidence relevant to early human dispersal out of Africa; early developments in seafaring, fishing, shell gathering, and sea-mammal hunting; earliest colonization of Australia and the Americas; the first penetration of de-glaciated territory at high latitudes especially in Northwest Europe; the growth of sedentary societies; the earliest dispersal of agricultural economies around coastal basins such as the Mediterranean; and the roots of some of the earliest civilizations such as those of Minoan Crete and Mesopotamia. All these developments took place when sea-level was lower than today; all probably depended to some degree on the resources and opportunities of coastal regions; many of the later developments in this sequence, beginning during the Last Glacial period if not earlier, almost certainly required skills in seafaring and exploitation of marine resources. Most of the evidence must, by definition, be sought in the vicinity of paleocoastlines and must therefore be located underwater (Bailey and Flemming, 2008).

Submerged settlements and landscapes

The discovery and interpretation of prehistoric archaeological evidence cannot be divorced from the reconstruction of the wider physical landscape in which people lived. Geological techniques are essential to reconstruct the likely configuration of topography, drainage channels, lakes, paleoshorelines, and other natural resources, to give precision to the locations which might have been persistently attractive places for human activity and the discard of archaeological material, and to identify the processes that are likely to have preserved, buried, exposed, or destroyed archaeological material before, during, and after inundation. Successful underwater archaeological research increasingly depends on multidisciplinary and multinational collaboration involving archaeologists, marine geologists, geophysicists, sedimentologists, tectonic geomorphologists, oceanographers, and paleoclimatologists.

The initial discovery of submerged archaeological sites usually begins with chance finds by sports divers or by commercial activities on the seabed such as trawler fishing or dredging of marine channels. Dredging operations in Denmark and northern Germany early in the twentieth century first revealed the large quantities of prehistoric artifacts present on the seabed in the western Baltic. Large quantities of ice age fauna such as mammoth tusks and teeth are regularly dredged up by Dutch trawlers working in the North Sea (Glimmerveen et al., 2004), and these catches sometimes include rare finds of Paleolithic stone tools and in one recent case part of a Neanderthal skull (Hublin et al., 2009). On the eastern seaboard of

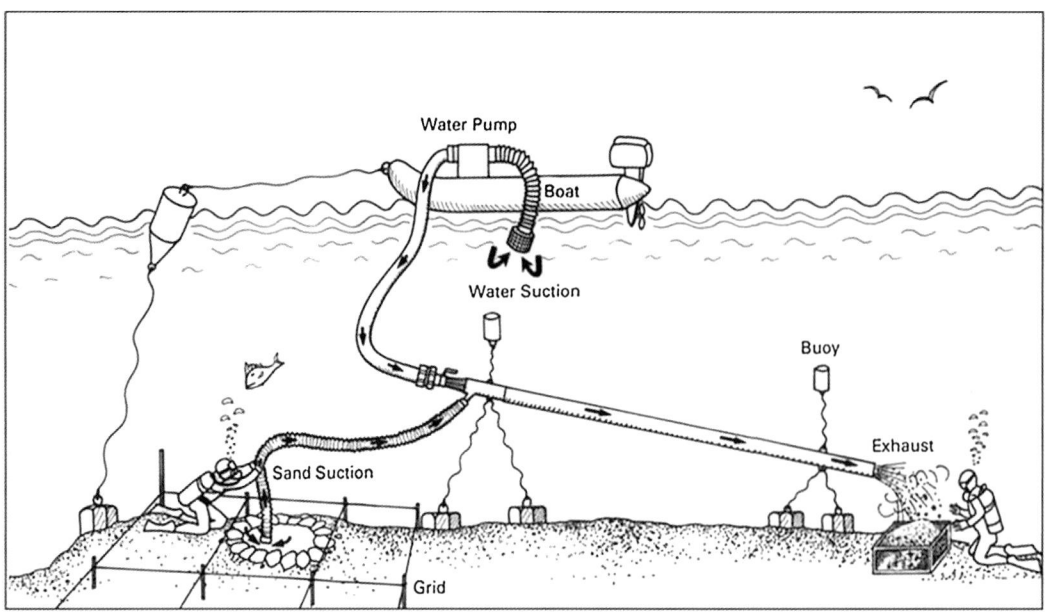

Underwater Archaeology, Figure 3 A diagrammatic representation, showing the techniques involved in the underwater excavation of Atlit Yam, Israel. After Galili et al. 1993, figure 2. Image supplied by Ehud Galili, courtesy of the Israel Antiquities Authority.

the USA, a mastodon tusk and a bifacially flaked leaf point were recovered from 70 m on the outer shelf in 1970 and languished in a local museum for over 30 years before their significance was recognized. Recent radiocarbon dating of the tusk at 23,000 years BP appears to extend human presence in the Americas by 10,000 years compared to the generally accepted date of earliest colonization (Stanford et al., 2014). The Atlit Yam site in Israel was discovered only by the coincidence of a storm that had temporarily removed the protective cover of sand and a local archaeologist who happened to be diving at that locality. The Cosquer Cave in southern France, with a magnificent gallery of Paleolithic cave art above modern sea-level, is only accessible by a single corridor with an entrance now some 40 m below sea-level, discovered by an intrepid sports diver (Clottes and Courtin, 1996).

Once materials and locations have been found, more systematic work can take place, either in the laboratory or through new work at the original seabed location using remote sensing, coring, and diving. Excavation work on shallow water sites down to depths of 10–20 m is logistically feasible, using SCUBA and normal air mixtures and the adaptation of conventional archaeological techniques, such as trench grids, hand tools, dredges, and sieves, for the processing of large quantities of sediments, and air lifts to take material to the surface (Figure 3). Good examples of such work are the Mesolithic site of Bouldnor Cliff in southern England (Momber, 2011; Momber et al., 2011); Tybrind Vig and other similar sites in Denmark (Fischer, 1997, 2004; Andersen, 1985, 2013; Skaarup and Grøn, 2004), all of Mesolithic date; Mesolithic and Neolithic submerged settlements in the Wismar Bay of northern Germany (Harff et al., 2007; Lübke et al., 2011); and Atlit Yam in Israel (Galili et al., 1993). Animal and human bones, remains of plant foods, and organic artifacts made from wood and plant fibers are often very well preserved because of burial in anaerobic sediments, providing detailed and well-dated information on paleoenvironment and paleoeconomy.

Sites at greater depth are more difficult to deal with and require divers trained in mixed gas techniques to ensure safe working at depth. Until now, very few examples have been reported other than experimental trials (e.g., Bailey et al., 2007). Other possibilities are the use of coring and dredging, but here the main challenges, in the case of coring, are the processing of sufficient volumes of sediment to locate and recover archaeological materials that may be present in relatively low densities and, in the case of dredging, the difficulties of achieving stratigraphic control. Nevertheless, recovery of cultural material including artifacts has been reported from coring or targeted grab sampling in several parts of the world up to depths of over 50 m below sea-level – on the shelf of the eastern USA (Gagliano et al., 1982), on the Pacific coast of British Columbia (Fedje and Josenhans, 2000), in the North Sea (Long et al., 1986), and in the Black Sea (V. Yanko-Hombach, personal communication, 2010).

Collaboration with industrial operators can be of significant help, especially where government legislation requires mitigation work by companies working on the seabed. In the UK a government tax levied on gravel extraction companies (MALSF – Marine Aggregates Levy Sustainability Fund) provided a multimillion pound (dollar) fund over a period of years that supported important archaeological projects. One was the computer analysis of seismic data provided by the hydrocarbon industry

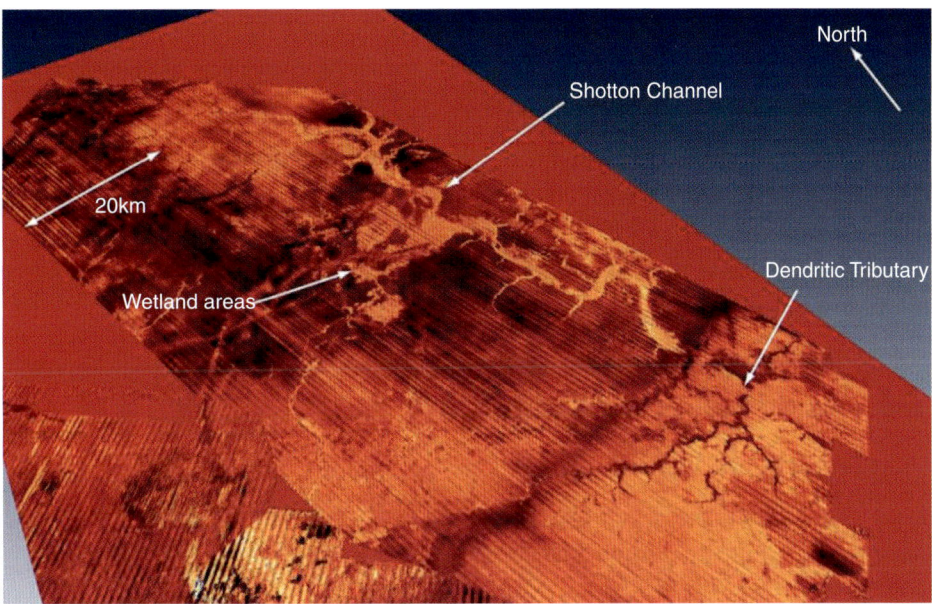

Underwater Archaeology, Figure 4 Part of the North Sea submerged landscape as reconstructed for the late Pleistocene period from seismic data. Image courtesy of Vince Gaffney and the Visual and Spatial Technology Centre (VISTA), University of Birmingham.

for a sector of the North Sea, which was able to reconstruct a 23,000 km² area of the pre-inundation landscape, showing river channels, paleoshorelines, marshes, and hills (Gaffney et al., 2007, 2009; Figure 4). In another case, at the A240 site, also in the North Sea, some 11 miles offshore of the East coast of England, companies processing large quantities of material for gravel extraction recovered Lower Paleolithic bifacial hand axes and cooperated with English Heritage (the government agency responsible for managing the underwater cultural heritage) and a commercial archaeological company to collect more material. Targeted acoustic survey and coring revealed a sequence of sediment-filled paleochannels that could be linked to the hand axe site, providing a date of 350–200,000 for the earliest material and information on the contemporaneous vegetation and environment (Tizzard et al., 2011; Bicket et al., 2014). That such material should have survived several sea-level cycles of inundation is not unique; a similar case of early-dated hand axes has been reported from South Africa (Werz and Flemming, 2001). In the Port of Rotterdam, active industrial collaboration using purpose-built dredges and industrial-scale sediment removal and sieving during the construction of new harbor facilities has resulted in the recovery of Mesolithic stone artifacts and animal-bone material in 17m depth of water (Weerts et al., 2012).

The total number of prehistoric archaeological sites discovered underwater is difficult to know at the present time, since the numbers are increasing all the time, and no systematic audit has yet been completed. But they certainly number in the thousands, including material of all ages from the Lower Paleolithic period (older than 250,000 years) up to the Bronze Age, at all depths from the outer edge of the continental shelf exposed at maximum sea-level regression up to the present-day intertidal zone, and in most continents, with the largest concentrations currently in Europe. At any rate, the number and range of discoveries is such as to refute emphatically the belief that nothing has survived on the seabed or can be recovered.

Purposeful research strategies aimed at discovering new archaeological sites are a much greater challenge than on land. Yet they are vital to the future development of the discipline, which cannot continue to rely on chance discoveries. Divers cannot traverse large areas of the seabed in the same way as with field-walking on land; remote sensing techniques of acoustic and seismic surveys are unlikely to identify archaeological sites unless they are very large and have a very distinctive acoustic signature, e.g., shell mounds, and are in any case expensive to deploy over large areas. Remotely operated vehicles with cameras, and manned submersibles, are other possibilities, but these require large support ships and considerable cost. Whatever the techniques available, it is essential to produce predictive models that can help target the likely location of archaeological material, and these models need to cater for two objectives. The first is the prediction of specific locations in the paleolandscape that were likely attractive places for human activity. The second is an understanding of the processes of landscape erosion or alteration associated with sea-level inundation. Both objectives require geological input, both to reconstruct terrestrial land surfaces and paleoshorelines as they would have existed prior to inundation and to understand the ongoing processes of submarine sedimentation and erosion that may have helped to preserve and make visible archaeological and terrestrial features of the original landscape. In Denmark, using Anders Fischer's (1995b) "fishing site model," shoreline locations most likely to be good fishing spots were searched by divers. This has resulted in the discovery of many hundreds of new sites and can be adapted for other regions (Benjamin, 2010). In the Gulf of Mexico and the Eastern seaboard of the USA, reconstruction of paleoriver channels and adjacent areas suitable for human settlement, combined with targeted coring and diver exploration, has resulted in successful location of archaeological sites (Faught and Gusick, 2011, Faught, 2014; Gagliano et al., 1982).

Examples of sea-level change

Archaeological settlements are often located at or very close to the ancient shoreline and can provide detailed information on shoreline depth below modern sea-level, chronology, and ecological conditions, all of significant value as sea-level index points. In the Baltic, archaeological data from a chronological sequence of submerged settlements, which underwent progressive inundation during the Holocene, has helped to give precision to the regional relative sea-level curve and the prediction of future shoreline change with continued sinking of the modern Baltic coast (Harff et al., 2007). In the Mediterranean, dates of shoreline features such as Roman fish tanks and loading areas for shipment of quarried stone, which are now submerged, provide sea-level reference points for modeling the complex interaction of tectonic and isostatic displacement (Antonioli et al., 2009; Florido et al., 2011; Furlani et al., 2013). Elsewhere, archaeologically defined problems about early human dispersal are providing the stimulus to more detailed modeling of changes in sea-level and the position of paleoshorelines (Lambeck et al., 2011).

Summary and conclusions

Underwater archaeology comprising the study of shipwrecks has been well established for over half a century. The study of formerly occupied submerged landscapes flooded by sea-level rise at the end of the last glaciation is a newer discipline. Only in the past decade has it become more widely established as a major focus for interdisciplinary research with the marine geosciences, and collaboration with offshore industries, and is opening up a completely new perspective on the contribution of prehistoric maritime activity and coastal settlement to the prehistory of human civilization.

Bibliography

Andersen, S. H., 1985. Tybrind Vig. A preliminary report on a submerged Ertebølle settlement on the west coast of Fyn. *Journal of Danish Archaeology*, **4**, 52–59.

Andersen, S. H., 2011. Ertebølle canoes and paddles from the submerged habitation site of Tybring Vig, Denmark. In Benjamin, J., Bonsall, C., Pickard, C., and Fischer, A. (eds.), Submerged Prehistory. Oxford: Oxbow Books, pp. 1–14.

Andersen, S. H., 2013. *Tybrind Vig: Submerged Mesolithic Settlements in Denmark*. Aarhus: Aarhus University Press. Jutland Archaeological Society Publications, Vol. 77.

Anderson, A., Barrett, J., and Boyle, K. (eds.), 2010. *The Global Origins and Development of Seafaring*. Cambridge: McDonald Institute for Archaeological Research.

Antonioli, F., Ferranti, L., Fontana, A., Amorosi, A., Bondesan, A., Braitenberg, C., Dutton, A., Fontolan, G., Furlani, S., Lambeck, K., Mastronuzzi, G., Monaco, C., Spada, G., and Stocchi, P., 2009. Holocene relative sea-level changes and vertical movements along the Italian and Istrian coastlines. *Quaternary International*, **206**, 103–133.

Bailey, G. N., and Flemming, N. C., 2008. Archaeology of the continental shelf: marine resources, submerged landscapes and underwater archaeology. *Quaternary Science Reviews*, **27** (23–24), 2153–2165.

Bailey, G. N., and Milner, N. J., 2002. Coastal hunters and gatherers and social evolution: marginal or central? *Before Farming: The Archaeology of Old World Hunter-Gatherers*, **3–4**(1), 1–15.

Bailey, G. N., Flemming, N., King, G. C. P., Lambeck, K., Momber, G., Moran, L., Al-Sharekh, A., and Vita-Finzi, C., 2007. Coastlines, submerged landscapes and human evolution: the Red Sea Basin and the Farasan Islands. *Journal of Island and Coastal Archaeology*, **2**(2), 127–160.

Bailey, G. N., Carrión, J. S., Fa, D. A., Finlayson, C., Finlayson, G., and Rodríguez-Vidal, J. (eds.), 2008. The Coastal Shelf of the Mediterranean and Beyond: Corridor and Refugium for Human Populations in the Pleistocene. *Quaternary Science Reviews*, **27**(Special Issue) (Vols 23–24), 2095–2270.

Bailey, G. N., Sakellariou, D., and members of the SPLASHCOS network, 2012. Submerged prehistoric archaeology & landscapes of the continental shelf. *Antiquity*, **86**, 334, http://antiquity.ac.uk/projgall/sakellariou334/.

Ballard, R. D. (ed.), 2008. *Archaeological Oceanography*. Princeton, NJ: Princeton University Press.

Bass, G., 1966. *Archaeology Underwater*. London: Thames and Hudson.

Benjamin, J., 2010. Submerged prehistoric landscapes and underwater site discovery: reevaluating the 'Danish model' for international practice. *Journal of Island and Coastal Archaeology*, **5**, 253–270.

Benjamin, J., Bonsall, C., Pickard, C., and Fischer, A. (eds.), 2011. *Submerged Prehistory*. Oxford: Oxbow Books.

Bicket, A., Firth, A., Tizzard, L., and Benjamin, J., 2014. Heritage management and submerged prehistory in the United Kingdom. In Evans, A., Flatman, J., and Flemming, N. C. (eds.), *Prehistoric Archaeology of the Continental Shelf: A Global Review*. New York: Springer, pp. 213–232.

Blot, J.-Y., 1996. *Underwater Archaeology: Exploring the World beneath the Sea*. New York: Abrams.

Clottes, J., and Courtin, J., 1996. *The Cave beneath the Sea: Paleolithic Images at Cosquer*. New York: Abrams.

Crumlin-Pedersen, O., 2010. Aspects of the origin of Atlantic and Baltic seafaring. In Anderson, A., Barrett, J., and Boyle, K. (eds.), *The Global Origins and Development of Seafaring*. Cambridge: McDonald Institute for Archaeological Research, pp. 109–127.

Cunliffe, B., 2010. Seafaring on the Atlantic seaboard. In Anderson, A., Barrett, J., and Boyle, K. (eds.), *The Global Origins and Development of Seafaring*. Cambridge: McDonald Institute for Archaeological Research, pp. 265–274.

Dawkins, W. B., 1870. On the discovery of flint and chert under a submerged forest in west Somerset. *The Journal of the Ethnological Society of London*, **2**(2), 141–146.

Delgado, J. (ed.), 1998. *Encyclopedia of Underwater and Maritime Archaeology*. New Haven, CO: Yale University Press.

Emery, K. O., and Edwards, R. L., 1966. Archaeological potential of the Atlantic continental Shelf. *American Antiquity*, **31**, 733.

Erlandson, J. M., 2001. The archaeology of aquatic adaptations: paradigms for a new millennium. *Journal of Archaeological Research*, **9**, 287–350.

Evans, A., Flatman, J., and Flemming, N. C. (eds.), 2014. *Prehistoric Archaeology on the Continental Shelf: A Global Review*. New York: Springer.

Faught, M. K., 2014. Remote sensing, target identification and testing for submerged prehistoric sites in Florida: process and protocol in underwater CRM projects. In Evans, A., Flatman, J., and Flemming, N. C. (eds.), *Prehistoric Archaeology of the Continental Shelf: A Global Review*. New York: Springer, pp. 37–52.

Faught, M. K., and Gusick, A., 2011. Submerged prehistory in the Americas. In Benjamin, J., Bonsall, C., Pickard, C., and Fischer, A. (eds.), *Submerged Prehistory*. Oxford: Oxbow Books, pp. 145–157.

Fedje, D. W., and Josenhans, H., 2000. Drowned forests and archaeology on the continental shelf of British Columbia, Canada. *Geology*, **28**(2), 99–102.

Fischer, A. (ed.), 1995a. *Man and Sea in the Mesolithic: Coastal Settlement above and below Present Sea Level*. Oxford: Oxbow.

Fischer, A., 1995b. An entrance to the Mesolithic world below the ocean. Status of ten years' work on the Danish sea floor. In Fischer, A. (ed.), *Man and Sea in the Mesolithic: Coastal Settlement Above and Below Present Sea Level*. Oxford: Oxbow, pp. 371–384.

Fischer, A., 1997. People and the sea – settlement and fishing along the Mesolithic coasts. In Pedersen, L., Fischer, A., and Aby, B. (eds.), *The Danish Storebaelt since the Ice Age – Man, Sea and Forest*. Copenhagen: Storebaelt Publications, pp. 63–77.

Fischer, A., 2004. Submerged Stone Age – Danish examples and North Sea potential. In Flemming, N. C. (ed.), *Submarine Prehistoric Archaeology of the North Sea*. York: Council for British Archaeology. CBA Research Report, Vol. 141, pp. 23–36.

Flemming, N. C., 1998. Archaeological evidence for vertical movement on the continental shelf during the Palaeolithic, Neolithic and Bronze Age periods. In Stewart, I. S., and Vita-Finzi, C. (eds.), *Coastal Tectonics*. London: Geological Society Special Publications, Vol. 146, pp. 129–146.

Flemming, N. C. (ed.), 2004. *Submarine Prehistoric Archaeology of the North Sea: Research Priorities and Collaboration with Industry*. York, London: Council for British Archaeology. CBA Research Report, Vol. 141.

Florido, E., Auriemma, R., Faivre, S., Radic Rossi, I., Antonioli, F., Furlani, S., and Spada, G., 2011. Istrian and Dalmatian fishtanks as sea-level markers. *Quaternary International*, **232**, 105–113.

Furlani, S., Antonioli, A., Biolchi, S., Gambin, T., Gauci, R., Lo Presti, V., Anzidei, M., Devoto, S., Palombo, M., and Sulli, A., 2013. Holocene sea level change in Malta. *Quaternary International*, **288**, 146–157.

Gaffney, V., Thomson, K., and Fitch, S. (eds.), 2007. *Mapping Doggerland: The Mesolithic Landscapes of the Southern North Sea*. Oxford: Archaeopress.

Gaffney, V., Fitch, S., and Smith, D. (eds.), 2009. *Europe's Lost World: The Rediscovery of Doggerland*. York: Council for British Archaeology. CBA Research Report, Vol. 160.

Gagliano, S. M., Pearson, C. E., Weinstein, R. A., Wiseman, D. E., and McClendon, C. M., 1982. *Sedimentary Studies of Prehistoric Archaeological Sites: Criteria for the Identification of Submerged Archaeological Sites on the Northern Gulf of Mexico Continental Shelf*. Washington, DC: National Park Service, U.S. Department of the Interior. Preservation Planning Series.

Galili, E., Weinstein-Evron, M., Hershkovitz, I., Gopher, A., Kislev, M., Lernau, O., Kolska-Horwitz, L., and Lernau, H., 1993. Atlit-Yam: a prehistoric site on the sea floor off the Israeli coast. *Journal of Field Archaeology*, **20**, 133–157.

Glimmerveen, J., Mol, D., Post, K., Reumer, J. W. F., Van der Plicht, H., De Vos, J., Van Geel, B., Van Reenen, G., and Pals, J. P., 2004. The North Sea Project: the first palaeontological, palynological, and archaeological results. In Flemming, N. C. (ed.), *Submarine Prehistoric Archaeology of the North Sea: Research Priorities and Collaboration with Industry*. Council for British Archaeology: York. CBA Research Report, Vol. 141, pp. 43–52.

Gould, R., 2000. *Archaeology and the Social History of Ships*. Cambridge: Cambridge University Press.

Harff, J., Lemke, W., Lampe, R., Lüth, F., Lübke, H., Meyer, M., Tauber, F., and Schmölcke, U., 2007. The Baltic Sea coast – a model of interrelations among geosphere, climate, and anthroposphere. *Geological Society of America, Special Paper*, **426**, 133–142.

Hublin, J.-J., Weston, D., Gunz, P., Richards, M., Roebroeks, W., Glimmerveen, J., and Anthonis, L., 2009. Out of the North Sea: the Zeeland Ridges Neandertal. *Journal of Human Evolution*, **57**, 777–785.

Lambeck, K., Purcell, A., Flemming, N. C., Vita-Finzi, C., Alsharekh, A., and Bailey, G. N., 2011. Sea level and shoreline reconstructions for the Red Sea: isostatic and tectonic considerations and implications for hominin migration out of Africa. *Quaternary Science Reviews*, **30**(25–26), 3542–74.

Long, D., Wickham-Jones, C. R., and Ruckley, N. A., 1986. A flint artefact from the northern North Sea. In Roe, D. A. (ed.), *Studies in the Upper Palaeolithic of Britain and Northwest Europe*. Oxford: British Archaeological Reports. British Archaeological Reports International Series, Vol. 296, pp. 55–62.

Lübke, H., Schmölcke, U., and Tauber, F., 2011. Mesolithic hunter-fishers in a changing world: a case study of submerged sites on the Jäckelberg, Wismar Bay, Northeastern Germany. In Benjamin, J., Bonsall, C., Pickard, C., and Fischer, A. (eds.), *Submerged Prehistory*. Oxford: Oxbow, pp. 21–37.

Masters, P. M., and Flemming, N. C. (eds.), 1983. *Quaternary Coastlines and Marine Archaeology*. London/New York: Academic.

Momber, G., 2011. Submerged landscape excavations in the Solent, southern Britain: climate change and cultural development. In Benjamin, J., Bonsall, C., Pickard, C., and Fischer, A. (eds.), *Submerged Prehistory*. Oxford: Oxbow Books, pp. 85–98.

Momber, G., Tomalin, D., Scaife, R., Satchell, J., and Gillespie, J. (eds.), 2011. *Bouldnor Cliff and the Submerged Mesolithic Landscape of the Solent*. York: Council for British Archaeology. CBA Monograph Series, Vol. 164.

Muckelroy, K., 1978. *Maritime Archaeology*. Cambridge: Cambridge University Press.

Ruppé, C. V., and Darstad, J. F. (eds.), 2002. *International Handbook of Underwater Archaeology*. New York: Kluwer Academic/Plenum.

Skaarup, J., and Grøn, O., 2004. *Møllegabet II: A Submerged Mesolithic Settlement in Southern Denmark*. Oxford: Archaeopress. BAR International Series, Vol. 1328.

Stanford, D., Lowery, D., Jodry, M., Bradley, B., Kay, M., Stafford, T. W., and Speakman, R. J., 2014. New evidence for a possible Paleolithic occupation of the Eastern North American continental shelf at the Last Glacial Maximum. In Evans, A., Flatman, J., and Flemming, N. C. (eds.), *Prehistoric Archaeology of the Continental Shelf: A Global Review*. New York: Springer, pp. 73–93.

Tizzard, L., Baggaley, P. A., and Firth, A. J., 2011. Seabed prehistory: investigating palaeolandscapes with Palaeolithic remains from the southern North Sea. In Benjamin, J., Bonsall, C., Pickard, C., and Fischer, A. (eds.), *Submerged Prehistory*. Oxford: Oxbow, pp. 65–74.

Weerts, H., Otte, A., Smit, B., Vos, P., Schiltmans, D., Waldus W., and Borst, W., 2012. Finding the needle in the haystack by using knowledge of Mesolithic human adaptation in a drowning delta. In Bebermeier, W., Hebenstreit, R., Kaiser, E., and Krause, J. (eds.), *Landscape Archaeology. Proceedings of the international conference held*, Berlin, June 6–8, Berlin: Etopoi, pp.17–24.

Journal for Ancient Studies, Special volume 3). Available at http://journal.topoi.org/index.php/etopoi.

Werz, B. E. J. S., and Flemming, N. C., 2001. Discovery in Table Bay of the oldest hand axes yet found underwater demonstrates preservation of hominid artefacts on the continental shelf. *South African Journal of Science*, **97**(5), 183–5.

Cross-references

Beach Processes
Coasts
Estuary, Estuarine Hydrodynamics
Eustasy
Glacio(hydro)-isostatic Adjustment
Marine Transgression
Relative Sea-level (RSL) Cycle
Sea-Level
Sediment Transport Models
Sedimentary Sequence
Technology in Marine Geosciences

UPWELLING

Colin Summerhayes
Scott Polar Research Institute (SPRI), Cambridge University, Cambridge, UK

Definition

Geologists have five good reasons for an interest in upwelling currents, which are highly productive and tend to be recorded for posterity through organic enrichment on continental margins. Firstly, massive accumulations of organic-rich sediment eventually form the source rocks for oil. Secondly, these accumulations may contain abundant remains of siliceous organisms like diatoms, providing commercially significant deposits of diatomite. Thirdly, the migration of phosphorus from organic-rich sediments into pore fluids or bottom waters may lead to the formation of phosphorite deposits important to the fertilizer industry. Fourthly, these organic rocks together or separately form part of the signature *triad* of organic-rich black shale, diatomite, and phosphorite that can be used in paleoclimatology as indicators of particular climate conditions including past upwelling (Parrish, 1982; Parrish and Curtis, 1982). And, finally, these rocks provide important clues to the operation of various biogeochemical processes like the carbon cycle, the silica cycle, and the phosphorus cycle.

Understanding the geological effects of upwelling requires familiarity with how upwelling works to affect the biogeochemistry of today's ocean and its relation to the oxygen depletion that may encourage the preservation of sedimentary organic matter. Comprehensive reviews of the nature of upwelling and its geological record can be found in Suess and Thiede (1983) and Thiede and Suess (1983), Summerhayes et al. (1992), Summerhayes et al. (1995a), and Summerhayes et al. (1995b) and are complemented by a review of the productivity of the ocean, past and present, by Berger et al. (1989a).

The winds play an extraordinary role in governing surface ocean properties. When they blow on the ocean surface, they force the surface waters in the same direction. The Earth's rotation deflects both winds and currents to the right in the northern hemisphere and to the left in the southern hemisphere. As pointed out by Swedish oceanographer Vagn Walfrid Ekman (1874–1954) in 1902, surface waters are deflected slightly from the direction of the wind acting upon them. Each layer within the water column deflects the next one down, and deflections continue downward until there is no more vertical transmission of energy. The result is a rotation of flow named the Ekman spiral, which causes the top 100 or so meters of the ocean to move at right angles to the direction of the prevailing wind. This moving layer is the Ekman layer. Its motion, known as Ekman transport, sets up a surface slope rising in the direction of transport. To compensate, cold, nutrient-rich water wells up from below; this is upwelling, referred to by some as Ekman pumping.

Persistent upwelling takes place where high-pressure cells over the ocean in midlatitudes force the winds to blow parallel to the arid coasts of northwest and southwest Africa, northwestern and southwestern America, and northeastern Africa and eastern Arabia, forcing the surface water offshore. Persistent upwelling also occurs in the open ocean, and these two locations are connected in the subsurface. The westerly winds of the Southern Ocean force the Antarctic Circumpolar Current to the east and drive the surface waters north, causing the upwelling of nutrient-rich Circumpolar Deep Water to feed the productive Southern Ocean ecosystem. This cold, northward-moving surface water eventually sinks beneath warmer and less dense surface waters at the Antarctic Convergence, or Polar Front, near 60°S, and spreads out to form the intermediate water layer of the world's oceans at depths of around 800–1,000 m. Somewhat less nutrient-rich Southern Ocean water also sinks at the Subantarctic Front to form Subantarctic Mode Water at depths of 200–300 m. The intermediate and mode water provide the deep source for upwelling water along continental margins, and mode water alone may be responsible for 33–75 % of global ocean productivity between 30°S and 30°N (Berger et al., 1989b; Palter et al., 2010). Persistent upwelling also characterizes the equatorial oceans, where the trade winds blow parallel to the equator, forcing surface waters away to the north and south (Shimmield and Jahnke, 1995). Because of their high productivity, upwelling zones form some of the world's richest fishing grounds. The extraordinary productivity of coastal upwelling may be further stimulated by the supply of iron-rich dust blown offshore from adjacent deserts – iron being a limiting nutrient for ocean productivity. Plumes of dust blowing to the west can commonly be seen in satellite photographs offshore from the Sahara and the Namib Desert (Summerhayes et al., 1995b). Impersistent and localized wind-driven upwelling also occurs beneath cyclones and anticyclones but for the most part leaves no geological imprint because these are transient phenomena.

Upwelling is also common above the edge of the continental shelf (Summerhayes et al., 1995b; Summerhayes et al., 1995c), where it may be stimulated by vertical mixing induced by the interaction of tides with the change in topography. Hence, it can be found both in areas of wind-driven coastal upwelling, for example, off Namibia, and away from them, for example, at the mouth of the English Channel. Namibia's shelf-edge upwelling is stimulated by wind stress curl, the location of the axis of maximum wind stress along the coast. There, the elevation of the hinterland along the coast forces that axis 200–300 km offshore, more or less over the shelf edge, where the southerly wind stress produces upwelling and divergence seaward of the axis and downwelling and convergence landward of the axis. Different plankton populations are found in the coastal and shelf-edge upwelling centers, because the coastal upwelling taps into nutrient-rich deep waters dragged up across the continental shelf, while the shelf-edge upwelling taps shallower waters with a lower nutrient content. The coastal upwelling waters tend to be rich in silica and support a rich diatom population, while shelf-edge waters tend to be depleted in silica.

Coastal upwelling typically takes place within a coastal current system, such as the California Current, the Humboldt Current (Peru-Chile), the Benguela Current (Namibia and South Africa), or the Somali Current (Somalia and Arabia). Beneath the thermocline in these large-scale systems, there is typically a poleward undercurrent with a core near 350–500 m. Where the continental shelf is deep, as off Namibia, this current may sweep the outer shelf clean of fine-grained recent sediment (Summerhayes et al., 1995b). Upwelling may also occur off the east coasts of ocean basins, for example, off the aptly named Cabo Frio, near Rio de Janeiro, and off Florida, due either to shelf-edge upwelling or to local wind patterns, but is usually less extensive in those locations. Although coastal upwelling might be expected off Western Australia, where winds commonly blow north along the coast, it does not usually occur there. Cold northward-moving water equivalent to the Benguela or Humboldt Currents does occur offshore in the West Australian Current, but at the coast, it is displaced by the warm, south-flowing Leeuwin Current, which is a branch of the Pacific South Equatorial Current and which generates coastal downwelling. In summer, coast-parallel winds induce local countercurrents (Ningaloo and Capes Currents) associated with localized coastal upwelling off Western Australia (Hanson et al., 2005). The West Australian example demonstrates that local ocean dynamics can work against received wisdom regarding the control of upwelling by basin-scale wind patterns.

Upwelling, the oxygen minimum zone, and organic matter accumulation

As plankton die and sink, they consume oxygen, enhancing the depletion of the oceanic oxygen minimum zone that lies between the ocean's oxygenated deep bottom water and its oxygenated surface waters. This zone is significantly depleted in oxygen along those continental margins where productivity is highest. It contains the least oxygen where the mid-depth waters of the ocean are the oldest, in the Pacific Ocean and the Arabian Sea, because sinking organic matter has stripped out oxygen by decomposition in these older water masses over long periods of time. There, the effect of coastal upwelling may be to make the oxygen minimum zone virtually devoid of oxygen – or "anoxic." Any organic remains falling to the seabed in such an environment would be completely preserved, aside from the effects of decomposition by anaerobic bacteria. For these reasons we find exceptionally organic-rich sediment on the continental slopes off Somalia, Arabia, and the eastern seaboard of the Americas, where there is strong upwelling at the surface and strong oxygen depletion at depth.

We tend not to find organic-rich sediments on the continental slopes of the Atlantic, regardless of upwelling, because the mid-depth waters of the Atlantic are relatively young and the oxygen minimum zone as a result is still quite rich in oxygen (Summerhayes, 1983). Nevertheless, there are commonly some residual biological signals of upwelling in the sediments there. For example, there is some organic enrichment on the continental slope of northwest Africa, south of latitude 22°S, and on the slope off Namibia. In the case of Namibia, this is at least in part because a poleward undercurrent brings oxygen-depleted water south to Namibia and South Africa from the Angola Basin along the upper continental slope (Summerhayes et al., 1995b; Summerhayes et al., 1995c). In the Pacific, ^{14}C studies of bomb carbon show that the extensive oxygen depletion in the oxygen minimum zone off Peru and Chile is driven at least in part by oxygen depletion of a similar subsurface countercurrent that makes its way there from New Zealand via the equator (Toggweiler and Carson, 1995).

For the most part, in the oceans of the Pleistocene and present times, organic enrichment beneath coastal upwelling centers is confined to the sediments of the continental slope, the notable exception being off Namibia where organic-rich sediments accumulate on the unusually deep continental shelf (Summerhayes et al., 1995b; Summerhayes et al., 1995c and references therein). It has commonly been supposed that the impingement of the oxygen minimum zone on the continental slope may account for the widespread occurrence of organic-rich sediments on continental margins. Calvert (1987 and elsewhere) has observed that slope sediments may be enriched because (a) nearer shore, the production of organic matter is high but rates of sedimentation of organic matter are low because of strong bottom currents on the continental shelf; (b) rates of sedimentation of all components, including organic matter, are high on the continental slope; and (c) further offshore, the rates of sedimentation of organic matter decline. Subsequent studies of the chemistry of pore waters by using cleverly designed oxygen and pH microelectrodes showed that organic matter does degrade more slowly under low-oxygen than high-oxygen conditions, the key to preservation being the duration of exposure to oxygen (Eglinton and Repeta, 2003). High rates of sedimentation on continental slopes do preserve organic matter preferentially, as Calvert suggested, but the impingement on the slope of an oxygen-deficient oxygen minimum zone adds to that preservation, as well as preserving annual laminations that would otherwise be destroyed by benthic organisms through bioturbation (Archer, 2003). Annual laminae are well preserved beneath productive surface upwelling in the anoxic basins of the sea off Southern California (Eglinton and Repeta, 2003).

Not all organic-rich sediments are associated with upwelling – they may form anywhere that the production of organic matter is moderately high, the supply of silt and clay low, and the water depleted in oxygen, for example, at the bottom of basins like today's Black Sea, in the Cariaco Trench of the continental shelf off Venezuela, and in the narrow deep basin of the Gulf of California behind the Baja peninsula.

Black organic-rich sapropels are also found in another closed basin, the Eastern Mediterranean. In Cretaceous times, black, organic-rich sediments were laid down in flooded continental margin basins and on the floor of the deep ocean during oceanic anoxic events (OAEs). They are commonly laminated and may have been laid down under anoxic conditions (for further discussion, see below). They may be associated with upwelling in some places.

Upwelling and phosphorite

Phosphorus is an essential element for life and abundant in the remains of marine plankton, fish bones, and shark teeth. Under appropriate conditions, phosphate-rich fluids can precipitate calcium fluorapatite ($Ca_5(PO_4)_3$ F) otherwise known as francolite. In marine sediments, francolite may also absorb carbonate ions (CO_3^{2-}) to form calcium carbonate-fluorapatite. The fluids can also phosphatize preexisting carbonate deposits (limestones or carbonate ooids). Primary or secondary deposits of phosphate-rich sediment are known as phosphorite, a rock with at least 15–20 % P_2O_5.

It was initially thought that phosphorites were not forming in the modern ocean, but the application of uranium series dating in the 1970s showed that phosphatic minerals are forming today beneath upwelling centers, at the edges of organic-rich mud deposits off Peru (Baturin et al., 1972; Burnett, 1977; Burnett et al., 1980). They have subsequently been found forming in the upwelling centers off Namibia (Thomson et al., 1984; Baturin, 2000), as well as off Baja California, Oman, and eastern Australia (Ruttenberg, 2005). Off Peru, phosphorite forms where conditions begin to become oxidizing. Pore-water gradients of phosphate and fluoride decrease downward in sediments as these ions are incorporated into authigenic carbonate-fluorapatite. The process can occur anywhere but tends to be most common where there is a high rate of supply of organic phosphorus, which is where organic carbon is also supplied, as in upwelling areas (Ruttenberg, 2005). Iron oxyhydroxides are known to play an important intermediate role in the transfer of organic phosphorus to the carbonate-fluorapatite mineral phase.

Phosphate minerals can also accumulate as rounded grains or ooids – sand grain-sized ovoid structures layered like onions in which each layer represents a period of mineral growth. Phosphorite is common as phosphatized limestone outcrops or nodules on the seabed in upwelling zones, for example, off South Africa, Namibia, California, and Morocco, where the phosphatic rocks may occur as rock outcrops or nodules (Burnett and Riggs, 1990). Phosphorite is also widespread on the Blake Plateau east of Florida and on the Chatham Rise east of New Zealand.

The association of submarine phosphorite with upwelling in middle latitudes was pointed out early on by McKelvey (1963). These nodules had been found off the coast of South Africa during the round-the-world oceanographic expedition of *H.M.S. Challenger* (1872–1876) (Murray and Renard, 1891), and a single sample had been dredged from a depth of 600 fathoms (1,097 m) on the continental slope west of the town of Safi, in Morocco, by the *S.S. Dacia*, in 1883, during a survey of the route for a submarine cable from Cadiz to the Canary Islands (Murray and Chumley, 1924). Upper Cretaceous (Maastrichtian) to Middle Eocene (Lutetian) phosphorite was known to be abundant onshore in Morocco and was subsequently found to be abundant on the continental shelf and upper continental slope off Morocco (Summerhayes et al., 1971). These phosphorites are part of a belt of phosphorite-rich deposits formed beneath upwelling currents on the southern margin of the narrow branch of the Tethyan Ocean which separated Europe and Africa in Cretaceous-Eocene times and which extends from western Morocco and the Western Sahara, through Algeria, Tunisia, and Egypt, to Israel, Jordan, and Syria.

Most ancient phosphorites formed as continental margins became flooded, so their control through time was not simply climatic (Cook and McElhinny, 1979); a confluence of different factors was important at different times (Ruttenberg, 2005). Nevertheless, large sustained sites of upwelling seem likely as a necessary condition for the supply of phosphorus.

The last major phosphorite-forming episode in the oceans took place in Miocene times. Miocene glauconitic phosphatic conglomerates are widespread beneath upwelling currents off South Africa, California, eastern New Zealand, and Morocco (e.g., Summerhayes and McArthur, 1990) but do not occur onshore in any of those places. The California phosphorites are of the same age as the organic- and silica-rich diatomaceous Monterey Formation of California (more on that below).

Upwelling and organic-rich sediments (Pleistocene to present)

Sediments rich in organic matter are common on the continental margins beneath the major coastal upwelling currents (California, Humboldt, Somali, and Benguela). They may typically contain as much as 5–6 % total organic carbon (Corg), equivalent to around 9–10 % nonskeletal organic matter (Corg × 1.8), but higher values prevail in the shallow shelf sediments off Walvis Bay, which contain up to 24 % organic matter (Bremner, 1983). They also commonly contain the remains of siliceous plankton, usually diatoms, which may be fragmented, as on the Namibian continental slope, where the rate of accumulation is modest (Summerhayes et al., 1995b), or whole, as on the Namibian continental shelf where deposition is rapid (Bremner, 1983). There is very little continental shelf off Peru and Chile, so most of the organic matters is dumped directly onto the continental slope there (Reimers and Suess, 1983; Jahnke and Shimmield, 1995). Off southern California, most of the organic matter ends up in the oxygen-poor basins of the California Borderland (Emery, 1960).

It is commonly thought that during glacial times, the equator-to-pole thermal gradient increased, so strengthening global winds and intensifying both upwelling and its associated productivity and, as a consequence, drawing down atmospheric CO_2 (Sarnthein et al., 1988). Under these circumstances, high rates of accumulation of organic matter would be expected on continental slopes in upwelling regions during glacial periods. But the real picture is not that simple, as pointed out, for instance, by Loubere et al. (2003). For example, off northwest Africa, the flux of organic matter rose slowly through Marine Isotope stages 4 and 3 to a maximum in stage 2, the Last Glacial Maximum (LGM) (Shimmield, 1992). There, the coldest surface water temperatures occurred not at the time of maximum ice volume (the LGM) at 21–23 ka but during Heinrich events just before and just after the LGM (Zhao et al., 1995). This cooling reflects not changes in upwelling but the introduction of cold surface water from the north during iceberg outbreaks. We shall see other examples of a divergence from the simplistic thesis that there is a straightforward relationship between productivity, upwelling, and wind strength. In large part, this reflects the advection into upwelling regions, either in surface currents or in oppositely directed undercurrents, of water from adjacent regions with different nutrient and/or oxygen contents.

Deep ocean drilling off northwest Africa indicates tight coupling between increases in African rainfall and decreases in the intensity of coastal upwelling (Adkins et al., 2006; Adkins et al., 2007). For instance, strong upwelling returned at 5.5 ka with the aridity at the end of the "African humid period," during which the Sahara had been both wet and green. Arid conditions with more upwelling also characterized the cold Younger Dryas period. These data contradict interpretations based on foraminifera and radiolaria (Haslett and Smart, 2006) but make more sense in terms of the observed fluxes of terrigenous material (Adkins et al., 2006; Adkins et al., 2007).

Interpretations of organic enrichment on continental slopes dating back into glacial times also have to consider the role of changing sea-level, which dropped by up to -130 m during the Last Glacial Maximum (LGM). This would have pushed the centers of coastal upwelling seaward from the continental shelf over the continental slope, in addition providing the possibility that erosion of former organic-rich shelf deposits could supply organic matter to the continental slope. Care must thus be taken in interpreting data from continental slope cores, as noted for NW Africa, for example, by Giraud and Paul (2010). They found that during the LGM, the subsurface waters there contained fewer nutrients, thus lowering productivity. The sedimentary record, in contrast, suggested an increase in productivity, which can be explained by the fall in sea-level and a seaward shift in the locus of upwelling.

Upwelling is also evident off southern Portugal. There, signs of heightened productivity (e.g., enhanced Ba content) reflecting pulses of upwelling were found at the transitions between glacial and interglacial periods (Thomson et al., 2000), probably reflecting the reorganization of ocean circulation and wind systems. Abreu et al. (2005) found evidence for upwelling and the associated deposition of organic matter increasing there as the climate deteriorated from the interglacial of Marine Isotope stage 11 (MIS-11) into the following glacial period. Upwelling events introducing colder surface waters could be differentiated from influxes of cold surface water associated with iceberg outbreaks (Heinrich events), by the association of organic matter with the former and characteristic minerals (e.g., dolomite) with the latter (Thomson et al., 1999; Abreu et al., 2005). Iberian margin sediments contain strong precessional signals associated with low-latitude wind-driven processes contributing to dust supply (high at times of precession minima = summer insolation maxima) and upwelling (high at times of precession maxima = summer insolation minima) (Hodell et al., 2013).

Over the continental slope off Namibia over the past 70 ka, sea surface temperatures were coldest and rates of accumulation of organic matter on the continental slope were highest in the interstadial Marine Isotope stage 3 (MIS-3) (Summerhayes et al., 1995b). Temperatures warmed steadily through late stage 3 and stage 2 to the present as the rate of accumulation of organic matter declined. This change occurs together with warming that began close to the Namibian coast just before and during the LGM with the incursion of warm subtropical surface water from the north (Kirst et al., 1999). Evidently, the wind systems were reorganizing and making conditions less favorable for upwelling. Isotope stage 4 was slightly warmer than MIS-3 (60–24 ka), tended to contain slightly less organic matter, and was associated with abundant terrigenous sediment, probably wind-blown dust, suggesting that winds at this time may have been more offshore than alongshore (Summerhayes, et al., 1995b). Thus, while in toto there was higher productivity off Namibia during glacial times, it did not occur throughout glacial times. The best conditions for upwelling seem to have been focused in stage 3, presumably because winds then were most persistently coast-parallel.

The upwelling signal off Namibia over the past 150 ka fluctuated with a period of around 22,000 years, suggesting primary control by the orbital precession signal, as off Portugal. During stage 3, the precession index and obliquity signal contributed to minimizing insolation, and southerly trade winds were at their strongest as the South Atlantic high-pressure cell shifted north. This pattern led to similar cooling and increased production at that time in the Angola Basin and in the Arabian Sea (Summerhayes et al., 1995b). Comparison of warming and cooling patterns off northwest and southwest Africa shows them to be in antiphase, with warmings in the north correlating with coolings in the south during the Ice Age (Little et al., 1997). This is typical of the seesaw pattern that connects ice core changes in Greenland to those in Antarctica.

Pichevin et al. (2005) used nitrogen isotopes ($\partial^{15}N$) to demonstrate that the upwelling structure off Namibia comprised two cells that decoupled nutrient dynamics on the shelf from those over the continental slope. As elsewhere, the lower slope cores displayed low $\partial^{15}N$ during cold periods and high $\partial^{15}N$ during warm periods, whereas the $\partial^{15}N$ signal of the coastal cell was relatively constant irrespective of wind strength, indicating that nitrate was never depleted in the surface water on the shelf. Over the lower slope, nitrate delivery to the photic zone was driven by the nutrient richness of South Atlantic Central Water, depending on Agulhas inflow, which depended in turn on the position of the Polar Front, rather than on atmospheric forcing.

In the California Current, offshore cores show a close association between cold SSTs and increases in ice volume, but there is only a small cooling associated with the LGM near the coast (Herbert, 2003). While there is some evidence that the coastal upwelling zones off California and Mexico were somewhat less productive during the recent glacial period (Sigman and Haug, 2003), fluctuations in organic carbon off Baja California, Santa Barbara, and Mexico were linked to global climate variability over the past 52 ka (Dean et al., 2006). Concentrations of biogenic proxies are highest in laminated sediments deposited during warm interstadials, including the Holocene, and lowest in bioturbated sediments deposited during cool periods. However, these fluctuations owe much to fluctuations in diluting terrigenous material, and the mass accumulation rate of organic carbon likely varied by no more than a factor of 2 over the past 52 ka (Dean et al., 2006). These variations may relate to changes in the nutrient content of the subsurface waters that form the source for upwelled water. During the last glaciation, the subsurface waters that filled the Santa Barbara Basin had nitrogen isotope ($\partial^{15}N$) values of 6–7 ‰, indicating a better oxygenated (less nutrient-rich) intermediate water mass, while the subsurface waters there today have $\partial^{15}N$ values of 8–9 ‰, indicative of low-oxygen intermediate waters with a higher nutrient content (Emmer and Thunell, 2000). A decrease in nitrogen values indicating a temporary return toward glacial conditions characterized the Younger Dryas at about 12 ka. Hendy et al. (2004) confirm that productivity was not simply linearly related to cold and warm cycles, but reflected the interplay between local winds and the strength and source of the California Undercurrent. For instance, warm conditions were more productive and cool conditions less so during MIS-3; the cold LGM was relatively productive, as was the Bolling-Allerod warm interval, and the cold Younger Dryas was less productive. Warm periods tended to be associated with a more intense oxygen minimum zone that gave rise to laminated sediments (Zheng et al., 2000).

Nitrogen isotopes show much the same picture of change in the northern part of the California Current, off Oregon (Kienast et al., 2002). Low values characterize glacial stages 2 and 4, and high values characterize the Holocene, interstadial stage 3 and the last interglacial (stage 5), when the California Undercurrent was strongest. The source of the nitrogen isotope signal is the oxygen-poor subsurface water of the eastern tropical North Pacific, which advects north into the California Current system via the poleward undercurrent. From the relationship between productivity and the $\partial^{15}N$ signal, Kienast et al. (2002) deduced that upwelling-favorable winds developed first in the NE after the last glaciation then propagated south along the coast. Hence, the response of the California Current and its associated upwelling was regionally diverse.

In the mid-Holocene (8–3 ka), there are indications that upwelling decreased and productivity fell off California, even though sea surface temperatures cooled (Diffenbaugh et al., 2003). This turns out to result from the decline in orbital insolation over the northern hemisphere, which led to a longer and less vigorous mid-Holocene upwelling season in the California Current, with decreased seasonal contrast. The cooling was brought about by the introduction of cooler source waters from the north, not by enhanced upwelling.

Upwelling off Peru was investigated on ODP Leg 112 (Heinze and Wefer, 1992). As off Namibia, Peru has a northward-flowing surface current, the Peru (or Humboldt) Current, beneath which is a southward-flowing Peru Countercurrent (or undercurrent). Sediments are rich in organic matter, except locally where there has been winnowing by the undercurrent, which is associated with slightly higher levels of oxygen than the typical bottom water of the oxygen minimum zone. As in many places, coastal waters cooled significantly during the Last Glacial Maximum, but half of the cooling here appears largely due to advection of cold water from high latitudes. The rest is due to a mixture of upwelling and uplifting of the pycnocline (Feldberg and Mix, 2003). Cooling began with increased upwelling and then continued as advected cold water arrived in the eastern boundary current. As off California and Mexico (see above), the coastal upwelling was apparently less productive during the LGM than before or since (Sigman and Haug, 2003). Loubere et al. (2003) confirm that productivity was least during cold Marine Isotope stages 2 and 4.

Upwelling off the west coast of South America is less well developed off Chile, where the surface waters belong to the Southern Ocean's high-nutrient, low-chlorophyll regime. Productivity there was significantly less than off Peru (Dezileau et al., 2004). Rain rates of opal and organic carbon there are in phase with the 20 ky precessional cycle, because during southern summer, insolation maxima increased river runoff, supplying large amounts of the limiting nutrient Fe to the coastal ocean, so enhancing productivity. Evidently, changes in upwelling induced by changes in the local wind field may not be the only controls on productivity along coasts where upwelling is prevalent. Mohtadi and Hebbeln (2004) point to latitudinal shirts of the Antarctic Circumpolar Current (ACC) as also being important in providing nutrients for the upwelling system. Close to 33°S, productivity off Chile was higher when the ACC was further north, during the LGM. Continental runoff was a supplementary source of nutrients north of 33°S; there, paleoproductivity values were higher before and after the LGM, when upwelling was stronger.

Off Somalia, in the Arabian Sea, where the Somali Current is driven by the southwest monsoon, sea surface temperatures were low in both the LGM and the Holocene, indicating the persistence of seasonal summer upwelling (Anand et al., 2008). Upwelling was maximal there during transitions between glacial and interglacial stages (e.g., isotope stages 2 to 1 and 15 to 10) (Caulet et al., 1992). It was also maximal, as off Walvis Bay (Summerhayes et al., 1995b), during isotope stage 3 (65–25 ka). At least during the past 60 ky, periods of increased upwelling in the Somalian, Arabian, and Peruvian upwelling systems were synchronous (Caulet et al., 1992).

Budziak et al. (2000) noticed that the organic carbon signal in the western Arabian Sea strongly followed the orbital signal of precession rather than simply changing from glacial to interglacial. Close to the coast, organic carbon values were more abundant in interglacials. Hence, while there is a general precessional control on paleoproductivity across the region, the strengthening of the southwest monsoon winds strengthened coastal upwelling and the supply of organic carbon especially during interglacials (Budziak et al., 2000). Productivity was evidently not driven by precession and associated upwelling alone, but also by changes in the strength of the northeast monsoon winds associated with the development of maximum ice volume in the northern hemisphere, which deepened mixing, so bringing more nutrients to the surface (Budziak et al., 2000).

Tamburini et al. (2003) recognized that productivity off Oman was high due to upwelling during the summer southwest monsoon but noticed that productivity remained high there during the winter northeast monsoon, when coastal upwelling switched off. They attributed the persistence of high productivity over the past 140 ka to the addition of deep mixing induced by the northeast monsoon. Bearing in mind that the northeast monsoonal conditions predominated during recent glacial periods, they surmised that the high productivity of glacial periods (isotope stages 2, 3, and 4) was probably caused by two factors. First, the northeast monsoon introduced more iron-rich dust, and second, wind-driven deep mixing stimulated the vertical supply of phosphorus from sediments to the water column at the top of an intensified oxygen minimum zone. In interglacial phases, by contrast, the coastal upwelling was strong, the winter mixing was diminished, the oxygen minimum zone was less intense, and the vertical supply of phosphorus from that source was cut off.

In the eastern Arabian Sea, off Goa, the sea surface temperature contrast, with more cold water in coastal waters than offshore, was greatest at the LGM, most likely due to increased winter upwelling driven by the NE monsoon along the west coast of India (Budziak et al., 2000; Anand et al., 2008). Productivity there was most abundant during glacial times but collapsed during northern hemisphere cold events (such as North Atlantic Heinrich events) (Singh et al., 2011). Upwelling there is weak today.

In the Southern Ocean, siliceous diatoms and radiolaria abound south of the Polar Front, where they are abundant due to wind-driven upwelling of Circumpolar Deep Water. Changes in their distribution down core show that during the LGM, the Polar Front was located some 7° further north of its present position, indicating a widening of the region subject to upwelling (Gersonde et al., 2005). This latitude also marks the approximate northern boundary of winter sea ice, which covered twice as much sea area then than now. Because of the northward displacement of the circum-Antarctic winds and their associated upwelling, which in turn reflected the growth of the sea ice field around the continent, the Antarctic sector was less productive in glacial times, while the subantarctic sector was more so. These changes are more prominent in the Atlantic and Indian Ocean sectors and less so in the Pacific sector, perhaps because the Pacific sector was the furthest from South American sources of Fe-rich dust that would have stimulated higher productivity.

Upwelling signals in older sediments

Sea surface temperature records from the continental margins of California and Peru show that prior to 3 Ma, the eastern Pacific was warmer during the Pliocene by 3–9 °C compared to today (Dekens et al., 2007). Evidently, the cold upwelling regions characterizing the modern Pacific did not exist in the Pliocene warm period. Even so, the eastern margins were productive, indicating that upwelling of nutrients did occur there. The warmth of the surface waters is thought to indicate the existence of a deeper, well-ventilated thermocline (i.e., a thicker column of warm water). Cooling toward the temperatures of the Pleistocene and Holocene began about 4 Ma ago, before the onset of northern hemisphere glaciation, presumably representing a gradual decrease in the thickness of the thermocline (Dekens et al., 2007).

California's siliceous and organic-rich Monterey Formation, a Miocene diatom ooze deposit (White et al., 1992), coincides with a substantial global increase in the ratio of $\partial^{13}C$ – which signifies excess trapping of ^{12}C-rich organic matter in sediments (Vincent and Berger, 1985). This was coincident with global cooling as indicated by mid-Miocene $\partial^{18}O$ values, leading Vincent and Berger (1985) to propose that the increased equator-to-pole thermal gradient caused by global cooling strengthened coastal winds that in turn enhanced the coastal upwelling, the productivity, the development of the oxygen minimum zone, and the sedimentation and entrapment of organic matter. Miocene expansion of the oxygen minimum zone in response to increased productivity is also called upon to preserve organic matter on the continental shelf and slope seaward of the California coastal basins (Summerhayes, 1981a), which in turn increased the prospect for the associated formation of phosphorite, as pointed out above. The main organic-rich rocks of the Monterey Formation are Middle to Late Miocene.

Initiation of the Benguela Current at about the same time (Siesser, 1980; Berger et al., 2002) confirms that these increases in upwelling and associated high productivity occurred in response to a global event – presumably the increased extent of the Antarctic ice sheet and associated changes in the circulation of deep water. While Dean et al. (1984) thought that the deposition of silica-rich sediments in the Benguela off Namibia was Late Pliocene-Early Pleistocene, more recently, Diester-Haass et al. (2004) found that organic enrichment began there at 12 Ma (mid-Miocene) and increased significantly (by factors of up to 7–9) at 7–6 Ma (Late Miocene). Oxygen concentrations in the oxygen minimum zone on the continental slope were lowest between 12 and 9 Ma. Diester-Haass et al. (2004) attributed the rise in productivity not to increased upwelling, but rather to ocean-wide increases in nutrient supply and delivery recognizable here and in the western and eastern equatorial Pacific and the Indian Ocean. Heinrich et al. (2011) used analyses of calcareous dinoflagellates to confirm the onset of modern upwelling at 11.8 Ma and to demonstrate pulses of upwelling at 11.5 and 10.5 Ma related to cooling events associated with increases in ice volume in Antarctica and a likely associated increase in southeasterly coastal winds. Permanent establishment of the upwelling regime with a source in Subantarctic Mode Water seems to date to about 10.4 Ma.

An increase in productivity related to an increase in the strength of upwelling in the Benguela Current peaked in the Pliocene, around 3 Ma ago. Data from Ocean Drilling Program 175 (Berger et al., 2002), collected along the entire southwest African margin from Cape Town to Angola, showed that a Namibian diatomaceous silica event in the Late Pliocene-Early Pleistocene covered the entire Namibian margin and that while silica production was in phase with organic matter production between 3 and 2 Ma, the two were out of phase between 2 and 1 Ma. The silica decrease at that time is attributed to gradual exhaustion of the silica supply from the upper ocean (essentially the thermocline) due to its precipitation in diatom-rich sediments. The deposition of organic matter remained high, reflecting a continued supply of nutrients. Evidently, the ecology changed from a productive environment favorable to diatoms to a productive environment unfavorable to them as global cooling continued.

Sampling by deep ocean drilling showed that laminated organic-rich black shales of Cretaceous age were widespread in the deep waters of the Atlantic, in surrounding basins on the continental margin, and in the equatorial Pacific (Schlanger and Jenkyns, 1976). Concentrations of these deposits at certain time intervals led to them becoming described as oceanic anoxic events (OAEs). OAEs recurred at intervals throughout the Cretaceous, and there was one in the Early Jurassic (Toarcian). Each lasted around 0.5–1.0 million years (Jenkyns, 2010). The Cretaceous black shales of the deep North Atlantic were richest in organic matter along the northwest African margin and off northeast Brazil and Guyana. The Atlantic was much smaller at the time, so these deposits were physically much closer together then than they are now. The organic-rich deposits of the margins of northwest Africa and Guyana were characterized by marine organic matter of planktonic origin, while elsewhere over much of the deep North Atlantic deposits of the same age were dominated by terrestrial organic matter, which were less organically enriched (Summerhayes, 1981b).

As with many open ocean sediments, it seemed highly probable that both the organic-rich and organic-poor layers in these cores were deposited either by turbidity currents, giving rise to thin structured *turbidites*, or from near-bottom suspensions of turbid water flowing down from the continental margin and across the basin floor to form *hemipelagic muds* (Summerhayes, 1987). Recent detailed sedimentological studies of black shales from the Cretaceous of both the North Atlantic and the South Atlantic confirm that many of the laminations in these organic-rich deposits are very thin-bedded fine-grained turbidites (Stow and Dean, 1984; Trabucho-Alexandre et al., 2011). The existence of thinly bedded organic-rich turbidite layers confirms that their source areas – the adjacent continental margins – were sites of deposition of organic-rich sediment, most probably originating from upwelling combined with an anoxic oxygen minimum zone impinging on the continental slope. In contrast, the open Atlantic was rather unproductive (Bralower and Thierstein, 1984). The arrival of organic-rich material on the deep Atlantic seabed via bottom currents, and its decay in situ, may have created anoxic conditions near or at the sediment-water interface, so preserving finely laminated structures from destruction by bioturbation.

Kuypers et al. (2002) found that evidence for anoxia in deep bottom waters during the Cenomanian-Turonian black shale event in the deep southern part of the proto-North Atlantic included "the occurrence of molecular fossils of anoxygenic photosynthetic green sulfur bacteria, lack of bioturbation, and high abundance of redox sensitive trace metals indicate sulfidic conditions, periodically reaching up into the photic zone." Indeed these conditions periodically occurred even at very shallow water depths of 15 m or less. However, the lack of change in bottom water conditions at some sites suggested that the increase in burial rates of organic carbon resulted from changes in productivity rather than increased anoxia.

In shallower water deposits of this age, for example, in the basins of Italy, the presence of delicate laminations along with molecular fossils derived from green sulfur bacteria indicates that bottom waters were anoxic and contained free hydrogen sulfide and that these events occurred when sea surface temperatures were exceptionally warm, 32–36 °C in the topics and 25 °C around the poles (Jenkyns, 2008). The global warming of these times "was accompanied by an accelerated hydrological cycle, increased continental weathering, enhanced nutrient discharge to oceans and lakes, intensified upwelling, and an increase in organic productivity" (Jenkyns, 2010). Jarvis et al. (2011) attribute the warming to an enhanced supply of CO_2 from an increase in volcanic degassing associated with emplacement of the Caribbean large igneous province and noted that it was associated with a decrease in oceanic oxygen concentration. Organic enrichment in the Cenomanian-Turonian was thus a function of increased nutrient supply and decreased oceanic oxygenation, which enhanced the usual effects of coastal upwelling.

Samples of Cenomanian-Turonian age were the richest in organic matter in the North Atlantic and surrounding areas, most of it marine and probably originating from continental margin upwelling (Summerhayes, 1981b; Summerhayes, 1987). It seems likely that this regional upwelling was characteristic of the entire Cretaceous. One possible model to explain the unusual enrichment of Cenomanian-Turonian sediments in organic matter across the North Atlantic basin involves the breaking apart of Africa and South America leading to an oceanic connection between the North and South Atlantic that allowed highly saline, oxygen-depleted, and nutrient-rich waters from the south to enter the North Atlantic, accentuating both productivity, through upwelling at the surface, and oxygen depletion and preservation of organic matter at depth (Summerhayes, 1987; Kuypers et al., 2002). The influx of this "new" deep water appears to have caused significant accumulation of organic-rich sediments until the new reserve of nutrients was exhausted or until continued widening of the connection to the south diminished the influx of highly saline bottom water. This event was associated with an exceptionally large positive excursion in $\partial^{13}C$ (Frakes et al., 1992), suggesting a significant increase in the burial of organic matter rich in ^{12}C.

An alternative source for subsurface nutrients in the North Atlantic was the development of an estuarine circulation pattern in which westward-moving nutrient-depleted Atlantic surface waters were replaced at depth by nutrient-rich Pacific waters moving eastward in an undercurrent through the gap between North and South America, which would have made the North Atlantic a nutrient trap (Summerhayes, 1981b; Trabucho-Alexandre et al., 2010; Topper et al., 2011). However, that circulation is likely to have characterized the entire Cretaceous, not just the Cenomanian-Turonian section.

Interestingly, the Cenomanian-Turonian black shale event along the nearby margins of northwest Africa, northeast Brazil, and Guyana was associated not with the sort of cooling one might expect from enhanced coastal upwelling, but with a rise in temperature of about 3 °C that made this the warmest time interval in the mid-Cretaceous (Forster et al., 2007). Forster et al. (2007) speculated that the rise in both temperature and sea-level at this time may have destabilized the water column, enhancing mixing that could have fed nutrients back into the photic zone to stimulate production. However, it still seems highly likely that upwelling currents played an important part in ensuring high productivity on the margins, as demonstrated by numerical models of the Cretaceous climate system (see below).

Hofmann and Wagner (2011) attribute repetitive fluctuations in productivity, ocean redox conditions, and clastic sediment supply on the Demerara Rise off Guyana at this time to shifts in the mean position of the Intertropical Convergence Zone (ITCZ), which connects the large-scale wind field patterns of the two hemispheres. Upwelling was strongest off Guyana, and burial of organic matter most pronounced, when the ITCZ was in its most southerly position, which maximized the impact of the NE trade winds, which fostered upwelling and organic carbon burial off NW Africa and NE Brazil. Fluctuations in the position of the ITCZ are likely to have been orbitally controlled.

Paleoclimate models of upwelling

Given the association of upwelling with the wind, it ought to be possible in principle to map upwelling from maps of paleo-wind direction. In 1977–1978, the Amoco International Oil Company asked Fred Ziegler of the University of Chicago to expand his *Paleogeographic Atlas Project* to apply to continental reconstruction maps the likely circulation patterns of the atmosphere and ocean, so as to determine the probable positions of oceanic upwelling zones that might be associated with the deposition of organic-rich rocks forming petroleum source beds (Ziegler, 1982). Judy Parrish was appointed to supervise that part of the *Atlas* project, and several other oil companies joined the consortium funding Ziegler's team. Parrish's first principle approach to paleoclimate mapping involved plotting on continental reconstruction maps the likely isobars of atmospheric pressure for summer and winter seasons and deducing from those crude qualitative pressure maps the likely directions of winds and from them the likely locations of wind-driven coastal upwelling currents back through time (Parrish, 1982; Parrish and Curtis, 1982). Parrish and her colleagues carried out this exercise for seven geological periods in the Mesozoic and Cenozoic (Earliest Triassic (Induan), Late-Early Jurassic (Pliensbachian), Latest Jurassic (Volgian), mid-Cretaceous (Cenomanian), Latest Cretaceous (Maastrichtian), mid-Eocene (Lutetian), and mid-Miocene (Vindobonian)) (Parrish and Curtis, 1982) and seven periods in the Paleozoic (Cambrian, Ordovician, Silurian, Devonian, Carboniferous (Mississippian), Carboniferous (Pennsylvanian), Permian) (Parrish et al., 1983). They found a general correlation between the positions of upwelling currents in their model and the actual locations of organic-rich rocks, phosphorites, and chert (representing silica derived from the dissolution of siliceous planktonic skeletons).

Scotese and Summerhayes (1986) quantified Parrish's conceptual and qualitative approach to paleoclimate modeling by applying numerical algorithms to recreate pressure isobars on maps of past continental positions, from which the location of coast-parallel winds – hence coastal upwelling – could be inferred for winter and summer seasons for the Early Triassic (Induan), Early Jurassic (Pliensbachian), Late Jurassic (Volgian), Middle Cretaceous (Cenomanian), Late Cretaceous (Maastrichtian), Middle Eocene (Lutetian), and Middle Miocene (Vindobonian). Taking the next natural step, Eric Barron applied a full-blown general circulation model to the warm world of the mid-Cretaceous and applied an algorithm to determine where upwelling should have taken place based on his model-generated wind directions (Barron, 1986). His maps of where upwelling was likely to have occurred correspond reasonably well with those hindcast by both Parrish and Curtis (1982) and Scotese and Summerhayes (1986). Paleoclimate maps for the Cenomanian-Turonian showed that upwelling should have been well developed then along the margin of northwest Africa and in the narrow gap between northwest Africa and Guyana and NE Brazil, where we find the most marine organic matter (Barron, 1986; Scotese and Summerhayes, 1986).

These modeling approaches are not ends in themselves, but guides to geological thinking. Just because a paleo-upwelling map suggests that upwelling may have characterized a particular continental margin does not mean that the rocks of the upwelling *triad* will be found there. Other requirements include a sedimentary basin, a limited supply of terrestrial silt and clay, a time of transgression, an oxygen minimum zone, and nutrient-enriched subsurface source waters.

Summary and conclusions

Upwelling currents are driven by the wind and increase in strength as the wind does. They bring up nutrients from below, stimulating productivity in surface waters where winds blow parallel to the coast and where winds blow persistently over the open ocean, especially at the equator and around Antarctica. Upwelling became more prominent in the Late Cenozoic as the world cooled and winds strengthened. It was prominent in the Miocene and has continued to the present, fluctuating from stronger in glacial times to weaker in interglacial times, with a tendency to increase at times of transition from glacial to interglacial. Signs of upwelling are found further offshore during times when sea-level was low. The oxygen minimum zone beneath upwelling centers may become significantly depleted in oxygen, permitting more preservation of organic matter; this is less true of well-oxygenated basins like the Atlantic and more true of poorly oxygenated basins like the North Pacific and Arabian Sea. Upwelling led to the deposition of abundant marine organic matter along the margins of northwest Africa and Guyana during the Cenomanian-Turonian, most likely due to an influx of nutrient-rich and oxygen-poor water from the South Atlantic as Africa and South America moved apart. The deep ocean black shales of oceanic anoxic events are not directly related to upwelling, but may contain abundant organic-rich turbidites derived from continental margins beneath upwelling currents. A by-product of decomposition beneath upwelling centers is the release of phosphate to precipitate as calcium carbonate-fluorapatite or to phosphatize preexisting carbonates, which may progress to the point of forming phosphorites. A further by-product is the deposition of sediments rich in the remains of siliceous diatoms, provided that surface waters have not become depleted in silica. Upwelling is thus frequently recognized in the geological record from an association between organic-rich shales or mudstones, diatomite, and phosphorite, the "signature triad." The absence of organic enrichment may not mean that upwelling was absent – it may reflect either low productivity, due to an inadequate supply of nutrients, or low preservation, due to either an abundance of oxygen or the absence of conditions suitable for the accumulation of fine-grained sediment. The richness of productivity associated with upwelling depends as well upon the source of nutrients in the intermediate depth waters of the adjacent ocean. Predictions of upwelling occurrence and organic enrichment based purely on past wind directions are likely to be overly simplistic.

Bibliography

Adkins, J., de Menocal, P., and Eshel, G., 2006. The "African humid period" and the record of marine upwelling from excess ^{230}Th in ocean drilling program hole 658C. *Paleoceanography*, **21**, PA4203, doi:10.1029/2005PA001200.

Adkins, J., de Menocal, P. B., and Eshel, G., 2007. Correction to The "African humid period" and the record of marine upwelling from excess ^{230}Th in ocean drilling program hole 658C. *Paleoceanography*, **22**, PA1206, doi:10.1029/2006PA001388.

Anand, P., Kroon, D., Singh, A. D., Ganeshram, R. S., Ganssen, G., and Elderfield, H., 2008. Coupled sea surface temperature-seawater ∂^{18}O reconstructions in the Arabian Sea at the millennial scale for the last 35 ka. *Paleoceanography*, **23**, PA4207, doi:10.1029/2007PA001564.

Anderson, R. F., 2003. Chemical tracers of particle transport. In Elderfield, H., Holland, H. D., and Turekian, K. K. (eds.), *The Oceans and Marine Geochemistry*. Oxford: Elsevier-Pergamon. Treatise on Geochemistry, Vol. 6, pp. 247–291.

Archer, D., 2003. Biological fluxes in the ocean and atmospheric pCO_2. In Elderfield, H., Holland, H. D., and Turekian, K. K. (eds.), *The Oceans and Marine Geochemistry*. Oxford: Elsevier-Pergamon. Treatise on Geochemistry, Vol. 6, pp. 275–291.

Barron, E. J., 1986. Mathematical climate models: insights into the relationship between climate and economic sedimentary deposits. In Parrish, J. T., and Barron, E. J. (eds.), *Paleoclimates and Economic Geology*. Tulsa: Society of Economic Paleontologists and Mineralogists. Lecture Notes for Short Course, Vol. 18, pp. 31–83.

Baturin, G. H., 2000. Formation and evolution of phosphorite grains and nodules on the Namibian Shelf, from recent to Pleistocene. In Glenn, C. R., Prevot-Lucas, L., and Lucas, J. (eds.), *Marine Authigenesis: From Global to Microbial*. Tulsa: Society of Economic Paleontologists and Mineralogists. SEPM Special Publication, Vol. 66, pp. 185–199.

Baturin, G. H., Merkulova, K. I., and Chalov, P. I., 1972. Radiometric evidence for recent formation of phosphatic nodules in marine shelf sediments. *Marine Geology*, **13**, M37–M41.

Berger, W. H., Smetacek, V. S., and Wefer, G. (eds.), 1989a. *Productivity of the Ocean: Present and Past. Dahlem Konferenzen*. Chichester: Wiley. 470pp.

Berger, W. H., Smetacek, V. S., and Wefer, G., 1989b. Ocean productivity – an overview. In Berger, W. H., Smetacek, V. S., and Wefer, G. (eds.), *Productivity of the Ocean: Present and Past. Dahlem Konferenzen*. Chichester: Wiley, pp. 1–34.

Berger, W. H., Lange, C. B., and Wefer, G., 2002. Upwelling history of the Benguela-Namibia system: a synthesis of Leg 1775 results. In Wefer, G., Berger, W. H., and Richter, C. (eds.), *Proceedings of ODP, Science Results, 175* (Online). Available from World Wide Web: http://www-odp.tamu.edu/publications/175_SR/synth/synth.htm

Bralower, T. J., and Thierstein, H. R., 1984. Low-productivity and slow deep-water circulation in mid-Cretaceous oceans. *Geology*, **12**, 614–618.

Bremner, J. M., 1983. Biogenic sediments of the South West African (Namibian) continental margin. In Thiede, J., and Suess, E. (eds.), *Coastal Upwelling, Its Sediment Record*. New York: Plenum, Vol. 2, pp. 73–103.

Budziak, D., Schneider, R., Rostek, F., Müller, P. J., Bard, E., and Wefer, G., 2000. Late Quaternary insolation forcing on total organic carbon and ^{37}C alkenone variations in the Arabian Sea. *Paleoceanography*, **15**(3), 307–321.

Burnett, W. C., 1977. Geochemistry and origin of phosphorite deposits from off Peru and Chile. *Geological Society of America Bulletin*, **88**, 813–823.

Burnett, W. C., and Riggs, S. R. (eds.), 1990. *Phosphate Deposits of the World, v3., Neogene to Modern Phosphorites*. Cambridge: Cambridge University Press.

Burnett, W. C., Veeh, H. H., and Soutar, A., 1980. U-Series, oceanographic and sedimentary evidence in support of recent formation of phosphate nodules off Peru. In Bentor, Y. K. (ed.), *Marine Phosphorites*. Tulsa: Society of Economics Paleontologists and Mineralogists. SEPM Special Publication, Vol. 29, pp. 61–71.

Calvert, S. E., 1987. Oceanographic controls on the accumulation of organic matter in marine sediments. In Brooks, J., and Fleet, A. J. (eds.), *Marine Petroleum Source Rocks*. Oxford: Blackwell. Geological Society of London, Special Publication, Vol. 26, pp. 137–151.

Caulet, J. P., Vénec-Peyré, M. T., Vergnaud-Grazzini, C., and Nigrini, C., 1992. Variation of South Somalian upwelling during the last 160 ka: radiolarian and foraminifera records in core MD 85674. In Summerhayes, C. P., Prell, W. L., and Emeis, K. C. (eds.), *Upwelling Systems: Evolution Since the Early Miocene*. London: Geological Society. Geological Society of London, Special Publication, Vol. 64, pp. 379–389.

Cook, P. J., and McElhinny, M. W., 1979. A reevaluation of the spatial and temporal distribution of sedimentary phosphate deposits in the light of plate tectonics. *Economic Geology*, **24**(2), 315–330.

de Abreu, L., Abrantes, F., Shackleton, N. J., Tzedakis, P. C., McManus, J. F., Oppo, D. W., and Hall, M. A., 2005. Ocean climate variability in the eastern North Atlantic during interglacial marine isotope stage 11: a partial analogue to the Holocene? *Paleoceanography*, **20**, PA3009, doi:10.1029/2004PA001091.

Dean, W. E., Hay, W. W., and Sibuet, J. C., 1984. Geologic evolution, sedimentation, and paleoenvironments of the Angola Basin and Adjacent Walvis ridge: synthesis of results of Deep Sea Drilling Project Leg 75. In Hay, W. W., Sibuet, J. C., et al. (eds.), *Initial Reports of the DSDP 75*. Washington, DC: U.S. Govt. Printing Office, pp. 509–544.

Dean, W. E., Zheng, Y., Ortiz, J. D., and van Geen, A., 2006. Sediment Cd and Mo accumulation in the oxygen-minimum zone off western Baja California linked to global climate over the past 52 kyr. *Paleoceanography*, **21**, PA4209, doi:10.1029/2005PA001239.

Dekens, P. S., Ravelo, A. C., and McCarthy, M. D., 2007. Warm upwelling regions in the Pliocene warm period. *Paleoceanography*, **22**, PA3211, doi:10.1029/2006PA001394.

Dezileau, L., Ulloa, O., Hebbeln, D., Lamy, F., Reyss, J.-L., and Fontugne, M., 2004. Iron control of past productivity in the coastal upwelling system off the Atacama Desert, Chile. *Paleoceanography*, **19**, PA3012, doi:10.1029/2004PA001006.

Diester-Haass, L., Meyers, P. A., and Rothe, P., 1992. The Benguela current and associated upwelling on the southwest African margin: a synthesis of the Neogene-Quaternary sedimentary record at DSDP sites 362 and 532. In Summerhayes, C. P., Prell, W. L., and Emeis, K. C. (eds.), *Upwelling Systems: Evolution Since the Early Miocene*. London: Geological Society. Geological Society of London, Special Publication, Vol. 64, pp. 331–342.

Diester-Haass, L., Meyers, P. A., and Bickert, T., 2004. Carbonate crash and biogenic bloom in the late Miocene: evidence from ODP Sites 1085, 1086, and 1087 in the Cape Basin, southeast Atlantic Ocean. *Paleoceanography*, **19**, PA1007, doi:10.1029/2003PA000933.

Diffenbaugh, N. S., Sloan, L. C., and Snyder, M. A., 2003. Orbital suppression of wind-driven upwelling in the California Current at 6 ka. *Paleoceanography*, **18**(2), 1051, doi:10.1029/2002PA000865.

Eglinton, T. I., and Repeta, D. J., 2003. Organic matter in the contemporary ocean. In Elderfield, H., Holland, H. D., and Turekian, K. K. (eds.), *The Oceans and Marine Geochemistry*. Oxford: Elsevier-Pergamon. Treatise on Geochemistry, Vol. 6, pp. 145–180.

Emery, K. O., 1960. *The Sea off Southern California: A Modern Habitat of Petroleum*. New York: Wiley. 366 pp.

Emmer, E., and Thunell, R. C., 2000. Nitrogen isotope variations in Santa Barbara Basin sediments: implications for denitrification in the eastern tropical North Pacific during the last 50,000 years. *Paleoceanography*, **15**(4), 377–387.

Feldberg, M. J., and Mix, A. C., 2003. Planktonic foraminifera, sea surface temperatures, and mechanisms of oceanic change in the Peru and south equatorial currents, 0–150 ka BP. *Paleoceanography*, **18**(1), 1016, doi:10.1029/2001PA000740.

Forster, A., Schouten, S., Moriya, K., Wilson, P. A., and Sinninghe Damsté, J. S., 2007. Tropical warming and intermittent cooling during the Cenomanian/Turonian Oceanic Anoxic Event 2: sea surface temperature records from the equatorial Atlantic. *Paleoceanography*, **22**, PA1219, doi:10.1029/2006PA001349.

Frakes, L. A., Francis, J. E., and Sytkus, J. I., 1992. *Climate Modes of the Phanerozoic – The History of the Earth's Climate Over the Past 600 Million Years*. Cambridge: Cambridge University Press. 274 pp.

Gersonde, R., Crosta, X., Abelmann, A., and Armand, L., 2005. Sea-surface temperature and sea ice distribution of the Southern Ocean at the EPILOG Last Glacial Maximum – a circum-Antarctic view based on siliceous microfossil records. *Quaternary Science Reviews*, **24**, 869–896.

Giraud, X., and Paul, A., 2010. Interpretation of the paleo-primary production record in the NW African coastal upwelling system as potentially biased by sea level change. *Paleoceanography*, **25**, PA4224, doi:10.1029/2009PA001795.

Hanson, C. E., Pattiaratchi, C., and Waite, A. M., 2005. Sporadic upwelling on a downwelling coast: phytoplankton responses to spatially variable nutrient dynamics off the Gascoyne region of Western Australia. *Continental Shelf Research*, **25**(12–13), 1561–1582.

Haslett, S. K., and Smart, C. W., 2006. Late Quaternary upwelling off tropical NW Africa: new micropalaeontological evidence from ODP Hole 658C. *Journal of Quaternary Science*, **21**, 259–269.

Heinrich, S., Zonneveld, K. A. F., Bickert, T., and Willems, H., 2011. The Benguela upwelling related to the Miocene cooling events and the development of the Antarctic Circumpolar Current: evidence from calcareous dinoflagellate cysts. *Paleoceanography*, **26**, PA3209, doi:10.1029/2010PA002065.

Heinze, P.-M., and Wefer, G., 1992. The history of coastal upwelling off Peru (11°S, ODP Leg 112, Site 680B) over the past 650,000 years. In Summerhayes, C. P., Prell, W. L., and Emeis, K. C. (eds.), *Upwelling Systems: Evolution Since the Early Miocene*. London: Geological Society. Geological Society of London, Special Publication, Vol. 64, pp. 451–462.

Hendy, I. L., Pedersen, T. F., Kennett, J. P., and Tada, R., 2004. Intermittent existence of a southern Californian upwelling cell during submillennial climate change of the last 60 kyr. *Paleoceanography*, **19**, PA3007, doi:10.1029/2003PA000965.

Herbert, T. D., 2003. Alkenone paleotemperature determinations. In Elderfield, H., Holland, H. D., and Turekian, K. K. (eds.), *The Oceans and Marine Geochemistry*. Oxford: Elsevier-Pergamon. Treatise on Geochemistry, Vol. 6, pp. 391–432.

Hodell, D., Crowhurst, S., Skinner, L., Tzedakis, P. C., Margari, V., Channell, J. E. T., Kamenov, G., Maclachlan, S., and Rothwell, G., 2013. Response of Iberian margin sediments to orbital and suborbital forcing over the past 420 ka. *Paleoceanography*, **28**, 185–199, doi:10.1002/palo.20017.

Hofmann, P., and Wagner, T., 2011. ITCZ controls on late cretaceous black shale sedimentation in the tropical Atlantic Ocean. *Paleoceanography*, **26**, PA4223, doi:10.1029/2011PA002154.

Jahnke, R. A., and Shimmield, G. B., 1995. Particle flux and its conversion to the sediment record: coastal ocean upwelling systems. In Summerhayes, C. P., Emeis, K. -C., Angel, M. V., Smith R. L., and Zeitschel, B., (eds.), *Upwelling in the Ocean – Modern Processes and Ancient Records*. Dahlem Workshop Reports, Environmental Sciences Report, 18. New York: Wiley, pp. 83–100.

Jarvis, I., Lignum, J. S., Gröcke, D. R., Jenkyns, H. C., and Pearce, M. A., 2011. Black shale deposition, atmospheric CO_2 drawdown, and cooling during the Cenomanian-Turonian Oceanic Anoxic Event. *Paleoceanography*, **26**, PA3201, doi:10.1029/2010PA002081.

Jenkyns, H. C., 2008. *Oceanic Anoxic Events: 30 Years On*. www.noc.soton.ac.uk/nocs/fridsemabs/Series_07-08/Hugh_Jenkyns_020508.pdf

Jenkyns, H. C., 2010. Geochemistry of oceanic anoxic events. *Geochemistry, Geophysics, Geosystems*, **11**, Q03004, doi:10.1029/2009GC002788.

Kienast, S. S., Calvert, S. E., and Pedersen, T. F., 2002. Nitrogen isotope and productivity variations along the northeast Pacific margin over the last 120 kyr: surface and subsurface paleoceanography. *Paleoceanography*, **17**(4), 1055, doi:10.1029/2001PA000650.

Kirst, G., Schneider, R. R., Muller, P. J., Von Storch, I., and Wefer, G., 1999. Late Quaternary temperature variability in the Benguela current system derived from alkenones. *Quaternary Research*, **52**, 92–103.

Kuypers, M. M. M., Pancost, R. D., Nijenhuis, I. A., and Sinninghe Damsté, J. S., 2002. Enhanced productivity led to increased organic carbon burial in the euxinic North Atlantic basin during the late Cenomanian oceanic anoxic event. *Paleoceanography*, **17**(4), 1051, doi:10.1029/2000PA000569.

Little, M. G., Schneider, R. R., Kroon, D., Price, N. B., Summerhayes, C. P., and Segl, M., 1997. Trade wind forcing of upwelling, seasonality, and Heinrich events as a response to sub-Milankovitch climate variability. *Paleoceanography*, **12**(4), 568–576.

Loubere, P., Fariduddin, M., and Murray, R. W., 2003. Patterns of export production in the eastern equatorial Pacific over the past 130,000 years. *Paleoceanography*, **18**(2), 1028, doi:10.1029/2001PA000658.

McKelvey, V. E., 1963. Successful new techniques in prospecting for phosphate deposits. In *US Department State, Natural Resources Contribution to UN Conference Application of Science and Technology for the Benefit of Less Developed Areas*. Geneva, Vol. 2, pp. 164–172.

Mohtadi, M., and Hebbeln, D., 2004. Mechanisms and variations of the paleoproductivity off northern Chile (24°S-33°S) during the last 40,000 years. *Paleoceanography*, **19**, PA2023, doi:10.1029/2004PA001003.

Murray, J., and Chumley, J., 1924. *The Deep Sea Deposits of the Atlantic Ocean*. Edinburgh: Robert Grant & Son. Transactions of the Royal Society of Edinburgh, Vol. 54, part 1, pp. 1–252.

Murray, J., and Renard, A. F., 1891. *Deep-Sea Deposits*. London: H. M.S.O.. Reports on the Scientific Results of the H.M.S. Challenger 1873–76.

Palter, J. S., Sarmiento, J. L., Gnanadesikan, A., Simeon, J., and Slater, D., 2010. Fueling primary productivity: nutrient return pathways from the deep ocean and their dependence on the meridional overturning circulation. *Biogeosciences Discussions*, **7**, 4045–4088.

Parrish, J. T., 1982. Upwelling and petroleum source beds, with reference to the Palaeozoic. *Bulletin of the American Association of Petroleum*, **66**, 750–754.

Parrish, J. T., and Curtis, R. L., 1982. Atmospheric circulation, upwelling, and organic-rich rocks in the Mesozoic and Cenozoic. *Palaeogeography Palaeoclimatology Palaeoecology*, **40**, 31–66.

Parrish, J. T., Ziegler, A. M., and Humphreville, R. G., 1983. Upwelling in the Paleozoic era. In Thiede, J., and Suess, E. (eds.), *Coastal Upwelling – Its Sediment Record, Part B, Sedimentary Records of Ancient Coastal Upwelling*. New York/London: Plenum. NATO Conference Series IV, Marine Sciences, pp. 553–578.

Pichevin, L., Martinez, P., Bertrand, P., Schneider, R., Giraudeau, J., and Emeis, K., 2005. Nitrogen cycling on the Namibian Shelf and slope over the last two climatic cycles: local and global forcings. *Paleoceanography*, **20**, PA2006, doi:10.1029/2004PA001001.

Reimers, C. E., and Suess, E., 1983. Late Quaternary fluctuations in the cycling of organic matter off central Peru: a proto-kerogen record. In Thiede, J., and Suess, E. (eds.), *Coastal Upwelling, Its Sediment Record. Part A: Responses of the Sedimentary Regime to Present Coastal Upwelling.* New York: Plenum, pp. 497–525.

Ruttenberg, K. C., 2005. The global phosphorus cycle. In Elderfield, H., Holland, H. D., and Turekian, K. K. (eds.), *The Oceans and Marine Geochemistry.* Amsterdam: Elsevier. Treatise on Geochemistry, Vol. 6, pp. 585–643.

Sarnthein, J. M., Winn, K., Duplessy, J.-C., and Fontugne, M., 1988. Global variations of surface ocean productivity in low and mid latitudes: influence on CO_2 reservoirs of the deep ocean and atmosphere during the last 21,000 years. *Paleoceanography,* **3**(4), 361–399.

Schlanger, S. O., and Jenkyns, H. C., 1976. Cretaceous oceanic anoxic events: causes and consequences. *Geologie en Mijnbouw,* **55**, 179–184.

Scotese, C. R., and Summerhayes, C. P., 1986. Computer model of paleoclimate predicts coastal upwelling in the Mesozoic and Cenozoic. *Geobyte,* **1**(3), 28–44 and 94.

Shimmield, G. B., 1992. Can sediment geochemistry record changes in coastal upwelling palaeoproductivity? Evidence from Northwest Africa and the Arabian Sea. In Summerhayes, C. P., Prell, W. L., and Emeis, K. C. (eds.), *Upwelling Systems: Evolution Since the Early Miocene.* London: Geological Society. Geological Society of London, Special Publication, Vol. 64, pp. 29–46.

Shimmield, G. B., and Jahnke, R. A., 1995. Particle flux and its conversion to the sediment record: open ocean upwelling systems. In Summerhayes, C. P., Emeis, K. C., Angel, M. V., Smith R. L., and Zeitschel, B. (eds.), *Upwelling in the Ocean – Modern Processes and Ancient Records.* Dahlem Workshop Reports, Environmental Sciences Report, 18. New York: Wiley, pp. 171–191.

Siesser, W. G., 1980. Late Miocene origin of the Benguela upwelling system off northern Namibia. *Science,* **208**, 283–285.

Sigman, D. M., and Haug, G. H., 2003. Biological pump in the past. In Elderfield, H., Holland, H. D., and Turekian, K. K. (eds.), *The Oceans and Marine Geochemistry.* Oxford: Elsevier-Pergamon. Treatise on Geochemistry, Vol. 6, pp. 491–528.

Singh, A. D., Jung, S. J. A., Darling, K., Ganeshram, R., Ivanochko, T., and Kroon, D., 2011. Productivity collapses in the Arabian Sea during glacial cold phases. *Paleoceanography,* **26**, PA3210, doi:10.1029/2009PA001923.

Stow, D. A. V., and Dean, W. E., 1984. Middle Cretaceous black sales at site 530 in the southeastern Angola Basin. In Hay, W. W., Sibuet, J.-C., et al. (eds.), *Initial Reports of the DSDP, 75.* Washington, DC: U.S. Govt. Printing Office, pp. 809–817.

Suess, E., and Thiede, J. (eds.), 1983. *Coastal Upwelling, Its Sediment Record. Part A: Responses of the Sedimentary Regime to Present Coastal Upwelling.* New York: Plenum. 604 pp.

Summerhayes, C. P., 1981a. Oceanographic controls on organic matter in the Miocene Monterey Formation, Offshore California. In Garrison, R. E., and Douglas, R. G. (eds.), *The Monterey Formation and Related Siliceous Rocks of California.* Los Angeles: Society of Economic Paleontologists and Mineralogists, pp. 213–219.

Summerhayes, C. P., 1981b. Organic facies of middle Cretaceous black shales in deep North Atlantic. *Bulletin of the American Association of Petroleum,* **65**(11), 2364–2380.

Summerhayes, C. P., 1983. Sedimentation of organic matter in upwelling regimes. In Thiede, J., and Suess, E. (eds.), *Coastal Upwelling, Its Sediment Record, Part B: Sedimentary Records of Ancient Coastal Upwelling.* New York: Plenum, pp. 29–72.

Summerhayes, C. P., 1987. Organic-rich Cretaceous sediments from the North Atlantic. In Brooks, J., and Fleet, A. J. (eds.), *Marine Petroleum Source Rocks.* Oxford: Blackwell. Geological Society of London, Special Publication, Vol. 26, pp. 301–316.

Summerhayes, C. P., and McArthur, J. M., 1990. Moroccan offshore phosphate deposits. In Burnett, W. C., and Riggs, S. R. (eds.), *Phosphate Deposits of the World.* Cambridge: Cambridge University Press, Vol. 3, pp. 159–166.

Summerhayes, C. P., Nutter, A. H., and Tooms, J. S., 1971. Geological structure and development of the continental margin of northwest Africa. *Marine Geology,* **11**, 1–25.

Summerhayes, C. P., Prell, W. L., and Emeis, K.-C., 1992. *Upwelling Systems: Evolution Since the Early Miocene.* London: Geological Society. Geological Society of London, Special Publication, Vol. 64. 519 pp.

Summerhayes, C. P., Emeis, K. -C., Angel, M. V., Smith R. L., and Zeitschel, B., (eds.), 1995a. *Upwelling in the Ocean – Modern Processes and Ancient Records.* Dahlem Workshop Reports, Environmental Sciences Report, 18. New York: Wiley, 418 pp.

Summerhayes, C. P., Kroon, D., Rosell-Melé, A., Jordan, R. W., Schrader, H.-J., Hearn, R., Villanueva, J., Grimalt, J. O., and Eglinton, G., 1995b. Variability in the Benguela Current upwelling system over the past 70,000 years. *Progress in Oceanography,* **35**, 207–251.

Summerhayes, C. P., Emeis, K. -C., Angel, M. V., Smith R. L., and Zeitschel, B., 1995c. Upwelling in the ocean – modern processes and ancient records. In Summerhayes, C. P., Emeis, K-C., Angel, M. V., Smith, R. L., and Zeitschel, B. (eds.), *Upwelling in the Ocean – Modern Processes and Ancient Records.* Dahlem Workshop Reports, Environmental Sciences Report, 18. New York: Wiley, pp. 1–37.

Tamburini, F., Föllmi, K. B., Adatte, T., Bernasconi, S. M., and Steinmann, P., 2003. Sedimentary phosphorus record from the Oman margin: new evidence of high productivity during glacial periods. *Paleoceanography,* **18**(1), 1015, doi:10.1029/2000PA000616.

Thiede, J., and Suess, E. (eds.), 1983. *Coastal Upwelling, Its Sediment Record. Part B: Sedimentary Records of Ancient Coastal Upwelling.* New York: Plenum. 610 pp.

Thomson, J., Calvert, S. E., Mukherjee, S., Burnett, W. C., and Bremner, J. M., 1984. Further studies of the nature, composition and ages of contemporary phosphorite from the Namibian Shelf. *Earth and Planetary Science Letters,* **69**, 341–353.

Thomson, J., Nixon, S., Summerhayes, C. P., Schonfeld, J., Zahn, R., and Grootes, P., 1999. Implications for sediment changes on the Iberian margin over the last two glacial/interglacial transitions from (^{230Th} excess)o systematics. *Earth and Planetary Science Letters,* **165**, 255–270.

Thomson, J., Nixon, S., Summerhayes, C. P., Rohling, E., Schonfeld, J., Zahn, R., Grootes, P., Abrantes, F., Gaspar, L., and Vaqueiro, S., 2000. Enhanced productivity on the Iberian margin during glacial/interglacial transitions revealed by barium and diatoms. *Journal of the Geological Society of London,* **157**, 667–677.

Toggweiler, J. R., and Carson, S., 1995. What are upwelling systems contributing to the ocean's carbon and nutrient budgets? In Summerhayes, C. P., Emeis, K. -C., Angel, M. V., Smith, R. L., and Zeitschel, B. (eds.), *Upwelling in the Ocean – Modern Processes and Ancient Records.* Dahlem Workshop Reports, Environmental Sciences Report, 18. New York: Wiley, pp. 337–360.

Topper, R. P. M., Trabucho-Alexandre, J., Tuenter, E., and Meijer, P. T., 2011. A regional ocean circulation model for the mid-Cretaceous North Atlantic Basin: implications for black shale formation. *Climate of the Past,* **7**, 277–297.

Trabucho Alexandre, J., Tuenter, E., Henstra, G. A., van der Zwan, K. J. R., van de Wal, S. W., Dijkstra, H. A., and de Boer, P. L., 2010. The mid-Cretaceous North Atlantic nutrient trap: black shales and OAEs. *Paleoceanography*, **25**, PA4201, doi:10.1029/2010PA001925.

Trabucho-Alexandre, J., Van Gilst, R. I., Rodriguez-Lopez, J. P., and De Boer, P. L., 2011. The sedimentary expression of oceanic anoxic event 1b in the North Atlantic. *Sedimentology*, **58**, 1217–1246.

Vincent, E., and Berger, W. H., 1985. Carbon dioxide and polar cooling in the Miocene: the monterey hypothesis. In Sundquist, E. T., and Broecker, W. S. (eds.), *The Carbon Cycle and Atmospheric CO_2: Natural Variations Archean to Present*. Washington, DC: American Geophysical Union. Geophysical Monograph, Vol. 32, pp. 455–468.

White, L. D., Garrison, R. E., and Barron, J. A., 1992. Miocene intensification of upwelling along the California margin as recorded in siliceous facies of the Monterey Formation and Offshore DSDP sites. In Summerhayes, C. P., Prell, W. L., and Emeis, K. C. (eds.), *Upwelling Systems: Evolution Since the Early Miocene*. London: Geological Society. Geological Society of London, Special Publication, Vol. 64, pp. 429–442.

Zhao, M., Beveridge, N. A. S., Shackleton, N. J., and Sarnthein, M., 1995. Molecular stratigraphy of cores off Northwest Africa: sea surface temperature history over the last 80 ka. *Paleoceanography*, **10**, 661–675.

Zheng, Y., van Geen, A., and Anderson, R. F., 2000. Intensification of the northeast Pacific oxygen minimum zone during the Bölling-Alleröd warm period. *Paleoceanography*, **15**(5), 528–536.

Ziegler, A. M., 1982. *The University of Chicago Paleogeographic Atlas Project: Background – Current Status – Future Plans*. Department of Geophysical Sciences, University of Chicago. Unpublished Ms. 19 pp.

Cross-references

Anoxic Oceans
Continental Slope
Dust in the Ocean
Energy Resources
Laminated Sediments
Marine Mineral Resources
Modelling Past Oceans
Organic Matter
Paleoproductivity
Phosphorites
Silica
Turbidites

V

VOLCANISM AND CLIMATE

Olav Eldholm[1] and Millard F. Coffin[2]
[1]Department of Earth Sciences, University of Bergen, Bergen, Norway
[2]Institute for Marine and Antarctic Studies, University of Tasmania, Hobart, TAS, Australia

Introduction

The impact of volcanism on weather and climate is unequivocally documented by temporal correlations in both historical and prehistorical records. Although volcanic events have been traced far back in time, only Quaternary and Holocene records provide the detail and resolution necessary to address many otherwise ambiguous connections. Both observations and modeling suggest inherent relationships among the geosphere, hydrosphere, cryosphere, and atmosphere, setting the stage for various climate responses commonly incorporating complex feedback loop systems and their elusive dependence on proximity to thresholds. Furthermore, both volcanism and climate may trigger secondary phenomena of sometimes catastrophic nature. The cause-and-effect sequence is particularly intriguing, i.e., whether volcanism forces climate or changing climate forces volcanism or if, in fact, either or both can occur in nature. This overview is based on a number of studies that comprehensively address the overall theme (e.g., Cas and Wright, 1987; Sigurdsson, 1999; Robock, 2000; Schmincke, 2004; Oppenheimer, 2011; McGuire, 2012; Cashman and Sparks, 2013; and references therein).

Volcanic impact

Volcanic impact, V_{imp}, and how it affects Earth system, climate, and human society depend on many factors of which timing, duration, and size are key variables. Qualitatively, it may be expressed as

$$V_{imp} \sim f\ (L,\ T,\ M,\ E,\ C,\ S), \qquad (1)$$

where L is the location, in particular relative to major fluid and gas circulation systems; T is the duration; M is the size, for example, given by the volcanic explosivity index (VEI); E is the mode of eruption, i.e., explosive or effusive; C is the ash, gas, and lava composition; and S is the state of societal development. However, relatively large uncertainties in some parameters may preclude assessing volcano-climate relationships at a high confidence level in the geological past. In addition to Eq. 1, the complex relationship of volcanism to environmental change is greatly contingent on the relative abundances of volcanic products (lava, ash, pyroclastics, volatiles, etc.) and their respective physical and chemical properties and on the eruption mode and setting, of which the explosive and effusive eruption processes and submarine and subaerial settings may be considered end-members, respectively.

Short-term volcanic climate forcing is well documented, and even relatively small events may not only cause severe local disturbance and devastation but also have regional and sometimes global impact. Two Icelandic events illustrate such forcing; the 2010 Eyjafjallajökull eruption, which did not have a climatic impact, caused regional havoc on modern air traffic, grounding more than 100,000 flights; the loss to aviation alone has been put at §250 million per day (Gudmundsson et al., 2010; Robock, 2013). The 8-month-long 1783–1784 Laki

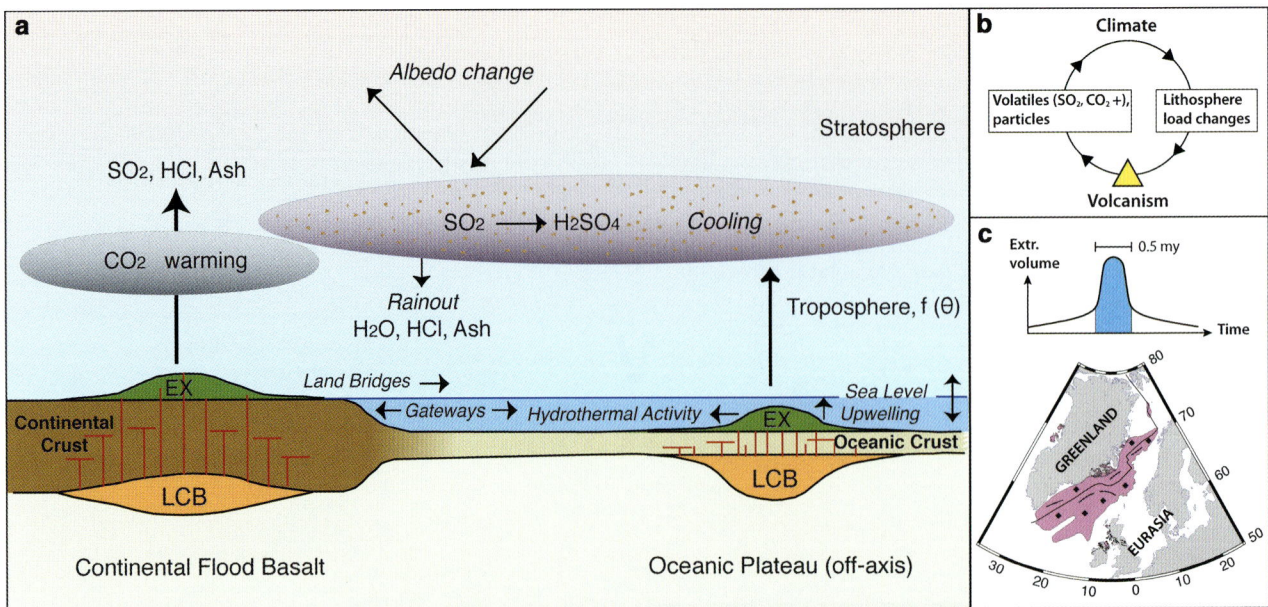

Volcanism and Climate, Figure 1 Schematic diagrams showing (**a**) key environmental effects of volcanism, shown in a LIP setting, indicating how eruptions may significantly perturb the Earth system (modified from McCormick et al. (1995) and Coffin et al. (2006)). Note that many present submarine LIPs are formed in subaerial and shallow marine environments. *EX* extrusive. (**b**) The classical chicken-and-egg conundrum. Note that both causal paths may act simultaneously. (**c**) Extent, shown in *red*, and rate of extrusive magma, in *blue*, erupted during formation of the North Atlantic LIP. Adjacent sedimentary basins also commonly contain lava, dike, and volatile expulsion features.

fissure basalt eruption caused a cold spell, thousands of deaths of both humans and animals, and famine also in Europe and beyond (Vasey, 1991). Moreover, the well-monitored 1991 silicic explosive eruption of Pinatubo in the Philippines provides clear evidence for its impact on global temperature, wind, and precipitation patterns. Globally, an average temperature anomaly of $-0.5\ °C$ is inferred that lasted for about 2 years (McCormick et al., 1995).

Eruptions may trigger secondary phenomena with links to climate as well, specifically by changing or modifying oceanic circulation patterns by forming and destroying marine barriers and gateways. Nonetheless, the greatest impact arises from atmospheric modifications caused by the largest eruptions. Climate forcing is inferred from historic eruptions, but they are dwarfed by some gigantic, cataclysmic effusive and explosive eruptions, from commonly denoted supervolcanoes, at irregular intervals farther back in the geological record. An example is the Toba eruption at ~74 ka (Rampino and Self, 1992). These huge events are relatively rare, approximately two events each 100 kyr since 2 Ma. The most recent one was Taupo in New Zealand, at 26.5 ka ago. Many supervolcanoes are associated with mantle plumes as are the large igneous provinces (LIPs) (Coffin and Eldholm, 1994). Thus, single eruptions of individual volcanoes (e.g., Eyjafjallajökull) and supervolcanoes (e.g., Toba) or cumulative multiple eruptions within a limited time period may force climate change.

Atmosphere processes

Climate is directly linked to the heat budget of the Earth's atmosphere and oceans which is governed by astronomical parameters, solar activity, insolation, and albedo. In turn, both natural and anthropogenic sources and processes modify the budget. Volcanic climate forcing depends largely on the amount and composition of volcanic gas, a minor magma component, injected into the atmosphere. The most important gases are SO_2 and other sulfur compounds and the "greenhouse" gas CO_2 (Figure 1a). In fact, much sulfur gas, in particular SO_2, has a volcanic origin, and present and past volcanic eruptions have had a significant impact on chemical and radiative properties and the energy budget of the atmosphere.

The volcanic record and measurements from space show that eruptions may affect areas far from the source, modifying climate for several years. Their impact depends primarily on the state and composition of the volcanic constituents, the amount of sulfur injected into the stratosphere, and to which atmospheric level the volcanic plume, carrying particulates and gases, extends. In particular, the stratosphere, above $>\sim10–20$ km and lowest at high latitudes, appears sensitive to the physical-chemical properties of volcanic products, affecting the insolation properties and thereby the atmospheric temperature. However, SO_2 and CO_2 modify climate in opposite ways, inducing cooling and warming, respectively. There is much uncertainty about highly variable volcanic CO_2 fluxes and how they affect atmospheric CO_2

concentrations, the role of positive and negative feedbacks, and the intricacies of modeling complex and coupled systems. The quantity of particulates, on the other hand, plays a minor role as they leave the stratosphere in a few months. Moreover, explosive eruptions may also induce changes in atmospheric and oceanic circulation patterns, in precipitation, as well as depletion of the stratospheric ozone layer.

Large events, such as the Toba eruption, may be an efficient mechanism for injecting SO_2 and other sulfur gases into the stratosphere where they interact with atmospheric water vapor yielding a veil of sulfuric acid aerosols that spread globally and affect solar radiation, thus cooling the Earth's lower atmosphere (McCormick et al., 1995). This process is supported by the correlation of eruption events and volcanic sulfate in ice cores (e.g., Abbott and Davies, 2012). Although particulates leave the stratosphere within some months, acid aerosols can persist for several years, spreading out over large parts of the Earth.

Basaltic and trachyandesitic magmas contain far more sulfur than highly differentiated magma, but the eruption rates are commonly too low for stratospheric injection, in contrast to large silicic explosive eruptions that can rapidly inject considerable amounts of SO_2 into the stratosphere. The net amounts depend on the eruption frequency and intensity. Nonetheless, large explosive eruptions, large fissure basalt eruptions, and smaller events of sulfur-rich magma may form ash and aerosol clouds at high altitudes (Self and Rampino, 1988). Although much interest has been directed toward the great silicic explosive eruptions, the Laki eruption and more abundant sulfur in mafic magmas suggest that the atmosphere is also affected by fissure basalt events generating convective plumes reaching the stratosphere, in some cases leading to a wide-ranging "volcanic winter" (Stothers et al., 1986).

The chicken-and-egg conundrum

That volcanism which can drive climate at irregular intervals is documented both by ice cores on Earth and measurements of volatile veils in the stratosphere. On the other hand, observational data also point toward a converse process forcing or modifying the volcanic behavior, i.e., an environmental driver. In fact, changes in lithospheric stress and melt potential and opening of melt pathways appear to govern the process. The postglacial period, in particular, provides evidence for feedbacks of this kind, the most obvious candidate being deformation of the lithosphere by on- and off-loading of large ice sheets (e.g., Maclennan et al., 2002; Schmidt et al., 2013). If eruptions change or modify climate, which in turn has the potential to initiate eruptions, the stage is set for chicken-and-egg scenarios where potential feedback loops and thresholds may yield complex responses (Figure 1b). In fact, only minor external environmental change may be required to trigger a volcanic system that is primed and ready to go.

This setting is typified by two opposite feedback mechanisms (McGuire, 2012). In a cooling world, falling sea-level may result in a positive feedback inducing explosive volcanism that furnishes the stratosphere with SO_2 droplets, thus accelerating cooling, which in turn may lead to new eruptions reinforcing the trend. Conversely, a negative feedback in a warming world with melting ice sheets and rising sea-level may also promote volcanism, also via SO_2 aerosols injected into the stratosphere, resulting in cooling opposing the warming trend.

On the other hand, Huybers and Langmuir (2009) proposed that volcanic events during major climate change inject large amounts of CO_2 into the atmosphere due to enhanced volcanic activity, counteracting the cooling of the SO_2 aerosols. They suggested that the two- to sixfold early Holocene global increase in subaerial volcanism increased atmospheric CO_2 resulting in a positive feedback accelerating warming during deglaciation. Nonetheless, the fact that the amount of CO_2 expelled during most eruptions is small or even insignificant appears to favor that cooling due to emitted SO_2 dominates CO_2-induced warming.

The environmental geosphere driver

As lithospheric stresses change, spatially and/or temporally, the geosphere may deform abruptly or gradually over time. Climate change is an instigator of deformation, principally by water and ice load changes. Data from different regions of active volcanism show enhanced volcanic activity during high rates of sea-level change (McGuire, 2012). Furthermore, removal of large ice masses may induce pressure-release melting and create magma pathways to the surface. The most important observation is that only a small change in loading rate may trigger an eruption, suggesting that the volcano system is approaching a threshold level. A similar, but more elusive, relationship appears to exist between load changes and seismicity.

Early Holocene time offers the opportunity to study lithospheric deformation and volcanic activity in Iceland when a warming climate shifted significant loads from continent (ice) to ocean (water), contributing to the ~130 m sea-level rise since the last glacial maximum at ca. 20 ka. Both data and numerical modeling support intense early Holocene volcanism resulting from glacial unloading of the lithosphere in conjunction with local tectonic features. Warming reduced the ice load on subglacial volcanoes and thereby decreased pressure in the magmatic systems, which encouraged both melting and magma chamber failure, i.e., a climate-volcanism link (Sigmundsson et al., 2010). Thus, mantle melting and magma upwelling are amplified by the pressure reduction due to glacial unloading. We also note that Kutterolf et al. (2013) suggest that circum-Pacific volcanism appears to lag slightly behind the obliquity band of Milankovitch cycles, consistent with a climate link.

Despite incomplete understanding of processes, it appears that deformation of the geosphere within the loaded/unloaded areas and surrounding lithospheric compensation comprises a common denominator for initiating earthquakes, as well as for allowing magma ascent and eruption, yielding volcanism that may contribute to climate change on multiple scales. Thus, the geosphere is a major player in the complex climate system.

The LIP contribution

While LIP formation clearly has affected local and regional environments, its potential global impact is less well understood despite temporal correlations between LIPs and environmental changes (Coffin et al., 2006) that suggest, but do not prove, causal relationships. Detailed studies have tended to focus on the continental flood basalt LIP category due to their relative accessibility, whereas submarine LIPs such as oceanic plateaus and volcanic margins have been less investigated despite their huge dimensions. A primary challenge is to estimate the spatial and temporal budget of volcanism within a given LIP, particularly eruption and emission rates. Some of the largest LIPs appear to have had a protracted history of volcanism from many volcanoes and fissures, but the most massive eruptive period was relatively short lived, possibly 0.5 Ma or less (Figure 1c), producing voluminous basaltic lavas erupted in predominantly subaerial and/or shallow marine environments.

The temporal correlations include eruption of the Siberian flood basalts at 250 Ma, which coincides with the largest biotic extinction in the geological record. During the 145–50 my period, LIPs have been linked to several changes in the global ocean, episodic formation of black shale during oceanic anoxic events being one example. Moreover, the huge submarine basaltic constructions of the North Atlantic LIP have led to suggestions of potential causal links to the early Eocene greenhouse (Eldholm and Thomas, 1993; Svensen et al., 2004; Storey et al., 2007).

A causal link to global climate should consider that (1) voluminous eruptions of basaltic magma release volatiles such as sulfur, chlorine, and fluorine (Figure 1a), and their amounts and relative proportions depend strongly on whether eruptions are subaerial or submarine. (2) The volatiles remain in the troposphere during most basaltic eruptions. However, convective plumes from large single events and the mainly basaltic magma may carry SO_2 and other volatiles into the stratosphere forming a mist of SO_2 aerosol droplets. Self et al. (2005) infer that the Deccan flood basalts emitted gas fluxes of ~1 Gt/year of SO_2 and CO_2 and that the subsequent atmospheric cooling from SO_2 aerosols of only one flood basalt eruption may have been severe, persisting for a decade or more. On the other hand, they considered CO_2-induced warming during an eruption as insignificant. Thus, LIPs may impact the global environment in a manner similar to more silica-rich eruptions, at least at low stratospheric levels at high latitudes. (3) The construction and the subsidence history of submarine LIPs may form barriers and/or open gateways that affect oceanic circulation which in turn may force climate.

Summary

Incontrovertible evidence for causal connections between volcanism and climate extends far back in Earth history. In particular, the link is well established for Holocene time. On the other hand, climate change or major transient climate events may also initiate episodic and protracted volcanism, commonly associated with lithospheric loading and unloading. Both mechanisms document an intimate interplay among the geosphere, hydrosphere, cryosphere, and atmosphere. Cause-and-effect relations incorporate complex feedback loops, not yet fully understood, but which may become increasingly efficient during an environmental threshold situation because active volcanoes are unstable systems quite sensitive to small changes in their external environment. The volcano-climate link may also initiate secondary events, some catastrophic. Depending on the state of societal development, volcanism may at times have inflicted severe climatic strain on human societies affecting food supply, living conditions, and, in cases, migration patterns.

Bibliography

Abbott, P. M., and Davies, S. M., 2012. Volcanism and the Greenland ice-cores: the tephra record. *Earth-Science Reviews*, **115**, 173–191.

Cas, R. A. F., and Wright, J. V., 1987. *Volcanic Successions – Modern and Ancient*. London: Allen & Unwin.

Cashman, K. V., and Sparks, R. S. J., 2013. How volcanoes work: a 25 year perspective. *Geological Society of America Bulletin*, **125**, 664–690.

Coffin, M. F., and Eldholm, O., 1994. Large igneous provinces: crustal structure, dimensions, and external consequences. *Reviews of Geophysics*, **32**, 1–36.

Coffin, M. F., Duncan, R. A., Eldholm, O., Fitton, J. G., Frey, F. A., Larsen, H. C., Mahoney, J. J., Saunders, A. D., Schlich, R., and Wallace, P. J., 2006. Large igneous provinces and scientific ocean drilling: status quo and a look ahead. *Oceanography*, **19**, 150–160.

Eldholm, O., and Thomas, E., 1993. Environmental impact of volcanic margin formation. *Earth and Planetary Science Letters*, **117**, 319–329.

Gudmundsson, M. T., Pedersen, R., Vogfjörd, K., Thorbjarnardóttir, B., Jakobsdóttir, S., and Roberts, M. J., 2010. Eruptions of Eyjafjallajökull Volcano, Iceland. *Eos*, **91**, 190–191.

Huybers, P. J., and Langmuir, C., 2009. Feedback between deglaciation, volcanism, and atmospheric CO_2. *Earth and Planetary Science Letters*, **286**, 479–491.

Kutterolf, S., Jegen, M., Mitrovica, J. X., Kwasnitschka, T., Freundt, A., and Huybers, P. J., 2013. A detection of Milankovitch frequencies in global volcanic activity. *Geology*, **41**, 227–230.

Maclennan, J., Jull, M., McKenzie, D., Slater, L., and Grönvold, K., 2002. The link between volcanism and deglaciation in Iceland.

Geochemistry, Geophysics, Geosystems, 3, 1062, doi:10.1029/2001GC000282.

McCormick, M. P., Thomason, L. W., and Trepte, C. R., 1995. Atmospheric effects of the Mt Pinatubo eruption. Nature, 373, 399–404.

McGuire, B., 2012. Waking the Giant. Oxford, UK: Oxford University Press.

Oppenheimer, C., 2011. Eruptions that Shook the World. Cambridge, UK: Cambridge University Press.

Rampino, M. R., and Self, S., 1992. Volcanic winter and accelerated glaciation following the Toba super-eruption. Nature, 359, 50–52.

Robock, A., 2000. Volcanic eruptions and climate. Reviews of Geophysics, 38, 191–219.

Robock, A., 2013. The latest on volcanic eruptions and climate. Eos, 94, 305–306.

Schmidt, P., Lund, B., Hieronymus, C., Maclennan, J., Arnadóttir, T., and Pagli, C., 2013. Effects of present-day deglaciation in Iceland on mantle melt production rates. Journal of Geophysical Research, 118, 1–14, doi:10.1002/jgrb50273.

Schmincke, H.-U., 2004. Volcanism. Berlin: Springer.

Self, S., and Rampino, M. R., 1988. The relationship between volcanic eruptions and climate change: still a conundrum? Eos, 69, 74–86.

Self, S., Thordarson, T., and Widdowson, M., 2005. Gas fluxes from flood basalt eruptions. Elements, 1, 283–287.

Sigmundsson, F., Pinel, V., Lund, B., Albino, F., Pagli, C., Geirsson, H., and Sturkell, E., 2010. Climate effects on volcanism: influence on magmatic systems of loading and unloading from ice mass variations, with examples from Iceland. Philosophical Transactions of the Royal Society, Series A, 368, 2519–2534.

Sigurdsson, H., 1999. Melting the Earth: The History of Ideas on Volcanic Eruptions. New York: Oxford University Press.

Storey, M., Duncan, R. A., and Swisher, C. C., 2007. Paleocene-Eocene thermal maximum and the opening of the North Atlantic. Science, 316, 587–589.

Stothers, R. B., Wolff, J. A., Self, S., and Rampino, M. R., 1986. Basaltic fissure eruptions, plume heights, and atmospheric aerosols. Geophysical Research Letters, 13, 725–728.

Svensen, H., Planke, S., Malthe-Sørenssen, A., Jamtveit, B., Myklebust, R., Rasmussen Eidem, T., and Rey, S. S., 2004. Release of methane from a volcanic basin as a mechanism for initial Eocene global warming. Nature, 429, 542–545.

Vasey, D. E., 1991. Population, agriculture, and famine: Iceland, 1784–1785. Human Ecology, 19, 323–350.

Cross-references

Astronomical Frequencies in Paleoclimates
Anoxic Oceans
Earthquakes
Events
Explosive Volcanism in the Deep Sea
Geohazards: Coastal Disasters
Hot Spots and Mantle Plumes
Intraplate Magmatism
Lithosphere: Structure and Composition
Ocean Drilling
Oceanic Plateaus
Paleoceanography
Tsunamis

VOLCANOGENIC MASSIVE SULFIDES

John W. Jamieson[1], Mark D. Hannington[1,2],
Sven Petersen[1] and Margaret K. Tivey[3]
[1]GEOMAR Helmholtz Centre for Ocean Research, Kiel, Germany
[2]Goldcorp Chair in Economic Geology, Department of Earth Sciences, University of Ottawa, Ottawa, ON, Canada
[3]Marine Chemistry & Geochemistry, Woods Hole Oceanographic Institution, Woods Hole, MA, USA

Synonyms

Seafloor Massive Sulfides

Definition

Volcanogenic Massive Sulfides. Accumulations of dominantly sulfide minerals that form at sites of focused hydrothermal discharge on the seafloor. Also refers to a class of ore deposit mined from ancient oceanic crust that is now exposed on land.

Introduction

Volcanogenic massive sulfide (VMS) deposits are mineral accumulations that form on or near the seafloor at sites of high-temperature hydrothermal vent fluid discharge. The deposits are formed by the precipitation of dominantly sulfide minerals around hydrothermal vents (e.g., Black and White Smokers) where high-temperature, metal- and sulfur-rich fluids mix with cold seawater. VMS deposits commonly form along submarine tectonic plate boundaries such as mid-ocean ridges and subduction-relating settings such as volcanic arcs and back-arc basins Although high-temperature "black smoker" chimneys are the most recognizable features, the deposits can take on a variety of forms, from individual chimneys to large mounds that host clusters of active chimneys. At the time of writing, more than 300 sites of high-temperature seafloor hydrothermal systems and related mineral deposits have been found (Figure 1).

Hydrothermal vents, first discovered on the East Pacific Rise spreading center (Francheteau et al., 1979), provided the first direct evidence of the formation of VMS deposits on the modern seafloor. However, the notion of sulfide deposits forming on the seafloor existed among economic geologists long before their discovery. By the late 1970s, many of the details of massive sulfide formation in the submarine environment had already been established from the study of fossil VMS deposits that are hosted in ancient oceanic crust now exposed on land. These ancient deposits have been mined since the time of the ancient Greeks and continue to be an important source for metals such as Cu, Zn, Pb, Au, and Ag. Volcanogenic massive sulfides that are forming on the modern seafloor may represent a

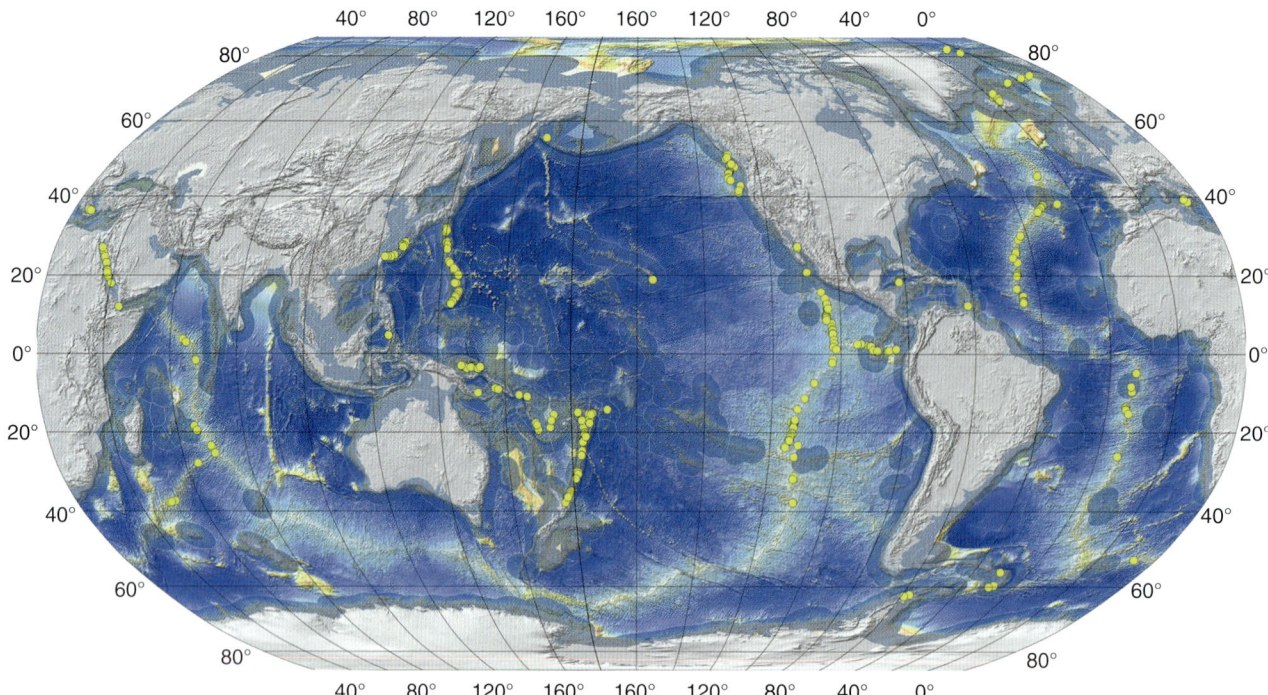

Volcanogenic Massive Sulfides, Figure 1 Global distribution of seafloor hydrothermal systems and related mineral deposits (from Beaulieu, 2013).

significant marine mineral resource for these metals that, unlike their ancient counterparts, have yet to be exploited.

Formation of volcanogenic massive sulfides

Volcanogenic massive sulfide formation is linked to circulation of hydrothermal vent fluids through oceanic crust at volcanically active areas on the seafloor. Large deposits typically consist of a consolidated basal mound of massive sulfide, overlain by hydrothermal crusts, sulfide talus from collapsed black and white smoker chimneys, and metalliferous sediment, and underlain by a metal-bearing subseafloor stockwork (vein and disseminated mineralization) within the underlying crust (Figure 2). The black and white smoker chimneys that often occur on the top and flanks of deposits only represent the uppermost part of a typical deposit and account for only a small fraction of the total amount of sulfide present on the seafloor. Different vents on the same mound are usually fed by the same high-temperature fluids at depth, with the venting temperatures mainly controlled by the degree of subseafloor mixing of the ascending high-temperature ($\sim$350 °C) fluids with local cold ($\sim$2 °C) seawater. When little to no subseafloor fluid mixing occurs, a large proportion of the dissolved metals in the venting fluids are lost to the hydrothermal plume in the overlying water column via high-temperature venting (Figure 2). Fluids that vent at lower temperatures due to subsurface mixing are commonly clear and lack the mineral-rich "black smoke" appearance often seen in high-temperature fluids. In this case, the decreased temperature has resulted in the precipitation of sulfide minerals prior to venting and the accumulation of sulfide material within the sulfide mound or below the seafloor.

The hydrothermal upflow zones below deposits typically contain intensely altered rocks with vertically extensive networks of sulfide veins and disseminated sulfides, referred to as the "stockwork zone" (Figure 2). Stockwork mineralization can contain as much as 30–40 % of the total metal in a deposit, confirming that a large proportion of the metal in the hydrothermal vent fluids never reaches the seafloor. The main alteration minerals are quartz, chlorite, sericite, and illite. In most cases, the alteration causes the rocks surrounding the discharge zone to become insulated, sealing the system against the ingress of seawater and allowing higher-temperature fluids to reach the seafloor without significant mixing with local seawater. Abundant anhydrite is also found within some feeder zones, precipitated from the progressive heating of seawater that is drawn into the mound beneath the hydrothermal vents. Lenses of massive sulfide can form, via subseafloor replacement, within permeable zones in the volcanic substrate, such as interflow breccias, hyaloclastite, or sediments.

The loss of metals to hydrothermal plumes means that the efficiency of metal deposition within deposits is very low in most cases. As a result, special conditions and/or very long periods of hydrothermal upflow are required to form large deposits. Radioisotope ages of hydrothermal

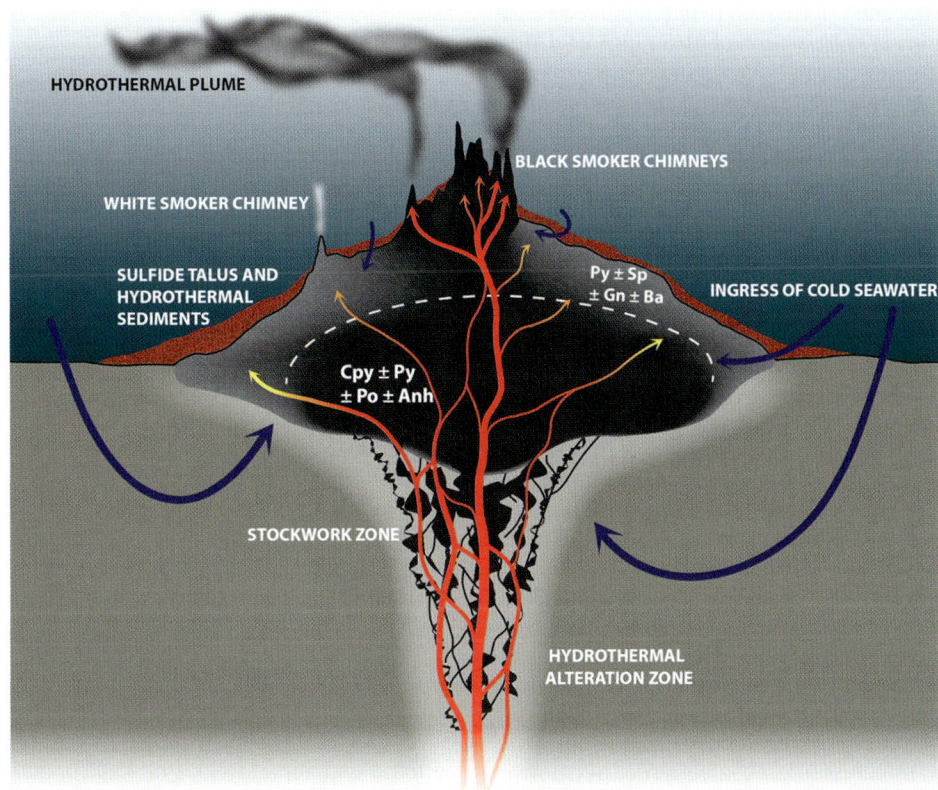

Volcanogenic Massive Sulfides, Figure 2 Cross-section through a typical active volcanogenic massive sulfide mound on the seafloor. As high-temperature hydrothermal fluid ascends to the seafloor, it mixes with local cold seawater, resulting in the precipitation of sulfide minerals and formation of chimneys and mounds on and below the seafloor. *Py* pyrite, *Cpy* chalcopyrite, *Sp* sphalerite, *Po* pyrrhotite, *Gn* galena, *Anh* anhydrite, *Ba* barite.

samples from some deposits, based on uranium-series disequilibrium measurements (e.g., using U/Th and ^{226}Ra/Ba), indicate that tens of thousands of years are required to form the largest accumulations of massive sulfide on the mid-ocean ridges.

Where do VMS deposits form?

Most hydrothermal activity and associated VMS formation are found at plate margins, where a strong spatial and temporal correlation exists between magmatism, seismicity, and high-temperature venting (Figure 1). The divergent plate boundaries include both mid-ocean ridges and intracontinental rifts such as the Red Sea. Convergent plate boundaries, which are characterized by volcanic arcs and back-arc basins that develop as a result of subduction of oceanic plates beneath oceanic crust or continental margins, are host to a smaller but still significant proportion of the known hydrothermal deposits.

On the mid-ocean ridges, the number and distribution of seafloor hydrothermal deposits is roughly proportional to spreading rate. Sulfide accumulations on the fast-spreading ridges tend to be abundant but small; fewer deposits are located on the slow-spreading ridges, but they are commonly larger and more robust. The lack of known vents along many stretches of mid-ocean ridge (e.g., the Polar Regions and the Southern Ocean, Figure 1) mainly reflects the difficulties of marine research at these latitudes. Recent discoveries of hydrothermal plumes and massive sulfide deposits in the high Arctic (Michael et al., 2003) and in Antarctica (e.g., Bransfield Strait) (Klinkhammer et al., 2001) confirm that seafloor hydrothermal activity in remote parts of the oceans is similar to that observed elsewhere. The distribution of deposits along volcanic arcs and back-arcs is largely similar to those along the mid-ocean ridges.

The composition of VMS deposits

The composition of VMS deposits is primarily controlled by the composition of the hydrothermal vent fluids from which they form. Differences in the composition of hydrothermal vent fluids largely reflect variations in the composition of the underlying volcanic substrates from which the dissolved metals and sulfur in the fluids are leached. Slow- and fast-spreading ridges are generally underlain by mid-ocean ridge basalt (MORB). In contrast, convergent margins are characterized by a range of different crustal thicknesses, heat flow regimes, and magma compositions, depending on the composition and geometry of the converging plates. As a result, the compositions of the volcanic rocks at convergent margins vary from MORB to more felsic lavas (andesite, dacite, and rhyolite), and this leads to major differences in the composition of the hydrothermal vent fluids, the isotopic systematics, and the mineralogy and bulk composition of the sulfide deposits. The composition of VMS deposits can also be influenced by direct magmatic volatile contributions to the hydrothermal fluids, especially in arc and back-arc settings (Yang and Scott, 1996).

The mineralogy of VMS deposits consists of a high-temperature (~350 °C) mineral assemblage of Cu, Fe, and Zn sulfide minerals including pyrite (±marcasite), pyrrhotite, chalcopyrite, Fe-rich sphalerite, as well as minor isocubanite and bornite and the sulfate mineral anhydrite (Figure 2). A lower-temperature (<300 °C) mineral assemblage is dominated by pyrite, marcasite, sphalerite, and minor galena and sulfosalts, together with abundant amorphous silica, barite, and minor anhydrite.

A strong mineralogical zonation within VMS deposits reflects the temperature-dependent solubilities of the sulfide minerals in the cooling hydrothermal vent fluids, with high-temperature minerals (e.g., chalcopyrite, pyrrhotite) occupying the core of a deposit and lower-temperature minerals (e.g., pyrite, sphalerite) occurring toward the top and fringes of a deposit. However, detailed examination reveals complex intergrowths, replacements, and recrystallization of the minerals within VMS deposits, resulting from a dynamic and locally complex environment of sulfide precipitation.

The depth at which a deposit forms can also have an effect on composition. At the average depth of known vent sites (2,600 m), most hydrothermal vent fluids at 350 °C are below the boiling temperature of seawater. However, about one third of known seafloor hydrothermal vents occur at depths of less than 1,500 m, where boiling of the fluids is likely to occur. The resulting phase separation can be responsible for important differences in the chemistry of hydrothermal vent fluids at different locations as different elements partition into either the vapor or brine phase, and these have a major impact on the mineralogy and chemistry of the sulfide deposits.

Average metal concentrations of deposits on the mid-ocean ridges are 5.9 wt% Cu, 6.1 wt% Zn, <0.1 wt% Pb, 1.6 ppm Au, and 89 ppm Ag (Table 1). Although anhydrite, barite, and amorphous silica are important constituents, on average Ca, Ba, and SiO2 account for <20 wt% of the samples analyzed. Hydrothermal sulfide deposits in volcanic arc and back-arc settings have elevated Zn, Pb, Au, and Ag concentrations, relative to mid-ocean ridge deposits, reflecting the typically more evolved magmatic compositions in arc settings and potential magmatic contributions.

Volcanogenic Massive Sulfides, Table 1 Average bulk compositions of 95 VMS deposits by tectonic setting (From Hannington et al., 2010)

Location	(n)	Cu wt%	Zn	Pb	Au ppm	Ag
Mid-ocean ridges[a]	2,071	5.9	6.1	<0.1	1.6	89
Sedimented ridges	173	1.1	3.6	0.5	0.5	84
Intraoceanic back-arc basins	668	3.9	16.4	0.9	6.6	210
Intraoceanic arcs	169	5.3	17.7	2.4	9.6	407
Transitional arcs	728	6.4	14.8	2.0	12.2	692
Continental margin arcs	60	3.1	20.3	10.0	2.3	953

n numbers of samples
[a]Data for the mid-ocean ridges are based on deposits for which there are representative suites of 50 samples or more

Metal distribution within individual deposits is directly linked to the temperature-dependent mineralogical controls described above. For example, Cu is typically enriched in the high-temperature, chalcopyrite-rich zones, whereas Zn and Pb are enriched in the lower-temperature sphalerite- and galena-rich zones (Figure 2). The distribution of trace elements in VMS deposits generally follows similar temperature-controlled distribution, with Co, Se, Bi, and Mo commonly enriched in high-temperature zones within a deposit and Cd, Pb, As, Sb, and Ag enriched in lower-temperature zones.

The size of VMS deposits

Volcanogenic massive sulfide deposits on the seafloor can range from several meters up to 100 s of meters in diameter. The size of a deposit is controlled by a combination of the length of time for which the hydrothermal system is active, the total flux of hydrothermal vent fluids through the system, the concentration of metals and sulfur in the fluids, and the efficiency of deposition (i.e., how much of the mobilized metals is retained within the deposit versus ejected as "smoke" into the overlying hydrothermal plume).

Several distinct deposit morphologies have been recognized. These morphological differences likely reflect the geometry of the hydrothermal upflow and especially the permeability of the crust in the upper few hundred meters below the seafloor (Figure 3). Large-diameter, low-relief mounds (e.g., the TAG site on the Mid-Atlantic Ridge or the Galapagos sulfide mound, in the eastern Pacific) are seafloor manifestations of widespread, branching feeder systems (Figure 3a), whereas tall, free-standing structures (e.g., up to 45 m in height within the Endeavour vent fields on the Juan de Fuca Ridge in the NE Pacific) reflect highly focused, high-temperature upflow (Figure 3b) (Embley et al., 1988; Robigou et al., 1993). Deposits hosted in ultramafic rocks on slow-spreading ridges can form

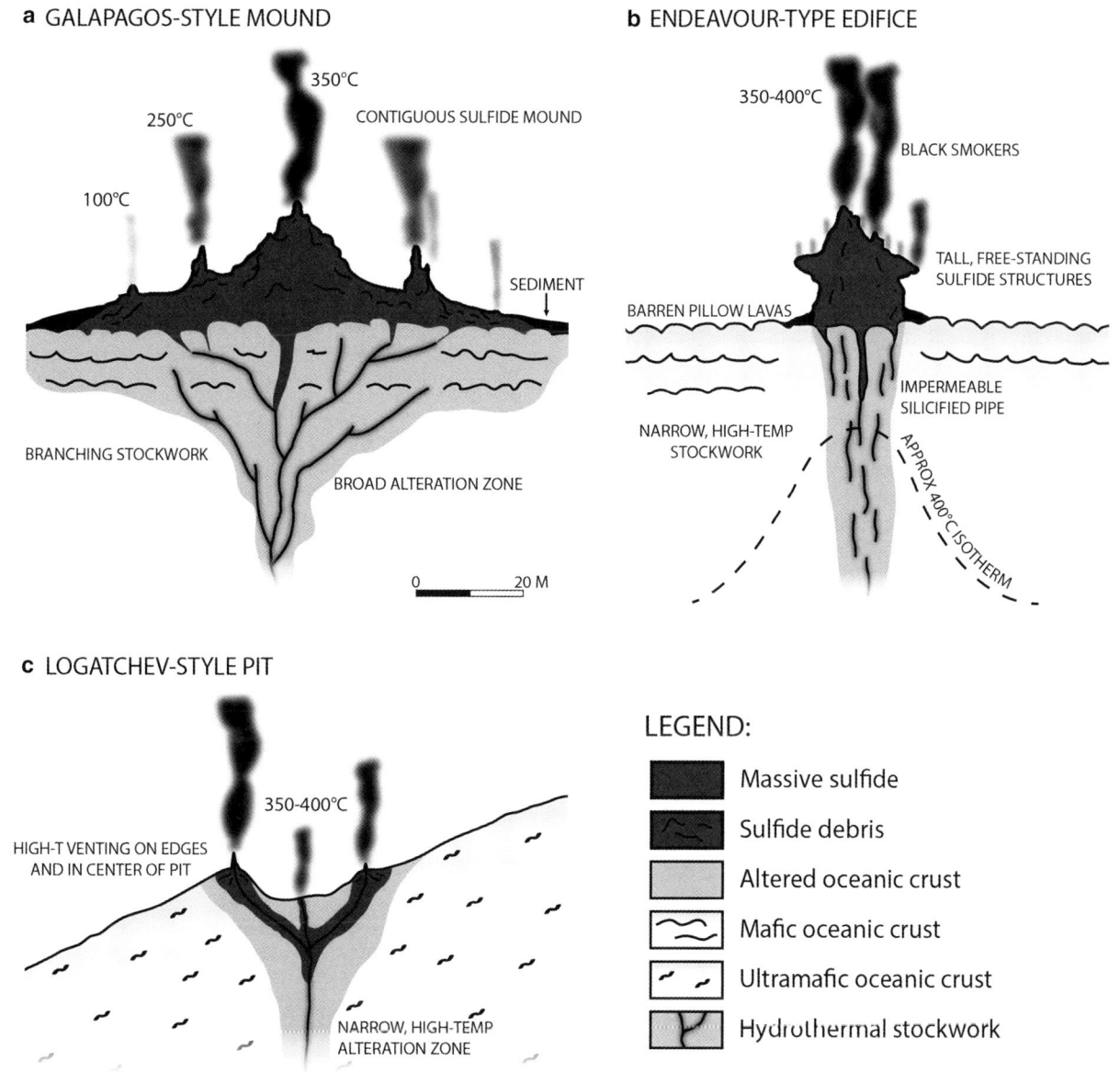

Volcanogenic Massive Sulfides, Figure 3 Illustration of contrasting volcanogenic massive sulfide deposit morphologies resulting from different permeabilities in the substrate. (a) Galapagos-type structures are low-relief mounds with a broad base. (b) Endeavour-type structures are steep sided and form over a narrow, restricted upflow zone. (c) Logatchev-style pits form in ultramafic settings. Donut-shaped structures have elevated rims consisting of a mixture of sulfide and host rock and host active high-temperature chimneys (modified from Hannington et al., 1995).

low-relief, donut-shaped depressions with elevated rims containing a mixture of sulfide and host rock (Figure 3c). A substrate with high porosity (e.g., volcanic breccias or sediments) may result in a significant amount of sulfide precipitation below the seafloor, forming stratabound sulfide lenses, and only a limited expression of hydrothermal venting at the seafloor.

The largest deposits so far discovered on the seafloor are up to several hundreds of meters in diameter and have total masses on the order of 1–5 million tonnes (Hannington et al., 2010). Reliable estimates of the total sulfide accumulation in large deposits have been possible in only a few cases where subseafloor drilling has provided information on deposit thicknesses (e.g., the TAG mound on the Mid-Atlantic Ridge, Middle Valley on the Juan de Fuca Ridge, and Solwara I in the Manus Basin), although developments in geophysical methods such as deep-towed magnetics and electromagnetic surveys are providing additional tools for measuring deposit sizes.

The largest deposits are all at least 100,000 years old (Hannington et al., 2010), implying that sustained hydrothermal venting over long periods is required to produce significant accumulations of massive sulfide at the seafloor. Hannington et al. (1998) estimated a growth rate for the main massive sulfide lens (2.7 million tonnes) at the TAG site on the Mid-Atlantic Ridge of about 500–1,000 tonnes/year. Similar growth rates have been estimated for other large deposits on the Mid-Atlantic Ridge (e.g., Logatchev, Ashadze, and Krasnov), based on the maximum ages and estimated tonnages of the deposits.

The relatively smaller sizes of most deposits discovered so far are likely related to the short-lived nature of their heat sources, which include narrow dike injections along the axial zones of the ridges. This is confirmed by uranium-series measurements indicating that hydrothermal discharge at fast-spreading centers is episodic on time scales of only 10–100 years (Lalou et al., 1993). By contrast, the protracted history of hydrothermal venting at places like TAG is a consequence of deep-seated magmatic activity followed by long periods of cooling and release of heat from depth. These deposits are situated in stable structural environments with relatively slow rates of spreading, far from the axis of the ridge. On sedimented ridges, like Middle Valley, long-term heat retention afforded by a thick sediment cover also may contribute to the large sizes of the deposits. Deposits on back-arc spreading ridges are thought to be comparable to deposits on mid-ocean ridges, with a similar relationship between size and spreading rate. Deposits on the summits of arc volcanoes are generally smaller, likely reflecting the small, shallow magma chambers and short-lived, episodic volcanic events that drive circulation (Hannington et al., 2005).

Summary

Volcanogenic massive sulfide deposits form on the seafloor as the result of mixing of hot hydrothermal fluids, rich in dissolved metals and sulfur, with cold seawater as the fluids ascend to the seafloor from deep within the underlying crust. The sulfide minerals that accumulate at these vent sites form spires (chimneys) or mounds that grow and coalesce over time. These sulfide edifices provide a substrate on which the macro- and microfauna that thrive on the hydrothermal fluids for sustenance live. VMS deposits are found along submarine tectonic boundaries, such as mid-ocean ridges, volcanic arcs, and back-arc basins. The deposits are often rich in economically valuable metals such as Cu, Zn, Au, and Ag. Although, at the time of writing, no direct exploitation of VMS deposits from the modern seafloor has occurred, with recent technological developments and elevated metal prices, the desire and feasibility to exploit seafloor deposits are such that exploitation of these deposits may occur in the near future.

Bibliography

Beaulieu, S. E., Baker, E. T., German, C. R., and Maffei, A., 2013. An authoritative global database for active submarine hydrothermal vent fields. *Geochemistry Geophysics Geosystems*, **14**, 4892–4905.

Davis, E. E., Mottl, M. J., and Fisher, A. T., et al., 1992. *Proceedings of the ODP, Initial Reports*, 139. College Station (Ocean Drilling Program). doi:10.2973/odp.proc.ir.139.1992

Embley, R. W., Jonasson, I. R., Perfit, M. R., Franklin, J. M., Tivey, M. A., Malahoff, A., Smith, M. F., and Francis, T. J. G., 1988. Submersible investigation of an extinct hydrothermal system on the Galapagos Ridge – Sulfide mounds, stockwork zone, and differentiated lavas. *Canadian Mineralogist*, **26**, 517–539.

Francheteau, J., Needham, H. D., Choukroune, P., Juteau, T., Seguret, M., Ballard, R. D., Fox, P. J., Normark, W., Carranza, A., Cordoba, D., Guerrero, J., Rangin, C., Bougault, H., Cambon, P., and Hekinian, R., 1979. Massive deep-sea sulfide ore-deposits discovered on the East Pacific Rise. *Nature*, **277**, 523–528.

Hannington, M., Jonasson, I., Herzig, P., and Petersen, S., 1995. Physical and chemical processes of seafloor mineralization at mid-ocean ridges. *Geophysical Monograph*, **11**, 115–157.

Hannington, M., Galley, A., Herzig, P., and Petersen, S., 1998. Comparison of the TAG mound and stockwork complex with Cyprus-type massive sulfide deposits. In Herzig, P. M., Humphris, S. E., Miller, D. J., and Zierenberg, R. A. (eds.), *Proceedings of ODP, Science Results*, 158, College Station, TX (Ocean Drilling Program), pp. 389–415.

Hannington, M., De Ronde, C., and Petersen, S., 2005. Sea-floor tectonics and submarine hydrothermal systems. In Hedenquist, J. W., Thompson, J. F. H., Goldfarb, R. J. and Richards, J. P. (eds.), *Economic Geology 100th Anniversary Volume*, Society of Economic Geologists, Littelton, Colorado, USA, pp. 111–141.

Hannington, M., Jamieson, J., Monecke, T., and Petersen, S., 2010. Modern sea-floor massive sulfides and base metal resources: towards an estimate of global sea-floor massive sulfide potential. *Society of Economic Geologists, Special Publication*, **15**, 111–141.

Klinkhammer, G. P., Chin, C. S., Keller, R. A., Dahlmann, A., Sahling, H., Sarthou, G., Petersen, S., and Smith, F., 2001. Discovery of new hydrothermal vent sites in Bransfield Strait, Antarctica. *Earth and Planetary Science Letters*, **193**, 395–407.

Lalou, C., Reyss, J. L., and Brichet, E., 1993. Actinide-series disequilibrium as a tool to establish the chronology of deep-sea hydrothermal activity. *Geochimica et Cosmochimica Acta*, **57**, 1221–1231.

Lipton, I., 2012. *Mineral resource estimate – Solwara Project, Bismarck Sea,* `. Canadian NI 43–101 Technical Report for Nautilus Minerals Inc., 240 p.

Michael, P. J., Langmuir, C. H., Dick, H. J. B., Snow, J. E., Goldstein, S. L., Graham, D. W., Lehnert, K., Kurras, G., Jokat, W., Muhe, R., and Edmonds, H. N., 2003. Magmatic and amagmatic seafloor generation at the ultraslow-spreading Gakkel ridge, Arctic Ocean. *Nature*, **423**, 956–U951.

Robigou, V., Delaney, J. R., and Stakes, D. S., 1993. Large massive sulfide deposits in a newly discovered active hydrothermal system, the High-rise Field, Endeavour Segment, Juan-de-Fuca Ridge. *Geophysical Research Letters*, **20**, 1887–1890.

Schmidt, R., and Schmincke, H. U., 2000. Seamounts and island building. In Sigurdsson, H. (ed.), *Encyclopedia of Volcanoes*. Sand Diego: Academic, pp. 383–402.

Yang, K. H., and Scott, S. D., 1996. Possible contribution of a metal-rich magmatic fluid to a sea-floor hydrothermal system. *Nature*, **383**, 420–423.

Zierenberg, R. A., Fouquet, Y., Miller, D. J., Bahr, J. M., Baker, P. A., Bjerkgard, T., Brunner, C. A., Duckworth, R. C., Gable, R., Gieskes, J., Goodfellow, W. D., Groschel-Becker, H. M., Guerin, G., Ishibashi, J., Iturrino, G., James, R. H., Lackschewitz, K. S., Marquez, L. L., Nehlig, P., Peter, J. M., Rigsby, C. A., Schultheiss, P., Shanks, W. C., Simoneit, B. R. T., Summit, M., Teagle, D. A. H., Urbat, M., and Zuffa, G. G., 1998. The deep structure of a sea-floor hydrothermal deposit. *Nature*, **392**, 485–488.

Cross-references

Black and White Smokers
Hydrothermal Plumes
Hydrothermal Vent Fluids (Seafloor)
Hydrothermalism
Intraplate Magmatism
Island Arc Volcanism, Volcanic Arcs
Marginal Seas
Marine Mineral Resources
Oceanic Spreading Centers
Ophiolites
Subduction

W

WADATI-BENIOFF-ZONE

Nina Kukowski
Institute of Geosciences, Friedrich-Schiller-University, Jena, Germany

Definition

Inclined thin planar zone of seismicity down to depths of about 700 km resulting from processes related to the subduction of an oceanic plate including interaction of descending and overriding plate.

Introduction

About three quarters of all tectonic earthquakes, both along all types of plate boundaries and in various intraplate environments, initiate at shallow depth of less than about 60 km. Among them are the largest earthquakes with M_w larger than 9, which solely occur in the seismogenic zone, the shallow portion of the plate interface in subduction zones. The remaining quarter of tectonic earthquakes occurs in depths deeper than about 60 km and as deep as close to 700 km. These deep earthquakes are nearly exclusively identified along inclined narrow zones of intraplate seismicity within the downgoing oceanic plate in subduction zones, the so-called Wadati-Benioff zones (Figure 1).

Downgoing oceanic slabs exhibit an unusually cold and complex thermal structure with steep temperature gradients to the surrounding lithosphere and uppermost mantle. Therefore, large portions of slabs remain beneath about 600 °C, a temperature, which seems to provide an upper limit for earthquakes to occur (e.g., Emmerson and McKenzie, 2007). Thus, e.g., the age and convergence rate-dependent thermal structure of subduction zones may explain the different cutoff depths of seismicity. However, deep earthquakes are not evenly distributed with depth, but bimodally. They strongly decrease in frequency down to about 300 km and then increase again in number with a maximum around 600-km depth and a relatively sharp lower boundary of their occurrence at about 700 km (Figure 1b). Further, many subduction zones show prominent gaps of deep seismicity. These observations suggest that earthquakes in different depth intervals may be due to different causes (e.g., Rayleigh and Paterson, 1965; Kirby et al., 1991; Kikuchi and Kanamori, 1994; Peacock, 2001; Jung et al., 2004; Kelemen and Hirth, 2007). Although detected in almost every subduction zone by now, deep earthquakes are not evenly distributed spatially. More than 60 % of all earthquakes deeper than 300 km have been recorded in the Tongan subduction zone, where they occur along the entire depth interval (e.g., Frohlich, 2006).

Deep earthquakes seem to be smaller in size than shallow earthquakes. The strongest instrumentally recorded deep earthquakes, the 1994 Bolivia (Kikuchi and Kanamori, 1994) and 2013 Okhotsk (Ye et al., 2013) deep events, had magnitudes of M_w 8.3. Since about 1950, there have only 13 $M_w \geq 7.5$- and three $M_w > 8$-deep seismic events been recorded.

Besides their larger depth, compared to shallow earthquakes, surface waves of deep earthquakes are considerably weaker or even absent, and often they do not have numerous aftershocks. However, stress drop associated with deep earthquakes mostly is considerably higher than for even very large shallow earthquakes (Frohlich, 2006). As deep earthquakes occur at high pressure and temperature, brittle failure most probably is not likely the cause. As the state of stress and ambient pressures and temperatures in the depth intervals of about 60–300 km, where intermediate-depth earthquakes occur, and deeper than about 350 km, where deep-focus earthquakes occur, differ significantly, seismic events in the two depth intervals probably have different causes. In an ongoing hot debate, since some decades, dehydration embrittlement (Rayleigh and Paterson, 1965) and phase transitions (e.g., Kirby

J. Harff et al. (eds.), *Encyclopedia of Marine Geosciences*, DOI 10.1007/978-94-007-6238-1,
© Springer Science+Business Media Dordrecht 2016

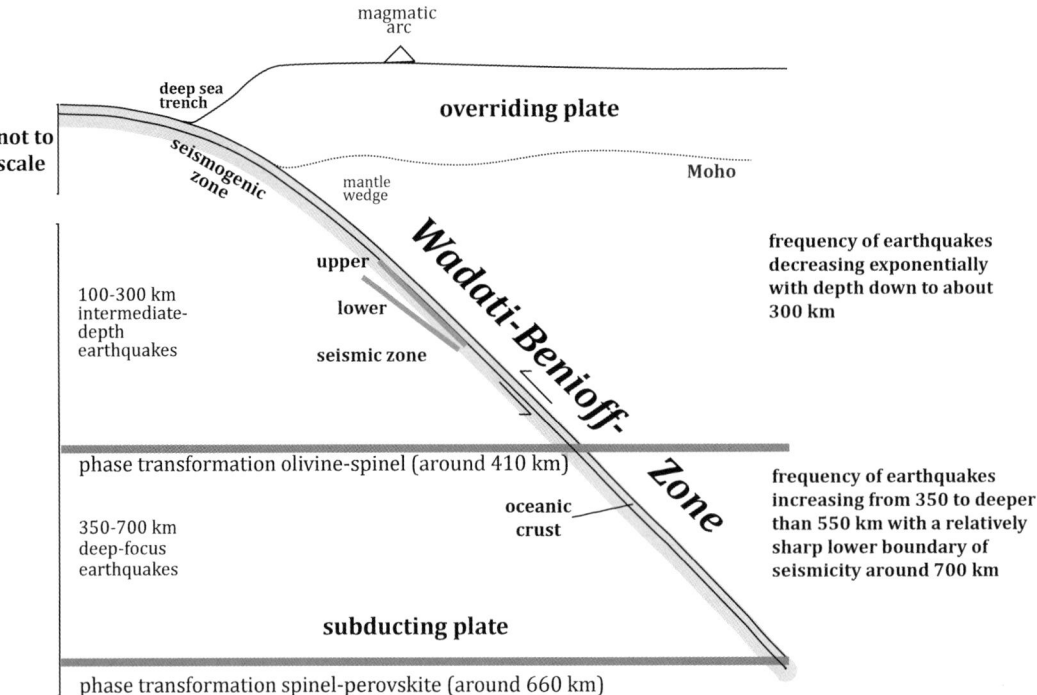

Wadati-Benioff-Zone, Figure 1 Unscaled schematic sketch of the Wadati-Benioff zone illustrating the environment of deep seismicity.

et al., 1991) are most widely discussed as major causes of deep earthquakes. However, stresses in a downgoing oceanic plate are also strongly influenced by bending and unbending. Therefore, these processes also may aid in generating intermediate-depth seismicity (Wang, 2002).

In the following, after a brief historical outline highlighting the important contributions of Kiyoo Wadati and Hugo Benioff to research on deep seismicity, first, the occurrence and potential causes of intermediate-depth earthquakes will be reviewed, followed by a review of deep-focus earthquakes. The Japanese subduction zones (Figure 2a) and Nazca plate subduction beneath South America (Figure 2b) serve as case studies both because Wadati and Benioff focused on these and they exert several of the most important features of Wadati-Benioff zones, South America also being the location of unusually many of the largest instrumentally recorded deep earthquakes. Tonga (Figure 2c), finally, results from fast subduction of the approximately 110-Myr-old Pacific plate and shows patterns of deep seismicity significantly different from those of the other subduction zones.

The historical contribution of Kiyoo Wadati and Hugo Benioff

In the 1920s, the discussion among leading geophysicists was centered on the question of the depth extent of earthquake occurrence and their characteristic seismic waves. Between 1928 and 1935, Kiyoo Wadati, a young scientist working at the Central Meteorological Observatory in Tokyo, published several papers (e.g., Wadati, 1928, 1931) in which he clearly showed the occurrence of earthquakes at all depths down to at least 500 km using the time interval between P- and S-waves arrivals. He then also suggested classifying earthquakes into shallow, intermediate-depth, and deep-focus events in the way this classification is still used today. Further, Wadati noted that deep earthquakes occur along inclined planes and might be related to continental drift (Wadati, 1935) long before plate tectonics was accepted and the presence of subduction zones was confirmed.

Hugo Benioff's main expertise was on the relation between earthquakes and motion along faults; however, using Gutenberg and Richter's seismic catalog, he synthesized the available information on Circum-Pacific deep earthquakes with a focus on South America. From this, he proposed the inclined thin zones of seismicity to represent boundaries between oceanic and continental material with their sharpness resulting from the different densities of these materials (Benioff, 1949). This hypothesis greatly influenced thinking leading to the formulation of the plate tectonics theory and its acceptance (Frohlich, 2006).

Since about the 1970s, the zones of intermediate-depth and deep-focus seismicity have been called Wadati-Benioff zones (WBZ) to acknowledge the important contributions of both scientists to our understanding of deep seismicity and also plate tectonics (Frohlich, 2006).

The characteristics and causes of deep earthquakes

Deep earthquakes with $M_w > 6$ mostly show durations of about 3–10 s and some very large events, like the 1970 M_w 8.1 Colombia earthquake or the 1994 M_w 8.3 Bolivia earthquake, even about half a minute. Rupture velocity mostly is between 0.3 and 0.9 v_s (Frohlich, 2006), inferring that deep earthquakes result from faulting, which, however, cannot resemble brittle failure. Although temperatures within subducting slabs are up to 1,000 K lower than at the same depth outside a slab, thermal models of subduction zones imply that temperatures in the coldest part of the slab in the depth interval, where intermediate-depth earthquakes occur, exceed about 300 °C even in fast-subducting, old slabs such as beneath northern Japan (e.g., Peacock, 2003).

Intermediate-depth earthquakes

A feature of intermediate-depth seismicity is a two-layered WBZ in a depth range from about 60 km to a maximum of approximately 200 km, more and more evidence for which has come especially from local seismic networks, which allow for the necessary precision of hypocenter localization (e.g., Fujita and Kanamori, 1981; Kawakatsu, 1986) since the 1970s, and about 30 years later, it was obvious that at least some segments of most subduction zones exert double seismic zones (DSZ) (Brudzinski et al., 2007). These layers are thin bands of seismicity, the upper one of which with more but smaller seismic events (e.g., Hasegawa et al., 1978) is located within the oceanic crust, whereas the lower one with fewer but larger seismic events (e.g., Fujita and Kanamori, 1981) occurs at variable depths within the uppermost oceanic mantle. Both seismic zones are separated by a seismically quiet layer of about roughly 10–30 km thickness that seems to be a weak function of the age of the descending plate (Brudzinski et al., 2007). The distance between the seismic layers is largest at their shallow end, and mostly both layers converge to the maximum depth extent of a DSZ. Temperatures and stress regimes are different in the upper and lower seismic zone, respectively, with stresses in the upper seismic zone often characterized by a down-dip compression, whereas in the lower one, down-dip tension is observed (Fujita and Kanamori, 1981; Wang, 2002). However, there are double seismic layers exhibiting the same tensional state of stress and related faulting, e.g., beneath the central Andes (Rietbrock and Waldhauser, 2004).

Mechanisms so far discussed for intermediate-depth seismicity mainly can be attributed to two processes: dehydration embrittlement or shear instabilities, e.g., involving thermal runaway or shear heating. Unbending also is suggested to play a role for intermediate-depth seismicity (Wang, 2002; Ranero et al., 2005).

Rayleigh and Paterson already proposed dehydration embrittlement as a potential cause of intermediate-depth

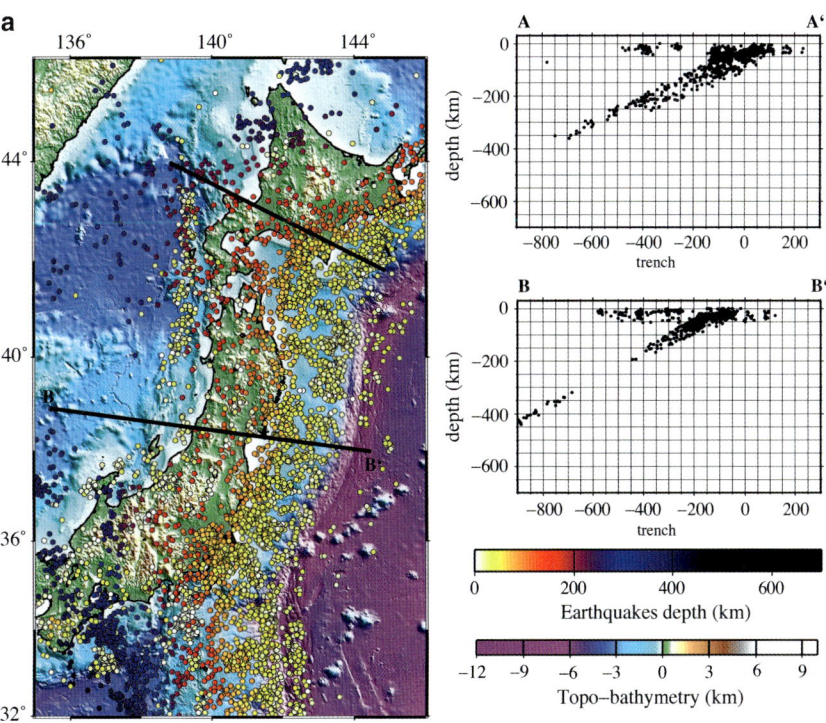

Wadati-Benioff-Zone, Figure 2 (continued)

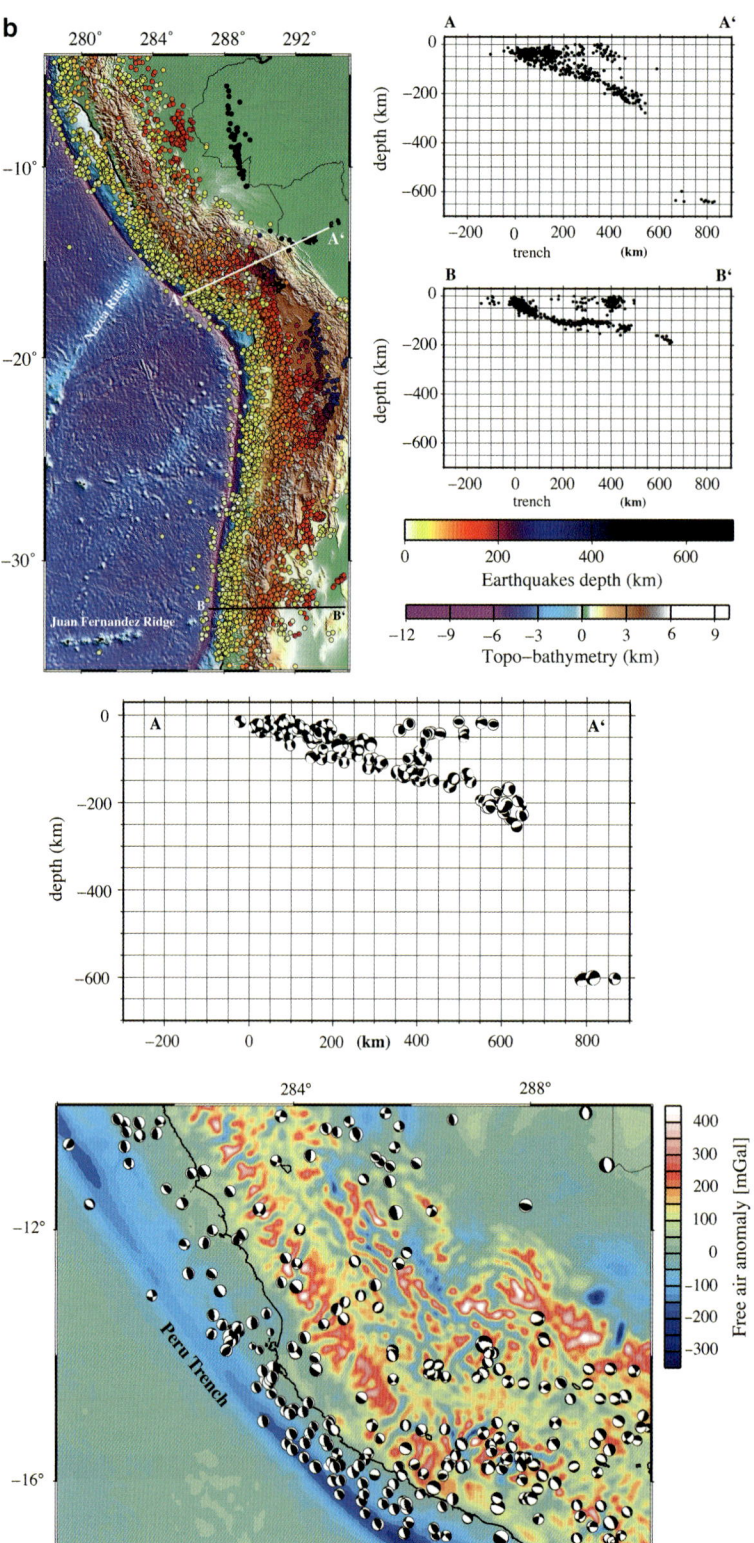

Wadati-Benioff-Zone, Figure 2 (continued)

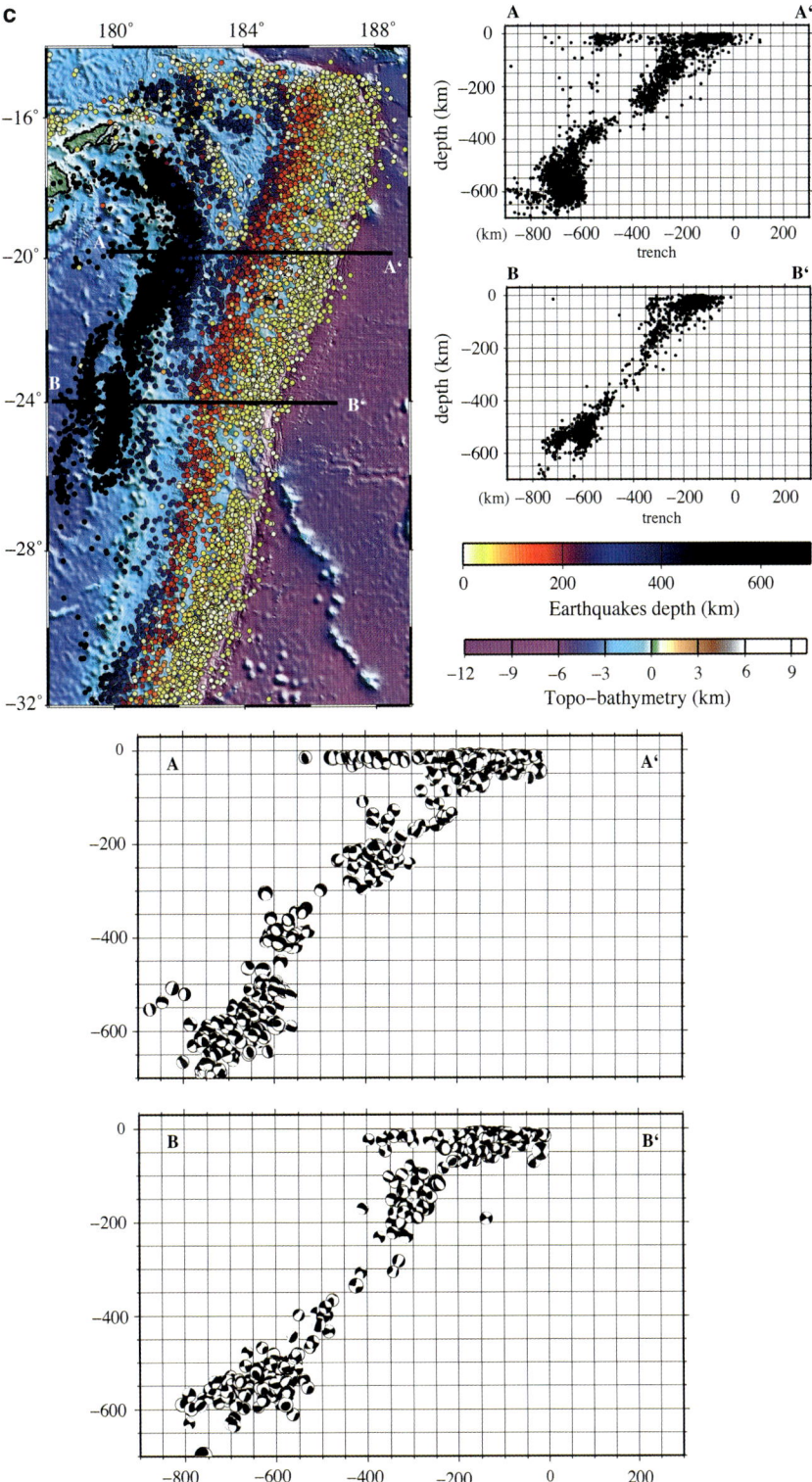

Wadati-Benioff-Zone, Figure 2 Maps and profiles illustrating seismicity characteristic for subduction zone segments exhibiting deep earthquakes: (**a**) northern and central Japan, (**b**) South America, and (**c**) Tonga. Profile locations are shown on the maps. Earthquakes hypocenters are from the EHB catalog, focal solutions are from the CMT catalog, and maps have been created using the submap 4.0 tool (Heuret and Lallemand, 2005). Note that focal solutions are only available for a subset of recorded earthquakes.

seismicity in 1965, and since then our understanding of this process has increased considerably through numerous studies (e.g., Jung et al., 2004; Green, 2007; Shiina et al., 2013). Newly created oceanic lithosphere does not have a significant water content which however is required for dehydration reactions to take place, implying that oceanic lithosphere must have been hydrated before it is subducted to where intermediate-depth earthquakes take place. Processes proposed for oceanic lithosphere hydration include faulting at outer rises through which seawater may percolate down to a depth of up to several 10s of km (Peacock, 2001), volatile supply through mantle plumes (Seno and Yamamaka, 1996), or advection in the upper fractured oceanic lithosphere before it enters a subduction zone, which leads to serpentinization along fractures (Wang, 2002). Indeed, evidence for the latter process comes from, e.g., the Nazca plate off northern Chile (Ranero and Sallares, 2004).

Hydrous minerals such as serpentine or antigorite are abundant in slabs at intermediate-depth and thus provide the fluids released through dehydration leading to embrittlement. The first hypotheses of dehydration embrittlement arose from laboratory studies on serpentine (Rayleigh and Paterson, 1965), which, however, has a negative volume change, which would make failure more difficult and thus shear instability would vanish (Jung et al., 2004). The latter workers therefore suggested that antigorite dehydration, which results in fine-grained shear zones under the entire pressure range present at intermediate-depth, is the more probable process causing intermediate-depth seismicity. Water released through dehydration reactions may assist in elevating fluid pressures especially together with the presence of aqueous fluids, which are inferred to be present in some Pacific slabs through unusually high vp/vs ratios (Shiina et al., 2013), thus reducing effective stress and abetting failure.

Viscous creep in thin, i.e., only cm-wide fine-grained shear zones may lead to further weakening a shear zone resulting in further slip. This process of thermal runaway (John et al., 2009) connected with a significant temperature increase of up to a few hundred Ks (Kelemen and Hirth, 2007) including even partial melting (Prieto et al., 2013) provides a failure mechanism without the need of a fluid phase present. However, fluids still may be invoked as previous hydration of the rocks facilitates the formation of fine-grained material prone to later failure (John et al., 2009). Plastic failure of anhydrous, anisotropic peridotite has been suggested as a mechanism to explain lower seismic zone seismicity without the need to invoke former hydration, which may be limited to the uppermost few km beneath the oceanic crust (Ranero et al., 2005); therefore, inferring the uppermost mantle in which the lower seismic zone is located may be dry (Reynard et al., 2010), but this failure then may generate the thin shear zones inevitable for thermal runaway.

Summarizing, observations from various subduction zones reveal that most likely different processes may explain intermediate-depth seismicity in the upper and lower seismic zones. Age and speed of the descending plate, the product of which is the so-called thermal parameter, seem to exert major control on the characteristics of intermediate-depth seismicity, and p-T-paths, i.e., the type of metamorphic reactions a slab is undergoing, the release of water as well as the state of stress when the plate is unbending are suggested to also do so (Abers et al., 2013).

Deep-focus earthquakes

Deep-focus seismicity mainly occurs well beneath the olivine-spinel phase boundary but above the spinel-perovskite boundary (Figure 1). Processes related to the breakdown of metastable olivine (Kirby et al., 1996), the most abundant mineral in the upper mantle, are most widely accepted now to explain deep-focus seismicity (Kirby et al., 1991). If so, slabs should be entirely dehydrated before reaching the depth interval in which deep-focus events occur (Green, 2007).

Deep-focus earthquakes take place in both warm and cold subduction zones, and in both with shallowly dipping slabs and steeply dipping slabs. However, in warm subduction zones like South America rupture velocity, seismic efficiency, and the number of aftershocks are low, whereas they are high in cold subduction zones such as Tonga. Therefore, temperature has been suggested to significantly influencing rupture characteristics (e.g., Tibi et al., 2003).

The increase in seismic moment around 550-km depth may be explained with an equilibrium phase change such as the transformation of garnet to perovskite, which is accompanied by a considerable dehydration (Estabrook, 2004), and therefore possible in a non-dehydrated slab, opposite to what would be required for the breakdown of metastable olivine.

Mechanisms discussed for the largest very deep earthquakes so far recorded, the 1994 637-km-deep Bolivia M_w 8.3 earthquake with a large stress drop of 25–110 MPa, which ruptured along a 30–50 km large subhorizontal plane (Kikuchi and Kanamori, 1994), and the 2013 609-km-deep Okhotsk M_w 8.3 earthquake with a stress drop of about 15 MPa, that had a rupture length of 180 km (Ye et al., 2013) include phase transition triggering but also shear instability. The Bolivia double-couple, highly anisotropic event is inferred to have resulted from volume reduction phase transition of olivine-rich peridotite (Kikuchi and Kanamori, 1994) leading to melting along a very thin (less than a meter) shear zone (Kanamori et al., 1998). This is in line with the slow rupture velocity of only about 1 kms^{-1} and the large stress drop. Rupture of the relatively fast Okhotsk earthquake propagated along two segments in different directions. Olivine to spinel phase transformation seems to be the cause for the NE segment of the rupture, while rupture characteristics to the SSE are better compatible with thermal shear instability (Meng et al., 2014), respectively.

Three case studies: northern Japan, Central Andes, and Tonga

Much of the early research on deep seismicity has been focused on northern Japan and South America, and the seminal contributions of Wadati and Benioff resulted from observations in these subduction zones. Tonga, on the other hand, is the subduction zone, where most deep earthquakes occur, but no event larger than M_w 7.6 was recorded so far. A comparison between these subduction zones reveals their very different characteristics of deep seismicity.

The Japanese Islands were among the first regions equipped with relatively dense seismic networks such that precise location and the estimate of focal mechanisms were already achieved early in the twentieth century. The old Pacific plate and the young Philippine Sea plate are subducting beneath the Japanese Islands at a high and relatively low speed, respectively. Thus, the thermal parameters of the Japanese subduction zones are very different from each other, whereas the release of water from the slabs is relatively uniform regionally (van Keken at al., 2011). Whereas beneath north and central Japan (Figure 2a), there is a pronounced zone of intermediate-depth seismicity, deep-focus seismicity is relatively rare, although present down to about 550 km, beneath the Nankai segment of southwest Japan, intraslab seismicity beneath about 65 km is lacking (e.g., Peacock, 2003). Along the entire Japanese Islands, there are pronounced gaps between intervals of seismicity.

The Nazca plate subducting beneath South America (Figure 2b) has a very variable geometry and exerts three segments with an almost horizontal slab. Oblique convergence and the varying geometry of the Nazca plate lead to a complicated pattern of seismicity (Figure 2b, lower panels). Many seismic events exhibit at least some strike-slip component. Intermediate seismicity occurs persistently from Ecuador to Chile. A double seismic zone identified beneath northern Chile is characterized by an unusually narrow seismically quiet interlayer of only 9-km width (Rietbrock and Waldhauser, 2004), most possibly because of the relative youth of the subducting slab. Contrastingly, the width between both seismic layers identified beneath central Chile approximately where Juan Fernandez Ridge subducts is about 30 km, while the descending Nazca plate is not older than 35 Ma here. This may result from an unusually cold slab relative to its age (Marot et al., 2013), which would hint to the cooling of the upper portion of the Nazca plate lithosphere through advecting seawater as proposed by Wang (2002). The young subducting slab makes South American deep seismicity still somewhat enigmatic: there is close to no seismicity between about 300 and 550 km (Figure 2b), but so far, since earthquakes are recorded instrumentally, more very large very deep events have occurred here than in any other subduciton zone.

Despite the high number of deep earthquakes in the cold Tonga subduction zone (Figure 2b), so far no M_w >7.6 deep-focus earthquakes have been recorded here (Frohlich, 2006). Tongan deep earthquakes much more resemble shallow earthquakes than deep earthquakes in other subduction zones. The Tongan slab is dipping significantly steeper than those beneath South America and Japan, and there is a remarkable clustering of earthquakes around 400-km depth. Displayed focal mechanisms (Figure 2c, lower panels) reveal that, whereas most intermediate-depth faults are rupturing subhorizontally, deep-focus seismicity occurs along variously oriented faults implying these are newly generated faults (Warren et al., 2007).

Summary and open questions

Most subduction zones exhibit double seismic layers at an intermediate-depth at least along some of their segments and also deep-focus seismicity. Processes mostly favored for intermediate-depth seismicity include dehydration embrittlement and shear instability, whereas those for deep-focus earthquakes include phase transitions and thermal runaway. This reveals that, although significantly less numerous than shallow earthquakes and restricted to one specific plate-boundary environment, their causes and failure mechanisms are much more variable. This may result from the very different p-T-t- conditions in the oceanic crust and uppermost mantle of descending slabs.

Temperature, faults created through plate bending and unbending, and water, both as aqueous fluids and bound in mineral phases, are thought to exert control on deep seismicity in WBZs. Especially debated is the role of water, as some proposed processes, i.e., thermal runaway or metastable olivine breakdown (Green, 2007; Kelemen and Hirth, 2007), do not require the presence of water. However, other workers have argued for the necessity of the presence of water to explain deep seismicity (e.g., Barcheck et al., 2012), and recent estimates of much higher percentages of water in subducting slabs (Garth and Rietbrock, 2014) hint to its even more prominent role than previously thought.

Bibliography

Abers, G. A., Nakajima, J., van Keken, P. E., Kita, S., and Hacker, B. R., 2013. Thermal-petrological controls on the location of earthquakes within subducting plates. *EPSL*, **369–370**, 178–187, doi:101016/j.epsl.2013.03.022.

Barcheck, C. G., Wiens, D. A., van Keken, P. E., and Hacker, B. R., 2012. The relationship of intermediate- and deep-focus seismicity to the hydration and dehydration of subducting slabs. *EPSL*, **349–350**, 153–160, doi:10.1016/epsl2012.06.055.

Benioff, H., 1949. Seismic evidence for the fault origin of oceanic deeps. *GSA Bulletin*, **60**, 1837–1866.

Brudzinski, M., Thurber, C. H., Hacker, B. R., and Engdahl, E. R., 2007. Global prevalence of double Benioff zones. *Science*, **316**, 1472–1474.

Chen, Y., Wen, L., and Ji, C., 2014. A cascading failure during the 24 May 2013 great Okhotsk deep earthquake. *Journal of Geophysical Research*, **119**, 3035–3049. doi:10.1002/2013JB010926

Emmerson, B., and McKenzie, D., 2007. Thermal structure and seismicity of subducting lithosphere. *Physics of the Earth and*

Planetary Interiors, **163**, 191–208, doi:10.1016/j.pepi.2007.05.007.

Estabrook, C. H., 2004. Seismic constraints on mechanisms of deep earthquake rupture. *Journal of Geophysical Research*, **109**, B02306, doi:10.1029/2003JB002449.

Frohlich, C., 2006. *Deep earthquakes*. Cambridge: Cambridge University Press, p. 573.

Fujita, K., and Kanamori, H., 1981. Double seismic zones and stresses of intermediate depth earthquakes. *Geophysical Journal of the Royal Astronomical Society*, **66**, 131–156.

Garth, T., and Rietbrock, A., 2014. Order of magnitude increase in subducted H2O due to hydrated normal faults within the wadati-benioff zone. *Geology*, **42**(3), 207–210, doi:10.1130/G34703.1.

Green, H. W., 2007. Shearing instabilities accompanying high-pressure phase transformations and the mechanics of deep earthquakes. *PNAS*, **104**, 9133–9138, doi:10.1073/pnas0608045104.

Hasegawa, A., Umino, N., and Horiuchi, S., 1978. Double-planed deep seismic zone and upper mantle structure in the northeastern Japan arc. *Royal Astr Social Geophys Journal*, **54**, 281–296.

Heuret, A., and Lallemand, S., 2005. Plate motions, slab dynamics and back-arc deformation. *PEPI*, **149**, 31–51.

John, T., Medvedev, S., Rüpke, L. H., Andersen, T. B., Podladchikov, Y. Y., and Austrheim, A., 2009. Generation of intermediate-depth earthquakes by self-localization thermal runaway. *Nature Geoscience*, **2**, 137–140.

Jung, H., Green, H. W., and Dobrzhinetskaya, L. F., 2004. Intermediate-depth earthquake faulting by dehydration embrittlement with negative volume change. *Nature*, **428**, 545–549.

Kanamori, H., Anderson, D. L., and Heaton, T. H., 1998. Frictional melting during the rupture of the 1994 Bolivian earthquake. *Science*, **279**, 839–842.

Kawakatsu, H., 1986. Double seismic zones: kinematics. *Journal of Geophysical Research*, **91**, 4811–4825.

Kelemen, P. B., and Hirth, G., 2007. A periodic shear-heating mechanism for intermediate-depth earthquakes in the mantle. *Nature*, **446**, 787–790.

Kikuchi, M., and Kanamori, H., 1994. The mechanism of the deep Bolivia earthquake of June 9, 1994. *Geophysical Research Letters*, **21**, 2341–2344.

Kirby, S. H., Durham, W. B., and Stern, L. A., 1991. Mantle phase changes and deep-earthquake faulting in subducting lithosphere. *Science*, **252**, 216–225.

Kirby, S. H., Stein, S., Okal, E. A., and Rubie, D. C., 1996. Metastable mantle phase transformations and deep earthquakes in subducting oceanic lithosphere. *Reviews of Geophysics*, **34**, 261–306.

Marot, M., Monfret, T., Pardo, M., Ranalli, G., and Nolet, G. 2013. A double seismic zone in the subducting Juan Fernandez Ridge of the Nazca plate (32°), central Chile. *Journal of Geophysical Research*, **118**, 3462–3475. doi:10.1002/jgrb.50240

Meng, L., Ampuero, J.-P., and Bürgmann, R., 2014. The Okhotsk deep-focus earthquake: rupture beyond the metastable olivine wedge and thermally controlled rise time near the edge of a slab. *Geophysical Research Letters*, **41**, 3779–3785, doi:10.1002/2014GL059968.

Peacock, S. M., 2001. Are the lower planes of double seismic zones caused by serpentine dehydration in subducting oceanic mantle? *Geology*, **29**, 299–301.

Peacock, S. M., 2003. Thermal structure and metamorphic evolution of subducting slabs. inside the subduction factory. *Geophysical Monograph*, **138**, 7–22.

Prieto, G., Florez, M., Barrett, S. A., Beroza, G. C., Pedraza, P., Blanco, J. F., and Poveda, E., 2013. Seismic evidence for thermal runaway during intermediate-depth earthquake rupture. *Geophysical Research Letters*, **40**, 6064–6068, doi:10.1002/2013GL058109.

Ranero, C. R., and Sallares, V., 2004. Geophysical evidence for hydration of the crust and mantle of the nazca plate during bending at the north Chile trench. *Geology*, **32**, 549–552, doi:10.1130/G20379.1.

Ranero, C. R., Villaseñor, A., Phipps Morgan, J., and Weinrebe, W., 2005. Relationship between bend-faulting at trenches and intermediate-depth seismicity. *Geochemistry, Geophysics, Geosystems*, **6**, Q12002. doi: 10.1029/2005GC000997.

Rayleigh, C. B., and Paterson, M. S., 1965. Experimental deformation of serpentinite and its tectonic implications. *Journal of Geophysical Research*, **70**, 3965–3985.

Reynard, B., Nakajima, J., and Kawakatsu, H., 2010. Earthquakes and plastic deformation of anhydrous slab mantle in double wadati-benioff zones. *Geophysical Research Letters*, **37**, L24309, doi:10.1029/2010GL045494.

Rietbrock, A., and Waldhauser, F. 2004. A narrowly spaced double-seismic zone in the subducting Nazca plate. *Geophysical Research Letters*, **31**, L10608, doi:10.1029/2004GL019610.

Seno, T., and Yamamaka, Y., 1996. Double seismic zones, compressional deep trench-outer rise events, and superplumes. In: Bebout GE (Ed.), subduction: Top to bottom. *Monograph*, **96**, 347–355.

Shiina, T., Nakajima, J., and Matsuzawa, T., 2013. Seismic evidence for high pore pressures in the oceanic crust: implications for fluid-related embrittlement. *Geophysical Research Letters*, **40**, 05–28, doi:10.1002/grl.50468.

Tibi, R., Bock, G., and Wiens, D. A., 2003. Source characteristics of large deep earthquakes: constraints on the faulting mechanisms at great depths. *Journal of Geophysical Research*, **108**, B22091, doi:10.1029/2002JB001948.

van Keken, P. E., Hacker, B. R., Syracuse, E. M., and Abers, G. A. 2011. Subduction factory: 4. Depth-dependent flux of H_2O from subducting slabs worldwide. *Journal of Geophysical Research*, **116**, B01401. doi:10.1029/2010JB007922

Wadati, K., 1928. On shallow and deep earthquakes. *Geophysical Magazine*, **1**, 162–202.

Wadati, K., 1931. Shallow and deep earthquakes, 3rd paper. *Geophysical Magazine*, **4**, 231–283.

Wadati, K., 1935. On the activity of deep-focus earthquakes in the Japan islands and neighboorhoods. *Geophysical Magazine*, **8**, 305–326.

Wang, K., 2002. Unbending combined with dehydration embrittlement as a cause for double and triple seismic zones. *Geophysical Research Letters*, **29**, 1889. doi:10.1029/2002GL015441

Warren, L. M., Hughes, A. N., and Silver, P. G., (2007). Earthquake mechanics and deformation in the tonga-Kermadec subduction zone from fault plane orientations of intermediate- and deep-focus earthquakes. *Journal of Geophysical Research*, **112**, B05314. doi:10.1029/2006JB004677

Wiens, D. A., and McGuire, J., 1995. The 1994 Bolivia and Tonga events: fundamentally different types of deep earthquakes? *Geophysical Research Letters*, **22**, 2245–2248.

Ye, L., Lay, T., Hanamori, H., and Koper, K. D., 2013. Energy release of the 2013 Mw 8.3 Sea of Okhotsk earthquake and deep slab stress heterogeneity. *Science*, **341**, 1380–1384.

Cross-references

Earthquakes
Epicenter, Hypocenter
Intraoceanic Subduction Zone
Seismogenic Zone
Subduction
Subduction Erosion

WADDEN SEA

Achim Wehrmann
Marine Research Department, Senckenberg am Meer, Wilhelmshaven, Germany

Synonyms

Vadehavet (Danish); Waddenzee (Dutch); Wattenmeer (German)

Definition

Tide influenced coastal zone of the southeastern North Sea. During low tide, the globally largest coherent system of sandy to muddy tidal flats emerge between the mainland and the barrier islands. The Wadden Sea evolved during late Holocene transgression and is a high productive ecosystem with strongly fluctuating environmental conditions. Besides the extended tidal flats, main morphological units are barrier islands, dunes, ebb tidal deltas, tidal inlets, tidal channels, salt marshes, and estuaries.

Introduction

The Wadden Sea forms the coastal margin of the southeastern North Sea, a broad and shallow shelf of the passive continental margin of Northwestern Europe. It extends from Den Helder (The Netherlands) to Blåvands Huk (Denmark) (Figure 1) over 490 km. The average width of the tidal flat area is 5–25 km. Toward the North Sea, the Wadden Sea is sheltered over long distances by a chain of barrier islands. Open tidal flats are restricted to the inner part of the German Bight where high tidal ranges and strong wave energy prevent the formation of barrier islands. Also, the larger estuaries and embayments of the rivers Ems, Jade, Weser, and Elbe are flanked by tidal flats. The Wadden Sea covers an area of approx. 9,800 km^2. About half of this (4,700 km^2) is situated in the intertidal environment represented by sand and mud flats (Common Wadden Sea Secretariat, 2008). The coherent tidal flats of the Wadden Sea are known as a conspicuous landscape of global relevance; however, they are part of a worldwide system of barrier islands, tidal flats, and lagoons bordering many coastal regions in different climatic settings.

Hydrodynamic regime

According to Davis (1964) and Hayes (1979), mean tidal range and mean wave height are the two principle energy sources which control depositional coastlines and therefore the presence and distribution of barriers. Those parts of the Wadden Sea coast with barrier islands and backbarrier tidal flats are typically mesotidal (i.e., tidal range 1.0–3.5 m sensu Hayes, 1979). As the tidal range continuously increases toward the inner part of the German Bight (Figure 1), low macrotidal conditions (tidal range > 3.5 m) occur in the larger estuaries of Jade (3.9 m), Weser (4.2 m), and Elbe (3.7 m) and the open tidal flats up north to the peninsula of Eiderstedt (3.5 m).

The tides are semidiurnal resulting in an in- and outflow of 15 km^3 of seawater through the tidal inlets during each tidal cycle, which is half of the entire water volume (30 km^3) of the area. Therefore, the water body of the Wadden Sea is well mixed.

Considering tidal range, wave action, and morphologic development, the East Frisian Islands reveal high-energy conditions. Accordingly, the West Frisian Islands and the Danish Wadden Sea both having considerably lower mean tidal ranges (1.4 m at Den Helder and 1.5 m at Esbjerg) (Figure 1) represent the lower energy spectrum of barrier island systems. Offshore, mean significant wave heights vary seasonally between 0.5 and 2 m (Bartholdy and Pejrup, 1994; Dobrynin et al., 2010) but can reach values of 8–11 m during storm surges.

In high-energy settings (high tidal ranges and/or high wave impact) ephemeral sand bank islands are developed. Temporarily, these islands may exhibit a poor salt marsh and dune vegetation which, however, has not the capacity to stabilize the islands permanently. The sand bank islands of the Wadden Sea are highly dynamic and transient sediment bodies (Wehrmann and Tilch, 2008; Hellwig and Stock, 2014).

The tidal flats are dissected by a multitude of main and tributary channels and creeks (Figure 2). Their morphology is mainly influenced by tidal currents. The main channels have mean current velocities of 0.7–1.3 m/s, whereas only approx. 0.3 m/s are recorded upon the tidal flats (Krögel, 1997; Gehm et al., 2000). In tidal inlets average maximum tidal velocities is 1–2 m/s (Bartholdy and Pejrup, 1994).

Oceanography

The salinity of the water body in the tidal basins ranges seasonally between 27 and 32, but can vary between 22 and 35 (Becker, 1998; monthly mean Kaiser and Niemeyer, 1999). The daily river discharge into the Wadden Sea is 1.64×10^8 m^3. However, no substantial fluvial input of sediment material into the Wadden Sea occurs (Bartholdy and Pejrup, 1994; Oost et al., 2012). Mean water temperature changes from 1 °C in winter to 21 °C in summers. In long-lasting freezing periods, the backbarrier zone is covered by drift ice (long-term average: 19 days per year).

The complex channel systems discharge the tidal basins during the ebb phase toward the tidal inlets. From the narrow and up to 42 m deep incising tidal inlets, the major channels began to furcate into minor channels and tidal creeks (Figure 2). The hydrodynamic energy in the channels is lowest close to the drainage divides separating the respective tidal basins. Therefore, these parts of the channels function as sediment traps which results in a pronounced accumulation of mud. In the West and East Frisian Wadden Sea, semi-consolidated mud beds are occasionally exposed along deep main channels. They may be related to former (historic) time periods when the muddy channel zones (of the watersheds) were located

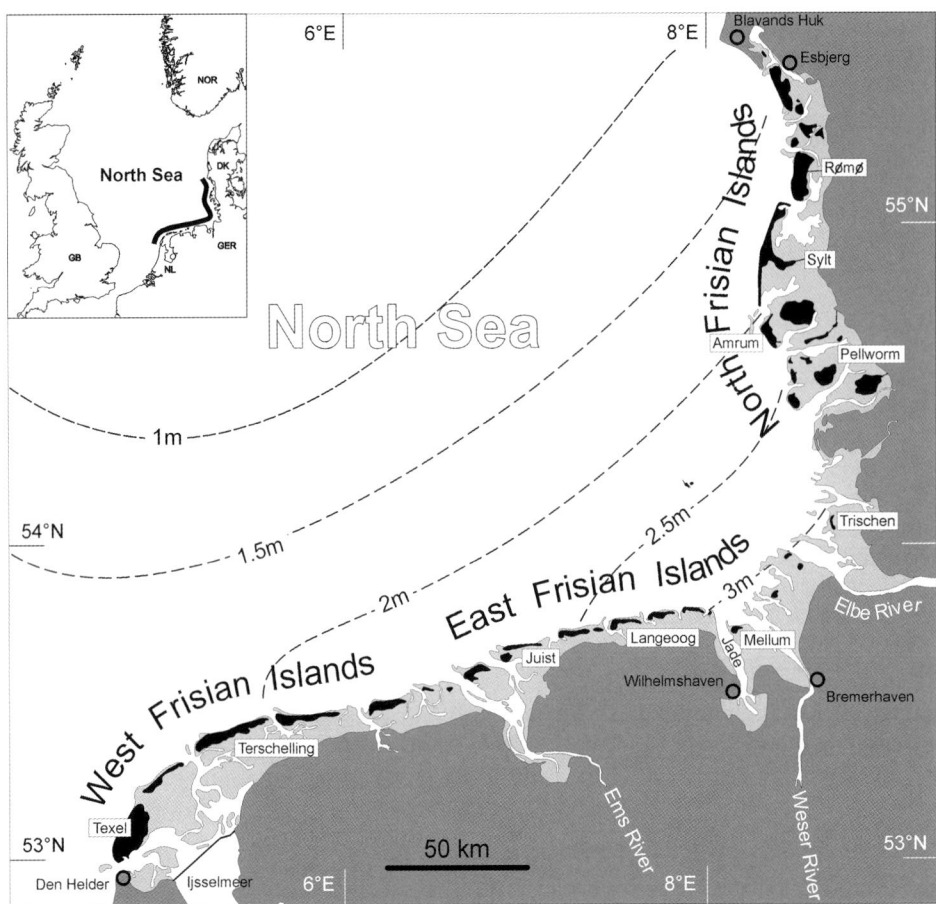

Wadden Sea, Figure 1 Geographic map of the Wadden Sea. The large coherent system of tidal flats (*bright gray*) is protected toward the open North Sea by a long chain of barrier islands. In the innermost part of the German Bight, which has macrotidal conditions (tidal range > 3.5 m), only small ephemeral sand bank islands and extended open tidal flats occur. *Dashed lines* indicate mean tidal ranges.

in the respective position, i.e., much more westerly than today (Homeier and Luck, 1969). These mud beds reflect the long-term shifting of the backbarrier watersheds toward southeast during the last centuries.

In total, the Wadden Sea consists of 39 tidal basins, of which seven have an estuarine character (Kraft et al., 2011).

Morphodynamics

From sediment trend analysis of the western tidal basins, McLaren et al. (1998) mentioned that some principle processes are similar for each tidal basin and their corresponding tidal inlet/tidal delta system. Inside the tidal basins, the sediment transport is generally directed from the tidal inlets toward the land. The sediment trends follow the branching network of channels and creeks with net accumulation on the tidal flats. Within the tidal basins, the hydrodynamic regime controls the grain size distribution of the siliciclastic sediments. As a general trend the sediment is fining toward the coast, from sand flats over mixed tidal flats to mud flats (Bartholomä and Flemming, 2007) (Figure 2). In detail, this principle transition in grain size distribution shows a much more complex pattern where the tidal flats are dissected by tidal channel systems (Hertweck, 1994; Flemming and Ziegler, 1995; Hertweck, 1998; Hertweck et al., 2005).

The ebb tidal deltas (Figure 2) are marked by a complex system of transport gyres originating from the deepest channel portion of the tidal inlets. Sediment transport results from both ebb tidal currents and wave action.

Offshore from the barrier islands, an eastward-directed long-shore transport is most obvious in the West and East Frisian Wadden Sea, whereas littoral drift is mostly directed southward in the Nord Frisian and Danish Wadden Sea (Bartholdy and Pejrup, 1994; Zeiler et al., 2000). At each tidal inlet this process is interrupted by the dynamics of the respective tidal basin. Perspectively, the rising sea-level, the loss of accommodation space, and the retrogradational development of sedimentary

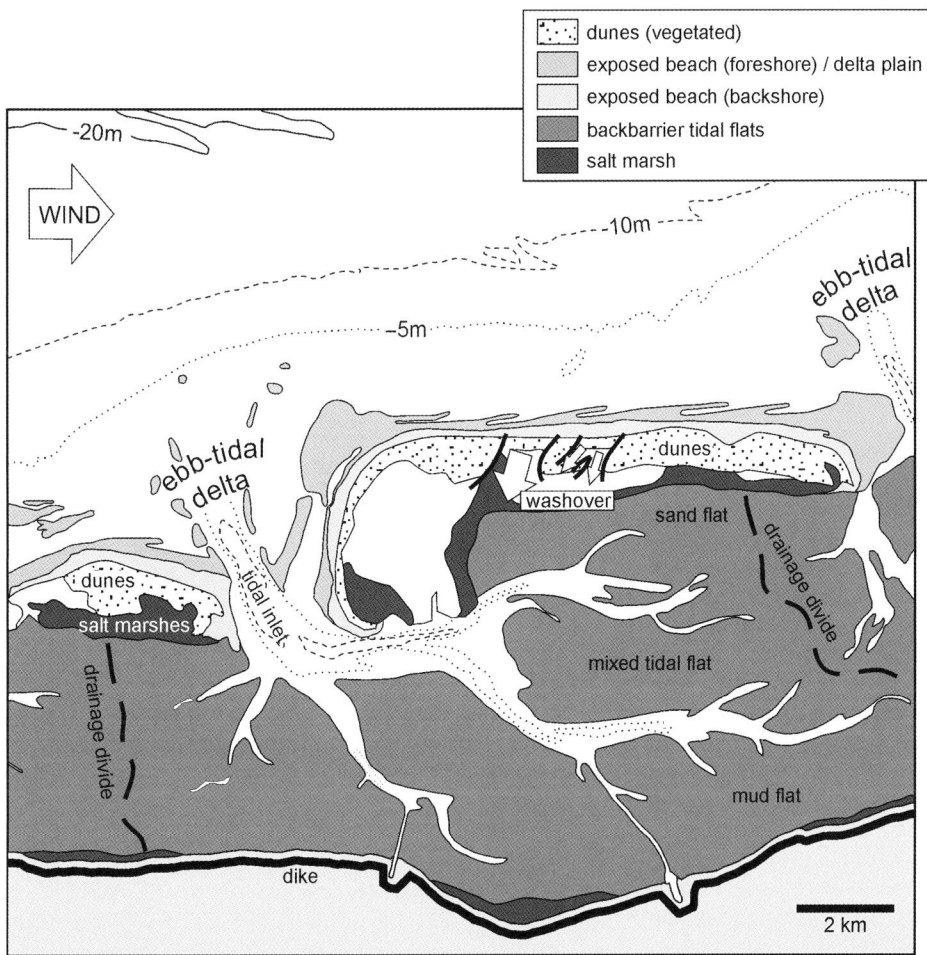

Wadden Sea, Figure 2 Main structural elements (morphological units) of the Wadden Sea. The tidal basins are separated by drainage divides. Water exchange between tidal basins and the open North Sea is supported by tidal inlets. The morphodynamic of the exposed beaches of the barrier islands is mainly controlled by wave action and undergoes seasonal modification. Salt marshes mark the transition from marine to terrestrial environment in sheltered positions. Washover fans document local dune erosion during severe storm surges.

facies will subsequently squeeze out the finest sediment particles, i.e., the mud flat facies (Mai and Bartholomä, 2000). However, extended mud flats are widely known from the Holocene sedimentary record of the coastal plain.

Sediment-organism interaction

The entire tidal flats are intensely populated by bivalves (e.g., *Macoma balthica*, *Cerastoderma edule*, *Mya arenaria*) and polychaetes (e.g., *Arenicola marina*, *Heteromastus filiformis*, *Hediste diversicolor*, and the tube-building polychaetes *Lanice conchilega* and *Pygospio elegans*). Another characteristic feature of the central tidal flats are extended mussel beds of *Mytilus edulis*, of which most of them were recently transformed into oyster reefs by bioinvasion of the Pacific oyster *Crassostrea gigas* (Wehrmann et al., 2000). As an ecosystem engineering species, the Pacific oyster has a strong impact to the Wadden Sea ecosystem causing significant changes in biodiversity. Additionally, by their filter feeding activity they produce enormous amounts of feces and pseudofeces accumulated as muddy organic-rich biodeposits within and around the reefs. As these biodeposits are stabilized by biofilms and microbial mats, they are not in equilibrium with the hydrodynamic regime.

All tidal flats display a similar structural pattern (Hertweck, 1998) as well as a typical distribution of bioturbation (Hertweck, 1994):

(a) Stable platforms with *Lanice conchilega* colonies and frequently with *Crassostrea* reefs or *Mytilus* beds in central position of the platforms.
(b) Sandy tidal flats in the marginal areas undergoing permanent physical reworking; the macrofauna is mostly reduced to the lugworm *Arenicola marina* and subordinately the cockle *Cerastoderma edule*.

(c) Mobile sand bodies in the point bars of channels and at the flanks and spits of the tidal flats which undergo intense reworking and are, therefore, nearly devoid of macrofauna.

At tidal flat margins and spits, which are highly exposed to tidal currents and wave action, sediment bodies are generated indicating intense physical sediment reworking. They consist of fine sand, partly laminated and partly with intercalations of thin mud flasers. Due to the lateral migration of smaller tidal channels and creeks, longitudinal cross-bedding and shell lag deposits are prominent sedimentary structures of tidal flat deposits.

Salt marshes

The salt marshes of the Wadden Sea represent the natural transition between the intertidal flats and the terrestrial environment (Figure 2) where average energy input is low. They are situated above the mean tidal high-water level. Thus, they may frequently be affected by flood events during extraordinary high-water levels and storm surges. Besides the abundant plant community (Dörjes, 1978; Streif, 1990), the salt marshes are also characterized by meandering and multifariously branched channels effecting drainage after extreme high-water events.

Where accretion prevails, the transition from the tidal flats to the salt marshes is marked by a silting-up zone (30–40 cm below the high-water line) which is characterized by the occurrence of the halophyte *Salicornia* sp., a pioneer colonizer stabilizing the sediment by its rootlets. Dense standing stocks of *Salicornia* favor baffling and accumulation of very-fine-grained sediment particles. In the same level, multilayered microbial mats are formed by diatom algae, filamentous green algae, cyanobacteria, and sulfate-reducing bacteria also efficiently stabilizing the sediment surface (Gerdes et al., 1985; Gerdes and Wehrmann, 2008). As these microbial mats with their typical lamination have a high preservation potential, they are a characteristic indicator for the proximate land-sea transition in low-energy environments. The transition from the *Salicornia* zone to the *Puccinellia* zone is situated around the high-water line. All species of the *Puccinellia* zone are genuine halophytes able to tolerate seawater flooding. The topographically highest zone of the salt marsh is the *Festuca* zone which is flooded only during highest spring tides and storm surges. The number of flooding events is less than 100 per year (Streif, 1990).

The sedimentologically most conspicuous feature of the salt marshes are the storm deposits. These are intercalations of silty sand and/or layers of shells and shell debris deposited during storm surges (Bouma, 1963; Hertweck et al., 2005). Flooding of the salt marshes under calm conditions results in the accumulation of muddy sediments.

Dunes

One of the main characteristics of the barrier islands are the dunes, separating the salt marshes of the islands from the shore environment (Figure 2). The dunes are generated at the upper limit of the backshore where sufficient desiccation allows the mobilization of sand by eolian processes.

The organic debris and algal material accumulated at the high-water lines cause a temporary enrichment of nutrients within the sand, resulting in the development of particular plant associations. Primary (embryonic) dunes are generated in leeward position of this pioneer vegetation. They may reach up to 2 m height several months after initial formation. The sediment baffling initiating the formation of primary dunes is caused by *Agropyron junceum* which is highly resistant to sand covering.

In higher parts of the primary dunes *Elymus arenarius* is the second characteristic plant. Both species are characterized by widespread rootlets, thus being highly adapted to sand transport, desiccation, and salt water spray. The fixation of the eolian sand by the plant community causes the formation of secondary dunes ("white dunes") of up to 20 m height. The influence of salt water within the sediment body strongly decreases. Simultaneously, salt water is diluted out of the dunes by precipitation. *Ammophila arenaria* is the characteristic plant species of the white dunes. Deep-anchored roots are a typical adaptation of this species to strong sand transport.

The disappearance of *Elymus arenarius* and *Ammophila arenaria* vegetation on the leeward dunes together with initial soil formation marks the beginning of the tertiary dune development ("gray" and "brown dunes").

Beaches

The upper shoreface and foreshore of the barrier islands are sandy beaches which undergo the hydrodynamic changes of ebb and flood tide but strongly differ to tidal flats. Tidal beaches are situated at the offshore-directed coasts of barrier islands and thus, are fully exposed to onshore winds and waves. During the tidal cycle the effect of wave action continuously undergoes periodical modifications.

Under fair weather conditions two or three coast-parallel ridges are formed in the shallow upper shoreface. These are continuously shifted toward the shore, whereas new ridges are formed at the lower limit of this zone. The same process takes place in the foreshore zone. Depending on the local tidal range, two or three beach ridges alternating with beach runnels occur. The most beachward runnel marks the transition to the backshore. The subparallel beach runnels are some 100 m long and show an extremely acute angle toward the respective waterline (Figure 2). Only at the mouth of a fully emerged runnel a short curved portion is visible effecting discharge in rectangular direction. Thus, the beach ridges and runnels are not parallel structures following the isohypses, but displaying a stepwise pattern along the foreshore zone (Reineck, 1960). The beach ridges are mainly formed by onshore wave action.

The dominating sediments in the beach area and the dune belt are fine-to-medium sands.

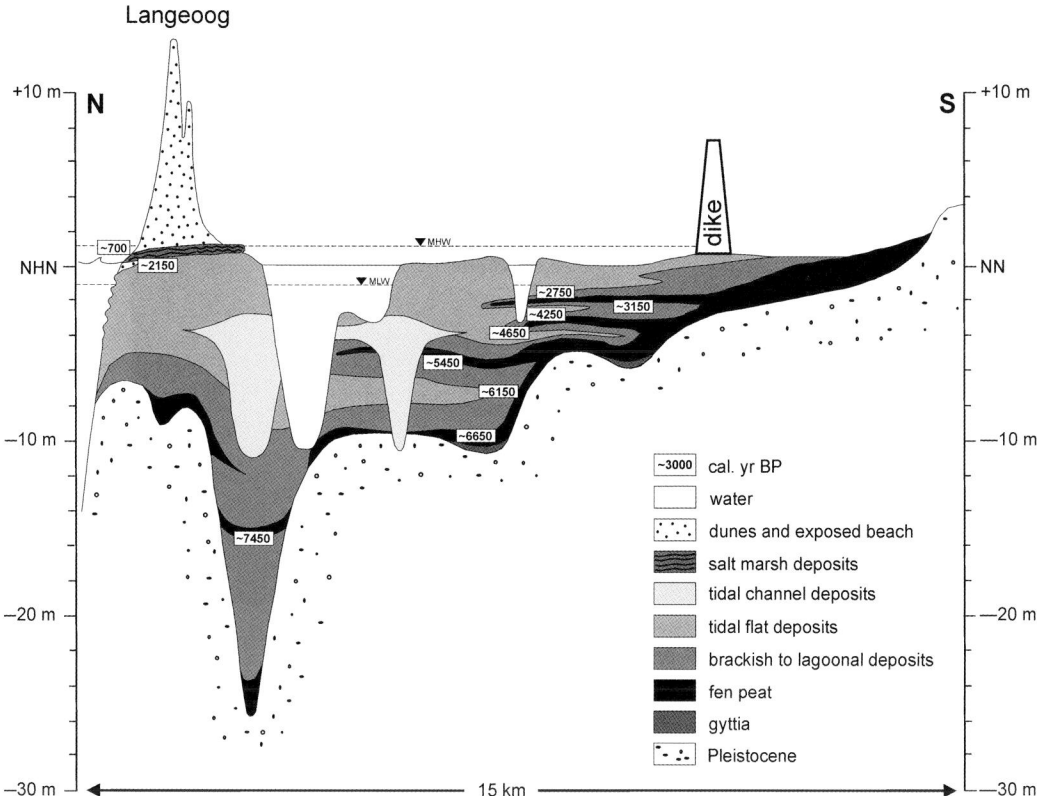

Wadden Sea, Figure 3 Geological cross section through the Central Wadden Sea perpendicular to the coastline (modified after Streif (2004) and Bungenstock and Schäfer (2009)). The relief of the early Holocene surface is based on core and seismic data. The narrow sediment wedge of sediments deposited during Holocene transgression represents a fourth-order TST and HST. Marked slowdown in sea-level rise favors the formation and progradation of distinct fen peat horizons. Since land reclamation and dike construction limit retrograde shift of sedimentary facies and gradual fade of the hydrodynamic energy, the finest grain sizes are exported out of the system. This will perspectively cause a depletion of the muddy facies. *MLW* mean low water line, *MHW* mean high-water line.

Holocene evolution of the Wadden Sea and development of the barrier islands

The Wadden Sea is a young sedimentary and ecological system which developed after the last glaciation. Most of the sediment material originates from local reworking of Pleistocene deposits during Holocene transgression, forming the narrow sediment wedge of the Wadden Sea (Figure 3). In the West and East Frisian Wadden Sea, this up to 35 m thick sediment wedge (Streif, 2004) represents fourth-order TST and HST (Bungenstock and Schäfer, 2009). Initially, the rate of sea-level rise was more than 1 m per century but strongly slowed down around 7,500–7,000 year BP (Kiden et al., 2002; Vink et al., 2007). At approx. 8,000–7,500 year BP, the Holocene transgression reached the area where the barrier islands were formed, as evident from ^{14}C datings of reed material (in lagoonal to brackish water deposits from −24 m below chart datum) sampled offshore of Wangerooge Island (Hanisch, 1980). Outcrops of fossil tidal flat deposits, mud beds and salt marshes at beaches, shoreface and offshore areas of the islands point to a shifting of these islands over their own tidal sediments, i.e., backbarrier areas (Figure 4). It has to be assumed that the West and East Frisian Islands were formed between 6,000 and 5,000 year BP several km offshore from its present position and, therefore, gradually shifted toward south-southeast during the Holocene transgression. For the last 2,000 years, displacement rates of more than 100 m per century may be stated (Barckhausen, 1969; Streif, 1990). In contrast, the data from the North Frisian Islands Rømø suggested that formation there started around 8,000 year BP (Madsen et al., 2010).

Both, seismic sub-bottom profiling and investigation of long sediment cores revealed a rugged Pleistocene relief below the Holocene deposits which only was preservable if an essential W-E displacement of the tidal inlets can be excluded. According to Streif (1990), the channel bases of the inlets are usually less than 3 km wide, exceptionally 5–6 km. In the coastal area the deep incised paleo-valleys of the rivers Ems, Weser, and Elbe were transformed into estuaries also around 8,000 year BP. During the late Holocene, the sea-level shows several fluctuations

Wadden Sea, Figure 4 Fossil semi-consolidated mud beds outcropping at the offshore exposed beaches of a barrier island at low tide indicates southward (retrograde) shift of the entire Holocene sediment wedge. The mud beds were formerly deposited in sheltered backbarrier environments of the respective island. From historical navigation charts migration rates of more than 100 m per century are documented.

(i.e., stillstand and/or slow down) with multifold progradation of the coastline as indicated by formation of fen peat intercalating the clastic sedimentary sequence (Figure 3).

The formation of the barrier islands of the Wadden Sea mainly results from two contrasting processes. The West and East Frisian Island (except Texel) (Figure 1), the so-called dune islands, were formed by the aquatic and eolian deposition of sandy material, mainly trapped and stabilized by vegetation.

All dune islands originate from shoals having developed by the interaction of hydrodynamic energy and eolian processes. The initial base of the shoals consists of ridges of the Pleistocene moraine landscape submerged by the transgression. The shoals emerging during low tide became more elevated by storm-related accumulation of sand and were, subsequently, flooded only exceptionally. In the leeward area sheltered by these durable shoals, tidal flats could develop favored by the low-energy conditions. Additionally, sand from the offshore exposed beaches of the shoals was mobilized by onshore winds. Thus, primary dunes were formed, subsequently stabilized by plants. During transgression, the barrier islands were shifted toward south passing over their own backbarrier tidal flats (Figure 4). Most of the islands consist of several isolated dune ridges cut by washover fans (Figure 2) which were generated by storm surges.

The North Frisian Islands on the other hand are the remnants of the marsh-like coastal plain (e.g., Pellworm, Nordstrand) or of Tertiary to Pleistocene deposits (e.g., Sylt, Amrum, Föhr), resulting from erosional processes during storm surges (Hoffmann, 2004). Nevertheless, most of the more exposed North Frisian Islands also show the principle sedimentary processes and morphological units of beach and dunes environments.

Summary

The Wadden Sea is a unique coastal landscape of the southeastern North Sea which has evolved during the late Holocene under rising sea-level conditions. Therefore, it is a very young system in terms of geomorphology and ecology. The Wadden Sea is characterized by a coherent system of extended tidal flats, barrier islands, and salt marshes which is one of the largest worldwide. The development of barrier island, which separates the intertidal sandy to muddy flats from the offshore shelf, is linked to two different processes: formation of dunes on supratidal shoals by eolian sediment transport and subsequent stabilization by vegetation (West and East Frisian Islands), and erosion of mainly Pleistocene headlands and/or early Holocene salt marshes (North Frisian Islands), respectively. Water exchange during semidiurnal tide between the tidal basins and the open North Sea is supported by the complex of tidal channels, tidal inlet, and ebb tidal delta.

Bibliography

Barckhausen, J., 1969. Entstehung und Entwicklung der Insel Langeoog. Beitrag zur Quartärgeologie und Paläogeographie eines ostfriesischen Küstenabschnittes. *Oldenburger Jahrbuch*, **68**, 239–281.

Bartholdy, J., and Pejrup, M., 1994. Holocene evolution of the Danish Wadden Sea. *Senckenbergiana Maritima*, **24**, 187–209.

Bartholomä, A., and Flemming, B. W., 2007. Progressive grain-size sorting along an intertidal energy gradient. *Sedimentary Geology*, **202**, 464–472.

Becker, G., 1998. *Der Salzgehalt im Wattenmeer*. Stuttgart: Ulmer. Umweltatlas Wattenmeer, Vol. 1, pp. 60–61.

Bouma, A. H., 1963. A graphic presentation of the facies model of salt marsh deposits. *Sedimentology*, **2**, 122–129.

Bungenstock, F., and Schäfer, A., 2009. The Holocene relative sea-level curve for the tidal basin of the barrier island Langeoog, German Bight, Southern North Sea. *Global and Planetary Change*, **66**, 34–51.

Common Wadden Sea Secretariat, 2008. *Nomination of the Dutch-German Wadden Sea as World Heritage Site*. Wadden Sea Ecosystem, Wilhelmshaven: CWSS, Vol. 24, 200 pp.

Davis, J. L., 1964. A morphogenetic approach to world shorelines. *Zeitschrift für Geomorphologie*, **8**, 27–42.

Dobrynin, M., Gayer, G., Pleskachevsky, A., and Guenther, H., 2010. Effect of waves and currents on the dynamics and seasonal variations of suspended particulate matter in the North Sea. *Journal of Marine Systems*, **82**(1–2), 1–20.

Dörjes, J., 1978. Das Watt als Lebensraum. In Reineck, H.-E. (ed.), *Das Watt – Ablagerungs- und Lebensraum*. Frankfurt a. M: Kramer, pp. 107–143.

Flemming, B. W., and Ziegler, K., 1995. High-resolution grain size distribution patterns and textural trends in the backbarrier environment of Spiekeroog Island (southern North Sea). *Senckenbergiana Maritima*, **26**, 1–24.

Gehm, G., Wilken, M., and Liebezeit, G., 2000. Temporal and spatial high resolution analysis of nutrient dynamics in the near bottom layer of the Wadden Sea: methodological aspects and first results. *Senckenbergiana Maritima*, **30**, 105–114.

Gerdes, G., and Wehrmann, A., 2008. Biofilms in surface sediments of the ephemeral sand bank island Kachelotplate (southern North Sea). *Senckenbergiana Maritima*, **38**(2), 173–183.

Gerdes, G., Krumbein, W. E., and Reineck, H.-E., 1985. The depositional record of sandy, versicolored tidal flats (Mellum Island, southern North Sea). *Journal of Sedimentary Petrology*, **55**, 265–278.

Hanisch, J., 1980. Neue Meeresspiegeldaten aus dem Raum Wangerooge. *Eiszeitalter und Gegenwart*, **30**, 221–228.

Hayes, M. O., 1979. Barrier island morphology as a function of tidal and wave regime. In Leatherman, S. P. (ed.), *Barrier Islands*. New York: Academic, pp. 1–27.

Hellwig, U., and Stock, M. (eds.), 2014. *Dynamic Islands in the Wadden Sea*. Wadden Sea Ecosystems, Wilhelmshaven: CWSS, Vol. 33, 132 pp.

Hertweck, G., 1994. Zonation of benthos and lebensspuren in the tidal flats of the Jade Bay, southern North Sea. *Senckenbergiana Maritima*, **24**(1/6), 157–170.

Hertweck, G., 1998. Facies characteristics of back-barrier tidal flats of the East Frisian Island of Spiekeroog, southern North Sea. I. In *Tidalites: Processes and Products*. SEPM Special Publication, 61, pp. 23–30

Hertweck, G., Wehrmann, A., Liebezeit, G., and Steffens, M., 2005. Ichnofabric zonation in modern tidal flats: palaeoenvironmental and palaeotrophic implications. *Senckenbergiana Maritima*, **35**(2), 189–201.

Hoffmann, D., 2004. Holocene landscape development in the marshes of the West Coast of Schleswig-Holstein, Germany. *Quaternary International*, **112**, 29–36.

Homeier, H., and Luck, G., 1969. Das Historische Kartenwerk 1:50 000 der Niedersächsischen Wasserwirtschaftsverwaltung als Ergebnis historisch-topographischer Untersuchungen und Grundlage zur kausalen Deutung der Hydrovorgänge im Küstengebiet. Veröffentlichungen Niedersächsisches Institut für Landeskunde und Landesentwicklung Universität Göttingen, Reihe A: Forschungen zur Landes- und Volkskunde. I. Natur, Wirtschaft, Siedlung u. Planung, 93. Göttingen: Wurm, 36 pp. + 28 pp. appendix

Kaiser, R., and Niemeyer, H. D., 1999. *Wasser-Beschaffenheit*. Stuttgart: Ulmer. Umweltatlas Wattenmeer, Vol. 2, pp. 32–33.

Kiden, P., Denys, L., and Johnston, P., 2002. Late Quaternary sea-level change and isostatic and tectonic land movements along the Belgian-Dutch North Sea coast: geological data and model results. *Journal of Quaternary Science*, **17**(5–6), 535–546.

Kraft, D., Folmer, E. O., Meyerdirks, J., and Stiehl, T. 2011. *Data inventory of the tidal basins in the trilateral Wadden Sea*. Programma Naar een Rijke Waddenzee Report, 42 pp.

Krögel, F., 1997. *Einfluß von Viskosität und Dichte des Seewassers auf Transport und Ablagerung von Wattsedimenten (Langeooger Rückseitenwatt, Südliche Nordsee)*. Bremen: Berichte FB Geowissenschaften Universität, Vol. 102, 169 pp.

Madsen, A. T., Murray, A. S., Andersen, T. J., and Pejrup, M., 2010. Luminescence dating of Holocene sedimentary deposits from Rømø, a barrier island in the Wadden Sea, Denmark. *The Holocene*, **20**(8), 1247–1256.

Mai, S., and Bartholomä, A., 2000. The missing mud flats of the wadden Sea: a reconstruction of sediments and accommodation space lost in the wake of land reclamation. In Flemming, B. W., Delafontaine, M. T., and Liebezeit, G. (eds.), *Muddy coast dynamics and resource management*. Amsterdam: Elsevier, pp. 257–272.

McLaren, P., Steyaert, F., and Powys, R., 1998. Sediment transport studies in the tidal basins of the Waddenzee. *Senckenbergiana Maritima*, **29**(1/2), 53–61.

Oost, A. P., Hoekstra, P., Wiersma, A., Flemming, B., Lammerts, E. J., Pejrup, M., Hofstede, J., van der Valk, B., Kiden, P., Bartholdy, J., van der Berg, M. W., Vos, P. C., de Vries, S., and Wang, Z. B., 2012. Barrier island management: lessons from the past and directions for the future. *Ocean and Coastal Management*, **68**, 18–38.

Reineck, H.-E, 1960. Über den Transport des Riffsandes. *Jahresbericht Forschungsstelle Norderney 1959*, **11**, 21–38.

Streif, H., 1990. *Das ostfriesische Küstengebiet. Nordsee, Inseln, Watten und Marschen*. Berlin: Borntraeger. Sammlung geologischer Führer, 57.

Streif, H., 2004. Sedimentary record of Pleistocene and Holocene marine inundations along the North Sea coast of Lower Saxony, Germany. *Quaternary International*, **112**, 3–28.

Vink, A., Steffen, H., Reinhardt, L., and Kaufmann, G., 2007. Holocene relative sea-level change, isostatic subsidence and the radial viscosity structure of the mantle of northwest Europe (Belgium, the Netherlands, Germany, southern North Sea). *Quaternary Science Reviews*, **26**, 3249–3275.

Wehrmann, A., and Tilch, E., 2008. Sedimentary dynamic of an ephemeral sand bank island (Kachelotplate, German Wadden Sea): an atlas of sedimentary structures. *Senckenbergiana Maritima*, **38**(2), 185–198.

Wehrmann, A., Herlyn, M., Bungenstock, F., Hertweck, G., and Millat, G., 2000. The distribution gap is closed – first record of naturally settled Pacific oysters *Crassostrea gigas* in the East Frisian Wadden Sea, North Sea. *Senckenbergiana Maritima*, **30**(3/6), 153–160.

Zeiler, M., Schulz-Ohlberg, J., and Figge, K., 2000. Mobile sand deposits and shoreface sediment dynamics in the inner German Bight (North Sea). *Marine Geology*, **170**, 363–380.

Cross-references

Beach Processes
Coasts
Dunes
Integrated Coastal Zone Management
Marginal Seas
Tidal Depositional Systems

WAVES

Tarmo Soomere
Institute of Cybernetics at Tallinn University of Technology, Tallinn, Estonia
Estonian Academy of Sciences, Tallinn, Estonia

Definition

Wave motion is the fundamental mode of energy propagation over large distances without substantial deformation or permanent displacement of the medium. Waves organize a part of energy and momentum supplied to the water masses and propagate as coherent motions that rapidly carry the energy to the coasts. Surface gravity waves with periods from about 1 to 25 s are called *ocean waves* below.

Introduction

The most important classes among a variety of wave motions in the ocean (LeBlond and Mysak, 1978) are surface waves and internal waves. The motions in surface waves are concentrated at sea surface, while internal waves (Miropolsky, 2001) propagate along a stratified medium, often along major density jumps in the water column. The basic restoring force for both is gravity and both involve a combination of longitudinal and transverse waves. They can be either progressive or standing and may exhibit all classical properties of waves such as dispersion, reflection, refraction, diffraction, or focusing.

The length of surface waves may range from millimeters (capillary and capillary-gravity waves for which the surface tension is substantial) to many kilometers (tsunamis, seiches, tides, for which the impact of Coriolis force becomes significant). Their periods are from fractions of seconds up to a few days. The presence of coasts substantially modifies large-scale waves such as tides and gives rise for asymmetric and/or coastally trapped waves (e.g., edge waves, Kelvin waves). Energy carried by ocean waves has many implications on beach processes and sediment dynamics, formation of bedforms, and driving of coastal erosion and is an important renewable energy source (Cruz, 2008).

Mathematical description of elementary surface waves

The simplest meaningful way to describe ocean waves is the *first-order*, *small-amplitude*, or *Airy* wave theory, called *linear wave theory*. Although it is based on a number of assumptions and approximations, its outcome replicates the behavior and impact of ocean waves with reasonable accuracy (Dean and Dalrymple, 1991; Craik, 2005; Holthuijsen, 2007).

The linear wave theory addresses nonstationary, inviscid, barotropic, irrotational flow in an incompressible medium. The wave-induced motion of water parcels thus always possesses a *velocity potential* – a function $\phi(x,y,z,t)$, such that velocity components $\vec{u} = (u,v,w)$ can be expressed as its partial derivatives $\vec{u} = (\partial\phi/\partial x, \partial\phi/\partial y, \partial\phi/\partial z)$. Ideal infinite plane (sine or elementary) wave trains correspond to the velocity potential $\phi = ag\omega^{-1}\cosh\kappa(z+d)(\cosh\kappa d)^{-1}\sin(kx+ly-\omega t)$ and produce a moving pattern of undulations $\eta(x,t) = a\cos(kx+ly-\omega t)$. The pattern is characterized by wave amplitude a or height (the vertical distance between the wave trough and crest) $h = 2a$, period T (the time interval between the arrival of two subsequent wave crests) or angular frequency $\omega = 2\pi/T$, and length L (Figure 1). The ratio h/L is called wave steepness. The wave vector $\vec{\kappa} = (k,l)$ represents the direction of wave propagation on the (x,y) plane along *wave rays* (lines normal to wave crests). Its length (wave number) is $\kappa = |\vec{\kappa}| = 2\pi/L$. For essentially one-dimensional long-crested waves propagating in the x-direction, the y-direction is often ignored by setting $l = 0$.

The wave height is arbitrary in the linear theory. The wave number and angular frequency of ocean waves satisfy the dispersion relation $\omega = \sqrt{g\kappa \tanh\kappa d}$, where g is gravity acceleration and d is the water depth.

Motions in the wave

Wave propagation involves three velocities: of water parcels, of wave crests, and of energy. The velocity $\vec{u}$ of water parcels is proportional to the wave amplitude. The maximum (orbital) speed $u_{maxs} = ag\kappa/\omega = ag/c_f$ occurs at the surface. The maximum speed at the ideal bottom is $u_{maxb} = a\sqrt{2g\kappa/\sinh 2\kappa d} = a\omega/\sinh\kappa d$. A bottom boundary layer is formed at the realistic seabed.

The wave *energy* is the energy of water parcels brought into motion by the wave. It is usually expressed as the energy density $E = a^2\rho g/2$ (ρ is the water density) averaged over the entire water column and over a wave period (or over a wavelength). The *phase speed* (also *celerity*, propagation speed of wave crests or any constant-phase point) is $\vec{c}_f = (\omega/k, \omega/l)$ and the *group speed* (propagation speed of energy) is $\vec{c}_g = (\partial\omega/\partial k, \partial\omega/\partial l)$ for any class of small-amplitude waves (Landau and Lifshits, 2000).

The dependence of many surface wave properties on the water depth is governed by the dispersion relation through the function $\tanh\kappa d$. As $\tanh\kappa d \approx 1$ for $\kappa d \geq 3$ and $\tanh\kappa d \approx \kappa d$ for $\kappa d \leq 1/3$ (Figure 2), waves with $L \leq 2\pi d/3 \approx 2d$ (*deepwater waves*) adequately obey the dispersion relation $\omega = \sqrt{g\kappa}$ and waves with $L \geq 6\pi d \approx 20d$ (*shallow-water waves*) satisfactorily conform to the dispersion relation $\omega = \kappa\sqrt{gd}$. Note that this classification relies on the relationship between the wavelength and the water depth rather than the water depth only (Table 1).

Both phase and group speeds for longer waves in deep and finite-depth water exceed that of shorter waves. Ocean waves thus exhibit *normal dispersion*. All shallow-water waves propagate with the same speed in a fixed depth and are *nondispersive*. Very short surface waves (capillary and

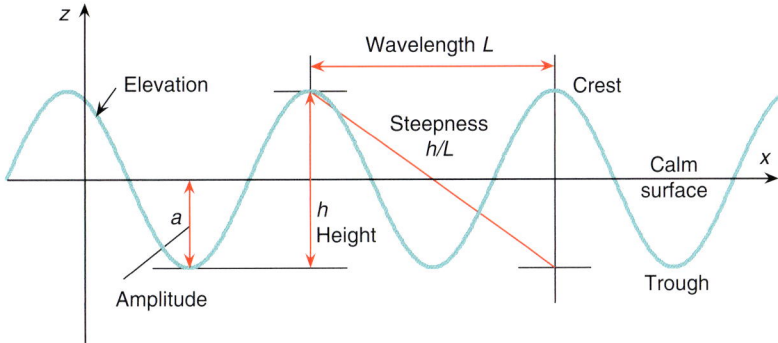

Waves, Figure 1 Elementary wave propagating in the x-direction. The z-axis is traditionally directed upward and $z = 0$ corresponds to the calm water surface.

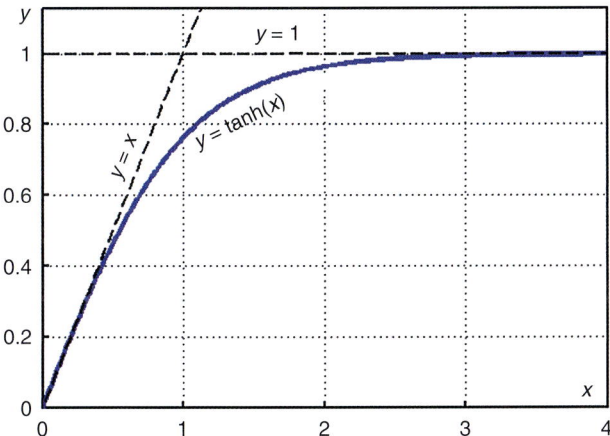

Waves, Figure 2 The function $\tanh x$.

capillary-gravity waves) possess *anomalous dispersion*: shorter waves propagate faster than longer ones.

Wave energy flux (also called *wave power*) $\vec{P} = E\vec{c}_g$ is the product of wave energy and its propagation speed. The flux of momentum created by surface wave motion is characterized by *radiation stress tensor* (Longuet-Higgins and Stewart, 1964). For elementary waves propagating along the x-axis, the x-component of radiation stress is $S_{xx} = E\left(\frac{1}{2} + \frac{2\kappa d}{\sin 2\kappa d}\right) = E\left(2\frac{c_g}{c_f} - \frac{1}{2}\right)$.

Wave transformation in the nearshore

A surface wave undergoes certain changes when it propagates from the open ocean (of large depth) into the nearshore (of smaller depth). If a wave propagates along a homogeneous (in the y-direction) underwater slope, modifications to the wave properties in the process of *shoaling* (during which waves become higher and shorter and thus steeper) are derived from the assumption that the angular frequency and energy flux of the wave remain constant. Changes to the wavelength are dictated by the dispersion relation. The wave height, originally h_0 at a certain depth and $h_1 = h_0 K_s$ at another location, follows the adaptation of the group speed from c_{g0} to c_{g1} through the shoaling coefficient $K_s = \sqrt{c_{g0}/c_{g1}}$.

A wave propagating obliquely with respect to isobaths experiences *refraction*, i.e., gradual turn in the wave propagation (equivalently, wave vector) direction. This process is quantified through Snell's law $\sin\theta/c_f = $ const that relates the angle θ between the wave crests and isobaths with the variation in the phase speed. Wave energy usually converges at shoals and headlands (and may be focused at the downwave side of shoals or small islands) and diverges along underwater canyons. Similar processes occur when waves propagate over a sea area that contains spatially variable currents. Additional changes in the wave properties occur in the nearshore owing to wave-bottom interaction and depth-induced breaking.

Reflection of a part of wave energy occurs from steeper slopes. *Standing waves* (called *seiches* in natural water bodies) may be formed when wave energy ideally reflects from a vertical wall. The period of the longest seiche in a basin of constant depth d and with a length L_b is expressed as a particular case of Merian's formula $T_{s0} = 2L_b/\sqrt{gd}$.

Wave *grouping* occurs when two sine waves with almost equal lengths (or periods) propagate in the same direction. Wave *diffraction* is the process of energy and momentum transfer in a direction other than the wave propagation direction, e.g., when a wave propagates through a gap or across an edge of a sheltered domain.

Realistic surface wave systems have much more complex nature than sine waves and exhibit many nonlinear properties (Osborne, 2010). Deepwater waves of considerable amplitude are often described as Stokes waves or using the nonlinear Schrödinger equation. Long waves in shallow domains often obtain a cnoidal shape or become similar to shallow-water solitons described by the Korteweg-de Vries equation.

Ocean waves

Ocean wave fields are traditionally divided into *windseas* (or *seas*, waves that are supported by the same wind that

Waves, Table 1 Phase and group speeds for deepwater, finite-depth, and shallow-water waves

Deep water: $\kappa d \geq 3$ or $L \leq 2d$	Finite depth: $1/3 \leq \kappa d \leq 3$ or $2d \leq L \leq 20d$	Shallow water: $\kappa d \leq 1/3$ or $L \geq 20d$
$c_f = \sqrt{\frac{g}{\kappa}} = 2c_g$	$c_f = \sqrt{\frac{g}{\kappa} \tanh \kappa d}$	$c_g = c_f = \sqrt{gd}$
$c_g = \frac{1}{2}\sqrt{\frac{g}{\kappa}} = \frac{1}{2}c_f$	$c_g = \frac{\omega}{2k}\left(1 + \frac{2kd}{\sinh 2kd}\right) = \frac{c_f}{2}\left(1 + \frac{2kd}{\sinh 2kd}\right)$	

has generated the field) and swell (usually longer waves that have propagated out of the generation area). Windseas are a superposition of many elementary waves with different periods, directions, and heights and are characterized by an extremely complicated shape of water surface. Swell is usually more regular, has well-defined long crests, and resembles propagation of elementary waves.

The analysis of wave heights originally relied on the identification of single waves from the record of surface elevation at a fixed point. For narrow-banded wave systems, the distribution of wave heights is close to a Rayleigh distribution. The distribution of wave periods in offshore conditions is close to a normal distribution (WMO, 1998), while in semi-sheltered domains and at coastal sites, it may be skewed toward shorter waves.

A widely used measure of ocean waves is the *significant wave height*. It has been historically defined as the average height $h_{1/3}$ of one-third of the highest waves (Sverdrup and Munk, 1947). For Rayleigh-distributed waves $h_{1/3} \approx 4.005\sigma$, where σ is the standard deviation of the water surface elevation. Contemporary numerical simulations (Komen et al., 1994) and wave measurement devices define the significant wave height as $h_s = 4\sigma$.

In a Rayleigh-distributed wave system, the probabilities of having waves higher than h_s and $2h_s$ are $e^{-2} \approx 1/7$ and $e^{-8} \approx 1/3000$, respectively. Waves higher than $2h_s$ (sometimes higher than $2.2h_s$) are called *freak* or *rogue waves* (Kharif et al., 2009). They are usually short-living phenomena (Osborne, 2010) but may have much longer lifetime (Peterson et al., 2003) and impact in shallow areas (Nikolkina and Didenkulova, 2012). Their exact prediction is probably not possible, but it is feasible to forecast the areas with enhanced probability of their occurrence (Janssen, 2008).

Wave spectrum and forecast

The practical analysis of ocean waves relies on the assumption that wave fields can be adequately described by their *energy spectrum*. The growth of a windsea usually leads to a wave system with a pronounced single peak at the *peak period*. The condition until which waves grow under a constant wind blowing long time over a large sea area is termed a *fully developed sea*, for which the Pierson-Moskowitz spectrum is an acceptable approximation. The peak period $T_p \approx 0.64 u_{10}$ and significant wave height $h_s \approx 0.022 u_{10}^2$ of such waves depend only on the wind speed (here u_{10} is the wind speed at a height of 10 m).

The dependence of wave properties on *fetch length* L_f is accounted for in the JONSWAP spectrum (Komen et al., 1994), for which the peak frequency is $\omega_p \approx 22 [g^2/(u_{10}L_f)]^{1/3}$ and $h_s \approx 0.0005 u_{10}\sqrt{L_f}$. These and similar relationships form the basis of *parametric wave models* (WMO, 1998). Contemporary numerical wave models (such as WAM, SWAN, WAVEWATCH, etc.) replicate all major features of wave generation, propagation, interactions, and decay by tracking the evolution of the wave energy spectrum (Komen et al., 1994).

Finite-amplitude and breaking waves

Steep waves ($h/L > 0.17$, Figure 1) may break at any depth, but most of wave breaking is concentrated in the surf zone at the coast. The maximum height limit above which the wave becomes unstable is called *breaking wave height* h_b. Generally it is a function of depth and wavelength. A commonly used assumption in coastal engineering is that waves approaching a natural beach break when their height h_b is about 78 % of the water depth d_b, equivalently, when the *breaking index* $\gamma_b = h_b/d_b \approx 0.78$ (Dean and Dalrymple, 1991; 2002). For strongly reflecting beaches, the breaking index may reach values up to ~1.5, while for domains with horizontal bed, it is in the range of 0.55–0.6. Breaking waves produce plumes of bubbles that may be dispersed in the water column to the depths of $4h_s$ (Thorpe, 1995).

Waves of finite amplitude generate a certain mass transport, estimated as $M = E/c_f$ in Starr (1947). Waves propagating over shallow areas may substantially affect the local water level. Nonbreaking waves produce *wave setdown* $\bar{\eta} = -a^2\kappa/(8\sinh 2\kappa d)$, with a maximum value of $\bar{\eta}_b = -(1/16)\gamma_b h_b$ at the seaward border of the surf zone. The gradual release of cross-shore momentum carried by breaking waves may lead to much higher *wave setup*, theoretically up to $\bar{\eta}_{max} = (5/16)\gamma_b h_b$. This process may provide 30–60 % of the extreme water levels (Dean and Bender, 2006). On realistic beaches, an acceptable approximation is $\bar{\eta}_{max} \approx 0.17 h_{s10}$, where h_{s10} is the significant wave height at a depth of 10 m (Guza and Thornton, 1987). The release of the alongshore component of momentum contributes to the formation of the alongshore current in the surf zone.

Wave climate and extreme wave heights

The characteristics of the offshore and onshore wave climate vary drastically over different domains of the ocean (Davies, 1980). The largest annual average significant wave heights (around 4.5 m) apparently occur within the "Roaring Forties" on the southern hemisphere. The highest 100-year return h_s (about 25 m) is likely to occur in the northern Atlantic according to a hindcast based on the ERA-40 reanalysis (Sterl and Caires, 2005). The largest instrumentally measured offshore value of h_s is 18.5 m near Rockall in the North Atlantic (Holliday et al., 2006), whereas $h_s \approx 24$ m and single waves over 32 m have been reported in specific conditions under Typhoon Krosa (Babanin et al., 2011). Extreme wave heights are much smaller in semi-sheltered basins such as the Mediterranean Sea or the Baltic Sea.

Although several variations in wave properties have been identified for different domains of the ocean based on relatively short data sets during the last decades, there is no consensus about the nature of recent changes in the wave climate. The temporal changes in the wave height are likely a part of multi-decadal fluctuations of storminess and seem to exhibit no obvious trend (Weisse and von Storch, 2010). The changes have been most pronounced in semi-sheltered seas where even small variations in the local storm activity may lead to considerable changes in the wave climate (Soomere and Räämet, 2011). Regional variations in the wave direction, changes in the joint occurrence of rough seas and high water levels, and the length of ice cover may have particularly large impact on the intensity of coastal processes in the Arctic and seasonally ice-covered seas.

Conclusions

Although the science of ocean waves is generally a mature field with strong physical foundations and highly developed mathematical gear, it still offers fascinating challenges and reveals the alluring secrets of nature. It has recently been one of few examples in Earth sciences that has opened a principally new research direction into rogue waves in various environments and has contributed into studies of the fundamentals of soliton interactions. The challenges have, however, largely moved from the examination of wave dynamics into the analysis of interaction of waves with various natural and anthropogenic structures. The changing marine environment calls for considerable efforts to understand and quantify the impact of waves in the conditions of rising sea-level and especially in relation with dramatic ice reduction in the Arctic.

Bibliography

Babanin, A. V., Hsu, T.-W., Roland, A., Ou, S.-H., Doong, D.-J., and Kao, C. C., 2011. Spectral wave modelling of Typhoon Krosa. *Natural Hazards and Earth System Sciences*, **11**, 501–511.

Craik, A. D. D., 2004. The origins of water wave theory. *Annual Review of Fluid Mechanics*, **36**, 1–28, doi:10.1146/annurev.fluid.36.050802.122118.

Cruz, J., 2008. *Ocean Wave Energy. Current Status and Future Perspectives*. Berlin/Heidelberg: Springer, p. 431.

Davies, J. L., 1980. *Geographical Variation in Coastal Development*, 2nd edn. London: Longman.

Dean, R. G., and Bender, C. J., 2006. Static wave setup with emphasis on damping effects by vegetation and bottom friction. *Coastal Engineering*, **53**(2–3), 149–165.

Dean, R. G., and Dalrymple, R. A., 1991. *Water Wave Mechanics for Engineers and Scientists*. Singapore/Teaneck: World Scientific Press, p. 292.

Dean, R. G., and Dalrymple, R. A., 2002. *Coastal Processes with Engineering Applications*. Cambridge: Cambridge University Press, p. 475.

Guza, R. T., and Thornton, E. B., 1987. Wave set-up on a natural beach. *Journal of Geophysical Research*, **96**, 4133–4137.

Holliday, N. P., Yelland, M. J., Pascal, R., Swail, V. R., Taylor, P. K., Griffiths, C. R., and Kent, E., 2006. Were extreme waves in the Rockall Trough the largest ever recorded? *Geophysical Research Letters*, **33**(5), L05613, doi:10.1029/2005GL025238.

Holthuijsen, L. H., 2007. *Waves in Oceanic and Coastal Waters*. Cambridge: Cambridge University Press, p. 389.

Janssen, P. A. E. M., 2008. Progress in ocean wave forecasting. *Journal of Computational Physics*, **227**(7), 3572–3594, doi:10.1016/j.jcp.2007.04.029.

Kharif, C., Pelinovsky, E., and Slunyaev, A., 2009. *Rogue Waves in the Ocean*. Berlin/Heidelberg: Springer, p. 216.

Komen, G. J., Cavaleri, L., Donelan, M., Hasselmann, K., Hasselmann, S., and Janssen, P. A. E. M., 1994. *Dynamics and Modelling of Ocean Waves*. Cambridge, NY: Cambridge University Press.

Landau, L. D., and Lifshits, E. M., 2000. *Fluid Mechanics*, 2nd edn. Oxford, UK: Butterworth-Heinemann. Course of Theoretical Physics, Vol. 6, p. 539.

LeBlond, P. H., and Mysak, L. A., 1978. *Waves in the Ocean*. Amsterdam: Elsevier, p. 602.

Longuet-Higgins, M. S., and Stewart, R. W., 1964. Radiation stresses in water waves: a physical discussion with applications. *Deep-Sea Research*, **11**, 529–562.

Miropolsky, Y. M., 2001. *Dynamics of Internal Gravity Waves in the Ocean*. Dordrecht/Boston: Kluwer, p. 424.

Nikolkina, I., and Didenkulova, I., 2012. Catalogue of rogue waves reported in media in 2006–2010. *Natural Hazards*, **61**(3), 989–1006.

Osborne, A. R., 2010. *Nonlinear Ocean Waves and the Inverse Scattering Transform*. Amsterdam: Elsevier, p. 917.

Peterson, P., Soomere, T., Engelbrecht, J., and van Groesen, E., 2003. Soliton interaction as a possible model for extreme waves in shallow water. *Nonlinear Processes in Geophysics*, **10**(6), 503–510.

Soomere, T., and Räämet, A., 2011. Spatial patterns of the wave climate in the Baltic Proper and the Gulf of Finland. *Oceanologia*, **53**(1-TI), 335–371.

Starr, V. P., 1947. A momentum integral for surface waves in deep water. *Journal of Marine Research*, **6**(2), 126–135.

Sterl, A., and Caires, S., 2005. Climatology, variability and extreme of ocean waves: the web-based KNMI/ERA-40 wave atlas. *International Journal of Climatology*, **25**, 963–977.

Sverdrup, H. U., and Munk, W. H., 1947. *Wind, Sea and Swell: Theory of Relations for Forecasting*. Washington, DC: US Navy Hydrographic Office. Publication 601.

Thorpe, S. A., 1995. Dynamical processes of transfer at the sea surface. *Progress in Oceanography*, **35**, 315–352.

Weisse, R., and von Storch, H., 2010. *Marine Climate and Climate Change*. Chichester: Springer Praxis, p. 219.

WMO, 1998. *Guide to Wave Analysis and Forecasting*. Geneva: World Meteorological Organisation. WMO-702, p. 159.

Cross-references

Beach
Beach Processes
Bed Forms

Coastal Engineering
Coasts
Currents
Energy Resources
Engineered Coasts
Hurricanes and Typhoons
Sea-Level
Sediment Dynamics
Tsunamis

WILSON CYCLE

Martin Meschede
Institute of Geography and Geology, Ernst-Moritz-Arndt University, Greifswald, Germany

Definition

A Wilson cycle describes a large tectonic cycle lasting for more than 100 million years. Such a cycle, named after J. Tuzo Wilson (1908–1993), one of the doyens of plate tectonics, starts with the breakup of a continent and growth of an ocean at a newly formed mid-oceanic rift system. Such oceans may remain limited in size or attain the dimensions of the Atlantic or even Pacific Ocean. Eventually, the ocean closes during continent-continent collision ending the cycle.

Stages of the Wilson cycle

Beginning with the splitting of a so far stable craton (Figure 1a), probably related to a rising mantle plume, the Wilson cycle continues with the formation of a graben system (Figure 1b). The graben is still underlain by continental lithosphere which, however, may be thinned because of extensional tectonics. With ongoing separation of the continental blocks, a new zone of production of oceanic lithosphere in between is formed. This rift zone remains in the middle of the new oceanic basin as a divergent plate boundary (Figure 1c, d). Spreading occurs along the mid-oceanic rift system producing new oceanic lithosphere, while the continents drift apart. The size of the ocean between the continental blocks will determine whether crustal growth by magmatism is large, small, or insignificant. Since young oceanic lithosphere is lighter than older one, a mid-oceanic ridge develops which may surmount the deep sea plains by several kilometers. Oceanic basins may reach widths of several thousand kilometers and depths down to more than 5 km. The transition

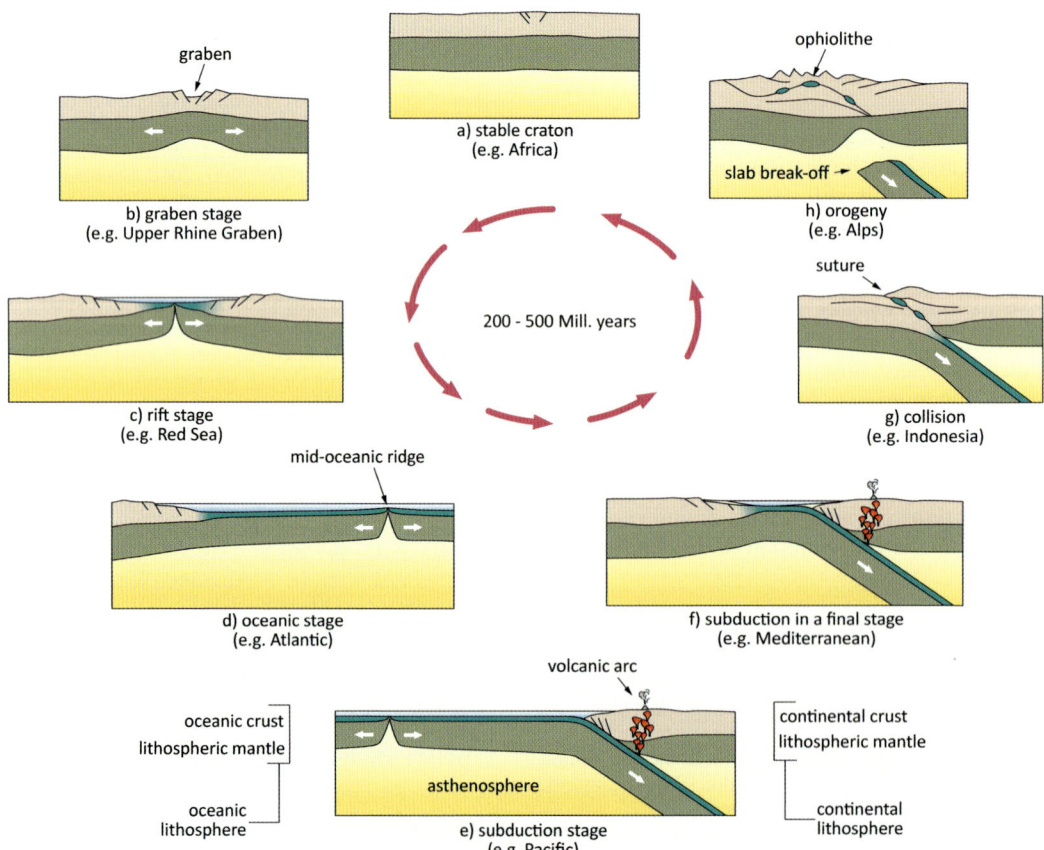

Wilson Cycle, Figure 1 Schematic representation of the different stages of a Wilson cycle (modified from Frisch et al., 2011).

from the oceanic basins to the continents is marked by the continental slope. It ends in the shelf area which is part of a passive continental margin dominated by extensional tectonics.

The turnabout in the Wilson cycle from divergent plate movement to continent convergence may be performed by "spontaneous subduction" of old oceanic lithosphere or, in the case of young intermediate oceanic realms, forced by the global plate drift pattern, by "forced subduction." Subduction (Figure 1e) occurs either along continental margins or intraoceanic island arcs. In both cases volcanic arcs develop above the subducting plate. Subduction ends when two continents move against each other and progressively reduce the width of the ocean in between (Figure 1f).

The process of orogeny becomes initiated when a continent-continent collision starts at a convergent plate margin. Continental crust may be thickened to more than 70 km, in comparison to normal continental crust that is 30–40 km thick, when the plates override. Mountain belts form as elongate zones of slightly elevated continental crust. The orogenic process leads to crustal thickening, deformation, and metamorphism (Figure 1g). At the end, the oceanic lithosphere breaks off (slab break off) leading to rapid uplift and increasing mountain formation (Figure 1h). This is called the Alpine-type orogeny in contrast to the Cordilleran-type orogeny, where subduction lasts over long periods with repeated episodes of collision that involve island arcs, oceanic plateaus, and sometimes even microcontinents which are accreted to the convergent plate margin. This orogenic style is exceptionally rich in volcanic and/or plutonic production.

During the final stage of the Wilson cycle, the mountain belt levels out to a normal crustal thickness and a new peneplain is formed. The next cycle can begin when the continental crust gets under extensional stress caused by active or passive rifting induced by warming or a rising mantle plume. The motor of the cyclicity is the subduction process which requires an oceanic basin floored by oceanic crust. This type of crust can be subducted, whereas continental crust cannot. The force of the slab pull is the most powerful force to drag down the oceanic lithosphere.

Bibliography

Frisch, W., Meschede, M., and Blakey, R. C., 2011. *Plate Tectonics*. Heidelberg/Dordrecht/London/New York: Springer, 212 pp.

Cross-references

Layering of Oceanic Crust
Lithosphere: Structure and Composition
Oceanic Rifts
Oceanic Spreading Centers
Orogeny
Paleophysiography of Ocean Basins
Plate Tectonics
Subduction

Author Index

A
Anderson, Leif G., 78
Andresen, Camilla S., 250

B
Bach, Wolfgang, 344, 779
Bailey, Geoffrey N., 893
Bąk, Marta, 467
Baker, Edward T., 335
Barckhausen, Udo, 299
Berger, André, 25
Berger, Wolfgang H., 71, 437
Binczewska, Anna, 251
Bjørklund, Kjell R., 700
Bohrmann, Gerhard, 24, 117, 433, 495
Boltovskoy, Demetrio, 255
Borrero, Jose C., 884
Brumsack, Hans-Jürgen, 20
Buchs, David M., 754
Bünz, Stefan, 62

C
Cambrollé, Jesús, 196
Canales, Juan P., 392
Coffin, Millard F., 372, 913
Colpron, Maurice, 835
Correa, Nancy, 255

D
de Vernal, Anne, 653
Dean, Walter E., 427
Devey, Colin, 328
Dickson, Alexander J., 610
Diepenbroek, Michael, 143
Dietrich, Reinhard, 238, 748
Drachev, Sergey S., 741
Duke, Norman C., 412
Dullo, Christian, 718, 747
Dypvik, Henning, 460

E
Edgar, Kirsty, 467
Eisenhauer, Anton, 699
Eldholm, Olav, 913
Elken, Jüri, 139

F
Farmer, Elizabeth J., 251
Finkl, Charles W., 41, 47, 103, 710
Fischer, Gerhard, 247
FitzGerald, Duncan, 171
Foley, Stephen F., 13
Forwick, Matthias, 490
Franke, Dieter, 217
Frisch, Wolfgang, 249, 523, 619

G
Gaedicke, Christoph, 217
Georgiou, Ioannis, 171
Gill, Jim, 379
Gischler, Eberhard, 302
Gohl, Karsten, 721
Götze, Hans-Jürgen, 138
Goutorbe, Bruno, 449
Gradstein, Felix M., 55, 283
Graf, Gerhard, 57
Grasemann, Bernhard, 602
Grevemeyer, Ingo, 754
Grootes, Pieter M., 695
Guieu, Cécile, 203
Gundlach, Erich, 587
Gürbüz, Alper, 687, 692

H
Hannington, Mark D., 917
Harding, Ian, 467
Harff, Jan, 437, 480, 489, 735
Hasterok, Derrick, 449
Hay, William W., 122, 124, 231, 239, 787, 807
Hein, James R., 113, 408
Hellebrand, Eric, 665
Helo, Christoph, 241
Henrich, Rüdiger, 397
Herrle, Jens O., 73
Heyde, Ingo, 299
Hillier, John, 449
Hoernle, Kaj, 754
Hoffmann, Gösta, 276
Hübscher, Christian, 721
Huet, Benjamin, 602
Hüneke, Heiko, 68, 127, 156, 888

I
Iglesia Llanos, María P., 407, 632
Ildefonse, Benoît, 133
Iversen, Morten, 247

J
Jakobsson, Martin, 365
Jamieson, John W., 344, 917
Jöns, Nils, 779

K
Kay, Shannon, 329
Kemp, Alan, 391
Kendall, Christopher G. St. C., 132, 768, 773
Kerr, Andrew C., 558
Kjellström, Erik, 540
Knutz, Paul, 359
Koepke, Jürgen, 137, 263
Kopf, Achim, 527
Körper, Janina, 216
Koschinsky, Andrea, 339
Kotilainen, Aarno, 448
Kowalewska-Kalkowska, Halina, 235
Kucera, Michal, 514
Kudrass, Hermann, 269, 666, 668
Kuijpers, Antoon., 250, 302, 359
Kukowski, Nina, 209, 231, 925
Kulp, Mark, 171
Kuznetsov, Vladislav, 271

L
Laberg, Jan S., 490
Lallemand, Serge, 9, 793
Lambeck, Kurt, 294
Lisitzin, Alexander P., 288
Lithgow-Bertelloni, Carolina, 193, 515
Loubere, Paul, 648
Lyle, Mitchell, 156

M
Mann, Paul, 481, 659
Marks, Roman, 235
März, Christian, 20
Maslin, Mark, 610
Matthiessen, Jens, 189

McCartney, Kevin, 467
Meschede, Martin, 6, 393, 676, 731, 803, 944
Mienert, Jürgen, 62
Miettinen, Arto, 185
Moros, Matthias, 359
Mulder, Thierry, 68, 156, 888
Müller, R. Dietmar, 638, 669
Muñoz-Vallés, Sara, 196

N
Negrete, Raquel, 449
Nelson, JoAnne, 835
Nicolas, Adolphe, 567, 571, 592
Nielsen, Tove, 302
Nürnberg, Dirk, 499

O
Osadczuk, Andrzej, 385

P
Pedersen, Thomas, 539
Peijnenburg, Katja T.C.A., 680
Perfit, Michael R., 33, 808
Petersen, Sven, 344, 475, 825, 917
Pierrot-Bults, Annelies C., 680
Polovodova Asteman, Irina, 251

R
Reicherter, Klaus, 276
Reijmer, John J. G., 80
Resio, Don, 329
Riebesell, Ulf, 541

Ring, Uwe, 491
Romankevich, Evgeny, 596
Rubin, Ken H., 501

S
Schenke, Hans W., 268
Schernewski, Gerald, 363
Schippers, Axel, 144
Schomacker, Anders, 519
Schreck, Michael, 189
Sclater, John G., 449
Seibold, Eugen, 437
Şengör, A. M. Celâl, 838, 859, 876
Seton, Maria, 638, 669
Shevchenko, Vladimir P., 203, 288
Shimmield, Graham, 56
Silver, Eli, 773
Singh, Satish C., 571
Soomere, Tarmo, 940
Soule, Samuel A., 33, 808
Stein, Rüdiger, 87, 746, 792
Stern, Robert J., 399
Stracke, Andreas, 182
Summerhayes, Colin, 900
Sündermann, Jürgen, 514
Suyehiro, Kiyoshi, 542

T
Thiede, Jörn, 437, 628
Thierstein, Hans R., 537
Tivey, Margaret K., 58
Torres, Marta E., 117, 433, 495

Treguer, Paul, 792
Tunnicliffe, Verena, 84

U
Ueda, Hayato, 311, 367

V
Vetrov, Alexander, 596
Voelker, David, 1, 821
von Huene, Roland, 817
Vorren, Tore O., 490

W
Wagner, Thomas, 73
Walker, H. Jesse, 226
Wallendorf, Louise, 99, 747
Wefer, Gerold, 247, 552, 622
Wehrmann, Achim, 849, 933
White, William M., 316
Whittaker, Joanne M., 372
Witkowski, Jakub, 467

Y
Yeo, Isobel, 36
Yokokawa, Miwa, 55

Z
Zhang, Jing, 93
Zhang, Wenyan, 67, 761, 764
Zhou, Di, 423

Subject Index

A

Absolute dynamic topography, 748, 749
Absolute plate motions, 669, 671, 674
Abukuma-type metamorphism, 620–622
Abyssal peridotite, 665, 785
Abyssal plains, 1–5
  bathymetry, 1–3
  life in, 4–5
  sedimentation, 3–4
Acantharia, 700, 702
Acanthocircus, 468
Acarinina, 468
ACC. *See* Antarctic Circumpolar Current (ACC)
Accelerator mass spectrometry (AMS), 695
Accreted oceanic plateaus, 559, 563–564
Accretionary wedges, 6–9
  earthquakes, 7
  formation, 6–7
  occurrences, 7–9
Acidification, ocean, 541–542
Acridine orange direct counts (AODC), 144
Active continental margins, 9–12, 678, 798
  birth, life and death, 10–11
  generalities, 9–10
  morphology, 10
  tectonic activity, 11–12
Active seismology, 827
Agglutinated foraminifers, 252
Airy-type isostasy, 758
Alpine-type orogeny, 604, 945
AMC. *See* Axial magma chamber (AMC)
AML. *See* Axial melt lens (AML)
Amorphous silica, 792
Amphimelissa setosa, 707
Anaerobic ammonium oxidation
    (anammox), 539
Anaerobic methanotrophic archaea (ANMEs),
    497, 498
Anaerobic oxidation of methane (AOM),
    497, 498
Anatectic magmas, 402
Anchoring force, 795–796
Ancient plate tectonics, 13–19. *See also* Plate
    tectonics Archean geology, 15
  temperature evolution, earth, 14–15
Angiosperms, 655, 656
Anhydrite, 918
Anhydrite plus halite, 428

Anoxic oceans, 20–23
  environmental controls, 21–22
  modern analogues, 21
  Phanerozoic, 21
  Proterozoic, 20
  in sedimentary archives, 22
Antarctic Circumpolar Current (ACC), 905
Apparent polar wander (APW) path, 632,
    634, 759
Archaea, 150
Archaeocenosphaera, 468
Archean geology, 15
Arc magmatism. *See* Convergent margin
    magmatism
Argentine, seismic profiles, 556
Artificial reefs, 710, 716–717
Asphalt volcanism, 24–25
  Chapopote Knoll, 24
  seepage *vs.* volcanism, 24
AST. *See* Axial summit trough (AST)
Asthenosphere, 423, 424, 572
Astronomical frequencies, paleoclimates,
    25–33
  characteristics, 25
  climatic precession, 29–31
  Earth's elliptical orbit, 26–27
  eccentricity, 29–31
  instability, astronomical periods, 31–32
  long-term variations, 28
  obliquity, 29–31
  orbital elements, planet, 27–28
Atlantic Ocean
  hemipelagic sediment components, 161
  margins, 553
  mid-ocean rifting, 570
  saturation depths, 79
  turbidites, 161
Atlit Yam site in Israel, 896
Atoll, 712–713, 719
  geomorphology, 304–305
  vs. guyots, 305–307
  occurrence, 304–305
  seamounts and, 823
  sediments, 304–305
Authigenic sediments, 761
Autonomous underwater vehicles (AUVs)
  marine geosciences, 826, 827, 829, 830
  topographic mapping, 754

Axial magma chamber (AMC), 504–505, 576
Axial melt lens (AML), 572, 577
Axial summit trough (AST), 33–36, 810
  development, 34
  volcanic-tectonic history,
      34–35
Axial volcanic ridges (AVRs), 36–38
  eruption style and volcanic
      architecture, 37
  formation and growth, 38

B

Back-arc
  oceanic spreading centers, 584
  spreading, 423
Back-stripping analysis, 775
Bacteria
  deep biosphere, 147–150
  diversity of, 86
  heterotrophic, 206
  oil-consuming, 222
  sulfate-reducing, 145
Bahamian coral reef complex, 714
Bar-built estuaries, 235
Barrier islands
  development, 937–938
  offshore from, 934
  and sandspits, 852
  segment of, 108
Barrier reef, 714–715, 719
Basal anhydrite, 428
Basalt, 401
Basaltic crust, 561
Basaltic magmas, 915
Batholiths, 403
Bathymetry, 1–3
Beach, 41–46
  artificial, 46
  chemical composition, 43–44
  classification, 45–46
  color, 43–44
  etymology, 42
  geographic occurrence, 42–43
  grain size, 43–44
  shape, 43–44
  types, 44–45
  usage, 42
  Wadden Sea, 936

Beach processes, 47–54
  bar formation and migration, 52
  large-scale morphodynamic, 52–53
  morphodynamic, 53
  plunge step formation, 51–52
  rip currents and channels, 52
  runoff channel, 51
  saltation, berm and backbeach, 50
  sapping-seepage erosion, 51
  sheetwash, 51
  small-scale, 49–50
  study of, 48–49
  swash run-up and rundown, 50–51
Bed forms, 55
Bed-load transport, 764
Bending moment, 795
Benthic boundary layer (BBL), 556. *See also* Bottom boundary layer (BBL)
Benthic fauna, 763
Bilocular foraminifers, 252
Biochronology, 55–56, 440–441
Biogenic barium, 56–57
Bioherms, 720
Biomass production, in marginal seas, 426–427
Biostratigraphy, 55–56, 440–441
  microfossils in, 471–472
  nannofossils coccoliths, 537
Biostromes, 720
Bioturbation, 57–58
Black shales
  deposits, 21
  formation, 21, 564–565
Black smokers, 58–61
  carbonate-rich chimneys, 61
  chimneys, 917
  deposits, 476
  discovery, 58
  formation, 58–59
Bond cycle, 441
Bond event, 441
Boninites, 370
Bottom boundary layer (BBL), 67–68. *See also* Benthic boundary layer (BBL)
Bottom-simulating seismic reflectors (BSR), 62–66, 225, 435
  diagenesis-related, 63
  dynamics, 66
  gas hydrate-related, 63–65
  seismic character, 65–66
Bouma sequence, 68–69
Box corers, 831
British Oceanographic Data Centre (BODC), 268
Brucite, 352, 782
BSR. *See* Bottom-simulating seismic reflectors (BSR)

C

Calcareous foraminifers, 252
Calcareous microfossils, 470
Calcidiscus leptoporus, 538
Calcifying organisms, 541
Calcite compensation depth (CCD), 71–73, 408, 549
  distribution pattern, 72
  fluctuations, 73
Calcite crystallites, 537
Calcite platelets, 537
Carbon, 695
Carbonate
  accretion, 720
  deposition, 789
  destruction, 720
  sequence stratigraphy, 775, 777
  tidal systems, 857–858

Carbonate compensation depth (CCD), 470
Carbonate dissolution, 78–80
  aragonite, 78
  calcite, 78
  ikaite, 79
  position, 79
Carbonate factories, 80–83
  C-factory, 82
  environmental parameters, 80–81
  M-factory, 82
  production profiles, 83
  stick, 83
  T-factory, 81–82
Carbonate ion concentrations, 611
Carboniferous to Cretaceous oceanic plateaus, 563
Carbon isotopes, 73–77
  applications, 73–74
  carbon sources, 74
  chemostratigraphy, 75
  CIE, 75–76
  ecology, 75
  Phanerozoic record, 76
  seawater masses tracing, 75
  shallow seawater, 74
  surface water productivity, 75
Caribbean, Great Arc, 485
Carnallite, 428
Cataclasis, 806
Cavolinia, 683
Cavolinia teschi, 685
CCZ. *See* Clarion-Clipperton Zone (CCZ)
Cellular Automata (CA) models, 767
Cementation, 720
Cenomanian–Turonian Boundary (CTB)
  environmental perturbation, 564
  event, 564
  oceanic plateau volcanism, 564
Cenomanian–Turonian Boundary Event (CTBE/OAE2), 21
Cenozoic ocean basins, 639–647
Cenozoic sediment record, 21
Challenger Deep, 732
Chapopote Knoll, 24
Chemical sediments, 761
Chemosynthetic ecosystems, 349
Chemosynthetic life, 84–87
  chemoautotrophic metabolism, 84–85
  habitats, 85–86
  microbial diversity and carbon fixation, 86–87
Chicxulub impact crater, 488
Chile-type subduction, 526–527, 794, 805
Chimney
  black smokers, 58–59, 917
  carbonate-rich, 61
  characteristics, 58
  white smokers, 58–61
Choked lagoons, 387
Cibicidoides, 468
Circulation Obviation Retrofit Kit (CORK) technologies, 551
Circum-Pacific subduction zones, 804
Clarion-Clipperton Zone (CCZ), 408–411, 476
Clastic systems, sequence stratigraphy, 775, 777
Clay minerals, 87–92
  determination, 88
  formation, 88
  significance, 88–92
  structure and composition, 88
Cliff collapse, 276
CLIMAP (Climate: Long-range Investigation, Mapping, and Prediction) project, 441, 630, 631

Climate
  change, 419
  coupling, 555
  forcing, volcanic, 913, 914
  gas hydrates and, 436
  modeling, 631
  plate tectonics, 629
Climate change, impacts of, 98, 363
Climate, volcanism, 913–917
  chicken-and-egg conundrum, 914, 915
  environmental geosphere driver, 915–916
  LIP contribution, 916
  volcanic impact, 913–914
Clio cuspidata, 684, 685
Clio pyramidata, 682, 686
Clione limacina, 681, 682, 686
Cluster effect, 723
Coastal bio-geochemical cycles, 93–98
  biogeochemical modeling and budgets, 96–98
  characteristics, 94
  dynamics, coastal ocean, 95–96
  earth system and human society, feedbacks, 98
  external driving forces, 94
  machinery, 95
Coastal disasters, geohazards, 276–281
  cliff collapse, 276
  extreme wave events, 276–278
  low elevation coastal zone, 276
  paleostorm deposits, 278–279
  paleotsunami, 279–281
  reconstruction, paleostorm and tsunami events, 278–281
  storm surges, 277
  tsunami, 277–278
Coastal engineering, 99–103
  available tools, 100
  beach fill/nourishment, 101
  berm breakwaters, 102
  breakwaters, 101–102
  construction resources, 101
  customer, 101
  design manuals, 100
  designs, 101
  floating breakwaters, 102
  groins, 101
  living shorelines, 101
  local knowledge, 101
  numerical tools, 100
  physical modeling, 100–101
  regulation, 101
  seawalls/revetments, 102–103
  self-closing flood control structures, 103
  setback/relocation, 103
  site environment, 100
Coastal lagoons, 389
Coastal Zone Management Act (CZMA) of 1972, 363
Coastal Zone Management Program (CZMP), 363
Coasts, 103–112
  beach and barrier, 107–108
  classification, 105–106
  coastal area, 104–105
  coastal zone, 104
  coastal zone management, 110–111
  coastline length, determination, 105
  ecosystems, 106
  ice (polar), 110
  landforms, 106
  muddy, 108–109
  and ocean jurisdictions, 111
  reef, 109–110
  rocky, 106–107
  types, 105–106

COB. *See* Continent-ocean boundary (COB)
Cobalt-rich ferromanganese crusts, 476–477
Cobalt-rich manganese crusts, 113–117, 476
　composition, 115–116
　distribution, 113–114
　formation, 114–115
　occurrence, 113
　paleoceanography, 116
　resource consideration, 117
Coccolithophorids, 537
Coccoliths, 537
　nannofossils (*see* Nannofossils coccoliths)
Cohesive sediments, 761–763
Cold seeps, 117–121
　authigenic minerals, 120–121
　basic process, 118
　locations, 118
　methane and biomes, 120
　ocean margin systems, 557
　paleoseeps, 121
　sources, 118
　transport, 119–120
Cold-water corals, ocean margin systems, 557
Collision-related basins, 482–487
Collosphaera huxleyi, 702
Conservative plate boundaries, 859
Continental crust, composition, 679
Continental drift, 13, 439, 442, 669
Continental margin, 423, 554, 660. *See also* Ocean margin systems
　active, 678. *See also* Active continental margins
　passive, 6, 124, 127, 156, 160–161, 219–226, 933, 945
　salt diapirism in, 741–745
Continental rifting, 494, 602
Continental rise, 122–124
　carbonate deposition, 123
　crust nature, 122
　currents, 123
　detrital sediment supply, 122–123
　geologic basis, 122
Continental shelf
　controls configuration of, 789, 791
　geologic basis, 788–789
　history, 787
　landslides, 818
　sea-level, 791
Continental slope, 124–127
　carbonate deposition, 125
　crustal thinning, 124–125
　currents, 125
　geologic basis, 124
　history, 124
　mass wasting, 125
　profiles, 125–127
　sea-level, 125
　waves, 125
Continental subduction, 793–794
Continent-ocean boundary (COB), 659
Contourites, 127–131
　bottom currents, 127–128
　drifts, 128–130
　ocean margin systems, 556
　sediment facies, 130–131
Controlled-source electromagnetics (CSEM), 829
Controlled-source seismology, 722
Conventional seafloor mapping, 826
Convergence texture, seismic reflector, 132
Convergent margin magmatism
　crustal processes, 401–403
　global map, 401
　mantle wedge processes, 399–401
　mineral deposits, 404–405

　subduction zone, 399–401
　sulfur, 404
　volatile-rich nature, 403
　volcanism, 403–404
　volcano-plutonic complex, 402
Convergent plate boundaries, 781
Coral mounds, 719–720
　Coral reef, 710–712. *See also* Reefs cold-water, 720
　with islands, 716
Cordilleran ophiolites, 593, 594
Cordilleran-type/arc-type orogens, 604
Cosmogenic isotopes, 699
Cosquer Cave in southern France, 896
Coupled geodynamic plate models, 671
Creseis virgula, 685
Cretaceous nannofossil chalk, 538
Cretaceous oceanic plateaus, Carboniferous to, 563
Cribroperidinium, 469
Crucella, 468
Crustal accretion, 133–135
　fast-spreading ridges, 133
　slow-spreading ridges, 133–135
Crustal processes, 401–403
Crystalline core zone, 605–606
CSEM. *See* Controlled-source electromagnetics (CSEM)
Cumulates, 137
Curie temperature, 138
Currents, 139–142
　changing ocean, 142
　observation techniques development, 139–140
　surface and deep currents, ocean, 141–142
　theoretical concepts, 140–141
Cuvierina atlantica, 682

D
Dansgaard-Oeschger cycles, 441
DART. *See* Deep-ocean Assessment and Reporting of Tsunamis (DART)
Darwin Point, 757
Data, 143
Deep biosphere, 144–151
　biomass, 144–145
　life in ocean crust, 149–151
　in marine sediments, 144–149
　microbial activity and biogeochemistry, 145–146
　microbial diversity and quantification, 146–149
Deep earthquakes, Wadati-Benioff-zone
　deep-focus earthquakes, 927, 929, 930
　intermediate-depth earthquakes, 927–930
Deep-focus seismicity, 930
Deep hot biosphere, 444–445
Deep-ocean Assessment and Reporting of Tsunamis (DART), 887
Deep ocean drilling, 442–444, 903, 906
Deep Sea Drilling Project (DSDP), 359, 428, 430–431, 433, 438, 469, 502, 707
　erosion, 231–232
Deep-sea fans, 156
Deep-sea mining
　environmental concerns of, 477
　exploration and, 479
　licenses, 479
Deep-sea sediments, 156–168
　aeolian and volcanic ash sediment components, 161–162
　authigenic, 166
　biogenic sediment components, 162–165
　cosmogenic sediment components, 166
　ferromanganese nodules, 166
　geophysical and geologic study, 158

　hydrothermal sediment components, 165–166
　near-bottom sediment movement and sediment focusing, 167
　seafloor processes, 166
　sediment dissolution and carbonate compensation depth, 167
　turbidites and hemipelagic sediment components, 160
　types, 158–160
Deep-sea trench, 525–526
Deep-sea turbidite systems, 889
　architectural elements, 890–891
　high-efficiency fan systems, 890
　low-efficiency fan systems, 890
Deep-sea vent communities, 342
Deepwater waves, 942
Deflandrea, 469
De Geer moraines, 519
Dehydration embrittlement, 927–928, 930
Delaware Basin, 429
Deltas, 171–180
　channel processes, 177–178
　components, 175
　lower delta plain, 176
　modern, 179–180
　morphology, 178–179
　processes, 177
　sediments, 176–177
　subaqueous delta plain, 176
　types and formation, 172–175
　upper delta plain, 175–176
Dendrochronology, 747
Denitrification
　in oxygen-minimum zones, 539
　in pore waters, 539
Depleted mantle, 182–184
　composition, 183–184
　Earth's mantle facts, 182–183
　formation and evolution, 183
Depleted MORB mantle (DMM), 665
Depositional sequences, stratigraphy, 768, 775
Deposits. *See also* Mineral deposit
　placer, 669
　tidal flats, 937
　VMS, 917
Desmopterus pacificus, 682
Deterministic models, sediment transport, 765–766
Detrital sediment supply, continental shelf, 789
DEViation in Axial Linearity (DEVAL), 575
Diacanthocapsa, 468
Diacavolinia, 683
Diacria schmidti, 685
Diatoms, 185–188
　applications, 185
　biostratigraphy, 185–186
　characteristics, 185
　Late Holocene SSTs, 186–187
　sea-ice reconstructions, 186
　sea surface temperature (SST) reconstructions, 186
Diazotrophy, 539
Dictyocoryne truncatum, 701
Didymocyrtis tetrathalamus, 706
Digital bathymetric model (DBM), 365–367
Digital terrain model (DTM), 365
Dinoflagellates, 189–191
　biostratigraphy, 191
　characteristics, 189
　cysts, 190
　ecology, 189–190
　extant, ecology, 190–191
　fossil record, 190
　paleoecology, 191

Dissolved inorganic carbon (DIC), 611
Dissolved organic carbon (DOC), 696
Dissolved organic matter (DOM), 597–599
Dissolved silica (DSi), 792
DMM. *See* Depleted MORB mantle (DMM)
DOM. *See* Dissolved organic matter (DOM)
Dominant gliding models, 744
Double seismic zones (DSZ), 927
Driving forces: slab pull, ridge push, 193–195
　historical overview, 193
　principles, 193–194
　state of current knowledge, 195
Drowned river valleys, 235
DSDP. *See* Deep Sea Drilling Project (DSDP)
DSZ. *See* Double seismic zones (DSZ)
Dunes, 196–202
　barchans, 197
　cliff-top, 198
　climbing, 197
　echo, 198
　as ecosystems, 200–201
　embryo, 197
　environmental gradients, 200
　environmental services and conservation, 201
　foredunes, 197
　formation, 198
　morphologies, 199
　occurrence, 196
　reversing, 197
　transgressive/precipitation, 197
　transverse, 197
　typology, 197
　Wadden Sea, 936
　wrap-around, 198
Dust in ocean, 203–206
　deposition, 203–204
　emissions, 203
　fate of, 203
　and ocean biogeochemistry, 204–206
　ocean sedimentation contribution, 204
　transport, 203

E

Earth Impact Database, 462
Earthquakes, 209–215. *See also* Deep earthquakes, Wadati-Benioff-zoneaccretionary wedges, 7
　fault-plane solutions (*see* Fault-plane solutions)
　global occurrence, 214
　seismic hazard, 215
　seismic waves (*see* Seismic waves, earthquakes)
　size and frequency, 213–214
　slow, 215
　subduction, 798
　swarms, 215
　tectonic activity, 11
　unusual, 215
Earth's elliptical orbit, astronomical frequencies, 26–27
East African rift system, 481, 483
East Pacific Rise (EPR), 574
　vs. Oman, ophiolite structure, 577–579
Ebb–tidal deltas, 851–852, 934
ECORD. *See* European Consortium for Ocean Research Drilling (ECORD)
Ekman transport, 900
Electromagnetics (EM), marine geosciences, 829
Elementary quantitative models, 820
Elementary surface waves, 940
Elements of Chemical and Physical Geology (Bischoff), 428

El Niño Southern Oscillation (ENSO), 216–217, 751
Emiliania huxleyi, 538
Energy resources, 217–226
　environmental impact of offshore petroleum production, 221
　example: gas hydrates: an unconventional energy resource, 225
　forward-moving energy, 50
　fossil energy resources, 218
　future targets for the petroleum exploration, 222
　gas hydrates and, 436
　offshore petroleum exploration and production, 219–226
　offshore renewable energy, 218
　petroleum classifications, reserves and resources, 218–219
Engineered coasts, 226–230
　harbors and ports, 228–229
　occupational history and coastal impact, 227
　shoreline protection, 229
　soft engineering, 229
　from tide pools to polders, 227–228
　types, 229–230
ENSO. *See* El Niño Southern Oscillation (ENSO)
Entactinaria, 702
Environment
　controls, anoxic oceans, 21–22
　effects of volcanism, 914
　impact of oceanic plateaus formation, 564–565
　perturbation, volcanism and, 564
Environmental sensitivity index (ESI), 589
Epicenter, 231
EPR. *See* East Pacific Rise (EPR)
Erosion, 231–234
　hiatuses, extent, 233
　history, 231–232
　lithospheric thermal, 569
　ocean floor, 233
　pattern, 232–233
Eruptions, 914
　style and volcanic architecture, 37
　Toba, 915
　volcanic, 679
ESI. *See* Environmental sensitivity index (ESI)
Estuarine hydrodynamics, 235–238
　investigations, 237
　tidal straining, 236–237
Estuarine lagoons, 387
Estuary, 235–238
　inverse/negative, 235
　neutral, 235
　origin, 235
　positive, 235
　principal forcing variables, 236
　salinity vertical structure and mixing processes, 236
　salt-wedge, 236
　strongly stratified (fjord-type), 236
　tide-dominated, 854–857
　water balance, 235–236
　weakly/partially stratified, 236
　well-mixed, 236
Euler's theorem, 670
European Consortium for Ocean Research Drilling (ECORD), 544
Eustasy, 238–239
Euthecosomata order, 681
Events, in ocean history, 239–241
Exclusive economic zone (EEZ)
　exploration activities in, 478–479
　manganese nodules, 409, 410

　marine mineral resources, 475
　outer limit of, 363
Exhumation tectonics, 314
Explosive volcanism,
　in deep sea, 241–246, 756
　classification, 242–243
　eruption products, 244–245
　magmatic volatiles, 245–246
　mechanism, 245
　occurrence, 243–244
　terminology, 242–243
Export production (EP), 247
　aggregation, 247
　deep ocean flux, 247
　fecal pellets, 247
　marine snow, 247
　primary production, 247
　sediment traps, 247
Extratropical transition cyclone, 332
Extreme wave events, geohazards, 276–278
Eye and eye wall, tropical cyclone, 332

F

FAST algorithm. *See* First-arrival seismic tomographic (FAST) algorithm
Fault-plane solutions, 249–250
　ambiguity, 250
　principle, 250
Felsic eruptions, 403
Felsic igneous rocks, 402
Fe–Mn crusts. *See* Cobalt-rich manganese crusts
Festuca zone, 936
Finite-depth waves, 942
First-arrival seismic tomographic (FAST) algorithm, 729
Fishing site model, 898
Fixism, 442
Fjords, 235, 250–251
Flake tectonics, 18
Flood–tidal deltas, 851
Fold-and-thrust belt, 605
Fold belts, 664–665
Foraminifera (benthic), 251–254, 471
　agglutinated forms, 252
　bilocular, 252
　biology, 253
　calcareous forms, 252
　ecology and geographic distribution, 253–254
　growth and reproduction, 253
　locomotion, 253
　morphology and taxonomy, 251–253
　multilocular, 252
　pseudopodia, 253
　shell geochemistry, 254
　trophic relationships, 253
　unilocular, 252
Foraminifera (planktonic), 255–260
　geographic patterns, 258–259
　reproduction and growth, 256–257
　shell and cell, 256
　stratigraphic and paleoecologic applications, 259–260
　taxonomy, 257–258
　trophic relationships, 257
Forearc intraoceanic subduction zone, 369
Forearcs, 399
Foreland basin, 604–605
Forward ray-tracing, 729
Fossil energy resources, 218
Fossil semi-consolidated mud beds, 938
Fracture zone, 370
Fringing reef, 713–714, 719
Frontal accretion process, 798

G

Gabbro, 263–267
    accretion within oceanic crust, 265–266
    characteristics, 264–265
    classification, 263–264
    formation, 266
    geochemistry, 267
Gakkel Ridge, 581, 583, 584
Galveston Seawall, 748
Gas hydrates 225, 433–434, 555
    cold seeps, 118
    energy resources, 224
    marine. *See* Marine gas hydrates
    ocean margin systems, 555
    stability in marine sediments, 433–434
Gas hydrate stability zone (GHSZ), 434
GDH1, 451–453
General Bathymetric Chart of the Ocean (GEBCO), 268–269, 365, 754, 788
    history, 268
    structure, 268–269
General ocean circulation, 269–270
    bottom currents, 269–270
    surface currents, 269
Generic Mapping Tools (GMT), 365
Genetic stratigraphic sequences, 768
Geochemical Ocean Section Study (GEOSECS), 337
Geochronology, 271–274
    $^{230}Th_{excess}$ and $^{231}Pa_{excess}$ dating methods, 271–272
    $^{230}Th/^{234}U$ and $^{231}Pa/^{235}U$ dating methods, 273–274
Geohazards, 276–281
    cliff collapse, 276
    extreme wave events, 276–278
    low elevation coastal zone, 276
    paleostorm deposits, 278–279
    paleotsunami, 279–281
    reconstruction, paleostorm and tsunami events, 278–281
    storm surges, 277
    tsunami, 277–278
Geoid, 748
Geological time
    marine microfossils distribution, 470–471
    plate motions through, 671
Geologic time scale (GTS), 283–287
    conventions and standards, 284
    GTS2012, 284–287
    historical overview, 284
    International Commission on Stratigraphy, 283
    stratigraphic charts and tables, 287
Geology of the Sea (Klenova), 439
Geomagnetism, 442
Geotextiles, 748
GHSZ. *See* Gas hydrate stability zone (GHSZ)
GI-Gun, 723
Glacial-marine sedimentation, 288–292
    from glacier ice, 288–290
    sea-ice sedimentation, 290–292
    subtypes, 288
Glacio(hydro)-isostatic adjustment, 294–299
    Earth's response, 294–295
    glacial rebound, 298–299
    inversion of sea-level data, 297–298
    pattern of global sea-level change, 295–297
Glacio-isostatic adjustments (GIA), 739
Global conveyor belt, 630
Global heat loss, 456–458
Global ice volume, 611
Global ocean carbon climatology, 697
Global plate motion, 674–675
Global Positioning System (GPS), 671

Gonyaulax, 469
Gorgospyris perplexus, 703
Grain size, 630, 761–762
Granites, 605–606
Granular sediments, 761, 762
Gravity corers, 831
Gravity field, 299–301
    International Union for Geodesy and Geophysics, 300
    marine gravity and tectonics, 301
    measurements, 300–301
    Newton's law, 299
Great Barrier Reef (GBR), 712
Greenhouse warming potential (GWP), 436
Greenstone belts, 15
Grounding line, 302
Guyot, 302–307, 754, 757, 787
    vs. atoll, 303–304
    geomorphology, 303–304
    occurrence, 303–304
    sediments, 303–304
GWP. *See* Greenhouse warming potential (GWP)
Gymnosomata order, 681, 683, 685
Gymnosomes, 681
Gymnosperms, pollen grains, 655–656
Gypsum, 428

H

Halesium, 468
Halokinesis, 741
Harbors
    engineered coasts, 228–229
    tsunamis in, 887
Hartsalz, 428
Harzburgite lherzolite ophiolite type, 593, 594
Harzburgite ophiolite type (HOT), 593, 594
Hawaiian–Emperor seamount chain, 756, 758–759, 823
Hawaiian mantle plume, 317
Haystacks, 715
Hazards, submarine landslides, 819–820
Heat emission, 676
Heat flow, 449
    measurements, 829
Heinrich Events, 441
Heliconoides mercinensis, 686
Hiatuses, extent of, 233
High-latitude eocene zonation, 471–472
High-pressure/low-temperature metamorphism, 311–315, 369
    exhumation tectonics, 314
    geodynamic settings, 311–312
    mineralogy and metamorphic facies, 312–313
    occurrences in the sea, 314–315
    pressure-temperature paths and dynamics, 313–314
    prograde and retrograde metamorphism, 313
    source rocks, 311–312
    subduction and collision zones, 311
High-pressure metamorphism, 619–620
Hinge line, 440
Holocene volcanism, 915
HOT. *See* Harzburgite ophiolite type (HOT)
Hotspot-ridge interaction, 328
Hot spots and mantle plumes, 316–324
    enumeration, 318
    Hawaiian mantle plume, 317
    intraplate volcanism, 317
    island arc volcanism, 317
    and large igneous provinces, 318
    lithological composition, 322
    melting point, 322–323
    Morgan's plume theory, 317

    movability, 318
    and oceanic island chains, 319
    petrology and geochemistry, 318–319
    plate tectonics, 316
    primitive and recycled materials, 319–322
    rejuvenescent volcanism, 323–324
Hotspot volcanism, 756
Human disturbance, mangrove coasts, 416–417
Hurricanes, 329–334
    terminology, 330
    tropical cyclone. *See* Tropical cyclone
Hydrates
    gas. *See* Marine gas hydrates
    methane, 555
Hydrocarbon (HC) exploration, 722. *See also* Energy resources
Hydrodynamic(s)
    energy, 933
    regime, 933
Hydrogenic sediments, 761
Hydrogen sulfate, 497
Hydrology, mangrove coasts, 417–418
Hydrothermal circulation, 572, 870
    at basaltic mid-ocean ridge, 348–349
    hydrothermal discharge, 349
    in oceanic crust, 451
    phase separation, 350–351
    recharge zones, 347–348
    on seafloor, 335
    stockwork zone, 349–350
    vent fluids, 340
Hydrothermal crusts, 918
Hydrothermalism, 344–354
    discharge, 349
    discovery, 345–347
    global distribution, 351–352
    high-temperature reaction zones, 348–349
    hydrothermal circulation, 347–351
    mid-ocean ridges, 351–352
    plumes, 353–354
    sedimentary environments, 353
    suprasubduction zones, 352–353
    ultramafic environments, 352
Hydrothermal plumes, 335–338, 918
    applications, 335–338
    characteristics, 335
    event plumes, 337–338
    exploration, 335–336
    global ocean tracers, 337
    hydrothermal flux measurements, 337
    ocean chemistry, 336–337
Hydrothermal systems, ultramafic-hosted, 784
Hydrothermal upflow zones, 918
Hydrothermal vent fluids (seafloor), 339–343, 917
    composition, 340–341
    deep-sea, 784
    fields, 572
    heat and energy transfer, 342
    occurrence, 340
    ore deposits formation, 342
    origin of life and deep-sea vent communities, 342
    radioisotope ages of, 918–919
Hydrous minerals, 930
Hypocenter, 231

I

Iceberg surging mechanisms, 361
Ice, in ocean, 629
Ice-marginal moraines, 519–521
Ice-rafted debris (IRD), 359–361, 441
    history of observations, 359
    iceberg surging mechanisms, 361

Ice-rafted debris (IRD) (*Continued*)
  processes and analysis, 360
  sources and distribution, 360–361
IHS. *See* Inclined heterolithic stratification (IHS), 852, 857
Inland flooding, tropical cyclone, 333
Instability, astronomical periods, 31–32
Integrated coastal zone management, 363–364. *See also* coastal zone management, 110–111
  background and history, 363
  state and challenges, 363–364
  success measurement, 364
Integrated Ocean Drilling Program (IODP), 544, 708, 812, 825
Intergovernmental Oceanographic Commission (IOC), 365
Intermediate-depth earthquakes
  seismicity, 927
  Wadati-Benioff-zone, 927–930
International Arctic Science Committee (IASC), 365
International Bathymetric Chart of the Arctic Ocean (IBCAO), 365–367
  digital bathymetric models, 365–367
  printed map, 367
International Commission on Stratigraphy (ICS), 283
International Hydrographic Organization (IHO), 269, 365
International Seabed Authority (ISA), 475, 478
International Union for Geodesy and Geophysics (IUGG), 300
Intertidal flats, 852
Intertropical Convergence Zone (ITCZ), 907
Intraoceanic subduction zone, 367–370
  architecture, 368
  back-arc basin, 367
  crustal growth, 370
  extension and weak seismic coupling, 369
  forearc, 369
  intraoceanic trench, 368–369
  oceanic crusts, 367
  subduction, earliest stage, 370
  subduction interface, 369
  volcanic arcs, 367
Intraoceanic trench, 368–369
Intraplate magmatism, 372–377
  causes, 374–376
  morphology and characteristics, 372–374
  nature and source, 376
  tectonic setting, 372
Intraplate volcanism, 317
Inverted barometer effect, 277
IODP. *See* Integrated Ocean Drilling Program (IODP)
IRD. *See* Ice-rafted debris (IRD)
ISA. *See* International Seabed Authority (ISA)
Island arcs, magmatic zones, 620
Island arc volcanism, 317, 379–382
  across-arc variations, 380
  along-arc variations, 380
  arc crust, 379–380
  characteristics, 379
  ore deposits, 382
  vs. subduction parameters, 380–381
  temporal evolution, 381–382
Island shelf, 787–789
Isotopes, nitrogen, 539
ITCZ. *See* Intertropical Convergence Zone (ITCZ)
Izu-Bonin-Mariana arc, 795

J
Joint Global Ocean Flux Study (JGOFS), 98
Juan de Fuca Ridge (JdFR), 508
Jurassic APW paths, 634
Jurassic paleogeography, 632–637
Jurassic rocks, magnetostratigraphy, 407

K
Karliova-type triple junctions, 882–883
Kieserite, 428
Kleithriasphaeridium, 469
Knolls, 754
Kuzey detachment, 492

L
Lagoons, 385–390
  choked, 387
  coastal, 389
  estuarine, 387
  formation and evolution, 387–390
  key features and variety, 385–387
  leaky, 387
  Lucke model, 389
  marine, 387
  open, 384
  partly closed, 384
  restricted, 387
  types, 387
  worldwide distribution, 386
Laguncularia racemosa, 854
Laminated sediments, 391–392
Land-based reserve base (LBRB), 411
Land-sea interactions, 423
Landslide-generated tsunamis, 885
Large igneous provinces (LIPs), 372, 549, 558, 639, 914, 916
  formation, 565
Last glacial maximum (LGM)
  climatic situation, 707
  and deglaciation period, 739
  transition, 630
  upwelling, 903–905
Late Holocene SSTs, 186–187
Laurentide Ice Sheet (LIS), 359
Lava flows
  intra-flow variations, 812–813
  lobate flows, 810, 811
  mid-ocean ridges, 808–809
  pillow flows, 809–810
  seawater vapor phase, 814
  sheet flows, 810–811
  variations in, 812
Lava lakes, 810
Layering, oceanic crust, 392–393
LBRB. *See* Land-based reserve base (LBRB)
Leaky lagoons, 387
Lesser Antilles island arc, 486
LGM. Last Glacial Maximum (LGM)
Lherzolite ophiolite type (LOT), 593, 594
LHOT. *See* Harzburgite lherzolite ophiolite type
Limacina, 683
  L. bulimoides, 685
  L. helicina, 685, 686
  L. retroversa, 682
Linear reefs, 714
Linear wave theory, 940
LIPs. *See* Large igneous provinces (LIPs)
Lithomitra lineata, 706
Lithosphere, 370, 393–397, 423
  composition, 395–396
  strength, 396–397
  structure, 394–395
Lithospheric material, triple junctions, 882–883
Lithospheric plates, 677
Lithospheric thermal erosion, 569
Lithostratigraphy, 397
Littoral beds, 438
LOT. *See* Lherzolite ophiolite type (LOT)
Louisville seamount chain, 757
Low elevation coastal zone (LECZ), 276
Low-latitude Eocene zonation, 471
Lucke model, lagoons, 389
Lucky Strike volcano, 582

M
Magma, 914
  basaltic and trachyandesitic, 915
  chamber, 572
  compositions, 505
  and mantle source compositions, 503
  at mid-ocean ridge, 503–505
  plutonic crust, 505
  processes, 503–505
Magmatic timescales, 508–509
Magmatic zone
  of island arcs, 620
  metamorphism in, 620
Magmatism
  intraplate magmatism, 373, 375
  at mid-ocean ridges, 501–510
  at subduction zones, 800
Magmatism, convergent plate boundaries, 399–406
  crustal processes, 401–403
  global map, 401
  mantle wedge processes, 399–401
  mineral deposits, 404–405
  subduction zone, 399–401
  sulfur, 404
  volatile-rich nature, 403
  volcanism, 403–404
  volcano-plutonic complex, 402
Magnetic measurements, 828–829
Magnetic polarity scales, 407
Magnetostratigraphy, Jurassic rocks, 407
Manganese nodules, 408–412, 476
  chemical composition, 411
  composition, 409–411
  cross-section, 409
  formation, 409
  occurrence and distribution, 408–409
  resource consideration, 411–412
Mangrove, 854
Mangrove coasts, 412–421
  bordering, 414
  climate change, 419
  factors influencing, 414–415
  human disturbance, 416–417
  hydrology and sediments, 417–418
  indicators of change, 415–416
  management, 414
  morphological and physiological adaptations, 413
  natural disturbance, 419–420
  pests and pathogens, 419
  pollutants, 418–419
  pressures on, 415
  salt marsh vegetation, 413
  tidal wetlands, 415
Mangrove forests, 412
Mangrove plants, 412
Mantle compositions, mantle heterogeneities, 509
Mantle drag, 795–796
Mantle heterogeneities, 503, 509
Mantle wedge processes, 399–401
Maraş-type triple junctions, 883
Marginal-marine evaporite deposition, 430

Marginal seas, 423–427
　back-arc spreading models, 425
　biomass production, 426–427
　distribution around Pacific Ocean, 424
　features, 423
　formation models, 423–424
　geographic distribution, 423
　history, 423
　importance of, 427
　primary productivity, 426–427
　sedimentation, 424–426
　system of, 426
　water circulation, 424
Mariana arc, 404
Mariana-type subduction, 526–527, 794, 805–806
Marine algae, in surface ocean, 649
Marine environment
　serpentinization, 780–784
　source/reservoir rocks, 793
Marine evaporites, 427–432
　minerals, 428
　Permian Castile, 429–430
　Permian Zechstein Formation, 428–429
　Salado formations, 429–430
　shallow-water evaporites, 430–431
Marine gas hydrates, 433–436
　and climate, 436
　distribution and dynamics, 435–436
　and energy, 436
　gas hydrate stability, 433–434
　historic background, 433
　imaging and quantification, 435
　observatories, 436
　remote sensing techniques, 434–435
Marine geosciences, 437–446
　deep biosphere, 444–445
　deep-sea drilling, 442–443
　electromagnetics, 829
　fallacy of quiet deep ocean, 440
　global seafloors, 437–438
　gravity data, 829
　heat flow measurements, 829
　magnetic measurements, 828–829
　pioneers, 438–440
　plate tectonics, 441–442
　research vessels, 825–826
　seafloor observation methods, 826–830
　seafloor sampling tools, 830–833
　seafloor spreading, 441–442
　seafloor treasures, 443–444
　sea-level and society, 445
　sediment physics, 829–830
　seismology, 827
　sensor platforms, 828
　sequence stratigraphy, 445
　ship-based multibeam bathymetry, 826
　sidescan sonar, 827
　single-beam echo sounders, 826
　stratigraphic changes, 440–441
　synthetic aperture sonar, 827
　technology in, 825–833
　Tethys in, 838–847
　visual imaging, 830
Marine geotope protection, 448
Marine heat flow, 449–458
　global heat flow estimates, 456–458
　global heat loss, 456–458
　heat flow data, 453–454
　heat flow vs. age, 454–456
　hydrothermal circulation, 451
　method of measurement, 449
　modern thermal models, 451–453
　plate creation, 451

　probes, 450
Marine impacts, 460–466
　asteroid and comet impacts, 460
　Chicxulub impact, 465
　consequences, 465–466
　cratering, 461–464
　Eltanin impact, 465
　excavation stage, 461
　layers of target sites, 461–462
　Mjølnir impact, 466
　modification stage, 461
　Pliocene event, 465
　shallow marine, 460, 464
　studying, 464
　submarine craters, 462
　water depth, 462
Marine lagoons, 387
Marine microfossils, 467–473, 629
　biostratigraphy in, 471
　distribution patterns, 470
　elemental composition, 472
　examples, 469
　high-latitude zonation, 471–472
　history of study, 467–470
　low-latitude Eocene zonation, 471
　organic-walled microfossils, 470
　paleoenvironmental utility, 472–473
　paleoproductivity estimates, 472–473
　scanning electron micrographs, 468
　siliceous microfossils, 472
　stable isotope records, 472
　through geological time, 470–471
Marine mineral resources, 475–479
　deep-sea mining, 477
　exploration contracts, 477–479
　legal aspects, 475–476
　seabed, 476–477
Marine regression, 480–481
Marine sedimentary basins, 481–489
　collision-related basins, 482–487
　continental and marine rift basins, 481–482
　failed rift basins, 482
　sedimentary transitions, 481, 482
　strike-slip basins, 489
　subdivision, 481
　subduction-related basins, 487–489
　submarine bolide impacts, 489
Marine transgression, 489–490
Maritime archaeology, 893
Massive sulfides, 476
Mass wasting, 490–491
Maximum flooding surface (MFS), 775
Mean sea-level pressure (MSLP), 540
Medlinia, 469
Meiourogonyaulax, 469
Melting, peridotites, 665
Mesozoic ocean basins, paleophysiography of, 639–647
Messinian Salinity Crisis (MSC), 430
Metamorphic core complexes, 491–494
　continental crust and extension, 493–494
　core complex, 491–493
　detachment faults, 491–492
　ductile-to-brittle deformation, 493
　extended lithosphere, 494
　footwalls of, 492
　high-temperature deformation, 492
　rift-type extension, 494
　shear-sense indicators, 493
　syn-extensional clastic sediments, 492–493
　thermal stratification, 493–494
Metamorphism
　Abukuma-type, 620

　high-pressure/low-temperature, 369
　high-pressure/subduction, 619–620
　in magmatic zone, 620
　prograde and retrograde, 313
Methane, 495–498
　acetoclastic fermentation, 496
　distribution, 495–496
　gas bubble ebullition, 498
　hydrates, 555
　hyperthermophilic methanogens, 496
　microbial, 496, 497
　production, 495
　sediment reservoirs, 496–497
　sinks, 497–498
　sources, 496
　stability, 434
　thermogenic, 496, 497
Methanogens, 496
Methoxymirabilis oxyfera, 497
Mg/Ca paleothermometry, 499–500
Microfossils
　marine, 629
　organic-walled, 658
Microtextures, serpentinization, 783
Mid-ocean ridge (MOR), 33, 450–451
　global distribution, 351–352
　lava flows on, 808–809
　ophiolites, 592, 594, 595
　spreading centers, 780–781
　subsidence of oceanic crust, 821, 822
Mid-ocean ridge basalt (MORB), 501, 756, 808, 920
　composition of, 502, 503
　heat supply model, 505
　migration of, 503
　spectral analysis of, 509
　trace constituents in, 503
Mid-ocean ridge magmatism and volcanism, 501–510
　compositional attributes of, 504
　decadal-scale studies, 506
　history, 501–502
　investigations and controversies, 508–510
　magma generation at, 503–505
　volcanic eruptions at, 505–508
Milankovitch frequency, 630–631
Mineral deposit
　geological characteristics of, 404
　global distribution of, 918
　phosphorites, 666, 668
　sources of, 405
　sulfide mineral deposits, 404
Mineralization, stockwork, 918
Mineralogy
　and metamorphic facies, 312–313
　of VMS deposits, 920
Minerals, hydrous, 930
Mobilism, 442
Modelling past oceans, 514
Modern analog technique (MAT), 514–515
Mohorovičić discontinuity (MOHO), 515–518
　active-source seismology, 516
　controversies and gaps, 517
　global crustal thickness model, 517
　historical overview, 515–516
　observational techniques, 516
　observations, 516
　seismic discontinuity at, 515
　velocity discontinuity, 515
Moho transition zone (MTZ), 578, 579, 583
Moraines, 519–521
　De Geer moraines, 520, 521
　end moraines, 520
　Holmströmbreen glacier, 520

Moraines (*Continued*)
  sedimentologic and geomorphologic features, 519
  swath-bathymetric data, 520
  tidewater glaciers, 519–521
MORB. *See* Mid-ocean ridge basalt (MORB)
Morgan's plume theory, 317
Morphodynamics, Wadden Sea, 934–935
Morphology across convergent plate boundaries, 523–527
  Mariana-and Chile-type subduction, 526–527
  structure and topography, 525–526
  types of, 523–525
Motions, in wave, 940–941
Mountain building, tectonic activity, 12
MTZ. *See* Moho transition zone (MTZ)
Mud beds, fossil semi-consolidated, 938
Mud mounds, 719–720
Mud volcano (MV), 527–534
  abundance of, 528
  causes for overpressuring, 531
  formation, occurrence, and habitats, 528–532
  future prospects, 534
  historical background, 528
  investigations, 533–534
  map of, 529
  mud diapir, 530
  negative topography, 532
  pockmark and, 533
  process, 532–533
  research avenues, 534
  sedimentary volcanism, 532
Multibeam bathymetry, 817
Multicorers, 831
Multilocular foraminifers, 252
MV. *See* Mud volcano (MV)

N

Namibia upwelling system, 666
Nannofossils coccoliths, 537–539
  biostratigraphy, 537
  evolution, 538–539
  paleoceanography, 537–538
  significance, 537
NAO. *See* North Atlantic Oscillation (NAO)
National Geophysical Data Center (NGDC), 268
National Physical Laboratory (NPL), 450
Natural disturbance
  change type and corresponding factors, 420
  factor indicators, 420
  mangrove coasts, 419–420
  predominant drivers, 419–420
Natural waters, sediment transport in, 764
Nautical archaeology, 893
Nautilus Minerals, 476–479
Naxos detachment, 492
Negative buoyancy, 424
Neo-Tethys
  evolution and end of, 845
  marine geosciences, 838–846
Neovolcanic zone, 572
Nitrogen fixation, 539
Nitrogen isotopes, 539
  paleoceanographic proxies, 624
  in sea, 539
  upwelling, 904
Non-biogenic reefs, 711
Non-cohesive sediments, 761–763
Nonvolcanic continental rifted margins, 780, 781
Normal mid-ocean ridge basalt (N-MORB), 38
Normal move-out (NMO), 725
North Atlantic Oscillation (NAO), 540
North Atlantic Oscillation Index (NAOI), 540

O

Ocean acidification, 541–542
Ocean basins
  paleophysiography of (Paleophysiography of ocean basins)
  physiography and age-area distribution, 647
Ocean-bottom hydrophones (OBH), 727–728
Ocean-bottom seismometers (OBS), 727–728, 827
Ocean circulation. *See* General ocean circulation
Ocean-continent transition (OCT), 871
Ocean drilling, 423, 542–551
  climate history, 548–549
  coring tools, 546
  drill core, 548
  fluids, 550
  history, 544–545, 547
  logistics and technology, 545–548
  microbiology, 549
  observatories, 550–551
  ocean igneous crust, 549–550
  platforms, 544, 545
  programs, 547
  solid earth process, 549
Ocean Drilling Program (ODP), 736, 807, 906
Ocean dynamics, 751
Ocean formations, uranium-series dating, 271–274
  $^{230}Th_{excess}$ and $^{231}Pa_{excess}$ dating methods, 271–272
  $^{230}Th/^{234}U$ and $^{231}Pa/^{235}U$ dating methods, 273–274
Oceanic anoxic events (OAEs), 564, 902, 906
Oceanic core complexes (OCC), 579, 581
Oceanic crust, 677–679
  hydrothermal circulation in, 451
  intraoceanic subduction zone, 367
  layering, 392–393
  model, 573
  subsidence of, 821–823
Oceanic fracture zones
  in Central Atlantic, 872
  finer geology of, 869–873
  in geological record, 873
  transform faults, 867, 869–873
Oceanic intraplate volcanoes, 756
Oceanic lithosphere, 602–603
Oceanic plateaus, 558–566
  accreted, 563–564
  and black shale formation, 564–565
  buoyancy and accretion of, 561–562
  Carboniferous to Cretaceous oceanic plateaus accreted in Japan, 563
  carboniferous to cretaceous, 563
  characteristics, 558–559
  compositional differences between groups, 561
  cross sections, 562
  duration of formation, 559
  environmental impact, 564–565
  geochemical features of, 560–561
  in geological record, 562–563
  Precambrian, 563–564
  and sediment cover, 639
  structure, 559
  temperature of source, 559–560
  volcanism and environmental perturbation, 564
  Wrangellia, 563
Oceanic ridges
  conceptual model for, 573–574
  ophiolites and geophysical data, 585
Oceanic rifts, 567–571
  historical review, 567–568
  Northern Red Sea, active rift, 568–569
  passive rifting, 567, 569

Oceanic spreading centers, 571–585
  back-arc spreading centers, 584
  deep detachment model, 583
  EPR *vs.* Oman ophiolite, 577–579
  fast spreading center, 574–576
  fast *vs.* slow spreading ridges, 579
  oceanic ridges model, 573–574
  slow spreading center, 579–581
  ultraslow spreading centers, 581–584
Oceanic subduction, 11, 793–794, 800, 801
Ocean igneous crust, 549–550
Ocean margin systems, 552–558
  Atlantic-type (passive) margins, 553
  benthic boundary layer, 556
  biology, 557
  climate coupling, 555
  cold seeps, 557
  cold-water corals, 557
  continental margin, 554–555
  contourites, 556
  deep biosphere, 557
  gas hydrates, 555
  human usage, 557–558
  margin types, 553
  morphology, 554–555
  organic matter deposition, 555
  Pacific-type (active) margins, 553–554
  sedimentation process, 553
  sedimentological features, 556–557
  shelf, 554
  turbidites, 556
Oceans
  salt diapirism in, 741–745
  topography, 826
  wave, 941–942
OCT. *See* Ocean-continent transition (OCT)
ODP. *See* Ocean Drilling Program (ODP)
Oestrupia, 469
Offshore petroleum exploration
  Arctic realm, 220–221
  deepwater petroleum, 219–220
  environmental impact, 221–222
  evolution, 218–219
  gas hydrates, 225
  passive rifted margins, 223
  simeulue forearc basin, 224–225
  volcanic rifted margins, 224
Offshore renewable energy, 217–218
Oil spill, 587–590
  applying geosciences to, 587–588
  coastal geomorphology, 589
  geoscience advise during response operations, 589–590
  modeling, 590
  oil composition and weathering, 587
  rock-dominated shorelines, 588–589
  sediment-dominated shorelines, 588
  sensitivity, 589
  shore zone coastal mapping, 590
  vegetated shorelines, 589
OM. *See* Organic matter (OM)
Oman ophiolite structure, 577–579
Open lagoons, 384
Open water
  embayment, 389
  sediment transport in, 765
Ophiolites, 592–596
  Cordilleran, 593, 594
  historical development, 592
  mid-ocean ridge, 592, 594
  modes of emplacement, 593–594
  MOR *vs.* SSZ origin, 595
  ridge structure and dynamics, 594
  serpentinization, 781–784

Tethyan, 593–595
typology, 592–593
Orbital elements, astronomical frequencies, 27–28
Ore-bearing fluids, 405
Ore deposits
  formation, 342
  island arc volcanism, 382
Organic carbon
  fluxes with depth in water, 600
  in World Ocean, 598
Organic matter (OM), 596–601
  accumulation, 901–902
  dissolved, 597–599
  forms of, 597
  ocean margin systems, 555
  particulate, 599–600
  of sediments, 600–601, 903–905
  sources of, 597
Organic-walled microfossils, 658
The Origin of Continents and Oceans (Wegener), 439
Orogenic belts, 602
Orogenic cycle, 602–603
Orogens, 602–607
  Alpine-type, 604
  anatomy, 604–606
  Cordilleran-type/arc-type, 604
  destruction, 603, 606–607
  types, 603–604
Oxygen isotopes, 610–616
  benthic foraminifera, 615
  carbonate ion concentrations, 611
  fractionation of, 629–630
  global ice volume, 611
  influence of diagenesis, 615
  in marine archives, 610–611
  paleosalinity, 612–614
  paleotemperature, 611–612
  planktonic foraminifera, 614
  sulfur cycle, 615–616
Oxygen-minimum zone (OMZ)
  cobalt-rich manganese crusts, 115
  Miocene expansion, 906
  oxygen concentrations, 906
  upwelling, 901–902, 908

P

Pacific Tsunami Warning Center (PTWC), 887
Pacific-type ocean margins, 553–554
Paired metamorphic belts, 619–622
  Abukuma-type metamorphism, 620
  high-pressure/subduction metamorphism, 619–620
  in Japan, 620
Paleoceanographic proxies, 622–627, 630
  circulation of ocean, 625
  CO_2 content of atmosphere, 625
  land vegetation and rainfall, 625
  productivity, 624
  salinity, 624
  sea-level, 625
  source of, 623
  stratigraphy, 625–627
  temperatures, 623–624
Paleoceanography, 440–441
  from infancy to maturity, 628–631
  nannofossils coccoliths, 537–538
Paleocene-Eocene Thermal Maximum (PETM), 21
Paleoclimate models
  astronomical frequencies, 25–33
  characteristics, 25
  earth's elliptical orbit, 26–27
  eccentricity, obliquity, and climatic precession, 29–31
  instability, astronomical periods, 31–32
  long-term variations, 28
  obliquity and precession, 29
  orbital elements, planet, 27–28
  upwelling, 907–908
Paleoenvironmental utility, microfossils, 472–473
Paleogeographic Atlas Project, 907
Paleogeography, Jurassic, 632–637
Paleomagnetism, 632–637
Paleophysiography of ocean basins, 629, 638–647
  history, 638
  Mesozoic and Cenozoic ocean basins, 639–647
  ocean gateways, 647
  reconstructing mid-ocean ridges and flanks, 638–639
Paleoproductivity, 648–652
  marine production, 649–652
  tracers of past biological production, 651
Paleosalinity
  in ocean, 629
  oxygen isotopes, 612–614
Paleostorm deposits, geohazards, 278–279
Paleotemperature in ocean, 629
Paleo-Tethys
  marine geosciences, 838–846
  Pangaea, 842, 843
  Permian extinctions, 846
Paleotsunami, geohazards, 279–281
Palynology, 653–658
  advantages and disadvantages, 658
  pollen analysis, 653–654
  spores and pollen grains, 654–657
Palynomorphs, sporopollenin of, 657
Parasequences, 769–771, 775
Paratethys, marine geosciences, 838–846
Particle-based modeling, 762
Particulate organic matter (POM), 599–600
Partly closed lagoons, 384
Passive plate margins, 659–665
  in cross section, 661–663
  fold belts, 664–665
  morphology and sedimentation, 663
  rifted margins underlie, 659
  rifting and, 663–664
  tectonic origins, 661
  volcanic *vs.* nonvolcanic margins, 663
Passive seismology, 827
Patch reefs, 715–716
Pathogens, mangrove coasts, 419
Peak to bubble ratio (PBR), 723
Pelagic beds, 438
Peracle bispinosa, 682
Peracle reticulata, 685
Peridotites, 665
  abyssal, 665
  melting process, 665
  subgroups, 665
Permanent Service for Mean Sea Level (PSMSL), 751
Permian Castile, 429–430
Permian Zechstein Formation, 428–429
Permo-Triassic extinction event, 720
Pests, mangrove coasts, 419
Petit spots, 754, 756
Petroleum
  classifications, 218–219
  offshore exploration (*see* Offshore petroleum exploration)
  reserves, 218–219
  resources, 218–219
Phaeodaria, 702
Phanerozoic, anoxic oceans, 21
Phase separation, 405
Phosphorites, 666–668
  authigenic precipitation, 666
  diagenetic phosphatization, 666
  mineral deposits, formation, 666, 668
  types and regional distribution, 666
  upwelling, 902–903
Pillow lavas, 809–810
Piston corers, 831
Placer deposits, 668–669
Plate driving forces, 193
Plate motions, 638, 669–675
  absolute, 671, 674
  Euler's theorem, 670
  history, 669–670
  models through time, 674–675
  plate tectonic hypothesis, 670
  present-day, 670–671
  relative, 670–671
  through geological time, 671
Plate stratigraphy, 629
Plate tectonics, 423, 439, 441–442, 676–679
  characteristics, 15–18
  and climate, 629
  history, 13
  hypothesis, 670
  mechanism, 13–14
  theory, 572
  types, 18
  upper, 798–800
Platform reefs, 719
Platinum, 410
Plinian eruptions, 404
Plumes, 353–354
Plutons, 403
Pneumodermopsis macrochira, 682
Polar wandering, 442, 759
Pole, 442
Pollen analysis, 653–654
Pollen grains
  angiosperms, 656
  gymnosperms, 655–656
  from marine sediment cores, 657
  stratigraphy, 656
  systematics, biology, and morphology, 654–656
  transport into marine sediments, 656–657
  use of, 657
Pollutants, mangrove coasts, 418–419
Polycystina, 702
Polyhalite, 428, 429
POM. *See* Particulate organic matter (POM)
Pore fluids, subduction erosion, 806
Porphyry Cu-Au deposits, 405
Ports
  engineered coasts, 228–229
  tsunamis, 887
Precambrian oceanic plateaus, 563–564
Present-day plate motions, 670–671
Pressure-temperature paths and dynamics, 313–314
Pre-stack depth migration (PSDM), 727
Primary productivity, in marginal seas, 426–427
Probabilistic models, sediment transport, 766–767
Prograde and retrograde metamorphism, 313
Project Mohole, 515–516
Proterozoic, anoxic oceans, 20
Pseudoceratium, 469
Pseudodictyomitra, 468
Pseudopodia, foraminifers, 253
Pseudothecosomata order, 681
PSMSL. *See* Permanent Service for Mean Sea Level (PSMSL)
Pteropods, 680–686

Pteropods (*Continued*)
  distribution, 685
  ecology, 685–686
  geology, 686
  Gymnosomata order, 681, 683
  hierarchical subdivisions, 681
  shelled, 681, 686
  Thecosomata order, 681
PTWC. *See* Pacific Tsunami Warning Center (PTWC)
Puccinellia zone, 936
Pull-apart basins, 687–690
  geometrical aspects, 688–689
  mechanical aspects, 688
  model, development, 687–688
  sedimentary aspects, 689–690
Pull-apart force, 424
Pure spreading models, salt tectonics, 744–745
Push-up blocks, 692–694
  geomorphic aspects, 693
  mechanical aspects, 693
  model, development, 692
  structure, 693

R
Radiation stress tensor, 941
Radioactive isotopes, 699
Radiocarbon, 695–698
  carbon cycle, 696–698
  climate-related redistribution, 696
  clock, 696
  dating, 696, 918–919
  decay, 695
  dispersion, 695
  measurement, 695–696
  mixing and reservoir age, 696–697
  ocean and climate, 697
  in oceanic circulation, 697
  plateau tuning, 697
  production, 695
  radiocarbon dating, 696
  tree-ring calibration, 696
Radiogenic tracers, 699
Radiolarians, 471, 700–709
  calymma, 700
  capsular membrane, 700
  cell organization, 700–702
  diversity, 705–706
  ectoplasm, 700
  endoplasm, 700
  lateral transportation, 705
  longevity, 703–704
  luminescence, 705
  microsphere, 702
  migration, 705
  myonemes, 706
  Neogene-Quaternary stratigraphy, 707–709
  nutrition, 705
  preservation in sediments, 706
  pseudopodia, 701–702
  reproduction, 704–705
  sarcodictyum, 701
  sarcomatrix, 700
  shell deposition, 702
  skeleton organization, 702–703
  Spongaster tetras, 703
  taxonomic position, 700
  vertical migration, 705
  zoogeography through time, 707–709
Radiometric dating, 508
Rare earth elements (REEs), 410
Rawhide Buckskin detachment, 492
RCB. *See* Rotary core barrel (RCB)
Realistic surface wave systems, 941

Red Sea oceanic rift, 568–569
Reef(s), 710, 718–721
  coral mounds, 719–720
  Devonian reef outcropping, 721
  in geological past, 720
  growth, 720
  linear, 714
  mud mounds, 719–720
  as reservoir rocks, 720–721
  tropical reefs, 718–719
  types, 719
Reef coasts, 710–717
  artificial reefs, 716–717
  atoll, 712–713
  barrier reef, 714–715
  classification, 712
  coral reefs, 711–712
  coral reefs with islands, 716
  etymology and usage, 711
  fringing reef, 713–714
  patch reefs, 715–716
  rock reefs and non-biogenic reefs, 711
REEs. *See* Rare earth elements (REEs)
Reflection/refraction seismology, 721–731
  acquisition schemes, 725–727
  data acquisition, 724–725, 727–728
  data processing, 724–725, 727, 728
  earth science, contribution to, 730
  energy sources, 723–724
  history, 722–723
  interpretation, 729–730
  inversion, 728–729
  modeling, 728–729
  principles, 722
  seismographs, 725
  signal enhancement, 728
  streamers, 724–725
  time-domain air gun specifications, 723, 724
  tomography, 728–729
  2D multichannel seismic acquisition scheme, 726
  velocity, 727
  Vines Branch experiment, 723
Reflux model, 429
Regional marine geology, 731–735
  ages of oceanic lithosphere, 734
  Amur and Okhotsk plates, 733
  Antarctic plate, 734
  Atlantic Plate, 734–735
  Caribbean Plate, 734
  Challenger Deep, 732
  Farallon Plate, 734
  Indo-Australian Plate, 733, 735
  Laurasian Plate, 734
  lithospheric plates on Earth, 732
  Nazca Plate, 734
  New Hebrides-Fiji Plate, 733
  Pacific Plate, 733, 734
  plate boundaries, 732
  Somalia Plate, 733
  Tonga-Kermadec subduction zone, 733
Rejuvenescent volcanism, 323–324
Relative plate motions, 670–671
Relative sea-level (RSL) cycle, 735–740, 748
  change, 481
  climate and, 738
  deglaciation process, 739
  duration, 736
  eustatic cycle, 736
  factors influencing, 736–738
  glacial and postglacial, 738–739
  global, 737
  order of, 736–737
  tectonic/isostatic cycle, 736

Remotely controlled robotic seabed drills, 831
Remotely operated vehicles (ROVs), 438, 826, 829, 830
Research vessels, 825–826
Reservoir rock, 720–721, 792–793
Restricted lagoons, 387
Revetments, 747–748
Revised local reference (RLR), 751
Rhine valley, 553
Rhizophora mangle, 854
Rhizosphaera sp., 703
Ribbon reef, 714–715
Ridge propagation, triple junctions, 880–882
Ridge push, 14, 193
Rifted margins underlie passive margins, 659
Rifting
  active, 568–569
  Atlantic Ocean, 570
  continental, 602
  passive, 567, 569
  symmetrical, 568
  theoretical models, 568
River deltas, tide-dominated, 854, 855
RLR. *See* Revised local reference (RLR)
Rocks
  reef, 711
  in subduction zones, 619
Rotary core barrel (RCB), 545
Rough seas, tropical cyclone, 333
ROVs. *See* Remotely operated vehicles (ROVs)
RSL cycle. *See* Relative sea-level (RSL) cycle

S
Sabkhas, 430, 857, 858
Sag basins, 663–664
Salado formations, 429–430
Salicornia zone, 936
Salt diapirism, 741–745
  classification, 741–742
  driving forces, 742–745
  geometry, 742
  oceans and continental margins, 741–745
  progressive development, 743
Salt marshes
  tidal, 413, 420
  vegetation, 853
  Wadden Sea, 936
Salt tectonics, 741–745
San Andreas Fault system, 487
Sandy tidal flats, 935
Sapropels, 746–747
Satellite altimetry
  sea-level, 749, 751, 754
  seamounts, 754
Sclater curve, 821
Sclerochronology, 747
SC on Regional Undersea Mapping (SCRUM), 268
SC on Undersea Feature Names (SCUFN), 268
Scripps Institution of Oceanography (SIO), 450
SCUBA. *See* Self-Contained Underwater Breathing Apparatus (SCUBA)
Seabed, 476–477
Seafloor
  depth measurements, 826
  electromagnetics, 829
  gravity data, 829
  heat flow measurements, 829
  hydrothermal system, 918
  magnetic measurements, 828–829
  observation methods, 826–830
  sampling tools, 830–833
  sediment physics, 829–830
  seismology, 827
  sidescan sonar, 827

spot-sampling instruments, 831
spreading, 368, 423, 439, 441–442, 571
subseafloor sampling instruments, 831–833
synthetic aperture sonar, 827
thermal subsidence, 822–823
topography, 826
towed sampling instruments, 831
visual imaging, 830
Seafloor massive sulfides (SMS), 444, 476
Sea-ice sedimentation, 290–292
Sea-level, 748–753
archaeological importance, 895
change, 424
continental shelf, 791
examples of, 898
geological records, 748
instrumental period, 748–749
long-term trends, 752
measurement, 748
present-day, 752
related quantities and, 749
record, 753
satellite altimetry, 749, 751, 754
seiches, 750
society and, 445
storm surge, 750, 751
with tide gauge, 748–749, 751
variability in time and space, 749–752
Sea-level equation (SLE), 739
Seamounts, 4, 754–759
architecture, 757–758
and atolls, 823
evolution and formation, 756–757
global distribution, 755
lithospheric flexure beneath, 758
on-/near-/off-axis, 754
origins, 754–756
satellite altimetry, 754
ship SONAR technique, 754
size–frequency relationship, 754
structure, 757–758
subduction, 804, 805
true polar wander, 759
volcanism, 756
Sea-surface height (SSH)
instantaneous, 748, 749
mean, 748, 749
Sea surface salinity (SSS), 612
Sea surface temperature (SST), 217, 499–500
diatoms, 186
Late Holocene, 186–187
Sea walls, 747–748
Sedimentary archives, anoxia in, 22
Sedimentary environments, 353
Sedimentary sequence, 768–772
parasequences, 769–771
systems tracts, 768–769
Sedimentation, 720
abyssal plains, 3–4
in marginal seas, 424–426
ocean margin systems, 553
sequence, 768–771
tsunamis, 886–887
Sediment cover, oceanic plateaus and, 639
Sediment dynamics, 761–763
biological impacts, 763
complexity, 761
origin of, 761
settling and recycling, 762–763
Stokes' law, 762
transport, 761–762
Sediment pore fluids, 409
Sediments
mangrove coasts, 417–418

organic matter of, 600–601
physical properties, 829–830
traps, 831
Sediment transport, 761–762, 764–767
bed-load, 764
Cellular Automata (CA) models, 767
carbonate tidal settings, 857
deterministic models, 765–766
inlet–ebb delta system, 852
natural waters, 764
open water, 765
probabilistic models, 766–767
suspended-load, 764
Seiches, 750
Seismicity
deep-focus, 930
intermediate-depth, 927
subduction earthquakes, 798
transform faults, 862, 863
Seismic reflection, 132, 572
Seismic refraction, 572
Seismic stratigraphy, 730
Seismic waves, earthquakes
compressional/P-waves, 212
generation, 209–211
recording, 209–211
records and fault plane solutions, 212–213
shear stress, 211
travel, 209–211
Seismogenic zone, 773, 795–796
Seismology
active, 827
marine geosciences, 827
passive, 827
Self-Contained Underwater Breathing Apparatus (SCUBA), 894, 896
Sequence stratigraphy, 445, 730, 773–779
back-stripping analysis, 775
bounding surfaces, 774–775
carbonate systems, 775, 777
clastic systems, 775, 777
depositional systems, 768, 775
framework, 768, 769
genetic stratigraphic sequences, 768
methodology, 768
over simplification of time, 778
stacking patterns and geometries, 776, 778
Steno's laws, 775–776
stratigraphic tools, 778
surfaces of, 775
systems tracts, 768–769
terminology, 778
transgressive-regressive sequences, 768
Walther's law, 775–776
Serpentinization, 311–312, 779–785
convergent plate boundaries, 781
geochemical changes during, 782
mantle peridotites, 783, 784
marine environment, 780–784
microtextures, 783
mid-oceanic ridge spreading centers, 780–781
mineralogical changes during, 782
nonvolcanic continental rifted margins, 781
ophiolite complexes, 781–784
P-T-fluid conditions, 782
related rock types, 784
societal relevance of, 785
ultramafic-hosted hydrothermal systems, 784
Shallow-water
coral reefs, 711–712
evaporites, 430–431
waves, 942
Shelf, 554, 787–791
continental, 788–791

history, 787–788
island, 788–789
oceanward slope, 788
width and depth to shelf break, 790
Shelled pteropods, 681, 686
Shoaling process, 941
Shoreline Cleanup Assessment Technique (SCAT), 589
Shoreline protection, 229
Sidescan sonar, 827
Silica, 792
with brucite, 782
Siliceous microfossils, 472
Silled basins, 392
Single-beam echo sounders, 826
Slab pull, 14, 193, 795–796
Slip partitioning process, 800
Slow earthquakes, 215
SMS. See Seafloor massive sulfides (SMS)
Soapstones, 784
Society of Exploration Geophysicists (SEG), 725
Solid earth process, 549
SOund Navigation And Ranging (SONAR), 787
Source rocks, 311–312, 792–793
Southwest Indian Ridge (SWIR), 581–583
Spatial scales, mantle heterogeneities, 509
Sphaerozoum fuscum, 700
Spindletop Dome, in Beaumont (Texas), 741
Spores
stratigraphy, 656
systematics, biology, and morphology of, 654–656
transport into marine sediments, 656–657
use of, 657
Sporopollenin, 656, 657
Spumellaria, 702
SSS. Sea surface salinity (SSS)
SSZ. Suprasubduction zone (SSZ)
Stacking patterns, sequence stratigraphy, 776, 778
Steno's laws, 775–776, 778
STEP. Subduction-Transform Edge Propagator (STEP)
Stilostimella, 468
Stockwork mineralization, 918
Stokes' law, 762
Storm surges, 750, 751
geohazards, 277
tropical cyclone, 332–333
Stratigraphy, 440–441
paleoceanographic proxies, 625–627
sequence. See Sequence stratigraphy
spores and pollen grains, 656
tools, 778
unconformity-bounded unit, 768
Strike-slip basins, 489
Strike-slip fault, 424
Strong winds, 333
Subaqueous eruptions, 403
Subduction, 793–801
basins formed by stretching, 487–489
channel, 796
Chilean-type, 794, 805
Circum-Pacific, 804
classification, 793–794
and collision zones, 311
distinctive characteristics of, 794–795
distribution on earth, 794
earthquakes, 798
interface, 796–799
Izu-Bonin-Mariana arc, 795
magmatism, 800
main driving and resistive forces in, 795–796
mantle wedge processes, 399–401

Subduction (*Continued*)
 Mariana-type, 794, 805–806
 metamorphism, 619–620
 migration, 805
 rocks in, 619
 slab shape and dynamics, 795
 submarine lava types, 813
 tectonic erosion in, 803–804
 upper plate tectonics, 798–800
Subduction erosion, 803–807
 controlling factors, 805–806
 Costa Rica, 806
 occurrences, 803
 pore fluids, 806
 record of, 806–807
 seamounts, 804, 805
Subduction-Transform Edge Propagator (STEP), 869
Submarine arc volcanism, 403
Submarine bolide impacts, 489
Submarine canyons, 807
Submarine landslides, 817–820
 on continental shelf, 818
 elementary quantitative models, 820
 environments, 817–819
 hazards, 819–820
 history, 817
 multibeam bathymetry, 817
 probabilistic analysis, 820
 slope instability/failure, 819
Submarine lava types, 808–814
 features, 814
 intra-flow variations, 812–813
 lava flows on mid-ocean ridges, 808–809
 morphology, 809–812
 subduction zone environments, 813
 variations, 808
Submerged landscapes, 895–898
Subseafloor stockwork mineralization, 349–350
Subsidence of oceanic crust, 821–823
 mid-ocean ridges, 821, 822
 morphological expression, 821–822
 seafloor, 822–823
 seamounts and atolls, 823
Sulfate-methane interface zone (SMI), 497–498
Sulfate-reducing bacteria (SRB), 145
Sulfur, 404
 cycle, 615–616
 pools, 404
Sumatra earthquakes, 798
Sunda Arc, 8
Supercontinent cycle. Orogenic cycle
Suprasubduction zone (SSZ), 352–353, 592, 594, 595
Surface ocean, marine algae in, 649
Surface waves, 940
Suspended-load transport, 764
Suture zone belt, 606
SWIR. Southwest Indian Ridge (SWIR)
Sylvanite, 428
Symbionts, 705
Synorogenic S-type granites, 605–606
Synthetic aperture sonar, 827
Systems tracts, 768–769
 internal architecture of, 769
 and parasequences, 769–771

T
Technical SC on Ocean Mapping (TSCOM), 268
Technology in marine geosciences, 825–833
 electromagnetics, 829
 gravity data, 829
 heat flow measurements, 829
 magnetic measurements, 828–829
 research vessels, 825–826
 seafloor observation methods, 826–830
 seafloor sampling tools, 830–833
 sediment physics, 829–830
 seismology, 827
 sensor platforms, 828
 ship-based multibeam bathymetry, 826
 sidescan sonar, 827
 single-beam echo sounders, 826
 synthetic aperture sonar, 827
 visual imaging, 830
Tectonic activity
 active continental margins, 11–12
 earthquakes, 11
Tectonically generated tsunamis, 885
Tectonics
 erosion, 798, 803–804
 estuaries, 235
 exhumation, 314
 origins of passive margins, 661
 plate, 669, 670
Temperature evolution, earth, 14–15
Terranes, 835–837
 Canadian-Alaskan Cordillera, 836
 collision/accretion, 12
 Cordilleran, 835, 837
 evolution, 835
 paleogeographic affinities, 837
Tethyan ophiolites, 593–595
Tethys in marine geosciences, 838–847
 Central Mediterranean during, 840
 Neo-Tethys, 842–845
 Paleo-Tethys, 842–846
 Pangaea, 842
 Paratethys, 845–846
 present usage, 841–842
Thecosomata order, 681
Thermal erosion, 569
Thermal subsidence, seafloor, 822–823
Tidal deltas, 850–852
 ebb, 851–852, 934
 flood–tidal deltas, 851
Tidal depositional systems, 849–858
 carbonate, 857–858
 classification, 850
 coastal wetlands, 852–854
 ebb–tidal deltas, 851–852
 estuaries, 854–857
 flood–tidal deltas, 851
 global distribution, 850
 intertidal flats, 852
 river deltas, 854, 855
Tidal flats, 852, 853, 933
 deposits, 937
 sandy, 935
Tidal inlets, 850–852
Tidal salt marsh, 413
Tidal salt pan, 413
Tidal wetlands, 413
Tide gauge, 748–749, 751
Tide pools to polders, engineered coasts, 227–228
Tidewater glaciers, 519–521
TMF. Trough mouth fans (TMF)
Toarcian Oceanic Anoxic Event (T-OAE), 21
Toba eruption, 915
Tohoku earthquakes, 798, 800
Tomography. Seismic refraction
Tonalite trondhjemite-granodiorite (TTG) suite, 15
Torculum, 468
TPW. True polar wander (TPW)
Trachyandesitic magmas, 915
Transform faults, 370, 442, 859–873
 behaviours of, 860
 conservative plate boundaries, 859
 displacement, 865
 large-scale geological structures, 861–869
 oceanic fracture zones, 867, 869–873
 Paleozoic and Mesozoic ages, 861
 plate boundary, 864, 866, 867
 seismicity, 862, 863
Transgressive-regressive (TR) sequences, 768
Transgressive surface (TS) forms, 775
Travel time inversion, 729
Travel time tomography, 729
Trench-arc system, 423
Triple junctions, 876–884
 centroid, 878
 Karlıova-type, 882–883
 lithospheric holes, 882–883
 Maraş-type, 883
 ridge propagation, 880–882
 ridge-ridge-ridge, 878–881
 stability, 876–880
 third spatial dimension, 882
 velocity space, 878
Tropical cyclone
 anatomy, 331–332
 extratropical transition, 332
 eye and eye wall, 332
 features within, 332
 formation, 330–331
 impacts, 332–334
 inland flooding, 333
 rain bands, 332
 rough seas, 333
 storm surge, 332–333
 strong winds, 333
Tropical reefs, 718–719
Trough mouth fans (TMF), 491
True polar wander (TPW), 633, 759
Tsunamis, 750–751, 884–888
 deposits, 886–887
 earthquake, 885
 events reconstruction, geohazards, 278–281
 events worldwide, 887
 generation, 885
 geohazards, 277–278
 inundation and run-up, 886
 landslide-generated, 885
 paleo-tsunami studies, 886–887
 in ports and harbors, 887
 propagation, 886
 tectonically generated, 885
Tsunami warning systems (TWS), 887
Turbidites, 4, 888–891
 beds, 891
 currents, 888–889
 deep-sea, 889–891
 ocean margin systems, 556
 related density flows, 888–889
Typhoons. Hurricanes

U
Ultramafic environments, 352
Ultra Short Base Line (USBL), 830
Ultraslow oceanic spreading centers, 581–584
Unconformity-bounded stratigraphic unit, 768
Underplating, 798
Underwater archaeology, 893–898
 Atlit Yam site in Israel, 896
 Cosquer Cave in southern France, 896
 fishing site model, 898
 geological techniques, 895
 geoscientific research, 894
 history, 894–895
 submerged settlements/landscapes, 895–898
Uniformitarianism, 439
Unilocular foraminifers, 252

United Nations Conference on Environment and Development (UNCED), 363
United Nations Convention on the Law of the Sea (UNCLOS), 365, 475
University of New Hampshire (UNH), 269
Unusual earthquakes, 215
Upwelling, 4, 900–908
   Cenomanian-Turonian black shale, 907, 908
   coastal, 901
   geological effects, 900
   nitrogen isotopes, 904
   organic matter accumulation, 901–902
   organic-rich sediments, 903–905
   oxygen minimum zone, 901–902
   paleoclimate models, 907–908
   phosphorite, 902–903
   signals in older sediments, 905–907
Uranium-series dating, ocean formations, 271–274
   $^{230}Th_{excess}$ and $^{231}Pa_{excess}$ dating methods, 271–272
   $^{230}Th/^{234}U$ and $^{231}Pa/^{235}U$ dating methods, 273–274
USBL. Ultra Short Base Line (USBL)

V

VEI. See Volcanic explosivity index (VEI)
Velocity space, 878
Very-long-baseline interferometry (VLBI), 671
Vessel-mounted drill systems, 831
Vibro corers, 831
Viscous creep, 930
Visual imaging, marine geosciences, 830
VMS deposits. Volcanogenic massive sulfide (VMS) deposits
Volcanic arcs, 379–382, 606
   across-arc variations, 380
   along-arc variations, 380
   arc crust, 379–380
   characteristics, 379
   island arc volcanism vs. subduction parameters, 380–381
   ore deposits, 382
   temporal evolution, 381–382
Volcanic climate forcing, 913, 914
Volcanic eruptions, 679
   deep-sea eruption, 507
   detection, 508
   dynamics, 506–508
   frequency and duration, 506
   at mid-ocean ridge, 505–508
   products, 506
Volcanic explosivity index (VEI), 403, 913
Volcanic impact (V_{imp}), 913–914
Volcanic-tectonic history, 34–35
Volcanism, 403–404, 501, 913–916
   environmental effects of, 914
   and environmental perturbation, 564
   explosive, 756
   Holocene, 915
   hotspot, 756
   rejuvenescent, 323–324
   seamounts, 756
   tectonic activity, 11–12
Volcanism and climate
   atmosphere process, 914–915
   chicken-and-egg conundrum, 914, 915
   environmental geosphere driver, 915–916
   LIP contribution, 916
   volcanic impact, 913–914
Volcanogenic massive sulfide (VMS) deposits, 345, 917–922
   composition of, 919–920
   formation, 918–919
   mineralogy of, 920
   morphologies, 921
   size of, 920–922

W

Wadati-Benioff zone (WBZ), 400, 925–931
   case study, 931
   deep-focus earthquakes, 927, 929, 930
   of deep seismicity, 926
   intermediate-depth earthquakes, 927–930
   Kiyoo Wadati and Hugo Benioff, 926
Wadden Sea, 933–938
   barrier islands development, 938
   beaches, 936
   dunes, 936
   geographic map, 934
   geological cross section, 937
   Holocene evolution of, 937–938
   hydrodynamic regime, 933
   morphodynamics, 934–935
   oceanography, 933–934
   salt marshes, 936
   sediment-organism interaction, 935–936
Walther's law, 775–776, 778
WARRP. See Wide-angle reflection/refraction profiling (WARRP)
Waste Isolation Pilot Plant (WIPP) site, 429
Water circulation patterns, in marginal seas, 424
Wave-field inversion, 729
Wavelet, 723
Waves, 940–943
   action in continental shelf, 789
   climate, 942–943
   diffraction, 941
   elementary surface, 940
   energy, 940
   energy flux (Waves, power)
   finite-amplitude and breaking, 942
   motion in, 940–941
   ocean, 941–942
   power, 941
   propagation, 940, 941
   spectrum and forecast, 942
   transformation in nearshore, 941
Weathering, 587
   chemical, 20, 21, 761, 792
   clay minerals formation, 88
   oxidative, 20
   semiconsolidated rocks, 276
   subaerial, 311
West Siberian rift basin, 484
Wetzelliella, 469
White smokers, 58–61
   carbonate-rich chimneys, 61
   discovery, 58
   formation, 59–61
Wide-angle reflection/refraction profiling (WARRP), 722
Wilson cycle, 602–603, 944–945
Woods Hole Oceanographic Institution (WHOI), 830
World Climate Research Program (WCRP), 697
World Ocean Circulation Experiment (WOCE), 337, 697
Wrangellia oceanic plateau, 563

Z

Zigzagiceras zigzag ammonite zone, 283

This encyclopedia includes no entries for X, Y & Z.

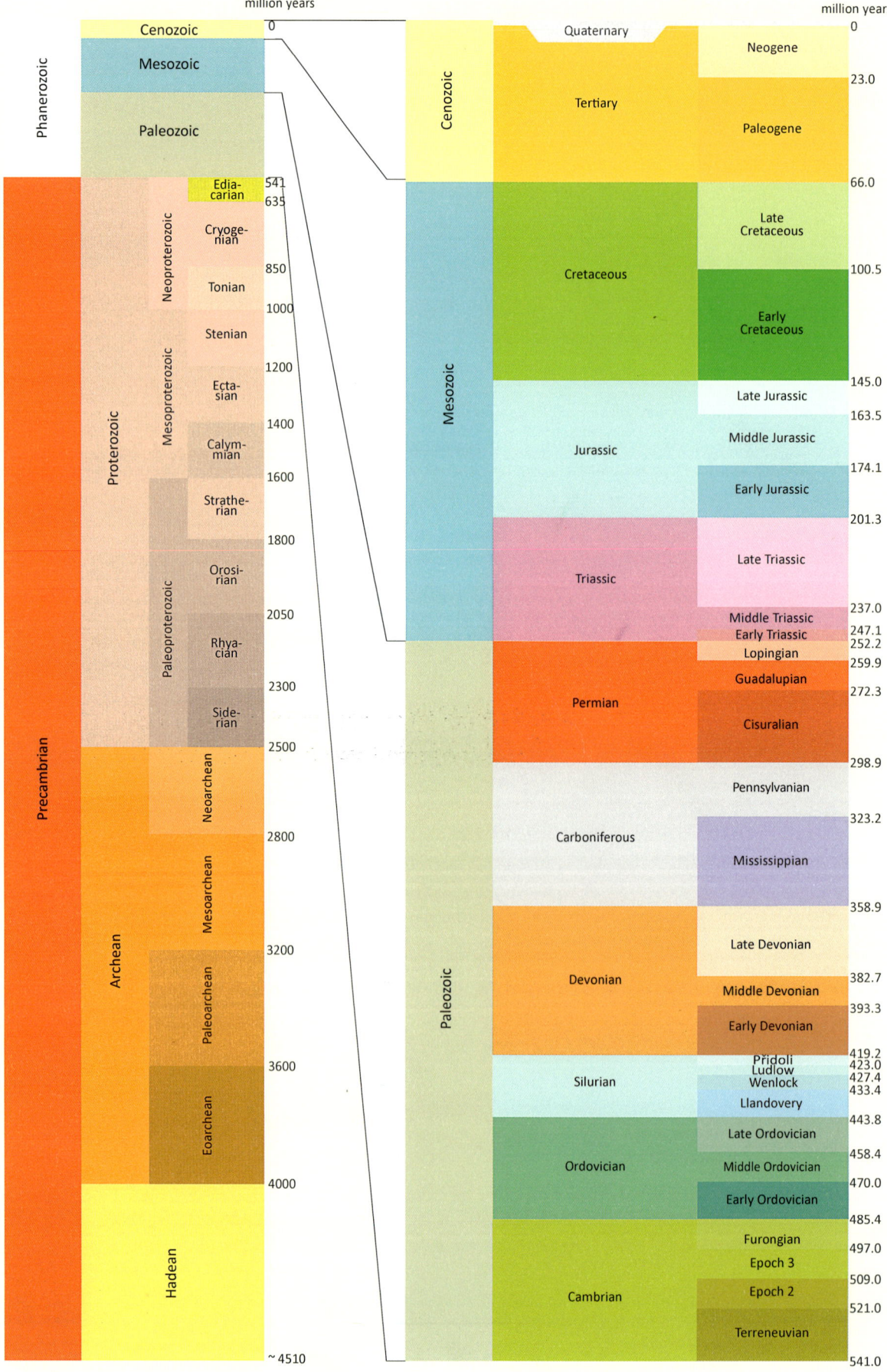